INTERNATIONAL TABLE OF ATOMIC WEIGHTS

Based on relative atomic mass of $^{12}C = 12$.

The following values apply to elements as they exist in materials of terrestrial origin and to certain artificial elements. Values in parentheses are the mass number of the isotope of longest half-life.

Name	Symbol	Atomic Number	Atomic Weight	Name	Symbol	Atomic Number	Atomic Weight	Name	Symbol	Atomic Number	Atomic Weight
Actinium[d,e]	Ac	89	(227)	Helium[g]	He	2	4.002602	Radium[d,e,g]	Ra	88	(226)
Aluminum[a]	Al	13	26.981539	Holmium[a,b]	Ho	67	164.93032	Radon[d,e]	Rn	86	(222)
Americium[d,e]	Am	95	(243)	Hydrogen[b,c,g]	H	1	1.00794	Rhenium	Re	75	186.207
Antimony	Sb	51	121.75	Indium	In	49	114.82	Rhodium[a]	Rh	45	102.90550
(Stibium)				Iodine[a]	I	53	126.90447	Rubidium[g]	Rb	37	85.4678
Argon[b,g]	Ar	18	39.948	Iridium	Ir	77	192.22	Ruthenium[g]	Ru	44	101.07
Arsenic[a]	As	33	74.92159	Iron	Fe	26	55.847	Samarium[g]	Sm	62	150.36
Astatine[d,e]	At	85	(210)	Krypton[c,g]	Kr	36	83.80	Scandium[a]	Sc	21	44.955910
Barium	Ba	56	137.327	Lanthanum[g]	La	57	138.9055	Selenium	Se	34	78.96
Berkelium[d,e]	Bk	97	(247)	Lawrencium[d,e]	Lr	103	(260)	Silicon[b]	Si	14	28.0855
Beryllium[a]	Be	4	9.012182	Lead[b,g]	Pb	82	207.2	Silver[g]	Ag	47	107.8682
Bismuth[a]	Bi	83	208.98037	Lithium[b,c,g]	Li	3	6.941	Sodium	Na	11	22.989768
Boron[b,c,g]	B	5	10.811	Lutetium[g]	Lu	71	174.967	(Natrium)[a]			
Bromine	Br	35	79.904	Magnesium	Mg	12	24.3050	Strontium[b,g]	Sr	38	87.62
Cadmium	Cd	48	112.411	Manganese[a]	Mn	25	54.93805	Sulfur[b]	S	16	32.066
Calcium[g]	Ca	20	40.078	Mendelevium[d,e]	Md	101	(258)	Tantalum	Ta	73	180.9479
Californium[d,e]	Cf	98	(251)	Mercury	Hg	80	200.59	Technetium[d,e]	Tc	43	(98)
Carbon	C	6	12.011	Molybdenum	Mo	42	95.94	Tellurium[g]	Te	52	127.60
Cerium[b,g]	Ce	58	140.115	Neodymium[g]	Nd	60	144.24	Terbium[a]	Tb	65	158.92534
Cesium[a]	Cs	55	132.90543	Neon[c,g]	Ne	10	20.1797	Thallium	Tl	81	204.3833
Chlorine	Cl	17	35.4527	Neptunium[d,e]	Np	93	(237)	Thorium[b,f,g]	Th	90	232.0381
Chromium	Cr	24	51.9961	Nickel	Ni	28	58.69	Thulium[a,b]	Tm	69	168.93421
Cobalt[a]	Co	27	58.93320	Niobium[a]	Nb	41	92.90638	Tin[g]	Sn	50	118.710
Copper[b]	Cu	29	63.546	Nitrogen[b,g]	N	7	14.00674	Titanium	Ti	22	47.88
Curium[b]	Cm	96	(247)	Nobelium[d,e]	No	102	(259)	Tungsten	W	74	183.85
Dysprosium[e,g]	Dy	66	162.50	Osmium[g]	Os	76	190.2	(Wolfram)			
Einsteinium[d,e]	Es	99	(252)	Oxygen[b,g]	O	8	15.9994	Unnilquadium[d]	Unq	104	(261)
Erbium[g]	Er	68	167.26	Palladium[g]	Pd	46	106.42	Unnilpentium[d]	Unp	105	(262)
Europium[g]	Eu	63	151.965	Phosphorus[a]	P	15	30.973762	Unnilhexium[d]	Unh	106	(263)
Fermium[d,e]	Fm	100	(257)	Platinum	Pt	78	195.08	Unnilseptium[d]	Uns	107	(262)
Fluorine[a]	F	9	18.9984032	Plutonium[d,e]	Pu	94	(244)	Uranium[c,f,g]	U	92	238.0289
Francium[d,e]	Fr	87	(223)	Polonium[d,e]	Po	84	(209)	Vanadium	V	23	50.9415
Gadolinium[g]	Gd	64	157.25	Potassium	K	19	39.0983	Xenon[a,c,g]	Xe	54	131.29
Gallium	Ga	31	69.723	(Kalium)				Ytterbium[g]	Yb	70	173.04
Germanium	Ge	32	72.61	Praseodymium[a]	Pr	59	140.90765	Yttrium[a]	Y	39	88.90585
Gold[a]	Au	79	196.96654	Promethium[d,e]	Pm	61	(145)	Zinc	Zn	30	65.39
Hafnium	Hf	72	178.49	Protactinium[f]	Pa	91	231.03588	Zirconium[g]	Zr	40	91.224

[a] *Elements with only one stable nuclide.*

[b] *Element for which known variation in isotopic abundance in terrestrial samples limits the precision of the atomic weight given.*

[c] *Element for which users are cautioned against the possibility of large variations in atomic weight due to inadvertent or undisclosed artificial separation in commercially available materials.*

[d] *Element has no stable nuclides.*

[e] *Radioactive element that lacks a characteristic terrestrial isotopic composition.*

[f] *An element, without stable nuclide(s), exhibiting a range of characteristic terrestrial compositions of long-lived radionuclide(s) such that a meaningful atomic weight can be given.*

[g] *In some geological specimens this element has an anomalous isotopic composition, corresponding to an atomic weight significantly different from that given.*

General Chemistry with Qualitative Analysis

Fifth Edition

General Chemistry with Qualitative Analysis

Fifth Edition

KENNETH W. WHITTEN
University of Georgia, Athens

RAYMOND E. DAVIS
University of Texas at Austin

M. LARRY PECK
Texas A&M University

with essay contributions from
Ronald A. DeLorenzo
Middle Georgia College

SAUNDERS COLLEGE PUBLISHING

Harcourt Brace College Publishers

Forth Worth   Philadelphia   San Diego   New York   Orlando   Austin
San Antonio   Toronto   Montreal   London   Sydney   Tokyo

Requests for permission to make copies of any part of the work should be mailed to: Permissions Department, Harcourt Brace & Company, 6277 Sea Harbor Drive, Orlando, Florida 32887-6777.

Text Typeface: Times Roman
Compositor: York Graphic Services, Inc.
Vice President, Publisher: John Vondeling
Associate Editor: Jennifer Bortel
Managing Editor: Carol Field
Project Editor: Beth Ahrens
Copy Editor: Martha Brown
Manager of Art and Design: Carol Bleistine
Art Directors: Anne Muldrow and Joan Wendt
Associate Art Director: Sue Kinney
Art & Design Coordinator: Kathleen Flanagan
Text Designer: Susan Blaker
Cover Designer: Melissa Walters
Text Artwork: Rolin Graphics, Inc.
Layout Artist: Anne Muldrow
Manager, Photos and Permissions: Dena Digilio-Betz
Vice President, Director of EDP: Tim Frelick
Production Manager: Charlene Squibb
Vice President, Marketing: Marjorie Waldron
Product Manager: Angus McDonald

Cover credit: Decomposition of red mercuric oxide. HgO(s) is heated. At a temperature of 400°C, mercuric oxide turns black and releases oxygen. Mercury begins to collect on the sides of the test tube. Richard Megna, Fundamental Photographs, New York.
Title page credit: Bob Evans/Peter Arnold, Inc.

Printed in the United States of America

GENERAL CHEMISTRY WITH QUALITATIVE ANALYSIS, Fifth Edition
0-03-006222-5

Library of Congress Catalog Card Number: 95-067684

678901234   048   10 98765432

To the memory of Kenneth Durwood Gailey

*G*eneral Chemistry and *General Chemistry with Qualitative Analysis,* fifth edition, are intended for use in the introductory chemistry course taken by students of chemistry, biology, geology, physics, engineering, and related subjects. Although some background in high school science is assumed, no specific knowledge of topics in chemistry is presupposed. These books are self-contained presentations of the fundamentals of chemistry. The aim is to convey to students the dynamic and changing aspects of chemistry in the modern world.

Richard Megna/Fundamental Photographs

CHANGES TO THE FIFTH EDITION

In revising *General Chemistry* and *General Chemistry with Qualitative Analysis,* we have incorporated many helpful suggestions that we received from professors who used earlier editions. The fifth edition is 15% smaller than the fourth edition because we have made every effort, based on reviewers' suggestions, to delete some material and to condense certain chapters.

The biggest change to this edition is the addition of new co-author Larry Peck of Texas A&M University. Larry Peck brings many new ideas and a fresh approach to this successful text. He has been instrumental in rearranging material to create a better flow of topics, and in adding many clear explanations and examples. Larry Peck's philosophy of teaching chemistry is similar to that of the other authors, so the integrity of the classic presentation and clear writing style is maintained.

This text provides students with an understanding of fundamental concepts of chemistry; their ability to solve problems is based on this understanding. Our goal in this revision is to provide students with the best possible tool for learning chemistry by adding features that enhance their understanding of concepts and guide them through the more challenging aspects of learning chemistry. Here are some of these new features:

- **Problem-Solving Tips:** Found in almost every chapter, these highlighted helpful hints show students how to avoid common mistakes and pitfalls, and guide them through more complex subject areas. These tips are based on the authors' experiences and sensitivity to difficulties that the students face.

- **"Building Your Knowledge" Exercises:** This is a new category of end-of-chapter questions that ask students to apply knowledge they learned in previous chapters to the current chapter. These questions help students retain previously learned information and show them that chemistry is an integrated science.

- **Titles have been added to each Example** so students can see more clearly what the Example is explaining. This is also useful for review purposes before exams.

- **A Glossary** has been added to the index, so students can look up a term at the back of the book as well as in the Key Terms at the end of the chapter.

- **"Chemistry in Use" boxes:** This successful feature from the previous edition has been improved in a number of ways: There are numerous new boxes (many by

Ronald A. DeLorenzo, Middle Georgia College) that provide interesting information and describe relevant applications of chemistry. The boxes have been divided into the following categories, each with its own icon, to show why topics are important:

 *The Environment*

 *The Development of Science*

 *Research & Technology*

 *Our Daily Lives*

The boxes help students understand that chemistry is not only facts and theory, but also has modern applications in the world around them.

We have also made changes in the order and depth of coverage, based on detailed comments by reviewers:

- The main features of atomic composition are summarized in a preliminary presentation in Section 2-1. This gives students a better foundation for understanding the basic ideas later in the chapter. Atomic structure is then discussed in more detail in Chapter 5.

- A new brief introduction to naming compounds is provided in Section 2-4 for instructors who want their students to be exposed to this material early. Systematic inorganic nomenclature is presented at the end of Chapter 4.

- Balancing equations for oxidation–reduction reactions has been moved from Chapter 4 to Chapter 11, where students will be better prepared for this challenging topic.

- Section 16-5 (Using Integrated Rate Equations to Determine Reaction Order) has been recast as an Enrichment section.

- The descriptive material on representative metals and transition metals has been condensed and combined into a single chapter (Chapter 23).

- Three previous chapters on nonmetallic elements have been condensed and combined into one chapter (Chapter 24).

- Cations that create serious disposal problems have been eliminated in the qualitative analysis chapters. Mercury, silver, lead, and most chromium have been removed.

- Some discussions of ionic equilibria have been moved into Chapter 36.

STRONG CLASSIC FEATURES

We have also continued to employ features that were well-received in earlier editions of the text:

- A *Chapter Outline* and a list of *Objectives* are provided at the beginning of each chapter. These allow students to preview the chapter prior to reading it.

- *Enrichment* sections provide more insight into selected topics for better-prepared students, but can be easily omitted without any loss of continuity. Enrichment sections are marked with a special icon ▼ .

- *Key Terms* are boldfaced in the text and are defined at the end of each chapter, immediately reinforcing terminology and concepts.

- The end-of-chapter *Exercises* have been carefully examined and revised: more than one-half of the problems are new or modified. All Exercises have been carefully reviewed for accuracy.

- *A note at the end of most Examples,* "You should now work Exercise X," encourages students to practice the appropriate end-of-chapter Exercise, and more closely ties illustrative Examples to related Exercises, thereby reinforcing concepts. Each Example also contains a *Plan* that explains the logic used to solve the problem.

- *Margin notes* are used to point out historical facts, provide additional information, emphasize further some important points, relate information to ideas developed earlier, and note the relevance of discussions.

- *Figures* have been redrawn as necessary to improve appearance and clarity, and many new *photographs* have been added to illustrate important points and provide visual interest.

We have also continued to use many ideas and teaching philosophies developed over the five editions of this text:

We have used color extensively to make the text easier to read and comprehend. A detailed description of our pedagogical use of color is given on page xxii in the "To the Student" section. Pedagogical use of color makes the text more clear, more accurate, and easier to understand.

We have kept in mind that chemistry is an experimental science and have emphasized the important role of theory in science. We have presented many of the classical experiments followed by interpretations and explanations of these milestones in the development of scientific thought.

We have defined each new term as accurately as possible and illustrated its meaning as early as is practical. We begin each chapter at a very fundamental level and then progress through carefully graded steps to a reasonable level of sophistication. *Numerous* illustrative Examples are provided throughout the text and keyed to end-of-chapter Exercises. The first Examples in each section are quite simple, the last considerably more complex. The unit-factor method has been emphasized where appropriate.

We have used a blend of SI and more traditional metric units, because many students plan careers in areas in which SI units are not yet widely used. The health-care fields, the biological sciences, textiles, and agriculture are typical examples. We have used the joule rather than the calorie in nearly all energy calculations.

ORGANIZATION

There are 28 chapters in *General Chemistry,* and eight additional chapters in *General Chemistry with Qualitative Analysis.*

We present stoichiometry (**Chapters 2 and 3**) before atomic structure and bonding (**Chapters 5–9**) to establish a sound foundation for a laboratory program as early as possible. These chapters are virtually self-contained to provide flexibility to those who wish to cover structure and bonding before stoichiometry.

Because much of chemistry involves chemical reactions, we have introduced chemical reactions in a simplified, systematic way early in the text **(Chapter 4).** A logical, orderly introduction to formula unit, total ionic, and net ionic equations is included so that this information can be used throughout the remainder of the text. Solubility rules are also presented in this chapter, so that students can use them in writing chemical equations in their laboratory work. Finally, naming inorganic compounds has been moved to Chapter 4 to give students earlier exposure to nomenclature.

Many students have difficulty systematizing and using information, so we have done our utmost to assist them. At many points throughout the text we summarize the results of recent discussions or illustrative examples in tabular form to help students see the "big picture." The basic ideas on chemical periodicity are introduced early **(Chapters 4 and 6)** and are used throughout the text. The simplified classification of acids and bases introduced in Chapter 4 is expanded in **Chapter 10:** Acids, Bases, and Salts, after the appropriate background on structure and bonding.

References are made to the classification of acids and bases and to the solubility rules throughout the text to emphasize the importance of systematizing and using previously covered information.

Chapter 11 covers solution stoichiometry for both acid–base and redox reactions.

After our excursion through Gases and the Kinetic–Molecular Theory **(Chapter 12),** Liquids and Solids **(Chapter 13),** and Solutions **(Chapter 14),** students have the appropriate background for a wide variety of laboratory experiments.

Comprehensive chapters are presented on Chemical Thermodynamics **(Chapter 15)** and Chemical Kinetics **(Chapter 16).** The distinction between the roles of standard and nonstandard Gibbs free energy change in predicting reaction spontaneity is clearly discussed. Chapter 16, Chemical Kinetics, provides early and consistent emphasis on the experimental basis of kinetics.

These chapters provide the necessary background for a strong introduction to Chemical Equilibrium in **Chapter 17.** This is followed by three chapters on Equilibria in Aqueous Solutions. A chapter on Electrochemistry **(Chapter 21)** completes the common core of the text except for Nuclear Chemistry **(Chapter 26),** which is self-contained and may be studied at any point in the course.

A group of basically descriptive chapters follow. For the fifth edition, we have condensed this section of the book. **Chapter 22,** Metals I: Metallurgy, and **Chapter 23,** Metals II: Properties and Reactions, give broad coverage to the chemistry of metals. **Chapter 24** covers Some Nonmetals, and **Chapter 25,** Coordination Compounds, is a sound introduction to that field. Throughout these chapters, we have been careful to include appropriate applications of the principles that have been developed in the first part of the text to explain descriptive chemistry.

Organic chemistry is covered in detail in Chapters 27 and 28. **Chapter 27** (Organic Chemistry I: Compounds) presents the classes of compounds, their structures, and nomenclature with major emphasis on the principal functional groups. **Chapter 28** (Organic Chemistry II: Molecular Geometry and Reactions) is a highly structured, concise, well-illustrated discussion of the geometries of organic molecules, the three fundamental classes of organic reactions, and some reactions of key functional groups. This material provides a broad overview for students who will not take a course in organic chemistry. It also provides the introduction to important concepts for those who will study organic chemistry.

Eight additional chapters are included in *General Chemistry with Qualitative Analysis.* In **Chapter 29,** the important properties of the metals of the cation groups are tabulated, their properties are discussed, the sources of the elements are listed, their metallurgies are described, and a few uses of each metal are given.

Chapter 30 is a detailed introduction to the laboratory procedures used in semimicro qualitative analysis.

Chapters 31–35 cover the analysis of the groups of cations. Each chapter includes a discussion of the important oxidation states of the metals, an introduction to the analytical procedures, and comprehensive discussions of the chemistry of each cation group. Detailed laboratory directions, set off in color, follow. Students are alerted to pitfalls in advance, and alternate confirmatory tests and "clean-up" procedures are described for troublesome cations. A set of Exercises accompanies each chapter.

In **Chapter 31,** the traditional Group I has been replaced by the traditional Group IIA (minus lead). Traditional Group IIB (minus mercury) constitutes the first part of **Chapter 32;** then Groups I and II (traditional IIA + IIB minus lead and mercury) make up the last part of Chapter 32. **Chapter 33** includes all of the usual Group III elements. **Chapter 34** covers Group IV and **Chapter 35** discusses Group V.

Chapter 36 contains a discussion of some of the more sophisticated ionic equilibria of qualitative analysis. The material is presented in a single chapter for the convenience of the instructor.

A FLEXIBLE PRESENTATION

We have exerted great effort to make the presentation as flexible as possible, to give instructors the freedom to choose the order in which they teach topics. Some examples follow:

1. We have clearly delineated the parts of Chapter 15, Chemical Thermodynamics, that can be moved forward for those who wish to cover thermochemistry (Sections 15-1 through 15-8) after stoichiometry (Chapters 2 and 3).

2. Chapter 4, Some Types of Chemical Reactions, is based on the periodic table and introduces chemical reactions just after stoichiometry. Reactions are classified as: (a) precipitation, (b) acid–base, (c) displacement, or (d) oxidation–reduction reactions. Chapter 4 can be moved to other positions later in the text, e.g., after structure and bonding, for those who prefer this order.

3. Some instructors prefer to discuss gases (Chapter 12) after stoichiometry (Chapters 2 and 3). Chapter 12 can be moved to that position.

4. Chapter 5 (The Structure of Atoms), Chapter 6 (Chemical Periodicity), and Chapter 7 (Chemical Bonding) provide comprehensive coverage of these key topics.

5. As in earlier editions, Molecular Structure and Covalent Bonding Theories (Chapter 8) includes parallel comprehensive VSEPR and VB descriptions of simple molecules. This approach has been widely accepted. However, some instructors prefer separate presentations of these theories of covalent bonding. The chapter has been carefully organized into numbered subdivisions to accommodate these professors; detailed suggestions are included at the beginning of the chapter.

6. Chapter 9 (Molecular Orbitals in Chemical Bonding) is a "stand-alone chapter" that may be omitted or moved with no loss in continuity.

7. Chapter 10 (Reactions in Aqueous Solutions I: Acids, Bases, and Salts) and Chapter 11 (Reactions in Aqueous Solution II: Calculations) include comprehensive discussions of acid–base and redox reactions in aqueous solutions, and solution stoichiometry calculations for acid–base and redox reactions.

INSTRUCTOR ANCILLARIES

Instructor's Solutions Manual by M. Larry Peck, Raymond E. Davis, and Kenneth W. Whitten. Contains answers and solutions to all odd-numbered end-of-chapter Exercises.

Test Bank by Frank L. Kolar, Texas A&M University. This brand new test bank is the best ever to accompany the Whitten texts. All questions have been reviewed for accuracy.

ExaMaster+[TM] **Computerized Test Bank** is a software version of the printed test bank, which enables instructors to add or edit problems and to create their own tests. Available in IBM Windows and Macintosh formats.

Overhead Transparencies A set of 150 full-color overhead transparency acetates provides a stunning visual accompaniment to the lecture. Figure and Table numbers for each acetate are displayed prominently.

Telegraph Colour Library

MULTIMEDIA MATERIALS

Cal Tech Video Animation Series in General Chemistry comprises five high-quality videos developed on state-of-the-art hardware and software and distributed exclusively for Saunders College Publishing. These videos offer professionally produced 20-minute lessons on the following topics:

1. Atomic Orbitals
2. Valence Shell Electron Pair Repulsion Theory (VSEPR)
3. Crystals
4. Periodic Trends
5. Molecular Orbitals

These videos are free to adopters of the text.

Shakhashiri Chemical Demonstration Videotapes feature well-known instructor Bassam Shakhashiri from the University of Wisconsin performing fifty 3- to 5-minute chemistry demonstrations. An accompanying **Instructor's Manual** describes each demonstration and includes discussion questions.

"Chemistry in Perspective" Videodisc contains over 110 minutes of motion footage (including chemical reactions, animations of chemistry principles, and videos demonstrating chemical principles at work in everyday life), 100 molecular model animations, and approximately 2000 still images, including images from Whitten/Davis/Peck.

"Chemistry of Life" Videodisc contains over 100 molecular model animations, chemical reaction videos, and approximately 2500 still images from a variety of Saunders' general chemistry texts.

Saunders General Chemistry Videodisc Version II This videodisc includes over 1400 images taken from ten of Saunders' chemistry texts, and also includes the Shakhashiri Chemical Demonstrations.

LectureActive™ Software comprises all video clip and still images from our videodiscs, allowing instructors to create custom lectures quickly and easily. Lectures can be read from the computer screen or printed with barcodes.

Saunders Interactive General Chemistry CD-ROM by John Kotz and William Vining, is a stand-alone multimedia presentation of general chemistry that can be used in conjunction with or as an alternative to a textbook. Integrated video, animation, and audio enhance the traditional chemistry presentation. Students also can perform simulations of laboratory experiments on their computers. Available for both Macintosh and Windows.

f(g) Scholar Spreadsheet/Graphing and Calculator/Graphing Software is a powerful scientific/engineering spreadsheet software program with over 300 built-in math functions, developed by Future Graph, Inc. It uniquely integrates graphing calculator, spreadsheet, and graphing applications into one, and allows for quick and easy movement among the applications. Students will find many uses for f(g) Scholar across their science, math, and engineering courses, including working through their laboratories from start to finished reports. Other features include programming language for defining math functions, curve fitting, three-dimensional graphing, and equation displaying. When bookstores order f(g) Scholar through Saunders College Publishing, they can pass on our exclusive low price to the student.

Cambridge Scientific ChemDraw and Chem3D are software packages that enable students to draw molecular structures. Users draw with ChemDraw; they can then transfer their work into Chem3D, which allows them to create and manipulate three-dimensional color models for a sharper image of a molecule's shape and reaction sites. Cambridge Scientific provides an accompanying User's Guide and Quick Reference Card, written exclusively for Saunders College Publishing. Available shrinkwrapped with the text for a price that students can afford.

STUDENT ANCILLARIES

Pocket Guide contains useful summaries of each text section as well as helpful problem-solving reminders and tips. The pocket guide, a $15 value, is available at a reduced price when adopted with the text.

Student Solutions Manual by Yi-Noo Tang and Wendy Keeney-Kennicutt, both of Texas A&M University. Contains answers and solutions to all even-numbered end-of-chapter Exercises. Solutions are divided by section for easy reference by students.

Student Study Guide by Raymond E. Davis. Includes chapter summaries, which highlight the main themes of each chapter; study goals with section references; a list of important terms; a preliminary test for each chapter, which provides an average of 80 drill and concept questions per chapter; and answers to the preliminary tests.

Problem Solving for General Chemistry by Leslie N. Kinsland, University of Southwestern Louisiana. Provides a valuable study tool for students who want more practice with problem solving. Contains a brief discussion of appropriate topics, a variety of examples to illustrate the topics, a set of exercises coded by topic, miscellaneous exercises, and answers to all exercises.

Student Lecture Outline by Richard M. Hedges, Texas A&M University, and Kenneth W. Whitten. Helps students organize material in the text and serves as a helpful classroom note-taking supplement, freeing students to pay more attention to the lecture. It provides great flexibility for the professor and makes more time available for special topics, increased drill, or whatever the professor chooses to do.

LABORATORY MANUAL

Standard and Microscale Experiments in General Chemistry 3/e by Carl Bishop and Muriel Bishop, Clemson University, and Kenneth W. Whitten. Approximately one-third of the experiments in this laboratory manual have been converted to microscale, wherever principles can be observed with small quantities of chemicals or when chemical waste disposal may be a serious problem. The lab manual also contains all-new pre- and post-laboratory worksheets to familiarize students with the basics of the experiment before they start and help them draw conclusions after the experiment. An **Instructor's Manual** also accompanies this title.

Acknowledgments

The list of individuals who contributed to the evolution of this book is long indeed. First, we would like to express our appreciation to the professors who contributed so much to our scientific education: Professors Arnold Gilbert, M. L. Bryant, the late W. N. Pirkle and Alta Sproull, C. N. Jones, S. F. Clark, R. S. Drago (KWW); the late Dorothy Vaughn, David Harker, and Calvin Vanderwerf, Professors Ralph N. Adams, F. S. Rowland, A. Tulinsky, and Wm. von E. Doering (RED); Professors R. O'Connor, G. L. Baker, W. B. Cook, G. J. Hunt, A. E. Martell, and M. Passer (MLP).

© Milton Heiberg/Photo Researchers, Inc.

The staff at Saunders College Publishing has contributed immeasurably to the evolution of this book. Our developmental editor, Jennifer Bortel, has great energy, a fertile imagination, and wisdom well beyond her years. She has provided more help than we can ever acknowledge. With ingenuity, persistence, and patience, our photo researcher, Dena Digilio-Betz, has gathered many excellent photographs. Anne Muldrow and Joan Wendt have given us high-quality design and artwork that contribute to the appearance and the substance of the book. Additionally, we have drawn from the excellent artwork of other Saunders texts. With remarkable good cheer, our project editor, Beth Ahrens, has handled innumerable details with skill, insight, and imagination. Her keen eye for detail has improved consistency and spared us many embarrassments. We express our continued deep appreciation to our publisher and friend, John Vondeling, the best editor in the business. John has guided us at every step in the development of each edition of these books, and our respect and admiration for him have grown with each passing day.

Our secretary, Martha Dove, has been patient and skillful through the many revisions. We are indeed grateful for her patience, her skill, and her dedication. Jim Morgenthaler (Athens), Charles Steele (Austin), and Charles D. Winters (Oneonta) did the original photography. We thank them for the tremendous contribution that their enthusiasm and their many hours of patient work and professional expertise have brought to these editions.

We are thankful to Professor Gary F. Riley of the St. Louis College of Pharmacy who has again worked all the end-of-chapter Exercises. Only the authors appreciate the magnitude of this important contribution.

Professor Donald C. Kleinfelter at the University of Tennessee at Knoxville has made a major contribution to this edition by a critical reading of the final manuscript.

Larry Brown at Texas A&M University helped tremendously in catching errors in the galley and page proof stages. His keen eye has helped create a text that is as close to error-free as possible.

We would also like to thank Ronald DeLorenzo of Middle Georgia College for writing the fine "Chemistry in Use" boxes that appear throughout the text. Professor DeLorenzo's writing has brought a good deal of interest, insight, and humor to our text.

Finally, we are deeply indebted to our families, Betty, Andy, and Kathryn Whitten; Sharon, Angela, and Brian Davis and Laura Kane; and Sandy Peck, Molly Levine, and Marci Culp. They have supported us during the many years we have worked on this project. Their understanding, encouragement, and moral support have "kept us going."

REVIEWERS OF THE FIFTH EDITION

The following individuals have reviewed the manuscripts for the fifth edition of *General Chemistry* and *General Chemistry with Qualitative Analysis;* several also provided detailed comments on earlier editions. Their suggestions have improved the texts significantly.

Major Charles Bass, *United States Military Academy*
James R. Blanton, *The Citadel*
Joseph Branch, *Central Alabama Community College*
Mark Cracolice, *University of Montana*
Julian Davies, *University of Toledo*
Harry Eick, *Michigan State University*
Dale Ensor, *Tennessee Technological University*
Mark Freilich, *Memphis State University*
Bruce Hoffman, *Lewis and Clark College*
Stephen W. John, *Lane Community College*
Andrew Jorgensen, *University of Toledo*
Margaret Kastner, *Bucknell University*
Bob Kowerski, *College of San Mateo*
Charles Kriley, *Purdue University, Calumet*
Larry Krannich, *University of Alabama at Birmingham*
Robert Lamb, *Ohio Northern University*
Ramon Lopez de la Vega, *Florida International University*
William E. McMullen, *Texas A&M University*
Deborah Nycz, *Broward Community College*
Richard A. Pierce, *Jefferson College*
Roland R. Roskos, *University of Wisconsin, La Crosse*
Mary Jane Schultz, *Tufts University*
Henry Tracy, *University of Southern Maine*
Victor Viola, *Indiana University*
Wendy S. Wolbach, *Illinois Wesleyan University*
Kevin L. Wolf, *Texas A&M University*
James Wood, *Palm Beach Community College*

REVIEWERS OF THE FIRST FOUR EDITIONS OF GENERAL CHEMISTRY

Edwin Abbott, *Montana State University*
Ed Acheson, *Millikin University*
David R. Adams, *North Shore Community College*
Carolyn Albrecht
Steven Albrecht, *Ball State University*
Ale Arrington, *South Dakota School of Mines*
George Atkinson, *Syracuse University*
Jerry Atwood, *University of Alabama*
William G. Bailey, *Broward Community College*
J. M. Bellama, *University of Maryland*
Carl B. Bishop, *Clemson University*

Muriel B. Bishop, *Clemson University*
George Bodner, *Purdue University*
Greg Brewer, *The Citadel*
Clark Bricker, *University of Kansas*
Robert Broman, *University of Missouri*
Robert F. Bryan, *University of Virginia*
L. A. Burns, *St. Clair County Commuity College*
James Carr, *University of Nebraska, Lincoln*
Elaine Carter, *Los Angeles City College*
Thomas Cassen, *University of North Carolina*
Evelyn A. Clarke, *Community College of Philadelphia*
Lawrence Conroy, *University of Minnesota*
John M. DeKorte, *Northern Arizona University*
George Eastland, Jr., *Saginaw Valley State University*
Lawrence Epstein, *University of Pittsburgh*
Sandra Etheridge, *Gulf Coast Community College*
Darrell Eyman, *University of Iowa*
Wade A. Freeman, *University of Illinois, Chicago Circle*
Richard Gaver, *San Jose State University*
Gary Gray, *University of Alabama, Birmingham*
Robert Hanrahan, *University of Florida*
Henry Heikkinen, *University of Maryland*
Forrest C. Hentz, *North Carolina State University*
R. K. Hill, *University of Georgia*
Larry Houck, *Memphis State University*
Arthur Hufnagel, *Erie Community College, North Campus*
Wilbert Hutton, *Iowa State University*
Albert Jache, *Marquette University*
William Jensen, *South Dakota State University*
M. D. Joesten, *Vanderbilt University*
Philip Kinsey, *University of Evansville*
Leslie N. Kinsland, *University of Southwestern Louisiana*
Marlene Kolz
James Krueger, *Oregon State University*
Norman Kulevsky, *University of North Dakota*
Alfred Lee, *City College of San Francisco*
Patricia Lee, *Bakersfield College*
William Litchman, *University of New Mexico*
Gilbert J. Mains, *Oklahoma State University*
Ronald Marks, *Indiana University of Pennsylvania*
William Masterton, *University of Connecticut*
Clinton Medbery, *The Citadel*
Joyce Miller, *San Jacinto College*
Joyce Neiburger, *Purdue University*
Barbara O'Brien, *Texas A&M University*
Christopher Ott, *Assumption College*
James L. Pauley, *Pittsburgh State University*
William Pietro, *University of Wisconsin, Madison*
Ronald O. Ragsdale, *University of Utah*
Susan Raynor, *Rutgers University*

Randal Remmel
Gary F. Riley, *St. Louis College of Pharmacy*
Eugene Rochow, *Harvard University*
John Ruff, *University of Georgia*
Don Roach, *Miami Dade Community College*
George Schenk, *Wayne State University*
William Scroggins, *El Camino College*
Curtis Sears, *Georgia State University*
Diane Sedney, *George Washington University*
Mahesh Sharma, *Columbus College*
C. H. Stammer, *University of Georgia*
Yi-Noo Tang, *Texas A&M University*
Margaret Tierney, *Prince George's Community College*
Janice Turner, *Augusta College*
James Valentini, *University of California, Irvine*
Douglas Vaughan
W. H. Waggoner, *University of Georgia*
Susan Weiner, *West Valley College*
David Winters, *Tidewater Community College*
Steve Zumdahl, *University of Illinois*

Kenneth W. Whitten
Raymond E. Davis
M. Larry Peck

To the Student

Charles D. Winters

We have written this text to assist you as you study chemistry. Chemistry is a fundamental science—some call it the central science. As you and your classmates pursue diverse career goals you will find that the vocabulary and ideas presented in this text will be useful in more places and in more ways than you may imagine now.

We begin with the most basic vocabulary and ideas. We then carefully evolve increasingly sophisticated ideas that are necessary and useful in all the other physical sciences, the biological sciences, and the applied sciences such as medicine, dentistry, engineering, agriculture, and home economics.

We have made the early chapters as nearly self-contained as possible. The material can be presented in the order considered most appropriate by your professor. Some professors will cover chapters in different orders or will omit some chapters completely—the text was designed to accommodate this.

Early in each section we have attempted to provide the experimental basis for the ideas we evolve. By *experimental basis* we mean the observations and experiments on the phenomena that have been most important in developing concepts. We then present an explanation of the experimental observations.

Chemistry is an experimental science. We know what we know because we (literally thousands of scientists) have observed it to be true. Theories have been evolved to explain experimental observations (facts). Successful theories explain observations fully and accurately. More importantly, they enable us to predict the results of experiments that have not yet been performed. Thus, we should always keep in mind the fact that experiment and theory go hand-in-hand. They are intimately related parts of our attempt to understand and explain natural phenomena.

"What is the best way to study chemistry?" is a question we are asked often by our students. While there is no single answer to this question, the following suggestions may be helpful. Your professor may provide additional suggestions. A number of supplementary materials accompany this text. All are designed to assist you as you study chemistry. Your professor may suggest that you use some of them.

Students often underestimate the importance of the act of *writing* as a tool for learning. Whenever you read, do not just highlight passages in the text, but also *take notes*. Whenever you work problems or answer questions *write yourself explanations* of why each step was done or how you reasoned out the answer. Keep a special section of your notebook for working out problems or answering questions. The very act of writing forces you to concentrate more on what you are doing, and you learn more. This is true even if you never go back to review what you wrote earlier. Of course, these notes will also help you to review for an examination.

You should always read over the assigned material before it is covered in class. This helps you to recognize the ideas as your professor discusses them. Take careful class notes. *At the first opportunity,* and certainly the same day, you should recopy your class notes. As you do this, fill in more detail where you can. Try to work the illustrative examples that your professor solved in class, without looking at the solution in your notes. If you must look at the solution, look at only one line (step), and then try to figure out the next step. Read the assigned material again and take notes, integrating these with your class notes. Reading should be much more informative the second time.

Review the "key terms" at the end of the chapter to be sure that you know the exact meaning of each. Work the illustrative examples in the text while covering the solutions with a sheet of paper. If you find it necessary to look at the solutions, look at only one line at a time and try to figure out the next step. Answers to illustrative examples are displayed on blue backgrounds. At the end of most examples, we suggest related questions from the end-of-chapter exercises. You should work these suggested exercises as you come to them. Make sure you read the Problem-Solving Tips; these will help you avoid common mistakes and understand more complex ideas.

This is a good time to work through the appropriate chapter in the STUDY GUIDE TO GENERAL CHEMISTRY. This will help you to see an overview of the chapter, to set specific study goals, and then to check and improve your grasp of basic vocabulary, concepts, and skills. Next, work the assigned exercises at the end of the chapter.

The Appendices contain much useful information. You should become familiar with them and their contents so that you may use them whenever necessary. Answers to all even-numbered numerical exercises are given at the end of the text so that you may check your work.

We heartily recommend the STUDY GUIDE TO GENERAL CHEMISTRY by Raymond E. Davis, the SOLUTIONS MANUAL by Professors Yi-Noo Tang and Wendy Keeney-Kennicutt, PROBLEM SOLVING FOR GENERAL CHEMISTRY by Leslie Kinsland, and the STUDENT LECTURE OUTLINE by Richard Hedges, all of which were written to accompany this text. The STUDY GUIDE provides an overview of each chapter and emphasizes the threads of continuity that run through chemistry. It lists study goals, tells you which ideas are most important and why they are important, and provides many forward and backward references. Additionally, the STUDY GUIDE contains many easy to moderately difficult questions that enable you to gauge your progress. These short questions provide excellent practice in preparing for examinations. Answers are provided for all questions, and many have explanations or references to appropriate sections in the text.

The SOLUTIONS MANUAL contains detailed solutions and answers to all even-numbered end-of-chapter exercises. It also has many helpful references to appropriate sections and illustrative examples in the text.

PROBLEM SOLVING FOR GENERAL CHEMISTRY provides a valuable study tool if you want more practice with problem solving. It contains a brief discussion of appropriate topics, a variety of examples to illustrate the topics, a set of exercises coded by topic, miscellaneous exercises, and answers to all exercises.

The STUDENT LECTURE OUTLINE helps you organize material in the text and serves as a helpful classroom note-taking supplement, so you can pay more attention to the lecture.

If you have suggestions for improving this text, please write to us and tell us about them.

KEYS FOR COLOR CODES

In addition to four-color photography and art, we have used color to help you identify and organize important ideas, techniques, and concepts as you study this book.

1. Important ideas, mathematical relationships, and summaries are displayed on tan screens, the width of the text.

There is no observable change in the quantity of matter during a chemical reaction or during a physical change.

2. Answers to examples are shown on blue screens. Intermediate steps (logic, guidance, and so on) are shown on gold screens.

EXAMPLE 1-9   *English–Metric Conversion*

Express 1.0 gallon in milliliters.

Plan

We ask $\underset{\sim}{?}$ mL = 1.0 gal and multiply by the appropriate factors.

$$\boxed{\text{gallons}} \longrightarrow \boxed{\text{quarts}} \longrightarrow \boxed{\text{liters}} \longrightarrow \boxed{\text{milliliters}}$$

Solution

$$\underset{\sim}{?}\ \text{mL} = 1.0\ \text{gal} \times \frac{4\ \text{qt}}{1\ \text{gal}} \times \frac{1\ \text{L}}{1.06\ \text{qt}} \times \frac{1000\ \text{mL}}{1\ \text{L}} = \boxed{3.8 \times 10^3\ \text{mL}}$$

You should now work Exercises 27 and 36.

---

3. Acidic and basic properties are contrasted by using pink and blue, respectively. Neutral solutions are indicated in pale purple.

Table 19-4 *Titration Data for 100.0 mL of 0.100 M HCl versus NaOH*

mL of 0.100 M NaOH Added	mmol NaOH Added	mmol Excess Acid or Base	pH
0.0	0.00	10.0 H_3O^+	1.00
20.0	2.00	8.0	1.17
50.0	5.00	5.0	1.48
90.0	9.00	1.0	2.28
99.0	9.90	0.10	3.30
99.5	9.95	0.05	3.60
100.0	10.00	0.00 (eq. pt.)	7.00
100.5	10.05	0.05 OH^-	10.40
110.0	11.00	1.00	11.68
120.0	12.00	2.00	11.96

4. Red and blue are used in oxidation–reduction reactions and electrochemistry.
   (a) Oxidation numbers are shown in red circles to avoid confusion with ionic charges. Oxidation is indicated by blue and reduction is indicated by red.

$$H^+(aq) + NO_3^-(aq) + Cu(s) \longrightarrow Cu^{2+}(aq) + NO(g) + H_2O(\ell)$$

with oxidation numbers: NO_3^- (+5), $Cu(s)$ (0), Cu^{2+} (+2), NO (+2)

$+2$

-3

(b) In electrochemistry (Chapter 21), we learn that oxidation occurs at the *anode;* we use blue to indicate the anode and its half-reaction. Similarly, reduction occurs at the *cathode;* we use red to indicate the cathode and its half-reaction.

$$2Cl^- \longrightarrow Cl_2(g) + 2e^- \quad \text{(oxidation, anode half-reaction)}$$
$$2[Na^+ + e^- \longrightarrow Na(\ell)] \quad \text{(reduction, cathode half-reaction)}$$
$$2Na^+ + 2Cl^- \longrightarrow 2Na(\ell) + Cl_2(g) \quad \text{(overall cell reaction)}$$
$$\underbrace{}_{2NaCl(\ell)}$$

5. Atomic orbitals are shown in tan. Bonding and antibonding molecular orbitals are shown in blue and red, respectively.

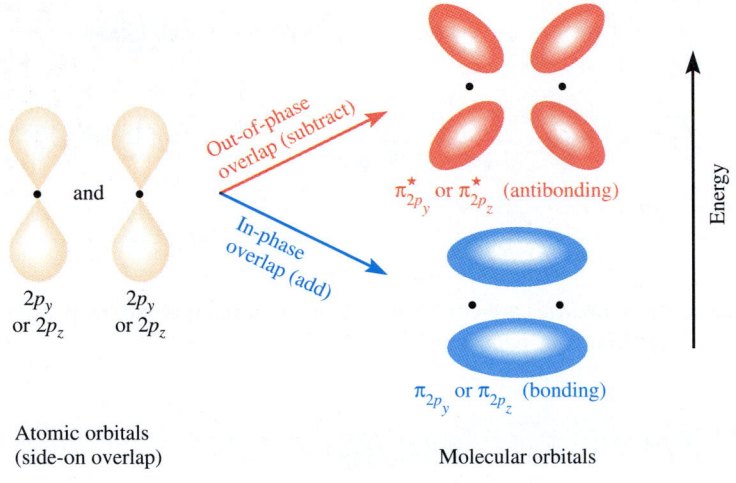

6. Hybridization schemes and hybrid orbitals are emphasized in green.

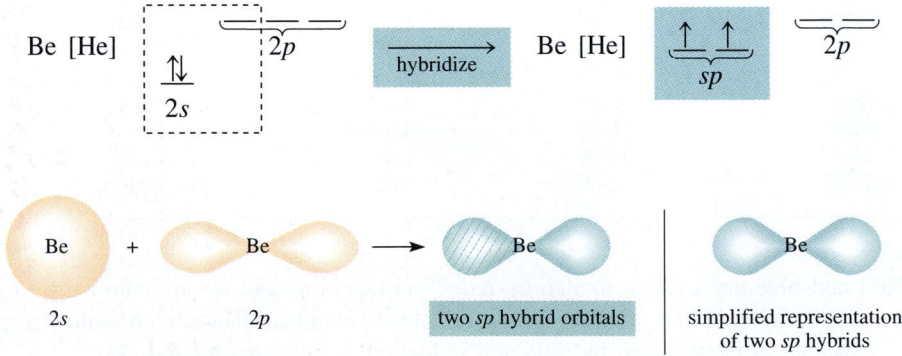

7. Color-coded periodic tables emphasize the classification of the elements as metals (blue), nonmetals (yellow), and metalloids (green). Please study the periodic table inside the front cover carefully so that you recognize this color scheme.

SAUNDERS INVITES YOU TO BE A PART OF OUR LATEST INNOVATION IN CHEMISTRY!

SAUNDERS COLLEGE PUBLISHING *a division of Harcourt Brace College Publishers* • The Public Ledger Building, Suite 1250 • 150 South Independence Mall West • Philadelphia, PA 19106

Contents Overview

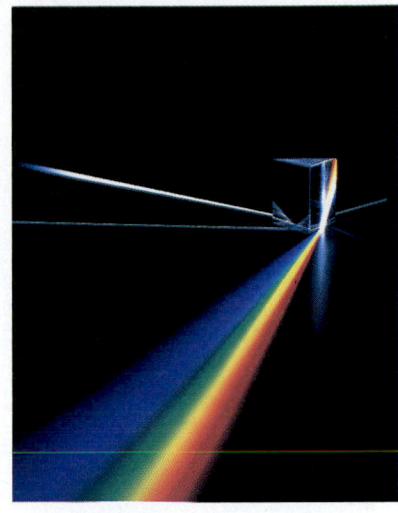

Alfred Pasieka/Peter Arnold, Inc.

Qualitative Analysis   1053

Table of Contents

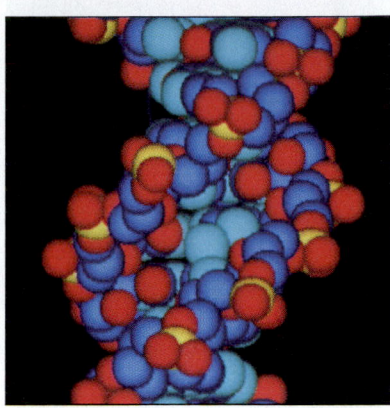

© *Peter Menzel*

Charles Steele

Steve Elmore/Tom Stack & Associates

Ralph Earlandson/Imagery

9   Molecular Orbitals in Chemical Bonding   316

10   Reactions in Aqueous Solutions I: Acids, Bases, and Salts   335

Larry Lefever/Grant Heilman, Inc.

Charles D. Winters

© Michael Lustbader/Photo Researchers, Inc.

Courtesty of Ashland Oil, Inc.

Courtesy of Niagara Mohawk Power Corp./Frank Warner

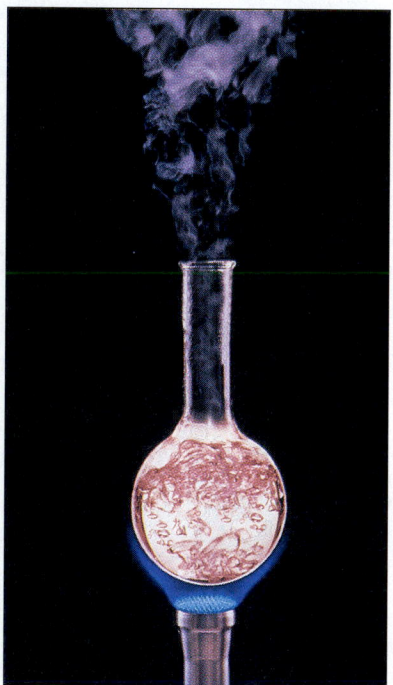

Lester Lefkowitz/Telegraph Colour Library

Gordon Garradd/Science Photo Library/Photo Researchers, Inc.

Brian Parker/Tom Stack & Associates

Courtesy of Am General Corporation

Charles Steele

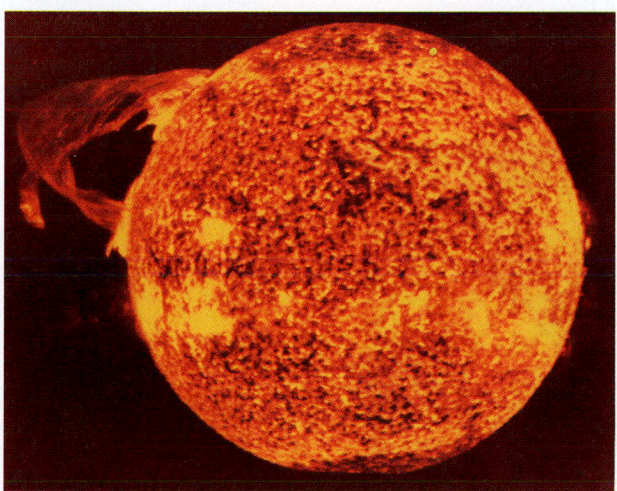

NASA

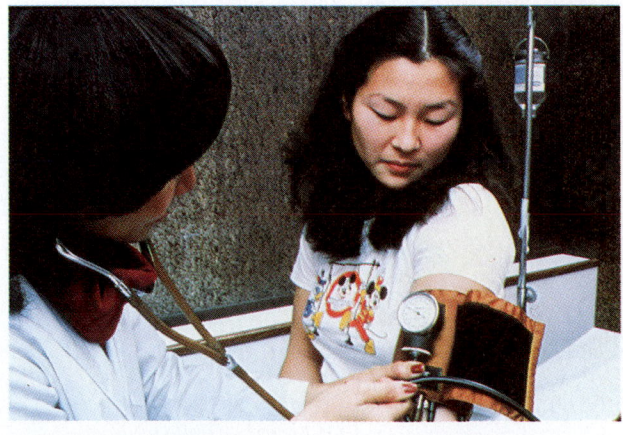

Science VU-WHC/Visuals Unlimited

Charles D. Winters

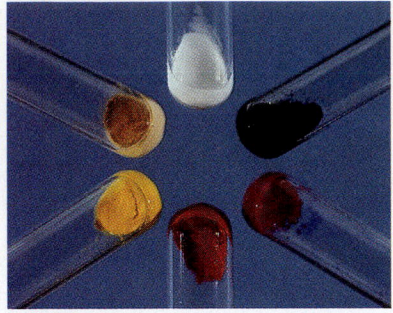

James Morgenthaler

Qualitative Analysis   1053

Charles Steele

34 **Analysis of Cation Group IV   1115**

35 **Analysis of Cation Group V   1126**

36 **Ionic Equilibria in Qualitative Analysis   1134**

from Levin

General Chemistry with Qualitative Analysis

Fifth Edition

The Foundations of Chemistry

1

A fireworks display involves many aspects of chemistry. These include the physical and chemical properties of the substances used; the physical and chemical changes that occur; the rates at which these changes occur; the elements and compounds that react and are formed; and the conversion of chemical energy to light, heat, and sound energy.

OBJECTIVES

As you study this chapter, you should learn to

- *Use the basic vocabulary of matter and energy*

- *Distinguish between chemical and physical properties and between chemical and physical changes*

- *Recognize various forms of matter: homogeneous and heterogeneous mixtures, substances, compounds, and elements*

- *Apply the concept of significant figures*

- *Apply appropriate units to describe the results of measurement*

- *Use the unit factor method to carry out conversions among units*

- *Describe temperature measurements on various common scales, and convert between these scales*

- *Carry out calculations relating temperature change to heat absorbed or liberated*

T housands of practical questions are studied by chemists. A few of them are

How can we modify a useful drug so as to improve its effectiveness while minimizing harmful or unpleasant side effects?

How can we develop better materials to be used as synthetic bone in transplants?

Which substances could help to avoid rejection of foreign tissue in organ transplants?

What improvements in fertilizers or pesticides can improve agricultural yields? How can this be done with minimal environmental danger?

How can we get the maximum work from a fuel while producing the least harmful emissions possible?

Which really poses the greater environmental threat—the burning of fossil fuels and its contribution to the greenhouse effect and climatic change, or the use of nuclear power and the related radiation and disposal problems?

How can we develop suitable materials for the semiconductor and microelectronics industry? Can we develop a battery that is cheaper, lighter, and more powerful?

What changes in structural materials could help to make aircraft lighter and more economical, yet at the same time stronger and safer?

What relation is there between the substances we eat, drink, or breathe and the possibility of developing cancer? How can we develop substances that are effective in killing cancer cells preferentially over normal cells?

Can we economically produce fresh water from sea water for irrigation or consumption?

How can we slow down unfavorable reactions, such as corrosion of metals, while speeding up favorable ones, such as the growth of foodstuffs?

Chemistry touches almost every aspect of our lives, our culture, and our environment. Its scope encompasses the air we breathe, the food we eat, the fluids we drink, our clothing, dwellings, transportation and fuel supplies, and our fellow creatures.

> Chemistry is the science that describes matter—its chemical and physical properties, the chemical and physical changes it undergoes, and the energy changes that accompany those processes.

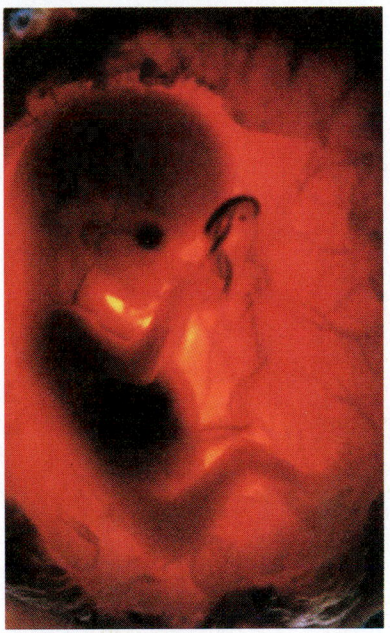

Enormous numbers of chemical reactions are necessary to produce a human embryo (here at 10 weeks, 6 cm long).

Matter includes everything that is tangible, from our bodies and the stuff of our everyday lives to the grandest objects in the universe. Some call chemistry the central science. It rests on the foundation of mathematics and physics and in turn underlies the life sciences—biology and medicine. To understand living systems fully, we must first understand the chemical reactions and chemical influences that operate within them. The chemicals of our bodies profoundly affect even the personal world of our thoughts and emotions.

No one can be expert in all aspects of such a broad science as chemistry. Sometimes we arbitrarily divide the study of chemistry into various branches. Carbon is very versatile in its bonding and behavior, and is a key element in many substances that are essential to life. All living matter contains carbon combined with hydrogen. The chemistry of compounds of carbon and hydrogen is called **organic chemistry.** (In the early days of chemistry, it was believed that living matter and inanimate matter were entirely different. We now know that many of the compounds found in living matter can be made from nonliving or ''inorganic'' sources. Thus, the terms ''organic'' and ''inorganic'' have different meanings than they did originally.) The study of substances that do not contain carbon combined with hydrogen is called **inorganic chemistry.** The branch of chemistry that is concerned with the detection or identification of substances present in a sample (*qualitative analysis*) or with the amount of each that is present (*quantitative analysis*) is called **analytical chemistry. Physical chemistry** applies the mathematical theories and methods of physics to the properties of matter and to the study of chemical processes and the accompanying energy changes. As its name suggests, **biochemistry** is the study of the chemistry of processes in living organisms. Such divisions are arbitrary and most chemical studies involve more than one of these traditional areas of chemistry. The principles you will learn in a general chemistry course are the foundation of all branches of chemistry.

We understand simple chemical systems well; they lie near chemistry's fuzzy boundary with physics. They can often be described exactly by mathematical equations. We fare less well with more complicated systems. Even where our understanding is fairly thorough, we must make approximations, and often our knowledge is far from complete. Each year researchers provide new insights into the nature of matter and its interactions. As chemists find answers to old questions, they learn to ask new ones. Our scientific knowledge has been described as an expanding sphere that, as it grows, encounters an ever-enlarging frontier.

In our search for understanding, we eventually must ask fundamental questions such as the following:

How do substances combine to form other substances? How are energy changes involved in chemical and physical changes?

How is matter constructed, in its intimate detail? How are atoms and the ways that they combine related to the properties of the matter that we can measure, such as color, hardness, chemical reactivity, and electrical conductivity?

What fundamental factors influence the stability of a substance? How can we force a desired (but energetically unfavorable) change to take place? What factors control the rate at which a chemical change takes place?

In your study of chemistry, you will learn about these and many other basic ideas that chemists have developed to help them describe and understand the behavior of matter. Along the way, we hope that you come to appreciate the development of this science, one of the grandest intellectual achievements of human endeavor. You will also learn how to apply these fundamental principles to solve real problems. One of your major goals in the study of chemistry should be to develop your ability to think critically and to solve problems (not just do numerical calculations!). In other words, you need to learn to manipulate not only numbers, but also quantitative ideas, words, and concepts.

In the first chapter, our main goals are (1) to begin to get an idea of what chemistry is about and the ways in which chemists view and describe the material world and (2) to acquire some skills that are useful and necessary in the understanding of chemistry, its contribution to science and engineering, and its role in our daily lives.

1-1 MATTER AND ENERGY

Matter is anything that has mass and occupies space. Mass is a measure of the quantity of matter in a sample of any material. The more massive an object is, the more force is required to put it in motion. All bodies consist of matter. Our senses of sight and touch usually tell us that an object occupies space. In the case of colorless, odorless, tasteless gases (such as air), our senses may fail us.

Energy is defined as the capacity to do work or to transfer heat. We are familiar with many forms of energy, including mechanical energy, light energy, electrical energy, and heat energy. Light energy from the sun is used by plants as they grow; electrical energy allows us to light a room by flicking a switch; and heat energy cooks our food and warms our homes. Energy can be classified into two principal types: kinetic energy and potential energy.

A body in motion, such as a rolling boulder, possesses energy because of its motion. Such energy is called **kinetic energy.** Kinetic energy represents the capacity for doing work directly. It is easily transferred between objects. **Potential energy** is the energy an object possesses because of its position, condition, or composition. Coal, for example, possesses chemical energy, a form of potential energy, because of its composition. Many electrical generating plants burn coal, producing heat, which is converted to electrical energy. A boulder located atop a mountain possesses potential energy because of its height. It can roll down the mountainside and convert its potential energy into kinetic

We might say that we can "touch" air when it blows in our faces, but we depend on other evidence to show that a still body of air fits our definition of matter.

The term comes from the Greek word *kinein,* meaning "to move." The word "cinema" is derived from the same word.

Nuclear energy is an important kind of potential energy.

CHEMISTRY IN USE

The Development of Science

Science and the Scientific Method

All living and nonliving entities, from single cells and complex animals to stars and galaxies, undergo change. They emerge, age, die, and continue to change even after death. *Science* is the careful inquiry into the many changes that occur within us, around us, and throughout our universe.

Many people are interested in and seek explanations for the changes they observe. Some become so interested and fascinated with transformations that they devote their lives to the study of changes. In so doing, they gain a deeper understanding of reality. Science is a vehicle used to study and understand changes; those who pursue this study are called *scientists*.

Some of the observations that we make about the properties and behavior of the matter around us are *qualitative* (one substance is a shiny solid that conducts electricity; another substance reacts violently with water) and others are *quantitative* (the pressure of a gas is measured to be 1 atmosphere; water is 11.1% hydrogen and 88.9% oxygen by mass). The result of an accurate observation is a *fact*. Often we find patterns in the facts we observe. A *law* is a concise verbal or mathematical statement of a generally observed pattern of behavior.

After many facts have been observed, it is human nature to try to discover or develop an explanation for the observations (facts). Scientists use their imaginations and reasoning powers to develop speculative explanations called *hypotheses* for their observations. They then perform additional experiments, make further observations to weed out false conjecture, and try to develop better explanations called *theories*. This approach used by scientists to develop, test, and refine explanations for observed changes is called the *scientific method*.

We can understand the scientific method by applying it to an everyday situation such as trying to determine why an air conditioner in a car isn't working properly. Let's assume that one day, while driving to school, you notice that the inside of your car is warmer and more humid than usual. You have observed a change. You remember that while you were stopped at a traffic light yesterday, another student drove into the side of your car. You speculate that the damage from the impact causes your car door to close improperly. If the door does not close properly, the cool air produced by the air conditioner can escape, and the air conditioner becomes less effective. This is your educated guess or hypothesis for the change in temperature and humidity that you observe.

You check the car door and find that it is damaged and doesn't close properly. Your educated guess has passed this additional observation. However, after having your door properly aligned at a body shop, you notice that the heat and humidity within the car are still uncomfortable. Your educated guess was wrong, or at least incomplete. Your tentative conjecture to explain the change in humidity and heat is not supported by reproducible observations or experiments. So you look further. You consider the possibility that the refrigerant level inside the air conditioner may be low, so you take your car to an air-conditioning service center. The mechanic corrects the problem, and your air conditioner then works properly. You now feel confident that your "refrigerant explanation" was correct. Your last educated guess (hypothesis) has now been elevated to the level of a theory.

Unknown to you, however, the mechanic had noticed that a piece of paper had blown against the cooling coils of your air conditioner. This paper prevented the air conditioner from working properly. The mechanic simply removed the paper and charged you $95 to "service your air conditioner," and you go through life mistakenly confident about your "refrigerant theory."

A good hypothesis or theory does more than correctly explain reality. A good hypothesis or theory must also be able to predict reality. When prehistoric men observed changes, they sometimes created superstitions to explain their observations. For example, when thunderstorms suddenly developed, the angry Thunder God explanation was invoked. Although this explanation provided our early ancestors with an explanation for sudden changes in weather, it did not give them the ability to make correct predictions about thunderstorms, a serious shortcoming. A hypothesis or theory gains status as its predictions are tested and verified.

Sometimes, both scientists and the public misuse the word "theory" to mean either an educated guess or a fact. A theory is neither a fact nor a simple educated guess. Facts are things that we can observe. Correctly made measurements are facts. A simple educated guess is a hypothesis. A theory may have begun as an educated guess, but the educated guess achieves the status of theory only after it survives repeated testing.

For any given phenomenon, it is possible to create many explanations. Only with time, continued observations, and confirmation by others can one explanation dominate all others and become accepted by the majority of scientists. Even so, the accepted *law* or model may still require modifications as more observations are made.

Ronald DeLorenzo
Middle Georgia College

energy. We discuss energy because all chemical processes are accompanied by energy changes. As some processes occur, energy is released to the surroundings, usually as heat energy. We call such processes **exothermic.** Any combustion (burning) reaction is exothermic. However, some chemical reactions and physical changes are **endothermic;** i.e., they absorb energy from their surroundings. An example of a physical change that is endothermic is the melting of ice.

The Law of Conservation of Matter

When we burn a sample of metallic magnesium in the air, the magnesium combines with oxygen from the air (Figure 1-1) to form magnesium oxide, a white powder. This chemical reaction is accompanied by the release of large amounts of heat energy and light energy. When we weigh the product of the reaction, magnesium oxide, we find that it is heavier than the original piece of magnesium. The increase in mass of the solid is due to the combination of oxygen with magnesium to form magnesium oxide. Many experiments have shown that the mass of the magnesium oxide is exactly the sum of the masses of magnesium and oxygen that combined to form it. Similar statements can be made for all chemical reactions. These observations are summarized in the **Law of Conservation of Matter:**

> There is no observable change in the quantity of matter during a chemical reaction or during a physical change.

This statement is an example of a **scientific (natural) law,** a general statement based on the observed behavior of matter to which no exceptions are known. A nuclear reaction is *not* a chemical reaction.

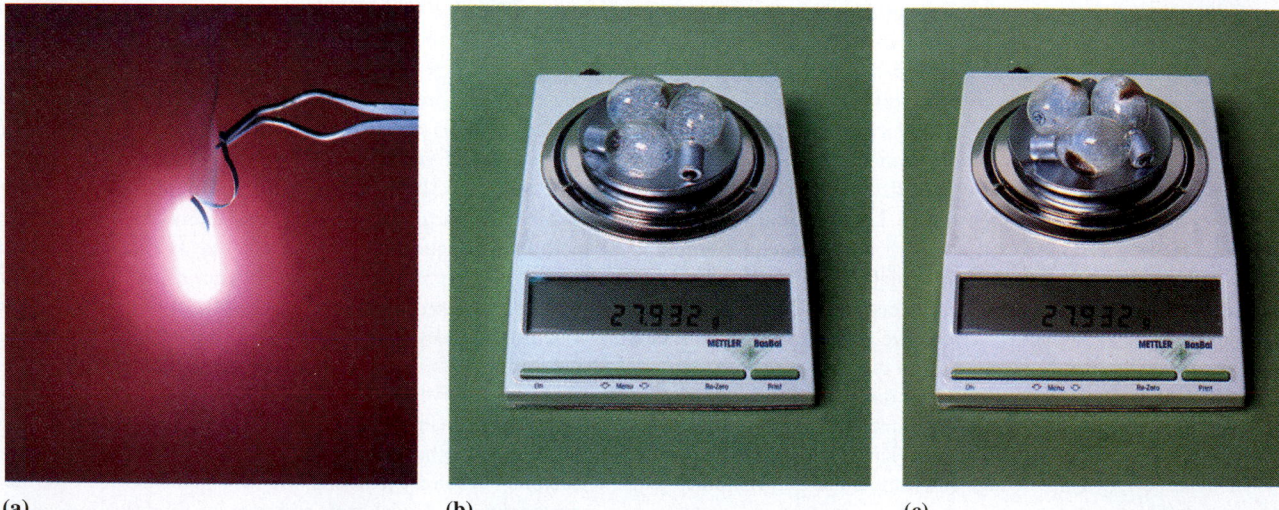

(a)   **(b)**   **(c)**

Figure 1-1   (a) Magnesium burns in the oxygen of the air to form magnesium oxide, a white solid. This reaction occurs in photographic flashbulbs (b, c). There is no gain or loss of mass as the reaction occurs.

The Law of Conservation of Energy

In exothermic chemical reactions, *chemical energy* is usually converted into *heat energy*. Some exothermic processes involve other kinds of energy changes. For example, some liberate light energy without heat, and others produce electrical energy without heat or light. In *endothermic* reactions, heat energy, light energy, or electrical energy is converted into chemical energy. Although chemical changes always involve energy changes, some energy transformations do not involve chemical changes at all. For example, heat energy may be converted into electrical energy or into mechanical energy without any simultaneous chemical changes. Many experiments have demonstrated that all of the energy involved in any chemical or physical change appears in some form after the change. These observations are summarized in the **Law of Conservation of Energy:**

> Energy cannot be created or destroyed in a chemical reaction or in a physical change. It can only be converted from one form to another.

Electricity is produced in hydroelectric plants by the conversion of mechanical energy (from flowing water) into electrical energy.

The Law of Conservation of Matter and Energy

With the dawn of the nuclear age in the 1940s, scientists, and then the world, became aware that matter can be converted into energy. In nuclear reactions (Chapter 26) matter is transformed into energy. The relationship between matter and energy is given by Albert Einstein's now famous equation

$$E = mc^2$$

This equation tells us that the amount of energy released when matter is transformed into energy is the product of the mass of matter transformed and the speed of light squared. At the present time, we have not (knowingly) observed the transformation of energy into matter on a large scale. It does, however, happen on an extremely small scale in "atom smashers," or particle accelerators, used to induce nuclear reactions. Now that the equivalence of matter and energy is recognized, the **Law of Conservation of Matter and Energy** can be stated in a single sentence:

> The combined amount of matter and energy in the universe is fixed.

Einstein formulated this equation in 1905 as a part of his theory of relativity. Its validity was demonstrated in 1939 with the first controlled nuclear reaction.

1-2 STATES OF MATTER

Matter can be classified into three states (Figure 1-2), although most of us can think of examples that do not fit neatly into any of the three categories. In the **solid** state, substances are rigid and have definite shapes. Volumes of solids do not vary much with changes in temperature and pressure. In many solids, called crystalline solids, the individual particles that make up the solid occupy definite positions in the crystal structure. The strengths of interaction between the individual particles determine how hard and how strong the crystals are. In the **liquid** state, the individual particles are confined to a given volume. A liquid flows and assumes the shape of its container up to the volume of the liquid. Liquids are very hard to compress. **Gases** are much less dense than liquids and solids. They occupy all parts of any vessel in which they are confined. Gases are capable of infinite expansion and are compressed easily. We conclude that they consist primarily of empty space; i.e., the individual particles are quite far apart.

(a)

(b)

(c)

Figure 1-2 (a) Iodine, a solid element. (b) Bromine, a liquid element. (c) Chlorine, a gaseous element.

The properties of a person include height, weight, sex, skin and hair color, and the many subtle features that constitute that person's general appearance.

One atmosphere of pressure is the average atmospheric pressure at sea level.

1-3 CHEMICAL AND PHYSICAL PROPERTIES

To distinguish among samples of different kinds of matter, we determine and compare their **properties.** We recognize different kinds of matter by their properties, which are broadly classified into chemical properties and physical properties.

Chemical properties are properties exhibited by matter as it undergoes changes in composition. These properties of substances are related to the kinds of chemical changes that the substances undergo. For instance, we have already described the combination of metallic magnesium with gaseous oxygen to form magnesium oxide, a white powder. A chemical property of magnesium is that it can combine with oxygen, releasing energy in the process. A chemical property of oxygen is that it can combine with magnesium.

All substances also exhibit **physical properties** that can be observed in the *absence of any change in composition.* Color, density, hardness, melting point, boiling point, and electrical and thermal conductivities are physical properties. Some physical properties of a substance depend on the conditions, such as temperature and pressure, under which they are measured. For instance, water is a solid (ice) at low temperatures but is a liquid at higher temperatures. At still higher temperatures, it is a gas (steam). As water is converted from one state to another, its composition is constant. Its chemical properties change very little. On the other hand, the physical properties of ice, liquid water, and steam are different (Figure 1-3a).

Properties of matter can be further classified according to whether or not they depend on the *amount* of substance present. The volume and the mass of a sample depend on, and are directly proportional to, the amount of matter in that sample. Such properties, which depend on the amount of material examined, are called **extensive properties.** By contrast, the color and the melting point of a substance are the same for a small sample and for a large one. Properties such as these, which are independent of the amount of material examined, are called **intensive properties.** All chemical properties are intensive properties.

Because no two substances have identical sets of chemical and physical properties under the same conditions, we are able to identify and distinguish among different substances. For instance, water is the only clear, colorless liquid that freezes at 0°C, boils at 100°C at one atmosphere of pressure, dissolves a wide variety of substances (such as copper(II) sulfate), and reacts violently with sodium (Figure 1-4). Table 1-1 compares several physical properties of a few substances. A sample of any of these substances can be distinguished from the others by observing their properties.

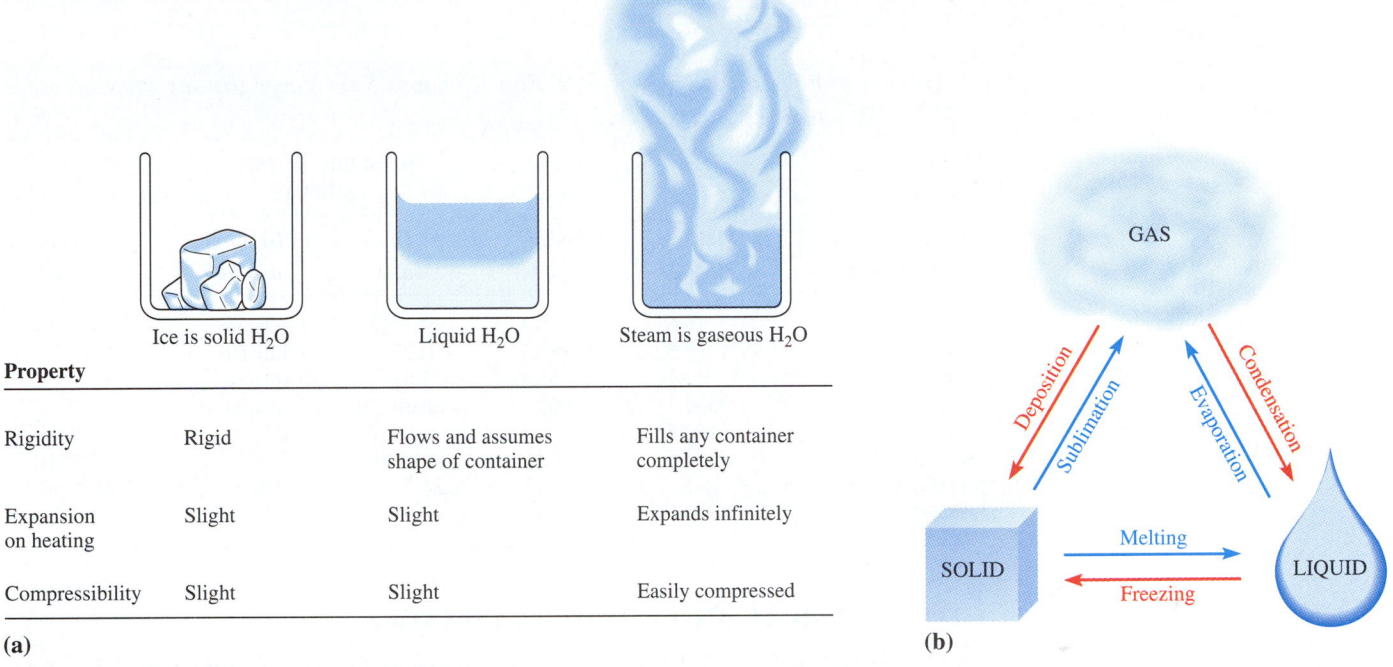

Figure 1-3 (a) A comparison of some physical properties of the three states of matter (for water). Steam, shown in blue in the artist's renditions here and in Figure 1-5, is invisible; the foggy appearance that we usually associate with steam is due to tiny droplets of liquid water. (b) Physical changes that occur among the three states of matter. *Sublimation* is the conversion of a solid directly to a gas without passing through the liquid state; the reverse of that process is called *deposition*. The changes shown in blue are endothermic (absorb heat); those shown in red are exothermic (release heat).

Figure 1-4 Some physical and chemical properties of water. *Physical:* (a) It melts at 0°C; (b) it boils at 100°C (at normal atmospheric pressure); (c) it dissolves a wide range of substances, including copper(II) sulfate, a blue solid. *Chemical:* (d) It reacts with sodium to form hydrogen gas and a solution of sodium hydroxide. The solution contains a little phenolphthalein, which is pink in the presence of sodium hydroxide.

Table 1-1 *Physical Properties of a Few Common Substances (at one atmosphere pressure)*

| Substance | Melting Pt. (°C) | Boiling Pt. (°C) | Solubility at 25°C (g/100 g) | | Density (g/cm³) |
			In water	In ethyl alcohol	
acetic acid	16.6	118.1	infinite	infinite	1.05
benzene	5.5	80.1	0.07	infinite	0.879
bromine	−7.1	58.8	3.51	infinite	3.12
iron	1530	3000	insoluble	insoluble	7.86
methane	−182.5	−161.5	0.0022	0.033	6.67×10^{-4}
oxygen	−218.8	−183.0	0.0040	0.037	1.33×10^{-3}
sodium chloride	801	1473	36.5	0.065	2.16
water	0	100	—	infinite	1.00

1-4 CHEMICAL AND PHYSICAL CHANGES

We described the reaction of magnesium as it burns in the oxygen of the air (Figure 1-1). This reaction is a *chemical change.* In any **chemical change,** (1) one or more substances are used up (at least partially), (2) one or more new substances are formed, and (3) energy is absorbed or released. As substances undergo chemical changes they demonstrate their chemical properties. A **physical change,** on the other hand, occurs with *no change in chemical composition.* Physical properties are usually altered significantly as matter undergoes physical changes (Figure 1-3b). In addition, a physical change *may* suggest that a chemical change has also taken place. For instance, a color change, a warming, or the formation of a solid when two solutions are mixed could indicate a chemical change.

Energy is always released or absorbed when chemical or physical changes occur. Energy is required to melt ice, and energy is required to boil water. Conversely, the condensation of steam to form liquid water always liberates energy, as does the freezing of liquid water to form ice. The changes in energy that accompany these physical changes for water are shown in Figure 1-5. At a pressure of one atmosphere, ice always melts at the same temperature (0°C) and pure water always boils at the same temperature (100°C).

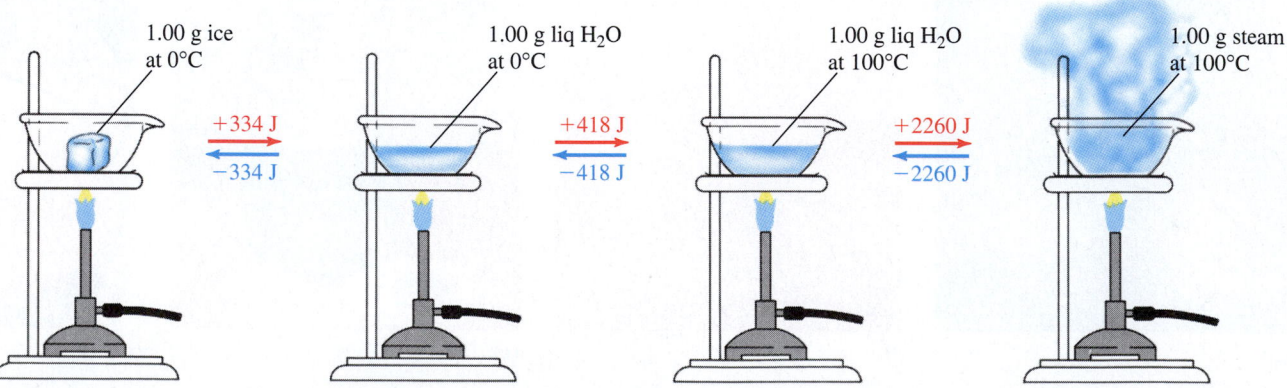

Figure 1-5 Changes in energy that accompany some physical changes for water. The energy unit joules (J) is defined in Section 1-13. The positive signs preceding joules above the arrows tell us that heat is *absorbed.* Negative signs below the arrows tell us that heat is *liberated.*

1-5 MIXTURES, SUBSTANCES, COMPOUNDS, AND ELEMENTS

Mixtures are combinations of two or more pure substances in which each substance retains its own composition and properties. Almost every sample of matter that we ordinarily encounter is a mixture. The most easily recognized type of mixture is one in which different portions of the sample have recognizably different properties. Such a mixture, which is not uniform throughout, is called **heterogeneous.** Examples include mixtures of salt and charcoal (in which two components with different colors can be distinguished readily from one another by sight), foggy air (which includes a suspended mist of water droplets), and vegetable soup. Another kind of mixture has uniform properties throughout; such a mixture is described as a **homogeneous mixture** and is also called a **solution.** Examples include saltwater; some **alloys,** which are homogeneous mixtures of metals in the solid state; and air (free of particulate matter or mists). Air is a mixture of gases. It is mainly nitrogen, oxygen, argon, carbon dioxide, and water vapor. There are only trace amounts of other substances in the atmosphere.

An important characteristic of all mixtures is that they can have variable composition. (For instance, we can make an infinite number of different mixtures of salt and sugar by varying the relative amounts of the two components used.) Consequently, repeating the same experiment on mixtures from different sources may give different results, whereas the same treatment of a pure sample will always give the same results. When the distinction between homogeneous mixtures and pure substances was realized and methods were developed (in the late 1700s) for separating mixtures and studying pure substances, consistent results could be obtained. This resulted in reproducible studies of chemical properties, which formed the basis of real progress in the development of chemical theory.

Mixtures can be separated by physical means because each component retains its properties (see Figures 1-6 and 1-7). For example, a mixture of salt and water can be separated by evaporating the water and leaving the solid salt behind. To separate a mixture of sand and salt, we could treat it with water to dissolve the salt, collect the sand by filtration, and then evaporate the water to reclaim the solid salt. Very fine iron powder can be mixed with powdered sulfur to give what appears to the naked eye to be a homogeneous mixture of the two. However, separation of the components of this mixture is easy. The iron may be removed by a magnet, or the sulfur may be dissolved in carbon disulfide, which does not dissolve iron (Figure 1-6).

By "composition of a mixture" we mean both the identities of the substances present and their relative amounts in the mixture.

The blue copper(II) sulfate solution in Figure 1-4d is a homogeneous mixture.

A heterogeneous mixture of the two minerals galena (black) and quartz (white).

In *any* mixture, (1) the composition can be varied and (2) each component of the mixture retains its own properties.

(a)

(b)

Figure 1-6 (a) A mixture of iron and sulfur is a *heterogeneous* mixture. (b) Like any mixture, it can be separated by physical means, such as removing the iron with a magnet.

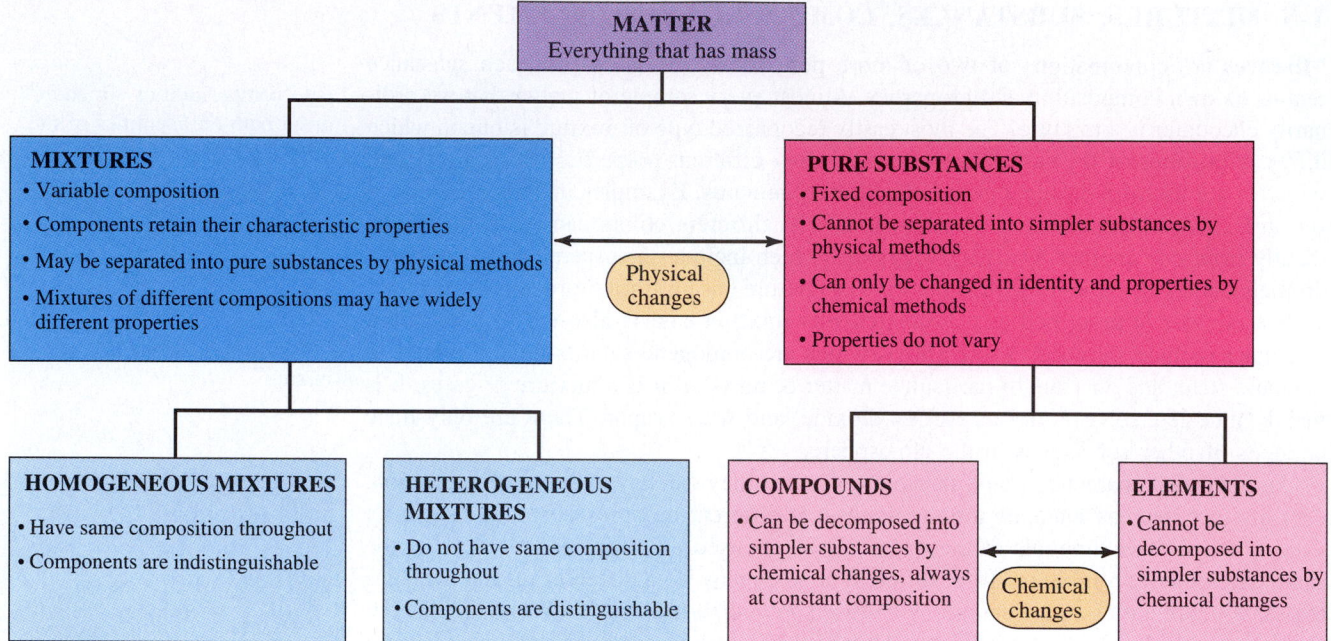

Figure 1-7 One scheme for classification of matter. Arrows indicate the general means by which matter can be separated.

Imagine that we have a sample of muddy river water (a heterogeneous mixture). We might first separate the suspended dirt from the liquid by filtration. Then we could remove dissolved air by warming the water. Dissolved solids might be removed by cooling the sample until some of it freezes, pouring off the liquid, and then melting the ice. Other dissolved components might be separated by distillation or other methods. Eventually we would obtain a sample of pure water that could not be further separated by any physical separation methods. No matter what the original source of the impure water—the ocean, the Mississippi River, a can of tomato juice, and so on—water samples obtained by purification all have identical composition and, under identical conditions, they all have identical properties. Any such sample is called a substance, or sometimes a pure substance.

The first ice that forms is quite pure. The dissolved solids tend to stay behind in the remaining liquid.

A **substance** cannot be further broken down or purified by physical means.

If we use the definition given here of a substance, the phrase pure substance may appear to be redundant.

Now suppose we decompose some water by passing electricity through it (Figure 1-8). (An *electrolysis* process is a chemical reaction.) We find that the water is converted into two simpler substances, hydrogen and oxygen; more significantly, hydrogen and oxygen are *always* present in the same ratio by mass, 11.1% to 88.9%. These observations allow us to identify water as a compound.

A **compound** is a substance that can be decomposed by chemical means into simpler substances, always in the same ratio by mass.

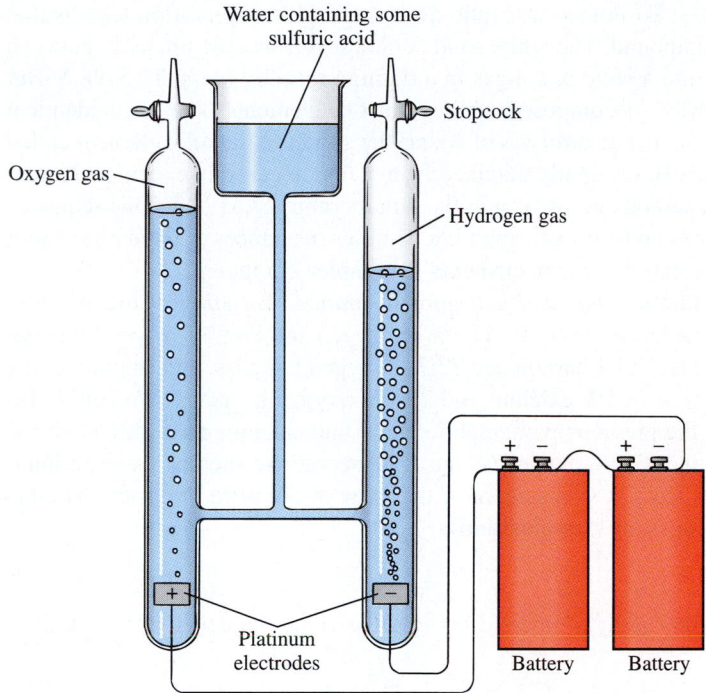

Figure 1-8 Electrolysis apparatus for small-scale decomposition of water by electrical energy. The volume of hydrogen produced (right) is twice that of oxygen (left). Some dilute sulfuric acid is added to increase the conductivity.

As we continue this process, starting with any substance, we eventually reach a stage at which the new substances formed cannot be further broken down by chemical means. The substances at the end of this chain are called elements.

An **element** is a substance that cannot be decomposed into simpler substances by chemical changes.

For instance, neither of the two gases obtained by the electrolysis of water—hydrogen and oxygen—can be further decomposed, so they are elements.

As another illustration (see Figure 1-9), pure calcium carbonate (a white solid present in limestone and seashells) can be broken down by heating to give another white solid

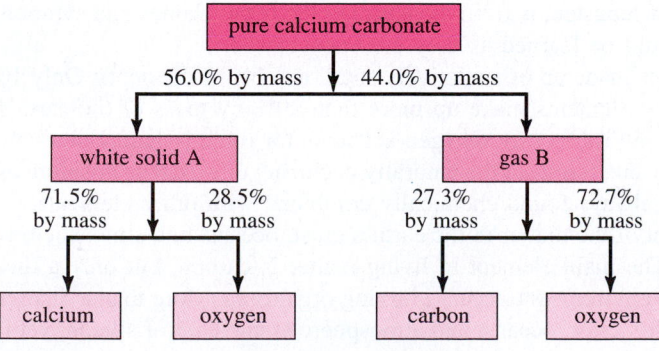

Figure 1-9 Diagram of the decomposition of calcium carbonate to give a white solid A (56.0% by mass) and a gas B (44.0% by mass). This decomposition into simpler substances at fixed ratio proves that calcium carbonate is a compound. The white solid A further decomposes to give the elements calcium (71.5% by mass) and oxygen (28.5% by mass). This proves that the white solid A is a compound; it is known as calcium oxide. The gas B also can be broken down to give the elements carbon (27.3% by mass) and oxygen (72.7% by mass). This establishes that gas B is a compound; it is known as carbon dioxide.

How would you know that the white solid we called A is a compound?

(call it A) and a gas (call it B) in the mass ratio 56.0:44.0. This observation tells us that calcium carbonate is a compound. The white solid A obtained from calcium carbonate can be further broken down into a solid and a gas in a definite ratio by mass, 71.5:28.5. But neither of these can be further decomposed, so they must be elements. The gas is identical to the oxygen obtained from the electrolysis of water; the solid is a metallic element called calcium. Similarly, the gas B, originally obtained from calcium carbonate, can be decomposed into two elements, carbon and oxygen in fixed mass ratio, 27.3:72.7. This sequence illustrates that a compound can be broken apart into simpler substances at fixed mass ratio; those simpler substances may be either elements or simpler compounds.

Further, we may say that *a compound is a pure substance consisting of two or more different elements in a fixed ratio.* Water is 11.1% hydrogen and 88.9% oxygen by mass. Similarly, carbon dioxide is 27.3% carbon and 72.7% oxygen by mass, and calcium oxide (the white solid A above) is 71.5% calcium and 28.5% oxygen by mass. We could also combine the numbers in the previous paragraph to show that calcium carbonate is 40.1% calcium, 12.0% carbon, and 47.9% oxygen by mass. Observations such as these on innumerable pure compounds led to the statement of the **Law of Definite Proportions** (also known as the **Law of Constant Composition**):

Different samples of any pure compound contain the same elements in the same proportions by mass.

Figure 1-10   The reaction of sodium and chlorine to produce table salt, sodium chloride.

The physical and chemical properties of a compound are different from the properties of its constituent elements. Sodium chloride is a white solid that we ordinarily use as table salt. This compound is produced by the combination of the element sodium (a soft, silvery white metal that reacts violently with water; Figure 1-4d) and the element chlorine (a pale green, corrosive, poisonous gas; Figure 1-2c). See Figure 1-10.

Recall that elements are substances that cannot be decomposed into simpler substances by chemical changes. Nitrogen, silver, aluminum, copper, gold, and sulfur are other examples of elements.

We use a set of **symbols** to represent the elements. These symbols can be written more quickly than names, and they occupy less space. The symbols for the first 103 elements consist of either a capital letter *or* a capital letter and a lowercase letter, such as C (carbon) or Ca (calcium). Symbols for elements beyond number 103 consist of three letters. A list of the known elements and their symbols is inside the front cover.

Elements beyond number 103 are presently given names and three-letter symbols based on a numerical system; however, such designations are temporary until the question of the right to name the newly-discovered elements is resolved.

A short list of symbols of common elements is given in Table 1-2. Learning this list will be helpful. Many symbols consist of the first one or two letters of the element's English name. Some are derived from the element's Latin name (indicated in parentheses in Table 1-2) and one, W for tungsten, is from the German *Wolfram.* Names and symbols for additional elements should be learned as they are encountered.

The other known elements have been made artificially in laboratories, as described in Chapter 26.

Most of the earth's crust is made up of a relatively small number of elements. Only 10 of the 88 naturally occurring elements make up more than 99% by mass of the earth's crust, oceans, and atmosphere (Table 1-3). Oxygen accounts for roughly half. Relatively few elements, approximately one fourth of the naturally occurring ones, occur in nature as free elements. The rest are always found chemically combined with other elements.

Only a very small amount of the matter in the earth's crust, oceans, and atmosphere is involved in living matter. The main element in living matter is carbon, but only a tiny fraction of the carbon in the environment occurs in living organisms. More than a quarter of the total mass of the earth's crust, oceans, and atmosphere is made up of silicon, yet it has almost no biological role.

Table 1-2 *Some Common Elements and Their Symbols*

Symbol	Element	Symbol	Element	Symbol	Element
Ag	silver (*argentum*)	F	fluorine	Ni	nickel
Al	aluminum	Fe	iron (*ferrum*)	O	oxygen
Au	gold (*aurum*)	H	hydrogen	P	phosphorus
B	boron	He	helium	Pb	lead (*plumbum*)
Ba	barium	Hg	mercury	Pt	platinum
Bi	bismuth	I	iodine	S	sulfur
Br	bromine	K	potassium (*kalium*)	Sb	antimony (*stibium*)
C	carbon	Kr	krypton	Si	silicon
Ca	calcium	Li	lithium	Sn	tin (*stannum*)
Cd	cadmium	Mg	magnesium	Sr	strontium
Cl	chlorine	Mn	manganese	Ti	titanium
Co	cobalt	N	nitrogen	U	uranium
Cr	chromium	Na	sodium (*natrium*)	W	tungsten (*Wolfram*)
Cu	copper (*cuprum*)	Ne	neon	Zn	zinc

Mercury is the only metal that is a liquid at room temperature.

Table 1-3 *Abundance of Elements in the Earth's Crust, Oceans, and Atmosphere*

Element	Symbol	% by Mass		Element	Symbol	% by Mass	
oxygen	O	49.5%		chlorine	Cl	0.19%	
silicon	Si	25.7		phosphorus	P	0.12	
aluminum	Al	7.5		manganese	Mn	0.09	
iron	Fe	4.7		carbon	C	0.08	
calcium	Ca	3.4	99.2%	sulfur	S	0.06	0.7%
sodium	Na	2.6		barium	Ba	0.04	
potassium	K	2.4		chromium	Cr	0.033	
magnesium	Mg	1.9		nitrogen	N	0.030	
hydrogen	H	0.87		fluorine	F	0.027	
titanium	Ti	0.58		zirconium	Zr	0.023	
				All others combined		~0.1%	

The stable form of sulfur at room temperature is a solid.

1-6 MEASUREMENTS IN CHEMISTRY

In the next section, we shall introduce the standards for basic units of measurement. These standards were selected because they allow us to make precise measurements and because they are reproducible and unchanging. The values of fundamental units are arbitrary.[1] In the United States, all units of measure are set by the National Institute of Standards and Technology, NIST (formerly the National Bureau of Standards, NBS). Measurements in the scientific world are usually expressed in the units of the metric system or its modernized successor, the International System of Units (SI). The SI, adopted by the National Bureau of Standards in 1964, is based on the seven fundamental units listed in Table 1-4. All other units of measurement are derived from them.

The abbreviation SI comes from the French *le Système International.*

---

[1] *Prior to the establishment of the U.S. Bureau of Standards in 1901, at least 50 different distances had been used as "1 foot" in measuring land within New York City. Thus the size of a 100-ft by 200-ft lot in New York City depended on the generosity of the seller and did not necessarily represent the expected dimensions.*

Table 1-4 *The Seven Fundamental Units of Measurement (SI)*

Physical Property	Name of Unit	Symbol
length	meter	m
mass	kilogram	kg
time	second	s
electric current	ampere	A
temperature	kelvin	K
luminous intensity	candela	cd
amount of substance	mole	mol

In this text we shall use both metric units and SI units. Conversions between non-SI and SI units are usually straightforward. Appendix C lists some important units of measurement and their relationships to each other. Appendix D lists several useful physical constants. The most frequently used of these appear inside the back cover.

The metric and SI systems are *decimal systems, in which prefixes are used to indicate fractions and multiples of ten*. The same prefixes are used with all units of measurement. The distances and masses in Table 1-5 illustrate the use of some common prefixes and the relationships among them.

1-7 UNITS OF MEASUREMENT

Mass and Weight

We distinguish between mass and weight. **Mass** is the measure of the quantity of matter a body contains (Section 1-1). The mass of a body does not vary as its position changes. On the other hand, the **weight** of a body is a measure of the gravitational attraction of the earth for the body, and this varies with distance from the center of the earth. An object weighs ever so slightly less high up on a mountain than at the bottom of a deep valley. Because the mass of a body does not vary with its position, the mass of a body is a more fundamental property than its weight. However, we have become accustomed to using the term "weight" when we mean mass, because weighing is one way of measuring mass (Figure

The prefixes used in the SI and metric systems may be thought of as *multipliers;* e.g., the prefix *kilo-* indicates multiplication by 1000 or 10^3, and *milli-* indicates multiplication by 0.001 or 10^{-3}.

Table 1-5 *Common Prefixes Used in the SI and Metric Systems*

Prefix	Abbreviation	Meaning	Example
mega-	M	10^6	1 megameter (Mm) = 1×10^6 m
kilo-	k	10^3	1 kilometer (km) = 1×10^3 m
deci-	d	10^{-1}	1 decimeter (dm) = 0.1 m
centi-	c	10^{-2}	1 centimeter (cm) = 0.01 m
milli-	m	10^{-3}	1 milligram (mg) = 0.001 g
micro-	μ*	10^{-6}	1 microgram (μg) = 1×10^{-6} g
nano-	n	10^{-9}	1 nanogram (ng) = 1×10^{-9} g
pico-	p	10^{-12}	1 picogram (pg) = 1×10^{-12} g

This is the Greek letter μ (pronounced "mew").

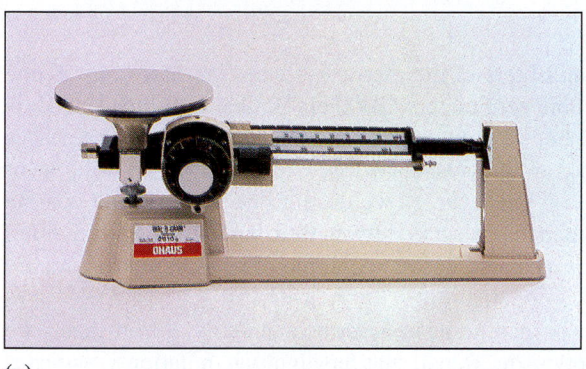

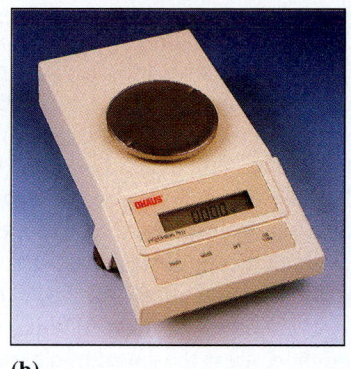

(a)   **(b)**

Figure 1-11 Three types of laboratory balances. (a) A triple-beam balance used for determining mass to about ±0.01 g. (b) A modern electronic top-loading balance that gives a direct readout of mass to ±0.001 g. (c) A modern analytical balance that can be used to determine mass to ±0.0001 g. Analytical balances are used when masses must be determined as precisely as possible.

(c)

1-11). Because we usually discuss chemical reactions at constant gravity, weight relationships are just as valid as mass relationships. We should keep in mind that the two are not identical.

The basic unit of mass in the SI system is the **kilogram** (Table 1-6). The kilogram is defined as the mass of a platinum–iridium cylinder stored in a vault in Sevres, near Paris, France. A one-pound object has a mass of 0.4536 kilogram. The basic mass unit in the earlier *metric system* was the gram. A U.S. five-cent coin (a "nickel") has a mass of about five grams.

Length

The **meter** is the standard unit of length (distance) in both SI and metric systems. The meter is *defined* as the distance light travels in a vacuum in 1/299,792,468 second. It is approximately 39.37 inches. In situations where the English system would use inches, the metric centimeter (1/100 meter) is convenient. The relationship between inches and centimeters is shown in Figure 1-12.

The meter was originally defined (1791) as one ten-millionth of the distance between the North Pole and the Equator.

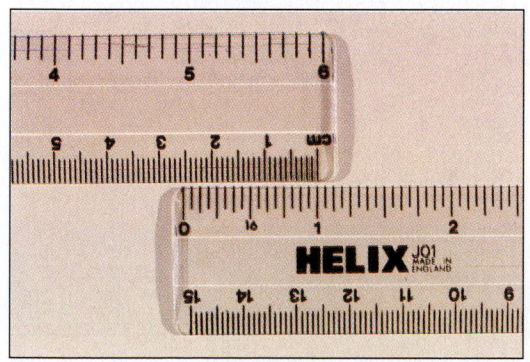

Figure 1-12 The relationship between inches and centimeters (1 in = 2.54 cm).

Table 1-6 *Some SI Units of Mass*	
*kilo*gram, kg	base unit
gram, g	1000 g = 1 kg
*milli*gram, mg	1000 mg = 1 g
*micro*gram, μg	1,000,000 μg = 1 g

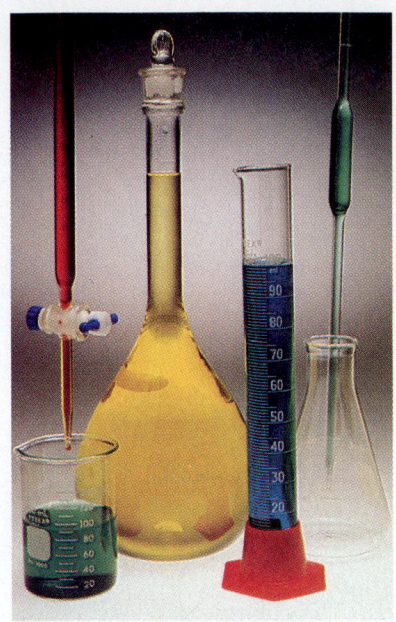

Figure 1-13 Some laboratory apparatus used to measure volumes of liquids: 150-mL beaker (bottom left, green liquid); 25-mL buret (top left, red); 1000-mL volumetric flask (center, yellow); 100-mL graduated cylinder (right front, blue); and 10-mL volumetric pipet (right rear, green).

In exponential form these numbers are

6.02×10^{23} gold atoms

3.27×10^{-22} gram

Volume

Volumes are measured in liters or milliliters in the metric system. One liter (1 L) is one cubic decimeter (1 dm^3), or 1000 cubic centimeters (1000 cm^3). One milliliter (1 mL) is 1 cm^3. In the SI the cubic meter is the basic volume unit, and the cubic decimeter replaces the metric unit, liter. Different kinds of glassware are used to measure the volume of liquids. The one we choose depends on the accuracy we desire. For example, the volume of a liquid dispensed can be measured more accurately with a buret than with a small graduated cylinder (Figure 1-13). Equivalences between common English units and metric units are summarized in Table 1-7.

Sometimes we must combine two or more units to describe a quantity. For instance, we might express the speed of a car as 60 mi/hr. Recall that the algebraic notation x^{-1} means $1/x$; applying this notation to units, we see that hr^{-1} means 1/hr, or "per hour." So the unit of speed could also be expressed as mi · hr^{-1}.

1-8 USE OF NUMBERS

In chemistry, we measure and calculate many things, so we must be sure we understand how to use numbers. In this section we discuss two aspects of the use of numbers: (1) the notation of very large and very small numbers and (2) an indication of how well we actually know the numbers we are using. You will carry out many calculations with calculators. Please refer to Appendix A for some instructions about the use of electronic calculators.

Scientific Notation

We use **scientific notation** when we deal with very large and very small numbers. For example, 197 grams of gold contains approximately

$$602,000,000,000,000,000,000,000 \text{ gold atoms}$$

The mass of one gold atom is approximately

$$0.000\ 000\ 000\ 000\ 000\ 000\ 000\ 327 \text{ gram}$$

Table 1-7 *Conversion Factors Relating Length, Volume, and Mass (weight) Units*

	Metric		English		Metric–English Equivalents	
Length	1 km	$= 10^3$ m	1 ft	$= 12$ in	2.54 cm	$= 1$ in
	1 cm	$= 10^{-2}$ m	1 yd	$= 3$ ft	39.37 in*	$= 1$ m
	1 mm	$= 10^{-3}$ m	1 mile	$= 5280$ ft	1.609 km*	$= 1$ mile
	1 nm	$= 10^{-9}$ m				
	1 Å	$= 10^{-10}$ m				
Volume	1 mL	$= 1$ cm$^3 = 10^{-3}$ L	1 gal	$= 4$ qt $= 8$ pt	1 L	$= 1.057$ qt*
	1 m^3	$= 10^6$ cm$^3 = 10^3$ L	1 qt	$= 57.75$ in^3*	28.32 L	$= 1$ ft^3
Mass	1 kg	$= 10^3$ g	1 lb	$= 16$ oz	453.6 g*	$= 1$ lb
	1 mg	$= 10^{-3}$ g			1 g	$= 0.03527$ oz*
	1 metric tonne	$= 10^3$ kg	1 short ton	$= 2000$ lb	1 metric tonne	$= 1.102$ short ton*

These conversion factors, unlike the others listed, are inexact. They are quoted to four significant figures, which is ordinarily more than sufficient.

In using such large and small numbers, it is inconvenient to write down all the zeroes. In scientific (exponential) notation, we place one nonzero digit to the left of the decimal.

$$4{,}300{,}000. = 4.3 \times 10^6$$

6 places to the left, ∴ exponent of 10 is 6

$$0.000348 = 3.48 \times 10^{-4}$$

4 places to the right, ∴ exponent of 10 is −4

The reverse process converts numbers from exponential to decimal form. See Appendix A for more detail, if necessary.

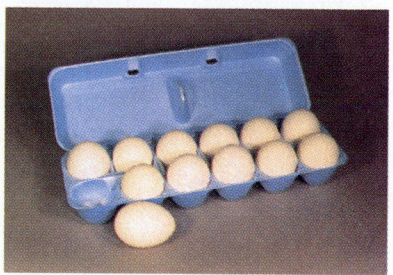
(a)

▼ PROBLEM-SOLVING TIP *Know How to Use Your Calculator*

Students sometimes make mistakes when they try to enter numbers into their calculators in scientific notation. Suppose you want to enter the number 4.36×10^{-2}. On most calculators, you would

(1) press 4.36

(2) press EE or EXP, which stands for "times ten to the"

(3) press 2 (the magnitude of the exponent) and then ± or CHS (to change its sign)

The calculator display might show the value as $\boxed{4.36 \quad -02}$ or as $\boxed{0.0436}$. Different calculators show different numbers of digits, which can sometimes be adjusted.

Other common errors include changing the sign of the exponent when the intent was to change the sign of the entire number (e.g., -3.48×10^4 entered as 3.48×10^{-4}). When in doubt, carry out a trial calculation for which you already know the answer. For instance, multiply 300 by 2 by entering the first value as 3.00×10^2 and then multiplying by 2; you know the answer should be 600, and if you get any other answer, you know you have done something wrong. If you cannot find (or understand) the printed instructions for *your calculator,* your instructor or a classmate might be able to help.

(b)

Figure 1-14 (a) A dozen eggs is exactly 12 eggs. (b) A swarm of honeybees contains an *exact* number of live bees (but it would be difficult to count them).

Significant Figures

There are two kinds of numbers. **Exact numbers** may be *counted* or *defined.* They are known to be absolutely accurate. For example, the exact number of people in a closed room can be counted, and there is no doubt about the number of people. A dozen eggs is defined as exactly 12 eggs, no more, no fewer (Figure 1-14).

Numbers obtained from measurements are not exact. Every measurement involves an estimate. For example, suppose you are asked to measure the length of this page to the nearest 0.1 mm. How do you do it? The smallest divisions (calibration lines) on a meter stick are 1 mm apart (Figure 1-12). An attempt to measure to 0.1 mm requires estimation. If three different people measure the length of the page to 0.1 mm, will they get the same answer? Probably not. We deal with this problem by using significant figures.

Significant figures are digits believed to be correct by the person who makes a measurement. We assume that the person is competent to use the measuring device. Suppose one measures a distance with a meter stick and reports the distance as 343.5 mm. What does this number mean? In this person's judgment, the distance is greater than 343.4 mm but less than 343.6 mm, and the best estimate is 343.5 mm. The number 343.5 mm contains four significant figures. The last digit, 5, is a *best estimate* and is therefore doubtful, but it is considered to be a significant figure. In reporting numbers obtained from measure-

An *exact* number may be thought of as containing an *infinite* number of significant figures.

There is some uncertainty in all measurements.

Significant figures indicate the *uncertainty* in measurements.

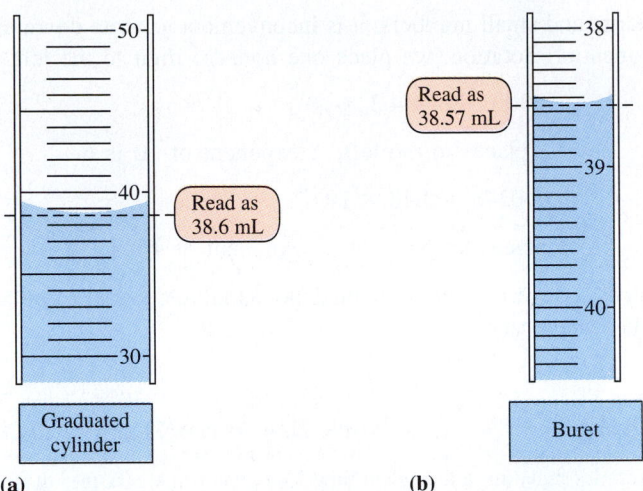

Figure 1-15   Measurement of the volume of water using two different pieces of volumetric glassware. For consistency, we always read the bottom of the meniscus (the curved surface of the water). (a) The level in a 50-mL graduated cylinder can be estimated to within 0.2 mL. The level here is 38.6 mL (three significant figures). (b) The level in a 50-mL buret can be read to within 0.02 mL. The level here is 38.57 mL (four significant figures).

ments, *we report one estimated digit, and no more.* Because the person making the measurement is not certain that the 5 is correct, it would be meaningless to report the distance as 343.53 mm.

To see more clearly the part significant figures play in reporting the results of measurements, consider Figure 1-15a. Graduated cylinders are used to measure volumes of liquids when a high degree of accuracy is not necessary. The calibration lines on a 50-mL graduated cylinder represent 1-mL increments. Estimation of the volume of liquid in a 50-mL cylinder to within 0.2 mL ($\frac{1}{5}$ of one calibration increment) with reasonable certainty is possible. We might measure a volume of liquid in such a cylinder and report the volume as 38.6 mL, i.e., to three significant figures.

Burets are used to measure volumes of liquids when higher accuracy is required. The calibration lines on a 50-mL buret represent 0.1-mL increments, allowing us to make estimates to within 0.02 mL ($\frac{1}{5}$ of one calibration increment) with reasonable certainty (Figure 1-15b). Experienced individuals estimate volumes in 50-mL burets to 0.01 mL with considerable reproducibility. For example, using a 50-mL buret, we can measure out 38.57 mL (four significant figures) of liquid with reasonable accuracy.

Accuracy refers to how closely a measured value agrees with the correct value. **Precision** refers to how closely individual measurements agree with each other. Ideally, all measurements should be both accurate and precise. Measurements may be quite precise, yet quite inaccurate, because of some *systematic error,* which is an error repeated in each measurement. (A faulty balance, for example, might produce a systematic error.) Very accurate measurements are seldom imprecise.

Measurements are frequently repeated to improve accuracy and precision. Average values obtained from several measurements are usually more reliable than individual measurements. Significant figures indicate how accurately measurements have been made (assuming the person who made the measurements was competent).

Some simple rules govern the use of significant figures in calculations.

> **1. Zeros used just to position the decimal point are not significant figures.**

The underlined zeros in the number 0.0000593 serve only to locate the decimal point, so they are *not* significant; this number has only three significant figures. The number could also be reported as 5.93×10^{-5} in scientific notation (Appendix A). When zeroes precede the decimal point, but come after other digits, we may have some difficulty in deciding whether the zeroes are significant figures or not. How many significant figures does the number 23,000 contain? We are given insufficient information to answer the question. If all three of the zeroes are used just to place the decimal point, the number should appear as 2.3×10^{4} (two significant figures). If only two of the zeroes are being used to place the decimal point, the number is 2.30×10^{4} (three significant figures). In the unlikely event that the number is actually known to be $23,000 \pm 1$, it should be written as 2.3000×10^{4} (five significant figures).

When we wish to specify that all of the zeroes in such a number *are* significant, we may indicate this by placing a decimal point after the number. For instance, 130. grams can represent a mass known to *three* significant figures, that is, 130 ± 1 gram.

> **2. In multiplication and division, an answer contains no more significant figures than the least number of significant figures used in the operation.**

EXAMPLE 1-1 *Significant Figures (Multiplication)*

What is the area of a rectangle 1.23 cm wide and 12.34 cm long?

Plan

The area of a rectangle is its length times its width. We should first check to see that the width and length are expressed in the same units. (They are, but if they were not, one would first have to be converted to the units of the other.) Then we multiply the width by the length. We then follow Rule 2 for significant figures to find the correct number of significant figures. The units for the result are equal to the product of the units for the individual terms in the multiplication.

Solution

$$A = \ell \times w = (12.34 \text{ cm})(1.23 \text{ cm}) = \boxed{15.2 \text{ cm}^2}$$

$$(\text{calculator result} = 15.1782)$$

Because three is the smallest number of significant figures used, the answer should contain only three significant figures. The number generated by an electronic calculator (15.1782) implies more accuracy than is justified; the result cannot be more accurate than the information that led to it. Calculators have no judgment, so you must exercise yours.

You should now work Exercise 21.

With many examples we suggest selected exercises from the end of the chapter. These exercises use the skills or concepts from that example. Now you should work Exercise 21 from the end of this chapter.

The step-by-step calculation in the margin demonstrates why the area is reported as 15.2 cm^2 rather than 15.1782 cm^2. The length, 12.34 cm, contains four significant figures, whereas the width, 1.23 cm, contains only three. If we underline each uncertain figure, as well as each figure obtained from an uncertain figure, the step-by-step multiplication gives the result reported in Example 1-1. We see that there are only two certain figures (15) in the result. We report the first doubtful figure (.2), but no more. Division is just the reverse of multiplication, and the same rules apply.

$$
\begin{array}{r}
12.3\underline{4} \text{ cm} \\
\times \quad 1.2\underline{3} \text{ cm} \\
\hline
3\,7\,0\,\underline{2} \\
2\,4\,6\,\underline{8} \\
12\,3\,4 \\
\hline
15.1\underline{7}\,\underline{8}\,\underline{2} \text{ cm}^2 = 15.2 \text{ cm}^2
\end{array}
$$

3. In addition and subtraction, the last digit retained in the sum or difference is determined by the position of the first doubtful digit.

EXAMPLE 1-2   *Significant Figures (Addition and Subtraction)*

(a) Add 37.24 mL and 10.3 mL. (b) Subtract 21.2342 g from 27.87 g.

Plan

Again we first check to see that the quantities to be added or subtracted are expressed in the same units. We carry out the addition or subtraction. Then we follow Rule 3 for significant figures to express the answer to the correct number of significant figures.

Solution

Doubtful digits are underlined in this example.

(a)

$$37.2\underline{4} \text{ mL}$$
$$+10.\underline{3} \text{ mL}$$

$$47.5\underline{4} \text{ mL is reported as } \boxed{47.5 \text{ mL}} \text{ (calculator gives 47.54)}$$

(b)

$$27.8\underline{7} \text{ g}$$
$$-21.234\underline{2} \text{ g}$$

$$6.6\underline{358} \text{ g is reported as } \boxed{6.64 \text{ g}} \text{ (calculator gives 6.6358)}$$

You should now work Exercise 24.

In the three simple arithmetic operations we have performed, the number combination generated by an electronic calculator is not the "answer" in a single case! However, the correct result of each calculation can be obtained by "rounding off." The rules of significant figures tell us where to round off.

In rounding off, certain conventions have been adopted. When the number to be dropped is less than 5, the preceding number is left unchanged (e.g., 7.34 rounds off to 7.3). When it is more than 5, the preceding number is increased by 1 (e.g., 7.37 rounds off to 7.4). When the number to be dropped is 5, the preceding number is set to the nearest *even* number (e.g., 7.45 rounds off to 7.4, and 7.35 rounds off to 7.4).

Rounding off to an even number is intended to reduce the accumulation of errors in chains of calculations.

1-9   THE UNIT FACTOR METHOD (DIMENSIONAL ANALYSIS)

Many chemical and physical processes can be described by numerical relationships. In fact, many of the most useful ideas in science must be treated mathematically. Let us devote a little time to reviewing problem-solving skills.

First, multiplication by unity (by one) does not change the value of an expression. If we represent "one" in a useful way, we can do many conversions by just "multiplying by one." This method of performing calculations is known as **dimensional analysis,** the **factor-label method,** or the **unit factor method.** Regardless of the name chosen, it is a very powerful mathematical tool that is almost foolproof.

Unit factors may be constructed from any two terms that describe the same or equivalent "amounts" of whatever we may consider. For example, 1 foot is equal to exactly

12 inches, by definition. We may write an equation to describe this equality:

$$1 \text{ ft} = 12 \text{ in}$$

Dividing both sides of the equation by 1 ft gives

$$\frac{1 \text{ ft}}{1 \text{ ft}} = \frac{12 \text{ in}}{1 \text{ ft}} \qquad \text{or} \qquad 1 = \frac{12 \text{ in}}{1 \text{ ft}}$$

The factor (fraction) 12 in/1 ft is a unit factor because the numerator and denominator describe the same distance. Dividing both sides of the original equation by 12 in gives 1 = 1 ft/12 in, a second unit factor that is the reciprocal of the first. *The reciprocal of any unit factor is also a unit factor.* Stated differently, division of an amount by the same amount always yields one!

In the English system we can write many unit factors, such as

$$\frac{1 \text{ yd}}{3 \text{ ft}}, \quad \frac{1 \text{ yd}}{36 \text{ in}}, \quad \frac{1 \text{ mile}}{5280 \text{ ft}}, \quad \frac{4 \text{ qt}}{1 \text{ gal}}, \quad \frac{2000 \text{ lb}}{1 \text{ ton}}$$

The reciprocal of each of these is also a unit factor. Items in retail stores are frequently priced with unit factors, such as 39¢/lb and $3.98/gal. When all the quantities in a unit factor come from definitions, the unit is known to an unlimited (infinite) number of significant figures. For instance, if you bought eight 1-gallon jugs of something priced at $3.98/gal, the total cost would be 8 × $3.98, or $31.84; the merchant would not round this to $31.80, let alone to $30.

In science, nearly all numbers have units. What does 12 mean? Usually we must supply appropriate units, such as 12 eggs or 12 people. In the unit factor method, the units guide us through calculations in a step-by-step process, because all units except those in the desired result cancel.

Unless otherwise indicated, a "ton" refers to a "short ton," 2000 lb. There are also the "long ton," which is 2240 lb, and the metric tonne, which is 1000 kg.

EXAMPLE 1-3 *Unit Factors*

Express 1.47 miles in inches.

Plan

First we write down the units of what we wish to know preceded by a question mark. Then we set it equal to whatever we are given:

$$\underline{?} \text{ in} = 1.47 \text{ miles}$$

Then we choose unit factors to convert the given units (miles) to the desired units (inches):

$$\boxed{\text{miles}} \longrightarrow \boxed{\text{feet}} \longrightarrow \boxed{\text{inches}}$$

We relate (a) miles to feet and then (b) feet to inches.

Solution

$$\underline{?} \text{ in} = 1.47 \text{ miles} \times \frac{5280 \text{ ft}}{1 \text{ mile}} \times \frac{12 \text{ in}}{1 \text{ ft}} = \boxed{9.31 \times 10^4 \text{ in}} \qquad \text{(calculator gives 93139.2)}$$

Note that both miles and feet cancel, leaving only inches, the desired unit. Thus, there is no ambiguity as to how the unit factors should be written. The answer contains three significant figures because there are three significant figures in 1.47 miles.

You should now work Exercise 26.

In the interest of clarity, cancellation of units will be omitted in the remainder of this book. You may find it useful to continue the cancellation of units.

> ▼ **PROBLEM-SOLVING TIP** *Significant Figures*
>
> *"How do defined quantities affect significant figures?"* Any quantity that comes from a *definition* is exact, that is, it is known to an unlimited number of significant figures. In Example 1-3, the quantities 5280 ft, 1 mile, 12 in, and 1 ft all come from definitions, so they do not limit the significant figures in the answer.

> ▼ **PROBLEM-SOLVING TIP** *Think About Your Answer!*
>
> It is often helpful to ask yourself, "Does the answer make sense?" In Example 1-3, the distance involved is more than a mile. We expect this distance to be many inches, so a large answer is not surprising. Suppose we had mistakenly multiplied by the unit factor $\dfrac{1 \text{ mile}}{5280 \text{ feet}}$ (and not noticed that the units did not cancel properly); we would have gotten the answer 3.34×10^{-3} inches (0.00334 inches), which we should have immediately recognized as nonsense!

Within the SI and metric systems, many measurements are related to each other by powers of ten.

EXAMPLE 1-4 *Unit Conversions*

The Ångstrom (Å) is a unit of length, 1×10^{-10} meter, that provides a convenient scale on which to express the radii of atoms. Radii of atoms are often expressed in nanometers. The radius of a phosphorus atom is 1.10 Å. What is the distance expressed in centimeters and nanometers?

Plan

We use the equalities 1 Å = 1×10^{-10} m, 1 cm = 1×10^{-2} m, and 1 nm = 1×10^{-9} m to construct the unit factors that convert 1.10 Å to the desired units.

$$\text{Å} \rightarrow \text{m} \rightarrow \text{cm}$$
$$\text{Å} \rightarrow \text{m} \rightarrow \text{nm}$$

Solution

$$\underline{?}\ \text{cm} = 1.10\ \text{Å} \times \frac{1 \times 10^{-10}\ \text{m}}{1\ \text{Å}} \times \frac{1\ \text{cm}}{1 \times 10^{-2}\ \text{m}} = \boxed{1.10 \times 10^{-8}\ \text{cm}}$$

$$\underline{?}\ \text{nm} = 1.10\ \text{Å} \times \frac{1.0 \times 10^{-10}\ \text{m}}{1\ \text{Å}} \times \frac{1\ \text{nm}}{1 \times 10^{-9}\ \text{m}} = \boxed{0.110\ \text{nm}}$$

All the unit factors used in this example contain only exact numbers.

You should now work Exercise 30.

EXAMPLE 1-5 *Volume Calculation*

Assuming a phosphorus atom is spherical, calculate its volume in Å^3, cm^3, and nm^3. The volume of a sphere is $V = (\frac{4}{3})\pi r^3$. Refer to Example 1-4.

Plan

We use the results of Example 1-4 to calculate the volume in each of the desired units.

$$1\ \text{Å} = 10^{-10}\ \text{m} = 10^{-8}\ \text{cm}$$

Solution

$$\underline{?}\,Å^3 = (\tfrac{4}{3})\pi(1.10\,Å)^3 = \boxed{5.58\,Å^3}$$

$$\underline{?}\,cm^3 = (\tfrac{4}{3})\pi(1.10 \times 10^{-8}\,cm)^3 = \boxed{5.58 \times 10^{-24}\,cm^3}$$

$$\underline{?}\,nm^3 = (\tfrac{4}{3})\pi(1.10 \times 10^{-1}\,nm)^3 = \boxed{5.58 \times 10^{-3}\,nm^3}$$

You should now work Exercise 38.

EXAMPLE 1-6   *Mass Conversion*

A sample of gold has a mass of 0.234 mg. What is its mass in g? in kg?

Plan

We use the relationships 1 g = 1000 mg and 1 kg = 1000 g to write the required unit factors.

Solution

$$\underline{?}\,g = 0.234\,mg \times \frac{1\,g}{1000\,mg} = \boxed{2.34 \times 10^{-4}\,g}$$

$$\underline{?}\,kg = 2.34 \times 10^{-4}\,g \times \frac{1\,kg}{1000\,g} = \boxed{2.34 \times 10^{-7}\,kg}$$

Again, this example includes unit factors that contain only exact numbers.

You should now work Exercise 39.

Unity raised to *any* power is one. *Any* unit factor raised to a power is still a unit factor, as the next example shows.

EXAMPLE 1-7   *Volume Conversion*

One liter is exactly 1000 cubic centimeters. How many cubic inches are there in 1000 cubic centimeters?

Plan

We would multiply by the unit factor $\dfrac{1\,in}{2.54\,cm}$ to convert cm to in. Here we require the *cube* of this unit factor.

Solution

$$\underline{?}\,in^3 = 1000\,cm^3 \times \left(\frac{1\,in}{2.54\,cm}\right)^3 = 1000\,cm^3 \times \frac{1\,in^3}{16.4\,cm^3} = \boxed{61.0\,in^3}$$

$$\left(\frac{1\,in}{2.54\,cm}\right)^3 = \frac{1\,in^3}{16.4\,cm^3} = 1$$

You should now work Exercise 29.

Example 1-7 shows that a unit factor *cubed* is still a unit factor.

EXAMPLE 1-8 *Energy Conversion*

A common unit of energy is the erg. Convert 3.74×10^{-2} erg to the SI units of energy, joules and kilojoules. One erg is exactly 1×10^{-7} joule.

Plan

The definition that relates ergs and joules is used to generate the needed unit factor. The second conversion uses a unit factor that is based on the definition of the prefix *kilo-*.

Solution

$$\underline{?}\ J = 3.74 \times 10^{-2}\ \text{erg} \times \frac{1 \times 10^{-7}\ J}{1\ \text{erg}} = \boxed{3.74 \times 10^{-9}\ J}$$

$$\underline{?}\ kJ = 3.74 \times 10^{-9}\ J \times \frac{1\ kJ}{1000\ J} = \boxed{3.74 \times 10^{-12}\ kJ}$$

Conversions between the English and SI (metric) systems are conveniently made by the unit factor method. Several conversion factors are listed in Table 1-7. It may be helpful to remember one each for

length	1 in = 2.54 cm (exact)
mass and weight	1 lb = 454 g (near sea level)
volume	1 qt = 0.946 L or 1 L = 1.06 qt

EXAMPLE 1-9 *English–Metric Conversion*

Express 1.0 gallon in milliliters.

Plan

We ask $\underline{?}$ mL = 1.0 gal and multiply by the appropriate factors.

$$\boxed{\text{gallons}} \longrightarrow \boxed{\text{quarts}} \longrightarrow \boxed{\text{liters}} \longrightarrow \boxed{\text{milliliters}}$$

We relate
(a) gallons to quarts, then
(b) quarts to liters, and then
(c) liters to milliliters.

Solution

$$\underline{?}\ mL = 1.0\ \text{gal} \times \frac{4\ \text{qt}}{1\ \text{gal}} \times \frac{1\ L}{1.06\ \text{qt}} \times \frac{1000\ mL}{1\ L} = \boxed{3.8 \times 10^3\ mL}$$

You should now work Exercises 27 and 36.

The fact that all other units cancel to give the desired unit, mL, shows that we used the correct unit factors. The factors 4 qt/gal and 1000 mL/L contain only exact numbers. The factor 1 L/1.06 qt contains three significant figures, while 1.0 gal contains only two. The answer contains only two significant figures.

Examples 1-1 through 1-9 show that multiplication by one or more unit factors changes the units and the number of units, but not the amount of whatever we are calculating.

1-10 PERCENTAGE

We often use percentages to describe quantitatively how a total is made up of its parts. In Table 1-3, we described the amounts of elements present in terms of the percentage of each element.

Percentages can be treated as unit factors. For any mixture containing substance A,

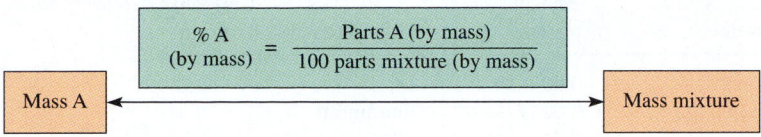

$$\frac{\% \text{ A}}{\text{(by mass)}} = \frac{\text{Parts A (by mass)}}{100 \text{ parts mixture (by mass)}}$$

Mass A ⟷ Mass mixture

If we say that a sample is 24.4% carbon by mass, we mean that out of every 100 parts (exactly) by mass of sample, 24.4 parts by mass are carbon. This relationship can be represented by whichever of the two unit factors we find useful:

$$\frac{24.4 \text{ parts carbon}}{100 \text{ parts sample}} \quad \text{or} \quad \frac{100 \text{ parts sample}}{24.4 \text{ parts carbon}}$$

This ratio can be expressed in terms of grams of carbon for every 100 grams of sample, pounds of carbon for every 100 pounds of sample, or any other mass or weight unit. The next example illustrates the use of dimensional analysis involving percentage.

EXAMPLE 1-10 *Percentage*

U.S. pennies made since 1982 consist of 97.6% zinc and 2.4% copper. The mass of a particular penny is measured to be 1.494 grams. How many grams of zinc does this penny contain?

Plan

From the percentage information given, we may write the required unit factor

$$\frac{97.6 \text{ g zinc}}{100 \text{ g sample}}$$

Solution

$$\underline{?} \text{ g Zn} = 1.494 \text{ grams sample} \times \frac{97.6 \text{ g zinc}}{100 \text{ g sample}} = \boxed{1.46 \text{ g zinc}}$$

The number of significant figures in the result is limited by the three significant figures in 97.6%. Because the definition of percentage involves *exactly* 100 parts, the number 100 is known to an infinite number of significant figures.

You should now work Exercises 31 and 32.

1-11 DENSITY AND SPECIFIC GRAVITY

In science we use many terms that involve combinations of different units. Such quantities may be thought of as unit factors that can be used to convert among these units. The **density** of a sample of matter is defined as the mass per unit volume:

$$\text{density} = \frac{\text{mass}}{\text{volume}} \quad \text{or} \quad D = \frac{m}{V}$$

Densities may be used to distinguish between two substances or to assist in identifying a particular substance. They are usually expressed as g/cm^3 or g/mL for liquids and solids

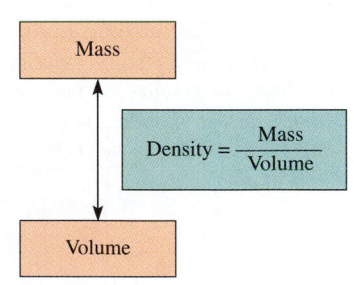

Mass ⟷ Volume

$$\text{Density} = \frac{\text{Mass}}{\text{Volume}}$$

The intensive property *density* relates the two extensive properties *mass* and *volume*.

These densities are given at room temperature and *one atmosphere* pressure, the average atmospheric pressure at sea level. Densities of solids and liquids change only slightly, but densities of gases change greatly, with changes in temperature and pressure.

Table 1-8 *Densities of Common Substances**			
Substance	**Density g/cm³**	**Substance**	**Density g/cm³**
hydrogen (gas)	0.000089	sand*	2.32
carbon dioxide (gas)	0.0019	aluminum	2.70
cork*	0.21	iron	7.86
oak wood*	0.71	copper	8.92
ethyl alcohol	0.789	silver	10.50
water	1.00	lead	11.34
magnesium	1.74	mercury	13.59
table salt	2.16	gold	19.3

Cork, oak wood, and sand are common materials that have been included to provide familiar reference points. They are not pure elements or compounds as are the other substances listed here.

Six materials with different densities. The liquid layers are gasoline (top), water (middle), and mercury (bottom). A cork floats on gasoline. A piece of oak wood sinks in gasoline, but floats on water. Brass sinks in water, but floats on mercury.

Observe that density gives two unit factors. In this case, they are $\dfrac{0.789 \text{ g}}{1 \text{ mL}}$ and $\dfrac{1 \text{ mL}}{0.789 \text{ g}}$.

and as g/L for gases. These units can also be expressed as $g \cdot cm^{-3}$, $g \cdot mL^{-1}$, and $g \cdot L^{-1}$, respectively. Densities of several substances are listed in Table 1-8.

EXAMPLE 1-11   Density, Mass, Volume

A 47.3-mL sample of ethyl alcohol (ethanol) has a mass of 37.32 g. What is its density?

Plan

We use the definition of density.

Solution

$$D = \frac{m}{V} = \frac{37.32 \text{ g}}{47.3 \text{ mL}} = \boxed{0.789 \text{ g/mL}}$$

You should now work Exercise 44.

EXAMPLE 1-12   Density, Mass, Volume

If 103 g of ethanol is needed for a chemical reaction, what volume of liquid would you use?

Plan

We determined the density of ethanol in Example 1-11. Here we are given the mass, m, of a sample of ethanol. So we know values for D and m in the relationship

$$D = \frac{m}{V}$$

We rearrange this relationship to solve for V, put in the known values, and carry out the calculation. Alternatively, we can use the unit factor method to solve the problem.

Solution

The density of ethanol is 0.789 g/mL (Table 1-8).

$$D = \frac{m}{V}, \quad \text{so} \quad V = \frac{m}{D} = \frac{103 \text{ g}}{0.789 \text{ g/mL}} = \boxed{130 \text{ mL}}$$

Alternatively,

$$\underline{?}\ \text{mL} = 103\ \text{g} \times \frac{1\ \text{mL}}{0.789\ \text{g}} = \boxed{130\ \text{mL}}$$

You should now work Exercise 47.

EXAMPLE 1-13 *Unit Conversion*

Express the density of mercury in lb/ft^3.

Plan

The density of mercury is 13.59 g/cm^3 (Table 1-8). To convert this value to the desired unit, we can use unit factors constructed from the conversion factors in Table 1-7.

Solution

$$\underline{?}\ \frac{\text{lb}}{\text{ft}^3} = 13.59\frac{\text{g}}{\text{cm}^3} \times \frac{1\ \text{lb}}{453.6\ \text{g}} \times \left(\frac{2.54\ \text{cm}}{1\ \text{in}}\right)^3 \times \left(\frac{12\ \text{in}}{1\ \text{ft}}\right)^3 = \boxed{848.4\ \text{lb/ft}^3}$$

It would take a very strong person to lift a cubic foot of mercury!

The **specific gravity** (Sp. Gr.) of a substance is the ratio of its density to the density of water, both at the same temperature.

$$\text{Sp. Gr.} = \frac{D_{\text{substance}}}{D_{\text{water}}}$$

Density and specific gravity are both intensive properties; i.e., they do not depend upon the size of the sample. Specific gravities are dimensionless numbers.

The density of water is 1.000 g/mL at 3.98°C, the temperature at which the density of water is greatest. However, variations in the density of water with changes in temperature are small enough that we may use 1.00 g/mL up to 25°C without introducing significant errors into our calculations.

EXAMPLE 1-14 *Density, Specific Gravity*

The density of table salt is 2.16 g/mL at 20°C. What is its specific gravity?

Plan

We use the definition of specific gravity given above. The numerator and denominator have the same units, so the result is dimensionless.

Solution

$$\text{Sp. Gr.} = \frac{D_{\text{salt}}}{D_{\text{water}}} = \frac{2.16\ \text{g/mL}}{1.00\ \text{g/mL}} = \boxed{2.16}$$

You should now work Exercise 51.

This example also demonstrates that the density and specific gravity of a substance are numerically equal near room temperature if density is expressed in g/mL (g/cm^3).

Labels on commercial solutions of acids give specific gravities and the percentage by mass of the acid present in the solution. From this information, the amount of acid present in a given volume of solution can be calculated.

CHEMISTRY IN USE

Our Daily Lives

Can Solids Be Lighter Than Air?

What would you think if you saw a bar of soap floating in midair? Is it a magician's trick? We know, and we can also see in Table 1-8, that the densities of solids are usually greater than the densities of gases. However, chemists have recently created solids called SEAgels* and aerogels that are less dense than air. Let's see why these solids are so light, how they are made, and how they may be used in your home.

SEAgels are prepared in much the same way as gelatin desserts (for example, Jell-O™). When making a gelatin dessert, we first dissolve gelatin (an animal protein) in hot water and then let it cool. Upon cooling, the mixture sets to form a flexible solid called a gel. The gel is made up of a thin-walled, honeycombed network of protein (the gelatin), whose spaces are filled with water. If we could evaporate the water from the gelatin dessert and leave the honeycombed protein network behind, we would have a very light-weight solid. However, if we tried to evaporate water from a gelatin dessert, the attractive forces between the water and the solid network would cause the gel to shrink as the water evaporates.

SEAgel is made by dissolving agar (a gelatinous material prepared from certain saltwater algae and used for thickening foods) in a water/organic solvent mixture and then letting it cool. Upon cooling, this mixture also sets much like Jell-O™. Also like Jell-O™, if we tried to evaporate the water from this gel, the solid agar network would shrink. However, we can freeze-dry the gel without shrinking it.

In freeze-drying, chemists first freeze the gel to lock its shape in place and then place the frozen gel into a combination freezer and vacuum chamber to evaporate the water. After the water evaporates, the delicate honeycombed agar structure remains with its spaces filled with air. Although it would be more correct to call this resulting substance a foam, most people continue to refer to such foams as aerogels.

The density of SEAgel is about 1.5 g/L and, as shown in Table 1-8, the density of carbon dioxide is about 1.9 g/L. This difference in densities allows us to do an interesting demonstra-

*SEA stands for Safe Emulsion Agar.

A piece of silica aerogel looks like a blue haze floating on beaten egg whites.

tion with a soap-bar-sized sample of SEAgel, an aquarium tank, and carbon dioxide gas. First, half fill the aquarium tank with carbon dioxide. Because carbon dioxide is more dense than air, carbon dioxide will displace the air in the tank and sink to cover the tank bottom. Second, drop the white soap-bar-sized sample of SEAgel into the tank and watch as it falls to the bottom of the tank and then floats to the top of the carbon dioxide layer.

Like all foams, solid aerogels are excellent insulators. In fact, aerogels have the best insulating properties of any known solid. Because of this, a California company has begun using aerogels as insulating material in refrigerators.

Ronald DeLorenzo
Middle Georgia College

EXAMPLE 1-15   *Specific Gravity, Volume, % by Mass*

Battery acid is 40.0% sulfuric acid, H_2SO_4, and 60.0% water by mass. Its specific gravity is 1.31. Calculate the mass of pure H_2SO_4 in 100.0 mL of battery acid.

Plan

The percentages are given on a mass basis, so we must first convert the 100.0 mL of acid solution to mass. To do this, we need a value for the density. We have demonstrated that density and specific gravity are numerically equal at 20°C because the density of water is 1.00 g/mL. We can use the density as a unit factor to convert the given volume of solution to mass of solution. Then we use the percentage by mass to convert the mass of solution to mass of acid.

Solution

From the given value for specific gravity, we may write

$$\text{density} = 1.31 \text{ g/mL}$$

The solution is 40.0% H_2SO_4 and 60.0% H_2O by mass. From this information we may construct the desired unit factor:

$$\frac{40.0 \text{ g } H_2SO_4}{100 \text{ g soln}} \longrightarrow \boxed{\text{because 100 g of solution contains 40.0 g of } H_2SO_4}$$

We can now solve the problem:

$$\underline{?}\ H_2SO_4 = 100.0 \text{ mL soln} \times \frac{1.31 \text{ g soln}}{1 \text{ mL soln}} \times \frac{40.0 \text{ g } H_2SO_4}{100 \text{ g soln}} = \boxed{52.4 \text{ g } H_2SO_4}$$

You should now work Exercise 53.

1-12 HEAT AND TEMPERATURE

In Section 1-1 you learned that heat is one form of energy. You also learned that the many different forms of energy can be interconverted and that in chemical processes, chemical energy is converted to heat energy or vice versa. The amount of heat a process uses (*endothermic*) or gives off (*exothermic*) can tell us a great deal about that process. For this reason it is important for us to be able to measure intensity of heat.

Temperature measures the intensity of heat, the "hotness" or "coldness" of a body. A piece of metal at 100°C feels hot to the touch, while an ice cube at 0°C feels cold. Why? Because the temperature of the metal is higher, and that of the ice cube lower, than body temperature. **Heat** is a form of energy that *always flows spontaneously from a hotter body to a colder body*—never in the reverse direction.

Temperatures can be measured with mercury-in-glass thermometers. A mercury thermometer consists of a reservoir of mercury at the base of a glass tube, open to a very thin (capillary) column extending upward. Mercury expands more than most other liquids as its temperature rises. As it expands, its movement up into the evacuated column can be seen.

Anders Celsius, a Swedish astronomer, developed the Celsius temperature scale, formerly called the centigrade temperature scale. When we place a Celsius thermometer in a beaker of crushed ice and water, the mercury level stands at exactly 0°C, the lower reference point. In a beaker of water boiling at one atmosphere pressure, the mercury level stands at 100°C, the higher reference point. There are 100 equal steps between these two mercury levels. They correspond to an interval of 100 degrees between the melting point of ice and the boiling point of water at one atmosphere. Figure 1-16 shows how temperature marks between the reference points are established.

In the United States, temperatures are frequently measured on the temperature scale devised by Gabriel Fahrenheit, a German instrument maker. On this scale the freezing and

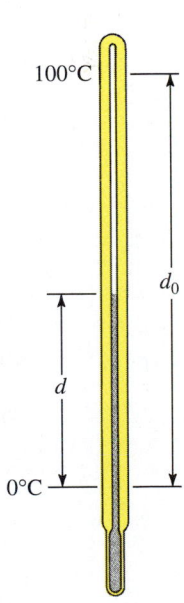

Figure 1-16 At 45°C, as read on a mercury-in-glass thermometer, d equals $0.45d_0$ where d_0 is the distance from the mercury level at 0°C to the level at 100°C.

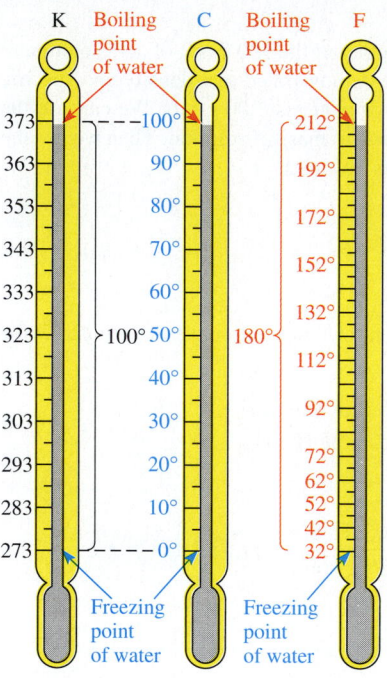

Figure 1-17 The relationships among the Kelvin, Celsius (centigrade), and Fahrenheit temperature scales.

boiling points of water are defined as 32°F and 212°F, respectively. In scientific work, temperatures are often expressed on the **Kelvin** (absolute) temperature scale. As we shall see in Section 12-5, the zero point of the Kelvin temperature scale is *derived* from the observed behavior of all matter.

Relationships among the three temperature scales are illustrated in Figure 1-17. Between the freezing point of water and the boiling point of water, there are 100 steps (degrees C or kelvins, respectively) on the Celsius and Kelvin scales. Thus the "degree" is the same size on the Celsius and Kelvin scales. But every Kelvin temperature is 273.15 units above the corresponding Celsius temperature. The relationship between these two scales is as follows:

We shall usually round 273.15 to 273.	$\underline{?}\,K = °C + 273.15°$ or $\underline{?}°C = K - 273.15°$

In the SI system, "degrees Kelvin" are abbreviated simply as K rather than °K and are called **kelvins.**

Please recognize that any temperature *change* has the same numerical value whether expressed on the Celsius scale or on the Kelvin scale. For example, a change from 25°C to 59°C represents a *change* of 34 Celsius degrees. Converting these to the Kelvin scale, the same change is expressed as (273 + 25) = 298 K to (59 + 273) = 332 K, or a *change* of 34 kelvins.

Comparing the Fahrenheit and Celsius scales, we find that the intervals between the same reference points are 180 Fahrenheit degrees and 100 Celsius degrees, respectively. Thus a Fahrenheit degree must be smaller than a Celsius degree. It takes 180 Fahrenheit degrees to cover the same temperature *interval* as 100 Celsius degrees. From this information, we can construct the unit factors for temperature *changes:*

$$\frac{180°F}{100°C} \quad \text{or} \quad \frac{1.8°F}{1.0°C} \quad \text{and} \quad \frac{100°C}{180°F} \quad \text{or} \quad \frac{1.0°C}{1.8°F}$$

But the starting points of the two scales are different, so we *cannot convert* a temperature on one scale to a temperature on the other just by multiplying by the unit factor. In converting from °F to °C, we must subtract 32 Fahrenheit degrees to reach the zero point on the Celsius scale (Figure 1-17).

$$\underline{?}°F = \left(x°C \times \frac{1.8°F}{1.0°C} \right) + 32°F \quad \text{and} \quad \underline{?}°C = \frac{1.0°C}{1.8°F}(x°F - 32°F)$$

These are often remembered in abbreviated form:

$$°F = 1.8°C + 32°$$

$$°C = \frac{(°F - 32°)}{1.8}$$

EXAMPLE 1-16 *Temperature Conversion*

When the temperature reaches "100.°F in the shade," it's hot. What is this temperature on the Celsius scale?

Plan

We use the relationship $\underline{?}°C = \frac{1.0°C}{1.8°F}(x°F - 32°F)$ to carry out the desired conversion.

Solution

$$\underline{?}°C = \frac{1.0°C}{1.8°F}(100.°F - 32°F) = \frac{1.0°C}{1.8°F}(68°F) = \boxed{38°C}$$

A temperature of 100.°F is 38°C.

EXAMPLE 1-17 *Temperature Conversion*

When the absolute temperature is 400 K, what is the Fahrenheit temperature?

Plan

We first use the relationship $\underline{?}°C = K - 273°$ to convert from kelvins to degrees Celsius; then we carry out the further conversion from degrees Celsius to degrees Fahrenheit.

Solution

$$\underline{?}°C = (400 \text{ K} - 273 \text{ K})\frac{1.0°C}{1.0 \text{ K}} = 127°C$$

$$\underline{?}°F = \left(127°C \times \frac{1.8°F}{1.0°C} \right) + 32°F = \boxed{261°F}$$

You should now work Exercise 56.

1-13 HEAT TRANSFER AND THE MEASUREMENT OF HEAT

Chemical reactions and physical changes occur with either the simultaneous evolution of heat (**exothermic processes**) or the absorption of heat (**endothermic processes**). The amount of heat transferred in a process is usually expressed in joules or in calories.

In English units this corresponds to a 4.4-pound object moving at 197 feet per minute, or 2.2 miles per hour. In terms of electrical energy, one joule is equal to one watt · second. Thus, one joule is enough energy to operate a 10-watt light bulb for $\frac{1}{10}$ second.

The calorie was originally defined as the amount of heat necessary to raise the temperature of one gram of water at one atmosphere from 14.5°C to 15.5°C.

The specific heat of a substance varies *slightly* with temperature and pressure. These variations can be ignored for calculations in this text.

In this example, we calculate the amount of heat needed to prepare a cup of hot tea.

The SI unit of energy and work is the **joule (J),** which is defined as $1 \text{ kg} \cdot \text{m}^2/\text{s}^2$. The kinetic energy (KE) of a body of mass m moving at speed v is given by $\frac{1}{2}mv^2$. A 2-kg object moving at one meter per second has $\text{KE} = \frac{1}{2}(2 \text{ kg})(1 \text{ m/s})^2 = 1 \text{ kg} \cdot \text{m}^2/\text{s}^2 = 1$ joule. You may find it more convenient to think in terms of the amount of heat required to raise the temperature of one gram of water from 14.5°C to 15.5°C, which is 4.184 joules.

One **calorie** is defined as exactly 4.184 joules. The so-called "large calorie," used to indicate the energy content of foods, is really one kilocalorie, i.e., 1000 calories. We shall do most calculations in joules.

The **specific heat** of a substance is the amount of heat required to raise the temperature of one gram of the substance one degree C (also one kelvin) with no change in phase. Changes in phase (physical state) absorb or liberate relatively large amounts of energy (Figure 1-5). The specific heat of each substance, a physical property, is different for the solid, liquid, and gaseous phases of the substance. For example, the specific heat of ice is 2.09 J/g · °C near 0°C; for liquid water it is 4.18 J/g · °C; and for steam it is 2.03 J/g · °C near 100°C. The specific heat for water is quite high. A table of specific heats is provided in Appendix E.

$$\text{specific heat} = \frac{(\text{amount of heat in J})}{(\text{mass of substance in g})(\text{temperature change in °C})}$$

The **heat capacity** of a body is the amount of heat required to raise its temperature 1°C. The heat capacity of a body is its mass in grams times its specific heat.

EXAMPLE 1-18   *Specific Heat*

How much heat, in joules, is required to raise the temperature of 205 g of water from 21.2°C to 91.4°C?

Plan

The specific heat of a substance is the amount of heat required to raise the temperature of 1 g of substance 1°C:

$$\text{specific heat} = \frac{(\text{amount of heat in J})}{(\text{mass of substance in g})(\text{temperature change in °C})}$$

We can rearrange the equation so that

$$(\text{amount of heat}) = (\text{mass of substance})(\text{specific heat})(\text{temperature change})$$

Alternatively, we can use the unit factor approach.

Solution

$$\text{amount of heat} = (205 \text{ g})(4.18 \text{ J/g} \cdot \text{°C})(70.2°\text{C}) = \boxed{6.02 \times 10^4 \text{ J}}$$

By the unit factor approach,

$$\text{amount of heat} = (205 \text{ g})\left(\frac{4.18 \text{ J}}{1 \text{ g} \cdot \text{°C}}\right)(70.2°\text{C}) = \boxed{6.02 \times 10^4 \text{ J}} \quad \text{or} \quad \boxed{60.2 \text{ kJ}}$$

All units except joules cancel. To cool 205 g of water from 91.4°C to 21.2°C, it would be necessary to remove exactly the same amount of heat, 60.2 kJ.

You should now work Exercise 61.

EXAMPLE 1-19 *Specific Heat*

How much heat, in calories, kilocalories, joules, and kilojoules, is required to raise the temperature of 205 g of iron from 294.2 K to 364.4 K? The specific heat of iron is 0.106 cal/g·°C, or 0.444 J/g·°C.

Plan

First we recall (Section 1-12) that a temperature *change* expressed in kelvins has the *same* numerical value expressed in degrees Celsius. Remembering that specific heat is in terms of temperature change, we can write the specific heat of iron as 0.106 cal/g·K or 0.444 J/g·K. Then we can solve this problem with the temperature change expressed in kelvins, and avoid the work of converting temperatures to °C.

This is the same temperature change for the same mass of iron that we used for water in Example 1-18, except that here the temperatures are given in kelvins. You may wish to check the result of this example by converting the starting and ending temperatures to °C and reworking the problem on the Celsius scale.

Solution

$$\text{temperature change} = 364.4 \text{ K} - 294.2 \text{ K} = 70.2 \text{ K}$$

$$\underline{?}\text{ cal} = (205 \text{ g})(0.106 \text{ cal/g·K})(70.2 \text{ K}) = \boxed{1.52 \times 10^3 \text{ cal}} \quad \text{or} \quad \boxed{1.52 \text{ kcal}}$$

$$\underline{?}\text{ J} = (205 \text{ g})(0.444 \text{ J/g·K})(70.2 \text{ K}) = \boxed{6.39 \times 10^3 \text{ J}} \quad \text{or} \quad \boxed{6.39 \text{ kJ}}$$

You should now work Exercises 62 and 63.

The specific heat of iron is much smaller than the specific heat of water.

$$\frac{\text{specific heat of iron}}{\text{specific heat of water}} = \frac{0.444 \text{ J/g·K}}{4.18 \text{ J/g·K}} = 0.106$$

As a result, the amount of heat required to raise the temperature of 205 g of iron by 70.2 K (70.2°C) is less than that required to do the same for 205 g of water, by the same ratio.

$$\frac{\text{amount of heat for iron}}{\text{amount of heat for water}} = \frac{6.39 \text{ kJ}}{60.2 \text{ kJ}} = 0.106$$

Key Terms

Accuracy How closely a measured value agrees with the correct value.

Calorie Defined as exactly 4.184 joules. Originally defined as the amount of heat required to raise the temperature of one gram of water from 14.5°C to 15.5°C.

Chemical change A change in which one or more new substances are formed.

Chemical property See *Properties*.

Compound A substance composed of two or more elements in fixed proportions. Compounds can be decomposed into their constituent elements.

Density Mass per unit volume, $D = m/V$.

Element A substance that cannot be decomposed into simpler substances by chemical means.

Endothermic Describes processes that absorb heat energy.

Energy The capacity to do work or transfer heat.

Exothermic Describes processes that release heat energy.

Extensive property A property that depends upon the amount of material in a sample.

Heat A form of energy that flows between two samples of matter because of their difference in temperature.

Heat capacity The amount of heat required to raise the temperature of a body (of whatever mass) one degree Celsius.

Heterogeneous mixture A mixture that does not have uniform composition and properties throughout.

Homogeneous mixture A mixture that has uniform composition and properties throughout.

Intensive property A property that is independent of the amount of material in a sample.

Joule A unit of energy in the SI system. One joule is $1 \text{ kg·m}^2/\text{s}^2$, which is also 0.2390 calorie.

Kinetic energy Energy that matter possesses by virtue of its motion.

Law of Conservation of Energy   Energy cannot be created or destroyed in a chemical reaction or in a physical change; it may be changed from one form to another.

Law of Conservation of Matter   There is no detectable change in the quantity of matter during a chemical reaction or during a physical change.

Law of Conservation of Matter and Energy   The combined amount of matter and energy available in the universe is fixed.

Law of Constant Composition   See *Law of Definite Proportions.*

Law of Definite Proportions   Different samples of any pure compound contain the same elements in the same proportions by mass; also known as the *Law of Constant Composition.*

Mass   A measure of the amount of matter in an object. Mass is usually measured in grams or kilograms.

Matter   Anything that has mass and occupies space.

Mixture   A sample of matter composed of variable amounts of two or more substances, each of which retains its identity and properties.

Physical change   A change in which a substance changes from one physical state to another, but no substances with different compositions are formed.

Physical property   See *Properties.*

Potential energy   Energy that matter possesses by virtue of its position, condition, or composition.

Precision   How closely repeated measurements of the same quantity agree with each other.

Properties   Characteristics that describe samples of matter. Chemical properties are exhibited as matter undergoes chemical changes. Physical properties are exhibited by matter with no changes in chemical composition.

Scientific (natural) law   A general statement based on the observed behavior of matter, to which no exceptions are known.

Significant figures   Digits that indicate the precision of measurements—digits of a measured number that have uncertainty only in the last digit.

Specific gravity   The ratio of the density of a substance to the density of water at the same temperature.

Specific heat   The amount of heat required to raise the temperature of one gram of a substance one degree Celsius.

Substance   Any kind of matter all specimens of which have the same chemical composition and physical properties.

Symbol   A letter or group of letters that represents (identifies) an element.

Temperature   A measure of the intensity of heat, i.e., the hotness or coldness of a sample or object.

Unit factor   A factor in which the numerator and denominator are expressed in different units but represent the same or equivalent amounts. Multiplying by a unit factor is the same as multiplying by one.

Weight   A measure of the gravitational attraction of the earth for a body.

Exercises

Asterisks are used to denote more challenging exercises.

Basic Ideas

1. Define the following terms and illustrate each with a specific example: (a) matter; (b) energy; (c) mass; (d) exothermic process.

2. Define the following terms and illustrate each with a specific example: (a) weight; (b) potential energy; (c) kinetic energy; (d) endothermic process.

3. State the following laws and illustrate each.
   (a) the Law of Conservation of Matter
   (b) the Law of Conservation of Energy
   (c) the Law of Conservation of Matter and Energy

4. List the three states of matter and some characteristics of each. How are they alike? different?

5. Distinguish between the following pairs of terms and give two specific examples of each: (a) chemical properties and physical properties; (b) intensive properties and extensive properties; (c) chemical changes and physical changes; (d) mass and weight.

6. Which of the following are chemical properties, and which are physical properties? (a) Baking powder gives off bubbles of carbon dioxide when added to water. (b) A particular type of steel consists of 95% iron, 4% carbon, and 1% miscellaneous other elements. (c) The density of gold is 19.3 g/mL. (d) Iron dissolves in hydrochloric acid with the evolution of hydrogen gas. (e) Fine steel wool burns in air.

7. Which of the following are chemical properties, and which are physical properties? (a) Metallic sodium is soft enough to be cut with a knife. (b) When sodium metal is cut, the surface is at first shiny; after a few seconds of exposure to air, it turns a dull gray. (c) The density of sodium is 0.97 g/mL. (d) Cork floats on water. (e) When sodium comes in contact with water, it melts, evolves a flammable gas, and eventually disappears altogether.

8. Describe each of the following as a chemical change, a physical change, or both. (a) A wet towel dries in the sun. (b) Lemon juice is added to tea, causing its color to change. (c) Hot air rises over a radiator. (d) Coffee is brewed by passing hot water through ground coffee.

9. Describe each of the following as a chemical change, a physical change, or both. (a) Powdered sulfur is heated, first melting and then burning. (b) Alcohol is evaporated by heating. (c) Transparent rock candy (pure sugar crystals) is finely ground into an opaque white powder. (d) Chlorine gas is bubbled through concentrated seawater, releasing liquid bromine. (e) Electricity is passed through water, resulting in the evolution of hydrogen and oxygen gases.

10. Which of the following processes are exothermic? endothermic? How can you tell? (a) combustion; (b) freezing water; (c) melting ice; (d) boiling water; (e) condensing steam.

11. Which of the following properties of a sample of matter are extensive? Which are intensive? (a) density; (b) melting point; (c) volume; (d) mass; (e) ability to conduct electricity; (f) temperature.

12. Define the following terms clearly and concisely. Give two illustrations of each. (a) substance; (b) mixture; (c) element; (d) compound.

13. Classify each of the following as an element, a compound, or a mixture. Justify your classification. (a) a soft drink; (b) water; (c) air; (d) chicken noodle soup; (e) table salt; (f) popcorn.

14. Classify each of the following as an element, a compound, or a mixture. Justify your classification. (a) coffee; (b) silver; (c) calcium carbonate; (d) ink from a ballpoint pen; (e) toothpaste.

15. What is a homogeneous mixture? Which of the following are homogeneous mixtures? Explain your answers. (a) sugar dissolved in water; (b) coffee; (c) french onion soup; (d) mud; (e) a clear liquid with no internal boundaries, consisting of corn oil and olive oil.

16. What is a heterogeneous mixture? Which of the following are heterogeneous mixtures? Explain your answers. (a) salt and sulfur; (b) milk; (c) clean air; (d) gasoline; (e) a chocolate chip cookie.

Scientific Notation and Significant Figures

17. Express the following numbers in scientific notation. (a) 6500.; (b) 0.00092; (c) 860 (assume that this number is measured to ± 10); (d) 860 (assume that this number is measured to ± 1); (e) 186,000; (f) 0.0516.

18. Express the following exponentials as ordinary numbers. (a) 5.26×10^4; (b) 4.10×10^{-6}; (c) 16.00×10^2; (d) 8.206×10^{-2}; (e) 9.346×10^3; (f) 9.346×10^{-3}.

19. Which of the following are likely to be exact numbers? Why? (a) 554 inches; (b) 7 computers; (c) $20,355.47; (d) 5 pounds of sugar; (e) 14.7 gallons of diesel fuel; (f) 5,446 ants.

20. To which of the quantities appearing in the following statements would the concept of significant figures apply? Where it would apply, indicate the number of significant figures. (a) The density of platinum at 20°C is 21.45 g/cm^3. (b) Wilbur Shaw won the Indianapolis 500-mile race in 1940 with an average speed of 114.277 mi/h. (c) A mile is defined as 5280 feet. (d) The International Committee for Weights and Measures "accepts that the curie be . . . retained as a unit of radioactivity, with the value 3.7×10^{10} s^{-1}." (This resolution was passed in 1964.)

21. The circumference of a circle is given by πd, where d is the diameter of the circle. Calculate the circumference of a circle with a diameter of 6.73 cm. Use the value of 3.141593 for π.

22. What is the total weight of 12 cars weighing an average of 1532.5 pounds?

In Exercises 23–24, perform the indicated operations, and round off your answers to the proper number of significant figures. Your answers should include appropriate units. Assume that all numbers were obtained from measurements.

23. (a) 2.68 ft + 11.4 ft; (b) 511 mi/2.2 h; (c) Three men work for 1.25 hours each. How many man-hours of work did they do?

24. (d) $(1.54 \times 10^2 \text{ cm})(2.336 \times 10^3 \text{ cm})$; (e) What is the area of a square 6.67 m on edge?

24. (a) 423.1 in + 0.256 in − 116 in; (b) The volume of a sphere is given by $\frac{4}{3}\pi r^3$, where r is the radius of the sphere. Note that the numbers 4 and 3 are exact numbers. Calculate the volume of a sphere with radius 3.31 in. Use the value of 3.141593 for π. (c) 6.057×10^3 m − 9.35 m; (d) $(8.54 \times 10^5 \text{ mi})/(22 \text{ days})$.

25. Indicate the multiple or fraction of 10 by which a quantity is multiplied when it is preceded by each of the following prefixes. (a) M; (b) m; (c) c; (d) d; (e) k; (f) μ.

Conversions and Dimensional Analysis

26. Carry out each of the following conversions. (a) 16.3 m to km; (b) 16.3 km to m; (c) 247 kg to g; (d) 4.32 L to mL; (e) 85.9 dL to L; (f) 7654 L to cm^3.

27. Express 13.5 yards in millimeters, centimeters, meters, and kilometers.

28. (a) Express 95 miles per hour in kilometers per hour. (b) If you are traveling at 95 miles per hour, how many feet are you traveling per second?

29. Express: (a) 12.00 gallons in cm^3; (b) 12.00 pints in milliliters; (c) 5.43 cubic yards in cubic inches; (d) 5.43 cubic feet in milliliters.

30. For each of the following pairs, determine which quantity is larger. (a) 24.0 mg or 24.0 cg; (b) 250 cm or 0.25 m; (c) 0.8 nm or 8 Å; (d) 10 L or 6.4 m^3.

31. A sample is marked as containing 22.8% calcium carbonate by mass. (a) How many grams of calcium carbonate are contained in 64.33 grams of the sample? (b) How many grams of the sample would contain 11.4 grams of calcium carbonate?

32. An iron ore is found to contain 9.24% hematite (a compound that contains iron). (a) How many tons of this ore would contain 8.40 tons of hematite? (b) How many kilograms of this ore would contain 8.40 kilograms of hematite?

*33. A foundry releases 5.0 tons of gas into the atmosphere each day. The gas contains 2.2% sulfur dioxide by mass. What mass of sulfur dioxide is released in one week?

*34. A certain chemical process requires 157 gallons of pure water each day. The available water contains 11 parts per million by mass of salt (i.e., for every 1,000,000 parts of available water, 11 parts of salt). What mass of salt must be removed each day? A gallon of water weighs 3.67 kg.

*35. The radius of a hydrogen atom is about 0.37 Å, and the average radius of the earth's orbit around the sun is about 1.5×10^8 km. Find the ratio of the average radius of the earth's orbit to the radius of the hydrogen atom.

36. If the price of gasoline is $1.159 per gallon, what is its price in cents per liter?

37. Suppose your automobile gas tank holds 16 gallons and the price of gasoline is $0.315 per liter. How much would it cost to fill your gas tank?

38. A particular medicine dropper delivers 22 drops of water to make 1.0 mL. (a) What is the volume of one drop in cubic centimeters? In microliters? (b) If the drop were spherical, what would be its diameter in millimeters? (Volume of a sphere = $\frac{4}{3}\pi r^3$.)

39. Express the following masses or weights in grams and in kilograms. (a) 8.7 ounces; (b) 3.15 short tons; (c) 3.6×10^6 milligrams; (d) 7.33×10^4 centigrams.

40. Express: (a) 326 kilograms in pounds; (b) 326 pounds in kilograms; (c) 326 ounces in centigrams.

*41. At a given point in its orbit, the earth is 92.98 million miles from the sun (center to center). The radius of the sun is 432,000 miles and the radius of the earth is 3960 miles. How long does it take for light from the surface of the sun to reach the earth's surface? The speed of light is 3.00×10^8 m/s.

*42. Cesium atoms are the largest naturally occurring atoms. The radius of a cesium atom is 2.62 Å. How many cesium atoms would have to be laid side by side to give a row of cesium atoms 1.00 inch long? Assume that the atoms are spherical.

Density and Specific Gravity

43. Which is more dense at 0°C, ice or water? Which has the higher specific gravity? How do you know?

44. What is the density of silicon, if 50.6 g occupies 21.72 mL?

45. What is the mass of a rectangular piece of copper 24.4 cm × 11.4 cm × 7.9 cm? The density of copper is 8.92 g/cm^3.

46. A small crystal of sucrose (table sugar) had a mass of 5.536 mg. The dimensions of the box-like crystal were $2.20 \times 1.36 \times 1.12$ mm. What is the density of sucrose expressed in g/cm^3?

47. Vinegar has a density of 1.0056 g/cm^3. What is the mass of two liters of vinegar?

*48. The radius of a neutron is approximately 1.5×10^{-15} m and its mass is 1.675×10^{-24} g. Find the density of a neutron, in g/cm^3.

*49. A container has a mass of 73.91 g when empty and 91.44 g when filled with water. The density of water is 1.0000 g/cm^3. (a) Calculate the volume of the container. (b) When filled with an unknown liquid, the container had a mass of 88.42 g. Calculate the density of the unknown liquid.

*50. The mass of an empty container is 77.664 g. The mass of the container filled with water is 99.646 g. (a) Calculate the volume of the container, using a density of 1.0000 g/cm^3 for water. (b) A piece of metal was added to the empty container and the combined mass was 85.308 g. Calculate the mass of the metal. (c) The container with the metal was filled with water and the mass of the entire system was 106.442 g. What mass of water was added? (d) What volume of water was added? (e) What is the volume of the piece of metal? (f) Calculate the density of the metal.

51. What is the specific gravity of a liquid if 325 mL of the liquid has the same mass as 396 mL of water?

52. The specific gravity of silver is 10.5. (a) What is the volume, in cm^3, of an ingot of silver with mass 0.765 kg? (b) If this sample of silver is a cube, how long is each edge in cm? (c) How long is the edge of this cube in inches?

53. The acid in an automobile battery is sulfuric acid. The density of a particular sulfuric acid solution is 1.36 g/mL. This solution is 46.3% H_2SO_4 by mass, the remainder being water.
  (a) 185 grams of the solution contains _____ grams of H_2SO_4 and _____ grams of water.
  (b) 185 mL of the solution contains _____ grams of H_2SO_4 and _____ grams of water.

Temperature Scales

54. Which represents a larger temperature interval: (a) a Celsius degree or a Fahrenheit degree? (b) a kelvin or a Fahrenheit degree?

55. Express: (a) 283°C in K; (b) 10.15 K in °C; (c) −32.0°C in °F; (d) 100.0°F in K.

56. Express: (a) 0°F in °C; (b) 98.6°F in K; (c) 298 K in °F; (d) 23.4°C in °F.

*57. Make each of the following temperature conversions: (a) 50°C to °F, (b) −50°C to °F, and (c) 100°F to °C.

*58. On the Réamur scale, which is no longer used, water freezes at 0°R and boils at 80°R. (a) Derive an equation that relates this to the Celsius scale. (b) Derive an equation that relates this to the Fahrenheit scale. (c) Mercury is a liquid metal at room temperature. It boils at 356.6°C (673.9°F). What is the boiling point of mercury on the Réamur scale?

59. Liquefied gases have boiling points well below room temperature. On the Kelvin scale the boiling points of the following gases are: He, 4.2 K; O_2, 90.2 K. Convert these temperatures to the Celsius and the Fahrenheit scales.

60. Convert the temperatures at which the following metals melt to the Celsius and Fahrenheit scales: Al, 933.6 K; Ag, 1235.1 K.

Heat and Heat Transfer

61. Calculate the amount of heat required to raise the temperature of 92.5 grams of water from 10.0°C to 35.0°C.

62. The specific heat of aluminum is 0.895 J/g · °C. Calculate the amount of heat required to raise the temperature of 22.1 grams of aluminum from 27.0°C to 44.3°C.

63. How much heat must be removed from 15.5 grams of water at 90.0°C to cool it to 23.0°C?

*64. In some solar-heated homes, heat from the sun is stored in rocks during the day, and then released during the cooler night. (a) Calculate the amount of heat required to raise the temperature of 78.7 kg of rocks from 25.0°C to 43.0°C. Assume that the rocks are limestone, which is essentially pure calcium carbonate. The specific heat of calcium carbonate is 0.818 J/g · °C. (b) Suppose that when the rocks in part (a) cool to 30.0°C, all the heat released goes to warm the 10,000 cubic feet (2.83×10^5 liters) of air in the house, originally at 10.0°C. To what final temperature would the air be heated? The specific heat of air is 1.004 J/g · °C, and its density is 1.20×10^{-3} g/mL.

*65. A small immersion heater is used to heat water for a cup of coffee. We wish to use it to heat 235 mL of water (about a teacupful) from 25°C to 90°C in 2.00 minutes. What must be the heat rating of the heater, in kJ/min, to accomplish this? Neglect the heat that goes to heat the teacup itself. The density of water is 0.997 g/mL.

Mixed Exercises

66. A student found the following list of properties of iodine in an encyclopedia: (a) grayish black granules, (b) metallic luster, (c) characteristic odor, (d) forms a purple vapor, (e) density = 4.93 g/cm^3, (f) melting point = 113.5°C, (g) soluble in alcohol, (h) insoluble in water, (i) noncombustible, (j) forms ions in aqueous solutions, (k) poisonous. Which of these properties are chemical properties?

67. Which of the properties listed in Exercise 66 are intensive properties?

*68. Use data from Table 1-8. (a) What would be the mass of a rectangular block of aluminum 2.11 × 6.25 in × 12.00 in? (b) Calculate the volume in cm^3 of 2.00 lb of mercury.

*69. At what temperature will a Fahrenheit thermometer give (a) the same reading as a Celsius thermometer? (b) a reading that is twice that on the Celsius thermometer? (c) a reading that is numerically the same but opposite in sign from that on the Celsius thermometer?

*70. The lethal dose of potassium cyanide (KCN) taken orally is 1.6 milligrams per kilogram of body weight. Calculate the lethal dose of potassium cyanide taken orally by a 165-pound person.

71. Suppose you ran a mile in 4.00 minutes. (a) What would be your average speed in km/h? (b) What would be your average speed in cm/s? (c) What would be your time (in minutes:seconds) for 1500 meters?

72. The distance light travels through space in one year is called one light-year. Using the speed of light in vacuum listed in Appendix D, and assuming that one year is 365 days, determine the distance of a light-year in kilometers and in miles.

73. Calculate the density of a metal, if a piece weighing 0.625 kg causes the water level in a graduated cylinder to rise from 10.3 mL to 80.2 mL.

74. Aluminum foil is sold in supermarkets in long rolls measuring 66⅔ yards by 12 inches, in a thickness of 0.00065 inch. If the specific gravity of aluminum at 22°C is 2.70, calculate the mass of a roll. The density of water at 22°C is 0.998 g/mL.

BUILDING YOUR KNOWLEDGE

75. Based on what you have learned during the study of this chapter, write a question that requires you to use chemical information but no mathematical calculations.

76. As you write out the answer to an end-of-chapter exercise, what chemical changes occur? Did your answer involve knowledge not covered in Chapter 1?

77. *Combustion* is discussed later in this textbook. However, you probably already know what the term means. (Look it up to be sure.) List two other chemical terms that were in your vocabulary before you read Chapter 1.

2 Chemical Formulas and Composition Stoichiometry

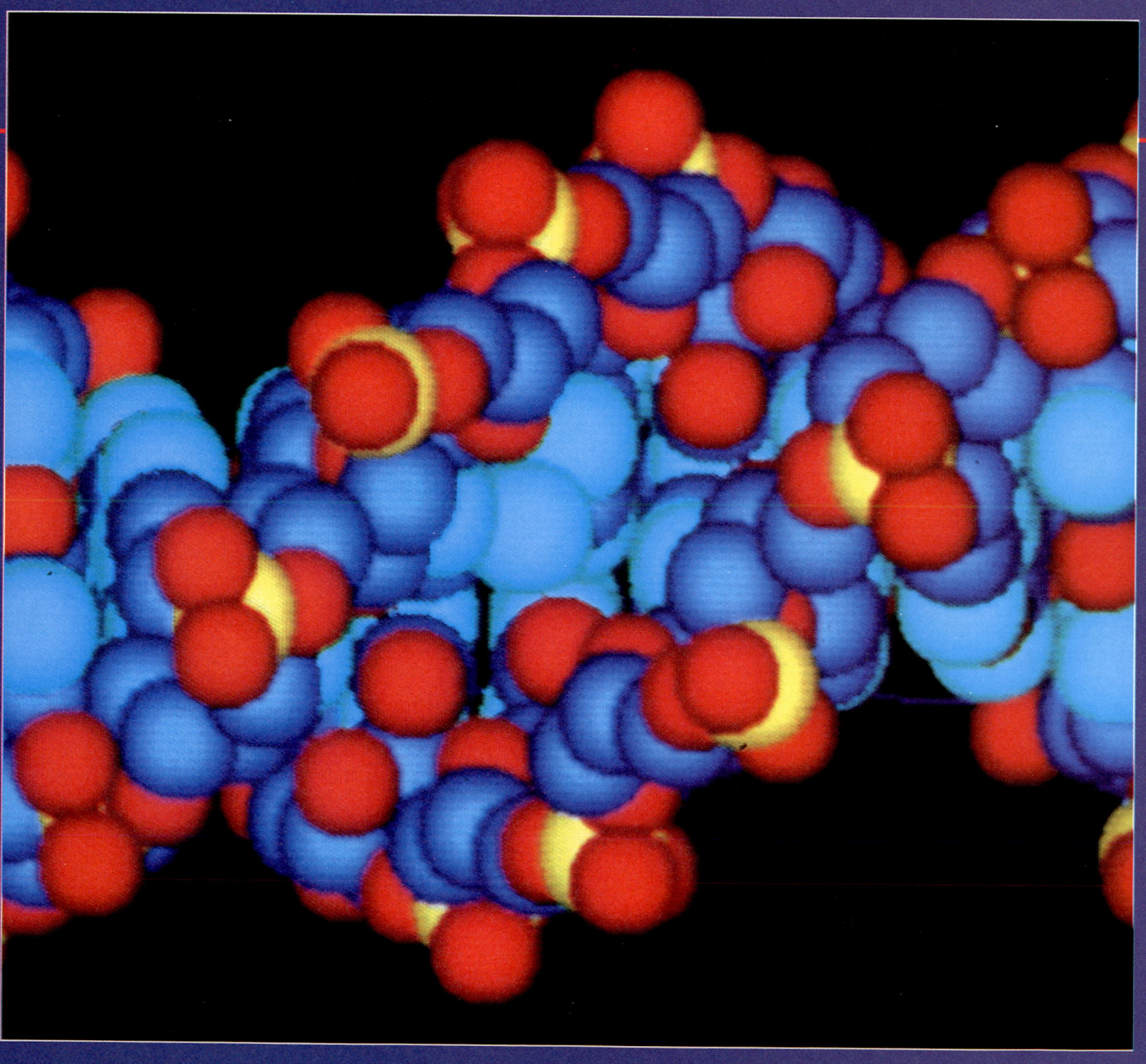

OBJECTIVES

As you study this chapter, you should learn to

• *Understand an early concept of atoms*

• *Use chemical formulas to solve various kinds of chemical problems*

• *Relate names to formulas and charges of simple ions*

• *Combine simple ions to write formulas and names of ionic compounds*

• *Recognize and use formula weights and mole relationships*

• *Interconvert masses, moles, and formulas in problems*

• *Determine percent compositions in compounds*

• *Determine formulas from composition*

• *Perform calculations about purity of substances*

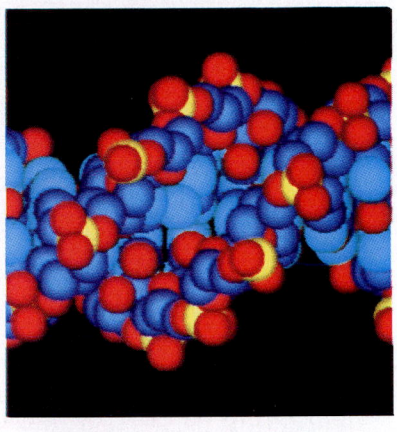

A space-filling model of a small portion of a molecule of deoxyribonucleic acid, DNA, the primary genetic material of all cells.

The language that chemists use to describe the forms of matter and the changes in its composition appears throughout the scientific world. Chemical symbols, formulas, and equations are used in such diverse areas as agriculture, home economics, engineering, geology, physics, biology, medicine, and dentistry. In this chapter we shall describe the simplest atomic theory. We shall use it as we represent chemical formulas of elements and compounds. Later this theory will be expanded when we discuss chemical changes.

It is important to learn this fundamental material well so that you can use it correctly and effectively.

The word "stoichiometry" is derived from the Greek *stoicheion,* which means "first principle or element," and *metron,* which means "measure." **Stoichiometry** describes the quantitative relationships among elements in compounds (composition stoichiometry) and among substances as they undergo chemical changes (reaction stoichiometry). In this chapter we shall be concerned with chemical formulas and composition stoichiometry. In Chapter 3 we shall discuss chemical equations and reaction stoichiometry.

2-1 ATOMS AND MOLECULES

Around 400 BC, the Greek philosopher Democritus suggested that all matter is composed of tiny, discrete, indivisible particles that he called atoms. His ideas, based entirely on philosophical speculation rather than experimental evidence, were rejected for 2000 years.

The term "atom" comes from the Greek language and means "not divided" or "indivisible."

By the late 1700s, scientists began to realize that the concept of atoms provided an explanation for many experimental observations about the nature of matter.

By the early 1800s, the Law of Conservation of Matter (Section 1-1) and the Law of Definite Proportions (Section 1-5) were both accepted as general descriptions of how matter behaves. John Dalton, an English schoolteacher, tried to explain why matter behaves in such simple and systematic ways as those expressed above. In 1808, he published the first "modern" ideas about the existence and nature of atoms. He summarized and expanded the nebulous concepts of early philosophers and scientists; more importantly, his ideas were based on *reproducible experimental results* of measurements by many scientists. Taken together, these ideas form the core of **Dalton's Atomic Theory,** one of the highlights of scientific thought. In condensed form, Dalton's ideas may be stated as follows:

> The radius of a calcium atom is only 1.97×10^{-8} cm, and its mass is 6.66×10^{-23} g.

1. An element is composed of extremely small indivisible particles called atoms.
2. All atoms of a given element have identical properties, which differ from those of other elements.
3. Atoms cannot be created, destroyed, or transformed into atoms of another element.
4. Compounds are formed when atoms of different elements combine with each other in small whole-number ratios.
5. The relative numbers and kinds of atoms are constant in a given compound.

> Statement 3 is true for *chemical* reactions. However, it is not true for *nuclear* reactions (Chapter 26).

Dalton believed that atoms were solid indivisible spheres, an idea we now reject. But he showed remarkable insight into the nature of matter and its interactions. Some of his ideas could not be verified (or refuted) experimentally at the time. They were based on the limited experimental observations of his day. Even with their shortcomings, Dalton's ideas provided a framework that could be modified and expanded by later scientists. Thus John Dalton is the father of modern atomic theory.

The smallest particle of an element that maintains its chemical identity through all chemical and physical changes is called an **atom** (Figure 2-1). In Chapter 5, we shall study the structure of the atom in detail; let us simply summarize here the main features of atomic composition. Atoms, and therefore *all* matter, consist principally of three **fundamental particles:** *electrons, protons,* and *neutrons.* These are the basic building blocks of all atoms. The masses and charges of the three fundamental particles are shown in Table 2-1. The masses of protons and neutrons are nearly equal, but the mass of an electron is much smaller. Neutrons carry no charge. The charge on a proton is equal in magnitude, but opposite in sign, to the charge on an electron. Because atoms are electrically neutral,

an atom contains equal numbers of electrons and protons.

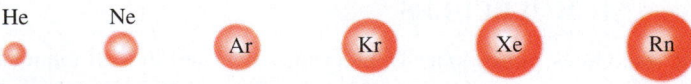

Figure 2-1   Relative sizes of monatomic molecules (single atoms) of the noble gases.

Table 2-1 *Fundamental Particles of Matter*		
Particle (symbol)	**Mass***	**Charge (relative scale)**
electron (e^-)	0.00054858 amu	1−
proton (p or p^+)	1.0073 amu	1+
neutron (n or n^0)	1.0087 amu	none

*1 amu = 1.6605×10^{-24} g

The **atomic number** (symbol is **Z**) of an element is defined as the number of protons in the nucleus. In the periodic table, elements are arranged in order of increasing atomic numbers. These are the red numbers above the symbols for the elements in the periodic table inside the front cover. For example, the atomic number of silver is 47.

A **molecule** is the smallest particle of an element or compound that can have a stable independent existence. In nearly all molecules, two or more atoms are bonded together in very small, discrete units (particles) that are electrically neutral.

Individual oxygen atoms are not stable at room temperature and atmospheric pressure. Single atoms of oxygen mixed under these conditions quickly combine to form pairs. The oxygen with which we are all familiar is made up of two atoms of oxygen; it is a *diatomic* molecule, O_2. Hydrogen, nitrogen, fluorine, chlorine, bromine, and iodine are other examples of diatomic molecules (Figure 2-2).

Some other elements exist as more complex molecules. Phosphorus molecules consist of four atoms, while sulfur exists as eight-atom molecules at ordinary temperatures and pressures. Molecules that contain two or more atoms are called *polyatomic* molecules. See Figure 2-3.

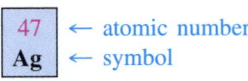
47 ← atomic number
Ag ← symbol

For Group 0 elements, the noble gases, a molecule contains only one atom and so an atom and a molecule are the same (Figure 2-1).

You should remember the common elements that occur as diatomic molecules: H_2, N_2, O_2, F_2, Cl_2, Br_2, I_2.

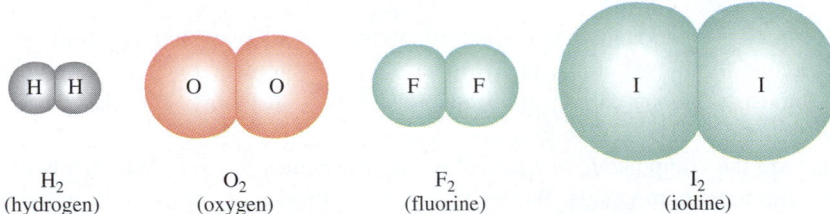

H H	O O	F F	I I
H_2 (hydrogen)	O_2 (oxygen)	F_2 (fluorine)	I_2 (iodine)

Figure 2-2 Models of diatomic molecules of some elements, approximately to scale.

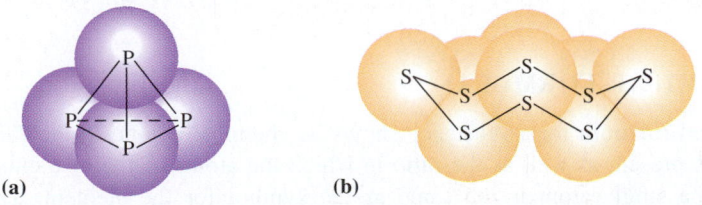

(a)   (b)

Figure 2-3 (a) A model of the P_4 molecule of white phosphorus. (b) A model of the S_8 ring found in rhombic sulfur.

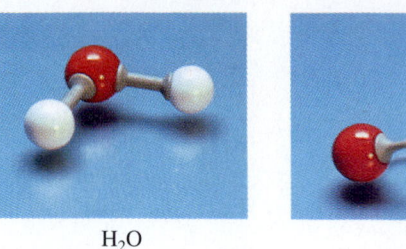
H₂O
(water)

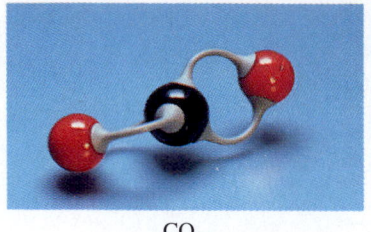
CO₂
(carbon dioxide)

CH₄
(methane)

C₂H₅OH
(ethyl alcohol)

Figure 2-4 Formulas and models for molecules of some compounds.

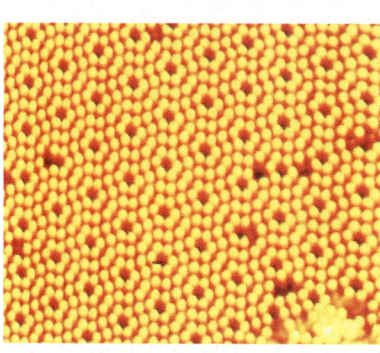

Figure 2-5 A computer reconstruction of the surface of a sample of silicon, as observed with a scanning tunnelling electron microscope (STM), reveals the regular pattern of individual silicon atoms. Many important reactions occur on the surfaces of solids. Observations of the atomic arrangements on surfaces help chemists understand such reactions. New information available with the STM will give many details about chemical bonding in solids.

In modern terminology, O_2 is named dioxygen, H_2 is dihydrogen, P_4 is tetraphosphorus, and so on. Even though such terminology is officially preferred, it has not yet gained wide acceptance. Most chemists still refer to O_2 as oxygen, H_2 as hydrogen, P_4 as phosphorus, and so on.

Molecules of compounds are composed of more than one kind of atom. A water molecule consists of two atoms of hydrogen and one atom of oxygen. A molecule of methane consists of one carbon atom and four hydrogen atoms. The shapes of a few molecules are shown in Figure 2-4.

Atoms are the components of molecules, and molecules are the components of many elements and most compounds. We are able to see samples of compounds and elements that consist of large numbers of atoms and molecules. With the scanning tunnelling microscope it is now possible to "see" atoms (Figure 2-5). It would take millions of atoms to make a row as long as the diameter of the period at the end of this sentence.

Methane is the principal component of natural gas.

2-2 CHEMICAL FORMULAS

The **chemical formula** for a substance shows its chemical composition. This represents the elements present as well as the ratio in which the atoms of the elements occur. The formula for a single atom is the same as the symbol for the element. Thus, Na can represent a single sodium atom. It is unusual to find such isolated atoms in nature, with the exception of the noble gases (He, Ne, Ar, Kr, Xe, and Rn). A subscript following the

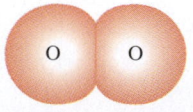
An O_2 molecule.

CHEMISTRY IN USE

The Development of Science

Do Atoms and Molecules Really Exist?

As we saw earlier, the idea that all matter is made up of tiny indivisible particles called atoms was first suggested about 2400 years ago. Some people believe that the "atoms" shown in textbooks really exist. The existence of atoms is rarely questioned because our atomic model does such a good job of explaining and predicting the behavior of matter. Sometimes, we forget that our model of the atom is only a theory or representation of reality. *Models are useful in explaining reality, but they are not reality.* To illustrate, consider a simple analogy using a toy car, a toy soldier, and a rock.

Suppose we place the large rock at the bottom of a small incline, the toy car at the top of the incline, and the toy soldier on the car. If we let the car roll down the incline, it strikes the rock, comes to a stop, and the toy soldier falls off. We can think of these objects and their behavior as models of reality. Our toy models could be used to predict correctly what would happen to a real soldier standing on top of a real car rolling down a hill and colliding with a large boulder. However, our toy car model is incomplete; it lacks details, such as an internal combustion engine, and so it tells us nothing about the atmospheric pollution from real cars. The toy car is an incomplete and imperfect model of reality.

In a similar way, consider the differences between the reality we call matter and our atomic model of matter. Gravity exists. We feel its effects every moment of our lives. Yet nothing in the atomic theory of matter explains gravity. Atomic theory cannot tell us why gravity exists or why it is an attractive rather than a repulsive force. As an example of our lack of understanding of gravity, consider what might happen if our sun were to vanish at exactly 12 noon tomorrow. How long would it take for the loss of gravity to affect the earth? Would it be several minutes before the earth felt the loss of the sun's gravity? Would earth feel this effect instantaneously at exactly 12 noon, or is the transmission of gravitational attraction limited to the speed of light or some other speed? The atomic theory of matter doesn't explain gravity and answer these questions. So we recognize that the atomic theory, like the toy car model, is not reality.

If someone develops another model of matter that appears to explain everything including gravity, we still couldn't call this new theory reality. To call any theory reality is to assume that we know with absolute certainty that we have observed every possible phenomenon. No number of experiments, no matter how large, can prove a theory conclusively; but only one set of observations that are inconsistent with the theory reveals its weakness. What should we do when scientists develop other models or theories that also explain and predict reality equally well? Is one of the models reality and the other just a model? We can't answer this question at this time.

A poet and a scientist have different ways of representing reality. One does it with metaphors and the other with models, and neither believes completely their representations because both recognize the limitations of models. Scientists have varying degrees of confidence in the ability of a particular theory to represent and explain some aspects of matter. Our history has shown us that as we make more and better observations, better theories and models are often "just around the corner."

Ronald DeLorenzo
Middle Georgia College

symbol of an element indicates the number of atoms in a molecule. For instance, F_2 indicates a molecule containing two fluorine atoms, and P_4 a molecule containing four phosphorus atoms.

Some elements exist in more than one form. Familiar examples include (1) oxygen, found as O_2 molecules, and ozone, found as O_3 molecules, and (2) two different crystalline forms of carbon—diamond and graphite (Figure 13-31). Different forms of the same element in the same physical state are called **allotropic modifications** or **allotropes.**

Compounds contain two or more elements in chemical combination in fixed proportions. Many compounds exist as molecules (Table 2-2). Hence, each molecule of hydrogen chloride, HCl, contains one atom of hydrogen and one atom of chlorine; each molecule of carbon tetrachloride, CCl_4, contains one carbon atom and four chlorine atoms. An aspirin molecule, $C_9H_8O_4$, contains nine carbon atoms, eight hydrogen atoms, and four oxygen atoms.

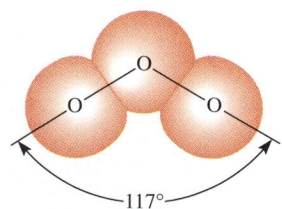

An O_3 molecule.

Table 2-2 *Names and Formulas of Some Common Molecular Compounds*

Name	Formula	Name	Formula	Name	Formula
water	H_2O	sulfur dioxide	SO_2	butane	C_4H_{10}
hydrogen peroxide	H_2O_2	sulfur trioxide	SO_3	pentane	C_5H_{12}
hydrogen chloride*	HCl	carbon monoxide	CO	benzene	C_6H_6
sulfuric acid	H_2SO_4	carbon dioxide	CO_2	methanol (methyl alcohol)	CH_3OH
nitric acid	HNO_3	methane	CH_4	ethanol (ethyl alcohol)	CH_3CH_2OH
acetic acid	CH_3COOH	ethane	C_2H_6	acetone	CH_3COCH_3
ammonia	NH_3	propane	C_3H_8	diethyl ether (ether)	$CH_3CH_2-O-CH_2CH_3$

Called hydrochloric acid if dissolved in water.

A space-filling model of a TNT molecule, $C_7H_5(NO_2)_3$.

We find many organic compounds in nature. **Organic compounds** contain C–C or C–H bonds or both. Eleven of the compounds listed in Table 2-2 are organic compounds (acetic acid and the last ten entries). All of the other compounds in the table are **inorganic compounds.**

Some groups of atoms behave chemically as single entities. For instance, one nitrogen atom and two oxygen atoms may combine to form a *nitro* group that is a part of a molecule. In formulas of compounds containing two or more of the same group, the group formula is enclosed in parentheses. Thus, 2,4,6-trinitrotoluene (often abbreviated TNT) contains three *nitro* groups, and its formula is $C_7H_5(NO_2)_3$ (see margin). When you count up the number of atoms in this molecule from its formula, you must multiply the numbers of nitrogen and oxygen atoms in the NO_2 group by 3. There are *seven* carbon atoms, *five* hydrogen atoms, *three* nitrogen atoms, and *six* oxygen atoms in a molecule of TNT.

Compounds were first recognized as distinct substances because of their different physical properties and because they could be separated from one another by physical methods. Once the concept of atoms and molecules was established, the reason for these differences in properties could be understood: Two compounds differ from one another because their molecules are different. Conversely, if two molecules contain the same number of the same kinds of atoms, arranged the same way, then both are molecules of the same compound. Thus, the atomic theory explains the **Law of Definite Proportions** (Section 1-5).

This law, also known as the **Law of Constant Composition,** can now be extended to include its interpretation in terms of atoms. It is so important for performing the calculations in this chapter that we restate it here:

> Different pure samples of a compound always contain the same elements in the same proportion by mass; this corresponds to atoms of these elements combined in fixed numerical ratios.

So we see that for a substance composed of molecules, the *chemical formula* gives the number of atoms of each type in the molecule. But this formula does not express the order in which the atoms in the molecules are bonded together. The **structural formula** shows the order in which atoms are connected. The lines connecting atomic symbols represent chemical bonds between atoms. The bonds are actually forces that tend to hold atoms at certain distances and angles from one another. For instance, the structural formula of propane shows that the three C atoms are linked in a chain, with three H atoms bonded to

Chemical Formula	Structural Formula	Ball-and-Stick Model	Space-Filling Model
H_2O, water	H—O—H		
H_2O_2, hydrogen peroxide	H—O—O—H		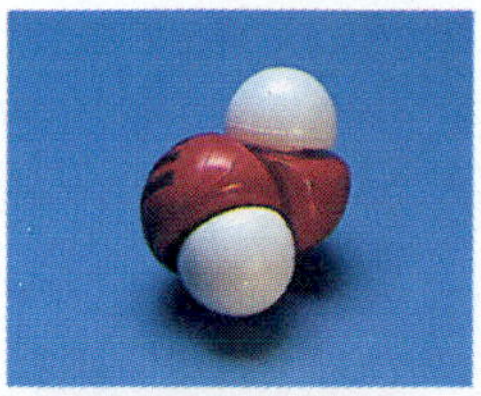
CCl_4, carbon tetrachloride	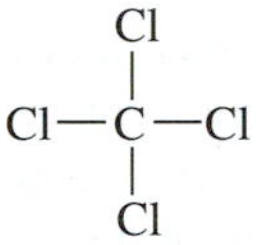 Cl | Cl—C—Cl | Cl		
C_3H_8, propane	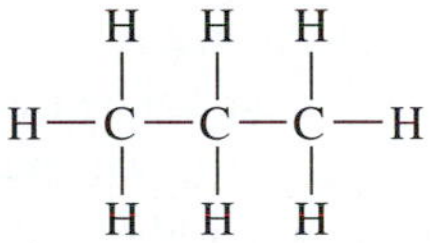 H H H | | | H—C—C—C—H | | | H H H	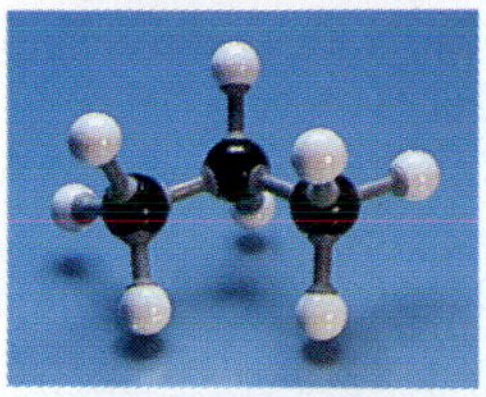	
C_2H_5OH, ethanol	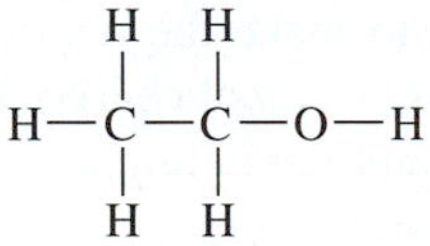 H H | | H—C—C—O—H | | H H		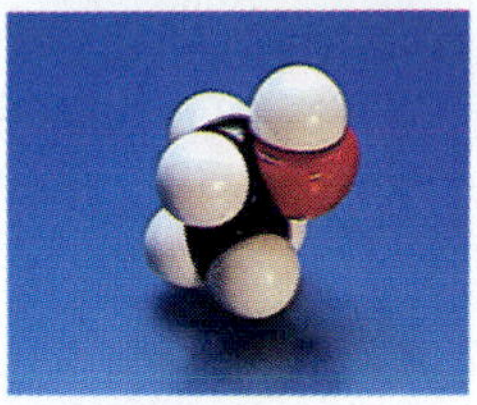

Figure 2-6 Formulas and models for some molecules. Structural formulas show the order in which atoms are connected, but do not represent true molecular shapes. Ball-and-stick models use balls of different colors to represent atoms and sticks to represent bonds; they show the three-dimensional shapes of molecules. Space-filling models show the (approximate) relative sizes of atoms and the shapes of molecules.

each of the end C atoms and two H atoms bonded to the center C. **Ball-and-stick** molecular models and **space-filling** molecular models help us to see the shapes and relative sizes of molecules. These four representations are shown in Figure 2-6. The ball-and-stick and space-filling models show (1) the *bonding sequence,* i.e., the order in which the atoms are connected to each other, and (2) the *geometrical arrangements* of the atoms in the molecule. As we shall see later, both are extremely important because they determine the properties of compounds.

2-3 IONS AND IONIC COMPOUNDS

So far we have discussed only compounds that exist as discrete molecules. Some compounds, such as sodium chloride, NaCl, consist of collections of large numbers of ions. An **ion** is an atom or group of atoms that carries an electrical charge. Ions that possess a *positive* charge, such as the sodium ion, Na^+, are called **cations.** Those carrying a *negative* charge, such as the chloride ion, Cl^-, are called **anions.** The charge on an ion *must* be included as a superscript on the right side of the chemical symbol(s) when we write the formula for the individual ion.

As we shall see in Chapter 5, an atom consists of a very small, very dense, positively charged *nucleus* surrounded by a diffuse distribution of negatively charged particles called *electrons*. The number of positive charges in the nucleus defines the identity of the element to which the atom corresponds. Electrically neutral atoms contain the same number of electrons outside the nucleus as positive charges (protons) within the nucleus. Ions are formed when neutral atoms lose or gain electrons. An Na^+ ion is formed when a sodium atom loses one electron, and a Cl^- ion is formed when a chlorine atom gains one electron.

The compound NaCl consists of an extended array of Na^+ and Cl^- ions (Figure 2-7). Within the crystal (though not on the surface) each Na^+ ion is surrounded at equal distances by six Cl^- ions, and each Cl^- ion is similarly surrounded by six Na^+ ions. *Any* compound, whether ionic or molecular, is electrically neutral; i.e., it has no net charge. In NaCl this means that the Na^+ and Cl^- ions are present in a 1 : 1 ratio, and this is indicated by the formula NaCl.

The general term "formula unit" applies to molecular or ionic compounds, whereas the more specific term "molecule" applies only to elements and compounds that exist as discrete molecules.

Because there are no "molecules" of ionic substances, we should not refer to "a molecule of sodium chloride, NaCl," for example. Instead, we refer to a **formula unit** (**FU**) of NaCl, which consists of one Na^+ ion and one Cl^- ion. Likewise, one formula unit of $CaCl_2$ consists of one Ca^{2+} ion and two Cl^- ions. Similarly, we speak of the formula unit of all ionic compounds. It is also acceptable to refer to a formula unit of a molecular compound. One formula unit of propane, C_3H_8, is the same as one molecule of C_3H_8; it contains three C atoms and eight H atoms bonded together into a group.

For the present, we shall tell you which substances are ionic and which are molecular when it is important to know. Later you will learn to make the distinction yourself.

In this text, we use the standard convention of representing multiple charges with the number before the sign, e.g., Ca^{2+}, not Ca^{+2} and SO_4^{2-}, not SO_4^{-2}.

Polyatomic ions are groups of atoms that bear an electrical charge. Examples include the ammonium ion, NH_4^+, the sulfate ion, SO_4^{2-}, and the nitrate ion, NO_3^-. Table 2-3 shows the formulas, ionic charges, and names of some common ions.

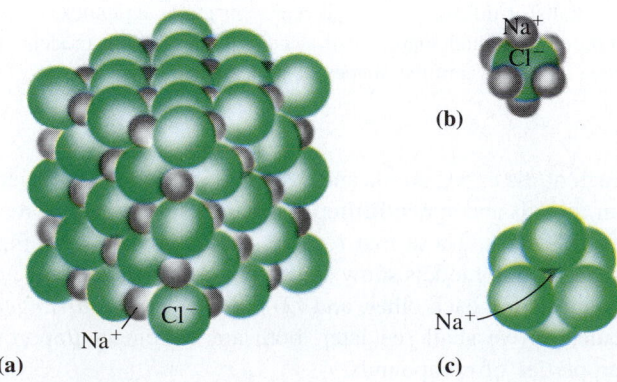

Figure 2-7   The arrangement of ions in NaCl. (a) A crystal of sodium chloride consists of an extended array that contains equal numbers of sodium ions (small spheres) and chloride ions (large spheres). Within the crystal, each chloride ion is surrounded by six sodium ions (b), and each sodium ion is surrounded by six chloride ions (c).

Table 2-3 *Formulas, Ionic Charges, and Names of Some Common Ions*

Common Cations (positive ions)			Common Anions (negative ions)		
Formula	*Charge*	*Name*	*Formula*	*Charge*	*Name*
Na^+	1+	sodium	F^-	1−	fluoride
K^+	1+	potassium	Cl^-	1−	chloride
NH_4^+	1+	ammonium	Br^-	1−	bromide
Ag^+	1+	silver	OH^-	1−	hydroxide
			CH_3COO^-	1−	acetate
Mg^{2+}	2+	magnesium	NO_3^-	1−	nitrate
Ca^{2+}	2+	calcium			
Zn^{2+}	2+	zinc	O^{2-}	2−	oxide
Cu^+	1+	copper(I) or cuprous	S^{2-}	2−	sulfide
Cu^{2+}	2+	copper(II) or cupric	SO_3^{2-}	2−	sulfite
Fe^{2+}	2+	iron(II) or ferrous	SO_4^{2-}	2−	sulfate
			CO_3^{2-}	2−	carbonate
Fe^{3+}	3+	iron(III) or ferric			
Al^{3+}	3+	aluminum	PO_4^{3-}	3−	phosphate

As we shall see, some metals can form more than one kind of ion with positive charge. For such metals, we specify which ion we mean with a Roman numeral—e.g., iron(II) or iron(III). Because zinc forms no stable ions other than Zn^{2+}, we do not need to use Roman numerals in its name.

2-4 AN INTRODUCTION TO NAMING COMPOUNDS

Throughout your study of chemistry you will have many occasions to refer to compounds by name. In this section, we shall see how to name a few compounds. More comprehensive rules for naming compounds will be presented at the appropriate places later in the text.

Table 2-2 includes examples of names for a few common molecular compounds. You should learn that short list. We shall name many more molecular compounds as we encounter them in later chapters.

The names of some common ions appear in Table 2-3. You should learn the names and formulas of these frequently encountered ions. They can be used to write the formulas and names of many ionic compounds. We write the formula of an ionic compound by adjusting the relative numbers of positive and negative ions so their total charges cancel (i.e., add to zero). The name of an ionic compound is formed by giving the names of the ions, with the positive ion named first.

Pure quartz crystals (silicon dioxide, SiO_2) are clear and colorless. Many gemstones consist of quartz that contains traces of certain metal ions. Amethyst, shown here, is quartz that contains small amounts of Fe^{3+} ions. Its color can range from pale lilac to royal purple depending on the amounts of Fe^{3+} ions present. Quartz is widely distributed over the earth's surface. White sand is relatively pure quartz that has been weathered into small pieces.

EXAMPLE 2-1   *Formulas for Ionic Compounds*

Write the formulas for the following ionic compounds: (a) sodium fluoride, (b) calcium fluoride, (c) iron(II) sulfate, (d) zinc phosphate.

Plan

In each case, we identify the chemical formulas of the ions from Table 2-3. These ions must be present in a ratio that gives the compound *no net charge*. The formulas and names of ionic compounds are written by giving the positively charged ion first.

Solution

(a) The formula for the sodium ion is Na^+ and the formula for the fluoride ion is F^- (Table 2-3). Because the charges on these two ions are equal in magnitude, the ions must be present in equal numbers, or in a 1:1 ratio. Thus, the formula for sodium fluoride is   NaF.

(b) The formula for the calcium ion is Ca^{2+} and the formula for the fluoride ion is F^-. Now each positive ion (Ca^{2+}) provides twice as much charge as each negative ion (F^-). So there must be twice as many F^- ions as Ca^{2+} ions to equalize the charge. This means that the ratio of calcium to fluoride ions is 1:2. So the formula for calcium fluoride is CaF_2.

(c) The iron(II) ion is Fe^{2+} and the sulfate ion is SO_4^{2-}. As in part (a), the equal magnitudes of positive and negative charges tell us that the ions must be present in equal numbers, or in a 1:1 ratio. The formula for iron(II) sulfate is $FeSO_4$.

(d) The zinc ion is Zn^{2+} and the phosphate ion is PO_4^{3-}. Now it will take *three* Zn^{2+} ions to account for as much charge (6+ total) as would be present in *two* PO_4^{3-} ions (6− total). So the formula for zinc phosphate is $Zn_3(PO_4)_2$.

You should now work Exercises 12 and 14.

EXAMPLE 2-2   *Names for Ionic Compounds*

Name the following ionic compounds: (a) $(NH_4)_2S$, (b) $Cu(NO_3)_2$, (c) $ZnCl_2$, (d) $Fe_2(CO_3)_3$.

Plan

In naming ionic compounds, it is helpful to inspect the formula for atoms or groups of atoms that we recognize as representing familiar ions.

Solution

(a) The presence of the polyatomic grouping NH_4 in the formula suggests to us the presence of the ammonium ion, NH_4^+. There are two of these, each accounting for 1+ in charge. To balance this, the single S must account for 2− in charge, or S^{2-}, which we recognize as the sulfide ion. Thus, the name of the compound is   ammonium sulfide.

(b) The NO_3 grouping in the formula tells us that the nitrate ion, NO_3^-, is present. Two of these nitrate ions account for $2 \times 1- = 2-$ in negative charge. To balance this, copper must account for 2+ charge and be the copper(II) ion. The name of the compound is   copper(II) nitrate   or, alternatively,   cupric nitrate.

(c) The positive ion present is zinc ion, Zn^{2+}, and the negative ion is chloride, Cl^-. The name of the compound is   zinc chloride.

(d) Each CO_3 grouping in the formula must represent the carbonate ion, CO_3^{2-}. The presence of *three* such ions accounts for a total of 6− in negative charge, so there must be a total of 6+ present in positive charge to balance this. It takes *two* iron ions to provide this 6+, so each ion must have a charge of 3+ and be Fe^{3+}, the iron(III) ion or ferric ion. The name of the compound is   iron(III) carbonate   or   ferric carbonate.

You should now work Exercises 11 and 13.

> We use the information that the carbonate ion has a 2− charge to find the charge on the iron ions. The total charges must add to zero.

A more extensive discussion on naming compounds appears in Sections 4-9 and 4-10.

2-5 ATOMIC WEIGHTS

As the chemists of the eighteenth and nineteenth centuries painstakingly sought information about the compositions of compounds and tried to systematize their knowledge, it became apparent that each element has a characteristic mass relative to every other element. Although these early scientists did not have the experimental means to measure the mass of each kind of atom, they succeeded in defining a *relative* scale of atomic masses.

An early observation was that carbon and hydrogen have relative atomic masses, also traditionally called **atomic weights, AW,** of approximately 12 and 1, respectively. Thousands of experiments on the compositions of compounds have resulted in the establishment of a scale of relative atomic weights based on the **atomic mass unit (amu)**, which is defined as *exactly $\frac{1}{12}$ of the mass of an atom of a particular kind of carbon atom, called carbon-12.*

On this scale, the atomic weight of hydrogen (H) is 1.00794 amu, that of sodium (Na) is 22.989768 amu, and that of magnesium (Mg) is 24.3050 amu. This tells us that Na atoms have nearly 23 times the mass of H atoms, while Mg atoms are about 24 times heavier than H atoms.

When you need values of atomic weights, consult the periodic table or the alphabetical listing of elements, both found on facing pages inside the front cover.

The term "atomic weight" is widely accepted because of its traditional use, although it is properly a mass rather than a weight. "Atomic mass" is often used.

2-6 THE MOLE

Even the smallest bit of matter that can be handled reliably contains an enormous number of atoms. So we must deal with large numbers of atoms in any real situation, and some unit for conveniently describing a large number of atoms is desirable. The idea of using a unit to describe a particular number (amount) of objects has been around for a long time. You are already familiar with the dozen (12 items) and the gross (144 items).

The SI unit for amount is the **mole,** abbreviated mol. It is *defined* as the amount of substance that contains as many entities (atoms, molecules, or other particles) as there are atoms in exactly 0.012 kg of pure carbon-12 atoms. Many experiments have refined the number, and the currently accepted value is

"Mole" is derived from the Latin word *moles,* which means "a mass." "Molecule" is the diminutive form of this word and means "a small mass."

$$1 \text{ mole} = 6.0221367 \times 10^{23} \text{ particles}$$

This number, often rounded off to 6.022×10^{23}, is called **Avogadro's number** in honor of Amedeo Avogadro (1776–1856), whose contributions to chemistry are discussed in Section 12-8.

According to its definition, the mole unit refers to a fixed number of entities, whose identities must be specified. Just as we speak of a dozen eggs or a pair of aces, we refer to a mole of atoms or a mole of molecules (or a mole of ions, electrons, or other particles). We could even think about a mole of eggs, although the size of the required carton staggers the imagination! Helium exists as discrete He atoms, so one mole of helium consists of 6.022×10^{23} He *atoms*. Hydrogen commonly exists as diatomic (two-atom) molecules, so one mole of hydrogen is 6.022×10^{23} H_2 *molecules* and $2(6.022 \times 10^{23})$ H atoms.

Every kind of atom, molecule, or ion has a definite characteristic mass. It follows that one mole of a given pure substance also has a definite mass, regardless of the source of the sample. This idea is of central importance in many calculations throughout the study of chemistry and the related sciences.

Because the mole is defined as the number of atoms in 0.012 kg (or 12 grams) of carbon-12, and the atomic mass unit is defined as $\frac{1}{12}$ of the mass of a carbon-12 atom, the following convenient relationship is true:

The mass of one mole of atoms of a pure element in grams is numerically equal to the atomic weight of that element in amu. This is also called the **molar mass** of the element; its units are grams/mole.

12 eggs
or
1 dozen eggs
or
24 ounces of eggs

6.022×10^{23} Fe atoms
or
1 mole of Fe atoms
or
55.847 grams of iron

Figure 2-8 Three different ways of representing amounts.

The atomic weight of iron (Fe) is 55.847 amu. Suppose that one dozen eggs weighs 24 ounces.

For instance, if you obtain a pure sample of the metallic element titanium (Ti), whose atomic weight is 47.88 amu, and measure out 47.88 grams of it, you will have one mole, or 6.022×10^{23} titanium atoms.

The symbol for an element can (1) identify the element, (2) represent one atom of the element, or (3) represent one mole of atoms of the element. The last interpretation will be extremely useful in calculations in the next chapter.

A quantity of a substance may be expressed in a variety of ways. For example, consider a dozen eggs and 55.847 grams of iron, or one mole of iron (Figure 2-8). We can express the amount of eggs or iron present in any of several different units. We can then construct unit factors to relate an amount of the substance expressed in one kind of unit to the same amount expressed in another unit.

Unit Factors for Eggs

$$\frac{12 \text{ eggs}}{1 \text{ doz eggs}}$$

$$\frac{12 \text{ eggs}}{24 \text{ ounces of eggs}}$$

and so on

Unit Factors for Iron

$$\frac{6.022 \times 10^{23} \text{ Fe atoms}}{1 \text{ mol Fe atoms}}$$

$$\frac{6.022 \times 10^{23} \text{ Fe atoms}}{55.847 \text{ g Fe}}$$

and so on

Table 2-4 *Mass of One Mole of Atoms of Some Common Elements*

Element	A Sample with a Mass of	Contains
carbon	12.011 g C	6.022×10^{23} C atoms or 1 mole of C atoms
titanium	47.88 g Ti	6.022×10^{23} Ti atoms or 1 mole of Ti atoms
gold	196.96654 g Au	6.022×10^{23} Au atoms or 1 mole of Au atoms
hydrogen	1.00794 g H$_2$	6.022×10^{23} H atoms or 1 mole of H atoms (3.011×10^{23} H$_2$ molecules or 1/2 mole of H$_2$ molecules)
sulfur	32.066 g S$_8$	6.022×10^{23} S atoms or 1 mole of S atoms (0.7528×10^{23} S$_8$ molecules or 1/8 mole of S$_8$ molecules)

Figure 2-9 One mole of atoms of some common elements. Back row (left to right): bromine, aluminum, mercury, copper. Front row (left to right): sulfur, zinc, iron.

As Table 2-4 suggests, the concept of a mole as applied to atoms is especially useful. It provides a convenient basis for comparing the masses of equal numbers of atoms of different elements.

Figure 2-9 shows what one mole of atoms of each of some common elements looks like. Each of the examples in Figure 2-9 represents 6.022×10^{23} *atoms* of the element.

The relationship between the mass of a sample of an element and the number of moles of atoms in the sample is illustrated in Example 2-3.

EXAMPLE 2-3   *Moles of Atoms*

How many moles of atoms does 245.2 g of iron metal contain?

Plan

The atomic weight of iron is 55.85 amu. This tells us that the molar mass of iron is 55.85 g/mol, or that one mole of iron atoms is 55.85 grams of iron. We can express this as either of two unit factors:

To the required four significant figures, 1 mol Fe atoms = 55.85 g Fe.

$$\frac{\text{1 mol Fe atoms}}{\text{55.85 g Fe}} \quad \text{or} \quad \frac{\text{55.85 g Fe}}{\text{1 mol Fe atoms}}$$

Because one mole of iron has a mass of 55.85 g, we expect that 245.2 g will be a fairly small number of moles (greater than one, but less than ten).

Solution

$$\underline{?}\ \text{mole Fe atoms} = 245.2\ \text{g Fe} \times \frac{\text{1 mol Fe atoms}}{\text{55.85 g Fe}} = \boxed{4.390\ \text{mol Fe atoms}}$$

You should now work Exercise 22.

Once the number of moles of atoms of an element is known, the number of atoms in the sample can be calculated, as Example 2-4 illustrates.

CHEMISTRY IN USE

The Development of Science

Avogadro's Number

If you think that the value of Avogadro's number, 6×10^{23}, is too large to be useful to anyone but chemists, look up into the sky on a clear night. You may be able to see about 3,000 stars with the naked eye, but the total number of stars swirling around you in the known universe is approximately equal to Avogadro's number. Just think, the known universe contains approximately one mole of stars! You don't have to leave Earth to encounter such large numbers. The water in the Pacific Ocean has a volume of about 6×10^{23} milliliters and a mass of about 6×10^{23} grams.

Avogadro's number is almost incomprehensibly large. For example, if one mole of dollars were given away at the rate of a million dollars per second beginning when the Earth first formed some 4.5 billion years ago, would any remain today? Surprisingly, about three fourths of the original mole of dollars would be left today; it would take about fourteen billion, five hundred million more years to give away the remaining money at one million dollars per second.

Computers can be used to provide another illustration of the magnitude of Avogadro's number. If a computer can count up to one billion in one second, it would take that computer about 20 million years to count up to 6×10^{23}. In contrast, recorded human history goes back only a few thousand years.

The impressively large size of Avogadro's number can give us very important insights into the very small sizes of individual molecules. Suppose one drop of water evaporates in one hour. There are about 20 drops in one milliliter of water, which weighs one gram. So one drop of water is about 0.05 gram of water. How many H_2O molecules evaporate per second?

$$\frac{\underline{?}\ H_2O\ \text{molecules}}{\text{second}} = \frac{0.05\ \text{g}\ H_2O}{1\ \text{hr}} \times \frac{1\ \text{mol}\ H_2O}{18\ \text{g}\ H_2O} \times$$

$$\frac{6 \times 10^{23}\ H_2O\ \text{molecules}}{1\ \text{mol}\ H_2O} \times \frac{1\ \text{hr}}{60\ \text{min}} \times \frac{1\ \text{min}}{60\ \text{s}}$$

$$= 5 \times 10^{17}\ H_2O\ \text{molecules/second}$$

5×10^{17} H_2O molecules evaporating per second is five hundred million billion H_2O molecules evaporating per second—a number that is beyond our comprehension! This calculation helps us to recognize that water molecules are incredibly small. There are approximately 1.7×10^{21} water molecules in a single drop of water.

By gaining some appreciation of the vastness of Avogadro's number, we gain a greater appreciation of the extremely tiny volumes occupied by individual atoms, molecules, and ions.

Ronald DeLorenzo
Middle Georgia College
Original concept by Larry Nordell,
J. Chem Educ., pending

EXAMPLE 2-4 *Numbers of Atoms*

How many atoms are contained in 4.390 moles of iron atoms?

Plan

One mole of atoms of an element contains Avogadro's number of atoms, or 6.022×10^{23} atoms. This lets us generate the two unit factors

$$\frac{6.022 \times 10^{23}\ \text{atoms}}{1\ \text{mol atoms}} \quad \text{and} \quad \frac{1\ \text{mol atoms}}{6.022 \times 10^{23}\ \text{atoms}}$$

Solution

Write this number with its zeroes and try to name it.

$$\underline{?}\ \text{Fe atoms} = 4.390\ \text{mol Fe atoms} \times \frac{6.022 \times 10^{23}\ \text{Fe atoms}}{1\ \text{mol Fe atoms}} = \boxed{2.644 \times 10^{24}\ \text{Fe atoms}}$$

We expected the number of atoms in more than four moles of atoms to be a very large number.

You should now work Exercise 28.

If we know the atomic weight of an element on the carbon-12 scale, we can use the mole concept and Avogadro's number to calculate the *average* mass of one atom of that element in grams (or any other mass unit we choose).

EXAMPLE 2-5 *Masses of Atoms*

Calculate the mass of one iron atom in grams.

Plan

We expect that the mass of a single atom in grams would be a *very* small number. We know that one mole of Fe atoms has a mass of 55.85 g and contains 6.022×10^{23} Fe atoms. We use this information to generate unit factors to carry out the desired conversion.

Solution

$$\frac{?\ \text{g Fe}}{\text{Fe atom}} = \frac{55.85\ \text{g Fe}}{1\ \text{mol Fe atoms}} \times \frac{1\ \text{mol Fe atoms}}{6.022 \times 10^{23}\ \text{Fe atoms}} = \boxed{9.274 \times 10^{-23}\ \text{g Fe/Fe atom}}$$

To gain some appreciation of how little this is, write 9.274×10^{-23} gram as a decimal fraction and try to name the fraction.

Thus, we see that the mass of one Fe atom is only 9.274×10^{-23} g.

You should now work Exercise 30.

Example 2-5 demonstrates how small atoms are and why it is necessary to use large numbers of atoms in practical work. The next example will help you to realize how large Avogadro's number is.

EXAMPLE 2-6 *Avogadro's Number*

A stack of 500 sheets of typing paper is 1.9 inches thick. Calculate the thickness, in inches and in miles, of a stack of typing paper that contains one mole (Avogadro's number) of sheets.

Plan

We construct unit factors from the data given, from conversion factors in Table 1-7, and from Avogadro's number.

Solution

$$?\ \text{in} = 1\ \text{mol sheets} \times \frac{6.022 \times 10^{23}\ \text{sheets}}{1\ \text{mol sheets}} \times \frac{1.9\ \text{in}}{500\ \text{sheets}} = \boxed{2.3 \times 10^{21}\ \text{in}}$$

$$?\ \text{mi} = 2.3 \times 10^{21}\ \text{in} \times \frac{1\ \text{ft}}{12\ \text{in}} \times \frac{1\ \text{mi}}{5280\ \text{ft}} = \boxed{3.6 \times 10^{16}\ \text{mi}}$$

Imagine the number of trees required to make this much paper!

By comparison, the sun is about 93 million miles from the earth. This stack of paper would make 387 million stacks that reach from the earth to the sun.

You should now work Exercise 18.

▼ **PROBLEM-SOLVING TIP** *When Do We Round Off?*

Even though the number 1.9 has two significant figures, we carry the other numbers in Example 2-6 to more significant figures. Then we round off at the end to the appropriate number of significant figures. The numbers in the distance conversions are exact numbers.

2-7 FORMULA WEIGHTS, MOLECULAR WEIGHTS, AND MOLES

> The **formula weight** (**FW**) of a substance is the sum of the atomic weights (AW) of the elements in the formula, each taken the number of times the element occurs. Hence a formula weight gives the mass of one formula unit in amu.

Formula weights, like the atomic weights on which they are based, are relative masses. The formula weight for sodium hydroxide, NaOH, (rounded off to the nearest 0.01 amu) is found as follows.

No. of Atoms of Stated Kind		× Mass of One Atom	= Mass Due to Element
1 × Na =	1	× 23.00 amu	= 23.00 amu of Na
1 × H =	1	× 1.01 amu	= 1.01 amu of H
1 × O =	1	× 16.00 amu	= 16.00 amu of O

Formula weight of NaOH = 40.01 amu

The term "formula weight" is correctly used for either ionic or molecular substances. When we refer specifically to molecular (nonionic) substances, i.e., substances that exist as discrete molecules, we often substitute the term **molecular weight** (**MW**).

EXAMPLE 2-7 *Formula Weights*

Calculate the formula weight (molecular weight) of 2,4,6-trinitrotoluene (TNT), $C_7H_5(NO_2)_3$, using the precisely known values for atomic weights given in the International Table of Atomic Weights inside the front cover of the text.

Plan

We add the atomic weights of the elements in the formula, each multiplied by the number of times the element occurs. Because the least precisely known atomic weight (12.011 amu for C) is known to three significant figures past the decimal point, the result is shown to only that number of significant figures.

Solution

No. of Atoms of Stated Kind		× Mass of One Atom	= Mass Due to Element
7 × C =	7	× 12.011 amu	= 84.077 amu of C
5 × H =	5	× 1.00794 amu	= 5.03970 amu of H
3 × N =	3	× 14.00674 amu	= 42.02022 amu of N
6 × O =	6	× 15.9994 amu	= 95.9964 amu of O

Formula weight (molecular weight) of 2,4,6-trinitrotoluene (TNT) = 227.133 amu

You should now work Exercise 24.

> The amount of substance that contains the mass in grams numerically equal to its formula weight in amu contains 6.022×10^{23} formula units, or *one mole* of the substance. This is sometimes called the **molar mass** of the substance. Molar mass is *numerically equal* to the formula weight of the substance (the atomic weight for atoms of elements), and has the units grams/mole.

Table 2-5 *One Mole of Some Common Molecular Substances*

Substance	Molecular Weight	A Sample with a Mass of	Contains
hydrogen	2.016	2.016 g H_2	6.022×10^{23} H_2 molecules or 1 mol of H_2 molecules (contains $2 \times 6.022 \times 10^{23}$ H atoms or 2 mol of H atoms)
oxygen	32.00	32.00 g O_2	6.022×10^{23} O_2 molecules or 1 mol of O_2 molecules (contains $2 \times 6.022 \times 10^{23}$ O atoms or 2 mol of O atoms)
methane	16.04	16.04 g CH_4	6.022×10^{23} CH_4 molecules or 1 mol of CH_4 molecules (contains 6.022×10^{23} C atoms and $4 \times 6.022 \times 10^{23}$ H atoms)
2,4,6-trinitro-toluene (TNT)	227.13	227.13 g $C_7H_5(NO_2)_3$	6.022×10^{23} $C_7H_5(NO_2)_3$ molecules or 1 mol of $C_7H_5(NO_2)_3$ molecules

One mole of sodium hydroxide is 40.01 g of NaOH, and one mole of TNT is 227.133 g of $C_7H_5(NO_2)_3$. One mole of any molecular substance contains 6.022×10^{23} molecules of the substance, as Table 2-5 illustrates.

The physical appearance of one mole of each of some compounds is illustrated in Figure 2-10. Two different forms of oxalic acid are shown. The formula unit (molecule) of oxalic acid is $(COOH)_2$ (FW = 90.04 amu; molar mass = 90.04 g/mol). However, when oxalic acid is obtained by crystallization from a water solution, two molecules of water are present for each molecule of oxalic acid, even though it appears dry. The formula of this **hydrate** is $(COOH)_2 \cdot 2H_2O$ (FW = 126.06 amu; molar mass = 126.06 g/mol). The dot shows that the crystals contain two H_2O molecules per $(COOH)_2$ molecule. The water can be driven out of the crystals by heating to leave **anhydrous** oxalic acid, $(COOH)_2$. Anhydrous means "without water." Copper(II) sulfate, an *ionic* compound, shows similar

Figure 2-10 One mole of some compounds. The colorless liquid is water, H_2O (1 mol = 18.0 g = 18.0 mL). The white solid (left) is *anhydrous* oxalic acid, $(COOH)_2$ (1 mol = 90.0 g). The second white solid is *hydrated* oxalic acid, $(COOH)_2 \cdot 2H_2O$ (1 mol = 126.0 g). The blue solid is hydrated copper(II) sulfate, $CuSO_4 \cdot 5H_2O$ (1 mol = 249.68 g). The red solid is mercury(II) oxide (1 mol = 216.59 g).

Table 2-6 *One Mole of Some Ionic Compounds*

Compound	Formula Weight	A Sample with a Mass of 1 Mole	Contains
sodium chloride	58.44	58.44 g NaCl	6.022×10^{23} Na$^+$ ions or 1 mole of Na$^+$ ions 6.022×10^{23} Cl$^-$ ions or 1 mole of Cl$^-$ ions
calcium chloride	111.0	111.0 g CaCl$_2$	6.022×10^{23} Ca^{2+} ions or 1 mole of Ca^{2+} ions $2(6.022 \times 10^{23})$ Cl$^-$ ions or 2 moles of Cl$^-$ ions
aluminum sulfate	342.1	342.1 g Al$_2$(SO$_4$)$_3$	$2(6.022 \times 10^{23})$ Al^{3+} ions or 2 moles of Al^{3+} ions $3(6.022 \times 10^{23})$ SO$_4$$^{2-}$ ions or 3 moles of SO$_4$$^{2-}$ ions

behavior. Anhydrous copper(II) sulfate (CuSO$_4$; FW = 159.60 amu; molar mass = 159.60 g/mol) is almost white. Hydrated copper(II) sulfate (CuSO$_4 \cdot 5H_2O$; FW = 249.68 amu; molar mass =249.68 g/mol) is deep blue.

Because there are no simple NaCl molecules at ordinary temperatures, it is inappropriate to refer to the "molecular weight" of NaCl or any ionic compound. One mole of an ionic compound contains 6.022×10^{23} *formula units* (FU) of the substance. Recall that one formula unit of sodium chloride consists of one sodium ion, Na$^+$, and one chloride ion, Cl$^-$. One mole, or 58.44 grams, of NaCl contains 6.022×10^{23} Na$^+$ ions and 6.022×10^{23} Cl$^-$ ions. See Table 2-6.

The mole concept, together with Avogadro's number, provides important connections among the extensive properties mass of substance, number of moles of substance, and number of molecules or ions. These are summarized as follows.

Heating blue CuSO$_4 \cdot 5H_2O$ forms anhydrous CuSO$_4$, which is white. Some blue CuSO$_4 \cdot 5H_2O$ is visible in the cooler center portion of the crucible.

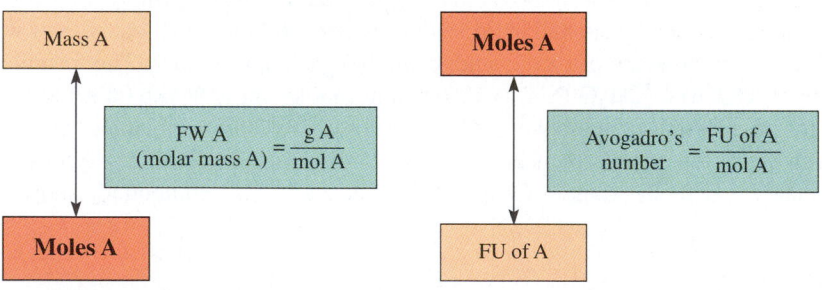

The following examples show the relations between numbers of molecules, atoms, or formula units and their masses.

EXAMPLE 2-8 *Masses of Molecules*

What is the mass in grams of 10.0 billion SO$_2$ molecules?

Plan

One mole of SO$_2$ contains 6.02×10^{23} SO$_2$ molecules and has a mass of 64.1 grams.

When fewer than four significant figures are used in calculations, Avogadro's number may be rounded off to 6.02×10^{23}.

Solution

$$\underline{?} \text{ g } SO_2 = 10.0 \times 10^9 \text{ } SO_2 \text{ molecules} \times \frac{64.1 \text{ g } SO_2}{6.02 \times 10^{23} \text{ } SO_2 \text{ molecules}}$$

$$= 1.06 \times 10^{-12} \text{ g } SO_2$$

Ten billion SO_2 molecules have a mass of only 0.00000000000106 gram. Commonly used analytical balances are capable of weighing to ± 0.0001 gram.

You should now work Exercise 34.

EXAMPLE 2-9 *Moles*

How many (a) moles of O_2, (b) O_2 molecules, and (c) O atoms are contained in 40.0 grams of oxygen gas (dioxygen) at 25°C?

Plan

We construct the needed unit factors from the following equalities: (a) the mass of one mole of O_2 is 32.0 g (molar mass O_2 = 32.0 g/mol); (b) one mole of O_2 contains 6.02×10^{23} O_2 molecules; (c) one O_2 molecule contains two O atoms.

Solution

One mole of O_2 contains 6.02×10^{23} O_2 molecules, and its mass is 32.0 g.

(a)
$$\underline{?} \text{ mol } O_2 = 40.0 \text{ g } O_2 \times \frac{1 \text{ mol } O_2}{32.0 \text{ g } O_2} = 1.25 \text{ mol } O_2$$

(b)
$$\underline{?} \text{ } O_2 \text{ molecules} = 40.0 \text{ g } O_2 \times \frac{6.02 \times 10^{23} \text{ } O_2 \text{ molecules}}{32.0 \text{ g } O_2} = 7.52 \times 10^{23} \text{ molecules}$$

Or, we can use the number of moles of O_2 calculated in part (a) to find the number of O_2 molecules.

$$\underline{?} \text{ } O_2 \text{ molecules} = 1.25 \text{ mol } O_2 \times \frac{6.02 \times 10^{23} \text{ } O_2 \text{ molecules}}{1 \text{ mol } O_2} = 7.52 \times 10^{23} \text{ } O_2 \text{ molecules}$$

(c)
$$\underline{?} \text{ O atoms} = 40.0 \text{ g } O_2 \times \frac{6.02 \times 10^{23} \text{ } O_2 \text{ molecules}}{32.0 \text{ g } O_2} \times \frac{2 \text{ O atoms}}{1 \text{ } O_2 \text{ molecule}}$$

$$= 1.50 \times 10^{24} \text{ O atoms}$$

You should now work Exercise 36.

EXAMPLE 2-10 *Numbers of atoms*

Calculate the number of hydrogen atoms in 39.6 grams of ammonium sulfate, $(NH_4)_2SO_4$.

Plan

One mole of $(NH_4)_2SO_4$ is 6.02×10^{23} formula units (FU) and has a mass of 132 g.

g of $(NH_4)_2SO_4$ $\longrightarrow$ mol of $(NH_4)_2SO_4$ $\longrightarrow$ Fu of $(NH_4)_2SO_4$ $\longrightarrow$ H atoms

In Example 2-10, we relate (a) g to mol, (b) mol to FU, and (c) FU to H atoms

Solution

$$\underset{-}{?} \text{ H atoms} = 39.6 \text{ g } (NH_4)_2SO_4 \times \frac{1 \text{ mol } (NH_4)_2SO_4}{132 \text{ g } (NH_4)_2SO_4}$$

$$\times \frac{6.02 \times 10^{23} \text{ FU } (NH_4)_2SO_4}{1 \text{ mol } (NH_4)_2SO_4} \times \frac{8 \text{ H atoms}}{1 \text{ FU } (NH_4)_2SO_4}$$

$$= \boxed{1.44 \times 10^{24} \text{ H atoms}}$$

You should now work Exercise 38.

$$FW = \frac{\text{no. of grams}}{\text{no. of moles}}$$

$$FW = \frac{\text{no. of milligrams}}{\text{no. of millimoles}}$$

The term "millimole" (mmol) is useful in laboratory work. As the prefix indicates, one **mmol** is 1/1000 of a mole. Small masses are frequently expressed in milligrams (mg) rather than grams. The relation between millimoles and milligrams is the same as that between moles and grams (Table 2-7).

EXAMPLE 2-11   *Millimoles*

Calculate the number of millimoles of sulfuric acid in 0.147 gram of H_2SO_4.

Plan

1 mol H_2SO_4 = 98.1 g H_2SO_4; 1 mmol H_2SO_4 = 98.1 mg H_2SO_4, or 0.0981 g H_2SO_4. We can use these equalities to solve this problem by either of two methods. Method 1: Express formula weight in g/mmol, then convert g H_2SO_4 to mmol H_2SO_4. Method 2: Convert g H_2SO_4 to mg H_2SO_4, then use the unit factor mg/mmol to convert to mmol H_2SO_4.

Solution
Method 1:

$$\underset{-}{?} \text{ mmol } H_2SO_4 = 0.147 \text{ g } H_2SO_4 \times \frac{1 \text{ mmol } H_2SO_4}{0.0981 \text{ g } H_2SO_4} = \boxed{1.50 \text{ mmol } H_2SO_4}$$

Method 2: Using 0.147 g H_2SO_4 = 147 mg H_2SO_4, we have

$$\underset{-}{?} \text{ mmol } H_2SO_4 = 147 \text{ mg } H_2SO_4 \times \frac{1 \text{ mmol } H_2SO_4}{98.1 \text{ mg } H_2SO_4} = \boxed{1.50 \text{ mmol } H_2SO_4}$$

You should now work Exercise 40.

Grams can be converted to milligrams by shifting the decimal three places to the right.

Table 2-7 *Comparison of Moles and Millimoles*		
Compound	**1 Mole**	**1 Millimole**
NaOH	40.0 g	40.0 mg or 0.0400 g
H_3PO_4	98.1 g	98.1 mg or 0.0981 g
SO_2	64.1 g	64.1 mg or 0.0641 g
C_3H_8	44.1 g	44.1 mg or 0.0441 g

2-8 PERCENT COMPOSITION AND FORMULAS OF COMPOUNDS

If the formula of a compound is known, its chemical composition can be expressed as the mass percent of each element in the compound. For example, one carbon dioxide molecule, CO_2, contains one C atom and two O atoms. Percentage is the part divided by the whole times 100 percent (or simply parts per 100), so we can represent the percent composition of carbon dioxide as follows:

$$\% \ C = \frac{\text{mass of C}}{\text{mass of } CO_2} \times 100\% = \frac{\text{AW of C}}{\text{MW of } CO_2} \times 100\% = \frac{12.0 \ \text{amu}}{44.0 \ \text{amu}} \times 100\% = \boxed{27.3\%}$$

$$\% \ O = \frac{\text{mass of O}}{\text{mass of } CO_2} \times 100\% = \frac{2 \times \text{AW of O}}{\text{MW of } CO_2} \times 100\% = \frac{2(16.0 \ \text{amu})}{44.0 \ \text{amu}} \times 100\% = \boxed{72.7\% \ O}$$

As a check, we see that the percentages add to 100%.

One *mole* of CO_2 (44.0 g) contains one *mole* of C atoms (12.0 g) and two *moles* of O atoms (32.0 g). Therefore, we could have used these masses in the preceding calculation. These numbers are the same as the ones used—only the units are different. In Example 2-12 we shall base our calculation on one *mole* rather than one *molecule*.

EXAMPLE 2-12 *Percent Composition*

Calculate the percent composition of HNO_3 by mass.

Plan

We first calculate the mass of one mole as in Example 2-7. Then we express the mass of each element as a percent of the total.

Solution

The molar mass of HNO_3 is calculated first.

No. of Mol of Atoms		× Mass of One Mol of Atoms	= Mass Due to Element
1 × H =	1	× 1.0 g	= 1.0 g of H
1 × N =	1	× 14.0 g	= 14.0 g of N
3 × O =	3	× 16.0 g	= 48.0 g of O

Mass of 1 mol of HNO_3 = 63.0 g

Now, its percent composition is

$$\% \ H = \frac{\text{mass of H}}{\text{mass of } HNO_3} \times 100\% = \frac{1.0 \ \text{g}}{63.0 \ \text{g}} \times 100\% = \boxed{1.6\% \ H}$$

$$\% \ N = \frac{\text{mass of N}}{\text{mass of } HNO_3} \times 100\% = \frac{14.0 \ \text{g}}{63.0 \ \text{g}} \times 100\% = \boxed{22.2\% \ N}$$

$$\% \ O = \frac{\text{mass of O}}{\text{mass of } HNO_3} \times 100\% = \frac{48.0 \ \text{g}}{63.0 \ \text{g}} \times 100\% = \boxed{76.2\% \ O}$$

Total = 100.0%

When chemists use the % notation, they mean percent by mass unless they specify otherwise.

You should now work Exercise 44.

CHEMISTRY IN USE

The Development of Science

Names of the Elements

If you were to discover a new element, how would you name it? Throughout history, scientists have answered this question in different ways. Most have chosen to honor a person or place or to describe the new substance.

Until the Middle Ages only nine elements were known: gold, silver, tin, mercury, copper, lead, iron, sulfur, and carbon. The metals' chemical symbols are taken from descriptive Latin names: *aurum* ("yellow"), *argentum* ("shining"), *stannum* ("dripping" or "easily melted"), *hydrargyrum* ("silvery water"), *cuprum* (Cyprus, where many copper mines were located), *plumbum* (exact meaning unknown—possibly "heavy"), and *ferrum* (also unknown). Mercury is named after the planet, one reminder that the ancients associated metals with gods and celestial bodies. In turn, both the planet, which moves rapidly across the sky, and the element, which is the only metal that is liquid at room temperature and thus flows rapidly, are named for the fleet god of messengers in Roman mythology. In English, mercury is nicknamed "quicksilver."

Prior to the reforms of Antoine Lavoisier, chemistry was a largely nonquantitative, unsystematic science in which experimenters had little contact with each other. In 1787 Lavoisier published his *Methode de Nomenclature Chimique,* which proposed, among other changes, that all new elements be named descriptively. For the next 125 years, most elements were given names that corresponded to their properties. Greek roots were one popular source, as evidenced by hydrogen (*hydros-gen,* "water-producing"), oxygen (*oksys-gen,* "acid-producing"), nitrogen (*nitron-gen,* "soda-producing"), bromine (*bromos,* "stink"), and argon (*a-er-gon,* "no reaction"). The discoverers of argon, Ramsay and Rayleigh, originally proposed the name *aeron* (from *aer* or air) but critics thought it was too close to the biblical name Aaron! Latin roots such as *radius* ("ray") were also used (radium and radon are both naturally radioactive elements that emit "rays"). Color was often the determining property, especially after the invention of the spectroscope in 1859, because different elements (or the light that they emit) have prominent characteristic colors. Cesium, indium, iodine, rubidium, and thallium were all named in this manner. Their respec-

tive Greek and Latin roots denote blue-gray, indigo, violet, red, and green (*thallus* means "tree sprout"). Because of the great variety of colors of its compounds, iridium takes its name from the Latin *iris,* meaning rainbow. Alternatively, an element name might suggest a mineral or the ore that contained it. One example is wolfram or tungsten (W), which was isolated from wolframite. Two other "inconsistent" elemental symbols, K and Na, arose from occurrence as well. *Kalium* was first obtained from the saltwort plant, *Salsola kali,* and *natrium* from niter. Their English names, potassium and sodium, are derived from the ores potash and soda.

Other elements, contrary to Lavoisier's suggestion, were named after planets, mythological figures, places, or superstitions. "Celestial elements" include helium (sun), tellurium (earth), selenium (moon—the element was discovered in close proximity to tellurium), cerium (the asteroid Ceres, which was discovered only two years before the element), and uranium (the planet Uranus, discovered a few years earlier). The first two transuranium elements (those *beyond* uranium) to be produced were named neptunium and plutonium for the next two planets, Neptune and Pluto. The names promethium (Prometheus, who stole fire from heaven), vanadium (Scandinavian goddess, Vanadis), titanium (Titans, the first sons of the earth), tantalum (Tantalos, father of Niobe), and thorium (Thor, Scandinavian god of war) all arise from Greek or Norse mythology.

"Geographical elements," shown on the map, sometimes honored the discoverer's native country or workplace. The Latin names for Russia (*ruthenium*), France (*gallium*), Paris (*lutetium*), and Germany (*germanium*) were among those used. Marie Sklodowska Curie named one of the elements that she discovered polonium, after her native Poland. Often the locale of discovery lends its name to the element; the record holder is certainly the Swedish village Ytterby, the site of ores from which the four elements terbium, erbium, ytterbium, and yttrium were isolated. Elements honoring important scientists include curium, einsteinium, nobelium, fermium, and lawrencium.

Most of the elements now known were given titles peacefully, but a few were not. Niobium, isolated in 1803 by Ekeberg

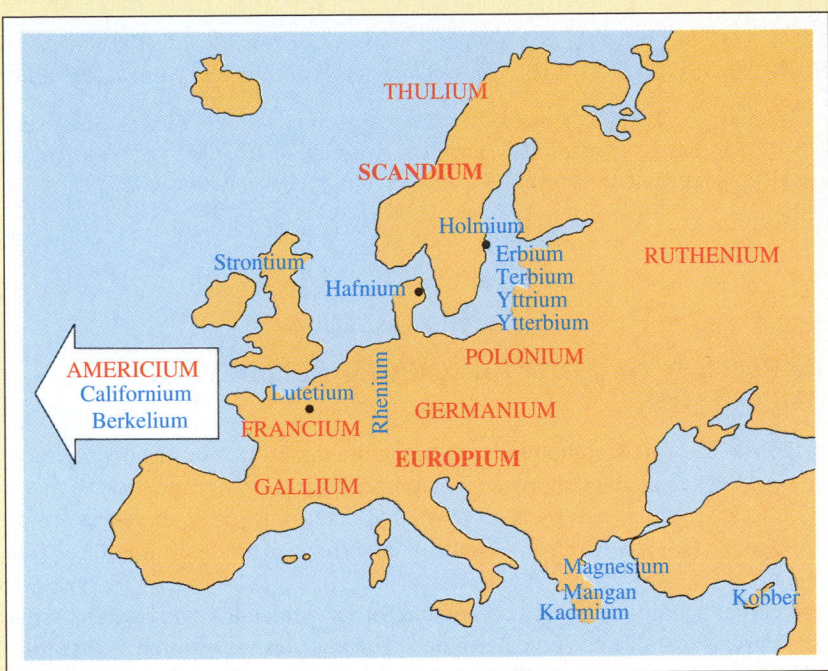

Many chemical elements were named after places.

from an ore that also contained tantalum, and named after the Greek goddess Niobe (daughter of Tantalus), was later found to be identical to an 1802 discovery of Hatchett, columbium. (Interestingly, Hatchett first found the element in an ore sample that had been sent to England more than a century earlier by John Winthrop, the first governor of Connecticut.) While "niobium" became the accepted designation in Europe, the Americans, not surprisingly, chose "columbium." It was not until 1949—when the International Union of Pure and Applied Chemistry (IUPAC) ended more than a century of controversy by ruling in favor of mythology—that element 41 received a unique name.

In 1978, the IUPAC recommended that, at least for now, the elements beyond element 103 be known by systematic names based on numerical roots; element 104 is unnilquadium (*un* for 1, *nil* for 0, *quad* for 4, plus the -*ium* ending), followed by unnilpentium, unnilhexium, and so on. Arguments over the proper names of elements 104 and 105 prompted the IUPAC to

begin hearing claims of priority to numbers 104 through 109, but names for those elements have not yet been officially approved by IUPAC. In November 1993, the Nomenclature Committee of the American Chemical Society (ACS) endorsed the use in the United States of the names rutherfordium (Rf) and hahnium (Ha) for elements 104 and 105; and nielsbohrium (Ns), hassium (Hs) and meitnerium (Mt) for 107, 108, and 109, respectively. The name seaborgium (Sg), proposed by ACS for element 106, has been rejected by IUPAC. You will recognize some of these as derived from the names of scientists prominent in the development of atomic theory; others are named for scientists who were involved in the discovery of these heavy elements.

Lisa L. Saunders
Chemistry graduate student
University of California, Berkeley

A ball-and-stick model of a molecule of water, H_2O.

A ball-and-stick model of a molecule of hydrogen peroxide, H_2O_2.

Nitric acid is 1.6% H, 22.2% N, and 76.2% O by mass. All samples of pure HNO_3 have this composition, according to the Law of Definite Proportions.

> ▼ **PROBLEM-SOLVING TIP**   *The Whole Is Equal to the Sum of Its Parts*
>
> Percentages must add to 100%. However, roundoff errors may not cancel, and totals such as 99.9% or 100.1% may be obtained in calculations. As an alternative method of calculation, if we know all of the percentages except one, we can subtract their sum from 100% to obtain the other value.

2-9 DERIVATION OF FORMULAS FROM ELEMENTAL COMPOSITION

Each year thousands of new compounds are made in laboratories or discovered in nature. One of the first steps in characterizing a new compound is the determination of its percent composition. A *qualitative* analysis is performed to determine *which* elements are present in the compound. Then a *quantitative* analysis is performed to determine the *amount* of each element.

Once the percent composition of a compound (or its elemental composition by mass) is known, the simplest formula can be determined. The **simplest** or **empirical formula** for a compound is the smallest whole-number ratio of atoms present. For molecular compounds the **molecular formula** indicates the *actual* numbers of atoms present in a molecule of the compound. It may be the same as the simplest formula or else some whole-number multiple of it. For example, the simplest and molecular formulas for water are both H_2O. However, for hydrogen peroxide, they are HO and H_2O_2, respectively.

EXAMPLE 2-13   *Simplest Formulas*

Compounds containing sulfur and oxygen are serious air pollutants; they represent the major cause of acid rain. Analysis of a sample of a pure compound reveals that it contains 50.1% sulfur and 49.9% oxygen by mass. What is the simplest formula of the compound?

Plan

One mole of atoms of any element is 6.022×10^{23} atoms, so the ratio of moles of atoms in any sample of a compound is the same as the ratio of atoms in that compound. This calculation is carried out in two steps. Step 1: Let's consider 100.0 grams of compound, which contains 50.1 grams of S and 49.9 grams of O. We calculate the number of moles of atoms of each. Step 2: We then obtain a whole-number ratio between these numbers that gives the ratio of atoms in the sample, and hence in the simplest formula for the compound.

Remember that *percent* means *parts per hundred.*

Solution

Step 1:
$$\underset{\_}{?} \text{ mol S atoms} = 50.1 \text{ g S} \times \frac{1 \text{ mol S atoms}}{32.1 \text{ g S}} = 1.56 \text{ mol S atoms}$$

$$\underset{\_}{?} \text{ mol O atoms} = 49.9 \text{ g O} \times \frac{1 \text{ mol O atoms}}{16.0 \text{ g O}} = 3.12 \text{ mol O atoms}$$

Step 2: Now we know that 100.0 grams of the compound contains 1.56 moles of S atoms and 3.12 moles of O atoms. We obtain a whole-number ratio between these numbers that gives the ratio of atoms in the simplest formula.

$$\frac{1.56}{1.56} = 1.00 \text{ S}$$

$$\frac{3.12}{1.56} = 2.00 \text{ O}$$

SO_2

A simple and useful way to obtain whole-number ratios among several numbers follows. (a) Divide each number by the smallest, and then, (b) if necessary, multiply all of the resulting numbers by the smallest whole number that will eliminate fractions.

You should now work Exercise 48.

The solution for Example 2-13 can be set up in tabular form.

Element	Relative Mass of Element	Relative Number of Atoms (divide mass by AW)	Divide by Smaller Number	Smallest Whole-Number Ratio of Atoms
S	50.1	$\dfrac{50.1}{32.1} = 1.56$	$\dfrac{1.56}{1.56} = 1.00$ S	
O	49.9	$\dfrac{49.9}{16.0} = 3.12$	$\dfrac{3.12}{1.56} = 2.00$ O	SO_2

This tabular format provides a convenient way to solve simplest-formula problems, as the next example illustrates.

The "Relative Mass" column is proportional to the mass of each element in grams. With this interpretation, the next column could be headed "Relative Number of *Moles* of Atoms." Then the last column would represent the smallest whole-number ratios of *moles* of atoms. But because a mole is always the same number of items (atoms), that ratio is the same as the smallest whole-number ratio of atoms.

EXAMPLE 2-14   *Simplest Formula*

A 20.882-gram sample of an ionic compound is found to contain 6.072 grams of Na, 8.474 grams of S, and 6.336 grams of O. What is its simplest formula?

Plan

We reason as in Example 2-13, calculating the number of moles of each element and the ratio among them. Here we use the tabular format that was introduced above.

Solution

Element	Relative Mass of Element	Relative Number of Atoms (divide mass by AW)	Divide by Smallest Number	Convert Fractions to Whole Numbers (multiply by integer)	Smallest Whole-Number Ratio of Atoms
Na	6.072	$\dfrac{6.072}{23.0} = 0.264$	$\dfrac{0.264}{0.264} = 1.00$	$1.00 \times 2 = 2$ Na	
S	8.474	$\dfrac{8.474}{32.1} = 0.264$	$\dfrac{0.264}{0.264} = 1.00$	$1.00 \times 2 = 2$ S	$Na_2S_2O_3$
O	6.336	$\dfrac{6.336}{16.0} = 0.396$	$\dfrac{0.396}{0.264} = 1.50$	$1.50 \times 2 = 3$ O	

The ratio of atoms in the simplest formula *must be a whole-number ratio* (by definition). To convert the ratio 1 : 1 : 1.5 to a whole-number ratio, each number in the ratio was multiplied by 2, which gave the simplest formula $Na_2S_2O_3$.

You should now work Exercise 49.

▼ **PROBLEM-SOLVING TIP**   *Know Common Fractions in Decimal Form*

As Example 2-14 illustrates, sometimes we must convert a fraction to a whole number by multiplication by the correct integer. But we must first recognize which fraction is represented by a nonzero part of a number. The decimal equivalents of the following fractions may be useful.

Fraction	Decimal Equivalent (to 2 places)	To Convert to Integer, Multiply by
1/2	0.50	2
1/3	0.33	3
2/3	0.67	3
1/4	0.25	4
3/4	0.75	4
1/5	0.20	5

The fractions 2/5, 3/5, 4/5 are equal to 0.40, 0.60, and 0.80, respectively; these should be multiplied by 5.

When we use the procedure given in this section, we often obtain numbers such as 0.99 and 1.52. Because there is usually some error in results obtained by analysis of samples (as well as roundoff errors), we would interpret 0.99 as 1.0 and 1.52 as 1.5.

2-10 DETERMINATION OF MOLECULAR FORMULAS

Percent composition data yield only simplest formulas. To determine the molecular formula for a molecular compound, *both* its simplest formula and its molecular weight must be known. Some methods for experimental determination of molecular weights are introduced in Chapters 12 and 14.

Millions of compounds are composed of carbon, hydrogen, and oxygen. Analyses for C and H can be performed in a C-H combustion system (Figure 2-11). An accurately known mass of a compound is burned in a furnace in a stream of oxygen. The carbon and

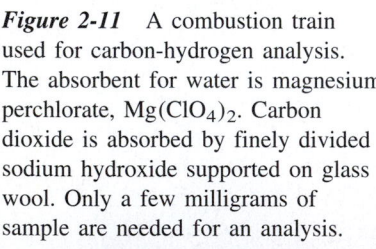

Figure 2-11   A combustion train used for carbon-hydrogen analysis. The absorbent for water is magnesium perchlorate, $Mg(ClO_4)_2$. Carbon dioxide is absorbed by finely divided sodium hydroxide supported on glass wool. Only a few milligrams of sample are needed for an analysis.

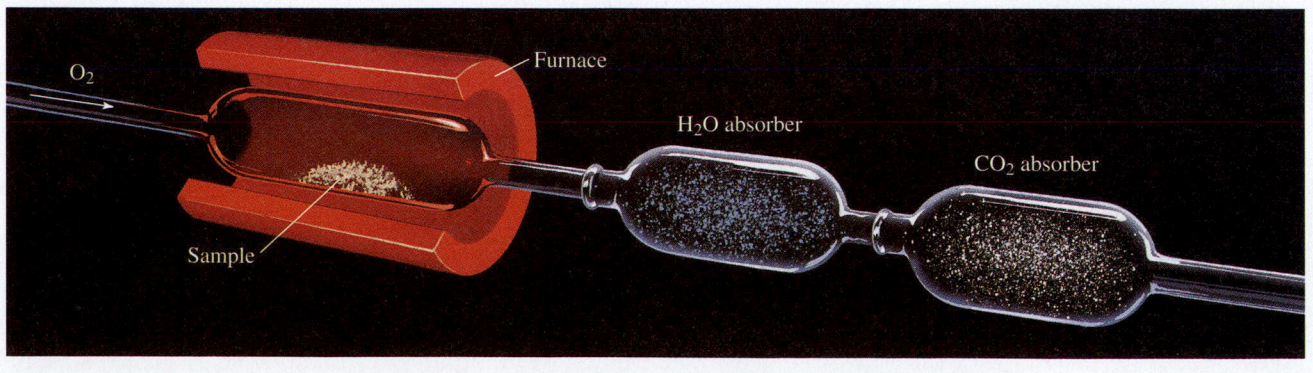

hydrogen in the sample are converted to carbon dioxide and water vapor, respectively. The resulting increases in masses of the CO_2 and H_2O absorbers can then be related to the masses and percentages of carbon and hydrogen in the original sample.

EXAMPLE 2-15 *Percent Composition*

Hydrocarbons are organic compounds composed entirely of hydrogen and carbon. A 0.1647-gram sample of a pure hydrocarbon was burned in a C-H combustion train to produce 0.4931 gram of CO_2 and 0.2691 gram of H_2O. Determine the masses of C and H in the sample and the percentages of these elements in this hydrocarbon.

Plan

Step 1: We use the observed mass of CO_2, 0.4931 grams, to determine the mass of carbon in the original sample. There is one mole of carbon atoms, 12.01 grams, in each mole of CO_2, 44.01 grams; we use this information to construct the unit factor

$$\frac{12.01 \text{ g C}}{44.01 \text{ g } CO_2}$$

Step 2: Likewise, we can use the observed mass of H_2O, 0.2691 grams, to calculate the amount of hydrogen in the original sample. We use the fact that there are two moles of hydrogen atoms, 2.016 grams, in each mole of H_2O, 18.02 grams, to construct the unit factor

$$\frac{2.016 \text{ g H}}{18.02 \text{ g } H_2O}$$

Step 3: Then we calculate the percentages by mass of each element in turn, using the relationship

$$\% \text{ element} = \frac{\text{g element}}{\text{g sample}} \times 100\%$$

Solution

Step 1: $\underset{\text{—}}{?} \text{ g C} = 0.4931 \text{ g } CO_2 \times \dfrac{12.01 \text{ g C}}{44.01 \text{ g } CO_2} = \boxed{0.1346 \text{ g C}}$

Step 2: $\underset{\text{—}}{?} \text{ g H} = 0.2691 \text{ g } H_2O \times \dfrac{2.016 \text{ g H}}{18.02 \text{ g } H_2O} = \boxed{0.03010 \text{ g H}}$

Step 3: $\% \text{ C} = \dfrac{0.1346 \text{ g C}}{0.1647 \text{ g sample}} \times 100\% = \boxed{81.72\% \text{ C}}$

$\% \text{ H} = \dfrac{0.03010 \text{ g H}}{0.1647 \text{ g sample}} \times 100\% = \boxed{18.28\% \text{ H}}$

Total = 100.00%

You should now work Exercise 52.

Hydrocarbons are obtained from coal and coal tar and from oil and gas wells. The main use of hydrocarbons is as fuels. The simplest hydrocarbons are

methane	CH_4
ethane	C_2H_6
propane	C_3H_8
butane	C_4H_{10}

We could calculate the mass of H by subtracting mass of C from mass of sample. However, it is good experimental practice, if possible, to base both on experimental measurements as we have done here. This would help to check for errors in the analysis or calculation.

Can you show that the hydrocarbon in Example 2-15 is propane, C_3H_8?

When the compound to be analyzed contains oxygen, the calculation of the amount or percentage of oxygen in the sample is somewhat different. Part of the oxygen that goes to form CO_2 and H_2O comes from the sample and part comes from the oxygen stream supplied. Therefore we cannot directly determine the amount of oxygen already in the sample. The approach is to analyze as we did in Example 2-15 for all elements *except* oxygen. Then we subtract the sum of their masses from the mass of the original sample to obtain the mass of oxygen. The next example illustrates such a calculation.

EXAMPLE 2-16   *Percent Composition*

A 0.1014-gram sample of purified glucose was burned in a C-H combustion train to produce 0.1486 gram of CO_2 and 0.0609 gram of H_2O. An elemental analysis showed that glucose contains only carbon, hydrogen, and oxygen. Determine the masses of C, H, and O in the sample and the percentages of these elements in glucose.

Plan

Steps 1 and 2: We first calculate the masses of carbon and hydrogen as we did in Example 2-15. Step 3: The rest of the sample must be oxygen because glucose has been shown to contain only C, H, and O. So we subtract the masses of C and H from the total mass of sample. Step 4: Then we calculate the percentage by mass for each element.

Solution

Step 1: $\underset{=}{?}$ g C $= 0.1486$ g $CO_2 \times \dfrac{12.01 \text{ g C}}{44.01 \text{ g } CO_2} = \boxed{0.04055 \text{ g C}}$

Step 2: $\underset{=}{?}$ g H $= 0.0609$ g $H_2O \times \dfrac{2.016 \text{ g H}}{18.02 \text{ g } H_2O} = \boxed{0.00681 \text{ g H}}$

We say that the mass of O in the sample is calculated by difference.

Step 3: $\underset{=}{?}$ g O $= 0.1014$ g sample $- [0.04055 \text{ g C} + 0.00681 \text{ g H}] = \boxed{0.0540 \text{ g O}}$

Step 4: Now we can calculate the percentages by mass for each element:

$$\% \text{ C} = \frac{0.04055 \text{ g C}}{0.1014 \text{ g}} \times 100\% = \boxed{39.99\% \text{ C}}$$

$$\% \text{ H} = \frac{0.00681 \text{ g H}}{0.1014 \text{ g}} \times 100\% = \boxed{6.72\% \text{ H}}$$

$$\% \text{ O} = \frac{0.0540 \text{ g O}}{0.1014 \text{ g}} \times 100\% = \boxed{53.2\% \text{ O}}$$

$$\text{Total} = 99.9\%$$

You should now work Exercise 50.

Glucose, a simple sugar, is the main component of intravenous feeding liquids. Its common name is dextrose. It is also one of the products of carbohydrate metabolism.

For many compounds the molecular formula is a multiple of the simplest formula. Consider butane, C_4H_{10}. The simplest formula for butane is C_2H_5, but the molecular formula contains twice as many atoms; i.e., $2 \times (C_2H_5) = C_4H_{10}$. Benzene, C_6H_6, is another example. The simplest formula for benzene is CH, but the molecular formula contains six times as many atoms; i.e., $6 \times (CH) = C_6H_6$.

The molecular formula for a compound is either the same as, or an *integer* multiple of, the simplest formula.

$$\text{molecular formula} = n \times \text{simplest formula}$$

So we can write

$$\text{molecular weight} = n \times \text{simplest-formula weight}$$

$$n = \frac{\text{molecular weight}}{\text{simplest-formula weight}}$$

The molecular formula is then obtained by multiplying the simplest formula by the integer, n.

EXAMPLE 2-17 *Molecular Formula*

In Example 2-16 we found the elemental composition of glucose. Other experiments show that its molecular weight is approximately 180 amu. Determine the simplest formula and the molecular formula of glucose.

Plan
Step 1: We first use the masses of C, H, and O found in Example 2-16 to determine the simplest formula. Step 2: We can use the simplest formula to calculate the simplest-formula weight. Because the molecular weight of glucose is known (approximately 180 amu), we can determine the molecular formula by dividing the molecular weight by the simplest-formula weight.

$$n = \frac{\text{molecular weight}}{\text{simplest-formula weight}}$$

The molecular weight is n times the simplest-formula weight, so the molecular formula of glucose is n times the simplest formula.

As an alternative, we could have used the percentages by mass from Example 2-16. Using the earliest available numbers helps to minimize the effects of rounding-off errors.

Solution
Step 1:

Element	Mass of Element	Moles of Element (divide mass by AW)	Divide by Smallest	Smallest Whole-Number Ratio of Atoms
C	0.04055 g	$\frac{0.04055}{12.01} = 0.003376$ mol	$\frac{0.003376}{0.003376} = 1.00$ C	
H	0.00681 g	$\frac{0.00681}{1.008} = 0.00676$ mol	$\frac{0.00676}{0.003376} = 2.00$ H	CH_2O
O	0.0540 g	$\frac{0.0540}{16.00} = 0.00338$ mol	$\frac{0.00338}{0.003376} = 1.00$ O	

Step 2: The simplest formula is CH_2O, which has a formula weight of 30.02 amu. Because the molecular weight of glucose is approximately 180 amu, we can determine the molecular formula by dividing the molecular weight by the simplest-formula weight.

$$n = \frac{180 \text{ amu}}{30.02 \text{ amu}} = 6.00$$

The molecular weight is six times the simplest-formula weight, $6 \times (CH_2O) = C_6H_{12}O_6$, so the molecular formula of glucose is $C_6H_{12}O_6$.

You should now work Exercise 56.

There are many sugars that are rich energy sources in our diet. The most familiar is ordinary table sugar, which is sucrose, $C_{12}H_{22}O_{11}$. An enzyme in our saliva readily splits sucrose into two simple sugars, glucose and fructose. The simplest formula for both glucose and fructose is $C_6H_{12}O_6$. However, they have different structures and different properties, so they are different compounds.

As we shall see when we discuss the composition of compounds in some detail, two (and sometimes more) elements may form more than one compound. The **Law of Multiple Proportions** summarizes many experiments on such compounds. It is usually stated: When two elements, A and B, form more than one compound, the ratio of the masses of element B that combine with a given mass of element A in each of the compounds can be expressed by small whole numbers. Water, H_2O, and hydrogen peroxide, H_2O_2, provide an example. The ratio of masses of oxygen that combine with a given mass of hydrogen is 1:2 in H_2O and H_2O_2. Many similar examples, such as CO and CO_2 (1:2 ratio) and SO_2 and SO_3 (2:3 ratio), are known. The Law of Multiple Proportions had been recognized from studies of elemental composition before the time of Dalton. It provided additional support for his atomic theory.

EXAMPLE 2-18 *Law of Multiple Proportions*

What is the ratio of the masses of oxygen that are combined with 1.00 gram of nitrogen in the compounds N_2O_3 and NO?

Plan

First we calculate the mass of O that combines with one gram of N in each compound. Then we determine the ratio of the values of $\dfrac{g\ O}{g\ N}$ for the two compounds.

Solution

In N_2O_3:
$$\frac{?\ g\ O}{g\ N} = \frac{48.0\ g\ O}{28.0\ g\ N} = 1.71\ g\ O/g\ N$$

In NO:
$$\frac{?\ g\ O}{g\ N} = \frac{16.0\ g\ O}{14.0\ g\ N} = 1.14\ g\ O/g\ N$$

The ratio is
$$
\begin{cases}
\dfrac{g\ O}{g\ N}\ (\text{in } N_2O_3) \\[2em]
\dfrac{g\ O}{g\ N}\ (\text{in NO})
\end{cases}
\quad \frac{1.71\ g\ O/g\ N}{1.14\ g\ O/g\ N} = \frac{1.5}{1.0} = \boxed{\frac{3}{2}}
$$

We see that the ratio is 3 mass units of O (in N_2O_3) to 2 mass units of O (in NO).

You should now work Exercises 59 and 60.

2-11 SOME OTHER INTERPRETATIONS OF CHEMICAL FORMULAS

Once we master the mole concept and the meaning of chemical formulas, we can use them in many other ways. The examples in this section illustrate a few additional kinds of information we can get from a chemical formula and the mole concept.

EXAMPLE 2-19 *Composition of Compounds*

What mass of chromium is contained in 35.8 grams of $(NH_4)_2Cr_2O_7$?

Plan

Let us first solve the problem in several steps. Step 1: The formula tells us that each mole of $(NH_4)_2Cr_2O_7$ contains two moles of Cr atoms, so we first find the number of moles of $(NH_4)_2Cr_2O_7$, using the unit factor

$$\frac{1\ mol\ (NH_4)_2Cr_2O_7}{252.0\ g\ (NH_4)_2Cr_2O_7}$$

Step 2: Then we convert the number of moles of $(NH_4)_2Cr_2O_7$ into the number of moles of Cr atoms it contains, using the unit factor

$$\frac{2\ mol\ Cr\ atoms}{1\ mol\ (NH_4)_2Cr_2O_7}$$

Step 3: We then use the atomic weight of Cr to convert the number of moles of chromium atoms to mass of chromium.

$$\text{Mass (NH}_4\text{)}_2\text{Cr}_2\text{O}_7 \longrightarrow \text{mol (NH}_4\text{)}_2\text{Cr}_2\text{O}_7 \longrightarrow \text{mol Cr} \longrightarrow \text{Mass Cr}$$

Solution

Step 1: $\underset{\text{—}}{?}$ mol $(NH_4)_2Cr_2O_7 = 35.8$ g $(NH_4)_2Cr_2O_7 \times \dfrac{1 \text{ mol } (NH_4)_2Cr_2O_7}{252.0 \text{ g } (NH_4)_2Cr_2O_7}$

$$= \boxed{0.142 \text{ mol } (NH_4)_2Cr_2O_7}$$

Step 2: $\underset{\text{—}}{?}$ mol Cr atoms $= 0.142$ mol $(NH_4)_2Cr_2O_7 \times \dfrac{2 \text{ mol Cr atoms}}{1 \text{ mol } (NH_4)_2Cr_2O_7}$

$$= \boxed{0.284 \text{ mol Cr atoms}}$$

Step 3: $\underset{\text{—}}{?}$ g Cr $= 0.284$ mol Cr atoms $\times \dfrac{52.0 \text{ g Cr}}{1 \text{ mol Cr atoms}} = \boxed{14.8 \text{ g Cr}}$

If you understand the reasoning in these conversions, you should be able to solve this problem in a single setup:

$$\underset{\text{—}}{?} \text{ g Cr} = 35.8 \text{ g } (NH_4)_2Cr_2O_7 \times \frac{1 \text{ mol } (NH_4)_2Cr_2O_7}{252.0 \text{ g } (NH_4)_2Cr_2O_7} \times \frac{2 \text{ mol Cr atoms}}{1 \text{ mol } (NH_4)_2Cr_2O_7} \times \frac{52.0 \text{ g Cr}}{1 \text{ mol Cr}} = \boxed{14.8 \text{ g Cr}}$$

You should now work Exercise 64.

EXAMPLE 2-20 *Composition of Compounds*

What mass of potassium chlorate, $KClO_3$, would contain 40.0 grams of oxygen?

Plan

The formula $KClO_3$ tells us that each mole of $KClO_3$ contains three moles of oxygen atoms. Each mole of oxygen atoms weighs 16.0 grams. So we can set up the solution to convert:

$$\text{Mass O} \longrightarrow \text{mol O} \longrightarrow \text{mol KClO}_3 \longrightarrow \text{Mass KClO}_3$$

Solution

$$\underset{\text{—}}{?} \text{ g KClO}_3 = 40.0 \text{ g O} \times \frac{1 \text{ mol O atoms}}{16.0 \text{ g O atoms}} \times \frac{1 \text{ mol KClO}_3}{3 \text{ mol O atoms}} \times \frac{122.6 \text{ g KClO}_3}{1 \text{ mol KClO}_3}$$

$$= \boxed{102 \text{ g KClO}_3}$$

You should now work Exercise 66.

▼ PROBLEM-SOLVING TIP *How Do We Know When . . . ?*

How do we know when to represent oxygen as O and when as O_2? A *compound* that contains oxygen *does not* contain O_2 molecules. So we solve problems such as Examples 2-20 and 2-21 in terms of moles of O atoms. Thus, we must use the formula weight for O, which is $\dfrac{16.0 \text{ g O atoms}}{1 \text{ mol O atoms}}$. Similar reasoning applies to compounds containing other elements that are polyatomic molecules in *pure elemental form*, such as H_2, Cl_2, or P_4.

EXAMPLE 2-21   *Composition of Compounds*

(a) What mass of sulfur dioxide, SO_2, would contain the same mass of oxygen as is contained in 33.7 g of arsenic pentoxide, As_2O_5? (b) What mass of calcium chloride, $CaCl_2$, would contain the same number of chloride ions as are contained in 48.6 g of sodium chloride, NaCl?

Plan

(a) We would find explicitly the number of grams of O in 33.7 g of As_2O_5, and then find the mass of SO_2 that contains that same number of grams of O. But this method includes some unnecessary calculation. We need only convert to *moles* of O (because this is the same amount of O regardless of its environment) and then to SO_2.

$$\boxed{\text{Mass } As_2O_5} \longrightarrow \boxed{\text{mol } As_2O_5} \longrightarrow \boxed{\text{mol O atoms}} \longrightarrow \boxed{\text{mol } SO_2} \longrightarrow \boxed{\text{Mass } SO_2}$$

(b) Because one mole always consists of the same number (Avogadro's number) of items, we can reason in terms of *moles* of Cl^- ions:

$$\boxed{\text{Mass NaCl}} \longrightarrow \boxed{\text{mol NaCl}} \longrightarrow \boxed{\text{mol } Cl^- \text{ ions}} \longrightarrow \boxed{\text{mol } CaCl_2} \longrightarrow \boxed{\text{Mass } CaCl_2}$$

Solution

(a)
$$\underset{=}{?} \text{ g } SO_2 = 33.7 \text{ g } As_2O_5 \times \frac{1 \text{ mol } As_2O_5}{229.8 \text{ g } As_2O_5} \times \frac{5 \text{ mol O atoms}}{1 \text{ mol } As_2O_5}$$

$$\times \frac{1 \text{ mol } SO_2}{2 \text{ mol O atoms}} \times \frac{64.1 \text{ g } SO_2}{1 \text{ mol } SO_2} = \boxed{23.5 \text{ g } SO_2}$$

(b)
$$\underset{=}{?} \text{ g } CaCl_2 = 48.6 \text{ NaCl} \times \frac{1 \text{ mol NaCl}}{58.4 \text{ g NaCl}} \times \frac{1 \text{ mol } Cl^-}{1 \text{ mol NaCl}}$$

$$\times \frac{1 \text{ mol } CaCl_2}{2 \text{ mol } Cl^-} \times \frac{111.0 \text{ g } CaCl_2}{1 \text{ mol } CaCl_2} = \boxed{46.2 \text{ g } CaCl_2}$$

You should now work Exercise 68.

We have already mentioned the existence of hydrates (for example, $(COOH)_2 \cdot 2H_2O$ and $CuSO_4 \cdot 5H_2O$ in Section 2-7). In such hydrates, two components, water and another compound are present in a definite integer ratio by moles. The following example illustrates how we might find and use the formula of a hydrate.

EXAMPLE 2-22   *Composition of Compounds*

A reaction requires pure anhydrous calcium sulfate, $CaSO_4$. Only an unidentified hydrate of calcium sulfate, $CaSO_4 \cdot xH_2O$, is available. (a) We heat 67.5 g of the unknown hydrate until all the water has been driven off. The resulting mass of pure $CaSO_4$ is 53.4 g. What is the formula of the hydrate, and what is its formula weight? (b) Suppose we wish to obtain enough of this hydrate to supply 95.5 grams of $CaSO_4$. How many grams should we weigh out?

Plan

(a) To find the formula of the hydrate, we must find the value of x in the formula $CaSO_4 \cdot xH_2O$. The mass of water removed from the sample is equal to the difference in the two masses given. The value of x is the number of moles of H_2O per mole of $CaSO_4$ in the hydrate.
(b) The formula weights of $CaSO_4$, 136.2 g/mol, and of $CaSO_4 \cdot xH_2O$, $(136.2 + x18.0)$ g/mol, allow us to write the conversion factor required for the calculation.

Solution

(a) $\underline{?}$ g water driven off = 67.5 g $CaSO_4 \cdot xH_2O$ − 53.4 g $CaSO_4$ = 14.1 g H_2O

$$x = \frac{\underline{?} \text{ mol } H_2O}{\text{mol } CaSO_4} = \frac{14.1 \text{ g } H_2O}{53.4 \text{ g } CaSO_4} \times \frac{1 \text{ mol } H_2O}{18.0 \text{ g } H_2O} \times \frac{136.2 \text{ g } CaSO_4}{1 \text{ mol } CaSO_4} = \frac{2.00 \text{ mol } H_2O}{\text{mol } CaSO_4}$$

Thus, the formula of the hydrate is $CaSO_4 \cdot 2H_2O$. Its formula weight is

$$FW = 1 \times (\text{formula weight } CaSO_4) + 2 \times (\text{formula weight } H_2O)$$

$$= 136.2 \text{ g/mol} + 2(18.0 \text{ g/mol}) = 172.2 \text{ g/mol}$$

(b) The formula weights of $CaSO_4$ (136.2 g/mol) and of $CaSO_4 \cdot 2H_2O$ (172.2 g/mol) allow us to write the unit factor

$$\frac{172.2 \text{ g } CaSO_4 \cdot 2H_2O}{136.2 \text{ g } CaSO_4}$$

We use this factor to perform the required conversion:

$$\underline{?} \text{ g } CaSO_4 \cdot 2H_2O = 95.5 \text{ g } CaSO_4 \text{ desired} \times \frac{172.2 \text{ g } CaSO_4 \cdot 2H_2O}{136.2 \text{ g } CaSO_4}$$

$$= 121 \text{ g } CaSO_4 \cdot 2H_2O$$

You should now work Exercise 72.

2-12 PURITY OF SAMPLES

Most substances obtained from laboratory reagent shelves are not 100% pure. When impure samples are used for precise work, account must be taken of impurities. The marginal photo shows the label from reagent-grade sodium hydroxide, NaOH, which is 98.2% pure by mass. From this information we know that total impurities represent 1.8% of the mass of this material. We can write several unit factors:

$$\frac{98.2 \text{ g NaOH}}{100 \text{ g sample}}, \qquad \frac{1.8 \text{ g impurities}}{100 \text{ g sample}}, \qquad \text{and} \qquad \frac{1.8 \text{ g impurities}}{98.2 \text{ g NaOH}}$$

The inverse of each of these gives us a total of six unit factors.

Impurities are not necessarily bad. For example, inclusion of 0.02% KI, potassium iodide, in ordinary table salt has nearly eliminated goiter in the United States. Goiter is a disorder of the thyroid gland caused by a deficiency of iodine. Mineral water tastes better than purer distilled water.

EXAMPLE 2-23 *Percent Purity*

Calculate the masses of NaOH and impurities in 45.2 g of 98.2% pure NaOH.

Plan

The percentage of NaOH in the sample gives the unit factor $\dfrac{98.2 \text{ g NaOH}}{100 \text{ g sample}}$. The remainder of the sample is 100% − 98.2% = 1.8% impurities; this gives the unit factor $\dfrac{1.8 \text{ g impurities}}{100 \text{ g sample}}$.

Solution

$$\underline{?} \text{ g NaOH} = 45.2 \text{ g sample} \times \frac{98.2 \text{ g NaOH}}{100 \text{ g sample}} = 44.4 \text{ g NaOH}$$

ACTUAL ANALYSIS, LOT G22931

Meets A.C.S. Specifications		
Assay (NaOH) (by acidimetry)	98.2	%
Sodium Carbonate (Na₂CO₃)	0.2	%
Chloride (Cl)	< 0.0005	%
Ammonium Hydroxide Precipitate	< 0.01	%
Heavy Metals (as Ag)	< 0.0005	%
Copper (Cu)	0.0003	%
Potassium (K) (by FES)	0.002	%
Trace Impurities (in ppm):		
Nitrogen Compounds (as N)	< 2	
Phosphate (PO₄)	< 1	
Sulfate (SO₄)	< 5	
Iron (Fe)	< 2	
Mercury (Hg) (by AAS)	< 0.003	
Nickel (Ni)	< 2	

A label from a bottle of sodium hydroxide, NaOH.

$$\underline{?}\text{ g impurities} = 45.2\text{ g sample} \times \frac{1.8\text{ g impurities}}{100\text{ g sample}} = \boxed{0.81\text{ g impurities}}$$

You should now work Exercises 73 and 76.

▼ **PROBLEM-SOLVING TIP**   *Utility of the Unit-Factor Method*

Observe the beauty of the unit-factor approach to problem solving! Such questions as "Do we multiply by 0.982 or divide by 0.982?" never arise. The units always point toward the correct answer because we use unit factors constructed so that units *always* divide out until we arrive at the desired unit.

Many important relationships have been introduced in this chapter. Some of the most important transformations you have seen in Chapters 1 and 2 are summarized in Figure 2-12.

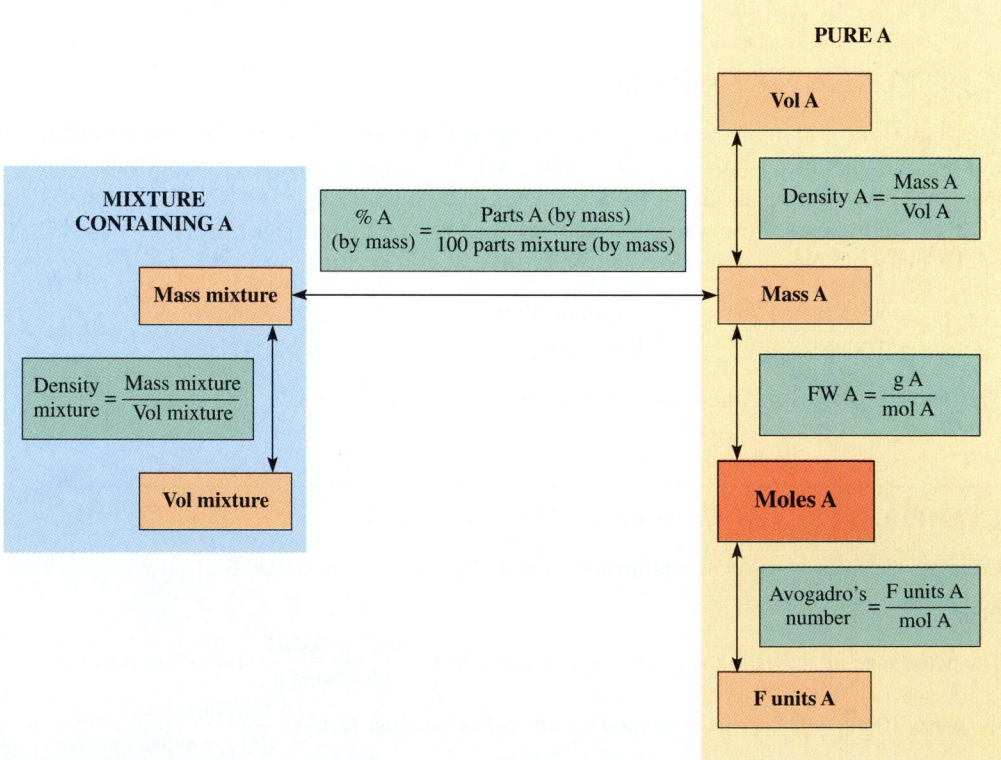

Figure 2-12   Some important relationships from Chapters 1 and 2. The relationships that provide unit factors are enclosed in green boxes.

Key Terms

Allotropic modifications (allotropes) Different forms of the same element in the same physical state.

Anhydrous Without water.

Anion A negative ion.

Atom The smallest particle of an element that maintains its chemical identity through all chemical and physical changes.

Atomic mass unit (amu) One twelfth of the mass of an atom of the carbon-12 isotope; a unit used for stating atomic and formula weights.

Atomic number The number of protons in the nucleus of an atom.

Atomic weight Weighted average of the masses of the constituent isotopes of an element; the relative mass of atoms of different elements.

Avogadro's number 6.022×10^{23} of the specified items. See *Mole*.

Cation A positive ion.

Chemical formula Combination of symbols that indicates the chemical composition of a substance.

Composition stoichiometry Describes the quantitative (mass) relationships among elements in compounds.

Empirical formula See *Simplest formula*.

Formula Combination of symbols that indicates the chemical composition of a substance.

Formula unit The smallest repeating unit of a substance—for nonionic substances, the molecule.

Formula weight The mass, in atomic mass units, of one formula unit of a substance. Numerically equal to the mass, in grams, of one mole of the substance (see *Molar mass*). This number is obtained by adding the atomic weights of the atoms specified in the formula.

Hydrate A crystalline sample that contains water, H_2O, and another compound in a fixed mole ratio. Examples include $CuSO_4 \cdot 5H_2O$ and $(COOH)_2 \cdot 2H_2O$.

Ion An atom or group of atoms that carries an electrical charge. A positive ion is a *cation;* a negative ion is an *anion.*

Ionic compound A compound that is composed of cations and anions. An example is sodium chloride, NaCl.

Law of Constant Composition See *Law of Definite Proportions*.

Law of Definite Proportions Different samples of a pure compound always contain the same elements in the same proportions by mass; this corresponds to atoms of these elements in fixed numerical ratios. Also known as the Law of Constant Composition.

Law of Multiple Proportions When two elements, A and B, form more than one compound, the ratio of the masses of element B that combine with a given mass of element A in each of the compounds can be expressed by small whole numbers.

Molar mass The mass of substance in one mole of the substance; numerically equal to the formula weight of the substance. See *Formula weight;* see *Molecular weight*.

Mole 6.022×10^{23} (Avogadro's number of) formula units (or molecules, for a molecular substance) of the substance under discussion. The mass of one mole, in grams, is numerically equal to the formula (molecular) weight of the substance.

Molecular formula A formula that indicates the actual number of atoms present in a molecule of a molecular substance. Compare with *Simplest formula*.

Molecule The smallest particle of an element or compound that can have a stable independent existence.

Molecular weight The mass, in atomic mass units, of one molecule of a nonionic (molecular) substance. Numerically equal to the mass, in grams, of one mole of such a substance. This number is obtained by adding the atomic weights of the atoms specified in the formula.

Monatomic Consisting of a single atom. The noble gases exist as monatomic molecules. Chloride ion, Cl^-, is a monatomic ion.

Percent composition The mass percentage of each element in a compound.

Percent purity The percentage of a specified compound or element in an impure sample.

Polyatomic Consisting of more than one atom. Elements such as Cl_2, P_4, and S_8 exist as polyatomic molecules. Examples of polyatomic ions are ammonium ion, NH_4^+, and sulfate ion, SO_4^{2-}

Simplest formula The smallest whole-number ratio of atoms present in a compound; also called empirical formula. Compare with *Molecular formula*.

Stoichiometry Description of the quantitative relationships among elements in compounds (composition stoichiometry) and among substances as they undergo chemical changes (reaction stoichiometry).

Structural formula A representation that shows how atoms are connected in a compound.

Exercises

Basic Ideas

1. (a) What is the origin of the word "stoichiometry"? (b) Distinguish between composition stoichiometry and reaction stoichiometry.
2. List the basic ideas of Dalton's atomic theory.
3. Give examples of molecules that contain (a) two atoms, (b) three atoms, (c) four atoms, and (d) eight atoms.
4. Give two examples of diatomic molecules and two examples of more complex molecules.
5. Write formulas for the following compounds: (a) sulfuric acid; (b) methyl alcohol; (c) sulfur dioxide; (d) acetic acid; (e) pentane.
6. Name the following compounds: (a) HNO_3; (b) C_2H_6; (c) NH_3; (d) CH_3COCH_3; (e) CH_3CH_2OH.

Ions and Ionic Compounds

7. Define and illustrate the following: (a) ion; (b) cation; (c) anion; (d) polyatomic ion; (e) molecule.

8. (a) There are no *molecules* in ionic compounds. Why not? (b) What is the difference between a formula unit of an ionic compound and a polyatomic molecule?

9. Name each of the following ions. Classify each as a monatomic or polyatomic ion. Classify each as a cation or an anion. (a) Na^+; (b) OH^-; (c) SO_4^{2-}; (d) S^{2-}; (e) Zn^{2+}.

10. Write the chemical symbol for each of the following ions. Classify each as a monatomic or polyatomic ion. Classify each as a cation or an anion. (a) potassium ion; (b) sulfate ion; (c) iron(III) ion; (d) ammonium ion; (e) carbonate ion.

11. Name each of the following compounds: (a) $CaCl_2$; (b) $Cu(NO_3)_2$; (c) K_2SO_4; (d) $Mg(OH)_2$; (e) $FeSO_4$.

12. Write the chemical formula for each of the following ionic compounds: (a) potassium acetate; (b) ammonium carbonate; (c) zinc phosphate; (d) calcium oxide; (e) aluminum nitrate.

13. Write the chemical formula for the ionic compound formed between each of the following pairs of ions. Name each compound. (a) Na^+ and S^{2-}; (b) Al^{3+} and SO_4^{2-}; (c) K^+ and PO_4^{3-}; (d) Mg^{2+} and NO_3^-; (e) Fe^{3+} and SO_4^{2-}.

14. Write the chemical formula for the ionic compound formed between each of the following pairs of ions. Name each compound. (a) Cu^{2+} and CO_3^{2-}; (b) Mg^{2+} and OH^-; (c) NH_4^+ and CO_3^{2-}; (d) Zn^{2+} and Cl^-; (e) Fe^{2+} and CH_3COO^-.

15. Convert each of the following into a correct formula represented with correct notation. (a) $AlOH_3$; (b) $Mg(CO_3)$; (c) $Zn(CO_3)_2$; (d) $(NH_4)^2SO_4$; (e) $Mg_2(SO_4)_2$.

Atomic Weights

16. What is the mass ratio (4 significant figures) of one atom of Ag to 1 atom of Cl?

17. 4.05 g of magnesium combines exactly with 6.33 g of fluorine, forming magnesium fluoride, MgF_2. Find the relative masses of the atoms of magnesium and fluorine. Check your answer using a table of atomic weights. If the formula were not known, could you still do this calculation?

The Mole Concept

18. The mass of one round-head wood screw is 4.11 g. Find the mass of one dozen, one gross, and one mole of these wood screws. Compare the latter value with the mass of the earth, 5.98×10^{24} kg.

19. Sulfur molecules exist under various conditions as S_8, S_6, S_4, S_2, and S. (a) Is the mass of one mole of each of these molecules the same? (b) Is the number of molecules in one mole of each of these molecules the same? (c) Is the mass of sulfur in one mole of each of these molecules the same? (d) Is the number of atoms of sulfur in one mole of each of these molecules the same?

20. Complete the following table. Refer to a table of atomic weights.

Element	Atomic Weight	Mass of One Mole of Atoms
(a) Ca	_____	_____
(b) _____	79.904 amu	_____
(c) P	_____	_____
(d) _____	_____	51.9961 g

21. Complete the following table. Refer to a table of atomic weights.

Element	Formula	Mass of One Mole of Molecules
(a) Br	Br_2	_____
(b) _____	H_2	_____
(c) _____	P_4	_____
(d) _____	_____	20.1797 g
(e) S	_____	256.528 g
(f) O	_____	_____

22. How many moles of Ni atoms are contained in 245.2 grams of nickel?

23. Two (2.000) moles of Ni atoms have the same mass as 1.223 moles of atoms of another element. What is the atomic weight of the other element? What is it?

24. Determine the formula weight of each of the following substances: (a) iodine, I_2; (b) water, H_2O; (c) saccharin, $C_7H_5NSO_3$; (d) sodium dichromate, $Na_2Cr_2O_7$.

25. Determine the formula weight of each of the following substances: (a) calcium sulfate, $CaSO_4$; (b) butane, C_4H_{10}; (c) the sulfa drug sulfanilamide, $C_6H_4SO_2(NH_2)_2$; (d) uranyl phosphate, $(UO_2)_3(PO_4)_2$.

26. How many moles of substance are contained in each of the following samples? (a) 16.8 g of NH_3; (b) 3.25 kg of ammonium bromide; (c) 5.6 g of PCl_5; (d) 126.5 g of Sn.

27. How many moles of substance are contained in each of the following samples? (a) 24.5 g of diethyl ether; (b) 12.62 g of calcium carbonate; (c) 33.5 g of acetic acid; (d) 24.3 g of ethanol.

28. How many atoms are contained in 4.178 moles of Ni atoms?

29. A sample of a metal contains 2.516×10^{23} atoms and has a mass of 82.29 grams. How many moles of metal atoms are present in the sample? What is the metal?

30. Calculate the mass of a nickel atom in grams.

31. An atom of an element has a mass ever-so-slightly greater than twice the mass of an Ni atom. Identify the element.

32. Calculate the number of Ni atoms in 1.0 millionth of a gram of nickel.

33. Calculate the number of Ni atoms in 1.0 trillionth of a gram of nickel.

34. What is the mass of 10.0 million methane, CH_4, molecules?

35. A sample of ethane, C_2H_6, has the same mass as 10.0 million molecules of methane, CH_4. How many C_2H_6 molecules does the sample contain?

36. How many molecules are in 14.0 g of each of the following substances? (a) CO; (b) N_2; (c) P_4; (d) P_2. (e) Do parts (c) and (d) contain the same number of atoms of phosphorus?

37. How many (a) moles of P_4, (b) P_4 molecules, and (c) P atoms are contained in 250 grams of phosphorus?

38. How many hydrogen atoms are contained in 75.0 grams of propane, C_3H_8?

39. How many atoms of C, H, and O are in each of the following? (a) 1.24 mol of glucose, $C_6H_{12}O_6$; (b) 3.31×10^{19} glucose ($C_6H_{12}O_6$) molecules; (c) 0.275 g of glucose.

40. An experiment requires 2.34 grams of H_2SO_4. How many millimoles of H_2SO_4 is this?

41. An experiment requires 10.5 millimoles of H_2SO_4. What mass of H_2SO_4 is this?

42. Complete the following table.

Moles of Compound	Moles of Cations	Moles of Anions
1 mol NaCl	_____	_____
2 mol Na_2SO_4	_____	_____
0.2 mol calcium nitrate	_____	_____
_____	0.50 mol NH_4^+	0.25 mol SO_4^{2-}

43. What mass, in grams, should be weighed for an experiment that requires 125 mmol of $(NH_4)_2HPO_4$?

Percent Composition and Simplest Formulas

44. Calculate the percent composition of each of the following compounds: (a) nicotine, $C_{10}H_{14}N_2$; (b) codeine, $C_{18}H_{21}NO_3$; (c) vanillin, $C_8H_8O_3$.

45. Calculate the percent composition of each of the following compounds: (a) diethyl ether, $C_4H_{10}O$; (b) carborundum, SiC; (c) aspirin, $CH_3COOC_6H_4COOH$.

*46. Copper is obtained from ores containing the following minerals: azurite, $Cu_3(CO_3)_2(OH)_2$; chalcocite, Cu_2S; chalcopyrite, $CuFeS_2$; covelite, CuS; cuprite, Cu_2O; and malachite, $Cu_2CO_3(OH)_2$. Which mineral has the highest copper content on a percent-by-mass basis?

47. Determine the simplest formula for each of the following compounds.
 (a) copper(II) tartrate: 30.03% Cu; 22.70% C; 1.91% H; 45.37% O.
 (b) nitrosyl fluoroborate: 11.99% N; 13.70% O; 9.25% B; 65.06% F.

48. The hormone epinephrine is released in the human body during stress and increases the body's metabolic rate. Like many biochemical compounds, epinephrine is composed of carbon, hydrogen, oxygen, and nitrogen. The percent composition of this hormone is 56.8% C, 6.56% H, 28.4% O, and 8.28% N. What is the simplest formula of epinephrine?

49. (a) A compound is found to contain 5.60 g N, 14.2 g Cl, and 0.800 g H. What is the simplest formula of this compound? (b) Another compound containing the same elements is found to be 26.2% N, 66.4% Cl, and 7.5% H. What is the simplest formula of this compound?

50. Combustion of 0.5707 mg of a hydrocarbon produces 1.790 mg of CO_2. What is the simplest formula of the hydrocarbon?

*51. Complicated chemical reactions occur at hot springs on the ocean floor. One compound obtained from such a hot spring consists of Mg, Si, H, and O. From a 0.334-gram sample, the Mg is recovered as 0.115 g of MgO; H is recovered as 25.7 mg of H_2O; and Si is recovered as 0.172 g of SiO_2. What is the simplest formula of this compound?

52. A 0.1647-gram sample of a pure hydrocarbon was burned in a C-H combustion train to produce 0.5694 gram of CO_2 and 0.0826 gram of H_2O. Determine the masses of C and H in the sample and the percentages of these elements in this hydrocarbon.

53. Naphthalene is a hydrocarbon that is used for mothballs. A 0.3204-gram sample of naphthalene was burned in a C-H combustion train to produce 1.100 grams of carbon dioxide and 0.1802 gram of water. What masses and percentages of C and H are present in naphthalene?

54. A 1.000-gram sample of an alcohol was burned in oxygen to produce 1.913 g of CO_2 and 1.174 g of H_2O. The alcohol contained only C, H, and O. What is the simplest formula of the alcohol?

Determination of Molecular Formulas

55. An alcohol is 64.81% C, 13.60% H, and 21.59% O by mass. Another experiment shows that its molecular weight is approximately 74 amu. What is the molecular formula of the alcohol?

56. Skatole is found in coal tar and in human feces. It contains three elements: C, H, and N. It is 82.40% C and 6.92% H by mass. Its simplest formula is its molecular formula. What are (a) the formula and (b) the molecular weight of skatole?

57. Testosterone, the male sex hormone, contains only C, H, and O. It is 79.12% C and 9.79% H by mass. Each molecule contains two O atoms. What are (a) the molecular weight and (b) the molecular formula for testosterone?

*58. The β-blocker drug, timolol, is expected to reduce the need for heart bypass surgery. Its composition by mass is 47.2% C, 6.55% H, 13.0% N, 25.9% O, and 7.43% S. The mass of 0.0100 mol of timolol is 4.32 g. (a) What is the simplest formula of timolol? (b) What is the molecular formula of timolol?

The Law of Multiple Proportions

59. Show that the compounds water, H_2O, and hydrogen peroxide, H_2O_2, obey the Law of Multiple Proportions.

60. Nitric oxide, NO, is produced in internal combustion engines. When NO comes in contact with air, it is quickly converted into nitrogen dioxide, NO_2, a very poisonous, corrosive gas. What mass of O is combined with 3.00 g of N in (a) NO and (b) NO_2? Show that NO and NO_2 obey the Law of Multiple Proportions.

61. Phosphorus forms two chlorides. A 30.00-gram sample of one chloride decomposes to give 4.35 g of P and 25.65 g of Cl. A 30.00-gram sample of the other chloride decomposes to give 6.61 g of P and 23.39 g of Cl. Show that these compounds obey the Law of Multiple Proportions.

62. What mass of oxygen is combined with 1.50 g of sulfur in (a) sulfur dioxide, SO_2, and in (b) sulfur trioxide, SO_3?

Interpretation of Chemical Formulas

63. One prominent ore of copper contains chalcopyrite, $CuFeS_2$. How many tons of copper are contained in 315 tons of pure $CuFeS_2$?

64. Mercury occurs as a sulfide ore called *cinnabar,* HgS. How many grams of mercury are contained in 225.0 g of pure HgS?

65. (a) How many grams of copper are contained in 175 g of $CuSO_4$? (b) How many grams of copper are contained in 175 g of $CuSO_4 \cdot 5H_2O$?

66. What mass of $KMnO_4$ would contain 10.5 g of manganese?

67. What mass of azurite, $Cu_3(CO_3)_2(OH)_2$, would contain 575 g of copper?

68. Two minerals that contain copper are chalcopyrite, $CuFeS_2$, and chalcocite, Cu_2S. What mass of chalcocite would contain the same mass of copper as is contained in 175 tons of chalcopyrite?

69. Tungsten is a very dense metal (19.3 g/cm^3) with extremely high melting and boiling points (3370°C and 5900°C). When a small amount of it is included in steel, the resulting alloy is far harder and stronger than ordinary steel. Two important ores of tungsten contain $FeWO_4$ and $CaWO_4$. How many grams of $CaWO_4$ would contain the same mass of tungsten that is present in 679 g of $FeWO_4$?

70. What mass of NaCl would contain the same *total* number of ions as 245 g of $MgCl_2$?

71. Suppose we have equal masses of potassium sulfate, K_2SO_4, and aluminum sulfate, $Al_2(SO_4)_3$. (a) What is the ratio of the numbers of sulfate ions contained in these two samples? (b) What is the ratio of the masses of sulfur contained in these two samples?

*72. When a mole of $CuSO_4 \cdot 5H_2O$ is heated to 110°C, it loses four moles of H_2O to form $CuSO_4 \cdot H_2O$. When it is heated to temperatures above 150°C, the other mole of H_2O is lost. (a) How many grams of $CuSO_4 \cdot H_2O$ could be obtained by heating 556 g of $CuSO_4 \cdot 5H_2O$ to 110°C? (b) How many grams of anhydrous $CuSO_4$ could be obtained by heating 556 g of $CuSO_4 \cdot 5H_2O$ to 180°C?

Percent Purity

73. (a) How many grams of sodium chloride, NaCl, are contained in 33.0 g of saline solution that is 5.0% NaCl by mass? (b) Vinegar is 5.0% acetic acid, CH_3COOH, by mass. How many grams of acetic acid are contained in 33.0 g of vinegar? (c) How many pounds of acetic acid are contained in 33.0 pounds of vinegar?

74. (a) What is the percent by mass of oxalic acid, $(COOH)_2$, in a sample of pure oxalic acid dihydrate, $(COOH)_2 \cdot 2H_2O$? (b) What is the percent by mass of $(COOH)_2$ in a sample that is 72.4% $(COOH)_2 \cdot 2H_2O$ by mass?

75. What masses of (a) Sr and (b) N are contained in 94.9 g of 88.2% pure $Sr(NO_3)_2$? Assume that the impurities do not contain the elements mentioned.

76. What weight of magnesium carbonate is contained in 671 pounds of an ore that is 27.7% magnesium carbonate by weight? (b) What weight of impurities is contained in the sample? (c) What weight of magnesium is contained in the sample? (Assume that no magnesium is present in the impurities.)

Mixed Examples

77. How many moles of bromine atoms are contained in each of the following? (a) 79.9×10^{23} Br atoms; (b) 79.9×10^{23} Br_2 molecules; (c) 79.9 g of bromine; (d) 79.9 mol of Br_2.

78. What is the *maximum* number of moles of CO_2 that could be obtained from the carbon in each of the following? (a) 2.00 mol of $Fe_2(CO_3)_3$; (b) 2.00 mol of $CaCO_3$; (c) 2.00 mol of $Ni(CO)_4$.

79. (a) How many formula units are contained in 127.3 g of K_2MoO_4? (b) How many potassium ions? (c) How many MoO_4^{2-} ions? (d) How many atoms of all kinds?

80. (a) How many moles of ozone molecules are contained in 94.0 g of ozone, O_3? (b) How many moles of oxygen atoms are contained in 94.0 g of ozone? (c) What mass of O_2 would contain the same number of oxygen atoms as 94.0 g of ozone? (d) What mass of oxygen gas, O_2, would contain the same number of molecules as 94.0 g of ozone?

81. Cocaine has the following percent composition by mass: 67.30% C, 6.930% H, 21.15% O, and 4.62% N. What is the simplest formula of cocaine?

82. Find the number of moles of Ag needed to form each of the following: (a) 0.555 mol Ag_2S; (b) 0.555 mol Ag_2O; (c) 0.555 g Ag_2S; (d) 5.55×10^{20} formula units of Ag_2S.

*83. (a) A sample contains 50.0% NaCl and 50.0% KCl by mass. What is the percent Cl, by mass, in this sample? (b) A second sample of NaCl and KCl contains 50.0% Cl by mass. What is the mass percent of NaCl in this sample?

*84. Analysis of a 20.0-mg sample of an organic compound for H, N, and C yields 1.99 mg H_2O, 1.25 mg NH_3, and 6.47 mg CO_2. The Cl and Br are recovered as a mixture of AgCl and AgBr with a mass of 48.8 mg. Finally, when all the AgBr is converted to AgCl (by adding chloride from an outside source), the mass of AgCl formed from the AgBr plus the mass of AgCl originally present in the mixture are 42.3 mg. Calculate the empirical formula of the compound.

85. A metal, M, forms an oxide having the empirical formula M_2O_3. This oxide contains 52.9% of the metal by mass. (a) Calculate the atomic weight of the metal. (b) Identify the metal.

86. Three samples of magnesium oxide were analyzed to determine the mass ratios O/Mg, giving the following results: $\dfrac{1.60 \text{ g O}}{2.43 \text{ g Mg}}$, $\dfrac{0.658 \text{ g O}}{1.00 \text{ g Mg}}$, $\dfrac{2.29 \text{ g O}}{3.48 \text{ g Mg}}$. Which law of chemical combination is illustrated by these data?

*87. The molecular weight of hemoglobin is about 65,000 g/mol. Hemoglobin contains 0.35% Fe by mass. How many iron atoms are in a hemoglobin molecule?

*88. We can drive off the water from copper sulfate pentahydrate, $CuSO_4 \cdot 5H_2O$, by heating to obtain anhydrous (meaning "without water") $CuSO_4$. An experiment calls for 10.0 g of $CuSO_4$. How many grams of $CuSO_4 \cdot 5H_2O$ should we use to supply 10.0 g of $CuSO_4$?

*89. During volcanic action, S_8 is converted to S, which is then converted to H_2S. In turn, the H_2S reacts with Fe, forming FeS_2. In water containing O_2, the FeS_2 reacts to form "mine acid,"

H_2SO_4. Find the maximum mass, in grams, of H_2SO_4 that can be formed from 0.717 mol of S_8.

90. One method of analyzing for the amount of Cr_2O_3 in a sample involves converting the chromium to $BaCrO_4$, and then weighing the amount of $BaCrO_4$ formed. Suppose that this process could be carried out with no loss of chromium. How many grams of Cr_2O_3 are present in the original sample for every gram of $BaCrO_4$ that could be isolated and weighed?

91. A 912-mg sample was found to contain 34.8% $PbSO_4$ (and no other source of lead). (a) How many milligrams of lead were contained in the sample? (b) What is the percent of lead in the sample?

*92. More than 1 billion pounds of adipic acid (MW 146.1 g/mol) is manufactured in the United States each year. Most of it is used to make synthetic fabrics. Adipic acid contains only C, H, and O. Combustion of a 1.6380-gram sample of adipic acid gives 2.960 g of CO_2 and 1.010 g of H_2O. (a) What is the simplest formula for adipic acid? (b) What is its molecular formula?

*93. Calcium lactate is used in the food and beverage industries. It has also been used medicinally for treatment of various allergies, for treatment of muscular leg cramps, and as an antidote for a variety of poisons, including lead, arsenicals, and carbon tetrachloride. A 0.8274-gram sample of anhydrous calcium lactate is found by analysis to contain 0.2732 g C, 0.0382 g H, 0.1520 g Ca, and 0.3640 g O. (a) Determine the simplest formula of calcium lactate. (b) Each mole of calcium lactate is found to contain one mole of calcium ions. Calculate the formula weight of calcium lactate. (c) When calcium lactate is recrystallized from water, it

forms the pentahydrate, for which we can represent the formula as "calcium lactate" $\cdot 5H_2O$. Suppose we dissolve 1.00 gram of anhydrous calcium lactate in water and then recover it all as crystalline calcium lactate pentahydrate. How many grams of calcium lactate pentahydrate would be recovered?

BUILDING YOUR KNOWLEDGE

NOTE *Beginning with this chapter, Exercises under the "Building Your Knowledge" heading will often require that you use skills, concepts, or information that you should have mastered in earlier chapters. This provides you an excellent opportunity to "tie things together" as you study.*

94. A 22-mL (19-g) sample of an unknown liquid is analyzed and the percent composition is found to be 53% C, 11% H, and 36% O. Is this compound likely to be ethanol, CH_3CH_2OH? Give two reasons for your answer. (*Hint:* See Table 1-8.)

95. Three allotropes of phosphorus are observed, with molecular weights of 62.0, 31.0, and 124.0 amu, respectively. Write the molecular formula for each allotrope.

96. Near room temperature the density of water is 1.00 g/mL and the density of ethanol (grain alcohol) is 0.789 g/mL. What volume of ethanol contains the same number of molecules as are present in 125 mL of H_2O?

97. Calculate the volume of one mole of mercury, a liquid metal. (*Hint:* See Table 1-8.)

3

Chemical Equations and Reaction Stoichiometry

OBJECTIVES

As you study this chapter, you should learn to

· *Write a balanced chemical equation to describe a chemical reaction*

· *Interpret a balanced chemical equation to calculate the* moles *of reactants and products involved in the reaction*

· *Interpret a balanced chemical equation to calculate the* masses *of reactants and products involved in the reaction*

· *Determine which is the limiting reactant*

· *Use the limiting reactant concept in calculations with chemical equations*

· *Compare the amount of substance actually formed in a reaction (actual yield) with the predicted amount (theoretical yield), and determine the percent yield*

· *Understand sequential reactions*

· *Use the terminology of solutions—solute, solvent, concentration*

· *Calculate concentrations of solutions when they are diluted*

· *Understand the concept of, and calculations for, titrations*

Potassium metal reacts vigorously with water

$$2K(s) + 2H_2O(\ell) \longrightarrow H_2(g) + 2KOH(aq)$$

Heat given off by the reaction causes unreacted potassium to give off a purple glow and ignites the hydrogen gas that is formed.

$$2H_2(g) + O_2(g) \longrightarrow 2H_2O(g)$$

The room was totally dark; the light was produced by the chemical reaction.

In Chapter 2 we studied composition stoichiometry, the quantitative relationships among elements in compounds. In this chapter we shall study reaction stoichiometry, the quantitative relationships among substances as they participate in chemical reactions. We ask several important questions. *How* can we describe the reaction of one substance with another? *How much* of one substance reacts with a given amount of another substance? *Which reactant* determines the amounts of products formed in a chemical reaction? *How* can we describe reactions in aqueous solutions?

Whether we are concerned with describing a reaction used in a chemical analysis, one used industrially in the production of a plastic, or one that occurs during metabolism in the body, we must describe it accurately. Chemical equations represent a very precise, yet a very versatile, language that describes chemical changes. We shall begin by studying chemical equations.

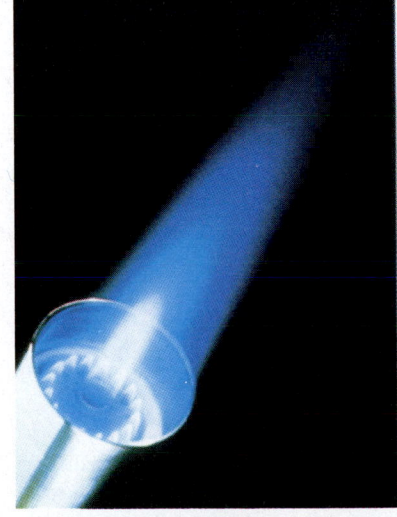

Methane, CH_4, is the main component of natural gas.

3-1 CHEMICAL EQUATIONS

Chemical reactions always involve changing one or more substances into one or more different substances. That is, they involve regrouping atoms or ions to form other substances.

Chemical equations are used to describe chemical reactions, and they show (1) *the substances that react,* called **reactants,** (2) *the substances formed,* called **products,** and (3) *the relative amounts of the substances involved.* As a typical example, let's consider the combustion (burning) of natural gas, a reaction used to heat buildings and cook foods. Natural gas is a mixture of several substances, but the principal component is methane, CH_4. The equation that describes the reaction of methane with excess oxygen is

$$\underbrace{CH_4 + 2O_2}_{\text{reactants}} \longrightarrow \underbrace{CO_2 + 2H_2O}_{\text{products}}$$

What does this equation tell us? In the simplest terms, it tells us that methane reacts with oxygen to produce carbon dioxide, CO_2, and water. More specifically, it says that for every CH_4 molecule that reacts, two molecules of O_2 also react, and that one CO_2 molecule and two H_2O molecules are formed. That is,

$$\underset{\text{1 molecule}}{CH_4} \; + \; \underset{\text{2 molecules}}{2O_2} \; \overset{\Delta}{\longrightarrow} \; \underset{\text{1 molecule}}{CO_2} \; + \; \underset{\text{2 molecules}}{2H_2O}$$

This description of the reaction of CH_4 with O_2 is based on *experimental observations.* By this we mean experiments have shown that when one CH_4 molecule reacts with two O_2 molecules, one CO_2 molecule and two H_2O molecules are formed. *Chemical equations are based on experimental observations.* Special conditions required for some reactions are indicated by notation over the arrow. Figure 3-1 gives a pictorial representation of the rearrangement of atoms described by this equation.

As we pointed out in Section 1-1, *there is no detectable change in the quantity of matter during an ordinary chemical reaction.* This guiding principle, the **Law of Conservation of Matter,** provides the basis for "balancing" chemical equations and for calculations based on those equations. Because matter is neither created nor destroyed during a chemical reaction,

a balanced chemical equation must always include the same number of each kind of atom on both sides of the equation.

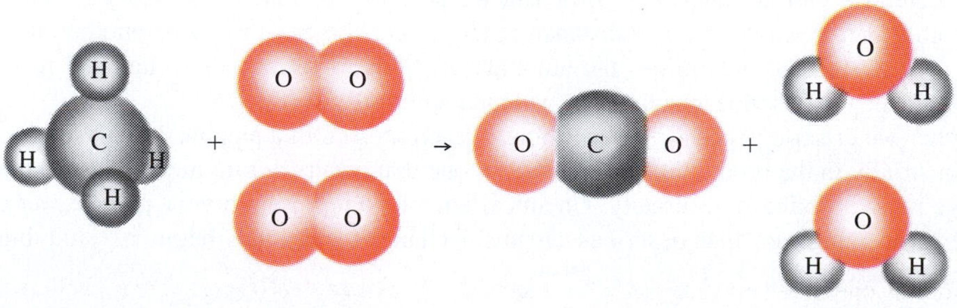

Sometimes it is not possible to represent a chemical change with a single chemical equation. For example, when too little O_2 is present, both CO_2 and CO are found as products, and a second chemical equation must be used to describe the process. In the present case (excess oxygen), only one equation is required.

The arrow may be read "yields." The capital Greek letter delta (Δ) means that heat is necessary to start this reaction.

Figure 3-1 Representation of the reaction of methane with oxygen to form carbon dioxide and water. Some chemical bonds are broken and some new ones are formed.

Chemists usually write equations with the smallest possible whole-number coefficients.

Before we attempt to balance an equation, all substances must be represented by formulas that describe them *as they exist*. For instance, we must write H_2 to represent diatomic hydrogen molecules—not H, which represents hydrogen atoms. Once the correct formulas are written, the subscripts in the formulas may not be changed. Different subscripts in formulas specify different compounds, so the equation would no longer describe the same reaction if formulas were changed.

Let's generate the balanced equation for the reaction of aluminum metal with hydrochloric acid to produce aluminum chloride and hydrogen. The unbalanced "equation" is

$$Al + HCl \longrightarrow AlCl_3 + H_2$$

Atoms of	In reactants	In products
Al	1	1
H	1	2
Cl	1	3
Equation is not balanced.		

As it now stands, the "equation" does not satisfy the Law of Conservation of Matter because there are two H atoms in the H_2 molecule and three Cl atoms in one formula unit of $AlCl_3$ (right side), but only one H atom and one Cl atom in the HCl molecule (left side).

Let us first balance chlorine by putting a coefficient of 3 in front of HCl.

$$Al + \boxed{3HCl} \longrightarrow AlCl_3 + H_2$$

Atoms of	In reactants	In products
Al	1	1
H	3	2
Cl	3	3
Equation is not balanced.		

Now there are 3H on the left and 2H on the right. The least common multiple of 3 and 2 is 6; to balance H, we multiply the 3HCl by 2 and the H_2 by 3.

$$Al + \boxed{6HCl} \longrightarrow AlCl_3 + \boxed{3H_2}$$

Atoms of	In reactants	In products
Al	1	1
H	6	6
Cl	6	3
Equation is still not balanced.		

Now Cl is again unbalanced (6Cl on the left, 3 on the right), but we can fix this by putting a coefficient of 2 in front of $AlCl_3$ on the right.

$$Al + 6HCl \longrightarrow \boxed{2AlCl_3} + 3H_2$$

Atoms of	In reactants	In products
Al	1	2
H	6	6
Cl	6	6
Equation is still not balanced.		

Now all elements except Al are balanced (1 on the left, 2 on the right); we complete the balancing by putting a coefficient of 2 in front of Al on the left.

$$\boxed{2Al} + 6HCl \longrightarrow 2AlCl_3 + 3H_2$$
aluminum   hydrochloric acid   aluminum chloride   hydrogen

Atoms of	In reactants	In products
Al	2	2
H	6	6
Cl	6	6
Now the equation is balanced.		

When we think that we have finished the balancing, we should *always* do a complete check for each element, as shown in red in the margin.

Dimethyl ether, C_2H_6O, burns in an excess of oxygen to give carbon dioxide and water. Let's balance the equation for this reaction. In unbalanced form,

$$C_2H_6O + O_2 \longrightarrow CO_2 + H_2O$$

Atoms of	In reactants	In products
C	2	1
H	6	2
O	3	3
Equation is not balanced.		

Carbon appears in only one compound on each side, and the same is true for hydrogen. We begin by balancing these elements:

$$C_2H_6O + O_2 \longrightarrow \boxed{2CO_2} + \boxed{3H_2O}$$

Atoms of	In reactants	In products
C	2	2
H	6	6
O	3	7
Equation is still not balanced.		

Now we have an odd number of atoms of O on each side. The single O in C_2H_6O balances one of the atoms of O on the right. We balance the other six by placing a coefficient of 3 before O_2 on the left.

$$C_2H_6O + \boxed{3O_2} \longrightarrow 2CO_2 + 3H_2O$$

Atoms of	In reactants	In products
C	2	2
H	6	6
O	7	7
Now the equation is balanced.		

Balancing chemical equations "by inspection" is a *trial-and-error* approach. It requires a great deal of practice, but it is *very important!* Remember that we use the smallest whole-number coefficients.

> ▼ **PROBLEM-SOLVING TIP**   *Balancing Chemical Equations*
>
> 1. When you are balancing chemical equations, look for elements that appear in only one place on each side of the equation and balance these elements first.
> 2. If free, uncombined elements appear on either side, balance them last.
>
> Notice how these suggestions worked in the procedures we have just illustrated.

3-2 CALCULATIONS BASED ON CHEMICAL EQUATIONS

As we indicated earlier, chemical equations represent a very precise and versatile language. We are now ready to use them to calculate the relative *amounts* of substances involved in chemical reactions. Let us again consider the combustion of methane in excess oxygen. The balanced chemical equation for that reaction is

$$CH_4 + 2O_2 \longrightarrow CO_2 + 2H_2O$$

On a quantitative basis, at the molecular level, the equation says

A balanced chemical equation may be interpreted on a molecular *basis.*

$$CH_4 \quad + \quad 2O_2 \quad \longrightarrow \quad CO_2 \quad + \quad 2H_2O$$

| 1 molecule | 2 molecules | 1 molecule | 2 molecules |
| of methane | of oxygen | of carbon dioxide | of water |

EXAMPLE 3-1   *Number of Molecules*

How many O_2 molecules would react with 47 CH_4 molecules according to the above equation?

Plan

The *balanced* equation tells us that *one* CH_4 molecule reacts with *two* O_2 molecules. We can construct two unit factors from this fact:

$$\frac{1 \text{ CH}_4 \text{ molecule}}{2 \text{ O}_2 \text{ molecules}} \quad \text{and} \quad \frac{2 \text{ O}_2 \text{ molecules}}{1 \text{ CH}_4 \text{ molecule}}$$

These are unit factors for *this* reaction because the numerator and denominator are *chemically equivalent.* In other words, the numerator and the denominator represent the same amount of reaction. We convert CH_4 molecules to O_2 molecules.

Solution

$$\underline{?} \text{ O}_2 \text{ molecules} = 47 \text{ CH}_4 \text{ molecules} \times \frac{2 \text{ O}_2 \text{ molecules}}{1 \text{ CH}_4 \text{ molecule}} = \boxed{94 \text{ O}_2 \text{ molecules}}$$

You should now work Exercise 8.

A chemical equation also indicates the relative amounts of each reactant and product in a given chemical reaction. We showed earlier that formulas can represent moles of substances. Suppose Avogadro's number of CH_4 molecules, rather than just one CH_4 molecule, undergo this reaction. Then the equation can be written

$$CH_4 \quad + \quad 2O_2 \quad \longrightarrow \quad CO_2 \quad + \quad 2H_2O$$

| 6.02×10^{23} molecules | $2(6.02 \times 10^{23}$ molecules) | 6.02×10^{23} molecules | $2(6.02 \times 10^{23}$ molecules) |
| 1 mol | 2 mol | 1 mol | 2 mol |

This interpretation tells us that *one* mole of methane reacts with *two* moles of oxygen to produce *one* mole of carbon dioxide and *two* moles of water.

A balanced chemical equation may be interpreted in terms of moles *of reactants and products.*

EXAMPLE 3-2   *Number of Moles Formed*

How many moles of water could be produced by the reaction of 3.5 moles of methane with excess oxygen?

Plan

The equation for the combustion of methane

$$CH_4 + 2O_2 \longrightarrow CO_2 + 2H_2O$$
$$\text{1 mol}\quad\text{2 mol}\qquad\text{1 mol}\quad\text{2 mol}$$

shows that one mole of methane reacts with two moles of oxygen to produce two moles of water. From this information we construct two *unit factors*:

$$\frac{\text{1 mol } CH_4}{\text{2 mol } H_2O} \quad \text{and} \quad \frac{\text{2 mol } H_2O}{\text{1 mol } CH_4}$$

We use the second factor in this calculation.

Please don't try to memorize unit factors for chemical reactions; rather, learn the general method *for constructing them from balanced chemical equations.*

Solution

$$\underline{?}\text{ mol } H_2O = 3.5\text{ mol } CH_4 \times \frac{\text{2 mol } H_2O}{\text{1 mol } CH_4} = \boxed{7.0\text{ mol } H_2O}$$

You should now work Exercises 12 and 14.

We know the mass of one mole of each of these substances, so we can also write

$$CH_4 + \ 2O_2 \longrightarrow CO_2 + 2H_2O$$

1 mol	2 mol	1 mol	2 mol
16 g	2(32 g)	44 g	2(18 g)
16 g	64 g	44 g	36 g

$$\underbrace{}_{\text{80 g reactants}} \qquad \underbrace{}_{\text{80 g products}}$$

The equation now tells us that 16 grams of CH_4 reacts with 64 grams of O_2 to form 44 grams of CO_2 and 36 grams of H_2O. The Law of Conservation of Matter is satisfied. Chemical equations describe **reaction ratios**, i.e., the *mole ratios* of reactants and products as well as the *relative masses* of reactants and products.

A balanced equation may be interpreted on a mass *basis. We have rounded off molecular weights to the nearest whole number of grams here.*

EXAMPLE 3-3   *Mass of a Reactant Required*

What mass of oxygen is required to react completely with 1.2 moles of CH_4?

Plan

The balanced equation

$$CH_4 + \ 2O_2 \longrightarrow CO_2 + 2H_2O$$

1 mol	2 mol	1 mol	2 mol
16 g	2(32 g)	44 g	2(18 g)

gives the relationships among moles and grams of reactants and products.

$$\boxed{\text{mol } CH_4} \longrightarrow \boxed{\text{mol } O_2} \longrightarrow \boxed{\text{g } O_2}$$

Solution

$$\underset{\text{—}}{?} \text{ g O}_2 = 1.2 \text{ mol CH}_4 \times \frac{2 \text{ mol O}_2}{1 \text{ mol CH}_4} \times \frac{32 \text{ g O}_2}{1 \text{ mol O}_2} = \boxed{77 \text{ g O}_2}$$

You should now work Exercise 10.

EXAMPLE 3-4   *Mass of a Reactant Required*

What mass of oxygen is required to react completely with 24 grams of CH_4?

Plan

Recall the balanced equation in Example 3-3.

$$CH_4 + 2O_2 \longrightarrow CO_2 + 2H_2O$$

| 1 mol | 2 mol | 1 mol | 2 mol |
| 16 g | 2(32) g | 44 g | 2(18) g |

This shows that 1 mole of CH_4 reacts with 2 moles of O_2. These two quantities are chemically equivalent, so we can construct *unit factors*.

Solution

$$CH_4 + 2O_2 \longrightarrow CO_2 + 2H_2O$$

| 1 mol | 2 mol | 1 mol | 2 mol |

Here we solve the problem in three steps and so we convert
1. g $CH_4 \rightarrow$ mol CH_4

$$\underset{\text{—}}{?} \text{ mol CH}_4 = 24 \text{ g CH}_4 \times \frac{1 \text{ mol CH}_4}{16 \text{ g CH}_4} = \underline{1.5 \text{ mol CH}_4}$$

2. mol $CH_4 \rightarrow$ mol O_2

$$\underset{\text{—}}{?} \text{ mol O}_2 = 1.5 \text{ mol CH}_4 \times \frac{2 \text{ mol O}_2}{1 \text{ mol CH}_4} = \underline{3.0 \text{ mol O}_2}$$

3. mol $O_2 \rightarrow$ g O_2

$$\underset{\text{—}}{?} \text{ g O}_2 = 3.0 \text{ mol O}_2 \times \frac{32 \text{ g O}_2}{1 \text{ mol O}_2} = \boxed{96 \text{ g O}_2}$$

All these steps could be combined into one setup in which we convert

$$\boxed{\text{g of CH}_4} \longrightarrow \boxed{\text{mol of CH}_4} \longrightarrow \boxed{\text{mol of O}_2} \longrightarrow \boxed{\text{g of O}_2}$$

$$\underset{\text{—}}{?} \text{ g O}_2 = 24 \text{ g CH}_4 \times \frac{1 \text{ mol CH}_4}{16 \text{ g CH}_4} \times \frac{2 \text{ mol O}_2}{1 \text{ mol CH}_4} \times \frac{32 \text{ g O}_2}{1 \text{ mol O}_2} = \boxed{96 \text{ g O}_2}$$

The same answer, 96 grams of O_2, is obtained by both methods.

You should now work Exercises 16 and 20.

The question posed in Example 3-4 may be reversed, as in Example 3-5.

EXAMPLE 3-5   *Mass of a Reactant Required*

What mass of CH_4, in grams, is required to react with 96 grams of O_2?

Plan

We recall that one mole of CH_4 reacts with two moles of O_2.

Solution

$$\underline{?} \text{ g CH}_4 = 96 \text{ g O}_2 \times \frac{1 \text{ mol O}_2}{32 \text{ g O}_2} \times \frac{1 \text{ mol CH}_4}{2 \text{ mol O}_2} \times \frac{16 \text{ g CH}_4}{1 \text{ mol CH}_4} = \boxed{24 \text{ g CH}_4}$$

These unit factors are the reciprocals of those used in Example 3-4.

You should now work Exercise 22.

This is the amount of CH_4 in Example 3-4 that reacted with 96 grams of O_2.

EXAMPLE 3-6 *Mass of a Product Formed*

Most combustion reactions occur in excess O_2, i.e., more than enough O_2 to burn the substance completely. Calculate the mass of CO_2, in grams, that can be produced by burning 6.0 moles of CH_4 in excess O_2.

It is important to recognize that the reaction must stop when the 6.0 mol of CH_4 has been used up. Some O_2 will remain unreacted.

Plan

The balanced equation tells us that one mole of CH_4 produces one mole of CO_2.

$$\text{CH}_4 + 2\text{O}_2 \longrightarrow \text{CO}_2 + 2\text{H}_2\text{O}$$

| 1 mol | 2 mol | 1 mol | 2 mol |
| 16 g | 2(32 g) | 44 g | 2(18 g) |

Solution

$$\underline{?} \text{ g CO}_2 = 6.0 \text{ mol CH}_4 \times \frac{1 \text{ mol CO}_2}{1 \text{ mol CH}_4} \times \frac{44 \text{ g CO}_2}{1 \text{ mol CO}_2} = \boxed{2.6 \times 10^2 \text{ g CO}_2}$$

You should now work Exercise 24.

From the mole interpretation of the chemical equation for the combustion of methane, we can see many chemically equivalent pairs of terms. Each pair gives a unit factor relating substances. Some of these factors are

$$\frac{1 \text{ mol CH}_4}{2 \text{ mol O}_2} \qquad \frac{1 \text{ mol CH}_4}{64 \text{ g O}_2} \qquad \frac{1 \text{ mol CH}_4}{2(6.02 \times 10^{23}) \text{ O}_2 \text{ molecules}}$$

$$\frac{16 \text{ g CH}_4}{2 \text{ mol O}_2} \qquad \frac{16 \text{ g CH}_4}{64 \text{ g O}_2} \qquad \frac{16 \text{ g CH}_4}{2(6.02 \times 10^{23}) \text{ O}_2 \text{ molecules}}$$

$$\frac{6.02 \times 10^{23} \text{ CH}_4 \text{ molecules}}{2 \text{ mol O}_2} \qquad \frac{6.02 \times 10^{23} \text{ CH}_4 \text{ molecules}}{64 \text{ g O}_2} \qquad \frac{6.02 \times 10^{23} \text{ CH}_4 \text{ molecules}}{2(6.02 \times 10^{23}) \text{ O}_2 \text{ molecules}}$$

We have written nine unit factors that relate CH_4 and O_2 for this particular reaction. The nine factors obtained by inverting each of these give a total of 18 unit factors relating CH_4 and O_2 for *this* reaction in three kinds of units. In fact, we can write down many factors relating *any* two substances involved in any chemical reaction! Try writing down some that involve reactants and products.

▼ **PROBLEM-SOLVING TIP** *Different Ways to Solve Problems*

There are many ways to solve stoichiometry problems, but all are based on the mole ratio (reaction ratio) from the balanced chemical equation. We

(1) convert the given units to moles,

(2) use the mole ratio, and

(3) if necessary, convert moles to the desired units.

Reaction stoichiometry usually involves interpreting a balanced chemical equation to relate a *given* bit of information to the *desired* bit of information.

EXAMPLE 3-7   *Mass of a Reactant Required*

What mass of CH_4 produces 3.01×10^{23} H_2O molecules when burned in excess O_2?

Plan

The balanced equation tells us that one mole of CH_4 produces two moles of H_2O.

$$CH_4 \ + \ 2O_2 \ \longrightarrow \ CO_2 \ + \ 2H_2O$$

1 mol   2 mol   1 mol   2 mol

$$\boxed{H_2O \text{ molecules}} \longrightarrow \boxed{\text{mol of } H_2O} \longrightarrow \boxed{\text{mol of } CH_4} \longrightarrow \boxed{\text{g of } CH_4}$$

Solution

$$\underline{?} \text{ g } CH_4 = 3.01 \times 10^{23} \text{ } H_2O \text{ molecules} \times \frac{1 \text{ mol } H_2O}{6.02 \times 10^{23} \text{ } H_2O \text{ molecules}} \times \frac{1 \text{ mol } CH_4}{2 \text{ mol } H_2O} \times \frac{16 \text{ g } CH_4}{1 \text{ mol } CH_4} = \boxed{4.0 \text{ g } CH_4}$$

You should now work Exercise 26.

The possibilities for this kind of problem solving go on and on. Before you continue, you should work Exercises 12–27 at the end of the chapter.

3-3   THE LIMITING REACTANT CONCEPT

In the problems we have worked thus far, the presence of an excess of one reactant was stated or implied. The calculations were based on the substance that was used up first, called the **limiting reactant.** Before we study the concept of the limiting reactant in stoichiometry, let's develop the basic idea by considering a simple but analogous non-chemical example.

Suppose you have four slices of ham and six slices of bread and you wish to make as many ham sandwiches as possible using only one slice of ham and two slices of bread per sandwich. Obviously, you can make only three sandwiches, at which point you run out of bread. (In a chemical reaction this would correspond to one of the reactants being used up—so the reaction would stop.) Therefore the bread is the "limiting reactant" and the extra slice of ham is the "excess reactant." The amount of product, ham sandwiches, is determined by the amount of the limiting reactant, bread in this case. The limiting reactant is not necessarily the reactant present in the smallest amount. We have four slices of ham, the smallest amount, and six slices of bread. But the "reaction ratio" is two slices of bread to one piece of ham, and so bread is the limiting reactant.

EXAMPLE 3-8   *Limiting Reactant*

What mass of CO_2 could be formed by the reaction of 16 g of CH_4 with 48 g of O_2?

Plan

The balanced equation tells us that *one* mole of CH_4 reacts with *two* moles of O_2.

$$CH_4 + 2O_2 \longrightarrow CO_2 + 2H_2O$$

| 1 mol | 2 mol | 1 mol | 2 mol |
| 16 g | 2(32 g) | 44 g | 2(18 g) |

We are given masses of both CH_4 and O_2, so we calculate the number of moles of each reactant, and then determine the number of moles of each reactant required to react with the other. From these calculations we can identify the limiting reactant. We base the calculation on it.

Solution

$$\underline{?}\ mol\ CH_4 = 16\ g\ CH_4 \times \frac{1\ mol\ CH_4}{16\ g\ CH_4} = \underline{1.0\ mol\ CH_4}$$

$$\underline{?}\ mol\ O_2 = 48\ g\ O_2 \times \frac{1\ mol\ O_2}{32\ g\ O_2} = \underline{1.5\ mol\ O_2}$$

Now we return to the balanced equation. First we calculate the number of moles of O_2 required to react with 1.0 mole of CH_4.

$$\underline{?}\ mol\ O_2 = 1.0\ mol\ CH_4 \times \frac{2\ mol\ O_2}{1\ mol\ CH_4} = 2\ mol\ O_2$$

We see that 2 moles of O_2 are required, but we have 1.5 moles of O_2, so O_2 is the limiting reactant. Alternatively, we can calculate the number of moles of CH_4 that would react with 1.5 moles of O_2.

$$\underline{?}\ mol\ CH_4 = 1.5\ mol\ O_2 \times \frac{1\ mol\ CH_4}{2\ mol\ O_2} = 0.75\ mol\ CH_4$$

This tells us that only 0.75 mole of CH_4 would be required to react with 1.5 moles of O_2. But we have 1.0 mole of CH_4, so we see again that O_2 is the limiting reactant. The reaction must stop when the limiting reactant, O_2, is used up, so we base the calculation on O_2.

$$\boxed{\text{g of } O_2} \longrightarrow \boxed{\text{mol of } O_2} \longrightarrow \boxed{\text{mol of } CO_2} \longrightarrow \boxed{\text{g of } CO_2}$$

$$\underline{?}\ g\ CO_2 = 48\ g\ O_2 \times \frac{1\ mol\ O_2}{32\ O_2} \times \frac{1\ mol\ CO_2}{2\ mol\ O_2} \times \frac{44\ g\ CO_2}{1\ mol\ CO_2} = \boxed{33\ g\ CO_2}$$

You should now work Exercise 28.

Thus, 33 grams of CO_2 is the most CO_2 that can be produced from 16 grams of CH_4 and 48 grams of O_2. If we had based our calculation on CH_4 rather than O_2, our answer would be too big (44 grams) and *wrong* because more O_2 than we have would be required.

Another approach to problems like Example 3-8 is to calculate the number of moles of each reactant:

$$\underline{?}\ mol\ CH_4 = 16\ g\ CH_4 \times \frac{1\ mol\ CH_4}{16\ g\ CH_4} = \underline{1.0\ mol\ CH_4}$$

$$\underline{?}\ mol\ O_2 = 48\ g\ O_2 \times \frac{1\ mol\ O_2}{32\ g\ O_2} = \underline{1.5\ mol\ O_2}$$

Then we return to the balanced equation. We first calculate the *required ratio* of reactants as indicated by the balanced chemical equation. We then calculate the *available ratio* of reactants and compare the two:

Required Ratio   **Available Ratio**

$$\frac{1 \text{ mol CH}_4}{2 \text{ mol O}_2} = \frac{0.50 \text{ mol CH}_4}{1.00 \text{ mol O}_2} \qquad \frac{1.0 \text{ mol CH}_4}{1.50 \text{ mol O}_2} = \frac{0.67 \text{ mol CH}_4}{1.00 \text{ mol O}_2}$$

We see that each mole of O_2 would require exactly 0.50 mole of CH_4 to be completely used up. But we have 0.67 mole of CH_4 for each mole of O_2, so there is *insufficient* O_2 to react with all of the available CH_4. The reaction must stop when the O_2 is gone; O_2 is the limiting reactant and we must base the calculation on it.

▼ **PROBLEM-SOLVING TIP**   *Choosing the Limiting Reactant*

If you can't decide how to solve a limiting reactant problem, as a last resort do the entire calculation based on each reactant. The *smallest* answer is always right.

EXAMPLE 3-9   *Limiting Reactant*

What is the maximum mass of $Ni(OH)_2$ that could be prepared by mixing two solutions that contain 25.9 grams of $NiCl_2$ and 10.0 grams of NaOH, respectively?

$$NiCl_2 + 2NaOH \longrightarrow Ni(OH)_2 + 2NaCl$$

Plan

Interpreting the balanced equation as usual, we have

$$NiCl_2 + 2NaOH \longrightarrow Ni(OH)_2 + 2NaCl$$

1 mol	2 mol	1 mol	2 mol
129.6 g	2(40.0 g)	92.7 g	2(58.4 g)

We determine the number of moles of $NiCl_2$ and NaOH present. Then we find the number of moles of each reactant required to react with the other reactant. These calculations identify the limiting reactant. We base the calculation on it.

Solution

$$\underset{=}{?} \text{ mol NiCl}_2 = 25.9 \text{ g NiCl}_2 \times \frac{1 \text{ mol NiCl}_2}{129.6 \text{ g NiCl}_2} = 0.200 \text{ mol NiCl}_2$$

$$\underset{=}{?} \text{ mol NaOH} = 10.0 \text{ g NaOH} \times \frac{1 \text{ mol NaOH}}{40.0 \text{ g NaOH}} = 0.250 \text{ mol NaOH}$$

We return to the balanced equation and calculate the number of moles of NaOH required to react with 0.200 mole of $NiCl_2$.

$$\underset{=}{?} \text{ mol NaOH} = 0.200 \text{ mol NiCl}_2 \times \frac{2 \text{ mol NaOH}}{1 \text{ mol NiCl}_2} = 0.400 \text{ mol NaOH}$$

But we have only 0.250 mole of NaOH, so NaOH is the limiting reactant.

g of NaOH $\longrightarrow$ mol of NaOH $\longrightarrow$ mol $Ni(OH)_2$ $\longrightarrow$ g of $Ni(OH)_2$

A precipitate of solid $Ni(OH)_2$ forms when colorless NaOH solution is added to green $NiCl_2$ solution.

Even though the reaction occurs in aqueous solution, this calculation is similar to earlier examples because we are given the amounts of both reactants.

$$\underline{?} \text{ g Ni(OH)}_2 = 10.0 \text{ g NaOH} \times \frac{1 \text{ mol NaOH}}{40.0 \text{ g NaOH}} \times \frac{1 \text{ mol Ni(OH)}_2}{2 \text{ mol NaOH}} \times \frac{92.7 \text{ g Ni(OH)}_2}{1 \text{ mol Ni(OH)}_2}$$

$$= \boxed{11.6 \text{ g Ni(OH)}_2}$$

You should now work Exercise 29.

3-4 PERCENT YIELDS FROM CHEMICAL REACTIONS

The **theoretical yield** from a chemical reaction is the yield calculated by assuming that the reaction goes to completion. In practice we often do not obtain as much product from a reaction mixture as is theoretically possible. There are several reasons. (1) Many reactions do not go to completion; i.e., the reactants are not completely converted to products. (2) In some cases, a particular set of reactants undergoes two or more reactions simultaneously, forming undesired products as well as desired products. Reactions other than the desired one are called "side reactions." (3) In some cases, separation of the desired product from the reaction mixture is so difficult that not all of the product formed is successfully isolated.

In the examples we have worked to this point, the amounts of products that we calculated were theoretical yields.

The term **percent yield** is used to indicate how much of a desired product is obtained from a reaction.

$$\text{percent yield} = \frac{\text{actual yield of product}}{\text{theoretical yield of product}} \times 100\%$$

Consider the preparation of nitrobenzene, $C_6H_5NO_2$, by the reaction of a limited amount of benzene, C_6H_6, with excess nitric acid, HNO_3. The balanced equation for the reaction may be written

$$C_6H_6 + HNO_3 \longrightarrow C_6H_5NO_2 + H_2O$$

| 1 mol | 1 mol | 1 mol | 1 mol |
| 78.1 g | 63.0 g | 123.1 g | 18.0 g |

EXAMPLE 3-10 *Percent Yield*

A 15.6-gram sample of C_6H_6 is mixed with excess HNO_3. We isolate 18.0 grams of $C_6H_5NO_2$. What is the percent yield of $C_6H_5NO_2$ in this reaction?

Plan

First we interpret the balanced chemical equation to calculate the theoretical yield of $C_6H_5NO_2$. Then we use the actual (isolated) yield with the definition given above to calculate the percent yield.

Solution

We calculate the theoretical yield of $C_6H_5NO_2$.

$$\underline{?} \text{ g } C_6H_5NO_2 = 15.6 \text{ g } C_6H_6 \times \frac{1 \text{ mol } C_6H_6}{78.1 \text{ g } C_6H_6} \times \frac{1 \text{ mol } C_6H_5NO_2}{1 \text{ mol } C_6H_6} \times \frac{123.1 \text{ g } C_6H_5NO_2}{1 \text{ mol } C_6H_5NO_2}$$

$$= 24.6 \text{ g } C_6H_5NO_2 \leftarrow \text{theoretical yield}$$

It is not necessary to know the mass of one mole of HNO_3 to solve this problem.

This tells us that if *all* the C_6H_6 were converted to $C_6H_5NO_2$ and isolated, we should obtain 24.6 grams of $C_6H_5NO_2$ (100% yield). However, we isolate only 18.0 grams of $C_6H_5NO_2$.

$$\text{percent yield} = \frac{\text{actual yield of product}}{\text{theoretical yield of product}} \times 100\% = \frac{18.0\ g}{24.6\ g} \times 100\%$$

$$= 73.2 \text{ percent yield}$$

You should now work Exercise 36.

The amount of nitrobenzene obtained *in this experiment* is 73.2% of the amount that would be expected *if* the reaction had gone to completion, *if* there were no side reactions, and *if* we could have recovered all of the product as pure substance.

In many important chemical processes, especially those encountered in the chemical industry, several equations are required to describe the chemical change. An analysis of the products often lets us describe the fraction of the change that occurs by each reaction.

3-5 SEQUENTIAL REACTIONS

Often more than one reaction is required to change starting materials into the desired product. This is true for many reactions that we carry out in the laboratory and for many industrial processes. These are called **sequential reactions.** The amount of desired product from each reaction is taken as the starting material for the next reaction.

EXAMPLE 3-11 *Sequential Reactions*

At high temperatures carbon reacts with water to produce a mixture of carbon monoxide, CO, and hydrogen, H_2.

$$C + H_2O \xrightarrow{\text{red heat}} CO + H_2$$

Carbon monoxide is separated from H_2 and then used to separate nickel from cobalt by forming a volatile compound, nickel tetracarbonyl, $Ni(CO)_4$.

$$Ni + 4CO \longrightarrow Ni(CO)_4$$

What mass of $Ni(CO)_4$ could be obtained from the CO produced by the reaction of 75.0 grams of carbon? Assume 100% reaction and 100% recovery in both steps.

Plan

We interpret both chemical equations in the usual way, and solve the problem in two steps. They tell us that one mole of C produces one mole of CO and that four moles of CO are required to produce one mole of $Ni(CO)_4$.

1. We determine the number of moles of CO formed in the first reaction.

2. From the number of moles of CO produced in the first reaction, we calculate the number of grams of $Ni(CO)_4$ that would be formed in the second reaction.

Solution

1.
$$\begin{array}{ccccccc} C & + & H_2O & \longrightarrow & CO & + & H_2 \\ 1\ mol & & 1\ mol & & 1\ mol & & 1\ mol \\ 12.0\ g & & & & & & \end{array}$$

$$\underline{?} \text{ mol CO} = 75.0 \text{ g C} \times \frac{1 \text{ mol C}}{12.0 \text{ g C}} \times \frac{1 \text{ mol CO}}{1 \text{ mol C}} = 6.25 \text{ mol CO}$$

2.
$$\begin{array}{cccc} \text{Ni} & + & 4\text{CO} & \longrightarrow & \text{Ni(CO)}_4 \\ 1 \text{ mol} & & 4 \text{ mol} & & 1 \text{ mol} \\ & & & & 171 \text{ g} \end{array}$$

$$\underline{?} \text{ g Ni(CO)}_4 = 6.25 \text{ mol CO} \times \frac{1 \text{ mol Ni(CO)}_4}{4 \text{ mol CO}} \times \frac{171 \text{ g Ni(CO)}_4}{1 \text{ mol Ni(CO)}_4}$$

$$= \boxed{267 \text{ g Ni(CO)}_4}$$

Alternatively, we can set up a series of unit factors based on the conversions in the reaction sequence and solve the problem in one setup.

$$\boxed{\text{g C}} \longrightarrow \boxed{\text{mol C}} \longrightarrow \boxed{\text{mol CO}} \longrightarrow \boxed{\text{mol Ni(CO)}_4} \longrightarrow \boxed{\text{g Ni(CO)}_4}$$

$$\underline{?} \text{ g Ni(CO)}_4 = 75.0 \text{ g C} \times \frac{1 \text{ mol C}}{12.0 \text{ g C}} \times \frac{1 \text{ mol CO}}{1 \text{ mol C}} \times \frac{1 \text{ mol Ni(CO)}_4}{4 \text{ mol CO}} \times \frac{171 \text{ g Ni(CO)}_4}{1 \text{ mol Ni(CO)}_4}$$

$$= \boxed{267 \text{ g Ni(CO)}_4}$$

You should now work Exercise 42.

EXAMPLE 3-12 *Sequential Reactions*

Phosphoric acid, H_3PO_4, is a very important compound used to make fertilizers. H_3PO_4 can be prepared in a two-step process.

$$\text{Rxn 1:} \quad P_4 + 5O_2 \longrightarrow P_4O_{10}$$

$$\text{Rxn 2:} \quad P_4O_{10} + 6H_2O \longrightarrow 4H_3PO_4$$

We allow 272 grams of phosphorus to react with excess oxygen, which forms tetraphosphorus decoxide, P_4O_{10}, in 89.5% yield. In the second step reaction, a 96.8% yield of H_3PO_4 is obtained. What mass of H_3PO_4 is obtained?

Approximately 100 lb of H_3PO_4 are required to support the lifestyle of each American for one year.

Plan

1. We interpret the first equation as usual and calculate the amount of P_4O_{10} *obtained.*

$$\begin{array}{cccc} P_4 & + & 5O_2 & \longrightarrow & P_4O_{10} \\ 1 \text{ mol} & & 5 \text{ mol} & & 1 \text{ mol} \\ 124 \text{ g} & & 5(32.0 \text{ g}) & & 284 \text{ g} \end{array}$$

$$\boxed{\text{g } P_4} \longrightarrow \boxed{\text{mol } P_4} \longrightarrow \boxed{\text{mol } P_4O_{10}} \longrightarrow \boxed{\text{g } P_4O_{10}}$$

2. Then we interpret the second equation and calculate the amount of H_3PO_4 *obtained* from the P_4O_{10} from the first step.

$$\begin{array}{cccc} P_4O_{10} & + & 6H_2O & \longrightarrow & 4H_3PO_4 \\ 1 \text{ mol} & & 6 \text{ mol} & & 4 \text{ mol} \\ 284 \text{ g} & & 6(18.0 \text{ g}) & & 4(98.0 \text{ g}) \end{array}$$

$$\boxed{\text{g } P_4O_{10}} \longrightarrow \boxed{\text{mol } P_4O_{10}} \longrightarrow \boxed{\text{mol } H_3PO_4} \longrightarrow \boxed{\text{g } H_3PO_4}$$

Solution

1. $$\underline{?} \text{ g } P_4O_{10} = 272 \text{ g } P_4 \times \frac{1 \text{ mol } P_4}{124 \text{ g } P_4} \times \frac{1 \text{ mol } P_4O_{10}}{1 \text{ mol } P_4} \times \frac{284 \text{ g } P_4O_{10} \text{ theor.}}{1 \text{ mol } P_4O_{10} \text{ theor.}}$$

$$\times \left(\frac{89.5 \text{ g } P_4O_{10} \text{ actual}}{100 \text{ g } P_4O_{10} \text{ theor.}} \right) = \boxed{558 \text{ g } P_4O_{10}}$$

The unit factors that account for less than 100% reaction and less than 100% recovery are included in parentheses.

The sodium hydroxide and aluminum in some drain cleaners do not react while they are stored in solid form. When water is added, the NaOH dissolves and begins to act on trapped grease. At the same time, NaOH and Al react to produce H_2 gas; the resulting turbulence helps to dislodge the blockage. Do you see why the container should be kept tightly closed?

In some solutions, such as a nearly equal mixture of ethyl alcohol and water, the distinction between solute *and* solvent *is arbitrary.*

2. $\underline{?}$ g H_3PO_4 = 558 g $P_4O_{10} \times \dfrac{1 \text{ mol } P_4O_{10}}{284 \text{ g } P_4O_{10}} \times \dfrac{4 \text{ mol } H_3PO_4}{1 \text{ mol } P_4O_{10}} \times \dfrac{98.0 \text{ g } H_3PO_4 \text{ theor.}}{1 \text{ mol } H_3PO_4 \text{ theor.}}$

$\times \left(\dfrac{96.8 \text{ g } H_3PO_4 \text{ actual}}{100 \text{ g } H_3PO_4 \text{ theor.}} \right) = \boxed{746 \text{ g } H_3PO_4}$

You should now work Exercise 44.

Chemists have determined the structures of many naturally occurring compounds. One way of proving the structure of such a compound is by synthesizing it from available starting materials. Professor Grieco, now at Indiana University, was assisted by Majetich and Ohfune in the synthesis of helenalin, a powerful cancer drug, in a forty-step process. This forty-step synthesis gave a remarkable average yield of about 90% for each step, which resulted in an overall yield of about 1.5%.

3-6 CONCENTRATIONS OF SOLUTIONS

Many chemical reactions are more conveniently carried out with the reactants mixed in solution rather than as pure substances. A **solution** is a homogeneous mixture, at the molecular level, of two or more substances. Simple solutions usually consist of one substance, the **solute,** dissolved in another substance, the **solvent.** The solutions used in the laboratory are usually liquids, and the solvent is often water. These are called **aqueous solutions.** For example, solutions of hydrochloric acid are prepared by dissolving hydrogen chloride (HCl, a gas at room temperature and atmospheric pressure) in water. Solutions of sodium hydroxide are prepared by dissolving solid NaOH in water.

We often use solutions to supply the reactants for chemical reactions. Solutions allow the most intimate mixing of the reacting substances at the molecular level, much more than would be possible in solid form. (A practical example is drain cleaner, shown in the photo.) Furthermore, the rate of the reaction can often be controlled by adjusting the concentrations of the solutions. In this section we shall study methods for expressing the quantities of the various components present in a given amount of solution.

Concentrations of solutions are expressed in terms of *either* the amount of solute present in a given mass or volume of *solution,* or the amount of solute dissolved in a given mass or volume of *solvent.*

Percent by Mass

Concentrations of solutions may be expressed in terms of percent by mass of solute, which gives the mass of solute per 100 mass units of solution. The gram is the usual mass unit.

$$\% \text{ solute} = \frac{\text{mass of solute}}{\text{mass of solution}} \times 100\%$$

A 10.0% solution of $Ca(C_6H_{11}O_7)_2$ is sometimes administered intravenously in emergency treatment for black widow spider bites.

Thus, a solution that is 10.0% calcium gluconate, $Ca(C_6H_{11}O_7)_2$, by mass contains 10.0 grams of calcium gluconate in 100.0 grams of *solution.* This could be described as 10.0 grams of calcium gluconate in 90.0 grams of water. The density of a 10.0% solution of calcium gluconate is 1.07 g/mL, so 100 mL of a 10.0% solution of calcium gluconate has a mass of 107 grams. Observe that 100 grams of a solution usually does *not* occupy 100 mL. Unless otherwise specified, percent means percent *by mass,* and water is the solvent.

EXAMPLE 3-13 *Percent of Solute*

Calculate the mass of nickel(II) sulfate, $NiSO_4$, contained in 200 grams of a 6.00% solution of $NiSO_4$.

Plan

The percentage information tells us that the solution contains 6.00 grams of $NiSO_4$ per 100 grams of solution. The desired information is the mass of $NiSO_4$ in 200 grams of solution. A unit factor is constructed by placing 6.00 grams of $NiSO_4$ over 100 grams of solution. Multiplication of the mass of the solution, 200 grams, by this unit factor gives the mass of $NiSO_4$ in the solution.

Solution

$$\underline{?} \text{ g NiSO}_4 = 200 \text{ g soln} \times \frac{6.00 \text{ g NiSO}_4}{100 \text{ g soln}} = \boxed{12.0 \text{ g NiSO}_4}$$

EXAMPLE 3-14 *Mass of Solution*

Calculate the mass of 6.00% $NiSO_4$ solution that contains 40.0 grams of $NiSO_4$.

Plan

Placing 100 grams of solution over 6.00 grams of $NiSO_4$ gives another unit factor.

Solution

$$\underline{?} \text{ g soln} = 40.0 \text{ g NiSO}_4 \times \frac{100 \text{ g soln}}{6.00 \text{ g NiSO}_4} = \boxed{667 \text{ g soln}}$$

You should now work Exercise 50.

EXAMPLE 3-15 *Mass of Solute*

Calculate the mass of $NiSO_4$ contained in 200 mL of a 6.00% solution of $NiSO_4$. The density of the solution is 1.06 g/mL at 25°C.

Plan

The volume of a solution multiplied by its density gives the mass of solution (Section 1-11). The mass of solution is then multiplied by the mass fraction due to $NiSO_4$ (6.00 g $NiSO_4$/100 g soln) to give the mass of $NiSO_4$ in 200 mL of solution.

Solution

$$\underline{?} \text{ g NiSO}_4 = \underbrace{200 \text{ mL soln} \times \frac{1.06 \text{ g soln}}{1.00 \text{ mL soln}}}_{212 \text{ g soln}} \times \frac{6.00 \text{ g NiSO}_4}{100 \text{ g soln}} = \boxed{12.7 \text{ g NiSO}_4}$$

Volume of solution × density of solution = mass of solution

You should now work Exercise 52.

EXAMPLE 3-16 *Percent Solute and Density*

What volume of a solution that is 15.0% iron(III) nitrate contains 30.0 grams of $Fe(NO_3)_3$? The density of the solution is 1.16 g/mL at 25°C.

Plan

Two unit factors relate mass of $Fe(NO_3)_3$ and mass of solution, 15.0 g $Fe(NO_3)_3$/100 g soln and 100 g soln/15.0 g $Fe(NO_3)_3$. The second factor converts grams of $Fe(NO_3)_3$ to grams of solution.

Solution

$$? \text{ mL soln} = \underbrace{30.0 \text{ g Fe(NO}_3)_3 \times \frac{100 \text{ g soln}}{15.0 \text{ g Fe(NO}_3)_3}}_{200 \text{ g soln}} \times \frac{1.00 \text{ mL soln}}{1.16 \text{ g soln}} = \boxed{172 \text{ mL}}$$

Note that the answer is not 200 mL but considerably less because 1.00 mL of solution has a mass of 1.16 grams. However, 172 mL of the solution has a mass of 200 grams.

You should now work Exercise 54.

Molarity (molar concentration)

The definition of molarity specifies the amount of solute *per unit volume of solution,* whereas percent specifies the amount of solute *per unit mass of solution.* Therefore, molarity depends on temperature and pressure, whereas percent by mass does not.

Molarity (M), or molar concentration, is a common unit for expressing the concentrations of solutions. **Molarity** is defined as the number of moles of solute per liter of solution:

$$\text{molarity} = \frac{\text{number of moles of solute}}{\text{number of liters of solution}}$$

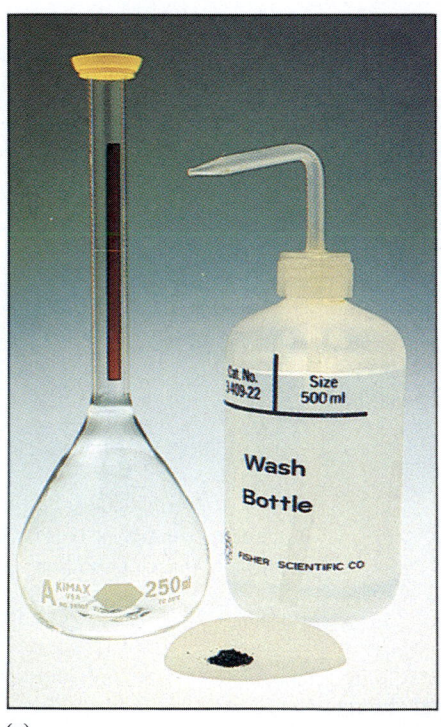

(a)

(b)

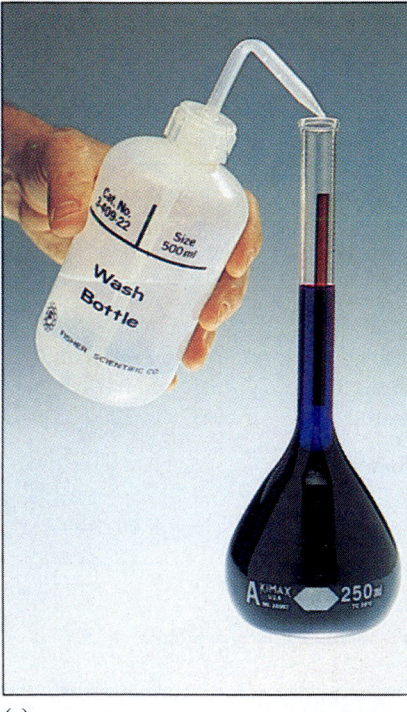

(c)

Figure 3-2 Preparation of 0.0100 M solution of KMnO$_4$, potassium permanganate. 250 mL of 0.0100 M KMnO$_4$ solution contains 0.395 g of KMnO$_4$ (1 mol = 158 g). (a) 0.395 g of KMnO$_4$ (0.00250 mole) is weighed out carefully and transferred into a 250-mL volumetric flask. (b) The KMnO$_4$ is dissolved in water. (c) Distilled H$_2$O is added to the volumetric flask until the volume of solution is 250 mL. The flask is then stoppered, and its contents are mixed thoroughly to give a homogeneous solution.

To prepare one liter of a one molar solution, one mole of solute is placed in a one-liter volumetric flask, enough solvent is added to dissolve the solute, and solvent is then added until the volume of the solution is exactly one liter. Students sometimes make the mistake of assuming that a one molar solution contains one mole of solute in a liter of solvent. This is *not* the case; one liter of solvent *plus* one mole of solute usually has a total volume of more than one liter. A 0.100 *M* solution contains 0.100 mole of solute per liter, and a 0.0100 *M* solution contains 0.0100 mole of solute per liter (Figure 3-2).

We often express the volume of a solution in milliliters rather than in liters. Likewise, we may express the amount of solute in millimoles (mmol) rather than in moles. Because one milliliter is 1/1000 of a liter and one millimole is 1/1000 of a mole, molarity also may be expressed as the number of millimoles of solute per milliliter of solution:

$$\text{molarity} = \frac{\text{number of millimoles of solute}}{\text{number of milliliters of solution}}$$

Water is the solvent in *most* of the solutions that we encounter. Unless otherwise indicated, we assume that water is the solvent. When the solvent is other than water, we state this explicitly.

EXAMPLE 3-17   *Molarity*

Calculate the molarity (*M*) of a solution that contains 3.65 grams of HCl in 2.00 liters of solution.

Plan
We are given the number of grams of HCl in 2.00 liters of solution. We apply the definition of molarity, remembering to convert grams of HCl to moles of HCl.

Solution

$$\frac{?\text{ mol HCl}}{\text{L}} = \frac{3.65\text{ g HCl}}{2.00\text{ L}} \times \frac{1\text{ mol HCl}}{36.5\text{ g HCl}} = \boxed{0.0500\text{ mol HCl/L}}$$

We place 3.65 g HCl over 2.00 L of solution, and then convert g HCl to mol HCl.

The concentration of the HCl solution is 0.0500 molar, and the solution is called 0.0500 *M* hydrochloric acid. One liter of the solution contains 0.0500 mole of HCl.

You should now work Exercises 56 and 58.

EXAMPLE 3-18   *Mass of Solute*

Calculate the mass of Ba(OH)$_2$ required to prepare 2.50 liters of a 0.06000 molar solution of barium hydroxide.

Plan
The volume of the solution, 2.50 liters, is multiplied by the concentration, 0.06000 mol Ba(OH)$_2$/L, to give the number of moles of Ba(OH)$_2$. The number of moles of Ba(OH)$_2$ is then multiplied by the mass of Ba(OH)$_2$ in one mole, 171.3 g Ba(OH)$_2$/mol Ba(OH)$_2$, to give the mass of Ba(OH)$_2$ in the solution.

Solution

$$\underset{?}{\text{g Ba(OH)}_2} = 2.50 \text{ L} \times \frac{0.06000 \text{ mol Ba(OH)}_2}{1 \text{ L}} \times \frac{171.3 \text{ g Ba(OH)}_2}{1 \text{ mol Ba(OH)}_2}$$

$$= \boxed{25.7 \text{ g Ba(OH)}_2}$$

You should now work Exercise 62.

The stock solutions of acids and bases that are sold commercially are too concentrated for most laboratory uses. We often dilute these solutions before we use them. We must know the molar concentration of a stock solution before it is diluted. This can be calculated from the specific gravity and the percentage data given on the label of the bottle.

ANALYSIS
Assay (H_2SO_4) W/W...Min. 95.0%--Max. 98.0%

MAXIMUM LIMITS OF IMPURITIES
Appearance ....Passes A.C.S. Test
Color (APHA) ....10 Max.
Residue after Ignition.................4 ppm
Chloride (Cl)...........................0.2 ppm
Nitrate (NO_3) ....0.5 ppm
Ammonium (NH_4).......................1 ppm
Substances Reducing $KMnO_4$ (limit about 2ppm as SO_2) ....Passes A.C.S. Test
Arsenic (As) ....0.004 ppm
Heavy Metals (as Pb)..............0.8 ppm
Iron (Fe) ....0.2 ppm
Mercury (Hg) ....5 ppb
Specific Gravity ....~1.84
Normality.....................................~36

Suitable for Mercury Determinations

A photo that shows the analysis of sulfuric acid.

EXAMPLE 3-19   *Molarity*

A sample of commercial sulfuric acid is 96.4% H_2SO_4 by mass, and its specific gravity is 1.84. Calculate the molarity of this sulfuric acid solution.

Plan

The specific gravity of a solution is numerically equal to its density, so the density of the solution is 1.84 g/mL. The solution is 96.4% H_2SO_4 by mass. Therefore 100 grams of solution contains 96.4 grams of *pure* H_2SO_4. From this information, we can find the molarity of the solution. First, we calculate the mass of one liter of solution.

Solution

$$\frac{\underset{?}{\text{g soln}}}{\text{L}} = \frac{1.84 \text{ g soln}}{\text{mL}} \times \frac{1000 \text{ mL}}{\text{L}} = 1840 \text{ g soln/L}$$

The solution is 96.4% H_2SO_4 by mass, so the mass of H_2SO_4 in one liter is

$$\frac{\underset{?}{\text{g H}_2\text{SO}_4}}{\text{L}} = \frac{1840 \text{ g soln}}{\text{L}} \times \frac{96.4 \text{ g H}_2\text{SO}_4}{100.0 \text{ g soln}} = 1.77 \times 10^3 \text{ g H}_2\text{SO}_4\text{/L}$$

The molarity is the number of moles of H_2SO_4 per liter of solution.

$$\frac{\underset{?}{\text{mol H}_2\text{SO}_4}}{\text{L}} = \frac{1.77 \times 10^3 \text{ g H}_2\text{SO}_4}{\text{L}} \times \frac{1 \text{ mol H}_2\text{SO}_4}{98.1 \text{ g H}_2\text{SO}_4} = \boxed{18.0 \text{ mol H}_2\text{SO}_4\text{/L}}$$

Thus, the solution is an 18.0 *M* H_2SO_4 solution. This problem can also be solved by using a series of three unit factors.

$$\frac{\underset{?}{\text{mol H}_2\text{SO}_4}}{\text{L}} = \frac{1.84 \text{ g soln}}{\text{mL}} \times \frac{96.4 \text{ g H}_2\text{SO}_4}{100 \text{ g soln}} \times \frac{1 \text{ mol H}_2\text{SO}_4}{98.1 \text{ g H}_2\text{SO}_4} \times \frac{1000 \text{ mL}}{\text{L}}$$

$$= 18.1 \text{ mol H}_2\text{SO}_4\text{/L} = \boxed{18.1 \text{ M H}_2\text{SO}_4}$$

The small difference is due to rounding.

You should now work Exercise 66.

3-7 DILUTION OF SOLUTIONS

Recall that the definition of molarity is the number of moles of solute divided by the volume of the solution in liters:

$$\text{molarity} = \frac{\text{number of moles of solute}}{\text{number of liters of solution}}$$

Multiplying both sides of the equation by the volume, we obtain

volume (in L)   × molarity = number of moles of solute

or   volume (in mL) × molarity = number of mmol of solute

> Multiplication of the volume of a solution by its molar concentration gives the amount of solute in the solution.

When we dilute a solution by mixing it with more solvent, the number of moles of solute present does not change. But the volume and the concentration of the solution *do* change. Because the same number of moles of solute is divided by a larger number of liters of solution, the molarity decreases. Using a subscript 1 to represent the original concentrated solution and a subscript 2 to represent the dilute solution, we obtain

volume$_1$ × molarity$_1$ = number of moles of solute = volume$_2$ × molarity$_2$

or

$$V_1 \times M_1 = V_2 \times M_2 \qquad \text{(for dilution only)}$$

This expression can be used to calculate any one of four quantities when the other three are known (Figure 3-3). We frequently need a certain volume of dilute solution of a given molarity for use in the laboratory, and we know the concentration of the stock solution available. Then we can calculate the amount of stock solution that must be used to make the dilute solution.

We could use any volume unit as long as we use the same unit on both sides of the equation. This relationship also applies when the concentration is changed by evaporating some solvent.

(a)

(b)

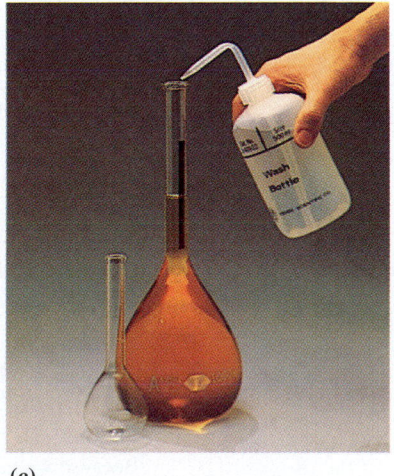

(c)

Figure 3-3 Dilution of solution. (a) A 100-mL volumetric flask is filled to the calibration line with 0.100 M potassium dichromate, $K_2Cr_2O_7$, solution. (b) The 0.100 M $K_2Cr_2O_7$ solution is transferred into a 1.00-L volumetric flask. The small flask is rinsed with distilled H_2O several times, and the rinse solutions are added to the larger flask. (c) Distilled water is added until the 1.00-L flask contains 1.00 L of solution. The flask is stoppered and its contents are mixed thoroughly. The new solution is 0.0100 M $K_2Cr_2O_7$. (100 mL of 0.100 M $K_2Cr_2O_7$ solution has been diluted to 1000 mL.)

> **C A U T I O N**
>
> Dilution of a concentrated solution, especially of a strong acid or base, frequently liberates a great deal of heat. This can vaporize drops of water as they hit the concentrated solution and can cause dangerous spattering. As a safety precaution, *concentrated solutions of acids or bases are always poured slowly into water,* allowing the heat to be absorbed by the larger quantity of water. Calculations are usually simpler to visualize by assuming that water is added to the concentrated solution.

EXAMPLE 3-20   *Dilution*

Calculate the volume of 18.0 M H_2SO_4 required to prepare 1.00 liter of a 0.900 M solution of H_2SO_4.

Plan

The volume (1.00 L) and molarity (0.900 M) of the final solution, as well as the molarity (18.0 M) of the original solution, are given. Therefore, the relation $V_1 \times M_1 = V_2 \times M_2$ can be used, with subscript 1 for the commercial acid solution and subscript 2 for the dilute solution. We solve

$$V_1 \times M_1 = V_2 \times M_2 \qquad \text{for } V_1$$

Solution

$$V_1 = \frac{V_2 \times M_2}{M_1} = \frac{1.00 \text{ L} \times 0.900 \, M}{18.0 \, M} = 0.0500 \text{ L} = \boxed{50.0 \text{ mL}}$$

The dilute solution contains 1.00 L × 0.900 M = 0.900 mol of H_2SO_4, so 0.900 mole of H_2SO_4 must be present in the original concentrated solution. Indeed, 0.0500 L × 18.0 M = 0.900 mol of H_2SO_4.

You should now work Exercises 70 and 72.

3-8 USING SOLUTIONS IN CHEMICAL REACTIONS

If we plan to carry out a reaction in a solution, we must calculate the amounts of solutions that we need. If we know the molarity of a solution, we can calculate the amount of solute contained in a specified volume of that solution. This is illustrated in Example 3-21.

EXAMPLE 3-21   *Amount of Solute*

Calculate (a) the number of moles of H_2SO_4, (b) the number of millimoles of H_2SO_4, and (c) the mass of H_2SO_4 in 500 mL of 0.324 M H_2SO_4 solution.

Plan

Because we have two parallel calculations in this example, we shall state the plan for each step just before the calculation is done.

Solution

(a) The volume of a solution in liters multiplied by its molarity gives the number of moles of solute, H_2SO_4 in this case.

500 mL is more conveniently expressed as 0.500 L in this problem. By now, you should be able to convert mL to L (and the reverse) without writing out the conversion.

$$\underline{?} \text{ mol } H_2SO_4 = 0.500 \text{ L soln} \times \frac{0.324 \text{ mol } H_2SO_4}{\text{L soln}} = \boxed{0.162 \text{ mol } H_2SO_4}$$

(b) The volume of a solution in milliliters multiplied by its molarity gives the number of millimoles of solute, H_2SO_4.

$$\underset{=}{?} \text{ mmol } H_2SO_4 = 500 \text{ mL soln} \times \frac{0.324 \text{ mmol } H_2SO_4}{\text{mL soln}} = \boxed{162 \text{ mmol } H_2SO_4}$$

(c) We may use the results of *either* part (a) or part (b) to calculate the mass of H_2SO_4 in the solution.

$$\underset{=}{?} \text{ g } H_2SO_4 = 0.162 \text{ mol } H_2SO_4 \times \frac{98.1 \text{ g } H_2SO_4}{1 \text{ mol } H_2SO_4} = \boxed{15.9 \text{ g } H_2SO_4}$$

A mole of H_2SO_4 is 98.1 g. A millimole is 98.1 mg.

or

$$\underset{=}{?} \text{ mg } H_2SO_4 = 162 \text{ mmol } H_2SO_4 \times \frac{98.1 \text{ mg } H_2SO_4}{1 \text{ mmol } H_2SO_4} = \boxed{1.59 \times 10^4 \text{ mg } H_2SO_4}$$

The mass of H_2SO_4 in the solution can be calculated without solving explicitly for the number of moles (or millimoles) of H_2SO_4.

Volume in liters times molarity gives moles of H_2SO_4. Molarity is a unit factor; i.e.,

$$\underset{=}{?} \text{ g } H_2SO_4 = 0.500 \text{ L soln} \times \frac{0.324 \text{ mol } H_2SO_4}{\text{L soln}} \times \frac{98.1 \text{ g } H_2SO_4}{1 \text{ mol } H_2SO_4} = \boxed{15.9 \text{ g } H_2SO_4}$$

$$\frac{\text{mol solute}}{\text{L soln}}$$

One of the most important uses of molarity relates the volume of a solution of known concentration of one reactant to the mass of the other reactant.

EXAMPLE 3-22 *Solution Stoichiometry*

Calculate the volume of a 0.324 *M* solution of sulfuric acid required to react completely with 2.792 grams of Na_2CO_3 according to the equation

$$H_2SO_4 + Na_2CO_3 \longrightarrow Na_2SO_4 + CO_2 + H_2O$$

The indicator methyl orange changes from yellow, its color in basic solutions, to orange, its color in acidic solutions, when the reaction in Example 3-22 reaches completion.

Plan

The balanced equation tells us that one mole of H_2SO_4 reacts with one mole of Na_2CO_3, and we can write

$$H_2SO_4 + Na_2CO_3 \longrightarrow Na_2SO_4 + CO_2 + H_2O$$
$$\text{1 mol} \qquad \text{1 mol} \qquad \qquad \text{1 mol} \qquad \text{1 mol} \qquad \text{1 mol}$$
$$\qquad \qquad \text{106.0 g}$$

We convert (1) grams of Na_2CO_3 to moles of Na_2CO_3, (2) moles of Na_2CO_3 to moles of H_2SO_4, and (3) moles of H_2SO_4 to liters of H_2SO_4 solution.

$$\boxed{\text{g } Na_2CO_3} \longrightarrow \boxed{\text{mol } Na_2CO_3} \longrightarrow \boxed{\text{mol } H_2SO_4} \longrightarrow \boxed{\text{L } H_2SO_4 \text{ soln}}$$

Solution

$$\underline{?}\text{ L } H_2SO_4 = 2.792 \text{ g } Na_2CO_3 \times \frac{1 \text{ mol } Na_2CO_3}{106.0 \text{ g } Na_2CO_3} \times \frac{1 \text{ mol } H_2SO_4}{1 \text{ mol } Na_2CO_3} \times \frac{1 \text{ L } H_2SO_4 \text{ soln}}{0.324 \text{ mol } H_2SO_4}$$

$$= \boxed{0.0813 \text{ L } H_2SO_4 \text{ soln}} \quad \text{or} \quad \boxed{81.3 \text{ mL } H_2SO_4 \text{ soln}}$$

You should now work Exercise 76.

Often we must calculate the volume of solution of known molarity that is required to react with a specified volume of another solution. We always examine the balanced chemical equation for the reaction to determine the *reaction ratio,* i.e., the relative numbers of moles (or millimoles) of reactants.

EXAMPLE 3-23   *Volume of Solution Required*

Find the volume of 0.505 *M* NaOH solution required to react with 40.0 mL of 0.505 *M* H_2SO_4 solution according to the reaction

$$H_2SO_4 + 2NaOH \longrightarrow Na_2SO_4 + 2H_2O$$

Plan

We shall work this example in several steps, stating the "plan," or reasoning, just before each step in the calculation. Then we shall use a single setup to solve the problem.

Solution

The balanced equation tells us that the reaction ratio is one mole of H_2SO_4 to two moles of NaOH.

$$H_2SO_4 + 2NaOH \longrightarrow Na_2SO_4 + 2H_2O$$
$$\text{1 mol} \quad \text{2 mol} \qquad \qquad \text{1 mol} \qquad \text{2 mol}$$

From the volume and the molarity of the H_2SO_4 solution, we can calculate the number of moles of H_2SO_4.

The volume of H_2SO_4 solution is expressed as 0.0400 L rather than 40.0 mL.

$$\underline{?}\text{ mol } H_2SO_4 = 0.0400 \text{ L } H_2SO_4 \text{ soln} \times \frac{0.505 \text{ mol } H_2SO_4}{\text{L soln}} = 0.0202 \text{ mol } H_2SO_4$$

The number of moles of H_2SO_4 is related to the number of moles of NaOH by the reaction ratio, 1 mol H_2SO_4/2 mol NaOH:

$$\underline{?}\text{ mol NaOH} = 0.0202 \text{ mol } H_2SO_4 \times \frac{2 \text{ mol NaOH}}{1 \text{ mol } H_2SO_4} = 0.0404 \text{ mol NaOH}$$

Now we can calculate the volume of 0.505 *M* NaOH solution that contains 0.0404 mole of NaOH:

$$\underline{?} \text{ L NaOH soln} = 0.0404 \text{ mol NaOH} \times \frac{1.00 \text{ L NaOH soln}}{0.505 \text{ mol NaOH}} = \boxed{0.0800 \text{ L NaOH soln}}$$

Again we see that molarity is a unit factor. In this case,

$$\frac{1.00 \text{ L NaOH soln}}{0.505 \text{ mol NaOH}}$$

which we usually call $\boxed{80.0 \text{ mL of NaOH solution.}}$

We have worked through the problem stepwise; let us solve it in a single setup.

$$\boxed{\begin{array}{c} \text{L } H_2SO_4 \text{ soln} \\ \text{available} \end{array}} \longrightarrow \boxed{\begin{array}{c} \text{mol } H_2SO_4 \\ \text{available} \end{array}} \longrightarrow \boxed{\begin{array}{c} \text{mol NaOH} \\ \text{soln needed} \end{array}} \longrightarrow \boxed{\begin{array}{c} \text{L NaOH} \\ \text{soln needed} \end{array}}$$

$$\underline{?} \text{ L NaOH soln} = 0.0400 \text{ L } H_2SO_4 \text{ soln} \times \frac{0.505 \text{ mol } H_2SO_4}{\text{L } H_2SO_4 \text{ soln}} \times \frac{2 \text{ mol NaOH}}{1 \text{ mol } H_2SO_4}$$

$$\times \frac{1.00 \text{ L NaOH soln}}{0.505 \text{ mol NaOH}}$$

$$= \boxed{0.0800 \text{ L NaOH soln or } 80.0 \text{ mL NaOH soln}}$$

You should now work Exercise 78.

We could have retained volumes in mL and expressed molarities as mmol/mL. Try working this example in those terms.

3-9 TITRATIONS

In Example 3-23, we calculated the volume of one solution that is required to react with a given volume of another solution, with the concentrations of *both* solutions given. In the laboratory we often measure the volume of one solution that is required to react with a given volume of another solution of known concentration. Then we calculate the concentration of the first solution. The process is called **titration** (Figure 3-4).

> Titration is the process in which a solution of one reactant, the titrant, is carefully added to a solution of another reactant, and the volume of titrant required for complete reaction is measured.

Solutions of accurately known concentrations are called **standard solutions.** Often we prepare solutions of such substances and then determine their concentrations by titration with a standard solution.

How does one know when to stop a titration—that is, when is the chemical reaction just complete? In one method, a few drops of an *indicator* solution are added to the solution to be titrated. An **indicator** is a substance that can exist in different forms, with different colors that depend upon the concentration of H^+ in the solution. At least one of these forms must be very intensely colored so that even very small amounts of it can be seen.

Suppose we titrate an acid solution of unknown concentration by adding a standardized solution of sodium hydroxide dropwise from a **buret** (Figure 3-4). A common buret is graduated in large intervals of 1 mL and in smaller intervals of 0.1 mL so that it is possible to estimate the volume of a solution dispensed to within at least ± 0.02 mL. (Experienced individuals can often read a buret to ± 0.01 mL.) The analyst tries to choose an indicator that changes color clearly at the point at which stoichiometrically equivalent amounts of

(a)

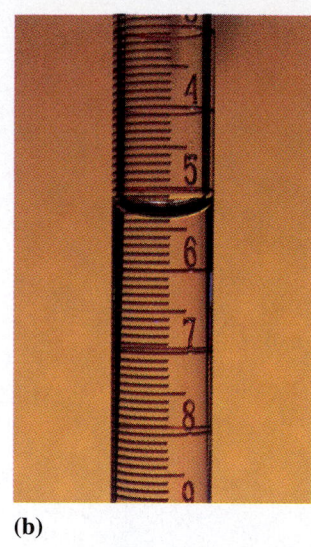

(b)

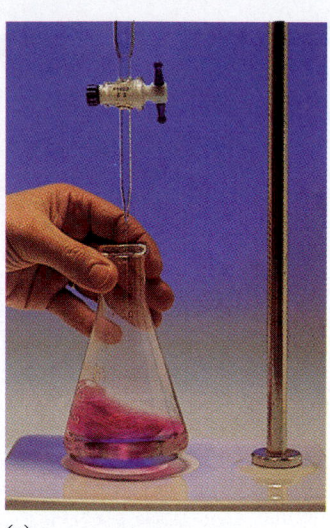

(c)

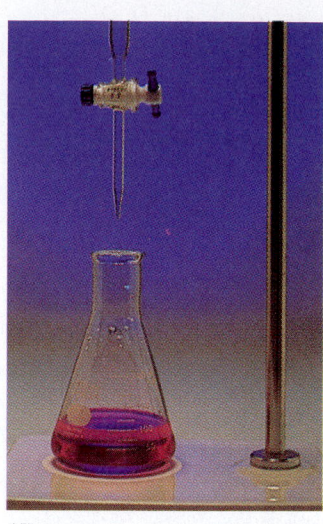

(d)

Figure 3-4 The titration process. (a) A typical setup for titration in a teaching laboratory. The solution to be titrated is placed in an Erlenmeyer flask, and a few drops of indicator are added. The buret is filled with a standard solution (or the solution to be standardized). The volume of solution in the buret is read carefully. (b) The meniscus describes the surface of the liquid in the buret. Aqueous solutions wet glass, so the meniscus of an aqueous solution is always concave. The position of the *bottom* of the meniscus is read and recorded. (c) The solution in the buret is added (dropwise near the end point), with stirring, to the Erlenmeyer flask until the end point is reached. (d) The end point is signaled by the appearance (or change) of color *throughout* the solution being titrated. (A very large excess of indicator was used to make this photograph.) The volume of the liquid is read again—the difference between the final and initial buret readings is the volume of the solution used.

acid and base have reacted, the **equivalence point.** The point at which the indicator changes color and the titration is stopped is called the **end point.** Ideally, the end point should coincide with the equivalence point. Phenolphthalein is colorless in acidic solution and reddish violet in basic solution. In a titration in which a base is added to an acid, phenolphthalein is often used as the indicator. The end point is signaled by the first appearance of a faint pink coloration that persists for at least 15 seconds as the solution is swirled.

EXAMPLE 3-24 *Titration*

What is the molarity of a hydrochloric acid solution if 36.7 mL of the HCl solution is required to react with 43.2 mL of 0.236 M sodium hydroxide solution?

$$HCl + NaOH \longrightarrow NaCl + H_2O$$

Plan

The balanced equation tells us that the reaction ratio is one mole of HCl to one mole of NaOH, which gives the unit factor, 1 mol HCl/1 mol NaOH.

$$HCl + NaOH \longrightarrow NaCl + H_2O$$
$$1 \text{ mol} \quad 1 \text{ mol} \quad \quad 1 \text{ mol} \quad 1 \text{ mol}$$

First we find the number of moles of NaOH. The reaction ratio is one mole of HCl to one mole of NaOH, so the HCl solution must contain the same number of moles of HCl. Then we can calculate the molarity of the HCl solution because we know its volume.

 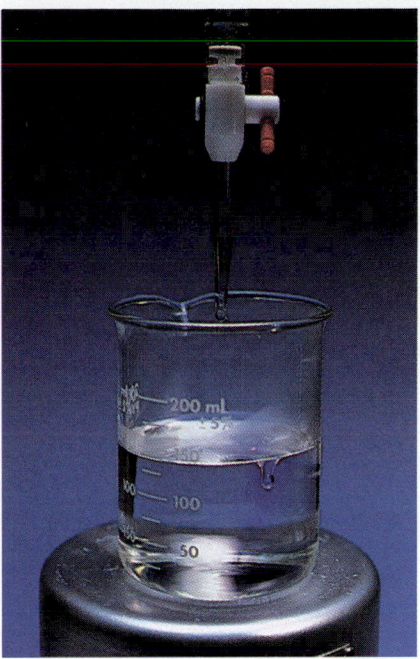 

The indicator phenolphthalein changes from colorless, its color in acidic solutions, to pink, its color in basic solutions, when the reaction in Example 3-24 reaches completion. Note the first appearance of a faint pink coloration in the middle beaker; this signals that the end point is near.

Solution

The volume of a solution (in liters) multiplied by its molarity gives the number of moles of solute.

$$\underset{\_}{?} \text{ mol NaOH} = 0.0432 \text{ L NaOH soln} \times \frac{0.236 \text{ mol NaOH}}{1 \text{ L NaOH soln}} = 0.0102 \text{ mol NaOH}$$

Because the reaction ratio is one mole of NaOH to one mole of HCl, the HCl solution must contain 0.0102 mole of HCl.

$$\underset{\_}{?} \text{ mol HCl} = 0.0102 \text{ mol NaOH} \times \frac{1 \text{ mol HCl}}{1 \text{ mol NaOH}} = 0.0102 \text{ mol HCl}$$

We know the volume of the HCl solution, so we can calculate its molarity.

$$\frac{\underset{\_}{?} \text{ mol HCl}}{\text{L HCl soln}} = \frac{0.0102 \text{ mol HCl}}{0.0367 \text{ L HCl soln}} = \boxed{0.278 \ M \text{ HCl}}$$

You should now work Exercise 88.

EXAMPLE 3-25   *Titration*

A 43.2-mL sample of 0.236 M sodium hydroxide solution reacts completely with 36.7 mL of a sulfuric acid solution. What is the molarity of the H_2SO_4 solution?

$$2\text{NaOH} + H_2SO_4 \longrightarrow Na_2SO_4 + H_2O$$

Plan

The balanced equation tells us that the reaction ratio is one mole of H_2SO_4 to two moles of NaOH, which gives the unit factor, 1 mol H_2SO_4/2 mol NaOH.

$$\begin{array}{ccccc} 2\text{NaOH} & + & H_2SO_4 & \longrightarrow & Na_2SO_4 & + & H_2O \\ 2 \text{ mol} & & 1 \text{ mol} & & 1 \text{ mol} & & 1 \text{ mol} \end{array}$$

First we find the number of moles of NaOH. The reaction ratio is one mole of H_2SO_4 to two moles of NaOH, so the number of moles of H_2SO_4 must be one half of the number of moles of NaOH. Then we can calculate the molarity of the H_2SO_4 solution because we know its volume.

Solution

The volume of a solution (in liters) multiplied by its molarity gives the number of moles of solute.

$$\underset{?}{\text{—}}\text{ mol NaOH} = 0.0432\text{ L NaOH soln} \times \frac{0.236\text{ mol NaOH}}{1\text{ L NaOH soln}} = 0.0102\text{ mol NaOH}$$

Because the reaction ratio is two moles of NaOH to one mole of H_2SO_4, the H_2SO_4 solution must contain 0.00510 mole of H_2SO_4.

$$\underset{?}{\text{—}}\text{ mol }H_2SO_4 = 0.0102\text{ mol NaOH} \times \frac{1\text{ mol }H_2SO_4}{2\text{ mol NaOH}} = 0.00510\text{ mol }H_2SO_4$$

We know the volume of the H_2SO_4 solution, so we can calculate its molarity.

$$\frac{\underset{?}{\text{—}}\text{ mol }H_2SO_4}{\text{L }H_2SO_4\text{ soln}} = \frac{0.00510\text{ mol }H_2SO_4}{0.0367\text{ L }H_2SO_4\text{ soln}} = \boxed{0.139\ M\ H_2SO_4}$$

Notice the similarity between Examples 3-24 and 3-25 in which 43.2 mL of 0.236 M NaOH solution is used. In Example 3-24 the reaction ratio is 1 mol acid/1 mol base, whereas in Example 3-25 the reaction ratio is 1 mol acid/2 mol base, and so the molarity of the HCl solution (0.278 M) is twice the molarity of the H_2SO_4 solution (0.139 M).

Key Terms

Actual yield   The amount of a specified pure product actually obtained from a given reaction. Compare with *Theoretical yield.*

Buret   A piece of volumetric glassware, usually graduated in 0.1-mL intervals, that is used in titrations to deliver solutions in a quantitative (dropwise) manner.

Chemical equation   Description of a chemical reaction by placing the formulas of reactants on the left and the formulas of products on the right of an arrow. A chemical equation must be balanced; i.e., it must have the same number of each kind of atom on both sides.

Concentration   The amount of solute per unit volume or mass of solvent or of solution.

Dilution   The process of reducing the concentration of a solute in solution, usually simply by mixing with more solvent.

End point   The point at which an indicator changes color and a titration is stopped.

Equivalence point   The point at which chemically equivalent amounts of reactants have reacted.

Indicators   For acid–base titrations, organic compounds that exhibit different colors in solutions of different acidities; used to determine the point at which reaction between two solutes is complete.

Limiting reactant   A substance that stoichiometrically limits the amount of product(s) that can be formed.

Molarity (M)   The number of moles of solute per liter of solution.

Percent by mass   100% multiplied by the mass of a solute divided by the mass of the solution in which it is contained.

Percent yield   100% times actual yield divided by theoretical yield.

Products   Substances produced in a chemical reaction.

Reactants   Substances consumed in a chemical reaction.

Reaction ratio   The relative amounts of reactants and products involved in a reaction; may be the ratio of moles, millimoles, or masses.

Reaction stoichiometry   Description of the quantitative relationships among substances as they participate in chemical reactions.

Sequential reaction   A chemical process in which several reaction steps are required to convert starting materials into products.

Solute   The dispersed (dissolved) phase of a solution.

Solution   A homogeneous mixture of two or more substances.

Solvent   The dispersing medium of a solution.

Standard solution   A solution of accurately known concentration.

Stoichiometry   Description of the quantitative relationships among elements and compounds as they undergo chemical changes.

Theoretical yield   The maximum amount of a specified product that could be obtained from specified amounts of reactants, assuming complete consumption of the limiting reactant according to only one reaction and complete recovery of the product. Compare with *Actual yield.*

Titration   The process by which the volume of a standard solution required to react with a specific amount of a substance is determined.

Exercises

1. What is a chemical equation? What information does it contain?
2. What fundamental law is the basis for balancing a chemical equation?
3. Use words to state explicitly the relationships among numbers of molecules of reactants and products in the equation for the combustion of hexane, C_6H_{14}.

$$2C_6H_{14} + 19O_2 \longrightarrow 12CO_2 + 14H_2O$$

Balance each "equation" in Exercises 4–7.

4. (a) $Al + O_2 \rightarrow Al_2O_3$
   (b) $N_2 + O_2 \rightarrow N_2O$
   (c) $K + KNO_3 \rightarrow K_2O + N_2$
   (d) $H_2O + KO_2 \rightarrow KOH + O_2$
   (e) $H_2SO_4 + NH_3 \rightarrow (NH_4)_2SO_4$
5. (a) $P_4 + O_2 \rightarrow P_4O_6$
   (b) $P_4 + O_2 \rightarrow P_4O_{10}$
   (c) $KClO_3 + H_2SO_4 \rightarrow HClO_3 + K_2SO_4$
   (d) $KOH + CO_2 \rightarrow K_2CO_3 + H_2O$
   (e) $KOH + CO_2 \rightarrow KHCO_3$
6. (a) $NaCl + H_2O \rightarrow NaOH + Cl_2 + H_2$
   (b) $RbOH + SO_2 \rightarrow Rb_2SO_3 + H_2O$
   (c) $Ba(OH)_2 + P_4O_{10} \rightarrow Ba_3(PO_4)_2 + H_2O$
   (d) $(NH_4)_2Cr_2O_7 \rightarrow N_2 + H_2O + Cr_2O_3$
   (e) $Al + Cr_2O_3 \rightarrow Al_2O_3 + Cr$
7. (a) $UO_2 + HF \rightarrow UF_4 + H_2O$
   (b) $CrCl_3 + NH_3 + H_2O \rightarrow Cr(OH)_3 + NH_4Cl$
   (c) $PbO + NH_3 \rightarrow Pb + N_2 + H_2O$
   (d) $C_3H_8 + O_2 \rightarrow CO_2 + H_2O$
   (e) $Cu + HNO_3 \rightarrow Cu(NO_3)_2 + NO + H_2O$

Calculations Based on Chemical Equations

In Exercises 8–11, (a) write the balanced chemical equation that represents the reaction described by words, and then perform calculations to answer parts (b) and (c).

8. (a) Nitrogen, N_2, combines with hydrogen, H_2, to form ammonia, NH_3.
   (b) How many hydrogen molecules would be required to react with 200 nitrogen molecules?
   (c) How many ammonia molecules would be formed in part (b)?
9. (a) Sulfur, S_8, combines with oxygen at elevated temperatures to form sulfur dioxide.
   (b) If 125 oxygen molecules are used up in this reaction, how many sulfur molecules reacted?
   (c) How many sulfur dioxide molecules were formed in part (b)?
10. (a) Lime, CaO, dissolves in muriatic acid, HCl, to form calcium chloride, $CaCl_2$, and water.
    (b) How many moles of HCl would be required to dissolve 8.8 mol of CaO?
    (c) How many moles of water would be formed in part (b)?
11. (a) Aluminum building materials have a hard, transparent, protective coating of aluminum oxide, Al_2O_3, formed by reaction with oxygen in the air. The sulfuric acid, H_2SO_4, in acid rain dissolves this protective coating and forms aluminum sulfate, $Al_2(SO_4)_3$, and water.
    (b) How many moles of H_2SO_4 are required to react with 4.0 mol of Al_2O_3?
    (c) How many moles of $Al_2(SO_4)_3$ are formed in part (b)?

12. How many moles of oxygen can be obtained by the decomposition of 10.0 mol of reactant in each of the following reactions?
    (a) $2KClO_3 \rightarrow 2KCl + 3O_2$

(b) $2H_2O_2 \rightarrow 2H_2O + O_2$
(c) $2HgO \rightarrow 2Hg + O_2$
(d) $2NaNO_3 \rightarrow 2NaNO_2 + O_2$
(e) $KClO_4 \rightarrow KCl + 2O_2$

13. For the formation of 5.00 mol of water, which reaction uses the most nitric acid?
    (a) $3Cu + 8HNO_3 \rightarrow 3Cu(NO_3)_2 + 2NO + 4H_2O$
    (b) $Al_2O_3 + 6HNO_3 \rightarrow 2Al(NO_3)_3 + 3H_2O$
    (c) $4Zn + 10HNO_3 \rightarrow 4Zn(NO_3)_2 + NH_4NO_3 + 3H_2O$
14. Consider the reaction

$$NH_3 + O_2 \xrightarrow{\text{not balanced}} NO + H_2O$$

For every 10.00 mol of NH_3, (a) how many moles of O_2 are required, (b) how many moles of NO are produced, and (c) how many moles of H_2O are produced?
15. Consider the reaction

$$2NO + Br_2 \longrightarrow 2NOBr$$

For every 5.00 mol of bromine that reacts, how many moles of (a) NO react and (b) NOBr are produced?
16. Find the mass of chlorine that will combine with 4.77 g of hydrogen to form hydrogen chloride.

$$H_2 + Cl_2 \longrightarrow 2HCl$$

17. What mass of ClO_2 is required to produce 8.36 kg of $HClO_3$ by the following reaction?

$$2ClO_2 + H_2O \longrightarrow HClO_3 + HClO_2$$

18. A sample of magnetic iron oxide, Fe_3O_4, reacted completely with hydrogen at red heat. The water vapor formed by the reaction

$$Fe_3O_4 + 4H_2 \xrightarrow{\Delta} 3Fe + 4H_2O$$

was condensed and found to weigh 11.25 g. Calculate the mass of Fe_3O_4 that reacted.
19. What masses of cobalt(II) chloride and of hydrogen fluoride are needed to prepare 12.0 moles of cobalt(II) fluoride by the following reaction?

$$CoCl_2 + 2HF \longrightarrow CoF_2 + 2HCl$$

20. We allow 32.0 g of methane, CH_4, to react as completely as possible with excess oxygen, O_2, to form CO_2 and water. Write the balanced equation for this reaction. What mass of oxygen will react?
21. We allow 48.0 g of propane, C_3H_8, to react as completely as possible with excess oxygen, O_2, to form CO_2 and water. Write the balanced equation for this reaction. What mass of oxygen will react?
22. Gaseous chlorine will displace bromide ion from an aqueous solution of potassium bromide to form aqueous potassium chloride and aqueous bromine. Write the chemical equation for this reaction. What mass of bromine will be produced if 0.361 g of chlorine undergoes reaction?

23. Solid zinc sulfide reacts with hydrochloric acid to form a mixture of aqueous zinc chloride and hydrogen sulfide, H_2S. Write the chemical equation for this reaction. What mass of zinc sulfide is needed to react with 12.10 g of HCl?
24. What is the maximum mass of water that can be produced by burning 26.0 grams of butane, C_4H_{10}, in excess oxygen, O_2?
25. Calculate the number of moles of CO_2 that can be produced by burning 86.0 grams of hexane, C_6H_{14}, in excess oxygen, O_2.
26. Calculate the number of molecules of propane, C_3H_8, that will produce 3.20 grams of water when burned in excess oxygen, O_2.
27. What mass of pentane, C_5H_{12}, produces 3.01×10^{22} CO_2 molecules when burned in excess oxygen, O_2?

Limiting Reactant

28. How many grams of NH_3 can be prepared from 85.5 grams of N_2 and 17.3 grams of H_2?

$$N_2 + 3H_2 \longrightarrow 2NH_3$$

29. Silver nitrate solution reacts with barium chloride solution according to the equation

$$2AgNO_3 + BaCl_2 \longrightarrow Ba(NO_3)_2 + 2AgCl$$

    All of the substances involved in this reaction are soluble in water except silver chloride, AgCl, which forms a solid (precipitate) at the bottom of the flask. Suppose we mix together a solution containing 24.8 g of $AgNO_3$ and one containing 18.4 g of $BaCl_2$. What mass of AgCl would be formed?
*30. "Superphosphate," a water-soluble fertilizer, is a mixture of $Ca(H_2PO_4)_2$ and $CaSO_4$ on a 1:2 *mole* basis. It is formed by the reaction

$$Ca_3(PO_4)_2 + 2H_2SO_4 \longrightarrow Ca(H_2PO_4)_2 + 2CaSO_4$$

    We treat 450 g of $Ca_3(PO_4)_2$ with 300 g of H_2SO_4. How many grams of superphosphate could be formed?

31. Silicon carbide, an abrasive, is made by the reaction of silicon dioxide with graphite.

$$SiO_2 + C \xrightarrow{\Delta} SiC + CO \qquad \text{(balanced?)}$$

    We mix 377 g of SiO_2 and 255 g of C. If the reaction proceeds as far as possible, which reactant will be left over? How much of this reactant will remain?

32. What mass of potassium can be produced by the reaction of 150.0 g of Na with 150.0 g of KCl?

$$Na + KCl \xrightarrow{\Delta} NaCl + K$$

33. A reaction mixture contains 55.0 g of PCl_3 and 35.0 g of PbF_2. What mass of $PbCl_2$ can be obtained from the following reaction?

$$3PbF_2 + 2PCl_3 \longrightarrow 2PF_3 + 3PbCl_2$$

    How much of which reactant will be left unchanged?

Percent Yield from Chemical Reactions

34. The percent yield for the reaction

$$PCl_3 + Cl_2 \longrightarrow PCl_5$$

    is 83.2%. What mass of PCl_5 would be expected from the reaction of 56.7 g of PCl_3 with excess chlorine?
35. The percent yield for the following reaction carried out in carbon tetrachloride solution is 67.0%.

$$Br_2 + Cl_2 \longrightarrow 2BrCl$$

    (a) What amount of BrCl would be formed from the reaction of 0.0250 mol Br_2 with 0.0250 mol Cl_2? (b) What amount of Br_2 is left unchanged?
36. Solid silver nitrate undergoes thermal decomposition to form silver metal, nitrogen dioxide, and oxygen. Write the chemical equation for this reaction. A 0.443-g sample of silver metal was obtained from the decomposition of a 0.722-g sample of $AgNO_3$. What is the percent yield of the reaction?
37. Gaseous nitrogen and hydrogen undergo a reaction to form gaseous ammonia (the Haber process). Write the chemical equation for this reaction. At a temperature of 400°C and a total pressure of 250 atm, 1.86 g of NH_3 was produced by the reaction of 5.85 g of N_2 with excess H_2. What is the percent yield of the reaction?
38. Ethylene oxide, C_2H_4O, a fumigant sometimes used by exterminators, is synthesized in 88.1% yield by reaction of ethylene bromohydrin, C_2H_5OBr, with sodium hydroxide:

$$C_2H_5OBr + NaOH \longrightarrow C_2H_4O + NaBr + H_2O$$

    How many grams of ethylene bromohydrin would be consumed in the production of 383 g of ethylene oxide, at 88.1% yield?
*39. How much 74.1% pure Na_2SO_4 could be produced from 245 g of 93.9% pure NaCl?

$$2NaCl + H_2SO_4 \longrightarrow Na_2SO_4 + 2HCl$$

*40. Calcium carbide is made in an electric furnace by the reaction

$$CaO + 3C \longrightarrow CaC_2 + CO$$

    The crude product is usually 85% CaC_2 and 15% unreacted CaO. (a) How much CaO should we start with to produce 450 kg of crude product? (b) How much CaC_2 would this crude product contain?

41. Ethylene glycol, $C_2H_6O_2$, is used as antifreeze in automobile radiators. A method of producing small amounts of ethylene glycol in the laboratory is by reaction of 1,2-dichloroethane with sodium carbonate in a water solution, followed by distillation of the reaction mixture to purify the ethylene glycol.

$$C_2H_4Cl_2 + Na_2CO_3 + H_2O \longrightarrow C_2H_6O_2 + 2NaCl + CO_2$$

When 31.5 g of 1,2-dichloroethane is used in this reaction, 11.3 g of ethylene glycol is obtained. (a) Calculate the theoretical yield of ethylene glycol. (b) What is the percent yield of ethylene glycol in this process? (c) What mass of Na_2CO_3 is consumed?

Sequential Reactions

42. Consider the two-step process for the formation of tellurous acid described by the following equations:

$$TeO_2 + 2OH^- \longrightarrow TeO_3^{2-} + H_2O$$

$$TeO_3^{2-} + 2H^+ \longrightarrow H_2TeO_3$$

What mass of H_2TeO_3 would be formed from 72.1 g of TeO_2, assuming 100% yield?

43. Consider the formation of cyanogen, C_2N_2, and its subsequent decomposition in water given by the equations

$$2Cu^{2+} + 6CN^- \longrightarrow 2[Cu(CN)_2]^- + C_2N_2$$

$$C_2N_2 + H_2O \longrightarrow HCN + HOCN$$

How much hydrocyanic acid, HCN, can be produced from 20.00 g of KCN, assuming 100% yield?

44. What mass of potassium chlorate would be required to supply the proper amount of oxygen needed to burn 33.2 g of methane, CH_4?

$$2KClO_3 \longrightarrow 2KCl + 3O_2$$

$$CH_4 + 2O_2 \longrightarrow CO_2 + 2H_2O$$

45. Hydrogen, obtained by the electrical decomposition of water, was combined with chlorine to produce 62.5 g of hydrogen chloride. Calculate the mass of water decomposed.

$$2H_2O \longrightarrow 2H_2 + O_2$$

$$H_2 + Cl_2 \longrightarrow 2HCl$$

*46. The Grignard reaction is a two-step reaction used to prepare pure hydrocarbons. Consider the preparation of pure ethane, CH_3CH_3, from ethyl chloride, CH_3CH_2Cl.

Step 1: $CH_3CH_2Cl + Mg \longrightarrow CH_3CH_2MgCl$

Step 2: $CH_3CH_2MgCl + H_2O \longrightarrow CH_3CH_3 + Mg(OH)Cl$

We allow 27.2 grams of CH_3CH_2Cl (64.4 g/mol) to react with excess magnesium. From the first step reaction, CH_3CH_2MgCl (88.7 g/mol) is obtained in 79.5% yield. In the second step reaction, a 78.8% yield of CH_3CH_3 (30.0 g/mol) is obtained. What mass of CH_3CH_3 is obtained?

*47. When sulfuric acid dissolves in water, the following reactions take place:

$$H_2SO_4 \longrightarrow H^+ + HSO_4^-$$

$$HSO_4^- \longrightarrow H^+ + SO_4^{2-}$$

The first reaction is 100.0% complete and the second reaction is 10.0% complete. Calculate the concentrations of the various ions in a 0.200 M aqueous solution of H_2SO_4.

*48. The chief ore of zinc is the sulfide, ZnS. The ore is concentrated by flotation and then heated in air, which converts the ZnS to ZnO.

$$2ZnS + 3O_2 \longrightarrow 2ZnO + 2SO_2$$

The ZnO is then treated with dilute H_2SO_4

$$ZnO + H_2SO_4 \longrightarrow ZnSO_4 + H_2O$$

to produce an aqueous solution containing the zinc as $ZnSO_4$. An electrical current is passed through the solution to produce the metal.

$$2ZnSO_4 + 2H_2O \longrightarrow 2Zn + 2H_2SO_4 + O_2$$

What mass of Zn will be obtained from an ore containing 225 kg of ZnS? Assume the flotation process to be 90.6% efficient, the electrolysis step to be 98.2% efficient, and the other steps to be 100% efficient.

*49. About half of the world's production of pigments for paints involves the formation of white TiO_2. In the United States, it is made on a large scale by the *chloride process,* starting with ores containing only small amounts of rutile, TiO_2. The ore is treated with chlorine and carbon (coke). This produces $TiCl_4$ and gaseous products.

$$2TiO_2 + 3C + 4Cl_2 \longrightarrow 2TiCl_4 + CO_2 + 2CO$$

The $TiCl_4$ is then converted into TiO_2 of high purity.

$$TiCl_4 + O_2 \longrightarrow TiO_2 + 2Cl_2$$

Suppose the first process can be carried out with 70.0% yield and the second with 93.0% yield. How many kg of TiO_2 could be produced starting with 1.00 metric ton (1.00×10^6 g) of an ore that is 0.75% rutile?

Concentrations of Solutions—Percent by Mass

50. (a) How many moles of solute are contained in 100 grams of a 0.250% aqueous solution of Na_2S? (b) How many grams of solute are contained in the solution of part (a)? (c) How many grams of water (the solvent) are contained in the solution of part (a)?

51. (a) How many moles of solute are contained in 500 grams of a 2.00% aqueous solution of $K_2Cr_2O_7$? (b) How many grams of solute are contained in the solution of part (a)? (c) How many grams of water (the solvent) are contained in the solution of part (a)?

52. The density of an 18.0% solution of ammonium sulfate, $(NH_4)_2SO_4$, is 1.10 g/mL. What mass of $(NH_4)_2SO_4$ would be required to prepare 425 mL of this solution?

53. The density of an 18.0% solution of ammonium chloride, NH_4Cl, solution is 1.05 g/mL. What mass of NH_4Cl does 425 mL of this solution contain?

54. What volume of the solution of $(NH_4)_2SO_4$ described in Exercise 52 contains 90.0 g of $(NH_4)_2SO_4$?

*55. A reaction requires 33.6 g of NH_4Cl. What volume of the solution described in Exercise 53 would you use if you wished to use a 25.0% excess of NH_4Cl?

Concentrations of Solutions—Molarity

56. What is the molarity of a solution that contains 650 g of phosphoric acid, H_3PO_4, in 3.00 L of solution?

57. What is the molarity of a solution that contains 3.37 g of sodium chloride in 40.0 mL of solution?

58. What is the molarity of a solution containing 0.335 mol H_3PO_4 in 250 mL of solution?

59. A solution contains 0.100 mole per liter of each of the following acids: HCl, H_2SO_4, H_3PO_4.
    (a) Is the molarity the same for each acid?
    (b) Is the number of molecules per liter the same for each acid?
    (c) Is the mass per liter the same for each acid?

60. A solution contains 2.25 g of a rubbing alcohol, $(CH_3)_2CHOH$, in 150 mL of solution. Find (a) the molarity of the solution and (b) the number of moles in 2.00 mL of solution.

61. (a) Calculate the molarity of caffeine in a 12-oz cola drink containing 50 mg caffeine, $C_8H_{10}N_4O_2$.
    (b) Cola drinks are usually 5.06×10^{-3} M with respect to H_3PO_4. How much of this acid is in a 250-mL drink? (1 oz = 29.6 mL)

62. How many grams of the cleansing agent Na_3PO_4 (a) are needed to prepare 250 mL of 0.40 M solution, and (b) are in 250 mL of 0.40 M solution?

63. How many kg of ethylene glycol, $C_2H_6O_2$, are needed to prepare a 9.50 M solution to protect a 14.0-L car radiator against freezing? What is the mass of $C_2H_6O_2$ in 14.0 L of 9.50 M solution?

64. A solution made by dissolving 16.0 g of $CaCl_2$ in 64.0 g of water has a density of 1.180 g/mL at 20°C.
    (a) What is the percent by mass of $CaCl_2$ in the solution?
    (b) What is the molarity of $CaCl_2$ in the solution?

65. Stock phosphoric acid solution is 85.0% H_3PO_4 and has a specific gravity of 1.70. What is the molarity of the solution?

66. Stock hydrofluoric acid solution is 49.0% HF and has a specific gravity of 1.17. What is the molarity of the solution?

67. What mass of sodium sulfate, Na_2SO_4, is contained in 650 mL of a 1.50 molar solution?

68. What is the molarity of a barium chloride solution prepared by dissolving 5.50 g of $BaCl_2 \cdot 2H_2O$ in enough water to make 600 mL of solution?

69. What mass of potassium benzoate trihydrate, $KC_7H_5O_2 \cdot 3H_2O$, is needed to prepare one liter of a 0.225 molar solution of potassium benzoate?

Dilution of Solutions

70. Commercial concentrated hydrochloric acid is 12.0 M HCl. What volume of concentrated hydrochloric acid is required to prepare 4.50 L of 1.80 M HCl solution?

71. Commercially available concentrated sulfuric acid is 18.0 M H_2SO_4. Calculate the volume of concentrated sulfuric acid required to prepare 4.50 L of 1.80 M H_2SO_4 solution.

72. Calculate the volume of 0.0600 M $Ba(OH)_2$ solution that contains the same number of moles of $Ba(OH)_2$ as 180 mL of 0.0900 M $Ba(OH)_2$ solution.

73. Calculate the volume of 4.00 M NaOH solution required to prepare 200 mL of a 0.750 M solution of NaOH.

*74. Calculate the resulting molarity when 145 mL of 6.00 M H_2SO_4 solution is mixed with 245 mL of 3.00 M H_2SO_4.

*75. Calculate the resulting molarity when 60.0 mL of 2.30 M NaCl solution is mixed with 90.0 mL of 1.40 M NaCl.

Using Solutions in Chemical Reactions

76. Calculate the volume of a 0.225 M solution of potassium hydroxide, KOH, required to react with 0.155 grams of acetic acid, CH_3COOH, according to the following reaction.

$$KOH + CH_3COOH \longrightarrow KCH_3COO + H_2O$$

77. Calculate the number of grams of carbon dioxide, CO_2, that can react with 175 mL of a 0.357 M solution of potassium hydroxide, KOH, according to the following reaction.

$$2KOH + CO_2 \longrightarrow K_2CO_3 + H_2O$$

78. What volume of 0.246 M HNO_3 solution is required to react completely with 32.0 mL of 0.0515 M $Ba(OH)_2$?

$$Ba(OH)_2 + 2HNO_3 \longrightarrow Ba(NO_3)_2 + 2H_2O$$

79. What volume of 0.60 M HBr is required to react completely with 0.80 mol of $Ca(OH)_2$?

$$2HBr + Ca(OH)_2 \longrightarrow CaBr_2 + 2H_2O$$

80. An excess of $AgNO_3$ reacts with 100.0 mL of an $AlCl_3$ solution to give 0.325 g of AgCl. What is the concentration, in mol/L, of the $AlCl_3$ solution?

$$AlCl_3 + 3AgNO_3 \longrightarrow 3AgCl + Al(NO_3)_3$$

81. An impure sample of solid Na_2CO_3 was allowed to react with 0.1225 M HCl.

$$Na_2CO_3 + 2HCl \longrightarrow 2NaCl + CO_2 + H_2O$$

A 0.1337-g sample of sodium carbonate required 15.55 mL of HCl solution. What is the purity of the sodium carbonate?

82. Calculate the theoretical yield of AgCl formed from the reaction of an aqueous solution containing excess $ZnCl_2$ with 45.0 mL of 0.425 M $AgNO_3$.

$$ZnCl_2 + 2AgNO_3 \longrightarrow Zn(NO_3)_2 + 2AgCl$$

Titrations

83. Define and illustrate the following terms clearly and concisely: (a) standard solution; (b) titration.

84. Distinguish between the *equivalence point* and *end point* of a titration.

85. What volume of 0.295 molar hydrochloric acid solution reacts with 42.4 mL of 0.150 molar sodium hydroxide solution?

86. What volume of 0.0496 M $HClO_4$ reacts with 25.0 mL of a 0.505 M KOH solution according to the following reaction?

$$KOH + HClO_4 \longrightarrow KClO_4 + H_2O$$

87. What is the molarity of a solution of sodium hydroxide, NaOH, if 36.9 mL of this solution is required to react with 35.2 mL of 0.101 M hydrochloric acid solution according to the following reaction?

$$HCl + NaOH \longrightarrow NaCl + H_2O$$

88. What is the molarity of a solution of sodium hydroxide, NaOH, if 36.2 mL of this solution is required to react with 25.0 mL of 0.0513 M nitric acid solution according to the following reaction?

$$HNO_3 + NaOH \longrightarrow NaNO_3 + H_2O$$

89. Benzoic acid, C_6H_5COOH, is sometimes used for the standardization of solutions of bases. A 1.922-g sample of the acid reacts with 29.47 mL of an NaOH solution. What is the molarity of the base solution?

$$C_6H_5COOH + NaOH \longrightarrow C_6H_5COONa + H_2O$$

90. An antacid tablet containing calcium carbonate as an active ingredient requires 26.7 mL of 0.0887 M HCl for complete reaction. What mass of $CaCO_3$ did the tablet contain?

$$2HCl + CaCO_3 \longrightarrow CaCl_2 + H_2O + CO_2$$

Mixed Exercises

*91. What mass of sulfuric acid can be obtained from 1.00 kg of sulfur by the following series of reactions?

$$S + O_2 \xrightarrow{\text{98\% yield}} SO_2$$

$$2SO_2 + O_2 \xrightarrow{\text{96\% yield}} 2SO_3$$

$$SO_3 + H_2SO_4 \xrightarrow{\text{100\% yield}} H_2S_2O_7$$

$$H_2S_2O_7 + H_2O \xrightarrow{\text{97\% yield}} 2H_2SO_4$$

*92. What is the total mass of products formed when 33.8 g of carbon disulfide is burned in air? What mass of carbon disulfide would have to be burned to produce a mixture of carbon dioxide and sulfur dioxide that has a mass of 54.2 g?

$$CS_2 + 3O_2 \xrightarrow{\Delta} CO_2 + 2SO_2$$

*93. Iron(II) chloride, $FeCl_2$, reacts with ammonia, NH_3, and water, H_2O, to produce iron(II) hydroxide, $Fe(OH)_2$, and ammonium chloride, NH_4Cl.
(a) Write the balanced equation for this reaction.
(b) We mix 78.5 g $FeCl_2$, 25.0 g NH_3, and 25.0 g H_2O, which then react as completely as possible. Which is the limiting reactant?

(c) How many grams of ammonium chloride, NH_4Cl, are formed?
(d) How many grams of each of the two leftover reactants remain at the completion of the reaction?

*94. An iron ore that contains Fe_3O_4 reacts according to the reaction

$$Fe_3O_4 + C \longrightarrow 3Fe + 2CO_2$$

We obtain 2.09 g of Fe from the reaction of 50.0 g of the ore. What is the percent Fe_3O_4 in the ore?

*95. If 86.3% of the iron can be recovered from an ore that is 43.2% magnetic iron oxide, Fe_3O_4, what mass of iron could be recovered from 2.00 kilograms of this ore? The reduction of magnetic iron oxide is a complex process that can be represented in simplified form as

$$Fe_3O_4 + 4CO \longrightarrow 3Fe + 4CO_2$$

BUILDING YOUR KNOWLEDGE

96. Acetic acid, CH_3COOH, reacts with ethanol, CH_3CH_2OH, to form ethyl acetate, $CH_3COOCH_2CH_3$, (density = 0.902 g/mL) by the following reaction.

$$CH_3COOH + CH_3CH_2OH \longrightarrow CH_3COOCH_2CH_3 + H_2O$$

We combine 20.2 mL of acetic acid with 20.1 mL of ethanol. (a) Which compound is the limiting reactant? (b) If 27.5 mL of pure ethyl acetate is produced, what is the percent yield? [*Hint:* See Tables 1-1 and 1-8.]

97. Gaseous chlorine and gaseous fluorine undergo a combination reaction to form the interhalogen compound ClF. (a) Write the chemical equation for this reaction. (b) Calculate the mass of fluorine needed to react with 3.47 grams of Cl_2. (c) How many grams of ClF are formed?

98. Suppose you are designing an experiment for the preparation of hydrogen. For the production of equal amounts of hydrogen, which metal, Zn or Al, is less expensive if Zn costs about half as much as Al on a mass basis?

$$Zn + 2HCl \longrightarrow ZnCl_2 + H_2$$

$$2Al + 6HCl \longrightarrow 2AlCl_3 + 3H_2$$

*99. Concentrated hydrochloric acid solution is 37.0% HCl and has a density of 1.19 g/mL. A dilute solution of HCl is prepared by diluting 2.00 mL of this concentrated HCl solution to 100.00 mL with water. Then 10.0 mL of this dilute HCl solution reacts with an $AgNO_3$ solution according to the following reaction.

$$HCl(aq) + AgNO_3(aq) \longrightarrow HNO_3(aq) + AgCl(s)$$

How many milliliters of 0.108 M $AgNO_3$ solution will be required to precipitate all of the chloride as AgCl(s)?

*100. In a particular experiment, 272 grams of phosphorus, P_4, reacted with excess oxygen to form tetraphosphorus decoxide, P_4O_{10}, in 89.5% yield. In the second step reaction, a 97.8% yield of H_3PO_4 was obtained. (a) Write the balanced equations for these two reaction steps. (b) What mass of H_3PO_4 was obtained?

4

Some Types of Chemical Reactions

When a copper wire is placed in a solution of silver nitrate, crystals of metallic silver are deposited on the wire.

OBJECTIVES

As you study this chapter, you should learn

• *About the periodic table and the classification of elements*

• *About reactions of solutes in aqueous solutions*

• *To recognize nonelectrolytes, strong electrolytes, and weak electrolytes*

• *The classification of acids, bases, and salts*

• *Which kinds of compounds are soluble and which kinds are insoluble in water*

• *How to describe reactions in aqueous solutions by writing formula unit equations as well as total ionic and net ionic equations*

• *About displacement reactions and the activity series*

• *About precipitation reactions*

• *About oxidation numbers*

• *To name common binary and ternary inorganic compounds*

W e observe that some elements form compounds with only a few other elements, while some, such as oxygen, combine with nearly every other element, and some, such as carbon, form millions of compounds. Some elements are metals, and others obviously lack metallic properties. Some substances are gases, others are liquids, and still others are solids. Some solids are soft (paraffin), others are quite hard (diamond), and others are quite strong (steel). Clearly, there are significant differences among the chemical bonds and other attractive forces that hold atoms together in different substances.

Compounds are formed when atoms of different elements are joined together by chemical bonds. Chemical bonds are broken when atoms are separated and the original compounds cease to exist. The attractive forces between atoms are electrical in nature, and chemical reactions between atoms involve *changes* in their electronic structures.

As we shall see in the next few chapters,

> the positions of elements in the periodic table are related to the arrangements of their electrons; these determine the chemical and physical properties of the elements.

Let us now turn our attention to the periodic table and to **chemical periodicity,** the variation in properties of elements with their positions in the periodic table.

4-1 THE PERIODIC TABLE: METALS, NONMETALS, AND METALLOIDS

Pronounced "men-del-*lay*-ev."

In 1869 the Russian chemist Dimitri Mendeleev and the German chemist Lothar Meyer independently published arrangements of known elements that are much like the periodic table in use today. Mendeleev's classification was based primarily on chemical properties of the elements, whereas Meyer's classification was based largely on physical properties. The tabulations were surprisingly similar. Both emphasized the *periodicity,* or regular periodic repetition, of properties with increasing atomic weight.

Mendeleev arranged the known elements in order of increasing atomic weight in successive sequences so that elements with similar chemical properties fell in the same column. He noted that both physical and chemical properties of the elements vary in a periodic fashion with atomic weight. His periodic table of 1872 contained the 62 known elements (Figure 4-1). Mendeleev placed H, Li, Na, and K in his table as "Gruppe I." These were known to combine with F, Cl, Br, and I of "Gruppe VII" to produce compounds that have similar formulas such as HF, LiCl, NaCl, and KI. All these compounds dissolve in water to produce solutions that conduct electricity. The "Gruppe II" elements were known to form compounds such as $BeCl_2$, $MgBr_2$, and $CaCl_2$, as well as compounds with O and S from "Gruppe VI" such as MgO, CaO, MgS, and CaS. These and other chemical properties led him to devise a table in which the elements were arranged by increasing atomic weights and grouped into vertical families.

In most areas of human endeavor progress is slow and faltering. However, there is an occasional individual who develops concepts and techniques that clarify confused situations. Mendeleev was such an individual. One of the brilliant successes of his periodic table was that it provided for elements that were unknown at the time. When he encountered "missing" elements, Mendeleev left blank spaces. Some appreciation of his genius in constructing the table as he did can be gained by comparing the predicted (1871) and observed properties of germanium, which was not discovered until 1886. Mendeleev

REIHEN	GRUPPE I — R^2O	GRUPPE II — RO	GRUPPE III — R^2O^3	GRUPPE IV RH^4 RO^2	GRUPPE V RH^3 R^2O^5	GRUPPE VI RH^2 RO^3	GRUPPE VII RH R^2O^7	GRUPPE VIII — RO^4
1	H = 1							
2	Li = 7	Be = 9,4	B = 11	C = 12	N = 14	O = 16	F = 19	
3	Na = 23	Mg = 24	Al = 27,3	Si = 28	P = 31	S = 32	Cl = 35,5	
4	K = 39	Ca = 40	– = 44	Ti = 48	V = 51	Cr = 52	Mn = 55	Fe = 56, Co = 59, Ni = 59, Cu = 63.
5	(Cu = 63)	Zn = 65	– = 68	– = 72	As = 75	Se = 78	Br = 80	
6	Rb = 85	Sr = 87	?Yt = 88	Zr = 90	Nb = 94	Mo = 96	– = 100	Ru = 104, Rh = 104, Pd = 106, Ag = 108.
7	(Ag = 108)	Cd = 112	In = 113	Sn = 118	Sb = 122	Te = 125	J = 127	
8	Cs = 133	Ba = 137	?Di = 138	?Ce = 140	–	–	–	– – – –
9	(–)	–	–	–	–	–	–	
10	–	–	?Er = 178	?La = 180	Ta = 182	W = 184	–	Os = 195, Ir = 197, Pt = 198, Au = 199.
11	(Au = 199)	Hg = 200	Tl = 204	Pb = 207	Bi = 208	–	–	
12	–	–	–	Th = 231	–	U = 240	–	– – – –

Figure 4-1 Mendeleev's early periodic table (1872). "J" is the German symbol for iodine.

called the undiscovered element eka-silicon because it fell below silicon in his table. He was familiar with the properties of germanium's neighboring elements. They served as the basis for his predictions of properties of germanium (Table 4-1). Some modern values for properties of germanium differ significantly from those reported in 1886. But many of the values upon which Mendeleev based his predictions were inaccurate.

Table 4-1 *Predicted and Observed Properties of Germanium*

Property	Eka-Silicon Predicted, 1871	Germanium Reported, 1886	Modern Values
Atomic weight	72	72.32	72.61
Atomic volume	13 cm^3	13.22 cm^3	13.5 cm^3
Specific gravity	5.5	5.47	5.35
Specific heat	0.073 cal/g°C	0.076 cal/g°C	0.074 cal/g°C
Maximum valence*	4	4	4
Color	Dark gray	Grayish white	Grayish white
Reaction with water	Will decompose steam with difficulty	Does not decompose water	Does not decompose water
Reactions with acids and alkalis	Slight with acids; more pronounced with alkalis	Not attacked by HCl or dilute aqueous NaOH; reacts vigorously with molten NaOH	Not dissolved by HCl or H_2SO_4 or dilute NaOH; dissolved by concentrated NaOH
Formula of oxide	EsO_2	GeO_2	GeO_2
Specific gravity of oxide	4.7	4.703	4.228
Specific gravity of tetrachloride	1.9 at 0°C	1.887 at 18°C	1.8443 at 30°C
Boiling point of tetrachloride	100°C	86°C	84°C
Boiling point of tetraethyl derivative	160°C	160°C	186°C

*"Valence" refers to the combining power of a specific element.

Because Mendeleev's arrangement of the elements was based on increasing *atomic weights,* several elements would have been out of place in his table. But Mendeleev put the controversial elements (Te and I, Co and Ni) in locations consistent with their properties. He thought the apparent reversal of atomic weights was due to inaccurate values for those weights. Careful redetermination showed that the values were correct. Explanation of the locations of these "out-of-place" elements had to await the development of the concept of *atomic number,* approximately 50 years after Mendeleev's work. The **atomic number** (Section 5-5) of an element is the number of protons in the nucleus of its atoms. (It is also the number of electrons in an atom of an element.) This quantity is fundamental to the identity of each element because it is related to the electrical make-up of atoms. Elements are arranged in the periodic table in order of increasing atomic number. With the development of this concept, the **periodic law** attained essentially its present form:

The properties of the elements are periodic functions of their atomic numbers.

Table 4-2 *The Periodic Table*

There are other systems for numbering the groups in the periodic table. We number the groups by the standard American system of A and B groups. An alternate system in which the groups are numbered 1 through 18 is shown in parentheses.

Silicon (top), germanium, and tin (bottom).

Three of the halogens: (left to right) chlorine, bromine, iodine.

The periodic law tells us that if we arrange the elements in order of increasing atomic number, we periodically encounter elements that have similar chemical and physical properties. The presently used "long form" of the periodic table (Table 4-2 and inside the front cover) is such an arrangement. The vertical columns are referred to as **groups** or **families,** and the horizontal rows are called **periods.** Elements in a *group* have similar chemical and physical properties, while those within a *period* have properties that change progressively across the table. Several groups of elements have common names that are used so frequently they should be learned. The Group IA elements, except H, are referred to as **alkali metals,** and the Group IIA elements are called the **alkaline earth metals.** The Group VIIA elements are called **halogens,** which means "salt formers," and the Group 0 elements are called **noble** (or **rare**) **gases.**

The general properties of metals and nonmetals are distinct. Physical and chemical properties that distinguish metals from nonmetals are summarized in Tables 4-3 and 4-4.

Alkaline means basic. The character of basic compounds is described in Section 10-4.

About 80% of the elements are metals.

Table 4-3 *Some Physical Properties of Metals and Nonmetals*

Metals	Nonmetals
1. High electrical conductivity that decreases with increasing temperature	1. Poor electrical conductivity (except carbon in the form of graphite)
2. High thermal conductivity	2. Good heat insulators (except carbon in the form of diamond)
3. Metallic gray or silver luster*	3. No metallic luster
4. Almost all are solids[†]	4. Solids, liquids, or gases
5. Malleable (can be hammered into sheets)	5. Brittle in solid state
6. Ductile (can be drawn into wires)	6. Nonductile

*Except copper and gold.

[†] Except mercury; cesium and gallium melt in protected hand.

Copper is drawn into wire, which is then collected into cables for use as an electrical conductor.

Table 4-4 *Some Chemical Properties of Metals and Nonmetals*	
Metals	**Nonmetals**
1. Outer shells contain few electrons— usually three or fewer	**1.** Outer shells contain four or more electrons*
2. Form cations (positive ions) by losing electrons	**2.** Form anions (negative ions) by gaining electrons[†]
3. Form ionic compounds with nonmetals	**3.** Form ionic compounds with metals[†] and molecular (covalent) compounds with other nonmetals
4. Solid state characterized by metallic bonding	**4.** Covalently bonded molecules; noble gases are monatomic

* *Except hydrogen and helium.*
[†] *Except the noble gases.*

Not all metals and nonmetals possess all these properties, but they share most of them to varying degrees. The physical properties of metals can be explained on the basis of metallic bonding in solids (Section 13-17).

Table 4-2, The Periodic Table, shows how we classify the known elements as *metals* (shown in blue), *nonmetals* (yellow), and *metalloids* (green). The elements to the left of those touching the heavy stairstep line are *metals* (except hydrogen), while those to the right are *nonmetals*. Such a classification is somewhat arbitrary, and several elements do not fit neatly into either class. Most elements adjacent to the heavy line are often called *metalloids* (or semimetals), because they are metallic (or nonmetallic) only to a limited degree.

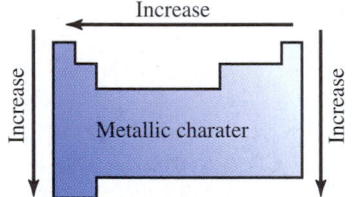

General trends in metallic character of A group elements with position in the periodic table.

Metallic character increases from top to bottom and from right to left with respect to position in the periodic table.

Cesium, atomic number 55, is the most active naturally-occurring metal. Francium and radium are radioactive and do not occur in nature in appreciable amounts. Noble gases seldom bond with other elements. They are unreactive, monatomic gases. The most active nonmetal is fluorine, atomic number 9.

Silicon, a metalloid, is widely used in the manufacture of electronic chips.

Nonmetallic character increases from bottom to top and from left to right in the periodic table.

Metalloids show some properties that are characteristic of both metals and nonmetals. Many of the metalloids, such as silicon, germanium, and antimony, act as semiconductors, which are important in solid-state electronic circuits. **Semiconductors** are insulators at lower temperatures, but become conductors at higher temperatures (Section 13-17). The conductivities of metals, by contrast, decrease with increasing temperature.

Aluminum is the least metallic of the metals and is sometimes classified as a metalloid. It is metallic in appearance, and an excellent conductor of electricity.

In this and later chapters we shall study some chemical reactions of elements and their compounds and relate the reactions to the locations of the elements in the periodic table.

Aluminum is the most abundant metal in the earth's crust (7.5% by mass).

The Periodic Table

The periodic table is one of the first things a student of chemistry encounters. It appears invariably in textbooks, in lecture halls, and in laboratories. Scientists consider it an indispensable reference. And yet, less than 150 years ago, the idea of arranging the elements by atomic weight or number was considered absurd. At an 1866 meeting of the Chemical Society at Burlington House, England, J. A. R. Newlands presented a theory he called the Law of Octaves. It stated that when the known elements were listed by increasing atomic weights, those that were eight places apart would be similar, much like notes on a piano keyboard. His colleagues' reactions are probably summed up best by the remark of a Professor Foster: "Have you thought of arranging the elements according to their initial letters? Maybe some better connections would come to light that way."

It is not surprising that poor Newlands was not taken seriously. In the 1860s, little information was available to illustrate relationships among the elements. Only 62 of them had been distinguished from more complex substances when Mendeleev first announced his discovery of the Periodic Law in 1869. However, as advances in atomic theory were made and as new experiments contributed to the understanding of chemical behavior, some scientists had begun to see similarities and patterns among the elements. In 1869 Lothar Meyer and Dmitri Mendeleev independently published similar versions of the now-famous periodic table.

Mendeleev's discovery was the result of many years of hard work. He gathered information on the elements from all corners of the earth—by corresponding with colleagues, studying books and papers, and redoing experiments to confirm data. He put the statistics of each element on a small card and pinned the cards to his laboratory wall, where he arranged and rearranged them many times until he was sure that they were in the right order. One especially farsighted feature of Mendeleev's accomplishment was his realization that some elements were missing from the table. He predicted the properties of these substances (gallium, scandium, and germanium). (It is important to remember that Mendeleev's periodic table organization was devised more than 50 years before the discovery and characterization of subatomic particles.)

Since its birth in 1869, the periodic table has been discussed and revised many times. Spectroscopic and other discoveries have filled in the blanks left by Mendeleev and added a new column consisting of the noble gases. As scientists learned more about atomic structure, the basis for ordering was changed from atomic weight to atomic number. The perplexing rare earths were sorted out and given a special place, along with many of the elements created by atomic bombardment. Even the form of the table has been experimented with, resulting in everything from spiral and circular tables to exotic shapes such as the one suggested by Charles Janet. A three-dimensional periodic table that takes into account valence-shell energies has been proposed by Professor Leland C. Allen of Princeton University.

During the past century, chemistry has become a fast-moving science in which methods and instruments are often outdated within a few years. But it is doubtful that our old friend, the periodic table, will ever become obsolete. It may be modified, but it will always stand as a statement of basic relationships in chemistry and as a monument to the wisdom and insight of its creator, Dmitri Mendeleev.

Lisa L. Saunders
Graduate student
University of California, Berkeley

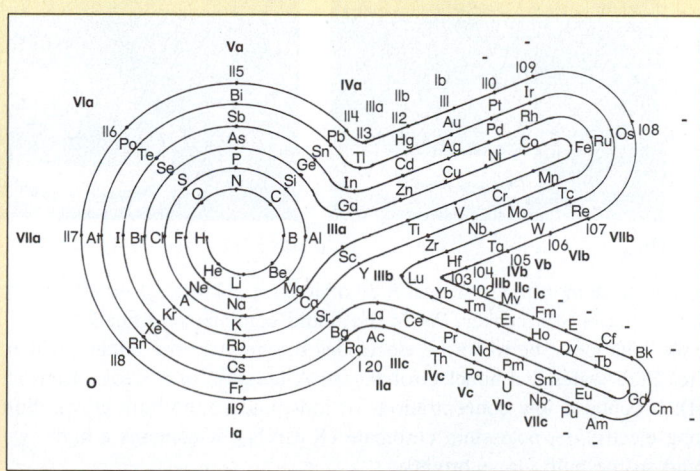

An alternative representation of the periodic table, as proposed by Charles Janet, 1928.

The White Cliffs of Dover, England, are mainly calcium carbonate ($CaCO_3$).

Recall that *ions* are charged particles. The movement of charged particles conducts electricity.

4-2 AQUEOUS SOLUTIONS—AN INTRODUCTION

Approximately three fourths of the earth's surface is covered with water. The body fluids of all plants and animals are mainly water. Thus we can see that many important chemical reactions occur in aqueous (water) solutions, or in contact with water. In Chapter 3, we introduced solutions and methods of expressing concentrations of solutions. It is useful to know the kinds of substances that are soluble in water, and the forms in which they exist, before we begin our systematic study of chemical reactions.

1 Electrolytes and Extent of Ionization

Solutes that are water-soluble can be classified as either electrolytes or nonelectrolytes. **Electrolytes** are substances whose aqueous solutions conduct electric current. **Strong electrolytes** are substances that conduct electricity well in dilute aqueous solution. **Weak electrolytes** conduct electricity poorly in dilute aqueous solution. Aqueous solutions of **nonelectrolytes** do not conduct electricity. Electric current is carried through aqueous solution by the movement of ions. The strength of an electrolyte depends upon the number of ions in solution and also on the charges on these ions (see Figure 4-2).

Dissociation refers to the process in which a solid *ionic compound*, such as NaCl, separates into its ions in solution:

$$NaCl \xrightarrow{H_2O} Na^+ + Cl^-$$

Molecular compounds, for example *pure* HCl, exist as discrete molecules and do not contain ions. However, many such compounds form ions in solution. **Ionization** refers to

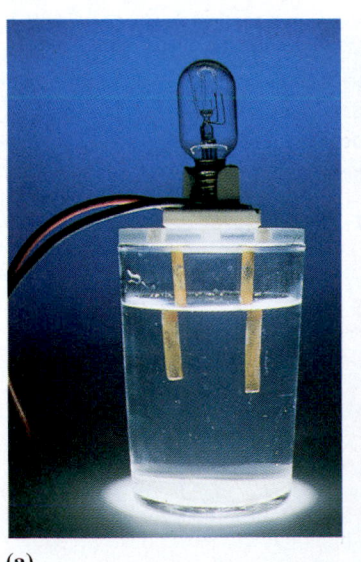
(a)

(b)

(c)

Figure 4-2 An experiment to demonstrate the presence of ions in solution. Two copper electrodes are dipped into a liquid in a beaker. When the liquid contains significant concentrations of ions, the ions move between the electrodes to complete the circuit (which includes a light bulb). (a) Pure water is a nonelectrolyte. (b) A solution of a weak electrolyte, acetic acid (CH_3COOH); it contains low concentrations of ions, and so the bulb glows dimly. (c) A solution of a strong electrolyte, potassium chromate (K_2CrO_4); it contains a high concentration of ions, and so the bulb glows brightly.

the process in which a *molecular compound* separates to form ions in solution:

$$HCl \xrightarrow{\text{H}_2\text{O}} H^+ + Cl^-$$

Three major classes of solutes are strong electrolytes: (1) strong acids, (2) strong soluble bases, and (3) most soluble salts. *These compounds are completely or nearly completely ionized (or dissociated) in dilute aqueous solutions,* and therefore are strong electrolytes.

Acids and bases are further identified in Subsections 2, 3, and 4. They are discussed in more detail in Chapter 10.

An **acid** can be defined as a substance that produces hydrogen ions, H^+, in aqueous solutions. We usually write the formulas of inorganic acids with hydrogen written first. Organic acids can often be recognized by the presence of the COOH group in the formula. A **base** is a substance that produces hydroxide ions, OH^-, in aqueous solutions. A **salt** is a compound that contains a cation other than H^+ and an anion other than hydroxide ion, OH^-, or oxide ion, O^{2-} (see Table 2-3 on page 49). As we shall see later in this chapter, salts are formed when acids react with bases.

Positively charged ions are called *cations* while negatively charged ions are called *anions* (Section 2-3). The formula for a salt may include H or OH, but it *must* contain another cation *and* another anion. For example, $NaHSO_4$ and $Al(OH)_2Cl$ are salts.

2 Strong and Weak Acids

As a matter of convenience we place acids into two classes: strong acids and weak acids. **Strong acids** ionize (separate into hydrogen ions and stable anions) completely, or very nearly completely, in dilute aqueous solution. The seven common strong acids and their anions are listed in Table 4-5. Please learn this short list. (A complete list of common weak acids would be very long.)

Because strong acids ionize completely or very nearly completely in dilute solutions, their solutions contain predominantly their ions rather than acid molecules. Consider the ionization of hydrochloric acid. Pure hydrogen chloride, HCl, is a molecular compound that is a gas at room temperature and atmospheric pressure. When it dissolves in water, it reacts nearly 100% to produce a solution that contains hydrogen ions and chloride ions:

$$HCl(g) \xrightarrow{\text{H}_2\text{O}} H^+(aq) + Cl^-(aq) \qquad \text{(to completion)}$$

Similar equations can be written for all strong acids.

Many properties of aqueous solutions of acids are due to $H^+(aq)$ ions. These are described in Section 10-4.

To give a more complete description of reactions, we indicate the physical states of reactants and products: (g) for gases, (ℓ) for liquids, and (s) for solids. The notation (aq) following ions indicates that they are hydrated in aqueous solution; that is, they interact with water molecules in solution. The complete ionization of a strong electrolyte is indicated by a single arrow ($\rightarrow$).

Table 4-5 *Common Strong Acids and Their Anions*

Common Strong Acids		Anions of These Strong Acids	
Formula	*Name*	*Formula*	*Name*
HCl	hydrochloric acid	Cl^-	chloride ion
HBr	hydrobromic acid	Br^-	bromide ion
HI	hydroiodic acid	I^-	iodide ion
HNO_3	nitric acid	NO_3^-	nitrate ion
$HClO_4$	perchloric acid	ClO_4^-	perchlorate ion
$HClO_3$	chloric acid	ClO_3^-	chlorate ion
H_2SO_4	sulfuric acid	$\begin{cases} HSO_4^- \\ SO_4^{2-} \end{cases}$	hydrogen sulfate ion sulfate ion

Citrus fruits contain citric acid, and so their juices are acidic. This is shown here by the color changes on the indicator paper. Acids taste sour.

Because there are so many weak acids, please learn the list of common strong acids (Table 4-5). Then assume that the other acids you encounter are weak.

Table 4-6 *Some Common Weak Acids and Their Anions*

Common Weak Acids		Anions of These Weak Acids	
Formula	*Name*	*Formula*	*Name*
HF*	hydrofluoric acid	F^-	fluoride ion
CH_3COOH	acetic acid	CH_3COO^-	acetate ion
HCN	hydrocyanic acid	CN^-	cyanide ion
HNO_2†	nitrous acid	NO_2^-	nitrite ion
H_2CO_3†	carbonic acid	$\begin{cases} HCO_3^- \\ CO_3^{2-} \end{cases}$	hydrogen carbonate ion carbonate ion
H_2SO_3†	sulfurous acid	$\begin{cases} HSO_3^- \\ SO_3^{2-} \end{cases}$	hydrogen sulfite ion sulfite ion
H_3PO_4	phosphoric acid	$\begin{cases} H_2PO_4^- \\ HPO_4^{2-} \\ PO_4^{3-} \end{cases}$	dihydrogen phosphate ion hydrogen phosphate ion phosphate ion
$(COOH)_2$	oxalic acid	$\begin{cases} H(COO)_2^- \\ (COO)_2^{2-} \end{cases}$	hydrogen oxalate ion oxalate ion

*HF is a weak acid, whereas HCl, HBr, and HI are strong acids.

†*Free acid molecules exist only in dilute aqueous solution or not at all. However, many salts of these acids are common, stable compounds.*

The species we have shown as H^+(aq) is sometimes shown as H_3O^+ or $[H(H_2O)^+]$ to emphasize hydration. However, it really exists in varying degrees of hydration, such as $H(H_2O)^+$, $H(H_2O)_2^+$, and $H(H_2O)_3^+$.

Weak acids ionize only slightly (usually less than 5%) in dilute aqueous solution. Some common weak acids are listed in Appendix F, and a few of them and their anions are given in Table 4-6.

The equation for the ionization of acetic acid, CH_3COOH, in water is typical of weak acids:

Acetic acid is the most familiar organic acid.

$$CH_3COOH(aq) \rightleftharpoons H^+(aq) + CH_3COO^-(aq) \qquad \text{(reversible)}$$

The double arrow ($\rightleftharpoons$) generally signifies that the reaction occurs in *both* directions and that the forward reaction does not go to completion. All of us are familiar with solutions of acetic acid. Vinegar is 5% acetic acid by mass. Our use of oil and vinegar as a salad dressing tells us that acetic acid is a weak acid. To be specific, acetic acid is 0.5% ionized (and 99.5% nonionized) in 5% solution.

Our stomachs have linings that are much more resistant to attack by acids than are our other tissues.

A multitude of organic acids occur in living systems. Organic acids contain the carboxylate grouping of atoms, —COOH. Most common organic acids are weak. They can ionize slightly by breaking the O—H bond, as shown here for acetic acid:

The carboxylate group —COOH is

$$-\overset{\displaystyle O}{\underset{\displaystyle O-H}{\overset{\|}{C}}}$$

$$H_3C-C\overset{\displaystyle O}{\underset{\displaystyle O-\fbox{H}}{\diagup}} \text{(aq)} \rightleftharpoons H_3C-C\overset{\displaystyle O}{\underset{\displaystyle O^-}{\diagup}} \text{(aq)} + \fbox{H^+(aq)}$$

Other organic acids have other groups in the position of the H_3C— group in acetic acid.

Organic acids are discussed in Chapter 27. Carbonic acid, H_2CO_3, and hydrocyanic acid, HCN(aq), are two common acids that contain carbon but that are considered to be *inorganic* acids. Inorganic acids are often called **mineral acids** because they are obtained primarily from nonliving sources.

Inorganic acids may be strong or weak.

EXAMPLE 4-1 *Strong and Weak Acids*

In the following lists of common acids, which are strong and which are weak? (a) H_3PO_4, HCl, H_2CO_3, HNO_3; (b) $HClO_4$, H_2SO_4, HClO, HF.

Plan

We recall that Table 4-5 lists the common strong acids. Other *common* acids are assumed to be weak.

Solution

(a) HCl and HNO_3 are strong acids; H_3PO_4 and H_2CO_3 are weak acids.
(b) $HClO_4$ and H_2SO_4 are strong acids; HClO and HF are weak acids.

You should now work Exercises 17 and 19.

Many common foods and household products are acidic or basic.

3 Reversible Reactions

Reactions that can occur in both directions are **reversible reactions.** We use a double arrow ($\rightleftharpoons$) to indicate that a reaction is *reversible*. What is the fundamental difference between reactions that go to completion and those that are reversible? We have seen that the ionization of HCl in water is nearly complete. Suppose we dissolve some table salt, NaCl, in water and then add some dilute nitric acid to it. The resulting solution contains Na^+ and Cl^- ions (from the dissociation of NaCl) as well as H^+ and NO_3^- (from the ionization of HNO_3). The H^+ and Cl^- ions do *not* react significantly to form nonionized HCl molecules; this would be the reverse of the ionization of HCl.

$$H^+(aq) + Cl^-(aq) \longrightarrow \text{no reaction}$$

In contrast, when a sample of sodium acetate, $NaCH_3COO$, is dissolved in H_2O and mixed with nitric acid, the resulting solution initially contains Na^+, CH_3COO^-, H^+, and NO_3^- ions. But most of the H^+ and CH_3COO^- ions combine to produce nonionized molecules of acetic acid, the reverse of the ionization of the acid. Thus, the ionization of acetic acid, like that of any other weak electrolyte, is reversible.

Na^+ and NO_3^- ions do not combine because $NaNO_3$ is a soluble ionic compound.

$$H^+(aq) + CH_3COO^-(aq) \rightleftharpoons CH_3COOH(aq) \qquad \text{(reversible)}$$

4 Strong Soluble Bases, Insoluble Bases, and Weak Bases

Most common bases are ionic metal hydroxides. **Strong soluble bases** are soluble in water and are dissociated completely in dilute aqueous solution. The common strong soluble bases are listed in Table 4-7. They are the hydroxides of the Group IA metals and the heavier members of Group IIA. The equation for the dissociation of sodium hydroxide in water is typical. Similar equations can be written for other strong soluble bases.

Solutions of bases have a set of common properties due to the OH^- ion. These are described in Section 10-4.

$$NaOH(s) \xrightarrow{H_2O} Na^+(aq) + OH^-(aq) \qquad \text{(to completion)}$$

Strong soluble bases are ionic compounds in the solid state.

Other metals form ionic hydroxides, but these are so sparingly soluble in water that they cannot produce strongly basic solutions. They are called **insoluble bases.** Typical examples include $Cu(OH)_2$, $Zn(OH)_2$, $Fe(OH)_2$, and $Fe(OH)_3$.

Table 4-7 *Common Strong Soluble Bases*		
Group IA		**Group IIA**
LiOH lithium hydroxide		
NaOH sodium hydroxide		
KOH potassium hydroxide	$Ca(OH)_2$ calcium hydroxide	
RbOH rubidium hydroxide	$Sr(OH)_2$ strontium hydroxide	
CsOH cesium hydroxide	$Ba(OH)_2$ barium hydroxide	

The weak bases are *molecular* substances that ionize only slightly in water; they are sometimes called molecular bases.

Common **weak bases** are soluble in water but ionize only slightly in solution. The most common weak base is ammonia, NH_3. Closely related N-containing compounds, the *amines,* such as methylamine, CH_3NH_2, and aniline, $C_6H_5NH_2$, are also weak bases.

$$NH_3(aq) + H_2O(\ell) \rightleftharpoons NH_4^+(aq) + OH^-(aq) \qquad \text{(reversible)}$$

EXAMPLE 4-2   *Classifying Bases*

From the following lists, choose (i) the strong soluble bases, (ii) the insoluble bases, and (iii) the weak bases. (a) NaOH, $Cu(OH)_2$, $Pb(OH)_2$, $Ba(OH)_2$; (b) $Fe(OH)_3$, KOH, $Mg(OH)_2$, $Sr(OH)_2$, NH_3.

Plan

(i) We recall that Table 4-7 lists the *common strong soluble bases.* (ii) Other common metal hydroxides are assumed to be *insoluble bases.* (iii) Ammonia and closely related nitrogen-containing compounds, the amines, are the common *weak bases.*

Solution

(a) (i)   The strong soluble bases are NaOH and $Ba(OH)_2$, so
    (ii)   the insoluble bases are $Cu(OH)_2$ and $Pb(OH)_2$.
(b) (i)   The strong soluble bases are KOH and $Sr(OH)_2$, so
    (ii)   the insoluble bases are $Fe(OH)_3$ and $Mg(OH)_2$, and
    (iii) the weak base is NH_3.

You should now work Exercises 20 and 22.

5 Solubility Rules for Compounds in Aqueous Solution

Solubility is a complex phenomenon, and it is not possible to give a simple summary of all of our observations. The following summary for solutes in aqueous solutions will be very useful. These generalizations are often called the *solubility rules.* Compounds whose solubility in water is less than about 0.02 mole per liter are usually classified as insoluble compounds, whereas those that are more soluble are classified as soluble compounds. No gaseous or solid substances are infinitely soluble in water. You may wish to review Tables 2-3 (page 49), 4-5, and 4-6. They list some common ions. Table 4-14 on page 146 contains a more comprehensive list.

There is no sharp dividing line between "soluble" and "insoluble" compounds. Compounds whose solubilities fall near the arbitrary dividing line are called "moderately soluble" compounds.

1. The common inorganic acids are soluble in water. Low molecular weight organic acids are soluble.

2. All common compounds of the Group IA metal ions (Li^+, Na^+, K^+, Rb^+, Cs^+) and the ammonium ion, NH_4^+, are soluble in water.

3. The common nitrates, NO_3^-; acetates, CH_3COO^-; chlorates, ClO_3^-; and perchlorates, ClO_4^-, are soluble in water.

4. (a) The common chlorides, Cl^-, are soluble in water except $AgCl$, Hg_2Cl_2, and $PbCl_2$.
   (b) The common bromides, Br^-, and iodides, I^-, show approximately the same solubility behavior as chlorides, but there are some exceptions. As these halide ions (Cl^-, Br^-, I^-) increase in size, the solubilities of their slightly soluble compounds decrease.

5. The common sulfates, SO_4^{2-}, are soluble in water except $PbSO_4$, $BaSO_4$, and $HgSO_4$; $CaSO_4$ and Ag_2SO_4 are moderately soluble.

6. The common metal hydroxides, OH^-, are *insoluble* in water except those of the Group IA metals and the heavier members of the Group IIA metals, beginning with $Ca(OH)_2$.

7. The common carbonates, CO_3^{2-}, phosphates, PO_4^{3-}, and arsenates, AsO_4^{3-}, are *insoluble* in water except those of the Group IA metals and NH_4^+. $MgCO_3$ is moderately soluble.

8. The common sulfides, S^{2-}, are *insoluble* in water except those of the Group IA and Group IIA metals and the ammonium ion.

Table 4-8 summarizes much of the information about the solubility rules.

Table 4-8 *Solubility of Common Ionic Compounds in Water*

Generally Soluble	Exceptions
Na^+, K^+, NH_4^+ compounds	No common exceptions
chlorides (Cl^-)	Insoluble: $AgCl$, Hg_2Cl_2 Soluble in hot water: $PbCl_2$
bromides (Br^-)	Insoluble: $AgBr$, Hg_2Br_2, $PbBr_2$ Moderately soluble: $HgBr_2$
iodides (I^-)	Insoluble: many heavy metal iodides
sulfates (SO_4^{2-})	Insoluble: $BaSO_4$, $PbSO_4$, $HgSO_4$ Moderately soluble: $CaSO_4$, $SrSO_4$, Ag_2SO_4
nitrates (NO_3^-), nitrites (NO_2^-)	Moderately soluble: $AgNO_2$
chlorates (ClO_3^-), perchlorates (ClO_4^-), permanganates (MnO_4^-)	Moderately soluble: $KClO_4$
acetates (CH_3COO^-)	Moderately soluble: $AgCH_3COO$

Generally Insoluble	Exceptions
sulfides (S^{2-})	Soluble: those of NH_4^+, Na^+, K^+, Mg^{2+}, Ca^{2+}
oxides (O^{2-}), hydroxides (OH^-)	Soluble: Li_2O*, $LiOH$, Na_2O*, $NaOH$, K_2O*, KOH, BaO*, $Ba(OH)_2$ Moderately soluble: CaO*, $Ca(OH)_2$, SrO*, $Sr(OH)_2$
carbonates (CO_3^{2-}), phosphates (PO_4^{3-}), arsenates (AsO_4^{3-})	Soluble: those of NH_4^+, Na^+, K^+

Dissolves with evolution of heat and formation of hydroxides.

We shall now discuss chemical reactions in some detail. Because millions of reactions are known, it is useful to group them into classes, or types, so that we can deal with such massive amounts of information systematically. We shall classify them as (1) precipitation reactions, (2) acid–base reactions, (3) displacement reactions, and (4) oxidation–reduction reactions. We shall also distinguish among (1) formula unit, (2) total ionic, and (3) net ionic equations for chemical reactions and indicate the advantages and disadvantages of these methods for representing chemical reactions. As we study different kinds of chemical reactions, we shall learn to predict the products of other similar reactions.

In Chapter 6 we shall describe typical reactions of hydrogen, oxygen, and their compounds. These reactions will illustrate periodic relationships with respect to chemical properties. Our system is not an attempt to transform nature so that it fits into small categories; rather, it is an attempt to give some order to our observations of nature. We shall see that many oxidation–reduction reactions also fit into other categories and that some reactions do not fit neatly into any of these categories.

We have distinguished between strong and weak electrolytes and between soluble and insoluble compounds. Let us now see how we can describe chemical reactions in aqueous solutions.

4-3   REACTIONS IN AQUEOUS SOLUTIONS

Many important chemical reactions occur in aqueous solutions. In this chapter you should learn to describe such aqueous reactions and to predict the products of many reactions.

Let us first look at how we write chemical equations that describe reactions in aqueous solutions. We use three kinds of chemical equations. Table 4-9 shows the kinds of infor-

Table 4-9 *Bonding, Solubility, Electrolyte Characteristics, and Predominant Forms of Solutes in Contact with Water*

	Acids		Bases			Salts	
	Strong acids	*Weak acids*	*Strong soluble bases*	*Insoluble bases*	*Weak bases*	*Soluble salts*	*Insoluble salts*
Examples	HCl HNO$_3$	CH$_3$COOH HF	NaOH Ca(OH)$_2$	Mg(OH)$_2$ Al(OH)$_3$	NH$_3$ CH$_3$NH$_2$	KCl, NaNO$_3$, NH$_4$Br	BaSO$_4$, AgCl, Ca$_3$(PO$_4$)$_2$
Pure compound ionic or molecular?	Molecular	Molecular	Ionic	Ionic	Molecular	Ionic	Ionic
Water soluble or insoluble?	Soluble*	Soluble*	Soluble	Insoluble	Soluble†	Soluble	Insoluble
~100% ionized or dissociated in dilute aqueous solution?	Yes	No	Yes	(footnote ‡)	No	Yes§	(footnote ‡)
Written in ionic equations as	Separate ions	Molecules	Separate ions	Complete formulas	Molecules	Separate ions	Complete formulas

*Most common inorganic acids and the low-molecular-weight organic acids (—COOH) are water soluble.

† The low-molecular-weight amines are water soluble.

‡ The very small concentrations of "insoluble" metal hydroxides and insoluble salts in saturated aqueous solutions are nearly completely dissociated.

§ There are a few exceptions. A few soluble salts are molecular (and not ionic) compounds.

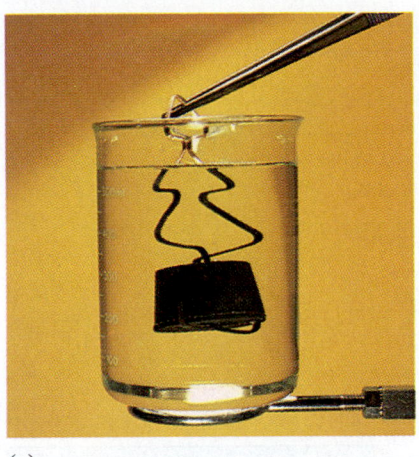 

(a) **(b)**

Figure 4-3 (a) Copper wire and a silver nitrate solution. (b) The copper wire has been placed in the solution and some finely divided silver has deposited on the wire. The solution is blue because it contains copper(II) nitrate.

mation about each substance that we use in writing equations for reactions in aqueous solutions. Some typical examples are included. Refer to Table 4-9 often as you study the following sections.

1. In **formula unit equations,** we show complete formulas for all compounds. When metallic copper is added to a solution of (colorless) silver nitrate, the more active metal—copper—displaces silver ions from the solution. The resulting solution contains blue copper(II) nitrate, and metallic silver forms as a finely divided solid (Figure 4-3):

$$2AgNO_3(aq) + Cu(s) \longrightarrow 2Ag(s) + Cu(NO_3)_2(aq)$$

Both silver nitrate and copper(II) nitrate are soluble ionic compounds (for solubility rules see page 124 and Table 4-8).

2. In **total ionic equations,** formulas are written to show the (predominant) form in which each substance exists when it is in contact with aqueous solution. We often use brackets in total ionic equations to show ions that have a common source or that remain in solution after the reaction is complete. The total ionic equation for this reaction is

$$2[Ag^+(aq) + NO_3^-(aq)] + Cu(s) \longrightarrow 2Ag(s) + [Cu^{2+}(aq) + 2NO_3^-(aq)]$$

Examination of the total ionic equation shows that NO_3^- ions do not participate in the reaction. Because they do not change, they are sometimes called "spectator" ions.

3. The **net ionic equation** shows only the species that react. The net ionic equation is obtained by eliminating the spectator ions and the brackets from the total ionic equation.

$$2Ag^+(aq) + Cu(s) \longrightarrow 2Ag(s) + Cu^{2+}(aq)$$

Net ionic equations allow us to focus on the *essence* of a chemical reaction in aqueous solutions. On the other hand, if we are dealing with stoichiometric calculations we frequently must deal with formula weights and therefore with the *complete* formulas of all species. In such cases, formula unit equations are more useful. Total ionic equations provide the bridge between the two.

Because we have not studied periodic trends in properties of transition metals, it would be difficult for you to predict that Cu is more active than Ag. The fact that this reaction occurs (see Figure 4-3) shows that it is.

Brackets are not used in net ionic equations.

This is why it is important to know how and when to construct net ionic equations from formula unit equations.

▼ **PROBLEM-SOLVING TIP**   *Writing Ionic Equations*

The following chart will help in deciding which formula units are to be written as separate ions in the total ionic equation and which ones are to be written as unchanged formula units. You must answer two questions about a substance to determine whether it should be written in ionic form or as a formula unit in the total and net ionic equations.

1. Does it dissolve in water? If not, write the full formula.

2. If it dissolves, does it either ionize or dissociate?

If both answers are yes, the substance is a soluble strong electrolyte, and its formula is written in ionic form.

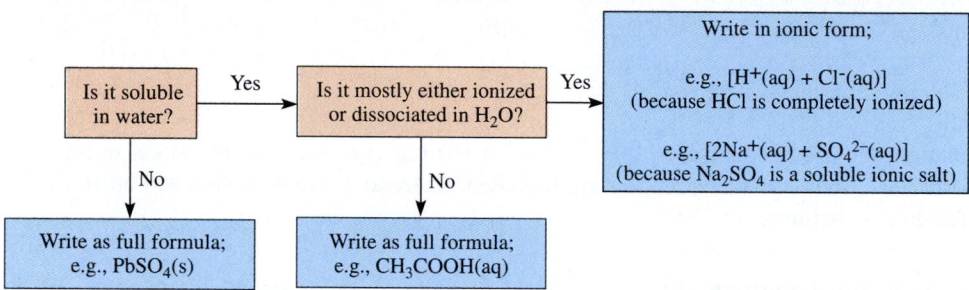

Recall the lists of strong acids (Table 4-5) and strong soluble bases (Table 4-7). These acids and bases are completely or almost completely ionized or dissociated in dilute aqueous solutions. Other common acids and bases are either insoluble or only slightly ionized or dissociated. In addition, the solubility rules (page 124 and Table 4-8) allow you to determine which salts are soluble in water. Most salts that are soluble in water are also strong electrolytes. Exceptions such as lead acetate, $Pb(CH_3COO)_2$, which is soluble but predominantly nonionized, will be noted as they are encountered.

> The only common substances that should be written in ionized or dissociated form in ionic equations are (1) strong acids, (2) strong soluble bases, and (3) soluble ionic salts.

Many reactions that occur in aqueous solution result in the removal of ions from solution. This happens in one of three ways: (1) formation of an insoluble solid product, (2) formation of predominantly nonionized molecules (weak or nonelectrolytes) in solution, or (3) formation of a gas that escapes. In the next two sections, we shall see many examples that illustrate these concepts.

4-4 PRECIPITATION REACTIONS

To understand the discussion of precipitation reactions, you *must* know the solubility rules (page 124) and Table 4-8.

In **precipitation reactions** an insoluble solid, a **precipitate,** forms and then settles out of solution. The driving force for these reactions is the strong attraction between cations and anions. This results in the removal of ions from solution by the formation of a precipitate.

Figure 4-4 A precipitation reaction. When K_2CrO_4 solution is added to aqueous $Pb(NO_3)_2$ solution, the yellow compound $PbCrO_4$ precipitates. The resulting solution contains K^+ and NO_3^- ions, the ions of KNO_3.

Our teeth and bones were formed by very slow precipitation reactions in which mostly calcium phosphate $Ca_3(PO_4)_2$ was deposited in the correct geometric arrangements.

An example of a precipitation reaction is the formation of bright yellow insoluble lead(II) chromate when we mix solutions of the soluble ionic compounds lead(II) nitrate and potassium chromate (Figure 4-4). The other product of the reaction is KNO_3, a soluble ionic salt.

The balanced formula unit, total ionic, and net ionic equations for this reaction follow.

$$Pb(NO_3)_2(aq) + K_2CrO_4(aq) \longrightarrow PbCrO_4(s) + 2KNO_3(aq)$$

$$[Pb^{2+}(aq) + 2\,NO_3^-(aq)] + [2K^+(aq) + CrO_4^{2-}(aq)] \longrightarrow$$
$$PbCrO_4(s) + 2[K^+(aq) + NO_3^-(aq)]$$

$$Pb^{2+}(aq) + CrO_4^{2-}(aq) \longrightarrow PbCrO_4(s)$$

Another important precipitation reaction involves the formation of insoluble carbonates (solubility rule 7). Limestone deposits are mostly calcium carbonate, $CaCO_3$, although many also contain significant amounts of magnesium carbonate, $MgCO_3$.

Suppose we mix together aqueous solutions of sodium carbonate, Na_2CO_3, and calcium chloride, $CaCl_2$. We recognize that *both* Na_2CO_3 and $CaCl_2$ (solubility rules 2, 4a, and 7) are soluble ionic compounds. At the instant of mixing, the resulting solution contains four ions:

$$Na^+(aq), \qquad CO_3^{2-}(aq), \qquad Ca^{2+}(aq), \qquad Cl^-(aq)$$

One pair of ions, Na^+ and Cl^-, *cannot* form an insoluble compound (solubility rules 2 and 4). We look for a pair of ions that could form an insoluble compound. Ca^{2+} ions and CO_3^{2-} ions are such a combination; they form insoluble $CaCO_3$ (solubility rule 7). The equations for the reaction follow.

$$CaCl_2(aq) + Na_2CO_3(aq) \longrightarrow CaCO_3(s) + 2\,NaCl(aq)$$

$$[Ca^{2+}(aq) + 2\,Cl^-(aq)] + [2Na^+(aq) + CO_3^{2-}(aq)] \longrightarrow$$
$$CaCO_3(s) + 2[Na^+(aq) + Cl^-(aq)]$$

$$Ca^{2+} + CO_3^{2-}(aq) \longrightarrow CaCO_3(s)$$

Seashells, which are formed in very slow precipitation reactions, are mostly calcium carbonate ($CaCO_3$), a white compound. Traces of transition metal ions give them color.

EXAMPLE 4-3   *Solubility Rules*

Will a precipitate form when aqueous solutions of $Ca(NO_3)_2$ and NaCl are mixed in reasonable concentrations?

Plan

We recognize that both $Ca(NO_3)_2$ (solubility rule 3) and NaCl (solubility rules 2 and 4) are soluble compounds. At the instant of mixing, the resulting solution contains four ions:

$$Ca^{2+}(aq), \quad NO_3^-(aq), \quad Na^+(aq), \quad Cl^-(aq)$$

New combinations of ions *could* be $CaCl_2$ and $NaNO_3$. Solubility rule 4 tells us that $CaCl_2$ is a soluble compound, while solubility rules 2 and 3 tell us that $NaNO_3$ is a soluble compound.

Solution

Therefore no precipitate forms in this solution.

You should now work Exercise 30.

EXAMPLE 4-4   *Writing Chemical Equations*

Will a precipitate form when aqueous solutions of $CaCl_2$ and K_3PO_4 are mixed in reasonable concentrations? Write the appropriate equations for any reaction.

Plan

Both $CaCl_2$ (solubility rule 4) and K_3PO_4 (solubility rule 2) are soluble compounds. At the instant of mixing, four ions are present in the solution:

$$Ca^{2+}(aq), \quad Cl^-(aq), \quad K^+(aq), \quad PO_4^{3-}(aq)$$

New combinations of these ions *could* be KCl and $Ca_3(PO_4)_2$. Solubility rules 2 and 4 tell us that potassium chloride, KCl, is a soluble compound.

Solution

Solubility rule 7 tells us that calcium phosphate, $Ca_3(PO_4)_2$, is an insoluble compound and so it forms a precipitate.

The equations for the formation of calcium phosphate follow.

$$3CaCl_2(aq) + 2K_3PO_4(aq) \longrightarrow Ca_3(PO_4)_2(s) + 6KCl(aq)$$

$$3[Ca^{2+}(aq) + 2\,Cl^-(aq)] + 2[3K^+(aq) + PO_4^{3-}(aq)] \longrightarrow$$
$$Ca_3(PO_4)_2(s) + 6[K^+(aq) + Cl^-(aq)]$$

$$3Ca^{2+}(aq) + 2PO_4^{3-}(aq) \longrightarrow Ca_3(PO_4)_2(s)$$

You should now work Exercise 38.

4-5   ACID–BASE REACTIONS

Acid–base reactions are among the most important kinds of chemical reactions. Many acid–base reactions occur in nature in both plants and animals. Many acids and bases are essential compounds in an industrialized society (see Table 4-10). For example, approximately 350 pounds of sulfuric acid, H_2SO_4, and approximately 135 pounds of ammonia, NH_3, are required to support the lifestyle of an average American for one year.

The manufacture of fertilizers consumes more H_2SO_4 *and* more NH_3 than any other single use.

Table 4-10 *1993 Production of Acids, Bases, and Salts in the United States*

Formula	Name	Billions of Pounds	Major Uses
H_2SO_4	sulfuric acid	80.31	Manufacture of fertilizers and other chemicals
CaO, $Ca(OH)_2$	lime (calcium oxide and calcium hydroxide)	36.80	Manufacture of other chemicals, steelmaking, water treatment
NH_3	ammonia	34.50	Fertilizer; manufacture of fertilizers and other chemicals
NaOH	sodium hydroxide	25.71	Manufacture of other chemicals, pulp and paper, soap and detergents, aluminum, textiles
H_3PO_4	phosphoric acid	23.04	Manufacture of fertilizers
Na_2CO_3	sodium carbonate (soda ash)	19.80	Manufacture of glass, other chemicals, detergents, pulp, and paper
HNO_3	nitric acid	17.07	Manufacture of fertilizers, explosives, plastics, and lacquers
NH_4NO_3	ammonium nitrate	16.79	Fertilizer and explosive
$C_6H_4(COOH)_2$*	terephthalic acid	7.84	Manufacture of fibers (polyesters), films, and bottles
HCl	hydrochloric acid	6.45	Manufacture of other chemicals and rubber; metal cleaning
$(NH_4)_2SO_4$	ammonium sulfate	4.80	Fertilizer
CH_3COOH*	acetic acid	3.66	Manufacture of acetate esters
KOH, K_2CO_3	potash	3.31	Manufacture of fertilizers
$Al_2(SO_4)_2$	aluminum sulfate	2.23	Water treatment; dyeing textiles
Na_2SiO_3	sodium silicate	1.97	Manufacture of detergents, cleaning agents, and adhesives
$C_4H_8(COOH)_2$*	adipic acid	1.56	Manufacture of Nylon 66
Na_2SO_4	sodium sulfate	1.44	Manufacture of paper, glass, and detergents
$CaCl_2$	calcium chloride	1.40	De-icing roads in winter, controlling dust in summer, concrete additive

Organic compound.

The reaction of an acid with a metal hydroxide base produces a salt and water. Such reactions are called **neutralization reactions** because the typical properties of acids and bases are neutralized.

In nearly all neutralization reactions, the driving force is the combination of $H^+(aq)$ from an acid and $OH^-(aq)$ from a base (or a base plus water) to form water molecules.

When a base such as ammonia or an amine reacts with an acid, a salt, but no water, is formed.

Consider the reaction of hydrochloric acid, HCl, with aqueous sodium hydroxide, NaOH. Table 4-5 tells us that HCl is a strong acid, and Table 4-7 tells us that NaOH is a strong soluble base. The salt sodium chloride, NaCl, is formed in this reaction. It contains the cation of its parent base, Na^+, and the anion of its parent acid, Cl^-. Solubility rules 2 and 4 tell us that NaCl is a soluble salt.

$$HCl(aq) + NaOH(aq) \longrightarrow H_2O(\ell) + NaCl(aq)$$

$$[H^+(aq) + Cl^-(aq)] + [Na^+(aq) + OH^-(aq)] \longrightarrow H_2O(\ell) + [Na^+(aq) + Cl^-(aq)]$$

$$H^+(aq) + OH^-(aq) \longrightarrow H_2O(\ell)$$

The net ionic equation for *all* reactions of strong acids with strong soluble bases that form soluble salts and water is

$$H^+(aq) + OH^-(aq) \longrightarrow H_2O(\ell)$$

▼ **PROBLEM-SOLVING TIP**   *Salt Formation*

If our goal were to obtain the salt from reaction of aqueous HCl with aqueous NaOH, we could evaporate the water and obtain solid NaCl.

EXAMPLE 4-5   *Neutralization Reactions*

Predict the products of the reaction between HI(aq) and $Ca(OH)_2$(aq). Write balanced formula unit, total ionic, and net ionic equations.

Plan

This is an acid–base neutralization reaction; the products are H_2O and the salt that contains the cation of the base, Ca^{2+}, and the anion of the acid, I^-; CaI_2 is a soluble salt (solubility rule 4). HI is a strong acid (Table 4-5), $Ca(OH)_2$ is a strong soluble base (Table 4-7), and CaI_2 is a soluble ionic salt, so all are written in ionic form.

Solution

$$2HI(aq) + Ca(OH)_2(aq) \longrightarrow CaI_2(aq) + 2H_2O(\ell)$$

$$2[H^+(aq) + I^-(aq)] + [Ca^{2+}(aq) + 2OH^-(aq)] \longrightarrow [Ca^{2+}(aq) + 2I^-(aq)] + 2H_2O(\ell)$$

We cancel the spectator ions.

$$2H^+(aq) + 2OH^-(aq) \longrightarrow 2H_2O(\ell)$$

Dividing by 2 gives the net ionic equation

$$H^+(aq) + OH^-(aq) \longrightarrow H_2O(\ell)$$

You should now work Exercise 41.

Recall that in balanced equations we show the smallest whole-number coefficients possible.

Reactions of *weak* acids with strong soluble bases also produce salts and water, but there is a significant difference in the balanced ionic equations because weak acids are only *slightly* ionized.

EXAMPLE 4-6   *Neutralization Reactions*

Write balanced formula unit, total ionic, and net ionic equations for the reaction of acetic acid with potassium hydroxide.

Plan

Neutralization reactions involving metal hydroxide bases produce a salt and water. CH_3COOH is a weak acid (Table 4-6) and so it is written as formula units. KOH is a strong soluble base (Table 4-7) and KCH_3COO is a soluble salt (solubility rules 2 and 3), and so both are written in ionic form.

Solution

$$CH_3COOH(aq) + KOH(aq) \longrightarrow KCH_3COO(aq) + H_2O(\ell)$$

$$CH_3COOH(aq) + [K^+(aq) + OH^-(aq)] \longrightarrow [K^+(aq) + CH_3COO^-(aq)] + H_2O(\ell)$$

The spectator ion is K^+, the cation of the strong soluble base, KOH.

$$CH_3COOH(aq) + OH^-(aq) \longrightarrow CH_3COO^-(aq) + H_2O(\ell)$$

Thus, we see that *this* net ionic equation includes *molecules* of the weak acid and *anions* of the weak acid.

You should now work Exercise 42.

The reactions of *weak monoprotic acids* with *strong soluble bases* that form *soluble salts* can be represented in general terms as

$$HA(aq) + OH^-(aq) \longrightarrow A^-(aq) + H_2O(\ell)$$

where HA represents the weak acid and A^- represents its anion.

A *monoprotic acid* contains one acidic H per formula unit.

EXAMPLE 4-7   *Salt Formation*

Write balanced formula unit, total ionic, and net ionic equations for an acid–base reaction that will produce the salt, barium chloride.

Plan

Neutralization reactions produce a salt. The salt contains the cation from the base and the anion from the acid. Therefore, the base must contain Ba^{2+}, i.e., $Ba(OH)_2$, and the acid must contain Cl^-, i.e., HCl. We write equations that represent the reaction between the base, $Ba(OH)_2$, and the acid, HCl.

Solution

$$2HCl(aq) + Ba(OH)_2(aq) \longrightarrow BaCl_2(aq) + 2H_2O(\ell)$$

$$2[H^+(aq) + Cl^-(aq)] + [Ba^{2+}(aq) + 2OH^-(aq)] \longrightarrow [Ba^{2+}(aq) + 2Cl^-(aq)] + 2H_2O(\ell)$$

We cancel the spectator ions.

$$2H^+(aq) + 2OH^-(aq) \longrightarrow 2H_2O(\ell)$$

Dividing by 2 gives the net ionic equation:

$$H^+(aq) + OH^-(aq) \longrightarrow H_2O(\ell)$$

You should now work Exercises 49 and 50.

The net ionic equation shows the driving force for this reaction. The formula unit equation shows the salt formed or that could be isolated if the water were evaporated.

Removal of Ions from Aqueous Solutions

Prior to Section 4-4, we listed three ways that a reaction can be driven by the removal of ions. They are (1) the formation of an insoluble solid product (precipitation reactions), (2) the formation of predominantly nonionized molecules in solution (acid–base reactions), or (3) the formation of a gas that escapes.

Blackboard chalk is mostly calcium carbonate, $CaCO_3$. Bubbles of carbon dioxide, CO_2, are clearly visible in this photograph of $CaCO_3$ dissolving in HCl.

Let us illustrate the third case. When an acid—for example, hydrochloric acid—is added to solid calcium carbonate, a reaction occurs in which carbonic acid, a weak acid, is produced.

$$2HCl(aq) + CaCO_3(s) \longrightarrow H_2CO_3(aq) + CaCl_2(aq)$$

$$2[H^+(aq) + Cl^-(aq)] + CaCO_3(s) \longrightarrow H_2CO_3(aq) + [Ca^{2+}(aq) + 2Cl^-(aq)]$$

$$2H^+(aq) + CaCO_3(s) \longrightarrow H_2CO_3(aq) + Ca^{2+}(aq)$$

The heat generated in the reaction causes thermal decomposition of carbonic acid to gaseous carbon dioxide and water:

$$H_2CO_3(aq) \longrightarrow CO_2(g) + H_2O(\ell)$$

Most of the CO_2 bubbles off and the reaction goes to completion (with respect to the limiting reactant). The net effect is the conversion of ionic species into nonionized molecules of a gas (CO_2) and water.

Table 4-11 *Common Oxidation Numbers (States) for Group A Elements in Compounds and Ions*

Element(s)	Common Ox. Nos.	Examples	Other Ox. Nos.
H	+1	H_2O, CH_4, NH_4Cl	-1 in metal hydrides, e.g., NaH, CaH_2
Group IA	+1	KCl, NaH, $RbNO_3$, K_2SO_4	None
Group IIA	+2	$CaCl_2$, MgH_2, $Ba(NO_3)_2$, $SrSO_4$	None
Group IIIA	+3	$AlCl_3$, BF_3, $Al(NO_3)_3$, GaI_3	None in common compounds
Group IVA	+2 +4	CO, PbO, $SnCl_2$, $Pb(NO_3)_2$ CCl_4, SiO_2, SiO_3^{2-}, $SnCl_4$	Many others are also seen for C and Si
Group VA	-3 in binary compounds with metals -3 in NH_4^+, binary compounds with H	Mg_3N_2, Na_3P, Cs_3As NH_3, PH_3, AsH_3, NH_4^+	$+3$, e.g., NO_2^-, PCl_3 $+5$, e.g., NO_3^-, PO_4^{3-}, AsF_5, P_4O_{10}
O	-2	H_2O, P_4O_{10}, Fe_2O_3, CaO, ClO_3^-	$+2$ in OF_2 -1 in peroxides, e.g., H_2O_2, Na_2O_2 $-\frac{1}{2}$ in superoxides, e.g., KO_2, RbO_2
Group VIA (other than O)	-2 in binary compounds with metals and H -2 in binary compounds with NH_4^+	H_2S, CaS, Fe_2S_3, Na_2Se $(NH_4)_2S$, $(NH_4)_2Se$	$+4$ with O and the lighter halogens, e.g., SO_2, SeO_2, Na_2SO_3, SO_3^{2-}, SF_4 $+6$ with O and the lighter halogens, e.g., SO_3, TeO_3, H_2SO_4, SO_4^{2-}, SF_6
Group VIIA	-1 in binary compounds with metals and H -1 in binary compounds with NH_4^+	MgF_2, KI, $ZnCl_2$, $FeBr_3$ NH_4Cl, NH_4Br	Cl, Br, or I with O or with a lighter halogen $+1$, e.g., BrF, ClO^-, BrO^- $+3$, e.g., ICl_3, ClO_2^-, BrO_2^- $+5$, e.g., BrF_5, ClO_3^-, BrO_3^- $+7$, e.g., IF_7, ClO_4^-, BrO_4^-

4-6 OXIDATION NUMBERS

Many reactions involve the transfer of electrons from one species to another. They are called **oxidation–reduction reactions** or simply **redox reactions.** We use oxidation numbers to keep track of electron transfers.

The **oxidation number,** or **oxidation state,** of an element in a simple *binary* ionic compound is the number of electrons gained or lost by an atom of that element when it forms the compound. In the case of a single-atom ion, it corresponds to the actual charge on the ion. In molecular compounds, oxidation numbers do not have the same significance they have in binary ionic compounds. However, oxidation numbers are very useful aids in writing formulas and in balancing equations. In molecular species, the oxidation numbers are assigned according to an arbitrary set of rules. The element farther to the right and higher up in the periodic table is assigned a negative oxidation number, and the element farther to the left and lower down in the periodic table is assigned a positive oxidation number.

Oxidation–reduction and displacement reactions are discussed in Sections 4-7 and 4-8.

Binary means two. Binary compounds contain two elements.

Table 4-12 *Some Nonzero Oxidation States (numbers)*

The terms "oxidation number" and "oxidation state" are used interchangeably.

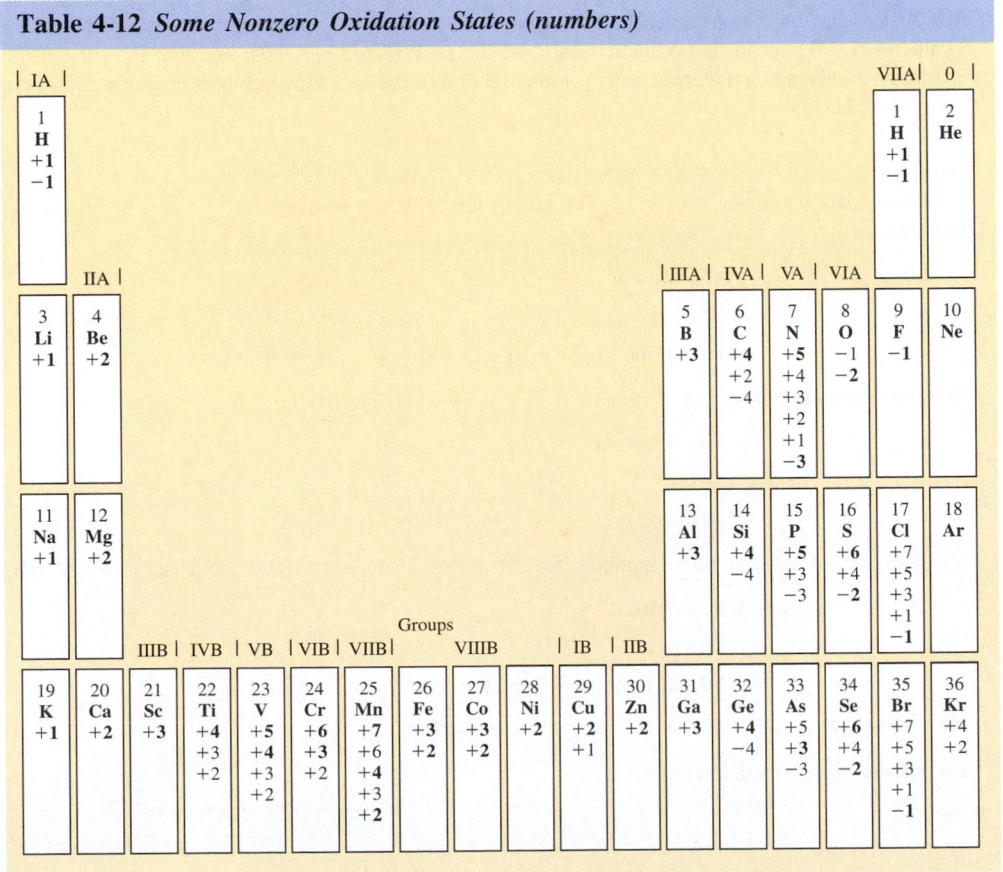

Some rules for assigning oxidation numbers follow. These rules are not comprehensive, but they cover most cases. *Oxidation numbers are always assigned on a per atom basis.*

Polyatomic elements have two or more atoms per molecule.

1. The oxidation number of any free, uncombined element is zero. This includes polyatomic elements such as H_2, O_2, O_3, and S_8.
2. The oxidation number of an element in a simple (monatomic) ion is the charge on the ion.
3. The sum of the oxidation numbers of all atoms in the compound is zero.
4. In a polyatomic ion, the sum of the oxidation numbers of the constituent atoms is equal to the charge on the ion.

Tables 4-11 and 4-12 (see pp. 134 and 135) list some common nonzero oxidation numbers.

Aqueous solutions of some compounds that contain chromium. Left to right: chromium(II) chloride ($CrCl_2$) is blue; chromium(III) chloride ($CrCl_3$) is green; potassium chromate (K_2CrO_4) is yellow; potassium dichromate ($K_2Cr_2O_7$) is orange.

EXAMPLE 4-8   *Oxidation Numbers*

Determine the oxidation numbers of nitrogen in the following species: (a) N_2O_4, (b) NH_3, (c) NO_3^-, (d) N_2.

Plan

We first assign oxidation numbers to elements that exhibit a single common oxidation number (Table 4-11). We recall that oxidation numbers are represented *per atom* and that the sum of the oxidation numbers in a molecule is zero, while the sum of the oxidation numbers in an ion equals the charge on the ion.

Solution

By convention, oxidation numbers are represented as $+n$ and $-n$, while ionic charges are represented as $n+$ and $n-$. We shall circle oxidation numbers associated with formulas and show them in red. Both oxidation numbers and ionic charges can be combined algebraically.

(a) The oxidation number of O is -2. The sum of the oxidation numbers for all atoms in a molecule must be zero:

ox. no./atom: $(x)\ (-2)$
$$N_2O_4$$

total ox. no.: $2x + 4(-2) = 0$ or $x = \boxed{+4}$

(b) The oxidation number of H is $+1$:

ox. no./atom: $(x)\ (+1)$
$$NH_3$$

total ox. no.: $x + 3(1) = 0$ or $x = \boxed{-3}$

Usually the element with the positive oxidation number is written first. However, for historic reasons, in compounds containing nitrogen and hydrogen, such as NH_3, and many compounds containing carbon and hydrogen, such as CH_4, hydrogen is written last, although it has a positive oxidation number.

(c) The sum of the oxidation numbers for all atoms in an ion equals the charge on the ion:

ox. no./atom: $(x)\ (-2)$
$$NO_3^-$$

total ox. no.: $x + 3(-2) = -1$ or $x = \boxed{+5}$

(d) The oxidation number of any free element is $\boxed{\text{zero.}}$

You should now work Exercise 52.

4-7 OXIDATION–REDUCTION REACTIONS—AN INTRODUCTION

The term "oxidation" originally referred to the combination of a substance with oxygen. This results in an increase in the oxidation number of an element in that substance. According to the original definition, the following reactions involve oxidation of the substance shown on the far left of each equation. Oxidation numbers are shown for *one* atom of the indicated kind.

1. The formation of rust, Fe_2O_3, iron(III) oxide:

   oxidation state of Fe

   $$4Fe(s) + 3O_2(g) \longrightarrow 2Fe_2O_3(s) \qquad\qquad 0 \longrightarrow +3$$

2. Combustion reactions:

   oxidation state of C

   $$C(s) + O_2(g) \longrightarrow CO_2(g) \qquad\qquad 0 \longrightarrow +4$$

   $$2CO(g) + O_2(g) \longrightarrow 2CO_2(g) \qquad\qquad +2 \longrightarrow +4$$

   $$C_3H_8(g) + 5O_2(g) \longrightarrow 3CO_2(g) + 4H_2O(g) \qquad -8/3 \longrightarrow +4$$

> Oxidation number is a formal concept adopted for our convenience. The numbers are determined by reliance upon rules. These rules can result in a fractional oxidation number, as shown here. This does not mean that electronic charges are split.

Originally, *reduction* described the removal of oxygen from a compound. Oxide ores are reduced to metals (a very real reduction in mass). For example, tungsten for use in light bulb filaments can be prepared by reduction of tungsten(VI) oxide with hydrogen at 1200°C:

oxidation number of W

$$WO_3(s) + 3H_2(g) \longrightarrow W(s) + 3H_2O(g) \qquad +6 \longrightarrow 0$$

Tungsten is reduced and its oxidation state decreases from +6 to zero. Hydrogen is oxidized from zero to the +1 oxidation state. The terms "oxidation" and "reduction" are now applied much more broadly.

Oxidation is an algebraic increase in oxidation number and corresponds to the loss, or apparent loss, of electrons. **Reduction** is an algebraic decrease in oxidation number and corresponds to a gain, or apparent gain, of electrons.

> In biological systems, *reduction* often corresponds to the addition of hydrogen to molecules or polyatomic ions and *oxidation* often corresponds to the removal of hydrogen.

Electrons are neither created nor destroyed in chemical reactions. So oxidation and reduction always occur simultaneously, and to the same extent, in ordinary chemical reactions. In the four equations cited previously as *examples of oxidation,* the oxidation numbers of iron and carbon atoms increase as they are oxidized. In each case oxygen is reduced as its oxidation number decreases from zero to −2.

Because oxidation and reduction occur simultaneously in all of these reactions, they are referred to as oxidation–reduction reactions. For brevity, we usually call them **redox** reactions. Redox reactions occur in nearly every area of chemistry and biochemistry. We need to be able to identify oxidizing agents and reducing agents and to balance oxidation–reduction equations. These skills are necessary for the study of electrochemistry in Chapter 21. Electrochemistry involves electron transfer between physically separated oxidizing and reducing agents and interconversions between chemical energy and electrical energy. These skills are also fundamental to the study of biology, biochemistry, environmental science, and materials science.

> Oxidizing agents are species that (1) oxidize other substances, (2) are reduced, and (3) gain (or appear to gain) electrons. Reducing agents are species that (1) reduce other substances, (2) are oxidized, and (3) lose (or appear to lose) electrons.

The equations below represent examples of redox reactions. Oxidation numbers are shown above the formulas, and oxidizing and reducing agents are indicated:

The following abbreviations are widely used:

> ox. no. = oxidation number
> ox. agt. = oxidizing agent
> red. agt. = reducing agent

$$\overset{0}{2Fe(s)} + \overset{0}{3Cl_2(g)} \longrightarrow \overset{+3 \; -1}{2FeCl_3(s)}$$
$$\text{red. agt.} \qquad \text{ox. agt.}$$

$$\overset{+3 \; -1}{2FeBr_3(aq)} + \overset{0}{3Cl_2(g)} \longrightarrow \overset{+3 \; -1}{2FeCl_3(aq)} + \overset{0}{3Br_2(\ell)}$$
$$\text{red. agt.} \qquad \text{ox. agt.}$$

Equations for redox reactions can also be written as total ionic and net ionic equations. For example, the previous equation may also be written as shown below. We distinguish between oxidation numbers and actual charges on ions by denoting oxidation numbers as $+n$ or $-n$ *in red circles just above the symbols of the elements,* and actual charges as $n+$ or $n-$ above and to the right of formulas of ions.

$$2[Fe^{3+}(aq) + 3Br^-(aq)] + 3Cl_2(g) \longrightarrow 2[Fe^{3+}(aq) + 3Cl^-(aq)] + 3Br_2(\ell)$$

The spectator ions, Fe^{3+}, do not participate in electron transfer. Their cancellation allows us to focus on the oxidizing agent, $Cl_2(g)$, and the reducing agent, $Br^-(aq)$.

$$2Br^-(aq) + Cl_2(g) \longrightarrow 2Cl^-(aq) + Br_2(\ell)$$

A **disproportionation reaction** is a redox reaction in which the same element is oxidized and reduced. An example is:

$$\overset{0}{Cl_2} + H_2O \longrightarrow \overset{-1}{HCl} + \overset{+1}{HClO}$$

Iron reacting with chlorine to form iron(III) chloride.

EXAMPLE 4-9   *Redox Reactions*

Write each of the following formula unit equations as a net ionic equation if the two differ. Which ones are redox reactions? For the redox reactions, identify the oxidizing agent, the reducing agent, the species oxidized, and the species reduced.

(a) $2AgNO_3(aq) + Cu(s) \longrightarrow Cu(NO_3)_2(aq) + 2Ag(s)$

(b) $4KClO_3(s) \overset{\Delta}{\longrightarrow} KCl(s) + 3KClO_4(s)$

(c) $3AgNO_3(aq) + K_3PO_4(aq) \longrightarrow Ag_3PO_4(s) + 3KNO_3(aq)$

Plan

To write ionic equations, we must recognize compounds that are (1) soluble in water and (2) ionized or dissociated in aqueous solutions. To determine which are oxidation–reduction reactions, we must assign an oxidation number to each element.

Solution

(a) According to the solubility rules (page 124), both silver nitrate, $AgNO_3$, and copper(II) nitrate, $Cu(NO_3)_2$, are water-soluble ionic compounds. The total ionic equation and oxidation numbers are

$$2[\overset{+1}{Ag^+}(aq) + \overset{+5\ -2}{NO_3^-}(aq)] + \overset{0}{Cu}(s) \longrightarrow [\overset{+2}{Cu^{2+}}(aq) + 2\overset{+5\ -2}{NO_3^-}(aq)] + 2\overset{0}{Ag}(s)$$

$$+2 \quad\quad -1$$

The nitrate ions, NO_3^-, are spectator ions. Canceling them from both sides gives the net ionic equation:

$$2\overset{+1}{Ag^+}(aq) + \overset{0}{Cu}(s) \longrightarrow \overset{+2}{Cu^{2+}}(aq) + 2\overset{0}{Ag}(s)$$

This is a redox equation. The oxidation number of silver decreases from +1 to zero; silver ion is reduced and is the oxidizing agent. The oxidation number of copper increases from zero to +2; copper is oxidized and is the reducing agent.

(b) This reaction involves three solids, so the formula unit and net ionic equations are identical. It is a redox reaction:

$$4\overset{+1\ +5\ -2}{KClO_3}(s) \longrightarrow \overset{+1\ -1}{KCl}(s) + 3\overset{+1\ +7\ -2}{KClO_4}(s)$$

$$-6 \quad\quad +2$$

Chlorine is reduced from +5 in $KClO_3$ to the −1 oxidation state in KCl; the oxidizing agent is $KClO_3$. Chlorine is oxidized from +5 in $KClO_3$ to the +7 oxidation state in $KClO_4$. $KClO_3$ is also the reducing agent. This is a disproportionation reaction. We see that $KClO_3$ is both the oxidizing agent and the reducing agent.

(c) The solubility rules indicate that all these salts are soluble and ionic except for silver phosphate, Ag_3PO_4. The total ionic equation is

$$3[Ag^+(aq) + NO_3^-(aq)] + [3K^+(aq) + PO_4^{3-}(aq)] \longrightarrow$$
$$Ag_3PO_4(s) + 3[K^+(aq) + NO_3^-(aq)]$$

Eliminating the spectator ions gives the net ionic equation:

$$3\overset{+1}{Ag^+}(aq) + \overset{+5\ -2}{PO_4^{3-}}(aq) \longrightarrow \overset{+1\ +5\ -2}{Ag_3PO_4}(s)$$

There are no changes in oxidation numbers; this is not a redox reaction.

You should now work Exercise 58.

In Chapter 11 we shall learn to balance redox equations and to carry out stoichiometric calculations using the balanced equations.

4-8 DISPLACEMENT REACTIONS

Reactions in which one element displaces another from a compound are called **displacement reactions.** These reactions are always redox reactions. The more readily a metal forms positive ions, the more active we say that it is.

Metallic silver formed by immersing a spiral of copper wire in a silver nitrate solution (Example 4-9a).

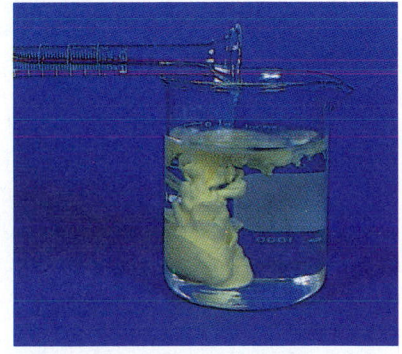

The reaction of $AgNO_3$(aq) and K_3PO_4(aq) is a precipitation reaction (Example 4-9c).

Table 4-13 *Activity Series of Some Elements*

Element	Common Reduced Form	Common Oxidized Forms
Li	Li	Li^+
K	K	K^+
Ca	Ca	Ca^{2+}
Na	Na	Na^+
Mg	Mg	Mg^{2+}
Al	Al	Al^{3+}
Mn	Mn	Mn^{2+}
Zn	Zn	Zn^{2+}
Cr	Cr	Cr^{3+}, Cr^{6+}
Fe	Fe	Fe^{2+}, Fe^{3+}
Cd	Cd	Cd^{2+}
Co	Co	Co^{2+}
Ni	Ni	Ni^{2+}
Sn	Sn	Sn^{2+}, Sn^{4+}
Pb	Pb	Pb^{2+}, Pb^{4+}
H (a nonmetal)	H_2	H^+
Sb (a metalloid)	Sb	Sb^{3+}
Cu	Cu	Cu^+, Cu^{2+}
Hg	Hg	Hg_2^{2+}, Hg^{2+}
Ag	Ag	Ag^+
Pt	Pt	Pt^{2+}, Pt^{4+}

Bracketed regions across the first column: Displace hydrogen from nonoxidizing acids (Li through Pb); Displace hydrogen from steam (Li through Fe); Displace hydrogen from cold water (Li through Na).

> Active metals displace less active metals or hydrogen from their compounds in aqueous solution to form the oxidized form of the more active metal and the reduced (free metal) form of the other metal or hydrogen.

In Table 4-13, the most active metals are listed at the top of the first column. These metals tend to react to form their oxidized forms (cations). Elements at the bottom of the activity series (the first column of Table 4-13) tend to remain in their reduced form. They are easily converted from their oxidized forms to their reduced forms.

$$1 \quad \begin{bmatrix} \textbf{More Active Metal +} \\ \textbf{Salt of Less Active Metal} \end{bmatrix} \longrightarrow \begin{bmatrix} \textbf{Less Active Metal +} \\ \textbf{Salt of More Active Metal} \end{bmatrix}$$

The reaction of copper with silver nitrate that was described in detail in Section 4-3 is typical. Please refer to it.

EXAMPLE 4-10 *Displacement Reaction*

A large piece of zinc metal is placed in a copper(II) sulfate, $CuSO_4$, solution. The blue solution becomes colorless as copper metal falls to the bottom of the container. The resulting solution contains zinc sulfate, $ZnSO_4$. Write balanced formula unit, total ionic, and net ionic equations for the reaction.

Plan

The metals zinc and copper are *not* ionized or dissociated in contact with H_2O. Both $CuSO_4$ and $ZnSO_4$ are soluble salts (solubility rule 5), and so they are written in ionic form.

Solution

$$CuSO_4(aq) + Zn(s) \longrightarrow Cu(s) + ZnSO_4(aq)$$

$$[Cu^{2+}(aq) + SO_4{}^{2-}(aq)] + Zn(s) \longrightarrow Cu(s) + [Zn^{2+}(aq) + SO_4{}^{2-}(aq)]$$

$$Cu^{2+}(aq) + Zn(s) \longrightarrow Cu(s) + Zn^{2+}(aq)$$

In this *displacement reaction,* the more active metal, zinc, displaces the ions of the less active metal, copper, from aqueous solution.

You should now work Exercise 60.

A strip of zinc was placed in a blue solution of copper(II) sulfate, $CuSO_4$. The copper has been displaced from solution and has fallen to the bottom of the beaker. The resulting zinc sulfate solution is colorless.

2 [Active Metal + Nonoxidizing Acid] ⟶ [Hydrogen + Salt of Acid]

A common method for the preparation of small amounts of hydrogen involves the reaction of active metals with nonoxidizing acids, such as HCl and H_2SO_4. For example, when zinc is dissolved in H_2SO_4, the reaction produces zinc sulfate; hydrogen is displaced from the acid, and it bubbles off as gaseous H_2. The formula unit equation for this reaction is

$$\underset{\text{strong acid}}{Zn(s) + H_2SO_4(aq)} \longrightarrow \underset{\text{soluble salt}}{ZnSO_4(aq)} + H_2(g)$$

Both sulfuric acid (in very dilute solution) and zinc sulfate exist primarily as ions; so the total ionic equation is

$$Zn(s) + [2H^+(aq) + SO_4{}^{2-}(aq)] \longrightarrow [Zn^{2+}(aq) + SO_4{}^{2-}(aq)] + H_2(g)$$

Elimination of unreacting species common to both sides of the total ionic equation gives the net ionic equation:

$$Zn(s) + 2H^+(aq) \longrightarrow Zn^{2+}(aq) + H_2(g)$$

Table 4-13 lists the **activity series.** When any metal listed above hydrogen in this series is added to a solution of a *nonoxidizing* acid such as hydrochloric acid, HCl, and sulfuric acid, H_2SO_4, the metal dissolves to produce hydrogen, and a salt is formed. HNO_3 is the common *oxidizing acid.* It reacts with active metals to produce oxides of nitrogen, but *not* hydrogen, H_2.

Zinc dissolves in dilute H_2SO_4 to produce H_2 and a solution that contains $ZnSO_4$.

EXAMPLE 4-11 *Displacement Reaction*

Which of the following metals can displace hydrogen from hydrochloric acid solution? Write appropriate equations for any reactions that can occur.

$$Al, \quad Cu, \quad Ag$$

Plan

The activity series of the metals, Table 4-13, tells us that copper and silver *do not* displace hydrogen from solutions of nonoxidizing acids. Aluminum is an active metal that can displace H_2 from HCl and form aluminum chloride (photo on page 372).

Figure 4-5 Potassium, like other Group IA metals, reacts vigorously with water. The room was completely dark, and all the light for this photograph was produced by dropping a small piece of potassium into a beaker of water.

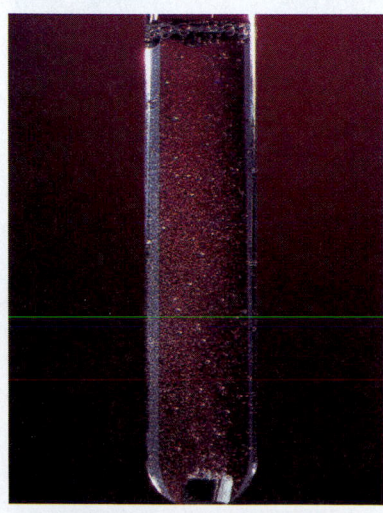

The reaction of calcium with water at room temperature produces a lazy stream of bubbles of hydrogen.

Solution

$$6HCl(aq) + 2Al(s) \longrightarrow 3H_2(g) + 2AlCl_3(aq)$$
$$6[H^+(aq) + Cl^-(aq)] + 2Al(s) \longrightarrow 3H_2(g) + 2[Al^{3+}(aq) + 3Cl^-(aq)]$$
$$6H^+(aq) + 2Al(s) \longrightarrow 3H_2(g) + 2Al^{3+}(aq)$$

You should now work Exercises 62 and 66.

Very active metals can even displace hydrogen from water. However, such reactions of very active metals of Group IA are dangerous because they generate enough heat to cause explosive ignition of the hydrogen (Figure 4-5). The reaction of potassium, or another metal of Group IA, with water is also a *displacement reaction:*

$$2K(s) + 2H_2O(\ell) \longrightarrow 2[K^+(aq) + OH^-(aq)] + H_2(g)$$

EXAMPLE 4-12 *Displacement Reaction*

Which of the following metals can displace hydrogen from water at room temperature? Write appropriate equations for any reactions that can occur.

$$Sn, \qquad Ca, \qquad Hg$$

Plan

The activity series, Table 4-13, tells us that tin and mercury *cannot* displace hydrogen from water. Calcium is a very active metal (Table 4-13) that displaces hydrogen from cold water and forms calcium hydroxide, a strong soluble base.

Solution

$$Ca(s) + 2H_2O(\ell) \longrightarrow H_2(g) + Ca(OH)_2(aq)$$
$$Ca(s) + 2H_2O(\ell) \longrightarrow H_2(g) + [Ca^{2+}(aq) + 2OH^-(aq)]$$
$$Ca(s) + 2H_2O(\ell) \longrightarrow H_2(g) + Ca^{2+}(aq) + 2OH^-(aq)$$

You should now work Exercise 67.

$$3 \left[\begin{matrix} \textbf{Active Nonmetal +} \\ \textbf{Salt of Less Active Nonmetal} \end{matrix} \right] \longrightarrow \left[\begin{matrix} \textbf{Less Active Nonmetal +} \\ \textbf{Salt of More Active Nonmetal} \end{matrix} \right]$$

Many *nonmetals* displace less active nonmetals from combination with a metal or other cation. For example, when chlorine is bubbled through a solution containing bromide ions (derived from a soluble ionic salt such as sodium bromide, NaBr), chlorine displaces bromide ions to form elemental bromine and chloride ions (as aqueous sodium chloride):

$$\underset{\text{chlorine}}{Cl_2(g)} + \underset{\text{sodium bromide}}{2[Na^+(aq) + Br^-(aq)]} \longrightarrow \underset{\text{sodium chloride}}{2[Na^+(aq) + Cl^-(aq)]} + \underset{\text{bromine}}{Br_2(\ell)}$$

Similarly, when bromine is added to a solution containing iodide ions, the iodide ions are displaced by bromine to form iodine and bromide ions:

$$\underset{\text{bromine}}{Br_2(\ell)} + \underset{\text{sodium iodide}}{2[Na^+(aq) + I^-(aq)]} \longrightarrow \underset{\text{sodium bromide}}{2[Na^+(aq) + Br^-(aq)]} + \underset{\text{iodine}}{I_2(s)}$$

Each halogen will displace less active (heavier) halogens from their binary salts; i.e., the order of increasing activities is

$$I_2 < Br_2 < Cl_2 < F_2$$

Activity of the halogens decreases going down the group in the periodic table.

Conversely, a halogen will *not* displace more active (lighter) members from their salts:

$$I_2(s) + 2F^- \longrightarrow \text{no reaction}$$

EXAMPLE 4-13   *Displacement Reactions*

Which of the following combinations would result in a displacement reaction? Write appropriate equations for any reactions that occur.
(a) $I_2(s)$ $+ NaBr(aq) \longrightarrow$
(b) $Cl_2(g) + NaI(aq)$ $\longrightarrow$
(c) $Br_2(\ell) + NaCl(aq) \longrightarrow$

Plan
The activity of the halogens decreases from top to bottom in the periodic table. We see (a) that Br is above I and (c) that Cl is above Br in the periodic table. Therefore neither combination (a) nor combination (c) could result in reaction. Cl is above I in the periodic table, and so combination (b) results in a displacement reaction.

Solution
The more active halogen, Cl_2, displaces the less active halogen, I_2, from its compounds.

$$2NaI(aq) + Cl_2(g) \longrightarrow I_2(s) + 2NaCl(aq)$$
$$2[Na^+(aq) + I^-(aq)] + Cl_2(g) \longrightarrow I_2(s) + 2[Na^+(aq) + Cl^-(aq)]$$
$$2I^-(aq) + Cl_2(g) \longrightarrow I_2(s) + 2Cl^-(aq)$$

You should now work Exercise 66.

NAMING INORGANIC COMPOUNDS

The rules for naming inorganic compounds were set down by the Committee on Inorganic Nomenclature of the International Union of Pure and Applied Chemistry (IUPAC). The names and formulas of several organic compounds were given in Table 2-2.

Because millions of compounds are known, it is important to be able to associate names and formulas in a systematic way.

4-9 NAMING BINARY COMPOUNDS

Binary compounds consist of two elements; they may be either ionic or molecular. The rule is to name the more metallic element first and the less metallic element second. The less metallic element is named by adding an "-ide" suffix to the element's *unambiguous* stem. Stems for the nonmetals follow.

The stem for each element is derived from the name of the element.

IIIA		IVA		VA		VIA		VIIA	
								H	hydr
B	bor	C	carb	N	nitr	O	ox	F	fluor
		Si	silic	P	phosph	S	sulf	Cl	chlor
				As	arsen	Se	selen	Br	brom
				Sb	antimon	Te	tellur	I	iod

Binary ionic compounds contain metal cations and nonmetal anions. The cation is named first and the anion second according to the rule described.

Formula	Name	Formula	Name
KBr	potassium bromide	Rb_2S	rubidium sulfide
$CaCl_2$	calcium chloride	Al_2Se_3	aluminum selenide
NaH	sodium hydride	SrO	strontium oxide

The preceding method is sufficient for naming binary ionic compounds containing metals that exhibit *only one oxidation number* other than zero (Section 4-6). Most transition metals and the metals of Groups IIIA (except Al), IVA, and VA, exhibit more than one oxidation number. These metals may form two or more binary compounds with the same nonmetal. To distinguish among all the possibilities, the oxidation number of the metal is indicated by a Roman numeral in parentheses following its name. This method can be applied to any binary compound of a metal and a nonmetal, whether the compound is ionic or molecular.

Roman numerals are *not* necessary for metals that commonly exhibit only one oxidation number in their compounds.

Formula	Ox. No. of Metal	Name	Formula	Ox. No. of Metal	Name
Cu_2O	+1	copper(I) oxide	$SnCl_2$	+2	tin(II) chloride
CuF_2	+2	copper(II) fluoride	$SnCl_4$	+4	tin(IV) chloride
FeS	+2	iron(II) sulfide	PbO	+2	lead(II) oxide
Fe_2O_3	+3	iron(III) oxide	PbO_2	+4	lead(IV) oxide

The advantage of the IUPAC system is that if you know the formula you can write the exact and unambiguous name; if you are given the name you can write the formula at once. An older method, still in use but not recommended by the IUPAC, uses ''-ous'' and ''-ic'' suffixes to indicate lower and higher oxidation numbers, respectively. This system can distinguish between only two different oxidation numbers for a metal. Therefore, it is not as useful as the Roman numeral system.

Familiarity with the older system is still necessary. It is still widely used in many scientific, engineering, and medical fields.

Formula	Ox. No. of Metal	Name	Formula	Ox. No. of Metal	Name
CuCl	+1	cuprous chloride	SnF_2	+2	stannous fluoride
$CuCl_2$	+2	cupric chloride	SnF_4	+4	stannic fluoride
FeO	+2	ferrous oxide	Hg_2Cl_2	+1	mercurous chloride
$FeBr_3$	+3	ferric bromide	$HgCl_2$	+2	mercuric chloride

The common pseudobinary anions
hydroxide OH^-
cyanide CN^-
thiocyanate SCN^-
The common pseudobinary cation
ammonium NH_4^+

Pseudobinary ionic compounds contain more than two elements. In these compounds one or more of the ions consist of more than one element but behave as simple ions. Some common examples of such anions are the hydroxide ion, OH^-; the cyanide ion, CN^-; and the thiocyanate ion, SCN^-. As before, the name of the anion ends in ''-ide.'' The ammonium ion, NH_4^+, is the common cation that behaves like a simple metal cation.

Formula	Name	Formula	Name
NH_4I	ammonium iodide	NH_4CN	ammonium cyanide
$Ca(CN)_2$	calcium cyanide	$Cu(OH)_2$	copper(II) hydroxide or cupric hydroxide
NaOH	sodium hydoxide	$Fe(OH)_3$	iron(III) hydroxide or ferric hydroxide

Nearly all **binary molecular compounds** involve two *nonmetals* bonded together. Although many nonmetals can exhibit different oxidation numbers, their oxidation numbers are *not* properly indicated by Roman numerals or suffixes. Instead, elemental proportions in binary covalent compounds are indicated by using a *prefix* system for both elements. The Greek and Latin prefixes used are mono, di, tri, tetra, penta, hexa, hepta, octa, nona, and deca. The prefix "mono-" is omitted for both elements except in the common name for CO, carbon monoxide. We use the minimum number of prefixes needed to name a compound unambiguously.

If you don't already know them, you should learn these common prefixes.

Number	Prefix
2	di
3	tri
4	tetra
5	penta
6	hexa
7	hepta
8	octa
9	nona
10	deca

Formula	Name	Formula	Name
SO_2	sulfur dioxide	Cl_2O_7	dichlorine heptoxide
SO_3	sulfur trioxide	CS_2	carbon disulfide
N_2O_4	dinitrogen tetroxide	As_4O_6	tetraarsenic hexoxide

Binary acids are compounds in which H is bonded to a Group VIA or VIIA element; they act as acids when dissolved in water. The pure compounds are named as typical binary compounds. Their aqueous solutions are named by modifying the characteristic stem of the nonmetal with the prefix "hydro-" and the suffix "-ic" followed by the word "acid." The stem for sulfur in this instance is "sulfur" rather than "sulf."

Formula	Name of Compound	Name of Aqueous Solution
HCl	hydrogen chloride	hydrochloric acid, HCl(aq)
HF	hydrogen fluoride	hydrofluoric acid, HF(aq)
H_2S	hydrogen sulfide	hydrosulfuric acid, H_2S(aq)
HCN	hydrogen cyanide	hydrocyanic acid, HCN(aq)

A list of common cations and anions appears in Table 4-14. It will enable you to name many of the ionic compounds you encounter. In later chapters we shall learn additional systematic rules for naming more complex compounds.

4-10 NAMING TERNARY ACIDS AND THEIR SALTS

A ternary compound consists of three elements. *Ternary acids* (**oxoacids**) are compounds of hydrogen, oxygen, and (usually) a nonmetal. Nonmetals that exhibit more than one oxidation state form more than one ternary acid. These ternary acids differ in the number of oxygen atoms they contain. The suffixes "-ous" and "-ic" following the stem name of the central element indicate lower and higher oxidation states, respectively. One common ternary acid of each nonmetal is (somewhat arbitrarily) designated as the "-ic" acid. That is, it is named *"stemic acid."* The common ternary "-ic acids" are shown below. There are no common "-ic" ternary acids for the omitted nonmetals.

It is important to learn the names and formulas of these acids, because the names of all other ternary acids and salts are derived from them.

The oxoacid with the central element in the highest oxidation state usually contains more O atoms. Oxoacids with their central elements in lower oxidation states usually have fewer O atoms.

Table 4-14 *Formulas, Ionic Charges, and Names for Some Common Ions*

Common Cations			Common Anions		
Formula	**Charge**	**Name**	**Formula**	**Charge**	**Name**
Li^+	1+	lithium ion	F^-	1−	fluoride ion
Na^+	1+	sodium ion	Cl^-	1−	chloride ion
K^+	1+	potassium ion	Br^-	1−	bromide ion
NH_4^+	1+	ammonium ion	I^-	1−	iodide ion
Ag^+	1+	silver ion	OH^-	1−	hydroxide ion
			CN^-	1−	cyanide ion
Mg^{2+}	2+	magnesium ion	ClO^-	1−	hypochlorite ion
Ca^{2+}	2+	calcium ion	ClO_2^-	1−	chlorite ion
Ba^{2+}	2+	barium ion	ClO_3^-	1−	chlorate ion
Cd^{2+}	2+	cadmium ion	ClO_4^-	1−	perchlorate ion
Zn^{2+}	2+	zinc ion	CH_3COO^-	1−	acetate ion
Cu^{2+}	2+	copper(II) ion or cupric ion	MnO_4^-	1−	permanganate ion
Hg_2^{2+}	2+	mercury(I) ion or mercurous ion	NO_2^-	1−	nitrite ion
Hg^{2+}	2+	mercury(II) ion or mercuric ion	NO_3^-	1−	nitrate ion
Mn^{2+}	2+	manganese(II) ion or manganous ion	SCN^-	1−	thiocyanate ion
Co^{2+}	2+	cobalt(II) ion or cobaltous ion			
Ni^{2+}	2+	nickel(II) ion or nickelous ion	O^{2-}	2−	oxide ion
Pb^{2+}	2+	lead(II) ion or plumbous ion	S^{2-}	2−	sulfide ion
Sn^{2+}	2+	tin(II) ion or stannous ion	HSO_3^-	1−	hydrogen sulfite ion or bisulfite ion
Fe^{2+}	2+	iron(II) ion or ferrous ion	SO_3^{2-}	2−	sulfite ion
			HSO_4^-	1−	hydrogen sulfate ion or bisulfate ion
Fe^{3+}	3+	iron(III) ion or ferric ion	SO_4^{2-}	2−	sulfate ion
Al^{3+}	3+	aluminum ion	HCO_3^-	1−	hydrogen carbonate ion or bicarbonate ion
Cr^{3+}	3+	chromium(III) ion or chromic ion	CO_3^{2-}	2−	carbonate ion
			CrO_4^{2-}	2−	chromate ion
			$Cr_2O_7^{2-}$	2−	dichromate ion
			PO_4^{3-}	3−	phosphate ion
			AsO_4^{3-}	3−	arsenate ion

Periodic Group of Central Elements

IIIA	IVA	VA	VIA	VIIA
(+3) H_3BO_3 boric acid	(+4) H_2CO_3 carbonic acid	(+5) HNO_3 nitric acid		
	(+4) H_4SiO_4 silicic acid	(+5) H_3PO_4 phosphoric acid	(+6) H_2SO_4 sulfuric acid	(+5) $HClO_3$ chloric acid
		(+5) H_3AsO_4 arsenic acid	(+6) H_2SeO_4 selenic acid	(+5) $HBrO_3$ bromic acid
			(+6) H_6TeO_6 telluric acid	(+5) HIO_3 iodic acid

Note that the oxidation state of the central atom is equal to its periodic group number, except for the halogens.

Acids containing *one fewer oxygen atom* per central atom are named in the same way except that the "-ic" suffix is changed to "-ous." The oxidation number of the central element is *lower by 2* in the "-ous" acid than in the "-ic" acid.

Formula	Ox. No.	Name	Formula	Ox. No.	Name
H_2SO_3	+4	sulfur*ous* acid	H_2SO_4	+6	sulfur*ic* acid
HNO_2	+3	nitr*ous* acid	HNO_3	+5	nitr*ic* acid
H_2SeO_3	+4	selen*ous* acid	H_2SeO_4	+6	selen*ic* acid
$HBrO_2$	+3	brom*ous* acid	$HBrO_3$	+5	brom*ic* acid

Ternary acids that have one fewer O atom than the "-ous" acids (two fewer O atoms than the "-ic" acids) are named using the prefix "hypo-" and the suffix "-ous." These are acids in which the oxidation state of the central nonmetal is lower *by 2* than that of the central nonmetal in the "-ous acids."

Formula	Ox. No.	Name
HClO	+1	*hypo*chlor*ous* acid
H_3PO_2	+1	*hypo*phosphor*ous* acid
HIO	+1	*hypo*iod*ous* acid
$H_2N_2O_2$	+1	*hypo*nitr*ous* acid

Notice that $H_2N_2O_2$ has a 1:1 ratio of nitrogen to oxygen, as would the hypothetical HNO.

Acids containing *one more oxygen atom* per central nonmetal atom than the normal "-ic acid" are named "*per*stem*ic*" acids.

Formula	Ox. No.	Name
$HClO_4$	+7	*per*chlor*ic* acid
$HBrO_4$	+7	*per*brom*ic* acid
HIO_4	+7	*per*iod*ic* acid

The oxoacids of chlorine follow.

Formula	Ox. No.	Name
HClO	+1	*hypo*chlor*ous* acid
$HClO_2$	+3	chlor*ous* acid
$HClO_3$	+5	chlor*ic* acid
$HClO_4$	+7	*per*chlor*ic* acid

Ternary salts are compounds that result from replacing the hydrogen in a ternary acid with another ion. They usually contain metal cations or the ammonium ion. As with binary compounds, the cation is named first. The name of the anion is based on the name of the ternary acid from which it is derived.

An anion derived from a ternary acid with an "-ic" ending is named by dropping the "-ic acid" and replacing it with "-ate." An anion derived from an "-ous acid" is named by replacing the suffix "-ous acid" with "-ite." The "per-" and "hypo-" prefixes are retained.

Formula	Name
$(NH_4)_2SO_4$	ammonium sulfate (SO_4^{2-}, from H_2SO_4)
KNO_3	potassium nitrate (NO_3^-, from HNO_3)
$Ca(NO_2)_2$	calcium nitrite (NO_2^-, from HNO_2)
$LiClO_4$	lithium perchlorate (ClO_4^-, from $HClO_4$)
$FePO_4$	iron(III) phosphate (PO_4^{3-}, from H_3PO_4)
$NaClO$	sodium hypochlorite (ClO^-, from $HClO$)

Acidic salts contain anions derived from ternary acids in which one or more acidic hydrogen atoms remain. These salts are named as if they were the usual type of ternary salt, with the word "hydrogen" or "dihydrogen" inserted after the name of the cation to show the number of acidic hydrogen atoms.

Formula	Name	Formula	Name
$NaHSO_4$	sodium hydrogen sulfate	KH_2PO_4	potassium dihydrogen phosphate
$NaHSO_3$	sodium hydrogen sulfite	K_2HPO_4	potassium hydrogen phosphate
		$NaHCO_3$	sodium hydrogen carbonate

An older, commonly used method (which is not recommended by the IUPAC, but which is widely used in commerce) involves the use of the prefix "bi-" attached to the name of the anion to indicate the presence of an acidic hydrogen. According to this system, $NaHSO_4$ is called sodium bisulfate and $NaHCO_3$ is called sodium bicarbonate.

▼ **PROBLEM-SOLVING TIP** *Naming Ternary Acids and Their Anions*

The following table might help you to remember the names of the ternary acids and their ions. First learn the formulas of the acids mentioned above that end with -ic acid. Then relate possible other acids to the following table. The stem (XXX) represents the stem of the name, e.g., "nitr," "sulfur," or "chlor."

	Ternary Acid	Anion	
Decreasing oxidation number of central atom →	*per*XXX*ic* acid	*per*XXX*ate*	Decreasing number of oxygen atoms on central atom →
	XXX*ic* acid	XXX*ate*	
	XXX*ous* acid	XXX*ite*	
	*hypo*XXX*ous* acid	*hypo*XXX*ite*	

Key Terms

Acid A substance that produces H^+(aq) ions in aqueous solution. Strong acids ionize completely or almost completely in dilute aqueous solution. Weak acids ionize only slightly.

Active metal A metal that readily loses electrons to form cations.

Activity series A listing of metals (and hydrogen) in order of decreasing activity.

Alkali metals Elements of Group IA in the periodic table, except hydrogen.

Alkaline earth metals Group IIA elements in the periodic table.

Atomic number The number of protons in the nucleus of an atom of an element.

Base A substance that produces OH⁻(aq) ions in aqueous solution. Strong soluble bases are soluble in water and are completely *dissociated*. Weak bases ionize only slightly.

Binary acid A binary compound in which H is bonded to one of the less metallic nonmetals.

Binary compound A compound consisting of two elements; may be ionic or molecular.

Chemical periodicity The variation in properties of elements with their positions in the periodic table.

Combustion reaction A highly exothermic reaction of a substance with oxygen, usually with a visible flame.

Displacement reaction A reaction in which one element displaces another from a compound.

Disproportionation reaction A redox reaction in which the oxidizing agent and the reducing agent are the same element.

Dissociation In aqueous solution, the process in which a solid *ionic compound* separates into its ions.

Electrolyte A substance whose aqueous solutions conduct electricity.

Formula unit equation An equation for a chemical reaction in which all formulas are written as complete formulas.

Group (family) The elements in a vertical column of the periodic table.

Halogens Group VIIA elements in the periodic table.

Ionization In aqueous solution, the process in which a *molecular compound* separates to form ions.

Metal An element below and to the left of the stepwise division (metalloids) in the upper right corner of the periodic table; about 80% of the known elements are metals.

Metalloids Elements with properties intermediate between metals and nonmetals: B, Si, Ge, As, Sb, Te, Po, and At.

Net ionic equation An equation that results from canceling spectator ions and eliminating brackets from a total ionic equation.

Neutralization The reaction of an acid with a base to form a salt. Often, the reaction of hydrogen ions with hydroxide ions to form water molecules.

Noble (rare) gases Elements of Group 0 in the periodic table.

Nonelectrolyte A substance whose aqueous solutions do not conduct electricity.

Nonmetals Elements above and to the right of the metalloids in the periodic table.

Oxidation An algebraic increase in oxidation number; corresponds to a loss of electrons.

Oxidation numbers Arbitrary numbers that can be used as mechanical aids in writing formulas and balancing equations; for single-atom ions they correspond to the charge on the ion; less metallic atoms are assigned negative oxidation numbers.

Oxidation states See *Oxidation numbers*.

Oxidation–reduction reaction A reaction in which oxidation and reduction occur; also called redox reactions.

Oxidizing agent The substance that oxidizes another substance and is reduced.

Period The elements in a horizontal row of the periodic table.

Periodicity Regular periodic variations of properties of elements with atomic number (and position in the periodic table).

Periodic law The properties of the elements are periodic functions of their atomic numbers.

Periodic table An arrangement of elements in order of increasing atomic number that also emphasizes periodicity.

Precipitate An insoluble solid that forms and separates from a solution.

Precipitation reaction A reaction in which a precipitate forms.

Pseudobinary ionic compound A compound that contains more than two elements but is named like a binary compound.

Redox reaction See *Oxidation–reduction reaction.*

Reducing agent The substance that reduces another substance and is oxidized.

Reduction An algebraic decrease in oxidation number; corresponds to a gain of electrons.

Reversible reaction A reaction that occurs in both directions; indicated by double arrows ($\rightleftharpoons$).

Salt A compound that contains a cation other than H⁺ and an anion other than OH⁻ or O^{2-}.

Semiconductor A substance that does not conduct electricity at low temperatures but does so at higher temperatures.

Spectator ions Ions in solution that do not participate in a chemical reaction.

Strong acid An acid that ionizes (separates into ions) completely, or very nearly completely, in dilute aqueous solution.

Strong electrolyte A substance that conducts electricity well in dilute aqueous solution.

Strong soluble base Metal hydroxide that is soluble in water and dissociates completely in dilute aqueous solution.

Ternary acid A ternary compound containing H, O, and another element, often a nonmetal.

Ternary compound A compound consisting of three elements; may be ionic or molecular.

Total ionic equation An equation for a chemical reaction written to show the predominant form of all species in aqueous solution or in contact with water.

Weak acid An acid that ionizes only slightly in dilute aqueous solution.

Weak base A molecular substance that ionizes only slightly in water to produce an alkaline (base) solution.

Weak electrolyte A substance that conducts electricity poorly in dilute aqueous solution.

Exercises

The Periodic Table

1. State the periodic law. What does it mean?
2. What was Mendeleev's contribution to the construction of the modern periodic table?
3. Consult a handbook of chemistry and look up melting points of the elements of periods 2 and 3. Show that melting point is a property that varies periodically for these elements.

*4. Mendeleev's periodic table was based on increasing atomic weight. Argon has a higher atomic weight than potassium, yet in the modern table argon appears before potassium. Explain how this can be.

5. Estimate the density of antimony from the following densities (g/cm^3): As, 5.72; Bi, 9.8; Sn, 7.30; Te, 6.24. Show how you arrived at your answer.

6. Given the following melting points in °C, estimate the value for CBr$_4$: CF$_4$, −184; CCl$_4$, −23; CI$_4$, 171 (decomposes).

7. Calcium and magnesium form the following compounds: CaCl$_2$, MgCl$_2$, CaO, MgO, Ca$_3$N$_2$, and Mg$_3$N$_2$. Predict the formula for a compound of (a) barium and sulfur, (b) strontium and iodine.

8. The formulas of some hydrides of second-period representative elements are as follows: BeH$_2$, BH$_3$, CH$_4$, NH$_3$, H$_2$O, HF. A famous test in criminology laboratories for the presence of arsenic (As) involves the formation of arsine, the hydride of arsenic. Predict the formula of arsine.

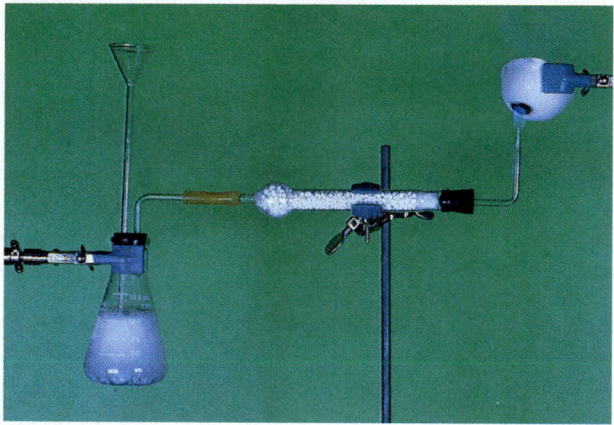

Arsine burns to form a dark spot.

9. Distinguish between the following terms clearly and concisely, and provide specific examples of each: groups (families) of elements, and periods of elements.

10. Write names and symbols for (a) the alkaline earth metals, (b) the Group IVA elements, (c) the Group VIB elements.

11. Write names and symbols for (a) the alkali metals, (b) the noble gases, (c) the Group IIIA elements.

12. Define and illustrate the following terms clearly and concisely: (a) metals, (b) nonmetals, (c) halogens.

Aqueous Solutions

13. Define and distinguish among (a) strong electrolytes, (b) weak electrolytes, and (c) nonelectrolytes.

14. Three common classes of compounds are electrolytes. Name them and give an example of each.

15. Define (a) acids, (b) bases, (c) salts, and (d) molecular compounds.

16. How can a salt be related to a particular acid and a particular base?

17. List the names and formulas of the common strong acids.

18. Write equations for the ionization of the following acids: (a) hydrochloric acid, (b) nitric acid, (c) chloric acid.

19. List names and formulas of five weak acids.

20. List names and formulas of the common strong soluble bases.

21. Write equations for the ionization of the following acids. Which ones ionize only slightly? (a) HF, (b) HNO$_2$, (c) CH$_3$COOH, (d) HNO$_3$.

22. The most common weak base is present in a common household chemical. Write the equation for the ionization of this weak base.

23. Summarize the electrical properties of strong electrolytes, weak electrolytes, and nonelectrolytes.

24. Write the formulas of two soluble and two insoluble chlorides, sulfates, and hydroxides.

25. Describe an experiment for classifying each of these compounds as a strong electrolyte, a weak electrolyte, or a nonelectrolyte: K$_2$CO$_3$, HCN, CH$_3$OH, H$_2$S, H$_2$SO$_4$, NH$_3$. Predict and explain the expected results.

26. (a) Which of these are acids? HBr, NH$_3$, H$_2$SeO$_4$, BF$_3$, H$_3$SbO$_4$, Al(OH)$_3$, H$_2$S, C$_6$H$_6$, CsOH, H$_3$BO$_3$, HCN. (b) Which of these are bases? NaOH, H$_2$Se, BCl$_3$, NH$_3$.

*27. Classify each substance as either an electrolyte or a nonelectrolyte: NH$_4$Cl, HI, C$_6$H$_6$, Zn(CH$_3$COO)$_2$, Cu(NO$_3$)$_2$, CH$_3$COOH, C$_{12}$H$_{22}$O$_{11}$ (sugar), LiOH, KHCO$_3$, CCl$_4$, La$_2$(SO$_4$)$_3$, I$_2$.

*28. Classify each substance as either a strong or weak electrolyte, and then list (a) the strong acids, (b) the strong bases, (c) the weak acids, and (d) the weak bases. NaCl, MgSO$_4$, HCl, H$_2$C$_2$O$_4$, Ba(NO$_3$)$_2$, H$_3$PO$_4$, RbOH, HNO$_3$, HI, Ba(OH)$_2$, LiOH, C$_2$H$_5$COOH, NH$_3$, KOH, Mg(CH$_3$COO)$_2$, HCN, HClO$_4$.

29. Based on the solubility rules given in Table 4-8, how would you write the formulas for the following substances in a net ionic equation? (a) PbSO$_4$, (b) Na(CH$_3$COO), (c) Na$_2$CO$_3$, (d) MnS, (e) BaCl$_2$.

30. Repeat Exercise 29 for the following: (a) (NH$_4$)$_2$SO$_4$, (b) NaBr, (c) Sr(OH)$_2$, (d) Mg(OH)$_2$, (e) K$_2$CO$_3$.

Precipitation Reactions

Refer to the solubility rules on page 124. Classify the compounds in Exercises 31 through 34 as soluble, moderately soluble, or insoluble in water.

31. (a) NaClO$_4$, (b) AgCl, (c) Pb(NO$_3$)$_2$, (d) KOH, (e) CaSO$_4$

32. (a) BaSO$_4$, (b) Al(NO$_3$)$_3$, (c) CuS, (d) Na$_2$S, (e) Ca(CH$_3$COO)$_2$

33. (a) Fe(NO$_3$)$_3$, (b) Hg(CH$_3$COO)$_2$, (c) BeCl$_2$, (d) CuSO$_4$, (e) CaCO$_3$

34. (a) KClO$_3$, (b) NH$_4$Cl, (c) NH$_3$, (d) HNO$_2$, (e) PbS

Exercises 35 and 36 describe precipitation reactions *in aqueous solutions*. For each, write balanced (i) formula unit, (ii) total ionic, and (iii) net ionic equations. Refer to the solubility rules as necessary.

35. (a) Black-and-white photographic film contains some silver bromide, which can be formed by the reaction of sodium bromide with silver nitrate.

(b) Barium sulfate is used when X-rays of the gastrointestinal tract are made. Barium sulfate can be prepared by reacting barium chloride with dilute sulfuric acid.

(c) In water purification, small solid particles are often "trapped" as aluminum hydroxide precipitates and falls to the bottom of the sedimentation pool. Aluminum sulfate reacts with calcium hydroxide (from lime) to form aluminum hydroxide and calcium sulfate.

*36. (a) Our bones are mostly calcium phosphate. Calcium chloride reacts with potassium phosphate to form calcium phosphate and potassium chloride.

(b) Mercury compounds are very poisonous. Mercury(II) nitrate reacts with sodium sulfide to form mercury(II) sulfide, which is very insoluble, and sodium nitrate.

(c) Chromium(III) ions are very poisonous. They can be removed from solution by precipitating very insoluble chromium(III) hydroxide. Chromium(III) chloride reacts with calcium hydroxide to form chromium(III) hydroxide and calcium chloride.

In Exercises 37 and 38, write balanced (i) formula unit, (ii) total ionic, and (iii) net ionic equations for the reactions that occur when *aqueous solutions* of the compounds are mixed.

37. (a) $Ba(NO_3)_2 + K_2CO_3 \rightarrow$
    (b) $NaOH + CoCl_2 \rightarrow$
    (c) $Al_2(SO_4)_3 + NaOH \rightarrow$
38. (a) $Cu(NO_3)_2 + Na_2S \rightarrow$
    (b) $CdSO_4 + H_2S \rightarrow$
    (c) $Bi_2(SO_4)_3 + (NH_4)_2S \rightarrow$

39. In each of the following, both compounds are water soluble. Predict whether a precipitate will form when solutions of the two are mixed, and, if so, identify the compound that precipitates. (a) $Pb(NO_3)_2$, NaI, (b) $Ba(NO_3)_2$, KCl, (c) $(NH_4)_2S$, $AgNO_3$

40. In each of the following, both compounds are water soluble. Predict whether a precipitate will form when solutions of the two are mixed, and, if so, identify the compound that precipitates. (a) NH_4Br, $Hg_2(NO_3)_2$, (b) KOH, Na_2S, (c) Cs_2SO_4, $MgCl_2$

Acid–Base Reactions

In Exercises 41 through 44, write balanced (i) formula unit, (ii) total ionic, and (iii) net ionic equations for the reactions that occur between the acid and the base. Assume that all reactions occur in water or in contact with water.

41. (a) hydrochloric acid + calcium hydroxide
    (b) dilute sulfuric acid + potassium hydroxide
    (c) perchloric acid + aqueous ammonia
42. (a) acetic acid + sodium hydroxide
    (b) sulfurous acid + sodium hydroxide
    (c) hydrofluoric acid + lithium hydroxide
*43. (a) potassium hydroxide + hydrosulfuric acid
    (b) barium hydroxide + hydrosulfuric acid
    (c) lead(II) hydroxide + hydrosulfuric acid
44. (a) sodium hydroxide + sulfuric acid
    (b) calcium hydroxide + phosphoric acid
    (c) copper(II) hydroxide + nitric acid

In Exercises 45 through 48, write balanced (i) formula unit, (ii) total ionic, and (iii) net ionic equations for the reaction of an acid and a base that will produce the indicated salts.

45. (a) sodium chloride, (b) sodium phosphate, (c) barium acetate
46. (a) calcium perchlorate, (b) ammonium sulfate, (c) copper(II) acetate
*47. (a) sodium carbonate, (b) barium carbonate, (c) nickel(II) nitrate
*48. (a) sodium sulfide, (b) aluminum phosphate, (c) lead(II) arsenate

49. Write a balanced equation for the preparation of each of the following salts by a neutralization reaction. SrC_2O_4 is insoluble in water. $Ca(NO_3)_2$, SrC_2O_4, $ZnSO_3$, Na_2CO_3.
50. Write the formulas for the acid and the base that could react to form each of the following *insoluble* salts. (a) $NiCO_3$, (b) Ag_2CrO_4, (c) $Hg_3(PO_4)_2$.

Oxidation Numbers

51. Assign oxidation numbers to the element specified in each group of compounds.
    (a) N in NO, N_2O_3, N_2O_4, NH_3, N_2H_4, NH_2OH, HNO_2, HNO_3
    (b) C in CO, CO_2, CH_2O, CH_4O, C_2H_6O, $(COOH)_2$, Na_2CO_3
    (c) S in S_8, H_2S, SO_2, SO_3, Na_2SO_3, H_2SO_4, K_2SO_4
52. Assign oxidation numbers to the element specified in each group of compounds.
    (a) P in PCl_3, P_4O_6, P_4O_{10}, HPO_3, H_3PO_3, $POCl_3$, $H_4P_2O_7$, $Mg_3(PO_4)_2$
    (b) Cl in Cl_2, HCl, HClO, $HClO_2$, $KClO_3$, Cl_2O_7, $Ca(ClO_4)_2$
    (c) Mn in MnO, MnO_2, $Mn(OH)_2$, K_2MnO_4, $KMnO_4$, Mn_2O_7
    (d) O in OF_2, Na_2O, Na_2O_2, KO_2
53. Assign oxidation numbers to the element specified in each group of ions.
    (a) S in S^{2-}, SO_3^{2-}, SO_4^{2-}, $S_2O_3^{2-}$, $S_4O_6^{2-}$, HS^-
    (b) Cr in CrO_2^-, $Cr(OH)_4^-$, CrO_4^{2-}, $Cr_2O_7^{2-}$
    (c) B in BO_2^-, BO_3^{3-}, $B_4O_7^{2-}$
54. Assign oxidation numbers to the element specified in each group of ions.
    (a) N in N^{3-}, NO_2^-, NO_3^-, N_3^-, NH_4^+
    (b) Br in Br^-, BrO^-, BrO_2^-, BrO_3^-, BrO_4^-

Oxidation–Reduction Reactions

55. Define and illustrate the following terms: (a) oxidation, (b) reduction, (c) oxidizing agent, (d) reducing agent.
56. Why must oxidation and reduction always occur simultaneously in chemical reactions?
57. Determine which of the following are oxidation–reduction reactions. For those that are, identify the oxidizing and reducing agents.
    (a) $3Zn(s) + 2CoCl_3(aq) \rightarrow 3ZnCl_2(aq) + 2Co(s)$
    (b) $ICl(s) + H_2O(\ell) \rightarrow HCl(aq) + HIO(aq)$
    (c) $3HCl(aq) + HNO_3(aq) \rightarrow Cl_2(g) + NOCl(g) + 2H_2O(\ell)$
    (d) $Fe_2O_3(s) + 3CO(g) \xrightarrow{\Delta} 2Fe(s) + 3CO_2(g)$
58. Determine which of the following are oxidation–reduction reactions. For those that are, identify the oxidizing and reducing agents.
    (a) $HgCl_2(aq) + 2KI(aq) \rightarrow HgI_2(s) + 2KCl(aq)$

(b) $4NH_3(g) + 3O_2(g) \rightarrow 2N_2(g) + 6H_2O(g)$

(c) $CaCO_3(s) + 2HNO_3(aq) \rightarrow$
$$Ca(NO_3)_2(aq) + CO_2(g) + H_2O(\ell)$$

(d) $PCl_3(\ell) + 3H_2O(\ell) \rightarrow 3HCl(aq) + H_3PO_3(aq)$

Displacement Reactions

59. Which of the following would displace hydrogen when a piece of the metal is dropped into dilute H_2SO_4 solution? Write balanced net ionic equations for the reactions: Zn, Cu, Fe, Al.

60. Which of the following metals would displace copper from an aqueous solution of copper(II) sulfate? Write balanced net ionic equations for the reactions: Hg, Zn, Fe, Pt.

61. Arrange the metals listed in Exercise 59 in order of increasing activity.

62. Arrange the metals listed in Exercise 60 in order of increasing activity.

63. Which of the following metals would displace hydrogen from cold water? Write balanced net ionic equations for the reactions: Zn, Na, Ca, Cr.

64. Arrange the metals listed in Exercise 63 in order of increasing activity.

65. What is the order of increasing activity of the halogens?

66. Of the possible displacement reactions shown, which one(s) could occur?

(a) $2Cl^-(aq) + Br_2(\ell) \rightarrow 2Br^-(aq) + Cl_2(g)$

(b) $2Br^-(aq) + F_2(g) \rightarrow 2F^-(aq) + Br_2(\ell)$

(c) $2I^-(aq) + Cl_2(g) \rightarrow 2Cl^-(aq) + I_2(s)$

(d) $2Br^-(aq) + Cl_2(g) \rightarrow 2Cl^-(aq) + Br_2(\ell)$

67. (a) Name two common metals—one that *does not* displace hydrogen from water, and one that *does not* displace hydrogen from water or acid solutions.

(b) Name two common metals—one that *does* displace hydrogen from water, and one that displaces hydrogen from acid solutions but not from water. Write net ionic equations for the reactions that occur.

68. Predict the products of each mixture. If a reaction occurs, write the net ionic equation. If no reaction occurs, write "no reaction."

(a) $Cd^{2+}(aq) + Al \rightarrow$

(b) $K + H_2O \rightarrow$

(c) $Ni + H_2O \rightarrow$

(d) $Hg + HCl(aq) \rightarrow$

(e) $Ni + H_2SO_4(aq) \rightarrow$

(f) $Fe + H_2SO_4(aq) \rightarrow$

69. Use the activity series to predict whether or not the following reactions will occur:

(a) $Cu(s) + Mg^{2+} \rightarrow Mg(s) + Cu^{2+}$

(b) $Ni(s) + Cu^{2+} \rightarrow Ni^{2+} + Cu(s)$

(c) $Cu(s) + 2H^+ \rightarrow Cu^{2+} + H_2(g)$

(d) $Mg(s) + H_2O(g) \rightarrow MgO(s) + H_2(g)$

70. Repeat Exercise 69 for

(a) $Sn(s) + Ca^{2+} \rightarrow Sn^{2+} + Ca(s)$

(b) $Al_2O_3(s) + 3H_2(g) \xrightarrow{\Delta} 2Al(s) + 3H_2O(g)$

(c) $Cu(s) + 2H^+ \rightarrow Cu^{2+} + H_2(g)$

(d) $Cu(s) + Pb^{2+} \rightarrow Cu^{2+} + Pb(s)$

Mixed Reactions

The following reactions apply to Exercises 71 through 77.

a. $H_2SO_4(aq) + 2KOH(aq) \rightarrow K_2SO_4(aq) + 2H_2O(\ell)$

b. $2Rb(s) + Br_2(\ell) \xrightarrow{\Delta} 2RbBr(s)$

c. $2KI(aq) + F_2(g) \rightarrow 2KF(aq) + I_2(s)$

d. $CaO(s) + SiO_2(s) \xrightarrow{\Delta} CaSiO_3(s)$

e. $S(s) + O_2(g) \xrightarrow{\Delta} SO_2(g)$

f. $BaCO_3(s) \xrightarrow{\Delta} BaO(s) + CO_2(g)$

g. $HgS(s) + O_2(g) \xrightarrow{\Delta} Hg(\ell) + SO_2(g)$

h. $AgNO_3(aq) + HCl(aq) \rightarrow AgCl(s) + HNO_3(aq)$

i. $Pb(s) + 2HBr(aq) \rightarrow PbBr_2(s) + H_2(g)$

j. $2HI(aq) + H_2O_2(aq) \rightarrow I_2(s) + 2H_2O(\ell)$

k. $RbOH(aq) + HNO_3(aq) \rightarrow RbNO_3(aq) + H_2O(\ell)$

l. $N_2O_5(s) + H_2O(\ell) \rightarrow 2HNO_3(aq)$

m. $H_2O(g) + CO(g) \xrightarrow{\Delta} H_2(g) + CO_2(g)$

n. $MgO(s) + H_2O(\ell) \rightarrow Mg(OH)_2(s)$

o. $PbSO_4(s) + PbS(s) \xrightarrow{\Delta} 2Pb(s) + 2SO_2(g)$

71. Identify the precipitation reactions.

72. Identify the acid–base reactions.

73. Identify the oxidation–reduction reactions.

74. Identify the oxidizing agent and reducing agent for each oxidation–reduction reaction.

75. Identify the oxidation–reduction reactions that are also displacement reactions.

76. Why can some reactions fit into more than one class?

77. Which of these reactions do not fit into any of our classes of reactions?

Naming Compounds

78. Name the following monatomic cations, using the IUPAC system of nomenclature. (a) Li^+, (b) Au^{3+}, (c) Ba^{2+}, (d) Zn^{2+}, (e) Ag^+.

79. Write the chemical symbol for each of the following: (a) sodium ion, (b) zinc ion, (c) silver ion, (d) mercury(II) ion, (e) bismuth(III) ion.

80. Write the chemical symbol for each of the following: (a) chloride ion, (b) sulfide ion, (c) telluride ion, (d) iodide ion, (e) oxide ion.

81. Name the following ionic compounds: (a) Li_2S, (b) SnO_2, (c) RbBr, (d) Li_2O, (e) Ba_3N_2.

82. Name the following ionic compounds: (a) NaI, (b) Hg_2S, (c) Li_3N, (d) $MnCl_2$, (e) $CuCO_3$, (f) FeO.

83. Write the chemical formula for each of the following compounds: (a) sodium fluoride, (b) zinc oxide, (c) barium oxide, (d) magnesium bromide, (e) hydrogen chloride, (f) copper(I) chloride.

84. Write the chemical formula for each of the following compounds: (a) copper(II) chlorate, (b) potassium nitrite, (c) barium phosphate, (d) copper(I) sulfate, (e) sodium carbonate.

85. What is the name of the acid with the formula H_2CO_3? Write the formulas of the two anions derived from it and name these ions.

86. What is the name of the acid with the formula H_3PO_3? What is the name of the $HPO_3{}^{2-}$ ion?

87. Name the following binary molecular compounds: (a) CO, (b) CO_2, (c) SF_6, (d) $SiCl_4$, (e) IF.

88. Name the following binary molecular compounds: (a) AsF_3, (b) Br_2O, (c) BrF_5, (d) CSe_2, (e) Cl_2O_7.

89. Write the chemical formula for each of the following compounds: (a) iodine bromide, (b) silicon dioxide, (c) phosphorus trichloride, (d) tetrasulfur dinitride, (e) bromine trifluoride, (f) hydrogen telluride, (g) xenon tetrafluoride.

90. Write the chemical formula for each of the following compounds: (a) diboron trioxide, (b) dinitrogen pentasulfide, (c) phosphorus triiodide, (d) sulfur tetrachloride, (e) silicon sulfide, (f) hydrogen sulfide, (g) tetraphosphorus hexoxide.

91. Write formulas for the compounds that are expected to be formed by the following pairs of ions:

	A. F^-	B. OH^-	C. SO_3^{2-}	D. PO_4^{3-}	E. NO_3^-
1. NH_4^+		Omit – see note			
2. K^+					
3. Mg^{2+}					
4. Cu^{2+}					
5. Fe^{3+}					
6. Ag^+					

NOTE: The compound NH_4OH does not exist. The solution commonly labeled "NH_4OH" is aqueous ammonia, $NH_3(aq)$.

92. Write the names for the compounds of Exercise 91.

93. Write balanced chemical equations for each of the following processes: (a) Calcium phosphate reacts with sulfuric acid to produce calcium sulfate and phosphoric acid. (b) Calcium phosphate reacts with water containing dissolved carbon dioxide to produce calcium hydrogen carbonate and calcium hydrogen phosphate.

94. Write balanced chemical equations for each of the following processes: (a) When heated, nitrogen and oxygen combine to form nitrogen oxide. (b) Heating a mixture of lead(II) sulfide and lead(II) sulfate produces metallic lead and sulfur dioxide.

BUILDING YOUR KNOWLEDGE

95. All of the following useful ammonium salts dissolve easily in water. Suggest the appropriate aqueous acid for preparing each salt and describe each preparation by a balanced "complete formula" equation. (a) Ammonium chloride: Used in medicine as an expectorant and in certain soldering operations as a "flux." (b) Ammonium nitrate: A valuable fertilizer, but potentially explosive; responsible for the famous Texas City disaster of 1947. (c) Ammonium sulfate: A common fertilizer often recommended for alkaline soils. (d) Ammonium phosphate: A particularly valuable fertilizer providing both nitrogen and phosphorus, essential elements for plant growth.

96. How many moles of oxygen can be obtained by the decomposition of 10.0 grams of reactant in each of the following reactions?
(a) $2KClO_3(s) \xrightarrow{\text{cat.}} 2KCl(s) + 3O_2(g)$
(b) $2H_2O_2(aq) \rightarrow 2H_2O(\ell) + O_2(g)$
(c) $2HgO(s) \xrightarrow{\Delta} 2Hg(\ell) + O_2(g)$

97. Magnesium oxide, marketed as tablets or as an aqueous slurry called "milk of magnesia," is a common commercial antacid. What volume, in milliliters, of fresh gastric juice, corresponding in acidity to 0.17 M HCl, could be neutralized by 125 mg of magnesium oxide?

$$MgO(s) + 2HCl(aq) \longrightarrow MgCl_2(aq) + H_2O(\ell)$$

98. What mass of Zn is needed to displace 14.5 grams of Cu from $CuSO_4 \cdot 5H_2O$?

5 The Structure of Atoms

Excited atoms of different elements give off light of different colors. The yellowish glow of some highway lamps is due to a sodium arc. Mercury lamps give a bluish glow.

OBJECTIVES

As you study this chapter, you should learn about

- *The evidence for the existence and properties of electrons, protons, and neutrons*

- *The arrangements of these particles in atoms*

- *Isotopes and their composition*

- *The relation between isotopic abundance and observed atomic weights*

- *The wave view of light and how wavelength, frequency, and speed are related*

- *The particle description of light, and how it is related to the wave description*

- *Atomic emission and absorption spectra, and how these were the basis for an important advance in atomic theory*

- *The quantum mechanical picture of the atom*

- *The four quantum numbers and possible combinations of their values*

- *The shapes of orbitals and the usual order of their relative energies*

- *Ways to determine electron configurations of atoms*

- *The relation between electron configurations and the positions of elements in the periodic table*

The Dalton theory of the atom and related ideas were the basis for our study of *composition stoichiometry* (Chapter 2) and *reaction stoichiometry* (Chapter 3). But that level of atomic theory leaves many questions unanswered. *Why* do atoms combine to form compounds? *Why* do they combine only in simple numerical ratios? *Why* are particular numerical ratios of atoms observed in compounds? *Why* do different elements have different properties—gases, liquids, solids, metals, nonmetals, and so on? *Why* do some groups of elements have similar properties, and form compounds with similar formulas? The answers to these and many other fascinating questions in chemistry are supplied by our modern understanding of the nature of atoms. But how can we study something as small as an atom?

Much of the development of modern atomic theory was based on two broad types of research carried out by dozens of scientists just before and after 1900. The first type dealt with the electrical nature of matter. These studies led scientists to recognize that atoms are composed of more fundamental particles, and helped them to describe the approximate arrangements of these particles in atoms. The second broad area of research dealt with the interaction of matter with energy in the form of light. Such research included studies of the colors of light that substances give off or absorb. These studies led to a much more detailed understanding of the arrangements of particles in atoms. It became clear that the arrangement of the particles determines the chemical and physical properties of each element. As we learn more about the structures of atoms, we are able to organize chemical facts in ways that help us to understand the behavior of matter.

We shall first study the particles that make up atoms and the basic structure of atoms. Then we shall trace the development of the quantum mechanical theory of atoms and see how this theory describes the arrangement of the electrons in atoms. Current atomic theory is considerably less than complete. Even so, it is a powerful tool that helps us describe the forces holding atoms in chemical combination with each other.

SUBATOMIC PARTICLES

5-1 FUNDAMENTAL PARTICLES

In our study of atomic theory, we look first at the **fundamental particles.** These are the basic building blocks of all atoms. Atoms, and hence *all* matter, consist principally of three fundamental particles: *electrons, protons,* and *neutrons.* Knowledge of the nature and functions of these particles is essential to understanding chemical interactions. The masses and charges of the three fundamental particles are shown in Table 5-1. The mass of an electron is very small compared with the mass of either a proton or a neutron. The charge on a proton is equal in magnitude, but opposite in sign, to the charge on an electron. Let's examine these particles in more detail.

Many other subatomic particles, such as quarks, positrons, neutrinos, pions, and muons, have also been discovered. It is not necessary to study their characteristics to learn the fundamentals of atomic structure that are important in chemical reactions.

Table 5-1 *Fundamental Particles of Matter*		
Particle	**Mass**	**Charge** (relative scale)
electron (e^-)	0.00054858 amu	1−
proton (p or p^+)	1.0073 amu	1+
neutron (n or n^0)	1.0087 amu	none

5-2 THE DISCOVERY OF ELECTRONS

Some of the earliest evidence about atomic structure was supplied in the early 1800s by the English chemist Humphrey Davy. He found that when he passed electrical current through some substances, the substances decomposed. This led him to propose that the elements of a chemical compound are held together by electrical forces. In 1832–33, Michael Faraday, Davy's protégé, determined the quantitative relationship between the amount of electricity used in electrolysis and the amount of chemical reaction that occurs. Studies of Faraday's work by George Stoney led him to suggest in 1874 that units of electrical charge are associated with atoms. In 1891 he suggested that they be named *electrons*.

The most convincing evidence for the existence of electrons came from experiments using *cathode ray tubes* (Figure 5-1). Two electrodes are sealed in a glass tube containing gas at a very low pressure. When a high voltage is applied, current flows and rays are

The process is called chemical electrolysis. *Lysis* means "splitting apart."

Study Figures 5-1 and 5-2 carefully as you read this section.

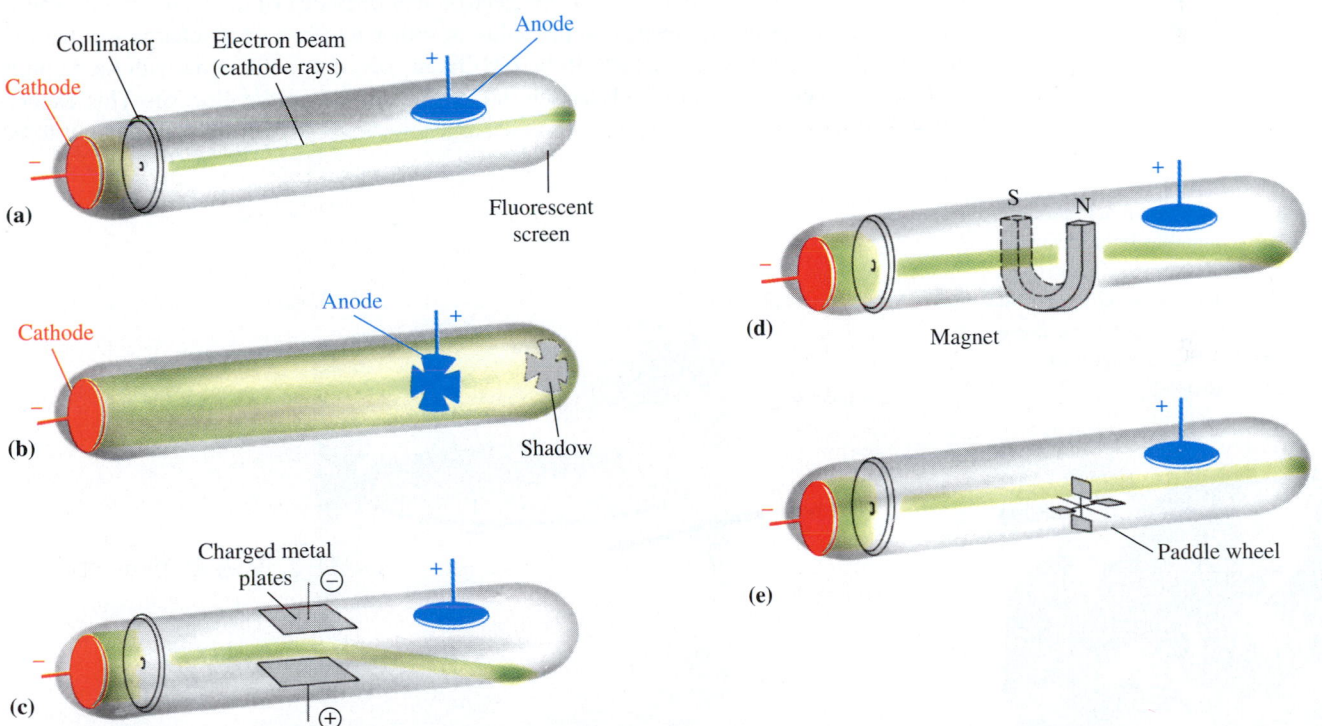

Figure 5-1 Some experiments with cathode ray tubes that show the nature of cathode rays. (a) A cathode ray (discharge) tube, showing the production of a beam of electrons (cathode rays). The beam is detected by observing the glow of a fluorescent screen. (b) A small object placed in a beam of cathode rays casts a shadow. This shows that cathode rays travel in straight lines. (c) Cathode rays have negative electrical charge, as demonstrated by their deflection in an electric field. (The electrically charged plates produce an electric field.) (d) Interaction of cathode rays with a magnetic field is also consistent with negative charge. The magnetic field goes from one pole to the other. (e) Cathode rays have mass, as shown by their ability to turn a small paddle wheel in their path.

given off by the cathode (negative electrode). These rays travel in straight lines toward the anode (positive electrode) and cause the walls opposite the cathode to glow. An object placed in the path of the cathode rays casts a shadow on a zinc sulfide screen placed near the anode. The shadow shows that the rays travel from the cathode toward the anode. Therefore the rays must be negatively charged. Additionally, they are deflected by magnetic and electrical fields in the directions expected for negatively charged particles.

In 1897 J. J. Thomson studied these negatively charged particles more carefully. He called them **electrons,** the name Stoney had suggested in 1891. By studying the degree of deflections of cathode rays in different magnetic and electric fields, Thomson determined the charge (e) to mass (m) ratio for electrons. The modern value for this ratio is

$$e/m = 1.75882 \times 10^8 \text{ coulomb (C) per gram}$$

This ratio is the same regardless of the type of gas in the tube, the composition of the electrodes, or the nature of the electrical power source. The clear implication of Thomson's work was that electrons are fundamental particles present in all atoms. We now know that this is true and that all atoms contain integral numbers of electrons.

Once the charge-to-mass ratio for the electron had been determined, additional experiments were necessary to determine the value of either its mass or its charge, so that the other could be calculated. In 1909 Robert Millikan solved this dilemma with his famous "oil-drop experiment," in which he determined the charge on the electron. This experiment is described in Figure 5-2. All of the charges measured by Millikan turned out to be

> The coulomb (C) is the standard unit of *quantity* of electrical charge. It is defined as the quantity of electricity transported in one second by a current of one ampere. It corresponds to the amount of electricity that will deposit 0.00111798 g of silver in an apparatus set up for plating silver.

> X-rays are radiations of much shorter wavelength than visible light (Section 5-10). They are sufficiently energetic to knock electrons out of the atoms in the air. In Millikan's experiment these free electrons became attached to some of the oil droplets.

Robert A. Millikan (left, 1868–1953) was an American physicist who was a professor at the University of Chicago and later director of the physics laboratory at the California Institute of Technology. For his investigations into photoelectric phenomena and the determination of the charge on the electron, he won the 1923 Nobel Prize in physics.

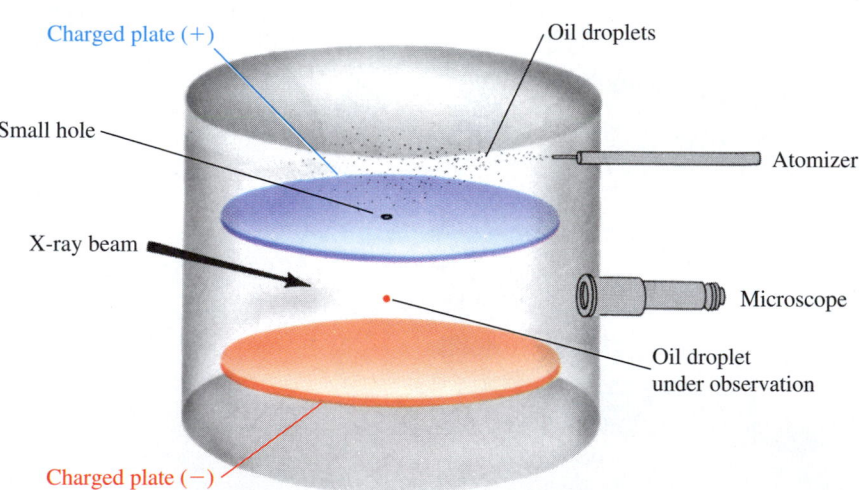

Figure 5-2   The Millikan oil-drop experiment. Tiny oil droplets are produced by an atomizer. A few of them fall through the hole in the upper plate. Irradiation with X-rays gives some of these oil droplets a negative charge. When the voltage between the plates is increased, a negatively charged drop falls more slowly because it is attracted by the positively charged upper plate and repelled by the negatively charged lower plate. At one particular voltage, the electrical force (up) and the gravitational force (down) on the drop are exactly balanced, and the drop remains stationary. If we know this voltage and the mass of the drop, we can calculate the charge on the drop. The mass of the spherical drop can be calculated from its volume (obtained from a measurement of the radius of the drop with a microscope) and the known density of the oil.

integral multiples of the same number. He assumed that this smallest charge was the charge on one electron. This value is 1.60218×10^{-19} coulomb (modern value).

The charge-to-mass ratio, $e/m = 1.75882 \times 10^8$ C/g, can be used in inverse form to calculate the mass of the electron:

$$m = \frac{1 \text{ g}}{1.75882 \times 10^8 \text{ C}} \times 1.60218 \times 10^{-19} \text{ C}$$

$$= 9.10940 \times 10^{-28} \text{ g per electron}$$

This is only about 1/1836 the mass of a hydrogen atom, the lightest of all atoms. Millikan's simple oil-drop experiment stands as one of the most clever, yet most fundamental, of all classic scientific experiments. It was the first experiment to suggest that atoms contain integral numbers of electrons; we now know this to be true.

The charge on one mole (Avogadro's number) of electrons is 96,485 coulombs.

The value of e/m obtained by Thomson and the values of e and m obtained by Millikan differ slightly from the modern values given in this text because early measurements were not as accurate as modern ones.

5-3 CANAL RAYS AND PROTONS

In 1886 Eugen Goldstein first observed that a cathode ray tube also generates a stream of positively charged particles that moves toward the cathode. These were called **canal rays** because they were observed occasionally to pass through a channel, or "canal," drilled in the negative electrode (Figure 5-3). These *positive rays,* or *positive ions,* are created when cathode rays knock electrons from the gaseous atoms in the tube, forming positive ions by processes such as

$$\text{atom} \longrightarrow \text{cation}^+ + e^- \qquad \text{or} \qquad \text{X} \longrightarrow \text{X}^+ + e^-$$

Different elements give positive ions with different e/m ratios. The regularity of the e/m values for different ions led to the idea that there is a unit of positive charge and that it resides in the **proton.** The proton is a fundamental particle with a charge equal in magnitude but opposite in sign to the charge on the electron. Its mass is almost 1836 times that of the electron.

The proton was observed by Rutherford and Chadwick in 1919 as a particle that is emitted by bombardment of certain atoms with alpha particles.

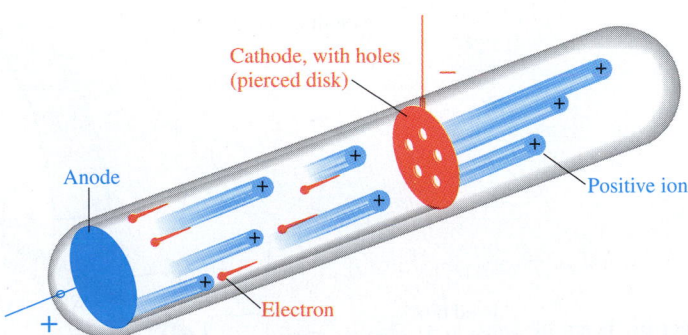

Cathode, with holes (pierced disk)

Anode

Positive ion

Electron

Figure 5-3 A cathode ray tube with a different design and with a perforated cathode. Such a tube was used to produce canal rays and to demonstrate that they travel toward the cathode. Like cathode rays, these *positive* rays are deflected by magnetic or electric fields, but in the opposite direction from cathode rays. Canal ray particles have e/m ratios many times smaller than those of electrons due to their much greater masses. When different elements are in the tube, positive ions with different e/m ratios are observed.

5-4 RUTHERFORD AND THE NUCLEAR ATOM

By the first decade of this century, it was clear that each atom contained regions of both positive and negative charge. The question was, how are these charges distributed? The dominant view of that time was summarized in J. J. Thomson's model of the atom, in which the positive charge was assumed to be distributed evenly throughout the atom. The negative charges were pictured as being imbedded in the atom like plums in a pudding (the "plum pudding model").

Soon after Thomson developed his model, tremendous insight into atomic structure was provided by one of Thomson's former students, Ernest Rutherford, who was the outstanding experimental physicist of his time.

By 1909 Ernest Rutherford had established that alpha (α) particles are positively charged particles. They can be emitted by some radioactive atoms, i.e., atoms that disintegrate spontaneously. In 1910 Rutherford's research group carried out a series of experiments that had enormous impact on the scientific world. They bombarded a very thin gold foil with α-particles from a radioactive source. A fluorescent zinc sulfide screen was placed behind the foil to indicate the scattering of the α-particles by the gold foil (Figure 5-4). Scintillations (flashes) on the screen, caused by the individual α-particles, were counted to determine the relative numbers of α-particles deflected at various angles. Alpha particles were known to be extremely dense, much denser than gold. Furthermore, they were known to be emitted at high kinetic energies.

If the Thomson model of the atom were correct, any α-particles passing through the foil would be expected to be deflected by very small angles. Quite unexpectedly, nearly all of the α-particles passed through the foil with little or no deflection. However, a few were deflected through large angles. A very few α-particles even returned from the gold foil in the direction from which they had come! Rutherford was astounded. In his own words,

> It was quite the most incredible event that has ever happened to me in my life. It was almost as if you fired a 15-inch shell into a piece of tissue paper and it came back and hit you.

Rutherford's mathematical analysis of his results showed that the scattering of positively charged α-particles was caused by repulsion from very dense regions of positive charge in the gold foil. He concluded that the mass of one of these regions is nearly equal to that of a gold atom, but that the diameter is no more than 1/10,000 that of an atom. Many

Alpha particles are now known to be helium atoms without their two electrons, or helium nuclei, which have 2+ charges (see Chapter 26).

Radioactivity is contrary to the Daltonian idea of the indivisibility of atoms.

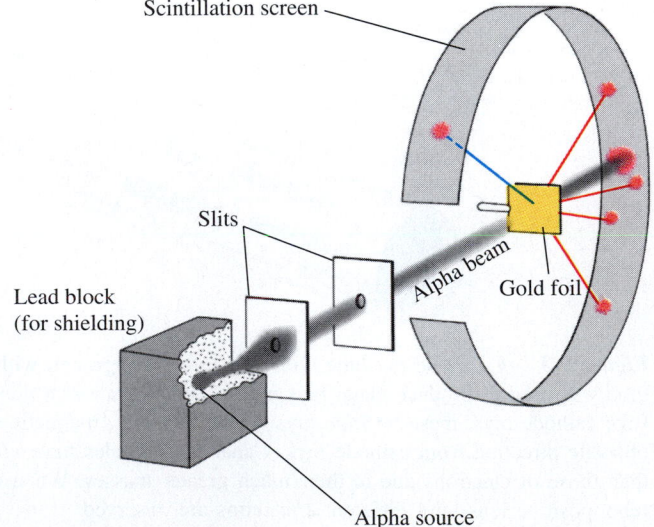

Figure 5-4 The Rutherford scattering experiment. A narrow beam of alpha particles (helium atoms stripped of their electrons) from a radioactive source was directed at a very thin gold foil. Most of the particles passed right through the gold foil (gray). Many were deflected through moderate angles (shown in red). These deflections were surprises, but the 0.001% of the total that were reflected at acute angles (shown in blue) were totally unexpected. Similar results were observed using foils of other metals.

Ernest Rutherford (1871–1937) was one of the giants in the development of our understanding of atomic structure. A native of New Zealand, Rutherford traveled to England in 1895 where he worked for much of his life. While working with J. J. Thomson at Cambridge University, he discovered α and β radiation. He spent the years 1899–1907 at McGill University in Canada where he proved the nature of these two radiations, for which he received the Nobel Prize in chemistry in 1908. He returned to England in 1908, and it was there, at Manchester University, that he and his coworkers Geiger and Marsden performed the famous gold foil experiments that led to the nuclear model of the atom. Not only did he perform much important research in physics and chemistry, but he also guided the work of ten future recipients of the Nobel Prize.

experiments with foils of different metals yielded similar results. Realizing that these observations were inconsistent with previous theories about atomic structure, Rutherford discarded the old theory and proposed a better one. He suggested that each atom contains a *tiny, positively charged, massive center* that he called an **atomic nucleus.** Most α-particles pass through metal foils undeflected because atoms are *primarily* empty space populated only by the very light electrons. The few particles that are deflected are the ones that come close to the heavy, highly charged metal nuclei (Figure 5-5).

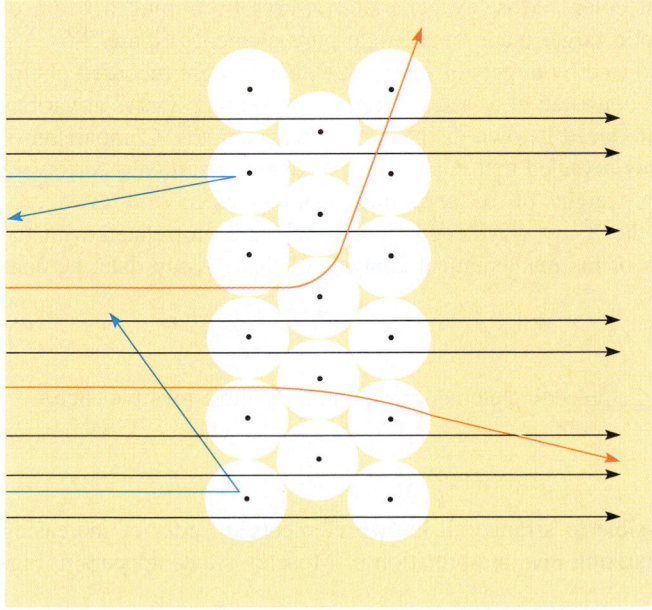

This representation is *not* to scale. If nuclei were as large as the black dots that represent them, each white region, which represents the size of an atom, would have a diameter of more than 30 feet!

Figure 5-5 An interpretation of the Rutherford scattering experiment. The atom is pictured as consisting mostly of "open" space. At the center is a tiny and extremely dense nucleus that contains all of the atom's positive charge and nearly all of the mass. The electrons are thinly distributed throughout the "open" space. Most of the positively charged alpha particles (black arrows) pass through the open space undeflected, not coming near any gold nuclei. The few that pass fairly close to a nucleus (red arrows) are repelled by electrostatic forces and thereby deflected. The very few particles that are on a "collision course" with gold nuclei are repelled backward at acute angles (blue arrows). Calculations based on the results of the experiment indicated that the diameter of the open-space portion of the atom is from 10,000 to 100,000 times greater than the diameter of the nucleus.

H. G. J. Moseley was one of the many remarkable scientists who worked with Ernest Rutherford. In 1913 Moseley found that the wavelengths of X-rays emitted by an element are related in a precise way to the atomic number of the element. This discovery led to the realization that atomic number, related to electrical properties of the atom, was more fundamental to determining the properties of the elements than atomic weight. This put the ideas of the periodic table on a more fundamental footing. Moseley's scientific career was very short. He was enlisted in the British army during World War I, and died in battle in the Gallipoli campaign in 1915. In subsequent wars, most countries have not allowed their promising scientists to take part in military combat.

In a modern technique known as "X-ray fluorescence spectroscopy," the wavelengths of X-rays given off by a sample target indicate which elements are present in the sample.

▼ **PROBLEM-SOLVING TIP**   *Analogy to the Rutherford Gold Foil Experiment*

As a somewhat similar experiment, imagine that we shoot pellets from a BB gun at a chain-link fence. Because the "particles" are much smaller than the "empty spaces" in such a fence, most of the pellets would go through the fence undeflected. However, some would glance off the wires of the fence and be deflected through moderate angles; a few would hit a wire directly enough to bounce back to our side of the fence. If we fired a very large number, say a million such "particles," at the fence, and counted how many went through undeflected and how many were deflected by various angles, we could perhaps calculate the fraction of the area of the fence that is open space, and even discover something about the size and distribution of the wires. (Gold foil is many atomic layers thick, so we could suppose that there were several such fences, one behind the other.)

Rutherford was able to determine the magnitudes of the positive charges on the atomic nuclei. The picture of atomic structure that he developed is called the Rutherford model of the atom.

Atoms consist of very small, very dense positively charged nuclei surrounded by clouds of electrons at relatively great distances from the nuclei.

5-5 ATOMIC NUMBER

Only a few years after Rutherford's scattering experiments, H. G. J. Moseley studied X-rays given off by various elements. Max von Laue had shown that X-rays could be diffracted by crystals into a spectrum in much the same way that visible light can be separated into its component colors. Moseley generated X-rays by aiming a beam of high-energy electrons at a solid target made of a single pure element (Figure 5-6).

The spectra of X-rays produced by targets of different elements were recorded photographically. Each photograph consisted of a series of lines representing X-rays at various wavelengths; each element produced its own distinct set of wavelengths. Comparison of results from different elements revealed that corresponding lines were displaced toward shorter wavelengths as atomic weights of the target materials increased, with few exceptions. Moseley showed that the X-ray wavelengths could be better correlated with the atomic number. On the basis of his mathematical analysis of these X-ray data, he concluded that

each element differs from the preceding element by having one more positive charge in its nucleus.

For the first time it was possible to arrange all known elements in order of increasing nuclear charge. A plot summarizing this interpretation of Moseley's data appears in Figure 5-7.

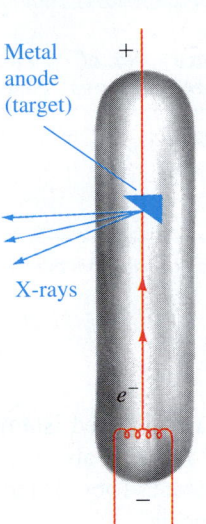

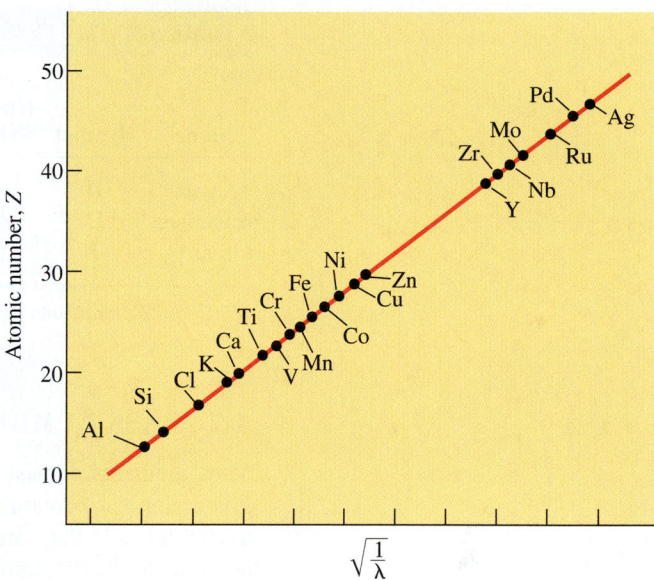

Figure 5-6 A simplified representation of the production of X-rays by bombardment of a solid target with a high-energy beam of electrons.

Figure 5-7 A plot of some of Moseley's X-ray data. The atomic number of an element is found to be directly proportional to the square root of the reciprocal of the wavelength of a particular X-ray spectral line. Wavelength (Section 5-10) is represented by λ.

We now know that every nucleus contains an integral number of protons exactly equal to the number of electrons in a neutral atom of the element. Every hydrogen atom contains one proton, every helium atom contains two protons, and every lithium atom contains three protons. The number of protons in the nucleus of an atom determines its identity; this number is known as the **atomic number** of that element.

5-6 NEUTRONS

The third fundamental particle, the neutron, eluded discovery until 1932. James Chadwick correctly interpreted experiments on the bombardment of beryllium with high-energy alpha particles. Later experiments showed that nearly all elements up to potassium, element 19, produce neutrons when they are bombarded with high-energy alpha particles. The **neutron** is an uncharged particle with a mass slightly greater than that of the proton.

This does not mean that elements above number 19 do not have neutrons, only that neutrons are not generally knocked out of atoms of higher atomic number by alpha particle bombardment.

> Atoms consist of very small, very dense nuclei surrounded by clouds of electrons at relatively great distances from the nuclei. All nuclei contain protons; nuclei of all atoms except the common form of hydrogen also contain neutrons.

Nuclear diameters are about 10^{-4} Ångstroms (10^{-5} nanometers); atomic diameters are about 1 Ångstrom (10^{-1} nanometers). To put this difference in perspective, suppose that you wish to build a model of an atom using a basketball (diameter about 9.5 inches) as the nucleus; on this scale, the atomic model would be nearly 6 miles across!

Table 5-2 *The Three Isotopes of Hydrogen*

Name	Symbol	Nuclide Symbol	Mass (amu)	Atomic Abundance in Nature	No. of Protons	No. of Neutrons	No. of Electrons (in neutral) atoms)
hydrogen	H	^1_1H	1.007825	99.985%	1	0	1
deuterium	D	^2_1H	2.0140	0.015%	1	1	1
tritium*	T	^3_1H	3.01605	0.000%	1	2	1

No known natural sources; produced by decomposition of artificial isotopes.

5-7 MASS NUMBER AND ISOTOPES

Most elements consist of atoms of different masses, called **isotopes.** The isotopes of a given element contain the same number of protons (and also the same number of electrons) because they are atoms of the same element. They differ in mass because they contain different numbers of neutrons in their nuclei.

For example, there are three distinct kinds of hydrogen atoms, commonly called hydrogen, deuterium, and tritium. (This is the only element for which we give each isotope a different name.) Each contains one proton in the atomic nucleus. The predominant form of hydrogen contains no neutrons, but each deuterium atom contains one neutron and each tritium atom contains two neutrons in its nucleus (Table 5-2). All three forms of hydrogen display very similar chemical properties.

The **mass number** of an atom is the sum of the number of protons and the number of neutrons in its nucleus; i.e.,

$$\text{mass number} = \text{number of protons} + \text{number of neutrons}$$
$$= \text{atomic number} \quad + \text{neutron number}$$

A mass number is a count of the number of protons plus neutrons present, so it must be a whole number. Because the masses of the proton and the neutron are both about 1 amu, the mass number is *approximately* equal to the actual mass of the isotope (which is not a whole number).

The mass number for normal hydrogen atoms is 1, for deuterium 2, and for tritium 3. The composition of a nucleus is indicated by its **nuclide symbol.** This consists of the symbol for the element (E), with the atomic number (Z) written as a subscript at the lower left and the mass number (A) as a superscript at the upper left, $^A_Z E$. By this system, the three isotopes of hydrogen are designated as ^1_1H, ^2_1H, and ^3_1H.

5-8 MASS SPECTROMETRY AND ISOTOPIC ABUNDANCE

Mass spectrometers are instruments that measure the charge-to-mass ratio of charged particles (Figure 5-8). A gas sample at very low pressure is bombarded with high-energy electrons. This causes electrons to be ejected from some of the gas molecules, creating positive ions. The positive ions are then focused into a very narrow beam and accelerated by an electric field toward a magnetic field. The magnetic field deflects the ions from their straight-line path. The extent to which the beam of ions is deflected depends upon four factors:

1. *Magnitude of the accelerating voltage (electric field strength).* Higher voltages result in beams of more rapidly moving particles that are deflected less than the beams of the more slowly moving particles produced by lower voltages.

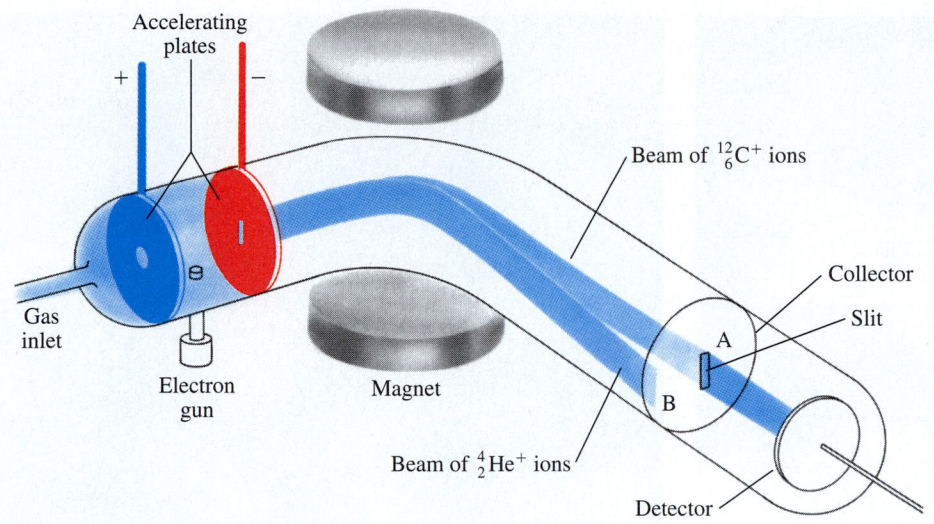

Figure 5-8 The mass spectrometer. In the mass spectrometer, gas molecules at low pressure are ionized and accelerated by an electric field. The ion beam is then passed through a magnetic field. In that field the beam is resolved into components, each containing particles of equal charge-to-mass ratio. Lighter particles are deflected more strongly than heavy ones with the same charge. In a beam containing $^{12}_{6}C^+$ and $^{4}_{2}He^+$ ions, the lighter $^{4}_{2}He^+$ ions would be deflected more than the heavier $^{12}_{6}C^+$ ions. The spectrometer shown is adjusted to detect the $^{12}_{6}C^+$ ions. By changing the magnitude of the magnetic or electric field, we can move the beam of $^{4}_{2}He^+$ ions striking the collector from B to A, where it would be detected.

2. *Magnetic field strength.* Stronger fields deflect a given beam more than weaker fields.

3. *Masses of the particles.* Because of their inertia, heavier particles are deflected less than lighter particles that carry the same charge.

4. *Charges on the particles.* Particles with higher charges interact more strongly with magnetic fields and are thus deflected more than particles of equal mass with smaller charges.

The mass spectrometer is used to measure masses of isotopes as well as isotopic abundances, i.e., the relative amounts of the isotopes. Helium occurs in nature almost exclusively as $^{4}_{2}He$. Let's see how its atomic mass is measured. To simplify the picture, we'll assume that only ions with 1+ charge are formed in the experiment illustrated in Figure 5-8.

A beam of Ne^+ ions in the mass spectrometer is split into three segments. The mass spectrum of these ions (a graph of the relative numbers of ions of each mass) is shown in Figure 5-9. This indicates that neon occurs in nature as three isotopes: $^{20}_{10}Ne$, $^{21}_{10}Ne$, and $^{22}_{10}Ne$. In Figure 5-9 we see that the isotope $^{20}_{10}Ne$, mass 19.99244 amu, is the most abundant isotope (has the tallest peak). It accounts for 90.5% of the atoms. $^{22}_{10}Ne$ accounts for 9.2% and $^{21}_{10}Ne$ for only 0.3% of the atoms.

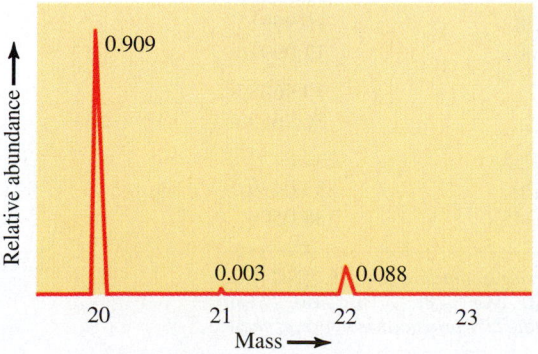

Figure 5-9 Mass spectrum of neon (1+ ions only). Neon consists of three isotopes, of which neon-20 is by far the most abundant (90.9%). The mass of that isotope, to five decimal places, is 19.99244 amu on the carbon-12 scale. The number by each peak represents the number of Ne^+ ions corresponding to that isotope, expressed as a fraction of all Ne^+ ions.

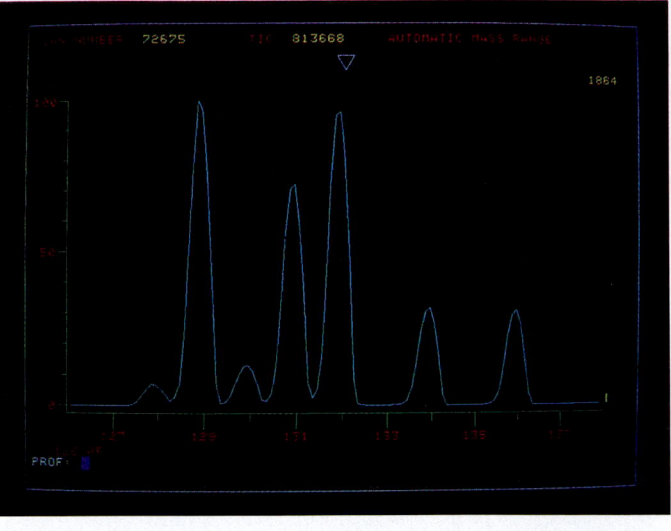

(a)   **(b)**

Figure 5-10   (a) A modern mass spectrometer, the Finnigan MAT TSQ 7002. (b) The mass spectrum of Xe^{1+} ions. The isotope ^{126}Xe is at too low an abundance (0.090%) to appear in this experiment.

Figure 5-10 shows a modern mass spectrometer and a typical spectrum of an element. In nature, some elements, such as fluorine and phosphorus, exist in only one form, but most elements occur as isotopic mixtures. Some examples of natural isotopic abundances are given in Table 5-3. The percentages are based on the numbers of naturally occurring atoms of each isotope, *not* on their masses.

The distribution of isotopic masses, while nearly constant, does vary somewhat depending on the source of the element. For example, the abundance of $^{13}_{6}C$ in atmospheric CO_2 is slightly different from that in seashells. The chemical history of a compound can be inferred from small differences in isotope ratios.

Table 5-3 *Some Naturally Occurring Isotopic Abundances*			
Element	**Isotope**	**% Natural Abundance**	**Mass (amu)**
boron	$^{10}_{5}B$	19.8	10.01294
	$^{11}_{5}B$	80.2	11.00931
oxygen	$^{16}_{8}O$	99.762	15.99492
	$^{17}_{8}O$	0.038	16.99913
	$^{18}_{8}O$	0.200	17.99916
chlorine	$^{35}_{17}Cl$	75.77	34.96885
	$^{37}_{17}Cl$	24.23	36.96590
uranium	$^{234}_{92}U$	0.0055	234.0409
	$^{235}_{92}U$	0.720	235.0439
	$^{238}_{92}U$	99.2745	238.0508

The 20 elements that have only one naturally occurring isotope are $^{9}_{4}Be$, $^{19}_{9}F$, $^{23}_{11}Na$, $^{27}_{13}Al$, $^{31}_{15}P$, $^{45}_{21}Sc$, $^{55}_{25}Mn$, $^{59}_{27}Co$, $^{75}_{33}As$, $^{89}_{39}Y$, $^{93}_{41}Nb$, $^{103}_{45}Rh$, $^{127}_{53}I$, $^{133}_{55}Cs$, $^{141}_{59}Pr$, $^{159}_{65}Tb$, $^{165}_{67}Ho$, $^{169}_{69}Tm$, $^{197}_{79}Au$, and $^{209}_{83}Bi$. However, there are other, artificially produced isotopes of these elements.

5-9 THE ATOMIC WEIGHT SCALE AND ATOMIC WEIGHTS

We said in Section 2-5 that the **atomic weight scale** is based on the mass of the carbon-12 isotope. As a result of action taken by the International Union of Pure and Applied Chemistry in 1962,

> one **amu** is exactly 1/12 of the mass of a carbon-12 atom.

Described another way, the mass of one atom of $^{12}_{6}C$ is *exactly* 12 amu.

This is approximately the mass of one atom of ^{1}H, the lightest isotope of the element with lowest mass.

In Section 2-6 we said that one mole of atoms contains 6.022×10^{23} atoms. The mass of one mole of atoms of any element, in grams, is numerically equal to the atomic weight of the element. Because the mass of one carbon-12 atom is exactly 12 amu, the mass of one mole of carbon-12 atoms is exactly 12 grams.

Let us now show the relationship between atomic mass units and grams.

$$\underline{?} \text{ amu} = 1 \text{ g } ^{12}_{6}C \text{ atoms} \times \frac{1 \text{ mol } ^{12}_{6}C}{12 \text{ g } ^{12}_{6}C \text{ atoms}} \times \frac{6.022 \times 10^{23} \; ^{12}_{6}C \text{ atoms}}{1 \text{ mol } ^{12}_{6}C \text{ atoms}} \times \frac{12 \text{ amu}}{^{12}_{6}C \text{ atom}}$$

$$= 6.022 \times 10^{23} \text{ amu}$$

Thus, we see that

> $$1 \text{ g} = 6.022 \times 10^{23} \text{ amu} \quad or \quad 1 \text{ amu} = 1.660 \times 10^{-24} \text{ g}$$

We saw in Chapter 2 that Avogadro's number is also the number of particles of a substance in one mole of that substance. We now see that Avogadro's number also represents the number of amu in one gram. You may wish to verify that the same result is obtained regardless of the element or isotope chosen.

At this point, we emphasize the differences among the following quantities:

1. The *atomic number, Z*, is an integer equal to the number of protons in the nucleus of an atom of the element. It is also the number of electrons in a neutral atom. It is the same for all atoms of an element.

2. The *mass number, A*, is an integer equal to the *sum* of the number of protons and the number of neutrons in the nucleus of an atom of a *particular isotope* of an element. It is different for different isotopes of the same element.

3. Many elements occur in nature as mixtures of isotopes. The *atomic weight* of such an element is the weighted average of the masses of its isotopes. Atomic weights are fractional numbers, not integers.

The atomic weight that we determine experimentally (for an element that consists of more than one isotope) is such a weighted average. The following example shows how an atomic weight can be calculated from measured isotopic abundances.

EXAMPLE 5-1 *Calculation of Atomic Weight*

Three isotopes of magnesium occur in nature. Their abundances and masses, determined by mass spectrometry, are listed below. Use this information to calculate the atomic weight of magnesium.

Isotope	% Abundance	Mass (amu)
$^{24}_{12}Mg$	78.99	23.98504
$^{25}_{12}Mg$	10.00	24.98584
$^{26}_{12}Mg$	11.01	25.98259

Plan

We multiply the fraction of each isotope by its mass and add these numbers to obtain the atomic weight of magnesium.

Solution

atomic weight = 0.7899(23.98504 amu) + 0.1000(24.98584 amu) + 0.1101(25.98259 amu)

$\qquad$ = 18.94 amu $\qquad$ + 2.498 amu $\qquad$ + 2.861 amu

$\qquad$ = 24.30 amu $\quad$ (to four significant figures)

The two heavier isotopes make small contributions to the atomic weight of magnesium, because most magnesium atoms are the lightest isotope.

You should now work Exercise 30.

▼ **PROBLEM-SOLVING TIP** *"Weighted" Averages*

Consider the following analogy to the calculation of atomic weights. Suppose you want to calculate the average weight of your classmates. Imagine that one half of them weigh 100 pounds each while the other one half weigh 200 pounds each. The average weight would be

$$\text{average weight} = \frac{1}{2}(100 \text{ lb}) + \frac{1}{2}(200 \text{ lb}) = 150 \text{ lb}$$

Imagine, however, that three quarters of the class members weigh 100 pounds each while the other one quarter weigh 200 pounds each. Now, the average weight would be

$$\text{average weight} = \frac{3}{4}(100 \text{ lb}) + \frac{1}{4}(200 \text{ lb}) = 125 \text{ lb}$$

We can express the fractions in this calculation in decimal form:

$$\text{average weight} = 0.750(100 \text{ lb}) + 0.250(200 \text{ lb}) = 125 \text{ lb}$$

In such a calculation, the value (in this case, the weight) of each thing (people, atoms) is multiplied by the fraction of things that have that value. In Example 5-1 we expressed each percentage as a decimal fraction, e.g.,

$$78.99\% = \frac{78.99 \text{ parts}}{100 \text{ parts total}} = 0.7899$$

Example 5-2 shows how the process can be reversed. Isotopic abundances can be calculated from isotopic masses and from the atomic weight of an element that occurs in nature as a mixture of only two isotopes.

EXAMPLE 5-2 *Calculation of Isotopic Abundance*

The atomic weight of gallium is 69.72 amu. The masses of the naturally occurring isotopes are 68.9257 amu for $^{69}_{31}\text{Ga}$ and 70.9249 amu for $^{71}_{31}\text{Ga}$. Calculate the percent abundance of each isotope.

Plan

We represent the fraction of each isotope algebraically. Atomic weight is the weighted average of the masses of the constituent isotopes. So the fraction of each isotope is multiplied by its mass and the sum of the results is equal to the atomic weight.

Solution

Let x = fraction of $_{31}^{69}Ga$. Then $(1 - x)$ = fraction of $_{31}^{71}Ga$.

$$x(68.9257 \text{ amu}) + (1 - x)(70.9249 \text{ amu}) = 69.72 \text{ amu}$$
$$68.9257x + 70.9249 - 70.9249x = 69.72$$
$$-1.9992x = -1.20$$
$$x = 0.600$$

$$x = 0.600 = \text{fraction of } _{31}^{69}Ga \quad \therefore \quad \boxed{60.0\% \ _{31}^{69}Ga}$$

$$(1 - x) = 0.400 = \text{fraction of } _{31}^{71}Ga \quad \therefore \quad \boxed{40.0\% \ _{31}^{71}Ga}$$

You should now work Exercise 24.

> When a quantity is represented by fractions, the sum of the fractions must always be unity. In this case, $x + (1 - x) = 1$.

THE ELECTRONIC STRUCTURES OF ATOMS

The Rutherford model of the atom is consistent with the evidence presented so far, but it has some serious limitations. It does not answer important questions such as the following. *Why* do different elements have such different chemical and physical properties? *Why* does chemical bonding occur at all? *Why* does each element form compounds with characteristic formulas? *How* can atoms of different elements give off or absorb light only of characteristic colors (as was known long before 1900)?

To improve our understanding, we must first learn more about the arrangements of electrons in atoms. The theory of these arrangements is based largely on the study of the light given off and absorbed by atoms. Then we shall develop a detailed picture of the *electron configurations* of different elements. A knowledge of these arrangements will help us to understand the periodic table and chemical bonding.

5-10 ELECTROMAGNETIC RADIATION

Our ideas about the arrangements of electrons in atoms have evolved slowly. Much of the information has been derived from **atomic emission spectra.** These are the lines, or bands, produced on photographic film by radiation that has passed through a refracting glass prism after being emitted from electrically or thermally excited atoms. To help us understand the nature of atomic spectra, we first describe electromagnetic radiation.

All types of electromagnetic radiation, or radiant energy, can be described in the terminology of waves. To help characterize any wave, we specify its *wavelength* (or its *frequency*). Let us use a familiar kind of wave, that on the surface of water (Figure 5-11), to illustrate these terms. The significant feature of wave motion is its repetitive nature. The **wavelength,** λ, is the distance between any two adjacent identical points of the wave, for instance, two adjacent crests. The **frequency** is the number of wave crests passing a given point per unit time; it is represented by the symbol ν (Greek letter "nu") and is usually expressed in cycles/second or, more commonly, simply as 1/s or s^{-1} with "cycles" understood. For a wave that is "traveling" at some speed, the wavelength and the frequency are related to each other by

$$\lambda\nu = \text{speed of propagation of the wave} \quad \text{or} \quad \lambda\nu = c$$

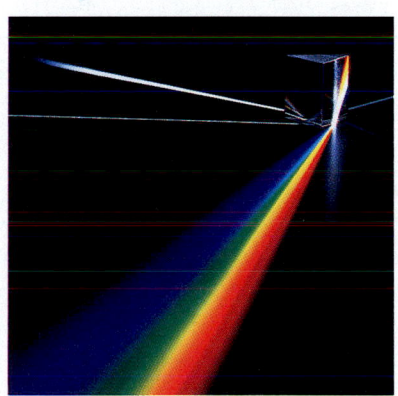

White light is dispersed by a prism into a *continuous* spectrum.

> One cycle per second is also called one *hertz* (Hz), after Heinrich Hertz. In 1887 Hertz discovered electromagnetic radiation outside the visible range and measured its speed and wavelengths.

The diffraction of white light by the closely spaced grooves of a compact disk spreads the light into its component colors. Diffraction is described as the constructive and destructive interference of light waves.

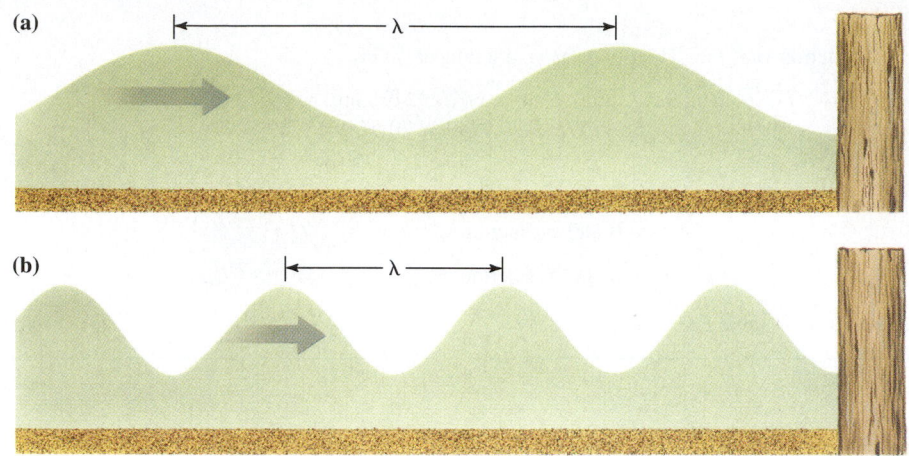

Figure 5-11 Illustrations of the wavelength and frequency of water waves. The distance between any two identical points, e.g., crests, is the wavelength, λ. We could measure the frequency, ν, of the wave by observing how often the level rises and falls at a fixed point in its path—for instance, at the post—or how often crests hit the post. (a) and (b) represent two waves that are traveling at the same speed. In (a) the wave has long wavelength and low frequency; in (b) the wave has shorter wavelength and higher frequency.

Sir Isaac Newton (1642–1727), one of the giants of science. You probably know of him from his theory of gravitation. In addition, he made enormous contributions to the understanding of many other aspects of physics, including the nature and behavior of light, optics, and the laws of motion. He is credited with the discoveries of differential calculus and of expansions into infinite series.

Thus, wavelength and frequency are inversely proportional to each other; for the same wave speed, shorter wavelengths correspond to higher frequencies.

For water waves, it is the surface of the water that changes repetitively; for a vibrating violin string, it is the displacement of any point on the string. Electromagnetic radiation is a form of energy that consists of electric and magnetic fields that vary repetitively. The electromagnetic radiation most obvious to us is visible light. It has wavelengths ranging from about 4×10^{-7} m (violet) to about 7×10^{-7} m (red). Expressed in frequencies, this range is about 7.5×10^{14} Hz (violet) to about 4.3×10^{14} Hz (red).

Isaac Newton first recorded the separation of sunlight into its component colors by allowing it to pass through a prism. Because sunlight (white light) contains all wavelengths of visible light, it gives the *continuous spectrum* observed in a rainbow (Figure 5-12a). Visible light represents only a tiny segment of the electromagnetic radiation spectrum (Figure 5-12b). In addition to all wavelengths of visible light, sunlight also contains shorter wavelength (ultraviolet) radiation as well as longer wavelength (infrared) radiation. Neither of these can be detected by the human eye. Both may be detected and recorded photographically or by detectors designed for that purpose. Many other familiar kinds of radiation are simply electromagnetic radiation of longer or shorter wavelengths.

In a vacuum, the speed of electromagnetic radiation, c, is the same for all wavelengths, 2.9979249×10^8 m/s. The relationship between the wavelength and frequency of electromagnetic radiation, with c rounded to three significant figures, is

$$\lambda\nu = c = 3.00 \times 10^8 \text{ m/s}$$

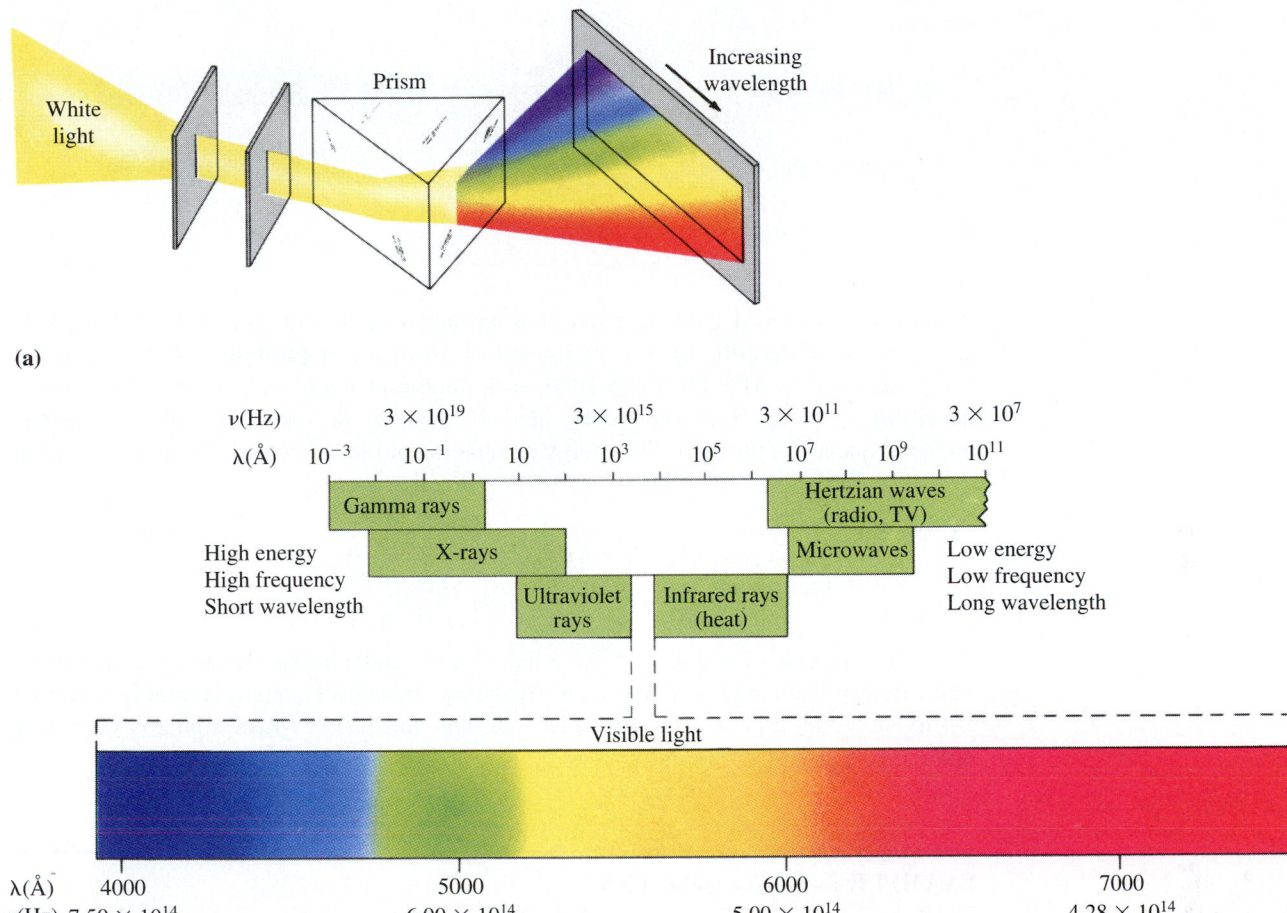

(a)

(b)

Figure 5-12 (a) Dispersion of visible light by a prism. Light from a source of white light is passed through a slit and then through a prism. It is spread into a continuous spectrum of all wavelengths of visible light. (b) Visible light is only a very small portion of the electromagnetic spectrum. Some radiant energy has longer or shorter wavelengths than our eyes can detect. The upper part shows the approximate ranges of the electromagnetic spectrum on a logarithmic scale. The lower part shows the visible region on an expanded scale. Note that wavelength increases as frequency decreases.

EXAMPLE 5-3 *Wavelength of Light*

Light near the middle of the ultraviolet region of the spectrum has a frequency of $2.73 \times 10^{16}\ \text{s}^{-1}$. Yellow light near the middle of the visible region of the spectrum has a frequency of $5.45 \times 10^{14}\ \text{s}^{-1}$. Calculate the wavelength that corresponds to each of these two frequencies of light.

Plan

Wavelength and frequency are inversely proportional to each other, $\lambda\nu = c$. We solve this relationship for λ and calculate the wavelengths.

Solution

$$\text{(uv light)} \quad \lambda = \frac{c}{\nu} = \frac{3.00 \times 10^8 \text{ m} \cdot \text{s}^{-1}}{2.73 \times 10^{16} \text{ s}^{-1}} = \boxed{1.10 \times 10^{-8} \text{ m } (1.10 \times 10^2 \text{ Å})}$$

$$\text{(yellow light)} \quad \lambda = \frac{c}{\nu} = \frac{3.00 \times 10^8 \text{ m} \cdot \text{s}^{-1}}{5.45 \times 10^{14} \text{ s}^{-1}} = \boxed{5.50 \times 10^{-7} \text{ m } (5.50 \times 10^3 \text{ Å})}$$

You should now work Exercise 33.

We have described light in terms of wave behavior. Under certain conditions, it is also possible to describe light as composed of *particles,* or **photons.** According to the ideas presented by Max Planck in 1900, each photon of light has a particular amount (a **quantum**) of energy. Furthermore, the amount of energy possessed by a photon depends on the frequency of the light. The energy of a photon of light is given by Planck's equation

$$E = h\nu \quad \text{or} \quad E = \frac{hc}{\lambda}$$

where h is Planck's constant, 6.6262×10^{-34} J $\cdot$ s, and ν is the frequency of the light. Thus, energy is directly proportional to frequency. Planck's equation is used in Example 5-4 to show that a photon of ultraviolet light has more energy than a photon of yellow light.

EXAMPLE 5-4   *Energy of Light*

In Example 5-3 we calculated the wavelengths of ultraviolet light of frequency of 2.73×10^{16} s^{-1} and of yellow light of frequency of 5.45×10^{14} s^{-1}. Calculate the energy, in joules, of an individual photon of each. Compare these photons by calculating the ratio of their energies.

Plan

We use each frequency to calculate the photon energy from the relationship $E = h\nu$. Then we calculate the required ratio.

Solution

$$\text{(uv light)} \quad E = h\nu = (6.63 \times 10^{-34} \text{ J} \cdot \text{s})(2.73 \times 10^{16} \text{ s}^{-1}) = \boxed{1.81 \times 10^{-17} \text{ J}}$$

$$\text{(yellow light)} \quad E = h\nu = (6.63 \times 10^{-34} \text{ J} \cdot \text{s})(5.45 \times 10^{14} \text{ s}^{-1}) = \boxed{3.61 \times 10^{-19} \text{ J}}$$

$\left(\text{You can check these answers by calculating the energies directly from the wavelengths, using the equation } E = \dfrac{hc}{\lambda}.\right)$

Now, we compare the energies.

$$\frac{E_{uv}}{E_{yellow}} = \frac{1.81 \times 10^{-17} \text{ J/photon}}{3.61 \times 10^{-19} \text{ J/photon}} = \boxed{50}$$

This is one reason why ultraviolet light damages your skin much more rapidly than visible light. Another reason is that many of the organic compounds in the skin absorb uv light more readily than visible light. The absorbed ultraviolet light breaks bonds in many biologically important molecules.

The energy of a photon of light near the middle of the ultraviolet region is 50 times greater than the energy of a photon of light near the middle of the visible region.

You should now work Exercise 35.

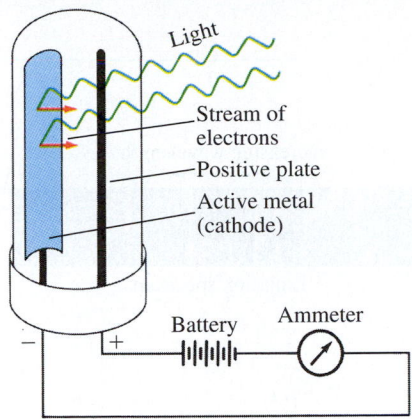

Light

Stream of electrons

Positive plate

Active metal (cathode)

Battery

Ammeter

$-$ $+$

Figure 5-13 The photoelectric effect. When electromagnetic radiation of sufficient minimum energy strikes the surface of a metal (negative electrode) inside an evacuated tube, electrons are stripped off the metal to create an electric current. The current increases with increasing radiation intensity.

5-11 THE PHOTOELECTRIC EFFECT

One experiment that had not been satisfactorily explained with the wave model of light was the **photoelectric effect.** The apparatus for the photoelectric effect is shown in Figure 5-13. The negative electrode in the evacuated tube is made of a pure metal such as cesium. When light of a sufficiently high energy strikes the metal, electrons are knocked off its surface. They then travel to the positive electrode and form a current flowing through the circuit. The important observations follow.

1. Electrons can be ejected only if the light is of sufficiently short wavelength (has sufficiently high energy), no matter how long or how brightly the light shines. This wavelength limit is different for different metals.

2. The current (the number of electrons emitted per second) increases with increasing *brightness* (intensity) of the light. However, it does not depend on the color of the light as long as the wavelength is short enough (has high enough energy).

The intensity of light is the brightness of the light. In wave terms, it is related to the amplitude of the light waves.

Classical theory said that even "low" energy light should cause current to flow if the metal is irradiated long enough. Electrons should accumulate energy and be released when they have enough energy to escape from the metal atoms. According to the old theory, if the light is made more energetic, then the current should increase even though the light intensity remains the same. Such is *not* the case.

The answer to the puzzle was provided by Albert Einstein. In 1905 he extended Planck's idea that light behaves as though it were composed of *photons,* each with a particular amount (a quantum) of energy. According to Einstein, each photon can transfer its energy to a single electron during a collision. When we say that the intensity of light is increased, we mean that the number of photons striking a given area per second is increased. The picture is now one of a particle of light striking an electron near the surface of the metal and giving up its energy to the electron. If that energy is equal to or greater than the amount needed to liberate the electron, it can escape to join the photoelectric current. For this explanation, Einstein received the 1921 Nobel Prize in physics.

The photoelectric effect is used in the photocells of automatic cameras. The photoelectric sensors that open some supermarket and elevator doors also utilize this effect.

5-12 ATOMIC SPECTRA AND THE BOHR ATOM

Incandescent ("red hot" or "white hot") solids, liquids, and high-pressure gases give continuous spectra. However, when an electric current is passed through a gas in a vacuum tube at very low pressures, the light that the gas emits is dispersed by a prism into distinct

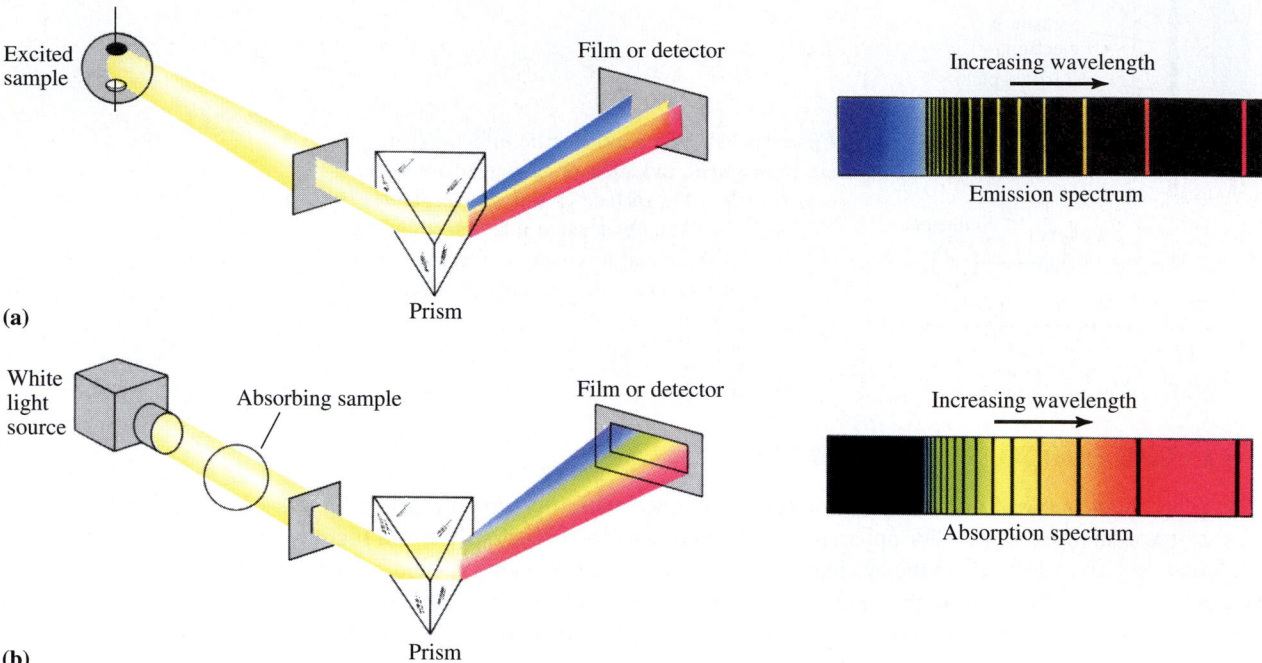

Figure 5-14 (a) *Atomic emission.* The light emitted by a sample of excited hydrogen atoms (or any other element) can be passed through a prism and separated into certain discrete wavelengths. Thus, an emission spectrum, which is a photographic recording of the separated wavelengths, is called a line spectrum. Any sample of reasonable size contains an enormous number of atoms. Although a single atom can be in only one excited state at a time, the collection of atoms contains all possible excited states. The light emitted as these atoms fall to lower energy states is responsible for the spectrum. (b) *Atomic absorption.* When white light is passed through unexcited hydrogen and then through a slit and a prism, the transmitted light is lacking in intensity at the same wavelengths as are emitted in (a). The recorded absorption spectrum is also a line spectrum and the photographic negative of the emission spectrum.

lines (Figure 5-14a). Such an **emission spectrum** is described as a *bright line spectrum*. The lines can be recorded photographically, and the wavelength of light that produced each line can be calculated from the position of that line on the photograph.

Similarly, we can shine a beam of white light (containing a continuous distribution of wavelengths) through a gas and analyze the beam that emerges. We find that only certain wavelengths have been absorbed (Figure 5-14b). The wavelengths that are absorbed in this **absorption spectrum** are the same as those given off in the emission experiment. Each spectral line corresponds to a specific wavelength of light and thus to a specific amount of energy that is either absorbed or emitted. An atom of each element displays its own characteristic set of lines in its emission or absorption spectrum (Figure 5-15). These spectra can serve as "fingerprints" that allow us to identify different elements present in a sample, even in trace amounts.

EXAMPLE 5-5 *Energy of Light*

A green line of wavelength 4.86×10^{-7} m is observed in the emission spectrum of hydrogen. Calculate the energy of one photon of this green light.

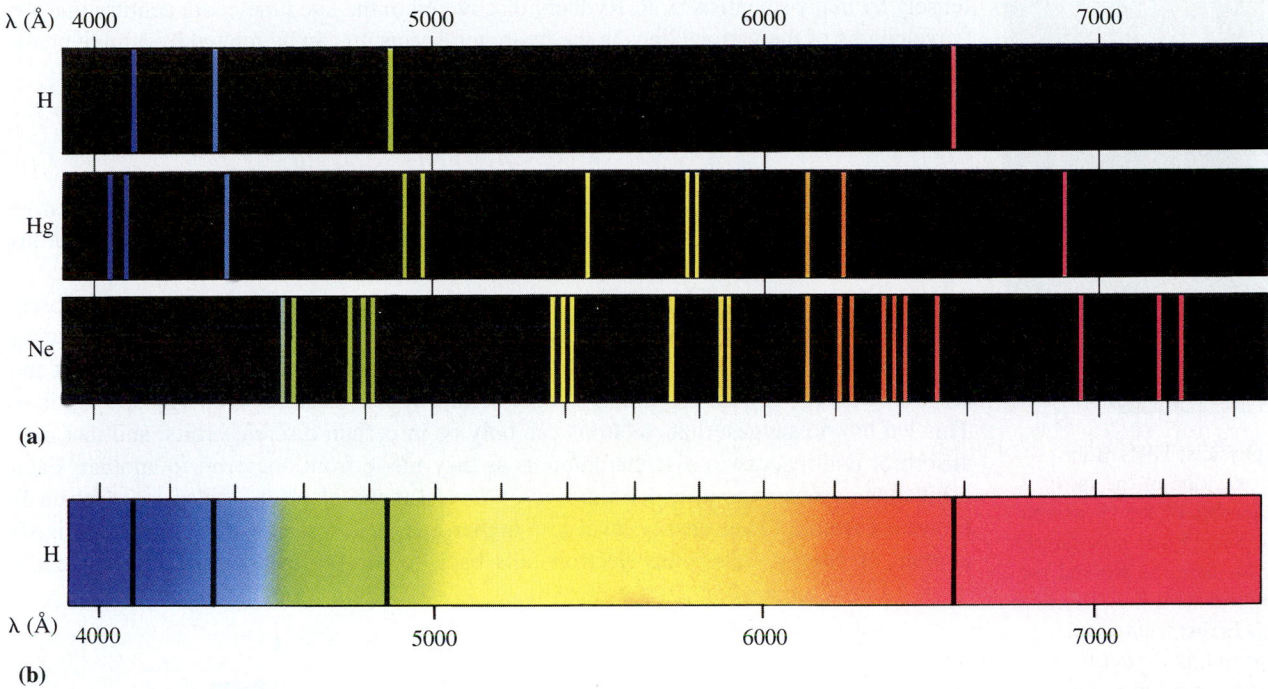

(a)

(b)

Figure 5-15 Atomic spectra in the visible region for some elements. Figure 5-14a shows how such spectra are produced. (a) Emission spectra for some elements. (b) Absorption spectrum for hydrogen. Compare the positions of these lines with those in the emission spectrum for H in (a).

Plan

We know the wavelength of the light, and we calculate its frequency so that we can then calculate the energy of each photon.

Solution

$$\lambda\nu = c \quad \text{so} \quad \nu = \frac{c}{\lambda} = \frac{3.00 \times 10^8 \text{ m/s}}{4.86 \times 10^{-7} \text{ m}} = 6.17 \times 10^{14} \text{ s}^{-1}$$

$$E = h\nu = (6.63 \times 10^{-34} \text{ J} \cdot \text{s})(6.17 \times 10^{14} \text{ s}^{-1}) = \boxed{4.09 \times 10^{-19} \text{ J/photon}}$$

To gain a better appreciation of the amount of energy involved, let's calculate the total energy, in kilojoules, emitted by one mole of atoms. (Each atom emits one photon.)

$$\frac{?\text{ kJ}}{\text{mol}} = 4.09 \times 10^{-19} \frac{\text{J}}{\text{atom}} \times \frac{1 \text{ kJ}}{1 \times 10^3 \text{ J}} \times \frac{6.02 \times 10^{23} \text{ atoms}}{\text{mol}}$$

$$= \boxed{2.46 \times 10^2 \text{ kJ/mol}}$$

This calculation shows that when each atom in one mole of hydrogen atoms emits light of wavelength 4.86×10^{-7} m, the mole of atoms loses 246 kJ of energy as green light. (This would be enough energy to operate a 100-watt light bulb for more than 40 minutes.)

You should now work Exercises 36 and 38.

When an electric current is passed through hydrogen gas at very low pressures, several series of lines in the spectrum of hydrogen are produced. These lines were studied in-

The lightning flashes produced in electrical storms and the light produced by neon gas in neon signs are two familiar examples of visible light produced by electronic transitions.

The Danish physicist Niels Bohr (1885–1962) was one of the most influential scientists of the twentieth century. Like many other now-famous physicists of his time, he worked for a time in England with J. J. Thomson and later with Ernest Rutherford. During this period, he began to develop the ideas that led to the publication of his explanation of atomic spectra and his theory of atomic structure, for which he received the Nobel Prize in 1922. After escaping from German-occupied Denmark to Sweden in 1943, he helped to arrange the escape of hundreds of Danish Jews from the Hitler regime. He later went to the United States, where, until 1945, he worked with other scientists at Los Alamos, New Mexico, on the development of the atomic bomb. From then until his death in 1962, he worked hard for the development and use of atomic energy for peaceful purposes.

tensely by many scientists. J. R. Rydberg discovered in the late nineteenth century that the wavelengths of the various lines in the hydrogen spectrum can be related by a mathematical equation:

$$\frac{1}{\lambda} = R\left(\frac{1}{n_1^2} - \frac{1}{n_2^2}\right)$$

Here R is 1.097×10^7 m^{-1} and is known as the Rydberg constant. The n's are positive integers, and n_1 is smaller than n_2. The Rydberg equation was derived from numerous observations, not theory. It is thus an empirical equation.

In 1913 Niels Bohr, a Danish physicist, provided an explanation for Rydberg's observations. He wrote equations that described the electron of a hydrogen atom as revolving around the nucleus of an atom in circular orbits. He included the assumption that the electronic energy is *quantized;* that is, only certain values of electron energy are possible. This led him to suggest that electrons can only be in certain discrete orbits, and that they absorb or emit energy in discrete amounts as they move from one orbit to another. Each orbit thus corresponds to a definite *energy level* for the electron. When an electron is promoted from a lower energy level to a higher one, it absorbs a definite (or quantized) amount of energy. When the electron falls back to the original energy level, it emits

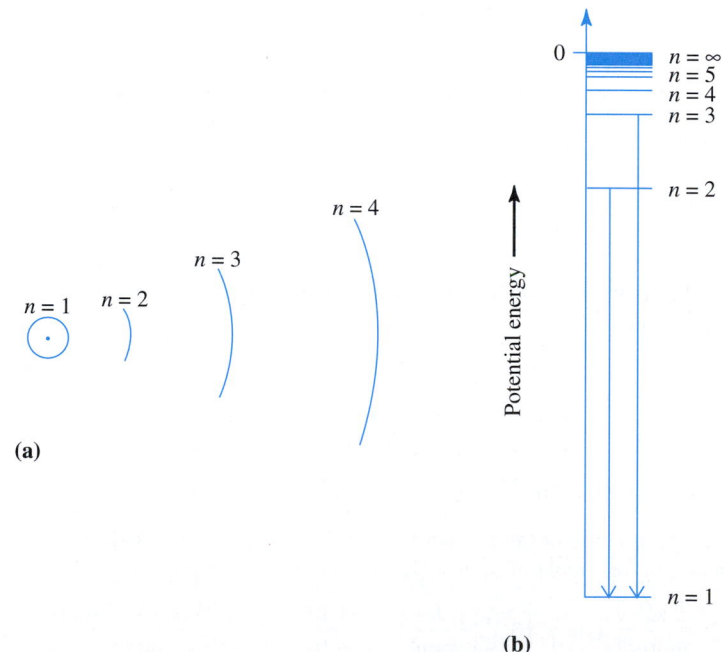

Figure 5-16 (a) The radii of the first four Bohr orbits for a hydrogen atom. The dot at the center represents the nuclear position. The radius of each orbit is proportional to n^2, so these four are in the ratio 1:4:9:16. (b) Relative values for the energies associated with the various energy levels in a hydrogen atom. The energies become closer together as n increases. They are so close together for large values of n that they form a continuum. By convention, potential energy is defined as zero when the electron is at an infinite distance from the atom. Any more stable arrangement would have a lower energy. Therefore, potential energies of electrons in atoms are always negative. Some possible electronic transitions corresponding to lines in the hydrogen emission spectrum are indicated by arrows. Transitions in the opposite directions account for lines in the absorption spectrum.

exactly the same amount of energy it absorbed in moving from the lower to the higher energy level. Figure 5-16 illustrates these transitions schematically. The values of n_1 and n_2 in the Rydberg equation identify the lower and higher levels, respectively, of these electronic transitions.

The Bohr Theory and the Rydberg Equation

ENRICHMENT

From mathematical equations describing the orbits for the hydrogen atoms, together with the assumption of quantization of energy, Bohr was able to determine two significant aspects of each allowed orbit:

1. *Where* (with respect to the nucleus) the electron can be—that is, the radius, r, of the circular orbit. This is given by

$$r = n^2 a_0$$

where n is a positive integer (1, 2, 3, ...) that tells which orbit is being described and a_0 is the *Bohr radius,* a constant with the dimensions of length. Bohr was able to calculate the value of a_0 from a combination of Planck's constant, the charge of the electron, and the mass of the electron as

$$a_0 = 5.292 \times 10^{-11} \text{ m} = 0.5292 \text{ Å}$$

2. *How stable* the electron would be in that orbit—that is, its potential energy, E. This is given by

$$E = -\frac{1}{n^2}\left(\frac{h^2}{8\pi^2 m a_0}\right) = -\frac{2.180 \times 10^{-18} \text{ J}}{n^2}$$

where h = Planck's constant, m = the mass of the electron, and the other symbols have the same meaning as before. E is always negative when the electron is in the atom; $E = 0$ when the atom is completely removed from the atom (n = infinity).

Results of evaluating these equations for some of the possible values of n (1, 2, 3, ...) are shown in Figure 5-17. The larger the value of n, the farther from the nucleus is the orbit being described, and the radius of this orbit increases as the *square of n* increases. As n increases, n^2 increases, $1/n^2$ decreases, and thus the electronic energy increases (becomes less negative and smaller in magnitude). For orbits farther from the nucleus, the electronic potential energy is higher (less negative—the electron is in a *higher* energy level or in a less stable state). Going away from the nucleus, the allowable orbits are farther apart in distance, but closer together in energy. Consider the two possible limits of these equations. One limit is when $n = 1$; this describes the electron at the smallest possible distance from the nucleus and at its lowest (most negative) energy. The other limit is for very large values of n, i.e., as n approaches infinity. As this limit is approached, the electron is very far from the nucleus, or effectively removed from the atom; the energy is as high as possible, approaching zero.

With these equations and the relationship that the Planck equation provides between energy and the frequency or wavelength, Bohr was able to predict the wavelengths observed in the hydrogen emission spectrum. Figure 5-17 illustrates the relationship between lines in the emission spectrum and the electronic transitions (changes of energy level) that occur in hydrogen atoms.

Note: r is proportional to n^2.

Note: E is proportional to $-\dfrac{1}{n^2}$.

We define the potential energy of a set of charged particles to be zero when the particles are infinitely far apart.

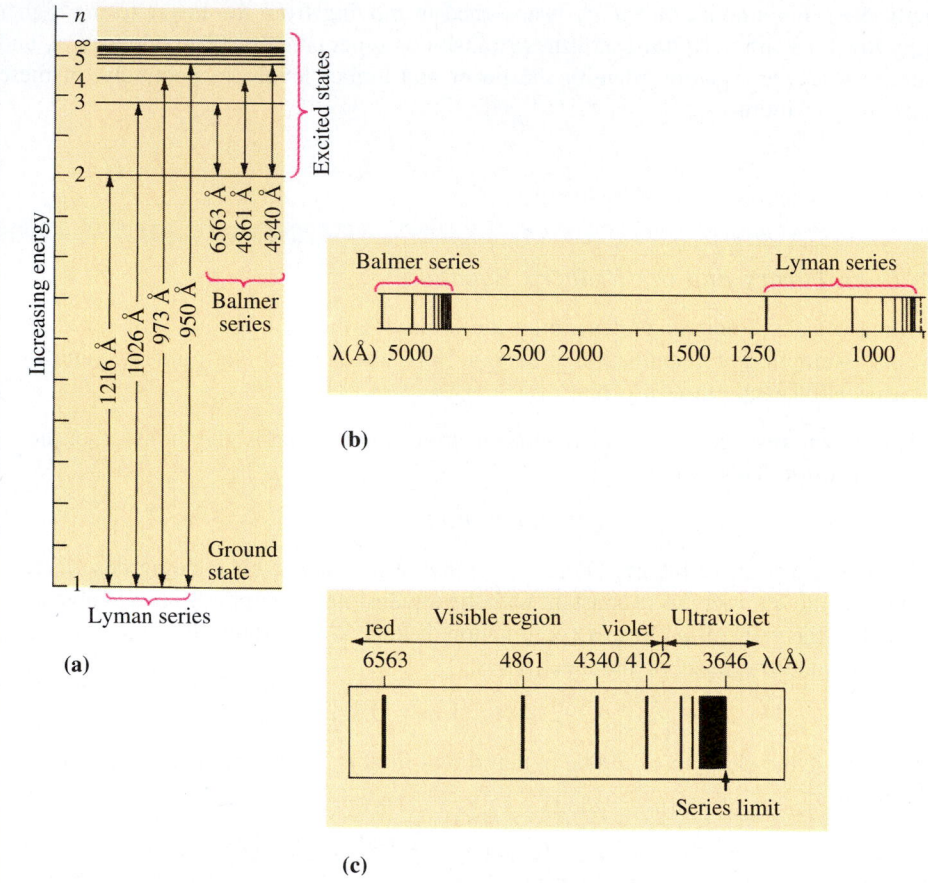

Figure 5-17   (a) The energy levels that the electron can occupy in a hydrogen atom and a few of the transitions that cause the emission spectrum of hydrogen. The numbers on the vertical lines show the wavelengths of light emitted when the electron falls to a lower energy level. (Light of the same wavelength is absorbed when the electron is promoted to the higher energy level.) The difference in energy between two given levels is exactly the same for all hydrogen atoms, so it corresponds to a specific wavelength and to a specific line in the emission spectrum of hydrogen. In a given sample, some hydrogen atoms could have their electrons excited to the $n = 2$ level. Some of these electrons could then fall to the $n = 1$ energy level, giving off the *difference* in energy in the form of light (the 1216-Å transition). Other hydrogen atoms might have their electrons excited to the $n = 3$ level; subsequently some could fall to the $n = 1$ level (the 1026-Å transition). Because higher energy levels become closer and closer in energy, *differences* in energy between successive transitions become smaller and smaller. The corresponding lines in the emission spectrum become closer together and eventually result in a continuum, a series of lines so close together that they are indistinguishable. (b) The emission spectrum of hydrogen. The series of lines produced by the electron falling to the $n = 1$ level is known as the *Lyman series;* it is in the ultraviolet region. A transition in which the electron falls to the $n = 2$ level gives rise to a similar set of lines in the visible region of the spectrum, known as the *Balmer series*. Not shown are series involving transitions to energy levels with higher values of n. (c) The Balmer series shown on an expanded scale. The line at 6563 Å (the $n = 3 \rightarrow n = 2$ transition) is much more intense than the line at 4861 Å (the $n = 4 \rightarrow n = 2$ transition) because the first transition occurs much more frequently than the second. Successive lines in the spectrum become less intense as the series limit is approached because the transitions that correspond to these lines are less probable.

Each line in the emission spectrum represents the *difference in energies* between two allowed energy levels for the electron. When the electron goes from energy level n_2 to energy level n_1, the difference in energy is given off as a single photon. The energy of this photon can be calculated from Bohr's equation for the energy, as follows.

$$E \text{ of photon} = E_2 - E_1 = -\frac{2.180 \times 10^{-18} \text{ J}}{n_2^{\ 2}} - \left(-\frac{2.180 \times 10^{-18} \text{ J}}{n_1^{\ 2}}\right)$$

Factoring out the constant 2.180×10^{-18} J and rearranging, we get

$$E \text{ of photon} = 2.180 \times 10^{-18} \text{ J}\left(\frac{1}{n_1^{\ 2}} - \frac{1}{n_2^{\ 2}}\right)$$

The Planck equation, $E = hc/\lambda$, relates the energy of the photon to the wavelength of the light, so

$$\frac{hc}{\lambda} = 2.180 \times 10^{-18} \text{ J}\left(\frac{1}{n_1^{\ 2}} - \frac{1}{n_2^{\ 2}}\right)$$

$h = 6.626 \times 10^{-34} \text{ J} \cdot \text{s}$
$c = 2.9979 \times 10^{8} \text{ m/s}$

Rearranging for $1/\lambda$, we obtain

$$\frac{1}{\lambda} = \frac{2.180 \times 10^{-18} \text{ J}}{hc}\left(\frac{1}{n_1^{\ 2}} - \frac{1}{n_2^{\ 2}}\right)$$

Comparing this to the Rydberg equation, Bohr showed that the Rydberg constant is equivalent to 2.180×10^{-18} J/hc. We could use the values for h and c to obtain the same value, $1.097 \times 10^{7} \text{ m}^{-1}$, that was obtained by Rydberg on a solely empirical basis. Further, Bohr showed the physical meaning of the two whole numbers n_1 and n_2; they represent the two energy states between which the transition takes place. Using this approach, Bohr was able to use fundamental constants to calculate the wavelengths of the observed lines in the hydrogen emission spectrum. Thus, Bohr's application of the idea of quantization of energy to the electron in an atom provided the answer to a half-century-old puzzle concerning the discrete colors given off in the spectrum.

We now accept the fact that electrons occupy only certain energy levels in atoms. In most atoms, some of the energy differences between levels correspond to the energy of visible light. Thus, colors associated with electronic transitions in such elements can be observed by the human eye.

Although the Bohr theory satisfactorily explained the spectra of hydrogen and of other species containing one electron (He^+, Li^{2+}, and so on) it could not calculate the wavelengths in the observed spectra of more complex species. Bohr's assumption of circular orbits was modified in 1916 by Sommerfeld, who assumed elliptical orbits. Even so, the Bohr approach was doomed to failure, because it modified classical mechanics to solve a problem that could not be solved by classical mechanics. It was a contrived solution. This failure of classical mechanics set the stage for the development of a new physics, quantum mechanics, to deal with small particles. However, the Bohr theory did introduce the ideas that only certain energy levels are possible, that these energy levels are described by quantum numbers that can have only certain allowed values, and that the quantum numbers indicate something about where and how stable the electrons are in these energy levels. The ideas of modern atomic theory have replaced Bohr's original theory. But his achievement in showing a link between electronic arrangements and Rydberg's empirical description of light absorption, and in establishing the quantization of electronic energy, was a very important step toward an understanding of atomic structure.

Two big questions remained about electrons in atoms: (1) How are electrons arranged in atoms? (2) How do these electrons behave? We now have the background to consider how modern atomic theory answers these questions.

5-13 THE WAVE NATURE OF THE ELECTRON

Be careful to distinguish between the letter v, which represents velocity and the Greek letter nu, ν, which represents frequency (Section 5-10).

The idea that light can exhibit both wave properties and particle properties suggested to Louis de Broglie that very small particles, such as electrons, might also display wave properties under the proper circumstances. In his doctoral thesis in 1925, de Broglie predicted that a particle with a mass m and velocity v should have a wavelength associated with it. The numerical value of this de Broglie wavelength is given by

$$\lambda = h/mv \quad \text{(where } h = \text{Planck's constant)}$$

Two years after de Broglie's prediction, C. Davisson and L. H. Germer at the Bell Telephone Laboratory demonstrated diffraction of electrons by a crystal of nickel. This behavior is an important characteristic of waves. It shows conclusively that electrons do have wave properties. Davisson and Germer found that the wavelength associated with electrons of known energy is exactly that predicted by de Broglie. Similar diffraction experiments have been successfully performed with other particles, such as neutrons.

EXAMPLE 5-6   de Broglie Equation

(a) Calculate the wavelength of an electron traveling at 1.24×10^7 m/s. The mass of an electron is 9.11×10^{-28} g. (b) Calculate the wavelength of a baseball of mass 5.25 oz traveling at 92.5 mi/h. Recall that $1 \text{ J} = 1 \text{ kg} \cdot \text{m}^2/\text{s}^2$.

Plan

For each calculation, we use the de Broglie equation

$$\lambda = \frac{h}{mv}$$

where

$$h \text{ (Planck's constant)} = 6.63 \times 10^{-34} \text{ J} \cdot \text{s} \times \frac{1 \dfrac{\text{kg} \cdot \text{m}^2}{\text{s}^2}}{1 \text{ J}}$$

$$= 6.63 \times 10^{-34} \frac{\text{kg} \cdot \text{m}^2}{\text{s}}$$

For consistency of units, mass must be expressed in kilograms. In part (b), we must also convert the speed to meters per second.

Solution

(a)
$$m = 9.11 \times 10^{-28} \text{ g} \times \frac{1 \text{ kg}}{1000 \text{ g}} = 9.11 \times 10^{-31} \text{ kg}$$

Substituting into the de Broglie equation,

$$\lambda = \frac{h}{mv} = \frac{6.63 \times 10^{-34} \dfrac{\text{kg} \cdot \text{m}^2}{\text{s}}}{(9.11 \times 10^{-31} \text{ kg})\left(1.24 \times 10^7 \dfrac{\text{m}}{\text{s}}\right)} = \boxed{5.87 \times 10^{-11} \text{ m}}$$

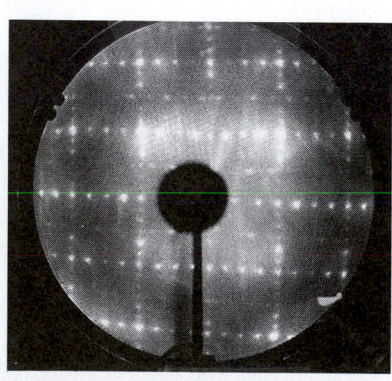

Materials scientists study electron diffraction patterns to learn about the surfaces of solids.

Though this seems like a very short wavelength, it is similar to the spacing between atoms in many crystals. A stream of such electrons hitting a crystal gives easily measurable diffraction patterns.

(b)
$$m = 5.25 \text{ oz} \times \frac{1 \text{ lb}}{16 \text{ oz}} \times \frac{1 \text{ kg}}{2.205 \text{ lb}} = 0.149 \text{ kg}$$

$$v = \frac{92.5 \text{ mi}}{\text{h}} \times \frac{1 \text{ h}}{3600 \text{ s}} \times \frac{1.609 \text{ km}}{1 \text{ mi}} \times \frac{1000 \text{ m}}{1 \text{ km}} = 41.3 \frac{\text{m}}{\text{s}}$$

Now, we substitute into the de Broglie equation.

$$\lambda = \frac{h}{mv} = \frac{6.63 \times 10^{-34} \frac{\text{kg} \cdot \text{m}^2}{\text{s}}}{(0.149 \text{ kg})\left(41.3 \frac{\text{m}}{\text{s}}\right)} = 1.08 \times 10^{-34} \text{ m}$$

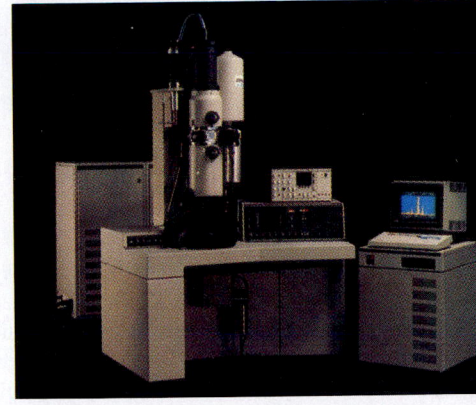

A modern electron microscope.

This wavelength is far too short to give any measurable effects. Recall that atomic diameters are of the order of 10^{-10} m, which is 24 powers of 10 greater than the baseball "wavelength."

You should now work Exercise 58.

As you can see from the results of Example 5-6, the particles of the subatomic world behave very differently from the macroscopic objects with which we are familiar. To talk about behavior of atoms and their particles, we must give up many of our long-held prejudices about the behavior of matter. We must be willing to visualize a world of new and unfamiliar properties, such as the ability to act in some ways like a particle and in other ways like a wave.

The wave behavior of electrons is exploited in the electron microscope. This instrument allows magnification of objects far too small to be seen with an ordinary light microscope.

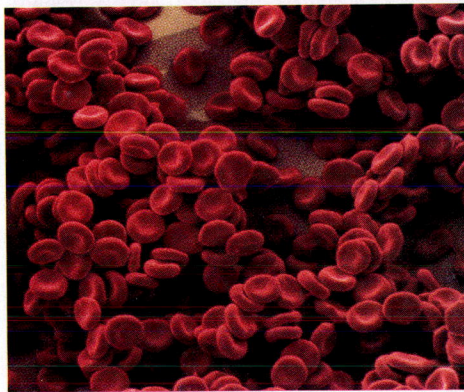

A color-enhanced scanning electron micrograph of human red blood cells, magnified 1200×.

5-14 THE QUANTUM MECHANICAL PICTURE OF THE ATOM

Through the work of de Broglie, Davisson and Germer, and others, we now know that electrons in atoms can be treated as waves more effectively than as small compact particles traveling in circular or elliptical orbits. Large objects such as golf balls and moving automobiles obey the laws of classical mechanics (Isaac Newton's laws), but very small particles such as electrons, atoms, and molecules do not. A different kind of mechanics, called **quantum mechanics,** which is based on the *wave* properties of matter, describes the behavior of very small particles much better. Quantization of energy is a consequence of these properties.

One of the underlying principles of quantum mechanics is that we cannot determine precisely the paths that electrons follow as they move about atomic nuclei. The **Heisenberg Uncertainty Principle,** stated in 1927 by Werner Heisenberg, is a theoretical assertion that is consistent with all experimental observations.

It is impossible to determine accurately both the momentum and the position of an electron (or any other very small particle) simultaneously.

CHEMISTRY IN USE

Research & Technology

Astrochemistry

Does life exist in other places in the universe, or are we unique and alone? Astrochemistry is a relatively new branch of chemistry that studies the chemical composition of interstellar space. The analytical techniques and instrumentation used by astrochemists provide insight into the search for life in the cosmos. To help us understand how astrochemists work, we must first learn a little more about electron and proton spins.

We have seen that electrons behave as though they were spinning like tops and that when two electrons occupy an orbital their spins must be opposite; that is, their spins must be paired. The most common kind of hydrogen atom is made up of one electron and one proton. Like electrons, protons behave as if they were spinning. Two electrons in the same orbital have paired spins. A hydrogen atom is more stable when the spins of its proton and electron are paired; astrochemists use this information to detect hydrogen atoms in space.

Much interstellar space contains hydrogen atoms at a concentration of about one hydrogen atom per mL of space. Most of these hydrogen atoms have paired proton and electron spins; these hydrogen atoms are said to be in their *ground states*. Occasionally an electron in a hydrogen atom absorbs energy from a nearby star and its electron begins spinning in the opposite direction. These hydrogen atoms are said to be *excited* or unstable. After a while, an excited hydrogen atom loses its extra energy, and the proton and electron spins again become paired. The energy that excited hydrogen atoms lose when they return to their ground states is emitted as radio waves; scientists use radio telescopes to pick up these signals to locate clouds of hydrogen. In a very real sense, astrochemists using radiotelescopes make discoveries as great as Galileo did when he first trained his newly-invented telescope on the heavens early in the 17th century.

Astrochemists have discovered more than one hundred molecules in space. One molecular substance can be distinguished from another because each molecule releases its own unique radio wave frequency when it goes from an excited state to a ground state. Many organic molecules (Chapters 27 and 28) have been discovered in space; some of them are precursors to molecules such as proteins, which are associated with life (Section 28-11). Because astrochemists have found some building blocks of life throughout interstellar space, many wonder if there might be life elsewhere in the universe.

The Rosette Nebula.

To answer this question, radio telescopes have been used in about 50 separate projects since 1959 to search for alien radio signals. In 1992, National Aeronautical and Space Administration (NASA) scientists began a project known as SETI (Search for Extra-Terrestrial Intelligence). This ten-year project utilizes two large radio telescopes and an assortment of computer software to analyze 14 million radio signals simultaneously. Within SETI's first few hours of operation, more signals were scanned than had been scanned in the thousands of hours of search time in earlier projects.

What will be found during the SETI project? Will astrochemists detect enough of the molecules of life to reach a conclusion? Will we receive intelligible radio signals? Will we be any closer to knowing if life exists elsewhere in the universe, or if we are unique and alone?

Ronald DeLorenzo
Middle Georgia College

Momentum is mass times velocity, *mv*. Because electrons are so small and move so rapidly, their motion is usually detected by electromagnetic radiation. Photons that interact with electrons have about the same energies as the electrons. Consequently, the interaction of a photon with an electron severely disturbs the motion of the electron. It is not possible to determine simultaneously both the position and the velocity of an electron, so we resort to a statistical approach and speak of the probability of finding an electron within specified regions in space.

This is like trying to locate the position of a moving automobile by driving another automobile into it.

With these ideas in mind, we list some basic ideas of quantum mechanics.

1. Atoms and molecules can exist only in certain energy states. In each energy state, the atom or molecule has a definite energy. When an atom or molecule changes its energy state, it must emit or absorb just enough energy to bring it to the new energy state (the quantum condition).

Atoms and molecules possess various forms of energy. Let us focus our attention on their *electronic energies*.

2. When atoms or molecules emit or absorb radiation (light), they change their energies. The energy change in the atom or molecule is related to the frequency or wavelength of the light emitted or absorbed by the equations:

$$\Delta E = h\nu \quad \text{or} \quad \Delta E = hc/\lambda$$

Recall that $\lambda\nu = c$, so $\nu = c/\lambda$.

This gives a relationship between the energy change, ΔE, and the wavelength, λ, of the radiation emitted or absorbed. *The energy lost (or gained) by an atom as it goes from higher to lower (or lower to higher) energy states is equal to the energy of the photon emitted (or absorbed) during the transition.*

3. The allowed energy states of atoms and molecules can be described by sets of numbers called *quantum numbers*.

The mathematical approach of quantum mechanics involves treating the electron in an atom as a *standing wave*. A standing wave is a wave that does not travel and therefore has at least one point at which it has zero amplitude, called a node. As an example, consider the various ways that a guitar string can vibrate when it is plucked (Figure 5-18). Because both ends are fixed (nodes), the string can vibrate only in ways in which there is a whole number of *half-wavelengths* in the length of the string (Figure 5-18a). Any possible motion of the string can be described as some combination of these allowed vibrations. In a similar way, we can imagine that the electron in the hydrogen atom behaves as a wave (recall the de Broglie relationship in the last section). The electron can be described by the

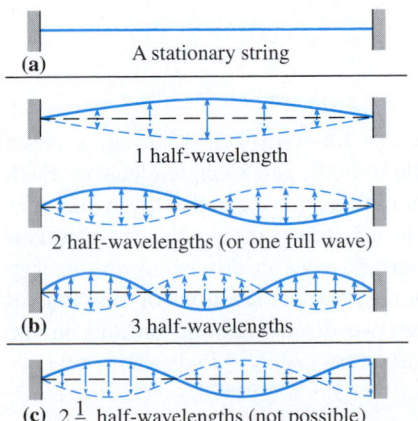

(a)   A stationary string

1 half-wavelength

2 half-wavelengths (or one full wave)

(b)   3 half-wavelengths

(c) $2\frac{1}{2}$ half-wavelengths (not possible)

Figure 5-18 When a string that is fixed at both ends—such as a guitar string (a)—is plucked, it has a number of natural patterns of vibration, called normal modes. Because the string is fixed at both ends, the ends must be stationary. Each different possible vibration is a standing wave, and can be described by a wave function. The only waves that are possible are those in which a whole number of half-wavelengths fits into the string length. These allowed waves comprise a harmonic series. Any total motion of the string is some combination of these allowed harmonics. (b) Some of the ways in which a plucked guitar string can vibrate. The position of the string at one extreme of each vibration is shown as a solid line, and at the other extreme as a dashed line. (c) An example of a vibration that is *not* possible for a plucked string. In such a vibration, an end of the string would move; this is not possible because the ends are fixed.

same kind of standing-wave mathematics that is applied to the vibrating guitar string. In this approach, the electron is characterized by a wave function, ψ. In a given space around the nucleus, only certain "waves" can exist. Each "allowed wave" corresponds to a stable energy state for the electron, and is described by a particular set of quantum numbers.

The quantum mechanical treatment of atoms and molecules is highly mathematical. The important point is that each solution of the Schrödinger wave equation (see Enrichment section) describes a possible energy state for the electrons in the atom. Each solution is described by a set of three **quantum numbers.** These numbers are in accord with those deduced from experiment and from empirical equations such as the Rydberg equation. Solutions of the Schrödinger equation also tell us about the shapes and orientations of the statistical probability distributions of the electrons. (The Heisenberg Principle implies that this is how we must describe the positions of the electrons.) These *atomic orbitals* (which are described in Section 5-16) are deduced from the solutions of the Schrödinger equation. The orbitals are directly related to the quantum numbers.

 **ENRICHMENT**

The Schrödinger Equation

In 1926 Erwin Schrödinger modified an existing equation that described a three-dimensional standing wave by imposing wavelength restrictions suggested by de Broglie's ideas. The modified equation allowed him to calculate the energy levels in the hydrogen atom. It is a differential equation that need not be memorized or even understood to read this book. A knowledge of differential calculus would be necessary.

$$-\frac{h^2}{8\pi^2 m}\left(\frac{\partial^2 \psi}{\partial x^2} + \frac{\partial^2 \psi}{\partial y^2} + \frac{\partial^2 \psi}{\partial z^2}\right) + V\psi = E\psi$$

This equation has been solved exactly only for one-electron species such as the hydrogen atom and the ions He^+ and Li^{2+}. Simplifying assumptions are necessary to solve the equation for more complex atoms and molecules. However, chemists and physicists have used their intuition and ingenuity (and modern computers) to apply this equation to more complex systems.

In 1928 Paul A. M. Dirac reformulated electron quantum mechanics to take into account the effects of relativity. This gave rise to a fourth quantum number.

5-15 QUANTUM NUMBERS

The solutions of the Schrödinger and Dirac equations for hydrogen atoms give wave functions, ψ, that describe the various states available to hydrogen's single electron. Each of these possible states is described by four quantum numbers. We can use these quantum numbers to designate the electronic arrangements in all atoms, their so-called **electron configurations.** These quantum numbers play important roles in describing the energy levels of electrons and the shapes of the orbitals that describe distributions of electrons in space. This interpretation will become clearer when we discuss atomic orbitals in the following section. For now, let's say that an **atomic orbital** is a region of space in which the probability of finding an electron is high.

We define each quantum number and describe the range of values it may take.

1. The **principal quantum number,** n, describes the *main energy level* an electron occupies. It may be any positive integer:

$$n = 1, 2, 3, 4, \ldots$$

2. The **subsidiary** (or **azimuthal**) **quantum number,** ℓ, designates the *shape of the region* in space an electron occupies. Within each energy level (defined by the value of n, the principal quantum number), ℓ may take integral values from 0 up to and including $(n-1)$:

$$\ell = 0, 1, 2, \ldots, (n-1)$$

Thus, the maximum value of ℓ is $(n-1)$. The subsidiary quantum number designates a **sublevel,** or a specific *kind* of atomic orbital, that an electron may occupy. We give a letter notation to each value of ℓ. Each letter corresponds to a different kind of atomic orbital.

$$\ell = 0, 1, 2, 3, \ldots, (n-1)$$
$$s \quad p \quad d \quad f$$

In the first energy level, the maximum value of ℓ is zero, which tells us that there is only an s sublevel and no p sublevel. In the second energy level, the permissible values of ℓ are 0 and 1, which tells us that there are only s and p sublevels.

3. The **magnetic quantum number,** m_ℓ, designates the spatial orientation of an atomic orbital. Within each sublevel, m_ℓ may take any integral values from $-\ell$ through zero up to and including $+\ell$:

$$m_\ell = (-\ell), \ldots, 0, \ldots, (+\ell)$$

The maximum value of m_ℓ depends on the value of ℓ. For example, when $\ell = 1$, which designates the p sublevel, there are three permissible values of m_ℓ: -1, 0, and $+1$. Thus, there are three distinct regions of space, called atomic orbitals, associated with a p sublevel. We refer to these orbitals as the p_x, p_y, and p_z orbitals (see the next section).

4. The **spin quantum number,** m_s, refers to the spin of an electron and the orientation of the magnetic field produced by this spin. For every set of n, ℓ, and m_ℓ values, m_s can take the value $+\frac{1}{2}$ or $-\frac{1}{2}$:

$$m_s = \pm\frac{1}{2}$$

The values of n, ℓ, and m_ℓ describe a particular atomic orbital. Each atomic orbital can accommodate no more than two electrons, one with $m_s = +\frac{1}{2}$ and another with $m_s = -\frac{1}{2}$.

Table 5-4 summarizes some permissible values for the four quantum numbers. Spectroscopic evidence confirms the quantum mechanical predictions about the number of atomic orbitals in each energy level.

The s, p, d, f designations arise from the characteristics of spectral emission lines produced by electrons occupying the orbitals: s (sharp), p (principal), d (diffuse), and f (fundamental).

Table 5-4 *Permissible Values of the Quantum Numbers Through n = 4*

n	ℓ	m_ℓ	m_s	Electron Capacity of Sublevel = $4\ell + 2$	Electron Capacity of Energy Level = $2n^2$
1	0 (1s)	0	$+\frac{1}{2}, -\frac{1}{2}$	2	2
2	0 (2s)	0	$+\frac{1}{2}, -\frac{1}{2}$	2	8
	1 (2p)	$-1, 0, +1$	$\pm\frac{1}{2}$ for each value of m_ℓ	6	
3	0 (3s)	0	$+\frac{1}{2}, -\frac{1}{2}$	2	18
	1 (3p)	$-1, 0, +1$	$\pm\frac{1}{2}$ for each value of m_ℓ	6	
	2 (3d)	$-2, -1, 0, +1, +2$	$\pm\frac{1}{2}$ for each value of m_ℓ	10	
4	0 (4s)	0	$+\frac{1}{2}, -\frac{1}{2}$	2	32
	1 (4p)	$-1, 0, +1$	$\pm\frac{1}{2}$ for each value of m_ℓ	6	
	2 (4d)	$-2, -1, 0, +1, +2$	$\pm\frac{1}{2}$ for each value of m_ℓ	10	
	4 (4f)	$-3, -2, -1, 0, +1, +2, +3$	$\pm\frac{1}{2}$ for each value of m_ℓ	14	

5-16 ATOMIC ORBITALS

As you study the next two sections, keep in mind that the wave function, ψ, for an orbital characterizes two features of an electron in that orbital: (1) *where* (the region in space) the probability of finding the electron is high and (2) *how stable* that electron is (its energy).

Let us now describe the distributions of electrons in atoms. For each neutral atom, we must account for a number of electrons equal to the number of protons in the nucleus, i.e., the atomic number of the atom. Each electron is said to occupy an atomic orbital defined by a set of quantum numbers n, ℓ, and m_ℓ. An orbital is the region of space in which there is a high probability of finding the electron. In any atom, each orbital can hold a maximum of two electrons. Within each atom, these atomic orbitals, taken together, can be represented as a diffuse cloud of electrons (Figure 5-19).

The energy level of each atomic orbital in an atom is indicated by the principal quantum number n (from the Schrödinger equation). As we have seen, the principal quantum number takes integral values: $n = 1, 2, 3, 4, \ldots$. The value $n = 1$ describes the first, or lowest, energy level. These energy levels have been referred to as electron shells. Successive shells are at increasingly greater distances from the nucleus. For example, the $n = 2$ shell is farther from the nucleus than the $n = 1$ shell. The electron capacity of each energy level is indicated in the right-hand column of Table 5-4. For a given n, the capacity is $2n^2$.

By the rules of Section 5-15, each energy level has one s sublevel (defined by $\ell = 0$) consisting of one s atomic orbital (defined by $m_\ell = 0$). We distinguish among orbitals in different principal shells (main energy levels) by using the principal quantum number as a coefficient; $1s$ indicates the s orbital in the first energy level, $2s$ is the s orbital in the second energy level, $2p$ is a p orbital in the second energy level, and so on (Table 5-4).

For each solution to the quantum mechanical equation, we can calculate the electron probability density (sometimes just called the electron density) at each point in the atom. This is the probability of finding an electron at that point. It can be shown that this electron density is proportional to $r^2\psi^2$, where r is the distance from the nucleus.

In the graphs in Figure 5-20, the electron probability density at a given distance from the nucleus is plotted against distance from the nucleus, for s orbitals. It is found that the electron probability density curve is the same regardless of the direction in the atom. We describe an s orbital as *spherically symmetrical;* i.e., it is round like a basketball (Figure 5-21). The electron clouds (electron densities) associated with the $1s$, $2s$, and $3s$ atomic orbitals are shown just below the plots. The electron clouds are three-dimensional, and only cross sections are shown here. The regions shown in some figures (Figures 5-21, 5-22, 5-23, 5-24, and 5-25) appear to have surfaces or skins only because they are arbitrar-

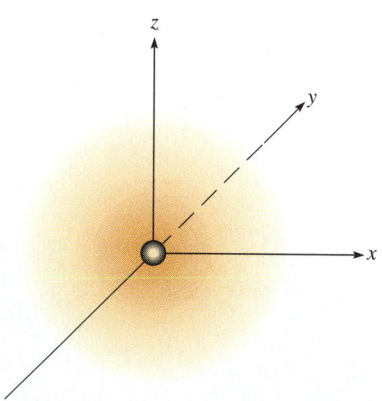

Figure 5-19 An electron cloud surrounding an atomic nucleus. The electron density drops off rapidly but smoothly as distance from the nucleus increases.

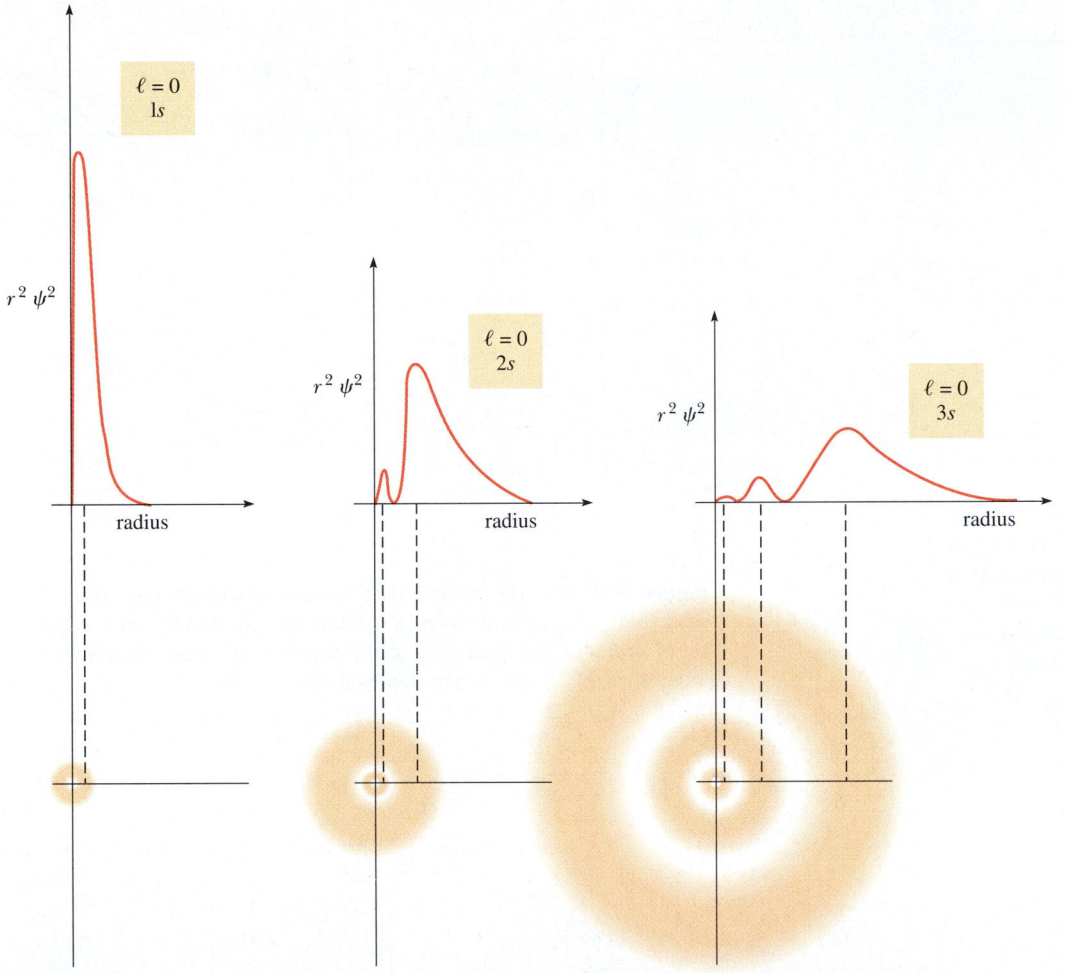

Figure 5-20 Plots of the electron density distributions associated with *s* orbitals. For any *s* orbital, this plot is the same in any direction (spherically symmetrical). The sketch below each plot shows a cross section, in the plane of the atomic nucleus, of the electron cloud associated with that orbital.

ily "cut off" so that there is a 90% probability of finding an electron occupying the orbital somewhere within the surfaces.

Beginning with the second energy level, each level contains a *p* sublevel, defined by $\ell = 1$. Each of these sublevels consists of a set of *three p* atomic orbitals, corresponding to the three allowed values of m_ℓ (-1, 0, and $+1$) when $\ell = 1$. The sets are referred to as 2*p*, 3*p*, 4*p*, 5*p*, . . . orbitals to indicate the main energy levels in which they occur. Each set of atomic *p* orbitals resembles three mutually perpendicular equal-arm dumbbells (Figure 5-22). The nucleus defines the origin of a set of Cartesian coordinates with the usual *x*, *y*, and *z* axes (Figure 5-23a). The subscript *x*, *y*, or *z* indicates the axis along which each of the three two-lobed orbitals is directed. A set of three *p* atomic orbitals may be represented as in Figure 5-23b.

Beginning at the third energy level, each level contains a third sublevel ($\ell = 2$) composed of a set of *five d* atomic orbitals ($m_\ell = -2, -1, 0, +1, +2$). They are designated

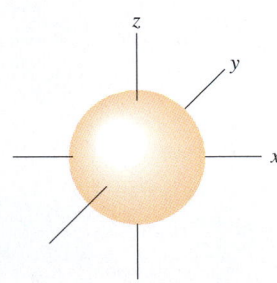

Figure 5-21 The shape of an *s* orbital.

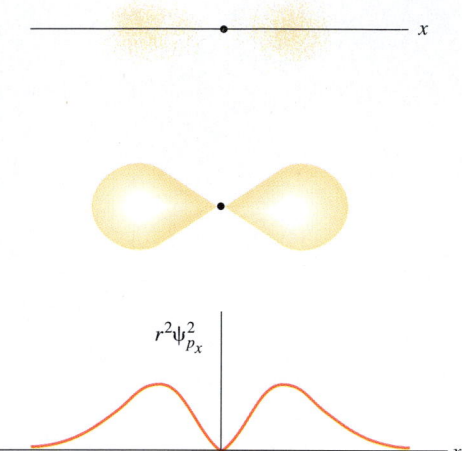

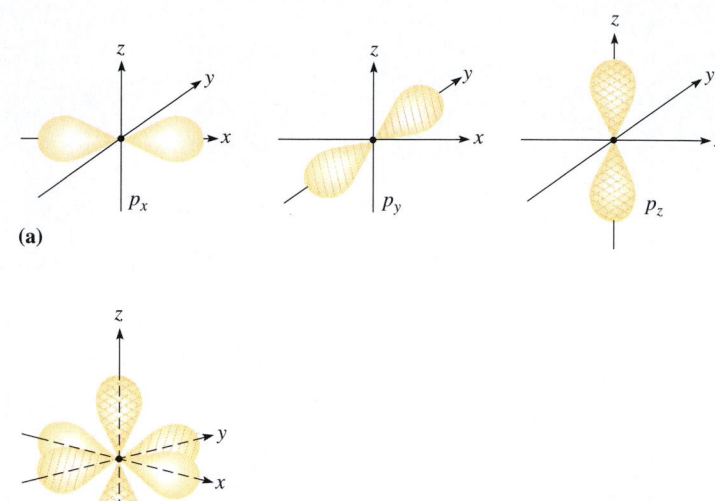

Figure 5-22 Three representations of the shape of a *p* orbital. The plot at the bottom is along the axis of maximum electron density for this orbital. A plot along any other direction would be different, because a *p* orbital is *not* spherically symmetrical.

Figure 5-23 (a) The relative directional character of a set of *p* orbitals. (b) A model of three *p* orbitals (p_x, p_y, and p_z) of a single set of orbitals. The nucleus is at the center. (The lobes are actually more diffuse ("fatter") than depicted. See Figure 5-26.)

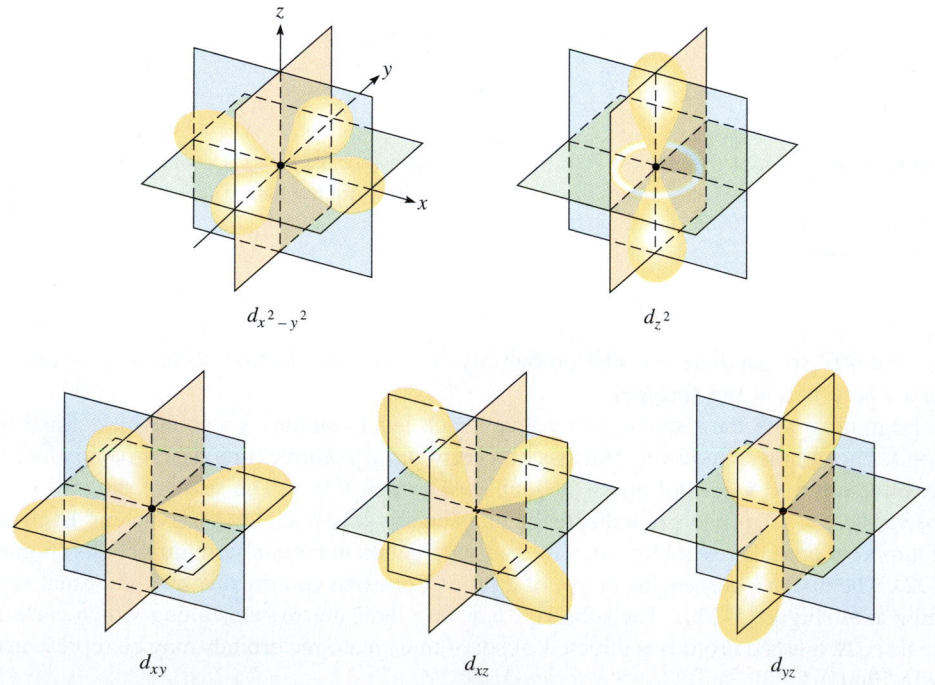

Figure 5-24 Spatial orientation of *d* orbitals. Note that the lobes of the $d_{x^2-y^2}$ and d_{z^2} orbitals lie along the axes, whereas the lobes of the others lie along diagonals between the axes.

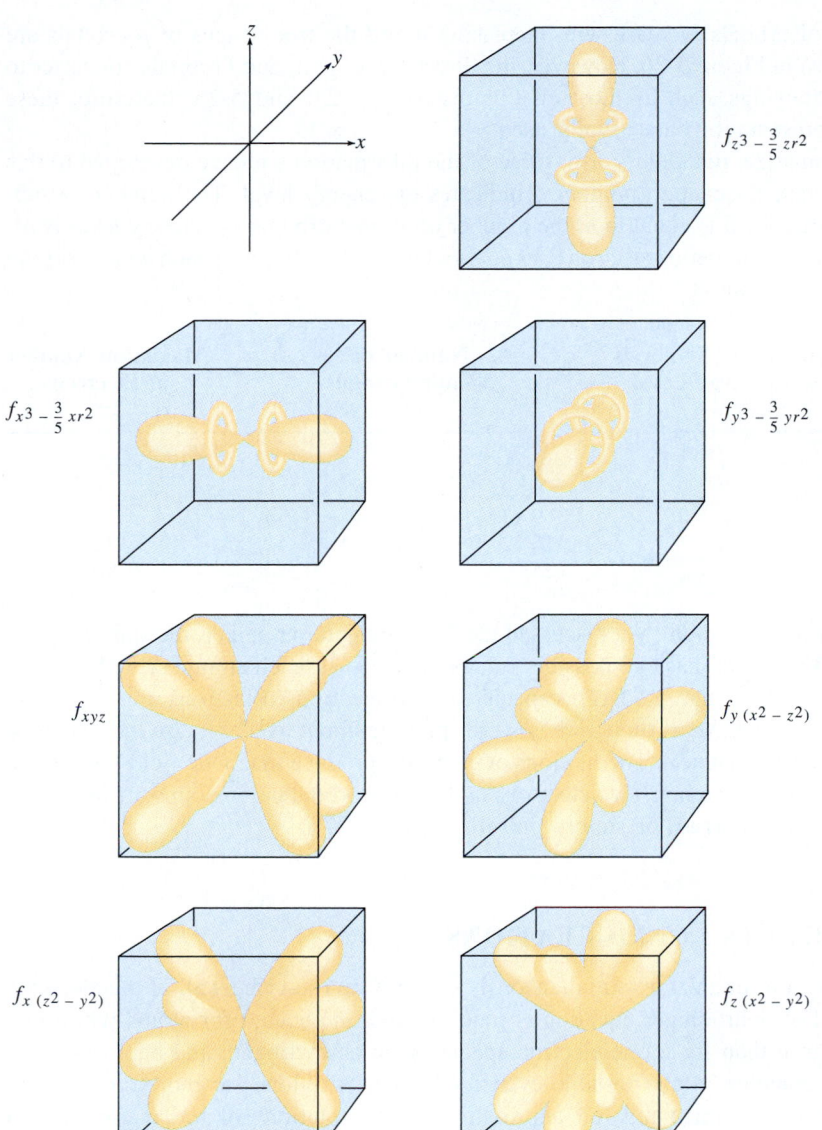

Figure 5-25 Relative directional character of f orbitals. The seven orbitals are shown within cubes as an aid to visualization.

$3d$, $4d$, $5d$, ... to indicate the energy level in which they are found. The shapes of the members of a set are indicated in Figure 5-24.

In each of the fourth and higher energy levels, there is also a fourth sublevel, containing a set of *seven f* atomic orbitals ($\ell = 3$, $m_\ell = -3$, -2, -1, 0, $+1$, $+2$, $+3$). These are shown in Figure 5-25.

Thus, we see the first energy level contains only the $1s$ orbital; the second energy level contains the $2s$ and three $2p$ orbitals; the third energy level contains the $3s$, three $3p$, and five $3d$ orbitals; and the fourth energy level consists of a $4s$, three $4p$, five $4d$, and seven $4f$ orbitals. All subsequent energy levels contain s, p, d, and f sublevels as well as others that are not occupied in any presently known elements in their lowest energy states.

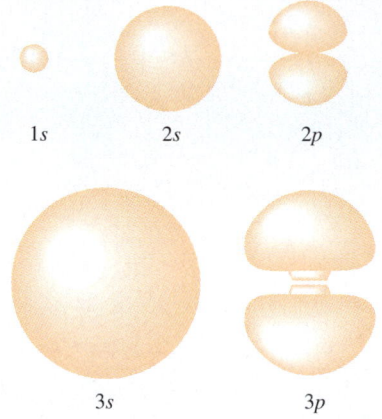

1s   2s   2p

3s   3p

Figure 5-26 Shapes and approximate relative sizes of several orbitals in an atom.

The sizes of orbitals increase with increasing n and the true shapes of p orbitals are "fat," as shown in Figure 5-26. However, the directions of p, d, and f orbitals are easier to visualize in drawings such as those in Figures 5-23, 5-24, and 5-25; therefore, these "slender" representations are usually used.

Let us summarize, in tabular form, some of the information we have developed to this point. The principal quantum number n indicates the energy level. The number of sublevels per energy level is equal to n, the number of atomic orbitals per energy level is n^2, and the maximum number of electrons per energy level is $2n^2$, because each atomic orbital can hold two electrons.

Energy Level n	Number of Sublevels per Energy Level n	Number of Atomic Orbitals n^2	Maximum Number of Electrons $2n^2$
1	1	1 (1s)	2
2	2	4 ($2s$, $2p_x$, $2p_y$, $2p_z$)	8
3	3	9 ($3s$, three $3p$'s, five $3d$'s)	18
4	4	16	32
5	5	25	50

In this section, we haven't yet discussed the fourth quantum number, the spin quantum number, m_s. Because m_s has two possible values, $+\frac{1}{2}$ and $-\frac{1}{2}$, each atomic orbital, defined by the values of n, ℓ, and m_ℓ, has a capacity of two electrons. Electrons are negatively charged, and they behave as though they were spinning about axes through their centers, so they act like tiny magnets. The motions of electrons produce magnetic fields, and these can interact with one another. Two electrons in the same orbital having opposite m_s values are said to be **spin-paired,** or simply **paired** (Figure 5-27).

5-17 ELECTRON CONFIGURATIONS

The wave function for an atom simultaneously depends on (describes) all of the electrons in the atom. The Schrödinger equation is much more complicated for atoms with more than one electron than for a one-electron species such as hydrogen, and an explicit solution to this equation is not possible even for helium, let alone for more complicated atoms. Therefore we must rely on approximations to solutions of the many-electron Schrödinger equation. We shall use one of the most common and useful, called the **orbital approximation.** In this approximation, the electron cloud of an atom is assumed to be

The great power of modern computers has allowed scientists to make numerical approximations to this solution to very high accuracy for simple atoms such as helium. However, as the number of electrons increases, even such numerical approaches become quite difficult to apply and interpret. For many purposes, more quantitative approximations are suitable.

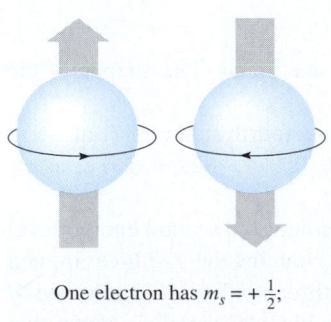

One electron has $m_s = +\frac{1}{2}$;

the other has $m_s = -\frac{1}{2}$.

Figure 5-27 Electron spin. Electrons act as though they spin about an axis through their centers. Because there are two directions in which an electron may spin, the spin quantum number has two possible values, $+\frac{1}{2}$ and $-\frac{1}{2}$. Each electron spin produces a magnetic field. When two electrons have opposite spins, the attraction due to their opposite magnetic fields (gray arrows) helps to overcome the repulsion of their like charges. This permits two electrons to occupy the same region (orbital).

the superposition of charge clouds, or orbitals, arising from the individual electrons; these orbitals resemble the atomic orbitals of hydrogen (for which exact solutions are known), which we described in some detail in the last section. Each electron is described by the same allowed combinations of quantum numbers (n, ℓ, m_ℓ, and m_s) that we used for the hydrogen atom; however, the order of energies of the orbitals is often different from that in hydrogen.

Let us now examine the electronic structures of atoms of different elements. The electronic arrangement that we shall describe for each atom is called the **ground state electron configuration.** This corresponds to the isolated atom in its lowest energy, or unexcited, state. We shall consider the elements in order of increasing atomic number, using as our guide the periodic table inside the front cover of this text.

In building up ground state electron configurations, the guiding idea is that the *total energy* of the atom is as low as possible. To determine these configurations, we use the **Aufbau Principle** as a guide:

The German verb *aufbauen* means "to build up."

Each atom is "built up" by (1) adding the appropriate numbers of protons and neutrons as specified by the atomic number and the mass number, and (2) adding the necessary number of electrons into orbitals in the way that gives the lowest *total* energy for the atom.

As we apply this principle, we shall focus on the difference in electronic arrangement between a given element and the element with an atomic number that is one lower. Though we do not always point it out, we *must* keep in mind that the atomic number (the charge on the nucleus) also differs. We also emphasize the particular electron that distinguishes each element from the previous one; however, we should remember that this distinction is artificial, because electrons are not really distinguishable.

The orbitals increase in energy with increasing value of the quantum number n. For a given value of n, energy increases with increasing value of ℓ. In other words, within a particular major energy level, the s sublevel is lowest in energy, the p sublevel is the next lowest, then the d, then the f, and so on. As a result of changes in the nuclear charge and interactions among the electrons in the atom, the order of energies of the orbitals can vary somewhat from atom to atom. The *usual* order of energies of the orbitals of an atom and a helpful device for remembering this order are shown in Figures 5-28 and 5-29.

The electronic structures of atoms are governed by the **Pauli Exclusion Principle:**

No two electrons in an atom may have identical sets of four quantum numbers.

An orbital is described by a particular allowed set of values for n, ℓ, and m_ℓ. Thus, two electrons can occupy the same orbital only if they have opposite spins, m_s. Two such electrons in the same orbital are *paired*. For simplicity, we shall indicate atomic orbitals as __ and show an unpaired electron as ↿ and spin-paired electrons as ↿⇂. By "unpaired electron" we mean an electron that occupies an orbital singly.

Row 1. The first energy level consists of only one atomic orbital, 1s. This can hold a maximum of two electrons. Hydrogen, as we have already noted, contains just one electron. Helium, a noble gas, has a filled first energy level (two electrons). The atom is so stable that no chemical reactions of helium are known.

Helium's electrons can be displaced only by electrical forces, as in excitation by high-voltage discharge.

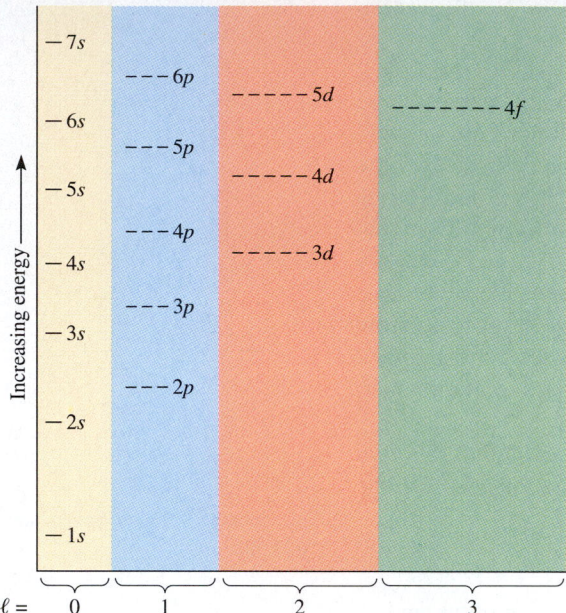

Figure 5-28 The usual order of filling (Aufbau order) of the orbitals of an atom. The energy scale varies for different elements, but the following main features should be noted: (1) The largest energy gap is between the $1s$ and $2s$ orbitals. (2) The energies of orbitals are generally closer together at higher energies. (3) The gap between np and $(n + 1)s$ (e.g., between $2p$ and $3s$ or between $3p$ and $4s$) is fairly large. (4) The gap between $(n - 1)d$ and ns (e.g., between $3d$ and $4s$) is quite small. (5) The gap between $(n - 2)f$ and ns (e.g., between $4f$ and $6s$) is even smaller.

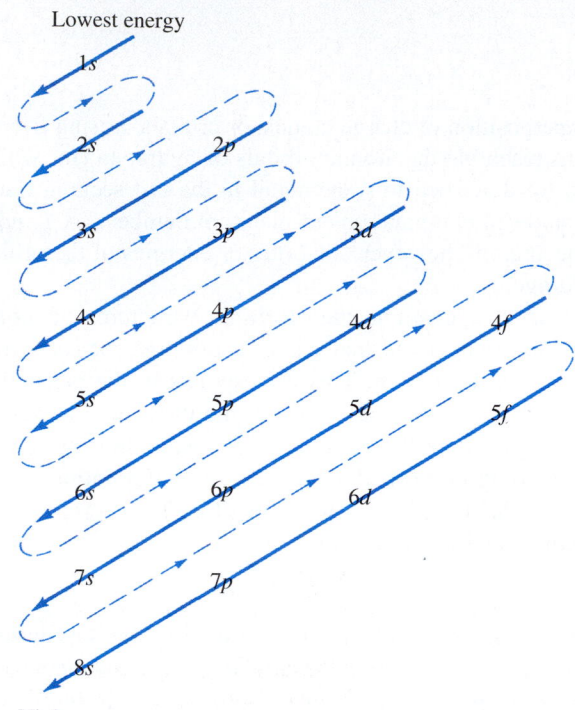

Higher energy

Figure 5-29 An aid to remembering the Aufbau order of atomic orbitals. Write all sublevels in the same major energy level on the same horizontal line. Write all like sublevels in the same vertical column. Draw parallel arrows diagonally from upper right to lower left. The arrows are read from top to bottom, and tail to head. The order is $1s$, $2s$, $2p$, $3s$, $3p$, $4s$, $3d$, $4p$, $5s$, $4d$, and so on.

| | **Orbital Notation** | | |
| | $1s$ | **Simplified Notation** | |

In the simplified notation, we indicate with superscripts the number of electrons in each sublevel.

	Orbital Notation $1s$	Simplified Notation
$_1$H	↑	$1s^1$
$_2$He	⇅	$1s^2$

Row 2. Elements of atomic numbers 3 through 10 occupy the second period, or horizontal row, in the periodic table. In neon atoms the second energy level is filled completely. Neon, a noble gas, is extremely stable. No reactions of it are known.

In writing electronic structures of atoms, we frequently simplify notations. The abbreviation [He] indicates that the $1s$ orbital is completely filled, $1s^2$, as in helium.

	Orbital Notation			**Simplified Notation**	
	$1s$	$2s$	$2p$		
$_3$Li	⇅	↑		$1s^2 2s^1$ or	[He] $2s^1$
$_4$Be	⇅	⇅		$1s^2 2s^2$	[He] $2s^2$
$_5$B	⇅	⇅	↑ _ _	$1s^2 2s^2 2p^1$	[He] $2s^2 2p^1$
$_6$C	⇅	⇅	↑ ↑ _	$1s^2 2s^2 2p^2$	[He] $2s^2 2p^2$
$_7$N	⇅	⇅	↑ ↑ ↑	$1s^2 2s^2 2p^3$	[He] $2s^2 2p^3$
$_8$O	⇅	⇅	⇅ ↑ ↑	$1s^2 2s^2 2p^4$	[He] $2s^2 2p^4$
$_9$F	⇅	⇅	⇅ ⇅ ↑	$1s^2 2s^2 2p^5$	[He] $2s^2 2p^5$
$_{10}$Ne	⇅	⇅	⇅ ⇅ ⇅	$1s^2 2s^2 2p^6$	[He] $2s^2 2p^6$

We see that some atoms have unpaired electrons in the same set of energetically equivalent, or **degenerate,** orbitals. We have already seen that two electrons can occupy a given atomic orbital (with the same values of n, ℓ, and m_ℓ) *only* if their spins are paired (have opposite values of m_s). Even with pairing of spins, however, two electrons that are in the same orbital repel each other more strongly than do two electrons in different (but equal-energy) orbitals. Thus, both theory and experimental observations (Enrichment section) lead to **Hund's Rule:**

As with helium, neon's electrons can be displaced by high-voltage electrical discharge, as is observed in neon signs.

> Electrons must occupy all the orbitals of a given sublevel singly before pairing begins. These unpaired electrons have parallel spins.

Thus, carbon has two unpaired electrons in its $2p$ orbitals, and nitrogen has three.

Paramagnetism and Diamagnetism

ENRICHMENT

Substances that contain unpaired electrons are weakly *attracted* into magnetic fields and are said to be **paramagnetic.** By contrast, those in which all electrons are paired are very weakly repelled by magnetic fields and are called **diamagnetic.** The magnetic effect can be measured by hanging a test tube full of a substance on a balance by a long thread and suspending it above the gap of an electromagnet (Figure 5-30). When the current is switched on, a paramagnetic substance such as copper(II) sulfate is pulled into the strong field. The paramagnetic attraction per mole of substance can be measured by weighing the sample before and after energizing the magnet. The paramagnetism per mole increases with increasing number of unpaired electrons per formula unit. Many transition metals and ions have one or more unpaired electrons and are paramagnetic.

Both paramagnetism and diamagnetism are hundreds to thousands of times weaker than *ferromagnetism,* the effect seen in iron bar magnets.

The metals of the iron triad (Fe, Co, and Ni) are the only *free* elements that exhibit **ferromagnetism.** This property is much stronger than paramagnetism; it allows a sub-

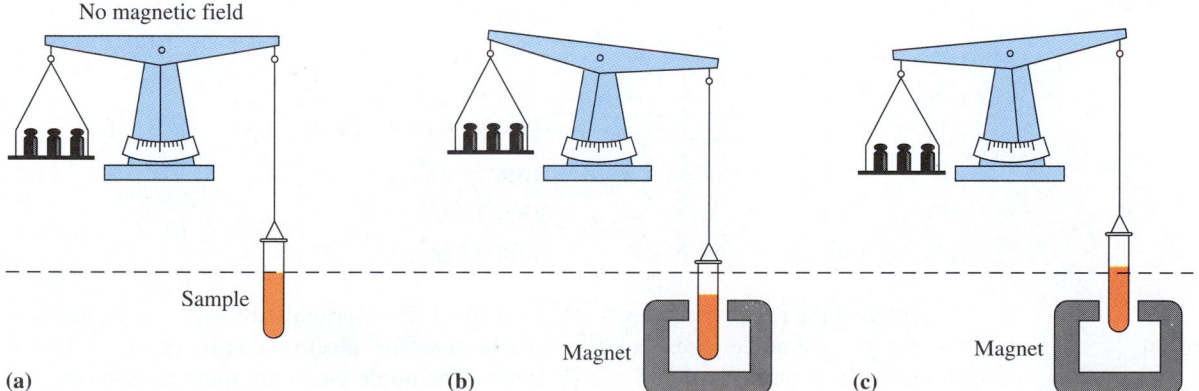

(a)   **(b)**   **(c)**

Figure 5-30   Diagram of an apparatus for measuring the paramagnetism of a substance. The tube contains a measured amount of the substance, often in solution. (a) Before the magnetic field is turned on, the position and mass of the sample are determined. (b) When the field is on, a paramagnetic substance is attracted *into* the field. (c) A diamagnetic substance would be repelled *very weakly* by the field.

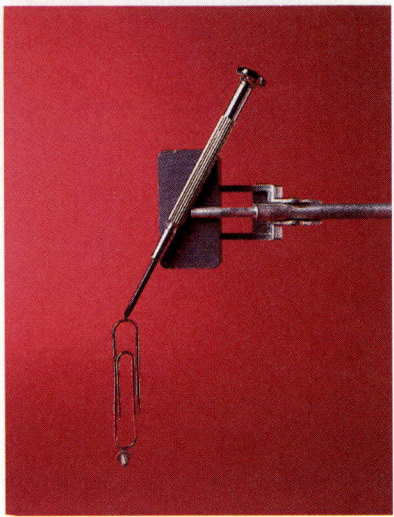

Iron displays ferromagnetism.

stance to become permanently magnetized when placed in a magnetic field. This happens as randomly oriented electron spins align themselves with an applied field. To exhibit ferromagnetism, the atoms must be within the proper range of sizes so that unpaired electrons on adjacent atoms can interact cooperatively with each other, but not to the extent that they pair. Experimental evidence suggests that in ferromagnets, atoms cluster together into *domains* that contain large numbers of atoms in fairly small volumes. The atoms within each domain interact cooperatively with each other.

Row 3. The next element beyond neon is sodium. Here we begin to add electrons to the third energy level. Elements 11 through 18 occupy the third period in the periodic table.

	Orbital Notation		
	3s	**3p**	**Simplified Notation**
$_{11}$Na	[Ne] ↑		[Ne] $3s^1$
$_{12}$Mg	[Ne] ↑↓		[Ne] $3s^2$
$_{13}$Al	[Ne] ↑↓	↑ __ __	[Ne] $3s^2 3p^1$
$_{14}$Si	[Ne] ↑↓	↑ ↑ __	[Ne] $3s^2 3p^2$
$_{15}$P	[Ne] ↑↓	↑ ↑ ↑	[Ne] $3s^2 3p^3$
$_{16}$S	[Ne] ↑↓	↑↓ ↑ ↑	[Ne] $3s^2 3p^4$
$_{17}$Cl	[Ne] ↑↓	↑↓ ↑↓ ↑	[Ne] $3s^2 3p^5$
$_{18}$Ar	[Ne] ↑↓	↑↓ ↑↓ ↑↓	[Ne] $3s^2 3p^6$

Although the third energy level is not yet filled (the *d* orbitals are still empty), argon is a noble gas. All noble gases except helium have $ns^2 np^6$ electron configurations (where *n* indicates the highest occupied energy level). The noble gases are quite unreactive.

Rows 4 and 5. It is an experimentally observed fact that *an electron occupies the available orbital that gives the atom the lowest total energy*. It is observed that filling the 4*s* orbitals before electrons enter the 3*d* orbitals *usually* leads to a lower total energy for the atom than some other arrangement. Therefore we fill the orbitals in this order (see Figure

5-28). According to the normal Aufbau order (recall Figures 5-28 and 5-29), $4s$ fills before $3d$. In general, *the $(n + 1)s$ orbital fills before the nd orbital*. This is sometimes referred to as the $(n + 1)$ *rule*.

After the $3d$ sublevel is filled to its capacity of 10 electrons, the $4p$ orbitals fill next, taking us to the noble gas krypton. Then the $5s$ orbital, the five $4d$ orbitals, and the three $5p$ orbitals fill to take us to xenon, a noble gas.

Let us now examine the electronic structures of the 18 elements in the fourth period in some detail. Some of these have electrons in d orbitals.

Orbital Notation

		3d	4s	4p	Simplified Notation
$_{19}$K	[Ar]		↑		[Ar] $4s^1$
$_{20}$Ca	[Ar]		↑↓		[Ar] $4s^2$
$_{21}$Sc	[Ar]	↑ _ _ _ _	↑↓		[Ar] $3d^1 4s^2$
$_{22}$Ti	[Ar]	↑ ↑ _ _ _	↑↓		[Ar] $3d^2 4s^2$
$_{23}$V	[Ar]	↑ ↑ ↑ _ _	↑↓		[Ar] $3d^3 4s^2$
$_{24}$Cr	[Ar]	↑ ↑ ↑ ↑ ↑	↑	[Ar] $3d^5 4s^1$
$_{25}$Mn	[Ar]	↑ ↑ ↑ ↑ ↑	↑↓		[Ar] $3d^5 4s^2$
$_{26}$Fe	[Ar]	↑↓ ↑ ↑ ↑ ↑	↑↓		[Ar] $3d^6 4s^2$
$_{27}$Co	[Ar]	↑↓ ↑↓ ↑ ↑ ↑	↑↓		[Ar] $3d^7 4s^2$
$_{28}$Ni	[Ar]	↑↓ ↑↓ ↑↓ ↑ ↑	↑↓		[Ar] $3d^8 4s^2$
$_{29}$Cu	[Ar]	↑↓ ↑↓ ↑↓ ↑↓ ↑↓	↑		[Ar] $3d^{10} 4s^1$
$_{30}$Zn	[Ar]	↑↓ ↑↓ ↑↓ ↑↓ ↑↓	↑↓		[Ar] $3d^{10} 4s^2$
$_{31}$Ga	[Ar]	↑↓ ↑↓ ↑↓ ↑↓ ↑↓	↑↓	↑ _ _	[Ar] $3d^{10} 4s^2 4p^1$
$_{32}$Ge	[Ar]	↑↓ ↑↓ ↑↓ ↑↓ ↑↓	↑↓	↑ ↑ _	[Ar] $3d^{10} 4s^2 4p^2$
$_{33}$As	[Ar]	↑↓ ↑↓ ↑↓ ↑↓ ↑↓	↑↓	↑ ↑ ↑	[Ar] $3d^{10} 4s^2 4p^3$
$_{34}$Se	[Ar]	↑↓ ↑↓ ↑↓ ↑↓ ↑↓	↑↓	↑↓ ↑ ↑	[Ar] $3d^{10} 4s^2 4p^4$
$_{35}$Br	[Ar]	↑↓ ↑↓ ↑↓ ↑↓ ↑↓	↑↓	↑↓ ↑↓ ↑	[Ar] $3d^{10} 4s^2 4p^5$
$_{36}$Kr	[Ar]	↑↓ ↑↓ ↑↓ ↑↓ ↑↓	↑↓	↑↓ ↑↓ ↑↓	[Ar] $3d^{10} 4s^2 4p^6$

As you study these electron configurations, you should be able to see how most of them are predicted from the Aufbau order. However, as we fill the $3d$ set of orbitals, from $_{21}$Sc to $_{30}$Zn, we see that these orbitals are not filled quite regularly. Some sets of orbitals are so close in energy (e.g., $4s$ and $3d$) that minor changes in their relative energies may occasionally change the order of filling.

Chemical and spectroscopic evidence indicates that the configurations of Cr and Cu have only one electron in the $4s$ orbital. Their $3d$ sets are half-filled and filled, respectively, in the ground state. Calculations from the quantum mechanical equations also indicate that *half-filled and filled sets of equivalent orbitals have a special stability*. In $_{24}$Cr, for example, this increased stability is apparently sufficient to make the *total* energy of [Ar] $3d$ ↑ ↑ ↑ ↑ ↑ $4s$ ↑ lower than that of [Ar] $3d$ ↑ ↑ ↑ ↑ _ $4s$ ↑↓. Similar reasoning helps us understand the apparent exception of the configuration of $_{29}$Cu from that predicted by the Aufbau Principle.

End-of-chapter Exercises 73–95 provide much valuable practice in writing electron configurations.

You may wonder why such an exception does not occur in, for example, $_{32}$Ge or $_{14}$Si, where we could have an s^1p^3 configuration that would have half-filled sets of s and p orbitals. It does not occur because of the very large energy gap between ns and np orbitals. We shall see evidence in Chapter 6 that does, however, illustrate the enhanced stability of half-filled sets of p orbitals.

▼ **PROBLEM-SOLVING TIP**   *Exceptions to the Aufbau Order*

In Appendix B, you will find a number of exceptions to the electron configurations predicted from the Aufbau Principle. You should realize that statements such as the Aufbau Principle and the $(n + 1)$ rule merely represent general guidelines and should not be viewed as hard-and-fast rules; the *total energy* of the atom is as low as possible. Some of the reasons for exceptions are

1. The Aufbau order of orbital energies is based on calculations for the hydrogen atom, which contains only one electron. The orbital energies also depend on additional factors such as the nuclear charge and interactions of electrons in different occupied orbitals.

2. The energy scale varies with the atomic number.

3. Some orbitals are very close together, so their order can change, depending on the occupancies of other orbitals.

Some types of exceptions to the Aufbau order are general enough to remember easily, for example, those based on the special stability of filled or half-filled sets of orbitals. Other exceptions are quite unpredictable. Your instructor may expect you to remember some of the exceptions.

Let us now write the quantum numbers to describe each electron in an atom of nitrogen. Keep in mind the fact that Hund's Rule must be obeyed. Thus, there is only one (unpaired) electron in each $2p$ orbital in a nitrogen atom.

EXAMPLE 5-7   *Electron Configurations and Quantum Numbers*

Write an acceptable set of four quantum numbers for each electron in a nitrogen atom.

Plan

Nitrogen has seven electrons, which occupy the lowest energy orbitals available. Two electrons can occupy the first energy level, $n = 1$, in which there is only one s orbital; when $n = 1$, then ℓ must be zero, and therefore $m_\ell = 0$. The two electrons differ only in spin quantum number, m_s. The next five electrons can all fit into the second energy level, for which $n = 2$ and ℓ may be either 0 or 1. The $\ell = 0$ (s) sublevel fills first, and the $\ell = 1$ (p) sublevel is occupied next.

Solution

Electrons are indistinguishable. We have numbered them 1, 2, 3, and so on as an aid to counting them.

In the lowest energy configurations, the three $2p$ electrons either all have $m_s = +\frac{1}{2}$ or all have $m_s = -\frac{1}{2}$.

Electron	n	ℓ	m_ℓ	m_s	e^- Configuration
1, 2	1	0	0	$+\frac{1}{2}$	$1s^2$
	1	0	0	$-\frac{1}{2}$	
3, 4	2	0	0	$+\frac{1}{2}$	$2s^2$
	2	0	0	$-\frac{1}{2}$	
5, 6, 7	2	1	-1	$+\frac{1}{2}$ or $-\frac{1}{2}$	$2p_x^1$
	2	1	0	$+\frac{1}{2}$ or $-\frac{1}{2}$	$2p_y^1$ or $2p^3$
	2	1	$+1$	$+\frac{1}{2}$ or $-\frac{1}{2}$	$2p_z^1$

EXAMPLE 5-8 *Electron Configurations and Quantum Numbers*

Write an acceptable set of four quantum numbers for each electron in a chlorine atom.

Plan

Chlorine is element number 17. Its first seven electrons have the same quantum numbers as those of nitrogen in Example 5-7. Electrons 8, 9, and 10 complete the filling of the $2p$ sublevel ($n = 2$, $\ell = 1$) and therefore also the second energy level. Electrons 11 through 17 fill the $3s$ sublevel ($n = 3$, $\ell = 0$) and partially fill the $3p$ sublevel ($n = 3$, $\ell = 1$).

Solution

Electron	n	ℓ	m_ℓ	m_s	e^- Configuration
1, 2	1	0	0	$\pm\frac{1}{2}$	$1s^2$
3, 4	2	0	0	$\pm\frac{1}{2}$	$2s^2$
5–10	$\begin{cases} 2 \\ 2 \\ 2 \end{cases}$	$\begin{matrix} 1 \\ 1 \\ 1 \end{matrix}$	$\begin{matrix} -1 \\ 0 \\ +1 \end{matrix}$	$\left.\begin{matrix} \pm\frac{1}{2} \\ \pm\frac{1}{2} \\ \pm\frac{1}{2} \end{matrix}\right\}$	$2p^6$
11, 12	3	0	0	$\pm\frac{1}{2}$	$3s^2$
13–17	$\begin{cases} 3 \\ 3 \\ 3 \end{cases}$	$\begin{matrix} 1 \\ 1 \\ 1 \end{matrix}$	$\begin{matrix} -1 \\ 0 \\ +1 \end{matrix}$	$\left.\begin{matrix} \pm\frac{1}{2} \\ \pm\frac{1}{2} \\ +\frac{1}{2} \text{ or } -\frac{1}{2}* \end{matrix}\right\}$	$3p^5$

*The 3p orbital with only a single electron can be any one of the set, not necessarily the one with $m_\ell = +1$.

You should now work Exercises 88 and 92.

5-18 THE PERIODIC TABLE AND ELECTRON CONFIGURATIONS

In this section, we view the *periodic table* (see inside front cover and Section 4-1) from a modern, much more useful perspective—as a systematic representation of the electron configurations of the elements. In the periodic table, elements are arranged in blocks based on the kinds of atomic orbitals that are being filled (Figure 5-31). The periodic tables in this text are divided into "A" and "B" groups. The A groups contain elements in which s and p orbitals are being filled. Elements within any particular A group have similar electron configurations and chemical properties, as we shall see in the next chapter. The B groups are those in which there are one or two electrons in the s orbital of the highest occupied energy level, and the d orbitals one energy level lower are being filled.

Lithium, sodium, and potassium, elements of the leftmost column of the periodic table (Group IA), have a single electron in their outermost s orbital (ns^1). Beryllium and magnesium, of Group IIA, have two electrons in their highest energy level, ns^2, while boron and aluminum (Group IIIA) have three electrons in their highest energy level, ns^2np^1. Similar observations can be made for each A group.

The electron configurations of the A group elements and the noble gases can be predicted reliably from Figures 5-28 and 5-29. However, there are some irregularities in the B groups below the fourth period that require special attention. In the heavier B group elements, the higher energy sublevels in different principal energy levels have energies that are very nearly equal (Figure 5-29). It is easy for an electron to jump from one orbital

H is shown in the $1s$ block in Figure 5-31. It is usually shown in Group IA.

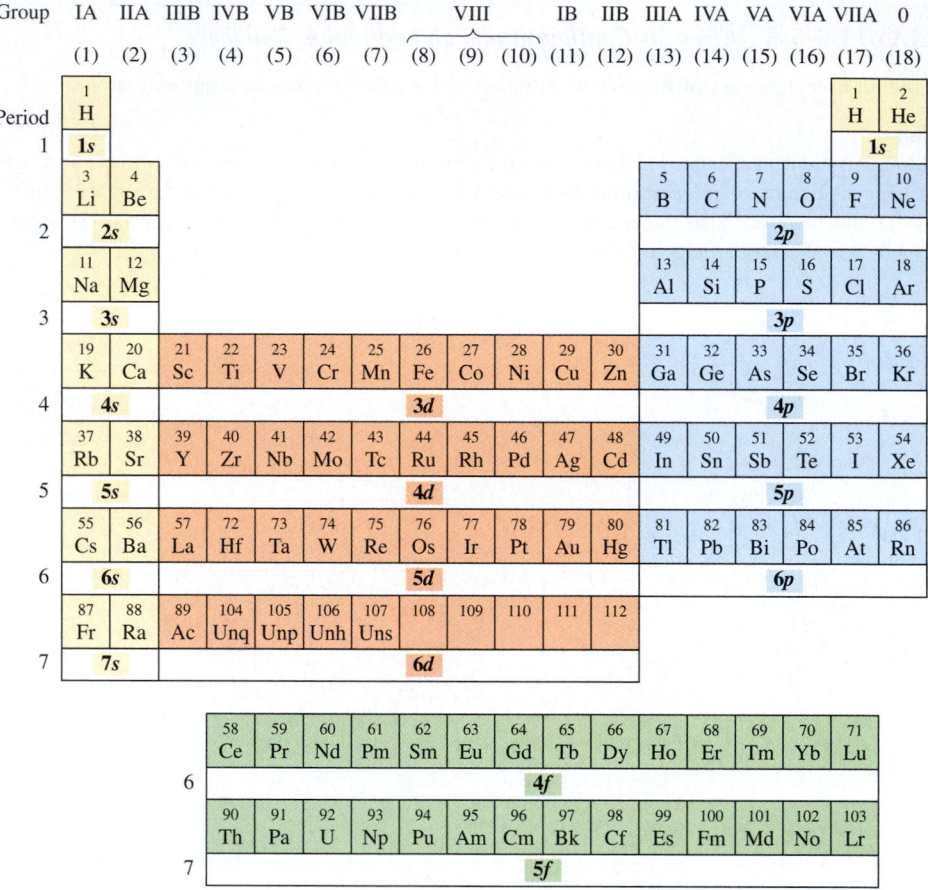

Figure 5-31   A periodic table colored to show the kinds of atomic orbitals (sublevels) being filled, below the symbols of blocks of elements. The electronic structures of the A group and 0 group elements are perfectly regular and can be predicted from their positions in the periodic table, but there are many exceptions in the *d* and *f* blocks. The colors in this figure are the same as those in Figure 5-28.

Hydrogen and helium are shown here in their usual positions in the periodic table. These may seem somewhat unusual based just on their electron configurations. However, we should remember that the first main energy level ($n = 1$) can hold a maximum of only two electrons. This shell is entirely filled in helium, so He behaves as a noble gas, and we put it in the column with the other noble gases (Group 0). Hydrogen has one electron that is easily lost, like the metals in Group IA, so we put it in Group IA even though it is not a metal. Furthermore, hydrogen is one electron short of a noble gas configuration (He), so we could also place it with the other such elements in Group VIIA.

to another of nearly the same energy in a different set. This is because the orbital energies are *perturbed* (change slightly) as the nuclear charge changes, and an extra electron is added in going from one element to the next. This phenomenon gives rise to other irregularities that are analogous to those of Cr and Cu, described earlier.

We can extend the information in Figure 5-31 to indicate the electron configurations that are represented by each *group* (column) of the periodic table. Table 5-5 shows this interpretation of the periodic table, along with the most important exceptions. (A more

Table 5-5 *The s, p, d, and f Blocks of the Periodic Table**

s orbital block; d orbital block; f orbital block; p orbital block

GROUPS

	IA	IIA	IIIB		IVB	VB	VIB	VIIB	VIII	VIII	VIII	IB	IIB	IIIA	IVA	VA	VIA	VIIA	0
	(1)	(2)	(3)	f orbital block	(4)	(5)	(6)	(7)	(8)	(9)	(10)	(11)	(12)	(13)	(14)	(15)	(16)	(17)	(18)
	s^1	s^2	d^1s^2		d^2s^2	d^3s^2	d^5s^1	d^5s^2	d^6s^2	d^7s^2	d^8s^2	$d^{10}s^1$	$d^{10}s^2$	s^2p^1	s^2p^2	s^2p^3	s^2p^4	s^2p^5	s^2p^6
$n = 1$	1 H																		2 He (s^2)
$n = 2$	3 Li	4 Be												5 B	6 C	7 N	8 O	9 F	10 Ne
$n = 3$	11 Na	12 Mg												13 Al	14 Si	15 P	16 S	17 Cl	18 Ar
$n = 4$	19 K	20 Ca	21 Sc		22 Ti	23 V	24 Cr	25 Mn	26 Fe	27 Co	28 Ni	29 Cu	30 Zn	31 Ga	32 Ge	33 As	34 Se	35 Br	36 Kr
$n = 5$	37 Rb	38 Sr	39 Y		40 Zr	41 Nb d^4s^1	42 Mo	43 Tc	44 Ru d^7s^1	45 Rh d^8s^1	46 Pd	47 Ag	48 Cd	49 In	50 Sn	51 Sb	52 Te	53 I	54 Xe
$n = 6$	55 Cs	56 Ba	57 La	58 Ce → 71 Lu	72 Hf	73 Ta	74 W d^4s^2	75 Re	76 Os	77 Ir	78 Pt d^9s^1	79 Au	80 Hg	81 Tl	82 Pb	83 Bi	84 Po	85 At	86 Rn
$n = 7$	87 Fr	88 Ra	89 Ac	90 Th → 103 Lr	104 Unq	105 Unp	106 Unh	107 Uns	108	109	110	111	112						

$n = 6$ LANTHANIDE SERIES	58 Ce	59 Pr	60 Nd	61 Pm	62 Sm	63 Eu	64 Gd	65 Tb	66 Dy	67 Ho	68 Er	69 Tm	70 Yb	71 Lu	4*f* subshell being filled
$n = 7$ ACTINIDE SERIES	90 Th	91 Pa	92 U	93 Np	94 Pu	95 Am	96 Cm	97 Bk	98 Cf	99 Es	100 Fm	101 Md	102 No	103 Lr	5*f* subshell being filled

*n is the principal quantum number. The d^1s^2, d^2s^2, . . . designations represent *known* configurations. They refer to $(n-1)d$ and ns orbitals. Several exceptions to the configurations indicated above each group are shown in gray.

complete listing of electron configurations is given in Appendix B.) We can use this interpretation of the periodic table to write, quickly and reliably, the electron configurations for elements.

EXAMPLE 5-9 *Electron Configurations*

Use Table 5-5 to determine the electron configurations of (a) germanium, Ge; (b) magnesium, Mg; and (c) molybdenum, Mo.

Plan

We use the electron configurations indicated in Table 5-5 for each group. Each *period* (row) begins filling a new shell (new value of n). Elements to the right of the d orbital block have the d orbitals in the $(n-1)$ shell already filled. We often find it convenient to collect all sets of orbitals with the same value of n together, to emphasize the number of electrons in the *outermost* shell, i.e., the shell with the highest value of n.

Solution

(a) Germanium, Ge, is in Group IVA, for which Table 5-5 shows the general configuration s^2p^2. It is in Period 4 (the 4th row), so we interpret this as $4s^24p^2$. The last filled noble gas configuration is that of argon, Ar, accounting for 18 electrons. In addition, Ge lies beyond the d orbital block, so we know that the $3d$ orbitals are completely filled. The electron configuration of Ge is [Ar] $4s^23d^{10}4p^2$ or   [Ar] $3d^{10}4s^24p^2$.

(b) Magnesium, Mg, is in Group IIA, which has the general configuration s^2; it is in Period 3 (3rd row). The last filled noble gas configuration is that of neon, or [Ne]. The electron configuration of Mg is [Ne] $3s^2$.

(c) Molybdenum, Mo, is in Group VIB, with general configuration d^5s^1; it is in Period 5, which begins with $5s$ and is beyond the noble gas krypton. The electron configuration of Mo is [Kr] $5s^14d^5$ or   [Kr] $4d^55s^1$. The electron configuration of molybdenum is analogous to that of chromium, Cr, the element just above it. The configuration of Cr was discussed in Section 5-17 as one of the exceptions to the Aufbau order of filling.

You should now work Exercise 90.

EXAMPLE 5-10   *Unpaired Electrons*

Determine the number of unpaired electrons in an atom of tellurium, Te.

Plan

Te is in Group VIA in the periodic table, which tells us that its configuration is s^2p^4. All other shells are completely filled, so they contain only paired electrons. We need only to find out how many unpaired electrons are represented by s^2p^4.

Solution

The notation s^2p^4 is a short representation for s ⇅ p ⇅ ↑ ↑. This shows that an atom of Te contains two unpaired electrons.

You should now work Exercises 84 and 86.

The periodic table has been described as "the chemist's best friend." In this chapter, we have seen that the fundamental basis of the periodic table is that it reflects similarities and trends in electron configurations. It is easy to use the periodic table to determine many important aspects of electron configurations of atoms. Practice until you can use the periodic table with confidence to answer many questions about electron configurations. As we continue our study of chemistry, we shall learn many other useful ways to interpret the periodic table. We should always keep in mind that the many trends in chemical and physical properties that we correlate with the periodic table are ultimately based on the trends in electron configurations.

Key Terms

Absorption spectrum   The spectrum associated with absorption of electromagnetic radiation by atoms (or other species) resulting from transitions from lower to higher energy states.

Alpha (α) particle   A helium ion with 2+ charge; an assembly of two protons and two neutrons.

amu   See *Atomic mass unit*.

Anode   In a cathode ray tube, the positive electrode.

Atomic mass unit   An arbitrary mass unit defined to be exactly one-twelfth the mass of the carbon-12 isotope.

Atomic number   The integral number of protons in the nucleus; defines the identity of an element.

Atomic orbital   The region or volume in space in which the probability of finding electrons is highest.

Aufbau ("building up") Principle   Describes the order in which electrons fill orbitals in atoms.

Canal ray   A stream of positively charged particles (cations) that moves toward the negative electrode in a cathode ray tube; observed to pass through canals in the negative electrode.

Cathode   In a cathode ray tube, the negative electrode.

Cathode ray   The beam of electrons going from the negative electrode toward the positive electrode in a cathode ray tube.

Cathode ray tube   A closed glass tube containing a gas under low pressure, with electrodes near the ends and a luminescent screen at the end near the positive electrode; produces cathode rays when high voltage is applied.

Continuous spectrum   The spectrum that contains all wavelengths in a specified region of the electromagnetic spectrum.

Degenerate   Of the same energy.

Diamagnetism   *Weak* repulsion by a magnetic field.

d **orbitals**   Beginning in the third energy level, a set of five degenerate orbitals per energy level, higher in energy than *s* and *p* orbitals of the same energy level.

Electromagnetic radiation   Energy that is propagated by means of electric and magnetic fields that oscillate in directions perpendicular to the direction of travel of the energy.

Electron   A subatomic particle having a mass of 0.00054858 amu and a charge of $1-$.

Electron configuration   The specific distribution of electrons in the atomic orbitals of atoms or ions.

Electronic transition   The transfer of an electron from one energy level to another.

Emission spectrum   The spectrum associated with emission of electromagnetic radiation by atoms (or other species) resulting from electronic transitions from higher to lower energy states.

Excited state   Any state other than the ground state of an atom or molecule.

f **orbitals**   Beginning in the fourth energy level, a set of seven degenerate orbitals per energy level, higher in energy than *s*, *p*, and *d* orbitals of the same energy level.

Ferromagnetism   The property that allows a substance to become permanently magnetized when placed in a magnetic field; exhibited by iron, cobalt, and nickel.

Frequency   The number of repeating corresponding points on a wave that pass a given point per unit time.

Fundamental particles   Subatomic particles of which all matter is composed; protons, electrons, and neutrons are fundamental particles.

Ground state   The lowest energy state or most stable state of an atom, molecule, or ion.

Group   A vertical column in the periodic table; also called a family.

Heisenberg Uncertainty Principle   It is impossible to determine accurately both the momentum and position of an electron simultaneously.

Hund's Rule   Each orbital of a given sublevel must be occupied by a single electron before pairing begins. See *Aufbau Principle.*

Isotopes   Two or more forms of atoms of the same element with different masses; atoms containing the same number of protons but different numbers of neutrons.

Line spectrum   An atomic emission or absorption spectrum.

Magnetic quantum number (m_ℓ)   Quantum mechanical solution to a wave equation that designates the particular orbital within a given set (s, p, d, f) in which an electron resides.

Mass number   The integral sum of the numbers of protons and neutrons in an atom.

Mass spectrometer   An instrument that measures the charge-to-mass ratios of charged particles.

Natural radioactivity   Spontaneous decomposition of an atom.

Neutron   A neutral subatomic nuclear particle having a mass of 1.0087 amu.

Nucleus   The very small, very dense, positively charged center of an atom containing protons and neutrons, except for $^1_1$H.

Nuclide symbol   The symbol for an atom, $^A_Z E$, in which E is the symbol for an element, Z is its atomic number, and A is its mass number.

Pairing of electrons   Favorable interaction of two electrons with opposite m_s values in the same orbital (⥮).

Paramagnetism   Attraction toward a magnetic field, stronger than diamagnetism, but still very weak compared with ferromagnetism.

Pauli Exclusion Principle   No two electrons in the same atom may have identical sets of four quantum numbers.

Period   A horizontal row in the periodic table.

Photoelectric effect   Emission of an electron from the surface of a metal, caused by impinging electromagnetic radiation of certain minimum energy; current increases with increasing intensity of radiation.

Photon   A "packet" of light or electromagnetic radiation; also called a quantum of light.

p **orbitals**   Beginning with the second energy level, a set of three degenerate mutually perpendicular, equal-arm, dumbbell-shaped atomic orbitals per energy level.

Principal quantum number (n)   The quantum mechanical solution to a wave equation that designates the major energy level, or shell, in which an electron resides.

Proton   A subatomic particle having a mass of 1.0073 amu and a charge of $+1$, found in the nuclei of atoms.

Quantum   A "packet" of energy. See *Photon.*

Quantum mechanics   A mathematical method of treating particles on the basis of quantum theory, which assumes that energy (of small particles) is not infinitely divisible.

Quantum numbers   Numbers that describe the energies of electrons in atoms; derived from quantum mechanical treatment.

Radiant energy   See *Electromagnetic radiation.*

Rydberg equation   An empirical equation that relates wavelengths in the hydrogen emission spectrum to integers.

s **orbital**   A spherically symmetrical atomic orbital; one per energy level.

Spectral line   Any of a number of lines corresponding to definite wavelengths in an atomic emission or absorption spectrum; represents the energy difference between two energy levels.

Spectrum   Display of component wavelengths (colors) of electromagnetic radiation.

Spin quantum number (m_s)   The quantum mechanical solution to a wave equation that indicates the relative spins of electrons.

Subsidiary quantum number (ℓ)   The quantum mechanical solution to a wave equation that designates the sublevel, or set of orbitals (s, p, d, f), within a given major energy level in which an electron resides.

Wavelength   The distance between two corresponding points of a wave.

Exercises

Particles and the Nuclear Atom

1. List the three fundamental particles of matter and indicate the mass and charge associated with each.
2. In the oil-drop experiment, how did Millikan know that none of the oil droplets he observed were ones that had a deficiency of electrons rather than an excess?
3. How many electrons carry a total charge of 1.00 coulomb?
4. (a) How do we know that canal rays have charges opposite in sign to cathode rays? What are canal rays? (b) Why are cathode rays from all samples of gases identical, whereas canal rays are not?
*5. The following data are measurements of the charges on oil droplets using an apparatus similar to that used by Millikan:

$$13.458 \times 10^{-19} \text{ C} \qquad 17.308 \times 10^{-19} \text{ C}$$
$$15.373 \times 10^{-19} \text{ C} \qquad 28.844 \times 10^{-19} \text{ C}$$
$$17.303 \times 10^{-19} \text{ C} \qquad 11.545 \times 10^{-19} \text{ C}$$
$$15.378 \times 10^{-19} \text{ C} \qquad 19.214 \times 10^{-19} \text{ C}$$

Each should be a whole-number ratio of some fundamental charge. Using these data, determine the value of the fundamental charge.
*6. Suppose we discover a new positively charged particle, which we call the "whizatron." We want to determine its charge. (a) What modifications would we have to make to the Millikan oil-drop apparatus to carry out the corresponding experiment on whizatrons?
(b) In such an experiment, we observe the following charges on five different droplets:

$$6.52 \times 10^{-19} \text{ C} \qquad 11.40 \times 10^{-19} \text{ C}$$
$$8.16 \times 10^{-19} \text{ C} \qquad 9.78 \times 10^{-19} \text{ C}$$
$$3.26 \times 10^{-19} \text{ C}$$

What is the charge on the whizatron?

7. What are alpha particles? Characterize them as to mass and charge.
8. The approximate radius of a hydrogen atom is 0.0529 nm, and that of a proton is 1.5×10^{-15} m. Assuming both the hydrogen atom and the proton to be spherical, calculate the fraction of the space in an atom of hydrogen that is occupied by the nucleus. $V = (4/3)\pi r^3$ for a sphere.
9. The approximate radius of a neutron is 1.5×10^{-15} m, and the mass is 1.675×10^{-27} kg. Calculate the density of a neutron. $V = (4/3)\pi r^3$ for a sphere.
10. Arrange the following in order of increasing ratio of charge to mass: $^{12}C^+$, $^{12}C^{2+}$, $^{13}C^+$, $^{13}C^{2+}$.
11. Refer to Exercise 10. Suppose all of these high energy ions are present in a mass spectrometer. For which one will its path be changed (a) the most and (b) the least by increasing the external magnetic field?

Atom Composition, Isotopes, and Atomic Weights

12. Estimate the percentage of the total mass of a $^{197}_{79}Au$ atom that is due to (a) electrons, (b) protons, and (c) neutrons by *assuming* that the mass of the atom is simply the sum of the masses of the appropriate numbers of subatomic particles.
13. (a) How are isotopic abundances determined experimentally? (b) How do the isotopes of a given element differ?
14. Define and illustrate the following terms clearly and concisely: (a) atomic number, (b) isotope, (c) mass number, (d) nuclear charge.
15. Write the composition of one atom of each of the three isotopes of neon: ^{20}Ne, ^{21}Ne, ^{22}Ne.
16. Write the composition of one atom of each of the four isotopes of strontium: ^{84}Sr, ^{86}Sr, ^{87}Sr, ^{88}Sr.
17. Complete Chart A for neutral atoms.
18. Complete Chart B for neutral atoms.

Chart A

Name of Element	Atomic Number	Mass Number	Isotope	Number of Protons	Number of Electrons	Number of Neutrons
			$^{40}_{20}Ca$			
potassium		39				
	14	28				
		202		80		

Chart B

Name of Element	Atomic Number	Mass Number	Isotope	Number of Protons	Number of Electrons	Number of Neutrons
cobalt						32
			$^{193}_{77}Ir$			
					25	30
		182			78	

*19. Prior to 1962 the atomic weight scale was based on the assignment of an atomic weight of exactly 16 amu to the *naturally occurring* mixture of oxygen. The atomic weight of cobalt is 58.9332 amu on the carbon-12 scale. What was it on the older scale?

20. Determine the number of protons, neutrons, and electrons in each of the following species: (a) $^{24}_{12}Mg$; (b) $^{45}_{21}Sc$; (c) $^{91}_{40}Zr$; (d) $^{27}_{13}Al^{3+}$; (e) $^{65}_{30}Zn^{2+}$; (f) $^{108}_{47}Ag^{+}$.

21. Determine the number of protons, neutrons, and electrons in each of the following species: (a) $^{52}_{24}Cr$; (b) $^{93}_{41}Nb$; (c) $^{137}_{56}Ba$; (d) $^{63}_{29}Cu^{+}$; (e) $^{56}_{26}Fe^{2+}$; (f) $^{55}_{26}Fe^{3+}$.

22. What is the symbol of the species composed of each of the following sets of subatomic particles? (a) $25p$, $30n$, $25e$; (b) $20p$, $20n$, $18e$; (c) $33p$, $42n$, $33e$; (d) $53p$, $74n$, $54e$.

23. What is the symbol of the species composed of each of the following sets of subatomic particles? (a) $94p$, $150n$, $94e$; (b) $79p$, $118n$, $76e$; (c) $34p$, $45n$, $36e$; (d) $54p$, $77n$, $54e$.

24. The atomic weight of lithium is 6.941 amu. The two naturally occurring isotopes of lithium have the following masses: ^{6}Li, 6.01512 amu; ^{7}Li, 7.01600 amu. Calculate the percent of ^{6}Li in naturally occurring lithium.

25. The atomic weight of rubidium is 85.4678 amu. The two naturally occurring isotopes of rubidium have the following masses: ^{85}Rb, 84.9118 amu; ^{87}Rb, 86.9092 amu. Calculate the percent of ^{85}Rb in naturally occurring rubidium.

26. Determine the charge on each of the following species: (a) Ca with 18 electrons; (b) Cu with 28 electrons; (c) Cu with 26 electrons; (d) Pt with 74 electrons; (e) F with 10 electrons; (f) C with 6 electrons; (g) C with 7 electrons; (h) C with 5 electrons.

27. The following is a mass spectrum of the 1+ charged ions of an element. Calculate the atomic weight of the element. What is the element?

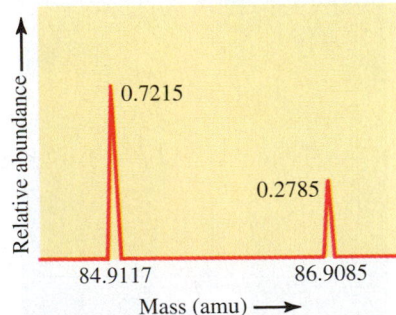

28. Suppose you measure the mass spectrum of the 1+ charged ions of germanium, atomic weight 72.61 amu. Unfortunately, the recorder on the mass spectrometer jams at the beginning and again at the end of your experiment. You obtain only the partial spectrum shown in the next column, which *may or may not be complete*. From the information given here, can you tell whether one of the germanium isotopes is missing? If one is missing, at which end of the plot should it appear?

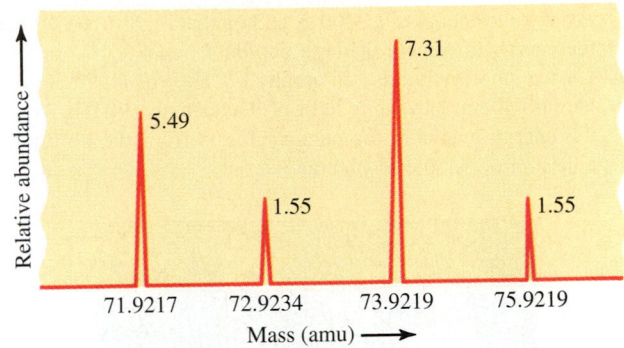

29. Calculate the atomic weight of silicon using the following data for the percent natural abundance and mass of each isotope: 92.23% ^{28}Si (27.9769 amu); 4.67% ^{29}Si (28.9765 amu); 3.10% ^{30}Si (29.9738 amu).

30. Calculate the atomic weight of chromium using the following data for the percent natural abundance and mass of each isotope: 4.35% ^{50}Cr (49.9461 amu); 83.79% ^{52}Cr (51.9405 amu); 9.50% ^{53}Cr (52.9406 amu); 2.36% ^{54}Cr (53.9389 amu).

31. Consider the ions $^{16}_{8}O^{+}$, $^{17}_{8}O^{+}$, $^{16}_{8}O^{2+}$, $^{17}_{8}O^{2+}$ produced in a mass spectrometer. Which ion's path would be deflected (a) most and (b) least by a magnetic field?

*32. In a suitable reference such as the Table of Isotopes in the *Handbook of Chemistry and Physics* (The Chemical Rubber Co.), look up the following information for germanium: (a) the total number of known naturally occurring isotopes, (b) the atomic mass, and (c) the percent natural abundance and mass of each of the stable isotopes. (d) Calculate the atomic weight of germanium.

Electromagnetic Radiation

33. Calculate the wavelengths, in meters, of radiation of the following frequencies:
    (a) 4.80×10^{15} s^{-1}
    (b) 1.18×10^{14} s^{-1}
    (c) 5.44×10^{12} s^{-1}

34. Calculate the frequency of radiation of each of the following wavelengths: (a) 9774 Å; (b) 492 nm; (c) 4.92 cm; (d) 4.92×10^{-9} cm.

35. What is the energy of a photon of each of the radiations in Exercise 33? Express your answer in joules per photon. In which regions of the electromagnetic spectrum do these radiations fall?

36. Excited lithium ions emit radiation at a wavelength of 670.8 nm in the visible range of the spectrum. (This characteristic color is often used as a qualitative analysis test for the presence of Li^{+}.) Calculate (a) the frequency and (b) the energy of a photon of this radiation. (c) What color is this light?

37. Find the energy of the photons corresponding to the red line, 6573 Å, in the spectrum of the Ca atom.

38. Ozone in the upper atmosphere absorbs ultraviolet radiation, which induces the following chemical reaction:

$$O_3(g) \longrightarrow O_2(g) + O(g)$$

What is the energy of a 3400-Å photon that is absorbed? What is the energy of a mole of these photons?

*39. During photosynthesis, chlorophyll-*a* absorbs light of wavelength 440 nm and emits light of wavelength 670 nm. What is the energy available for photosynthesis from the absorption–emission of a mole of photons?

Photosynthesis.

*40. Assume that 10^{-17} J of light energy is needed by the interior of the human eye to "see" an object. How many photons of green light (wavelength = 495 nm) are needed to generate this minimum energy?

*41. The human eye receives a 2.500×10^{-14} J-signal consisting of photons of orange light, $\lambda = 6150$ Å. How many photons reach the eye?

*42. Water absorbs microwave radiation of wavelength 3 mm. How many photons are needed to raise the temperature of a cup of water (250 g) from 25°C to 75°C in a microwave oven, using this radiation? The specific heat of water is 4.184 J/g · °C.

The Photoelectric Effect

43. What evidence supports the idea that electromagnetic radiation is (a) wave-like; (b) particle-like?

44. Describe the influence of frequency and intensity of electromagnetic radiation on the current in the photoelectric effect.

*45. Cesium is often used in "electric eyes" for self-opening doors in an application of the photoelectric effect. The amount of energy required to ionize (remove an electron from) a cesium atom is 3.89 electron volts (1 eV = 1.60×10^{-19} J). Show by calculation whether a beam of yellow light with wavelength 5830 Å would ionize a cesium atom.

*46. Refer to Exercise 45. What would be the wavelength, in nanometers, of light with just sufficient energy to ionize a cesium atom? What color would this light be?

Atomic Spectra and the Bohr Theory

47. (a) Distinguish between an atomic emission spectrum and an atomic absorption spectrum. (b) Distinguish between a continuous spectrum and a line spectrum.

48. Prepare a sketch similar to Figure 5-16b that shows a ground energy state and three excited energy states. Using vertical arrows, indicate the transitions that would correspond to the absorption spectrum for this system.

49. What is the Rydberg equation? Why is it called an empirical equation?

50. Hydrogen atoms absorb energy so that the electrons are excited to the energy level $n = 7$. Electrons then undergo these transitions: (1) $n = 7 \rightarrow n = 1$; (2) $n = 7 \rightarrow n = 6$; (3) $n = 2 \rightarrow n = 1$. Which of these transitions will produce the photon with (a) the smallest energy; (b) the highest frequency; (c) the shortest wavelength? (d) What is the frequency of a photon resulting from the transition $n = 6 \rightarrow n = 1$?

*51. Five energy levels of the He atom are given in J/atom above an *arbitrary* reference energy: (1) 6.000×10^{-19}; (2) 8.812×10^{-19}; (3) 9.381×10^{-19}; (4) 10.443×10^{-19}; (5) 10.934×10^{-19}. Construct an energy level diagram for He and find the energy of the photon (a) absorbed for the electron transition from level 1 to level 5 and (b) emitted for the electron transition from level 4 to level 1.

52. The following are prominent lines in the visible region of the emission spectra of the elements listed. The lines can be used to identify the elements. What color is the light responsible for each line? (a) lithium, 4603 Å; (b) neon, 540.0 nm; (c) calcium, 6573 Å; (d) cesium, $\nu = 3.45 \times 10^{14}$ Hz; (e) potassium, $\nu = 3.90 \times 10^{14}$ Hz.

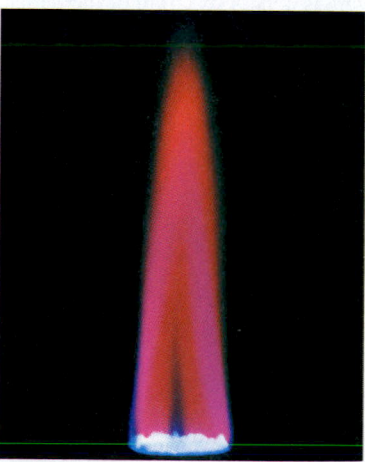

53. Hydrogen atoms have an absorption line at 1026 Å. What is the frequency of the photons absorbed, and what is the energy difference, in joules, between the ground state and this excited state of the atom?

*54. If each atom in one mole of atoms emits a photon of wavelength 6.64×10^3 Å, how much energy is lost? Express the answer in kJ/mol. As a reference point, burning one mole (16 g) of CH_4 produces 819 kJ of heat.

*55. Suppose we could excite all of the electrons in a sample of hydrogen atoms to the $n = 6$ level. They would then emit light as they relaxed to lower energy states. Some atoms might undergo the transition $n = 6$ to $n = 1$, while others might go from $n = 6$ to $n = 5$, then from $n = 5$ to $n = 4$, and so on. How many lines would we expect to observe in the resulting emission spectrum?

*56. An argon laser emits blue light with a wavelength of 488.0 nm. How many photons are emitted by this laser in 2.00 seconds, operating at a power of 515 milliwatts? One watt (a unit of power) is equal to 1 joule/second.

Lasers.

The Wave–Particle View of Matter

57. (a) What evidence supports the idea that electrons are particle-like? (b) What evidence supports the idea that electrons are wave-like?

58. (a) What is the de Broglie wavelength of a proton moving at a speed of 2.50×10^7 m/s? The proton mass is 1.67×10^{-24} g. (b) What is the de Broglie wavelength of a stone with a mass of 30.0 g moving at 2.00×10^3 m/h (≈ 100 mi/h)? (c) How do the wavelengths in parts (a) and (b) compare with the typical radii of atoms? (See the atomic radii in Figure 6-1.)

59. What is the wavelength corresponding to a neutron of mass 1.67×10^{-27} kg moving at 2360 m/s?

60. What is the velocity of an α-particle (a helium nucleus) that has a de Broglie wavelength of 0.529 Å?

Quantum Numbers and Atomic Orbitals

61. (a) What is a quantum number? What is an atomic orbital? (b) How many quantum numbers are required to specify a single atomic orbital? What are they?

62. How are the possible values for the subsidiary quantum number for a given electron restricted by the value of n?

63. Without giving the ranges of possible values of the four quantum numbers, n, ℓ, m_ℓ and m_s, describe briefly what information each one gives.

64. (a) How are the values of m_ℓ for a particular electron restricted by the value of ℓ? (b) What are the letter designations for the values $n = 1, 2, 3, 4$? (c) What are the letter designations for the values $\ell = 0, 1, 2, 3$?

65. What are the values of n and ℓ for the following sublevels? (a) $1s$; (b) $4d$; (c) $3p$; (d) $5s$; (e) $4f$.

66. How many individual orbitals are there in the third major energy level? Write out n, ℓ and m_ℓ quantum numbers for each one and label each set by the s, p, d, f designations.

67. (a) Write the possible values of ℓ when $n = 5$. (b) Write the allowed number of orbitals (1) with the quantum numbers $n = 4$, $\ell = 3$; (2) with the quantum number $n = 4$; (3) with the quantum numbers $n = 7$, $\ell = 6$, $m_\ell = 6$; (4) with quantum numbers $n = 6$, $\ell = 5$.

68. Write the subshell notations that correspond to (a) $n = 3$, $\ell = 0$; (b) $n = 3$, $\ell = 1$; (c) $n = 7$, $\ell = 0$; (d) $n = 4$, $\ell = 3$.

69. What values can m_ℓ take for (a) a $4d$ orbital, (b) a $1s$ orbital, and (c) a $3p$ orbital?

70. How many orbitals in any atom can have the given quantum number or designation? (a) $4p$; (b) $3p$; (c) $3p_x$; (d) $n = 5$; (e) $6d$; (f) $5d$; (g) $5f$; (h) $7s$.

71. The following incorrect sets of quantum numbers in the order n, ℓ, m_ℓ, m_s are written for paired electrons or for one electron in an orbital. Correct them, assuming n values are correct. (a) 1, 0, 0, $+\frac{1}{2}$, $+\frac{1}{2}$; (b) 2, 2, 1, $\pm\frac{1}{2}$; (c) 3, 2, 3, $\pm\frac{1}{2}$; (d) 3, 1, 2, $+\frac{1}{2}$; (e) 2, 1, -1, 0; (f) 3, 0, -1, $-\frac{1}{2}$.

72. (a) How are a $1s$ orbital and a $2s$ orbital in an atom similar? How do they differ? (b) How are a $2p_x$ orbital and a $2p_y$ orbital in an atom similar? How do they differ?

Electron Configurations and the Periodic Table

You should be able to use the positions of elements in the periodic table to answer the exercises in this section.

73. Draw representations of ground state electron configurations using the orbital notation ($\underline{\uparrow\downarrow}$) for the following elements. (a) N; (b) Fe; (c) S; (d) Rh.

74. Draw representations of ground state electron configurations using the orbital notation ($\underline{\uparrow\downarrow}$) for the following elements. (a) P; (b) Ni; (c) Ca; (d) Zr.

75. Give the ground state electron configurations for the elements of Exercise 73 using shorthand notation—that is, $1s^2 2s^2 2p^6$, and so on.

76. Give the ground state electron configurations for the elements of Exercise 74 using shorthand notation—that is, $1s^2 2s^2 2p^6$, and so on.

77. State the Pauli Exclusion Principle. Would any of the following electron configurations violate this rule: (a) $1s^2$; (b) $1s^2 2p^1$; (c) $1s^3$? Explain.

78. State Hund's Rule. Would any of the following electron configurations violate this rule: (a) $1s^2$; (b) $1s^2 2s^2 2p_x^2$; (c) $1s^2 2s^2 2p_x^1 2p_y^1$; (d) $1s^2 2s^2 2p_x^1 2p_z^1$; (e) $1s^2 2s^2 2p_x^2 2p_y^1 2p_z^1$? Explain.

*79. Classify each of the following atomic electron configurations as (i) a ground state, (ii) an excited state, or (iii) a forbidden state: (a) $1s^2 2s^2 2p^5 3s^1$; (b) [Kr] $4d^{10} 5s^3$; (c) $1s^2 2s^2 2p^6 3s^2 3p^6 3d^8 4s^2$; (d) $1s^2 2s^2 2p^6 3s^2 3p^6 3d^1$; (e) $1s^2 2s^2 2p^{10} 3s^2 3p^5$.

80. Which elements are represented by the following electron configurations?
    (a) $1s^2 2s^2 2p^6 3s^2 3p^6 3d^{10} 4s^2 4p^5$
    (b) [Kr] $4d^{10} 4f^{14} 5s^2 5p^6 5d^{10} 5f^{14} 6s^2 6p^6 6d^6 7s^2$
    (c) [Kr] $4d^{10} 4f^{14} 5s^2 5p^6 5d^{10} 6s^2 6p^5$
    (d) [Kr] $4d^5 5s^2$
    (e) $1s^2 2s^2 2p^6 3s^2 3p^6 3d^3 4s^2$
81. Repeat Exercise 80 for
    (a) $1s^2 2s^2 2p^6 3s^2 3p^6 3d^5 4s^1$
    (b) [Kr] $4d^{10} 4f^{14} 5s^2 5p^6 5d^{10} 6s^2 6p^3$
    (c) $1s^2 2s^2 2p^6 3s^2 3p^3$
    (d) [Kr] $4d^{10} 4f^{14} 5s^2 5p^6 5d^{10} 6s^2 6p^6 7s^2$
82. Find the total number of s, p, and d electrons in each of the following: (a) Si; (b) Ar; (c) Ni; (d) Zn; (e) Rb.
83. (a) Distinguish between the terms "diamagnetic" and "paramagnetic," and provide an example that illustrates the meaning of each. (b) How is paramagnetism measured experimentally?
84. How many unpaired electrons are in atoms of Na, Ne, B, Be, Se, and Ti?
85. Which of the following ions or atoms possess paramagnetic properties? (a) Cl^-; (b) Na^+; (c) Co; (d) Ar^-; (e) P.
86. Which of the following ions or atoms possess paramagnetic properties? (a) F; (b) Ne; (c) Ne^+; (d) Zn; (e) S^{2-}.
87. Write the electron configurations of the Group IA elements Li, Na, and K (see inside front cover). What similarity do you observe?
88. Construct a table in which you list a possible set of values for the four quantum numbers for each electron in the following atoms in their ground states. (a) N; (b) S; (c) Zn.
89. Construct a table in which you list a possible set of values for the four quantum numbers for each electron in the following atoms in their ground states. (a) B; (b) Cl; (c) Cu.
90. Draw general electronic structures for the A group elements using the ⇅ notation, where n is the principal quantum number for the highest occupied energy level.

	ns	np
IA	—	— — —
IIA	—	— — —
and so on		

91. Repeat Exercise 90 using $ns^x np^y$ notation.
92. List n, ℓ, and m_ℓ quantum numbers for the highest energy electron (or one of the highest energy electrons if there are more than one) in the following atoms in their ground states (a) P; (b) Nb; (c) Cl; (d) Pr.
93. List n, ℓ, and m_ℓ quantum numbers for the highest energy electron (or one of the highest energy electrons if there are more than one) in the following atoms in their ground states. (a) As; (b) Zn; (c) Mg; (d) Pu.
94. Write the ground state electron configurations for elements A–E.

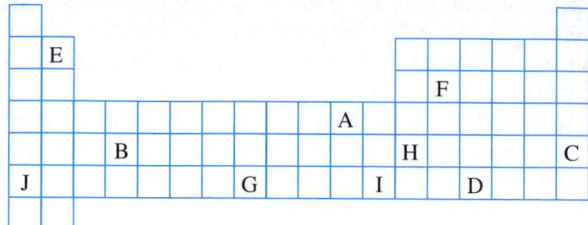

95. Repeat Exercise 94 for elements F–J.

BUILDING YOUR KNOWLEDGE

96. There are two naturally occurring isotopes of hydrogen (^{1}H, >99%, and ^{2}H, <1%) and two of chlorine (^{35}Cl, 76%, and ^{37}Cl, 24%). (a) How many different masses of HCl molecules can be formed from these isotopes? (b) What is the approximate mass of each of the molecules, expressed in atomic mass units? (Use atomic weights rounded to the nearest whole number.) (c) List these HCl molecules in order of decreasing relative abundance.
97. CH_4 is methane. If ^{1}H, ^{2}H, ^{12}C, and ^{13}C were the only isotopes present in a given sample of methane, show the different formulas and formula weights that might exist in that sample. (Use atomic weights rounded to the nearest whole number.)
98. Classical music radio station KMFA in Austin broadcasts at a frequency of 89.5 MHz. What is the wavelength of its signal in meters?
99. A helium atom (^{4}He) contains 2 protons, 2 neutrons, and 2 electrons. Using the masses listed in Table 5-1, calculate the mass of a mole of helium atoms. Compare the calculated value to the listed atomic weight of helium. From your calculated value, which isotopes of helium might you find in a natural sample of helium? (This question ignores binding energy, a topic discussed later in the chapter on nuclear chemistry.)
100. When compounds of barium are heated in a flame, green light of wavelength 554 nm is emitted. How much energy is lost when one mole of barium atoms each emit one photon of this wavelength?
101. A 60-watt light bulb consumes energy at the rate of 60 $J \cdot s^{-1}$. Much of the light is emitted in the infrared region, and less than 5% of the energy appears as visible light. Calculate the number of visible photons emitted per second. Make the simplifying assumptions that 5.0% of the light is visible and that all visible light has a wavelength of 550 nm (yellow/green).

Chemical Periodicity

Steel wool (iron) burns vigorously in pure oxygen.

OBJECTIVES

As you study this chapter, you should learn

· *More about the usefulness of the periodic table*

· *About chemical periodicity in physical properties:*
    Atomic radii
    Ionization energy
    Electron affinity
    Ionic radii
    Electronegativity

· *About chemical periodicity in the reactions of*
    Hydrogen
    Oxygen

· *About chemical periodicity in the compounds of*
    Hydrogen
    Oxygen

The properties of elements are correlated with their positions in the periodic table. Chemists use the periodic table as an invaluable guide in their search for new, useful materials. A barium sodium niobate crystal can convert infrared laser light into visible green light. This harmonic generation or "frequency doubling" is very important in chemical research using lasers and in the telecommunications industry.

6-1 MORE ABOUT THE PERIODIC TABLE

In Chapter 4 we described the development of the periodic table, some terminology for it, and its guiding principle, the *periodic law*.

> The properties of the elements are periodic functions of their atomic numbers.

In Chapter 5 we described electron configurations of the elements. In the long form of the periodic table, elements are arranged in blocks based on the kinds of atomic orbitals being filled. (Please review Table 5-5 and Figure 5-31 carefully.) We saw that electron configurations of elements in the A groups and in Group 0 are entirely predictable from their positions in the periodic table. There are some irregularities within the B groups.

Now we classify the elements according to their electron configurations, which is a very useful system.

Some transition metals (left to right): Ti, V, Cr, Mn, Fe, Co, Ni, Cu.

Noble Gases. For many years the Group 0 elements—the noble gases—were called inert gases because no chemical reactions were known for them. We now know that the heavier members do form compounds, mostly with fluorine and oxygen. Except for helium, each of these elements has eight electrons in its highest occupied energy level. Their structures may be represented as . . . ns^2np^6.

$[He] = 1s^2$

Representative Elements. The A group elements in the periodic table are called representative elements. They have partially occupied highest energy levels. Their "last" electron was added to an s or p orbital. These elements show distinct and fairly regular variations in their properties with changes in atomic number.

d-Transition Elements. Elements in the B groups (except IIB) in the periodic table are known as the d-transition elements or, more simply, as transition elements or transition metals. They were considered to be transitions between the alkaline elements (base-formers) on the left and the acid-formers on the right. All are metals and are characterized by electrons being added to d orbitals. Stated differently, the d-transition elements contain electrons in both the ns and $(n-1)d$ orbitals, but not in the np orbitals. The first transition series, Sc through Zn, has electrons in the $4s$ and $3d$ orbitals, but not in the $4p$ orbitals. They are referred to as

First transition series:	$_{21}$Sc through $_{30}$Zn
Second transition series:	$_{39}$Y through $_{48}$Cd
Third transition series:	$_{57}$La and $_{72}$Hf through $_{80}$Hg
Fourth transition series:	$_{89}$Ac and elements 104 through 112
	(not complete)

f-Transition Elements. Sometimes known as *inner transition elements*, these are elements in which electrons are being added to f orbitals. In these elements, the second from the highest occupied energy level is building from 18 to 32 electrons. All are metals. The f-transition elements are located between Groups IIIB and IVB in the periodic table. They are

First f-transition series (lanthanides):	$_{58}$Ce through $_{71}$Lu
Second f-transition series (actinides):	$_{90}$Th through $_{103}$Lr

The elements of Period 3. Properties progress (left to right) from solids (Na, Mg, Al, Si, P, S) to gases (Cl, Ar) and from the most metallic (Na) to the most nonmetallic (Ar).

The A and B designations for groups of elements in the periodic table are arbitrary, and they are reversed in some periodic tables. In another designation, the groups are numbered 1 through 18. The system used in this text is the one commonly used in the United States. Elements with the same group numbers, but with different letters, have relatively few similar properties. The origin of the A and B designations is the fact that some compounds of elements with the same group numbers have similar formulas but quite different properties, e.g., NaCl (IA) and AgCl (IB), $MgCl_2$ (IIA) and $ZnCl_2$ (IIB). As we shall see, variations in the properties of the B groups across a row are not nearly as regular and dramatic as the variations observed across a row of A group elements. Elements in the B groups contain electrons in the $(n - 1)d$ and ns orbitals but not in the np orbitals, where n represents the highest energy level that contains electrons.

In any atom the *outermost* electrons are those that have the highest value of the principal quantum number, n.

> The *outermost* electrons have the greatest influence on the properties of elements. Adding an electron to an *inner d* orbital results in less striking changes in properties than adding an electron to an *outer s* or *p* orbital.

PERIODIC PROPERTIES OF THE ELEMENTS

We shall now investigate the nature of periodicity. Knowledge of periodicity is valuable in understanding bonding in simple compounds. Many physical properties, such as melting points, boiling points, and atomic volumes, show periodic variations. For now, we describe the variations that are most useful in predicting chemical behavior. The variations in these properties depend on electron configurations, especially the configurations in the outermost occupied shell, and on how far away that shell is from the nucleus.

6-2 ATOMIC RADII

In Section 5-16 we described individual atomic orbitals in terms of probabilities of distributions of electrons over certain regions in space. Similarly, we can visualize the total electron cloud that surrounds an atomic nucleus as somewhat indefinite. Further, we cannot isolate a single atom and measure its diameter the way we can measure the diameter of a golf ball. For all practical purposes, the size of an individual atom cannot be uniquely defined. An indirect approach is required. The size of an atom is determined by its immediate environment, especially its interaction with surrounding atoms. By analogy, suppose we arrange some golf balls in an orderly array in a box. If we know how the balls are positioned, the number of balls, and the dimensions of the box, we can calculate the diameter of an individual ball. Application of this reasoning to solids and their densities leads us to values for the atomic sizes of many elements. In other cases, we derive atomic radii from the observed distances between atoms that are combined with each other. For example, the distance between atomic centers (nuclei) in the Cl_2 molecule is measured to be 1.98 Å. We take the radius of *each* Cl atom to be half the interatomic distance, or 0.99 Å. We collect the data obtained from many such measurements to indicate the *relative* sizes of individual atoms.

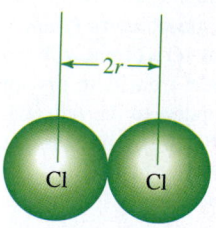

The radius of an atom, r, is taken as half of the distance between nuclei in *homonuclear* molecules such as Cl_2.

The top of Figure 6-1 displays the relative sizes of atoms of the representative elements and the noble gases. It shows the periodicity in atomic radii. (The ionic radii at the bottom of Figure 6-1 are discussed in Section 6-5.)

The **effective nuclear charge,** Z_{eff}, experienced by an electron in an outer energy level is less than the actual nuclear charge, Z. This is because the *attraction* of outer shell electrons by the nucleus is partly counteracted by the *repulsion* of these outer shell electrons by electrons in filled inner shells. We say that the electrons in filled sets of energy levels *screen* or *shield* electrons in outer energy levels from the full effect of the nuclear charge. This **screening,** or **shielding, effect** helps us understand many periodic trends in atomic properties.

Consider the Group IA metals, Li–Cs. Lithium, element number 3, has two electrons in a filled energy level, $1s^2$, and one electron in the $2s$ orbital, $2s^1$. The electron in the $2s$ orbital is fairly effectively screened from the nucleus by the two electrons in the filled $1s$ orbital, the He configuration. The electron in the $2s$ orbital "feels" an effective nuclear charge of about $1+$, rather than the full nuclear charge of $3+$. Sodium, element number 11, has ten electrons in filled sets of orbitals, $1s^2 2s^2 2p^6$, the Ne configuration. These ten electrons in a noble gas configuration fairly effectively screen (shield) its outer electron ($3s^1$) from the nucleus. Thus, the $3s$ electron in sodium also "feels" an effective nuclear charge of about $1+$ rather than $11+$. A sodium atom is larger than a lithium atom because (1) the effective nuclear charge is *approximately* the same for both atoms, and (2) the "outer" electron in a sodium atom is in the third energy level. A similar argument explains why potassium atoms are larger than sodium atoms.

> Within a group of representative elements, atomic radii *increase* from top to bottom as electrons are added to higher energy levels.

As we move *across* the periodic table, atoms become smaller due to increasing effective nuclear charges. Consider the elements B ($Z = 5$, $1s^2 2s^2 2p^1$) to F ($Z = 9$, $1s^2 2s^2 2p^5$). In B there are two electrons in a noble gas configuration, $1s^2$, and three electrons in the second energy level, $2s^2 2p^1$. The two electrons in the noble gas configuration fairly effectively screen out the effect of two protons in the nucleus. So the electrons in the second energy level of B "feel" an effective nuclear charge of approximately $3+$. By similar arguments, we see that in carbon ($Z = 6$, $1s^2 2s^2 2p^2$) the electrons in the second energy level "feel" an effective nuclear charge of approximately $4+$. So we

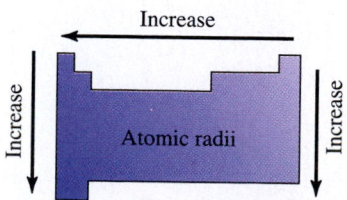

General trends in atomic radii of A group elements with position in the periodic table.

Atomic radii are often stated in **angstroms** (1 Å = 10^{-10} m) or in the SI units **nanometers** (1 nm = 10^{-9} m) or **picometers** (1 pm = 10^{-12} m). To convert from Å to nm, move the decimal point to the left one place (1 Å = 0.1 nm). For example, the atomic radius of Li is 1.52 Å, or 0.152 nm.

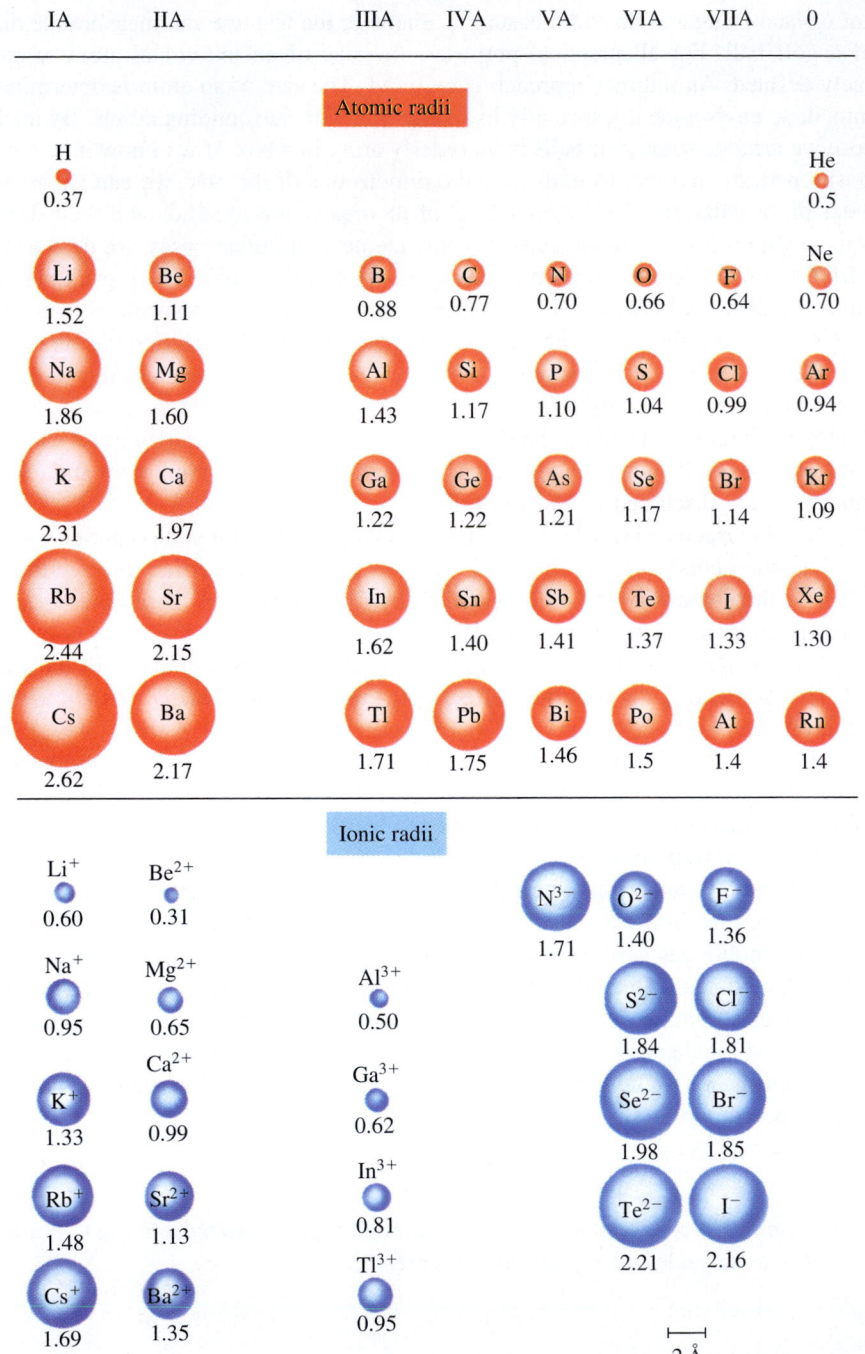

Figure 6-1 (Top) Atomic radii of the A group (representative) elements and the noble gases, in angstroms, Å (Section 6-2). Atomic radii *increase as a group is descended* because electrons are being added to shells farther from the nucleus. Atomic radii *decrease from left to right within a given period* owing to increasing effective nuclear charge. Hydrogen atoms are the smallest and cesium atoms are the largest naturally occurring atoms.

(Bottom) Sizes of ions of the A group elements, in angstroms (Section 6-5). Positive ions (cations) are always *smaller* than the neutral atoms from which they are formed. Negative ions (anions) are always *larger* than the neutral atoms from which they are formed.

expect C atoms to be smaller than B atoms, and they are. In nitrogen ($Z = 7$, $1s^2 2s^2 2p^3$) the electrons in the second energy level "feel" an effective nuclear charge of approximately 5+, and so N atoms are smaller than C atoms.

> As we move from left to right *across a period* in the periodic table, atomic radii of representative elements *decrease* as a proton is added to the nucleus and an electron is added to a particular energy level.

For the transition elements, the variations are not so regular because electrons are being added to an inner shell. All transition elements have smaller radii than the preceding Group IA and IIA elements in the same period.

EXAMPLE 6-1   *Trends in Atomic Radii*

Arrange the following elements in order of increasing atomic radii. Justify your order.

$$\text{Cs, \quad F, \quad K, \quad Cl}$$

Plan

Both K and Cs are Group IA metals, whereas F and Cl are halogens (VIIA nonmetals). Figure 6-1 shows that atomic radii increase as a group is descended, so K < Cs and F < Cl. Atomic radii decrease from left to right.

Solution

The order of increasing atomic radii is   F < Cl < K < Cs.

You should now work Exercises 14 and 16.

6-3 IONIZATION ENERGY

The **first ionization energy (IE_1),** also called *first ionization potential,* is

> the minimum amount of energy required to remove the most loosely bound electron from an isolated gaseous atom to form an ion with a 1+ charge.

For calcium, for example, the first ionization energy, IE_1, is 590 kJ/mol:

$$\text{Ca(g)} + 590 \text{ kJ} \longrightarrow \text{Ca}^+(\text{g}) + e^-$$

The **second ionization energy (IE_2)** is the amount of energy required to remove the second electron. For calcium, it may be represented as

$$\text{Ca}^+(\text{g}) + 1145 \text{ kJ} \longrightarrow \text{Ca}^{2+}(\text{g}) + e^-$$

For a given element, IE_2 *is always greater than IE_1* because it is always more difficult to remove an electron from a positively charged ion than from the corresponding neutral atom. Table 6-1 gives first ionization energies.

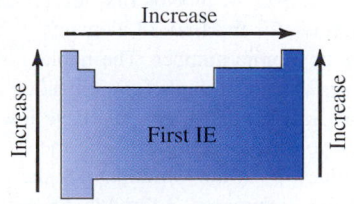

General trends in first ionization energies of A group elements with position in the periodic table. Exceptions occur at Groups IIIA and VIA.

Table 6-1 *First Ionization Energies (kJ/mol of atoms) of Some Elements*

H 1312																	He 2372
Li 520	Be 899											B 801	C 1086	N 1402	O 1314	F 1681	Ne 2081
Na 497	Mg 738											Al 578	Si 786	P 1012	S 1000	Cl 1251	Ar 1521
K 419	Ca 590	Sc 631	Ti 658	V 650	Cr 653	Mn 717	Fe 759	Co 758	Ni 737	Cu 745	Zn 906	Ga 579	Ge 762	As 947	Se 941	Br 1140	Kr 1351
Rb 403	Sr 549	Y 616	Zr 660	Nb 664	Mo 685	Tc 702	Ru 711	Rh 720	Pd 805	Ag 731	Cd 868	In 558	Sn 709	Sb 834	Te 869	I 1008	Xe 1170
Cs 376	Ba 503	La 538	Hf 675	Ta 761	W 770	Re 760	Os 840	Ir 878	Pt 870	Au 890	Hg 1007	Tl 589	Pb 716	Bi 703	Po 812	At 920	Rn 1037

Ionization energies measure how tightly electrons are bound to atoms. Ionization always requires energy to remove an electron from the attractive force of the nucleus. Low ionization energies indicate ease of removal of electrons, and hence ease of positive ion (cation) formation. Figure 6-2 shows a plot of first ionization energy versus atomic number for several elements.

Elements with low ionization energies (IE) lose electrons easily to form cations.

We see that in each period of Figure 6-2, the noble gases have the highest first ionization energies. This should not be surprising, because the noble gases are known to be very unreactive elements. It requires more energy to remove an electron from a helium atom (slightly less than 4.0×10^{-18} J/atom or 2372 kJ/mol) than to remove one from a neutral atom of any other element.

$$\text{He}(g) + 2372 \text{ kJ} \longrightarrow \text{He}^{+}(g) + e^{-}$$

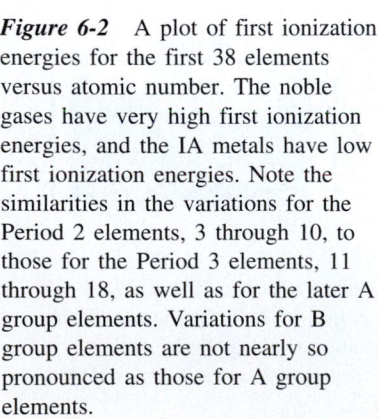

Figure 6-2 A plot of first ionization energies for the first 38 elements versus atomic number. The noble gases have very high first ionization energies, and the IA metals have low first ionization energies. Note the similarities in the variations for the Period 2 elements, 3 through 10, to those for the Period 3 elements, 11 through 18, as well as for the later A group elements. Variations for B group elements are not nearly so pronounced as those for A group elements.

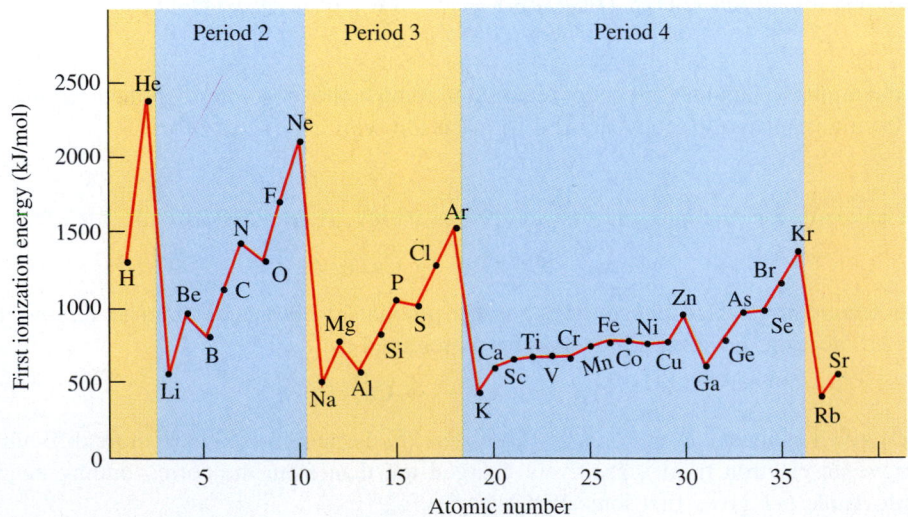

The Group IA metals (Li, Na, K, Rb, Cs) have very low first ionization energies. Each of these elements has only one electron in its highest energy level ($\ldots ns^1$), and they are the largest atoms in their periods. The first electron added to a principal energy level is easily removed to form a noble gas configuration. As we move down the group, the first ionization energies become smaller. The force of attraction of the positively charged nucleus for electrons decreases as the square of the distance between them increases. So as atomic radii increase in a given group, first ionization energies decrease because the outermost electrons are farther from the nucleus. The shielding effect of the electrons in filled inner shells, and the decreased Z_{eff} as the group is descended, further weaken the attraction for the outer shell electrons.

By Coulomb's Law, $F \propto \dfrac{(q^+)(q^-)}{d^2}$, the attraction for the outer shell electrons is directly proportional to the *effective* charges and inversely proportional to the *square* of the distance between the charges.

The first ionization energies of the Group IIA elements (Be, Mg, Ca, Sr, Ba) are significantly higher than those of the Group IA elements in the same periods. This is because the Group IIA elements have smaller atomic radii and higher Z_{eff} values. Thus, their outermost electrons are held more tightly than those of the neighboring IA metals. It is harder to remove an electron from a pair in the filled outermost *s* orbitals of the Group IIA elements than to remove the single electron from the half-filled outermost *s* orbitals of the Group IA elements.

The first ionization energies for the Group IIIA elements (B, Al, Ga, In, Tl) are exceptions to the general horizontal trends. They are *lower* than those of the IIA elements in the same periods because the IIIA elements have only a single electron in their outermost *p* orbitals. It requires less energy to remove the first *p* than to remove the second *s* electron from the same principal energy level because an *ns* orbital is lower in energy (more stable) than an *np* orbital.

The second peak for each period in the ionization energy curve occurs at the Group VA elements (N, P, As, Sb, Bi). These elements have three unpaired electrons in the three outermost *p* orbitals, that is, ns ⇅ np ↑ ↑ ↑, a half-filled set of *p* orbitals. The Group VIA elements (O, S, Se, Te, Po), like the IIIA elements, are exceptions to the horizontal trend. They have slightly *lower* first ionization energies than the VA elements in the same periods. This tells us that it takes slightly less energy to remove a paired electron from a VIA element than to remove an unpaired *p* electron from a VA element in the same period. This is due to the relative stability of the half-filled set of *p* orbitals in the VA elements. Removal of one electron from the VIA elements gives a half-filled set of *p* orbitals.

We have seen other consequences of the special stability of half-filled sets of equivalent orbitals, i.e., exceptions to the Aufbau Principle (Section 5-17).

Knowledge of the relative values of ionization energies assists us in predicting whether an element is likely to form ionic or molecular (covalent) compounds. Elements with low ionization energies form ionic compounds by losing electrons to form **cations** (positively charged ions). Elements with intermediate ionization energies generally form molecular compounds by sharing electrons with other elements. Elements with very high ionization energies, e.g., Groups VIA and VIIA, often gain electrons to form **anions** (negatively charged ions).

Here is one reason why trends in ionization energies are important.

One factor that favors an atom of a *representative* element forming a monatomic ion in a compound is the formation of a stable noble gas configuration. Energy considerations are consistent with this observation. For example, as one mole of Li from Group IA forms one mole of Li^+ ions, it absorbs 520 kJ per mole of Li atoms. The IE_2 value is 14 times greater, 7298 kJ/mol, and is prohibitively large for the formation of Li^{2+} ions under ordinary conditions. For Li^{2+} ions to form, an electron would have to be removed from the filled first energy level. We recognize that this is unlikely. The other alkali metals behave in the same way, for the same reason.

Noble gas configurations are stable only for ions in *compounds*. In fact, $Li^+(g)$ is less stable than $Li(g)$ by 520 kJ/mol.

The first two ionization energies of Be (Group IIA) are 899 and 1757 kJ/mol, but IE_3 is more than eight times larger, 14,849 kJ/mol. So Be forms Be^{2+} ions, but not Be^{3+} ions. The other alkaline earth metals—Mg, Ca, Sr, Ba, and Ra—behave in a similar way.

Due to the high energy required, *simple monatomic cations with charges greater than 3+ do not form under ordinary circumstances.*

Only the lower members of Group IIIA, beginning with Al, form 3+ ions. Bi and some *d*- and *f*-transition metals do so, too. We see that the magnitudes of successive ionization energies support the ideas of electron configurations discussed in Chapter 5.

EXAMPLE 6-2   *Trends in First IEs*

Arrange the following elements in order of increasing first ionization energy. Justify your order.

$$Na, \quad Mg, \quad Al, \quad Si$$

Plan
Table 6-1 shows that first ionization energies generally increase from left to right in the periodic table, but there are exceptions at Groups IIIA and VIA. Al is a IIIA element with only one electron in its outer *p* orbitals, $1s^2 2s^2 2p^1$.

Solution
There is a slight dip at Group IIIA in the plot of first IE versus atomic number. The order of increasing first ionization energy is Na < Al < Mg < Si.

You should now work Exercises 20–22.

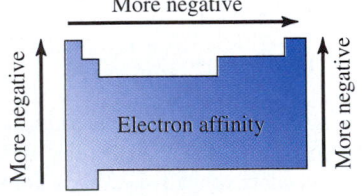

General trends in electron affinities of A group elements with position in the periodic table. There are many exceptions.

6-4 ELECTRON AFFINITY

The **electron affinity (EA)** of an element is defined as

the amount of energy *absorbed* when an electron is added to an isolated gaseous atom to form an ion with a 1− charge.

This is consistent with thermodynamic convention.

The convention is to assign a positive value when energy is absorbed and a negative value when energy is released. For most elements, energy is absorbed. We can represent the electron affinities of beryllium and chlorine as

$$Be(g) + e^- + 241 \text{ kJ} \longrightarrow Be^-(g) \qquad\qquad EA = \ \ 241 \text{ kJ/mol}$$

$$Cl(g) + e^- \longrightarrow Cl^-(g) + 348 \text{ kJ} \qquad\qquad EA = -348 \text{ kJ/mol}$$

The value of EA for Cl can also be represented as -5.78×10^{-19} J/atom or -3.61 eV/atom. The electron volt (eV) is a unit of energy $(1 \text{ eV} = 1.6022 \times 10^{-19} \text{ J})$.

The first equation tells us that when one mole of gaseous beryllium atoms gain one electron each to form gaseous Be⁻ ions, 241 kJ/mol of ions is *absorbed (endothermic)*. The second equation tells us that when one mole of gaseous chlorine atoms gain one electron each to form gaseous chloride ions, 348 kJ of energy is *released (exothermic)*. Figure 6-3 shows a plot of electron affinity versus atomic number for several elements.

Electron affinity involves the *addition* of an electron to a neutral gaseous atom. The process by which a neutral atom X gains an electron (EA),

$$X(g) + e^- \longrightarrow X^-(g) \qquad (EA)$$

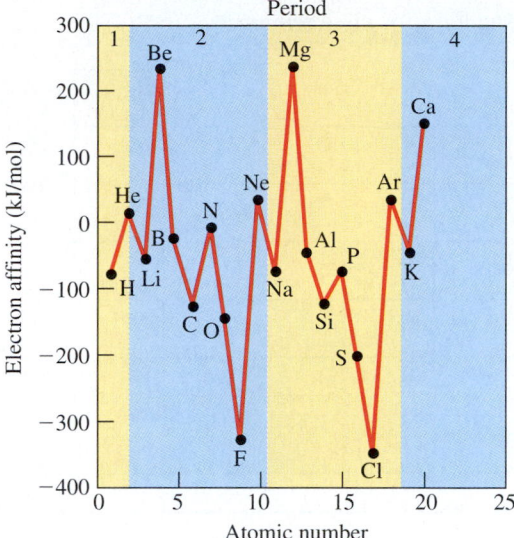

Figure 6-3 A plot of electron affinity versus atomic number for the first 20 elements. The *general* horizontal trend is that electron affinities become more negative (more energy is released as an extra electron is added) from Group IA through Group VIIA for a given period. Exceptions occur at the IIA and VA elements.

is *not* the reverse of the ionization process,

$$X^+(g) + e^- \longrightarrow X(g) \qquad \text{(reverse of IE}_1)$$

The first process begins with a neutral atom, whereas the second begins with a positive ion. Thus, IE_1 and EA are *not* simply equal in value with the signs reversed. We see from Figure 6-3 that electron affinities generally become more negative from left to right across a row in the periodic table (excluding the noble gases). This means that most representative elements in Groups IA to VIIA show a greater attraction for an extra electron from left to right. The halogens, which have the outer electron configuration ns^2np^5, have the most negative electron affinities. They form stable anions with noble gas configurations, . . . ns^2np^6, by gaining one electron.

Elements with very negative electron affinities gain electrons easily to form negative ions (anions).

"Electron affinity" is a precise and quantitative term, like "ionization energy," but it is difficult to measure. Table 6-2 shows electron affinities for the representative elements.

For many reasons, the variations in electron affinities are not regular across a period. The general trend is: the electron affinities of the elements become more negative from left to right in each period. Noteworthy exceptions are the elements of Groups IIA and VA, which have less negative (more positive) values than the trends suggest (Figure 6-3). It is very difficult to add an electron to a IIA metal atom because its outer *s* subshell is filled. The values for the VA elements are slightly less negative than expected because they apply to the addition of an electron to a relatively stable half-filled set of *np* orbitals ($ns^2np^3 \rightarrow ns^2np^4$).

Energy is always required to bring a negative charge (electron) closer to another negative charge (anion). So the addition of a second electron to a $1-$ anion to form an ion with a $2-$ charge is always endothermic. Thus electron affinities of *anions* are always positive.

Table 6-2 *Electron Affinity Values (kJ/mol) of Some Elements**

	IA	IIA		IB	IIIA	IVA	VA	VIA	VIIA	
1	H −72									He (21)
2	Li −60	Be (241)			B −23	C −122	N 0	O −142	F −322	Ne (29)
3	Na −53	Mg (231)		Cu −123	Al −44	Si −119	P −74	S −200	Cl −348	Ar (35)
4	K −48	Ca (156)		Ag −125	Ga (−36)	Ge −116	As −77	Se −194	Br −323	Kr (39)
5	Rb −47	Sr (119)		Au −222	In (−34)	Sn −120	Sb −101	Te −190	I −295	Xe (40)
6	Cs −45	Ba (52)			Tl (−48)	Pb −101	Bi −101	Po (−173)	At (−270)	Rn (40)
7	Fr (−44)									

Estimated values are in parentheses.

EXAMPLE 6-3   *Trends in EAs*

Arrange the following elements in order of increasing values of electron affinity, i.e., from most negative to most positive.

$$\text{K, Br, Cs, Cl}$$

Plan

Table 6-2 shows that electron affinity values generally become more negative from left to right across a period with major exceptions at Groups IIA (Be) and VA (N). They generally become more negative from bottom to top.

Solution

The order of increasing values of electron affinity is

(most negative EA)   Cl < Br < K < Cs   (least negative EA)

You should now work Exercise 31.

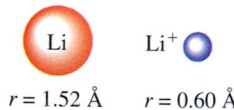

Li
r = 1.52 Å   Li$^+$ r = 0.60 Å

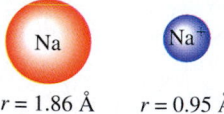

Na
r = 1.86 Å   Na$^+$ r = 0.95 Å

The nuclear charge remains constant when the ion is formed.

6-5   IONIC RADII

Many elements on the left side of the periodic table react with other elements by *losing* electrons to form positively charged ions. Each of the Group IA elements (Li, Na, K, Rb, Cs) has only one electron in its highest energy level (electron configuration . . . ns^1). These elements react with other elements by losing one electron to attain noble gas configurations. They form the ions Li$^+$, Na$^+$, K$^+$, Rb$^+$, and Cs$^+$. A neutral lithium atom, Li, contains three protons in its nucleus and three electrons, with its outermost electron in the 2s orbital. However, a lithium ion, Li$^+$, contains three protons in its nucleus but only two electrons, both in the 1s orbital. So a Li$^+$ ion is much smaller than a neutral Li atom (see margin). Likewise, a sodium ion, Na$^+$, is considerably smaller than a sodium atom, Na. The relative sizes of atoms and common ions of some representative elements are shown in Figure 6-1.

Isoelectronic species have the same number of electrons. We see that the ions formed by the Group IIA elements (Be^{2+}, Mg^{2+}, Ca^{2+}, Sr^{2+}, Ba^{2+}) are significantly smaller than the *isoelectronic* ions formed by the Group IA elements in the same period. The radius of the Li^+ ion is 0.60 Å, while the radius of the Be^{2+} ion is only 0.31 Å. This is what we expect. A beryllium ion, Be^{2+}, is formed when a beryllium atom, Be, loses both of its $2s$ electrons while the 4+ nuclear charge remains constant. We expect the 4+ nuclear charge in Be^{2+} to attract the remaining two electrons quite strongly. Comparison of the ionic radii of the IIA elements with their atomic radii indicates the validity of our reasoning. Similar reasoning indicates that the ions of the Group IIIA metals (Al^{3+}, Ga^{3+}, In^{3+}, Tl^{3+}) should be even smaller than the ions of Group IA and Group IIA elements in the same periods.

Now consider the Group VIIA elements (F, Cl, Br, I). These have the outermost electron configuration . . . ns^2np^5. These elements can completely fill their outermost p orbitals by *gaining* one electron each to attain noble gas configurations. Thus, when a fluorine atom (with seven electrons in its highest energy level) gains one electron, it becomes a fluoride ion, F^-, with eight electrons in its highest energy level. These eight electrons repel one another more strongly than the original seven, so the electron cloud expands. The F^- ion is much larger than the neutral F atom (see margin). Similar reasoning indicates that a chloride ion, Cl^-, should be larger than a neutral chlorine atom, Cl. Observed ionic radii (see Figure 6-1) verify this prediction.

Comparing the sizes of an oxygen atom (Group VIA) and an oxide ion, O^{2-}, again we find that the negatively charged ion is larger than the neutral atom. The oxide ion is also larger than the isoelectronic fluoride ion because the oxide ion contains ten electrons held by a nuclear charge of only 8+, whereas the fluoride ion has ten electrons held by a nuclear charge of 9+. In summary,

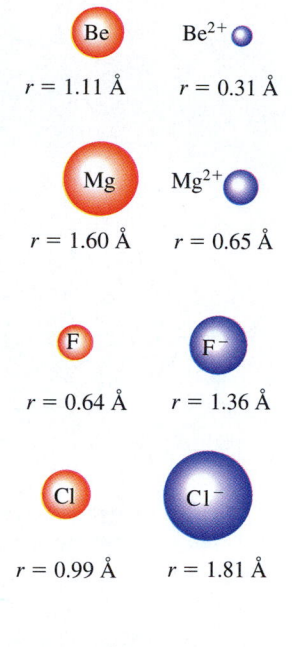

Be	Be^{2+}
$r = 1.11$ Å	$r = 0.31$ Å
Mg	Mg^{2+}
$r = 1.60$ Å	$r = 0.65$ Å
F	F^-
$r = 0.64$ Å	$r = 1.36$ Å
Cl	Cl^-
$r = 0.99$ Å	$r = 1.81$ Å

General trends in ionic radii of A group elements with position in the periodic table.

1. Simple positively charged ions (cations) are always smaller than the neutral atoms from which they are formed.

2. Simple negatively charged ions (anions) are always larger than the neutral atoms from which they are formed.

3. Within an isoelectronic series of ions, ionic radii decrease with increasing atomic number because of increasing nuclear charge.

An isoelectronic series of ions

	N^{3-}	O^{2-}	F^-	Na^+	Mg^{2+}	Al^{3+}
Ionic radius (Å)	1.71	1.40	1.36	0.95	0.65	0.50
No. of electrons	10	10	10	10	10	10
Nuclear charge	+7	+8	+9	+11	+12	+13

EXAMPLE 6-4 *Trends in Ionic Radii*

Arrange the following ions in order of increasing ionic radii: (a) Ca^{2+}, K^+, Al^{3+}; (b) S^{2-}, Cl^-, Te^{2-}.

Plan

Figure 6-1 shows that ionic radii increase from right to left across a period and from top to bottom within a group.

Solution

(a) Cations are always smaller than the neutral atoms from which they are formed.

$$Al^{3+} < Ca^{2+} < K^+$$

(b) Anions are always larger than the neutral atoms from which they are formed.

$$Cl^- < S^{2-} < Te^{2-}$$

You should now work Exercises 36 and 38.

6-6 ELECTRONEGATIVITY

Because the noble gases form few compounds, they are not included in this discussion.

The **electronegativity (EN)** of an element is a measure of the relative tendency of an atom to attract electrons to itself *when it is chemically combined with another atom.*

Elements with high electronegativities (nonmetals) often gain electrons to form anions. Elements with low electronegativities (metals) often lose electrons to form cations.

Table 6-3 *Electronegativity Values of the Elements**

	IA																	0
1	1 H 2.1	IIA		Metals									IIIA	IVA	VA	VIA	VIIA	2 He
2	3 Li 1.0	4 Be 1.5		Nonmetals									5 B 2.0	6 C 2.5	7 N 3.0	8 O 3.5	9 F 4.0	10 Ne
3	11 Na 1.0	12 Mg 1.2	IIIB	Metalloids	IVB	VB	VIB	VIIB	VIIIB		IB	IIB	13 Al 1.5	14 Si 1.8	15 P 2.1	16 S 2.5	17 Cl 3.0	18 Ar
4	19 K 0.9	20 Ca 1.0	21 Sc 1.3	22 Ti 1.4	23 V 1.5	24 Cr 1.6	25 Mn 1.6	26 Fe 1.7	27 Co 1.7	28 Ni 1.8	29 Cu 1.8	30 Zn 1.6	31 Ga 1.7	32 Ge 1.9	33 As 2.1	34 Se 2.4	35 Br 2.8	36 Kr
5	37 Rb 0.9	38 Sr 1.0	39 Y 1.2	40 Zr 1.3	41 Nb 1.5	42 Mo 1.6	43 Tc 1.7	44 Ru 1.8	45 Rh 1.8	46 Pd 1.8	47 Ag 1.6	48 Cd 1.6	49 In 1.6	50 Sn 1.8	51 Sb 1.9	52 Te 2.1	53 I 2.5	54 Xe
6	55 Cs 0.8	56 Ba 1.0	57 * La 1.1	72 Hf 1.3	73 Ta 1.4	74 W 1.5	75 Re 1.7	76 Os 1.9	77 Ir 1.9	78 Pt 1.8	79 Au 1.9	80 Hg 1.7	81 Tl 1.6	82 Pb 1.7	83 Bi 1.8	84 Po 1.9	85 At 2.1	86 Rn
7	87 Fr 0.8	88 Ra 1.0	89 † Ac 1.1															

*	58 Ce 1.1	59 Pr 1.1	60 Nd 1.1	61 Pm 1.1	62 Sm 1.1	63 Eu 1.1	64 Gd 1.1	65 Tb 1.1	66 Dy 1.1	67 Ho 1.1	68 Er 1.1	69 Tm 1.1	70 Yb 1.0	71 Lu 1.2
†	90 Th 1.2	91 Pa 1.3	92 U 1.5	93 Np 1.3	94 Pu 1.3	95 Am 1.3	96 Cm 1.3	97 Bk 1.3	98 Cf 1.3	99 Es 1.3	100 Fm 1.3	101 Md 1.3	102 No 1.3	103 Lr 1.5

*Electronegativity values are given at the bottoms of the boxes. The noble gases are not included in this discussion.

Electronegativities of the elements are expressed on a somewhat arbitrary scale, called the Pauling scale (Table 6-3). The electronegativity of fluorine (4.0) is higher than that of any other element. This tells us that when fluorine is chemically bonded to other elements, it has a greater tendency to attract electron density to itself than does any other element. Oxygen is the second most electronegative element.

> For the representative elements, electronegativities usually increase from left to right across periods and from bottom to top within groups.

Variations among the transition metals are not as regular. In general, both ionization energies and electronegativities are low for elements at the lower left of the periodic table and high for those at the upper right.

General trends in electronegativities of A group elements with position in the periodic table.

EXAMPLE 6-5   *Trends in ENs*

Arrange the following elements in order of increasing electronegativity.

$$B, \quad Na, \quad F, \quad O$$

Plan
Table 6-3 shows that electronegativities increase from left to right across a period and from bottom to top within a group.

Solution
The order of increasing electronegativity is $Na < B < O < F.$

You should now work Exercise 42.

Although the electronegativity scale is somewhat arbitrary, we can use it with reasonable confidence to make predictions about bonding. Two elements with quite different electronegativities (a metal and a nonmetal) tend to react with each other to form ionic compounds. The less electronegative element gives up its electron(s) to the more electronegative element. Two nonmetals with similar electronegativities tend to form covalent bonds with each other. That is, they share their electrons. In this sharing, the more electronegative element attains a greater share. This is discussed in detail in Chapters 7 and 8.

Ionization energy (Section 6-3) and electron affinity (Section 6-4) are precise quantitative concepts. However, we find that the more qualitative concept of *electronegativity* is more useful in describing chemical bonding.

CHEMICAL REACTIONS AND PERIODICITY

Now we shall illustrate the periodicity of chemical properties by considering some reactions of hydrogen, oxygen, and their compounds. We choose to discuss hydrogen and oxygen because, of all the elements, they form the most kinds of compounds with other elements. Additionally, compounds of hydrogen and oxygen are very important in such diverse phenomena as all life processes and most corrosion processes.

6-7 HYDROGEN AND THE HYDRIDES

1 Hydrogen

Elemental hydrogen is a colorless, odorless, tasteless diatomic gas with the lowest atomic weight and density of any known substance. Discovery of the element is attributed to the

The name "hydrogen" means "water former."

Englishman Henry Cavendish, who prepared it in 1766 by passing steam through a red-hot gun barrel (mostly iron) and by the reaction of acids with active metals. The latter is still the method commonly used for the preparation of small amounts of H_2 in the laboratory. In each case, H_2 is liberated by a displacement (and redox) reaction, of the kind described in Section 4-8. (See the activity series, Table 4-13.)

$$3Fe(s) + 4H_2O(g) \xrightarrow{\Delta} Fe_3O_4(s) + 4H_2(g)$$

$$Zn(s) + 2HCl(aq) \longrightarrow ZnCl_2(aq) + H_2(g)$$

Can you write the net ionic equation for the reaction of Zn with HCl(aq)?

Hydrogen also can be prepared by electrolysis of water.

$$2H_2O(\ell) \xrightarrow{\text{electricity}} 2H_2(g) + O_2(g)$$

In the future, if it becomes economical to convert solar energy into electrical energy that can be used to electrolyze water, H_2 could become an important fuel (although the dangers of storage and transportation would have to be overcome). The *combustion* of H_2 liberates a great deal of heat. **Combustion** is the highly exothermic combination of a substance with oxygen, usually with a flame. (See Section 6-8, part 3.)

This is the reverse of the decomposition of H_2O.

$$2H_2(g) + O_2(g) \xrightarrow[\text{or } \Delta]{\text{spark}} 2H_2O(\ell) + \text{energy}$$

Hydrogen is no longer used in blimps and dirigibles. It has been replaced by helium, which is slightly denser, nonflammable, and much safer.

Hydrogen is very flammable; it was responsible for the Hindenburg airship disaster in 1937. A spark is all it takes to initiate the **combustion reaction,** which is exothermic enough to provide the heat necessary to sustain the reaction.

Hydrogen is prepared by the "water gas reaction," which results from the passage of steam over white-hot coke (impure carbon, a nonmetal) at 1500°C. The mixture of products commonly called "water gas" is used industrially as a fuel. Both components, CO and H_2, undergo combustion.

$$\underset{\text{in coke}}{C(s)} + \underset{\text{steam}}{H_2O(g)} \longrightarrow \underbrace{CO(g) + H_2(g)}_{\text{"water gas"}}$$

Vast quantities of hydrogen are produced commercially each year by a process called *steam cracking*. Methane reacts with steam at 830°C in the presence of a nickel catalyst.

$$CH_4(g) + H_2O(g) \xrightarrow[\text{Ni}]{\Delta} CO(g) + 3H_2(g)$$

2 Reactions of Hydrogen and Hydrides

The use of the term "hydride" does not necessarily imply the presence of the hydride ion, H^-.

Now let us turn to some reactions of hydrogen with metals and other nonmetals to form binary compounds called **hydrides.** We shall also discuss characteristic reactions of the hydrides. Atomic hydrogen has the $1s^1$ electron configuration. It can form (1) **ionic hydrides** containing hydride ions, H^-, by gaining one electron per atom from an active metal; or (2) **molecular hydrides** by sharing its electrons with an atom of another nonmetal.

The ionic or molecular character of the binary compounds of hydrogen depends upon the position of the other element in the periodic table (Figure 6-4). The reactions of H_2 with the *alkali* (IA) and the heavier (more active) *alkaline earth* (IIA) *metals* result in solid

IA	IIA	IIIA	IVA	VA	VIA	VIIA
LiH	BeH_2	B_2H_6	CH_4	NH_3	H_2O	HF
NaH	MgH_2	$(AlH_3)_x$	SiH_4	PH_3	H_2S	HCl
KH	CaH_2	Ga_2H_6	GeH_4	AsH_3	H_2Se	HBr
RbH	SrH_2	InH_3	SnH_4	SbH_3	H_2Te	HI
CsH	BaH_2	TlH	PbH_4	BiH_3	H_2Po	HAt

Figure 6-4 Common hydrides of the representative elements. The ionic hydrides are shaded blue, molecular hydrides are shaded red, and those of intermediate character are shaded purple.

ionic hydrides, often called *saline* or *salt-like hydrides.* The reaction with the molten (liquid) IA metals may be represented in general terms as

$$2M(\ell) + H_2(g) \xrightarrow[\text{high pressures of } H_2]{\text{high temperatures}} 2(M^+, H^-)(s) \qquad M = \text{Li, Na, K, Rb, Cs}$$

Thus, hydrogen combines with lithium to form lithium hydride and with sodium to form sodium hydride.

> The ionic hydrides are named by naming the metal first, followed by "hydride."

$$2Li(\ell) + H_2(g) \longrightarrow 2LiH(s) \qquad \text{lithium hydride (mp 680°C)}$$

$$2Na(\ell) + H_2(g) \longrightarrow 2NaH(s) \qquad \text{sodium hydride (mp 800°C)}$$

In general terms, the reactions of the heavier (more active) IIA metals may be represented as

$$M(\ell) + H_2(g) \longrightarrow (M^{2+}, 2H^-)(s) \qquad M = \text{Ca, Sr, Ba}$$

Thus, calcium combines with hydrogen to form calcium hydride.

$$Ca(\ell) + H_2(g) \longrightarrow CaH_2(s) \qquad \text{calcium hydride (mp 816°C)}$$

These *ionic hydrides are all basic* because hydride ions react with water to form hydroxide ions. When water is added by drops to lithium hydride, for example, lithium hydroxide and hydrogen are produced. The reaction of calcium hydride is similar.

> Ionic hydrides can serve as sources of hydrogen. However, they must be stored in environments free of moisture and O_2.
>
> We show LiOH and $Ca(OH)_2$ as solids here because not enough water is available to act as a solvent.

$$LiH(s) + H_2O(\ell) \longrightarrow LiOH(s) + H_2(g)$$

$$CaH_2(s) + 2H_2O(\ell) \longrightarrow Ca(OH)_2(s) + 2H_2(g)$$

Hydrogen reacts with *nonmetals* to form binary *molecular hydrides.* For example, H_2 combines with the halogens to form colorless, gaseous hydrogen halides (Figure 6-5):

$$H_2(g) + X_2 \longrightarrow 2HX(g) \qquad X = \text{F, Cl, Br, I}$$
$$\text{hydrogen halides}$$

> The hydrogen halides are named by the word "hydrogen" followed by the stem for the halogen with an "-ide" ending.

Specifically, hydrogen reacts with fluorine to form hydrogen fluoride and with chlorine to form hydrogen chloride:

$$H_2(g) + F_2(g) \longrightarrow 2HF(g) \qquad \text{hydrogen fluoride}$$

$$H_2(g) + Cl_2(g) \longrightarrow 2HCl(g) \qquad \text{hydrogen chloride}$$

Hydrogen combines with Group VIA elements to form molecular compounds:

$$2H_2(g) + O_2(g) \xrightarrow{\Delta} 2H_2O(g)$$

Figure 6-5 Hydrogen, H_2, burns in an atmosphere of pure chlorine, Cl_2, to produce hydrogen chloride.

$$H_2 + Cl_2 \longrightarrow 2HCl$$

These compounds are named:
  H_2O, hydrogen oxide (water)
  H_2S, hydrogen sulfide
  H_2Se, hydrogen selenide
  H_2Te, hydrogen telluride
All except H_2O are *very* toxic.

As we shall see in Chapter 10, even H_2O is weakly acidic.

Figure 6-6 Ammonia may be applied directly to the soil as a fertilizer.

The heavier members of this family also combine with hydrogen to form binary compounds that are gases at room temperature. Their formulas resemble that of water.

The primary industrial use of H_2 is in the synthesis of ammonia, a molecular hydride, by the Haber process (Section 17-6). Most of the NH_3 is used in liquid form as a fertilizer (Figure 6-6) or to make other fertilizers, such as ammonium nitrate, NH_4NO_3, and ammonium sulfate, $(NH_4)_2SO_4$:

$$N_2(g) + 3H_2(g) \xrightarrow[\Delta,\text{ high pressure}]{\text{catalysts}} 2NH_3(g)$$

Many of the molecular (nonmetal) hydrides are acidic; their aqueous solutions produce hydrogen ions. These include HF, HCl, HBr, HI, H_2S, H_2Se, and H_2Te.

EXAMPLE 6-6 *Predicting Products of Reactions*

Predict the products of the reactions involving the reactants shown. Write a balanced formula unit equation for each.

(a) $H_2(g) + I_2(g) \xrightarrow{\Delta}$

(b) $K(\ell) + H_2(g) \xrightarrow{\Delta}$

(c) $NaH(s) + H_2O(\ell)$ (excess) $\longrightarrow$

Plan

(a) Hydrogen reacts with the halogens (Group VIIA) to form hydrogen halides—in this example, HI.

(b) Hydrogen reacts with active metals to produce hydrides—in this case, KH.

(c) Active metal hydrides react with water to produce a metal hydroxide and H_2.

Remember that hydride ions, H^-, react with (reduce) water to produce OH^- ions and $H_2(g)$.

Solution

(a) $H_2(g) + I_2(g) \xrightarrow{\Delta} 2HI(g)$

(b) $2K(\ell) + H_2(g) \xrightarrow{\Delta} 2KH(s)$

(c) $NaH(s) + H_2O(\ell) \longrightarrow NaOH(aq) + H_2(g)$

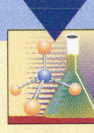
Fill 'er Up with Hydrides?

The science fiction writer Jules Verne had an uncanny knack of predicting the future. In 1874 he wrote, "I believe that water will one day be employed as fuel, that hydrogen and oxygen which constitute it, used singly or together, will furnish an inexhaustible source of heat and light." Many now believe this prophecy may be fulfilled in the twenty-first century when hydrogen may become a universal fuel that can be burned in colossal power plants and in small stoves. At that time, hydrogen may be the fuel used to power automobiles. Now, conventional automobile engines can burn hydrogen with only minor adjustments to the carburetor; hydride fuel tanks (using hydrides as a source of hydrogen) have been used in experimental vehicles for over a decade. Hydrogen is attractive as a fuel because its only combustion product is water; it produces no soot, toxic gases, smoke, or undesirable greenhouse gases.

Hydrogen's lack of undesirable combustion products is currently causing automobile makers to consider it as a fuel. California requires that by 1998, 2% of all cars sold there must be "zero-polluting"; by 2003, this figure will increase to 10%. New York and Massachusetts have passed similar laws that become effective in 1999, and ten other states, plus Washington, D.C., are considering comparable measures. Although some automobile manufacturers may turn to electric cars initially to reduce toxic vehicle emissions, total poisonous emissions may actually increase with battery-powered electric vehicles. The main toxic emissions are those given off by power plants that supply the electricity needed to recharge the very large batteries. These toxic emissions will be higher per mile driven by battery-powered cars than by cars that burn gasoline. However, the toxic emissions would be *at* power plants, not *on* city streets.

As with any new energy source, problems must be solved before hydrogen can take its place as a universal fuel. One problem is storage. Hydrogen could be stored as a compressed gas, cooled and compressed into a liquid and stored at low temperatures, or stored in "hydride compounds." Many transition metals and metal alloys such as magnesium-nickel can absorb large quantities of hydrogen. When hydrogen is absorbed by such metals, the resulting materials are called *interstitial hydrides*—they are not true compounds. Gentle heating of these hydrides, for example by automobile engine exhausts, releases the stored hydrogen. In fact, experimental vehicles with hydride fuel tanks have been in operation since the early 1980s. There are advantages of using metals to store hydrogen in the form of interstitial hydrides. The need for high pressures and low temperatures is eliminated. Some metals have such a great ability to absorb hydrogen that interstitial hydrides can store twice as much hydrogen in the space required by compressed hydrogen or liquid hydrogen tanks. Also, because these hydrides release hydrogen slowly when heated, hydrogen explosions are less likely. Interestingly, hydrogen stored as a compressed gas or in liquid form can be safer than gasoline. Because hydrogen is much less dense than air and disperses quickly into the atmosphere, leaking hydrogen would be less likely to explode than leaking gasoline.

As Jules Verne suggested, hydrogen can be produced by decomposing water in electrolytic cells (Chapter 21). However, all techniques used to produce hydrogen for use as a fuel are currently prohibitively expensive. Research chemists are developing artificial chloroplasts, similar to the natural chloroplasts found in green plant cells, that split hydrogen from water. These artificial chloroplasts sever hydrogen from water molecules and release gaseous hydrogen. In a second approach, researchers are searching for ways to improve the efficiency of photoelectric cells. The electricity produced by photoelectric cells could be used to decompose water economically into hydrogen and oxygen.

There may be a better approach than simply burning hydrogen as a fuel. An alternative oxidizes hydrogen in fuel cells (Chapter 21) that directly convert hydrogen into electricity; water is the only product of this reaction. Fuel cells produce energy much more efficiently than burning hydrogen and then converting the heat energy into mechanical power.

For the immediate future, you can expect to use gasoline to power your car. But within a few decades, historians may be impressed by the wisdom of Jules Verne.

Ronald DeLorenzo
Middle Georgia College

EXAMPLE 6-7 *Ionic and Molecular Properties*

Predict the ionic or molecular character of the products in Example 6-6.

Plan

We refer to Figure 6-4, which displays the nature of hydrides.

Solution

Reaction (a) is a reaction between hydrogen and another nonmetal. The product, HI, must be molecular. Reaction (b) is the reaction of hydrogen with an active Group IA metal. Thus, KH must be ionic. The products of reaction (c) are molecular $H_2(g)$ and the strong soluble base, NaOH, which is ionic.

You should now work Exercises 56–58.

6-8 OXYGEN AND THE OXIDES

1 Oxygen and Ozone

The name "oxygen" means "acid former."

Oxygen was discovered in 1774 by an English minister and scientist, Joseph Priestley. He observed the thermal decomposition of mercury(II) oxide, a red powder:

$$2HgO(s) \xrightarrow{\Delta} 2Hg(\ell) + O_2(g)$$

That part of the earth we see—land, water, and air—is approximately 50% oxygen by mass. About two thirds of the mass of the human body is due to oxygen in H_2O. Elemental oxygen, O_2, is an odorless and colorless gas that makes up about 21% by volume of dry air. In the liquid and solid states it is pale blue. Oxygen is very slightly soluble in water; only about 0.04 gram dissolves in 1 liter of water at 25°C. This is sufficient to sustain fish and other marine organisms. Oxygen is obtained commercially by the fractional distillation of liquid air. The greatest single industrial use of O_2 is for oxygen-enrichment in blast furnaces for the conversion of pig iron to steel.

Liquid O_2 is used as an oxidizer for rocket fuels. O_2 also is used in the health fields for oxygen-enriched air.

Allotropes are different forms of the same element in the same physical state (Section 2-2).

Oxygen also exists in a second allotropic form, ozone, O_3. Ozone is an unstable, pale blue gas at room temperature. It is formed by passing an electrical discharge through gaseous oxygen. Its unique, pungent odor is often noticed during electrical storms and in the vicinity of electrical equipment. Not surprisingly, its density is about $1\frac{1}{2}$ times that of O_2. At −112°C it condenses to a deep blue liquid. It is a very strong oxidizing agent. As a concentrated gas or a liquid, ozone can easily decompose explosively:

$$2O_3(g) \longrightarrow 3O_2(g)$$

A radical is a species containing one or more unpaired electrons; many radicals are very reactive.

Oxygen atoms, or **radicals,** are intermediates in this exothermic decomposition of O_3 to O_2. They act as strong oxidizing agents in such applications as destroying bacteria in water purification.

The ozone molecule is angular and diamagnetic (page 45). Both oxygen—oxygen bond lengths (1.28 Å) are identical and are intermediate between typical single and double bond lengths.

2 Reactions of Oxygen and the Oxides

Oxygen forms oxides by direct combination with all other elements except the noble gases and noble (unreactive) metals (Au, Pd, Pt). **Oxides** are binary compounds that contain oxygen. Although such reactions are generally very exothermic, many proceed quite slowly and require heating to supply the energy necessary to break the strong bonds in O_2 molecules. Once these reactions are initiated, most release more than enough energy to be self-sustaining and sometimes become "red hot."

Reactions of O_2 with Metals

In general, metallic oxides (including peroxides and superoxides) are ionic solids. The Group IA metals combine with oxygen to form three kinds of solid ionic products called oxides, peroxides, and superoxides. Lithium combines with oxygen to form lithium oxide.

$$4Li(s) + O_2(g) \longrightarrow 2Li_2O(s) \qquad \text{lithium oxide (mp > 1700°C)}$$

By contrast, sodium reacts with an excess of oxygen to form sodium peroxide, Na_2O_2, rather than sodium oxide, Na_2O, as the *major* product.

$$2Na(s) + O_2(g) \longrightarrow Na_2O_2(g) \qquad \text{sodium peroxide (decomposes at 460°C)}$$

Peroxides contain the $O-O^{2-}$, O_2^{2-} group, in which the oxidation number of oxygen is -1, whereas *normal oxides* such as lithium oxide, Li_2O, contain oxide ions, O^{2-}. The heavier members of the family (K, Rb, Cs) react with excess oxygen to form **superoxides.** These contain the superoxide ion, O_2^-, in which the oxidation number of oxygen is $-\frac{1}{2}$. The reaction with K is

$$K(s) + O_2(g) \longrightarrow KO_2(s) \qquad \text{potassium superoxide (mp 430°C)}$$

The tendency of the Group IA metals to form oxygen-rich compounds increases as the group is descended. This is because cation radii increase going down the group. A similar trend is observed in the reactions of the Group IIA metals with oxygen. You can recognize these classes of compounds as

Class	Contains Ions	Oxidation No. of Oxygen
normal oxides	O^{2-}	-2
peroxides	O_2^{2-}	-1
superoxides	O_2^-	$-\frac{1}{2}$

With the exception of Be, the Group IIA metals react with oxygen at moderate temperatures to form normal ionic oxides, MO, and at high pressures of oxygen the heavier ones form ionic peroxides, MO_2 (Table 6-4).

Beryllium reacts with oxygen only at elevated temperatures and forms only the normal oxide, BeO.

$$2M(s) + O_2(g) \longrightarrow 2(M^{2+}, O^{2-})(s) \qquad M = Be, Mg, Ca, Sr, Ba$$

$$M(s) + O_2(g) \longrightarrow (M^{2+}, O_2^{2-})(s) \qquad M = Ca, Sr, Ba$$

Table 6-4 *Oxygen Compounds of the IA and IIA Metals**

	IA					IIA				
	Li	Na	K	Rb	Cs	Be	Mg	Ca	Sr	Ba
normal oxides	Li_2O	Na_2O	K_2O	Rb_2O	Cs_2O	BeO	MgO	CaO	SrO	BaO
peroxides	Li_2O_2	Na_2O_2	K_2O_2	Rb_2O_2	Cs_2O_2			CaO_2	SrO_2	BaO_2
superoxides		NaO_2	KO_2	RbO_2	CsO_2					

*The shaded compounds represent the principal products of the direct reaction of the metal with oxygen.

For example, the equations for the reactions of calcium and oxygen are

$$2Ca(s) + O_2(g) \longrightarrow 2CaO(s) \qquad \text{calcium oxide (mp 2580°C)}$$

$$Ca(s) + O_2(g) \longrightarrow CaO_2(s) \qquad \text{calcium peroxide (decomposes at 275°C)}$$

The other metals, with the exceptions noted previously (Au, Pd, and Pt), react with oxygen to form solid metal oxides. Many metals to the right of Group IIA show variable oxidation states, so they may form several oxides. For example, iron combines with oxygen in the following series of reactions to form three different oxides (Figure 6-7).

$$2Fe(s) + O_2(g) \xrightarrow{\Delta} 2FeO(s) \qquad \text{iron(II) oxide } or \text{ ferrous oxide}$$

$$6FeO(s) + O_2(g) \xrightarrow{\Delta} 2Fe_3O_4(s) \qquad \text{magnetic iron oxide (a mixed oxide)}$$

$$4Fe_3O_4(s) + O_2(g) \xrightarrow{\Delta} 6Fe_2O_3(s) \qquad \text{iron(III) oxide } or \text{ ferric oxide}$$

Copper reacts with a limited amount of oxygen to form red Cu_2O, whereas with excess oxygen it forms black CuO.

$$4Cu(s) + O_2(g) \xrightarrow{\Delta} 2Cu_2O(s) \qquad \text{copper(I) oxide } or \text{ cuprous oxide}$$

$$2Cu(s) + O_2(g) \xrightarrow{\Delta} 2CuO(s) \qquad \text{copper(II) oxide } or \text{ cupric oxide}$$

Figure 6-7 Iron powder burns brilliantly to form iron(III) oxide, Fe_2O_3.

> Metals that exhibit variable oxidation states react with a limited amount of oxygen to give lower oxidation state oxides (such as FeO and Cu_2O). They react with an excess of oxygen to give higher oxidation state oxides (such as Fe_2O_3 and CuO).

Reactions of Metal Oxides with Water

Oxides of metals are called **basic anhydrides** because many of them combine with water to form bases with no change in oxidation state of the metal (Figure 6-8). "Anhydride" means "without water"; in a sense, the metal oxide is a hydroxide base with the water

Increasing acidic character ⟶

Increasing base character ↓

IA	IIA	IIIA	IVA	VA	VIA	VIIA
Li_2O	BeO	B_2O_3	CO_2	N_2O_5		F_2O
Na_2O	MgO	Al_2O_3	SiO_2	P_4O_{10}	SO_3	Cl_2O_7
K_2O	CaO	Ga_2O_3	GeO_2	As_2O_5	SeO_3	Br_2O_7
Rb_2O	SrO	In_2O_3	SnO_2	Sb_2O_5	TeO_3	I_2O_7
Cs_2O	BaO	Tl_2O_3	PbO_2	Bi_2O_5	PoO_3	At_2O_7

Figure 6-8 The normal oxides of the representative elements in their maximum oxidation states. Acidic oxides (acid anhydrides) are shaded red, amphoteric oxides are shaded purple, and basic oxides (basic anhydrides) are shaded blue. An **amphoteric** oxide is one that shows some acidic and some basic properties.

"removed." Metal oxides that are soluble in water react to produce the corresponding hydroxides.

Metal Oxide + Water ⟶ Metal Hydroxide (base)

sodium oxide	$Na_2O(s)$	$+ H_2O(\ell) \longrightarrow 2\,NaOH(aq)$	sodium hydroxide
calcium oxide	$CaO(s)$	$+ H_2O(\ell) \longrightarrow Ca(OH)_2(aq)$	calcium hydroxide
barium oxide	$BaO(s)$	$+ H_2O(\ell) \longrightarrow Ba(OH)_2(aq)$	barium hydroxide

The oxides of the Group IA metals and the heavier Group IIA metals dissolve in water to give solutions of strong soluble bases. Most other metal oxides are insoluble in water.

Reactions of O_2 with Nonmetals

Oxygen combines with many nonmetals to form molecular oxides. For example, carbon burns in oxygen to form carbon monoxide or carbon dioxide, depending on the relative amounts of carbon and oxygen.

$$2C(s) + O_2(g) \longrightarrow 2\overset{+2}{C}O(s) \qquad \text{(excess C and limited } O_2)$$

$$C(s) + O_2(g) \longrightarrow \overset{+4}{C}O_2(g) \qquad \text{(limited C and excess } O_2)$$

Recall that oxidation states are indicated by red numbers in red circles.

When hydrocarbons burn in the air, similar reactions occur. For example, burning octane, C_8H_{18}, in an excess of oxygen (plenty of air) produces carbon dioxide and water. There are many similar compounds in gasoline and diesel fuels.

$$2C_8H_{18}(\ell) + \underset{\text{excess}}{25O_2(g)} \longrightarrow 16CO_2(g) + 18H_2O(\ell)$$

Carbon monoxide is produced by the incomplete burning of carbon-containing compounds in a limited amount of oxygen.

$$2C_8H_{18}(\ell) + \underset{\text{limited amount}}{17O_2(g)} \longrightarrow 16CO(g) + 18H_2O(\ell)$$

In severely limited oxygen, carbon (soot) is produced by partially burning hydrocarbons. For octane,

$$2C_8H_{18}(\ell) + \underset{\text{severely limited amount}}{9O_2(g)} \longrightarrow 16C(s) + 18H_2O(\ell)$$

When you see blue or black smoke (carbon) coming from an internal combustion engine, (or smell unburned fuel in the air) you may be quite sure that lots of carbon monoxide is also being produced and released into the air.

"White smoke" from auto exhaust systems is tiny droplets of water condensing in the cooler air.

Carbon monoxide is a very poisonous gas because it forms a stronger bond with the iron atom in hemoglobin than does an oxygen molecule. Attachment of the CO molecule to the iron atom destroys the ability of hemoglobin to pick up oxygen in the lungs and carry it to the brain and muscle tissue. Carbon monoxide poisoning is particularly insidious because the gas has no odor and because the victim first becomes drowsy.

Unlike carbon monoxide, carbon dioxide is not toxic. It is one of the products of the respiratory process. It is used to make carbonated beverages, which are mostly saturated solutions of carbon dioxide in water; a small amount of the carbon dioxide combines with the water to form carbonic acid (H_2CO_3), a very weak acid.

Figure 6-9 Sulfur burns in oxygen to form sulfur dioxide.

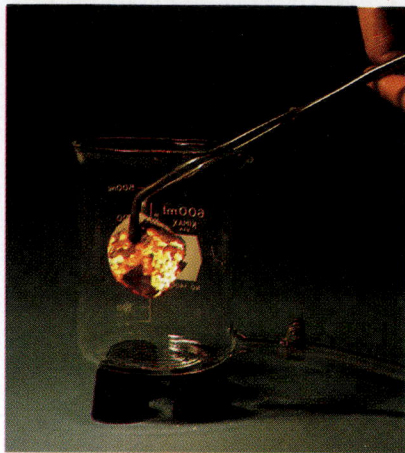

Carbon burns brilliantly in pure O_2 to form CO_2.

Phosphorus reacts with a limited amount of oxygen to form tetraphosphorus hexoxide, P_4O_6,

$$P_4(s) + 3O_2(g) \xrightarrow{\overset{+3}{}} P_4O_6(s) \qquad \text{tetraphosphorus hexoxide}$$

whereas reaction with an excess of oxygen gives tetraphosphorus decoxide, P_4O_{10}.

$$P_4(s) + 5O_2(g) \xrightarrow{\overset{+5}{}} P_4O_{10} \qquad \text{tetraphosphorus decoxide}$$

Sulfur burns in oxygen to form primarily sulfur dioxide (Figure 6-9) and only very small amounts of sulfur trioxide.

$$S_8(s) + 8O_2(g) \xrightarrow{\overset{+4}{}} 8SO_2(g) \qquad \text{sulfur dioxide (mp } -73°C)$$

The production of SO_3 at a reasonable rate requires the presence of a catalyst.

$$S_8(s) + 12O_2(g) \xrightarrow{\overset{+6}{}} 8SO_3(g) \qquad \text{sulfur trioxide (mp } 32.5°C)$$

Oxidation States of Nonmetals

Nearly all nonmetals exhibit more than one oxidation state in their compounds. In general, the *most* common oxidation states of a nonmetal are (1) its periodic group number, (2) its periodic group number minus two, and (3) its periodic group number minus eight. The reactions of nonmetals with a limited amount of oxygen usually give products that contain the nonmetals (other than oxygen) in lower oxidation states, usually case (2). Reactions with excess oxygen give products in which the nonmetals exhibit higher oxidation states, case (1). The examples we have cited are CO and CO_2, P_4O_6 and P_4O_{10}, and SO_2 and SO_3. The molecular formulas of the oxides are sometimes not easily predictable, but the *simplest* formulas are. For example, the two most common oxidation states of phosphorus in molecular compounds are $+3$ and $+5$. The simplest formulas for the corresponding phosphorus oxides therefore are P_2O_3 and P_2O_5, respectively. The molecular (true) formulas are twice these, P_4O_6 and P_4O_{10}.

Reactions of Nonmetal Oxides with Water

Nonmetal oxides are called **acid anhydrides** because many of them dissolve in water to form acids *with no change in oxidation state of the nonmetal* (Figure 6-8). Several **ternary**

acids can be prepared by reaction of the appropriate nonmetal oxides with water. Ternary acids contain three elements, usually H, O, and another nonmetal.

$$\begin{array}{c}\textbf{Nonmetal} \\ \textbf{Oxide}\end{array} + \textbf{Water} \longrightarrow \textbf{Ternary Acid}$$

carbon dioxide $\quad\quad \overset{(+4)}{CO_2}(g) + H_2O(\ell) \longrightarrow \overset{(+4)}{H_2}CO_3(aq)$ $\quad$ carbonic acid

sulfur dioxide $\quad\quad \overset{(+4)}{SO_2}(g) + H_2O(\ell) \longrightarrow \overset{(+4)}{H_2}SO_3(aq)$ $\quad$ sulfurous acid

sulfur trioxide $\quad\quad \overset{(+6)}{SO_3}(g) + H_2O(\ell) \longrightarrow \overset{(+6)}{H_2}SO_4(aq)$ $\quad$ sulfuric acid

dinitrogen pentoxide $\quad \overset{(+5)}{N_2}O_5(s) + H_2O(\ell) \longrightarrow 2\overset{(+5)}{H}NO_3(aq)$ $\quad$ nitric acid

tetraphosphorus decoxide $\quad \overset{(+5)}{P_4}O_{10}(s) + 6H_2O(\ell) \longrightarrow 4\overset{(+5)}{H_3}PO_4(aq)$ $\quad$ phosphoric acid

Nearly all oxides of nonmetals react with water to give solutions of ternary acids. The oxides of boron and silicon, which are insoluble, are two exceptions.

Reactions of Metal Oxides with Nonmetal Oxides

Another common kind of reaction of oxides is the *combination of metal oxides (basic anhydrides) with nonmetal oxides (acid anhydrides), with no change in oxidation states, to form salts.*

$$\begin{array}{c}\textbf{Metal} \\ \textbf{Oxide}\end{array} + \begin{array}{c}\textbf{Nonmetal} \\ \textbf{Oxide}\end{array} \longrightarrow \textbf{Salt}$$

calcium oxide + sulfur trioxide $\quad \overset{(+2)}{Ca}O(s) + \overset{(+6)}{SO_3}(g) \longrightarrow \overset{(+2)}{Ca}\overset{(+6)}{SO_4}(s)$ $\quad$ calcium sulfate

magnesium oxide + carbon dioxide $\quad \overset{(+2)}{Mg}O(s) + \overset{(+4)}{CO_2}(g) \longrightarrow \overset{(+2)}{Mg}\overset{(+4)}{CO_3}(s)$ $\quad$ magnesium carbonate

sodium oxide + tetraphosphorus decoxide $\quad 6\overset{(+1)}{Na_2}O(s) + \overset{(+5)}{P_4}O_{10}(s) \longrightarrow 4\overset{(+1)}{Na_3}\overset{(+5)}{P}O_4(s)$ $\quad$ sodium phosphate

EXAMPLE 6-8 *Acidic Character of Oxides*

Arrange the following oxides in order of increasing molecular (acidic) character: SO_3, Cl_2O_7, CaO, and PbO_2.

Plan

Molecular (acidic) character of oxides increases as nonmetallic character of the element that is combined with oxygen increases (Figure 6-8).

increasing nonmetallic character

$$Ca < Pb < S < Cl$$

Periodic group: IIA   IVA   VIA   VIIA

Solution

increasing molecular character
$\longrightarrow$

Thus, the order is $CaO < PbO_2 < SO_3 < Cl_2O_7$

EXAMPLE 6-9   *Basic Character of Oxides*

Arrange the oxides in Example 6-8 in order of increasing basicity.

Plan

The greater the molecular character of an oxide, the more acidic it is. Thus, the most basic oxides have the least molecular (most ionic) character (Figure 6-8).

Solution

increasing basic character
$\longrightarrow$

molecular $Cl_2O_7 < SO_3 < PbO_2 < CaO$ ionic

EXAMPLE 6-10   *Predicting Reaction Products*

Predict the products of the reactions involving the following reactants. Write a balanced equation for each reaction.

(a) $Cl_2O_7(\ell) + H_2O(\ell) \longrightarrow$

(b) $As_4(s) + O_2(g)$ (excess) $\xrightarrow{\Delta}$

(c) $Mg(s) + O_2(g)$ (low pressure) $\xrightarrow{\Delta}$

Plan

(a) The reaction of a nonmetal oxide (acid anhydride) with water forms a ternary acid in which the nonmetal (Cl) has the same oxidation state (+7) as in the oxide. Thus, the acid is perchloric acid, $HClO_4$.
(b) Arsenic, a nonmetal of Group VA, exhibits common oxidation states of +5 and +5 − 2 = +3. Reaction of arsenic with *excess* oxygen produces the higher-oxidation-state oxide, As_2O_5. By analogy with the oxide of phosphorus in the +5 oxidation state, P_4O_{10}, we might write the formula as As_4O_{10}, but this oxide is usually represented as As_2O_5.
(c) The reaction of a Group IIA metal with oxygen (at low pressure) produces the normal metal oxide—MgO in this case.

Solution

(a) $Cl_2O_7(\ell) + H_2O(\ell) \longrightarrow 2HClO_4(aq)$

(b) $As_4(s) + 5O_2(g) \xrightarrow{\Delta} 2As_2O_5(s)$

(c) $2Mg(s) + O_2(g) \xrightarrow{\Delta} 2MgO(s)$

EXAMPLE 6-11   *Predicting Reaction Products*

Predict the products of the following pairs of reactants. Write a balanced equation for each reaction.

(a) $CaO(s) + H_2O(\ell) \longrightarrow$

(b) $Li_2O(s) + SO_3(g) \longrightarrow$

Plan

(a) The reaction of a metal oxide with water produces the metal hydroxide.

(b) The reaction of a metal oxide with a nonmetal oxide produces a salt containing the cation of the metal oxide and the anion of the acid for which the nonmetal oxide is the anhydride. SO_3 is the acid anhydride of sulfuric acid, H_2SO_4.

Solution

(a) Calcium oxide reacts with water to form calcium hydroxide.

$$CaO(s) + H_2O(\ell) \longrightarrow Ca(OH)_2(aq)$$

(b) Lithium oxide reacts with sulfur trioxide to form lithium sulfate.

$$Li_2O(s) + SO_3(g) \longrightarrow Li_2SO_4(s)$$

You should now work Exercises 68–71.

CaO is called quicklime. $Ca(OH)_2$ is called slaked lime.

3 Combustion Reactions

Combustion, or burning, is an oxidation–reduction reaction in which oxygen combines rapidly with oxidizable materials in highly exothermic reactions, usually with a visible flame. The complete combustion of **hydrocarbons,** in fossil fuels for example, produces carbon dioxide and water (steam) as the major products.

Hydrocarbons are compounds that contain only hydrogen and carbon.

$$CH_4(g) + 2O_2(g) \xrightarrow{\Delta} CO_2(g) + 2H_2O(g) + heat$$

excess

$$C_6H_{12}(g) + 9O_2(g) \xrightarrow{\Delta} 6CO_2(g) + 6H_2O(g) + heat$$

cyclohexane   excess

As we have seen, the origin of the term "oxidation" lies in just such reactions, in which oxygen "oxidizes" another species.

4 Combustion of Fossil Fuels and Air Pollution

Fossil fuels are mixtures of variable composition that consist primarily of hydrocarbons. We burn them because they release energy, rather than to obtain chemical products (Figure 6-10). We have seen that the incomplete combustion of hydrocarbons yields undesirable products, carbon monoxide and elemental carbon (soot), which pollute the air. Unfor-

Carbon in the form of soot is one of many kinds of *particulate matter* in polluted air.

Figure 6-10 Georgia Power Company's Plant Bowen at Taylorsville, Georgia. In 1993 it burned 8,039,586 tons of coal and produced 21,373,720 megawatt-hours of electricity.

Figure 6-11 The luxuriant growth of vegetation that occurred during the carboniferous age is the source of our coal deposits.

Table 6-5 *Some Typical Coal Compositions in Percent (dry, ash-free)*					
	C	**H**	**O**	**N**	**S**
lignite	70.59	4.47	23.13	1.04	0.74
subbituminous	77.2	5.01	15.92	1.30	0.51
bituminous	80.2	5.80	7.53	1.39	5.11
anthracite	92.7	2.80	2.70	1.00	0.90

tunately, all fossil fuels—natural gas, coal, gasoline, kerosene, oil, and so on—also have undesirable nonhydrocarbon impurities that burn to produce oxides that act as additional air pollutants. At this time it is not economically feasible to remove all of these impurities.

Fossil fuels result from the decay of animal and vegetable matter (Figure 6-11). All living matter contains some sulfur and nitrogen, so fossil fuels also contain sulfur and nitrogen impurities to varying degrees. Table 6-5 gives composition data for some common kinds of coal.

Combustion of sulfur produces sulfur dioxide, SO_2, probably the single most harmful pollutant.

$$\overset{0}{S_8}(s) + 8O_2(g) \xrightarrow{\Delta} 8\overset{+4}{SO_2}(g)$$

Large amounts of SO_2 are produced by the burning of sulfur-containing coal.

Many metals occur in nature as sulfides. The process of extracting the free (elemental) metals involves **roasting**—heating an ore in the presence of air. For many metal sulfides this produces a metal oxide and SO_2. The metal oxides are then reduced to the free metals. Consider lead sulfide, PbS, as an example.

$$2PbS(s) + 3O_2(g) \longrightarrow 2PbO(s) + 2SO_2(g)$$

Sulfur dioxide is corrosive; it damages plants, structural materials, and humans. It is a nasal, throat, and lung irritant. Sulfur dioxide is slowly oxidized to sulfur trioxide, SO_3, by oxygen in air.

$$2SO_2(g) + O_2(g) \longrightarrow 2SO_3(\ell)$$

Sulfur trioxide combines with moisture in the air to form the strong, corrosive acid, sulfuric acid.

$$SO_3 + H_2O(\ell) \longrightarrow H_2SO_4(\ell)$$

Oxides of sulfur are the main cause of acid rain.

Compounds of nitrogen are also impurities in fossil fuels; they burn to form nitric oxide, NO. However, most of the nitrogen in the NO in exhaust gases from furnaces, automobiles, airplanes, and so on comes from the air that is mixed with the fuel.

$$\overset{0}{N_2}(g) + O_2(g) \longrightarrow 2\overset{+2}{NO}(g)$$

NO can be further oxidized by oxygen to nitrogen dioxide, NO_2; this reaction is enhanced in the presence of ultraviolet light from the sun.

Remember that "clean air" is *about* 80% N_2 and 20% O_2 by mass. This reaction does *not* occur at room temperature but does occur at the high temperatures of furnaces, internal combustion engines, and jet engines.

$$2\overset{+2}{NO}(g) + O_2(g) \xrightarrow[\text{light}]{uv} 2\overset{+4}{NO_2}(g) \qquad \text{(a reddish-brown gas)}$$

(Text continues on p. 238.)

CHEMISTRY IN USE

The Environment

Acid Rain

During the 1980s, the phenomenon of acid deposition, commonly known as acid rain, gained considerable attention from the public and the scientific community. It is the subject of intense research for thousands of environmental scientists. It has also been a source of political conflict and debate within and between nations in North America and Europe.

The term "acid rain" is used to describe all naturally occurring precipitation, including rain, snow, sleet, and hail, that has become acidified. The *acidity* or *alkalinity* of a substance is determined by the relative concentrations of hydrogen ions and hydroxide ions that it contains. Acidity is expressed on a logarithmic scale called the pH scale. Pure water, which has equal concentrations of hydrogen ions and hydroxide ions, has a pH of 7.0 and is said to be *neutral*. A substance with a higher concentration of hydrogen ions than hydroxide ions is acidic and has a value less than 7.0 on the pH scale. Conversely, a substance with a higher concentration of hydroxide ions than hydrogen ions is alkaline, or *basic,* and has a pH value greater than 7.0.

The farther a reading is from 7.0, below or above, the more acidic or basic (respectively) the substance is. Each full pH unit decrease represents a tenfold increase in acidity. For example, a solution whose pH value is 6.0 contains ten times more hydrogen ions than pure water and is ten times more acidic. A substance with a pH of 5.0 is 100 times more acidic than pure water.

Normal, uncontaminated precipitation is naturally slightly acidic, having a pH value of about 5.6. This natural acidity is the result of the combination of carbon dioxide in the atmosphere with water vapor to form carbonic acid, H_2CO_3, a very weak acid. Any rainfall with pH lower than 5.6 is considered excessively acidic. Unfortunately, as a result of human activities such as burning fossil fuels and smelting ores, reports of rain with pH values of 3.8 to 4.5 are common. Some of the most acidic rainfall recorded in the United States had a pH value of 1.5; this is 12,600 times more acidic than normal uncontaminated rain, and almost one third as acidic as battery acid, a very strong acid. Widespread acid rain has been known in northern Europe and

Most of the oxides of sulfur that contribute to acid rain in North America come from midwestern states. Prevailing winds carry the resulting acid droplets to the north and east, as far as Canada. Oxides of nitrogen also contribute to acid rain formation.

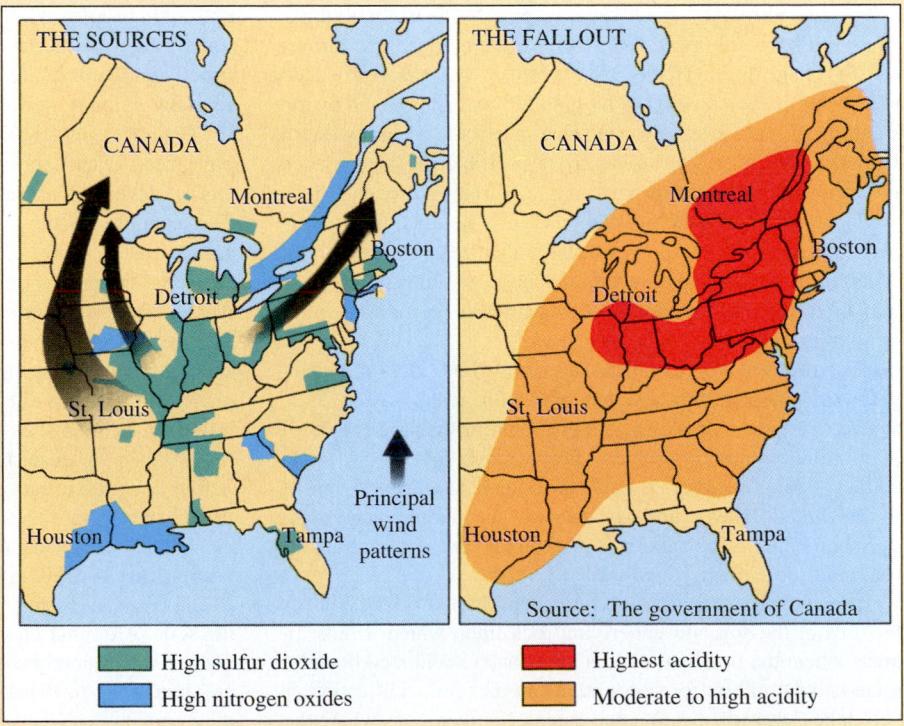

THE SOURCES

CANADA

Montreal
Boston
Detroit
St. Louis
Houston   Tampa

Principal wind patterns

THE FALLOUT

CANADA

Montreal
Boston
Detroit
St. Louis
Houston   Tampa

Source: The government of Canada

High sulfur dioxide
High nitrogen oxides

Highest acidity
Moderate to high acidity

Effects of acid rain on evergreen forests. The two photographs were taken at Camel's Hump, Vermont, 15 years apart.

eastern North America for some time. More recent work has led to the discovery of acid rain in western North America, Japan, China, the Soviet Union, and South America.

The main compounds that produce acid rain are the oxides of sulfur and nitrogen, which quickly react with water to form acids such as sulfurous (H_2SO_3), sulfuric (H_2SO_4), nitrous (HNO_2), and nitric (HNO_3) acids. Sulfur oxides are produced mainly by the combustion of high-sulfur coal and oil and by the smelting of metal ores. Coal-burning utilities in the midwestern United States appear to be largely responsible for the presence of sulfur oxides in North American acid rain. Nitrogen oxides are by-products of gasoline combustion in automobile and airplane engines and of some processes that generate electricity. After these compounds are emitted into the atmosphere, they may be transported thousands of miles before returning to earth in solution with rain, sleet, or snow.

As a result of the great distances traveled by sulfur and nitrogen oxides before they result in acid rain, what was once assumed to be a harmless dispersal of contaminants has become an international pollution problem. For instance, about 50% of the sulfates falling in eastern Canada are believed to have originated in the United States. Similarly, much of the acidic precipitation in Scandinavia originates in industrialized areas of central Europe and the United Kingdom.

The consequences of acid rain depend in part on the characteristics of the soil and underlying rock upon which it falls. In areas where the principal rock is limestone (calcium carbonate), there is a natural buffer system that can prevent acidification of soil, lakes, and streams to some extent. In other areas where the

soil and bodies of water do not contain such a natural buffer, the pH drops gradually as a result of acid precipitation. At times, it drops quite suddenly—for instance, when the spring melt causes all of the acids that have accumulated in the winter snows and ice to be released into soils and waters over a short period of time. Although the low pH resulting from the spring melt is usually temporary, it can be devastating to aquatic plant life and to many animals, such as small fish and frogs.

The effects of acid rain on aquatic and terrestrial ecosystems have been extensively documented. Many lakes are virtually sterile, devoid of fish as well as most plants and invertebrates. For example, in the Adirondacks region of the northeastern United States, more than 200 lakes have been rendered totally sterile by the effects of acid rain. Of the 100,000 lakes in Sweden, 4000 have become fishless. In acidified lakes in which fish still exist, the diversity of species has decreased and the life spans of the fish are greatly reduced. Some food chains have also been shortened due to the elimination of sensitive species at intermediate points in the chain. The disappearance of lower organisms may cause starvation of large predatory animals well before direct toxic action of hydrogen ions is evident. In terrestrial systems, trees and plants are suffering from nutrient deficiency, reduced efficiency of photosynthesis, and lowered resistance to disease. This is thought to be due mainly to the leaching of important nutrients, such as calcium and magnesium, from the soils. Aluminum and trace metals are leached from the soil and are subsequently washed into lakes and streams, where they are quite toxic to fishes and other organisms. The acidification of ground water also causes public water supplies to become

Effects of acid rain on statues. The photo on the left was taken at the Lincoln Cathedral in England in 1910; the one on the right was taken 74 years later.

The effects of acid rain on pH can be offset by adding calcium carbonate, CaCO₃, to react with excess hydrogen ions. Here a helicopter sprays finely ground limestone (mostly CaCO₃) over forests affected by acid rain.

acidified. Acidic water dissolves metals from plumbing, creating a drinking-water hazard.

Acid rain has also done considerable damage to structures. Rates of corrosion of metal structures are greatly increased by acid rain. Exfoliation (flaking) of marble and limestone monuments and buildings is common, as is the pitting of granitic stonework.

Acid rain is a serious worldwide pollution problem. Its potential consequences for humans are enormous: lowered crop yields, decreased timber production, and loss of important fishing and recreational areas as well as public water supplies. It is interesting to note that 21 European countries agreed in 1985 to lower their sulfur dioxide emissions by 30% or more over a 10-year period. By 1989, more than half of those countries had already reached that goal. In 1988 the Canadian government announced a goal of lowering its sulfur dioxide emissions by half by 1994. In the United States, progress has not been so rapid.

Measures to counteract the effects of acid rain have only limited effectiveness. The available processes that remove sulfur and nitrogen oxides *at the source,* before they enter the atmosphere to form acid rain, are expensive. But the monetary and ecological costs of allowing the conditions that create acid rain to continue are also very great. Scientists can provide important factual information, but the decisions about alleviating the acid rain problem will be made in the political arena.

Beth A. Trust
Graduate student in chemistry
University of Texas Marine Sciences Institute

Figure 6-12 Photochemical pollution (a brown haze) enveloping a city.

NO_2 is responsible for the reddish-brown haze that hangs over many cities on sunny afternoons (Figure 6-12) and probably for most of the respiratory problems associated with this kind of air pollution. It can react to produce other oxides of nitrogen and other secondary pollutants.

In addition to being a pollutant itself, nitrogen dioxide reacts with water in the air to form nitric acid, another major contributor to acid rain.

$$3NO_2(g) + H_2O(\ell) \longrightarrow 2HNO_3(aq) + NO(g)$$

Key Terms

Acid anhydride   A nonmetal oxide that reacts with water to form an acid.

Acidic oxide   See *Acid anhydride.*

Actinides   Elements 90 through 103 (after *actinium*).

Amphoteric oxide   An oxide that shows some acidic and some basic properties.

Amphoterism   The ability to react with both acids and bases.

Angstrom (Å) 10^{-10} meter.

Atomic radius   The radius of an atom.

Basic anhydride   A metal oxide that reacts with water to form a base.

Basic oxide   See *Basic anhydride.*

Catalyst   A substance that speeds up a chemical reaction without itself being consumed in the reaction.

Combustion reaction   The reaction of a substance with oxygen in a highly exothermic reaction, usually with a visible flame.

***d*-Transition elements (metals)**   The B group elements in the periodic table; sometimes called simply transition elements.

Effective nuclear charge (Z_{eff})   The nuclear charge experienced by the outermost electrons of an atom; the actual nuclear charge minus the effects of shielding due to inner shell electrons.

Electron affinity   The amount of energy absorbed in the process in which an electron is added to a neutral isolated gaseous atom to form a gaseous ion with a $1-$ charge; has a negative value if energy is released.

Electronegativity   A measure of the relative tendency of an atom to attract electrons to itself when chemically combined with another atom.

***f*-Transition elements (metals)**   Elements 58 through 71 and 90 through 103; also called inner transition elements (metals).

Hydride   A binary compound of hydrogen.

Inner transition elements   See *f-Transition elements.*

Ionic radius   The radius of an ion.

Ionization energy   The minimum amount of energy required to remove the most loosely held electron of an isolated gaseous atom or ion.

Isoelectronic   Having the same number of electrons.

Lanthanides   Elements 58 through 71 (after *lanthanum*).

Noble gases   Elements of periodic Group 0; also called rare gases; formerly called inert gases.

Noble gas configuration   The stable electron configuration of a noble gas.

Normal oxide A metal oxide containing the oxide ion, O^{2-} (oxygen in the -2 oxidation state).

Nuclear shielding See *Shielding effect.*

Oxide A binary compound of oxygen.

Periodicity Regular periodic variations of properties of elements with atomic number and position in the periodic table.

Periodic law The properties of the elements are periodic functions of their atomic numbers.

Peroxide A compound containing oxygen in the -1 oxidation state. Metal peroxides contain the peroxide ion, O_2^{2-}.

Radical A species containing one or more unpaired electrons; many radicals are very reactive.

Rare earths Inner transition elements.

Representative elements The A group elements in the periodic table.

Roasting Heating an ore of an element in the presence of air.

Shielding effect Electrons in filled sets of s and p orbitals between the nucleus and outer shell electrons shield the outer shell electrons somewhat from the effect of protons in the nucleus; also called screening effect.

Superoxide A compound containing the superoxide ion, O_2^- (oxygen in the $-\frac{1}{2}$ oxidation state).

Ternary acid An acid containing three elements, H, O, and (usually) another nonmetal.

Exercises

Classification of the Elements

1. Define and illustrate the following terms clearly and concisely: (a) representative elements, (b) *d*-transition elements, (c) inner transition elements.

2. Explain why Period 1 contains two elements and Period 2 contains eight elements.

3. Explain why Period 4 contains 18 elements.

4. The third major energy level ($n = 3$) has s, p, and d sublevels. Why does Period 3 contain only eight elements?

5. Account for the number of elements in Period 6.

*6. What would be the atomic number of the as-yet-undiscovered alkaline earth element of Period 8?

7. Identify the group, family, and/or other periodic table location of each element with the outer electron configuration
   (a) ns^2np^3,
   (b) ns^1,
   (c) $ns^2(n-1)d^{0-2}(n-2)f^{1-14}$.

8. Repeat Exercise 7 for
   (a) ns^2np^5
   (b) ns^2
   (c) $ns^2(n-1)d^{1-10}$
   (d) ns^2np^1.

9. Write the outer electron configurations for the (a) alkaline earth metals, (b) *d*-transition metals, and (c) halogens.

10. Which of the elements in the following periodic table is (are) (a) alkali metals, (b) an element with the outer configuration of d^8s^2, (c) lanthanides, (d) *p*-block representative elements, (e) elements with incompletely filled *f*-subshells, (f) halogens, (g) *s*-block representative elements, (h) actinides, (i) *d*-transition elements, (j) noble gases?

11. Identify the elements and the part of the periodic table in which the elements with the following configurations are found.
    (a) $1s^22s^22p^63s^23p^64s^1$
    (b) $[Kr]4d^85s^2$
    (c) $[Xe]4f^{14}5d^66s^2$
    (d) $[Xe]4f^{12}6s^2$
    (e) $[Kr]4d^{10}5s^25p^5$
    (f) $[Kr]4d^{10}4f^{14}5s^25p^65d^{10}6s^26p^2$

Atomic Radii

12. What is meant by nuclear shielding? What effect does it have on trends in atomic radii?

13. Why do atomic radii decrease from left to right within a period in the periodic table?

14. Why do atomic radii increase from top to bottom within a group in the periodic table?

15. Variations in the atomic radii of the transition elements are not so pronounced as those of the representative elements. Why?

16. Arrange each of the following sets of atoms in order of increasing atomic radii: (a) the alkaline earth elements; (b) the noble gases; (c) the elements in the second period; (d) N, Te, B, Sr, and Sb.

17. Arrange each of the following sets of atoms in order of increasing atomic volume: (a) O, Mg, Al, Si; (b) O, S, Se, Te; (c) Ca, Sr, Ga, In.

Ionization Energy

18. Define (a) first ionization energy and (b) second ionization energy.

19. Why is the second ionization energy for a given element always greater than the first ionization energy?

20. What is the usual relationship between atomic radius and first ionization energy, other factors being equal?

21. What is the usual relationship between nuclear charge and first ionization energy, other factors being equal?

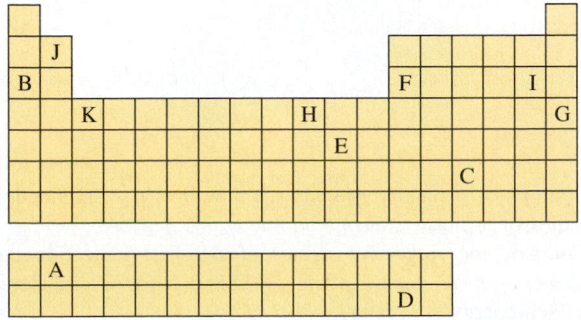

22. Arrange the members of each of the following sets of elements in order of increasing first ionization energies: (a) the alkali metals; (b) the halogens; (c) the elements in the second period; (d) Br, F, B, Ga, Cs, and H.

23. Explain why there is a general increase in first ionization energy across each period.

24. Explain the trend in first ionization energy upon descending a periodic group.

25. In a plot of first ionization energy versus atomic number for Periods 2 and 3, "dips" occur at the IIIA and VIA elements. Account for these dips.

26. What is the general relationship between the sizes of the atoms of Period 2 and their first ionization energies? Rationalize the relationship.

27. Why must a zinc atom absorb more energy than a calcium atom to ionize a $4s$ electron?

28. On the basis of electron configurations, would you expect a Ca^{3+} ion to exist in compounds? Why or why not? How about Al^{3+}?

29. How much energy, in kJ, must be absorbed by 1.00 mol of gaseous lithium atoms to convert all of them to gaseous Li^+ ions?

Electron Affinity

30. Arrange the following elements in order of increasing negative values of electron affinity: O, S, Cl, and Br.

31. Arrange the members of each of the following sets of elements in order of increasingly negative electron affinities: (a) the Group IA metals; (b) the Group VIIA elements; (c) the elements in the second period; (d) Li, K, C, F, and Cl.

32. The electron affinities of the halogens are much more negative than those of the Group VIA elements. Why is this so?

33. The addition of a second electron to form an ion with a $2-$ charge is always endothermic. Why is this so?

34. Write the equation for the change described by each of the following, and write the electron configuration for each atom or ion shown: (a) the electron affinity of oxygen, (b) the electron affinity of chlorine, (c) the electron affinity of calcium.

Ionic Radii

35. Compare the sizes of cations and the neutral atoms from which they are formed by citing three specific examples.

36. Arrange the members of each of the following sets of cations in order of increasing ionic radii: (a) K^+, Ca^{2+}, Ga^{3+}; (b) Ca^{2+}, Be^{2+}, Ba^{2+}, Mg^{2+}; (c) Al^{3+}, Sr^{2+}, Rb^+, K^+.

37. Compare the sizes of anions and the neutral atoms from which they are formed by citing three specific examples.

38. Arrange the following sets of anions in order of increasing ionic radii: (a) Cl^-, S^{2-}, P^{3-}; (b) O^{2-}, S^{2-}, Se^{2-}; (c) N^{3-}, S^{2-}, Br^-, P^{3-}.

39. Compare and explain the relative sizes of H^+, H, and H^-.

40. Most transition metals can form more than one simple positive ion. For example, iron forms both Fe^{2+} and Fe^{3+} ions, and copper forms both Cu^+ and Cu^{2+} ions. Which is the smaller ion of each pair, and why?

Electronegativity

41. What is electronegativity?

42. Arrange the members of each of the following sets of elements in order of increasing electronegativities: (a) B, Ga, Al, In; (b) S, Na, Mg, Cl; (c) P, N, Sb, Bi; (d) Se, Ba, F, Si, Sc.

43. Which of the following statements is better? Why?
(a) Magnesium has a weak attraction for electrons in a chemical bond because it has a low electronegativity.
(b) The electronegativity of magnesium is low because magnesium has a weak attraction for electrons in a chemical bond.

44. Comment on the validity of the following statement: "Chlorine has a high electronegativity because it forms chloride ions, Cl^-, readily."

*45. Do you think an element might be discovered (or made) with a lower electronegativity than francium? What would its electron configuration be?

*46. Some of the second-period elements show a similarity to the element one column to the right and one row down. For instance, Li is similar in many respects to Mg, and Be is similar to Al. This has been attributed to the charge density on the stable ions (Li^+ vs. Mg^{2+}; Be^{2+} vs. Al^{3+}). From the values of electronic charge (Chapter 5) and ionic radii (Chapter 6), calculate the charge density for these four ions, in coulombs/$Å^3$.

Additional Exercises on the Periodic Table

47. The P—Cl bond length in PCl_3 is 2.04 Å. The bond length in Cl_2 is 1.98 Å. Calculate the atomic radius for phosphorus. Using the atomic radius for F given in Figure 6-1, predict the P—F bond length in PF_3.

48. The bond lengths in F_2 and Cl_2 molecules are 1.42 Å and 1.98 Å, respectively. Calculate the atomic radii for these elements. Predict the Cl—F bond length. (The actual Cl—F bond length is 1.64 Å.)

*49. The atoms in crystalline nickel are arranged so that they are touching each other in a plane as shown in the sketch:

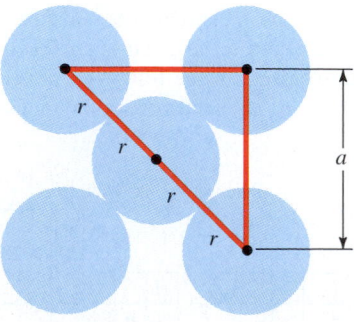

From plane geometry, we can see that $4r = a\sqrt{2}$. Calculate the radius of a nickel atom given that $a = 3.5238$ Å.

50. Compare the respective values of the first ionization energy (Table 6-1) and electron affinity (Table 6-2) for several elements. Which energy is greater? Why?

51. Compare the respective values of the first ionization energy (Table 6-1) and electron affinity (Table 6-2) for nitrogen to those for carbon and oxygen. Explain why the nitrogen values are considerably different.

*52. Based on general trends, the electron affinity of fluorine would be expected to be greater than that of chlorine; however, the value is less and is similar to the value for bromine. Explain.

*53. The first ionization energy of oxygen is 1313.9 kJ/mol at 0 K. The second ionization energy is 3388.1 kJ/mol. Why is this second value much larger than the first?

Hydrogen and the Hydrides

54. Summarize the physical properties of hydrogen.

55. Write balanced formula unit equations for (a) the reaction of iron with steam, (b) the reaction of calcium with hydrochloric acid, (c) the electrolysis of water, and (d) the "water gas" reaction.

56. Write a balanced formula unit equation for the preparation of (a) an ionic hydride and (b) a molecular hydride.

57. Classify the following hydrides as molecular or ionic: (a) NaH, (b) H_2S, (c) BaH_2, (d) RbH, (e) NH_3.

58. Explain why NaH and H_2S are different kinds of hydrides.

59. Write formula unit equations for the reactions of (a) NaH and (b) BaH_2 with water.

60. Name the following (pure) compounds: (a) H_2S, (b) HF, (c) KH, (d) NH_3, (e) H_2Se, (f) MgH_2, (g) CaH_2.

Oxygen and the Oxides

*61. How are O_2 and O_3 similar? Different?

62. Briefly compare and contrast the properties of oxygen with those of hydrogen.

63. Write molecular equations to show how oxygen can be prepared from (a) mercury(II) oxide, HgO, (b) hydrogen peroxide, H_2O_2, and (c) potassium chlorate, $KClO_3$.

*64. Which of the following elements form normal oxides as the *major* products of reactions with oxygen? (a) Li, (b) Na, (c) Rb, (d) Mg, (e) Zn (exhibits only one common oxidation state), (f) Al.

65. Write formula unit equations for the reactions of the following elements with an *excess* of oxygen: (a) Sr, (b) Fe, (c) Mn, (d) Cu.

66. Write formula unit equations for the reactions of the following elements with a *limited* amount of oxygen: (a) C, (b) As_4, (c) Ge.

67. Write formula unit equations for the reactions of the following elements with an *excess* of oxygen: (a) C, (b) As_4, (c) Ge.

68. Distinguish among normal oxides, peroxides, and superoxides. What is the oxidation state of oxygen in each case?

*69. Which of the following can be classified as basic anhydrides? (a) SO_2, (b) Li_2O, (c) SeO_3, (d) CaO, (e) N_2O_5.

70. Write balanced formula unit equations for the following reactions and name the products:
(a) sulfur dioxide, SO_2, with water
(b) sulfur trioxide, SO_3, with water
(c) selenium trioxide, SeO_3, with water
(d) dinitrogen pentoxide, N_2O_5, with water
(e) dichlorine heptoxide, Cl_2O_7, with water

71. Write balanced formula unit equations for the following reactions and name the products:
(a) sodium oxide, Na_2O, with water
(b) calcium oxide, CaO, with water
(c) lithium oxide, Li_2O, with water
(d) magnesium oxide, MgO, with sulfur dioxide, SO_2
(e) barium oxide, BaO, with carbon dioxide, CO_2

*72. Identify the acid anhydrides of the following ternary acids: (a) H_2SO_4, (b) H_2CO_3, (c) H_2SO_3, (d) H_3AsO_4, (e) HNO_2.

73. Identify the basic anhydrides of the following metal hydroxides: (a) NaOH, (b) $Ca(OH)_2$, (c) $Fe(OH)_2$, (d) $Al(OH)_3$.

Combustion Reactions

74. Define combustion. Why are all combustion reactions also redox reactions?

75. Write equations for the complete combustion of the following compounds: (a) ethane, $C_2H_6(g)$; (b) propane, $C_3H_8(g)$; (c) ethanol, $C_2H_5OH(\ell)$.

76. Write equations for the *incomplete* combustion of the following compounds to produce carbon monoxide: (a) ethane, $C_2H_6(g)$; (b) propane, $C_3H_8(g)$.

As we have seen, two substances may react to form different products when they are mixed in different proportions under different conditions. In Exercises 77 and 78, write balanced equations for the reactions described. Assign oxidation numbers.

77. (a) Methane burns in excess air to form carbon dioxide and water.
(b) Methane burns in a limited amount of air to form carbon monoxide and water.
(c) Methane burns (poorly) in a very limited amount of air to form elemental carbon and water.

78. (a) Butane (C_4H_{10}) burns in excess air to form carbon dioxide and water.
(b) Butane burns in a limited amount of air to form carbon monoxide and water.
(c) When heated in the presence of *very little* air, butane "cracks" to form acetylene, C_2H_2; carbon monoxide; and hydrogen.

*79. (a) How much SO_2 would be formed by burning 1.00 ton of bituminous coal that is 5.15% sulfur by mass? Assume that all of the sulfur is converted to SO_2.
(b) If 27.0% of the SO_2 escaped into the atmosphere and 84.2% of it were converted to H_2SO_4, how many grams of H_2SO_4 would be produced in the atmosphere?

*80. Write equations for the complete combustion of the following compounds. Assume that sulfur is converted to SO_2 and nitrogen is converted to NO. (a) $C_6H_5N(\ell)$, (b) $C_2H_5SH(\ell)$, (c) $C_7H_{10}NO_2S(\ell)$.

81. Describe the formation of the reddish-brown haze of some cities experiencing this kind of air pollution.

82. Show how two common oxides can account for the occurrence of acid rain.

BUILDING YOUR KNOWLEDGE

83. The only chemically stable ion of rubidium is Rb^+. The most stable monatomic ion of bromine is Br^-. Krypton (Kr) is among the least reactive of all elements. Compare the electron configurations of Rb^+, Br^-, and Kr. Then predict the most stable monatomic ions of strontium (Sr) and selenium (Se).

84. The first ionization energy of potassium, K, is 419 kJ/mol. What is the minimum frequency of light required to ionize gaseous potassium atoms?

85. Potassium and argon would be anomalies in a periodic table in which elements are arranged in order of increasing atomic weights. Identify two other elements among the transition elements whose positions in the periodic table would have been reversed in a "weight-sequence" arrangement. Which pair of elements would most obviously be out of place on the basis of their chemical behavior? Explain your answer in terms of the current atomic model, showing electron configurations for these elements.

86. The second ionization energy for magnesium is 1451 kJ/mol. How much energy, in kJ, must be absorbed by 1.00 g of gaseous magnesium atoms to convert them to gaseous Mg^{2+} ions?

Chemical Bonding

7

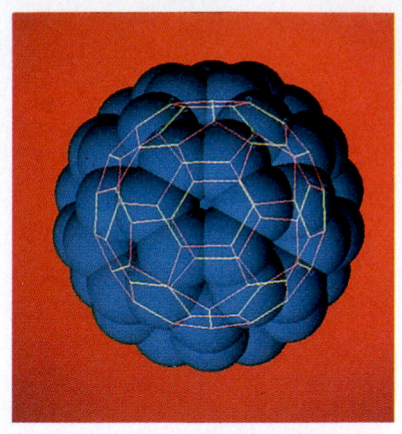

A representative of a third form of carbon, a new class of molecules called the fullerenes. One of the most important is C_{60}, buckminsterfullerene or "buckyball." This new form of carbon is very different from the two previously known forms, graphite and diamond. Scientists are trying to understand its unusual properties, such as its ability to trap other atoms and its superconductivity.

OBJECTIVES

As you study this chapter, you should learn

- *To write Lewis dot representations of atoms*

- *To predict whether bonding will be primarily ionic, covalent, or polar covalent*

- *To compare and contrast ionic and covalent compounds*

- *How the properties of compounds depend on their bonding*

- *How elements bond by electron transfer (ionic bonding)*

- *To predict the formulas of ionic compounds*

- *How elements bond by sharing electrons (covalent bonding)*

- *To write Lewis dot formulas for molecules and polyatomic ions*

- *To recognize exceptions to the octet rule*

- *To relate the nature of the bonding to electronegativity differences*

- *About resonance, when to write resonance structures, and how to do so*

- *How to write formal charges for atoms in covalent structures*

As you study Chapters 7 and 8, keep in mind the periodic similarities that you learned in Chapters 4 and 6. What you learn about the bonding of an element usually applies to the other elements in the same column of the periodic table, with minor variations.

Chemical bonding refers to the attractive forces that hold atoms together in compounds. There are two major classes of bonding. (1) **Ionic bonding** results from electrostatic interactions among ions, which often results from the net *transfer* of one or more electrons from one atom or group of atoms to another. (2) **Covalent bonding** results from *sharing* one or more electron pairs between two atoms. These two classes represent two extremes; all bonds between atoms of different elements have at least some degree of both ionic and covalent character. Compounds containing predominantly ionic bonding are called **ionic compounds.** Those that are held together mainly by covalent bonds are called **covalent compounds.** Some of the properties associated with many simple ionic and covalent compounds in the extreme cases are summarized below. The differences in properties can be accounted for by the differences in bonding between the atoms or ions.

Table 7-1 *Lewis Dot Formulas for Representative Elements*

Group	IA	IIA	IIIA	IVA	VA	VIA	VIIA	0
Number of electrons in valence shell	1	2	3	4	5	6	7	8 (except He)
Period 1	H·							He:
Period 2	Li·	Be:	B·	C·	·N·	·O:	·F:	:Ne:
Period 3	Na·	Mg:	Al·	Si·	·P·	·S:	·Cl:	:Ar:
Period 4	K·	Ca:	Ga·	Ge·	·As·	·Se:	·Br:	:Kr:
Period 5	Rb·	Sr:	In·	Sn·	·Sb·	·Te:	·I:	:Xe:
Period 6	Cs·	Ba:	Tl·	Pb·	·Bi·	·Po:	·At:	:Rn:
Period 7	Fr·	Ra:						

For the Group A elements, the number of valence electrons (dots in the Lewis formula) for the *neutral* atom is equal to the group number. Exceptions: H (one valence electron), He (two valence electrons), and the other noble gases (eight valence electrons).

Ionic Compounds

1. They are solids with high melting points (typically >400°C).

2. Many are soluble in polar solvents such as water.

3. Most are insoluble in nonpolar solvents, such as hexane, C_6H_{14}.

4. Molten compounds conduct electricity well because they contain mobile charged particles (ions).

5. Aqueous solutions conduct electricity well because they contain mobile charged particles (ions).

Covalent Compounds

1. They are gases, liquids, or solids with low melting points (typically <300°C).

2. Many are insoluble in polar solvents.

3. Most are soluble in nonpolar solvents, such as hexane, C_6H_{14}.

4. Liquid and molten compounds do not conduct electricity.

5. Aqueous solutions are *usually* poor conductors of electricity because most do not contain charged particles.

The distinction between polar and nonpolar molecules is made in Section 7-8.

As we saw in Section 4-2, aqueous solutions of some covalent compounds do conduct electricity.

7-1 LEWIS DOT FORMULAS OF ATOMS

The number and arrangements of electrons in the outermost shells of atoms determine the chemical and physical properties of the elements as well as the kinds of chemical bonds they form. We write **Lewis dot formulas** (or **Lewis dot representations**) as a convenient bookkeeping method for keeping track of these "chemically important electrons." We now introduce this method for atoms of elements; in our discussion of chemical bonding in subsequent sections, we shall frequently use such formulas for atoms, molecules, and ions.

Chemical bonding usually involves only the outermost electrons of atoms, also called **valence electrons.** In Lewis dot representations, only the electrons in the outermost occupied *s* and *p* orbitals are shown as dots. Paired and unpaired electrons are also indicated. Table 7-1 shows Lewis dot formulas for the representative elements. All elements in a given group have the same outer shell electron configuration.

Because of the large numbers of dots, such formulas are not as useful for the transition and inner transition elements.

IONIC BONDING

7-2 FORMATION OF IONIC COMPOUNDS

The chemical and physical properties of an ion are quite different from those of the atom from which the ion is derived. For example, an atom of Na and an Na^+ ion are quite different.

The first kind of chemical bonding we shall describe is **ionic bonding.** We recall (Section 2-3) that an **ion** is an atom or a group of atoms that carries an electrical charge. An ion in which the atom or group of atoms has more electrons than protons is negatively charged, and is called an **anion;** one that has fewer electrons than protons is positively charged, and is called a **cation.** An ion that consists of only one atom is described as a **monatomic ion.** Examples include the chloride ion, Cl^-, and the magnesium ion, Mg^{2+}. An ion that contains more than one atom is called a **polyatomic ion.** Examples include the ammonium ion, NH_4^+; the hydroxide ion, OH^-; and the sulfate ion, SO_4^{2-}. The atoms of a polyatomic ion are held together by covalent bonds. In this section we shall discuss how ions can be formed from individual atoms; polyatomic ions will be discussed along with other covalently bonded species.

> **Ionic bonding** is the attraction of oppositely charged ions (cations and anions) in large numbers to form a solid. Such a solid compound is called an *ionic solid.*

As our previous discussions of ionization energy, electronegativity, and electron affinity would suggest, ionic bonding can occur easily when elements that have low ionization energies (metals) react with elements having high electronegativities and very negative electron affinities (nonmetals). Many metals are easily *oxidized*—that is, they lose electrons to form cations; and many nonmetals are readily *reduced*—that is, they gain electrons to form anions.

> When the electronegativity difference, ΔEN, between two elements is large, the elements are likely to form a compound by ionic bonding (transfer of electrons).

Freshly cut sodium has a metallic luster. A little while after being cut, the sodium metal surface turns white as it reacts with the air.

Let us describe some combinations of metals with nonmetals to form ionic compounds.

Group IA Metals and Group VIIA Nonmetals

Consider the reaction of sodium (a Group IA metal) with chlorine (a Group VIIA nonmetal). Sodium is a soft silvery metal (mp 98°C), and chlorine is a yellowish-green corrosive gas at room temperature. Both sodium and chlorine react with water, sodium vigorously. By contrast, sodium chloride is a white solid (mp 801°C) that dissolves in water with no reaction and with the absorption of just a little heat. We can represent the reaction for its formation as

$$2Na(s) + Cl_2(g) \longrightarrow 2NaCl(s)$$
$$\text{sodium} \qquad \text{chlorine} \qquad \text{sodium chloride}$$

We can understand this reaction better by showing electron configurations for all species. We represent chlorine as individual atoms rather than molecules, for simplicity.

Halite crystals (naturally occurring NaCl).

		3s	**3p**			**3s**	**3p**	
$_{11}$Na	[Ne]	↑		} → {	Na^+ [Ne]			$1e^-$ lost
$_{17}$Cl	[Ne]	⇅	⇅ ⇅ ↑		Cl^- [Ne]	⇅	⇅ ⇅ ⇅	$1e^-$ gained

In this reaction, Na atoms lose one electron each to form Na^+ ions, which contain only ten electrons, the same number as the *preceding* noble gas, neon. We say that sodium ions have the neon electronic structure; Na^+ is *isoelectronic* with Ne. In contrast, Cl atoms gain one electron each to form Cl^- ions, which contain 18 electrons. This is the same number as the *following* noble gas, argon; Cl^- is *isoelectronic* with Ar. These processes can be represented compactly as

$$Na \longrightarrow Na^+ + e^- \qquad \text{and} \qquad Cl + e^- \longrightarrow Cl^-$$

The loss of electrons is *oxidation* (Section 4-7). Na atoms are *oxidized* to form Na^+ ions.

The gain of electrons is *reduction* (Section 4-7). Cl atoms are *reduced* to form Cl^- ions.

Similar observations apply to most ionic compounds formed by reactions between *representative metals and representative nonmetals*.

We can use Lewis dot formulas (Section 7-1) to represent the reaction.

$$Na\cdot \; + \; :\overset{..}{\underset{..}{Cl}}\cdot \; \longrightarrow \; Na^+[:\overset{..}{\underset{..}{Cl}}:]^-$$

The formula for sodium chloride, NaCl, indicates that the compound contains Na and Cl atoms in a $1:1$ ratio. This is the formula we predict based on the fact that each Na atom contains only one electron in its highest energy level and each Cl atom needs only one electron to fill completely its outermost p orbitals.

The chemical formula NaCl does not explicitly indicate the ionic nature of the compound, only the ratio of atoms. So we must learn to recognize, from positions of elements in the periodic table and known trends in electronegativity, when the difference in electronegativity is large enough to favor ionic bonding.

> The farther apart across the periodic table two Group A elements are, the more ionic their bonding will be.

The noble gases are excluded from this generalization.

The greatest difference in electronegativity occurs from lower left to upper right, so CsF is more ionic than LiI.

All the Group IA metals (Li, Na, K, Rb, Cs) will react with the Group VIIA elements (F, Cl, Br, I) to form ionic compounds of the same general formula, MX. All the resulting ions, M^+ and X^-, have noble gas configurations. Once we understand the bonding of one member of a group (column) in the periodic table, we know a great deal about the others in the family. Combining each of the five common alkali metals with each of the four common halogens gives $5 \times 4 = 20$ possible compounds. The discussion of NaCl presented here applies also to the other 19 such compounds.

We can represent the general reaction of the IA metals with the VIIA elements as follows:

$$2M(s) + X_2 \longrightarrow 2MX(s) \qquad M = Li, Na, K, Rb, Cs; \; X = F, Cl, Br, I$$

The Lewis dot representation for the generalized reaction is

$$2M\cdot \; + \; :\overset{..}{\underset{..}{X}}:\overset{..}{\underset{..}{X}}: \; \longrightarrow \; 2\,(M^+[:\overset{..}{\underset{..}{X}}:]^-)$$

Formulas for some of these are

LiF	NaF	KF
LiCl	NaCl	KCl
LiBr	NaBr	KBr
LiI	NaI	KI

Because of the opposite charges on Na^+ and Cl^-, an attractive force is developed. According to Coulomb's Law, the force of attraction, F, between two oppositely charged particles of charge magnitudes q^+ and q^- is directly proportional to the product of the charges and inversely proportional to the square of the distance separating their centers, d. Thus, the greater the charges on the ions and the smaller the ions are, the stronger the resulting ionic bonding. Of course, like-charged ions repel each other, so the distances separating ions in solids are those at which the attractions exceed the repulsions by the

Coulomb's Law is $F \propto \dfrac{q^+ q^-}{d^2}$. The symbol $\propto$ means "is proportional to."

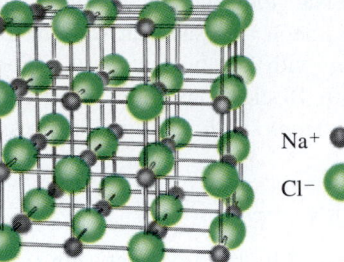

Figure 7-1 A representation of the crystal structure of NaCl. Each Cl⁻ ion (green) is surrounded by six sodium ions, and each Na⁺ ion (gray) is surrounded by six chloride ions. Any NaCl crystal includes billions of ions in the pattern shown. Adjacent ions actually are in contact with each other; in this drawing, the structure has been expanded to show the spatial arrangement of ions. The lines *do not* represent covalent bonds. Compare with Figure 2-7, a space-filling drawing of the NaCl structure.

greatest amount. The structure of common table salt, sodium chloride (NaCl), is shown in Figure 7-1. Like other simple ionic compounds, NaCl(s) exists in a regular, extended array of positive and negative ions, Na^+ and Cl^-.

Distinct molecules of solid ionic substances do not exist, so we must refer to *formula units* (Section 2-3) instead of molecules. The sum of all the forces that hold all the particles in an ionic solid is quite large. This explains why such substances have quite high melting and boiling points (a topic that we will discuss more fully in Chapter 13). When an ionic compound is melted or dissolved in water, its charged particles are free to move in an electric field, so such a liquid shows high electrical conductivity (Section 4-2, part 1).

Group IA Metals and Group VIA Nonmetals

Next, consider the reaction of lithium (Group IA) with oxygen (Group VIA) to form lithium oxide, a solid ionic compound (mp >1700°C). We may represent the reaction as

$$4Li(s) + O_2(g) \longrightarrow 2Li_2O(s)$$

lithium   oxygen   lithium oxide

The formula for lithium oxide, Li_2O, indicates that two atoms of lithium combine with one atom of oxygen. If we examine the structures of the atoms before reaction, we can see why two lithium atoms react with one oxygen atom.

Although the oxides of the other Group IA metals are prepared by different methods, similar descriptions apply to compounds between the Group IA metals (Li, Na, K, Rb, Cs) and the Group VIA nonmetals (O, S, Se, Te, Po).

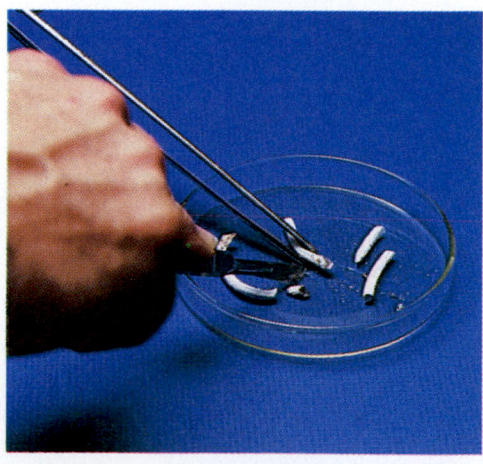

Lithium is a metal, as the shiny surface of freshly cut Li shows. Where it has been exposed to air, the surface is covered with lithium oxide.

	1s	**2s**	**2p**
$_3$Li	⇅	↑	
$_3$Li	⇅	↑	
$_8$O	⇅	⇅	⇅ ↑ ↑

$\longrightarrow$

	1s	**2s**	**2p**	
Li$^+$	⇅	—		1 e^- lost
Li$^+$	⇅	—		1 e^- lost
O^{2-}	⇅	⇅	⇅ ⇅ ⇅	2 e^- gained

Each Li atom has 1 e^- in its valence shell. Each O atom has 6 e^- in its valence shell and needs 2 e^- more to attain a noble gas configuration. The Li$^+$ ions are formed by oxidation of Li atoms, and the O^{2-} ions are formed by reduction of O atoms.

In a compact representation,

$$2[\text{Li} \longrightarrow \text{Li}^+ + e^-] \quad \text{and} \quad \text{O} + 2e^- \longrightarrow \text{O}^{2-}$$

The Lewis dot formulas for the atoms and ions are

$$2\text{Li}\cdot + :\overset{..}{\underset{.}{\text{O}}}\cdot \longrightarrow 2\text{Li}^+[:\overset{..}{\underset{..}{\text{O}}}:]^{2-}$$

Lithium ions, Li$^+$, are isoelectronic with helium atoms. Oxide ions, O^{2-}, are isoelectronic with neon atoms (10 e^-).

The very small size of the Li$^+$ ion gives it a much higher *charge density* (ratio of charge to size) than that of the larger Na$^+$ ion (Figure 6-1). Similarly, the O^{2-} ion is smaller than the Cl$^-$ ion, so its doubly negative charge gives it a much higher charge density. These more concentrated charges and smaller sizes bring the Li$^+$ and O^{2-} ions closer together in Li$_2$O than the Na$^+$ and Cl$^-$ ions are in NaCl. Consequently, the q^+q^- product in the numerator of Coulomb's Law is greater in Li$_2$O, and the d^2 term in the denominator is smaller. The net result is that the ionic bonding is much stronger in Li$_2$O than in NaCl. This accounts for the higher melting temperature of Li$_2$O (>1700°C) compared to NaCl (801°C).

Group IIA Metals and Group VIA Nonmetals

Calcium (Group IIA) reacts with oxygen (Group VIA) to form calcium oxide, a white solid ionic compound with a very high melting point, 2580°C.

$$2\text{Ca(s)} + \text{O}_2\text{(g)} \longrightarrow 2\text{CaO(s)}$$

calcium   oxygen   calcium oxide

This discussion applies to other ionic compounds between any Group IIA metal (Be, Mg, Ca, Sr, Ba) and any Group VIA nonmetal (O, S, Se, Te).

Again, we show the electronic structures of the atoms and ions, representing the inner electrons by the symbol of the preceding noble gas.

$_{20}$Ca	[Ar]	⇅ (4s)
$_8$O	[He]	⇅ (2s) ⇅ ↑ ↑ (2p)

$\longrightarrow$

Ca^{2+}	[Ar]	— (4s)	2 e^- lost
O^{2-}	[He]	⇅ (2s) ⇅ ⇅ ⇅ (2p)	2 e^- gained

In writing equations in which electrons in different energy levels in different atoms are involved, atomic orbitals are more conveniently labeled *under* the lines that represent them.

The Lewis dot notation for the atoms and ions is

$$\text{Ca}: + :\overset{..}{\underset{.}{\text{O}}}\cdot \longrightarrow \text{Ca}^{2+}[:\overset{..}{\underset{..}{\text{O}}}:]^{2-}$$

Calcium ions, Ca^{2+}, are isoelectronic with argon (18 e^-), the preceding noble gas. Oxide ions, O^{2-}, are isoelectronic with neon (10 e^-), the following noble gas.

Ca^{2+} is about the same size as Na$^+$ (Figure 6-1) but carries twice the charge, so its charge density is higher. Because the attraction between the two small, highly charged ions Ca^{2+} and O^{2-} is quite high, the ionic bonding is very strong, accounting for the very high melting point of CaO, 2580°C.

Group IIA Metals and Group VA Nonmetals

As our final example of ionic bonding, consider the reaction of magnesium (Group IIA) with nitrogen (Group VA). At elevated temperatures, they form magnesium nitride, Mg_3N_2, a white, solid ionic compound that decomposes at 800°C. We can represent the reaction as

$$3Mg(s) + N_2(g) \xrightarrow{\Delta} Mg_3N_2(s)$$

magnesium   nitrogen   magnesium nitride

The formula Mg_3N_2 indicates that magnesium and nitrogen atoms combine in a 3:2 ratio. We could predict this from the fact that Mg atoms contain two electrons in their highest energy level, whereas N atoms need three electrons to fill theirs completely. Thus, three Mg atoms lose two electrons each for a total of 6 e^-, and two N atoms gain three electrons each, also for a total of 6 e^-. As before, examination of the electronic structures makes the picture clearer.

$$\left. \begin{array}{l} 3 \times \left({}_{12}Mg \quad [Ne] \quad \frac{\uparrow\downarrow}{3s} \right) \\[2em] 2 \times \left({}_{7}N \quad [He] \quad \frac{\uparrow\downarrow}{2s} \; \frac{\uparrow}{} \; \frac{\uparrow}{} \; \frac{\uparrow}{2p} \right) \end{array} \right\} \xrightarrow{\Delta} \begin{cases} 3 \times \left(Mg^{2+} \quad [Ne] \quad \frac{}{3s} \right) & (2 \, e^- \text{ lost}) \times 3 \\[2em] 2 \times \left(N^{3-} \quad [He] \quad \frac{\uparrow\downarrow}{2s} \; \frac{\uparrow\downarrow}{} \; \frac{\uparrow\downarrow}{} \; \frac{\uparrow\downarrow}{2p} \right) & (3 \, e^- \text{ gained}) \times 2 \end{cases}$$

Using Lewis dot formulas for the atoms and ions, we have

$$3Mg: \; + \; 2 \cdot \overset{\cdot\cdot}{\underset{\cdot}{N}} \cdot \; \longrightarrow \; 3Mg^{2+}, \; 2[:\overset{\cdot\cdot}{\underset{\cdot\cdot}{N}}:]^{3-}$$

Both Mg^{2+} and N^{3-} ions are isoelectronic with neon (10 e^-).

Binary compounds contain two *elements.*

Table 7-2 summarizes the general formulas of binary ionic compounds formed by the representative elements. "M" represents metals and "X" represents nonmetals from the indicated groups. In these examples of ionic bonding, each of the metal atoms has lost one, two, or three electrons and each of the nonmetal atoms has gained one, two, or three electrons. *Simple (monatomic) ions rarely have charges greater than 3+ or 3−.* Ions with

Table 7-2 *Simple Binary Ionic Compounds*

Metal		Nonmetal		General Formula	Ions Present	Example	mp (°C)
IA*	+	VIIA	$\longrightarrow$	MX	(M^+, X^-)	LiBr	547
IIA	+	VIIA	$\longrightarrow$	MX_2	$(M^{2+}, 2X^-)$	$MgCl_2$	708
IIIA	+	VIIA	$\longrightarrow$	MX_3	$(M^{3+}, 3X^-)$	GaF_3	800 (subl)
IA*†	+	VIA	$--\rightarrow$	M_2X	$(2M^+, X^{2-})$	Li_2O	>1700
IIA	+	VIA	$\longrightarrow$	MX	(M^{2+}, X^{2-})	CaO	2580
IIIA	+	VIA	$\longrightarrow$	M_2X_3	$(2M^{3+}, 3X^{2-})$	Al_2O_3	2045
IA*	+	VA	$\longrightarrow$	M_3X	$(3M^+, X^{3-})$	Li_3N	840
IIA	+	VA	$\longrightarrow$	M_3X_2	$(3M^{2+}, 2X^{3-})$	Ca_3P_2	~1600
IIIA	+	VA	$\longrightarrow$	MX	(M^{3+}, X^{3-})	AlP	

Hydrogen is considered a nonmetal. All binary compounds of hydrogen are covalent except certain metal hydrides such as NaH and CaH₂, which contain hydride, H^-, ions.

†*As we saw in Section 6-8, part 2, the metals in Groups IA and IIA also commonly form peroxides (containing the O_2^{2-} ion) or superoxides (containing the O_2^- ion). See Table 6-4. The peroxide and superoxide ions contain atoms that are covalently bonded to one another.*

greater charges would interact so strongly with the electron clouds of other ions that the electron clouds would be very distorted, and considerable covalent character in the bonds would result.

The *d*- and *f*-transition elements form many compounds that are essentially ionic in character. Many simple ions of transition metals do not have noble gas configurations.

The distortion of the electron cloud of an anion by a small, highly charged cation is called polarization.

Introduction to Energy Relationships in Ionic Bonding

E N R I C H M E N T

The following discussion may help you to understand why ionic bonding occurs between elements with low ionization energies and those with high electronegativities. There is a general tendency in nature to achieve stability. One way to do this is by lowering potential energy; *lower* energies generally represent *more stable* arrangements.

Let us use energy relationships to describe why the ionic solid NaCl is more stable than a mixture of individual Na and Cl atoms. Consider a gaseous mixture of one mole of sodium atoms and one mole of chlorine atoms, $Na(g) + Cl(g)$. The energy change associated with the loss of one mole of electrons by one mole of Na atoms to form one mole of Na^+ ions (step 1 in Figure 7-2) is given by the *first ionization energy* of Na (Section 6-3).

$$Na(g) \longrightarrow Na^+(g) + e^- \qquad \text{first ionization energy} = 496 \text{ kJ/mol}$$

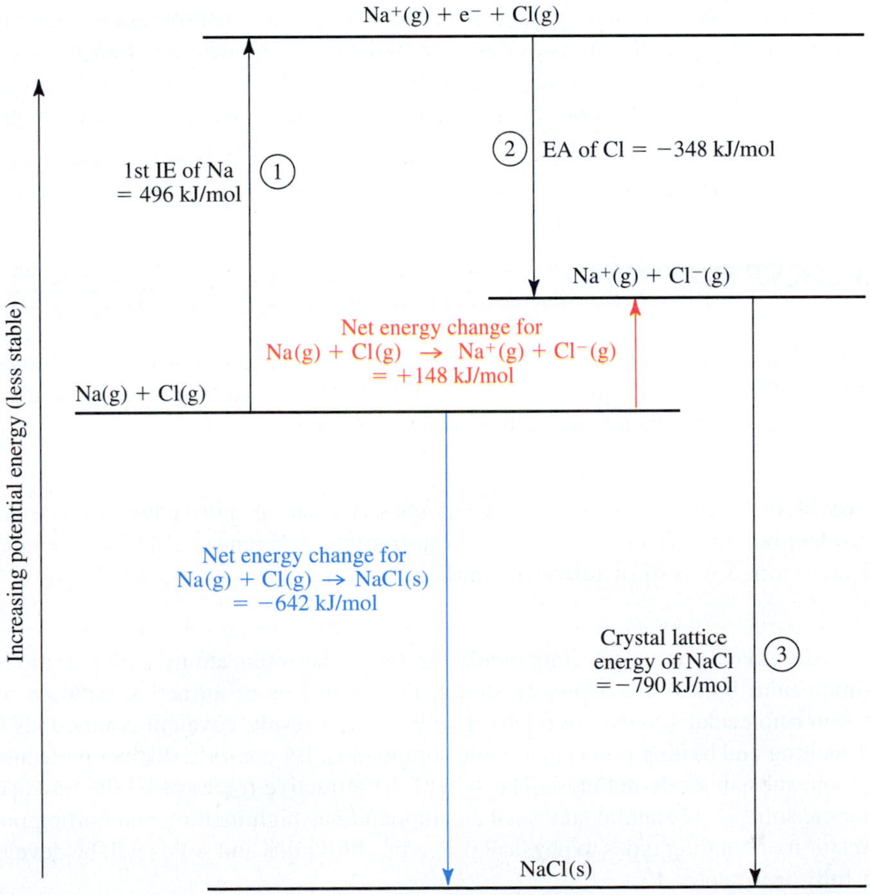

$Na^+(g) + e^- + Cl(g)$

Increasing potential energy (less stable)

1st IE of Na ①
= 496 kJ/mol

② EA of Cl = −348 kJ/mol

$Na^+(g) + Cl^-(g)$

Net energy change for
$Na(g) + Cl(g) \rightarrow Na^+(g) + Cl^-(g)$
= +148 kJ/mol

$Na(g) + Cl(g)$

Net energy change for
$Na(g) + Cl(g) \rightarrow NaCl(s)$
= −642 kJ/mol

Crystal lattice
energy of NaCl ③
= −790 kJ/mol

$NaCl(s)$

Figure 7-2 A schematic representation of the energy changes that accompany the process $Na^+(g) + Cl^-(g) \rightarrow NaCl(s)$. The red arrow represents the *positive* energy change (unfavorable) for the process of ion formation, $Na(g) + Cl(g) \rightarrow Na^+(g) + Cl^-(g)$. The blue arrow represents the *negative* energy change (favorable) for the overall process, including the formation of the ionic solid.

This is a positive value, so the mixture $Na^+(g) + e^- + Cl(g)$ is 496 kJ/mol higher in energy than the original mixture of atoms (the mixture $Na^+ + e^- + Cl$ is *less stable* than the mixture of atoms).

The energy change for the gain of one mole of electrons by one mole of Cl atoms to form one mole of Cl^- ions (step 2) is given by the *electron affinity* of Cl (Section 6-4).

$$Cl(g) + e^- \longrightarrow Cl^-(g) \qquad \text{electron affinity} = -348 \text{ kJ/mol}$$

This negative value, -348 kJ/mol, lowers the energy of the mixture, but the mixture of separated ions, $Na^+ + Cl^-$, is still *higher* in energy by $(496 - 348)$ kJ/mol = 148 kJ/mol than the original mixture of atoms (the red arrow in Figure 7-2). Thus, just the formation of ions does not explain why the process occurs. The strong attractive force between ions of opposite charge draws the ions together into the regular array shown in Figure 7-1. The energy associated with this attraction (step 3) is the *crystal lattice energy* of NaCl, -790 kJ/mol.

$$Na^+(g) + Cl^-(g) \longrightarrow NaCl(s) \qquad \text{crystal lattice energy} = -790 \text{ kJ/mol}$$

The crystal (solid) formation thus further *lowers* the energy to $(148 - 790)$ kJ/mol = -642 kJ/mol. The overall result is that one mole of NaCl(s) is 642 kJ/mol lower in energy (more stable) than the original mixture of atoms (the blue arrow in Figure 7-2). Thus, we see that a major driving force for the formation of ionic compounds is the large electrostatic stabilization due to the attraction of the ionic charges (step 3).

In this discussion we have not taken into account the fact that sodium is a solid metal or that chlorine actually exists as diatomic molecules. The additional energy changes involved when these are changed to gaseous Na and Cl atoms, respectively, are small enough that the overall energy change starting from Na(s) and $Cl_2(g)$ is still negative.

COVALENT BONDING

Ionic bonding cannot result from a reaction between two nonmetals, because their electronegativity difference is not great enough for electron transfer to take place. Instead, reactions between two nonmetals result in *covalent bonding*.

> A **covalent bond** is formed when two atoms share one or more pairs of electrons. Covalent bonding occurs when the electronegativity difference, ΔEN, between elements (atoms) is zero or relatively small.

In predominantly covalent compounds the bonds between atoms *within* a molecule (*intra*molecular bonds) are relatively strong, but the forces of attraction *between* molecules (*inter*molecular forces) are relatively weak. As a result, covalent compounds have lower melting and boiling points than ionic compounds. By contrast, distinct molecules of solid ionic substances do not exist. The sum of the attractive forces of all the interactions in an ionic solid is substantial, and such a compound has high melting and boiling points. The relation of bonding types to physical properties of liquids and solids will be developed more fully in Chapter 13.

7-3 FORMATION OF COVALENT BONDS

Let us look at a simple case of covalent bonding, the reaction of two hydrogen atoms to form the diatomic molecule H_2. As you recall, an isolated hydrogen atom has the ground state electron configuration $1s^1$, with the probability density for this one electron spherically distributed about the hydrogen nucleus (Figure 7-3a). As two hydrogen atoms approach one another, the electron of each hydrogen atom is attracted by the nucleus of the *other* hydrogen atom as well as by its own nucleus (Figure 7-3b). If these two electrons have opposite spins so that they can occupy the same region (orbital), both electrons can now preferentially occupy the region *between* the two nuclei (Figure 7-3c), because they are attracted by both nuclei. The electrons are *shared* between the two hydrogen atoms, and a single covalent bond is formed. We say that the $1s$ orbitals *overlap* so that both electrons are now in the orbitals of both hydrogen atoms. The closer together the atoms come, the more nearly this is true. In that sense, each hydrogen atom now has the helium configuration, $1s^2$.

The bonded atoms are at lower energy (more stable) than the separated atoms. This is shown in the plot of energy versus distance in Figure 7-4. However, as the two atoms get closer together, the two nuclei, being positively charged, exert an increasing repulsion on one another. At some distance, a minimum energy, -435 kJ/mol, is reached; it corresponds to the most stable arrangement and occurs at 0.74 Å, the actual distance between two hydrogen nuclei in an H_2 molecule. At greater internuclear separation, the repulsive forces diminish, but the attractive forces decrease even faster. At smaller separations, repulsive forces increase more rapidly than attractive forces.

Other pairs of nonmetal atoms share electron pairs to form covalent bonds. The result of this sharing is that each atom attains a more stable electron configuration—frequently the same as that of the nearest noble gas. (This is discussed in Section 7-5.) Most covalent bonds involve sharing of two, four, or six electrons—that is, one, two, or three *pairs* of electrons. Two atoms form a **single covalent bond** when they share one pair of electrons, a **double covalent bond** when they share two electron pairs, and a **triple covalent bond** when they share three electron pairs. These are usually called simply *single*, *double*, and

Of course this one electron is unpaired.

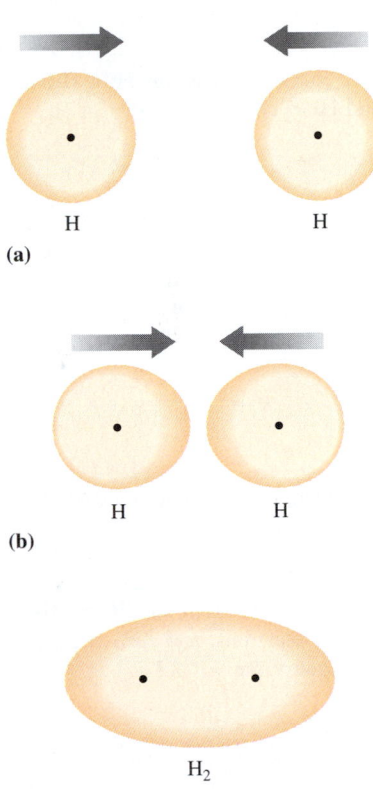

(a)

(b)

(c)

Figure 7-3 A representation of the formation of a covalent bond between two hydrogen atoms. The position of each positively charged nucleus is represented by a black dot. Electron density is indicated by the depth of shading. (a) Two hydrogen atoms separated by a large distance (essentially isolated). (b) As the atoms approach one another, the electron of each atom is attracted by the positively charged nucleus of the other atom, so the electron density begins to shift. (c) The two electrons can both occupy the region where the two $1s$ orbitals overlap; the electron density is highest in the region between the two atoms.

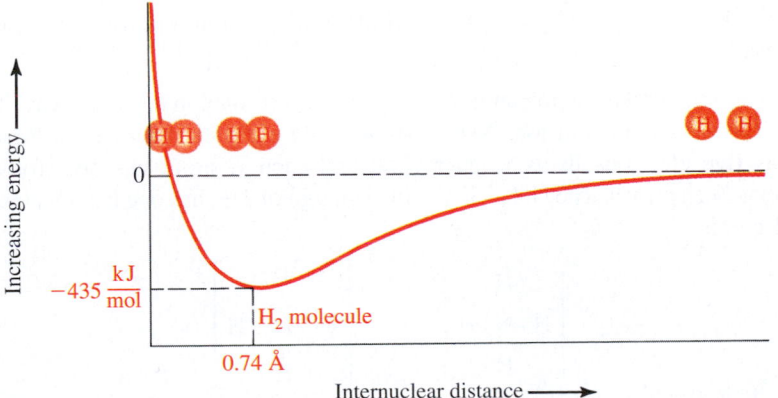

Figure 7-4 The potential energy of the H_2 molecule as a function of the distance between the two nuclei. The lowest point in the curve, -435 kJ/mol, corresponds to the internuclear distance actually observed in the H_2 molecule, 0.74 Å. (The minimum potential energy, -435 kJ/mol, corresponds to a value of -7.23×10^{-19} joule per H_2 molecule.) Energy is compared with that of two separated hydrogen atoms.

triple bonds. Covalent bonds that involve sharing of one and three electrons are known, but are relatively rare.

In a Lewis formula, we represent a covalent bond by writing *each shared electron pair* either as a pair of two dots between the two atom symbols or as a dash connecting them. Thus, the formation of H_2 from two H atoms could be represented as

$$H \cdot \; + \; \cdot H \longrightarrow H:H \qquad \text{or} \qquad H\!-\!H$$

where the dash represents a single bond. Similarly, the combination of a hydrogen atom and a fluorine atom to form a hydrogen fluoride (HF) molecule can be shown as

$$H \cdot \; + \; \cdot \ddot{\underset{\cdot\cdot}{F}}: \longrightarrow H:\ddot{\underset{\cdot\cdot}{F}}: \qquad \text{or} \qquad H\!-\!\ddot{\underset{\cdot\cdot}{F}}:$$

We shall see many more examples of this representation.

> The bonding in gaseous HCl, HBr, and HI is analogous to that in HF.

In our discussion, we have postulated that bonds form by the **overlap** of two atomic orbitals. This is the essence of the **valence bond theory,** which we will describe in more detail in the next chapter. Another theory, **molecular orbital theory,** is discussed in Chapter 9. For now, let us concentrate on the *number* of electron pairs shared and defer the discussion of *which* orbitals are involved in the sharing until the next chapter.

7-4 LEWIS FORMULAS FOR MOLECULES AND POLYATOMIC IONS

> A polyatomic ion is an ion that contains more than one atom.

In Sections 7-1 and 7-2 we drew *Lewis formulas* for atoms and monatomic ions. In Section 7-3, we used Lewis formulas to show the *valence electrons* in two simple molecules. A water molecule can be represented by either of the following diagrams.

> In H_2O, the O atom contributes six valence electrons, and each H atom contributes one.

$$\begin{array}{c} H:\ddot{O}: \\ H \end{array} \qquad \text{or} \qquad \begin{array}{c} H\!-\!\ddot{O}: \\ | \\ H \end{array}$$

dot formula   dash formula

An H_2O molecule has two shared electron pairs, i.e., two single covalent bonds. The O atom also has two unshared pairs.

In *dash formulas,* a shared pair of electrons is indicated by a dash. There are two *double* bonds in carbon dioxide, and its Lewis formula is

> In CO_2, the C atom contributes four valence electrons, and each O atom contributes six.

$$\ddot{\underset{\cdot\cdot}{O}}::C::\ddot{\underset{\cdot\cdot}{O}} \qquad \text{or} \qquad \ddot{\underset{\cdot\cdot}{O}}\!=\!C\!=\!\ddot{\underset{\cdot\cdot}{O}}$$

dot formula   dash formula

A CO_2 molecule has four shared electron pairs, i.e., two double bonds. The central atom (C) has no unshared pairs.

> The NH_3 molecule, like the NH_4^+ ion, has eight valence electrons about the N atom.
>
> $$\begin{array}{c} H:\ddot{N}:H \\ H \end{array} \qquad \text{or} \qquad \begin{array}{c} H\!-\!N\!-\!H \\ | \\ H \end{array}$$

The covalent bonds in a polyatomic ion can be represented in the same way. The Lewis formula for the ammonium ion, NH_4^+, shows only eight electrons, even though the N atom has five electrons in its valence shell and each H atom has one, for a total of $5 + 4(1) = 9$ electrons. The NH_4^+ ion, with a charge of $1+$, has one less electron than the original atoms.

$$\left[\begin{array}{c} H \\ \ddot{} \\ H:N:H \\ \ddot{} \\ H \end{array} \right]^+ \qquad \left[\begin{array}{c} H \\ | \\ H\!-\!N\!-\!H \\ | \\ H \end{array} \right]^+$$

dot formula   dash formula

The writing of Lewis formulas is an electron bookkeeping method that is useful as a first approximation to suggest bonding schemes. It is important to remember that Lewis dot formulas only show the number of valence electrons, the number and kinds of bonds, and the order in which the atoms are connected. *They are not intended to show the three-*

dimensional shapes of molecules and polyatomic ions. We shall see in Chapter 8, however, that the three-dimensional geometry of a molecule can be predicted from its Lewis formula.

7-5 THE OCTET RULE

Representative elements usually attain stable noble gas electron configurations when they share electrons. In the water molecule the O has a share in eight outer shell electrons, the neon configuration, while H shares two electrons, the helium configuration. Likewise, the C and O of CO_2 and the N of NH_3 and the NH_4^+ ion each have a share in eight electrons in their outer shells. The H atoms in NH_3 and NH_4^+ each share two electrons. Many Lewis formulas are based on the idea that

> in *most* of their compounds, the representative elements achieve noble gas configurations.

This statement is usually called the **octet rule,** because the noble gas configurations have $8\ e^-$ in their outermost shells (except for He, which has $2\ e^-$).

   For now, we shall restrict our discussion to compounds of the *representative elements.* The octet rule alone does not let us write Lewis formulas. We still must decide how to place the electrons around the bonded atoms—that is, how many of the available valence electrons are **bonding electrons** (shared) and how many are **unshared electrons** (associated with only one atom). A pair of unshared electrons in the same orbital is called a **lone pair.** A simple mathematical relationship is helpful here:

> $$S = N - A$$
>
> S is the total number of electrons *shared* in the molecule or polyatomic ion.
> N is the number of valence shell electrons *needed* by all the atoms in the molecule or ion to achieve noble gas configurations ($N = 8 \times$ number of atoms not including H, plus $2 \times$ number of H atoms).
> A is the number of electrons *available* in the valence shells of all of the (representative) atoms. This is equal to the sum of their periodic group numbers.

   For example, in CO_2, A for each O atom is 6 and A for the carbon atom is 4, so A for CO_2 is $4 + 2(6) = 16$. The following general steps describe the use of this relationship in constructing dot formulas for molecules and polyatomic ions.

Writing Lewis Formulas

1. Select a reasonable (symmetrical) "skeleton" for the molecule or polyatomic ion.
   a. The *least electronegative element* is usually the central element, except that H is never the central element, because it forms only one bond. The least electronegative element is usually the one that needs the most electrons to fill its octet. Example: CS_2 has the skeleton S C S.
   b. Oxygen atoms do not bond to each other except in (1) O_2 and O_3 molecules; (2) the

In some compounds, the central atom does not achieve a noble gas configuration. Such exceptions to the octet rule are discussed later in this chapter.

The representative elements are those in the A groups of the periodic table.

peroxides, which contain the O_2^{2-} group; and (3) the rare superoxides, which contain the O_2^- group. Example: The sulfate ion, SO_4^{2-}, has the skeleton

$$\begin{bmatrix} & O & \\ O & S & O \\ & O & \end{bmatrix}^{2-}$$

A ternary acid contains *three* elements—H, O, and another element, often a nonmetal.

c. In *ternary acids* (oxoacids), hydrogen usually bonds to an O atom, *not* to the central atom. Example: nitrous acid, HNO_2, has the skeleton H O N O. However, there are a few exceptions to this guideline, such as H_3PO_3 and H_3PO_2.

d. For ions or molecules that have more than one central atom, the most symmetrical skeletons possible are used. Examples: C_2H_4 and $P_2O_7^{4-}$ have the following skeletons:

$$\begin{array}{cc} H & H \\ C & C \\ H & H \end{array} \quad \text{and} \quad \begin{bmatrix} O & & O \\ O & P & O & P & O \\ O & & O \end{bmatrix}^{4-}$$

For compounds containing only representative elements, N is equal to $8 \times$ number of atoms *not* including H, plus $2 \times$ number of H atoms.

2. Calculate N, *the number of valence (outer) shell electrons* needed by all atoms in the molecule or ion to achieve noble gas configurations. Examples:

For H_2SO_4,

$$N = 1 \times 8 \text{ (S atom)} + 4 \times 8 \text{ (O atoms)} + 2 \times 2 \text{ (H atoms)}$$
$$= 8 + 32 + 4 = 44 \ e^- \text{ needed}$$

For SO_4^{2-},

$$N = 8 + 32 = 40 \ e^- \text{ needed}$$

For the representative elements, the number of valence shell electrons in an atom is equal to its periodic group number. Exceptions: 1 for an H atom and 8 for a noble gas (except 2 for He).

3. Calculate A, *the number of electrons available* in the valence (outer) shells of all the atoms. For negatively charged ions, add to the total the number of electrons equal to the charge on the anion; for positively charged ions, subtract the number of electrons equal to the charge on the cation. Examples:

For H_2SO_4,

$$A = 2 \times 1 \text{ (H atoms)} + 1 \times 6 \text{ (S atom)} + 4 \times 6 \text{ (O atoms)}$$
$$= 2 + 6 + 24 = 32 \ e^- \text{ available}$$

For SO_4^{2-},

$$A = 1 \times 6 \text{ (S atom)} + 4 \times 6 \text{ (O atoms)} + 2 \text{ (for 2– charge)}$$
$$= 6 + 24 + 2 = 32 \ e^- \text{ available}$$

4. Calculate S, *total number of electrons shared* in the molecule or ion, using the relationship $S = N - A$. Examples:

For H_2SO_4,

$$S = N - A = 44 - 32$$
$$= 12 \text{ electrons shared (6 pairs of } e^- \text{ shared)}$$

For SO_4^{2-},

$$S = N - A = 40 - 32$$
$$= 8 \text{ electrons shared (4 pairs of } e^- \text{ shared)}$$

5. Place the S electrons into the skeleton as *shared pairs*. Use double and triple bonds only when necessary. Lewis formulas may be shown as either dot formulas or dash formulas.

C, N, and O often form double and triple bonds. S and Se can form double bonds with C, N, and O.

Formula	Skeleton	Dot Formula ("bonds" in place, but incomplete)	Dash Formula ("bonds" in place, but incomplete)
H_2SO_4	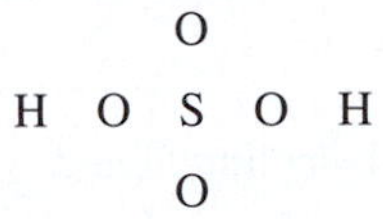 H O S O H (with O above and below S)	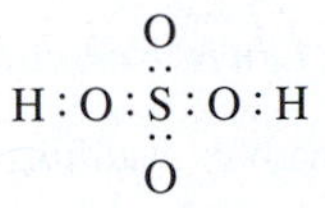 H:O:S:O:H (with O above and below S)	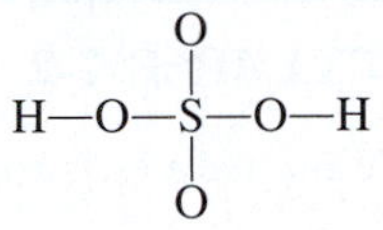 H—O—S—O—H (with O above and below S)
SO_4^{2-}	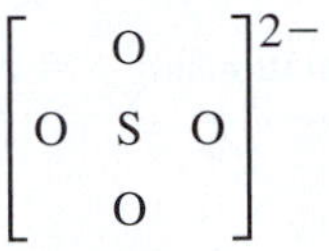 $[O\ S\ O]^{2-}$ (with O above and below S)	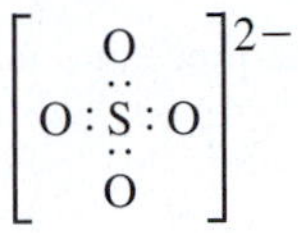 $[O:S:O]^{2-}$ (with O above and below S)	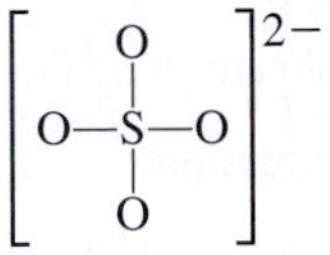 $[O—S—O]^{2-}$ (with O above and below S)

6. Place the additional electrons into the skeleton as *unshared (lone) pairs* to fill the octet of every A group element (except H, which can share only $2\ e^-$). Check that the total number of electrons is equal to A, from Step 3. Examples:

For H_2SO_4,

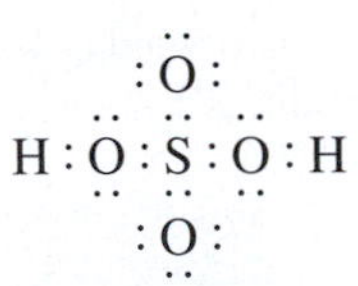

$$H:\overset{..}{\underset{..}{O}}:\overset{:\overset{..}{O}:}{\underset{:\overset{..}{O}:}{S}}:\overset{..}{\underset{..}{O}}:H \quad \text{or} \quad H-\overset{..}{\underset{..}{O}}-\overset{:\overset{..}{O}:}{\underset{:\overset{..}{O}:}{S}}-\overset{..}{\underset{..}{O}}-H$$

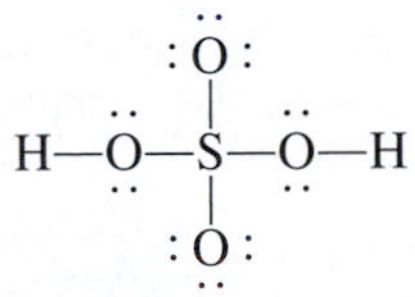

Check: 16 pairs of e^- have been used. $2 \times 16 = 32\ e^-$ available.

For SO_4^{2-},

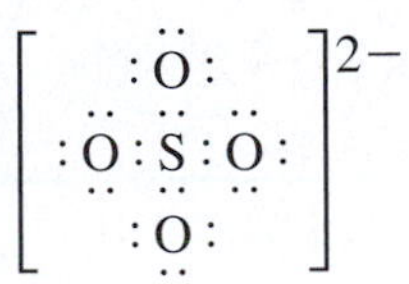

$$\left[:\overset{..}{\underset{..}{O}}:\overset{:\overset{..}{O}:}{\underset{:\overset{..}{O}:}{S}}:\overset{..}{\underset{..}{O}}: \right]^{2-} \quad \text{or} \quad \left[:\overset{..}{\underset{..}{O}}-\overset{:\overset{..}{O}:}{\underset{:\overset{..}{O}:}{S}}-\overset{..}{\underset{..}{O}}: \right]^{2-}$$

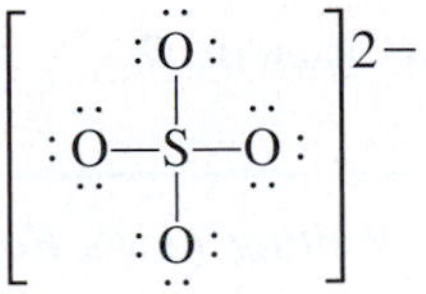

Check: 16 pairs of e^- have been used. $2 \times 16 = 32\ e^-$ available.

Please note that a Lewis formula *does not* represent the geometry of a molecule or an ion. We will discuss the actual geometries of molecules and ions in Chapter 8.

EXAMPLE 7-1 *Writing Lewis Formulas*

Write the Lewis formula for the nitrogen molecule, N_2.

Plan

We follow the stepwise procedure that was just presented for writing Lewis formulas.

Solution

Step 1: The skeleton is N   N.

Step 2: $N = 2 \times 8 = 16\ e^-$ needed (total) by both atoms

Step 3: $A = 2 \times 5 = 10\ e^-$ available (total) for both atoms

Step 4: $S = N - A = 16\ e^- - 10\ e^- = 6\ e^-$ shared

Step 5: N:::N $6\ e^-$ (3 pairs) are shared; a *triple* bond.

Step 6: The additional $4\,e^-$ are accounted for by a lone pair on each N. The complete Lewis formula is

$$:N:::N: \qquad \text{or} \qquad :N{\equiv}N:$$

Check: $10\,e^-$ (5 pairs) have been used.

You should now work Exercises 34 and 36.

EXAMPLE 7-2 *Writing Lewis Formulas*

Write the Lewis formula for carbon disulfide, CS_2, an ill-smelling liquid.

Plan

Again, we follow the stepwise procedure to apply the relationship $S = N - A$.

Solution

C is the central atom, or the element in the middle of the molecule. It needs four more electrons to acquire an octet, while each S atom needs only two more electrons.

Step 1: The skeleton is S   C   S.

Step 2: $N = 1 \times 8$ (for C) $+ 2 \times 8$ (for S) $= 24\,e^-$ needed by all atoms

Step 3: $A = 1 \times 4$ (for C) $+ 2 \times 6$ (for S) $= 16\,e^-$ available

Step 4: $S = N - A = 24\,e^- - 16\,e^- = 8\,e^-$ shared

Step 5: S : : C : : S $8\,e^-$ (4 pairs) are shared; two *double* bonds.

Step 6: C already has an octet, so the remaining $8e^-$ are distributed as lone pairs on the S atoms to give each S an octet. The complete Lewis formula is

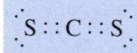 or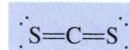

Check: $16\,e^-$ (8 pairs) have been used. The bonding picture is similar to that of CO_2; this is not surprising, because S is below O in Group VIA.

You should now work Exercise 38.

EXAMPLE 7-3 *Writing Lewis Formulas*

Write the Lewis formula for the carbonate ion, CO_3^{2-}.

Plan

The same stepwise procedure can be applied to ions. We must remember to adjust A, the total number of electrons, to account for the charge shown on the ion.

Solution

A number of minerals contain the carbonate ion. A very common one is calcium carbonate, $CaCO_3$, the main constituent of limestone and of stalactites and stalagmites.

$$\text{O}^{\ 2-}$$

Step 1: The skeleton is O   C   O

Step 2: $N = 1 \times 8$ (for C) $+ 3 \times 8$ (for O) $= 8 + 24 = 32\,e^-$ needed by all atoms

Step 3: $A = 1 \times 4$ (for C) $+ 3 \times 6$ (for O) $+ 2$ (for the $2-$ charge)
$= 4 + 18 + 2 = 24\,e^-$ available

Step 4: $S = N - A = 32\,e^- - 24\,e^- = 8\,e^-$ (4 pairs) shared

$$\text{O}^{\ 2-}$$

Step 5: O : C : : O   (Four pairs are shared. At this point it doesn't matter which O is doubly bonded.)

Step 6: The Lewis formula is

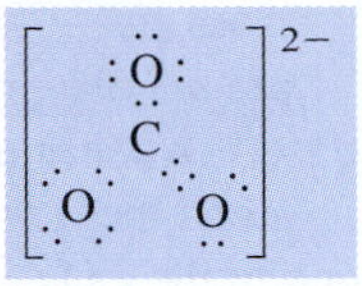

$$\left[\begin{array}{c} :\ddot{O}: \\ C \\ :\ddot{O}\quad\ddot{O}: \end{array}\right]^{2-} \quad \text{or} \quad \left[\begin{array}{c} :\ddot{O}: \\ C \\ :\ddot{O}\quad\ddot{O}: \end{array}=\right]^{2-}$$

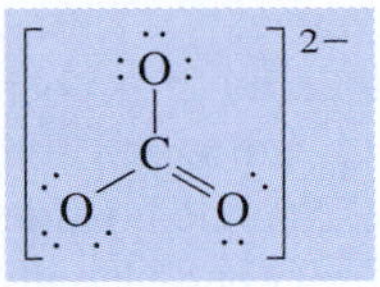

Check: 24 e^- (12 pairs) have been used.

You should now work Exercise 39.

You should practice writing many Lewis formulas. A few common types of organic compounds and their Lewis formulas are shown here. Each follows the octet rule. Methane, CH_4, is the simplest of a huge family of organic compounds called *hydrocarbons* (composed solely of hydrogen and carbon). Propane, C_3H_8, is another hydrocarbon that contains only single bonds. Ethylene, C_2H_4, has a carbon–carbon double bond.

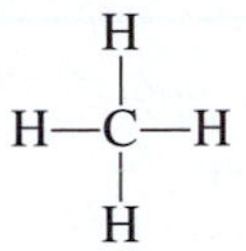

methane, CH_4 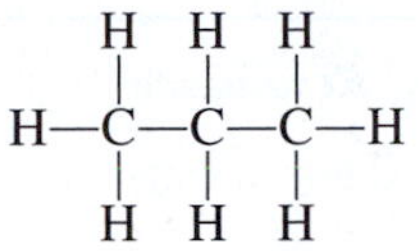   propane, C_3H_8 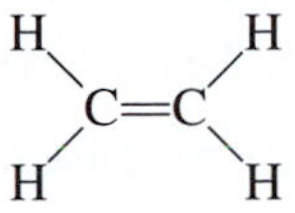   ethylene, C_2H_4

Halogen atoms can appear in place of hydrogen atoms in many organic compounds, because both hydrogen and halogen atoms need one more electron to attain noble gas configurations. An example is chloroform, $CHCl_3$. Alcohols contain the group C—O—H; the simplest alcohol is methanol, CH_3OH. An organic compound that contains a carbon–oxygen double bond is formaldehyde, H_2CO.

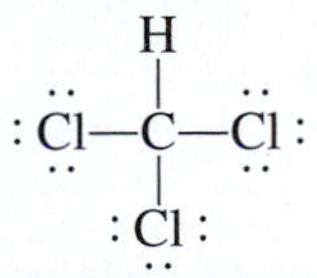

chloroform, $CHCl_3$ 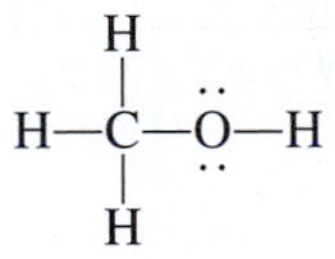   methanol, CH_3OH 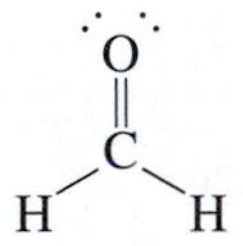   formaldehyde, H_2CO

7-6 RESONANCE

In addition to the Lewis formula shown in Example 7-3, two others with the same skeleton for the $CO_3{}^{2-}$ ion are equally acceptable. In these formulas, 4 e^- could be shared between the carbon atom and either of the other two oxygen atoms.

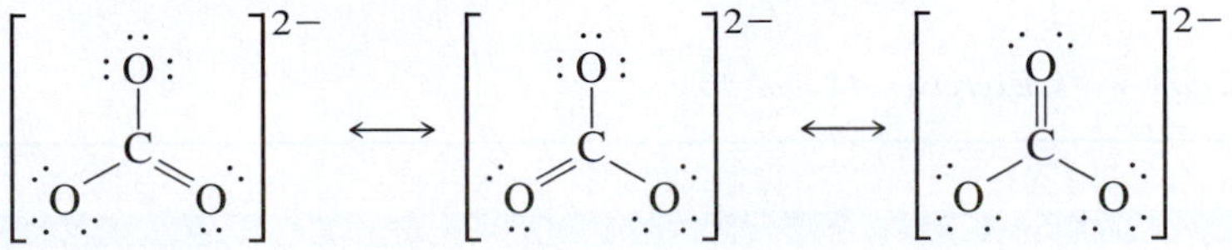

A molecule or polyatomic ion for which two or more Lewis formulas with the same arrangements of atoms can be drawn to describe the bonding is said to exhibit **resonance.** The three structures above are **resonance structures** of the carbonate ion. The relationship among them is indicated by the double-headed arrows, ↔. This symbol *does not mean* that the ion flips back and forth among these three structures. The true structure is like an average of the three.

Experiments show that the C—O bonds in $CO_3{}^{2-}$ are *neither* double nor single bonds, but are intermediate in bond length and strength. Based on measurements in many compounds, the typical C—O single bond length is 1.43 Å, and the typical C=O double bond length is 1.22 Å. The C—O bond length for each bond in the $CO_3{}^{2-}$ ion is intermediate at 1.29 Å. Another way to represent this situation is by **delocalization** of bonding electrons:

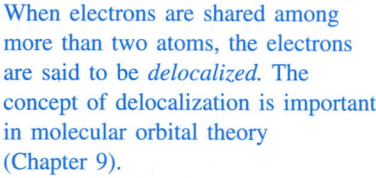

(lone pair on O atoms not shown)

The dashed lines indicate that some of the electrons shared between C and O atoms are *delocalized* among all four atoms; that is, the four pairs of shared electrons are equally distributed among three C—O bonds.

> When electrons are shared among more than two atoms, the electrons are said to be *delocalized*. The concept of delocalization is important in molecular orbital theory (Chapter 9).

EXAMPLE 7-4   *Lewis Formulas, Resonance*

Draw two resonance structures for the sulfur dioxide molecule, SO_2.

Plan

The stepwise procedure (Section 7-5) can be used to write each resonance structure.

Solution

$$N = 1(8) + 2(8) = 24\ e^-$$
$$A = 1(6) + 2(6) = 18\ e^-$$
$$S = N - A = 6\ e^-\ \text{shared}$$

The resonance structures are

or

We could show delocalization of electrons as follows:

O==S==O   (lone pairs on O not shown)

Remember that Lewis formulas *do not necessarily show shapes*. SO_2 molecules are angular, not linear.

You should now work Exercises 44 and 46.

The rose on the left is in an atmosphere of sulfur dioxide, SO_2. Gaseous SO_2 and its aqueous solutions are used as bleaching agents. A similar process is used to bleach wood pulp before it is converted to paper.

 **E N R I C H M E N T**

Formal Charges

An experimental determination of the structure of a molecule or polyatomic ion is necessary to establish unequivocally its correct structure. However, we often do not have these results available. **Formal charge** is the charge on an atom *in a molecule or* *polyatomic ion;* to find the formal charge, we count bonding electrons as though they were equally shared between the two bonded atoms. The concept of formal charges helps us to write correct Lewis formulas in most cases. The most energetically favorable formula for a molecule is usually one in which the formal charge on each atom is zero or as near zero as possible.

Consider the reaction of NH_3 with hydrogen ion, H^+, to form the ammonium ion, NH_4^+.

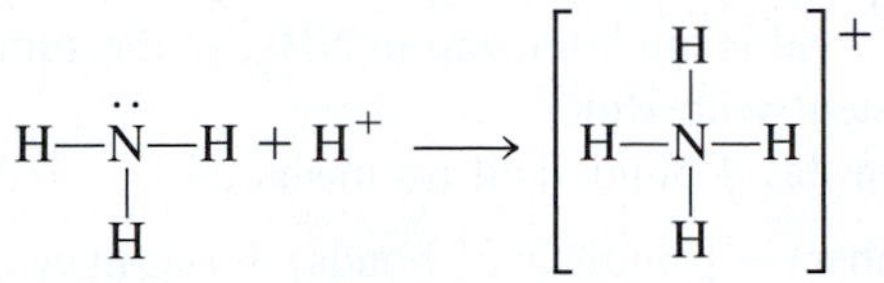

$$H-\overset{..}{\underset{\underset{H}{|}}{N}}-H + H^+ \longrightarrow \left[H-\underset{\underset{H}{|}}{\overset{\overset{H}{|}}{N}}-H \right]^+$$

The unshared pair of electrons on the N atom in the NH_3 molecule is shared with the H^+ ion to form the NH_4^+ ion, in which the N atom has four covalent bonds. Because N is a Group VA element, we expect it to form three covalent bonds to complete its octet. How can we describe the fact that N has four covalent bonds in species like NH_4^+? The answer is obtained by calculating the *formal charge* on each atom in NH_4^+ by the following rules:

Rules for Assigning Formal Charges to Atoms of A Group Elements

1. a. In a molecule, the sum of the formal charges is zero.
   b. In a polyatomic ion, the sum of the formal charges is equal to the charge.

2. The formal charge, abbreviated FC, on an atom in a Lewis formula is given by the relationship

   $$FC = (\text{group number}) - [(\text{number of bonds}) + (\text{number of unshared } e^-)]$$

   The group number of the noble gases is taken as VIIIA, rather than zero, in calculating formal charges. Formal charges are represented by $\oplus$ and $\ominus$ to distinguish them from real charges on ions.

3. In a Lewis formula, an atom that has the same number of bonds as its periodic group number has a formal charge of zero.

   This bookkeeping system helps us to select reasonable Lewis formulas, according to the following guidelines:

a. The most likely formula for a molecule or ion is usually one in which the formal charge on each atom is zero or as near zero as possible.

b. Negative formal charges are more likely to occur on the more electronegative elements.

c. Lewis formulas in which adjacent atoms have formal charges of the same sign are usually *not* accurate representations (the *adjacent charge rule*).

Let us apply these rules to the ammonia molecule, NH_3, and to the ammonium ion, NH_4^+. Because N is a Group VA element, its group number is 5.

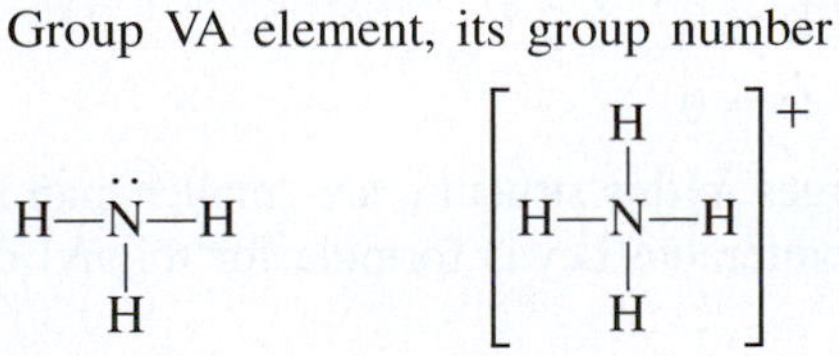

$$H-\overset{..}{\underset{\underset{H}{|}}{N}}-H \qquad \left[H-\underset{\underset{H}{|}}{\overset{\overset{H}{|}}{N}}-H \right]^+$$

In NH_3 the N atom has 3 bonds and 2 unshared e^-, and so for N,

> FC = (group number) − [(number of bonds) + (number of unshared e^-)]
> = 5 − (3 + 2) = 0 (for N)

For H,

> FC = (group number) − [(number of bonds) + (number of unshared e^-)]
> = 1 − (1 + 0) = 0 (for H)

The formal charges of N and H are both zero in NH_3, so the sum of the formal charges is $0 + 3(0) = 0$, consistent with Rule 1a.

In NH_4^+ the N atom has 4 bonds and no unshared e^-, and so for N,

> FC = (group number) − [(number of bonds) + (number of unshared e^-)]
> = 5 − (4 + 0) = 1+ (for N)

Calculation of the FC for H atoms gives zero, as above. The sum of the formal charges in NH_4^+ is $(1+) + 4(0) = 1+$. This is consistent with Rule 1b.

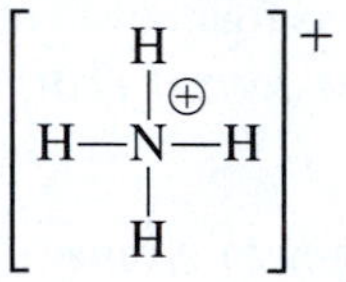

Thus, we see that the octet rule is obeyed in both NH_3 and NH_4^+. The sum of the formal charges in each case is that predicted by Rule 1, even though nitrogen has four covalent bonds in the NH_4^+ ion.

Let us now write a Lewis formula for, and assign formal charges to, the atoms in thionyl chloride, $SOCl_2$, a compound often used in organic synthesis. Both Cl atoms and the O atom are bonded to the S atom. A Lewis formula that satisfies the octet rule is

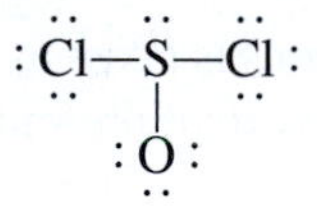

Formal charges on the atoms are calculated by the usual relationship.

> FC = (group number) − [(number of bonds) + (number of unshared e^-)]

For Cl: FC = 7 − (1 + 6) = 0

For S: FC = 6 − (3 + 2) = 1+

For O: FC = 6 − (1 + 6) = 1−

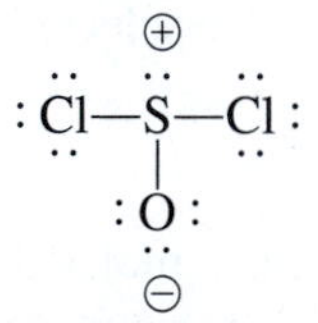

Another possible Lewis structure for thionyl chloride is

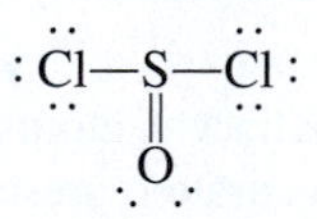

Let us calculate formal charges on the atoms in this structure.

For Cl: FC = 7 − (1 + 6) = 0

For S: FC = 6 − (4 + 2) = 0

For O: FC = 6 − (2 + 4) = 0

Because the formal charges in this structure are smaller than in the previous one, we conclude that this is a preferable Lewis formula for thionyl chloride.

FCs are indicated by ⊕ and ⊖. The sum of the formal charges in a polyatomic ion is equal to the charge on the ion—1+ in NH_4^+.

As we shall see in Section 7-7, sulfur has more than eight valence electrons in some of its compounds.

7-7 LIMITATIONS OF THE OCTET RULE FOR LEWIS FORMULAS

Recall that representative elements achieve noble gas electron configurations in *most* of their compounds. But when the octet rule is not applicable, the relationship $S = N - A$ is not valid without modification. The following are general cases for which the procedure in Section 7-5 *must be modified*—i.e., cases in which there are limitations of the octet rule.

1. Most covalent compounds of beryllium, Be. Because Be contains only two valence shell electrons, it usually forms only two covalent bonds when it bonds to two other atoms. Therefore, we use *four electrons* as the number *needed* by Be in Step 2, Section 7-5. In Steps 5 and 6 we use only two pairs of electrons for Be.

2. Most covalent compounds of the Group IIIA elements, especially boron, B. The IIIA elements contain only three valence shell electrons, so they often form three covalent bonds when they bond to three other atoms. Therefore, we use *six electrons* as the number *needed* by the IIIA elements in Step 2; and in Steps 5 and 6 we use only three pairs of electrons for the IIIA elements.

3. Compounds or ions containing an odd number of electrons. Examples are NO, with 11 valence shell electrons, and NO_2, with 17 valence shell electrons.

4. Compounds or ions in which the central element needs a share in more than eight valence shell electrons to hold all the available electrons, A. Extra rules are added to Steps 4 and 6 when this is encountered.

   Step 4a: If S, the number of electrons shared, is less than the number needed to bond all atoms to the central atom, then S is increased to the number of electrons needed.

   Step 6a: If S must be increased in Step 4a, then the octets of all the atoms might be satisfied before all A of the electrons have been added. Place the extra electrons on the central element.

Lewis formulas are not normally written for compounds containing d- and f-transition metals. The d- and f-transition metals utilize d and/or f orbitals in bonding as well as s and p orbitals. Thus, they can accommodate more than eight valence electrons.

Many species that violate the octet rule are quite reactive. For instance, compounds containing atoms with only four valence electrons (limitation 1 above) or six valence electrons (limitation 2 above) frequently react with other species that supply electron pairs. Compounds such as these that accept a share in a pair of electrons are called *Lewis acids*; a *Lewis base* is a species that makes available a share in a pair of electrons. (This kind of behavior will be discussed in detail in Section 10-10.) Molecules with an odd number of electrons often *dimerize* (combine in pairs) to give products that do satisfy the octet rule. Examples are the dimerization of NO to form N_2O_2 (Section 24-15) and of NO_2 to form N_2O_4 (Section 24-15). Examples 7-5 through 7-8 illustrate some limitations and show how such Lewis formulas are constructed.

EXAMPLE 7-5 *Limitations of Octet Rule*

Write the Lewis formula for gaseous beryllium chloride, $BeCl_2$, a covalent compound.

Plan

This is an example of limitation 1. So, as we follow the steps in writing the Lewis formula, we must remember to use *four electrons* as the number *needed* by Be in Step 2. Steps 5 and 6 should show only two pairs of electrons for Be.

Solution

Step 1: The skeleton is Cl   Be   Cl.

see limitation 1
↓

Step 2: $N = 2 \times 8$ (for Cl) $+ 1 \times 4$ (for Be) $= 20\,e^-$ needed

Step 3: $A = 2 \times 7$ (for Cl) $+ 1 \times 2$ (for Be) $= 16\,e^-$ available

Step 4: $S = N - A = 20\,e^- - 16\,e^- = 4\,e^-$ shared

Step 5: Cl : Be : Cl

Step 6:   :Cl : Be : Cl :   or   :Cl—Be—Cl :

Calculation of formal charges shows that

$$\text{for Be, FC} = 2 - (2 + 0) = 0 \qquad \text{and} \qquad \text{for Cl, FC} = 7 - (1 + 6) = 0$$

You should now work Exercise 48.

In $BeCl_2$, the chlorine atoms achieve the argon configuration, [Ar], while the beryllium atom has a share of only four electrons. Compounds such as $BeCl_2$, in which the central atom shares fewer than $8\,e^-$, are sometimes referred to as **electron deficient** compounds. This "deficiency" refers only to satisfying the octet rule for the central atom. The term does not imply that there are fewer electrons than there are protons in the nuclei, as in the case of a cation, because the molecule is neutral.

A Lewis formula can be written for $BeCl_2$ that *does* satisfy the octet rule (see margin). Let us evaluate the formal charges for that formula:

$$\text{for Be, FC} = 2 - (4 + 0) = 2- \qquad \text{and} \qquad \text{for Cl, FC} = 7 - (2 + 4) = 1+$$

As mentioned earlier, the most favorable structure for a molecule is one in which the formal charge on each atom is zero, if possible. In case some atoms did have nonzero formal charges, we would expect that the more electronegative atoms (Cl) would be the ones with lowest formal charge. Thus, we prefer the Lewis structure shown in Example 7-5 over the one in the margin.

One might expect a similar situation for compounds of the other IIA metals, Mg, Ca, Sr, Ba, and Ra. However, these elements have *lower ionization energies* and *larger radii* than Be, so they usually form ions by losing two electrons.

EXAMPLE 7-6   *Limitations of Octet Rule*

Write the Lewis formula for boron trichloride, BCl_3, a covalent compound.

Plan

This covalent compound of boron is an example of limitation 2. As we follow the steps in writing the Lewis formula, we use *six electrons* as the number *needed* by B in Step 2. Steps 5 and 6 should show only three pairs of electrons for B.

Solution

                              Cl

Step 1: The skeleton is Cl   B   Cl.

see limitation 2
↓

Step 2: $N = 3 \times 8$ (for Cl) $+ 1 \times 6$ (for B) $= 30\,e^-$ needed

BF$_3$ and BCl$_3$ are gases at room temperature. Liquid BBr$_3$ and solid BI$_3$ are shown here.

Step 3: $A = 3 \times 7$ (for Cl) $+ 1 \times 3$ (for B) $= 24 \, e^-$ available

Step 4: $S = N - A = 30 \, e^- - 24 \, e^- = 6 \, e^-$ shared

Step 5: Cl
      Cl : B : Cl

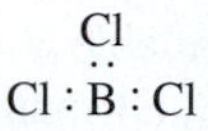

Step 6: 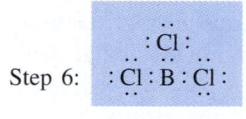 or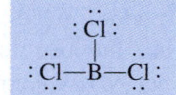

Each chlorine atom achieves the Ne configuration. The boron (central) atom acquires a share of only six valence shell electrons. Calculation of formal charges shows that

for B, FC $= 3 - (3 + 0) = 0$ and   for Cl, FC $= 7 - (1 + 6) = 0$

You should now work Exercise 50.

EXAMPLE 7-7 *Limitations of Octet Rule*

Write the Lewis formula for phosphorus pentafluoride, PF_5, a covalent compound.

Plan

We apply the usual stepwise procedure to write the Lewis formula. In PF_5, all five F atoms are bonded to P. This requires the sharing of a minimum of $10 \, e^-$, so this is an example of limitation 4. Therefore we add the extra Step 4a, and increase S from the calculated value of $8 \, e^-$ to $10 \, e^-$.

Solution

Step 1: The skeleton is
          F F
        F P F
            F

Step 2: $N = 5 \times 8$ (for F) $+ 1 \times 8$ (for P) $= 48 \, e^-$ needed

Step 3: $A = 5 \times 7$ (for F) $+ 1 \times 5$ (for P) $= 40 \, e^-$ available

Step 4: $S = N - A = 8 \, e^-$ shared
Five F atoms are bonded to P. This requires the sharing of a minimum of $10 \, e^-$. But only $8 \, e^-$ have been calculated in Step 4. Therefore, this is an example of limitation 4.

Step 4a: Increase S from $8 \, e^-$ to $10 \, e^-$. The number of electrons available, 40, does not change.

Step 5:
        F   F
          P
      F   F
         F

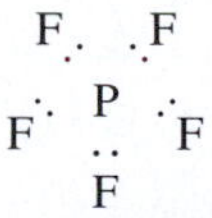

Step 6: 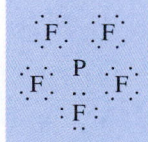 or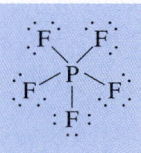

When the octets of the five F atoms have been satisfied, all 40 of the available electrons have been added. The phosphorus (central) atom has a share of ten electrons.

Calculation of formal charges shows that

for P, FC $= 5 - (5 + 0) = 0$ and   for F, FC $= 7 - (1 + 6) = 0$

You should now work Exercise 52.

This is sometimes referred to as *hypervalence.*

When an atom has a share of more than eight electrons, as does P in PF_5, we say that it exhibits an *expanded valence shell.* The electronic basis of the octet rule is that one s and three p orbitals in the valence shell of an atom can accommodate a maximum of eight electrons. The valence shell of phosphorus has $n = 3$, so it also has available d orbitals that can be involved in bonding. It is for this reason that phosphorus (and many other representative elements of period 3 and beyond) can exhibit expansion of valence. By contrast, elements in the *second row* of the periodic table can *never* exceed eight electrons in their valence shells, because each atom has only one s and three p orbitals in that shell. Thus, we understand why NF_3 can exist but NH_5 cannot.

EXAMPLE 7-8   *Limitations of Octet Rule*

Write the Lewis formula for the triiodide ion, I_3^-.

Plan

We apply the usual stepwise procedure. The calculation of $S = N - A$ in Step 4 shows only $2\ e^-$ shared, but a minimum of $4\ e^-$ are required to bond two I atoms to the central I. Limitation 4 applies, and we proceed accordingly.

Solution

Step 1:   The skeleton is $[I \quad I \quad I]^-$.

Step 2: $N = 3 \times 8$ (for I) $= 24\ e^-$ needed

Step 3: $A = 3 \times 7$ (for I) $+ 1$ (for the $1-$ charge) $= 22\ e^-$ available

Step 4: $S = N - A = 2\ e^-$ shared. Two I atoms are bonded to the central I. This requires a minimum of $4\ e^-$, but only $2\ e^-$ have been calculated in Step 4. Therefore, this is an example of limitation 4.

Step 4a: Increase S from $2\ e^-$ to $4\ e^-$.

Step 5: $[I : I : I]^-$

Step 6: $[: \ddot{I} : \ddot{I} : \ddot{I} :]^-$

Step 6a: Now we have satisfied the octets of all atoms using only 20 of the $22\ e^-$ available. We place the other two electrons on the central I atom.

The central iodine atom in I_3^- has an expanded valence shell.

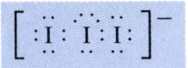   or

Calculation of formal charges shows that

for I on ends, $FC = 7 - (1 + 6) = 0$
for I in middle, $FC = 7 - (2 + 6) = 1-$

You should now work Exercise 54.

We have seen that *atoms attached to the central atom nearly always attain noble gas configurations,* even when the central atom does not.

7-8 POLAR AND NONPOLAR COVALENT BONDS

Covalent bonds may be either *polar* or *nonpolar.* In a **nonpolar bond** such as that in the hydrogen molecule, H_2, the electron pair is *shared equally* between the two hydrogen

$H:H$ or $H—H$

nuclei. Recall (Section 6-6) that we defined electronegativity as the tendency of an atom to attract electrons to itself in a chemical bond. Both H atoms have the same electronegativity. This means that the shared electrons are equally attracted to both hydrogen nuclei and therefore spend equal amounts of time near each nucleus. In this nonpolar covalent bond, the **electron density** is symmetrical about a plane that is perpendicular to a line between the two nuclei. This is true for all homonuclear *diatomic molecules,* such as H_2, O_2, N_2, F_2, and Cl_2, because the two identical atoms have identical electronegativities. We can generalize:

> The covalent bonds in all homonuclear diatomic molecules must be nonpolar.

Let us now consider *heteronuclear diatomic molecules.* Start with the fact that hydrogen fluoride, HF, is a gas at room temperature. This tells us that it is a covalent compound. We also know that the H—F bond has some degree of polarity because H and F are not identical atoms and therefore do not attract the electrons equally. But how polar will this bond be?

The electronegativity of hydrogen is 2.1, and that of fluorine is 4.0 (Table 6-3). Clearly, the F atom, with its higher electronegativity, attracts the shared electron pair much more strongly than does H. We represent the structure of HF as shown in the margin. Notice the unsymmetrical distribution of electron density; the electron density is distorted in the direction of the more electronegative F atom. This small shift of electron density leaves H somewhat positive.

Covalent bonds, such as the one in HF, in which the *electron pairs are shared unequally* are called **polar covalent bonds.** Two kinds of notation used to indicate polar bonds are shown in the margin.

The $\delta-$ over the F atom indicates a "partial negative charge." This means that the F end of the molecule is somewhat more negative than the H end. The $\delta+$ over the H atom indicates a "partial positive charge," or that the H end of the molecule is positive *with respect to* the F end. We are *not* saying that H has a charge of 1+ or that F has a charge of 1−! A second way to indicate the polarity is to draw an arrow so that the head points toward the negative end (F) of the bond and the crossed tail indicates the positive end (H).

The separation of charge in a polar covalent bond creates an electric **dipole.** We expect the dipoles in the covalent molecules HF, HCl, HBr, and HI to be different because F, Cl, Br, and I have different electronegativities. This tells us that atoms of these elements have different tendencies to attract an electron pair that they share with hydrogen. We indicate this difference as shown below, where $\Delta(EN)$ is the difference in electronegativity between two atoms that are bonded together.

A **homonuclear** molecule contains only one kind of atom. A molecule that contains two or more kinds of atoms is described as **heteronuclear.**

Remember that ionic compounds are solids at room temperature.

$$H : \overset{\cdot\cdot}{\underset{\cdot\cdot}{F}} :$$

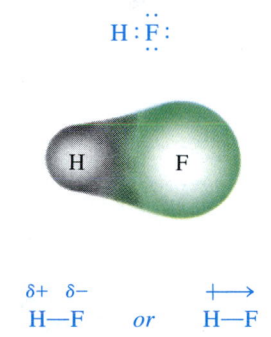

$$\overset{\delta+ \quad \delta-}{H—F} \quad or \quad \overset{\longmapsto}{H—F}$$

The word "dipole" means "two poles." Here it refers to the positive and negative poles that result from the separation of charge within a molecule.

The values of electronegativity are obtained from Table 6-3.

	Most polar			Least polar
	$\longmapsto$	$\longmapsto$	$\longmapsto$	$\longmapsto$
	H—F	H—Cl	H—Br	H—I
EN:	2.1 4.0	2.1 3.0	2.1 2.8	2.1 2.5
$\Delta(EN)$	1.9	0.9	0.7	0.4

The longest arrow indicates the largest dipole, or greatest separation of electron density in the molecule (see Table 7-3). For comparison, the $\Delta(EN)$ values for some typical 1:1 ionic compounds are NaBr, 1.8; RbF, 3.1; and KCl, 2.1.

Table 7-3 $\Delta(EN)$ Values and Dipole Moments for Some Pure (Gaseous) Substances		
Substance	**Δ(EN)**	**Dipole Moment (μ)***
HF	1.9	1.91 D
HCl	0.9	1.03 D
HBr	0.7	0.79 D
HI	0.4	0.38 D
H—H	0	0 D

*The magnitude of a dipole moment is given by the product of charge × distance of separation. Molecular dipole moments are usually expressed in debyes (D).

$\mu = d \times q$

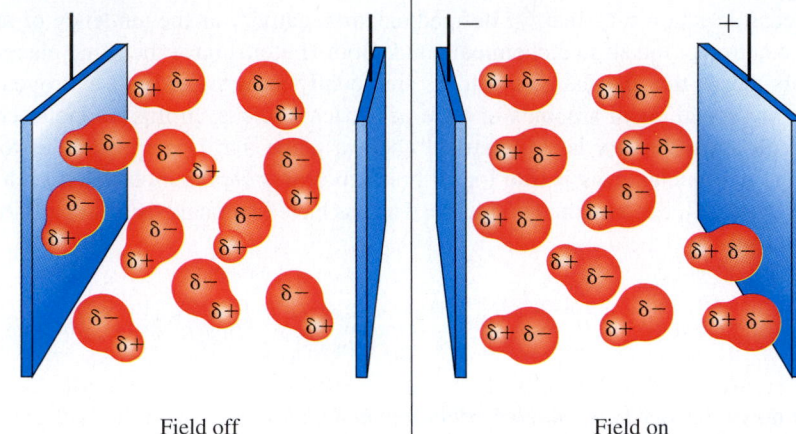

Field off   Field on

Figure 7-5   If polar molecules, such as HF, are subjected to an electric field, they tend to line up very slightly in a direction opposite to that of the field. This minimizes the electrostatic energy of the molecules. Nonpolar molecules are not oriented by an electric field. The effect is greatly exaggerated in this drawing.

7-9 DIPOLE MOMENTS

It is convenient to express bond polarities on a numerical scale. We indicate the polarity of a molecule by its dipole moment, which measures the separation of charge within the molecule. The **dipole moment,** μ, is defined as the product of the distance, d, separating charges of equal magnitude and opposite sign, and the magnitude of the charge, q. A dipole moment is measured by placing a sample of the substance between two plates and applying a voltage. This causes a small shift in electron density of any molecule, so the applied voltage is diminished very slightly. However, diatomic molecules that contain polar bonds, such as HF, HCl, and CO, tend to orient themselves in the electric field (Figure 7-5). This causes the measured voltage between the plates to decrease more markedly for these substances, and so we say that these molecules are *polar.* Molecules such as F_2 or N_2 do not reorient, so the change in voltage between the plates remains slight; we say that these molecules are *nonpolar.*

Generally, as electronegativity differences increase in diatomic molecules, the measured dipole moments increase. This can be seen clearly from the data for the hydrogen halides (Table 7-3).

Unfortunately, the dipole moments associated with *individual bonds* can be measured only in simple diatomic molecules. *Entire molecules* rather than selected pairs of atoms must be subjected to measurement. Measured values of dipole moments reflect the *overall* polarities of molecules. For polyatomic molecules they are the result of all the bond dipoles in the molecules. In Chapter 8, we shall see that structural features, such as molecular geometry and the presence of lone (unshared) pairs of electrons, also affect the polarity of a molecule.

7-10 THE CONTINUOUS RANGE OF BONDING TYPES

Let us now clarify our classification of bonding types. The degree of electron sharing or transfer depends on the electronegativity difference between the bonding atoms. Nonpolar covalent bonding (involving *equal sharing* of electron pairs) is one extreme, occurring when the atoms are identical (ΔEN is zero). Ionic bonding (involving *complete transfer* of electrons) represents the other extreme, and occurs when two elements with very different electronegativities interact (ΔEN is large).

Polar covalent bonds may be thought of as intermediate between pure (nonpolar) covalent bonds and pure ionic bonds. In fact, bond polarity is sometimes described in terms of **partial ionic character.** This usually increases with increasing difference in electronegativity between bonded atoms. Calculations based on the measured dipole moment of gaseous HCl indicate about 17% "ionic character."

When cations and anions interact strongly, some amount of electron sharing takes place; in such cases we can consider the ionic compound as having some **partial covalent character.** For instance, the high charge density of the very small Li^+ ion causes it to distort large anions that it approaches. The distortion attracts electron density from the anion to the region between it and the Li^+ ion, giving lithium compounds a higher degree of covalent character than in other alkali metal compounds.

Almost all bonds have both ionic and covalent character. By experimental means, a given type of bond can usually be identified as being "closer" to one or the other extreme type. We find it useful and convenient to use the labels for the major classes of bonds to describe simple substances, keeping in mind that they represent ranges of behavior.

Above all, we must recognize that any classification of a compound that we might suggest based on electronic properties *must* be consistent with the physical properties of ionic and covalent substances described at the beginning of the chapter. For instance, HCl has a rather large electronegativity difference (0.9), and its aqueous solutions conduct electricity. But we know that we cannot view it as an ionic compound because it is a gas, and not a solid, at room temperature. Liquid HCl is a nonconductor.

HCl *ionizes* in aqueous solution.

Let us point out another aspect of the classification of compounds as ionic or covalent. Not all ions consist of single charged atoms. Many are small groups of atoms that are covalently bonded together, yet they still have excess positive or negative charge. Examples of such *polyatomic ions* are ammonium ion, NH_4^+, sulfate ion, SO_4^{2-}, and nitrate ion, NO_3^-. A compound such as potassium sulfate, K_2SO_4, contains potassium ions, K^+, and sulfate ions, SO_4^{2-}, in a 2:1 ratio. We should recognize that this compound contains both covalent bonding (electron sharing *within* each sulfate ion) and ionic bonding (electrostatic attractions *between* potassium and sulfate ions). However, we classify this compound as *ionic,* because it is a high-melting solid (mp 1069°C), it conducts electricity both in molten form and in aqueous solution, and it displays the properties that we generally associate with ionic compounds. Put another way, covalent bonding holds each sulfate ion together, but the forces that hold the *entire* substance together are ionic.

In summary, we can describe chemical bonding as a continuum that may be represented as

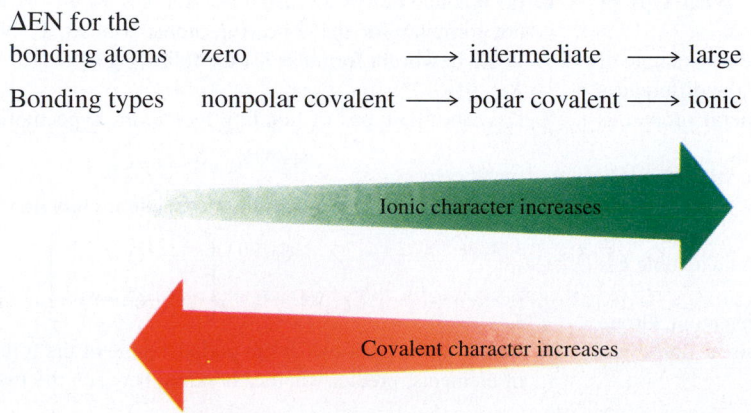

ΔEN for the
bonding atoms   zero $\longrightarrow$ intermediate $\longrightarrow$ large

Bonding types   nonpolar covalent $\longrightarrow$ polar covalent $\longrightarrow$ ionic

Ionic character increases

Covalent character increases

Key Terms

Anion A negatively charged ion; that is, an ion in which the atom or group of atoms has more electrons than protons.

Binary compound A compound consisting of two elements; may be ionic or covalent.

Bonding pair A pair of electrons involved in a covalent bond. Also called *shared pair*.

Cation A positively charged ion; that is, an ion in which the atom or group of atoms has fewer electrons than protons.

Chemical bonds Attractive forces that hold atoms together in elements and compounds.

Covalent bond A chemical bond formed by the sharing of one or more electron pairs between two atoms.

Covalent compound A compound containing predominantly covalent bonds.

Debye The unit used to express dipole moments.

Delocalization of electrons Refers to bonding electrons distributed among more than two atoms that are bonded together; occurs in species that exhibit resonance.

Dipole Refers to the separation of charge between two covalently bonded atoms.

Dipole moment (μ) The product of the distance separating opposite charges of equal magnitude and the magnitude of the charge; a measure of the polarity of a bond or molecule. A measured dipole moment refers to the dipole moment of an entire molecule.

Double bond A covalent bond resulting from the sharing of four electrons (two pairs) between two atoms.

Electron deficient compound A compound containing at least one atom (other than H) that shares fewer than eight electrons.

Formal charge The charge on an atom in a covalently bonded molecule or ion; bonding electrons are counted as though they were shared equally between the two bonded atoms.

Heteronuclear Consisting of different elements.

Homonuclear Consisting of only one element.

Ion An atom or a group of atoms that carries an electrical charge.

Ionic bonding The attraction of oppositely charged ions (cations and anions) in large numbers to form a solid. Ions result from the transfer of one or more electrons from one atom or group of atoms to another.

Ionic compound A compound containing predominantly ionic bonding.

Isoelectronic Having the same number of electrons.

Lewis acid A substance that accepts a share in a pair of electrons from another species.

Lewis base A substance that makes available a share in an electron pair.

Lewis formula The representation of a molecule, ion, or formula unit by showing atomic symbols and only outer shell electrons; does not show shape.

Lone pair A pair of electrons residing on one atom and not shared by other atoms; unshared pair.

Monatomic ion An ion that consists of only one atom.

Nonpolar bond A covalent bond in which electron density is symmetrically distributed.

Octet rule Many representative elements attain at least a share of eight electrons in their valence shells when they form molecular or ionic compounds; there are some limitations.

Polyatomic ion An ion that consists of more than one atom.

Polar bond A covalent bond in which there is an unsymmetrical distribution of electron density.

Resonance A concept in which two or more equivalent Lewis formulas for the same arrangement of atoms (resonance structures) are necessary to describe the bonding in a molecule or ion.

Single bond A covalent bond resulting from the sharing of two electrons (one pair) between two atoms.

Triple bond A covalent bond resulting from the sharing of six electrons (three pairs) between two atoms.

Unshared pair See *Lone pair*.

Valence electrons The electrons in the outermost shell of an atom.

Exercises

Chemical Bonding—Basic Ideas

1. Give a suitable definition of chemical bonding. What type of force is responsible for chemical bonding?

2. List the two basic types of chemical bonding. Give an example of a substance with each type of bonding. What are the differences between these types? What are some of the general properties associated with the two main types of bonding?

3. Why are covalent bonds called directional bonds, whereas ionic bonding is termed nondirectional?

4. The outermost electron configuration of any alkali metal atom, Group IA, is ns^1. How can an alkali metal atom attain a noble gas configuration?

5. The outermost electron configuration of any halogen atom, Group VIIA, is ns^2np^5. How can a halogen atom attain a noble gas configuration?

6. (a) What do Lewis dot formulas for atoms show? (b) Write Lewis dot formulas for the following atoms: He; Si; P; Ne; Mg; Cl.

7. Write Lewis dot formulas for the following atoms: Li; B; As; Na; Xe; Al.

8. Describe the types of bonding in sodium hypochlorite, NaOCl.

$$\text{Na}^+ \qquad [\,:\ddot{\text{O}}{-}\ddot{\ddot{\text{Cl}}}:\,]^-$$

9. Describe the types of bonding in calcium chlorate, $Ca(ClO_3)_2$.

$$\text{Ca}^{2+} \qquad 2\left[\begin{array}{c} :\ddot{\text{O}}: \\ | \\ :\ddot{\text{O}}{-}\ddot{\text{Cl}}{-}\ddot{\text{O}}: \end{array}\right]^-$$

10. Based on the positions in the periodic table of the following pairs of elements, predict whether bonding between the two would be

primarily ionic or covalent. Justify your answers. (a) Ba and Cl; (b) P and O; (c) Br and I; (d) Li and I; (e) Si and Br; (f) Ca and F.

11. Predict whether the bonding between the following pairs of elements would be ionic or covalent. Justify your answers. (a) K and Cl; (b) N and O; (c) Ca and Cl; (d) P and S; (e) C and F; (f) K and N.

12. Classify the following compounds as ionic or covalent: (a) $CaSO_4$; (b) SO_2; (c) KNO_3; (d) $NiCl_2$; (e) H_2CO_3; (f) PCl_3; (g) Li_2O; (h) N_2H_4; (i) $SOCl_2$.

Ionic Bonding

13. Describe what happens to the valence electron(s) as a metal atom and a nonmetal atom combine to form an ionic compound.

*14. Describe an ionic crystal. What factors might determine the geometrical arrangement of the ions?

15. Why are solid ionic compounds rather poor conductors of electricity? Why does conductivity increase when an ionic compound is melted or dissolved in water?

16. Write the formula for the ionic compound that forms between each of the following pairs of elements: (a) Cs and Br_2; (b) Ba and S; (c) K and Cl_2.

17. Write the formula for the ionic compound that forms between each of the following pairs of elements: (a) Ca and F_2; (b) Sr and Cl_2; (c) K and Se.

18. When a d-transition metal ionizes, it loses its outer s electrons before it loses any d electrons. Using [noble gas]$(n - 1)d^x$ representations, write the outer electron configurations for the following ions: (a) Cr^{3+}; (b) Mn^{2+}; (c) Ag^+; (d) Fe^{3+}; (e) Cu^{2+}; (f) Sc^{3+}; (g) Cu^+.

19. Which of the following do not accurately represent stable binary ionic compounds? Why? $BaCl_2$; NaS; AlF_4; SrS_2; Ca_2O_3; $NaBr_2$; Li_2S.

20. Which of the following do not accurately represent stable binary ionic compounds? Why? MgI; $Al(OH)_2$; InF_2; CO_2; $RbCl_2$; CsSe; Be_2O.

21. (a) Write Lewis formulas for the positive and negative ions in these compounds: $SrBr_2$; K_2O; Ca_3P_2; SnF_2; Bi_2O_3. (b) Which ions do not have a noble gas configuration?

22. (a) What are isoelectronic species? List three pairs of isoelectronic species. (b) All but one of the following species are isoelectronic. Which one is not isoelectronic with the others? Ne; Al^{3+}; O^{2-}; Na^+; Mg^{2+}; F.

23. All but one of the following species are isoelectronic. Which one is not isoelectronic with the others? S^{2-}; Ga^{2+}; Ar; K^+; Ca^{2+}; Sc^{3+}.

24. Write formulas for two cations and two anions that are isoelectronic with argon.

25. Write formulas for two cations and two anions that are isoelectronic with krypton.

26. Write formulas for two cations that have the following electron configurations in their *highest* occupied energy level: (a) $3s^2 3p^6$; (b) $4s^2 4p^6$.

27. Write formulas for two anions that have the electron configurations listed in Exercise 26.

Covalent Bonding—General Concepts

28. What does Figure 7-4 tell us about the attractive and repulsive forces in a hydrogen molecule?

29. Distinguish between heteronuclear and homonuclear diatomic molecules.

30. How many electrons are shared between two atoms in (a) a single covalent bond, (b) a double covalent bond, and (c) a triple covalent bond?

31. What is the maximum number of covalent bonds that a second-period element could form? How can the representative elements beyond the second period form more than this number of covalent bonds?

Lewis Formulas for Molecules and Polyatomic Ions

32. What information about chemical bonding can a Lewis formula for a compound or ion give? What information about bonding is not directly represented by a Lewis formula?

33. Write Lewis formulas for the following: H_2; N_2; Cl_2; HCl; HBr.

34. Write Lewis formulas for the following: H_2O; NH_3; OH^-; F^-.

35. Use Lewis formulas to represent the covalent molecules formed by these pairs of elements. Write only structures that satisfy the octet rule. (a) P and H; (b) Se and Br; (c) C and Cl; (d) Si and Cl.

36. Use Lewis formulas to represent the covalent molecules formed by these pairs of elements. Write only structures that satisfy the octet rule. (a) S and Cl; (b) As and F; (c) I and Cl; (d) N and Cl.

37. Find the total number of valence electrons for (a) $SnCl_4$; (b) NH_2^-; (c) CH_5O^+; (d) ClO_3^-; (e) CN_2H_2; (f) SF_4.

38. Write Lewis formulas for the following covalent molecules: (a) H_2S; (b) PCl_3; (c) BCl_3; (d) SiH_4; (e) NOCl.

39. Write Lewis formulas for (a) ClO_4^-; (b) NOF; (c) SeF_4; (d) $COCl_2$; (e) ClF_3; (f) C_2H_6S (two possibilities).

40. Write Lewis structures for (a) H_2O_2; (b) IO_4^-; (c) BeH_2; (d) NCl_3; (e) HClO; (f) XeF_4; (g) C_3H_4 (three possibilities).

*41. (a) Write the Lewis formula for $AlCl_3$, a molecular compound. Note that in $AlCl_3$, the aluminum atom is an exception to the octet rule. (b) In the gaseous phase, two molecules of $AlCl_3$ join together (dimerize) to form Al_2Cl_6. (The two molecules are joined by two "bridging" Al—Cl—Al bonds.) Write the Lewis formula for this molecule.

*42. Write the Lewis formula for molecular ClO_2. There is a single unshared electron on the chlorine atom in this molecule.

43. What do we mean by the term "resonance"? Do the resonance structures that we draw actually represent the bonding in the substance? Explain your answer.

*44. We can write two resonance structures for toluene, $C_6H_5CH_3$:

How would you expect the carbon–carbon bond lengths in the six-membered ring to compare with the carbon–carbon bond length between the CH_3 group and the carbon atom on the ring?

45. Write resonance structures for the formate ion, $HCOO^-$.

46. Write resonance structures for the nitrate ion, NO_3^-.

47. Write Lewis formulas for (a) H_2NOH (i.e., one H bonded to O); (b) S_8 (a ring of eight atoms); (c) SiH_4; (d) F_2O_2 (O atoms in center, F atoms on outside); (e) CO; (f) $SeCl_6$.

48. Write the Lewis formula for each of the following covalent compounds. Which ones contain at least one atom with a share in less than an octet of valence electrons? (a) $BeBr_2$; (b) BBr_3; (c) BCl_3; (d) $AlCl_3$.

49. Which of the following species contain at least one atom that violates the octet rule?

(a) $:\!\ddot{F}\!-\!\ddot{C}l\!:$ (b) $:\!\ddot{O}\!-\!\ddot{C}l\!-\!\ddot{O}\!:$ (c) $:\!\ddot{F}\!-\!Xe\!-\!\ddot{F}\!:$

(d) $\left[:\!\ddot{O}\!-\!\underset{\underset{:\ddot{O}:}{|}}{\overset{\overset{:\ddot{O}:}{|}}{S}}\!-\!\ddot{O}\!: \right]^{2-}$

50. Write the Lewis formula for each of the following molecules or ions. Which ones contain at least one atom with a share in less than an octet of valence electrons? (a) $CHCl_3$; (b) BF_3; (c) BCl_4^-; (d) AlF_4^-.

*51. None of the following is known to exist. What is wrong with each one?

(a) $:\!\ddot{C}l\!-\!\ddot{S}\!=\!\dot{\ddot{O}}$ (c) $:O\!\equiv\!N\!-\!\ddot{O}\!:^-$

(b) $H\!-\!H\!-\!\ddot{O}\!-\!\underset{\underset{:\ddot{C}l:}{|}}{P}\!-\!\ddot{C}l\!:$ (d) $Na\!-\!\underset{\underset{Na}{|}}{\ddot{O}}\!:$

52. Write the Lewis formula for each of the following molecules or ions. Which ones contain at least one atom with a share in more than an octet of valence electrons? (a) $AlCl_4^-$; (b) $AsCl_5$; (c) SF_4; (d) C_2H_6.

*53. "El" is the general symbol for a representative element. In each case, in which periodic group is El located? Justify your answers and cite a specific example for each one.

(a) $\left[:\!\ddot{O}\!-\!\underset{\underset{:\ddot{O}:}{|}}{El}\!-\!\ddot{O}\!: \right]^{-}$ (c) $H\!-\!\ddot{O}\!-\!El\!=\!\dot{\ddot{O}}$

(b) $\left[:\!\ddot{O}\!-\!\underset{\underset{:\ddot{O}:}{|}}{\overset{\overset{:\ddot{O}:}{|}}{El}}\!-\!\ddot{O}\!: \right]^{2-}$ (d) $H\!-\!\ddot{O}\!-\!\underset{\underset{H}{|}}{El}\!-\!H$

54. Write the Lewis formula for each of the following molecules or ions. Which ones contain at least one atom with a share in more than an octet of valence electrons? (a) KCl_3; (b) KrF_2; (c) BrF_5; (d) PF_6^-.

55. Many common stains, such as those of chocolate and other fatty foods, can be removed by dry-cleaning solvents such as tetrachloroethylene, C_2Cl_4. Is C_2Cl_4 ionic or covalent? Write its Lewis formula.

*56. Draw acceptable Lewis formulas for the following common air pollutants: (a) SO_2; (b) NO_2; (c) CO; (d) O_3 (ozone); (e) SO_3; (f) $(NH_4)_2SO_4$. Which one is a solid? Which ones exhibit resonance? Which ones violate the octet rule?

Polluted air over a city.

Formal Charges

57. Assign a formal charge to each atom in the following:

(a) $:\!\ddot{C}l\!-\!\underset{\underset{:\ddot{C}l:}{|}}{\ddot{O}}\!:$ (d) $\left[\ddot{O}\!=\!\underset{\underset{:\ddot{O}:}{|}}{C}\!-\!\ddot{O}\!: \right]^{2-}$

(b) $:\!\ddot{O}\!-\!\ddot{S}\!=\!\dot{\ddot{O}}$

(c) $:\!\ddot{O}\!-\!\underset{\underset{:\ddot{O}:}{|}}{\overset{\overset{:O:}{|}}{\ddot{C}l}}\!-\!\ddot{O}\!-\!\underset{\underset{:\ddot{O}:}{|}}{\ddot{C}l}\!-\!\ddot{O}\!:$ (e) $\left[:\!\ddot{O}\!-\!\underset{\underset{:\ddot{O}:}{|}}{\overset{\overset{:\ddot{O}:}{|}}{\ddot{C}l}}\!-\!\ddot{O}\!: \right]^{-}$

58. Assign a formal charge to each atom in the following:

(a) $:\!\ddot{F}\!-\!\underset{\underset{:\ddot{F}:}{|}}{As}\!-\!\ddot{F}\!:$ (b) $\underset{\ddot{F}}{\overset{:\ddot{F}}{\diagdown}}\underset{\underset{:\ddot{F}:}{|}}{P}\underset{\ddot{F}}{\overset{\ddot{F}:}{\diagup}}$ (c) $\ddot{O}\!=\!C\!=\!\ddot{O}$

(d) $\left[\ddot{O}\!=\!N\!=\!\ddot{O} \right]^{+}$ (e) $\left[:\!\ddot{C}l\!-\!\underset{\underset{:\ddot{C}l:}{|}}{\overset{\overset{:\ddot{C}l:}{|}}{Al}}\!-\!\ddot{C}l\!: \right]^{-}$

*59. With the aid of formal charges, explain which Lewis formula is more likely to be correct for each given molecule.

(a) For Cl_2O, $:\!\ddot{C}l\!-\!\ddot{O}\!-\!\ddot{C}l\!:$ or $:\!\ddot{C}l\!-\!\ddot{C}l\!-\!\ddot{O}\!:$

(b) For HN_3, $H—\overset{..}{N}=N=\overset{..}{N}$ or $H—N\equiv N—\overset{..}{\underset{..}{N}}:$

(c) For N_2O, $\overset{..}{N}=O=\overset{..}{N}$ or $:N\equiv N—\overset{..}{\underset{..}{O}}:$

60. Write Lewis formulas for three different atomic arrangements with the molecular formula HCNO. Indicate all formal charges. Predict which arrangement is likely to be the least stable and justify your selection.

Ionic versus Covalent Character and Bond Polarities

61. Distinguish between polar and nonpolar covalent bonds.
62. Why is an HBr molecule polar while a Br_2 molecule is nonpolar?
63. Why do we show only partial charges, and not full charges, on the atoms of a polar molecule?
64. (a) Which two of the following pairs of elements are most likely to form ionic bonds? Te and H; C and F; Ba and F; N and F; K and O. (b) Of the remaining three pairs, which one forms the least polar, and which the most polar, covalent bond?
65. (a) List three reasonable nonpolar covalent bonds between dissimilar atoms. (b) List three pairs of elements whose compounds should exhibit extreme ionic character.
66. Classify the bonding between the following pairs of atoms as ionic, polar covalent, or nonpolar covalent. (a) Li and O; (b) Br and I; (c) Na and H; (d) O and O; (e) H and O.
67. The following properties can be found in a handbook of chemistry:

> *camphor*, $C_{10}H_{16}O$—colorless crystals; specific gravity 0.990 at 25°C; sublimes 204°C; insoluble in water; very soluble in alcohol and ether.

> *praseodymium chloride*, $PrCl_3$—blue-green needle crystals; specific gravity 4.02; melting point 786°C; boiling point 1700°C; solubility in cold water, 103.9 g/100 mL H_2O; very soluble in hot water.

Would you describe each of these as ionic or covalent? Why?

68. Look up the properties of NaCl and PCl_3 in a handbook of chemistry. Why do we describe NaCl as an ionic compound and PCl_3 as a covalent compound?
69. (a) How many moles of electrons are transferred when 10.0 grams of sodium react as completely as possible with 10.0 g of fluorine to form NaF? (b) How many electrons is this? (c) Look up the charge on the electron in coulombs. What is the total charge, in coulombs, that is transferred?

BUILDING YOUR KNOWLEDGE

70. Write the formula for the compound that forms between (a) calcium and nitrogen, (b) aluminum and oxygen, (c) potassium and selenium, and (d) strontium and chlorine. Classify each compound as covalent or ionic.
*71. Write the Lewis formulas for the nitric acid molecule (HNO_3) that are consistent with the following bond length data: 1.405 Å for the bond between the nitrogen atom and the oxygen atom that is attached to the hydrogen atom; 1.206 Å for the bonds between the nitrogen atom and each of the other oxygen atoms.
72. Write the total ionic and net ionic equations for the reaction between each of the following pairs of compounds in aqueous solution. Then give the Lewis formula for each species in these equations. (a) HCN and NaOH; (b) HCl and NaOH; (c) $CaCl_2$ and Na_2CO_3.
73. Sketch a portion of an aqueous solution of NaCl. Show the Lewis formulas of the solute and solvent species. Suggest the relative location of each species with respect to the others.
74. Sketch a portion of an aqueous solution of CH_3COOH. Show the Lewis formulas of the solute and solvent species. Suggest the relative location of each species with respect to the others.
75. Determine the oxidation number of each element in the following compounds: (a) CO_2; (b) CH_4; (c) PF_3; (d) PF_5; (e) Na_2O; (f) Na_2O_2.

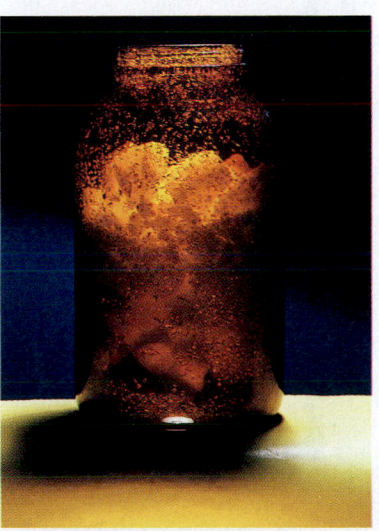

Camphor sublimes readily.

8 Molecular Structure and Covalent Bonding Theories

A model of the structure of diamond.

OBJECTIVES

As you study this chapter, you should learn

- *The basic ideas of the valence shell electron pair repulsion (VSEPR) theory*

- *To use the VSEPR theory to predict electronic geometry of polyatomic molecules and ions*

- *To use the VSEPR theory to predict molecular geometry of polyatomic molecules and ions*

- *The relationships between molecular shapes and molecular polarities*

- *To predict whether a molecule is polar or nonpolar*

- *The basic ideas of the valence bond (VB) theory*

- *To analyze the hybrid orbitals used in bonding in polyatomic molecules and ions*

- *To use hybrid orbitals to describe the bonding in double and triple bonds*

We know a great deal about the molecular structures of many thousands of compounds, all based on reliable experiments. In our discussion of theories of covalent bonding, we must keep in mind that the theories represent *an attempt to explain and organize experimental observations.* For bonding theories to be valid, they must be consis-

275

tent with the large body of experimental observations about molecular structure. In this chapter we shall study two theories of covalent bonding, which usually allow us to predict correct structures and properties. Like any simplified theories, they are not entirely satisfactory in describing *every* known structure; however, their successful application to many thousands of structures justifies their continued use.

8-1 AN OVERVIEW OF THE CHAPTER

The electrons in the outer shell, or **valence shell,** of an atom are the electrons involved in bonding. In most of our discussion of covalent bonding, we will focus attention on these electrons. Valence shell electrons are those that were not present in the *preceding* noble gas, ignoring *filled* sets of *d* and *f* orbitals. Lewis formulas show the number of valence shell electrons in a polyatomic molecule or ion (Sections 7-4 through 7-7). We shall write Lewis formulas for each molecule or polyatomic ion we discuss. The theories introduced in this chapter apply equally well to polyatomic molecules and to ions.

Two theories go hand in hand in a discussion of covalent bonding. The *valence shell electron pair repulsion (VSEPR) theory* helps us to understand and predict the spatial arrangement of atoms in a polyatomic molecule or ion. However, it does not explain *how* bonding occurs, just *where* it occurs, as well as where unshared pairs of valence shell electrons are directed. The *valence bond (VB) theory* describes *how* the bonding takes place, in terms of *overlapping atomic orbitals.* In this theory, the atomic orbitals discussed in Chapter 5 are often "mixed," or *hybridized,* to form new orbitals with different spatial orientations. Used together, these two simple ideas enable us to understand the bonding, molecular shapes, and properties of a wide variety of polyatomic molecules and ions.

We shall first discuss the basic ideas and application of these two theories. Then we shall learn how an important molecular property, *polarity,* depends on molecular shape. Most of this chapter will then be devoted to studying how these ideas are applied to various types of polyatomic molecules and ions.

> In Chapter 7 we used valence bond terminology to discuss the bonding in H_2 although we did not name the theory there.

Important Note

Different instructors prefer to cover these two theories in different ways. Your instructor will tell you the order in which you should study the material in this chapter. However you study this chapter, Tables 8-1, 8-2, 8-3, and 8-4 are important summaries, and you should refer to them often.

A. One approach is to discuss both the VSEPR theory and the VB theory together, emphasizing how they complement one another. If your instructor prefers this parallel approach, you should study the chapter in the order in which it is presented.

B. An alternative approach is to first master the VSEPR theory and the related topic of molecular polarity for different structures, and then learn how the VB theory describes the overlap of bonding orbitals in these structures. If your instructor takes this approach, you should study this chapter in the following order:
   1. Read the summary material under the main heading "Molecular Shapes and Bonding" preceding Section 8-5.
   2. *VSEPR theory, molecular polarity.* Study Sections 8-2 and 8-3; then in Sections 8-5 through 8-12, study only the subsections marked A and B.
   3. *VB theory.* Study Section 8-4; then in Sections 8-5 through 8-12, study the valence bond subsections, marked C; then study Sections 8-13 and 8-14.

No matter which order your instructor prefers, the following procedure will help you analyze the structure and bonding in any compound.

1. Write the Lewis formula for the molecule or polyatomic ion, and identify a *central atom*—an atom that is bonded to more than one other atom (Section 8-2).

2. Count the *number of regions of high electron density* on the central atom (Section 8-2).

3. Apply the VSEPR theory to determine the arrangement of the *regions of high electron density* (the *electronic geometry*) about the central atom (Section 8-2; Tables 8-1 and 8-4).

4. Using the Lewis formula as a guide, determine the arrangement of the *bonded atoms* (the *molecular geometry*) about the central atom, as well as the location of the unshared valence electron pairs on that atom (parts B of Sections 8-5 through 8-12; Tables 8-3 and 8-4). This description includes predicted bond angles.

5. If there are unshared (lone) pairs of electrons on the central atom, consider how their presence might modify somewhat the *ideal* molecular geometry and bond angles deduced in Step 4 (Section 8-2; parts B of Sections 8-8 through 8-12).

6. Use the VB theory to determine the *hybrid orbitals* utilized by the central atom; describe the overlap of these orbitals to form bonds; describe the orbitals that contain unshared valence shell electron pairs on the central atom (parts C of Sections 8-5 through 8-12; Sections 8-13; 8-14; Tables 8-2 and 8-4).

7. If more than one atom can be identified as a central atom, repeat Steps 2 through 6 for each central atom, to build up a picture of the geometry and bonding in the entire molecule.

8. When all central atoms have been accounted for, use the entire molecular geometry, electronegativity differences, and the presence of lone pairs of valence shell electrons on the central atom to predict *molecular polarity* (Section 8-3; parts B of Sections 8-5 through 8-12).

Never skip to Step 5 until you have done Step 4. The electronic geometry and the molecular geometry may or may not be the same; knowing the electronic geometry first will enable you to find the correct molecular geometry.

Many chemists use the terms *lone pair* and *unshared pair* interchangeably, as we shall do throughout this discussion.

The following diagram summarizes this procedure.

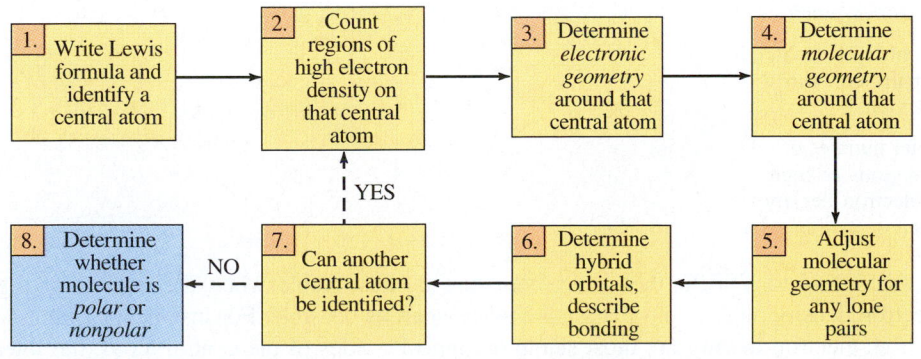

Learn this procedure and use it as a mental "checklist." Trying to do this reasoning in a different order often leads to confusion or wrong answers.

In Section 7-4 we showed that Lewis formulas of polyatomic ions can be constructed in the same way as those of neutral molecules. Once the Lewis formula of an ion is known, we use the VSEPR and VB theories to deduce its electronic geometry, shape, and hybridization, just as for neutral molecules.

Recall that we must take into account the "extra" electrons on anions and the "missing" electrons of cations.

8-2 VALENCE SHELL ELECTRON PAIR REPULSION (VSEPR) THEORY

The basic ideas of the **valence shell electron pair repulsion (VSEPR) theory** are:

> Each set of valence shell electrons on a central atom is significant. The sets of valence shell electrons on the *central atom* repel one another. They are arranged about the *central atom* so that repulsions among them are as small as possible.

This results in maximum separation of the regions of high electron density about the central atom.

A **central atom** is any atom that is bonded to more than one other atom. In some molecules, more than one central atom may be present. In such cases, we determine the arrangement around each in turn, to build up a picture of the overall shape of the entire molecule or ion. We first count the number of **regions of high electron density** around the *central atom,* as follows:

> 1. Each bonded atom is counted as *one* region of high electron density, *whether the bonding is single, double, or triple.*
>
> 2. Each unshared pair of valence electrons on the central atom is counted as *one* region of high electron density.

Consider the following molecules and polyatomic ions as examples.

Formula:	CH_4	NH_3	CO_2	SO_4^{2-}
Lewis dot formula:	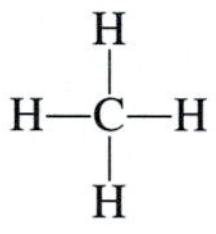 H—C—H with H above and H below	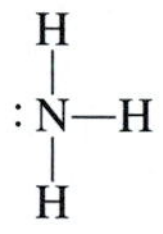 :N—H with H above and H below	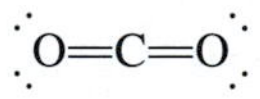 :O=C=O:	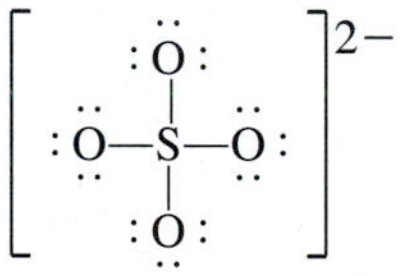 $\begin{bmatrix} :O:\ :O—S—O:\ :O: \end{bmatrix}^{2-}$
Central atom:	C	N	C	S
Number of atoms bonded to *central atom:*	4	3	2	4
Number of unshared pairs on *central atom:*	0	1	0	0
Total number of regions of high electron density on *central atom:*	4	4	2	4

According to VSEPR theory, the structure is most stable when the regions of high electron density on the central atom are as far apart as possible. For instance, two regions of high electron density are most stable on opposite sides of the central atom (the linear arrangement). Three regions are most stable when they are arranged at the corners of an equilateral triangle (the trigonal planar arrangement). The arrangement of these *regions of high electron density* around the central atom is referred to as the **electronic geometry** of the central atom.

CHEMISTRY IN USE

Research & Technology

The War Against the Super Germs

Just a few years ago, medical practitioners thought that their rapidly expanding list of antibiotics would eventually conquer most disease-spreading microbes. That confidence has disappeared with the emergence of some recent epidemics. Nearly every known disease organism is now resistant to at least one antibiotic, and many diseases are immune to several antibiotics. Scourges such as AIDS appear to defy all treatment. In addition, once-conquered microorganisms such as tuberculosis bacteria are making a comeback as new antibiotic-resistant strains called *super germs*.

Over the last few years, newspapers have carried many stories about super germs. For example, in 1993, super germs struck a 1200-student high school in California. One third of the pupils at this school tested positive for tuberculosis. At least a dozen of these students were infected with strains of a tuberculosis bacterium that do not respond to any of the antibiotics administered.

That same year, an epidemic of whooping cough (pertussis) struck children in Cincinnati. Between 1979 and 1992, only 542 cases of whooping cough were reported in the United States. In 1993 alone, there were 352 reported cases. Even more alarming is the realization that an unusually hardy strain of the pertussis bacterium is emerging, one that is very difficult to treat with standard antibiotics.

The cholera epidemic that killed 50,000 people in Rwandan refugee camps in 1994 involved a strain of the cholera bacterium that cannot be treated with standard antibiotics.

One of medicine's worst nightmares has recently come true in the form of a drug-resistant strain of severe invasive strep A, the so-called flesh-eating bacteria. Physicians worry that as strep A infections increase, their treatment with antibiotics will further increase the drug-resistance of these microbes.

Bacteria become drug-resistant in a variety of ways, one of which is by producing special chemicals called *enzymes* (Section 16.9). Bacteria use enzyme molecules to render inactive the drug molecules (antibiotics) that were once so effective in killing bacteria. The precise way in which the molecular shapes of enzyme molecules match antibiotic molecules allows these enzymes to trigger the deactivation of antibiotic molecules.

Chemists are trying to learn the molecular shapes of the enzyme molecules used by drug-resistant bacteria. Then chemists may be able to design "fighter molecules" with the molecular shapes needed to fit precisely onto these enzyme molecules and

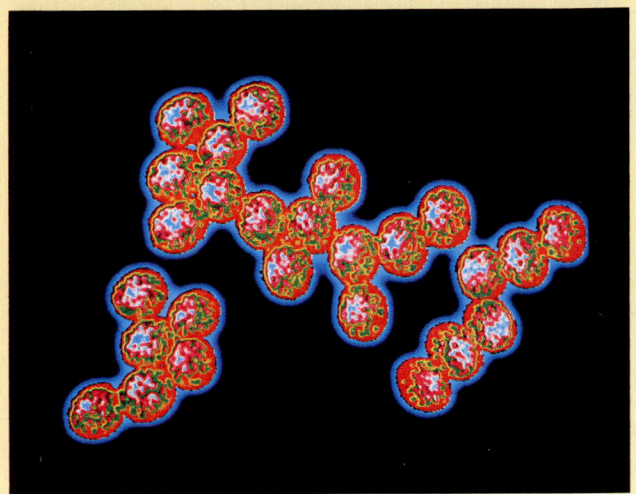

Staphylococcus bacteria (7850×).

chemically deactivate them. Once enzyme molecules are deactivated by fighter molecules, the bacteria would be deprived of a crucial element of their defense. These bacteria would then become susceptible again to the drugs that were lethal to them originally.

Knowing the structure of molecules is critically important to chemists because this knowledge allows chemists to understand how molecules react. Models such as Lewis structures and VSEPR theory are vitally important because they yield information about molecular structure. Research chemists use even more sophisticated computer models to study and develop drugs. Computer software allows chemists to see structures of molecules on computer screens before the chemicals are made. By using models to eliminate chemicals that are unlikely drug prospects, chemists can reduce the time required to develop disease-fighting drugs by many years.

If we didn't have useful theories such as VSEPR and Lewis structures, we would have to memorize the individual properties of the many chemicals we know about. Using VSEPR theory to predict molecular structures, you can mimic research chemists as you predict molecular characteristics such as molecular shape, polarity, hydrogen bonding, and solubility.

Ronald DeLorenzo
Middle Georgia College

Table 8-1 ***Number of Regions of High Electron Density about a Central Atom***

Number of Regions of High Electron Density	Electronic Geometry*		Angles†
2	linear	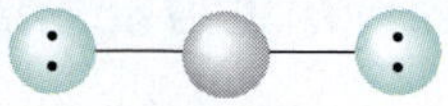	180°
3	trigonal planar	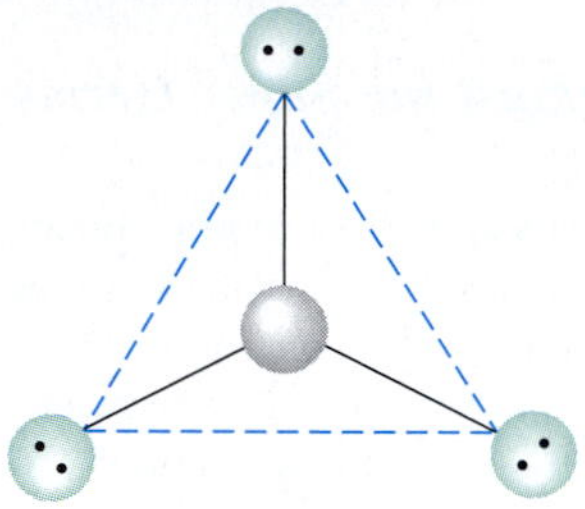	120°
4	tetrahedral	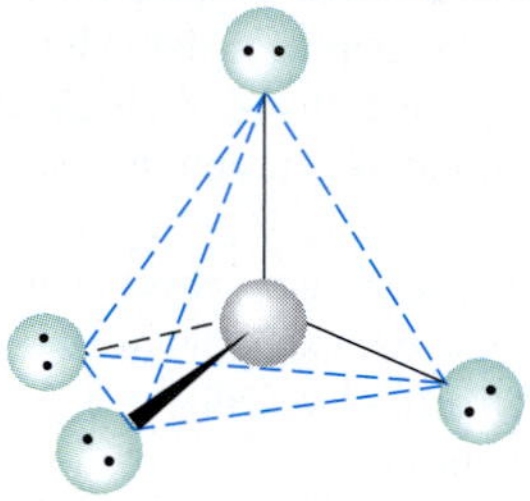	109.5°
5	trigonal bipyramidal	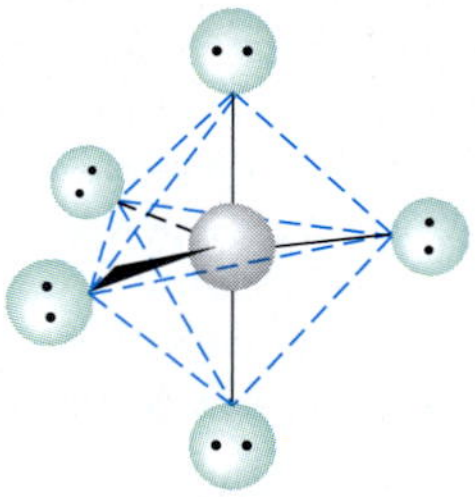	90°, 120°, 180°
6	octahedral	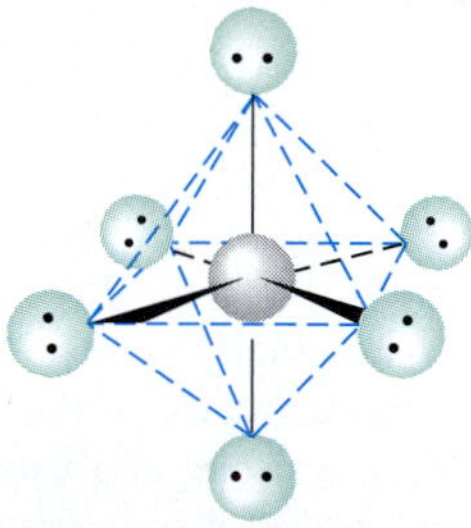	90°, 180°

**Electronic geometries are illustrated here using only single pairs of electrons as regions of high electron density. The symbol (:) represents the regions of high electron density about the central atom ●. By convention, a line in the plane of the drawing is represented by a solid line ——, a line behind this plane is shown as a dashed line ---, and a line in front of this plane is shown as a wedge ◄ with the fat end of the wedge nearest the viewer. Each shape is outlined in blue dashed lines to help you visualize it.*

†Angles made by imaginary lines through the nucleus and the centers of regions of high electron density.

Table 8-1 shows the relationship between the common numbers of regions of high electron density and their corresponding electronic geometries. After we know the electronic geometry (and *only then*), we consider how many of these regions of high electron density connect (bond) the central atom to other atoms. This lets us deduce the arrangement of *atoms* around the central atom, called the **molecular geometry.** If necessary, we repeat this procedure for each central atom in the molecule or ion. These procedures are illustrated in parts B of Sections 8-5 through 8-12.

Although the terminology is not as precise as we might wish, we use "molecular geometry" to describe the arrangement of atoms in polyatomic ions as well as in molecules.

8-3 POLAR MOLECULES—THE INFLUENCE OF MOLECULAR GEOMETRY

In Chapter 7 we saw that the unequal sharing of electrons between two atoms with different electronegativities, $\Delta EN > 0$, results in a *polar bond.* For heteronuclear diatomic molecules such as HF, this bond polarity results in a *polar molecule.* Then the entire molecule acts as a dipole, and we would find that the molecule has a measurable *dipole moment,* i.e., greater than zero.

When a molecule consists of more than two atoms joined by polar bonds, we must also take into account the *arrangement* of the resulting bond dipoles in deciding whether or not a molecule is polar. For such a case, we first use VSEPR theory to deduce the molecular geometry (arrangement of atoms), as described in the preceding section and exemplified in parts A and B of Sections 8-5 through 8-12. Then we determine whether the bond dipoles are arranged in such a way that they cancel (so that the resulting molecule is *nonpolar*) or do not cancel (so that the resulting molecule is *polar*).

In this section we shall discuss the ideas of cancellation of dipoles in general terms, using general atomic symbols X and Y. Then we will apply these ideas to specific molecular geometries and molecular polarities in parts B of Sections 8-5 through 8-12.

Let us consider a heteronuclear triatomic molecule with the formula XY_2 (X is the central atom). Such a molecule must have one of the following two molecular geometries:

$$Y—X—Y \qquad or \qquad Y—X$$
$$\diagdown Y$$

linear   angular

The angular form could have different angles, but either the molecule is linear or it is not. The angular arrangement is sometimes called V-shaped or bent.

Suppose that atom Y has a higher electronegativity than atom X. Then each X—Y bond is polar, with the negative end of the bond dipole pointing toward Y. We can view each bond dipole as an *electronic vector,* with a *magnitude* and a *direction.* In the linear XY_2 arrangement, the two bond dipoles are *equal* in magnitude and *opposite* in direction. Therefore, they cancel to give a nonpolar molecule (dipole moment equal to zero).

$$\overleftarrow{}\!+\!\overrightarrow{}$$
$$Y—X—Y$$

Net dipole = 0
(nonpolar molecule)

In the case of the angular arrangement, the two equal dipoles *would not cancel,* but would add to give a dipole moment greater than zero. The angular molecular arrangement represents a polar molecule.

$$Y—X$$

Net dipole > 0
(polar molecule)

If the electronegativity differences were reversed in this Y—X—Y molecule—that is, if X were more electronegative than Y—the directions of all bond polarities would be reversed. But the bond polarities would still cancel in the linear arrangement, to give a nonpolar molecule. In the angular arrangement, bond polarities would still add to give a polar molecule, but with the net dipole pointing in the opposite direction from that described above.

We can make similar arguments based on addition of bond dipoles for other arrangements. As we shall see in Section 8-8, unshared pairs on the central atom also affect the direction and the magnitude of the net molecular dipole, so the presence of unshared pairs on the central atom must always be taken into account.

> For a molecule to be polar, *two* conditions must be met:
>
> 1. There must be at least one polar bond or one unshared (lone) pair on the central atom.
>   *and*
>
> 2. a. The polar bonds, if there are more than one, must not be symmetrically arranged so that their polarities (bond dipoles) cancel.
>   *or*
>   b. If there are two or more unshared (lone) pairs on the central atom, they must not be symmetrically arranged so that their polarities cancel.

Put another way, if there are no polar bonds or unshared pairs of electrons on the central atom, the molecule *cannot* be polar. Even if polar bonds or unshared pairs are present, they may be arranged so that their polarities cancel one another, resulting in a nonpolar molecule.

For instance, carbon dioxide, CO_2, is a three-atom molecule in which each carbon–oxygen bond is *polar* because of the electronegativity difference between C and O. But the molecule *as a whole* is shown by experiment to be nonpolar. This tells us that the polar bonds are arranged in such a way that the bond polarities cancel. Water, H_2O, on the other hand, is a very polar molecule; this tells us that the H—O bond polarities do not cancel

$$\overset{\longleftarrow\;+\;\longrightarrow}{\ddot{\text{O}}=\text{C}=\ddot{\text{O}}}$$

linear molecule;
bond dipoles cancel;
molecule is nonpolar

$$\text{H}\!\rightleftharpoons\!\ddot{\text{O}}\overset{}{\underset{\searrow}{}}\text{H}$$

angular molecule;
bond dipoles do not cancel;
molecule is polar

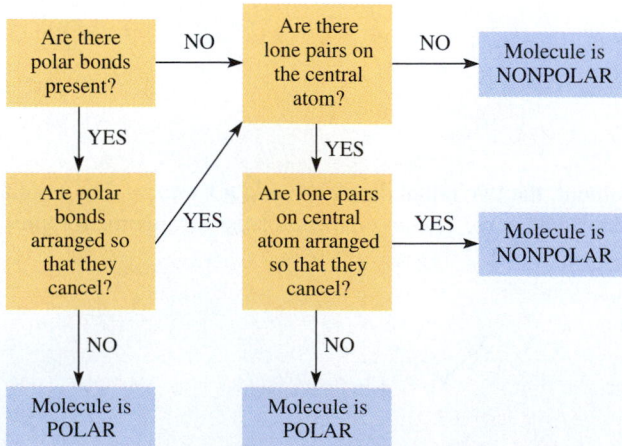

Figure 8-1 A guide to determining whether a polyatomic molecule is polar or nonpolar. Study the more detailed presentation in the text.

one another. Molecular shapes clearly play a crucial role in determining molecular dipole moments. We shall develop a better understanding of molecular shapes in order to understand molecular polarities.

The logic used in deducing whether a molecule is polar or nonpolar is outlined in Figure 8-1. The approach described in this section will be applied to various electronic and molecular geometries in parts B of Sections 8-5 through 8-12.

8-4 VALENCE BOND (VB) THEORY

In Chapter 7 we described covalent bonding as electron pair sharing that results from the overlap of orbitals from two atoms. This is the basic idea of the **valence bond (VB) theory**—it describes *how* bonding occurs. In many examples throughout this chapter, we first use the VSEPR theory to describe the *orientations* of the regions of high electron density. Then we use the VB theory to describe the atomic orbitals that overlap to produce the bonding with that geometry. We also assume that each unshared pair occupies a separate orbital. Thus, the two theories work together to give a fuller description of the bonding.

We learned in Chapter 5 that an isolated atom has its electrons arranged in orbitals in the way that leads to the lowest total energy for the atom. But usually these "pure atomic" orbitals do not have the correct energies or orientations to describe where the electrons are when an atom is bonded to other atoms. When other atoms are nearby as in a molecule or ion, an atom can combine its valence shell orbitals to form a new set of orbitals that is at a lower total energy in the presence of the other atoms than the pure atomic orbitals would be. This process is called **hybridization,** and the new orbitals that are formed are called **hybrid orbitals.** Such hybrid orbitals *usually* give an improved description of the experimentally observed geometry of the molecule or ion.

The designation (label) given to a set of hybridized orbitals reflects the *number and kind* of atomic orbitals that hybridize to produce the set (Table 8-2). Further details about hybridization and hybrid orbitals appear in the following sections. Throughout the text, hybrid orbitals are shaded in green.

VSEPR theory describes the locations of bonded atoms around the central atom, as well as where its lone pairs of valence shell electrons are directed.

We can describe hybridization as the mathematical combination of the waves that represent the orbitals of the atom. This is analogous to the formation of new waves on the surface of water when different waves interact.

MOLECULAR SHAPES AND BONDING

We are now ready to study the structures of some simple molecules. We often refer to generalized chemical formulas in which "A" represents the central atom and "B" represents an atom bonded to A. We first give the known (experimentally determined) facts about polarity and shape. Then we follow the eight steps of analysis outlined in Section 8-1 and write the Lewis formula (part A of each section). Then we explain these facts in terms of the VSEPR and VB theories. The simpler VSEPR theory will be used to explain (or predict) first the *electronic geometry* and then the *molecular geometry* in the molecule (part B). We then show how the molecular polarity of a molecule is a result of bond polarities, unshared pairs, and molecular geometry. Finally, we use the VB theory to describe the bonding in molecules in more detail, usually using hybrid orbitals (part C). As you study each section, refer frequently to the summaries that appear in Table 8-4.

See the "Important Note" in Section 8-1 and consult your instructor for guidance on the order in which you should study Sections 8-5 through 8-12.

8-5 LINEAR ELECTRONIC GEOMETRY—AB$_2$ SPECIES (NO UNSHARED PAIRS OF ELECTRONS ON A)

A. Experimental Facts and Lewis Formulas

Several linear molecules consist of a central atom plus two atoms of another element, abbreviated as AB$_2$. These compounds include gaseous BeCl$_2$, BeBr$_2$, and BeI$_2$, as well as CdX$_2$ and HgX$_2$, where X = Cl, Br, or I. All of these are known to be linear (bond angle = 180°), nonpolar, covalent compounds.

> The high melting point of BeCl$_2$ is due to its polymeric nature in the solid state.

Let's focus on *gaseous* BeCl$_2$ molecules (mp 405°C). The electronic structures of Be and Cl *atoms* in their ground states are

	1s	2s			1s	2s	2p	3s	3p
Be	↑↓	↑↓	and	Cl	↑↓	↑↓	↑↓ ↑↓ ↑↓	↑↓	↑↓ ↑↓ ↑

We wrote the Lewis formula for BeCl$_2$ in Example 7-5. It shows two single covalent bonds, with Be and Cl each contributing one electron to each bond.

In many of its compounds, Be does not satisfy the octet rule (Section 7-7).

$$: \ddot{\text{Cl}} : \text{Be} : \ddot{\text{Cl}} : \qquad \text{or} \qquad : \ddot{\text{Cl}}—\text{Be}—\ddot{\text{Cl}} :$$

B. VSEPR Theory

> VSEPR theory assumes that regions of high electron density (electron pairs) on the central atom will be as far from one another as possible.

Valence shell electron pair repulsion theory places the two electron pairs on Be 180° apart, i.e., with **linear** *electronic geometry.* Both electron pairs are bonding pairs, so VSEPR predicts a linear atomic arrangement, or *linear molecular geometry,* for BeCl$_2$.

If we examine the bond dipoles, we see that the electronegativity difference (see Table 6-3) is large (1.5 units) and each bond is quite polar:

A model of a linear AB$_2$ molecule, e.g., BeCl$_2$.

$$\text{Cl}—\text{Be}—\text{Cl}$$
$$\text{EN} = 3.0 \quad 1.5 \quad 3.0$$
$$\Delta(\text{EN}) = \underbrace{1.5}\ \underbrace{1.5}$$

$$: \ddot{\text{Cl}}—\text{Be}—\ddot{\text{Cl}} :$$
Net dipole = 0

> It is important to distinguish between *nonpolar bonds* and *nonpolar molecules.*

The two bond dipoles are *equal* in magnitude and *opposite* in direction. Therefore, they cancel to give nonpolar molecules.

The difference in electronegativity between Be and Cl is so large that we might expect ionic bonding. However, the radius of Be^{2+} is so small (0.31 Å) and its **charge density** (ratio of charge to size) is so high that most simple beryllium compounds are covalent rather than ionic. The high charge density of Be^{2+} causes it to attract and distort the electron cloud of monatomic anions of all but the most electronegative elements. As a result, the bonds in BeCl$_2$ are polar covalent rather than being ionic. Two exceptions are BeF$_2$ and BeO. They are ionic compounds because they contain very electronegative elements bonded to Be.

> We say that the Be^{2+} ion *polarizes* the anions, Cl$^-$.

C. Valence Bond Theory

Consider the ground state electron configuration of Be. There are two electrons in the 1s orbital, but these nonvalence (inner) electrons are *not* involved in bonding. There are two more electrons *paired* in the 2s orbital. How, then, will two Cl atoms bond to Be? The Be atom must somehow make available one orbital for each bonding Cl electron (the unpaired *p* electrons). The following *ground state* electron configuration for Be is the configuration for an isolated Be atom. Another configuration may be more stable when the Be atom is covalently bonded. Suppose that the Be atom "promoted" one of the paired 2s electrons to one of the 2p orbitals, the next higher energy orbitals.

Be [He] $\underset{2s}{\uparrow\downarrow}$ $\overline{}\,\overline{}\,\overline{}\atop 2p$ $\xrightarrow[\text{promote}]{}$ Be [He] $\underset{2s}{\uparrow}$ $\underset{2p}{\uparrow\;\overline{}\;\overline{}}$

Then there would be two Be orbitals available for bonding. However, this description is still not fully consistent with experimental fact. The Be 2s and 2p orbitals could not overlap a Cl 3p orbital with equal effectiveness; that is, this "promoted pure atomic" arrangement would predict two *nonequivalent* Be—Cl bonds. Yet we observe experimentally that the Be—Cl bonds are *identical* in bond length and bond strength.

For these two orbitals on Be to become equivalent, they must *hybridize* to give two orbitals intermediate between the *s* and *p* orbitals. These are called ***sp* hybrid orbitals.** Consistent with Hund's Rule, the two valence electrons of Be would occupy each of these equivalent hybrid orbitals individually.

Be [He] $\underset{2s}{\uparrow\downarrow}$ $\overline{}\,\overline{}\,\overline{}\atop 2p$ $\xrightarrow[\text{hybridize}]{}$ Be [He] $\underset{sp}{\uparrow\;\uparrow}$ $\overline{}\,\overline{}\atop 2p$

The *sp* hybrid orbitals are described as *linear orbitals,* and we say that Be has *linear electronic geometry.*

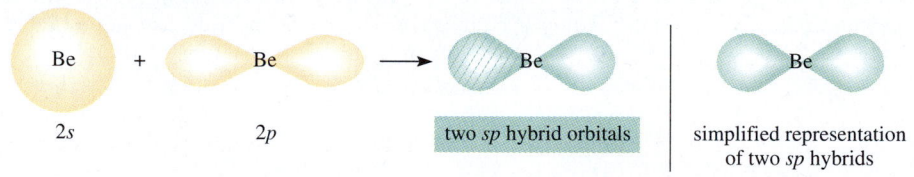

| Be 2s | + | Be 2p | → | Be two *sp* hybrid orbitals | Be simplified representation of two *sp* hybrids |

We can imagine that there is one electron in each of these hybrid orbitals on the Be atom. Recall that each Cl atom has a half-filled 3p orbital that can overlap with a half-filled *sp* hybrid of Be. We picture the bonding in BeCl₂ in the following diagram, in which only the bonding electrons are represented.

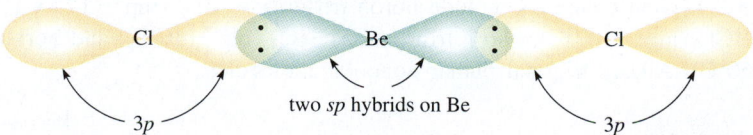

Cl ⋯ Be ⋯ Cl
3p — two *sp* hybrids on Be — 3p

Thus, the Be and two Cl nuclei would lie on a straight line. *This is consistent with the experimental observation that the molecule is linear.*

Cl ground state configuration:

[Ne] $\underset{3s}{\uparrow\downarrow}$ $\underset{3p}{\uparrow\downarrow\;\;\uparrow\downarrow\;\;\uparrow}$

Hund's Rule is discussed in Section 5-17.

As we did for pure atomic orbitals, we often draw hybrid orbitals more slender than they actually are. Such drawings are intended to remind us of the orientations and general shapes of orbitals.

Unshared pairs of e⁻ on Cl atoms are not shown. The hybrid orbitals on the central atom are shown in green in this and subsequent drawings.

Table 8-2 *Relation Between Electronic Geometries and Hybridization*

Regions of High Electron Density	Electronic Geometry	Atomic Orbitals Mixed from Valence Shell of Central Atom	Hybridization
2	linear	one s, one p	sp
3	trigonal planar	one s, two p's	sp^2
4	tetrahedral	one s, three p's	sp^3
5	trigonal bipyramidal	one s, three p's, one d	sp^3d
6	octahedral	one s, three p's, two d's	sp^3d^2

The structures of beryllium bromide, $BeBr_2$, and beryllium iodide, BeI_2, are similar to that of $BeCl_2$. The chlorides, bromides, and iodides of cadmium, CdX_2, and mercury, HgX_2, are also linear, covalent molecules (where X = Cl, Br, or I).

The two X's within one structure are identical.

> sp hybridization occurs at the central atom whenever there are two regions of high electron density around the central atom. AB_2 molecules and ions with no unshared pairs on the central atom have linear electronic geometry, linear molecular geometry, and sp hybridization on the central atom.

▼ **PROBLEM-SOLVING TIP** *Number and Kind of Hybrid Orbitals*

One additional idea about hybridization is worth special emphasis:

The number of hybrid orbitals is always equal to the number of atomic orbitals that hybridize.

Hybrid orbitals are named by indicating the *number and kind* of atomic orbitals hybridized. Hybridization of *one s* orbital and *one p* orbital gives *two sp hybrid orbitals*. We shall see presently that hybridization of *one s* and *two p* orbitals gives *three* sp^2 hybrid orbitals; hybridization of *one s* orbital and *three p* orbitals gives *four* sp^3 hybrids, and so on (Table 8-2).

Hybridization usually involves orbitals from the same main shell (same n).

8-6 TRIGONAL PLANAR ELECTRONIC GEOMETRY—AB₃ SPECIES (NO UNSHARED PAIRS OF ELECTRONS ON A)

A. Experimental Facts and Lewis Formulas

Boron is a Group IIIA element that forms many covalent compounds by bonding to three other atoms. Typical examples include boron trifluoride, BF_3 (mp −127°C); boron trichloride, BCl_3 (mp −107°C); boron tribromide, BBr_3 (mp −46°C); and boron triiodide, BI_3 (mp 50°C). All are trigonal planar nonpolar molecules.

A trigonal planar molecule is a flat molecule in which all three bond angles are 120°.

The solid lines represent bonds between B and F atoms. The dashed blue lines emphasize the shape of the molecule.

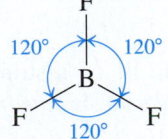

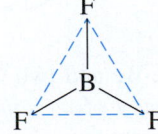

The Lewis formula for BF$_3$ is derived from the following: (a) each B atom has three electrons in its valence shell and (b) each B atom is bonded to three F (or Cl, Br, I) atoms. In Example 7-6 we wrote the Lewis formula for BCl$_3$. Both F and Cl are members of Group VIIA, and so the Lewis formulas for BF$_3$ and BCl$_3$ should be similar.

We see that BF$_3$ and other similar molecules have a central element that does *not* satisfy the octet rule. Boron shares only six electrons.

A model of a trigonal planar AB$_3$ molecule, e.g., BF$_3$.

B. VSEPR Theory

Boron, the central atom, has three regions of high electron density (three bonded atoms, no unshared pairs on B). VSEPR theory predicts **trigonal planar** *electronic geometry* for molecules such as BF$_3$ because this structure gives maximum separation among the three regions of high electron density. There are no unshared pairs of electrons associated with the boron atom, so a fluorine atom is at each corner of the equilateral triangle, and the *molecular geometry* is also trigonal planar. The maximum separation of any three items (electron pairs) around a fourth item (B atom) is at 120° angles in a single plane. All four atoms are in the same plane. The three F atoms are at the corners of an equilateral triangle, with the B atom in the center. The structures of BCl$_3$, BBr$_3$, and BI$_3$ are similar.

Examination of the bond dipoles of BF$_3$ shows that the electronegativity difference (Table 6-3) is very large (2.0 units) and that the bonds are very polar.

The B^{3+} ion is so small (radius = 0.20 Å) that boron does not form simple ionic compounds.

$$
\begin{array}{c}
\text{B--F} \\
\text{EN} \;=\; \underbrace{2.0 \quad 4.0} \\
\Delta(\text{EN}) = \quad 2.0
\end{array}
$$

Net molecular dipole = 0

However, the three bond dipoles are symmetrical, so they cancel to give nonpolar molecules.

C. Valence Bond Theory

To be consistent with experimental findings and the predictions of VSEPR theory, the VB theory must explain three *equivalent* B—F bonds. Again we use the idea of hybridization. Now the 2s orbital and two of the 2p orbitals of B hybridize to form a set of three degenerate *sp*^{2} **hybrid orbitals.**

"Degenerate" refers to orbitals of the same energy.

B [He]   2p $\xrightarrow{\text{hybridize}}$ B [He] $\uparrow\ \uparrow\ \uparrow$ sp^2 $\overline{\quad}$ 2p

2s

Three *sp*^{2} hybrid orbitals point toward the corners of an equilateral triangle:

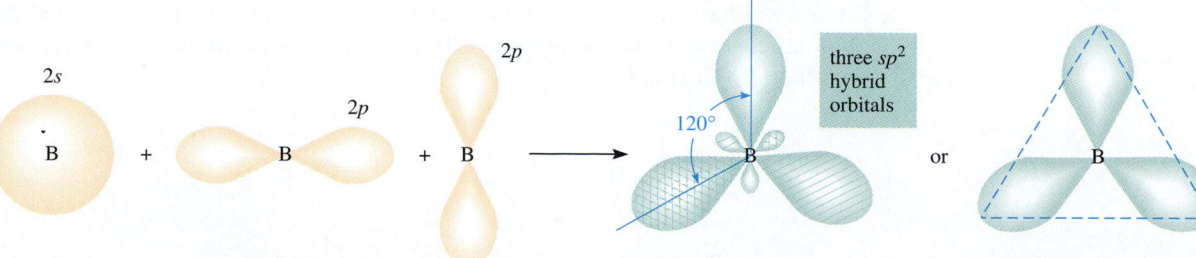

We can imagine that there is one electron in each of these hybrid orbitals. Each of the three F atoms has a $2p$ orbital with one unpaired electron. The $2p$ orbitals can overlap the three sp^2 hybrid orbitals on B. Three electron pairs are shared among one B and three F atoms:

Unshared pairs of e^- are not shown for the F atoms.

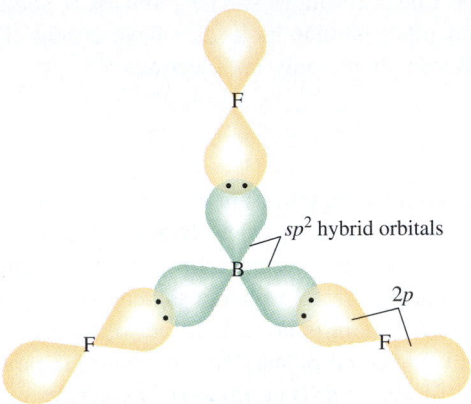

sp^2 hybridization occurs at the central atom whenever there are three regions of high electron density around the central atom. AB_3 molecules and ions with no unshared pairs on the central atom have trigonal planar electronic geometry, trigonal planar molecular geometry, and sp^2 hybridization on the central atom.

A molecule that has an incomplete valence shell on the central atom frequently reacts by accepting a share in an electron pair from another species. A substance that behaves in this way is called a **Lewis acid** (Section 10-10). Both beryllium chloride, $BeCl_2$, and boron trichloride, BCl_3, can react as Lewis acids. The fact that both compounds so readily take a share of additional pairs of electrons tells us that Be and B atoms do not have octets of electrons in these gaseous compounds.

8-7 TETRAHEDRAL ELECTRONIC GEOMETRY—AB_4 SPECIES (NO UNSHARED PAIRS OF ELECTRONS ON A)

A. Experimental Facts and Lewis Formulas

Each Group IVA element has four electrons in its highest occupied energy level. The Group IVA elements form many covalent compounds by sharing those four electrons with four other atoms. Typical examples include CH_4 (mp $-182°C$), CF_4 (mp $-184°C$), CCl_4 (mp $-23°C$), SiH_4 (mp $-185°C$), and SiF_4 (mp $-90°C$). All are tetrahedral, nonpolar molecules (bond angles $= 109.5°$). In each, the Group IVA atom is located in the center of a regular tetrahedron. The other four atoms are located at the four corners of the tetrahedron.

The Group IVA atom contributes four electrons in a tetrahedral AB_4 molecule, and the other four atoms contribute one electron each. The Lewis formulas for methane, CH_4, and carbon tetrafluoride, CF_4, are typical.

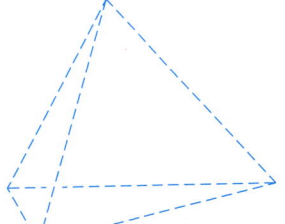

The names of many solid figures are based on the numbers of plane faces they have. A *regular* tetrahedron is a three-dimensional figure with four equal-sized equilateral triangular faces (the prefix *tetra-* means "four").

CH_4, methane CF_4, carbon tetrafluoride

Ammonium ion, NH_4^+, and sulfate ion, SO_4^{2-}, are familiar examples of polyatomic ions of this type. In each of these ions, the central atom is located at the center of a regular tetrahedron with the other atoms at the corners (H—N—H and O—S—O bond angles = 109.5°).

NH_4^+, ammonium ion

SO_4^{2-}, sulfate ion

B. VSEPR Theory

VSEPR theory predicts that four valence shell electron pairs are directed toward the corners of a regular tetrahedron. That shape gives the maximum separation for four electron pairs around one atom. Thus, VSEPR theory predicts **tetrahedral** *electronic geometry* for an AB₄ molecule that has no unshared electrons on A. Because there are no unshared pairs of electrons on the central atom, another atom is at each corner of the tetrahedron. VSEPR theory predicts a *tetrahedral molecular geometry* of each of these molecules.

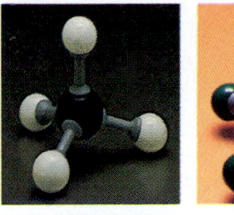

Models of two tetrahedral AB₄ molecules: CH_4 (left) and CF_4 (right).

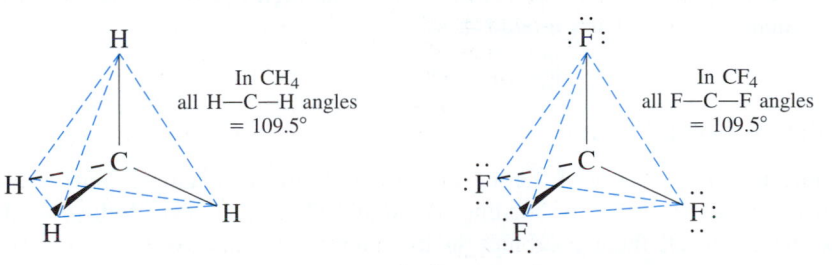

In CH₄ all H—C—H angles = 109.5°

In CF₄ all F—C—F angles = 109.5°

> When a molecule or polyatomic ion has no unshared pairs of valence electrons on the central atom, the *electronic geometry* and the *molecular geometry* are the same.

Examination of bond dipoles shows that in CH₄ the individual bonds are only slightly polar, whereas in CF₄ the bonds are quite polar. In CH₄ the bond dipoles are directed toward carbon, but in CF₄ they are directed away from carbon. Both molecules are very symmetrical, so the bond dipoles cancel, and both molecules are nonpolar. This is true for all AB₄ molecules in which there are *no unshared electron pairs on the central element* and all four B atoms are identical.

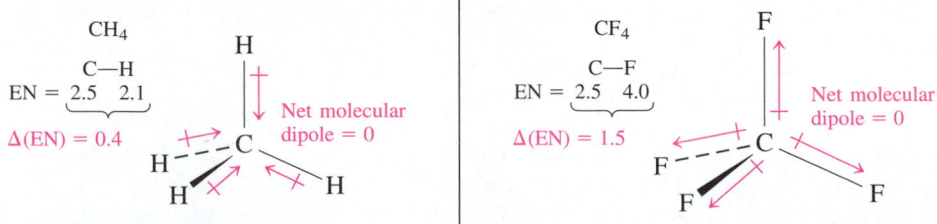

In some tetrahedral molecules, the atoms bonded to the central atom are not all the same. Such molecules are usually polar; the degree of polarity depends on the relative sizes of the bond dipoles present. In CH_3F or CH_2F_2, for example, the addition of unequal dipoles makes the molecule polar.

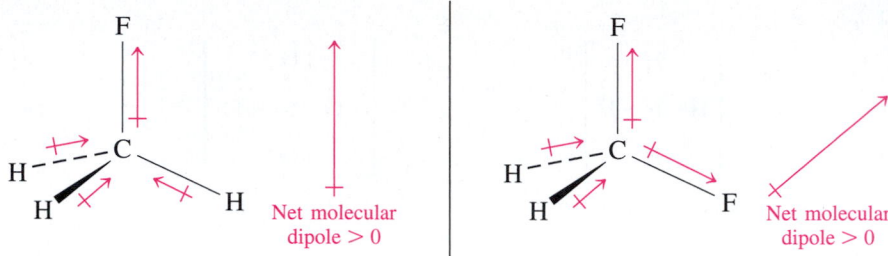

VSEPR theory also predicts that NH_4^+ and SO_4^{2-} ions have tetrahedral electronic geometry. Each region of high electron density bonds the central atom to another atom (H in NH_4^+, O in SO_4^{2-}) at the corner of the tetrahedral arrangement. We describe the molecular geometry of each of these ions as tetrahedral.

You may wonder whether square planar AB_4 molecules exist. They do, in compounds of some of the transition metals. All *simple* square planar AB_4 molecules have unshared electron pairs on A. The bond angles in square planar molecules are only 90°. Most AB_4 molecules are tetrahedral, however, with larger bond angles (109.5°) and greater separation of valence electron pairs around A.

C. Valence Bond Theory

According to VB theory, each Group IVA atom (C in our example) must make four equivalent orbitals available for bonding. To do this, C forms four *sp*³ **hybrid orbitals** by mixing the *s* and all three *p* orbitals in its outer shell. This results in four unpaired electrons.

These *sp*³ hybrid orbitals are directed toward the corners of a regular tetrahedron, which has a 109.5° angle from any corner to the center to any other corner.

Each of the four atoms that bond to C has a half-filled atomic orbital; these can overlap the half-filled sp^3 hybrid orbitals, as is illustrated for CH_4 and CF_4.

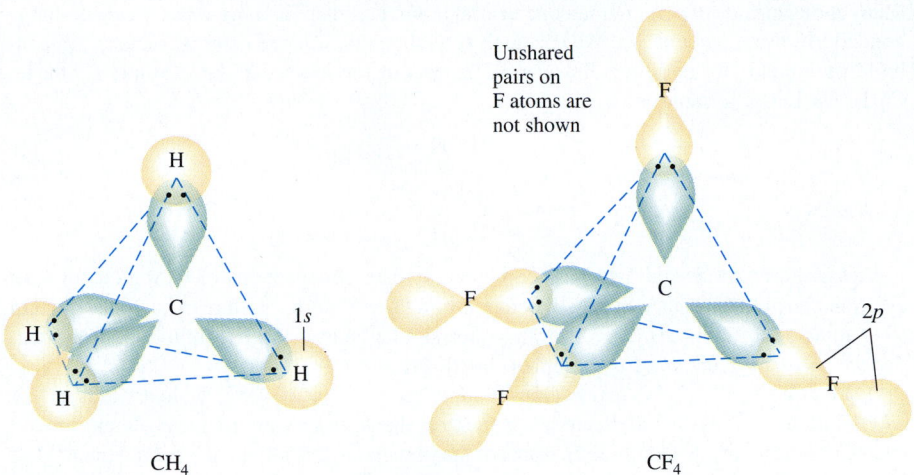

CH$_4$ CF$_4$

We can give the same VB description for the hybridization of the central atoms in polyatomic ions. In NH_4^+ and SO_4^{2-}, the N and S atoms, respectively, form four sp^3 hybrid orbitals directed toward the corners of a regular tetrahedron. Each of these sp^3 hybrid orbitals overlaps with an orbital on a neighboring atom (H in NH_4^+, O in SO_4^{2-}) to form a bond.

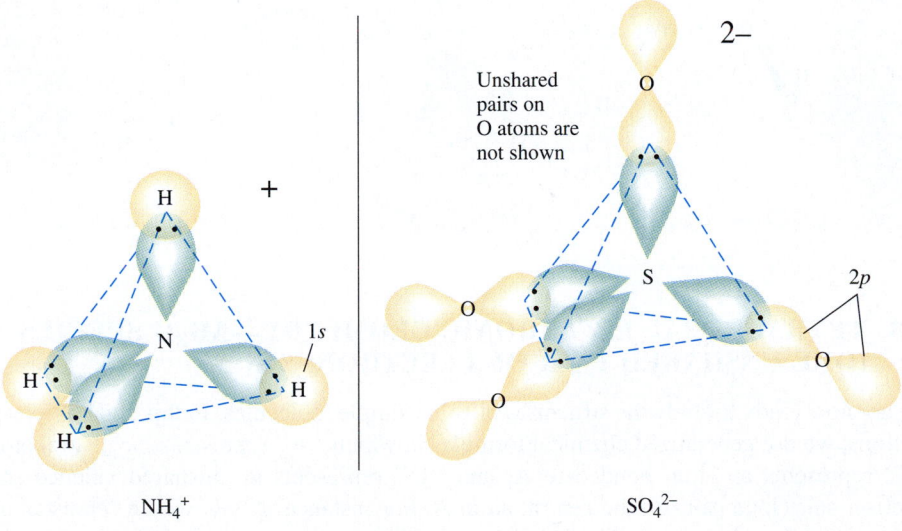

NH$_4^+$ SO$_4^{2-}$

sp^3 hybridization occurs at the central atom whenever there are four regions of high electron density around the central atom. AB$_4$ molecules and ions with no unshared pairs on the central atom have tetrahedral electronic geometry, tetrahedral molecular geometry, and sp^3 hybridization on the central atom.

▼ **PROBLEM-SOLVING TIP**   *Often There Is More Than One Central Atom*

Many molecules contain more than one central atom, i.e., there is more than one atom that is bonded to several other atoms. We can analyze such molecules one central atom at a time, to build up a picture of the three-dimensional aspects of the molecule. An example is ethane, C_2H_6; its Lewis formula is

$$\begin{array}{c}
\quad\;\;\text{H}\;\;\text{H}\\
\quad\;\;|\quad\;\;|\\
\text{H}-\text{C}-\text{C}-\text{H}\\
\quad\;\;|\quad\;\;|\\
\quad\;\;\text{H}\;\;\text{H}
\end{array}$$

Let us first consider the left-hand carbon atom. The arrangements of its regions of high electron density allows us to locate the other C and three H atoms with respect to that C atom (the atoms outlined in red). Then we carry out a similar analysis for the right-hand C atom to deduce the arrangements of *its* neighbors (outlined in blue).

Each C atom in C_2H_6 has four regions of high electron density. VSEPR theory tells us that each C atom has tetrahedral electronic geometry; the resulting atomic arrangement around each C atom has one C and three H atoms at the corners of this tetrahedral arrangement. The VB interpretation is that each C atom is sp^3 hybridized. The C—C bond is formed by overlap of a half-filled sp^3 hybrid orbital of one C atom with a half-filled sp^3 hybrid orbital of the other C atom. Each C—H bond is formed by the overlap of a half-filled sp^3 hybrid orbital on C with the half-filled $1s$ orbital of an H atom.

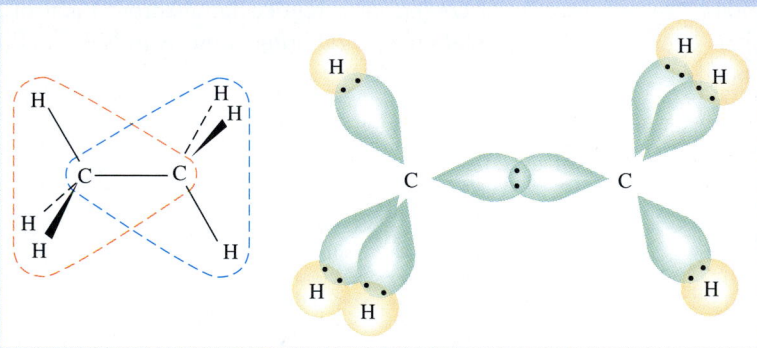

8-8   TETRAHEDRAL ELECTRONIC GEOMETRY—AB₃U SPECIES (ONE UNSHARED PAIR OF ELECTRONS ON A)

We are now ready to study the structures of some simple molecules. In this and subsequent sections, we use generalized chemical formulas in which "A" represents the central atom, "B" represents an atom bonded to A, and "U" represents an unshared valence shell electron pair (lone pair) on the central atom A. For instance, AB_3U would represent any molecule with three B atoms bonded to a central atom A, with one unshared valence pair on A.

A. Experimental Facts and Lewis Formulas

Each Group VA element has five electrons in its valence shell. The Group VA elements form some covalent compounds by sharing three of those electrons with three other atoms.

Some Group VA elements also form covalent compounds by sharing all five valence electrons (Sections 7-7 and 8-11).

Let us describe two examples: ammonia, NH_3, and nitrogen trifluoride, NF_3. Each is a trigonal pyramidal, polar molecule with an unshared pair on the nitrogen atom. Each has a nitrogen atom at the apex and the other three atoms at the corners of the triangular base of the pyramid.

The Lewis dot formulas for NH_3 and NF_3 are

$$\text{H : N : H} \qquad \qquad \text{:F : N : F:}$$
$$\text{H} \qquad \qquad \qquad \text{:F:}$$

$$NH_3 \qquad \qquad \qquad NF_3$$

Sulfite ion, $SO_3{}^{2-}$, is an example of a polyatomic ion of the AB_3U type. It is a trigonal pyramidal ion with an unshared pair on the sulfur atom.

$$\left[\text{ :O—S—O: } \right]^{2-}$$
$$\text{:O:}$$

B. VSEPR Theory

As in Section 8-7, VSEPR theory predicts that the *four* regions of high electron density around a central atom will be directed toward the corners of a tetrahedron, because this gives maximum separation. So N has tetrahedral electronic geometry in each molecule.

Let us reemphasize the distinction between electronic geometry and molecular geometry. *Electronic geometry* refers to the geometric arrangement of *regions of electron density* around the central atom. *Molecular geometry* excludes the unshared pairs on the central atom and refers only to the arrangement of *atoms* (that is, nuclei) around the central atom. For example, CH_4, CF_4, NH_3, and NF_3 all have tetrahedral electronic geometry. But CH_4 and CF_4 have tetrahedral molecular geometry, whereas NH_3 and NF_3 have trigonal pyramidal molecular geometry.

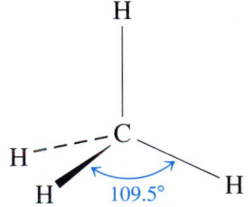

> In molecules or polyatomic ions that contain unshared pairs of valence electrons on the central atom, the *electronic geometry* and the *molecular geometry* cannot be the same.

The term "lone pair" refers to a pair of valence electrons that is associated with only one nucleus. The known geometries of many molecules and polyatomic ions, based on measurements of bond angles, show that *lone pairs of electrons occupy more space than bonding pairs*. A lone pair has only one atom exerting strong attractive forces on it, so it resides closer to the nucleus than do bonding electrons. The relative magnitudes of the repulsive forces between pairs of electrons on an atom are

$$lp/lp \gg lp/bp > bp/bp$$

where *lp* refers to lone (unshared) pairs and *bp* refers to bonding pairs of valence shell electrons. We are most concerned with the repulsions among the electrons in the valence shell of the *central atom* of a molecule or polyatomic ion. The angles at which repulsive forces among valence shell electron pairs are minimized are the angles at which the bonding pairs and unshared pairs (and therefore nuclei) are found in covalently bonded molecules and polyatomic ions. Due to *lp/bp* repulsions in NH_3 and NF_3, their bond angles are *less* than the angles of 109.5° we observed in CH_4 and CF_4 molecules.

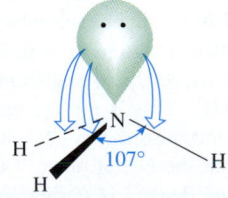

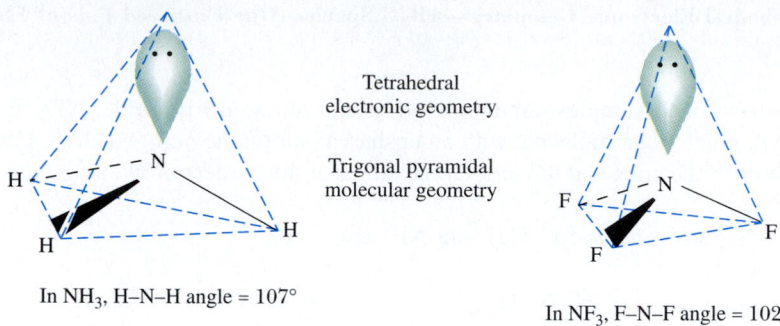

Tetrahedral
electronic geometry

Trigonal pyramidal
molecular geometry

In NH₃, H–N–H angle = 107°

In NF₃, F–N–F angle = 102°

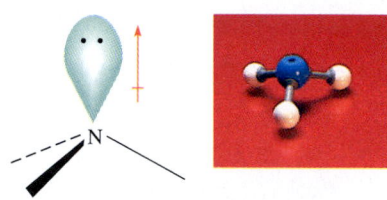

A drawing and a model of a trigonal pyramidal molecule (AB₃U).

The formulas are frequently written as $:NH_3$ and $:NF_3$ to emphasize the unshared pairs of electrons. The unshared pairs must be considered as the polarities of these molecules are examined; they are extremely important in chemical reactions. This is why NH_3 is a base, as we saw in Section 4-2.4 and as we shall discuss more fully in Chapter 10. The contribution of each unshared pair to polarity can be depicted as shown in the margin.

The electronegativity differences in NH_3 and NF_3 are nearly equal, *but* the resulting nearly equal bond polarities are in opposite directions.

$$\begin{array}{ccc} & \text{N—H} & \longleftarrow \\ \text{EN} = & \underbrace{3.0 \quad 2.1} & \text{N—H} \\ \Delta(\text{EN}) = & 0.9 \end{array} \qquad \begin{array}{ccc} & \text{N—F} & \longmapsto \\ \text{EN} = & \underbrace{3.0 \quad 4.0} & \text{N—F} \\ \Delta(\text{EN}) = & 1.0 \end{array}$$

Thus, we have

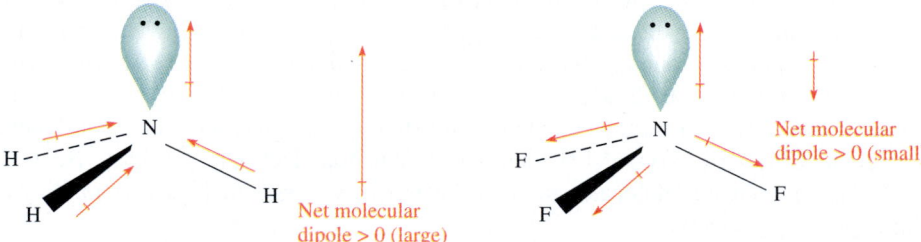

In NH_3 the bond dipoles *reinforce* the effect of the unshared pair, so NH_3 is very polar ($\mu = 1.5$ D). In NF_3 the bond dipoles *oppose* the effect of the unshared pair, so NF_3 is only slightly polar ($\mu = 0.2$ D).

We can now use this information to explain the bond angles observed in NF_3 and NH_3. Because of the direction of the bond dipoles in NH_3, the electron-rich end of each N—H bond is at the central atom, N. On the other hand, the fluorine end of each bond in NF_3 is the electron-rich end. As a result, the lone pair can more closely approach the N in NF_3 than in NH_3. Therefore, in NF_3 the lone pair exerts greater repulsion toward the bonded pairs than in NH_3. In addition, the longer N—F bond length makes the *bp–bp* distance greater in NF_3 than in NH_3, so that the *bp/bp* repulsion in NF_3 is less than that in NH_3. The net effect is that the bond angles are reduced more in NF_3. We can represent this situation as:

We might expect the larger F atoms ($r = 0.64$ Å) to repel each other more strongly than H atoms ($r = 0.37$ Å), leading to larger bond angles in NF_3 than in NH_3. This is not the case, however, because the N—F bond is longer than the N—H bond. The N—F bond density is farther from the N than the N—H bond density.

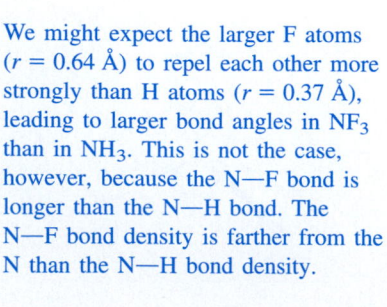

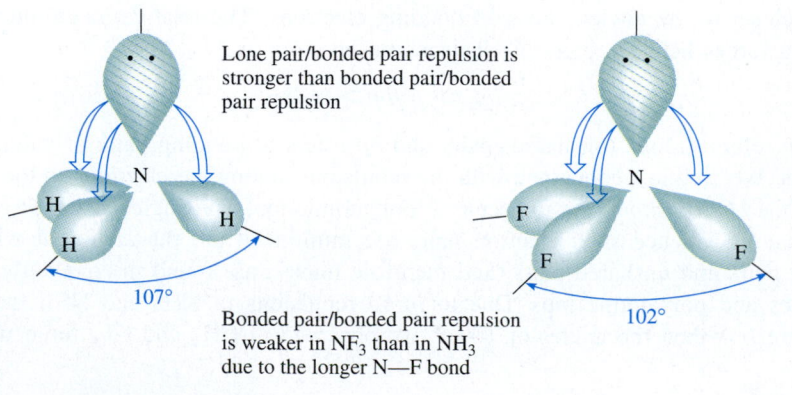

Lone pair/bonded pair repulsion is stronger than bonded pair/bonded pair repulsion

Bonded pair/bonded pair repulsion is weaker in NF_3 than in NH_3 due to the longer N—F bond

107°

102°

With the same kind of reasoning, VSEPR theory predicts that sulfite ion, SO_3^{2-}, has tetrahedral electronic geometry. One of these tetrahedral locations is occupied by the sulfur unshared pair, and oxygen atoms are at the other three locations. The molecular geometry of this ion is trigonal pyramidal, the same as for other AB₃U species.

C. Valence Bond Theory

Experimental results suggest four nearly equivalent orbitals (three involved in bonding, a fourth to accommodate the unshared pair), so we again need four sp^3 hybrid orbitals.

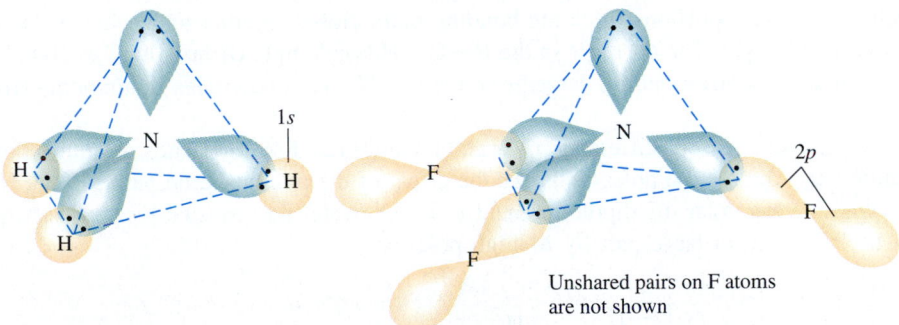

In both NH₃ and NF₃ the unshared pair of electrons occupies one of the sp^3 hybrid orbitals. Each of the other three sp^3 orbitals participates in bonding by sharing electrons with another atom. They overlap with half-filled H $1s$ orbitals and F $2p$ orbitals in NH₃ and NF₃, respectively.

Unshared pairs on F atoms
are not shown

As we shall see in Section 10-10, many compounds that have an unshared electron pair on the central element are called Lewis bases. **Lewis bases** react by making available an electron pair that can be shared by other species.

The sulfur atom in the sulfite ion, SO_3^{2-}, can be described as sp^3 hybridized. One of these hybrid orbitals contains the sulfur lone pair, and the remaining three overlap with oxygen orbitals to form bonds.

> AB₃U molecules and ions, each having four regions of high electron density around the central atom, *usually* have tetrahedral electronic geometry, trigonal pyramidal molecular geometry, and sp^3 hybridization on the central atom.

We must remember that *theory* (and its application) depends on fact, not the other way around. Sometimes the experimental facts are not consistent with the existence of hybrid orbitals. In such cases, we just use the "pure" atomic orbitals rather than hybrid orbitals to describe the bonding. In PH₃ and AsH₃, each H—P—H bond angle is 93.7°, and each H—As—H bond angle is 91.8°. These angles very nearly correspond to three p orbitals at 90° to each other. Thus, there appears to be no need to use hybridization to describe the bonding in these molecules.

A model of H_2O, an angular molecule (AB_2U_2).

8-9 TETRAHEDRAL ELECTRONIC GEOMETRY—AB_2U_2 SPECIES (TWO UNSHARED PAIRS OF ELECTRONS ON A)

A. Experimental Facts and Lewis Formulas

Each Group VIA element has six electrons in its valence shell. The Group VIA elements form many covalent compounds by acquiring a share in two additional electrons from two other atoms. Typical examples are H_2O, H_2S, and Cl_2O. All are angular, polar molecules. The Lewis formulas for these molecules are

$$H_2O \quad H-\overset{\cdot\cdot}{\underset{|}{O}}: \qquad H_2S \quad H-\overset{\cdot\cdot}{\underset{|}{S}}: \qquad Cl_2O \quad :\overset{\cdot\cdot}{\underset{\cdot\cdot}{Cl}}-\overset{\cdot\cdot}{\underset{|}{O}}:$$
$$\qquad\qquad H \qquad\qquad\qquad H \qquad\qquad\qquad\qquad :\overset{}{\underset{\cdot\cdot}{Cl}}:$$

Let us consider the structure of water in detail. The bond angle in water is 104.5°, and the molecule is very polar.

B. VSEPR Theory

VSEPR theory predicts that the four electron pairs around the oxygen atom in H_2O should be 109.5° apart in a tetrahedral arrangement. The observed H—O—H bond angle is 104.5°. The two lone (unshared) pairs strongly repel each other and the bonding pairs of electrons. These repulsions force the bonding pairs closer together and result in the decreased bond angle. The decrease in the H—O—H bond angle (from 109.5° to 104.5°) is greater than the corresponding decrease in the H—N—H bond angles in ammonia (from 109.5° to 107°).

The electronegativity difference is large (1.4 units) and so the bonds are quite polar. Additionally, the bond dipoles *reinforce* the effect of the two unshared pairs, so the H_2O molecule is very polar. Its dipole moment is 1.8 D. Water has unusual properties, which can be explained in large part by its high polarity.

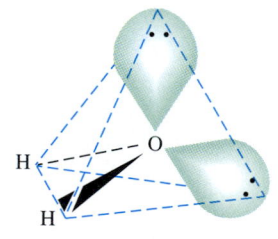

$$\begin{array}{c} O-H \\ EN = \underbrace{3.5 \quad 2.1} \\ \Delta(EN) = \quad 1.4 \end{array}$$

Molecular dipole; includes effect of two unshared electron pairs.

C. Valence Bond Theory

The bond angle in H_2O (104.5°) is closer to the tetrahedral value (109.5°) than to the 90° angle that would result from bonding by pure $2p$ atomic orbitals on O. Therefore valence bond theory postulates four sp^3 hybrid orbitals centered on the O atom: two to participate in bonding and two to hold the two unshared pairs.

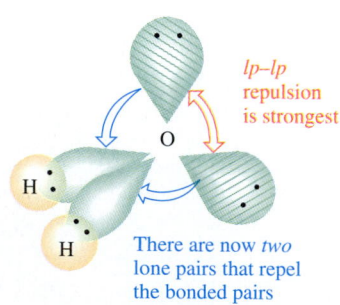

lp–lp repulsion is strongest

There are now *two* lone pairs that repel the bonded pairs

> AB_2U_2 molecules and ions, each having four regions of high electron density around the central atom, *usually* have tetrahedral electronic geometry, angular molecular geometry, and sp^3 hybridization on the central atom.

Hydrogen sulfide, H_2S, is also an angular molecule, but the H—S—H bond angle is 92.2°. This is very close to the 90° angles between two unhybridized $3p$ orbitals of S. Therefore, we *do not* propose hybrid orbitals to describe the bonding in H_2S. The two H atoms are able to exist at approximately right angles to each other when they are bonded to the larger S atom. The bond angles in H$_2$Se and H$_2$Te are 91° and 89.5°, respectively.

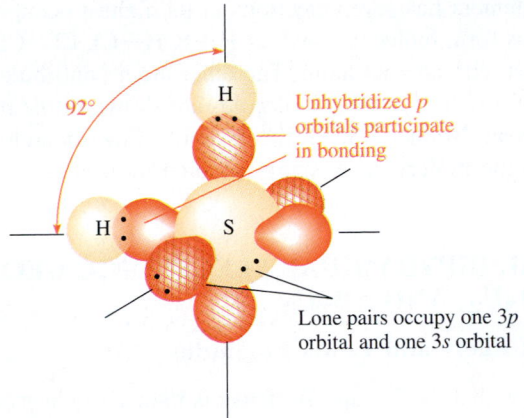

92°

H

Unhybridized *p* orbitals participate in bonding

H   S

Lone pairs occupy one 3*p* orbital and one 3*s* orbital

▼ **PROBLEM-SOLVING TIP**   *Some Molecules Have Two Central Atoms and Unshared Pairs*

Methanol, CH$_3$OH, is a simple molecule that has more than one central atom *and* two unshared pairs. It is the simplest member of a family of organic molecules called *alcohols;* all alcohols contain the atom grouping C—O—H. The Lewis formula for methanol is

$$
\begin{array}{c}
\text{H} \\
| \\
\text{H}\!-\!\text{C}\!-\!\ddot{\text{O}}\!-\!\text{H} \\
| \\
\text{H}
\end{array}
$$

Again, we consider the arrangements around the two central atoms in sequence. The carbon atom (outlined in red) has four regions of high electron density. VSEPR theory tells us that this atom has tetrahedral electronic geometry; the resulting atomic arrangement around the C atom has the O and three H atoms at the corners of this tetrahedral arrangement. The oxygen atom (outlined in blue) has four regions of high electron density, so it, too, has tetrahedral electronic geometry. Thus, the C—O—H arrangement is angular and there are two unshared pairs on O. The VB interpretation is that both atoms are sp^3 hybridized. The C—O bond is formed by overlap of a half-filled sp^3 hybrid orbital on the C atom with a half-filled sp^3 hybrid orbital on the O atom. Each covalent bond to an H is formed by the overlap of a half-filled sp^3 hybrid orbital with the half-filled $1s$ orbital on an H atom. Each unshared pair of electrons on O is in an sp^3 hybrid orbital.

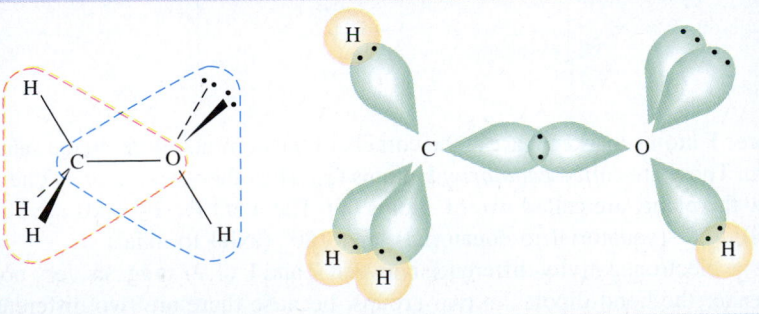

8-10 TETRAHEDRAL ELECTRONIC GEOMETRY—ABU₃ SPECIES (THREE UNSHARED PAIRS OF ELECTRONS ON A)

We represent the halogen as X.

$$HX, \quad H\!:\!\ddot{\underset{\cdot\cdot}{X}}\!: \qquad X_2, \quad :\!\ddot{\underset{\cdot\cdot}{X}}\!:\!\ddot{\underset{\cdot\cdot}{X}}\!:$$

In the latter case, either halogen may be considered the "A" atom of AB.

Each Group VIIA element has seven electrons in its highest occupied energy level. The Group VIIA elements form molecules such as H—F, H—Cl, Cl—Cl, and I—I by sharing one of their electrons with another atom. The other atom contributes one electron to the bonding. Lewis dot formulas for these molecules are shown in the margin. Any diatomic molecule must be linear. Neither VSEPR theory nor VB theory adds anything to what we already know about the molecular geometry of such molecules.

8-11 TRIGONAL BIPYRAMIDAL ELECTRONIC GEOMETRY—AB₅, AB₄U, AB₃U₂, AND AB₂U₃

A. Experimental Facts and Lewis Formulas

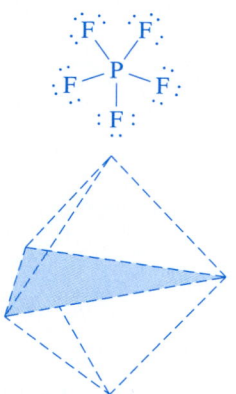

A trigonal bipyramid. The triangular base common to the two pyramids is shaded.

In Section 8-8 we saw that the Group VA elements have five electrons in their outermost occupied shells and form some molecules by sharing only three of these electrons with other atoms (for example, NH_3 and NF_3). Other Group VA elements (P, As, and Sb) form some covalent compounds by sharing all five of their valence electrons with five other atoms (Section 7-7). Phosphorus pentafluoride, PF_5 (mp $-83°C$), is such a compound. Each P atom has five valence electrons to share with five F atoms. The Lewis formula for PF_5 (Example 7-7) is shown in the margin. PF_5 molecules are *trigonal bipyramidal* nonpolar molecules. A **trigonal bipyramid** is a six-sided polyhedron consisting of two pyramids joined at a common triangular (trigonal) base.

A model of a trigonal bipyramidal AB₅ molecule, e.g., PF_5.

B. VSEPR Theory

VSEPR theory predicts that the five regions of high electron density around the phosphorus atom in PF_5 should be as far apart as possible. Maximum separation of five items around a sixth item is achieved when the five items (bonding pairs) are placed at the corners and the sixth item (P atom) is placed in the center of a trigonal bipyramid. This is in agreement with experimental observation.

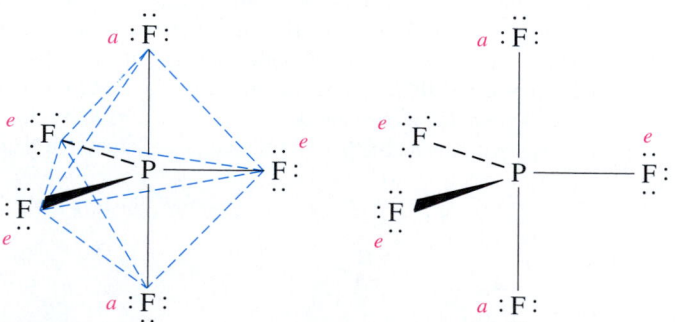

As an exercise in geometry, in how many different ways can five fluorine atoms be arranged *symmetrically* around a phosphorus atom? Compare the hypothetical bond angles in such arrangements with those in a trigonal bipyramidal arrangement.

The three F atoms marked *e* are at the corners of the common base, in the same plane as the P atom. These are called *equatorial* F atoms (*e*). The other two F atoms, one above and one below the plane, are called *axial* F atoms (*a*). The axial P—F bonds are 90° (axial to equatorial), 120° (equatorial to equatorial), and 180° (axial to axial).

The large electronegativity difference between P and F (1.9) suggests very polar bonds. Let's consider the bond dipoles in two groups, because there are two different kinds of P—F bonds in PF_5 molecules, axial and equatorial.

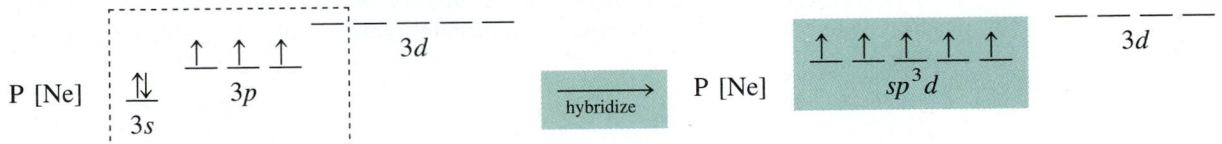

$$P—F$$
$$EN = 2.1 \quad 4.0$$
$$\Delta(EN) = 1.9$$

axial bonds   equatorial bonds

The two axial bond dipoles cancel each other, and the three equatorial bond dipoles cancel, so PF$_5$ molecules are nonpolar.

C. Valence Bond Theory

Because phosphorus is the central element in PF$_5$ molecules, it must have available five half-filled orbitals to form bonds with five F atoms. Hybridization involves one d orbital from the vacant set of 3d orbitals along with the 3s and 3p orbitals of the P atom.

P [Ne]   3s 3p 3d $\xrightarrow{\text{hybridize}}$ P [Ne] sp^3d 3d

The five **sp^3d hybrid orbitals** point toward the corners of a trigonal bipyramid. Each is overlapped by a singly occupied 2p orbital of an F atom. The resulting pairing of P and F electrons forms five covalent bonds.

Imagine that there is one electron in each of the phosphorus hybrid orbitals before it bonds to the five F atoms.

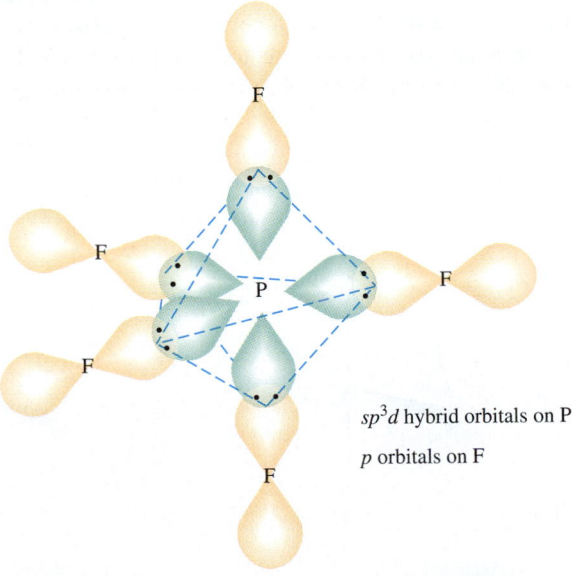

sp^3d hybrid orbitals on P

p orbitals on F

sp^3d hybridization occurs at the central atom whenever there are five regions of high electron density around the central atom. AB$_5$ molecules and ions with no unshared pairs on the central atom have trigonal bipyramidal electronic geometry, trigonal bipyramidal molecular geometry, and sp^3d hybridization on the central atom.

We see that sp^3d hybridization uses an available d orbital in the outermost occupied shell of the central atom, P. The heavier Group VA elements—P, As, and Sb—can form five covalent bonds using this hybridization. But nitrogen, also in Group VA, cannot form five covalent bonds, because the valence shell of N has only one s and three p orbitals (and no d orbitals). The set of s and p orbitals in a given energy level (and therefore any set of hybrids composed only of s and p orbitals) can accommodate a *maximum* of eight electrons and participate in a *maximum* of four covalent bonds. The same is true of all elements of the second period, because they have only s and p orbitals in their valence shells. No atoms in the first and second periods can exhibit expanded valence.

The P atom is said to have an *expanded valence shell* (Section 7-7).

D. Unshared Valence Electron Pairs in Trigonal Bipyramidal Electronic Geometry

Relative magnitudes of repulsive forces:

$$lp/lp \gg lp/bp > bp/bp$$

As we saw in Sections 8-8 and 8-9, lone pairs of electrons occupy more space than bonding pairs, resulting in increased repulsions from lone pairs. What happens when one or more of the five regions of high electron density on the central atom are unshared electron pairs? Let us first consider a molecule such as SF_4, for which the Lewis formula is

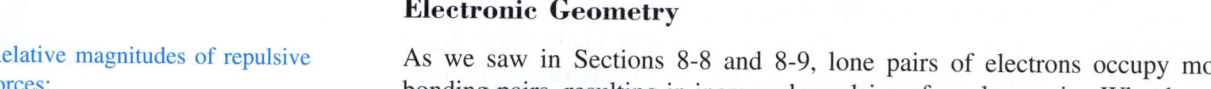

The central atom, S, is bonded to four atoms and has one unshared valence electron pair. This is an example of the general formula AB_4U. Sulfur has five regions of high electron density, so we know that the electronic geometry is trigonal bipyramidal and that the bonding orbitals are sp^3d hybrids. But now a new question arises: Is the unshared (lone) pair more stable in an axial (a) or in an equatorial (e) position? If it were in an axial position, it would be 90° from the *three* closest other pairs (the pairs bonding three F

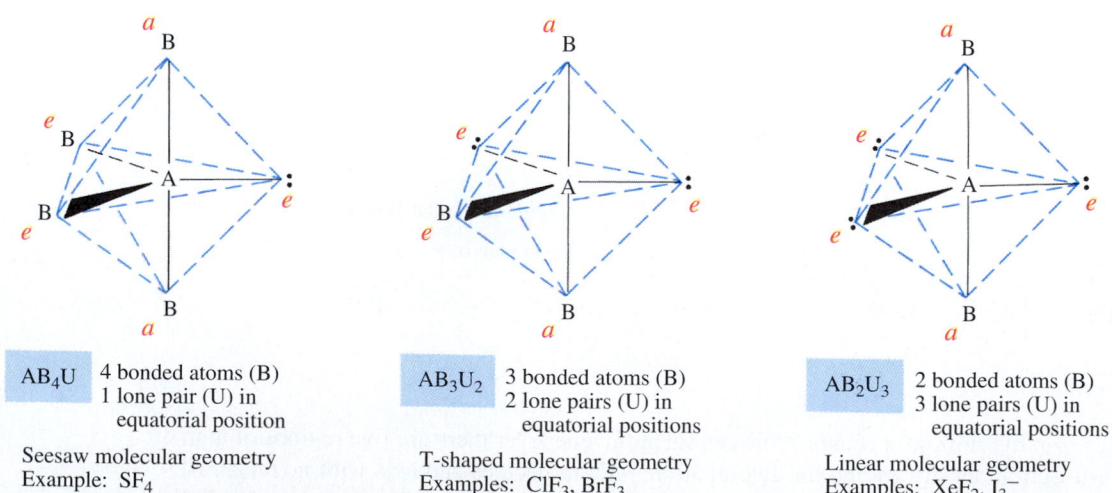

AB_4U	4 bonded atoms (B) 1 lone pair (U) in equatorial position	AB_3U_2	3 bonded atoms (B) 2 lone pairs (U) in equatorial positions	AB_2U_3	2 bonded atoms (B) 3 lone pairs (U) in equatorial positions

Seesaw molecular geometry
Example: SF_4

T-shaped molecular geometry
Examples: ClF_3, BrF_3

Linear molecular geometry
Examples: XeF_2, I_3^-

Figure 8-2 Arrangements of bonded atoms and lone pairs (five regions of high electron density—trigonal bipyramidal electronic geometry).

atoms in equatorial positions) and 180° from the other axial pair. If it were in an equatorial position, only the *two* axial pairs would be at 90°, while the other two equatorial pairs would be less crowded at 120° apart.

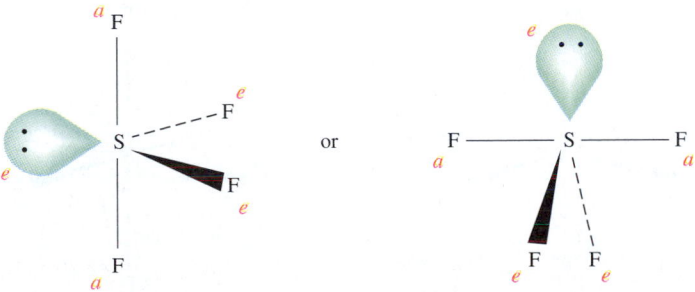

Trigonal bipyramidal electronic geometry

Lone pair in axial position

vs.

Lone pair in equatorial position (preferred arrangement)

We see that the lone pair would be less crowded in an *equatorial* position. The four F atoms then occupy the remaining four positions. We describe the resulting arrangement of *atoms* as a **seesaw arrangement.**

Imagine rotating the arrangement so that the line joining the two axial positions is the board on which the two seesaw riders sit, and the two bonded equatorial positions are the pivot of the seesaw.

As we saw in Sections 8-8 and 8-9, the differing magnitudes of repulsions involving lone pairs and bonding pairs often result in observed bond angles that are slightly different from idealized values. For instance, *bp/lp* repulsion in the seesaw molecule SF_4 causes distortion of the axial S—F bonds away from the lone pair, to an angle of 177°; the two equatorial S—F bonds, ideally at 120°, move much closer together to an angle of 101.6°.

By the same reasoning, we understand why additional lone pairs also take equatorial positions (AB_3U_2 with both lone pairs equatorial or AB_2U_3 with all three lone pairs equatorial). These arrangements are summarized in Figure 8-2.

8-12 OCTAHEDRAL ELECTRONIC GEOMETRY—AB_6, AB_5U, AND AB_4U_2

A. Experimental Facts and Lewis Formulas

The heavier Group VIA elements form some covalent compounds of the AB_6 type by sharing their six valence electrons with six other atoms. Sulfur hexafluoride SF_6 (mp −51°C), an unreactive gas, is an example. Sulfur hexafluoride molecules are non-polar octahedral molecules. The hexafluorophosphate ion, PF_6^-, is an example of a poly-atomic ion of the type AB_6.

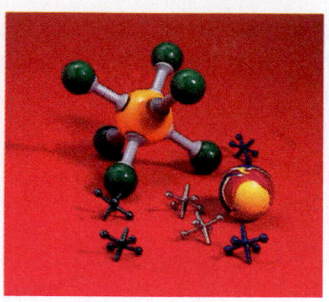

(Left) A model of an octahedral AB_6 molecule, e.g., SF_6. (Right) Some familiar octahedral toys.

B. VSEPR Theory

In an SF_6 molecule we have six valence electron pairs and six F atoms surrounding one S atom. Because there are no lone pairs in the valence shell of sulfur, the electronic and molecular geometries are identical. The maximum separation possible for six electron pairs around one S atom is achieved when the electron pairs are at the corners and the S atom is at the center of a regular octahedron. Thus, VSEPR theory is consistent with the observation that SF_6 molecules are **octahedral.**

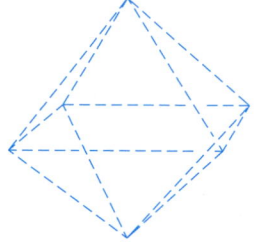

In a regular octahedron, each of the eight faces is an equilateral triangle.

In this octahedral molecule the F—S—F bond angles are 90° and 180°. Each S—F bond is quite polar, but each bond dipole is cancelled by an equal dipole at 180° from it. So the large bond dipoles cancel and the SF_6 molecule is nonpolar.

By similar reasoning, VSEPR theory predicts octahedral electronic geometry and octahedral molecular geometry for the PF_6^- ion, which has six valence electron pairs and six F atoms surrounding one P atom.

C. Valence Bond Theory

Sulfur atoms can use one $3s$, three $3p$, and two $3d$ orbitals to form six hybrid orbitals that accommodate six electron pairs:

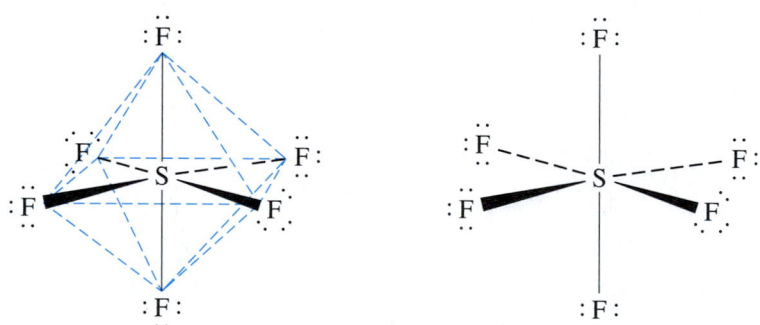

Se and Te, in the same group, form analogous compounds. O cannot do so, for the same reasons as discussed earlier for N (Section 8-11).

The six sp^3d^2 **hybrid orbitals** are directed toward the corners of a regular octahedron. Each sp^3d^2 hybrid orbital is overlapped by a half-filled $2p$ orbital from fluorine, to form a total of six covalent bonds.

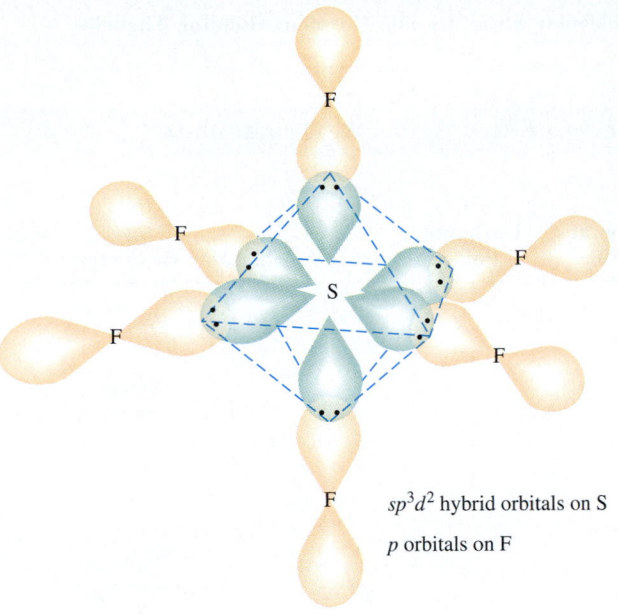

sp^3d^2 hybrid orbitals on S

p orbitals on F

An analogous picture could be drawn for the PF_6^- ion.

sp^3d^2 hybridization occurs at the central atom whenever there are six regions of high electron density around the central atom. AB_6 molecules and ions with no lone pairs on the central atom have octahedral electronic geometry, octahedral molecular geometry, and sp^3d^2 hybridization on the central atom.

D. Unshared Valence Electron Pairs in Octahedral Electronic Geometry

We can reason along the lines of part D in Section 8-10 to predict the preferred locations of unshared electron pairs on the central atom in octahedral electronic geometry. Because of the high symmetry of the octahedral arrangement, all six positions are equivalent, so it does not matter in which position in the drawing we put the first lone pair. AB_5U molecules and ions are described as having **square pyramidal** molecular geometry. When a second lone pair is present, the most stable arrangement has the two lone pairs in two octahedral positions at 180° angles from one another. This leads to a **square planar** molecular geometry for AB_4U_2 species. These arrangements are shown in Figure 8-3. Table 8-3 summarizes a great deal of information—study this table carefully.

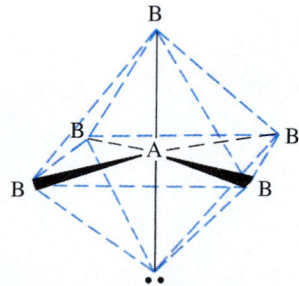

AB_5U 5 bonded atoms (B)
1 lone pair (U)

Square pyramidal molecular geometry
Examples: IF_5, BrF_5

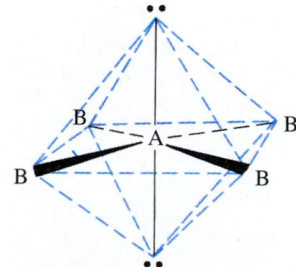

AB_4U_2 4 bonded atoms (B)
2 lone pairs (U)

Square planar molecular geometry
Examples: XeF_4, IF_4^-

Figure 8-3 Arrangements of bonded atoms and lone pairs (six regions of high electron density—octahedral electronic geometry).

Table 8-3 *Molecular Geometry of Species with Unshared Pairs (U) on the Central Atom*

General Formula	Regions of High Electron Density	Electronic Geometry	Hybridization at Central Atom	Unshared Pairs	Molecular Geometry	Examples
AB_2U	3	trigonal planar	sp^2	1	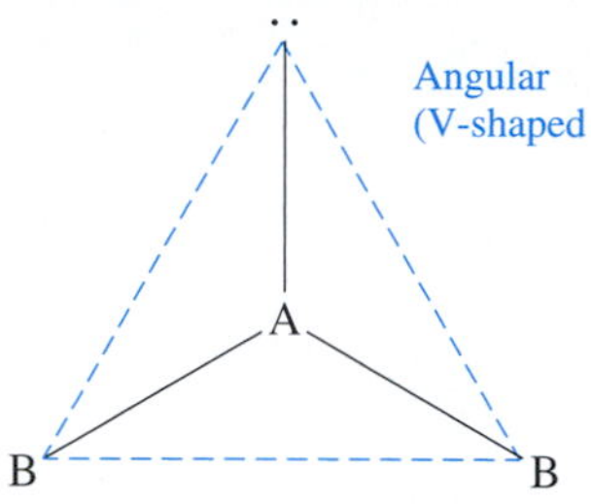 Angular (V-shaped or bent)	O_3, NO_2^-
AB_3U	4	tetrahedral	sp^3	1	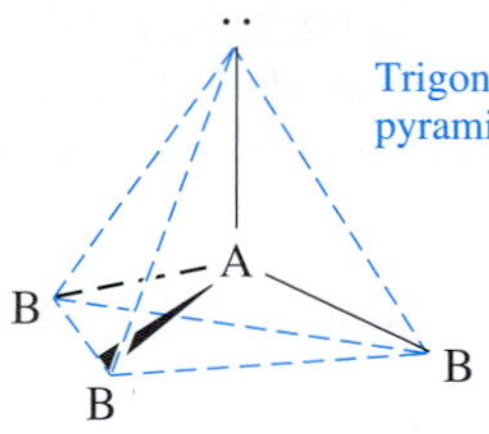 Trigonal pyramidal	NH_3, SO_3^{2-}
AB_2U_2	4	tetrahedral	sp^3	2	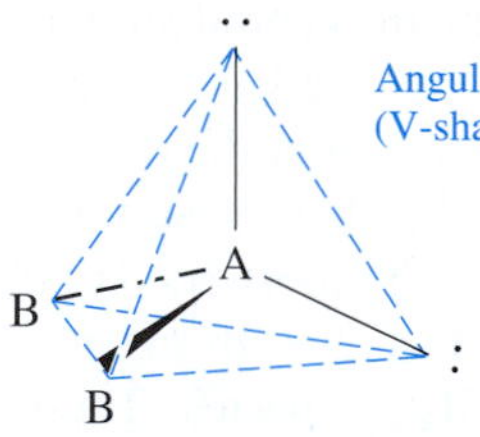 Angular (V-shaped or bent)	H_2O
AB_4U	5	trigonal bipyramidal	sp^3d	1	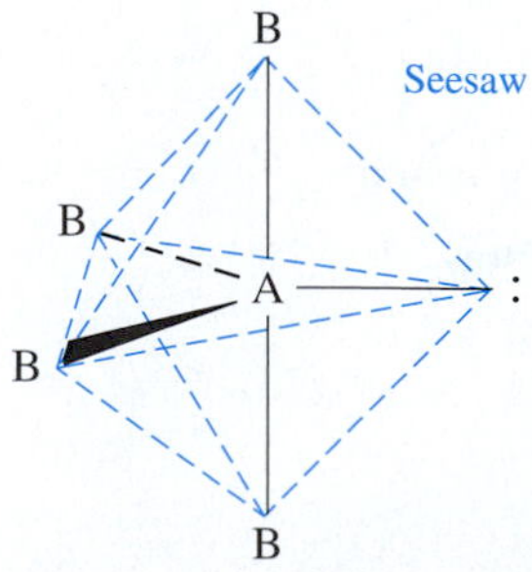 Seesaw	SF_4

Table 8-3 (*continued*)

General Formula	Regions of High Electron Density	Electronic Geometry	Hybridization at Central Atom	Unshared Pairs	Molecular Geometry	Examples
AB_3U_2	5	trigonal bipyramidal	sp^3d	2	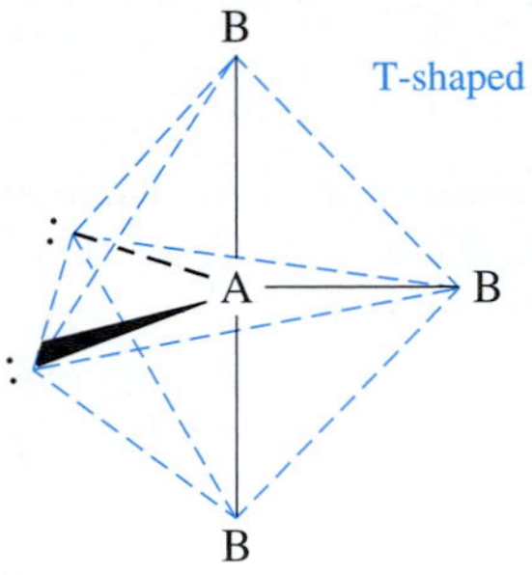 T-shaped	ICl_3, ClF_3
AB_2U_3	5	trigonal bipyramidal	sp^3d	3	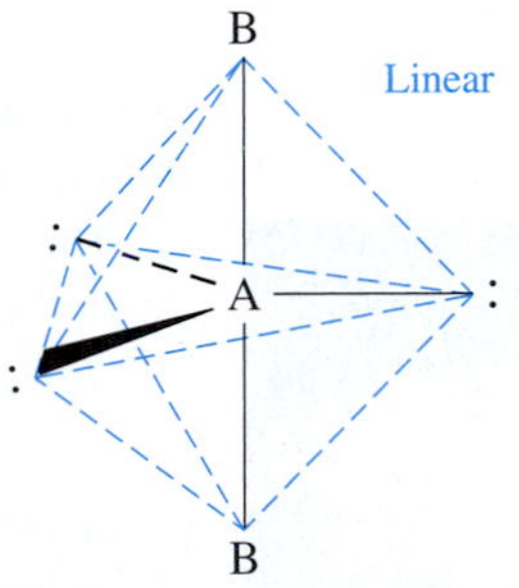 Linear	XeF_2, I_3^-
AB_5U	6	octahedral	sp^3d^2	1	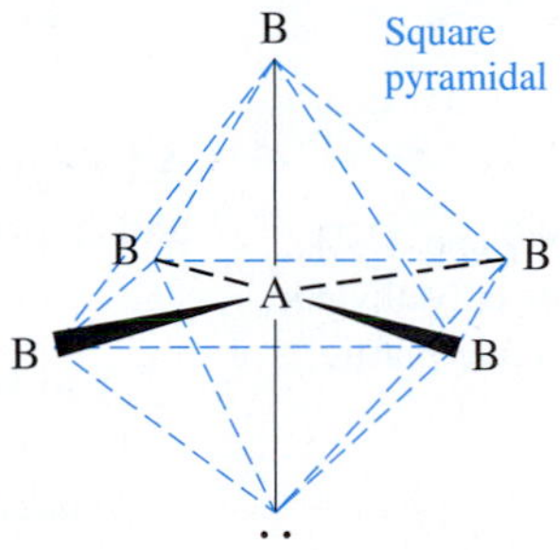 Square pyramidal	IF_5, BrF_5
AB_4U_2	6	octahedral	sp^3d^2	2	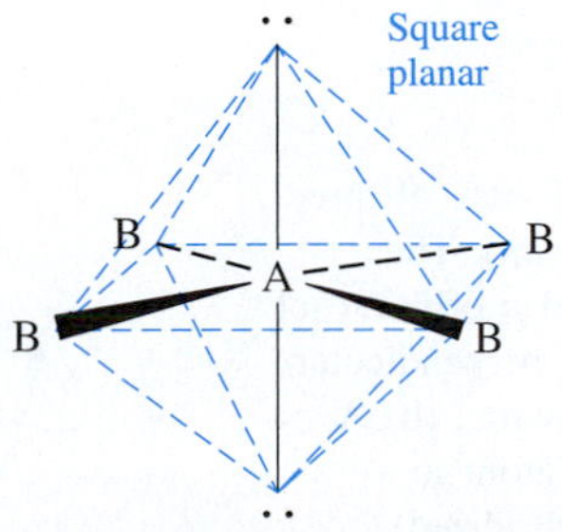 Square planar	XeF_4, IF_4^-

8-13 COMPOUNDS CONTAINING DOUBLE BONDS

In Chapter 7 we constructed Lewis formulas for some molecules and polyatomic ions that contain double and triple bonds. We have not yet considered bonding and shapes for such species. Let us consider ethylene (ethene), C_2H_4, as a specific example. Its dot formula is

$$S = N - A$$
$$= 24 - 12 = \underline{12e^-} \text{ shared}$$

Here each C atom is considered a central atom. Remember that each bonded atom counts as *one* region of high electron density.

There are three regions of high electron density around each C atom. VSEPR theory tells us that each C atom is at the center of a trigonal plane.

Valence bond theory pictures each doubly bonded carbon atom as sp^2 hybridized with one electron in each sp^2 hybrid orbital and one electron in the unhybridized $2p$ orbital. This $2p$ orbital is perpendicular to the plane of the three sp^2 hybrid orbitals:

Relative energies of pure atomic orbitals and hybridized orbitals are not indicated here.

C [He]   ⇅ 2s   ↑ ↑ 2p   →hybridize→   C [He]   ↑ ↑ ↑ sp^2 ↑ $2p_z$

Recall that sp^2 hybrid orbitals are directed toward the corners of an equilateral triangle. Figure 8-4 shows top and side views of these hybrid orbitals.

The two C atoms interact by head-on (end-to-end) overlap of sp^2 hybrids pointing toward each other to form a *sigma* (σ) *bond* and by side-on overlap of the unhybridized $2p$ orbitals to form a *pi* (π) *bond*.

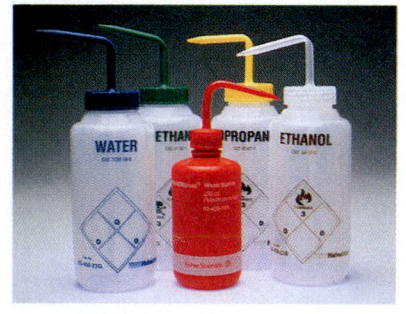

An important use of ethylene, C_2H_4, is in the manufacture of polyethylene, a nonbreakable, nonreactive plastic.

> A **sigma bond** is a bond resulting from head-on overlap of atomic orbitals. *The region of electron sharing is along and cylindrically around the imaginary line connecting the bonded atoms.*

All single bonds are sigma bonds. We have seen that many kinds of pure atomic orbitals and hybridized orbitals can be involved in sigma bond formation.

> A **pi bond** is a bond resulting from side-on overlap of atomic orbitals. *The regions of electron sharing are on opposite sides of the imaginary line connecting the bonded atoms and parallel to this line.*

Figure 8-4 (a) A top view of three sp^2 hybrid orbitals (green). The remaining unhybridized p orbital (not shown in this view) is perpendicular to the plane of the drawing. (b) A side view of a carbon atom in a trigonal planar (sp^2-hybridized) environment, showing the remaining p orbital (tan). This p orbital is perpendicular to the plane of the three sp^2 hybrid orbitals.

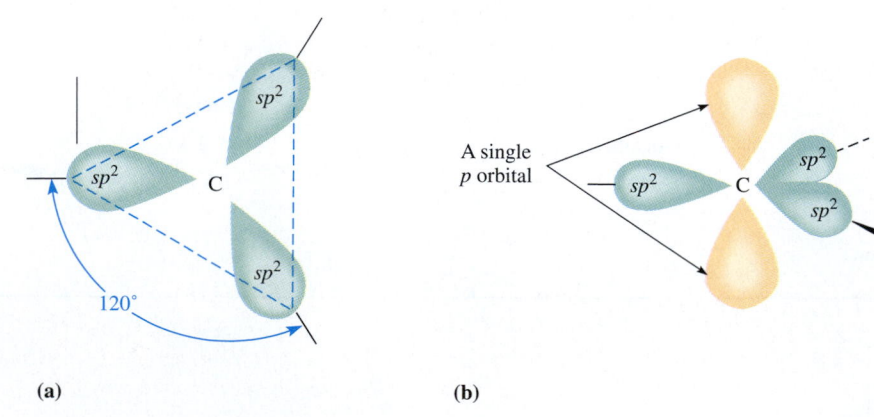

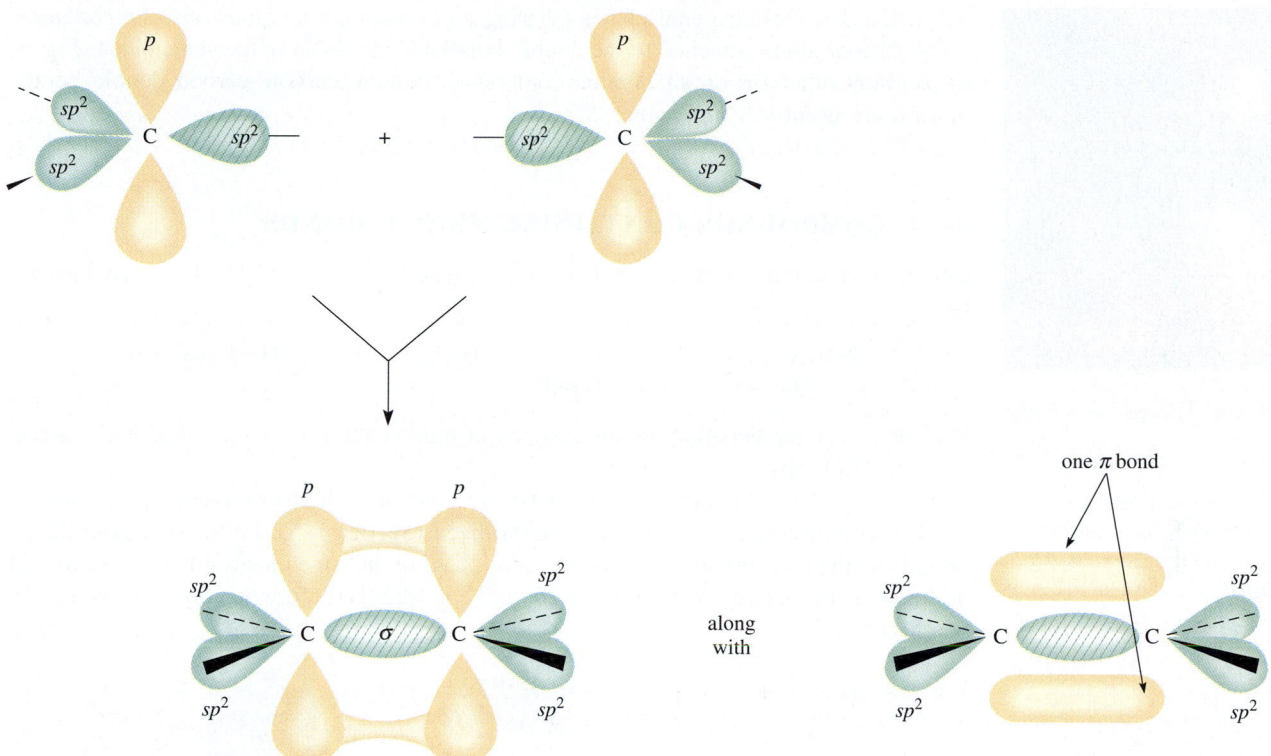

Figure 8-5 A schematic representation of the formation of a carbon–carbon double bond. Two sp^2-hybridized carbon atoms form a sigma (σ) bond by overlap of two sp^2 orbitals (green, hatched) and a pi (π) bond by overlap of properly aligned p orbitals (tan). All orbitals are fatter than shown here.

A pi bond can form *only* if there is *also* a sigma bond between the same two atoms. The sigma and pi bonds together make a double bond (Figure 8-5). The $1s$ orbitals (with one e^- each) of four hydrogen atoms overlap the remaining four sp^2 orbitals (with one e^- each) on the carbon atoms to form four C—H sigma bonds (Figure 8-6).

> A double bond consists of one sigma bond and one pi bond.

As a consequence of the sp^2 hybridization of C atoms in carbon–carbon double bonds, each carbon atom is at the center of a trigonal plane. The p orbitals that overlap to form the π bond must be parallel to each other for effective overlap to occur. This adds the further

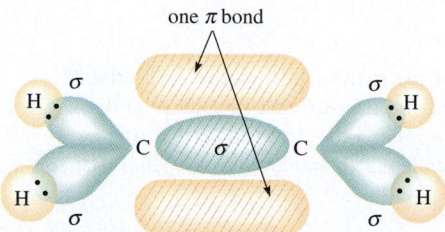

Figure 8-6 Four C—H σ bonds, one C—C σ bond (green, hatched), and one C—C π bond (tan, hatched) in the planar C_2H_4 molecule.

Acetylene (ethyne) is used in welding.

restriction that these trigonal planes (sharing a common corner) must also be *coplanar*. Thus, all four atoms attached to the doubly bonded C atoms lie in the same plane (Figure 8-6). Many other important organic compounds contain carbon–carbon double bonds. Several are described in Chapter 27.

8-14 COMPOUNDS CONTAINING TRIPLE BONDS

One compound that contains a triple bond is ethyne (acetylene), C_2H_2. Its Lewis formula is

$$S = N - A \qquad\qquad H:C:::C:H \qquad H-C\equiv C-H$$
$$= 20 - 10 = \underline{10e^-} \text{ shared}$$

VSEPR theory predicts that the two regions of high electron density around each carbon atom are 180° apart.

Each triple-bonded carbon atom has two regions of high electron density, so valence bond theory postulates that each is *sp*-hybridized (Section 8-5). Let us designate the p_x orbitals as the ones involved in hybridization. Carbon has one electron in each *sp* hybrid orbital and one electron in each of the $2p_y$ and $2p_z$ orbitals (before bonding is considered). See Figure 8-7.

The three *p* orbitals in a set are indistinguishable. We can label the one involved in hybridization as "p_x" to help us visualize the orientations of the two unhybridized *p* orbitals on carbon.

The two carbon atoms form one sigma bond by head-on overlap of the *sp* hybrid orbitals; each C atom also forms a sigma bond with one H atom. The *sp* hybrids on each atom are 180° apart. Thus, the entire molecule must be linear.

The unhybridized atomic $2p_y$ and $2p_z$ orbitals are perpendicular to each other and to the line through the centers of the two *sp* hybrid orbitals (Figure 8-8). The side-on overlap of the $2p_y$ orbitals on the two C atoms forms one pi bond; the side-on overlap of the $2p_z$ orbitals forms another pi bond.

> A triple bond consists of one sigma bond and two pi bonds.

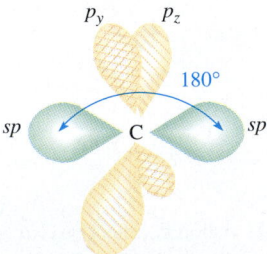

Figure 8-7 Diagram of the two linear hybridized *sp* orbitals (green) of an atom. These lie in a straight line, and the two unhybridized *p* orbitals p_y (tan, cross-hatched) and p_z (tan, hatched) lie in the perpendicular plane and are perpendicular to one another.

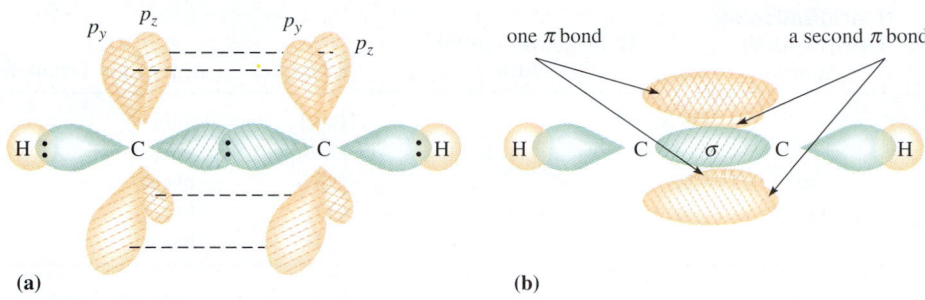

Figure 8-8 The acetylene molecule, C_2H_2. (a) The overlap diagram of two sp-hybridized carbon atoms and two s orbitals from two hydrogen atoms. The hybridized sp orbitals on each C are shown in green and the unhybridized p orbitals are shown in tan. The dashed lines, each connecting two lobes, indicate the side-by-side overlap of the four unhybridized p orbitals to form two π bonds. There are two C—H σ bonds, one C—C σ bond (green hatched), and two C—C π bonds (hatched and cross-hatched). This makes the net carbon–carbon bond a triple bond. (b) The π bonding orbitals (tan) are positioned with one above and below the line of the σ bonds (green) and the other behind and in front of the line of the σ bonds.

Other molecules containing triply bonded atoms are nitrogen, $:N{\equiv}N:$, hydrogen cyanide, $H{-}C{\equiv}N:$, and propyne, $CH_3{-}C{\equiv}C{-}H$. In each case, both atoms involved in the triple bonds are sp hybridized. In the triple bond, each atom participates in one sigma and two pi bonds. The C atom in carbon dioxide, $:O{=}C{=}O:$, must participate in two pi bonds (to two different O atoms). It also participates in two sigma bonds, so it is also sp hybridized and the molecule is linear.

In propyne, the C atom in the CH_3 group is sp^3 hybridized and at the center of a tetrahedral arrangement.

8-15 A SUMMARY OF ELECTRONIC AND MOLECULAR GEOMETRIES

We have discussed several common types of polyatomic molecules and ions, and provided a reasonable explanation for the observed structures and polarities of these species. Table 8-4 provides a summation.

Our discussion of covalent bonding illustrates two important points:

1. Molecules and polyatomic ions have definite shapes.
2. The properties of molecules and polyatomic ions are determined to a great extent by their shapes. Incompletely filled electron shells and unshared pairs of electrons on the central element are very important.

Our ideas about chemical bonding have developed over many years. As experimental techniques for determining the *structures* of molecules have improved, our understanding of chemical bonding has improved also. Experimental observations on molecular geometry support our ideas about chemical bonding. The ultimate test for any theory is this: Can it correctly predict the results of experiments before they are performed? When the answer is *yes*, we have confidence in the theory. When the answer is *no*, the theory must be modified. Current theories of chemical bonding enable us to make predictions that are usually accurate.

Table 8-4 A Summary of Electronic and Molecular Geometries of Polyatomic Molecules and Ions

Regions of High Electron Density[a]	Electronic Geometry	Hybridization at Central Atom (Angles)	Hybridized Orbital Orientation	Examples	Molecular Geometry
2	 linear	sp (180°)		$BeCl_2$ $HgBr_2$ CdI_2 CO_2[b] C_2H_2[c]	linear linear linear linear linear
3	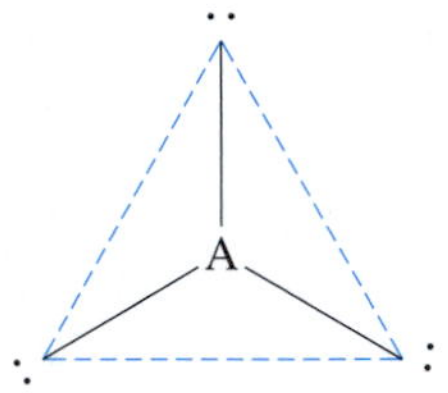 trigonal planar	sp^2 (120°)	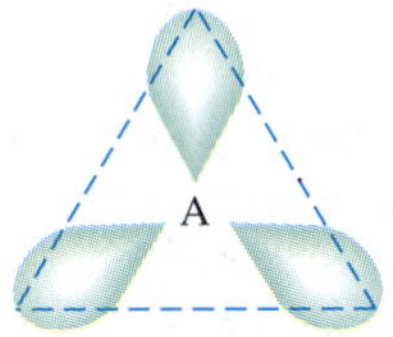	BF_3 BCl_3 $NO_3{}^{-e}$ SO_2[d,e] $NO_2{}^{-d,e}$ C_2H_4[f]	trigonal planar trigonal planar trigonal planar angular (AB_2U) angular (AB_2U) planar (trig. planar at each C)
4	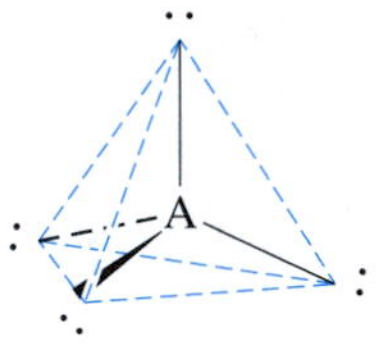 tetrahedral	sp^3 (109.5°)	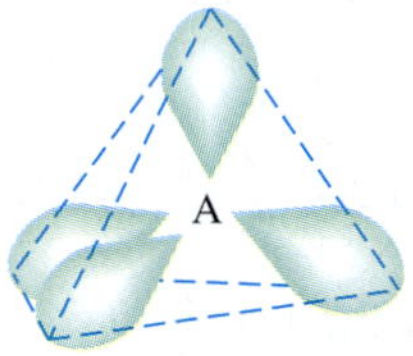	CH_4 CCl_4 $NH_4{}^+$ $SO_4{}^{2-}$ $CHCl_3$ NH_3[d] $SO_3{}^{2-d}$ H_3O^{+d} H_2O[d]	tetrahedral tetrahedral tetrahedral tetrahedral distorted tet. pyramidal (AB_3U) pyramidal (AB_3U) pyramidal (AB_3U) angular (AB_2U_2)
5	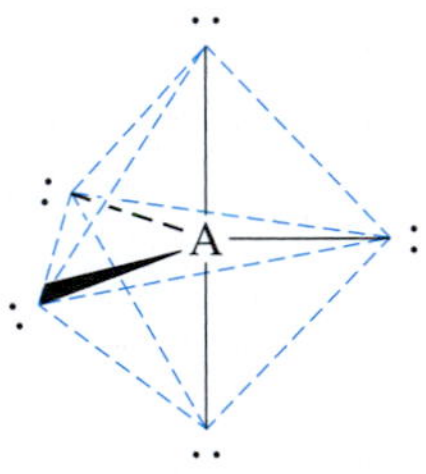 trigonal bipyramidal	sp^3d or dsp^3 (90°, 120°, 180°)	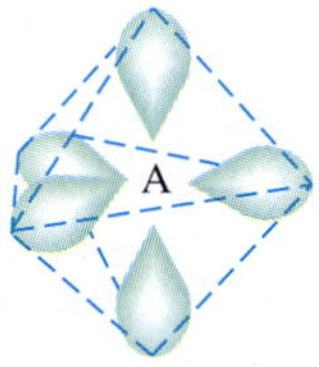	PF_5 $SbCl_5$ SF_4[d] ClF_3[d] XeF_2[d] $I_3{}^{-d}$	trigonal bipyramidal trigonal bipyramidal seesaw (AB_4U) T-shaped (AB_3U_2) linear (AB_2U_3) linear (AB_2U_3)
6	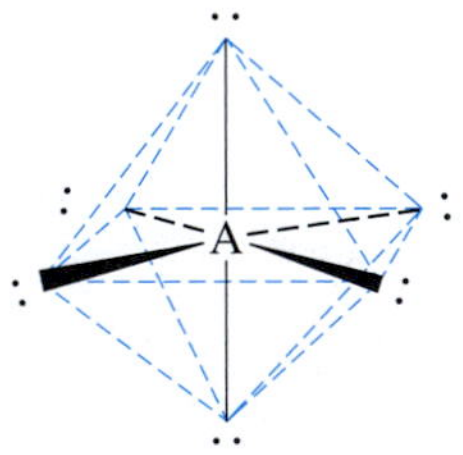 octahedral	sp^3d^2 or d^2sp^3 (90°, 180°)	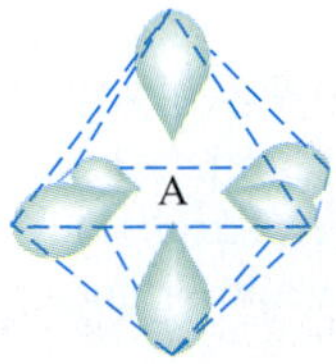	SF_6 SeF_6 $PF_6{}^-$ BrF_5[d] XeF_4[d]	octahedral octahedral octahedral square pyramidal (AB_5U) square planar (AB_4U_2)

[a] The number of locations of high electron density around the central atom. A region of high electron density may be a single bond, a double bond, a triple bond, or an unshared pair. These determine the electronic geometry, and thus the hybridization of the central atom.

[b] Contains two double bonds.

[c] Contains a triple bond.

[d] Central atom in molecule or ion has unshared pair(s) of electrons.

[e] Bonding involves resonance.

[f] Contains one double bond.

Key Terms

Angular A term used to describe the molecular geometry of a molecule that has two atoms bonded to a central atom and one or more unshared pairs on the central atom (AB_2U or AB_2U_2). Also called *V-shaped* or *bent*.

Central atom An atom in a molecule or polyatomic ion that is bonded to more than one other atom.

Electronic geometry The geometric arrangement of orbitals containing the shared and unshared electron pairs surrounding the central atom of a molecule or polyatomic ion.

Hybridization The mixing of a set of atomic orbitals on an atom to form a new set of hybrid orbitals with the same total electron capacity and with properties and energies intermediate between those of the original unhybridized orbitals.

Hybrid orbitals Orbitals formed on an atom by the process of hybridization.

Ionic geometry The arrangement of atoms (not unshared pairs of electrons) about the central atom of a polyatomic ion.

Lewis acid A substance that accepts a share in a pair of electrons from another species.

Lewis base A substance that makes available a share in an electron pair.

Lewis formula A method of representing a molecule or formula unit by showing atoms and only outer shell electrons; does not show shape.

Linear A term used to describe the electronic geometry around a central atom that has two regions of high electron density. Also used to describe the molecular geometry of a molecule or polyatomic ion that has one atom in the center bonded to two atoms on opposite sides (180°) of the central atom (AB_2 or AB_2U_3).

Molecular geometry The arrangement of atoms (*not* unshared pairs of electrons) around a central atom of a molecule or polyatomic ion.

Octahedral A term used to describe the electronic geometry around a central atom that has six regions of high electron density. Also used to describe the molecular geometry of a molecule or polyatomic ion that has one atom in the center bonded to six atoms at the corners of an octahedron (AB_6).

Octahedron A polyhedron with eight equal-sized, equilateral triangular faces and six apices (corners).

Overlap of orbitals The interaction of orbitals on different atoms in the same region of space.

Pi (π) bond A bond resulting from the side-on overlap of atomic orbitals, in which the regions of electron sharing are on opposite sides of and parallel to an imaginary line connecting the bonded atoms.

Seesaw A term used to describe the molecular geometry of a molecule or polyatomic ion that has four atoms bonded to a central atom and one unshared pair on the central atom (AB_4U).

Sigma (σ) bond A bond resulting from the head-on overlap of atomic orbitals, in which the region of electron sharing is along and (cylindrically) symmetrical to the imaginary line connecting the bonded atoms.

Square planar A term used to describe molecules and polyatomic ions that have one atom in the center and four atoms at the corners of a square.

Square pyramidal A term used to describe the molecular geometry of a molecule or polyatomic ion that has five atoms bonded to a central atom and one unshared pair on the central atom (AB_5U).

Tetrahedral A term used to describe the electronic geometry around a central atom that has four regions of high electron density. Also used to describe the molecular geometry of a molecule or polyatomic ion that has one atom in the center bonded to four atoms at the corners of a tetrahedron (AB_4).

Tetrahedron A polyhedron with four equal-sized, equilateral triangular faces and four apices (corners).

Trigonal bipyramid A six-sided polyhedron with five apices (corners), consisting of two pyramids sharing a common triangular base.

Trigonal bipyramidal A term used to describe the electronic geometry around a central atom that has five regions of high electron density. Also used to describe the molecular geometry of a molecule or polyatomic ion that has one atom in the center bonded to five atoms at the corners of a trigonal bipyramid (AB_5).

Trigonal planar A term used to describe the electronic geometry around a central atom that has three regions of high electron density. Also used to describe the molecular geometry of a molecule or polyatomic ion that has one atom in the center bonded to three atoms at the corners of an equilateral triangle (AB_3).

Trigonal pyramidal A term used to describe the molecular geometry of a molecule or polyatomic ion that has three atoms bonded to a central atom and one unshared pair on the central atom (AB_3U).

T-shaped A term used to describe the molecular geometry of a molecule or polyatomic ion that has three atoms bonded to a central atom and two unshared pairs on the central atom (AB_3U_2).

Valence bond (VB) theory Assumes that covalent bonds are formed when atomic orbitals on different atoms overlap and electrons are shared.

Valence shell The outermost occupied electron shell of an atom.

Valence shell electron pair repulsion (VSEPR) theory Assumes that valence electron pairs are arranged around the central element of a molecule or polyatomic ion so that there is maximum separation (and minimum repulsion) among regions of high electron density.

Exercises

VSEPR Theory—General Concepts

1. State in your own words the basic idea of the VSEPR theory.
2. (a) Distinguish between "lone pairs" and "bonding pairs" of electrons. (b) Which has the greater spatial requirement? How do we know this? (c) Indicate the order of increasing repulsions among lone pairs and bonding pairs of electrons.
3. Distinguish between electronic geometry and molecular geometry.
4. Under what conditions is molecular (or ionic) geometry identical to electronic geometry about a central atom?
5. What two shapes can a triatomic species have? How would the electronic geometries for the two shapes differ?
6. How are double and triple bonds treated when VSEPR theory is used to predict molecular geometry? How is a single unshared electron treated?
7. Sketch the three different possible arrangements of the three B atoms around the central atom A for the molecule AB_3U_2. Which of these structures correctly describes the molecular geometry? Why? What are the predicted ideal bond angles? How would observed bond angles deviate from these values?
8. Sketch the three different possible arrangements of the two B atoms around the central atom A for the molecule AB_2U_3. Which of these structures correctly describes the molecular geometry? Why?

Valence Bond Theory—General Concepts

9. What are hybridized atomic orbitals? How is the theory of hybridized orbitals useful?
10. Prepare sketches of the overlaps of the following atomic orbitals: (a) s with s; (b) s with p along the bond axis; (c) p with p along the bond axis (head-on overlap); (d) p with p perpendicular to the bond axis (side-on overlap).
11. Prepare a sketch of the cross-section (through the atomic centers) taken between two atoms that have formed (a) a single σ bond, (b) a double bond consisting of a σ bond and a π bond, and (c) a triple bond consisting of a σ bond and two π bonds.
12. Prepare sketches of the orbitals around atoms that are (a) sp, (b) sp^2, (c) sp^3, (d) sp^3d, and (e) sp^3d^2 hybridized. Show in the sketches any unhybridized p orbitals that might participate in multiple bonding.
13. What hybridization is associated with these electronic geometries: trigonal planar; linear; tetrahedral; octahedral; trigonal bipyramidal?
14. What angles are associated with orbitals in the following sets of hybrid orbitals? (a) sp; (b) sp^2; (c) sp^3; (d) sp^3d; (e) sp^3d^2. Sketch each.
15. What types of hybridization would you predict for molecules having the following general formulas? (a) AB_4; (b) AB_2U_2; (c) AB_3U; (d) ABU_4; (e) ABU_3.
16. Repeat Exercise 15 for (a) ABU_5; (b) AB_2U_4; (c) AB_2; (d) AB_3U_2; (e) AB_5.
17. What are the primary factors upon which we base a decision on whether the bonding in a molecule is better described in terms of simple orbital overlap or overlap involving hybridized atomic orbitals?

Electronic and Molecular Geometry

18. Write a Lewis formula for each of the following species. Indicate the number of regions of high electron density and describe the electronic and molecular geometries. (a) NCl_3; (b) molecular $AlCl_3$; (c) SiH_4; (d) SF_6; (e) IO_4^-.
19. Write a Lewis formula for each of the following species. Indicate the number of regions of high electron density and describe the electronic and molecular or ionic geometries. (a) IF_4^-; (b) CO_2; (c) AlH_4^-; (d) NH_4^+; (e) PCl_3; (f) ClO_3^-.
20. Write a Lewis formula for each of the following species. Indicate the number of regions of high electron density and describe the electronic and molecular or ionic geometries. (a) SeF_6; (b) ONCl (N is the central atom); (c) Cl_2CO; (d) $AsCl_3$; (e) BCl_3; (f) ClO_4^-.
21. Write a Lewis formula for each of the following species. Indicate the number of regions of high electron density and describe the electronic and molecular geometries. (a) molecular $MgCl_2$; (b) ClO_2; (c) XeF_2; (d) CBr_4; (e) $CdCl_2$; (f) AsF_5.
22. Write a Lewis formula for each of the following species. Indicate the number of regions of high electron density and the electronic and molecular or ionic geometries. (a) H_2O; (b) $SnCl_4$; (c) BrF_3; (d) SbF_6^-.
23. Write a Lewis dot formula for each of the following species. Indicate the number of regions of high electron density and the electronic and molecular or ionic geometries. (a) BF_3; (b) NF_3; (c) IO_3^-; (d) $SiCl_4$; (e) SeF_6.
24. (a) What would be the ideal bond angles in the species in Exercise 22, ignoring unshared pair effects? (b) How do these differ, if at all, from the actual values? Why?
25. (a) What would be the ideal bond angles in the molecules or ions in Exercise 23, ignoring unshared pair effects? (b) Are these values greater than, less than, or equal to the actual values? Why?
26. Carbon forms two common oxides, CO and CO_2, both of which are linear. It forms a third (very uncommon) oxide, carbon suboxide, C_3O_2, which is also linear. The structure has terminal oxygen atoms on both ends. Write the Lewis formula for C_3O_2. How many regions of high electron density are there about each of the three carbon atoms?
27. Pick the member of each pair that you would expect to have the smaller bond angles, if different, and explain why. (a) SF_2 and SO_2; (b) BF_3 and BCl_3; (c) CF_4 and SF_4; (d) NF_3 and OF_2.
28. Draw a Lewis formula, sketch the three-dimensional shape, and name the electronic and ionic geometries for the following polyatomic ions. (a) $AsCl_4^-$; (b) PCl_6^-; (c) PCl_4^-; (d) $SbCl_4^+$.
29. As the name implies, the interhalogens are compounds that contain two halogens. Write Lewis formulas and three-dimensional structures for the following. Name the electronic and molecular geometries of each. (a) IF_3; (b) IBr; (c) IF_5.

30. A number of ions derived from the interhalogens are known. Write Lewis formulas and three-dimensional structures for the following ions. Name the electronic and ionic geometries of each. (a) IF_4^+; (b) ICl_4^-; (c) BrF_4^-.

*31. (a) Write a Lewis formula for each of the following molecules: BF_3; NF_3; BrF_3. (b) Contrast the molecular geometries of these three molecules. Account for differences in terms of the VSEPR theory.

*32. (a) Write a Lewis formula for each of the following molecules: GeF_4; SF_4; XeF_4. (b) Contrast the molecular geometries of these three molecules. Account for differences in terms of the VSEPR theory.

33. Write the Lewis formulas and predict the shapes of these very reactive carbon-containing species: H_3C^+ (a carbocation); $H_3C:^-$ (a carbanion); and $:CH_2$ (a carbene whose unshared electrons are paired).

34. Write the Lewis formulas and predict the shapes of (a) ICl_2^-; (b) $TeCl_4$; (c) XeO_3; (d) $NOBr$ (N is the central atom); (e) NO_2Cl (N is the central atom); (f) $SOCl_2$ (S is the central atom).

35. Describe the shapes of these polyatomic ions: (a) BO_3^{3-}; (b) AsO_4^{3-}; (c) SO_3^{2-}; (d) NO_2^-.

36. Describe the shapes of these polyatomic ions: (a) H_3O^+; (b) GeF_3^-; (c) ClF_3^{2-}; (d) $IO_2F_2^-$.

37. Which of the following molecules are polar? Why? (a) CH_4; (b) CH_3Br; (c) CH_2Br_2; (d) $CHBr_3$; (e) CBr_4.

38. Which of the following molecules are polar? Why? (a) CdI_2; (b) BCl_3; (c) NF_3; (d) H_2O; (e) SF_6.

39. Which of the following molecules are nonpolar? Justify your answer. (a) SO_3; (b) IF; (c) Cl_2O; (d) $AsCl_3$; (e) $CHCl_3$.

40. The PF_2Cl_3 molecule is nonpolar. Use this information to sketch its three-dimensional shape. Justify your choice.

*41. In what two major ways does the presence of unshared pairs of valence electrons affect the polarity of a molecule? Describe two molecules for which the presence of unshared pairs on the central atom helps to make the molecules polar. Can you think of a bonding arrangement that has unshared pairs of valence electrons on the central atom but that is nonpolar?

Valence Bond Theory

42. What is the hybridization on the central atom in each of the following? (a) NCl_3; (b) molecular $AlCl_3$; (c) SiH_4; (d) SF_6; (e) IO_4^-.

43. What is the hybridization on the central atom in each of the following? (a) IF_4^-; (b) CO_2; (c) AlH_4^-; (d) NH_4^+; (e) PCl_3; (f) ClO_3^-.

44. What is the hybridization on the central atom in each of the following? (a) SeF_6; (b) $ONCl$ (N is the central atom); (c) Cl_2CO; (d) $AsCl_3$; (e) BCl_3; (f) ClO_4^-.

45. What is the hybridization on the central atom in each of the following? (a) molecular $MgCl_2$; (b) ClO_2; (c) XeF_2; (d) CBr_4; (e) $CdCl_2$; (f) AsF_5.

46. (a) Describe the hybridization of the central atom in each of these covalent species. (1) $CHCl_3$; (2) CH_2Cl_2; (3) NF_3; (4) ClO_4^-; (5) IF_6^+; (6) SiF_6^{2-}. (b) Give the shape of each species.

47. Describe the hybridization of the underlined atoms in $\underline{C}_2F_4$, $\underline{C}_2F_2$, $\underline{N}_2F_4$, and $(H_2\underline{N})_2\underline{C}O$.

48. (a) Describe the hybridization of N in NO_2^+ and NO_2^-. (b) Predict the bond angle in each case.

*49. After comparing experimental and calculated dipole moments, Charles A. Coulson suggested that the Cl atom in HCl is sp hybridized. (a) Give the orbital electronic structure for an sp hybridized Cl atom. (b) Which HCl molecule would have a larger dipole moment—one in which the chlorine uses pure p orbitals for bonding with the H atom or one in which sp hybrid orbitals are used?

*50. Predict the hybridization at each carbon atom in each of the following molecules.

(a) ethanol (ethyl alcohol or grain alcohol)

(b) glycine (an amino acid)

(c) tetracyanoethylene

(d) chloroprene (used to make neoprene, a synthetic rubber)

(e) 3-penten-1-yne

*51. Predict the hybridization at the numbered atoms (①, ②, and so on) in the following molecules and predict the approximate bond angles at those atoms.

(a) diethyl ether, an anesthetic

(b) caffeine, a stimulant in coffee and in many over-the-counter medicinals*

(c) acetylsalicylic acid (aspirin)*

(d) nicotine*

(e) ephedrine, a nasal decongestant*

52. How many sigma and how many pi bonds are there in each of the following molecules?

*In these kinds of structural drawings, each intersection of lines represents a C atom; sometimes H atoms are not shown at all these intersections.

*53. How many sigma and how many pi bonds are there in each of the following molecules?

54. Describe the bonding in the N_2 molecule with a three-dimensional VB structure. Show the orbital overlap and label the orbitals.

55. Write Lewis formulas for molecular oxygen and ozone. Assuming that all of the valence electrons in the oxygen atoms are in hybrid orbitals, what will be the hybridization of the oxygen atoms in each substance? Prepare sketches of the molecules.

*56. A water solution of cadmium bromide, $CdBr_2$, contains Cd^{2+}, Br^-, $CdBr^+$, $CdBr_2$, $CdBr_3^-$, and $CdBr_4^{2-}$ ions. Describe the type of hybrid orbital used by Cd in each polyatomic species and describe the shape of the species.

*57. In their crystalline states, PCl_5 exists as $(PCl_4)^+(PCl_6)^-$, and PBr_5 exists as $PBr_4^+Br^-$. (a) Predict the shapes of all the polyatomic ions. (b) Indicate the hybrid orbital structure for the P atom in each of its different types of ions.

*58. Draw a Lewis formula and a three-dimensional structure for each of the following polycentered molecules. Indicate hybridizations and bond angles at each carbon atom. (a) butane, C_4H_{10}; (b) propene, $H_2C{=}CHCH_3$; (c) 1-butyne, $HC{\equiv}CCH_2CH_3$; (d) acetaldehyde, CH_3CHO.

59. How many σ bonds and how many π bonds are there in each of the molecules of Exercise 58?

60. (a) Describe the hybridization of N in each of these species. (1) $:NH_3$; (2) NH_4^+; (3) $HN{=}NH$; (4) $HC{\equiv}N:$; (5) $H_2N{-}NH_2$.
(b) Give an orbital description for each species, specifying the location of any unshared pairs and the orbitals used for the multiple bonds.

61. Write the Lewis formulas and predict the orbital types and the shapes of these polyatomic ions and covalent molecules: (a) $HgCl_2$; (b) BF_3; (c) BF_4^-; (d) SeF_2; (e) $SbCl_5$; (f) SbF_6^-.

62. (a) What is the hybridization of each C in these molecules? (1) $H_2C{=}O$; (2) $HC{\equiv}N$; (3) $CH_3CH_2CH_3$; (4) ketene, $H_2C{=}C{=}O$. (b) Describe the shape of each molecule.

*63. The following fluorides of xenon have been well characterized: XeF_2, XeF_4, and XeF_6. (a) Write Lewis formulas for these substances and decide what type of hybridization of the Xe atomic orbitals has taken place. (b) Draw all of the possible atomic arrangements of XeF_2 and discuss your choice of molecular geometry. (c) What shape do you predict for XeF_4?

*64. Iodine and fluorine form a series of interhalogen molecules and ions. Among these are IF (minute quantities observed spectroscopically), IF_3, IF_4^-, IF_5, IF_6^-, and IF_7. (a) Write Lewis formulas for each of these species. (b) Identify the type of hybridization that the orbitals of the iodine atom have undergone in each substance. (c) Identify the shape of the molecule or ion.

Mixed Exercises

65. In the pyrophosphate ion, $P_2O_7^{4-}$, one oxygen atom is bonded to both phosphorus atoms. Write a Lewis formula and sketch the three-dimensional shape of the ion. Describe the ionic geometry with respect to the central O atom and with respect to each P atom.

66. Briefly discuss the bond angles in the hydroxylamine molecule in terms of the ideal geometry and the small changes caused by electron pair repulsions.

$$H—\overset{..}{\underset{|}{N}}—\overset{..}{\underset{..}{O}}—H$$
$$\overset{|}{H}$$

67. Repeat Exercise 66 for nitric acid.

$$H—\overset{..}{\underset{..}{O}}—\overset{\overset{..}{O}}{\underset{||}{N}}—\overset{..}{\underset{..}{O}}: \longleftrightarrow H—\overset{..}{\underset{..}{O}}—N=\overset{..}{\underset{..}{O}}:$$

*68. The methyl free radical $\cdot CH_3$ has bond angles of about 120°, whereas the methyl carbanion $:CH_3^-$ has bond angles of about 109°. What can you infer from these facts about the repulsive force exerted by an unpaired, unshared electron as compared to that exerted by an unshared pair of electrons?

*69. Two Lewis structures can be written for the square planar molecule $PtCl_2Br_2$:

$$\begin{array}{ccc} Br & & Cl \\ & \diagdown Pt \diagup & \\ Br & & Cl \end{array} \quad and \quad \begin{array}{ccc} Br & & Cl \\ & \diagdown Pt \diagup & \\ Cl & & Br \end{array}$$

Show how a difference in dipole moments can distinguish between these two possible structures.

70. Prepare a sketch of the molecule $CH_3CH=CH_2$ showing orbital overlaps. Identify the type of hybridization of atomic orbitals for each carbon atom.

*71. The skeleton for the nitrous acid molecule, HNO_2, is shown below. Draw the Lewis formula. What are the hybridizations at the middle O and N atoms?

$$\begin{array}{c} H \\ 0.98\ \text{Å} \diagdown \quad 104° \\ O \longrightarrow N \quad 1.45\ \text{Å} \\ 116° \searrow \\ O \end{array}$$

72. Describe the change in hybridization that occurs at the central atom of the reactant at the left in each of the following reactions.
(a) $PF_5 + F^- \rightarrow PF_6^-$
(b) $2CO + O_2 \rightarrow 2CO_2$

(c) $AlI_3 + I^- \rightarrow AlI_4^-$
(d) What change in hybridization occurs in the following reaction?

$$:NH_3 + BF_3 \longrightarrow H_3N:BF_3$$

*73. Consider the following proposed Lewis formulas for ozone (O_3):

(i) $:\overset{..}{O}—\overset{..}{O}=\overset{..}{O}: \longleftrightarrow :\overset{..}{O}=\overset{..}{O}—\overset{..}{O}:$ (ii) $\begin{array}{c} :\overset{..}{O}: \\ \diagup \diagdown \\ :\overset{..}{O} \quad \overset{..}{\underset{..}{O}}: \end{array}$

(iii) $:\overset{..}{O}—\overset{..}{O}—\overset{..}{O}:$

(a) Which of these correspond to a polar molecule? (b) Which of these predict covalent bonds of equal lengths and strengths? (c) Which of these predict a diamagnetic molecule? (d) The properties listed in parts (a), (b), and (c) are those observed for ozone. Which structure correctly predicts all three? (e) Which of these contain a considerable amount of "strain"? Explain.

74. What hybridizations are predicted for the central atoms in molecules having the formulas AB_2U_2 and AB_3U? What are the predicted bond angles for these molecules? The observed bond angles for representative substances are

H_2O	104.5°	NH_3	106.7°
H_2S	92.2°	PH_3	93.7°
H_2Se	91.0°	AsH_3	91.8°
He_2Te	89.5°	SbH_3	91.3°

What would be the predicted bond angle if no hybridization occurred? What conclusion can you draw concerning the importance of hybridization for molecules of compounds involving elements with higher atomic numbers?

75. Describe the orbitals (s, p, sp^2, and so on) of the central atom used for bond orbitals and unshared pairs in (a) H_2S (bond angle 91°), (b) CH_3OCH_3 (C—O—C angle 110°), (c) S_8 (S—S—S angle 105°), (d) $(CH_3)_2Sn^{2+}$ (C—Sn—C angle 180°), (e) $(CH_3)_3Sn^+$ (planar with respect to the Sn and 3 C atoms), and (f) $SnCl_2$ (angle 95°).

BUILDING YOUR KNOWLEDGE

76. Write the equation for the reaction between gaseous NH_3 and HCl to form solid NH_4Cl. Rewrite the equation using a Lewis formula for each species in the equation. Write the equation a third time, but this time use sketches to represent the molecular geometry of each species in the equation.

77. Sketch three-dimensional representations of the following molecules and indicate the direction of any net dipole for each molecule. (a) CH_4; (b) CH_3Cl; (c) CH_2Cl_2; (d) $CHCl_3$; (e) CCl_4.

78. Write the Lewis formula and identify the electronic geometry and molecular geometry for each polyatomic species in the following equations:
(a) $H^+ + H_2O \rightarrow H_3O^+$
(b) $NH_3 + H^+ \rightarrow NH_4^+$

9 Molecular Orbitals in Chemical Bonding

OBJECTIVES

As you study this chapter, you should learn to

- *Understand the basic ideas of molecular orbital theory*

- *Relate the shapes and overlap of atomic orbitals to the shapes and energies of the resulting molecular orbitals*

- *Distinguish between bonding and antibonding orbitals*

- *Apply the Aufbau Principle to find molecular orbital descriptions for homonuclear diatomic molecules and ions*

- *Apply the Aufbau Principle to find molecular orbital descriptions for heteronuclear diatomic molecules and ions*

- *Find the bond order in diatomic molecules*

- *Relate bond order to bond stability*

- *Use the MO concept of delocalization for molecules in which VB theory would postulate resonance*

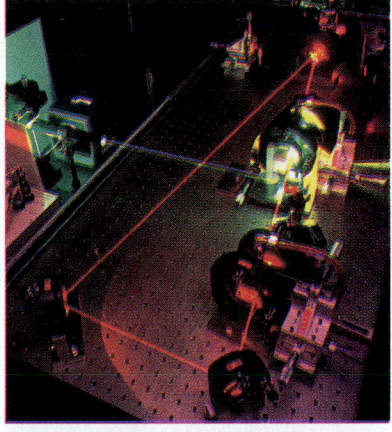

Lasers are used to study the structures and reactions of molecules.

W e have described bonding and molecular geometry in terms of valence bond theory. In valence bond theory we postulate that bonds result from the sharing of electrons in overlapping orbitals of different atoms. These orbitals may be *pure atomic orbitals* or *hybridized atomic orbitals* of *individual* atoms. We describe electrons in overlapping orbitals of different atoms as being localized in the bonds between the two atoms involved. We then invoke hybridization when it helps to account for the geometry of a molecule.

In **molecular orbital theory,** we postulate

the combination of atomic orbitals on different atoms forms **molecular orbitals** (MOs), so that electrons in them belong to the molecule as a whole.

Valence bond and molecular orbital theories are alternate descriptions of chemical bonding. They have strengths and weaknesses, so they are complementary. Valence bond theory is descriptively attractive and it lends itself well to visualization. Molecular orbital (MO) theory gives better descriptions of electron cloud distributions, bond energies, and magnetic properties, but its results are not as easy to visualize.

In some polyatomic molecules, a molecular orbital may extend over only a fraction of the molecule.

Polyatomic ions such as CO_3^{2-}, SO_4^{2-}, and NH_4^+ can be described by the molecular orbital approach.

317

An early triumph of molecular orbital theory was its ability to account for the observed paramagnetism of oxygen, O_2. According to earlier theories, O_2 was expected to be diamagnetic, that is, to have only paired electrons.

The valence bond picture of bonding in the O_2 molecule involves a double bond.

$$\ddot{O} :: \ddot{O}$$

This shows no unpaired electrons, so it predicts that O_2 is diamagnetic. However, experiments show that O_2 is paramagnetic; therefore, it has unpaired electrons. Thus, the VB structure is inconsistent with experiment and cannot be accepted as a description of the bonding. Molecular orbital theory *predicts* that O_2 has two unpaired electrons. This ability of MO theory to explain the paramagnetism of O_2 gave it credibility as a major theory of bonding. We shall develop some of the ideas of MO theory and apply them to some molecules and polyatomic ions.

9-1 MOLECULAR ORBITALS

We learned in Chapter 5 that each solution to the Schrödinger equation, called a wave function, represents an atomic orbital. The mathematical pictures of hybrid orbitals in valence bond theory can be generated by combining the wave functions that describe two or more atomic orbitals on a *single* atom. Similarly, combining wave functions that describe atomic orbitals on *separate* atoms generates mathematical descriptions of molecular orbitals.

An orbital has physical meaning only when we square its wave function to describe the electron density. Thus, the overall sign on the wave function that describes an atomic orbital is arbitrary. But when we *combine* two orbitals, the signs relative to one another matter. When waves are combined, they may interact either constructively or destructively (Figure 9-1). Likewise, when two atomic orbitals overlap, they can be in phase (added) or out of phase (subtracted). When they overlap in phase, constructive interaction occurs in the region between the nuclei, and a **bonding orbital** is produced. The energy of the bonding orbital is always lower (more stable) than the energies of the combining orbitals. When they overlap out of phase, destructive interaction reduces the probability of finding electrons in the region between the nuclei, and an **antibonding orbital** is produced. This is higher in energy (less stable) than the original atomic orbitals. The overlap of two atomic orbitals always produces two MOs—one bonding and one antibonding.

We can illustrate this basic principle by considering the combination of the 1s atomic orbitals *on two different atoms* (Figure 9-2). When these orbitals are occupied by electrons, the shapes of the orbitals are plots of electron density. These plots show the regions in molecules where the probabilities of finding electrons are greatest.

Figure 9-1 An illustration of constructive and destructive interference of waves. (a) If the two identical waves shown at the left are added, they interfere constructively to produce the more intense wave at the right. (b) Conversely, if they are subtracted, it is as if the phases (signs) of one wave were reversed and added to the first wave. This causes destructive interference, resulting in the wave at the right with zero amplitude; i.e., a straight line.

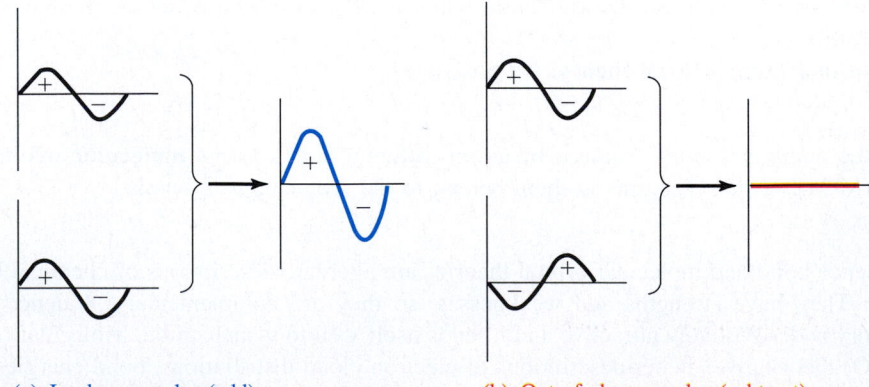

(a) In-phase overlap (add)   (b) Out-of-phase overlap (subtract)

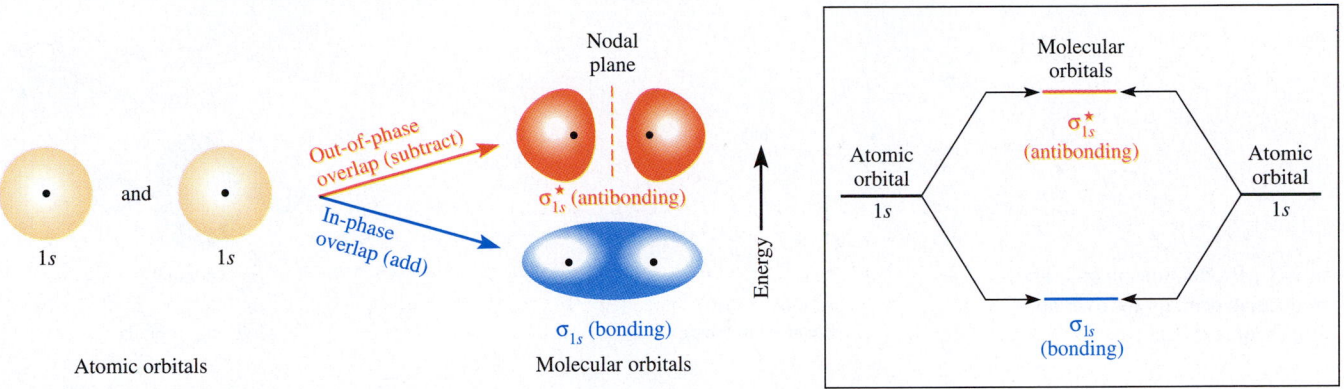

Figure 9-2 Molecular orbital (MO) diagram for the combination of the $1s$ atomic orbitals on two identical atoms (at the left) to form two MOs. One is a *bonding* orbital, σ_{1s} (blue), resulting from addition of the wave functions of the $1s$ orbitals. The other is an *antibonding* orbital, $\sigma_{1s}^{\star}$ (red), at higher energy resulting from subtraction of the waves that describe the combining $1s$ orbitals. In all σ-type MOs, the electron density is symmetrical about an imaginary line connecting the two nuclei. The terms "subtraction of waves," "out of phase," and "destructive interference in the region between the nuclei" all refer to the formation of an antibonding MO. Nuclei are represented by dots.

In the bonding orbital, the two $1s$ orbitals have reinforced one another in the region between the two nuclei by in-phase overlap, or addition of their electron waves. In the antibonding orbital, they have canceled one another in this region by out-of-phase overlap, or subtraction of their electron waves. We designate both molecular orbitals as **sigma (σ) molecular orbitals** (which indicates that they are cylindrically symmetrical about the internuclear axis). We indicate with subscripts the atomic orbitals that have been combined. The star denotes an antibonding orbital. Thus, two $1s$ orbitals produce a σ_{1s} (read "sigma-$1s$") bonding orbital and a $\sigma_{1s}^{\star}$ (read "sigma-$1s$-star") antibonding orbital. The right-hand side of Figure 9-2 shows the relative energy levels of these orbitals. All sigma antibonding orbitals have nodal planes bisecting the internuclear axis. A **node** or **nodal plane** is a region in which the probability of finding electrons is zero.

Another way of viewing the relative stabilities of these orbitals follows. In a bonding molecular orbital, there is high electron density *between* the two atoms, where it stabilizes the arrangement by exerting a strong attraction for both nuclei. By contrast, an antibonding orbital has a node (a region of zero electron density) between the nuclei; this allows a strong net repulsion between the nuclei, which makes the arrangement less stable. Electrons are *more* stable (have lower energy) in bonding molecular orbitals than in the individual atoms. Placing electrons in antibonding orbitals, on the other hand, requires an increase in their energy, which makes them *less* stable than in the individual atoms.

For any two sets of p orbitals on two different atoms, corresponding orbitals such as p_x orbitals can overlap *head-on*. This gives σ_p and $\sigma_p^{\star}$ orbitals, as shown in Figure 9-3 for the head-on overlap of $2p_x$ orbitals on the two atoms. If the remaining p orbitals overlap (p_y with p_y and p_z with p_z), they must do so sideways, or *side-on*, forming *pi (π) molecular orbitals*. Depending on whether all p orbitals overlap, there can be as many as two π_p and two $\pi_p^{\star}$ orbitals. Figure 9-4 illustrates the overlap of two corresponding $2p$ orbitals on two atoms to form π_{2p} and $\pi_{2p}^{\star}$ molecular orbitals. There is a nodal plane along the internuclear axis for all pi molecular orbitals. If one views a sigma molecular orbital along the

How we name the axes is arbitrary. We shall designate the internuclear axis as the x direction.

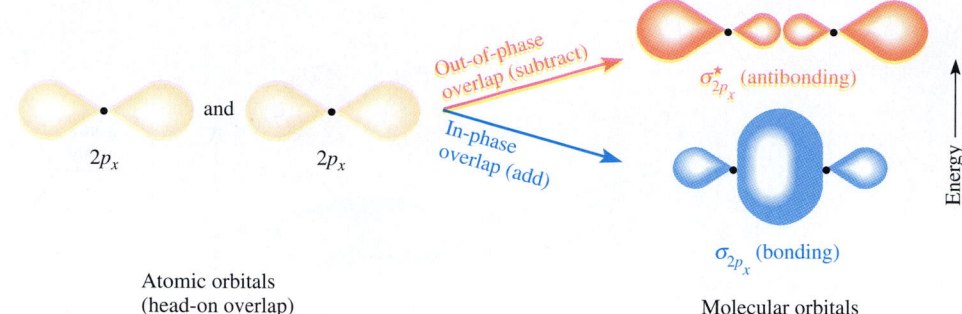

Figure 9-3 Production of σ_{2p_x} and $\sigma^\star_{2p_x}$ molecular orbitals by overlap of $2p_x$ orbitals on two atoms.

Atomic orbitals
(head-on overlap)

Molecular orbitals

This would involve rotating Figures 9-2, 9-3, and 9-4 by 90° so that the internuclear axes are perpendicular to the plane of the pages.

internuclear axis, it appears to be symmetrical around the axis like a pure *s* atomic orbital. A similar cross-sectional view of a pi molecular orbital looks like a pure *p* atomic orbital, with a node along the internuclear axis.

> The number of molecular orbitals (MOs) formed is equal to the number of atomic orbitals that are combined. When two atomic orbitals are combined, one of the resulting MOs is at a *lower* energy than the original atomic orbitals; this is a *bonding* orbital. The other MO is at a *higher* energy than the original atomic orbitals; this is an *antibonding* orbital.

9-2 MOLECULAR ORBITAL ENERGY-LEVEL DIAGRAMS

In the same way that atomic orbitals can be arranged by increasing energy into an energy-level diagram (Figure 5-28), we can draw molecular orbital energy-level diagrams for simple molecules. The simplest examples, shown in Figure 9-5, apply to homonuclear diatomic molecules of elements in the first and second periods. Each diagram is an extension of the right-hand diagram in Figure 9-2, to which we have added the molecular orbitals formed from $2s$ and $2p$ atomic orbitals.

If we had chosen the *z* axis as the axis of head-on overlap of the $2p$ orbitals in Figure 9-3, side-on overlap of the $2p_x$–$2p_x$ and $2p_y$–$2p_y$ orbitals would form the π-type molecular orbitals.

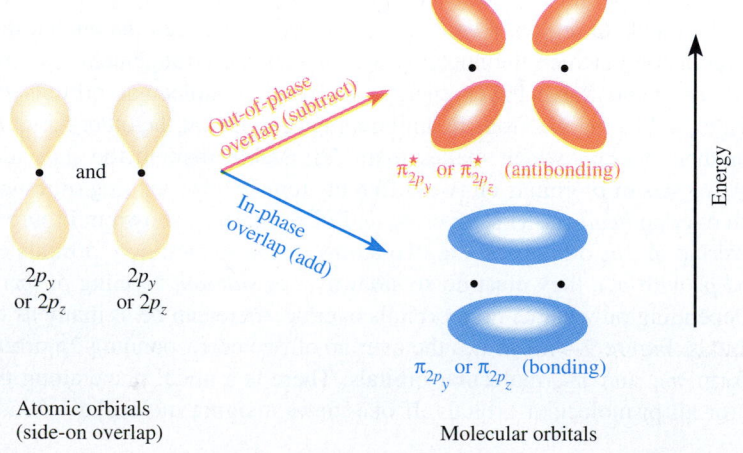

Figure 9-4 The π_{2p} and $\pi^\star_{2p}$ molecular orbitals from overlap of one pair of $2p$ atomic orbitals (for instance, $2p_y$ orbitals). There can be an identical pair of molecular orbitals at right angles to these, formed by another pair of *p* orbitals on the same two atoms (in this case, $2p_z$ orbitals).

Atomic orbitals
(side-on overlap)

Molecular orbitals

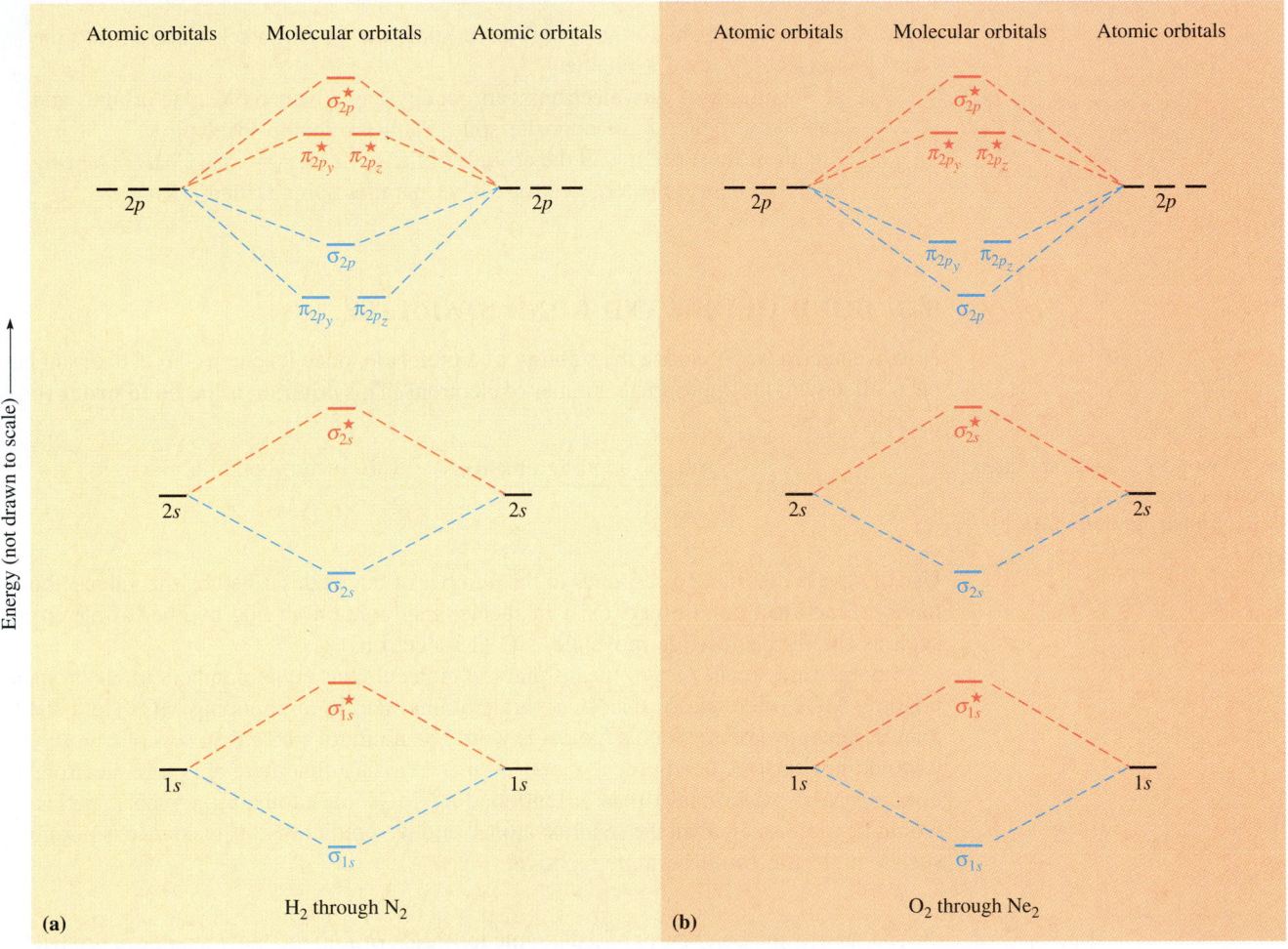

Figure 9-5 Energy-level diagrams for first and second period homonuclear diatomic molecules and ions (not drawn to scale). The solid lines represent the relative energies of the indicated atomic and molecular orbitals. (a) The diagram for H_2, He_2, Li_2, Be_2, B_2, C_2, and N_2 molecules and their ions. (b) The diagram for O_2, F_2, and Ne_2 molecules and their ions.

For the cases shown in Figure 9-5a, the two π_{2p} orbitals are lower in energy than the σ_{2p} orbital. However, molecular orbital calculations indicate that for O_2, F_2, and hypothetical Ne_2 molecules, the σ_{2p} orbital is lower in energy than the π_{2p} orbitals (Figure 9-5b).

Diagrams such as these are used to describe the bonding in a molecule in MO terms. Electrons occupy MOs according to the same rules developed for atomic orbitals; they follow the Aufbau Principle, the Pauli Exclusion Principle, and Hund's Rule. (See Section 5-17.)

Spectroscopic data support these orders.

1. Draw (or select) the appropriate molecular orbital energy-level diagram.
2. Determine the *total* number of electrons in the molecule. Note that in applying MO theory, we usually account for *all* electrons. This includes both the inner shell electrons and the valence electrons.

3. Add these electrons to the energy-level diagram, putting each electron into the lowest energy level available.
   a. A maximum of *two* electrons can occupy any given molecular orbital, and then only if they have opposite spin (Pauli Exclusion Principle).
   b. Electrons must occupy all the orbitals of the same energy singly before pairing begins. These unpaired electrons have parallel spins (Hund's Rule).

9-3 BOND ORDER AND BOND STABILITY

Now we need a way to judge the stability of a molecule, once its energy-level diagram has been filled with the appropriate number of electrons. This criterion is the **bond order** (bo):

Electrons in bonding orbitals are often called **bonding electrons;** electrons in antibonding orbitals, **antibonding electrons.**

$$\text{bond order} = \frac{(\text{no. of bonding electrons}) - (\text{no. of antibonding electrons})}{2}$$

Usually the bond order corresponds to the number of bonds described by the valence bond theory. Fractional bond orders exist in species that contain an odd number of electrons, such as the nitrogen oxide molecule, NO (15 electrons).

A bond order *equal to zero* means that the molecule has equal numbers of electrons in bonding MOs (more stable than in separate atoms) and in antibonding MOs (less stable than in separate atoms). Such a molecule would be no more stable than separate atoms, so it would not exist. A bond order *greater than zero* means that there are more electrons in bonding MOs (stabilizing) than in antibonding MOs (destabilizing). Such a molecule would be more stable than the separate atoms, and we predict that its existence is possible; such a molecule could be quite reactive.

The greater the bond order of a diatomic molecule or ion, the more stable we predict it to be. Likewise, for a bond between two given atoms, the greater the bond order, the shorter the bond length and the greater the bond energy.

The **bond energy** is the amount of energy necessary to break a mole of bonds (Section 15-9). Bond energy is therefore a measure of bond strength.

▼ **PROBLEM-SOLVING TIP**   *Working with MO Theory*

MO theory is often the best model to predict the bond order, bond stability, and/or magnetic properties of a molecule or ion. We:

1. draw (or select) the appropriate MO energy-level diagram.
2. count the total number of electrons in the molecule or ion.
3. follow the Pauli Exclusion Principle and Hund's Rule to add the electrons to the MO diagram.
4. calculate the bond order: $\text{bo} = \left(\dfrac{\text{bonding } e\text{'s} - \text{antibonding } e\text{'s}}{2} \right)$.
5. use the bond order to evaluate stability.
6. look for the presence of unpaired electrons to determine if a species is paramagnetic.

Table 9-1 *Molecular Orbitals for First and Second Period (row) Diatomic Molecules*[a]

	H_2	He_2[c]	Li_2[b]	Be_2[c]	B_2[b]	C_2[b]	N_2		O_2	F_2	Ne_2[c]
$\sigma^{\star}_{2p}$	—	—	—	—	—	—	—		—	—	↑↓
$\pi^{\star}_{2p_y},\ \pi^{\star}_{2p_z}$	— —	— —	— —	— —	— —	— —	— —		↑ ↑	↑↓ ↑↓	↑↓ ↑↓
σ_{2p}	—	—	—	—	—	—	↑↓	$\pi_{2p_y},\ \pi_{2p_z}$	↑↓ ↑↓	↑↓ ↑↓	↑↓ ↑↓
$\pi_{2p_y},\ \pi_{2p_z}$	— —	— —	— —	— —	↑ ↑	↑↓ ↑↓	↑↓ ↑↓	σ_{2p}	↑↓	↑↓	↑↓
$\sigma^{\star}_{2s}$	—	—	—	↑↓	↑↓	↑↓	↑↓		↑↓	↑↓	↑↓
σ_{2s}	—	—	↑↓	↑↓	↑↓	↑↓	↑↓		↑↓	↑↓	↑↓
$\sigma^{\star}_{1s}$	—	↑↓	↑↓	↑↓	↑↓	↑↓	↑↓		↑↓	↑↓	↑↓
σ_{1s}	↑↓	↑↓	↑↓	↑↓	↑↓	↑↓	↑↓		↑↓	↑↓	↑↓
Paramagnetic?	no	no	no	no	yes	no	no		yes	no	no
Bond order	1	0	1	0	1	2	3		2	1	0
Observed bond length (Å)	0.74	—	2.67	—	1.59	1.31	1.09		1.21	1.43	—
Observed bond energy (kJ/mol)	435	—	110	9	~270	602	946		498	159	—

[a] *Electron distribution in molecular orbitals, bond order, bond length, and bond energy of homonuclear diatomic molecules of the first and second row elements. Note that nitrogen molecules, N_2, have the highest bond energies listed; they have a bond order of three. The species C_2 and O_2, with a bond order of two, have the next highest bond energies.*

[b] *Exists only in the vapor state at elevated temperatures.*

[c] *Unknown species.*

9-4 HOMONUCLEAR DIATOMIC MOLECULES

The electron distributions for the homonuclear diatomic molecules of the first and second periods are shown in Table 9-1 together with their bond orders, bond lengths, and bond energies.

> "Homonuclear" means consisting only of atoms of the same element. "Diatomic" means consisting of two atoms.

The Hydrogen Molecule, H_2

The overlap of the $1s$ orbitals of two hydrogen atoms produces σ_{1s} and $\sigma^{\star}_{1s}$ molecular orbitals. The two electrons of the molecule occupy the lower energy σ_{1s} orbital (Figure 9-6a).

Because the two electrons in an H_2 molecule are in a bonding orbital, the bond order is one. We conclude that the H_2 molecule would be stable, and it is. The energy associated

$$H_2 \text{ bo} = \frac{2 - 0}{2} = 1$$

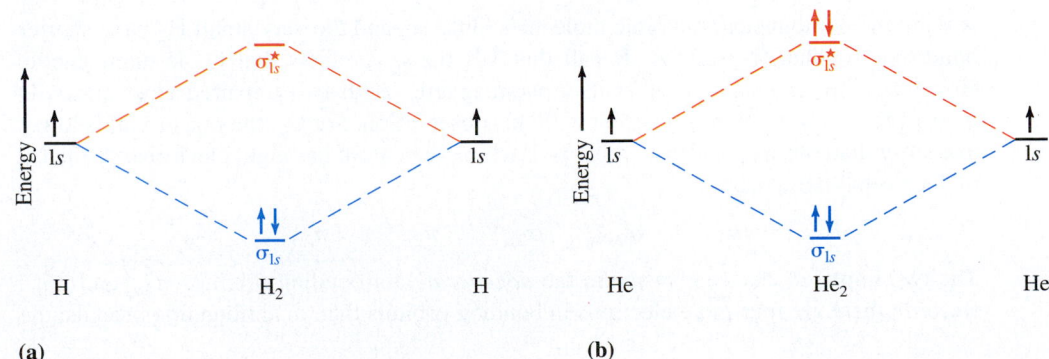

Figure 9-6 Molecular orbital diagrams for (a) H_2 and (b) He_2.

(a)   (b)

with two electrons in the H_2 molecule is lower than that associated with the same two electrons in separate $1s$ atomic orbitals. The lower the energy of a system, the more stable it is.

The Helium Molecule (hypothetical), He₂

$$He_2 \ bo = \frac{2-2}{2} = 0$$

The energy-level diagram for He_2 is similar to that for H_2 except that it has two more electrons. These occupy the antibonding $\sigma_{1s}^{\star}$ orbital (Figures 9-5a and 9-6b and Table 9-1), giving He_2 a bond order of zero. That is, the two electrons in the bonding orbital of He_2 would be *more stable* than in the separate atoms. But the two electrons in the antibonding orbital would be *less stable* than in the separate atoms by the same amount. These effects cancel, so the molecule would be no more stable than the separate atoms and would not exist. In fact, He_2 is not known.

The Boron Molecule, B₂

$$B_2 \ bo = \frac{6-4}{2} = 1$$

The boron atom has the configuration $1s^2 2s^2 2p^1$. Here p electrons participate in the bonding. Figure 9-5a and Table 9-1 show that the π_{p_y} and π_{p_z} molecular orbitals are lower in energy than the σ_{2p} for B_2. Thus, the electron configuration is

$$\sigma_{1s}^2 \quad \sigma_{1s}^{\star 2} \quad \sigma_{2s}^2 \quad \sigma_{2s}^{\star 2} \quad \pi_{2p_y}^1 \quad \pi_{2p_z}^1$$

Orbitals of equal energy are called *degenerate* orbitals. Hund's Rule for filling degenerate orbitals was discussed in Section 5-17.

The unpaired electrons are consistent with the observed paramagnetism of B_2. Here we illustrate Hund's Rule in molecular orbital theory. The π_{2p_y} and π_{2p_z} orbitals are equal in energy and contain a total of two electrons. As a result, one electron occupies each orbital. The bond order is one. Experiments verify that B_2 molecules exist in the vapor state.

The Nitrogen Molecule, N₂

$$N_2 \ bo = \frac{10-4}{2} = 3$$

In the valence bond representation, N_2 is shown as N≡N, with a triple bond.

Experimental thermodynamic data show that the N_2 molecule is stable, is diamagnetic, and has a very high bond energy, 946 kJ/mol. This is consistent with molecular orbital theory. Each nitrogen atom has seven electrons, so the diamagnetic N_2 molecule has 14 electrons.

$$\sigma_{1s}^2 \quad \sigma_{1s}^{\star 2} \quad \sigma_{2s}^2 \quad \sigma_{2s}^{\star 2} \quad \pi_{2p_y}^2 \quad \pi_{2p_z}^2 \quad \sigma_{2p}^2$$

There are six more electrons in bonding orbitals than in antibonding orbitals, so the bond order is three. We see (Table 9-1) that N_2 has a very short bond length, only 1.09 Å, the shortest of any diatomic species except H_2.

The Oxygen Molecule, O₂

Among the homonuclear diatomic molecules, only N_2 and the very small H_2 have shorter bond lengths than O_2, 1.21 Å. Recall that VB theory predicts that O_2 is diamagnetic. However, experiments show that it is paramagnetic, with two unpaired electrons. MO theory predicts a structure consistent with this observation. For O_2, the σ_{2p} orbital is lower in energy than the π_{2p_y} and π_{2p_z} orbitals. Each oxygen atom has eight electrons, so the O_2 molecule has 16 electrons.

$$\sigma_{1s}^2 \quad \sigma_{1s}^{\star 2} \quad \sigma_{2s}^2 \quad \sigma_{2s}^{\star 2} \quad \sigma_{2p}^2 \quad \pi_{2p_y}^2 \quad \pi_{2p_z}^2 \quad \pi_{2p_y}^{\star 1} \quad \pi_{2p_z}^{\star 1}$$

$$O_2 \ bo = \frac{10-6}{2} = 2$$

The two unpaired electrons reside in the *degenerate* antibonding orbitals, $\pi_{2p_y}^{\star}$ and $\pi_{2p_z}^{\star}$. Because there are four more electrons in bonding orbitals than in antibonding orbitals, the

CHEMISTRY IN USE

The Environment

The Blaze of the Century

To many it was the "blaze of the century." Temperatures were so high that even the sand melted into glass while dazzling orange flames billowed high into the air. The firefighters who battled the extreme heat had another name for this Gulf War aftermath—"Operation Desert Hell."

The saga began when Iraqi President Saddam Hussein sent his military forces to invade Kuwait in 1991. After the United States and other nations intervened, retreating Iraqi soldiers set fire to more than 600 oil wells. Kuwaiti officials estimated that about 6 million barrels of oil per day went up in smoke, at a cost of over $1,000 a second. Air pollution from the fires was a serious environmental concern with smoke plumes reaching countries more than 1,200 miles away. Firefighters from around the world converged on Kuwait to help extinguish the burning oil wells and to minimize the resulting environmental damage.

Oil well fires cannot be safely extinguished with water. Frequently, oil fires are snuffed out by exploding dynamite above burning wells, depriving the fires of oxygen needed to support combustion. This method requires large reservoirs of water for two important reasons. First, water is needed to keep the explosives cool until they are detonated. Second, because the explosives can extinguish an oil well fire only briefly, additional water is needed to prevent the hot structural metal from reigniting the gas and oil.

When large reservoirs of water are unavailable, oil-well firefighters use nitrogen to help put out the blaze; using nitrogen requires much less water. To use nitrogen, firefighters attach a 25-foot steel tube with a 2.5-foot diameter to a long pole extending from a crane. Workers then connect one end of a hose to the base of the 25-foot steel tube and the other end to a nearby truck containing liquid nitrogen. Liquid nitrogen is heated to change it into a gas before it reaches the steel tube. The crane places the steel tube over the spout of burning oil so the well propels its burning oil through the tube and burns at the top. The burning oil at the top of the steel tube resembles an enormous Bunsen burner. Now nitrogen is pumped into the base of the steel tube. Because nitrogen does not support combustion, the fire is quickly smothered as nitrogen displaces oxygen at the source of the fire.

An oil well fire.

In addition to its inability to support combustion, another important property of nitrogen makes it particularly useful in fighting oil well fires. As we have seen, molecular orbital theory predicts that nitrogen molecules have a bond order of three, the maximum bond order possible for a diatomic molecule. The greater the bond order of a diatomic molecule, the more stable we expect the molecule to be. Because of nitrogen's great stability, it can withstand the intense heat produced by the burning oil.

Throughout the fire-fighting process, water is pumped onto the metal tube to keep it from melting. After the flames shooting from the top of the steel tube are extinguished, in their place is a huge fountain of hot but nonburning oil. The dangerous part has just begun because hot oil rains down covering machinery and workers alike; the slightest spark could ignite the oil again. To protect the firefighters, a water cannon continuously shoots streams of water over the firefighters until they shut off the oil flow.

Ronald DeLorenzo
Middle Georgia College

bond order is two (Figure 9-5b and Table 9-1). We see why the molecule is much more stable than two free O atoms.

Molecular orbital theory can also be used to predict the structures and stabilities of ions, as Example 9-1 shows.

EXAMPLE 9-1 *Predicting Stabilities and Bond Orders*

Predict the stabilities and bond orders of the ions (a) O_2^+ and (b) O_2^-.

Plan

(a) The O_2^+ ion is formed by removing one electron from the O_2 molecule, given above. The electrons that are withdrawn most easily are those in the highest energy orbitals. (b) The superoxide ion, O_2^-, results from adding an electron to the O_2 molecule.

Solution

(a) We remove one of the $\pi_{2p}^\star$ electrons of O_2 to find the configuration of O_2^+:

$$\sigma_{1s}^2 \qquad \sigma_{1s}^{\star\,2} \qquad \sigma_{2s}^2 \qquad \sigma_{2s}^{\star\,2} \qquad \sigma_{2p}^2 \qquad \pi_{2p_y}^2 \qquad \pi_{2p_z}^2 \qquad \pi_{2p_y}^{\star\,1}$$

There are five more electrons in bonding orbitals than in antibonding orbitals, so the bond order is 2.5. We conclude that the ion would be reasonably stable relative to other diatomic ions, and it does exist.

In fact, the unusual ionic compound $[O_2^+][PtF_6^-]$ played an important role in the discovery of the first noble gas compound, $XePtF_6$ (Section 24-2).

(b) We add one electron to the appropriate orbital of O_2 to find the configuration of O_2^-. Following Hund's Rule, we add this electron into the $\pi_{2p}^\star$ orbital to form a pair:

$$\sigma_{1s}^2 \quad \sigma_{1s}^{\star\,2} \quad \sigma_{2s}^2 \quad \sigma_{2s}^{\star\,2} \quad \sigma_{2p}^2 \quad \pi_{2p_y}^2 \quad \pi_{2p_z}^2 \quad \pi_{2p_y}^{\star\,2} \quad \pi_{2p_z}^{\star\,1}$$

There are three more bonding electrons than antibonding electrons, so the bond order is 1.5. We conclude that the ion should exist but be less stable than O_2.

The known superoxides of the heavier Group IA elements—KO_2, RbO_2, and CsO_2—contain the superoxide ion, O_2^-. These compounds are formed by combination of the free metals with oxygen (Section 6-8, part 2).

You should now work Exercise 30.

The Fluorine Molecule, F_2

Each fluorine atom has nine electrons, so there are 18 electrons in F_2.

$$\sigma_{1s}^2 \quad \sigma_{1s}^{\star\,2} \quad \sigma_{2s}^2 \quad \sigma_{2s}^{\star\,2} \quad \sigma_{2p}^2 \quad \pi_{2p_y}^2 \quad \pi_{2p_z}^2 \quad \pi_{2p_y}^{\star\,2} \quad \pi_{2p_z}^{\star\,2}$$

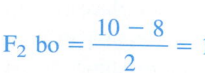

$$F_2 \text{ bo} = \frac{10 - 8}{2} = 1$$

The bond order is one. As you know, F_2 exists. The F—F bond distance is longer (1.43 Å) than the bond distances in O_2 (1.21 Å) or N_2 (1.09 Å) molecules. The bond order in F_2 (one) is less than that in O_2 (two) or N_2 (three). The bond energy of the F_2 molecules is quite low (159 kJ/mol), so F_2 molecules are very reactive.

Heavier Homonuclear Diatomic Molecules

It might appear reasonable to use the same types of molecular orbital diagrams to predict the stability or existence of homonuclear diatomic molecules of the third and subsequent periods. However, the heavier halogens, Cl_2, Br_2, and I_2, which contain only sigma (single) bonds, are the only well-characterized examples at room temperature. We would predict from both molecular orbital theory and valence bond theory that the other (non-halogen) homonuclear diatomic molecules from below the second period would exhibit pi bonding and therefore multiple bonding.

Some heavier elements exist as diatomic species, such as S_2, in the vapor phase at elevated temperatures. These species are neither common nor very stable. The instability is related to the inability of the heavier elements to form strong pi bonds *with each other*. For larger atoms, the sigma bond length is too great to allow the atomic *p* orbitals on different atoms to overlap very effectively. Therefore, the strength of pi bonding decreases rapidly with increasing atomic size. For example, N_2 is *much* more stable than P_2. This is because the 3*p* orbitals on one P atom do not overlap side by side in a pi-bonding manner with corresponding 3*p* orbitals on another P atom nearly as effectively as do the corresponding 2*p* orbitals on the smaller N atoms. MO theory does not predict multiple bonding for Cl_2, Br_2, or I_2, each of which has a bond order of one.

9-5 HETERONUCLEAR DIATOMIC MOLECULES

Heteronuclear Diatomic Molecules of Second Period Elements

Corresponding atomic orbitals of two different elements, such as the 2*s* orbitals of carbon and oxygen atoms, have different energies because their nuclei have different charges and therefore different attractions for electrons. The atomic orbitals of the *more electronegative element* are *lower* in energy than the corresponding orbitals of the less electronegative element. As a result, a molecular orbital diagram such as Figure 9-5 is inappropriate for *heteronuclear* diatomic molecules. If the two elements are similar (as in CO, NO, or CN molecules, for example), we can modify the diagram of Figure 9-5 by skewing it slightly. Figure 9-7 shows the energy-level diagram and electron configuration for carbon monoxide, CO.

NOTE: CN is a reactive molecule, not the stable cyanide ion, CN^-.

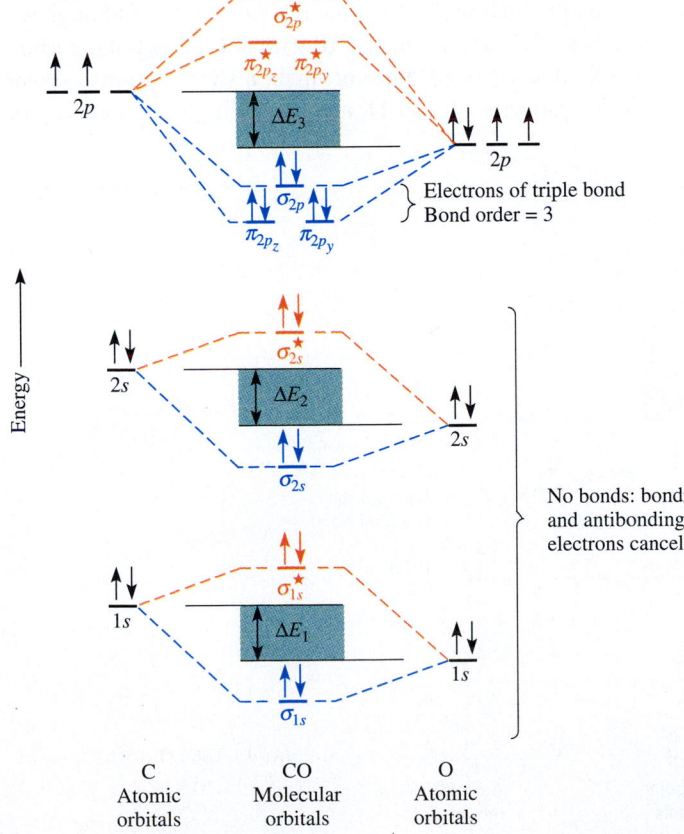

Figure 9-7 MO energy-level diagram for carbon monoxide, CO, a slightly polar heteronuclear diatomic molecule ($\mu =$ 0.11 D). The atomic orbitals of oxygen, the more electronegative element, are a little lower in energy than the corresponding atomic orbitals of carbon, the less electronegative element. For this molecule, the energy differences ΔE_1, ΔE_2, and ΔE_3 are not very large; the molecule is not very polar.

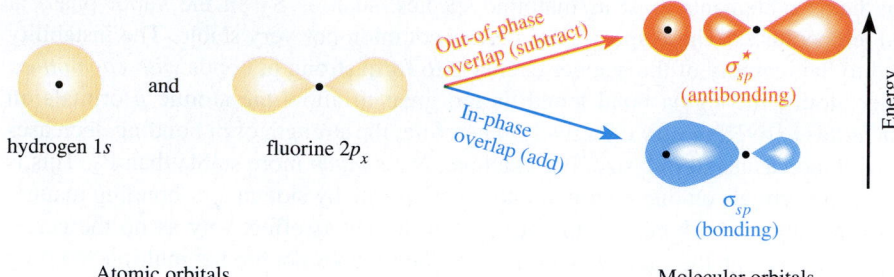

Figure 9-8   Formation of σ_{sp} and $\sigma_{sp}^{\star}$ molecular orbitals in HF by overlap of the $1s$ orbital of H with a $2p$ orbital of F.

The closer the energy of a molecular orbital is to the energy of one of the atomic orbitals from which it is formed, the more of the character of that atomic orbital it shows. Thus, as we see in Figure 9-7, the bonding MOs in the CO molecule have more oxygen-like atomic orbital character, and the antibonding orbitals have more carbon-like atomic orbital character.

In general the energy differences ΔE_1, ΔE_2, and ΔE_3 (green backgrounds in Figure 9-7) depend on the difference in electronegativities between the two atoms. The greater these energy differences, the more polar is the bond joining the atoms and the greater is its ionic character. On the other hand, the energy differences reflect the degree of overlap between atomic orbitals; the smaller these differences, the more the orbitals can overlap, and the greater is the covalent character of the bond.

We see that CO has a total of 14 electrons, making it isoelectronic with the stable N_2 molecule. Therefore, the distribution of electrons is the same in CO as in N_2, although we expect the energy levels of the MOs to be different. In accord with our predictions, carbon monoxide is a very stable molecule. It has a bond order of three, a short carbon–oxygen bond length of 1.13 Å, a low dipole moment of 0.11 D, and a very high bond energy of 1071 kJ/mol.

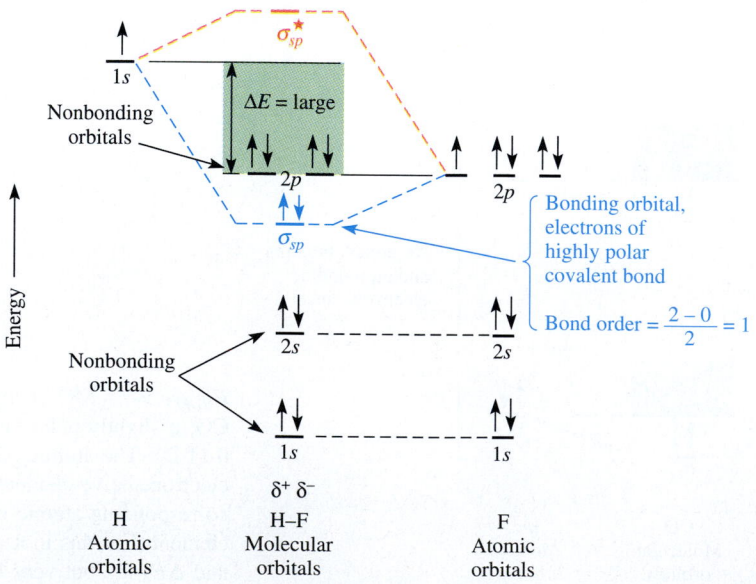

Figure 9-9   MO energy-level diagram for hydrogen fluoride, HF, a very polar molecule ($\mu = 1.91$ D). ΔE is large because the electronegativity difference is large.

The Hydrogen Fluoride Molecule, HF

The electronegativity difference between hydrogen (EN = 2.1) and fluorine (EN = 4.0) is very large (ΔEN = 1.9). The hydrogen fluoride molecule contains a very polar bond (μ = 1.9 D). The bond in HF involves the $1s$ electron of H and an unpaired electron from a $2p$ orbital of F. Figure 9-8 shows the overlap of the $1s$ orbital of H with a $2p$ orbital of F to form σ_{sp} and $\sigma_{sp}^{\star}$ molecular orbitals. The remaining two $2p$ orbitals of F have no net overlap with H orbitals. They are called **nonbonding** orbitals. The same is true for the F $2s$ and $1s$ orbitals. These nonbonding orbitals retain the characteristics of the F atomic orbitals from which they are formed. The MO diagram of HF is shown in Figure 9-9.

9-6 DELOCALIZATION AND THE SHAPES OF MOLECULAR ORBITALS

In Section 7-6 we described resonance formulas for molecules and polyatomic ions. Resonance is said to exist when two or more equivalent Lewis formulas can be written for the same species and a single such formula does not account for the properties of a substance. In molecular orbital terminology, a more appropriate description involves *delocalization* of electrons. The shapes of molecular orbitals for species in which electron delocalization occurs can be predicted by combining all the contributing atomic orbitals.

The Carbonate Ion, $CO_3{}^{2-}$

Consider the trigonal planar carbonate ion, $CO_3{}^{2-}$, as an example. All the carbon–oxygen bonds in the ion have the same bond length and the same energy, intermediate between those of typical C—O and C=O bonds. Valence bond theory describes the ion in terms of three contributing resonance structures (Figure 9-10a). No one of the three resonance forms adequately describes the bonding.

> The average carbon–oxygen bond order in the $CO_3{}^{2-}$ ion is $1\frac{1}{3}$.

According to valence bond theory, the C atom is described as sp^2 hybridized, and it forms one sigma bond with each of the three O atoms. This leaves one unhybridized $2p$ atomic orbital on the C atom, say the $2p_z$ orbital. This orbital is capable of overlapping and mixing with the $2p_z$ orbital of any of the three O atoms. The sharing of two electrons in the resulting localized pi orbital would form a pi bond. Thus, three equivalent resonance structures can be drawn in valence bond terms (Figure 9-10b). We emphasize that there is *no evidence* for the existence of these separate resonance structures.

The MO description of the pi bonding involves the simultaneous overlap and mixing of the carbon $2p_z$ orbital with the $2p_z$ orbitals of all three oxygen atoms. This forms a delocalized bonding pi molecular orbital system extending above and below the plane of the sigma system, as well as an antibonding pi orbital system. Electrons are said to occupy the entire set of bonding pi MOs, as depicted in Figure 9-10c. The shape is obtained by averaging the contributing valence bond resonance structures. The bonding in such species as nitrate ion, $NO_3{}^{-}$, and sulfur dioxide, SO_2, can be described in a similar manner.

The Benzene Molecule, C_6H_6

Now let us consider the benzene molecule, C_6H_6, whose two valence bond resonance forms are shown in Figure 9-11a. The valence bond description involves sp^2 hybridization at each C atom. Each C atom is at the center of a trigonal plane, and the entire molecule is known to be planar. There are sigma bonds from each C atom to the two adjacent C atoms

> There is no evidence for the existence of either of these forms of benzene. The MO description of benzene is far better than the VB description.

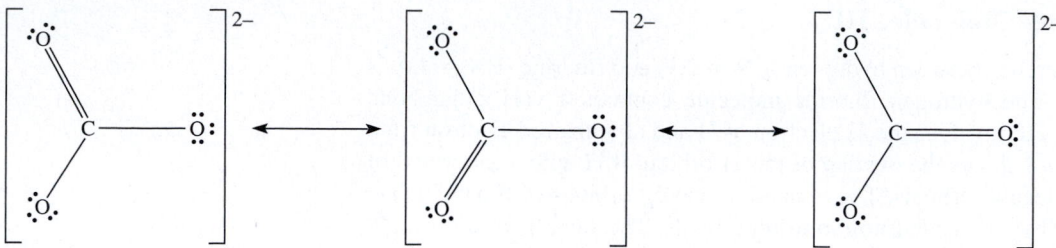

(a) Lewis formulas for valence bond resonance structures

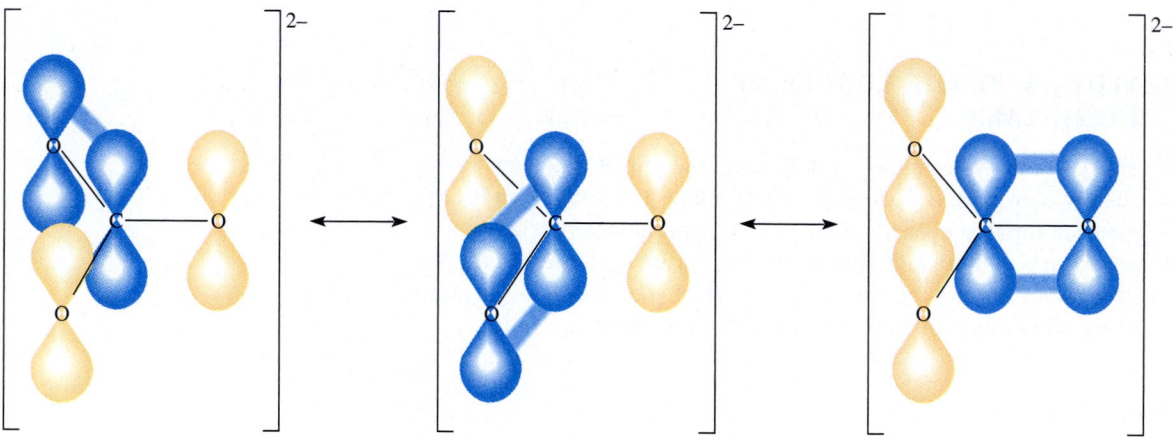

(b) *p*-orbital overlap in valence bond resonance structures

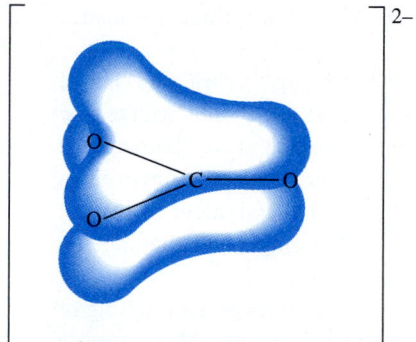

(c) Delocalized MO representation

Figure 9-10 Alternative representations of the bonding in the carbonate ion, CO_3^{2-}. (a) Lewis formulas of the three valence bond resonance structures. (b) Representation of the *p* orbital overlap in the valence bond resonance structures. In each resonance form, the *p* orbitals on two atoms would overlap to form the π components of the hypothetical double bonds. Each O atom has two additional sp^2 orbitals (not shown) in the plane of the nuclei. Each of these additional sp^2 orbitals contains an oxygen unshared pair. (c) In the MO description, the electrons in the π-bonded region are spread out, or *delocalized,* over all four atoms of the CO_3^{2-} ion. This MO description is more consistent with the experimental observation of equal bond lengths and energies than are the valence bond pictures in (a) and (b).

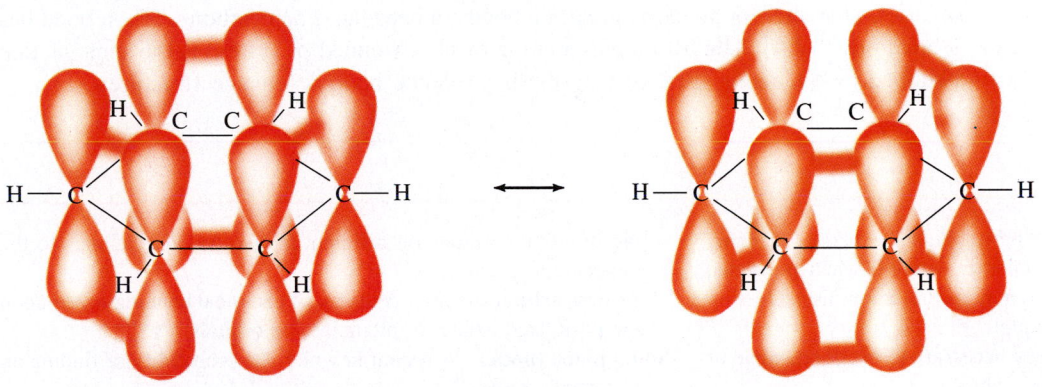

(a) Lewis formulas for valence bond resonance structures

(b) *p*-orbital overlap in valence bond resonance structures

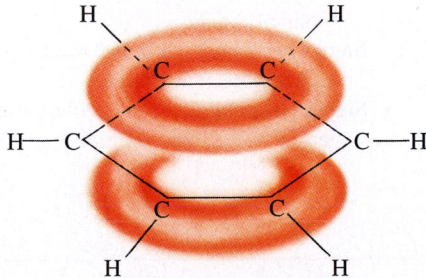

(c) Delocalized MO representation

Figure 9-11 Representations of the bonding in the benzene molecule, C_6H_6. (a) Lewis formulas of the two valence bond resonance structures. (b) The six *p* orbitals of the benzene ring, shown overlapping to form the (hypothetical) double bonds of the two resonance forms of valence bond theory. (c) In the MO description the six electrons in the pi-bonded region are *delocalized;* i.e., they occupy an extended pi-bonding region above and below the plane of the six C atoms.

and to one H atom. This leaves one unhybridized $2p_z$ orbital on each C atom and one remaining valence electron for each. According to valence bond theory, adjacent pairs of $2p_z$ orbitals and the six remaining electrons occupy the regions of overlap to form a total of three pi bonds in either of the two ways shown in Figure 9-11b.

Experimental studies of the C_6H_6 structure prove that it does *not* contain alternating single and double carbon–carbon bonds. The usual C—C single bond length is 1.54 Å, and the usual C=C double bond length is 1.34 Å. All six of the carbon–carbon bonds in benzene are the same length, 1.39 Å, intermediate between those of single and double bonds.

This is well explained by the MO theory, which predicts that the six $2p_z$ orbitals of the C atoms overlap and mix to form three pi-bonding and three pi-antibonding molecular orbitals. For instance, the most strongly bonding pi molecular orbital in the benzene pi–MO system is that in Figure 9-11c. The six pi electrons occupy three bonding MOs of this extended (delocalized) system. Thus, they are distributed throughout the molecule as a whole, above and below the plane of the sigma-bonded framework. This results in identical character for all carbon–carbon bonds in benzene. Each carbon–carbon bond has a bond order of $1\frac{1}{2}$. The MO representation of the extended pi system is the same as that obtained by averaging the two contributing valence bond resonance structures.

Key Terms

Antibonding orbital   A molecular orbital higher in energy than any of the atomic orbitals from which it is derived; when populated with electrons, lends instability to a molecule or ion. Denoted with a star (★) superscript on its symbol.

Bond energy   The amount of energy necessary to break one mole of bonds of a given kind (in the gas phase).

Bond order   Half the number of electrons in bonding orbitals minus half the number of electrons in antibonding orbitals.

Bonding orbital   A molecular orbital lower in energy than any of the atomic orbitals from which it is derived; when populated with electrons, lends stability to a molecule or ion.

Delocalization   The formation of a set of molecular orbitals that extend over more than two atoms; important in species that valence bond theory describes in terms of *resonance.*

Heteronuclear   Consisting of different elements.

Homonuclear   Consisting of only one element.

Molecular orbital (MO)   An orbital resulting from overlap and mixing of atomic orbitals on different atoms. An MO belongs to the molecule as a whole.

Molecular orbital theory   A theory of chemical bonding based upon the postulated existence of molecular orbitals.

Nodal plane (node)   A region in which the probability of finding an electron is zero.

Nonbonding orbital   A molecular orbital derived only from an atomic orbital of one atom; lends neither stability nor instability to a molecule or ion when populated with electrons.

Pi (π) bond   A bond resulting from electron occupation of a pi molecular orbital.

Pi (π) orbital   A molecular orbital resulting from side-on overlap of atomic orbitals.

Sigma (σ) bond   A bond resulting from electron occupation of a sigma molecular orbital.

Sigma (σ) orbital   A molecular orbital resulting from head-on overlap of two atomic orbitals.

Exercises

MO Theory—General Concepts

1. Describe the main differences between the valence bond theory and the molecular orbital theory.
2. What is a molecular orbital? What two types of information can be obtained from molecular orbital calculations? How do we use such information to describe the bonding within a molecule?
3. What is the relationship between the maximum number of electrons that can be accommodated by a set of molecular orbitals and the maximum number that can be accommodated by the atomic orbitals from which the MOs are formed? What is the maximum number of electrons that one MO can hold?
4. Answer Exercise 3 after replacing "molecular orbitals" with "hybridized atomic orbitals."
5. What differences and similarities exist among (a) atomic orbitals, (b) localized hybridized atomic orbitals according to valence bond theory, and (c) molecular orbitals?
6. What is the relationship between the energy of a bonding molecu-

lar orbital and the energies of the original atomic orbitals? What is the relationship between the energy of an antibonding molecular orbital and the energies of the original atomic orbitals?

7. Compare and contrast the following three concepts: (a) bonding orbitals; (b) antibonding orbitals; (c) nonbonding orbitals.

8. Describe the shapes, including the locations of the atoms, of σ and $\sigma^{\star}$ orbitals.

9. Describe the shapes, including the locations of the atoms, of π and $\pi^{\star}$ orbitals.

10. State the three rules for placing electrons in molecular orbitals.

11. What is meant by the term "bond order"? How is the value of the bond order calculated?

12. Compare and illustrate the differences between (a) atomic orbitals and molecular orbitals, (b) bonding and antibonding molecular orbitals, (c) σ bonds and π bonds, and (d) localized and delocalized molecular orbitals.

13. Is it possible for a molecule or complex ion in its ground state to have a negative bond order? Why?

Homonuclear Diatomic Species

14. What do we mean when we say that a molecule or ion is (a) *homonuclear,* (b) *heteronuclear,* or (c) *diatomic?*

15. Use the appropriate molecular orbital energy diagram to write the electron configuration for each of the following; calculate the bond order of each, and predict which would exist. (a) H_2^{+}; (b) H_2; (c) H_2^{-}; (d) H_2^{2-}.

16. Repeat Exercise 15 for (a) He_2^{+} and (b) He_2.

17. Repeat Exercise 15 for (a) N_2, (b) Ne_2, and (c) C_2.

18. Repeat Exercise 15 for (a) Li_2, (b) Li_2^{+}, and (c) C_2^{-}.

19. Determine the electron configurations of the following molecules and ions: (a) Be_2, Be_2^{+}, Be_2^{-}; (b) B_2, B_2^{+}, B_2^{-}; (c) CO, CO^{+}.

20. What is the bond order of each of the species in Exercise 19?

21. Which of the species in Exercise 19 are diamagnetic (D) and which are paramagnetic (P)?

22. Use MO theory to predict relative stabilities of the species in Exercise 19. Comment on the validity of these predictions. What else *must* be considered in addition to electron occupancy of MOs?

*23. Which homonuclear diatomic molecules or ions of the second period have the following electron distributions in MOs? In other words, identify X in each.

(a) X_2 $\sigma_{1s}^{2} \sigma_{1s}^{\star 2} \sigma_{2s}^{2} \sigma_{2s}^{\star 2} \pi_{2p_y}^{2} \pi_{2p_z}^{2}$

(b) X_2^{+} $\sigma_{1s}^{2} \sigma_{1s}^{\star 2} \sigma_{2s}^{2} \sigma_{2s}^{\star 2} \sigma_{2p}^{2} \pi_{2p_y}^{2} \pi_{2p_z}^{2} \pi_{2p_y}^{\star 2} \pi_{2p_z}^{\star 1}$

(c) X_2^{-} $\sigma_{1s}^{2} \sigma_{1s}^{\star 2} \sigma_{2s}^{2} \sigma_{2s}^{\star 2} \pi_{2p_y}^{2} \pi_{2p_z}^{2} \sigma_{2p}^{2} \pi_{2p_y}^{\star 1}$

24. What is the bond order of each of the species in Exercise 23?

25. (a) Give the MO designations for O_2, O_2^{-}, O_2^{2-}, O_2^{+}, and O_2^{2+}. (b) Give the bond order in each case. (c) Match these species with the following observed bond lengths: 1.04 Å; 1.12 Å; 1.21 Å; 1.33 Å; and 1.49 Å.

26. (a) Give the MO designations for N_2, N_2^{-}, and N_2^{+}. (b) Give the bond order in each case. (c) Rank these three species by increasing predicted bond length.

27. Assuming that the σ_{2p} MO is lower in energy than the π_{2p_y} and π_{2p_z} MOs for the following species, write out electron configurations for all of them. (a) F_2, F_2^{+}, F_2^{-}; (b) NO, NO^{+}.

28. (a) What is the bond order of each species in Exercise 27? (b) Are they diamagnetic or paramagnetic? (c) What would MO theory predict about the stabilities of these species?

Heteronuclear Diatomic Species

The following is the molecular orbital energy-level diagram for a heteronuclear diatomic molecule, XY, in which both X and Y are from Period 2 and Y is the more electronegative element. This diagram may be useful in answering questions in this section.

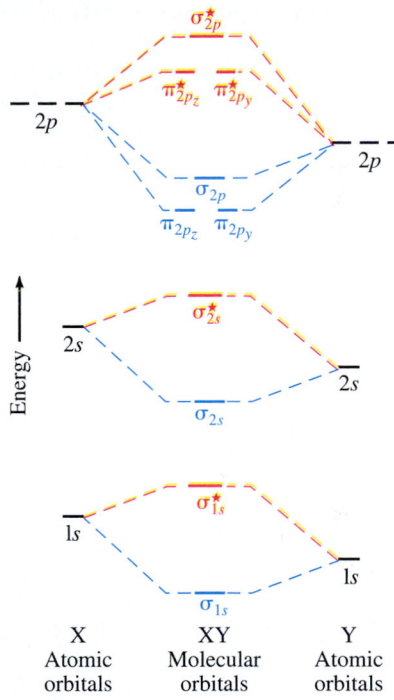

X	XY	Y
Atomic orbitals	Molecular orbitals	Atomic orbitals

29. Use the preceding diagram to fill in an MO diagram for nitrogen oxide, NO. What is the bond order of NO? Is it paramagnetic? How would you assess its stability?

30. Assuming that the preceding MO diagram is valid for CN, CN^{+}, and CN^{-}, write the MO descriptions for these species. Which would be most stable? Why?

31. For each of the two species OF and OF^{+}: (a) Draw MO energy-level diagrams. (b) Write out electron configurations. (c) Determine bond orders and predict relative stabilities. (d) Predict diamagnetism or paramagnetism. Refer to the preceding diagram. Assume that the σ_{2p} MO is lower in energy than the π_{2p_y} and π_{2p_z} MOs.

32. For each of the two species NF and NF^{-}: (a) Draw MO energy-level diagrams. (b) Write out electron configurations. (c) Determine bond orders and predict relative stabilities. (d) Predict diamagnetism or paramagnetism. Refer to the preceding diagram. Assume that the σ_{2p} MO is lower in energy than the π_{2p_y} and π_{2p_z} MOs.

33. Considering the shapes of MO energy-level diagrams for non-polar covalent and polar covalent molecules, what would you predict about MO diagrams, and therefore about overlap of atomic orbitals, for ionic compounds?
34. To increase the strength of the bonding in the hypothetical compound BO, would you add or subtract an electron? Explain your answer with the aid of an MO electron structure.

Delocalization

35. Use Lewis formulas to depict the resonance structures of the following species from the valence bond point of view, and then draw MOs for the delocalized π systems. (a) SO_2, sulfur dioxide; (b) HCO_3^-, hydrogen carbonate ion (H is bonded to O); (c) NO_2^-, nitrite ion.
36. Use Lewis formulas to depict the resonance structure of the following species from the valence bond point of view, and then draw MOs for the delocalized π systems: (a) SO_3, sulfur trioxide; (b) O_3 ozone; (c) HCO_2^-, formate ion (H is bonded to C).

Mixed Exercises

37. Draw and label the complete MO energy-level diagrams for the following species. For each, determine the bond order, predict the stability of the species, and predict whether the species will be paramagnetic. (a) He_2^+; (b) CN; (c) HeH^+.

38. Draw and label the complete MO energy-level diagrams for the following species. For each, determine the bond order, predict the stability of the species, and predict whether the species will be paramagnetic. (a) O_2^{2+}; (b) HO^-; (c) HCl.
39. Which of these species would you expect to be paramagnetic? (a) He_2^-; (b) NO; (c) NO^+; (d) N_2^{2+}; (e) CO^{2-}; (f) F_2^+.

BUILDING YOUR KNOWLEDGE

40. Rationalize the following observations in terms of the stabilities of σ and π bonds: (a) The most common form of nitrogen is N_2, whereas the most common form of phosphorus is P_4 (see the structure in Figure 2-3); (b) The most common forms of oxygen are O_2 and (less common) O_3, whereas the most common form of sulfur is S_8.
41. When carbon vaporizes at extremely high temperatures, among the species present in the vapor is the diatomic molecule C_2. Write a Lewis formula for C_2. Does your Lewis formula of C_2 obey the octet rule? (C_2 does not contain a quadruple bond.) Does C_2 contain a single, a double, or a triple bond? Is it paramagnetic or diamagnetic? Show how molecular orbital theory can be used to predict the answers to questions left unanswered by valence bond theory.

Reactions in Aqueous Solutions I: Acids, Bases, and Salts

10

Acids, bases, and salts occur in many foods.

OBJECTIVES

As you study this chapter, you should learn

- *To understand the Arrhenius theory*

- *About hydrated hydrogen ions*

- *To understand the Brønsted–Lowry theory*

- *The properties of aqueous solutions of acids*

- *The properties of aqueous solutions of bases*

- *How to predict the strengths of binary acids*

- *To understand the strengths of ternary acids*

- *About acid–base reactions*

- *About acidic and basic salts*

- *About amphoterism*

- *How to prepare acids*

- *To understand the Lewis theory*

You will encounter many of these in your laboratory work.

In highly developed societies, acids, bases, and salts are indispensable compounds. Table 4-10 lists the 18 such compounds that were included in the top 50 chemicals produced in the United States in 1993. The production of H_2SO_4 (number 1) was more than twice as great as the production of lime (number 2). Sixty-five percent of the H_2SO_4 is used in the production of fertilizers.

Many acids, bases, and salts occur in nature and serve a wide variety of purposes. For instance, your "digestive juice" contains approximately 0.10 mole of hydrochloric acid per liter. Human blood and the aqueous components of most cells are mildly acidic. The liquid in your automobile battery is approximately 40% H_2SO_4 by mass. Sodium hydroxide, a base, is used in the manufacture of soaps, paper, and many other chemicals. "Drāno" is solid NaOH that contains some aluminum turnings. Sodium chloride is used to season food and as a food preservative. Calcium chloride is used to melt ice on highways and in the emergency treatment of cardiac arrest. Several ammonium salts are used as fertilizers.

Many organic acids (carboxylic acids) and their derivatives occur in nature. Amino acids are the building blocks of proteins, which are important materials in the bodies of animals, including humans. Amino acids are carboxylic acids that also contain basic

Many common household liquids are acidic, including soft drinks, vinegar, and fruit juices. Most cleaning materials are basic.

groups derived from ammonia. The pleasant odors and flavors of ripe fruit are due in large part to the presence of esters (Section 27-13), which are formed from the acids in unripe fruit.

10-1 PROPERTIES OF AQUEOUS SOLUTIONS OF ACIDS AND BASES

Aqueous solutions of most **protonic acids** (those containing acidic hydrogen atoms) exhibit certain properties, which are properties of hydrated hydrogen ions.

1. They have a sour taste. Pickles are usually preserved in vinegar, a 5% solution of acetic acid. Many pickled condiments contain large amounts of sugar so that the taste of acetic acid is not so pronounced. Lemons contain citric acid, which is responsible for their characteristic sour taste.

2. They change the colors of many indicators (highly colored dyes). Acids turn blue litmus red, and cause bromthymol blue to change from blue to yellow.

3. Nonoxidizing acids react with metals above hydrogen in the activity series (Section 4-8, part 2) to liberate hydrogen gas, H_2. (Recall that HNO_3 is a common oxidizing acid.)

4. They react with (neutralize) metal oxides and metal hydroxides to form salts and water (Section 4-5).

5. They react with salts of weaker or more volatile acids to form the weaker or more volatile acid and a new salt.

6. Aqueous solutions of protonic acids conduct an electric current because they are wholly or partly ionized.

Aqueous solutions of most bases also exhibit certain properties. These are due to the hydrated hydroxide ions that are present in aqueous solutions of bases.

1. They have a bitter taste.

2. They have a slippery feeling. Soaps are common examples; they are mildly basic. A solution of household bleach feels very slippery because it is strongly basic.

The indicator bromthymol blue is yellow in acidic solution and blue in basic solution.

Lifesaving Sulfuric Acid

Faced with problems such as soil depletion, animal fertilizer quality, and population increases, farmers around the world are finding it difficult to grow sufficient supplies of food. As a result, within the next hour about 1,100 people (mostly children) will starve to death, and another 50,000 undernourished people will suffer permanent mental or physical injuries.

For thousands of years, civilizations such as the Roman empire used animal fertilizers (manure) to help grow crops. Then, around 1750, Benjamin Franklin invented chemical fertilizers that were man-made salts. Ammonium phosphate and potassium nitrate are two examples. Because approximately one third of all food grown today is dependent upon these types of compounds, it is easy to understand why so many people owe their lives and well-being to man-made chemical fertilizers. With an increasing worldwide demand for fertilizers, it is not surprising that many of the top dozen chemicals produced in the United States are used in the production of fertilizers. Most sulfuric acid goes into the production of fertilizers. Most of it is used to treat a mineral called "phosphate rock," which is primarily calcium phosphate, $Ca_3(PO_4)_2$, and calcium triphosphate fluoride,[1] $Ca_5(PO_4)_3F$.

Phosphate rock is the primary source of phosphorus in man-made fertilizers; Florida sits on one of the richest deposits of phosphate rock. However, without sulfuric acid, the phosphate rock is essentially useless to plant life, because it is not very soluble in water. Plant nutrients must be soluble in water so that plants can obtain the nutrients through their roots. Treating phosphate rock[2] with sulfuric acid produces soluble calcium dihydrogen phosphate, $Ca(H_2PO_4)_2$, and phosphoric acid, H_3PO_4.

$$2Ca_5(PO_4)_3F + 7H_2SO_4 + 17H_2O \longrightarrow$$
$$7Ca(SO_4)_2 \cdot 2H_2O + 3Ca(H_2PO_4)_2 \cdot H_2O + 2HF$$

[1] *Calcium triphosphate fluoride is commonly called fluorapatite or apatite. It is the chief component of the bony structure of teeth.*

[2] *Hydrogen fluoride, which is produced by the reaction shown in the first equation, is very toxic. It causes bone deformities in animals and interferes with plant growth. HF is trapped to prevent it from escaping into the atmosphere.*

Farming is often made economical by the use of fertilizers produced from minerals.

$$Ca_3(PO_4)_2 + 3H_2SO_4 + 6H_2O \longrightarrow$$
$$2H_3PO_4 + 3Ca(SO_4)_2 \cdot 2H_2O$$

Phosphoric acid, produced by the second reaction, is used in a variety of ways to produce other fertilizers. For example, ammonium phosphate is produced when phosphoric acid reacts with ammonia.

$$3NH_3 + H_3PO_4 \longrightarrow (NH_4)_3PO_4$$

The chemical industry's primary products are fertilizers or chemicals such as sulfuric acid and phosphoric acid, which are used to make fertilizers. Thanks to the insight of Benjamin Franklin more than two hundred years ago, we have been able to nourish more people than would have been possible otherwise.

Ronald DeLorenzo
Middle Georgia College

3. They change the colors of many indicators: litmus changes from red to blue, and bromthymol blue changes from yellow to blue, in bases.

4. They react with (neutralize) protonic acids to form salts and water.

5. Their aqueous solutions conduct an electric current because they are dissociated or ionized.

10-2 THE ARRHENIUS THEORY

In 1680 Robert Boyle noted that acids (1) dissolve many substances, (2) change the colors of some natural dyes (indicators), and (3) lose their characteristic properties when mixed with alkalis (bases). By 1814 J. Gay-Lussac concluded that acids *neutralize* bases and that the two classes of substances should be defined in terms of their reactions with each other.

This is an extremely important idea.

In 1884 Svante Arrhenius presented his theory of electrolytic dissociation, which resulted in the Arrhenius theory of acid–base reactions. In his view,

> an **acid** is a substance that contains hydrogen and produces H^+ in aqueous solution. A **base** is a substance that contains the OH group and produces hydroxide ions, OH^-, in aqueous solution.

Neutralization is defined as the combination of H^+ ions with OH^- ions to form H_2O molecules.

$$H^+(aq) + OH^-(aq) \longrightarrow H_2O(\ell) \qquad \text{(neutralization)}$$

We now know that all ions are hydrated in aqueous solution.

The Arrhenius theory of acid–base behavior satisfactorily explained reactions of *protonic acids* with metal hydroxides (hydroxy bases). It was a significant contribution to chemical thought and theory in the latter part of the nineteenth century. We used this theory in introducing acids and bases and discussing some of their reactions. The Arrhenius model of acids and bases, although limited in scope, led to the development of more general theories of acid–base behavior. They will be considered in later sections.

Review Sections 4-2, 4-5, 6-7, and 6-8.

10-3 THE HYDRATED HYDROGEN ION

Although Arrhenius described H^+ ions in water as bare ions (protons), we now know that they are hydrated in aqueous solution and exist as $H^+(H_2O)_n$ in which n is some small integer. This is due to the attraction of the H^+ ions, or protons, for the oxygen end ($\delta-$) of water molecules. While we do not know the extent of hydration of H^+ in most solutions, we often represent the hydrated hydrogen ion as the hydronium ion, H_3O^+, or $H^+(H_2O)_n$ in which $n = 1$.

The most common isotope of hydrogen, ^1_1H, has no neutrons. Thus, $^1_1H^+$ is a bare proton.

> The hydrated hydrogen ion is the species that gives aqueous solutions of acids their characteristic acidic properties.

Whether we use the designation $H^+(aq)$ or H_3O^+, we always refer to the hydrogen ion.

$$H^+ + \ :\!\overset{..}{O}\!-\!H \longrightarrow H\!-\!\overset{..}{O}\!-\!H^+$$
$$\qquad\quad |\qquad\qquad\qquad\ |$$
$$\qquad\quad H\qquad\qquad\qquad H$$

10-4 THE BRØNSTED–LOWRY THEORY

In 1923 J. N. Brønsted and T. M. Lowry independently presented logical extensions of the Arrhenius theory. Brønsted's contribution was more thorough than Lowry's, and the result is known as the **Brønsted theory** or the **Brønsted–Lowry theory.**

The Brønsted–Lowry theory is especially useful for reactions in aqueous solutions. It is widely used in medicine and in the biological sciences.

> An **acid** is defined as a *proton donor*, H^+, and a **base** is defined as a *proton acceptor*.

These definitions are sufficiently broad that any hydrogen-containing molecule or ion capable of releasing a proton, H^+, is an acid, whereas any molecule or ion that can accept a proton is a base.

> An acid–base reaction is the transfer of a proton from an acid to a base.

Thus, the complete ionization of hydrogen chloride, HCl, a *strong* acid, in water is an acid–base reaction in which water acts as a base, a proton acceptor.

Step 1: $HCl(aq) \longrightarrow H^+(aq) + Cl^-(aq)$ (Arrhenius description)
Step 2: $H^+(aq) + H_2O(\ell) \longrightarrow H_3O^+$

Overall: $HCl(aq) + H_2O(\ell) \longrightarrow H_3O^+ + Cl^-(aq)$ (Brønsted–Lowry description)

Remember that in this text we use red to indicate acids and blue to indicate bases. We use rectangles to indicate one conjugate acid–base pair and ovals to indicate the other pair.

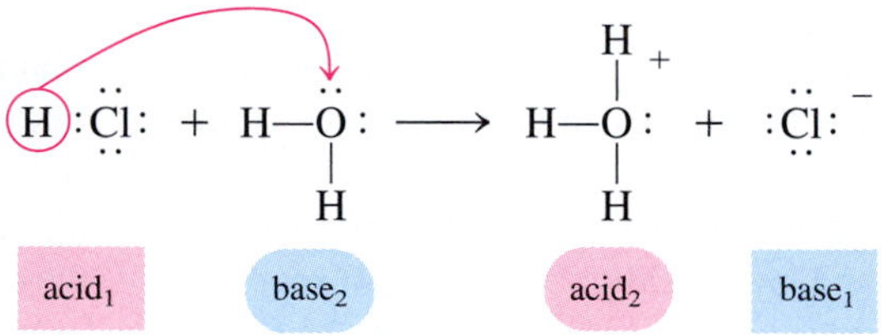

The ionization of hydrogen fluoride, a *weak* acid, is similar, but it occurs to only a slight extent. So we use a double arrow to indicate that it is reversible.

Various measurements (electrical conductivity, freezing point depression, and so on) indicate that HF is only *slightly* ionized in water.

The double arrow is used to indicate that the reaction occurs in both the forward and the reverse directions.

$$HF(g) + H_2O(\ell) \rightleftharpoons H_3O^+ + F^-(aq)$$

We can describe Brønsted–Lowry acid–base reactions in terms of **conjugate acid–base pairs.** These are species that differ by a proton. In the preceding equation, HF ($acid_1$) and F^- ($base_1$) are one conjugate acid–base pair, and H_2O ($base_2$) and H_3O^+ ($acid_2$) are the other pair. The members of each conjugate pair are designated by the same numerical subscript. In the forward reaction, HF and H_2O act as acid and base, respectively. In the reverse reaction, H_3O^+ acts as the acid, or proton donor, and F^- acts as the base, or proton acceptor.

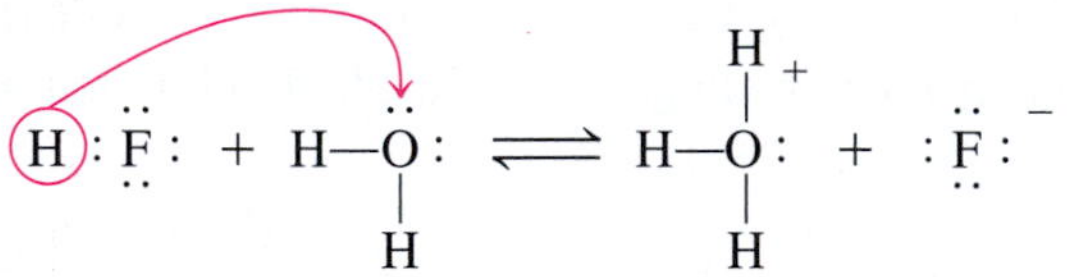

It makes no difference which conjugate acid–base pair, HF and F^- or H_3O^+ and H_2O, is assigned the subscripts 1 and 2.

When the *weak* acid, HF, dissolves in water, the HF molecules give up some H^+ ions that can be accepted by either of two bases, F^- or H_2O. The fact that HF is only slightly ionized tells us that F^- is a stronger base than H_2O. When the *strong* acid, HCl, dissolves in water, the HCl molecules give up H^+ ions that can be accepted by either of two bases, Cl^- or H_2O. The fact that HCl is completely ionized in dilute aqueous solution tells us that Cl^- is a weaker base than H_2O. Thus, the weaker acid, HF, has the stronger conjugate base, F^-. The stronger acid, HCl, has the weaker conjugate base, Cl^-. We can generalize:

F^- is a stronger base than H_2O. H_2O is a stronger base than Cl^-. Therefore, F^- is a stronger base than Cl^-.

The weaker an acid is, the greater is the base strength of its conjugate base. Likewise, the weaker a base is, the stronger is its conjugate acid.

"Strong" and "weak," like many other adjectives, are used in a relative sense. We do not mean to imply that the fluoride ion, F^-, is a strong base compared with species such as the hydroxide ion, OH^-. We mean that *relative to the anions of strong acids, which are very weak bases,* F^- is a much stronger base.

Ammonia acts as a weak Brønsted–Lowry base, and water acts as an acid in the ionization of aqueous ammonia.

Be careful to avoid confusing solubility in water *and* extent of ionization. *They are not necessarily related. Ammonia is very* soluble *in water (~15 mol/L at 25°C). In 0.10 M solution, NH_3 is only 1.3% ionized and 98.7% nonionized.*

$$NH_3(aq) + H_2O(\ell) \rightleftharpoons NH_4^+(aq) + OH^-(aq)$$

base$_1$ acid$_2$ acid$_1$ base$_2$

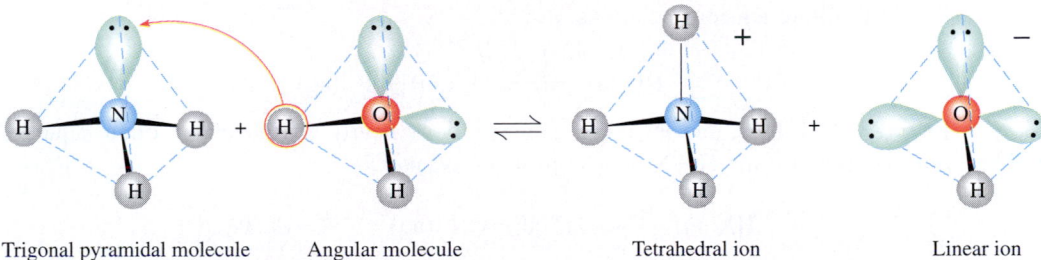

As we see in the reverse reaction, ammonium ion, NH_4^+, is the conjugate acid of NH_3. The hydroxide ion, OH^-, is the conjugate base of water. In three dimensions, the molecular structures are

Trigonal pyramidal molecule   Angular molecule   Tetrahedral ion   Linear ion

Water acts as an acid (H^+ donor) in its reaction with NH_3, whereas it acts as a base (H^+ acceptor) in its reaction with HCl and HF.

Whether water acts as an acid or as a base depends on the other species present.

Careful measurements show that pure water ionizes ever so slightly to produce equal numbers of hydrated hydrogen ions and hydroxide ions.

Base$_1$ Acid$_2$ Acid$_1$ Base$_2$

In simplified notation, we represent this reaction as

$$H_2O(\ell) \rightleftharpoons H^+(aq) + OH^-(aq)$$

This **autoionization** (self-ionization) of water is an acid–base reaction according to the Brønsted–Lowry theory. One H_2O molecule (the acid) donates a proton to another H_2O molecule (the base). The H_2O molecule that donates a proton becomes an OH^- ion, the conjugate base of water. The H_2O molecule that accepts a proton becomes an H_3O^+ ion. Examination of the reverse reaction (right to left) shows that H_3O^+ (an acid) donates a proton to OH^- (a base) to form two H_2O molecules. One H_2O molecule behaves as an acid and the other acts as a base in the autoionization of water. Water is said to be **amphiprotic;** that is, H_2O molecules can accept and donate protons.

> The prefix "amphi-" means "of both kinds." "Amphiprotism" refers to amphoterism by accepting and donating a proton in different reactions.

As we saw in Section 4-5, H_3O^+ and OH^- ions combine to form nonionized water molecules when strong acids and strong soluble bases react to form soluble salts and water. The reverse reaction, the autoionization of water, occurs only slightly, as expected.

10-5 STRENGTHS OF ACIDS

Binary Acids

The ease of ionization of binary protonic acids depends on both (1) the ease of breaking H—X bonds and (2) the stability of the resulting ions in solution. Let us consider the relative strengths of the Group VIIA hydrohalic acids. Hydrogen fluoride ionizes only slightly in dilute aqueous solutions.

> A *weak* acid may be very reactive. For example, HF dissolves sand and glass. The equation for its reaction with sand is
>
> $SiO_2(s) + 4HF(g) \longrightarrow$
> $\qquad SiF_4(g) + 2H_2O(\ell)$
>
> The reaction with glass and other silicates is similar. These reactions are *not* related to acid strength.

$$HF(aq) \xrightarrow{H_2O} H^+(aq) + F^-(aq)$$

However, HCl, HBr, and HI ionize completely or nearly completely in dilute aqueous solutions because the H—X bonds are much weaker.

$$HX(aq) \xrightarrow{H_2O} H^+(aq) + X^-(aq) \qquad X = Cl, \ Br, \ I$$

The order of *bond strengths* for the hydrogen halides is

$$HF \gg HCl > HBr > HI$$

> Bond strength is shown by the bond energies introduced in Chapter 7 and tabulated in Section 15-9. The strength of the H—F bond is due largely to the very small size of the F atom.

To understand why HF is so much weaker an acid than the other hydrogen halides, let us consider the following factors.

1. In HF the electronegativity difference is 1.9, compared with 0.9 in HCl, 0.7 in HBr, and 0.4 in HI (Section 7-8). We might expect the very polar H—F bond in HF to ionize easily. The fact that HF is the *weakest* of these acids suggests that this effect must be of minor importance.

2. The bond strength is considerably greater in HF than in the other three molecules. This tells us that the H—F bond is harder to break than the H—Cl, H—Br, and H—I bonds.

3. The small, highly charged F^- ion, formed when HF ionizes, causes increased ordering of the water molecules. This increase is unfavorable to the process of ionization.

The net result of all factors is that the order of *acid strengths* is

$$HF \ll HCl < HBr < HI$$

In dilute aqueous solutions, hydrochloric, hydrobromic, and hydroiodic acids are completely ionized and all show the same apparent acid strength. Water is sufficiently basic that it does not distinguish among the acid strengths of HCl, HBr, and HI, and therefore it is referred to as a **leveling solvent** for these acids. It is not possible to determine the order of the strengths of these three acids in water because they are so nearly completely ionized.

However, when these compounds dissolve in anhydrous acetic acid or other solvents less basic than water, significant differences in their acid strengths are observed.

$$HCl < HBr < HI \qquad \text{(strongest acid)}$$

As a result of the leveling effect of water,

the hydronium ion is the strongest acid that can exist in aqueous solution.

Acids stronger than $H_3O^+(aq)$ react with water to produce $H_3O^+(aq)$ and their conjugate bases. For example, $HClO_4$ (see Table 10-1) reacts completely with H_2O to form $H^+(aq)$ and $ClO_4^-(aq)$.

$$HClO_4(\ell) \xrightarrow{H_2O} H^+(aq) + ClO_4^-(aq)$$

Similar observations have been made for aqueous solutions of strong soluble bases such as NaOH and KOH. Both are completely dissociated in dilute aqueous solutions.

$$Na^+OH^-(aq) \xrightarrow{H_2O} Na^+(aq) + OH^-(aq)$$

The hydroxide ion is the strongest base that can exist in aqueous solution.

Table 10-1 *Relative Strengths of Conjugate Acid–Base Pairs*

Acid		Base
$\begin{cases} HClO_4 \\ HI \\ HBr \\ HCl \\ HNO_3 \end{cases}$ 100% ionized in dilute aq. soln. No molecules of nonionized acid.	Negligible base strength in water.	$\begin{cases} ClO_4^- \\ I^- \\ Br^- \\ Cl^- \\ NO_3^- \end{cases}$

Acid strength increases → (left arrow, upward)

Base strength increases → (right arrow, downward)

$$\underset{+H^+}{\overset{-H^+}{\rightleftharpoons}}$$

Acid		Base
$\begin{cases} H_3O^+ \\ HF \\ CH_3COOH \\ HCN \\ NH_4^+ \\ H_2O \\ NH_3 \end{cases}$ Equilibrium mixture of nonionized molecules of acid, conjugate base, and $H^+(aq)$.	Reacts completely with H_2O; cannot exist in aqueous solution.	$\begin{cases} H_2O \\ F^- \\ CH_3COO^- \\ CN^- \\ NH_3 \\ OH^- \\ NH_2^- \end{cases}$

The amide ion, NH_2^-, is a stronger base than OH^-.

Bases stronger than OH^- react with H_2O to produce OH^- and their conjugate acids. When metal amides such as sodium amide, $NaNH_2$, are placed in H_2O, the amide ion, NH_2^-, reacts with H_2O completely.

$$NH_2^- + H_2O \longrightarrow NH_3(aq) + OH^-(aq)$$

Thus, we see that H_2O is a leveling solvent for all bases stronger than OH^-.

The trends in binary acid strengths *across* a period (e.g., $CH_4 < NH_3 < H_2O < HF$) are *not* those predicted from trends in bond energies and electronegativity differences. The correlations used for *vertical* trends cannot be used for *horizontal* trends. This is because a "horizontal" series of compounds has different stoichiometries and different numbers of unshared pairs of electrons on the central atoms.

Acid strengths for other *vertical* series of binary acids vary in the same way as those of the VIIA elements. The order of bond strengths for the VIA hydrides is

$$H_2O \gg H_2S > H_2Se > H_2Te$$

H—O bonds are much stronger than the other H—El bonds. As we might expect, the order of acid strengths for these hydrides is just the reverse of the order of bond strengths.

(weakest acid) $H_2O \ll H_2S < H_2Se < H_2Te$ (strongest acid)

Table 10-1 displays relative acid and base strengths of a number of conjugate acid–base pairs.

Ternary Acids

Most ternary acids are *hydroxides of nonmetals* (oxoacids) that ionize to produce $H^+(aq)$. The formula for nitric acid is commonly written HNO_3 to emphasize the presence of an acidic hydrogen atom, but it could also be written as $NO_2(OH)$, as its structure shows (see margin).

Bond that breaks to form H^+ and NO_3^-

Hydroxyl group

We usually reserve the term "hydroxide" for substances that produce basic solutions, and call the other "hydroxides" acids because they ionize to produce $H^+(aq)$. In most ternary acids the hydroxyl oxygen is bonded to a fairly electronegative nonmetal. In nitric acid the nitrogen draws the electrons of the N—O (hydroxyl) bond closer to itself than would a less electronegative element such as sodium. The oxygen pulls the electrons of the O—H bond close enough so that the hydrogen atom ionizes as H^+, leaving NO_3^-.

$$HNO_3(aq) \longrightarrow H^+(aq) + NO_3^-(aq)$$

In contrast, let us consider hydroxides of metals. Oxygen is much more electronegative than most metals, such as sodium. It draws the electrons of the sodium–oxygen bond in NaOH (a strong soluble base) so close to itself that the bonding is ionic. Therefore, NaOH exists as Na^+ and OH^- ions, even in the solid state, and dissociates into Na^+ and OH^- ions when it dissolves in H_2O.

$$Na^+OH^-(s) \xrightarrow{H_2O} Na^+(aq) + OH^-(aq)$$

We usually write the formula for sulfuric acid as H_2SO_4 to emphasize that it is an acid. However, the formula can also be written as $SO_2(OH)_2$, because the structure of sulfuric acid (see margin) shows clearly that H_2SO_4 contains two —O—H groups bound to a sulfur atom. Because the O—H bonds are easier to break than the S—O bonds, sulfuric acid ionizes as an acid.

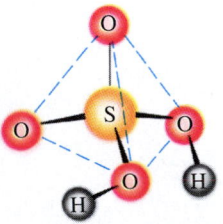

Sulfuric acid is called a polyprotic acid because it has more than one ionizable hydrogen atom per molecule. It is the only *common* polyprotic acid that is also a strong acid.

Step 1: $H_2SO_4(aq) \longrightarrow H^+(aq) + HSO_4^-(aq)$

Step 2: $HSO_4^-(aq) \rightleftharpoons H^+(aq) + SO_4^{2-}(aq)$

The first step in the ionization of H_2SO_4 is complete in dilute aqueous solution. The second step is nearly complete only in very dilute solutions. The first step in the ionization of a polyprotic acid occurs to a greater extent than the second step.

Sulfurous acid, H_2SO_3, is a polyprotic acid that contains the same elements as H_2SO_4. However, H_2SO_3 is a weak acid, which tells us that the H—O bonds in H_2SO_3 are stronger than those in H_2SO_4.

Comparison of the acid strengths of nitric acid, HNO_3, and nitrous acid, HNO_2, shows that HNO_3 is a much stronger acid than HNO_2.

Acid strengths of most ternary acids containing the same central element increase with increasing oxidation state of the central element and with increasing numbers of oxygen atoms.

The following orders of increasing acid strength are typical.

$H_2SO_3 < H_2SO_4$
$HNO_2 < HNO_3$ (strongest acids are on the right side)
$HClO < HClO_2 < HClO_3 < HClO_4$

For most ternary acids containing different elements in the same oxidation state from the same group in the periodic table, acid strengths increase with increasing electronegativity of the central element.

$H_2SeO_4 < H_2SO_4$ $H_2SeO_3 < H_2SO_3$
$H_3PO_4 < HNO_3$
$HBrO_4 < HClO_4$ $HBrO_3 < HClO_3$

Contrary to what we might expect, H_3PO_3 is a stronger acid than HNO_2. Care must be exercised to compare acids that have *similar structures*. For example, H_3PO_2, which has two H atoms bonded to the P atom, is a stronger acid than H_3PO_3, which has one H atom bonded to the P atom. H_3PO_3 is a stronger acid than H_3PO_4, which has no H atoms bonded to the P atom.

In *most* ternary inorganic acids, all H atoms are bonded to O.

10-6 REACTIONS OF ACIDS AND BASES

In Section 4-5 we introduced classical acid–base reactions. We defined neutralization as the reaction of an acid with a base to form a salt and (in most cases) water. Most *salts* are ionic compounds that contain a cation other than H^+ and an anion other than OH^- or O^{2-}. The common *strong acids* and common *strong soluble bases* are listed in the margin. Recall that other common acids may be assumed to be weak. The other common metal hydroxides (bases) are insoluble in water.

Arrhenius and Brønsted–Lowry acid–base neutralization reactions all have one thing in common. They involve the reaction of an acid with a base to form a salt that contains the cation characteristic of the base and the anion characteristic of the acid. Water is also usually formed. This is indicated in the formula unit equation. However, the general form of the net ionic equation and the essence of the reaction are different for different acid–base reactions. They depend upon the solubility and extent of ionization or dissociation of each reactant and product.

In writing ionic equations, we always write the formulas of the predominant forms of the compounds in, or in contact with, aqueous solution. Writing ionic equations from

Common Strong Acids	
Binary	**Ternary**
HCl	$HClO_4$
HBr	$HClO_3$
HI	HNO_3
	H_2SO_4

Strong Soluble Bases	
LiOH	
NaOH	
KOH	$Ca(OH)_2$
RbOH	$Sr(OH)_2$
CsOH	$Ba(OH)_2$

formula unit equations requires a knowledge of the lists of strong acids and strong soluble bases, as well as of the generalizations on solubilities of inorganic compounds. Please review carefully all of Sections 4-2 and 4-3. Study Tables 4-9 and 4-10 carefully because they summarize much information that you are about to use again.

In Section 4-5 we examined some reactions of strong acids with strong soluble bases to form soluble salts. Let us illustrate one additional example. Perchloric acid, $HClO_4$, reacts with sodium hydroxide to produce sodium perchlorate, $NaClO_4$, a soluble ionic salt.

$$HClO_4(aq) + NaOH(aq) \longrightarrow NaClO_4(aq) + H_2O(\ell)$$

The total ionic equation for this reaction is

$$[H^+(aq) + ClO_4^-(aq)] + [Na^+(aq) + OH^-(aq)] \longrightarrow [Na^+(aq) + ClO_4^-(aq)] + H_2O(\ell)$$

Eliminating the spectator ions, Na^+ and ClO_4^-, gives the net ionic equation

This is considered to be the same as

$$H_3O^+ + OH^- \rightarrow 2H_2O$$

$$H^+(aq) + OH^-(aq) \longrightarrow H_2O(\ell)$$

This is the net ionic equation for the reaction of all strong acids with strong soluble bases to form soluble salts and water.

Many weak acids react with strong soluble bases to form soluble salts and water. For example, acetic acid, CH_3COOH, reacts with sodium hydroxide, $NaOH$, to produce sodium acetate, $NaCH_3COO$.

$$CH_3COOH(aq) + NaOH(aq) \longrightarrow NaCH_3COO(aq) + H_2O(\ell)$$

The total ionic equation for this reaction is

$$CH_3COOH(aq) + [Na^+(aq) + OH^-(aq)] \longrightarrow [Na^+(aq) + CH_3COO^-(aq)] + H_2O(\ell)$$

Elimination of Na^+ from both sides gives the net ionic equation

Monoprotic acids contain one, diprotic acids contain two, and triprotic acids contain three acidic (ionizable) hydrogen atoms per formula unit. Polyprotic acids (those that contain more than one ionizable hydrogen atom) are discussed in detail in Chapter 18.

$$CH_3COOH(aq) + OH^-(aq) \longrightarrow CH_3COO^-(aq) + H_2O(\ell)$$

In general terms, the reaction of a *weak monoprotic acid* with a *strong soluble base* to form a *soluble salt* may be represented as

$$HA(aq) + OH^-(aq) \longrightarrow A^-(aq) + H_2O(aq) \qquad \text{(net ionic equation)}$$

EXAMPLE 10-1 *Equations for Acid–Base Reactions*

Write (a) formula unit, (b) total ionic, and (c) net ionic equations for the complete neutralization of phosphoric acid, H_3PO_4, with calcium hydroxide, $Ca(OH)_2$.

Plan

(a) The salt contains the cation of the base, Ca^{2+}, and the anion of the acid, PO_4^{3-}. The salt is $Ca_3(PO_4)_2$.

(b) H_3PO_4 is a weak acid—it is not written in ionic form. $Ca(OH)_2$ is a strong soluble base, and so it is written in ionic form. $Ca_3(PO_4)_2$ is an *insoluble salt,* and so it is written as a complete formula unit.

(c) There are no species common to both sides of the equation, and so there are no spectator ions. The net ionic equation is the same as the total ionic equation except that there are no brackets.

Solution

(a) $2H_3PO_4(aq) + 3Ca(OH)_2(aq) \longrightarrow Ca_3(PO_4)_2(s) + 6H_2O(\ell)$

(b) $2H_3PO_4(aq) + 3[Ca^{2+}(aq) + 2OH^-(aq)] \longrightarrow Ca_3(PO_4)_2(s) + 6H_2O(\ell)$

(c) $2H_3PO_4(aq) + 3Ca^{2+}(aq) + 6OH^-(aq) \longrightarrow Ca_3(PO_4)_2(s) + 6H_2O(\ell)$

You should now work Exercise 60.

EXAMPLE 10-2 *Equations for Acid–Base Reactions*

Write (a) formula unit, (b) total ionic, and (c) net ionic equations for the neutralization of aqueous ammonia with nitric acid.

Plan

(a) The salt contains the cation of the base, NH_4^+, and the anion of the acid, NO_3^-. The salt is NH_4NO_3.
(b) HNO_3 is a strong acid—we write it in ionic form. Ammonia is a weak base. NH_4NO_3 is a soluble salt that is completely dissociated—we write it in ionic form.
(c) We cancel the spectator ions, NO_3^-, and obtain the net ionic equation.

Solution

(a) $HNO_3(aq) + NH_3(aq) \longrightarrow NH_4NO_3(aq)$

(b) $[H^+(aq) + NO_3^-(aq)] + NH_3(aq) \longrightarrow [NH_4^+(aq) + NO_3^-(aq)]$

(c) $H^+(aq) + NH_3(aq) \longrightarrow NH_4^+(aq)$

You should now work Exercises 56 and 58.

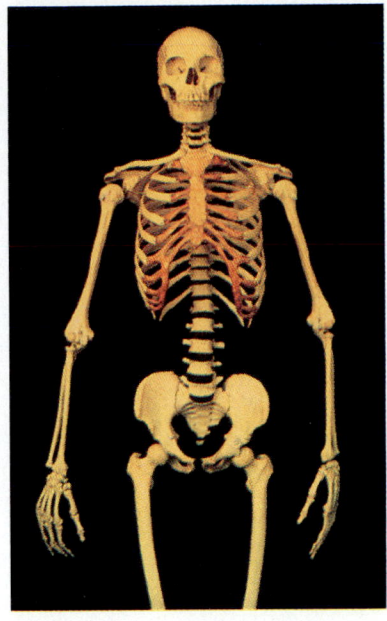

Our bones are mostly calcium phosphate, $Ca_3(PO_4)_2$, an insoluble compound. They are formed by reactions of calcium ions, Ca^{2+}, with phosphate ions, PO_4^{3-}. Calcium and phosphorus are required for normal growth and maintenance of bones and teeth. Thus, they are essential elements in human nutrition.

10-7 ACIDIC SALTS AND BASIC SALTS

To this point we have examined acid–base reactions in which stoichiometric amounts of acids and bases were mixed to form *normal salts*. As the name implies, **normal salts** contain no unreacted H^+ or OH^- ions.

If less than stoichiometric amounts of bases react with *polyprotic* acids, the resulting salts are known as **acidic salts** because they are still capable of neutralizing bases. The reaction of phosphoric acid, H_3PO_4, a weak acid, with strong bases can produce three different salts, depending on the relative amounts of acid and base used.

$$H_3PO_4(aq) + NaOH(aq) \longrightarrow NaH_2PO_4(aq) + H_2O(\ell)$$
    1 mole   1 mole   sodium dihydrogen phosphate,
                             an acidic salt

$$H_3PO_4(aq) + 2NaOH(aq) \longrightarrow Na_2HPO_4(aq) + 2H_2O(\ell)$$
    1 mole   2 moles   sodium hydrogen phosphate,
                             an acidic salt

$$H_3PO_4(aq) + 3NaOH(aq) \longrightarrow Na_3PO_4(aq) + 3H_2O(\ell)$$
    1 mole   3 moles   sodium phosphate,
                             a normal salt

Sodium hydrogen carbonate, baking soda, is the most familiar example of an acidic salt. It can neutralize strong bases, but its aqueous solutions are slightly basic, as the color of the indicator bromthymol blue shows.

There are many additional examples of acidic salts. Sodium hydrogen carbonate, $NaHCO_3$, commonly called sodium bicarbonate, is classified as an acidic salt. However, it is the acidic salt of an extremely weak acid—carbonic acid, H_2CO_3—and solutions of sodium bicarbonate are slightly basic, as are solutions of salts of other extremely weak acids.

Polyhydroxy bases (bases that contain more than one OH per formula unit) react with less than stoichiometric amounts of acids to form **basic salts,** i.e., salts that contain unreacted OH groups. For example, the reaction of aluminum hydroxide with hydrochloric acid can produce three different salts.

> These basic aluminum salts are called "aluminum chlorohydrate." They are components of some deodorants.

$$Al(OH)_3(s) + HCl(aq) \longrightarrow Al(OH)_2Cl(s) + H_2O(\ell)$$
$$\text{1 mole} \qquad \text{1 mole} \qquad \text{aluminum dihydroxide chloride,}$$
$$\text{a basic salt}$$

$$Al(OH)_3(s) + 2HCl(aq) \longrightarrow Al(OH)Cl_2(s) + 2H_2O(\ell)$$
$$\text{1 mole} \qquad \text{2 moles} \qquad \text{aluminum hydroxide dichloride,}$$
$$\text{a basic salt}$$

$$Al(OH)_3(s) + 3HCl(aq) \longrightarrow AlCl_3(aq) + 3H_2O(\ell)$$
$$\text{1 mole} \qquad \text{3 moles} \qquad \text{aluminum chloride,}$$
$$\text{a normal salt}$$

Aqueous solutions of basic salts are not necessarily basic, but they can neutralize acids. Most basic salts are rather insoluble in water.

10-8 AMPHOTERISM

As we have seen, whether a particular substance behaves as an acid or as a base depends on its environment. In Section 10-4 we described the amphiprotic nature of water. **Amphoterism** is the general term that describes the ability of a substance to react either as an acid or as a base. *Amphiprotic behavior* describes the cases in which substances exhibit amphoterism by accepting and by donating a proton, H^+. Several *insoluble* metal hydroxides are amphoteric; i.e., they react with acids to form salts and water, but they also dissolve in and react with excess strong soluble bases.

> All hydroxides containing small, highly charged metal ions are insoluble in water.

Aluminum hydroxide is a typical amphoteric metal hydroxide. Its behavior as a *base* is illustrated by its reaction with nitric acid to form a *normal salt.*

CHEMISTRY IN USE

Our Daily Lives

Everyday Salts of Ternary Acids

You may have encountered some salts of ternary acids without even being aware of them. For example, did you know that cigarette tobacco and paper are treated with potassium nitrate, KNO_3, so that smokers do not have to relight their cigarettes as often as cigar and pipe smokers do? When heated by a burning cigarette, potassium nitrate slowly decomposes into potassium nitrite (KNO_2) and oxygen (O_2). This oxygen supports combustion and keeps cigarettes from becoming extinguished. Unfortunately, the luxury of having cigarettes remaining lighted comes at a high price. When people fall asleep while smoking cigarettes in bed, house fires often result. The odds of a tobacco-induced fire are much less for pipe and cigar smokers because their tobacco has not been treated with potassium nitrate. Thousands of lives and billions of dollars are lost annually from cigarette-initiated fires. Fires kill more Americans than are killed by all major natural disasters such as hurricanes, earthquakes, tornados, and floods. In fact, the United States leads the industrialized world in the number of fire deaths on a per capita basis, and cigarettes are the leading cause of home fires in this country.

Turning to your pantry, the iron in many of your breakfast cereals and breads may have been added in the form of iron(II) sulfate, $FeSO_4$, or iron(II) phosphate, $Fe_3(PO_4)_2$; the calcium in these foods often comes from the addition of calcium carbonate, $CaCO_3$. Fruits and vegetables keep fresh longer after an application of sodium sulfite, Na_2SO_3, and sodium hydrogen sulfite, $NaHSO_3$. Restaurants also use these two sulfites to keep their salad bars more appetizing. The red color of fresh meat is maintained for much longer by the additives sodium nitrate, $NaNO_3$, and sodium nitrite, $NaNO_2$. Sodium phosphate, Na_3PO_4, is used to prevent metal ion flavors and to control acidity in some canned goods.

Many salts of ternary acids are used in medicine. Lithium carbonate, Li_2CO_3, has been used successfully to combat severe jet lag. Lithium carbonate is also useful in the treatment of mania, depression, alcoholism, and schizophrenia. Magnesium

The tips of "strike anywhere" matches contain tetraphosphorus trisulfide, red phosphorus, and potassium chlorate. Friction converts kinetic energy into heat, which initiates a spontaneous reaction.

$$P_4S_3(s) + 8O_2 \longrightarrow P_4O_{10}(s) + 3SO_2(g)$$

The thermal decomposition of $KClO_3$ provides additional oxygen for this reaction.

sulfate, $MgSO_4$, sometimes helps to prevent convulsions during pregnancy and to reduce the solubility of toxic barium sulfate in internally administered preparations consumed before gastrointestinal X-rays are taken.

Other salts of ternary acids that you may find in your home include potassium chlorate, $KClO_3$, in matches as an oxidizing agent and oxygen source; sodium hypochlorite, $NaClO$, in bleaches and mildew removers; and ammonium carbonate, $(NH_4)_2CO_3$, which is the primary ingredient in smelling salts. Limestone and marble are calcium carbonate; gypsum and plaster of Paris are primarily calcium sulfate, $CaSO_4$. Fireworks get their brilliant colors from salts such as barium nitrate, $Ba(NO_3)_2$, which imparts a green color; strontium carbonate, $SrCO_3$, which gives a red color; and copper(II) sulfate, $CuSO_4$, which produces a blue color. Should your fireworks get out of hand and accidentally start a fire, the ammonium phosphate, $(NH_4)_3PO_4$, sodium hydrogen carbonate, $NaHCO_3$, and potassium hydrogen carbonate, $KHCO_3$, in your ABC dry fire extinguisher will come in handy.

An unexpected place to find ternary acid salts is in your long distance phone bills; nitrates, NO_3^-, are cheaper than day rates.

Ronald DeLorenzo
Middle Georgia College

$$Al(OH)_3(s) + 3HNO_3(aq) \longrightarrow Al(NO_3)_3(aq) + 3H_2O(\ell)$$

$$Al(OH)_3(s) + 3[H^+(aq) + NO_3^-(aq)] \longrightarrow [Al^{3+}(aq) + 3NO_3^-(aq)] + 3H_2O(\ell)$$

$$Al(OH)_3(s) + 3H^+(aq) \longrightarrow Al^{3+}(aq) + 3H_2O(\ell)$$

Table 10-2 *Amphoteric Hydroxides*

Metal or Metalloid ions	Insoluble Amphoteric Hydroxide	Complex Ion Formed in an Excess of a Strong Soluble Base
Be^{2+}	$Be(OH)_2$	$[Be(OH)_4]^{2-}$
Al^{3+}	$Al(OH)_3$	$[Al(OH)_4]^-$
Cr^{3+}	$Cr(OH)_3$	$[Cr(OH)_4]^-$
Zn^{2+}	$Zn(OH)_2$	$[Zn(OH)_4]^{2-}$
Sn^{2+}	$Sn(OH)_2$	$[Sn(OH)_3]^-$
Sn^{4+}	$Sn(OH)_4$	$[Sn(OH)_6]^{2-}$
Pb^{2+}	$Pb(OH)_2$	$[Pb(OH)_4]^{2-}$
As^{3+}	$As(OH)_3$	$[As(OH)_4]^-$
Sb^{3+}	$Sb(OH)_3$	$[Sb(OH)_4]^-$
Si^{4+}	$Si(OH)_4$	SiO_4^{4-} and SiO_3^{2-}
Co^{2+}	$Co(OH)_2$	$[Co(OH)_4]^{2-}$
Cu^{2+}	$Cu(OH)_2$	$[Cu(OH)_4]^{2-}$

When an excess of a solution of any strong soluble base, e.g., NaOH, is added to solid aluminum hydroxide, the $Al(OH)_3$ dissolves. The equation for the reaction is usually written

$$Al(OH)_3(s) + NaOH(aq) \longrightarrow NaAl(OH)_4(aq)$$

an acid   a base   sodium aluminate, a soluble compound

The total ionic and net ionic equations are

$$Al(OH)_3(s) + [Na^+(aq) + OH^-(aq)] \longrightarrow [Na^+(aq) + Al(OH)_4^-(aq)]$$

$$Al(OH)_3(s) + OH^-(aq) \longrightarrow Al(OH)_4^-(aq)$$

Other amphoteric metal hydroxides undergo similar reactions.

Table 10-2 contains a list of the common amphoteric hydroxides. Three are hydroxides of metalloids, the elements located along the line that divides metals and nonmetals in the periodic table.

Generally, elements of intermediate electronegativity form amphoteric hydroxides. Those of high and low electronegativity form acidic and basic "hydroxides," respectively.

10-9 THE PREPARATION OF ACIDS

Binary acids may be prepared by combination of appropriate elements with hydrogen (Section 6-7, part 2).

Small quantities of the hydrogen halides (their solutions are called hydrohalic acids) and other *volatile acids* are usually prepared by dropping concentrated nonvolatile acids onto the appropriate salts. (Sulfuric and phosphoric acids are classified as *nonvolatile acids* because they have much higher boiling points than other common acids.) The reactions of concentrated sulfuric acid with solid sodium fluoride and sodium chloride produce gaseous hydrogen fluoride and hydrogen chloride, respectively.

$$H_2SO_4(\ell) + NaF(s) \longrightarrow NaHSO_4(s) + HF(g)$$

sulfuric acid   sodium fluoride   sodium hydrogen   hydrogen fluoride
bp = 338°C   sulfate   bp = 19.59°C

$$H_2SO_4(\ell) + NaCl(s) \longrightarrow NaHSO_4(s) + HCl(g)$$

sodium chloride   hydrogen chloride
bp = −84.9°C

Because concentrated sulfuric acid is a fairly strong oxidizing agent, it cannot be used to prepare hydrogen bromide or hydrogen iodide; instead, the free halogens are produced. Phosphoric acid, a nonoxidizing acid, is dropped onto solid sodium bromide or sodium iodide to produce hydrogen bromide or hydrogen iodide, as the following equations show.

$$H_3PO_4(\ell) + NaBr(s) \xrightarrow{\Delta} NaH_2PO_4(s) + HBr(g)$$

phosphoric acid   sodium bromide   sodium dihydrogen   hydrogen bromide
bp = 213°C   phosphate   bp = −67.0°C

$$H_3PO_4(\ell) + NaI(s) \xrightarrow{\Delta} NaH_2PO_4(s) + HI(g)$$

sodium iodide   hydrogen iodide
bp = −35°C

This kind of reaction may be generalized as

$$\begin{matrix} \text{nonvolatile} \\ \text{acid} \end{matrix} + \begin{matrix} \text{salt of} \\ \text{volatile acid} \end{matrix} \longrightarrow \text{volatile acid} + \begin{matrix} \text{salt of} \\ \text{nonvolatile acid} \end{matrix}$$

In Section 6-8, part 2 we saw that many nonmetal oxides, called acid anhydrides, react with water to form *ternary acids* with no changes in oxidation numbers. For example, dichlorine heptoxide, Cl_2O_7, forms perchloric acid when it dissolves in water.

$$\overset{+7}{Cl_2}O_7(\ell) + H_2O(\ell) \longrightarrow 2[H^+(aq) + \overset{+7}{Cl}O_4^-(aq)]$$

Some *high oxidation state transition metal oxides* are acidic oxides; that is, they dissolve in water to give solutions of ternary acids. Manganese(VII) oxide, Mn_2O_7, and chromium(VI) oxide, CrO_3, are the most common examples.

$$\overset{+7}{Mn_2}O_7(\ell) + H_2O(\ell) \longrightarrow 2[H^+(aq) + \overset{+7}{Mn}O_4^-(aq)]$$

manganese(VII) oxide   permanganic acid

$$2\overset{+6}{Cr}O_3(s) + H_2O(\ell) \longrightarrow [2H^+(aq) + \overset{+6}{Cr_2}O_7^{2-}(aq)]$$

chromium(VI) oxide   dichromic acid

Neither permanganic acid nor dichromic acid has been isolated in pure form. Many stable salts of both are well known.

The halides and oxyhalides of some nonmetals hydrolyze (react with water) to produce two acids—a (binary) hydrohalic acid and a (ternary) oxyacid of the nonmetal. Phosphorus trihalides react with water to produce the corresponding hydrohalic acids and phosphorous acid, a weak diprotic acid, while phosphorus pentahalides give phosphoric acid and the corresponding hydrohalic acid.

$$\overset{+3}{P}X_3 + 3H_2O(\ell) \longrightarrow \overset{+3}{H_3}PO_3(aq) + 3HX(aq)$$

$$\overset{+5}{P}X_5 + 4H_2O(\ell) \longrightarrow \overset{+5}{H_3}PO_4(aq) + 5HX(aq)$$

The volatile acid HCl can be made by dropping concentrated H_2SO_4 onto solid NaCl. Gaseous HCl is liberated. HCl(g) dissolves in the water on a piece of filter paper. The indicator methyl red on the paper turns red, its color in acidic solution.

A solution of dichromic acid, $H_2Cr_2O_7$, is deep red.

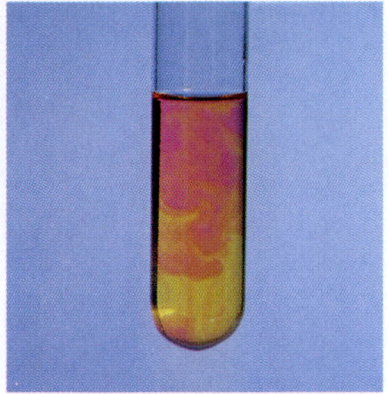

A drop of PCl_3 is added to water that contains the indicator methyl orange. As PCl_3 reacts with water to form HCl and H_3PO_3, the indicator turns red, its color in acidic solution.

This is the same Lewis who made many contributions to our understanding of chemical bonding.

There are no changes in oxidation numbers in these reactions. Consider the reactions of PCl_3 and PCl_5 with H_2O.

$$\overset{+3}{P}Cl_3(\ell) + 3H_2O(\ell) \longrightarrow H_3\overset{+3}{P}O_3(aq) + 3[H^+(aq) + Cl^-(aq)]$$

phosphorus trichloride   phosphorous acid

$$\overset{+5}{P}Cl_5(s) + 4H_2O(\ell) \longrightarrow H_3\overset{+5}{P}O_4(aq) + 5[H^+(aq) + Cl^-(aq)]$$

phosphorus pentachloride   phosphoric acid

10-10 THE LEWIS THEORY

In 1923 Professor G. N. Lewis presented the most comprehensive of the classic acid–base theories. The Lewis definitions follow.

> An **acid** is any species that can accept a share in an electron pair. A **base** is any species that can make available, or "donate," a share in an electron pair.

These definitions do *not* specify that an electron pair must be transferred from one atom to another—only that an electron pair, residing originally on one atom, must be shared between two atoms. *Neutralization* is defined as **coordinate covalent bond formation.** This results in a covalent bond in which both electrons were furnished by one atom.

The reaction of boron trichloride with ammonia is a typical Lewis acid–base reaction.

$$BCl_3(g) + NH_3(g) \longrightarrow Cl_3B:NH_3$$

acid   base   product

The Brønsted–Lowry theory is more general than the Arrhenius theory. It includes all Arrhenius acid–base reactions *plus* many others. The Lewis theory is the most general. It includes all Brønsted–Lowry acid–base reactions *plus* a great many others.

The Lewis theory is sufficiently general that it covers *all* acid–base reactions that the other theories include, plus many additional reactions such as complex formation (Chapter 25).

The autoionization of water (Section 10-4) was described in terms of Brønsted–Lowry theory. In Lewis theory terminology, this is also an acid–base reaction. The acceptance of a proton, H^+, by a base involves the formation of a *coordinate covalent bond*.

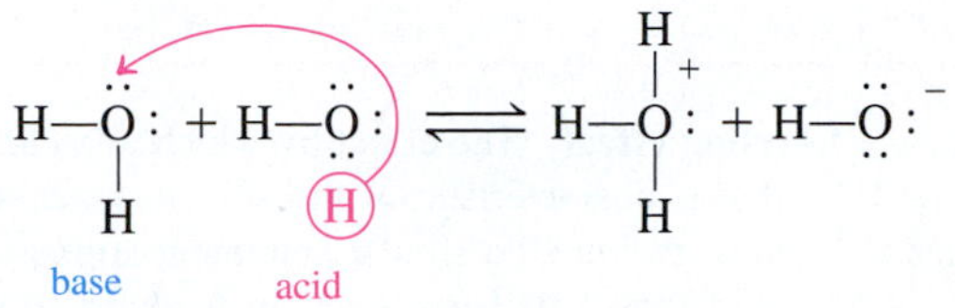

$$H—\overset{..}{\underset{\underset{H}{|}}{O}}: + H—\overset{..}{\underset{\underset{(H)}{|}}{O}}: \;\rightleftharpoons\; H—\overset{\overset{H}{|}}{\underset{\underset{H}{|}}{O}}:^{+} + H—\overset{..}{\underset{..}{O}}:^{-}$$

base   acid

Theoretically, any species that contains an unshared electron pair could act as a base. In fact, most ions and molecules that contain unshared electron pairs undergo some reactions by sharing their electron pairs. Conversely, many Lewis acids contain only six electrons in the highest occupied energy level of the central element. They react by accepting a share in an additional pair of electrons. These species are said to have an **open sextet.** Many compounds of the Group IIIA elements are Lewis acids, as illustrated by the reaction of boron trichloride with ammonia, presented earlier.

Anhydrous aluminum chloride is a common Lewis acid that is used to catalyze many organic reactions. $AlCl_3$ acts as a Lewis acid when it dissolves in hydrochloric acid to give a solution that contains $AlCl_4^-$ ions.

$$AlCl_3(s) + Cl^-(aq) \longrightarrow AlCl_4^-(aq)$$

acid   base   product

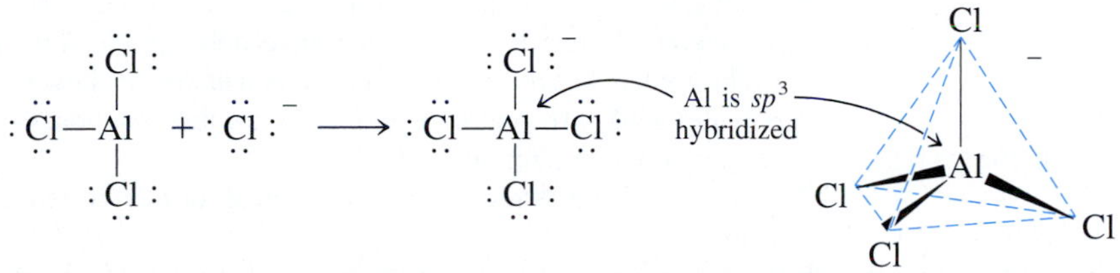

Other ions and molecules behave as Lewis acids by expansion of the valence shell of the central element. Anhydrous tin(IV) chloride, often called stannic chloride, is a colorless liquid that also is frequently used as a Lewis acid catalyst. The tin atom (Group IVA) can expand its valence shell by utilizing vacant d orbitals. It can accept shares in two additional electron pairs, as its reaction with hydrochloric acid illustrates.

$$SnCl_4(\ell) + 2Cl^-(aq) \longrightarrow SnCl_6^{2-}(aq)$$

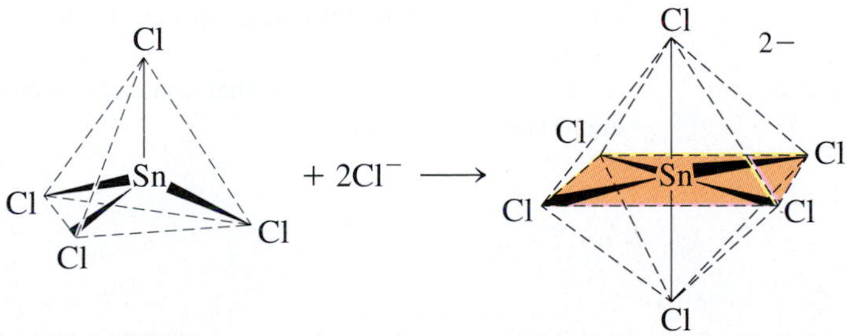

acid   base

Sn is sp^3 hybridized   Sn is sp^3d^2 hybridized
(tetrahedral)   (octahedral)

Many organic and biological reactions are acid–base reactions that do not fit within the Arrhenius or Brønsted–Lowry theories. Experienced chemists find the Lewis theory to be very useful because so many more chemical reactions are covered by it. The less experienced sometimes find the theory less useful, but as their knowledge expands so does its utility.

Key Terms

Acid (Arrhenius or Brønsted–Lowry)   A substance that produces $H^+(aq)$ ions in aqueous solution. Strong acids ionize completely or almost completely in dilute aqueous solution; weak acids ionize only slightly.

Acid anhydride   The oxide of a nonmetal that reacts with water to form an acid.

Acidic salt   A salt that contains an ionizable hydrogen atom; does not necessarily produce acidic solutions.

Amphiprotism   The ability of a substance to exhibit amphoterism by accepting or donating protons.

Amphoterism   Ability of a substance to act as either an acid or a base.

Anhydrous   Without water.

Autoionization   An ionization reaction between identical molecules.

Base (Arrhenius)   A substance that produces $OH^-(aq)$ ions in aqueous solution. Strong soluble bases are soluble in water and are completely *dissociated*. Weak bases ionize only slightly.

Basic anhydride   The oxide of a metal that reacts with water to form a base.

Basic salt   A salt containing a basic OH group.

Brønsted–Lowry acid   A proton donor.

Brønsted–Lowry base   A proton acceptor.

Conjugate acid–base pair   In Brønsted–Lowry terminology, a reactant and product that differ by a proton, H^+.

Coordinate covalent bond   A covalent bond in which both shared electrons are furnished by the same species; the bond between a Lewis acid and a Lewis base.

Dissociation   In aqueous solution, the process in which a *solid ionic compound* separates into its ions.

Electrolyte   A substance whose aqueous solutions conduct electricity.

Formula unit equation   A chemical equation in which all compounds are represented by complete formulas.

Hydration   The process by which water molecules bind to ions or molecules in the solid state or in solution.

Hydride   A binary compound of hydrogen.

Hydrolysis   Reaction of a substance with water.

Ionization   In aqueous solution, the process in which a *molecular* compound reacts with water to form ions.

Leveling effect   The effect by which all acids stronger than the acid that is characteristic of the solvent react with the solvent to produce that acid; a similar statement applies to bases. The strongest acid (base) that can exist in a given solvent is the acid (base) characteristic of that solvent.

Lewis acid   Any species that can accept a share in an electron pair to form a coordinate covalent bond.

Lewis base   Any species that can make available a share in an electron pair to form a coordinate covalent bond.

Net ionic equation   The equation that results from canceling spectator ions and eliminating brackets from a total ionic equation.

Neutralization   The reaction of an acid with a base to form a salt and (usually) water; usually, the reaction of hydrogen ions with hydroxide ions to form water molecules.

Nonelectrolyte   A substance whose aqueous solutions do not conduct electricity.

Normal oxide   A metal oxide containing the oxide ion, O^{2-} (oxygen in the -2 oxidation state).

Normal salt   A salt containing no ionizable H atoms or OH groups.

Open sextet   Refers to species that have only six electrons in the highest energy level of the central element (many Lewis acids).

Polyprotic acid   An acid that contains more than one ionizable hydrogen atom per formula unit.

Protonic acid   An Arrhenius (classical) acid, or a Brønsted–Lowry acid.

Salt   A compound that contains a cation other than H^+ and an anion other than OH^- or O^{2-}.

Spectator ions   Ions in solution that do not participate in a chemical reaction.

Strong electrolyte   A substance that conducts electricity well in dilute aqueous solution.

Ternary acid   An acid that contains three elements—usually H, O, and another nonmetal.

Ternary compound   A compound that contains three elements.

Total ionic equation   The equation for a chemical reaction written to show the predominant form of all species in aqueous solution or in contact with water.

Weak electrolyte   A substance that conducts electricity poorly in dilute aqueous solution.

Exercises

Basic Ideas

1. Robert Boyle observed that acids have certain properties. Which ones did he observe?

2. Gay-Lussac reached an important conclusion about acids and bases. What was it?

3. Define the following terms. You may wish to refer to Chapter 4 to check the definitions. (a) acid; (b) neutralization; (c) ionization.

The Arrhenius Theory

4. Outline Arrhenius' ideas about acids and bases. (a) How did he define the following terms: acid, base, neutralization? (b) Give an example that illustrates each term.

5. Define and illustrate the following terms clearly and concisely. Give an example of each. (a) strong electrolyte; (b) weak electrolyte; (c) nonelectrolyte; (d) strong acid; (e) strong soluble base;

(f) weak acid; (g) weak base; (h) insoluble base.

6. Distinguish between the following pairs of terms and provide a specific example of each. (a) strong acid and weak acid; (b) strong soluble base and weak base; (c) strong soluble base and insoluble base.

7. Write formulas and names for (a) the common strong acids; (b) three weak acids; (c) the common strong soluble bases; (d) the most common weak base; (e) four soluble ionic salts; (f) four insoluble salts.

8. Describe an experiment for classifying each of these compounds as a strong electrolyte, a weak electrolyte, or a nonelectrolyte. Classify each. K_2CO_3; HCN; C_2H_5COOH; CH_3OH; H_2S; H_2SO_4; HCOOH; NH_3.

9. Summarize the electrical properties of strong electrolytes, weak electrolytes, and nonelectrolytes.

The Hydrated Hydrogen Ion

10. Write the formula of a hydrated hydrogen ion that contains only one water of hydration. Give another name for the hydrated hydrogen ion.

11. Why is the hydrated hydrogen ion important?

12. Criticize the following statement: "The hydrated hydrogen ion should always be represented as H_3O^+."

Brønsted–Lowry Theory

13. State the basic ideas of the Brønsted–Lowry theory.

14. Use Brønsted–Lowry terminology to define the following terms. Illustrate each with a specific example. (a) acid; (b) conjugate base; (c) base; (d) conjugate acid; (e) conjugate acid–base pair.

15. Write balanced equations that describe the ionization of the following acids in dilute aqueous solution. Use a single arrow ($\rightarrow$) to represent complete, or nearly complete, ionization and a double arrow ($\rightleftharpoons$) to represent a small extent of ionization. (a) HNO_3; (b) CH_3COOH; (c) H_2S; (d) HCN; (e) HF; (f) $HClO_4$.

16. Use words and equations to describe how ammonia can act as a base in (a) aqueous solution and (b) the pure state, i.e., as gaseous ammonia molecules when it reacts with gaseous hydrogen chloride or a similar anhydrous acid.

17. What does autoionization mean? How can the autoionization of water be described as an acid–base reaction?

18. What do we mean when we say that water is amphiprotic? (a) Can we also describe water as amphoteric? Why? (b) Illustrate the amphiprotic nature of water by writing two equations for reactions in which water exhibits this property.

19. What structural features must a compound have to be able to undergo autoionization?

20. Illustrate, with appropriate equations, the fact that these species are bases in water: NH_3; HS^-; CH_3COO^-; O^{2-}.

21. In terms of Brønsted–Lowry theory, state the differences between (a) a strong and a weak base and (b) a strong and a weak acid.

22. Give the products in the following acid–base reactions. Identify the conjugate acid–base pairs.
    (a) $NH_4^+ + CN^-$
    (b) $HS^- + HSO_4^-$

(c) $HClO_4 + [H_2NNH_3]^+$
(d) $NH_2^- + H_2O$

23. List the conjugate acids of H_2O, OH^-, I^-, HCl, AsO_4^{3-}, NH_2^-, HPO_4^{2-}, and SO_4^{2-}.

24. List the conjugate bases of H_2O, HSe^-, HCl, PH_4^+, and $HOCH_3$.

25. Identify the Brønsted–Lowry acids and bases in these reactions and group them into conjugate acid–base pairs.
    (a) $NH_3 + HBr \rightleftharpoons NH_4^+ + Br^-$
    (b) $NH_4^+ + HS^- \rightleftharpoons NH_3 + H_2S$
    (c) $H_3O^+ + PO_4^{3-} \rightleftharpoons HPO_4^{2-} + H_2O$
    (d) $HSO_3^- + CN^- \rightleftharpoons HCN + SO_3^{2-}$

26. Identify all species in the following reactions as either an acid or a base, in the Brønsted–Lowry sense.
    (a) $CN^- + H_2O \rightleftharpoons HCN + OH^-$
    (b) $HCO_3^- + H_2SO_4 \rightleftharpoons HSO_4^- + H_2CO_3$
    (c) $H_2C_2O_4 + NO_2^- \rightleftharpoons HNO_2 + HC_2O_4^-$

27. Identify all species in the following reactions as either an acid or a base, in the Brønsted–Lowry sense.
    (a) $NH_4^+ + HCO_3^- \rightleftharpoons NH_3 + H_2CO_3$
    (b) $NH_2^- + H_2O \rightleftharpoons NH_3 + OH^-$
    (c) $O^{2-} + H_2O \rightleftharpoons OH^- + OH^-$

28. Arrange the species in the reactions of Exercise 26 as Brønsted–Lowry conjugate pairs.

29. Arrange the species in the reactions of Exercise 27 as Brønsted–Lowry conjugate pairs.

Properties of Aqueous Solutions of Acids and Bases

30. Write equations and designate conjugate pairs for the stepwise reactions in water of (a) sulfuric acid, H_2SO_4, and (b) H_2SO_3.

31. List six properties of aqueous solutions of protonic acids.

32. List five properties of bases in aqueous solution. Does aqueous ammonia exhibit these properties? Why?

33. We say that strong acids, weak acids, and weak bases *ionize* in water, while strong soluble bases *dissociate* in water. What is the difference between ionization and dissociation?

34. Distinguish between solubility in water and extent of ionization in water. Provide specific examples that illustrate the meanings of both terms.

35. Write three general statements that describe the extents to which acids, bases, and salts are ionized in dilute aqueous solutions.

Strengths of Acids

36. Classify each of the hydrides NaH, BeH_2, BH_3, CH_4, NH_3, H_2O, and HF as a Brønsted–Lowry base, a Brønsted–Lowry acid, or neither.

37. What does "acid strength" mean?

38. What does "base strength" mean?

39. Which of the following substances are (a) strong soluble bases, (b) insoluble bases, (c) strong acids, or (d) weak acids? LiOH; HCl; $Ca(OH)_2$; $Fe(OH)_2$; H_2S; H_2CO_3; H_2SO_4; $Zn(OH)_2$.

40. (a) What are binary protonic acids? (b) Write names and formulas for four binary protonic acids.

41. (a) How can the order of increasing acid strength in a series of similar binary protonic acids be explained? (b) Illustrate your answer for the series HF, HCl, HBr, and HI. (c) What is the order of

increasing base strength of the conjugate bases of the acids in (b)? Why? (d) Is your explanation applicable to the series H_2O, H_2S, H_2Se, and H_2Te? Why?

42. What does the term "leveling effect" mean? Illustrate your answer with three specific examples.

43. (a) Which is the stronger acid of each pair? (1) NH_4^+, NH_3; (2) H_2O, H_3O^+; (3) HS^-, H_2S; (4) HSO_3^-, H_2SO_3. (b) How are acidity and charge related?

44. Arrange the members of each group in order of decreasing acidity: (a) H_2O, H_2Se, H_2S; (b) HI, HCl, HF, HBr; (c) H_2S, S^{2-}, HS^-.

45. Illustrate the leveling effect of water by writing equations for the reactions of HCl and HNO_3 with water.

Ternary Acids

46. Why can we describe nitric and sulfuric acids as "hydroxides" of nonmetals?

47. What are ternary acids? Write names and formulas for four of them.

48. Write proton-transfer autoionization equations for the following amphiprotic solvents. (a) NH_3, (b) NH_2OH, (c) H_2SO_4.

49. Explain the order of increasing acid strength for the following groups of acids and the order of increasing base strength for their conjugate bases. (a) H_2SO_3, H_2SO_4; (b) HNO_2, HNO_3; (c) H_3PO_3, H_3PO_4; (d) HClO, $HClO_2$, $HClO_3$, $HClO_4$.

50. (a) Write a generalization that describes the order of acid strengths for a series of ternary acids that contain different elements in the same oxidation state from the same group in the periodic table. (b) Indicate the order of acid strengths for the following: (1) HNO_3, H_3PO_4; (2) H_3PO_4, H_3AsO_4; (3) H_2SO_4, H_2SeO_4; (4) $HClO_3$, $HBrO_3$, HIO_3.

*51. List the following acids in order of increasing strength: (a) sulfuric, phosphoric, and perchloric; (b) HIO_3, HIO_2, HIO, and HIO_4; (c) selenous, sulfurous, and tellurous acids; (d) hydrosulfuric, hydroselenic, and hydrotelluric acids; (e) H_2CrO_4, H_2CrO_2, $HCrO_3$, and H_3CrO_3.

Reactions of Acids and Bases

52. Why are acid–base reactions described as neutralization reactions?

53. Distinguish among (a) formula unit equations, (b) total ionic equations, and (c) net ionic equations. What are the advantages and limitations of each?

54. Classify each substance as either an electrolyte or a nonelectrolyte: NH_4Cl; HI; C_6H_6; RaF_2; $Zn(CH_3COO)_2$; $Cu(NO_3)_2$; CH_3COOH; $C_{12}H_{22}O_{11}$ (table sugar); LiOH; $KHCO_3$; CCl_4; $La_2(SO_4)_3$; I_2.

55. Classify each substance as either a strong or a weak electrolyte, and then list (a) the strong acids, (b) the strong bases, (c) the weak acids, and (d) the weak bases. NaCl; $MgSO_4$; HCl; $(COOH)_2$; $Ba(NO_3)_2$; H_3PO_4; $Sr(OH)_2$; HNO_3; HI; $Ba(OH)_2$; LiOH; C_3H_5COOH; NH_3; CH_3NH_2; KOH; HCN; $HClO_4$.

For Exercises 56–58, write balanced (1) formula unit, (2) total ionic, and (3) net ionic equations for reactions between the acid–base pairs. Name all compounds except water. Assume complete neutralization.

56. (a) HNO_3 + KOH →
    (b) H_2SO_4 + NaOH →
    (c) HCl + $Ca(OH)_2$ →
    (d) CH_3COOH + KOH →
    (e) HI + NaOH →

57. (a) H_2CO_3 + $Sr(OH)_2$ →
    (b) H_2SO_4 + $Ca(OH)_2$ →
    (c) H_3PO_4 + $Ba(OH)_2$ →
    (d) H_2S + KOH →
    (e) H_3AsO_4 + KOH →

58. (a) $HClO_4$ + $Ba(OH)_2$ →
    (b) HBr + NH_3 →
    (c) HNO_3 + NH_3 →
    (d) H_2SO_4 + $Fe(OH)_3$ →
    (e) H_3PO_4 + $Ba(OH)_2$ →

59. Complete these equations by writing the formulas of the omitted compounds.
    (a) $Ba(OH)_2$ + ? → $BaSO_4(s)$ + H_2O
    (b) FeO(s) + ? → $Fe(NO_3)_2(aq)$ + H_2O
    (c) HCl(aq) + ? → $ZnCl_2(aq)$ + ?
    (d) Na_2O + ? → 2NaOH(aq)
    (e) NaOH + ? → $Na_2HPO_4(aq)$ + ?
    (two possible answers)

60. Although many salts may be formed by a variety of reactions, salts are usually thought of as being derived from the reaction of an acid with a base. For each of the salts listed below, choose the acid and base that would react with each other to form the salt. Write the (1) formula unit, (2) total ionic, and (3) net ionic equations for the formation of each salt. (a) $Pb(NO_3)_2$; (b) $AlCl_3$; (c) $(NH_4)_2CO_3$; (d) $Ca(ClO_4)_2$; (e) $Al_2(SO_4)_3$.

Acidic and Basic Salts

61. What are polyprotic acids? Write names and formulas for five polyprotic acids.

62. What are acidic salts? Write balanced equations to show how the following acidic salts can be prepared from the appropriate acid and base: $NaHSO_3$; $KHCO_3$; NaH_2PO_4; Na_2HPO_4.

63. Indicate the molar ratio of acid and base required in each case in Exercise 62.

64. The following salts are components of fertilizers. They are made by reacting gaseous NH_3 with concentrated solutions of acids. The heat produced by the reactions evaporates most of the water. Write balanced formula unit equations that show the formation of each. (a) NH_4NO_3; (b) $NH_4H_2PO_4$; (c) $(NH_4)_2HPO_4$; (d) $(NH_4)_3PO_4$; (e) $(NH_4)_2SO_4$.

65. What are polyhydroxy bases? Write names and formulas for five polyhydroxy bases.

66. What are basic salts? (a) Write balanced equations to show how each of the following basic salts can be prepared from the appropriate acid and base: Ba(OH)Cl; $Al(OH)_2Cl$; $Al(OH)Cl_2$. (b) Indicate the molar ratio of acid and base required in each case.

67. What are amphoteric metal hydroxides? (a) Are they bases? (b) Write the names and formulas for four amphoteric metal hydroxides.

68. Chromium(III) hydroxide and lead(II) hydroxide are typical amphoteric hydroxides. (a) Write the formula unit, total ionic, and net ionic equations for the complete reaction of each hydroxide with nitric acid. (b) Write the same kinds of equations for the reaction of each hydroxide with an excess of potassium hydroxide solution. Reference to Table 10-2 may be helpful.

69. Write the chemical equations for the stepwise ionization of oxalic acid, $(COOH)_2$, a diprotic acid.

70. Write the chemical equations for the stepwise ionization of citric acid, $C_3H_5O(COOH)_3$, a triprotic acid.

Preparation of Acids

71. A volatile acid such as nitric acid, HNO_3, can be prepared by adding concentrated H_2SO_4 to a salt of the acid. (a) Write the chemical equation for the reaction of H_2SO_4 with sodium nitrate (called Chile saltpeter). (b) A dilute aqueous solution of H_2SO_4 cannot be used. Why?

72. Outline a method of preparing each of the following acids and write appropriate balanced equations for each preparation: (a) H_2S; (b) HCl; (c) CH_3COOH.

73. Repeat Exercise 72 for (a) carbonic acid, (b) perchloric acid, (c) permanganic acid, and (d) phosphoric acid (two methods).

74. Give the formula for an example chosen from the representative elements for (a) an acidic oxide, (b) an amphoteric oxide, and (c) a basic oxide.

The Lewis Theory

75. Define and illustrate the following terms clearly and concisely. Write an equation to illustrate the meaning of each term. (a) Lewis acid; (b) Lewis base; (c) neutralization according to Lewis theory.

76. What are the advantages and limitations of the Brønsted–Lowry theory?

77. What are the advantages and limitations of the Lewis theory?

78. Write a Lewis formula for each species in the following equations. Label the acids and bases using Lewis theory terminology.
(a) $H_2O + H_2O \rightleftharpoons H_3O^+ + OH^-$
(b) $HCl(g) + H_2O \rightarrow H_3O^+ + Cl^-$
(c) $NH_3(g) + H_2O \rightleftharpoons NH_4^+ + OH^-$
(d) $NH_3(g) + HCl(g) \rightarrow NH_4Cl(s)$

79. What is the term for a single covalent bond in which both electrons in the shared pair come from the same atom? Identify the Lewis acid and base and the donor and acceptor atoms in the following.

$$\underset{\underset{H}{|}}{\overset{\overset{H}{|}}{H-N:}} \; + \; \underset{\underset{:\ddot{F}:}{|}}{\overset{\overset{:\ddot{F}:}{|}}{B-\ddot{F}:}} \rightarrow \underset{\underset{H}{|}}{\overset{\overset{H}{|}}{H-N}}-\underset{\underset{:\ddot{F}:}{|}}{\overset{\overset{:\ddot{F}:}{|}}{B-\ddot{F}:}}$$

80. Identify the Lewis acid and base and the donor and acceptor atoms in each of the following.

(a) $H-\ddot{O}: + H^+ \rightarrow \left[H-\overset{\overset{H}{|}}{O}-H \right]^+$

(b) $6 \left[:\ddot{C}l: \right]^- + Pt^{4+} \rightarrow \left[\begin{array}{c} :\ddot{C}l \quad :\ddot{C}l: \\ :\ddot{C}l-Pt-\ddot{C}l: \\ :\ddot{C}l: \quad :\ddot{C}l: \end{array} \right]^{2-}$

81. Iodine, I_2, is much more soluble in a water solution of potassium iodide, KI, than it is in H_2O. The anion found in the solution is I_3^-. Write an equation for the reaction that forms I_3^-, indicating the Lewis acid and the Lewis base.

82. A group of very strong acids are the fluoroacids, H_mXF_n. Two such acids are formed by Lewis acid–base reactions. (a) Identify the Lewis acid and the Lewis base.

$$HF + SbF_5 \longrightarrow H(SbF_6) \quad \text{(called a ''super'' acid, hexafluoroantimonic acid)}$$

$$HF + BF_3 \longrightarrow H(BF_4) \quad \text{(tetrafluoroboric acid)}$$

(b) To which atom is the H of the product bonded? How is the H bonded?

Mixed Exercises

83. Identify each of the following as (i) acidic, (ii) basic, or (iii) amphoteric. Assume all oxides are dissolved in or are in contact with water. Do not be intimidated by the way in which the formula of the compound is written. (a) Cs_2O; (b) N_2O_5; (c) HCl; (d) $SO_2(OH)_2$; (e) HNO_2; (f) Al_2O_3; (g) BaO; (h) H_2O; (i) CO_2; (j) SO_2.

84. Indicate which of the following substances—(a) H_2S; (b) $PO(OH)_3$; (c) H_2CaO_2; (d) $ClO_3(OH)$; (e) $Sb(OH)_3$—can act as (i) an acid, (ii) a base, or (iii) both according to the Arrhenius (classical) theory and/or the Brønsted–Lowry theory. Do not be confused by the way in which the formulas are written.

85. (a) Write equations for the reactions (1) $HCO_3^- + H_3O^+$ and (2) $HCO_3^- + OH^-$, and indicate the conjugate acid–base pairs in each case. (b) A substance such as HCO_3^- that reacts with both H_3O^+ and OH^- is said to be _____. (Fill in the missing word.)

86. (a) List the conjugate bases of (1) H_3PO_4, (2) NH_4^+, and (3) OH^- and the conjugate acids of (4) HSO_4^-, (5) PH_3, and (6) PO_4^{3-}. (b) Given that NO_2^- is a stronger base than NO_3^-, which is the stronger acid—nitric acid, HNO_3, or nitrous acid, HNO_2?

*87. A 0.1 M solution of copper(II) chloride, $CuCl_2$, causes the light bulb in Figure 4-2 to glow brightly. When hydrogen sulfide, H_2S, a very weak acid, is added to the solution, a black precipitate of copper(II) sulfide, CuS, forms and the bulb still glows brightly. The experiment is repeated with a 0.1 M solution of copper(II) acetate, $Cu(CH_3COO)_2$, which also causes the bulb to glow brightly. Again, CuS forms, but this time the bulb glows dimly. With the aid of ionic equations, explain the difference in behavior of the $CuCl_2$ and $Cu(CH_3COO)_2$ solutions.

Conductivity experiment.

88. Referring again to Figure 4-2, explain the following results of a conductivity experiment (use ionic equations). (a) Individual solutions of NaOH and HCl cause the bulb to glow brightly. When the solutions are mixed, the bulb still glows brightly but not as brightly as before. (b) Individual solutions of NH_3 and CH_3COOH cause the bulb to glow dimly. When the solutions are mixed, the bulb glows brightly.

89. Which statements are true? Rewrite any false statement so that it is correct. (a) Strong acids and bases are virtually 100% ionized or dissociated in dilute aqueous solutions. (b) The leveling effect is the seemingly identical strengths of all acids and bases in aqueous solutions. (c) A conjugate acid is a molecule or ion formed by the addition of a proton to a base. (d) Amphoterism and amphiprotism are the same in aqueous solution.

BUILDING YOUR KNOWLEDGE

90. Autoionization can occur when an ion other than an H^+ is transferred, as exemplified by the transfer of a Cl^- ion from one PCl_5 molecule to another. Write the equation for this reaction. What are the shapes of the two ions that are formed?

91. Limestone, $CaCO_3$, is a water-insoluble material, whereas $Ca(HCO_3)_2$ is soluble. Caves are formed when rainwater containing dissolved CO_2 passes over limestone for long periods of time. Write a chemical equation for the acid–base reaction.

92. Acids react with metal carbonates and hydrogen carbonates to form carbon dioxide and water. (a) Write the balanced equation for the reaction that occurs when baking soda, $NaHCO_3$, and vinegar, 5% acetic acid, are mixed. What causes the "fizz"? (b) Lactic acid, $CH_3CH(OH)COOH$, is found in sour milk and in buttermilk. Many of its reactions are very similar to those of acetic acid. Write the balanced equation for the reaction of baking soda, $NaHCO_3$, with lactic acid. Explain why bread "rises" during the baking process.

93. Some of the acid formed in tissues is excreted through the kidneys. One of the bases removing the acid is HPO_4^{2-}. Write the equation for the reaction. Could Cl^- serve this function?

94. One of the chemical products of muscle contraction is lactic acid $(CH_3CH(OH)CO_2H)$, a monoprotic acid whose structure is

$$\begin{array}{ccc} & H & OH & \\ & | & | & \diagup O \\ H - & C - C - & C \\ & | & | & \diagdown O-H \\ & H & H \end{array}$$

Prolonged exercise can temporarily overload the body's capacity for elimination of this substance, and the resulting increase in lactic acid concentration in the muscles causes pain and stiffness. (a) Lactic acid has six H atoms, yet it acts as a monoprotic acid in an aqueous environment. Which of the H atoms is ionizable? (b) Draw the structural formula of the conjugate base. (c) Write a net ionic equation that illustrates the ionization of lactic acid in water. (d) Describe the geometry around each of the carbon atoms in lactic acid.

Reactions in Aqueous Solutions II: Calculations

A concentrated NaOH solution added to a CH_3COOH solution that contains a large amount of phenolphthalein indicator.

OBJECTIVES

As you study this chapter, you should

• *Review molarity calculations and expand your understanding of molarity*

• *Learn about acid–base titrations, standardization, and the associated calculations using*
    The mole method and molarity
    Equivalent weights and normality

• *Learn to balance equations for oxidation–reduction reactions*

• *Learn about redox titrations and the associated calculations*

AQUEOUS ACID–BASE REACTIONS

Digestive juice is the acidic fluid secreted by glands in the lining of the stomach.

Hydrochloric acid, HCl, is called "stomach acid" because it is the main acid (~0.10 M) in our digestive juices. When the concentration of HCl is too high in humans, problems result. These problems may range from "heartburn" to ulcers, which can eat through the lining of the stomach wall. Snakes have very high concentrations of HCl in their digestive juices so that they can digest whole small animals and birds.

Automobile batteries contain 40% H_2SO_4 by mass. When the battery has "run down," the concentration of H_2SO_4 is significantly lower than 40%. A technician checks an automobile battery by drawing some battery acid into a hydrometer, which indicates the density of the solution. This density is related to the concentration of H_2SO_4.

There are many practical applications of acid–base chemistry in which we must know the concentration of a solution of an acid or a base.

11-1 CALCULATIONS INVOLVING MOLARITY

In Sections 3-6 through 3-9 we introduced methods for expressing concentrations of solutions and discussed some related calculations. Review of those sections will be helpful as we learn more about acid–base reactions in solutions.

In *some cases,* one mole of an acid reacts with one mole of a base.

$$HCl + NaOH \longrightarrow NaCl + H_2O$$
$$HNO_3 + KOH \longrightarrow KNO_3 + H_2O$$

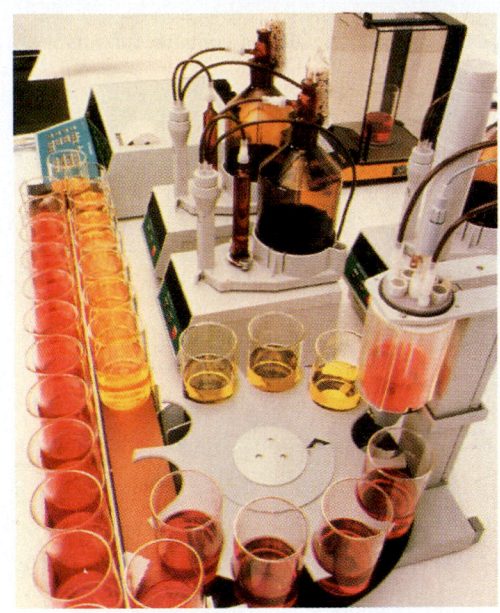

Automatic titrators are used in modern analytical laboratories. Such titrators rely on electrical properties of the solutions. Methyl red indicator changes from yellow to red at the end point of this titration.

Because one mole of each acid reacts with one mole of each base in these examples, *one liter of a one-molar solution of either of these acids* reacts with *one liter of a one-molar solution of either of these bases.* These acids have only one acidic hydrogen per formula unit, and these bases have one hydroxide ion per formula unit, so one formula unit of base reacts with one hydrogen ion.

The *reaction ratio* is the relative numbers of moles of reactants and products shown in the balanced equation.

EXAMPLE 11-1 *Acid–Base Reactions*

If 100 mL of 1.00 *M* HCl solution and 100 mL of 1.00 *M* NaOH are mixed, what is the molarity of the salt in the resulting solution? You may assume that the volumes are additive.

Plan

We first write the balanced equation for the acid–base reaction, and then construct the reaction summary that shows the amounts (millimoles) of NaOH and HCl. We determine the amount of salt formed from the reaction summary. The final (total) volume is the sum of the volumes mixed. Then we can calculate the molarity of the salt.

Solution

The following tabulation shows that equal numbers of millimoles or moles of HCl and NaOH are mixed, and therefore all of the HCl and NaOH react. The resulting solution contains only NaCl, the salt formed by the reaction, and water

	NaOH	+	HCl	$\longrightarrow$	NaCl	+	H$_2$O
Rxn ratio:	1 mmol		1 mmol		1 mmol		1 mmol
Start:	$\left[100 \text{ mL} \left(\dfrac{1.00 \text{ mmol}}{\text{mL}}\right)\right]$		$\left[100 \text{ mL} \left(\dfrac{1.00 \text{ mmol}}{\text{mL}}\right)\right]$		0 mmol		
	= 100 mmol NaOH		= 100 mmol HCl				
Change:	−100 mmol		−100 mmol		+100 mmol		
After rxn:	0 mmol		0 mmol		100 mmol		

Experiments have shown that volumes of dilute aqueous solutions are very nearly additive. No significant error is introduced by making this assumption. 100 mL of NaOH solution mixed with 100 mL of HCl solution gives 200 mL of solution.

The millimole (mmol) was introduced at the end of Section 2-7. Please review Example 2-11. Recall that

$$1 \text{ mol} = 1000 \text{ mmol}$$
$$1 \text{ L} = 1000 \text{ mL}$$

$$\text{molarity} = \frac{\text{no. mol}}{\text{L}} = \frac{\text{mmol}}{\text{mL}}$$

The HCl and NaOH neutralize each other exactly, and the resulting solution contains 100 mmol of NaCl in 200 mL of solution. Its molarity is

$$\underset{?}{}\frac{\text{mmol NaCl}}{\text{mL}} = \frac{100 \text{ mmol NaCl}}{200 \text{ mL}} = \boxed{0.500 \ M \text{ NaCl}}$$

You should now work Exercise 8.

EXAMPLE 11-2   *Acid–Base Reactions*

If 100 mL of 1.00 M HCl and 100 mL of 0.75 M NaOH solutions are mixed, what are the molarities of the solutes in the resulting solution?

Plan

We proceed as we did in Example 11-1. This reaction summary shows that NaOH is the limiting reactant and that we have excess HCl.

Solution

	HCl	+	NaOH	$\longrightarrow$	NaCl	+	H$_2$O
Rxn ratio:	1 mmol		1 mmol		1 mmol		1 mmol
Start:	100 mmol		75 mmol		0 mmol		
Change:	−75 mmol		−75 mmol		+75 mmol		
After rxn:	25 mmol		0 mmol		75 mmol		

Because two solutes are present in the solution after reaction, we must calculate the concentrations of both.

$$\underset{?}{}\frac{\text{mmol HCl}}{\text{mL}} = \frac{25 \text{ mmol HCl}}{200 \text{ mL}} = \boxed{0.12 \ M \text{ HCl}}$$

$$\underset{?}{}\frac{\text{mmol NaCl}}{\text{mL}} = \frac{75 \text{ mmol NaCl}}{200 \text{ mL}} = \boxed{0.38 \ M \text{ NaCl}}$$

Both HCl and NaCl are strong electrolytes, so the solution is 0.12 M in H$^+$(aq), (0.12 + 0.38) M = 0.50 M in Cl$^-$, and 0.38 M in Na$^+$ ions.

You should now work Exercise 10.

▼ **PROBLEM-SOLVING TIP**   *Review Limiting Reactant Calculations*

To solve many of the problems in this chapter, you will need to apply the limiting reactant concept (Section 3-3). In Example 11-1, we confirm that the two reactants are initially present in the mole ratio required by the balanced chemical equation; they both react completely, so there is no excess of either one. In Example 11-2, we need to determine which reactant limits the reaction. Before you proceed, be sure you understand how the ideas of Section 3-3 are used in these examples.

In many cases more than one mole of a base will be required to neutralize completely one mole of an acid, or more than one mole of an acid will be required to neutralize completely one mole of a base.

$$\text{H}_2\text{SO}_4 + 2\text{NaOH} \longrightarrow \text{Na}_2\text{SO}_4 + 2\text{H}_2\text{O}$$
$$\text{1 mol} \qquad \text{2 mol} \qquad \qquad \text{1 mol}$$

$$2HCl + Ca(OH)_2 \longrightarrow CaCl_2 + 2H_2O$$
$$\quad \text{2 mol} \qquad \text{1 mol} \qquad\qquad \text{1 mol}$$

The first equation shows that one mole of H_2SO_4 reacts with two moles of NaOH. Thus, *two* liters of 1 M NaOH solution are required to neutralize one liter of 1 M H_2SO_4 solution. The second equation shows that two moles of HCl react with one mole of $Ca(OH)_2$. Thus, *two* liters of HCl solution are required to neutralize one liter of $Ca(OH)_2$ solution of equal molarity.

EXAMPLE 11-3   *Volume of Acid to Neutralize Base*

What volume of 0.00300 M HCl solution would just neutralize 30.0 mL of 0.00100 M $Ca(OH)_2$ solution?

Plan

We write the balanced equation for the reaction to determine the reaction ratio. Then we convert (1) milliliters of $Ca(OH)_2$ solution to moles of $Ca(OH)_2$ using molarity as a unit factor, 0.00100 mol $Ca(OH)_2$/1000 mL $Ca(OH)_2$ solution; (2) moles of $Ca(OH)_2$ to moles of HCl using the unit factor, 2 mol HCl/1 mol $Ca(OH)_2$ (from the balanced equation); and (3) moles of HCl to milliliters of HCl solution using the unit factor, 1000 mL HCl/0.00300 mol HCl, that is, molarity inverted.

mL Ca(OH)$_2$ soln	$\longrightarrow$	mol Ca(OH)$_2$ present	$\longrightarrow$	mol HCl needed	$\longrightarrow$	mL HCl(aq) needed

Solution

The balanced equation for the reaction is

$$2HCl + Ca(OH)_2 \longrightarrow CaCl_2 + 2H_2O$$
$$\quad \text{2 mol} \qquad \text{1 mol} \qquad\qquad \text{1 mol} \qquad \text{2 mol}$$

$$\underline{?}\ \text{mL HCl} = 30.0\ \text{mL Ca(OH)}_2 \times \frac{0.00100\ \text{mol Ca(OH)}_2}{1000\ \text{mL Ca(OH)}_2} \times \frac{2\ \text{mol HCl}}{1\ \text{mol Ca(OH)}_2} \times \frac{1000\ \text{mL HCl}}{0.00300\ \text{mol HCl}}$$

$$= \boxed{20.0\ \text{mL HCl}}$$

You should now work Exercise 12.

In the preceding example we used the unit factor, 2 mol HCl/1 mol $Ca(OH)_2$, to convert moles of $Ca(OH)_2$ to moles of HCl because the balanced equation shows that two moles of HCl are required to neutralize one mole of $Ca(OH)_2$. We must always write the balanced equation and determine the *reaction ratio*.

▼ **PROBLEM-SOLVING TIP**   *There Is More Than One Way to Solve Some Problems*

In many problems more than one "plan" can be followed. In Example 11-3 a particular plan was used successfully. Many students can more easily visualize the solution by following a plan like that in Examples 11-1 and 11-4. We suggest that you use the plan that you find most understandable.

EXAMPLE 11-4   *Acid–Base Reactions*

If 100 mL of 1.00 M H_2SO_4 solution is mixed with 200 mL of 1.00 M KOH, what salt is produced, and what is its molarity?

Plan

We proceed as we did in Example 11-1. We note that the reaction ratio is 1 mmol of H_2SO_4 to 2 mmol of KOH to 1 mmol of K_2SO_4.

Solution

$$H_2SO_4 \quad + \quad 2KOH \quad \longrightarrow \quad K_2SO_4 \quad + H_2O$$

	H_2SO_4	$2KOH$	K_2SO_4
Rxn ratio:	1 mmol	2 mmol	1 mmol
Start:	$\left[100 \text{ mL} \left(\dfrac{1.00 \text{ mmol}}{\text{mL}}\right)\right]$ $= 100$ mmol	$\left[200 \text{ mL} \left(\dfrac{1.00 \text{ mmol}}{\text{mL}}\right)\right]$ $= 200$ mmol	0 mmol
Change:	-100 mmol	-200 mmol	$+100$ mmol
After rxn:	0 mmol	0 mmol	100 mmol

The reaction produces 100 mmol of potassium sulfate. This is contained in 300 mL of solution, and so the concentration is

$$\underset{?}{\_}\frac{\text{mmol } K_2SO_4}{\text{mL}} = \frac{100 \text{ mmol } K_2SO_4}{300 \text{ mL}} = \boxed{0.333 \text{ } M \text{ } K_2SO_4}$$

You should now work Exercise 16.

Because K_2SO_4 is a strong electrolyte, this corresponds to 0.666 M K^+ and 0.333 M SO_4^{2-}.

ACID–BASE TITRATIONS

In Section 3-9 we discussed *titrations* of solutions of acids and bases and introduced the terminology used to describe titrations.

Standardization is the process by which one determines the concentration of a solution by measuring accurately the volume of the solution required to react with an exactly known amount of a **primary standard.** The standardized solution is then known as a **secondary standard** and is used in the analysis of unknowns.

The properties of an ideal *primary standard* include the following.

CO_2, H_2O, and O_2 are present in the atmosphere. They react with many substances.

1. It must not react with or absorb the components of the atmosphere, such as water vapor, oxygen, and carbon dioxide.

2. It must react according to one invariable reaction.

3. It must have a high percentage purity.

4. It should have a high formula weight to minimize the effect of error in weighing.

5. It must be soluble in the solvent of interest.

6. It should be nontoxic.

7. It should be readily available (inexpensive).

8. It should be environmentally friendly.

The first five of these characteristics are essential to minimize the errors involved in analytical methods. The last three characteristics are just as important as the first five in most analytical laboratories. Because primary standards are often costly and difficult to prepare, secondary standards are often used in day-to-day work.

11-2 THE MOLE METHOD AND MOLARITY

Let us now describe the use of a few primary standards for acids and bases. One primary standard for solutions of acids is sodium carbonate, Na_2CO_3, a solid compound.

Refer to the Brønsted–Lowry theory (Section 10-4).

$$H_2SO_4 + Na_2CO_3 \longrightarrow Na_2SO_4 + CO_2 + H_2O$$

$$\text{1 mol} \qquad \text{1 mol} \qquad \text{1 mol} \qquad \text{1 mol} \qquad \text{1 mol}$$

$$\text{1 mol } Na_2CO_3 = 106.0 \text{ g} \qquad \text{and} \qquad \text{1 mmol } Na_2CO_3 = 0.1060 \text{ g}$$

Sodium carbonate is a salt. However, because a base can be broadly defined as a substance that reacts with hydrogen ions, in *this* reaction Na_2CO_3 can be thought of as a base.

EXAMPLE 11-5 *Standardization of an Acid Solution*

Calculate the molarity of a solution of H_2SO_4 if 40.0 mL of the solution neutralizes 0.364 gram of Na_2CO_3.

Plan

We know from the balanced equation that 1 mol of H_2SO_4 reacts with 1 mol of Na_2CO_3, 106.0 g. This provides the unit factors that convert 0.364 g of Na_2CO_3 to the corresponding number of moles of H_2SO_4, from which we can calculate molarity.

$$\boxed{\begin{array}{c} \text{g } Na_2CO_3 \\ \text{available} \end{array}} \longrightarrow \boxed{\begin{array}{c} \text{mol } Na_2CO_3 \\ \text{present} \end{array}} \longrightarrow \boxed{\begin{array}{c} \text{mol } H_2SO_4 \\ \text{used} \end{array}} \longrightarrow \boxed{\begin{array}{c} \text{molarity} \\ \text{of } H_2SO_4 \end{array}}$$

Solution

$$\underset{\text{?}}{} \text{ mol } H_2SO_4 = 0.364 \text{ g } Na_2CO_3 \times \frac{1 \text{ mol } Na_2CO_3}{106.0 \text{ g } Na_2CO_3} \times \frac{1 \text{ mol } H_2SO_4}{1 \text{ mol } Na_2CO_3}$$

$$= 0.00343 \text{ mol } H_2SO_4 \qquad \text{(present in 40.0 mL of solution)}$$

Now we calculate the molarity of the H_2SO_4 solution

$$\frac{\text{? mol } H_2SO_4}{L} = \frac{0.00343 \text{ mol } H_2SO_4}{0.0400 \text{ L}} = \boxed{0.0858 \text{ } M \text{ } H_2SO_4}$$

You should now work Exercise 22.

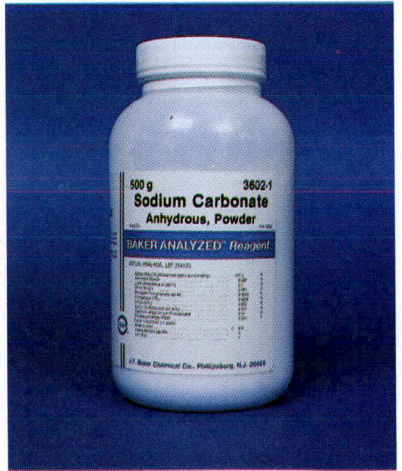

Sodium carbonate is often used as a primary standard for acids.

Most inorganic bases are metal hydroxides, all of which are solids. However, even in the solid state, most inorganic bases react rapidly with CO_2 (an acid anhydride) from the atmosphere. Most metal hydroxides also absorb H_2O from the air. These properties make it *very* difficult to accurately weigh out samples of pure metal hydroxides. Chemists obtain solutions of bases of accurately known concentration by standardizing the solutions

The "ph" in *phthalate* is silent. Phthalate is pronounced "thalate."

against an acidic salt, potassium hydrogen phthalate, $KC_6H_4(COO)(COOH)$. This is produced by neutralization of one of the two ionizable hydrogens of an organic acid, phthalic acid.

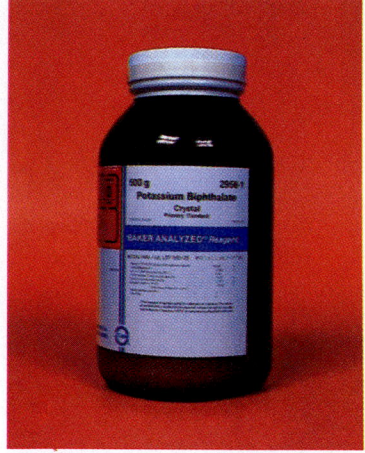

Very pure KHP is readily available.

$$C_6H_4(COOH)_2 \text{ phthalic acid} + KOH \longrightarrow H_2O + KC_6H_4(COO)(COOH) \text{ potassium hydrogen phthalate (KHP)}$$

This acidic salt, known as KHP, has one acidic hydrogen (highlighted) that reacts with bases. KHP is easily obtained in a high state of purity, and is soluble in water. It is used as a primary standard for bases.

EXAMPLE 11-6   *Standardization of Base Solution*

A 20.00-mL sample of a solution of NaOH reacts with 0.3641 gram of KHP. Calculate the molarity of the basic solution.

Plan

We first write the balanced equation for the reaction between NaOH and KHP. We then calculate the number of moles of NaOH in 20.00 mL of solution from the amount of KHP that reacts with it. Then we can calculate the molarity of the NaOH solution.

$$\boxed{\text{g KHP available}} \longrightarrow \boxed{\text{mol KHP available}} \longrightarrow \boxed{\text{mol NaOH required}} \longrightarrow \boxed{\text{molarity of NaOH}}$$

The P in KHP stands for the phthalate ion, $C_6H_4(COO)_2^{2-}$, *not* phosphorus.

Solution

$$NaOH + KHP \longrightarrow NaKP + H_2O$$
$$\text{1 mol} \quad \text{1 mol} \quad \quad \text{1 mol} \quad \text{1 mol}$$

We see that NaOH and KHP react in a 1:1 mole ratio. One mole of KHP is 204.2 g.

$$\underline{?} \text{ mol NaOH} = 0.3641 \text{ g KHP} \times \frac{1 \text{ mol KHP}}{204.2 \text{ g KHP}} \times \frac{1 \text{ mol NaOH}}{1 \text{ mol KHP}} = 0.001783 \text{ mol NaOH}$$

Then we calculate the molarity of the NaOH solution.

$$\frac{\underline{?} \text{ mol NaOH}}{L} = \frac{0.001783 \text{ mol NaOH}}{0.02000 \text{ L}} = \boxed{0.08915 \text{ } M \text{ NaOH}}$$

You should now work Exercise 24.

Impure samples of acids can be titrated with standard solutions of bases. The results can be used to determine percentage purity of the samples.

EXAMPLE 11-7 *Determination of Percent Acid*

Oxalic acid, $(COOH)_2$, is used to remove iron stains and some ink stains from fabrics. A 0.1743-gram sample of *impure* oxalic acid required 39.82 mL of 0.08915 M NaOH solution for complete neutralization. No acidic impurities were present. Calculate the percentage purity of the $(COOH)_2$.

Plan

We write the balanced equation for the reaction and calculate the number of moles of NaOH in the standard solution. Then we calculate the mass of $(COOH)_2$ in the sample, which gives us the information we need to calculate percentage purity.

Solution

The equation for the reaction of NaOH with $(COOH)_2$ is

$$2NaOH + (COOH)_2 \longrightarrow Na_2(COO)_2 + 2H_2O$$
$$\quad 2 \text{ mol} \qquad 1 \text{ mol} \qquad\qquad 1 \text{ mol} \qquad 2 \text{ mol}$$

Two moles of NaOH neutralize completely one mole of $(COOH)_2$. The number of moles of NaOH that react is the volume times the molarity of the solution.

$$\underline{?} \text{ mol NaOH} = 0.03982 \text{ L} \times \frac{0.08915 \text{ mol NaOH}}{L} = 0.003550 \text{ mol NaOH}$$

Now we calculate the mass of $(COOH)_2$ that reacts with 0.003550 mol NaOH.

$$\underline{?} \text{ g } (COOH)_2 = 0.003550 \text{ mol NaOH} \times \frac{1 \text{ mol } (COOH)_2}{2 \text{ mol NaOH}} \times \frac{90.04 \text{ g } (COOH)_2}{1 \text{ mol } (COOH)_2}$$

$$= 0.1598 \text{ g } (COOH)_2$$

The sample contained 0.1598 g of $(COOH)_2$, and its percentage purity was

$$\% \text{ purity} = \frac{0.1598 \text{ g } (COOH)_2}{0.1743 \text{ g sample}} \times 100\% = \boxed{91.68\% \text{ pure } (COOH)_2}$$

You should now work Exercise 28.

Each molecule of $(COOH)_2$ contains two acidic H's.

$$H-O-\overset{\overset{\displaystyle O}{\|}}{C}-\overset{\overset{\displaystyle O}{\|}}{C}-O-H$$

1 mol = 90.04 g

11-3 EQUIVALENT WEIGHTS AND NORMALITY

Because one mole of an acid does not necessarily neutralize one mole of a base, some chemists prefer a method of expressing concentration other than molarity to retain a one-to-one relationship. Concentrations of solutions of acids and bases are frequently expressed as *normality* (N). The **normality** of a solution is defined as the number of equivalent weights, or simply equivalents (eq), of solute per liter of solution. Normality may be represented symbolically as

$$\text{normality} = \frac{\text{number of equivalent weights of solute}}{\text{liter of solution}} = \frac{\text{no. eq}}{L}$$

By definition there are 1000 milliequivalent weights (meq) in one equivalent weight of an acid or base. Normality may also be represented as

$$\text{normality} = \frac{\text{number of milliequivalent weights of solute}}{\text{milliliter of solution}} = \frac{\text{no. meq}}{mL}$$

Any calculation that can be carried out with equivalent weights and normality can also be done by the mole method using molarity. However, the methods of this section are widely used in health-related fields and in many industrial laboratories.

An **equivalent weight** is often referred to simply as an **equivalent** (eq).

A **milliequivalent weight** is often referred to simply as a **milliequivalent** (meq).

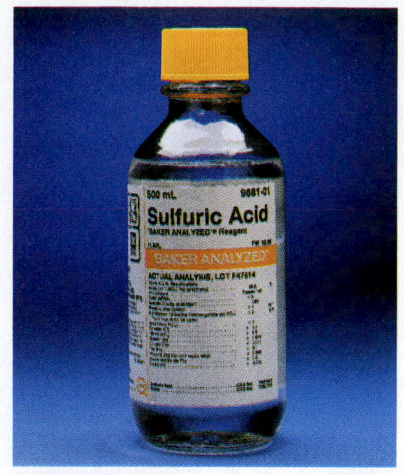

One mole of H_2SO_4 is two equivalent weights of H_2SO_4.

In acid–base reactions, one **equivalent weight,** or **equivalent (eq), of an acid** is defined as the mass of the acid (expressed in grams) that will furnish 6.022×10^{23} hydrogen ions (1 mol) or that will react with 6.022×10^{23} hydroxide ions (1 mol). One mole of an acid contains 6.022×10^{23} formula units of the acid. Consider hydrochloric acid as a typical monoprotic acid.

$$HCl \xrightarrow{H_2O} H^+(aq) + Cl^-(aq)$$

1 mol	1 mol	1 mol
36.46 g	1.008 g	35.45 g
6.022×10^{23} FU	6.022×10^{23} FU	6.022×10^{23} FU

We see that one mole of HCl can produce 6.022×10^{23} H^+, and so *one mole of HCl is one equivalent.* The same is true for all monoprotic acids.

Sulfuric acid is a diprotic acid. One molecule of H_2SO_4 can furnish $2H^+$ ions.

$$H_2SO_4 \xrightarrow{H_2O} 2H^+(aq) + SO_4^{2-}(aq)$$

1 mol	2 mol	1 mol
98.08 g	2(1.008 g)	96.06 g
6.022×10^{23} FU	$2(6.022 \times 10^{23})$ FU	6.022×10^{23} FU

This equation shows that one mole of H_2SO_4 can produce $2(6.022 \times 10^{23})$ H^+; therefore, one mole of H_2SO_4 is *two* equivalent weights in all reactions in which *both* acidic hydrogen atoms react.

One **equivalent weight of a base** is defined as the mass of the base (expressed in grams) that will furnish 6.022×10^{23} hydroxide ions or the mass of the base that will react with 6.022×10^{23} hydrogen ions.

The equivalent weight of an *acid* is obtained by dividing its formula weight in grams either by the number of acidic hydrogens furnished by one formula unit of the acid *or* by the number of hydroxide ions with which one formula unit of the acid reacts. The equivalent weight of a *base* is obtained by dividing its formula weight in grams either by the number of hydroxide ions furnished by one formula unit *or* by the number of hydrogen ions with which one formula unit of the base reacts. Equivalent weights of some common acids and bases are given in Table 11-1.

Table 11-1 *Equivalent Weights* of Some Acids and Bases*

Acids		Bases	
Symbolic representation	*One eq*	*Symbolic representation*	*One eq*
$\dfrac{HNO_3}{1}$	$= \dfrac{63.02 \text{ g}}{1} = 63.02$ g HNO_3	$\dfrac{NaOH}{1}$	$= \dfrac{40.00 \text{ g}}{1} = 40.00$ g NaOH
$\dfrac{CH_3COO\underline{H}}{1}$	$= \dfrac{60.03 \text{ g}}{1} = 60.03$ g $CH_3COO\underline{H}$	$\dfrac{NH_3}{1}$	$= \dfrac{17.04 \text{ g}}{1} = 17.04$ g NH_3
$\dfrac{KC_6H_4(COO)(COO\underline{H})}{1}$	$= \dfrac{204.2 \text{ g}}{1} = 204.2$ g $KC_6H_4(COO)(COO\underline{H})$	$\dfrac{Ca(OH)_2}{2}$	$= \dfrac{74.10 \text{ g}}{2} = 37.05$ g $Ca(OH)_2$
$\dfrac{H_2SO_4}{2}$	$= \dfrac{98.08 \text{ g}}{2} = 49.04$ g H_2SO_4	$\dfrac{Ba(OH)_2}{2}$	$= \dfrac{171.36 \text{ g}}{2} = 85.68$ g $Ba(OH)_2$

*Complete neutralization is assumed.

EXAMPLE 11-8 *Concentration of a Solution*

Calculate the normality of a solution of 4.202 grams of HNO_3 in 600 mL of solution.

Plan

We convert grams of HNO_3 to moles of HNO_3 and then to equivalents of HNO_3, which lets us calculate the normality.

$$\frac{g\ HNO_3}{L} \longrightarrow \frac{mol\ HNO_3}{L} \longrightarrow \frac{eq\ HNO_3}{L} = N\ HNO_3$$

Solution

$$N = \frac{no.\ eq\ HNO_3}{L}$$

$$\underbrace{\frac{?\ eq\ HNO_3}{L} = \frac{4.202\ g\ HNO_3}{0.600\ L} \times \frac{1\ mol\ HNO_3}{63.02\ g\ HNO_3}}_{M_{HNO_3}} \times \frac{1\ eq\ HNO_3}{mol\ HNO_3} = \boxed{0.111\ N\ HNO_3}$$

You should now work Exercise 32.

Because normality is equal to molarity times the number of equivalents per mole of solute, a solution's normality is always equal to or greater than its molarity.

$$normality = molarity \times \frac{no.\ eq}{mol} \quad or \quad N = M \times \frac{no.\ eq}{mol}$$

EXAMPLE 11-9 *Concentration of a Solution*

Calculate (a) the molarity and (b) the normality of a solution that contains 9.50 grams of barium hydroxide in 2000 mL of solution.

Plan

(a) We use the same kind of logic we used in Example 11-8.

(b) Because each mole of $Ba(OH)_2$ produces 2 moles of OH^- ions, 1 mole of $Ba(OH)_2$ is 2 equivalents. Thus,

$$N = M \times \frac{2\ eq}{mol} \quad or \quad M = \frac{N}{2\ eq/mol}$$

Because each formula unit of $Ba(OH)_2$ contains two OH^- ions,

$$1\ mol\ Ba(OH)_2 = 2\ eq\ Ba(OH)_2$$

Thus, molarity is one half of normality for $Ba(OH)_2$ solutions.

Solution

(a) $$\frac{?\ mol\ Ba(OH)_2}{L} = \frac{9.50\ g\ Ba(OH)_2}{2.00\ L} \times \frac{1\ mol\ Ba(OH)_2}{171.4\ g\ Ba(OH)_2} = \boxed{0.0277\ M\ Ba(OH)_2}$$

(b) $$\frac{?\ eq\ Ba(OH)_2}{L} = \frac{0.0277\ mol\ Ba(OH)_2}{L} \times \frac{2\ eq\ Ba(OH)_2}{1\ mol\ Ba(OH)_2} = \boxed{0.0554\ N\ Ba(OH)_2}$$

You should now work Exercise 34.

From the definitions of one equivalent of an acid and of a base, we see that *one equivalent of an acid reacts with one equivalent of any base.* It is *not* true that one mole of any acid reacts with one mole of any base in a specific chemical reaction. As a consequence of the definition of equivalents, 1 eq acid ≅ 1 eq base. We may write the following for *all* acid–base reactions that go to completion.

> no. eq acid = no. eq base   *or*   no. meq acid = no. meq base

The product of the volume of a solution, in liters, and its normality is equal to the number of equivalents of solute contained in the solution. For a solution of an acid,

Remember that the product of volume and concentration equals the amount of solute.

$$L_{acid} \times N_{acid} = L_{acid} \times \frac{eq\ acid}{L_{acid}} = eq\ acid$$

Alternatively,

$$mL_{acid} \times N_{acid} = mL_{acid} \times \frac{meq\ acid}{mL_{acid}} = meq\ acid$$

Similar relationships can be written for a solution of a base. Because 1 eq of acid *always* reacts with 1 eq of base, we may write

$$no.\ eq\ acid = no.\ eq\ base$$

so

> $$L_{acid} \times N_{acid} = L_{base} \times N_{base}\quad or\quad mL_{acid} \times N_{acid} = mL_{base} \times N_{base}$$

EXAMPLE 11-10   *Volume Required for Neutralization*

What volume of 0.100 N HNO$_3$ solution is required to neutralize completely 50.0 mL of a 0.150 N solution of Ba(OH)$_2$?

Plan

We know three of the four variables in the relationship

$$mL_{acid} \times N_{acid} = mL_{base} \times N_{base}, \text{ and so we solve for } mL_{acid}.$$

Solution

$$\underline{?}\ mL_{acid} = \frac{mL_{base} \times N_{base}}{N_{acid}} = \frac{50.0\ mL \times 0.150\ N}{0.100\ N} = \boxed{75.0\ mL\ of\ HNO_3\ solution}$$

You should now work Exercise 36.

In Example 11-11 let us again solve Example 11-5, this time using normality rather than molarity. The balanced equation for the reaction of H$_2$SO$_4$ with Na$_2$CO$_3$, interpreted in terms of equivalent weights, is

By definition, there must be equal numbers of equivalents of all reactants and products in a balanced chemical equation.

$$\begin{array}{ccccccc}
H_2SO_4 & + & Na_2CO_3 & \longrightarrow & Na_2SO_4 & + & CO_2 & + & H_2O \\
1\ mol & & 1\ mol & & 1\ mol & & 1\ mol & & 1\ mol \\
2\ eq & & 2\ eq & & 2\ eq & & 2\ eq & & 2\ eq \\
98.08\ g & & 106.0\ g & & & & &
\end{array}$$

$$1\ eq\ Na_2CO_3 = 53.0\ g \quad and \quad 1\ meq\ Na_2CO_3 = 0.0530\ g$$

EXAMPLE 11-11   *Standardization of Acid Solution*

Calculate the normality of a solution of H_2SO_4 if 40.0 mL of the solution reacts completely with 0.364 gram of Na_2CO_3.

Plan

We refer to the balanced equation. We are given the mass of Na_2CO_3, so we convert grams of Na_2CO_3 to milliequivalents of Na_2CO_3, then to milliequivalents of H_2SO_4, which lets us calculate the normality of the H_2SO_4 solution.

$$\boxed{\begin{array}{c} \text{g } Na_2CO_3 \\ \text{present} \end{array}} \longrightarrow \boxed{\begin{array}{c} \text{meq } Na_2CO_3 \\ \text{present} \end{array}} \longrightarrow \boxed{\begin{array}{c} \text{meq } H_2SO_4 \\ \text{needed} \end{array}} \longrightarrow \boxed{\dfrac{\text{meq } H_2SO_4}{\text{mL}}}$$

Solution

First we calculate the number of milliequivalents of Na_2CO_3 in the sample.

$$\text{no. meq } Na_2CO_3 = 0.364 \text{ g } Na_2CO_3 \times \frac{1 \text{ meq } Na_2CO_3}{0.0530 \text{ g } Na_2CO_3} = 6.87 \text{ meq } Na_2CO_3$$

Because no. meq H_2SO_4 = no. meq Na_2CO_3, we may write

$$mL_{H_2SO_4} \times N_{H_2SO_4} = 6.87 \text{ meq } H_2SO_4$$

$$N_{H_2SO_4} = \frac{6.87 \text{ meq } H_2SO_4}{mL_{H_2SO_4}} = \frac{6.87 \text{ meq } H_2SO_4}{40.0 \text{ mL}} = \boxed{0.172 \text{ } N \text{ } H_2SO_4}$$

You should now work Exercise 38.

The normality of this H_2SO_4 solution is twice the molarity obtained in Example 11-5 because 1 mol of H_2SO_4 is 2 eq.

OXIDATION–REDUCTION REACTIONS

Our rules for assigning oxidation numbers are constructed so that in all redox reactions

the total increase in oxidation numbers must equal the total decrease in oxidation numbers.

This equivalence provides the basis for balancing redox equations. Although there is no single "best method" for balancing all redox equations, two methods are particularly useful: (1) the change-in-oxidation-number method and (2) the half-reaction method, which is used extensively in electrochemistry (Chapter 21).

Most redox equations can be balanced by both methods, but in some instances one may be easier to use than the other.

11-4   CHANGE-IN-OXIDATION-NUMBER METHOD

The next few examples illustrate this method, which is based on *equal total increases and decreases in oxidation numbers*. While many redox equations can be balanced by simple inspection, you should learn the method because it can be used to balance difficult equations. The general procedure follows.

For brevity, we will refer to this as the CON method.

1. Write as much of the overall *unbalanced* equation as possible.

2. Assign oxidation numbers (Section 4-6) to find the elements that undergo changes in oxidation numbers.

3. a. Draw a bracket to connect atoms of the element that are oxidized. Show the increase in oxidation number *per atom*. Draw a bracket to connect atoms of the element that are reduced. Show the decrease in oxidation number *per atom*.

   b. Determine the factors that will make the *total* increase and decrease in oxidation numbers equal.

4. Insert coefficients into the equation to make the total increase and decrease in oxidation numbers equal.

5. Balance the other atoms by inspection.

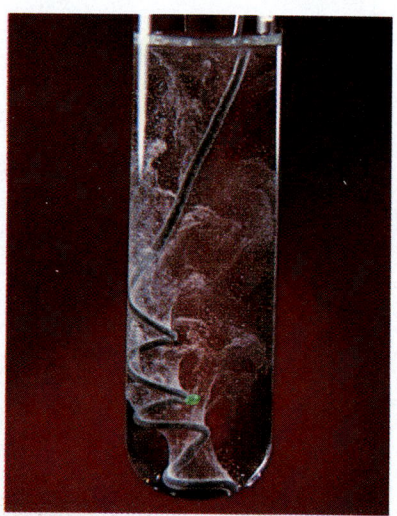

Aluminum wire reacting with hydrochloric acid.

EXAMPLE 11-12   *Balancing Redox Equations (CON method)*

Aluminum reacts with hydrochloric acid to form aqueous aluminum chloride and gaseous hydrogen. Balance the formula unit equation and identify the oxidizing and reducing agents.

Plan
We follow the five-step procedure, one step at a time.

Solution
The unbalanced formula unit equation and oxidation numbers (Steps 1 and 2) are

$$\overset{+1\ -1}{HCl}(aq) + \overset{0}{Al}(s) \longrightarrow \overset{+3\ -1}{AlCl_3}(aq) + \overset{0}{H_2}(g)$$

The oxidation number of Al increases from 0 to +3. Al is the reducing agent; it is oxidized. The oxidation number of H decreases from +1 to 0. HCl is the oxidizing agent; it is reduced.

$$\overset{+1}{HCl} + \overset{0}{Al} \longrightarrow \overset{+3}{AlCl_3} + \overset{0}{H_2} \qquad (Step\ 3a)$$
$$+3\ \uparrow$$
$$-1$$

We make the *total* increase and decrease in oxidation numbers equal (Step 3b).

Oxidation Numbers	Change/Atom	Equalizing Changes Gives
Al = 0 $\longrightarrow$ Al = +3	+3	$1(+3) = +3$
H = +1 $\longrightarrow$ H = 0	−1	$3(-1) = -3$

Each change must be multiplied by two because there are 2 H in each H_2.

$$2(+3) = +6 \text{ (total increase)} \qquad 2(-3) = -6 \text{ (total decrease)}$$

We need 2 Al and 6 H on each side of the equation (Step 4).

$$6HCl(aq) + 2Al(s) \longrightarrow 2AlCl_3(aq) + 3H_2(g)$$

You should now work Exercise 40.

All balanced equations must satisfy two criteria.

1. There must be mass balance. That is, the same number of atoms of each kind must be shown as reactants and products.

2. There must be charge balance. The sums of actual charges on the left and right sides of the equation must be equal.

When balancing redox equations, we often find it convenient to omit the spectator ions (Section 4-3) so that we can focus on the oxidation and reduction processes. We use the methods presented in this chapter to balance the net ionic equation. If necessary we add the spectator ions and combine species to write the balanced formula unit equation. Example 11-13 illustrates this approach.

> In Section 4-3, we first wrote the *formula unit equation.* We separated any ionized or dissociated species into ions to obtain the *total ionic equation.* Then we eliminated the spectator ions to obtain the *net ionic equation.* Now we reverse the procedure.

EXAMPLE 11-13   *Balancing Redox Equations (CON method)*

Copper is a widely used metal. Before it is welded (brazed), copper is cleaned by dipping it into nitric acid. HNO_3 oxidizes Cu to Cu^{2+} ions and is reduced to NO. The other product is H_2O. Write the balanced net ionic and formula unit equations for the reaction. Excess HNO_3 is present.

Plan

In writing ionic equations, we recall that strong acids, strong soluble bases, and most soluble salts are strong electrolytes. Then we apply our five-step procedure for redox equations.

Solution

We write the unbalanced net ionic equation and assign oxidation numbers. HNO_3 is a strong acid.

$$\overset{+5}{H^+}(aq) + \overset{0}{NO_3^-}(aq) + \overset{+2}{Cu}(s) \longrightarrow \overset{+2}{Cu^{2+}}(aq) + NO(g) + H_2O(\ell)$$

We see that copper is oxidized; it is the reducing agent. Nitrate ions are reduced; they are the oxidizing agent.

$$\overset{+5}{H^+}(aq) + \overset{0}{NO_3^-}(aq) + \overset{+2}{Cu}(s) \longrightarrow \overset{+2}{Cu^{2+}}(aq) + NO(g) + H_2O(\ell)$$

$$+2$$
$$-3$$

We make the *total* increase and decrease in oxidation numbers equal.

Oxidation Numbers	Change/Atom	Equalizing Changes Gives
Cu = 0 $\longrightarrow$ Cu = +2	+2	3(+2) = +6
N = +5 $\longrightarrow$ N = +2	−3	2(−3) = −6

Now we balance the *redox part* of the reaction.

$$H^+ + 2NO_3^- + 3Cu \longrightarrow 3Cu^{2+} + 2NO + H_2O$$

There are 6 O on the left in NO_3^- ions. A coefficient of 4 before H_2O balances O and gives 8 H on the right. So we need 8 H^+ ions on the left to balance the net ionic equation.

$$8H^+(aq) + 2NO_3^-(aq) + 3Cu(s) \longrightarrow 3Cu^{2+}(aq) + 2NO(g) + 4H_2O(\ell)$$

This solution contains excess HNO_3, so NO_3^- is the only anion present in significant concentration. Therefore, we add six more NO_3^- ions to each side to give the total ionic equation.

$$8[H^+(aq) + NO_3^-(aq)] + 3Cu(s) \longrightarrow 3[Cu^{2+}(aq) + 2NO_3^-(aq)] + 2NO(g) + 4H_2O(\ell)$$

Now we can write the balanced formula unit equation.

$$8HNO_3(aq) + 3Cu(s) \longrightarrow 3Cu(NO_3)_2(aq) + 2NO(g) + 4H_2O(\ell)$$

You should now work Exercise 42.

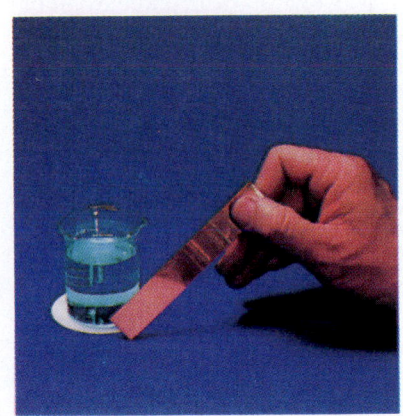

Copper is cleaned by dipping it into nitric acid.

> ▼ **PROBLEM-SOLVING TIP** *Converting Ionic to Formula Unit Equations*
>
> We learned in Section 4-3 how to convert the formula unit equation to the net ionic equation. To do this, we convert the formulas for all *strong electrolytes* into their ions, and then cancel *spectator ions* from both sides of the equation. In the last step of Example 11-13 we reverse this procedure. The balanced net ionic equation contains an excess 6+ charge on each side. To balance this excess charge, we must add negatively charged spectator ions to bring the total charge on each side to zero. Considering the contents of the solution, we see that nitric acid, a strong electrolyte, is present as a source of NO_3^- ions. We add six NO_3^- ions to bring the total charge on the left-hand side to zero; to maintain balance, we must add the same number of NO_3^- ions on the right. Then we combine species to give complete formula units.

11-5 ADDING H^+, OH^-, OR H_2O TO BALANCE OXYGEN OR HYDROGEN

Frequently we need more oxygen or hydrogen to complete the mass balance for a reaction in aqueous solution. However, we must be careful not to introduce other changes in oxidation number or to use species that could not actually be present in the solution. We cannot add H_2 or O_2 to equations because these species are not present in aqueous solutions. Acidic solutions do not contain significant concentrations of OH^- ions. Basic solutions do not contain significant concentrations of H^+ ions.

> In acidic solution: We add only H^+ or H_2O (*not* OH^- in acidic solution).
> In basic solution: We add only OH^- or H_2O (*not* H^+ in basic solution).

The following chart shows how to balance hydrogen and oxygen.

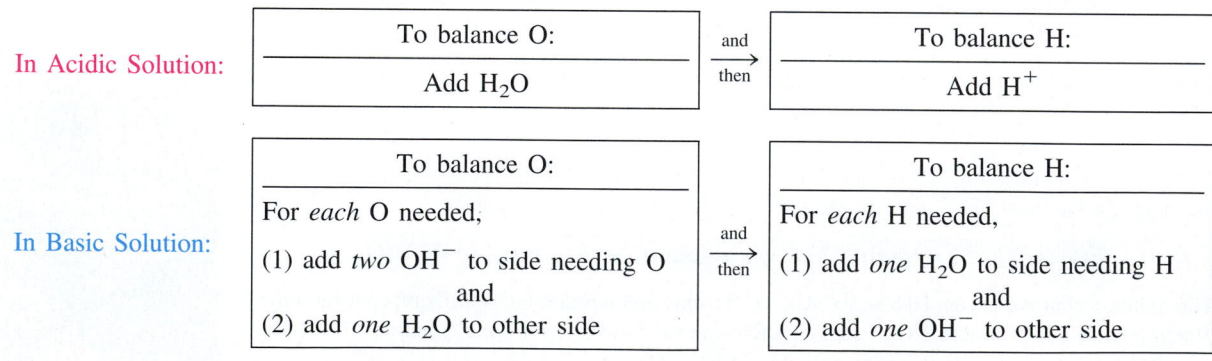

EXAMPLE 11-14 *Balancing Redox Equations (CON method)*

"Drāno" drain cleaner is solid sodium hydroxide that contains some aluminum turnings. When Drāno is added to water, the NaOH dissolves rapidly with the evolution of a lot of heat. The Al reduces H_2O in the basic solution to produce $[Al(OH)_4]^-$ ions and H_2 gas, which gives the bubbling action. Write the balanced net ionic and formula unit equations for this reaction.

Plan

We are given formulas for reactants and products. Recall that NaOH is a strong soluble base (OH^- and H_2O can be added to either side as needed). We apply our five-step procedure.

Solution

We write the unbalanced net ionic equation and assign oxidation numbers.

$$\overset{0}{OH^-}(aq) + \overset{+1}{Al}(s) + \overset{+3}{H_2O}(\ell) \longrightarrow \overset{0}{[Al(OH)_4]^-}(aq) + \overset{0}{H_2}(g)$$

Aluminum is oxidized; it is the reducing agent. H_2O is reduced; it is the oxidizing agent.

$$\overset{+1}{OH^-}(aq) + \overset{0}{H_2O}(\ell) + \overset{+3}{Al}(s) \longrightarrow \overset{0}{[Al(OH)_4]^-}(aq) + H_2(g)$$

$+3$

-1

We make the *total* increase and decrease in oxidation numbers equal.

Oxidation Numbers	Change/Atom	Equalizing Changes Gives
Al = 0 ⟶ Al = +3	+3	1(+3) = +3
H = +1 ⟶ H = 0	−1	3(−1) = −3

Each change must be multiplied by two because there are 2 H in each H_2.

$$2(+3) = +6 \text{ (total increase)} \qquad 2(-3) = -6 \text{ (total decrease)}$$

Now we balance the redox part of the equation. We need 2 Al on each side. Because only one H in each H_2O molecule is reduced (the other H is in OH^-), we show 6 H_2O on the left and 3 H_2 on the right.

$$OH^- + 6H_2O + 2Al \longrightarrow 2[Al(OH)_4]^- + 3H_2$$

The net charge on the right is 2−, and so we need 2 OH^- on the left to balance the net ionic equation.

$$2OH^-(aq) + 6H_2O(\ell) + 2Al(s) \longrightarrow 2[Al(OH)_4]^-(aq) + 3H_2(g)$$

This reaction occurs in excess NaOH solution. We need two $Na^+(aq)$ on each side to balance the negative charges.

$$2NaOH(aq) + 6H_2O(\ell) + 2Al(s) \longrightarrow 2Na[Al(OH)_4](aq) + 3H_2(g)$$

You should now work Exercise 44.

The Drāno reaction.

EXAMPLE 11-15   *Balancing Redox Equations (CON method)*

The breathalyzer detects the presence of ethanol (ethyl alcohol) in the breath of persons suspected of drunken driving. It utilizes the oxidation of ethanol to acetaldehyde by dichromate ions in acidic solution. The $Cr_2O_7{}^{2-}(aq)$ ion is orange (see page 382). The $Cr^{3+}(aq)$ ion is green. The appearance of a green color signals alcohol in the breath that exceeds the legal limit. Balance the net ionic equation for this reaction.

$$H^+(aq) + Cr_2O_7{}^{2-}(aq) + C_2H_5OH(\ell) \longrightarrow Cr^{3+}(aq) + C_2H_4O(\ell) + H_2O(\ell)$$

Plan

We are given the unbalanced equation, which includes H^+. This tells us that the reaction occurs in acidic solution. We apply our five-step procedure.

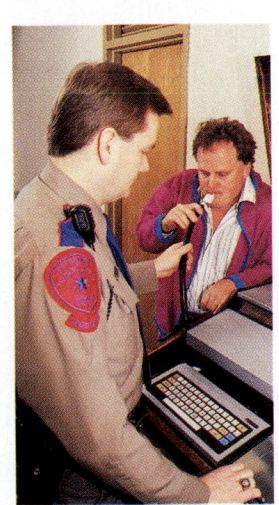

Many breathalyzers have been developed over the years. This model has proved to be effective.

Solution

We first assign oxidation numbers to the elements that change.

$$\overset{(-2)}{}\quad\overset{(+6)}{}\qquad\overset{(+3)}{}\quad\overset{(-1)}{}$$
$$H^+ + C_2H_5OH + Cr_2O_7{}^{2-} \longrightarrow Cr^{3+} + C_2H_4O + H_2O$$

We see that ethanol is oxidized; it is the reducing agent. $Cr_2O_7{}^{2-}$ ions are reduced; they are the oxidizing agent.

$$\overset{(-2)}{}\quad\overset{(+6)}{}\qquad\overset{(+3)}{}\quad\overset{(-1)}{}$$
$$H^+ + C_2H_5OH + Cr_2O_7{}^{2-} \longrightarrow Cr^{3+} + C_2H_4O + H_2O$$
$$-3$$
$$+1$$

Oxidation Numbers	Change/Atom	Equalizing Changes Gives
$Cr = +6 \longrightarrow Cr = +3$	-3	$1(-3) = -3$
$C = -2 \longrightarrow C = -1$	$+1$	$3(+1) = +3$

Each change must be multiplied by two because there are 2 Cr in each $Cr_2O_7{}^{2-}$ and 2 C in C_2H_5OH.

$$2(-3) = -6 \text{ (total decrease)} \qquad 2(+3) = +6 \text{ (total increase)}$$

We need 2 Cr and 6 C on each side of the equation to balance the redox part.

$$H^+ + 3C_2H_5OH + Cr_2O_7{}^{2-} \longrightarrow 2Cr^{3+} + 3C_2H_4O + H_2O$$

Now we balance H and O using our chart. There are 10 O on the left and only 4 O on the right. So we add 6 *more* H_2O molecules on the right.

$$H^+ + 3C_2H_5OH + Cr_2O_7{}^{2-} \longrightarrow 2Cr^{3+} + 3C_2H_4O + 7H_2O$$

Now there are 26 H on the right and only 19 on the left. So we add 7 *more* H^+ ions on the left to give the balanced net ionic equation.

$$8H^+(aq) + 3C_2H_5OH(\ell) + Cr_2O_7{}^{2-}(aq) \longrightarrow 2Cr^{3+}(aq) + 3C_2H_4O(\ell) + 7H_2O(\ell)$$

You should now work Exercise 42.

Every balanced equation must have both mass balance and charge balance. Once the redox part of an equation has been balanced, we may count *either* atoms or charges. After we balanced the redox part in Example 11-15, we had

$$H^+ + 3C_2H_5OH + Cr_2O_7{}^{2-} \longrightarrow 2Cr^{3+} + 3C_2H_4O + H_2O$$

The net charge on the left side is $(1 + 2-) = 1-$. On the right, it is $2(3+) = 6+$. Because H^+ is the *only charged species whose coefficient isn't known*, we add 7 *more* H^+ to give a net charge of $6+$ on both sides.

$$8H^+ + 3C_2H_5OH + Cr_2O_7{}^{2-} \longrightarrow 2Cr^{3+} + 3C_2H_4O + H_2O$$

Now we have 10 O on the left and only 4 O on the right. We add six *more* H_2O molecules to give the balanced net ionic equation.

$$8H^+(aq) + 3C_2H_5OH(\ell) + Cr_2O_7{}^{2-}(aq) \longrightarrow 2Cr^{3+}(aq) + 3C_2H_4O(\ell) + 7H_2O(\ell)$$

How can you tell whether to balance atoms or charges first? Look at the equation *after you have balanced the redox part.* Decide which is simpler, and do that. In the preceding equation, it is easier to balance charges than to balance atoms.

11-6 THE HALF-REACTION METHOD

In the half-reaction method we separate and completely balance equations describing oxidation and reduction **half-reactions.** Then we equalize the numbers of electrons gained and lost in each. Finally, we add the resulting half-reactions to give the overall balanced equation. The general procedure follows.

1. Write as much of the overall unbalanced equation as possible, omitting spectator ions.

2. Construct unbalanced oxidation and reduction half-reactions (these are usually incomplete as well as unbalanced). Show complete formulas for polyatomic ions and molecules.

3. Balance by inspection all elements in each half-reaction, except H and O. Then use the chart in Section 11-5 to balance H and O in each half-reaction.

4. Balance the charge in each half-reaction by adding electrons as "products" or "reactants."

5. Balance the electron transfer by multiplying the balanced half-reactions by appropriate integers.

6. Add the resulting half-reactions and eliminate any common terms.

For brevity, we will refer to this as the HR method. It is sometimes referred to as the ion–electron method.

EXAMPLE 11-16   *Balancing Redox Equations (HR method)*

A useful analytical procedure involves the oxidation of iodide ions to free iodine. The free iodine is then titrated with a standard solution of sodium thiosulfate, $Na_2S_2O_3$. Iodine oxidizes $S_2O_3^{2-}$ ions to tetrathionate ions, $S_4O_6^{2-}$, and is reduced to I^- ions. Write the balanced net ionic equation for this reaction.

Plan

We are given the formulas for two reactants and two products. We use these to write as much of the equations as possible. We construct and balance the appropriate half-reactions using the rules just described. Then we add the half-reactions and eliminate common terms.

Solution

$$I_2 + S_2O_3^{2-} \longrightarrow I^- + S_4O_6^{2-}$$

$$I_2 \longrightarrow I^- \qquad \text{(red. half-reaction)}$$

$$I_2 \longrightarrow 2I^-$$

$$I_2 + 2e^- \longrightarrow 2I^- \qquad \text{(balanced red. half-reaction)}$$

$$S_2O_3^{2-} \longrightarrow S_4O_6^{2-} \qquad \text{(ox. half-reaction)}$$

$$2S_2O_3^{2-} \longrightarrow S_4O_6^{2-}$$

$$2S_2O_3^{2-} \longrightarrow S_4O_6^{2-} + 2e^- \qquad \text{(balanced ox. half-reaction)}$$

Each I_2 gains $2e^-$. I_2 is reduced; it is the oxidizing agent.

Each $S_2O_3^{2-}$ ion loses an e^-. $S_2O_3^{2-}$ is oxidized; it is the reducing agent.

Each half-reaction has two electrons. We add these half-reactions and cancel the electrons.

$$I_2 + 2e^- \longrightarrow 2I^-$$
$$\underline{2S_2O_3^{2-} \longrightarrow S_4O_6^{2-} + 2e^-}$$
$$I_2(s) + 2S_2O_3^{2-}(aq) \longrightarrow 2I^-(aq) + S_4O_6^{2-}(aq)$$

You should now work Exercise 44.

These common household chemicals should never be mixed because they react to form chloramine (NH_2Cl), a very poisonous volatile compound.

$$NH_3(aq) + ClO^-(aq) \longrightarrow NH_2Cl(aq) + OH^-(aq)$$

Bleaches sold under trade names such as Clorox and Purex are 5% solutions of sodium hypochlorite. The hypochlorite ion is a very strong oxidizing agent in basic solution. It oxidizes many stains to colorless substances.

EXAMPLE 11-17 *Balancing Redox Equations (HR method)*

In basic solution, hypochlorite ions, ClO^-, oxidize chromite ions, CrO_2^-, to chromate ions, CrO_4^{2-}, and are reduced to chloride ions. Write the balanced net ionic equation for this reaction.

Plan

We are given the formulas for two reactants and two products; we write as much of the equation as possible. The reaction occurs in basic solution; we can add OH^- and H_2O as needed. We construct and balance the appropriate half-reactions, equalize the electron transfer, add the half-reactions, and eliminate common terms.

Solution

$$CrO_2^- + ClO^- \longrightarrow CrO_4^{2-} + Cl^-$$

$$CrO_2^- \longrightarrow CrO_4^{2-} \qquad \text{(ox. half-rxn)}$$

$$CrO_2^- + 4OH^- \longrightarrow CrO_4^{2-} + 2H_2O$$

$$CrO_2^- + 4OH^- \longrightarrow CrO_4^{2-} + 2H_2O + 3e^- \qquad \text{(balanced ox. half-rxn)}$$

$$ClO^- \longrightarrow Cl^- \qquad \text{(red. half-rxn)}$$

$$ClO^- + H_2O \longrightarrow Cl^- + 2OH^-$$

$$ClO^- + H_2O + 2e^- \longrightarrow Cl^- + 2OH^- \qquad \text{(balanced red. half-rxn)}$$

One half-reaction involves three electrons and the other involves two electrons. We balance the electron transfer and add the half-reactions term by term.

$$2(CrO_2^- + 4OH^- \longrightarrow CrO_4^{2-} + 2H_2O + 3e^-)$$

$$\underline{3(ClO^- + H_2O + 2e^- \longrightarrow Cl^- + 2OH^-)}$$

$$2CrO_2^- + 8OH^- + 3ClO^- + 3H_2O \longrightarrow 2CrO_4^{2-} + 4H_2O + 3Cl^- + 6OH^-$$

We see 6 OH^- and 3 H_2O that can be eliminated from both sides to give the balanced net ionic equation.

$$2CrO_2^-(aq) + 2OH^-(aq) + 3ClO^-(aq) \longrightarrow 2CrO_4^{2-}(aq) + H_2O(\ell) + 3Cl^-(aq)$$

You should now work Exercise 46.

EXAMPLE 11-18 *Net Ionic and Formula Unit Equations*

Permanganate ions oxidize iron(II) to iron(III) in sulfuric acid solution. Permanganate ions are reduced to manganese(II) ions. (a) Write the balanced net ionic equation for this reaction. (b) Using $KMnO_4$, $FeSO_4$, and H_2SO_4 as reactants, write the balanced formula unit equation for this reaction.

Plan

We use the given information to write as much of the equation as possible. The reaction occurs in H_2SO_4 solution; we can add H^+ and H_2O as needed to construct and balance the appropriate half-reactions. Then we proceed as in earlier examples.

These common household chemicals should never be mixed because they react to form chlorine, a very poisonous gas.

$$2H^+(aq) + ClO^-(aq) + Cl^-(aq) \longrightarrow Cl_2(g) + H_2O(\ell)$$

Solution

$$Fe^{2+} + MnO_4^- \longrightarrow Fe^{3+} + Mn^{2+}$$

$$Fe^{2+} \longrightarrow Fe^{3+} \qquad \text{(ox. half-reaction)}$$

$$Fe^{2+} \longrightarrow Fe^{3+} + 1e^- \qquad \text{(balanced ox. half-reaction)}$$

$$MnO_4^- \longrightarrow Mn^{2+} \qquad \text{(red. half-reaction)}$$

$$MnO_4^- + 8H^+ \longrightarrow Mn^{2+} + 4H_2O$$

$$MnO_4^- + 8H^+ + 5e^- \longrightarrow Mn^{2+} + 4H_2O \qquad \text{(balanced red. half-reaction)}$$

One half-reaction involves one electron and the other involves five electrons. Now we balance the electron transfer and then add the two equations term by term. This gives the balanced net ionic equation.

$$5(Fe^{2+} \longrightarrow Fe^{3+} + 1e^-)$$

$$\underline{1(MnO_4^- + 8H^+ + 5e^- \longrightarrow Mn^{2+} + 4H_2O)}$$

$$5Fe^{2+}(aq) + MnO_4^-(aq) + 8H^+(aq) \longrightarrow 5Fe^{3+}(aq) + Mn^{2+}(aq) + 4H_2O(\ell)$$

The reaction occurs in H_2SO_4 solution. The SO_4^{2-} ion is the counter anion in the formula unit equation. Because the Fe^{3+} ion occurs twice in $Fe_2(SO_4)_3$, there must be an even number of Fe. So the net ionic equation is multiplied by two. We add 18 SO_4^{2-} to each side to give complete formulas in the balanced formula unit equation.

$$10FeSO_4(aq) + 2KMnO_4(aq) + 8H_2SO_4(aq) \longrightarrow$$
$$5Fe_2(SO_4)_3(aq) + 2MnSO_4(aq) + K_2SO_4(aq) + 8H_2O(\ell)$$

You should now work Exercise 48.

11-7 STOICHIOMETRY OF REDOX REACTIONS

One method of analyzing samples quantitatively for the presence of *oxidizable* or *reducible* substances is by **redox titration.** In such analyses, the concentration of a solution is determined by allowing it to react with a carefully measured amount of a *standard* solution of an oxidizing or reducing agent.

As in other kinds of chemical reactions, we must pay particular attention to the mole ratio in which oxidizing agents and reducing agents react.

Potassium permanganate, $KMnO_4$, is a strong oxidizing agent. Through the years it has been the "workhorse" of redox titrations. For example, in acidic solution, $KMnO_4$ reacts with iron(II) sulfate, $FeSO_4$, according to the balanced equations below. A strong acid, such as H_2SO_4, is used in such titrations (Example 11-18).

A word about terminology. The reaction involves MnO_4^- ions and Fe^{2+} ions in acidic solution. The source of MnO_4^- ions usually is the soluble ionic compound $KMnO_4$. We often refer to "permanganate solutions." Such solutions also contain cations—in this case, K^+. Likewise, we often refer to "iron(II) solutions" without specifying what the anion is.

Because it has an intense purple color, $KMnO_4$ acts as its own indicator. One drop of 0.020 M $KMnO_4$ solution imparts a pink color to a liter of pure water. When $KMnO_4$ solution is added to a solution of a reducing agent, the end point in the titration is taken as the point at which a pale pink color appears in the solution being titrated and persists for at least 30 seconds.

CHEMISTRY IN USE

The Development of Science

Matches, Cap Guns, Fireworks, and Space Travel

What do matches, cap guns, fireworks, and space travel have in common? The chemistry of the combustion reactions associated with each is similar; it all began 1000 years ago when the Chinese developed gunpowder. Contrary to popular belief, gunpowder wasn't originally used in guns; it was first used in fireworks.

Because modern gunpowder differs in composition from the early Chinese recipe, many people refer to the original mixture as black powder. Black powder consists of a mixture of potassium nitrate (an oxidizing agent), carbon (a fuel), and sulfur (another fuel). In the proper proportions, these substances undergo very rapid combustion in an enclosed, airless space. In a tiny fraction of a second, the oxidation of the carbon and sulfur produces gases, such as carbon dioxide and sulfur dioxide, which occupy a volume thousands of times greater than the original black powder. The Chinese found that this very rapid increase in volume produced a considerable thrust when the black powder was placed into a narrow tube and then ignited.

Although black powder is no longer used for most of its original uses, the fireworks industry continues to use the old Chinese recipe, consuming about one-half million pounds per year. Rapidly burning black powder provides the thrust needed to launch fireworks into the sky and the energy for the brilliant explosive display that follows. Americans spend about 100 million dollars each year to observe this dazzling spectacle.

Other chemicals are added to fireworks to give brilliant color effects. For example, adding strontium nitrate produces red flames, barium nitrate produces green flames, and copper carbonate produces blue flames. Adding iron filings produces gold sparks; adding titanium produces white sparks. To obtain slightly different results, some fireworks manufacturers use other oxidizing agents, such as ammonium perchlorate, and other fuels, such as powdered aluminum.

A mixture of ammonium perchlorate and aluminum is used in the solid rocket boosters that launch American space shuttles. About 1.5 million pounds of ammonium perchlorate are consumed every launch; the reaction in the solid rocket booster provides about 75% of the thrust needed to lift the shuttle off the ground.

Like the mixtures used in black powder and in rocket boosters, safety matches also rely on an oxidizer, potassium chlorate, and fuel, which is a blend of carbon, sulfur, and antimony sulfide. The striking surface on a matchbook cover contains red

phosphorus, which is easily ignited by the heat generated by friction. The heat from this reaction in turn ignites the mixture inside the match head. Nonsafety matches, the kind that light when struck against rough surfaces, have tetraphosphorus trisulfide, P_4S_3, at the tips of their heads. P_4S_3 is easily ignited by the heat generated by friction.

The caps used in toy guns contain a mixture of potassium chlorate, red phosphorus, and sand. When struck by the hammer of a cap gun, the sand provides friction to produce the heat that ignites the red phosphorus. Heat produced by burning red phosphorus initiates the decomposition of potassium chlorate which in turn supplies additional oxygen to the burning red phosphorus.

Interestingly, a discovery made over 1000 years ago gives us instant fire, enhances our patriotic feelings on national holidays, and helps us explore the rest of the universe.

Ronald DeLorenzo
Middle Georgia College

EXAMPLE 11-19   *Redox Titration*

What volume of 0.0200 *M* KMnO₄ solution is required to oxidize 40.0 mL of 0.100 *M* FeSO₄ in sulfuric acid solution (Figure 11-1)?

Plan

We refer to the balanced equation in Example 11-18 to find the reaction ratio, 1 mol MnO_4^-/5 mol Fe^{2+}. Then we calculate the number of moles of Fe^{2+} to be titrated, which lets us find the number of moles of MnO_4^- required *and* the volume in which this number of moles of KMnO₄ is contained.

One mole of KMnO₄ contains one mole of MnO_4^- ions. Therefore, the number of moles of KMnO₄ is *always* equal to the number of moles of MnO_4^- ions required in a reaction. Similarly, one mole of FeSO₄ contains 1 mole of Fe^{2+} ions.

Solution

The reaction ratio is

$$MnO_4^-(aq) + 8H^+(aq) + 5Fe^{2+}(aq) \longrightarrow 5Fe^{3+}(aq) + Mn^{2+}(aq) + 4H_2O$$

rxn ratio: 1 mol   5 mol

The number of moles of Fe^{2+} to be titrated is

$$\underline{?}\text{ mol Fe}^{2+} = 40.0\text{ mL} \times \frac{0.100\text{ mol Fe}^{2+}}{1000\text{ mL}} = 4.00 \times 10^{-3}\text{ mol Fe}^{2+}$$

We use the balanced equation to find the number of moles of MnO_4^- required.

$$\underline{?}\text{ mol MnO}_4^- = 4.00 \times 10^{-3}\text{ mol Fe}^{2+} \times \frac{1\text{ mol MnO}_4^-}{5\text{ mol Fe}^{2+}} = 8.00 \times 10^{-4}\text{ mol MnO}_4^-$$

Each formula unit of KMnO₄ contains one MnO_4^- ion, and so

$$1\text{ mol KMnO}_4 \cong 1\text{ mol MnO}_4^-$$

The volume of 0.0200 *M* KMnO₄ solution that contains 8.00×10^{-4} mol of KMnO₄ is

$$\underline{?}\text{ mL KMnO}_4\text{ soln} = 8.00 \times 10^{-4}\text{ mol KMnO}_4 \times \frac{1000\text{ mL KMnO}_4\text{ soln}}{0.0200\text{ mol KMnO}_4}$$

$$= 40.0\text{ mL KMnO}_4\text{ soln}$$

You should now work Exercise 52.

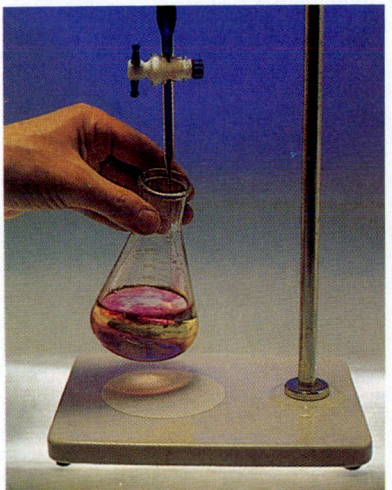

(a)

(b)

Figure 11-1 (a) Nearly colorless FeSO₄ solution is titrated with deep-purple KMnO₄. (b) The end point is the point at which the solution becomes pink, owing to a *very small* excess of KMnO₄. Here a considerable excess of KMnO₄ was added so that the pink color could be reproduced photographically.

K$_2$Cr$_2$O$_7$ is orange in acidic solution. Cr$_2$(SO$_4$)$_3$ is green in acidic solution.

Potassium dichromate, K$_2$Cr$_2$O$_7$, is another frequently used oxidizing agent. However, an indicator must be used when reducing agents are titrated with dichromate solutions. K$_2$Cr$_2$O$_7$ is orange, and its reduction product, Cr^{3+}, is green.

Consider the oxidation of sulfite ions, SO$_3^{2-}$, to sulfate ions, SO$_4^{2-}$, by Cr$_2$O$_7^{2-}$ ions in the presence of a strong acid such as sulfuric acid. We shall balance the equation by the half-reaction method.

$$Cr_2O_7^{2-} \longrightarrow Cr^{3+} \qquad \text{(red. half-rxn)}$$

$$Cr_2O_7^{2-} \longrightarrow \boxed{2Cr^{3+}}$$

$$\boxed{14H^+} + Cr_2O_7^{2-} \longrightarrow 2Cr^{3+} + \boxed{7H_2O}$$

$$\boxed{6e^-} + 14H^+ + Cr_2O_7^{2-} \longrightarrow 2Cr^{3+} + 7H_2O \qquad \text{(balanced red. half-rxn)}$$

$$SO_3^{2-} \longrightarrow SO_4^{2-} \qquad \text{(ox. half-rxn)}$$

$$SO_3^{2-} + \boxed{H_2O} \longrightarrow SO_4^{2-} + \boxed{2H^+}$$

$$SO_3^{2-} + H_2O \longrightarrow SO_4^{2-} + 2H^+ + \boxed{2e^-} \qquad \text{(balanced ox. half-rxn)}$$

We now equalize the electron transfer, add the balanced half-reactions, and eliminate common terms.

$$(6e^- + 14H^+ + Cr_2O_7^{2-} \longrightarrow 2Cr^{3+} + 7H_2O) \qquad \text{(reduction)}$$

$$\underline{3(SO_3^{2-} + H_2O \longrightarrow SO_4^{2-} + 2H^+ + 2e^-)} \qquad \text{(oxidation)}$$

$$8H^+(aq) + Cr_2O_7^{2-}(aq) + 3SO_3^{2-}(aq) \longrightarrow 2Cr^{3+}(aq) + 3SO_4^{2-}(aq) + 4H_2O(\ell)$$

The balanced equation tells us that the reaction ratio is 3 mol SO$_3^{2-}$/mol Cr$_2$O$_7^{2-}$ or 1 mol Cr$_2$O$_7^{2-}$/3 mol SO$_3^{2-}$. Potassium dichromate is the usual source of Cr$_2$O$_7^{2-}$ ions, and Na$_2$SO$_3$ is the usual source of SO$_3^{2-}$ ions. Thus, the preceding reaction ratio could also be expressed as 1 mol K$_2$Cr$_2$O$_7$/3 mol Na$_2$SO$_3$.

EXAMPLE 11-20   *Redox Titration*

A 20.00-mL sample of Na$_2$SO$_3$ was titrated with 36.30 mL of 0.05130 *M* K$_2$Cr$_2$O$_7$ solution in the presence of H$_2$SO$_4$. Calculate the molarity of the Na$_2$SO$_3$ solution.

Plan

We can calculate the number of moles of Cr$_2$O$_7^{2-}$ in the standard solution. Then we refer to the balanced equation in the preceding discussion, which gives us the reaction ratio, 3 mol SO$_3^{2-}$/1 mol Cr$_2$O$_7^{2-}$. The reaction ratio lets us calculate the number of moles of SO$_3^{2-}$(Na$_2$SO$_3$) that reacted and the molarity of the solution.

$$\boxed{\text{L Cr}_2\text{O}_7^{2-} \text{ soln}} \longrightarrow \boxed{\text{mol Cr}_2\text{O}_7^{2-}} \longrightarrow \boxed{\text{mol SO}_3^{2-}} \longrightarrow \boxed{M \text{ SO}_3^{2-} \text{ soln}}$$

Solution

From the preceding discussion we know the balanced equation and the reaction ratio.

$$3SO_3^{2-} + Cr_2O_7^{2-} + 8H^+ \longrightarrow 3SO_4^{2-} + 2Cr^{3+} + 4H_2O$$
$$\quad\; 3 \text{ mol} \qquad\; 1 \text{ mol}$$

The number of moles of Cr$_2$O$_7^{2-}$ used is

$$\underline{?} \text{ mol Cr}_2\text{O}_7^{2-} = 0.03630 \text{ L} \times \frac{0.05130 \text{ mol Cr}_2\text{O}_7^{2-}}{\text{L}} = 0.001862 \text{ mol Cr}_2\text{O}_7^{2-}$$

The number of moles of SO_3^{2-} that reacted with 0.001862 mol of $Cr_2O_7^{2-}$ is

$$\underline{?}\ mol\ SO_3^{2-} = 0.001862\ mol\ Cr_2O_7^{2-} \times \frac{3\ mol\ SO_3^{2-}}{1\ mol\ Cr_2O_7^{2-}} = 0.005586\ mol\ SO_3^{2-}$$

The Na_2SO_3 solution contained 0.005586 mol of SO_3^{2-} (or 0.005586 mol of Na_2SO_3). Its molarity is

$$\underline{?}\ \frac{mol\ Na_2SO_3}{L} = \frac{0.005586\ mol\ Na_2SO_3}{0.02000\ L} = 0.2793\ M\ Na_2SO_3$$

You should now work Exercise 56.

Key Terms

Buret   A piece of volumetric glassware, usually graduated in 0.1-mL intervals, that is used in titrations to deliver solutions in a quantitative (dropwise) manner.

End point   The point at which an indicator changes color and a titration is stopped.

Equivalence point   The point at which chemically equivalent amounts of reactants have reacted.

Equivalent weight in acid–base reactions   The mass of an acid or base that furnishes or reacts with 6.022×10^{23} H_3O^+ or OH^- ions.

Half-reaction   Either the oxidation part or the reduction part of a redox reaction.

Indicator   For acid–base titrations, an organic compound that exhibits different colors in solutions of different acidities; used to determine the point at which the reaction between two solutes is complete.

Normality   The number of equivalent weights (equivalents) of solute per liter of solution.

Oxidation   An algebraic increase in oxidation number; may correspond to a loss of electrons.

Oxidation–reduction reaction   A reaction in which oxidation and reduction occur; also called redox reaction.

Oxidizing agent   The substance that oxidizes another substance and is reduced.

Primary standard   A substance of a known high degree of purity that undergoes one invariable reaction with the other reactant of interest.

Redox reaction   An oxidation–reduction reaction.

Redox titration   The quantitative analysis of the amount or concentration of an oxidizing or reducing agent in a sample by observing its reaction with a known amount or concentration of a reducing or oxidizing agent.

Reducing agent   The substance that reduces another substance and is oxidized.

Reduction   An algebraic decrease in oxidation number; may correspond to a gain of electrons.

Secondary standard   A solution that has been titrated against a primary standard. A standard solution is a secondary standard.

Standard solution   A solution of accurately known concentration.

Standardization   The process by which the concentration of a solution is accurately determined by titrating it against an accurately known amount of a primary standard.

Titration   The process by which the volume of a standard solution required to react with a specific amount of a substance is determined.

Exercises

Molarity

1. Why can we describe molarity as a "method of convenience" for expressing concentrations of solutions?

2. Why can molarity be expressed in mol/L *and* in mmol/mL?

3. Calculate the molarities of solutions that contain the following masses of solute in the indicated volumes: (a) 45 g of H_3AsO_4 in 500 mL of solution; (b) 8.3 g of $(COOH)_2$ in 600 mL of solution; (c) 10.5 g of $(COOH)_2 \cdot 2H_2O$ in 750 mL of solution.

4. What mass of $(NH_4)_2SO_4$ is required to prepare 2.75 L of 0.188 M $(NH_4)_2SO_4$ solution?

5. Calculate the mass of NaOH required to prepare 2.50 L of 3.15 M NaOH solution.

6. Calculate the molarity of a solution that is 39.77% H_2SO_4 by mass. The specific gravity of the solution is 1.305.

7. Calculate the molarity of a solution that is 19.0% HNO_3 by mass. The specific gravity of the solution is 1.11.

8. If 150 mL of 4.32 M HCl solution is added to 300 mL of 2.16 M NaOH solution, the resulting solution will be _____ molar in NaCl.

9. If 450 mL of 1.05 M NaOH solution and 225 mL of 2.10 M acetic acid solution are mixed, what is the molarity of the salt in the resulting solution?

10. If 225 mL of 3.68 M H_3PO_4 solution is added to 775 mL of 3.68 M NaOH solution, the resulting solution will be _____ molar in Na_3PO_4 and _____ molar in _____.

11. If 400 mL of 0.200 M HCl solution is added to 800 mL of 0.0400 M $Ba(OH)_2$ solution, the resulting solution will be _____ molar in $BaCl_2$ and _____ molar in _____.

12. What volume of 0.0150 M acetic acid solution would just neutralize 25.2 mL of 0.0105 M $Ba(OH)_2$ solution?

13. What volume of 0.300 M potassium hydroxide solution would just neutralize 25.0 mL of 0.100 M H_2SO_4 solution?

14. A vinegar solution is 5.11% acetic acid. Its density is 1.007 g/mL. What is its molarity?

15. A household ammonia solution is 5.03% ammonia. Its density is 0.979 g/mL. What is its molarity?

16. What volumes of 1.50 M NaOH and 3.00 M H_3PO_4 solutions would be required to form 1.00 mol of Na_3PO_4?

Standardization and Acid–Base Titrations—Mole Method

17. Define and illustrate the following terms clearly and concisely: (a) standard solution; (b) titration; (c) primary standard; (d) secondary standard.

18. Describe the preparation of a standard solution of NaOH, a compound that absorbs both CO_2 and H_2O from the air.

19. Distinguish between the *net ionic equation* and the *formula unit equation.*

20. (a) What is potassium hydrogen phthalate? (b) For what is it used?

21. Why can sodium carbonate be used as a primary standard for solutions of acids?

22. Calculate the molarity of a solution of HNO_3 if 35.72 mL of the solution neutralizes 0.364 g of Na_2CO_3.

23. If 41.38 mL of a sulfuric acid solution reacts with 0.2888 g of Na_2CO_3, what is the molarity of the sulfuric acid solution?

24. A solution of sodium hydroxide is standardized against potassium hydrogen phthalate. From the following data, calculate the molarity of the NaOH solution.

mass of KHP	0.8407 g
buret reading before titration	0.23 mL
buret reading after titration	38.78 mL

25. Calculate the molarity of a KOH solution if 31.53 mL of the KOH solution reacted with 0.4084 g of potassium hydrogen phthalate.

26. (a) What are the properties of an ideal primary standard? (b) What is the importance of each property?

27. The secondary standard solution of NaOH of Exercise 24 was used to titrate a solution of unknown concentration of HCl. A 30.00-mL sample of the HCl solution required 34.21 mL of the NaOH solution for complete neutralization. What is the molarity of the HCl solution?

*28. An impure sample of $(COOH)_2 \cdot 2H_2O$ that had a mass of 2.00 g was dissolved in water and titrated with standard NaOH solution. The titration required 38.32 mL of 0.198 M NaOH solution. Calculate the percentage of $(COOH)_2 \cdot 2H_2O$ in the sample.

*29. A 50.0-mL sample of 0.0500 M $Ca(OH)_2$ is added to 10.0 mL of 0.200 M HNO_3. (a) Is the resulting solution acidic or basic? (b) How many moles of excess acid or base are present? (c) How many mL of 0.0500 M $Ca(OH)_2$ or 0.0500 M HNO_3 would be required to neutralize the solution?

*30. An antacid tablet containing calcium carbonate as an active ingredient required 24.5 mL of 0.0932 M HCl for complete neutralization. What mass of $CaCO_3$ did the tablet contain?

*31. Butyric acid, whose empirical formula is C_2H_4O, is the acid responsible for the odor of rancid butter. The acid has one ionizable hydrogen per molecule. A 1.000-g sample of butyric acid is neutralized by 54.42 mL of 0.2088 M NaOH solution. What are (a) the molecular weight and (b) the molecular formula of butyric acid?

Standardization and Acid–Base Titrations—Equivalent Weight Method

In answering Exercises 32–35, assume that the acids and bases will be completely neutralized.

32. What is the normality of a solution that contains 7.08 g of H_3PO_4 in 185 mL of solution?

33. Calculate the molarity and the normality of a solution that was prepared by dissolving 24.2 g of barium hydroxide in enough water to make 4000 mL of solution.

34. Calculate the molarity and the normality of a solution that contains 19.6 g of arsenic acid, H_3AsO_4, in enough water to make 500 mL of solution.

35. What are the molarity and normality of a sulfuric acid solution that is 19.6% H_2SO_4 by mass? The density of the solution is 1.14 g/mL.

36. A 25.0-mL sample of 0.206 normal nitric acid solution required 35.2 mL of barium hydroxide solution for neutralization. Calculate the molarity of the barium hydroxide solution.

37. Vinegar is an aqueous solution of acetic acid, CH_3COOH. Suppose you titrate a 25.00-mL sample of vinegar with 17.62 mL of a standardized 0.1060 N solution of NaOH. (a) What is the normality of acetic acid in this vinegar? (b) What is the mass of acetic acid contained in 1.000 L of vinegar?

38. Calculate the normality and molarity of an HCl solution if 43.1 mL of the solution reacts with 0.318 g of Na_2CO_3.

$$2HCl + Na_2CO_3 \longrightarrow 2NaCl + CO_2 + H_2O$$

39. Calculate the normality and molarity of an H_2SO_4 solution if 40.0 mL of the solution reacts with 0.424 g of Na_2CO_3.

$$H_2SO_4 + Na_2CO_3 \longrightarrow Na_2SO_4 + CO_2 + H_2O$$

Balancing Redox Equations

In Exercises 40 and 41, write balanced formula unit equations for the reactions described by words.

*40. (a) Iron reacts with hydrochloric acid to form aqueous iron(II) chloride and gaseous hydrogen. (b) Chromium reacts with sulfuric acid to form aqueous chromium(III) sulfate and gaseous hydrogen. (c) Tin reacts with concentrated nitric acid to form tin(IV) oxide, nitrogen dioxide, and water.

41. (a) Carbon reacts with hot concentrated nitric acid to form carbon dioxide, nitrogen dioxide, and water. (b) Sodium reacts with water to form aqueous sodium hydroxide and gaseous hydrogen. (c) Zinc reacts with sodium hydroxide solution to form aqueous sodium tetrahydroxozincate and gaseous hydrogen. (The tetrahydroxozincate ion is $[Zn(OH)_4]^{2-}$.)

42. Balance the following ionic equations.
(a) $MnO_4^-(aq) + H^+(aq) + Br^-(aq) \rightarrow$
$Mn^{2+}(aq) + Br_2(\ell) + H_2O(\ell)$
(b) $Cr_2O_7^{2-}(aq) + H^+(aq) + I^-(aq) \rightarrow$
$Cr^{3+}(aq) + I_2(s) + H_2O(\ell)$
(c) $MnO_4^-(aq) + SO_3^{2-}(aq) + H^+(aq) \rightarrow$
$Mn^{2+}(aq) + SO_4^{2-}(aq) + H_2O(\ell)$
(d) $Cr_2O_7^{2-}(aq) + Fe^{2+}(aq) + H^+(aq) \rightarrow$
$Cr^{3+}(aq) + Fe^{3+}(aq) + H_2O(\ell)$

43. Balance the following ionic equations.
(a) $Cr(OH)_4^-(aq) + OH^-(aq) + H_2O_2(aq) \rightarrow$
$CrO_4^{2-}(aq) + H_2O(\ell)$
(b) $MnO_2(s) + H^+(aq) + NO_2^-(aq) \rightarrow$
$NO_3^-(aq) + Mn^{2+}(aq) + H_2O(\ell)$
(c) $Sn(OH)_3^-(aq) + Bi(OH)_3(s) + OH^-(aq) \rightarrow$
$Sn(OH)_6^{2-}(aq) + Bi(s)$

44. Balance the following ionic equations.
(a) $Al(s) + NO_3^-(aq) + OH^-(aq) + H_2O \rightarrow$
$Al(OH)_4^-(aq) + NH_3(g)$
(b) $NO_2(g) + OH^-(aq) \rightarrow NO_3^-(aq) + NO_2^-(aq) + H_2O(\ell)$
(c) $MnO_4^-(aq) + H_2O(\ell) + NO_2^-(aq) \rightarrow$
$MnO_2(s) + NO_3^-(aq) + OH^-(aq)$
(d) $I^-(aq) + H^+(aq) + NO_2^-(aq) \rightarrow NO(g) + H_2O(\ell) + I_2(s)$
(e) $Hg_2Cl_2(s) + NH_3(aq) \rightarrow$
$Hg(\ell) + HgNH_2Cl(s) + NH_4^+(aq) + Cl^-(aq)$

45. Balance the following ionic equations.
(a) $CrO_4^{2-}(aq) + H_2O(\ell) + HSnO_2^-(aq) \rightarrow$
$CrO_2^-(aq) + OH^-(aq) + HSnO_3^-(aq)$
(b) $C_2H_4(g) + MnO_4^-(aq) + H^+(aq) \rightarrow$
$CO_2(g) + Mn^{2+}(aq) + H_2O(\ell)$
(c) $H_2S(aq) + H^+(aq) + Cr_2O_7^{2-}(aq) \rightarrow$
$Cr^{3+}(aq) + S(s) + H_2O(\ell)$
(d) $ClO_3^-(aq) + H_2O(\ell) + I_2(s) \rightarrow$
$IO_3^-(aq) + Cl^-(aq) + H^+(aq)$
(e) $Cu(s) + H^+(aq) + SO_4^{2-}(aq) \rightarrow$
$Cu^{2+}(aq) + H_2O(\ell) + SO_2(g)$

46. Balance the following ionic equations for reactions in acidic solution. H^+ or H_2O (but not OH^-) may be added as necessary.
(a) $Fe^{2+}(aq) + MnO_4^-(aq) \rightarrow Fe^{3+}(aq) + Mn^{2+}(aq)$
(b) $Br_2(\ell) + SO_2(g) \rightarrow Br^-(aq) + SO_4^{2-}(aq)$
(c) $Cu(s) + NO_3^-(aq) \rightarrow Cu^{2+}(aq) + NO_2(g)$
(d) $PbO_2(s) + Cl^-(aq) \rightarrow PbCl_2(s) + Cl_2(g)$
(e) $Zn(s) + NO_3^-(aq) \rightarrow Zn^{2+}(aq) + N_2(g)$

47. Balance the following ionic equations for reactions in acidic solution. H^+ or H_2O (but not OH^-) may be added as necessary.
(a) $P_4(s) + NO_3^-(aq) \rightarrow H_3PO_4(aq) + NO(g)$

(b) $H_2O_2(aq) + MnO_4^-(aq) \rightarrow Mn^{2+}(aq) + O_2(g)$
(c) $HgS(s) + Cl^-(aq) + NO_3^-(aq) \rightarrow$
$HgCl_4^{2-}(aq) + NO_2(g) + S(s)$
(d) $HBrO(aq) \rightarrow Br^-(aq) + O_2(g)$
(e) $Cl_2(g) \rightarrow ClO_3^-(aq) + Cl^-(aq)$

48. Write the balanced net ionic equations for the reactions given. Using the reactants shown in parentheses, write balanced formula unit equations for the ionic equations that you have just written.
(a) $MnO_4^- + C_2O_4^{2-} + H^+ \rightarrow Mn^{2+} + CO_2 + H_2O$
($KMnO_4$, HCl, and $K_2C_2O_4$)
(b) $Zn + NO_3^- + H^+ \rightarrow Zn^{2+} + NH_4^+ + H_2O$
($Zn(s)$ and HNO_3)

49. Write the balanced net ionic equations for the reactions given. Using the reactants shown in parentheses, write balanced formula unit equations for the ionic equations that you have just written.
(a) $I_2 + S_2O_3^{2-} \rightarrow I^- + S_4O_6^{2-}$
(I_2 and $Na_2S_2O_3$)
(b) $IO_3^- + N_2H_4 + Cl^- + H^+ \rightarrow N_2 + ICl_2^- + H_2O$
($NaIO_3 + N_2H_4$, and HCl)

50. Write the balanced net ionic equations for the reactions given. Using the reactants shown in parentheses, write balanced formula unit equations for the ionic equations that you have just written.
(a) $Zn + Cu^{2+} \rightarrow Cu + Zn^{2+}$
(Zn and $CuSO_4$)
(b) $Cr + H^+ \rightarrow Cr^{3+} + H_2$
(Cr and H_2SO_4)

51. Write the balanced net ionic equations for the reactions given. Using the reactants shown in parentheses, write balanced formula unit equations for the ionic equations that you have just written.
(a) $Cl_2 + OH^- \rightarrow ClO_3^- + Cl^- + H_2O$
(Cl_2 and hot NaOH)
(b) $Pb + H^+ + Br^- \rightarrow PbBr_2(s) + H_2$
($Pb(s)$ and HBr)

Redox Titrations—Mole Method and Molarity

52. What volume of 0.150 M $KMnO_4$ would be required to oxidize 20.0 mL of 0.100 M $FeSO_4$ in acidic solution? Refer to Example 11-19.

53. What volume of 0.100 M $K_2Cr_2O_7$ would be required to oxidize 70.0 mL of 0.100 M Na_2SO_3 in acidic solution? The products include Cr^{3+} and SO_4^{2-} ions. Refer to Example 11-20.

54. What volume of 0.200 M $KMnO_4$ would be required to oxidize 50.0 mL of 0.100 M KI in acidic solution? Products include Mn^{2+} and I_2.

55. What volume of 0.200 M $K_2Cr_2O_7$ would be required to oxidize 50.0 mL of 0.100 M KI in acidic solution? Products include Cr^{3+} and I_2.

56. (a) A solution of sodium thiosulfate, $Na_2S_2O_3$, is 0.1455 M. 40.00 mL of this solution reacts with 26.36 mL of I_2 solution. Calculate the molarity of the I_2 solution.

$$2Na_2S_2O_3 + I_2 \longrightarrow Na_2S_4O_6 + 2NaI$$

(b) 25.32 mL of the I_2 solution is required to titrate a sample containing As_2O_3. Calculate the mass of As_2O_3 (197.8 g/mol) in the sample.

$$As_2O_3 + 5H_2O + 2I_2 \longrightarrow 2H_3AsO_4 + 4HI$$

57. Copper(II) ions, Cu^{2+}, can be determined by the net reaction

$$2Cu^{2+} + 2I^- + 2S_2O_3^{2-} \longrightarrow 2CuI(s) + S_4O_6^{2-}$$

A 2.075-g sample containing $CuSO_4$ and excess KI is titrated with 32.55 mL of 0.1214 M solution of $Na_2S_2O_3$. What is the percentage of $CuSO_4$ (159.6 g/mol) in the sample?

58. Find the volume of 0.250 M HI solution required to titrate
(a) 25.0 mL of 0.100 M NaOH
(b) 5.03 g of $AgNO_3$ ($Ag^+ + I^- \rightarrow AgI(s)$)
(c) 0.621 g $CuSO_4$ ($2Cu^{2+} + 4I^- \rightarrow 2CuI(s) + I_2(s)$)

*59. The iron in a sample containing some Fe_2O_3 is reduced to Fe^{2+}. The Fe^{2+} is titrated with 12.02 mL of 0.1467 M $K_2Cr_2O_7$ in an acid solution.

$$6Fe^{2+} + Cr_2O_7^{2-} + 14H^+ \longrightarrow 6Fe^{3+} + 2Cr^{3+} + 7H_2O$$

Find (a) the mass of Fe and (b) the percentage of Fe in a 5.675-g sample.

60. Calculate the molarity of a solution that contains 12.6 g of $KMnO_4$ in 500 mL of solution to be used in the reaction that produces MnO_4^{2-} ions as the reduction product.

*61. A 0.783-g sample of an ore of iron is dissolved in acid and converted to Fe(II). The sample is oxidized by 38.50 mL of 0.161 M ceric sulfate, $Ce(SO_4)_2$, solution; the ceric ion, Ce^{4+}, is reduced to Ce^{3+} ion. (a) Write a balanced equation for the reaction. (b) What is the percentage of iron in the ore?

Mixed Exercises

62. Calculate the molarity of a hydrochloric acid solution if 32.75 mL of it reacts with 0.2013 g of sodium carbonate.

63. Calculate the molarity and the normality of a sulfuric acid solution if 32.75 mL of it reacts with 0.2013 g of sodium carbonate.

64. Find the number of mmols of HCl that reacts with 25.5 mL of 0.110 M NaOH. What volume of 0.205 M HCl is needed to furnish this amount of HCl?

65. What is the composition of the final solution when 25.5 mL of 0.110 M NaOH and 25.5 mL of 0.205 M HCl solutions are mixed?

66. What volume of 0.1123 M HCl is needed to neutralize 1.58 g of $Ca(OH)_2$?

67. What mass of NaOH is needed to neutralize 45.50 mL of 0.1036 M HCl? If the NaOH is available as a 0.1021 M aqueous solution, what volume will be required?

68. What volume of 0.203 M H_2SO_4 solution would be required to neutralize completely 39.4 mL of 0.302 M KOH solution?

69. What volume of 0.333 N H_2SO_4 solution would be required to neutralize completely 37.4 mL of 0.302 N KOH solution?

70. What volume of 0.2045 normal sodium hydroxide would be required to neutralize completely 38.38 mL of 0.1023 normal H_2SO_4 solution?

71. Benzoic acid, C_6H_5COOH, is sometimes used as a primary standard for the standardization of solutions of bases. A 1.862-g sample of the acid is neutralized by 33.00 mL of an NaOH solution. What is the molarity of the base solution?

$$C_6H_5COOH(s) + NaOH(aq) \longrightarrow$$
$$C_6H_5COONa(aq) + H_2O(\ell)$$

BUILDING YOUR KNOWLEDGE

72. Limonite is an ore of iron that contains $2Fe_2O_3 \cdot 3H_2O$. A 0.5166-g sample of limonite is dissolved in acid and treated so that all the iron is converted to ferrous ions, Fe^{2+}. This sample requires 42.96 mL of 0.02130 M sodium dichromate solution, $Na_2Cr_2O_7$, for titration. Fe^{2+} is oxidized to Fe^{3+} and $Cr_2O_7^{2-}$ is reduced to Cr^{3+}. What is the percentage of iron in the limonite? If your answer had been over 100% limonite, what conclusion could you make, presuming that the analytical data are correct?

73. For the formation of 1.00 mol of water, which reaction uses the most nitric acid?
(a) $3 Cu(s) + 8HNO_3(aq) \rightarrow$
$$3Cu(NO_3)_2(aq) + 2NO(g) + 4H_2O(\ell)$$
(b) $Al_2O_3(s) + 6HNO_3(aq) \rightarrow 2Al(NO_3)_3(aq) + 3H_2O(\ell)$
(c) $4Zn(s) + 10HNO_3(aq) \rightarrow$
$$4Zn(NO_3)_2(aq) + NH_4NO_3(aq) + 3H_2O(\ell)$$

74. The typical concentration of HCl in stomach acid (digestive juice) is a concentration of about 8.0×10^{-2} M. One experiences "acid stomach" when the stomach contents reach about 1.0×10^{-1} M HCl. One Rolaids® tablet (an antacid) contains 334 mg of active ingredient, $NaAl(OH)_2CO_3$. Assume that you have acid stomach and that your stomach contains 800 mL of 1.0×10^{-1} M HCl. Calculate the number of mmol of HCl in the stomach and the number of mmol of HCl that the tablet *can* neutralize. Which is greater? (The neutralization reaction produces NaCl, $AlCl_3$, CO_2, and H_2O.)

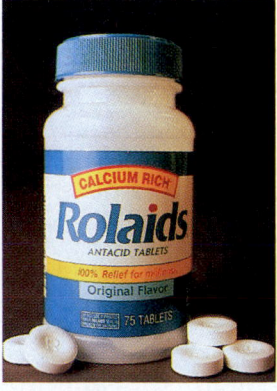

75. Refer to Exercises 14 and 15. Notice that the percentage by mass of solute is nearly the same for both solutions. How many moles of solute are present per liter of each solution? Are the moles of solute per liter also nearly equal? Why or why not?

76. The etching of glass by hydrofluoric acid may be represented by the simplified reaction of silica with HF.

$$SiO_2(s) + HF(aq) \longrightarrow H_2SiF_6(aq) + H_2O(\ell)$$

This is an acid–base reaction in which a weak acid is used to produce an even weaker acid; is it also an oxidation–reduction reaction? Balance the equation.

77. Write a Lewis formula for the anion SiF_6^{2-} that would be produced from the weak acid H_2SiF_6. Use the VSEPR theory to predict the shape of SiF_6^{2-}.

78. Ascorbic acid (vitamin C), along with many other reputed properties, acts as an antioxidant. The following equation illustrates its antioxidant properties.

$$H_2C_6H_6O_6 \longrightarrow C_6H_6O_6 + H_2$$

What is an antioxidant? Assign an oxidation number to each atom. Is vitamin C oxidized or reduced in this reaction?

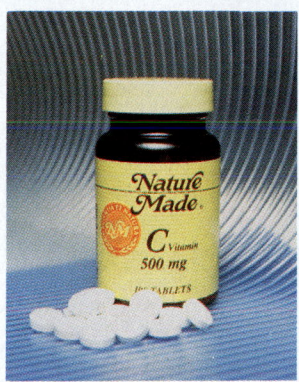

79. Oxalic acid, a poisonous compound, is found in certain vegetables such as spinach and rhubarb, but in concentrations well below toxic limits. The manufacturers of a spinach juice concentrate routinely test their product using an oxalic acid analysis to avoid any problems from an unexpectedly high concentration of this chemical. A titration with potassium permanganate is used for the oxalic acid assay, according to the following net equation.

$$5H_2C_2O_4 + 2MnO_4^- + 6H^+ \longrightarrow 10CO_2 + 2Mn^{2+} + 8H_2O$$

Calculate the molarity of an oxalic acid solution requiring 23.2 mL of 0.127 M permanganate for a 25.0 mL aliquot of the solution.

80. Baking soda, $NaHCO_3$, used to be a common remedy for "acid stomach." What weight of baking soda would be required to neutralize 85 mL of digestive juice, corresponding in acidity to 0.17 M HCl?

12 Gases and the Kinetic–Molecular Theory

One dish contains concentrated aqueous NH_3 solution and the other contains concentrated aqueous HCl solution. Gaseous NH_3 and gaseous HCl from these solutions react to form solid NH_4Cl, the white smoke.

OBJECTIVES

As you study this chapter, you should learn

• *About the properties of gases and how they differ from liquids and solids*

• *How we measure pressure*

• *To understand the absolute (Kelvin) temperature scale*

• *About the relationships among pressure, volume, temperature, and amount of gas (Boyle's Law, Charles' Law, Avogadro's Law, and the Combined Gas Law, and the limitations of each one)*

• *To use Boyle's Law, Charles' Law, Avogadro's Law, and the Combined Gas Law, as appropriate, to calculate changes in pressure, volume, temperature, and amount of gas*

• *To do calculations about gas densities and the standard molar volume*

• *About the ideal gas equation, and how to use it to do calculations about a sample of gas*

• *To determine molecular weights and formulas of gaseous substances from measured properties of gases*

• *To describe how mixtures of gases behave and to predict their properties (Dalton's Law of Partial Pressures)*

• *About the kinetic–molecular theory of gases, and how this theory is consistent with the observed gas laws*

• *About molecular motion, diffusion, and effusion of gases*

• *What molecular features are responsible for nonideal behavior of real gases, and when this nonideal behavior is important*

• *To carry out calculations about the gases involved in chemical reactions*

Table 12-1 *Densities and Molar Volumes of Three Substances at Atmospheric Pressure**

Substance	Solid		Liquid (20°C)		Gas (100°C)	
	Density (g/mL)	*Molar volume (mL/mol)*	*Density (g/mL)*	*Molar volume (mL/mol)*	*Density (g/mL)*	*Molar volume (mL/mol)*
water (H_2O)	0.917 (0°C)	19.6	0.998	18.0	0.000588	30,600
benzene (C_6H_6)	0.899 (0°C)	86.9	0.876	89.2	0.00255	30,600
carbon tetrachloride (CCl_4)	1.70 (−25°C)	90.5	1.59	96.8	0.00503	30,600

**The molar volume of a substance is the volume occupied by one mole of that substance.*

12-1 COMPARISON OF SOLIDS, LIQUIDS, AND GASES

Matter exists in three physical states: solids, liquids, and gases. In the solid state water is known as ice, in the liquid state it is called water, and in the gaseous state it is known as steam or water vapor. Most, but not all, substances can exist in all three states. Most solids change to liquids and most liquids change to gases as they are heated. Liquids and gases are known as **fluids** because they flow freely. Solids and liquids are referred to as **condensed states** because they have much higher densities than gases. Table 12-1 displays the densities of a few common substances in different states.

As the data in Table 12-1 indicate, solids and liquids are many times more dense than gases. The molecules must be very far apart in gases and much closer together in liquids and solids. For example, the volume of one mole of liquid water is about 18 milliliters, whereas one mole of steam occupies about 30,600 milliliters at 100°C and at atmospheric pressure. Gases are easily compressed, and they completely fill any container in which they are present. This tells us that the molecules in a gas are far apart compared to their sizes and that interactions among them are weak. The possibilities for interaction among gaseous molecules would be minimal (because they are so far apart) were it not for their rapid motion.

All substances that are gases at room temperature may be liquefied by cooling and compressing them. Volatile liquids are easily converted to gases at room temperature or slightly above. The term **"vapor"** refers to a gas that is formed by evaporation of a liquid or sublimation of a solid; it is commonly used when some of the liquid or solid remains in contact with the gas.

Some compounds decompose before melting or boiling.

Ice is less dense than liquid water. This behavior is quite unusual; most substances are more dense in the solid state than in the liquid state.

Volatile liquids evaporate readily. They have low boiling points.

Table 12-2 *Composition of Dry Air*

Gas	% by Volume
N_2	78.09
O_2	20.94
Ar	0.93
CO_2	0.03*
He, Ne, Kr, Xe	0.002
CH_4	0.00015*
H_2	0.00005
All others combined[†]	<0.00004

**Variable.*

[†]*Atmospheric moisture varies.*

12-2 COMPOSITION OF THE ATMOSPHERE AND SOME COMMON PROPERTIES OF GASES

Many important chemical substances are gases. The earth's atmosphere is a mixture of gases and particles of liquids and solids (Table 12-2). The major gaseous components are N_2 (bp −195.79°C) and O_2 (bp −182.98°C), with smaller concentrations of other gases. All gases are *miscible;* that is, they mix completely *unless* they react with each other.

Several scientists, notably Torricelli (1643), Boyle (1660), Charles (1787), and Graham (1831), laid an experimental foundation upon which our present understanding of gases is based. For example, their investigations showed that

(a)

(b)

Diffusion of bromine vapor in air. Some liquid bromine (dark reddish-brown) was placed in the small inner bottle. As the liquid evaporated, the resulting reddish-brown gas diffused. In (a) the diffusion has occurred for a shorter time than in (b).

1. Gases can be compressed into smaller volumes; that is, their densities can be increased by applying increased pressure.

2. Gases exert pressure on their surroundings; in turn, pressure must be exerted to confine gases.

3. Gases expand without limits, and so gas samples completely and uniformly occupy the volume of any container.

4. Gases diffuse into each other, and so samples of gas placed in the same container mix completely. Conversely, different gases in a mixture do not separate on standing.

5. The amounts and properties of gases are described in terms of temperature, pressure, the volume occupied, and the number of molecules present. For example, a sample of gas occupies a greater volume hot than it does when cold at the same pressure, but the number of molecules does not change.

Investigating four variables at once is difficult. In Sections 12-4 through 12-8 we shall see how to study these variables two at a time. Section 12-9 will consolidate these descriptions into a single relationship, the ideal gas equation.

12-3 PRESSURE

Pressure is defined as force per unit area—for example, lb/in², commonly known as *psi*. Pressure may be expressed in many different units, as we shall see. The mercury **barometer** is a simple device for measuring atmospheric pressures. Figure 12-1a illustrates the "heart" of the mercury barometer. A glass tube (about 800 mm long) is sealed at one end, filled with mercury, and then carefully inverted into a dish of mercury without air being allowed to enter. The mercury in the tube falls to the level at which the pressure of the air on the surface of the mercury in the dish equals the gravitational pull downward on the mercury in the tube. The air pressure is measured in terms of the height of the mercury column, i.e., the vertical distance between the surface of the mercury in the open dish and that inside the closed tube. The pressure exerted by the atmosphere is equal to the pressure exerted by the column of mercury.

(Text continues on p. 394)

CHEMISTRY IN USE

The Environment

The Greenhouse Effect

During the last century, the great increase in our use of fossil fuels has caused a significant rise in the concentration of carbon dioxide, CO_2, in the atmosphere. Scientists believe that the concentration of atmospheric CO_2 could double by early in the 21st century, compared with its level just before the Industrial Revolution. During the last 100 to 200 years, the CO_2 concentration has increased by 25%; nearly a fifth of this rise took place just in the decade from 1975 to 1985. The curve in Figure (a) shows the recent steady rise in atmospheric CO_2 concentration.

Energy from the sun reaches the earth in the form of light: Neither CO_2 nor H_2O vapor absorbs the visible light in sunlight, so they do not prevent it from reaching the surface of the earth. However, the energy given off by the earth in the form of lower-energy infrared (heat) radiation is readily absorbed by both CO_2 and H_2O (as it is by the glass or plastic of greenhouses). Thus, some of the heat the earth must lose to stay in thermal equilibrium can become trapped in the atmosphere, causing the temperature to rise (Figure b). This phenomenon, called the **greenhouse effect,** has been the subject of much discussion among scientists and of many articles in the popular press. The anticipated rise in average global temperature by the year 2050 due to increased CO_2 concentration is predicted to be 2 to 5°C. Indeed, eight of the ten warmest years on record in the Northern Hemisphere were between 1980 and 1989.

An increase of 2 to 5°C may not seem like much. However, this is thought to be enough to cause a dramatic change in climate, transforming now productive land into desert and altering the habitats of many animals and plants beyond their ability to adapt. Another drastic consequence of even this small temperature rise would be the partial melting of the polar ice caps. The resulting rise in sea level, though only a few feet, would mean that water would inundate coastal cities such as Los Angeles, New York, and Houston, and low-lying coastal areas such as southern Florida and Louisiana. On a global scale, the effects would be devastating.

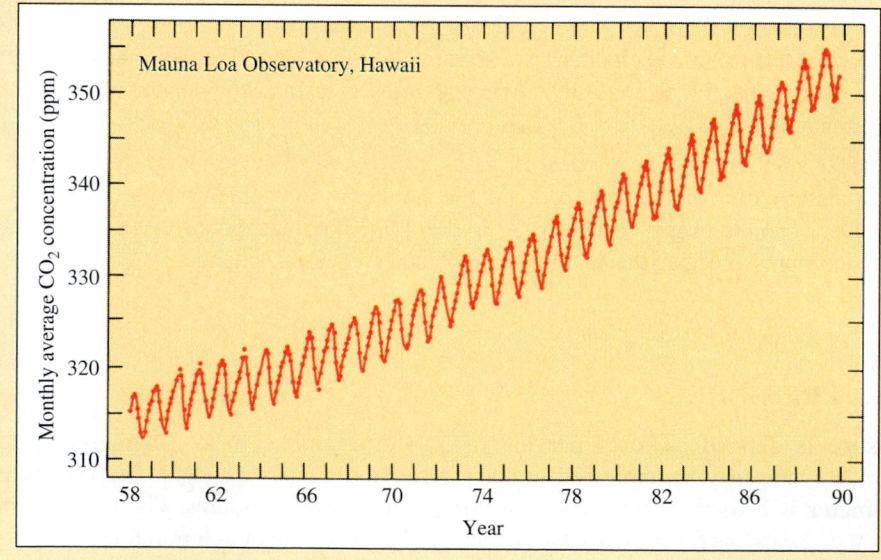

(a) A plot of the monthly average CO_2 concentration in parts per million (ppm), measured at Mauna Loa Observatory, Hawaii, far from significant sources of CO_2 from human activities. Annual fluctuations occur because plants in the Northern Hemisphere absorb CO_2 in the spring and release it as they decay in the fall.

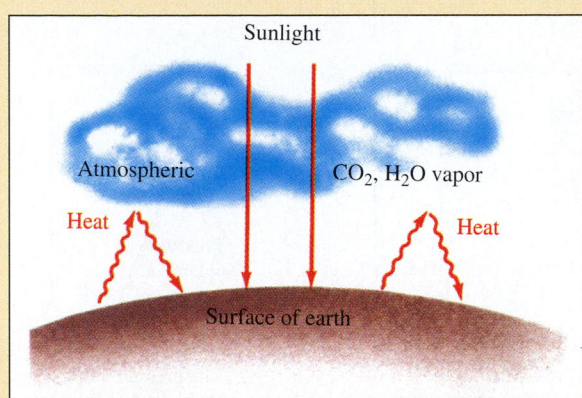

Sunlight

Atmospheric

CO_2, H_2O vapor

Heat

Heat

Surface of earth

(b) The greenhouse effect. Visible light passes through atmospheric H_2O and CO_2, but heat radiated from the surface of the earth is absorbed by these gases.

Tropical rain forests are important in maintaining the balance of CO_2 and O_2 in the earth's atmosphere. In recent years a portion of the South American forests (by far the world's largest) larger than France has been destroyed, either by flooding caused by hydroelectric dams or by clearing of forest land for agricultural or ranching use. Such destruction continues at a rate of more than 20,000 square kilometers per year. If current trends continue, many of the world's rain forests will be severely reduced or even obliterated by the year 2000. The fundamental question—"What are the long-term consequences of the destruction of tropical rain forests?"—remains unanswered.

The earth's forests and jungles play a crucial role in maintaining the balance of gases in the atmosphere, removing CO_2 and supplying O_2. The massive destruction, for economic reasons, of heavily forested areas such as the Amazon rain forest in South America is cited as another long-term contributor to global environmental problems. Worldwide, more than three million square miles of once-forested land are now barren for some reason; at least 60% of this land is now unused. Environmental scientists estimate that if even one quarter of this land could be reforested, the vegetation would absorb 1.1 billion tons of CO_2 annually.

Some scientists are more skeptical about the role of human-produced CO_2 in climate changes and, indeed, about whether global warming is a significant phenomenon or simply another of the recognized warm–cold cycles that have occurred throughout the earth's history. Such skeptics point out an unexplained increase in atmospheric CO_2 during an extended period in the 17th century, and an even higher and more prolonged peak about 130,000 years ago. However, even the most skeptical observers seem to agree that responsible stewardship of the planet requires that we do something in a reasoned fashion to reduce production of greenhouse gases, primarily CO_2, and that this will involve decreasing our dependence on energy from fossil fuels. Despite the technical and political problems of waste dis-posal, an essentially all-electric economy based on nuclear power is viewed by some as the most reasonable course in the long run.

Much CO_2 is eventually absorbed by the vast amount of water in the oceans, where the carbonate–bicarbonate buffer system almost entirely counteracts any adverse effects of ocean water acidity. Ironically, there is also evidence to suggest that other types of air pollution in the form of particulate matter may partially counteract the greenhouse effect. The particles reflect visible (sun) radiation rather than absorbing it, blocking some light from entering the atmosphere. However, it seems foolish to depend on one form of pollution to help rescue us from the effects of another! Real solutions to current environmental problems such as the greenhouse effect are not subject to quick fixes, but depend on long-term cooperative international efforts that are based on the firm knowledge resulting from scientific research.

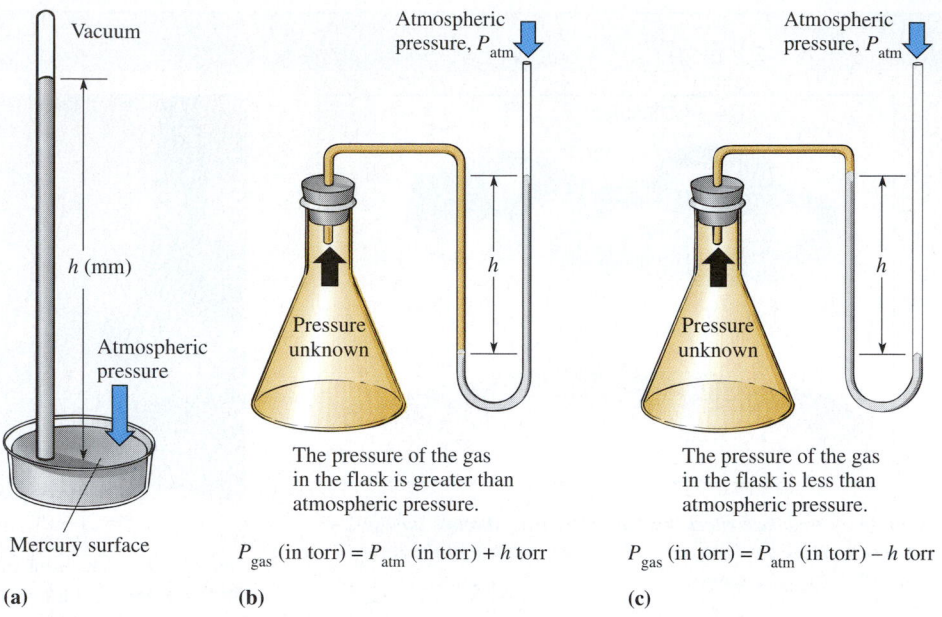

P_{gas} (in torr) = P_{atm} (in torr) + h torr

P_{gas} (in torr) = P_{atm} (in torr) − h torr

(a)   **(b)**   **(c)**

Figure 12-1   Some laboratory devices for measuring pressure. (a) Schematic diagram of a closed-end barometer. At the level of the lower mercury surface, the pressure both inside and outside the tube must be equal to that of the atmosphere. There is no air inside the tube, so the pressure is exerted only by the mercury column h mm high. Hence, the atmospheric pressure must equal the pressure exerted by h mm Hg, or h torr. (b) The two-arm mercury barometer is called a manometer. In this sample, the pressure of the gas is *greater than* the external atmospheric pressure. At the level of the lower mercury surface, the total pressure on the mercury in the left arm must equal the total pressure on the

mercury in the right arm. The pressure exerted by the gas is equal to the external pressure *plus* the pressure exerted by the mercury column of height h mm, or P_{gas} (in torr) = P_{atm} (in torr) + h torr. (c) When the gas pressure measured by the manometer is *less than* the external atmospheric pressure, the pressure exerted by the atmosphere is equal to the gas pressure *plus* the pressure exerted by the mercury column, or P_{atm} = P_{gas} + h. We can rearrange this to write P_{gas} (in torr) = P_{atm} (in torr) − h torr.

The unit *torr* was named for Evangelista Torricelli (1608–1647), who invented the mercury barometer.

Mercury barometers are simple and well known, so gas pressures are frequently expressed in terms of millimeters of mercury (mm Hg, or just mm). In recent years the unit **torr** has been used to indicate pressure; it is defined as 1 torr = 1 mm Hg.

A mercury **manometer** consists of a glass U-tube partially filled with mercury. One arm is open to the atmosphere and the other is connected to a container of gas (Figure 12-1b,c).

Atmospheric pressure varies with atmospheric conditions and distance above sea level. It is lower at high elevations because the air at low levels is compressed by the air above it. Approximately one half of the matter in the atmosphere is less than 20,000 feet above sea level. Thus, atmospheric pressure is only about one-half as great at 20,000 feet as it is at sea level. Mountain climbers and pilots use portable barometers to determine their altitudes (Figure 12-2). At sea level, at a latitude of 45°, the average atmospheric pressure supports a column of mercury 760 mm high in a simple mercury barometer when the mercury is at 0°C. This average sea-level pressure of 760 mm Hg is called **one atmosphere of pressure.**

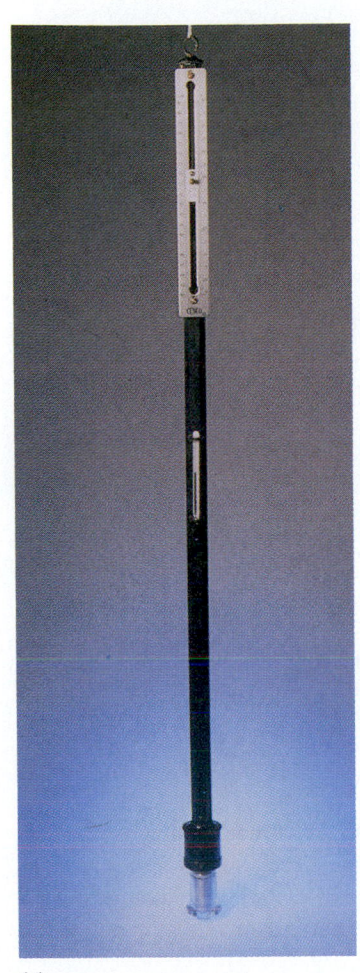

(a)

(b)

(c)

Figure 12-2   Some commercial pressure-measuring devices. (a) A commercial mercury barometer. (b) Portable barometers. This type is called an *aneroid* ("not wet") barometer. Some of the air has been removed from the airtight box, which is made of thin, flexible metal. When the pressure of the atmosphere changes, the remaining air in the box expands or contracts (Boyle's Law), moving the flexible box surface and an attached pointer along a scale. (c) A tire gauge. This kind of gauge registers "relative" pressure, i.e., the *difference* between internal pressure and the external atmospheric pressure. For instance, when the gauge reads 30 psi (pounds per square inch), the total gas pressure in the tire is 30 psi + 1 atm, or about 45 psi. In engineering terminology, this is termed "psig" (g = gauge).

1 atmosphere (atm) = 760 mm Hg at 0°C = 760 torr

The SI unit of pressure is the **pascal** (Pa), defined as the pressure exerted by a force of one newton acting on an area of one square meter. By definition, one **newton** (N) is the force required to give a mass of one kilogram an acceleration of one meter/second per second. Symbolically we represent one newton as

Acceleration is the change in velocity (m/s) per unit time (s), m/s^2.

$$1 \text{ N} = \frac{1 \text{ kg} \cdot \text{m}}{\text{s}^2} \quad \text{so} \quad 1 \text{ Pa} = \frac{1 \text{ N}}{\text{m}^2} = \frac{1 \text{ kg}}{\text{m} \cdot \text{s}^2}$$

One atmosphere of pressure = 1.01325×10^5 Pa, or 101.325 kPa.

12-4 BOYLE'S LAW: THE RELATION OF VOLUME TO PRESSURE

Early experiments on the behavior of gases were carried out by Robert Boyle in the 17th century. In a typical experiment (Figure 12-3), a sample of a gas was trapped in a U-tube and allowed to come to constant temperature. Then its volume and the difference in the

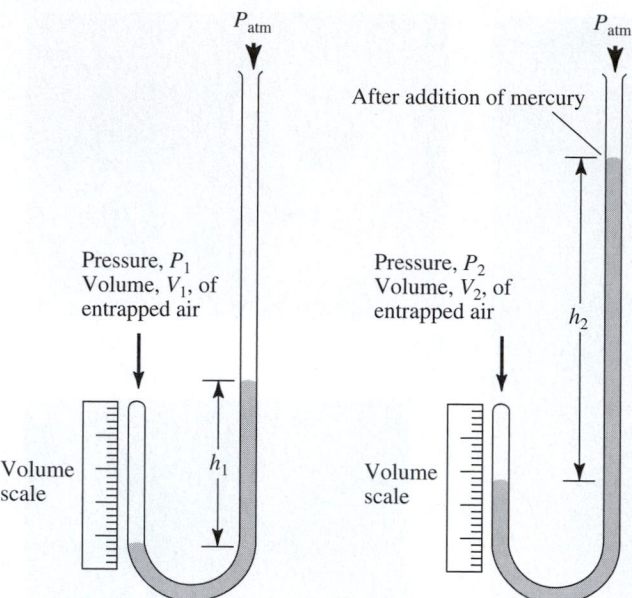

Figure 12-3 A representation of Boyle's experiment. A sample of air is trapped in a tube in such a way that the pressure on the air can be changed and the volume of the air measured. P_{atm} is the atmospheric pressure, measured with a barometer. $P_1 = h_1 + P_{atm}$, $P_2 = h_2 + P_{atm}$.

heights of the two mercury columns were recorded. This difference in height plus the pressure of the atmosphere represents the pressure on the gas. Addition of more mercury to the tube increases the pressure by changing the height of the mercury column. As a result, the gas volume decreases. The results of several such experiments are tabulated in Figure 12-4a.

Boyle showed that for a given sample of gas at constant temperature, the product of pressure and volume, $P \times V$, was always the same number.

P	V	$P \times V$	$1/P$
5.0	40.0	200	0.20
10.0	20.0	200	0.10
15.0	13.3	200	0.067
17.0	11.8	201	0.059
20.0	10.0	200	0.050
22.0	9.10	200	0.045
30.0	6.70	201	0.033
40.0	5.00	200	0.025

(a)

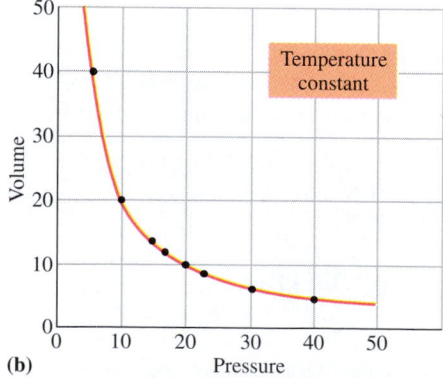

(b)

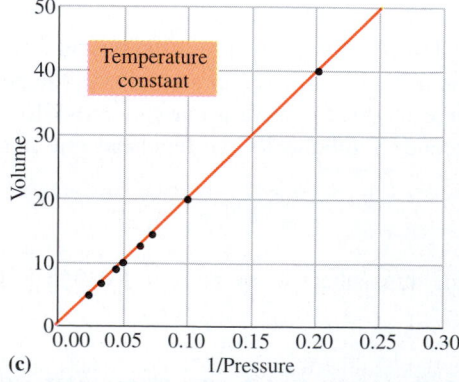

(c)

Figure 12-4 (a) Some typical data from an experiment such as that in Figure 12-3. Measured values of P and V are presented in the first two columns, on an arbitrary scale. (b, c) Graphical representations of Boyle's Law, using the data of part (a). (b) V versus P. (c) V versus $1/P$.

At a given temperature, the product of pressure and volume of a definite mass of gas is constant.

$$PV = k \qquad \text{(constant } n, T\text{)}$$

This relationship is **Boyle's Law.** The value of k depends on the amount (number of moles, n) of gas present and on the temperature, T. Units for k are determined by the units used to express the volume (V) and pressure (P).

When the volume of a gas is plotted against its pressure at constant temperature, the resulting curve is one branch of a hyperbola. Figure 12-4b is a graphic illustration of this inverse relationship. When volume is plotted versus the reciprocal of the pressure, $1/P$, a straight line results (Figure 12-4c). In 1662 Boyle summarized the results of his experiments on various samples of gases in an alternative statement of Boyle's Law:

Because neither V nor P of a sample of a gas can be less than zero, the other branch of the hyperbola (third quadrant) has no physical significance.

At constant temperature the volume, V, occupied by a definite mass of a gas is inversely proportional to the applied pressure, P.

$$V \propto \frac{1}{P} \qquad \text{or} \qquad V = k\left(\frac{1}{P}\right) \qquad \text{(constant } n, T\text{)}$$

The symbol $\propto$ reads "is proportional to." A proportionality is converted into an equality by introducing a proportionality constant, k.

At normal temperatures and pressure, most gases obey Boyle's Law rather well. We call this *ideal behavior.* Deviations from ideality are discussed in Section 12-14.

Let us think about a fixed mass of gas at constant temperature, but at two different conditions of pressure and volume (Figure 12-3). For the first condition we can write

$$P_1 V_1 = k \qquad \text{(constant } n, T\text{)}$$

and for the second condition we can write

$$P_2 V_2 = k \qquad \text{(constant } n, T\text{)}$$

Because the right-hand sides of these two equations are the same, the left-hand sides must be equal, or

$$P_1 V_1 = P_2 V_2 \qquad \text{(for a given amount of a gas at constant temperature)}$$

This form of Boyle's Law is useful for calculations involving pressure and volume changes, as the following examples demonstrate.

EXAMPLE 12-1 *Boyle's Law Calculation*

A sample of gas occupies 12 liters under a pressure of 1.2 atm. What would its volume be if the pressure were increased to 2.4 atm? Assume that the temperature of the gas sample does not change.

Plan

We know the volume at one pressure and wish to find the volume at another pressure (constant temperature). We can solve Boyle's Law for the second volume and substitute.

Pressure and volume are inversely proportional. Doubling the pressure halves the volume of a sample of gas at constant temperature.

Solution

We have $P_1 = 1.2$ atm; $V_1 = 12$ L; $P_2 = 2.4$ atm; $V_2 = \underline{?}$. Solving Boyle's Law, $P_1V_1 = P_2V_2$, for V_2 and substituting yields

$$V_2 = \frac{P_1V_1}{P_2} = \frac{(1.2 \text{ atm})(12 \text{ L})}{2.4 \text{ atm}} = \boxed{6.0 \text{ L}}$$

> It is often helpful to tabulate what is given and what is asked for in a problem.

▼ **PROBLEM-SOLVING TIP**　*Units in Boyle's Law Calculations*

Which units for volume and pressure are appropriate for Boyle's Law calculations? Boyle's Law in the form $P_1V_1 = P_2V_2$ can be written as $V_1/V_2 = P_2/P_1$. This involves a ratio of volumes, so they can be expressed in any volume units—liters, milliliters, cubic feet—as long as the *same* units are used for both volumes. Likewise, because Boyle's Law involves a ratio of pressures, you can use any units for pressures—atmospheres, torr, pascals—as long as the *same* units are used for both pressures.

EXAMPLE 12-2　*Boyle's Law Calculation*

A sample of oxygen occupies 10.0 liters under a pressure of 790 torr (105 kPa). At what pressure would it occupy 13.4 liters if the temperature did not change?

Plan

We know the pressure at one volume and wish to find the pressure at another volume (at constant temperature). We can solve Boyle's Law for the second pressure and substitute.

Solution

We have $P_1 = 790$ torr; $V_1 = 10.0$ L; $P_2 = \underline{?}$; $V_2 = 13.4$ L. Solving Boyle's Law, $P_1V_1 = P_2V_2$, for P_2 and substituting yields

> The problem tells us that *V increases* from 10.0 L to 13.4 L (at constant temperature). Therefore *P* must *decrease.*

$$P_2 = \frac{P_1V_1}{V_2} = \frac{(790 \text{ torr})(10.0 \text{ L})}{13.4 \text{ L}} = \boxed{590 \text{ torr}} \quad \left(\times \frac{101.3 \text{ kPa}}{760 \text{ torr}} = \boxed{78.6 \text{ kPa}} \right)$$

You should now work Exercises 18 and 19.

▼ **PROBLEM-SOLVING TIP**　*Use What You Can Predict about the Answer*

In Example 12-1 the calculated volume decrease is consistent with the increase in pressure. We can use this reasoning in another method for solving that problem, that is, by setting up a "Boyle's Law factor" to change the volume in the direction required by the pressure change. We reason that the pressure increases from 1.2 atm to 2.4 atm, so the volume *decreases* by the factor (1.2 atm/2.4 atm). The solution then becomes

$$\underline{?} \text{ L} = 12 \text{ L} \times (\text{Boyle's Law factor that would decrease the volume})$$

$$= 12 \text{ L} \times \left(\frac{1.2 \text{ atm}}{2.4 \text{ atm}} \right) = \boxed{6.0 \text{ L}}$$

Perhaps you can solve Example 12-2 using a Boyle's Law factor.

12-5 CHARLES' LAW: THE RELATION OF VOLUME TO TEMPERATURE; THE ABSOLUTE TEMPERATURE SCALE

In his pressure–volume studies on gases, Robert Boyle noticed that heating a sample of gas caused some volume change, but he did not follow up on this observation. About 1800, two French scientists—Jacques Charles and Joseph Gay-Lussac, pioneer balloonists at the time—began studying the expansion of gases with increasing temperature. Their studies showed that the rate of expansion with increased temperature was constant and was the same for all the gases they studied as long as pressure remained constant. The implications of their discovery were not fully recognized until nearly a century later. Then scientists used this behavior of gases as the basis of a new temperature scale, the absolute temperature scale.

The change of volume with temperature, at constant pressure, is illustrated in Figure 12-5. From the table of typical data in Figure 12-5b, we see that volume (V, mL) increases as temperature (t, °C) increases, but the quantitative relationship is not yet obvious. These data are plotted in Figure 12-5c (line A), together with similar data for the same gas sample at different pressures (lines B and C).

Lord Kelvin, a British physicist, noticed that an extension of the different temperature–volume lines back to zero volume (dashed line) yields a common intercept at −273.15°C on the temperature axis. Kelvin named this temperature **absolute zero**. The degrees are the same size over the entire scale, so 0°C becomes 273.15 degrees above absolute zero. In honor of Lord Kelvin's work, this scale is called the Kelvin temperature scale. As pointed out in Section 1-12, the relationship between the Celsius and Kelvin temperature scales is K = °C + 273.15°.

An artist's representation of Jacques Charles' first ascent in a hydrogen balloon at the Tuileries, Paris, December 1, 1783.

Lord Kelvin (1842–1907) was born William Thompson. At the age of ten he was admitted to Glasgow University. Because its new appliance was based on Kelvin's theories, a refrigerator company named its product the Kelvinator.

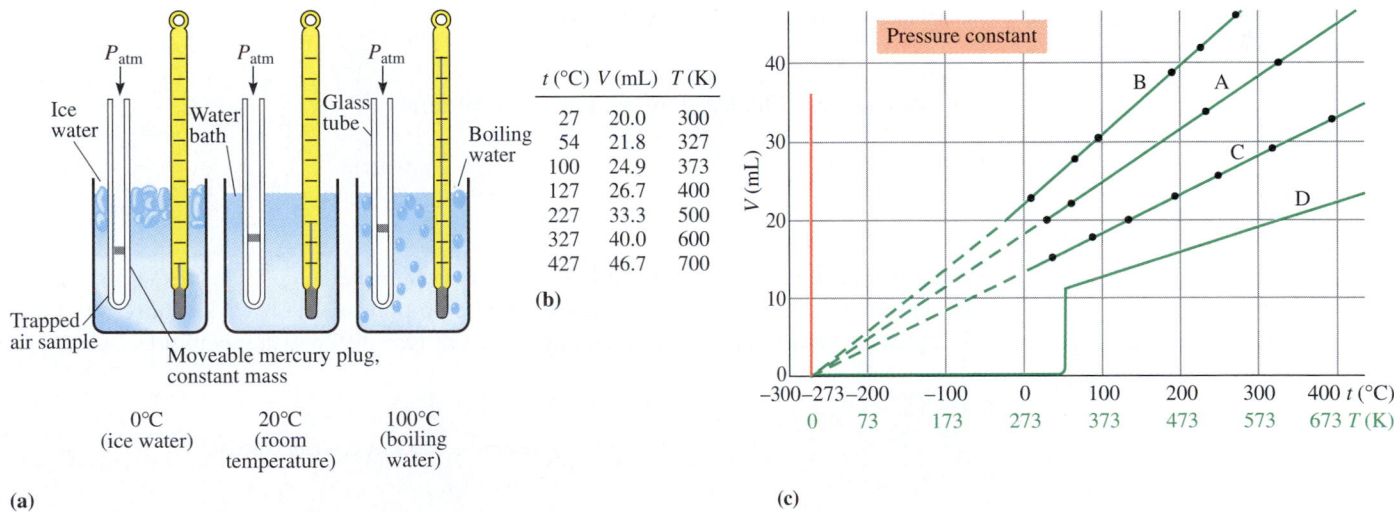

t (°C)	V (mL)	T (K)
27	20.0	300
54	21.8	327
100	24.9	373
127	26.7	400
227	33.3	500
327	40.0	600
427	46.7	700

(b)

(a)

(c)

Figure 12-5 An experiment showing that the volume of an ideal gas increases as the temperature is increased at constant pressure. (a) A mercury plug of constant weight, plus atmospheric pressure, maintains a constant pressure on the trapped air. (b) Some representative volume–temperature data at constant pressure. The relationship becomes clear when t (°C) is converted to T (K) by adding 273°. (c) A graph in which volume is plotted versus temperature on two different scales. Lines A, B, and C represent the same mass of the same ideal gas at different pressures. Line A represents the data tabulated in part (b). Graph D shows the behavior of a gas that condenses to form a liquid (in this case, at 50°C) as it is cooled.

When balloons filled with air are cooled in liquid nitrogen (bp −196°C), each shrinks to a small fraction of its original volume. Because the boiling points of the other components of air, except He and Ne, are lower than −196°C, they condense to form liquids. When the balloons are removed from the liquid nitrogen, the liquids vaporize to form gases again. As the air warms to room temperature, the balloons expand to their original volume (Charles' Law).

Recall that temperatures on the Kelvin scale are expressed in kelvins (not degrees Kelvin) and represented by K, not °K.

Absolute zero may be thought of as the limit of thermal contraction for an ideal gas.

If we convert temperatures (°C) to absolute temperatures (K), the green scale in Figure 12-5c, the volume–temperature relationship becomes obvious. This relationship is known as **Charles' Law.**

> At constant pressure, the volume occupied by a definite mass of a gas is directly proportional to its absolute temperature.

We can express Charles' Law in mathematical terms as

$$V \propto T \qquad \text{or} \qquad V = kT \qquad \text{(constant } n, P\text{)}$$

Rearranging the expression gives $V/T = k$, a concise statement of Charles' Law. As the temperature increases, the volume must increase proportionally. If we let subscripts 1 and 2 represent values for the same sample of gas at two different temperatures, we obtain

$$\frac{V_1}{T_1} = \frac{V_2}{T_2} \qquad \text{(for a definite mass of gas at constant pressure)}$$

which is the more useful form of Charles' Law. This relationship is valid *only* when temperature, T, is expressed on an absolute (usually the Kelvin) scale.

EXAMPLE 12-3   *Charles' Law Calculation*

A sample of nitrogen occupies 117 mL at 100°C. At what temperature would it occupy 234 milliliters if the pressure did not change?

Plan

We know the volume of the sample at one temperature and wish to know its temperature corresponding to a second volume (constant pressure). We can solve Charles' Law for the second temperature. We must remember to carry out calculations with all temperatures expressed on the Kelvin scale, converting to or from Celsius as necessary.

Solution

$$V_1 = 117 \text{ mL} \qquad\qquad V_2 = 234 \text{ mL}$$

$$T_1 = 100°C + 273° = 373 \text{ K} \qquad T_2 = \underline{?}$$

$$\frac{V_1}{T_1} = \frac{V_2}{T_2} \quad \text{and} \quad T_2 = \frac{V_2 T_1}{V_1} = \frac{(234 \text{ mL})(373 \text{ K})}{(117 \text{ mL})} = \boxed{746 \text{ K}}$$

$$°C = 746 \text{ K} - 273° = \boxed{473°C}$$

The temperature doubles on the Kelvin scale, from 373 K to 746 K, so the volume doubles.

You should now work Exercise 28.

▼ **PROBLEM-SOLVING TIP**   *Be Careful of Units in Charles' Law Calculations*

Which units for volume and temperature are appropriate for Charles' Law calculations? The equation $\frac{V_1}{T_1} = \frac{V_2}{T_2}$ can be written as $\frac{V_1}{V_2} = \frac{T_1}{T_2}$. This involves a ratio of volumes, so they can be expressed in any volume units—liters, milliliters, cubic feet—as long as the *same* units are used for both volumes. But the relationship *does not apply at all* unless the temperatures are both expressed on an absolute scale. Remember to express all temperatures in kelvins for Charles' Law calculations.

12-6 STANDARD TEMPERATURE AND PRESSURE

We have seen that both temperature and pressure affect the volumes (and therefore the densities) of gases. It is often convenient to choose some "standard" temperature and pressure as a reference point for discussing gases. **Standard temperature and pressure (STP)** are, by international agreement, 0°C (273.15 K) and one atmosphere of pressure (760 torr).

12-7 THE COMBINED GAS LAW EQUATION

Boyle's Law relates the pressures and volumes of a sample of gas at constant temperature, $P_1 V_1 = P_2 V_2$. Charles' Law relates the volumes and temperatures at constant pressure $V_1/T_1 = V_2/T_2$. Combination of Boyle's Law and Charles' Law into a single expression gives the **combined gas law equation.**

$$\frac{P_1 V_1}{T_1} = \frac{P_2 V_2}{T_2} \qquad \text{(constant } n\text{)}$$

Notice that the combined gas law equation becomes
1. $P_1 V_1 = P_2 V_2$ (Boyle's Law) when T is constant;
2. $\dfrac{V_1}{T_1} = \dfrac{V_2}{T_2}$ (Charles' Law) when P is constant; and
3. $\dfrac{P_1}{T_1} = \dfrac{P_2}{T_2}$ when V is constant.

CHEMISTRY IN USE

Research & Technology

Can Refrigerators Stay Cool with Rock 'N' Roll?

Some refrigerators keep their cool by using sound emitted by loudspeakers. Scientists at Los Alamos (New Mexico) National Laboratory made such a refrigerator and named it STAR (Space ThermoAcoustic Refrigerator) in the early 1980s. STAR was first used to cool electronics aboard the space shuttle Discovery in January 1992. Within the next five years, refrigerators and freezers whose design is based on STAR may be used in our homes.

To understand how this new breed of sonic refrigerators works, we recall that around 1800, Joseph Gay-Lussac observed that for a sample of gas in a container of constant volume, the pressure exerted by the gas is directly proportional to the absolute temperature.

$$P \propto T$$

He also found that the rate of increase in pressure with increasing temperature was independent of the gas studied. Thus, doubling the absolute temperature of a sample of gas also doubles its pressure at constant volume.

The pressure inside automobile tires increases as tire temperature increases from road friction and contact with hot pavement. Measuring tire pressures after a drive could give abnormally high pressure readings, causing owners to deflate their heated tires. Then, after the deflated tires cooled they would be under-inflated. Under-inflated tires wear out more quickly than properly inflated tires.

At the other extreme, when temperatures plunged to −86°F (−66°C) in Siberia in January 1989, automobile tire pressures decreased to 16 psi. To make matters worse, the rubber of the flattened tires froze, and it became impossible to drive the cars because each tire had a "flat side."

You may have noticed that the temperature of a fixed volume of gas depends upon its pressure when a bike tire got warm as you pumped air into it. Likewise, you may have noticed a cooling effect when discharging the gas from a pressurized aerosol can or releasing the air from a fully inflated bicycle tire. As the pressure is decreased in the aerosol can or the bike tire, they become cooler.

The sonic refrigerator uses sound waves from two four-inch speakers to exert pressure on a mixture of argon and helium confined in a specially shaped container. The mixture of gases heats up as the sound exerts a pressure on the gases, then cools down when the pressure decreases in the absence of sound. This sound is between 10,000 and 100,000 times louder than that of a typical rock concert, but the sound is contained within rigid, nonvibrating plastic walls. As long as the walls don't vibrate, no sound escapes from the refrigerator. In fact, the unit is so quiet that a stethoscope is sometimes used to check to see if the unit is operating properly.

A heat exchanger, similar to a car's radiator, removes excess heat from the pressurized gas sample. Then, as the gas mixture moves away from the speakers, its pressure decreases. As the pressure of the gas mixture decreases, so does its temperature. The cooled gas mixture then passes by and cools a liquid in a surrounding chamber; the cool liquid is circulated around a storage compartment that contains food (or electronic devices as discussed earlier).

Efforts are currently being directed at developing a home refrigerator called the ThermoAcoustic Life Sciences Refrigerator (TALSR); it has the same operating system as STAR but contains improvements that make it more suitable for home use. Many environmentalists are excited about TALSR because thermoacoustic refrigerators don't use chlorofluorocarbons (CFCs), the principal contributors to the destruction of earth's protective ozone layer. The principles of TALSR may also be utilized in a new generation of air conditioners that will be more environmentally friendly; their use would help to eliminate CFCs and a variety of CFC-replacement chemicals from the earth's atmosphere.

Ronald DeLorenzo
Middle Georgia College

When any five of the variables in the equation are known, the sixth variable can be calculated.

EXAMPLE 12-4 *Combined Gas Law Calculation*

A sample of neon occupies 105 liters at 27°C under a pressure of 985 torr. What volume would it occupy at standard temperature and pressure (STP)?

Plan

A sample of gas is changing in all three quantities P, V, and T. This suggests that we use the combined gas law equation. We tabulate what is known and what is asked for, solve the combined gas law equation for the unknown quantity, V_2, and substitute known values.

Solution

$$V_1 = 105 \text{ L} \qquad P_1 = 985 \text{ torr} \qquad T_1 = 27°C + 273° = 300 \text{ K}$$
$$V_2 = \underline{?} \qquad P_2 = 760 \text{ torr} \qquad T_2 = 273 \text{ K}$$

Solving for V_2,

$$\frac{P_1 V_1}{T_1} = \frac{P_2 V_2}{T_2} \qquad \text{so} \qquad V_2 = \frac{P_1 V_1 T_2}{P_2 T_1} = \frac{(985 \text{ torr})(105 \text{ L})(273 \text{ K})}{(760 \text{ torr})(300 \text{ K})} = \boxed{124 \text{ L}}$$

Alternatively, we can multiply the original volume by a Boyle's Law factor and a Charles' Law factor. As the pressure decreases from 985 torr to 760 torr, the volume increases, so the Boyle's Law factor is 985 torr/760 torr. As the temperature decreases from 300 K to 273 K, the volume decreases, so the Charles' Law factor is 273 K/300 K. Multiplication of the original volume by these factors gives the same result.

$$\underline{?} \text{ L} = 105 \text{ L} \times \frac{985 \text{ torr}}{760 \text{ torr}} \times \frac{273 \text{ K}}{300 \text{ K}} = \boxed{124 \text{ L}}$$

The temperature decrease (from 300 K to 273 K) alone would give only a small *decrease* in the volume of neon. The pressure decrease (from 985 torr to 760 torr) alone would result in a greater *increase* in the volume. The result of the two changes is that the volume increases from 105 liters to 124 liters.

EXAMPLE 12-5 *Combined Gas Law Calculation*

A sample of gas occupies 10.0 liters at 240°C under a pressure of 80.0 kPa. At what temperature would the gas occupy 20.0 liters if we increased the pressure to 107 kPa?

As a reference, one atm is 101.3 kPa.

Plan

The approach is the same as for Example 12-4 except that the unknown quantity is the temperature, T_2.

Solution

$$V_1 = 10.0 \text{ L} \qquad P_1 = 80.0 \text{ kPa} \qquad T_1 = 240°C + 273° = 513 \text{ K}$$
$$V_2 = 20.0 \text{ L} \qquad P_2 = 107 \text{ kPa} \qquad T_2 = \underline{?}$$

We solve the combined gas law equation for T_2.

$$\frac{P_1 V_1}{T_1} = \frac{P_2 V_2}{T_2} \qquad \text{so} \qquad T_2 = \frac{P_2 V_2 T_1}{P_1 V_1} = \frac{(107 \text{ kPa})(20.0 \text{ L})(513 \text{ K})}{(80.0 \text{ kPa})(10.0 \text{ L})} = \boxed{1.37 \times 10^3 \text{ K}}$$

$$K = °C + 273° \qquad \text{so} \qquad °C = 1.37 \times 10^3 \text{ K} - 273° = \boxed{1.10 \times 10^3 \text{ °C}}$$

You should now work Exercises 36 and 38.

▼ **PROBLEM-SOLVING TIP**   *Units in Combined Gas Law Calculations*

The combined gas law equation is derived by combining Boyle's and Charles' Laws, so the comments in earlier Problem-Solving Tips also apply to this equation. Remember to express all temperatures in kelvins. Volumes can be expressed in any units as long as both are in the same units. Similarly, any pressure units can be used, so long as both are in the same units. Example 12-4 uses torr for both pressures; Example 12-5 uses kPa for both pressures.

12-8 AVOGADRO'S LAW AND THE STANDARD MOLAR VOLUME

In 1811 Amedeo Avogadro postulated that

at the same temperature and pressure, equal volumes of all gases contain the same number of molecules.

Many experiments have demonstrated that Avogadro's hypothesis is accurate to within about ±2%, and the statement is now known as **Avogadro's Law.**

Avogadro's Law can also be stated as follows.

At constant temperature and pressure, the volume occupied by a gas sample is directly proportional to the number of moles of gas.

$$V \propto n \qquad \text{or} \qquad V = kn \qquad \text{or} \qquad \frac{V}{n} = k \qquad \text{(constant } P, T\text{)}$$

For two different samples of gas at the same temperature and pressure, the relation between volumes and numbers of moles can be represented as

$$\frac{V_1}{n_1} = \frac{V_2}{n_2} \qquad \text{(same } T, P\text{)}$$

The volume percentages given in Table 12-2 are also equal to mole percentages.

The volume occupied by a mole of gas at *standard temperature and pressure,* STP, is referred to as the standard molar volume. It is nearly constant for all gases (Table 12-3).

The **standard molar volume** of an ideal gas is taken to be 22.414 liters per mole at STP.

Density is defined as mass per unit volume.

Gas densities depend on pressure and temperature. However, the number of moles of gas in a given sample does not change with temperature or pressure. Pressure changes affect volumes of gases according to Boyle's Law, while temperature changes affect volumes of gases according to Charles' Law. We can use these laws to convert gas densities at various temperatures and pressures to *standard temperature and pressure.* Table 12-3 gives the experimentally determined densities of several gases at standard temperature and pressure.

Table 12-3 *Standard Molar Volumes and Densities of Some Gases*

Gas	Formula	(g/mol)	Standard Molar Volume (L/mol)	Density at STP (g/L)
hydrogen	H_2	2.02	22.428	0.090
helium	He	4.003	22.426	0.178
neon	Ne	20.18	22.425	0.900
nitrogen	N_2	28.01	22.404	1.250
oxygen	O_2	32.00	22.394	1.429
argon	Ar	39.95	22.393	1.784
carbon dioxide	CO_2	44.01	22.256	1.977
ammonia	NH_3	17.03	22.094	0.771
chlorine	Cl_2	70.91	22.063	3.214

Deviations in standard molar volume indicate that gases do not behave ideally.

EXAMPLE 12-6 *Molecular Weight, Density*

One (1.00) mole of a gas occupies 27.0 liters, and its density is 1.41 g/L at a particular temperature and pressure. What is its molecular weight? What is the density of the gas at STP?

Plan

We can use dimensional analysis to convert the density, 1.41 g/L, to molecular weight, g/mol. To calculate the density at STP, we recall that the volume occupied by one mole would be 22.4 L.

Solution

We multiply the density under the original conditions by the unit factor 27.0 L/1.00 mol to generate the appropriate units, g/mol.

$$\frac{?\ \text{g}}{\text{mol}} = \frac{1.41\ \text{g}}{\text{L}} \times \frac{27.0\ \text{L}}{\text{mol}} = \boxed{38.1\ \text{g/mol}}$$

At STP, 1.00 mol of the gas, 38.1 g, would occupy 22.4 L, and its density would be

$$\text{density} = \frac{38.1\ \text{g}}{1\ \text{mol}} \times \frac{1\ \text{mol}}{22.4\ \text{L}} = \boxed{1.70\ \text{g/L}}\ \text{at STP}$$

You should now work Exercise 44.

12-9 SUMMARY OF GAS LAWS—THE IDEAL GAS EQUATION

Let us summarize what we have learned about gases. Any sample of gas can be described in terms of its pressure, temperature (Kelvin), volume, and the number of moles, n, present. Any three of these variables determine the fourth. The gas laws we have studied give several relationships among these variables. An **ideal gas** is one that obeys these gas laws exactly. Many real gases show slight deviations from ideality, but at normal temperatures and pressures the deviations are usually small enough to be ignored. We shall do so for the present and discuss deviations later.

We can summarize the behavior of ideal gases as follows.

$$\text{Boyle's Law} \qquad V \propto \frac{1}{P} \qquad \text{(at constant } T \text{ and } n\text{)}$$

Charles' Law	$V \propto T$	(at constant P and n)
Avogadro's Law	$V \propto n$	(at constant T and P)
Summarizing	$V \propto \dfrac{nT}{P}$	(no restrictions)

As before, a proportionality can be written as an equality by introducing a proportionality constant, for which we'll use the symbol R. This gives

$$V = R\left(\frac{nT}{P}\right) \quad \text{or, rearranging,} \quad \boxed{PV = nRT}$$

This equation takes into account the values of n, T, P, and V. Therefore, restrictions that apply to the individual gas laws are not needed for the ideal gas equation.

This relationship is called the **ideal gas equation** or the *ideal gas law*. The numerical value of R, the **universal gas constant,** depends on the choices of the units for P, V, and T. One mole of an ideal gas occupies 22.414 liters at 1.00 atmosphere and 273.15 K (STP). Solving the ideal gas law for R gives

$$R = \frac{PV}{nT} = \frac{(1.00 \text{ atm})(22.414 \text{ L})}{(1.00 \text{ mol})(273.15 \text{ K})} = 0.082057 \frac{\text{L} \cdot \text{atm}}{\text{mol} \cdot \text{K}}$$

In working problems, we often round R to 0.0821 L·atm/mol·K. R may be expressed in other units, as it is inside the back cover.

EXAMPLE 12-7   *Units of R*

R can have any *energy* units per mole per kelvin. Calculate R in terms of joules per mole per kelvin and in SI units of kPa·dm³/mol·K.

Plan

We apply dimensional analysis to convert to the required units.

Solution

Appendix C shows that 1 L·atm = 101.325 joules.

$$R = \frac{0.082057 \text{ L} \cdot \text{atm}}{\text{mol} \cdot \text{K}} \times \frac{101.325 \text{ J}}{1 \text{ L} \cdot \text{atm}} = \boxed{8.3144 \text{ J/mol} \cdot \text{K}}$$

Recall that 1 dm³ = 1 L.

Now evaluate R in SI units. One atmosphere pressure is 101.325 kilopascals, and the molar volume at STP is 22.414 dm³.

$$R = \frac{PV}{nT} = \frac{101.325 \text{ kPa} \times 22.414 \text{ dm}^3}{1 \text{ mol} \times 273.15 \text{ K}} = \boxed{8.3145 \frac{\text{kPa} \cdot \text{dm}^3}{\text{mol} \cdot \text{K}}}$$

You should now work Exercise 49.

We can now express R, the universal gas constant, to four digits in three different sets of units.

$$R = 0.08206 \frac{\text{L} \cdot \text{atm}}{\text{mol} \cdot \text{K}} = \frac{8.314 \text{ J}}{\text{mol} \cdot \text{K}} = 8.314 \frac{\text{kPa} \cdot \text{dm}^3}{\text{mol} \cdot \text{K}}$$

The usefulness of the ideal gas equation is that it relates the four variables, P, V, n, and T, that describe a sample of gas at *one set of conditions*. If any three of these variables are known, the fourth can be calculated.

EXAMPLE 12-8   *Ideal Gas Equation*

What is the volume of a gas balloon filled with 4.00 moles of He when the atmospheric pressure is 748 torr and the temperature is 30°C?

Plan

We first list the variables with the proper units. Then we solve the ideal gas equation for *V* and substitute values.

Solution

$$P = 748 \text{ torr} \times \frac{1 \text{ atm}}{760 \text{ torr}} = 0.984 \text{ atm} \qquad n = 4.00 \text{ mol}$$

$$T = 30°C + 273° = 303 \text{ K} \qquad V = \underline{?}$$

Solving $PV = nRT$ for *V* and substituting gives

$$PV = nRT; \qquad V = \frac{nRT}{P} = \frac{(4.00 \text{ mol})\left(0.0821 \dfrac{\text{L} \cdot \text{atm}}{\text{mol} \cdot \text{K}}\right)(303 \text{ K})}{0.984 \text{ atm}} = \boxed{101 \text{ L}}$$

You should now work Exercise 54.

You may wonder why pressures are given in torr or mm Hg and temperatures in °C. This is because pressures are often measured with mercury barometers, and temperatures are measured with Celsius thermometers.

EXAMPLE 12-9   *Ideal Gas Equation*

A helium-filled weather balloon with a diameter of 24.0 feet has a volume of 7240 cubic feet. How many grams of helium would be required to inflate this balloon to a pressure of 745 torr at 21°C? (1 ft^3 = 28.3 L)

Plan

We shall use the ideal gas equation to find *n*, the number of moles required, and then convert to grams. We must convert each quantity to one of the units stated for *R*. ($R = 0.0821$ L · atm/mol · K)

Solution

$$P = 745 \text{ torr} \times \frac{1 \text{ atm}}{760 \text{ torr}} = 0.980 \text{ atm} \qquad T = 21°C + 273 = 294 \text{ K}$$

$$V = 7240 \text{ ft}^3 \times \frac{28.3 \text{ L}}{1 \text{ ft}^3} = 2.05 \times 10^5 \text{ L} \qquad n = \underline{?}$$

Solving $PV = nRT$ for *n* and substituting gives

$$n = \frac{PV}{RT} = \frac{(0.980 \text{ atm})(2.05 \times 10^5 \text{ L})}{\left(0.0821 \dfrac{\text{L} \cdot \text{atm}}{\text{mol} \cdot \text{K}}\right)(294 \text{ K})} = 8.32 \times 10^3 \text{ mol He}$$

$$\underline{?} \text{ g He} = (8.32 \times 10^3 \text{ mol He})\left(4.00 \dfrac{\text{g}}{\text{mol}}\right) = \boxed{3.33 \times 10^4 \text{ g He}}$$

A helium-filled weather balloon.

EXAMPLE 12-10 *Ideal Gas Equation*

What pressure, in kPa, is exerted by 54.0 grams of Xe in a 1.00-liter flask at 20°C?

Plan

We list the variables with the proper units. Then we solve the ideal gas equation for P and substitute values.

Solution

$$V = 1.00 \text{ L} = 1.00 \text{ dm}^3 \qquad n = 54.0 \text{ g Xe} \times \frac{1 \text{ mol}}{131.3 \text{ g Xe}} = 0.411 \text{ mol}$$

$$T = 20°C + 273° = 293 \text{ K} \qquad P = \underline{?}$$

Solving $PV = nRT$ for P and substituting gives

$$P = \frac{nRT}{V} = \frac{(0.411 \text{ mol})\left(\dfrac{8.31 \text{ kPa} \cdot \text{dm}^3}{\text{mol} \cdot \text{K}}\right)(293 \text{ K})}{1.00 \text{ dm}^3} = \boxed{1.00 \times 10^3 \text{ kPa}} \quad (9.87 \text{ atm})$$

You should now work Exercise 46.

▼ **PROBLEM-SOLVING TIP**   *Watch Out for Units in Ideal Gas Law Calculations*

The units of R that are appropriate for ideal gas law calculations are those that involve units of volume, pressure, moles, and temperature—i.e., $R = 0.08206 \text{ L} \cdot \text{atm/mol} \cdot \text{K}$. When you use this value, remember to express all quantities in a calculation in these units. Pressures should be expressed in atmospheres, volumes in liters, temperature in kelvins, and amount of gas in moles. In Examples 12-8 and 12-9 we converted pressures from torr to atm. In Example 12-9 the volume was converted from ft^3 to L.

Summary of the ideal gas laws

1. The individual gas laws are usually used to calculate the *changes* in conditions for a sample of gas (subscripts can be thought of as "before" and "after").

   Boyle's Law $\qquad\qquad P_1V_1 = P_2V_2 \qquad$ (for a given amount of a gas at constant temperature)

   Charles' Law $\qquad\qquad \dfrac{V_1}{T_1} = \dfrac{V_2}{T_2} \qquad$ (for a given amount of a gas at constant pressure)

   Combined gas law $\qquad \dfrac{P_1V_1}{T_1} = \dfrac{P_2V_2}{T_2} \qquad$ (for a given amount of a gas)

   Avogadro's Law $\qquad\quad \dfrac{V_1}{n_1} = \dfrac{V_2}{n_2} \qquad$ (for gas samples at the same temperature and pressure)

2. The ideal gas equation is used to calculate one of the four variables P, V, n, and T, which describe a sample of gas at *any single set of conditions*.

   $$PV = nRT$$

The ideal gas equation can also be used to calculate the densities of gases.

EXAMPLE 12-11 *Ideal Gas Equation*

Nitric acid, a very important industrial chemical, is made by dissolving the gas nitrogen dioxide, NO_2, in water. Calculate the density of NO_2 gas, in g/L, at 1.24 atm and 50°C.

Plan

We use the ideal gas equation to find the number of moles, n, in any volume, V, at the specified pressure and temperature. Then we convert moles to grams. Because we want to express density in g/L, we choose a volume of one liter.

Solution

$$V = 1.00 \text{ L} \qquad\qquad n = \underline{\frac{?}{}}$$
$$T = 50°C + 273° = 323 \text{ K} \qquad P = 1.24 \text{ atm}$$

Solving $PV = nRT$ for n and substituting gives

$$n = \frac{PV}{RT} = \frac{(1.24 \text{ atm})(1.00 \text{ L})}{\left(0.0821 \dfrac{\text{L} \cdot \text{atm}}{\text{mol} \cdot \text{K}}\right)(323 \text{ K})} = 0.0468 \text{ mol}$$

So there is 0.0468 mol NO_2/L at the specified P and T. Converting this to grams of NO_2,

$$\text{density} = \frac{? \text{ g}}{\text{L}} = \frac{0.0468 \text{ mol } NO_2}{\text{L}} \times \frac{46.0 \text{ g } NO_2}{\text{mol } NO_2} = \boxed{2.15 \text{ g/L}}$$

You should now work Exercise 52.

12-10 DETERMINATION OF MOLECULAR WEIGHTS AND MOLECULAR FORMULAS OF GASEOUS SUBSTANCES

In Section 2-10 we distinguished between simplest and molecular formulas of compounds. We showed how simplest formulas can be calculated from percent compositions of compounds. The molecular weight must be known to determine the molecular formula of a compound. For compounds that are gases at convenient temperatures and pressures, the ideal gas law provides a basis for determining molecular weights.

EXAMPLE 12-12 *Molecular Weight*

A 0.109-gram sample of a pure gaseous compound occupies 112 mL at 100°C and 750 torr. What is the molecular weight of the compound?

Plan

We first use the ideal gas law, $PV = nRT$, to find the number of moles of gas. Then, knowing the mass of that number of moles of gas, we calculate the mass of one mole, the molecular weight.

Solution

$$V = 0.112 \text{ L}; \ T = 100°C + 273 = 373 \text{ K}; \ P = 750 \text{ torr} \times \frac{1 \text{ atm}}{760 \text{ torr}} = 0.987 \text{ atm}$$

$$n = \frac{PV}{RT} = \frac{(0.987 \text{ atm})(0.112 \text{ L})}{\left(0.0821 \dfrac{\text{L} \cdot \text{atm}}{\text{mol} \cdot \text{K}}\right)(373 \text{ K})} = 0.00361 \text{ mol}$$

The mass of 0.00361 mole of this gas is 0.109 g, so the mass of one mole is

The molecular weight of the gas is 30.2 amu. The gas could be ethane, C_2H_6, MW = 30.1 amu. Can you think of other possibilities?

$$\text{molecular weight} = \frac{?\ g}{mol} = \frac{0.109\ g}{0.00361\ mol} = \boxed{30.2\ g/mol}$$

You should now work Exercise 58.

EXAMPLE 12-13   *Molecular Formula*

Analysis of a sample of a gaseous compound shows that it contains 85.7% carbon and 14.3% hydrogen by mass. At standard temperature and pressure, 100 mL of the compound has a mass of 0.188 gram. What is its molecular formula?

Plan

We first find the simplest formula for the compound as we did in Section 2-9 (Examples 2-13 and 2-14). We use the gas volume and mass data to find the molecular weight as in Example 12-12. To find the molecular formula, we reason as in Example 2-17. We use the experimentally known molecular weight to find the ratio

$$n = \frac{\text{molecular weight}}{\text{simplest-formula weight}}$$

The molecular weight is n times the simplest-formula weight, so the molecular formula is n times the simplest formula.

Solution

We first determine the simplest formula.

Element	Relative Mass of Element	Relative Number of Atoms (divide mass by AW)	Divide by Smallest Number	Smallest Whole-Number Ratio of Atoms
C	85.7	$\dfrac{85.7}{12.0} = 7.14$	$\dfrac{7.14}{7.14} = 1.00$	1
H	14.3	$\dfrac{14.3}{1.01} = 14.2$	$\dfrac{14.2}{7.14} = 1.99$	2

CH_2

The simplest formula is CH_2 (simplest-formula weight = 14.0 amu), so the molecular formula of the compound must be some multiple of CH_2. We use the ideal gas law, $PV = nRT$, to find the number of moles in the sample.

The gas in this sample is at STP, so we could use the standard molar volume, 22.4 L/mol, to determine the number of moles of gas. The method shown here is applicable at all temperatures and pressures.

$$V = 0.100\ L \qquad P = 1.00\ atm \qquad T = 273\ K$$

$$n = \frac{PV}{RT} = \frac{(1.00\ atm)(0.100\ L)}{\left(0.0821\ \dfrac{L \cdot atm}{mol \cdot K}\right)(273\ K)} = 0.00446\ mol$$

This result tells us that 0.00446 mole of this gas has a mass of 0.188 g, which enables us to calculate its molecular weight.

$$\frac{?\ g}{mol} = \frac{0.188\ g}{0.00446\ mol} = \boxed{42.2\ g/mol}$$

We know that one mole of the compound has a mass of 42.2 grams. To determine its molecular formula, we divide the molecular weight of the compound (which corresponds to its molecular formula) by the simplest-formula weight to obtain the nearest integer.

$$\frac{\text{molecular weight}}{\text{simplest-formula weight}} = \frac{42.2 \text{ amu}}{14 \text{ amu}} = 3$$

This tells us that the molecular formula is three times the simplest formula. Therefore, the molecular formula is $(CH_2)_3 = \boxed{C_3H_6}$. The gas could be either propene or cyclopropane, both of which have the formula C_3H_6.

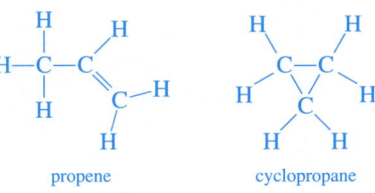

propene   cyclopropane

You should now work Exercise 61.

This kind of calculation of molecular weights can be extended to volatile liquids. In the **Dumas method** a sample of volatile liquid is placed in a previously weighed flask, and the flask is placed in a boiling-water bath. The liquid is allowed to evaporate, and the excess vapor escapes into the air (Figure 12-6). When the last bit of liquid evaporates, the container is filled completely with vapor of the volatile liquid at 100°C. The container is then removed from the boiling-water bath, cooled quickly, and weighed again. Rapid cooling condenses the vapor that filled the flask at 100°C, and air reenters the flask. Recall that the volume of a liquid is extremely small in comparison with the volume occupied by the same mass of the gaseous compound at atmospheric pressure. The difference between the mass of the flask containing the condensed liquid and the mass of the empty flask is the mass of the condensed liquid. It is also equal to the mass of the vapor that filled the flask at 100°C and atmospheric pressure. The "empty" flask contains very nearly the same amount of air both times it is weighed, so the mass of air does not affect the result.

Although the compound is a liquid at room temperature, it is a gas at 100°C, and therefore we can apply the gas laws. If the water surrounding the flask is replaced by a liquid with a higher boiling point, the Dumas method can be applied to liquids that vaporize at or below the boiling point of the liquid surrounding the flask. Example 12-14 illustrates the determination of the molecular weight of a volatile liquid by the Dumas method.

Because the resulting vapor is only slightly above its liquefaction point (the boiling point of the liquid), it does not behave very ideally. Hence, errors of a few percent are common in results of the Dumas method. Such errors are acceptable in determining molecular weights and molecular formulas (Examples 12-14 and 12-15).

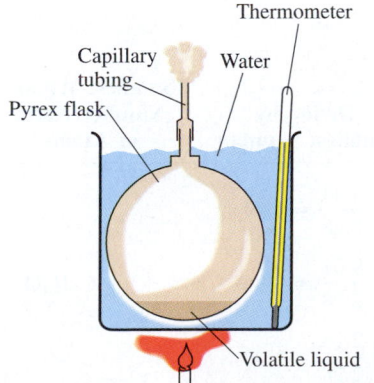

Figure 12-6 Determination of molecular weight by the Dumas method. A volatile liquid is vaporized in the boiling-water bath, and excess vapor is driven from the flask. The remaining vapor, at known *P*, *T*, and *V*, is condensed and weighed.

EXAMPLE 12-14 *Molecular Weight, Dumas Method*

The molecular weight of a volatile liquid was determined by the Dumas method. A 120-mL flask contained 0.345 gram of vapor at 100°C and 1.00 atm pressure. What is the molecular weight of the compound?

Plan

We use the ideal gas law, $PV = nRT$, to determine the number of moles of vapor that filled the flask. Then, knowing the mass of this number of moles, we can calculate the mass of one mole.

Solution

$$V = 0.120 \text{ L} \qquad P = 1.00 \text{ atm} \qquad T = 100°C + 273 = 373 \text{ K}$$

$$n = \frac{PV}{RT} = \frac{(1.00 \text{ atm})(0.120 \text{ L})}{\left(0.0821 \dfrac{\text{L} \cdot \text{atm}}{\text{mol} \cdot \text{K}}\right)(373 \text{ K})} = 0.00392 \text{ mol}$$

The mass of 0.00392 mol of vapor is 0.345 g, so the mass of one mole is

$$\frac{? \text{ g}}{\text{mol}} = \frac{0.345 \text{ g}}{0.00392 \text{ mol}} = \boxed{88.0 \text{ g/mol}}$$

The molecular weight of the vapor, and therefore of the volatile liquid, is 88.0 amu.

You should now work Exercise 56.

Let's carry the calculation one step further in the next example.

EXAMPLE 12-15 *Molecular Formula*

Analysis of the volatile liquid in Example 12-14 showed that it contained 54.5% carbon, 9.10% hydrogen, and 36.4% oxygen by mass. What is its molecular formula?

Plan

We reason just as we did in Example 12-13. We first determine the simplest formula for the compound. Then we use the molecular weight that we determined in Example 12-14 to find the molecular formula.

Solution

Element	Relative Mass of Element	Relative Number of Atoms (divide mass by AW)	Divide by Smallest Number	Smallest Whole-Number Ratio of Atoms
C	54.5	$\dfrac{54.5}{12.0} = 4.54$	$\dfrac{4.54}{2.28} = 1.99$	2
H	9.10	$\dfrac{9.10}{1.01} = 9.01$	$\dfrac{9.01}{2.28} = 3.95$	4 C_2H_4O
O	36.4	$\dfrac{36.4}{16.0} = 2.28$	$\dfrac{2.28}{2.28} = 1.00$	1

The simplest formula is C_2H_4O; the simplest-formula weight is 44 amu.

Division of the molecular weight by the simplest-formula weight gives

$$\frac{\text{molecular weight}}{\text{simplest-formula weight}} = \frac{88 \text{ amu}}{44 \text{ amu}} = 2$$

Therefore, the molecular formula is $2 \times (C_2H_4O) =$ $C_4H_8O_2$.

One compound that has this molecular formula is ethyl acetate, a common solvent used in nail polishes and polish removers. Another is butyric acid, the compound responsible for the odor of rancid butter.

12-11 DALTON'S LAW OF PARTIAL PRESSURES

Many gas samples, including our atmosphere, are mixtures that consist of different kinds of gases. The total number of moles in a mixture of gases is

$$n_{\text{total}} = n_A + n_B + n_C + \cdots$$

where n_A, n_B, and so on represent the number of moles of each kind of gas present. Rearranging the ideal gas equation, $P_{\text{total}}V = n_{\text{total}}RT$, for the total pressure, P_{total}, and then substituting for n_{total} gives

$$P_{\text{total}} = \frac{n_{\text{total}}RT}{V} = \frac{(n_A + n_B + n_C + \cdots)RT}{V}$$

Multiplying out the right-hand side gives

$$P_{\text{total}} = \frac{n_A RT}{V} + \frac{n_B RT}{V} + \frac{n_C RT}{V} + \cdots$$

Now $n_A RT/V$ is the *partial pressure* P_A that the n_A moles of gas A alone would exert in the container at temperature T; similarly, $n_B RT/V = P_B$, and so on. Substituting these into the equation for P_{total}, we obtain **Dalton's Law of Partial Pressures** (Figure 12-7).

John Dalton was the first to notice this effect. He did so in 1807 while studying the compositions of moist and dry air. The pressure that each gas exerts in a mixture is called its **partial pressure.** No way has been devised to measure the pressure of an individual gas in a mixture; it must be calculated from other quantities.

$$P_{\text{total}} = P_A + P_B + P_C + \cdots \qquad \text{(constant } V, T\text{)}$$

The total pressure exerted by a mixture of ideal gases is the sum of the partial pressures of those gases.

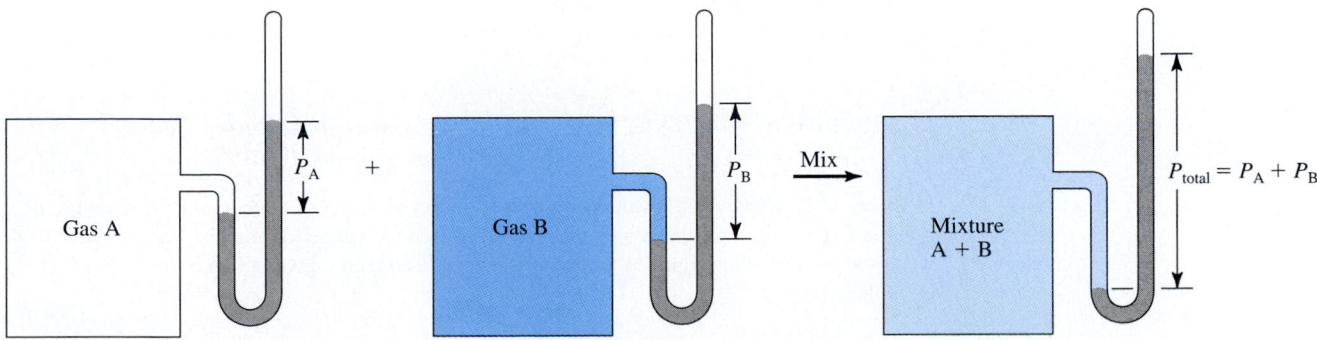

Figure 12-7 An illustration of Dalton's Law. When the two gases A and B are mixed in the same container at the same temperature, they exert a total pressure equal to the sum of their partial pressures.

Dalton's Law is useful in describing real gaseous mixtures at moderate pressures because it allows us to relate total measured pressures to the composition of mixtures.

EXAMPLE 12-16   *Mixture of Gases*

A 10.0-liter flask contains 0.200 mole of methane, 0.300 mole of hydrogen, and 0.400 mole of nitrogen at 25°C. (a) What is the pressure, in atmospheres, inside the flask? (b) What is the partial pressure of each component of the mixture of gases?

Plan

(a) We are given the number of moles of each component. The ideal gas law is then used to calculate the total pressure from the total number of moles. (b) The partial pressure of each gas in the mixture can be calculated by substituting the number of moles of each gas into $PV = nRT$ individually.

Solution

(a) $n = 0.200$ mol CH_4 + 0.300 mol H_2 + 0.400 mol N_2 = 0.900 mol of gas
     $V = 10.0$ L $T = 25°C + 273° = 298$ K

Solving $PV = nRT$ for P gives $P = nRT/V$. Substitution gives

$$P = \frac{(0.900 \text{ mol})\left(0.0821 \ \dfrac{L \cdot atm}{mol \cdot K}\right)(298 \text{ K})}{10.0 \text{ L}} = \boxed{2.20 \text{ atm}}$$

(b) Now we find the partial pressures. For CH_4, $n = 0.200$ mol, and the values for V and T are the same as above.

$$P_{CH_4} = \frac{(n_{CH_4})RT}{V} = \frac{(0.200 \text{ mol})\left(0.0821 \ \dfrac{L \cdot atm}{mol \cdot K}\right)(298 \text{ K})}{10.0 \text{ L}} = \boxed{0.489 \text{ atm}}$$

Similar calculations for the partial pressures of hydrogen and nitrogen give

$$P_{H_2} = \boxed{0.734 \text{ atm}} \quad \text{and} \quad P_{N_2} = \boxed{0.979 \text{ atm}}$$

As a check, we use Dalton's Law: $P_{total} = P_A + P_B + P_C + \cdots$. Addition of the partial pressures in this mixture gives the total pressure.

$$P_{total} = P_{CH_4} + P_{H_2} + P_{N_2} = (0.489 + 0.734 + 0.979) \text{ atm} = \boxed{2.20 \text{ atm}}$$

The total pressure exerted by the mixture of gases is 2.20 atmospheres.

You should now work Exercises 63 and 64.

▼ **PROBLEM-SOLVING TIP**   *Amounts of Gases in Mixtures Can Be Expressed in Various Units*

In Example 12-16 we were given the number of moles of each gas. Sometimes the amount of a gas is expressed in other units that can be converted to numbers of moles. For instance, if we know the formula weight (or the formula), we can convert a given mass of gas to number of moles.

We can describe the composition of any mixture in terms of the mole fraction of each component. The **mole fraction,** X_A, of component A in a mixture is defined as

$$X_A = \frac{\text{no. mol } A}{\text{total no. mol of all components}}$$

Like any other fraction, mole fraction is a dimensionless quantity. For each component in a mixture, the mole fraction is

$$X_A = \frac{\text{no. mol } A}{\text{no. mol } A + \text{no. mol } B + \cdots},$$

$$X_B = \frac{\text{no. mol } B}{\text{no. mol } A + \text{no. mol } B + \cdots}, \quad \text{and so on}$$

The sum of all mole fractions in a mixture is equal to 1.

$$X_A + X_B + \cdots = 1 \text{ for any mixture}$$

We can use this relationship to check mole fraction calculations or to find a remaining mole fraction if we know all the others.

For a gaseous mixture, we can relate the mole fraction of each component to its partial pressure as follows. From the ideal gas equation, the number of moles of each component can be written as

$$n_A = P_A V/RT, \qquad n_B = P_B V/RT, \qquad \text{and so on}$$

and the total number of moles is

$$n_{\text{total}} = P_{\text{total}} V/RT$$

Substituting into the definition of X_A,

$$X_A = \frac{n_A}{n_A + n_B + \cdots} = \frac{P_A V/RT}{P_{\text{total}} V/RT}$$

The quantities V, R, and T cancel to give

$$X_A = \frac{P_A}{P_{\text{total}}}; \quad \text{similarly, } X_B = \frac{P_B}{P_{\text{total}}}; \quad \text{and so on}$$

We can rearrange these equations to give another statement of Dalton's Law of Partial Pressures.

$$P_A = X_A \times P_{\text{total}}; \qquad P_B = X_B \times P_{\text{total}}; \qquad \text{and so on}$$

The partial pressure of each gas is equal to its mole fraction in the gaseous mixture times the total pressure of the mixture.

EXAMPLE 12-17   *Mole Fraction, Partial Pressure*

Calculate the mole fractions of the three gases in Example 12-16.

Plan

One way to solve this problem is to use the numbers of moles given in the problem. Alternatively, we could use the partial pressures and the total pressure from Example 12-16.

In Example 12-17 we see that, for a gas mixture, relative numbers of moles of components are the same as relative pressures of the components.

Solution

Using the moles given in Example 12-16,

$$X_{CH_4} = \frac{n_{CH_4}}{n_{total}} = \frac{0.200 \text{ mol}}{0.900 \text{ mol}} = \boxed{0.222}$$

$$X_{H_2} = \frac{n_{H_2}}{n_{total}} = \frac{0.300 \text{ mol}}{0.900 \text{ mol}} = \boxed{0.333}$$

$$X_{N_2} = \frac{n_{N_2}}{n_{total}} = \frac{0.400 \text{ mol}}{0.900 \text{ mol}} = \boxed{0.444}$$

Using the partial and total pressures calculated in Example 12-16,

The difference between the two calculated results is due to rounding.

$$X_{CH_4} = \frac{P_{CH_4}}{P_{total}} = \frac{0.489 \text{ atm}}{2.20 \text{ atm}} = \boxed{0.222}$$

$$X_{H_2} = \frac{P_{H_2}}{P_{total}} = \frac{0.734 \text{ atm}}{2.20 \text{ atm}} = \boxed{0.334}$$

$$X_{N_2} = \frac{P_{N_2}}{P_{total}} = \frac{0.979 \text{ atm}}{2.20 \text{ atm}} = \boxed{0.445}$$

You should now work Exercise 66.

EXAMPLE 12-18   *Partial Pressure, Mole Fraction*

The mole fraction of oxygen in the atmosphere is 0.2094. Calculate the partial pressure of O_2 in air when the atmospheric pressure is 760. torr.

Plan

The partial pressure of each gas in a mixture is equal to its mole fraction in the mixture times the total pressure of the mixture.

Solution

$$P_{O_2} = X_{O_2} \times P_{total}$$

$$P_{O_2} = 0.2094 \times 760. \text{ torr} = \boxed{159 \text{ torr}}$$

Dalton's Law can be used in combination with other gas laws, as the following example shows.

EXAMPLE 12-19   *Mixture of Gases*

Two tanks are connected by a closed valve. Each tank is filled with gas as shown, and both tanks are held at the same temperature. We open the valve and allow the gases to mix.

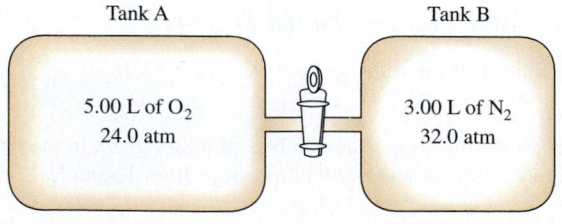

Tank A — 5.00 L of O_2, 24.0 atm   Tank B — 3.00 L of N_2, 32.0 atm

(a) After the gases mix, what is the partial pressure of each gas, and what is the total pressure?
(b) What is the mole fraction of each gas in the mixture?

Plan

(a) Each gas expands to fill the entire 8.00 liters. We can use Boyle's Law to calculate the partial pressure that each gas would exert after expansion. The total pressure is equal to the sum of the partial pressures of the two gases. (b) The mole fractions can be calculated from the ratio of the partial pressure of each gas to the total pressure.

Solution

(a) For O_2,

$$P_1V_1 = P_2V_2 \quad \text{or} \quad P_{2,O_2} = \frac{P_1V_1}{V_2} = \frac{24.0 \text{ atm} \times 5.00 \text{ L}}{8.00 \text{ L}} = \boxed{15.0 \text{ atm}}$$

For N_2,

$$P_1V_1 = P_2V_2 \quad \text{or} \quad P_{2,N_2} = \frac{P_1V_1}{V_2} = \frac{32.0 \text{ atm} \times 3.00 \text{ L}}{8.00 \text{ L}} = \boxed{12.0 \text{ atm}}$$

The total pressure is the sum of the partial pressures.

$$P_{total} = P_{2,O_2} + P_{2,N_2} = 15.0 \text{ atm} + 12.0 \text{ atm} = \boxed{27.0 \text{ atm}}$$

(b)

$$X_{O_2} = \frac{P_{2,O_2}}{P_{total}} = \frac{15.0 \text{ atm}}{27.0 \text{ atm}} = \boxed{0.556}$$

$$X_{N_2} = \frac{P_{2,N_2}}{P_{total}} = \frac{12.0 \text{ atm}}{27.0 \text{ atm}} = \boxed{0.444}$$

As a check, the sum of the mole fractions is 1.

You should now work Exercise 68.

Notice that this problem has been solved without calculating the number of moles of either gas.

Some gases can be collected over water. Figure 12-8 illustrates the collection of a sample of hydrogen over water. A gas produced in a reaction displaces the more dense water from the inverted water-filled jars. The pressure on the gas inside the collection vessel can be made equal to atmospheric pressure by raising or lowering the vessel until the water level inside is the same as that outside.

Gases that are soluble in water or that react with water cannot be collected by this method. Other liquids can be used.

Figure 12-8 Apparatus for preparing hydrogen from zinc and sulfuric acid.

$$Zn(s) + 2H^+(aq) \longrightarrow Zn^{2+}(aq) + H_2(g)$$

The hydrogen is collected over water.

CHEMISTRY IN USE

The Environment

Radon and Smoking

Lung cancer kills more than 120,000 Americans each year. The leading cause of lung cancer is well known; the Surgeon General of the United States has warned us about the dangers of tobacco smoke since 1964. The second leading cause of lung cancer is radon, a naturally occurring radioactive gas. Because of its widespread presence, many Americans test their homes for radon, which may seep into foundations from the surrounding soil.

Radon is *relatively* harmless. If you inhale a radon atom, you will probably exhale that radon atom with no complications. The danger arises when radon decays to produce polonium, which is dangerously radioactive. If this happens while radon atoms are in your lungs, your risk of lung cancer increases.

Why is exposure to very small concentrations of this elemental gas dangerous? Assume that every liter of air in your home contains two radon atoms. If 1000 people were exposed to this concentration of radon in their homes over average lifetimes, as many as fifty would probably get lung cancer who otherwise wouldn't.

Let's calculate the molarity of radon in a 1-L air sample that contains two radon atoms and explain why this very low concentration of radon is dangerous.

$$\frac{2 \text{ Rn atoms}}{\text{L air}} \times \frac{1 \text{ mol Rn}}{6 \times 10^{23} \text{ Rn atoms}} = \frac{3 \times 10^{-24} \text{ mol Rn}}{\text{L air}}$$

The concentration of this radon solution (air) is 3×10^{-24} molar. For most circumstances, this would represent a negligibly small solute concentration. Chemists usually work with solutions whose concentrations are many orders of magnitude larger.

Although radon and tobacco smoke each present a health hazard, together they form a combination that is more deadly than radon gas alone—up to 1500% more deadly! Here's why.

Radon gas is transformed into polonium by a nuclear reaction. The polonium particles attach themselves to surfaces such as walls and furniture. However, when there is tobacco smoke in the air, much of the polonium is attracted to the particles of smoke. Cigarette smoke can remain in the air inside a building for a long time. A single cigarette burned in a room raises the concentration of polonium in the air by 25%, after about five hours. This increased concentration of polonium then remains nearly constant for about nine hours before trailing off. When twenty cigarettes are burned over a 24-hour period, simulating a pack-a-day smoker, there is a 300% increase in the concentration of polonium and other dangerous radioactive particles in the air. Breathing this air results in a risk of radon-induced lung cancer that is 15 times greater than breathing smoke-free air with the same radon concentration. The combination of radon with tobacco smoke affects everyone who spends time in such a building.

Ronald DeLorenzo
Middle Georgia College

The partial pressure exerted by the vapor above a liquid is called the vapor pressure of that liquid. A more extensive table of the vapor pressure of water appears in Appendix E.

One complication arises, however. A gas in contact with water soon becomes saturated with water vapor. The pressure inside the vessel is the sum of the partial pressure of the gas itself *plus* the partial pressure exerted by the water vapor in the gas mixture (the **vapor pressure** of water). Every liquid shows a characteristic vapor pressure that varies only with temperature, and *not* with the volume of vapor present, so long as both liquid and vapor are present. Table 12-4 displays the vapor pressure of water near room temperature.

The relevant point here is that a gas collected over water is "moist"; that is, it is saturated with water vapor. Measuring the atmospheric pressure at which the gas is collected, we can write

$$P_{atm} = P_{gas} + P_{H_2O} \quad \text{or} \quad P_{gas} = P_{atm} - P_{H_2O}$$

Example 12-20 provides a detailed illustration.

Table 12-4 Vapor Pressure of Water Near Room Temperature

Temperature (°C)	Vapor Pressure of Water (torr)
19	16.48
20	17.54
21	18.65
22	19.83
23	21.07
24	22.38
25	23.76
26	25.21
27	26.74
28	28.35

EXAMPLE 12-20 Gas Collected over Water

A 300-mL sample of hydrogen was collected over water (Figure 12-8) at 21°C on a day when the atmospheric pressure was 748 torr. (a) How many moles of H_2 were present? (b) How many moles of water vapor were present in the moist gas mixture? (c) What is the mole fraction of hydrogen in the moist gas mixture? (d) What would be the mass of the gas sample if it were dry?

Plan

(a) The vapor pressure of H_2O, $P_{H_2O} = 19$ torr at 21°C, is obtained from Table 12-4. Applying Dalton's Law, $P_{H_2} = P_{atm} - P_{H_2O}$. We then use the partial pressure of H_2 in the ideal gas equation to find the number of moles of H_2 present. (b) The partial pressure of water vapor (the vapor pressure of water at the stated temperature) is used in the ideal gas equation to find the number of moles of water vapor present. (c) The mole fraction of H_2 is the ratio of its partial pressure to the total pressure. (d) The number of moles found in part (a) can be converted to mass of H_2.

Solution

(a) $P_{H_2} = P_{atm} - P_{H_2O} = (748 - 19) \text{ torr} = 729 \text{ torr} \times \dfrac{1 \text{ atm}}{760 \text{ torr}} = 0.959 \text{ atm}$

We also know

$$V = 300 \text{ mL} = 0.300 \text{ L} \quad \text{and} \quad T = 21°C + 273 = 294 \text{ K}$$

Solving the ideal gas equation for n_{H_2} gives

$$n_{H_2} = \frac{P_{H_2}V}{RT} = \frac{(0.959 \text{ atm})(0.300 \text{ L})}{\left(0.0821 \dfrac{\text{L} \cdot \text{atm}}{\text{mol} \cdot \text{K}}\right)(294 \text{ K})} = 1.19 \times 10^{-2} \text{ mol } H_2$$

Remember that each gas occupies the total volume of the container.

At STP, this dry hydrogen would occupy 266 mL. Can you work this out?

(b) $P_{H_2O} = 19 \text{ torr} \times \dfrac{1 \text{ atm}}{760 \text{ torr}} = 0.025 \text{ atm}$

V and T have the same values as in part (a).

$$n_{H_2O} = \frac{P_{H_2O}V}{RT} = \frac{(0.025 \text{ atm})(0.300 \text{ L})}{\left(0.0821 \dfrac{\text{L} \cdot \text{atm}}{\text{mol} \cdot \text{K}}\right)(294 \text{ K})} = 3.1 \times 10^{-4} \text{ mol } H_2O \text{ vapor}$$

(c) $X_{H_2} = \dfrac{P_{H_2}}{P_{total}} = \dfrac{729 \text{ torr}}{748 \text{ torr}} = 0.974$

(d) $\underline{?} \text{ g } H_2 = 1.19 \times 10^{-2} \text{ mol} \times \dfrac{2.02 \text{ g}}{1 \text{ mol}} = 2.40 \times 10^{-2} \text{ g } H_2$

You should now work Exercise 72.

12-12 THE KINETIC–MOLECULAR THEORY

As early as 1738, Daniel Bernoulli envisioned gaseous molecules in ceaseless motion striking the walls of their container and thereby exerting pressure. In 1857 Rudolf Clausius published a theory that attempted to explain various experimental observations that had been summarized by Boyle's, Dalton's, Charles', and Avogadro's laws. The basic assumptions of the **kinetic–molecular theory** for an ideal gas follow.

1. The observation that gases can be easily compressed indicates that the molecules are far apart. At ordinary temperatures and pressures, the gas molecules themselves occupy an insignificant fraction of the total volume of the container.

2. Near temperatures and pressures at which a gas liquefies, the gas does not behave ideally (Section 12-14), and attractions or repulsions among gas molecules *are* significant.

3. At any given instant, only a small fraction of the molecules are involved in collisions.

1. Gases consist of discrete molecules. The individual molecules are very small and are very far apart compared to their own sizes.

2. The gas molecules are in continuous, random, straight-line motion with varying velocities (see Figure 12-9).

3. The collisions between gas molecules and with the walls of the container are elastic; the total energy is conserved during a collision; i.e., there is no net energy gain or loss.

4. Between collisions, the molecules exert no attractive or repulsive forces on one another; instead, each molecule travels in a straight line with a constant velocity.

Kinetic energy is the energy a body possesses by virtue of its motion. It is $\frac{1}{2}mu^2$, where m, the body's mass, can be expressed in grams and u, its velocity, can be expressed in meters per second, m/s. The assumptions of the kinetic–molecular theory can be used to relate temperature and molecular kinetic energy (see the Enrichment section, pages 423–425).

The average kinetic energy of gaseous molecules is directly proportional to the absolute temperature of the sample. The average kinetic energies of molecules of different gases are equal at a given temperature.

For instance, in samples of H_2, He, CO_2, and SO_2 at the same temperature, all the molecules have the same average kinetic energies. But the lighter molecules, H_2 and He, have much higher average velocities than do the heavier molecules, CO_2 and SO_2, at the same temperature.

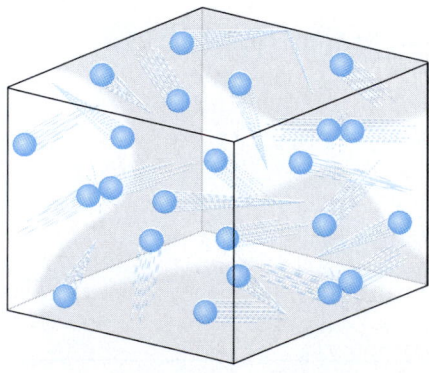

Figure 12-9 A representation of molecular motion. Gaseous molecules, in constant motion, undergo collisions with one another and with the walls of the container.

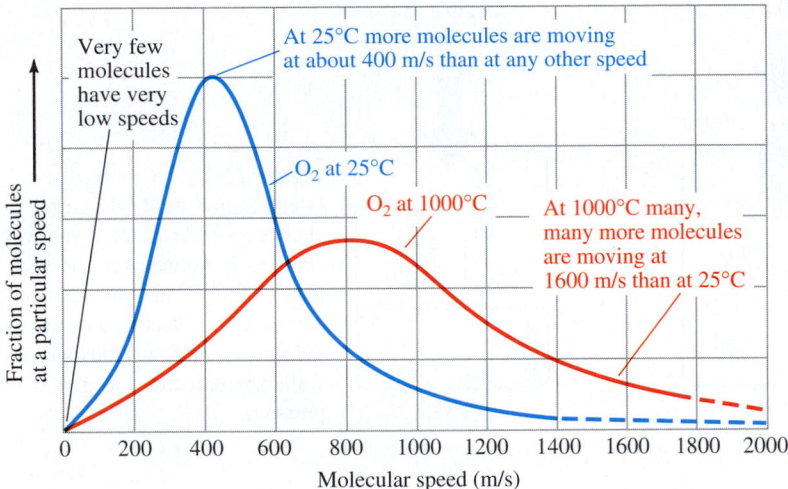

Figure 12-10 The Maxwellian distribution function for molecular speeds. This graph shows the relative numbers of O_2 molecules having a given speed at 25°C and at 1000°C. At 25°C, most O_2 molecules have speeds between 200 and 600 m/s (450 to 1350 miles per hour). The graph approaches the horizontal axis, but does not reach it; some of the molecules have very high speeds.

We can summarize this very important result from the kinetic–molecular theory.

$$\text{average molecular } KE = \overline{KE} \propto T$$

or

$$\text{average molecular speed} = \overline{u} \propto \sqrt{\frac{T}{\text{molecular weight}}}$$

A bar over a quantity denotes an *average* of that quantity.

Molecular kinetic energies of gases increase with increasing temperature and decrease with decreasing temperature. We have referred only to the *average* kinetic energy; in a given sample, some molecules may be moving quite rapidly while others are moving more slowly. Figure 12-10 shows the distribution of speeds of gaseous molecules at two temperatures.

The kinetic–molecular theory satisfactorily explains most of the observed behavior of gases in terms of molecular behavior. Let's look at the gas laws in light of the kinetic–molecular theory.

Boyle's Law

The pressure exerted by a gas upon the walls of its container is caused by gas molecules striking the walls. Clearly, pressure depends upon two factors: (1) the number of molecules striking the walls per unit time and (2) how vigorously the molecules strike the walls. If the temperature is held constant, the mean speed and the force of the collisions remain the same. But halving the volume of a sample of gas doubles the pressure because

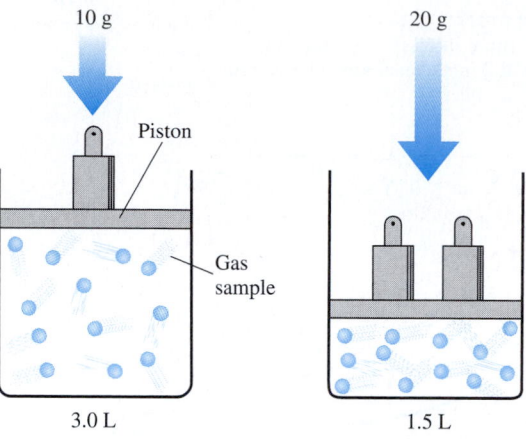

Figure 12-11 A molecular interpretation of Boyle's Law—the change in pressure of a gas with changes in volume (at constant temperature). The entire apparatus is enclosed in a vacuum. In the smaller volume, more molecules strike the walls per unit time, to give a higher pressure.

twice as many molecules strike a given area on the walls per unit time. Likewise, doubling the volume of a sample of gas halves the pressure because only half as many gas molecules strike a given area on the walls per unit time (Figure 12-11).

Dalton's Law

In a gas sample the molecules are very far apart and do not attract one another significantly. Each kind of gas molecule acts independently of the presence of the other kind. The molecules of each gas thus collide with the walls with a frequency and vigor that do not change even if other molecules are present (Figure 12-12). As a result, each gas exerts a partial pressure that is independent of the presence of the other gas, and the total pressure is due to the sum of all the molecule–wall collisions.

Charles' Law

Recall that average kinetic energy is directly proportional to the absolute temperature. Doubling the *absolute* temperature of a sample of gas doubles the average kinetic energy of the gaseous molecules, and the increased force of the collisions of molecules with the

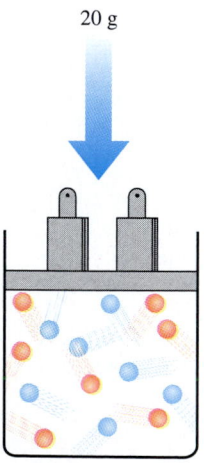

Figure 12-12 A molecular interpretation of Dalton's Law. The molecules act independently, so each gas exerts its own partial pressure due to its molecular collisions with the walls.

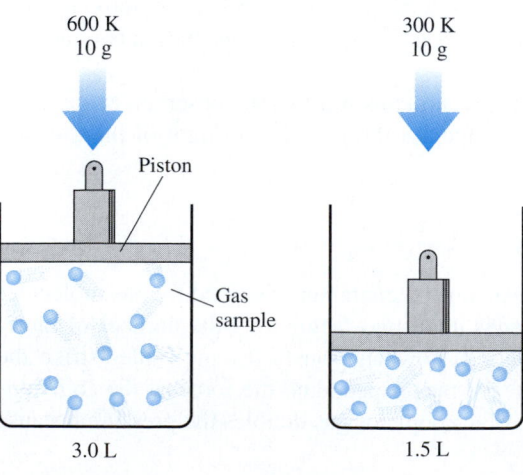

Figure 12-13 A molecular interpretation of Charles' Law—the change in volume of a gas with changes in temperature (at constant pressure). At the lower temperature, molecules strike the walls less often and less vigorously. Thus, the volume must be less to maintain the same pressure.

walls doubles the volume at constant pressure. Similarly, halving the absolute temperature decreases kinetic energy to one-half its original value; at constant pressure, the volume decreases by one half because of reduced vigor of collision of gaseous molecules with the container walls (Figure 12-13).

Kinetic–Molecular Theory, the Ideal Gas Equation, and Molecular Speeds

ENRICHMENT

In 1738 Daniel Bernoulli derived Boyle's Law from Newton's laws of motion applied to gas molecules. This derivation was the basis for an extensive mathematical development of the kinetic–molecular theory more than a century later by Clausius, Maxwell, Boltzmann, and others. Although we do not need to study the detailed mathematical presentation of this theory, we can gain some insight into its concepts from the reasoning behind Bernoulli's derivation. Here we present that reasoning based on proportionality arguments.

In the kinetic–molecular theory pressure is viewed as the result of collisions of gas molecules with the walls of the container. As each molecule strikes a wall, it exerts a small impulse. The pressure is the total force thus exerted on the walls divided by the area of the walls. The total force on the walls (and thus the pressure) is proportional to two factors: (1) the impulse exerted by each collision and (2) the rate of collisions (number of collisions in a given time interval).

$$P \propto (\text{impulse per collision}) \times (\text{rate of collisions})$$

Let us represent the mass of an individual molecule by m and its speed by u. The heavier the molecule is (greater m) and the faster it is moving (greater u), the harder it pushes on the wall when it collides. The impulse due to each molecule is proportional to its *momentum, mu.*

$$\text{impulse per collision} \propto mu$$

Recall that momentum is mass × speed.

The rate of collisions, in turn, is proportional to two factors. First, the rate of collision must be proportional to the molecular speed; the faster the molecules move, the more often they reach the wall to collide. Second, this collision rate must be proportional to the number of molecules per unit volume, N/V. The greater the number of molecules, N, in a given volume, the more molecules collide in a given time interval.

$$\text{rate of collisions} \propto (\text{molecules per unit volume}) \times (\text{molecular speed})$$

or

$$\text{rate of collisions} \propto \left(\frac{N}{V}\right) \times (u)$$

We can introduce these proportionalities into the one describing pressure, to conclude that

$$P \propto (mu) \times u \times \frac{N}{V} \quad \text{or} \quad P \propto \frac{Nmu^2}{V} \quad \text{or} \quad PV \propto Nmu^2$$

At any instant not all molecules are moving at the same speed, u. We should reason in terms of the *average* behavior of the molecules, and express the quantity u^2 in average terms as $\overline{u^2}$, the **mean-square speed.**

$$PV \propto Nm\overline{u^2}$$

$\overline{u^2}$ is the average of the squares of the molecular speeds. It is proportional to the square of the average speed, but the two quantities are not equal.

Not all molecules collide with the walls at right angles, so we must average (using calculus) over all the trajectories. This gives a proportionality constant of $\frac{1}{3}$, and

$$PV = \frac{1}{3}Nm\overline{u^2}$$

This describes the quantity PV (pressure × volume) in terms of *molecular quantities*—number of molecules, molecular masses, and molecular speeds. The number of molecules, N, is given by the number of moles, n, times Avogadro's number, N_{Av}, or $N = nN_{Av}$. Making this substitution, we obtain

$$PV = \frac{1}{3}nN_{Av}m\overline{u^2}$$

The ideal gas equation describes (pressure × volume) in terms of *measurable quantities*—number of moles and absolute temperature.

$$PV = nRT$$

So we see that the ideas of the kinetic–molecular theory lead to an equation of the same form as the macroscopic ideal gas equation. Thus, the molecular picture of the theory is consistent with the ideal gas equation and gives support to the theory. Equating the right-hand sides of these last two equations and cancelling n gives

$$\frac{1}{3}N_{Av}m\overline{u^2} = RT$$

This equation can also be written as

$$\frac{1}{3}N_{Av} \times (2 \times \frac{1}{2}m\overline{u^2}) = RT$$

From physics we know that the *kinetic energy* of a particle of mass m moving at speed u is $\frac{1}{2}mu^2$. So we can write

$$\frac{2}{3}N_{Av} \times (\text{avg } KE \text{ per molecule}) = RT$$

or

$$N_{Av} \times (\text{avg } KE \text{ per molecule}) = \frac{3}{2}RT$$

This equation shows that the absolute temperature is directly proportional to the average molecular kinetic energy, as postulated by the kinetic–molecular theory. Because there are N_{Av} molecules in a mole, the left-hand side of this equation is equal to the total kinetic energy of a mole of molecules.

$$\text{total kinetic energy per mole of gas} = \frac{3}{2}RT$$

With this interpretation, the total molecular kinetic energy of a mole of gas depends *only* on the temperature, and not on the mass of the molecules or the gas density.

We can also obtain some useful equations for molecular speeds from the above reasoning. Solving the equation

$$\frac{1}{3}N_{Av}m\overline{u^2} = RT$$

for root-mean-square speed, $u_{rms} = \sqrt{\overline{u^2}}$, we obtain

$$u_{rms} = \sqrt{\frac{3RT}{N_{Av}m}}$$

We recall that m is the mass of a single molecule. So $N_{Av}m$ is the mass of Avogadro's number of molecules, or one mole of substance; this is equal to the *molecular weight, M,* of the gas.

$$u_{rms} = \sqrt{\frac{3RT}{M}}$$

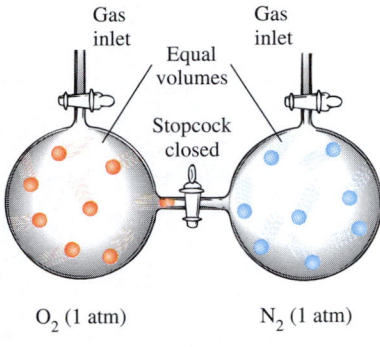

O_2 (1 atm) N_2 (1 atm)

(a)

EXAMPLE 12-21   *Molecular Speed*

Calculate the root-mean-square speed of H_2 molecules in meters per second at 20°C. Recall that $1 \text{ J} = 1 \dfrac{\text{kg} \cdot \text{m}^2}{\text{s}^2}$.

Plan

We substitute the appropriate values into the equation relating u_{rms} to temperature and molecular weight. Remember that R must be expressed in the appropriate units.

$$R = 8.314 \frac{\text{J}}{\text{mol} \cdot \text{K}} = 8.314 \frac{\text{kg} \cdot \text{m}^2}{\text{mol} \cdot \text{K} \cdot \text{s}^2}$$

Solution

$$u_{rms} = \sqrt{\frac{3RT}{M}} = \sqrt{\frac{3 \times 8.314 \dfrac{\text{kg} \cdot \text{m}^2}{\text{mol} \cdot \text{K} \cdot \text{s}^2} \times 293 \text{ K}}{2.016 \dfrac{\text{g}}{\text{mol}} \times \dfrac{1 \text{ kg}}{1000 \text{ g}}}}$$

$$u_{rms} = \sqrt{3.62 \times 10^6 \text{ m}^2/\text{s}^2} = \boxed{1.90 \times 10^3 \text{ m/s}} \qquad \text{(about 4250 mi/hr)}$$

You should now work Exercise 78.

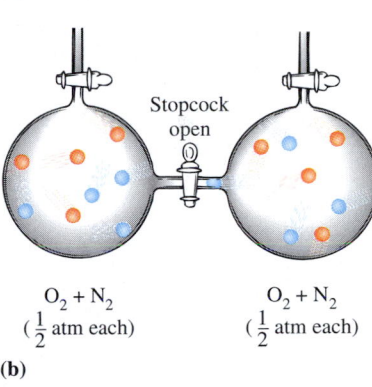

$O_2 + N_2$ $O_2 + N_2$
($\frac{1}{2}$ atm each)   ($\frac{1}{2}$ atm each)

(b)

Figure 12-14   A representation of diffusion of gases. The space between the molecules allows for ease of mixing one gas with another. Collisions of molecules with the walls of the container are responsible for the pressure of the gas.

12-13   DIFFUSION AND EFFUSION OF GASES

Because gas molecules are in constant, rapid, random motion, they diffuse quickly throughout any container (Figure 12-14). For example, if hydrogen sulfide (the essence of rotten eggs) is released in a large room, the odor can eventually be detected throughout the room. If a mixture of gases is placed in a container with thin porous walls, the molecules effuse through the walls. Because they move faster, lighter gas molecules effuse through the tiny openings of porous materials faster than heavier molecules (Figure 12-15).

Although they are the most abundant elements in the universe, hydrogen and helium occur as gases only in trace amounts in our atmosphere. This is due to the high average molecular speeds resulting from their low molecular weights. At temperatures in our atmosphere, these molecules reach speeds exceeding the escape velocity required for them to break out of the earth's gravitational pull and diffuse into interplanetary space. Thus, most of the gaseous atmospheric hydrogen and helium that were probably present in large concentrations in the earth's early atmosphere have long since diffused away. The same is true of other small planets in our solar system, especially those with higher average

Scientists use the word "effusion" to describe the escape of a gas through a tiny hole, and the word "diffusion" to describe movement of a gas into a space or the mixing of one gas with another. The distinction made by chemists is somewhat sharper than that found in the dictionary.

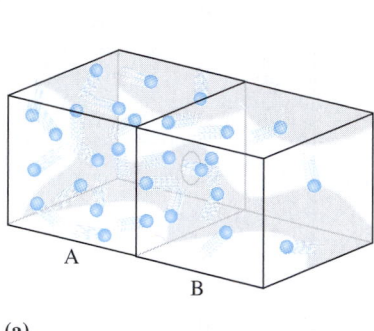

(a)

(b)

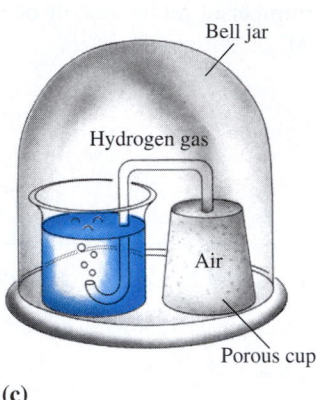

(c)

Figure 12-15   Effusion of gases. (a) A molecular interpretation of effusion. Molecules are in constant motion; occasionally they strike the opening and escape. (b) Rubber balloons were filled with the same volume of He (yellow), N_2 (blue), and O_2 (red). Lighter molecules, such as He, effuse through the tiny pores of these balloons more rapidly than does N_2 or O_2. The silver party balloon is made of a metal-coated polymer with pores that are too small to allow rapid He effusion. (c) If a bell jar full of hydrogen is brought down over a porous cup full of air, rapidly moving hydrogen diffuses into the cup faster than the oxygen and nitrogen in the air can effuse out of the cup. This causes an increase in pressure in the cup sufficient to produce bubbles in the water in the beaker.

NH_3 gas (left) and HCl gas (right) escape from concentrated aqueous solutions. The white smoke (solid NH_4Cl) shows where the gases mix and react.

$$NH_3(g) + HCl(g) \longrightarrow NH_4Cl(s)$$

temperatures than ours (Mercury and Venus). The Mariner 10 spacecraft in 1974 revealed measurable amounts of He in the atmosphere of Mercury; the source of this helium is unknown. Massive bodies such as stars (including our own sun) are mainly H and He.

12-14   REAL GASES—DEVIATIONS FROM IDEALITY

Our discussions up to now have dealt with *ideal* behavior of gases. By this we mean that the identity of a gas does not affect how it behaves, and the same equations should work equally well for all gases. Under ordinary conditions most *real* gases do behave ideally; their *P* and *V* are predicted by the ideal gas laws, so they do obey the postulates of the kinetic–molecular theory. According to the kinetic–molecular model, (1) all but a negligible volume of a gas sample is empty space, and (2) the molecules of *ideal* gases do not attract each other because they are so far apart compared to their own sizes.

However, under some conditions most gases can have pressures and/or volumes that are *not* accurately predicted by the ideal gas laws. This tells us that they are not behaving entirely as postulated by the kinetic–molecular theory.

Nonideal gas behavior (deviation from the predictions of the ideal gas laws) is most significant at *high pressures* and/or *low temperatures,* i.e., near the conditions under which the gas liquefies.

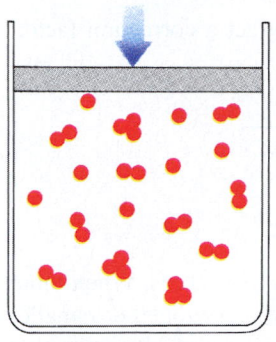 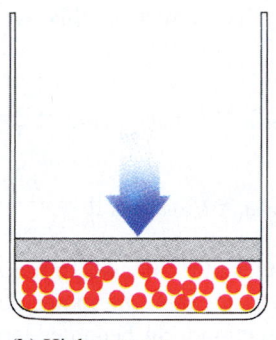

(a) Low temperature   (b) High pressure

Figure 12-16 A molecular interpretation of deviations from ideal behavior. (a) A sample of gas at a low temperature. Each sphere represents a molecule. Because of their low kinetic energies, attractive forces between molecules can now cause a few molecules to "stick together." (b) A sample of gas under high pressure. The molecules are quite close together. The free volume is now a much smaller fraction of the total volume.

Johannes van der Waals studied deviations of real gases from ideal behavior. In 1867 he empirically adjusted the ideal gas equation

$$P_{ideal}V_{ideal} = nRT$$

to take into account two complicating factors.

1. According to the kinetic–molecular theory, the molecules are so small, compared to the total volume of the gas, that each molecule can move through virtually the entire *measured volume* of the container, $V_{measured}$ (Figure 12-16a). But under high pressures, a gas is compressed so that the volume of the molecules themselves becomes a significant fraction of the total volume occupied by the gas. As a result, the *available volume, $V_{available}$,* for any molecule to move in is less than the *measured volume* by an amount that depends on the volume excluded by the presence of the other molecules (Figure 12-16b). To account for this, we subtract a correction factor, *nb*.

$$V_{ideally\ available} = V_{measured} - nb$$

The factor *nb* corrects for the volume occupied by the molecules themselves. Larger molecules have greater values of *b*, and the greater the number of molecules in a sample (higher *n*), the larger is the volume correction. However, the correction term becomes negligibly small when the volume is large.

2. The kinetic–molecular theory describes pressure as resulting from molecular collisions with the walls of the container; this theory assumes that attractive forces between molecules are not significant. For any real gas, the molecules can attract one another. But at higher temperatures, the potential energy due to intermolecular attractions is negligibly small compared to the high kinetic energy due to the rapid motion of the molecules and to the great distances between them. When the temperature is quite low (low kinetic energy), the molecules move so slowly that the potential energy due to even small attractive forces *does* become important. This perturbation becomes even more important when the molecules are very close together (at high pressure). As a result, the molecules deviate from their straight-line paths and take longer to reach the walls, so fewer collisions take place in a given time interval. Further, for a molecule about to collide with the wall, the attraction by its neighbors causes the collision to be less energetic than it would otherwise be (Figure 12-17). As a consequence, the pressure that the gas exerts, $P_{measured}$, is less than the pressure it would exert if attractions

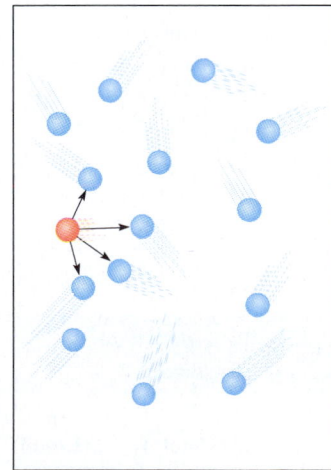

Figure 12-17 A gas molecule strikes the walls of a container with diminished force. The attractive forces between a molecule and its neighbors are significant.

were truly negligible, $P_{\text{ideally exerted}}$. To correct for this, we subtract a correction factor, n^2a/V^2, from the ideal pressure.

$$P_{\text{measured}} = P_{\text{ideally exerted}} - \frac{n^2a}{V_{\text{measured}}^2}$$

or

$$P_{\text{ideally exerted}} = P_{\text{measured}} + \frac{n^2a}{V_{\text{measured}}^2}$$

In this correction term, large values of a indicate strong attractive forces. When more molecules are present (greater n) and when the molecules are close together (smaller V^2 in the denominator), the correction term becomes larger. However, the correction term becomes negligibly small when the volume is large.

When we substitute these two expressions for corrections into the ideal gas equation, we obtain the equation

$$\left(P_{\text{measured}} + \frac{n^2a}{V_{\text{measured}}^2}\right)(V_{\text{measured}} - nb) = nRT$$

or

The van der Waals equation, like the ideal gas equation, is known as an *equation of state,* i.e., an equation that describes a state of matter.

$$\left(P + \frac{n^2a}{V^2}\right)(V - nb) = nRT$$

This is the **van der Waals equation.** In this equation, P, V, T, and n represent the *measured* values of pressure, volume, temperature (expressed on the absolute scale), and number of moles, respectively, just as in the ideal gas equation. The quantities a and b are experimentally derived constants that differ for different gases (Table 12-5). When a and b are both zero, the van der Waals equation reduces to the ideal gas equation.

We can understand the relative values of a and b in Table 12-5 in terms of molecular properties. Note that a for helium is very small. This is the case for all noble gases and many other nonpolar molecules, because only very weak attractive forces, called London forces, exist between them. **London forces** result from short-lived electrical dipoles produced by the attraction of one atom's nucleus for an adjacent atom's electrons. These forces exist for all molecules but are especially important for nonpolar molecules, which would never liquefy if London forces did not exist. Polar molecules such as ammonia, NH_3, have permanent charge separations (dipoles), so they exhibit greater forces of attraction for each other. This explains the high value of a for ammonia. London forces and permanent dipole forces of attraction are discussed in more detail in the next chapter.

Larger molecules have greater values of b. For instance, H_2, a first-row diatomic molecule, has a greater b value than the first-row monatomic He. The b value for CO_2, which contains three second-row atoms, is greater than that for N_2, which contains only two second-row atoms.

The following example illustrates the deviation of methane, CH_4, from ideal gas behavior under high pressure.

Table 12-5 *van der Waals Constants*

Gas	a ($L^2 \cdot atm/mol^2$)	b (L/mol)
H_2	0.244	0.0266
He	0.034	0.0237
N_2	1.39	0.0391
NH_3	4.17	0.0371
CO_2	3.59	0.0427
CH_4	2.25	0.0428

EXAMPLE 12-22   *van der Waals Equation*

Calculate the pressure exerted by 1.00 mole of methane, CH_4, in a 500-mL vessel at 25.0°C assuming (a) ideal behavior and (b) nonideal behavior.

Plan

(a) Ideal gases obey the ideal gas equation. We can solve this equation for P.

(b) To describe methane as a nonideal gas, we use the van der Waals equation and solve for P.

Solution

(a) Using the ideal gas equation to describe ideal gas behavior,

$$PV = nRT$$

$$P = \frac{nRT}{V} = \frac{(1.00 \text{ mol})\left(0.0821 \dfrac{\text{L} \cdot \text{atm}}{\text{mol} \cdot \text{K}}\right)(298 \text{ K})}{0.500 \text{ L}} = \boxed{48.9 \text{ atm}}$$

(b) Using the van der Waals equation to describe nonideal gas behavior,

$$\left(P + \frac{n^2 a}{V^2}\right)(V - nb) = nRT$$

For CH_4, $a = 2.25 \text{ L}^2 \cdot \text{atm/mol}^2$ and $b = 0.0428 \text{ L/mol}$ (Table 12-5).

$$\left[P + \frac{(1.00 \text{ mol})^2(2.25 \text{ L}^2 \cdot \text{atm/mol}^2)}{(0.500 \text{ L})^2}\right]\left[0.500 \text{ L} - (1.00 \text{ mol})\left(0.0428 \frac{\text{L}}{\text{mol}}\right)\right]$$

$$= (1.00 \text{ mol})\left(0.0821 \frac{\text{L} \cdot \text{atm}}{\text{mol} \cdot \text{K}}\right)(298 \text{ K})$$

Combining terms and canceling units, we get

$$[P + 9.00 \text{ atm}][0.457 \text{ L}] = 24.5 \text{ L} \cdot \text{atm}$$

$$P + 9.00 \text{ atm} = 53.6 \text{ atm}$$

$$P = \boxed{44.6 \text{ atm}}$$

You should now work Exercises 86 and 87.

The pressure is 4.3 atm (8.8%) less than that calculated from the ideal gas law. A significant error would be introduced by assuming ideal behavior at this high pressure.

Repeating the calculations of Example 12-22 with the volume doubled ($V = 10.0 \text{ L}$) gives ideal and nonideal pressures, respectively, of 2.45 and 2.44 atm, a difference of only 0.4%.

Many other equations have been developed to describe the behavior of real gases. Each of these contains quantities that must be empirically derived for each gas.

12-15 MASS–VOLUME RELATIONSHIPS IN REACTIONS INVOLVING GASES

Many chemical reactions produce gases. For instance, the combustion of a hydrocarbon in excess oxygen at high temperatures produces both carbon dioxide and water as gases, as illustrated for octane.

$$2C_8H_{18}(g) + 25O_2(g) \longrightarrow 16CO_2(g) + 18H_2O(g)$$

The N_2 gas produced by the very rapid decomposition of sodium azide, $NaN_3(s)$, inflates air bags used as safety devices in automobiles.

We know that one mole of gas, measured at STP, occupies 22.4 liters; we can use the ideal gas equation to find the volume of a mole of gas at any other conditions. This information can be utilized in stoichiometry calculations (Section 3-2).

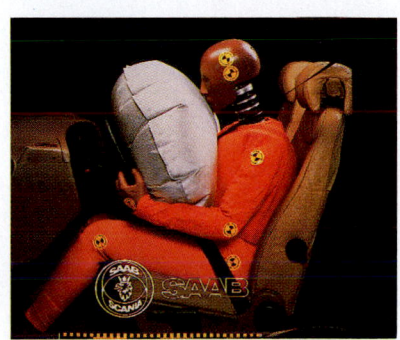

The nitrogen gas formed in the rapid reaction

$$2NaN_3(s) \longrightarrow 2Na(s) + 3N_2(g)$$

fills an automobile air bag during a collision. The air bag fills within 1/20th of a second after a front collision.

Production of a gas by a reaction.

$$2NaOH(g) + 2Al(s) + 6H_2O(\ell) \longrightarrow$$
$$2Na[Al(OH)_4](aq) + 3H_2(g)$$

This reaction is used in some solid drain cleaners.

Small amounts of oxygen can be produced in the laboratory by heating solid potassium chlorate, $KClO_3$, in the presence of a catalyst, manganese(IV) oxide, MnO_2. Solid potassium chloride, KCl, is also produced. (CAUTION: Heating $KClO_3$ can be dangerous.)

$$2KClO_3(s) \xrightarrow[\Delta]{MnO_2} 2KCl(s) + 3O_2(g)$$

2 mol	2 mol	3 mol
2(122.6 g)	2(74.6 g)	3(22.4 L_{STP})

Unit factors can be constructed using any two of these quantities.

EXAMPLE 12-23   Gas Volume in a Chemical Reaction

What volume of O_2 (STP) can be produced by heating 112 grams of $KClO_3$?

Plan

The preceding equation shows that two moles of $KClO_3$ produce three moles of O_2. We construct appropriate unit factors from the balanced equation and the standard molar volume of oxygen to solve the problem.

Solution

$$\underline{?}\, L_{STP}\, O_2 = 112\ g\ KClO_3 \times \frac{1\ mol\ KClO_3}{122.6\ g\ KClO_3} \times \frac{3\ mol\ O_2}{2\ mol\ KClO_3} \times \frac{22.4\ L_{STP}\ O_2}{1\ mol\ O_2}$$
$$= \boxed{30.7\ L_{STP}\ O_2}$$

We could have solved this problem using a single unit factor, $3(22.4\ L_{STP})\ O_2/2(122.6\ g)\ KClO_3$, read directly from the equation.

$$\underline{?}\, L_{STP}\, O_2 = 112\ g\ KClO_3 \times \frac{3(22.4\ L_{STP})\ O_2}{2(122.6\ g)\ KClO_3} = \boxed{30.7\ L_{STP}\ O_2}$$

This calculation shows that the thermal decomposition of 112 grams of $KClO_3$ produces 30.7 liters of oxygen measured at standard conditions.

You should now work Exercise 96.

EXAMPLE 12-24   Gas Volume in a Chemical Reaction

A 1.80-gram mixture of potassium chlorate, $KClO_3$, and potassium chloride, KCl, was heated until all of the $KClO_3$ had decomposed. After being dried, the liberated oxygen occupied 405 mL at 25°C when the barometric pressure was 745 torr. (a) How many moles of O_2 were produced? (b) What percentage of the mixture was $KClO_3$? KCl?

Plan

(a) The number of moles of O_2 produced can be calculated from the ideal gas equation. (b) Then we use the balanced chemical equation to relate the known number of moles of O_2 formed and the mass of $KClO_3$ that decomposed to produce it.

Solution

(a) $V = 405\ mL = 0.405\ L;\ P = 745\ torr \times \dfrac{1\ atm}{760\ torr} = 0.980\ atm$

$T = 25°C + 273 = 298\ K$

Solving the ideal gas equation for n and evaluating gives

$$n = \frac{PV}{RT} = \frac{(0.980 \text{ atm})(0.405 \text{ L})}{\left(0.0821 \dfrac{\text{L} \cdot \text{atm}}{\text{mol} \cdot \text{K}} (298 \text{ K})\right)} = \boxed{0.0162 \text{ mol } O_2}$$

(b) $\underline{?} \text{ g KClO}_3 = 0.0162 \text{ mol } O_2 \times \dfrac{2 \text{ mol KClO}_3}{3 \text{ mol } O_2} \times \dfrac{122.6 \text{ g KClO}_3}{1 \text{ mol KClO}_3} = 1.32 \text{ g KClO}_3$

The sample contained 1.32 grams of $KClO_3$. The percent of $KClO_3$ in the sample is

$$\% \text{ KClO}_3 = \frac{\text{g KClO}_3}{\text{g sample}} \times 100\% = \frac{1.32 \text{ g}}{1.80 \text{ g}} \times 100\% = \boxed{73.3\% \text{ KClO}_3}$$

The sample contains 73.3% $KClO_3$ and $(100.0 - 73.3)\% = \boxed{26.7\% \text{ KCl}}$.

You should now work Exercise 98.

Key Terms

Absolute zero The zero point on the absolute temperature scale; $-273.15°C$ or 0 K; theoretically, the temperature at which molecular motion is a minimum.

Atmosphere (atm) A unit of pressure; the pressure that will support a column of mercury 760 mm high at 0°C; 760 torr.

Avogadro's Law At the same temperature and pressure, equal volumes of all gases contain the same number of molecules.

Barometer A device for measuring atmospheric pressure. See Figures 12-1 and 12-2. The liquid is usually mercury.

Boyle's Law At constant temperature, the volume occupied by a given mass of a gas is inversely proportional to the applied pressure.

Charles' Law At constant pressure, the volume occupied by a definite mass of a gas is directly proportional to its absolute temperature.

Condensed states The solid and liquid states.

Dalton's Law See *Law of Partial Pressures.*

Diffusion The movement of a substance (e.g., a gas) into a space or the mixing of one substance (e.g., a gas) with another.

Dumas method A method used to determine the molecular weights of volatile liquids. See Figure 12-6.

Effusion The escape of a gas through a tiny hole or a thin porous wall.

Equation of state An equation that describes the behavior of matter in a given state; e.g., the van der Waals equation describes the behavior of the gaseous state.

Fluids Substances that flow freely; gases and liquids.

Ideal gas A hypothetical gas that obeys exactly all postulates of the kinetic–molecular theory.

Ideal Gas Equation The product of the pressure and volume of an ideal gas is directly proportional to the number of moles of the gas and the absolute temperature.

Kinetic–molecular theory A theory that attempts to explain macroscopic observations on gases in microscopic or molecular terms.

Law of Partial Pressures The total pressure exerted by a mixture of gases is the sum of the partial pressures of the individual gases; also called Dalton's Law.

Manometer A two-armed barometer. See Figure 12-1.

Mole fraction The number of moles of a component of a mixture divided by the total number of moles in the mixture.

Partial pressure The pressure exerted by one gas in a mixture of gases.

Pascal (Pa) The SI unit of pressure; defined as the pressure exerted by a force of one newton acting on an area of one square meter.

Pressure Force per unit area.

Real gases Gases that deviate from ideal gas behavior.

Root-mean-square speed, u_{rms} The square root of the mean-square speed, $\sqrt{\overline{u^2}}$. This is equal to $\sqrt{\dfrac{3RT}{M}}$ for an ideal gas. The root-mean-square speed is slightly different from the average speed, but the two quantities are proportional.

Standard temperature and pressure (STP) Standard temperature, 0°C (273.15 K), and standard pressure, one atmosphere, are standard conditions for gases.

Standard molar volume The volume occupied by one mole of an ideal gas under standard conditions; 22.414 liters.

Torr A unit of pressure; the pressure that will support a column of mercury one mm high at 0°C.

Universal gas constant R, the proportionality constant in the ideal gas equation, $PV = nRT$.

van der Waals equation An equation of state that extends the ideal gas law to real gases by inclusion of two empirically determined parameters, which are different for different gases.

Vapor A gas formed by boiling or evaporation of a liquid or sublimation of a solid; a term commonly used when some of the liquid or solid remains in contact with the gas.

Vapor pressure The pressure exerted by a vapor in equilibrium with its liquid or solid.

Exercises

You may assume *ideal gas behavior* unless otherwise indicated.

Basic Ideas

1. What are the three states of matter? Compare and contrast them.
2. State whether each property is characteristic of all gases, some gases, or no gas: (a) transparent to light; (b) colorless; (c) unable to pass through filter paper; (d) more difficult to compress than liquid water; (e) odorless; (f) settles on standing.
3. Suppose you were asked to supply a particular mass of a specified gas in a container of fixed volume at a specified pressure and temperature. Is it likely that you could fulfill the request? Explain.
4. State whether each of the following samples of matter is a gas. If the information is insufficient for you to decide, write "insufficient information."
   (a) A material is in a steel tank at 100 atm pressure. When the tank is opened to the atmosphere, the material immediately expands, increasing its volume many-fold.
   (b) A material, on being emitted from an industrial smokestack, rises about 10 m into the air. Viewed against a clear sky, it has a white appearance.
   (c) 1.0 mL of material weighs 8.2 g.
   (d) When a material is released from a point 30 ft below the level of a lake at sea level (equivalent in pressure to about 76 cm of mercury), it rises rapidly to the surface, at the same time doubling its volume.
   (e) A material is transparent and pale green in color.
   (f) One cubic meter of a material contains as many molecules as 1 m^3 of air at the same temperature and pressure.

Pressure

5. Define pressure. Give a precise scientific definition—one that can be understood by someone without any scientific training.
6. What are some of the units used to measure pressure? How do they relate to one another? What is the value of the average pressure of the atmosphere in these various units? Why do we have more than one unit for pressure?
7. Describe the mercury barometer. How does it work?
8. What is a manometer? How does it work?
9. Express a pressure of 685 torr in the following units: (a) mm Hg; (b) atm; (c) Pa; (d) kPa.
10. A typical laboratory atmospheric pressure reading is 755 torr. Convert this value to (a) psi, (b) cm Hg, (c) inches Hg, (d) kPa, (e) atm, and (f) ft H_2O.
11. Complete the following table.

	atm	torr	Pa	kPa
Standard atmosphere	1			
Partial pressure of nitrogen in the atmosphere		593		

	atm	torr	Pa	kPa
A tank of compressed hydrogen			1.61×10^5	
Atmospheric pressure at the summit of Mt. Everest				33.7

*12. Consider a container of mercury open to the atmosphere on a day when the barometric pressure is 754 torr. Calculate the total pressure, in torr and atmospheres, within the mercury at depths of (a) 100 mm and (b) 5.04 cm.

*13. The densities of mercury and corn oil are 13.5 g/mL and 0.92 g/mL, respectively. If corn oil were used in a barometer, what would be the height of the column, in meters, at standard atmospheric pressure? (The vapor pressure of the oil is negligible.)

14. Steel tanks for storage of gases are capable of withstanding pressures greater than 150 atm. Express this pressure in psi.

15. Automobile tires are normally inflated to a pressure of 28 psi as measured by a tire gauge. (a) Express this pressure in atmospheres. (b) Assuming standard atmospheric pressure, calculate the absolute pressure inside a tire.

Boyle's Law: The Pressure–Volume Relationship

16. (a) On what kinds of observations (measurements) is Boyle's Law based? State the law. (b) Use the statement of Boyle's Law to derive a simple mathematical expression for Boyle's Law.

17. Could the words "a fixed number of moles" be substituted for "a definite mass" in the statement of Boyle's Law? Explain.

18. What pressure is needed to confine an ideal gas to 105 L after it has expanded from 35 L and 1.50 atm at constant temperature?

19. A sample of krypton gas occupies 75.0 mL at 0.400 atm. If the temperature remained constant, what volume would the krypton occupy at (a) 4.00 atm, (b) 0.00400 atm, (c) 765 torr, (d) 4.00 torr, and (e) 3.5×10^{-2} torr?

20. A 50-L sample of gas collected in the upper atmosphere at a pressure of 18.3 torr is compressed into a 150-mL container at the same temperature. (a) What is the new pressure, in atmospheres? (b) To what volume would the original sample have had to be compressed to exert a pressure of 10.0 atm?

21. Assume that, for some set of conditions, the value of k in $PV = k$ is 18. (a) Plot the graph of P (x axis) versus V (y axis) for the values $V = 1, 2, 3, 4, 6, 12, 15,$ and 18. What is the shape of the curve? (b) Plot the graph of $1/P$ (x axis) versus V (y axis) for the same values of V. What plot is obtained?

*22. A cylinder containing 29 L of helium gas at a pressure of 165 atm is to be used to fill toy balloons to a pressure of 1.1 atm. Each inflated balloon has a volume of 2.0 L. What is the maximum number of balloons that can be inflated? (Remember that 29 L of helium at 1.1 atm will remain in the "exhausted" cylinder.)

The Absolute Temperature Scale

23. (a) Can an absolute temperature scale based on Fahrenheit rather than Celsius degrees be evolved? Why? (b) Can an absolute temperature scale that is based on a "degree" twice as large as a Celsius degree be developed? Why?

24. (a) What does "absolute temperature scale" mean? (b) Describe the experiments that led to the evolution of the absolute temperature scale. What is the relationship between the Celsius and Kelvin temperature scales? (c) What does "absolute zero" mean?

25. Complete the table by making the required temperature conversions. Pay attention to significant figures.

	Temperature	
	K	**°C**
Normal boiling point of water		100
Standard for thermodynamic data	298.15	
Dry ice becomes a gas at atmospheric pressure		−78.5
The center of the sun (more or less)	1.53×10^7	

Charles' Law: The Volume–Temperature Relationship

26. (a) Why is a plot of volume versus temperature at constant pressure a straight line (Figure 12-5)? (b) On what kind of observations (measurements) is Charles' Law based? State the law.

27. A gas occupies a volume of 31.0 L at 17.0°C. If the gas temperature rises to 34.0°C at constant pressure, (a) would you expect the volume to double to 62.0 L? Explain. Calculate the new volume (b) at 34.0°C, (c) at 400 K, and (d) at −34.0°C.

28. A sample of helium occupies 90.0 mL at 25.0°C. To what temperature, in °C, must it be cooled, at constant pressure, to reduce its volume to 30.0 mL?

29. Which of the following statements are true? Which are false? Why is each true or false? *Assume constant pressure in each case.*

(a) If a sample of gas is heated from 100°C to 200°C, the volume will double.

(b) If a sample of gas is heated from 0°C to 273°C, the volume will double.

(c) If a sample of gas is cooled from 1273°C to 500°C, the volume will decrease by a factor of two.

(d) If a sample of gas is cooled from 1000°C to 200°C, the volume will decrease by a factor of five.

(e) If a sample of gas is heated from 473°C to 1219°C, the volume will increase by a factor of two.

*30. The device shown below is a **gas thermometer**. (a) At the ice point, the gas volume is 1.400 L. What would be the new volume if the gas temperature were raised from the ice point to 8.0°C? (b) Assume the cross-sectional area of the graduated arm is 1.0 cm². What would be the difference in height if the gas temperature changed from 0°C to 8.0°C? (c) What modifications could be made to increase the sensitivity of the thermometer?

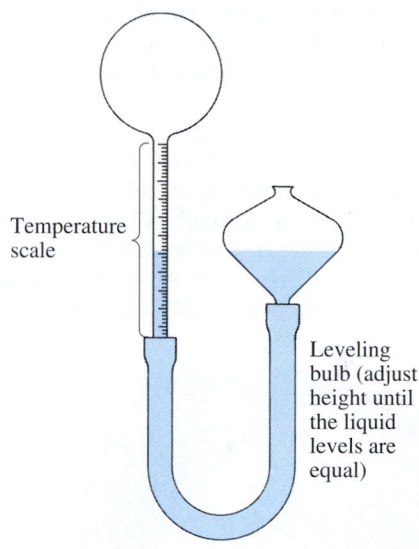

Temperature scale

Leveling bulb (adjust height until the liquid levels are equal)

31. What volume of gas should we use at 25°C and 1 atm if we wish to have the sample occupy 12.3 L at 185°C and one atm?

32. Calculate the volume of an ideal gas at the temperatures of dry ice (−78.5°C), liquid N_2 (−195.8°C), and liquid He (−268.9°C) if it occupies 5.00 L at 25.0°C. Assume constant pressure. Plot your results and extrapolate to zero volume. At what temperature would zero volume be reached?

The Combined Gas Law

33. Classify the relationship between the variables (a) P and V, (b) V and T, and (c) P and T as either (i) directly proportional or (ii) inversely proportional.

34. Prepare sketches of plots of (a) P vs. V, (b) P vs. $1/V$, (c) V vs. T, and (d) P vs. T for an ideal gas.

35. A sample of gas occupies 400 mL at STP. Under what pressure would this sample occupy 200 mL if the temperature were increased to 819°C?

36. A sample of hydrogen occupies 375 mL at STP. If the temperature were increased to 773°C, what final pressure would be necessary to keep the volume constant at 375 mL?

37. A 247-mL sample of a gas exerts a pressure of 3.33 atm at 16.0°C. What volume would it occupy at 100°C and 1.00 atm?

38. A 280-ml sample of neon exerts a pressure of 660 torr at 26°C. At what temperature in °C would it exert a pressure of 880 torr in a volume of 440 mL?

39. What temperature would be necessary to double the volume of an ideal gas initially at STP if the pressure decreased by 25.0%?

STP, Standard Molar Volume, and Gas Densities

40. (a) What is Avogadro's Law? What does it mean? (b) What does "standard molar volume" mean?

41. How many molecules of an ideal gas are contained in a 1.00-L flask at STP?

42. The limit of sensitivity for the analysis of carbon monoxide, CO, in air is 1 ppb (ppb = parts per billion) by volume. What is the smallest number of CO molecules that can be detected in 10 L of air at STP?

43. Sodium vapor has been detected recently as a major component of the thin atmosphere of Mercury using a ground-based telescope and a spectrometer. Its concentration is estimated to be about 1.0×10^5 atoms per cm^3. (a) Express this in moles per liter. (b) The maximum temperature of the atmosphere was measured by Mariner 10 to be about 970°C. What is the approximate partial pressure of sodium vapor at that temperature?

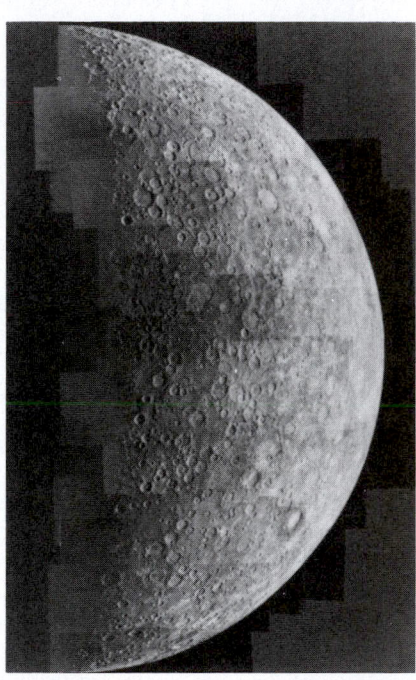

The surface of the planet Mercury.

44. Ethylene dibromide (EDB) was formerly used as a fumigant for fruits and grains, but now is banned because of its potential carcinogenicity. EDB is a liquid that boils at 109°C. Its molecular weight is 188 g/mol. Calculate the density of its vapor at 180°C and 1.00 atm.

45. A student's lab instructor asked him to calculate the number of moles of gas contained in a 320-mL bulb in the laboratory. The student determined that the pressure and temperature of the gas were 732 torr and 23.3°C. His calculations follow.

$$\underline{?}\ \text{mol} = 0.320\ \text{L} \times \frac{1\ \text{mol}}{22.4\ \text{L}} = 0.0143\ \text{mol}$$

Was he correct? Why?

46. A laboratory technician forgot what the color coding on some commercial cylinders of gas meant, but remembered that each of two specific tanks contained one of the following gases: He, Ne, Ar, or Kr. Measurements at STP made on samples of the gases from the two cylinders showed the gas densities to be 3.74 g/L and 0.178 g/L. Determine by calculation which of these gases was present in each tank.

*47. A 503-mL flask contains 0.0243 mol of an ideal gas at a given temperature and pressure. Another flask contains 0.0188 mol of the gas at the same temperature and pressure. What is the volume of the second flask?

The Ideal Gas Equation

48. (a) What is an ideal gas? (b) What is the ideal gas equation? (c) Outline the logic used to obtain the ideal gas equation. (d) What is R? How is it obtained?

49. Calculate R in L · atm/mol · K, in kPa · dm^3/mol · K, in J/mol · K, and in kJ/mol · K.

50. Calculate the pressure needed to contain 2.44 mol of an ideal gas at 45°C in a volume of 3.45 L.

51. (a) A chemist is preparing to carry out a reaction at high pressure that requires 36.0 mol of hydrogen gas. The chemist pumps the hydrogen into a 12.3-L rigid steel vessel at 25°C. To what pressure (in atmospheres) must the hydrogen be compressed? (b) What would be the density of the high-pressure hydrogen?

52. A 5.00-mol sample of neon is confined in a 3.14-L vessel. (a) What is the temperature if the pressure is 2.50 atm? (b) What is the density of the sample?

53. How many gaseous molecules are in a 1.00-L container if the pressure is 2.50×10^{-9} torr and the temperature is 1225 K?

*54. A barge containing 580 tons of liquid chlorine was involved in an accident. What volume would this amount of chlorine occupy if it were all converted to a gas at 750 torr and 18°C? Assume that the chlorine is confined to a width of 0.50 mile and an average depth of 60 ft. What would be the length, in feet, of this chlorine "cloud"?

Molecular Weights and Formulas for Gaseous Compounds

55. A sample of a liquid with a boiling point of 56.5°C was vaporized in a Dumas apparatus (Figure 12-6) as the 295-mL bulb was immersed into boiling water. The barometric pressure was 733 torr. The boiling point of water at this pressure is 99°C. Filled only with air, the bulb had a mass of 124.773 g; filled with the vapor, it had a mass of 125.437 g. Determine the molecular weight of the liquid.

*56. A student was given a container of ethane, C_2H_6, that had been closed at STP. By making appropriate measurements, she found that the mass of the sample of ethane was 0.244 g and the volume of the container was 185 mL. Use her data to calculate the molecular weight of ethane. What percent error is obtained? Suggest some possible sources of the error.

57. Analysis of a volatile liquid shows that it contains 37.23% carbon, 7.81% hydrogen, and 54.96% chlorine by mass. At 150°C and 1.00 atm, 500 mL of the vapor has a mass of 0.922 g.

(a) What is the molecular weight of the compound? (b) What is its molecular formula?

58. What is the molecular weight of an ideal gas if 0.480 g of the gas occupies 590 mL at 365 torr and 45°C?

59. The Dumas method was used to determine the molecular weight of a liquid. The vapor occupied a 122-mL volume at 98°C and 708 torr. The condensed vapor has a mass of 0.757 g. Calculate the molecular weight of the liquid.

*60. A highly volatile liquid was allowed to vaporize completely into a 250-mL flask immersed in boiling water. From the following data, calculate the molecular weight of the liquid. Mass of empty flask = 65.347 g; mass of flask filled with water at room temperature = 327.4 g; mass of flask and condensed liquid = 65.739 g; atmospheric pressure = 743.3 torr; temperature of boiling water = 99.8°C; density of water at room temperature = 0.997 g/mL.

61. A pure gas contains 85.63% carbon and 14.37% hydrogen by mass. Its density is 2.50 g/L at STP. What is its molecular formula?

Gas Mixtures and Dalton's Law

62. (a) What are partial pressures of gases? (b) State Dalton's Law. Express it symbolically.

63. A sample of oxygen of mass 30.0 g is confined in a vessel at 0°C and 1000 torr. Then 8.00 g of hydrogen is pumped into the vessel at constant temperature. What will be the final pressure in the vessel (assuming only mixing with no reaction)?

64. A gaseous mixture contains 5.23 g of chloroform, $CHCl_3$, and 1.66 g of methane, CH_4. What pressure is exerted by the mixture inside a 50.0-mL metal bomb at 345°C? What pressure is contributed by the $CHCl_3$?

65. A cyclopropane–oxygen mixture can be used as an anesthetic. If the partial pressures of cyclopropane and oxygen are 140 torr and 560 torr, respectively, what is the ratio of the number of moles of cyclopropane to the number of moles of oxygen in this mixture?

66. What is the mole fraction of each gas in a mixture that contains 0.267 atm of He, 0.317 atm of Ar, and 0.277 atm of Xe?

*67. Assume that unpolluted air has the composition shown in Table 12-2. (a) Calculate the number of molecules of N_2, of O_2, and of Ar in 1.00 L of air at 21°C and 1.00 atm. (b) Calculate the mole fractions of N_2, O_2, and Ar in the air.

68. Individual samples of O_2, N_2, and He are present in three 2.50-L vessels. Each exerts a pressure of 1.50 atm. (a) If all three gases are forced into the same 1.00-L container with no change in temperature, what will be the resulting pressure? (b) What is the partial pressure of O_2 in the mixture? (c) What are the partial pressures of N_2 and He?

69. A sample of hydrogen was collected over water at 20°C and 755 torr. The volume of the gas was 23.5 mL. What volume would the dry hydrogen occupy at STP?

70. A sample of dry nitrogen occupies 447 mL at STP. What would its volume be if it were collected over water at 26°C and 750 torr?

*71. A study of climbers who reached the summit of Mt. Everest without supplemental oxygen revealed that the partial pressures of O_2 and CO_2 in their lungs were 35 torr and 7.5 torr, respectively. The barometric pressure at the summit was 253 torr. Assume that the lung gases are saturated with moisture at a body temperature of 37°C. Calculate the partial pressure of inert gas (mostly nitrogen) in the climbers' lungs.

72. A 6.00-L flask containing He at 6.00 atm is connected to a 3.00-L flask containing N_2 at 3.00 atm and the gases are allowed to mix. (a) Find the partial pressures of each gas after they are allowed to mix. (b) Find the total pressure of the mixture. (c) What is the mole fraction of helium?

The Kinetic–Molecular Theory and Molecular Speeds

73. Outline the kinetic–molecular theory.

74. The radius of a typical molecule of a gas is 2.0 Å. (a) Find the volume of a molecule assuming it to be spherical. ($V = 4/3\pi r^3$ for a sphere.) (b) Calculate the volume actually occupied by 1.0 mol of these molecules. (c) If 1.0 mol of this gas occupies 22.4 L, find the fraction of the volume actually occupied by the molecules. (d) Comment on your answer to (c) in view of the first statement summarizing the kinetic–molecular theory of an ideal gas.

75. How does the kinetic–molecular theory explain (a) Boyle's Law? (b) Dalton's Law? (c) Charles' Law?

76. SiH_4 molecules are heavier than CH_4 molecules; yet, according to kinetic–molecular theory, the average kinetic energies of the two gases at the same temperature are equal. How can this be?

*77. At 22°C, Cl_2 molecules have some rms speed (which we need not calculate). At what temperature would the rms speed of F_2 molecules be the same?

*78. (a) How do average speeds of gaseous molecules vary with temperature? (b) Calculate the ratio of the rms speed of N_2 molecules at 100°C to the rms speed of the same molecules at 0°C.

Real Gases and Deviations from Ideality

79. (a) How do "real" and "ideal" gases differ? (b) Under what kinds of conditions are deviations from ideality most important? Why?

80. Which of the following gases would be expected to behave most nearly ideally under the same conditions? H_2, F_2, HF. Which one would be expected to deviate from ideal behavior the most? Explain both answers.

81. Does the effect of intermolecular attraction on the properties of a gas become more significant or less significant if (a) the gas is compressed to a smaller volume at constant temperature? (b) more gas is forced into the same volume at the same temperature? (c) the temperature of the gas is raised at constant pressure?

82. Does the effect of molecular volume on the properties of a gas become more significant or less significant if (a) the gas is compressed to a smaller volume at constant temperature? (b) more gas is forced into the same volume at the same temperature? (c) the temperature of the gas is raised at constant pressure?

83. A sample of gas has a molar volume of 10.3 L at a pressure of 745 torr and a temperature of $-138°C$. Is the gas behaving ideally?

84. Calculate the compressibility factor, $(P_{real})(V_{real})/RT$, for a 1.00-mol sample of NH_3 under the following conditions: in a 500-mL vessel at $-10.0°C$ it exerts a pressure of 30.0 atm. What would be the *ideal* pressure for 1.00 mol of NH_3 at $-10.0°C$ in a 500-mL vessel? Compare this with the real pressure and account for the difference.

85. What is the van der Waals equation? How does it differ from the ideal gas equation?

86. Find the pressure of a sample of carbon tetrachloride, CCl_4, if 1.00 mol occupies 35.0 L at $77.0°C$ (slightly above its normal boiling point). Assume that CCl_4 obeys (a) the ideal gas law; (b) the van der Waals equation. The van der Waals constants for CCl_4 are $a = 20.39$ $L^2 \cdot atm/mol^2$ and $b = 0.1383$ L/mol.

87. Repeat the calculations of Exercise 86 using a 3.10-mol gas sample confined to 6.15 L at $135°C$.

Stoichiometry in Reactions Involving Gases

88. During a collision, automobile air bags are inflated by the N_2 gas formed by the explosive decomposition of sodium azide, NaN_3.

$$2NaN_3 \longrightarrow 2Na + 3N_2$$

What mass of sodium azide would be needed to inflate a 30.0-L bag to a pressure of 1.40 atm at $25°C$?

89. Assuming the volumes of all gases in the reaction are measured at the same temperature and pressure, calculate the volume of water vapor obtainable by the explosive reaction of a mixture of 440 mL of hydrogen gas and 220 mL of oxygen gas.

*90. One liter of sulfur vapor at $600°C$ and 1.00 atm is burned in excess pure oxygen to give sulfur dioxide gas, SO_2, measured at the same temperature and pressure. What mass of SO_2 gas is obtained?

91. Calculate the volume of methane, CH_4, measured at 300 K and 770 torr, that can be produced by the bacterial breakdown of 1.00 kg of a simple sugar.

$$C_6H_{12}O_6 \longrightarrow 3CH_4 + 3CO_2$$

*92. A common laboratory preparation of oxygen is

$$2KClO_3(s) \xrightarrow[\Delta]{MnO_2} 2KCl(s) + 3O_2(g)$$

If you were designing an experiment to generate four bottles (each containing 250 mL) of O_2 at $25°C$ and 741 torr and allowing for 25% waste, what mass of potassium chlorate would be required?

93. Many campers use small propane stoves to cook meals. What volume of air (Table 12-2) will be required to burn 10.0 L of propane, C_3H_8? Assume all gas volumes are measured at the same temperature and pressure.

$$C_3H_8(g) + 5O_2(g) \longrightarrow 3CO_2(g) + 4H_2O(g)$$

*94. If 1.67 L of nitrogen and 4.42 L of hydrogen were allowed to react, how many grams of $NH_3(g)$ can form? Assume all gases are at the same temperature and pressure, and that the limiting reactant is used up.

$$N_2(g) + 3H_2(g) \longrightarrow 2NH_3(g)$$

*95. We burn 11.31 L of ammonia in 16.14 L of oxygen at $500°C$. What volume of nitric oxide, NO, gas can form? What volume of steam, $H_2O(g)$, is formed? Assume all gases are at the same temperature and pressure, and that the limiting reactant is used up.

$$4NH_3(g) + 5O_2(g) \longrightarrow 4NO(g) + 6H_2O(g)$$

96. What mass of KNO_3 would have to be decomposed to produce 21.1 L of oxygen measured at STP?

$$2KNO_3(s) \xrightarrow{\Delta} 2KNO_2(s) + O_2(g)$$

97. Refer to Exercise 96. An impure sample of KNO_3 that had a mass of 50.3 g was heated until all of the KNO_3 had decomposed. The liberated oxygen occupied 4.22 L at STP. What percentage of the sample was KNO_3?

*98. Heating a 5.913-g sample of an ore containing a metal sulfide, in the presence of excess oxygen, produces 1.177 L of dry SO_2, measured at $35.0°C$ and 755 torr. Calculate the percentage by mass of sulfur in the ore.

*99. The following reactions occur in a gas mask (self-contained breathing apparatus) sometimes used by underground miners. The H_2O and CO_2 come from exhaled air, and O_2 is inhaled as it is produced. KO_2 is potassium superoxide. The CO_2 is converted to the solid salt $KHCO_3$, potassium hydrogen carbonate, so that CO_2 is not inhaled in significant amounts.

$$4KO_2(s) + 2H_2O(\ell) \longrightarrow 4KOH(s) + 3O_2(g)$$

$$CO_2(g) + KOH(s) \longrightarrow KHCO_3(s)$$

(a) What volume of O_2, measured at STP, is produced by the complete reaction of 1.00 g of KO_2? (b) What is this volume at body temperature, $37°C$, and 1.00 atm? (c) What mass of KOH is produced in (a)? (d) What volume of CO_2, measured at STP, will react with the mass of KOH of (c)? (e) What is the volume of CO_2 of (d) measured at $37°C$ and 1.00 atm?

*100. Let us represent gasoline as octane, C_8H_{18}. When hydrocarbon fuels burn in the presence of sufficient oxygen, CO_2 is formed.

$$\text{Reaction A:} \quad 2C_8H_{18} + 25O_2 \longrightarrow 16CO_2 + 18H_2O$$

But when the supply of oxygen is limited, the poisonous gas carbon monoxide, CO, is formed.

$$\text{Reaction B:} \quad 2C_8H_{18} + 17O_2 \longrightarrow 16CO + 18H_2O$$

Any automobile engine, no matter how well tuned, burns its fuel by some combination of these two reactions. Suppose an automobile engine is running at idle speed in a closed garage with air volume 97.5 m³. This engine burns 95.0% of its fuel by reaction A, and the remainder by reaction B. (a) How many liters of octane, density 0.702 g/mL, must be burned for the CO to reach a concentration of 2.00 g/m³? (b) If the engine running at idle speed burns fuel at the rate of 1.00 gal/h (0.0631 L/min), how long does it take to reach the CO concentration in (a)?

Mixed Exercises

101. A tilting McLeod gauge is used to measure very low pressures of gases in glass vacuum lines in the laboratory. It operates by compressing a large volume of gas at low pressure to a much smaller volume so that the pressure is more easily measured. What is the pressure of a gas in a vacuum line if a 53.3-mL volume of the gas, when compressed to 0.133 mL, supports a 16.9-mm column of mercury?

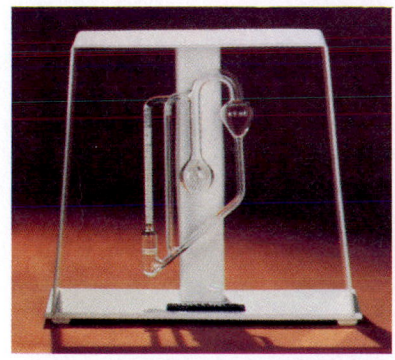

A McLeod gauge.

102. Imagine that you live in a cabin with an interior volume of 150 m³. On a cold morning your indoor air temperature is 10°C, but by the afternoon the sun has warmed the cabin air to 18°C. The cabin is not sealed; therefore, the pressure inside is the same as it is outdoors. Assume that the pressure remains constant during the day. How many cubic meters of air would have been forced out of the cabin by the sun's warming? How many liters?

103. A particular tank can safely hold gas up to a pressure of 44.3 atm. When the tank contains 38.1 g of N_2 at 25°C, the gas exerts a pressure of 10.1 atm. What is the highest temperature to which the gas sample can be heated safely?

104. Find the molecular weight of Freon-12 (a chlorofluoromethane) if 8.29 L of vapor at 200°C and 790 torr has a mass of 26.8 g.

105. A flask of unknown volume was filled with air to a pressure of 3.25 atm. This flask was then attached to an evacuated flask with a known volume of 5.00 L, and the air was allowed to expand into the flask. The final pressure of the air (in both flasks) was 2.50 atm. Calculate the volume of the first flask.

*106. Relative humidity is the ratio of the pressure of water vapor in the air to the pressure of water vapor in air that is saturated with water vapor at the same temperature.

$$\text{relative humidity} = \frac{\text{actual partial pressure of } H_2O \text{ vapor}}{\text{partial pressure of } H_2O \text{ vapor if sat'd}}$$

Often this quantity is multiplied by 100 to give the percent relative humidity. Suppose the percent relative humidity is 80.0% at 91.4°F (33.0°C) in a house with volume 245 m³. Then an air conditioner is turned on. Due to the condensation of water vapor on the cold coils of the air conditioner, water vapor is also removed from the air as it cools. After the air temperature has reached 77.0°F (25.0°C), the percent relative humidity is measured to be 15.0%. (a) What mass of water has been removed from the air in the house? (*Reminder:* take into account the difference in saturated water vapor pressure at the two temperatures.) (b) What volume would this liquid water occupy at 25°C? (Density of liquid water at 25.0°C = 0.997 g/cm³.)

107. A 315-mL flask contains 0.350 g of nitrogen gas at a pressure of 744 torr. Are these data sufficient to allow you to calculate the temperature of the gas? If not, what is missing? If so, what is the temperature in °C?

108. Use both the ideal gas law and the van der Waals equation to calculate the pressure exerted by a 10.0-mol sample of ammonia in a 60.0-L container at 100°C. By what percentage do the two results differ?

109. What volume of hydrogen fluoride at 743 torr and 24°C will be released by the reaction of 38.3 g of xenon difluoride with a stoichiometric amount of water? The *unbalanced* equation is

$$XeF_2(s) + H_2O(\ell) \longrightarrow Xe(g) + O_2(g) + HF(g)$$

What volumes of oxygen and xenon will be released under these conditions?

110. Cyanogen is 46.2% carbon and 53.8% nitrogen by mass. At a temperature of 25°C and a pressure of 750 torr, 1.00 g of cyanogen gas occupies 0.476 L. Determine the empirical formula and the molecular formula of cyanogen.

111. When the flight attendant delivers your sealed bag of in-flight peanuts, you notice that the bag is "swollen" and that it makes a distinct "psss" sound when you tear it open. Explain.

112. A manufacturer of automobile tires suggests that you check the pressure of the air in the tires when the car has not just been driven at high speeds. Why do you suppose this suggestion is made?

113. Incandescent light bulbs contain inert gases, such as argon, so that the filament will last longer. The approximate volume of a 100-watt bulb is 130 cm³, and the bulb contains 0.125 g of argon. How many grams of argon would be contained in a 150-watt bulb under the same pressure and temperature conditions if the volume of the larger wattage bulb is 185 cm³?

BUILDING YOUR KNOWLEDGE

114. A 5.00-L reaction vessel contains hydrogen at a partial pressure of 0.588 atm and oxygen gas at a partial pressure of 0.302 atm. Which element is the limiting reactant?

$$2H_2(g) + O_2(g) \longrightarrow 2H_2O(g)$$

115. Suppose the gas mixture in Exercise 114 is ignited and the reaction produces the theoretical yield of the product. What would be the partial pressure of each substance present in the final mixture?

116. A 0.361-g sample of pentane, C_5H_{12}, is placed in a 4.00-L reaction vessel with excess O_2. The mixture is then ignited and the sample burns with complete combustion. What will be the partial pressures of CO_2 and of $H_2O(g)$ in the reaction vessel if the final temperature is 300°C?

117. When magnesium carbonate, $MgCO_3$, is heated to a high temperature, it decomposes.

$$MgCO_3(s) \xrightarrow{\text{heat}} MgO(s) + CO_2(g)$$

A 20.29-gram sample of impure magnesium carbonate is completely decomposed at 1000°C in a previously evacuated 2.00-L reaction vessel. After the reaction was complete, the solid residue (consisting only of MgO and the original impurities) had a mass of 15.90 grams. Assume that no other constituent of the sample produced a gas and that the volume of any solid was negligible compared to the gas volume. (a) How many grams of CO_2 were produced? (b) What was the pressure of the CO_2 produced? (c) What percent of the original sample was magnesium carbonate?

118. One natural source of atmospheric carbon dioxide is precipitation reactions such as the precipitation of silicates in the oceans.

$$Mg^{2+}(aq) + SiO_2(\text{dispersed}) + 2HCO_3^-(aq) \longrightarrow$$
$$MgSiO_3(s) + 2CO_2(g) + H_2O(\ell)$$

How many grams of magnesium silicate would be precipitated during the formation of 100 liters of carbon dioxide at 30°C and 775 torr?

Liquids and Solids

The dew on this spider web was formed by condensation of water vapor from the air.

OBJECTIVES

As you study this chapter, you should learn

• *About the properties of liquids and solids and how they differ from gases*

• *To understand the kinetic–molecular description of liquids and solids, and how this description differs from that for gases*

• *To use the terminology of phase changes*

• *To understand various kinds of intermolecular attractions and how they are related to physical properties such as vapor pressure, viscosity, melting point, boiling point, and so on*

• *To describe evaporation, condensation, and boiling in molecular terms*

• *To calculate the heat transfer involved in warming or cooling without change of phase*

• *To calculate the heat transfer involved in phase changes*

• *To describe melting, solidification, sublimation, and deposition in molecular terms*

• *To interpret P vs. T phase diagrams*

• *About the regular structure of crystalline solids*

• *About the various types of solids*

• *To relate the properties of different types of solids to the bonding or interactions among particles in these solids*

• *To visualize some common simple arrangements of atoms in solids*

• *To carry out calculations relating atomic arrangement, density, unit cell size, and ionic or atomic radii in some simple crystalline arrangements*

• *About the bonding in metals*

• *Why some substances are conductors, some are insulators, and others are semiconductors*

T he molecules of most gases are so widely separated at ordinary temperatures and pressures that they do not interact with each other significantly. The physical properties of gases are reasonably well described by the simple relationships in Chapter 12. In liquids and solids, the so-called **condensed phases,** the particles are close together so they interact strongly. Although the properties of liquids and solids can be described, they cannot be adequately explained by simple mathematical relationships. Table 13-1 and Figure 13-1 summarize some of the characteristics of gases, liquids, and solids.

Table 13-1 *Some Characteristics of Gases, Liquids, and Solids*

Gases	Liquids	Solids
1. Have no definite shape (fill containers completely)	1. Have no definite shape (assume shapes of containers)	1. Have definite shape (resist deformation)
2. Are compressible	2. Have definite volume (are only very slightly compressible)	2. Are nearly incompressible
3. Have low density	3. Have high density	3. Usually have higher density than liquids
4. Are fluid	4. Are fluid	4. Are not fluid
5. Diffuse rapidly	5. Diffuse through other liquids	5. Diffuse only very slowly through solids
6. Consist of extremely disordered particles and much empty space; particles have rapid, random motion in three dimensions	6. Consist of disordered clusters of particles that are quite close together; particles have random motion in three dimensions	6. Have an ordered arrangement of particles that are very close together; particles have vibrational motion only

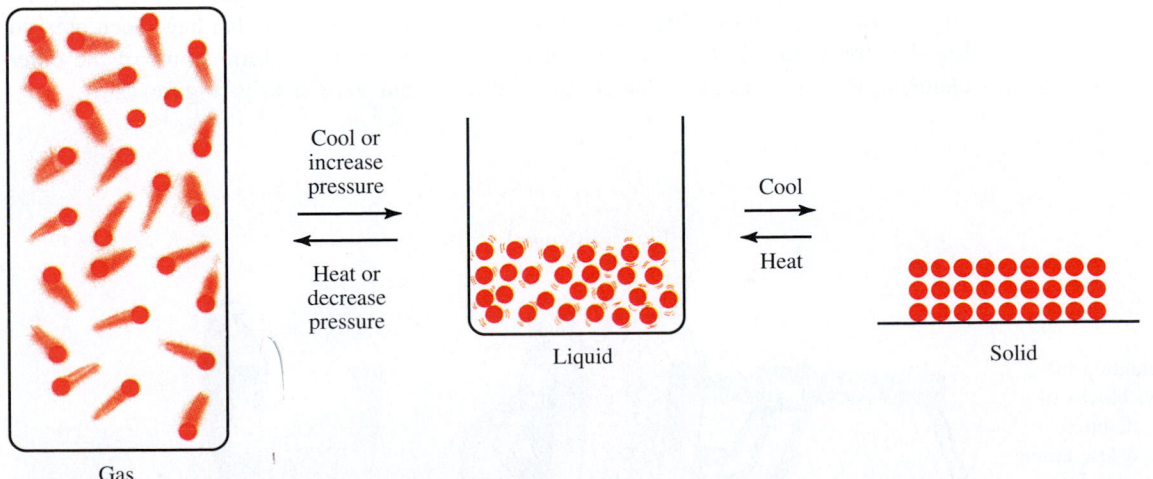

Figure 13-1 Representations of the kinetic–molecular interpretation of the three phases of matter.

13-1 KINETIC–MOLECULAR DESCRIPTION OF LIQUIDS AND SOLIDS

The properties listed in Table 13-1 can be qualitatively explained in terms of the kinetic–molecular theory of Chapter 12. We saw in Section 12-12 that the average kinetic energy of a collection of gas molecules decreases as the temperature is lowered. As a sample of gas is cooled and compressed, the rapid, random motion of gaseous molecules decreases. The molecules approach each other and the intermolecular attractions increase. Eventually these increasing intermolecular attractions overcome the reduced kinetic energies. At this point condensation (liquefaction) occurs. The temperatures and pressures required for condensation vary from gas to gas, because different kinds of molecules have different attractive forces.

In the liquid state the forces of attraction among particles are great enough that disordered clustering occurs. The particles are so close together that very little of the volume occupied by a liquid is empty space. As a result, it is very hard to compress a liquid. Particles in liquids have sufficient energy of motion to overcome partially the attractive forces among them. They are able to slide past each other so that liquids assume the shapes of their containers up to the volume of the liquid.

Liquids diffuse into other liquids with which they are *miscible*. For example, when a drop of red food coloring is added to a glass of water, the water becomes red throughout after diffusion is complete. The natural diffusion rate is slow at normal temperatures. Because the average separations among particles in liquids are far less than those in gases, the densities of liquids are much higher than the densities of gases (Table 12-1).

Cooling a liquid lowers its molecular kinetic energy and causes its molecules to slow down even more. If the temperature is lowered sufficiently, at ordinary pressures, stronger but shorter-range attractive interactions overcome the reduced kinetic energies of the molecules to cause *solidification*. The temperature required for *crystallization* at a given pressure depends on the nature of short-range interactions among the particles and is characteristic of each substance.

Most solids have ordered arrangements of particles with a very restricted range of motion. Particles in the solid state cannot move freely past one another so they only vibrate about fixed positions. Consequently, solids have definite shapes and volumes. Because the particles are so close together, solids are nearly incompressible and are very dense relative to gases. Solid particles do not diffuse readily into other solids. However, analysis of two blocks of different solids, such as copper and lead, that have been pressed together for a period of years shows that each block contains some atoms of the other element. This demonstrates that solids do diffuse, but very slowly (Figure 13-2).

*Inter*molecular attractions are those between different molecules or ions. *Intra*molecular attractions are those between atoms within a single molecule or ion.

The *miscibility* of two liquids refers to their ability to mix and produce a homogeneous solution.

Solidification and *crystallization* refer to the process in which a liquid changes to a solid.

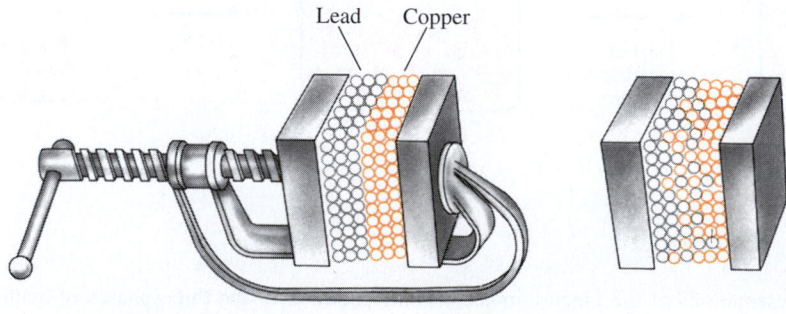

Lead   Copper

Figure 13-2   A representation of diffusion in solids. When blocks of two different metals are clamped together for a long time, a few atoms of each metal diffuse into the other metal.

13-2 INTERMOLECULAR ATTRACTIONS AND PHASE CHANGES

We have seen (Section 12-14) how the presence of strong attractive forces between gas molecules can cause gas behavior to become nonideal when the molecules get close together. In liquids and solids the molecules are much closer together than in gases. Therefore, properties of liquids, such as boiling point, vapor pressure, viscosity, and heat of vaporization, depend markedly on the strengths of the intermolecular attractive forces. These forces are also directly related to the properties of solids, such as melting point and heat of fusion. Let us preface our study of these condensed phases with a discussion of the types of attractive forces that can exist between molecules and ions.

*Inter*molecular forces refer to the forces *between* individual particles (atoms, molecules, ions) of a substance. These forces are quite weak relative to *intra*molecular forces, i.e., covalent and ionic bonds *within* compounds. For example, 920 kJ of energy is required to decompose one mole of water vapor into H and O atoms. This reflects the strength of intramolecular forces (chemical bonds). But only 40.7 kJ is required to convert one mole of liquid water into steam at 100°C. This reflects the strength of the intermolecular forces of attraction between the water molecules, mainly *hydrogen bonding.*

If it were not for the existence of intermolecular attractions, condensed phases (liquids and solids) could not exist. These are the forces that hold the particles close to one another in liquids and solids. As we shall see, the effects of these attractions on melting points of solids parallel those on boiling points of liquids. High boiling points are associated with compounds that have strong intermolecular attractions. Let us consider the effects of the general types of forces that exist among ionic, covalent, and monatomic species.

Ion–Ion Interactions

According to Coulomb's Law, the *force of attraction* between two oppositely charged ions is directly proportional to the charges on the ions, q^+ and q^-, and inversely proportional to the square of the distance between them, d.

$$F \propto \frac{q^+ q^-}{d^2}$$

Energy has the units of force × distance, $F \times d$, so the *energy of attraction* between two oppositely charged ions is directly proportional to the charges on the ions and inversely proportional to the distance of separation.

$$E \propto \frac{q^+ q^-}{d}$$

When oppositely charged ions are close together, d (the denominator) is small, so E, the energy of attraction, is large.

Ionic bonding may be thought of as both *inter-* and *intramolecular* bonding.

Ionic compounds such as NaCl, $CaBr_2$, and K_2SO_4 exist as extended arrays of discrete ions in the solid state. As we shall see in Section 13-16, the oppositely charged ions in these arrays are quite close together. As a result of these small distances, d, the energies of attraction in these solids are substantial. Most ionic bonding is strong, and as a result most ionic compounds have relatively high melting points (Table 13-2). At high enough temperatures, ionic solids melt as the added heat energy overcomes the potential energy associated with the attraction of oppositely charged ions. The ions in the resulting *molten* samples are free to move about, which accounts for the excellent electrical conductivity of molten ionic compounds.

For most substances, the liquid is less dense than the solid, but H_2O is one of the rare exceptions. Melting a solid nearly always produces greater average separations among the particles. This means that the forces (and energies) of attractions among the ions in an

Table 13-2 *Melting Points of Some Ionic Compounds*					
Compound	**mp (°C)**	**Compound**	**mp (°C)**	**Compound**	**mp (°C)**
NaF	993	CaF_2	1423	MgO	2800
NaCl	801	Na_2S	1180	CaO	2580
NaBr	747	K_2S	840	BaO	1923
KCl	770				

ionic liquid are less than in the solid state because the average d is greater in the melt. However, these energies of attraction are still much greater in magnitude than the energies of attraction among neutral species (molecules or atoms).

The product q^+q^- increases as the charges on ions increase. Ionic substances containing multiply charged ions such as Al^{3+}, Mg^{2+}, O^{2-}, and S^{2-} ions *usually* have higher melting and boiling points than ionic compounds containing only singly charged ions such as Na^+, K^+, F^-, and Cl^-. For a series of ions of similar charges, the closer approach of smaller ions results in stronger interionic attractive forces and higher melting points (compare NaF, NaCl, and NaBr in Table 13-2).

Dipole–Dipole Interactions

> Permanent dipole–dipole interactions occur between polar covalent molecules because of the attraction of the $\delta+$ atoms of one molecule to the $\delta-$ atoms of another molecule (Section 7-9).

Electrostatic forces between two ions decrease by the factor $1/d^2$ as their separation, d, increases. But dipole–dipole forces vary as $1/d^4$. Because of the higher power of d in the denominator, $1/d^4$ diminishes with increasing d much more rapidly than does $1/d^2$. Therefore, dipole forces are effective only over very short distances. Furthermore, for dipole–dipole forces, q^+ and q^- represent only "partial charges," so these forces are weaker than ion–ion forces. Average dipole–dipole interaction energies are approximately 4 kJ per mole of bonds. They are much weaker than ionic and covalent bonds, which have typical energies of about 400 kJ per mole of bonds. Substances in which permanent dipole–dipole interactions affect physical properties include bromine fluoride, BrF, and sulfur dioxide, SO_2. Dipole–dipole interactions are illustrated in Figure 13-3. All dipole–dipole interactions, including hydrogen bonding (discussed in the following section), are somewhat directional. An increase in temperature causes an increase in translational, rotational, and vibrational motion of molecules. This produces more randomness of orientation of molecules relative to each other. Consequently, the strength of dipole–dipole interactions decreases as temperature increases. All these factors make compounds having dipole–dipole interactions more volatile than ionic compounds.

Hydrogen Bonding

Hydrogen bonds are a special case of very strong dipole–dipole interaction. They are not really chemical bonds in the formal sense.

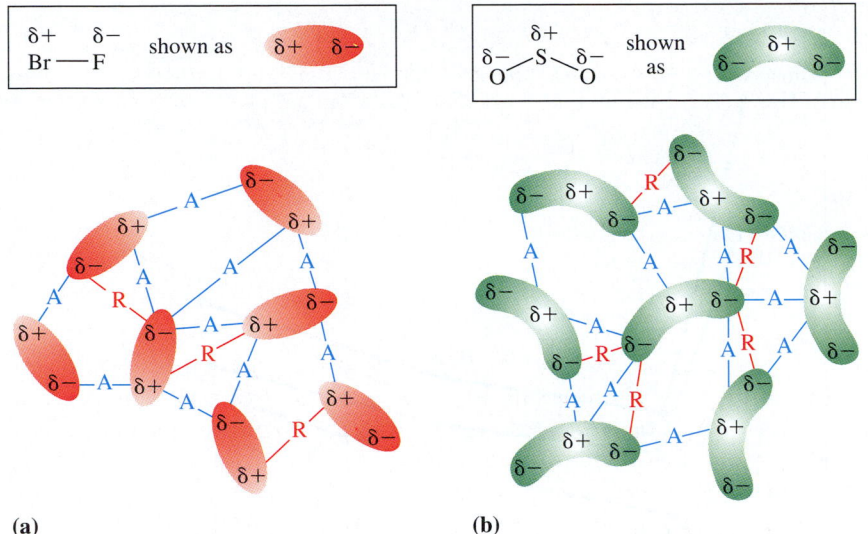

(a)   **(b)**

Figure 13-3 Dipole–dipole interactions among polar molecules. Each polar molecule is shaded with regions of highest negative charge ($\delta-$) darkest and regions of highest positive charge ($\delta+$) lightest. Attractive forces are shown as —A—, and repulsive forces are shown as —R—. Molecules tend to arrange themselves to maximize attractions by bringing regions of opposite charge together while minimizing repulsions by separating regions of like charge. (a) Bromine fluoride, BrF. (b) Sulfur dioxide, SO_2.

> Strong hydrogen bonding occurs among polar covalent molecules containing H and one of the three small, highly electronegative elements—F, O, or N.

Like ordinary dipole–dipole interactions, hydrogen bonds result from the attractions between $\delta+$ atoms of one molecule, in this case H atoms, and the $\delta-$ atoms of another molecule. The small sizes of the F, O, and N atoms, combined with their high electronegativities, concentrate the electrons of these molecules around these $\delta-$ atoms. This causes an H atom bonded to one of these quite electronegative elements to behave somewhat like a bare proton. The $\delta+$ H atom is attracted to an unshared pair of electrons on an F, O, or N atom other than the atom to which it is covalently bonded (Figure 13-4).

Typical hydrogen-bond energies are in the range 15 to 20 kJ/mol, which is four to five times greater than the energies of other dipole–dipole interactions. As a result, hydrogen

Most chemists usually restrict usage of the term "hydrogen bonding" to compounds in which H is covalently bonded to F, O, or N. Recently, careful studies of the arrangements of molecules in solids have led to the conclusion that the same kind of attraction occurs (although more weakly) when H is bonded to carbon, that is, very weak C—H---O "hydrogen bonds" exist. Similar observations suggest the existence of weak hydrogen bonds to chlorine atoms, e.g., O—H---Cl.

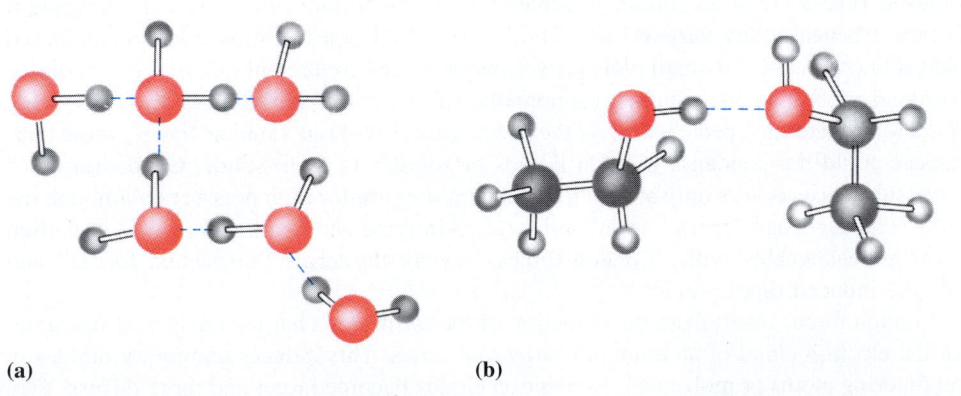

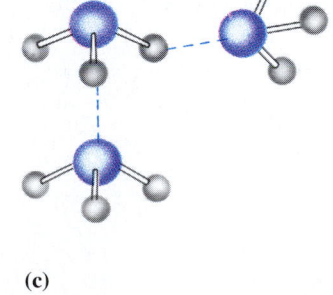

(a)   **(b)**   **(c)**

Figure 13-4 Hydrogen bonding (indicated by dashed lines) in (a) water, (b) ethyl alcohol, and (c) ammonia. Hydrogen bonding is a special case of very strong dipole–dipole interaction.

Figure 13-5 Boiling points of some hydrides as a function of molecular weight. The unusually high boiling points of NH_3, H_2O, and HF compared with those of other hydrides of the same groups are due to hydrogen bonding. The electronegativity difference between H and C is small and there are no unshared pairs on C; thus, CH_4 is not hydrogen bonded. Increasing molecular weight corresponds to increasing number of electrons; this makes the electron clouds easier to deform and causes increased London forces, accounting for the increase in boiling points for the nonhydrogen-bonded members of each series.

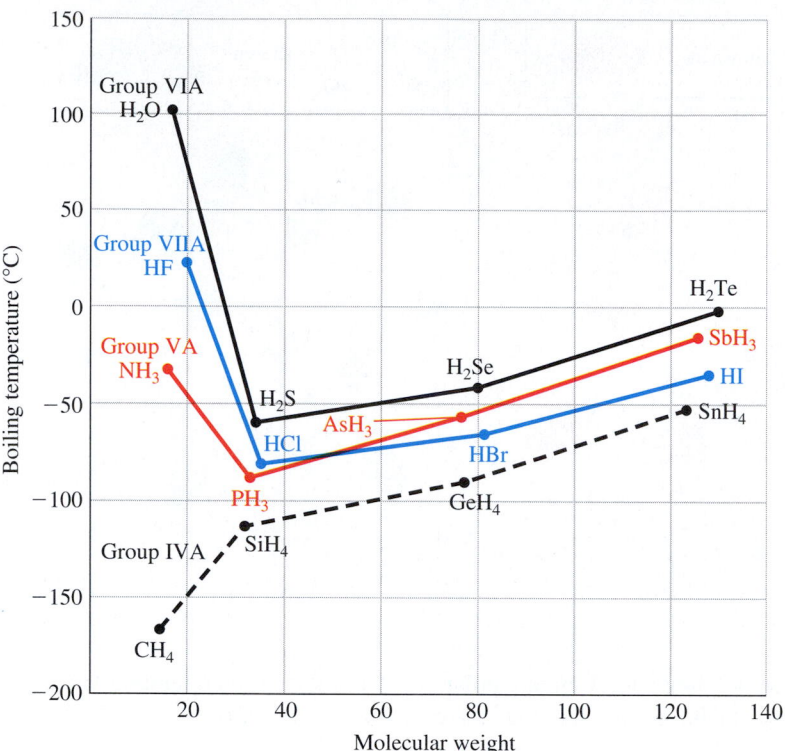

bonds exert a considerable influence on the properties of substances. Hydrogen bonding is responsible for the unusually high melting and boiling points of compounds such as water, ethyl alcohol, and ammonia compared to other compounds of similar molecular weight and molecular geometry (Figure 13-5). Hydrogen bonding between carboxylate groups ($—CO_2—$) and amino groups ($—NH_2$) of amino acid subunits is very important in establishing the three-dimensional structures of proteins.

London Forces

London forces are named after the German-born physicist Fritz London. He initially postulated their existence in 1930, on the basis of quantum theory.

London forces are weak attractive forces that are important only over *extremely* short distances because they vary as $1/d^7$. They exist for all types of molecules in condensed phases but are weak for small molecules. London forces are the only kind of intermolecular forces present among symmetrical nonpolar substances such as SO_3, CO_2, O_2, N_2, Br_2, H_2, and monatomic species such as the noble gases. Without London forces, these substances could not condense to form liquids or solidify to form solids. Condensation of some substances occurs only at very low temperatures and/or high pressures. Although the term "van der Waals forces" usually refers to all intermolecular attractions, it is also often used interchangeably with "London forces," as are the terms "dispersion forces" and "dipole-induced dipole forces."

London forces result from the attraction of the positively charged nucleus of one atom for the electron cloud of an atom in nearby molecules. This induces temporary dipoles in neighboring atoms or molecules. As electron clouds become larger and more diffuse, they are attracted less strongly by their own (positively charged) nuclei. Thus, they are more easily distorted, or *polarized,* by adjacent nuclei.

Polarizability increases with increasing sizes of molecules and therefore with increasing numbers of electrons. Therefore, London forces are generally stronger for molecules that are larger or have more electrons.

London forces are depicted in Figure 13-6. They exist in all substances.

Figure 13-5 shows that polar covalent compounds with hydrogen bonding (H_2O, HF, NH_3) boil at higher temperatures than analogous polar compounds without hydrogen bonding (H_2S, HCl, PH_3). Symmetrical, nonpolar compounds (CH_4, SiH_4) of comparable molecular weight boil at lower temperatures. In the absence of hydrogen bonding, boiling points of analogous substances (CH_4, SiH_4, GeH_4, SnH_4) increase fairly regularly with increasing number of electrons and molecular size (molecular weight). This is due to increasing effectiveness of London forces of attraction in the larger molecules and occurs even in the case of some polar covalent molecules. The increasing effectiveness of London forces, for example, accounts for the increase in boiling points in the sequences HCl < HBr < HI and H_2S < H_2Se < H_2Te, which involve nonhydrogen-bonded polar covalent molecules. The differences in electronegativities between hydrogen and other nonmetals *decrease* in these sequences, and the increasing London forces override the decreasing permanent dipole–dipole forces. Therefore, the *permanent* dipole–dipole interactions have very little effect on the boiling points of these compounds.

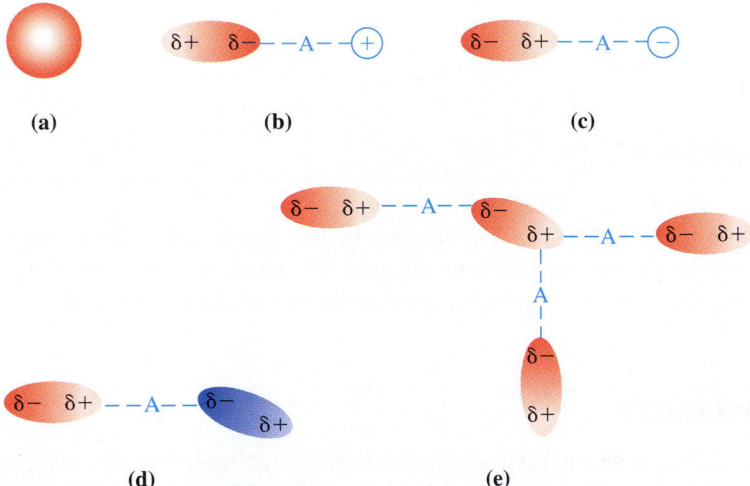

Figure 13-6 An illustration of how a temporary dipole can be induced in an atom. (a) An isolated argon atom, with spherical charge distribution (no dipole). (b) When a cation approaches the argon atom, the outer portion of the electron cloud is weakly attracted by the ion's positive charge. This induces a weak *temporary* dipole in the argon atom. (c) A temporary dipole can also be induced if the argon atom is approached by an anion. (d) The approach of a molecule with a permanent dipole (for instance, HF) could also temporarily polarize the argon atom. (e) Even in pure argon, the close approach of one argon atom to another results in temporary dipole formation in both atoms as each atom's electron cloud is attracted by the nucleus of the other atom or is repelled by the other atom's electron cloud. The resulting temporary dipoles cause weak attractions among the argon atoms. Molecules are even more easily polarized than isolated atoms.

Table 13-3 *Approximate Contributions to the Total Energy of Interaction Between Molecules, in kJ/mol*

Molecule	Permanent Dipole Moment (D)	Permanent Dipole–Dipole Energy	London Energy	Total Energy	Molar Heat of Vaporization (kJ/mol)
Ar	0	0	8.5	8.5	6.7
CO	0.1	~0	8.7	8.7	8.0
HCl	1.03	3.3	17.8	21	16.2
NH_3	1.47	13*	16.3	29	27.4
H_2O	1.85	36*	10.9	47	40.7

*Hydrogen-bonded.

Let us compare the magnitudes of the various contributions to the total energy of interactions in a group of simple molecules. Table 13-3 shows the permanent dipole moments and the energy contributions for five simple molecules. The contribution from London forces is substantial in all cases. The permanent dipole–dipole energy is greatest for substances in which hydrogen bonding occurs. The variations of these total energies of interaction are closely related to molar heats of vaporization. As we shall see in Section 13-9, the heat of vaporization measures the amount of energy required to overcome the attractive forces that hold the molecules together in a liquid.

Honey is a very viscous liquid.

OH OH OH
| | |
H—C—C—C—H
| | |
H   H   H

glycerine

THE LIQUID STATE

We shall briefly describe several properties of the liquid state. These properties vary markedly among various liquids, depending on the nature and strength of the attractive forces among the particles (atoms, molecules, ions) making up the liquid.

13-3 VISCOSITY

Viscosity is the resistance to flow of a liquid. Honey has a high viscosity at room temperature, and freely flowing gasoline has a low viscosity. The viscosity of a liquid can be measured with a viscometer such as the one in Figure 13-7.

For a liquid to flow, the molecules must be able to slide past one another. In general, the stronger the intermolecular forces of attraction, the more viscous the liquid is. Substances that have a great ability to form hydrogen bonds, especially involving several hydrogen-bonding sites per molecule, such as glycerine (margin), usually have high viscosities. Increasing the size and surface area of molecules generally results in increased viscosity, due to the increased London forces. For example, the shorter-chain hydrocarbon pentane (a free-flowing liquid at room temperature) is less viscous than dodecane (an oily liquid at room temperature). The longer the molecules are, the more they can get "tangled up" in the liquid, and the harder it is for them to flow.

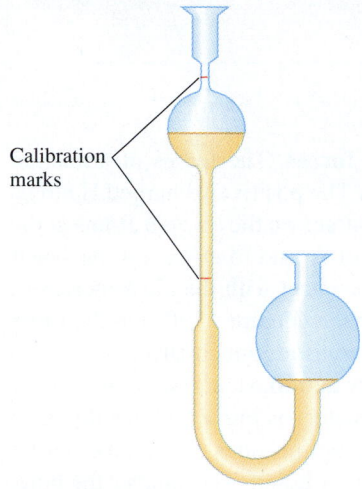

Calibration
marks

Figure 13-7 The Ostwald viscometer, a device used to measure viscosity of liquids. The time it takes for a known volume of a liquid to flow through a small neck of known size is measured. Liquids with low viscosities flow rapidly.

pentane, C_5H_{12}
viscosity = 0.24 centipoise

$$\begin{array}{c} \text{H} & \text{H} & \text{H} & \text{H} & \text{H} \\ | & | & | & | & | \\ \text{H}-\text{C}-\text{C}-\text{C}-\text{C}-\text{C}-\text{H} \\ | & | & | & | & | \\ \text{H} & \text{H} & \text{H} & \text{H} & \text{H} \end{array}$$

dodecane, $C_{12}H_{26}$
viscosity = 1.35 centipoise

$$\begin{array}{c} \text{H} & \text{H} & \text{H} & \text{H} & \text{H} & \text{H} & \text{H} & \text{H} & \text{H} & \text{H} & \text{H} & \text{H} \\ | & | & | & | & | & | & | & | & | & | & | & | \\ \text{H}-\text{C}-\text{C}-\text{C}-\text{C}-\text{C}-\text{C}-\text{C}-\text{C}-\text{C}-\text{C}-\text{C}-\text{C}-\text{H} \\ | & | & | & | & | & | & | & | & | & | & | & | \\ \text{H} & \text{H} & \text{H} & \text{H} & \text{H} & \text{H} & \text{H} & \text{H} & \text{H} & \text{H} & \text{H} & \text{H} \end{array}$$

The *poise* is the unit used to express viscosity. The viscosity of water at 25°C is 0.89 centipoise.

As temperature increases and the molecules move more rapidly, their kinetic energies are better able to overcome intermolecular attractions. Thus, viscosity decreases with increasing temperature, as long as no changes in composition occur.

13-4 SURFACE TENSION

Molecules below the surface of a liquid are influenced by intermolecular attractions from all directions. Those on the surface are attracted only toward the interior (Figure 13-8); these attractions pull the surface layer toward the center. The most stable situation is one in which the surface area is minimal. For a given volume, a sphere has the least possible surface area, so drops of liquid tend to assume spherical shapes. **Surface tension** is a measure of the inward forces that must be overcome to expand the surface area of a liquid.

The shapes of soap bubbles are due to surface tension, an important physical property of liquids. White light striking the bubbles gives brightly colored interference patterns.

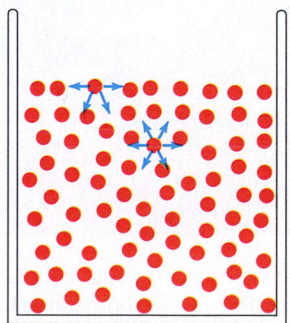

Figure 13-8 A molecular-level view of the attractive forces experienced by molecules at and below the surface of a liquid.

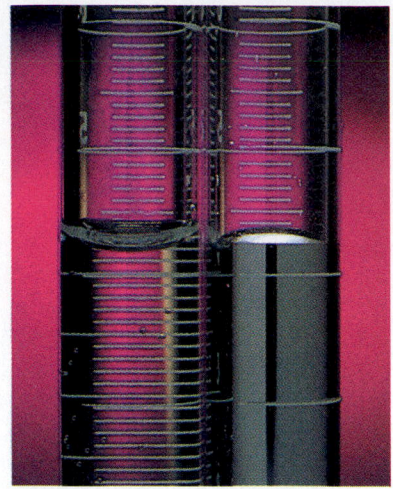

Figure 13-9 The meniscus, as observed in glass tubes with water and with mercury.

The surface tension of water supports this water strider. The nonpolar surfaces of its feet also help to repel the water.

Droplets of mercury lying on a glass surface. The small droplets are almost spherical, whereas the larger droplets are flattened due to the effects of gravity. This shows that surface tension has more influence on the shape of the small (lighter) droplets.

13-5 CAPILLARY ACTION

All forces holding a liquid together are called **cohesive forces.** The forces of attraction between a liquid and another surface are **adhesive forces.** The positively charged H atoms of water hydrogen bond strongly to the partial negative charges on the oxygen atoms at the surface of the glass. As a result, water *adheres* to glass, or is said to *wet* glass. As water creeps up the side of the glass tube, its favorable area of contact with the glass increases. The surface of the water, its **meniscus,** has a concave shape (Figure 13-9). On the other hand, mercury does not wet glass because its cohesive forces are much stronger than its attraction to glass. Thus, its meniscus is convex. **Capillary action** occurs when one end of a capillary tube, a glass tube with a small bore (inside diameter), is immersed in a liquid. If adhesive forces exceed cohesive forces, the liquid creeps up the sides of the tube until a balance is reached between adhesive forces and the weight of liquid. The smaller the bore, the higher the liquid climbs. Capillary action helps plant roots take up water and dissolved nutrients from the soil and transmit them up the stems. The roots, like glass, exhibit strong adhesive forces for water. Osmotic pressure (Section 14-15) also plays a major role in this process.

13-6 EVAPORATION

Evaporation, or **vaporization,** is the process by which molecules on the surface of a liquid break away and go into the gas phase (Figure 13-10). Kinetic energies of molecules in liquids depend on temperature in the same way as they do in gases. The distribution of kinetic energies among liquid molecules at two different temperatures is shown in Figure 13-11. To break away, the molecules must possess at least some minimum kinetic energy. Figure 13-11 shows that at a higher temperature, a greater fraction of molecules possess at least that minimum energy. The rate of evaporation increases as temperature increases.

Only the higher-energy molecules can escape from the liquid phase. The average molecular kinetic energy of the molecules remaining in the liquid state is thereby lowered, resulting in a lower temperature in the liquid. The liquid would then be cooler than its surroundings, so it absorbs heat from its surroundings. The cooling of your body by evaporation of perspiration is a familiar example of the cooling of the surroundings by evaporation of a liquid. This is called "cooling by evaporation."

Coating glass with a silicone polymer greatly reduces the adhesion of water to the glass. The left side of each glass has been treated with Rain-X, which contains a silicone polymer. Water on the treated side forms droplets that are easily swept away.

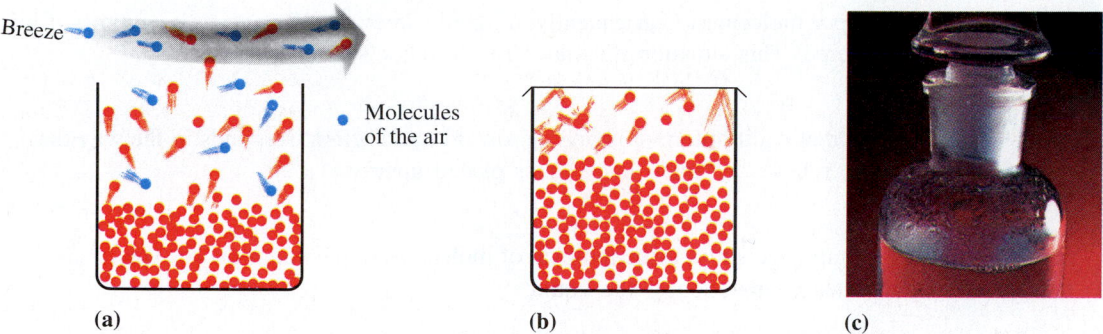

Breeze

Molecules of the air

(a)   (b)   (c)

Figure 13-10   (a) Liquid continuously evaporates from an open vessel. (b) Equilibrium between liquid and vapor is established in a closed container in which molecules return to the liquid at the same rate as they leave it. (c) A bottle in which liquid–vapor equilibrium has been established. Note that droplets have condensed.

A molecule in the vapor may strike the liquid surface and be captured there. This process, the reverse of evaporation, is called **condensation.** As evaporation occurs in a closed container, the volume of liquid decreases and the number of gas molecules above the surface increases. Because more gas phase molecules can collide with the surface, the rate of condensation increases. The system composed of the liquid and gas molecules of the same substance eventually achieves a **dynamic equilibrium** in which the rate of evaporation equals the rate of condensation in the closed container.

$$\text{liquid} \underset{\text{condensation}}{\overset{\text{evaporation}}{\rightleftharpoons}} \text{vapor}$$

The two opposing rates are not zero, but are equal to one another—hence we call this "dynamic," rather than "static," equilibrium. Even though evaporation and condensation are both continuously occurring, *no net change occurs* because the rates are equal.

However, if the vessel were left open to the air, this equilibrium could not be reached. Molecules would diffuse away and slight air currents would also sweep some gas molecules away from the liquid surface. This would allow more evaporation to occur to replace

As an analogy, suppose that 45 students per minute leave a classroom, moving into the closed hallway outside, and 45 students per minute enter it. The total number of students in the room would remain constant, as would the total number of students outside the room.

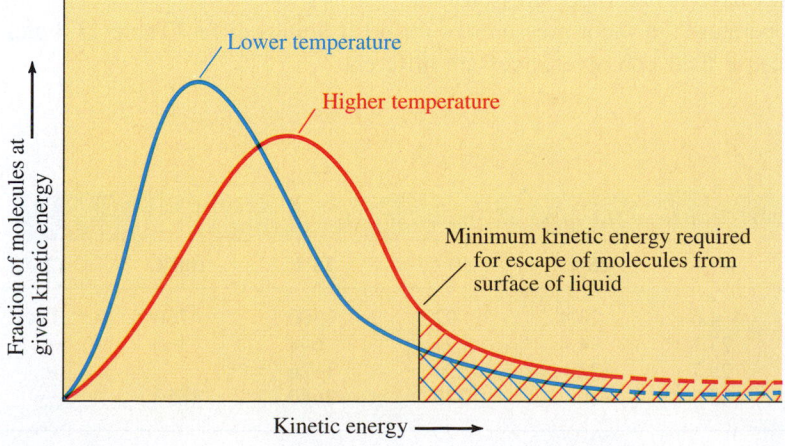

Lower temperature

Higher temperature

Minimum kinetic energy required for escape of molecules from surface of liquid

Fraction of molecules at given kinetic energy

Kinetic energy

Figure 13-11   Distribution of kinetic energies of molecules in a liquid at different temperatures. At the lower temperature, a smaller fraction of the molecules have the energy required to escape from the liquid, so evaporation is slower and the equilibrium vapor pressure (Section 13-7) is lower.

the lost vapor molecules. Consequently, a liquid can eventually evaporate entirely if it is left uncovered. This situation illustrates **LeChatelier's Principle:**

> A system at equilibrium, or changing toward equilibrium, responds in the way that tends to relieve or "undo" any stress placed upon it.

In this example the stress is the removal of molecules in the vapor phase. The response is the continued evaporation of the liquid.

This is one of the guiding principles that allows us to understand chemical equilibrium. It is discussed further in Chapter 17.

13-7 VAPOR PRESSURE

Vapor molecules cannot escape when vaporization of a liquid occurs in a closed container. As more molecules leave the liquid, more gaseous molecules collide with the walls of the container, with each other, and with the liquid surface, so more condensation occurs. This is responsible for the formation of liquid droplets that adhere to the sides of the vessel above a liquid surface and for the eventual establishment of equilibrium between liquid and vapor (Figure 13-10b and c).

> The partial pressure of vapor molecules above the surface of a liquid at equilibrium at a given temperature is the **vapor pressure (vp)** of the liquid at that temperature. Because the rate of evaporation increases and the rate of condensation decreases with increasing temperature, vapor pressures of liquids *always* increase as temperature increases.

As long as some liquid remains in contact with the vapor, the pressure does not depend on the volume or surface area of the liquid.

Easily vaporized liquids are called **volatile** liquids, and they have relatively high vapor pressures. The most volatile liquid in Table 13-4 is diethyl ether. Water is the least volatile. Vapor pressures can be measured with manometers (Figure 13-12).

Stronger cohesive forces tend to hold molecules in the liquid state. Methyl alcohol molecules are strongly linked by hydrogen bonding, whereas diethyl ether molecules are not, so methyl alcohol has a lower vapor pressure than diethyl ether. The very strong hydrogen bonding in water accounts for its unusually low vapor pressure (Table 13-4). London forces generally increase with increasing molecular size, so substances composed of larger molecules have lower vapor pressures.

At a given temperature the vapor pressures of different liquids differ (Table 13-4 and Figure 13-13) because their cohesive forces are different.

Table 13-4 *Vapor Pressures (in torr) of Some Liquids*						
	0°C	**25°C**	**50°C**	**75°C**	**100°C**	**125°C**
water	4.6	23.8	92.5	300	760	1741
benzene	27.1	94.4	271	644	1360	
methyl alcohol	29.7	122	404	1126		
diethyl ether	185	470	1325	2680	4859	

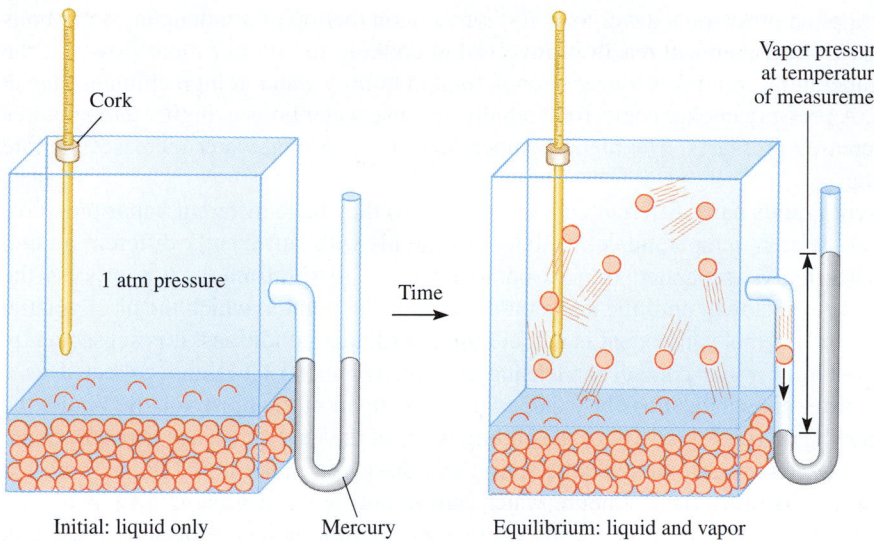

Figure 13-12 A simplified representation of the measurement of vapor pressure of a liquid at a given temperature. At the instant the liquid is added to the container, the space above the liquid is occupied by air only. Some of the liquid then vaporizes until equilibrium is established. The increase in height of the mercury column is a measure of the vapor pressure of the liquid at that temperature.

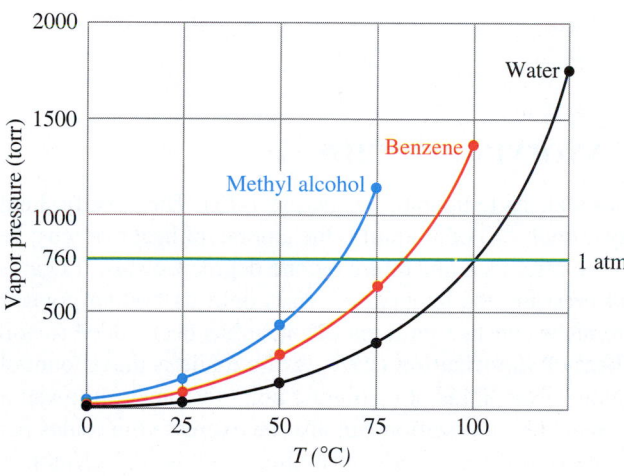

Figure 13-13 Plots of the vapor pressures of three of the liquids in Table 13-4. The *normal* boiling point of a liquid is the temperature at which its vapor pressure is equal to one atmosphere. Normal boiling points are: water, 100°C, benzene, 80.1°C; and methyl alcohol, 65.0°C. Notice that the increase in vapor pressure is *not* linear with temperature.

13-8 BOILING POINTS AND DISTILLATION

When heat energy is added to a liquid, it increases the kinetic energy of the molecules, and the temperature of the liquid increases. Heating a liquid always increases its vapor pressure. When a liquid is heated to a sufficiently high temperature under a given applied (usually atmospheric) pressure, bubbles of vapor begin to form below the surface. They rise to the surface and burst, releasing the vapor into the air. This process is called *boiling* and is different from evaporation. If the vapor pressure inside the bubbles is less than the applied pressure on the surface of the liquid, the bubbles collapse as soon as they form and no boiling occurs. The **boiling point** of a liquid is the temperature at which its vapor pressure equals the external pressure. The **normal boiling point** is the temperature at which the vapor pressure of a liquid is equal to exactly one atmosphere (760 torr). The vapor pressure of water is 760 torr at 100°C, its normal boiling point. As heat energy is added to a pure liquid *at its boiling point,* the temperature remains constant, because the energy is used to overcome the cohesive forces in the liquid to form vapor.

As water is being heated, but before it boils, small bubbles may appear in the container. This is not boiling, but rather the formation of bubbles of dissolved gases such as CO_2 and O_2, whose solubilities in water decrease with increasing temperature.

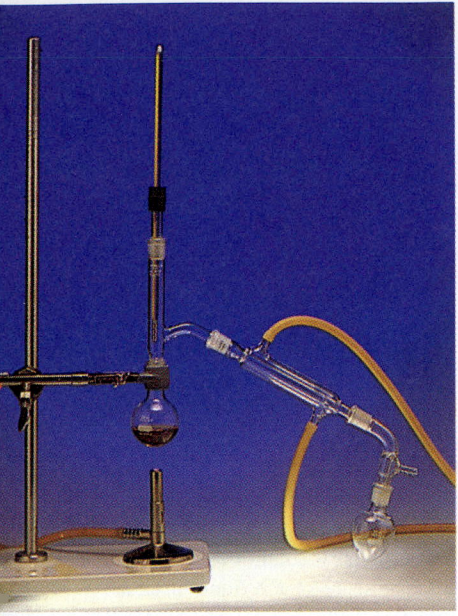

Figure 13-14 A laboratory setup for distillation. During distillation of an impure liquid, nonvolatile substances remain in the distilling flask. The liquid is vaporized and condensed before being collected in the receiving flask.

The specific heat and heat capacity of a substance change somewhat with its temperature. For most substances, this variation is small enough to ignore.

Molar heats of vaporization are often expressed in kilojoules rather than joules. The units of heat of vaporization do *not* include temperature. This is because boiling occurs with *no change in temperature.*

Heat of vaporization, like other heats of transition, is an *extensive* property of a substance, because the amount of heat absorbed depends on the amount of liquid that vaporizes.

If the applied pressure is lower than 760 torr, say on the top of a mountain, water boils below 100°C. The chemical reactions involved in cooking food occur more slowly at the lower temperature, so it takes longer to cook food in boiling water at high altitudes than at sea level. A pressure cooker cooks food rapidly because water boils at higher temperatures under increased pressures. The higher temperature of the boiling water increases the rate of cooking.

Different liquids have different cohesive forces, so they have different vapor pressures and boil at different temperatures. A mixture of liquids with sufficiently different boiling points can often be separated into its components by **distillation.** In this process the mixture is heated slowly until the temperature reaches the point at which the most volatile liquid boils off. If this component is a liquid under ordinary conditions, it is subsequently recondensed in a water-cooled condensing column (Figure 13-14) and collected as a distillate. After enough heat has been added to vaporize all of the most volatile liquid, the temperature again rises slowly until the boiling point of the next substance is reached, and the process continues. Any nonvolatile substances dissolved in the liquid do not boil, but remain in the distilling flask. Impure water can be purified and separated from its dissolved salts by distillation. Compounds with similar boiling points, especially those that interact very strongly with each other, are not effectively separated by simple distillation but require a modification called fractional distillation (Section 14-10).

13-9 HEAT TRANSFER INVOLVING LIQUIDS

Heat must be added to a liquid to raise its temperature (Section 1-13). The **specific heat** (J/g · °C) or **molar heat capacity** (J/mol · °C) of a liquid is the amount of heat that must be added to the stated mass of liquid to raise its temperature by one degree Celsius. If heat is added to a liquid under constant pressure, the temperature rises until its boiling point is reached. Then the temperature remains constant until enough heat has been added to boil away all the liquid. The **molar heat of vaporization (ΔH_{vap})** of a liquid is the amount of heat that must be added to one mole of the liquid at its boiling point to convert it to vapor with no change in temperature. Heats of vaporization can also be expressed in joules per gram. For example, the heat of vaporization for water at its boiling point is 40.7 kJ/mol, or 2.26×10^3 J/g.

$$\frac{? \text{ J}}{\text{g}} = \frac{40.7 \text{ kJ}}{\text{mol}} \times \frac{1000 \text{ J}}{\text{kJ}} \times \frac{1 \text{ mol}}{18.0 \text{ g}} = 2.26 \times 10^3 \text{ J/g}$$

Like many other properties of liquids, heats of vaporization reflect the strengths of intermolecular forces. Heats of vaporization generally increase as boiling points and intermolecular forces increase and as vapor pressures decrease. Table 13-5 illustrates this. The high heats of vaporization of water, ethylene glycol, and ethyl alcohol are due mainly to the strong hydrogen-bonding interactions in these liquids (Section 13-2). The very high value for water makes it very effective as a coolant and, in the form of steam, as a source of heat.

Liquids can evaporate even below their boiling points. The water in perspiration is an effective coolant for our bodies. As it evaporates, each gram absorbs 2.41 kJ of heat from the body and carries it into the air in the form of water vapor. We feel even cooler in a breeze because perspiration evaporates faster, so heat is removed more rapidly.

Table 13-5 *Heats of Vaporization, Boiling Points, and Vapor Pressures of Some Common Liquids*

Liquid	Vapor Pressure (torr at 20°C)	Boiling Point at 1 atm (°C)	Heat of Vaporization at Boiling Point	
			J/g	kJ/mol
water (MW 18.0)	17.5	100	2260	40.7
ethyl alcohol (MW 46.1)	43.9	78.3	858	39.3
benzene (MW 78.1)	74.6	80.1	395	30.8
diethyl ether (MW 74.1)	442	34.6	351	26.0
carbon tetrachloride (MW 153.8)	85.6	76.8	213	32.8
ethylene glycol (MW 67.1)	0.1	197.3	984	58.9

The heat of vaporization of water is higher at 37°C (normal body temperature) than at 100°C (2.41 kJ/g compared to 2.26 kJ/g).

Condensation is the reverse of evaporation. The amount of heat that must be removed from a vapor to condense it (without change in temperature) is called the **heat of condensation.**

$$\text{liquid} + \text{heat} \underset{\text{condensation}}{\overset{\text{evaporation}}{\rightleftharpoons}} \text{vapor}$$

The heat of condensation of a liquid is equal in magnitude, but opposite in sign, to the heat of vaporization. It is released by the vapor during condensation.

Because 2.26 kJ must be absorbed to vaporize one gram of water at 100°C, that same amount of heat must be released to the environment when one gram of steam at 100°C condenses to form liquid water at 100°C. In steam-heated radiators, steam condenses and releases 2.26 kJ of heat per gram as its molecules collide with the cooler radiator walls and condense there. The metallic walls conduct heat well. They transfer the heat to the air in contact with the outside walls of the radiator. The heats of condensation and vaporization of nonhydrogen-bonded liquids, such as benzene, have smaller magnitudes than those of hydrogen-bonded liquids (Table 13-5). Therefore, they are much less effective as heating and cooling agents.

Because of the large amount of heat released by steam as it condenses, burns caused by steam at 100°C are much more severe than burns caused by liquid water at 100°C.

EXAMPLE 13-1 *Heat of Vaporization*

Calculate the amount of heat, in joules, required to convert 180 grams of water at 10.0°C to steam at 105.0°C.

Plan

The total amount of heat absorbed is the sum of the amounts required to (1) warm the liquid water from 10.0°C to 100.0°C, (2) convert the liquid water to steam at 100°C, and (3) warm the steam from 100.0°C to 105.0°C. Steps 1 and 3 involve the specific heats of water and steam, 4.18 J/g·°C and 2.03 J/g·°C, respectively (Appendix E), whereas step 2 involves the heat of vaporization of water $(2.26 \times 10^3 \text{ J/g})$.

Steps 1 and 3 of this example involve warming with no *phase change. Such calculations were introduced in Section 1-13.*

Solution

1. $\underline{?} \text{ J} = 180 \text{ g} \times \dfrac{4.18 \text{ J}}{\text{g} \cdot °\text{C}} \times (100.0°\text{C} - 10.0°\text{C}) = 6.77 \times 10^4 \text{ J} = 0.677 \times 10^5 \text{ J}$

2. $\underline{?}$ J $= 180$ g $\times \dfrac{2.26 \times 10^3 \text{ J}}{\text{g}}$ $= 4.07 \times 10^5$ J

3. $\underline{?}$ J $= 180$ g $\times \dfrac{2.03 \text{ J}}{\text{g} \cdot {}^\circ\text{C}} \times (105.0{}^\circ\text{C} - 100.0{}^\circ\text{C}) = 1.8 \times 10^3$ J $= 0.018 \times 10^5$ J

Total amount of heat absorbed $= \boxed{4.76 \times 10^5 \text{ J}}$

You should now work Exercise 47.

EXAMPLE 13-2   *Heat of Vaporization*

Compare the amount of "cooling" experienced by an individual who drinks 400 mL of ice water (0°C) with the amount of "cooling" experienced by an individual who "sweats out" 400 mL of water. Assume that all the sweat evaporates. The density of water is very nearly 1.00 g/mL at both 0°C and 37.0°C, average body temperature. The heat of vaporization of water is 2.41 kJ/g at 37.0°C.

Plan

In the case of drinking ice water, the body is cooled by the amount of heat required to raise the temperature of 400 mL (400 g) of water from 0.0°C to 37.0°C. The amount of heat lost by perspiration is equal to the amount of heat required to vaporize 400 g of water at 37.0°C.

Solution

Raising the temperature of 400 g of water from 0.0°C to 37.0°C requires

$$\underline{?}\text{ J} = (400\text{ g})(4.18\text{ J/g}\cdot{}^\circ\text{C})(37.0{}^\circ\text{C}) = 6.19 \times 10^4 \text{ J, or } \boxed{61.9\text{ kJ}}$$

Evaporating (i.e., "sweating out") 400 mL of water at 37°C requires

$$\underline{?}\text{ J} = (400\text{ g})(2.41 \times 10^3 \text{ J/g}) = 9.64 \times 10^5 \text{ J, or } \boxed{964\text{ kJ}}$$

For health reasons, it is important to replace the water lost by perspiration.

Thus, we see that "sweating out" 400 mL of water removes 964 kJ of heat from one's body, whereas drinking 400 mL of ice water cools it by only 61.9 kJ. Stated differently, sweating removes $(964/61.9) = 15.6$ times more heat than drinking ice water!

You should now work Exercise 51.

 **E N R I C H M E N T**

The Clausius–Clapeyron equation is used for three types of calculations: (1) to predict the vapor pressure of a liquid at a specified temperature, as in Example 13-3; (2) to determine the temperature at which a liquid has a specified vapor pressure; and (3) to calculate ΔH_{vap} from measurement of vapor pressures at different temperatures.

The Clausius–Clapeyron Equation

We have seen (Figure 13-13) that vapor pressure increases with increasing temperature. Let us now discuss the quantitative expression of this relationship.

When the temperature of a liquid is changed from T_1 to T_2, the vapor pressure of the liquid changes from P_1 to P_2. These changes are related to the molar heat of vaporization, ΔH_{vap}, for the liquid by the **Clausius–Clapeyron equation.**

$$\ln\left(\frac{P_2}{P_1}\right) = \frac{\Delta H_{vap}}{R}\left(\frac{1}{T_1} - \frac{1}{T_2}\right) \quad \text{or} \quad \log\left(\frac{P_2}{P_1}\right) = \frac{\Delta H_{vap}}{2.303R}\left(\frac{1}{T_1} - \frac{1}{T_2}\right)$$

Although ΔH_{vap} changes somewhat with temperature, it is usually adequate to use the value tabulated at the normal boiling point of the liquid (Appendix E) unless more precise values are available. The units of R must be consistent with those of ΔH_{vap}.

EXAMPLE 13-3   *Vapor Pressure versus Temperature*

The normal boiling point of ethanol, C_2H_5OH, is 78.3°C and its molar heat of vaporization is 39.3 kJ/mol (Appendix E). What would be the vapor pressure, in torr, of ethanol at 50.0°C?

Plan

The normal boiling point of a liquid is the temperature at which its vapor pressure is 760 torr, so we designate this as one of the conditions (subscript 1). We wish to find the vapor pressure at another temperature (subscript 2), and we know the molar heat of vaporization. We use the Clausius–Clapeyron equation to solve for P_2.

Solution

$$P_1 = 760 \text{ torr} \quad \text{at} \quad T_1 = 78.3°C + 273.2 = 351.5 \text{ K}$$
$$P_2 = \underline{?} \quad \text{at} \quad T_2 = 50.0°C + 273.2 = 323.2 \text{ K}$$
$$\Delta H_{vap} = 39.3 \text{ kJ/mol} \quad \text{or} \quad 3.93 \times 10^4 \text{ J/mol}$$

We solve for P_2.

$$\ln\left(\frac{P_2}{760 \text{ torr}}\right) = \frac{3.93 \times 10^4 \text{ J/mol}}{\left(8.314 \dfrac{\text{J}}{\text{mol} \cdot \text{K}}\right)}\left(\frac{1}{351.5 \text{ K}} - \frac{1}{323.2 \text{ K}}\right)$$

$$\ln\left(\frac{P_2}{760 \text{ torr}}\right) = -1.18 \quad \text{so} \quad \left(\frac{P_2}{760 \text{ torr}}\right) = e^{-1.18} = 0.307$$

$$P_2 = 0.307(760 \text{ torr}) = \boxed{233 \text{ torr}} \quad \begin{array}{l}\text{(lower vapor pressure} \\ \text{at lower temperature)}\end{array}$$

You should now work Exercises 36 and 37.

The normal boiling point of acetone is 56.2°C, and its ΔH_{vap} is 32.0 kJ/mol. It boils at lower temperatures under reduced pressure. Can you use the Clausius–Clapeyron equation to calculate the pressure at which acetone boils at 14°C, as in this photograph?

We have described many properties of liquids and discussed how they depend on intermolecular forces of attraction. The general effects of these attractions on the physical properties of liquids are summarized in Table 13-6. "High" and "low" are relative terms. Table 13-6 is intended to show only very general trends. Example 13-4 illustrates the use of intermolecular attractions to predict boiling points.

Table 13-6 *General Effects of Intermolecular Attractions on Physical Properties of Liquids*

Property	Volatile Liquids (weak intermolecular attractions)	Nonvolatile Liquids (strong intermolecular attractions)
Cohesive forces	Low	High
Viscosity	Low	High
Surface tension	Low	High
Specific heat	Low	High
Vapor pressure	High	Low
Rate of evaporation	High	Low
Boiling point	Low	High
Heat of vaporization	Low	High

EXAMPLE 13-4 *Boiling Points versus Intermolecular Forces*

Predict the order of increasing boiling points for the following: H_2S; H_2O; CH_4; H_2; KBr.

Plan

We analyze the polarity and size of each substance to determine the kinds of intermolecular forces that are present. In general the stronger the intermolecular forces, the higher the boiling point of the substance.

Solution

KBr is ionic, so it boils at the highest temperature. Water exhibits hydrogen bonding and boils at the next highest temperature. Hydrogen sulfide is the only other polar covalent substance in the list, so it boils below H_2O but above the other two substances. Both CH_4 and H_2 are nonpolar. The larger CH_4 molecule is more easily polarized than the very small H_2 molecule, so London forces are stronger in CH_4. Thus, CH_4 boils at a higher temperature than H_2.

$$H_2 < CH_4 < H_2S < H_2O < KBr$$
increasing boiling points →

You should now work Exercise 20.

THE SOLID STATE

13-10 MELTING POINT

The **melting point (freezing point)** of a substance is the temperature at which its solid and liquid phases coexist in equilibrium.

$$\text{liquid} \underset{\text{melting}}{\overset{\text{freezing}}{\rightleftharpoons}} \text{solid}$$

The *melting point* of a solid is the same as the *freezing point* of its liquid. It is the temperature at which the rate of melting of a solid is the same as the rate of freezing of its liquid under a given applied pressure.

The **normal melting point** of a substance is its melting point at one atmosphere pressure. Changes in pressure have very small effects on melting points; they have large effects on boiling points.

13-11 HEAT TRANSFER INVOLVING SOLIDS

When heat is added to a solid below its melting point, its temperature rises. After enough heat has been added to bring the solid to its melting point, additional heat is required to convert the solid to liquid. During this melting process, the temperature remains constant at the melting point until all of the substance has melted. After melting is complete, the continued addition of heat results in an increase in the temperature of the liquid, until the boiling point is reached. This is illustrated graphically in the first three segments of the heating curve in Figure 13-15.

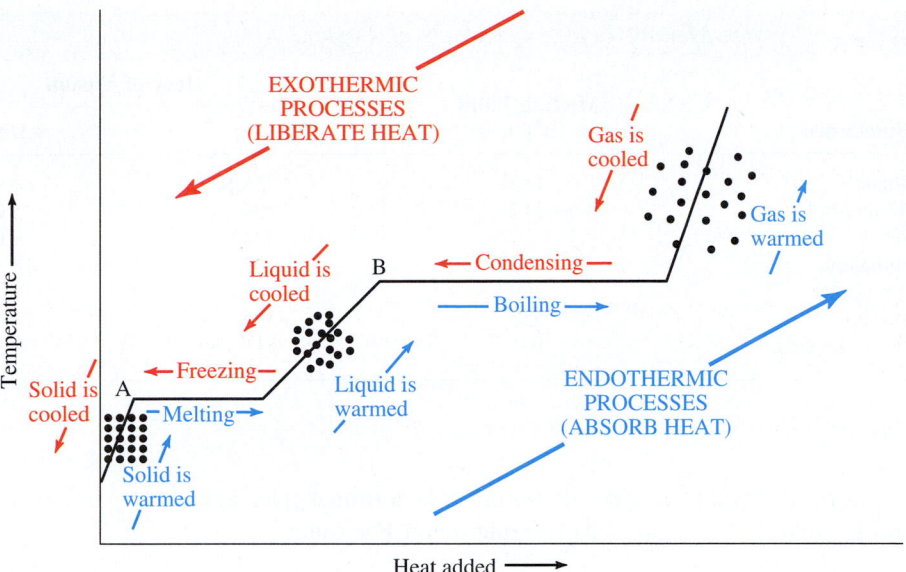

Figure 13-15 A typical heating curve at constant pressure. When heat energy is added to a solid below its melting point, the temperature of the solid rises until its melting point is reached (point A). In this region of the plot the slope is rather steep because of the low specific heats of solids [e.g., 2.09 J/g · °C for $H_2O(s)$]. If the solid is heated at its melting point (A), its temperature remains constant until the solid has melted, because the melting process requires energy. The length of this horizontal line is proportional to the heat of fusion of the substance—the higher the heat of fusion, the longer the line. When all of the solid has melted, heating the liquid raises its temperature until its boiling point is reached (point B). The slope of this line is less steep than that for warming the solid, because the specific heat of the liquid phase [e.g., 4.18 J/g · °C for $H_2O(\ell)$] is usually greater than that of the corresponding solid. If heat is added to the liquid at its boiling point (B), the added heat energy is absorbed as the liquid boils. This horizontal line is longer than the previous one, because the heat of vaporization of a substance is always higher than its heat of fusion. When all of the liquid has been converted to a gas (vapor), the addition of more heat raises the temperature of the gas. This segment of the plot has a steep slope because of the relatively low specific heat of the gas phase [e.g., 2.03 J/g · °C for $H_2O(g)$]. Each step in the process can be reversed by removing the same amount of heat.

The **molar heat of fusion (ΔH_{fus}; kJ/mol)** is the amount of heat required to melt one mole of a solid at its melting point. Heats of fusion can also be expressed on a per gram basis. The heat of fusion depends on the *inter*molecular forces of attraction in the solid state. These forces "hold the molecules together" as a solid. Heats of fusion are *usually* higher for substances with higher melting points. Values for some common compounds are shown in Table 13-7. Appendix E has more values.

The **heat of solidification, ΔH_{sol},** of a liquid is equal in magnitude, but opposite in sign, to the heat of fusion, ΔH_{fus}.

$$\Delta H_{solidification} = -\Delta H_{fusion}$$

Melting is always endothermic. The term "fusion" means "melting."

| | Melting Point | Heat of Fusion | |
Substance	(°C)	J/g	kJ/mol
methane	−182	58.6	0.92
ethyl alcohol	−117	109	5.02
water	0	334	6.02
naphthalene	80.2	147	18.8
silver nitrate	209	67.8	11.5
aluminum	658	395	10.6
sodium chloride	801	519	30.3

Table 13-7 *Some Melting Points and Heats of Fusion*

It represents removal of a sufficient amount of heat from a given amount (1 g or 1 mol) of liquid to solidify the liquid at its freezing point. For water,

$$\text{water} \underset{\substack{+6.02\ \text{kJ/mol} \\ \text{or } +334\ \text{J/g}}}{\overset{\substack{-6.02\ \text{kJ/mol} \\ \text{or } -334\ \text{J/g}}}{\rightleftharpoons}} \text{ice} \qquad (\text{at } 0°C)$$

EXAMPLE 13-5   *Heat of Fusion*

The molar heat of fusion, ΔH_{fus}, of Na is 2.6 kJ/mol at its melting point, 97.5°C. How much heat must be absorbed by 5.0 g of solid Na at 97.5°C to melt it?

Plan

The molar heat of fusion tells us that every mole of Na, 23 grams, absorbs 2.6 kJ of heat at 97.5°C during the melting process. We want to know the amount of heat that 5.0 grams would absorb. We use the appropriate unit factors, constructed from the atomic weight and ΔH_{fus}, to find the amount of heat absorbed.

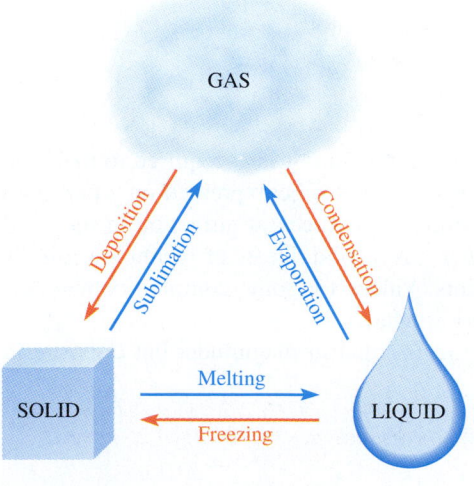

Transitions among the three states of matter. The transitions shown in blue are endothermic (absorb heat); those shown in red are exothermic (release heat).

Solution

$$\underline{?}\ kJ = 5.0\ g\ Na \times \frac{1\ mol\ Na}{23\ g\ Na} \times \frac{2.6\ kJ}{1\ mol\ Na} = \boxed{0.57\ kJ}$$

EXAMPLE 13-6 *Heat of Fusion*

Calculate the amount of heat that must be absorbed by 50.0 grams of ice at $-12.0°C$ to convert it to water at $20.0°C$. Refer to Appendix E.

Plan

We must determine the amount of heat absorbed during three steps: (1) warming 50.0 g of ice from $-12.0°C$ to its melting point, $0.0°C$ (we use the specific heat of ice, $2.09\ J/g \cdot °C$); (2) melting the ice with no change in temperature (we use the heat of fusion of ice at $0.0°C$, 334 J/g); and (3) warming the resulting water from $0.0°C$ to $20.0°C$ (we use the specific heat of water, $4.18\ J/g \cdot °C$).

Ice is very efficient for cooling because considerable heat is required to melt a given mass of it. However, ΔH_{vap} is generally much greater than ΔH_{fusion}, so evaporative cooling is preferable when possible.

Solution

1. $50.0\ g \times \dfrac{2.09\ J}{g \cdot °C} \times [0.0 - (-12.0)]°C = 1.25 \times 10^3\ J = 0.125 \times 10^4\ J$

2. $50.0\ g \times \dfrac{334\ J}{g} \hspace{5.5cm} = 1.67 \times 10^4\ J$

3. $50.0\ g \times \dfrac{4.18\ J}{g \cdot °C} \times (20.0 - 0.0)°C = 4.18 \times 10^3\ J \hspace{0.5cm} = 0.418 \times 10^4\ J$

$$\text{Total amount of heat absorbed} = \boxed{2.21 \times 10^4\ J = 22.1\ kJ}$$

Note that most of the heat was absorbed in step 2, melting the ice.

You should now work Exercise 52.

13-12 SUBLIMATION AND THE VAPOR PRESSURE OF SOLIDS

Some solids, such as iodine and carbon dioxide, vaporize at atmospheric pressure without passing through the liquid state. We say that they **sublime.** Solids exhibit vapor pressures just as liquids do, but they generally have much lower vapor pressures. Solids with high vapor pressures sublime easily. The characteristic odors of the common household solids naphthalene (mothballs) and *para*-dichlorobenzene (bathroom deodorizer) are due to sublimation. The reverse process, by which a vapor solidifies without passing through the liquid phase, is called **deposition.**

$$\text{solid} \underset{\text{deposition}}{\overset{\text{sublimation}}{\rightleftharpoons}} \text{gas}$$

Some impure solids can be purified by sublimation and subsequent deposition of the vapor (as a solid) onto a cooler surface. Purification of iodine by sublimation is illustrated in Figure 13-16.

Figure 13-16 Sublimation can be used to purify volatile solids. The high vapor pressure of the solid substance causes it to sublime when heated. Crystals of purified substance are formed when the vapor is deposited as solid on the cooler (inner) portion of the apparatus. Iodine, I_2, sublimes readily. I_2 vapor is purple.

13-13 PHASE DIAGRAMS (*P* VERSUS *T*)

Now that we have discussed the general properties of the three phases of matter, we can describe **phase diagrams.** They show the equilibrium pressure–temperature relationships among the different phases of a given pure substance in a closed system. Our discussion of phase diagrams applies only to *closed systems* (e.g., a sample in a sealed container), in which matter does not escape into the surroundings. This limitation is especially important when the vapor phase is involved. Figure 13-17 shows a portion of the phase diagrams for water and carbon dioxide. The curves are not drawn to scale. The distortion allows us to describe the changes of state accompanying different pressure or temperature changes using one diagram.

The curved line from *A* to *C* in Figure 13-17a is a vapor pressure curve obtained experimentally by measuring the vapor pressures of water at various temperatures (Table

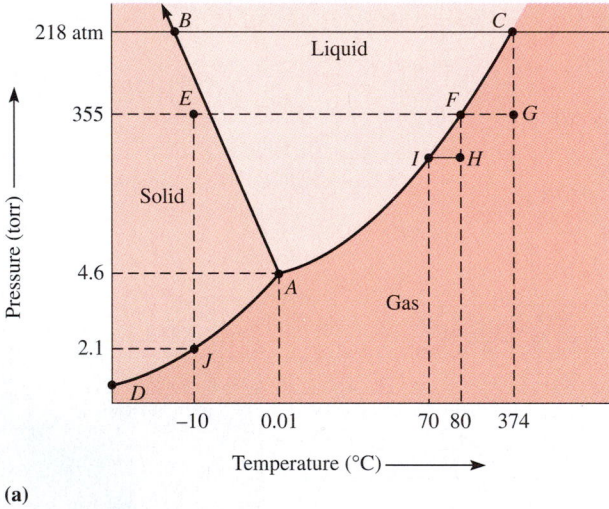

(a)

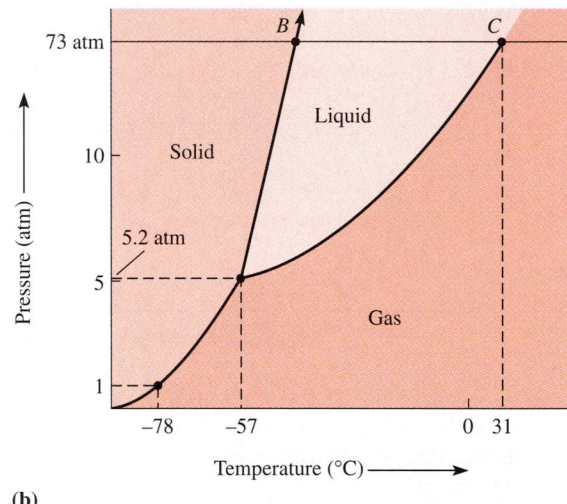

(b)

Figure 13-17 Phase diagrams (not to scale). (a) Diagram for water. For water and a few other substances for which the solid is less dense than the liquid, the solid–liquid equilibrium line (*AB*) has negative slope, i.e., up and to the left. (b) Diagram for carbon dioxide, a substance for which the solid is more dense than the liquid. Note that the solid–liquid equilibrium line has positive slope, i.e., up and to the right. This is true for most substances. (c) Two paths by which a gas can be liquefied. (1) Below the critical temperature. Compressing the sample at *constant* temperature is represented by the vertical line *WZ*. Where this line crosses the vapor pressure curve *AC*, the gas liquefies; at that set of conditions, *two distinct phases,* gas and liquid, are present in equilibrium with one another. These two phases have different properties—e.g., different densities. Raising the pressure further results in a completely liquid sample at point *Z*. (2) Above the critical temperature. Suppose that we instead first warm the gas at constant pressure from *W* to *X*, a temperature above its critical temperature. Then, holding the temperature constant, we increase the pressure to point *Y*. Along this path, the sample increases *smoothly* in density, with no sharp transition between phases. From *Y*, we then decrease the temperature to reach final point *Z*, where the sample is clearly a liquid.

(c)

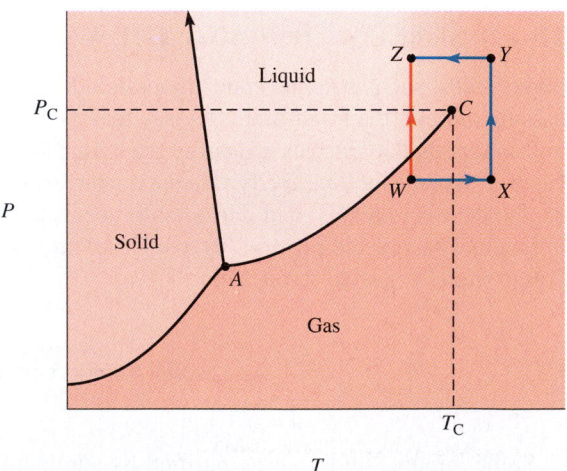

temperature (°C)	−10	0	20	30	50	70	90	95	100	101
Table 13-8 *Points on the Vapor Pressure Curve for Water*										
vapor pressure (torr)	2.1	4.6	17.5	31.8	92.5	234	526	634	760	788

Camphor, which is used in inhalers, has a high vapor pressure. When stored in a bottle, camphor sublimes and then deposits elsewhere in the bottle.

13-8). Points along this curve represent the temperature–pressure combinations for which liquid and gas (vapor) coexist in equilibrium. At points above *AC*, the stable form of water is liquid; below the curve, it is vapor.

Line *AB* represents the liquid–solid equilibrium conditions. We see that it has a negative slope. Water is one of the very few substances for which this is the case. The negative slope (up and to the left) indicates that increasing the pressure sufficiently on the surface of ice causes it to melt. This is because ice is *less dense* than liquid water in the vicinity of the liquid–solid equilibrium. The network of hydrogen bonding in ice is more extensive than that in liquid water and requires a greater separation of H_2O molecules. This causes ice to float in liquid water. Almost all other solids are more dense than their corresponding liquids; they would have positive slopes associated with line *AB*. The stable form of water at points to the left of *AB* is solid (ice). Thus *AB* is called a *melting curve*.

There is only one point, *A*, at which all three phases of a substance—solid, liquid, and gas—can coexist at equilibrium. This is called the **triple point.** For water it occurs at 4.6 torr and 0.01°C.

At pressures below the triple-point pressure, the liquid phase does not exist; rather, the substance goes directly from solid to gas (sublimes) or the reverse happens (crystals deposit from the gas). At pressures and temperatures along *AD*, the *sublimation curve*, solid and vapor are in equilibrium.

Consider CO_2 (Figure 13-17b). The triple point is at 5.2 atmospheres and −57°C. This pressure is *above* normal atmospheric pressure so liquid CO_2 cannot exist at atmospheric pressure. Dry ice (solid CO_2) sublimes and does not melt at atmospheric pressure.

The **critical temperature** is the temperature above which a gas cannot be liquefied, i.e., the temperature above which the liquid and gas do not exist as distinct phases. A substance at a temperature above its critical temperature is called a *supercritical fluid.* The

The CO_2 in common fire extinguishers is liquid. As you can see from Figure 13-17b, the liquid must be at some pressure greater than 10 atm for temperatures above 0°C. It is ordinarily at about 65 atm (more than 900 lb/in^2), so these cylinders must be handled with care.

Benzene is *more* dense as a solid than as a liquid, so the solid sinks in the liquid (right). This is the behavior shown by nearly all known substances except water (left).

Compare the description here with that accompanying Figure 13-15. Each horizontal line on a phase diagram contains the information in one heating curve obtained at a particular pressure. Each horizontal line in the heating curve (Figure 13-15) corresponds to crossing one of the phase equilibrium lines in the phase diagram (Figure 13-17) at a particular pressure. Phase diagrams are obtained by combining the results of heating curves measured experimentally at different pressures.

critical pressure is the pressure required to liquefy a gas (vapor) *at* its critical temperature. The combination of critical temperature and critical pressure is called the **critical point** (*C* in Figure 13-17). For H_2O, the critical point is 374°C and 218 atmospheres; for CO_2, 31°C and 73 atmospheres. However, there is no such upper limit to the solid/liquid line, as emphasized by the arrowhead at the top of that line.

To illustrate the use of a phase diagram in determining the physical state or states of a system under different sets of pressures and temperatures, let's consider a sample of water at point *E* in Figure 13-17a (355 torr and −10°C). At this point all the water is in the form of ice. Suppose that we hold the pressure constant and gradually increase the temperature—in other words, trace a path from left to right along *EG*. At the temperature at which *EG* intersects *AB*, the melting curve, some of the ice melts. If we stopped here, equilibrium between ice and liquid water would eventually be established, and both phases would be present. If we added more heat, all the ice would melt with no temperature change. Remember that all phase changes of pure substances occur at constant temperature.

Once the ice is completely melted, additional heat causes the temperature to rise. Eventually, at point *F* (355 torr and 80°C), some of the liquid begins to boil; liquid and vapor are in equilibrium. Adding more heat at constant pressure vaporizes the rest of the water with no temperature change. Adding still more heat warms the gas from *F* to *G*. Complete vaporization would also occur if, at point *F* and before all the liquid had vaporized, the temperature were held constant and the pressure were decreased to, say, 234 torr at point *H*. If we wished to hold the pressure at 234 torr and condense some of the vapor, it would be necessary to cool the vapor to 70°C, point *I*, which lies on the vapor pressure curve, *AC*. To state this in another way, the vapor pressure of water at 70°C is 234 torr.

Suppose we move back to ice at point *E* (355 torr and −10°C). If we now hold the temperature at −10°C and reduce the pressure, we move vertically down along *EJ*. At a pressure of 2.1 torr we reach the sublimation curve, at which point the solid passes directly into the gas phase (sublimes) until all the ice has sublimed. An important application of this phenomenon is in the freeze-drying of foods. In this process a water-containing food

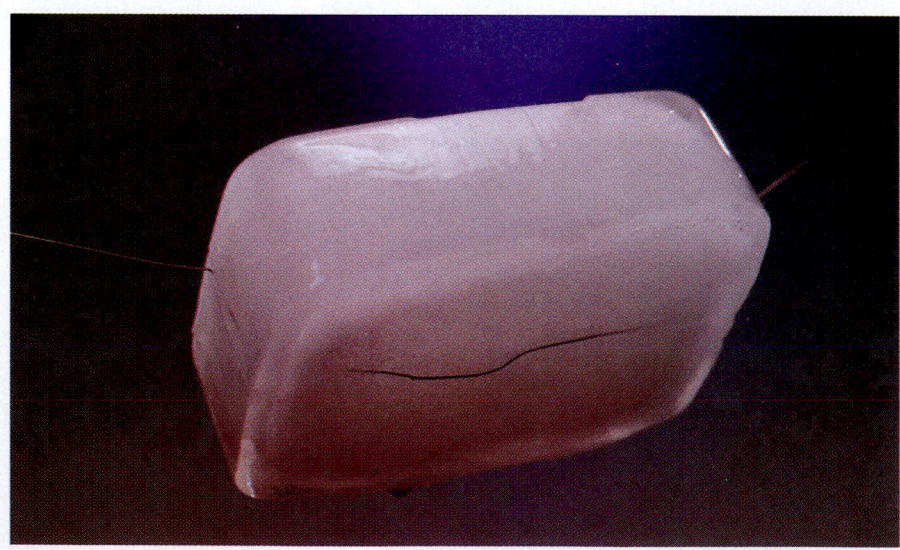

A weighted wire cuts through a block of ice. The ice melts under the high pressure of the wire, and then refreezes behind the wire.

is cooled below the freezing point of water to form ice, which is then removed as a vapor by decreasing the pressure.

Let us clarify the nature of the fluid phases (liquid and gas) and of the critical point by describing two different ways by which a gas can be liquefied. A sample at point W in the phase diagram of Figure 13-17c is in the vapor (gas) phase, below its critical temperature. Suppose we compress the sample at constant T from point W to point Z. We can identify a definite pressure (the intersection of line WZ with the vapor pressure curve AC) where the transition from gas to liquid takes place. However, if we go *around* the critical point by the path $WXYZ$, no such clear-cut transition takes place. By this second path, the density and other properties of the sample vary in a continuous manner; there is no definite point at which we can say that the sample changes from gas to liquid. We use the term *fluid* to describe either a gas or a liquid.

A fluid *below* its critical temperature may properly be identified as a liquid or as a gas. *Above* the critical temperature, we should use the term "fluid." The somewhat unusual properties of supercritical fluids lead to some novel uses. For instance, supercritical CO_2 is a useful solvent that can be removed much more completely than more conventional solvents.

13-14 AMORPHOUS SOLIDS AND CRYSTALLINE SOLIDS

We have already seen that solids have definite shapes and volumes, are not very compressible, are dense, and diffuse only very slowly into other solids. They are generally characterized by compact, ordered arrangements of particles that vibrate about fixed positions in their structures.

However, some noncrystalline solids, called **amorphous solids,** have no well-defined, ordered structure. Examples include rubber, some kinds of plastics, and amorphous sulfur.

Some amorphous solids are called glasses because like liquids, they flow, although *very* slowly. The irregular structures of glasses are intermediate between those of freely flowing liquids and those of crystalline solids; there is only short-range order. Crystalline solids such as ice and sodium chloride have well-defined, sharp melting temperatures. Particles in amorphous solids are irregularly arranged, so intermolecular forces among their particles vary in strength within a sample. Melting occurs at different temperatures for various portions of the same sample as the intermolecular forces are overcome. Unlike crystalline solids, glasses and other amorphous solids do not exhibit sharp melting points, but soften over a temperature range.

One test for the purity of a crystalline solid is the sharpness of its melting point. Impurities disrupt the intermolecular forces and cause melting to occur over a considerable temperature range.

The shattering of a crystalline solid produces fragments having the same (or related) interfacial angles and structural characteristics as the original sample. The shattering of a cube of rock salt produces several smaller cubes of rock salt. This cleaving occurs preferentially along crystal lattice planes between which the interionic or intermolecular forces of attraction are weakest. Amorphous solids with irregular structures, such as glasses, shatter irregularly to yield pieces with curved edges and irregular angles.

The lattice planes are planes within the crystal containing ordered arrangements of particles.

The regular external shape of a crystal is the result of regular internal arrangements of atoms, molecules, or ions. Crystals of the ionic solid sodium chloride, NaCl, from a kitchen saltshaker have the same shape as the large crystal shown here.

X-Ray Diffraction

Atoms, molecules, and ions are much too small to be seen with the eye. The arrangements of particles in crystalline solids are determined indirectly by X-ray diffraction (scattering). In 1912 the German physicist Max von Laue showed that any crystal could serve as a three-dimensional diffraction grating for incident electromagnetic radiation with wavelengths approximating the internuclear separations of atoms in the crystal. Such radiation is in the X-ray region of the electromagnetic spectrum. Using an apparatus such as that shown in Figure 13-18, a monochromatic (single-wavelength) X-ray beam is defined by a system of slits and directed onto a crystal. The crystal is rotated to vary the angle of incidence θ. At various angles, strong beams of deflected X-rays hit a photographic plate. Upon development, the plate shows a set of symmetrically arranged spots due to deflected X-rays. Different crystals produce different arrangements of spots.

In 1913 the English scientists William and Lawrence Bragg found that diffraction photographs are more easily interpreted by considering the crystal as a reflection grating rather than a diffraction grating. The analysis of the spots is somewhat complicated, but an experienced crystallographer can determine the separations between atoms within identical layers and the distances between layers of atoms. The more electrons an atom has, the more strongly it scatters X-rays, so it is also possible to determine the identities of individual atoms.

Figure 13-19 illustrates the determination of spacings between layers of atoms. The X-ray beam strikes parallel layers of atoms in the crystal at an angle θ. Those rays colliding with atoms in the first layer are reflected at the same angle θ. Those passing through the first layer may be reflected from the second layer, third layer, and so forth. A reflected beam results only if all rays are in phase.

William and Lawrence Bragg are the only father and son to receive the Nobel Prize, which they shared in physics in 1915.

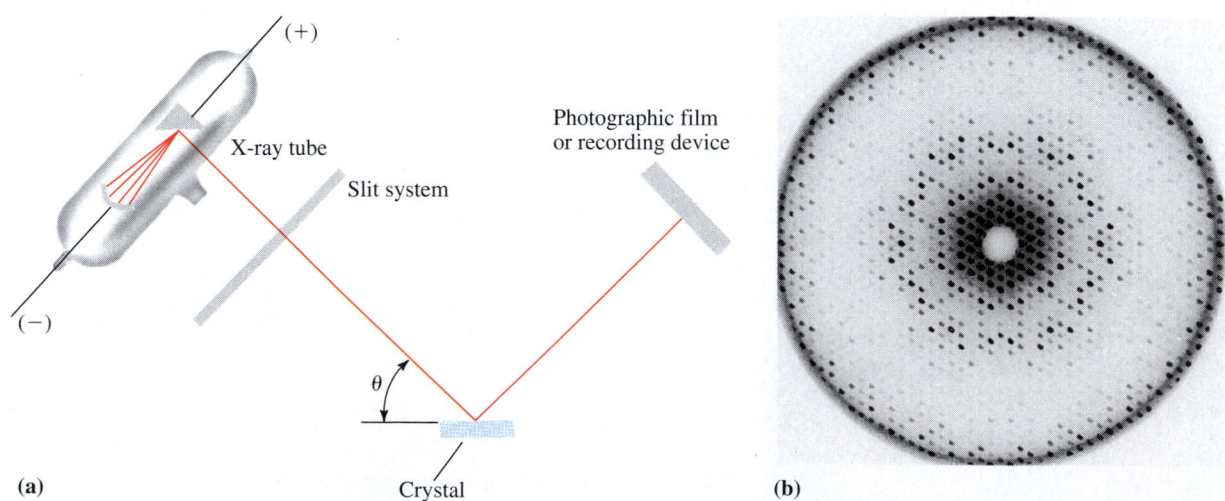

Figure 13-18   (a) X-ray diffraction by crystals (schematic). (b) A photograph of the X-ray diffraction pattern from a crystal of the enzyme histidine decarboxylase (MW ~37,000 amu). The crystal was rotated so that many different lattice planes with different spacings were moved in succession into diffracting position (Figure 13-19).

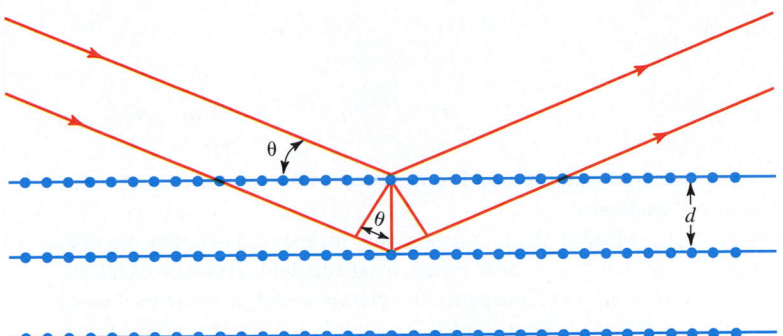

Figure 13-19 Reflection of a monochromatic beam of X-rays by two lattice planes (layers of atoms) of a crystal.

For the waves to be in phase (interact constructively), the difference in path length must be equal to the wavelength, λ, times an integer, n. This leads to the condition known as the **Bragg equation.**

$$n\lambda = 2d \sin \theta \qquad \text{or} \qquad \sin \theta = \frac{n\lambda}{2d}$$

It tells us that for X-rays of a given wavelength λ, atoms in planes separated by distances d give rise to reflections at angles of incidence θ. The reflection angles increase with increasing order, $n = 1, 2, 3, \ldots$.

13-15 STRUCTURES OF CRYSTALS

All crystals contain regularly repeating arrays of atoms, molecules, or ions. They are analogous (but in three dimensions) to a wallpaper pattern (Figure 13-20). Once we discover the pattern of a wallpaper, we can repeat it in two dimensions to cover a wall. To describe such a repeating pattern we must specify two things: (1) the size and shape of the repeating unit and (2) the contents of this unit. In the wallpaper pattern of Figure 13-20a, several choices of the repeating unit are outlined. Repeating unit A contains one complete cat; unit B, with the same area, contains parts of several different cats, but these still add

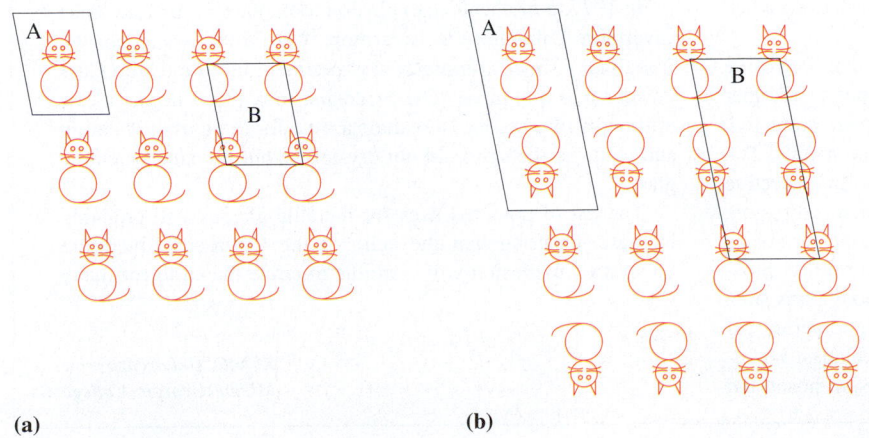

(a)   (b)

Figure 13-20 Patterns that repeat in two dimensions. Such patterns might be used to make wallpaper. We must imagine that the pattern extends indefinitely (to the end of the wall). In each pattern two of the many possible choices of unit cells are outlined. Once we identify a unit cell and its contents, repetition by translating this unit generates the entire pattern. In (a) the unit cell contains only one cat. In (b) each cell contains two cats related to one another by a 180° rotation. Any crystal is an analogous pattern in which the contents of the three-dimensional unit cell consist of atoms, molecules, or ions. The pattern extends in *three* dimensions to the boundaries of the crystal, usually including many thousands of unit cells.

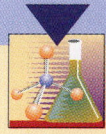
Quick-Frozen Metals

What do replacement parts for human bodies have in common with the walls of fusion reactors? We may see dramatic improvements in both due to a relatively new discovery called *metallic glass*. Metallic glass is different from the crystalline metals we commonly use. In the crystalline state, metal atoms are arranged in a regularly repeating three-dimensional array produced by relatively slow cooling of the molten metal. Today it is possible to cool a mixture of molten metals so quickly that their atoms do not have time to arrange themselves into an orderly pattern. The result is an amorphous solid or glass that may be the strongest, most corrosion-resistant, and most easily magnetized material we know. Because of their low production and raw material costs, metallic glasses are among the least expensive forms of metals.

To cool a molten metal mixture quickly enough to produce metallic glass, the molten mixture is shot in a continuous jet onto the cold rim of a rapidly rotating wheel. When the molten metal hits the rim of the wheel, the metal, initially at 1200°C, cools to room temperature in a thousandth of a second. This represents a cooling rate of about one million degrees per second; the result is a ribbon of metallic glass that flies off the rim of the spinning wheel at about 6000 feet per minute.

Although the ribbon looks, feels, and conducts electricity like a metal, its atoms are randomly dispersed and not arranged in distinct crystalline patterns. This unique atomic arrangement gives it unusual properties, such as great strength.

This improvement in strength results from a lack of crystal boundaries where metal fractures usually start. As a result, a metallic glass alloy is about 15 times stronger than the same alloy mixture that has been cooled more slowly. Minor cracks, nicks, and chips don't weaken metallic glasses as they do regular metals and window glass. However, unlike conventional metals and window glass, metallic glass has "flexible" atomic bonding, which allows it to bend without shattering. In fact, metallic glass is so pliable that it does not fracture even when bent back on itself by 180 degrees.

Another advantage of metallic glasses is that corrosive chemicals do not penetrate them easily. This property is further enhanced because metallic glasses form their own protective films, as do aluminum and a few other crystalline metals. The lack of crystalline boundaries and the ability to form protective films cause metallic glasses to be very resistant to corrosion. Their unique combination of strength and corrosion resistance appears to make them nearly ideal for the three million prosthetic devices used each year to replace human body parts such as hip joints. Other potential applications include undersea cables, sutures to close surgical wounds, and razor blades. Metallic glass may also be used to reinforce high-pressure hoses and automobile tires, and to construct corrosion-resistant chemical equipment.

Metallic glasses are so easily magnetized that they are magnetized by the earth's weak magnetic field. This ease of magnetization could make metallic glasses useful in computers, which rely heavily upon quickly magnetizing materials, such as floppy disks, to store information. Ease of magnetization may also provide financial rewards in building transformers. (Is there one at the top of a nearby pole outside your home?) The iron cores inside these transformers are magnetized and demagnetized 60 times every second, and some energy is lost during each cycle. This energy loss adds about 500 million dollars each year to our electric bills. Because metallic glasses require less energy to become magnetized and demagnetized, much of this waste might be eliminated.

Unfortunately, metallic glass comes in only a few shapes such as ribbons, wires, and powders. Metallic glass would be even more useful if it could be made into sheets, fabrics, tubes, cylinders, and plates. It may be possible to weave, braid, helically wrap, or laminate metallic glass ribbons to construct these shapes. Even so, the number of shapes could remain somewhat limited. Work is under way to treat metallic glass powders with shock compression to increase the variety of possible shapes of objects.

Since 1956, shock compression has been used to change powders into larger shapes. In shock compression, explosives are placed around a powder, and all the explosives are set off at the same instant. The explosive forces push the powder together from all sides, thereby creating a solid object. However, using this technique becomes difficult with glassy powders because the explosion must not be too intense and melt the glass, which would then become crystalline. On the other hand, if the explosion were too weak, the powder wouldn't be pushed together with enough force to form new shapes.

In 1977, a momentous explosion took place at the Lawrence Livermore Laboratory in Livermore, California, where, for the first time, a shock compression experiment fused metallic glass powder into one piece. The explosion took place in just a few millionths of a second, too short a time for the glass to heat up and melt. So the solid did not crystallize but remained metallic glass.

The list of potential uses for metallic glasses will probably increase, ensuring that the echo of the Lawrence Livermore Laboratory explosion will continue to circle the earth for many years.

Ronald DeLorenzo
Middle Georgia College

up to one complete cat. From whichever unit we choose, we can obtain the entire pattern by repeatedly translating the contents of that unit in two dimensions.

In a crystal the repeating unit is three-dimensional; its contents consist of atoms, molecules, or ions. The smallest unit of volume of a crystal that shows all the characteristics of the crystal's pattern is a **unit cell.** We note that the unit cell is just the fundamental *box* that describes the arrangement. The unit cell is described by the lengths of its edges—a, b, c (which are related to the spacings between layers, d)—and the angles between the edges— α, β, γ (Figure 13-21). Unit cells are stacked in three dimensions to build a lattice, the three-dimensional *arrangement* corresponding to the crystal. It can be proven that unit cells must fit into one of the seven crystal systems (Table 13-9). Each crystal system is distinguished by the relations between the unit cell lengths and angles *and* by the symmetry of the resulting three-dimensional patterns. Crystals have the same symmetry as their constituent unit cells because all crystals are repetitive multiples of such cells.

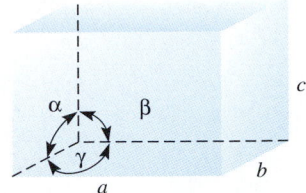

Figure 13-21 A representation of a unit cell.

Let us replace each repeat unit in the crystal by a point (called a *lattice point*) placed at the same place in the unit. All such points have the same environment and are indistinguishable from each other. The resulting three-dimensional array of points is called a **lattice.** It is a simple but complete description of the way in which a crystal structure is built up.

The unit cells shown in Figure 13-22a are the simple, or primitive, unit cells corresponding to the seven crystal systems listed in Table 13-9. Each of these unit cells corresponds to *one* lattice point. As a two-dimensional representation of the reasoning behind this statement, look at the unit cell marked "B" in Figure 13-20a. Each corner of the unit cell is a lattice point, and can be imagined to represent one cat. The cat at each corner is shared among four unit cells (remember—we are working in two dimensions here). The unit cell has four corners, and in the corners of the unit cell are enough pieces to make one complete cat. Thus, unit cell B contains one cat, the same as the alternative unit cell choice marked "A." Now imagine that each lattice point in a three-dimensional crystal represents an object (a molecule, an atom, and so on). Such an object at a corner (Figure 13-23a) is shared by the eight unit cells that meet at that corner. Because each unit cell has eight corners, it contains eight "pieces" of the object, so it contains $8(\frac{1}{8}) = 1$ object. Similarly, an object on an edge, but not at a corner, is shared by four unit cells (Figure 13-23b), and an object on a face is shared by two unit cells (Figure 13-23c).

Each unit cell contains atoms, molecules, or ions in a definite arrangement. Often the unit cell contents are related by some additional symmetry. (For instance, the unit cell in Figure 13-20b contains *two* cats, related to one another by a rotation of 180°.) Different

Table 13-9 *The Unit Cell Relationships for the Seven Crystal Systems**

	Unit Cell		
System	*Lengths*	*Angles*	**Example (common name)**
cubic	$a = b = c$	$\alpha = \beta = \gamma = 90°$	NaCl (rock salt)
tetragonal	$a = b \neq c$	$\alpha = \beta = \gamma = 90°$	TiO_2 (rutile)
orthorhombic	$a \neq b \neq c$	$\alpha = \beta = \gamma = 90°$	$MgSO_4 \cdot 7H_2O$ (epsomite)
monoclinic	$a \neq b \neq c$	$\alpha = \gamma = 90°$; $\beta \neq 90°$	$CaSO_4 \cdot 2H_2O$ (gypsum)
triclinic	$a \neq b \neq c$	$\alpha \neq \beta \neq \gamma \neq 90°$	$K_2Cr_2O_7$ (potassium dichromate)
hexagonal	$a = b \neq c$	$\alpha = \beta = 90°$; $\gamma = 120°$	SiO_2 (silica)
rhombohedral	$a = b = c$	$\alpha = \beta = \gamma \neq 90°$	$CaCO_3$ (calcite)

**In these definitions, the sign $\neq$ means "is not necessarily equal to."*

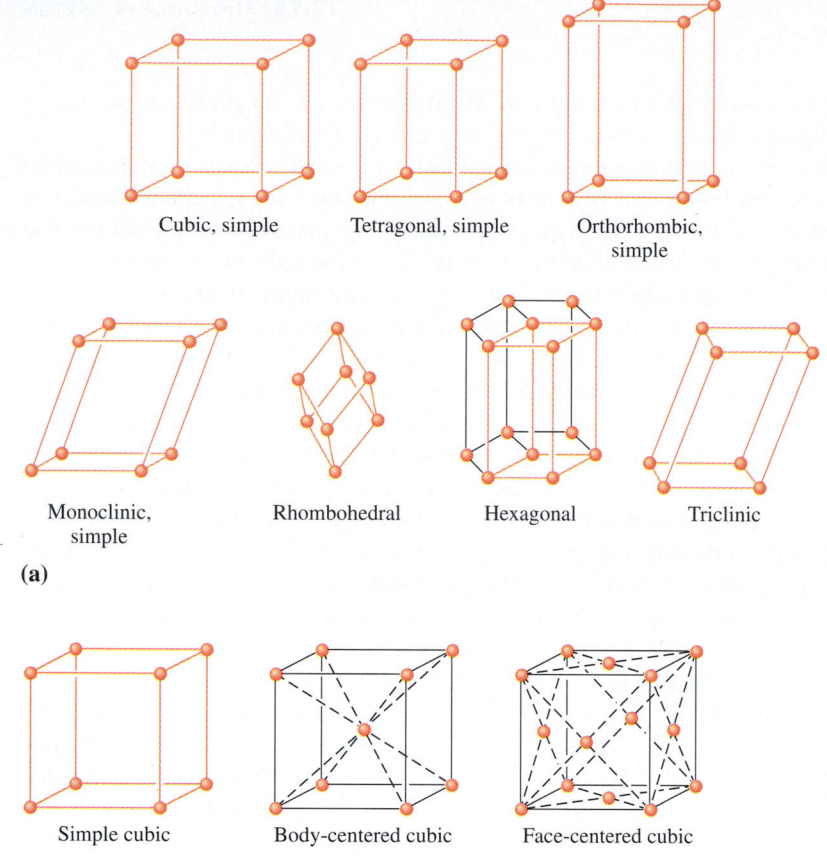

Figure 13-22 (a) Simple (primitive) unit cells of the seven crystal systems. (b) Simple cubic, body-centered cubic, and face-centered cubic unit cells.

Cubic, simple   Tetragonal, simple   Orthorhombic, simple

Monoclinic, simple   Rhombohedral   Hexagonal   Triclinic

(a)

Simple cubic   Body-centered cubic   Face-centered cubic

(b)

A crystal of one form of manganese metal has Mn atoms at the corners of a simple cubic unit cell that is 6.30 Å on edge (Example 13-7).

Each object in a face is shared between two unit cells, so it is counted $\frac{1}{2}$ in each; there are six faces in each unit cell.

substances that crystallize in the same type of lattice with the same atomic arrangement are said to be **isomorphous.** A single substance that can crystallize in more than one arrangement is said to be **polymorphous.**

In a *simple,* or *primitive,* lattice, only the eight corners of the unit cell are equivalent. In other types of crystals, objects equivalent to those forming the outline of the unit cell may occupy extra positions within the unit cell. (In this context, "equivalent" means that the same atoms, molecules, or ions appear in *identical environments and orientations* at the eight corners of the cell and, when applicable, at other locations in the unit cell.) This results in additional lattices besides the simple ones in Figure 13-22a. Two of these are shown in Figure 13-22b. A *body-centered* lattice has equivalent points at the eight unit cell corners *and* at the center of the unit cell (Figure 13-22b). Iron, chromium, and many other metals crystallize in a body-centered cubic (bcc) arrangement. The unit cell of such a metal contains $8(\frac{1}{8}) = 1$ atom at the corners of the cell *plus* one atom at the center of the cell (and therefore entirely in this cell); this makes a total of *two* atoms per unit cell. A *face-centered* structure involves the eight points at the corners and six more equivalent points, one in the middle of each of the six square faces of the cell. A metal (calcium and silver are cubic examples) that crystallizes in this arrangement has $8(\frac{1}{8}) = 1$ atom at the corners *plus* $6(\frac{1}{2}) = 3$ more in the faces, for a total of *four* atoms per unit cell. In more complicated crystals, each lattice site may represent several atoms or an entire molecule.

We have discussed some simple structures that are easy to visualize. More complex compounds crystallize in structures with unit cells that can be more difficult to describe. Experimental determination of the crystal structures of such solids is correspondingly

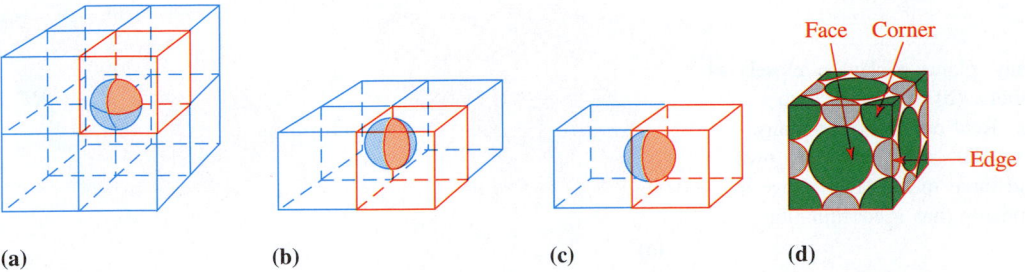

(a) (b) (c) (d)

Figure 13-23 Representation of the sharing of an object (an atom, ion, or molecule) among unit cells. The fraction of each sphere that "belongs" to a single unit cell is shown in red. (a) The sharing of an object at a corner by eight unit cells. (b) The sharing of an object on an edge by four unit cells. (c) The sharing of an object in a face by two unit cells. (d) A representation of a unit cell that illustrates the portions of atoms presented in more detail in Figure 13-27. The green ion at each corner is shared by eight unit cells, as in (a). The gray ion at each edge is shared by four unit cells, as in (b). The green ion in each face is shared by two unit cells, as in (c).

more complex. However, modern computer-controlled instrumentation can collect and analyze the large amounts of X-ray diffraction data used in such studies. This now allows analysis of structures ranging from simple metals to complex biological molecules such as proteins and nucleic acids. Most of our knowledge about the three-dimensional arrangements of atoms depends on crystal structure studies.

13-16 BONDING IN SOLIDS

We classify crystalline solids into categories according to the types of particles in the crystal and the bonding or interactions among them. The four categories are (1) metallic solids, (2) ionic solids, (3) molecular solids, and (4) covalent solids. Table 13-10 summarizes these categories of solids and their typical properties.

Table 13-10 *Characteristics of Types of Solids*

	Metallic	Ionic	Molecular	Covalent
Particles of unit cell	Metal ions in "electron gas"	Anions, cations	Molecules (or atoms)	Atoms
Strongest interparticle forces	Metallic bonds (attraction between cations and e^-'s)	Electrostatic	London, dipole–dipole, and/or hydrogen bonds	Covalent bonds
Properties	Soft to very hard; good thermal and electrical conductors; wide range of melting points (-39 to $3400°C$)	Hard; brittle; poor thermal and electrical conductors; high melting points (400 to $3000°C$)	Soft; poor thermal and electrical conductors; low melting points (-272 to $400°C$)	Very hard; poor thermal and electrical conductors;* high melting points (1200 to $4000°C$)
Examples	Li, K, Ca, Cu, Cr, Ni (metals)	NaCl, $CaBr_2$, K_2SO_4 (typical salts)	CH_4 (methane), P_4, O_2, Ar, CO_2, H_2O, S_8	C (diamond), SiO_2 (quartz)

Exceptions: diamond is a good conductor of heat; graphite is soft and conducts electricity well.

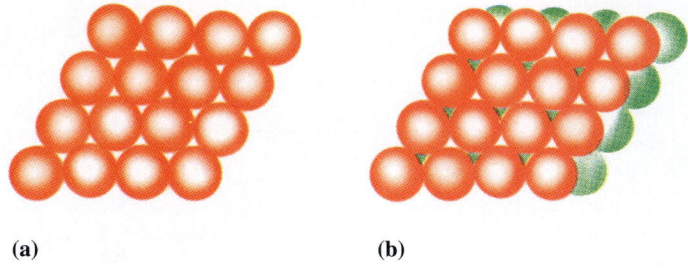

Figure 13-24   (a) Spheres in the same plane, packed as closely as possible. Each sphere touches six others. (b) Spheres in two planes, packed as closely as possible. Real crystals have many more than two planes. Each sphere touches six others in its own layer, three in the layer below it, and three in the layer above it; i.e., it contacts a total of 12 other spheres (has a coordination number of 12).

(a)   **(b)**

Metallic Solids

Metals crystallize as solids in which metal ions may be thought to occupy the lattice sites and are embedded in a cloud of delocalized valence electrons. Nearly all metals crystallize in one of three types of lattices: (1) body-centered cubic (bcc), (2) face-centered cubic (fcc; also called cubic close-packed), and (3) hexagonal close-packed. The latter two types are called close-packed structures because the particles (in this case metal atoms) are packed together as closely as possible. The differences between the two close-packed structures are illustrated in Figures 13-24 and 13-25. Let spheres of equal size represent identical metal atoms, or any other particles, that form close-packed structures. Consider a layer of spheres packed in a plane, *A*, as closely as possible (Figure 13-25a). An identical

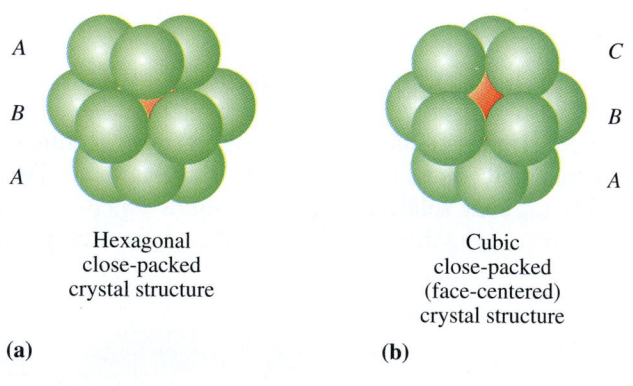

A
B
A

Hexagonal
close-packed
crystal structure

(a)

C
B
A

Cubic
close-packed
(face-centered)
crystal structure

(b)

Figure 13-25   There are two crystal structures in which atoms are packed together as compactly as possible. The diagrams show the structures expanded to clarify the difference between them. (a) In the hexagonal close-packed structure, the first and third layers are oriented in the same direction, so that each atom in the third layer (*A*) lies directly above an atom in the first layer (*A*). (b) In the cubic close-packed structure, the first and third layers are oriented in opposite directions, so that no atom in the third layer (*C*) is directly above an atom in either of the first two layers (*A* and *B*). In both cases, every atom is surrounded by 12 other atoms if the structure is extended indefinitely, so each atom has a coordination number of 12. Although it is not obvious from this figure, the cubic close-packed structure is face-centered cubic. To see this, we would have to include additional atoms and tilt the resulting cluster of atoms.

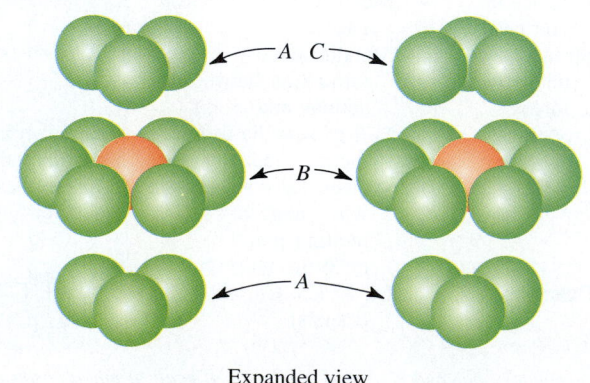

A   C

B

A

Expanded view

plane of spheres, *B*, is placed in the depressions of plane *A*. If the third plane is placed with its spheres directly above those in plane *A*, the *ABA* arrangement results. This is the hexagonal close-packed structure (Figure 13-25a). The extended pattern of arrangement of planes is *ABABAB . . . .* If the third layer is placed in the alternate set of depressions in the second layer so that spheres in the first and third layers are *not* directly above and below each other, the cubic close-packed structure, *ABCABCABC . . .* , results (Figure 13-25b). In close-packed structures each sphere has a *coordination number* of 12, i.e., 12 nearest neighbors. In ideal close-packed structures 74% of a given volume is due to spheres and 26% is empty space. The body-centered cubic structure is less efficient in packing; each sphere has only eight nearest neighbors, and there is more empty space.

The term ''coordination number'' is used in crystallography in a somewhat different sense from that in coordination chemistry (Section 25-3). Here it refers to the number of near neighbors.

EXAMPLE 13-7 *Nearest Neighbors*

In the simple cubic form of manganese there are Mn atoms at the corners of a simple cubic unit cell that is 6.30 Å on edge. (a) What is the shortest distance between neighboring Mn atoms? (b) How many nearest neighbors does each atom have?

Plan
We visualize the simple cubic cell.

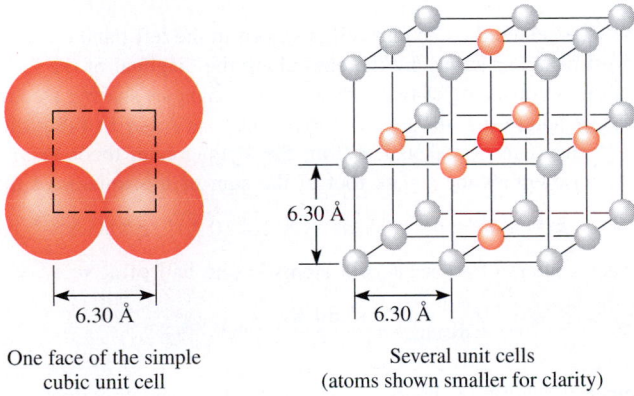

One face of the simple
cubic unit cell

Several unit cells
(atoms shown smaller for clarity)

Solution
(a) One face of the cubic unit cell is shown in the left-hand drawing, with the atoms touching. The nearest-neighbor atoms are separated by one unit cell edge, at the distance 6.30 Å.
(b) A three-dimensional representation of eight unit cells is also shown. In that drawing the atoms are shown smaller for clarity. Some atoms are represented with different colors to aid in visualizing the arrangement, but *all atoms are identical.* Consider the atom shown in red at the center (at the intersection of the eight unit cells). Its nearest neighbors in all of the unit cell directions are shown as light red atoms. As we can see, there are six nearest neighbors. The same would be true of any other atom in the structure.

EXAMPLE 13-8 *Nearest Neighbors*

Gold crystals are face-centered cubic, with a cell edge of 4.10 Å. (a) What is the distance between centers of the two closest Au atoms? (b) How many nearest neighbors does each atom have?

Plan

We reason as in Example 13-7, except that now the two atoms closest to one another are those along the face diagonal.

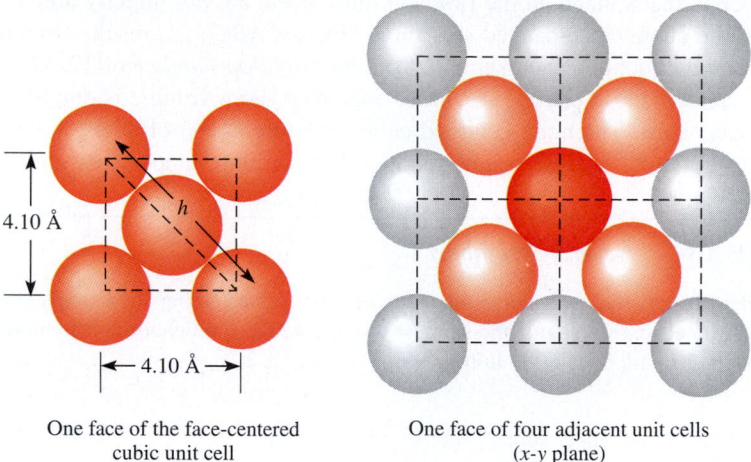

One face of the face-centered   One face of four adjacent unit cells
cubic unit cell   (x-y plane)

Solution

(a) One face of the face-centered cubic unit cell is shown in the left-hand drawing, with the atoms touching. The nearest-neighbor atoms are the ones along the diagonal of the face of the cube. We may visualize the face as consisting of two right isosceles triangles sharing a common hypotenuse, h, and having sides of length $a = 4.10$ Å. The hypotenuse is equal to *twice* the center-to-center distance. The hypotenuse can be calculated from the Pythagorean theorem, $h^2 = a^2 + a^2$. The length of the hypotenuse equals the square root of the sum of the squares of the sides.

$$h = \sqrt{a^2 + a^2} = \sqrt{2a^2} = \sqrt{2(4.10 \text{ Å})^2} = 5.80 \text{ Å}$$

The distance between centers of adjacent gold atoms is one half of h, so

$$\text{distance} = \frac{5.80 \text{ Å}}{2} = \boxed{2.90 \text{ Å}}$$

(b) To see the number of nearest neighbors, we expand the left-hand drawing to include several unit cells, as shown in the right-hand drawing. Suppose that this is the $x-y$ plane. The atom shown in red has four nearest neighbors in this plane. There are four more such neighbors in the $x-z$ plane (perpendicular to the $x-y$ plane), and four additional neighbors in the $y-z$ plane (also perpendicular to the $x-y$ plane). This gives a total of $\boxed{12 \text{ nearest neighbors.}}$

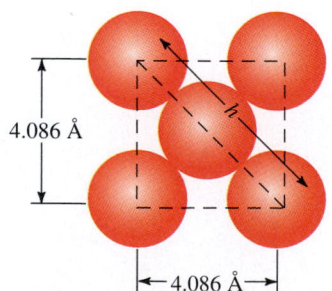

Figure 13-26   One face of the face-centered cubic unit cell of metallic silver.

EXAMPLE 13-9   *Packing Efficiency*

The unit cell of metallic silver is face-centered cubic (or cubic close-packed) with $a = b = c = 4.086$ Å. Calculate (a) the radius of an Ag atom; (b) the volume of an Ag atom in cm^3, given that, for a sphere, $V = (\frac{4}{3})\pi r^3$; (c) the percentage of the volume of a unit cell that is occupied by Ag atoms; and (d) the percentage that is empty space.

Plan

(a) One face of the fcc unit cell can be visualized (Figure 13-26) as consisting of (parts of) five Ag atoms describing two right triangles that share a hypotenuse. The hypotenuse, h, is four times the radius of the silver atom. We can use the Pythagorean theorem to evaluate r_{Ag}. (b) The volume of a sphere is $(\frac{4}{3})\pi r^3$. (c) We can use the length of the edge of the unit cell to find its total volume. The volume occupied by the Ag atoms is equal to the number of atoms per unit cell times the volume of each atom. These two values can be used to calculate the required percentage. (d) The percentage that is empty space is 100% minus the percentage that is occupied.

Solution

(a) $h = \sqrt{a^2 + a^2} = \sqrt{2a^2} = \sqrt{2(4.086 \text{ Å})^2} = 5.778 \text{ Å} = 4r_{Ag}$

$$r_{Ag} = \frac{5.778 \text{ Å}}{4} = \boxed{1.444 \text{ Å}}$$

(b) The volume of an Ag atom is $(\frac{4}{3})\pi r^3_{Ag}$.

$$V = (\tfrac{4}{3})\pi(1.444 \text{ Å})^3 = 12.61 \text{ Å}^3$$

We now convert Å^3 to cm^3.

$$\underline{?} \text{ cm}^3 = 12.61 \text{ Å}^3 \times \left(\frac{10^{-8} \text{ cm}}{\text{Å}}\right)^3 = \boxed{1.261 \times 10^{-23} \text{ cm}^3}$$

(c) The entire fcc unit cell contains $8(\frac{1}{8}) + 6(\frac{1}{2}) = 4$ Ag atoms. Each Ag atom has a volume of $1.261 \times 10^{-23} \text{ cm}^3$. Thus, the volume of four Ag atoms is

> The volume of the unit cell is the volume of the atoms *plus* the volume of the empty space *between* the atoms.

$$V_{Ag \text{ atoms}} = 4 \text{ Ag atoms} \times \frac{1.261 \times 10^{-23} \text{ cm}^3}{\text{Ag atom}} = 5.044 \times 10^{-23} \text{ cm}^3$$

Because the unit cell is cubic, its volume is a^3.

$$V_{\text{unit cell}} = (4.086 \text{ Å})^3 = 68.22 \text{ Å}^3 \times \left(\frac{10^{-8} \text{ cm}}{\text{Å}}\right)^3 = 6.822 \times 10^{-23} \text{ cm}^3$$

The percentage of the volume of the unit cell occupied by the silver atoms is

$$\% \text{ Ag}_{\text{volume}} = \frac{V_{Ag \text{ atoms}}}{V_{\text{unit cell}}} \times 100\% = \frac{5.044 \times 10^{-23} \text{ cm}^3}{6.822 \times 10^{-23} \text{ cm}^3} \times 100\% = \boxed{73.9\%}$$

(d) % empty space $= (100.0 - 73.9)\% = \boxed{26.1\%}$

> This result agrees well with our earlier observation that ideal close-packed structures are about 26% empty space.

You should now work Exercise 104.

EXAMPLE 13-10   *Density and Cell Volume*

From data in Example 13-9, calculate the density of metallic silver.

Plan

We first determine the mass of a unit cell, i.e., the mass of four atoms of silver. The density of the unit cell, and therefore of silver, is its mass divided by its volume.

Solution

$$\underline{?} \text{ g Ag per unit cell} = \frac{4 \text{ Ag atoms}}{\text{unit cell}} \times \frac{1 \text{ mol Ag}}{6.022 \times 10^{23} \text{ Ag atoms}} \times \frac{107.87 \text{ g Ag}}{1 \text{ mol Ag}}$$

$$= 7.165 \times 10^{-22} \text{ g Ag/unit cell}$$

$$\text{density} = \frac{7.165 \times 10^{-22} \text{ g Ag/unit cell}}{6.822 \times 10^{-23} \text{ cm}^3/\text{unit cell}} = \boxed{10.50 \text{ g/cm}^3}$$

A handbook gives the density of silver as 10.5 g/cm^3 at 20°C.

You should now work Exercise 80.

Data obtained from crystal structures and observed densities give us information from which we can calculate the value of Avogadro's number. The next example illustrates these calculations.

EXAMPLE 13-11   *Density, Cell Volume, and Avogadro's Number*

Titanium crystallizes in a body-centered cubic unit cell with an edge length of 3.306 Å. The density of titanium is observed to be 4.401 g/cm^3. Use these data to calculate Avogadro's number.

Plan

We relate the density and the volume of the unit cell to find the total mass contained in one unit cell. Knowing the number of atoms per unit cell, we can then find the mass of one atom. Comparing this to the known atomic weight, which is the mass of one mole (Avogadro's number) of atoms, we can evaluate Avogadro's number.

Solution

We first determine the volume of the unit cell.

$$V_{cell} = (3.306 \text{ Å})^3 = 36.13 \text{ Å}^3$$

We now convert Å^3 to cm^3.

$$\underline{?} \text{ cm}^3 = 36.13 \text{ Å}^3 \times \left(\frac{10^{-8} \text{ cm}}{\text{Å}}\right)^3 = 3.613 \times 10^{-23} \text{ cm}^3$$

The mass of the unit cell is its volume times the measured density.

$$\text{mass of unit cell} = 3.613 \times 10^{-23} \text{ cm}^3 \times \frac{4.401 \text{ g}}{cm^3} = 1.590 \times 10^{-22} \text{ g}$$

The bcc unit cell contains $8(\frac{1}{8}) + 1 = 2$ Ti atoms, so this represents the mass of two Ti atoms. The mass of a single Ti atom is

$$\text{mass of atom} = \frac{1.590 \times 10^{-22} \text{ g}}{2 \text{ atoms}} = 7.950 \times 10^{-23} \text{ g/atom}$$

From the known atomic weight of Ti (47.88), we know that the mass of one mole of Ti is 47.88 g/mol. Avogadro's number represents the number of atoms per mole, and can be calculated as

$$N_{Av} = \frac{47.88 \text{ g}}{\text{mol}} \times \frac{1 \text{ atom}}{7.950 \times 10^{-23} \text{ g}} = \boxed{6.023 \times 10^{23} \text{ atoms/mol}}$$

You should now work Exercise 82.

▼ PROBLEM-SOLVING TIP   *Nearest Neighbors*

In simple cubic structures the nearest neighbors are along the cell edge. In face-centered cubic structures the nearest neighbors are along the face diagonal; in body-centered cubic structures they are along the body diagonal.

Ionic Solids

Most salts crystallize as ionic solids with ions occupying the unit cell. Sodium chloride (Figure 13-27) is an example. Many other salts crystallize in the sodium chloride (face-centered cubic) arrangement. Examples are the halides of Li^+, K^+, and Rb^+, and

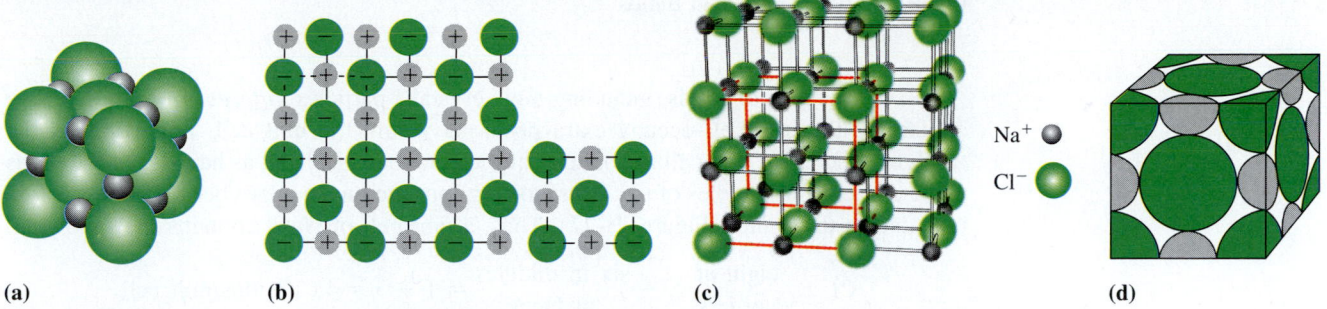

(a)   (b)   (c) Na^+ ● Cl^- ●   (d)

Figure 13-27 Some representations of the crystal structure of sodium chloride, NaCl. Sodium ions are shown in gray and chloride ions are shown in green. (a) One unit cell of the crystal structure of sodium chloride. (b) A cross section of the space lattice of NaCl, showing the repeating pattern of its unit cell at the right. The dashed lines outline an alternative choice of the unit cell. The entire pattern is generated by repeating either unit cell (and its contents) in all three directions. Several such choices of unit cells are usually possible. (c) The three-dimensional representation of part (b). One unit cell is outlined in red. (d) A representation of the unit cell of sodium chloride that indicates the relative sizes of the Na^+ and Cl^- ions as well as how ions are shared between unit cells. Particles at the corners, edges, and faces of unit cells are shared by other unit cells. Remember that there is an additional Na^+ ion at the center of the cube.

$M^{2+}X^{2-}$ oxides and sulfides such as MgO, CaO, CaS, and MnO. Two other common ionic structures are those of cesium chloride, CsCl (simple cubic lattice), and zincblende, ZnS (face-centered cubic lattice), shown in Figure 13-28. Salts that are isomorphous with the CsCl structure include CsBr, CsI, NH_4Cl, TlCl, TlBr, and TlI. The sulfides of Be^{2+}, Cd^{2+}, and Hg^{2+}, together with CuBr, CuI, AgI, and ZnO, are isomorphous with the zincblende structure (Figure 13-28c).

The ions in an ionic solid can vibrate only about their fixed positions, so ionic solids are poor electrical and thermal conductors. However, molten ionic compounds are excellent conductors because their ions are freely mobile.

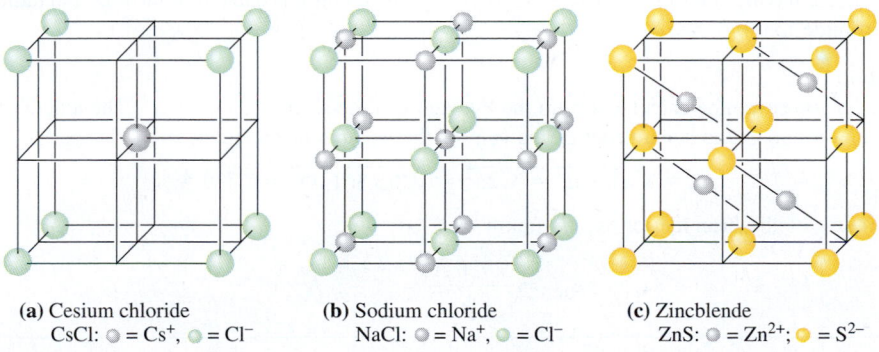

(a) Cesium chloride
CsCl: ● = Cs^+, ● = Cl^-

(b) Sodium chloride
NaCl: ● = Na^+, ● = Cl^-

(c) Zincblende
ZnS: ● = Zn^{2+}; ● = S^{2-}

Figure 13-28 Crystal structures of some ionic compounds of the MX type. The gray circles represent cations. One unit cell of each structure is shown. (a) The structure of cesium chloride, CsCl, is simple cubic. It is *not* body-centered, because the point at the center of the cell (Cs^+) is not the same as the point at a corner of the cell (Cl^-). (b) Sodium chloride, NaCl, is face-centered cubic. (c) Zincblende, ZnS, is face-centered cubic, with four Zn^{2+} and four S^{2-} ions per unit cell. The Zn^{2+} ions are related by the same translations as the S^{2-} ions.

In certain types of solids, including ionic crystals, particles *different* from those at the corners of the unit cell occupy extra positions within the unit cell. For example, the face-centered cubic unit cell of sodium chloride can be visualized as having chloride ions at the corners and middles of the faces; sodium ions are on the edges between the chloride ions and in the center (Figure 13-27). Thus, a unit cell of NaCl contains the following.

$$Cl^-: \quad \begin{matrix} \text{eight at} \\ \text{corners} \\ 8(\frac{1}{8}) \end{matrix} \quad + \quad \begin{matrix} \text{six in middles} \\ \text{of faces} \\ 6(\frac{1}{2}) \end{matrix} \quad = 1 + 3 = 4 \; Cl^- \; \text{ions/unit cell}$$

$$Na^+: \quad \begin{matrix} \text{twelve on} \\ \text{edges} \\ 12(\frac{1}{4}) \end{matrix} \quad + \quad \begin{matrix} \text{one in center} \\ \\ 1 \end{matrix} \quad = 3 + 1 = 4 \; Na^+ \; \text{ions/unit cell}$$

The unit cell contains equal numbers of Na^+ and Cl^- ions, as required by its chemical formula. Alternatively, we could translate the unit cell by half its length in any axial direction within the lattice, and visualize the unit cell in which sodium and chloride ions have exchanged positions. Such an exchange is not always possible. You should confirm that this alternative description also gives four chloride ions and four sodium ions per unit cell.

Ionic radii such as those in Figure 6-1 and Table 14-1 are obtained from X-ray crystallographic determinations of unit cell dimensions, assuming that adjacent ions are in contact with each other.

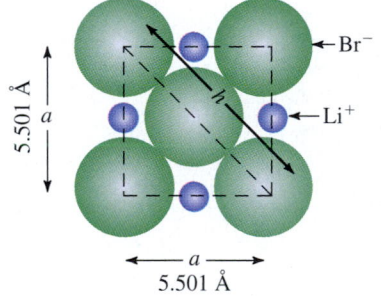

Figure 13-29 One face of the face-centered cubic unit cell of lithium bromide (Example 13-12).

EXAMPLE 13-12   *Ionic Radii versus Crystal Data*

Lithium bromide, LiBr, crystallizes in the NaCl face-centered cubic structure with a unit cell edge length of $a = b = c = 5.501$ Å. Assume that the Br^- ions at the corners of the unit cell are in contact with those at the centers of the faces. Determine the ionic radius of the Br^- ion. One face of the unit cell is depicted in Figure 13-29.

Plan

We may visualize the face as consisting of two right isosceles triangles sharing a common hypotenuse, h, and having sides of length $a = 5.501$ Å. The hypotenuse is equal to four times the radius of the bromide ion, $h = 4r_{Br^-}$.

Solution

The hypotenuse can be calculated from the Pythagorean theorem, $h^2 = a^2 + a^2$. The length of the hypotenuse equals the square root of the sum of the squares of the sides.

$$h = \sqrt{a^2 + a^2} = \sqrt{2a^2} = \sqrt{2(5.501 \text{ Å})^2} = 7.780 \text{ Å}$$

The radius of the bromide ion is one fourth of h, so

$$r_{Br^-} = \frac{7.780 \text{ Å}}{4} = \boxed{1.945 \text{ Å}}$$

This value is fairly close to the ionic radius of the Br^- ion listed in Figure 6-1. However, the tabulated value is the *average* value obtained from a number of crystal structures of compounds containing Br^- ions. Calculations of ionic radii assume that anion–anion contact exists. This is not always true. Therefore, calculated radii vary from structure to structure. We should not place too much emphasis on a value of an ionic radius obtained from any *single* structure determination.

EXAMPLE 13-13   *Ionic Radii versus Crystal Data*

Refer to Example 13-12. Calculate the ionic radius of Li^+ in LiBr, assuming anion–cation contact along an edge of the unit cell.

Plan

The edge length, $a = 5.501$ Å, is twice the radius of the Br^- ion plus twice the radius of the Li^+ ion. We know from Example 13-12 that the radius for the Br^- ion is 1.945 Å.

Solution

$$5.501 \text{ Å} = 2 \, r_{Br^-} + 2 \, r_{Li^+}$$

$$2 \, r_{Li^+} = 5.501 \text{ Å} - 2(1.945 \text{ Å}) = 1.611 \text{ Å}$$

$$r_{Li^+} = \boxed{0.806 \text{ Å}}$$

You should now work Exercises 74 and 78.

The radius of Li^+ is calculated to be 0.806 Å, but Figure 6-1 lists it as 0.60 Å. The discrepancy results from the assumption that the Li^+ ion is in contact with both Br^- ions simultaneously. It is too small and is free to vibrate about a fixed-center position between two large Br^- ions. We now see that there is some difficulty in determining precise values of ionic radii. Similar difficulties can arise in the determination of atomic radii from molecular and covalent solids or of metallic radii from solid metals.

Molecular Solids

The lattice positions that describe unit cells of molecular solids represent molecules or monatomic elements (sometimes referred to as monatomic molecules). Figure 13-30 shows the unit cells of two simple molecular crystals. Although the bonds *within* mole-

(Text continues on p. 482)

Figure 13-30 The packing arrangement in a molecular crystal depends on the shape of the molecule as well as on the electrostatic interactions of any regions of excess positive and negative charge in the molecules. The arrangements in some molecular crystals are shown here, with one unit cell outlined: (a) carbon dioxide, CO_2; (b) benzene, C_6H_6.

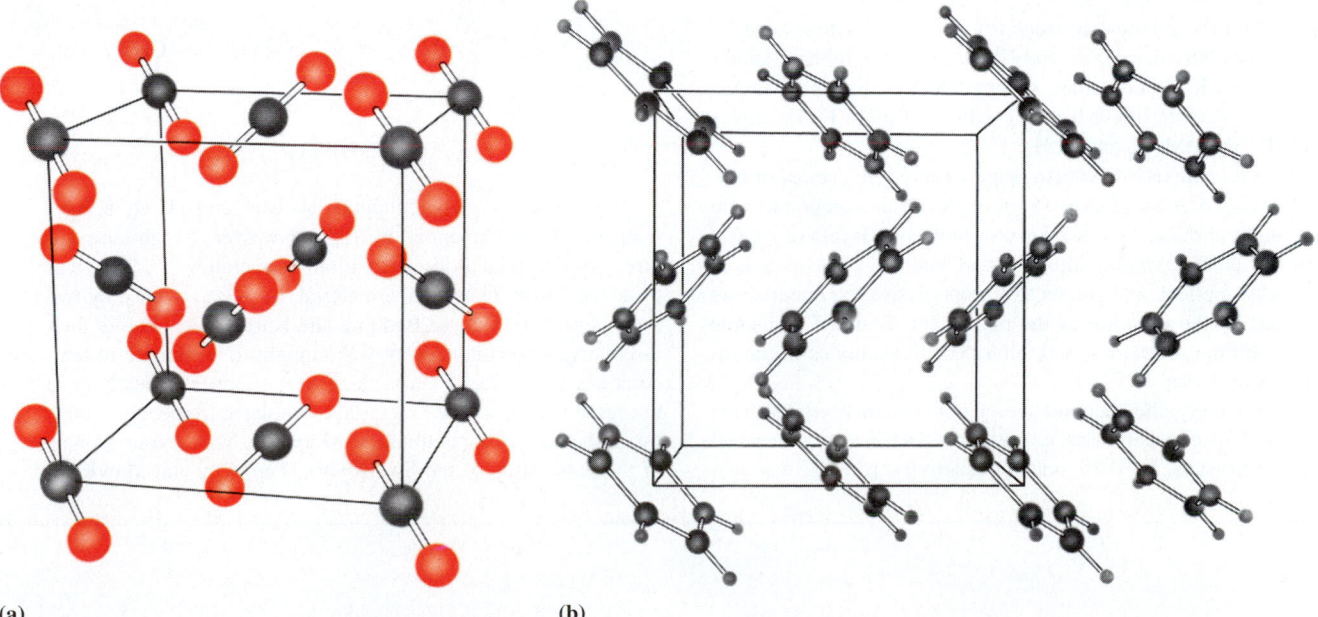

(a)   (b)

CHEMISTRY IN USE

Research & Technology

Superconductivity

Superconductivity was discovered by a Dutch scientist named Heike K. Onnes. In 1908 he was the first to accomplish the liquefaction of helium. Using liquid helium as a coolant, he then began to study the low-temperature conductive properties of metals. In 1911 he observed that the resistivity of mercury displays a remarkable behavior as it is cooled to a temperature approaching absolute zero. At about 4 K, all electrical resistance of mercury is suddenly lost within a temperature range of 0.01 K. In 1913 Onnes concluded that "mercury has passed into a new state that, on account of its remarkable electrical properties, may be called the superconducting state." The temperature at which a material enters the superconducting state is now called its *superconducting transition temperature, T_c*.

Soon after its initial discovery, superconductivity was observed in lead at 7.2 K and in tin at 3.7 K. Work done in the past 80 years has shown that superconductivity is a widespread phenomenon. Today about half the metallic elements have been shown to be superconductors. The highest superconducting transition temperature for a pure metal is that observed for niobium, $T_c = 9.5$ K. A number of compounds, as well as a large number of alloys, also exhibit superconductivity. Many of them display transition temperatures higher than those displayed by the pure metals.

Among the more familiar superconductors are those described as A15 superconductors ("A15" is a crystallographic symbol for the β-tungsten structure). This class includes materials such as Nb_3Al, Nb_3Ge, and Nb_3Sn, which exhibit T_c values up to 23.3 K. Many other superconductors have cubic β-1 (NaCl) structures; these have T_c values up to 18 K. Examples include NbN, MoN, and PdH.

Superconductivity has also been observed in a series of materials called *intercalation compounds*. Such compounds are formed by the insertion of atomic or molecular layers of a guest species, an *intercalant*, into the host material. The electrical, magnetic, optical, and conductive properties of the material are affected by the presence of the intercalant. Graphite and some metal sulfides, selenides, and tellurides are examples of the intercalation hosts.

Surprisingly, the phenomenon of superconductivity has been observed in unusual cases in materials such as polymers and organic crystals. In 1975, superconductivity at 0.3 K was dis-

covered in a material called polythiazyl, $(SN)_x$ (the subscript x indicates a large number of variable size). This observation provided the first, and still the only, example of superconductivity in a polymeric system, as well as the first example of superconductivity at ambient pressure in a material containing no metallic elements. Superconducting materials often contain an organic species or a metal in more than one oxidation state (sometimes called a mixed-valence species). Superconductivity ($T_c = 0.9$ K) in such an organic material (containing no metallic elements) was discovered in 1980 in the selenium-based salt $(TMTSF)_2PF_6$. TMTSF is tetramethyltetraselenafulvalene.

TMTSF

Since this initial discovery, a number of additional (TMTSF)X derivatives have been synthesized and found to be superconductors. In these, X is TaF_6^-, AsF_6^-, ReO_4^-, FSO_3^-, or ClO_4^-. More recently, superconductivity has been discovered in similar salts of the sulfur-based system $(BEDT-TTF)_2X$, where BEDT-TTF is bis-ethylenedithiotetrathiafulvalene.

BEDT-TTF

Until recently, superconductivity was considered a low-temperature phenomenon. In 1986, however, "high-temperature" superconductivity (transition temperature ≈ 35 K) was discovered in a copper oxide material, $La_{1.8}Ba_{0.2}CuO_4$, by two Swiss scientists, George Bednorz and Karl Muller. Before their discovery, superconductivity was thought to be limited to temperatures below 23.3 K. The discovery of superconductivity in the ceramic copper oxide materials stimulated tremendous interest in the scientific community. Moreover, within four months of the publication of the Swiss work, Paul Chu and MawKuen

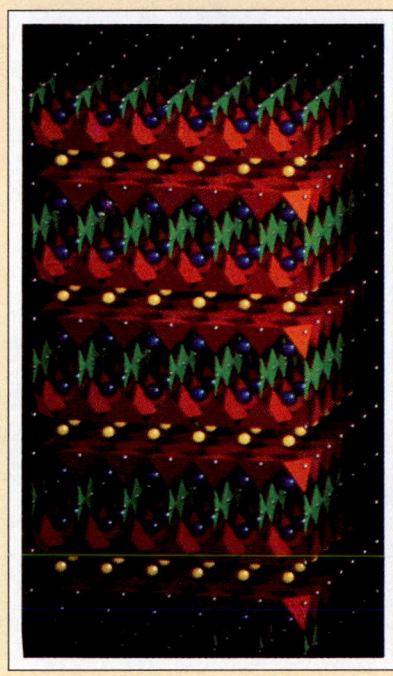

Until recently, only small, coin-size magnets could be floated above the new high-temperature superconductors. This is due to limited particle size in the ceramic superconductor. Attempts to increase the particle size caused the brittle ceramic to crack. A team of workers at the International Superconductivity Technical Center in Tokyo added a small amount of silver during the preparation of a superconductor. The result was a great increase in the weight that could be raised. Here, a goldfish in a 4.5-pound tank resting on a ring magnet is levitated.

A representation of the structure of the $YBa_2Cu_3O_7$ superconductor.

Wu found that a similar material, $YBa_2Cu_3O_7$, is superconducting at temperatures up to 92 K. While 92 K is not warm by most standards, it is well above the boiling point of liquid nitrogen (77 K). Liquid nitrogen is an inexpensive coolant that costs about 25 cents per liter. Liquid helium, on the other hand, which was required to cool the previous generations of low-temperature superconductors, is an inefficient refrigerant and costs about $4 per liter. Thus, there could be a tremendous economic advantage to using the new copper oxide superconductors in practical applications.

Development of liquid-nitrogen-cooled superconductors could have dramatic effects on research, industry, electronics, transportation, medicine, and utilities. Currently, nearly one third of the electric power transmitted over long distances is lost because of resistive heating in the power lines. Superconductive transmission lines could carry current for hundreds of miles without such dissipative losses; hence, power plants such as nuclear generators could be located far from population centers. Smaller, more powerful computers could be made with superconductive microchips. The strong magnetic fields that are generated with superconducting coils could lead to many applica-

tions. High-speed trains could be levitated above the tracks by superconductor-generated magnetic fields, thereby eliminating friction at the track surface. Magnetic resonance imaging (MRI) scanners, used in diagnostic medicine to make images of body tissue, could be built more cheaply and simply, making them accessible to many more hospitals and clinics. Furthermore, the strong magnetic fields produced by superconductors could aid high-energy physics research. Currently, the plasma produced in fusion reactions cannot be contained by any known material, but with the use of superconducting magnets the plasma might be contained by strong magnetic fields.

Much work must still be done to translate laboratory results into such revolutionary technological applications. Intensive research continues to develop such applications, to search for other superconductive materials, and to illuminate the underlying mechanism that is responsible for high-temperature superconductivity.

Professor John T. McDevitt
University of Texas at Austin

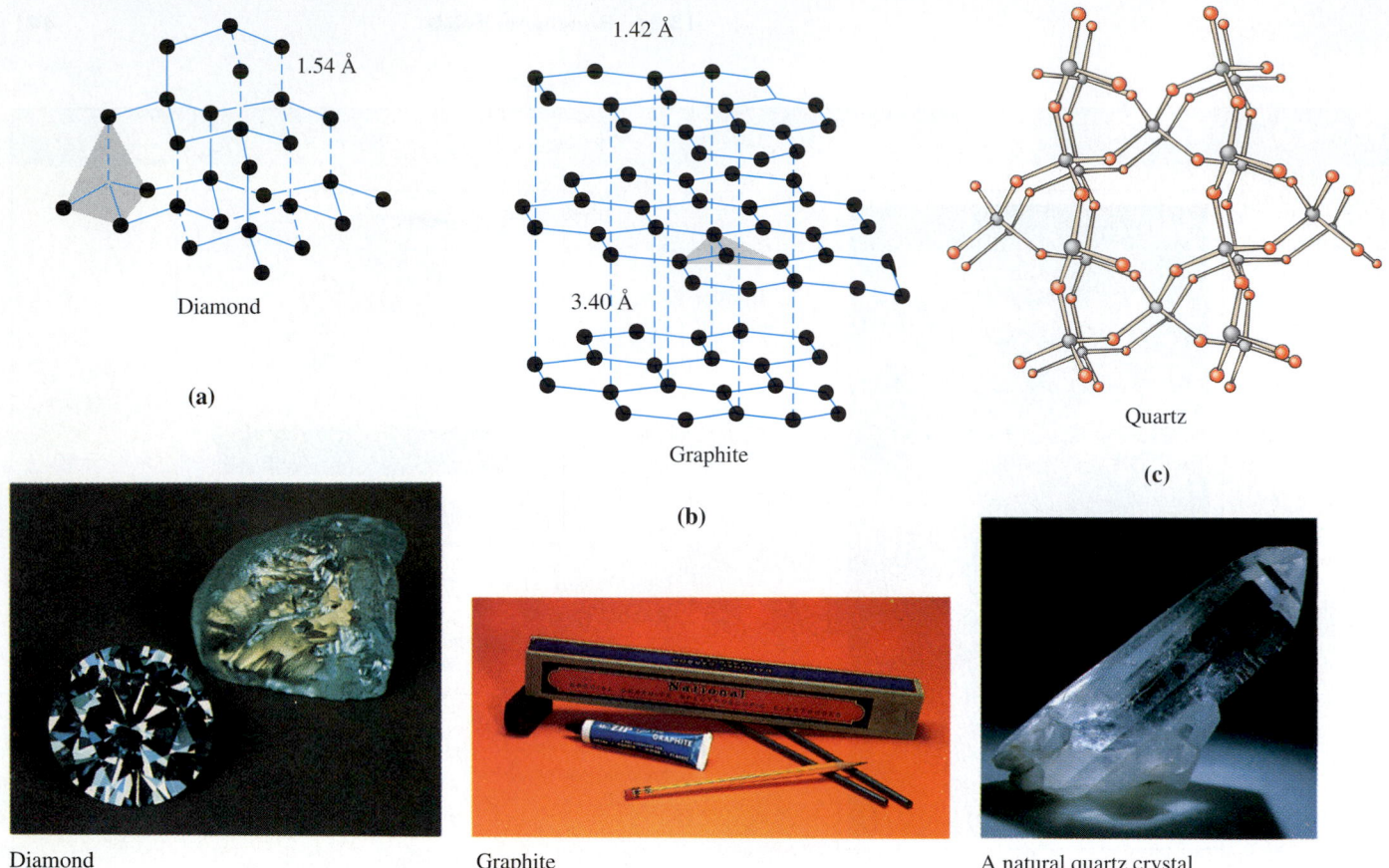

1.54 Å

Diamond

(a)

1.42 Å

3.40 Å

Graphite

(b)

Quartz

(c)

Diamond

Graphite

A natural quartz crystal

Figure 13-31   Portions of the atomic arrangements in three covalent solids. (a) Diamond. Each C is bonded tetrahedrally to four others through sp^3-sp^3 σ-bonds (1.54 Å). (b) Graphite. C atoms are linked (1.42 Å) in planes by sp^2-sp^2 σ-bonds. Electrons move freely through the π-bonding network in these planes, but they do not jump between planes easily. The crystal is soft, owing to the weakness of the attractions between planes (3.40 Å). (c) Quartz (SiO_2). Each Si atom (gray) is bonded tetrahedrally to four O atoms (red).

cules are covalent and strong, the forces of attraction *between* molecules are much weaker. They range from hydrogen bonds and weaker dipole–dipole interactions in polar molecules such as H_2O and SO_2 to very weak London forces in symmetrical, nonpolar molecules such as CH_4, CO_2, and O_2 and monatomic elements, e.g., the noble gases. Because of the relatively weak intermolecular forces of attraction, molecules can be easily displaced. Thus, molecular solids are usually soft substances with low melting points. Because electrons do not move from one molecule to another under ordinary conditions, molecular solids are poor electrical conductors and good insulators.

London forces are also present among polar molecules.

Covalent Solids

Covalent solids (or "network solids") can be considered giant molecules that consist of covalently bonded atoms in an extended, rigid crystalline network. Diamond (one crystalline form of carbon) and quartz are examples of covalent solids (Figure 13-31). Because of their rigid, strongly bonded structures, *most* covalent solids are very hard and melt at high temperatures. Because electrons are localized in covalent bonds, they are not freely mobile. As a result, covalent solids are *usually* poor thermal and electrical conductors at ordinary temperatures. (However, diamond is a good conductor of heat; jewelers use this property to distinguish diamonds from imitations.)

An important exception to these generalizations about properties is *graphite,* an allotropic form of carbon. It has the layer structure shown in Figure 13-31c. The overlap of an extended π-electron network in each plane makes graphite an excellent conductor. The very weak attraction between layers allows these layers to slide over one another easily. Graphite is used as a lubricant, as an additive for motor oil, and in pencil "lead" (combined with clay and other fillers to control hardness).

It is interesting to note that the allotropes of carbon include one very hard substance and one very soft substance. They differ only in the arrangement and bonding of the C atoms.

13-17 BAND THEORY OF METALS

As described in the previous section, most metals crystallize in close-packed structures. The ability of metals to conduct electricity and heat must result from strong electronic interactions among the 8 to 12 nearest neighbors. This is somewhat difficult to rationalize if we recall that each Group IA and Group IIA metal atom has only one or two valence electrons available for bonding. This is too few to participate in bonds localized between it and each of its nearest neighbors.

Bonding in metals is called **metallic bonding.** It results from the electrical attractions among positively charged metal ions and mobile, delocalized electrons belonging to the crystal as a whole. The properties associated with metals—metallic luster, high thermal and electrical conductivity, and so on—can be explained by the **band theory** of metals, which we now describe.

The interaction of two atomic orbitals, say the 3s orbitals of two sodium atoms, produces two molecular orbitals, one bonding orbital and one antibonding orbital (Chapter 9). If N atomic orbitals interact, N molecular orbitals are formed. In a single metallic crystal containing one mole of sodium atoms, for example, the interaction of 6.022×10^{23} 3s atomic orbitals produces 6.022×10^{23} molecular orbitals. Atoms interact more strongly with nearby atoms than with those farther away. As discussed in Chapter 9, the energy that separates bonding and antibonding molecular orbitals resulting from two given atomic orbitals decreases as the interaction (overlap) between the atomic orbitals decreases. When we consider all possible interactions among the mole of Na atoms, there results a series of very closely spaced molecular orbitals (formally σ_{3s} and $\sigma_{3s}^{\star}$). These comprise a nearly continuous **band** of orbitals belonging to the crystal as a whole. One mole of Na atoms contributes 6.022×10^{23} valence electrons (Figure 13-32a), so the 6.022×10^{23} orbitals in the band are half-filled.

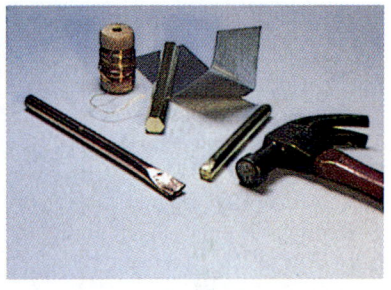

Metals can be formed into many shapes because of their malleability and ductility.

Figure 13-32 (a) The band of orbitals resulting from interaction of the 3s orbitals in a crystal of sodium. (b) Overlapping of a half-filled "3s" band (black) with an empty "3p" band (red) of Na_N crystal.

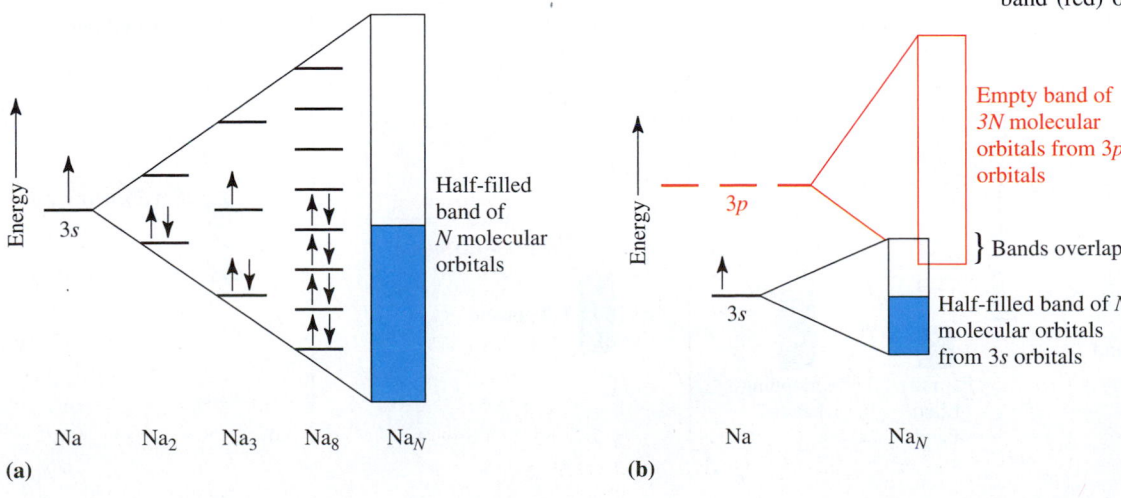

The empty $3p$ atomic orbitals of the Na atoms also interact to form a wide band of $3 \times 6.022 \times 10^{23}$ orbitals. The $3s$ and $3p$ atomic orbitals are quite close in energy, so the fanned-out bands of molecular orbitals overlap, as shown in Figure 13-32b. The two overlapping bands contain $4 \times 6.022 \times 10^{23}$ orbitals and only 6.022×10^{23} electrons. Because each orbital can hold two electrons, the resulting combination of bands is only one-eighth full.

The ability of metallic Na to conduct electricity is due to the ability of any of the highest-energy electrons in the "$3s$" band to jump to a slightly higher-energy vacant orbital in the same band when an electric field is applied. The resulting net flow of electrons through the crystal is in the direction of the applied field.

Overlap of "$3s$" and "$3p$" bands is not necessary to explain the ability of Na or of any other Group IA metal to conduct electricity. It could do so utilizing only the half-filled "$3s$" band. In the Group IIA metals, however, such overlap is important. Consider a crystal of magnesium as an example. The $3s$ atomic orbital of an isolated Mg atom is filled with two electrons. Thus, without this overlap, the "$3s$" band in a crystal of Mg is also filled. Mg is a good conductor at room temperature because the highest-energy electrons are able to move readily into vacant orbitals in the "$3p$" band (Figure 13-33).

According to band theory, the highest-energy electrons of metallic crystals occupy either a partially filled band or a filled band that overlaps an empty band. A band within which (or into which) electrons must move to allow electrical conduction is called a **conduction band.** The electrical conductivity of a metal decreases as temperature increases. The increase in temperature causes thermal agitation of the metal ions. This impedes the flow of electrons when an electric field is applied.

Crystalline nonmetals, such as diamond and phosphorus, are **insulators**—they do not conduct electricity. The reason is that their highest-energy electrons occupy filled bands of molecular orbitals that are separated from the lowest empty band (conduction band) by an energy difference called the **band gap.** In an insulator, this band gap is an energy difference that is too large for electrons to jump to get to the conduction band (Figure 13-34).

Elements that are **semiconductors** have filled bands that are only slightly below, but do not overlap with, empty bands. They do not conduct electricity at low temperatures, but a small increase in temperature is sufficient to excite some of the highest-energy electrons into the empty conduction band.

Let us now explain some of the physical properties of metals in terms of the band theory of metallic bonding.

The alkali metals are those of Group IA; the alkaline earth metals are those of Group IIA.

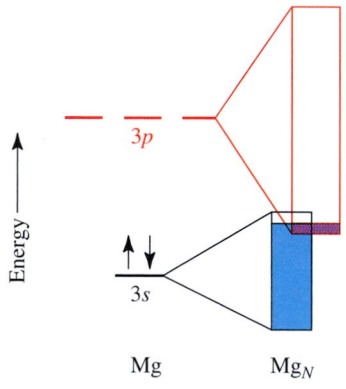

Figure 13-33 Overlapping of a filled "$3s$" band (blue) with an empty "$3p$" band of Mg$_N$ crystal. The higher-energy electrons are able to move into the "$3p$" band (red) as a result of this overlap.

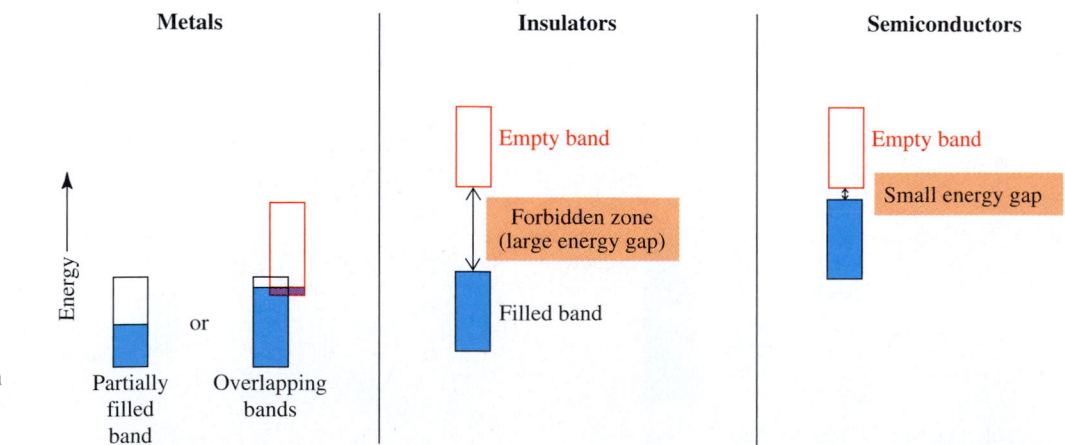

Figure 13-34 Distinction among metals, insulators, and semiconductors. In each case an unshaded area represents a conduction band.

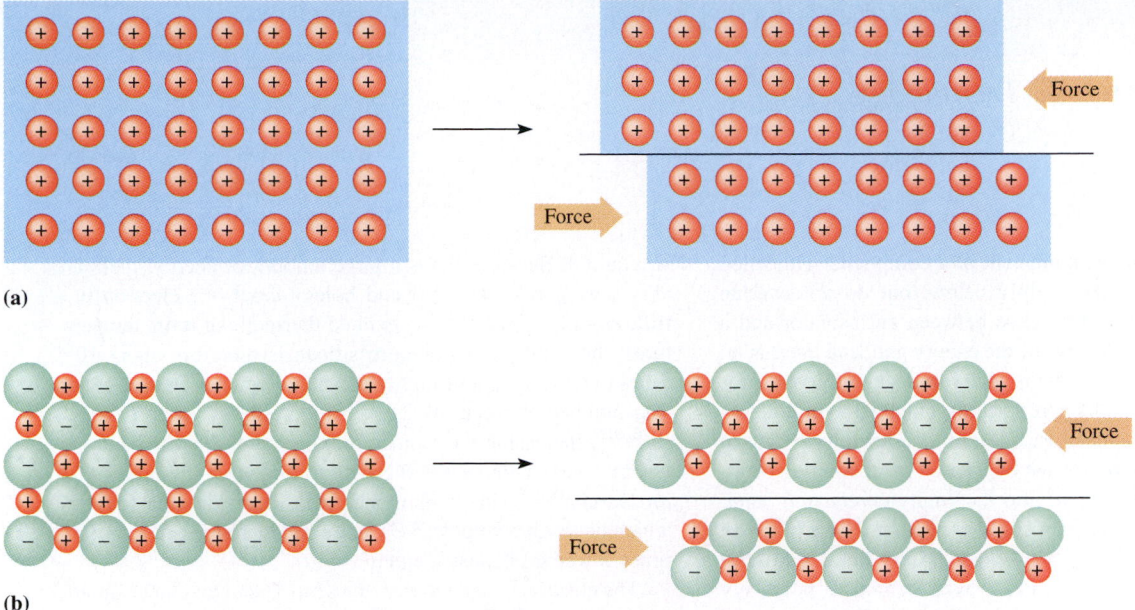

(a)

(b)

Figure 13-35 (a) In a metal, the positively charged metal ions are immersed in a "sea of electrons." When the metal is distorted (e.g., rolled into sheets or drawn into wires), the environment around the metal atoms is essentially unchanged, and no new repulsive forces occur. This explains why metal sheets and wires remain intact. (b) By contrast, when an ionic crystal is subjected to a force that causes it to slip along a plane, the increased repulsive forces between like-charged ions cause the crystal to break.

1. We have just accounted for the *ability of metals to conduct electricity.*

2. Metals are also *conductors of heat.* They can absorb heat as electrons become thermally excited to low-lying vacant orbitals in a conduction band. The reverse process accompanies the release of heat.

3. Metals have a *lustrous appearance* because the mobile electrons can absorb a wide range of wavelengths of radiant energy as they jump to higher energy levels. Then they emit photons of visible light and fall back to lower levels within the conduction band.

4. Metals are *malleable and/or ductile.* A crystal of a metal is easily deformed when a mechanical stress is applied to it. All of the metal ions are identical, and they are imbedded in a "sea of electrons." As bonds are broken, new ones are readily formed with adjacent metal ions. The features of the lattice remain unchanged, and the environment of each metal ion is the same as before the deformation occurred (Figure 13-35). The breakage of bonds involves the promotion of electrons to higher-energy levels. The formation of bonds is accompanied by the return of the electrons to the original energy levels.

A **malleable** substance can be rolled or pounded into sheets. A **ductile** substance can be drawn into wires.

A 6-inch wafer of ultrapure silicon, a semiconductor (left). Gallium phosphide, GaP, is a semiconducting compound (right). The analogous semiconducting compound gallium arsenide, GaAs, is used in many solid-state electronic devices.

Semiconductors

A **semiconductor** is an element or a compound with filled bands that are only slightly below, but do not overlap with, empty bands. The difference between an insulator and a semiconductor is only the size of the energy gap, and there is no sharp distinction between them. An **intrinsic** semiconductor (i.e., a semiconductor in its pure form) is a much poorer conductor of electricity than a metal because, for conduction to occur in a semiconductor, electrons must be excited from bonding orbitals in the filled *valence band* into the empty *conduction band*. Figure (a) shows how this happens. An electron that is given an excitation energy greater than or equal to the **band gap (E_g)** enters the conduction band and leaves behind a positively charged **hole** (h^+, the absence of a bonding electron) in the valence band. Both the electron and the hole reside in *delocalized* orbitals, and both can move in an electric field, much as electrons move in a metal. (Holes can migrate because an electron in a nearby orbital can move to fill in the hole, thereby creating a new hole in the nearby orbital.) Electrons and holes move in opposite directions in an electric field.

Silicon, a semiconductor of great importance in electronics, has a band gap of 1.94×10^{-22} kJ, or 1.21 *electron volts* (eV). This is the energy needed to create one electron and one hole or, put another way, the energy needed to break one Si—Si bond. This energy can be supplied either thermally or by using light with a photon energy greater than the band gap. To excite one *mole* of electrons from the valence band to the conduction band, an energy of

$$\frac{6.022 \times 10^{23} \text{ electrons}}{\text{mol}} \times \frac{1.94 \times 10^{-22} \text{ kJ}}{\text{electron}} = 117 \text{ kJ/mol}$$

is required. Because this is a large amount of energy, there are very few mobile electrons and holes (about one electron in a trillion—i.e., 1 in 10^{12}—is excited thermally at room temperature); the conductivity of pure silicon is therefore about 10^{11} times lower than that of highly conductive metals such as silver. The number of electrons excited thermally is proportional to $e^{-E_g/RT}$. Increasing the temperature or decreasing the band gap energy leads to higher conductivity for an intrinsic semiconductor. Insulators such as diamond and silicon dioxide (quartz), which have large values of E_g, have conductivities 10^{15} to 10^{20} times lower than most metals.

The electrical conductivity of a semiconductor can be greatly increased by **doping** with impurities. For example, silicon, a Group IVA element, can be doped by adding small amounts of a Group VA element, such as phosphorus, or a Group IIIA element, such as boron. Figure (b) shows the effect of substituting phosphorus for silicon in the crystal structure (silicon has the same structure as diamond, Figure 13-31a). There are exactly enough valence band orbitals to accommodate four of the valence electrons from the phosphorus atom. However, the phosphorus atom has one more electron (and one more proton in its nucleus) than does silicon. The fifth electron enters a higher energy orbital that is localized in the lattice near the phosphorus atom; the energy of this orbital, called a **donor level,** is just below the conduction band, within the energy gap. An electron in this orbital can easily become *delocalized* when a small amount of thermal energy promotes it into the conduction band. Because the phosphorus-doped silicon contains mobile, *negatively* charged carriers (electrons), it is said to be doped **n-type.** Doping the silicon crystal with boron produces a related, but opposite, effect. Each boron atom contributes only three valence

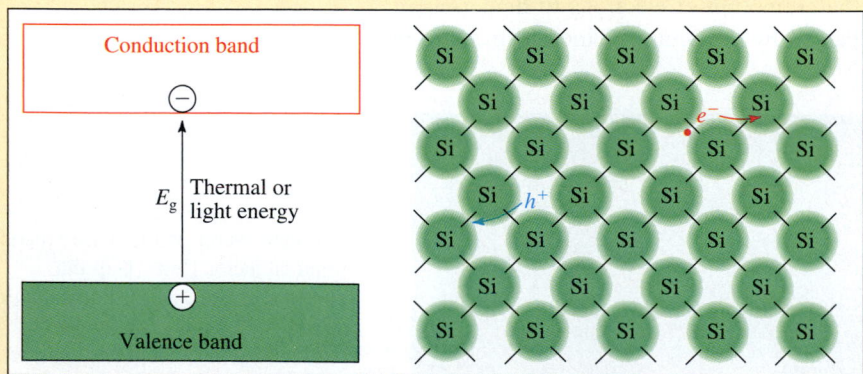

(a)   Generation of an electron–hole pair in silicon, an intrinsic semiconductor. The electron (e^-) and hole (h^+) have opposite charges, and so move in opposite directions in an electric field.

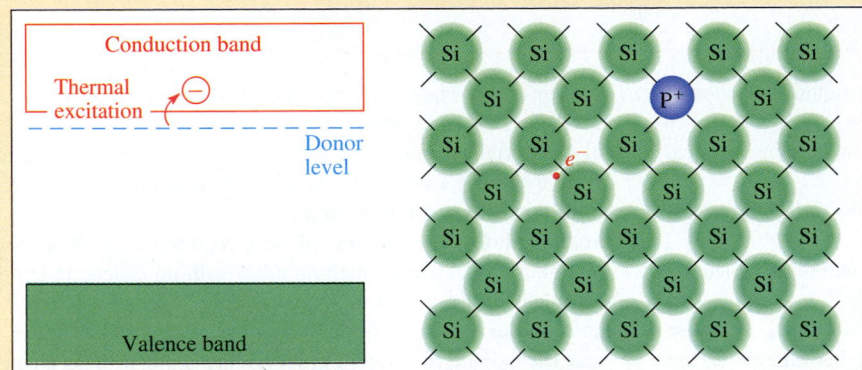

(b) n-type doping of silicon by phosphorus. The extra valence electron from a phosphorus atom is thermally excited into the conduction band, leaving a fixed positive charge on the phosphorus atom.

The colors of semiconductors are determined by the band gap energy E_g. Only photons with energy greater than E_g can be absorbed. From the Planck radiation formula ($E = h\nu$) and $\lambda\nu = c$, we calculate that the wavelength, λ, of an absorbed photon must be less than hc/E_g. Gallium arsenide (GaAs; $E_g = 1.4$ eV) absorbs photons of wavelengths shorter than 890 nm, which is in the near infrared region. Because it absorbs all wavelengths of visible light, gallium arsenide appears black to the eye. Iron oxide (Fe_2O_3; $E_g = 2.2$ eV) absorbs light of wavelengths shorter than 570 nm; it absorbs both yellow and blue light, and therefore appears red. Cadmium sulfide (CdS; $E_g = 2.6$ eV), which absorbs blue light ($\lambda \le 470$ nm), appears yellow. Strontium titanate ($SrTiO_3$; $E_g = 3.2$ eV) absorbs only in the ultraviolet ($\lambda \le 390$ nm) and therefore appears white to the eye.

electrons to bonding orbitals in the valence band, and therefore a *hole* is localized near each boron atom. Thermal energy is enough to separate the negatively charged boron atom from the hole, delocalizing the latter. In this case the charge carriers are *positive* holes, and the crystal is doped **p-type.** In both *p-* and *n*-type doping, an extremely small concentration of dopants (as little as one part per billion) is enough to cause a significant increase in conductivity. For this reason, great pains are taken to purify the semiconductors used in electronic devices.

Even in a doped semiconductor, mobile electrons and holes are both present, although one carrier type is predominant. For example, in a sample of silicon doped with arsenic (*n*-type doping), the concentrations of mobile electrons are slightly less than the concentration of arsenic atoms (usually expressed in terms of atoms/cm^3), and the concentrations of mobile holes are extremely low. Interestingly, the concentrations of electrons and holes always follow an equilibrium expression that is entirely analogous to that for the autodissociation of water into H^+ and OH^- ions (Chapter 18); that is,

$$[e^-][h^+] = K_{eq}$$

where the equilibrium constant K_{eq} depends only on the identity of the semiconductor and the absolute temperature. For silicon at room temperature, $K_{eq} = 4.9 \times 10^{19}$ carriers^2/cm^6.

Doped semiconductors are extremely important in electronic applications. A **p–n junction** is formed by joining *p-* and *n*-type semiconductors. At the junction, free electrons and holes combine, annihilating each other and leaving positively and negatively charged dopant atoms on opposite sides. The unequal charge distribution on the two sides of the junction causes an electric field to develop and gives rise to current rectification (electrons can flow, with a small applied voltage, only from the *n* side to the *p* side of the junction; holes flow only in the reverse direction). Devices such as **diodes** and bipolar **transistors,** which form the bases of most analog and digital electronic circuits, are composed of *p–n* junctions.

Professor Thomas A. Mallouk
University of Texas at Austin

Key Terms

Adhesive force   Force of attraction between a liquid and another surface.

Allotropes   Different forms of the same element in the same physical state.

Amorphous solid   A noncrystalline solid with no well-defined, ordered structure.

Band   A series of very closely spaced, nearly continuous molecular orbitals that belong to the crystal as a whole.

Band gap   An energy separation between an insulator's highest filled electron energy band and the next higher-energy vacant band.

Band theory of metals   A theory that accounts for the bonding and properties of metallic solids.

Boiling point   The temperature at which the vapor pressure of a liquid is equal to the external pressure; also the condensation point.

Capillary action   The drawing of a liquid up the inside of a small-bore tube when adhesive forces exceed cohesive forces, or the depression of the surface of the liquid when cohesive forces exceed adhesive forces.

Cohesive forces   All the forces of attraction among particles of a liquid.

Condensation   Liquefaction of vapor.

Condensed phases   The liquid and solid phases; phases in which particles interact strongly.

Conduction band   A partially filled band or a band of vacant energy levels just higher in energy than a filled band; a band within which, or into which, electrons must be promoted to allow electrical conduction to occur in a solid.

Coordination number   In describing crystals, the number of nearest neighbors of an atom or ion.

Critical point   The combination of critical temperature and critical pressure of a substance.

Critical pressure   The pressure required to liquefy a gas (vapor) at its critical temperature.

Critical temperature   The temperature above which a gas cannot be liquefied; the temperature above which a substance cannot exhibit distinct gas and liquid phases.

Crystal lattice   The pattern of arrangement of particles in a crystal.

Crystalline solid   A solid characterized by a regular, ordered arrangement of particles.

Deposition   The direct solidification of a vapor by cooling; the reverse of sublimation.

Dipole–dipole interactions   Interactions between polar molecules, i.e., between molecules with permanent dipoles.

Dipole-induced dipole interaction   See *London forces.*

Distillation   The separation of a liquid mixture into its components on the basis of differences in boiling points.

Dynamic equilibrium   A situation in which two (or more) processes occur at the same rate so that no net change occurs.

Evaporation   Vaporization of a liquid below its boiling point.

Heat of condensation   The amount of heat that must be removed from a specific amount of a vapor at its condensation point to condense the vapor with no change in temperature; usually expressed in J/g. See *Molar heat of condensation.*

Heat of crystallization   The amount of heat that must be removed from a specific amount of a liquid at its freezing point to freeze it with no change in temperature.

Heat of fusion   The amount of heat required to melt a specific amount of a solid at its melting point with no change in temperature; usually expressed in J/g. See *Molar heat of fusion.*

Heat of vaporization   The amount of heat required to vaporize a specific amount of a liquid at its boiling point with no change in temperature; usually expressed in J/g.

Hydrogen bond   A fairly strong dipole–dipole interaction (but still considerably weaker than covalent or ionic bonds) between molecules containing hydrogen directly bonded to a small, highly electronegative atom, such as N, O, or F.

Insulator   A poor conductor of electricity and heat.

Intermolecular forces   Forces *between* individual particles (atoms, molecules, ions) of a substance.

Intramolecular forces   Forces between atoms (or ions) *within* molecules (or formula units).

Isomorphous   Refers to crystals having the same atomic arrangement.

LeChatelier's Principle   A system at equilibrium, or striving to attain equilibrium, responds in such a way as to counteract any stress placed upon it.

London forces   Very weak and very short-range attractive forces between short-lived temporary (induced) dipoles; also called dispersion forces.

Melting point   The temperature at which liquid and solid coexist in equilibrium; also the freezing point.

Meniscus   The upper surface of a liquid in a cylindrical container.

Metallic bonding   Bonding within metals due to the electrical attraction of positively charged metal ions for mobile electrons that belong to the crystal as a whole.

Molar heat capacity   The amount of heat necessary to raise the temperature of one mole of a substance one degree Celsius with no change in state; usually expressed in kJ/mol. See *Specific heat.*

Molar heat of condensation   The amount of heat that must be removed from one mole of a vapor at its condensation point to condense the vapor with no change in temperature; usually expressed in kJ/mol. See *Heat of condensation.*

Molar heat of fusion   The amount of heat required to melt one mole of a solid at its melting point with no change in temperature; usually expressed in kJ/mol. See *Heat of fusion.*

Molar heat of vaporization   The amount of heat required to vaporize one mole of a liquid at its boiling point with no change in temperature; usually expressed in kJ/mol. See *Heat of vaporization.*

Normal boiling point   The temperature at which the vapor pressure of a liquid is equal to one atmosphere pressure.

Normal melting point   The melting (freezing) point at one atmosphere pressure.

Phase diagram A diagram that shows equilibrium temperature–pressure relationships for different phases of a substance.

Polymorphous Refers to substances that crystallize in more than one crystalline arrangement.

Semiconductor A substance that does not conduct electricity well at low temperatures but that does at higher temperatures.

Specific heat The amount of heat necessary to raise the temperature of a specific amount of a substance one degree Celsius with no change in state; usually expressed in J/g°C. See *Molar heat capacity.*

Sublimation The direct vaporization of a solid by heating without passing through the liquid state.

Supercritical fluid A substance at a temperature above its critical temperature.

Surface tension A measure of the inward intermolecular forces of attraction among liquid particles that must be overcome to expand the surface area.

Triple point The point on a phase diagram that corresponds to the only pressure and temperature at which three phases (solid, liquid, and gas) of a substance can coexist at equilibrium.

Unit cell The smallest repeating unit showing all the structural characteristics of a crystal.

Vapor pressure The partial pressure of a vapor at the surface of its parent liquid.

Viscosity The tendency of a liquid to resist flow; the inverse of its fluidity.

Volatility The ease with which a liquid vaporizes.

Exercises

General Concepts

1. Explain why liquids are nearly incompressible and gases are very compressible.

2. Why does a gas completely fill its container, a liquid spread to take the shape of its container, and a solid retain its shape?

3. What causes London forces? What factors determine the strengths of London forces between molecules?

4. What is hydrogen bonding? Under what conditions can strong hydrogen bonds be formed?

5. Which of the following substances have permanent dipole–dipole forces? (a) SiH_4, (b) molecular $MgCl_2$, (c) NCl_3, (d) F_2O.

6. Which of the following substances have permanent dipole–dipole forces? (a) Molecular $AlCl_3$, (b) SF_6, (c) NO, (d) SeF_4.

7. For which of the substances in Exercise 5 are London forces the only important forces in determining boiling points?

8. For which of the substances in Exercise 6 are London forces the only important forces in determining boiling points?

9. For each of the following pairs of compounds, predict which compound would exhibit stronger hydrogen bonding. Justify your prediction. It may help to write a Lewis formula for each. (a) water, H_2O, or hydrogen sulfide, H_2S; (b) difluoromethane, CH_2F_2, or fluoroamine, NH_2F; (c) acetone, C_3H_6O (contains a C=O double bond) or ethyl alcohol, C_2H_6O (contains one C—O single bond).

10. For each of the following pairs of compounds, predict which would exhibit stronger hydrogen bonding. Justify your prediction. It may help to write a Lewis formula for each. (a) ammonia, NH_3, or phosphine, PH_3; (b) ethylene, C_2H_4, or hydrazine, N_2H_4; (c) hydrogen fluoride, HF, or hydrogen chloride, HCl.

11. Describe the intermolecular forces that are present in each of the following compounds. Which kind of force would have the greatest influence on the properties of each compound? (a) bromine pentafluoride, BrF_5; (b) acetone, C_3H_6O (contains a central C=O double bond); (c) carbonyl fluoride, F_2CO.

12. Describe the intermolecular forces that are present in each of the following compounds. Which kind of force would have the greatest influence on the properties of each compound? (a) ethyl alcohol, C_2H_6O (contains one C—O single bond); (b) phosphine, PH_3; (c) sulfur hexafluoride, SF_6.

13. Give the correct names for these changes in state: (a) Crystals of *para*-dichlorobenzene, used as a moth repellent, gradually become vapor without passing through the liquid phase. (b) As you enter a warm room from the outdoors on a cold winter day, your eyeglasses become fogged with a film of moisture. (c) On the same (windy) winter day, a pan of water is left outdoors. Some of it turns to vapor, the rest to ice.

14. The normal boiling point of trichlorofluoromethane, CCl_3F, is 24°C, and its freezing point is −111°C. Complete these sentences by supplying the proper terms. (a) At standard temperature and pressure, CCl_3F is a _____. (b) In an arctic winter at −40°C and 1 atm pressure, CCl_3F is a _____. If it is cooled to −120°C, the molecules arrange themselves in an orderly lattice, the CCl_3F _____ and becomes a _____. (c) If crystalline CCl_3F is held at a temperature of −120°C while a stream of helium gas is blown over it, the crystals will gradually disappear by the process of _____. If liquid CCl_3F is boiled at atmospheric pressure, it is converted to a _____ at a temperature of _____.

15. Why does HF have a lower boiling point and lower heat of vaporization than H_2O, even though their molecular weights are nearly the same and the hydrogen bonds between molecules of HF are stronger?

*16. Many carboxylic acids form dimers in which two molecules "stick together." These dimers result from the formation of *two* hydrogen bonds between the two molecules. Use acetic acid to draw a likely structure for this kind of hydrogen-bonded dimer.

$$CH_3-\overset{\overset{\displaystyle ..O..}{\|}}{C}-\overset{..}{\underset{..}{O}}-H$$

The Liquid State

17. Use the kinetic–molecular theory to describe the behavior of liquids with changing temperature. Why are liquids more dense than gases?

18. Distinguish between evaporation and boiling. Use the kinetic–molecular theory to explain the dependence of rate of evaporation on temperature.

19. Support or criticize the statement that liquids with high normal boiling points have low vapor pressures. Give examples of three common liquids that have relatively high vapor pressures at 25°C and three that have low vapor pressures at 25°C.

20. Within each group, assign each of the boiling points to the appropriate substance on the basis of intermolecular forces. (a) Ne, Ar, Kr: −246°C, −186°C, −152°C; (b) NH_3, H_2O, HF: −33°C, 20°C, 100°C.

21. Within each group, assign each of the boiling points to the respective substances on the basis of intermolecular forces. (a) N_2, HCN, C_2H_6: −196°C, −89°C, 26°C; (b) H_2, HCl, Cl_2: −35°C, −259°C, −85°C.

22. (a) What is the definition of the normal boiling point? (b) Why is it necessary to specify the atmospheric pressure over a liquid when measuring a boiling point?

23. What factors determine how viscous a liquid is? How does viscosity change with increasing temperature?

24. What is the surface tension of a liquid? What causes this property? How does surface tension change with increasing temperature?

25. What happens inside a capillary tube when a liquid "wets" the tube? What happens when a liquid does not "wet" the tube?

26. Choose from each pair the substance that, in the liquid state, would have the greater vapor pressure at a given temperature. Base your choice on predicted strengths of intermolecular forces. (a) $BiBr_3$ or $BiCl_3$, (b) CO or CO_2, (c) N_2 or NO, (d) CH_3COOH or $HCOOCH_3$.

27. Repeat Exercise 26 for (a) C_6H_6 or C_6Cl_6, (b) $H_2C=O$ or CH_3OH, (c) He or H_2.

28. The temperatures at which the vapor pressures of the following liquids are all 100 torr are given. Predict the order of increasing boiling points of the liquids. Normal butane, C_4H_{10}, −44.2°C; 1-butanol, $C_4H_{10}O$, 70.1°C; diethyl ether, $C_4H_{10}O$, −11.5°C.

29. The vapor pressure of liquid bromine at room temperature is 168 torr. Suppose that bromine is introduced drop by drop into a closed system containing air at 745 torr and room temperature. (The volume of liquid bromine is negligible compared to the gas volume.) If the bromine is added until no more vaporizes and a few drops of liquid are present in the flask, what would be the total pressure? What would be the total pressure if the volume of this closed system were decreased to one-half its original value at the same temperature?

30. A closed flask contains water at 75.0°C. The total pressure of the air-and-water-vapor mixture is 633.5 torr. The vapor pressure of water at this temperature is given in Appendix E as 289.1 torr. What is the partial pressure of the air in the flask?

*31. ΔH_{vap} is usually greater than ΔH_{fusion} for a substance, yet the nature of interactions that must be overcome in the vaporization and fusion processes are similar. Why should ΔH_{vap} be greater?

*32. The heat of vaporization of water at 100°C is 2.26 kJ/g; at 37°C (body temperature) it is 2.41 kJ/g.
(a) Convert the latter value to standard molar heat of vaporization, $\Delta H^0{}_{vap}$, at 37°C.
(b) Why is the heat of vaporization greater at 37°C than at 100°C?

33. Plot a vapor pressure curve for $C_2Cl_2F_4$ from the following vapor pressures. Determine the boiling point of $C_2Cl_2F_4$ under a pressure of 300 torr from the plot:

t (°C)	−95.4	−72.3	−53.7	−39.1	−12.0	3.5
vp (torr)	1	10	40	100	400	760

34. Plot a vapor pressure curve for $C_2H_4F_2$ from the following vapor pressures. From the plot, determine the boiling point of $C_2H_4F_2$ under a pressure of 200 torr.

t (°C)	−77.2	−51.2	−31.1	−15.0	14.8	31.7
vp (torr)	1	10	40	100	400	760

Clausius–Clapeyron Equation

35. Toluene, $C_6H_5CH_3$, is a liquid used in the manufacture of TNT. Its normal boiling point is 111.0°C, and its molar heat of vaporization is 35.9 kJ/mol. What would be the vapor pressure (torr) of toluene at 75.00°C?

36. At their normal boiling points, the heat of vaporization of water (100°C) is 40,656 J/mol and that of heavy water (101.41°C) is 41,606 J/mol. Use these data to calculate the vapor pressure of each liquid at 80.00°C.

37. (a) Use the Clausius–Clapeyron equation to calculate the temperature (°C) at which pure water would boil at a pressure of 400.0 torr. (b) Compare this result with the temperature read from Figure 13-13. (c) Compare the results of (a) and (b) with a value obtained from Appendix E.

*38. Show that the Clausius–Clapeyron equation can be written as

$$\log P = \frac{-\Delta H_{vap}}{2.303\,RT} + B$$

where B is a constant that has different values for different substances. This is an equation for a straight line. (a) What is the expression for the slope of this line? (b) Using the following vapor pressure data, plot $\log P$ vs. $1/T$ for ethyl acetate, $CH_3COOC_2H_5$, a common organic solvent used in nail polish removers.

t (°C)	−43.4	−23.5	−13.5	−3.0	+9.1
vp (torr)	1	5	10	20	40

t (°C)	16.6	27.0	42.0	59.3	
vp (torr)	60	100	200	400	

(c) From the plot, estimate ΔH_{vap} for ethyl acetate. (d) From the plot, estimate the normal boiling point of ethyl acetate.

*39. Repeat Exercise 38(c) for mercury, using the following data for liquid mercury. Then compare this value with the one in Appendix E.

t (°C)	126.2	184.0	228.8	261.7	323.0
vp (torr)	1	10	40	100	400

Phase Changes and Associated Heat Transfer

The following values will be useful in some exercises in this section:

Specific heat of ice	2.09 J/g · °C
Heat of fusion of ice at 0°C	334 J/g
Specific heat of liquid H_2O	4.18 J/g · °C
Heat of vaporization of liquid H_2O at 100°C	2.26×10^3 J/g
Specific heat of steam	2.03 J/g · °C

40. Is the equilibrium that is established between two physical states of matter an example of static or dynamic equilibrium? Explain your answer.

41. Which of the following changes of state are exothermic? (a) fusion, (b) liquefaction, (c) sublimation, (d) deposition. Explain.

42. Suppose that heat was added to a 21.8-g sample of solid zinc at the rate of 9.84 J/s. After the temperature reached the normal melting point of zinc, 420°C, it remained constant for 3.60 minutes. Calculate ΔH^0_{fusion} at 420°C, in J/mol, for zinc.

43. The specific heat of silver is 0.237 J/g · °C. Its melting point is 961°C. Its heat of fusion is 11 J/g. How much heat is needed to change 7.50 g of silver from solid at 25°C to liquid at 961°C?

44. The heat of fusion of thallium is 21 J/g, and its heat of vaporization is 795 J/g. The melting and boiling points are 304°C and 1457°C. The specific heat of liquid thallium is 0.13 J/g · °C. How much heat is needed to change 225 g of solid thallium at 304°C to vapor at 1457°C and 1 atm?

45. Calculate the amount of heat required to convert 50.0 g of ice at 0°C to liquid water at 100°C.

46. Calculate the amount of heat required to convert 75.0 g of ice at −15.0°C to steam at 125.0°C.

47. Use data in Appendix E to calculate the amount of heat required to warm 175 g of mercury from 25°C to its boiling point and then to vaporize it.

48. If 275 g of liquid water at 100°C and 475 g of water at 30.0°C are mixed in an insulated container, what is the final temperature?

49. If 25.0 g of ice at −10.0°C and 25.0 g of liquid water at 100°C are mixed in an insulated container, what will the final temperature be?

50. If 175 g of liquid water at 0°C and 17.5 g of steam at 110°C are mixed in an insulated container, what will the final temperature be?

51. Water can be cooled in hot climates by the evaporation of water from the surfaces of canvas bags. What mass of water can be cooled from 35.0°C to 20.0°C by the evaporation of one gram of water? Assume that ΔH_{vap} does not change with temperature.

52. (a) How much heat must be removed to prepare 15.0 g of ice at 0°C from 15.0 g of water at 25.0°C? (b) Suppose this heat is to be absorbed by vaporization of Freon-12 (a common household refrigerant, CCl_2F_2). What mass of refrigerant must be vaporized? The heat of vaporization of Freon-12 is 165.1 J/g.

Phase Diagrams

53. How many phases exist at a triple point? Describe what would happen if a small amount of heat were added under constant-volume conditions to a sample of water at the triple point. Assume a negligible volume change during fusion.

54. What is the critical point? Will a substance always be a liquid below the critical temperature? Why or why not?

Refer to the phase diagram of CO_2 in Figure 13-17b to answer Exercises 55–58.

55. What phase of CO_2 exists at 2 atm pressure and a temperature of −90°C? −60°C? 0°C?

56. What phases of CO_2 are present (a) at a temperature of −78°C and a pressure of 1 atm? (b) at −57°C and a pressure of 5.2 atm?

57. List the phases that would be observed if a sample of CO_2 at 8 atm pressure were heated from −80°C to 40°C.

58. How does the melting point of CO_2 change with pressure? What does this indicate about the relative density of solid CO_2 versus liquid CO_2?

*59. You are given the following data for ethanol, C_2H_5OH.

Normal melting point	−114°C
Normal boiling point	78.5°C
Critical temperature	243°C
Critical pressure	63.0 atm

Assume that the triple point is slightly lower in temperature than the melting point and that the vapor pressure at the triple point is about 10^{-5} torr. (a) Sketch a phase diagram for ethanol. (b) Ethanol at 1 atm and 140°C is compressed to 70 atm. Are two phases present at any time during this process? (c) Ethanol at 1 atm and 270°C is compressed to 70 atm. Are two phases present at any time during this process?

*60. You are given the following data for butane, C_4H_{10}.

Normal melting point	−138°C
Normal boiling point	0°C
Critical temperature	152°C
Critical pressure	38 atm

Assume that the triple point is slightly lower in temperature than the melting point and that the vapor pressure at the triple point is 3×10^{-5} torr. (a) Sketch a phase diagram for butane. (b) Butane at 1 atm and 140°C is compressed to 40 atm. Are two phases present at any time during this process? (c) Butane at 1 atm and 200°C is compressed to 40 atm. Are two phases present at any time during this process?

Exercises 61 and 62 refer to the phase diagram for sulfur on the next page. (The vertical axis is on a logarithmic scale.) Sulfur has two *solid* forms, monoclinic and rhombic.

*61. (a) How many triple points are there for sulfur? (b) Indicate the approximate pressure and temperature at each triple point. (c) Which phases are in equilibrium at each triple point?

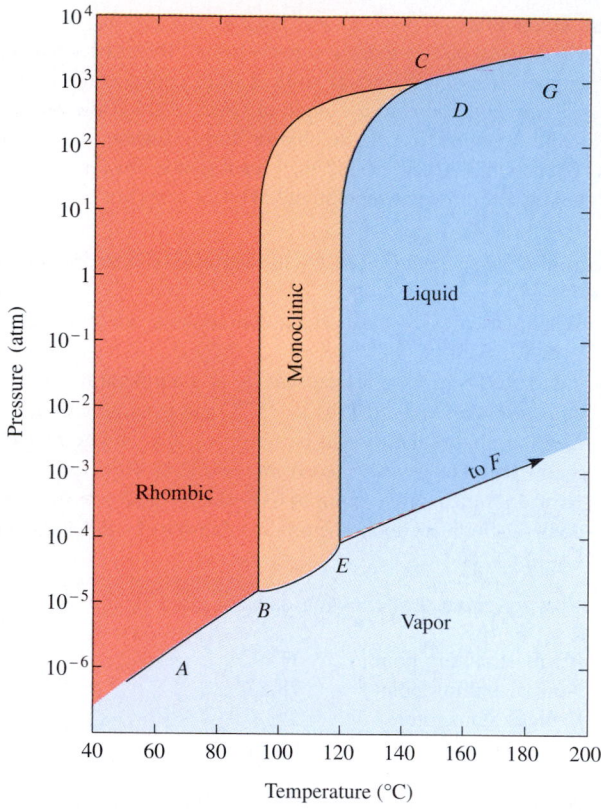

The Solid State

*62. Which physical states should be present at equilibrium under the following conditions? (a) 10^{-1} atm and 140°C, (b) 10^{-5} atm and 80°C, (c) 5×10^3 atm and 160°C, (d) 10^{-1} atm and 110°C, (e) 10^{-5} atm and 140°C, (f) 1 atm and 140°C.

63. Comment on the following statement: "The only perfectly ordered state of matter is the crystalline state."

64. Ice floats in water. Why? Would you expect solid mercury to float in liquid mercury at its freezing point? Explain.

65. Distinguish among and compare the characteristics of molecular, covalent, ionic, and metallic solids. Give two examples of each kind of solid.

66. Classify each of the following substances, in the solid state, as a molecular, ionic, covalent (network), or metallic solid.

	Melting Point (°C)	Boiling Point (°C)	Electrical Conductor	
			Solid	Liquid
SiO$_2$ (crystobalite)	1713	2230	no	no
Na$_2$S	1180	—	no	yes
Cr(CO)$_6$	110	210 (dec.)	no	no
Ti	1660	3287	yes	yes

67. Classify each of the following substances, in the solid state, as a molecular, ionic, covalent (network), or metallic solid.

	Melting Point (°C)	Boiling Point (°C)	Electrical Conductor	
			Solid	Liquid
NH$_4$NO$_3$	167	210	no	yes
Mg	649	1090	yes	yes
GeO$_2$	1115	—	no	no
S$_8$ (rhombic)	113	445	no	no

68. Based only on their formulas, classify each of the following in the solid state as a molecular, ionic, covalent (network), or metallic solid: (a) SO$_2$F, (b) MgF$_2$, (c) W, (d) Pb, (e) PF$_5$.

69. Based only on their formulas, classify each of the following in the solid state as a molecular, ionic, covalent (network), or metallic solid: (a) Au, (b) NO$_2$, (c) CaF$_2$, (d) SF$_4$, (e) C$_{diamond}$.

70. Arrange the following solids in order of increasing melting points and account for the order: NaF, MgF$_2$, AlF$_3$.

71. Arrange the following solids in order of increasing melting points and account for the order: MgO, CaO, SrO, BaO.

72. Distinguish among and sketch simple cubic, body-centered cubic (bcc), and face-centered cubic (fcc) lattices. Use CsCl, sodium, and nickel as examples of solids existing in simple cubic, bcc, and fcc lattices, respectively.

73. Describe a unit cell as precisely as you can.

74. Consider a unit cell that consists of a cube in which there is a cation at each corner and an anion at the center of each face. (a) Sketch the unit cell. How many (b) cations and (c) anions are present? (d) The simplest formula of the compound is of the type _____.

75. Consider a unit cell that consists of a cube in which there is an anion at each corner and one at the center of the unit cell, and a cation at the center of each face. How many cations and anions make up the unit cell? The simplest formula of this compound is of the type _____.

76. Choose two different unit cells in the two-dimensional lattice shown. Make one as simple as possible and the other orthogonal (containing right angles). Which of these is simpler to use to calculate area and so on?

Unit Cell Data; Atomic and Ionic Sizes

77. Refer to Figure 13-28a. (a) If the unit cell edge is represented as a, what is the distance (center to center) from Cs$^+$ to its nearest

neighbor? (b) How many equidistant nearest neighbors does each Cs^+ ion have? What are the identities of these nearest neighbors? (c) What is the distance (center to center), in terms of a, from a Cs^+ ion to the nearest Cs^+ ion? (d) How many equidistant nearest neighbors does each Cl^- ion have? What are their identities?

78. Refer to Figure 13-28b. (a) If the unit cell edge is represented as a, what is the distance (center to center) from Na^+ to its nearest neighbor? (b) How many equidistant nearest neighbors does each Na^+ ion have? What are the identities of these nearest neighbors? (c) What is the distance (center to center), in terms of a, from an Na^+ ion to the nearest Na^+ ion? (d) How many equidistant nearest neighbors does each Cl^- ion have? What are their identities?

79. Polonium crystallizes in a simple cubic unit cell with an edge length of 3.36 Å. (a) What is the mass of the unit cell? (b) What is the volume of the unit cell? (c) What is the theoretical density of Po?

80. Calculate the density of Na metal. The length of the body-centered cubic unit cell is 4.24 Å.

81. Tungsten has a density of 19.3 g/cm^3 and crystallizes in a cubic lattice whose unit cell edge length is 3.16 Å. Which type of cubic unit cell is it?

82. A Group IVA element with a density of 11.35 g/cm^3 crystallizes in a face-centered cubic lattice whose unit cell edge length is 4.95 Å. Calculate its atomic weight. What is the element?

83. The crystal structure of CO_2 is cubic, with a cell edge length of 5.540 Å. A diagram of the cell is shown in Figure 13-30a. (a) What is the number of molecules of CO_2 per unit cell? (b) Is this structure face-centered cubic? How can you tell? (c) What is the density of solid CO_2 at this temperature?

84. The structure of diamond is shown below, with each sphere representing a carbon atom. (a) How many carbon atoms are there per unit cell in the diamond structure? (b) Verify, by extending the drawing if necessary, that each carbon atom has four nearest

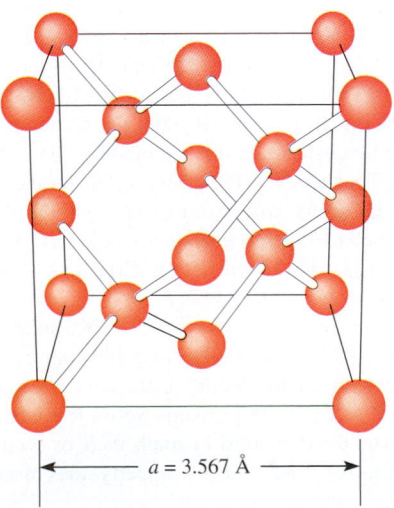

$a = 3.567$ Å

neighbors. What is the arrangement of these nearest neighbors? (c) What is the distance (center to center) from any carbon atom to its nearest neighbor, expressed in terms of a, the unit cell edge? (d) The observed unit cell edge length in diamond is 3.567 Å. What is the C—C single bond length in diamond? (e) Calculate the density of diamond.

85. Crystalline silicon has the same structure as diamond, with a unit cell edge length of 5.430 Å. (a) What is the Si—Si distance in this crystal? (b) Calculate the density of crystalline silicon.

*86. (a) What types of electromagnetic radiation are suitable for diffraction studies of crystals? (b) Describe the X-ray diffraction experiment. (c) What must be the relationship between the wavelength of incident radiation and the spacing of the particles in a crystal for diffraction to occur?

*87. (a) Write the Bragg equation. Identify each symbol. (b) X-rays from a palladium source ($\lambda = 0.576$ Å) were reflected by a sample of copper at an angle of 9.40°. This reflection corresponds to the unit cell length ($d = a$) with $n = 2$ in the Bragg equation. Calculate the length of the copper unit cell.

88. The spacing between successive planes of platinum atoms parallel to the cubic unit cell face is 2.256 Å. When X-radiation emitted by copper strikes a crystal of platinum metal, the minimum diffraction angle of X-rays is 19.98°. What is the wavelength of the Cu radiation?

89. Gold crystallizes in an fcc structure. When X-radiation of 0.70926 Å wavelength from molybdenum is used to determine the structure of metallic gold, the minimum diffraction angle of X-rays by the gold is 8.683°. Calculate the spacing between parallel layers of gold atoms.

Metallic Bonding and Semiconductors

90. Describe metallic bonding. Nonmetals do not form metallic bonds. Why not?

91. Compare the temperature dependence of electrical conductivity of a metal with that of a typical metalloid. Explain the difference.

92. In general, metallic solids are ductile and malleable, whereas ionic salts are brittle and shatter readily (although they are hard). Explain this observation.

93. What single factor accounts for the ability of metals to conduct both heat and electricity in the solid state? Why are ionic solids poor conductors of heat and electricity even though they are composed of charged particles?

Mixed Exercises

94. The three major components of air are N_2 (bp $-196°C$), O_2 (bp $-183°C$), and Ar (bp $-186°C$). Suppose we have a sample of liquid air at $-200°C$. In what order will these gases evaporate as the temperature is raised?

*95. A 10.0-g sample of liquid ethanol, C_2H_5OH, absorbs 3.42×10^3 J of heat at its normal boiling point, 78.5°C. The molar enthalpy of vaporization of ethanol, ΔH_{vap}, is 39.3 kJ/mol.
(a) What volume of C_2H_5OH vapor is produced? The volume is measured at 78.5°C and 1.00 atm pressure.
(b) What mass of C_2H_5OH remains in the liquid state?

*96. What is the pressure predicted by the ideal gas law for one mole of steam in 31.0 L at 100°C? What is the pressure predicted by the van der Waals equation (Section 12-14) given that $a = 5.464 \text{ L}^2 \cdot \text{atm/mol}^2$ and $b = 0.03049$ L/mol? What is the percent difference between these values? Does steam deviate from ideality significantly at 100°C? Why?

*97. The boiling points of HCl, HBr, and HI increase with increasing molecular weight. Yet the melting and boiling points of the sodium halides, NaCl, NaBr, and NaI, decrease with increasing formula weight. Explain why the trends are opposite.

98. The structures for three molecules having the formula $C_2H_2Cl_2$ are

Describe the intermolecular forces present in each of these compounds and predict which has the lowest boiling point.

99. The vapor pressure of $CH_3CH_2CH_2Cl$ at room temperature is 385 torr. The total pressure of $CH_3CH_2CH_2Cl$ and air in a container is 745 torr. What will happen to the pressure if the volume of the container is doubled at constant temperature? Assume that a small amount of liquid $CH_3CH_2CH_2Cl$ is present in the container at all times.

100. Tantalum (181 g/mol, density 16.7 g/cm^3) crystallizes in a cubic lattice whose unit cell edge length is 3.32 Å. How many atoms are in one unit cell? What is the type of cubic crystal lattice?

101. Isopropyl alcohol, C_3H_8O, is marketed as "rubbing alcohol." Its vapor pressure is 100 torr at 39.5°C and 400 torr at 67.8°C. (a) Estimate the molar heat of vaporization of isopropyl alcohol. (b) Predict the normal boiling point of isopropyl alcohol.

*102. The heat of vaporization of cyclohexane, C_6H_{12}, is 390 J/g at its boiling point, 80.7°C. For how many minutes would heat have to be supplied at a rate of 10.0 J/s to a 1.00 mol sample of liquid cyclohexane at 80.7°C to vaporize all of it?

103. The van der Waals constants (Section 12-14) are $a = 19.01 \text{ L}^2 \cdot \text{atm/mol}^2$, $b = 0.1460$ L/mol for pentane, and $a = 18.05 \text{ L}^2 \cdot \text{atm/mol}^2$, $b = 0.1417$ L/mol for isopentane.

(a) Basing your reasoning on intermolecular forces, why would you expect a for pentane to be greater? (b) Basing your reasoning on molecular size, why would you expect b for pentane to be greater?

104. The density of solid lead is 11.288 g/cm^3 at 20°C; that of liquid lead is 10.43 g/cm^3 at 500°C; and that of gaseous lead is 1.110 g/L at 2000°C and 1 atm pressure. (a) Calculate the volume occupied by one mole of lead in each state. The radius of a lead atom is 1.75 Å. Calculate (b) the volume actually occupied by one mole of lead atoms and (c) the fraction of volume of each state actually occupied by the atoms.

105. Refer to the sulfur phase diagram that accompanies Exercises 61 and 62. (a) Can rhombic sulfur be sublimed? If so, under what conditions? (b) Can monoclinic sulfur be sublimed? If so, under what conditions? (c) Describe what happens if rhombic sulfur is slowly heated from 80°C to 140°C at constant 1-atm pressure. (d) What happens if rhombic sulfur is heated from 80°C to 140°C under constant pressure of 5×10^{-6} atm?

106. The normal boiling point of ammonia, NH_3, is −33°C, and its freezing point is −78°C. Fill in the blanks. (a) At STP (0°C, 1 atm pressure), NH_3 is a _____. (b) If the temperature drops to −40°C, the ammonia will _____ and become a _____. (c) If the temperature drops further to −80°C and the molecules arrange themselves in an orderly pattern, the ammonia will _____ and become a _____. (d) If crystals of ammonia are left on the planet Mars at a temperature of −100°C, they will gradually disappear by the process of _____ and form a _____.

BUILDING YOUR KNOWLEDGE

107. In a flask containing dry air and some *liquid* silicon tetrachloride, $SiCl_4$, the total pressure was 988 torr at 225°C. Halving the volume of the flask increased the pressure to 1742 torr at constant temperature. What is the vapor pressure of $SiCl_4$ in this flask at 225°C?

108. A friend comes to you with this problem: "I looked up the vapor pressure of water in a table; it is 26.7 torr at 300 K and 92,826 torr at 600 K. That means that the vapor pressure increases by a factor of 3477 when the absolute temperature doubles over this temperature range. But I thought the pressure was proportional to the absolute temperature, $P = nRT/V$. The pressure doesn't just double. Why?" How would you help the friend?

109. Using as many as six drawings (frames) for each, depict the changes that occur at the molecular level during each of the following physical changes: (a) melting an ice cube, (b) sublimation of an ice cube below 4.6°C and 4.6 torr, and (c) evaporation of a droplet of water at room temperature and pressure.

110. Write the Lewis formula of each member of each of the following pairs. Then use VSEPR theory to predict the geometry about each central atom, and describe any features that lead you to decide which member of each pair would have the lower boiling point. (a) CH_3COOH and $HCOOCH_3$, (b) NHF_2 and BH_2Cl, (c) CH_3CH_2OH and CH_3OCH_3.

111. At its normal melting point of 271.3°C, solid bismuth has a density of 9.73 g/cm^3 and liquid bismuth has a density of 10.05 g/cm^3. A mixture of liquid and solid bismuth is in equilibrium at 271.3°C. If the pressure were increased from 1 atm to 10 atm, would more solid bismuth melt or would more liquid bismuth freeze? What unusual property does bismuth share with water?

112. More than 150 years ago Dulong and Petit discovered a *rule of thumb* that the heat capacity of one mole of a pure solid element is about 6.0 calories per °C (i.e., about 25 J/°C). A 100.2-g sample of an unknown metal at 99.9°C is placed in 50.6 g of water at 24.8°C. The temperature is 36.6°C when the system comes to equilibrium. Assume all heat lost by the metal is absorbed by the water. What is the likely identity of this metal?

113. The doping of silicon with boron to an atomic concentration of 0.0010% boron atoms vs. silicon atoms increases its conductivity by a factor of 10^3 at room temperature. How many atoms of boron would be needed to dope 12.5 g of silicon? What mass of boron is this?

14 Solutions

A petroleum refinery uses many distillation towers. In such towers, petroleum (a very complex mixture) is separated into many useful components.

OBJECTIVES

As you study this chapter, you should learn

• *About the factors that favor the dissolution process*

• *About the dissolution of solids in liquids, liquids in liquids, and gases in liquids*

• *How temperature and pressure affect solubility*

• *To express concentrations of solutions in terms of molality and mole fractions*

• *About colligative properties of solutions: lowering of vapor pressure (Raoult's Law), boiling point elevation, and freezing point depression*

• *To use colligative properties to determine molecular weights of compounds*

• *About dissociation and ionization of compounds, and their effects on colligative properties*

• *About osmotic pressure and some of its applications*

• *About colloids: the Tyndall effect, the adsorption phenomenon, hydrophilic and hydrophobic colloids*

S olutions are common in nature and are extremely important in all life processes, in all scientific areas, and in many industrial processes. The body fluids of all forms of life are solutions. Variations in their concentrations, especially those of blood and urine, give physicians valuable clues about a person's health.

Solutions include many different combinations in which a solid, liquid, or gas acts as either solvent or solute. Usually the solvent is a liquid. For instance, sea water is an aqueous solution of many salts and some gases such as carbon dioxide and oxygen. Carbonated water is a saturated solution of carbon dioxide in water. Examples of solutions in which the solvent is not a liquid also are common. Air is a solution of gases with variable composition. Dental fillings are solid amalgams, or solutions of liquid mercury dissolved in solid metals. Alloys are solid solutions of solids dissolved in a metal.

A solution is defined as a *homogeneous mixture* of substances in which no settling occurs. A solution consists of a solvent and one or more solutes, whose proportions vary from one solution to another. By contrast, a pure substance has fixed composition. The *solvent* is the medium in which the *solutes* are dissolved. The fundamental units of solutes are usually ions or molecules.

It is usually obvious which of the components of a solution is the solvent and which is (are) the solute(s): The solvent is usually the most abundant species present. In a cup of instant coffee, the coffee and any added sugar are considered solutes, and the hot water is the solvent. If we mix 10 grams of alcohol with 90 grams of water, alcohol is the solute. If we mix 10 grams of water with 90 grams of alcohol, water is the solute. But which is the solute and which is the solvent in a solution of 50 grams of water and 50 grams of alcohol? In such cases, the terminology is arbitrary and, in fact, unimportant.

Many naturally occurring fluids contain particulate matter suspended in a solution. For example, blood contains a solution (plasma) with suspended blood cells. Sea water contains dissolved substances as well as suspended solids.

THE DISSOLUTION PROCESS

14-1 SPONTANEITY OF THE DISSOLUTION PROCESS

In Section 4-2, part 5, we listed the solubility rules for aqueous solutions. Now we investigate the major factors that influence solubility.

A substance may dissolve with or without reaction with the solvent. For example, metallic sodium "dissolves" in water with the evolution of bubbles of hydrogen and a great deal of heat. A chemical change occurs in which H_2 and soluble ionic sodium hydroxide, NaOH, are produced.

$$2Na(s) + 2H_2O \longrightarrow 2[Na^+(aq) + OH^-(aq)] + H_2(g)$$

If the resulting solution is evaporated to dryness, solid sodium hydroxide, NaOH, is obtained rather than metallic sodium. This, along with the production of bubbles of hydrogen, is evidence of a reaction with the solvent.

Solid sodium chloride, NaCl, on the other hand, dissolves in water with no evidence of chemical reaction.

$$NaCl(s) \xrightarrow{H_2O} Na^+(aq) + Cl^-(aq)$$

Ionic solutes that do not react with the solvent undergo solvation. This is a kind of reaction in which molecules of solvent are attached in oriented clusters to the solute particles.

Evaporation of the water from the sodium chloride solution yields the original NaCl. We focus on dissolution of this type, in which no irreversible reaction occurs between components.

The ease of dissolution of a solute depends upon two factors: (1) the change in energy and (2) the change in disorder (called entropy change) that accompanies the process. In the next chapter we shall study both these factors in detail for many kinds of physical and chemical changes. For now, we point out that a process is *favored* by (1) a *decrease in the energy* of the system, which corresponds to an *exothermic process,* and (2) an *increase in the disorder,* or randomness, of the system.

Let us look at the first of these factors. If a solution gets hotter as a substance dissolves, this means that energy is being released in the form of heat. This energy change is called the **heat of solution, $\Delta H_{\text{solution}}$.** It depends mainly on how strongly solute and solvent particles interact. A negative value of $\Delta H_{\text{solution}}$ designates the release of heat. More negative (less positive) values of $\Delta H_{\text{solution}}$ favor the dissolution process.

In a pure liquid all the intermolecular forces are between like molecules; when the liquid and a solute are mixed, each molecule then interacts with molecules (or ions) unlike it as well as with like molecules. The relative strengths of these interactions help to determine the extent of solubility of a solute in a solvent. The main interactions that affect the dissolution of a solute in a solvent follow.

1. Strong solvent–solute attractions favor solubility.
2. Weak solvent–solvent attractions favor solubility.
3. Weak solute–solute attractions favor solubility.

Figure 14-1 illustrates the interplay of these factors. The intermolecular or interionic attractions among solute particles in the pure solute must be overcome (Step a) to dissolve the solute. This part of the process requires an *input* of energy (endothermic). Separating the solvent molecules from each other (Step b) to "make room" for the solute particles also requires the *input* of energy (endothermic). However, energy is *released* as the solute particles and solvent molecules interact in the solution (Step c, exothermic). The overall dissolution process is exothermic (and favored) if the amount of heat absorbed in hypothetical Steps a and b is less than the amount of heat released in Step c. The process is

We can consider the energy changes separately, even though the actual process cannot be carried out in these separate steps.

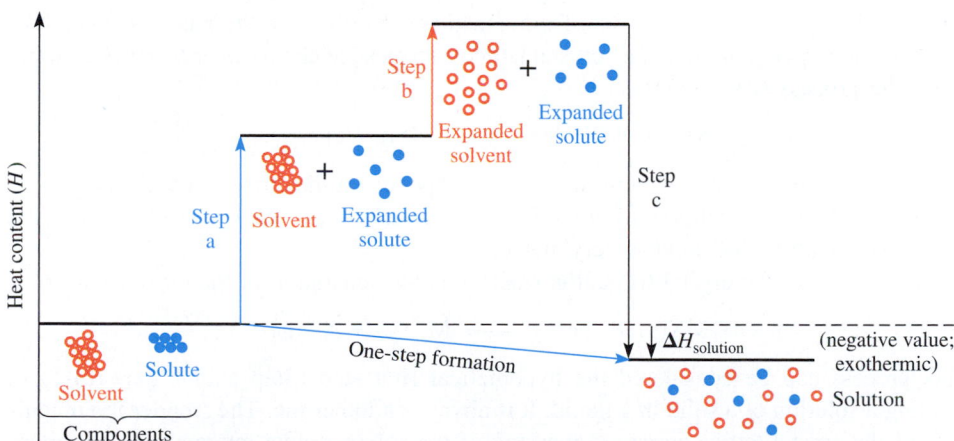

Figure 14-1 A diagram representing the changes in heat content associated with the hypothetical three-step sequence in a dissolution process—in this case, for a solid solute dissolving in a liquid solvent. (Similar considerations would apply to other combinations.) An *exothermic* process is depicted here. The amount of heat absorbed in Steps a and b is *less* than the amount of heat released in Step c, so the heat of the solution is favorable. In an *endothermic* process (not shown), the heat content of the solution would be *higher* than that of the original solvent plus solute. Thus, the amount of heat absorbed in Steps a and b would be *greater* than the amount of heat released in Step c, so heat of solution would be unfavorable.

endothermic (and disfavored) if the amount of heat absorbed in Steps a and b is greater than the amount of heat released in Step c.

However, many solids do dissolve in liquids by *endothermic* processes. The reason such processes can occur is that the endothermicity can be outweighed by a large increase in disorder of the solute during the dissolution process. The solute particles are highly ordered in a solid crystal, but are free to move about randomly in liquid solutions. Likewise, the degree of disorder in the solvent increases as the solution is formed, because solvent molecules are then in a more random environment. They are surrounded by a mixture of solvent and solute particles.

Nearly all dissolution processes are accompanied by an increase in the disorder of both solvent and solute. Thus, the disorder factor is usually *favorable* to solubility. The determining factor, then, is whether the heat of solution also favors dissolution or, if it does not, whether it is small enough to be outweighed by the favorable effects of the increasing disorder. In gases, for instance, the molecules are so far apart that intermolecular forces are quite weak. Thus, when gases are mixed, changes in the intermolecular forces are very slight. So the very favorable increase in disorder that accompanies mixing is always more important than possible changes in intermolecular attractions (energy). Hence, gases that do not react with each other can always be mixed with each other in any proportion.

The most common types of solutions are those in which the solvent is a liquid. In the next several sections we consider these in more detail.

One of very few exceptions is the dissolution of NaF. The water molecules become more ordered around the small F$^-$ ions. This is due to the strong hydrogen bonding between H$_2$O molecules and F$^-$ ions.

$$\underset{H}{\overset{|}{O}}-H\text{---}F^-\text{---}H-\underset{H}{\overset{|}{O}}$$

However, the amount of heat released on mixing outweighs the disadvantage of this ordering.

14-2 DISSOLUTION OF SOLIDS IN LIQUIDS

The ability of a solid to go into solution depends most strongly on its crystal lattice energy, or the strength of attractions among the particles making up the solid. The **crystal lattice energy** is defined as the energy change accompanying the formation of one mole of formula units in the crystalline state from constituent particles in the gaseous state. This process is always exothermic; i.e., crystal lattice energies are always *negative*. For an ionic solid, the process is written as

A very negative crystal lattice energy indicates very strong attractions within the solid.

$$M^+(g) + X^-(g) \longrightarrow MX(s) + \text{energy}$$

The amount of energy involved in this process depends on the attraction between ions in the solid. When these attractions are strong, a large amount of energy is released as the solid forms, and so the solid is very stable.

The reverse of the crystal formation reaction is the separation of the crystal into ions.

$$MX(s) + \text{energy} \longrightarrow M^+(g) + X^-(g)$$

This process can be considered the hypothetical first step (Step a in Figure 14-1) in forming a solution of a solid in a liquid. It is always endothermic. The smaller the magnitude of the crystal lattice energy (a measure of the solute–solute interactions), the more readily dissolution occurs. Less energy must be supplied to start the dissolution process.

If the solvent is water, the energy that must be supplied to expand the solvent (Step b in Figure 14-1) includes that required to break up some of the hydrogen bonding between water molecules.

The third major factor contributing to the heat of solution is the extent to which solvent molecules interact with particles of the solid. The process in which solvent molecules surround and interact with solute ions or molecules is called **solvation.** When the solvent is water, the more specific term is **hydration. Hydration energy** (equal to the sum of

*Hydration energy is also referred to as the **heat of hydration.***

Steps b and c in Figure 14-1) is defined as the energy change involved in the (exothermic) hydration of one mole of gaseous ions.

$$M^{n+}(g) + xH_2O \longrightarrow M(OH_2)_x^{n+} + \text{energy} \qquad \text{(for cation)}$$
$$X^{y-}(g) + rH_2O \longrightarrow X(H_2O)_r^{y-} + \text{energy} \qquad \text{(for anion)}$$

Hydration is usually highly exothermic for ionic or polar covalent compounds, because the polar water molecules interact very strongly with ions and polar molecules. In fact, the only solutes that are appreciably soluble in water either undergo dissociation or ionization or are able to form hydrogen bonds with water.

The overall heat of solution for a solid dissolving in a liquid is equal to the heat of solvation minus the crystal lattice energy.

$$\Delta H_{\text{solution}} = (\text{heat of solvation}) - (\text{crystal lattice energy})$$

Remember that both terms on the right are always negative.

Nonpolar solids such as naphthalene, $C_{10}H_8$, do not dissolve appreciably in polar solvents such as water because the two substances do not attract each other significantly. This is true despite the fact that crystal lattice energies of solids consisting of nonpolar molecules are much less negative (smaller in magnitude) than those of ionic solids. Naphthalene dissolves readily in nonpolar solvents such as benzene because there are no strong attractive forces between solute molecules or between solvent molecules. In such cases, the increase in disorder controls the process. These facts help explain the observation that "like dissolves like."

Consider what happens when a piece of sodium chloride, a typical ionic solid, is placed in water. The $\delta+$ ends of water molecules attract the negative chloride ions on the surface of the solid NaCl, as shown in Figure 14-2. Likewise, the $\delta-$ ends of H_2O molecules (O atoms) orient themselves toward the Na^+ ions and solvate them. These attractions help to overcome the forces holding the ions in the crystal, and NaCl dissolves in the H_2O.

$$NaCl(s) \xrightarrow{H_2O} Na^+(aq) + Cl^-(aq)$$

A cube of sugar is slowly lifted through a solution.

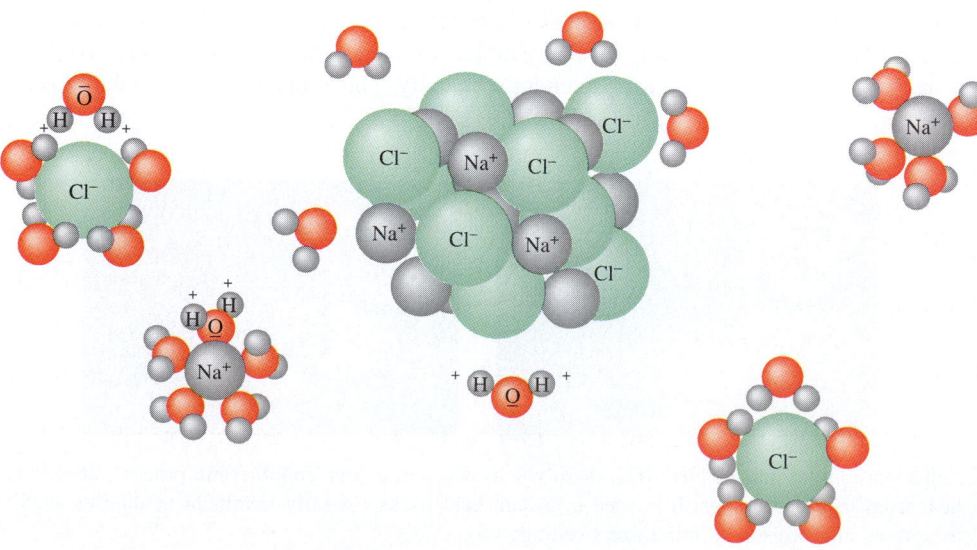

Figure 14-2 The role of electrostatic attractions in the dissolution of NaCl in water. The $\delta+$ H of the polar H_2O molecule helps to attract Cl^- away from the crystal. Likewise, Na^+ is attracted by the $\delta-$ O. Once they are separated from the crystal, both kinds of ions are surrounded by water molecules, to complete the hydration process.

The charge/radius ratio is the ionic charge divided by the ionic radius in angstroms. This is a measure of the *charge density* around the ion. A negative value for heat of hydration indicates that heat is *released* during hydration.

Table 14-1 *Ionic Radii, Charge/Radius Ratios, and Heats of Hydration for Some Cations*

Ion	Ionic Radius (Å)	Charge/Radius Ratio	Heat of Hydration (kJ/mol)
K^+	1.33	0.75	-351
Na^+	0.95	1.05	-435
Li^+	0.60	1.67	-544
Ca^{2+}	0.99	2.02	-1650
Fe^{2+}	0.76	2.63	-1980
Zn^{2+}	0.74	2.70	-2100
Cu^{2+}	0.72	2.78	-2160
Fe^{3+}	0.64	4.69	-4340
Cr^{3+}	0.62	4.84	-4370
Al^{3+}	0.50	6.00	-4750

For simplicity, we often omit the (aq) designations from dissolved ions. Remember that all ions are hydrated in aqueous solution, whether this is indicated or not.

Review the sizes of ions in Figure 6-1 carefully.

When we write Na^+(aq) and Cl^-(aq), we refer to hydrated ions. The number of H_2O molecules attached to an ion differs with different ions. Sodium ions are thought to be hexahydrated; that is, Na^+(aq) probably represents $[Na(OH_2)_6]^+$. Most cations in aqueous solution are surrounded by four to nine H_2O molecules, with six being the most common. Generally, larger cations can accommodate more H_2O molecules than smaller cations.

Many solids that are appreciably soluble in water are ionic compounds. Magnitudes of crystal lattice energies generally increase with increasing charge and decreasing size of ions. That is, the size of the lattice energy increases as the ionic charge densities increase and, therefore, as the strength of electrostatic attractions within the crystal increases. Hydration energies vary in the same order (Table 14-1). As we indicated earlier, crystal lattice energies and hydration energies are generally much smaller in magnitude for molecular solids than for ionic solids.

Hydration and the effects of attractions in a crystal oppose each other in the dissolution process. Hydration energies and lattice energies are usually of about the same magnitude for low-charge species, so they often nearly cancel each other. As a result, the dissolution process is slightly endothermic for many ionic substances. Ammonium nitrate, NH_4NO_3, is an example of a salt that dissolves endothermically. This property is used in the "instant

Solid ammonium nitrate, NH_4NO_3, dissolves in water in a very endothermic process, absorbing heat from its surroundings. It is used in instant cold packs for early treatment of injuries such as sprains and bruises, to minimize swelling.

cold packs" used to treat sprains and other minor injuries. Ammonium nitrate and water are packaged in a plastic bag in which they are kept separate by a partition that is easily broken when squeezed. As the NH_4NO_3 dissolves in the H_2O, the mixture absorbs heat from its surroundings and the bag becomes cold to the touch.

Some ionic solids dissolve with the release of heat. Examples are anhydrous sodium sulfate, Na_2SO_4; calcium acetate, $Ca(CH_3COO)_2$; calcium chloride, $CaCl_2$; and lithium sulfate hydrate, $Li_2SO_4 \cdot H_2O$.

As the charge-to-size ratio (charge density) increases for ions in ionic solids, the magnitude of the crystal lattice energy usually increases more than the hydration energy. This makes dissolution of solids that contain highly charged ions—such as aluminum fluoride, AlF_3; magnesium oxide, MgO; and chromium(III) oxide, Cr_2O_3—very endothermic. As a result, these compounds are not very soluble in water.

14-3 DISSOLUTION OF LIQUIDS IN LIQUIDS (MISCIBILITY)

Miscibility is the ability of one liquid to dissolve in another. The three kinds of attractive interactions (solvent–solute, solvent–solvent, and solute–solute) must be considered for liquid–liquid solutions just as they were for solid–liquid solutions. Because solute–solute attractions are usually much lower for liquid solutes than for solids, this factor is less important and so the mixing process is often exothermic for miscible liquids. Polar liquids tend to interact strongly with and dissolve readily in other polar liquids. Methanol, CH_3OH; ethanol, CH_3CH_2OH; acetonitrile, CH_3CN; and sulfuric acid, H_2SO_4, are all polar liquids that are soluble in most polar solvents (such as water). The hydrogen bonding between methanol and water molecules and the dipolar interaction between acetonitrile and water molecules are depicted in Figure 14-3.

Because hydrogen bonding is so strong between sulfuric acid, H_2SO_4, and water, large amounts of heat are released when concentrated H_2SO_4 is diluted with water (Figure 14-4). This can cause the solution to boil and spatter. If the major component of the mixture is water, this heat can be absorbed with less increase in temperature because of the

Hydrogen bonding and dipolar interactions were discussed in Section 13-2.

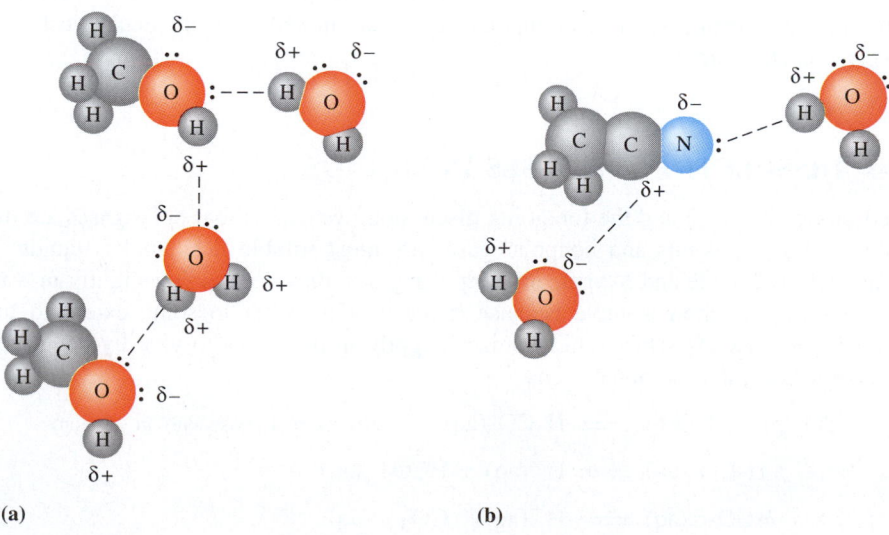

(a) **(b)**

Figure 14-3 (a) Hydrogen bonding in methanol–water solution. (b) Dipolar interaction in acetonitrile–water solution.

The nonpolar molecules in oil do not attract polar water molecules, so oil and water are immiscible. The polar water molecules attract each other strongly—they "squeeze out" the nonpolar molecules in the oil. Oil is less dense than water, so it floats on water.

Figure 14-4 The heat released by pouring 50 mL of sulfuric acid, H_2SO_4, into 50 mL of water increases the temperature by 100°C (from 21°C to 121°C)!

When water is added to concentrated acid, the danger is due more to the spattering of the acid itself than to the steam from boiling water.

unusually high specific heat of H_2O. For this reason, *sulfuric acid (as well as other mineral acids) is always diluted by adding the acid slowly and carefully to water. Water is never added to the acid.* If spattering does occur when the acid is added to water, it is mainly water that spatters, not the corrosive concentrated acid.

Nonpolar liquids that do not react with the solvent generally are not very soluble in polar liquids because of the mismatch of forces of interaction. Nonpolar liquids are, however, usually quite soluble in other nonpolar liquids. Between nonpolar molecules (whether alike or different) there are only London forces, which are weak and easily overcome. As a result, when two nonpolar liquids are mixed, their molecules just "slide between" each other.

14-4 DISSOLUTION OF GASES IN LIQUIDS

Based on Section 13-2 and the foregoing discussion, we expect that polar gases are most soluble in polar solvents and nonpolar gases are most soluble in nonpolar liquids. Although carbon dioxide and oxygen are nonpolar gases, they do dissolve slightly in water. CO_2 is somewhat more soluble because it reacts with water to some extent to form carbonic acid, H_2CO_3. This in turn ionizes slightly in two steps to give hydrogen ions, bicarbonate ions, and carbonate ions.

Carbon dioxide is called an acid anhydride, i.e., an "acid without water." As noted in Section 6-8, part 2, many other oxides of nonmetals, such as N_2O_5, SO_3, and P_4O_{10}, are also acid anhydrides.

$$CO_2(g) + H_2O(\ell) \rightleftharpoons H_2CO_3(aq) \quad \text{carbonic acid (exists only in solution)}$$

$$H_2CO_3(aq) \rightleftharpoons H^+(aq) + HCO_3^-(aq)$$

$$HCO_3^-(aq) \rightleftharpoons H^+(aq) + CO_3^{2-}(aq)$$

CHEMISTRY IN USE

Our Daily Lives

Solubility and the Spread of AIDS

Many people know that oil and water do not mix, but ignorance of the underlying principles contributes to the spread of AIDS and other sexually transmitted diseases. To understand this, we recall that nonpolar compounds, such as oils, are insoluble in polar solvents, such as water. Most nonpolar liquids dissolve in each other, and many polar compounds dissolve in polar liquids. Chemists summarize these observations in the adage, "Like dissolves like."

Lack of familiarity with this adage was made obvious in a 1990 report by the Kinsey Institute for Research in Sex, Gender, and Reproduction. The use of an appropriate lubricant with a condom adds a measure of safety because lubrication decreases the chance that the condom will break. The Kinsey Institute reported that half of Americans 18 years and older do not know that oil-based lotions and creams should not be used with latex condoms or latex diaphragms. These include baby oil, petroleum jelly, and many hand lotions; all contain nonpolar compounds. Latex is nonpolar and such oils, creams, and jellies can dissolve latex in as little as one minute, which may cause ruptures and tears. Physical damage to condoms from lubricating oils and creams is often invisible to the naked eye, but the latex may be punctured with tiny holes that are large enough to allow passage of the HIV virus or sperm. Because this damage cannot be visually detected, the practice of using nonpolar oil-based lubricants continues; this increases the odds of unwanted pregnancies and the spread of diseases. We can only speculate about how many of the 2 million unwanted pregnancies and approximately 1 million new cases of AIDS in the United States each year result from such ignorance.

Even those who know better than to use oil-based lotions as lubricants often confuse "water-based" and "water soluble" lotions. Some products such as Vaseline Intensive Care Lotion can be washed away easily with water; they also contain mineral

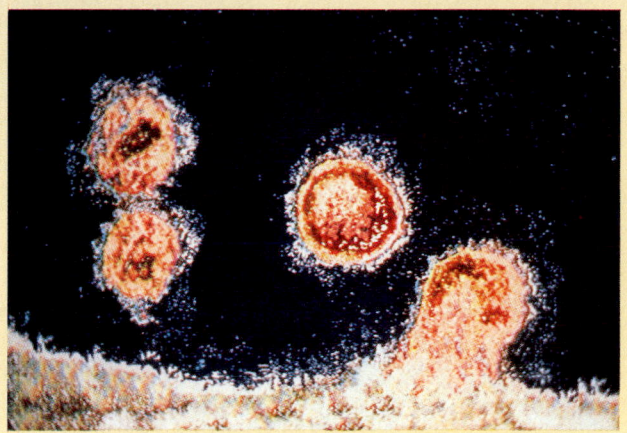

AIDS virus particle as it emerges from an infected lymphocyte.

oil. The results of this confusion became obvious in 1992 when another survey showed that more than 90% of the men who knew better than to use oil-based lotions, and who experienced frequent condom failure, were unaware that their "water soluble" lubricant may have been the cause of the failure.

Lotion manufacturers add *surfactants* to oil-based lotions to make the lotions water soluble. Surfactants are special chemicals that are soluble in both polar and nonpolar substances. Soap is an example of a surfactant. Soap has the ability to mix with oil and water to produce an *emulsion* of soap, water, and oil. The presence of surfactants in oil-based lotions allows these lotions to be rinsed away with plain water, just as soap allows us to wash away oily grime with water. The fact that some other substances are present in "water-based" lotions doesn't interfere with the ability of mineral oil and the other nonpolar substances to dissolve latex!

Ronald DeLorenzo
Middle Georgia College

Approximately 1.45 grams of CO_2 (0.033 mole) dissolves in a liter of water at 25°C and one atmosphere pressure.

Oxygen, O_2, is less soluble in water than CO_2, but it does dissolve to a noticeable extent due to London forces (induced dipoles, Section 13-2). Only about 0.041 gram of O_2 (1.3×10^{-3} mole) dissolves in a liter of water at 25°C and one atm pressure; yet this is sufficient to support aquatic life.

The hydrogen halides, HF, HCl, HBr, and HI, are all polar covalent gases. In the gas phase the interactions among the widely separated molecules are not very strong, so

Aqueous HCl, HBr, and HI are strong acids (Sections 4-2, part 2 and 10-5). Aqueous HF is a weak acid.

Oxygen gas is sufficiently soluble in water to support a wide variety of aquatic life.

solute–solute attractions are minimal, and the dissolution processes in water are exothermic. The resulting solutions, called hydrohalic acids, contain predominantly ionized HX (X = Cl, Br, I). The ionization involves *protonation* of a water molecule by HX to form a hydrated hydrogen ion and halide ion X$^-$ (also hydrated). HCl is used as an example.

$$\text{H} : \overset{..}{\underset{..}{\text{Cl}}} : + \text{H}-\overset{..}{\underset{|}{\underset{\text{H}}{\text{O}}}} : \longrightarrow : \overset{..}{\underset{..}{\text{Cl}}} : ^- + \left[\text{H}-\overset{\text{H}}{\underset{|}{\underset{\text{H}}{\overset{|}{\text{O}}}}} : \right]^+$$

HF is only slightly ionized in aqueous solution because of the strong covalent bond between highly electronegative fluorine and hydrogen atoms. In addition, the more polar bond between H and the small F atoms in HF causes very strong hydrogen bonding between H_2O and the largely intact HF molecules.

$$\overset{\delta+ \quad \delta-}{\text{H}-\overset{..}{\underset{|}{\underset{\underset{\delta+}{\text{H}}}{\text{O}}}} :} --- \overset{\delta+ \quad \delta-}{\text{H}-\overset{..}{\underset{..}{\text{F}}} :} \qquad \text{as well as} \qquad \overset{\delta+ \quad \delta-}{\text{H}-\overset{..}{\underset{..}{\text{F}}} :} --- \overset{\delta+ \quad \delta-}{\text{H}-\overset{..}{\underset{|}{\underset{\underset{\delta+}{\text{H}}}{\text{O}}}} :}$$

The only gases that dissolve appreciably in water are those that are capable of hydrogen bonding (such as HF), those that ionize (such as HCl, HBr, and HI), and those that react with water (such as CO_2).

Figure 14-5 A mortar and pestle are used for grinding solids.

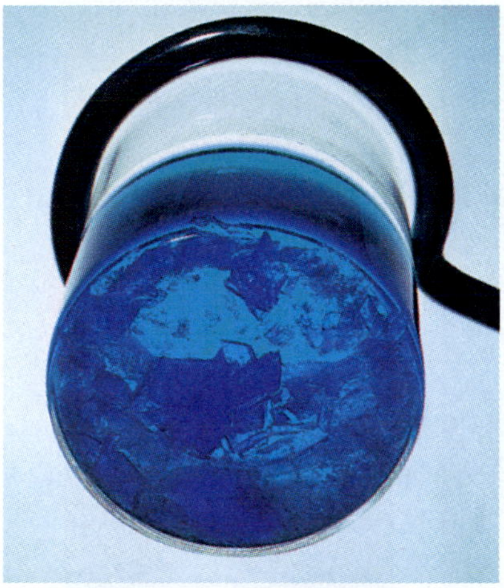

A saturated solution of copper(II) sulfate, $CuSO_4$, in water. As H_2O evaporates, crystals of blue $CuSO_4 \cdot 5H_2O$ form. They are in dynamic equilibrium with the saturated solution.

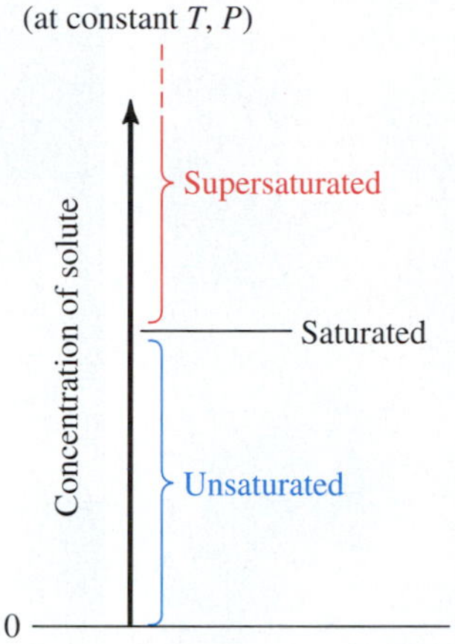

A solution that contains less than the amount of solute necessary for saturation is said to be unsaturated.

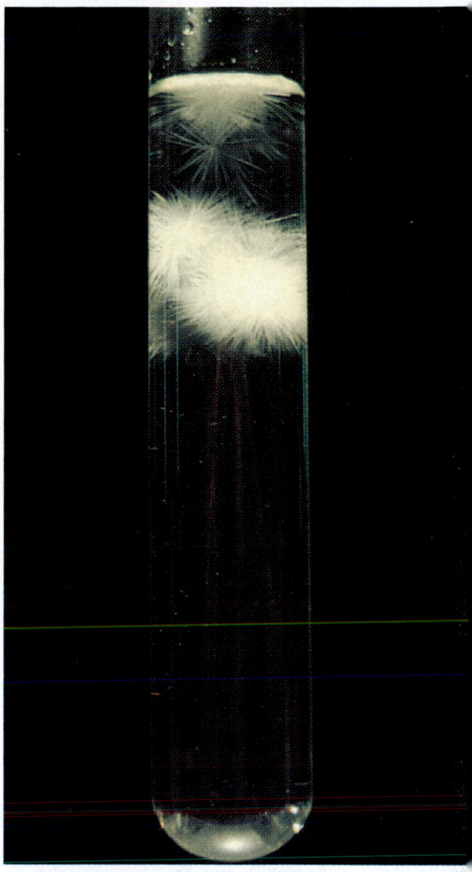

A tiny crystal of sodium acetate, $NaCH_3COO$, was added to a clear, colorless, supersaturated solution of $NaCH_3COO$. This photo shows solid $NaCH_3COO$ just beginning to crystallize in a very rapid process.

14-5 RATES OF DISSOLUTION AND SATURATION

At a given temperature, the rate of dissolution of a solid increases if large crystals are ground to a powder (Figure 14-5). Grinding increases the surface area, which in turn increases the number of solute ions or molecules in contact with the solvent. When a solid is placed in water, some of its particles solvate and dissolve. The rate of this process slows as time passes because the surface area of the crystals gets smaller and smaller. At the same time, the number of solute particles in solution increases, so they collide with the solid more frequently. Some of these collisions result in recrystallization. The rates of the two opposing processes become equal after some time. The solid and dissolved ions are then in equilibrium with each other.

$$\text{solid} \underset{\text{crystallization}}{\overset{\text{dissolution}}{\rightleftharpoons}} \text{dissolved ions}$$

Such a solution is said to be **saturated.** Saturation occurs at very low concentrations of dissolved species for slightly soluble substances and at high concentrations for very soluble substances. When imperfect crystals are placed in saturated solutions of their ions, surface defects on the crystals are slowly "patched" with no net increase in mass of solid. Often, after some time has passed, we see fewer but larger crystals. These observations provide evidence of the dynamic nature of the solubility equilibrium. After equilibrium is established, no more solid dissolves without the simultaneous crystallization of an equal mass of dissolved ions.

Dynamic equilibria occur in all saturated solutions; for instance, there is a continuous exchange of oxygen molecules across the surface of water in an open container. This is fortunate for fish, which "breathe" dissolved oxygen.

The solubilities of many solids increase at higher temperatures. **Supersaturated solutions** contain higher-than-saturated concentrations of solute. They can sometimes be prepared by saturating a solution at a high temperature. The saturated solution is cooled slowly, without agitation, to a temperature at which the solute is less soluble. At this point, the resulting supersaturated solution is *metastable.* This may be thought of as a state of pseudoequilibrium in which the system is at a higher energy than in its most stable state.

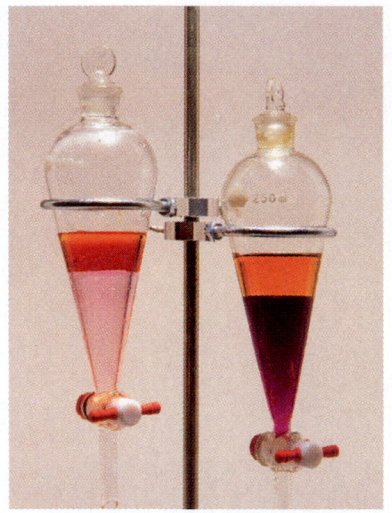

Solid iodine, I_2, dissolves to a limited extent in water to give an orange solution. This aqueous solution does not mix with nonpolar carbon tetrachloride, CCl_4 (left). Iodine is much more soluble in the nonpolar carbon tetrachloride. After the funnel is shaken and the liquids are allowed to separate (right), the upper aqueous phase is lighter orange and the lower CCl_4 layer is much more highly colored. This is because iodine is much more soluble in the nonpolar carbon tetrachloride than in water; much of the iodine dissolves preferentially in the lower (CCl_4) phase. The design of the separatory funnel allows the lower (more dense) layer to be drained off. Fresh CCl_4 could be added and the process repeated. This method of separation is called *extraction*. It takes advantage of the different solubilities of a solute in two immiscible liquids.

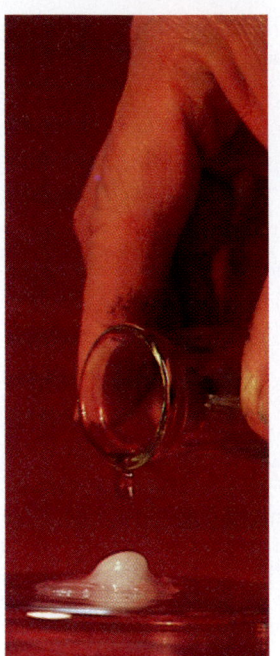  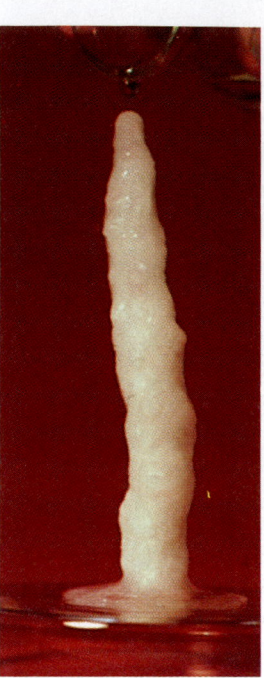

Figure 14-6 Another method of seeding a supersaturated solution is by pouring it very slowly onto a seed crystal. A supersaturated sodium acetate solution was used in these photographs.

In such a case, the solute has not yet become sufficiently organized for crystallization to begin. A supersaturated solution produces crystals rapidly if it is slightly disturbed or if it is "seeded" with a dust particle or a tiny crystal. Under such conditions enough solid crystallizes to leave a saturated solution (Figure 14-6).

14-6 EFFECT OF TEMPERATURE ON SOLUBILITY

In Section 13-6 we introduced LeChatelier's Principle, which states that *when a stress is applied to a system at equilibrium, the system responds in a way that best relieves the stress.* Recall that exothermic processes release heat and endothermic processes absorb heat.

$$\text{Exothermic:} \qquad \text{reactants} \longrightarrow \text{products} + \text{heat}$$

$$\text{Endothermic:} \qquad \text{reactants} + \text{heat} \longrightarrow \text{products}$$

Many ionic solids dissolve by endothermic processes. Their solubilities in water usually *increase* as heat is added and the temperature increases. For example, KCl dissolves endothermically.

$$KCl(s) + 17.2 \text{ kJ} \xrightarrow{\text{H}_2\text{O}} K^+(aq) + Cl^-(aq)$$

Figure 14-7 shows that the solubility of KCl increases as the temperature increases because more heat is available to drive the dissolving process. Raising the temperature (adding *heat*) causes a stress on the solubility equilibrium. This stress favors the process that *consumes* heat. In this case, more KCl dissolves.

Calcium acetate, $Ca(CH_3COO)_2$, is more soluble in cold water than in hot water. When a cold, concentrated solution of calcium acetate is heated, solid calcium acetate precipitates.

Some solids, such as anhydrous Na_2SO_4, and many liquids and gases dissolve by exothermic processes. Their solubilities usually decrease as temperature increases. The solubility of O_2 in water decreases (by 22%) from 0.041 gram per liter of water at 25°C to 0.03 gram per liter at 50°C. Raising the temperature of rivers and lakes by dumping heated waste water from industrial plants and nuclear power plants is called **thermal pollution.** A slight increase in the temperature of the water causes a small but significant decrease in the concentration of dissolved oxygen. As a result, the water can no longer support the marine life it ordinarily could.

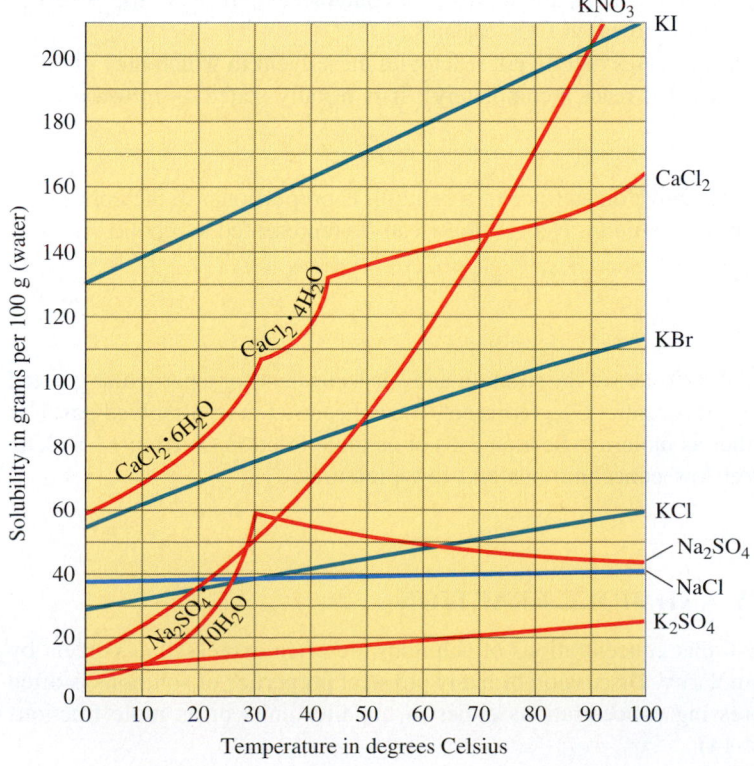

Figure 14-7 A graph that illustrates the effect of temperature on the solubilities of some salts. Some compounds exist either as nonhydrated crystalline substances or as hydrated crystals. Hydrated and nonhydrated crystal forms of the same compounds often have different solubilities, because of the different total forces of attraction in the solids. The discontinuities in the solubility curves for $CaCl_2$ and Na_2SO_4 are due to transitions between hydrated and nonhydrated crystal forms.

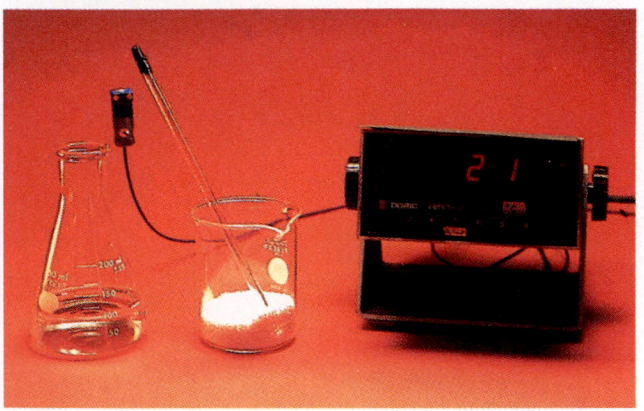
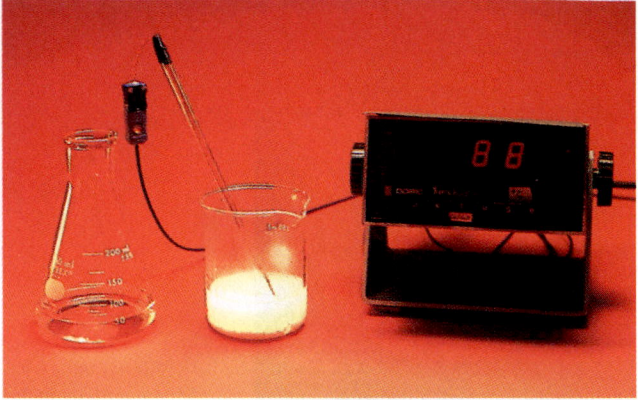

The dissolution of anhydrous calcium chloride, $CaCl_2$, in water is quite exothermic. This dissolution process is utilized in commercial instant hot packs for quick treatment of injuries requiring heat.

14-7 EFFECT OF PRESSURE ON SOLUBILITY

Changing the pressure has no appreciable effect on the solubilities of either solids or liquids in liquids. However, the solubilities of gases in all solvents increase as the partial pressures of the gases increase (Figure 14-8). Carbonated water is a saturated solution of carbon dioxide in water under pressure. When a can or bottle of a carbonated beverage is opened, the pressure on the surface of the beverage is reduced to atmospheric pressure, and much of the CO_2 bubbles out of solution. If the container is left open, the beverage becomes "flat" because the released CO_2 escapes.

Henry's Law applies to gases that do not react with the solvent in which they dissolve (or, in some cases, gases that react incompletely). It is usually stated as follows.

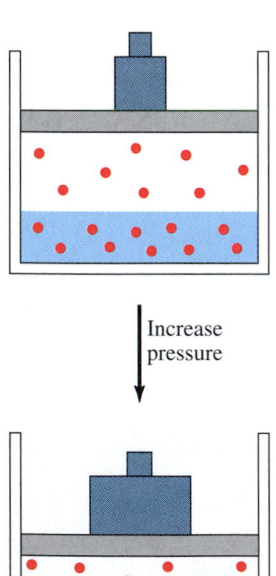

The pressure of a gas above the surface of a solution is proportional to the concentration of the gas in the solution. Henry's Law can be represented symbolically as

$$P_{gas} = kC_{gas}$$

P_{gas} is the pressure of the gas above the solution and k is a constant for a particular gas and solvent at a particular temperature. C_{gas} represents the concentration of dissolved gas; it is usually expressed either as molarity (Section 3-6) or as mole fraction (Section 14-8). The relationship is valid at low concentrations and low pressures.

Figure 14-8 An illustration of Henry's Law. The solubility of a gas (that does not react completely with the solvent) increases with increasing pressure of the gas above the solution.

14-8 MOLALITY AND MOLE FRACTION

We saw in Section 3-6 that concentrations of solutions are often expressed as percent by mass of solute or as molarity. Discussion of many physical properties of solutions is often made easier by expressing concentrations either in molality units or as mole fractions (Sections 14-9 to 14-14).

Molality

The **molality**, *m*, of a solute in solution is the number of moles of solute *per kilogram of solvent.*

$$\text{molality} = \frac{\text{number of moles solute}}{\text{number of kilograms solvent}}$$

Molality is based on the amount of *solvent not solution.*

EXAMPLE 14-1   *Molality*

What is the molality of a solution that contains 128 g of CH_3OH in 108 g of water?

Plan

We convert the amount of solute (CH_3OH) to moles, express the amount of solvent (water) in kilograms, and apply the definition of molality.

$$\boxed{\text{g } CH_3OH} \longrightarrow \boxed{\text{mol } CH_3OH}$$
$$\boxed{\text{g } H_2O} \longrightarrow \boxed{\text{kg } H_2O} \longrightarrow \boxed{\text{molality}}$$

Solution

$$\frac{?\ \text{mol } CH_3OH}{\text{kg } H_2O} = \frac{128 \text{ g } CH_3OH}{0.108 \text{ kg } H_2O} \times \frac{1 \text{ mol } CH_3OH}{32.0 \text{ g } CH_3OH} = \frac{37.0 \text{ mol } CH_3OH}{\text{kg } H_2O}$$

$$= \boxed{37.0\ m\ CH_3OH}$$

You should now work Exercise 28.

EXAMPLE 14-2   *Molality*

How many grams of H_2O must be used to dissolve 50.0 grams of sucrose to prepare a 1.25 *m* solution of sucrose, $C_{12}H_{22}O_{11}$?

Plan

We convert the amount of solute ($C_{12}H_{22}O_{11}$) to moles, solve the molality expression for kilograms of solvent (water), and then express the result in grams.

Solution

$$?\ \text{mol } C_{12}H_{22}O_{11} = 50.0 \text{ g } C_{12}H_{22}O_{11} \times \frac{1 \text{ mol } C_{12}H_{22}O_{11}}{342 \text{ g } C_{12}H_{22}O_{11}} = 0.146 \text{ mol } C_{12}H_{22}O_{11}$$

$$\text{molality of solution} = \frac{\text{mol } C_{12}H_{22}O_{11}}{\text{kg } H_2O}$$

Rearranging gives

$$\text{kg } H_2O = \frac{\text{mol } C_{12}H_{22}O_{11}}{\text{molality of solution}} = \frac{0.146 \text{ mol } C_{12}H_{22}O_{11}}{1.25 \text{ mol } C_{12}H_{22}O_{11}/\text{kg } H_2O}$$

$$= 0.117 \text{ kg } H_2O = \boxed{117 \text{ g } H_2O}$$

You should now work Exercise 32(a).

Each beaker holds the amount of a crystalline ionic compound that will dissolve in 100 grams of water at 100°C. The compounds are (top row, left to right) 39 grams of sodium chloride (NaCl, white), 102 grams of potassium dichromate ($K_2Cr_2O_7$, red-orange), 341 grams of nickel sulfate hexahydrate ($NiSO_4 \cdot 6H_2O$, green); (bottom row; left to right) 79 grams of potassium chromate (K_2CrO_4, yellow), 191 grams of cobalt(II) chloride hexahydrate ($CoCl_2 \cdot 6H_2O$, dark red), and 203 grams of copper sulfate pentahydrate ($CuSO_4 \cdot 5H_2O$, blue).

In other examples later in this chapter we shall calculate several properties of the solution in Example 14-2.

Mole Fraction

Recall that in Chapter 12 the **mole fractions,** X_A and X_B, of each component in a mixture containing components A and B were defined as

$$X_A = \frac{\text{no. mol A}}{\text{no. mol A + no. mol B}} \quad \text{and} \quad X_B = \frac{\text{no. mol B}}{\text{no. mol A + no. mol B}}$$

Mole fraction is a dimensionless quantity, i.e., it has no units.

EXAMPLE 14-3 *Mole Fraction*

What are the mole fractions of CH_3OH and H_2O in the solution described in Example 14-1? It contains 128 grams of CH_3OH and 108 grams of H_2O.

Plan
We express the amount of both components in moles, and then apply the definition of mole fraction.

Solution

$$\underline{?}\text{ mol CH}_3\text{OH} = 128\text{ g CH}_3\text{OH} \times \frac{1\text{ mol CH}_3\text{OH}}{32.0\text{ g CH}_3\text{OH}} = 4.00\text{ mol CH}_3\text{OH}$$

$$\underline{?}\text{ mol H}_2\text{O} = 108\text{ g H}_2\text{O} \times \frac{1\text{ mol H}_2\text{O}}{18.0\text{ g H}_2\text{O}} = 6.00\text{ mol H}_2\text{O}$$

Now we calculate the mole fraction of each component.

$$X_{CH_3OH} = \frac{\text{no. mol CH}_3\text{OH}}{\text{no. mol CH}_3\text{OH + no. mol H}_2\text{O}} = \frac{4.00\text{ mol}}{(4.00 + 6.00)\text{ mol}} = \boxed{0.400}$$

$$X_{H_2O} = \frac{\text{no. mol H}_2\text{O}}{\text{no. mol CH}_3\text{OH + no. mol H}_2\text{O}} = \frac{6.00\text{ mol}}{(4.00 + 6.00)\text{ mol}} = \boxed{0.600}$$

You should now work Exercises 32(b) and 38.

In any mixture the sum of the mole fractions must be 1:

$$0.400 + 0.600 = 1$$

COLLIGATIVE PROPERTIES OF SOLUTIONS

Colligative means "tied together."

Physical properties of solutions that depend upon the *number,* but not the *kind,* of solute particles in a given amount of solvent are called **colligative properties.** There are four important colligative properties of a solution that are directly proportional to the number of solute particles present. They are (1) vapor pressure lowering, (2) boiling point elevation, (3) freezing point depression, and (4) osmotic pressure. These properties of a solution depend on the *total concentration of all solute particles,* regardless of their ionic or molecular nature, charge, or size. For most of this chapter, we shall consider *nonelectrolyte* solutes (Section 4-2, part 1); these substances dissolve to give one mole of dissolved particles for each mole of solute. In Section 14-14 we shall learn to modify our predictions of colligative properties to account for ion formation in electrolyte solutions.

14-9 LOWERING OF VAPOR PRESSURE AND RAOULT'S LAW

Many experiments have shown that a solution containing a *nonvolatile* liquid or a solid as a solute always has a lower vapor pressure than the pure solvent (Figure 14-9). The vapor pressure of a liquid depends on the ease with which the molecules are able to escape from the surface of the liquid. When a solute is dissolved in a liquid, some of the total volume of the solution is occupied by solute molecules, and so there are fewer solvent molecules *per unit area* at the surface. As a result, solvent molecules vaporize at a slower rate than if no solute were present. The increase in disorder that accompanies evaporation is also a significant factor. Because a solution is already more disordered ("mixed up") than a pure solvent, the evaporation of the pure solvent involves a larger increase in disorder, and is thus more favorable. Hence, the pure solvent exhibits a higher vapor pressure than does the solution. The lowering of the vapor pressure of the solution is a colligative property. It is a function of the number, and not the kind, of solute particles in solution. We emphasize that solutions of gases or low-boiling (volatile) liquids can have *higher* total vapor pressures than the pure solvents, so this discussion does not apply to them.

The lowering of the vapor pressure of a solvent due to the presence of *nonvolatile, nonionizing* solutes is summarized by **Raoult's Law.**

> The vapor pressure of a solvent in an ideal solution is directly proportional to the mole fraction of the solvent in the solution.

The relationship can be expressed mathematically as

$$P_{solvent} = X_{solvent}P^0_{solvent}$$

$X_{solvent}$ represents the mole fraction of the solvent in a solution, $P^0_{solvent}$ is the vapor pressure of the *pure* solvent, and $P_{solvent}$ is the vapor pressure of the solvent *in the solution* (see Figure 14-10). If the solute is nonvolatile, the vapor pressure of the solution is entirely due to the vapor pressure of the solvent, $P_{solution} = P_{solvent}$.

The *lowering* of the vapor pressure, $\Delta P_{solvent}$, is defined as

$$\Delta P_{solvent} = P^0_{solvent} - P_{solvent}$$

Thus,

$$\Delta P_{solvent} = P^0_{solvent} - (X_{solvent}P^0_{solvent}) = (1 - X_{solvent})P^0_{solvent}$$

Now $X_{solvent} + X_{solute} = 1$, so $1 - X_{solvent} = X_{solute}$. We can express the *lowering* of the vapor pressure in terms of the mole fraction of solute.

$$\Delta P_{solvent} = X_{solute}P^0_{solvent}$$

Solutions that obey this relationship exactly are called **ideal solutions.** The vapor pressures of many solutions do not behave ideally.

EXAMPLE 14-4 *Vapor Pressure of a Solution of Nonvolatile Solute*

Sucrose is a nonvolatile, nonionizing solute in water. Determine the vapor pressure lowering, at 25°C, of the 1.25 *m* sucrose solution in Example 14-2. Assume that the solution behaves ideally. The vapor pressure of pure water at 25°C is 23.8 torr (Appendix E).

A *vapor* is a gas formed by the boiling or evaporation of a liquid or sublimation of a solid. The *vapor pressure* of a liquid is the pressure (partial pressure) exerted by a vapor in equilibrium with its liquid.

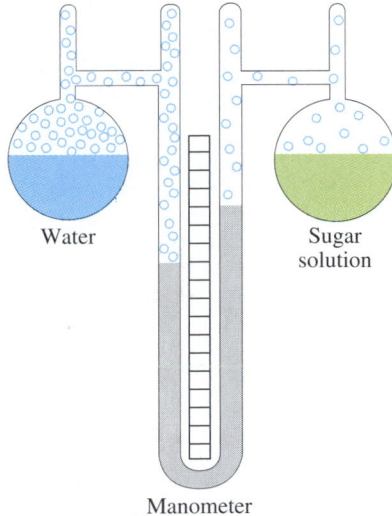

Figure 14-9   Lowering of vapor pressure. If no air is present in the apparatus, the pressure above each liquid is due to water vapor. This pressure is less over the solution of sugar and water.

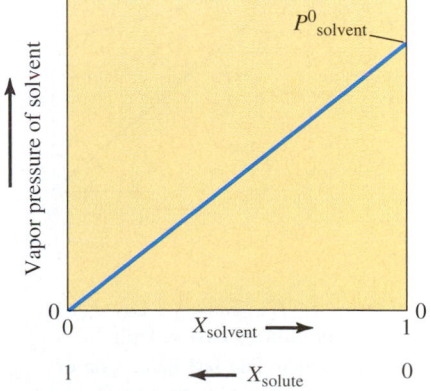

Figure 14-10   Raoult's Law for an ideal solution of a solute in a volatile liquid. The vapor pressure exerted by the liquid is proportional to *its* mole fraction in the solution.

Plan

The solution in Example 14-2 was made by dissolving 50.0 grams of sucrose (0.146 mol) in 117 grams of water (6.50 mol). We calculate the mole fraction of solute in the solution. Then we apply Raoult's Law to find the vapor pressure lowering, $\Delta P_{solvent}$.

Solution

The vapor pressure of water in the solution is $(23.8 - 0.524)$ torr $= 23.3$ torr. We could calculate this vapor pressure directly from the mole fraction of the solvent (water) in the solution, using the relationship $P_{solvent} = (X_{solvent})(P^0_{solvent})$.

$$X_{sucrose} = \frac{0.146 \text{ mol}}{0.146 \text{ mol} + 6.50 \text{ mol}} = 0.0220$$

Applying Raoult's Law in terms of the vapor pressure lowering,

$$\Delta P_{solvent} = (X_{solute})(P^0_{solvent}) = (0.0220)(23.8 \text{ torr}) = \boxed{0.524 \text{ torr}}$$

You should now work Exercise 44.

When a solution consists of two components that are very similar, each component behaves essentially as it would if it were pure. For example, the two liquids heptane, C_7H_{16}, and octane, C_8H_{18}, are so similar that each heptane molecule experiences nearly the same intermolecular forces whether it is near another heptane molecule or near an octane molecule, and similarly for each octane molecule. The properties of such a solution can be predicted from a knowledge of its composition and the properties of each component. Such a solution is very nearly ideal.

If component B were nonvolatile, then P^0_B would be zero, and this description would be the same as that given earlier for a solution of a nonvolatile, nonionizing solute in a volatile solvent, $P_{total} = P_{solvent}$.

Consider an ideal solution of two volatile components, A and B. The vapor pressure of each component above the solution is proportional to its mole fraction in the solution.

$$P_A = X_A P^0_A \qquad \text{and} \qquad P_B = X_B P^0_B$$

The total vapor pressure of the solution is, by Dalton's Law of Partial Pressures (Section 12-11), equal to the sum of the vapor pressures of the two components.

$$P_{total} = P_A + P_B \qquad \text{or} \qquad P_{total} = X_A P^0_A + X_B P^0_B$$

This is shown graphically in Figure 14-11. We can use these relationships to predict the vapor pressures of an ideal solution, as Example 14-5 illustrates.

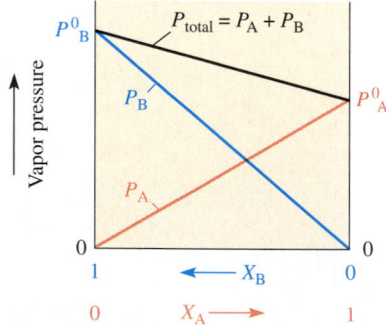

Figure 14-11   Raoult's Law for an ideal solution of two volatile components. The left-hand side of the plot corresponds to pure B ($X_A = 0$, $X_B = 1$), and the right-hand side corresponds to pure A ($X_A = 1$, $X_B = 0$). Of these hypothetical liquids, B is more volatile than A ($P^0_B > P^0_A$).

EXAMPLE 14-5   *Vapor Pressure of a Solution of Volatile Components*

At 40°C, the vapor pressure of pure heptane is 92.0 torr and the vapor pressure of pure octane is 31.0 torr. Consider a solution that contains 1.00 mole of heptane and 4.00 moles of octane. Calculate the vapor pressure of each component and the total vapor pressure above the solution.

Plan

We first calculate the mole fraction of each component in the liquid solution. Then we apply Raoult's Law to each of the two volatile components. The total vapor pressure is the sum of the vapor pressures of the components.

Solution

We first calculate the mole fraction of each component in the liquid solution.

$$X_{heptane} = \frac{1.00 \text{ mol heptane}}{(1.00 \text{ mol heptane}) + (4.00 \text{ mol octane})} = 0.200$$

$$X_{octane} = 1 - X_{heptane} = 0.800$$

Then, applying Raoult's Law for volatile components,

$$P_{heptane} = X_{heptane} P^0_{heptane} = (0.200)(92.0 \text{ torr}) = \boxed{18.4 \text{ torr}}$$

$$P_{\text{octane}} = X_{\text{octane}}P^0_{\text{octane}} = (0.800)(31.0 \text{ torr}) = \boxed{24.8 \text{ torr}}$$

$$P_{\text{total}} = P_{\text{heptane}} + P_{\text{octane}} = 18.4 \text{ torr} + 24.8 \text{ torr} = \boxed{43.2 \text{ torr}}$$

You should now work Exercise 46.

The vapor in equilibrium with a liquid solution of two or more volatile components is *richer than the liquid solution* in the more volatile component.

EXAMPLE 14-6 *Composition of Vapor*

Calculate the mole fractions of heptane and octane in the vapor that is in equilibrium with the solution in Example 14-5.

Plan
We learned in Section 12-11 that the mole fraction of a component in a gaseous mixture equals the ratio of its partial pressure to the total pressure. In Example 14-5 we calculated the partial pressure of each component in the vapor and the total vapor pressure.

Solution
In the *vapor*

$$X_{\text{heptane}} = \frac{P_{\text{heptane}}}{P_{\text{total}}} = \frac{18.4 \text{ torr}}{43.2 \text{ torr}} = \boxed{0.426}$$

$$X_{\text{octane}} = \frac{P_{\text{octane}}}{P_{\text{total}}} = \frac{24.8 \text{ torr}}{43.2 \text{ torr}} = \boxed{0.574}$$

You should now work Exercise 48.

Heptane (pure vapor pressure = 92.0 torr at 40°C) is a more volatile liquid than octane (pure vapor pressure = 31.0 torr at 40°C). Its mole fraction in the vapor, 0.426, is higher than its mole fraction in the liquid, 0.200.

Many dilute solutions behave ideally. Some solutions do not behave ideally over the entire concentration range. For some solutions, the observed vapor pressure is greater than that predicted by Raoult's Law (Figure 14-12a). This kind of deviation, known as a

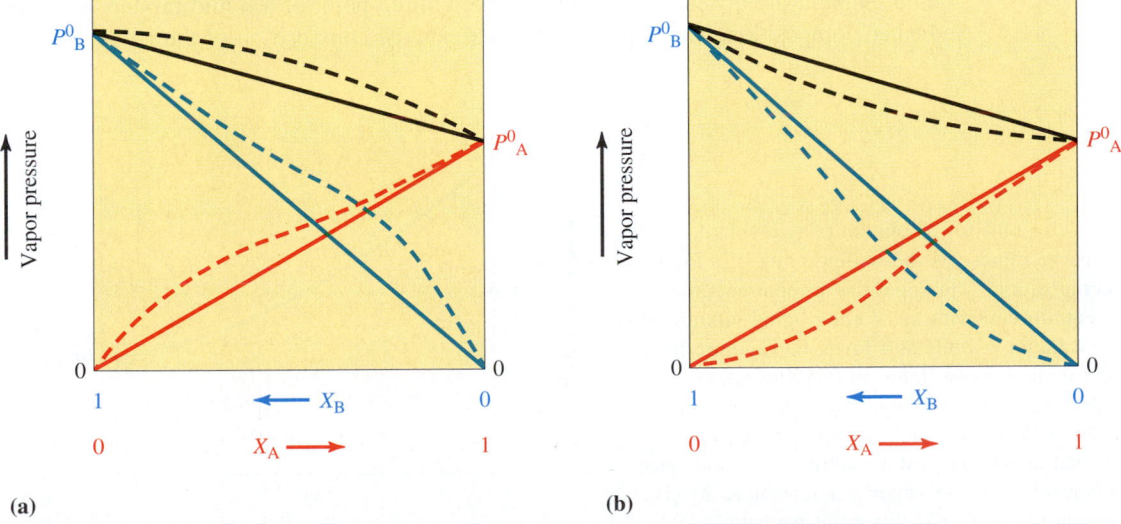

(a)   **(b)**

Figure 14-12 Deviations from Raoult's Law for two volatile components. (a) Positive deviation. (b) Negative deviation.

positive deviation, is due to differences in polarity of the two components. On the molecular level, the two substances do not mix entirely randomly, so there is self-association of each component with local regions enriched in one type of molecule or the other. In a region enriched in A molecules, substance A acts as though its mole fraction were greater than it is in the solution as a whole, and the vapor pressure due to A is greater than if the solution were ideal. A similar description applies to component B. The total vapor pressure is then greater than it would be if the solution were behaving ideally. A solution of acetone and carbon disulfide is an example of a solution that shows a positive deviation from Raoult's Law.

Another, more common type of deviation occurs when the total vapor pressure is less than that predicted (Figure 14-12b). This is called a *negative deviation.* Such an effect is due to unusually strong attractions (such as hydrogen bonding) between *unlike* molecules. As a result, unlike species hold one another especially tightly in the liquid phase, so fewer molecules escape to the vapor phase. The observed vapor pressure of each component is thus less than ideally predicted. An acetone–chloroform solution and an ethanol–water solution are two examples that show negative deviations from Raoult's Law.

14-10 FRACTIONAL DISTILLATION

In Section 13-8 we described *simple* distillation as a process in which a liquid solution can be separated into volatile and nonvolatile components. But separation of volatile components is not very efficient by this method. Consider the simple distillation of a liquid solution consisting of two volatile components. If the temperature is slowly raised, the solution begins to boil when the sum of the vapor pressures of the components reaches the applied pressure on the surface of the solution. Both components exert vapor pressures, so both are carried away as a vapor. The resulting distillate is richer than the original liquid in the more volatile component (Example 14-6).

As a mixture of volatile liquids is distilled, the compositions of both the liquid and the vapor, as well as the boiling point of the solution, change continuously. *At constant pressure,* we can represent these quantities in a **boiling point diagram,** Figure 14-13. In such a diagram the lower curve represents the boiling point of a liquid mixture with the indicated composition. The upper curve represents the composition of the *vapor* in equi-

The applied pressure is often atmospheric pressure.

acetone $CH_3-\overset{\overset{\displaystyle O}{\|}}{C}-CH_3$

carbon disulfide $S=C=S$

chloroform $CHCl_3$

ethanol CH_3CH_2OH

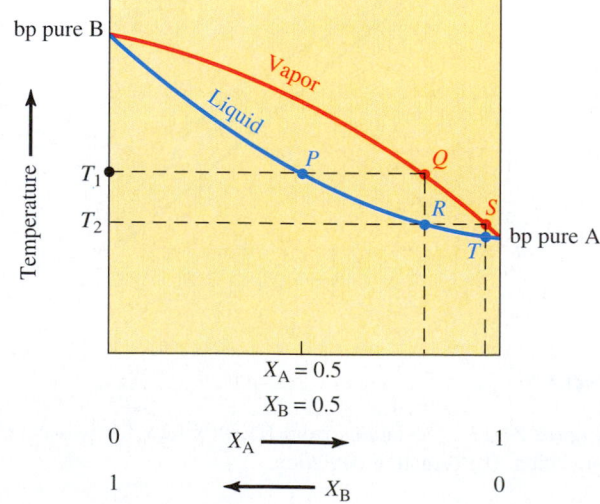

Figure 14-13 A boiling point diagram for a solution of two volatile liquids, A and B. The lower curve represents the boiling point of a liquid mixture with the indicated composition. The upper curve represents the composition of the *vapor* in equilibrium with the boiling liquid mixture at the indicated temperature. Pure liquid A boils at a lower temperature than pure liquid B; hence, A is the more volatile liquid in this illustration. Suppose we begin with an ideal equimolar mixture ($X_A = X_B = 0.5$) of liquids A and B. The point P represents the temperature at which this solution boils, T_1. The vapor that is present at this equilibrium is indicated by point Q ($X_A \approx 0.8$). Condensation of that vapor at temperature T_2 gives a liquid of the same composition (point R). At this point we have described one step of simple distillation. The boiling liquid at point R is in equilibrium with the vapor of composition indicated by point S ($X_A > 0.95$), and so on.

librium with the boiling liquid mixture at the indicated temperature. The intercepts at the two vertical axes show the boiling points of the two pure liquids. The distillation of the two liquids is described in the legend to Figure 14-13.

From the boiling point diagram in Figure 14-13, we see that two or more volatile liquids cannot be completely separated from each other by a single distillation step. The vapor collected at any boiling temperature is always enriched in the more volatile component (A); however, at any temperature it is still some mixture of the two vapors. A *series* of simple distillations would provide distillates increasingly richer in the more volatile component, but the repeated distillations would be very tedious.

Repeated distillations may be avoided by using **fractional distillation.** A *fractionating column* is inserted above the solution and attached to the condenser, as shown in Figure 14-14a. The column is constructed so that it has a large surface area or is packed with many small glass beads or another material with a large surface area. These provide surfaces upon which condensation can occur. Contact between the vapor and the packing favors condensation of the less volatile component. The column is cooler at the top than at the bottom. By the time the vapor reaches the top of the column, practically all of the less volatile component has condensed and fallen back down the column. The more volatile component goes into the condenser, where it is liquefied and delivered as a highly enriched distillate into the collection flask. The longer the column or the more packing, the better the separation.

Distillation under vacuum lowers the applied pressure. This allows boiling at lower temperatures than under atmospheric pressure. This technique allows distillation of some substances that would decompose at higher temperatures.

Many fractions such as gasoline, kerosene, fuel oil, paraffin, and asphalt are separated from crude oil by fractional distillation.

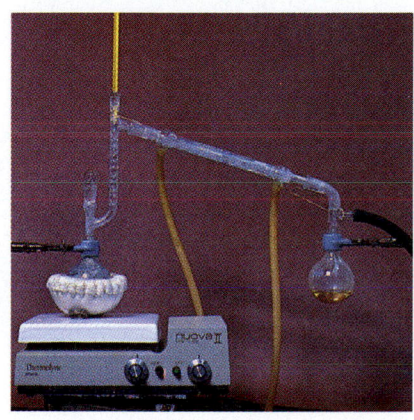

(a)

Figure 14-14 (a) A fractional distillation apparatus. The vapor phase rising in the column is in equilibrium with the liquid phase that has condensed out and is flowing slowly back down the column. (b) Fractional distillation is used for separations in many industrial processes. In this Pennsylvania plant, atmospheric air is liquefied by cooling and compression and then is separated by distillation in towers. This plant produces more than 1000 tons daily of gases from air (nitrogen, oxygen, and argon).

(b)

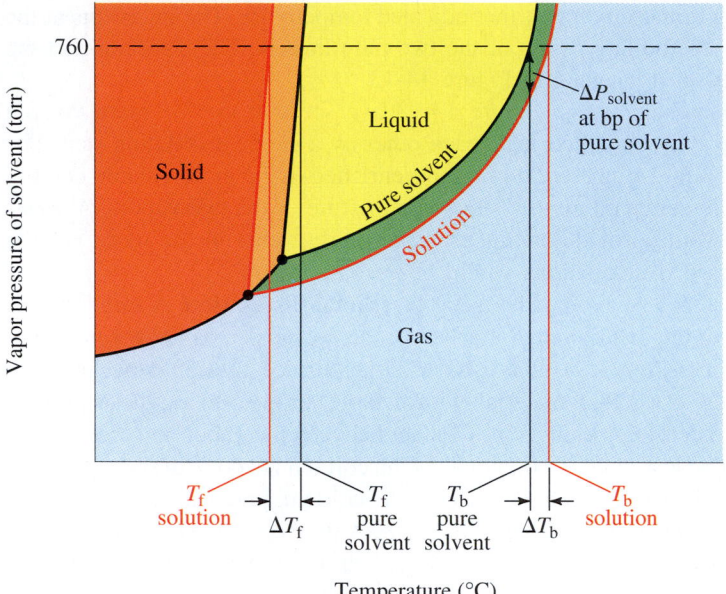

Figure 14-15 Because a *nonvolatile* solute lowers the vapor pressure of a solvent, the boiling point of a solution is higher and the freezing point lower than the corresponding points for the pure solvent. The magnitude of boiling point elevation, ΔT_b, is less than the magnitude of freezing point depression, ΔT_f.

14-11 BOILING POINT ELEVATION

Recall that the boiling point of a liquid is the temperature at which its vapor pressure equals the applied pressure on its surface. For liquids in open containers, this is atmospheric pressure. We have seen that the vapor pressure of a solvent at a given temperature is lowered by the presence in it of a *nonvolatile* solute. Such a solution must be heated to a higher temperature than the pure solvent to cause the vapor pressure of the solvent to equal atmospheric pressure (Figure 14-15). In accord with Raoult's Law, the elevation of the boiling point of a solvent caused by the presence of a nonvolatile, nonionized solute is proportional to the number of moles of solute dissolved in a given mass of solvent. Mathematically, this is expressed as

> When the solute is nonvolatile, only the *solvent* distills from the solution.

$$\Delta T_b = K_b m$$

The term ΔT_b represents the elevation of the boiling point of the solvent, i.e., the boiling point of the solution minus the boiling point of the pure solvent. The m is the molality of the solute, and K_b is a proportionality constant called the **molal boiling point elevation constant.** This constant is different for different solvents and does not depend on the solute (Table 14-2).

> $\Delta T_b = T_{b(soln)} - T_{b(solvent)}$. The boiling points of solutions that contain nonvolatile solutes are always higher than the boiling points of the pure solvents.

> K_b corresponds to the change in boiling point produced by a one-molal *ideal* solution of a nonvolatile nonelectrolyte. The units of K_b are °C/m.

Elevations of boiling points and depressions of freezing points, which will be discussed later, are usually quite small for solutions of typical concentrations. However, they can be measured with specially constructed differential thermometers that measure small temperature changes accurately to the nearest 0.001°C.

Table 14-2 *Some Properties of Common Solvents*				
Solvent	bp (pure)	K_b (°C/m)	fp (pure)	K_f (°C/m)
water	100*	0.512	0*	1.86
benzene	80.1	2.53	5.48	5.12
acetic acid	118.1	3.07	16.6	3.90
nitrobenzene	210.88	5.24	5.7	7.00
phenol	182	3.56	43	7.40
camphor	207.42	5.61	178.40	40.0

Exact values.

EXAMPLE 14-7 *Boiling Point Elevation*

What is the normal boiling point of the 1.25 *m* sucrose solution in Example 14-2?

Plan
We first find the *increase* in boiling point from the relationship $\Delta T_b = K_b m$. The boiling point is *higher* by this amount than the normal boiling point of pure water.

Solution
From Table 14-2, K_b for $H_2O = 0.512$°C/m, so

$$\Delta T_b = (0.512°C/m)(1.25\ m) = 0.640°C$$

The solution would boil at a temperature that is 0.640°C higher than pure water would boil. The normal boiling point of pure water is exactly 100°C, so at 1.00 atm this solution boils at 100°C + 0.640°C = 100.640°C.

You should now work Exercise 52.

14-12 FREEZING POINT DEPRESSION

Molecules of liquids move more slowly and approach each other more closely as the temperature is lowered. The freezing point of a liquid is the temperature at which the forces of attraction among molecules are just great enough to overcome their kinetic energies and thus cause a phase change from the liquid to the solid state. Strictly speaking, the freezing (melting) point of a substance is the temperature at which the liquid and solid phases are in equilibrium. When a dilute solution freezes, it is the *solvent* that begins to solidify first, leaving the solute in a more concentrated solution. Solvent molecules in a solution are somewhat more separated from each other (because of solute particles) than they are in the pure solvent. Consequently, the temperature of a solution must be lowered below the freezing point of the pure solvent to freeze it.

The freezing point depressions of solutions of nonelectrolytes have been found to be equal to the molality of the solute times a proportionality constant called the **molal freezing point depression constant, K_f.**

$$\Delta T_f = K_f m$$

ΔT_f is the *depression* of freezing point. It is defined as

$$\Delta T_f = T_{f(solvent)} - T_{f(soln)}$$

so it is always *positive*.

Lime, CaO, is added to molten iron ore during the manufacture of pig iron. It lowers the melting point of the mixture. The metallurgy of iron is discussed in more detail in Chapter 22.

The values of K_f for a few solvents are given in Table 14-2. Each is numerically equal to the freezing point depression of a one-molal *ideal* solution of a nonelectrolyte in that solvent.

EXAMPLE 14-8   *Freezing Point Depression*

When 15.0 grams of ethyl alcohol, C_2H_5OH, is dissolved in 750 grams of formic acid, the freezing point of the solution is 7.20°C. The freezing point of pure formic acid is 8.40°C. Evaluate K_f for formic acid.

Plan

The molality and the depression of the freezing point are calculated first. Then we solve the equation $\Delta T_f = K_f m$ for K_f and substitute values for m and ΔT_f.

$$K_f = \frac{\Delta T_f}{m}$$

Solution

$$\frac{?\ \text{mol } C_2H_5OH}{\text{kg formic acid}} = \frac{15.0\ \text{g } C_2H_5OH}{0.750\ \text{kg formic acid}} \times \frac{1\ \text{mol } C_2H_5OH}{46.0\ \text{g } C_2H_5OH} = 0.435\ m$$

$$\Delta T_f = (T_{f[\text{formic acid}]}) - (T_{f[\text{solution}]}) = 8.40°C - 7.20°C = 1.20°C \qquad (\text{depression})$$

Then $K_f = \dfrac{\Delta T_f}{m} = \dfrac{1.20°C}{0.435\ m} = \boxed{2.76°C/m}$ for formic acid.

You should now work Exercise 54.

EXAMPLE 14-9   *Freezing Point Depression*

What is the freezing point of the 1.25 m sucrose solution in Example 14-2?

Plan

We first find the *decrease* in freezing point from the relationship $\Delta T_f = K_f m$. The temperature at which the solution freezes is *lower* than the freezing point of pure water by this amount.

Solution

From Table 14-2, K_f for H_2O = 1.86°C/m, so

$$\Delta T_f = (1.86°C/m)(1.25\ m) = 2.32°C$$

The temperature at which the solution freezes is 2.32°C *below* the freezing point of pure water, or

$$T_{f(\text{solution})} = 0.00°C - 2.32°C = \boxed{-2.32°C}$$

You should now work Exercise 58.

The total concentration of all dissolved solute species determines the colligative properties. As we shall emphasize in Section 14-14, we must take into account the extent of ion formation in solutions of ionic solutes.

You may be familiar with several examples of the effects we have studied. Sea water does not freeze on some days when fresh water does, because sea water contains higher concentrations of solutes, mostly ionic solutes. Spreading soluble salts such as sodium chloride, NaCl, or calcium chloride, $CaCl_2$, on an icy road lowers the freezing point of the ice, causing the ice to melt.

A familiar application is the addition of "permanent" antifreeze, mostly ethylene glycol, $HOCH_2CH_2OH$, to the water in an automobile radiator. Because the boiling point of the solution is elevated, addition of a solute as a winter antifreeze also helps to protect against loss of the coolant by summer "boil-over." The amounts by which the freezing and boiling points change depend on the concentration of the ethylene glycol solution. However, the addition of too much ethylene glycol is counterproductive. The freezing point of pure ethylene glycol is about −12°C. A solution that is mostly ethylene glycol would have a somewhat lower freezing point due to the presence of water as a solute. Suppose you graph the freezing point depression of water below 0°C as ethylene glycol is added, and also graph the freezing point depression of ethylene glycol below −12°C as water is added. Obviously, these two curves would intersect at some temperature, indicating the limit of lowering that can occur. (At these high concentrations, the solutions do not behave ideally, so the temperatures could not be accurately predicted by the equations we have introduced in this chapter, but the main ideas still apply.) Most antifreeze labels recommend a 50:50 mixture by volume (fp −34°F, bp 265°F with a 15-psi pressure cap on the radiator), and cite the limit of possible protection with a 70:30 mixture by volume of antifreeze:water (fp −84°F, bp 276°F with a 15-psi pressure cap).

Ethylene glycol, $HOCH_2CH_2OH$, is the major component of "permanent" antifreeze. It depresses the freezing point of water in an automobile radiator and also raises its boiling point. The solution remains in the liquid phase over a wider temperature range than does pure water. This protects against both freezing and boil-over.

14-13 DETERMINATION OF MOLECULAR WEIGHT BY FREEZING POINT DEPRESSION OR BOILING POINT ELEVATION

The colligative properties of freezing point depression and, to a lesser extent, boiling point elevation are useful in the determination of molecular weights of solutes. The solutes *must* be nonvolatile in the temperature range of the investigation if boiling point elevations are to be determined. We will restrict our discussion of determination of molecular weight to nonelectrolytes.

EXAMPLE 14-10   *Molecular Weight from a Colligative Property*

A 1.20-gram sample of an unknown covalent compound is dissolved in 50.0 grams of benzene. The solution freezes at 4.92°C. Calculate the molecular weight of the compound.

Plan

To calculate the molecular weight of the unknown compound, we find the number of moles that is represented by the 1.20 grams of unknown compound. We first use the freezing point data to find the molality of the solution. The molality relates the number of moles of solute and the mass of solvent (known), so this allows us to calculate the number of moles of unknown.

Solution

From Table 14-2, the freezing point of pure benzene is 5.48°C and K_f is 5.12°C/*m*.

$$\Delta T_f = 5.48°C - 4.92°C = 0.56°C$$

$$m = \frac{\Delta T_f}{K_f} = \frac{0.56°C}{5.12°C/m} = 0.11 \ m$$

The molality is the number of moles of solute per kilogram of benzene, so the number of moles of solute in 50.0 g (0.0500 kg) of benzene can be calculated.

$$0.11 \ m = \frac{\underline{?} \ \text{mol solute}}{0.0500 \ \text{kg benzene}}$$

$$\underline{?} \ \text{mol solute} = (0.11 \ m)(0.0500 \ \text{kg}) = 0.0055 \ \text{mol solute}$$

$$\text{mass of 1.0 mol} = \frac{\text{no. of g solute}}{\text{no. of mol solute}} = \frac{1.20 \ \text{g solute}}{0.0055 \ \text{mol solute}} = 2.2 \times 10^2 \ \text{g/mol}$$

$$\text{molecular weight} = \boxed{2.2 \times 10^2 \ \text{amu}}$$

You should now work Exercise 61.

EXAMPLE 14-11   *Molecular Weight from a Colligative Property*

Either camphor ($C_{10}H_{16}O$, molecular weight = 152 g/mol) or naphthalene ($C_{10}H_8$, molecular weight 128 g/mol) can be used to make mothballs. A 5.2-gram sample of mothballs was dissolved in 100.0 grams of ethyl alcohol, and the resulting solution had a boiling point of 78.90°C. Were the mothballs made of camphor or naphthalene? Pure ethyl alcohol has a boiling point of 78.41°C; its K_b = 1.22°C/m.

Plan

We can distinguish between the two possibilities by determining the molecular weight of the unknown solute. We do this by the method shown in Example 14-10, except that now we use the observed boiling point data.

Solution

The observed boiling point elevation is

$$\Delta T_b = T_{b(solution)} - T_{b(solvent)} = (78.90 - 78.41)°C = 0.49°C$$

Using ΔT_b = 0.49°C and K_b = 1.22°C/m, we can find the molality of the solution.

$$\text{molality} = \frac{\Delta T_b}{K_b} = \frac{0.49°C}{1.22°C/m} = 0.40\ m$$

The number of moles of solute in the 100.0 g (0.1000 kg) of solvent used is

$$\left(0.40\ \frac{\text{mol solute}}{\text{kg solvent}}\right)(0.1000\ \text{kg solvent}) = 0.040\ \text{mol solute}$$

The molecular weight of the solute is its mass divided by the number of moles.

$$\frac{?\ g}{\text{mol}} = \frac{5.2\ g}{0.040\ \text{mol}} = 130\ g/\text{mol}$$

The value 130 g/mol for the molecular weight indicates that   naphthalene   was used to make these mothballs.

You should now work Exercise 63.

14-14   COLLIGATIVE PROPERTIES AND DISSOCIATION OF ELECTROLYTES

As we have emphasized, colligative properties depend on the *number* of solute particles in a given mass of solvent. A 0.100 molal *aqueous* solution of a covalent compound that does not ionize gives a freezing point depression of 0.186°C. If dissociation were complete, 0.100 m KBr would have an *effective* molality of 0.200 m (that is, 0.100 m K^+ + 0.100 m Br^-). So we might predict that a 0.100 molal solution of this 1:1 strong electrolyte would have a freezing point depression of 2 × 0.186°C, or 0.372°C. In fact, the *observed* depression is only 0.349°C. This value for ΔT_f is about 6% less than we would expect for an effective molality of 0.200 m.

In an ionic solution the solute particles are not randomly distributed. Rather, each positive ion has more negative than positive ions near it. The resulting electrical interactions cause the solution to behave nonideally. Some of the ions undergo **association** in solution (Figure 14-16). At any given instant, some K^+ and Br^- ions collide and "stick together." During the brief time that they are in contact, they behave as a single particle. This tends to reduce the effective molality. Therefore, the freezing point depression is reduced (as well as the boiling point elevation and the lowering of vapor pressure).

A (more concentrated) 1.00 m solution of KBr might be expected to have a freezing point depression of 2 × 1.86°C = 3.72°C, but the observed depression is only 3.29°C.

$\Delta T_f = K_f m = (1.86°C/m)(0.100\ m)$
$= 0.186°C$

Ionic solutions are elegantly described by the Debye–Hückel theory, which is beyond the scope of this text.

This value for ΔT_f is about 11% less than we would expect. We see a greater deviation from the depression predicted (neglecting ionic association) in the more concentrated solution. This is because the ions are closer together and collide more often in the more concentrated solution. Consequently, the ionic association is greater.

One measure of the extent of dissociation (or ionization) of an electrolyte in water is the **van't Hoff factor, i,** for the solution. This is the ratio of the *actual* colligative property to the value that *would* be observed *if no dissociation occurred*.

$$i = \frac{\Delta T_{f(actual)}}{\Delta T_f \text{ (if nonelectrolyte)}} = \frac{K_f m_{effective}}{K_f m_{stated}} = \frac{m_{effective}}{m_{stated}}$$

The ideal, or limiting, value of i for a solution of KBr would be 2, and the value for a 2:1 electrolyte such as Na_2SO_4 would be 3; these values would apply to infinitely dilute solutions in which no ion association occurs. For 0.10 m and 1.0 m solutions of KBr, i is *less than* 2.

For 0.10 m: $i = \dfrac{0.349°C}{0.186°C} = 1.88$ For 1.0 m: $i = \dfrac{3.29°C}{01.86°C} = 1.77$

Table 14-3 lists actual and ideal values of i for solutions of some strong electrolytes, based on measurements of freezing point depressions.

Many weak electrolytes are quite soluble in water, but they ionize only slightly. The percent ionization and i value for a weak electrolyte in solution can also be determined from freezing point depression data (Example 14-12).

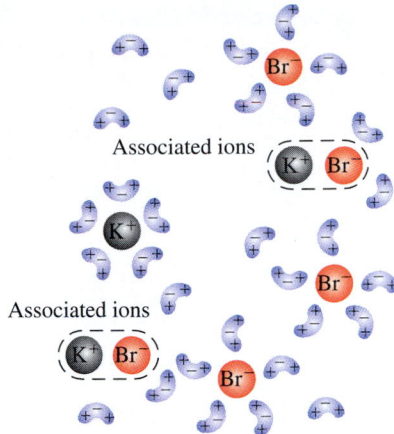

Figure 14-16   Diagrammatic representation of the various species thought to be present in a solution of KBr in water. This would explain unexpected values for its colligative properties, such as freezing point depression.

Weak acids and weak bases (Section 4-2) are weak electrolytes.

▼ **PROBLEM-SOLVING TIP**   *Selection of a van't Hoff Factor*

Use an ideal value for the van't Hoff factor unless the question clearly indicates to do otherwise, as in the following example and in some of the end-of-chapter exercises. For a strong electrolyte dissolved in water the ideal value for its van't Hoff factor is listed in Table 14-3. For nonelectrolytes dissolved in water or any solute dissolved in common nonaqueous solvents, the van't Hoff factor is considered to be 1. For weak electrolytes dissolved in water, the van't Hoff factor is a little greater than 1.

Table 14-3 *Actual and Ideal van't Hoff Factors, i, for Aqueous Solutions of Nonelectrolytes and Strong Electrolytes*

Compound	i for 1.0 m Solution	i for 0.10 m Solution
nonelectrolytes	1.00	1.00
sucrose, $C_{12}H_{22}O_{11}$	1.00 (ideal)	1.00 (ideal)
If 2 ions in solution/formula unit	2.00 (ideal)	2.00 (ideal)
KBr	1.77	1.88
NaCl	1.83	1.87
If 3 ions in solution/formula unit	3.00 (ideal)	3.00 (ideal)
K_2CO_3	2.39	2.45
K_2CrO_4	1.95	2.39
If 4 ions in solution/formula unit	4.00 (ideal)	4.00 (ideal)
$K_3[Fe(CN)_6]$	—	2.85

Percent Ionization and i Value from Freezing Point Depression Data

EXAMPLE 14-12   *Colligative Property and Weak Electrolytes*

Lactic acid, $C_2H_4(OH)COOH$, is found in sour milk. It is also formed in muscles during intense physical activity and is responsible for the pain felt during strenuous exercise. It is a weak monoprotic acid and therefore a weak electrolyte. The freezing point of a $0.0100\ m$ aqueous solution of lactic acid is $-0.0206°C$. Calculate (a) the i value and (b) the percent ionization in the solution.

Plan for (a)

To evaluate the van't Hoff factor, i, we first calculate $m_{\text{effective}}$ from the observed freezing point depression and K_f for water; we then compare $m_{\text{effective}}$ and m_{stated} to find i.

Solution for (a)

$$m_{\text{effective}} = \frac{\Delta T_f}{K_f} = \frac{0.0206°C}{1.86°C/m} = 0.0111\ m$$

$$i = \frac{m_{\text{effective}}}{m_{\text{stated}}} = \frac{0.0111\ m}{0.0100\ m} = \boxed{1.11}$$

Plan for (b)

The percent ionization is given by

$$\% \text{ ionization} = \frac{m_{\text{ionized}}}{m_{\text{original}}} \times 100\% \qquad (\text{where } m_{\text{original}} = m_{\text{stated}} = 0.0100\ m)$$

The freezing point depression is caused by the $m_{\text{effective}}$, the total *concentration of all dissolved species*—in this case, the sum of the concentrations of HA, H^+, and A^-. We know the value of $m_{\text{effective}}$ from part (a). Thus, we need to construct an expression for the effective molality in terms of the amount of lactic acid that ionizes. We represent the molality of lactic acid that ionizes as an unknown, x, and write the concentrations of all species in terms of this unknown.

Solution for (b)

In many calculations, it is helpful to write down (1) the values, or symbols for the values, of initial concentrations; (2) changes in concentrations due to reaction; and (3) final concentrations, as shown below. The coefficients of the equation are all ones, so the reaction ratio must be 1:1:1.

Let x = molality of lactic acid that ionizes; then

x = molality of H^+ and lactate ions that have been formed

	HA	$\longrightarrow$	H^+	+	A^-
Start	$0.0100\ m$		0		0
Change	$-x\ m$		$+x\ m$		$+x\ m$
Final	$(0.0100 - x)\ m$		$x\ m$		$x\ m$

The $m_{\text{effective}}$ is equal to the sum of the molalities of all the solute particles.

$$m_{\text{effective}} = m_{\text{HA}} \qquad + m_{H^+} + m_{A^-}$$
$$= (0.0100 - x)\ m + x\ m \ + x\ m = (0.0100 + x)\ m$$

This must equal the value for $m_{\text{effective}}$ calculated earlier, $0.0111\ m$.

$$0.0111\ m = (0.0100 + x)\ m$$

$$x = 0.0011\ m = \text{molality of the acid that ionizes}$$

To simplify the notation, we denote the weak acid as HA and its anion as A^-. The reaction summary used here to analyze the extent of reaction was introduced in Chapter 11.

We can now calculate the percent ionization.

$$\% \text{ ionization} = \frac{m_{\text{ionized}}}{m_{\text{original}}} \times 100\% = \frac{0.0011\ m}{0.0100\ m} \times 100\% = \boxed{11\%}$$

This experiment shows that in 0.0100 m solution, only 11% of the lactic acid has been converted into H^+ and $C_2H_4(OH)COO^-$ ions. The remainder, 89%, exists as nonionized molecules.

You should now work Exercises 74 and 76.

14-15 OSMOTIC PRESSURE

Osmosis is the spontaneous process by which the solvent molecules pass through a semipermeable membrane from a solution of lower concentration of solute into a solution of higher concentration of solute. A **semipermeable membrane** (such as cellophane) separates two solutions. Solvent molecules may pass through the membrane in either direction, but the rate at which they pass into the more concentrated solution is found to be greater than the rate in the opposite direction. The initial difference between the two rates is directly proportional to the difference in concentration between the two solutions. Solvent particles continue to pass through the membrane (Figure 14-17a). The column of liquid continues to rise until the hydrostatic pressure due to the weight of the solution in the column is sufficient to force solvent molecules back through the membrane at the same rate at which they enter from the dilute side. The pressure exerted under this condition is called the **osmotic pressure** of the solution.

Osmosis is one of the main ways in which water molecules move into and out of living cells. The membranes and cell walls in living organisms allow solvent to pass through. Some of these also selectively permit passage of ions and other small solute particles.

Osmotic pressure depends on the number, and not the kind, of solute particles in solution; it is therefore a colligative property.

The osmotic pressure of a given aqueous solution can be measured with an apparatus such as that depicted in Figure 14-17a. The solution of interest is placed inside an inverted thistle tube that has a membrane firmly fastened across the bottom. This part of the thistle tube and its membrane are then immersed in a container of pure water. As time passes, the height of the solution in the neck rises until the pressure it exerts just counterbalances the osmotic pressure.

Alternatively, we can view osmotic pressure as the external pressure exactly sufficient to prevent osmosis. The pressure required (Figure 14-18) is equal to the osmotic pressure of the solution.

The greater the number of solute particles, the greater the height to which the column rises, and the greater the osmotic pressure.

Like molecules of an ideal gas, solute particles are widely separated in very dilute solutions and do not interact significantly with each other. For very dilute solutions, osmotic pressure, π, is found to follow the equation

$$\pi = \frac{nRT}{V}$$

In this equation n is the number of moles of solute in volume V (in liters) of the solution. The other quantities have the same meaning as in the ideal gas law. The term n/V is a concentration term. In terms of molarity (M),

$$\pi = MRT$$

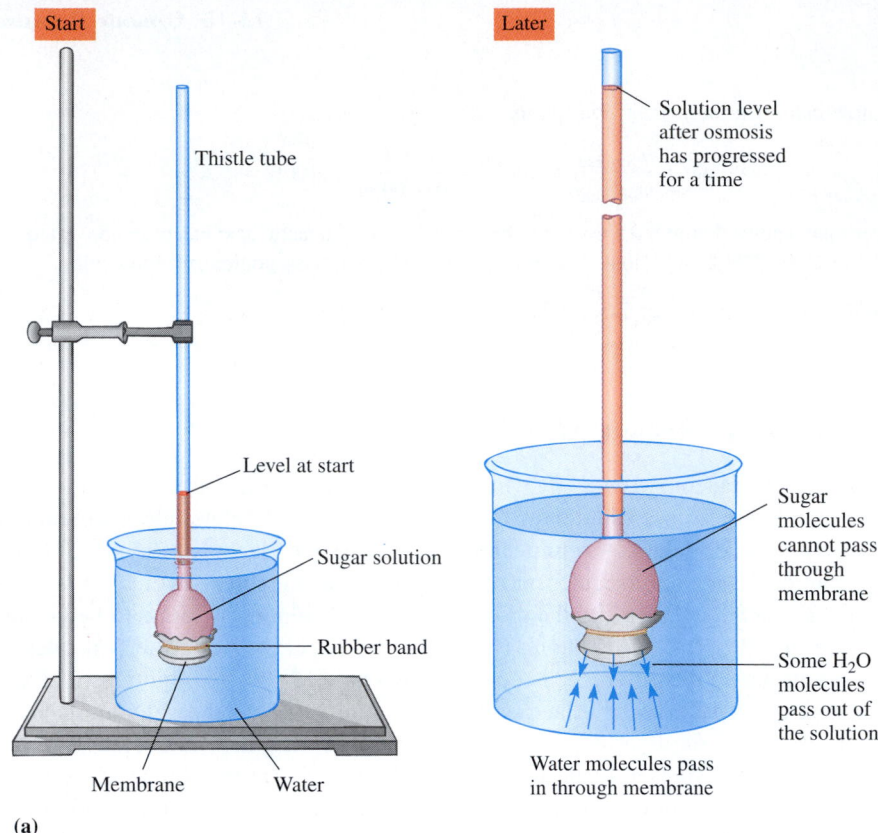

Thistle tube

Solution level after osmosis has progressed for a time

Level at start

Sugar molecules cannot pass through membrane

Sugar solution

Rubber band

Some H_2O molecules pass out of the solution

Membrane   Water

Water molecules pass in through membrane

(a)

Figure 14-17 (a) Laboratory apparatus for demonstrating osmosis. The picture at the right gives some details of the process, which is analogous to the transfer of solvent into a solution through the space above them (b). In (a) the solute particles cannot pass through the semipermeable membrane. In (b) the solute particles cannot pass through the vapor phase because they are nonvolatile.

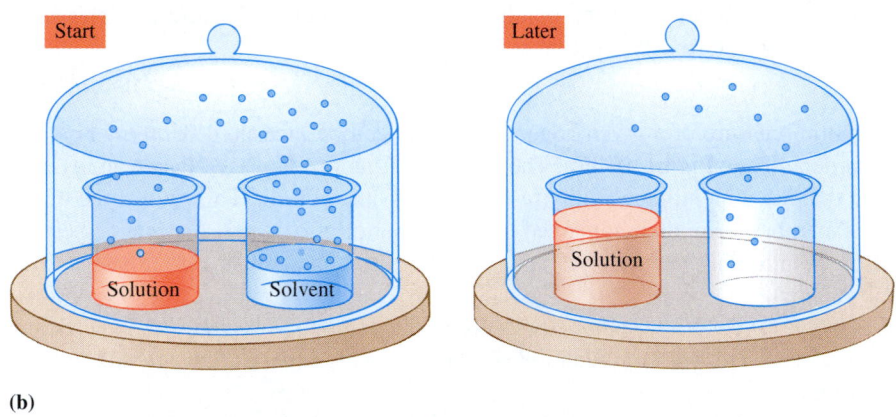

Solution   Solvent

Solution

(b)

Osmotic pressure increases with increasing temperature because T affects the number of solvent–membrane collisions per unit time. It also increases with increasing molarity because M affects the difference in the numbers of solvent molecules hitting the membrane from the two sides, and because a higher M leads to a stronger drive to equalize the concentration difference by dilution and to increase disorder in the solution. For *dilute aqueous solutions,* the molarity is approximately equal to the molality (because the density is nearly 1 kg/L), so

For a solution of an electrolyte, $\pi = m_{effective}RT$.

$$\pi = mRT \qquad \text{(dilute aqueous solutions)}$$

Osmotic pressures represent very significant forces. For example, a 1.0 molal solution of a nonelectrolyte in water at 0°C produces an equilibrium osmotic pressure of approximately 22.4 atmospheres (~330 psi).

EXAMPLE 14-13   *Osmotic Pressure Calculation*

What osmotic pressure would the 1.25 *m* sucrose solution in Example 14-2 exhibit at 25°C? The density of this solution is 1.34 g/mL.

Plan

We note that the approximation $M \approx m$ is not very good for this solution, because the density of this solution is quite different from 1 g/mL or kg/L. Thus, we must first find the molarity of sucrose, and then use the relationship $\pi = MRT$.

Solution

Recall that 167 g of solution contains 50.0 g of sucrose (0.146 mol) in 117 g of H_2O. The volume of this solution is

$$\text{vol solution} = 167 \text{ g} \times \frac{1 \text{ mL}}{1.34 \text{ g}} = 125 \text{ mL, or } 0.125 \text{ L}$$

Thus, the molarity of sucrose in the solution is

$$M_{\text{sucrose}} = \frac{0.146 \text{ mol}}{0.125 \text{ L}} = 1.17 \text{ mol/L}$$

Now we can calculate the osmotic pressure.

$$\pi = MRT = (1.17 \text{ mol/L})\left(0.0821 \, \frac{\text{L} \cdot \text{atm}}{\text{mol} \cdot \text{K}}\right)(298 \text{ K}) = \boxed{28.6 \text{ atm}}$$

You should now work Exercise 80.

Let's compare the magnitudes of the four colligative properties for this 1.25 *m* sucrose solution.

vapor pressure lowering	= 0.524 torr	(Example 14-4)
boiling point elevation	= 0.640°C	(Example 14-7)
freezing point depression	= 2.32°C	(Example 14-9)
osmotic pressure	= 28.6 atm	(Example 14-13)

The first of these is so small that it would be hard to measure precisely. Even this small lowering of the vapor pressure is sufficient to raise the boiling point by an amount that could be measured, although with difficulty. The freezing point depression is greater, but still could not be measured very precisely without special apparatus. The osmotic pressure, on the other hand, is so large that it could be measured much more precisely. Thus, osmotic pressure is often the most easily measured of the four colligative properties, especially when very dilute solutions are used.

The use of measurements of osmotic pressure for the determination of molecular weights has several advantages. Even very dilute solutions give easily measurable osmotic pressures. This method therefore is useful in determination of the molecular weights of (1) very expensive substances, (2) substances that can be prepared only in very small amounts, and (3) substances of very high molecular weight. Because many high-molec-

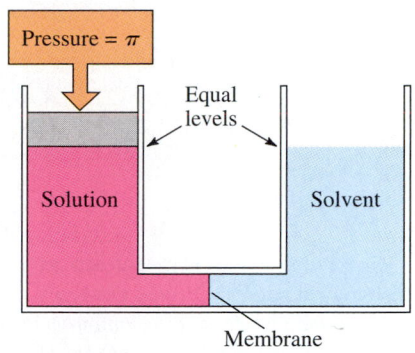

Figure 14-18 The pressure that is just sufficient to prevent solvent flow from the pure solvent side through the semipermeable membrane to the solution side is a measure of the osmotic pressure of the solution.

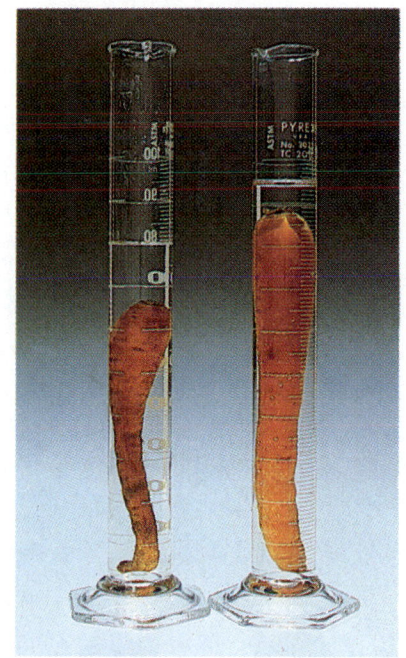

An illustration of osmosis. When a carrot is soaked in a concentrated salt solution, water flows out of the plant cells by osmosis. A carrot soaked overnight in salt solution (left) has lost much water and become limp. A carrot soaked overnight in pure water (right) is little affected.

ular-weight materials are difficult, and in some cases impossible, to obtain in a high state of purity, determinations of their molecular weights are not as accurate as we might like. Nonetheless, osmotic pressures provide a very useful method of estimating molecular weights.

EXAMPLE 14-14   *Molecular Weight from Osmotic Pressure*

An enzyme is a protein that acts as a biological catalyst. Pepsin catalyzes the metabolic cleavage of amino acid chains (called peptide chains) in other proteins.

Pepsin is an enzyme present in the human digestive tract. A solution of a 0.500-gram sample of purified pepsin in 30.0 mL of aqueous solution exhibits an osmotic pressure of 8.92 torr at 27.0°C. Estimate the molecular weight of pepsin.

Plan

As we did in earlier molecular weight determinations (Section 14-13), we must first find n, the number of moles that 0.500 grams of pepsin represents. We use the relationship $\pi = MRT$. The molarity of pepsin is equal to the number of moles of pepsin per liter of solution, n/V. We substitute this for M and solve for n.

Solution

$$\pi = MRT = \left(\frac{n}{V}\right)RT \qquad \text{or} \qquad n = \frac{\pi V}{RT}$$

We convert 8.92 torr to atmospheres to be consistent with the units of R.

$$n = \frac{\pi V}{RT} = \frac{\left(8.92 \text{ torr} \times \dfrac{1 \text{ atm}}{760 \text{ torr}}\right)(0.0300 \text{ L})}{\left(0.0821 \dfrac{\text{L} \cdot \text{atm}}{\text{mol} \cdot \text{K}}\right)(300 \text{ K})} = 1.43 \times 10^{-5} \text{ mol pepsin}$$

Thus, 0.500 g of pepsin is 1.43×10^{-5} mol. We now estimate its molecular weight.

The freezing point depression of this very dilute solution would be only about 0.0009°C, which would be difficult to measure accurately. The osmotic pressure of 8.92 torr, on the other hand, is easily measured.

$$\underline{?} \text{ g/mol} = \frac{0.500 \text{ g}}{1.43 \times 10^{-5} \text{ mol}} = \boxed{3.50 \times 10^4 \text{ g/mol}}$$

The molecular weight of pepsin is approximately 35,000 amu. This is typical for medium-size proteins.

You should now work Exercise 86.

▼ **PROBLEM-SOLVING TIP**   *Units in Osmotic Pressure Calculations*

Strictly speaking, the equation for osmotic pressure is presented in terms of molarity, $\pi = MRT$. Osmotic pressure, π, has the units of pressure (atmospheres); M (mol/L); and T (kelvins). Therefore, the appropriate value of R is $0.0821 \dfrac{\text{L} \cdot \text{atm}}{\text{mol} \cdot \text{K}}$. We can balance the units in this equation as follows.

$$\text{atm} = \left(\frac{\text{mol}}{\text{L}}\right)\left(\frac{\text{L} \cdot \text{atm}}{\text{mol} \cdot \text{K}}\right)(\text{K})$$

When we use the approximation that $M \approx m$, we might think that the units do not balance. For very dilute aqueous solutions, we can think of M as being *numerically* equal to m, but having the units mol/L, as shown above.

COLLOIDS

A solution is a homogeneous mixture in which no settling occurs and in which solute particles are at the molecular or ionic state of subdivision. This represents one extreme of mixtures. The other extreme is a suspension, a clearly heterogeneous mixture in which solute-like particles settle out after mixing with a solvent-like phase. Such a situation results when a handful of sand is stirred into water. **Colloids (colloidal dispersions)** represent an intermediate kind of mixture in which the solute-like particles, or **dispersed phase,** are suspended in the solvent-like phase, or **dispersing medium.** The particles of the dispersed phase are so small that settling is negligible. However, they are large enough to make the mixture appear cloudy or even opaque, because light is scattered as it passes through the colloid.

Table 14-4 indicates that all combinations of solids, liquids, and gases can form colloids except mixtures of nonreacting gases (all of which are homogeneous and, therefore, true solutions). Whether a given mixture forms a solution, a colloidal dispersion, or a suspension depends upon the size of the solute-like particles (Table 14-5), as well as solubility and miscibility.

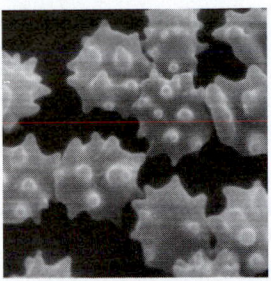

Cells shrink in solution of greater solute concentration (a hypertonic solution)

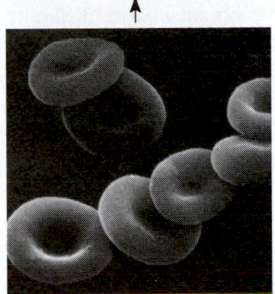

Normal cells in isotonic solution

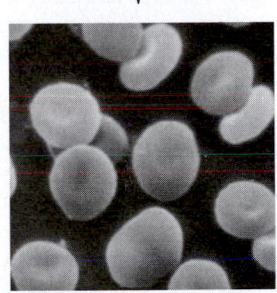

Cells expand in solution of lower solute concentration (a hypotonic solution)

Living cells contain solutions. When living cells are put in contact with solutions having different total solute concentrations, the resulting osmotic pressures can cause solvent to flow out of (top) or into (bottom) the cells.

Table 14-4 *Types of Colloids*

Dispersed (solute-like) Phase		Dispersing (solvent-like) Medium	Common Name	Examples
Solid	in	Solid	Solid sol	Many alloys (such as steel and duralumin), some colored gems, reinforced rubber, porcelain, pigmented plastics
Liquid	in	Solid	Solid emulsion	Cheese, butter, jellies
Gas	in	Solid	Solid foam	Sponge, rubber, pumice, Styrofoam
Solid	in	Liquid	Sols and gels	Milk of magnesia, paints, mud, puddings
Liquid	in	Liquid	Emulsion	Milk, face cream, salad dressings, mayonnaise
Gas	in	Liquid	Foam	Shaving cream, whipped cream, foam on beer
Solid	in	Gas	Solid aerosol	Smoke, airborne viruses and particulate matter, auto exhaust
Liquid	in	Gas	Liquid aerosol	Fog, mist, aerosol spray, clouds

Table 14-5 *Approximate Sizes of Dispersed Particles*

Mixture	Example	Approximate Particle Size
Suspension	Sand in water	Larger than 10,000 Å
Colloidal dispersion	Starch in water	10–10,000 Å
Solution	Sugar in water	1–10 Å

CHEMISTRY IN USE

Our Daily Lives

Water Purification and Hemodialysis

Semipermeable membranes play important roles in the normal functioning of many living systems. In addition, they are used in a wide variety of industrial and medical applications. Membranes with different permeability characteristics have been developed for many different purposes. One of these is the purification of water by reverse osmosis.

Suppose we place a semipermeable membrane between a saline (salt) solution and pure water. If the saline solution is pressurized under a greater pressure than its osmotic pressure, the direction of flow can be reversed. That is, the net flow of water molecules will be from the saline solution through the membrane into the pure water. This process is called **reverse osmosis.** The membrane usually consists of cellulose acetate or hollow fibers of a material structurally similar to Nylon. This method has been used for the purification of brackish (mildly saline) water. It has the economic advantages of low cost, ease of apparatus construction, and simplicity of operation. Because this method of water purification requires no heat, it has a great advantage over distillation.

The city of Sarasota, Florida, has built a large reverse osmosis plant to purify drinking water. It processes more than 4 million gallons of water per day from local wells. Total dissolved solids are reduced in concentration from 1744 parts per million (0.1744% by mass) to 90 parts per million (ppm). This water is mixed with additional well water purified by an ion exchange system. The final product is more than 10 million gallons of water per day containing less than 500 ppm of total dissolved

A reverse osmosis unit used to provide all the fresh water (82,500 gallons per day) for the steamship Norway.

solids, the standard for drinking water set by the World Health Organization. The Kuwaiti and Saudi water purification plants that were of strategic concern in the Persian Gulf war use reverse osmosis in one of their primary stages.

Human kidneys carry out many important functions. One of the most crucial is the removal of metabolic waste products (such as creatinine, urea, and uric acid) from the blood without removal of substances needed by the body (such as glucose, electrolytes, and amino acids). The process by which this is accomplished in the kidney involves *dialysis*, a phenomenon in which the membrane allows transfer of both solvent molecules

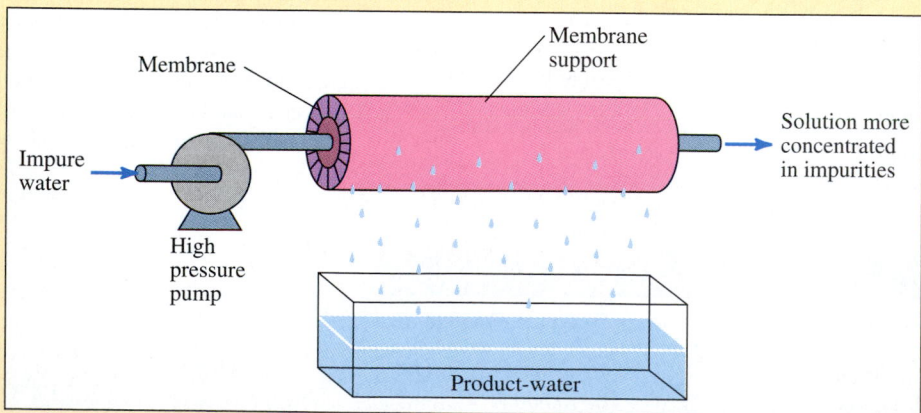

The reverse osmosis method of water purification.

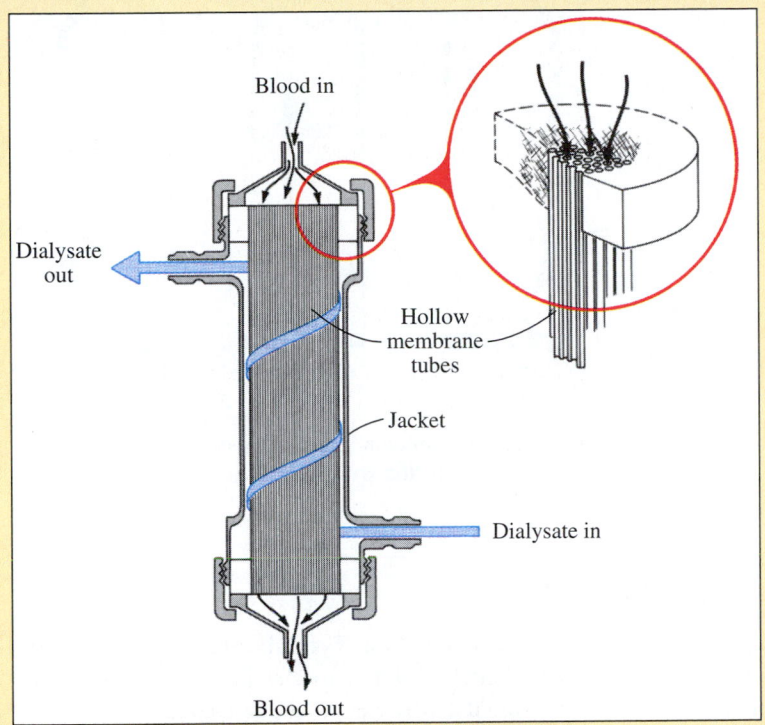

A schematic diagram of the hollow fiber (or capillary) dialyzer, the most commonly used artificial kidney. The blood flows through many small tubes constructed of semipermeable membrane; these tubes are bathed in the dialyzing solution.

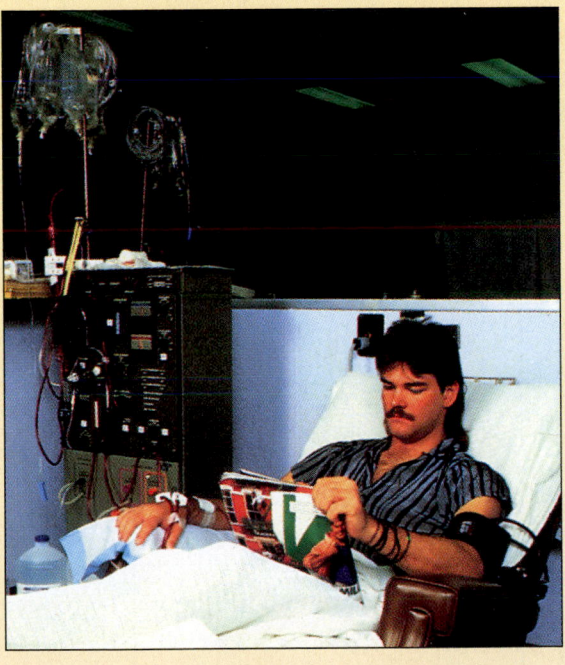

and certain solute molecules and ions, usually small ones. A patient whose kidneys have failed can often have this dialysis performed by an artificial kidney machine. In this mechanical procedure, called *hemodialysis,* the blood is withdrawn from the body and passed in contact with a semipermeable membrane.

The membrane separates the blood from a dialyzing solution, or *dialysate,* that is similar to blood plasma in its concentration of needed substances (such as electrolytes and amino acids) but contains none of the waste products. Because the concentrations of undesirable substances are thus higher in the blood than in the dialysate, they flow preferentially out of the blood and are washed away. The concentrations of *needed* substances are the same on both sides of the membrane, so these substances are maintained at the proper concentrations in the blood. The small pore size of the membrane prevents passage of blood cells. However, Na^+ and Cl^- ions and some small molecules do pass through the membrane. A patient with total kidney failure may require up to four hemodialysis sessions per week, at three to four hours per session. To help hold down the cost of such treatment, the dialysate solution is later purified by a combination of filtration, distillation, and reverse osmosis and is then re-used.

Freshly made wines are often cloudy because of colloidal particles. Removing these colloidal particles clarifies the wine.

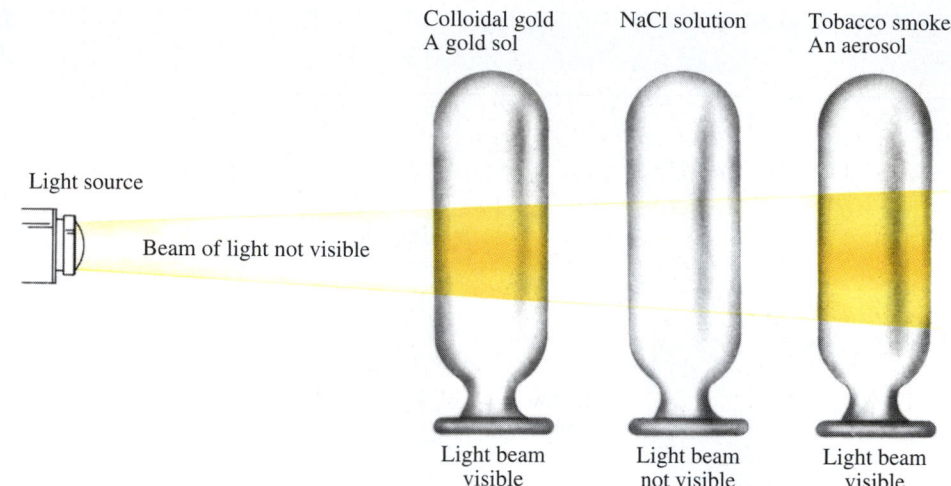

Figure 14-19   The dispersion of a beam of light by colloidal particles is called the Tyndall effect. The presence of colloidal particles is easily detected with the aid of a light beam.

14-16 THE TYNDALL EFFECT

The scattering of light by colloidal particles is called the **Tyndall effect** (Figure 14-19). Particles cannot scatter light if they are too small. Solute particles in solutions are below this limit. The maximum dimension of colloidal particles is about 10,000 Å.

The scattering of light from automobile headlights by fogs and mists is an example of the Tyndall effect, as is the scattering of a light beam in a laser show by dust particles in the air in a darkened room.

Figure 14-20   Stabilization of a colloid (Fe_2O_3 sol) by electrostatic forces. Each colloidal particle of this red sol is a cluster of many formula units of hydrated Fe_2O_3. Each attracts positively charged Fe^{3+} ions to its surface. (Fe^{3+} ions fit readily into the crystal structure, so they are preferentially adsorbed rather than Cl^-.) Each particle is then surrounded by a shell of positively charged ions, so the particles repel one another and cannot combine to the extent necessary to cause actual precipitation. The suspended particles scatter light, making the path of the light beam through the suspension visible.

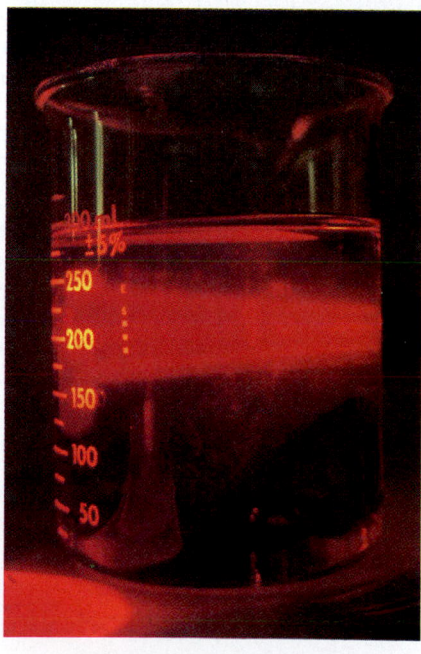

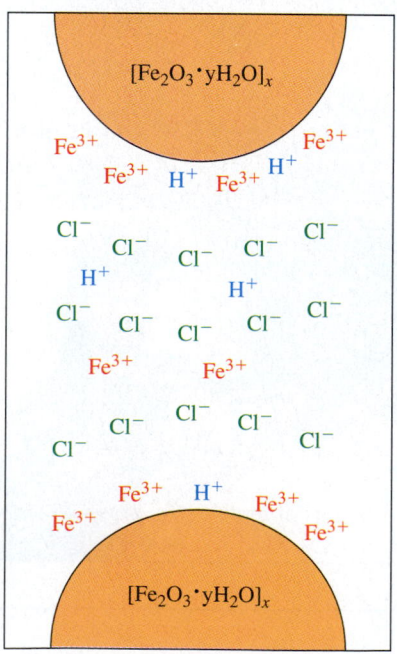

14-17 THE ADSORPTION PHENOMENON

Much of the chemistry of everyday life is the chemistry of colloids, as one can tell from the examples in Table 14-4. Because colloidal particles are so finely divided, they have tremendously high total surface area in relation to their volume. It is not surprising, therefore, that an understanding of colloidal behavior requires an understanding of surface phenomena.

Atoms on the surface of a colloidal particle are bonded only to other atoms of the particle on and below the surface. These atoms interact with whatever comes in contact with the surface. Colloidal particles often adsorb ions or other charged particles, as well as gases and liquids. The process of **adsorption** involves adhesion of any such species onto the surfaces of particles. For example, a bright red **sol** (solid dispersed in liquid) is formed by mixing hot water with a concentrated aqueous solution of iron(III) chloride (Figure 14-20).

$$2x[Fe^{3+}(aq) + 3Cl^-(aq)] + x(3 + y)H_2O \longrightarrow [Fe_2O_3 \cdot yH_2O]_x(s) + 6x[H^+ + Cl^-]$$

yellow solution   bright red sol

Each colloidal particle of this sol is a cluster of many formula units of hydrated Fe_2O_3. Each attracts positively charged Fe^{3+} ions to its surface. Because each particle is then surrounded by a shell of positively charged ions, the particles repel each other and cannot combine to the extent necessary to cause precipitation.

Figure 14-21 Examples of hydrophilic groups at the surface of a giant molecule (macromolecule) that help keep the macromolecule suspended in water.

14-18 HYDROPHILIC AND HYDROPHOBIC COLLOIDS

Colloids are classified as **hydrophilic** ("water loving") or **hydrophobic** ("water hating") based on the surface characteristics of the dispersed particles.

Hydrophilic Colloids

Proteins such as the oxygen-carrier hemoglobin form hydrophilic sols when they are suspended in saline aqueous body fluids, e.g., blood plasma. Such proteins are macromolecules (giant molecules) that fold and twist in an aqueous environment so that polar groups are exposed to the fluid whereas nonpolar groups are encased (Figure 14-21). Protoplasm and human cells are examples of **gels,** which are special types of sols in which the solid particles (in this case mainly proteins and carbohydrates) join together in a semirigid network structure that encloses the dispersing medium. Other examples of gels are gelatin, jellies, and gelatinous precipitates such as $Al(OH)_3$ and $Fe(OH)_3$.

Hydrophobic Colloids

Hydrophobic colloids cannot exist in polar solvents without the presence of **emulsifying agents,** or **emulsifiers.** These agents coat the particles of the dispersed phase to prevent their coagulation into a separate phase. Milk and mayonnaise are examples of hydrophobic colloids (milk fat in milk, vegetable oil in mayonnaise) that stay suspended with the aid of emulsifying agents (casein in milk and egg yolk in mayonnaise).

Consider the mixture resulting from vigorous shaking of salad oil (nonpolar) and vinegar (polar). Droplets of hydrophobic oil are temporarily suspended in the water. However, in a short time, the very polar water molecules, which attract each other strongly, squeeze out the nonpolar oil molecules. The oil then coalesces and floats to the top. If we add an emulsifying agent, such as egg yolk, and shake or beat the mixture, a stable emulsion (mayonnaise) results.

$Fe(OH)_3$ is a gelatinous precipitate (a gel).

"Dry" cleaning, on the other hand, does not involve water. The solvents that are used in dry cleaning dissolve grease to form true solutions.

Oil and grease are mostly long-chain hydrocarbons that are nonpolar. Our most common solvent is water, a polar substance that does not dissolve nonpolar substances. To use water to wash soiled fabrics, greasy dishes, or our bodies, we must enable the water to suspend and remove nonpolar substances. Soaps and detergents are emulsifying agents that accomplish this. Their function is controlled by the intermolecular interactions that result from their structures.

Solid soaps are usually sodium salts of long-chain organic acids called fatty acids. They have a polar "head" and a nonpolar "hydrocarbon tail." Sodium stearate, a typical soap, is shown below.

$$\left[H-\underset{\underset{H}{|}}{\overset{\overset{H}{|}}{C}}-\underset{\underset{H}{|}}{\overset{\overset{H}{|}}{C}}-\underset{\underset{H}{|}}{\overset{\overset{H}{|}}{C}}-\underset{\underset{H}{|}}{\overset{\overset{H}{|}}{C}}-\underset{\underset{H}{|}}{\overset{\overset{H}{|}}{C}}-\underset{\underset{H}{|}}{\overset{\overset{H}{|}}{C}}-\underset{\underset{H}{|}}{\overset{\overset{H}{|}}{C}}-\underset{\underset{H}{|}}{\overset{\overset{H}{|}}{C}}-\underset{\underset{H}{|}}{\overset{\overset{H}{|}}{C}}-\underset{\underset{H}{|}}{\overset{\overset{H}{|}}{C}}-\underset{\underset{H}{|}}{\overset{\overset{H}{|}}{C}}-\underset{\underset{H}{|}}{\overset{\overset{H}{|}}{C}}-\underset{\underset{H}{|}}{\overset{\overset{H}{|}}{C}}-\underset{\underset{H}{|}}{\overset{\overset{H}{|}}{C}}-\underset{\underset{H}{|}}{\overset{\overset{H}{|}}{C}}-\underset{\underset{H}{|}}{\overset{\overset{H}{|}}{C}}-\underset{\underset{O}{\|}}{\overset{\overset{O}{}}{C}} \right]^{-} Na^{+}$$

$\underbrace{\qquad\qquad\qquad\qquad\qquad\qquad\qquad}_{\text{hydrocarbon tail (soluble in oil)}}$ $\underbrace{\quad}_{\substack{\text{polar head}\\ \text{(soluble in } H_2O)}}$

sodium stearate (a soap)

Sodium stearate is also a major component of some stick deodorants.

The stearate ion is typical of the anions in soaps. It has a polar carboxylate head, $-\overset{\overset{O}{\|}}{C}-O^{-}$, and a long nonpolar tail, $CH_3(CH_2)_{16}-$. The head of the stearate ion is compatible with ("soluble in") water, whereas the hydrocarbon tail is compatible with ("soluble in") oil and grease. Groups of such ions can be dispersed in water because they form **micelles** (Figure 14-22a). Their "water-insoluble" tails are in the interior of a micelle and their polar heads on the outside where they can interact with the polar water molecules. When sodium stearate is stirred into water, the result is not a true solution. Instead it contains negatively charged micelles of stearate ions, surrounded by the positively charged Na^{+} ions. The result is a suspension of micelles in water. These micelles are large enough to scatter light, so a soap–water mixture appears cloudy. Oil and grease

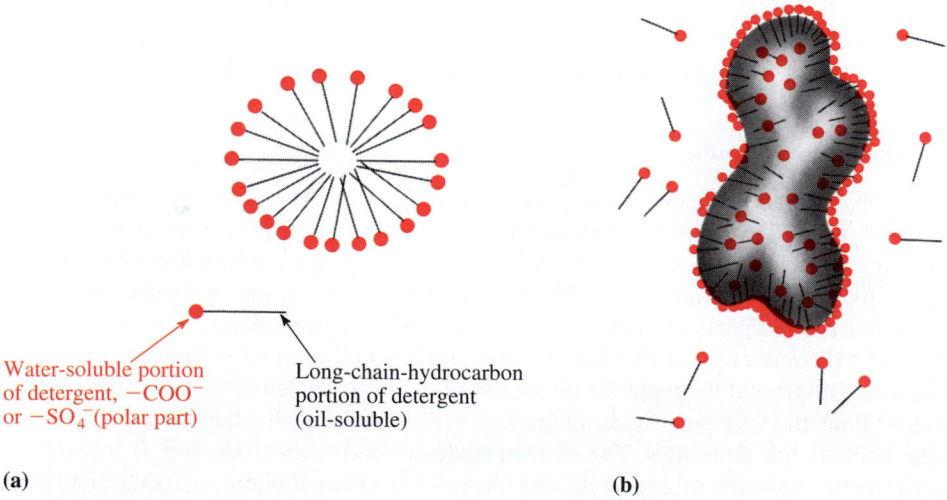

Figure 14-22 (a) A representation of a micelle. The nonpolar tails "dissolve" in one another in the center of the cluster and the polar heads on the outside interact favorably with the polar water molecules. (b) Attachment of soap or detergent molecules to a droplet of oily dirt to suspend it in water.

Oil droplet (dirt) suspended in water

Water-soluble portion of detergent, $-COO^{-}$ or $-SO_4^{-}$ (polar part)

Long-chain-hydrocarbon portion of detergent (oil-soluble)

(a)   (b)

CHEMISTRY IN USE

Our Daily Lives

Why Does Red Wine Go with Red Meat?

Choosing the appropriate wine to go with dinner is a problem for some diners. However, experts have offered a simple rule for generations, "serve red wine with red meat and white wine with fish." Are these cuisine choices just traditions, or are there fundamental reasons for them?

Red wine is usually drunk with red meat because of a desirable matching of the chemicals found in each. The most influential ingredient in red meat is fat; it gives red meats their desirable flavor. As you chew a piece of red meat, the fat from the meat coats your tongue and palate, which desensitizes your taste buds. As a result, your second bite of red meat is less tasty than the first. Your steak would taste better if you washed your mouth between mouthfuls; there is an easy way to wash away the fat deposits.

Red wine contains a surfactant that cleanses your mouth, removing fat deposits, re-exposing your taste buds, and allowing you to savor the next bite of red meat almost as well as the first bite. The tannic acid (also called tannin) in red wine provides a soap-like action. Like soap, tannic acid consists of both a non-polar complex hydrocarbon part as well as a polar one. The polar part of tannic acid dissolves in polar saliva, while the nonpolar part dissolves in the fat film that coats your palate. When you sip red wine, a suspension of micelles forms in the saliva. This micelle emulsion has the fat molecules in its interior; the fat is washed away by swallowing the red wine.

White wines go poorly with red meats because they lack the tannic acid needed to cleanse the palate. In fact, the presence or absence of tannic acid distinguishes red wines from white wines. Grapes fermented with their skins produce red wines; grapes fermented without their skins produce white wines.

Because fish has less fat than red meats, fish can be enjoyed without surfactants to cleanse the palate. Also, tannic acid has a strong flavor that can overpower the delicate flavor of many fish. The absence of tannic acid in white wines gives them a lighter flavor than red wines, and many people prefer this lighter flavor with their fish dinners.

Ronald DeLorenzo
Middle Georgia College

"dissolve" in soapy water because the nonpolar oil and grease are taken into the nonpolar interior of micelles (Figure 14-22b). Micelles form a true emulsion in water, so the oil and grease can be washed away. Sodium stearate is called a **surfactant** (meaning "surface-active agent") or wetting agent because it has the ability to suspend and wash away oil and grease. Other soaps and detergents behave similarly.

"Hard" water contains Fe^{3+}, Ca^{2+}, and/or Mg^{2+} ions, all of which displace Na^+ from soaps to form precipitates. This removes the soap from the water and puts an undesirable coating on the bathtub or on the fabric being laundered. **Synthetic detergents** are soap-like emulsifiers that contain sulfonate, $-SO_3^-$, or sulfate, $-OSO_3^-$, instead of carboxylate groups, $-COO^-$. They do not precipitate the ions of hard water, so they can be used in hard water as soap substitutes without forming undesirable scum.

Phosphates were added to commercial detergents for various purposes. They complexed the metal ions that contribute to water hardness and kept them dissolved, controlled acidity, and influenced micelle formation. The use of detergents containing phosphates is now discouraged because they cause **eutrophication** in rivers and streams that receive sewage. This is a condition (not related to colloids) in which there is an overgrowth of vegetation caused by the high concentration of phosphorus, a plant nutrient. This overgrowth and the subsequent decay of the dead plants lead to decreased dissolved O_2 in the water, which causes the gradual elimination of marine life. There is also a foaming prob-

Some edible colloids.

The 14th-century Saint Martin's Bridge over the Tajo River in Toledo, Spain. The presence of nonbiodegradable detergents is responsible for the foam.

lem associated with branched alkylbenzenesulfonate (ABS) detergents in streams and in pipes, tanks, and pumps of sewage treatment plants. Such detergents are not **biodegradable;** that is, they cannot be broken down by bacteria.

$$\left[\begin{array}{c} \overset{\displaystyle CH_3 \quad CH_3 \quad CH_3 \quad CH_3}{\underset{\displaystyle CH_3CHCH_2CHCH_2CHCH_2CH}{\big| \quad\quad \big| \quad\quad \big| \quad\quad \big|}} \!\!-\!\! \bigcirc \!\!-\!\! SO_3 \end{array} \right]^{-} Na^{+}$$

a sodium branched alkylbenzenesulfonate (ABS)—a nonbiodegradable detergent

Currently used linear-chain alkylbenzenesulfonate (LAS) detergents are biodegradable and do not cause such foaming.

The two detergents shown here each have a $C_{12}H_{25}$ tail but the branched one is not biodegradable.

$$\left[CH_3CH_2CH_2CH_2CH_2CH_2CH_2CH_2CH_2CH_2CH_2CH_2 \!\!-\!\! \bigcirc \!\!-\!\! SO_3 \right]^{-} Na^{+}$$

sodium lauryl benzenesulfonate
a linear alkylbenzenesulfonate (LAS)—a biodegradable detergent

Key Terms

The following terms were defined at the end of Chapter 3: **concentration, dilution, molarity, percent by mass, solute, solution,** and **solvent.**

Adsorption   Adhesion of species onto surfaces of particles.

Associated ions   Short-lived species formed by the collision of dissolved ions of opposite charge.

Biodegradability   The ability of a substance to be broken down into simpler substances by bacteria.

Boiling point elevation   The increase in the boiling point of a solvent caused by dissolution of a nonvolatile solute.

Boiling point elevation constant, K_b   A constant that corresponds to the change in boiling point produced by a one-molal *ideal* solution of a nonvolatile nonelectrolyte.

Colligative properties   Physical properties of solutions that depend on the number but not the kind of solute particles present.

Colloid   A heterogeneous mixture in which solute-like particles do not settle out; also called *colloidal dispersion.*

Crystal lattice energy   The energy change when one mole of formula units of a crystalline solid is formed from its ions, atoms, or molecules in the gas phase; always negative.

Detergent   A soap-like emulsifier that contains a sulfonate, $-SO_3^-$, or sulfate, $-OSO_3^-$, group instead of a carboxylate, $-COO^-$, group.

Dispersed phase   The solute-like species in a colloid.

Dispersing medium   The solvent-like phase in a colloid.

Dispersion   See *Colloid.*

Distillation   The process in which components of a mixture are separated by boiling away the more volatile liquid.

Effective molality   The sum of the molalities of all solute particles in solution.

Emulsifier   See *Emulsifying agent.*

Emulsifying agent   A substance that coats the particles of a dispersed phase and prevents coagulation of colloidal particles; an emulsifier.

Emulsion   A colloidal dispersion of a liquid in a liquid.

Eutrophication   The undesirable overgrowth of a vegetation caused by high concentrations of plant nutrients in bodies of water.

Foam   A colloidal dispersion of a gas in a liquid.

Fractional distillation   The process in which a fractionating column is used in a distillation apparatus to separate components of a liquid mixture that have different boiling points.

Freezing point depression   The decrease in the freezing point of a solvent caused by the presence of a solute.

Freezing point depression constant, K_f   A constant that corresponds to the change in freezing point produced by a one-molal *ideal* solution of a nonvolatile nonelectrolyte.

Gel   A colloidal dispersion of a solid in a liquid; a semirigid sol.

Hard water   Water containing Fe^{3+}, Ca^{2+}, and/or Mg^{2+} ions, which form precipitates with soaps.

Heat of solution (molar)   The amount of heat absorbed in the formation of a solution that contains one mole of solute; the value is positive if heat is absorbed (endothermic) and negative if heat is released (exothermic).

Henry's Law   The pressure of the gas above a solution is proportional to the concentration of the gas in the solution.

Hydration   The interaction (surrounding) of a solute particle with water molecules.

Hydration energy (molar) of an ion   The energy change accompanying the hydration of a mole of gaseous ions.

Hydrophilic colloids   Colloidal particles that attract water molecules.

Hydrophobic colloids   Colloidal particles that repel water molecules.

Ideal solution   A solution that obeys Raoult's Law exactly.

Liquid aerosol   A colloidal dispersion of a liquid in gas.

Micelle   A cluster of a large number of soap or detergent molecules or ions, assembled with their hydrophobic tails directed toward the center and their hydrophilic head directed outward.

Miscibility   The ability of one liquid to mix with (dissolve in) another liquid.

Molality (m)   Concentration expressed as number of moles of solute per kilogram of solvent.

Mole fraction of a component in solution   The number of moles of the component divided by the total number of moles of all components.

Osmosis   The process by which solvent molecules pass through a semipermeable membrane from a dilute solution into a more concentrated solution.

Osmotic pressure   The hydrostatic pressure produced on the surface of a semipermeable membrane by osmosis.

Percent ionization of weak electrolytes   The percent of the weak electrolyte that ionizes in a solution of given concentration.

Raoult's Law   The vapor pressure of a solvent in an ideal solution is directly proportional to the mole fraction of the solvent in the solution.

Reverse osmosis   The forced flow of solvent molecules through a semipermeable membrane from a concentrated solution into a dilute solution. This is accomplished by application of hydrostatic pressure on the concentrated side greater than the osmotic pressure that is opposing it.

Saturated solution   A solution in which no more solute will dissolve.

Semipermeable membrane   A thin partition between two solutions through which certain molecules can pass but others cannot.

Soap   An emulsifier that can disperse nonpolar substances in water; the sodium salt of a long chain organic acid; consists of a long hydrocarbon chain attached to a carboxylate group, $-CO_2^-Na^+$.

Sol   A colloidal dispersion of a solid in a liquid.

Solid aerosol   A colloidal dispersion of a solid in a gas.

Solid emulsion   A colloidal dispersion of a liquid in a solid.

Solid foam   A colloidal dispersion of a gas in a solid.

Solid sol   A colloidal dispersion of a solid in a solid.

Solvation   The process by which solvent molecules surround and interact with solute ions or molecules.

Supersaturated solution   A (metastable) solution that contains a higher-than-saturation concentration of solute; slight disturbance or seeding causes crystallization of excess solute.

Surfactant   A "surface-active agent"; a substance that has the ability to emulsify and wash away oil and grease in an aqueous suspension.

Thermal pollution   Introduction of heated waste water into natural waters.

Tyndall effect   The scattering of light by colloidal particles.

van't Hoff factor, i   A number that indicates the extent of dissociation or ionization of a solute; equal to the actual colligative property divided by the colligative property calculated assuming no ionization or dissociation.

Exercises

General Concepts—The Dissolving Process

1. Support or criticize the statement "Solutions and mixtures are the same thing."

2. Give an example of a solution that contains each of the following: (a) a solid dissolved in a liquid; (b) a gas dissolved in a gas; (c) a gas dissolved in a liquid; (d) a liquid dissolved in a liquid; (e) a solid dissolved in a solid. Identify the solvent and the solute in each case.

3. There are no *true* solutions in which the solvent is gaseous and the solute is either liquid or solid. Why?

4. Explain why (a) solute–solute, (b) solvent–solvent, and (c) solute–solvent interactions are important in determining the extent to which a solute dissolves in a solvent.

5. What is the relative importance of each factor listed in Exercise 4 when (a) solids, (b) liquids, and (c) gases dissolve in water?

6. Why is the dissolving of many ionic solids in water an endothermic process, whereas the mixing of most miscible liquids is an exothermic process?

7. Define and distinguish between dissolution, solvation, and hydration.

8. The amount of heat released or absorbed in the dissolution process is important in determining whether the dissolution process is spontaneous, i.e., whether it can occur. What is the other important factor? How does it influence solubility?

9. An old saying is that "oil and water don't mix." Explain, on a molecular basis, why this saying is true.

10. Two liquids, A and B, do not react chemically and are completely miscible. What would be observed as one is poured into the other? What would be observed in the case of two completely immiscible liquids and in the case of two partially miscible liquids?

11. Consider the following solutions. In each case, predict whether the solubility of the solute should be high or low. Justify your answers. (a) KCl in hexane, C_6H_{14}; (b) $CaCl_2$ in H_2O; (c) C_6H_{14} in H_2O; (d) CCl_4 in C_6H_{14}; (e) C_6H_{14} in CCl_4.

12. Consider the following solutions. In each case predict whether the solubility of the solute should be high or low. Justify your answers. (a) HCl in H_2O; (b) HF in H_2O; (c) Al_2O_3 in H_2O; (d) S_8 in H_2O; (e) $NaNO_3$ in hexane, C_6H_{14}.

13. For those solutions in Exercise 11 that can be prepared in "reasonable" concentrations, classify the solutes as nonelectrolytes, weak electrolytes, or strong electrolytes.

14. For those solutions in Exercise 12 that can be prepared in "reasonable" concentrations, classify the solutes as nonelectrolytes, weak electrolytes, or strong electrolytes.

15. Both methanol, CH_3OH, and ethanol, CH_3CH_2OH, are completely miscible with water at room temperature because of strong solvent–solute intermolecular forces. Predict the trend in solubility in water for 1-propanol, $CH_3CH_2CH_2OH$; 1-butanol, $CH_3CH_2CH_2CH_2OH$; and 1-pentanol, $CH_3CH_2CH_2CH_2CH_2OH$.

16. (a) Does the solubility of a solid in a liquid exhibit an appreciable dependence on pressure? (b) Is the same true for the solu-

bility of a liquid in a liquid? Why?

17. Describe the effect of increasing pressure on the solubilities of gases in liquids.

18. Describe the effect of increasing temperature on the solubilities of most gases in water.

*19. A handbook lists the value of the Henry's Law constant as 3.02×10^4 atm for ethane, C_2H_6, dissolved in water at 25°C. The absence of concentration units on k means that the constant is meant to be used with concentration expressed as a mole fraction. Calculate the mole fraction of ethane in water at an ethane pressure of 0.15 atm.

*20. The mole fraction of methane, CH_4, dissolved in water can be calculated from the Henry's Law constants of 4.13×10^4 atm at 25°C and 5.77×10^4 atm at 50°C. Calculate the solubility of methane at these temperatures for a methane pressure of 10 atm above the solution. Does the solubility increase or decrease with increasing temperature? (See Exercise 19 for interpretation of units.)

21. Choose the ionic compound from each pair for which the crystal lattice energy should be the most negative. Justify your choice. (a) LiF or LiBr; (b) KF or CaF_2; (c) FeF_2 or FeF_3; (d) NaF or KF.

22. Choose the ion from each pair that should be more strongly hydrated in aqueous solution. Justify your choice. (a) Na^+ or Rb^+; (b) Cl^- or Br^-; (c) Fe^{3+} or Fe^{2+}; (d) Na^+ or Mg^{2+}.

*23. The crystal lattice energy, ΔH_{xtal}, for LiBr(s) is -818.6 kJ/mol at 25°C. The hydration energy of the ions of LiBr is -867.4 kJ/mol at 25°C (for infinite dilution). (a) What is the heat of solution of LiBr(s) at 25°C (for infinite dilution)? (b) The heat of hydration of $Li^+(g)$ is -544 kJ/mol at 25°C. What is ΔH_{hyd} for $Br^-(g)$ at 25°C?

24. Describe and explain the effect of changing the temperature on the solubility of (a) a solid that dissolves by an exothermic process, and (b) a solid that dissolves by an endothermic process.

Concentrations of Solutions

25. Under what conditions are the molarity and molality of a solution nearly the same? Which concentration unit is more useful when measuring volume with burets, pipets, and volumetric flasks in the laboratory? Why?

26. Many handbooks list solubilities in units of (g solute/100 g H_2O). How would you convert from this unit to mass percent?

27. A 60.0-mL sample of diethyl ether, $(C_2H_5)_2O$, is dissolved in enough methanol, CH_3OH, to make 300.0 mL of solution. The density of the ether is 0.714 g/mL. What is the molarity of this solution?

*28. Describe how to prepare 1.000 L of 0.225 m NaCl. The density of this solution is 1.01 g/mL.

29. Urea, $(NH_2)_2CO$, is a product of metabolism of proteins. An aqueous solution is 32.0% urea by mass and has a density of 1.087 g/mL. Calculate the molality of urea in the solution.

30. What masses of NaCl and H_2O are present in 210 g of a 15.0% aqueous solution of NaCl?

31. Sodium fluoride has a solubility of 4.22 g in 100.0 g of water at 18°C. Express the solute concentration in terms of (a) mass percent, (b) mole fraction, and (c) molality.

32. The solubility of $K_2[ZrF_6]$ at 100°C is 25 g/100 g H_2O. Express the concentration of the solute in terms of (a) mass percent, (b) mole fraction, and (c) molality.

*33. A piece of jewelry is marked "14 carat gold," meaning that on a mass basis the jewelry is 14/24 pure gold. What is the molality of this alloy—considering the other metal as the solvent?

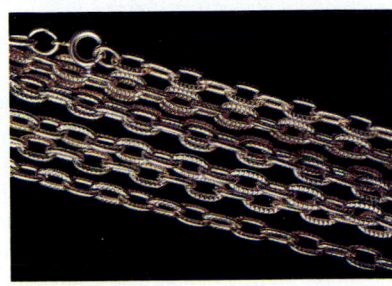

34. Calculate the molality of a solution that contains 90.0 g of benzoic acid, C_6H_5COOH, in 350 mL of ethanol, C_2H_5OH. The density of ethanol is 0.789 g/mL.

35. A solution contains 15.0 g of urea, $(NH_2)_2CO$, 10.0 g of fructose, $C_6H_{12}O_6$, and 75.0 g H_2O. Calculate the mole fraction of water.

*36. A solution that is 24.0% fructose, $C_6H_{12}O_6$, in water has a density of 1.10 g/mL at 20°C. (a) What is the molality of fructose in this solution? (b) At a higher temperature, the density is lower. Would the molality be less than, greater than, or the same as the molality at 20°C? Explain.

37. The density of a sulfuric acid solution taken from a car battery is 1.225 g/cm^3. This corresponds to a 3.75 M solution. Express the concentration of this solution in terms of molality, mole fraction of H_2SO_4, and % of water by mass.

38. What are the mole fractions of ethanol, C_2H_5OH, and water in a solution prepared by mixing 75.0 g of ethanol with 35.0 g of water?

39. What are the mole fractions of ethanol, C_2H_5OH, and water in a solution prepared by mixing 75.0 mL of ethanol with 35.0 mL of water at 25°C? The density of ethanol is 0.789 g/mL, and that of water is 1.00 g/mL.

40. The density of an aqueous solution containing 12.50 g K_2SO_4 in 100.00 g solution is 1.083 g/mL. Calculate the concentration of this solution in terms of molarity, molality, percent of K_2SO_4, and mole fraction of solvent.

*41. (a) What is the mass % ethanol in an aqueous solution in which the mole fraction of each component is 0.500? (b) What is the molality of ethanol in such a solution?

Raoult's Law and Vapor Pressure

42. In your own words, explain briefly *why* the vapor pressure of a solvent is lowered by dissolving a nonvolatile solute in it.

43. (a) Calculate the vapor pressure lowering associated with dissolving 35.5 g of table sugar, $C_{12}H_{22}O_{11}$, in 400 g of water at 25.0°C. (b) What is the vapor pressure of the solution? Assume that the solution is ideal. The vapor pressure of pure water at 25°C is 23.76 torr. (c) What is the vapor pressure of the solution at 100°C?

44. Calculate (a) the lowering of vapor pressure and (b) the vapor pressure of a solution prepared by dissolving 35.5 g of naphthalene, $C_{10}H_8$ (a nonvolatile nonelectrolyte), in 150.0 g of benzene, C_6H_6, at 20°C. Assume that the solution is ideal. The vapor pressure of pure benzene is 74.6 torr at 20°C.

45. At -100°C ethane, CH_3CH_3, and propane, $CH_3CH_2CH_3$, are liquids. At that temperature, the vapor pressure of pure ethane is 394 torr and that of pure propane is 22 torr. What is the vapor pressure at -100°C over a solution containing equal molar amounts of these substances?

46. Using Raoult's Law, predict the partial pressures in the vapor above a solution containing 0.250 mol acetone ($P^0 = 345$ torr) and 0.300 mol chloroform ($P^0 = 295$ torr).

47. What is the composition of the vapor above the solution described in Exercise 45?

48. What is the composition of the vapor above the solution described in Exercise 46?

49. Use the following vapor pressure diagram to estimate (a) the partial pressure of chloroform, (b) the partial pressure of ace-

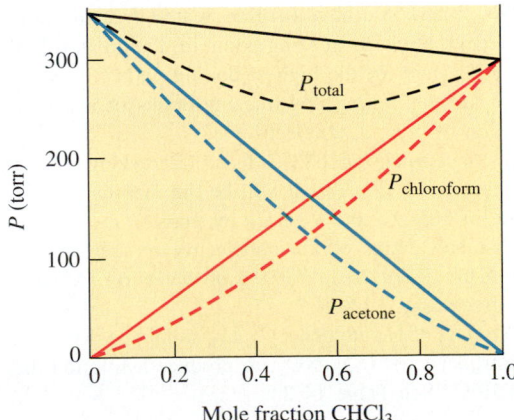

tone, and (c) the total vapor pressure of a solution in which the mole fraction of $CHCl_3$ is 0.3, assuming *ideal* behavior.

50. Answer Exercise 49 for the *real* solution of acetone and chloroform.

51. A solution is prepared by mixing 60.0 g of dichloromethane, CH_2Cl_2, and 30.0 g of dibromomethane, CH_2Br_2, at 0°C. The vapor pressure at 0°C of pure CH_2Cl_2 is 0.175 atm, and that of CH_2Br_2 is 0.0150 atm. (a) Assuming ideal behavior, calculate the total vapor pressure of the solution. (b) Calculate the mole fractions of CH_2Cl_2 and CH_2Br_2 in the *vapor* above the liquid. Assume that both the vapor and the solution behave ideally.

Boiling Point Elevation and Freezing Point Depression—Solutions of Nonelectrolytes

52. What is the boiling point of a 2.15 *m* aqueous solution of ethylene glycol, a nonvolatile nonelectrolyte?

Ethylene glycol solutions.

53. A solution is prepared by dissolving 6.41 g of ordinary sugar (sucrose, $C_{12}H_{22}O_{11}$, 342 g/mol) in 32.0 g of water. Calculate the boiling point of the solution. Sucrose is a nonvolatile nonelectrolyte.

54. What is the freezing point of the solution described in Exercise 52?

55. What is the freezing point of the solution described in Exercise 53?

56. Refer to Table 14-2. Suppose you had a 0.150 *m* solution of a nonvolatile nonelectrolyte in each of the solvents listed there. Which one would have (a) the greatest freezing point depression, (b) the lowest freezing point, (c) the greatest boiling point elevation, and (d) the highest boiling point?

57. What are the boiling and freezing points of a 2.75 *m* aqueous solution of urea, a nonelectrolyte?

58. Pure copper melts at 1083°C. Its molal freezing point depression constant is 23°C/*m*. What will be the melting point of a brass made of 10% Zn and 90% Cu by mass?

59. How many grams of the nonelectrolyte sucrose, $C_{12}H_{22}O_{11}$, should be dissolved in 700.0 g of water to produce a solution that freezes at −2.50°C?

60. What mass of naphthalene, $C_{10}H_8$, should be dissolved in 400 g of nitrobenzene, $C_6H_5NO_2$, to produce a solution that boils at 214.20°C? See Table 14-2.

61. A solution was made by dissolving 3.75 g of a nonvolatile solute in 108.7 g of acetone. The solution boiled at 56.58°C. The boiling point of pure acetone is 55.95°C, and $K_b = 1.71$°C/*m*. Calculate the molecular weight of the solute.

62. The molecular weight of an organic compound was determined by measuring the freezing point depression of a benzene solution. A 0.500-g sample was dissolved in 50.0 g of benzene, and the resulting depression was 0.42°C. What is the approximate molecular weight? The compound gave the following elemental analysis: 40.0% C, 6.67% H, 53.3% O by mass. Determine the formula and exact molecular weight of the substance.

63. When 0.154 g of sulfur is finely ground and melted with 4.38 g of camphor, the freezing point of the camphor is lowered by 5.47°C. What is the molecular weight of sulfur? What is its molecular formula?

*64. (a) Suppose we dissolve a 6.00-g sample of a mixture of naphthalene, $C_{10}H_8$, and anthracene, $C_{14}H_{10}$, in 360 g of benzene. The solution is observed to freeze at 4.85°C. Find the percent composition (by mass) of the sample. (b) At what temperature would the solution boil? Assume that naphthalene and anthracene are nonvolatile nonelectrolytes.

Boiling Point Elevation and Freezing Point Depression—Solutions of Electrolytes

65. What is ion association in solution? Can you suggest why the term "ion pairing" is sometimes used to describe this phenomenon?

66. You have separate 0.10 *M* aqueous solutions of the following salts: $LiNO_3$, $Ca(NO_3)_2$, and $Al(NO_3)_3$. In which one would you expect to find the highest particle concentration? Which solution would you expect to conduct electricity most strongly? Explain your reasoning.

67. What is the significance of the van't Hoff factor, *i*?

68. What is the value of the van't Hoff factor, *i*, for the following strong electrolytes at infinite dilution? (a) Na_2SO_4; (b) KOH; (c) $Al_2(SO_4)_3$; (d) $SrSO_4$.

69. Compare the number of solute particles that are present in solutions of equal concentrations of strong electrolytes, weak electrolytes, and nonelectrolytes.

70. Four beakers contain 0.010 *m* aqueous solutions of CH_3OH, $KClO_3$, $CaCl_2$, and CH_3COOH, respectively. Which of these solutions has the lowest freezing point?

*71. A 0.050 *m* aqueous solution of $K_3[Fe(CN)_6]$ has a freezing point of −0.2800°C. Calculate the total concentration of solute particles in this solution and interpret your results.

*72. One gram each of NaCl, NaBr, and NaI was dissolved in 100.0 g water. What is the vapor pressure above the solution at 100°C? Assume complete dissociation of the three salts.

73. Formic acid, HCOOH, ionizes slightly in water.

$$HCOOH(aq) \rightleftharpoons H^+(aq) + HCOO^-(aq)$$

A 0.0100 *m* formic acid solution freezes at −0.02092°C. Calculate the percent ionization of HCOOH in this solution.

74. A 0.100 *m* acetic acid solution in water freezes at −0.1884°C. Calculate the percent ionization of CH_3COOH in this solution.

*75. In a home ice cream freezer, we lower the freezing point of the water bath surrounding the ice cream can by dissolving NaCl in water to make a brine solution. A 15.0% brine solution is observed to freeze at $-10.888°C$. What is the van't Hoff factor, i, for this solution?

76. CsCl dissolves in water according to

$$CsCl(s) \xrightarrow{H_2O} Cs^+(aq) + Cl^-(aq)$$

A 0.121 m solution of CsCl freezes at $-0.403°C$. Calculate i and the apparent percent dissociation of CsCl in this solution.

Osmotic Pressure

77. What are osmosis and osmotic pressure?

*78. Show numerically that the molality and molarity of 1.00×10^{-4} M aqueous sodium chloride are nearly equal. Why is this true? Would this be true if another solvent, say acetonitrile, CH_3CN, replaced water? Why or why not? The density of CH_3CN is 0.786 g/mL at 20°C.

79. Show how the expression $\pi = MRT$, where π is osmotic pressure, is similar to the ideal gas law. Rationalize qualitatively why this should be so.

80. What is the osmotic pressure associated with a 0.0111 M aqueous solution of a nonvolatile nonelectrolyte solute at 75°C?

81. The osmotic pressure of an aqueous solution of a nonvolatile nonelectrolyte solute is 1.17 atm at 0°C. What is the molarity of the solution?

82. Calculate the freezing point depression and boiling point elevation associated with the solution in Exercise 81.

83. Estimate the osmotic pressure associated with 24.5 g of an enzyme of molecular weight 4.21×10^6 dissolved in 1740 mL of ethyl acetate solution at 38.0°C.

84. Calculate the osmotic pressure at 25°C of 0.10 m K_2CrO_4 in water, taking ion association into account. Refer to Table 14-3.

85. Estimate the osmotic pressure at 25°C of 0.10 m K_2CrO_4 in water, assuming no ion association.

*86. Many biological compounds are isolated and purified in very small amounts. We dissolve 11.0 mg of a biological macromolecule with molecular weight of 2.00×10^4 in 10.0 g of water. (a) Calculate the freezing point of the solution. (b) Calculate the osmotic pressure of the solution at 25°C. (c) Suppose we are trying to use freezing point measurements to *determine* the molecular weight of this substance and that we make an error of only 0.001°C in the temperature measurement. What percent error would this cause in the calculated molecular weight? (d) Suppose we could measure the osmotic pressure with an error of only 0.1 torr (not a very difficult experiment). What percent error would this cause in the calculated molecular weight?

Colloids

87. How does a colloidal dispersion differ from a true solution?

88. Distinguish among (a) sol, (b) gel, (c) emulsion, (d) foam, (e) solid sol, (f) solid emulsion, (g) solid foam, (h) solid aerosol, and (i) liquid aerosol. Try to give an example of each that is not listed in Table 14-4.

89. What is the Tyndall effect, and how is it caused?

90. Distinguish between hydrophilic and hydrophobic colloids.

91. What is an emulsifier?

92. Distinguish between soaps and detergents. How do they interact with hard water? Write an equation to show the interaction between a soap and hard water that contains Ca^{2+} ions.

93. What is the disadvantage of branched alkylbenzenesulfonate (ABS) detergents compared to linear alkylbenzenesulfonate (LAS) detergents?

Mixed Exercises

*94. The heat of solution (for infinite dilution) of KF(s) is -17.7 kJ/mol at 25°C. The crystal lattice energy, ΔH_{xtal}, is -825.9 kJ/mol at 25°C. What is the hydration energy of KF for infinite dilution at 25°C? [Here we refer to the sum of the hydration energies of $K^+(g)$ and $F^-(g)$.]

95. Dry air contains 20.94% O_2 by volume. The solubility of O_2 in water at 25°C is 0.041 gram O_2 per liter of water. How many liters of water would dissolve the O_2 in one liter of dry air at 25°C and 1.00 atm?

96. (a) The freezing point of a 1.00% aqueous solution of acetic acid, CH_3COOH, is $-0.310°C$. What is the approximate formula weight of acetic acid in water? (b) A 1.00% solution of acetic acid in benzene has a freezing point depression of 0.441°C. What is the formula weight of acetic acid in this solvent? Explain the difference.

97. An aqueous ammonium chloride solution contains 6.50 mass % NH_4Cl. The density of the solution is 1.0201 g/mL. Express the concentration of this solution in molarity, molality, and mole fraction of solute.

98. Starch contains C—C, C—H, C—O, and O—H bonds. Hydrocarbons contain only C—C and C—H bonds. Both starch and hydrocarbon oils can form colloidal dispersions in water. (a) Which dispersion is classified as hydrophobic? (b) Which is hydrophilic? (c) Which dispersion would be easier to make and maintain?

*99. Suppose we put some one-celled microorganisms in various aqueous NaCl solutions. We observe that the cells remain unperturbed in 0.7% NaCl, whereas they shrink in more concentrated solutions and expand in more dilute solutions. Assume that 0.7% NaCl behaves as an *ideal* 1:1 electrolyte. Calculate the osmotic pressure of the aqueous fluid within the cells at 25°C.

*100. A sample of a drug ($C_{21}H_{23}O_5N$, molecular weight = 369 g/mol) mixed with lactose (a sugar, $C_{12}H_{22}O_{11}$, molecular weight = 342 g/mol) was analyzed by osmotic pressure to determine the amount of sugar present. If 100.00 mL of solution containing 1.00 g of the drug–sugar mixture has an osmotic pressure of 527 torr at 25°C, what is the percent sugar present?

101. A solution containing 4.22 g of a nonelectrolyte polymer per liter of benzene solution has an osmotic pressure of 0.646 torr at 20.0°C. (a) Calculate the molecular weight of the polymer. (b) Assume that the density of the dilute solution is the same as that of benzene, 0.879 g/mL. What would be the freezing point depression for this solution? (c) Why are boiling point elevations and freezing point depressions difficult to use to measure molecular weights of polymers?

102. On what basis would you choose the components to prepare an ideal solution of a molecular solute? Which of the following combinations would you expect to act most nearly ideally? (a) $CH_4(\ell)$ and $CH_3OH(\ell)$; (b) $CH_3OH(\ell)$ and $NaCl(s)$; (c) $CH_4(\ell)$ and $CH_3CH_3(\ell)$.

*103. At what temperature would a 1.00 *M* aqueous solution of sugar have an osmotic pressure of 1.00 atm? Is this answer reasonable?

*104. In the Signer method for estimating molecular weights, separate solutions of two compounds are placed at constant temperature in an evacuated system similar to that in Figure 14-17b. The same solvent is used in each solution. In one solution is dissolved a known mass of a compound of known molecular weight, while the other solution contains a known mass of the substance of unknown molecular weight. The solvent evaporates preferentially from the more dilute solution and condenses preferentially in the more concentrated solution. Eventually the mole fractions of solute in both solutions become equal, at which point the vapor pressures of the two solutions are equal (Ra-oult's Law); at this point, equilibrium is reached. If both solutions are sufficiently dilute, two approximations can be made: (1) The volume due to each solute is negligible compared to the volume due to the solvent, and (2) the number of moles of each solute is negligible compared to the number of moles of solvent. The volumes of the two solutions are measured, and the molecular weight of the unknown compound is calculated. Suppose we dissolve 40.6 mg of an unknown compound to form one solution and 45.1 mg of azobenzene (molecular weight 168.2) to form the other solution. After four days at 50°C, when equilibrium has been reached, the volume of the azobenzene solution is 1.80 mL and the volume of the "unknown" solution is 1.41 mL. Calculate the molecular weight of the unknown compound.

BUILDING YOUR KNOWLEDGE

105. Draw Figure 14-1, but instead of using colored circles to represent the solvent and solute molecules, use Lewis formulas to represent water as the solvent and acetone, CH_3COCH_3, as the solute. Use dashed lines to show hydrogen bonds. Twelve water molecules and two acetone molecules should be sufficient to illustrate the interaction between these two kinds of molecules.

106. A sugar maple tree grows to a height of 45 feet, and its roots are in contact with water in the soil. What must be the concentration of the sugar in its sap so that osmotic pressure forces the sap to the top of the tree at 0°C? The density of mercury is 13.6 g/cm^3, and the density of the sap can be considered to be 1.00 g/cm^3.

107. Many metal ions become hydrated in solution by forming coordinate covalent bonds with the unshared pair of electrons from the water molecules to form "AB_6" ions. Because of their sizes, these hydrated ions are unable to pass through the semipermeable membrane described in Section 14-15, while water as a trimer, $(H_2O)_3$, or a tetramer, $(H_2O)_4$, can pass through. Anions also hydrate but not via coordinate covalent bonds. Using the VSEPR theory, prepare 3-dimensional drawings of $Cu(H_2O)_6^{2+}$ and a possible $(H_2O)_3$ that show their relative shapes and sizes.

Chemical Thermodynamics

15

A continual exchange of energy is required to build the complex molecules and maintain the molecular order that are essential for life.

OBJECTIVES

As you study this chapter, you should learn to

- *Pay close attention to the terminology of thermodynamics, especially the significance of the signs of changes*

- *Use the concept of state functions*

- *Carry out calculations of calorimetry to determine changes in energy and enthalpy*

- *Use Hess' Law to find the enthalpy change, ΔH, for a reaction by combining thermochemical equations with known ΔH values*

- *Use the First Law of Thermodynamics to relate heat, work, and energy changes*

- *Relate the work done on or by a system to changes in its volume*

- *Use Hess' Law to find the enthalpy change, ΔH, for a reaction by using tabulated values of standard molar enthalpies of formation*

- *Use bond energies to estimate heats of reaction for gas phase reactions; use ΔH values for gas phase reactions to find bond energies*

- *Understand what is meant by the spontaneity of a process*

- *Understand the relationship of entropy to the order/disorder of a system*

- *Understand how the spontaneity of a process is related to entropy changes— the Second Law of Thermodynamics*

- *Use tabulated values of absolute entropies to calculate the entropy change, ΔS, for a process*

- *Calculate changes in Gibbs free energy, ΔG, (a) from values of ΔH and ΔS and (b) from tabulated values of standard molar free energies of formation; know when to use each type of calculation*

- *Use ΔG to predict the spontaneity of a process at constant T and P*

- *Understand how changes in temperature can affect the spontaneity of a process*

E nergy is very important in every aspect of our daily lives. The food we eat supplies the energy to sustain life with all of its activities and concerns. The availability of relatively inexpensive energy is an important factor in our technological society. This is seen in the costs of fuel, heating and cooling our homes and workplaces, and the electricity to power our lights, appliances, and computers. It is also seen in the costs of the goods and services we purchase, because a substantial part of the cost of production is for energy in one form or another. We must understand the storage and use of energy on a scientific basis to learn how to decrease our dependence on consumable oil and natural gas as our main energy sources. Such understanding has profound ramifications, ranging from our daily lifestyles to international relations.

The concept of energy is at the very heart of science. All physical and chemical processes are accompanied by the transfer of energy. Because energy cannot be created or destroyed, we must understand how to do the "accounting" of energy transfers from one body or one substance to another or from one form of energy to another.

Some forms of energy are potential, kinetic, electrical, heat, and light.

In **thermodynamics** we study the energy changes that accompany physical and chemical processes. Usually these energy changes involve *heat*—hence the "thermo-" part of the term. In this chapter we study the two main aspects of thermodynamics. The first is **thermochemistry.** This practical subject is concerned with how we *observe, measure,* and *predict* energy changes for both physical changes and chemical reactions. The second part of the chapter addresses a more fundamental aspect of thermodynamics. There we will learn to use energy changes to tell whether or not a given process can occur under specified conditions, and how to make a process more (or less) favorable.

HEAT CHANGES AND THERMOCHEMISTRY

15-1 THE FIRST LAW OF THERMODYNAMICS

We can define energy as follows.

> Energy is the capacity to do work or to transfer heat.

We classify energy into two general types—kinetic and potential. **Kinetic energy** is the energy of motion. The kinetic energy of an object is equal to one-half its mass, m, times the square of its velocity, v.

$$E_{kinetic} = \tfrac{1}{2}mv^2$$

The heavier a hammer is and the more rapidly it moves, the greater its kinetic energy and the more work it can accomplish.

Potential energy is the energy that a system possesses by virtue of its position or composition. The work that we do to lift an object is stored in the object as energy; we describe this as potential energy. If we drop a hammer, its potential energy is converted into kinetic energy as it falls, and it could do work on something it hits—for example, drive a nail or break a piece of glass. Similarly, an electron in an atom has potential energy because of the electrostatic force between it and the positively charged nucleus. Energy can take many other forms: electrical energy, radiant energy (light), nuclear energy, and chemical energy. At the atomic or molecular level, we can think of each of these as either kinetic or potential energy.

As matter falls from a higher to a lower level, its gravitational potential energy is converted into kinetic energy. A hydroelectric power plant converts the kinetic energy of falling water into electrical (potential) energy.

Figure 15-1 The difference between the heat content of the reactants—one mole of $CH_4(g)$ and two moles of $O_2(g)$—and that of the products—one mole of $CO_2(g)$ and two moles of $H_2O(\ell)$—is the amount of heat evolved in this *exothermic* reaction at constant pressure. For this reaction, it is 890 kJ/mol of reaction. Some initial activation, for example by heat, is needed to get the reaction started. In the absence of such activation energy, a mixture of CH_4 and O_2 can be kept at room temperature for a long time without reacting. For an *endothermic* reaction, the final level is higher than the initial level.

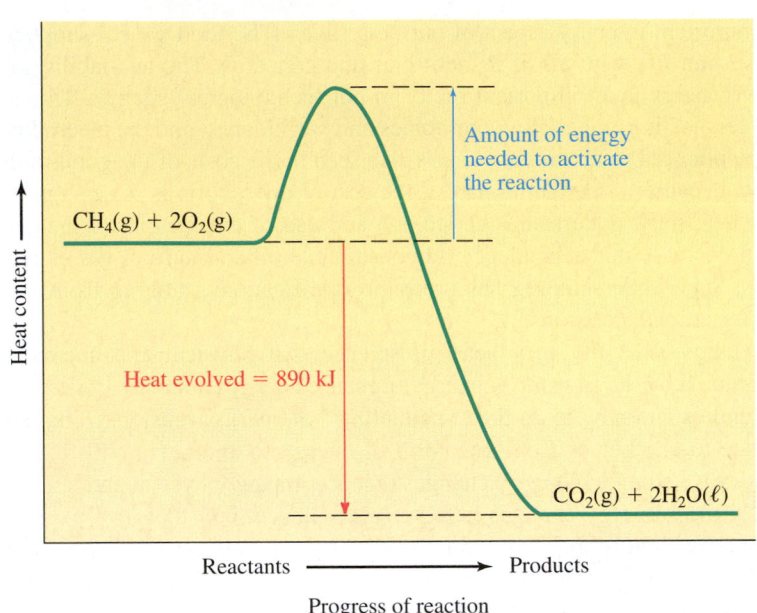

The chemical energy in a fuel or food can be viewed as potential energy stored in the electrons and nuclei due to their arrangements in the molecules. This stored chemical energy can be released when compounds undergo chemical changes, such as those that occur in combustion and metabolism. Reactions that release energy in the form of heat are called **exothermic** reactions.

A hydrocarbon is a binary compound of hydrogen and carbon. Hydrocarbons may be gaseous, liquid, or solid. All burn.

Combustion reactions of fossil fuels are familiar examples of exothermic reactions. Hydrocarbons—including methane, the main component of natural gas, and octane, a minor component of gasoline—undergo combustion with an excess of O_2 to yield CO_2 and H_2O. These reactions release heat energy. The amounts of heat energy released at constant pressure are shown for the reactions of one mole of methane and of two moles of octane.

$$CH_4(g) + 2O_2(g) \longrightarrow CO_2(g) + 2H_2O(\ell) + 890 \text{ kJ}$$

$$2C_8H_{18}(\ell) + 25O_2(g) \longrightarrow 16CO_2(g) + 18H_2O(\ell) + 1.090 \times 10^4 \text{ kJ}$$

The amount of heat shown in such an equation always refers to the reaction for the number of moles of reactants and products specified by the coefficients. We call this *one mole of reaction*. It is important to specify the physical states of all substances, because different physical states have different energy contents.

In such reactions, the total energy of the products is lower than that of the reactants by the amount of energy released, most of which is heat. Some initial activation (e.g., by heat) is needed to get these reactions started. This is shown for CH_4 in Figure 15-1. However, this activation energy *plus* 890 kJ is released as one mole of $CO_2(g)$ and two moles of $H_2O(\ell)$ are formed. A process that absorbs energy from its surroundings is called **endothermic.** One such process is shown in Figure 15-2.

Energy changes accompany physical changes, too (Chapter 13). For example, the melting of one mole of ice at 0°C at constant pressure must be accompanied by the absorption of 6.02 kJ of energy.

$$H_2O(s) + 6.02 \text{ kJ} \longrightarrow H_2O(\ell)$$

This tells us that the total energy of the water is raised by 6.02 kJ in the form of heat during the phase change.

(a)   **(b)**

Figure 15-2 An endothermic process. (a) When solid hydrated barium hydroxide [Ba(OH)$_2$·8H$_2$O] and *excess* solid ammonium nitrate [NH$_4$NO$_3$] are mixed, a reaction occurs.

$$Ba(OH)_2 \cdot 8H_2O(s) + 2NH_4NO_3(s) \longrightarrow Ba(NO_3)_2(s) + 2NH_3(g) + 10H_2O(\ell)$$

The excess ammonium nitrate dissolves in the water produced in the reaction. (b) The dissolution process is very endothermic. If the flask is placed on a wet wooden block, the water freezes and attaches the block to the flask.

Some important ideas about energy are summarized in the **First Law of Thermodynamics.**

> The total amount of energy in the universe is constant.

The **Law of Conservation of Energy** is just another statement of the First Law of Thermodynamics.

> Energy is neither created nor destroyed in ordinary chemical reactions and physical changes.

In Chapter 1 we pointed out the equivalence of matter and energy. The word "energy" is understood to include the energy equivalent of all matter in the universe. Stated differently, the total amount of mass and energy in the universe is constant.

15-2 SOME THERMODYNAMIC TERMS

The substances involved in the chemical and physical changes of interest are called the **system.** Everything in the system's environment constitutes its **surroundings.** The **universe** is the system plus its surroundings. The system may be thought of as the part of the universe under investigation. The First Law of Thermodynamics tells us that energy is neither created nor destroyed; it is only transferred between the system and its surroundings.

The **thermodynamic state of a system** is defined by a set of conditions that completely specifies all the properties of the system. This set commonly includes the temperature, pressure, composition (identity and number of moles of each component), and physical state (gas, liquid, or solid) of each part of the system. Once the state has been specified, all other properties—both physical and chemical—are fixed.

The properties of a system—e.g., P, V, T—are called **state functions.** The *value* of a state function depends *only* on the state of the system and not on the way in which the system came to be in that state. A *change* in a state function describes a *difference* between the two states. It is independent of the process or pathway by which the change occurs.

For instance, consider a sample of one mole of pure liquid water at 30°C and 1 atm pressure. If at some later time the temperature of the sample is 22°C at the same pressure, then it is in a different thermodynamic state. We can tell that the *net* temperature change is −8°C. It does not matter whether (1) the cooling took place directly (either slowly or rapidly) from 30°C to 22°C, or (2) the sample was first heated to 36°C, then cooled to 10°C, and finally warmed to 22°C, or (3) any other conceivable path was followed from the initial state to the final state. The change in other properties (e.g., the pressure) of the sample is likewise independent of path.

The most important use of state functions in thermodynamics is to describe *changes*. We describe the difference in any quantity, X, as

$$\Delta X = X_{\text{final}} - X_{\text{initial}}$$

When X increases, the final value is greater than the initial value, so ΔX is *positive;* a decrease in X makes ΔX a *negative* value.

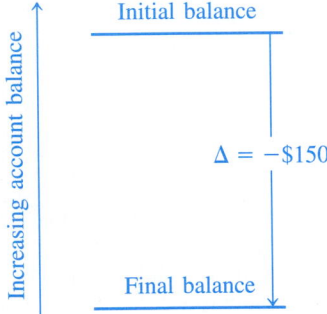

Here is a graphical representation of a $150 decrease in your bank balance. We express the change in your bank balance as $\Delta\$ = \$_{\text{final}} - \$_{\text{initial}}$. Your final balance is *less* than your initial balance, so the result is *negative,* indicating a *decrease*. There are many ways to get this same net change—one large withdrawal or some combination of deposits, withdrawals, interest earned, and service charges. All of the Δ values we shall see in this chapter can be thought of in this way.

▼ **PROBLEM-SOLVING TIP**   *Your Bank Balance Is a State Function*

You can consider a state function as analogous to a bank account. With a bank account, at any time you can measure the amount of money in your account (your balance) in convenient terms—dollars and cents. Changes in this balance can occur for several reasons, such as deposit of your paycheck, writing of checks, or service charges assessed by the bank. In our analogy these transactions are *not* state functions, but they do cause *changes in* the state function (the balance in the account). You can think of the bank balance on a vertical scale; a deposit of $150 changes the balance by +150 dollars, no matter what it was at the start, just as a withdrawal of $150 would change the balance by −150 dollars. Similarly, we shall see that the energy of a system is a state function that can be changed—for instance, by an energy "deposit" of heat absorbed or work done on the system, or by an energy "withdrawal" of heat given off or work done by the system.

We can describe *differences* between levels of a state function, regardless of where the zero level happens to be located. In the case of a bank balance, the "natural" zero level is obviously the point at which we open the account, before any deposits or withdrawals. In contrast, the zero levels on most temperature scales are set arbitrarily. When we say that the temperature of an ice–water mixture is "zero degrees Celsius," we are not saying that the mixture contains no temperature! We have simply chosen to describe this point on the temperature scale by the number *zero;* conditions of higher temperature are described by positive temperature values, and those of lower temperature have negative values, "below zero." The phrase "15 degrees cooler" has the same meaning anywhere on the scale. Many of the scales that we use in thermodynamics are arbitrarily defined in this way.

Arbitrary scales are useful when we are interested only in *changes* in the quantity being described.

Any property of a system that depends only on the values of its state functions is also a state function. For instance, the volume of a sample of water depends only on temperature, pressure, and physical state; volume is a state function. We shall encounter other thermodynamic state functions.

15-3 ENTHALPY CHANGES

Most chemical reactions and physical changes occur at constant (usually atmospheric) pressure.

> The change in heat content of a system that accompanies a process at constant pressure, q_p, is defined as the **enthalpy change, ΔH,** of the process.

We use q to represent the amount of heat absorbed by the system. The subscript p indicates a constant-pressure process.

An enthalpy change is sometimes loosely referred to as a *heat change* or a *heat of reaction*. The enthalpy change is equal to the enthalpy or "heat content," H, of the substances produced minus the enthalpy of the substances consumed.

$$\Delta H = H_{\text{final}} - H_{\text{initial}} \qquad \text{or} \qquad \Delta H = H_{\text{substances produced}} - H_{\text{substances consumed}}$$

It is impossible to know the absolute enthalpy (heat content) of a system. However, *enthalpy is a state function,* and it is the *change in enthalpy* in which we are interested; this can be measured for many processes. In the next several sections we focus on chemical reactions and the enthalpy changes that occur in these processes. We first discuss the experimental determination of enthalpy changes.

15-4 CALORIMETRY

We can determine the energy change associated with a chemical or physical process by using an experimental technique called **calorimetry.** This technique is based on observing the temperature change when a system absorbs or releases energy in the form of heat. The experiment is carried out in a device called a **calorimeter,** in which the temperature change of a known amount of substance (often water) of known specific heat is measured. The temperature change is caused by the absorption or release of heat by the chemical or physical process under study. A review of calculations involved with heat transfer (Sections 1-13, 13-9, and 13-11) may be helpful for understanding this section.

A "coffee-cup" calorimeter (Figure 15-3) is often used in laboratory classes to measure "heats of reaction" at constant pressure, q_p, in aqueous solutions. Reactions are chosen so that there are no gaseous reactants or products. Thus, all reactants and products remain in the vessel throughout the experiment. Such a calorimeter could be used to measure the amount of heat absorbed or released when a reaction takes place in aqueous solution. For an exothermic reaction, the amount of heat evolved by the reaction can be calculated from the amount by which it causes the temperature of the system to rise. The heat can be visualized as divided into two parts.

The polystyrene insulation of the coffee-cup calorimeter ensures that little or no heat passes between the system and the surroundings.

$$\left(\begin{array}{c}\text{amount of heat}\\\text{released by reaction}\end{array}\right) = \left(\begin{array}{c}\text{amount of heat gained}\\\text{by calorimeter}\end{array}\right) + \left(\begin{array}{c}\text{amount of heat}\\\text{gained by solution}\end{array}\right)$$

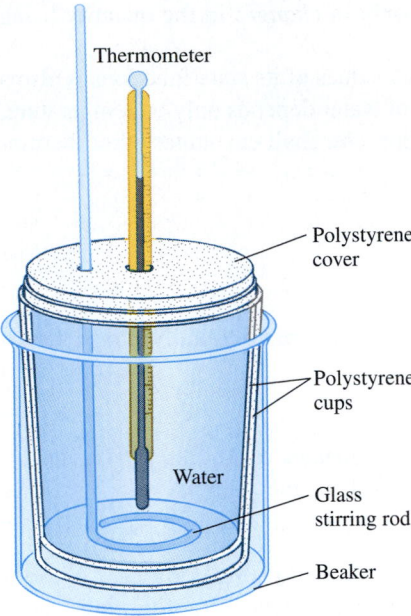

Figure 15-3   A coffee-cup calorimeter. The stirring rod is moved up and down to ensure thorough mixing and uniform heating of the solution during reaction. The polystyrene walls and top provide insulation so that very little heat escapes. This kind of calorimeter measures q_p, the heat transfer due to a reaction occurring at constant *pressure*.

The amount of heat absorbed by a calorimeter is sometimes expressed as the heat capacity of the calorimeter, in joules per degree.

The heat capacity of a calorimeter is determined by adding a known amount of heat and measuring the rise in temperature of the calorimeter and of the solution it contains. This heat capacity of a calorimeter is sometimes called its *calorimeter constant*. It depends on the materials of which the calorimeter is constructed and the mass of the calorimeter.

EXAMPLE 15-1   *Heat Capacity of Calorimeter*

One way to add heat is to use an electric heater.

We add 3.358 kJ of heat to a calorimeter that contains 50.00 g of water. The temperature of the water and the calorimeter, originally at 22.34°C, increases to 36.74°C. Calculate the heat capacity of the calorimeter in J/°C. The specific heat of water is 4.184 J/g·°C.

Plan

We first calculate the amount of heat gained by the water in the calorimeter. The rest of the heat must have been gained by the calorimeter, so we can determine the heat capacity of the calorimeter.

Solution

$$50.00 \text{ g H}_2\text{O at } 22.34°\text{C} \longrightarrow 50.00 \text{ g H}_2\text{O at } 36.74°\text{C}$$

The temperature change is $(36.74 - 22.34)°\text{C} = 14.40°\text{C}$.

$$\underline{?}\, \text{J} = 50.00 \text{ g} \times \frac{4.184 \text{ J}}{\text{g}\cdot°\text{C}} \times 14.40°\text{C} = 3.012 \times 10^3 \text{ J}$$

The total amount of heat added was 3.358 kJ or 3.358×10^3 J. The difference between these heat values is the amount of heat absorbed by the calorimeter.

$$\underline{?}\, \text{J} = 3.358 \times 10^3 \text{ J} - 3.012 \times 10^3 \text{ J} = 0.346 \times 10^3 \text{ J} \text{ or } 346 \text{ J absorbed by calorimeter}$$

To obtain the heat capacity of the calorimeter, we divide the amount of heat absorbed by the calorimeter, 346 J, by its temperature change.

$$\underline{?}\, \frac{\text{J}}{°\text{C}} = \frac{346 \text{ J}}{14.40°\text{C}} = \boxed{24.0 \text{ J/°C}}$$

The calorimeter absorbs 24.0 J of heat for each degree Celsius increase in its temperature.

You should now work Exercise 50.

EXAMPLE 15-2 *Heat Measurements Using a Calorimeter*

A 50.0-mL sample of 0.400 M copper(II) sulfate solution at 23.35°C is mixed with 50.0 mL of 0.600 M sodium hydroxide solution, also at 23.35°C, in the coffee-cup calorimeter of Example 15-1. After the reaction occurs, the temperature of the resulting mixture is measured to be 26.65°C. The density of the final solution is 1.02 g/mL. Calculate the amount of heat evolved. Assume that the specific heat of the solution is the same as that of pure water, 4.184 J/g·°C.

$$CuSO_4(aq) + 2NaOH(aq) \longrightarrow Cu(OH)_2(s) + Na_2SO_4(aq)$$

Plan

The amount of heat released by the reaction is absorbed by the solution *and* the calorimeter (assuming negligible loss to the surroundings). To find the amount of heat absorbed by the solution, we must know the mass of solution; to find that, we assume that the volume of the reaction mixture is the sum of volumes of the original solutions.

When *dilute aqueous solutions* are mixed, their volumes are very nearly additive.

Solution

The mass of solution is

$$\underline{?}\ g\ soln = (50.0 + 50.0)\ mL \times \frac{1.02\ g\ soln}{mL} = 102\ g\ soln$$

The amount of heat absorbed by the solution *plus* the calorimeter is

$$\underline{?}\ J = \overbrace{102\ g \times \frac{4.18\ J}{g \cdot °C} \times (26.65 - 23.35)°C}^{\substack{\text{amount of heat} \\ \text{absorbed by solution}}} + \overbrace{\frac{24.0\ J}{°C} \times (26.65 - 23.35)°C}^{\substack{\text{amount of heat} \\ \text{absorbed by calorimeter}}}$$

$$= 1.41 \times 10^3\ J + 79\ J = 1.49 \times 10^3\ J \text{ absorbed by solution plus calorimeter}$$

Thus, the reaction must have liberated 1.49×10^3 J, or 1.49 kJ, of heat.

You should now work Exercise 52(a).

15-5 THERMOCHEMICAL EQUATIONS

A balanced chemical equation, with its value of ΔH, is called a **thermochemical equation.** For example,

$$\underset{1\ mol}{C_2H_5OH(\ell)} + \underset{3\ mol}{3O_2(g)} \longrightarrow \underset{2\ mol}{2CO_2(g)} + \underset{3\ mol}{3H_2O(\ell)} + 1367\ kJ$$

is a thermochemical equation that describes the combustion (burning) of one mole of liquid ethanol at a particular temperature and pressure. The coefficients in such a description *must* be interpreted as *numbers of moles.* Thus, 1367 kJ of heat is released when *one* mole of $C_2H_5OH(\ell)$ reacts with *three* moles of $O_2(g)$ to give *two* moles of $CO_2(g)$ and *three* moles of $H_2O(\ell)$. We can refer to this amount of reaction as one **mole of reaction,** which we abbreviate "mol rxn." This interpretation allows us to write various unit factors as desired.

$$\frac{1\ mol\ C_2H_5OH(\ell)}{1\ mol\ rxn}, \quad \frac{2\ mol\ O_2(g)}{1\ mol\ rxn}, \quad \text{and so on}$$

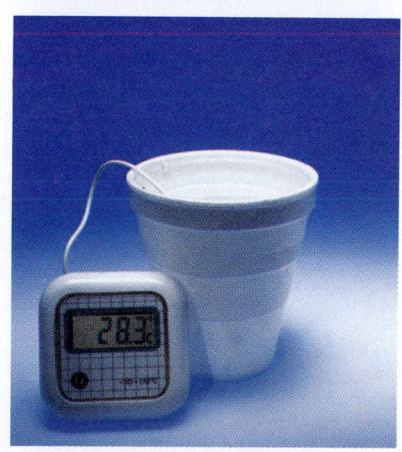

The heat released by the reaction of HCl(aq) with NaOH(aq) causes the temperature of the solution to rise.

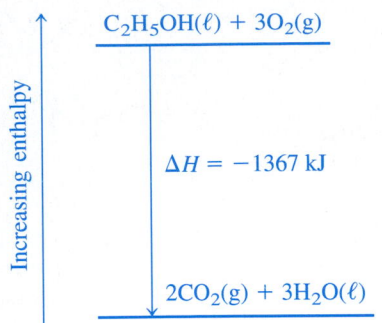

Before we know what a value of ΔH means, we must know the balanced chemical equation to which it refers.

We can also write the thermochemical equation as

$$C_2H_5OH(\ell) + 3O_2(g) \longrightarrow 2CO_2(g) + 3H_2O(\ell) \qquad \Delta H = -1367 \text{ kJ/mol rxn}$$

The negative sign indicates that this is an *exothermic* reaction (it *gives off* heat).

> We always interpret ΔH as the enthalpy change for the reaction as written; that is, as (enthalpy change)/(mole of reaction), where the denominator means "for the number of moles of each substance shown in the balanced equation."

We can then use several unit factors to interpret this thermochemical equation.

$$\frac{1367 \text{ kJ given off}}{\text{mol of reaction}} = \frac{1367 \text{ kJ given off}}{\text{mol } C_2H_5OH(\ell) \text{ consumed}} = \frac{1367 \text{ kJ given off}}{3 \text{ mol } O_2(g) \text{ consumed}}$$

$$= \frac{1367 \text{ kJ given off}}{2 \text{ mol } CO_2(g) \text{ formed}} = \frac{1367 \text{ kJ given off}}{3 \text{ mol } H_2O(\ell) \text{ formed}}$$

The reverse reaction would require the absorption of 1367 kJ under the same conditions; i.e., it is *endothermic* with $\Delta H = +1367$ kJ.

$$2CO_2(g) + 3H_2O(\ell) \longrightarrow C_2H_5OH(\ell) + 3O_2(g) \qquad \Delta H = +1367 \text{ kJ/mol rxn}$$

It is important to remember the following conventions regarding thermochemical equations:

> 1. The coefficients in a balanced thermochemical equation refer to the numbers of *moles* of reactants and products involved. In the thermodynamic interpretation of equations we *never* interpret the coefficients as *numbers of molecules*. Thus, it is acceptable to write coefficients as fractions rather than as integers, when necessary.
>
> 2. The numerical value of ΔH (or any other thermodynamic change) refers to the *number of moles* of substances specified by the equation. This amount of change of substances is called *one mole of reaction*, so we can express ΔH in units of energy/mol rxn. For brevity, the units of ΔH are sometimes written kJ/mol or even just kJ. No matter what units are used, be sure that you interpret the thermodynamic change *per mole of reaction for the balanced chemical equation to which it refers*. If a different amount of material is involved in the reaction, then the ΔH (or other change) must be scaled accordingly.
>
> 3. The physical states of all species are important and must be specified.
>
> 4. The value of ΔH usually does not change significantly with moderate changes in temperature.

EXAMPLE 15-3   *Thermochemical Equations*

Write the thermochemical equation for the reaction in Example 15-2.

Plan

We must determine *how much* reaction occurred—that is, how many moles of reactants were consumed. We first multiply the volume, in liters, of each solution by its concentration in mol/L (molarity), to determine the number of moles of each reactant mixed. Then we identify the limiting reactant. We scale the amount of heat released in the experiment to correspond to the number of moles of that reactant shown in the balanced equation.

Solution
Using the data from Example 15-2,

$$\underline{?}\ \text{mol CuSO}_4 = 0.0500\ \text{L} \times \frac{0.400\ \text{mol CuSO}_4}{1.00\ \text{L}} = 0.0200\ \text{mol CuSO}_4$$

$$\underline{?}\ \text{mol NaOH} = 0.0500\ \text{L} \times \frac{0.600\ \text{mol NaOH}}{1.00\ \text{L}} = 0.0300\ \text{mol NaOH}$$

We determine which is the limiting reactant (Section 3-3).

Required Ratio	**Available Ratio**

$$\frac{1\ \text{mol CuSO}_4}{2\ \text{mol NaOH}} = \frac{0.50\ \text{mol CuSO}_4}{1.00\ \text{mol NaOH}} \qquad \frac{0.0200\ \text{mol CuSO}_4}{0.0300\ \text{mol NaOH}} = \frac{0.667\ \text{mol CuSO}_4}{1.00\ \text{mol NaOH}}$$

More $CuSO_4$ is available than is required to react with the NaOH. Thus, 1.49 kJ of heat was released during the consumption of 0.0300 mol of NaOH. The amount of heat released per "mole of reaction" is

NaOH is the limiting reactant.

$$\frac{\underline{?}\ \text{kJ released}}{\text{mol of rxn}} = \frac{1.49\ \text{kJ released}}{0.0300\ \text{mol NaOH}} \times \frac{2\ \text{mol NaOH}}{\text{mol of rxn}} = \frac{99.3\ \text{kJ released}}{\text{mol of rxn}}$$

Thus, when the reaction occurs *to the extent indicated by the balanced chemical equation*, 99.3 kJ is released. Remembering that exothermic reactions have negative values of ΔH_{rxn}, we write

Heat is released, so this is an exothermic reaction.

$$CuSO_4(aq) + 2NaOH(aq) \longrightarrow Cu(OH)_2(s) + Na_2SO_4(aq) \qquad \Delta H_{rxn} = -99.3\ \text{kJ/mol rxn}$$

You should now work Exercise 52(b).

EXAMPLE 15-4   *Amount of Heat versus Extent of Reaction*

When aluminum metal is exposed to atmospheric oxygen (as in aluminum doors and windows), it is oxidized to form aluminum oxide. How much heat is released by the complete oxidation of 24.2 grams of aluminum at 25°C and 1 atm? The thermochemical equation is

$$4Al(s) + 3O_2(g) \longrightarrow 2Al_2O_3(s) \qquad \Delta H = -3352\ \text{kJ/mol rxn}$$

Plan
The thermochemical equation tells us that 3352 kJ of heat is released for every mole of reaction, that is, for every 4 moles of Al that reacts. We convert 24.2 g of Al to moles, and then calculate the number of kilojoules corresponding to that number of moles of Al, using the unit factors

$$\frac{-3352\ \text{kJ}}{\text{mol rxn}} \quad \text{and} \quad \frac{1\ \text{mol rxn}}{4\ \text{mol Al}}$$

Solution
For 24.2 g Al,

$$\underline{?}\ \text{kJ} = 24.2\ \text{g Al} \times \frac{1\ \text{mol Al}}{27.0\ \text{g Al}} \times \frac{1\ \text{mol rxn}}{4\ \text{mol Al}} \times \frac{-3352\ \text{kJ}}{\text{mol rxn}} = -751\ \text{kJ}$$

This tells us that 751 kJ of heat is released to the surroundings during the oxidation of 24.2 grams of aluminum.

You should now work Exercises 10 and 11.

The sign tells us that heat was released, but it would be grammatical nonsense to say in words that "−751 kJ of heat was released." As an analogy, suppose you give your friend \$5. Your $\Delta\$$ is −\$5, but in describing your action you would not say "I gave her minus five dollars," but rather "I gave her five dollars."

▼ **PROBLEM-SOLVING TIP** *Mole of Reaction*

Remember that a thermochemical equation can imply *different* numbers of moles of *different* reactants or products. In Example 15-4 one mole of reaction also corresponds to 3 moles of $O_2(g)$ and to 2 moles of $Al_2O_3(s)$.

15-6 STANDARD STATES AND STANDARD ENTHALPY CHANGES

The **thermochemical standard state** of a substance is its most stable state under standard pressure (one atmosphere) and at some specific temperature (25°C or 298 K unless otherwise specified). Examples of elements in their standard states at 25°C are hydrogen, gaseous diatomic molecules, $H_2(g)$; mercury, a silver-colored liquid metal, $Hg(\ell)$; sodium, a silvery-white solid metal, $Na(s)$; and carbon, a grayish-black solid called graphite, $C(graphite)$. We use $C(graphite)$ instead of $C(s)$ to distinguish it from another form of carbon, $C(diamond)$. Examples of standard states of compounds include ethanol (ethyl alcohol or grain alcohol), a liquid, $C_2H_5OH(\ell)$; water, a liquid, $H_2O(\ell)$; calcium carbonate, a solid, $CaCO_3(s)$; and carbon dioxide, a gas, $CO_2(g)$. Keep in mind the following conventions for thermochemical standard states.

A temperature of 25°C is 77°F. This is slightly above typical room temperature.

1. For a *pure* substance in the liquid or solid phase, the standard state is the pure liquid or solid.
2. For a gas, the standard state is the gas at a pressure of *one atmosphere;* in a mixture of gases, its partial pressure must be one atmosphere.
3. For a substance in solution, the standard state refers to *one-molar* concentration.

For ease of comparison and tabulation, we often refer to thermochemical or thermodynamic changes "at standard states" or, more simply, to a *standard change*. To indicate a change at standard pressure, we add a superscript zero. If some temperature other than standard temperature of 25°C (298 K) is specified, we indicate it with a subscript; if no subscript appears, a temperature of 25°C (298 K) is implied. For instance, the **standard enthalpy change, ΔH^0_{rxn},** for reaction

This is sometimes referred to as the *standard heat of reaction*. It is sometimes represented as ΔH^0_{rxn}.

$$\text{reactants} \longrightarrow \text{products}$$

refers to the ΔH when the specified number of moles of reactants, all at standard states, are converted *completely* to the specified number of moles of products, all at standard states. We allow the reaction to take place, with changes in temperature or pressure if necessary; when the reaction is complete, we return the products to the same conditions of temperature and pressure that we started with, *keeping track of energy or enthalpy changes* as we do so. When we describe a process as taking place "at constant T and P," we mean that the initial and final conditions are the same. Because we are dealing with changes in state functions, the net change is the same as the change we would have obtained hypothetically with T and P actually held constant.

15-7 STANDARD MOLAR ENTHALPIES OF FORMATION, ΔH^0_f

It is not possible to determine the total enthalpy content of a substance on an absolute scale. However, we need to describe only *changes* in this state function, so we can define an *arbitrary scale* as follows.

The **standard molar enthalpy of formation, ΔH_f^0,** of a substance is the enthalpy change for the reaction in which *one mole* of the substance in a specified state is formed from its elements in their standard states.

Standard molar enthalpy of formation is often called **standard molar heat of formation** or, more simply, **heat of formation.** The superscript zero in ΔH_f^0 signifies standard pressure, 1 atmosphere. Negative values for ΔH_f^0 describe exothermic formation reactions, whereas positive values for ΔH_f^0 describe endothermic formation reactions.

By the definition of ΔH_f^0, the formation of an element in its standard state from that element in its standard state is "no reaction." Thus,

The ΔH_f^0 value for any *element in its standard state* is zero.

The enthalpy change for a balanced equation may not give directly a molar enthalpy of formation for the compound formed. Consider the following exothermic reaction at standard conditions.

$$H_2(g) + Br_2(\ell) \longrightarrow 2HBr(g) \qquad \Delta H_{rxn}^0 = -72.8 \text{ kJ/mol rxn}$$

We see that *two* moles of HBr(g) are formed in the reaction as written. Half as much energy, 36.4 kJ, is liberated when *one mole* of HBr(g) is produced from its constituent elements in their standard states. For HBr(g), $\Delta H_f^0 = -36.4$ kJ/mol. This can be shown by dividing all coefficients in the balanced equation by 2.

$$\tfrac{1}{2}H_2(g) + \tfrac{1}{2}Br_2(\ell) \longrightarrow HBr(g) \qquad \Delta H_{rxn}^0 = -36.4 \text{ kJ/mol rxn}$$

$$\Delta H_{f\,HBr(g)}^0 = -36.4 \text{ kJ/mol HBr(g)}$$

Standard heats of formation of some common substances are tabulated in Table 15-1. Appendix K contains a larger listing.

When referring to a thermodynamic quantity for a *substance,* we often omit the description of the substance from the units. Units for tabulated ΔH_f^0 values are given as

We can think of ΔH_f^0 as the enthalpy content of each substance, in its standard state, relative to the enthalpy content of the elements, in their standard states.

The coefficients $\tfrac{1}{2}$ preceding $H_2(g)$ and $Br_2(\ell)$ do *not* imply half a molecule of each. In thermochemical equations, the coefficients always refer to the number of *moles* under consideration.

Table 15-1 *Selected Standard Molar Enthalpies of Formation at 298 K*

Substance	ΔH_f^0 (kJ/mol)	Substance	ΔH_f^0 (kJ/mol)
$Br_2(\ell)$	0	HgS(s) red	−58.2
$Br_2(g)$	30.91	$H_2(g)$	0
C(diamond)	1.897	HBr(g)	−36.4
C(graphite)	0	$H_2O(\ell)$	−285.8
$CH_4(g)$	−74.81	$H_2O(g)$	−241.8
$C_2H_4(g)$	52.26	NO(g)	90.25
$C_6H_6(\ell)$	49.03	Na(s)	0
$C_2H_5OH(\ell)$	−277.7	NaCl(s)	−411.0
CO(g)	−110.52	$O_2(g)$	0
$CO_2(g)$	−393.51	$SO_2(g)$	−296.8
CaO(s)	−635.5	$SiH_4(g)$	34
$CaCO_3(s)$	−1207	$SiCl_4(g)$	−657.0
$Cl_2(g)$	0	$SiO_2(s)$	−910.9

The ΔH_f^0 values of $Br_2(g)$ and C(diamond) are *not equal to 0* at 298 K. The standard states of these elements are $Br_2(\ell)$ and C(graphite), respectively.

"kJ/mol"; we must interpret this as "per mole of the substance in the specified state." For instance, for HBr(g) the tabulated ΔH_f^0 value of -36.4 kJ/mol should be interpreted as

$$\frac{-36.4 \text{ kJ}}{\text{mol HBr(g)}}.$$

EXAMPLE 15-5   *Interpretation of* ΔH_f^0

The standard molar enthalpy of formation of ethanol, $C_2H_5OH(\ell)$, is -277.7 kJ/mol. Write the thermochemical equation for the reaction for which $\Delta H_{rxn}^0 = -277.7$ kJ/mol rxn.

Plan

The balanced equation (whole number coefficients) shows the formation of two moles of C_2H_5OH.

$$4C(\text{graphite}) + 6H_2(g) + O_2(g) \longrightarrow 2C_2H_5OH(\ell)$$

ΔH_f^0 values apply to the formation of *one* mole of the substance, so we divide the coefficients in the chemical equation by 2.

Solution

$$2C(\text{graphite}) + 3H_2(g) + \tfrac{1}{2}O_2(g) \longrightarrow C_2H_5OH(\ell) \qquad \Delta H = -277.7 \text{ kJ/mol rxn}$$

You should now work Exercise 20.

15-8   HESS' LAW

In 1840, G. H. Hess published his **law of heat summation,** which he derived on the basis of numerous thermochemical observations.

> The enthalpy change for a reaction is the same whether it occurs by one step or by any series of steps.

As an analogy, consider traveling from Kansas City (elevation 884 ft above sea level) to Denver (elevation 5280 ft). The change in elevation is $(5280 - 884)$ ft $= 4396$ ft, regardless of the route taken.

Enthalpy is a state function. Therefore, its *change* is independent of the pathway by which a reaction occurs. We do not need to know whether the reaction *does,* or even *can,* occur by the series of steps used in the calculation. The steps must (if only "on paper") result in the overall reaction. Hess' Law lets us calculate enthalpy changes for reactions for which the changes could be measured only with difficulty, if at all. In general terms, Hess' Law of heat summation may be represented as

> $$\Delta H_{rxn}^0 = \Delta H_a^0 + \Delta H_b^0 + \Delta H_c^0 + \cdots$$

Here a, b, c, . . . refer to balanced thermochemical equations that can be summed to give the equation for the desired reaction.

Consider the following reaction.

$$C(\text{graphite}) + \tfrac{1}{2}O_2(g) \longrightarrow CO(g) \qquad \Delta H_{rxn}^0 = \underline{?}$$

The enthalpy change for this reaction cannot be measured directly. Even though CO(g) is the predominant product of the reaction of graphite with a *limited* amount of $O_2(g)$, some $CO_2(g)$ is always produced as well. However, the following reactions do go to completion with excess $O_2(g)$. Therefore, ΔH^0 values have been measured experimentally for them. (Pure $CO(g)$ is readily available.)

$$C(graphite) + O_2(g) \longrightarrow CO_2(g) \qquad \Delta H^0_{rxn} = -393.5 \text{ kJ/mol rxn} \qquad (1)$$

$$CO(g) + \tfrac{1}{2}O_2(g) \longrightarrow CO_2(g) \qquad \Delta H^0_{rxn} = -283.0 \text{ kJ/mol rxn} \qquad (2)$$

We can "work backwards" to find out how to combine these two known equations to obtain the desired equation. We want one mole of CO on the right, so we reverse equation (2); heat is then absorbed instead of released, so we must change the sign of its ΔH^0 value. Then we add it to equation (1), canceling equal numbers of moles of the same species on each side. This gives the equation for the reaction we want. Adding the corresponding enthalpy changes gives the enthalpy change we seek.

You are familiar with the addition and subtraction of algebraic equations. This method of combining thermochemical equations is analogous.

$$\Delta H^0$$

$C(graphite) + O_2(g) \longrightarrow CO_2(g)$	-393.5 kJ/mol rxn	(1)
$CO_2(g) \longrightarrow CO(g) + \tfrac{1}{2}O_2(g)$	$-(-283.0$ kJ/mol rxn)	(-2)
$C(graphite) + \tfrac{1}{2}O_2(g) \longrightarrow CO(g)$	$\Delta H^0_{rxn} = -110.5$ kJ/mol rxn	

This equation shows the formation of one mole of $CO(g)$ in its standard state from the elements in their standard states. In this way, we determine that ΔH^0_f for $CO(g)$ is -110.5 kJ/mol.

A schematic representation of the enthalpy changes for the reaction $C(graphite) + \tfrac{1}{2}O_2(g) \rightarrow CO(g)$. The ΔH value for each step refers to the number of moles of each substance indicated.

EXAMPLE 15-6 *Combining Thermochemical Equations—Hess' Law*

Use the thermochemical equations given below to determine ΔH^0_{rxn} at 25°C for the following reaction.

$$C(graphite) + 2H_2(g) \longrightarrow CH_4(g)$$

$$\Delta H^0$$

$C(graphite) + O_2(g) \longrightarrow CO_2(g)$	-393.5 kJ/mol rxn	(1)
$H_2(g) + \tfrac{1}{2}O_2(g) \longrightarrow H_2O(\ell)$	-285.8 kJ/mol rxn	(2)
$CH_4(g) + 2O_2(g) \longrightarrow CO_2(g) + 2H_2O(\ell)$	-890.3 kJ/mol rxn	(3)

These are combustion reactions, for which ΔH^0_{rxn} values can be readily determined.

Plan

(i) We want one mole of C(graphite) as reactant, so we write down equation (1).

(ii) We want two moles of $H_2(g)$ as reactants, so we multiply equation (2) by 2 [designated below as $2 \times (2)$].

(iii) We want one mole of $CH_4(g)$ as product, so we reverse equation (3) to give (-3).

(iv) We do the same operations on each ΔH^0 value.

(v) Then we add these equations term by term. The result is the desired thermochemical equation; all unwanted substances cancel. The sum of the ΔH^0 values is the ΔH^0 for the desired reaction.

Solution

$$\Delta H^0$$

$C(graphite) + O_2(g) \longrightarrow CO_2(g)$	-393.5 kJ/mol rxn	(1)
$2H_2(g) + O_2(g) \longrightarrow 2H_2O(\ell)$	$2(-285.8$ kJ/mol rxn)	$2 \times (2)$
$CO_2(g) + 2H_2O(\ell) \longrightarrow CH_4(g) + 2O_2(g)$	$+890.3$ kJ/mol rxn	(-3)
$C(graphite) + 2H_2(g) \longrightarrow CH_4(g)$	$\Delta H^0_{rxn} = -74.8$ kJ/mol rxn	

$CH_4(g)$ cannot be formed directly from C(graphite) and $H_2(g)$, so its ΔH^0_f value cannot be measured directly. The result of Example 15-6 tells us that this value is -74.8 kJ/mol.

We have used a series of reactions for which ΔH^0 values can be easily measured to calculate ΔH^0 for a reaction that cannot be carried out.

EXAMPLE 15-7   *Combining Thermochemical Equations—Hess' Law*

Given the following thermochemical equations, calculate the heat of reaction at 298 K for the reaction of ethylene with water to form ethanol.

$$C_2H_4(g) + H_2O(\ell) \longrightarrow C_2H_5OH(\ell)$$

$$\Delta H^0$$

$C_2H_5OH(\ell) + 3O_2(g) \longrightarrow 2CO_2(g) + 3H_2O(\ell)$	-1367 kJ/mol rxn	(1)
$C_2H_4(g) + 3O_2(g) \longrightarrow 2CO_2(g) + 2H_2O(\ell)$	-1411 kJ/mol rxn	(2)

Plan

We reverse equation (1) to give (-1); when the equation is reversed, the sign of ΔH^0 is changed because the reverse of an exothermic reaction is endothermic. Then we add it to equation (2).

Solution

$$\Delta H^0$$

$2CO_2(g) + 3H_2O(\ell) \longrightarrow C_2H_5OH(\ell) + 3O_2(g)$	$+1367$ kJ/mol rxn	(-1)
$C_2H_4(g) + 3O_2(g) \longrightarrow 2CO_2(g) + 2H_2O(\ell)$	-1411 kJ/mol rxn	(2)
$C_2H_4(g) + H_2O(\ell) \longrightarrow C_2H_5OH(\ell)$	$\Delta H^0_{rxn} = -44$ kJ/mol rxn	

You should now work Exercises 24 and 26.

▼ **PROBLEM-SOLVING TIP** ΔH^0_f **Refers to Specific Reaction**

The ΔH^0 for the reaction in Example 15-7 is -44 kJ for each mole of $C_2H_5OH(\ell)$ formed. However, this reaction does not involve formation of $C_2H_5OH(\ell)$ from its constituent elements. Therefore, ΔH^0_{rxn} is *not* ΔH^0_f for $C_2H_5OH(\ell)$.

Another interpretation of Hess' Law lets us use tables of ΔH^0_f values to calculate the enthalpy change accompanying a reaction. Let us reconsider the reaction of Example 15-7.

$$C_2H_4(g) + H_2O(\ell) \longrightarrow C_2H_5OH(\ell)$$

A table of ΔH^0_f values (Appendix K) gives $\Delta H^0_{f\,C_2H_5OH(\ell)} = -277.7$ kJ/mol, $\Delta H^0_{f\,C_2H_4(g)} = 52.3$ kJ/mol, and $\Delta H^0_{f\,H_2O(\ell)} = -285.8$ kJ/mol. We may express this information in the form of the following thermochemical equations.

$$\Delta H^0$$

$2C(graphite) + 3H_2(g) + \frac{1}{2}O_2(g) \longrightarrow C_2H_5OH(\ell)$	-277.7 kJ/mol rxn	(1)
$2C(graphite) + 2H_2(g) \longrightarrow C_2H_4(g)$	52.3 kJ/mol rxn	(2)
$H_2(g) + \frac{1}{2}O_2(g) \longrightarrow H_2O(\ell)$	-285.8 kJ/mol rxn	(3)

We may generate the equation for the desired net reaction by adding equation (1) to the reverse of equations (2) and (3). The value of ΔH^0 for the desired reaction is then the sum of the corresponding ΔH^0 values.

$$\Delta H^0$$

$2C(graphite) + 3H_2(g) + \frac{1}{2}O_2(g) \longrightarrow C_2H_5OH(\ell)$	-277.7 kJ/mol rxn	(1)
$C_2H_4(g) \longrightarrow 2C(graphite) + 2H_2(g)$	-52.3 kJ/mol rxn	(-2)
$H_2O(\ell) \longrightarrow H_2(g) + \frac{1}{2}O_2(g)$	$+285.8$ kJ/mol rxn	(-3)
net rxn: $C_2H_4(g) + H_2O(\ell) \longrightarrow C_2H_5OH(\ell)$	$\Delta H^0_{rxn} = -44.2$ kJ/mol rxn	

We see that ΔH^0 for this reaction is given by

$$\Delta H^0_{rxn} = \Delta H^0_{(1)} + \Delta H^0_{(-2)} + \Delta H^0_{(-3)}$$

Or by

$$\Delta H^0_{rxn} = \Delta H^0_{f\ C_2H_5OH(\ell)} - [\Delta H^0_{f\ C_2H_4(g)} + \Delta H^0_{f\ H_2O(\ell)}]$$

with labels: *product* pointing to $\Delta H^0_{f\ C_2H_5OH(\ell)}$ and *reactants* pointing to the bracketed terms.

In general terms this is a very useful form of Hess' Law.

$$\Delta H^0_{rxn} = \Sigma\, n\Delta H^0_{f\ products} - \Sigma\, n\Delta H^0_{f\ reactants}$$

The standard enthalpy change of a reaction is equal to the sum of the standard molar enthalpies of formation of the products, each multiplied by its coefficient, n, in *the balanced equation,* minus the corresponding sum of the standard molar enthalpies of formation of the reactants.

The capital Greek letter sigma (Σ) is read "the sum of." The $\Sigma\, n$ means that the ΔH^0_f value of each product and reactant must be multiplied by its coefficient, n, in the balanced equation. The resulting values are then added.

In effect this form of Hess' Law supposes that the reaction occurs by converting reactants to the elements in their standard states, then converting these to products (Figure 15-4). Few, if any, reactions actually occur by such a pathway. Nevertheless, the ΔH^0 for this *hypothetical* pathway for *reactants* → *products* would be the same as that for any other pathway—including the one by which the reaction actually occurs.

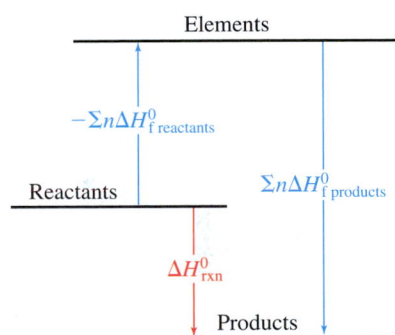

EXAMPLE 15-8 *Using ΔH^0_f Values—Hess' Law*

Calculate ΔH^0_{rxn} for the following reaction at 298 K.

$$SiH_4(g) + 2O_2(g) \longrightarrow SiO_2(s) + 2H_2O(\ell)$$

Plan

We apply Hess' Law in the form $\Delta H^0_{rxn} = \Sigma\, n\Delta H^0_{f\ products} - \Sigma\, n\Delta H^0_{f\ reactants}$, so we use the ΔH^0_f values tabulated in Appendix K.

Solution

$$\Delta H^0_{rxn} = \Sigma\, n\Delta H^0_{f\ products} - \Sigma\, n\Delta H^0_{f\ reactants}$$

$$\Delta H^0_{rxn} = [\Delta H^0_{f\ SiO_2(s)} + 2\Delta H^0_{f\ H_2O(\ell)}] - [\Delta H^0_{f\ SiH_4(g)} + 2\Delta H^0_{f\ O_2(g)}]$$

$$\Delta H^0_{rxn} = \left[\frac{1\ mol\ SiO_2(s)}{mol\ rxn} \times \frac{-910.9\ kJ}{mol\ SiO_2(s)} + \frac{2\ mol\ H_2O(\ell)}{mol\ rxn} \times \frac{-285.8\ kJ}{mol\ H_2O(\ell)} \right]$$
$$- \left[\frac{1\ mol\ SiH_4(g)}{mol\ rxn} \times \frac{+34\ kJ}{mol\ SiH_4(g)} + \frac{2\ mol\ O_2(g)}{mol\ rxn} \times \frac{0\ kJ}{mol\ O_2(g)} \right]$$

$$\Delta H^0_{rxn} = -1516\ kJ/mol\ rxn$$

You should now work Exercise 30.

Figure 15-4 A schematic representation of Hess' Law. The red arrow represents the *direct* path from reactants to products. The series of blue arrows is a path (hypothetical) in which reactants are converted to elements, and they in turn are converted to products—all in their standard states.

$O_2(g)$ is an element in its standard state, so its ΔH^0_f is zero.

Each term in the sums on the right-hand side of the solution in Example 15-8 has the units

$$\frac{mol\ substance}{mol\ rxn} \times \frac{kJ}{mol\ substance} \quad or \quad \frac{kJ}{mol\ rxn}$$

For brevity, we shall omit units in the intermediate steps of calculations of this type, and just assign the proper units to the answer. Be sure that you understand how these units arise.

Suppose we measure ΔH^0_{rxn} at 298 K and know all but one of the ΔH^0_f values for reactants and products. We can then calculate the unknown ΔH^0_f value.

EXAMPLE 15-9   *Using ΔH^0_f Values—Hess' Law*

We will consult Appendix K only after working the problem, to check the answer.

Use the following information to determine ΔH^0_f for PbO(s, yellow).

$$PbO(s, yellow) + CO(g) \longrightarrow Pb(s) + CO_2(g) \qquad \Delta H^0_{rxn} = -65.69 \text{ kJ}$$

$$\Delta H^0_f \text{ for } CO_2(g) = -393.5 \text{ kJ/mol} \qquad \text{and} \qquad \Delta H^0_f \text{ for } CO(g) = -110.5 \text{ kJ/mol}$$

Plan

We again use Hess' Law in the form $\Delta H^0_{rxn} = \Sigma\, n\Delta H^0_{f\,products} - \Sigma\, n\Delta H^0_{f\,reactants}$. Now we are given ΔH^0_{rxn} and the ΔH^0_f values for all substances *except* PbO(s, yellow). We can solve for this unknown.

Solution

$$\Delta H^0_{rxn} = \Sigma\, n\Delta H^0_{f\,products} \qquad\qquad - \Sigma\, n\Delta H^0_{f\,reactants}$$

$$\Delta H^0_{rxn} = \Delta H^0_{f\,Pb(s)} + \Delta H^0_{f\,CO_2(g)} - [\Delta H^0_{f\,PbO(s,\,yellow)} + \Delta H^0_{f\,CO(g)}]$$

$$-65.69 = 0 + (-393.5) \qquad\quad - [\Delta H^0_{f\,PbO(s,\,yellow)} + (-110.5)]$$

Rearranging to solve for $\Delta H^0_{f\,PbO(s,\,yellow)}$, we have

$$\Delta H^0_{f\,PbO(s,\,yellow)} = 65.69 - 393.5 + 110.5 = \boxed{-217.3 \text{ kJ/mol of PbO}}$$

You should now work Exercise 34.

15-9   BOND ENERGIES

Chemical reactions involve the breaking and making of chemical bonds. Energy is always required to break a chemical bond. Often this energy is supplied in the form of heat.

For all practical purposes, the bond energy is the same as bond enthalpy. Tabulated values of average bond energies are actually average bond enthalpies. We use the term "bond *energy*" rather than "bond *enthalpy*" because it is common practice to do so.

The **bond energy (B.E.)** is the amount of energy necessary to break *one mole* of bonds in a gaseous covalent substance to form products in the gaseous state at constant temperature and pressure.

We have discussed these changes in terms of absorption or release of heat. Another way of breaking bonds is by absorption of light energy (Chapter 5). Bond energies can be determined from the energies of the photons that cause bond dissociation.

Consider the following reaction.

$$H_2(g) \longrightarrow 2H(g) \qquad \Delta H^0_{rxn} = \Delta H_{H—H} = +435 \text{ kJ/mol H—H bonds}$$

The bond energy of the hydrogen–hydrogen bond is 435 kJ/mol of bonds. This endothermic reaction (ΔH^0_{rxn} is positive) can be written

$$H_2(g) + 435 \text{ kJ} \longrightarrow 2H(g)$$

Some average bond energies are listed in Tables 15-2 and 15-3.

Table 15-2 *Some Average Single Bond Energies in kJ/mol of Bonds*

H	C	N	O	F	Si	P	S	Cl	Br	I	
435	414	389	464	569	293	318	339	431	368	297	H
	347	293	351	439	289	264	259	330	276	238	C
		159	201	272	—	209	—	201	243?	—	N
			138	184	368	351	—	205	—	201	O
				159	540	490	327	255	197?	—	F
					176	213	226	360	289	213	Si
						213	230	331	272	213	P
							213	251	213	—	S
								243	218	209	Cl
									192	180	Br
										151	I

Let us consider more complex reactions. For example,

$$CH_4(g) \longrightarrow C(g) + 4H(g) \qquad \Delta H^0_{rxn} = 1.66 \times 10^3 \text{ kJ/mol rxn}$$

The four hydrogen atoms are identical, so all the C—H bonds are identical in bond length and energy *in methane molecules*. However, the energies required to break the individual C—H bonds differ for successively broken bonds, as shown below.

$CH_4(g) \longrightarrow CH_3(g) + H(g)$	$\Delta H^0 =$	$+ \ 427$ kJ/mol rxn
$CH_3(g) \longrightarrow CH_2(g) + H(g)$	$\Delta H^0 =$	$+ \ 439$ kJ/mol rxn
$CH_2(g) \longrightarrow CH(g) \ + H(g)$	$\Delta H^0 =$	$+ \ 452$ kJ/mol rxn
$CH(g) \longrightarrow C(g) \ \ \ + H(g)$	$\Delta H^0 =$	$+ \ 347$ kJ/mol rxn
$CH_4(g) \longrightarrow C(g) \ \ \ + 4H(g)$	$\Delta H^0 =$	$+ 1665$ kJ/mol rxn

The hydrogen atoms are indistinguishable, so it makes no difference which one is removed first.

"average" C—H bond energy in $CH_4 = \Delta H^0/4$ mol bonds
$= 416$ kJ/mol bonds

We see that the *average C—H bond energy in methane* is 416 kJ/mol of bonds. No mole of single C—H bonds is actually broken by absorption of exactly that amount of energy. Average C—H bond energies differ slightly from compound to compound, as in CH_4, CH_3Cl, CH_3NO_2, and so on. Nevertheless, they are sufficiently constant to be useful in estimating thermodynamic data that are not readily available for another approach. Values of ΔH^0_{rxn} estimated in this way are not as reliable as those obtained from ΔH^0_f values for the substances involved in the reaction.

A special case of Hess' Law involves the use of bond energies to *estimate* heats of reaction. Consider the enthalpy diagrams in Figure 15-5. In general terms, ΔH^0_{rxn} is related to the bond energies of the reactants and products *in gas phase reactions* by the following version of Hess' Law.

Bond energies for double and triple bonds are not simply two or three times those for the corresponding single bonds. A single bond is a σ bond, whereas double and triple bonds involve a combination of σ and π bonding. The bond energy measures the difficulty of overcoming the orbital overlap, and we should not expect the stability of a π bond to be the same as that of a σ bond between the same two atoms.

Table 15-3 *Some Average Multiple Bond Energies in kJ/mol of Bonds*

N=N	418	C=C	611	O=O	498
N≡N	946	C≡C	837		
C=N	615	C=O	741		
C≡N	891	C≡O	1070		

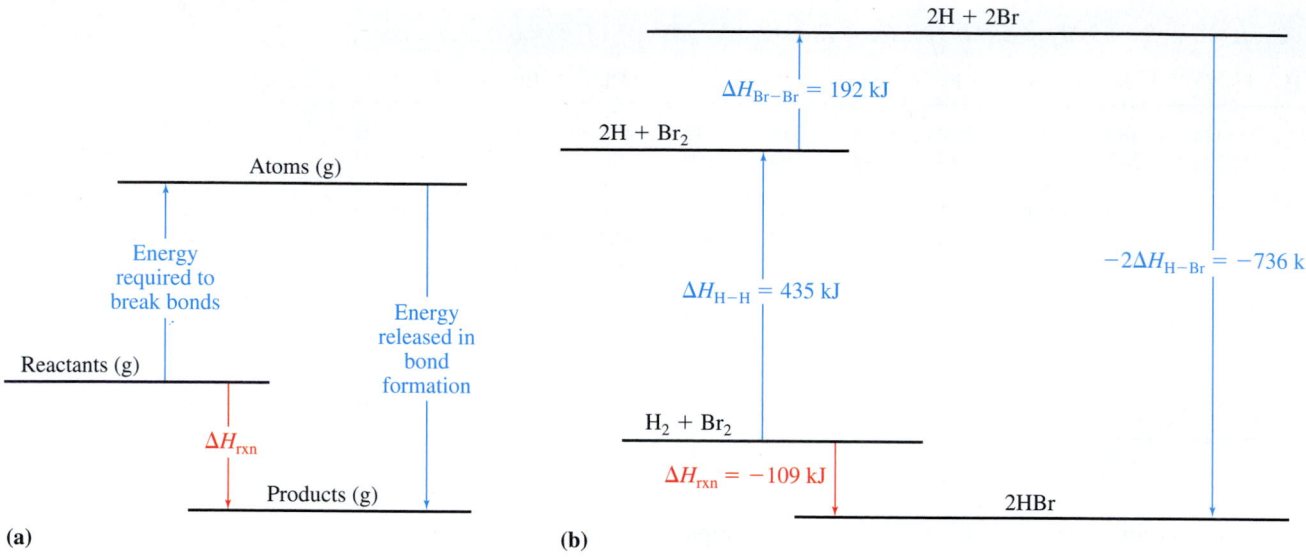

Figure 15-5   A schematic representation of the relationship between bond energies and ΔH_{rxn} for gas phase reactions. (a) For a general reaction (exothermic). (b) For the gas phase reaction

$$H_2(g) + Br_2(g) \longrightarrow 2HBr(g)$$

As usual for such diagrams, the value shown for each change refers to the number of moles of substances or bonds indicated in the diagram.

Note that this equation involves bond energies of *reactants* minus bond energies of *products*.

$$\Delta H^0_{rxn} = \Sigma \text{ B.E.}_{reactants} - \Sigma \text{ B.E.}_{products} \qquad \text{in gas phase reactions only}$$

The net enthalpy change of a reaction is the amount of energy required to break all the bonds in reactant molecules *minus* the amount of energy required to break all the bonds in product molecules. Stated in another way, the amount of energy released when a bond is formed is equal to the amount absorbed when the same bond is broken. The heat of reaction for a gas phase reaction can be described as the amount of energy released in forming all the bonds in the products minus the amount of energy released in forming all the bonds in the reactants (Figure 15-5).

The definition of bond energies is limited to the bond-breaking process *only*, and does not include any provision for changes of state. Thus, it is valid only for substances in the gaseous state. Therefore, the calculations of this section apply *only* when all substances in the reaction are gases. If liquids or solids were involved, then additional information such as heats of vaporization and fusion would be needed to account for phase changes.

EXAMPLE 15-10   *Bond Energies*

Use the bond energies listed in Table 15-2 to estimate the heat of reaction at 298 K for the following reaction. All bonds are single bonds.

$$Br_2(g) + 3F_2(g) \longrightarrow 2BrF_3(g)$$

Plan

Each BrF_3 molecule contains three Br—F bonds, so two moles of BrF_3 contain six moles of Br—F bonds. Three moles of F_2 contain a total of three moles of F—F bonds, and one mole of Br_2 contains one mole of Br—Br bonds. Using the bond energy form of Hess' Law,

$$F—\overset{..}{\underset{|}{Br}}\!\!\!\overset{..}{}—F$$
$$\underset{F}{|}$$

Solution

$$\Delta H^0_{rxn} = [\Delta H_{Br—Br} + 3\Delta H_{F—F}] - [6\Delta H_{Br—F}]$$

$$= 192 + 3(159) - 6(197) = \boxed{-513 \text{ kJ/mol rxn}}$$

You should now work Exercise 38.

For each term in the sum, the units are

$$\frac{\text{mol bonds}}{\text{mol rxn}} \times \frac{\text{kJ}}{\text{mol bonds}}$$

EXAMPLE 15-11 *Bond Energies*

Use the bond energies listed in Table 15-2 to estimate the heat of reaction at 298 K for the following reaction.

$$C_3H_8(g) \qquad + Cl_2(g) \longrightarrow \qquad C_3H_7Cl(g) \quad + HCl(g)$$

$$\underset{\underset{H}{|}}{\overset{\overset{H}{|}}{H—C}}\!\!-\!\!\underset{\underset{H}{|}}{\overset{\overset{H}{|}}{C}}\!\!-\!\!\underset{\underset{H}{|}}{\overset{\overset{H}{|}}{C}}\!\!-\!\!H + Cl—Cl \longrightarrow \underset{\underset{H}{|}}{\overset{\overset{H}{|}}{H—C}}\!\!-\!\!\underset{\underset{H}{|}}{\overset{\overset{H}{|}}{C}}\!\!-\!\!\underset{\underset{H}{|}}{\overset{\overset{H}{|}}{C}}\!\!-\!\!Cl + H—Cl$$

Plan

Two moles of C—C bonds and seven moles of C—H bonds are the same before and after reaction, so we do not need to include them in the bond energy calculation. The only reactant bonds that are broken are one mole of C—H bonds and one mole of Cl—Cl bonds. On the product side, the only new bonds formed are one mole of C—Cl bonds and one mole of H—Cl bonds. We need to take into account only the bonds that are different on the two sides of the equation. As before, we add and subtract the appropriate bond energies, using values from Table 15-2.

We would get the same value for ΔH^0_{rxn} if we used the full bond energy form of Hess' Law and assumed that *all* bonds in reactants were broken and then *all* bonds in products were formed. In such a calculation the bond energies for the unchanged bonds would cancel. Why? Try it!

Solution

$$\Delta H^0_{rxn} = [\Delta H_{C—H} + \Delta H_{Cl—Cl}] - [\Delta H_{C—Cl} + \Delta H_{H—Cl}]$$

$$= [414 + 243] - [330 + 431] = \boxed{-104 \text{ kJ/mol rxn}}$$

You should now work Exercises 40 and 41.

15-10 CHANGES IN INTERNAL ENERGY, ΔE

The **internal energy,** E, of a specific amount of a substance represents all the energy contained within the substance. It includes such forms as kinetic energies of the molecules; energies of attraction and repulsion among subatomic particles, atoms, ions, or molecules; and other forms of energy. The internal energy of a collection of molecules is a state function. The difference between the internal energy of the products and the internal energy of the reactants of a chemical reaction or physical change, ΔE, is given by the equation

Internal energy is a state function, so it is represented by a capital letter.

In some older texts, this equation was written in the form $\Delta E = q - w$, using the reverse convention for the sign of w. In this text we use the convention of physical chemistry. Heat added to the system or work done on the system would increase its energy.

$$\Delta E = E_{final} - E_{initial} = E_{products} - E_{reactants} = q + w$$

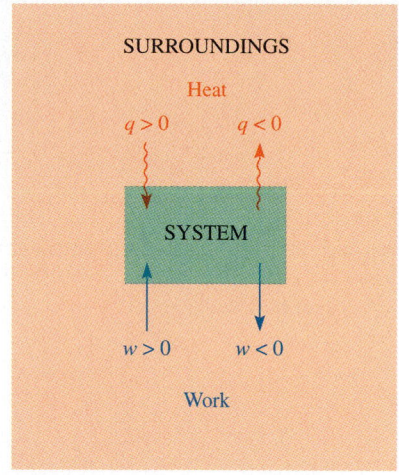

Sign conventions for q and w.

At 25°C the change in internal energy for the combustion of methane is -887 kJ/mol CH_4. The change in heat content is -890 kJ/mol CH_4 (Section 15-1). The small difference is due to work done on the system as it is compressed by the atmosphere.

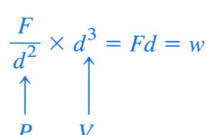

The terms q and w represent heat and work, respectively. These are two ways in which energy can flow into or out of a system. **Work** involves a change of energy in which a body is moved through a distance, d, against some force, f; that is, $w = fd$.

$$\Delta E = (\text{amount of heat absorbed by system}) + (\text{amount of work done on system})$$

The following conventions apply to the signs of q and w.

q is positive:	Heat is *absorbed* by the system from the surroundings.
q is negative:	Heat is *released* by the system to the surroundings.
w is positive:	Work is done *on* the system by the surroundings.
w is negative:	Work is done *by* the system on the surroundings.

In other words, whenever energy is added to a system, either as heat or as work, the energy of the system increases.

When energy is released by a reacting system, ΔE is negative; energy can be written as a product in the equation for the reaction. When the system absorbs energy from the surroundings, ΔE is positive; energy can be written as a reactant in the equation.

For example, the complete combustion of CH_4 at constant volume at 25°C *releases* energy.

$$CH_4(g) + 2O_2(g) \longrightarrow CO_2(g) + 2H_2O(\ell) + 887 \text{ kJ}$$

indicates release of energy

We can write the *change in energy* that accompanies this reaction as

$$CH_4(g) + 2O_2(g) \longrightarrow CO_2(g) + 2H_2O(\ell) \qquad \Delta E = -887 \text{ kJ/mol rxn}$$

As discussed in Section 15-2, the negative sign indicates a *decrease* in energy of the system, or a *release* of energy by the system.

The reverse of this reaction *absorbs* energy. It can be written as

$$CO_2(g) + 2H_2O(\ell) + 887 \text{ kJ} \longrightarrow CH_4(g) + 2O_2(g)$$

indicates absorption of energy

or

$$CO_2(g) + 2H_2O(\ell) \longrightarrow CH_4(g) + 2O_2(g) \qquad \Delta E = +887 \text{ kJ/mol rxn}$$

If the latter reaction could be forced to occur, the system would have to absorb 887 kJ of energy per mole of reaction from its surroundings.

The only type of work involved in most chemical and physical changes is pressure–volume work. From dimensional analysis we can see that the product of pressure and volume is work. Pressure is the force exerted per unit area, where area is distance squared, d^2; volume is distance cubed, d^3. Thus, the product of pressure and volume is force times distance, which is work. When a gas is produced against constant external pressure, such as in an open vessel at atmospheric pressure, the gas does work as it expands against the pressure of the atmosphere. If no heat is absorbed, this results in a decrease in the internal energy of the system. On the other hand, when a gas is consumed in a reaction, the atmosphere does work on the reacting system.

Let us illustrate the latter case. Consider the complete reaction of a 2:1 mole ratio of H_2 and O_2 to produce steam at some constant temperature above 100°C and at one atmosphere pressure (Figure 15-6).

$$2H_2(g) + O_2(g) \longrightarrow 2H_2O(g) + \text{heat}$$

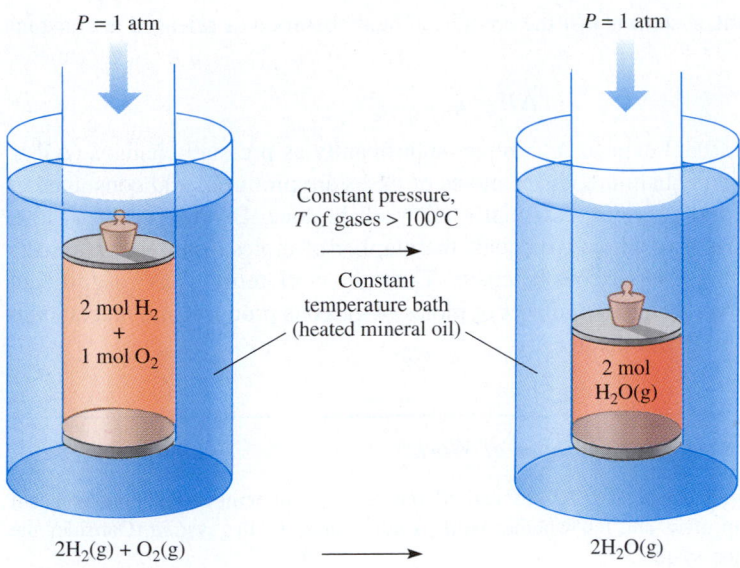

$2H_2(g) + O_2(g) \longrightarrow 2H_2O(g)$

Figure 15-6 An illustration of the one-third decrease in volume that accompanies the reaction of H_2 with O_2 at constant temperature. The temperature is above 100°C.

Assume that the constant-temperature bath surrounding the reaction vessel completely absorbs all the evolved heat so that the temperature of the gases does not change. The volume of the system decreases by one third (3 mol gaseous reactants → 2 mol gaseous products). The surroundings exert a constant pressure of one atmosphere and do work on the system by compressing it. The internal energy of the system increases by an amount equal to the amount of work done on it.

The work done on or by a system depends on the *external* pressure and the volume. When the external pressure is constant during a change, the amount of work done is equal to this pressure times the change in volume. The work done *on* a system equals $-P\Delta V$ or $-P(V_2 - V_1)$.

V_2 is the final volume and V_1 is the initial volume.

Compression (volume decreases)	Expansion (volume increases)
Work is done *by* the surroundings *on* the system, so the sign of w is positive	Work is done *by* the system *on* the surroundings, so the sign of w is negative
V_2 is less than V_1, so $\Delta V = (V_2 - V_1)$ is negative	V_2 is greater than V_1, so $\Delta V = (V_2 - V_1)$ is positive
$w = -P\Delta V$ is positive $(-) \times (+) \times (-) = +$	$w = -P\Delta V$ is negative $(-) \times (+) \times (+) = -$
Can be due to a *decrease* in number of moles of gas (Δn negative)	Can be due to an *increase* in number of moles of gas (Δn positive)

We substitute $-P\Delta V$ for w in the equation $\Delta E = q + w$ to obtain

$$\Delta E = q - P\Delta V$$

In constant-volume reactions, no $P\Delta V$ work is done. The absence of a change in volume means that nothing "moves through a distance," so $d = 0$ and $fd = 0$. The change

Do not make the error of setting work equal to $V\Delta P$.

in internal energy of the system is just the amount of heat absorbed or released at constant volume, q_v.

$$\Delta E = q_v$$

A subscript v indicates a constant-volume process; a subscript p indicates a constant-pressure process, and so on.

Solids and liquids do not expand or contract significantly as pressure changes ($\Delta V \approx 0$). In reactions in which equal numbers of moles of gases are produced and consumed at constant temperature and pressure, essentially no work is done. By the ideal gas law, $P\Delta V = (\Delta n)RT$ and $\Delta n = 0$, where Δn equals the number of moles of gaseous products minus the number of moles of gaseous reactants. Thus, the work term w has a significant value only when there are different numbers of moles of gaseous products and reactants so that the volume of the system changes.

EXAMPLE 15-12 *Predicting the Sign of Work*

For each of the following chemical reactions carried out at constant temperature and constant pressure, predict the sign of w, and tell whether work is done *on* or *by* the system. Consider the reaction mixture to be the system.

(a) Ammonium nitrate, commonly used as a fertilizer, decomposes explosively.

$$2NH_4NO_3(s) \longrightarrow 2N_2(g) + 4H_2O(g) + O_2(g)$$

This reaction was responsible for an explosion in 1947 that destroyed nearly the entire port of Texas City, Texas, and killed 576 people.

(b) The combination of hydrogen and chlorine forms hydrogen chloride gas.

$$H_2(g) + Cl_2(g) \longrightarrow 2HCl(g)$$

(c) The oxidation of sulfur dioxide to sulfur trioxide is one step in the production of sulfuric acid.

$$2SO_2(g) + O_2(g) \longrightarrow 2SO_3(g)$$

Plan

Δn refers to the balanced equation.

For a process at constant pressure, $w = -P\Delta V = -(\Delta n)RT$. For each reaction, we evaluate Δn, the change in the number of moles of *gaseous* substances in the reaction.

$$\Delta n = \text{no. of moles of gaseous products} - \text{no. of moles of gaseous reactants}$$

Because both R and T (on the Kelvin scale) are positive quantities, the sign of w is opposite from that of Δn; it tells us whether the work is done *on* ($w = +$) or *by* ($w = -$) the system.

Solution

Here there are no gaseous reactants.

(a) $\Delta n = [2 \text{ mol } N_2(g) + 4 \text{ mol } H_2O(g) + 1 \text{ mol } O_2(g)] - 0 \text{ mol}$
    $= 7 \text{ mol} - 0 \text{ mol} = +7 \text{ mol}$

Δn is positive, so **w is negative.** This tells us that **work is done *by* the system.** The large amount of gas formed by the reaction pushes against the surroundings (as happened with devastating effect in the Texas City disaster).

(b) $\Delta n = [2 \text{ mol } HCl(g)] - [1 \text{ mol } H_2(g) + 1 \text{ mol } Cl_2(g)]$
    $= 2 \text{ mol} - 2 \text{ mol} = 0 \text{ mol}$

Thus, **$w = 0$,** and **no work is done** as the reaction proceeds. We can see from the balanced equation that for every two moles (total) of gas that react, two moles of gas are formed, so the volume neither expands nor contracts as the reaction occurs.

(c) $\Delta n = [2 \text{ mol } SO_3(g)] - [2 \text{ mol } SO_2(g) + 1 \text{ mol } O_2(g)]$
    $= 2 \text{ mol} - 3 \text{ mol} = -1 \text{ mol}$

Δn is negative, so w is positive. This tells us that work is done *on* the system as the reaction proceeds. The surroundings push against the diminishing volume of gas.

You should now work Exercises 61 and 62.

A bomb calorimeter is a device that measures the amount of heat evolved or absorbed by a reaction occurring at constant volume (Figure 15-7). A strong steel vessel (the bomb) is immersed in a large volume of water. As heat is produced or absorbed by a reaction inside the steel vessel, the heat is transferred to or from the large volume of water. Thus, only rather small temperature changes occur. For all practical purposes, the energy changes associated with the reactions are measured at constant volume and constant temperature. No work is done when a reaction is carried out in a bomb calorimeter, even if gases are involved, because $\Delta V = 0$. Therefore,

$$\Delta E = q_v \qquad \text{(constant volume)}$$

The "calorie content" of a food can be determined by burning it in excess oxygen inside a bomb calorimeter and determining the heat released.

EXAMPLE 15-13 *Bomb Calorimeter*

A 1.000-gram sample of ethanol, C_2H_5OH, was burned in a bomb calorimeter whose heat capacity had been determined to be 2.71 kJ/°C. The temperature of 3000 grams of water rose from 24.284°C to 26.225°C. Determine ΔE for the reaction in joules per gram of ethanol, then in kilojoules per mole of ethanol. The specific heat of water is 4.184 J/g · °C. The combustion reaction is

$$C_2H_5OH(\ell) + 3O_2(g) \longrightarrow 2CO_2(g) + 3H_2O(\ell)$$

Benzoic acid, C_6H_5COOH, is often used to determine the heat capacity of a calorimeter. It is a solid that can be compressed into pellets. Its heat of combustion is accurately known: 3227 kJ/mol benzoic acid, or 26.46 kJ/g benzoic acid. Another way to measure the heat capacity of a calorimeter is to add a known amount of heat electrically.

Plan

The amount of heat given off by the system (in the sealed compartment) raises the temperature of the calorimeter and its water. The amount of heat absorbed by the water can be calculated using the specific heat of water; similarly, we use the heat capacity of the calorimeter to find the amount of heat absorbed by the calorimeter. The sum of these two amounts of heat is the total amount of heat released by the combustion of 1.000 gram of ethanol. We must then scale that result to correspond to one mole of ethanol.

Solution

The increase in temperature is

$$\underline{?}°C = 26.225°C - 24.284°C = 1.941°C \text{ rise}$$

The amount of heat responsible for this increase in temperature of 3000 grams of water is

$$\text{heat to warm water} = 1.941°C \times \frac{4.184 \text{ J}}{\text{g} \cdot °C} \times 3000 \text{ g} = 2.436 \times 10^4 \text{ J} = 24.36 \text{ kJ}$$

The amount of heat responsible for the warming of the calorimeter is

$$\text{heat to warm calorimeter} = 1.941°C \times \frac{2.71 \text{ kJ}}{°C} = 5.26 \text{ kJ}$$

The total amount of heat absorbed by the calorimeter *and* the water is

$$\text{total amount of heat} = 24.36 \text{ kJ} + 5.26 \text{ kJ} = 29.62 \text{ kJ}$$

Combustion of one gram of C_2H_5OH liberates 29.62 kJ of energy in the form of heat, i.e.,

$$\Delta E = q_v = \boxed{-29.62 \text{ kJ/g ethanol}}$$

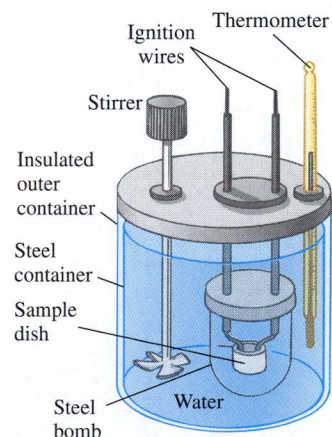

Figure 15-7 A bomb calorimeter measures q_v, the amount of heat given off or absorbed by a reaction occurring at constant *volume*. The amount of energy introduced via the ignition wires is measured and taken into account.

The negative sign indicates that energy is released by the system to the surroundings. Now we may evaluate ΔE in kJ/mol of ethanol by converting grams of C_2H_5OH to moles.

$$\frac{?\ kJ}{mol\ ethanol} = \frac{-29.62\ kJ}{g} \times \frac{46.07\ g\ C_2H_5OH}{1\ mol\ C_2H_5OH} = -1365\ kJ/mol\ ethanol$$

$$\Delta E = \boxed{-1365\ kJ/mol\ ethanol}$$

This calculation shows that for the combustion of ethanol at constant temperature and constant volume, the change in internal energy is -1365 kJ/mol ethanol.

You should now work Exercises 54 and 55.

The balanced chemical equation involves one mole of ethanol, so we can write the unit factor $\dfrac{1\ mol\ ethanol}{1\ mol\ rxn}$. Then we express the result of Example 15-13 as

$$\Delta E = \frac{-1365\ kJ}{mol\ ethanol} \times \frac{1\ mol\ ethanol}{1\ mol\ rxn} = -1365\ kJ/mol\ rxn$$

15-11 RELATIONSHIP BETWEEN ΔH AND ΔE

The fundamental definition of enthalpy, H, is

$$H = E + PV$$

For a process at constant temperature and pressure,

$$\Delta H = \Delta E + P\Delta V \qquad \text{(constant } T \text{ and } P\text{)}$$

From Section 15-10, we know that $\Delta E = q + w$, so

$$\Delta H = q + w + P\Delta V \qquad \text{(constant } T \text{ and } P\text{)}$$

At constant pressure, $w = -P\Delta V$, so

$$\Delta H = q + (-P\Delta V) + P\Delta V$$

$$\Delta H = q_p \qquad \text{(constant } T \text{ and } P\text{)}$$

The difference between ΔE and ΔH is the amount of expansion work ($P\Delta V$ work) that the system can do. Unless there is a change in the number of moles of gas present, this difference is extremely small and can usually be neglected. For an ideal gas, $PV = nRT$. At constant temperature and constant pressure, $P\Delta V = (\Delta n)RT$, a work term. Substituting gives

As usual, Δn refers to the number of moles of *gaseous products* minus the number of moles of *gaseous reactants*.

$$\Delta H = \Delta E + (\Delta n)RT \qquad \text{or} \qquad \Delta E = \Delta H - (\Delta n)RT \qquad \text{(constant } T \text{ and } P\text{)}$$

In Example 15-13 we found that the change in internal energy, ΔE, for the combustion of ethanol is -1365 kJ/mol ethanol at 298 K. Combustion of one mole of ethanol at 298 K and constant pressure releases 1367 kJ of heat. Therefore (Section 15-5)

$$\Delta H = -1367 \, \frac{\text{kJ}}{\text{mol ethanol}}$$

The difference between ΔH and ΔE is due to the work term, $-P\Delta V$ or $-(\Delta n)RT$. In this reaction there are fewer moles of gaseous products than of gaseous reactants: $\Delta n = 2 - 3 = -1$.

$$C_2H_5OH(\ell) + 3O_2(g) \longrightarrow 2CO_2(g) + 3H_2O(\ell)$$

Thus, the atmosphere does work on the system (compresses it). Let us find the work done on the system per mole of reaction.

$$w = -P\Delta V = -(\Delta n)RT$$

$$w = -(\Delta n)RT = -(-1 \text{ mol})\left(\frac{8.314 \text{ J}}{\text{mol} \cdot \text{K}}\right)(298 \text{ K}) = +2.48 \times 10^3 \text{ J}$$

$$w = +2.48 \text{ kJ} \qquad \text{or} \qquad (\Delta n)RT = -2.48 \text{ kJ}$$

We can now calculate ΔE for the reaction from ΔH and $(\Delta n)RT$ values.

$$\Delta E = \Delta H - (\Delta n)RT = [-1367 - (-2.48)] = -1365 \text{ kJ/mol rxn}$$

The positive sign is consistent with the fact that work is done on the system. The balanced equation involves one mole of ethanol, so this is the amount of work done when one mole of ethanol undergoes combustion.

This value agrees with the result that we obtained in Example 15-13. The size of the work term ($+2.48$ kJ) is very small compared with ΔH (-1367 kJ/mol rxn). This is true for many reactions. Of course, if $\Delta n = 0$, then $\Delta H = \Delta E$, and the same amount of heat would be absorbed or given off by the reaction whether it is carried out at constant pressure or at constant volume.

SPONTANEITY OF PHYSICAL AND CHEMICAL CHANGES

Another major concern of thermodynamics is predicting *whether* a particular process can occur under specified conditions. We may summarize this concern in the question "Which would be more stable at the given conditions—the reactants or the products?" A change for which the collection of products is thermodynamically *more stable* than the collection of reactants under the given conditions is said to be **spontaneous** under those conditions. A change for which the products are thermodynamically *less stable* than the reactants under the given conditions is described as **nonspontaneous** under those conditions. Some changes are spontaneous under all conditions; others are nonspontaneous under all conditions. The great majority of changes, however, are spontaneous under some conditions but not under others. We use thermodynamics to predict conditions for which the latter type of reactions can occur.

The concept of spontaneity has a very specific interpretation in thermodynamics. A spontaneous chemical reaction or physical change is one that can happen without any continuing outside influence. Any spontaneous change has a natural direction, like the rusting of a piece of iron, the burning of a piece of paper, or the melting of ice at room temperature. We can think of a spontaneous process as one for which products are favored over reactants at the specified conditions. Although a spontaneous reaction *might* occur

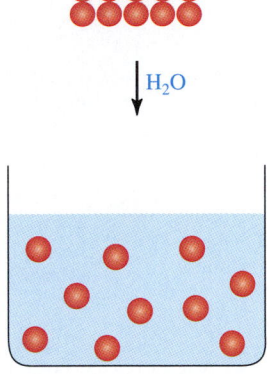

Figure 15-8 As particles leave a crystal to go into solution, they become more disordered. This increase in disorder favors the dissolution of the crystal.

rapidly, thermodynamic spontaneity is not related to speed. The fact that a process is spontaneous does not mean that it will occur at an observable rate. It may occur rapidly, at a moderate rate, or very slowly. The rate at which a spontaneous reaction occurs is addressed by kinetics (Chapter 16). We now study the factors that influence spontaneity of a physical or chemical change.

15-12 THE TWO PARTS OF SPONTANEITY

Many spontaneous reactions are exothermic. For instance, the combustion (burning) reactions of hydrocarbons such as methane and octane are all exothermic and spontaneous. The enthalpy contents of the products are lower than those of the reactants. However, not all exothermic changes are spontaneous, nor are all spontaneous changes exothermic. As an example, consider the freezing of water, which is an exothermic process (heat is released). This process is spontaneous at temperatures below 0°C, but it certainly is not spontaneous at temperatures above 0°C. Likewise, we can find conditions at which the melting of ice, an endothermic process, is spontaneous. Spontaneity is *favored* but not required when heat is released during a chemical reaction or a physical change.

Another factor, related to the disorder of reactants and products, also plays a role in determining spontaneity. The dissolution of ammonium nitrate, NH_4NO_3, in water is spontaneous. Yet a beaker in which this process occurs becomes colder (Figure 15-2). The system (consisting of the water, the solid NH_4NO_3, and the resulting hydrated NH_4^+ and NO_3^- ions) absorbs heat from the surroundings as the endothermic process occurs. Nevertheless, the process is spontaneous because the system becomes more disordered as the regularly arranged ions of crystalline ammonium nitrate become more randomly distributed hydrated ions in solution (Figure 15-8). An increase in disorder in the system favors the spontaneity of a reaction. In this particular case, the increase in disorder overrides the effect of endothermicity. The balance of these two effects is considered in Section 15-15.

15-13 THE SECOND LAW OF THERMODYNAMICS

We now know that two factors determine whether a reaction is spontaneous under a given set of conditions. The effect of one factor, the enthalpy change, is that spontaneity is favored (but not required) by exothermicity, and nonspontaneity is favored (but not required) by endothermicity. The effect of the other factor is summarized by the **Second Law of Thermodynamics.**

Suppose we shake a beaker containing marbles. A disordered arrangement (left) is more likely than an ordered arrangement (right) in which all marbles of the same color remain together.

In spontaneous changes the universe tends toward a state of greater disorder.

The Second Law of Thermodynamics is based on our experiences. Some examples illustrate this law in the macroscopic world. When a mirror is dropped, it can shatter. When a drop of food coloring is added to a glass of water, it diffuses until a homogeneously colored solution results. When a truck is driven down the street, it consumes fuel and oxygen, producing carbon dioxide, water vapor, and other emitted substances.

The reverse of any spontaneous change is nonspontaneous, because if it did occur, the universe would tend toward a state of greater order. This is contrary to our experience. We would be very surprised if we dropped some pieces of silvered glass on the floor and a mirror spontaneously assembled. A truck cannot be driven along the street, even in reverse gear, so that it sucks up CO_2, water vapor, and other substances and produces fuel and oxygen.

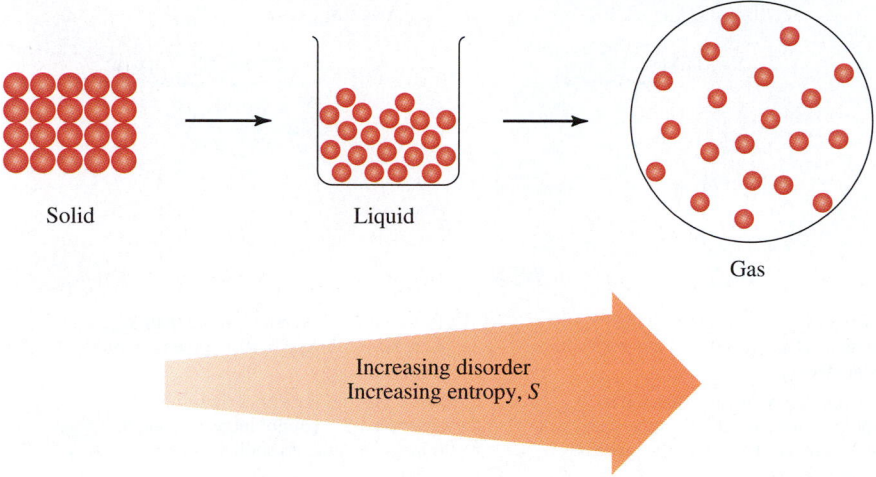

Solid

Liquid

Gas

Increasing disorder
Increasing entropy, S

Figure 15-9 As a sample changes from solid to liquid to gas, its particles become increasingly less ordered (more disordered), so its entropy increases.

15-14 ENTROPY, S

The thermodynamic state function **entropy, S,** is a measure of the disorder of the system. The greater the disorder of a system, the higher its entropy. For any substance, the particles are more highly ordered in the solid state than in the liquid state. These, in turn, are more highly ordered than in the gaseous state. Thus, the entropy of any substance increases as the substance goes from solid to liquid to gas (Figure 15-9).

If the entropy of a system increases during a process, the spontaneity of the process is favored but not required. The Second Law of Thermodynamics says that the entropy of the *universe* (not the system) increases during a spontaneous process, i.e.,

$$\Delta S_{universe} = \Delta S_{system} + \Delta S_{surroundings} > 0 \qquad \text{(spontaneous process)}$$

Of the two ideal gas samples in Figure 15-10, the more ordered arrangement (Figure 15-10a) has lower entropy than the randomly mixed arrangement with the same volume (Figure 15-10b). Because these ideal gas samples mix without absorbing or releasing heat and without a change in total volume, they do not interact with the surroundings, so the entropy of the surroundings does not change. In this case

$$\Delta S_{universe} = \Delta S_{system}$$

If we open the stopcock between the two bulbs in Figure 15-10a, we expect the gases to mix spontaneously, with an increase in the disorder of the system, that is, ΔS_{system} is positive.

unmixed gases $\longrightarrow$ mixed gases $\qquad \Delta S_{universe} = \Delta S_{system} > 0$

We do not expect the more homogeneous sample in Figure 15-10b to spontaneously "unmix" to give the arrangement in Figure 15-10a (which would correspond to a decrease in ΔS_{system}).

mixed gases $\longrightarrow$ unmixed gases $\qquad \Delta S_{universe} = \Delta S_{system} < 0$

The ideas of entropy, order, and disorder are related to probability. The more ways an event can happen, the more probable that event is. In Figure 15-10b each individual red molecule is equally likely to be in either container, as is each individual blue molecule. As a result, there are many ways in which the mixed arrangement of Figure 15-10b can occur,

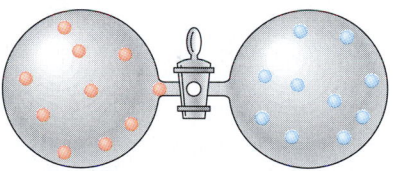

(a) Stopcock closed

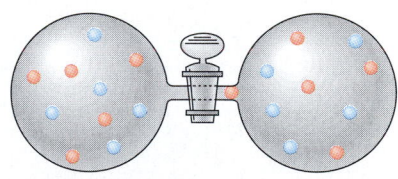

(b) Stopcock open

Figure 15-10 (a) A sample of gas in which all molecules of one gas are in one bulb and all molecules of the other gas are in the other bulb. (b) A sample of gas that contains the same number of each kind of molecule as in (a), but with the two kinds randomly mixed in the two bulbs. Sample (b) has greater disorder (higher entropy), and is thus more probable.

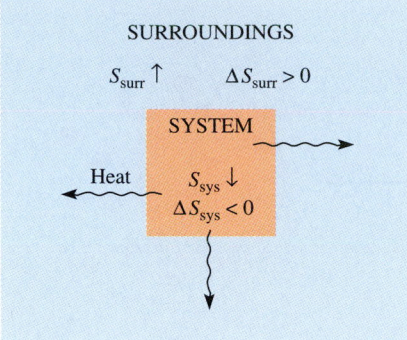

(a) Freezing below mp

SURROUNDINGS

$S_{surr} \uparrow$ $\Delta S_{surr} > 0$

SYSTEM

Heat $S_{sys} \downarrow$
$\Delta S_{sys} < 0$

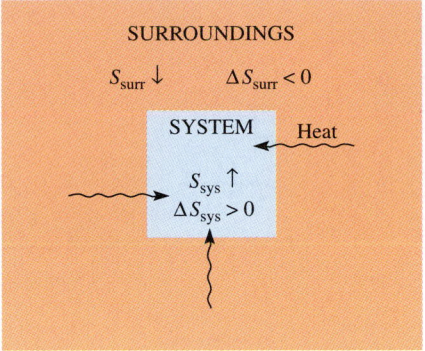

(b) Melting above mp

SURROUNDINGS

$S_{surr} \downarrow$ $\Delta S_{surr} < 0$

SYSTEM   Heat

$S_{sys} \uparrow$
$\Delta S_{sys} > 0$

Below mp, S_{surr} *increases* more than S_{sys} decreases, so $\Delta S_{univ} > 0$, and the process is spontaneous.

Above mp, S_{surr} is already high, so it cannot increase enough to compensate for the decrease in S_{sys}, and the process is not spontaneous.

Above mp, S_{sys} *increases* more than S_{surr} decreases, so $\Delta S_{univ} > 0$, and the process is spontaneous.

Below mp, too little heat flows into the system, so S_{sys} cannot increase enough to make the process spontaneous.

Figure 15-11 A schematic representation of heat flow and entropy changes for (a) freezing and (b) melting of a pure substance.

so the probability of its occurrence is high, and so its entropy is high. In contrast, there is only one way the unmixed arrangement in Figure 15-10a can occur. The resulting probability is extremely low, and the entropy of this arrangement is low.

The entropy of a system can decrease during a spontaneous process or increase during a nonspontaneous process, depending on the accompanying $\Delta S_{surroundings}$. If ΔS_{sys} is negative (decrease in disorder), then ΔS_{univ} may still be positive (overall increase in disorder) *if* ΔS_{surr} is more positive than ΔS_{sys} is negative. A refrigerator provides an illustration. It removes heat from inside the box (the system) and ejects that heat, *plus* the heat generated by the compressor, into the room (the surroundings). The entropy of the system decreases because the air molecules inside the box move more slowly. However, the increase in the entropy of the surroundings more than makes up for that, so the entropy of the universe (refrigerator + room) increases.

Similarly, if ΔS_{sys} is positive but ΔS_{surr} is even more negative, then ΔS_{univ} is still negative. Such a process will be nonspontaneous.

Let's consider the entropy changes that occur when a liquid solidifies at a temperature *below* its freezing (melting) point (Figure 15-11). ΔS_{sys} is negative because a solid forms from its liquid, yet we know that this is a spontaneous process. A liquid releases heat to its surroundings (atmosphere) as it crystallizes. The released heat increases the motion (disorder) of the molecules of the surroundings, so ΔS_{surr} is positive. As the temperature decreases, the ΔS_{surr} contribution becomes more important. When the temperature is low enough (below the freezing point), the positive ΔS_{surr} outweighs the negative ΔS_{sys}. Then ΔS_{univ} becomes positive and the freezing process becomes spontaneous.

The situation is reversed when a liquid is boiled or a solid is melted (Figure 15-11b). For example, at temperatures above its melting point, a solid spontaneously melts, and ΔS_{sys} is positive. The heat absorbed when the solid (system) melts comes from its surroundings. This decreases the motion of the molecules of the surroundings. Thus, ΔS_{surr} is negative (the surroundings become less disordered). However, the positive ΔS_{sys} is greater in magnitude than the negative ΔS_{surr}, so ΔS_{univ} is positive and the process is spontaneous.

Above the melting point, ΔS_{univ} is positive for melting. Below the melting point, ΔS_{univ} is positive for freezing. At the melting point, ΔS_{surr} is equal in magnitude and opposite in sign to ΔS_{sys}. Then ΔS_{univ} is zero for both melting and freezing; the system is at *equilibrium.* Table 15-4 lists the entropy effects for these changes of physical state.

We shall abbreviate these subscripts as follows: system = sys, surroundings = surr, and universe = univ.

Similar arguments apply for condensing a gas at its condensation point (boiling point).

Table 15-4 *Entropy Effects Associated with Melting and Freezing*

Change	Temperature	Sign of ΔS_{sys}	Sign of ΔS_{surr}	(Magnitude of ΔS_{sys}) Compared with (Magnitude of ΔS_{surr})	$\Delta S_{univ} = \Delta S_{sys} + \Delta S_{surr}$	Spontaneity
1. Melting (solid → liquid)	(a) >mp	+	−	>	>0	Spontaneous
	(b) =mp	+	−	=	=0	Equilibrium
	(c) <mp	+	−	<	<0	Nonspontaneous
2. Freezing (liquid → solid)	(a) >mp	−	+	>	<0	Nonspontaneous
	(b) =mp	−	+	=	=0	Equilibrium
	(c) <mp	−	+	<	>0	Spontaneous

We have said that ΔS_{univ} is positive for all spontaneous processes. Unfortunately, it is not possible to make direct measurements of ΔS_{univ}. Consequently, entropy changes accompanying physical and chemical changes are reported in terms of ΔS_{sys}. The subscript "system" is usually omitted. The symbol ΔS refers to the change in the entropy of the reacting system, just as ΔH refers to the change in enthalpy of the reacting system. The **Third Law of Thermodynamics** establishes the zero of the entropy scale.

Can you develop a comparable table for boiling (liquid → gas) and condensation (gas → liquid)? (Study Table 15-4 carefully.)

The entropy of a pure, perfect crystalline substance (perfectly ordered) is zero at absolute zero (0 K).

As the temperature of a substance increases, the particles vibrate more vigorously, so the entropy increases (Figure 15-12). Further heat input causes either increased temperature (still higher entropy) or phase transitions (melting, sublimation, or boiling) that also

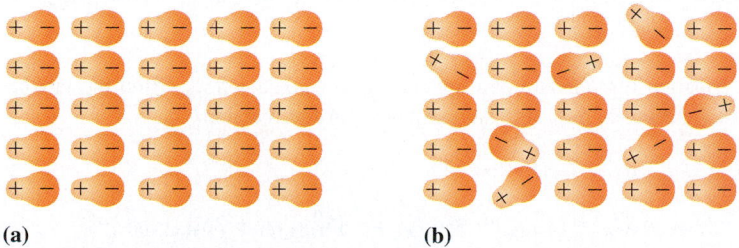

(a) (b)

Figure 15-12 (a) A simplified representation of a side view of a "perfect" crystal of a polar substance at 0 K. Note the perfect alignment of the dipoles in all molecules in a perfect crystal. This causes its entropy to be zero at 0 K. However, there are no perfect crystals, because even the purest substances that scientists have prepared are contaminated by traces of impurities that occupy a few of the positions in the crystal structure. Additionally, there are some vacancies in the crystal structures of even very highly purified substances such as those used in semiconductors (Section 13-17). (b) A simplified representation of the same "perfect" crystal at a temperature above 0 K. Vibrations of the individual molecules within the crystal cause some dipoles to be oriented in directions other than those in a perfect arrangement. The entropy of such a crystalline solid is greater than zero, because there is disorder in the crystal.

Enthalpies are measured only as *differences* with respect to an arbitrary standard state. Entropies, in contrast, are defined relative to an absolute zero level. In either case, the *per mole* designation means *per mole of substance in the specified state.*

result in higher entropy. The entropy of a substance at any condition is its **absolute entropy,** also called standard molar entropy. Consider the absolute entropies at 298 K listed in Table 15-5. At 298 K, *any* substance is more disordered than if it were in a perfect crystalline state at absolute zero, so tabulated S_{298}^0 values are *always positive*. Notice especially that S_{298}^0 of an element, unlike its ΔH_f^0, is *not* equal to zero. The reference state for absolute entropy is specified by the Third Law of Thermodynamics. It is different from the reference state for ΔH_f^0 (Section 15-7). The absolute entropies, S_{298}^0, of various substances under standard conditions are tabulated in Appendix K.

The **standard entropy change, ΔS^0,** of a reaction can be determined from the absolute entropies of reactants and products. The relationship is analogous to Hess' Law.

The $\Sigma\, n$ means that each S^0 value must be multiplied by the appropriate coefficient, n, from the balanced equation. These values are then added.

$$\Delta S_{rxn}^0 = \Sigma\, n S_{products}^0 - \Sigma\, n S_{reactants}^0$$

S^0 values are tabulated in units of *J/mol · K* rather than the larger units involving kilojoules that are used for enthalpy changes. The "mol" term in the units for a *substance* refers to a mole of the substance, whereas for a *reaction* it refers to a mole of reaction. Each term in the sums on the right-hand side of the equation has the units

$$\frac{\text{mol substance}}{\text{mol rxn}} \times \frac{J}{(\text{mol substance}) \cdot K} = \frac{J}{(\text{mol rxn}) \cdot K}$$

The result is usually abbreviated as J/mol · K, or sometimes even as J/K. As before, we shall usually omit units in intermediate steps and then apply appropriate units to the result.

EXAMPLE 15-14   *Calculation of ΔS_{rxn}^0*

Use the values of standard molar entropies in Appendix K to calculate the entropy change at 25°C and one atmosphere pressure for the reaction of hydrazine with hydrogen peroxide. This explosive reaction has been used for rocket propulsion. Do you think the reaction is spontaneous? The balanced equation for the reaction is

$$N_2H_4(\ell) + 2H_2O_2(\ell) \longrightarrow N_2(g) + 4H_2O(g) \qquad \Delta H_{rxn}^0 = -642.2 \text{ kJ/mol reaction}$$

Plan

We use the equation for standard entropy change to calculate ΔS_{rxn}^0 from the tabulated values of standard molar entropies, S_{298}^0, for the substances in the reaction.

Solution

$$\Delta S_{rxn}^0 = [S_{N_2(g)}^0 + 4S_{H_2O(g)}^0] - [S_{N_2H_4(\ell)}^0 + 2S_{H_2O_2(\ell)}^0]$$
$$= [1\,(191.5) + 4\,(188.7)] - [1\,(121.2) + 2\,(109.6)]$$
$$\Delta S_{rxn}^0 = +605.9 \text{ J/mol} \cdot K$$

The "mol" designation for ΔS_{rxn}^0 refers to a mole of reaction, that is, one mole of $N_2H_4(\ell)$, two moles of $H_2O_2(\ell)$, and so on. Although it may not appear to be, +605.9 J/mol · K is a relatively large value of ΔS_{sys}^0. The positive entropy change favors spontaneity. This reaction is also exothermic (ΔH^0 is negative). As we shall see, this reaction *must* be spontaneous, because ΔS^0 is positive and ΔH^0 is negative.

You should now work Exercise 76.

Small booster rockets adjust the course of a satellite in orbit. Some of these small rockets are powered by the N_2H_4/H_2O_2 reaction.

Because changes in the thermodynamic quantity *entropy* may be understood in terms of changes in *molecular disorder,* we can often predict the sign of ΔS_{sys}. The following illustrations emphasize several common types of processes that result in predictable entropy changes.

(a) *Phase changes.* When melting occurs, the molecules or ions are taken from their quite ordered crystalline arrangement to a more disordered one in which they are able to move past one another in the liquid. Thus, a melting process is always accompanied by an entropy increase ($\Delta S_{sys} > 0$). Likewise, vaporization and sublimation both take place with large increases in disorder, and hence with increases in entropy. For the reverse processes of freezing, condensation, and deposition, entropy decreases because order increases.

(b) *Temperature changes*—for example, warming a gas from 25°C to 50°C. As any sample is warmed, the molecules undergo more (random) motion; hence entropy increases ($\Delta S_{sys} > 0$) as temperature increases. Likewise, as we raise the temperature of a solid, the particles vibrate more vigorously about their positions in the crystal, so that at any instant there is a larger average displacement from their mean positions; this results in an increase in entropy.

(c) *Volume changes.* When the volume of a sample of gas increases, the molecules can occupy more positions, and hence are more randomly arranged than when they are closer together in a smaller volume. Hence, an expansion is accompanied by an increase in entropy ($\Delta S_{sys} > 0$). Conversely, as a sample is compressed, the molecules are more restricted in their locations, and a situation of greater order (lower entropy) results.

(d) *Mixing of substances,* even without chemical reaction. Situations in which the molecules are more "mixed up" are more disordered, and hence are at higher entropy. We pointed out that the mixed gases of Figure 15-10b were more disordered than the separated gases of Figure 15-10a, and that the former was a situation of higher entropy. We see, then, that mixing of gases by diffusion is a process for which $\Delta S_{sys} > 0$; we know from experience that it is always spontaneous. We have already pointed out (Section 14-2) that the increase in disorder (entropy increase) that accompanies mixing often provides the driving force for solubility of one substance in another. For example, when one mole of solid NaCl dissolves in water, NaCl(s) → NaCl(aq), the entropy (Appendix K) increases from 72.4 J/mol·K to 115.5 J/mol·K, or $\Delta S^0 = +43.1$ J/mol·K. The term "mixing" can be interpreted rather liberally. For example, the reaction $H_2(g) + Cl_2(g) → 2HCl(g)$ has $\Delta S^0 > 0$; in the reactants, each atom is bonded to an identical atom, a less "mixed-up" situation than in the products, where unlike atoms are bonded together.

(e) *Increase in the number of particles,* as in the dissociation of a diatomic gas such as $F_2(g) → 2F(g)$. Any process in which the number of particles increases results in an increase in entropy, $\Delta S_{sys} > 0$. Values of ΔS^0 calculated for several reactions of this type are given in Table 15-6. As you can see, the ΔS^0 values for the dissociation process $X_2 → 2X$ are all similar for X = H, F, Cl, and N. Why is the value given in Table 15-6 so much larger for X = Br? This process starts with *liquid* Br_2. The total process $Br_2(\ell) → 2Br(g)$, for which $\Delta S^0 = 197.5$ J/mol·K, can be treated as the result of *two* processes. The first of these is *vaporization,* $Br_2(\ell) → Br_2(g)$, for which $\Delta S^0 = 93.1$ J/mol·K. The second step is the dissociation of gaseous bromine, $Br_2(g) → 2Br(g)$, for which $\Delta S^0 = 104.4$ J/mol·K; this entropy increase is about the same as for the other processes that involve *only* dissociation of a gaseous diatomic species. Can you rationalize the even higher value given in the table for $I_2(s) → 2I(g)$?

(f) *Changes in the number of moles of gaseous substances.* Processes that result in an increase in the number of moles of gaseous substances have $\Delta S_{sys} > 0$. Example 15-14 illustrates this. There are no gaseous reactants, but the products include five moles of gas.

Table 15-5 *Absolute Entropies at 298 K for a Few Common Substances*

	S^0 (J/mol·K)
C(diamond)	2.38
C(g)	158.0
$H_2O(\ell)$	69.91
$H_2O(g)$	188.7
$I_2(s)$	116.1
$I_2(g)$	260.6

Table 15-6 *Entropy Changes for Some Processes $X_2 → 2X$*

Reaction	ΔS^0 (J/mol·K)
$H_2(g) \longrightarrow 2H(g)$	98.0
$N_2(g) \longrightarrow 2N(g)$	114.9
$F_2(g) \longrightarrow 2F(g)$	114.5
$Cl_2(g) \longrightarrow 2Cl(g)$	107.2
$Br_2(\ell) \longrightarrow 2Br(g)$	197.5
$I_2(s) \longrightarrow 2I(g)$	245.3

Conversely, the process $2H_2(g) + O_2(g) \rightarrow 2H_2O(g)$ has a negative ΔS^0 value; here, three moles of gas are consumed while only two moles are produced, for a net decrease in the number of moles in the gas phase. You should be able to calculate the value of ΔS^0 for this reaction from the values in Appendix K.

Do you think that the reaction

$$2H_2(g) + O_2(g) \longrightarrow 2H_2O(\ell)$$

would have a higher or lower value of ΔS^0 than when the water is in the gas phase? Confirm by calculation.

15-15 FREE ENERGY CHANGE, ΔG, AND SPONTANEITY

Energy is the capacity to do work. If heat is released in a chemical reaction (ΔH is negative), *some* of the heat may be converted into useful work. Some of it may be expended to increase the order of the system (if ΔS is negative). However, if a system becomes more disordered ($\Delta S > 0$), more useful energy becomes available than indicated by ΔH alone. J. Willard Gibbs, a prominent 19th-century American professor of mathematics and physics, formulated the relationship between enthalpy and entropy in terms of another state function that we now call **Gibbs free energy, G.** It is defined as

$$G = H - TS$$

The **Gibbs free energy change, ΔG,** at constant temperature and pressure, is

This is often called simply the Gibbs energy change or the free energy change.

$$\Delta G = \Delta H - T\Delta S \qquad \text{(constant } T \text{ and } P\text{)}$$

The amount by which the Gibbs free energy decreases is the *maximum useful energy* obtainable in the form of work from a given process at constant temperature and pressure. It is also the *indicator of spontaneity of a reaction or physical change* at constant T and P. If there is a net decrease of useful energy, ΔG is negative and the process is spontaneous. We see from the equation that ΔG becomes more negative as (1) ΔH becomes more negative (the process gives off more heat) and (2) ΔS becomes more positive (the process results in greater disorder). If there is a net increase in free energy of the system during a process, ΔG is positive and the process is nonspontaneous. This means that the reverse

(a) The entropy of an organism decreases (unfavorable) when new cells are formed. The energy to sustain animal life is provided by the metabolism of food. This energy is released when the chemical bonds in the food are broken. Exhalation of gases and excretion of waste materials increase the entropy of the surroundings enough that the entropy of the universe increases and the overall process can occur. (b) Stored chemical energy can later be transformed by the organism to the mechanical energy for muscle contraction, to the electrical energy for brain function, or to another needed form.

(a)

(b)

process is spontaneous under the given conditions. When $\Delta G = 0$, there is no net transfer of free energy; both the forward and reverse processes are equally favorable. Thus, $\Delta G = 0$ describes a system at *equilibrium.*

The relationship between ΔG and spontaneity may be summarized as follows.

ΔG	Spontaneity of Reaction (constant T and P)
ΔG is positive	Reaction is nonspontaneous
ΔG is zero	System is at equilibrium
ΔG is negative	Reaction is spontaneous

The free energy content of a system depends on temperature and pressure (and, for mixtures, on concentrations). The value of ΔG for a process depends on the states and the concentrations of the various substances involved. It also depends strongly on temperature, because the equation $\Delta G = \Delta H - T\Delta S$ includes temperature. Just as for other thermodynamic variables, we choose some set of conditions as a standard state reference. The standard state for $\Delta G°$ is the same as for $\Delta H°$—1 atm and the specified temperature, usually 25°C (298 K). Values of standard molar free energy of formation, ΔG_f^0, for many substances are tabulated in Appendix K. For *elements* in their standard states, $\Delta G_f^0 = 0$. The values of ΔG_f^0 may be used to calculate the standard free energy change of a reaction *at 298 K* by using the following relationship.

$$\Delta G_{rxn}^0 = \Sigma\, n\Delta G_{f\,products}^0 - \Sigma\, n\Delta G_{f\,reactants}^0 \qquad \text{(1 atm and 298 K } \textit{only}\text{)}$$

The value of ΔG_{rxn}^0 allows us to predict the spontaneity of a very special hypothetical reaction that we call the *standard reaction.*

In the **standard reaction,** the numbers of moles of reactants shown in the balanced equation, all in their standard states, are *completely* converted to the numbers of moles of products shown in the balanced equation, all in their standard states.

In other words, are the *reactants* or the *products* more stable *in their standard states*?

We must remember that it is ΔG, and not ΔG^0, that is the general criterion for spontaneity. ΔG depends on concentrations of reactants and products in the mixture. For most reactions, there is an *equilibrium mixture* of reactants and products that is more stable than either all reactants or all products. In Chapter 17 we shall study the concept of equilibrium and see how to find ΔG for mixtures.

EXAMPLE 15-15 *Spontaneity of Standard Reaction*

Diatomic nitrogen and oxygen molecules make up about 99% of all the molecules in reasonably "unpolluted" dry air. Evaluate ΔG^0 for the following reaction at 298 K, using ΔG_f^0 values from Appendix K. Is the standard reaction spontaneous?

$$N_2(g) + O_2(g) \longrightarrow 2NO(g) \qquad \text{(nitrogen oxide)}$$

For the reverse reaction at 298 K, $\Delta G^0_{rxn} = -173.1$ kJ/mol. It is spontaneous, but very slow at room temperature. The NO formed in automobile engines is oxidized to even more harmful NO_2 much more rapidly than it spontaneously decomposes to N_2 and O_2. Thermodynamic spontaneity does not guarantee that a process occurs at an observable rate. The oxides of nitrogen in the atmosphere represent a major environmental problem.

Plan

The reaction conditions are 1 atm and 298 K, so we can use the tabulated values of ΔG^0_f for each substance in Appendix K to evaluate ΔG^0_{rxn} in the equation given above. The treatment of units for calculation of ΔG^0 is the same as that for ΔH^0 in Example 15-8.

Solution

$$\Delta G^0_{rxn} = 2\Delta G^0_{f\, NO(g)} - [\Delta G^0_{f\, N_2(g)} + \Delta G^0_{f\, O_2(g)}]$$

$$= (2(86.57) \quad - [0 \quad\quad + 0])$$

$$\boxed{\Delta G^0_{rxn} = +173.1 \text{ kJ/mol rxn}} \quad\quad \text{for the reaction as written}$$

Because ΔG^0 is positive, the reaction is nonspontaneous at 298 K under standard state conditions.

You should now work Exercise 83.

ΔG^0 can also be calculated by the equation

$$\Delta G^0 = \Delta H^0 - T\Delta S^0 \qquad \text{(constant } T \text{ and } P\text{)}$$

Strictly, this last equation applies to standard conditions. However, ΔH^0 and ΔS^0 often do not vary much with temperature, so it can often be used to *estimate* free energy changes at other temperatures.

EXAMPLE 15-16   *Spontaneity of Standard Reaction*

Make the same determination as in Example 15-15, using heats of formation and absolute entropies rather than free energies of formation.

Plan

First we calculate ΔH^0_{rxn} and ΔS^0_{rxn}. We use the relationship $\Delta G^0 = \Delta H^0 - T\Delta S^0$ to evaluate the free energy change under standard state conditions at 298 K.

Solution

$$\Delta H^0_{rxn} = \Sigma\, n\Delta H^0_{f\, products} - \Sigma\, n\Delta H^0_{f\, reactants}$$

$$= 2\Delta H^0_{f\, NO(g)} \quad - [\Delta H^0_{f\, N_2(g)} + \Delta H^0_{f\, O_2(g)}]$$

$$= (2[90.25] \quad - [0 + 0]) = 180.5 \text{ kJ/mol}$$

$$\Delta S^0_{rxn} = \Sigma\, nS^0_{products} \quad - \Sigma\, nS^0_{reactants}$$

$$= 2S^0_{NO(g)} \quad - [S^0_{N_2(g)} + S^0_{O_2(g)}]$$

$$= (2[210.7] \quad - [191.5 + 205.0]) = 24.9 \text{ J/mol} \cdot \text{K} = 0.0249 \text{ kJ/mol} \cdot \text{K}$$

Now we use the relationship $\Delta G^0 = \Delta H^0 - T\Delta S^0$, with $T = 298$ K, to evaluate the free energy change under standard state conditions at 298 K.

$$\Delta G^0_{rxn} = \Delta H^0_{rxn} \quad - T\Delta S^0_{rxn}$$

$$= 180.5 \text{ kJ/mol} - (298 \text{ K})(0.0249 \text{ kJ/mol} \cdot \text{K})$$

$$= 180.5 \text{ kJ/mol} - 7.42 \text{ kJ/mol}$$

$$\boxed{\Delta G^0_{rxn} = +173.1 \text{ kJ/mol rxn,}} \quad \text{the same value obtained in Example 15-15}$$

You should now work Exercise 84.

15-16 THE TEMPERATURE DEPENDENCE OF SPONTANEITY

The methods developed in Section 15-15 can also be used to estimate the temperature at which a process is in equilibrium.

EXAMPLE 15-17 *Estimation of Boiling Point*

Use the thermodynamic data in Appendix K to estimate the normal boiling point of bromine, Br_2. Assume that ΔH and ΔS do not change with temperature.

Plan

The process we must consider is

$$Br_2(\ell) \longrightarrow Br_2(g)$$

By definition, the normal boiling point of a liquid is the temperature at which pure liquid and pure gas coexist in equilibrium at 1 atm. Therefore, $\Delta G = 0$. We assume that $\Delta H_{rxn} = \Delta H_{rxn}^0$ and $\Delta S_{rxn} = \Delta S_{rxn}^0$. We can evaluate these two quantities, substitute them in the relationship $\Delta G = \Delta H - T\Delta S$, and then solve for T.

Solution

$$\Delta H_{rxn} = \Delta H_{f\,Br_2(g)}^0 - \Delta H_{f\,Br_2(\ell)}^0$$
$$= 30.91 \quad - 0 = 30.91 \text{ kJ/mol}$$
$$\Delta S_{rxn} = S_{Br_2(g)}^0 \quad - S_{Br_2(\ell)}^0$$
$$= (245.4 \quad - 152.2) = 93.2 \text{ J/mol} \cdot \text{K} = 0.0932 \text{ kJ/mol} \cdot \text{K}$$

We can now solve for the temperature at which the system is in equilibrium, i.e., the boiling point of Br_2.

$$\Delta G_{rxn} = \Delta H_{rxn} - T\Delta S_{rxn} = 0 \quad \text{so} \quad \Delta H_{rxn} = T\Delta S_{rxn}$$

$$T = \frac{\Delta H_{rxn}}{\Delta S_{rxn}} = \frac{30.91 \text{ kJ/mol}}{0.0932 \text{ kJ/mol} \cdot \text{K}} = \boxed{332 \text{ K } (59°\text{C})}$$

This is the temperature at which the system is in equilibrium, i.e., the boiling point of Br_2. The value listed in a handbook of chemistry and physics is 58.78°C.

You should now work Exercise 92.

> Actually, both ΔH_{rxn}^0 and ΔS_{rxn}^0 vary with temperature, but usually not enough to introduce significant errors for modest temperature changes. The value of ΔG_{rxn}^0, on the other hand, is strongly dependent on the temperature.

The free energy change and spontaneity of a reaction depend upon both enthalpy and entropy changes. Both ΔH and ΔS may be either positive or negative, so we can classify reactions into four categories with respect to spontaneity (Figure 15-13).

$\Delta G = \Delta H - T\Delta S$	*(constant temperature and pressure)*
1. $\Delta H = -$, $\Delta S = +$	Rxns are spontaneous at all temperatures.
2. $\Delta H = -$, $\Delta S = -$	Rxns become spontaneous below a definite temperature.
3. $\Delta H = +$, $\Delta S = +$	Rxns become spontaneous above a definite temperature.
4. $\Delta H = +$, $\Delta S = -$	Rxns are nonspontaneous at all temperatures.

Table 15-7 gives examples of reactions in each category, as well as temperatures at which the changeover from spontaneous to nonspontaneous occurs, where appropriate. The numbering system corresponds to that above.

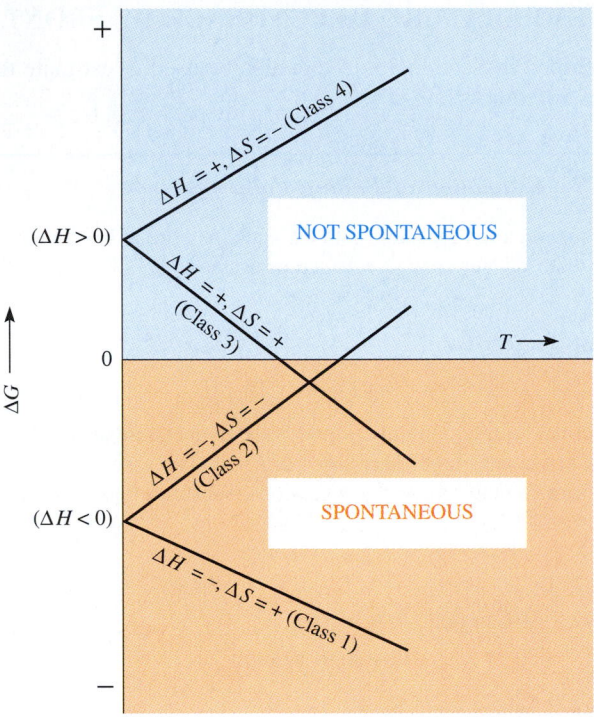

Figure 15-13 A graphical representation of the dependence of ΔG and spontaneity on temperature for each of the four classes of reactions listed in the text and in Table 15-7.

We can estimate the temperature range over which a chemical reaction is spontaneous by evaluating ΔH_{rxn}^0 and ΔS_{rxn}^0 from tabulated data. The temperature at which $\Delta G_{rxn}^0 = 0$ is the temperature limit of spontaneity. The sign of ΔS_{rxn}^0 tells us whether the reaction is spontaneous *below* or *above* this limit (Classes 2 and 3 in Table 15-7).

Table 15-7 *Thermodynamic Classes of Reactions*

Class	Examples	ΔH (kJ/mol)	ΔS (J/mol · K)	Temperature Range of Spontaneity
1	$2H_2O_2(\ell) \longrightarrow 2H_2O(\ell) + O_2(g)$	-196	$+126$	All temperatures
	$H_2(g) + Br_2(\ell) \longrightarrow 2HBr(g)$	-72.8	$+114$	All temperatures
2	$NH_3(g) + HCl(g) \longrightarrow NH_4Cl(s)$	-176	-285	Lower temperatures (<619 K)
	$2H_2S(g) + SO_2(g) \longrightarrow 3S(s) + 2H_2O(\ell)$	-233	-424	Lower temperatures (<550 K)
3	$NH_4Cl(s) \longrightarrow NH_3(g) + HCl(g)$	$+176$	$+285$	Higher temperatures (>619 K)
	$CCl_4(\ell) \longrightarrow C(graphite) + 2Cl_2(g)$	$+136$	$+235$	Higher temperatures (>517 K)
4	$2H_2O(\ell) + O_2(g) \longrightarrow 2H_2O_2(\ell)$	$+196$	-126	Nonspontaneous, all temperatures
	$3O_2(g) \longrightarrow 2O_3(g)$	$+285$	-137	Nonspontaneous, all temperatures

EXAMPLE 15-18 *Temperature Range of Spontaneity*

Mercury(II) sulfide is a dark red mineral called cinnabar. Metallic mercury is obtained by roasting the sulfide in a limited amount of air. Estimate the temperature range in which the *standard* reaction is spontaneous.

$$HgS(s) + O_2(g) \longrightarrow Hg(\ell) + SO_2(g)$$

Plan

We evaluate ΔH^0_{rxn} and ΔS^0_{rxn} and assume that their values are independent of temperature. We find that both factors are favorable to spontaneity.

Solution

$$\Delta H^0_{rxn} = \Delta H^0_{f\,Hg(\ell)} + \Delta H^0_{f\,SO_2(g)} - [\Delta H^0_{f\,HgS(s)} + \Delta H^0_{f\,O_2(g)}]$$

$$= (0 - 296.8 + 58.2 - 0) = -238.6 \text{ kJ/mol}$$

$$\Delta S^0_{rxn} = S^0_{Hg(\ell)} + S^0_{SO_2(g)} - [S^0_{HgS(s)} + S^0_{O_2(g)}]$$

$$= (76.02 + 248.1 - 82.4 - 205.0) = +36.7 \text{ J/mol} \cdot \text{K}$$

ΔH^0_{rxn} is negative and ΔS^0_{rxn} is positive, so the reaction is spontaneous at all temperatures. The reverse reaction is, therefore, nonspontaneous at all temperatures.

Heating red HgS in air produces liquid Hg. The gaseous SO_2 escapes. Cinnabar, an important ore of mercury, contains HgS.

 The fact that a reaction is spontaneous at all temperatures does not mean that the reaction occurs fast enough to be useful at all temperatures. As a matter of fact, $Hg(\ell)$ can be obtained from HgS(s) by this reaction at a reasonable rate only at high temperatures.

EXAMPLE 15-19 *Temperature Range of Spontaneity*

Estimate the temperature range for which the following standard reaction is spontaneous.

$$SiO_2(s) + 2C(graphite) + 2Cl_2(g) \longrightarrow SiCl_4(g) + 2CO(g)$$

Plan

When we proceed as in Example 15-18, we find that ΔS^0_{rxn} is favorable to spontaneity, whereas ΔH^0_{rxn} is unfavorable. Thus, we know that the reaction becomes spontaneous *above* some temperature. We can set ΔG^0 equal to zero in the equation $\Delta G^0 = \Delta H^0 - T\Delta S^0$ and solve for the temperature at which the system is *at equilibrium*. This will represent the temperature above which the reaction would be spontaneous.

Solution

$$\Delta H^0_{rxn} = [\Delta H^0_{f\,SiCl_4(g)} + 2\Delta H^0_{f\,CO(g)}] - [\Delta H^0_{f\,SiO_2(s)} + 2\Delta H^0_{f\,C(graphite)} + 2\Delta H^0_{f\,Cl_2(g)}]$$

$$= [(-657.0) + 2(-110.5)] - [(-910.9) + 2(0) + 2(0)]$$

$$= +32.9 \text{ kJ/mol}$$

$$\Delta S^0_{rxn} = S^0_{SiCl_4(g)} + 2S^0_{CO(g)} - [S^0_{SiO_2(s)} + 2S^0_{C(graphite)} + 2S^0_{Cl_2(g)}]$$

$$= [330.6 + 2(197.6)] - [41.84 + 2(5.740) + 2(223.0)]$$

$$= 226.5 \text{ J/mol} \cdot \text{K} = 0.2265 \text{ kJ/mol} \cdot \text{K}$$

When $\Delta G^0 = 0$, neither the forward nor the reverse reaction is spontaneous. Let's find the temperature at which $\Delta G^0 = 0$ and the system is at equilibrium.

$$\Delta G^0 = \Delta H^0 - T\Delta S^0 = 0$$

$$\Delta H^0 = T\Delta S^0$$

$$T = \frac{\Delta H^0}{\Delta S^0} = \frac{+32.9 \text{ kJ/mol}}{+0.2265 \text{ kJ/mol} \cdot \text{K}}$$

$$T = 145 \text{ K}$$

At temperatures above 145 K, the $T\Delta S^0$ would be greater ($-T\Delta S^0$ would be more negative) than the ΔH^0 term, which would make ΔG^0 negative; so the reaction would be spontaneous above 145 K. At temperatures below 145 K, the $T\Delta S^0$ term would be smaller than the ΔH^0 term, which would make ΔG^0 positive; so the reaction would be nonspontaneous below 145 K.

However, 145 K ($-128°C$) is a very low temperature. For all practical purposes, the reaction is spontaneous at all but very low temperatures. In practice, it is carried out at 800 to 1000°C because of the greater reaction rate at these higher temperatures. This gives a useful and economical rate of production of $SiCl_4$, an important industrial chemical.

You should now work Exercises 89 and 90.

Key Terms

Absolute entropy (of a substance) The entropy of a substance relative to its entropy in a perfectly ordered crystalline form at 0 K (where its entropy is zero).

Bomb calorimeter A device used to measure the heat transfer between system and surroundings at constant volume.

Bond energy The amount of energy necessary to break one mole of bonds in a gaseous substance, to form gaseous products at the same temperature and pressure.

Calorimeter A device used to measure the heat transfer between system and surroundings.

Endothermic process A process that absorbs heat.

Enthalpy, H The heat content of a specific amount of a substance; defined as $E + PV$.

Enthalpy change, ΔH The change in heat content of a system that accompanies a process at constant pressure, q_p.

Entropy, S A thermodynamic state property that measures the degree of disorder or randomness of a system; $T\Delta S$ represents a form of energy.

Equilibrium A state of dynamic balance in which the rates of forward and reverse processes (reactions) are equal; the state of a system when neither the forward nor the reverse process is thermodynamically favored.

Exothermic process A process that gives off heat.

First Law of Thermodynamics The total amount of energy in the universe is constant (also known as the Law of Conservation of Energy); energy is neither created nor destroyed in ordinary chemical reactions and physical changes.

Gibbs free energy, G The thermodynamic state function of a system that indicates the amount of energy available for the system to do useful work at constant T and P. $G = H - TS$.

Gibbs free energy change, ΔG The indicator of spontaneity of a process at constant T and P. $\Delta G = \Delta H - T\Delta S$. If ΔG is negative, the process is spontaneous.

Hess' Law of heat summation The enthalpy change for a reaction is the same whether it occurs in one step or a series of steps.

Internal energy, E All forms of energy associated with a specific amount of a substance.

Mole of reaction (mol rxn) The amount of reaction that corresponds to the number of moles of each substance shown in the balanced equation.

Pressure–volume work Work done by a gas when it expands against an external pressure or work done on a system as gases are compressed or consumed in the presence of an external pressure.

Second Law of Thermodynamics The universe tends toward a state of greater disorder in spontaneous processes.

Spontaneity Of a process, its property of being energetically favorable and therefore capable of proceeding in the forward direction (but not necessarily at an observable rate).

Standard enthalpy change, ΔH^0 The enthalpy change when the number of moles of reactants specified in the balanced chemical equation, all at standard states, are converted completely to the specified number of moles of products, all at standard states.

Standard entropy, S^0 (of a substance) The absolute entropy of a substance in its standard state at 298 K.

Standard entropy change, ΔS^0 The entropy change when the number of moles of reactants specified in the balanced chemical equation, all at standard states, are converted completely to the specified number of moles of products, all at standard states.

Standard molar enthalpy of formation, ΔH_f^0 (of a substance) The enthalpy change for the formation of one mole of a substance in a specified state from its elements in their standard states.

Standard reaction   A reaction in which the numbers of moles of reactants shown in the balanced equation, all in their standard states, are *completely* converted to the numbers of moles of products shown in the balanced equation, also all at their standard states.

State function   A variable that defines the state of a system; a function that is independent of the pathway by which a process occurs.

Surroundings   Everything in the environment of the system.

System   The substances of interest in a process; the part of the universe under investigation.

Thermochemical equation   A balanced chemical equation together with a designation of the corresponding value of ΔH_{rxn}. Sometimes used with changes in other thermodynamic quantities.

Thermochemical standard state   The state in which a substance exists at 25°C (298 K) and one atmosphere.

Thermochemistry   The observation, measurement, and prediction of energy changes that accompany physical changes or chemical reactions.

Thermodynamics   The study of the energy transfers accompanying physical and chemical processes.

Thermodynamic state of a system   A set of conditions that completely specifies all of the properties of the system.

Thermodynamic standard state of a substance   The most stable state of the substance at one atmosphere pressure and at some specific temperature (25°C unless otherwise specified).

Third Law of Thermodynamics   The entropy of a hypothetical pure, perfect, crystalline substance at absolute zero temperature is zero.

Universe   The system plus the surroundings.

Work   The application of a force through a distance; for physical changes or chemical reactions at constant external pressure, the work done on the system is $-P\Delta V$.

Exercises

General Concepts

1. State precisely the meaning of each of the following terms. You may need to review Chapter 1 to refresh your memory concerning terms introduced there. (a) Energy; (b) kinetic energy; (c) potential energy; (d) calorie; (e) joule.

2. State precisely the meaning of each of the following terms. You may need to review Chapter 1 to refresh your memory about terms introduced there. (a) Heat; (b) temperature; (c) system; (d) surroundings; (e) thermodynamic state of system; (f) work.

3. (a) Give an example of the conversion of heat into work. (b) Give an example of the conversion of work into heat.

4 Distinguish between endothermic and exothermic processes. If we know that a reaction is endothermic in one direction, what can be said about the reaction in the reverse direction?

5. What is a state function? Would Hess' Law be a law if enthalpy were not a state function?

Enthalpy and Changes in Enthalpy

6. (a) Distinguish between ΔH and ΔH^0 for a reaction. (b) Distinguish between ΔH_{rxn}^0 and ΔH_f^0.

7. A reaction is characterized by $\Delta H_{rxn} = -500$ kJ/mol. Does it absorb heat from the surroundings or release heat to them?

8. For each of these reactions, (a) does the enthalpy increase or decrease; (b) is $H_{reactant} > H_{product}$ or is $H_{product} > H_{reactant}$; (c) is ΔH positive or negative?
   (1) $Al_2O_3(s) \longrightarrow 2Al(s) + 1\frac{1}{2}O_2(g)$ (endothermic)
   (2) $Sn(s) + Cl_2(g) \longrightarrow SnCl_2(s)$ (exothermic)

9. (a) The combustion of 0.0222 g of isooctane vapor, C_8H_{18}, at constant pressure raises the temperature of a calorimeter 0.400°C. The heat capacity of the calorimeter and water combined is 2.48 kJ/°C. Find the molar heat of combustion of gaseous isooctane.

$$C_8H_{18}(g) + 12\frac{1}{2}O_2(g) \longrightarrow 8CO_2(g) + 9H_2O(\ell)$$

(b) How many grams of $C_8H_{18}(g)$ must be burned to obtain 225 kJ of heat energy?

10. Methanol, CH_3OH, is an efficient fuel with a high octane rating that can be produced from coal and hydrogen.

$$CH_3OH(g) + 1\frac{1}{2}O_2(g) \longrightarrow CO_2(g) + 2H_2O(\ell)$$
$$\Delta H = -762 \text{ kJ/mol rxn}$$

(a) Find the heat evolved when 90.0 g $CH_3OH(g)$ burns in excess oxygen. (b) What mass of O_2 is consumed when 825 kJ of heat is given out?

11. How much heat is liberated when 0.0426 moles of sodium reacts with excess water according to the following equation?

$$2Na(s) + 2H_2O(\ell) \longrightarrow H_2(g) + 2NaOH(aq)$$
$$\Delta H = -368 \text{ kJ/mol rxn}$$

12. What is ΔH of the reaction

$$PbO(s) + C(s) \longrightarrow Pb(s) + CO(g)$$

if 23.8 kJ must be supplied to convert 49.7 g lead(II) oxide to lead?

13. Methylhydrazine is burned with dinitrogen tetroxide in the altitude-control engines of the space shuttle.

$$CH_6N_2(\ell) + \tfrac{5}{4}N_2O_4(\ell) \longrightarrow CO_2(g) + 3H_2O(\ell) + \tfrac{9}{4}N_2(g)$$

The two substances ignite instantly on contact, producing a flame temperature of 3000 K. The energy liberated per 0.100 g of CH_6N_2 at constant atmospheric pressure after the products are cooled back to 25°C is 750 J. (a) Find ΔH for the reaction as written. (b) How many kilojoules are liberated when 44.0 g of N_2 is produced?

The space shuttle.

14. Which is more exothermic, the combustion of one mole of methane to form $CO_2(g)$ and liquid water or the combustion of one mole of methane to form $CO_2(g)$ and steam? Why? (No calculations are necessary.)

15. Which is more exothermic, the combustion of one mole of gaseous benzene, C_6H_6, or the combustion of one mole of liquid benzene? Why? (No calculations are necessary.)

Thermochemical Equations, ΔH_f^0, and Hess' Law

16. Explain the meaning of the term "thermochemical standard state of a substance."

17. Explain the significance of each word in the term "standard molar enthalpy of formation."

18. From the data in Appendix K, determine the form that represents the standard state for each of the following elements: (a) chlorine; (b) iron; (c) nitrogen; (d) iodine; (e) sulfur.

19. From the data in Appendix K, determine the form that represents the standard state for each of the following elements: (a) oxygen; (b) carbon; (c) phosphorus; (d) tin; (e) calcium.

20. Write the balanced chemical equation whose ΔH_{rxn}^0 value is equal to ΔH_f^0 for each of the following substances: (a) calcium hydroxide, $Ca(OH)_2(s)$; (b) ethane, $C_2H_6(g)$; (c) sodium carbonate, $Na_2CO_3(s)$; (d) calcium fluoride, $CaF_2(s)$; (e) phosphine, $PH_3(g)$; (f) propane, $C_3H_8(g)$; (g) atomic sulfur, $S(g)$.

21. Write the balanced chemical equation whose ΔH_{rxn}^0 value is equal to ΔH_f^0 for each of the following substances: (a) hydrogen sulfide, $H_2S(g)$; (b) magnesium oxide, $MgO(s)$; (c) atomic oxygen, $O(g)$; (d) benzoic acid, $C_6H_5COOH(s)$; (e) hydrogen peroxide, $H_2O_2(\ell)$; (f) dinitrogen tetroxide, $N_2O_4(g)$.

*22. We burn 24.5 g of lithium in excess oxygen at constant atmospheric pressure to form Li_2O. Then we bring the reaction mixture back to 25°C. In this process 1029 kJ of heat is given off. What is the standard molar enthalpy of formation of Li_2O?

*23. We burn 24.5 g of magnesium in excess nitrogen at constant atmospheric pressure to form Mg_3N_2. Then we bring the reaction mixture back to 25°C. In this process 232.5 kJ of heat is given off. What is the standard molar enthalpy of formation of Mg_3N_2?

24. From the following enthalpies of reaction,

$$4HCl(g) + O_2(g) \longrightarrow 2H_2O(\ell) + 2Cl_2(g)$$
$$\Delta H = -148.4 \text{ kJ/mol rxn}$$

$$\tfrac{1}{2}H_2(g) + \tfrac{1}{2}F_2(g) \longrightarrow HF(\ell) \qquad \Delta H = -600.0 \text{ kJ/mol rxn}$$

$$H_2(g) + \tfrac{1}{2}O_2(g) \longrightarrow H_2O(\ell) \qquad \Delta H = -285.8 \text{ kJ/mol rxn}$$

find ΔH_{rxn} for $2HCl(g) + F_2(g) \longrightarrow 2HF(\ell) + Cl_2(g)$.

25. From the following heats of reaction,

$$CaCO_3(s) \longrightarrow CaO(s) + CO_2(g) \qquad \Delta H = +178.1 \text{ kJ/mol rxn}$$

$$CaO(s) + H_2O(\ell) \longrightarrow Ca(OH)_2(s) \qquad \Delta H = -64.8 \text{ kJ/mol rxn}$$

$$Ca(OH)_2(s) \longrightarrow Ca^{2+}(aq) + 2OH^-(aq)$$
$$\Delta H = -16.2 \text{ kJ/mol rxn}$$

calculate ΔH_{rxn} for

$$Ca^{2+}(aq) + 2OH^-(aq) + CO_2(g) \longrightarrow CaCO_3(s) + H_2O(\ell)$$

26. Given that

$$S(s) + O_2(g) \longrightarrow SO_2(g) \qquad \Delta H = -296.8 \text{ kJ/mol}$$

$$S(s) + \tfrac{3}{2}O_2(g) \longrightarrow SO_3(g) \qquad \Delta H = -395.7 \text{ kJ/mol}$$

determine the enthalpy change for the decomposition reaction $2SO_3(g) \rightarrow 2SO_2(g) + O_2(g)$.

27. Aluminum reacts vigorously with many oxidizing agents. For example,

$$4Al(s) + 3O_2(g) \longrightarrow 2Al_2O_3(s) \qquad \Delta H = -3352 \text{ kJ/mol}$$

$$4Al(s) + 3MnO_2(s) \longrightarrow 3Mn(s) + 2Al_2O_3(s)$$
$$\Delta H = -1792 \text{ kJ/mol}$$

Use this information to determine the enthalpy of formation of $MnO_2(s)$.

28. Given that

$$2H_2(g) + O_2(g) \longrightarrow 2H_2O(\ell) \qquad \Delta H = -571.7 \text{ kJ/mol}$$

$$C_3H_4(g) + 4O_2(g) \longrightarrow 3CO_2(g) + 2H_2O(\ell)$$
$$\Delta H = -1941 \text{ kJ/mol}$$

$$C_3H_8(g) + 5O_2(g) \longrightarrow 3CO_2(g) + 4H_2O(\ell)$$
$$\Delta H = -2220 \text{ kJ/mol}$$

determine the heat of the hydrogenation reaction

$$C_3H_4(g) + 2H_2(g) \longrightarrow C_3H_8(g)$$

29. Determine the heat of formation of liquid hydrogen peroxide at 25°C from the following thermochemical equations.

$$H_2(g) + \tfrac{1}{2}O_2(g) \longrightarrow H_2O(g) \qquad \Delta H^0 = -241.82 \text{ kJ/mol}$$

$$2H(g) + O(g) \longrightarrow H_2O(g) \qquad \Delta H^0 = -926.92 \text{ kJ/mol}$$

$$2H(g) + 2O(g) \longrightarrow H_2O_2(g) \qquad \Delta H^0 = -1070.60 \text{ kJ/mol}$$

$$2O(g) \longrightarrow O_2(g) \qquad \Delta H^0 = -498.34 \text{ kJ/mol}$$

$$H_2O_2(\ell) \longrightarrow H_2O_2(g) \qquad \Delta H^0 = 51.46 \text{ kJ/mol}$$

30. Use data in Appendix K to find the enthalpy of reaction for
    (a) $NH_4NO_3(s) \longrightarrow N_2O(g) + 2H_2O(g)$
    (b) $2FeS_2(s) + \tfrac{11}{2}O_2(g) \longrightarrow Fe_2O_3(s) + 4SO_2(g)$
    (c) $SiO_2(s) + 3C(s) \longrightarrow SiC(s) + 2CO(g)$

31. Repeat Exercise 30 for
    (a) $CaCO_3(s) \longrightarrow CaO(s) + CO_2(g)$
    (b) $2HI(g) + F_2(g) \longrightarrow 2HF(g) + I_2(s)$
    (c) $SF_6(g) + 3H_2O(\ell) \longrightarrow 6HF(g) + SO_3(g)$

32. The thermite reaction, used for welding iron, is the reaction of Fe_3O_4 with Al.

$$8Al(s) + 3Fe_3O_4(s) \longrightarrow 4Al_2O_3(s) + 9Fe(s)$$
$$\Delta H^0 = -3347.6 \text{ kJ/mol rxn}$$

Because this large amount of heat cannot be rapidly dissipated to the surroundings, the reacting mass may reach temperatures near 3000°C. How much heat is released by the reaction of 19.2 g of Al with 48.0 g of Fe_3O_4?

The thermite reaction.

33. When a welder uses an acetylene torch, the combustion of acetylene liberates the intense heat needed for welding metals together. The equation for this combustion reaction is

$$2C_2H_2(g) + 5O_2(g) \longrightarrow 4CO_2(g) + 2H_2O(g)$$

The heat of combustion of acetylene is -1300 kJ/mol of C_2H_2. How much heat is liberated when 3.462 kg of C_2H_2 is burned?

34. Silicon carbide, or carborundum, SiC, is one of the hardest substances known and is used as an abrasive. It has the structure of diamond with half of the carbons replaced by silicon. It is prepared industrially by reduction of sand (SiO_2) with coke (C) in an electric furnace.

$$SiO_2(s) + 3C(s) \longrightarrow SiC(s) + 2CO(g)$$

ΔH^0 for this reaction is 624.6 kJ, and the ΔH_f^0 for $SiO_2(s)$ and $CO(g)$ are -910.9 kJ/mol and -110.5 kJ/mol, respectively. Calculate ΔH_f^0 for silicon carbide.

35. Natural gas is mainly methane, $CH_4(g)$. Assume that gasoline is octane, $C_8H_{18}(\ell)$, and that kerosene is $C_{10}H_{22}(\ell)$. (a) Write the balanced equations for the combustion of each of these three hydrocarbons in excess O_2. The products are $CO_2(g)$ and $H_2O(\ell)$. (b) Calculate ΔH_{rxn}^0 at 25°C for each combustion reaction. ΔH_f^0 for $C_{10}H_{22}$ is -249.6 kJ/mol. (c) When burned at standard conditions, which of these three fuels would produce the most heat per mole? (d) When burned at standard conditions, which of the three would produce the most heat per gram?

Bond Energies

36. (a) How is the heat released or absorbed in a *gas phase reaction* related to bond energies of products and reactants? (b) Hess' Law states that

$$\Delta H_{rxn}^0 = \Sigma \, n\Delta H_{f\,products}^0 - \Sigma \, n\Delta H_{f\,reactants}^0$$

The relationship between ΔH_{rxn}^0 and bond energies for a *gas phase reaction* is

$$\Delta H_{rxn}^0 = \Sigma \text{ bond energies}_{reactants} - \Sigma \text{ bond energies}_{products}$$

It is *not* true, in general, that ΔH_f^0 for a substance is equal to the negative of the sum of the bond energies of the substance. Why?

37. (a) Suggest a reason for the fact that different amounts of energy are required for the successive removal of the three hydrogen atoms of an ammonia molecule, even though all N—H bonds in ammonia are equivalent. (b) Suggest why the N—H bonds in different compounds such as ammonia, NH_3; methylamine, CH_3NH_2; and ethylamine, $C_2H_5NH_2$, have slightly different bond energies.

38. Use tabulated bond energies to estimate the enthalpy of reaction for each of the following gas phase reactions.
    (a) $H_2C{=}CH_2 + Br_2 \longrightarrow BrH_2C{-}CH_2Br$
    (b) $H_2O_2 \longrightarrow H_2O + \tfrac{1}{2}O_2$

39. Use tabulated bond energies to estimate the enthalpy of reaction for each of the following gas phase reactions.
    (a) $N_2 + 3H_2 \longrightarrow 2NH_3$
    (b) $CH_4 + Cl_2 \longrightarrow CH_3Cl + HCl$
    (c) $CO + H_2O \longrightarrow CO_2 + H_2$

40. Use the bond energies in Table 15-2 to estimate the heat of reaction for

$$\underset{\underset{F}{|}}{\overset{\overset{Cl}{|}}{Cl-C-F}}(g) + F-F(g) \longrightarrow \underset{\underset{F}{|}}{\overset{\overset{F}{|}}{F-C-F}}(g) + Cl-Cl(g)$$

41. Estimate ΔH for the burning of one mole of butane, using the bond energies in Tables 15-2 and 15-3.

$$\underset{\underset{H}{|}\ \underset{H}{|}\ \underset{H}{|}\ \underset{H}{|}}{\overset{\overset{H}{|}\ \overset{H}{|}\ \overset{H}{|}\ \overset{H}{|}}{H-C-C-C-C-H}}(g) + \tfrac{13}{2}O=O(g) \longrightarrow$$

$$4O=C=O(g) + 5H-O-H(g)$$

Butane is used as a fuel in lighters.

42. Using data in Appendix K, calculate the average O—F bond energy in $OF_2(g)$.

43. Using data in Appendix K, calculate the average H—S bond energy in $H_2S(g)$.

44. Using data in Appendix K, calculate the average S—F bond energy in $SF_6(g)$.

*45. Methane undergoes several different exothermic reactions with gaseous chlorine. One of these forms chloroform, $CHCl_3(g)$.

$$CH_4(g) + 3Cl_2(g) \longrightarrow CHCl_3(g) + 3HCl(g)$$
$$\Delta H^0_{rxn} = -277 \text{ kJ/mol rxn}$$

Average bond energies per mole of bonds are: C—H = 414 kJ; Cl—Cl = 243 kJ; H—Cl = 431 kJ. Use these to calculate the average C—Cl bond energy in chloroform. Compare this with the value in Table 15-2.

*46. Ethylamine undergoes an endothermic gas phase dissociation to produce ethylene (or ethene) and ammonia.

$$\underset{\underset{H}{|}\ \underset{H}{|}}{\overset{\overset{H}{|}\ \overset{H}{|}}{H-C-C-N}}\overset{H}{\underset{H}{:}} \overset{\Delta}{\longrightarrow} \underset{\underset{H}{}\ \ \underset{H}{}}{\overset{\overset{H}{}\ \ \overset{H}{}}{C=C}} + \ddot{N}\underset{H\ H}{\overset{|}{\underset{|}{H}}}$$

$$\Delta H^0_{rxn} = +54.68 \text{ kJ/mol rxn}$$

The following average bond energies per mole of bonds are given: C—H = 414 kJ; C—C = 347 kJ; C=C = 611 kJ; N—H = 389 kJ. Calculate the C—N bond energy in ethylamine. Compare this with the value in Table 15-2.

Calorimetry

47. Define each of the following quantities, and give appropriate units. (a) heat capacity; (b) specific heat.

48. What is a bomb calorimeter? How do bomb calorimeters give us useful data?

49. What is a coffee-cup calorimeter? How do coffee-cup calorimeters give us useful information?

50. A calorimeter contained 75.0 g of water at 16.95°C. A 93.3-g sample of iron at 65.58°C was placed in it, giving a final temperature of 19.68°C for the system. Calculate the heat capacity of the calorimeter. Specific heats are 4.184 J/g·°C for H_2O and 0.444 J/g·°C for Fe.

51. A student wishes to determine the heat capacity of a coffee-cup calorimeter. After she mixes 100.0 g of water at 58.5°C with 100.0 g of water, already in the calorimeter, at 22.8°C, the final temperature of the water is 39.7°C. (a) Calculate the heat capacity of the calorimeter in J/°C. Use 4.18 J/g·°C as the specific heat of water. (b) Why is it more useful to express the value in J/°C rather than units of J/(g calorimeter·°C)?

52. A coffee-cup calorimeter having a heat capacity of 472 J/°C is used to measure the heat evolved when the following aqueous solutions, both initially at 22.6°C, are mixed: 100 g of solution containing 6.62 g of lead(II) nitrate, $Pb(NO_3)_2$, and 100 g of solution containing 6.00 g of sodium iodide, NaI. The final temperature is 24.2°C. Assume that the specific heat of the mixture is the same as that for water, 4.18 J/g·°C. The reaction is

$$Pb(NO_3)_2(aq) + 2NaI(aq) \longrightarrow PbI_2(s) + 2NaNO_3(aq)$$

(a) Calculate the heat evolved in the reaction. (b) Calculate the ΔH for the reaction [per mole of $Pb(NO_3)_2(aq)$ consumed] under the conditions of the experiment.

53. A coffee-cup calorimeter is used to determine the heat of reaction for the acid–base neutralization

$$CH_3COOH(aq) + NaOH(aq) \longrightarrow NaCH_3COO(aq) + H_2O(\ell)$$

When we add 20.00 mL of 0.625 M NaOH at 21.400°C to 30.00 mL of 0.500 M CH_3COOH already in the calorimeter at the same temperature, the resulting temperature is observed to be 24.347°C. The heat capacity of the calorimeter has previously been determined to be 27.8 J/°C. Assume that the specific heat of the mixture is the same as that of water, 4.18 J/g·°C and that the density of the mixture is 1.02 g/mL. (a) Calculate the amount of heat given off in the reaction. (b) Determine ΔH for the reaction under the conditions of the experiment.

54. In a bomb calorimeter compartment surrounded by 945 g of water, the combustion of 1.048 g of benzene, $C_6H_6(\ell)$, raised the temperature of the water from 23.640°C to 32.692°C. The heat capacity of the calorimeter is 891 J/°C. (a) Write the balanced equation for the combustion reaction, assuming that $CO_2(g)$ and $H_2O(\ell)$ are the only products. (b) Use the calorimetric data to calculate ΔE for the combustion of benzene in kJ/g and in kJ/mol.

55. A 2.00-g sample of hydrazine, N_2H_4, is burned in a bomb calorimeter that contains 6.40×10^3 g of H_2O, and the temperature increases from 25.00°C to 26.17°C. The heat capacity of the calorimeter is 3.76 kJ/°C. Calculate ΔE for the combustion of N_2H_4 in kJ/g and in kJ/mol.

Internal Energy and Changes in Internal Energy

56. (a) What are the sign conventions for q, the amount of heat added to or removed from a system? (b) What are the sign conventions for w, the amount of work done on or by a system?

57. What happens to ΔE for a system during a process in which (a) $q < 0$ and $w < 0$, (b) $q = 0$ and $w > 0$, and (c) $q > 0$ and $w < 0$?

58. What happens to ΔE for a system during a process in which (a) $q > 0$ and $w > 0$, (b) $q = w = 0$, and (c) $q < 0$ and $w > 0$?

59. A system performs 720 L·atm of pressure–volume work (1 L·atm = 101.325 J) on its surroundings and absorbs 6300 J of heat from its surroundings. What is the change in internal energy of the system?

60. A system receives 73 J of electrical work, delivers 227 J of pressure–volume work, and releases 212 J of heat. What is the change in internal energy of the system?

61. For each of the following chemical and physical changes carried out at constant pressure, state whether work is done by the system on the surroundings or by the surroundings on the system, or whether the amount of work is negligible.
(a) $C_6H_6(\ell) \longrightarrow C_6H_6(g)$
(b) $\frac{1}{2}N_2(g) + \frac{3}{2}H_2(g) \longrightarrow NH_3(g)$
(c) $SiO_2(s) + 3C(s) \longrightarrow SiC(s) + 2CO(g)$

62. Repeat Exercise 61 for
(a) $2SO_2(g) + O_2(g) \longrightarrow 2SO_3(g)$
(b) $CaCO_3(s) \longrightarrow CaO(s) + CO_2(g)$
(c) $CO_2(g) + H_2O(\ell) + CaCO_3(s) \longrightarrow$
$$Ca^{2+}(aq) + 2HCO_3{}^-(aq)$$

63. Assuming that the gases are ideal, calculate the amount of work done (in joules) in each of the following reactions. In each case, is the work done *on* or *by* the system? (a) A reaction in the Mond process for purifying nickel that involves formation of the gas, nickel(0) tetracarbonyl, at 50 to 100°C. Assume one mole of nickel is used and a constant temperature of 75°C is maintained.

$$Ni(s) + 4CO(g) \longrightarrow Ni(CO)_4(g)$$

(b) The conversion of one mole of brown nitrogen dioxide into colorless dinitrogen tetroxide at 10.0°C.

$$2NO_2(g) \longrightarrow N_2O_4(g)$$

64. Assuming that the gases are ideal, calculate the amount of work done (in joules) in each of the following reactions. In each case, is the work done *on* or *by* the system? (a) The oxidation of one mole of $HCl(g)$ at 200°C.

$$4HCl(g) + O_2(g) \longrightarrow 2Cl_2(g) + 2H_2O(g)$$

(b) The decomposition of one mole of nitric oxide (an air pollutant) at 300°C.

$$2NO(g) \longrightarrow N_2(g) + O_2(g)$$

*65. When an ideal gas expands at *constant temperature,* there is no change in molecular kinetic energy (kinetic energy is proportional to temperature), and there is no change in potential energy due to intermolecular attractions (these are zero for an ideal gas). Thus, for the isothermal (constant temperature) expansion of an ideal gas, $\Delta E = 0$. Suppose we allow an ideal gas to expand isothermally from 2.00 L to 5.00 L in two steps: (i) against a constant external pressure of 3.00 atm until equilibrium is reached, then (ii) against a constant external pressure of 2.00 atm until equilibrium is reached. Calculate q and w for this two-step expansion.

Entropy and Entropy Changes

66. What property of a system is described by entropy?

67. State the Second Law of Thermodynamics. We cannot use ΔS_{univ} directly as a measure of the spontaneity of a reaction. Why?

68. State the Third Law of Thermodynamics. What does it mean?

69. When solid sodium chloride is cooled from 25°C to 0°C, the entropy change is -4.4 J/mol·K. Is this an increase or decrease in randomness? Explain this entropy change in terms of what happens in the solid at the molecular level.

70. When a one-mole sample of argon gas at 0°C is compressed to one-half its original volume, the entropy change is -5.76 J/mol·K. Is this an increase or a decrease in randomness? Explain this entropy change in terms of what happens in the gas at the molecular level.

71. Which of the following processes are accompanied by an increase in entropy of the system? (No calculation is necessary.) (a) A building is constructed from bricks, mortar, lumber, and nails. (b) A building collapses into bricks, mortar, lumber, and nails. (c) Iodine sublimes, $I_2(s) \to I_2(g)$. (d) White silver sulfate, Ag_2SO_4, precipitates from a solution containing silver ions and sulfate ions. (e) A marching band is gathered into formation. (f) A partition is removed to allow two gases to mix.

72. Which of the following processes are accompanied by an increase in entropy of the system? (No calculation is necessary.) (a) Thirty-five pennies are removed from a bag and placed heads up on a table. (b) The pennies of part (a) are swept off the table and back into the bag. (c) Water freezes. (d) Carbon tetrachlo-

ride, CCl_4, evaporates.
(e) The reaction $PCl_5(g) \rightarrow PCl_3(g) + Cl_2(g)$ occurs.
(f) The reaction $PCl_3(g) + Cl_2(g) \rightarrow PCl_5(g)$ occurs.

73. For each of the following processes, tell whether the entropy of the *universe* increases, decreases, or remains constant. (a) melting one mole of ice to water at 0°C (b) freezing one mole of water to ice at 0°C (c) freezing one mole of water to ice at −10°C (d) freezing one mole of water to ice at 0°C and then cooling it to −10°C

*74. Consider the boiling of a pure liquid at constant pressure. Is each of the following greater than, less than, or equal to zero? (a) ΔS_{sys}; (b) ΔH_{sys}; (c) ΔT_{sys}

75. Use S^0 data from Appendix K to calculate the value of ΔS_{298}^0 for each of the following reactions. Compare the signs and magnitudes for these ΔS_{298}^0 values and explain your observations.
(a) $2NO(g) + H_2(g) \longrightarrow N_2O(g) + H_2O(g)$
(b) $2N_2O_5(g) \longrightarrow 4NO_2(g) + O_2(g)$
(c) $2NH_4NO_3(s) \longrightarrow 2N_2(g) + 4H_2O(g) + O_2(g)$

76. Use S^0 data from Appendix K to calculate the value of ΔS_{298}^0 for each of the following reactions. Compare the signs and magnitudes for these ΔS_{298}^0 values and explain your observations.
(a) $4HCl(g) + O_2(g) \longrightarrow 2Cl_2(g) + 2H_2O(g)$
(b) $PCl_3(g) + Cl_2(g) \longrightarrow PCl_5(g)$
(c) $2NO(g) \longrightarrow N_2(g) + O_2(g)$

Gibbs Free Energy Changes and Spontaneity

77. (a) What are the two factors that favor spontaneity of a process? (b) What is Gibbs free energy? What is change in Gibbs free energy? (c) Most spontaneous reactions are exothermic, but some are not. Explain. (d) Explain how the signs and magnitudes of ΔH and ΔS are related to the spontaneity of a process and how they affect it.

78. Which of the following conditions would predict a process that is (a) always spontaneous, (b) always nonspontaneous, or (c) spontaneous or nonspontaneous depending on the temperature and magnitudes of ΔH and ΔS? (i) $\Delta H > 0$, $\Delta S > 0$; (ii) $\Delta H > 0$, $\Delta S < 0$; (iii) $\Delta H < 0$, $\Delta S > 0$; (iv) $\Delta H < 0$, $\Delta S < 0$.

79. For the decomposition of $O_3(g)$ to $O_2(g)$

$$2O_3(g) \longrightarrow 3O_2(g)$$

$\Delta H^0 = -285.4$ kJ/mol and $\Delta S^0 = 137.55$ J/mol·K at 25°C. Calculate $\Delta G°$ for the reaction. Is the reaction spontaneous? Is either or both of the driving forces (ΔH^0 and ΔS^0) for the reaction favorable?

80. Calculate ΔG^0 at 25°C for the reaction

$$2NO_2(g) \longrightarrow N_2O_4(g)$$

given $\Delta H^0 = -57.20$ kJ/mol and $\Delta S^0 = -175.83$ J/mol·K. Is this reaction spontaneous? What is the driving force for spontaneity?

81. The standard Gibbs free energy of formation is −286.06 kJ/mol for NaI(s), −261.90 kJ/mol for Na$^+$(aq), and −51.57 kJ/mol for

I$^-$(aq) at 25°C. Calculate ΔG^0 for the reaction

$$NaI(s) \xrightarrow{\text{H}_2\text{O}} Na^+(aq) + I^-(aq)$$

*82. Use the following equations to find the ΔG^0 of formation of HBr(g) at 25°C.

$Br_2(\ell) \longrightarrow Br_2(g)$	$\Delta G^0 =$	3.110 kJ/mol
$HBr(g) \longrightarrow H(g) + Br(g)$	$\Delta G^0 =$	339.09 kJ/mol
$Br_2(g) \longrightarrow 2Br(g)$	$\Delta G^0 =$	164.792 kJ/mol
$H_2(g) \longrightarrow 2H(g)$	$\Delta G^0 =$	406.494 kJ/mol

83. Use values of standard free energy of formation, ΔG_f^0, from Appendix K, to calculate the standard free energy change for each of the following reactions at 25°C and 1 atm.
(a) $3NO_2(g) + H_2O(\ell) \longrightarrow 2HNO_3(\ell) + NO(g)$
(b) $SnO_2(s) + 2CO(g) \longrightarrow 2CO_2(g) + Sn(s)$
(c) $2Na(s) + 2H_2O(\ell) \longrightarrow 2NaOH(aq) + H_2(g)$

84. Make the same calculations as in Exercise 83, using values of standard enthalpy of formation and absolute entropy instead of values of ΔG_f^0.

85. Calculate ΔG^0 for the reduction of the oxides of iron and copper by carbon at 700 K represented by the equations

$$2Fe_2O_3(s) + 3C(graphite) \longrightarrow 4Fe(s) + 3CO_2(g)$$

$$2CuO(s) + C(graphite) \longrightarrow 2Cu(s) + CO_2(g)$$

Standard free energies of formation are −92 kJ/mol for CuO(s), −637 kJ/mol for Fe$_2$O$_3$(s), and −395 kJ/mol for CO$_2$(g). Which oxide can be reduced using carbon in a wood fire (which has a temperature of about 700 K), assuming standard state conditions?

Temperature Range of Spontaneity

86. Are the following statements true or false? Justify your answer. (a) An exothermic reaction is spontaneous. (b) If ΔH and ΔS are both positive, then ΔG will decrease when the temperature increases. (c) A reaction for which ΔS_{sys} is positive is spontaneous.

87. For the reaction

$$C(s) + O_2(g) \longrightarrow CO_2(g)$$

$\Delta H^0 = -393.51$ kJ/mol and $\Delta S^0 = 2.86$ J/mol·K at 25°C. Does this reaction become more or less favorable as the temperature increases? (b) For the reaction

$$C(s) + \tfrac{1}{2}O_2(g) \longrightarrow CO(g)$$

$\Delta H^0 = -110.52$ kJ/mol and $\Delta S^0 = 89.36$ J/mol·K at 25°C. Does this reaction become more or less favorable as the temperature increases? (c) Compare the temperature dependencies of these reactions.

88. How does the value of ΔG^0 change for the reaction

$$CCl_4(\ell) + H_2(g) \longrightarrow HCl(g) + CHCl_3(\ell)$$

if the reaction is carried out at 95°C rather than at 25°C? At 25°C, $\Delta G^0 = -103.75$ kJ/mol and $\Delta H^0 = -91.34$ kJ/mol for the reaction.

89. At 25°C, the enthalpy of reaction under standard state conditions for the combustion of CO,

$$CO(g) + \tfrac{1}{2}O_2(g) \longrightarrow CO_2(g)$$

is -282.98 kJ/mol, and ΔG^0 for the reaction is -257.19 kJ/mol. At what temperature will this reaction no longer be spontaneous under standard state conditions?

90. (a) Calculate ΔH^0, ΔG^0, and ΔS^0 for the reaction

$$2H_2O_2(\ell) \longrightarrow 2H_2O(\ell) + O_2(g)$$

at 25°C. (b) Is there any temperature at which $H_2O_2(\ell)$ is stable at 1 atm?

91. Estimate the normal boiling point of carbon tetrachloride, CCl_4, at 1 atm pressure, using Appendix K.

92. (a) Estimate the normal boiling point of water, at 1 atm pressure, using Appendix K. (b) Compare the temperature obtained with the known boiling point of water. Can you explain the discrepancy?

93. Sublimation and subsequent deposition onto a cold surface are a common method of purification of I_2 and other solids that sublime readily. Estimate the sublimation temperature (solid to vapor) of the dark-violet solid iodine, I_2, at 1 atm pressure, using the data of Appendix K.

Sublimation and deposition of I_2.

*94. (a) Is the reaction C(diamond) → C(graphite) spontaneous at 25°C and 1 atm? (b) Now are you worried about your diamonds turning to graphite? Why or why not? (c) Is there a temperature at which diamond and graphite are in equilibrium? If so, what is

this temperature? (d) How do you account for the formation of diamonds in the first place? (*Hint:* Diamond has a higher density than graphite.)

Mixed Exercises

*95. An ice calorimeter, shown below, can be used to measure the amount of heat released or absorbed by a reaction that is carried out at a constant temperature of 0°C. If heat is transferred from the system to the bath, some of the ice melts. A given mass of liquid water has a smaller volume than the same mass of ice, so the total volume of the ice/water mixture decreases. Measuring the volume decrease using the scale at the left indicates the amount of heat released by the reacting system. As long as some ice remains in the bath, the temperature remains at 0°C. In Example 15-2 we saw that the reaction

$$CuSO_4(aq) + 2NaOH(aq) \longrightarrow Cu(OH)_2(s) + Na_2SO_4(aq)$$

releases 1.49 kJ of heat at constant temperature and pressure when 50.0 mL of 0.400 M $CuSO_4$ solution and 50.0 mL of 0.600 M NaOH solution are allowed to react. (Because no gases are involved in the reaction, the volume change of the reaction mixture is negligible.) Calculate the change in volume of the ice/water mixture that would be observed if we carried out the same experiment in an ice calorimeter. The density of $H_2O(\ell)$ at 0°C is 0.99987 g/mL and that of ice is 0.917 g/mL. The heat of fusion of ice at 0°C is 334 J/g.

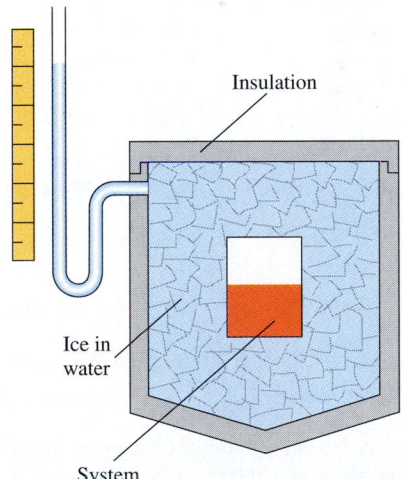

*96. Energy to power muscular work is produced from stored carbohydrate (glycogen) or fat (triglycerides). Metabolic consumption and production of energy are described with the nutritional "Calorie," which is equal to 1 kilocalorie. Average energy output per minute for various activities follows: sitting (1.7 kcal); walking, level, 3.5 mph (5.5 kcal); cycling, level, 13 mph (10 kcal); swimming (8.4 kcal); running 10 mph (19 kcal). Approximate energy values of some common foods are also given: large apple (100 kcal); 8-oz cola drink (105 kcal); malted milkshake (8 oz milk, 500 kcal); $\frac{3}{4}$ cup pasta with tomato sauce and cheese (195 kcal); hamburger on bun with sauce (350 kcal); 10-oz sirloin steak, including fat (1000 kcal). To maintain body weight, fuel intake should balance energy output. Prepare a table showing (a) each given food, (b) its fuel value, and (c) the minutes of each activity that would balance the kcal of each food.

97. Write the chemical equation for the reaction between ethene, C_2H_4, and hydrogen to form ethane, C_2H_6. Find the enthalpy change using the following thermochemical equations.

$$C_2H_4(g) + 3O_2(g) \longrightarrow 2CO_2(g) + 2H_2O(\ell)$$
$$\Delta H^0 = -1410.9 \text{ kJ/mol}$$

$$C_2H_6(g) + \tfrac{7}{2}O_2(g) \longrightarrow 2CO_2(g) + 3H_2O(\ell)$$
$$\Delta H^0 = -1559.8 \text{ kJ/mol}$$

$$H_2(g) + \tfrac{1}{2}O_2(g) \longrightarrow H_2O(\ell) \qquad \Delta H^0 = -285.8 \text{ kJ/mol}$$

98. It is difficult to prepare many compounds directly from their elements, so ΔH_f^0 values for these compounds cannot be measured directly. For many organic compounds, it is easier to measure the standard enthalpy of combustion by reaction of the compound with excess $O_2(g)$ to form $CO_2(g)$ and $H_2O(\ell)$. From the following standard enthalpies of combustion at 25°C, determine ΔH_f^0 for the compound. (a) cyclohexane, $C_6H_{12}(\ell)$, a useful organic solvent: $\Delta H_{combustion}^0 = -3920$ kJ/mol; (b) phenol, $C_6H_5OH(s)$, used as a disinfectant and in the production of thermo-setting plastics: $\Delta H_{combustion}^0 = -3053$ kJ/mol

*99. Standard entropy changes cannot be measured directly in the laboratory. They are calculated from experimentally obtained values of ΔG^0 and ΔH^0. From the data given below, calculate ΔS^0 at 298 K for each of the following reactions.

(a) $OF_2(g) + H_2O(g) \longrightarrow O_2(g) + 2HF(g)$
oxygen difluoride

$$\Delta H^0 = -323 \text{ kJ/mol} \qquad \Delta G^0 = -358.4 \text{ kJ/mol}$$

(b) $CaC_2(s) + 2H_2O(\ell) \longrightarrow Ca(OH)_2(s) + C_2H_2(g)$
calcium carbide   acetylene

$$\Delta H^0 = -125.4 \text{ kJ/mol} \qquad \Delta G^0 = -145.4 \text{ kJ/mol}$$

$CaO(s) + H_2O(\ell) \longrightarrow Ca(OH)_2(aq) \ \Delta H^0 = 81.5$ kJ/mol rxn

$$\Delta H^0 = -830.8 \text{ kJ/mol} \qquad \Delta G^0 = -780.9 \text{ kJ/mol}$$

*100. (a) A student heated a sample of a metal weighing 32.6 g to 99.83°C and put it into 100.0 g of water at 23.62°C in a calorimeter. The final temperature was 24.41°C. The student calculated the specific heat of the metal, but neglected to use the heat capacity of the calorimeter. The specific heat of water is 4.184 J/g · °C. What was his answer? The metal was known to be chromium, molybdenum, or tungsten. By comparing the value of the specific heat to those of the metals (Cr, 0.460; Mo, 0.250; W, 0.135 J/g · °C), the student identified the metal. What was the metal? (b) A student at the next laboratory bench did the same experiment, obtained the same data, and used the heat capacity

of the calorimeter in his calculations. The heat capacity of the calorimeter was 410 J/°C. Was his identification of the metal different?

101. Find the heat of formation of hydrogen peroxide, $H_2O_2(\ell)$, from

$$2H_2O_2(\ell) \longrightarrow 2H_2O(\ell) + O_2(g) \qquad \Delta H = -196.0 \text{ kJ/mol}$$

and $\Delta H_f^0 \ H_2O(\ell) = -285.8$ kJ/mol.

*102. Calculate q, w, and ΔE for the vaporization of 10.0 g of liquid ethanol (C_2H_5OH) at 1.00 atm at 78.0°C, to form gaseous ethanol at 1.00 atm at 78.0°C. Make the following simplifying assumptions: (i) the density of liquid ethanol at 78.0°C is 0.789 g/mL, and (ii) gaseous ethanol is adequately described by the ideal gas equation. The heat of vaporization of ethanol is 854 J/g.

103. We add 0.100 g of CaO(s) to 125 g H_2O at 23.6°C in a coffee-cup calorimeter. The following reaction occurs. What will be the final temperature of the solution?

$$CaO(s) + H_2O(\ell) \longrightarrow Ca(OH)_2(aq) \ \Delta H^0 = 81.5 \text{ kJ/mol rxn}$$

104. (a) The accurately known molar heat of combustion of naphthalene, $C_{10}H_8(s)$, $\Delta H = -5156.8$ kJ/mol $C_{10}H_8$, is used to calibrate calorimeters. The complete combustion of 0.01520 g of $C_{10}H_8$ at constant pressure raises the temperature of a calorimeter by 0.212°C. Find the heat capacity of the calorimeter. (b) The initial temperature of the calorimeter (part a) is 22.102°C; 0.1040 g of $C_8H_{18}(\ell)$, octane (molar heat of combustion $\Delta H = -5450$ kJ/mol C_8H_{18}), is completely burned in the calorimeter. Find the final temperature of the calorimeter.

105. When a gas expands suddenly, it may not have time to absorb a significant amount of heat: $q = 0$. Assume that 1.00 mol N_2 expands suddenly, doing 3000 J of work. (a) What is ΔE for the process? (b) The heat capacity of N_2 is 20.9 J/mol · °C. How much does its temperature fall during this expansion? (This is the principle of most snow-making machines, which use compressed air mixed with water vapor.)

BUILDING YOUR KNOWLEDGE

106. As a rubber band is stretched, it gets warmer; when released, it gets cooler. To obtain the more nearly linear arrangement of the rubber band's polymeric material from the more random relaxed rubber band requires that there be rotation about carbon–carbon single bonds. Based upon these data, give the sign of ΔG, ΔH, and ΔS for the stretching of a rubber band and for the relaxing of a stretched rubber band. What drives the spontaneous process?

107. From its heat of fusion, calculate the entropy change associated with the melting of one mole of ice at its melting point. From its heat of vaporization, calculate the entropy change associated with the boiling of one mole of water at its boiling point. Are your calculated values consistent with the simple model that we use to describe order in solids, liquids, and gases?

108. The energy content of dietary fat is 39 kJ/g, while for protein and carbohydrate it is 17 and 16 kJ/g, respectively. A 70.0 kg (155 lb) person utilizes 335 kJ/hr while resting and 1250 kJ/hr while walking 6 km/hr. How many hours would the person need to walk per day instead of resting if he or she consumed 100 g (about $\frac{1}{4}$ lb) of fat instead of 100 g of protein?

109. The enthalpy change for melting one mole of water at 273 K is $\Delta H^0_{273} = 6010$ J/mol, whereas that for vaporizing a mole of water at 373 K is $\Delta H^0_{373} = 40,660$ J/mol. Why is the second value so much larger?

110. A 436-g chunk of lead was removed from a beaker of boiling water, quickly dried, and dropped into a Styrofoam cup containing 50.0 g of water at 25.0°C. As the system reached equilibrium, the water temperature rose to 40.8°C. Calculate the heat capacity and the specific heat of the lead.

111. Methane, $CH_4(g)$, is the main constituent of natural gas. In excess oxygen, methane burns to $CO_2(g)$ and $H_2O(\ell)$, whereas in limited oxygen, the products are $CO(g)$ and $H_2O(\ell)$. Which would result in a higher temperature—a gas–air flame or a gas–oxygen flame? How can you tell?

112. A 0.483-g sample of butter was burned in a bomb calorimeter whose heat capacity was 4572 J/°C, and the temperature was observed to rise from 24.76 to 27.93°C. Calculate the fuel value of butter in (a) kJ/g; (b) nutritional Calories/g (one nutritional Calorie is equal to one kilocalorie); (c) nutritional Calories/5-gram pat.

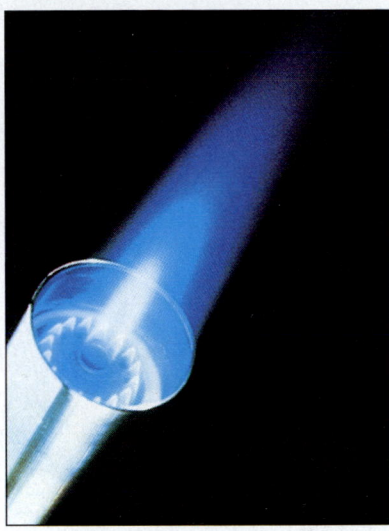

A methane flame.

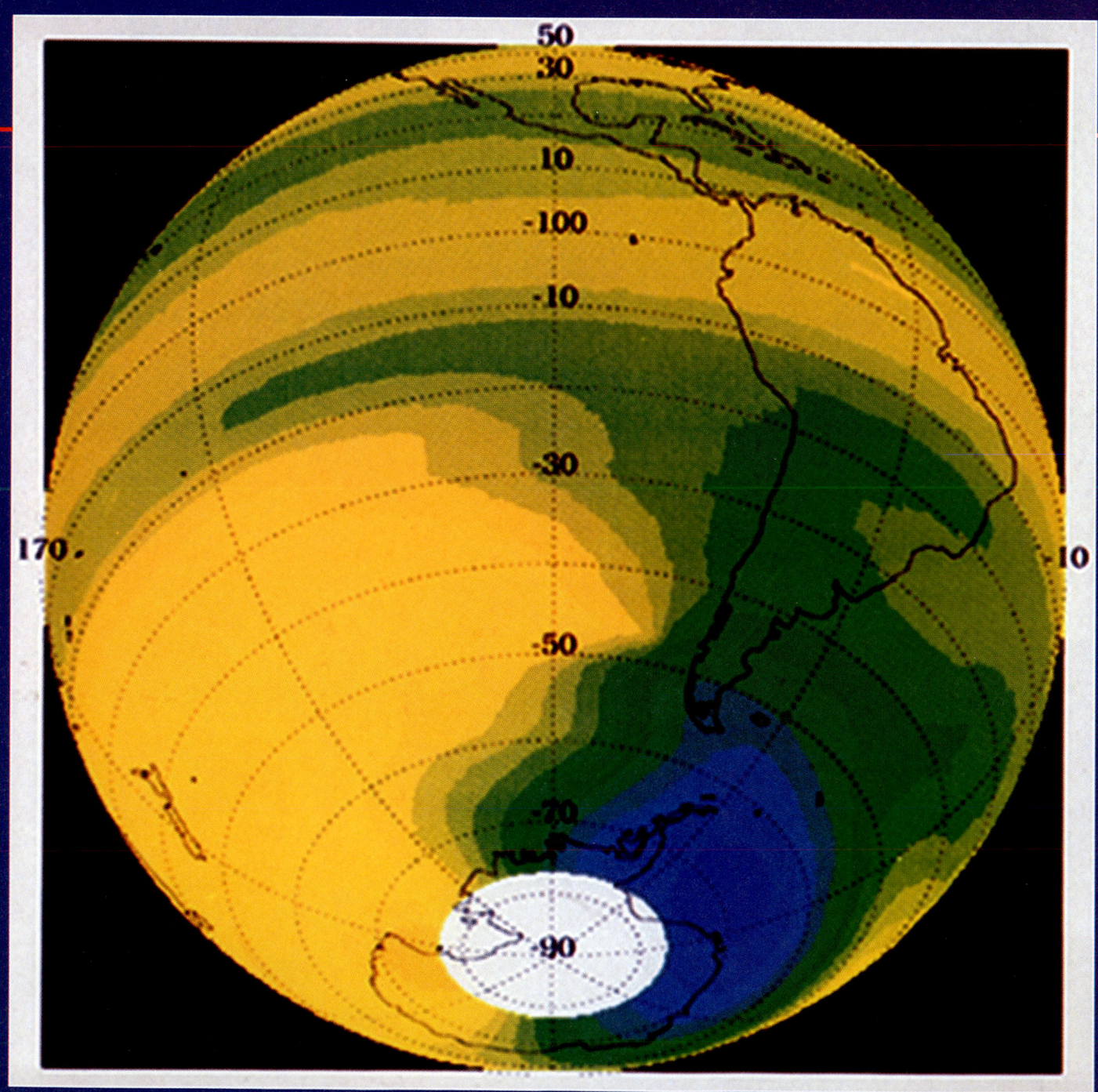

OBJECTIVES

As you study this chapter, you should learn

- *How to express the rate of a chemical reaction in terms of changes in concentrations of reactants and products with time*

- *About the experimental factors that affect the rates of chemical reactions*

- *How to use the rate-law expression for a reaction—the relationship between concentration and rate*

- *How to use the concept of order of a reaction*

- *To apply the method of initial rates to find the rate-law expression for a reaction*

- *How to use the integrated rate-law expression for a reaction—the relationship between concentration and time*

- *How to analyze concentration-versus-time data to determine the order of a reaction*

- *About the collision theory of reaction rates*

- *The main aspects of transition state theory and the role of activation energy in determining the rate of a reaction*

- *How the mechanism of a reaction is related to its rate-law expression*

- *How temperature affects rates of reactions*

- *How to use the Arrhenius equation to relate the activation energy for a reaction to changes in its rate constant with changing temperature*

- *How a catalyst changes the rate of a reaction*

- *About homogeneous catalysis and heterogeneous catalysis*

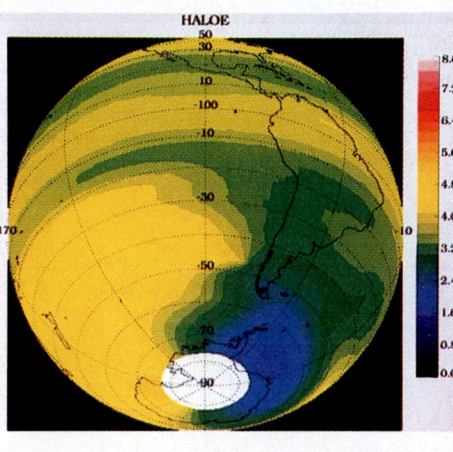

A computer map of ozone concentrations of the southern hemisphere during the period September 21 to October 15, 1992. The depletion of the ozone layer by man-made pollutants is a major environmental concern. This map shows the hole (dark blue) in the ozone layer over Antarctica.

W e are all familiar with processes in which some quantity changes with time—an automobile travels at 40 miles/hour, a faucet delivers water at 3 gallons/minute, or a factory produces 32,000 tires/day. Each of these ratios is called a rate. The **rate of a reaction** describes how fast reactants are used up and products are formed. **Chemical kinetics** is the study of *rates* of chemical reactions and the *mechanisms* (the series of steps) by which they occur.

Our experience tells us that different chemical reactions occur at very different rates. For instance, combustion reactions—such as the burning of methane, CH_4, in natural gas and the combustion of isooctane, C_8H_{18}, in gasoline—proceed very rapidly, sometimes even explosively.

$$CH_4(g) + 2O_2(g) \longrightarrow CO_2(g) + 2H_2O(g)$$

$$2C_8H_{18}(g) + 25O_2(g) \longrightarrow 16CO_2(g) + 18H_2O(g)$$

On the other hand, the rusting of iron occurs only very slowly.

In our study of thermodynamics we learned to assess whether a particular reaction was favorable. The question of whether substantial reaction would occur in a certain time period is addressed by kinetics. If a reaction is not thermodynamically favored, it will not occur appreciably under the given conditions. Even though a reaction is thermodynamically favored, it can occur, but not necessarily at a measurable rate.

The reactions of strong acids with strong bases are thermodynamically favored *and* occur at very rapid rates. Consider, for example, the reaction of hydrochloric acid solution with solid magnesium hydroxide. It is thermodynamically spontaneous at standard state conditions, as indicated by the negative ΔG^0 value. It also occurs rapidly.

$$2HCl(aq) + Mg(OH)_2(s) \longrightarrow MgCl_2(aq) + 2H_2O(\ell) \qquad \Delta G^0 = -97 \text{ kJ/mol}$$

The reaction of diamond with oxygen is also spontaneous.

$$C(\text{diamond}) + O_2(g) \longrightarrow CO_2(g) \qquad \Delta G^0 = -397 \text{ kJ/mol}$$

However, we know from experience that diamonds exposed to air, even over long periods, do not react to form carbon dioxide. The reaction does not occur at an observable rate near room temperature. This observation is explained by kinetics, not thermodynamics.

16-1 THE RATE OF A REACTION

Rates of reactions are usually expressed as moles per liter per unit time. If we know the chemical equation for a reaction, its rate can be determined by following the change in concentration of any product or reactant that can be detected quantitatively.

To describe the rate of a reaction, we must determine the concentration of a reactant or product at various times as the reaction proceeds. Devising effective methods for this is a continuing challenge for chemists who study chemical kinetics. If a reaction is slow enough, we can take samples from the reaction mixture after successive time intervals and then analyze them. For instance, if one reaction product is an acid, its concentration can be determined by titration (Section 11-2) after each time interval. The reaction of ethyl acetate with water in the presence of a small amount of strong acid produces acetic acid. The extent of the reaction at any desired time can be determined by titration of the acetic acid.

$$\underset{\text{ethyl acetate}}{CH_3\overset{O}{\overset{\|}{C}}-OCH_2CH_3(aq)} + H_2O \xrightarrow{H^+} \underset{\text{acetic acid}}{CH_3\overset{O}{\overset{\|}{C}}-OH(aq)} + \underset{\text{ethanol}}{CH_3CH_2OH(aq)}$$

This approach is suitable only if the reaction is sufficiently slow that the time elapsed during withdrawal and analysis of the sample is negligible. Sometimes the sample that is withdrawn is quickly cooled ("quenched"). This slows the reaction (Section 16-8) so much that the desired concentration does not change significantly while the analysis is performed.

The rusting of iron is a complicated process. It can be represented in simplified form as

$$4Fe(s) + 3O_2(g) \longrightarrow 2Fe_2O_3(s).$$

This is one of the reactions that occurs in a human digestive system when an antacid containing relatively insoluble magnesium hydroxide neutralizes excess stomach acid.

Recall that the units kJ/mol refer to the numbers of moles of reactants and products in the balanced equation.

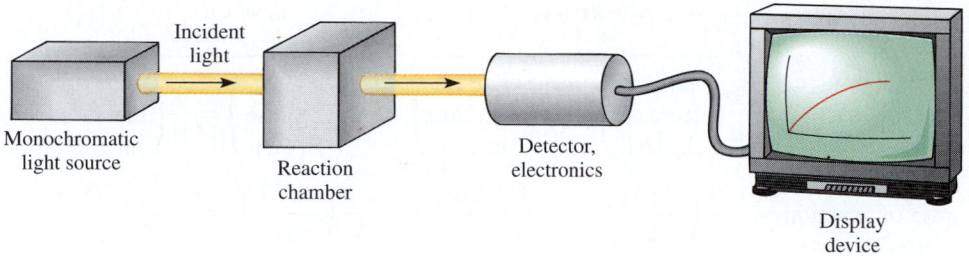

Figure 16-1 A spectroscopic method for determining reaction rates. Light of a wavelength that is absorbed by some substance whose concentration is changing is passed through a reaction chamber. Recording the change in light intensity gives a measure of the changing concentration of a reactant or product as the reaction progresses.

It is more convenient, especially when a reaction is rapid, to use a technique that continually monitors the change in some physical property of the system. If one of the reactants or products is colored, the increase (or decrease) in intensity of its color might be used to measure a decrease or increase in its concentration. Such an experiment is a special case of *spectroscopic* methods. These methods involve passing light (visible, infrared, or ultraviolet) through the sample. The light should have a wavelength that is absorbed by some substance whose concentration is changing (Figure 16-1). An appropriate light-sensing apparatus provides a signal that depends on the concentration of the absorbing substance. Modern techniques that use computer-controlled pulsing and sensing of lasers have enabled scientists to sample concentrations at intervals on the order of picoseconds (1 picosecond = 10^{-12} second) or even femtoseconds (1 femtosecond = 10^{-15} second).

If the progress of a reaction causes a change in the total number of moles of gas present, the change in pressure of the reaction mixture (held at constant temperature and constant volume) lets us measure how far the reaction has gone. For instance, the decomposition of dinitrogen pentoxide, $N_2O_5(g)$, has been studied by this method.

$$2N_2O_5(g) \longrightarrow 4NO_2(g) + O_2(g)$$

For every 2 moles of N_2O_5 gas that react, a total of 5 moles of gas is formed (4 moles of NO_2 and 1 mole of O_2). The resulting increase in pressure can be related by the ideal gas equation to the total number of moles of gas present. This indicates the extent to which the reaction has occurred.

In reactions involving gases, rates of reactions may be related to rates of change of partial pressures. Pressures of gases and concentrations of gases are directly proportional.

Once we have measured the changes in concentrations of reactants or products with time, how do we describe the rate of a reaction? Consider a hypothetical reaction.

$$aA + bB \longrightarrow cC + dD$$

The amount of each substance present can be given by its concentration, usually expressed as molarity (mol/L) and designated by brackets. The rate at which the reaction proceeds can be described in terms of the rate at which one of the reactants disappears, $-\Delta[A]/\Delta t$ or $-\Delta[B]/\Delta t$, or the rate at which one of the products appears, $\Delta[C]/\Delta t$ or $\Delta[D]/\Delta t$. The reaction rate must be positive because it describes the forward (left-to-right) reaction, which consumes A and B. The concentrations of reactants A and B decrease in the time interval Δt. Thus, $\Delta[A]/\Delta t$ and $\Delta[B]/\Delta t$ would be *negative* quantities. The purpose of a negative sign in the definition of a rate of reaction is to make the rate a positive quantity.

If no other reaction takes place, the changes in concentration are related to each other. For every a mol/L that [A] decreases, [B] must decrease by b mol/L, [C] must increase by c mol/L, and so on. We wish to describe the rate of reaction on a basis that is the same regardless of which reactant or product we choose to measure. Therefore, we divide each change by its coefficient in the balanced equation. This allows us to write the rate of reaction based on the rate of change of concentration of each species.

$$
\begin{array}{c}
\overbrace{}^{\text{in terms of reactants}} \qquad \overbrace{}^{\text{in terms of products}} \\[4pt]
\text{rate of reaction} = \frac{1}{a}\left(\begin{array}{c}\text{rate of} \\ \text{decrease} \\ \text{in [A]}\end{array}\right) = \frac{1}{b}\left(\begin{array}{c}\text{rate of} \\ \text{decrease} \\ \text{in [B]}\end{array}\right) = \frac{1}{c}\left(\begin{array}{c}\text{rate of} \\ \text{increase} \\ \text{in [C]}\end{array}\right) = \frac{1}{d}\left(\begin{array}{c}\text{rate of} \\ \text{increase} \\ \text{in [D]}\end{array}\right)
\end{array}
$$

$$
\text{rate of reaction} = -\frac{1}{a}\left(\frac{\Delta[A]}{\Delta t}\right) = -\frac{1}{b}\left(\frac{\Delta[B]}{\Delta t}\right) = \frac{1}{c}\left(\frac{\Delta[C]}{\Delta t}\right) = \frac{1}{d}\left(\frac{\Delta[D]}{\Delta t}\right)
$$

This representation gives several equalities, any one of which can be used to relate changes in observed concentrations to the rate of reaction.

The expressions just given describe the *average* rate over a time period Δt. The rigorous expressions for the rate at any instant involve the derivatives of concentrations with respect to time.

$$
-\frac{1}{a}\left(\frac{d[A]}{dt}\right), \ \frac{1}{c}\left(\frac{d[C]}{dt}\right), \ \text{and so on.}
$$

The shorter the time period, the closer $\dfrac{\Delta(\text{concentration})}{\Delta t}$ is to the corresponding derivative.

▼ **PROBLEM-SOLVING TIP**   *Signs and Divisors in Expressions for Rate*

As an analogy to these expressions, suppose we make sardine sandwiches by the following procedure:

$$
2 \text{ bread slices} + 3 \text{ sardines} + 1 \text{ pickle} \longrightarrow 1 \text{ sandwich}
$$

The number of sandwiches increases, so $\Delta(\text{sandwiches})$ is positive; the rate of the process is given by $\Delta(\text{sandwiches})/\Delta(\text{time})$. Alternatively, we could count the decreasing number of pickles at various times. Because $\Delta(\text{pickles})$ is negative, we must multiply by (-1) to make the rate positive; rate $= -\Delta(\text{pickles})/\Delta(\text{time})$. If we choose to measure the rate by counting slices of bread, we must also take into account that bread slices are consumed *twice as fast* as sandwiches are produced, so rate $= -\frac{1}{2}(\Delta(\text{bread})/\Delta(\text{time}))$. Four different ways of describing the rate all have the same numerical value.

$$
\text{rate} = \left(\frac{\Delta(\text{sandwiches})}{\Delta t}\right) = -\frac{1}{2}\left(\frac{\Delta(\text{bread})}{\Delta t}\right) = -\frac{1}{3}\left(\frac{\Delta(\text{sardines})}{\Delta t}\right) = -\left(\frac{\Delta(\text{pickles})}{\Delta t}\right)
$$

Consider as a specific chemical example the gas phase reaction that occurs when we mix 1.000 mole of hydrogen and 2.000 moles of iodine chloride at 230°C in a closed 1.000-liter container.

$$
H_2(g) + 2ICl(g) \longrightarrow I_2(g) + 2HCl(g)
$$

The coefficients tell us that 1 mole of H_2 disappears for every 2 moles of ICl that disappear and for every 1 mole of I_2 and 2 moles of HCl that are formed. In other terms, the rate of disappearance of moles of H_2 is one-half the rate of disappearance of moles of ICl, and so on. So we write the rate of reaction as

As we see from this discussion, the units of the rate of a reaction are $\dfrac{\text{mol rxn}}{L \cdot \text{time}}$. We often abbreviate this as mol/L $\cdot$ time or $M \cdot \text{time}^{-1}$.

$$
\text{rate of reaction} = \left(\begin{array}{c}\text{rate of} \\ \text{decrease} \\ \text{in [H}_2\text{]}\end{array}\right) = \frac{1}{2}\left(\begin{array}{c}\text{rate of} \\ \text{decrease} \\ \text{in [ICl]}\end{array}\right) = \left(\begin{array}{c}\text{rate of} \\ \text{increase} \\ \text{in [I}_2\text{]}\end{array}\right) = \frac{1}{2}\left(\begin{array}{c}\text{rate of} \\ \text{increase} \\ \text{in [HCl]}\end{array}\right)
$$

$$
\text{rate of reaction} = -\left(\frac{\Delta[H_2]}{\Delta t}\right) = -\frac{1}{2}\left(\frac{\Delta[ICl]}{\Delta t}\right) = \left(\frac{\Delta[I_2]}{\Delta t}\right) = \frac{1}{2}\left(\frac{\Delta[HCl]}{\Delta t}\right)
$$

Table 16-1 *Concentration and Rate Data for Reaction of 2.000 M ICl and 1.000 M H₂ at 230°C*

[ICl] (mol/L)	[H₂] (mol/L)	Average Rate During One Time Interval $= -\dfrac{\Delta[H_2]}{\Delta t}$ $(M \cdot s^{-1})$	Time (t) (seconds)
2.000	1.000		0
		0.326	
1.348	0.674		1
		0.148	
1.052	0.526		2
		0.090	
0.872	0.436		3
		0.062	
0.748	0.374		4
		0.046	
0.656	0.328		5
		0.035	
0.586	0.293		6
		0.028	
0.530	0.265		7
		0.023	
0.484	0.242		8

For example, the *average* rate over the interval from 1 to 2 seconds can be calculated as

$$-\frac{\Delta[H_2]}{\Delta t} =$$
$$-\frac{(0.526 - 0.674)\ \text{mol} \cdot \text{L}^{-1}}{(2 - 1)\ \text{s}}$$
$$= 0.148\ \text{mol} \cdot \text{L}^{-1} \cdot \text{s}^{-1}$$
$$= 0.148\ M \cdot \text{s}^{-1}$$

This does *not* mean that the reaction proceeds at this rate during the entire interval.

Table 16-1 lists the concentrations of reactants remaining at 1-second intervals, beginning with the time of mixing ($t = 0$ seconds). The *average* rate of reaction over each 1-second interval is indicated in terms of the rate of decrease in concentration of hydrogen. Verify for yourself that the rate of loss of ICl is twice that of H₂. Therefore, the rate of reaction could also be expressed as rate $= -\dfrac{1}{2}(\Delta[ICl]/\Delta t)$. Increases in concentrations of products could be used instead. Figure 16-2 shows graphically the rates of change of concentrations of all reactants and products.

Figure 16-3 is a plot of the hydrogen concentration versus time, using data of Table 16-1. The initial rate, or the rate at the instant of mixing the reactants, is the negative of the slope at $t = 0$. The *instantaneous* rate of reaction at time t (2.0 seconds, for example)

Suppose a driver goes 40 miles in an hour; we describe his average speed (rate) as 40 mi/h. This does not necessarily mean that he traveled at a steady speed. He might have stopped at a few traffic signals, made a fuel stop, driven sometimes faster, sometimes slower—his *instantaneous rate* (the rate at which he was traveling at any instant) was quite changeable.

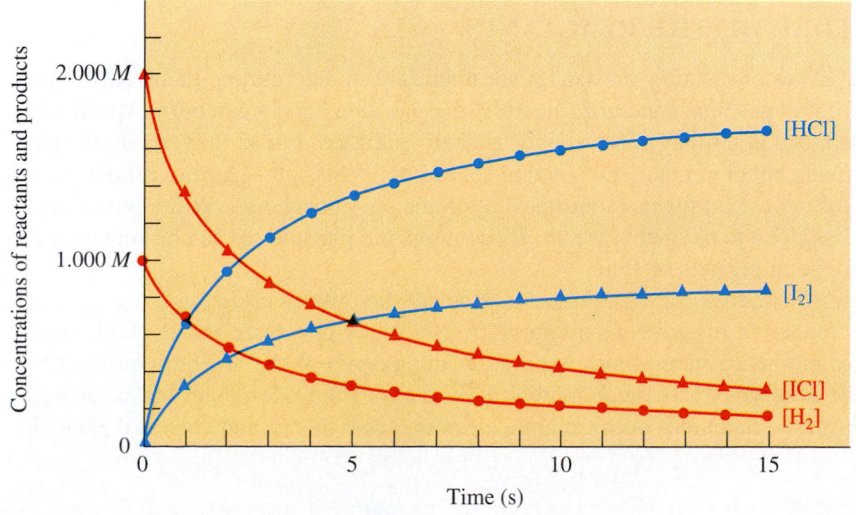

Figure 16-2 Plot of concentrations of all reactants and products versus time in the reaction of 2.000 *M* ICl with 1.000 *M* H₂ at 230°C, from data in Table 16-1 (and a few more points).

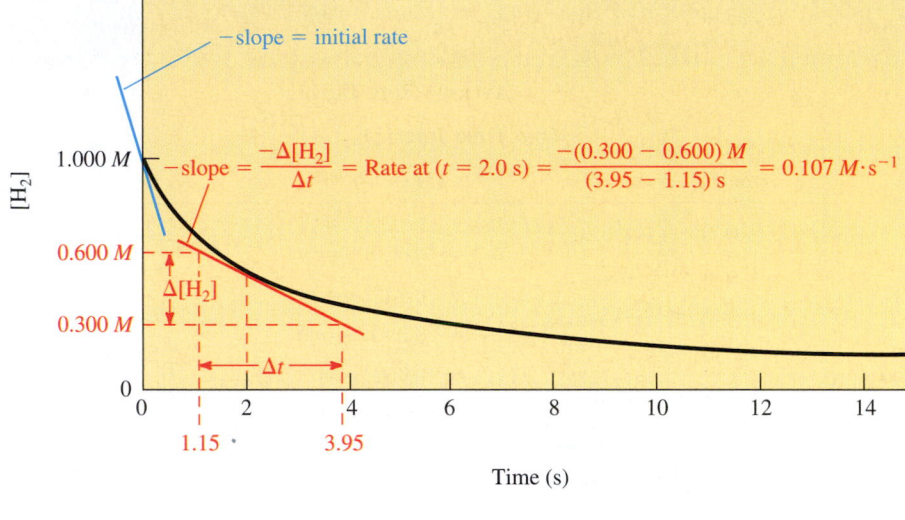

Figure 16-3   Plot of H_2 concentration versus time for the reaction of 2.000 M ICl with 1.000 M H_2. The instantaneous rate of reaction at any time, t, equals the negative of the slope of the tangent to this curve at time t. The initial rate of the reaction is equal to the negative of the initial slope ($t = 0$). The determination of the instantaneous rate at $t = 2$ seconds is illustrated. (If you do not recall how to find the slope of a straight line, refer to Figure 16-5.)

is the negative of the slope of the tangent to the curve at time t. We see that the rate decreases with time; lower concentrations of H_2 and ICl result in slower reaction. Had we plotted concentration of a product versus time, the rate would have been related to the *positive* slope of the tangent at time t.

FACTORS THAT AFFECT REACTION RATES

Four factors have marked effects on the rates of chemical reactions. They are (1) nature of the reactants, (2) concentrations of the reactants, (3) temperature, and (4) the presence of a catalyst. Understanding their effects can help us control the rates of reactions. The study of these factors gives important insight into the details of the processes by which a reaction occurs. This kind of study is the basis for developing theories of chemical kinetics. Now we study these factors and the related theories—collision theory and transition state theory.

16-2   NATURE OF THE REACTANTS

The physical states of reacting substances are important in determining their reactivities. A puddle of liquid gasoline can burn smoothly, but gasoline vapors can burn explosively. Two immiscible liquids may react slowly at their interface, but if they are intimately mixed to provide better contact, the reaction speeds up. White phosphorus and red phosphorus are different solid forms (allotropes) of elemental phosphorus. White phosphorus ignites when exposed to oxygen in the air. By contrast, red phosphorus can be kept in open containers for long periods of time.

The extent of subdivision of solids or liquids can be crucial in determining reaction rates. Large chunks of most metals do not burn. But many powdered metals, with larger surface areas and hence more atoms exposed to the oxygen of the air, burn easily. One pound of fine iron wire rusts much more rapidly than a solid one-pound chunk of iron. Violent explosions sometimes occur in grain elevators, coal mines, and chemical plants in

A burning building is an example of a rapid, highly exothermic reaction. Firefighters use basic principles of chemical kinetics. When water is sprayed onto a fire, its evaporation absorbs a large amount of energy; this lowers the temperature and slows the reaction. Other common methods for extinguishing fires include covering them with CO_2 (as with most household extinguishers), which decreases the supply of oxygen, and backburning (for grass and forest fires), which removes combustible material. In both cases, the removal of a reactant slows (or stops) the reaction.

which large amounts of powdered substances are produced. These explosions are examples of the effect of large surface areas on rates of reaction. The rate of reaction depends on the surface area or degree of subdivision. The ultimate degree of subdivision would make all reactant molecules (or ions or atoms) accessible to react at any given time. This situation is achieved when the reactants are in the gaseous state or in solution.

Chemical identities of elements and compounds affect reaction rates. Metallic sodium, with its low ionization energy, reacts rapidly with water at room temperature; metallic calcium has a higher ionization energy and reacts only slowly with water at room temperature. Solutions of a strong acid and a strong base react rapidly when they are mixed because the interactions involve mainly electrostatic attractions between ions in solution. Reactions that involve the breaking of covalent bonds are usually slower.

Samples of dry solid potassium sulfate, K_2SO_4, and dry solid barium nitrate, $Ba(NO_3)_2$, can be mixed with no appreciable reaction occurring for several years. But if aqueous solutions of the two are mixed, a reaction occurs rapidly, forming a white precipitate of barium sulfate.

$$Ba^{2+}(aq) + SO_4{}^{2-}(aq) \longrightarrow BaSO_4(s)$$

Two allotropes of phosphorus. White phosphorus (above) ignites and burns rapidly when exposed to oxygen in the air, so it is stored under water. Red phosphorus (below) reacts with air much more slowly, and can be stored in contact with air.

16-3 CONCENTRATIONS OF REACTANTS: THE RATE-LAW EXPRESSION

As the concentrations of reactants change at constant temperature, the rate of reaction changes. We write the **rate-law expression** (often called simply the **rate law**) for a reaction to describe how its rate depends on concentrations; this rate law is experimentally deduced for each reaction from a study of how its rate varies with concentration.

> The rate-law expression for a reaction in which A, B, . . . are reactants has the general form
> $$\text{rate} = k[A]^x[B]^y \cdots$$

The constant k is called the **specific rate constant** (or just the **rate constant**) for the reaction at a particular temperature. The values of the exponents, x and y, and of the rate constant, k, bear no necessary relationship to the coefficients in the *balanced chemical equation* for the overall reaction and must be determined *experimentally*.

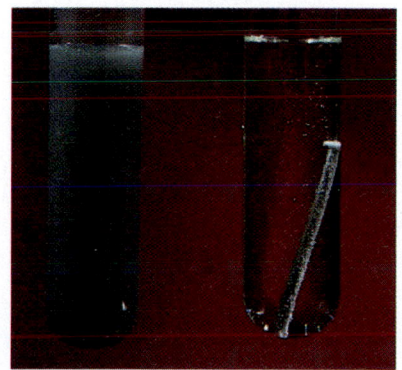

Powdered iron reacts rapidly with dilute sulfuric acid because it has a large total surface area. An iron nail reacts much more slowly.

The powers to which the concentrations are raised, x and y, are usually integers or zero but are occasionally fractional or even negative. A power of *one* means that the rate is directly proportional to the concentration of that reactant. A power of *two* means that the rate is directly proportional to the *square* of that concentration. A power of *zero* means that the rate does not depend on the concentration of that reactant, so long as some of the reactant is present. The value of x is said to be the **order** of the reaction with respect to A, and y is the order of the reaction with respect to B. The overall order of the reaction is $x + y$. Examples of observed rate laws for three reactions follow.

1. $3NO(g) \longrightarrow N_2O(g) + NO_2(g)$

$$\text{rate} = k[NO]^2 \qquad \text{second order in NO; second order overall}$$

2. $2NO_2(g) + F_2(g) \longrightarrow 2NO_2F(g)$

$$\text{rate} = k[NO_2][F_2] \qquad \text{first order in } NO_2 \text{ and first order in } F_2; \text{ second order overall}$$

There is *no way* to predict reaction orders from the balanced overall chemical equation. They must be determined experimentally.

Any number raised to the zero power is one.

3. $H_2O_2(aq) + 3I^-(aq) + 2H^+(aq) \longrightarrow 2H_2O(\ell) + I_3^-(aq)$

$$rate = k[H_2O_2][I^-]$$

first order in H_2O_2 and first order in I^-; zero order in H^+; second order overall

More details about values and units of k will be discussed in later sections.

It is important to remember the following points about the specific rate constant, k.

1. Its value is for a *specific reaction*, represented by a balanced equation.

2. Its units depend on the *overall order* of the reaction.

3. Its value does not change with concentrations of either reactants or products.

4. Its value does not change with time (Section 16-4).

5. Its value refers to the reaction *at a particular temperature* and changes if we change the temperature (Section 16-8).

6. Its value depends on whether a *catalyst* is present (Section 16-9).

7. Its value must be determined experimentally for the reaction at appropriate conditions.

We can use the **method of initial rates** to deduce the rate law from experimentally measured rate data. Usually we know the concentrations of all reactants at the beginning of the reaction. We can then measure the *initial rate* of the reaction, corresponding to these initial concentrations. The following tabulated data refer to the hypothetical reaction

$$A + 2B \longrightarrow AB_2$$

at a specific temperature. The brackets indicate the concentrations of the reacting species *at the beginning* of each experimental run listed in the first column—i.e., the initial concentrations for each experiment.

In such an experiment we often keep some initial concentrations the same and vary others by simple factors, such as 2 or 3. This makes it easier to assess the effect of each change on the rate.

Experiment	Initial [A]	Initial [B]	Initial Rate of Formation of AB_2
1	$1.0 \times 10^{-2} M$	$1.0 \times 10^{-2} M$	$1.5 \times 10^{-4} M \cdot s^{-1}$
2	$1.0 \times 10^{-2} M$	$2.0 \times 10^{-2} M$	$1.5 \times 10^{-4} M \cdot s^{-1}$
3	$2.0 \times 10^{-2} M$	$3.0 \times 10^{-2} M$	$6.0 \times 10^{-4} M \cdot s^{-1}$

Because we are describing the same reaction in each experiment, each is governed by the same rate-law expression. This expression has the form

$$rate = k[A]^x[B]^y$$

Let's compare the initial rates of formation of product (reaction rates) for different experimental runs to see how changes in concentrations of reactants affect the rate of reaction. This lets us evaluate x and y, and then k.

We see that the initial concentration of A is the same in experiments 1 and 2; for these trials, any change in reaction rate would be due to different initial concentrations of B. However, the initial rate of formation of product AB_2 is the same in these two experiments. So we conclude that the rate is independent of [B], or the exponent $y = 0$. Thus far, we know that the rate expression is

Anything raised to the zero power is 1. Here, $[B]^0 = 1$.

$$rate = k[A]^x[B]^0 \qquad \text{or} \qquad rate = k[A]^x$$

To evaluate x, we observe that the concentrations of [A] are different in experiments 1 and 3. The [B] concentrations also change, but we have already deduced that the rate does not depend on [B]. So we know that any change in observed rate between experiments 1 and 3 must be due *only* to the changed [A] concentration. Comparing these two experiments, we see that

[A] has been *multiplied* by a factor of

$$\frac{2.0 \times 10^{-2}}{1.0 \times 10^{-2}} = 2.0 = \text{[A] ratio}$$

Rate increases by a factor of

$$\frac{6.0 \times 10^{-4}}{1.5 \times 10^{-4}} = 4.0 = \text{rate ratio}$$

The exponent x can be deduced from

$$\text{rate ratio} = (\text{[A] ratio})^x$$
$$4.0 = (2.0)^x \quad \text{so} \quad x = 2$$

The reaction is second order in [A]. We can now write its rate-law expression.

$$\text{rate} = k[A]^2$$

The specific rate constant, k, can be evaluated by substituting any of the three sets of data into the rate-law expression. Using the data from experiment 1 gives

$$\text{rate}_1 = k[A]_1^{\ 2} \quad \text{or} \quad k = \frac{\text{rate}_1}{[A]_1^{\ 2}}$$

$$k = \frac{1.5 \times 10^{-4}\,M \cdot s^{-1}}{(1.0 \times 10^{-2}\,M)^2} = 1.5\,M^{-1} \cdot s^{-1}$$

At the temperature at which the measurements were made, the rate-law expression for this reaction is

$$\text{rate} = k[A]^2 \quad \text{or} \quad \text{rate} = 1.5\,M^{-1} \cdot s^{-1}\,[A]^2$$

We can check our result by evaluating k from one of the other sets of data.

> Alternatively, we could have used data from experiments 2 and 3 to deduce the value of x.

> Remember that the specific rate constant k does *not* change with concentration. Only a temperature change or the introduction of a catalyst can change the value of k.

> The units of k depend on the overall order of the reaction, consistent with converting the product of concentrations on the right to concentration/time on the left. For any reaction that is second order overall, the units of k are $M^{-1} \cdot \text{time}^{-1}$ or L/(mol · time).

▼ **PROBLEM-SOLVING TIP** *Be Sure to Use the Rate of* **Reaction**

The rate-law expression should always give the dependence of the *rate of reaction* on concentrations. The data for the preceding calculation describe the rate of formation of the product AB_2; the coefficient of AB_2 in the balanced equation is one, so the rate of reaction is equal to the rate of formation of AB_2. If the coefficient of the measured substance had been two, then we should have divided each value of the "initial rate of formation" by two to obtain the initial rate of reaction. For instance, suppose we measure the rate of formation of AB in the reaction

$$A_2 + B_2 \longrightarrow 2AB$$

Then

$$\text{rate of reaction} = \frac{1}{2}\left(\frac{\Delta[AB]}{\Delta t}\right) = \frac{1}{2}(\text{rate of formation of AB})$$

Then we would analyze how this reaction rate changes as we change the concentrations of reactants.

When heated in air, steel wool glows but does not burn rapidly, due to the low O_2 concentration in air (about 21%). When the glowing steel wool is put into pure oxygen, it burns vigorously because of the much greater accessibility of O_2 reactant molecules.

An Alternative Method

We can also use a simple algebraic approach to find the exponents in a rate-law expression. Consider the set of rate data given earlier for the hypothetical reaction

$$A + 2B \longrightarrow AB_2$$

Experiment	Initial [A]	Initial [B]	Initial Rate of Formation of AB_2 $(M \cdot s^{-1})$
1	$1.0 \times 10^{-2} M$	$1.0 \times 10^{-2} M$	1.5×10^{-4}
2	$1.0 \times 10^{-2} M$	$2.0 \times 10^{-2} M$	1.5×10^{-4}
3	$2.0 \times 10^{-2} M$	$3.0 \times 10^{-2} M$	6.0×10^{-4}

Because we are describing the same reaction in each experiment, all the experiments are governed by the same rate-law expression,

$$\text{rate} = k[A]^x[B]^y$$

The initial concentration of A is the same in experiments 1 and 2, so any change in the initial rates for these experiments would be due to different initial concentrations of B. To evaluate y, we solve the ratio of the rate-law expressions of these two experiments for y. We divide the first rate-law expression by the corresponding terms in the second rate-law expression.

$$\frac{\text{rate}_1}{\text{rate}_2} = \frac{k[A]_1{}^x[B]_1{}^y}{k[A]_2{}^x[B]_2{}^y}$$

The value of k always cancels from such a ratio because it is constant at a particular temperature. The initial concentrations of A are equal, so they too cancel. Thus, the expression simplifies to

$$\frac{\text{rate}_1}{\text{rate}_2} = \left(\frac{[B]_1}{[B]_2}\right)^y$$

The only unknown in this equation is y. We substitute data from experiments 1 and 2 into the equation, which gives us

$$\frac{1.5 \times 10^{-4} \, M \cdot s^{-1}}{1.5 \times 10^{-4} \, M \cdot s^{-1}} = \left(\frac{1.0 \times 10^{-2} \, M}{2.0 \times 10^{-2} \, M}\right)^y$$

$$1.0 = (0.5)^y \qquad \text{so} \qquad y = 0$$

Because the units of rate_1 and rate_2 are identical, they cancel. The units of $[B]_1$ and $[B]_2$ are identical, and they too cancel. Thus far, we know that the rate-law expression is

$$\text{rate} = k[A]^x[B]^0 \qquad \text{or} \qquad \text{rate} = k[A]^x$$

Next we evaluate x. In experiments 1 and 3, the initial concentration of A is doubled and the rate increases by a factor of four. We neglect the initial concentration of B because we have just shown that it does *not* affect the rate of the reaction.

$$\frac{\text{rate}_3}{\text{rate}_1} = \frac{k[A]_3{}^x}{k[A]_1{}^x} = \left(\frac{[A]_3}{[A]_1}\right)^x$$

$$\frac{6.0 \times 10^{-4}\ M \cdot s^{-1}}{1.5 \times 10^{-4}\ M \cdot s^{-1}} = \left(\frac{2.0 \times 10^{-2}\ M}{1.0 \times 10^{-2}\ M}\right)^x$$

$$4.0 = (2.0)^x \qquad \text{so} \qquad x = 2$$

The power to which [A] is raised in the rate-law expression is 2, so the rate-law expression for this reaction is the same as that obtained earlier.

$$\text{rate} = k[A]^2[B]^0 \qquad \text{or} \qquad \text{rate} = k[A]^2$$

EXAMPLE 16-1 *Method of Initial Rates*

Given the following data, determine the rate-law expression for the reaction

$$2A + B_2 + C \longrightarrow A_2B + BC$$

Experiment	Initial [A]	Initial [B$_2$]	Initial [C]	Initial Rate of Formation of BC
1	0.20 M	0.20 M	0.20 M	$2.4 \times 10^{-6}\ M \cdot min^{-1}$
2	0.40 M	0.30 M	0.20 M	$9.6 \times 10^{-6}\ M \cdot min^{-1}$
3	0.20 M	0.30 M	0.20 M	$2.4 \times 10^{-6}\ M \cdot min^{-1}$
4	0.20 M	0.40 M	0.60 M	$7.2 \times 10^{-6}\ M \cdot min^{-1}$

The coefficient of BC in the balanced equation is one, so the rate of reaction is equal to the rate of formation of BC.

Plan

The rate law is of the form Rate = $k[A]^x[B_2]^y[C]^z$. We must evaluate x, y, z, and k. We use the reasoning outlined earlier; in this presentation the first method is used.

The alternative algebraic method outlined above can also be used.

Solution

Dependence on [B$_2$]: In experiments 1 and 3, the initial concentrations of A and C are the same. Thus, any change in the rate would be due to the change in concentration of B$_2$. But we see that the rate is the same in experiments 1 and 3. Thus, the reaction rate is independent of [B$_2$], so $y = 0$. We can neglect changes in [B$_2$] in the subsequent reasoning. The rate law must be

$[B_2]^0 = 1$

$$\text{rate} = k[A]^x[C]^z$$

Dependence on [C]: Experiments 1 and 4 involve the same initial concentration of A; thus, the observed change in rate must be due entirely to the changed [C]. So we compare experiments 1 and 4 to find z.

$$\text{[C] has been } \textit{multiplied} \text{ by a factor of } \frac{0.60}{0.20} = 3.0 = \text{[C] ratio}$$

The effect on rate is that

$$\text{rate changes by a factor of } \frac{7.2 \times 10^{-6}}{2.4 \times 10^{-6}} = 3.0 = \text{rate ratio}$$

The exponent z can be deduced from

$$\text{rate ratio} = (\text{[C] ratio})^z$$

$$3.0 = (3.0)^z \qquad \text{so} \qquad z = 1 \qquad \text{The reaction is first order in [C].}$$

Now we know that the rate law is of the form

$$\text{rate} = k[A]^x[C]$$

Dependence on [A]: We use experiments 1 and 2 to evaluate x, because [A] is changed, [B$_2$] does not matter, and [C] is unaltered. The observed rate change is due *only* to the changed [A].

$$[A] \text{ has been } multiplied \text{ by a factor of } \frac{0.40}{0.20} = 2.0 = [A] \text{ ratio}$$

The effect on rate is that

$$\text{rate changes by a factor of } \frac{9.6 \times 10^{-6}}{2.4 \times 10^{-6}} = 4.0 = \text{rate ratio}$$

The exponent x can be deduced from

$$\text{rate ratio} = ([A] \text{ ratio})^x$$

$$4.0 = (2.0)^x \qquad \text{so} \qquad \boxed{x = 2 \qquad \text{The reaction is second order in [A].}}$$

From these results we can write the complete rate-law expression.

$$\text{rate} = k[A]^2[B_2]^0[C]^1 \qquad \text{or} \qquad \text{rate} = k[A]^2[C]$$

We can evaluate the specific rate constant, k, by substituting any of the four sets of data into the rate-law expression we have just derived. Data from experiment 2 give

$$\text{rate}_2 = k[A]_2{}^2[C]_2$$

$$k = \frac{\text{rate}}{[A]^2[C]} = \frac{9.6 \times 10^{-6} \, M \cdot min^{-1}}{(0.40 \, M)^2(0.20 \, M)} = \boxed{3.0 \times 10^{-4} \, M^{-2} \cdot min^{-1}}$$

The rate-law expression, rate $= k[A]^2[C]$, can now be written

$$\boxed{\text{rate} = 3.0 \times 10^{-4} \, M^{-2} \cdot min^{-1} \, [A]^2[C]}$$

This expression allows us to calculate the rate at which this reaction occurs with any known concentrations of A and C (so long as some B_2 is present). As we shall see presently, changes in temperature change reaction rates. This value of k is valid *only* at the temperature at which the data were collected.

EXAMPLE 16-2   *Method of Initial Rates*

Use the following initial rate data to determine the form of the rate-law expression for the reaction $3A + 2B \rightarrow 2C + D$.

Experiment	Initial [A]	Initial [B]	Initial Rate of Formation of D
1	$1.00 \times 10^{-2} \, M$	$1.00 \times 10^{-2} \, M$	$6.00 \times 10^{-3} \, M \cdot min^{-1}$
2	$2.00 \times 10^{-2} \, M$	$3.00 \times 10^{-2} \, M$	$1.44 \times 10^{-1} \, M \cdot min^{-1}$
3	$1.00 \times 10^{-2} \, M$	$2.00 \times 10^{-2} \, M$	$1.20 \times 10^{-2} \, M \cdot min^{-1}$

Plan

The rate law is of the form rate $= k[A]^x[B]^y$. Let's use the alternative method presented earlier to evaluate x and y.

Solution

The initial concentration of A is the same in experiments 1 and 3. We divide the third rate-law expression by the corresponding terms in the first one.

$$\frac{\text{rate}_3}{\text{rate}_1} = \frac{k[A]_3{}^x[B]_3{}^y}{k[A]_1{}^x[B]_1{}^y}$$

The initial concentrations of A are equal, so they cancel, as does k. Simplifying and then substituting known values of rates and [B],

$$\frac{\text{rate}_3}{\text{rate}_1} = \frac{[B]_3{}^y}{[B]_1{}^y} \qquad \text{or} \qquad \frac{1.20 \times 10^{-2} \, M \cdot min}{6.00 \times 10^{-3} \, M \cdot min} = \left(\frac{2.00 \times 10^{-2} \, M}{1.00 \times 10^{-2} \, M} \right)^y$$

$$2.0 = (2.0)^y \quad \text{or} \quad \boxed{y = 1} \quad \text{The reaction is first order in [B].}$$

No two of the experimental runs have the same concentrations of B, so we must proceed somewhat differently. Let us compare experiments 1 and 2. The observed change in rate must be due to the *combination* of the changes in [A] and [B]. We can divide the second rate-law expression by the corresponding terms in the first one, cancel the equal k values, and collect terms.

$$\frac{\text{rate}_2}{\text{rate}_1} = \frac{k[A]_2{}^x[B]_2{}^y}{k[A]_1{}^x[B]_1{}^y} = \left(\frac{[A]_2}{[A]_1}\right)^x \left(\frac{[B]_2}{[B]_1}\right)^y$$

Now let's insert the known values for rates and concentrations and the known [B] exponent of 1.

$$\frac{1.44 \times 10^{-1}\,M \cdot \text{min}^{-1}}{6.00 \times 10^{-3}\,M \cdot \text{min}^{-1}} = \left(\frac{2.00 \times 10^{-2}\,M}{1.00 \times 10^{-2}\,M}\right)^x \left(\frac{3.00 \times 10^{-2}\,M}{1.00 \times 10^{-2}\,M}\right)^1$$

$$24.0 = (2.00)^x(3.00)$$

$$8.00 = (2.00)^x \quad \text{or} \quad \boxed{x = 3} \quad \text{The reaction is third order in [A].}$$

The rate-law expression has the form $\boxed{\text{rate} = k[A]^3[B].}$

You should now work Exercises 14, 16, and 20.

16-4 CONCENTRATION VERSUS TIME: THE INTEGRATED RATE EQUATION

Often we wish to know the concentration of a reactant that would remain after some specified time, or how long it would take for some amount of the reactants to be used up.

> The equation that relates *concentration* and *time* is the **integrated rate equation.** We can also use it to calculate the **half-life, $t_{1/2}$,** of a reactant—the time it takes for half of that reactant to be converted into product. The integrated rate equation and the half-life are different for reactions of different order.

We shall look at relationships for some simple cases. (If you know calculus, you may be interested in the derivation of the integrated rate equations. This development is presented in the Enrichment at the end of this section.)

First-Order Reactions

For reactions involving $a\text{A} \rightarrow$ products that are *first order in* A and *first order overall,* the integrated rate equation is

a represents the coefficient of reactant A in the balanced overall equation.

$$\ln\left(\frac{[A]_0}{[A]}\right) = akt \quad \text{or} \quad \log\left(\frac{[A]_0}{[A]}\right) = \frac{akt}{2.303} \quad \text{(first order)}$$

$[A]_0$ is the initial concentration of reactant A, and [A] is its concentration at some time, t, after the reaction begins. Solving this relationship for t gives

$$t = \frac{1}{ak}\ln\left(\frac{[A]_0}{[A]}\right) \quad \text{or} \quad t = \frac{2.303}{ak}\log\left(\frac{[A]_0}{[A]}\right)$$

The same expression for $t_{1/2}$ would
be obtained using the base-10 log
form above.

By definition, $[A] = \frac{1}{2}[A]_0$ at $t = t_{1/2}$. Thus (using the ln form),

$$t_{1/2} = \frac{1}{ak} \ln \frac{[A]_0}{\frac{1}{2}[A]_0} = \frac{1}{ak} \ln (2)$$

$$t_{1/2} = \frac{\ln 2}{ak} = \frac{0.693}{ak} \qquad \text{(first order)}$$

Nuclear decay (Chapter 26) is a very
important first-order process.
Exercises at the end of this chapter
involve calculations of nuclear decay
rates.

This relates the half-life of a reactant in a *first-order reaction* and its rate constant, k. In such reactions, the half-life *does not depend* on the initial concentration of A. This is not true for reactions having overall orders other than first order.

EXAMPLE 16-3 *Half-Life—First-Order Reaction*

Compound A decomposes to form B and C in a reaction that is first order with respect to A and first order overall. At 25°C, the specific rate constant for the reaction is 0.0450 s^{-1}. What is the half-life of A at 25°C?

$$A \longrightarrow B + C$$

Plan

We use the equation given above for $t_{1/2}$ for a first-order reaction. The value of k is given in the problem; the coefficient of reactant A is $a = 1$.

Solution

$$t_{1/2} = \frac{\ln 2}{ak} = \frac{0.693}{1(0.0450 \text{ s}^{-1})} = \boxed{15.4 \text{ s}}$$

After 15.4 seconds of reaction, half of the original reactant remains, so that $[A] = \frac{1}{2}[A]_0$.

You should now work Exercise 33.

EXAMPLE 16-4 *Concentration vs. Time—First-Order Reaction*

The reaction $2N_2O_5(g) \rightarrow 2N_2O_4(g) + O_2(g)$ obeys the rate law Rate $= k[N_2O_5]$, in which the specific rate constant is 0.00840 s^{-1} at a certain temperature. If 2.50 moles of N_2O_5 were placed in a 5.00-liter container at that temperature, how many moles of N_2O_5 would remain after 1.00 minute?

Plan

We apply the first-order integrated rate equation.

$$\ln \left(\frac{[N_2O_5]_0}{[N_2O_5]} \right) = akt$$

Alternatively, we can use the base-10
log form of the first-order integrated
rate equation.

First we must determine $[N_2O_5]_0$, the original molar concentration of N_2O_5. Then we solve for $[N_2O_5]$, the molar concentration after 1.00 minute. We must remember to express k and t using the same time units. Finally, we convert molar concentration of N_2O_5 to moles remaining.

Solution

The original concentration of N_2O_5 is

$$[N_2O_5]_0 = \frac{2.50 \text{ mol}}{5.00 \text{ L}} = 0.500 \ M$$

The other quantities are

$$a = 2 \qquad k = 0.00840 \text{ s}^{-1} \qquad t = 1.00 \text{ min} = 60.0 \text{ s} \qquad [N_2O_5] = \underline{?}$$

1.00 minute = 60.0 seconds

The only unknown in the integrated rate equation is $[N_2O_5]$ after 1.00 minute. Let us solve for the unknown. Because $\ln x/y = \ln x - \ln y$,

$$\ln \frac{[N_2O_5]_0}{[N_2O_5]} = \ln [N_2O_5]_0 - \ln [N_2O_5] = akt$$

$$\ln [N_2O_5] = \ln [N_2O_5]_0 - akt$$

$$= \ln (0.500) - (2)(0.00840 \text{ s}^{-1})(60.0 \text{ s}) = -0.693 - 1.008$$

$$\ln [N_2O_5] = -1.701$$

Taking the inverse natural logarithm of both sides gives

inv $\ln x = e^{\ln x}$

$$[N_2O_5] = 1.82 \times 10^{-1} \, M$$

Thus, after 1.00 minute of reaction, the concentration of N_2O_5 is 0.182 M. The number of moles of N_2O_5 left in the 5.00-L container is

$$\underline{?} \text{ mol } N_2O_5 = 5.00 \text{ L} \times \frac{0.182 \text{ mol}}{\text{L}} = \boxed{0.910 \text{ mol } N_2O_5}$$

You should now work Exercises 30 and 34.

Second-Order Reactions

For reactions involving $a\text{A} \rightarrow$ products that are *second order with respect to* A *and second order overall,* the integrated rate equation is

$$\frac{1}{[A]} - \frac{1}{[A]_0} = akt \qquad \left(\begin{array}{l}\text{second order in A,} \\ \text{second order overall}\end{array}\right)$$

For $t = t_{1/2}$, we have $[A] = \frac{1}{2}[A]_0$, so

$$\frac{1}{\frac{1}{2}[A]_0} - \frac{1}{[A]_0} = akt_{1/2}$$

Simplifying and solving for $t_{1/2}$, we obtain the relationship between the rate constant and $t_{1/2}$.

$$t_{1/2} = \frac{1}{ak[A]_0} \qquad \left(\begin{array}{l}\text{second order in A,} \\ \text{second order overall}\end{array}\right)$$

You should carry out the algebraic steps to solve for $t_{1/2}$.

In this case $t_{1/2}$ *depends on the initial concentration of* A. Figure 16-4 illustrates the different behavior of half-life for first- and second-order reactions.

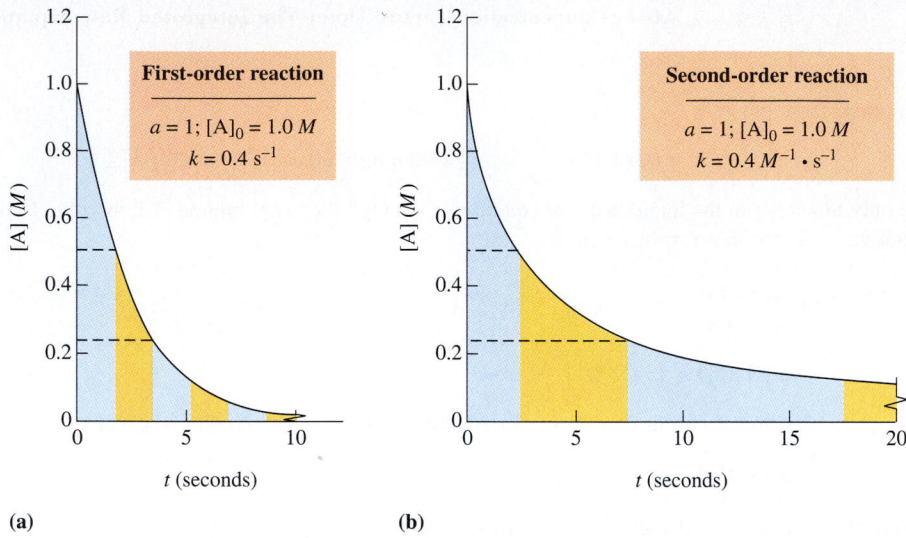

Figure 16-4 (a) Plot of concentration versus time for a first-order reaction. During the first half-life, 1.73 seconds, the concentration of A falls from 1.00 M to 0.50 M. An additional 1.73 seconds are required for the concentration to fall by half again, from 0.50 M to 0.25 M, and so on. For a first-order reaction, $t_{1/2} = \dfrac{\ln 2}{ak} = \dfrac{0.693}{ak}$; $t_{1/2}$ does not depend on the concentration at the beginning of that time period. (b) Plot of concentration versus time for a second-order reaction. The same values are used for a, $[A]_0$, and k as in part (a). During the first half-life, 2.50 seconds, the concentration of A falls from 1.00 M to 0.50 M. The concentration falls by half again from 2.50 to 7.50 seconds, so the second half-life is 5.00 seconds. The half-life beginning at 0.25 M is 10.00 seconds. For a second-order reaction,

$$t_{1/2} = \frac{1}{ak[A]_0}$$

$t_{1/2}$ is inversely proportional to the concentration at the beginning of that time period.

EXAMPLE 16-5   Half-Life—Second-Order Reaction

Compounds A and B react to form C and D in a reaction that was found to be second order overall and second order in A. The rate constant at 30°C is 0.622 liter per mole per minute. What is the half-life of A when $4.10 \times 10^{-2}\, M$ A is mixed with excess B?

$$A + B \longrightarrow C + D$$

Plan

As long as some B is present, only the concentration of A affects the rate. The reaction is second order in [A] and second order overall, so we use the appropriate equation for the half-life.

Solution

$$t_{1/2} = \frac{1}{ak[A]_0} = \frac{1}{(1)(0.622\, M^{-1} \cdot \text{min}^{-1})(4.10 \times 10^{-2}\, M)} = \boxed{39.2\ \text{min}}$$

EXAMPLE 16-6   Concentration vs. Time—Second-Order Reaction

The gas phase decomposition of NOBr is second order in [NOBr], with $k = 0.810\, M^{-1} \cdot \text{s}^{-1}$ at

10°C. We start with 4.00×10^{-3} M NOBr in a flask at 10°C. How many seconds does it take to use up 1.50×10^{-3} M of this NOBr?

$$2NOBr(g) \longrightarrow 2NO(g) + Br_2(g) \qquad Rate = k[NOBr]^2$$

Plan

We first determine the concentration of NOBr that remains after 1.50×10^{-3} M is used up. Then we use the second-order integrated rate equation to determine the time required to reach that concentration.

Solution

$$\underline{?} \text{ M NOBr remaining} = (0.00400 - 0.00150) \text{ M} = 0.00250 \text{ M} = [NOBr]$$

We solve the integrated rate equation $\dfrac{1}{[NOBr]} - \dfrac{1}{[NOBr]_0} = akt$ for t.

$$t = \frac{1}{ak}\left(\frac{1}{[NOBr]} - \frac{1}{[NOBr]_0}\right) = \frac{1}{(2)(0.810 \ M^{-1}\cdot s^{-1})}\left(\frac{1}{0.00250 \ M} - \frac{1}{0.00400 \ M}\right)$$

The coefficient of NOBr is $a = 2$.

$$= \frac{1}{1.62 \ M^{-1}\cdot s^{-1}}(400 \ M^{-1} - 250 \ M^{-1})$$

$$= \boxed{92.6 \text{ s } (1.54 \text{ min})}$$

You should now work Exercise 28.

EXAMPLE 16-7 *Concentration vs. Time—Second-Order Reaction*

Consider the reaction of Example 16-6 at 10°C. What concentration of NOBr will remain after 5.00 minutes of reaction, starting with 2.40×10^{-3} M of NOBr?

Plan

We use the second-order integrated rate equation to solve for the concentration of NOBr remaining at $t = 5.00$ minutes.

Solution

Again, we start with the expression $\dfrac{1}{[NOBr]} - \dfrac{1}{[NOBr]_0} = akt$. Let's put in the known values and solve for [NOBr].

$$\frac{1}{[NOBr]} - \frac{1}{2.40 \times 10^{-3} \ M} = (2)(0.810 \ M^{-1}\cdot s^{-1})(5.00 \text{ min})\left(\frac{60 \text{ s}}{1 \text{ min}}\right)$$

You may prefer to rearrange the original equation to solve for [NOBr] before substituting values.

$$\frac{1}{[NOBr]} - 4.17 \times 10^2 \ M^{-1} = 486 \ M^{-1}$$

$$\frac{1}{[NOBr]} = 486 \ M^{-1} + 417 \ M^{-1} = 903 \ M^{-1}$$

$$[NOBr] = \boxed{1.11 \times 10^{-3} \ M} \qquad (46\% \text{ remains unreacted})$$

Thus, about 54% of the original concentration of NOBr reacts within the first 5 minutes. This is reasonable because, as you may easily verify, the reaction has an initial half-life of 257 seconds, or 4.29 minutes.

You should now work Exercises 29 and 32.

Zero-Order Reaction

For a reaction $a\text{A} \rightarrow$ products that is zero order, we can write the rate-law expression as

$$-\frac{1}{a}\left(\frac{\Delta[\text{A}]}{\Delta t}\right) = k$$

The corresponding integrated rate equation is

$$[\text{A}] = [\text{A}]_0 - akt \qquad \text{(zero order)}$$

and the half-life is

$$t_{1/2} = \frac{[\text{A}]_0}{2ak} \qquad \text{(zero order)}$$

Table 16-2 summarizes the relationships that we have presented in Sections 16-3 and 16-4.

▼ **PROBLEM-SOLVING TIP**   *Which Equation Should Be Used?*

How can you tell which equation to use to solve a particular problem?

1. You must decide whether to use the rate-law expression or the integrated rate equation. Remember that

     the *rate-law expression* relates *rate and concentration*

   whereas

     the *integrated rate equation* relates *time and concentration.*

   When you need to find the *rate* that corresponds to particular concentrations, or the concentrations needed to give a desired rate, you should use the rate-law expression. When the problem at hand involves the *time* required to reach a stated concentration, or the concentration that remains or that has been consumed after a stated time, you should use the integrated rate equation.

2. You must choose the form of the rate-law expression or the integrated rate equation—zero, first, or second order—that is appropriate to the order of the reaction. These are summarized in Table 16-2. One of the following usually helps you decide.
   a. The statement of the problem may state explicitly what the order of the reaction is.
   b. The rate-law expression may be given, from which you can tell the order of the reaction from the exponents in that expression.
   c. The units of the specific rate constant, k, may be given; you can interpret these stated units to tell you the order of the reaction. The units of k for a zero-order reaction are $M \cdot \text{time}^{-1}$; those for a first-order reaction are time^{-1}; those for a second-order reaction are $M^{-1} \cdot \text{time}^{-1}$.

You should test this method using the concentration-versus-time data of Example 16-8, plotted in Figure 16-8a.

One method of assessing reaction order is based on comparing successive half-lives. As we have seen, $t_{1/2}$ for a first-order reaction does not depend on initial concentration. We can measure the time required for different concentrations of a reactant to fall to half of

Table 16-2 *Summary of Relationships for Various Orders of the Reaction aA → Products*

	Order		
	Zero	*First*	*Second*
Rate-law expression	rate $= k$	rate $= k[A]$	rate $= k[A]^2$
Integrated rate equation	$[A] = [A]_0 - akt$	$\ln \dfrac{[A]_0}{[A]} = akt$ or $\log \dfrac{[A]_0}{[A]} = \dfrac{akt}{2.303}$	$\dfrac{1}{[A]} - \dfrac{1}{[A]_0} = akt$
Half-life, $t_{1/2}$	$\dfrac{[A]_0}{2ak}$	$\dfrac{\ln 2}{ak} = \dfrac{0.693}{ak}$	$\dfrac{1}{ak[A]_0}$

their original values. If this time remains constant, it is an indication that the reaction is first order for that reactant and first order overall (Figure 16-4a). By contrast, for other orders of reaction, $t_{1/2}$ would change depending on initial concentration. For a second-order reaction, successively measured $t_{1/2}$ values would increase as $[A]_0$ decreases (Figure 16-4b). $[A]_0$ is measured at the *beginning of each particular measurement period*.

Calculus Derivation of Integrated Rate Equations

ENRICHMENT

The derivation of the integrated rate equation is an example of the use of calculus in chemistry. The following derivation is for a reaction that is assumed to be first order in a reactant A and first order overall. If you do not know calculus, you can still use the results of this derivation, as we have already shown in this section. For the reaction

$$aA \longrightarrow \text{products}$$

the rate is expressed as

$$\text{rate} = -\frac{1}{a}\left(\frac{\Delta[A]}{\Delta t}\right)$$

For a first-order reaction, the rate is proportional to the first power of $[A]$.

$$-\frac{1}{a}\left(\frac{\Delta[A]}{\Delta t}\right) = k[A]$$

In calculus terms, we express the change during an infinitesimally short time dt as the derivative of $[A]$ with respect to time.

$$-\frac{1}{a}\frac{d[A]}{dt} = k[A]$$

Separating variables, we obtain

$$-\frac{d[A]}{[A]} = (ak)dt$$

$-\dfrac{1}{a}\left(\dfrac{\Delta[A]}{\Delta t}\right)$ represents the *average* rate over a finite time interval Δt.
$-\dfrac{1}{a}\left(\dfrac{d[A]}{dt}\right)$ involves a change over an infinitesimally short time interval dt, so it represents the *instantaneous* rate.

We integrate this equation to get

$$-\ln [A] = akt + C$$

We evaluate C, the constant of integration, by setting $t = 0$. This refers to the beginning of the reaction. At that time, the concentration of A is $[A]_0$, its initial value, so $C = -\ln [A]_0$.

$$-\ln [A] = akt - \ln [A]_0$$

We now rearrange the equation, remembering that $\ln (x) - \ln (y) = \ln (x/y)$.

$$\ln [A]_0 - \ln [A] = akt$$

$$\ln \frac{[A]_0}{[A]} = akt \qquad \text{(first order)}$$

This is the integrated rate equation for a reaction that is first order in reactant A and first order overall.

Integrated rate equations can be derived similarly from other simple rate laws. For a reaction $a\text{A} \rightarrow$ products that is second order in reactant A and second order overall, we can write the rate equation as

$$-\frac{d[A]}{adt} = k[A]^2$$

Again, using the methods of calculus, we can separate variables, integrate, and rearrange to obtain the corresponding second-order integrated rate equation.

$$\frac{1}{[A]} - \frac{1}{[A]_0} = akt \qquad \text{(second order)}$$

For a reaction $a\text{A} \rightarrow$ products that is zero order overall, we can write the rate equation as

$$-\frac{d[A]}{adt} = k$$

In this case, the calculus derivation already described leads to the zero-order integrated rate equation

$$[A] = [A]_0 - akt \qquad \text{(zero order)}$$

ENRICHMENT

Using Integrated Rate Equations to Determine Reaction Order

The integrated rate equation can help us to analyze concentration-versus-time data to determine reaction order. A graphical approach is often used. We can rearrange the

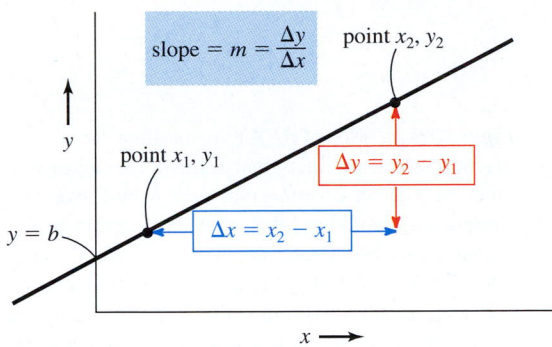

Figure 16-5 Plot of the equation $y = mx + b$, where m and b are constant. The slope of the line (positive in this case) is equal to m; the intercept on the y axis is equal to b.

first-order integrated rate equation

$$\ln \frac{[A]_0}{[A]} = akt$$

as follows. Remembering that the logarithm of a quotient, $\ln (x/y)$, is equal to the difference of the logarithms, $\ln x - \ln y$, we can write

$$\ln [A]_0 - \ln [A] = akt \qquad \text{or} \qquad \ln [A] = -akt + \ln [A]_0$$

Recall that the equation for a straight line may be written as

$$y = mx + b$$

where y is the variable plotted along the ordinate (vertical axis), x is the variable plotted along the abscissa (horizontal axis), m is the slope of the line, and b is the intercept of the line with the y axis (Figure 16-5). If we compare the last two equations, we find that $\ln [A]$ can be interpreted as y, and t as x.

$$\underbrace{\ln [A]}_{y} = \underbrace{-ak}_{m}\underbrace{t}_{x} + \underbrace{\ln [A]_0}_{b}$$

The quantity $-ak$ is a constant as the reaction proceeds, so it can be interpreted as m. The initial concentration of A is fixed, so $\ln [A]_0$ is a constant for each experiment, and $\ln [A]_0$ can be interpreted as b. Thus, a plot of $\ln [A]$ versus time for a first-order reaction would be expected to give a straight line (Figure 16-6) with the slope of the line equal to $-ak$ and the intercept equal to $\ln [A]_0$.

In similar fashion we can work with the integrated rate equation for a reaction that is second order in A and second order overall. We rearrange

$$\frac{1}{[A]} - \frac{1}{[A]_0} = akt \qquad \text{to read} \qquad \frac{1}{[A]} = akt + \frac{1}{[A]_0}$$

Again comparing this with the equation for a straight line, we see that a plot of $1/[A]$ versus time would be expected to give a straight line (Figure 16-7). The line would have a slope equal to ak and an intercept equal to $1/[A]_0$.

For a zero-order reaction, we can rearrange the integrated rate equation

$$[A]_0 - [A] = akt \qquad \text{to} \qquad [A] = -akt + [A]_0$$

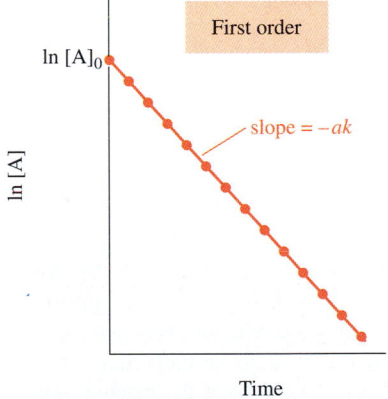

Figure 16-6 Plot of $\ln [A]$ versus time for a reaction $aA \rightarrow$ products that follows first-order kinetics. The observation that such a plot gives a straight line would confirm that the reaction is first order in [A] and first order overall, i.e., rate = $k[A]$. The slope is equal to $-ak$. Because a and k are positive numbers, the slope of the line is always negative. Logarithms are dimensionless, so the slope has the units $(\text{time})^{-1}$. A plot of (base-10) log [A] for a first-order reaction would give a straight line with a slope equal to $-ak/2.303$. The logarithm of a quantity less than 1 is negative, so data points for concentrations less than 1 molar would appear below the time-axis.

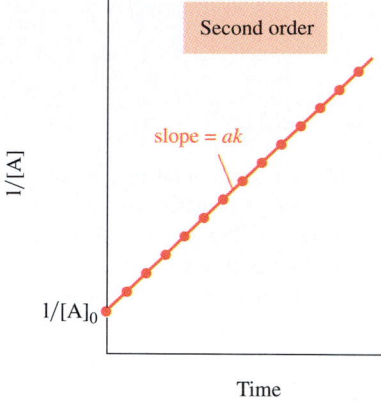

Figure 16-7 Plot of 1/[A] versus time for a reaction $aA \rightarrow$ products that follows second-order kinetics. The observation that such a plot gives a straight line would confirm that the reaction is second order in [A] and second order overall, i.e., rate $= k[A]^2$. The slope is equal to ak. Because a and k are positive numbers, the slope of the line is always positive. Because concentrations cannot be negative, 1/[A] is always positive, and the line is always above the time-axis.

Comparing this with the equation for a straight line, we see that a straight-line plot would be obtained by plotting concentration versus time, [A] versus t. The slope of this line is $-ak$, and the intercept is $[A]_0$.

This discussion suggests another way to deduce an unknown rate-law expression from experimental concentration data. The following approach is particularly useful for any decomposition reaction, one that involves only one reactant.

$$aA \longrightarrow products$$

We plot the data in various ways as suggested above. *If* the reaction followed zero-order kinetics, *then* a plot of [A] versus t would give a straight line. But *if* the reaction followed first-order kinetics, *then* a plot of ln [A] versus t would give a straight line whose slope could be interpreted to derive a value of k. *If* the reaction were second order in A and second order overall, *then* neither of these plots would give a straight line, but a plot of 1/[A] versus t would. If none of these plots gave a straight line (within expected scatter due to experimental error), we would know that none of these is the correct order (rate law) for the reaction. Plots to test for other orders can be devised, as can graphical tests for rate-law expressions involving more than one reactant, but those are subjects for more advanced texts. The graphical approach that we have described is illustrated in the following example.

It is not possible for all of the plots suggested here to yield straight lines for a given reaction. However, the nonlinearity of the plots may not become obvious if the reaction times used are too short.

Time (min)	[A] (mol/L)
0.00	2.000
1.00	1.488
2.00	1.107
3.00	0.823
4.00	0.612
5.00	0.455
6.00	0.338
7.00	0.252
8.00	0.187
9.00	0.139
10.00	0.103

EXAMPLE 16-8 *Graphical Determination of Reaction Order*

We carry out the reaction $A \rightarrow B + C$ at a particular temperature. As the reaction proceeds, we measure the molarity of the reactant, [A], at various times. The observed data are tabulated in the margin. (a) Plot [A] versus time. (b) Plot ln [A] versus time. (c) Plot 1/[A] versus time. (d) What is the order of the reaction? (e) Write the rate-law expression for the reaction. (f) What is the value of k at this temperature?

Plan

For parts (a)–(c), we use the observed data to make the required plots, calculating related values as necessary. (d) We can determine the order of the reaction by observing which of these plots gives a straight line. (e) Knowing the order of the reaction, we can write the rate-law expression. (f) The value of k can be determined from the slope of the straight-line plot.

Solution

(a) The plot of [A] versus time is given in Figure 16-8a.

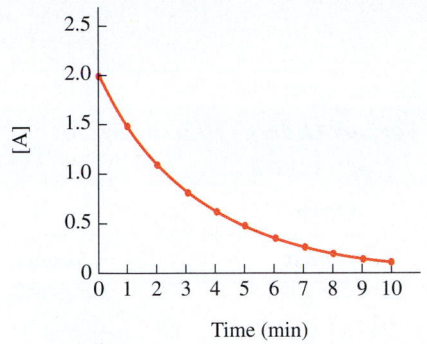

(a) Example 16-8(a).

Time (min)	[A]	ln [A]	1/[A]
0.00	2.000	0.693	0.5000
1.00	1.488	0.397	0.6720
2.00	1.107	0.102	0.9033
3.00	0.823	−0.195	1.22
4.00	0.612	−0.491	1.63
5.00	0.455	−0.787	2.20
6.00	0.338	−1.085	2.95
7.00	0.252	−1.378	3.97
8.00	0.187	−1.677	5.35
9.00	0.139	−1.973	7.19
10.00	0.103	−2.273	9.71

(b) Data for Example 16-8.

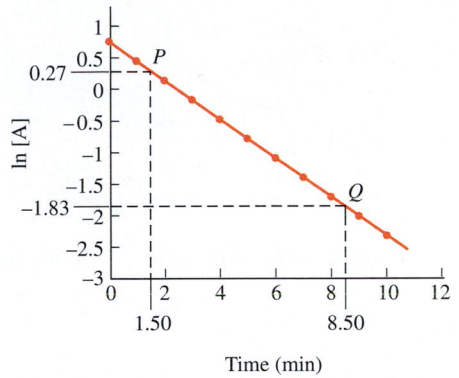

(c) Example 16-8(b).

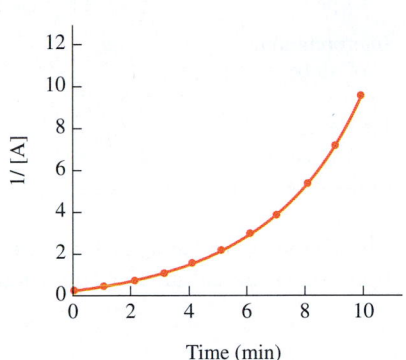

(d) Example 16-8(c).

Figure 16-8 Plots and data conversion for Example 16-8. (a) Plot of [A] versus time using the data given. (b) The data are used to calculate the two columns ln [A] and 1/[A]. (c) Plot of ln [A] versus time. The observation that this plot gives a straight line indicates that the reaction follows first-order kinetics. (d) Plot of 1/[A] versus time. If the reaction had followed second-order kinetics, this plot would have resulted in a straight line, and the plot in Figure 16-8c would not.

(b) We first use the given data to calculate the ln [A] column in Figure 16-8b. These data are then used to plot ln [A] versus time, as shown in Figure 16-8c.

(c) The given data are used to calculate the 1/[A] column in Figure 16-8b. Then we plot 1/[A] versus time, as shown in Figure 16-8d.

(d) It is clear from the answer to part (b) that the plot of ln [A] versus time gives a straight line. This tells us that the reaction is first order in [A].

(e) Putting our answer to part (d) in the form of a rate-law expression, we write rate $= k[A]$.

(f) We use the straight-line plot in Figure 16-8c to find the value of the rate constant for this first-order reaction from the relationship

$$\text{slope} = -ak \quad \text{or} \quad k = -\frac{\text{slope}}{a}$$

To determine the slope of the straight line, we pick any two points, such as P and Q, on the line. From their coordinates, we calculate

$$\text{slope} = \frac{\text{change in ordinate}}{\text{change in abscissa}} = \frac{(-1.83) - (0.27)}{(8.50 - 1.50) \text{ min}} = -0.300 \text{ min}^{-1}$$

$$k = -\frac{\text{slope}}{a} = -\frac{-0.300 \text{ min}^{-1}}{1} = 0.300 \text{ min}^{-1}$$

You may wish to confirm that a straight line would also be obtained by plotting log [A] versus time and that the proper interpretation of its slope would give the same value.

You should now work Exercises 38 and 39.

Remember that the ordinate is the vertical axis and the abscissa is the horizontal one. If you are not careful to keep the points in the same order in the numerator and denominator, you will get the wrong sign for the slope.

Table 16-3 *Graphical Interpretations for Various Orders of the Reaction aA → Products*

	Order		
	Zero	*First*	*Second*
Plot that gives straight line	[A] vs. t	ln [A] vs. t	$\dfrac{1}{[A]}$ vs. t
Direction of straight-line slope	down with time	down with time	up with time
Interpretation of slope	$-ak$	$-ak$	ak
Interpretation of intercept	$[A]_0$	ln $[A]_0$	$\dfrac{1}{[A]_0}$

The graphical interpretations of concentration-versus-time data for some common reaction orders are summarized in Table 16-3.

16-5 COLLISION THEORY OF REACTION RATES

The fundamental notion of the **collision theory of reaction rates** is that for reaction to occur, molecules, atoms, or ions must first collide. Increased concentrations of reacting species result in greater numbers of collisions per unit time. However, not all collisions result in reaction; i.e., not all collisions are **effective collisions.** For a collision to be effective, the reacting species must (1) possess at least a certain minimum energy necessary to rearrange outer electrons in breaking bonds and forming new ones and (2) have the proper orientations toward each other at the time of collision.

> Collisions must occur in order for a chemical reaction to proceed, but they do not guarantee that a reaction will occur.

A collision between atoms, molecules, or ions is not like one between two hard billiard balls. Whether or not chemical species "collide" depends on the distance at which they can interact with one another. For instance, the gas phase ion–molecule reaction $CH_4^+ + CH_4 \rightarrow CH_5^+ + CH_3$ can occur with a fairly long-range contact. This is because the interactions between ions and induced dipoles are effective over a relatively long distance. By contrast, the reacting species in the gas reaction $CH_3 + CH_3 \rightarrow C_2H_6$ are both neutral. They interact appreciably only through very short-range forces between induced dipoles, so they must approach one another very closely before we could say that they "collide."

Recall (Chapter 12) that the average kinetic energy of a collection of molecules is proportional to the absolute temperature. At higher temperatures, more of the molecules possess sufficient energy to react (Section 16-8).

If colliding molecules have improper orientations, they do not react even though they may possess sufficient energy. Figure 16-9 depicts collisions between molecules of NO

$$Zn(s) + 2H^+(aq) \longrightarrow Zn^{2+}(aq) + H_2(g)$$

Dilute sulfuric acid reacts slowly with zinc metal (left), whereas more concentrated acid reacts rapidly (right). The $H^+(aq)$ concentration is higher in the more concentrated acid, and so more $H^+(aq)$ ions collide with Zn per unit time.

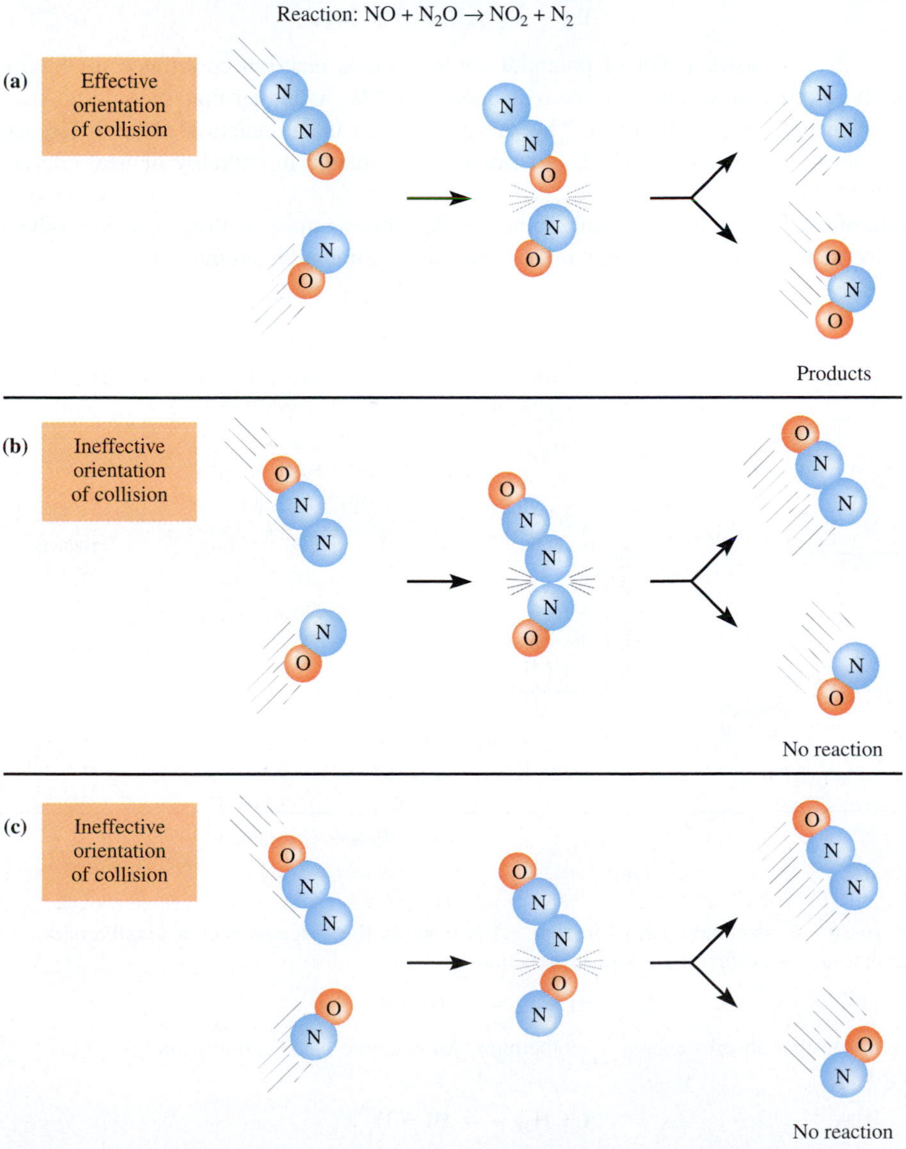

Reaction: $NO + N_2O \rightarrow NO_2 + N_2$

(a) Effective orientation of collision

Products

(b) Ineffective orientation of collision

No reaction

(c) Ineffective orientation of collision

No reaction

Figure 16-9 Some possible collisions between N_2O and NO molecules in the gas phase. (a) A collision that could be effective in producing the reaction. (b, c) Collision would be ineffective. The molecules must have the proper orientations relative to each other *and* have sufficient energy to react.

and N_2O. We assume that each possesses sufficient energy to react according to

$$NO + N_2O \longrightarrow NO_2 + N_2$$

For many reactions, a heterogeneous catalyst (Section 16-9) can increase the fraction of colliding molecules that have the proper orientations.

16-6 TRANSITION STATE THEORY

Chemical reactions involve the making and breaking of chemical bonds. The energy associated with a chemical bond is a form of potential energy. Reactions are accompanied by changes in potential energy. Consider the following hypothetical, one-step *exothermic* reaction at a certain temperature.

$$A + B_2 \longrightarrow AB + B + heat$$

Figure 16-10 shows a plot of potential energy versus reaction coordinate. In Figure 16-10a the ground state energy of the reactants, A and B_2, is higher than the ground state energy of the products, AB and B. The energy released in the reaction is the difference between these two energies, ΔE. It is related to the change in enthalpy or heat content (Section 15-11).

The "reaction coordinate" represents the *progress along the pathway* leading from reactants to products. This coordinate is sometimes labeled "progress of reaction."

Quite often, for reaction to occur, some covalent bonds must be broken so that others can be formed. This can occur only if the molecules collide *with enough kinetic energy to*

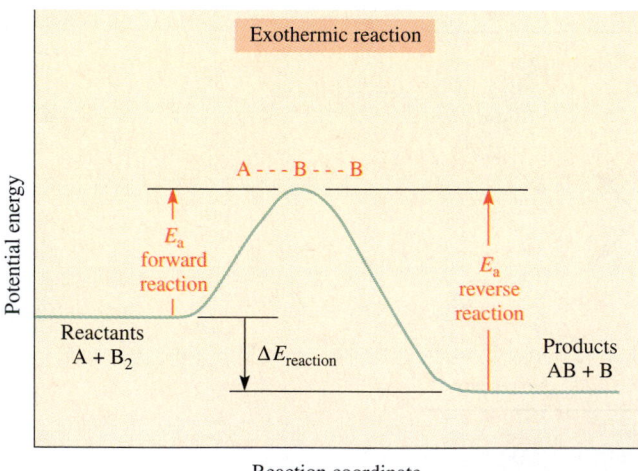

(a)   (b)

Figure 16-10 A potential energy diagram. (a) A reaction that releases energy (exothermic). An example of an exothermic gas-phase reaction is

$$H + I_2 \longrightarrow HI + I$$

(b) A reaction that absorbs energy (endothermic). An example of an endothermic gas-phase reaction is

$$I + H_2 \longrightarrow HI + H$$

overcome the potential energy stabilization of the bonds. According to the **transition state theory,** the reactants pass through a short-lived, high-energy intermediate state, called a **transition state,** before the products are formed.

$$A + B\text{—}B \longrightarrow A\text{---}B\text{---}B \longrightarrow A\text{—}B + B$$

reactants	transition state	products
$A + B_2$	AB_2	$AB + B$

The **activation energy, E_a,** is the additional energy that must be absorbed by the reactants in their ground states to allow them to reach the transition state. If A and B_2 molecules do not possess the necessary amount of energy, E_a, above their ground states when they collide, reaction cannot occur. If they do possess sufficient energy to "climb the energy barrier" to reach the transition state, the reaction can proceed. When the atoms go from the transition state arrangement to the product molecules, energy is *released.* If the reaction results in a net *release* of energy (Figure 16-10a), *more* energy than the activation energy is returned to the surroundings and the reaction is exothermic. If the reaction results in a *net absorption* of energy (Figure 16-10b), an amount less than E_a is given off when the transition state is converted to products and the reaction is endothermic. Thus, the activation energy must be supplied to the system from its environment, but some of that energy is subsequently released to the surroundings. The *net* release of energy is ΔE.

When the reverse reaction occurs, an increase in energy equal to the reverse activation energy, $E_{a\text{ reverse}}$, is required to convert the AB product molecules to the transition state. As you can see from the potential energy diagrams in Figure 16-10,

$$E_{a\text{ forward}} - E_{a\text{ reverse}} = \Delta E_{\text{reaction}}$$

As we shall see, increasing the temperature changes the rate by altering the fraction of molecules that can get over a given energy barrier (Section 16-8). Introducing a catalyst changes the rate by lowering the barrier (Section 16-9).

As a specific example that illustrates the ideas of collision theory and transition state theory, consider the reaction of iodide ions with methyl chloride.

$$I^- + CH_3Cl \longrightarrow CH_3I + Cl^-$$

Many studies have established that this reaction proceeds as shown in Figure 16-11a. The I^- ion must approach the CH_3Cl from the "back side" of the C—Cl bond, through the middle of the three hydrogen atoms. A collision of an I^- ion with a CH_3Cl molecule from any other angle would not lead to reaction. But a collision with the appropriate orientation could allow the new I—C bond to form at the same time that the C—Cl bond is breaking. This collection of atoms, which we represent as

$$I\text{---}\overset{\displaystyle H \quad H}{\underset{\displaystyle H}{C}}\text{---}Cl$$

is what we call the transition state of this reaction (Figure 16-11b). From this state, either of two things could happen: (1) the I—C bond could finish forming and the C—Cl bond could finish breaking with Cl^- leaving, leading to products, or (2) the I—C bond could fall apart with I^- leaving, and the C—Cl bond could re-form, leading back to reactants.

Remember that ΔE relates product energy to reactant energy, regardless of the pathway. ΔE is negative when energy is given off; ΔE is positive when energy is absorbed from the surroundings.

The CH_3Cl and CH_3I molecules are (distorted) tetrahedra.

We can view this transition state as though carbon is only partially bonded to I and only partially bonded to Cl.

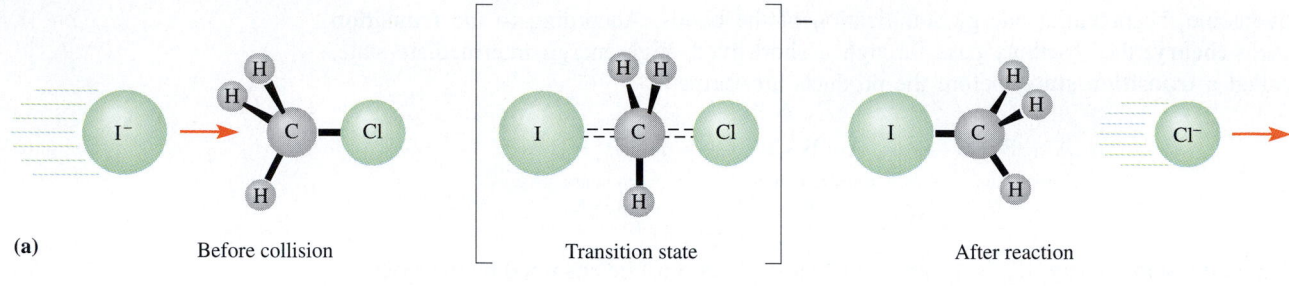

(a)   Before collision   Transition state   After reaction

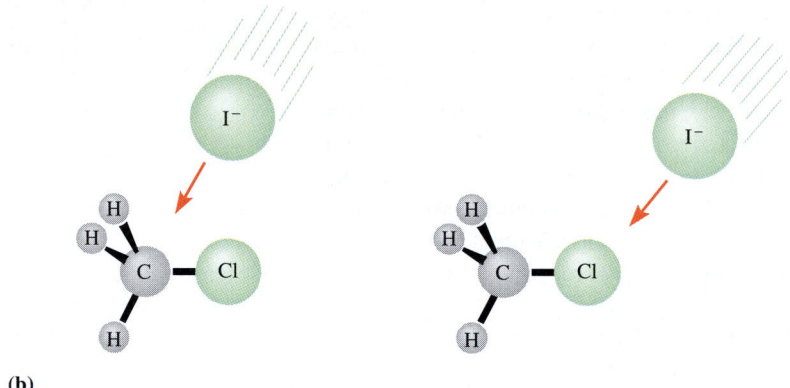

(b)

Figure 16-11   (a) A collision that could lead to reaction of I^- + CH_3Cl to give CH_3I + Cl^-. The I^- must approach along the "back side" of the C—Cl bond. (b) Two collisions that are not in the "correct" orientation.

16-7 REACTION MECHANISMS AND THE RATE-LAW EXPRESSION

The step-by-step pathway by which a reaction occurs is called its **mechanism.** Some reactions take place in a single step, but most reactions occur in a series of **elementary steps.**

> The reaction orders *for any single elementary step* are equal to the coefficients for that step.

In many mechanisms, however, one step is much slower than the others.

> A reaction can never occur faster than its slowest step.

This slow step is called the **rate-determining step.** The speed at which the slow step occurs limits the rate at which the overall reaction occurs.

The balanced equation for the overall reaction is equal to the sum of *all* the individual steps, including any steps that might follow the rate-determining step. We emphasize again that the rate-law exponents *do not necessarily match* the coefficients of the overall balanced equation.

For the general overall reaction

$$aA + bB \longrightarrow cC + dD$$

the experimentally determined rate-law expression has the form

$$\text{rate} = k[A]^x[B]^y$$

The values of x and y are related to the coefficients of the reactants in the slowest step, influenced in some cases by earlier steps.

Using a combination of experimental data and chemical intuition, we can *postulate* a mechanism by which a reaction could occur. We can never prove absolutely that a proposed mechanism is correct. All we can do is postulate a mechanism that is *consistent* with experimental data. We might later detect reaction-intermediate species that are not part of the proposed mechanism. We must then modify the mechanism or discard it and propose a new one.

As an example, the reaction of nitrogen dioxide and carbon monoxide has been found to be second order with respect to NO_2 and zero order with respect to CO below 225°C.

$$NO_2(g) + CO(g) \longrightarrow NO(g) + CO_2(g) \qquad \text{rate} = k[NO_2]^2$$

The balanced equation for the overall reaction shows the stoichiometry but *does not necessarily mean* that the reaction simply occurs by one molecule of NO_2 colliding with one molecule of CO. If the reaction really took place in *that* one step, then the rate would be first order in NO_2 and first order in CO, or rate = $k[NO_2][CO]$. The fact that the experimentally determined orders do not match the coefficients in the overall balanced equation tells us that *the reaction does not take place in one step.*

The following proposed two-step mechanism is consistent with the observed rate-law expression.

(1)	$NO_2 + NO_2 \longrightarrow N_2O_4$	(slow)
(2)	$N_2O_4 + CO \longrightarrow NO + CO_2 + NO_2$	(fast)
	$NO_2 + CO \longrightarrow NO + CO_2$	overall

The slowest step is always referred to as the slow step.

The rate-determining step of this mechanism involves a *bimolecular* collision between two NO_2 molecules. This is consistent with the rate expression involving $[NO_2]^2$. Because the CO is involved only after the slow step has occurred, the reaction rate would not depend on [CO] (that is, the reaction would be zero order in CO) if this were the actual mechanism. In this proposed mechanism, N_2O_4 is formed in one step and is completely consumed in a later step. Such a species is called a **reaction intermediate.**

However, in other studies of this reaction, nitrogen trioxide, NO_3, has been detected as a transient (short-lived) intermediate. The mechanism now thought to be correct is

Some reaction intermediates are so unstable that it is very difficult to prove experimentally that they exist.

(1)	$NO_2 + NO_2 \longrightarrow NO_3 + NO$	(slow)
(2)	$NO_3 + CO \longrightarrow NO_2 + CO_2$	(fast)
	$NO_2 + CO \longrightarrow NO + CO_2$	overall

In this proposed mechanism two molecules of NO_2 collide to produce one molecule each of NO_3 and NO. The reaction intermediate NO_3 then collides with one molecule of CO and reacts very rapidly to produce one molecule each of NO_2 and CO_2. Even though two NO_2 molecules are consumed in the first step, one is produced in the second step. The net result is that only the NO_2 molecule is consumed in the overall reaction.

Each of these proposed mechanisms meets both criteria for a plausible mechanism: (1) The steps add to give the equation for the overall reaction, and (2) the mechanism is consistent with the experimentally determined rate-law expression (in that two NO_2 molecules and no CO molecules are reactants in the slow step). The NO_3 that has been detected is evidence in favor of the second mechanism, but this does not unequivocally *prove* that mechanism; it may be possible to think of other mechanisms that would involve NO_3 as an intermediate and would also be consistent with the observed rate law.

The gas phase reaction of nitrogen oxide and bromine is known to be second order in NO and first order in Br_2.

$$2NO(g) + Br_2(g) \longrightarrow 2NOBr(g) \qquad rate = k[NO]^2[Br_2]$$

A one-step collision involving two NO molecules and one Br_2 molecule would be consistent with the experimentally determined rate-law expression. However, the likelihood of all *three* molecules colliding simultaneously is far less than the likelihood of two colliding. *Routes involving only bimolecular collisions or unimolecular decompositions are thought to be more favorable in reaction mechanisms.* The mechanism is believed to be

> *Think how unlikely it is for three moving billiard balls to collide simultaneously.*

> *Any fast step that precedes a slow step reaches equilibrium.*

$$
\begin{array}{lll}
(1) & NO + Br_2 \rightleftharpoons NOBr_2 & \text{(fast, equilibrium)} \\
(2) & NOBr_2 + NO \longrightarrow 2NOBr & \text{(slow)} \\
\hline
& 2NO + Br_2 \longrightarrow 2NOBr & \text{overall}
\end{array}
$$

The first step involves the collision of one NO and one Br_2 to produce the intermediate species $NOBr_2$. However, $NOBr_2$ can react rapidly to re-form NO and Br_2. We say that this is an *equilibrium step*. One of the $NOBr_2$ molecules can then react relatively slowly with one NO molecule to produce two NOBr molecules.

Denoting the rate constant for Step 2 as k_2, we could express the rate of this step as

$$rate = k_2[NOBr_2][NO]$$

> *The rate-law expression of Step 2 (the rate-determining step) determines the rate law for the overall reaction.*

However, $NOBr_2$ is a reaction intermediate, so its concentration at the beginning of the second step cannot be measured directly. Because $NOBr_2$ is formed in a fast equilibrium step, we can relate its concentration to the concentrations of the original reactants. When a reaction or reaction step is at *equilibrium,* its forward (f) and reverse (r) rates are equal.

$$rate_{1f} = rate_{1r}$$

Because this is an elementary step, we can write the rate expression for both directions from the equation for the elementary step

$$k_{1f}[NO][Br_2] = k_{1r}[NOBr_2]$$

and then rearrange for $[NOBr_2]$.

$$[NOBr_2] = \frac{k_{1f}}{k_{1r}}[NO][Br_2]$$

When we substitute the right side of this equation for $[NOBr_2]$ in the rate expression for the rate-determining step, $rate = k_2[NOBr_2][NO]$, we arrive at the experimentally determined rate-law expression.

> *The product and quotient of constants k_2, k_{1f}, and k_{1r} is another constant, k.*

$$rate = k_2\left(\frac{k_{1f}}{k_{1r}}[NO][Br_2]\right)[NO] \qquad or \qquad rate = k[NO]^2[Br_2]$$

Similar interpretations apply to most other overall third- or higher-order reactions, as well as many lower-order reactions. However, when several steps are about equally slow,

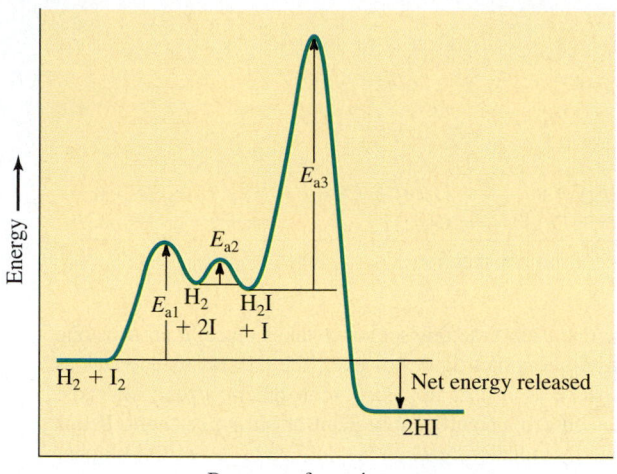

Figure 16-12 A graphical representation of the relative energies of activation for a postulated mechanism for the gas phase reaction

$$H_2 + I_2 \longrightarrow 2HI$$

the analysis of experimental data is much more complex. Fractional or negative reaction orders result from complex multistep mechanisms.

One of the earliest kinetic studies involved the gas phase reaction of hydrogen and iodine to form hydrogen iodide. The reaction was found to be first order in both hydrogen and iodine.

$$H_2(g) + I_2(g) \longrightarrow 2HI(g) \qquad \text{rate} = k[H_2][I_2]$$

The mechanism that was accepted for many years involved collision of single molecules of H_2 and I_2 in a simple one-step reaction. However, current evidence indicates a more complex process. Most kineticists now accept the following mechanism.

$$
\begin{array}{llll}
(1) & I_2 & \rightleftharpoons 2I & \text{(fast, equilibrium)} \\
(2) & I + H_2 & \rightleftharpoons H_2I & \text{(fast, equilibrium)} \\
(3) & H_2I + I & \longrightarrow 2HI & \text{(slow)} \\
\hline
& H_2 + I_2 & \longrightarrow 2HI & \text{overall}
\end{array}
$$

In this case neither original reactant appears in the rate-determining step, but both appear in the rate-law expression. Each step is a reaction in itself. Transition state theory tells us that each step has its own activation energy. Because Step 3 is the slowest, we know that its activation energy is the highest, as shown in Figure 16-12.

In summary

Apply the algebraic approach described above to show that this mechanism is consistent with observed rate-law expression.

> The experimentally determined reaction orders of reactants indicate the number of molecules of those reactants involved in (1) the slow step only, if it occurs first, or (2) the slow step *and* any fast equilibrium steps preceding the slow step.

16-8 TEMPERATURE: THE ARRHENIUS EQUATION

The average kinetic energy of a collection of molecules is proportional to the absolute temperature. At a particular temperature, T_1, a definite fraction of the reactant molecules

(Text continues on p. 626.)

Ozone

Ozone, O_3, is such a powerful oxidizing agent that in significant concentrations it destroys many plastics, metals, and rubber, as well as both plant and animal tissues. Therefore we try to minimize exposure to ozone in our immediate environment. However, ozone in the upper atmosphere plays a very important role in the absorption of harmful radiation from the sun. Maintaining appropriate concentrations of ozone—minimizing its production where ozone is harmful and preventing its destruction where ozone is helpful—is an important challenge in environmental chemistry.

Ozone is formed in the upper atmosphere as some O_2 molecules absorb high energy electromagnetic radiation from the sun and dissociate into oxygen atoms; these then combine with other O_2 molecules to form ozone.

$$O_2(g) + uv \longrightarrow 2O(g) \qquad \text{(step 1—occurs once)}$$
$$O_2(g) + O(g) \longrightarrow O_3(g) \qquad \text{(step 2—occurs twice)}$$
$$\overline{3O_2(g) + uv \longrightarrow 2O_3(g)} \qquad \text{(net reaction)}$$

Although it decomposes in the upper atmosphere, the ozone supply is continuously replenished by this process. Its concentration in the stratosphere (~7 to 31 miles above the earth's surface) is about 10 ppm (parts per million), whereas it is only about 0.04 ppm near the earth's surface.

The high-altitude ozone layer is responsible for absorbing much of the dangerous ultraviolet light from the sun in the 20–30 Å wavelength range.

$$O_3(g) + uv \longrightarrow O_2(g) + O(g) \qquad \text{(step 1—occurs once)}$$
$$O_2(g) + O(g) \longrightarrow O_3(g) \qquad \text{(step 2—occurs once)}$$
$$\text{no net reaction}$$

We see that each time this sequence takes place, it absorbs one photon of ultraviolet light; however, the process regenerates as much ozone as it uses up. Each stratospheric ozone molecule can thus absorb a significant amount of ultraviolet light. If this high-energy radiation reached the surface of the earth in higher intensity, it would be very harmful to plants and animals (including humans). It has been estimated that the incidence of skin cancer would increase by 2% for every 1% decrease in the concentration of ozone in the stratosphere.

Chlorofluorocarbons (CFCs) are chemically inert, nonflammable, nontoxic compounds that are superb solvents and have been used in many industrial processes; they are excellent coolants for air conditioners and refrigerators. Two CFCs that have been widely used are Freon-11 and Freon-12 (Freon is a DuPont trade name).

Freon-11 is CCl_3F Freon-12 is CCl_2F_2

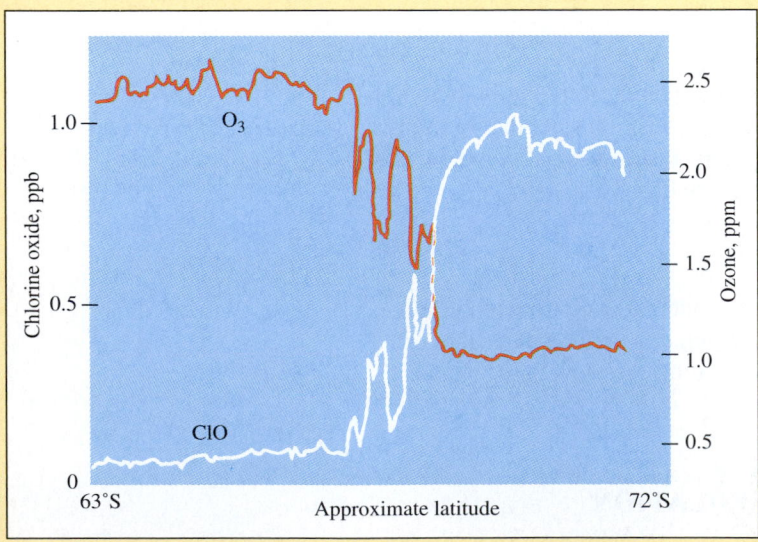

A plot that shows the decrease in $[O_3]$ as $[ClO]$ increases over Antarctica.

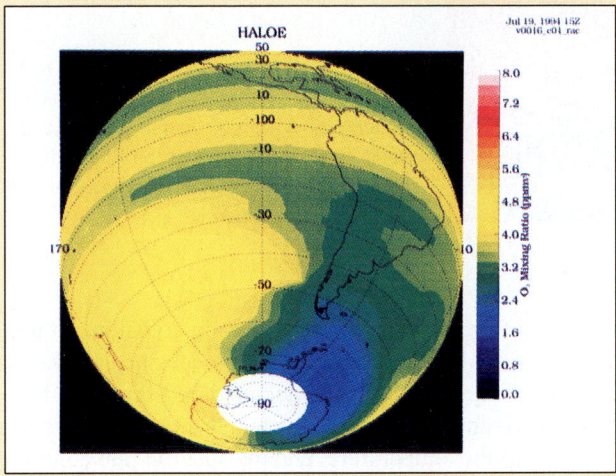

HALOE
Jul 19, 1994 15Z
v0016_c01_nat

O₃ Mixing Ratio (ppmv)

8.0
7.2
6.4
5.6
4.8
4.0
3.2
2.4
1.6
0.8
0.0

Halogen Occultation Experiment (HALOE) measurements of ozone were made by NASA during the Antarctic Spring, September 21 to October 15, 1992. The results show the "ozone hole," i.e., low levels of ozone at 15.5 miles over large areas of the Southern Hemisphere as well as the Antarctic region. Normal levels of ozone (yellow) are about 4.8 parts of ozone per million by volume (ppmv). The lowest levels of ozone are over the eastern part of Antarctica. In this computer-generated image, the dark blue region represents about 1.4 ppmv of ozone. Relatively low ozone levels also extend over the southern tip of South America (deep green, about 3.0 ppmv). Decreased levels extend over South America and into the mid-latitudes and the tropics (light green, about 4.0 ppmv). Measurements were not made over the white area at the South Pole.

The CFCs are so unreactive that they do not readily decompose, i.e., break down into simpler compounds, when they are released into the atmosphere. Over time the CFCs are carried into the stratosphere by air currents, where they are exposed to large amounts of ultraviolet radiation.

In 1974, Molinda and Rowland of the University of California–Irvine demonstrated in their laboratory that when CFCs are exposed to ultraviolet radiation they break down to form chlorine *radicals* ($:\overset{..}{\underset{..}{Cl}}\cdot$).

$$\underset{\underset{Cl}{|}}{\overset{\overset{F}{|}}{F-C-Cl}} \xrightarrow{uv} \underset{\underset{Cl}{|}}{\overset{\overset{F}{|}}{F-C\cdot}} + :\overset{..}{\underset{..}{Cl}}\cdot \quad \text{(a chlorine radical)}$$

Molinda and Rowland predicted that these very reactive radicals could cause problems by catalyzing the destruction of ozone in the stratosphere.

Each spring since 1979, researchers have observed a thinning of the ozone layer over Antarctica. Each spring (autumn in the Northern Hemisphere) beginning in 1983, satellite images have shown a "hole" in the ozone layer over the South Pole. During August and September 1987, a NASA research team flew a plane equipped with sophisticated analytical instruments into the ozone hole 25 times. Their measurements demonstrated that as the concentration of the chlorine oxide radicals, $Cl-\overset{..}{\underset{..}{O}}\cdot$, increased, the concentration of ozone decreased.

By September 1992 this **ozone hole** was nearly three times the area of the United States. In December 1994, three years of data from NASA's Upper Atmosphere Research Satellite (UARS) provided conclusive evidence that chlorofluorocarbons (CFCs) are primarily responsible for this destruction of the ozone layer. Considerable thinning of the ozone layer in the Northern Hemisphere has also been observed.

The following is a simplified representation of the **chain reaction** that is now believed to account for most of the ozone destruction in the stratosphere.

$$:\overset{..}{\underset{..}{Cl}}\cdot + O_3 \longrightarrow Cl-\overset{..}{\underset{..}{O}}\cdot + O_2 \quad \text{(step 1)}$$

$$Cl-\overset{..}{\underset{..}{O}}\cdot + O \longrightarrow :\overset{..}{\underset{..}{Cl}}\cdot + O_2 \quad \text{(step 2)}$$

$$\overline{O_3(g) + O \longrightarrow 2O_2(g)} \quad \text{(net reaction)}$$

A sufficient supply of oxygen atoms, O, is available in the upper atmosphere for the second step to occur. The net reaction results in the destruction of a molecule of ozone. However, the chlorine radical that initiates the first step of this reaction sequence is regenerated in the second step, and so a single chlorine radical can act as a catalyst to destroy many thousands of O_3 molecules. There are other well-known reactions that destroy ozone in the stratosphere, but the evidence shows conclusively that the CFCs are the principal culprits.

Since January 1978, the use of chlorofluorocarbons in aerosol cans in the United States has been banned; increasingly strict laws prohibit the release into the atmosphere of chlorofluorocarbons from sources such as automobile air conditioners and discarded refrigerators. Many other countries have joined to limit the use and release of chlorofluorocarbons. Efforts to develop suitable replacement substances and controls for existing chlorofluorocarbons continue. The good news is that scientists expect the ozone hole to decrease and possibly disappear during the 21st century *if* current international treaties remain in effect and *if* they are implemented throughout the world. These are two very large *ifs*.

Which egg will cook faster—the one in ice water (left) or the one in boiling water (right)? Why—in terms of what you have learned in this chapter?

have sufficient energy, E_a, to react to form product molecules upon collision. At a higher temperature, T_2, a greater fraction of the molecules possess the necessary activation energy, and the reaction proceeds at a faster rate. This is depicted in Figure 16-13.

From experimental observations, Svante Arrhenius developed the mathematical relationship among activation energy, absolute temperature, and the specific rate constant of a reaction, k, at that temperature. The relationship, called the Arrhenius equation, is

$e = 2.718$ is the base of *natural* logarithms (ln).

$$k = Ae^{-E_a/RT}$$

or, in logarithmic form,

$$\ln k = \ln A - \frac{E_a}{RT} \qquad \text{or} \qquad \log k = \log A - \frac{E_a}{2.303\,RT}$$

In this expression, A is a constant having the same units as the rate constant. It is proportional to the frequency of collisions between reacting molecules. R is the universal gas constant, expressed with the same energy units in its numerator as are used for E_a. For instance, when E_a is known in J/mol, the value $R = 8.314$ J/mol · K is appropriate. Here the unit "mol" is interpreted as "mole of reaction," as described in Chapter 15. One important point is the following: The greater the value of E_a, the smaller the value of k and the slower the reaction rate (other factors being equal). This is because fewer collisions take place with sufficient energy to get over a high energy barrier.

The Arrhenius equation predicts that increasing T results in faster reaction for the same E_a and concentrations.

The area between the distribution curve and the horizontal axis in Figure 16-13 is proportional to the total number of molecules present. The total area is the same at T_1 and T_2. The shaded areas represent the number of particles that exceed the energy of activation, E_a.

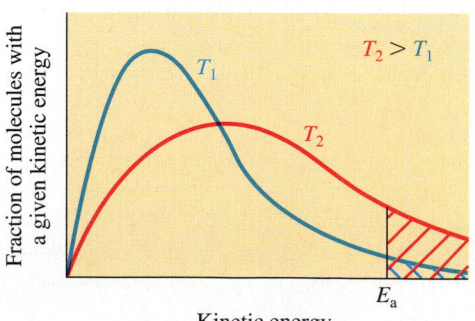

Figure 16-13   The effect of temperature on the number of molecules that have kinetic energies greater than E_a. At T_2, a higher fraction of molecules possess at least E_a, the activation energy.

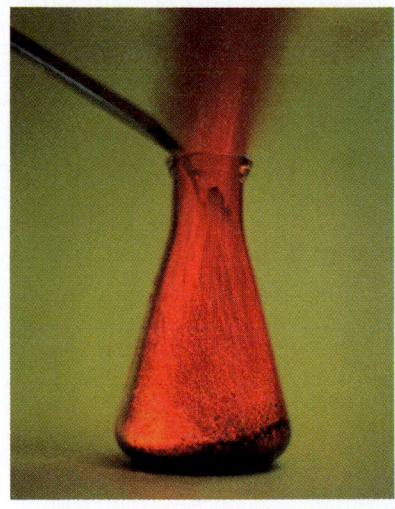 

Antimony powder reacts with bromine more rapidly at 75°C (left) than at 25°C (right).

| If T increases | $\Rightarrow$ | E_a/RT decreases | $\Rightarrow$ | $-E_a/RT$ increases | $\Rightarrow$ | $e^{-E_a/RT}$ increases | $\Rightarrow$ | k increases | $\Rightarrow$ | Reaction speeds up |

Let's look at how the rate constant varies with temperature for a given single reaction. Assume that the activation energy and the factor A do not vary with temperature. We can write the Arrhenius equation for two different temperatures. Then we subtract one equation from the other, and rearrange the result to obtain, in natural logarithm (ln) form,

$$\ln \frac{k_2}{k_1} = \frac{E_a}{R}\left(\frac{1}{T_1} - \frac{1}{T_2}\right)$$

In base-10 logarithm (log) form this equation is written as

$$\log \frac{k_2}{k_1} = \frac{E_a}{2.303R}\left(\frac{1}{T_1} - \frac{1}{T_2}\right)$$

Let's substitute some typical values into this equation. The activation energy for many reactions that occur near room temperature is about 50 kJ/mol (or 12 kcal/mol). For such a reaction, a temperature increase from 300 K to 310 K would result in

$$\ln \frac{k_2}{k_1} = \frac{50,000 \text{ J/mol}}{(8.314 \text{ J/mol} \cdot \text{K})}\left(\frac{1}{300 \text{ K}} - \frac{1}{310 \text{ K}}\right) = 0.647$$

$$\frac{k_2}{k_1} = 1.91 \approx 2$$

Chemists sometimes use the rule of thumb that near room temperature the rate of a reaction approximately doubles with a 10°C rise in temperature. Such a "rule" must be used with care, however, because it obviously depends on the activation energy.

EXAMPLE 16-9   *Arrhenius Equation*

The specific rate constant, k, for the following first-order reaction is 9.16×10^{-3} s^{-1} at 0.0°C. The activation energy of this reaction is 88.0 kJ/mol. Determine the value of k at 2.0°C.

$$N_2O_5 \longrightarrow NO_2 + NO_3$$

Plan

First we tabulate the values, remembering to convert temperature to the Kelvin scale.

$$E_a = 88{,}000 \text{ J/mol} \qquad\qquad R = 8.314 \text{ J/mol} \cdot \text{K}$$
$$k_1 = 9.16 \times 10^{-3} \text{ s}^{-1} \quad \text{at} \quad T_1 = 0.0°\text{C} + 273 = 273 \text{ K}$$
$$k_2 = \underline{?} \qquad\qquad \text{at} \quad T_2 = 2.0°\text{C} + 273 = 275 \text{ K}$$

We use these values in the "two-temperature" form of the Arrhenius equation.

Solution

For illustration, we use the *ln* version of the Arrhenius equation; if you prefer, you should establish that you get the same result with the *log* version.

$$\ln \frac{k_2}{k_1} = \frac{E_a}{R}\left(\frac{1}{T_1} - \frac{1}{T_2} \right)$$

$$\ln \left(\frac{k_2}{9.16 \times 10^{-3} \text{ s}^{-1}} \right) = \frac{88{,}000 \text{ J/mol}}{8.314 \dfrac{\text{J}}{\text{mol} \cdot \text{K}}} \left(\frac{1}{273 \text{ K}} - \frac{1}{275 \text{ K}} \right) = 0.282$$

Taking inverse (natural) logarithms of both sides,

$$\frac{k_2}{9.16 \times 10^{-3} \text{ s}^{-1}} = 1.32$$

$$k_2 = 1.32(9.16 \times 10^{-3} \text{ s}^{-1}) = \boxed{1.21 \times 10^{-2} \text{ s}^{-1}}$$

We see that a very small temperature difference, only 2°C, causes an increase in the rate constant (and hence in the reaction rate for the same concentrations) of about 32%. Such sensitivity of rate to temperature change makes the control and measurement of temperature extremely important in chemical reactions.

You should now work Exercise 45.

EXAMPLE 16-10   *Activation Energy*

The gas-phase decomposition of ethyl iodide to give ethylene and hydrogen iodide is a first-order reaction.

$$C_2H_5I \longrightarrow C_2H_4 + HI$$

At 600 K, the value of k is 1.60×10^{-5} s^{-1}. When the temperature is raised to 700 K, the value of k increases to 6.36×10^{-3} s^{-1}. What is the activation energy for this reaction?

Plan

We know k at two different temperatures. We solve the two-temperature form of the Arrhenius equation for E_a and evaluate.

Solution

$$k_1 = 1.60 \times 10^{-5} \text{ s}^{-1} \text{ at } T_1 = 600 \text{ K} \qquad k_2 = 6.36 \times 10^{-3} \text{ s}^{-1} \text{ at } T_2 = 700 \text{ K}$$
$$R = 8.314 \text{ J/mol} \cdot \text{K} \qquad\qquad\qquad E_a = \underline{?}$$

We arrange the Arrhenius equation for E_a.

$$\ln \frac{k_2}{k_1} = \frac{E_a}{R}\left(\frac{1}{T_1} - \frac{1}{T_2}\right) \qquad \text{so} \qquad E_a = \frac{R \ln \frac{k_2}{k_1}}{\left(\frac{1}{T_1} - \frac{1}{T_2}\right)}$$

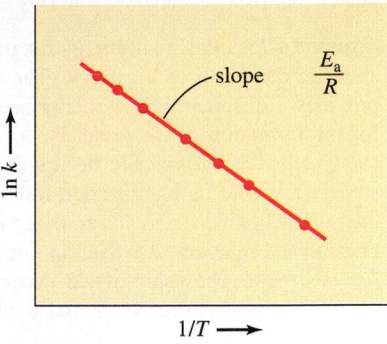

Substituting,

$$E_a = \frac{\left(8.314 \, \frac{J}{mol \cdot K}\right) \ln \left(\frac{6.36 \times 10^{-3} \, s^{-1}}{1.60 \times 10^{-5} \, s^{-1}}\right)}{\left(\frac{1}{600 \, K} - \frac{1}{700 \, K}\right)} = \frac{\left(8.314 \, \frac{J}{mol \cdot K}\right)(5.98)}{2.38 \times 10^{-4} \, K^{-1}}$$

$$E_a = \boxed{2.09 \times 10^5 \, \text{J/mol} \qquad \text{or} \qquad 209 \, \text{kJ/mol}}$$

You should now work Exercise 46.

Figure 16-14 A graphical method for determining activation energy, E_a. At each of several different temperatures, the rate constant, k, is determined by methods such as those in Sections 16-3 and 16-4. A plot of $\ln k$ versus $1/T$ gives a straight line with negative slope. The slope of this straight line is $-E_a/R$. Alternatively, a plot of $\log k$ versus $1/T$ gives a straight line whose slope is $-E_a/2.303R$. Use of this graphical method is often desirable, because it partially compensates for experimental errors in individual k and T values.

The determination of E_a in the manner illustrated in Example 16-10 may be subject to considerable error, because it depends on the measurement of k at only two temperatures. Any error in either of these k values would greatly affect the resulting value of E_a. A more reliable method that uses many measured values for the same reaction is based on a graphical approach. Let us rearrange the single-temperature logarithmic form of the Arrhenius equation and compare it with the equation for a straight line.

$$\underbrace{\ln k}_{y} = \underbrace{-\left(\frac{E_a}{R}\right)}_{m}\underbrace{\left(\frac{1}{T}\right)}_{x} + \underbrace{\ln A}_{b}$$

The value of the collision frequency factor, A, is very nearly constant over moderate temperature changes. Thus, $\ln A$ can be interpreted as the constant term in the equation (the intercept). The slope of the straight line obtained by plotting $\ln k$ versus $1/T$ can be interpreted as $-E_a/R$. This allows us to determine the value of the activation energy from the slope (Figure 16-14). Exercises 49 and 50 use this method.

Compare this approach to that described in the earlier Enrichment section for determining k.

16-9 CATALYSTS

Catalysts are substances that can be added to reacting systems to increase the rate of reaction. They allow reactions to occur via alternative pathways that increase reaction rates by lowering activation energies.

The activation energy is lowered in all catalyzed reactions, as depicted in Figures 16-15 and 16-16. A catalyst does take part in the reaction, but all of it is regenerated in later steps. Thus a catalyst does not appear in the balanced equation for the reaction.

For constant T and the same concentrations,

| If E_a decreases | $\Rightarrow$ | E_a/RT decreases | $\Rightarrow$ | $-E_a/RT$ increases | $\Rightarrow$ | $e^{-E_a/RT}$ increases | $\Rightarrow$ | k increases | $\Rightarrow$ | Reaction speeds up |

Figure 16-15 Potential energy diagrams showing the effect of a catalyst. The catalyst provides a different, lower-energy mechanism for the formation of the products. A catalyzed reaction typically occurs in several steps, each with its own barrier, but the overall energy barrier is lower than the uncatalyzed reaction. ΔE has the same value for each path. The value of ΔE depends only on the states of the reactants and products.

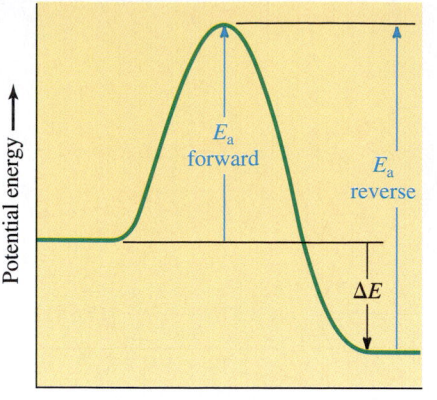

Reaction coordinate for uncatalyzed reaction

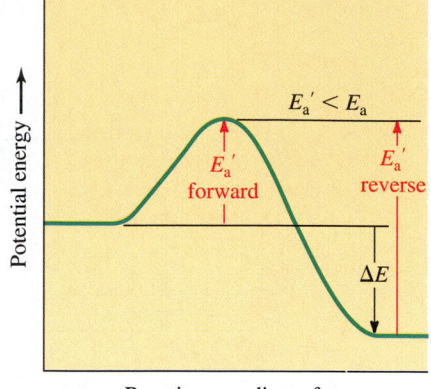

Reaction coordinate for catalyzed reaction

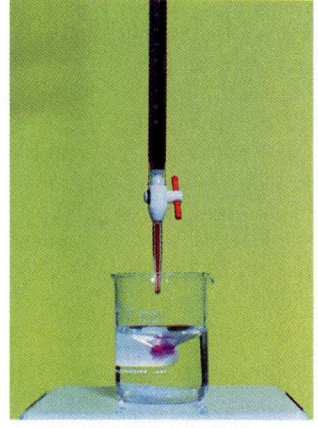

Figure 16-16 When a catalyst is present, the energy barrier is lowered. Thus, more molecules possess the minimum kinetic energy necessary for reaction. This is analogous to allowing more students to pass a course by lowering the requirements.

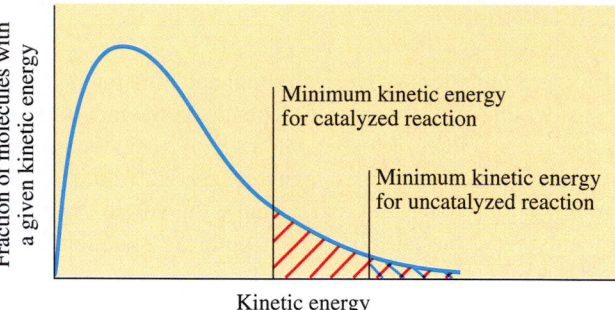

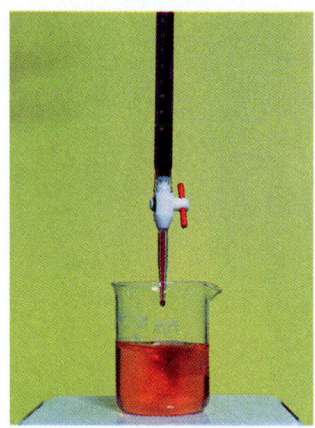

Oxalic acid, $(COOH)_2$, reduces intensely colored permanganate ions, MnO_4^-, to nearly colorless Mn^{2+} ions in acidic solution (top). The reaction is catalyzed by Mn^{2+} ions. If no Mn^{2+} ions are present, the reaction occurs *very* slowly, and MnO_4^- ions are reduced to brown solid MnO_2 (bottom). When we titrate $(COOH)_2$ with $KMnO_4$, we should always add the first few drops of $KMnO_4$ solution *very slowly* to produce some Mn^{2+} ions that can then catalyze the desired reaction.

We can describe two categories of catalysts: (1) homogeneous catalysts and (2) heterogeneous catalysts.

Homogeneous Catalysis

A **homogeneous catalyst** exists in the same phase as the reactants. Ceric ion, Ce^{4+}, was an important laboratory oxidizing agent that was used in many redox titrations (Section 11-7). For example, Ce^{4+} oxidizes thallium(I) ions in solution; this reaction is catalyzed by the addition of a very small amount of a soluble salt containing manganous ions, Mn^{2+}. The Mn^{2+} acts as a homogeneous catalyst.

$$2Ce^{4+} + Tl^+ \xrightarrow{Mn^{2+}} 2Ce^{3+} + Tl^{3+}$$

This reaction is thought to proceed by the following sequence of elementary steps.

$Ce^{4+} + Mn^{2+} \longrightarrow Ce^{3+} + Mn^{3+}$	step 1
$Ce^{4+} + Mn^{3+} \longrightarrow Ce^{3+} + Mn^{4+}$	step 2
$Mn^{4+} + Tl^+ \longrightarrow Mn^{2+} + Tl^{3+}$	step 3
$2Ce^{4+} + Tl^+ \longrightarrow 2Ce^{3+} + Tl^{3+}$	overall

Some of the Mn^{2+} catalyst reacts in step 1, but an equal amount is regenerated in step 3 and is thus available to react again. The two ions shown in blue, Mn^{3+} and Mn^{4+}, are *reaction intermediates*. Mn^{3+} ions are formed in step 1 and consumed in an equal amount in step 2; similarly, Mn^{4+} ions are formed in step 2 and consumed in an equal amount in step 3.

Be sure you can distinguish among the various species that appear in a reaction mechanism.

1. *Reactant:* more is consumed than is formed.

2. *Product:* more is formed than is consumed.

3. *Reaction intermediate:* formed in earlier steps, then consumed in an equal amount in later steps.

4. *Catalyst:* consumed in earlier steps, then regenerated in an equal amount in later steps.

Strong acids function as homogeneous catalysts in the acid-catalyzed hydrolysis of esters (a class of organic compounds—Section 27-13). Using ethyl acetate (a component of nail polish removers) as an example of an ester, we can write the overall reaction as follows.

"Hydrolysis" means reaction with water.

$$CH_3\overset{O}{\overset{\|}{C}}-OCH_2CH_3(aq) + H_2O \xrightarrow{H^+} CH_3\overset{O}{\overset{\|}{C}}-OH(aq) + CH_3CH_2OH(aq)$$

ethyl acetate   acetic acid   ethanol

This is a thermodynamically favored reaction, but because of its high energy of activation, it occurs only very, very slowly when no catalyst is present. In the presence of strong acids, however, the reaction occurs more rapidly. In this acid-catalyzed hydrolysis, different intermediates with lower activation energies are formed. The sequence of steps in the *postulated* mechanism follows.

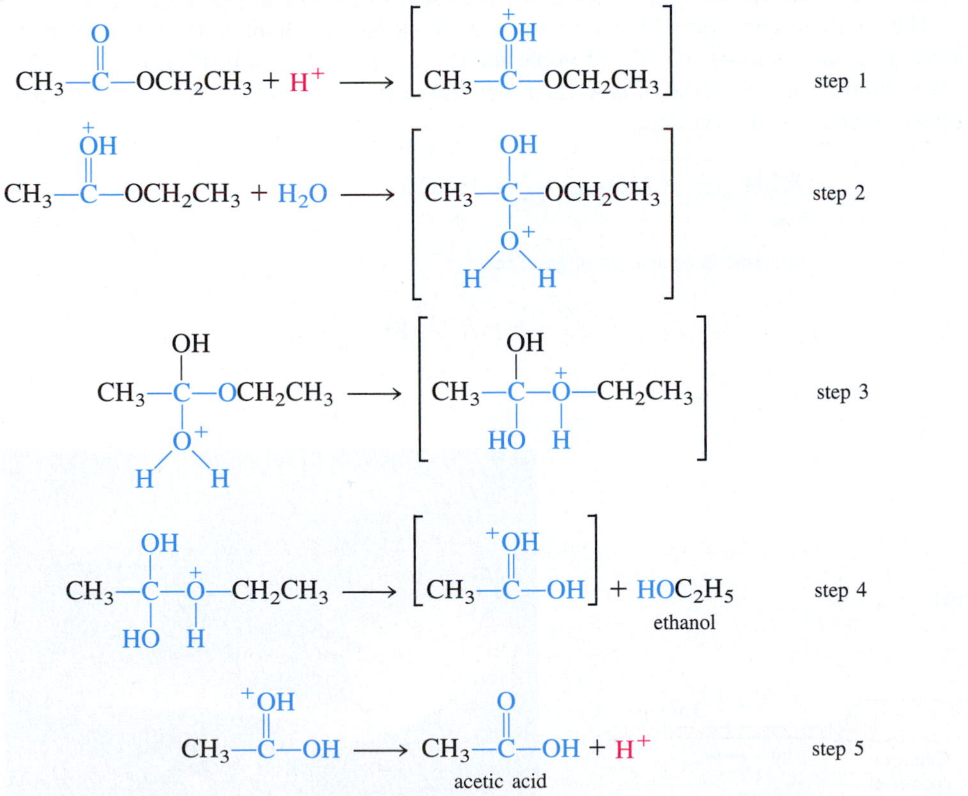

step 1

step 2

step 3

step 4

step 5

overall

Groups of atoms that are involved in the change in each step are shown in blue.

All intermediates in this sequence of elementary steps are charged species, but this is not always the case.

The petroleum industry uses numerous heterogeneous catalysts. Many of them contain highly colored compounds of transition metal ions. Several are shown here.

We see that H^+ is a reactant in step 1, but it is completely regenerated in step 5. Therefore, H^+ is a catalyst. The species shown in brackets in steps 1 through 4 are *reaction intermediates.* Ethyl acetate and water are the reactants, and acetic acid and ethanol are the products of the overall catalyzed reaction.

Heterogeneous Catalysis

A **heterogeneous catalyst** (also known as a **contact catalyst**) is present in a different phase than the reactants. They are usually solids, and they lower activation energies by providing surfaces on which reactions can occur. The first step in the catalytic process is usually *adsorption,* in which one or more of the reactants become attached to the solid surface. Some reactant molecules may be held in particular orientations; in other molecules, some bonds may be broken to form atoms or smaller molecular fragments. This causes *activation* of the reactants. As a result, *reaction* occurs more readily than would otherwise be possible. In a final step, *desorption,* the product molecules leave the surface, freeing reaction sites to be used again. Most contact catalysts are more effective as small particles, because they have relatively large surface areas.

Transition metals and their compounds function as effective catalysts in many homogeneous and heterogeneous reactions. Vacant *d* orbitals in many transition metal ions can accept electrons from reactants to form intermediates. These subsequently decompose to form products. Three transition metals, Pt, Pd, and Ni, are often used as finely divided solids to provide surfaces on which heterogeneous reactions can occur.

The catalytic converters built into automobile exhaust systems contain two types of heterogeneous catalysts, powdered noble metals and powdered transition metal oxides. They catalyze the oxidation of unburned fuel and of partial combustion products such as carbon monoxide (Figure 16-17).

$$2C_8H_{18}(g) + 25O_2(g) \xrightarrow[\text{NiO}]{\text{Pt}} 16CO_2(g) + 18H_2O(g)$$
↑
isooctane (a component of gasoline)

$$2CO(g) + O_2(g) \xrightarrow[\text{NiO}]{\text{Pt}} 2CO_2(g)$$

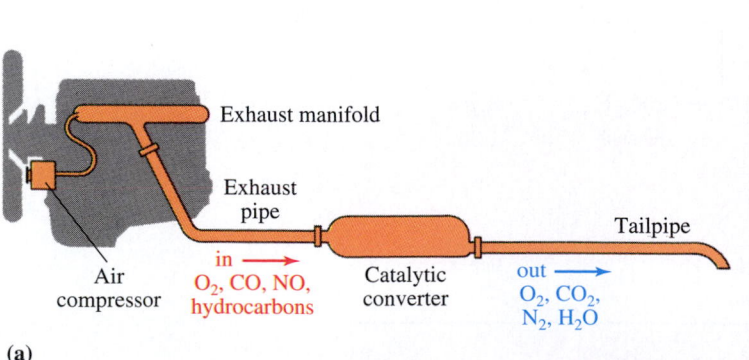

Exhaust manifold

Exhaust pipe

Tailpipe

Air compressor

in →
O_2, CO, NO, hydrocarbons

Catalytic converter

out →
O_2, CO_2, N_2, H_2O

(a)

(b)

Figure 16-17   (a) The arrangement of a catalytic converter in an automobile. (b) A cutaway view of a catalytic converter, showing the pellets of catalyst.

It is desirable to carry out these reactions in automobile exhaust systems. Carbon monoxide is very poisonous. The latter reaction is so slow that a mixture of CO and O_2 gas at the exhaust temperature would remain unreacted for thousands of years in the absence of a catalyst! Yet the addition of only a small amount of a solid, finely divided transition metal catalyst promotes the production of up to a mole of CO_2 per minute. Because this reaction is a very simple but important one, it has been studied extensively by surface chemists. It is one of the best understood heterogeneously catalyzed reactions. The major features of the catalytic process are shown in Figure 16-18.

(a) *Adsorption*: CO and O_2 reactant molecules become bound to the surface:

$$CO(g) \longrightarrow CO(surface) \quad \text{and} \quad O_2(g) \longrightarrow O_2(surface)$$

The CO molecules are linked through their C atoms to one or more metal atoms on the surface. The O_2 molecules are more weakly bound.

(b) *Activation*: The O_2 molecules dissociate into O atoms, which are held in place more tightly:

$$O_2(surface) \longrightarrow 2O(surface)$$

The CO molecules stick to the surface, but they migrate easily across the surface.

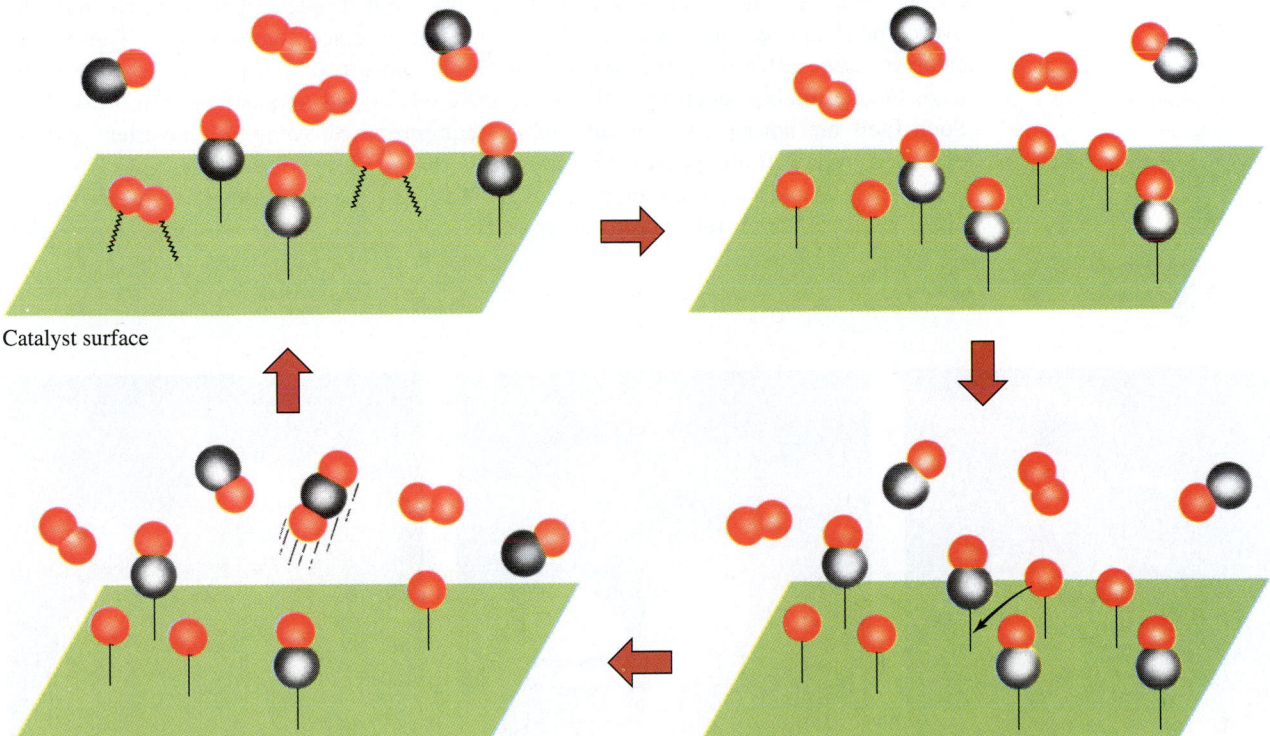

Catalyst surface

(d) *Desorption*: The CO_2 product molecules leave the surface:

$$CO_2(surface) \longrightarrow CO_2(g)$$

Fresh reactant molecules can then replace them to start the cycle again [back to step (a)].

(c) *Reaction*: O atoms react with bound CO molecules, to form CO_2 molecules:

$$CO(surface) + O(surface) \longrightarrow CO_2(surface)$$

The resulting CO_2 molecules bind to the surface *very poorly*.

Figure 16-18 A simplified representation of the catalysis of the reaction

$$2CO(g) + O_2(g) \longrightarrow 2CO_2(g)$$

on a metallic surface.

The same catalysts also catalyze another reaction, the decomposition of nitrogen oxide, NO, into harmless N_2 and O_2.

$$2NO(g) \xrightarrow[\text{NiO}]{\text{Pt}} N_2(g) + O_2(g)$$

At the high temperatures of the combustion of any fuel in air, nitrogen and oxygen combine to form nitrogen oxide.

Nitrogen oxide is a serious air pollutant because it is oxidized to nitrogen dioxide, NO_2. This reacts with water to form nitric acid and with alcohols to form nitrates, which are eye irritants.

These three reactions, catalyzed in catalytic converters, are all exothermic and thermodynamically favored. Unfortunately, other energetically favored reactions are also accelerated by the mixed catalysts. All fossil fuels contain sulfur compounds, which are oxidized to sulfur dioxide during combustion. Sulfur dioxide, itself an air pollutant, undergoes further oxidation to form sulfur trioxide as it passes through the catalytic bed.

$$2SO_2(g) + O_2(g) \xrightarrow[\text{NiO}]{\text{Pt}} 2SO_3(g)$$

Maintaining the continued efficiency of all three reactions in a "three-way" catalytic converter is a delicate matter. It requires control of such factors as the O_2 supply pressure and the order in which the reactants reach the catalyst. Some modern automobile engines use microcomputer chips, based on an O_2 sensor in the exhaust stream, to control air valves.

Sulfur trioxide is probably a worse pollutant than sulfur dioxide, because SO_3 is the acid anhydride of strong, corrosive sulfuric acid. Sulfur trioxide reacts with water vapor in the air, as well as in auto exhausts, to form sulfuric acid droplets. This problem must be overcome if the current type of catalytic converter is to see continued use. These same catalysts also suffer from the problem of being "poisoned"—i.e., made inactive—by lead. Leaded fuels contain tetraethyl lead, $Pb(C_2H_5)_4$, and tetramethyl lead, $Pb(CH_3)_4$. Such fuels are not suitable for automobiles equipped with catalytic converters and are excluded by law from use in such cars.

Reactions that occur in the presence of a solid catalyst, as on a metal surface (heterogeneous catalysis) often follow zero-order kinetics. For instance, the rate of decomposition

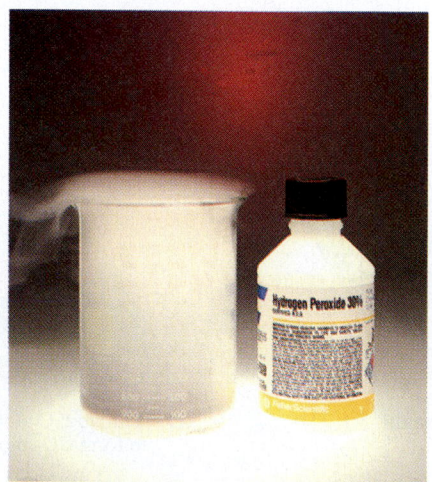

The decomposition of a 30% hydrogen peroxide solution is catalyzed by very small amounts of transition metal oxides. This catalyzed reaction is rapid, so the exothermic reaction quickly heats the solution to the boiling point of water, forming steam. The temperature increase further accelerates the decomposition. One should never use a syringe with a metal tip to withdraw a sample from a 30% hydrogen peroxide solution.

of $NO_2(g)$ at higher pressures on a platinum metal surface does not change if we add more NO_2. This is because only the NO_2 molecules on the surface can react. If the metal surface is completely covered with NO_2 molecules, no additional molecules can be *adsorbed* until the ones already there have reacted and the products have *desorbed*. Thus, the rate of the reaction is controlled only by the availability of reaction sites on the Pt surface, and not by the total number of NO_2 molecules available.

Some other important reactions that are catalyzed by transition metals and their oxides follow.

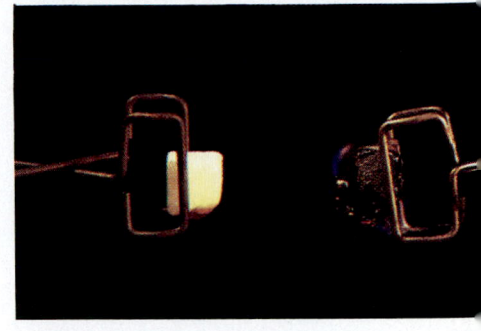

1. The Haber process for the production of ammonia (Section 17-6).

$$N_2 + 3H_2 \xrightarrow[\text{high T, P}]{\text{Fe, Fe oxides}} 2NH_3$$

2. The contact process for the production of sulfur trioxide in the manufacture of sulfuric acid (Section 24-11).

$$2SO_2 + O_2 \xrightarrow[400°C]{V_2O_5} 2SO_3$$

When heated, a sugar cube (sucrose, melting point 185°C) melts but does not burn. A sugar cube rubbed in cigarette ash burns before it melts. The cigarette ash contains trace amounts of metal compounds that catalyze the combustion of sugar.

3. The bromination of benzene (Section 28-4).

$$\underset{\text{benzene}}{C_6H_6} + Br_2 \xrightarrow{FeBr_3} \underset{\text{bromobenzene}}{C_6H_5Br} + HBr$$

4. The hydrogenation of unsaturated hydrocarbons (Section 28-5).

$$RCH{=}CH_2 + H_2 \xrightarrow{Pt} RCH_2CH_3 \qquad R = \text{organic groups}$$

Enzymes as Biological Catalysts

Enzymes are proteins that act as catalysts for specific biochemical reactions in living systems. The reactants in enzyme-catalyzed reactions are called **substrates.** Thousands of vital processes in our bodies are catalyzed by as many distinct enzymes. For instance, the enzyme carbonic anhydrase catalyzes the combination of CO_2 and water (the substrates), facilitating most of the transport of carbon dioxide in the blood. This combination reaction, ordinarily uselessly slow, proceeds rapidly in the presence of carbonic anhydrase; a single molecule of this enzyme can promote the conversion of more than 1 million molecules of carbon dioxide each second. Each enzyme is extremely specific, catalyzing only a few closely related reactions—or, in many cases, only one particular reaction—for only certain substrates. Modern theories of enzyme action attribute this to the requirement of very specific matching of shapes (molecular geometries) for a particular substrate to bind to a particular enzyme (Figure 16-19).

In the presence of a hot copper catalyst, acetone, CH_3COCH_3, is converted into ketene, CH_2CO, and methane, CH_4, in an exothermic reaction.

$$CH_3COCH_3(g) \xrightarrow[\Delta]{Cu} CH_2CO(g) + CH_4(g) + \text{heat}$$

Hot copper pennies are suspended over acetone in the flask. Acetone vapor reacts on the hot copper surface. The heat liberated by the reaction causes the pennies to glow brightly.

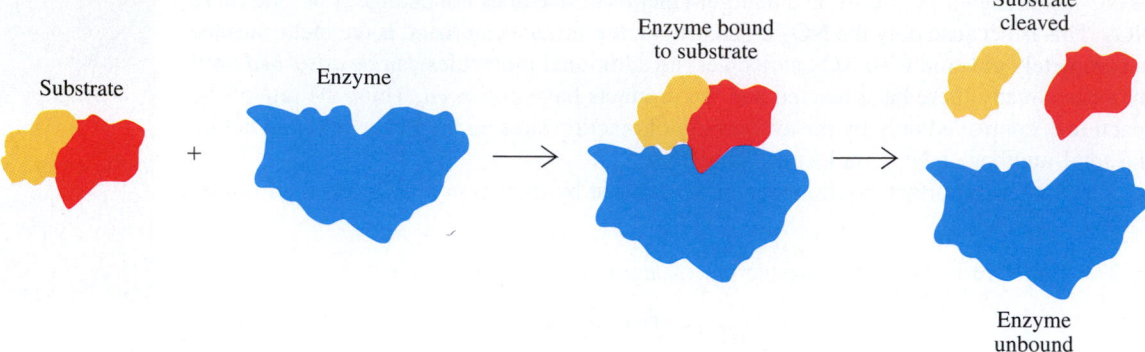

Figure 16-19 A schematic representation of a simplified mechanism (lock-and-key) for enzyme reaction. The substrates (reactants) fit the active sites of the enzyme molecule much as keys fit locks. When the reaction is complete, the products do not fit the active sites as well as the reactants did. They separate from the enzyme, leaving it free to catalyze the reaction of additional reactant molecules. The enzyme is not permanently changed by the process. The illustration here is for a process in which a complex reactant molecule is split to form two simpler product molecules. The formation of simple sugars from complex carbohydrates is a similar reaction. Some enzymes catalyze the combination of simple molecules to form more complex ones.

Enzyme-catalyzed reactions are important examples of zero-order reactions; that is, the rate of each of them is independent of the concentration of the substrate (provided *some* substrate is present).

$$rate = k$$

The active site on an enzyme can bind to only one substrate molecule at a time (or one pair, if the reaction links two reactant molecules), no matter how many other substrate molecules are available in the vicinity.

Ammonia is a very important industrial chemical. The Haber process for its preparation involves the use of iron as a *catalyst* at 450°C to 500°C and high pressures. The reaction is thermodynamically spontaneous.

$$N_2(g) + 3H_2(g) \xrightarrow{\text{Fe}} 2NH_3(g) \qquad \Delta G^0 = -194.7 \text{ kJ/mol (at 500°C)}$$

but very slow. In the absence of a catalyst, the reaction does not occur at an observable rate at room temperature. Even with a catalyst, it is not efficient at atmospheric pressure.

Most of the essential nutrients for both plants and animals contain nitrogen. The reaction between N_2 and H_2 to form NH_3 is catalyzed at room temperature and atmospheric pressure by a class of enzymes, called nitrogenases, that are present in some bacteria. Legumes are plants that support these bacteria; they are able to obtain nitrogen as N_2 from the atmosphere and convert it to ammonia.

In comparison with manufactured catalysts, most enzymes are tremendously efficient under very mild conditions. If chemists and biochemists could develop catalysts with a small fraction of the efficiency of enzymes, they could be a great boon to the world's health and economy. One of the most active areas of current chemical research involves attempts to discover or synthesize catalysts that can mimic the efficiency of naturally occurring enzymes such as nitrogenases. Such a development would be important in industry. It would eliminate the costs of the high temperature and high pressure that are

Transition metal ions are present in the active sites of some enzymes.

The process is called nitrogen fixation. The ammonia can be used in the synthesis of many nitrogen-containing biological compounds such as proteins and nucleic acids.

necessary in the Haber process. This could decrease the cost of food grown with the aid of ammonia-based fertilizers. Ultimately this would help greatly to feed the world's growing population.

Key Terms

Activation energy   The amount of energy that must be absorbed by reactants in their ground states to convert them to the transition state so that a reaction can occur.

Arrhenius equation   An equation that relates the specific rate constant to activation energy and temperature.

Catalyst   A substance that alters (usually increases) the rate at which a reaction occurs.

Chemical kinetics   The study of rates and mechanisms of chemical reactions and of the factors on which they depend.

Collision theory   A theory of reaction rates that states that effective collisions between reactant molecules must take place for reaction to occur.

Contact catalyst   See *Heterogeneous catalyst.*

Effective collision   A collision between molecules that results in reaction; one in which molecules collide with proper orientations and with sufficient energy to react.

Elementary step   An individual step in the mechanism by which a reaction occurs. For each elementary step, the reaction orders *do* match the reactant coefficients in that step.

Enzyme   A protein that acts as a catalyst in a biological system.

Half-life of a reactant   The time required for half of that reactant to be converted into product(s).

Heterogeneous catalyst   A catalyst that exists in a different phase (solid, liquid, or gas) from the reactants; a contact catalyst.

Homogeneous catalyst   A catalyst that exists in the same phase (solid, liquid, or gas) as the reactants.

Integrated rate equation   An equation that gives the concentration of a reactant remaining after a specified time; has different mathematical forms for different orders of reaction.

Method of initial rates   A method of determining the rate-law expression by carrying out a reaction with different initial concentrations and analyzing the resulting changes in initial rates.

Order of a reactant   The power to which the reactant's concentration is raised in the rate-law expression.

Order of a reaction   The sum of the powers to which all concentrations are raised in the rate-law expression; also called overall order of a reaction.

Rate-determining step   The slowest step in a reaction mechanism; the step that limits the overall rate of reaction.

Rate-law expression (also called **rate law**)   An equation that relates the rate of a reaction to the concentrations of the reactants and the specific rate constant; rate $= k[A]^x[B]^y$. The exponents of reactant concentrations *do not necessarily* match the coefficients in the overall balanced chemical equation. The rate-law expression must be determined from experimental data.

Rate of reaction   The change in concentration of a reactant or product per unit time.

Reaction coordinate   The progress along the pathway from reactants to products; sometimes called "progress of reaction."

Reaction intermediate   A species that is produced and then entirely consumed during a reaction; usually short-lived.

Reaction mechanism   The sequence of steps by which reactants are converted into products.

Specific rate constant (also called **rate constant**)   An experimentally determined proportionality constant that is different for different reactions and that, for a given reaction, changes only with temperature or the presence of a catalyst; k in the rate-law expression Rate $= k[A]^x[B]^y$.

Substrate   A reactant in an enzyme-catalyzed reaction.

Thermodynamically favorable (spontaneous) reaction   A reaction that occurs with a net release of free energy, G; a reaction for which ΔG is negative (see Section 15-15).

Transition state   A relatively high-energy state in which bonds in reactant molecules are partially broken and new ones are partially formed.

Transition state theory   A theory of reaction rates that states that reactants pass through high-energy transition states before forming products.

Exercises

General Concepts

1. Briefly summarize the effects of each of the four factors that affect rates of reactions.
2. Describe the basic features of collision theory and transition state theory.
3. What is a rate-law expression? Describe how it is determined for a particular reaction.
4. Distinguish between reactions that are thermodynamically favor-able and reactions that are kinetically favorable. What can be said about relationships between the two?
5. What is meant by the order of a reaction?
6. What, if anything, can be said about the relationship between the coefficients of the balanced *overall* equation for a reaction and the powers to which concentrations are raised in the rate-law expression? To what are these powers related?

7. Express the rate of reaction in terms of the rate of change of each reactant and each product in the following reactions.
   (a) $H_2O_2(aq) + 2H^+(aq) + 2I^-(aq) \longrightarrow I_2(aq) + 2H_2O(\ell)$
   (b) $2NO(g) + Br_2(g) \longrightarrow 2NOBr(g)$
   (c) $CH_3COOH(aq) + OH^-(aq) \longrightarrow CH_3COO^-(aq) + H_2O(\ell)$
8. Express the rate of reaction in terms of the rate of change of each reactant and each product in the following.
   (a) $3ClO^-(aq) \longrightarrow ClO_3^-(aq) + 2Cl^-(aq)$
   (b) $2SO_2(g) + O_2(g) \longrightarrow 2SO_3(g)$
   (c) $C_2H_4(g) + Br_2(g) \longrightarrow C_2H_4Br_2(g)$
9. At the instant when N_2 is reacting at a rate of 0.30 M/min, what is the rate at which H_2 is disappearing, and what is the rate at which NH_3 is forming?

$$N_2 + 3H_2 \longrightarrow 2NH_3$$

10. At the instant when NH_3 is reacting at a rate of 1.20 M/min, what is the rate at which the other reactant is disappearing, and what is the rate at which each product is being formed?

$$4NH_3 + 5O_2 \longrightarrow 4NO + 6H_2O$$

Rate-Law Expression

11. If doubling the initial concentration of a reactant doubles the initial rate of reaction, what is the order of the reaction with respect to the reactant? If the rate increases by a factor of eight, what is the order? If the rate remains the same, what is the order?
12. Use times expressed in seconds to give the units of the rate constant for reactions that are overall (a) first order; (b) second order; (c) third order; (d) of order $1\frac{1}{2}$.
13. Rate data were obtained at 25°C for the following reaction. What is the rate-law expression for this reaction?

$$A + 2B \longrightarrow C + 2D$$

Expt.	[A] [mol/L]	[B] [mol/L]	Initial Rate of Formation of C
1	0.10	0.10	$3.0 \times 10^{-4} M \cdot min^{-1}$
2	0.30	0.30	$9.0 \times 10^{-4} M \cdot min^{-1}$
3	0.30	0.10	$3.0 \times 10^{-4} M \cdot min^{-1}$
4	0.40	0.20	$6.0 \times 10^{-4} M \cdot min^{-1}$

14. Rate data were obtained for the following reaction at 25°C. What is the rate-law expression for the reaction?

$$2A + B + 2C \longrightarrow D + 2E$$

Expt.	Initial [A]	Initial [B]	Initial [C]	Initial Rate of Formation of D
1	0.10 M	0.20 M	0.10 M	$5.0 \times 10^{-4} M \cdot min^{-1}$
2	0.20 M	0.20 M	0.30 M	$1.5 \times 10^{-3} M \cdot min^{-1}$
3	0.30 M	0.20 M	0.10 M	$5.0 \times 10^{-4} M \cdot min^{-1}$
4	0.40 M	0.60 M	0.30 M	$4.5 \times 10^{-3} M \cdot min^{-1}$

15. The reaction $2NO + 2H_2 \rightarrow N_2 + 2H_2O$ gives the following initial rates at a high temperature.

Expt.	[NO]$_0$ (mol/L)	[H$_2$]$_0$ (mol/L)	Initial Rate of Formation of N$_2$
1	0.420	0.122	0.124 $M \cdot s^{-1}$
2	0.210	0.122	0.0310 $M \cdot s^{-1}$
3	0.210	0.244	0.0620 $M \cdot s^{-1}$
4	0.105	0.488	0.0310 $M \cdot s^{-1}$

(a) Write the rate-law expression for this reaction. (b) Find the rate of formation of N_2 at the instant when [NO] = 0.450 M and [H$_2$] = 0.161 M.

*16. The reaction $2NO + O_2 \rightarrow 2NO_2$ gives the following initial rates.

Expt.	[NO]$_0$ (mol/L)	[O$_2$]$_0$ (mol/L)	Initial Rate of Disappearance of NO
1	0.020	0.010	$1.5 \times 10^{-4} M \cdot s^{-1}$
2	0.040	0.010	$6.0 \times 10^{-4} M \cdot s^{-1}$
3	0.020	0.040	$6.0 \times 10^{-4} M \cdot s^{-1}$

(*Reminder:* Express the rate of *reaction* in terms of the rate of change of concentration of each reactant and each product.)
(a) Write the rate-law expression for this reaction. (b) Find the rate of reaction at the instant when [NO] = 0.055 M and [O$_2$] = 0.035 M. (c) At the time described in part (b), what will be the rate of disappearance of NO? the rate of disappearance of O_2? the rate of formation of NO_2?

17. (a) A certain reaction is zero order in reactant A and second order in reactant B. If the concentrations of both reactants are doubled, what happens to the reaction rate? (b) What would happen to the reaction rate if the reaction in part (a) were first order in A and first order in B?
18. The rate expression for the following reaction is rate = $k[A]^2[B_2]$. If, during a reaction, the concentrations of both A and B_2 were suddenly halved, the rate of the reaction would _____ by a factor of _____.

$$A + B_2 \longrightarrow products$$

19. Rate data were collected for the following reaction at a particular temperature.

$$A + B \longrightarrow products$$

Expt.	[A]$_0$ (mol/L)	[B]$_0$ (mol/L)	Initial Rate of Reaction
1	0.10	0.10	0.0090 $M \cdot s^{-1}$
2	0.20	0.10	0.036 $M \cdot s^{-1}$
3	0.10	0.20	0.018 $M \cdot s^{-1}$
4	0.10	0.30	0.027 $M \cdot s^{-1}$

(a) What is the rate-law expression for this reaction? (b) Describe the order of the reaction with respect to each reactant and to the overall order.

20. Rate data were collected for the following reaction at a particular temperature.

$$2ClO_2(aq) + 2OH^-(aq) \longrightarrow$$
$$ClO_3^-(aq) + ClO_2^-(aq) + H_2O(\ell)$$

Expt.	$[ClO_2]_0$ (mol/L)	$[OH^-]_0$ (mol/L)	Initial Rate of Reaction
1	0.012	0.012	$2.07 \times 10^{-4} \, M \cdot s^{-1}$
2	0.024	0.012	$8.28 \times 10^{-4} \, M \cdot s^{-1}$
3	0.012	0.024	$4.14 \times 10^{-4} \, M \cdot s^{-1}$
4	0.024	0.024	$1.66 \times 10^{-3} \, M \cdot s^{-1}$

(a) What is the rate-law expression for this reaction? (b) Describe the order of the reaction with respect to each reactant and to the overall order.

21. The reaction $(C_2H_5)_2(NH)_2 + I_2 \rightarrow (C_2H_5)_2N_2 + 2HI$ gives the following initial rates.

Expt.	$[(C_2H_5)_2(NH)_2]_0$ (mol/L)	$[I_2]_0$ (mol/L)	Initial Rate of Formation of $(C_2H_5)_2N_2$
1	0.015	0.015	$3.15 \, M \cdot s^{-1}$
2	0.015	0.045	$9.45 \, M \cdot s^{-1}$
3	0.030	0.045	$18.9 \, M \cdot s^{-1}$

Write the rate-law expression.

22. Given the following data for the reaction $A + B \rightarrow C$, write the rate-law expression.

Expt.	Initial [A]	Initial [B]	Initial Rate of Formation of C
1	0.20 M	0.10 M	$5.0 \times 10^{-6} \, M \cdot s^{-1}$
2	0.30 M	0.10 M	$7.5 \times 10^{-6} \, M \cdot s^{-1}$
3	0.40 M	0.20 M	$4.0 \times 10^{-5} \, M \cdot s^{-1}$

*23. Consider a chemical reaction between compounds A and B that is first order in A and first order in B. From the information given below, fill in the blanks.

Expt.	Rate ($M \cdot s^{-1}$)	[A]	[B]
1	0.24	0.20 M	0.050 M
2	0.20	____ M	0.030 M
3	0.80	0.40 M	____ M

*24. Consider a chemical reaction of compounds A and B that was found to be first order in A and second order in B. From the following information, fill in the blanks.

Expt.	Rate ($M \cdot s^{-1}$)	[A]	[B]
1	0.150	1.00 M	0.200 M
2	____	2.00 M	0.200 M
3	____	2.00 M	0.400 M

*25. The decomposition of NO_2 by the following reaction at some temperature proceeds at a rate 5.4×10^{-5} mol NO_2/L·s when $[NO_2] = 0.0100$ mol/L.

$$2NO_2(g) \longrightarrow 2NO(g) + O_2(g)$$

(a) Assume that the rate law is rate $= k[NO_2]$. What rate of disappearance of NO_2 would be predicted when $[NO_2] = 0.00500$ mol/L? (b) Now assume that the rate law is rate $= k[NO_2]^2$. What rate of disappearance of NO_2 would be predicted when $[NO_2] = 0.00500$ mol/L? (c) The observed rate when $[NO_2] = 0.00500$ mol/L is observed to be 1.4×10^{-5} mol NO_2/L·s. Which rate law is correct? (d) Calculate the rate constant. (*Reminder:* Express the rate of reaction in terms of rate of disappearance of NO_2.)

Integrated Rate Equations and Half-Life

26. What is meant by the half-life of a reactant?

27. The rate law for the reaction of sucrose in water,

$$C_{12}H_{22}O_{11} + H_2O \longrightarrow 2C_6H_{12}O_6$$

is rate $= k[C_{12}H_{22}O_{11}]$. After 2.57 hours at 25°C, 6.00 g/L of $C_{12}O_{22}O_{11}$ has decreased to 5.40 g/L. Evaluate k for this reaction at 25°C.

28. The rate constant for the decomposition of nitrogen dioxide

$$2NO_2 \longrightarrow 2NO + O_2$$

with a laser beam is $1.70 \, M^{-1} \cdot min^{-1}$. Find the time, in seconds, needed to decrease 2.00 mol/L of NO_2 to 1.25 mol/L.

29. The second-order rate constant for the following gas phase reaction is $0.0442 \, M^{-1} \cdot s^{-1}$. We start with 0.130 mol C_2F_4 in a 2.00-liter container, with no C_4F_8 initially present.

$$2C_2F_4 \longrightarrow C_4F_8$$

(a) What will be the concentration of C_2F_4 after 1.00 hour? (b) What will be the concentration of C_4F_8 after 1.00 hour? (c) What is the half-life of the reaction for the initial C_2F_4 concentration given in part (a)? (d) How long will it take for half of the C_2F_4 that remains after 1.00 hour to disappear?

30. The rate constant for the decomposition of gaseous azomethane

$$CH_3N{=}NCH_3 \longrightarrow N_2 + C_2H_6$$

is $40.8 \, min^{-1}$ at 425°C. (a) Find the number of moles of $CH_3N{=}NCH_3$ and N_2 in a flask 0.0500 min after 2.00 g $CH_3N{=}NCH_3$ is introduced. (b) How many grams of $CH_3N{=}NCH_3$ remain after 12.0 seconds?

*31. The first-order rate constant for the conversion of cyclobutane to ethylene at 1000°C is 87 s^{-1}.

cyclobutane   ethylene

(a) What is the half-life of this reaction at 1000°C? (b) If one started with 2.00 g of cyclobutane, how long would it take to consume 1.50 g of it? (*Hint:* Write the ratio of concentrations, $[A]_0/[A]$, in terms of mass, molecular weight, and volume.) (c) How much of an initial 1.00-g sample of cyclobutane would remain after 1.00 s?

*32. For the reaction

$$2NO_2 \longrightarrow 2NO + O_2$$

the rate equation is

$$\text{rate} = 1.4 \times 10^{-10} \, M^{-1} \cdot s^{-1} [NO_2]^2 \quad \text{at } 25°C$$

(a) If 3.00 mol of NO_2 is initially present in a sealed 2.00-L vessel at 25°C, what is the half-life of the reaction? (b) Refer to part (a). What concentration and how many grams of NO_2 remain after 115 years? (c) Refer to part (b). What concentration of NO would have been produced during the same period of time?

33. The first-order rate constant for the radioactive decay of radium-223 is 0.0606 day^{-1}. What is the half-life of radium-223?

34. Cyclopropane rearranges to form propene

cyclopropane   propene

in a reaction that follows first-order kinetics. At 800 K, the specific rate constant for this reaction is 2.74×10^{-3} s^{-1}. Suppose we start with a cyclopropane concentration of 0.290 M. How long will it take for 99.0% of the cyclopropane to disappear according to this reaction?

35. The rate constant for the first-order reaction

$$N_2O_5 \longrightarrow 2N_2O_4 + \tfrac{1}{2}O_2$$

is 1.20×10^{-2} s^{-1} at 45°C, and the initial concentration of N_2O_5 is 0.00500 M. (a) How long will it take for the concentration to decrease to 0.00100 M? (b) How much longer will it take for a further decrease to 0.000900 M?

36. It is found that 47.0 minutes is required for the concentration of substance A to decrease from 0.75 M to 0.25 M. What is the rate constant for this first-order decomposition?

$$A \longrightarrow B + C$$

37. The thermal decomposition of ammonia at high temperatures was studied in the presence of inert gases. Data at 2000 K are given for a single experiment.

$$NH_3 \longrightarrow NH_2 + H$$

Time (hours)	[NH$_3$] (mol/L)
0	8.000×10^{-7}
25	6.75×10^{-7}
50	5.84×10^{-7}
75	5.15×10^{-7}

Plot the appropriate concentration expressions against time to find the order of the reaction. Find the rate constant of the reaction from the slope of the line. Use the given data and the appropriate integrated rate equation to check your answer.

38. The following data were obtained from a study of the decomposition of a sample of HI on the surface of a gold wire. (a) Plot the data to find the order of the reaction, the rate constant, and the rate equation. (b) Calculate the HI concentration in mmol/L at 600 sec.

t (seconds)	[HI] (mmol/L)
0	5.46
250	4.10
500	2.73
750	1.37

39. The decomposition of SO_2Cl_2 in the gas phase,

$$SO_2Cl_2 \longrightarrow SO_2 + Cl_2$$

can be studied by measuring the concentration of Cl_2 gas as the reaction proceeds. We begin with $[SO_2Cl_2]_0 = 0.250 \, M$. Holding the temperature constant at 320°C, we monitor the Cl_2 concentration, with the following results.

Time (hours)	[Cl$_2$] (mol/L)
0.00	0.000
2.00	0.037
4.00	0.068
6.00	0.095
8.00	0.117
10.00	0.137
12.00	0.153
14.00	0.168
16.00	0.180
18.00	0.190
20.00	0.199

(a) Plot [Cl$_2$] versus t. (b) Plot [SO$_2$Cl$_2$] versus t. (c) Determine the rate law for this reaction. (d) What is the value, with units, for the specific rate constant at 320°C? (e) How long would it take for 95% of the original SO_2Cl_2 to react?

*40. At some temperature, the rate constant for the decomposition of HI on a gold surface is $0.080 \ M \cdot s^{-1}$.

$$2HI(g) \longrightarrow H_2(g) + I_2(g)$$

(a) What is the order of the reaction? (b) How long will it take for the concentration of HI to drop from $1.50 \ M$ to $0.15 \ M$?

Activation Energy, Temperature, and Catalysts

41. Draw typical reaction-energy diagrams for one-step reactions that release energy and that absorb energy. Distinguish between the net energy change, ΔE, for each kind of reaction and the activation energy. Indicate potential energies of products and reactants for both kinds of reactions.

42. Describe and illustrate with graphs how the presence of a catalyst can affect the rate of a reaction.

43. How do homogeneous catalysts and heterogeneous catalysts differ?

44. The energy of activation for a hypothetical reaction $A + B \rightarrow C$ is 285 kJ/mol for the uncatalyzed reaction and 135 kJ/mol for the reaction catalyzed on a metal surface. ΔE for this reaction is -40 kJ/mol of reaction. Draw and label a reaction coordinate diagram similar to Figure 16-15 for this reaction.

45. For a gas phase reaction, $E_a = 103$ kJ/mol and the rate constant is $0.0850 \ min^{-1}$ at 273 K. Find the rate constant at 373 K.

46. The rate constant of a reaction is tripled when the temperature is increased from 298 K to 308 K. Find E_a.

47. The rate constant for the decomposition of N_2O

$$2N_2O(g) \longrightarrow 2N_2(g) + O_2(g)$$

is $2.6 \times 10^{-11} \ s^{-1}$ at 300°C and $2.1 \times 10^{-10} \ s^{-1}$ at 330°C. Calculate the activation energy for this reaction. Prepare a reaction coordinate diagram like Figure 16-10 using -164.1 kJ/mol as the ΔE_{rxn}.

48. For a particular reaction, $\Delta E^0 = 51.51$ kJ/mol, $k = 8.0 \times 10^{-7}$ s^{-1} at 0.0°C, and $k = 8.9 \times 10^{-4} \ s^{-1}$ at 50.0°C. Prepare a reaction coordinate diagram like Figure 16-10 for this reaction.

*49. You are given the rate constant as a function of temperature for the exchange reaction

$$Mn(CO)_5(CH_3CN)^+ + NC_5H_5 \longrightarrow$$
$$Mn(CO)_5(NC_5H_5)^+ + CH_3CN$$

T (K)	k (min^{-1})
298	0.0409
308	0.0818
318	0.157

(a) Calculate E_a from a plot of log k versus $1/T$. (b) Use the graph to predict the value of k at 311 K. (c) What is the numerical value of the collision frequency factor, A, in the Arrhenius equation?

*50. The rearrangement of cyclopropane to propene described in Exercise 34 has been studied at various temperatures. The following values for the specific rate constant have been determined experimentally.

T (K)	k (s^{-1})
600	3.30×10^{-9}
650	2.19×10^{-7}
700	7.96×10^{-6}
750	1.80×10^{-4}
800	2.74×10^{-3}
850	3.04×10^{-2}
900	2.58×10^{-1}

(a) From the appropriate plot of these data, determine the value of the activation energy for this reaction. (b) Use the graph to estimate the value of k at 500 K. (c) Use the graph to estimate the temperature at which the value of k would be equal to $5.00 \times 10^{-5} \ s^{-1}$.

51. Biological reactions nearly always occur in the presence of enzymes as catalysts. The enzyme catalase, which acts on peroxides, reduces the E_a for the reaction from 72 kJ/mol (uncatalyzed) to 28 kJ/mol (catalyzed). By what factor does the reaction rate increase at normal body temperature, 37.0°C, for the same reactant (peroxide) concentration? Assume that the collision factor, A, remains constant.

*52. The enzyme carbonic anhydrase catalyzes the hydration of carbon dioxide.

$$CO_2 + H_2O \longrightarrow H_2CO_3$$

This reaction is involved in the transfer of CO_2 from tissues to the lung via the bloodstream. One enzyme molecule hydrates 10^6 molecules of CO_2 per second. How many kilograms of CO_2 are hydrated in one hour in one L by $1.0 \times 10^{-6} \ M$ enzyme?

53. The following gas phase reaction follows first-order kinetics.

$$ClO_2F \longrightarrow ClOF + O$$

The activation energy of this reaction is found to be 186 kJ/mol. The value of k at 322°C is found to be $6.76 \times 10^{-4} \ s^{-1}$. (a) What would be the value of k for this reaction at 25°C? (b) At what temperature would this reaction have a k value of $3.00 \times 10^{-2} \ s^{-1}$?

54. The following gas phase reaction is first order.

$$N_2O_5 \longrightarrow NO_2 + NO_3$$

The activation energy of this reaction is found to be 88 kJ/mol. The value of k at 0°C is found to be $9.16 \times 10^{-3} \ s^{-1}$. (a) What would be the value of k for this reaction at room temperature, 25°C? (b) At what temperature would this reaction have a k value of $3.00 \times 10^{-2} \ s^{-1}$?

Reaction Mechanisms

55. Define reaction mechanism. Why do we believe that only bimolecular collisions and unimolecular decompositions are important in most reaction mechanisms?

56. The rate equation for the reaction

$$Cl_2(aq) + H_2S(aq) \longrightarrow S(s) + 2HCl(aq)$$

is found to be rate = $k[Cl_2][H_2S]$. Which of the following mechanisms are consistent with the rate law?

(a)
$$Cl_2 \longrightarrow Cl^+ + Cl^- \quad \text{(slow)}$$
$$Cl^- + H_2S \longrightarrow HCl + HS^- \quad \text{(fast)}$$
$$Cl^+ + HS^- \longrightarrow HCl + S \quad \text{(fast)}$$
$$\overline{Cl_2 + H_2S \longrightarrow S + 2HCl \quad \text{overall}}$$

(b)
$$Cl_2 + H_2S \longrightarrow HCl + Cl^+ + HS^- \quad \text{(slow)}$$
$$Cl^+ + HS^- \longrightarrow HCl + S \quad \text{(fast)}$$
$$\overline{Cl_2 + H_2S \longrightarrow S + 2HCl \quad \text{overall}}$$

(c)
$$Cl_2 \rightleftharpoons Cl + Cl \quad \text{(fast, equilibrium)}$$
$$Cl + H_2S \rightleftharpoons HCl + HS \quad \text{(fast, equilibrium)}$$
$$HS + Cl \longrightarrow HCl + S \quad \text{(slow)}$$
$$\overline{Cl_2 + H_2S \longrightarrow S + 2HCl \quad \text{overall}}$$

57. The ozone, O_3, of the stratosphere can be decomposed by reaction with nitrogen oxide (commonly called nitric oxide), NO, from high-flying jet aircraft.

$$O_3(g) + NO(g) \longrightarrow NO_2(g) + O_2(g)$$

The rate expression is rate = $k[O_3][NO]$. Which of the following mechanisms are consistent with the observed rate expression?

(a)
$$NO + O_3 \longrightarrow NO_3 + O \quad \text{(slow)}$$
$$NO_3 + O \longrightarrow NO_2 + O_2 \quad \text{(fast)}$$
$$\overline{O_3 + NO \longrightarrow NO_2 + O_2 \quad \text{overall}}$$

(b)
$$NO + O_3 \longrightarrow NO_2 + O_2 \quad \text{(slow)}$$
$$\quad \text{(one step)}$$

(c)
$$O_3 \longrightarrow O_2 + O \quad \text{(slow)}$$
$$O + NO \longrightarrow NO_2 \quad \text{(fast)}$$
$$\overline{O_3 + NO \longrightarrow NO_2 + O_2 \quad \text{overall}}$$

(d)
$$NO \longrightarrow N + O \quad \text{(slow)}$$
$$O + O_3 \longrightarrow 2O_2 \quad \text{(fast)}$$
$$O_2 + N \longrightarrow NO_2 \quad \text{(fast)}$$
$$\overline{O_3 + NO \longrightarrow NO_2 + O_2 \quad \text{overall}}$$

(e)
$$NO \rightleftharpoons N + O \quad \text{(fast, equilibrium)}$$
$$O + O_3 \longrightarrow 2O_2 \quad \text{(slow)}$$
$$O_2 + N \longrightarrow NO_2 \quad \text{(fast)}$$
$$\overline{O_3 + NO \longrightarrow NO_2 + O_2 \quad \text{overall}}$$

58. A proposed mechanism for the decomposition of ozone, $2O_3 \rightarrow 3O_2$, is

$$O_3 \rightleftharpoons O_2 + O \quad \text{(fast, equilibrium)}$$
$$O + O_3 \longrightarrow 2O_2 \quad \text{(slow)}$$

Derive the rate equation for the net reaction.

59. A mechanism for the gas phase reaction

$$H_2 + I_2 \longrightarrow 2HI$$

was discussed in the chapter. (a) Show that this mechanism predicts the correct rate law, rate = $k[H_2][I_2]$.

$$I_2 \rightleftharpoons 2I \quad \text{(fast, equilibrium)}$$
$$I + H_2 \rightleftharpoons H_2I \quad \text{(fast, equilibrium)}$$
$$H_2I + I \longrightarrow 2HI \quad \text{(slow)}$$

(b) Identify any reaction intermediates in this proposed mechanism.

60. The combination of Cl atoms is catalyzed by $N_2(g)$. The following mechanism is suggested.

$$N_2 + Cl \rightleftharpoons N_2Cl \quad \text{(fast, equilibrium)}$$
$$N_2Cl + Cl \longrightarrow Cl_2 + N_2 \quad \text{(slow)}$$

(a) Identify any reaction intermediates in this proposed mechanism. (b) Is this mechanism consistent with the experimental rate law, rate = $k[N_2][Cl]^2$?

61. The reaction between NO and Br_2 was discussed in Section 16-7. The following mechanism has also been proposed.

$$2NO \rightleftharpoons N_2O_2 \quad \text{(fast, equilibrium)}$$
$$N_2O_2 + Br_2 \longrightarrow 2NOBr \quad \text{(slow)}$$

Is this mechanism consistent with the observation that the reaction is second order in NO and first order in Br_2?

*62. The following mechanism for the reaction between H_2 and CO to form formaldehyde, H_2CO, has been proposed.

$$H_2 \rightleftharpoons 2H \quad \text{(fast, equilibrium)}$$
$$H + CO \longrightarrow HCO \quad \text{(slow)}$$
$$H + HCO \longrightarrow H_2CO \quad \text{(fast)}$$

(a) Write the balanced equation for the overall reaction. (b) The observed rate dependence is found to be one-half order in H_2 and first order in CO. Is this proposed reaction mechanism consistent with the observed rate dependence?

Mixed Exercises

63. The following explanation of the operation of a pressure cooker appears in a cookbook: "Boiling water in the presence of air can never produce a temperature higher than 212°F, no matter how high the heat source. But in a pressure cooker, the air is withdrawn first, so the boiling water can be maintained at higher temperatures." Support or criticize this explanation.

*64. A cookbook gives the following general guideline for use of a pressure cooker: "For steaming vegetables, cooking time at a gauge pressure of 15 pounds per square inch (psi) is $\frac{1}{3}$ that at atmospheric pressure." Remember that gauge pressure is measured relative to the external atmospheric pressure, which is 15 psi at sea level. From this information, estimate the activation energy for the process of steaming vegetables. (*Hint:* Clausius and Clapeyron may be able to help you.)

65. The reaction between nitrogen dioxide and ozone,

$$2NO_2 + O_3 \longrightarrow N_2O_5 + O_2$$

has been studied at 231 K. The experimental rate equation is rate = $k[NO_2][O_3]$. (a) What is the order of the reaction? (b) Is either of the following proposed mechanisms consistent with the given kinetic data? Show how you arrived at your answer.

(a) $NO_2 + NO_2 \rightleftharpoons N_2O_4$ (fast, equilibrium)
$\quad N_2O_4 + O_3 \longrightarrow N_2O_5 + O_2$ (slow)
(b) $NO_2 + O_3 \longrightarrow NO_3 + O_2$ (slow)
$\quad NO_3 + NO_2 \longrightarrow N_2O_5$ (fast)

66. Data are given for the decomposition of N_2O_5 in CCl_4 solution at $45°C$.

$$2N_2O_5(sol) \longrightarrow 4NO_2(sol) + O_2(g)$$

The rate is determined by measuring the volume of O_2 produced.

$[N_2O_5]_0$ (M)	Initial Rate $M_{O_2} \cdot s^{-1}$
0.900	5.58×10^{-4}
0.450	2.79×10^{-4}
0.150	9.30×10^{-5}

Write the rate-law expression.

67. Refer to the reaction and data in Exercise 53. We begin with 3.60 mol of ClO_2F in a 3.00-L container. (a) How many moles of ClO_2F would remain after 1.00 min at $25°C$? (b) How much time would be required for 99.0% of the ClO_2F to decompose at $25°C$?

68. Refer to the reaction and data in Exercise 54. We begin with 3.60 mol of N_2O_5 in a 3.00-L container. (a) How many moles of N_2O_5 would remain after 1.00 min at $25°C$? (b) How much time would be required for 99.0% of the N_2O_5 to decompose at $25°C$?

69. The decomposition of gaseous dimethyl ether

$$CH_3OCH_3 \longrightarrow CH_4 + CO + H_2$$

follows first-order kinetics. Its half-life is 25.0 min at $500°C$. (a) Starting with 10.00 g of dimethyl ether at $500°C$, how many grams would remain after 120 min? (b) In part (a), how many grams would remain after 150 min? (c) In part (b), what fraction remains and what fraction reacts? (d) Calculate the time, in minutes, required to decrease 9.20 mg of dimethyl ether to 3.60 mg.

70. The rate of the hemoglobin (Hb)–carbon monoxide reaction,

$$4Hb + 3CO \longrightarrow Hb_4(CO)_3$$

has been studied at $20°C$. Concentrations are expressed in micromoles per liter ($\mu mol/L$).

Concentration ($\mu mol/L$)		Rate of Disappearance of Hb ($\mu mol \cdot L^{-1} \cdot s^{-1}$)
[Hb]	[CO]	
3.36	1.00	0.941
6.72	1.00	1.88
6.72	3.00	5.64

(a) Write the rate equation for the reaction. (b) Calculate the rate constant for the reaction. (c) Calculate the rate, at the instant when $[Hb] = 1.50$ and $[CO] = 0.600$ $\mu mol/L$.

*71. In the chapter it was calculated that the rate constant, k, for a reaction whose activation energy is 50 kJ/mol approximately doubles (factor of 1.9) when the temperature is increased from 300 K to 310 K. Over this same temperature range, determine the factor by which k increases for a reaction whose activation energy is 75 kJ/mol. Can you rationalize this result?

BUILDING YOUR KNOWLEDGE

72. Concentrations of reactants and products in a chemical reaction are measured in moles per liter and time is measured in seconds. (a) What are the units of the rate constant for a first-order reaction? (b) for a second-order reaction? (c) for an nth-order reaction?

73. For most reactions that involve an enzyme, the rate of product formation vs. reactant concentration increases as reactant concentration increases until a maximum value is obtained, after which further increases do not yield increased rates. Using a description like that in Figure 16-19, describe how the reaction may be first order with respect to substrate but the amount of enzyme can also be a determining factor.

74. Using the mechanism and energy values shown in Figure 16-12, prepare Lewis formulas that illustrate the species that are likely to be present at each of the peaks and troughs in the graphical representation given in Figure 16-12. (*Hint:* You may need to label some bonds as being weaker, stretched, in the process of being formed, and so on.)

75. The activation energy for the reaction

$$2HI(g) \longrightarrow H_2(g) + I_2(g)$$

is 179 kJ/mol. Construct a diagram similar to Figure 16-10 for this reaction. (*Hint:* Calculate ΔH^0 from values in Appendix K. How does ΔH^0 compare to ΔE^0 for this reaction?)

*76. The activation energy for the reaction between O_3 and NO is 9.6 kJ/mol.

$$O_3(g) + NO(g) \longrightarrow NO_2(g) + O_2(g)$$

(a) Use the thermodynamic quantities in Appendix K to calculate ΔH^0 for this reaction. (b) Prepare an activation energy plot similar to Figure 16-10 for this reaction. (*Hint:* How does ΔH^0 compare to ΔE^0 for this reaction?)

77. Write Lewis formulas for each species in Exercise 58.

17 Chemical Equilibrium

OBJECTIVES

As you study this chapter, you should learn

- *The basic ideas of chemical equilibrium*

- *What an equilibrium constant is and what it tells us*

- *What a reaction quotient is and what it tells us*

- *To use equilibrium constants to describe systems at equilibrium*

- *To recognize the factors that affect equilibria and predict the resulting effects*

- *About equilibrium constants expressed in terms of partial pressures (K_P's) and how these are related to K_c's*

- *About heterogeneous equilibria*

- *About relationships between thermodynamics and equilibrium*

- *How to estimate equilibrium constants at different temperatures*

A nighttime photo of a large plant for the commercial production of ammonia, NH_3. Such an installation can produce up to 7000 metric tons of ammonia per day. There are nearly a hundred such plants in the world.

17-1 BASIC CONCEPTS

Most chemical reactions do not go to completion. That is, when reactants are mixed in stoichiometric quantities, they are not completely converted to products. Reactions that do not go to completion *and* that can occur in either direction are called **reversible reactions.**

Reversible reactions can be represented in general terms as follows, where the capital letters represent formulas and the lowercase letters represent the stoichiometric coefficients in the balanced equation.

$$a\text{A} + b\text{B} \rightleftharpoons c\text{C} + d\text{D}$$

The double arrow ($\rightleftharpoons$) indicates that the reaction is reversible—i.e., both the forward and reverse reactions occur simultaneously. When A and B react to form C and D at the same rate at which C and D react to form A and B, the system is at *equilibrium.*

Chemical equilibrium exists when two opposing reactions occur simultaneously at the same rate.

Figure 17-1 Variation in the concentrations of species present in the $A + B \rightleftharpoons C + D$ system as equilibrium is approached, beginning with equal concentrations of A and B only. For this reaction, production of products is favored. As we shall see, this corresponds to a value of the equilibrium constant greater than 1.

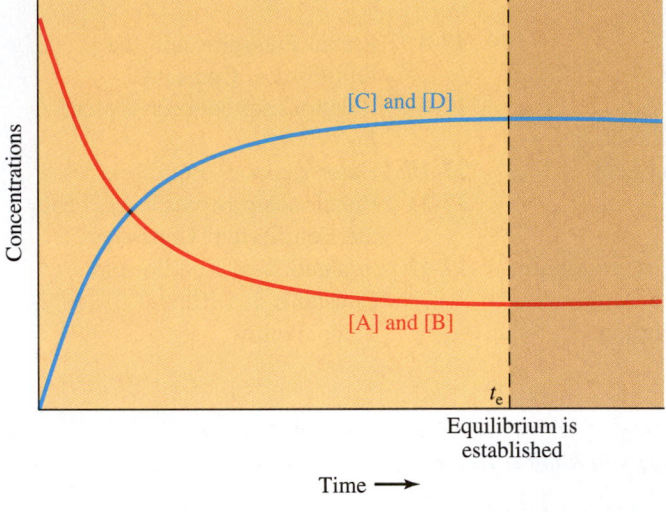

Equilibrium is established

Time $\longrightarrow$

Brackets, [ ], represent the concentration, in moles per liter, of the species enclosed within them.

The dynamic nature of chemical equilibrium can be proved experimentally by "tagging" a small percentage of molecules with radioactive atoms and following them through the reaction. Even when the initial mixture is at equilibrium, radioactive atoms eventually appear in both reactant and product molecules.

Chemical equilibria are **dynamic equilibria;** that is, individual molecules are continually reacting, even though the overall composition of the reaction mixture does not change. In a system at equilibrium, the equilibrium is said to lie toward the right if more C and D are present than A and B, and to lie toward the left if more A and B are present.

Consider a case in which the coefficients in the equation for a reaction are all 1. When substances A and B react, the rate of the forward reaction decreases as time passes because the concentrations of A and B decrease.

$$A + B \longrightarrow C + D \tag{1}$$

As the concentrations of C and D build up, they start to form A and B.

$$C + D \longrightarrow A + B \tag{2}$$

As more C and D molecules are formed, more can react, and so the rate of reaction between C and D increases with time. Eventually, the two reactions occur at the same rate, and the system is at equilibrium (Figure 17-1).

$$A + B \rightleftharpoons C + D$$

If a reaction begins with only C and D present, the rate of reaction (2) decreases with time, and the rate of reaction (1) increases with time until the two rates are equal.

The SO_2–O_2–SO_3 System

Consider the reversible reaction of sulfur dioxide with oxygen to form sulfur trioxide at 1500 K.

$$2SO_2(g) + O_2(g) \rightleftharpoons 2SO_3(g)$$

The numbers in this discussion were determined experimentally.

Suppose 0.400 mole of SO_2 and 0.200 mole of O_2 are injected into a closed 1.00-liter container. When equilibrium is established (at time t_e, Figure 17-2a), we find that 0.056 mole of SO_3 has formed and that 0.344 mole of SO_2 and 0.172 mole of O_2 remain unreacted. The reaction does not go to completion. These changes are summarized below,

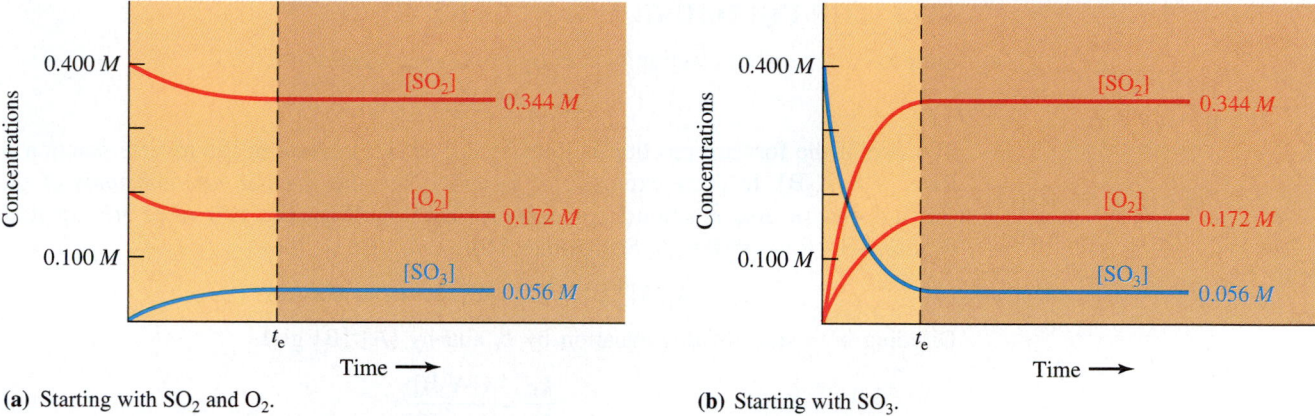

(a) Starting with SO_2 and O_2.   (b) Starting with SO_3.

Figure 17-2 Establishment of equilibrium in the $2SO_2 + O_2 \rightleftharpoons 2SO_3$ system. (a) Beginning with stoichiometric amounts of SO_2 and O_2 and no SO_3. (b) Beginning with only SO_3 and no SO_2 or O_2. Greater changes in concentrations occur to establish equilibrium when starting with SO_3 than when starting with SO_2 and O_2. The equilibrium favors SO_2 and O_2.

using molarity units rather than moles. (They are numerically identical here because the volume of the reaction vessel is 1.00 liter.) The *net reaction* is represented by the *changes* in concentrations.

	$2SO_2(g)$	+	$O_2(g)$	$\rightleftharpoons$	$2SO_3(g)$
initial conc'n	0.400 M		0.200 M		0
change due to rxn	−0.056 M		−0.028 M		+0.056 M
equilibrium conc'n	0.344 M		0.172 M		0.056 M

The ratio in the "change due to rxn" line is determined by the coefficients in the balanced equation.

In another experiment, 0.400 mole of SO_3 is introduced alone into a closed 1.00-liter container. (The same total numbers of sulfur and oxygen atoms are present as in the previous experiment.) When equilibrium is established (at time t_e, Figure 17-2b), 0.056 mole of SO_3, 0.172 mole of O_2, and 0.344 mole of SO_2 are present. (These are the same amounts found at equilibrium in the previous case.) This time the net reaction proceeds from *right to left* as the equation is written. The changes in concentration are in the same $2:1:2$ ratio as in the previous case, as required by the coefficients of the balanced equation. The time required to reach equilibrium may be longer or shorter.

	$2SO_2(g)$	+	$O_2(g)$	$\rightleftharpoons$	$2SO_3(g)$
initial conc'n	0		0		0.400 M
change due to rxn	+0.344 M		+0.172 M		−0.344 M
equilibrium conc'n	0.344 M		0.172 M		0.056 M

A setup such as this is called a "reaction summary."

The results of these experiments are summarized below and in Figure 17-2.

	Initial Concentrations			Equilibrium Concentrations		
	[SO_2]	[O_2]	[SO_3]	[SO_2]	[O_2]	[SO_3]
Experiment 1	0.400 M	0.200 M	0 M	0.344 M	0.172 M	0.056 M
Experiment 2	0 M	0 M	0.400 M	0.344 M	0.172 M	0.056 M

17-2 THE EQUILIBRIUM CONSTANT

Suppose a reversible reaction occurs by a *one-step mechanism.*

$$2A + B \rightleftharpoons A_2B$$

The rate of the forward reaction is $\text{Rate}_f = k_f[A]^2[B]$; the rate of the reverse reaction is $\text{Rate}_r = k_r[A_2B]$. In these expressions, k_f and k_r are the *specific rate constants* of the forward and reverse reactions, respectively. By definition, the two rates are equal *at equilibrium* ($\text{Rate}_f = \text{Rate}_r$). So we may write

$$k_f[A]^2[B] = k_r[A_2B] \qquad \text{(at equilibrium)}$$

Dividing both sides of this equation by k_r and by $[A]^2[B]$ gives

$$\frac{k_f}{k_r} = \frac{[A_2B]}{[A]^2[B]}$$

At any specific temperature, both k_f and k_r are constants, so k_f/k_r is a constant.

This ratio is given a special name and symbol—the **equilibrium constant, K_c** or simply **K.**

The subscript c refers to concentrations. The brackets, [ ], in this expression indicate *equilibrium* concentrations in moles per liter.

$$K_c = \frac{[A_2B]}{[A]^2[B]} \qquad \text{(at equilibrium)}$$

We have described a one-step reaction. Suppose, instead, that this reaction involves a *two-step mechanism* with the following rate constants.

$$\text{(step 1)} \quad 2A \underset{k_{1r}}{\overset{k_{1f}}{\rightleftharpoons}} A_2 \qquad \text{followed by} \qquad \text{(step 2)} \quad A_2 + B \underset{k_{2r}}{\overset{k_{2f}}{\rightleftharpoons}} A_2B$$

We add (1) and (2) to obtain the overall reaction, $2A + B \rightleftharpoons A_2B$. An equilibrium constant expression can be written for each step.

$$K_1 = \frac{k_{1f}}{k_{1r}} = \frac{[A_2]}{[A]^2} \qquad \text{and} \qquad K_2 = \frac{k_{2f}}{k_{2r}} = \frac{[A_2B]}{[A_2][B]}$$

If we multiply K_1 by K_2, the $[A_2]$ term can be eliminated. Because K_1 and K_2 are both constants, their product is also a constant, and we obtain

$$K_1 \times K_2 = \frac{[A_2]}{[A]^2} \times \frac{[A_2B]}{[A_2][B]} = \frac{[A_2B]}{[A]^2[B]} = K_c$$

Regardless of the mechanism by which this reaction occurs, the concentrations of reaction intermediates cancel out and the equilibrium constant expression has the same form. For a reaction in general terms, the equilibrium constant can always be written as follows.

For $\underbrace{aA + bB}_{\text{reactants}} \rightleftharpoons \underbrace{cC + dD,}_{\text{products}} \qquad K_c = \frac{[C]^c[D]^d}{[A]^a[B]^b} \quad \begin{matrix} \leftarrow \text{products} \\ \leftarrow \text{reactants} \end{matrix}$

The equilibrium constant, K_c, is defined as the product of the *equilibrium concentrations* (moles per liter) of the products, each raised to the power that corresponds to its coefficient in the balanced equation, divided by the product of the *equilibrium concentrations* of reactants, each raised to the power that corresponds to its coefficient in the balanced equation.

In general numerical values for K_c can come only from experiments. Some equilibrium constant expressions and their numerical values at 25°C are

$$N_2(g) + O_2(g) \rightleftharpoons 2NO(g) \qquad K_c = \frac{[NO]^2}{[N_2][O_2]} = 4.5 \times 10^{-31}$$

$$CH_4(g) + Cl_2(g) \rightleftharpoons CH_3Cl(g) + HCl(g) \qquad K_c = \frac{[CH_3Cl][HCl]}{[CH_4][Cl_2]} = 1.2 \times 10^{18}$$

$$N_2(g) + 3H_2(g) \rightleftharpoons 2NH_3(g) \qquad K_c = \frac{[NH_3]^2}{[N_2][H_2]^3} = 3.6 \times 10^8$$

The thermodynamic definition of the equilibrium constant involves activities rather than concentrations. The **activity** of a component of an ideal mixture is the ratio of its concentration or partial pressure to a standard concentration (1 M) or pressure (1 atm). For now, we can consider the activity of each species to be a dimensionless quantity whose numerical value can be determined as follows. (a) For pure liquids and solids, the activity is taken as 1. (b) For components of ideal solutions, the activity of each component is taken to be equal to its molar concentration. (c) For gases in an ideal mixture, the activity of each component is taken to be equal to its partial pressure in atmospheres. Because of the use of activities, all units cancel and *the equilibrium constant has no units; the values we put into K_c are numerically equal to molar concentrations, but are dimensionless,* i.e., have no units.

The magnitude of K_c is a measure of the extent to which reaction occurs. For any reaction, the value of K_c

1. varies only with temperature
2. is constant at a given temperature
3. is independent of the initial concentrations

These are three very important ideas.

A value of K_c *much* greater than 1 indicates that the "numerator concentrations" (products) would be much greater than the "denominator concentrations" (reactants); this means that at equilibrium most of the reactants would be converted into products. On the other hand, if K_c is quite small, equilibrium is established when most of the reactants remain unreacted and only small amounts of products are formed.

For a given chemical reaction at a specific temperature, the product of the concentrations of the products formed by the reaction, each raised to the appropriate power, divided by the product of the concentrations of the reactants, each raised to the appropriate power, always has the same value, i.e., K_c. This does *not* mean that the individual equilibrium concentrations for a given reaction are always the same, but it does mean that this particular numerical combination of their values (K_c) is constant.

Consider again the SO_2–O_2–SO_3 equilibrium described earlier in this chapter. The equilibrium concentrations were the same in the two experiments. We can use these equilibrium concentrations to calculate the value of the equilibrium constant for this reaction at 1500 K.

$$2SO_2(g) + O_2(g) \rightleftharpoons 2SO_3(g)$$

equil conc'n   0.344 M 0.172 M 0.056 M

Substituting the numerical values (without units) into the equilibrium expression gives the value of the equilibrium constant.

$$K_c = \frac{[SO_3]^2}{[SO_2]^2[O_2]} = \frac{(0.056)^2}{(0.344)^2(0.172)} = 1.5 \times 10^{-1}$$

For the reversible reaction written *as it is*, K_c is 0.15 at 1500 K.

EXAMPLE 17-1   *Calculation of K_c*

We place some nitrogen and hydrogen in an empty 5.00-liter vessel at 500°C. When equilibrium is established, 3.01 mol of N_2, 2.10 mol of H_2, and 0.565 mol of NH_3 are present. Evaluate K_c for the following reaction at 500°C.

$$N_2(g) + 3H_2(g) \rightleftharpoons 2NH_3(g)$$

Plan

The *equilibrium concentrations* are obtained by dividing the number of moles of each reactant and product by the volume, 5.00 liters. Then we substitute these equilibrium concentrations into the equilibrium constant expression.

Solution

The equilibrium concentrations are

$$[N_2] = 3.01 \text{ mol}/5.00 \text{ L} \ = 0.602 \ M$$
$$[H_2] = 2.10 \text{ mol}/5.00 \text{ L} \ = 0.420 \ M$$
$$[NH_3] = 0.565 \text{ mol}/5.00 \text{ L} = 0.113 \ M$$

We substitute these numerical values into the expression for K_c.

$$K_c = \frac{[NH_3]^2}{[N_2][H_2]^3} = \frac{(0.113)^2}{(0.602)(0.420)^3} = 0.286$$

Thus, for the reaction of H_2 and N_2 to form NH_3 at 500°C, we can write

$$K_c = \frac{[NH_3]^2}{[N_2][H_2]^3} = \boxed{0.286}$$

The small value of K_c indicates that the equilibrium lies to the left.

You should now work Exercises 12 and 14.

▼ PROBLEM-SOLVING TIP   *There Are No Units in Equilibrium Constants*

Students often wonder about units in equilibrium constants. In Section 17-2, we learned that the fundamental definition of K involves dimensionless quantities called *activities*. For ideal mixtures, these are numerically equal to the concentrations (for components in solutions) or the partial pressures in atmospheres (for gases). So strictly speaking, the numbers we put into the expression for K_c are numerically equal to the molar concentrations or the partial pressures, but are dimensionless. Thus, K has no units.

17-3 VARIATION OF K_c WITH THE FORM OF THE BALANCED EQUATION

The value of K_c depends on the form of the balanced equation for the reaction. We wrote the equation for the reaction of SO_2 and O_2 to form SO_3 and its equilibrium constant expression as

$$2SO_2(g) + O_2(g) \rightleftharpoons 2SO_3(g) \quad \text{and} \quad K_c = \frac{[SO_3]^2}{[SO_2]^2[O_2]} = 0.15$$

Suppose we write the equation for the same reaction in reverse. The equation and its equilibrium constant, written this way, are

$$2SO_3(g) \rightleftharpoons 2SO_2(g) + O_2(g) \quad \text{and} \quad K_c' = \frac{[SO_2]^2[O_2]}{[SO_3]^2} = \frac{1}{K_c} = 6.5$$

Reversing an equation is the same as multiplying all coefficients by -1. This reverses the roles of "reactants" and "products."

We see that K_c', the equilibrium constant for the reaction written in reverse, is the *reciprocal* of K_c, the equilibrium constant for the original reaction.

If the equation for the reaction were written as

$$SO_2(g) + \tfrac{1}{2}O_2(g) \rightleftharpoons SO_3(g) \quad K_c'' = \frac{[SO_3]}{[SO_2][O_2]^{1/2}} = K_c^{1/2} = 0.39$$

We see that K_c'' is the square root of K_c. $K_c^{1/2}$ means the square root of K_c.

If an *equation* for a reaction is *multiplied* by a positive or negative number, n, then the *original value of K_c is raised to the nth power*. Thus, we must always write the balanced chemical equation, as well as the value of K_c, for a chemical reaction.

EXAMPLE 17-2 *Variation of the Form of K_c*

You are given the following reaction and its equilibrium constant at a given temperature.

$$2HBr(g) + Cl_2(g) \rightleftharpoons 2HCl(g) + Br_2(g) \quad K_c = 4.0 \times 10^4$$

Write the expression for, and calculate the numerical value of, the equilibrium constant for each of the following at the same temperature.

(a) $\qquad\qquad\qquad 4HBr(g) + 2Cl_2(g) \rightleftharpoons 4HCl(g) + 2Br_2(g)$

(b) $\qquad\qquad\qquad HBr(g) + \tfrac{1}{2}Cl_2(g) \rightleftharpoons HCl(g) + \tfrac{1}{2}Br_2(g)$

A coefficient of $\tfrac{1}{2}$ refers to $\tfrac{1}{2}$ of a mole, *not* $\tfrac{1}{2}$ of a molecule.

Plan

We recall the definition of the equilibrium constant. For the original equation,

$$K_c = \frac{[HCl]^2[Br_2]}{[HBr]^2[Cl_2]} = 4.0 \times 10^4$$

Solution

(a) The original equation has been multiplied by 2, so K_c must be squared.

$$K_c' = \frac{[HCl]^4[Br_2]^2}{[HBr]^4[Cl_2]^2} \qquad K_c' = (K_c)^2 = (4.0 \times 10^4)^2 = \boxed{1.6 \times 10^9}$$

(b) The original equation has been multiplied by $\frac{1}{2}$ (divided by 2), so K_c must be raised to the $\frac{1}{2}$ power. This is the same as extracting the square root of K_c.

$$K_c'' = \frac{[HCl][Br_2]^{1/2}}{[HBr][Cl_2]^{1/2}} = \sqrt{K_c} = \sqrt{4.0 \times 10^4} = 2.0 \times 10^2$$

You should now work Exercise 16.

17-4 THE REACTION QUOTIENT

The reaction quotient is sometimes called the **mass action expression.**

The **reaction quotient, Q,** for the general reaction is given as follows.

For $aA + bB \rightleftharpoons cC + dD$, $Q = \dfrac{[C]^c[D]^d}{[A]^a[B]^b}$ ← not necessarily ← equilibrium concentrations

The reaction quotient has the same *form* as the equilibrium constant, but it involves specific values that are not *necessarily* equilibrium concentrations. If they *are* equilibrium concentrations, then $Q = K_c$. The concept of the reaction quotient is very useful. We can compare the magnitude of Q with that of K for a reaction under given conditions to decide whether a net forward or a net reverse reaction must occur to establish equilibrium.

When the forward reaction occurs to a greater extent than the reverse reaction, we say that a *net* forward reaction has occurred.

If at any time $Q < K$, the forward reaction must occur to a greater extent than the reverse reaction for equilibrium to be established. This is because when $Q < K$, the numerator of Q is too small and the denominator is too large. To increase the numerator and to reduce the denominator, A and B must react to produce C and D. Conversely, if $Q > K$, the reverse reaction must occur to a greater extent than the forward reaction for equilibrium to be reached. When the value of Q reaches the value of K, the system is at equilibrium, so no further *net* reaction occurs.

$Q < K$ Forward reaction predominates until equilibrium is established.

$Q = K$ System is at equilibrium.

$Q > K$ Reverse reaction predominates until equilibrium is established.

In Example 17-3 we calculate the value for Q and compare it with the *known* value of K_c to predict the direction of the net reaction that leads to equilibrium.

EXAMPLE 17-3   *The Reaction Quotient*

At a very high temperature, $K_c = 1.0 \times 10^{-13}$ for the following reaction.

$$2HF(g) \rightleftharpoons H_2(g) + F_2(g)$$

At a certain time the following concentrations were detected. Is the system at equilibrium? If not, what must occur for equilibrium to be established?

These concentrations could be present if we started with a mixture of HF, H_2, and F_2.

$$[HF] = 0.500\ M,\quad [H_2] = 1.00 \times 10^{-3}\ M,\quad \text{and}\ [F_2] = 4.00 \times 10^{-3}\ M$$

Plan

We substitute these concentrations into the expression for the reaction quotient to calculate Q. Then we compare Q with K_c to see whether the system is at equilibrium.

Solution

$$Q = \frac{[H_2][F_2]}{[HF]^2} = \frac{(1.00 \times 10^{-3})(4.00 \times 10^{-3})}{(0.500)^2} = 1.60 \times 10^{-5}$$

But $K_c = 1.0 \times 10^{-13}$, so $Q > K_c$. The system is *not* at equilibrium. For equilibrium to be established, the value of Q must decrease until it equals K_c. This can occur only if the numerator decreases and the denominator increases. Thus, the right-to-left reaction must occur to a greater extent than the forward reaction; i.e., H_2 and F_2 must react to form more HF to reach equilibrium.

You should now work Exercises 24, 26, and 28.

17-5 USES OF THE EQUILIBRIUM CONSTANT, K_c

We have seen (Section 17-2) how to calculate the value of K_c from one set of equilibrium concentrations. Once that value has been obtained, the process can be turned around to calculate equilibrium *concentrations* from the equilibrium *constant*.

The equilibrium constant is a "constant" only if the temperature does not change.

EXAMPLE 17-4 *Finding Equilibrium Concentrations*

The equation for the following reaction and the value of K_c at a given temperature are given. An equilibrium mixture in a 1.00-liter flask contains 0.25 mol of PCl_5 and 0.16 mol of PCl_3. What equilibrium concentration of Cl_2 must be present?

$$PCl_3(g) + Cl_2(g) \rightleftharpoons PCl_5(g) \qquad K_c = 1.9$$

Plan

We write the equilibrium constant expression and its value. Only one term, $[Cl_2]$, is unknown. We solve for it.

Solution

The equilibrium constant expression and its numeric value are

$$K_c = \frac{[PCl_5]}{[PCl_3][Cl_2]} = 1.9$$

$$[Cl_2] = \frac{[PCl_5]}{K_c[PCl_3]} = \frac{(0.25)}{(1.9)(0.16)} = 0.82\ M$$

In a one-liter flask the number of moles of each component equals the molarity of that component.

Often we know the starting concentrations and want to know how much of each reactant and each product would be present at equilibrium.

EXAMPLE 17-5 *Finding Equilibrium Concentrations*

For the following reaction, the equilibrium constant is 49.0 at a certain temperature. If 0.400 mol each of A and B are placed in a 2.00-liter container at that temperature, what concentrations of all species are present at equilibrium?

$$A + B \rightleftharpoons C + D$$

We have represented chemical formulas by single letters to simplify the notation in these calculations.

Plan

First we find the initial concentrations. Then we write the reaction summary and represent the equilibrium concentrations algebraically. Finally we substitute the representations of equilibrium concentrations into the K_c expression and find the equilibrium concentrations.

Solution
The initial concentrations are

$$[A] = \frac{0.400 \text{ mol}}{2.00 \text{ L}} = 0.200 \ M \qquad\qquad [C] = 0 \ M$$

$$[B] = \frac{0.400 \text{ mol}}{2.00 \text{ L}} = 0.200 \ M \qquad\qquad [D] = 0 \ M$$

We know that the reaction can only proceed to the right because only "reactants" are present. The reaction summary includes the values, or symbols for the values, of (1) initial concentrations, (2) changes in concentrations, and (3) concentrations at equilibrium.

> The coefficients in the equation are all 1's, so the reaction ratio must be 1:1:1:1.

Let x = moles per liter of A that react; then x = moles per liter of B that react and x = moles per liter of C and D that are formed.

	A	+	B	$\rightleftharpoons$	C	+	D
initial	0.200 M		0.200 M		0 M		0 M
change due to rxn	$-x\ M$		$-x\ M$		$+x\ M$		$+x\ M$
at equilibrium	$(0.200 - x)\ M$		$(0.200 - x)\ M$		$x\ M$		$x\ M$

Now K_c is known but concentrations are not. However, the equilibrium concentrations have all been expressed in terms of the single variable x. We substitute the equilibrium concentrations (*not* the initial ones) into the K_c expression and solve for x.

$$K_c = \frac{[C][D]}{[A][B]} = 49.0$$

$$\frac{(x)(x)}{(0.200 - x)(0.200 - x)} = \frac{x^2}{(0.200 - x)^2} = 49.0$$

This quadratic equation has a perfect square on both sides. We solve it by taking the square roots of both sides of the equation and then rearranging for x.

$$\frac{x}{0.200 - x} = 7.00$$

$$x = 1.40 - 7.00x \qquad 8.00x = 1.40 \qquad x = \frac{1.40}{8.00} = 0.175$$

Now we know the value of x, so the equilibrium concentrations are

> We see that the equilibrium concentrations of products are much greater than those of reactants because K_c is much greater than 1.

$$[A] = (0.200 - x)\ M = \boxed{0.025\ M}\ ; \qquad [C] = x\ M = \boxed{0.175\ M}$$

$$[B] = (0.200 - x)\ M = \boxed{0.025\ M}\ ; \qquad [D] = x\ M = \boxed{0.175\ M}$$

To check our answers we use *the equilibrium concentrations* to calculate Q_e and verify that its value is equal to K_c.

$$Q_e = \frac{[C][D]}{[A][B]} = \frac{(0.175)(0.175)}{(0.025)(0.025)} = 49.0 \qquad \text{Recall that } K_c = 49.0$$

The ideas developed in Example 17-5 may be applied to cases in which the reactants are mixed in nonstoichiometric amounts.

EXAMPLE 17-6 *Finding Equilibrium Concentrations*

Consider the same system as in Example 17-5 at the same temperature. If 0.600 mol of A and 0.200 mol of B are mixed in a 2.00-liter container and allowed to reach equilibrium, what are the equilibrium concentrations of all species?

Plan

We proceed as we did in Example 17-5. The only difference is that now we have *nonstoichiometric* amounts of reactants.

Solution

Let x = mol/L of A that react; then x = mol/L of B that react, and x = mol/L of C and D formed.

	A	+	B	$\rightleftharpoons$	C	+	D
initial	0.300 M		0.100 M		0 M		0 M
change due to rxn	$-x$ M		$-x$ M		$+x$ M		$+x$ M
equilibrium	$(0.300 - x)$ M		$(0.100 - x)$ M		x M		x M

The initial concentrations are governed by the amounts of reactants mixed together. But *changes in concentrations* due to reaction must occur in the 1:1:1:1 ratio required by the coefficients in the balanced equation.

$$K_c = \frac{[C][D]}{[A][B]} = 49.0 \quad \text{so} \quad \frac{(x)(x)}{(0.300 - x)(0.100 - x)} = 49.0$$

The left side of this equation is *not* a perfect square.

We can arrange this quadratic equation into the standard form.

$$\frac{x^2}{0.0300 - 0.400x + x^2} = 49.0$$

$$x^2 = 1.47 - 19.6x + 49.0x^2$$

$$48.0x^2 - 19.6x + 1.47 = 0$$

Quadratic equations can be solved by use of the quadratic formula.

$$x = \frac{-b \pm \sqrt{b^2 - 4ac}}{2a}$$

In this case $a = 48.0$, $b = -19.6$, and $c = 1.47$. Substituting these values gives

$$x = \frac{-(-19.6) \pm \sqrt{(-19.6)^2 - 4(48.0)(1.47)}}{2(48.0)} = \frac{19.6 \pm \sqrt{384 - 282}}{96.0}$$

$$= \frac{19.6 \pm \sqrt{102}}{96.0} = \frac{19.6 \pm 10.1}{96.0} = 0.309 \quad \text{or} \quad 0.099$$

Solving a quadratic equation always yields two roots. One root (the answer) has physical meaning. The other root, while mathematically correct, is extraneous; i.e., it has no physical meaning. The value of x is defined as the number of moles of A per liter that react and the number of moles of B per liter that react. No more B can be consumed than was initially present (0.100 M), so $x = 0.309$ is the extraneous root. Thus, $x = 0.099$ is the root in which we are interested, and the extraneous root is 0.309. The equilibrium concentrations are

$$[A] = (0.300 - x) M = \boxed{0.201 \ M} \ ; \quad [B] = (0.100 - x) M = \boxed{0.001 \ M} \ ;$$

$$[C] = [D] = x M = \boxed{0.099 \ M}$$

You should now work Exercises 30, 32, and 34.

Check Example 17-5:

$$Q = \frac{[C][D]}{[A][B]} = \frac{(0.175)(0.175)}{(0.025)(0.025)}$$

$$Q = 49 = K_c$$

Check Example 17-6:

$$Q = \frac{(0.099)(0.099)}{(0.201)(0.001)}$$

$$Q = 49 = K_c$$

The following table summarizes Examples 17-5 and 17-6.

	Initial Concentrations (*M*)				Equilibrium Concentrations (*M*)			
	[A]	**[B]**	**[C]**	**[D]**	**[A]**	**[B]**	**[C]**	**[D]**
Example 17-5	0.200	0.200	0	0	0.025	0.025	0.175	0.175
Example 17-6	0.300	0.100	0	0	0.201	0.001	0.099	0.099

The data from the table can be substituted into the reaction quotient expression, *Q*, as a check. Even though the reaction is initiated by different relative amounts of reactants in the two cases, the ratios of equilibrium concentrations of products to reactants (each raised to the first power) agree within roundoff error.

> ▼ **PROBLEM-SOLVING TIP**   *Solving Quadratic Equations*
>
> Quadratic equations can be rearranged into standard form.
>
> $$ax^2 + bx + c = 0$$
>
> All can be solved by the quadratic formula, which is
>
> $$x = \frac{-b \pm \sqrt{b^2 - 4ac}}{2a} \qquad \text{(Appendix A)}$$

17-6 FACTORS THAT AFFECT EQUILIBRIA

Once a reacting system has reached equilibrium, it remains at equilibrium until it is disturbed by some change of conditions. The guiding principle is known as **LeChatelier's Principle** (Section 13-6).

Remember that the *value* of an equilibrium constant changes only with temperature.

> If a change of conditions (stress) is applied to a system at equilibrium, the system responds in the way that best tends to reduce the stress in reaching a new state of equilibrium.

The reaction quotient, *Q*, helps us predict the direction of this response. There are four types of changes to consider.

1. Concentration changes
2. Pressure changes (volume changes for gas phase reactions)
3. Temperature changes
4. Introduction of catalysts

For reactions involving gases at constant temperature, changes in volume cause changes in pressure, and vice versa.

We now study the effects of these types of stresses from a qualitative, or descriptive, point of view. In Section 17-7 we expand our discussion with quantitative examples.

Changes in Concentration

Consider the following system *starting at equilibrium.*

$$A + B \rightleftharpoons C + D \qquad K_c = \frac{[C][D]}{[A][B]}$$

If more of any reactant or product is *added* to the system, the stress is relieved by shifting the equilibrium in the direction that consumes some of the added substance. Let us compare the mass action expressions for Q and K. If more A or B is added, then $Q < K$, and the forward reaction occurs to a greater extent than the reverse reaction until equilibrium is reestablished. If more C or D is added, $Q > K$, and the reverse reaction occurs to a greater extent until equilibrium is reestablished.

We can understand LeChatelier's Principle in the kinetic terms we used to introduce equilibrium. The rate of the forward reaction is proportional to the reactant concentrations raised to some powers,

$$\text{Rate}_f = k_f[A]^x[B]^y$$

When we add more A to an equilibrium mixture, this rate increases so that it no longer matches the rate of the reverse reaction. As the reaction proceeds to the right, consuming some A and B and forming more C and D, the forward rate diminishes and the reverse rate increases until they are again equal. At that point, a new equilibrium condition has been reached, with more C and D than were present in the original equilibrium mixture.

If a reactant or product is *removed* from a system at equilibrium, the reaction that produces *that* substance occurs to a greater extent than its reverse. If some C or D is removed, then $Q < K$, and the forward reaction is favored until equilibrium is reestablished. If some A or B is removed, the reverse reaction is favored.

Stress	Q	Direction of Shift of $A + B \rightleftharpoons C + D$
Increase concentration of A or B	$Q < K$	$\longrightarrow$ right
Increase concentration of C or D	$Q > K$	left $\longleftarrow$
Decrease concentration of A or B	$Q > K$	left $\longleftarrow$
Decrease concentration of C or D	$Q < K$	$\longrightarrow$ right

When a "new equilibrium" is established, (1) the rates of the forward and reverse reactions are equal again, and (2) K_c is again satisfied by the concentrations of reactants and products.

Practical applications of changes of this type are of great economic importance. Removing a product of a reversible reaction forces the reaction to produce more product than could be obtained if the reaction were simply allowed to reach equilibrium.

Changes in Volume and Pressure

Changes in pressure have little effect on the concentrations of solids or liquids because they are only slightly compressible. However, changes in pressure do cause significant changes in concentrations of gases. Therefore, such changes affect the value of Q for reactions in which the number of moles of gaseous reactants differs from the number of moles of gaseous products. For an ideal gas,

$$PV = nRT \qquad \text{or} \qquad P = (n/V)(RT)$$

The terminology used here is not as precise as we might like, but it is widely used. When we say that the equilibrium is "shifted to the left," we mean that the reaction to the left occurs to a greater extent than the reaction to the right.

This tabulation summarizes a lot of useful information. Study it carefully.

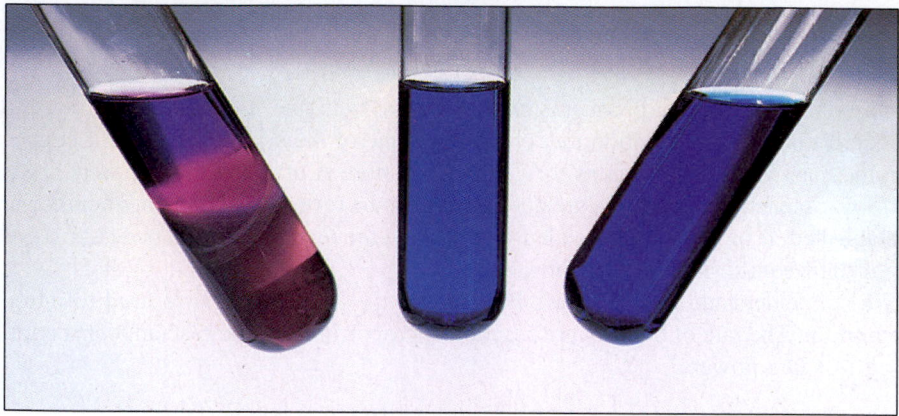

Effects of changes in concentration on the equilibrium

$$[Co(OH_2)_6]^{2+} + 4Cl^- \rightleftharpoons [CoCl_4]^{2-} + 6H_2O$$

A solution of $CoCl_2 \cdot 6H_2O$ in isopropyl alcohol is blue due to the $[CoCl_4]^{2-}$ ion. Addition of H_2O favors the reaction to the left to form a mixture of the pink and blue complexes, which is purple (middle test tube). When we add concentrated HCl, the excess Cl^- shifts the reaction to the right (blue, right). Adding $AgNO_3(aq)$ removes some Cl^- by precipitation of AgCl(s) and favors the reaction to the left (produces $[Co(OH_2)_6]^{2+}$). The resulting solution is pink (left).

The term (n/V) represents concentration, i.e., mol/L. At constant temperature n, R, and T are constants. Thus, if the volume occupied by a gas decreases, its partial pressure increases and its concentration (n/V) increases. If the volume of a gas increases, both its partial pressure and its concentration decrease.

Consider the following gaseous system at equilibrium.

$$A(g) \rightleftharpoons 2D(g) \qquad K = \frac{[D]^2}{[A]}$$

At constant temperature, a decrease in volume (increase in pressure) increases the concentrations of both A and D. In the expression for Q, the concentration of D is squared and the concentration of A is raised to the first power. As a result, the numerator of Q increases more than the denominator as pressure increases. Thus, $Q > K$, and this equilibrium shifts to the left. Conversely, an increase in volume (decrease in pressure) shifts this reaction to the right until equilibrium is reestablished, because $Q < K$. We can summarize the effect of pressure (volume) changes on *this* gas phase system at equilibrium.

Stress	$Q*$	Direction of Shift of $A(g) \rightleftharpoons 2D(g)$
Volume decrease, pressure increase	$Q > K$	Toward smaller number of moles of gas (left for *this* reaction)
Volume increase, pressure decrease	$Q < K$	Toward larger number of moles of gas (right for *this* reaction)

Study this tabulation carefully. How would these conclusions change for a reaction in which there are more moles of gaseous reactants than moles of gaseous products?

*In Q for *this* reaction, there are more moles of gaseous product than gaseous reactant.

A decrease in volume (increase in pressure) shifts the reaction in the direction that produces the smaller number of moles of gas.

An increase in volume (decrease in pressure) shifts the reaction in the direction that produces the larger number of moles of gas.

If there is no change in the number of moles of gases in a reaction, a volume (pressure) change does not affect the position of equilibrium.

One practical application of these ideas is illustrated later in this section by the Haber process.

The foregoing argument applies only when pressure changes are due to volume changes. It *does not apply* if the total pressure of a gaseous system is raised by merely pumping in an inert gas, e.g., He. In such a situation, the *partial* pressure of each reacting gas remains constant, so the system remains at equilibrium.

Changes in Temperature

Consider the following system at equilibrium:

$$A + B \rightleftharpoons C + D + \text{heat} \qquad (\Delta H \text{ is negative})$$

Heat is produced by the forward (exothermic) reaction. Suppose we increase the temperature at constant pressure by adding heat to the system. This favors the reaction to the left, removing some of the extra heat. Lowering the temperature favors the reaction to the right as the system replaces some of the heat that was removed.

By contrast, for the endothermic reaction at equilibrium,

$$W + X + \text{heat} \rightleftharpoons Y + Z \qquad (\Delta H \text{ is positive})$$

an increase in temperature at constant pressure favors the reaction to the right. A decrease in temperature favors the reaction to the left.

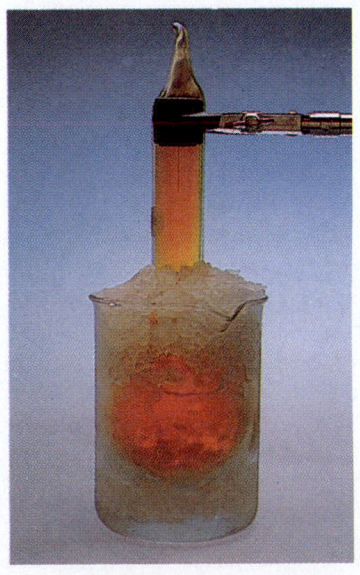 

The gas phase equilibrium for the *exothermic* reaction

$$2NO_2(g) \rightleftharpoons N_2O_4(g)$$

The two flasks contain the same *total* amounts of gas. NO_2 is brown, whereas N_2O_4 is colorless. The higher temperature (50°C) of the flask on the right favors the reverse reaction; this mixture is more highly colored because it contains more NO_2. The flask on the left, at the temperature of ice water, contains less brown NO_2 gas.

Effect of temperature changes on the equilibrium

$$[Co(OH_2)_6]^{2+} + 4Cl^- + heat \rightleftharpoons [CoCl_4]^{2-} + 6H_2O$$

We begin with a purple equilibrium mixture of the pink and blue complexes (middle test tube). In hot water the forward reaction is favored, so the solution is blue (right). At 0°C the reverse reaction is favored, so the solution is pink (left).

> An increase in temperature favors endothermic reactions.
>
> A decrease in temperature favors exothermic reactions.

The *values* of equilibrium constants change as temperature changes. The K's of exothermic reactions decrease with increasing T, and the K's of endothermic reactions increase with increasing T, as we shall see in Section 17-12. No other stresses affect the value of K.

Introduction of a Catalyst

Can you use the Arrhenius equation (Section 16-8) to show that lowering the activation energy barrier increases forward and reverse rates by the same factor?

Adding a catalyst to a system changes the rate of the reaction (Section 16-9), but this *cannot* shift the equilibrium in favor of either products or reactants. Because a catalyst affects the activation energy of *both* forward and reverse reactions equally, it changes both rates equally. Equilibrium is established more quickly in the presence of a catalyst.

Not all reactions attain equilibrium; they may occur too slowly, or else products or reactants may be continually added or removed. Such is the case with most reactions in biological systems. On the other hand, some reactions, such as typical acid–base neutralizations, achieve equilibrium very rapidly.

EXAMPLE 17-7 *Applying a Stress to a System at Equilibrium*

Given the following reaction at equilibrium in a closed container at 500°C, predict the effect of each of the following changes on the amount of NH_3: (a) increasing the temperature, (b) lowering the temperature, (c) increasing the pressure by decreasing the volume, (d) introducing some platinum catalyst, (e) forcing more H_2 into the system, and (f) removing some NH_3 from the system.

$$N_2(g) + 3H_2(g) \rightleftharpoons 2NH_3(g) \qquad \Delta H^0 = -92 \text{ kJ/mol rxn}$$

Plan

We apply LeChatelier's Principle to each part of the question individually.

Solution

(a) The negative value for ΔH tells us that the forward reaction is exothermic. Increasing the temperature favors the endothermic reaction (reverse in this case).

> Some NH_3 is used up.

(b) Lowering the temperature favors the exothermic reaction (forward in this case).

> More NH_3 is formed.

(c) Increasing the pressure favors the reaction that produces the smaller number of moles of gas (forward in this case).

> More NH_3 is formed.

(d) A catalyst does not favor either reaction.

> It would have no effect on the amount of NH_3.

(e) Adding a substance favors the reaction that uses up that substance (forward in this case).

> More NH_3 is formed.

(f) Removing a substance favors the reaction that produces that substance (forward in this case).

> More NH_3 is formed.

You should now work Exercises 42, 43, and 44.

Now we shall illustrate the commercial importance of these changes.

The Haber Process

Nitrogen, N_2, is very unreactive. The Haber process is the economically important industrial process by which atmospheric N_2 is converted to ammonia, NH_3, a soluble, reactive compound. Innumerable dyes, plastics, explosives, fertilizers, and synthetic fibers are made from ammonia. The Haber process provides insight into kinetic and thermodynamic factors that influence reaction rates and the positions of equilibria. In this process the reaction between N_2 and H_2 to produce NH_3 is never allowed to reach equilibrium, but moves toward it.

$$N_2(g) + 3H_2(g) \rightleftharpoons 2NH_3(g) \qquad \Delta H^0 = -92 \text{ kJ/mol}$$

$$K_c = \frac{[NH_3]^2}{[N_2][H_2]^3} = 3.6 \times 10^8 \qquad \text{(at 25°C)}$$

Approximately 135 pounds of NH_3 are required for each person per year in the United States. Haber developed the process to provide a cheaper and more reliable source of explosives as Germany prepared for World War I. (Britain controlled the seas and access to the natural nitrates in India and Chile.) The current use of the process is more humanitarian: most NH_3 is used to produce fertilizers.

The process is diagrammed in Figure 17-3. The reaction is run at about 450°C under pressures ranging from 200 to 1000 atmospheres. Hydrogen is obtained from coal gas or petroleum refining and nitrogen from liquefied air.

The value of K_c is 3.6×10^8 at 25°C. This very large value of K_c indicates that *at equilibrium* virtually all of the N_2 and H_2 (mixed in a 1:3 mole ratio) would be converted

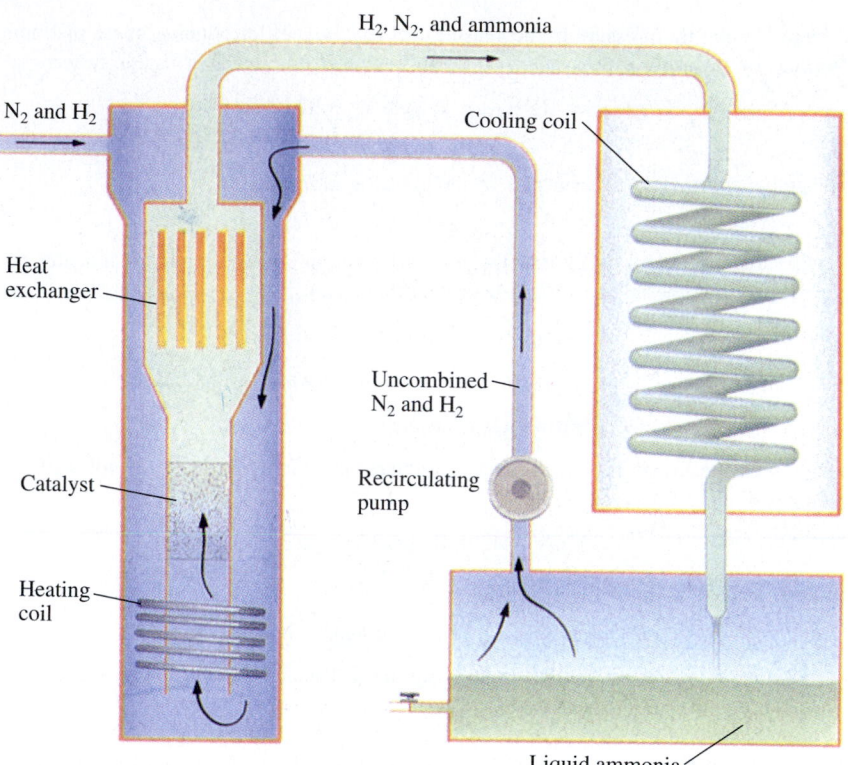

Figure 17-3 A simplified representation of the Haber process for synthesizing ammonia.

In practice the mixed reactants are compressed by special pumps and injected into the heated reaction vessel.

into NH_3. However, at 25°C the reaction occurs so slowly that no measurable amount of NH_3 is produced within a reasonable time. Thus, the large equilibrium constant (a thermodynamic factor) indicates that the reaction proceeds toward the right almost completely. However, it tells us *nothing* about how fast the reaction occurs (a kinetic factor).

There are four moles of gases on the left side of the equation and only two moles of gas on the right, so increasing the pressure favors the production of NH_3. Therefore, the Haber process is carried out at very high pressures, as high as the equipment will safely stand.

The reaction is exothermic (ΔH^0_{rxn} is negative), so increasing the temperature favors the *decomposition* of NH_3 (the reverse reaction). But, the rates of both forward and reverse reactions increase as temperature increases.

The addition of a catalyst of finely divided iron and small amounts of selected oxides also speeds up both the forward and reverse reactions. This allows NH_3 to be produced not only faster but at a lower temperature, which increases the yield of NH_3 and extends the life of the equipment.

Table 17-1 *Effect of T and P on Yield of Ammonia*				
		Mole % NH$_3$ in Equilibrium Mixture		
°C	**K_c**	**10 atm**	**100 atm**	**1000 atm**
209	650	51	82	98
467	0.5	4	25	80
758	0.014	0.5	5	13

A plant for the commercial production of ammonia. The automobile and truck help you comprehend the size of this plant.

Table 17-1 shows the effects of increases in temperature and pressure on the equilibrium yield of NH_3, starting with $1:3$ mole ratios of $N_2:H_2$. K_c decreases by more than ten orders of magnitude, from 3.6×10^8 at 25°C to only 1.4×10^{-2} at 758°C. This tells us that the reaction proceeds *very far to the left* at high temperatures. Casual examination of the data might suggest that the reaction should be run at lower temperatures, because a higher percentage of the N_2 and H_2 is converted into NH_3. However, the reaction occurs so slowly, even in the presence of a catalyst, that it cannot be run economically at temperatures below about 450°C.

The emerging reaction mixture is cooled down, and NH_3 (bp = -33.43°C) is removed as a liquid. This favors the forward reaction. The unreacted N_2 and H_2 are recycled. Excess N_2 is used to favor the reaction to the right.

Ten orders of magnitude is 10^{10}, i.e., 10 billion.

$$1 \times 10^{10} = 10,000,000,000$$

17-7 APPLICATION OF STRESS TO A SYSTEM AT EQUILIBRIUM

We can use equilibrium constants to determine new equilibrium concentrations that result from adding one or more species to, or removing one or more species from, a system at equilibrium.

EXAMPLE 17-8 *Applying a Stress to a System at Equilibrium*

Some hydrogen and iodine are mixed at 229°C in a 1.00-liter container. When equilibrium is established, the following concentrations are present: [HI] = 0.490 M, [H_2] = 0.080 M, and [I_2] = 0.060 M. If an additional 0.300 mol of HI is then added, what concentrations will be present when the new equilibrium is established?

$$H_2(g) + I_2(g) \rightleftharpoons 2HI(g)$$

Plan

We use the initial equilibrium concentrations to calculate the value of K_c. Then we determine the new concentrations after some HI has been added and calculate Q. The value of Q tells us which reaction is favored. Then we can represent the new equilibrium concentrations. We substitute these representations into the K_c expression and solve for the new equilibrium concentrations.

Solution

Calculate the value of K_c from the first set of equilibrium concentrations.

$$K_c = \frac{[HI]^2}{[H_2][I_2]} = \frac{(0.490)^2}{(0.080)(0.060)} = 50$$

When we add 0.300 mol of HI to the 1.00-liter container, the [HI] instantaneously increases by 0.300 M.

	$H_2(g)$	+	$I_2(g)$	$\rightleftharpoons$	$2HI(g)$
equilibrium	0.080 M		0.060 M		0.490 M
mol/L added	0 M		0 M		+0.300 M
new initial conc'n	0.080 M		0.060 M		0.790 M

> Here it is obvious that adding some HI favors the reaction to the left. In many cases we cannot tell which reaction will be favored. Calculating Q always lets us make the decision.

Substitution of these *new initial* concentrations into the reaction quotient gives

$$Q = \frac{[HI]^2}{[H_2][I_2]} = \frac{(0.790)^2}{(0.080)(0.060)} = 130$$

Because $Q > K$, the reaction proceeds to the left to establish a new equilibrium. The new equilibrium concentrations can be determined as follows. Let $x = $ mol/L of H_2 formed; so $x = $ mol/L of I_2 formed, and $2x = $ mol/L of HI consumed.

	$H_2(g)$	+	$I_2(g)$	$\rightleftharpoons$	$2HI(g)$
new initial	0.080 M		0.060 M		0.790 M
change due to rxn	$+x$ M		$+x$ M		$-2x$ M
new equilibrium	$(0.080 + x)$ M		$(0.060 + x)$ M		$(0.790 - 2x)$ M

> Equal concentrations of H_2 and I_2 must be formed by the *reaction*.

Substitution of these values into K_c allows us to evaluate x.

$$K_c = 50 = \frac{(0.790 - 2x)^2}{(0.080 + x)(0.060 + x)} = \frac{0.624 - 3.16x + 4x^2}{0.0048 + 0.14x + x^2}$$

$$0.24 + 7.0x + 50x^2 = 0.624 - 3.16x + 4x^2$$

$$46x^2 + 10.2x - 0.38 = 0$$

Solution by the quadratic formula gives $x = 0.032$ and -0.25.

Clearly, $x = -0.25$ is the extraneous root, because x cannot be less than zero. This reaction does not consume a negative quantity of HI, because the reaction is proceeding toward the left. Thus, $x = 0.032$ is the root with physical meaning, so the new equilibrium concentrations are

> To "consume a negative quantity of HI" would be to form HI. The value $x = -0.25$ would lead to $[H_2] = (0.080 + x)$ $M = (0.080 - 0.25)$ $M = -0.17$ M. A negative concentration is impossible, so $x = -0.25$ is the extraneous root.

$$[H_2] = (0.080 + x) M = (0.080 + 0.032) M = \boxed{0.112\ M}$$

$$[I_2] = (0.060 + x) M = (0.060 + 0.032) M = \boxed{0.092\ M}$$

$$[HI] = (0.790 - 2x) M = (0.790 - 0.064) M = \boxed{0.726\ M}$$

In summary,

Original Equilibrium	Stress Applied	New Equilibrium
$[H_2] = 0.080$ M		$[H_2] = 0.112$ M
$[I_2] = 0.060$ M	Add 0.300 M HI	$[I_2] = 0.092$ M
$[HI] = 0.490$ M		$[HI] = 0.726$ M

We see that some of the additional HI is consumed, but not all of it. More HI remains after the new equilibrium is established than was present before the stress was imposed. However, the new equilibrium $[H_2]$ and $[I_2]$ are substantially greater than the original equilibrium concentrations.

You should now work Exercise 52.

We can also use the equilibrium constant to calculate new equilibrium concentrations that result from decreasing the volume (increasing the pressure) of a gaseous system that was initially at equilibrium.

EXAMPLE 17-9 *Applying a Stress to a System at Equilibrium*

At 22°C the equilibrium constant, K_c, for the following reaction is 4.66×10^{-3}. (a) If 0.800 mol of N_2O_4 were injected into a closed 1.00-liter container at 22°C, how many moles of each gas would be present at equilibrium? (b) If the volume were halved (to 0.500 L) at constant temperature, how many moles of each gas would be present after the new equilibrium has been established?

$$N_2O_4(g) \rightleftharpoons 2NO_2(g) \qquad K_c = \frac{[NO_2]^2}{[N_2O_4]} = 4.66 \times 10^{-3}$$

Plan

(a) We are given the value for K_c and the initial concentration of N_2O_4. We write the reaction summary, which gives the representation of the equilibrium concentrations. Then we substitute these into the K_c expression and solve.
(b) We repeat the process we used in part (a).

Solution

(a) Let x = mol/L of N_2O_4 consumed and $2x$ = mol/L of NO_2 formed.

	$N_2O_4(g)$	$\rightleftharpoons$	$2NO_2(g)$
initial	0.800 *M*		0 *M*
change due to rxn	$-x$ *M*		$+2x$ *M*
equilibrium	$(0.800 - x)$ *M*		$2x$ *M*

$$K_c = \frac{[NO_2]^2}{[N_2O_4]} = 4.66 \times 10^{-3} = \frac{(2x)^2}{0.800 - x} = \frac{4x^2}{0.800 - x}$$

$$3.73 \times 10^{-3} - 4.66 \times 10^{-3}\, x = 4x^2$$

$$4x^2 + 4.66 \times 10^{-3}\, x - 3.73 \times 10^{-3} = 0$$

Solving by the quadratic formula gives $x = 3.00 \times 10^{-2}$ and $x = -3.11 \times 10^{-2}$. We use $x = 3.00 \times 10^{-2}$.

The value of x is the number of moles per liter of N_2O_4 that react. So x must be positive and cannot be greater than 0.800 *M*.

$$0 < x < 0.800\ M$$

The original equilibrium concentrations are

$$[NO_2] = 2x\ M = 6.00 \times 10^{-2}\ M$$
$$N_2O_4 = (0.800 - x)\ M = (0.800 - 3.00 \times 10^{-2})\ M = 0.770\ M$$

$$\underset{\_}{?}\ \text{mol NO}_2 = 1.00\ \text{L} \times \frac{6.00 \times 10^{-2}\ \text{mol NO}_2}{\text{L}} = \boxed{6.00 \times 10^{-2}\ \text{mol NO}_2}$$

$$\underset{\_}{?}\ \text{mol N}_2\text{O}_4 = 1.00\ \text{L} \times \frac{0.770\ \text{mol N}_2\text{O}_4}{\text{L}} = \boxed{0.770\ \text{mol N}_2\text{O}_4}$$

LeChatelier's Principle tells us that a decrease in volume (increase in pressure) favors the production of N_2O_4.

$$Q = \frac{(0.120)^2}{1.54} = 9.35 \times 10^{-3}$$

$$Q > K$$

$$\therefore \overset{\text{shift left}}{\longleftarrow}$$

(b) When the volume of the reaction vessel is halved, the concentrations are doubled, so the new *initial* concentrations of N_2O_4 and NO_2 are $2(0.770\,M) = 1.54\,M$ and $2(6.00 \times 10^{-2}\,M) = 0.120\,M$, respectively.

	$N_2O_4(g)$	$\rightleftharpoons$	$2NO_2(g)$
new initial	1.54 M		0.120 M
change due to rxn	$+x\,M$		$-2x\,M$
new equilibrium	$(1.54 + x)\,M$		$(0.120 - 2x)\,M$

$$K_c = \frac{[NO_2]^2}{[N_2O_4]} = 4.66 \times 10^{-3} = \frac{(0.120 - 2x)^2}{1.54 + x}$$

Rearranging into the standard form of a quadratic equation gives

$$x^2 - 0.121x + 1.81 \times 10^{-3} = 0$$

Solving as before gives $x = 0.104$ and $x = 0.017$.

The root $x = 0.104$ would give a *negative* concentration for NO_2, which is impossible.

The maximum value of x is $0.060\,M$, because $2x$ may not exceed the concentration of NO_2 that was present after the volume was halved. Thus, $x = 0.017\,M$ is the root with physical significance. The new equilibrium concentrations in the 0.500-liter container are

$$[NO_2] = (0.120 - 2x)\,M = (0.120 - 0.034)\,M = 0.086\,M$$
$$[N_2O_4] = (1.54 + x)\,M \quad = (1.54 + 0.017)\,M \quad = 1.56\,M$$

$$\underset{\cdot}{?}\ \text{mol } NO_2 = 0.500\ L \times \frac{0.086\ \text{mol } NO_2}{L} = \boxed{0.043\ \text{mol } NO_2}$$

$$\underset{\cdot}{?}\ \text{mol } N_2O_4 = 0.500\ L \times \frac{1.56\ \text{mol } N_2O_4}{L} = \boxed{0.78\ \text{mol } N_2O_4}$$

In summary,

First Equilibrium	Stress	New Equilibrium
0.770 mol of N_2O_4 0.0600 mol of NO_2	Decrease volume from 1.00 L to 0.500 L	0.78 mol of N_2O_4 0.043 mol of NO_2

Both $[N_2O_4]$ and $[NO_2]$ *increase* because of the large decrease in volume. You may wish to verify that the *number of moles* of N_2O_4 increases, while the *number of moles* of NO_2 decreases. We predict this from LeChatelier's Principle.

You should now work Exercise 54.

17-8 PARTIAL PRESSURES AND THE EQUILIBRIUM CONSTANT

It is often more convenient to measure pressures rather than concentrations of gases. Solving the ideal gas equation, $PV = nRT$, for pressure gives

$$P = \frac{n}{V}(RT) \quad \text{or} \quad P = M(RT)$$

The pressure of a gas is directly proportional to its concentration (n/V). For reactions in which all substances that appear in the equilibrium constant expression are gases, we sometimes prefer to express the equilibrium constant in terms of partial pressures *in atmospheres* (K_P) rather than in terms of concentrations (K_c).

In general for a reaction involving gases,

$$aA(g) + bB(g) \rightleftharpoons cC(g) + dD(g) \qquad K_P = \frac{(P_C)^c(P_D)^d}{(P_A)^a(P_B)^b}$$

K_P has no units for the same reasons that K_c has no units.

For instance, for the following reversible reaction,

$$N_2(g) + 3H_2(g) \rightleftharpoons 2NH_3(g) \qquad K_P = \frac{(P_{NH_3})^2}{(P_{N_2})(P_{H_2})^3}$$

EXAMPLE 17-10 *Calculation of K_P*

In an equilibrium mixture at 500°C, we find $P_{NH_3} = 0.147$ atm, $P_{N_2} = 1.41$ atm, and $P_{H_2} = 6.00$ atm. Evaluate K_P at 500°C for the following reaction.

$$N_2(g) + 3H_2(g) \rightleftharpoons 2NH_3(g)$$

Plan

We are given equilibrium partial pressures of all reactants and products. So we write the expression for K_P and substitute partial pressures in atmospheres into it.

Solution

$$K_P = \frac{(P_{NH_3})^2}{(P_{N_2})(P_{H_2})^3} = \frac{(0.147)^2}{(1.41)(6.00)^3} = \boxed{7.10 \times 10^{-5}}$$

You should now work Exercise 58.

17-9 RELATIONSHIP BETWEEN K_P AND K_c

If the ideal gas equation is rearranged, the molar concentration of a gas is

$$\left(\frac{n}{V}\right) = \frac{P}{RT} \qquad \text{or} \qquad M = \frac{P}{RT}$$

$\left(\dfrac{n}{V}\right)$ is $\left(\dfrac{\text{no. mol}}{L}\right)$

Substituting P/RT for n/V in the K_c expression for the N_2–H_2–NH_3 equilibrium gives the relationship between K_c and K_P for *this* reaction.

$$K_c = \frac{[NH_3]^2}{[N_2][H_2]^3} = \frac{\left(\dfrac{P_{NH_3}}{RT}\right)^2}{\left(\dfrac{P_{N_2}}{RT}\right)\left(\dfrac{P_{H_2}}{RT}\right)^3} = \frac{(P_{NH_3})^2}{(P_{N_2})(P_{H_2})^3} \times \frac{\left(\dfrac{1}{RT}\right)^2}{\left(\dfrac{1}{RT}\right)^4}$$

$$K_c = K_P (RT)^2 \qquad \text{and} \qquad K_P = K_c (RT)^{-2}$$

In general the relationship between K_c and K_P is

$$K_P = K_c(RT)^{\Delta n} \qquad \text{or} \qquad K_c = K_P(RT)^{-\Delta n} \qquad \Delta n = (n_{\text{gas prod}}) - (n_{\text{gas react}})$$

Δn refers to the numbers of moles of gaseous substances in the balanced equation, *not* in the reaction vessel.

Because K_c refers to mol/L and K_P refers to atmospheres, R must be expressed in L · atm/mol · K. For reactions in which there are equal numbers of moles of gases on both sides of the equation, $\Delta n = 0$ and $K_P = K_c$.

For the ammonia equilibrium,

$$N_2(g) + 3H_2(g) \rightleftharpoons 2NH_3(g) \qquad \Delta n = 2 - 4 = -2$$

$K_P = 7.10 \times 10^{-5}$ at 500°C (773 K). So we have

$$K_c = K_P(RT)^{-\Delta n} = (7.10 \times 10^{-5})[(0.0821)(773)]^{-(-2)} = 0.286$$

This agrees with the value from Example 17-1.

For *gas phase reactions,* we can calculate the amounts of substances present at equilibrium using either K_P or K_c. The results are the same by either method (when they are expressed in the same terms). To illustrate, let's solve the following problem by both methods.

EXAMPLE 17-11   *Using K_c and K_P*

We place 10.0 grams of $SbCl_5$ in a 5.00-liter container at 448°C, and allow the reaction to attain equilibrium. How many grams of $SbCl_5$ are present at equilibrium? Solve this problem (a) using K_c and molar concentrations and (b) using K_P and partial pressures.

$$SbCl_5(g) \rightleftharpoons SbCl_3(g) + Cl_2(g) \qquad \text{at 448°C, } K_c = 2.51 \times 10^{-2} \qquad \text{and} \qquad K_P = 1.48$$

(a) Plan (using K_c)

We calculate the initial concentration of $SbCl_5$, write the reaction summary and represent the equilibrium concentrations; then we substitute into the K_c expression to obtain the equilibrium concentrations.

(a) Solution (using K_c)

Because we are given K_c, we use concentrations. The initial concentration of $SbCl_5$ is

$$[SbCl_5] = \frac{10.0 \text{ g } SbCl_5}{5.00 \text{ L}} \times \frac{1 \text{ mol}}{299 \text{ g}} = 0.00669 \text{ } M \text{ } SbCl_5$$

Let x = mol/L of $SbCl_5$ that react. In terms of molar concentrations, the reaction summary is

	$SbCl_5$	$\rightleftharpoons$	$SbCl_3$ +	Cl_2
initial	0.00669 M		0	0
change due to rxn	$-x$ M		$+x$ M	$+x$ M
equilibrium	(0.00669 − x) M		x M	x M

$$K_c = \frac{[SbCl_3][Cl_2]}{[SbCl_5]}$$

$$2.51 \times 10^{-2} = \frac{(x)(x)}{0.00669 - x}$$

$$x^2 = 1.68 \times 10^{-4} - 2.51 \times 10^{-2} x$$

$$x^2 + 2.51 \times 10^{-2} x - 1.68 \times 10^{-4} = 0$$

Solving by the quadratic formula gives

$$x = 5.49 \times 10^{-3} \qquad \text{and} \qquad -3.06 \times 10^{-2} \text{ (extraneous root)}$$

$$[SbCl_5] = (0.00669 - x) \text{ } M = (0.00669 - 0.00549) \text{ } M = 1.20 \times 10^{-3} \text{ } M$$

$$\underline{?}\ g\ SbCl_5 = 5.00\ L \times \frac{1.20 \times 10^{-3}\ mol}{L} \times \frac{299\ g}{mol} = \boxed{1.79\ g\ SbCl_5}$$

Let us now solve the same problem *using K_P and partial pressures.*

(b) Plan (using K_P)

Calculate the initial partial pressure of $SbCl_5$ and write the reaction summary. Substitution of the representation of the equilibrium partial pressures into K_P gives their values.

(b) Solution (using K_P)

We calculate the initial *pressure* of $SbCl_5$ in atmospheres, using $PV = nRT$.

$$P_{SbCl_5} = \frac{nRT}{V} = \frac{\left[(10.0\ g)\left(\dfrac{1\ mol}{299\ g}\right)\right]\left(0.0821\ \dfrac{L \cdot atm}{mol \cdot K}\right)(721\ K)}{5.00\ L} = 0.396\ atm$$

Clearly, $P_{SbCl_3} = 0$ and $P_{Cl_2} = 0$ because only PCl_5 is present initially. We write the reaction summary in terms of partial pressures in atmospheres, because K_P refers to pressures in atmospheres.

Let $y =$ decrease in pressure (atm) of $SbCl_5$ due to reaction. In terms of partial pressures, the reaction summary is

The partial pressure of each substance is proportional to the number of moles of that substance.

	$SbCl_5$	$\rightleftharpoons$	$SbCl_3$	+	Cl_2
initial	0.396 atm		0		0
change due to rxn	$-y$ atm		$+y$ atm		$+y$ atm
equilibrium	$(0.396 - y)$ atm		y atm		y atm

$$K_P = \frac{(P_{SbCl_3})(P_{Cl_2})}{P_{SbCl_5}} = 1.48 = \frac{(y)(y)}{0.396 - y}$$

If we did not know the value of K_P, we could calculate it from the known value of K_c. $K_P = K_c(RT)^{\Delta n}$.

$$0.586 - 1.48y = y^2 \qquad y^2 + 1.48y - 0.586 = 0$$

Solving by the quadratic formula gives

$$y = 0.325 \qquad and \qquad -1.80\ \text{(extraneous root)}$$

$$P_{SbCl_5} = (0.396 - y) = (0.396 - 0.325) = 0.071\ atm$$

We use the ideal gas law, $PV = nRT$, to calculate the number of moles of $SbCl_5$.

$$n = \frac{PV}{RT} = \frac{(0.071\ atm)(5.00\ L)}{\left(0.0821\ \dfrac{L \cdot atm}{mol \cdot K}\right)(721\ K)} = 0.0060\ mol\ SbCl_5$$

$$\underline{?}\ g\ SbCl_5 = 0.0060\ mol \times \frac{299\ g}{mol} = \boxed{1.8\ g\ SbCl_5}$$

We see that within roundoff range the same result is obtained by both methods.

You should now work Exercise 64.

▼ PROBLEM-SOLVING TIP   *In K_P Calculations Gas Pressures Must Be Expressed in Atmospheres*

One error that students sometimes make when solving K_P problems is to express pressures in torr. Remember that these pressures must be expressed in atmospheres.

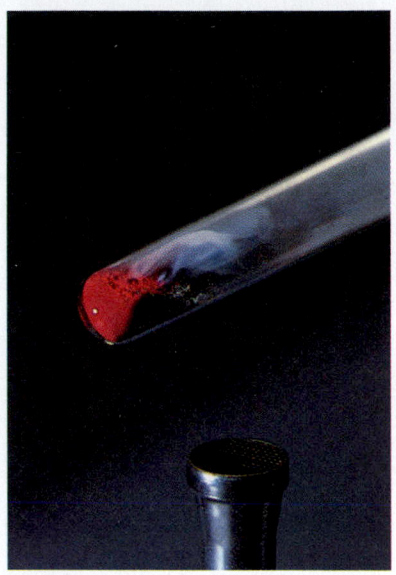

A photograph of the reaction

$$2HgO(s) \rightleftharpoons 2Hg(\ell) + O_2(g)$$

The reaction is not at equilibrium here, because O_2 gas has been allowed to escape.

17-10 HETEROGENEOUS EQUILIBRIA

Thus far, we have considered only equilibria involving species in a single phase, i.e., homogeneous equilibria. **Heterogeneous equilibria** involve species in more than one phase. Consider the following reversible reaction at 25°C.

$$2HgO(s) \rightleftharpoons 2Hg(\ell) + O_2(g)$$

When equilibrium is established for this system, a solid, a liquid, and a gas are present. Neither solids nor liquids are significantly affected by changes in pressure. The fundamental definition of the equilibrium constant in thermodynamics is in terms of the activities of the substances involved.

> For pure solids or liquids, the activity is taken as 1 (Section 17-2), so terms for pure liquids and pure solids *do not* appear in the K expressions for heterogeneous equilibria.

Thus, for the reaction

$$2HgO(s) \rightleftharpoons 2Hg(\ell) + O_2(g) \qquad K_c = [O_2] \quad \text{and} \quad K_P = P_{O_2}$$

These equilibrium constant expressions indicate that equilibrium exists at a given temperature for *one and only one* concentration and one partial pressure of oxygen in contact with liquid mercury and solid mercury(II) oxide.

EXAMPLE 17-12 K_c and K_P for Heterogeneous Equilibrium

Write both K_c and K_P expressions for the following reversible reactions.
(a) $2SO_2(g) + O_2(g) \rightleftharpoons 2SO_3(g)$
(b) $2NH_3(g) + H_2SO_4(\ell) \rightleftharpoons (NH_4)_2SO_4(s)$
(c) $S(s) + H_2SO_3(aq) \rightleftharpoons H_2S_2O_3(aq)$

Plan
We apply the definitions of K_c and K_P to each reaction.

Solution

(a) $\quad K_c = \dfrac{[SO_3]^2}{[SO_2]^2[O_2]} \qquad\qquad K_P = \dfrac{(P_{SO_3})^2}{(P_{SO_2})^2(P_{O_2})}$

(b) $\quad K_c = \dfrac{1}{[NH_3]^2} = [NH_3]^{-2} \quad K_P = \dfrac{1}{(P_{NH_3})^2} = (P_{NH_3})^{-2}$

(c) $\quad K_c = \dfrac{[H_2S_2O_3]}{[H_2SO_3]} \qquad\qquad K_P$ undefined; no gases involved

You should now work Exercise 69.

EXAMPLE 17-13   Heterogeneous Equilibria

The value of K_P is 27 for the thermal decomposition of potassium chlorate at a given temperature. What is the partial pressure of oxygen in a closed container in which the following system is at equilibrium at the given temperature? (This can be a dangerous reaction.)

$$2KClO_3(s) \xrightarrow{\Delta} 2KCl(s) + 3O_2(g)$$

Plan

Because one gas, O_2, and two solids, $KClO_3$ and KCl, are involved, we see that K_P involves only the partial pressure of O_2, i.e., $K_P = (P_{O_2})^3$.

Solution

We are given

$$K_P = (P_{O_2})^3 = 27$$

Let x atm $= P_{O_2}$ at equilibrium. Then we have

$$(P_{O_2})^3 = 27 = x^3 \qquad \boxed{x = 3.0 \text{ atm}}$$

This tells us that the partial pressure of oxygen is 3.0 atm at equilibrium.

17-11 RELATIONSHIP BETWEEN ΔG^0 AND THE EQUILIBRIUM CONSTANT

In thermodynamic terms let us consider what may happen when two substances are mixed together at constant temperature and pressure. First, as a result of mixing, there is usually an increase in entropy due to the increase in disorder. If the two substances can react with each other, the chemical reaction begins, heat is released or absorbed, and the concentrations of the substances in the mixture change. An additional change in entropy, which depends upon changes in the nature of the reactants and products, also begins to occur. The evolution or absorption of heat energy, the changes in entropy, and the changes in concentrations all continue until equilibrium is established. Equilibrium may be reached with large amounts of products formed, with virtually all of the reactants remaining unchanged, or at *any* intermediate combination of concentrations.

The standard free energy change for a reaction is ΔG^0. This is the free energy change that would accompany *complete* conversion of *all* reactants initially present in their standard states to *all* products in their standard states—the standard reaction (Section 15-15). The free energy change for any other concentrations or pressures is ΔG (no superscript zero). The two quantities are related by the equation

Thermodynamic standard states are (1) pure solids or pure liquids at 1 atm, (2) solutions of one-molar concentrations, and (3) gases at partial pressures of 1 atm.

$$\Delta G = \Delta G^0 + RT \ln Q \qquad \text{or} \qquad \Delta G = \Delta G^0 + 2.303RT \log Q$$

R is the universal gas constant, T is the absolute temperature, and Q is the reaction quotient (Section 17-4). When a system is *at equilibrium,* $\Delta G = 0$ (Section 15-15) and $Q = K$ (Section 17-4). Recall that the reaction quotient may represent nonequilibrium concentrations (or partial pressures) of products and reactants. As reaction occurs, the free energy of the mixture and the concentrations change until at equilibrium $\Delta G = 0$, and the concentrations of reactants and products satisfy the equilibrium constant. At that point, Q becomes equal to K (Section 17-4). Then

$$0 = \Delta G^0 + RT \ln K \qquad \text{or} \qquad 0 = \Delta G^0 + 2.303RT \log K \qquad \text{(at equilibrium)}$$

Rearranging gives

The energy units of R *must* match those of ΔG^0. We often use

$$R = 8.314 \text{ J/mol} \cdot \text{K}$$

$$\Delta G^0 = -RT \ln K \qquad \text{or} \qquad \Delta G^0 = -2.303RT \log K$$

This equation shows the relationship between the standard free energy change and the **thermodynamic equilibrium constant.**

For the following generalized reaction, the thermodynamic equilibrium constant is defined in terms of the activities of the species involved.

$$a\text{A} + b\text{B} \rightleftharpoons c\text{C} + d\text{D} \qquad K = \frac{(a_\text{C})^c (a_\text{D})^d}{(a_\text{A})^a (a_\text{B})^b}$$

where a_A is the activity of substance A, and so on. The mass action expression to which it is related involves concentration terms for species in solution and partial pressures for gases.

> For equilibria that involve only gases, the thermodynamic equilibrium constant (related to ΔG^0) is K_P. For those that involve species in solution, it is equal to K_c.

Figure 17-4 displays the relationships between free energy and equilibrium. The *left* end of each curve represents the total free energy of the reactants and the *right* end of each

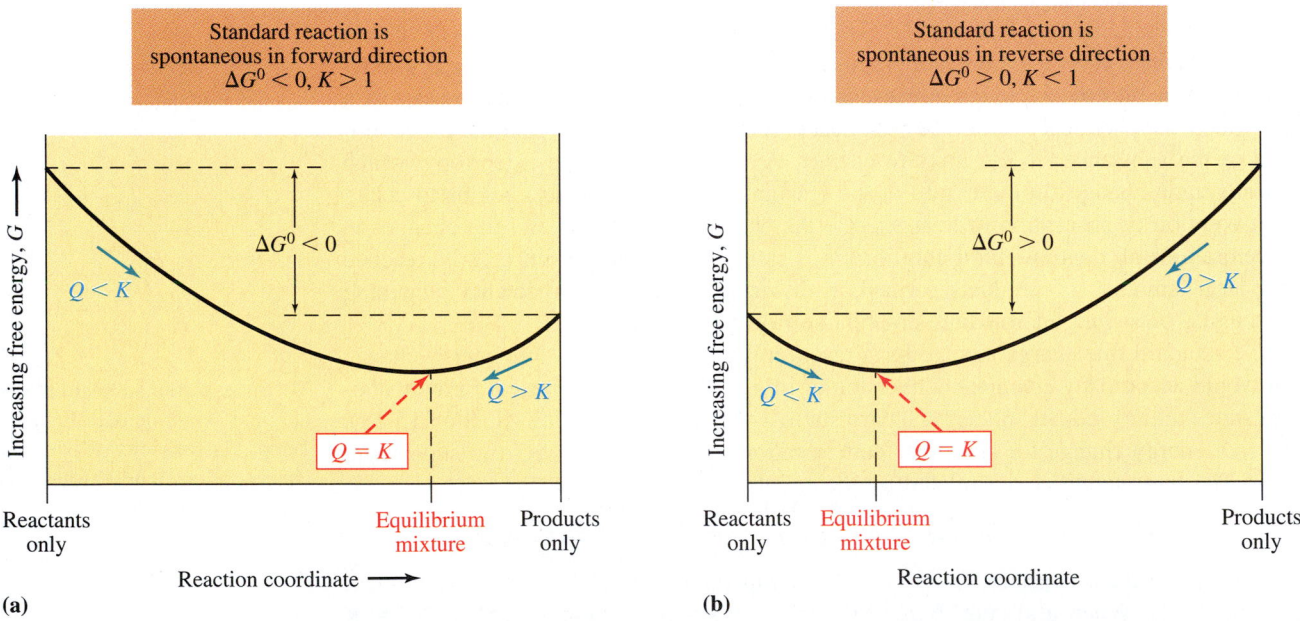

Figure 17-4 Variation in total free energy for a reversible reaction carried out at constant *T*. The *standard* free energy change, ΔG^0, represents the free energy change for the *standard reaction*—the *complete* conversion of reactants into products. In (a) this change is negative, indicating that the standard reaction is spontaneous; the collection of products would be more stable than the collection of reactants. However, the mixture of reactants and products corresponding to the minimum of the curve is even more stable, and represents the equilibrium mixture. Because ΔG^0 is negative, $K > 1$, and the equilibrium mixture contains more products than reactants. At any point on the curve, comparing Q and K indicates the direction in which the reaction must proceed to approach equilibrium, that is, which way is "downhill" in free energy. The plot in (b) is for positive ΔG^0 (the reverse standard reaction is thermodynamically spontaneous). In this case $K < 1$, and the equilibrium mixture contains more reactants than products.

curve represents the total free energy of the products at standard state conditions. The difference between them is ΔG^0; like K, ΔG^0 depends only on temperature and is a constant for any given reaction.

From the preceding equation that relates ΔG^0 and K, we see that when ΔG^0 is negative, K is greater than 1 (ln K *must* be positive). This tells us that products are favored over reactants at equilibrium. This case is illustrated in Figure 17-4a. When ΔG^0 is positive, K is less than 1 (ln K *must* be negative). This tells us that reactants are favored over products at equilibrium (Figure 17-4b). In the *rare* case of a chemical reaction for which $\Delta G^0 = 0$, then $K = 1$ and the numerator and the denominator must be equal in the equilibrium constant expression, (i.e., $[C]^c[D]^d \ldots = [A]^a[B]^b \ldots$). These relationships are summarized as follows.

ΔG^0	K	Product Formation
$\Delta G^0 < 0$	$K > 1$	Products favored over reactants at equilibrium
$\Delta G^0 = 0$	$K = 1$	At equilibrium when $[C]^c[D]^d \ldots = [A]^a[B]^b \ldots$ (very rare)
$\Delta G^0 > 0$	$K < 1$	Reactants favored over products at equilibrium

The direction of approach to equilibrium and the actual free energy change (ΔG) are *not* constants. They vary with the conditions and the initial concentrations. If the initial concentrations correspond to $Q < K$, equilibrium is approached from left to right on the curves in Figure 17-4, and the forward reaction is spontaneous. If $Q > K$, equilibrium is approached from right to left, and the reaction is spontaneous in the reverse direction.

The magnitude of ΔG^0 indicates the *extent* to which a chemical reaction occurs under standard state conditions, i.e., how far the reaction goes toward the formation of products before equilibrium is reached. The more negative the ΔG^0 value, the larger the value of K and the more favorable the formation of products. We think of some reactions as going to "completion." These generally have very negative ΔG^0 values. The more positive the ΔG^0 value, the smaller the value of K and the less favorable the formation of products.

EXAMPLE 17-14 *K versus* ΔG^0

Use the data in Appendix K to evaluate K_P for the following reaction at 25°C.

$$2C_2H_2(g) + 5O_2(g) \rightleftharpoons 4CO_2(g) + 2H_2O(g) \qquad (C_2H_2 \text{ is acetylene})$$

Plan

The temperature is 25°C, so we evaluate ΔG^0 for the reaction from ΔG_f^0 values in Appendix K. The reaction involves only gases so K is K_P. This means that $\Delta G^0 = -RT \ln K_P$ or $\Delta G^0 = -2.303RT \log K_P$. We solve for K_P.

Solution

$$\Delta G_{rxn}^0 = 4\Delta G_{f\,CO_2(g)}^0 + 2\Delta G_{f\,H_2O(g)}^0 - [2\Delta G_{f\,C_2H_2(g)}^0 + 5\Delta G_{f\,O_2(g)}^0]$$

$$= [4(-394.4) + 2(-228.6)] - [2(209.2) + 5(0)]$$

$$= -2.45 \times 10^3 \text{ kJ/mol} \qquad \text{or} \qquad -2.45 \times 10^6 \text{ J/mol}$$

This is a gas phase reaction, so ΔG_{rxn}^0 is related to K_P by

$$\Delta G_{rxn}^0 = -2.303RT \log K_P$$

Units cancel when we express ΔG^0 in joules per mole. We interpret this as meaning ''per mole of reaction''—that is, for the number of moles of each substance shown in the balanced equation.

$$\log K_P = \frac{\Delta G^0_{rxn}}{-2.303RT}$$

$$\log K_P = \frac{-2.45 \times 10^6 \text{ J/mol}}{-2.303(8.314 \text{ J/mol} \cdot \text{K})(298 \text{ K})} = 429$$

$$K_P = 10^{429}$$

The very large value for K_P tells us that the equilibrium lies *very* far to the right.

EXAMPLE 17-15 *K versus* ΔG^0

In Examples 15-16 and 15-17 we evaluated ΔG^0 for the following reaction at 25°C and found it to be +173.1 kJ/mol. Calculate K_P at 25°C for this reaction.

$$N_2(g) + O_2(g) \rightleftharpoons 2NO(g)$$

Plan

In this example we use $\Delta G^0 = -RT \ln K_P$.

Solution

$$\Delta G^0 = -RT \ln K_P$$

$$\ln K_P = \frac{\Delta G^0}{-RT} = \frac{1.731 \times 10^5 \text{ J/mol}}{(-8.314 \text{ J/mol} \cdot \text{K})(298 \text{ K})}$$

$$\ln K_P = -69.87$$

$$K_P = e^{-69.87} = 4.5 \times 10^{-31}$$

On some calculators, we evaluate e^x as follows: Enter the value of x, then press $\boxed{\text{INV}}$ followed by $\boxed{\ln x}$.

This very small number indicates that at equilibrium almost no N_2 and O_2 are converted to NO at 25°C. For all practical purposes, the reaction does not occur at 25°C.

You should now work Exercise 76.

A very important application of the relationships in this section is the use of measured K values to calculate ΔG^0.

EXAMPLE 17-16 *K versus* ΔG^0

The equilibrium constant, K_P, for the following reaction is 5.04×10^{17} at 25°C. Calculate ΔG^0_{298} for the hydrogenation of ethylene to form ethane.

$$C_2H_4(g) + H_2(g) \rightleftharpoons C_2H_6(g)$$

Plan

We use the relationship between ΔG^0 and K_P.

Solution

$$\begin{aligned} \Delta G^0_{298} &= -RT \ln K_P \\ &= -(8.314)(298) \ln (5.04 \times 10^{17}) \\ &= -(2478)(40.761) \\ &= -1.010 \times 10^5 \text{ J/mol} \end{aligned}$$

$$\Delta G^0_{298} = -101 \text{ kJ/mol}$$

You should now work Exercise 79.

17-12 EVALUATION OF EQUILIBRIUM CONSTANTS AT DIFFERENT TEMPERATURES

Chemists have determined equilibrium constants for thousands of reactions. It would be an impossibly huge task to catalog such constants at every temperature of interest for each reaction. Fortunately, there is no need to do this. If we determine the equilibrium constant, K_{T_1}, for a reaction at one temperature, T_1, and also its ΔH^0, we can then *estimate* the equilibrium constant at a second temperature, T_2, using the **van't Hoff equation.**

$$\ln\left(\frac{K_{T_2}}{K_{T_1}}\right) = \frac{\Delta H^0}{R}\left(\frac{1}{T_1} - \frac{1}{T_2}\right) \quad or \quad \log\left(\frac{K_{T_2}}{K_{T_1}}\right) = \frac{\Delta H^0}{2.303R}\left(\frac{1}{T_1} - \frac{1}{T_2}\right)$$

As a memory aid, compare the form of this equation to the Arrhenius equation (Section 16-8) and to the Clausius–Clapeyron equation (Section 13-9).

Thus, if we know ΔH^0 for a reaction and K at a given temperature (say 298 K), we can use the van't Hoff equation to calculate the value of K at any other temperature.

EXAMPLE 17-17 *Evaluation of K_P at Different Temperatures*

We found in Example 17-15 that $K_P = 4.6 \times 10^{-31}$ at 25°C (298 K) for the following reaction. $\Delta H^0 = 180.5$ kJ/mol for this reaction. Evaluate K_P at 2400 K, and then compare K_{2400} with K_{298}.

$$N_2(g) + O_2(g) \rightleftharpoons 2NO(g)$$

2400 K is a typical temperature inside the combustion chambers of automobile engines. Large quantities of N_2 and O_2 are present during gasoline combustion, because the gasoline is mixed with air.

Plan

We are given K_P at one temperature, 25°C, and the value for ΔH^0. We are given the second temperature, 2400 K. These data allow us to evaluate the right side of the van't Hoff equation, which gives us $\ln(K_{T_2}/K_{T_1})$. Because we know K_{T_1}, we can find the value for K_{T_2}.

Solution

Let $T_1 = 298$ K and $T_2 = 2400$ K. Then

$$\ln\left(\frac{K_{T_2}}{K_{T_1}}\right) = \frac{\Delta H^0}{R}\left(\frac{1}{T_1} - \frac{1}{T_2}\right)$$

Let us first evaluate the right side of the equation.

$$\ln\left(\frac{K_{T_2}}{K_{T_1}}\right) = \frac{1.805 \times 10^5 \text{ J/mol}}{8.314 \text{ J/mol}\cdot\text{K}}\left(\frac{1}{298 \text{ K}} - \frac{1}{2400 \text{ K}}\right) = 63.80$$

Now take the inverse logarithm of both sides.

$$\frac{K_{T_2}}{K_{T_1}} = e^{63.80} = 5.1 \times 10^{27}$$

Solving for K_{T_2} and substituting the known value of K_{T_1}, we obtain

$$K_{T_2} = (5.1 \times 10^{27})(K_{T_1}) = (5.1 \times 10^{27})(4.6 \times 10^{-31}) = \boxed{2.3 \times 10^{-3}} \text{ at 2400 K}$$

You should now work Exercise 80.

In Example 17-17 we see that K_{T_2} (K_P at 2400 K) is quite small, which tells us that the equilibrium favors N_2 and O_2 rather than NO. However, K_{T_2} is very much larger than K_{T_1}, which is 4.6×10^{-31}. At 2400 K, significantly more NO is present at equilibrium, relative to N_2 and O_2, than at 298 K. So automobiles emit small amounts of NO into the atmo-

The K_p value could be converted to K_c using the relationship $K_c = K_p (RT)^{-\Delta n}$ (Section 17-9).

sphere, which are sufficient to cause severe air pollution problems. Catalytic converters (Section 16-9) are designed to catalyze the breakdown of NO into N_2 and O_2.

$$2NO(g) \rightleftharpoons N_2(g) + O_2(g)$$

This reaction is spontaneous. Catalysts do not shift the position of equilibrium. They favor neither consumption nor production of NO. They merely allow the system to reach equilibrium more rapidly. The time factor is very important because the NO stays in the exhaust system for only a very short time.

Key Terms

Activity (of a component of an ideal mixture)   A dimensionless quantity whose magnitude is equal to molar concentration in an ideal solution, equal to partial pressure (in atmospheres) in an ideal gas mixture, and defined as 1 for pure solids or liquids.

Chemical equilibrium   A state of dynamic balance in which the rates of forward and reverse reactions are equal; there is no net change in concentrations of reactants or products while a system is at equilibrium.

Dynamic equilibrium   An equilibrium in which processes occur continuously, with no *net* change.

Equilibrium constant, K   A quantity that indicates the extent to which a reversible reaction occurs; its magnitude is equal to the mass action expression at equilibrium. K varies with temperature.

Heterogeneous equilibria   Equilibria involving species in more than one phase.

Homogeneous equilibria   Equilibria involving only species in a single phase, i.e., all gases, all liquids, or all solids.

LeChatelier's Principle   If a stress (change of conditions) is applied to a system at equilibrium, the system shifts in the direction that reduces the stress.

Mass action expression   For a reversible reaction,

$$aA + bB \rightleftharpoons cC + dD$$

the product of the concentrations of the products (species on the right), each raised to the power that corresponds to its coefficient in the balanced chemical equation, divided by the product of the concentrations of the reactants (species on the left), each raised to the power that corresponds to its coefficient in the balanced chemical equation. At equilibrium the mass action expression is equal to K; at other conditions, it is Q.

$$\frac{[C]^c[D]^d}{[A]^a[B]^b} = Q \text{ or, at equilibrium, } K$$

Reaction quotient, Q   The mass action expression under any set of conditions (not necessarily equilibrium); its magnitude relative to K determines the direction in which reaction must occur to establish equilibrium.

Reversible reactions   Reactions that do not go to completion and occur in both the forward and reverse directions.

van't Hoff equation   The relationship between ΔH^0 for a reaction and its equilibrium constants at two different temperatures.

Exercises

Basic Concepts

1. Define and illustrate the following terms: (a) reversible reaction, (b) chemical equilibrium, (c) equilibrium constant, K.
2. What can be said about the magnitude of the equilibrium constant in a reaction whose equilibrium lies far to the right? To the left?
3. What is the relationship between equilibrium and the rates of opposing processes?
4. What does the value of an equilibrium constant tell us about the time required for the reaction to reach equilibrium?

The Equilibrium Constant

5. Write the equilibrium constant expression, K_c, for each of the following reactions.
   (a) $2CO(g) + O_2(g) \rightleftharpoons 2CO_2(g)$
   (b) $2SO_2(g) + O_2(g) \rightleftharpoons 2SO_3(g)$
   (c) $2O_3(g) \rightleftharpoons 3O_2(g)$

6. Write the expression for K_c for each of the following reactions.
   (a) $CO_2(g) + H_2(g) \rightleftharpoons CO(g) + H_2O(g)$
   (b) $2NO_2(g) \rightleftharpoons 2NO(g) + O_2(g)$
   (c) $SrCO_3(s) \rightleftharpoons SrO(s) + CO_2(g)$
   (d) $2HBr(g) \rightleftharpoons H_2(g) + Br_2(\ell)$
   (e) $P_4(g) + 3O_2(g) \rightleftharpoons P_4O_6(s)$
   (f) $2CO(g) + O_2(g) \rightleftharpoons 2CO_2(g)$

7. Write the equilibrium constant expression for the following chemical reaction.

$$H_3PO_4(aq) \rightleftharpoons H^+(aq) + H_2PO_4^-(aq)$$

Describe what happens at the molecular level when H^+ ions from a strong acid such as hydrochloric acid are added to a system in which the H_3PO_4, H^+ ion, and $H_2PO_4^-$ ion have reached equilibrium.

8. Write the equilibrium constant expression for the following chemical reaction.

$$Cl_2(aq) + 2Br^-(aq) \rightleftharpoons 2Cl^-(aq) + Br_2(aq)$$

Describe what happens at the molecular level as solid NaBr, a soluble ionic compound, is added to an aqueous solution in which the four species have reached equilibrium.

9. Explain the significance of (a) a very large value of K, (b) a very small value of K, and (c) a value of K of about 1.0.

Calculation and Uses of K

10. On the basis of the equilibrium constant values, choose the reactions in which the *products* are favored.
(a) $NH_3(aq) + H_2O(\ell) \rightleftharpoons NH_4^+(aq) + OH^-(aq)$
$$K = 1.8 \times 10^{-5}$$
(b) $Au^+(aq) + 2CN^-(aq) \rightleftharpoons [Au(CN)_2]^-(aq)$
$$K = 2 \times 10^{38}$$
(c) $PbC_2O_4(s) \rightleftharpoons Pb^{2+}(aq) + C_2O_4^{2-}(aq)$ $K = 4.8$
(d) $HS^-(aq) + H^+(aq) \rightleftharpoons H_2S(aq)$ $K = 1.0 \times 10^7$

11. On the basis of the equilibrium constant values, choose the reactions in which the *reactants* are favored.
(a) $H_2O(\ell) \rightleftharpoons H^+(aq) + OH^-(aq)$ $K = 1.0 \times 10^{-14}$
(b) $[AlF_6]^{3-}(aq) \rightleftharpoons Al^{3+}(aq) + 6F^-(aq)$
$$K = 2 \times 10^{-24}$$
(c) $Ca_3(PO_4)_2(s) \rightleftharpoons 3Ca^{2+}(aq) + 2PO_4^{3-}(aq)$
$$K = 10^{-25}$$
(d) $2Fe^{3+}(aq) + 3S^{2-}(aq) \rightleftharpoons Fe_2S_3(s)$ $K = 1 \times 10^{88}$

12. The reaction between nitrogen and oxygen to form NO(g) is represented by the chemical equation

$$N_2(g) + O_2(g) \rightleftharpoons 2NO(g)$$

Equilibrium concentrations of the gases at 1500 K are 1.7×10^{-3} mol/L for O_2, 6.4×10^{-3} mol/L for N_2, and 1.1×10^{-5} mol/L for NO. Calculate the value of K_c at 1500 K from these data.

13. At elevated temperatures, BrF_5 establishes the following equilibrium.

$$2BrF_5(g) \rightleftharpoons Br_2(g) + 5F_2(g)$$

The equilibrium concentrations of the gases at 1500 K are 0.0064 mol/L for BrF_5, 0.0018 mol/L for Br_2, and 0.0090 mol/L for F_2. Calculate the value of K_c.

14. At some temperature the reaction

$$PCl_3(g) + Cl_2(g) \rightleftharpoons PCl_5(g)$$

is at equilibrium when the concentrations of PCl_3, Cl_2, and PCl_5 are 10, 9, and 12 mol/L, respectively. Calculate the value of K_c for this reaction at that temperature.

15. Nitrogen reacts with hydrogen to form ammonia.

$$N_2(g) + 3H_2(g) \rightleftharpoons 2NH_3(g)$$

An equilibrium mixture at a given temperature is found to contain 0.31 mol/L N_2, 0.50 mol/L H_2, and 0.14 mol/L NH_3. Calculate the value of K_c at the given temperature.

16. The following equation is given. At 500 K, $K_c = 7.9 \times 10^{11}$.

$$H_2(g) + Br_2(g) \rightleftharpoons 2HBr(g)$$

(a) $\frac{1}{2}H_2(g) + \frac{1}{2}Br_2(g) \rightleftharpoons HBr(g)$ $K_c = ?$
(b) $2HBr(g) \rightleftharpoons H_2(g) + Br_2(g)$ $K_c = ?$
(c) $4HBr(g) \rightleftharpoons 2H_2(g) + 2Br_2(g)$ $K_c = ?$

17. The equilibrium constant for the reaction

$$2SO_2 + O_2 \rightleftharpoons 2SO_3$$

is $K_c = 279$ at a given high temperature. What is the value of the equilibrium constant for each of the following reactions at this temperature?
(a) $2SO_3 \rightleftharpoons 2SO_2 + O_2$
(b) $SO_2 + \frac{1}{2}O_2 \rightleftharpoons SO_3$

18. Pick the more suitable reaction for the preparation of hydrogen chloride. Explain your choice.
(a) $NaCl(s) + H_3PO_4(aq) \rightleftharpoons NaH_2PO_4(s) + HCl(g)$
$$K = 5.23 \times 10^{-9}$$
(b) $NaCl(s) + H_2SO_4(aq) \rightleftharpoons NaHSO_4(s) + HCl(g)$
$$K = 7.23 \times 10^{-5}$$

19. NO and O_2 are mixed in a container of fixed volume kept at 1000 K. Their initial concentrations are 0.0200 mol/L and 0.0300 mol/L, respectively. When the reaction

$$2NO(g) + O_2(g) \rightleftharpoons 2NO_2(g)$$

has come to equilibrium, the concentration of NO_2 is 2.2×10^{-3} mol/L. Calculate (a) the concentration of NO at equilibrium, (b) the concentration of O_2 at equilibrium, and (c) the equilibrium constant K_c for the reaction.

20. A sealed tube initially contains 9.84×10^{-4} mol H_2 and 1.38×10^{-3} mol I_2. It is kept at 350°C until the reaction

$$H_2(g) + I_2(g) \rightleftharpoons 2HI(g)$$

comes to equilibrium. At equilibrium, 4.73×10^{-4} mol I_2 is present. Calculate (a) the numbers of moles of H_2 and HI present at equilibrium; (b) the equilibrium constant of the reaction.

21. The reversible reaction

$$2SO_2(g) + O_2(g) \rightleftharpoons 2SO_3(g)$$

has come to equilibrium in a vessel of specific volume and at a given temperature. Before the reaction, the concentrations of the reactants were 0.060 mol/L of SO_2 and 0.050 mol/L of O_2. No SO_3 was present. After equilibrium was reached, the concentration of SO_3 was 0.040 mol/L. What is the concentration of O_2 at equilibrium? What is the value of K_c?

22. One mole of A and one mole of B are placed in a 0.400-L container. After equilibrium has been established, 0.20 mol of C is present in the container. Calculate the equilibrium constant for the reaction.

$$A(g) + B(g) \rightleftharpoons C(g) + 2D(g)$$

The Reaction Quotient, Q
23. Define the reaction quotient Q. Distinguish between Q and K.

24. Why is it useful to compare Q with K? What is the situation when (a) $Q = K$? (b) $Q < K$? (c) $Q > K$?

25. How does the form of the reaction quotient compare with that of the equilibrium constant? What is the difference between these two expressions?

26. If the reaction quotient is larger than the equilibrium constant, what will happen to the reaction? What will happen if $Q < K$?

27. $K_c = 19.9$ for the reaction

$$Cl_2(g) + F_2(g) \rightleftharpoons 2ClF(g)$$

What will happen in a reaction mixture originally containing $[Cl_2] = 0.4$ mol/L, $[F_2] = 0.2$ mol/L, and $[ClF] = 7.3$ mol/L?

28. The concentration equilibrium constant for the gas-phase reaction

$$H_2CO \rightleftharpoons H_2 + CO$$

has the numerical value 0.50 at a given temperature. A mixture of H_2CO, H_2, and CO is introduced into a flask at this temperature. After a short time, analysis of a small sample of the reaction mixture shows the concentrations to be $[H_2CO] = 0.50\ M$, $[H_2] = 1.50\ M$, and $[CO] = 0.25\ M$. Classify each of the following statements about this reaction mixture as true or false.
(a) The reaction mixture is at equilibrium.
(b) The reaction mixture is not at equilibrium, but no further reaction will occur.
(c) The reaction mixture is not at equilibrium, but will move toward equilibrium by using up more H_2CO.
(d) The forward rate of this reaction is the same as the reverse rate.

29. The value of K_c at 25°C for

$$C(graphite) + CO_2(g) \rightleftharpoons 2CO(g)$$

is 3.7×10^{-23}. Describe what will happen if 2.0 mol of CO and 2.0 mol of CO_2 are mixed in a 1.0-L container with a suitable catalyst to make the reaction "go" at this temperature.

Uses of the Equilibrium Constant, K_c

30. For the reaction described by the equation

$$N_2(g) + C_2H_2(g) \rightleftharpoons 2HCN(g)$$

$K_c = 2.3 \times 10^{-4}$ at 300°C. What is the equilibrium concentration of hydrogen cyanide if the initial concentrations of N_2 and acetylene (C_2H_2) were 2.5 mol/L and 1.0 mol/L, respectively?

31. The equilibrium constant for the reaction

$$Br_2(g) + F_2(g) \rightleftharpoons 2BrF(g)$$

is 54.7. What are the equilibrium concentrations of all these gases if the initial concentrations of bromine and fluorine were both 0.250 mol/L?

32. Antimony pentachloride decomposes in a gas-phase reaction at 448°C.

$$SbCl_5(g) \rightleftharpoons SbCl_3(g) + Cl_2(g)$$

An equilibrium mixture in a 5.00-L vessel is found to contain 3.84 g of $SbCl_5$, 9.14 g of $SbCl_3$, and 2.84 g of Cl_2. Evaluate K_c at 448°C.

33. $K_c = 5.85 \times 10^{-3}$ at 25°C for the reaction

$$N_2O_4(g) \rightleftharpoons 2NO_2(g)$$

Fifteen (15.0) grams of N_2O_4 is confined in a 5.00-L flask at 25°C. Calculate (a) the number of moles of NO_2 present at equilibrium and (b) the percentage of the original N_2O_4 that is dissociated.

34. $K_c = 96.2$ at 400 K for the reaction

$$PCl_3(g) + Cl_2(g) \rightleftharpoons PCl_5(g)$$

What is the concentration of Cl_2 at equilibrium if the initial concentrations were 0.20 mol/L for PCl_3 and 7.0 mol/L for Cl_2?

35. If 12.5 g of $SbCl_5$ are placed in a 5.00-L container at 448°C and allowed to establish equilibrium as in Exercise 32, what will be the equilibrium concentrations of all species? For this reaction, $K_c = 2.51 \times 10^{-2}$.

36. If 7.50 g of $SbCl_5$ and 5.00 g of $SbCl_3$ are placed in a 7.52-L container and allowed to establish equilibrium as in Exercise 32, what will the equilibrium concentrations be? For this reaction, $K_c = 2.51 \times 10^{-2}$.

37. The following equilibrium is established in the presence of water and a particular enzyme (catalyst).

$$\underset{\text{fumaric acid}}{H_2C_4H_2O_4(aq)} + H_2O \rightleftharpoons \underset{\text{malic acid}}{H_2C_4H_4O_5(aq)}$$

One (1.00) liter of solution was prepared from water and 0.300 mol of pure malic acid. At equilibrium, the solution contained 0.067 mol of fumaric acid, in addition to some unchanged malic acid. Find the equilibrium constant for the reaction.

*38. At some temperature, the reaction

$$N_2(g) + 3H_2(g) \rightleftharpoons 2NH_3(g)$$

has an equilibrium constant K_c numerically equal to 1. State whether each of the following is true or false, and explain why.
(a) An equilibrium mixture must have the H_2 concentration three times that of N_2 and the NH_3 concentration twice that of H_2.
(b) An equilibrium mixture must have the H_2 concentration three times that of N_2.
(c) A mixture in which the H_2 concentration is three times that of N_2 *and* the NH_3 concentration is twice that of N_2 could be an equilibrium mixture.
(d) A mixture in which the concentration of each reactant and each product is $1\ M$ is an equilibrium mixture.
(e) Any mixture in which the concentrations of all reactants and products are equal is an equilibrium mixture.
(f) An equilibrium mixture must have equal concentrations of all reactants and products.

*39. Bromine chloride, BrCl, a reddish covalent gas with properties similar to those of Cl_2, may eventually replace Cl_2 as a water disinfectant. One mole of chlorine and one mole of bromine are enclosed in a 5.00-L flask and allowed to reach equilibrium at a certain temperature.

$$Cl_2(g) + Br_2(g) \rightleftharpoons 2BrCl(g) \qquad K_c = 4.7 \times 10^{-2}$$

(a) What percentage of the chlorine has reacted at equilibrium?
(b) What mass of each species is present at equilibrium?
(c) How would a decrease in volume shift the position of equilibrium, if at all?

*40. At a certain temperature, K_c for the following reaction is 0.450. If 0.600 mol of $POCl_3$ is placed in a closed 2.00-L vessel at this temperature, what percentage of it will be dissociated when equilibrium is established?

$$POCl_3(g) \rightleftharpoons POCl(g) + Cl_2(g)$$

Factors That Influence Equilibrium

41. State LeChatelier's Principle. Which factors have an effect on a system at equilibrium? How does the presence of a catalyst affect a system at chemical equilibrium? Explain your answer.

42. What will be the effect of increasing the total pressure on the equilibrium conditions for (a) a reaction that has more moles of gaseous products than gaseous reactants, (b) a reaction that has more moles of gaseous reactants than gaseous products, (c) a reaction that has the same number of moles of gaseous reactants and gaseous products, and (d) a reaction in which all reactants and products are either pure solids, pure liquids, or in aqueous solution?

43. What would be the effect on the equilibrium position of an equilibrium mixture of Br_2, F_2, and BrF_5 if the total pressure of the system were increased?

$$2BrF_5(g) \rightleftharpoons Br_2(g) + 5F_2(g)$$

44. What would be the effect on the equilibrium position of an equilibrium mixture of carbon, oxygen, and carbon monoxide if the total pressure of the system were increased?

$$2C(s) + O_2(g) \rightleftharpoons 2CO(g)$$

45. A weather indicator can be made with a hydrate of cobalt(II) chloride, which changes color as a result of the following reaction.

$$[Co(H_2O)_6]Cl_2(s) \rightleftharpoons [Co(H_2O)_4]Cl_2(s) + 2H_2O(g)$$
$$\text{pink} \qquad\qquad\qquad \text{blue}$$

Does a blue color indicate "moist" or "dry" air? Explain.

46. Consider the reaction

$$CaCO_3(s) \rightleftharpoons CaO(s) + CO_2(g)$$

Will the mass of $CaCO_3$ at equilibrium (i) increase, (ii) decrease, or (iii) remain the same if (a) CO_2 is added to the equilibrium system? (b) the pressure is decreased? (c) solid CaO is removed?

47. The reaction between NO and O_2 is exothermic.

$$2NO(g) + O_2(g) \rightleftharpoons 2NO_2(g)$$

Will the concentration of NO_2 at equilibrium (i) increase, (ii) decrease, or (iii) remain the same if (a) additional O_2 is introduced? (b) additional NO is introduced? (c) the total pressure is increased? (d) the temperature is increased?

48. Predict whether the equilibrium for the photosynthesis reaction described by the equation

$$6CO_2(g) + 6H_2O(\ell) \rightleftharpoons C_6H_{12}O_6(s) + 6O_2(g)$$
$$\Delta H^0 = 2801.69 \text{ kJ/mol}$$

would (i) shift to the right, (ii) shift to the left, or (iii) remain unchanged if (a) $[CO_2]$ were increased; (b) P_{O_2} were increased; (c) one half of the $C_6H_{12}O_6$ were removed; (d) the total pressure were decreased; (e) the temperature were increased; (f) a catalyst were added.

49. What would be the effect of decreasing the temperature on each of the following systems at equilibrium?
(a) $H_2(g) + I_2(g) \rightleftharpoons 2HI(g)$; $\Delta H^0 = -9.45$ kJ/mol
(b) $PCl_5(g) + 92.5$ kJ $\rightleftharpoons PCl_3(g) + Cl_2(g)$;
$$\Delta H^0 = 92.5 \text{ kJ/mol}$$
(c) $2SO_2(g) + O_2(g) \rightleftharpoons 2SO_3(g)$; $\Delta H^0 = -198$ kJ/mol
(d) $2NOCl(g) \rightleftharpoons 2NO(g) + Cl_2(g)$; $\Delta H^0 = 75$ kJ/mol
(e) $C(s) + H_2O(g) \rightleftharpoons CO(g) + H_2(g)$; $\Delta H^0 = 131$ kJ/mol

50. What would be the effect of decreasing the pressure by increasing the volume on each of the following systems at equilibrium?
(a) $2CO(g) + O_2(g) \rightleftharpoons 2CO_2(g)$
(b) $2NO(g) \rightleftharpoons N_2(g) + O_2(g)$
(c) $N_2O_4(g) \rightleftharpoons 2NO_2(g)$
(d) $Ni(s) + 4CO(g) \rightleftharpoons Ni(CO)_4(g)$
(e) $N_2(g) + 3H_2(g) \rightleftharpoons 2NH_3(g)$

51. The value of K_c is 0.020 at 2870°C for the reaction

$$N_2(g) + O_2(g) \rightleftharpoons 2NO(g)$$

There are 0.800 mole of N_2, 0.500 mole of O_2, and 0.400 mole of NO in a 1.00-liter container at 2870°C. Is the system at equilibrium or must the forward or reverse reaction occur to a greater extent to bring the system to equilibrium?

52. Given: $A(g) + B(g) \rightleftharpoons C(g) + D(g)$
(a) At equilibrium a 1.00-liter container was found to contain 1.60 moles of C, 1.60 moles of D, 0.40 mole of A, and 0.40 mole of B. Calculate the equilibrium constant for this reaction.
(b) If 0.20 mole of B and 0.20 mole of C are added to this system, what will the new *equilibrium* concentration of A be?

53. Given: $A(g) + B(g) \rightleftharpoons C(g) + D(g)$
When one mole of A and one mole of B are mixed and allowed to reach equilibrium at room temperature, the mixture is found to contain $\frac{2}{3}$ mole of C.
(a) Calculate the equilibrium constant.
(b) If two moles of A were mixed with two moles of B and allowed to reach equilibrium, how many moles of C would be present at equilibrium?

54. Given: $A(g) \rightleftharpoons B(g) + C(g)$
(a) When the system is at equilibrium at 200°C, the concentrations are found to be: $[A] = 0.30\ M$, $[B] = [C] = 0.20\ M$. Calculate K_c.
(b) If the volume of the container in which the system is at equilibrium is suddenly doubled at 200°C, what will the new equilibrium concentrations be?
(c) Refer back to part (a). If the volume of the container is suddenly halved at 200°C, what will the new equilibrium concentrations be?

*55. The equilibrium constant, K_c, for the dissociation of phosphorus pentachloride is 4.2×10^{-2} at 250°C. How many moles and grams of PCl_5 must be added to a 2.0-liter flask to obtain a Cl_2 concentration of 0.15 M?

$$PCl_5(g) \rightleftharpoons PCl_3(g) + Cl_2(g)$$

56. At 25°C, K_c is 5.84×10^{-3} for the dissociation of dinitrogen tetroxide to nitrogen dioxide.

$$N_2O_4(g) \rightleftharpoons 2NO_2(g)$$

(a) Calculate the equilibrium concentrations of both gases when 3.50 grams of N_2O_4 are placed in a 2.00-liter flask at 25°C.
(b) What will be the new equilibrium concentrations if the volume of the system is suddenly increased to 4.00 liters at 25°C?
(c) What will be the new equilibrium concentrations if the volume is decreased to 1.00 liter at 25°C?

K in Terms of Partial Pressures

57. Write the K_P expression for each reaction in Exercise 5.
58. Under what conditions are K_c and K_P for a reaction numerically equal?
59. 0.0100 mol of NH_4Cl and 0.0100 mol of NH_3 are placed in a closed 2.00-L container and heated to 603 K. At this temperature, all the NH_4Cl vaporizes. When the reaction

$$NH_4Cl(g) \rightleftharpoons NH_3(g) + HCl(g)$$

has come to equilibrium, 5.8×10^{-3} mol of HCl is present. Calculate (a) K_c and (b) K_P for this reaction at 603 K.
60. CO_2 is passed over graphite at 500 K. The emerging gas stream contains 4.0×10^{-3} mol percent CO. The total pressure is 1.00 atm. Assume that equilibrium is attained. Find K_P for the reaction

$$C(graphite) + CO_2(g) \rightleftharpoons 2CO(g)$$

61. At 425°C, the equilibrium partial pressures of H_2, I_2, and HI are 0.06443 atm, 0.06540 atm, and 0.4821 atm, respectively. Calculate K_P for the following reaction at this temperature.

$$2HI(g) \rightleftharpoons H_2(g) + I_2(g)$$

62. The following equilibrium partial pressures were measured at 750°C: $P_{H_2} = 0.387$ atm, $P_{CO_2} = 0.152$ atm, $P_{CO} = 0.180$ atm, and $P_{H_2O} = 0.252$ atm. What is the value of the equilibrium constant, K, for the reaction?

$$H_2 + CO_2 \rightleftharpoons CO + H_2O$$

63. For the reaction

$$H_2(g) + Cl_2(g) \rightleftharpoons 2HCl(g)$$

$K_c = 193$ at 2500 K. What is the value of K_P?
64. For the reaction

$$Br_2(g) \rightleftharpoons 2Br(g)$$

$K_P = 2550$ at 4000 K. What is the value of K_c?
*65. For the reaction

$$H_2(g) + I_2(g) \rightleftharpoons 2HI(g)$$

at 450°C, the equilibrium constant is 65. Calculate the equilibrium pressures of all species in a vessel that initially contained 0.500 atm HI.
*66. Ammonium carbamate decomposes according to

$$NH_4NH_2CO_2(s) \rightleftharpoons 2NH_3(g) + CO_2(g)$$

If the partial pressure of CO_2 is 0.300 atm after a sample of pure ammonium carbamate reaches equilibrium in a closed container, what is the value of K_P?
67. A stream of gas containing H_2 at an initial partial pressure of 0.200 atm is passed through a tube in which CuO is kept at 500 K. The reaction

$$CuO(s) + H_2(g) \rightleftharpoons Cu(s) + H_2O(g)$$

comes to equilibrium. For this reaction, $K_P = 1.6 \times 10^9$. What is the partial pressure of H_2 in the gas leaving the tube? Assume that the total pressure of the stream is unchanged.
68. In the distant future, when hydrogen may be cheaper than coal, steel mills may make iron by the reaction

$$Fe_2O_3(s) + 3H_2(g) \rightleftharpoons 2Fe(s) + 3H_2O(g)$$

For this reaction, $\Delta H = 96$ kJ/mol and $K_c = 8.11$ at 1000 K. (a) What percentage of the H_2 remains unreacted after the reaction has come to equilibrium at 1000 K? (b) Is this percentage greater or less if the temperature is decreased to below 1000 K?

Heterogeneous Equilibrium

69. At -10°C, the solid compound $Cl_2(H_2O)_8$ is in equilibrium with gaseous chlorine, water vapor, and ice. The partial pressures of the two gases in equilibrium with a mixture of $Cl_2(H_2O)_8$ and ice are 0.20 atm for Cl_2 and 0.00262 atm for water vapor. Find the equilibrium constant K_P for each of these reactions.
(a) $Cl_2(H_2O)_8(s) \rightleftharpoons Cl_2(g) + 8H_2O(g)$
(b) $Cl_2(H_2O)_8(s) \rightleftharpoons Cl_2(g) + 8H_2O(s)$
Why are your two answers so different?
70. A flask contains $NH_4Cl(s)$ in equilibrium with its decomposition products.

$$NH_4Cl(s) \rightleftharpoons NH_3(g) + HCl(g)$$

For this reaction, $\Delta H = 176$ kJ/mol. How is the mass of NH_3 in the flask affected by each of the following disturbances? (a) The temperature is decreased. (b) NH_3 is added. (c) HCl is added. (d) NH_4Cl is added, with no appreciable change in the gas volume. (e) A large amount of NH_4Cl is added, decreasing the volume available to the gases.
71. The equilibrium constant for the reaction

$$H_2(g) + Br_2(\ell) \rightleftharpoons 2HBr(g)$$

is $K_P = 4.5 \times 10^{18}$ at 25°C. The vapor pressure of liquid Br_2 at this temperature is 0.28 atm. (a) Find K_P at 25°C for the reaction

$$H_2(g) + Br_2(g) \rightleftharpoons 2HBr(g)$$

(b) How will the equilibrium in part (a) be shifted by an increase in the volume of the container if (1) liquid Br_2 is absent; (2) liquid Br_2 is present? Explain why the effect is different in these two cases.

72. At 700°C, K_P is 1.50 for the reaction

$$C(s) + CO_2(g) \rightleftharpoons 2CO(g)$$

Suppose the total gas pressure at equilibrium is 1.00 atm. What are the partial pressures of CO and CO_2?

Relationships Among K, ΔG^0, ΔH^0, and T

73. What kind of equilibrium constant can be calculated from a ΔG^0 value for a reaction involving only gases?

74. What must be true of the value of ΔG^0 for a reaction if (a) $K \gg 1$; (b) $K = 1$; (c) $K \ll 1$?

75. The equilibrium constant K_c of the reaction

$$H_2(g) + Br_2(g) \rightleftharpoons 2HBr(g)$$

is 1.6×10^5 at 1297 K and 3.5×10^4 at 1495 K. (a) Is ΔH^0 for this reaction positive or negative? (b) Find K_c for the reaction

$$\tfrac{1}{2}H_2(g) + \tfrac{1}{2}Br_2(g) \rightleftharpoons HBr(g)$$

at 1297 K. (c) Pure HBr is placed in a container of constant volume and heated to 1297 K. What percentage of the HBr is decomposed to H_2 and Br_2 at equilibrium?

76. The air pollutant sulfur dioxide can be partially removed from stack gases in industrial processes and converted to sulfur trioxide, the acid anhydride of commercially important sulfuric acid. Write the equation for the reaction, using the smallest whole-number coefficients. Calculate the value of the equilibrium constant for this reaction at 25°C, from values of ΔG_f^0 in Appendix K.

77. The value of ΔH^0 for the reaction in Exercise 76 is -197.6 kJ/mol. (a) Predict qualitatively (i.e., without calculation) whether the value of K_P for this reaction at 500°C would be greater than, the same as, or less than the value at room temperature (25°C). (b) Now calculate the value of K_P at 500°C.

78. Given that K_P is 4.6×10^{-14} at 25°C for the reaction

$$2Cl_2(g) + 2H_2O(g) \rightleftharpoons 4HCl(g) + O_2(g)$$
$$\Delta H^0 = +115 \text{ kJ/mol}$$

Calculate K_P and K_c for the reaction at 400°C and at 800°C.

79. A mixture of 3.00 mol of Cl_2 and 3.00 mol of CO is enclosed in a 5.00-L flask at 600°C. At equilibrium, 3.3% of the Cl_2 has been consumed.

$$CO(g) + Cl_2(g) \rightleftharpoons COCl_2(g)$$

(a) Calculate K_c for the reaction at 600°C.
(b) Calculate ΔG^0 for the reaction at this temperature.

80. (a) Use the tabulated thermodynamic values of ΔH_f^0 and S^0 to calculate the value of K_P at 25°C for the gas-phase reaction

$$CO + H_2O \rightleftharpoons CO_2 + H_2$$

(b) Calculate the value of K_P for this reaction at 200°C, by the same method as in part (a).
(c) Repeat the calculation of part (a), using tabulated values of ΔG_f^0.

81. At sufficiently high temperatures, chlorine gas dissociates, according to

$$Cl_2(g) \rightleftharpoons 2Cl(g)$$

At 800°C, K_P for this reaction is 5.63×10^{-7}.
(a) A sample originally contained Cl_2 at 1 atm and 800°C. Calculate the percentage dissociation of Cl_2 when this reaction has reached equilibrium.
(b) At what temperature would Cl_2 (originally at 1 atm pressure) be 1% dissociated into Cl atoms?

*82. The following is an example of an alkylation reaction that is important in the production of isooctane (2,2,4-trimethylpentane) from two components of crude oil, isobutane and isobutene. Isooctane is an antiknock additive for gasoline.

The thermodynamic equilibrium constant, K, for this reaction at 25°C is 4.3×10^6, and ΔH^0 is -78.58 kJ/mol.
(a) Calculate ΔG^0 at 25°C.
(b) Calculate K at 800°C.
(c) Calculate ΔG^0 at 800°C.
(d) How does the spontaneity of the forward reaction at 800°C compare with that at 25°C?
(e) Why do you think the reaction mixture is heated in the industrial preparation of isooctane?
(f) What is the purpose of the catalyst? Does it affect the forward reaction more than the reverse reaction?

83. $K_c = 19.9$ for the reaction

$$Cl_2(g) + F_2(g) \rightleftharpoons 2ClF(g)$$

What will happen in a reaction mixture originally containing $[Cl_2] = 0.200$ mol/L, $[F_2] = 0.100$ mol/L, and $[ClF] = 0.365$ mol/L?

Mixed Exercises

84. The value of $K_P = 1.11 \times 10^{21}$ at 25°C for

$$2CO(g) \rightleftharpoons C(graphite) + CO_2(g)$$

What is the value of K_c? Describe what will happen if 3 mol of CO and 2 mol of CO_2 are mixed in a 1-L container with a suitable catalyst to make the reaction "go" at this temperature.

85. $K_c = 0.21$ at 350°C for the reaction

$$2NO(g) + Br_2(g) \rightleftharpoons 2BrNO(g)$$

A 20.0-mL tube contains 1.0×10^{-4} mol NO, 2.0×10^{-4} mol Br_2, and 2.0×10^{-4} mol BrNO at 350°C. (a) Is the mixture in equilibrium? If it is not, which way can the reaction go? (Do not confuse number of moles and concentration.) (b) At equilibrium, the tube contains 2.8×10^{-4} mol NO and 1.6×10^{-5} mol BrNO. How many moles of Br_2 does it contain? (c) Do you need to know the volume of the tube to do this problem?

86. Given the reaction

$$H_2(g) + Br_2(g) \rightleftharpoons 2HBr(g)$$

(a) At 500 K, $K_c = 7.9 \times 10^{11}$. What is K_P?
(b) $\frac{1}{2}H_2(g) + \frac{1}{2}Br_2(g) \rightleftharpoons HBr(g)$ $K_P = ?$
(c) $2HBr(g) \rightleftharpoons H_2(g) + Br_2(g)$ $K_P = ?$
(d) $4HBr(g) \rightleftharpoons 2H_2(g) + 2Br_2(g)$ $K_P = ?$

87. What must be the pressure of hydrogen so that the reaction

$$WCl_6(g) + 3H_2(g) \rightleftharpoons W(s) + 6HCl(g)$$

will occur when $P_{WCl_6} = 0.012$ atm and $P_{HCl} = 0.10$ atm? $K_P = 1.37 \times 10^{21}$ at 900 K.

88. A mixture of CO, H_2, CH_4, and H_2O is kept at 1133 K until the reaction

$$CO(g) + 3H_2(g) \rightleftharpoons CH_4(g) + H_2O(g)$$

has come to equilibrium. The volume of the container is 0.100 L. The equilibrium mixture contains 1.21×10^{-4} mol CO, 2.47×10^{-4} mol H_2, 1.21×10^{-4} mol CH_4, and 5.63×10^{-8} mol H_2O. Calculate K_P for this reaction at 1133 K.

BUILDING YOUR KNOWLEDGE

89. Hemoglobin, Hb, has four Fe atoms per molecule that, on the average, pick up roughly 3 molecules of O_2.

$$Hb(aq) + 3O_2(g) \rightleftharpoons Hb(O_2)_3(aq)$$

Discuss mountain or space sickness in terms of this equilibrium.

90. At room temperature, the equilibrium constant for the reaction

$$2SO_2 + O_2 \rightleftharpoons 2SO_3$$

is 6.98×10^{24}. Calculate ΔG_{rxn}^0 and $\Delta G_f^0(SO_3)$, given the additional information that $\Delta G_f^0(SO_2) = -300.194$ kJ/mol. Check your answer by looking up $\Delta G_f^0(SO_3)$ in Appendix K.

91. At 25°C, 550.0 g of deuterium oxide, D_2O (20.0 g/mole; density 1.10 g/mL), and 498.5 g of H_2O (18.0 g/mole; density 0.997 g/mL) are mixed. The volumes are additive. 47.0% of the H_2O reacts to form HDO. Calculate K_c at 25°C for the reaction

$$H_2O + D_2O \rightleftharpoons 2HDO$$

92. At its normal boiling point of 100°C, the heat of vaporization of water is 40.66 kJ/mol. What is the equilibrium vapor pressure of water at 50°C? (You may wish to review Example 15-17.)

93. Use the data in the preceding exercise to calculate the temperature at which the vapor pressure of water is 2.00 atm.

Ionic Equilibria I: Acids and Bases

18

The red anthocyanin pigment in the common geranium is a naturally occurring acid–base indicator.

OBJECTIVES

As you study this chapter, you should learn

- *To recognize strong electrolytes and calculate concentrations of their ions*

- *To understand the autoionization of water*

- *To understand the pH and pOH scales and how they are used*

- *About ionization constants for weak monoprotic acids and bases*

- *To use ionization constants*

- *What acid–base indicators are and how they function*

- *About the common ion effect*

- *About buffer solutions, how they are prepared, and how they function*

- *How to prepare buffer solutions of a given pH*

- *How polyprotic acids ionize in steps and how to calculate concentrations of all species in these solutions*

Aqueous solutions are very important. Nearly three fourths of the earth's surface is covered with water. Enormous numbers of chemical reactions occur in the oceans and smaller bodies of water. Body fluids of plants and animals are mostly water. Life processes (chemical reactions) of all plants and animals occur in aqueous solutions or in contact with water. All of us developed in sacs filled with aqueous solutions, which protected and nurtured us until we had developed to the point that we could live in the atmosphere.

In previous discussions we have seen that water-soluble compounds may be classified as either electrolytes or nonelectrolytes. **Electrolytes** are compounds that ionize (or dissociate into their constituent ions) to produce aqueous solutions that conduct an electric current. **Nonelectrolytes** exist as molecules in aqueous solution, and such solutions do not conduct an electric current.

18-1 A BRIEF REVIEW OF STRONG ELECTROLYTES

Strong electrolytes are ionized or dissociated completely, or very nearly completely, in dilute aqueous solutions. Strong electrolytes include strong acids, strong soluble bases, and most soluble salts. You should review the discussions of these substances in Sections 4-2 and 10-6. The common strong acids and strong soluble bases are listed again in Table 18-1. See Section 4-2, part 5, for the solubility rules for ionic compounds.

Concentrations of ions in aqueous solutions of strong electrolytes can be calculated directly from the molarity of the strong electrolyte, as the next two examples illustrate.

Table 18-1 *Common Strong Acids and Strong Soluble Bases*	
Strong Acids	
HCl	HNO_3
HBr	$HClO_4$
HI	$HClO_3$
	H_2SO_4
Strong Soluble Bases	
LiOH	
NaOH	
KOH	
RbOH	$Ca(OH)_2$
CsOH	$Sr(OH)_2$
	$Ba(OH)_2$

EXAMPLE 18-1   *Calculation of Concentrations of Ions*

Calculate the molar concentrations of Ba^{2+} and OH^- ions in 0.030 M barium hydroxide.

Plan

Write the equation for the dissociation of $Ba(OH)_2$ and construct the reaction summary. $Ba(OH)_2$ is a strong soluble base that is completely dissociated.

Solution

From the equation for the dissociation of barium hydroxide, we see that *one* mole of $Ba(OH)_2$ produces *one* mole of Ba^{2+} ions and *two* moles of OH^- ions.

(strong base)	$Ba(OH)_2(s) \longrightarrow$	$Ba^{2+}(aq) +$	$2OH^-(aq)$
initial	0.030 M		
change due to rxn	$-0.030\ M$	$+0.030\ M$	$+2(0.030)\ M$
final	0 M	0.030 M	0.060 M

$$[Ba^{2+}] = 0.030\ M \quad \text{and} \quad [OH^-] = 0.060\ M$$

You should now work Exercise 2.

Recall that we use a single arrow to indicate that a reaction goes to completion, or nearly to completion, in the indicated direction.

EXAMPLE 18-2   *Calculation of Concentrations of Ions*

Calculate the concentrations of Mg^{2+} and Br^- ions in a solution that contains 0.92 gram of $MgBr_2$ in 500 mL of solution.

Plan

Calculate the concentration of $MgBr_2$ in moles per liter of solution. Then write the equation for the dissociation of $MgBr_2$ and construct the reaction summary. The solubility rules tell us that $MgBr_2$ is a soluble ionic salt, so it is completely dissociated.

Solution

$$\frac{?\ \text{mol MgBr}_2}{\text{L soln}} = \frac{0.92\ \text{g MgBr}_2}{0.500\ \text{L}} \times \frac{1\ \text{mol MgBr}_2}{184\ \text{g MgBr}_2} = 0.010\ \text{mol MgBr}_2/\text{L}$$

The solution is 0.010 M in $MgBr_2$. The equation for the dissociation of $MgBr_2$ shows that *one* mole of $MgBr_2$ produces *one* mole of Mg^{2+} and *two* moles of Br^-.

(soluble salt)	$MgBr_2(s) \longrightarrow$	$Mg^{2+}(aq) +$	$2Br^-(aq)$
initial	0.010 M		
change due to rxn	$-0.010\ M$	$+0.010\ M$	$+2(0.010)\ M$
final	0 M	0.010 M	0.020 M

$$[Mg^{2+}] = 0.010\ M \quad \text{and} \quad [Br^-] = 0.020\ M$$

You should now work Exercises 4 and 6.

This does not imply the existence of "molecules" of $MgBr_2$ in solution.

18-2 THE AUTOIONIZATION OF WATER

Careful experiments on its electrical conductivity have shown that pure water ionizes to a very slight extent.

$$H_2O(\ell) + H_2O(\ell) \rightleftharpoons H_3O^+(aq) + OH^-(aq)$$

Because the H_2O is pure, its activity is 1, so we do not include its concentration in the equilibrium constant expression. This equilibrium constant is known as the **ion product for water** and is usually represented as K_w.

$$K_w = [H_3O^+][OH^-]$$

The formation of an H_3O^+ ion by the ionization of water is always accompanied by the formation of an OH^- ion. Thus, in *pure* water the concentration of H_3O^+ is *always* equal to the concentration of OH^-. Careful measurements show that, in pure water at 25°C,

$$[H_3O^+] = [OH^-] = 1.0 \times 10^{-7} \text{ mol/L}$$

Substituting these concentrations into the K_w expression gives

$$K_w = [H_3O^+][OH^-] = (1.0 \times 10^{-7})(1.0 \times 10^{-7})$$
$$= 1.0 \times 10^{-14} \qquad (\text{at 25°C})$$

Although the expression $K_w = [H_3O^+][OH^-] = 1.0 \times 10^{-14}$ was obtained for pure water, *it is also valid for dilute aqueous solutions at 25°C*. This is one of the most useful relationships chemists have discovered. It gives a simple relationship between H_3O^+ and OH^- concentrations in *all* dilute aqueous solutions.

The *value* of K_w is different at different temperatures (Table 18-2), but the *relationship* $K_w = [H_3O^+][OH^-]$ is still valid.

> In this text, we shall assume a temperature of 25°C for all calculations involving aqueous solutions unless we specify another temperature.

Solutions in which the concentration of solute is less than about 1 mol/L are usually called dilute solutions.

EXAMPLE 18-3 *Calculation of Concentration of Ions*

Calculate the concentrations of H_3O^+ and OH^- ions in a 0.050 *M* HNO_3 solution.

Plan

Write the equation for the ionization of HNO_3, a strong acid, and construct the reaction summary, which gives the concentration of H_3O^+ ions directly. Then use the relationship $K_w = [H_3O^+][OH^-] = 1.0 \times 10^{-14}$ to find the concentration of OH^- ions.

Solution

The reaction summary for the ionization of HNO_3, a strong acid, is

(*strong acid*)	HNO_3	$+ H_2O \longrightarrow$	H_3O^+	$+ NO_3^-$
initial	0.050 *M*		~0 *M*	0 *M*
change due to rxn	−0.050 *M*		+0.050 *M*	+0.050 *M*
at equil	0 *M*		0.050 *M*	0.050 *M*

$$[H_3O^+] = [NO_3^-] = 0.050 \ M$$

Table 18-2 K_w *at Some Temperatures*

Temperature (°C)	K_w
0	1.13×10^{-15}
10	2.92×10^{-15}
25	1.00×10^{-14}
37*	2.38×10^{-14}
45	4.02×10^{-14}
60	9.61×10^{-14}

*Normal human body temperature.

The [OH$^-$] is determined from the equation for the autoionization of water.

	$2H_2O \rightleftharpoons$	$H_3O^+(aq)$	$+ OH^-$
initial		0.050 M	
change due to rxn	$-2x\, M$	$+x\, M$	$+x\, M$
at equil		$(0.050 + x)\, M$	$x\, M$

$$K_w = [H_3O^+][OH^-]$$

$$1.0 \times 10^{-14} = (0.050 + x)(x)$$

Because the product $(0.050 + x)(x)$ is such a small number, we know that x must be very small. Thus, it will not matter (much) whether we add x to 0.050; that is, we assume that $(0.050 + x) \approx$ 0.050. We substitute this approximation into the equation and solve.

$$1.0 \times 10^{-14} = (0.050)(x) \qquad \text{or} \qquad x = \frac{1.0 \times 10^{-14}}{0.050} = \boxed{2.0 \times 10^{-13}\, M = [OH^-]}$$

Recall that

$$[OH^-]_{\text{from } H_2O} = [H_3O^+]_{\text{from } H_2O}$$

in *all* aqueous solutions. So we know that $[H_3O^+]_{\text{from } H_2O}$ must be $2.0 \times 10^{-13}\, M$ also.

We see that the assumption that x is much smaller than 0.050 was a good one.

You should now work Exercise 14.

In solving Example 18-3 we assumed that *all* of the H_3O^+ (0.050 M) came from the ionization of HNO_3 and neglected the H_3O^+ formed by the ionization of H_2O. The ionization of H_2O produces only $2.0 \times 10^{-13}\, M\, H_3O^+$ and $2.0 \times 10^{-13}\, M\, OH^-$ *in this solution.* Thus, we were justified in assuming that the $[H_3O^+]$ is derived solely from the strong acid. A more concise way to carry out the calculation to find the [OH$^-$] concentration is to write directly.

$$K_w = [H_3O^+][OH^-] = 1.0 \times 10^{-14} \qquad \text{or} \qquad [OH^-] = \frac{1.0 \times 10^{-14}}{[H_3O^+]}$$

Then we substitute to obtain

$$[OH^-] = \frac{1.0 \times 10^{-14}}{0.050} = 2.0 \times 10^{-13}\, M$$

From now on, we shall use this more direct approach for such calculations.

When nitric acid is added to water, large numbers of H_3O^+ ions are produced. The large increase in $[H_3O^+]$ shifts the water equilibrium far to the left (LeChatelier's Principle), and the [OH$^-$] decreases.

$$H_2O(\ell) + H_2O(\ell) \rightleftharpoons H_3O^+(aq) + OH^-(aq)$$

In acidic solutions the H_3O^+ concentration is always greater than the OH$^-$ concentration. We should not conclude that acidic solutions contain no OH$^-$ ions. Rather, the [OH$^-$] is always less than $1.0 \times 10^{-7}\, M$ in such solutions. The reverse is true for basic solutions, in which the [OH$^-$] is always greater than $1.0 \times 10^{-7}\, M$. By definition, "neutral" aqueous solutions at 25°C are solutions in which $[H_3O^+] = [OH^-] = 1.0 \times 10^{-7}\, M$.

Solution	General Condition	At 25°C	
acidic	$[H_3O^+] > [OH^-]$	$[H_3O^+] > 1.0 \times 10^{-7}$	$[OH^-] < 1.0 \times 10^{-7}$
neutral	$[H_3O^+] = [OH^-]$	$[H_3O^+] = 1.0 \times 10^{-7}$	$[OH^-] = 1.0 \times 10^{-7}$
basic	$[H_3O^+] < [OH^-]$	$[H_3O^+] < 1.0 \times 10^{-7}$	$[OH^-] > 1.0 \times 10^{-7}$

18-3 THE pH AND pOH SCALES

The pH scale provides a convenient way to express the acidity and basicity of dilute aqueous solutions. The **pH** of a solution is defined as

$$pH = \log \frac{1}{[H_3O^+]} \quad \text{or} \quad pH = -\log [H_3O^+]$$

When dealing with pH, we always use the base-10 (common) logarithm, *not* the base-*e* (natural) logarithm. This is because pH is *defined* using base-10 logarithms.

Note that we use pH rather than pH_3O. At the time the pH concept was developed, H_3O^+ was represented as H^+. Various "p" terms are used. In general a lowercase "**p**" before a symbol means "negative logarithm of the symbol." Thus, pH is the negative logarithm of H_3O^+ concentration, **pOH** is the negative logarithm of OH^- concentration, and **p***K* refers to the negative logarithm of an equilibrium constant.

$$pH = -\log [H_3O^+] \quad \text{or} \quad [H_3O^+] = 10^{-pH}$$
$$pOH = -\log [OH^-] \quad \text{or} \quad [OH^-] = 10^{-pOH}$$

EXAMPLE 18-4 *Calculation of pH*

Calculate the pH of a solution in which the H_3O^+ concentration is 0.050 mol/L.

Plan

We are given the value for $[H_3O^+]$, and so we take the negative logarithm of this value.

Solution

$$[H_3O^+] = 0.050\ M = 5.0 \times 10^{-2}\ M$$
$$pH = -\log [H_3O^+] = -\log [5.0 \times 10^{-2}] = -(-1.30) = \boxed{1.30}$$

This answer contains only *two* significant figures. The "1" in 1.30 is *not* a significant figure; it comes from the power of ten.

You should now work Exercise 20.

EXAMPLE 18-5 *Calculation of Concentration from pH*

The pH of a solution is 3.301. What is the concentration of H_3O^+ in this solution?

Plan

By definition, $pH = -\log [H_3O^+]$. We are given the pH, so we solve for $[H_3O^+]$.

Solution

From the definition of pH, we write

$$-\log [H_3O^+] = 3.301$$

Multiplying through by -1 gives

$$\log [H_3O^+] = -3.301$$

Taking the inverse logarithm (antilog) of both sides of the equation gives

$$[H_3O^+] = 10^{-3.301} \quad \text{so} \quad \boxed{[H_3O^+] = 5.00 \times 10^{-4}\ M}$$

You should now work Exercises 22 and 24.

The pH of some common substances is shown by a universal indicator. Refer to Figure 18-2 to interpret the indicator colors.

pH Range for a Few Common Substances	
Substance	**pH Range**
Gastric contents (human)	1.6–3.0
Soft drinks	2.0–4.0
Lemons	2.2–2.4
Vinegar	2.4–3.4
Tomatoes	4.0–4.4
Beer	4.0–5.0
Urine (human)	4.8–8.4
Milk (cow's)	6.3–6.6
Saliva (human)	6.5–7.5
Blood plasma (human)	7.3–7.5
Egg white	7.6–8.0
Milk of magnesia	10.5
Household ammonia	11–12

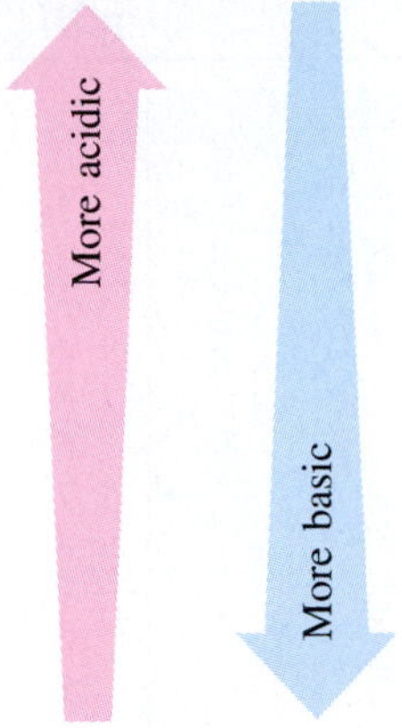

More acidic

More basic

A convenient relationship between pH and pOH in *all dilute solutions at 25°C* can be derived easily. We start with the K_w expression.

$$[H_3O^+][OH]^-] = 1.0 \times 10^{-14}$$

Taking the logarithm of both sides of this equation gives

$$\log [H_3O^+] + \log [OH^-] = \log (1.0 \times 10^{-14})$$

Multiplying both sides of this equation by -1 gives

$$(-\log [H_3O^+]) + (-\log [OH^-]) = -\log (1.0 \times 10^{-14})$$

or
$$pH + pOH = 14.00$$

We can now relate $[H_3O^+]$ and $[OH^-]$ as well as pH and pOH.

At any temperature, pH + pOH = pK_w.

$$[H_3O^+][OH^-] = 1.0 \times 10^{-14} \quad \text{and} \quad pH + pOH = 14.00 \quad \text{(at 25°C)}$$

Remember these relationships!

From this relationship, we see that pH and pOH can *both* be positive only if *both* are less than 14. If either pH or pOH is greater than 14, the other is obviously negative.

Please study the following summary carefully. It will be helpful.

Solution	General Condition	At 25°C
acidic	$[H_3O^+] > [OH^-]$ pH < pOH	$[H_3O^+] > 1.0 \times 10^{-7} M > [OH^-]$ pH < 7.00 < pOH
neutral	$[H_3O^+] = [OH^-]$ pH = pOH	$[H_3O^+] = 1.0 \times 10^{-7} M = [OH^-]$ pH = 7.00 = pOH
basic	$[H_3O^+] < [OH^-]$ pH > pOH	$[H_3O^+] < 1.0 \times 10^{-7} M < [OH^-]$ pH > 7.00 > pOH

To develop familiarity with the pH and pOH scales, consider a series of solutions in which $[H_3O^+]$ varies from $10 M$ to $1.0 \times 10^{-15} M$. Obviously, $[OH^-]$ will vary from $1.0 \times 10^{-15} M$ to $10 M$ in these solutions. Table 18-3 summarizes these scales.

Table 18-3 *Relationships Among $[H_3O^+]$, pH, pOH, and $[OH^-]$*

$[H_3O^+]$	pH		pOH	$[OH^-]$
10^{-15}	15		-1	10^1
10^{-14}	14		0	1
10^{-13}	13		1	10^{-1}
10^{-12}	12		2	10^{-2}
10^{-11}	11		3	10^{-3}
10^{-10}	10		4	10^{-4}
10^{-9}	9		5	10^{-5}
10^{-8}	8		6	10^{-6}
10^{-7}	7		7	10^{-7}
10^{-6}	6		8	10^{-8}
10^{-5}	5		9	10^{-9}
10^{-4}	4		10	10^{-10}
10^{-3}	3		11	10^{-11}
10^{-2}	2		12	10^{-12}
10^{-1}	1		13	10^{-13}
1	0		14	10^{-14}
10^1	-1		15	10^{-15}

H_3O^+ concentration ... OH^- concentration

Increasing basicity — Neutral — Increasing acidity

EXAMPLE 18-6   *Calculations Involving pH and pOH*

Calculate $[H_3O^+]$, pH, $[OH^-]$, and pOH for 0.015 M HNO_3 solution.

Plan

We write the equation for the ionization of the strong acid HNO_3, which gives us $[H_3O^+]$. Then we calculate pH. We use the relationships pH + pOH = 14.00 and $[H_3O^+][OH^-] = 1.0 \times 10^{-14}$ to find pOH and $[OH^-]$.

Solution

All ions are hydrated in aqueous solution. We often omit the designations (ℓ), (g), (s), (aq), and so on.

$$HNO_3 + H_2O \longrightarrow H_3O^+ + NO_3^-$$

Because nitric acid is a strong acid (it ionizes completely), we know that

$$[H_3O^+] = \boxed{0.015\ M}$$

$$pH = -\log[H_3O^+] = -\log(0.015) = -(-1.82) = \boxed{1.82}$$

We also know that pH + pOH = 14.00. Therefore,

$$pOH = 14.00 - pH = 14.00 - 1.82 = \boxed{12.18}$$

Because $[H_3O^+][OH^-] = 1.0 \times 10^{-14}$, $[OH^-]$ is easily calculated.

$$[OH^-] = \frac{1.0 \times 10^{-14}}{[H_3O^+]} = \frac{1.0 \times 10^{-14}}{0.015} = \boxed{6.7 \times 10^{-13}\ M}$$

You should now work Exercise 26.

An indicator is a compound that changes color as pH changes.

The pH of a solution can be determined using a pH meter (Figure 18-1) or by the indicator method. Acid–base **indicators** are intensely colored complex organic compounds that have different colors in solutions of different pH (Section 18-5). Many are weak acids or weak bases that are useful over rather narrow ranges of pH values. *Universal indicators* are mixtures of several indicators; they show several color changes over a wide range of pH values.

Figure 18-1 A pH meter gives the pH of the solution directly. When the electrode is dipped into a solution, the meter displays the pH. The pH of this solution is 7.01 at 25°C. The pH meter is based on the glass electrode. This sensing device generates a voltage that is proportional to the pH of the solution in which the electrode is placed. The instrument has an electrical circuit to amplify the voltage from the electrode and a meter that relates the voltage to the pH of the solution. Before being used, a pH meter must be calibrated with a series of solutions of known pH.

Figure 18-2 Solutions containing a universal indicator. A universal indicator shows a wide range of colors as pH varies. The pH values are given by the black numbers. These solutions range from quite acidic (upper left) to quite basic (lower right).

In the indicator method we prepare a series of solutions of known pH (standard solutions). We add a universal indicator to each; solutions with different pH have different colors (Figure 18-2). We then add the same universal indicator to the unknown solution and compare its color to those of the standard solutions. Solutions with the same pH have the same color.

Universal indicator papers can also be used to determine pH. A drop of solution is placed on a piece of paper or a piece of the paper is dipped into a solution. The color of the paper is then compared with a color chart on the container to establish the pH of the solution.

18-4 IONIZATION CONSTANTS FOR WEAK MONOPROTIC ACIDS AND BASES

We have discussed strong acids and strong soluble bases. There are relatively few of these. Weak acids are much more numerous than strong acids. For this reason you were asked to learn the list of common strong acids (Table 18-1). You may assume that nearly all other

CHEMISTRY IN USE

Our Daily Lives

Does Boiling Water Make It Acidic?

If we place a pH meter electrode into a beaker of pure water at 25°C, we find that the water has a pH of 7.00. Then, if we heat the water, the meter shows a decrease in pH. When the water begins to boil at 100°C, the pH meter reads 6.12, if the meter and electrode work as predicted. We find ourselves face to face with a paradox. On the one hand, we know that the water sample is neutral because pure water is neither acidic nor basic. On the other hand, the pH meter reading seems to indicate that the water sample has become acidic because the boiling water has a pH that is less than 7.00. We sometimes hear that all aqueous solutions with a pH of less than 7.00 are acidic (Table 18-3). How do we resolve this inconsistency?

First, let's calculate the concentration of H_3O^+ in boiling water. From the definition of pH, we write

$$pH = -\log [H_3O^+] = 6.12$$

Multiplying through by -1 gives

$$\log [H_3O^+] = -6.12$$

Taking the inverse logarithm (antilog) of both sides of the equation gives

$$[H_3O^+] = 10^{-6.12}$$

Therefore,

$$[H_3O^+] = 7.6 \times 10^{-7}$$

By boiling pure water, we have increased its acidity ($[H_3O^+]$) by a factor of 7.6 (at 25°C $[H_3O^+] = 1.00 \times 10^{-7}$).

Next, let's calculate the concentration of hydroxide ion, $[OH^-]$, in the boiling water. Consider the equilibrium for the ionization of pure water.

$$H_2O(\ell) + H_2O(\ell) \rightleftharpoons H_3O^+ + OH^-$$

From this equilibrium, we see that H_3O^+ and OH^- are always produced in equal amounts, so their concentrations must always be the same in pure water. The concentration of H_3O^+ in boiling water is 7.6×10^{-7}, so the concentration of OH^- must also be 7.6×10^{-7}. By definition, neutral solutions are those in which H_3O^+ equals OH^-; the boiling water is indeed neutral as we first suspected.

The beaker of pure water is neutral because it has equal concentrations of H_3O^+ and OH^-, yet it has a pH of 6.12. Table 18-2 shows that K_w increases with temperature. K_w increases with an increase in temperature because the ionization of water is endothermic (absorbs heat), and an increase in temperature favors endothermic reactions (Section 17-6). An increase in temperature favors the ionization of water, so higher concentrations of both H_3O^+ and OH^- are produced. The higher the concentration of H_3O^+, the lower the value of the pH.

We now realize that not all neutral aqueous solutions have a pH of 7.00. At 25°C neutral aqueous solutions do have a pH of 7.00, but at higher temperatures neutral aqueous solutions have pH values less than 7.00.

Just as an increase in temperature favors endothermic reactions, a decrease in temperature favors exothermic reactions. The lower the concentration of H_3O^+, the higher the value of the pH. As a result, neutral aqueous solutions at temperatures less than 25°C have pH values that are greater than 7.00. In fact, pure water at 0°C has a pH of 7.47, and water at 0°C is neutral, not basic.

Realizing that water remains neutral when boiled or frozen, you may continue to enjoy your hot beverage or frozen dessert without the fear of stomach upset or stomach damage by aqueous solutions that have suddenly turned acidic or basic.

Ronald DeLorenzo
Middle Georgia College

Table 18-4 *Ionization Constants and pK$_a$ Values for Some Weak Monoprotic Acids*

Acid	Ionization Reaction		K_a at 25°C	pK$_a$
hydrofluoric acid	$HF + H_2O$	$\rightleftharpoons H_3O^+ + F^-$	7.2×10^{-4}	3.14
nitrous acid	$HNO_2 + H_2O$	$\rightleftharpoons H_3O^+ + NO_2^-$	4.5×10^{-4}	3.35
acetic acid	$CH_3COOH + H_2O$	$\rightleftharpoons H_3O^+ + CH_3COO^-$	1.8×10^{-5}	4.74
hypochlorous acid	$HOCl + H_2O$	$\rightleftharpoons H_3O^+ + OCl^-$	3.5×10^{-8}	7.45
hydrocyanic acid	$HCN + H_2O$	$\rightleftharpoons H_3O^+ + CN^-$	4.0×10^{-10}	9.40

acids you encounter in this text will be weak acids. Table 18-4 contains names, formulas, ionization constants, and pK$_a$ values for a few common weak acids; Appendix F contains a longer list of K_a values. *Weak* acids ionize only slightly in dilute aqueous solution. Our classification of acids as strong or weak is based on the *extent to which they ionize in dilute aqueous solution.*

Recall that
pK$_a$ means $-\log K_a$.

Several weak acids are familiar to us. Vinegar is a 5% solution of acetic acid, CH_3COOH. Carbonated beverages are saturated solutions of carbon dioxide in water, which produces carbonic acid.

How can you tell that carbonated beverages are *saturated* CO_2 solutions?

$$CO_2 + H_2O \rightleftharpoons H_2CO_3$$

Citrus fruits contain citric acid, $C_3H_5O(COOH)_3$. Some ointments and powders used for medicinal purposes contain boric acid, H_3BO_3. These everyday uses of weak acids suggest that there is a significant difference between strong and weak acids. The difference is that *strong acids ionize completely in dilute aqueous solution, whereas weak acids ionize only slightly.*

Would you think of using sulfuric or nitric acid for any of these purposes?

Let us consider the reaction that occurs when a weak acid, such as acetic acid, is dissolved in water. The equation for the ionization of acetic acid is

$$CH_3COOH(aq) + H_2O(\ell) \rightleftharpoons H_3O^+(aq) + CH_3COO^-(aq)$$

The equilibrium constant for this reaction could be represented as

$$K_c = \frac{[H_3O^+][CH_3COO^-]}{[CH_3COOH][H_2O]}$$

This expression contains the concentration of water. If we restrict our discussion to *dilute* aqueous solutions, the concentration of water is very high. There are 55.5 moles of water in one liter of pure water. In dilute aqueous solutions the concentration of water is *essentially* constant; if we assume that it *is* constant, we can rearrange the preceding expression to give

$$K_c[H_2O] = \frac{[H_3O^+][CH_3COO^-]}{[CH_3COOH]}$$

Because both K_c and $[H_2O]$ are constant (at a specified temperature) the **ionization constant** of a weak acid is the product $K_c[H_2O]$; i.e., $K_a = K_c[H_2O]$. For acetic acid,

$$K_a = \frac{[H_3O^+][CH_3COO^-]}{[CH_3COOH]} = 1.8 \times 10^{-5}$$

The thermodynamic approach is that the activity of the (nearly) pure H_2O is essentially 1. The activity of each dissolved species is numerically equal to its molar concentration.

This expression tells us that in dilute aqueous solutions of acetic acid, the concentration of

H_3O^+ multiplied by the concentration of CH_3COO^- and then divided by the concentration of *nonionized* acetic acid is equal to 1.8×10^{-5}.

Ionization constants for weak acids (and bases) must be calculated from *experimentally determined data*. Measurements of pH, conductivity, or depression of freezing point provide data from which these constants can be calculated.

EXAMPLE 18-7 *Calculation of K_a from Equilibrium Concentrations*

The structure of nicotinic acid is

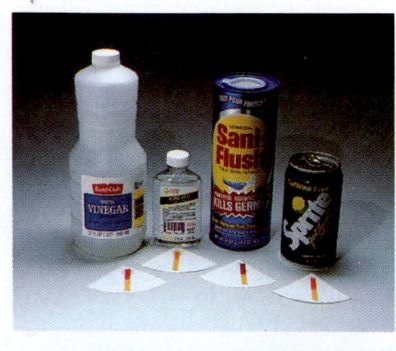

Nicotinic acid, also called niacin, is a necessary vitamin in our diets. It is not physiologically related to nicotine.

Nicotinic acid is a weak monoprotic organic acid that we can represent as HA.

$$HA + H_2O \rightleftharpoons H_3O^+ + A^-$$

A dilute solution of nicotinic acid was found to contain the following concentrations at equilibrium at 25°C. What is the value of K_a? $[HA] = 0.049\ M$; $[H^+] = [A^-] = 8.4 \times 10^{-4}\ M$.

Plan

We are given *equilibrium* concentrations, and so we substitute these into the expression for K_a.

Solution

$$HA + H_2O \rightleftharpoons H_3O^+ + A^- \qquad K_a = \frac{[H_3O^+][A^-]}{[HA]}$$

$$K_a = \frac{(8.4 \times 10^{-4})(8.4 \times 10^{-4})}{(0.049)} = 1.4 \times 10^{-5}$$

The equilibrium constant expression is

$$K_a = \frac{[H_3O^+][A^-]}{[HA]} = 1.4 \times 10^{-5}$$

You should now work Exercise 28.

EXAMPLE 18-8 *Calculation of K_a from Percent Ionization*

In $0.0100\ M$ solution, acetic acid is 4.2% ionized. Calculate its ionization constant.

Plan

We write the equation for the ionization of acetic acid and its equilibrium constant expression. Next we use the percent ionization to complete the reaction summary and then substitute into the K_a expression.

Solution

The equations for the ionization of CH_3COOH and its ionization constant are

$$CH_3COOH + H_2O \rightleftharpoons H_3O^+ + CH_3COO^- \qquad \text{and} \qquad K_a = \frac{[H_3O^+][CH_3COO^-]}{[CH_3COOH]}$$

Because 4.2% of the CH_3COOH ionizes,

$$M_{CH_3COOH} \text{ that ionizes} = 0.042 \times 0.0100\ M = 4.2 \times 10^{-4}\ M$$

Each mole of CH_3COOH that ionizes forms one mole of H_3O^+ and one mole of CH_3COO^-. We represent this in the reaction summary.

	$CH_3COOH + H_2O \rightleftharpoons$	H_3O^+	+	CH_3COO^-
initial	$0.0100\ M$	$\sim 0\ M$		$0\ M$
change	$-4.2 \times 10^{-4}\ M$	$+4.2 \times 10^{-4}\ M$		$+4.2 \times 10^{-4}\ M$
at equil	$9.58 \times 10^{-3}\ M$	$4.2 \times 10^{-4}\ M$		$4.2 \times 10^{-4}\ M$

Some common household weak acids. A strip of paper impregnated with a universal indicator is convenient for estimating the pH of a solution.

Substitution of these values into the K_a expression gives the value for K_a.

$$K_a = \frac{[H_3O^+][CH_3COO^-]}{[CH_3COOH]} = \frac{(4.2 \times 10^{-4})(4.2 \times 10^{-4})}{9.58 \times 10^{-3}} = \boxed{1.8 \times 10^{-5}}$$

You should now work Exercise 34.

EXAMPLE 18-9   *Calculation of K_a from pH*

The pH of a 0.115 M solution of chloroacetic acid, $ClCH_2COOH$, is measured to be 1.92. Calculate K_a for this weak monoprotic acid.

Plan

For simplicity, we represent $ClCH_2COOH$ as HA. We write the ionization equation and the expression for K_a. Next we calculate $[H_3O^+]$ for the given pH and complete the reaction summary. Finally, we substitute into the K_a expression.

Solution

The ionization of this weak monoprotic acid and its ionization constant expression may be represented as

$$HA + H_2O \rightleftharpoons H_3O^+ + A^- \quad \text{and} \quad K_a = \frac{[H_3O^+][A^-]}{[HA]}$$

We calculate $[H_3O^+]$ from the definition of pH.

$$pH = -\log [H_3O^+]$$

$$[H_3O^+] = 10^{-pH} = 10^{-1.92} = 0.012 \ M$$

We use the usual reaction summary as follows. At this point, we know the *original* [HA] and the *equilibrium* $[H_3O^+]$. From this information, we can fill out the "change" line and then deduce the other equilibrium values.

	HA + H₂O ⇌	H₃O⁺	+ A⁻
initial	0.115 M	~0 M	0 M
change due to rxn	−0.012 M	+0.012 M	0.012 M
at equil	0.103 M	0.012 M	0.012 M

Now that all concentrations are known, K_a can be calculated.

$$K_a = \frac{[H_3O^+][A^-]}{[HA]} = \frac{(0.012)(0.012)}{0.103} = \boxed{1.4 \times 10^{-3}}$$

You should now work Exercise 36.

▼ PROBLEM-SOLVING TIP   *Filling in Reaction Summaries*

In Examples 18-8 and 18-9 the value of an equilibrium concentration was used to determine the change in concentration. You should become proficient at using a variety of data to determine values that are related via a chemical equation. Let's review what we did in Example 18-9. Only the equilibrium expression, initial concentrations, and the equilibrium concentration of H_3O^+ were known when we started the reaction summary. As shown on the next page, we filled in the remaining values in the order indicated by the numbered red arrows.

1. $[H_3O^+]_{equil} = 0.012 \ M$, so we record this value

2. $[H_3O^+]_{initial} \sim 0$, so change in $[H_3O^+]$ due to rxn must be $+0.012 \ M$

3. Formation of 0.012 M H_3O^+ consumes 0.012 M HA, so the change in [HA] = $-0.012 \ M$

4. $[HA]_{equil} = [HA]_{orig} + [HA]_{chg} = 0.115\ M + (-0.012\ M) = 0.103\ M$

5. Formation of $0.012\ M\ H_3O^+$ also gives $0.012\ M\ A^-$

6. $[A^-]_{equil} = [A^-]_{orig} + [A^-]_{chg} = 0\ M + 0.012\ M = 0.012\ M$

At equilibrium, $[H_3O^+] = 0.012\ M$ so

	HA	$+\ H_2O \rightleftharpoons$	H_3O^+	$+$	A^-
initial	$0.115\ M$		$\sim0\ M$		$0\ M$
change due to rxn	$-0.012\ M$		$+0.012\ M$		$0.012\ M$
at equil	$0.103\ M$		$0.012\ M$		$0.012\ M$

Because ionization constants are equilibrium constants for ionization reactions, their values indicate the extents to which weak electrolytes ionize. Acids with larger ionization constants ionize to greater extents (and are stronger acids) than acids with smaller ionization constants. From Table 18-4, we see that the order of decreasing acid strength for these five weak acids is

$$HF > HNO_2 > CH_3COOH > HClO > HCN$$

Recall that in Brønsted–Lowry terminology, an acid forms its conjugate base by losing H^+.

Conversely, in Brønsted–Lowry terminology (Section 10-4), the order of increasing base strength of the anions of these acids is

$$F^- < NO_2^- < CH_3COO^- < ClO^- < CN^-$$

If we know the value of the ionization constant for a weak acid, we can calculate the concentrations of the species in solutions of known concentrations.

EXAMPLE 18-10   *Calculation of Concentrations from K_a*

We have written the formula for hypochlorous acid as HOCl rather than HClO to emphasize that its structure is H—O—Cl.

Calculate the concentrations of the various species in $0.10\ M$ hypochlorous acid, HOCl. For HOCl, $K_a = 3.5 \times 10^{-8}$.

Plan

We write the equation for the ionization of the weak acid and its K_a expression. Then, represent the *equilibrium* concentrations algebraically and substitute into the K_a expression.

Solution

The equation for the ionization of HOCl and its K_a expression are

$$HOCl + H_2O \rightleftharpoons H_3O^+ + OCl^- \quad \text{and} \quad K_a = \frac{[H_3O^+][OCl^-]}{[HOCl]} = 3.5 \times 10^{-8}$$

We would like to know the concentrations of H_3O^+, OCl^-, and nonionized HOCl in solution. An algebraic representation of concentrations is required, because there is no other obvious way to obtain the concentrations.

We neglect the 1.0×10^{-7} mol/L of H_3O^+ produced by the ionization of *pure* water. Recall (Section 18-2) that the addition of an acid to water suppresses the ionization of H_2O, so $[H_3O^+]$ from H_2O is even less than $1.0 \times 10^{-7}\ M$.

Let x = mol/L of HOCl that ionizes. Then, write the "change" line and complete the reaction summary.

	$HOCl$	$+\ H_2O \rightleftharpoons$	H_3O^+	$+$	OCl^-
initial	$0.10\ M$		$\sim0\ M$		$0\ M$
change due to rxn	$-x\ M$		$+x\ M$		$+x\ M$
at equil	$(0.10 - x)\ M$		$x\ M$		$x\ M$

Substituting these algebraic representations into the K_a expression gives

$$K_a = \frac{[H_3O^+][OCl^-]}{[HOCl]} = \frac{(x)(x)}{(0.10 - x)} = 3.5 \times 10^{-8}$$

This is a quadratic equation, but it is not necessary to solve it by the quadratic formula. If we assume that $(0.10 - x)$ is very nearly equal to 0.10, the equation becomes

$$\frac{x^2}{0.10} \approx 3.5 \times 10^{-8} \qquad x^2 \approx 3.5 \times 10^{-9} \qquad \text{so} \qquad x \approx 5.9 \times 10^{-5}$$

In our algebraic representation we let

$$[H_3O^+] = x\,M = \boxed{5.9 \times 10^{-5}\,M} \qquad [OCl^-] = x\,M = \boxed{5.9 \times 10^{-5}\,M}$$

$$[HOCl] = (0.10 - x)\,M = (0.10 - 0.000059)\,M = \boxed{0.10\,M}$$

You should now work Exercise 38.

▼ PROBLEM-SOLVING TIP   *Simplifying Quadratic Equations*

We often encounter quadratic or higher-order equations in equilibrium calculations. With modern programmable calculators, solving such problems by iterative methods is often feasible. But frequently a problem can be made much simpler by using some mathematical common sense.

When the linear variable (x) in a *quadratic* equation is added to or subtracted from a much larger number, it can often be disregarded if it is sufficiently small. A reasonable rule of thumb for determining whether the variable can be disregarded in equilibrium calculations is this: If the exponent of 10 in the K value is -4 or less (-5, -6, -7, and so on), then the variable may be small enough to disregard when it is added to or subtracted from a number greater than 0.05. Solve the problem neglecting x; then compare the value of x with the number it would have been added to (or subtracted from). If x is more than 5% of that number, the assumption was *not* justified, and you should solve the equation using the quadratic formula.

Let's examine the assumption as it applies to Example 18-10. Our quadratic equation is

$$\frac{(x)(x)}{(0.10 - x)} = 3.5 \times 10^{-8}$$

Because 3.5×10^{-8} is a very small K_a value, we know that the acid ionizes only slightly. Thus, x must be a very small number compared to 0.10, so we can write $(0.10 - x) \approx 0.10$. The equation then becomes $\dfrac{x^2}{0.10} \approx 3.5 \times 10^{-8}$. To solve this, we rearrange and take the square roots of both sides. To check, we see that the result, $x = 5.9 \times 10^{-5}$, is only 0.059% of 0.10. This error is much less than 5%, so our assumption is justified. You may also wish to use the quadratic formula to verify that the answer obtained this way is correct to within roundoff error.

The preceding argument is purely algebraic. We could use our chemical intuition to reach the same conclusion. A small K_a value (10^{-4} or less) tells us that the extent of ionization is very small. Therefore, nearly all of the weak acid exists as nonionized molecules. The amount that ionizes is not significant compared to the concentration of nonionized weak acid.

From our calculations we can draw some conclusions. In a solution containing *only a weak monoprotic acid,* the concentration of H_3O^+ is equal to the concentration of the anion of the acid. Unless the solution is *very* dilute, say less than 0.050 M, the concentration of nonionized acid is approximately equal to the molarity of the solution. When the value of K_a for the weak acid is greater than $\sim 10^{-4}$, then the extent of ionization will be large enough to make a significant difference between the concentration of nonionized acid and the molarity of the solution. In such cases we *cannot* make the simplifying assumption.

EXAMPLE 18-11   *Percent Ionization*

Calculate the percent ionization of a 0.10 M solution of acetic acid.

Plan

Write the ionization equation and the expression for K_a. Next, follow the procedure used in Example 18-10 to find the concentration of acid that ionized. Then, substitute the concentration of acid that ionized into the expression for percent ionization.

Solution

The equations for the ionization of CH_3COOH and its K_a are

$$CH_3COOH + H_2O \rightleftharpoons H_3O^+ + CH_3COO^- \qquad K_a = \frac{[H_3O^+][CH_3COO^-]}{[CH_3COOH]} = 1.8 \times 10^{-5}$$

Percentage is defined as (part/whole) $\times$ 100%, so the percent ionization is

$$\% \text{ ionization} = \frac{[CH_3COOH]_{ionized}}{[CH_3COOH]_{total}} \times 100\%$$

We proceed as we did in Example 18-10. Let $x = [CH_3COOH]_{ionized}$.

	$CH_3COOH + H_2O \rightleftharpoons$	$H_3O^+ +$	CH_3COO^-
initial	0.10 M	$\sim 0\ M$	0 M
change due to rxn	$-x\ M$	$+x\ M$	$+x\ M$
at equil	$(0.10 - x)\ M$	$x\ M$	$x\ M$

> We could write the original $[H_3O^+]$ as $1.0 \times 10^{-7}\ M$. In very *dilute* solutions of weak acids, we might have to take this into account. In this acid solution, $(1.0 \times 10^{-7} + x) \approx x$.

Substituting into the ionization constant expression gives

$$K_a = \frac{[H_3O^+][CH_3COO^-]}{[CH_3COOH]} = \frac{(x)(x)}{(0.10 - x)} = 1.8 \times 10^{-5}$$

If we make the simplifying assumption that $(0.10 - x) \approx 0.10$, we have

$$\frac{x^2}{0.10} = 1.8 \times 10^{-5} \qquad x^2 = 1.8 \times 10^{-6} \qquad x = 1.3 \times 10^{-3}$$

This gives $[CH_3COOH]_{ionized} = x = 1.3 \times 10^{-3}\ M$. Now we can calculate the percent ionization for 0.10 M CH_3COOH solution.

> Note that we need not solve explicitly for the equilibrium concentrations $[H_3O^+]$ and $[CH_3COO^-]$ to answer the question. From the setup, we see that these are both $1.3 \times 10^{-3}\ M$. The pH of the solution is 2.89.

$$\% \text{ ionization} = \frac{[CH_3COOH]_{ionized}}{[CH_3COOH]_{total}} \times 100\% = \frac{1.3 \times 10^{-3}\ M}{0.10\ M} \times 100\% = 1.3\%$$

Our assumption that $(0.10 - x)$ is approximately 0.10 is reasonable because $(0.10 - x) = (0.10 - 0.0013)$. This is only about 1% different than 0.10. However, when K_a for a weak acid is significantly greater than 10^{-4}, this assumption would introduce considerable error.

You should now work Exercise 40.

In dilute solutions, acetic acid exists primarily as nonionized molecules, as do all weak acids; there are relatively few hydronium and acetate ions. In 0.10 M solution, CH_3COOH

Table 18-5 *Comparison of Extents of Ionization of Some Acids*

Acid Solution	Ionization Constant	$[H_3O^+]$	pH	Percent Ionization
0.10 M HCl	very large	0.10 M	1.00	~100
0.10 M CH$_3$COOH	1.8×10^{-5}	0.0013 M	2.89	1.3
0.10 M HOCl	3.5×10^{-8}	0.000059 M	4.23	0.059

An inert solid has been suspended in the liquid to improve the quality of this photograph of a pH meter.

is 1.3% ionized; for each 1000 molecules of CH_3COOH originally placed in the solution, there are 13 H_3O^+ ions, 13 CH_3COO^- ions, and 987 nonionized CH_3COOH molecules. For weaker acids, the number of molecules of nonionized acid would be even larger.

By now we should have gained some "feel" for the strength of an acid by looking at its K_a value. Consider 0.10 M solutions of HCl (a strong acid), CH_3COOH (Example 18-11), and HOCl (Example 18-10). If we calculate the percent ionization for 0.10 M HOCl (as we did for 0.10 M CH_3COOH in Example 18-11), we find that it is 0.059% ionized. In 0.10 M solution, HCl is very nearly completely ionized. The data in Table 18-5 show that the $[H_3O^+]$ in 0.10 M HCl is approximately 77 times greater than that in 0.10 M CH_3COOH and approximately 1700 times greater than that in 0.10 M HOCl.

Many scientists prefer to use pK_a values rather than K_a values for weak acids. Recall that in general, "p" terms refer to negative logarithms. The pK_a value for a weak acid is just the negative logarithm of its K_a value.

EXAMPLE 18-12 pK_a Values

The K_a values for acetic acid and hydrofluoric acid are 1.8×10^{-5} and 7.2×10^{-4}, respectively. What are their pK_a values?

Plan

pK_a is defined as the negative logarithm of K_a (i.e., $pK_a = -\log K_a$) so we take the negative logarithm of each K_a.

Solution

For CH_3COOH,

$$pK_a = -\log K_a = -\log (1.8 \times 10^{-5}) = -(-4.74) = \boxed{4.74}$$

For HF,

$$pK_a = -\log K_a = -\log (7.2 \times 10^{-4}) = -(-3.14) = \boxed{3.14}$$

You should now work Exercise 42.

From Example 18-12, we see that the stronger acid (HF in this case) has the larger K_a value and the smaller pK_a value. Conversely, the weaker acid (CH_3COOH in this case) has the smaller K_a value and the larger pK_a value. The generalization is

A similar statement is true for weak bases; i.e., a stronger base has the greater K_b value and the smaller pK_b value.

The larger the value of K_a, the smaller the value of pK_a, and the stronger the acid.

EXAMPLE 18-13   *Acid Strengths and K_a Values*

Given the following list of weak acids and their K_a values, arrange the acids in order of (a) increasing acid strength and (b) increasing pK_a values.

Acid	K_a
HOCl	3.5×10^{-8}
HCN	4.0×10^{-10}
HNO$_2$	4.5×10^{-4}

Plan

(a) We see that HNO$_2$ is the strongest acid in this group because it has the largest K_a value. HCN is the weakest because it has the smallest K_a value.
(b) We do not need to calculate pK_a values to answer the question. We recall that the weakest acid has the largest pK_a value and the strongest acid has the smallest pK_a value, so the order of increasing pK_a values is just the reverse of the order in part (a).

Solution

(a)   Increasing acid strength: HCN < HOCl < HNO$_2$
(b)   Increasing pK_a values: HNO$_2$ < HOCl < HCN

You should now work Exercise 45.

You may know that hydrofluoric acid dissolves glass. But HF is *not* a strong acid. The reaction of glass with hydrofluoric acid occurs because silicates react with HF to produce silicon tetrafluoride, SiF$_4$, a very volatile compound. This reaction tells us nothing about the acid strength of hydrofluoric acid.

Thus far we have focused our attention on acids. Very few common weak bases are soluble in water. Aqueous ammonia is the most frequently encountered example. From our earlier discussion of bonding in covalent compounds (Section 8-8), we recall that there is one unshared pair of electrons on the nitrogen atom in NH$_3$. When ammonia dissolves in water, it accepts H$^+$ from a water molecule in a reversible reaction (Section 10-4). We say that NH$_3$ ionizes slightly when it undergoes this reaction. Aqueous solutions of NH$_3$ are basic because OH$^-$ ions are produced.

$$:NH_3 + H_2O \rightleftharpoons NH_4^+ + OH^-$$

Amines are derivatives of NH$_3$ in which one or more H atoms have been replaced by organic groups, as the following structures indicate.

H—N: (H above, H below)	H$_3$C—N: (H above, H below)	H$_3$C—N: (H$_3$C above, H below)	H$_3$C—N: (H$_3$C above, H$_3$C below)
ammonia NH$_3$	methylamine CH$_3$NH$_2$	dimethylamine (CH$_3$)$_2$NH	trimethylamine (CH$_3$)$_3$N

Thousands of amines are known, and many are very important in biochemical processes. Low-molecular-weight amines are soluble weak bases. The ionization of trimethylamine, for example, forms trimethylammonium ions and OH$^-$ ions.

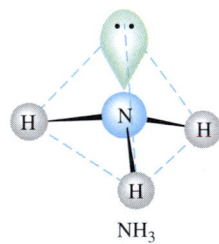

NH$_3$

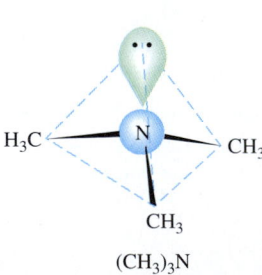

(CH$_3$)$_3$N

Table 18-6 *Ionization Constants and pK_b Values for Some Weak Bases*

Base	Ionization Reaction	K_b at 25°C	pK_b
ammonia	$NH_3 + H_2O \rightleftharpoons NH_4^+ + OH^-$	1.8×10^{-5}	4.74
methylamine	$(CH_3)NH_2 + H_2O \rightleftharpoons (CH_3)NH_3^+ + OH^-$	5.0×10^{-4}	3.30
dimethylamine	$(CH_3)_2NH + H_2O \rightleftharpoons (CH_3)_2NH_2^+ + OH^-$	7.4×10^{-4}	3.13
trimethylamine	$(CH_3)_3N + H_2O \rightleftharpoons (CH_3)_3NH^+ + OH^-$	7.4×10^{-5}	4.13
pyridine	$C_5H_5N + H_2O \rightleftharpoons C_5H_5NH^+ + OH^-$	1.5×10^{-9}	8.82

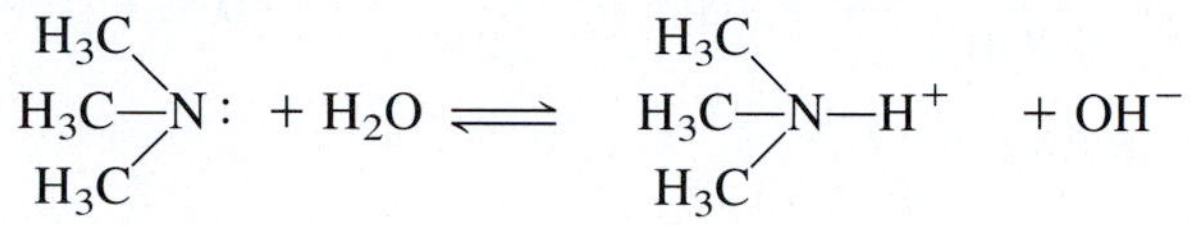

$$\begin{array}{ccc} H_3C & & H_3C \\ H_3C-N: + H_2O \rightleftharpoons & H_3C-N-H^+ & + OH^- \\ H_3C & & H_3C \end{array}$$

trimethylamine   trimethylammonium ion
$(CH_3)_3N$ $(CH_3)_3NH^+$

The structures of the ammonium and trimethylammonium ions are similar.

Now let us consider the behavior of ammonia in aqueous solutions. The reaction of ammonia with water and its ionization constant expression are

$$NH_3 + H_2O \rightleftharpoons NH_4^+ + OH^-$$

and

$$K_b = \frac{[NH_4^-][OH^-]}{[NH_3]} = 1.8 \times 10^{-5}$$

The subscript "b" indicates that the substance ionizes as a base.

The fact that K_b for aqueous NH_3 has the same value as K_a for CH_3COOH is pure coincidence. It does tell us that in aqueous solutions of the same concentration, CH_3COOH and NH_3 are ionized to the same extent. Table 18-6 lists K_b and pK_b values for a few common weak bases. Appendix G includes a longer list of K_b values.

We use K_b's for weak bases in the same way we used K_a's for weak acids and pK_b values for weak bases in the same way we used pK_a values for weak acids.

EXAMPLE 18-14 *pH of a Weak Base Solution*

Calculate the $[OH^-]$, pH, and percent ionization for $0.20\ M$ aqueous NH_3.

Plan

Write the equation for the ionization of aqueous NH_3 and represent the equilibrium concentrations algebraically. Then, substitute into the K_b expression and solve for $[OH^-]$ and $[NH_3]_{ionized}$.

Solution

The equation for the ionization of aqueous ammonia and the algebraic representations of equilibrium concentrations follow. Let $x = [NH_3]_{ionized}$.

	NH_3	$+ H_2O \rightleftharpoons$	NH_4^+	$+ OH^-$
initial	$0.20\ M$		$0\ M$	$\sim 0\ M$
change due to rxn	$-x\ M$		$+x\ M$	$+x\ M$
at equil	$(0.20 - x)\ M$		$x\ M$	$x\ M$

Measurement of the pH of a 0.20 M NH_3 solution.

Substitution into the ionization constant expression gives

$$K_b = \frac{[NH_4^+][OH^-]}{[NH_3]} = 1.8 \times 10^{-5} = \frac{(x)(x)}{(0.20 - x)}$$

If we assume that $(0.20 - x) \approx 0.20$, we have

$$\frac{x^2}{0.20} = 1.8 \times 10^{-5} \qquad x^2 = 3.6 \times 10^{-6} \qquad x = 1.9 \times 10^{-3}\ M$$

Then $[OH^-] = x = \boxed{1.9 \times 10^{-3}\ M,}$ pOH = 2.72, and pH = $\boxed{11.28.}$

$[NH_3]_{ionized} = x$, so the percent ionization may be calculated.

$$\% \text{ ionization} = \frac{[NH_3]_{ionized}}{[NH_3]_{total}} \times 100\% = \frac{1.9 \times 10^{-3}}{0.20} \times 100\% = \boxed{0.95\% \text{ ionized}}$$

You should now work Exercise 44.

EXAMPLE 18-15   *Household Ammonia*

The pH of a household ammonia solution is 11.50. What is its molarity?

Plan

We are given the pH of an aqueous NH_3 solution. Use pH + pOH = 14.00 to find pOH, which we can convert to $[OH^-]$. Then, complete the reaction summary and substitute the representations of equilibrium concentrations into the K_b expression.

Solution

At equilibrium pH = 11.50; we know that pOH = 2.50, so $[OH^-] = 10^{-2.50} = 3.2 \times 10^{-3}\ M$. This $[OH^-]$ results from the reaction, so we can write the change line. Then, letting x represent the *initial* concentration of NH_3, we can complete the reaction summary.

Measurement of the pH of a solution of household ammonia.

At equilibrium $[OH^-] = 3.2 \times 10^{-3}\ M$, so ⓵

	NH_3	$+ H_2O \rightleftharpoons$	NH_4^+	$+$	OH^-
initial	$x\ M$		$0\ M$ ③		$\sim 0\ M$
change	$-3.2 \times 10^{-3}\ M$ ④		$+3.2 \times 10^{-3}\ M$		$+3.2 \times 10^{-3}\ M$ ⑦
at equil	$(x - 3.2 \times 10^{-3})M$ ⑥	⑤	$3.2 \times 10^{-3}\ M$	②	$3.2 \times 10^{-3}\ M$

Substituting these values into the K_b expression for aqueous NH_3 gives

$$K_b = \frac{[NH_4^+][OH^-]}{[NH_3]} = \frac{(3.2 \times 10^{-3})(3.2 \times 10^{-3})}{(x - 3.2 \times 10^{-3})} = 1.8 \times 10^{-5}$$

This suggests that $(x - 3.2 \times 10^{-3}) \approx x$. So we can approximate.

$$\frac{(3.2 \times 10^{-3})(3.2 \times 10^{-3})}{x} = 1.8 \times 10^{-5} \qquad \text{and} \qquad x = \boxed{0.57\ M\ NH_3}$$

The solution is 0.57 M NH_3. Our assumption that $(x - 3.2 \times 10^{-3}) \approx x$ was justified.

You should now work Exercises 48, 50, and 52.

18-5   ACID–BASE INDICATORS

In Section 3-9 we described acid–base titrations and the use of indicators to tell us when to stop a titration. Detection of the end point in an acid–base titration is only one of the important uses of indicators.

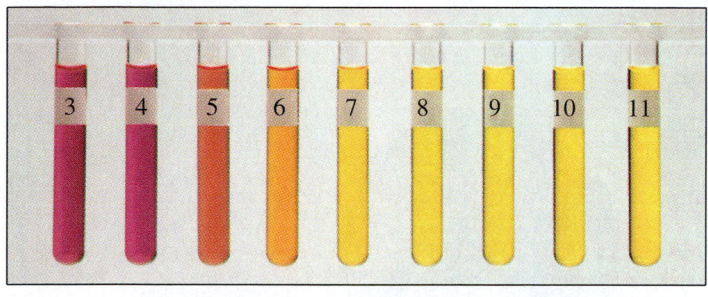

(a)

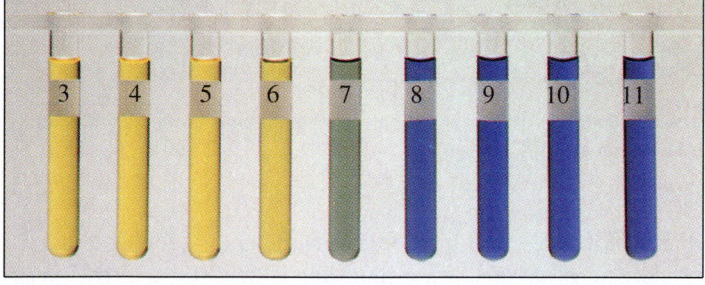

(b)

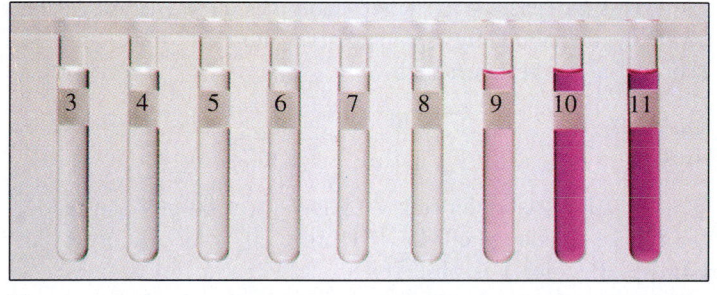

(c)

Figure 18-3 Three common indicators in solutions that cover the pH range 3 to 11 (the black numbers). (a) Methyl red is red at pH 4 and below; it is yellow at pH 7 and above. Between pH 4 and pH 7 it changes from red to red-orange, to orange, to yellow. (b) Bromthymol blue is yellow at pH 6 and below; it is blue at pH 8 and above. Between pH 6 and 8 it changes from yellow to yellow-green, to green, to blue-green, to blue. (c) Phenolphthalein is colorless below pH 8 and red above pH 10. It changes from colorless to pale pink, to pink, to red.

Phenolphthalein is also the active component of the laxative Ex-Lax. It is sometimes added to laboratory ethyl alcohol to discourage consumption.

An indicator is an organic dye; its color depends on the concentration of H_3O^+ ions, or pH, in the solution. By the color an indicator displays, it "indicates" the acidity or basicity of a solution. Figure 18-3 displays solutions that contain three common indicators in solutions over the pH range 3 to 11. Study Figure 18-3 and its legend carefully.

The first indicators used were vegetable dyes. Litmus is a familiar example. Most of the indicators that we use in the laboratory today are synthetic compounds; i.e., they have been made in laboratories by chemists. Phenolphthalein is the most common acid–base indicator. It is colorless in solutions of pH less than 8 ($[H_3O^+] > 10^{-8} M$) and turns red as pH approaches 10.

Many acid–base indicators are weak organic acids, HIn, where "In" represents complex organic groups. Bromthymol blue is such an indicator. Its ionization constant is 7.9×10^{-8}. We can represent its ionization in dilute aqueous solution and its ionization constant expression as

$$HIn + H_2O \rightleftharpoons H_3O^+ + In^-$$

$$K_a = \frac{[H_3O^+][In^-]}{[HIn]} = 7.9 \times 10^{-8}$$

color 1 yellow   ← for bromthymol blue →   color 2 blue

Bromthymol blue indicator is yellow in acidic solutions and blue in basic solutions.

Figure 18-4 The juice of the red (purple) cabbage is a naturally occurring universal indicator. From left to right are solutions of pH 1, 4, 7, 10, and 13.

HIn represents nonionized acid molecules, and In⁻ represents the anion (conjugate base) of HIn. The essential characteristic of an acid–base indicator is that HIn and In⁻ *must* have quite different colors. The relative amounts of the two species determine the color of the solution. Adding an acid favors the reaction to the left and gives more HIn molecules (color 1). Adding a base favors the reaction to the right and gives more In⁻ ions (color 2). The ionization constant expression can be rearranged.

$$\frac{[H_3O^+][In^-]}{[HIn]} = K_a \qquad \text{so} \qquad \frac{[In^-]}{[HIn]} = \frac{K_a}{[H_3O^+]}$$

This shows clearly how the $[In^-]/[HIn]$ ratio depends on $[H_3O^+]$ (or on pH) and the K_a value for the indicator. As a rule-of-thumb, when $[In^-]/[HIn] \geq 10$, color 2 is observed; conversely, when $[In^-]/[HIn] \leq \frac{1}{10}$, color 1 is observed.

Universal indicators are mixtures of several acid–base indicators that display a continuous range of colors over a wide range of pH values. Figure 18-2 shows concentrated solutions of a universal indicator in flat dishes so that the colors are very intense. The juice of red (purple) cabbage is a universal indicator. Figure 18-4 shows the color of red cabbage juice in solutions within the pH range 1 to 13.

One important use of universal indicators is in commercial indicator papers, which are small strips of paper impregnated with solutions of universal indicators. A strip of the paper is dipped into the solution of interest, and the color of the indicator on the paper indicates the pH of the solution. The photographs on page 688 (solutions of universal indicators) and page 694 (an indicator paper) illustrate the use of universal indicators to estimate pH. We shall describe the use of indicators in titrations more fully in Sections 19-6 and 19-7.

18-6 THE COMMON ION EFFECT AND BUFFER SOLUTIONS

In laboratory reactions, in industrial processes, and in the bodies of plants and animals, it is often necessary to keep the pH nearly constant despite the addition of acids or bases. The oxygen-carrying capacity of the hemoglobin in your blood and the activity of the enzymes in your cells are very sensitive to the pH of your body fluids. Our bodies use a combination of compounds known as a *buffer system* to keep the pH within a narrow range. The operation of a buffer solution depends on the *common ion effect,* a special case of LeChatelier's Principle.

Buffer systems resist changes in pH.

The Common Ion Effect

The term **common ion effect** is used to describe the behavior of a solution in which the same ion is produced by two different compounds. Many types of solutions exhibit this effect. Two of the most frequently encountered kinds are

1. A solution of a weak acid *plus* a soluble ionic salt of the weak acid
2. A solution of a weak base *plus* a soluble ionic salt of the weak base

Weak Acids Plus Salts of Weak Acids

Consider a solution that contains acetic acid *and* sodium acetate, a soluble ionic salt of CH_3COOH. The $NaCH_3COO$ is completely dissociated into its constituent ions, but CH_3COOH is only slightly ionized.

$$NaCH_3COO \xrightarrow{H_2O} Na^+ + CH_3COO^- \quad \text{(to completion)}$$

$$CH_3COOH + H_2O \rightleftharpoons H_3O^+ + CH_3COO^- \quad \text{(reversible)}$$

Both CH_3COOH and $NaCH_3COO$ are sources of CH_3COO^- ions. The completely dissociated $NaCH_3COO$ provides a high $[CH_3COO^-]$. This shifts the ionization equilibrium of CH_3COOH far to the left as CH_3COO^- combines with H_3O^+ to form nonionized CH_3COOH and H_2O. The result is a drastic decrease in $[H_3O^+]$ in the solution.

LeChatelier's Principle (Section 17-6) is applicable to equilibria in aqueous solution.

> Solutions that contain a weak acid plus a salt of the weak acid are always less acidic than solutions that contain the same concentration of the weak acid alone.

EXAMPLE 18-16 *Weak Acid/Salt of Weak Acid Buffers*

Calculate the concentration of H_3O^+ and the pH of a solution that is 0.10 *M* in CH_3COOH and 0.20 *M* in $NaCH_3COO$.

Plan

Write the appropriate equations for *both* $NaCH_3COO$ *and* CH_3COOH and the ionization constant expression for CH_3COOH. Then, represent the *equilibrium* concentrations algebraically and substitute into the K_a expression.

Solution

The appropriate equations and ionization constant expression are

$$NaCH_3COO \longrightarrow Na^+ + CH_3COO^- \quad \text{(to completion)}$$

$$CH_3COOH + H_2O \rightleftharpoons H_3O^+ + CH_3COO^- \quad \text{(reversible)}$$

$$K_a = \frac{[H_3O^+][CH_3COO^-]}{[CH_3COOH]} = 1.8 \times 10^{-5}$$

This K_a expression is valid for *all solutions* that contain CH_3COOH. In solutions that contain both CH_3COOH and $NaCH_3COO$, CH_3COO^- ions come from two sources. The ionization constant is satisfied by the *total* CH_3COO^- concentration.

The two solutions of Table 18-7, in the presence of universal indicator. The CH_3COOH solution is on the left.

Because $NaCH_3COO$ is completely dissociated, the $[CH_3COO^-]$ *from $NaCH_3COO$* will be 0.20 mol/L. Let $x = [CH_3COOH]$ that ionizes; then x is also equal to $[H_3O^+]$ *and* equal to $[CH_3COO^-]$ *from CH_3COOH*. The *total* concentration of CH_3COO^- is $(0.20 + x)$ M. The concentration of nonionized CH_3COOH is $(0.10 - x)$ M.

$$NaCH_3COO \longrightarrow Na^+ + CH_3COO^-$$
$$0.20\ M \rightleftharpoons 0.20\ M \quad 0.20\ M$$

$$CH_3COOH + H_2O \rightleftharpoons H_3O^+ + CH_3COO^-$$
$$(0.10 - x)\ M \qquad x\ M \qquad x\ M$$

$\longrightarrow$ Total $[CH_3COO^-] = (0.20 + x)\ M$

Substitution into the ionization constant expression for CH_3COOH gives

$$K_a = \frac{[H_3O^+][CH_3COO^-]}{[CH_3COOH]} = \frac{(x)(0.20 + x)}{(0.10 - x)} = 1.8 \times 10^{-5}$$

This equation suggests that x is very small. We can make two assumptions.

$$(0.20 + x) \approx 0.20 \qquad \text{and} \qquad (0.10 - x) \approx 0.10$$

Introducing these assumptions gives

$$\frac{0.20\ x}{0.10} = 1.8 \times 10^{-5} \qquad \text{and} \qquad x = 9.0 \times 10^{-6}$$

$$x\ M = [H_3O^+] = 9.0 \times 10^{-6}\ M \qquad \text{so} \qquad pH = 5.05$$

You should now work Exercise 60.

You can verify the validity of the assumption by substituting the value for x, 9.0×10^{-6}, into the original equation.

To see how much the acidity of the 0.10 M CH_3COOH solution is reduced by making it also 0.20 M in $NaCH_3COO$, refer back to Example 18-11. There we found that in 0.10 M CH_3COOH the H_3O^+ concentration is 1.3×10^{-3} mol/L (pH = 2.89).

Let us calculate the percent ionization in the solution of Example 18-16.

$$\% \text{ ionization} = \frac{[CH_3COOH]_{ionized}}{[CH_3COOH]_{original}} \times 100\%$$

$$= \frac{9.0 \times 10^{-6}\ M}{0.10\ M} \times 100\% = 0.0090\% \text{ ionized}$$

This compares with 1.3% ionization in 0.10 M CH_3COOH (Example 18-11). Table 18-7 compares these solutions. The third column shows that $[H_3O^+]$ is about 140 times greater in 0.10 M CH_3COOH than in the solution to which 0.20 mol/L $NaCH_3COO$ has been added (LeChatelier's Principle).

Table 18-7 *Comparison of $[H_3O^+]$ and pH in Acetic Acid and Sodium Acetate–Acetic Acid Solutions*

Solution	% CH_3COOH Ionized	$[H_3O^+]$	pH	
0.10 M CH_3COOH	1.3%	$1.3 \times 10^{-3}\ M$	2.89	
0.10 M CH_3COOH and 0.20 M $NaCH_3COO$	0.009%	$9.0 \times 10^{-6}\ M$	5.05	$\Delta pH = 2.16$

The calculation of $[H_3O^+]$ in solutions containing both a weak acid and a salt of the weak acid can be simplified greatly. Let us write the equation for the ionization of a *weak monoprotic acid* and its K_a in the following way.

$$HA + H_2O \rightleftharpoons H_3O^+ + A^- \qquad \text{and} \qquad \frac{[H_3O^+][A^-]}{[HA]} = K_a$$

HA and A^- represent the weak acid and its conjugate base, respectively.

Solving this expression for $[H_3O^+]$ gives

$$[H_3O^+] = \frac{[HA]}{[A^-]} \times K_a$$

We now impose two conditions: (1) The concentrations of both the weak acid and its salt are some reasonable values, say greater than 0.050 M, and (2) the salt contains a univalent cation. Under these conditions the concentration of the anion, $[A^-]$, in the solution can be assumed to be the same as the concentration of the salt. With these restrictions, the preceding expression for $[H_3O^+]$ becomes

These are the kinds of assumptions we made in Example 18-16.

$$[H_3O^+] = \frac{[\text{acid}]}{[\text{salt}]} \times K_a$$

[acid] is the concentration of nonionized weak acid (in most cases, this is the total acid concentration), and [salt] is the concentration of its salt.

If we take the logarithm of both sides of the preceding equation, we obtain

$$\log [H_3O^+] = \log \frac{[\text{acid}]}{[\text{salt}]} + \log K_a$$

Multiplying by -1 gives

$$-\log [H_3O^+] = -\log \frac{[\text{acid}]}{[\text{salt}]} - \log K_a$$

and rearrangement gives

$$pH = pK_a + \log \frac{[\text{salt}]}{[\text{acid}]} \qquad \text{where} \qquad pK_a = -\log K_a \qquad \text{(acid/salt buffer)}$$

The relationship is valid *only* for solutions that contain a weak *monoprotic* acid and a soluble, ionic salt of the weak acid with a *univalent* cation, both in reasonable concentrations.

This equation is known as the **Henderson–Hasselbalch equation.** Workers in the biological sciences use it frequently. In general terms, we can also write

$$pH = pK_a + \log \frac{[\text{conjugate base}]}{[\text{acid}]} \qquad \text{(acid/salt buffer)}$$

Weak Bases Plus Salts of Weak Bases

Let us consider the second common kind of buffer solution, containing a weak base and its salt. A solution that contains aqueous NH_3 and ammonium chloride, NH_4Cl, a soluble ionic salt of NH_3, is typical. The NH_4Cl is completely dissociated, but aqueous NH_3 is only slightly ionized.

$$NH_4Cl \xrightarrow{H_2O} \boxed{NH_4^+} + Cl^- \qquad \text{(to completion)}$$

$$NH_3 + H_2O \rightleftharpoons \boxed{NH_4^+} + \boxed{OH^-} \qquad \text{(reversible)}$$

Both NH_4Cl and aqueous NH_3 produce NH_4^+ ions. The completely dissociated NH_4Cl provides a high $[NH_4^+]$. This shifts the ionization equilibrium of aqueous NH_3 far to the left, as NH_4^+ ions combine with OH^- ions to form nonionized NH_3 and H_2O. Thus, $[OH^-]$ is decreased significantly.

> Solutions that contain a weak base plus a salt of the weak base are always less basic than solutions that contain the same concentration of the weak base alone.

EXAMPLE 18-17 *Weak Base/Salt of Weak Base Buffers*

Calculate the concentration of OH^- and the pH of a solution that is 0.20 M in aqueous NH_3 *and* 0.10 M in NH_4Cl.

Plan

Write the appropriate equations for *both* NH_4Cl and NH_3 and the ionization constant expression for NH_3. Then, represent the *equilibrium* concentrations algebraically and substitute into the K_b expression.

Solution

The appropriate equations and algebraic representations of concentrations are

$$\begin{array}{ccccc}
NH_4Cl & \longrightarrow & \boxed{NH_4^+} & + & Cl^- \\
0.10\ M & & 0.10\ M & & 0.10\ M \\[1em]
NH_3 & + H_2O \rightleftharpoons & \boxed{NH_4^+} & + & OH^- \\
(0.20 - x)\ M & & x\ M & & x\ M
\end{array}$$

$\longrightarrow$ Total $[NH_4^+] = (0.10 + x)\ M$

Substitution into the K_b expression for aqueous NH_3 gives

$$K_b = \frac{[NH_4^+][OH^-]}{[NH_3]} = 1.8 \times 10^{-5} = \frac{(0.10 + x)(x)}{(0.20 - x)}$$

Because K_b is small, we can assume that $(0.10 + x) \approx 0.10$ and $(0.20 - x) \approx 0.20$.

$$\frac{0.10x}{0.20} = 1.8 \times 10^{-5}\ M \qquad \text{and} \qquad x = 3.6 \times 10^{-5}\ M$$

$$x\ M = \boxed{[OH^-] = 3.6 \times 10^{-5}\ M} \qquad \text{so} \qquad pOH = 4.44 \qquad \text{and} \qquad \boxed{pH = 9.56}$$

You should now work Exercise 62.

In Example 18-14 we calculated $[OH^-]$ and pH for 0.20 M aqueous NH_3. Compare those results with the values obtained in Example 18-17 (Table 18-8). The concentration of OH^- is 53 times greater in the solution containing only 0.20 M aqueous NH_3 than in the solution to which 0.10 mol/L NH_4Cl has been added. This is another demonstration of Le Chatelier's Principle.

Table 18-8 *Comparison of [OH⁻] and pH in Ammonia and Ammonium Chloride—Ammonia Solutions*

Solution	% NH_3 Ionized	$[OH^-]$	pH	
0.20 M aq NH_3	0.95%	1.9×10^{-3} M	11.28	$\left.\begin{array}{l}\\\\\end{array}\right\}\Delta pH = -1.72$
0.20 M aq NH_3 and 0.10 M aq NH_4Cl	0.0018%	3.6×10^{-5} M	9.56	

The two solutions in Table 18-8, in the presence of universal indicator. The NH_3 solution is on the left. Can you calculate the percentage of NH_3 that is ionized in these two solutions?

We can derive a relationship for $[OH^-]$ in solutions containing weak bases *plus* salts of the weak bases, just as we did for weak acids. In general terms the equation for the ionization of a monoprotic weak base and its K_b expression are

$$\text{Base} + H_2O \rightleftharpoons (\text{base})H^+ + OH^- \qquad \text{and} \qquad \frac{[(\text{base})H^+][OH^-]}{[\text{base}]} = K_b$$

Solving the K_b expression for $[OH^-]$ gives

$$[OH^-] = \frac{[\text{base}]}{[(\text{base})H^+]} \times K_b$$

base and *(base)H⁺* represent the weak base and its conjugate acid, respectively—e.g., NH_3 and NH_4^+.

Taking the logarithm of both sides of the equation gives

$$\log [OH^-] = \log \frac{[\text{base}]}{[(\text{base})H^+]} + \log K_b$$

For salts of weak bases that contain *univalent* anions, $[(\text{base})H^+] = [\text{salt}]$. Multiplication by -1 and rearrangement gives another form of the *Henderson–Hasselbalch equation* for solutions containing a weak base plus a salt of the weak base.

$$pOH = pK_b + \log \frac{[\text{salt}]}{[\text{base}]} \qquad \text{where} \qquad pK_b = -\log K_b \qquad \text{(base/salt buffer)}$$

For salts such as $(NH_4)_2SO_4$ that contain divalent anions, $[(\text{base})H^+] = 2[\text{salt}]$.

The Henderson–Hasselbalch equation is valid for solutions of weak bases plus salts of weak bases with univalent anions in reasonable concentrations. In general terms we can also write this equation as

$$pOH = pK_b + \log \frac{[\text{conjugate acid}]}{[\text{base}]} \qquad \text{(base/salt buffer)}$$

18-7 BUFFERING ACTION

The two common kinds of buffer solutions are the ones we have just discussed—namely, solutions containing (1) a weak acid plus a soluble ionic salt of the weak acid and (2) a weak base plus a soluble ionic salt of the weak base.

A buffer solution contains a conjugate acid–base pair with both the acid and base in reasonable concentrations. The acidic component reacts with added strong bases. The basic component reacts with added acids.

> A buffer solution is able to react with either H_3O^+ or OH^- ions, whichever is added.

Thus, a buffer solution resists changes in pH. When we add a modest amount of a strong base or a strong acid to a buffer solution, the pH changes very little.

Solutions of a Weak Acid and a Salt of the Weak Acid

A solution containing acetic acid, CH_3COOH, and sodium acetate, $NaCH_3COO$, is an example of this kind of buffer solution. The acidic component is CH_3COOH. The basic component is $NaCH_3COO$ because the CH_3COO^- ion is the conjugate base of CH_3COOH. The operation of this buffer depends on the equilibrium

$$CH_3COOH + H_2O \rightleftharpoons H_3O^+ + CH_3COO^-$$
high conc   high conc (from salt)

If we add a strong acid such as HCl to this solution, it produces H_3O^+. As a result of the added H_3O^+, the reaction occurs to the *left,* to use up most of the added H_3O^+ and reestablish equilibrium. Because the $[CH_3COO^-]$ in the buffer solution is high, this can occur to a great extent. The net reaction is

$$H_3O^+ + CH_3COO^- \longrightarrow CH_3COOH + H_2O \qquad (\sim 100\%)$$

or, as a formula unit equation,

$$HCl + NaCH_3COO \longrightarrow CH_3COOH + NaCl \qquad (\sim 100\%)$$
added acid   base   weak acid   salt

This reaction goes nearly to completion because CH_3COOH is a *weak* acid; even when mixed from separate sources, its ions have a strong tendency to form nonionized CH_3COOH molecules rather than remain separate.

When a strong soluble base, such as NaOH, is added to the CH_3COOH–$NaCH_3COO$ buffer solution, it is consumed by the acidic component, CH_3COOH. This occurs in the following way. The additional OH^- causes the water autoionization reaction to proceed to the *left.*

$$2H_2O \rightleftharpoons H_3O^+ + OH^- \qquad \text{(shifts } left)$$

This uses up some H_3O^+, causing more CH_3COOH to ionize.

$$CH_3COOH + H_2O \rightleftharpoons CH_3COO^- + H_3O^+ \qquad \text{(shifts } right)$$

Because the $[CH_3COOH]$ is high, this can occur to a great extent. The net result is the neutralization of OH^- by CH_3COOH.

$$OH^- + CH_3COOH \longrightarrow CH_3COO^- + H_2O \qquad (\sim 100\%)$$

or, as a formula unit equation,

$$NaOH + CH_3COOH \longrightarrow NaCH_3COO + H_2O \qquad (\sim 100\%)$$
added base   acid   salt   water

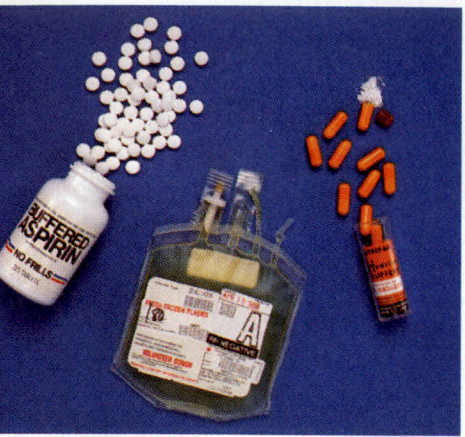

Three common examples of buffers. Many medications are buffered to minimize digestive upset. Most body fluids, including blood plasma, contain very efficient natural buffer systems. Buffer capsules are used in laboratories to prepare solutions of specified pH.

The net effect is to neutralize most of the H_3O^+ from HCl by forming nonionized CH_3COOH molecules. This slightly decreases the ratio $[CH_3COO^-]/[CH_3COOH]$, which governs the pH of the solution.

The net effect is to neutralize most of the OH^- from NaOH. This slightly increases the ratio $[CH_3COO^-]/[CH_3COOH]$, which governs the pH of the solution.

EXAMPLE 18-18   *Buffered Solutions*

If we add 0.010 mol of solid NaOH to 1.0 liter of a buffer solution that is 0.10 M in CH_3COOH and 0.10 M in $NaCH_3COO$, how much will $[H_3O^+]$ and pH change? Assume that there is no volume change due to the addition of solid NaOH.

Plan

Calculate $[H_3O^+]$ and pH for the original buffer solution. Then, write the reaction summary that shows how much of the CH_3COOH is neutralized by NaOH. Calculate $[H_3O^+]$ and pH for the resulting buffer solution. Finally, calculate ΔpH.

Solution

For the 0.10 M CH_3COOH and 0.10 M $NaCH_3COO$ solution, we can write

$$[H_3O^+] = \frac{[acid]}{[salt]} \times K_a = \frac{0.10}{0.10} \times 1.8 \times 10^{-5} = \underline{1.8 \times 10^{-5}\ M};\ \underline{pH = 4.74}$$

When solid NaOH is added, it reacts with CH_3COOH to form more $NaCH_3COO$.

	NaOH	+ CH_3COOH	$\longrightarrow$	$NaCH_3COO$ + H_2O
start	0.01 mol	0.10 mol		0.10 mol
change due to rxn	−0.01 mol	−0.01 mol		+0.01 mol
after rxn	0 mol	0.09 mol		0.11 mol

The volume of the solution is 1.0 liter, so we now have a solution that is 0.09 M in CH_3COOH and 0.11 M in $NaCH_3COO$. In this solution,

$$[H_3O^+] = \frac{[acid]}{[salt]} \times K_a = \frac{0.09}{0.11} \times 1.8 \times 10^{-5} = \underline{1.5 \times 10^{-5}\ M};\ \underline{pH = 4.82}$$

> The addition of 0.010 mol of solid NaOH to 1.0 liter of this buffer solution decreases $[H_3O^+]$ from $1.8 \times 10^{-5}\ M$ to $1.5 \times 10^{-5}\ M$ and increases pH from 4.74 to 4.82, a change of 0.08 pH unit, which is a very slight change.

This is enough NaOH to neutralize 10% of the acid.

You should now work Exercise 70.

Addition of 0.010 mole of solid NaOH to one liter of 0.10 M CH_3COOH (pH = 2.89 from Table 18-7) would give a solution that is 0.09 M in CH_3COOH and 0.01 M in $NaCH_3COO$. The pH of this solution is 3.80, which is 0.91 pH unit higher than that of the 0.10 M CH_3COOH solution.

By contrast, adding 0.010 mole of NaOH to enough pure H_2O to give one liter produces a 0.010 M solution of NaOH: $[OH^-] = 1.0 \times 10^{-2}\ M$ and pOH = 2.00. The pH of this solution is 12.00, an increase of 5 pH units above that of pure H_2O. In summary, 0.010 mole of NaOH

pH + pOH = 14

added to the $CH_3COOH/NaCH_3COO$ buffer, pH 4.74 $\longrightarrow$ 4.82

added to 0.10 M CH_3COOH, pH 2.89 $\longrightarrow$ 3.80

added to pure H_2O, pH 7.00 $\longrightarrow$ 12.00

In similar fashion we could calculate the effects of adding 0.010 mole of pure HCl(g) instead of pure NaOH to 1.00 liter of each of these three solutions. This would result in the following changes in pH.

	When We Add 0.010 mol NaOH(s)		When We Add 0.010 mol HCl(g)	
We Have 1.00 L of Original Solution	**pH Increases by**	**$[H_3O^+]$ Decreases by a Factor of**	**pH Decreases by**	**$[H_3O^+]$ Increases by a Factor of**
buffer solution (0.10 M NaCH$_3$COO and 0.10 M CH$_3$COOH)	+0.08 pH unit	1.2	−0.08 pH unit	1.2
0.10 M CH$_3$COOH	+0.91	8.1	−0.89	7.8
pure H$_2$O	+5.00	100,000	−5.00	100,000

Table 18-9 *Changes in pH Caused by Addition of Pure Acid or Base to One Liter of Solution*

added to the $CH_3COOH/NaCH_3COO$ buffer, pH 4.74 $\longrightarrow$ 4.66

added to 0.10 M CH$_3$COOH, pH 2.89 $\longrightarrow$ 2.00

added to pure H$_2$O, pH 7.00 $\longrightarrow$ 2.00

The results of adding NaOH or HCl to these solutions (Table 18-9) demonstrate the efficiency of the buffer solution. We recall that each change of 1 pH unit means that the $[H_3O^+]$ and $[OH^-]$ change by a *factor* of 10. In these terms, the effectiveness of the buffer solution in controlling pH is even more dramatic.

Solutions of a Weak Base and a Salt of the Weak Base

An example of this type of buffer solution is one that contains the weak base ammonia, NH_3, and its soluble ionic salt ammonium chloride, NH_4Cl. The reactions responsible for the operation of this buffer are

$$NH_4Cl \xrightarrow{H_2O} NH_4^+ + Cl^- \qquad \text{(to completion)}$$

$$\underset{\text{high conc}}{NH_3} + H_2O \rightleftharpoons \underset{\substack{\text{high conc} \\ \text{from salt}}}{NH_4^+} + OH^- \qquad \text{(reversible)}$$

If a strong acid such as HCl is added to this buffer solution, the resulting H_3O^+ shifts the equilibrium reaction

$$2H_2O \rightleftharpoons H_3O^+ + OH^- \qquad \text{(shifts *left*)}$$

strongly to the *left*. As a result of the diminished OH^- concentration, the reaction

$$NH_3 + H_2O \rightleftharpoons NH_4^+ + OH^- \qquad \text{(shifts *right*)}$$

shifts markedly to the *right*. Because the $[NH_3]$ in the buffer solution is high, this can occur to a great extent. The net reaction is

$$H_3O^+ + NH_3 \longrightarrow NH_4^+ + H_2O \qquad (\sim 100\%)$$

The net effect is to neutralize most of the H_3O^+ from HCl. This slightly increases the ratio $[NH_4^+]/[NH_3]$, which governs the pH of the solution.

or, as a formula unit equation,

CHEMISTRY IN USE

Our Daily Lives

Fun with Carbonates

Carbonates react with acids to produce carbon dioxide. This property of carbonates has been exploited in many ways, both serious and silly.

One of the giddiest applications of this behavior of carbonates is in Mad Dawg™, a foaming bubble gum developed in the early 1990s. If you chew a piece of this gum, large quantities of foam are produced so that it is difficult to keep the colorful lather from oozing out of your mouth. The froth begins to form as your teeth mix saliva with the gum's ingredients (sodium hydrogen carbonate, citric acid, malic acid, food coloring, and flavoring).

How is this foam produced? When citric acid and malic acid dissolve in saliva, they produce hydrogen ions which decompose the sodium hydrogen carbonate (baking soda) to produce carbon dioxide, a gas. These bubbles of carbon dioxide produce the foam. Large quantities of foam are produced because citric and malic acids taste sour, which stimulates salivation.

A common medical recipe for a similar combination of ingredients is found in Alka Seltzer™ tablets; these contain sodium hydrogen carbonate, citric acid, and aspirin. The acid and carbonate react in water to produce carbon dioxide, which gives the familiar fizz of Alka Seltzer™.

Makeup artists add baking soda to cosmetics to produce monster-flesh makeup. When the hero throws acid (which is actually vinegar, a dilute solution of acetic acid) into the monster's face, the acetic acid reacts with sodium hydrogen carbonate to produce the disgustingly familiar scenes of "dissolving flesh" that we see in horror movies. The ability of baking soda to produce carbon dioxide delights children of all ages as it creates monsters in the movies.

Many early fire extinguishers utilized the reaction of sodium hydrogen carbonate with acids. A metal cylinder was filled with a solution of sodium hydrogen carbonate and water; a bottle filled with sulfuric acid was placed above the water layer. In-

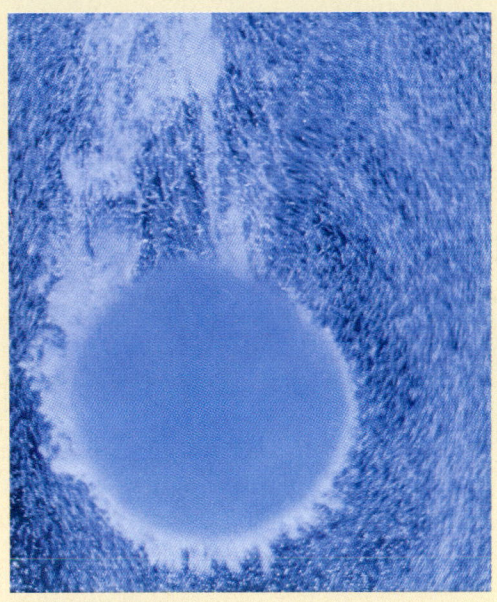

Alka Seltzer™.

verting the extinguisher activated it by causing the acid to spill into the carbonate solution. The pressure produced by gaseous carbon dioxide gas pushed the liquid contents out through a small hose.

Kitchen oven fires can usually be extinguished by throwing baking soda onto the flame. When heated, carbonates decompose to produce carbon dioxide, which smothers fires by depriving them of oxygen.

Chefs frequently use the heat-sensitive nature of carbonates to test the freshness of a box of baking soda. Pouring some boiling water over a little fresh baking soda results in active bubbling. Less active bubbling means the baking soda is unlikely to work well in a baking recipe.

Ronald DeLorenzo
Middle Georgia College

$$HCl + NH_3 \longrightarrow NH_4Cl \quad (\sim 100\%)$$

added acid   base   salt

When a strong soluble base such as NaOH is added to the *original* buffer solution, it is neutralized by the more acidic component, NH_4Cl, or NH_4^+, the conjugate acid of ammonia.

$$NH_3 + H_2O \rightleftharpoons NH_4^+ + OH^- \quad \text{(shifts } \textit{left}\text{)}$$

Because the $[NH_4^+]$ is high, this can occur to a great extent. The result is the neutralization of OH^- by NH_4^+.

$$OH^- + NH_4^+ \longrightarrow NH_3 + H_2O \qquad (\sim 100\%)$$

or, as a formula unit equation,

$$\underset{\text{added base}}{NaOH} + \underset{\text{acid}}{NH_4Cl} \longrightarrow \underset{\text{weak base}}{NH_3} + \underset{\text{water}}{H_2O} + NaCl \qquad (\sim 100\%)$$

The net effect is to neutralize most of the OH^- from NaOH. This slightly decreases the ratio $[NH_4^+]/[NH_3]$, which governs the pH of the solution.

Summary   Changes in pH are minimized in buffer solutions because the basic component can react with H_3O^+ ions and the acidic component can react with OH^- ions.

18-8 PREPARATION OF BUFFER SOLUTIONS

Buffer solutions can be prepared by mixing other solutions. When solutions are mixed, the volume in which each solute is contained increases, so solute concentrations change. These changes in concentration must be considered. If the solutions are dilute, we may assume that their volumes are additive.

EXAMPLE 18-19   *Preparation of Buffer Solutions*

Calculate the concentration of H_3O^+ in a buffer solution prepared by mixing 200 mL of 0.10 M NaF and 100 mL of 0.050 M HF. $K_a = 7.2 \times 10^{-4}$ for HF.

Plan

Calculate the number of millimoles (or moles) of NaF and HF and then the molarity of each solute in the solution after mixing. Write the appropriate equations for both NaF and HF, represent the equilibrium concentrations algebraically, and substitute into the K_a expression for HF.

Solution

When two dilute solutions are mixed, we assume that their volumes are additive. The volume of the new solution will be 300 mL. Mixing a solution of a weak acid with a solution of its salt does not form any new species. So we have a straightforward buffer calculation. We calculate the number of millimoles (or moles) of each compound and the molarities in the new solution.

Recall that

$$M = \frac{\text{\# mol of solute}}{\text{L}}$$

or

$$M = \frac{\text{\# mmol of solute}}{\text{mL}}$$

$$\underset{-}{?} \text{ mmol NaF} = 200 \text{ mL} \times \frac{0.10 \text{ mmol NaF}}{\text{mL}} = 20 \text{ mmol NaF}$$

$$\underset{-}{?} \text{ mmol HF} = 100 \text{ mL} \times \frac{0.050 \text{ mmol HF}}{\text{mL}} = 5.0 \text{ mmol HF}$$

$\left. \right\}$ in 300 mL

The molarities of NaF and HF in the solution are

$$\frac{20 \text{ mmol NaF}}{300 \text{ mL}} = 0.067 \, M \text{ NaF} \qquad \text{and} \qquad \frac{5.0 \text{ mmol HF}}{300 \text{ mL}} = 0.017 \, M \text{ HF}$$

The appropriate equations and algebraic representations of concentrations are

$$\begin{array}{ccccc}
\text{NaF} & \longrightarrow & \text{Na}^+ & + & \text{F}^- \\
0.067 \, M & & 0.067 \, M & & 0.067 \, M \\
\\
\text{HF} & + H_2O \rightleftharpoons & H_3O^+ & + & \text{F}^- \\
(0.017 - x) \, M & & x \, M & & x \, M
\end{array}$$

$\dashrightarrow$ Total $[F^-] = (0.067 + x) \, M$

Substituting into the K_a expression for hydrofluoric acid gives

$$K_a = \frac{[H_3O^+][F^-]}{[HF]} = \frac{(x)(0.067 + x)}{(0.017 - x)} = 7.2 \times 10^{-4}$$

Can we assume that x is negligible compared with 0.067 and 0.017 in this expression? When in doubt, solve the equation using the simplifying assumption. Then decide if the assumption was valid. Assume that $(0.067 + x) \approx 0.067$ and $(0.017 - x) \approx 0.017$.

$$\frac{0.067x}{0.017} = 7.2 \times 10^{-4} \qquad x = \boxed{1.8 \times 10^{-4} \, M = [H_3O^+]}$$

Our assumption is valid.

You should now work Exercise 82.

We often need a buffer solution of a given pH. One method by which such solutions can be prepared involves adding a salt of a weak base (or weak acid) to a solution of the weak base (or weak acid).

EXAMPLE 18-20 *Preparation of Buffer Solutions*

Calculate the numbers of moles and grams of NH_4Cl that must be used to prepare 500 mL of a buffer solution that is 0.10 M in aqueous NH_3 and has a pH of 9.15.

Plan

Convert the given pH to the desired $[OH^-]$ by the usual procedure. Write the appropriate equations for the reactions of NH_4Cl and NH_3 and represent the equilibrium concentrations. Then, substitute into the K_b expression and solve for the concentration of NH_4Cl required.

Solution

Because the desired pH = 9.15, pOH = 14.00 − 9.15 = 4.85.
So $[OH^-] = 10^{-pOH} = 10^{-4.85} = 1.4 \times 10^{-5} \, M \, OH^-$ desired.
Let x mol/L be the necessary molarity of NH_4Cl. Because $[OH^-] = 1.4 \times 10^{-5} \, M$, this must be the $[OH^-]$ produced by ionization of NH_3. The equations and representations of equilibrium concentrations follow.

$$
\begin{array}{ccccc}
NH_4Cl & \xrightarrow{100\%} & NH_4^+ & + & Cl^- \\
x \, M & \Longrightarrow & x \, M & & x \, M \\
\\
NH_3 \quad + \quad H_2O & \Longrightarrow & NH_4^+ & + & OH^- \\
(0.10 - 1.4 \times 10^{-5}) \, M & & 1.4 \times 10^{-5} \, M & & 1.4 \times 10^{-5} \, M
\end{array}
$$

$$\text{Total } [NH_4^+] = (x + 1.4 \times 10^{-5}) \, M$$

Substitution into the K_b expression for aqueous ammonia gives

$$K_b = \frac{[NH_4^+][OH^-]}{[NH_3]} = 1.8 \times 10^{-5} = \frac{(x + 1.4 \times 10^{-5})(1.4 \times 10^{-5})}{0.10 - 1.4 \times 10^{-5}}$$

NH_4Cl is 100% dissociated, so $x \gg 1.4 \times 10^{-5}$. Then $(x + 1.4 \times 10^{-5}) \approx x$.

$$\frac{(x)(1.4 \times 10^{-5})}{0.10} = 1.8 \times 10^{-5} \qquad x = 0.13 \, M = [NH_4^+] = M_{NH_4Cl}$$

Here x does *not* represent a *change* in concentration, but rather the initial concentration of NH_4Cl. We do *not* assume that $x \ll 1.4 \times 10^{-5}$, but rather the reverse.

Now we calculate the number of moles of NH_4Cl that must be added to prepare 500 mL (0.500 L) of buffer solution.

$$\underline{?} \text{ mol } NH_4Cl = 0.500 \text{ L} \times \frac{0.13 \text{ mol } NH_4Cl}{L} = 0.065 \text{ mol } NH_4Cl \quad (3.5 \text{ g } NH_4Cl)$$

You should now work Exercise 85.

18-9 POLYPROTIC ACIDS

Thus far we have considered only *monoprotic* weak acids. Acids that can furnish *two or more* hydronium ions per molecule are called **polyprotic acids.** The ionizations of polyprotic acids occur stepwise, i.e., one proton at a time. An ionization constant expression can be written for each step, as the following example illustrates. Consider phosphoric acid as a typical polyprotic acid. It contains three acidic hydrogen atoms and ionizes in three steps.

$$H_3PO_4 + H_2O \rightleftharpoons H_3O^+ + H_2PO_4^- \qquad K_1 = \frac{[H_3O^+][H_2PO_4^-]}{[H_3PO_4]} = 7.5 \times 10^{-3}$$

$$H_2PO_4^- + H_2O \rightleftharpoons H_3O^+ + HPO_4^{2-} \qquad K_2 = \frac{[H_3O^+][HPO_4^{2-}]}{[H_2PO_4^-]} = 6.2 \times 10^{-8}$$

$$HPO_4^{2-} + H_2O \rightleftharpoons H_3O^+ + PO_4^{3-} \qquad K_3 = \frac{[H_3O^+][PO_4^{3-}]}{[HPO_4^{2-}]} = 3.6 \times 10^{-13}$$

Each K expression includes $[H_3O^+]$, so each K expression must be satisfied by the *total* concentration of H_3O^+ in the solution.

We see that K_1 is much greater than K_2 and that K_2 is much greater than K_3. This is generally true for polyprotic *inorganic* acids (Appendix F). Successive ionization constants often decrease by a factor of approximately 10^4 to 10^6, although some differences are outside this range. Large decreases in the values of successive ionization constants mean that each step in the ionization of a polyprotic acid occurs to a much lesser extent than the previous step. Thus, the $[H_3O^+]$ produced in the first step is very large compared

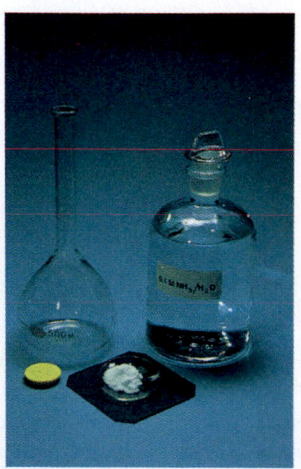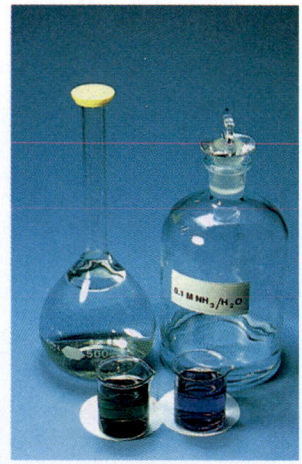

Preparation of the buffer solution in Example 18-20. We add 3.5 grams of NH_4Cl to a 500-mL volumetric flask and dissolve it in a little of the 0.10 M NH_3 solution. We then dilute to 500 mL with the 0.10 M NH_3 solution.

with the $[H_3O^+]$ produced in the second and third steps. As we shall see, except in extremely dilute solutions of H_3PO_4, the concentration of H_3O^+ may be assumed to be that furnished by the first step in the ionization alone.

EXAMPLE 18-21 *Solutions of Weak Polyprotic Acid*

Calculate the concentrations of all species present in 0.100 M H_3PO_4.

Plan
Because H_3PO_4 contains three acidic hydrogens per formula unit, we show its ionization in three steps. For each step, write the appropriate ionization equation, with its K_a expression and value. Then, represent the equilibrium concentrations from the *first*-step ionization and substitute into the K_1 expression. Repeat the procedure for the second and third steps *in order*.

Solution
First we calculate the concentrations of all species formed in the first-step ionization. Let $x = $ mol/L of H_3PO_4 that ionize; then $x = [H_3O^+]_{1st} = [H_2PO_4^-]$.

$$H_3PO_4 \quad + H_2O \rightleftharpoons H_3O^+ + H_2PO_4^-$$
$$(0.100 - x) \, M \qquad\qquad x \, M \qquad x \, M$$

Substitution into the expression for K_1 gives

$$\frac{[H_3O^+][H_2PO_4^-]}{[H_3PO_4]} = 7.5 \times 10^{-3} = \frac{(x)(x)}{(0.100 - x)}$$

This equation must be solved by the quadratic formula because K is too large to neglect x relative to 0.100 M. Solving gives the *positive* root $x = 2.4 \times 10^{-2}$. Thus, from the first step in the ionization of H_3PO_4,

$x = -3.1 \times 10^{-2}$ is the extraneous root of the quadratic equation.

$$x \, M = \quad [H_3O^+]_{1st} = [H_2PO_4^-] = 2.4 \times 10^{-2} \, M$$
$$(0.100 - x) \, M = \quad [H_3PO_4] = 7.6 \times 10^{-2} \, M$$

For the second step, we use the $[H_3O^+]$ and $[H_2PO_4^-]$ from the first step. Let $y = $ mol/L of $H_2PO_4^{2-}$ that ionize; then $y = [H_3O^+]_{2nd} = [HPO_4^{2-}]$.

$$H_2PO_4^- \quad + H_2O \rightleftharpoons \quad H_3O^+ \quad + HPO_4^{2-}$$
$$(2.4 \times 10^{-2} - y) \, M \qquad (2.4 \times 10^{-2} + y) \, M \qquad y \, M$$

from 1st step   from 2nd step

Substitution into the expression for K_2 gives

$$\frac{[H_3O^+][HPO_4^{2-}]}{[H_2PO_4^-]} = 6.2 \times 10^{-8} = \frac{(2.4 \times 10^{-2} + y)(y)}{(2.4 \times 10^{-2} - y)}$$

Examination of this equation suggests that $y \ll 2.4 \times 10^{-2}$, so

$$\frac{(2.4 \times 10^{-2})(y)}{(2.4 \times 10^{-2})} = 6.2 \times 10^{-8} \qquad y = \quad 6.2 \times 10^{-8} \, M = [HPO_4^{2-}] \quad = [H_3O^+]_{2nd}$$

We see that $[HPO_4^{2-}] = K_2$ and $[H_3O^+]_{2nd} \ll [H_3O^+]_{1st}$. In general, in solutions of reasonable concentration of weak polyprotic acids for which $K_1 \gg K_2$ and that contain no other electrolytes, *the concentration of the anion produced in the second-step ionization is always equal to* K_2.

For the third step, we use $[H_3O^+]$ from the *first* step and $[HPO_4^{2-}]$ from the *second* step. Let $z = $ mol/L of HPO_4^{2-} that ionize; then $z = [H_3O^+]_{3rd} = [PO_4^{3-}]$.

The pH of solutions of most polyprotic acids is governed by the first-step ionization.

$$HPO_4^{2-} + H_2O \rightleftharpoons H_3O^+ + PO_4^{3-}$$
$$(6.2 \times 10^{-8} - z)\, M \qquad (2.4 \times 10^{-2} + z)\, M \qquad z\, M$$

from 2nd step   from 3rd step   from 1st step   from 3rd step

$$\frac{[H_3O^+][PO_4^{3-}]}{[HPO_4^{2-}]} = 3.6 \times 10^{-13} = \frac{(2.4 \times 10^{-2} + z)(z)}{(6.2 \times 10^{-8} - z)}$$

We make the usual simplifying assumption, and find that

$$z\, M = \boxed{9.3 \times 10^{-19}\, M} = [PO_4^{3-}] = [H_3O^+]_{3rd}$$

$y = 6.2 \times 10^{-8}$ was disregarded in the second step and is also disregarded here.

You should now work Exercise 86.

We have calculated the concentrations of the species formed by the ionization of 0.100 M H_3PO_4. These concentrations are compared in Table 18-10. The concentration of [OH$^-$] in 0.100 M H_3PO_4 is included. It was calculated from the known [H$_3$O$^+$] using the ion product for water, $[H_3O^+][OH^-] = 1.0 \times 10^{-14}$.

Nonionized H_3PO_4 is present in greater concentration than any other species in 0.100 M H_3PO_4 solution. The only other species present in significant concentrations are H_3O^+ and $H_2PO_4^-$. Similar statements can be made for other weak polyprotic acids for which the last K is very small.

Phosphoric acid is a typical *weak* polyprotic acid. Let us now describe solutions of sulfuric acid, a *very strong* polyprotic acid.

EXAMPLE 18-22 *Solutions of Strong Polyprotic Acid*

Calculate concentrations of the species in 0.10 M H_2SO_4. $K_2 = 1.2 \times 10^{-2}$.

Plan

Because the first-step ionization of H_2SO_4 is complete, we read the concentrations for the first step from the balanced equation. The second-step ionization is *not* complete, and so we write the ionization equation, the K_2 expression, and the algebraic representations of equilibrium concentrations. Then we substitute into K_2 for H_2SO_4.

Solution

As we pointed out, the first-step ionization of H_2SO_4 is complete.

$$H_2SO_4 + H_2O \xrightarrow{100\%} H_3O^+ + HSO_4^-$$
$$0.10\, M \rightleftharpoons 0.10\, M \quad 0.10\, M$$

However, the second-step ionization is not complete.

$$HSO_4^- + H_2O \rightleftharpoons H_3O^+ + SO_4^{2-} \quad \text{and} \quad K_2 = \frac{[H_3O^+][SO_4^{2-}]}{[HSO_4^-]} = 1.2 \times 10^{-2}$$

Let $x = [HSO_4^-]$ that ionizes. [H$_3$O$^+$] is the sum of the concentrations produced in the first and second steps. So we represent the equilibrium concentrations as

$$HSO_4^- + H_2O \rightleftharpoons H_3O^+ + SO_4^{2-}$$
$$(0.10 - x)\, M \qquad (0.10 + x)\, M \quad x\, M$$

from 1st step   from 2nd step

Substitution into the ionization constant expression for K_2 gives

$$K_2 = \frac{[H_3O^+][SO_4^{2-}]}{[HSO_4^-]} = 1.2 \times 10^{-2} = \frac{(0.10 + x)(x)}{0.10 - x}$$

Clearly, x cannot be disregarded because K is too large. This equation must be solved by the quadratic formula, which gives $x = 0.010$ and $x = -0.12$ (extraneous). So $[H_3O^+]_{2nd} = [SO_4^{2-}] = 0.010\ M$. The concentrations of species in $0.10\ M\ H_2SO_4$ are

$$[H_2SO_4] \approx 0\ M \qquad [HSO_4^-] = (0.10 - x)\ M = 0.09\ M \qquad [SO_4^{2-}] = 0.010\ M$$

$$[H_3O^+] = (0.10 + x)\ M = 0.11\ M$$

$$[OH^-] = \frac{K_w}{[H_3O^+]} = \frac{1.0 \times 10^{-14}}{0.11} = 9.1 \times 10^{-14}\ M$$

In $0.10\ M\ H_2SO_4$ solution, the extent of the second-step ionization is 10%.

You should now work Exercise 88.

In Table 18-11 we compare $0.10\ M$ solutions of these two polyprotic acids. Their acidities are very different.

Table 18-10 *Concentrations of the Species in $0.10\ M\ H_3PO_4$*

Species	Concentration (mol/L)
H_3PO_4	$7.6 \times 10^{-2} = 0.076$
H_3O^+	$2.4 \times 10^{-2} = 0.024$
$H_2PO_4^-$	$2.4 \times 10^{-2} = 0.024$
HPO_4^{2-}	$6.2 \times 10^{-8} = 0.000000062$
OH^-	$4.2 \times 10^{-13} = 0.00000000000042$
PO_4^{3-}	$9.3 \times 10^{-19} = 0.00000000000000000093$

Table 18-11 *Comparison of $0.10\ M$ Solutions of Two Polyprotic Acids*

	$0.10\ M$ H_3PO_4	$0.10\ M$ H_2SO_4
K_1	7.5×10^{-3}	very large
K_2	6.2×10^{-8}	1.2×10^{-2}
K_3	3.6×10^{-13}	
$[H_3O^+]$	$2.4 \times 10^{-2}\ M$	$0.11\ M$
[acid molecules]	$7.6 \times 10^{-2}\ M$	$\sim 0\ M$

The two solutions in Table 18-11.

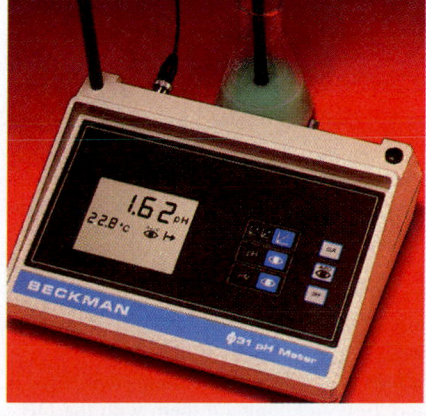

$0.10\ M\ H_3PO_4$
(Example 18-21)

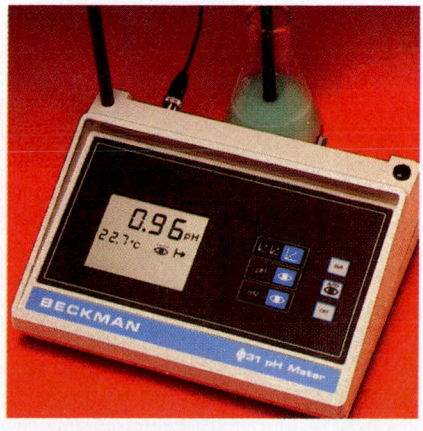

$0.10\ M\ H_2SO_4$
(Example 18-22)

Key Terms

Amines   Derivatives of ammonia in which one or more hydrogen atoms have been replaced by organic groups.

Buffer solution   A solution that resists changes in pH when acids or bases are added. A buffer solution contains an acid and its conjugate base, so it can react with added base or acid. Common buffer solutions contain either (1) a weak acid and a soluble ionic salt of the weak acid *or* (2) a weak base and a soluble ionic salt of the weak base.

Common ion effect   Suppression of ionization of a weak electrolyte by the presence in the same solution of a strong electrolyte containing one of the same ions as the weak electrolyte.

Henderson–Hasselbalch equation   An equation that enables us to calculate the pH or pOH of a buffer solution directly.

For acid/salt buffer $pH = pK_a + \log \dfrac{[salt]}{[acid]}$

For base/salt buffer $pOH = pK_b + \log \dfrac{[salt]}{[base]}$

Indicator   An organic compound that exhibits different colors in solutions of different acidities. Many indicators are weak acids,

HIn, in which HIn and In⁻ have different colors. Indicators may be used to estimate the pH of a solution.

Ion product for water   An equilibrium constant for the ionization of water,

$$K_w = [H_3O^+][OH^-] = 1.00 \times 10^{-14} \text{ at } 25°C$$

Ionization constant   An equilibrium constant for the ionization of a weak electrolyte.

Monoprotic acid   An acid that can form only one hydronium ion per molecule; may be strong or weak.

p[ ]   The negative logarithm of the concentration (mol/L) of the indicated species.

pH   The negative logarithm of the concentration (mol/L) of the $H_3O^+[H^+]$ ion; the commonly used scale ranges from 0 to 14.

pK_a   The negative logarithm of K_a, the ionization constant for a weak acid.

pK_b   The negative logarithm of K_b, the ionization constant for a weak base.

pOH   The negative logarithm of the concentration (mol/L) of the OH⁻ ion; the commonly used scale ranges from 14 to 0.

Polyprotic acid   An acid that can form two or more hydronium ions per molecule; often at least one step of the ionization is weak.

Exercises

NOTE *All exercises in this chapter assume a temperature of 25°C unless they specify otherwise. All logarithms are common (base 10).*

A Brief Review of Strong Electrolytes

1. List names and formulas for (a) the common strong acids; (b) six weak bases; (c) the common strong soluble bases; (d) ten soluble ionic salts.

2. Calculate the molarity of each of the following solutions. (a) 17.52 g of NaCl in 125 mL of solution; (b) 55.0 g of H_2SO_4 in 575 mL of solution; (c) 0.135 g of phenol, C_6H_5OH, in 1.5 L of solution.

3. Calculate the molarity of each of the following solutions. (a) 0.038 g of barium chloride in 50.0 mL of solution; (b) 0.050 g of iodine in 175 mL of solution; (c) 0.00055 g of HCl in 1.00 L of solution.

4. Calculate the concentrations of the constituent ions in solutions of the following compounds in the indicated concentrations. (a) 0.25 *M* HBr; (b) 0.055 *M* KOH; (c) 0.0020 *M* $CaCl_2$.

5. Calculate the concentrations of the constituent ions in solutions of the following compounds in the indicated concentrations. (a) 0.025 *M* $Sr(OH)_2$; (b) 0.00035 *M* $HClO_3$; (c) 0.0040 *M* K_2SO_4.

6. Calculate the concentrations of the constituent ions in the following solutions. (a) 2.5 g of KOH in 1.50 L of solution; (b) 0.72 g of $Ba(OH)_2$ in 250 mL of solution; (c) 2.64 g of $Ca(NO_3)_2$ in 100 mL of solution.

7. Calculate the concentrations of the constituent ions in the following solutions. (a) 1.77 g of $Al_2(SO_4)_3$ in 400 mL of solution; (b) 75.8 g of $CaCl_2 \cdot 6H_2O$ in 8.00 L of solution; (c) 18.4 g of HBr in 450 mL of solution.

The Autoionization of Water

8. (a) Write a chemical equation showing the ionization of water. (b) Write the equilibrium constant expression for this equation. (c) What is the special symbol used for this equilibrium constant? (d) What is the relationship between $[H^+]$ and $[OH^-]$ in pure water? (e) How can this relationship be used to define the terms "acidic" and "basic"?

9. Write mathematical definitions for pH and pOH. What is the relationship between pH and pOH? How can pH be used to define the terms "acidic" and "basic"?

10. (a) Why is the concentration of OH⁻ produced by the ionization of water neglected in calculating the concentration of OH⁻ in a 0.10 *M* solution of NaOH? (b) Demonstrate that it (OH⁻ from H_2O) may be neglected.

11. Calculate the concentrations of OH⁻ in the solutions described in Exercises 4(a), 5(b), and 7(c), and compare them with the OH⁻ concentration in pure water.

12. Calculate the concentrations of H_3O^+ in the solutions described in Exercises 4(b), 5(a), and 6(b), and compare them with the H_3O^+ concentration in pure water.

13. Calculate $[OH^-]$ that is in equilibrium with
    (a) $[H_3O^+] = 1.8 \times 10^{-4}$ mol/L
    (b) $[H_3O^+] = 8 \times 10^{-9}$ mol/L

14. Calculate $[H_3O^+]$ that is in equilibrium with

    $$[OH^-] = 6.32 \times 10^{-6} \text{ mol/L}$$

15. The equilibrium constant of the following reaction is 1.35×10^{-15} at 25°C.

    $$2D_2O \rightleftharpoons D_3O^+ + OD^-$$

    D is deuterium, ^2H. Calculate the pD of pure deuterium oxide (heavy water) at 25°C. What is the relationship between $[D_3O^+]$ and $[OD^-]$ in pure D_2O? Is pure D_2O acidic, basic, or neutral?

16. What are the logarithms of the following numbers? (a) 0.000052; (b) 5.7; (c) 5.8×10^{-12}; (d) 4.9×10^{-7}.

17. Refer to Exercise 11 and calculate the pOH of each solution.

18. Refer to Exercise 12 and calculate the pH of each solution.

19. Calculate the pH of a 1.5×10^{-4} M solution of $HClO_4$, a strong acid, at 25°C.

20. Calculate the pH of the following solutions. (a) 5.00×10^{-1} M HCl; (b) 0.030 M HNO_3; (c) 0.75 g·L^{-1} $HClO_4$.

21. Calculate the pH of the following solutions. (a) 7.5×10^{-2} M HBr; (b) 0.0062 M HI; (c) 2.84 g HNO_3 in 250 mL solution.

22. Calculate the pH of a 2.5×10^{-4} M solution of NaOH at 25°C.

23. Calculate the pH of a 1.5×10^{-11} M solution of HCl at 25°C.

24. A solution of HNO_3 has pH 3.52. What is the molarity of the solution?

25. Complete the following table. Is there an obvious relationship between pH and pOH? What is it?

Solution	$[H_3O^+]$	$[OH^-]$	pH	pOH
0.15 M HI	___	___	___	___
0.040 M RbOH	___	___	___	___
0.020 M Ba(OH)$_2$	___	___	___	___
0.00030 M HClO$_4$	___	___	___	___

26. Calculate the following values for each solution.

Solution	$[H_3O^+]$	$[OH^-]$	pH	pOH
(a) 0.055 M NaOH	___	___	___	___
(b) 0.055 M HCl	___	___	___	___
(c) 0.055 M Ca(OH)$_2$	___	___	___	___

27. Complete the following table by appropriate calculations.

$[H_3O^+]$	pH	$[OH^-]$	pOH
(a) ___	3.84	___	___
(b) ___	12.61	___	___
(c) ___	___	___	1.34
(d) ___	___	___	9.47

Ionization Constants for Weak Monoprotic Acids and Bases

28. Write a chemical equation that represents the ionization of a weak acid HA. Write the equilibrium constant expression for this reaction. What is the special symbol used for this equilibrium constant?

29. What is the relationship between the strength of an acid and the numerical value of K_a? What is the relationship between the acid strength and the value of pK_a?

30. (a) What is the pH of pure water at body temperature, 37°C? Refer to Table 18-2. (b) Is this acidic, basic, or neutral? Why?

31. Fill in the blanks in this table for given solutions a, b, c, and d.

Sol'n	Temp. (°C)	Concentration (mol/L)		pH
		$[H^+]$	$[OH^-]$	
(a)	25	1.0×10^{-6}	___	___
(b)	0	___	___	5.69
(c)	60	___	___	7.00
(d)	25	___	4.5×10^{-9}	___

32. Write a chemical equation that represents the equilibrium between water and a weak base B. Write the equilibrium constant expression for this reaction. What is the special symbol used for this equilibrium constant?

33. What is the relationship between base strength and the value of K_b? What is the relationship between base strength and the value of pK_b?

34. A 0.083 M solution of a monoprotic acid is known to be 1.07% ionized. What is the pH of the solution? Calculate the value of K_a for this acid.

35. A 0.045 M aqueous solution of a weak, monoprotic acid is 0.85% ionized. Calculate the ionization constant for this acid.

36. The pH of a 0.025 M solution of butanoic acid, C_3H_7COOH, is 3.21. What is the value of the ionization constant for butanoic acid?

37. The pH of a 0.35 M solution of uric acid is 2.17. What is the value of K_a for uric acid, a monoprotic acid?

38. Calculate the concentrations of all the species present in a 0.35 M benzoic acid solution. (See Appendix F.)

39. Find the concentrations of the various species present in a 0.25 M solution of hydrofluoric acid, HF. What is the pH of the solution?

40. What is the percent ionization in 0.0500 M solution of formic acid, HCOOH?

41. What is the percent ionization in (a) 0.100 M CH$_3$COOH solution and (b) 0.0100 M CH$_3$COOH solution?

42. The K_a values for benzoic acid and hydrocyanic acid are 6.3×10^{-5} and 4.0×10^{-10}, respectively. What are their pK_a values?

43. What is the concentration of OI^- in equilibrium with $[H^+] = 0.035$ mol/L and $[HOI] = 0.427$ mol/L? $K_a = 2.3 \times 10^{-11}$ for HOI.

44. Answer the following questions for 0.10 M solutions of the weak bases listed in Table 18-6. (a) In which solution is (i) the pH highest; (ii) the pH lowest; (iii) the pOH highest; (iv) the pOH lowest? (b) Which solution contains (i) the highest concentration

of the cation of the weak base; (ii) the lowest concentration of the cation of the weak base?

45. Answer the following questions for 0.10 M solutions of the weak acids listed in Table 18-4. Which solution contains (a) the highest concentration of H_3O^+; (b) the highest concentration of OH^-; (c) the lowest concentration of H_3O^+; (d) the lowest concentration of OH^-; (e) the highest concentration of nonionized acid molecules; (f) the lowest concentration of nonionized acid molecules?

46. Pyridine is 0.053% ionized in 0.00500 M solution. What is the pK_b of this monobasic compound?

47. A 0.068 M solution of benzamide has a pOH of 2.91. What is the value of pK_b for this monobasic compound?

48. In a 0.0100 M aqueous solution of methylamine, CH_3NH_2, the equilibrium concentrations of the species are $[CH_3NH_2] = 0.0080$ mol/L and $[CH_3NH_3^+] = [OH^-] = 2.0 \times 10^{-3}$ mol/L. Calculate K_b for this weak base.

$$CH_3NH_2(aq) + H_2O(\ell) \rightleftharpoons CH_3NH_3^+ + OH^-$$

49. What is the concentration of NH_3 in equilibrium with $[NH_4^+] = 0.015$ mol/L and $[OH^-] = 1.2 \times 10^{-5}$ mol/L?

50. Calculate $[OH^-]$, percent ionization, and pH for (a) 0.10 M aqueous ammonia, and (b) 0.10 M methylamine solution.

51. Calculate $[H_3O^+]$, $[OH^-]$, pH, pOH, and percent ionization for 0.16 M aqueous ammonia solution.

52. Because K_b is larger for triethylamine

$$(C_2H_5)_3N(aq) + H_2O(\ell) \rightleftharpoons (C_2H_5)_3NH^+ + OH^-$$
$$K_b = 5.2 \times 10^{-4}$$

than for trimethylamine

$$(CH_3)_3N(aq) + H_2O(\ell) \rightleftharpoons (CH_3)_3NH^+ + OH^-$$
$$K_b = 7.4 \times 10^{-5}$$

an aqueous solution of triethylamine should have a larger concentration of OH^- ion than an aqueous solution of trimethylamine of the same concentration. Confirm this statement by calculating the $[OH^-]$ for 0.015 M solutions of both weak bases.

53. The odor of cooked fish is due to the presence of amines. This odor is lessened by adding lemon juice, which contains citric acid. Why does this work?

Acid–Base Indicators

54. What are acid–base indicators? (b) What are the essential characteristics of acid–base indicators? (c) What determines the color of an acid–base indicator in an aqueous solution?

55. K_a is 7.9×10^{-8} for bromthymol blue, an indicator that can be represented as HIn. HIn molecules are yellow, and In^- ions are blue. What color will bromthymol blue be in a solution in which (a) $[H_3O^+] = 1.0 \times 10^{-4} M$ and (b) pH = 10.30?

*56. The indicator metacresol purple changes from yellow to purple at pH 8.2. At this point it exists in equal concentrations as the conjugate acid and the conjugate base. What are K_a and pK_a for metacresol purple, a weak acid represented as HIn?

57. Demonstrate mathematically that neutral red is red in solutions of pH 3.00, whereas it is yellow in solutions of pH 10.00. HIn is red, and In^- is yellow. K_a is 2.0×10^{-7}.

The Common Ion Effect and Buffer Solutions

58. A solution is 0.40 M in HCl and 1.00 M in formic acid, $HCHO_2$. Calculate the concentration of CHO_2^- ions in the solution.

59. Suppose that you have a solution that is 0.50 M in methylamine, CH_3NH_2, and 0.00050 M in the salt methylammonium chloride, CH_3NH_3Cl. Would you expect this to be an effective buffer solution? Why or why not?

60. Calculate pH for each of the following buffer solutions. (a) 0.15 M HF and 0.25 M KF; (b) 0.050 M CH_3COOH and 0.025 M $Ba(CH_3COO)_2$.

61. The pK_a of HOCl is 7.45. Calculate the pH of a solution that is 0.0493 M HOCl and 0.0872 M NaOCl.

62. Calculate the concentration of OH^- and the pH for the following buffer solutions. (a) 0.30 M $NH_3(aq)$ and 0.20 M NH_4NO_3; (b) 0.15 M $NH_3(aq)$ and 0.20 M $(NH_4)_2SO_4$.

63. Calculate the concentration of OH^- and the pH for the following solutions. (a) 0.45 M $NH_3(aq)$ and 0.20 M NH_4NO_3; (b) 0.10 M aniline and 0.20 M anilinium chloride, $C_6H_5NH_3Cl$.

*64. Buffer solutions are especially important in our body fluids and metabolism. Write net ionic equations to illustrate the buffering action of (a) the $H_2CO_3/NaHCO_3$ buffer system in blood and (b) the NaH_2PO_4/Na_2HPO_4 buffer system inside cells.

65. Calculate the ratio of concentrations $[NH_3]/[NH_4^+]$ that gives (a) solutions of pH = 9.75 and (b) solutions of pH = 9.10.

Buffering Action

66. Consider the ionization of formic acid, HCOOH.

$$HCOOH + H_2O \rightleftharpoons HCOO^- + H_3O^+$$

What effect does the addition of sodium formate (NaHCOO) have on the fraction of formic acid molecules that undergo ionization in aqueous solution?

67. Briefly describe why the pH of a buffer solution remains nearly constant when small amounts of acid or base are added. Over what pH range do we observe the best buffering action (nearly constant pH)?

68. What is the pH of a solution that is 0.10 M in $HClO_4$ and 0.10 M $KClO_4$? Is this a buffer solution? $HClO_4$ is a strong acid.

69. (a) Find the pH of a solution 0.50 M in formic acid and 0.40 M in sodium formate. (b) Find the pH of the solution after 0.050 mol HCl/L has been added to it.

70. One liter of 0.400 M NH_3 solution also contains 12.78 g of NH_4Cl. How much will the pH of this solution change if 0.155 mole of gaseous HCl is bubbled into it?

71. (a) Find the pH of a solution that is 0.12 M in nitrous acid and 0.15 M in sodium nitrite. (b) Find the pH after 0.15 mol NaOH has been added to 1.0 liter of the solution.

72. (a) Find the pH of a solution that is 1.00 M in NH_3 and 0.80 M in NH_4Cl. (b) Find the pH of the solution after 0.10 mol HCl/L has been added to it. (c) A solution of pH 9.34 is prepared by adding

NaOH to pure water. Find the pH of this solution after 0.10 mol HCl/L has been added to it.

73. (a) Calculate the concentrations of CH_3COOH and CH_3COO^- in a solution in which their total concentration is 0.20 mol/L and the pH is 4.50. (b) If 0.0100 mol of solid NaOH is added to 1.00 L of this solution, how much does pH change?

74. A solution contains bromoacetic acid and sodium bromoacetate with a total concentration of 0.20 mol/L. If the pH is 3.10, what are the concentrations of the acid and the salt? $K_a = 2.0 \times 10^{-3}$ for $BrCH_2COOH$.

75. Calculate the concentration of propionate ion, $CH_3CH_2COO^-$, in equilibrium with 0.020 M CH_3CH_2COOH (propionic acid) and 0.10 M H^+ from hydrochloric acid. $K_a = 1.3 \times 10^{-5}$ for CH_3CH_2COOH.

76. Calculate the concentration of $C_2H_5NH_3^+$ in equilibrium with 0.015 M $C_2H_5NH_2$ (ethylamine) and 0.0010 M OH^- ion from sodium hydroxide. $K_b = 4.7 \times 10^{-4}$ for ethylamine.

77. When chlorine gas is dissolved in water to make "chlorine water," HCl (a strong acid) and HOCl (a weak acid) are produced in equal amounts.

$$Cl_2(g) + H_2O(\ell) \longrightarrow HCl(aq) + HOCl(aq)$$

What is the concentration of OCl^- ion in a solution containing 0.010 mol of each acid in 1.00 L of solution?

Preparation of Buffer Solutions

78. A buffer solution of pH 5.30 is to be prepared from propionic acid and sodium propionate. The concentration of sodium propionate must be 0.50 mol/L. What should be the concentration of the acid? K_a is 1.3×10^{-5} for CH_3CH_2COOH.

79. We need a buffer with pH 9.00. It can be prepared from NH_3 and NH_4Cl. What must be the $[NH_4^+]/[NH_3]$ ratio?

80. What volumes of 0.1500 M acetic acid and 0.1000 M NaOH solutions must be mixed to prepare 1.000 L of a buffer solution of pH 4.50 at 25°C?

81. One liter of a buffer solution is prepared by dissolving 0.150 mol of $NaNO_2$ and 0.050 mol of HCl in water. What is the pH of this solution? If the solution is diluted twofold with water, what is the pH?

82. One liter of a buffer solution is made by mixing exactly 500 mL of 1.25 M acetic acid and 500 mL of 0.500 M calcium acetate solutions. What is the concentration of each of the following in the buffer solution? (a) CH_3COOH; (b) Ca^{2+}; (c) CH_3COO^-; (d) H^+. (e) What is the pH?

83. What must be the concentration of benzoate ion, $C_6H_5COO^-$, in a 0.015 M benzoic acid, C_6H_5COOH, solution so that the pH is 5.00?

84. What must be the concentration of chloroacetic acid, $ClCH_2COOH$, in a 0.015 M $NaCH_2ClCOO$ solution so that the pH is 3.00? $K_a = 1.4 \times 10^{-3}$ for $ClCH_2COOH$.

85. What must be the concentration of NH_4^+ in a 0.075 M NH_3 solution so that the pH is 8.80?

Polyprotic Acids

86. Calculate the concentrations of the various species in 0.200 M H_3AsO_4 solution. Compare the concentrations with those of the analogous species in 0.100 M H_3PO_4 solution (Example 18-21 and Table 18-10).

87. Citric acid, the acid in lemons and other citrus fruits, has the structure

$$
\begin{array}{c}
CH_2COOH \\
| \\
HO-C-COOH \\
| \\
CH_2COOH
\end{array}
$$

which we may abbreviate as $C_3H_5O(COOH)_3$ or H_3A. It is a triprotic acid. Write the chemical equations for the three stages in the ionization of citric acid with the appropriate K_a values.

88. Calculate the concentrations of H_3O^+, OH^-, $HSeO_4^-$, and SeO_4^{2-} in 0.10 M H_2SeO_4, selenic acid, solution.

89. Some kidney stones are crystalline deposits of calcium oxalate, a salt of oxalic acid, $(COOH)_2$. Calculate the concentrations of H_3O^+, OH^-, $COOCOO^-$, and $(COO^-)_2$ in 0.045 M $(COOH)_2$. Compare the concentrations with those obtained in Exercise 88. How can you explain the difference between the concentrations of $HSeO_4^-$ and $COOCOO^-$? between SeO_4^{2-} and $(COO^-)_2$?

Mixed Exercises

90. A solution of oxalic acid, $(COOH)_2$, is 0.20 M. (a) Calculate the pH. (b) What is the concentration of the oxalate ion, $(COO)_2^{2-}$?

91. Calculate the pH at 25°C of the following solutions. (a) 0.0050 M $Ca(OH)_2$; (b) 0.20 M chloroacetic acid, $ClCH_2COOH$, $K_a = 1.4 \times 10^{-3}$; (c) 0.040 M pyridine, C_5H_5N.

*92. What is the pH of a solution that is a mixture of HOCl and HOI, each at 0.12 M concentration? For HOI, $K_a = 2.3 \times 10^{-11}$.

BUILDING YOUR KNOWLEDGE

93. Arrange the following common kitchen samples from most acidic to most basic.

carrot juice, pH 5.1   blackberry juice, pH 3.4

soap, pH 11.0   red wine, pH 3.7

egg white, pH 7.8   milk of magnesia, pH 10.5

sauerkraut, pH 3.5   lime juice, pH 2.0

*94. The buildup of lactic acid in muscles causes pain during extreme physical exertion. The K_a for lactic acid, C_2H_5OCOOH, is 8.4×10^{-4}. Calculate the pH of a 0.100 M solution of lactic acid. Can you make a simplifying assumption in this case?

95. A 0.0100 molal solution of acetic acid ($CH_3COOH +$

$H_2O \rightleftharpoons H_3O^+ + CH_3COO^-$) freezes at $-0.01938°C$. Use this information to calculate the ionization constant for acetic acid. A 0.0100 molal solution is sufficiently dilute that it may be assumed to be 0.0100 molar without introducing a significant error.

96. Ascorbic acid, $C_5H_7O_4COOH$, also known as vitamin C, is an essential vitamin for all mammals. Among mammals, only humans, monkeys, and guinea pigs cannot synthesize it in their bodies. K_a for ascorbic acid is 7.9×10^{-5}. Calculate $[H_3O^+]$ and pH in a 0.100 M solution of ascorbic acid.

Ionic Equilibria II: Hydrolysis and Titrations

Many familiar mouthwashes contain salts of the weak organic base pyridine, $K_b = 1.5 \times 10^{-9}$. These salts produce weakly acidic solutions that are antibacterial agents.

OBJECTIVES

As you study this chapter, you should learn

• *About the concepts of solvolysis and hydrolysis*

• *To apply these concepts to salts of strong bases and weak acids*

• *To apply these concepts to salts of weak bases and strong acids*

• *To apply these concepts to salts of weak bases and weak acids*

• *To apply these concepts to salts of small, highly charged cations*

• *About titration curves for (a) strong acids and strong bases and (b) weak acids and strong bases*

Solvolysis is the reaction of a substance with the solvent in which it is dissolved. The solvolysis reactions that we shall consider in this chapter occur in aqueous solutions so they are called *hydrolysis* reactions. **Hydrolysis** is the reaction of a substance with water. Some hydrolysis reactions involve reaction with H_3O^+ or OH^- ions. One common kind of hydrolysis involves reaction of the anion of a *weak acid* with water to form *nonionized acid* molecules and OH^- ions. This upsets the H_3O^+/OH^- balance in water and produces basic solutions. This reaction is usually represented as

As in Chapter 18 we shall usually omit (aq) from molecules and ions in aqueous solution.

$$\underset{\substack{\text{anion of} \\ \text{weak acid}}}{A^-} + H_2O \rightleftharpoons \underset{\text{weak acid}}{HA} + OH^- \qquad \text{(excess } OH^-\text{, so solution is basic)}$$

Recall that in

neutral solutions	$[H_3O^+] = [OH^-] = 1.0 \times 10^{-7}\,M$
basic solutions	$[H_3O^+] < [OH^-]$ or $[OH^-] > 1.0 \times 10^{-7}\,M$
acidic solutions	$[H_3O^+] > [OH^-]$ or $[H_3O^+] > 1.0 \times 10^{-7}\,M$

In Brønsted–Lowry terminology anions of strong acids are extremely weak bases, whereas anions of weak acids are stronger bases (Section 10-4). To refresh your memory, consider the following examples.

Nitric acid, a common strong acid, is essentially completely ionized in dilute aqueous solution. *Dilute* aqueous solutions of HNO_3 contain equal concentrations of H_3O^+ and NO_3^- ions. In dilute aqueous solution nitrate ions show almost no tendency to react with H_3O^+ ions to form nonionized HNO_3; thus, NO_3^- is a very weak base.

$$HNO_3 + H_2O \xrightarrow{100\%} H_3O^+ + NO_3^-$$

On the other hand, acetic acid (a weak acid) is only slightly ionized in dilute aqueous solution. Acetate ions have a strong tendency to react with H_3O^+ to form CH_3COOH molecules. Acetic acid ionizes only slightly.

$$CH_3COOH + H_2O \rightleftharpoons H_3O^+ + CH_3COO^-$$

Hence, the CH_3COO^- ion is a much stronger base than the NO_3^- ion.

In dilute solutions, strong acids and strong soluble bases are completely ionized or dissociated. Let us now consider dilute aqueous solutions of salts. Based on our classification of acids and bases, we can identify four different kinds of salts.

1. Salts of strong soluble bases and strong acids
2. Salts of strong soluble bases and weak acids
3. Salts of weak bases and strong acids
4. Salts of weak bases and weak acids

Examples of conjugate acid–base pairs.

Acid	Conjugate Base
strong (HCl) $\longrightarrow$	weak (Cl^-)
weak (HCN) $\longrightarrow$	strong (CN^-)

Base	Conjugate Acid
strong (OH^-) $\longrightarrow$	weak (H_2O)
weak (NH_3) $\longrightarrow$	strong (NH_4^+)

19-1 SALTS OF STRONG SOLUBLE BASES AND STRONG ACIDS

We could also describe these as salts that contain the cation of a strong soluble base and the anion of a strong acid. Salts derived from strong soluble bases and strong acids give *neutral* solutions because neither the cation nor the anion reacts with H_2O. Consider an aqueous solution of NaCl, which is the salt of the strong soluble base NaOH and the strong acid HCl. Sodium chloride is ionic even in the solid state. It dissociates into hydrated ions in H_2O. H_2O ionizes slightly to produce equal concentrations of H_3O^+ and OH^- ions.

$$NaCl(solid) \xrightarrow[100\%]{H_2O} Na^+ + Cl^-$$

$$H_2O + H_2O \rightleftharpoons OH^- + H_3O^+$$

We see that aqueous solutions of NaCl contain four ions, Na^+, Cl^-, H_3O^+ and OH^-. The cation of the salt, Na^+, is such a weak acid that it does not react with the anion of water, OH^-. The anion of the salt, Cl^-, is such a weak base that it does not react with the cation of water, H_3O^+. Therefore, solutions of salts of strong bases and strong acids are *neutral* because neither ion of such a salt reacts to upset the H_3O^+/OH^- balance in water.

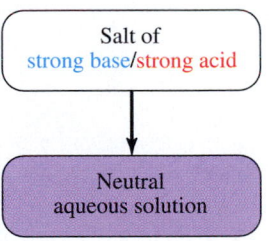

Salt of
strong base/strong acid

↓

Neutral
aqueous solution

19-2 SALTS OF STRONG SOLUBLE BASES AND WEAK ACIDS

When salts derived from strong soluble bases and weak acids are dissolved in water, the resulting solutions are always basic. This is because anions of weak acids react with water to form hydroxide ions. Consider a solution of sodium acetate, $NaCH_3COO$, which is the salt of the strong soluble base NaOH and the weak acid CH_3COOH. It is soluble and dissociates completely in water.

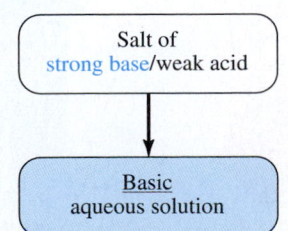

Salt of
strong base/weak acid

↓

Basic
aqueous solution

$$NaCH_3COO(solid) \xrightarrow[100\%]{H_2O} Na^+ + \boxed{CH_3COO^-}$$

$$\underbrace{H_2O + H_2O}_{\text{Equilibrium is shifted}} \rightleftharpoons OH^- + \boxed{H_3O^+}$$

(result is excess OH^-)

$$\Updownarrow$$

$$CH_3COOH + H_2O$$

LeChatelier's Principle applies to equilibria in aqueous solution. It enables us to make accurate predictions.

Consider the similarity of this reaction to the ionization of aqueous ammonia.

$$NH_3 + H_2O \rightleftharpoons NH_4^+ + OH^-$$

Both bases remove H^+ ions from H_2O molecules and form OH^- ions.

Acetate ion is the conjugate base of a *weak* acid, CH_3COOH. Thus, it combines with H_3O^+ to form CH_3COOH. As H_3O^+ is removed from the solution, causing more H_2O to ionize, an excess of OH^- builds up. So the solution becomes basic. The preceding equations can be combined into a single equation.

$$CH_3COO^- + H_2O \rightleftharpoons CH_3COOH + OH^-$$

The equilibrium constant for this reaction is called a (base) hydrolysis constant, or K_b for CH_3COO^-.

$$K_b = \frac{[CH_3COOH][OH^-]}{[CH_3COO^-]} \qquad (K_b \text{ for } CH_3COO^-)$$

We can evaluate this equilibrium constant from other known expressions. We multiply the preceding expression by $[H_3O^+]/[H_3O^+]$ to give

$$K_b = \frac{[CH_3COOH][OH^-]}{[CH_3COO^-]} \times \frac{[H_3O^+]}{[H_3O^+]} = \frac{[CH_3COOH]}{[H_3O^+][CH_3COO^-]} \times \frac{[H_3O^+][OH^-]}{1}$$

We recognize that

$$K_b = \frac{1}{K_{a\,(CH_3COOH)}} \times \frac{K_w}{1} = \frac{K_w}{K_{a\,(CH_3COOH)}} = \frac{1.0 \times 10^{-14}}{1.8 \times 10^{-5}}$$

which gives

$$K_b = \frac{[CH_3COOH][OH^-]}{[CH_3COO^-]} = 5.6 \times 10^{-10}$$

Hydrolysis constants, K_b's, for anions of weak acids can be determined experimentally. The values obtained from experiments agree with the calculated values. Please note that this K_b refers to a reaction in which the anion of a weak acid acts as a base.

We have calculated K_b, the hydrolysis constant for the acetate ion, CH_3COO^-.

We can do the same kind of calculations for the anion of any weak monoprotic acid and find that $K_b = K_w/K_a$ where K_a refers to the ionization constant for the weak monoprotic acid from which the anion is derived.

This equation can be rearranged to

$$K_w = K_a K_b \qquad \text{(valid for } \textit{any conjugate acid–base pair} \text{ in aqueous solution)}$$

If either K_a or K_b is known, the other can be calculated.

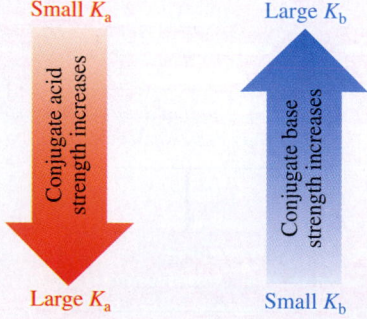

Small K_a — Large K_b

Conjugate acid strength increases

Conjugate base strength increases

Large K_a — Small K_b

EXAMPLE 19-1   *K_b for the Anion of a Weak Acid*

(a) Write the equation for the reaction of the base CN^- with water. (b) The value of the ionization constant for hydrocyanic acid, HCN, is 4.0×10^{-10}. What is the value of K_b for the cyanide ion, CN^-?

Plan

(a) The base CN^- accepts H^+ from H_2O to form the weak acid HCN and OH^- ions. (b) We know that $K_aK_b = K_w$. So we solve for K_b and substitute into the equation.

Solution

(a) $CN^- + H_2O \rightleftharpoons HCN + OH^-$

(b) We are given $K_a = 4.0 \times 10^{-10}$ for HCN, and we know that $K_w = 1.0 \times 10^{-14}$.

$$K_b = \frac{K_w}{K_a} = \frac{1.0 \times 10^{-14}}{4.0 \times 10^{-10}} = \boxed{2.5 \times 10^{-5}}$$

You should now work Exercise 12.

EXAMPLE 19-2 *Calculations Based on Hydrolysis*

Calculate $[OH^-]$, pH, and the percent hydrolysis for 0.10 *M* solutions of (a) $NaCH_3COO$, sodium acetate, and (b) NaCN, sodium cyanide. Both $NaCH_3COO$ and NaCN are soluble ionic salts that are completely dissociated in H_2O. From the text, K_b for $CH_3COO^- = 5.6 \times 10^{-10}$; from Example 19-1, K_b for $CN^- = 2.5 \times 10^{-5}$.

If we did not know K_b, we could use $K_{a(HCN)}$ to find $K_{b(CN^-)}$.

Plan

We recognize that both $NaCH_3COO$ and NaCN are salts of strong soluble bases and weak acids. The anions in such salts hydrolyze to give basic solutions. As we did in Chapters 17 and 18, we write the appropriate chemical equation and equilibrium constant expression; we complete the reaction summary; and then we substitute the algebraic representations of equilibrium concentrations into the equilibrium constant expression and solve for the unknown concentration(s).

Solution

(a) The overall equation for the reaction of CH_3COO^- with H_2O and its equilibrium constant expression are

$$CH_3COO^- + H_2O \rightleftharpoons CH_3COOH + OH^- \qquad K_b = \frac{[CH_3COOH][OH^-]}{[CH_3COO^-]} = 5.6 \times 10^{-10}$$

Let $x = $ mol/L of CH_3COO^- that hydrolyzes. Then $x = [CH_3COOH] = [OH^-]$.

	$CH_3COO^- + H_2O \longrightarrow$	$CH_3COOH +$	OH^-
initial	0.10 *M*		
change due to rxn	$-x\ M$	$+x\ M$	$+x\ M$
at equil	$(0.10 - x)\ M$	$x\ M$	$x\ M$

Substitution into the equilibrium constant expression gives

$$\frac{[CH_3COOH][OH^-]}{[CH_3COO^-]} = 5.6 \times 10^{-10} = \frac{(x)(x)}{(0.10 - x)} \qquad \text{so} \qquad x = 7.5 \times 10^{-6}$$

$$x = \boxed{7.5 \times 10^{-6}\ M = [OH^-]} \qquad pOH = 5.12 \qquad \text{and} \qquad \boxed{pH = 8.88}$$

The 0.10 *M* $NaCH_3COO$ solution is distinctly basic.

$$\% \text{ hydrolysis} = \frac{[CH_3COO^-]_{\text{hydrolyzed}}}{[CH_3COO^-]_{\text{initial}}} \times 100\% = \frac{7.5 \times 10^{-6}\ M}{0.10\ M} \times 100\%$$

$$= \boxed{0.0075\% \text{ hydrolysis}}$$

(b) Perform the same kind of calculation for 0.10 *M* NaCN. Let $y = $ mol/L of CN^- that hydrolyzes. Then $y = [HCN] = [OH^-]$.

The pH of 0.10 M NaCH$_3$COO is 8.88. The pH of 0.10 M NaCN is 11.20. An inert solid has been suspended in the liquids to improve the quality of photographs of pH meters.

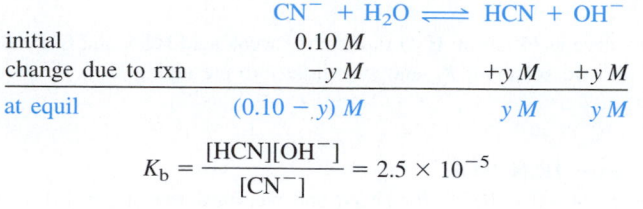

	CN$^-$ + H$_2$O $\rightleftharpoons$ HCN + OH$^-$		
initial	0.10 M		
change due to rxn	$-y\,M$	$+y\,M$	$+y\,M$
at equil	$(0.10 - y)\,M$	$y\,M$	$y\,M$

$$K_b = \frac{[\text{HCN}][\text{OH}^-]}{[\text{CN}^-]} = 2.5 \times 10^{-5}$$

Substitution into this expression gives

$$\frac{(y)(y)}{(0.10 - y)} = 2.5 \times 10^{-5} \qquad \text{so} \qquad y = 1.6 \times 10^{-3}\,M$$

$$y = \boxed{[\text{OH}^-] = 1.6 \times 10^{-3}\,M} \qquad \text{pOH} = 2.80 \qquad \text{and} \qquad \boxed{\text{pH} = 11.20}$$

The 0.10 M NaCN solution is even more basic than the 0.10 M NaCH$_3$COO solution in part (a).

$$\% \text{ hydrolysis} = \frac{[\text{CN}^-]_{\text{hydrolyzed}}}{[\text{CN}^-]_{\text{initial}}} \times 100\% = \frac{1.6 \times 10^{-3}\,M}{0.10\,M} \times 100\%$$

$$= \boxed{1.6\% \text{ hydrolysis}}$$

You should now work Exercises 15 and 16.

The 0.10 M solution of NaCN is much more basic than the 0.10 M solution of NaCH$_3$COO because CN$^-$ is a much stronger base than CH$_3$COO$^-$. This is expected because HCN is a much weaker acid than CH$_3$COOH.

The percent hydrolysis for 0.10 M NaCN (1.6%) is about 210 times greater than the percent hydrolysis for 0.10 M NaCH$_3$COO (0.0075%). In Table 19-1 we compare 0.10 M solutions of CH$_3$COO$^-$, CN$^-$, and NH$_3$ (the familiar weak base). We see that CH$_3$COO$^-$ is a much weaker base than NH$_3$, whereas CN$^-$ is a slightly stronger base than NH$_3$.

Table 19-1 *Data for 0.10 M Solutions of NaCH$_3$COO, NaCN, and NH$_3$*

	0.10 M NaCH$_3$COO	0.10 M NaCN	0.10 M aq NH$_3$
K_a for parent acid	1.8×10^{-5}	4.0×10^{-10}	
K_b for anion	5.6×10^{-10}	2.5×10^{-5}	K_b for NH$_3$ = 1.8×10^{-5}
[OH$^-$]	$7.5 \times 10^{-6}\,M$	$1.6 \times 10^{-3}\,M$	$1.3 \times 10^{-3}\,M$
% hydrolysis	0.0075%	1.6%	1.3% ionized
pH	8.88	11.20	11.11

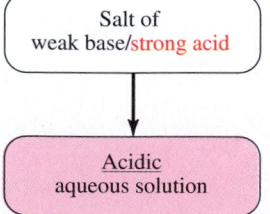

Salt of weak base/strong acid

↓

Acidic aqueous solution

19-3 SALTS OF WEAK BASES AND STRONG ACIDS

The second common kind of hydrolysis reaction involves the reaction of the cation of a weak base with OH$^-$ from water to form nonionized molecules of the weak base. The removal of OH$^-$ upsets the H$_3$O$^+$/OH$^-$ balance in water and gives an excess of H$_3$O$^+$. This makes such solutions *acidic*. Consider a solution of ammonium chloride, the salt of aqueous NH$_3$ and HCl.

$$NH_4Cl(solid) \xrightarrow[100\%]{H_2O} \boxed{NH_4^+} + Cl^-$$
$$H_2O + H_2O \rightleftharpoons \boxed{OH^- + H_3O^+} \quad \text{(result is excess } H_3O^+\text{)}$$

Equilibrium is shifted

$$\Updownarrow$$

$$NH_3 + H_2O$$

Ammonium chloride is an ionic salt that is soluble in water.

Ammonium ions from NH_4Cl react with OH^- to form nonionized NH_3 and H_2O molecules. This reaction removes OH^- from the system, so it causes more H_2O to ionize to produce an excess of H_3O^+. We combine the preceding equations into a single equation and write its equilibrium constant expression.

$$NH_4^+ + H_2O \rightleftharpoons NH_3 + H_3O^+ \qquad K_a = \frac{[NH_3][H_3O^+]}{[NH_4^+]}$$

Similar equations can be written for cations derived from other weak bases such as $CH_3NH_3^+$ and $(CH_3)_2NH_2^+$.

The expression $K_w = K_aK_b$ is valid for *any* conjugate acid–base pair in aqueous solution. We use it for the NH_4^+/NH_3 pair.

$$K_{a\,(NH_4^+)} = \frac{K_w}{K_{b\,(NH_3)}} = \frac{1.0 \times 10^{-14}}{1.8 \times 10^{-5}} = 5.6 \times 10^{-10} = \frac{[NH_3][H_3O^+]}{[NH_4^+]}$$

To derive $K_w = K_aK_b$ for this case, multiply the K_a expression by $[OH^-]/[OH^-]$ and simplify.

The fact that K_a for the ammonium ion, NH_4^+, is the same as K_b for the acetate ion, should not be surprising. Recall that ionization constants for CH_3COOH and aqueous NH_3 are equal (by coincidence). Thus, we expect CH_3COO^- to hydrolyze to the same extent as NH_4^+ does.

EXAMPLE 19-3 *pH of a Soluble Salt*

Calculate the pH of a $0.20\,M$ solution of ammonium nitrate, NH_4NO_3. K_a for $NH_4^+ = 5.6 \times 10^{-10}$.

Ammonium nitrate is widely used as a fertilizer because of its high nitrogen content. It contributes to soil acidity.

Plan

We recognize that NH_4NO_3 is the salt of a weak base, NH_3, and a strong acid, HNO_3, and that the cations of such salts hydrolyze to give acidic solutions. We proceed as we did in Example 19-2.

Solution

The cation of the weak base reacts with H_2O. Let $x = $ mol/L of NH_4^+ that hydrolyzes. Then $x = [NH_3] = [H_3O^+]$.

	$NH_4^+ + H_2O \rightleftharpoons$	NH_3	$+ H_3O^+$
initial	$0.20\,M$		
change due to rxn	$-x\,M$	$+x\,M$	$+x\,M$
at equil	$(0.20 - x)\,M$	$x\,M$	$x\,M$

Substituting into the K_a expression gives

$$K_a = \frac{[NH_3][H_3O^+]}{[NH_4^+]} = \frac{(x)(x)}{(0.20 - x)} = 5.6 \times 10^{-10}$$

Making the usual simplifying assumption gives $x = 1.1 \times 10^{-5}\,M = [H_3O^+]$ and $\boxed{pH = 4.96.}$ The $0.20\,M$ NH_4NO_3 solution is distinctly acidic.

You should now work Exercise 22.

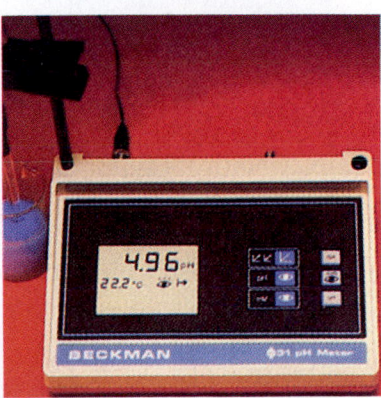

The pH of $0.20\,M$ NH_4NO_3 solution is 4.96.

19-4 SALTS OF WEAK BASES AND WEAK ACIDS

Salts of weak bases and weak acids are the fourth class of salts. Most are soluble. Salts of weak bases and weak acids contain cations that would give acidic solutions and anions that would give basic solutions. Will solutions of such salts be neutral, basic, or acidic? They may be any one of the three depending on the relative strengths of the weak molecular acid and weak molecular base from which each salt is derived. Thus, salts of this class may be divided into three types that depend on the relative strengths of their parent weak bases and weak acids.

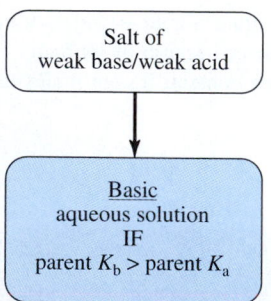

Salts of Weak Bases and Weak Acids for Which $K_b = K_a$

The common example of a salt of this type is ammonium acetate, NH_4CH_3COO, the salt of aqueous NH_3 and CH_3COOH. The ionization constants for both aqueous NH_3 and CH_3COOH are 1.8×10^{-5}. We know that ammonium ions react with water to produce H_3O^+.

$$NH_4^+ + H_2O \rightleftharpoons NH_3 + H_3O^+ \qquad K_a = \frac{[NH_3][H_3O^+]}{[NH_4^+]} = 5.6 \times 10^{-10}$$

We also recall that acetate ions react with water to produce OH^-.

$$CH_3COO^- + H_2O \rightleftharpoons CH_3COOH + OH^- \qquad K_b = \frac{[CH_3COOH][OH^-]}{[CH_3COO^-]} = 5.6 \times 10^{-10}$$

Because these K values are equal, the NH_4^+ produces just as many H_3O^+ ions as the CH_3COO^- produces OH^- ions. Thus, we predict that ammonium acetate solutions are neutral, and they are.

Salts of Weak Bases and Weak Acids for Which $K_b > K_a$

Salts of weak bases and weak acids for which K_b is greater than K_a are always basic because the anion of the weaker acid hydrolyzes to a greater extent than the cation of the stronger base.

Consider NH_4CN, ammonium cyanide. K_a for HCN (4.0×10^{-10}) is much smaller than K_b for NH_3 (1.8×10^{-5}), so K_b for CN^- (2.5×10^{-5}) is much larger than K_a for NH_4^+ (5.6×10^{-10}). This tells us that the CN^- ions hydrolyze to a much greater extent than do NH_4^+ ions, and so ammonium cyanide solutions are distinctly basic. Stated differently, CN^- is much stronger as a base than NH_4^+ is as an acid.

$$NH_4^+ + H_2O \rightleftharpoons NH_3 + H_3O^+$$
$$CN^- + H_2O \rightleftharpoons HCN + OH^-$$
$$\rightarrow 2H_2O$$

The second reaction occurs to greater extent ∴ the solution is basic.

Salts of Weak Bases and Weak Acids for Which $K_b < K_a$

Salts of weak bases and weak acids for which K_b is less than K_a are acidic because the cation of the weaker base hydrolyzes to a greater extent than the anion of the stronger acid. Consider ammonium fluoride, NH_4F, the salt of aqueous ammonia and hydrofluoric acid.

K_b for aqueous NH_3 is 1.8×10^{-5} and K_a for HF is 7.2×10^{-4}. So K_a value for NH_4^+ (5.6×10^{-10}) is slightly larger than the K_b value for F^- (1.4×10^{-11}). This tells us that

NH_4^+ ions hydrolyze to a slightly greater extent than F^- ions. In other words, NH_4^+ is slightly stronger as an acid than F^- is as a base. Ammonium fluoride solutions are slightly acidic.

$$NH_4^+ + H_2O \rightleftharpoons NH_3 + {\color{red}H_3O^+}$$
$$F^- + H_2O \rightleftharpoons HF + OH^-$$

$\Big\} \longrightarrow 2H_2O$

{\color{red}The first reaction occurs to greater extent ∴ the solution is acidic.}

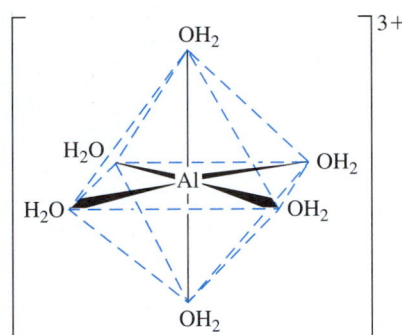

19-5 SALTS THAT CONTAIN SMALL, HIGHLY CHARGED CATIONS

Solutions of certain common salts of strong acids are acidic. For this reason, many home-owners apply iron(II) sulfate, $FeSO_4 \cdot 7H_2O$, or aluminum sulfate, $Al_2(SO_4)_3 \cdot 18H_2O$, to the soil around "acid-loving" plants such as azaleas, camelias, and hollies. You are probably familiar with the sour, "acid" taste of alum, $KAl(SO_4)_2 \cdot 12H_2O$, a substance that is frequently added to pickles.

Each of these salts contains a small, highly charged cation and the anion of a strong acid. Solutions of such salts are acidic because these cations hydrolyze to produce excess hydronium ions. Consider aluminum chloride, $AlCl_3$, as a typical example. When solid anhydrous $AlCl_3$ is added to water, the water becomes very warm as the Al^{3+} ions become hydrated in solution. In many cases, the interaction between positively charged ions and the negative ends of polar water molecules is so strong that salts crystallized from aqueous solution contain definite numbers of water molecules. Salts containing Al^{3+}, Fe^{2+}, Fe^{3+}, and Cr^{3+} ions usually crystallize from aqueous solutions with six water molecules bonded to each metal ion. These salts contain the hydrated cations $[Al(OH_2)_6]^{3+}$, $[Fe(OH_2)_6]^{2+}$, $[Fe(OH_2)_6]^{3+}$, and $[Cr(OH_2)_6]^{3+}$, respectively, in the solid state. Such species also exist in aqueous solutions. Each of these species is octahedral; i.e., the metal ion (M^{n+}) is located at the center of a regular octahedron, and the O atoms in six H_2O molecules are located at the corners (Figure 19-1). In the metal–oxygen bonds of the hydrated cation, electron density is decreased around the O end of each H_2O molecule by the positively charged metal ion. This weakens the H—O bonds in coordinated H_2O molecules relative to the H—O bonds in noncoordinated H_2O molecules. Consequently, the coordinated H_2O molecules can donate H^+ to solvent H_2O molecules to form H_3O^+ ions. This produces acidic solutions (Figure 19-2).

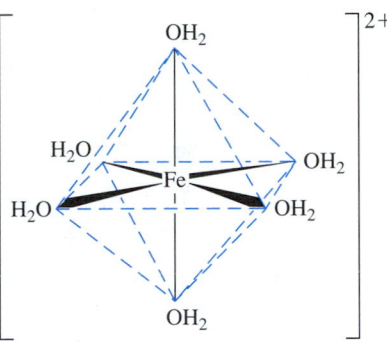

Figure 19-1 Structures of hydrated aluminum ions, $[Al(OH_2)_6]^{3+}$, and hydrated iron(II) ions, $[Fe(OH_2)_6]^{2+}$.

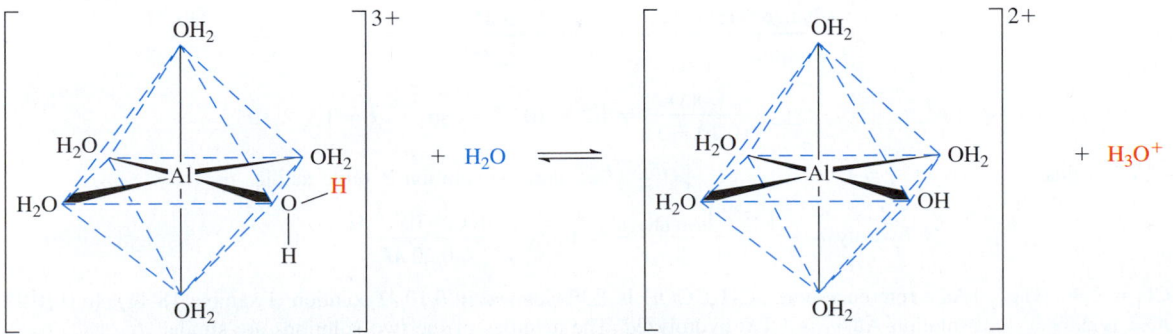

Figure 19-2 Hydrolysis of hydrated aluminum ions to produce H_3O^+—that is, the removal of a proton from a coordinated H_2O molecule by a noncoordinated H_2O molecule.

The equation for the hydrolysis of Al^{3+} is written as follows.

$$[Al(OH_2)_6]^{3+} + H_2O \rightleftharpoons [Al(OH)(OH_2)_5]^{2+} + H_3O^+$$

$$K_a = \frac{[[Al(OH)(OH_2)_5]^{2+}][H_3O^+]}{[[Al(OH_2)_6]^{3+}]} = 1.2 \times 10^{-5}$$

Removing an H^+ converts a coordinated water molecule to a coordinated hydroxide ion and decreases the positive charge on the hydrated species. This can be written in more abbreviated form as

$$Al^{3+} + 2H_2O \rightleftharpoons Al(OH)^{2+} + H_3O^+$$

$$K_a = \frac{[Al(OH)^{2+}][H_3O^+]}{[Al^{3+}]} = 1.2 \times 10^{-5}$$

Hydrolysis of small, highly charged cations may occur beyond the first step. In many cases these reactions are quite complex. They may involve two or more cations reacting with each other to form large polymeric species. For most common cations, consideration of the first hydrolysis constant is adequate for our calculations.

EXAMPLE 19-4   *Percent Hydrolysis*

Calculate the pH and percent hydrolysis in $0.10\,M$ $AlCl_3$ solution. $K_a = 1.2 \times 10^{-5}$ for $[Al(OH_2)_6]^{3+}$ (often abbreviated Al^{3+}).

Plan
We recognize that $AlCl_3$ produces a small, highly charged cation that hydrolyzes to give an acidic solution. We represent the equilibrium concentrations and proceed as we did in earlier examples.

Solution
The equation for the reaction and its hydrolysis constant can be represented as

$$Al^{3+} + 2H_2O \rightleftharpoons Al(OH)^{2+} + H_3O^+ \qquad K_a = \frac{[Al(OH)^{2+}][H_3O^+]}{[Al^{3+}]} = 1.2 \times 10^{-5}$$

Let $x = $ mol/L of Al^{3+} that hydrolyzes. Then $x = [Al(OH)^{2+}] = [H_3O^+]$.

	Al^{3+}	$+ 2H_2O \rightleftharpoons$	$Al(OH)^{2+}$	$+ H_3O^+$
initial	$0.10\,M$			
change due to rxn	$-x\,M$		$+x\,M$	$+x\,M$
at equil	$(0.10 - x)\,M$		$x\,M$	$x\,M$

$$\frac{(x)(x)}{(0.10 - x)} = 1.2 \times 10^{-5} \qquad so \qquad x = 1.1 \times 10^{-3}$$

Recall that we let $x = [Al^{3+}]$ that hydrolyzes.

$[H_3O^+] = 1.1 \times 10^{-3}\,M$, pH = 2.96, and the solution is quite acidic.

$$\% \text{ hydrolysis} = \frac{[Al^{3+}]_{\text{hydrolyzed}}}{[Al^{3+}]_{\text{total}}} \times 100\% = \frac{1.1 \times 10^{-3}\,M}{0.10\,M} \times 100\% = 1.1\% \text{ hydrolyzed}$$

The pH of $0.10\,M$ $AlCl_3$ is 2.96. The pH of $0.10\,M$ CH_3COOH is 2.89.

As a reference point, CH_3COOH is 1.3% ionized in $0.10\,M$ solution (Example 18-11). In $0.10\,M$ solution $AlCl_3$ is 1.1% hydrolyzed. The acidities of the two solutions are similar.

You should now work Exercise 30.

CHEMISTRY IN USE

Our Daily Lives

Taming Acids with Harmless Salts

From yellowing paper in old books and newsprint to heartburn and environmental spills, many Americans encounter unwanted or excessive amounts of acid. Neutralizing unwanted acids with hydroxide bases might appear to be a good way to combat these acids. But even more effective chemicals exist that can neutralize acids without the risks posed by hydroxide bases. These acid-neutralizing chemicals are salts of weak acids and strong bases. Such salts can neutralize acids because hydrolysis makes their aqueous solutions basic. More importantly, there is a significant advantage in using relatively harmless salts such as sodium hydrogen carbonate (baking soda) rather than stronger bases such as sodium hydroxide (lye). For example, if we used too much sodium hydroxide to neutralize sulfuric acid spilled from a car battery, any excess lye left behind would pose an environmental and human health threat about equal to that of the spilled sulfuric acid. (Lye is the major ingredient in such commercial products as oven cleaners and Drāno.) However, we would not be concerned if a little baking soda were left on the ground after the sulfuric acid from the car battery had been neutralized.

The same principle applies to acid indigestion. Rather than swallow lye (ugh!) or some other strong base to neutralize excess stomach acid, most people take antacids. Antacids typically contain salts such as calcium carbonate, sodium hydrogen carbonate (sodium bicarbonate), and magnesium carbonate, all of which are salts of weak acids. These salts hydrolyze to form hydroxide ions that reduce the degree of acidity in the stomach. Physicians also prescribe these and similar salts to treat peptic ulcers. The repeated use of antacids should always be under the supervision of a physician.

Salts of weak acids and strong bases can be used effectively against a major acid spill in much the same way they are used against sulfuric acid from a car battery or excess stomach acid. In a recent major acid spill, a tank car filled with nitric acid was punctured by the coupling of another rail car, spilling 22,000 gallons of concentrated nitric acid onto the ground. Many thousand residents living near the spill were evacuated. There were no fatalities or serious injuries, and there was no major environmental damage; resident fire fighters neutralized the concentrated nitric acid by using airport snow blowers to spread *relatively* harmless sodium carbonate (washing soda) over the contaminated area.

Salts of weak acids and strong bases are also being used to combat the destructive aging process of paper. Think how seri-

Dalton's Scale of Relative Atomic Weights.

ous this problem is for the Library of Congress, which loses 70,000 books each year to the decomposition of aging paper. Many of the twenty million books in the Library of Congress have a life expectancy of only 25 to 40 years. Paper ages because of the hydrolysis of aluminum sulfate. Aluminum sulfate has been used in the paper manufacturing process since the 1850s because it is an inexpensive sizing compound (it keeps ink from spreading out on paper). Aluminum sulfate is the salt of an insoluble weak base and a strong acid; it hydrolyzes in the water in paper (typically 4–7% H_2O) to give an acidic environment. The acid eats away at cellulose fibers; this causes the paper to turn yellow and eventually disintegrate. To combat this aging, the Library of Congress individually treats its collections with solutions of salts of weak acids and strong bases at great cost. Meanwhile, the paper industry is fighting this aging process by increasing its output of alkaline paper. Some alkaline paper contains calcium carbonate, the same salt found in several brands of antacids. Calcium carbonate increases the pH of paper to between 7.5 and 8.5. Special manufacturing techniques produce calcium carbonate that is very fine and that has uniform particle size. Alkaline papers are expected to last about 300 years, in contrast to the average 25- to 40-year life expectancy of standard acidic paper.

Salts that hydrolyze to produce basic solutions can settle upset stomachs, prevent yellowing pages, and neutralize major and minor acid spills. A knowledge of hydrolysis is very useful and has many applications.

Ronald DeLorenzo
Middle Georgia College

Table 19-2 *Ionic Radii and Hydrolysis Constants for Some Cations*

Cation	Ionic Radius (Å)	Hydrated Cation	K_a
Li^+	0.60	$[Li(OH_2)_4]^+$	1×10^{-14}
Be^{2+}	0.31	$[Be(OH_2)_4]^{2+}$	1.0×10^{-5}
Na^+	0.95	$[Na(OH_2)_6]^+$ (?)	10^{-14}
Mg^{2+}	0.65	$[Mg(OH_2)_6]^{2+}$	3.0×10^{-12}
Al^{3+}	0.50	$[Al(OH_2)_6]^{3+}$	1.2×10^{-5}
Fe^{2+}	0.76	$[Fe(OH_2)_6]^{2+}$	3.0×10^{-10}
Fe^{3+}	0.64	$[Fe(OH_2)_6]^{3+}$	4.0×10^{-3}
Co^{2+}	0.74	$[Co(OH_2)_6]^{2+}$	5.0×10^{-10}
Co^{3+}	0.63	$[Co(OH_2)_6]^{3+}$	1.7×10^{-2}
Cu^{2+}	0.96	$[Cu(OH_2)_6]^{2+}$	1.0×10^{-8}
Zn^{2+}	0.74	$[Zn(OH_2)_6]^{2+}$	2.5×10^{-10}
Hg^{2+}	1.10	$[Hg(OH_2)_6]^{2+}$	8.3×10^{-7}
Bi^{3+}	0.74	$[Bi(OH_2)_6]^{3+}$	1.0×10^{-2}

Pepto-Bismol contains $BiO(HOC_4H_6COO)$, bismuth subsalicylate, a *hydrolyzed* bismuth salt. Such salts "coat" polar surfaces such as glass and the lining of the stomach.

Smaller, more highly charged cations are stronger acids than larger, less highly charged cations (Table 19-2). This is because the smaller, more highly charged cations interact with coordinated water molecules more strongly.

For isoelectronic cations in the same period in the periodic table, the smaller, more highly charged cation is the stronger acid. (Compare K_a values for Li^+ and Be^{2+} and for Na^+, Mg^{2+}, and Al^{3+}.) For cations with the same charge from the same group in the periodic table, the smaller cation hydrolyzes to a greater extent. (Compare K_a values for Be^{2+} and Mg^{2+}.) If we compare cations of the same element in different oxidation states, the smaller, more highly charged cation is the stronger acid. (Compare K_a values for Fe^{2+} and Fe^{3+} and for Co^{2+} and Co^{3+}.)

TITRATION CURVES

19-6 STRONG ACID–STRONG BASE TITRATION CURVES

A **titration curve** is a plot of pH versus the amount (usually volume) of acid or base added. It displays graphically the change in pH as acid or base is added to a solution and shows how pH changes near the equivalence point.

The point at which the color of an indicator changes in a titration is known as the **end point.** It is determined by the K_a value for the indicator (Section 18-5). Table 19-3 shows a few acid–base indicators and the pH ranges over which their colors change. Typically, color changes occur over a range of 1.5 to 2.0 pH units.

The **equivalence point** is the point at which chemically equivalent amounts of acid and base have reacted.

Ideally, the end point and the equivalence point in a titration should coincide.

Table 19-3 *Range and Color Changes of Some Common Acid–Base Indicators*

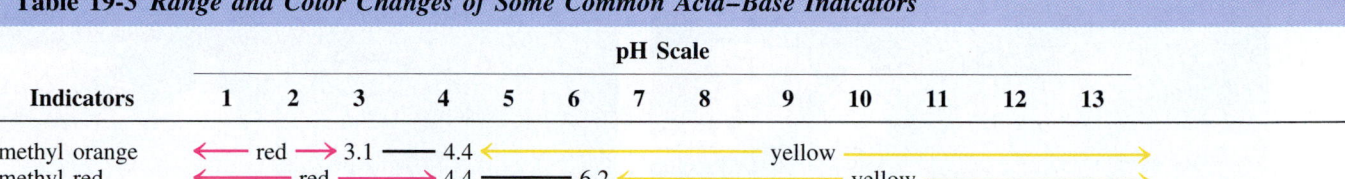

Indicators	pH Scale
	1 2 3 4 5 6 7 8 9 10 11 12 13
methyl orange	← red → 3.1 — 4.4 ← yellow →
methyl red	← red → 4.4 — 6.2 ← yellow →
bromthymol blue	← yellow → 6.2 — 7.6 ← blue →
neutral red	← red → 6.8 — 8.0 ← yellow →
phenolphthalein	← colorless → 8.0 — 10.0 ← red → colorless beyond 13.0

In practice, we try to select an indicator whose range of color change includes the equivalence point. We use the same procedures in both standardization and analysis to minimize any error arising from a difference between end point and equivalence point.

Review Section 3-9 before you consider the titration of 100.0 mL of a 0.100 M solution of HCl with a 0.100 M solution of NaOH. As we know, NaOH and HCl react in a 1:1 ratio. We calculate the pH of the solution at several stages as NaOH is added.

Titrations are usually done with 50-mL or smaller burets. We have used 100 mL of solution in this example to simplify the arithmetic.

1. Before any NaOH is added to the 0.100 M HCl solution:

$$HCl + H_2O \xrightarrow{100\%} H_3O^+ + Cl^-$$

$[H_3O^+] = 0.10\ M$ so   pH = 1.00

2. After 20.0 mL of 0.100 M NaOH has been added:

	HCl	+	NaOH	⟶	NaCl	+	H₂O
start	10.0 mmol		2.0 mmol		0 mmol		
change	−2.0 mmol		−2.0 mmol		+2.0 mmol		
after rxn	8.0 mmol		0 mmol		2.0 mmol		

The concentration of unreacted HCl in the total volume of 120 mL is

$$M_{HCl} = \frac{8.00\ \text{mmol HCl}}{120\ \text{mL}} = 0.067\ M\ \text{HCl}$$

$[H_3O^+] = 6.7 \times 10^{-2}\ M$ so   pH = 1.17

3. After 50.0 mL of 0.100 M NaOH has been added (midpoint of the titration):

	HCl	+	NaOH	⟶	NaCl	+	H₂O
start	10.0 mmol		5.0 mmol		0 mmol		
change	−5.0 mmol		−5.0 mmol		+5.0 mmol		
after rxn	5.0 mmol		0 mmol		5.0 mmol		

$$M_{HCl} = \frac{5.00\ \text{mmol HCl}}{150\ \text{mL}} = 0.033\ M\ \text{HCl}$$

$[H_3O^+] = 3.3 \times 10^{-2}\ M$ so   pH = 1.48

4. After 100 mL of 0.100 M NaOH has been added:

	HCl	+	NaOH	⟶	NaCl	+	H₂O
start	10.0 mmol		10.0 mmol		0 mmol		
change	−10.0 mmol		−10.0 mmol		+10.0 mmol		
after rxn	0 mmol		0 mmol		10.0 mmol		

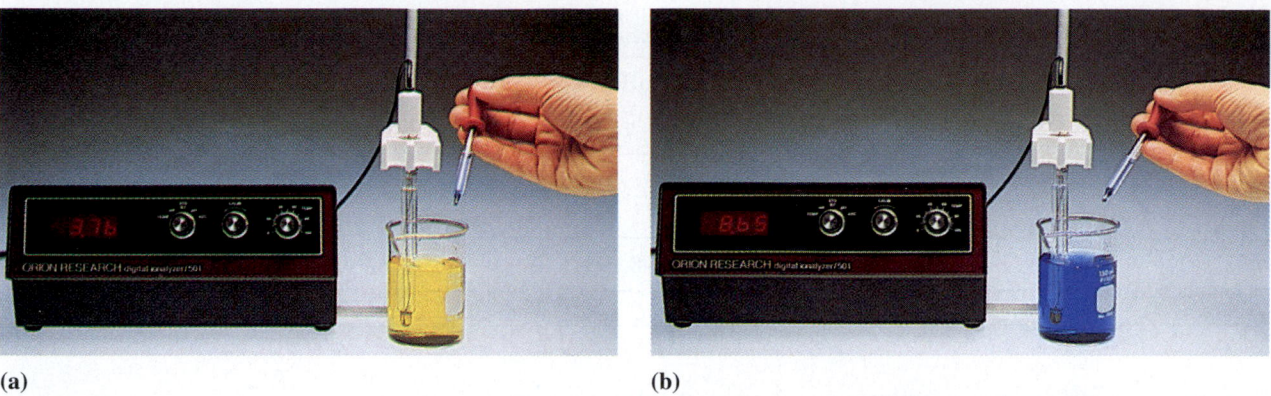

(a)   **(b)**

The end point of the titration of 0.100 M HCl with 0.100 M NaOH using another indicator, bromthymol blue.

We have added enough NaOH to neutralize the HCl exactly so this is the equivalence point. A strong acid and a strong base react to give a neutral salt solution so pH = 7.00.

5. After 110.0 mL of 0.100 M NaOH has been added:

	HCl	+	NaOH	$\longrightarrow$	NaCl	+	H$_2$O
start	10.0 mmol		11.0 mmol		0 mmol		
change	−10.0 mmol		−10.0 mmol		+10.0 mmol		
after rxn	0 mmol		1.0 mmol		10.0 mmol		

The pH is determined by the excess NaOH.

$$M_{\text{NaOH}} = \frac{1.0 \text{ mmol NaOH}}{210 \text{ mL}} = 0.0048 \ M \text{ NaOH}$$

$$[\text{OH}^-] = 4.8 \times 10^{-3} \ M \quad \text{so} \quad \text{pOH} = 2.32 \quad \text{and} \quad \text{pH} = 11.68$$

Table 19-4 displays the data for the titration of 100.0 mL of 0.100 M HCl by 0.100 M NaOH solution. A few additional points have been included to show the shape of the curve better. These data are plotted in Figure 19-3a. This titration curve has a long "vertical section" over which the pH changes very rapidly with the addition of very

Table 19-4 *Titration Data for 100.0 mL of 0.100 M HCl versus NaOH*

mL of 0.100 M NaOH Added	mmol NaOH Added	mmol Excess Acid or Base	pH
0.0	0.00	10.0 H$_3$O$^+$	1.00
20.0	2.00	8.0	1.17
50.0	5.00	5.0	1.48
90.0	9.00	1.0	2.28
99.0	9.90	0.10	3.30
99.5	9.95	0.05	3.60
100.0	10.00	0.00 (eq. pt.)	7.00
100.5	10.05	0.05 OH$^-$	10.40
110.0	11.00	1.00	11.68
120.0	12.00	2.00	11.96

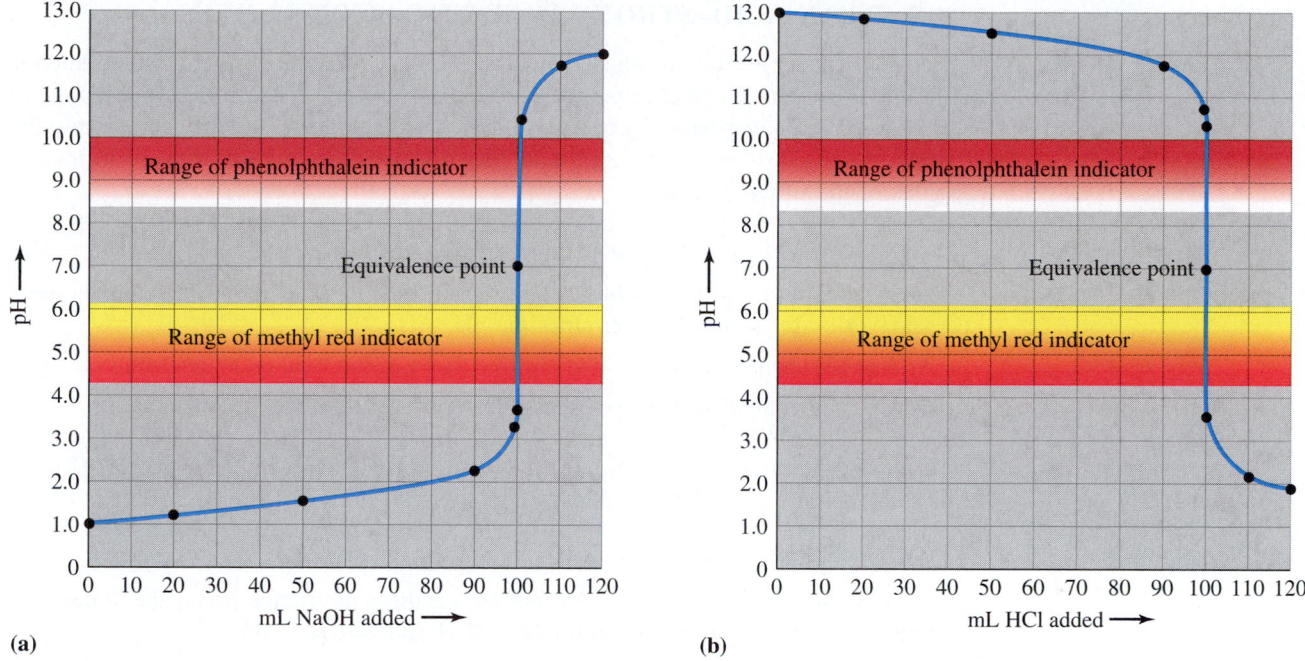

Figure 19-3 (a) The titration curve for 100 mL of 0.100 M HCl with 0.100 M NaOH. Note that the "vertical" section of the curve is quite long. The titration curves for other strong acids and bases are identical with this one *if* the same concentrations of acid and bases are used *and if* both are monoprotic. (b) The titration curve for 100 mL of 0.100 M NaOH with 0.100 M HCl. This curve is similar to that in part (a), but inverted.

small amounts of base. The pH changes from 3.60 (99.5 mL NaOH added) to 10.40 (100.5 mL of NaOH added) in the vicinity of the equivalence point (100.0 mL NaOH added). The midpoint of the vertical section (pH = 7.00) is the equivalence point. We can separate the calculations on this kind of titration into four distinct types, which correspond to four regions of the titration curves.

1. Before any strong base is added, the pH depends on the strong acid alone.

2. After some strong base has been added, but before the equivalence point, the remaining (excess) strong acid determines the pH.

3. At the equivalence point, the solution is neutral.

4. Beyond the equivalence point, excess strong base determines the pH.

Ideally, the indicator color change should occur at pH = 7.00. For practical purposes, indicators with color changes in the pH range 4 to 10 can be used in the titration of strong acids and strong bases because the vertical portion of the titration curve is so long. Figure 19-3 shows the color ranges for methyl red and phenolphthalein, two widely used indicators. Both fall within the vertical section of the NaOH/HCl titration curve. When a strong acid is added to a solution of a strong base, the titration curve is inverted, but its essential characteristics are the same (Figure 19-3b).

In Figure 19-3a we see that the curve rises very slowly before the equivalence point. It then rises very rapidly near the equivalence point because there is no hydrolysis. The curve becomes almost flat beyond the equivalence point.

19-7 WEAK ACID–STRONG BASE TITRATION CURVES

When a weak acid is titrated with a strong base, the curve is quite different. The solution is buffered *before* the equivalence point. It is basic *at* the equivalence point because salts of weak acids and strong bases hydrolyze to give basic solutions. So, we can separate the calculations on this kind of titration into four distinct types, which correspond to four regions of the titration curves.

1. Before any base is added, the pH depends on the weak acid alone.

2. After some base has been added, but before the equivalence point, a series of weak acid/salt buffer solutions determines the pH.

3. At the equivalence point, hydrolysis of the anion of the weak acid determines the pH.

4. Beyond the equivalence point, excess strong base determines the pH.

Consider the titration of 100.0 mL of 0.100 M CH_3COOH with 0.100 M NaOH solution. (The strong electrolyte is added to the weak electrolyte.)

1. Before any base is added, the pH is 2.89 (Example 18-11).

2. As soon as some NaOH is added, but before the equivalence point, the solution is buffered because it contains both CH_3COOH and $NaCH_3COO$.

$$NaOH + CH_3COOH \longrightarrow NaCH_3COO + H_2O$$
$$\text{lim amt} \qquad \text{excess}$$

For instance, after 20.0 mL of NaOH solution has been added, we have

	NaOH +	CH_3COOH $\longrightarrow$	$NaCH_3COO$ + H_2O
start	2.00 mmol	10.00 mmol	0 mmol
change	−2.00 mmol	−2.00 mmol	+2.00 mmol
after rxn	0 mmol	8.00 mmol	2.00 mmol

These amounts are present in 120 mL of solution, so the concentrations are

$$M_{CH_3COOH} = \frac{8.00 \text{ mmol } CH_3COOH}{120 \text{ mL}} = 0.0667 \, M \, CH_3COOH$$

$$M_{NaCH_3COO} = \frac{2.00 \text{ mmol } NaCH_3COO}{120 \text{ mL}} = 0.0167 \, M \, NaCH_3COO$$

The pH of this solution is calculated as in Example 18-16.

$$\frac{[H_3O^+][CH_3COO^-]}{[CH_3COOH]} = 1.8 \times 10^{-5}$$

$$[H_3O^+] = 1.8 \times 10^{-5} \times \frac{[CH_3COOH]}{[CH_3COO^-]} = 1.8 \times 10^{-5} \times \frac{0.0667}{0.0167}$$

$$[H_3O^+] = 7.2 \times 10^{-5} \, M \qquad \text{and} \qquad pH = 4.14$$

Just *before* the equivalence point, the solution contains relatively high concentrations of $NaCH_3COO$ and relatively low concentrations of CH_3COOH. Just *after* the equivalence point, the solution contains relatively high concentrations of $NaCH_3COO$ and relatively low concentrations of NaOH, both basic components. In both regions our calculations are only approximations. Exact calculations of pH in these regions are beyond the scope of this text.

After some NaOH has been added, the solution contains both $NaCH_3COO$ and CH_3COOH, and so it is buffered until the equivalence point is reached. All points before the equivalence point are calculated in the same way. You could also use the Henderson–Hasselbalch equation for these calculations.

3. At the equivalence point, the solution is 0.0500 M in $NaCH_3COO$.

Table 19-5 *Titration Data for 100.0 mL of 0.100 M CH$_3$COOH with 0.100 M NaOH*

mL 0.100 M NaOH Added		mmol Base Added	mmol Excess Acid or Base	pH
0.0 mL		0	10.0 CH$_3$COOH	2.89
20.0 mL		2.00	8.00	4.14
50.0 mL		5.00	5.00	4.74
75.0 mL	buffered	7.50	2.50	5.22
90.0 mL	region	9.00	1.00	5.70
95.0 mL		9.50	0.50	6.02
99.0 mL		9.90	0.10	6.74
100.0 mL		10.0	0 (equivalence point)	8.72
101.0 mL		10.1	0.10 OH$^-$	10.70
110.0 mL		11.0	1.0	11.68
120.0 mL		12.0	2.0	11.96

$$\text{NaOH} \quad + \quad \text{CH}_3\text{COOH} \longrightarrow \text{NaCH}_3\text{COO} + \text{H}_2\text{O}$$

start	10.0 mmol	10.0 mmol	
change	-10.0 mmol	-10.0 mmol	$+10.0$ mmol
after rxn	0	0	10.0 mmol

$$M_{\text{NaCH}_3\text{COO}} = \frac{10.0 \text{ mmol NaCH}_3\text{COO}}{200 \text{ mL}} = 0.0500 \ M \text{ NaCH}_3\text{COO}$$

The pH of a 0.0500 M solution of NaCH$_3$COO is 8.72 (Example 19-2 shows a similar calculation). The solution is distinctly basic at the equivalence point because of the hydrolysis of the acetate ion.

4. Beyond the equivalence point, the concentration of the excess NaOH determines the pH of the solution just as it did in the titration of a strong acid.

Table 19-5 lists several points on the titration curve, and Figure 19-4 shows the titration curve for 100.0 mL of 0.100 M CH$_3$COOH titrated with a 0.100 M solution of NaOH.

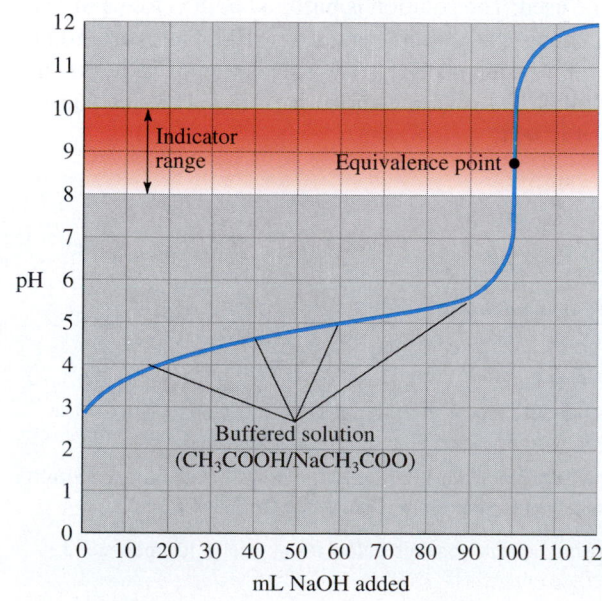

Figure 19-4 The titration curve for 100 mL of 0.100 M CH$_3$COOH with 0.100 M NaOH. The "vertical" section of this curve is much shorter than those in Figure 19-3 because the solution is buffered before the equivalence point.

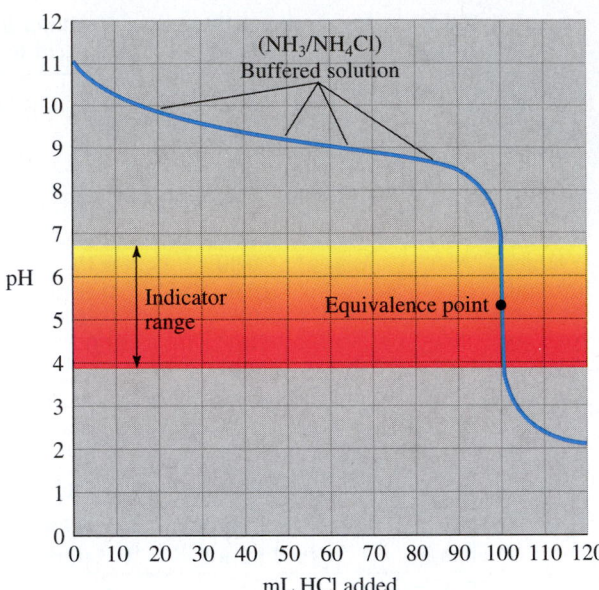

Figure 19-5 The titration curve for 100 mL of 0.100 M aqueous ammonia with 0.100 M HCl. The vertical section of the curve is relatively short because the solution is buffered before the equivalence point. The curve is very similar to that in Figure 19-4, but inverted.

This titration curve has a short vertical section (pH ≈ 7 to 10), and the indicator range is limited. Phenolphthalein is the indicator commonly used to titrate weak acids with strong bases (Table 19-3).

The titration curves for weak bases and strong acids are similar to those for weak acids and strong bases except that they are inverted (recall that strong is added to weak). Figure 19-5 displays the titration curve for 100.0 mL of 0.100 M aqueous ammonia titrated with 0.100 M HCl solution.

19-8 WEAK ACID–WEAK BASE TITRATION CURVES

In titration curves for weak acids and weak bases, pH changes near the equivalence point are too small for color indicators to be used. The solution is buffered both before and after the equivalence point. Figure 19-6 shows the titration curve for 100.0 mL of 0.100 M CH_3COOH solution titrated with 0.100 M aqueous NH_3. The calculation of values on the curve in Figure 19-6 other than the initial pH and an estimation of the pH at the equivalence point is beyond the scope of this text.

▼ **PROBLEM-SOLVING TIP** *Titration Curves*

You can consider a titration curve in four parts.

1. **Initial solution (before any titrant is added).**
2. **Region before the equivalence point.** This may or may not be buffered. The solution is buffered in this region if the substance being titrated is a weak acid or weak base.
3. **Equivalence point.** Its location depends upon the strengths of the acid and the base reacting.
4. **Region beyond the equivalence point.** This becomes nearly flat as more and more excess reactant is added. We often calculate only one or two points in this region.

Recognizing the four regions of a titration curve makes calculations easier to understand.

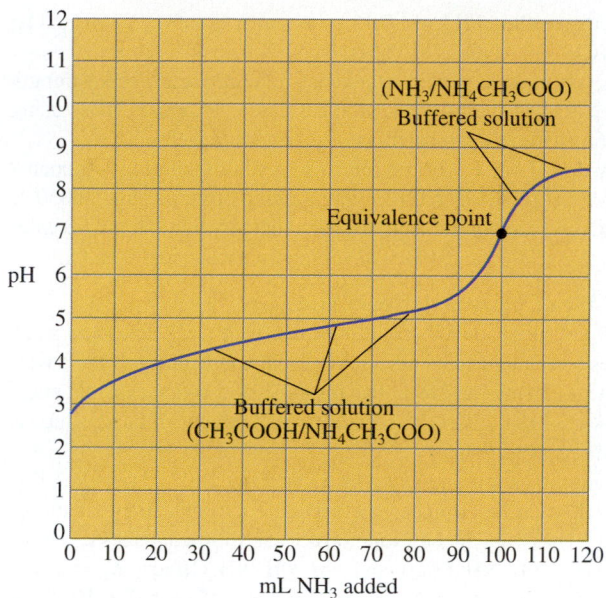

Figure 19-6 The titration curve for 100 mL of 0.100 M CH_3COOH with 0.100 M aqueous NH_3. Because the solution is buffered before and after the equivalence point, the vertical section of the curve is too short to be noticed. Therefore, color indicators cannot be used in such titrations. Physical methods such as conductivity measurements can be used to detect the end point.

Key Terms

End point The point at which an indicator changes color and a titration is stopped.

Equivalence point The point at which chemically equivalent amounts of reactants have reacted.

Hydrolysis The reaction of a substance with water or its ions.

Hydrolysis constant An equilibrium constant for a hydrolysis reaction.

Indicator (for acid–base titrations) An organic compound that exhibits different colors in solutions of different acidities; used to indicate the point at which reaction between an acid and a base is complete.

Solvolysis The reaction of a substance with the solvent in which it is dissolved.

Titration A procedure in which one solution is added to another solution until the chemical reaction between the two solutes is complete; usually the concentration of one solution is known and that of the other is unknown.

Titration curve (for acid–base titration) A plot of pH versus volume of acid or base solution added.

Exercises

Basic Ideas

1. How do we know that 0.10 M solutions of HCl and HNO_3 contain essentially no molecules of nonionized acid?
2. How do we know that 0.10 M solutions of HF and HNO_2 contain relatively few ions?
3. How can salts be classified conveniently into four classes? For each class, write the name and formula of a salt that fits into that category. Use examples other than those used in illustrations in this chapter.
4. Define and illustrate the following terms clearly and concisely: (a) solvolysis; (b) hydrolysis.

Salts of Strong Soluble Bases and Strong Acids

5. Some anions, when dissolved, undergo no significant reaction with water molecules. What is the relative base strength of such an anion compared to water? What effect will dissolution of such anions have on the pH of the solution?
6. Some cations in aqueous solution undergo no significant reactions with water molecules. What is the relative acid strength of such a cation compared to water? What effect will dissolution of such cations have on the pH of the solution?
7. Why do salts of strong soluble bases and strong acids give neutral aqueous solutions? Use KNO_3 to illustrate. Write names and formulas for three other salts of strong soluble bases and strong acids.
8. What determines whether the aqueous solution of a salt is acidic, neutral, or alkaline?

Salts of Strong Bases and Weak Acids

9. Some anions react with water to upset the H_3O^+/OH^- balance. What is the relative base strength of such an anion compared to water? What effect will dissolution of such anions have on the pH of the solution?

10. Why do salts of strong soluble bases and weak acids give basic aqueous solutions? Use sodium hypochlorite, NaOCl, to illustrate. (Clorox, Purex, and other "chlorine bleaches" are 5% NaOCl.)

11. Write names and formulas for three salts of strong soluble bases and weak acids other than NaOCl.

12. Calculate the equilibrium constant for the reaction of azide ions, N_3^-, with water.

13. Calculate the equilibrium constant for the reaction of hypobromite ions with water. $K_a = 2.5 \times 10^{-9}$ for HOBr.

14. Calculate hydrolysis constants for the following anions of weak acids: (a) NO_2^-; (b) OCl^-; (c) $HCOO^-$. What is the relationship between K_a, the ionization constant for a weak acid, and K_b, the hydrolysis constant for the anion of the weak acid? (See Appendix F.)

15. Calculate the pH of 0.15 M solutions of the following salts: (a) $NaNO_2$; (b) NaOCl; (c) NaHCOO. (Refer to Exercise 14 for K values.)

16. What is the percent hydrolysis in each of the solutions in Exercise 15?

17. What is the pH of a 0.15 M solution of KOI? $K_a = 2.3 \times 10^{-11}$ for HOI.

18. What is the pH of a 0.15 M solution of KF?

Salts of Weak Bases and Strong Acids

19. Why do salts of weak bases and strong acids give acidic aqueous solutions? Illustrate with NH_4NO_3, a common fertilizer.

20. Write names and formulas for four salts of weak bases and strong acids.

21. Use values found in Appendix G to calculate hydrolysis constants for the following cations of weak bases: (a) NH_4^+; (b) $CH_3NH_3^+$, methylammonium ion; (c) $C_6H_5NH_3^+$, anilinium ion.

22. Calculate the pH of 0.15 M solutions of (a) NH_4NO_3; (b) $(CH_3)NH_3NO_3$; (c) $C_6H_5NH_3NO_3$.

23. Can you make a general statement relating parent base strength and extent of hydrolysis of the cations of Exercise 21 by using hydrolysis constants calculated in that exercise?

24. How do pH values for the following pairs of solutions compare? Why? (a) 0.050 M NH_4Br, ammonium bromide, and 0.050 M NH_4NO_3, ammonium nitrate; (b) 0.010 M ammonium perchlorate, NH_4ClO_4, and 0.010 M ammonium chloride, NH_4Cl.

Salts of Weak Bases and Weak Acids

25. Why are some aqueous solutions of salts of weak acids and weak bases neutral, whereas others are acidic and still others are basic?

26. Write the names and formulas for three salts of a weak acid and a weak base that give (a) neutral, (b) acidic, and (c) basic aqueous solutions.

27. If both the cation and anion of a salt react with water when dissolved, what determines whether the solution will be acidic or basic? Classify aqueous solutions of the following salts as acidic or basic. (a) NH_4F(aq) and (b) CH_3NH_3OI(aq), $K_b = 1.8 \times 10^{-5}$ for NH_3, $K_a = 7.2 \times 10^{-4}$ for HF, $K_b = 5.0 \times 10^{-4}$ for CH_3NH_2 and $K_a = 2.3 \times 10^{-11}$ for HOI.

Salts That Contain Small, Highly Charged Cations

28. Choose the cations that react with water to give acidic solutions. (a) K^+; (b) $[Be(H_2O)_4]^{2+}$; (c) $[Al(H_2O)_6]^{3+}$; (d) $[Fe(H_2O)_6]^{3+}$; (e) $[Cu(H_2O)_6]^{2+}$. Write chemical equations for the reactions.

29. Why do salts that contain cations related to insoluble bases (metal hydroxides) and anions related to strong acids give acidic aqueous solutions? Use $Fe(NO_3)_3$ to illustrate.

30. Calculate pH and percent hydrolysis for the following (Table 19-2). (a) 0.15 M $Al(NO_3)_3$, aluminum nitrate; (b) 0.075 M $Co(ClO_4)_2$, cobalt(II) perchlorate; (c) 0.15 M $MgCl_2$, magnesium chloride.

*31. Given pH values for solutions of the following concentrations, calculate hydrolysis constants for the cations: (a) 0.00050 M $CeCl_3$, cerium(III) chloride, pH = 5.99; (b) 0.10 M $Cu(NO_3)_2$, copper(II) nitrate, pH = 4.50; (c) 0.10 M $Sc(ClO_4)_3$, scandium perchlorate, pH = 3.44.

Strong Acid–Strong Base Titration Curves

32. Make a rough sketch of the titration curve expected for the titration of a strong acid with a strong base. What determines the pH of the solution at the following points? (a) No base added; (b) half-equivalence point; (c) equivalence point; (d) excess base added. Compare your curve with Figure 19-3.

For Exercises 33, 36, 37, and 44, calculate and tabulate $[H_3O^+]$, $[OH^-]$, pH, and pOH at the indicated points as we did in Table 19-4. In each case assume that pure acid (or base) is added to exactly 1 L of a 0.0100 molar solution of the indicated base (or acid). This simplifies the arithmetic because we may assume that the volume of each solution is constant throughout the titration. Plot each titration curve with pH on the vertical axis and moles of base (or acid) added on the horizontal axis.

33. Solid NaOH is added to 1 L of 0.0500 *M* HCl solution. Number of moles of NaOH added: (a) none; (b) 0.00500; (c) 0.01500; (d) 0.02500 (50% titrated); (e) 0.03500; (f) 0.04500; (g) 0.04750; (h) 0.0500 (100% titrated); (i) 0.0525; (j) 0.0600; (k) 0.0750 (50% excess NaOH). Consult Table 19-3 and list the indicators that could be used in this titration.

34. A 25.0-mL sample of 0.125 *M* HNO_3 is titrated with 0.100 *M* NaOH. Calculate the pH of the solution (a) before the addition of NaOH and after the addition of (b) 5.0 mL, (c) 12.5 mL, (d) 25.0 mL, (e) 31.25 mL, (f) 37.5 mL of NaOH.

Weak Acid–Strong Base Titration Curves

35. Make a rough sketch of the titration curve expected for the titration of a weak monoprotic acid with a strong base. What determines the pH of the solution at the following points? (a) No base added; (b) half-equivalence point; (c) equivalence point; (d) excess base added. Compare your curve to Figure 19-4.

36. Solid NaOH is added to 1 L of 0.0200 *M* CH_3COOH solution. Number of moles NaOH added: (a) none; (b) 0.00400; (c) 0.00800; (d) 0.01000 (50% titrated); (e) 0.01400; (f) 0.01800; (g) 0.01900; (h) 0.0200 (100% titrated); (i) 0.0210; (j) 0.0240; (k) 0.0300 (50% excess NaOH). Consult Table 19-3 and list the indicators that could be used in this titration.

*37. Gaseous HCl is added to 1 L of 0.0100 *M* aqueous ammonia solution. Number of moles HCl added (a) none; (b) 0.00100; (c) 0.00300; (d) 0.00500 (50% titrated); (e) 0.00700; (f) 0.00900; (g) 0.00950; (h) 0.0100 (100% titrated); (i) 0.0105; (j) 0.0120; (k) 0.0150 (50% excess HCl). Consult Table 19-3 and list the indicators that could be used in this titration.

38. Calculate the pH at the equivalence point of the titration of 100.0 mL of each of the following with 0.150 *M* KOH: (a) 1.000 *M* acetic acid; (b) 0.100 *M* acetic acid; (c) 0.0100 *M* acetic acid.

39. A solution contains an unknown weak monoprotic acid HA. It takes 46.24 mL of NaOH solution to titrate 50.00 mL of the HA solution to the equivalence point. To another 50.00-mL sample of the same HA solution, 23.12 mL of the same NaOH solution is added. The pH of the resulting solution in the second experiment is 5.14. What are K_a and pK_a of HA?

Mixed Exercises

40. Classify aqueous solutions of the following salts as acidic, basic, or essentially neutral. Justify your choice. (a) $(NH_4)HSO_4$; (b) $(NH_4)_2SO_4$; (c) KCl; (d) LiBrO; (e) $Al(NO_3)_3$.

41. Repeat Exercise 40 for (a) $NaClO_4$; (b) $K_2C_2O_4$; (c) NH_4HTeO_3; (d) NaH_2PO_4; (e) NH_4CN. (See Appendix F.)

42. In aqueous solution some cations react with water to upset the H_3O^+/OH^- balance. What is the relative acid strength of such a cation compared to water? What effect will dissolution of these cations have on the pH of the solution?

43. Calculate the pH at the equivalence point for the titration of a solution containing 150.0 mg of ethylamine, $C_2H_5NH_2$, with 0.1000 *M* HCl solution. The volume of the solution at the equivalence point is 250 mL. Select a suitable indicator.

*44. Gaseous NH_3 is added to 1 L of 0.0100 *M* HNO_3 solution. Number of moles NH_3 added: (a) none; (b) 0.00100; (c) 0.00400; (d) 0.00500 (50% titrated); (e) 0.00900; (f) 0.00950; (g) 0.0100 (100% titrated); (h) 0.0105; (i) 0.0130. What is the major difference between the titration curve for the reaction of HNO_3 and NH_3 and the other curves you have plotted? Consult Table 19-3. Can you suggest a satisfactory indicator for this titration?

45. Use Table 19-3 to choose one or more indicators that could be used to "signal" reaching a pH of (a) 3.5; (b) 7; (c) 10.3; (d) 8.0.

46. A solution of 0.020 *M* acetic acid is to be titrated with a 0.015 *M* NaOH solution. What is pH at the equivalence point? Choose an appropriate indicator for the titration.

47. Some plants require acidic soils for healthy growth. Which of the following could be added to the soil around such plants to increase the acidity of the soil? Write equations to justify your answers. (a) $FeSO_4$; (b) Na_2SO_4; (c) $Al_2(SO_4)_3$; (d) $Fe_2(SO_4)_3$; (e) $BaSO_4$. Arrange the salts that give acidic solutions in order of increasing acidity.

48. Some of the following salts are used in detergents and other cleaning materials because they produce basic solutions. Which of the following could *not* be used for this purpose? Write equations to justify your answers. (a) Na_2CO_3; (b) Na_2SO_4; (c) $(NH_4)_2SO_4$; (d) Na_3PO_4.

BUILDING YOUR KNOWLEDGE

49. Acetylsalicylic acid, the active ingredient in aspirin, has a K_a value of 3.0×10^{-4}. We dissolve 0.0100 mole of acetylsalicylic acid in sufficient water to make 1.00 L of solution and then titrate it with 0.500 *M* NaOH solution. What is the pH at each of these points in the titration? (a) before any of the NaOH solution is added; (b) at the equivalence point; (c) when a volume of NaOH solution has been added that is equal to half the amount required to reach the equivalence point.

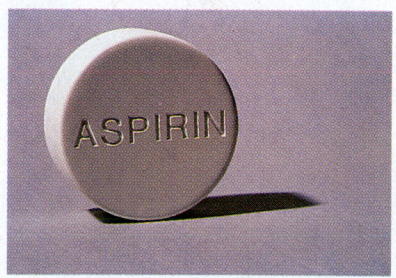

50. Phenol is a weak acid that is sometimes used as a disinfectant. What is the pH of a disinfectant solution that is 0.0100 *M* phenol? $K_a = 1.3 \times 10^{-10}$.

51. The pH of an equal molar acetic acid/sodium acetate buffer is 4.74. Draw a molecular representation of a small portion of this buffer solution. (You may omit the water molecules.) Draw another molecular representation of the solution after a very small amount of NaOH has been added.

20

Ionic Equilibria III: The Solubility Product Principle

OBJECTIVES

As you study this chapter, you should learn

- *To recognize common slightly soluble compounds*

- *To write solubility product constant expressions*

- *How K_{sp}'s are determined*

- *To use K_{sp}'s*

- *About fractional precipitation and its importance*

- *How simultaneous equilibria are used to control solubility*

- *Some methods for dissolving precipitates*

Pouring ammonium sulfide solution into a solution of cadmium nitrate gives a precipitate of cadmium sulfide.

$$(NH_4)_2S + Cd(NO_3)_2 \longrightarrow CdS(s) + 2NH_4NO_3$$

Cadmium sulfide is used as a pigment in artists' oil-based paints.

So far we have discussed mainly compounds that are quite soluble in water. Although most compounds dissolve in water to some extent, many are so slightly soluble that they are called "insoluble compounds." We shall now consider those that are only very slightly soluble. As a rough rule of thumb, compounds that dissolve in water to the extent of 0.020 mole/liter or more are classified as soluble. Refer to the solubility rules (Table 4-8) as necessary.

Slightly soluble compounds are important in many natural phenomena. Our bones and teeth are mostly calcium phosphate, $Ca_3(PO_4)_2$, a slightly soluble compound. Also, many natural deposits of $Ca_3(PO_4)_2$ rock are mined and converted into agricultural fertilizer. Limestone caves have been formed by acidic water slowly dissolving away calcium carbonate, $CaCO_3$. Sinkholes are created when acidic water dissolves away most of the underlying $CaCO_3$. The remaining limestone can no longer support the weight above it, so it collapses, and a sinkhole is formed. More examples can be found in Chapters 22 and 23.

20-1 SOLUBILITY PRODUCT CONSTANTS

Suppose we add one gram of solid barium sulfate, $BaSO_4$, to 1.0 liter of water at 25°C and stir until the solution is *saturated*. Very little $BaSO_4$ dissolves. Only 0.0025 gram of $BaSO_4$ dissolves in 1.0 liter of water, no matter how much more $BaSO_4$ is added. The $BaSO_4$ that does dissolve is completely dissociated into its constituent ions.

$$BaSO_4(s) \xrightleftharpoons{H_2O} Ba^{2+}(aq) + SO_4^{2-}(aq)$$

We usually omit H_2O over the arrows in equations like the one above.

Sinkholes are formed when the underlying limestone, mostly $CaCO_3$, is dissolved away by acidic water.

Barium sulfate is the "insoluble" substance taken orally before stomach X-rays are made because the barium atoms absorb X-rays well. Even though barium ions are quite toxic, barium sulfate can still be taken orally without danger. The compound is so insoluble that it passes through the digestive system essentially unchanged.

In equilibria that involve slightly soluble compounds in water, the equilibrium constant is called a **solubility product constant, K_{sp}**. The activity of the solid $BaSO_4$ is one (Section 17-10, Heterogeneous Equilibrium). Hence, the concentration of the solid is not included in the equilibrium constant expression. For a saturated solution of $BaSO_4$ in contact with solid $BaSO_4$, we write

$$BaSO_4(s) \rightleftharpoons Ba^{2+}(aq) + SO_4^{2-}(aq) \quad \text{and} \quad K_{sp} = [Ba^{2+}][SO_4^{2-}]$$

The solubility product constant for $BaSO_4$ is the product of the concentrations of its constituent ions in a saturated solution.

> In general, the **solubility product expression** for a compound is the product of the concentrations of its constituent ions, each raised to the power that corresponds to the number of ions in one formula unit of the compound. The quantity is constant at constant temperature for a saturated solution of the compound. This statement is the **solubility product principle.**

The existence of a substance in the solid state is indicated several ways. For example, $BaSO_4(s)$, $\underline{BaSO_4}$, and $BaSO_4\downarrow$ are sometimes used to represent solid $BaSO_4$. We use the (s) notation for formulas of solid substances in equilibrium with their saturated aqueous solutions.

Consider dissolving slightly soluble calcium fluoride, CaF_2, in H_2O.

$$CaF_2(s) \rightleftharpoons Ca^{2+}(aq) + 2F^-(aq) \quad K_{sp} = [Ca^{2+}][F^-]^2 = 3.9 \times 10^{-11}$$

Dissolving solid bismuth sulfide, Bi_2S_3, in H_2O gives two bismuth ions and three sulfide ions per formula unit.

$$Bi_2S_3(s) \rightleftharpoons 2Bi^{3+}(aq) + 3S^{2-}(aq) \quad K_{sp} = [Bi^{3+}]^2[S^{2-}]^3 = 1.6 \times 10^{-72}$$

Generally, we may represent the dissolution of a slightly soluble compound and its K_{sp} expression as

$$M_yX_z(s) \rightleftharpoons yM^{z+}(aq) + zX^{y-}(aq) \quad \text{and} \quad K_{sp} = [M^{z+}]^y[X^{y-}]^z$$

In some cases a compound contains more than two kinds of ions. Dissolution of the slightly soluble compound magnesium ammonium phosphate, $MgNH_4PO_4$, in water and its solubility product expression are represented as

$$MgNH_4PO_4(s) \rightleftharpoons Mg^{2+}(aq) + NH_4^+(aq) + PO_4^{3-}(aq)$$

$$K_{sp} = [Mg^{2+}][NH_4^+][PO_4^{3-}] = 2.5 \times 10^{-12}$$

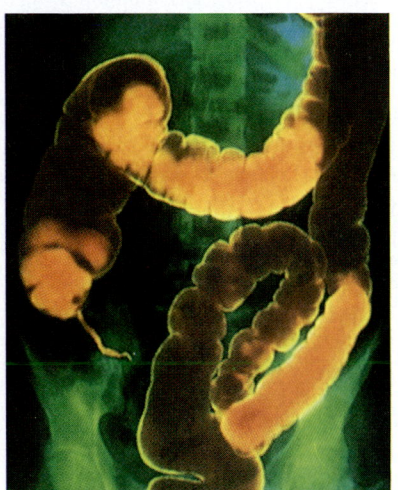

An X-ray photo of the gastrointestinal tract. The barium ions in $BaSO_4$ absorb X-radiation well.

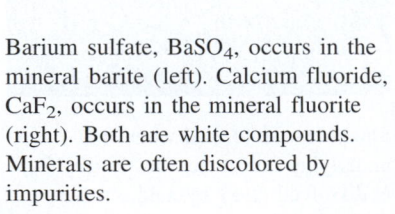

Barium sulfate, $BaSO_4$, occurs in the mineral barite (left). Calcium fluoride, CaF_2, occurs in the mineral fluorite (right). Both are white compounds. Minerals are often discolored by impurities.

We often shorten the term "solubility product constant" to "solubility product." Thus, the solubility products for barium sulfate, $BaSO_4$, and for calcium fluoride, CaF_2, are written as

$$K_{sp} = [Ba^{2+}][SO_4^{2-}] = 1.1 \times 10^{-10} \qquad K_{sp} = [Ca^{2+}][F^-]^2 = 3.9 \times 10^{-11}$$

The **molar solubility** of a compound is the number of moles that dissolve to give one liter of saturated solution.

20-2 DETERMINATION OF SOLUBILITY PRODUCT CONSTANTS

Solubility product constants can be determined in a number of ways. For example, careful measurements of conductivity show that one liter of a saturated solution of barium sulfate contains 0.0025 gram of dissolved $BaSO_4$. If the solubility of a compound is known, the value of its solubility product can be calculated.

Unless otherwise indicated, solubility product constants and solubility data are given for 25°C (Appendix H).

EXAMPLE 20-1 *Solubility Product Constants*

One (1.0) liter of saturated barium sulfate solution contains 0.0025 gram of dissolved $BaSO_4$. Calculate the solubility product constant for $BaSO_4$.

Plan

We write the equation for the dissolution of $BaSO_4$ and the expression for its solubility product constant, K_{sp}. From the solubility of $BaSO_4$ in H_2O, we calculate its molar solubility and the concentrations of the ions. This lets us calculate K_{sp}.

We frequently use statements such as "The solution contains 0.0025 gram of $BaSO_4$." What we mean is that 0.0025 gram of solid $BaSO_4$ *dissolves* to give a solution that contains equal concentrations of Ba^{2+} and SO_4^{2-} ions.

Solution

In saturated solutions, equilibrium exists between solid and dissolved solute. The equation for the dissolution of barium sulfate in water and its solubility product expression are

$$BaSO_4(s) \rightleftharpoons Ba^{2+}(aq) + SO_4^{2-}(aq) \qquad K_{sp} = [Ba^{2+}][SO_4^{2-}]$$

From the given solubility of $BaSO_4$ in H_2O we can calculate its *molar solubility*.

$$\frac{? \text{ mol } BaSO_4}{L} = \frac{2.5 \times 10^{-3} \text{ g } BaSO_4}{1.0 \text{ L}} \times \frac{1 \text{ mol } BaSO_4}{233 \text{ g } BaSO_4} = 1.1 \times 10^{-5} \text{ mol } BaSO_4/L$$
$$\text{(dissolved)}$$

1.1×10^{-5} mole of solid $BaSO_4$ dissolves to give a liter of saturated solution.

We know the molar solubility of $BaSO_4$. The dissolution equation shows that each formula unit of $BaSO_4$ that dissolves produces one Ba^{2+} ion and one SO_4^{2-} ion.

$$
\begin{array}{ccccc}
BaSO_4(s) & \rightleftharpoons & Ba^{2+}(aq) & + & SO_4^{2-}(aq) \\
1.1 \times 10^{-5} \text{ mol/L} & \Longrightarrow & 1.1 \times 10^{-5} M & & 1.1 \times 10^{-5} M \\
\text{(dissolved)} & & & &
\end{array}
$$

In a saturated solution $[Ba^{2+}] = [SO_4^{2-}] = 1.1 \times 10^{-5} M$. Substituting these values into the K_{sp} expression for $BaSO_4$ gives the value of K_{sp}.

$$K_{sp} = [Ba^{2+}][SO_4^{2-}] = (1.1 \times 10^{-5})(1.1 \times 10^{-5}) = \boxed{1.2 \times 10^{-10}}$$

You should now work Exercise 6.

The value calculated here is 1.2×10^{-10}, whereas the tabulated value is 1.1×10^{-10}. Roundoff error is responsible for the difference.

This expression is useful because it applies *to all saturated solutions of $BaSO_4$ at 25°C.* The *origin* of Ba^{2+} and SO_4^{2-} ions is not relevant. For example, suppose a solution of $BaCl_2$ and a solution of Na_2SO_4 are mixed at 25°C. (Both of these compounds are soluble and ionic.) When the concentration of Ba^{2+} ions multiplied by the concentration of SO_4^{2-} ions reaches 1.1×10^{-10}, the solution is saturated. At higher concentrations,

$BaSO_4$ precipitates and does so until the solution is just saturated. Or, if solid $BaSO_4$ is placed in water at 25°C, it dissolves until the $[Ba^{2+}]$ multiplied by the $[SO_4^{2-}]$ just equals 1.1×10^{-10}.

EXAMPLE 20-2   *Solubility Product Constant*

One (1.0) liter of a saturated solution of silver chromate at 25°C contains 0.0435 gram of Ag_2CrO_4. Calculate its solubility product constant.

Plan

We proceed as in Example 20-1.

Solution

The equation for the dissolution of silver chromate in water and its solubility product expression are

$$Ag_2CrO_4(s) \rightleftharpoons 2Ag^+(aq) + CrO_4^{2-}(aq) \qquad \text{and} \qquad K_{sp} = [Ag^+]^2[CrO_4^{2-}]$$

The molar solubility of silver chromate is calculated first.

1.31 × 10⁻⁴ mole of solid Ag_2CrO_4 dissolves to give a liter of saturated solution.

$$\frac{\underset{?}{\text{mol }Ag_2CrO_4}}{L} = \frac{0.0435 \text{ g } Ag_2CrO_4}{1.0 \text{ L}} \times \frac{1 \text{ mol } Ag_2CrO_4}{332 \text{ g } Ag_2CrO_4} = 1.31 \times 10^{-4} \text{ mol/L}$$
$$\text{(dissolved)}$$

The equation for dissolution of Ag_2CrO_4 and its molar solubility give the concentrations of Ag^+ and CrO_4^{2-} ions in the saturated solution.

$$\begin{array}{ccccc} Ag_2CrO_4(s) & \rightleftharpoons & 2Ag^+(aq) & + & CrO_4^{2-}(aq) \\ 1.31 \times 10^{-4} \text{ mol/L} & \Longrightarrow & 2.62 \times 10^{-4} M & & 1.31 \times 10^{-4} M \\ \text{dissolved} & & & & \end{array}$$

Substitution into the K_{sp} expression for Ag_2CrO_4 gives the value of K_{sp}.

$$K_{sp} = [Ag^+]^2[CrO_4^{2-}] = (2.62 \times 10^{-4})^2(1.31 \times 10^{-4}) = \boxed{8.99 \times 10^{-12}}$$

Solubility products are usually given to only two significant digits.

The K_{sp} of Ag_2CrO_4 is 9.0×10^{-12}.

You should now work Exercise 8.

The values for $BaSO_4$ and Ag_2CrO_4 are compared in Table 20-1. These data show that the molar solubility of Ag_2CrO_4 is greater than that of $BaSO_4$. However, K_{sp} for Ag_2CrO_4 is less than K_{sp} for $BaSO_4$ because the expression for Ag_2CrO_4 contains a *squared* term, $[Ag^+]^2$.

$K_{sp(AgCl)} = 1.8 \times 10^{-10}$
$K_{sp(BaSO_4)} = 1.1 \times 10^{-10}$

The molar solubility of AgCl is only slightly higher than that of $BaSO_4$.

If we compare K_{sp} values for two 1:1 compounds, e.g., AgCl and $BaSO_4$, the compound with the larger K_{sp} value has the higher molar solubility. The same is true for *any* two compounds that have the same ion ratio, e.g., the 1:2 compounds CaF_2 and $Mg(OH)_2$ and the 2:1 compound Ag_2S.

Appendix H lists some K_{sp} values. Refer to it as needed.

Table 20-1 *Comparison of Solubilities of $BaSO_4$ and Ag_2CrO_4*

Compound	Molar Solubility	K_{sp}
$BaSO_4$	1.1×10^{-5} mol/L	$[Ba^{2+}][SO_4^{2-}] = 1.1 \times 10^{-10}$
Ag_2CrO_4	1.3×10^{-4} mol/L	$[Ag^+]^2[CrO_4^{2-}] = 9.0 \times 10^{-12}$

The Effects of Hydrolysis on Solubility

In Section 19-2 we discussed the hydrolysis of anions of weak acids. For example, we found that for CH_3COO^- and CN^- ions,

$$CH_3COO^- + H_2O \rightleftharpoons CH_3COOH + OH^- \qquad K_b = \frac{[CH_3COOH][OH^-]}{[CH_3COO^-]} = 5.6 \times 10^{-10}$$

$$CN^- + H_2O \rightleftharpoons HCN + OH^- \qquad K_b = \frac{[HCN][OH^-]}{[CN^-]} = 2.5 \times 10^{-5}$$

We see that K_b for CN^-, the anion of a *very* weak acid, is much larger than K_b for CH_3COO^-, the anion of a much stronger acid. This tells us that in solutions of the same concentration, CN^- ions hydrolyze to a much greater extent than do CH_3COO^- ions. So we might expect that hydrolysis would have a much greater effect on the solubilities of cyanides such as AgCN than on the solubilities of acetates such as $AgCH_3COO$. It does.

Hydrolysis reduces the concentrations of anions of weak acids, so its effect must be taken into account when we do very precise solubility calculations. However, taking into account the effect of hydrolysis on solubilities of slightly soluble compounds is beyond the scope of this chapter.

20-3 USES OF SOLUBILITY PRODUCT CONSTANTS

When the solubility product for a compound is known, the solubility of the compound in H_2O at 25°C can be calculated as Example 20-3 illustrates.

EXAMPLE 20-3   *Molar Solubilities from K_{sp} Values*

Calculate the molar solubilities, concentrations of the constituent ions, and solubilities in grams per liter for (a) silver chloride, AgCl ($K_{sp} = 1.8 \times 10^{-10}$), and (b) zinc hydroxide, $Zn(OH)_2$ ($K_{sp} = 4.5 \times 10^{-17}$).

The values of K_{sp} are obtained from Appendix H.

Plan

We are given the value for each solubility product constant. In each case we write the appropriate equation, represent the equilibrium concentrations, and then substitute into the K_{sp} expression.

Solution

(a) The equation for the dissolution of silver chloride and its solubility product expression are

$$AgCl(s) \rightleftharpoons Ag^+(aq) + Cl^-(aq) \qquad K_{sp} = [Ag^+][Cl^-] = 1.8 \times 10^{-10}$$

Each formula unit of AgCl that dissolves produces one Ag^+ and one Cl^-. We let $x = $ mol/L of AgCl that dissolves, i.e., the molar solubility.

$$\begin{array}{ccc} AgCl(s) & \rightleftharpoons & Ag^+(aq) + Cl^-(aq) \\ x \text{ mol/L} \Longrightarrow & & x\,M \qquad\quad x\,M \end{array}$$

Substitution into the solubility product expression gives

$$[Ag^+][Cl^-] = (x)(x) = 1.8 \times 10^{-10} \qquad x^2 = 1.8 \times 10^{-10} \qquad x = 1.3 \times 10^{-5}$$

$$x = \text{molar solubility of AgCl} = 1.3 \times 10^{-5} \text{ mol/L}$$

One liter of saturated AgCl contains 1.3×10^{-5} mole of dissolved AgCl at 25°C. From the balanced equation we know the concentrations of the constituent ions.

$$x = [Ag^+] = [Cl^-] = 1.3 \times 10^{-5} \text{ mol/L} = 1.3 \times 10^{-5} \, M$$

Now we can calculate the mass of dissolved AgCl in one liter of saturated solution.

$$\frac{\underline{?} \text{ g AgCl}}{L} = \frac{1.3 \times 10^{-5} \text{ mol AgCl}}{L} \times \frac{143 \text{ g AgCl}}{1 \text{ mol AgCl}} = 1.9 \times 10^{-3} \text{ g AgCl/L}$$

A liter of saturated AgCl solution contains only 0.0019 g of dissolved AgCl.

(b) The dissolution of zinc hydroxide, $Zn(OH)_2$, in water and its solubility product expression are

$$Zn(OH)_2(s) \rightleftharpoons Zn^{2+}(aq) + 2OH^-(aq) \qquad K_{sp} = [Zn^{2+}][OH^-]^2 = 4.5 \times 10^{-17}$$

We let x = molar solubility, so $[Zn^{2+}] = x$ and $[OH^-] = 2x$, and we have

$$Zn(OH)_2(s) \rightleftharpoons Zn^{2+}(aq) + 2OH^-(aq)$$
$$x \text{ mol/L} \rightleftharpoons \quad x \, M \qquad \quad 2x \, M$$

Substitution into the solubility product expression gives

$$[Zn^{2+}][OH^-]^2 = (x)(2x)^2 = 4.5 \times 10^{-17}$$

$$4x^3 = 4.5 \times 10^{-17} \qquad x^3 = 11 \times 10^{-18} \qquad x = 2.2 \times 10^{-6}$$

$$x = \text{molar solubility of } Zn(OH)_2 = 2.2 \times 10^{-6} \text{ mol } Zn(OH)_2/L$$

$$x = [Zn^{2+}] = 2.2 \times 10^{-6} \, M \qquad \text{and} \qquad 2x = [OH^-] = 4.4 \times 10^{-6} \, M$$

> The $[OH^-]$ is twice the molar solubility of $Zn(OH)_2$ because each formula unit of $Zn(OH)_2$ produces two OH^-.

We can now calculate the mass of dissolved $Zn(OH)_2$ in one liter of saturated solution.

$$\frac{\underline{?} \text{ g } Zn(OH)_2}{L} = \frac{2.2 \times 10^{-6} \text{ mol } Zn(OH)_2}{L} \times \frac{99 \text{ g } Zn(OH)_2}{1 \text{ mol } Zn(OH)_2} = 2.2 \times 10^{-4} \text{ g } Zn(OH)_2/L$$

A liter of saturated $Zn(OH)_2$ solution contains only 0.00022 g of dissolved $Zn(OH)_2$.

You should now work Exercise 12.

In part (b) of Example 20-3 we found that $[OH^-] = 4.4 \times 10^{-6} \, M$ in a *saturated* $Zn(OH)_2$ solution. From this we find pOH = 5.36 and pH = 8.64. A saturated $Zn(OH)_2$ solution is not very basic because $Zn(OH)_2$ is not very soluble in H_2O. The $[OH^-]$ is 44 times greater than it is in pure water.

The Common Ion Effect in Solubility Calculations

The common ion effect applies to solubility equilibria just as it does to other ionic equilibria. The solubility of a compound is less in a solution that contains an ion common to the compound than in pure water (as long as no other reaction is caused by the presence of the common ion).

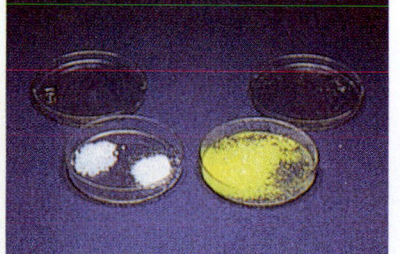

When white solid potassium iodide, KI, and white solid lead(II) nitrate, $Pb(NO_3)_2$, are stirred together, some yellow lead(II) iodide, PbI_2, forms. This reaction occurs in the small amount of water present in these solids.

EXAMPLE 20-4 *Molar Solubilities*

The molar solubility of magnesium fluoride, MgF_2, is 1.2×10^{-3} mol/L in pure water at 25°C. Calculate the molar solubility of MgF_2 in 0.10 M sodium fluoride, NaF, solution at 25°C. $K_{sp} = 6.4 \times 10^{-9}$ for MgF_2.

Plan

We recognize that NaF is a soluble ionic compound that is completely dissociated into its ions. MgF_2 is a slightly soluble compound. Both compounds produce F^- ions so this is a common ion effect problem. We write the appropriate chemical equations and solubility product expression, represent the equilibrium concentrations, and then substitute into the solubility product expression.

Solution

NaF is a soluble ionic salt and, therefore, 0.10 M F^- is produced by

$$NaF(s) \xrightarrow{H_2O} Na^+(aq) + F^-(aq) \quad \text{(complete)}$$
$$0.10\,M \implies 0.10\,M \quad 0.10\,M$$

We let x = molar solubility for MgF_2, a slightly soluble salt.

$$MgF_2(s) \rightleftharpoons Mg^{2+}(aq) + 2F^-(aq) \quad \text{(reversible)}$$
$$x\,\text{mol/L} \implies x\,M \quad 2x\,M$$

The total $[F^-]$ is 0.10 M from NaF *plus* $2x\,M$ from MgF_2, or $(0.10 + 2x)\,M$.

$$K_{sp} = [Mg^{2+}][F^-]^2 = 6.4 \times 10^{-9}$$
$$(x)(0.10 + 2x)^2 = 6.4 \times 10^{-9}$$

We make the simplifying assumption that $2x \ll 0.10$, so

$$(x)(0.10)^2 = 6.4 \times 10^{-9} \quad \text{and} \quad x = 6.4 \times 10^{-7}$$

The assumption is shown to be valid.

$$6.4 \times 10^{-7}\,M = \text{molar solubility of } MgF_2 \text{ in } 0.10\,M \text{ NaF}$$

The molar solubility of MgF_2 in 0.10 M NaF ($6.4 \times 10^{-7}\,M$) is nearly 1900 times less than in pure water ($1.2 \times 10^{-3}\,M$).

You should now work Exercises 16 and 18.

Ratio of solubility of MgF_2 in 0.10 M NaF to its solubility in water =
$$\frac{6.4 \times 10^{-7}}{1.2 \times 10^{-3}} = \frac{1}{1900}$$

The Reaction Quotient in Precipitation Reactions

Another application of the solubility product principle is the calculation of the maximum concentrations of ions that can coexist in solution. From these calculations we can determine whether a precipitate will form in a given solution. We compare Q_{sp} (Section 17-4) with K_{sp}.

If $Q_{sp} < K_{sp}$	Forward process is favored No precipitation occurs; more solid can dissolve
$Q_{sp} = K_{sp}$	Solution is *just* saturated Neither forward nor reverse process is favored
$Q_{sp} > K_{sp}$	Reverse process is favored; precipitation occurs

EXAMPLE 20-5 *Predicting Precipitate Formation*

If 100 mL of 0.00075 M sodium sulfate, Na_2SO_4, is mixed with 50.0 mL of 0.015 M barium chloride, $BaCl_2$, will a precipitate form?

Plan

We are mixing solutions of two soluble ionic salts. First we find the amount of each solute at the instant of mixing. Next we find the molarity of each solute *at the instant of mixing*. Then we find the concentration of each ion in the *new* solution. Now we ask the question "*Could* any combination of the ions in this solution form a slightly soluble compound?" The answer is "yes, Ba^{2+} and SO_4^{2-} *could* form $BaSO_4$," so we calculate Q_{sp} and compare it with K_{sp}.

Solution

We find the *amount* of each solute at the instant of mixing.

$$\underset{?}{\text{?}}\text{ mmol Na}_2\text{SO}_4 = 100\text{ mL} \times \frac{0.00075\text{ mmol Na}_2\text{SO}_4}{\text{mL}} = 0.075\text{ mmol Na}_2\text{SO}_4$$

$$\underset{?}{\text{?}}\text{ mmol BaCl}_2 = 50.0\text{ mL} \times \frac{0.015\text{ mmol BaCl}_2}{\text{mL}} = 0.75\text{ mmol BaCl}_2$$

Then we find the *molarity of each solute at the instant of mixing.*

$$M_{\text{Na}_2\text{SO}_4} = \frac{0.075\text{ mmol Na}_2\text{SO}_4}{150\text{ mL}} = 0.00050\ M\text{ Na}_2\text{SO}_4$$

$$M_{\text{BaCl}_2} = \frac{0.75\text{ mmol BaCl}_2}{150\text{ mL}} = 0.0050\ M\text{ BaCl}_2$$

> When *dilute* aqueous solutions are mixed, their volumes can be added to give the volume of the resulting solution.

Now we find the *concentration of each ion* in the new solution.

$$\text{Na}_2\text{SO}_4(s) \xrightarrow{100\%} 2\text{Na}^+(aq) + \text{SO}_4^{2-}(aq) \qquad \text{(to completion)}$$

$$0.00050\ M \implies \boxed{0.0010\ M} \qquad \boxed{0.00050\ M}$$

$$\text{BaCl}_2(s) \xrightarrow{100\%} \text{Ba}^{2+}(aq) + 2\text{Cl}^-(aq) \qquad \text{(to completion)}$$

$$0.0050\ M \implies \boxed{0.0050\ M} \qquad \boxed{0.010\ M}$$

We consider the kinds of compounds mixed and determine whether a reaction could occur. Both Na_2SO_4 and BaCl_2 are soluble ionic salts. At the moment of mixing, the new solution contains a mixture of Na^+, SO_4^{2-}, Ba^{2+}, and Cl^- ions. We must consider the possibility of forming two new compounds, NaCl and BaSO_4. Sodium chloride is a soluble ionic compound so Na^+ and Cl^- do not combine in dilute aqueous solutions. However, BaSO_4 is only very slightly soluble, and solid BaSO_4 will precipitate from the solution *if* $Q_{sp} > K_{sp}$ for BaSO_4. K_{sp} for BaSO_4 is 1.1×10^{-10}. Substituting $[\text{Ba}^{2+}] = 0.0050\ M$ and $[\text{SO}_4^{2-}] = 0.00050\ M$ into the Q_{sp} expression for BaSO_4, we get

> Recall that Q has the same form as the equilibrium constant, in this case K_{sp}, but the concentrations are not necessarily equilibrium concentrations.

$$Q_{sp} = [\text{Ba}^{2+}][\text{SO}_4^{2-}] = (5.0 \times 10^{-3})(5.0 \times 10^{-4}) = 2.5 \times 10^{-6} \qquad (Q_{sp} > K_{sp})$$

Because $Q_{sp} > K_{sp}$ solid BaSO_4 will precipitate until $[\text{Ba}^{2+}][\text{SO}_4^{2-}]$ just equals K_{sp} for BaSO_4.

You should now work Exercises 22 and 24.

▼ **PROBLEM-SOLVING TIP**   *Detection of Precipitates*

The human eye is not a very sensitive detection device. As a rule of thumb, a precipitate can be seen with the naked eye if $Q_{sp} > K_{sp}$ by a factor of 1000. In Example 20-5, Q_{sp} exceeds K_{sp} by a factor of $\dfrac{2.5 \times 10^{-6}}{1.1 \times 10^{-10}} = 2.3 \times 10^4 = 23{,}000$. We expect to be able to see the BaSO_4 precipitate that is formed. Modern techniques enable us to detect smaller amounts of precipitates.

EXAMPLE 20-6   *Initiation of Precipitation*

What $[\text{Ba}^{2+}]$ is necessary to start the precipitation of BaSO_4 in a solution that is $0.0015\ M$ in Na_2SO_4? Assume that the Ba^{2+} comes from addition of a solid soluble ionic compound such as BaCl_2. For BaSO_4, $K_{sp} = 1.1 \times 10^{-10}$.

Plan

These are the compounds in Example 20-5. We recognize that Na_2SO_4 is a soluble ionic compound and that the molarity of SO_4^{2-} is equal to the molarity of the Na_2SO_4 solution. We are given K_{sp} for $BaSO_4$, so we solve for $[Ba^{2+}]$.

Solution

Because Na_2SO_4 is a soluble ionic compound, we know that $[SO_4^{2-}] = 0.0015\ M$. We can use K_{sp} for $BaSO_4$ to calculate the $[Ba^{2+}]$ required for Q_{sp} to just equal K_{sp}.

$$[Ba^{2+}][SO_4^{2-}] = 1.1 \times 10^{-10}$$

$$[Ba^{2+}] = \frac{1.1 \times 10^{-10}}{[SO_4^{2-}]} = \frac{1.1 \times 10^{-10}}{1.5 \times 10^{-3}} = 7.3 \times 10^{-8}\ M$$

Addition of enough $BaCl_2$ to give a barium ion concentration of $7.3 \times 10^{-8}\ M$ *just satisfies* K_{sp} for $BaSO_4$; i.e., $Q_{sp} = K_{sp}$. Ever so slightly more $BaCl_2$ would be required for Q_{sp} to exceed K_{sp} and for precipitation of $BaSO_4$ to occur. Therefore

$$[Ba^{2+}] > 7.3 \times 10^{-8}\ M \qquad \text{(to initiate precipitation of } BaSO_4)$$

You should now work Exercise 26.

Often we wish to remove an ion from solution by forming an insoluble compound (as in water purification). We use K_{sp} values to calculate the concentrations of ions remaining in solution *after* precipitation has occurred.

EXAMPLE 20-7 *Concentration of Common Ion*

Suppose we wish to recover silver from an aqueous solution that contains a soluble silver compound such as $AgNO_3$ by precipitating insoluble silver chloride, $AgCl$. A soluble ionic compound such as $NaCl$ can be used as a source of Cl^-. What is the minimum concentration of chloride ion needed to reduce the dissolved silver ion concentration to a maximum of $1.0 \times 10^{-9}\ M$? For $AgCl$, $K_{sp} = 1.8 \times 10^{-10}$.

Plan

We are given K_{sp} for $AgCl$ and the required equilibrium $[Ag^+]$, so we solve for $[Cl^-]$.

Solution

The equation for the reaction of interest and the K_{sp} for $AgCl$ are

$$AgCl(s) \rightleftharpoons Ag^+(aq) + Cl^-(aq) \qquad \text{and} \qquad [Ag^+][Cl^-] = 1.8 \times 10^{-10}$$

To determine the $[Cl^-]$ required to reduce the $[Ag^+]$ to $1.0 \times 10^{-9}\ M$, we solve the K_{sp} expression for $[Cl^-]$.

$$[Cl^-] = \frac{1.8 \times 10^{-10}}{[Ag^+]} = \frac{1.8 \times 10^{-10}}{1.0 \times 10^{-9}} = 0.18\ M\ Cl^-$$

To reduce the $[Ag^+]$ to $1.0 \times 10^{-9}\ M$ (0.00000011 g Ag^+/L), $NaCl$ would be added until $[Cl^-] = 0.18\ M$ in the solution.

You should now work Exercise 28.

The recovery of silver from the solutions used in developing and fixing photographic film and prints presents just such a problem. Silver is an expensive metal, and the recovery is profitable. Moreover, if not recovered, the silver ions would constitute an undesirable pollutant in water supplies.

Silver chloride precipitates when chloride ions are added to a solution containing silver ions.

CHEMISTRY IN USE

Research & Technology

Growing Concrete

One day we may use a new form of concrete that is less expensive and stronger than traditional concrete. The new concrete is available in unlimited amounts; its widespread use would decrease environmental damage caused by current production and transportation methods. This may sound like a pipe dream, but since the early 1980s researchers have been using the oceans to successfully grow "concrete" structures such as construction panels, dams, jetties, reefs, pipes, and even entire buildings.

The technique used to grow concrete structures is relatively straightforward. First metal mesh is shaped into the form of a desired structure, and then the mesh is inserted into an ocean. A direct electric current is applied to the mesh. As magnesium and calcium ions migrate toward the negatively charged wire mesh, the concentrations of these ions increase until $Q_{sp} > K_{sp}$ for both magnesium hydroxide and calcium carbonate. When $Q_{sp} > K_{sp}$ for these slightly soluble compounds, they begin to precipitate onto the wire mesh where they form a thin coating. As time passes these precipitates accumulate and fill the spaces in the wire mesh until a solid concrete-like object develops. The form is that of the original wire mesh. The growth rate is about one inch every six weeks.

Let's compare this new ocean-grown concrete to traditional concrete. Traditional concrete is a mixture of gravel, pebbles, broken stone, slag, and cement. Cement is a construction adhesive that contains mostly powdered rock and clay; these form a paste with water and then set as a solid mass. In contrast, the new kind of concrete is made up primarily of magnesium hydroxide and calcium carbonate, depending on growing conditions, along with smaller quantities of other materials. New concrete has about the same color, texture, and density as traditional

concrete but is at least 30% stronger and much less expensive to make and transport. In fact, when new concrete is grown *at* construction sites, transportation costs vanish.

To better understand some of the details of the growing process, we calculate the concentration of magnesium ions needed around the wire mesh to produce a saturated solution of magnesium hydroxide, i.e., to make $Q_{sp} = K_{sp}$. Sea water has a pH of approximately 11; K_{sp} for magnesium hydroxide is 2×10^{-11}.

$$pH = 11; \quad K_{sp} = [Mg^{2+}][OH^-]^2 = 2 \times 10^{-11};$$
$$[H^+] = 10^{-11} M; [OH^-] = 10^{-3} M$$

$$[Mg^{2+}] = \frac{2 \times 10^{-11}}{[OH^-]^2} = \frac{2 \times 10^{-11}}{1 \times 10^{-6}} = 2 \times 10^{-5} M$$

When the electric current attracts enough magnesium ions near the wire mesh so that $[Mg^{2+}] = 2 \times 10^{-5} M$, we have a saturated solution of magnesium hydroxide, and $Q_{sp} = K_{sp}$. As the electric current causes the concentration of magnesium ions to increase, magnesium hydroxide begins to precipitate. Because the electrically charged metal mesh continuously attracts magnesium ions toward itself, the small Mg^{2+} concentration needed for saturation, $2 \times 10^{-5} M$, can easily, quickly, and perpetually be exceeded.

For millions of years, corals and clams have made their homes of calcium carbonate from ocean water. Perhaps humans can learn to do similar kinds of things. We need a few photovoltaic panels to convert solar energy into a direct electric current, some wire mesh, the sea, and some imagination.

Ronald DeLorenzo
Middle Georgia College

20-4 FRACTIONAL PRECIPITATION

We sometimes wish to remove some ions from solution while leaving others with similar properties in solution. This separation process is called **fractional precipitation.** Consider a solution that contains Cl^-, Br^-, and I^- ions. These halide ions are anions of elements in the same family in the periodic table. We expect them to have similar properties. But we also expect some differences in properties, and that is what we find. Consider the solubility products for these silver halides.

Clearly cations such as Na^+ or K^+ must also be present in this solution.

Compound	Solubility Product
AgCl	1.8×10^{-10}
AgBr	3.3×10^{-13}
AgI	1.5×10^{-16}

These K_{sp} values show that AgI is less soluble than AgBr and that AgBr is less soluble than AgCl. This is because the I^- ion is larger and more polarizable than the Br^- ion, which is larger and more polarizable than the Cl^- ion. Silver fluoride is quite soluble in water because the small F^- ion is not easily polarized.

> The distortion of the electron cloud of an ion by another ion is called *polarization* (Section 13-2). Greater polarization increases the attraction of the ions for one another and makes the salt less soluble.

EXAMPLE 20-8 *Concentration Required to Initiate Precipitation*

Solid silver nitrate is slowly added to a solution that is 0.0010 M each in NaCl, NaBr, and NaI. Calculate the $[Ag^+]$ required to initiate the precipitation of each of the silver halides. For AgI, $K_{sp} = 1.5 \times 10^{-16}$; for AgBr, $K_{sp} = 3.3 \times 10^{-13}$; for AgCl, $K_{sp} = 1.8 \times 10^{-10}$.

> NaCl, NaBr, NaI, $AgNO_3$, and $NaNO_3$ are soluble compounds that are completely dissociated in dilute aqueous solution.

Plan

We are given a solution that contains equal concentrations of Cl^-, Br^-, and I^- ions; all of which form insoluble silver salts. Then we slowly add Ag^+ ions. We use each K_{sp} to determine the $[Ag^+]$ that must be exceeded to initiate precipitation of each salt as we did in Example 20-6.

> We ignore the extremely small change in volume caused by addition of solid $AgNO_3$.

Solution

We calculate the $[Ag^+]$ that is necessary to begin to precipitate each of the silver halides. The solubility product for AgI is

$$[Ag^+][I^-] = 1.5 \times 10^{-16}$$

$[I^-] = 1.0 \times 10^{-3} M$, so the $[Ag^+]$ that must be exceeded to start precipitation of AgI is

$$[Ag^+] = \frac{1.5 \times 10^{-16}}{[I^-]} = \frac{1.5 \times 10^{-16}}{1.0 \times 10^{-3}} = 1.5 \times 10^{-13} M$$

Therefore, AgI will begin to precipitate when $[Ag^+] > 1.5 \times 10^{-13} M$.

Repeating this kind of calculation for silver bromide gives

$$[Ag^+][Br^-] = 3.3 \times 10^{-13}$$

$$[Ag^+] = \frac{3.3 \times 10^{-13}}{[Br^-]} = \frac{3.3 \times 10^{-13}}{1.0 \times 10^{-3}} = 3.3 \times 10^{-10} M$$

Thus, $[Ag^+] > 3.3 \times 10^{-10} M$ is needed to start precipitation of AgBr.

For the precipitation of silver chloride to begin,

$$[Ag^+][Cl^-] = 1.8 \times 10^{-10}$$

> We are not suggesting that a $1.5 \times 10^{-13} M$ solution of $AgNO_3$ be added. We are pointing out the fact that when sufficient $AgNO_3$ has been added to the solution to make $[Ag^+] > 1.5 \times 10^{-13} M$, AgI begins to precipitate.

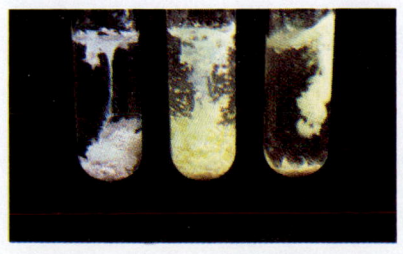

Freshly precipitated AgCl is white (left), AgBr is very pale yellow (center), and AgI is yellow (right). Polarizabilities of these halide ions increase in the order $Cl^- < Br^- < I^-$. Colors of the silver halides become more intense in the same direction. Solubilities of the silver halides increase in the opposite direction.

$$[Ag^+] = \frac{1.8 \times 10^{-10}}{[Cl^-]} = \frac{1.8 \times 10^{-10}}{1.0 \times 10^{-3}} = 1.8 \times 10^{-7} \, M$$

To precipitate AgCl, we must have $[Ag^+] > 1.8 \times 10^{-7} \, M$.

We have shown that

to precipitate AgI,	$[Ag^+] > 1.5 \times 10^{-13} \, M$
to precipitate AgBr,	$[Ag^+] > 3.3 \times 10^{-10} \, M$
to precipitate AgCl,	$[Ag^+] > 1.8 \times 10^{-7} \, M$

You should now work Exercise 32.

As $AgNO_3$ is added to the solution containing Cl^-, Br^-, and I^- ions, some AgBr and AgCl may precipitate *locally*. As the solution is stirred, AgBr and AgCl redissolve as long as $[Ag^+]$ is not large enough to exceed their K_{sp} values in the *bulk* of the solution.

This calculation tells us that when $AgNO_3$ is added slowly to a solution that is 0.0010 M in each of NaI, NaBr, and NaCl, AgI precipitates first, AgBr precipitates second, and AgCl precipitates last. We can also calculate the amount of I^- precipitated before Br^- begins to precipitate and the amounts of I^- and Br^- precipitated before Cl^- begins to precipitate (Example 20-9).

EXAMPLE 20-9 *Fractional Precipitation*

Refer to Example 20-8. (a) Calculate the percentage of I^- precipitated before AgBr precipitates. (b) Calculate the percentages of I^- and Br^- precipitated before Cl^- precipitates.

Plan

From Example 20-8 we know the $[Ag^+]$ that must be exceeded to initiate precipitation of each of three silver halides, AgI, AgBr, and AgCl. We use each of these values of $[Ag^+]$ with the appropriate K_{sp} expression, in turn, to find the concentration of each halide ion that remains in solution (unprecipitated). We express these halide ion concentrations as percent unprecipitated. Then we subtract each from exactly 100% to find the percentage of each halide that precipitates.

Solution

In Example 20-7 we did a similar calculation, but we did not express the result in terms of the percentage of an ion precipitated.

(a) In Example 20-8 we found that AgBr begins to precipitate when $[Ag^+] > 3.3 \times 10^{-10} \, M$. This value for $[Ag^+]$ can be substituted into the K_{sp} expression for AgI to determine $[I^-]$ remaining *unprecipitated* when AgBr begins to precipitate.

$$[Ag^+][I^-] = 1.5 \times 10^{-16}$$

$$[I^-]_{unppt'd} = \frac{1.5 \times 10^{-16}}{[Ag^+]} = \frac{1.5 \times 10^{-16}}{3.3 \times 10^{-10}} = 4.5 \times 10^{-7} \, M$$

The percentage of I^- unprecipitated is

$$\% \, I^-_{\,unppt'd} = \frac{[I^-]_{unppt'd}}{[I^-]_{orig}} \times 100\% = \frac{4.5 \times 10^{-7} \, M}{1.0 \times 10^{-3} \, M} \times 100\%$$

$$= 0.045\% \, I^- \, \text{unprecipitated}$$

We have subtracted 0.045% from *exactly* 100%. Therefore, we have *not* violated the rules for significant figures.

Therefore, 99.955% of the I^- precipitates *before* AgBr begins to precipitate.

(b) Similar calculations show that *just before* AgCl begins to precipitate, $[Ag^+] = 1.8 \times 10^{-7} \, M$, and the $[I^-]$ unprecipitated is calculated as in part (a).

$$[Ag^+][I^-] = 1.5 \times 10^{-16}$$

$$[I^-]_{unppt'd} = \frac{1.5 \times 10^{-16}}{[Ag^+]} = \frac{1.5 \times 10^{-16}}{1.8 \times 10^{-7}} = 8.3 \times 10^{-10} \, M$$

The percentage of I^- unprecipitated just before AgCl precipitates is

$$\% \ I^-_{\text{unppt'd}} = \frac{[I^-]_{\text{unppt'd}}}{[I^-]_{\text{orig}}} \times 100\% = \frac{8.3 \times 10^{-10} \, M}{1.0 \times 10^{-3} \, M} \times 100\%$$

$$= 0.000083\% \ I^- \ \text{unprecipitated}$$

Therefore, 99.999917% of the I^- precipitates before AgCl begins to precipitate.

A similar calculation for the amount of Br^- precipitated just before AgCl begins to precipitate gives

$$[Ag^+][Br^-] = 3.3 \times 10^{-13}$$

$$[Br^-]_{\text{unppt'd}} = \frac{3.3 \times 10^{-13}}{[Ag^+]} = \frac{3.3 \times 10^{-13}}{1.8 \times 10^{-7}} = 1.8 \times 10^{-6} \, M$$

$$\% \ Br^-_{\text{unppt'd}} = \frac{[Br^-]_{\text{unppt'd}}}{[Br^-]_{\text{orig}}} \times 100\% = \frac{1.8 \times 10^{-6} \, M}{1.0 \times 10^{-3} \, M} \times 100\%$$

$$= 0.18\% \ Br^- \ \text{unprecipitated}$$

Thus, 99.82% of the Br^- precipitates before AgCl begins to precipitate.

You should now work Exercises 32 and 34.

We have described the series of reactions that occurs when solid $AgNO_3$ is added slowly to a solution that is 0.0010 M in Cl^-, Br^-, and I^-. Silver iodide begins to precipitate first; 99.955% of the I^- precipitates before any solid AgBr is formed. Silver bromide begins to precipitate next; 99.82% of the Br^- and 99.999917% of the I^- precipitate before any solid AgCl forms. This shows that we can separate these ions very effectively by fractional precipitation.

20-5 SIMULTANEOUS EQUILIBRIA INVOLVING SLIGHTLY SOLUBLE COMPOUNDS

Many weak acids and bases react with many metal ions to form insoluble compounds. In such cases, we must take into account the weak acid or weak base equilibrium as well as the solubility equilibrium. The reaction of a metal ion with aqueous ammonia forms an insoluble metal hydroxide.

EXAMPLE 20-10 *Simultaneous Equilibria*

If a solution is made 0.10 M in magnesium nitrate, $Mg(NO_3)_2$, *and* 0.10 M in aqueous ammonia, a weak base, will magnesium hydroxide, $Mg(OH)_2$, precipitate? K_{sp} for $Mg(OH)_2$ is 1.5×10^{-11}, and K_b for aqueous NH_3 is 1.8×10^{-5}.

Plan

We first write equations for the two *reversible* reactions and their equilibrium constant expressions. We note that $[OH^-]$ appears in both equilibrium constant expressions. From the statement of the problem we know the concentration of Mg^{2+}. We use the K_b expression for aqueous NH_3 to find $[OH^-]$. Then we can calculate Q_{sp} for $Mg(OH)_2$ and compare it with its K_{sp}.

Solution

Two equilibria and their equilibrium constant expressions must be considered.

$$Mg(OH)_2(s) \rightleftharpoons Mg^{2+}(aq) + 2OH^-(aq) \qquad K_{sp} = [Mg^{2+}][OH^-]^2 = 1.5 \times 10^{-11}$$

$$NH_3(aq) + H_2O(\ell) \rightleftharpoons NH_4^+(aq) + OH^-(aq) \qquad K_b = \frac{[NH_4^+][OH^-]}{[NH_3]} = 1.8 \times 10^{-5}$$

The $[OH^-]$ in 0.10 M aqueous NH_3 is calculated as in Example 18-14.

$$\underset{(0.10 - x)\,M}{NH_3(aq)} + H_2O \rightleftharpoons \underset{x\,M}{NH_4^+(aq)} + \underset{x\,M}{OH^-(aq)}$$

$$\frac{[NH_4^+][OH^-]}{[NH_3]} = 1.8 \times 10^{-5} = \frac{(x)(x)}{(0.10 - x)} \qquad x = 1.3 \times 10^{-3}\,M = [OH^-]$$

Magnesium nitrate is a soluble ionic compound, so $[Mg^{2+}] = 0.10\,M$. Now that both $[Mg^{2+}]$ and $[OH^-]$ are known, we calculate Q_{sp} for $Mg(OH)_2$.

$$[Mg^{2+}][OH^-]^2 = Q_{sp}$$
$$(0.10)(1.3 \times 10^{-3})^2 = 1.7 \times 10^{-7} = Q_{sp}$$

We have calculated the $[OH^-]$ produced by the ionization of 0.10 M aqueous NH_3. This is the equilibrium concentration of OH^- in this solution. There is no reason to double this value!

$K_{sp} = 1.5 \times 10^{-11}$, so we see that $Q_{sp} > K_{sp}$.

Therefore, $\boxed{Mg(OH)_2 \text{ would precipitate until } Q_{sp} = K_{sp}.}$

You should now work Exercise 40.

Example 20-11 shows how we can calculate the concentration of weak base that is required to initiate precipitation of an insoluble metal hydroxide.

EXAMPLE 20-11 *Simultaneous Equilibria*

What concentration of aqueous ammonia is necessary to just start precipitation of $Mg(OH)_2$ from a 0.10 M solution of $Mg(NO_3)_2$? Refer to Example 20-10.

Plan

We have the same reactions and equilibrium constant expressions as in Example 20-10. We are given $[Mg^{2+}]$, and so we solve the K_{sp} expression for $Mg(OH)_2$ for the $[OH^-]$ that is necessary to initiate precipitation. Then we find the molarity of aqueous NH_3 solution that would furnish the desired $[OH^-]$.

Solution

Two equilibria and their equilibrium constant expressions must be considered.

$$Mg(OH)_2(s) \rightleftharpoons Mg^{2+}(aq) + 2OH^-(aq) \qquad K_{sp} = 1.5 \times 10^{-11}$$

$$NH_3(aq) + H_2O(\ell) \rightleftharpoons NH_4^+(aq) + OH^-(aq) \qquad K_b = 1.8 \times 10^{-5}$$

We find the $[OH^-]$ necessary to initiate precipitation of $Mg(OH)_2$ when $[Mg^{2+}] = 0.10\,M$.

$$K_{sp} = [Mg^{2+}][OH^-]^2 = 1.5 \times 10^{-11}$$

$$[OH^-]^2 = \frac{1.5 \times 10^{-11}}{[Mg^{2+}]} = \frac{1.5 \times 10^{-11}}{0.10} = 1.5 \times 10^{-10} \qquad [OH^-] = 1.2 \times 10^{-5}\,M$$

$\therefore [OH^-] > 1.2 \times 10^{-5}\,M$ to initiate precipitation of $Mg(OH)_2$

Now we find the $[NH_3]$ that will produce $1.2 \times 10^{-5}\,M$ OH^-. We let x be the original $[NH_3]$.

$$\underset{(x - 1.2 \times 10^{-5})\,M}{NH_3(aq)} + H_2O \rightleftharpoons \underset{1.2 \times 10^{-5}\,M}{NH_4^+(aq)} + \underset{1.2 \times 10^{-5}\,M}{OH^-(aq)}$$

$$K_b = \frac{[NH_4^+][OH^-]}{[NH_3]} = 1.8 \times 10^{-5} = \frac{(1.2 \times 10^{-5})(1.2 \times 10^{-5})}{(x - 1.2 \times 10^{-5})}$$

$$1.8 \times 10^{-5}\,x - 2.16 \times 10^{-10} = 1.44 \times 10^{-10}$$

$$1.8 \times 10^{-5}\,x = 3.6 \times 10^{-10} \qquad \text{so} \qquad x = 2.0 \times 10^{-5}\,M = [NH_3]_{original}$$

Because 1.2×10^{-5} and x are of comparable magnitude, neither can be neglected in the term $(x - 1.2 \times 10^{-5})$.

The solution must be ever so slightly greater than $2.0 \times 10^{-5}\,M$ in NH_3 to initiate precipitation of $Mg(OH)_2$ in a $0.10\,M$ solution of $Mg(NO_3)_2$.

A solution that contains a weak base can be buffered (by addition of a salt of the weak base) to decrease its basicity. Significant concentrations of some metal ions that form insoluble hydroxides can be kept in such solutions.

EXAMPLE 20-12 *Simultaneous Equilibria*

What minimum number of moles of NH_4Cl must be added to 1.0 liter of solution that is $0.10\,M$ in $Mg(NO_3)_2$ *and* $0.10\,M$ in NH_3 to prevent precipitation of $Mg(OH)_2$?

Plan

These are the same compounds, in the same concentrations, that we used in Example 20-10. Because we know $[Mg^{2+}]$, we must find the maximum $[OH^-]$ that can exist in the solution *without exceeding* K_{sp} for $Mg(OH)_2$. Then we find the minimum concentration of NH_4Cl that is necessary to buffer the NH_3 solution to keep the $[OH^-]$ below the calculated value.

Solution

The buffering action of NH_4Cl in the presence of NH_3 decreases the concentration of OH^-. Again we have two equilibria.

In Example 20-10 we found that $Mg(OH)_2$ will precipitate from a solution that is $0.10\,M$ in $Mg(NO_3)_2$ and $0.10\,M$ in NH_3.

$$Mg(OH)_2(s) \rightleftharpoons Mg^{2+}(aq) + 2OH^-(aq) \qquad K_{sp} = 1.5 \times 10^{-11}$$

$$NH_3(aq) + H_2O(\ell) \rightleftharpoons NH_4^+(aq) + OH^-(aq) \qquad K_b = 1.8 \times 10^{-5}$$

To find the *maximum $[OH^-]$ that can exist in solution without causing precipitation,* we substitute $[Mg^{2+}]$ into K_{sp} for $Mg(OH)_2$.

$$[Mg^{2+}][OH^-]^2 = 1.5 \times 10^{-11}$$

$$[OH^-]^2 = \frac{1.5 \times 10^{-11}}{[Mg^{2+}]} = \frac{1.5 \times 10^{-11}}{0.10} = 1.5 \times 10^{-10}$$

$$[OH^-] = 1.2 \times 10^{-5}\,M \qquad \text{(maximum } [OH^-] \text{ possible)}$$

To prevent precipitation of $Mg(OH)_2$ in *this* solution, $[OH^-]$ must be equal to or less than $1.2 \times 10^{-5}\,M$. K_b for aqueous NH_3 is used to calculate the number of moles of NH_4Cl necessary to buffer 1.0 L of $0.10\,M$ aqueous NH_3 so that $[OH^-] = 1.2 \times 10^{-5}\,M$. Let x = number of mol/L of NH_4Cl required.

You may wish to refer to Section 18-6 to refresh your memory on buffer solutions.

$$
\begin{array}{cccc}
NH_4Cl(aq) & \longrightarrow & NH_4^+(aq) & + Cl^-(aq) \qquad \text{(to completion)} \\
x\,M & \rightleftharpoons & x\,M & x\,M \\
\\
NH_3(aq) & + H_2O \rightleftharpoons & NH_4^+(aq) & + OH^-(aq) \\
(0.10 - 1.2 \times 10^{-5})\,M & & 1.2 \times 10^{-5}\,M & 1.2 \times 10^{-5}\,M
\end{array}
$$

We can assume that $(x + 1.2 \times 10^{-5}) \approx x$ and $(0.10 - 1.2 \times 10^{-5}) \approx 0.10$.

$$\frac{(x)(1.2 \times 10^{-5})}{0.10} = 1.8 \times 10^{-5}$$

$$x = 0.15 \text{ mol of } NH_4^+ \text{ per liter of solution}$$

Addition of 0.15 mol of NH_4Cl to 1.0 L of 0.10 M aqueous NH_3 decreases $[OH^-]$ to 1.2×10^{-5} M. Then K_{sp} for $Mg(OH)_2$ is not exceeded in this solution and so no precipitate would form.

You should now work Exercises 36 and 38.

All relevant equilibria must be satisfied when more than one equilibrium is required to describe a solution (Examples 20-10, 20-11, and 20-12).

20-6 DISSOLVING PRECIPITATES

A precipitate dissolves when the concentrations of its ions are reduced so that K_{sp} is no longer exceeded, i.e., when $Q_{sp} < K_{sp}$. The precipitate then dissolves until $Q_{sp} = K_{sp}$. Precipitates can be dissolved by the following three types of reactions. All involve removing ions from solution.

Converting an Ion to a Weak Electrolyte

Three specific illustrations follow.

1. Insoluble $Al(OH)_3$ dissolves in acids. H^+ ions react with OH^- ions (from the saturated $Al(OH)_3$ solution) to form the weak electrolyte H_2O. This makes $[Al^{3+}][OH^-]^3 < K_{sp}$, so that the dissolution equilibrium shifts to the right and $Al(OH)_3$ dissolves.

$$Al(OH)_3(s) \rightleftharpoons Al^{3+}(aq) + 3OH^-(aq)$$
$$3H^+(aq) + 3OH^-(aq) \longrightarrow 3H_2O(\ell)$$

overall rxn: $Al(OH)_3(s) + 3H^+(aq) \longrightarrow Al^{3+}(aq) + 3H_2O(\ell)$

2. Ammonium ions, from a salt such as NH_4Cl, dissolve insoluble $Mg(OH)_2$. The NH_4^+ ions combine with OH^- ions in the saturated $Mg(OH)_2$ solution. This forms the weak electrolytes NH_3 and H_2O. The result is $[Mg^{2+}][OH^-]^2 < K_{sp}$, and so the $Mg(OH)_2$ dissolves.

$$Mg(OH)_2(s) \rightleftharpoons Mg^{2+}(aq) + 2OH^-(aq)$$
$$2NH_4^+(aq) + 2OH^-(aq) \longrightarrow 2NH_3(aq) + 2H_2O(\ell)$$

overall rxn: $Mg(OH)_2(s) + 2NH_4^+(aq) \longrightarrow Mg^{2+}(aq) + 2NH_3(aq) + 2H_2O(\ell)$

This process, dissolution of $Mg(OH)_2$ in an NH_4Cl solution, is the reverse of the reaction we considered in Example 20-10. There, $Mg(OH)_2$ precipitated from a solution of aqueous NH_3.

3. Nonoxidizing acids dissolve most insoluble metal sulfides. For example, 6 M HCl dissolves MnS. The H^+ ions combine with S^{2-} ions to form H_2S, a gas that bubbles out of the solution. The result is $[Mn^{2+}][S^{2-}] < K_{sp}$, and so the MnS dissolves.

$$MnS(s) \rightleftharpoons Mn^{2+}(aq) + S^{2-}(aq)$$
$$2H^+(aq) + S^{2-}(aq) \longrightarrow H_2S(g)$$

overall rxn: $MnS(s) + 2H^+(aq) \longrightarrow Mn^{2+}(aq) + H_2S(g)$

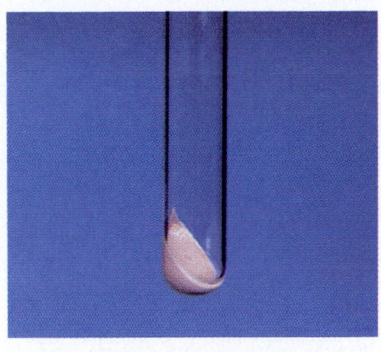 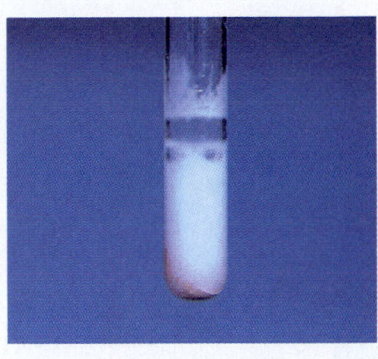 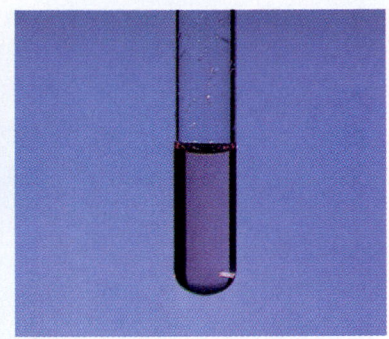

Manganese(II) sulfide, MnS, is salmon-colored. MnS dissolves in 6 M HCl. The resulting solution of MnCl$_2$ is pale pink.

Converting an Ion to Another Species by a Redox Reaction

Most insoluble metal sulfides dissolve in hot dilute HNO$_3$ because NO$_3^-$ ions oxidize S^{2-} ions to elemental sulfur. This removes S^{2-} ions from the solution and promotes the dissolving of more of the metal sulfide.

$$3S^{2-}(aq) + 2NO_3^-(aq) + 8H^+(aq) \longrightarrow 3S(s) + 2NO(g) + 4H_2O(\ell)$$

Consider copper(II) sulfide, CuS, in equilibrium with its ions. This equilibrium lies far to the left; $K_{sp} = 8.7 \times 10^{-36}$. Removal of the S^{2-} ions by oxidation to elemental sulfur favors the reaction to the right, and so CuS(s) dissolves in hot dilute HNO$_3$.

$$CuS(s) \rightleftharpoons Cu^{2+}(aq) + S^{2-}(aq)$$

$$3S^{2-}(aq) + 2NO_3^-(aq) + 8H^+(aq) \longrightarrow 3S(s) + 2NO(g) + 4H_2O(\ell)$$

We multiply the first equation by 3, add the two equations, and cancel like terms. This gives the net ionic equation for dissolving CuS(s) in hot dilute HNO$_3$.

$$3CuS(s) + 2NO_3^-(aq) + 8H^+(aq) \longrightarrow 3Cu^{2+}(aq) + 3S(s) + 2NO(g) + 4H_2O(\ell)$$

Complex Ion Formation

The cations in many slightly soluble compounds can form complex ions. This often results in dissolution of the slightly soluble compound. Some metal ions share electron pairs donated by molecules and ions such as NH$_3$, CN$^-$, OH$^-$, F$^-$, Cl$^-$, Br$^-$, and I$^-$. Coordinate covalent bonds are formed as these ligands replace H$_2$O molecules from hydrated metal ions. The decrease in the concentration of the hydrated metal ion shifts the solubility equilibrium to the right.

"Ligand" is the name given to an atom or a group of atoms bonded to the central element in complex ions. Ligands are Lewis bases.

 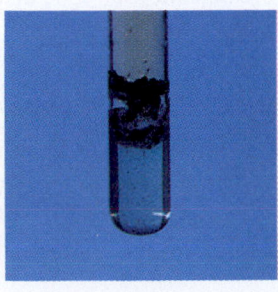 

Copper(II) sulfide, CuS, is black. As CuS dissolves in 6 M HNO$_3$, some NO is oxidized to brown NO$_2$ by O$_2$ in the air. The resulting solution of Cu(NO$_3$)$_2$ is blue.

Concentrated aqueous NH_3 was added *slowly* to a solution of copper(II) sulfate, $CuSO_4$. Unreacted blue copper(II) sulfate solution remains in the bottom part of the test tube. The light-blue precipitate in the middle is copper(II) hydroxide, $Cu(OH)_2$. The top layer contains deep-blue $[Cu(NH_3)_4]^{2+}$ ions that were formed as some $Cu(OH)_2$ dissolved in excess aqueous NH_3.

Many copper(II) compounds react with excess aqueous NH_3 to form the deep-blue complex ion $[Cu(NH_3)_4]^{2+}$.

$$Cu^{2+} + 4NH_3 \rightleftharpoons [Cu(NH_3)_4]^{2+}$$

The dissociation of this complex ion is represented as

$$[Cu(NH_3)_4]^{2+}(aq) \rightleftharpoons Cu^{2+}(aq) + 4NH_3(aq)$$

$$K_d = \frac{[Cu^{2+}][NH_3]^4}{[[Cu(NH_3)_4]^{2+}]} = 8.5 \times 10^{-13}$$

As before, the outer brackets mean molar concentrations. The inner brackets are part of the formula of the complex ion.

Recall that $Cu^{2+}(aq)$ is really a hydrated ion, $[Cu(H_2O)_6]^{2+}$. The preceding reaction and its K_d expression are represented more accurately as

$$[Cu(NH_3)_4]^{2+} + 6H_2O \rightleftharpoons [Cu(H_2O)_6]^{2+} + 4NH_3$$

$$K_d = \frac{[[Cu(H_2O)_6]^{2+}][NH_3]^4}{[[Cu(NH_3)_4]^{2+}]} = 8.5 \times 10^{-13}$$

The more effectively a ligand competes with H_2O for a coordination site on the metal ions, the smaller K_d is. This tells us that in a comparison of complexes with the same number of ligands, the smaller the K_d value, the more stable the complex ion. Some complex ions and their dissociation constants, K_d, are listed in Appendix I.

Copper(II) hydroxide dissolves in an excess of aqueous NH_3 to form the deep-blue complex ion $[Cu(NH_3)_4]^{2+}$. This decreases the $[Cu^{2+}]$ so that $[Cu^{2+}][OH^-]^2 < K_{sp}$, and so the $Cu(OH)_2$ dissolves.

For brevity we shall omit H_2O from formulas of hydrated ions. For example, we write $[Cu(H_2O)_6]^{2+}$ as Cu^{2+}.

$$Cu(OH)_2(s) \rightleftharpoons Cu^{2+}(aq) \qquad + 2OH^-(aq)$$
$$\underline{Cu^{2+}(aq) + 4NH_3(aq) \rightleftharpoons [Cu(NH_3)_4]^{2+}(aq)}$$
overall rxn: $Cu(OH)_2(s) + 4NH_3(aq) \rightleftharpoons [Cu(NH_3)_4]^{2+}(aq) + 2OH^-(aq)$

Similarly, $Zn(OH)_2$ dissolves in excess NH_3 to form $[Zn(NH_3)_4]^{2+}$ ions.

$$Zn(OH)_2(s) + 4NH_3(aq) \rightleftharpoons [Zn(NH_3)_4]^{2+}(aq) + 2OH^-$$

Amphoteric hydroxides such as $Zn(OH)_2$ dissolve in excess strong soluble base by forming complex ions (Section 10-8).

$$Zn(OH)_2(s) + 2OH^-(aq) \rightleftharpoons [Zn(OH)_4]^{2-}(aq)$$

We see that we are able to shift equilibria (in this case, dissolve $Zn(OH)_2$) by taking advantage of complex ion formation.

Key Terms

Complex ions Ions resulting from the formation of coordinate covalent bonds between simple ions and other ions or molecules.

Dissociation constant The equilibrium constant that applies to the dissociation of a complex ion into a simple ion and coordinating species (ligands).

Fractional precipitation Removal of some ions from solution by precipitation while leaving other ions, with similar properties, in solution.

Molar solubility The number of moles of a solute that dissolve to produce a liter of saturated solution.

Precipitate A solid formed by mixing in solution the constituent ions of a slightly soluble compound.

Solubility product constant, K_{sp} The equilibrium constant that applies to the dissolution of a slightly soluble compound.

Solubility product principle The solubility product constant expression for a slightly soluble compound is the product of the concentrations of the constituent ions, each raised to the power that corresponds to the number of ions in one formula unit.

Exercises

Consult Appendix H for solubility product constant values and Appendix I for complex ion dissociation constants, as needed.

Solubility Product

1. (a) Are "insoluble" substances really insoluble? (b) What do we mean when we refer to insoluble substances?

2. State the solubility product principle. What is its significance?

3. (a) Why are solubility product constant expressions written as products of concentrations of ions raised to appropriate powers? (b) Why do we not include a term for the solid in a solubility product expression?

4. What do we mean when we refer to the molar solubility of a compound?

5. Write the solubility product expression for the following salts. (a) BaF_2; (b) $Bi_2(SO_4)_3$; (c) $CuBr$; (d) $BaCO_3$.

6. Write the solubility product expression for the following salts. (a) $Co_3(AsO_4)_2$; (b) Hg_2I_2 [contains mercury(I) ions, Hg_2^{2+}]; (c) HgI_2; (d) $(Ag)_2CO_3$.

Experimental Determination of K_{sp}

7. From the solubility data given for the following compounds, calculate their solubility product constants. Your calculated values may not agree exactly with the solubility products given in Appendix H, because roundoff errors can be large in calculations to two significant figures.
   (a) $CuBr$, copper(I) bromide, 1.0×10^{-3} g/L
   (b) AgI, silver iodide, 2.8×10^{-8} g/10 mL
   (c) $Pb_3(PO_4)_2$, lead(II) phosphate, 6.2×10^{-7} g/L
   (d) Ag_2SO_4, silver sulfate, 5.0 mg/mL

8. From the solubility data given for the following compounds, calculate their solubility product constants. Your calculated values may not agree exactly with the solubility products given in Appendix H, because roundoff errors can be large in calculations to two significant figures.
   (a) $SrCrO_4$, strontium chromate, 1.2 mg/mL
   (b) BiI_3, bismuth iodide, 7.7×10^{-3} g/L
   (c) $Fe(OH)_2$, iron(II) hydroxide, 1.1×10^{-3} g/L
   (d) SnI_2, tin(II) iodide, 10.9 g/L

9. Construct a table like Table 20-1 for the compounds listed in Exercise 7. Which compound has (a) the highest molar solubility; (b) the lowest molar solubility; (c) the largest K_{sp}; (d) the smallest K_{sp}?

10. Construct a table like Table 20-1 for the compounds listed in Exercise 8. Which compound has (a) the highest molar solubility; (b) the lowest molar solubility; (c) the largest K_{sp}; (d) the smallest K_{sp}?

Uses of Solubility Product Constants

11. Calculate molar solubilities, concentrations of constituent ions, and solubilities in grams per liter for the following compounds at 25°C: (a) $Zn(CN)_2$, zinc cyanide; (b) MgF_2, magnesium fluoride; (c) $Pb_3(AsO_4)_2$, lead(II) arsenate; (d) Hg_2CO_3, mercury(I) carbonate [the formula for the mercury(I) ion is Hg_2^{2+}].

12. Calculate molar solubilities, concentrations of constituent ions, and solubilities in grams per liter for the following compounds at 25°C: (a) CuI, copper(I) iodide; (b) $Ba_3(PO_4)_2$, barium phosphate; (c) PbF_2, lead(II) fluoride; (d) $Pb_3(PO_4)_2$, lead(II) phosphate.

13. Construct a table similar to Table 20-1 for the compounds listed in Exercise 11. Which compound has (a) the highest molar solubility; (b) the lowest molar solubility; (c) the highest solubility, expressed in grams per liter; (d) the lowest solubility, expressed in grams per liter?

14. Construct a table similar to Table 20-1 for the compounds listed in Exercise 12. Which compound has (a) the highest molar solubility; (b) the lowest molar solubility; (c) the highest solubility, expressed in grams per liter; (d) the lowest solubility, expressed in grams per liter?

15. Calculate the molar solubility of AgBr in 0.015 M KBr solution.

16. Calculate the molar solubility of Ag_2SO_4 in 0.15 M K_2SO_4 solution.

17. Calculate the molar solubility of Ag_2CrO_4 (a) in pure water, (b) in 0.015 M $AgNO_3$, and (c) in 0.015 M K_2CrO_4.

18. Milk of magnesia is a suspension of the slightly soluble compound $Mg(OH)_2$ in water. (a) What is the molar solubility of $Mg(OH)_2$ in a 0.015 M NaOH solution? (b) What is the molar solubility of $Mg(OH)_2$ in a 0.015 M $MgCl_2$ solution?

19. How many moles of $Cr(OH)_3$ will dissolve in 1.00 L of a solution with a pH of 5.00?

20. Which is more soluble in 0.20 M K_2CrO_4 solution—$BaCrO_4$ or Ag_2CrO_4?

21. Will a precipitate form when 1.00 g of $AgNO_3$ is added to 50.0 mL of 0.050 M NaCl? If so, would you expect the precipitate to be visible?

22. Will a precipitate of $PbCl_2$ form when 5.0 g of solid $Pb(NO_3)_2$ is added to 1.00 L of 0.010 M NaCl? Assume that volume change is negligible.

23. Sodium bromide and lead nitrate are soluble in water. Will lead bromide precipitate when 1.03 g of NaBr and 0.332 g of $Pb(NO_3)_2$ are dissolved in sufficient water to make 1.00 L of solution?

24. Will a precipitate of $Cu(OH)_2$ form when 10.0 mL of 0.010 M NaOH is added to 1.00 L of 0.010 M $CuCl_2$?

25. Suppose you have three beakers that contain, respectively, 100 mL each of the following solutions: (i) 0.0015 M KOH; (ii) 0.0015 M K_2CO_3; (iii) 0.0015 M K_2S. (a) If solid lead nitrate, $Pb(NO_3)_2$, were added slowly to each beaker, what concentration of Pb^{2+} would be required to initiate precipitation? (b) If solid lead nitrate were added to each beaker until $[Pb^{2+}]$ = 0.0015 M, what concentrations of OH^-, CO_3^{2-}, and S^{2-} would remain in solution, i.e., unprecipitated? Neglect any volume change when solute is added.

26. Suppose you have three beakers that contain, respectively, 100 mL of each of the following solutions: (i) 0.0015 M KOH; (ii) 0.0015 M K_2CO_3; (iii) 0.0015 M K_2S. (a) If solid zinc nitrate, $Zn(NO_3)_2$, were added slowly to each beaker, what concentration of Zn^{2+} would be required to initiate precipitation? (b) If solid zinc nitrate were added to each beaker until $[Zn^{2+}]$ = 0.0015 M, what concentrations of OH^-, CO_3^{2-}, and S^{2-} would remain in solution, i.e., unprecipitated? Neglect any volume change when solid is added.

27. A solution is 0.0100 M in Pb^{2+} ions. If 0.103 mol of solid Na_2SO_4 is added to 1.00 L of this solution (with negligible volume change), what percentage of the Pb^{2+} ions remain in solution?

28. A solution is 0.0100 M in Pb^{2+} ions. If 0.103 mol of solid NaI is added to 1.00 L of this solution (with negligible volume change), what percentage of the Pb^{2+} ions remain in solution?

*29. A solution is 0.0100 M in $Ba(NO_3)_2$. If 0.103 mol of solid Na_3PO_4 is added to 1.00 L of this solution (with negligible volume change), what percentage of the Ba^{2+} ions remain in solution?

Fractional Precipitation

30. What is fractional precipitation?

31. Solid Na_2SO_4 is added slowly to a solution that is 0.10 M in $Pb(NO_3)_2$ and 0.10 M in $Ba(NO_3)_2$. In what order will solid $PbSO_4$ and $BaSO_4$ form? Calculate the percentage of Ba^{2+} that precipitates just before $PbSO_4$ begins to precipitate.

32. To a solution that is 0.10 M in Cu^+, 0.10 M in Ag^+, and 0.10 M in Au^+, *solid* NaCl is added slowly. Assume that there is no volume change due to the addition of solid NaCl. (a) Which compound will begin to precipitate first? (b) Calculate $[Au^+]$ when AgCl just begins to precipitate. What percentage of the Au^+ has precipitated at this point? (c) Calculate $[Au^+]$ and $[Ag^+]$ when CuCl just begins to precipitate.

33. A solution is 0.015 M in Pb^{2+} and 0.015 M in Ag^+. As Cl^- is introduced to the solution by the addition of solid NaCl, determine (a) which substance will precipitate first, AgCl or $PbCl_2$, and (b) the fraction of the metal ion in the first precipitate that remains in solution at the moment the precipitation of the second compound begins.

34. A solution is 0.050 M in K_2SO_4 and 0.050 M in K_2CrO_4. A solution of $Pb(NO_3)_2$ is added slowly without changing the volume appreciably. (a) Which salt, $PbSO_4$ or $PbCrO_4$, will precipitate first? (b) What is $[Pb^{2+}]$ when the salt in part (a) begins to precipitate? (c) What is $[Pb^{2+}]$ when the other lead salt begins to precipitate? (d) What are $[SO_4^{2-}]$ and $[CrO_4^{2-}]$ when the lead salt in part (c) begins to precipitate?

35. Solid $Pb(NO_3)_2$ is added slowly to a solution that is 0.015 M each in NaOH, K_2CO_3, and Na_2SO_4. (a) In what order will solid $Pb(OH)_2$, $PbCO_3$, and $PbSO_4$ begin to precipitate? (b) Calculate the percentages of OH^- and CO_3^{2-} that have precipitated when $PbSO_4$ begins to precipitate.

Simultaneous Equilibria

36. If a solution is made 0.080 M in $Mg(NO_3)_2$, 0.075 M in aqueous ammonia, and 3.5 M in NH_4NO_3, will $Mg(OH)_2$ precipitate? What is the pH of this solution?

37. If a solution is made 0.090 M in $Mg(NO_3)_2$, 0.090 M in aqueous ammonia, and 0.080 M in NH_4NO_3, will $Mg(OH)_2$ precipitate? What is the pH of this solution?

*38. Calculate the solubility of CaF_2 in a solution that is buffered at $[H^+]$ = 0.0050 M with [HF] = 0.10 M.

*39. Calculate the solubility of AgCN in a solution that is buffered at $[H^+]$ = 0.000200 M, with [HCN] = 0.01 M.

40. If a solution is 2.0×10^{-5} M in $Mn(NO_3)_2$ and 1.0×10^{-3} M in aqueous ammonia, will $Mn(OH)_2$ precipitate?

41. If a solution is 0.040 M in manganese(II) nitrate, $Mn(NO_3)_2$, and 0.080 M in aqueous ammonia, will manganese(II) hydroxide, $Mn(OH)_2$, precipitate?

*42. Find the minimum pH at which MnS will precipitate from a solution that is 0.100 M with respect to H_2S and 0.0100 M with respect to $MnCl_2$.

*43. What concentration of NH_4NO_3 is necessary to prevent precipitation of $Mn(OH)_2$ in the solution of Exercise 41?

44. (a) What is the pH of a saturated solution of $Mn(OH)_2$? (b) What is the solubility in g $Mn(OH)_2$/100 mL of solution?

45. (a) What is the pH of a saturated solution of $Mg(OH)_2$? (b) What is the solubility in g $Mg(OH)_2$/100 mL of solution?

Dissolution of Precipitates and Complex Ion Formation

46. Explain, by writing appropriate equations, how the following insoluble compounds can be dissolved by the addition of a solution of nitric acid. (Carbonates dissolve in strong acids to form carbon dioxide, which is evolved as a gas, and water.) What is the "driving force" for each reaction? (a) $Cu(OH)_2$; (b) $Al(OH)_3$; (c) $MnCO_3$; (d) $(PbOH)_2CO_3$.

47. Explain, by writing equations, how the following insoluble compounds can be dissolved by the addition of a solution of ammonium nitrate or ammonium chloride. (a) $Mg(OH)_2$; (b) $Mn(OH)_2$; (c) $Ni(OH)_2$.

48. The following insoluble sulfides can be dissolved in 3 M hydrochloric acid. Explain how this is possible and write the appropriate equations. (a) MnS; (b) CuS.

49. The following sulfides are less soluble than those listed in Exercise 48 and can be dissolved in hot 6 M nitric acid, an oxidizing acid. Explain how, and write the appropriate balanced equations. (a) PbS; (b) CuS; (c) Bi_2S_3.

50. Why would MnS be expected to be more soluble in 0.10 M HCl solution than in water? Would the same be true for $Mn(NO_3)_2$?

*51. For each pair, choose the salt that would be expected to be more soluble in acidic solution than in pure water and justify your choice: (a) $Hg_2(CH_3COO)_2$ or Hg_2Br_2; (b) $Pb(OH)_2$ or PbI_2; (c) AgI or $AgNO_2$.

Mixed Exercises

52. We mix 25.0 mL of a 0.0030 M solution of $BaCl_2$ and 50.0 mL of a 0.050 M solution of NaF. (a) Find $[Ba^{2+}]$ and $[F^-]$ in the mixed solution at the instant of mixing (before any possible reaction occurs). (b) Would BaF_2 precipitate?

*53. A concentrated, strong acid is added to a solid mixture of 0.015-mol samples of $Fe(OH)_2$ and $Cu(OH)_2$ placed in 1.0 L of water. At what values of pH will the dissolution of each hydroxide be complete? (Assume negligible volume change.)

54. A solution is 0.015 M in I^- ions and 0.015 M in Br^- ions. Ag^+ ions are introduced to the solution by the addition of solid $AgNO_3$. Determine (a) which compound will precipitate first, AgI or AgBr, and (b) the percentage of the halide ion in the first precipitate that is removed from solution before the precipitation of the second compound begins.

BUILDING YOUR KNOWLEDGE

55. Draw a picture of a portion of a saturated silver chloride solution at the molecular level. Show a small amount of solid plus some dissociated ions. You need not show water or waters of hydration. Prepare a second drawing that includes the same volume of solution but twice as much solid. Should your drawing include more, less, or the same number of silver ions?

56. A fluoridated water supply contains 1 mg/L of F^-. What is the maximum amount of Ca^{2+}, expressed in grams per liter, that can exist in this water supply?

57. Many industrial operations require very large amounts of water as a coolant in heat exchange processes. Muddy or cloudy water is usually unsatisfactory because the dispersed solids may clog filters or deposit sediment in pipes and pumps. Murky water can be clarified on a large scale by adding agents to coagulate colloidal material, and then allowing the precipitate to settle out in holding tanks or ponds before the clarified water is sent to plant intakes. Recent methods employ the addition of both calcium hydroxide and magnesium carbonate. If 56 g of $Ca(OH)_2$ and 45 g of $MgCO_3$ were added to 520 liters of water, would these compounds form a precipitate?

58. Magnesium carbonate is used in the manufacture of a high density *magnesite* brick. This material is not well suited to general exterior use because the magnesium carbonate easily erodes. What percent of 28 grams of surface-exposed $MgCO_3$ would be lost through the solvent action of 15 liters of water? Assume sufficient contact time for the water to become saturated with $MgCO_3$.

21 Electrochemistry

OUTLINE

Gaseous H_2 is produced from H_2O at an illuminated photoelectrode. Light from the sun may soon be used to produce hydrogen, the ultimate clean-burning fuel.

OBJECTIVES

As you study this chapter, you should learn

- *How to use the terminology of electrochemistry (terms such as cell, electrode, cathode, anode)*

- *About the differences between electrolytic cells and voltaic (galvanic) cells*

- *To recognize oxidation and reduction half-reactions, and to know at which electrode each occurs*

- *To write half-reactions and overall cell reactions for electrolysis processes*

- *To use Faraday's Law of Electrolysis to calculate amounts of products formed, amounts of current passed, time elapsed, and oxidation state*

- *About the refining and plating of metals by electrolytic methods*

- *To describe the construction of simple voltaic cells from half-cells and a salt bridge, and to understand the function of each component*

- *To write half-reactions and overall cell reactions for voltaic cells*

- *To compare various voltaic cells to determine the relative strengths of oxidizing and reducing agents*

- *About standard electrode potentials*
- *To use standard electrode potentials, E^0, to calculate the potential of a standard voltaic cell, E^0_{cell}*
- *To use standard electrode potentials to identify the cathode and the anode in a standard cell*
- *To use standard electrode potentials to predict the spontaneity of a redox reaction*
- *To use standard potentials to identify oxidizing and reducing agents in a cell or in a redox reaction*
- *About some processes of corrosion and some methods for its prevention*
- *To use the Nernst equation to relate electrode potentials and cell potentials to different concentrations and partial pressures*
- *To apply relationships among the standard cell potential (E^0_{cell}), the standard Gibbs free energy change (ΔG^0), and the equilibrium constant (K)*
- *To distinguish between primary and secondary voltaic cells*
- *To describe the compositions and reactions of some useful primary and secondary cells (batteries)*
- *To describe the electrochemical processes involved in discharging and recharging a lead storage (automobile) battery*

Electrochemistry deals with the chemical changes produced by electric current and with the production of electricity by chemical reactions. Many metals are purified or are plated onto jewelry by electrochemical methods. Digital watches, automobile starters, calculators, and pacemakers are just a few devices that depend on electrochemically produced power. Corrosion of metals is an electrochemical process.

We learn much about chemical reactions from the study of electrochemistry. The amount of electrical energy consumed or produced can be measured quite accurately. All electrochemical reactions involve the transfer of electrons and are therefore *oxidation–reduction* reactions. The sites of oxidation and reduction are separated physically so that oxidation occurs at one location while reduction occurs at the other. Electrochemical processes require some method of introducing a stream of electrons into a reacting chemical system and some means of withdrawing electrons. In most applications the reacting system is contained in a **cell,** and an electric current enters or exits by **electrodes.**

We classify electrochemical cells into two types.

1. **Electrolytic cells** are those in which electrical energy from an external source causes *nonspontaneous* chemical reactions to occur.
2. **Voltaic cells** are those in which *spontaneous* chemical reactions produce electricity and supply it to an external circuit.

We shall discuss several electrochemical cells. From experimental observations we deduce the electrode reactions and the overall reactions. We then construct simplified diagrams of the cells.

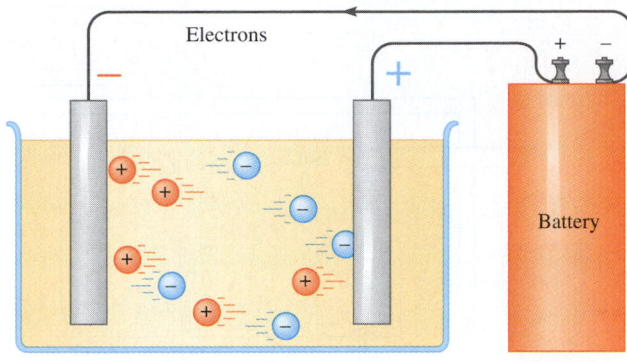

Figure 21-1 The motion of ions through a solution is an electric current. This accounts for ionic (electrolytic) conduction. Positively charged ions migrate toward the negative electrode, and negatively charged ions migrate toward the positive electrode. Here the rate of migration is greatly exaggerated for clarity. The ionic velocities are actually only slightly greater than random molecular speeds.

21-1 ELECTRICAL CONDUCTION

Electric current represents transfer of charge. Charge can be conducted through metals and through pure liquid electrolytes or solutions containing electrolytes. The former type of conduction is called **metallic conduction.** It involves the flow of electrons with no similar movement of the atoms of the metal and no obvious changes in the metal (Section 13-17). **Ionic** or **electrolytic conduction** is the conduction of electrical current by the motion of ions through a solution or a pure liquid. Positively charged ions migrate toward the negative electrode while negatively charged ions move toward the positive electrode. Both kinds of conduction, ionic and metallic, occur in electrochemical cells (Figure 21-1).

21-2 ELECTRODES

Electrodes are surfaces upon which oxidation or reduction half-reactions occur. They may or may not participate in the reactions. Those that do not react are called **inert electrodes.** Regardless of the kind of cell, electrolytic or voltaic, the electrodes are identified as follows.

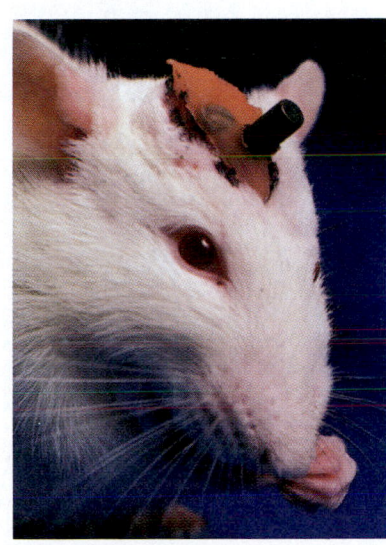

Many reactions that take place in living cells are redox reactions. These reactions can be studied with miniature electrodes.

> The **cathode** is defined as the electrode at which *reduction* occurs as electrons are gained by some species. The **anode** is the electrode at which *oxidation* occurs as electrons are lost by some species.

Each of these can be either the positive or the negative electrode.

ELECTROLYTIC CELLS

In some electrochemical cells *nonspontaneous* chemical reactions are forced to occur by the input of electrical energy. This process is called **electrolysis.** An electrolytic cell consists of a container for the reaction material with electrodes immersed in the reaction material and connected to a source of direct current. Inert electrodes are often used so that they do not react.

Lysis means "splitting apart." In many electrolytic cells compounds are split into their constituent elements.

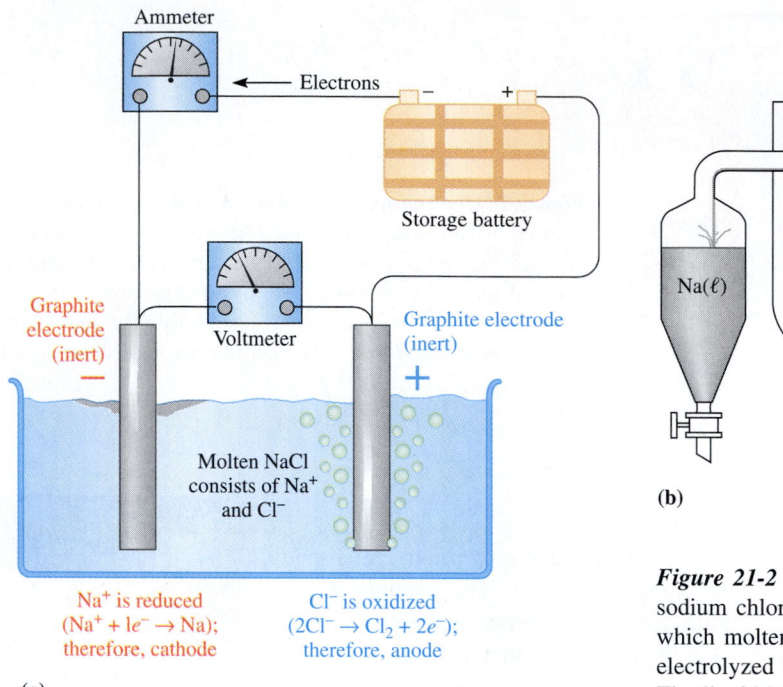

Na⁺ is reduced
$(Na^+ + 1e^- \rightarrow Na)$;
therefore, cathode

Cl⁻ is oxidized
$(2Cl^- \rightarrow Cl_2 + 2e^-)$;
therefore, anode

(a)

Figure 21-2   (a) Apparatus for electrolysis of molten sodium chloride. (b) The Downs cell, the apparatus in which molten sodium chloride is commercially electrolyzed to produce sodium metal and chlorine gas. The liquid Na floats on the more dense molten NaCl.

21-3 THE ELECTROLYSIS OF MOLTEN SODIUM CHLORIDE (DOWNS CELL)

Molten NaCl, melting point 801°C, is a clear, colorless liquid that looks like water.

Solid sodium chloride does not conduct electricity. Its ions vibrate about fixed positions, but they are not free to move throughout the crystal. However, molten (melted) NaCl is an excellent conductor because its ions are freely mobile. Consider a cell in which a source of direct current is connected by wires to two inert graphite electrodes (Figure 21-2a). They are immersed in a container of molten sodium chloride. When the current flows, we observe the following.

1. A pale green gas, which is chlorine, Cl_2, is liberated at one electrode.

The metal remains liquid because its melting point is only 97.8°C. It floats because it is less dense than the molten NaCl.

2. Molten, silvery-white metallic sodium, Na, forms at the other electrode and floats on top of the molten sodium chloride.

From these observations we can deduce the processes of the cell. Chlorine must be produced by oxidation of Cl^- ions, and the electrode at which this happens must be the anode. Metallic sodium is produced by reduction of Na^+ ions at the cathode, where electrons are being forced into the cell.

In this chapter, as in Chapters 4 and 11, we often use red type to emphasize reduction and blue type to emphasize oxidation.

$$
\begin{array}{ll}
2Cl^- \longrightarrow Cl_2(g) + 2e^- & \text{(oxidation, anode half-reaction)} \\
\underline{2[Na^+ + e^- \longrightarrow Na(\ell)]} & \text{(reduction, cathode half-reaction)} \\
\underbrace{2Na^+ + 2Cl^-}_{2NaCl(\ell)} \longrightarrow 2Na(\ell) + Cl_2(g) & \text{(overall cell reaction)}
\end{array}
$$

The formation of metallic Na and gaseous Cl_2 from NaCl is *nonspontaneous* except at temperatures very much higher than 801°C. The direct current (dc) source must supply electrical energy to force this reaction to occur. Electrons are used in the cathode half-reaction (reduction) and produced in the anode half-reaction (oxidation). Therefore, they travel through the wire from *anode* to *cathode*. The dc source forces electrons to flow nonspontaneously from the positive electrode to the negative electrode. The anode is the positive electrode and the cathode the negative electrode *in all electrolytic cells*. Figure 21-2a is a simplified diagram of the cell.

The direction of *spontaneous* flow for negatively charged particles is from negative to positive.

Sodium and chlorine must not be allowed to come in contact with each other because they react spontaneously, rapidly, and explosively to form sodium chloride. Figure 21-2b shows the Downs cell that is used for the industrial electrolysis of sodium chloride. The Downs cell is expensive to run, mainly because of the cost of construction, the cost of the electricity, and the cost of heating the NaCl to melt it. However, electrolysis of a molten sodium salt is the most practical means by which metallic Na can be obtained, owing to its extremely high reactivity. Once liberated by the electrolysis, the liquid Na metal is drained off, cooled, and cast into blocks. These must be stored in an inert environment (e.g., in mineral oil) to prevent reaction with O_2 or other components of the atmosphere.

Electrolysis of molten compounds is also the common method of obtaining other Group IA metals, IIA metals (except barium), and aluminum (Chapter 22). The Cl_2 gas produced in the Downs cell is cooled, compressed, and marketed. This partially offsets the expense of producing metallic sodium. But most chlorine is produced by the cheaper electrolysis of aqueous NaCl.

21-4 THE ELECTROLYSIS OF AQUEOUS SODIUM CHLORIDE

Consider the electrolysis of a moderately concentrated solution of NaCl in water, using inert electrodes. The following experimental observations are made when a sufficiently high voltage is applied across the electrodes of a suitable cell.

1. H_2 gas is liberated at one electrode. The solution becomes basic in that vicinity.

2. Cl_2 gas is liberated at the other electrode.

Chloride ions are obviously being oxidized to Cl_2 in this cell, as they were in the electrolysis of molten NaCl. But Na^+ ions are not reduced to metallic Na. Instead, gaseous H_2 and aqueous OH^- ions are produced by reduction of H_2O molecules at the cathode. Water is more easily reduced than Na^+ ions. This is primarily because the reduction of Na^+ would produce the very active metal Na, whereas the reduction of H_2O produces the more stable products $H_2(g)$ and $OH^-(aq)$. The active metals Li, K, Ca, and Na (Table 4-13) displace H_2 from aqueous solutions, so we do not expect these metals to be produced in aqueous solution. Later in this chapter (Section 21-15) we learn the quantitative basis for predicting which of several possible oxidations or reductions is favored. The half-reactions and overall cell reaction for this electrolysis are

$$2Cl^- \longrightarrow Cl_2 + 2e^- \qquad \text{(oxidation, anode)}$$
$$\underline{2H_2O + 2e^- \longrightarrow 2OH^- + H_2} \qquad \text{(reduction, cathode)}$$
$$2H_2O + 2Cl^- \longrightarrow 2OH^- + H_2 + Cl_2 \qquad \text{(overall cell reaction)}$$
$$\underbrace{+ 2Na^+}_{2NaCl} \longrightarrow \underbrace{+ 2Na^+}_{2NaOH} \qquad \text{(spectator ions)}$$

We shall omit the notation that indicates states of substances—(s), (ℓ), (g), and (aq)—except where states are not obvious. This abbreviates writing equations.

The cell is illustrated in Figure 21-3. As before, the electrons flow from the anode (+) through the wire to the cathode (−).

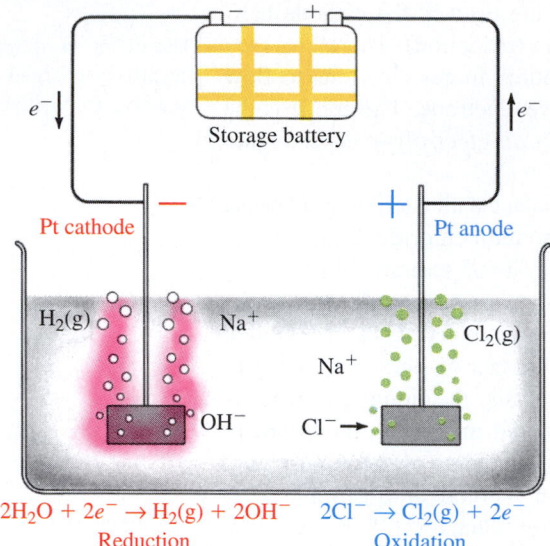

$$2H_2O + 2e^- \rightarrow H_2(g) + 2OH^- \qquad 2Cl^- \rightarrow Cl_2(g) + 2e^-$$

Reduction   Oxidation

Figure 21-3 Electrolysis of aqueous NaCl solution. Although several reactions occur at both the anode and the cathode, the net result is the production of $H_2(g)$ and NaOH at the cathode and $Cl_2(g)$ at the anode. A few drops of phenolphthalein indicator were added to the solution. The solution turns pink at the cathode, where OH^- ions are formed.

The electrolysis of the aqueous solution of KI, another Group IA–Group VIIA salt. At the cathode (left), water is reduced to $H_2(g)$ and OH^- ions, turning the phenolphthalein indicator pink. The characteristic reddish color of I_2 appears at the anode (right).

Not surprisingly, the fluctuations in commercial prices of these widely used industrial products—H_2, Cl_2, and NaOH—have often paralleled one another.

The overall cell reaction produces gaseous H_2 and Cl_2 and an aqueous solution of NaOH, called caustic soda. Solid NaOH is then obtained by evaporation of the residual solution. This is the most important commercial preparation of each of these substances. It is much less expensive than the electrolysis of molten NaCl, because it is not necessary to heat the solution.

21-5 THE ELECTROLYSIS OF AQUEOUS SODIUM SULFATE

In the electrolysis of aqueous sodium sulfate using inert electrodes, we observe the following.

1. Gaseous H_2 is produced at one electrode. The solution becomes basic around that electrode.

2. Gaseous O_2 is produced at the other electrode. The solution becomes acidic around that electrode.

As in the previous example, water is reduced in preference to Na^+ at the cathode. Observation 2 suggests that water is also preferentially oxidized relative to the sulfate ion, SO_4^{2-}, at the anode (Figure 21-4).

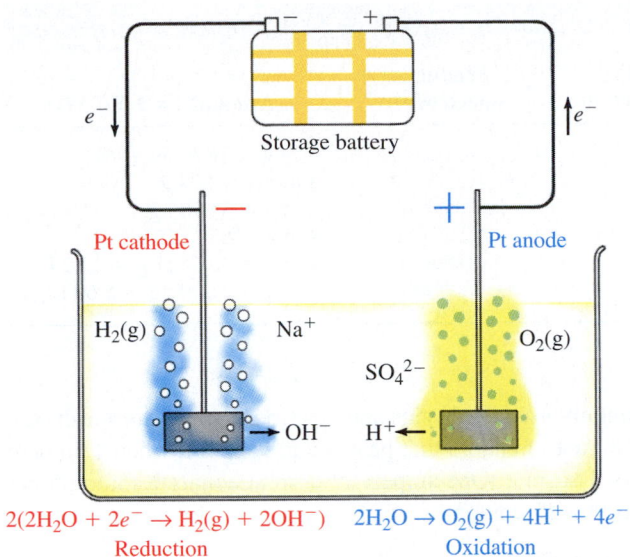

2($2H_2O + 2e^- \rightarrow H_2(g) + 2OH^-$) $2H_2O \rightarrow O_2(g) + 4H^+ + 4e^-$
Reduction   Oxidation

Figure 21-4 The electrolysis of aqueous Na_2SO_4 produces $H_2(g)$ at the cathode and O_2 at the anode. Bromthymol blue indicator has been added to the solution. This indicator turns blue in the basic solution near the cathode (where OH^- is produced), and yellow in the acidic solution near the anode (where H^+ is formed).

$$2(2H_2O + 2e^- \longrightarrow H_2 + 2OH^-) \quad \text{(reduction, cathode)}$$
$$2H_2O \longrightarrow O_2 + 4H^+ + 4e^- \quad \text{(oxidation, anode)}$$
$$6H_2O \longrightarrow 2H_2 + O_2 + \underbrace{4H^+ + 4OH^-}_{4H_2O} \quad \text{(overall cell reaction)}$$
$$2H_2O \longrightarrow 2H_2 + O_2 \quad \text{(net reaction)}$$

The net result is the electrolysis of water. This occurs because H_2O is more readily reduced than Na^+ and more readily oxidized than SO_4^{2-}. The ions of Na_2SO_4 conduct the current through the solution, but they take no part in the reaction.

21-6 FARADAY'S LAW OF ELECTROLYSIS

In 1832–33 Michael Faraday's studies of electrolysis led to this conclusion.

> The amount of substance that undergoes oxidation or reduction at each electrode during electrolysis is directly proportional to the amount of electricity that passes through the cell.

This is **Faraday's Law of Electrolysis.** A quantitative unit of electricity is now called the faraday.

> One **faraday** is the amount of electricity that reduces one equivalent weight of a substance at the cathode and oxidizes one equivalent weight of a substance at the anode. This corresponds to the gain or loss, and therefore the passage, of 6.022×10^{23} electrons. Thus, *one equivalent weight* of any substance is the amount of that substance that supplies or consumes *one mole* of electrons.

Michael Faraday (1791–1867) is considered the greatest experimental scientist of the 19th century. As a bookbinder's apprentice, he educated himself by extensive reading. Intrigued by his self-study of chemistry and by a lecture given by Sir Humphrey Davy, the leading chemist of the day, Faraday applied for a position with Davy at the Royal Institution. He subsequently became director of that laboratory. His public lectures on science were very popular.

One faraday of electricity corresponds to the passage of one mole of electrons.

Table 21-1 *Amounts of Elements Produced at One Electrode in Electrolysis by 1 Faraday of Electricity*

Half-Reaction	Number of e^- in Half-Reaction	Product (electrode)	Amount (= 1 Eq Wt)
$Ag^+(aq) + e^- \longrightarrow Ag(s)$	1	Ag (cathode)	1 mol = 107.868 g
$2H^+(aq) + 2e^- \longrightarrow H_2(g)$	2	H_2 (cathode)	$\frac{1}{2}$ mol = 1.008 g
$Cu^{2+}(aq) + 2e^- \longrightarrow Cu(s)$	2	Cu (cathode)	$\frac{1}{2}$ mol = 31.773 g
$Au^{3+}(aq) + 3e^- \longrightarrow Au(s)$	3	Au (cathode)	$\frac{1}{3}$ mol = 65.6555 g
$2Cl^- \longrightarrow Cl_2(g) + 2e^-$	2	Cl_2 (anode)	$\frac{1}{2}$ mol = 35.4527 g = 11.2 L_{STP}
$2H_2O(\ell) \longrightarrow O_2(g) + 4H^+(aq) + 4e^-$	4	O_2 (anode)	$\frac{1}{4}$ mol = 7.9997 g = 5.60 L_{STP}

A smaller electrical unit commonly used in physics and electronics is the **coulomb (C).** One coulomb is defined as the amount of charge that passes a given point when 1 ampere (A) of electrical current flows for 1 second. One ampere of current equals 1 coulomb per second. One faraday is equal to 96,485 coulombs of charge.

For comparison, a 100-watt household light bulb uses a current of about 0.8 ampere.

$$1 \text{ ampere} = 1 \frac{\text{coulomb}}{\text{second}} \quad \text{or} \quad 1 \text{ A} = 1 \text{ C/s}$$

$$1 \text{ faraday} = 6.022 \times 10^{23} \ e^- = 96,485 \text{ C}$$

We may restate Faraday's Law in a very useful form.

> In an electrochemical process, 1 faraday of electricity (96,485 coulombs = 1 mole of electrons) reduces and oxidizes, respectively, one equivalent weight each of the oxidizing and reducing agents.

Table 21-1 shows the amounts of several elements produced during electrolysis by the passage of 1 faraday of electricity.

The amount of electricity in Examples 21-1 and 21-2 would be sufficient to light a 100-watt household light bulb for about 150 minutes, or 2.5 hours.

EXAMPLE 21-1   *Electrolysis*

Calculate the mass of copper metal produced during the passage of 2.50 amperes of current through a solution of copper(II) sulfate for 50.0 minutes.

Plan

When the number of significant figures in the calculation warrants, the value 96,485 coulombs is usually rounded off to 96,500 coulombs.

The half-reaction that describes the reduction of copper(II) ions tells us the number of moles of electrons required to produce one mole of copper metal. Each mole of electrons corresponds to 1 faraday, or 96,500 coulombs, of charge. The product of current and time gives the number of coulombs.

$$\boxed{\begin{array}{c}\text{current} \\ \times \text{ time}\end{array}} \longrightarrow \boxed{\begin{array}{c}\text{no. of} \\ \text{coulombs}\end{array}} \longrightarrow \boxed{\begin{array}{c}\text{mol of } e^- \\ \text{passed}\end{array}} \longrightarrow \boxed{\begin{array}{c}\text{mass} \\ \text{of Cu}\end{array}}$$

Solution

The equation for the reduction of copper(II) ions to copper metal is

$$\begin{array}{ccccc} Cu^{2+} & + & 2e^- & \longrightarrow & Cu \qquad \text{(reduction, cathode)} \\ 1 \text{ mol} & & 2(6.02 \times 10^{23})e^- & & 1 \text{ mol} \\ 63.5 \text{ g} & & 2(96,500 \text{ C}) & & 63.5 \text{ g} \end{array}$$

We see that 63.5 grams of copper "plate out" for every 2 moles of electrons, or for every 2(96,500 coulombs) of charge. We first calculate the number of coulombs passing through the cell.

$$\underline{?}\ C = 50.0\ \text{min} \times \frac{60\ \text{s}}{1\ \text{min}} \times \frac{2.50\ \text{C}}{\text{s}} = 7.50 \times 10^3\ \text{C}$$

2.50A = 2.50 C/s

We calculate the mass of copper produced by the passage of 7.50×10^3 coulombs.

$$\underline{?}\ \text{g Cu} = 7.50 \times 10^3\ \text{C} \times \frac{1\ \text{mol}\ e^-}{96,500\ \text{C}} \times \frac{63.5\ \text{g Cu}}{2\ \text{mol}\ e^-}$$

$$= \boxed{2.47\ \text{g Cu}} \qquad \text{(about the mass of a copper penny)}$$

Notice how little copper is deposited by this considerable current in 50 minutes.

You should now work Exercises 24 and 28.

EXAMPLE 21-2 *Electrolysis*

What volume of oxygen gas (measured at STP) is produced by the oxidation of water in the electrolysis of copper(II) sulfate in Example 21-1?

Plan

We use the same approach as in Example 21-1. Here we relate the amount of charge passed to the number of moles, and hence the volume of O_2 gas produced at STP.

$$\boxed{\begin{array}{c}\text{current} \\ \times\ \text{time}\end{array}} \longrightarrow \boxed{\begin{array}{c}\text{no. of} \\ \text{coulombs}\end{array}} \longrightarrow \boxed{\begin{array}{c}\text{mol of}\ e^- \\ \text{passed}\end{array}} \longrightarrow \boxed{\begin{array}{c}L_{STP} \\ \text{of}\ O_2\end{array}}$$

Solution

The equation for the oxidation of water and the equivalence between the number of coulombs and the volume of oxygen produced at STP are

$$2H_2O \longrightarrow \underset{\substack{1\ \text{mol} \\ 22.4\ L_{STP}}}{O_2} + 4H^+ + \underset{\substack{4(6.02 \times 10^{23})e^- \\ 4(96,500\ \text{C})}}{4e^-} \qquad \text{(oxidation, anode)}$$

The number of coulombs passing through the cell is 7.50×10^3 C. For every 4(96,500 coulombs) passing through the cell, 22.4 L of O_2 at STP is produced.

$$\underline{?}\ L_{STP}\ O_2 = 7.50 \times 10^3\ \text{C} \times \frac{1\ \text{mol}\ e^-}{96,500\ \text{C}} \times \frac{22.4\ L_{STP}\ O_2}{4\ \text{mol}\ e^-} = \boxed{0.435\ L_{STP}\ O_2}$$

You should now work Exercise 26.

Notice how little product is formed by what seems to be a lot of electricity. This suggests why electrolytic production of gases and metals is so costly.

21-7 DETERMINATION OF OXIDATION STATE (CHARGE ON AN ION) BY ELECTROLYSIS

Faraday's Law can be used to determine oxidation states. We electrolyze a solution that contains an ion. We then relate the amount of an element produced to the number of moles of electrons passing through the cell.

CHEMISTRY IN USE

Our Daily Lives

A Spectacular View of One Mole of Electrons

Early in our study of chemistry, we saw that atoms are made up of protons, neutrons, and electrons. We also discussed the incredibly large size of Avogadro's number, 6.022×10^{23}. Although individual atoms and molecules are invisible to the naked eye, one mole of atoms or molecules is easily detected. Because protons and neutrons are even smaller than atoms and also invisible, you might never expect to see individual electrons, which have only about 1/2000th the mass of a proton and a neutron. However, let's consider the possibility of seeing a faraday of charge. A faraday of charge contains Avogadro's number of electrons. Would this collection of 6.022×10^{23} electrons be visible? If so, what might it look like? It would look quite spectacular!

Throughout the 1980s, scientists carefully studied data collected during five million lightning flashes along the eastern United States. The data were collected by thirty-six instruments that were collectively known as the National Lightning Detection Network. The investigating scientists found that the electrical currents in lightning flashes over northern Florida measured about 45,000 amps, about double the 25,000 amp currents in lightning flashes over the New England states. This study showed that the amount of current flowing during lightning flashes was inversely proportional to the latitude (distance from the equator) of the storm.

One coulomb is the amount of charge that passes a point when a one-ampere current flows for one second (1 coulomb = 1 ampere · second). Thus, a current of 96,500 amps flowing for one second contains Avogadro's number of electrons, or one faraday of charge.

Measurements taken in northern Florida show that a typical two-second lightning strike over that section of the country would transfer approximately Avogadro's number of electrons between the clouds and the earth. So, for those living in northern Florida, a spectacular mental view of one mole of electrons can be obtained by visualizing a two-second lightning strike. Keep in mind that the average lightning strike lasts only a small fraction of a second, and that we can only have a mental view of a two-second lightning strike by extrapolation of what is seen in nature. Because New England lightning strikes produce only about half the current of lightning strikes over northern Florida, people in New England must try to imagine a four-second lightning strike.

Ronald DeLorenzo
Middle Georgia College

EXAMPLE 21-3 *Determination of Oxidation States*

An aqueous solution of an unknown complex salt of chromium is electrolyzed by a current of 3.00 amperes for 1.00 hour. This produces 1.94 grams of chromium metal at the cathode. What is the oxidation state of chromium in this solution?

Plan

The magnitude of the oxidation state is equal to the number of moles of electrons required to oxidize or reduce one mole of any substance to the elemental state (i.e., to an oxidation state of zero).

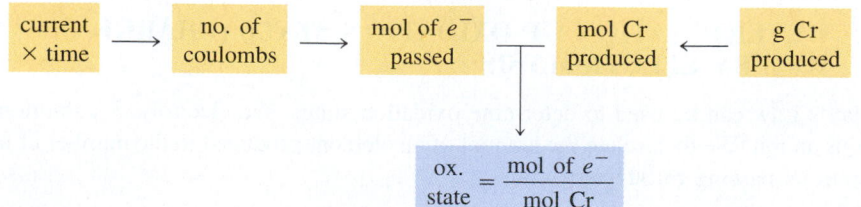

Solution

As before, we first find the amount of charge passing through the cell.

$$\underset{?}{\_}\,C = 1.00 \text{ hr} \times \frac{60 \text{ min}}{1 \text{ hr}} \times \frac{60 \text{ s}}{1 \text{ min}} \times \frac{3.00 \text{ C}}{1 \text{ s}} = 1.08 \times 10^4 \text{ C}$$

We then find the number of moles of electrons passed.

$$\underset{?}{\_}\,\text{mol } e^- = 1.08 \times 10^4 \text{ C} \times \frac{1 \text{ mol } e^-}{96{,}500 \text{ C}} = 0.112 \text{ mol } e^-$$

We now find the number of moles of Cr formed.

$$\underset{?}{\_}\,\text{mol Cr formed} = 1.94 \text{ g Cr} \times \frac{1 \text{ mol Cr}}{52.0 \text{ g Cr}} = 0.0373 \text{ mol Cr}$$

$$\text{ox. state} = \frac{\text{mol of } e^-}{\text{mol Cr formed}} = \frac{0.112 \text{ mol } e^-}{0.0373 \text{ mol Cr formed}} = \frac{3 \text{ mol } e^-}{\text{mol Cr formed}}$$

Because three moles of electrons are required to reduce one mole of chromium ions to Cr metal, the chromium in solution must be Cr(III).

You should now work Exercises 30 and 31.

21-8 ELECTROLYTIC REFINING AND ELECTROPLATING OF METALS

Electrolytic reduction is the most practical means by which many active metals can be obtained (Section 22-3). Less active metals can be obtained from their ores by less expensive chemical reduction, and some metals occur in nature in the uncombined state. But these are frequently purified or refined by electrolysis. The electrolytic method for refining metals is also called *electroplating* when used to plate a metal onto a surface. For example, impure metallic copper obtained from the chemical reduction of Cu_2S and CuS (Section 22-8) is purified using an electrolytic cell like the one shown in Figure 21-5. Thin sheets of very pure copper are made to act as cathodes by connecting them to the negative terminal of a dc generator. Chunks of impure copper connected to the positive terminal function as anodes. The electrodes are immersed in a solution of copper(II) sulfate and

A sludge called anode mud collects under the anodes. It contains such valuable and difficult-to-oxidize elements as Au, Pt, Ag, Se, and Te. The separation, purification, and sale of these elements reduce the cost of refined copper.

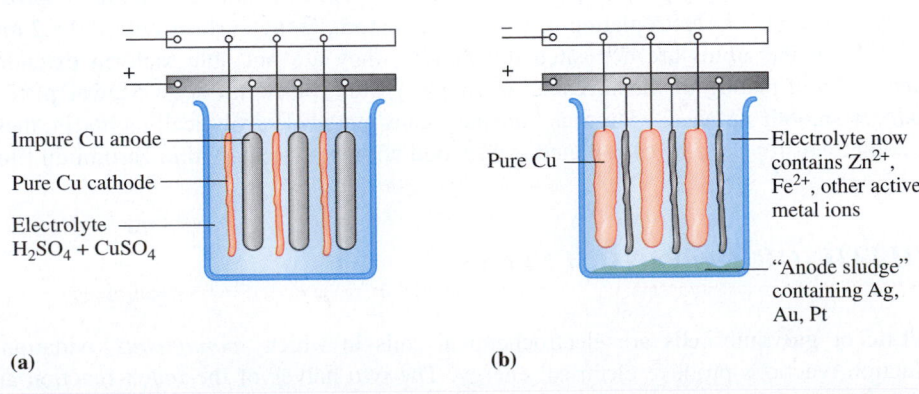

(a) (b)

(c)

Figure 21-5 A schematic diagram of the electrolytic cell used for refining copper (a) before electrolysis and (b) after electrolysis. (c) Commercial electrolysis cells for refining copper.

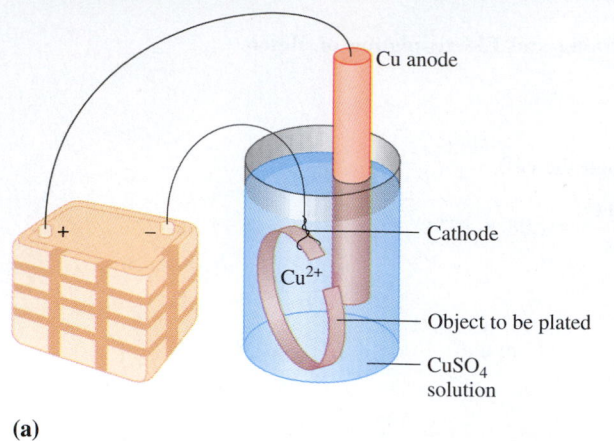

(a)

(b)

Figure 21-6 Electroplating with copper. (a) The anode is made of pure copper, which dissolves during the electroplating process. This replenishes the Cu^{2+} ions that are removed from the solution as Cu plates out on the cathode. (b) A family memento that has been electroplated with copper. To aid in electroplating onto nonconductors such as shoes, the material is first soaked in a concentrated electrolyte solution to make it conductive.

sulfuric acid. When the cell operates, Cu from the impure anodes is oxidized and goes into solution as Cu^{2+} ions; Cu^{2+} ions from the solution are reduced and plate out as metallic Cu on the pure Cu cathodes. Other active metals from the impure bars also go into solution after oxidation. They do not plate out onto the cathode bars of pure Cu because of the far greater concentration of the more easily reduced Cu^{2+} ions that are already in solution. Overall, there is no net reaction, merely a simultaneous transfer of Cu from anode to solution and from solution to cathode.

$$\text{(impure) Cu} \longrightarrow Cu^{2+} + 2e^- \qquad \text{(oxidation, anode)}$$
$$\underline{Cu^{2+} + 2e^- \longrightarrow \text{Cu (pure)}} \qquad \text{(reduction, cathode)}$$
$$\text{Cu (impure)} \longrightarrow \text{Cu (pure)} \qquad \text{(no net reaction)}$$

Nevertheless, the net effect is that small bars of very pure Cu and large bars of impure Cu are converted into large bars of very pure Cu and small bars of impure Cu. The energy provided by the electric generator forces a decrease in the entropy of the system by separating the Cu from its impurities in the impure bars. Copper can be plated onto other objects by the same mechanism (Figure 21-6).

Metal-plated articles are common in our society. Jewelry and tableware are often plated with silver. Gold is plated onto jewelry and electrical contacts. Some automobiles have steel bumpers plated with thin films of chromium. A chrome bumper requires approximately 3 seconds of electroplating to produce a smooth, shiny surface only 0.0002 mm thick. When the atoms are deposited too rapidly, they are not able to form extended lattices. Rapid plating of metal results in rough, grainy, black surfaces. Slower plating produces smooth surfaces. "Tin cans" are steel cans plated electrolytically with tin; these are sometimes replaced by cans plated in $\frac{1}{3}$ second with an extremely thin chromium film.

VOLTAIC OR GALVANIC CELLS

These are named for Allesandro Volta and Luigi Galvani, two Italian physicists of the 18th century.

Voltaic or **galvanic cells** are electrochemical cells in which *spontaneous* oxidation–reduction reactions produce electrical energy. The two halves of the redox reaction are separated, requiring electron transfer to occur through an external circuit. In this way, useful electrical energy is obtained. Everyone is familiar with some voltaic cells. The dry cells commonly used in flashlights, transistor radios, photographic equipment, and many toys and appliances are voltaic cells. Automobile batteries consist of voltaic cells connected in series so that their voltages add. We shall first consider some simple laboratory cells used to measure the potential difference, or voltage, of a reaction under study. We shall then look at some common voltaic cells.

21-9 THE CONSTRUCTION OF SIMPLE VOLTAIC CELLS

A **half-cell** contains the oxidized and reduced forms of an element, or other more complex species, in contact with each other. A common kind of half-cell consists of a piece of metal (the electrode) immersed in a solution of its ions. Consider two such half-cells in separate beakers (Figure 21-7). The electrodes are connected by a wire. A voltmeter can be inserted into the circuit to measure the potential difference between the two electrodes, or an ammeter can be inserted to measure the current flow. The electrical current is the result of the spontaneous redox reaction that occurs. We measure the potential of the cell.

Neither of these meters generates electrical energy.

The circuit between the two solutions is completed by a **salt bridge.** This can be any medium through which ions can slowly pass. A salt bridge can be made by bending a piece of glass tubing into the shape of a "U," filling it with a hot saturated salt/5% agar solution, and allowing it to cool. The cooled mixture "sets" to the consistency of firm gelatin. As a result, the solution does not run out when the tube is inverted (see Figure 21-7), but the ions in the gel are still able to move. A salt bridge serves three functions.

Agar is a gelatinous material obtained from algae.

1. It allows electrical contact between the two solutions.
2. It prevents mixing of the electrode solutions.
3. It maintains the electrical neutrality in each half-cell as ions flow into and out of the salt bridge.

A cell in which all reactants and products are in their thermodynamic standard states (1 M for dissolved species and 1 atm partial pressure for gases) is called a **standard cell.**

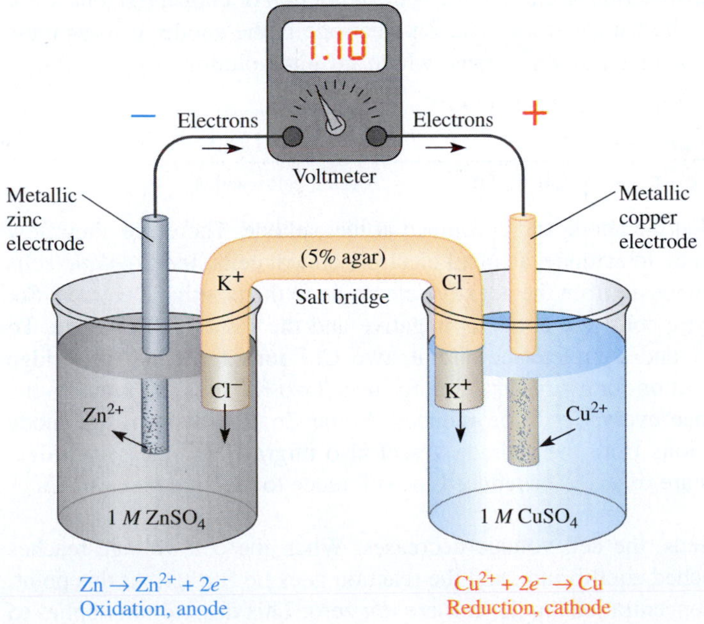

Figure 21-7 The zinc–copper voltaic cell utilizes the reaction

$$Zn(s) + Cu^{2+}(aq) \longrightarrow Zn^{2+}(aq) + Cu(s)$$

The standard potential of this cell is 1.10 volts. The cell can be represented as $Zn|Zn^{2+}(1.0\ M)\|Cu^{2+}(1.0\ M)|Cu$.

Voltmeter — reading: 1.10 — Electrons, Electrons, $-$, $+$

Metallic zinc electrode — Metallic copper electrode

(5% agar) — Salt bridge — K^+ — Cl^-

Zn^{2+} — Cl^- — K^+ — Cu^{2+}

1 M ZnSO$_4$ — 1 M CuSO$_4$

$Zn \rightarrow Zn^{2+} + 2e^-$
Oxidation, anode

$Cu^{2+} + 2e^- \rightarrow Cu$
Reduction, cathode

(Left) A strip of zinc was placed in a blue solution of copper(II) sulfate, $CuSO_4$. The copper has been displaced from solution and has fallen to the bottom of the beaker. The resulting zinc sulfate solution is colorless. (Right) No reaction occurs when copper wire is placed in a colorless zinc sulfate solution.

21-10 THE ZINC–COPPER CELL

Consider a standard cell made up of two half-cells, one a strip of metallic Cu immersed in 1.0 M copper(II) sulfate solution and the other a strip of Zn immersed in 1.0 M zinc sulfate solution (Figure 21-7). This cell is called the Daniell cell. The following experimental observations have been made about this cell.

1. The initial voltage is 1.10 volts.
2. The mass of the zinc electrode decreases. The concentration of Zn^{2+} increases in the solution around the zinc electrode as the cell operates.
3. The mass of the copper electrode increases. The concentration of Cu^{2+} decreases in the solution around this electrode as the cell operates.

We deduce that the half-reaction at the cathode is the reduction of copper(II) ions to Cu metal. This plates out on the Cu electrode. The Zn electrode is the anode. It loses mass because the Zn metal is oxidized to Zn^{2+} ions, which go into solution.

$$\begin{array}{ll} Zn \longrightarrow Zn^{2+} + 2e^- & \text{(oxidation, anode)} \\ \underline{Cu^{2+} + 2e^- \longrightarrow Cu} & \text{(reduction, cathode)} \\ Cu^{2+} + Zn \longrightarrow Cu + Zn^{2+} & \text{(overall cell reaction)} \end{array}$$

Electrons are released at the anode and consumed at the cathode. Therefore, they flow through the wire from anode to cathode, as in all electrochemical cells. In all *voltaic* cells the electrons flow spontaneously from the negative electrode to the positive electrode. So, in contrast with electrolytic cells, the anode is negative and the cathode is positive. To maintain electroneutrality and complete the circuit, two Cl^- ions from the salt bridge migrate into the anode solution for every Zn^{2+} ion formed. Two K^+ ions migrate into the cathode solution to replace every Cu^{2+} ion reduced. Some Zn^{2+} ions from the anode vessel and some SO_4^{2-} ions from the cathode vessel also migrate into the salt bridge. Neither Cl^- nor K^+ ions are oxidized or reduced in preference to the zinc metal or Cu^{2+} ions.

As the reaction proceeds, the cell voltage decreases. When the cell voltage reaches zero, the reaction has reached equilibrium, and the reaction goes no further. At this point, however, the metal ion concentrations in the cell are *not* zero. This description applies to any voltaic cell.

Compare the $-/+$, anode/cathode, and oxidation/reduction labels and the directions of electron flow in Figures 21-2a and 21-7.

Voltaic cells can be represented as follows for the zinc–copper cell.

salt bridge
$$Zn|Zn^{2+} \ (1.0 \ M)\|Cu^{2+} \ (1.0 \ M)|Cu$$
species (and concentrations)
in contact with electrode surfaces

In this representation, a single line (|) represents an interface at which a potential develops, i.e., an electrode. It is conventional to write the anode half-cell on the left in this notation.

The same reaction occurs when a piece of Zn is dropped into a solution of $CuSO_4$. The Zn dissolves and the blue color of Cu^{2+} ions disappears. Copper forms on the Zn and then settles to the bottom of the container. But no electricity flows in an external circuit, because the two half-reactions are *not* physically separated.

▼ **PROBLEM-SOLVING TIP**   *How to Tell the Anode from the Cathode*

The correspondence between the names *anode* and *cathode* and the charge on the electrode is *different* for electrolytic cells than for voltaic (galvanic) cells. Students sometimes get confused by trying to remember which is which. Check the definitions of these two terms in Section 21-2. The surest way to name these electrodes is to determine what process takes place at each one.

$$\text{anode} \Longleftrightarrow \text{oxidation} \quad \text{and} \quad \text{cathode} \Longleftrightarrow \text{reduction}$$

As a memory aid, both anode and oxidation begin with a vowel, whereas both cathode and reduction begin with a consonant.

21-11 THE COPPER–SILVER CELL

Now consider a similar standard voltaic cell consisting of a strip of Cu immersed in 1.0 M $CuSO_4$ solution and a strip of Ag immersed in 1.0 M $AgNO_3$ solution. A wire and a salt bridge complete the circuit. The following observations have been made.

1. The initial voltage of the cell is 0.46 volt.

(Left) A spiral of copper wire was placed in a colorless solution of silver nitrate, $AgNO_3$. The silver has been displaced from solution and adheres to the wire. The resulting copper nitrate solution is blue. (Right) No reaction occurs when silver wire is placed in a blue copper sulfate solution.

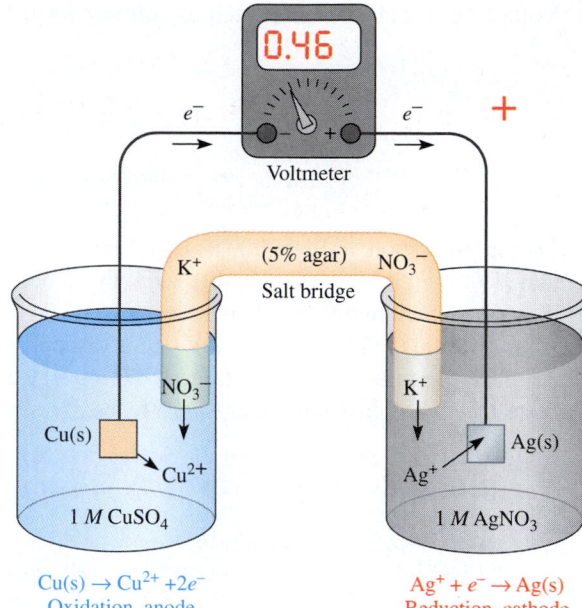

Figure 21-8   The copper–silver voltaic cell utilizes the reaction

$$Cu(s) + 2Ag^+(aq) \longrightarrow Cu^{2+}(aq) + 2Ag(s)$$

The standard potential of this cell is 0.46 volt. This cell can be represented as $Cu|Cu^{2+}(1.0\ M)\|Ag^+(1.0\ M)|Ag$.

$Cu(s) \rightarrow Cu^{2+} + 2e^-$
Oxidation, anode

$Ag^+ + e^- \rightarrow Ag(s)$
Reduction, cathode

2. The mass of the copper electrode decreases. The Cu^{2+} ion concentration increases in the solution around the copper electrode.

3. The mass of the silver electrode increases. The Ag^+ ion concentration decreases in the solution around the silver electrode.

In this cell the Cu electrode is the anode because Cu metal is oxidized to Cu^{2+} ions. The Ag electrode is the cathode because Ag^+ ions are reduced to metallic Ag (Figure 21-8).

$$
\begin{array}{ll}
Cu \longrightarrow Cu^{2+} + 2e^- & \text{(oxidation, anode)} \\
2(Ag^+ + e^- \longrightarrow Ag) & \text{(reduction, cathode)} \\
\hline
Cu + 2Ag^+ \longrightarrow Cu^{2+} + 2Ag & \text{(overall cell reaction)}
\end{array}
$$

As before, ions from the salt bridge migrate to maintain electroneutrality. Some NO_3^- ions (from the cathode vessel) and some Cu^{2+} ions (from the anode vessel) also migrate into the salt bridge.

Recall that in the zinc–copper cell the copper electrode is the *cathode;* now in the copper–silver cell the copper electrode is the *anode.*

> Whether a particular electrode acts as an anode or a cathode depends on what the other electrode of the cell is.

The two cells we have described show that the Cu^{2+} ion is more easily reduced (is a stronger oxidizing agent) than Zn^{2+}, so Cu^{2+} oxidizes metallic zinc to Zn^{2+}. By contrast, Ag^+ ion is more easily reduced (is a stronger oxidizing agent) than Cu^{2+} ion, so Ag^+ oxidizes Cu atoms to Cu^{2+}. Conversely, metallic Zn is a stronger reducing agent than metallic Cu, and metallic Cu is a stronger reducing agent than metallic Ag. We can now

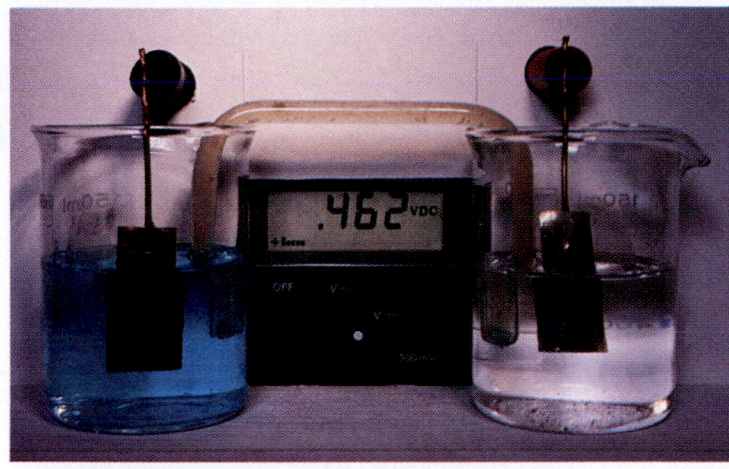

The standard $Cu|Cu^{2+}(1.0\ M)\|Ag^+\ (1.0\ M)|Ag$ cell.

arrange the species we have studied in order of increasing strength as oxidizing agents and as reducing agents.

$$Zn^{2+} < Cu^{2+} < Ag^+ \qquad Zn > Cu > Ag$$

$\xrightarrow{\hspace{2cm}}$ Increasing strength as oxidizing agents

$\xleftarrow{\hspace{2cm}}$ Increasing strength as reducing agents

STANDARD ELECTRODE POTENTIALS

The potentials of the standard zinc–copper and copper–silver voltaic cells are 1.10 volts and 0.46 volts, respectively. The magnitude of a cell's potential measures the spontaneity of its redox reaction. *Higher (more positive) cell potentials indicate greater driving force for the reaction as written.* Under standard conditions, the oxidation of metallic Zn by Cu^{2+} ions has a greater tendency to go toward completion than does the oxidation of metallic Cu by Ag^+ ions. It is convenient to separate the total cell potential into the individual contributions of the two half-reactions. This lets us determine the relative tendencies of particular oxidation or reduction half-reactions to occur. Such information gives us a quantitative basis for specifying strengths of oxidizing and reducing agents. In the next several sections we shall see how this is done for standard half-cells. Then we shall learn to correct for changes in temperature, concentration, and pressure (Section 21-20).

21-12 THE STANDARD HYDROGEN ELECTRODE

Every oxidation must be accompanied by a reduction (that is, the electrons must have somewhere to go). So it is impossible to determine experimentally the potential of any *single* electrode. Therefore, we establish an arbitrary standard. The reference electrode is the **standard hydrogen electrode (SHE).** This electrode contains a piece of metal electrolytically coated with a grainy black surface of inert platinum metal, immersed in a 1.0 M H^+ solution. Hydrogen, H_2, is bubbled at 1 atm pressure through a glass envelope over the platinized electrode (Figure 21-9).

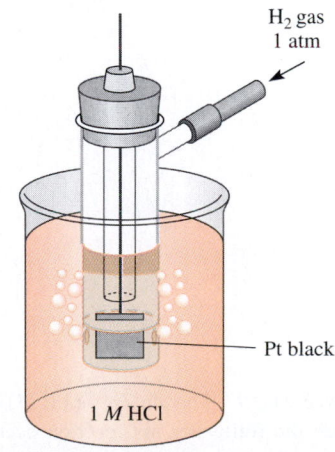

H_2 gas
1 atm

Pt black

1 M HCl

Figure 21-9 The standard hydrogen electrode (SHE).

By international agreement, the standard hydrogen electrode is arbitrarily assigned a potential of *exactly* 0.0000 . . . volt.

The superscript in E^0 indicates standard electrochemical conditions.

SHE Half-Reaction	E^0 (standard electrode potential)	
$H_2 \longrightarrow 2H^+ + 2e^-$	exactly 0.0000 . . . V	(SHE as anode)
$2H^+ + 2e^- \longrightarrow H_2$	exactly 0.0000 . . . V	(SHE as cathode)

We then construct a standard cell consisting of a standard hydrogen electrode and some other standard electrode (half-cell). Because the defined electrode potential of the SHE contributes exactly 0 volt to the sum, the voltage of the overall cell then lets us determine the **standard electrode potential** of the other half-cell. This is its potential with respect to the standard hydrogen electrode, measured at 25°C when the concentration of each ion in the solution is 1 M and the pressure of any gas involved is 1 atm.

By agreement, we always present the standard cell potential for each half-cell as a *reduction* process.

21-13 THE ZINC–SHE CELL

This cell consists of an SHE in one beaker and a strip of zinc immersed in 1.0 M zinc sulfate solution in another beaker (Figure 21-10). A wire and a salt bridge complete the circuit. When the circuit is closed, the following observations can be made.

1. The initial potential of the cell is 0.763 volt.

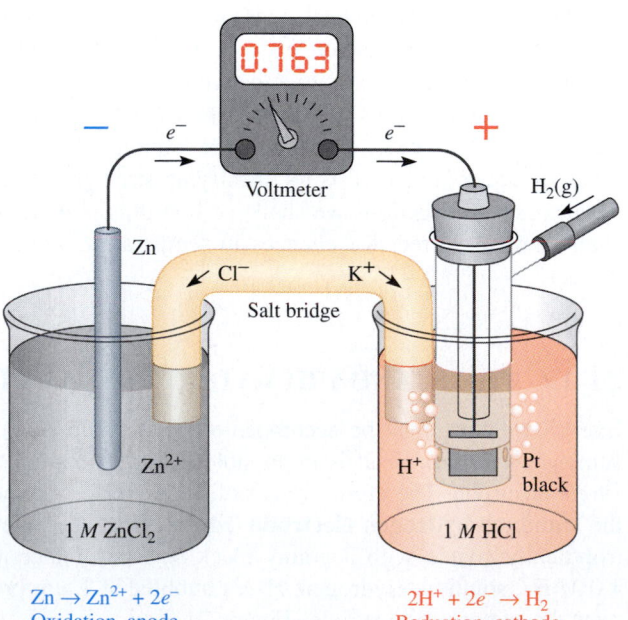

Figure 21-10 The $Zn|Zn^{2+}(1.0\ M)\|H^+(1.0\ M);\ H_2(1\ atm)|Pt$ cell, in which the following net reaction occurs.

$$Zn(s) + 2H^+(aq) \longrightarrow Zn^{2+}(aq) + H_2(g)$$

In this cell the standard hydrogen electrode functions as the cathode.

Voltmeter

$H_2(g)$

Zn

Cl^- K^+

Salt bridge

Zn^{2+}

H^+ Pt black

1 M ZnCl$_2$

1 M HCl

$Zn \rightarrow Zn^{2+} + 2e^-$
Oxidation, anode

$2H^+ + 2e^- \rightarrow H_2$
Reduction, cathode

2. As the cell operates, the mass of the zinc electrode decreases. The concentration of Zn^{2+} ions increases in the solution around the zinc electrode.

3. The H^+ concentration decreases in the SHE. Gaseous H_2 is produced.

We can conclude from these observations that the following half-reactions and cell reaction occur.

		E^0	
(oxidation, anode)	$Zn \longrightarrow Zn^{2+} + 2e^-$	0.763 V	
(reduction, cathode)	$2H^+ + 2e^- \longrightarrow H_2$	0.000 V	(by definition)
(cell reaction)	$Zn + 2H^+ \longrightarrow Zn^{2+} + H_2$	$E^0_{cell} = 0.763$ V	(measured)

The standard potential at the anode *plus* the standard potential at the cathode gives the standard cell potential. The potential of the SHE is 0.000 volt, and the standard cell potential is found to be 0.763 volt. So the standard potential of the zinc anode must be 0.763 volt. The $Zn|Zn^{2+}(1.0\ M)\|H^+(1.0\ M), H_2(1\ atm)|Pt$ cell is depicted in Figure 21-10.

Note that in *this* cell the SHE is the *cathode,* and metallic zinc reduces H^+ to H_2. The zinc electrode is the *anode* in this cell.

The idea of "electron pressure" helps to understand this process. The negative reduction potential for the half-reaction

$$Zn^{2+} + 2e^- \longrightarrow Zn \qquad E^0 = -0.763\ V$$

says that this reaction is *less favorable* than the corresponding reduction to H_2,

$$2H^+ + 2e^- \longrightarrow H_2 \qquad E^0 = 0.000\ V$$

Before they are connected, each half-cell builds up a supply of electrons waiting to be released, thus generating an electron pressure. Let us compare these electron pressures by reversing the two half-reactions to show production of electrons (and changing the signs of their E^0 values).

$$Zn \longrightarrow Zn^{2+} + 2e^- \qquad E^0_{oxidation} = +0.763\ V$$

$$H_2 \longrightarrow 2H^+ + 2e^- \qquad E^0_{oxidation} = 0.000\ V$$

The process with the more *positive* E^0 value is favored, so we reason that the electron pressure generated at the Zn electrode is greater than that at the H_2 electrode. As a result, when the cell is connected, electrons flow through the wire *from the Zn electrode to the H_2 electrode.* Oxidation occurs at the zinc electrode (anode), and reduction occurs at the hydrogen electrode (cathode).

What we have informally called "electron pressure" is the tendency to undergo oxidation. This is formally expressed as an *oxidation potential;* however, we usually tabulate reduction potentials (Table 21-2).

21-14 THE COPPER–SHE CELL

Another cell consists of an SHE in one beaker and a strip of Cu metal immersed in 1.0 M copper(II) sulfate solution in another beaker. A wire and a salt bridge complete the circuit. For this cell, we observe the following (Figure 21-11).

1. The initial cell potential is 0.337 volt.

2. Gaseous hydrogen is used up. The H^+ concentration increases in the solution of the SHE.

3. The mass of the copper electrode increases. The concentration of Cu^{2+} ions decreases in the solution around the copper electrode.

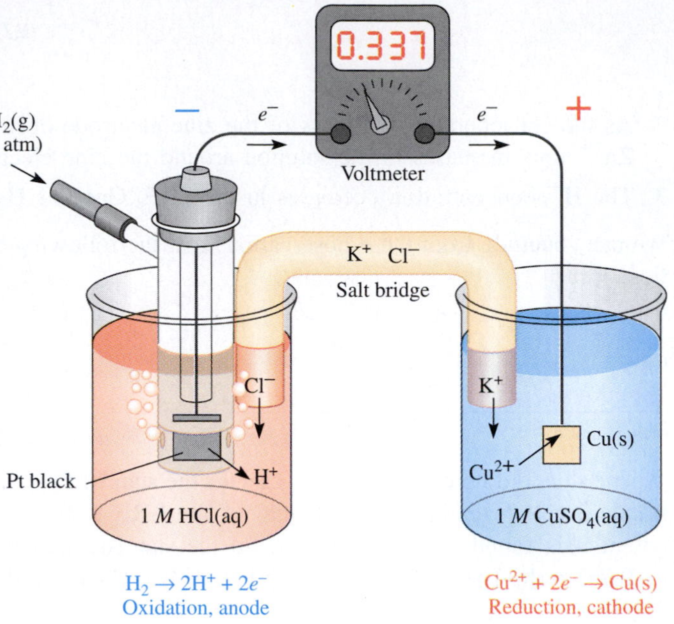

Figure 21-11 The standard copper–SHE cell,

$$Pt|H^+(1.0\ M);\ H_2(1\ atm)\|Cu^{2+}(1.0\ M)|Cu$$

In this cell, the standard hydrogen electrode functions as the anode. The net reaction is

$$H_2(g) + Cu^{2+}(aq) \longrightarrow 2H^+(aq) + Cu(s)$$

$H_2 \rightarrow 2H^+ + 2e^-$
Oxidation, anode

$Cu^{2+} + 2e^- \rightarrow Cu(s)$
Reduction, cathode

Thus, the following half-reactions and cell reaction occur.

		E^0	
(oxidation, anode)	$H_2 \longrightarrow 2H^+ + 2e^-$	0.000 V	(by definition)
(reduction, cathode)	$Cu^{2+} + 2e^- \longrightarrow Cu$	0.337 V	
(cell reaction)	$H_2 + Cu^{2+} \longrightarrow 2H^+ + Cu$	$E^0_{cell} = 0.337$ V	(measured)

Recall that in the Zn–SHE cell the SHE was the cathode.

The SHE functions as the *anode* in this cell, and Cu^{2+} ions oxidize H_2 to H^+ ions. The standard electrode potential of the copper half-cell is 0.337 volt as a *cathode* in the Cu–SHE cell.

Again, we can think of $E^0_{oxidation}$ in the two half-cells as "electron pressures."

$$Cu \longrightarrow Cu^{2+} + 2e^- \qquad E^0_{oxidation} = -0.337\ V$$

$$H_2 \longrightarrow 2H^+ + 2e^- \qquad E^0_{oxidation} = 0.000\ V$$

Now the hydrogen electrode has the higher electron pressure. When the cell is connected, electrons flow through the wire from the hydrogen electrode to the copper electrode. H_2 is oxidized to $2H^+$ (anode), and Cu^{2+} is reduced to Cu (cathode).

21-15 STANDARD ELECTRODE POTENTIALS

The activity series (Table 4-13) is based on standard electrode potentials.

We can develop a series of standard electrode potentials by measuring the potentials of other standard electrodes versus the SHE in the way we describe for the standard Zn–SHE and standard Cu–SHE voltaic cells. Many electrodes involve metals or nonmetals in contact with their ions. We saw (Section 21-13) that the standard Zn electrode behaves as the anode versus the SHE and that the standard *oxidation* potential for the Zn half-cell is 0.763 volt.

		$E^0_{oxidation}$
(as anode)	$Zn \longrightarrow Zn^{2+} + 2e^-$	+0.763 V
	reduced form $\longrightarrow$ oxidized form $+ ne^-$	(standard *oxidation* potential)

Therefore, the *reduction* potential for the standard zinc electrode (to act as a *cathode* relative to the SHE) is the negative of this, or -0.763 volt.

$$E^0_{reduction}$$

(as cathode) $\quad Zn^{2+} + 2e^- \longrightarrow Zn \qquad\qquad -0.763\ V$

$\qquad$ oxidized form $+ ne^- \longrightarrow$ reduced form $\qquad$ (standard *reduction* potential)

By international convention, the standard potentials of electrodes are tabulated for *reduction half-reactions*. These indicate the tendencies of the electrodes to behave as cathodes toward the SHE. Electrodes with positive E^0 values for reduction half-reactions act as *cathodes* versus the SHE. Those with negative E^0 values for reduction half-reactions act as anodes versus the SHE.

Different conventions have been used for writing half-reactions and the signs for their potentials. To avoid confusion, use the convention presented here consistently.

Electrodes with *Positive* $E^0_{reduction}$	Electrodes with *Negative* $E^0_{reduction}$
Reduction occurs *more readily* than the reduction of $2H^+$ to H_2.	Reduction is *more difficult* than the reduction of $2H^+$ to H_2.
Electrode acts as a *cathode* versus the SHE.	Electrode acts as an *anode* versus the SHE.

The more positive the E^0 value for a half-reaction, the greater the tendency for the half-reaction to occur in the forward direction as written. Conversely, the more negative the E^0 value for a half-reaction, the greater the tendency for the half-reaction to occur in the reverse direction as written.

Table 21-2 lists standard electrode potentials for a few elements.

Table 21-2 *Standard Aqueous Electrode Potentials at 25°C—The Electromotive Series*

Element	Reduction Half-Reaction	Standard Reduction Potential E^0, volts
Li	$Li^+ + e^- \longrightarrow Li$	−3.045
K	$K^+ + e^- \longrightarrow K$	−2.925
Ca	$Ca^{2+} + 2e^- \longrightarrow Ca$	−2.87
Na	$Na^+ + e^- \longrightarrow Na$	−2.714
Mg	$Mg^{2+} + 2e^- \longrightarrow Mg$	−2.37
Al	$Al^{3+} + 3e^- \longrightarrow Al$	−1.66
Zn	$Zn^{2+} + 2e^- \longrightarrow Zn$	−0.7628
Cr	$Cr^{3+} + 3e^- \longrightarrow Cr$	−0.74
Fe	$Fe^{2+} + 2e^- \longrightarrow Fe$	−0.44
Cd	$Cd^{2+} + 2e^- \longrightarrow Cd$	−0.403
Ni	$Ni^{2+} + 2e^- \longrightarrow Ni$	−0.25
Sn	$Sn^{2+} + 2e^- \longrightarrow Sn$	−0.14
Pb	$Pb^{2} + 2e^- \longrightarrow Pb$	−0.126
H_2	$2H^+ + 2e^- \longrightarrow H_2$	0.000 (reference electrode)
Cu	$Cu^{2+} + 2e^- \longrightarrow Cu$	+0.337
I_2	$I_2 + 2e^- \longrightarrow 2I^-$	+0.535
Hg	$Hg^{2} + 2e^- \longrightarrow Hg$	+0.789
Ag	$Ag^+ + e^- \longrightarrow Ag$	+0.7994
Br_2	$Br_2 + 2e^- \longrightarrow 2Br^-$	+1.08
Cl_2	$Cl_2 + 2e^- \longrightarrow 2Cl^-$	+1.360
Au	$Au^{3+} + 3e^- \longrightarrow Au$	+1.50
F_2	$F_2 + 2e^- \longrightarrow 2F^-$	+2.87

Increasing strength as oxidizing agent

Increasing strength as reducing agent

The *oxidizing agent* is *reduced*.

1. The species on the *left* side are all either cations of metals, hydrogen ions, or elemental nonmetals. These are all *oxidizing agents* (*oxidized forms* of the elements). Their strengths as oxidizing agents increase from top to bottom, i.e., as the $E^0_{\text{reduction}}$ values become more positive. Fluorine is the strongest oxidizing agent, and Li^+ is a very weak oxidizing agent.

The *reducing agent* is *oxidized*.

2. The species on the *right* side are free metals, hydrogen, or anions of nonmetals. These are all *reducing agents* (*reduced forms* of the elements). Their strengths as reducing agents increase from bottom to top, i.e., as the $E^0_{\text{reduction}}$ values become more negative. Metallic Li is a very strong reducing agent, and F^- is a very weak reducing agent.

The more positive the reduction potential, the stronger the species on the left is as an oxidizing agent and the weaker the species on the right is as a reducing agent. The elements with high ionization energies and highly negative electron affinities have the greatest tendencies to exist as anions. The elements with low ionization energies and positive or slightly negative electron affinities have the greatest tendencies to exist as cations.

21-16 USES OF STANDARD ELECTRODE POTENTIALS

The most important application of electrode potentials is the prediction of the spontaneity of redox reactions. Standard electrode potentials can be used to determine the spontaneity of redox reactions in general, whether or not the reactions can take place in electrochemical cells.

Suppose we ask a question such as this: At standard conditions, will Cu^{2+} ions oxidize metallic Zn to Zn^{2+} ions, or will Zn^{2+} ions oxidize metallic copper to Cu^{2+}? One of the two possible reactions is spontaneous, and the reverse reaction is nonspontaneous. We must determine which one is spontaneous. We already know the answer to this question from experimental results (Section 21-10), but let us demonstrate the procedure for predicting the spontaneous reaction.

1. Choose the appropriate half-reactions from a table of standard reduction potentials.
2. Write the equation for the half-reaction with the more positive (or less negative) E^0 value *for reduction* first, along with its potential.
3. Then write the equation for the other half-reaction *as an oxidation* and write its *oxidation potential;* to do this, reverse the tabulated reduction half-reaction and change the sign of E^0. (Reversing a half-reaction or a complete reaction also changes the sign of its potential.)
4. Balance the electron transfer. *We do not multiply the potentials by the numbers used to balance the electron transfer!* The reason is that each potential represents a *tendency* for a reaction process to occur relative to the SHE; this does not depend on *how many times* it occurs. An electrical potential is an *intensive property.*
5. Add the reduction and oxidation half-reactions, and add the reduction and oxidation potentials. E^0_{cell} will be *positive* for the resulting overall cell reaction. *This indicates that the forward reaction is spontaneous.* (A negative E^0_{cell} value indicates a nonspontaneous process.)

For the cell described above, the Cu^{2+}/Cu couple has the more positive reduction potential, so we keep it as the reduction half-reaction and reverse the other half-reaction. Following the steps outlined, we obtain the equation for the spontaneous reaction.

$$1(Cu^{2+} + 2e^- \longrightarrow Cu) \qquad\qquad +0.337 \text{ V} \longleftarrow \text{reduction potential}$$
$$\underline{1(Zn \longrightarrow Zn^{2+} + 2e^-) \qquad\qquad +0.763 \text{ V} \longleftarrow \text{oxidation potential}}$$
$$Cu^{2+} + Zn \longrightarrow Cu + Zn^{2+} \qquad E^0_{cell} = +1.100 \text{ V}$$

The positive E^0_{cell} value tells us that the forward reaction is spontaneous at standard conditions. So we conclude that copper(II) ions oxidize metallic zinc to Zn^{2+} ions as they are reduced to metallic copper. (Section 21-10 shows that the potential of the standard zinc–copper voltaic cell is 1.10 volts. This is the spontaneous reaction that occurs.)

The reverse reaction has a negative E^0 and is nonspontaneous.

nonspontaneous
reaction: $\qquad Cu + Zn^{2+} \longrightarrow Cu^{2+} + Zn \qquad E^0_{cell} = -1.10 \text{ volts}$

To make it occur, we would have to supply electrical energy with a potential difference greater than 1.10 volts. That is, this nonspontaneous reaction would have to be carried out in an *electrolytic cell.*

EXAMPLE 21-4 *Predicting the Direction of Reactions*

At standard conditions, will chromium(III) ions, Cr^{3+}, oxidize metallic copper to copper(II) ions, Cu^{2+}, or will Cu^{2+} oxidize metallic chromium to Cr^{3+} ions?

Plan

We refer to the table of standard reduction potentials and choose the two appropriate half-reactions. The copper half-reaction has the more positive reduction potential, so we write it first. Then we write the chromium half-reaction as an oxidation, balance the electron transfer, and add the two half-reactions and their potentials.

Solution

$$\begin{array}{lll} & & E^0 \\ 3(Cu^{2+} + 2e^- \longrightarrow Cu) & \text{(reduction)} & +0.337 \text{ V} \\ \underline{2(Cr \longrightarrow Cr^{3+} + 3e^-)} & \text{(oxidation)} & +0.74 \text{ V} \\ 2Cr + 3Cu^{2+} \longrightarrow 2Cr^{3+} + 3Cu & E^0_{cell} = & +1.08 \text{ V} \end{array}$$

E^0_{cell} is positive, so we know this is the spontaneous reaction.

Cu^{2+} ions spontaneously oxidize metallic Cr to Cr^{3+} ions and are reduced to metallic Cu.

You should now work Exercise 50a.

▼ **PROBLEM-SOLVING TIP** *The Sign of E^0 Indicates Spontaneity*

For a reaction that is spontaneous at *standard conditions*, E^0 must be positive. A negative value of E^0_{cell} indicates that the reverse of the reaction written would be spontaneous at standard conditions.

An alternative approach to Example 21-4 would be to write the reaction in whichever direction we like, and then calculate E^0_{cell} for that reaction. If E^0_{cell} is positive, we have chosen the correct direction for the reaction; if it is negative, we must reverse the equation to get the spontaneous reaction. Suppose we ask whether Cr^{3+} ions oxidize metallic copper to Cu^{2+} ions at standard conditions. First we write the half-reaction for reduction of Cr^{3+} to Cr; then we write the half-reaction for oxidation of Cu to Cu^{2+} by reversing the appropriate reduction half-reaction. We balance the electron transfer and then add the two half-reactions and their potentials.

$$E^0$$

$$
\begin{array}{llr}
2(Cr^{3+} + 3e^- \longrightarrow Cr) & \text{(reduction)} & -0.74 \text{ V} \\
3(Cu \longrightarrow Cu^{2+} + 2e^-) & \text{(oxidation)} & -0.337 \text{ V} \\
\hline
2Cr^{3+} + 3Cu \longrightarrow 2Cr + 3Cu^{2+} & E^0_{cell} = & -1.08 \text{ V}
\end{array}
$$

The negative sign of E^0_{cell} tells us that the reaction as written is *not spontaneous*, but that the reverse reaction is. This is the same answer we obtained in Example 21-4.

21-17 STANDARD ELECTRODE POTENTIALS FOR OTHER HALF-REACTIONS

Platinum metal is often used as the inert electrode material. These two standard half-cells could be shown in shorthand notation as

$Pt/Fe^{3+}(1.0\ M),\ Fe^{2+}(1.0\ M)$ and
$Pt/Cr_2O_7^{2-}(1.0\ M),\ Cr^{3+}(1.0\ M)$

There are some half-cells in which oxidized and reduced species are in solution as ions in contact with inert electrodes. For example, the standard iron(III) ion/iron(II) ion half-cell contains 1.0 M concentrations of the two ions. It involves the following half-reaction.

$$Fe^{3+} + e^- \longrightarrow Fe^{2+} \qquad E^0 = +0.771 \text{ V}$$

The standard dichromate ($Cr_2O_7^{2-}$) ion/chromium(III) ion half-cell consists of a 1.0 M concentration of each of the two ions in contact with an inert electrode. The balanced half-reaction in acidic solution (1.0 M H^+) is

$$Cr_2O_7^{2-} + 14H^+ + 6e^- \longrightarrow 2Cr^{3+} + 7H_2O \qquad E^0 = +1.33 \text{ V}$$

Standard electrode potentials for some other reactions are given in Table 21-3 and in Appendix J. These potentials can be used like those of the electromotive series.

Table 21-3 *Standard Electrode Potentials for Selected Half-Cells*

Reduction Half-Reaction		Standard Electrode Potential E^0 (volts)
$Zn(OH)_4^{2-} + 2e^-$	$\longrightarrow Zn + 4OH^-$	-1.22
$Fe(OH)_2 + 2e^-$	$\longrightarrow Fe + 2OH^-$	-0.877
$2H_2O + 2e^-$	$\longrightarrow H_2 + 2OH^-$	-0.8277
$PbSO_4 + 2e^-$	$\longrightarrow Pb + SO_4^{2-}$	-0.356
$NO_3^- + H_2O + 2e^-$	$\longrightarrow NO_2^- + 2OH^-$	$+0.01$
$Sn^{4+} + 2e^-$	$\longrightarrow Sn^{2+}$	$+0.15$
$AgCl + e^-$	$\longrightarrow Ag + Cl^-$	$+0.222$
$Hg_2Cl_2 + 2e^-$	$\longrightarrow 2Hg + 2Cl^-$	$+0.27$
$O_2 + 2H_2O + 4e^-$	$\longrightarrow 4OH^-$	$+0.40$
$NiO_2 + 2H_2O + 2e^-$	$\longrightarrow Ni(OH)_2 + 2OH^-$	$+0.49$
$H_3AsO_4 + 2H^+ + 2e^-$	$\longrightarrow H_3AsO_3 + H_2O$	$+0.58$
$Fe^{3+} + e^-$	$\longrightarrow Fe^{2+}$	$+0.771$
$ClO^- + H_2O + 2e^-$	$\longrightarrow Cl^- + 2OH^-$	$+0.89$
$NO_3^- + 4H^+ + 3e^-$	$\longrightarrow NO + 2H_2O$	$+0.96$
$O_2 + 4H^+ + 4e^-$	$\longrightarrow 2H_2O$	$+1.23$
$Cr_2O_7^{2-} + 14H^+ + 6e^-$	$\longrightarrow 2Cr^{3+} + 7H_2O$	$+1.33$
$MnO_4^- + 8H^+ + 5e^-$	$\longrightarrow Mn^{2+} + 4H_2O$	$+1.51$
$PbO_2 + SO_4^{2-} + 4H^+ + 2e^-$	$\longrightarrow PbSO_4 + 2H_2O$	$+1.685$

EXAMPLE 21-5 *Predicting the Direction of Reactions*

In an acidic solution at standard conditions, will tin(IV) ions, Sn^{4+}, oxidize gaseous nitrogen oxide, NO, to nitrate ions, NO_3^-, or will NO_3^- oxidize Sn^{2+} to Sn^{4+} ions?

Plan

The NO_3^-/NO reduction half-reaction has the more positive E^0 value, so we write the Sn^{4+}/Sn^{2+} half-reaction as an oxidation. We balance the electron transfer and add the two half-reactions to obtain the equation for the *spontaneous* reaction. Then we add the half-reaction potentials to obtain the overall cell potential.

Solution

$$
\begin{array}{ll}
 & E^0 \\
\hline
2(NO_3^- + 4H^+ + 3e^- \longrightarrow NO + 2H_2O) & +0.96 \text{ V} \\
3(Sn^{2+} \longrightarrow Sn^{4+} + 2e^-) & -0.15 \text{ V} \\
\hline
2NO_3^- + 8H^+ + 3Sn^{2+} \longrightarrow 2NO + 4H_2O + 3Sn^{4+} \quad E^0_{cell} = +0.81 \text{ V}
\end{array}
$$

E^0_{cell} is positive for this reaction, so

nitrate ions spontaneously oxidize tin(II) ions to tin(IV) ions and are reduced to nitrogen oxide in acidic solution.

You should now work Exercises 51, 52, and 53.

▼ **PROBLEM-SOLVING TIP** *A Common Error in E^0_{cell} Calculations*

Remember the italicized warning in step 4 of the procedure set out in Section 21-16. *We do not multiply the potentials by the numbers used to balance the electron transfer!* This is a very common error.

▼ **PROBLEM-SOLVING TIP** *Remember What We Mean by Standard Conditions*

A reaction at *standard conditions* involves the following.

1. Any solution that takes part in the reaction is at a concentration of exactly 1 M.

2. Any gas that takes part in the reaction is at a pressure of exactly 1 atm.

3. Any other substance that takes part in the reaction is *pure*.

4. The temperature is standard thermodynamic temperature, 25°C, unless stated otherwise.

When we say "takes part in the reaction" we mean either as a reactant or as a product. When one or more of these conditions is not satisfied, we must adjust our calculations for nonstandard conditions. We shall learn how to do this in Section 21-20.

Now that we know how to use standard reduction potentials, let us use them to explain the reaction that occurs in the electrolysis of aqueous NaCl. The first two electrolytic cells we considered involved *molten* NaCl and *aqueous* NaCl (Sections 21-3 and 21-4). There was no doubt that in molten NaCl metallic Na would be produced by reduction of Na^+, and gaseous Cl_2 would be produced by oxidation of Cl^-. But we found that in aqueous

NaCl, H_2O, rather than Na^+, was reduced. This is consistent with the less negative reduction potential of H_2O, compared with Na^+.

$$
\begin{array}{ll}
 & E^0 \\
\hline
2H_2O + 2e^- \longrightarrow H_2 + 2OH^- & -0.828 \text{ V} \\
Na^+ + e^- \longrightarrow Na & -2.714 \text{ V}
\end{array}
$$

The more easily reduced species, H_2O, is reduced.

Electrode potentials measure only the relative *thermodynamic* likelihood for various half-reactions. In practice kinetic factors can complicate matters. For instance, sometimes the electrode process is limited by the rate of diffusion of dissolved species to or from the electrode surface. At some cathodes, the rate of electron transfer from the electrode to a reactant is the rate-limiting step, and a higher voltage (called *overvoltage*) must be applied to accomplish the reduction. As a result of these factors, a half-reaction that is *thermodynamically* more favorable than some other process still might not occur at a significant rate. In the electrolysis of NaCl(aq), Cl^- is oxidized to Cl_2 gas (-1.360 V), instead of H_2O being oxidized to form O_2 gas (-1.229 V), because of the overvoltage of O_2 on Pt, the inert electrode.

21-18 CORROSION

Ordinary **corrosion** is the redox process by which metals are oxidized by oxygen, O_2, in the presence of moisture. There are other kinds, but this is the most common. The problem of corrosion and its prevention are of both theoretical and practical interest. Corrosion is responsible for the loss of billions of dollars annually in metal products. The mechanism of corrosion has been studied extensively. It is now known that the oxidation of metals occurs most readily at points of strain (where the metals are most "active"). Thus, a steel nail, which is mostly iron (Section 22-7), first corrodes at the tip and head (Figure 21-12). A bent nail corrodes most readily at the bend.

A point of strain in a steel object acts as an anode where the iron is oxidized to iron(II) ions, and pits are formed (Figure 21-13).

$$ Fe \longrightarrow Fe^{2+} + 2e^- \qquad \text{(oxidation, anode)} $$

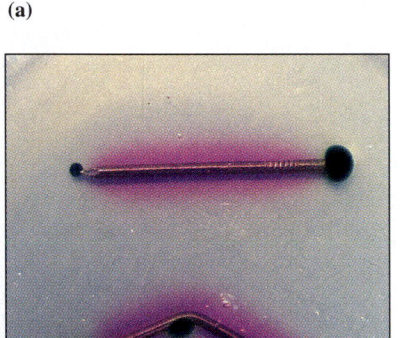

(a)

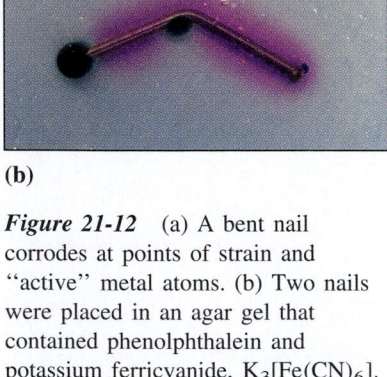

(b)

Figure 21-12 (a) A bent nail corrodes at points of strain and "active" metal atoms. (b) Two nails were placed in an agar gel that contained phenolphthalein and potassium ferricyanide, $K_3[Fe(CN)_6]$. As the nails corroded they produced Fe^{2+} ions at each end and at the bend. Fe^{2+} ions react with $[Fe(CN)_6]^{3-}$ ions to form $Fe_3[Fe(CN)_6]_2$, an intensely blue-colored compound. The rest of each nail is the cathode, at which water is reduced to H_2 and OH^- ions. The OH^- ions turn phenolphthalein pink.

Figure 21-13 The corrosion of iron. Pitting appears at the anodic region, where iron metal is oxidized to Fe^{2+}. Rust appears at the cathodic region.

Overall process: $2Fe(s) + \frac{3}{2}O_2(aq) + xH_2O(\ell) \rightarrow Fe_2O_3 \cdot xH_2O(s)$

$Fe \rightarrow Fe^{2} + 2e^-$
Oxidation

$O_2 + 2H_2O + 4e^- \rightarrow 4OH^-$
Reduction

The electrons produced then flow through the nail to areas exposed to O_2. These act as cathodes where oxygen is reduced to hydroxide ions, OH^-.

$$O_2 + 2H_2O + 4e^- \longrightarrow 4OH^- \qquad \text{(reduction, cathode)}$$

At the same time, the Fe^{2+} ions migrate through the moisture to the surface. The overall reaction is obtained by balancing the electron transfer and adding the two half-reactions.

$$\begin{array}{ll} 2(Fe \longrightarrow Fe^{2+} + 2e^-) & \text{(oxidation, anode)} \\ O_2 + 2H_2O + 4e^- \longrightarrow 4OH^- & \text{(reduction, cathode)} \\ \hline 2Fe + O_2 + 2H_2O \longrightarrow 2Fe^{2+} + 4OH^- & \text{(net reaction)} \end{array}$$

The Fe^{2+} ions can migrate from the anode through the solution toward the cathode region, where they combine with OH^- ions to form iron(II) hydroxide. Iron is further oxidized by O_2 to the $+3$ oxidation state. The material we call rust is a complex hydrated form of iron(III) oxides and hydroxides with variable water composition; it can be represented as $Fe_2O_3 \cdot xH_2O$. The overall reaction for the rusting of iron is

$$2Fe(s) + \tfrac{3}{2}O_2(aq) + xH_2O(\ell) \longrightarrow Fe_2O_3 \cdot xH_2O(s)$$

21-19 CORROSION PROTECTION

There are several methods for protecting metals against corrosion. The most widely used are

1. Plating the metal with a thin layer of a less easily oxidized metal

2. Connecting the metal directly to a "sacrificial anode," a piece of another metal that is more active and therefore preferentially oxidized

3. Allowing a protective film, such as a metal oxide, to form naturally on the surface of the metal

4. Galvanizing, or coating steel with zinc, a more active metal

5. Applying a protective coating such as paint

Rust is a serious economic problem.

Compare the potentials for the *oxidation* half-reactions (the reverse of tabulated reduction half-reactions) to see which metal is more easily oxidized.

	$E^0_{\text{oxidation}}$
$Fe \longrightarrow Fe^{2+} + 2e^-$	$+0.44$ V
$Sn \longrightarrow Sn^{2+} + 2e^-$	$+0.14$ V
$Cu \longrightarrow Cu^{2+} + 2e^-$	-0.337 V

Protection against corrosion. (Left) Galvanized objects are steel coated with zinc. (Right) Steel is plated with chromium for appearance as well as protection against corrosion.

CHEMISTRY IN USE

Our Daily Lives

Electrochemical Errors

The deterioration of the Statue of Liberty and the damage done at the Three Mile Island and Chernobyl nuclear facilities are just a few of the major problems that have resulted from ignorance about chemical reactivity.

When originally constructed over one hundred years ago, the Statue of Liberty had a 200,000-pound outer copper skin supported by a framework of 2000 iron bars. First, oxygen in the air oxidized the copper skin to form copper oxide. In a series of reactions, iron (the more active metal) then reduced the Cu^{2+} ions in copper oxide.

$$2Fe + 3Cu^{2+} \longrightarrow 2Fe^{3+} + 3Cu$$

Over the years, the supporting iron frame was reduced to less than half its original thickness; this made necessary the repairs done to the statue before the celebration of its 100th birthday on July 4, 1986.

Two major nuclear power plant accidents, one at Three Mile Island near Harrisburg, Pennsylvania, in 1979, and the other at Chernobyl in Ukraine in 1986, were also unexpected consequences of chemical reactivity. In each case, cooling pump failures sent temperatures soaring above 340°C. Like aluminum, zirconium (used in building the reactors) forms a protective oxide coating that protects it from further reactions. However, that protective coating breaks down at high temperatures. Without its protective coating, zirconium reacts with steam.

$$Zr(s) + 2H_2O(g) \longrightarrow ZrO_2(s) + 2H_2(g)$$

At Three Mile Island, this displacement reaction produced a 1000-cubic foot bubble of hydrogen gas. Because hydrogen is easily ignited by a spark, the nuclear power plant was in real danger of a complete meltdown until the hydrogen could be removed.

During the Middle Ages (~400–1400 AD), another displacement reaction completely misled alchemists into foolishly pursuing a philosopher's stone that was believed to have the power to turn base metals such as iron and lead into more precious metals such as silver and gold. The alchemists' ignorance of standard electrode potentials led them to believe that they had turned iron into a more precious metal when they inserted an iron rod into a blue copper(II) sulfate solution. In fact, the following displacement reaction had occurred, plating shiny copper metal onto the iron rod.

$$Fe(s) + 3Cu^{2+}(aq) \longrightarrow 2Fe^{3+}(aq) + Cu(s)$$

In the 1960s and 1970s, some automobile manufacturers showed their ignorance of basic electrochemistry by building cars with aluminum water pumps and aluminum engine heads attached to cast-iron engine blocks. These water pumps often leaked and the engine heads quickly deteriorated. These problems occurred as the more active aluminum reacted with iron(II) oxide (formed when the iron engine reacted with atmospheric oxygen).

$$Al + Fe^{3+} \longrightarrow Al^{3+} + Fe$$

Some dentists have made similar mistakes by placing gold caps over teeth that are adjacent to existing fillings. The slightly oxidized gold can react with a dental amalgam filling (an alloy of silver, tin, copper, and mercury). As the dental amalgam is oxidized, it dissolves in saliva to produce a persistent metallic taste in the patient's mouth.

When plumbers connect galvanized pipes (iron pipes coated with zinc) to copper pipes, copper ions oxidize the zinc coating and expose the underlying iron, allowing it to rust. The displacement reaction that occurs is

$$Zn + Cu^{2+} \longrightarrow Zn^{2+} + Cu$$

Once the zinc coating has been punctured on an iron pipe, oxidation of the iron pipes occurs rapidly because iron is a more active metal than copper.

It is important to keep in mind that a variety of other reactions probably take place in addition to the above displacement reactions. For example, less active metals (such as copper) can conduct electrons from the metals being oxidized to oxidizing agents (such as oxygen or the oxide of nitrogen and sulfur) that are present in the atmosphere. Oxygen plays an important role in all of these displacement examples.

Ronald DeLorenzo
Middle Georgia College

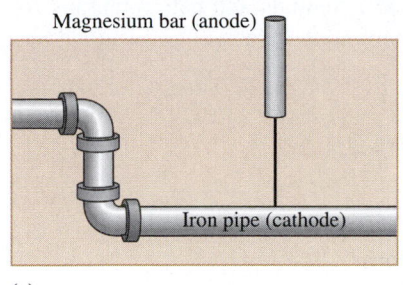

Magnesium bar (anode)

Iron pipe (cathode)

(a)

(b)

Figure 21-14 (a) Cathodic protection of buried iron pipe. A magnesium or zinc bar is oxidized instead of the iron. The "sacrificial" anode eventually must be replaced. (b) Cathodic protection of a ship's hull. The small yellow horizontal strips are blocks of titanium (coated with platinum) that are attached to the ship's hull. The hull is steel (mostly iron). When the ship is in salt water, the titanium blocks become the anode, and the hull the cathode, in a voltaic cell. Because oxidation always occurs at the anode, the ship's hull (the cathode) is protected from oxidation (corrosion).

The thin layer of tin on tin-plated steel cans is less easily oxidized than iron, and it protects the steel underneath from corrosion. It is deposited either by dipping the can into molten tin or by electroplating. Copper is also less active than iron (see Table 21-2). It is sometimes deposited by electroplating to protect metals when food is not involved. Whenever the layer of tin or copper is breached, the iron beneath it corrodes even more rapidly than it would without the coating, because of the adverse electrochemical cell that is set up.

Figure 21-14a shows an iron pipe connected to a strip of magnesium, a more active metal, to protect the iron from oxidation. The magnesium is preferentially oxidized. It is called a "sacrificial anode." Similar methods are used to protect bridges and ships' hulls from corrosion. Other active metals, such as zinc, are also used as sacrificial anodes.

Aluminum, a very active metal, reacts rapidly with O_2 from the air to form a surface layer of aluminum oxide, Al_2O_3, that is so thin that it is transparent. This very tough, hard substance is inert toward oxygen, water, and most other corrosive agents in the environment. In this way, objects made of aluminum form their own protective layers and need not be treated further to inhibit corrosion.

Compare the potentials for the *oxidation* half-reactions to see which metal is more easily oxidized.

		$E^0_{\text{oxidation}}$
Fe $\longrightarrow$ Fe^{2+} + 2e^-		+0.44 V
Zn $\longrightarrow$ Zn^{2+} + 2e^-		+0.763 V
Mg $\longrightarrow$ Mg^{2+} + 2e^-		+2.37 V

Acid rain endangers structural aluminum by dissolving this Al_2O_3 coating.

EFFECT OF CONCENTRATIONS (OR PARTIAL PRESSURES) ON ELECTRODE POTENTIALS

21-20 THE NERNST EQUATION

Standard electrode potentials, designated E^0, refer to standard state conditions. These standard state conditions are one-molar solutions for ions, one atmosphere pressure for gases, and all solids and liquids in their standard states at 25°C. (Remember that we refer to *thermodynamic* standard state conditions, and not standard temperature and pressure as in gas law calculations.) As any of the standard cells described earlier operates, and concentrations or pressures of reactants change, the observed cell voltage drops. Similarly, cells constructed with solution concentrations different from one molar, or gas pressures different from one atmosphere, cause the corresponding potentials to deviate from standard electrode potentials.

Walther Nernst (1864–1941) was a German scientist who made extensive contributions in thermodynamics and electrochemistry.

The **Nernst equation** is used to calculate electrode potentials and cell potentials for concentrations and partial pressures other than standard state values.

In this equation the expression following the minus sign represents how much the *nonstandard* conditions cause the electrode potential to deviate from its standard value, E^0. The Nernst equation is normally presented in terms of base-10 logarithms.

$$E = E^0 - \frac{2.303\,RT}{nF} \log Q$$

where

E = potential under the **nonstandard** conditions
E^0 = **standard** potential
R = gas constant, 8.314 J/mol $\cdot$ K
T = absolute temperature in K
n = number of moles of electrons transferred in the reaction
F = faraday, $96{,}485$ C/mol e^- $\times$ 1 J/(V $\cdot$ C)
   = $96{,}485$ J/V $\cdot$ mol e^-
Q = reaction quotient

The reaction quotient, Q, was introduced in Section 17-4. It involves a ratio of concentrations or pressures of products to those of reactants, each raised to the power indicated by the coefficient in the balanced equation. The Q expression that is used in the Nernst equation is the thermodynamic reaction quotient; it can include *both* concentrations and pressures. Substituting these values into the Nernst equation at 25°C gives

At 25°C, the value of $\dfrac{2.303\,RT}{F}$ is 0.0592; at any other temperature, this term must be recalculated. Can you show that this term has the units V $\cdot$ mol?

$$E = E^0 - \frac{0.0592}{n} \log Q \qquad \text{(\textit{Note:} in terms of base-10 log)}$$

In general half-reactions for standard reduction potentials are written

$$x\,\text{Ox} + ne^- \longrightarrow y\,\text{Red}$$

"Red" refers to the reduced species, and "Ox" to the oxidized species. The Nernst equation for any *cathode* half-cell (*reduction* half-reaction) is

$$E = E^0 - \frac{0.0592}{n} \log \frac{[\text{Red}]^y}{[\text{Ox}]^x} \qquad \text{(reduction half-reaction)}$$

For the familiar half-reaction involving metallic zinc and zinc ions,

$$\text{Zn}^{2+} + 2e^- \rightleftharpoons \text{Zn} \qquad E^0 = -0.763 \text{ V}$$

Metallic Zn is a pure solid, so its concentration does not appear in Q.

the corresponding Nernst equation is

$$E = E^0 - \frac{0.0592}{2} \log \frac{1}{[\text{Zn}^{2+}]} \qquad \text{(for reduction)}$$

We substitute the E^0 value into the equation to obtain

$$E = -0.763 \text{ V} - \frac{00592}{2} \log \frac{1}{[\text{Zn}^{2+}]}$$

EXAMPLE 21-6 *The Nernst Equation*

Calculate the potential, E, for the Fe^{3+}/Fe^{2+} electrode when the concentration of Fe^{2+} is five times that of Fe^{3+}

Plan

The Nernst equation lets us calculate potentials for concentrations other than one molar. The tabulation of standard reduction potentials gives us the value of E^0 for the reduction half-reaction. We use the balanced half-reaction and the given concentration ratio to calculate the value of Q. Then we substitute this into the Nernst equation with n equal to the number of moles of electrons involved in the half-reaction.

Solution

The reduction half-reaction is

$$Fe^{3+} + e^- \longrightarrow Fe^{2+} \qquad E^0 = +0.771 \text{ V}$$

We are told that the concentration of Fe^{2+} is five times that of Fe^{3+}, or $[Fe^{2+}] = 5[Fe^{3+}]$. Calculating the value of Q,

$$Q = \frac{[\text{Red}]^y}{[\text{Ox}]^x} = \frac{[Fe^{2+}]}{[Fe^{3+}]} = \frac{5[Fe^{3+}]}{[Fe^{3+}]} = 5$$

The balanced half-reaction shows one mole of electrons, or $n = 1$. Putting values into the Nernst equation,

$$E = E^0 - \frac{0.0592}{n} \log Q = +0.771 - \frac{0.0592}{1} \log 5 = (+0.771 - 0.041) \text{ V}$$

$$E = +0.730 \text{ V}$$

EXAMPLE 21-7 *The Nernst Equation*

Calculate E for the Fe^{3+}/Fe^{2+} electrode when the Fe^{3+} concentration is five times the Fe^{2+} concentration.

Plan

The approach is the same as for Example 21-6.

Solution

$$Q = \frac{[\text{Red}]^y}{[\text{Ox}]^x} = \frac{[Fe^{2+}]}{[Fe^{3+}]} = \frac{[Fe^{2+}]}{5[Fe^{2+}]} = \frac{1}{5} = 0.200$$

$$E = +0.771 - \frac{0.0592}{1} \log (0.200) = +0.771 - 0.0592(-0.699)$$

$$= (+0.771 + 0.041) \text{ V}$$

$$= +0.812 \text{ V} \qquad \text{(for reduction)}$$

We see that the correction factor, +0.041 volt, differs from that in Example 21-6 only in sign.

EXAMPLE 21-8 *The Nernst Equation*

Calculate the potential of the chlorine–chloride ion, Cl_2/Cl^-, electrode when the partial pressure of Cl_2 is 10.0 atm and $[Cl^-] = 1.00 \times 10^{-3} M$.

Plan

The approach is the same as for the preceding two examples. Now the Nernst equation involves the molar concentration of a substance in solution (Cl^-) and the pressure of a gaseous component (Cl_2); for this half-reaction, $n = 2$.

Solution

The half-reaction and standard reduction potential are

$$Cl_2 + 2e^- \longrightarrow 2Cl^- \qquad E^0 = +1.360 \text{ V}$$

The appropriate Nernst equation is

$$E = +1.360 - \frac{0.0592}{2} \log \frac{[Cl^-]^2}{P_{Cl_2}}$$

Substituting $[Cl^-] = 1.00 \times 10^{-3} M$ and $P_{Cl_2} = 10.0$ atm into the equation gives

> Remember that, in evaluating Q in the Nernst equation, (1) molar concentrations are used for dissolved species, and (2) partial pressures of gases are expressed in atmospheres.

$$E = +1.360 - \frac{0.0592}{2} \log \frac{(1.00 \times 10^{-3})^2}{10.0} = +1.360 - \frac{0.0592}{2} \log (1.00 \times 10^{-7})$$

$$= +1.360 - \frac{0.0592}{2}(-7.00) = [+1.360 - (-0.207)] \text{ V} = \boxed{+1.567 \text{ V}}$$

You should now work Exercises 69 and 70.

The Nernst equation can be applied to balanced equations for redox reactions, just as we applied it to equations for half-reactions.

EXAMPLE 21-9 *The Nernst Equation*

A cell is constructed at 25°C as follows. One half-cell consists of the Fe^{3+}/Fe^{2+} couple in which $[Fe^{3+}] = 1.00 M$ and $[Fe^{2+}] = 0.100 M$; the other involves the MnO_4^-/Mn^{2+} couple in acidic solution in which $[MnO_4^-] = 1.00 \times 10^{-2} M$, $[Mn^{2+}] = 1.00 \times 10^{-4} M$, and $[H^+] = 1.00 \times 10^{-3} M$. (a) Find the electrode potential for each half-cell with these concentrations, and (b) calculate the overall cell potential.

Plan

(a) We can apply the Nernst equation to find the reduction potential of each half-cell with the stated concentrations. (b) As in Section 21-16, we write the half-reaction with the more positive potential (*after* correction) along with its potential. We reverse the other half-reaction and change the sign of its E value. We balance the electron transfer, and then add the half-reactions and their potentials to find the overall cell potential.

Solution

(a) For the MnO_4^-/Mn^{2+} half-cell *as a reduction,*

$$MnO_4^- + 8H^+ + 5e^- \longrightarrow Mn^{2+} + H_2O \qquad E^0 = +1.51 \text{ V}$$

$$E = E^0 - \frac{0.0592}{n} \log \frac{[Mn^{2+}]}{[MnO_4^-][H^+]^8}$$

$$= +1.51 \text{ V} - \frac{0.0592}{5} \log \frac{1.00 \times 10^{-4}}{(1.00 \times 10^{-2})(1.00 \times 10^{-3})^8}$$

$$= +1.51 \text{ V} - \frac{0.0592}{5} \log (1.00 \times 10^{22}) = +1.51 \text{ V} - \frac{0.0592}{5}(22.0)$$

$$= \boxed{+1.25 \text{ V}}$$

For the Fe^{3+}/Fe^{2+} half-cell *as a reduction,*

$$Fe^{3+} + e^- \longrightarrow Fe^{2+} \qquad E^0 = +0.771 \text{ V}$$

$$E = E^0 - \frac{0.0592}{n} \log \frac{[Fe^{2+}]}{[Fe^{3+}]} = +0.771 \text{ V} - \frac{0.0592}{1} \log \frac{0.100}{1.00}$$

$$= +0.771 \text{ V} - \frac{0.0592}{1} \log (0.100) = +0.771 \text{ V} - \frac{0.0592}{1}(-1.0)$$

$$= \boxed{+0.83 \text{ V}}$$

The corrected potential for the MnO_4^-/Mn^{2+} half-cell is greater than that for the Fe^{3+}/Fe^{2+} half-cell, so we reverse the latter, balance the electron transfer, and add.

	E (corrected)
$MnO_4^- + 8H^+ + 5e^- \longrightarrow Mn^{2+} + 4H_2O$	$+1.25$ V
$5(Fe^{2+} \longrightarrow Fe^{3+} + e^-)$	-0.83 V
$MnO_4^- + 8H^+ + 5Fe^{2+} \longrightarrow Mn^{2+} + 4H_2O + 5Fe^{3+}$	$E_{cell} = 0.42$ V

If we prefer, we can find the cell potential for the nonstandard cell by first finding E^0_{cell} for the overall standard cell reaction, and then using the Nernst equation to correct for nonstandard concentrations.

EXAMPLE 21-10 *The Nernst Equation*

Apply the Nernst equation to the *overall* cell reaction to determine the cell potential for the cell described in Example 21-9.

Plan
First we determine the overall reaction and its *standard* cell potential, E^0_{cell}, as usual. Then we apply the Nernst equation to the overall cell.

Solution
Determining the standard cell potential, E^0_{cell},

	E^0
$MnO_4^- + 8H^+ + 5e^- \longrightarrow Mn^{2+} + 4H_2O$	$+1.51$ V
$5(Fe^{2+} \longrightarrow Fe^{3+} + e^-)$	$-(+0.771)$ V
$MnO_4^- + 8H^+ + 5Fe^{2+} \longrightarrow Mn^{2+} + 4H_2O + 5Fe^{3+}$	$E^0_{cell} = +0.74$ V

In the overall reaction $n = 5$; putting values into the Nernst equation,

$$E_{cell} = E^0_{cell} - \frac{0.0592}{5} \log \frac{[Mn^{2+}][Fe^{3+}]^5}{[MnO_4^-][H^+]^8[Fe^{2+}]^5}$$

$$= +0.74 \text{ V} - \frac{0.0592}{5} \log \frac{(1.00 \times 10^{-4})(1.00)^5}{(1.00 \times 10^{-2})(1.00 \times 10^{-3})^8(0.100)^5}$$

$$= +0.74 \text{ V} - \frac{0.0592}{5} \log (1.00 \times 10^{+27})$$

$$= +0.74 \text{ V} - \frac{0.0592}{5}(27.0) = (+0.74 - 0.32) \text{ V} = \boxed{+0.42 \text{ V}}$$

You should now work Exercises 72 and 76.

The method illustrated in Example 21-10, applying the Nernst equation to the *overall* cell reaction, usually involves less calculation than correcting the separate half-reactions as in Example 21-9. We interpret our results as follows: The positive cell potential in Examples 21-9 and 21-10 tells us that the cell reaction is spontaneous in the direction written, for the concentrations given. If the resulting cell potential were negative, the *reverse* reaction would be spontaneous at those concentrations. We could then reverse the equation for the overall cell reaction and change the sign of its potential to describe the spontaneous operation of the cell.

The reaction in Examples 21-9 and 21-10 is thermodynamically favored (spontaneous) under the stated conditions, with a potential of +0.42 volt *when the cell starts operation.* As the cell discharges and current flows, the product concentrations, $[Mn^{2+}]$ and $[Fe^{3+}]$, increase. At the same time, reactant concentrations, $[MnO_4^-]$, $[H^+]$, and $[Fe^{2+}]$, decrease. This increases $\log Q$, so the correction factor becomes more negative. Thus, the overall E_{cell} *decreases* (the reaction becomes less favorable). Eventually the cell potential approaches zero (equilibrium), and the cell "runs down."

> ▼ **PROBLEM-SOLVING TIP**   *Be Careful of the Value of n*
>
> How do you know what value of n to use? Remember that n must be the number of moles of electrons transferred in the *balanced* equation for the process to which you apply the Nernst equation.
>
> 1. For a *half-reaction, n* represents the number of moles of electrons in that half-reaction. In Example 21-9 we applied the Nernst equation to each half-reaction separately, so we used $n = 5$ for the half-reaction
>
> $$MnO_4^- + 8H^+ + 5e^- \longrightarrow Mn^{2+} + H_2O$$
>
> and we used $n = 1$ for the half-reaction
>
> $$Fe^{3+} + e^- \longrightarrow Fe^{2+}$$
>
> 2. For an *overall reaction, n* represents the total number of moles of electrons transferred. In Example 21-10 we applied the Nernst equation to an *overall* reaction in which 5 moles of electrons were transferred from five moles of Fe^{2+} to one mole of MnO_4^-, so we used the value $n = 5$.

Using Electrochemical Cells to Determine Concentrations

We can apply the ideas of this section to *measure* the voltage of a cell and then use the Nernst equation to solve for an unknown concentration. The following example illustrates such an application.

EXAMPLE 21-11   *The Nernst Equation*

This cell is similar to the zinc–hydrogen cell that we discussed in Section 21-13, except that the hydrogen concentration is not (necessarily) 1.00 *M*.

We construct an electrochemical cell at 25°C as follows. One half-cell is a standard Zn^{2+}/Zn cell, i.e., a strip of zinc immersed in a 1.00 *M* Zn^{2+} solution; the other is a *nonstandard* hydrogen electrode in which a platinum electrode is immersed in a solution of *unknown* hydrogen ion concentration with gaseous hydrogen bubbling through it at a pressure of 1.000 atm. The observed cell voltage is 0.522 V. (a) Calculate the value of the reaction quotient Q. (b) Calculate $[H^+]$ in the second half-cell. (c) Determine the pH of the solution in the second half-cell.

Plan

We saw in Section 21-13 that the zinc–hydrogen cell operated with oxidation at the zinc electrode and reduction at the hydrogen electrode, with a *standard* cell potential of 0.763 V.

$$\text{overall:} \quad Zn + 2H^+ \longrightarrow Zn^{2+} + H_2 \qquad E^0_{cell} = 0.763 \text{ V}$$

(a) We rearrange the Nernst equation to solve for the reaction quotient, Q, from the measured cell voltage and $n = 2$. (b) We substitute concentrations and partial pressures in the expression for Q. Then we can solve for the only unknown, $[H^+]$. (c) The pH can be determined from the $[H^+]$ determined in part (a).

Solution

(a)
$$E_{cell} = E^0_{cell} - \frac{0.0592}{n} \log Q$$

Substituting and solving for Q,

$$0.522 \text{ V} = 0.763 \text{ V} - \frac{0.0592}{2} \log Q$$

$$\frac{0.0592 \text{ V}}{2} \log Q = (0.763 - 0.522) \text{ V} = 0.241 \text{ V}$$

$$\log Q = \frac{(2)(0.241 \text{ V})}{0.0592 \text{ V}} = 8.14$$

$$Q = 10^{8.14} = \boxed{1.4 \times 10^8}$$

(b) We write the expression for Q from the balanced overall equation, and solve for $[H^+]$.

$$Q = \frac{[Zn^{2+}]P_{H_2}}{[H^+]^2}$$

$$[H^+]^2 = \frac{[Zn^{2+}]P_{H_2}}{Q} = \frac{(1.00)(1.00)}{1.4 \times 10^8} = 7.1 \times 10^{-9}$$

$$[H^+] = \boxed{8.4 \times 10^{-5} M}$$

(c)
$$pH = -\log [H^+] = -\log (8.4 \times 10^{-5}) = \boxed{4.08}$$

You should now work Exercises 73 and 74.

Concentration Cells

As we have seen, different concentrations of ions in a half-cell result in different half-cell potentials. We can use this idea to construct a **concentration cell,** in which both half-cells are composed of the same species, but in different ion concentrations. Suppose we set up such a cell using the Cu^{2+}/Cu half-cell that we introduced in Section 21-10. We put copper electrodes into two aqueous solutions, one that is $0.10 \ M$ $CuSO_4$ and another that is $1.00 \ M \ CuSO_4$. To complete the cell construction, we connect the two electrodes with a wire and join the two solutions with a salt bridge as usual (Figure 21-15). Now the relevant reduction half-reaction in either half-cell is

$$Cu^{2+} + 2e^- \longrightarrow Cu \qquad E^0 = +0.337 \text{ V}$$

The Nernst equation lets us evaluate the reduction potential for the possible half-reaction in each half-cell.

In the Half-Cell with $[Cu^{2+}] = 1.00 \ M$

The concentration is *standard,* so the reduction potential of this half-cell is equal to E^0.

$$Cu^{2+}(1.00 \ M) + 2e^- \longrightarrow Cu \qquad E = E^0 = +0.337 \text{ V}$$

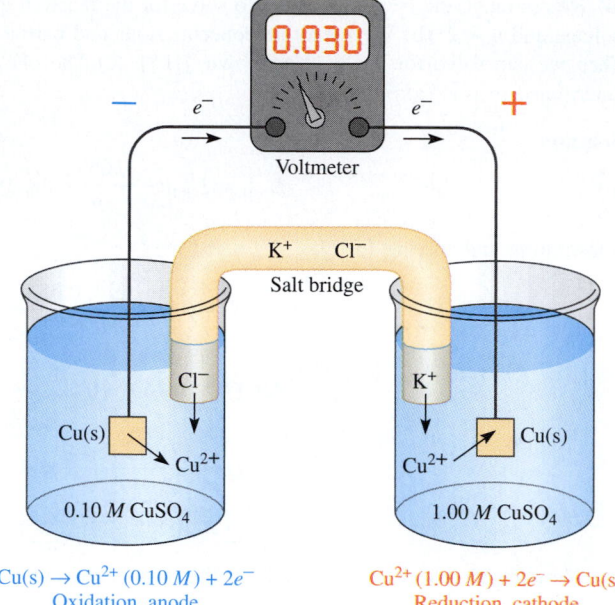

Figure 21-15   The concentration cell $Cu|Cu^{2+}(0.10\ M)\|Cu^{2+}(1.00\ M)|Cu$. The overall reaction lowers the $[Cu^{2+}]$ concentration in the more concentrated solution and increases it in the more dilute solution.

$Cu(s) \rightarrow Cu^{2+}(0.10\ M) + 2e^-$
Oxidation, anode

$Cu^{2+}(1.00\ M) + 2e^- \rightarrow Cu(s)$
Reduction, cathode

In the Half-Cell with $[Cu^{2+}] = 0.10\ M$

$$E = E^0 - \frac{0.0592}{n} \log \frac{1}{[Cu^{2+}]} = +0.337 - \frac{0.0592}{2} \log \frac{1}{0.10} = +0.307\ V$$

$$Cu^{2+}(0.10\ M) + 2e^- \longrightarrow Cu \qquad E = +0.307\ V$$

The more concentrated half-cell has the higher potential, so its half-reaction proceeds as written, i.e., as a reduction. In the more dilute half-cell, Cu is oxidized to Cu^{2+}.

		E	
$Cu^{2+}(1.00\ M) + 2e^- \longrightarrow Cu$		$+0.337\ V$	(red., cathode)
$Cu \longrightarrow Cu^{2+}(0.10\ M) + 2e^-$		$-0.307\ V$	(ox., anode)
$Cu^{2+}(1.00\ M) \longrightarrow Cu^{2+}(0.10\ M)$		$E_{cell} = 0.030\ V$	(overall rxn)

The overall cell potential could be calculated more directly by applying the Nernst equation to the overall cell reaction. We must first find E^0, the standard cell potential *at standard concentrations;* because the same electrode and the same type of ions are involved in both half-cells, this E^0 is always zero. Thus,

$$E_{cell} = E^0 - \frac{0.0592}{n} \log \frac{[\text{dilute solution}]}{[\text{concentrated solution}]}$$

$$= 0 - \frac{0.0592}{2} \log \frac{0.10}{1.00} = 0.30\ V$$

As the reaction proceeds, $[Cu^{2+}]$ decreases in the more concentrated half-cell and increases in the more dilute half-cell until the two concentrations are equal; at that point $E_{cell} = 0$ and equilibrium has been reached. This equilibrium $[Cu^{2+}]$ is the same concentration that would have been formed if we had just mixed the two solutions directly to obtain a solution of intermediate concentration.

In any concentration cell the spontaneous reaction is always from more concentrated solution to more dilute solution.

Electrochemical procedures that use the principles illustrated here provide a convenient method for making many concentration measurements.

21-21 THE RELATIONSHIP OF E^0_{cell} TO ΔG^0 AND K

In Section 17-11 we studied the relationship between the standard Gibbs free energy change, ΔG^0, and the thermodynamic equilibrium constant, K.

$$\Delta G^0 = -RT \ln K \quad \text{or} \quad \Delta G^0 = -2.303RT \log K$$

There is also a simple relationship between ΔG^0 and the standard cell potential, E^0_{cell}, for a redox reaction (reactants and products in standard states).

$$\Delta G^0 = -nFE^0_{\text{cell}}$$

ΔG^0 can be thought of as the *negative of the maximum electrical work* that can be obtained from a redox reaction. In this equation, n is the number of moles of electrons transferred in the overall process (mol e^-/mol rxn), and F is the faraday, 96,485 J/V · mol e^-.

Combining these relationships for ΔG^0 gives the relationship between E^0_{cell} values and equilibrium constants.

$$\underbrace{-nFE^0_{\text{cell}}}_{\Delta G^0} = \underbrace{-2.303RT \log K}_{\Delta G^0} \quad \text{or} \quad \underbrace{-nFE^0_{\text{cell}}}_{\Delta G^0} = \underbrace{-RT \ln K}_{\Delta G^0}$$

After multiplying by -1, we can rearrange.

	Rearranging for E^0_{cell}	**Rearranging for K**
$nFE^0_{\text{cell}} = RT \ln K$	$E^0_{\text{cell}} = \dfrac{RT \ln K}{nF}$	$\ln K = \dfrac{nFE^0_{\text{cell}}}{RT}$
$nFE^0_{\text{cell}} = 2.303RT \log K$	$E^0_{\text{cell}} = \dfrac{2.303RT \log K}{nF}$	$\log K = \dfrac{nFE^0_{\text{cell}}}{2.303RT}$

If any one of the three quantities ΔG^0, K, and E^0_{cell} is known, the other two can be calculated using these equations. It is usually much easier to determine K for a redox reaction from electrochemical measurements than by measuring equilibrium concentrations directly, as described in Chapter 17. Keep in mind the following for all redox reactions *under standard state conditions*.

Forward Reaction	ΔG^0	K	E^0_{cell}	
spontaneous	$-$	>1	$+$	
at equilibrium	0	1	0	(all substances at *standard state conditions*)
nonspontaneous	$+$	<1	$-$	

EXAMPLE 21-12   *Calculation of* ΔG^0

Calculate the standard Gibbs free energy change, ΔG^0, in J/mol at 25°C for the following reaction from standard electrode potentials.

$$3Sn^{4+} + 2Cr \longrightarrow 3Sn^{2+} + 2Cr^{3+}$$

Plan

We evaluate the standard cell potential as before. Then we apply the relationship $\Delta G^0 = -nFE^0_{cell}$.

Solution

The standard reduction potential for the Sn^{4+}/Sn^{2+} couple is $+0.15$ volt; that for the Cr^{3+}/Cr couple is -0.74 volt. The equation for the reaction shows Cr being oxidized to Cr^{3+}, so the sign of the E^0 value for the Cr^{3+}/Cr couple is reversed. The overall reaction, the sum of the two half-reactions, has a cell potential equal to the sum of the two half-reaction potentials.

	E^0
$3(Sn^{4+} + 2e^- \longrightarrow Sn^{2+})$	$+0.15$ V
$2(Cr \longrightarrow Cr^{3+} + 3e^-)$	$-(-0.74$ V$)$
$3Sn^{4+} + 2Cr \longrightarrow 3Sn^{2+} + 2Cr^{3+}$	$E^0_{cell} = +0.89$ V

The positive value of E^0_{cell} indicates that the forward reaction is spontaneous.

$$\Delta G^0 = -nFE^0_{cell} = -\left(\frac{6 \text{ mol } e^-}{\text{mol rxn}}\right)\left(96{,}500 \frac{J}{V \cdot \text{mol } e^-}\right)(+0.89 \text{ V})$$

$$\Delta G^0 = \boxed{-5.2 \times 10^5 \text{ J/mol rxn} \quad \text{or} \quad -5.2 \times 10^2 \text{ kJ/mol rxn}}$$

You should now work Exercises 85 and 90.

EXAMPLE 21-13   *Calculation of* K

Calculate the value of the equilibrium constant, K, for the reaction in Example 21-12 at 25°C (a) by relating it to ΔG^0 and (b) by relating it to E^0_{cell}.

Plan

We substitute the values of ΔG^0 and E^0_{cell} from Example 21-12 into the following relationships and solve for K.

$$\Delta G^0 = -RT \ln K \quad \text{and} \quad \ln K = \frac{nFE^0_{cell}}{RT}$$

Solution

(a)
$$\Delta G^0 = -RT \ln K$$

$$-5.2 \times 10^5 \text{ J/mol} = -\left(8.314 \frac{J}{\text{mol} \cdot K}\right)(298 \text{ K}) \ln K$$

$$\ln K = \frac{-5.2 \times 10^5}{-(8.314)(298)} = +210$$

$$K = \boxed{2 \times 10^{91}}$$

(b)
$$\ln K = \frac{nFE^0_{cell}}{KT}$$

Substituting the known values gives a numerical expression for K.

Recall from Chapter 15 that ΔG^0 can be expressed in joules per *mole of reaction*. Here we ask for the number of joules of free energy change that corresponds to the reaction of 2 moles of chromium with 3 moles of tin(IV) to give 3 moles of tin(II) ions and 2 moles of chromium(III) ions.

The very large value of K is also consistent with the fact that the forward reaction is spontaneous. This tells us nothing about the speed with which the reaction would occur.

$$\ln K = \frac{(6)\left(96{,}500\ \dfrac{J}{V\cdot mol}\right)(+0.89\ V)}{\left(8.314\ \dfrac{J}{mol\cdot K}\right)(298\ K)} = +210$$

$$K = \boxed{2 \times 10^{91}}$$

This very large value of K tells us that the reaction goes essentially to completion. At equilibrium, very little Sn^{4+} is present.

$$K = \frac{[Cr^{3+}]^2[Sn^{2+}]^3}{[Sn^{4+}]^3} = 2 \times 10^{91}$$

EXAMPLE 21-14 Calculation of K

Calculate the value of the equilibrium constant, K, at 25°C for the following reaction.

$$2Cu + PtCl_6{}^{2-} \longrightarrow 2Cu^+ + PtCl_4{}^{2-} + 2Cl^-$$

Plan

This example combines the ideas we used in Examples 21-12 and 21-13.

Solution

First we find the appropriate half-reactions. Cu is oxidized to Cu^+, so we write the Cu^+/Cu couple as an oxidation and reverse the sign of its tabulated E^0 value. We balance the electron transfer and then add the half-reactions. The resulting E^0_{cell} value can be used to calculate the equilibrium constant, K, for the reaction *as written*.

$2(Cu \longrightarrow Cu^+ + e^-)$	$-(+0.521\ V)$
$PtCl_6{}^{2-} + 2e^- \longrightarrow PtCl_4{}^{2-} + 2Cl^-$	$+0.68\ V$
$2Cu + PtCl_6{}^{2-} \longrightarrow 2Cu^+ + PtCl_4{}^{2-} + 2Cl^-$	$E^0_{cell} = +0.16\ V$

As the problem is stated, we must keep the equation as written. Therefore, we must accept either a positive or a negative value of E^0_{cell}. A negative value of E^0_{cell} would lead to $K < 1$.

Then we calculate K.

$$\ln K = \frac{nFE^0_{cell}}{RT} = \frac{(2)(96.5 \times 10^3\ J/V\cdot mol)(+0.16\ V)}{(8.314\ J/mol\cdot K)(298\ K)} = 12.4$$

$$K = e^{12.4} = \boxed{2.4 \times 10^5}$$

At equilibrium, $K = \dfrac{[Cu^+]^2[PtCl_4{}^{2-}][Cl^-]^2}{[PtCl_6{}^{2-}]} = 2.4 \times 10^5.$

The forward reaction is spontaneous, and the equilibrium lies far to the right.

You should now work Exercises 86 and 88.

Under nonstandard conditions, the relationship between the Gibbs free energy change, ΔG, and the cell potential, E_{cell}, is similar to that at standard conditions.

$$\Delta G = -nFE_{cell} \qquad \text{(at nonstandard conditions)}$$

From this equation, we can see that when E_{cell} reaches zero, ΔG is zero. From Section 17-11, we know that this is the condition of *equilibrium*. At that condition, the concentrations are *not* zero; they must have values that satisfy the equilibrium constant, K.

Remember the distinction between ΔG^0 and ΔG.

$$\Delta G^0 = -RT \ln K = -nFE^0_{cell}$$

but

$$\Delta G = -nFE_{cell}$$

EXAMPLE 21-15   *Calculation of* ΔG

Calculate ΔG at 25°C for the reaction in Example 21-14 at the concentrations indicated below.

$$[PtCl_6{}^{2-}] = 1.00 \times 10^{-2}\,M \qquad\qquad [Cu^+] = 1.00 \times 10^{-3}\,M$$

$$[Cl^-] = 1.00 \times 10^{-3}\,M \qquad\qquad [PtCl_4{}^{2-}] = 2.00 \times 10^{-5}\,M$$

Plan

The Gibbs free energy change, ΔG, is related to E_{cell} (not E_{cell}^0) by the relationship

$$\Delta G = -nFE_{cell}$$

We use the Nernst equation to evaluate E_{cell} as shown in Example 21-10.

Solution

From the solution to Example 21-4, E_{cell}^0 for this reaction—i.e., at standard conditions—is $+0.16$ V. For the overall cell reaction, $n = 2$ moles of electrons transferred.

Here we use the Nernst equation to find E_{cell} for the overall reaction, as in Example 21-10. Alternatively, we could find the potential for each half-reaction and then add these to obtain E_{cell}, as in Example 21-9.

$$E_{cell} = E_{cell}^0 - \frac{0.0592}{n} \log \frac{[Cu^+]^2[PtCl_4{}^{2-}][Cl^-]^2}{[PtCl_6{}^{2-}]}$$

$$= +0.16\text{ V} - \frac{0.0592}{2} \log \frac{(1.00 \times 10^{-3})^2(2.00 \times 10^{-5})(1.00 \times 10^{-3})^2}{(1.00 \times 10^{-2})}$$

$$= +0.16\text{ V} - \frac{0.0592}{2} \log (2.00 \times 10^{-15})$$

$$= +0.16\text{ V} - \frac{0.0592}{2}(-14.699) = \boxed{0.60\text{ V}}$$

We saw (Example 21-14) that the reaction

$$2Cu + PtCl_6{}^{2-} \longrightarrow 2Cu^+ + PtCl_4{}^{2-} + 2Cl^-$$

is spontaneous when all ions are present in 1.0 M concentrations ($E_{cell}^0 = +0.16$ V). We see that E_{cell} is even more positive ($+0.60$ V) under the stated conditions. The Gibbs free energy change can now be calculated.

$$\Delta G = -nFE_{cell} = -\left(\frac{2\text{ mol }e^-}{\text{mol rxn}}\right)\left(\frac{96.5 \times 10^3\text{ J}}{\text{V} \cdot \text{mol }e^-}\right)(0.60\text{ V}) = \boxed{-1.2 \times 10^5\text{ J/mol}}$$

You should now work Exercise 89.

PRIMARY VOLTAIC CELLS

As any voltaic cell produces current (**discharges**), chemicals are consumed. **Primary voltaic cells** cannot be "recharged." Once the chemicals have been consumed, further chemical action is not possible. The electrolytes and/or electrodes cannot be regenerated by reversing the current flow through the cell using an external direct current source. The most familiar examples of primary voltaic cells are the ordinary "dry" cells that are used as energy sources in flashlights and other small appliances.

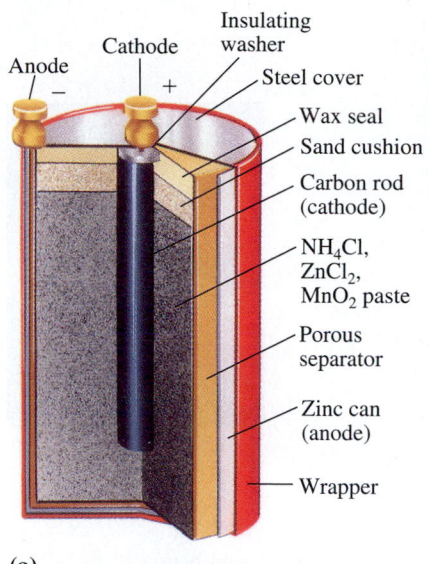

(a)

(b)

Figure 21-16 (a) The Leclanché cell is a dry cell that generates a potential difference of about 1.6 volts. (b) Some commercial alkaline dry cells.

21-22 THE DRY CELL

The first dry cell was patented by Georges Leclanché in 1866 (Figure 21-16). The container of this dry cell, made of zinc, also serves as one of the electrodes. The other electrode is a carbon rod in the center of the cell. The zinc container is lined with porous paper to separate it from the other materials of the cell. The rest of the cell is filled with a moist mixture (the cell is *not* really dry) of ammonium chloride (NH_4Cl), manganese(IV) oxide (MnO_2), zinc chloride ($ZnCl_2$), and a porous, inert filler. Dry cells are sealed to keep the moisture from evaporating. As the cell operates (the electrodes must be connected externally), the metallic Zn is oxidized to Zn^{2+}, and the liberated electrons flow along the container to the external circuit. Thus, the zinc electrode is the anode (negative electrode).

$$Zn \longrightarrow Zn^{2+} + 2e^- \quad \text{(oxidation, anode)}$$

The carbon rod is the cathode, at which ammonium ions are reduced.

$$2NH_4^+ + 2e^- \longrightarrow 2NH_3 + H_2 \quad \text{(reduction, cathode)}$$

Addition of the half-reactions gives the overall cell reaction

$$Zn + 2NH_4^+ \longrightarrow Zn^{2+} + 2NH_3 + H_2 \quad E_{cell} = 1.6 \text{ V}$$

As H_2 is formed, it is oxidized by MnO_2 in the cell. This prevents collection of H_2 gas on the cathode, which would stop the reaction.

$$H_2 + 2MnO_2 \longrightarrow 2MnO(OH)$$

The ammonia produced at the cathode combines with zinc ions and forms a soluble compound containing the complex ions, $[Zn(NH_3)_4]^{2+}$.

$$Zn^{2+} + 4NH_3 \longrightarrow [Zn(NH_3)_4]^{2+}$$

The buildup of reaction products at an electrode is called *polarization* of the electrode.

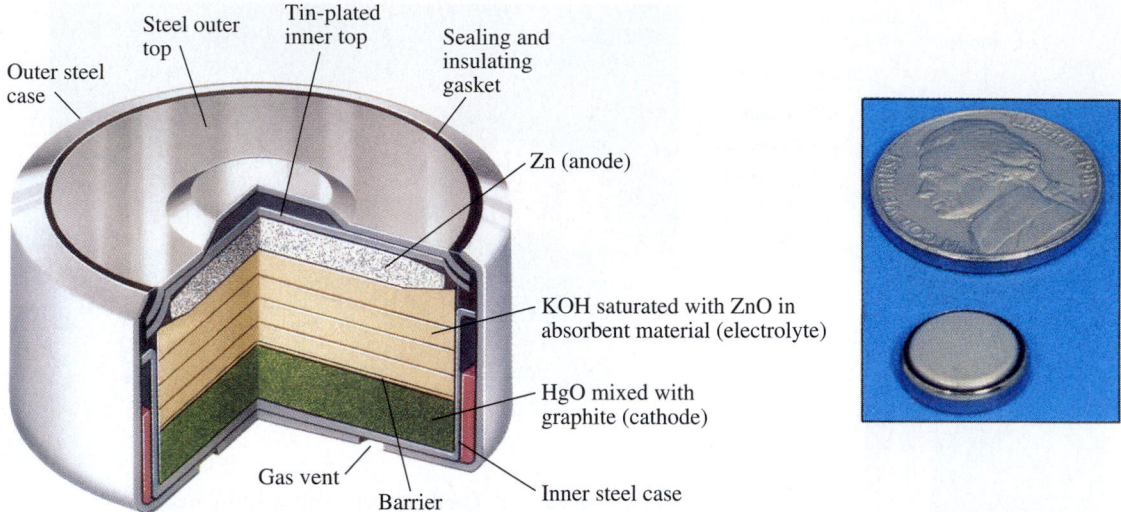

Outer steel case
Steel outer top
Tin-plated inner top
Sealing and insulating gasket
Zn (anode)
KOH saturated with ZnO in absorbent material (electrolyte)
HgO mixed with graphite (cathode)
Gas vent
Barrier
Inner steel case

The mercury battery of the type frequently used in watches, calculators, and hearing aids is a primary cell. Although mercury in the water supply is known to cause health problems, no conclusive evidence has been found that the disposal of household batteries contributes to such problems. Nevertheless, manufacturers are working to decrease the amount of mercury in batteries. In recent years, the amount of mercury in alkaline batteries decreased markedly; at the same time, the life of such batteries has increased dramatically.

This reaction prevents polarization due to the buildup of ammonia, and it prevents the concentration of Zn^{2+} from increasing substantially, which would decrease the cell potential.

Alkaline dry cells are similar to Leclanché dry cells except that (1) the electrolyte is basic (alkaline) because it contains KOH, and (2) the interior surface of the Zn container is rough; this gives a larger surface area. Alkaline cells have a longer shelf life than ordinary dry cells, and they stand up better under heavy use. The voltage of an alkaline cell is about 1.5 volts. During discharge, the alkaline dry cell reactions are

$$Zn(s) + 2OH^-(aq) \longrightarrow Zn(OH)_2(s) + 2e^- \qquad \text{(anode)}$$
$$\underline{2MnO_2(s) + 2H_2O(\ell) + 2e^- \longrightarrow 2MnO(OH)(s) + 2OH^-(aq)} \qquad \text{(cathode)}$$
$$Zn(s) + 2MnO_2(s) + 2H_2O(\ell) \longrightarrow Zn(OH)_2(s) + 2MnO(OH)(s) \qquad \text{(overall)}$$

SECONDARY VOLTAIC CELLS

In **secondary voltaic cells,** or *reversible cells,* the original reactants can be regenerated. This is done by passing a direct current through the cell in the direction opposite to the discharge current flow. This process is referred to as *charging,* or recharging, a cell or battery. The most common example of a secondary voltaic cell is the lead storage battery, used in most automobiles.

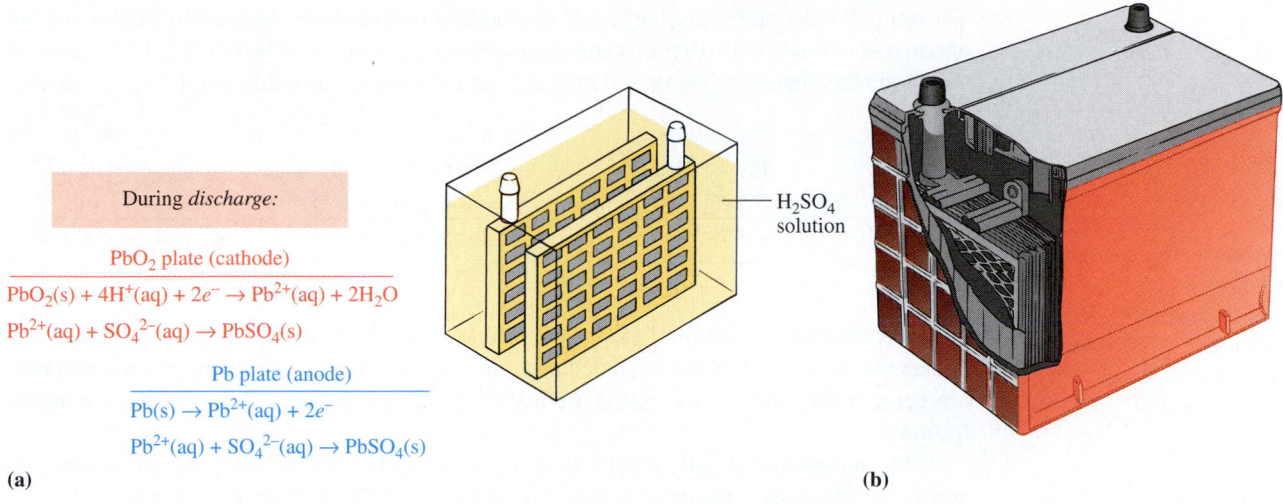

During *discharge:*

PbO$_2$ plate (cathode)

PbO$_2$(s) + 4H$^+$(aq) + 2e^- → Pb^{2+}(aq) + 2H$_2$O

Pb^{2+}(aq) + SO$_4{}^{2-}$(aq) → PbSO$_4$(s)

Pb plate (anode)

Pb(s) → Pb^{2+}(aq) + 2e^-

Pb^{2+}(aq) + SO$_4{}^{2-}$(aq) → PbSO$_4$(s)

— H$_2$SO$_4$ solution

(a)   **(b)**

Figure 21-17   (a) A schematic representation of one cell of a lead storage battery. The reactions shown are those taking place during the *discharge* of the cell. Alternate lead grids are packed with spongy lead and lead(IV) oxide. The grids are immersed in a solution of sulfuric acid, which serves as the electrolyte. To provide a large reacting surface, each cell contains several connected grids of each type as in (b); for clarity, only one of each is shown in (a). Such a cell generates a voltage of about 2 volts. Six of these cells are connected together in series, so that their voltages add, to make a 12-volt battery.

21-23 THE LEAD STORAGE BATTERY

The lead storage battery is depicted in Figure 21-17. One group of lead plates contains compressed spongy lead. These alternate with a group of lead plates that contain lead(IV) oxide, PbO$_2$. The electrodes are immersed in a solution of about 40% sulfuric acid. When the cell discharges, the spongy lead is oxidized to lead ions, and the lead plates accumulate a negative charge.

$$Pb \longrightarrow Pb^{2+} + 2e^- \qquad \text{(oxidation)}$$

The lead ions then combine with sulfate ions from the sulfuric acid to form insoluble lead(II) sulfate. This begins to coat the lead electrode.

$$Pb^{2+} + SO_4{}^{2-} \longrightarrow PbSO_4(s) \qquad \text{(precipitation)}$$

Thus, the net process at the anode *during discharge* is

$$Pb + SO_4{}^{2-} \longrightarrow PbSO_4(s) + 2e^- \qquad \text{(anode during discharge)}$$

The electrons travel through the external circuit and re-enter the cell at the PbO$_2$ electrode, which is the cathode during discharge. Here, in the presence of hydrogen ions, the lead(IV) oxide is reduced to lead(II) ions, Pb^{2+}. These ions also combine with SO$_4{}^{2-}$ ions from the H$_2$SO$_4$ to form an insoluble PbSO$_4$ coating on the lead(IV) oxide electrode.

$$PbO_2 + 4H^+ + 2e^- \longrightarrow Pb^{2+} + 2H_2O \qquad \text{(reduction)}$$
$$Pb^{2+} + SO_4{}^{2-} \longrightarrow PbSO_4(s) \qquad \text{(precipitation)}$$
$$\overline{PbO_2 + 4H^+ + SO_4{}^{2-} + 2e^- \longrightarrow PbSO_4(s) + 2H_2O} \qquad \text{(cathode during discharge)}$$

The net cell reaction for discharge and its standard potential are obtained by adding the net anode and cathode half-reactions and their tabulated potentials. The tabulated E^0 value for the anode half-reaction is reversed in sign because it occurs as oxidation during discharge.

$$
\begin{array}{lr}
 & E^0 \\
\hline
\text{Pb} + \text{SO}_4^{2-} \longrightarrow \text{PbSO}_4(s) + 2e^- & -(-0.356 \text{ V}) \\
\text{PbO}_2 + 4\text{H}^+ + \text{SO}_4^{2-} + 2e^- \longrightarrow \text{PbSO}_4(s) + 2\text{H}_2\text{O} & +1.685 \text{ V} \\
\hline
\text{Pb} + \text{PbO}_2 + \underbrace{4\text{H}^+ + 2\text{SO}_4^{2-}}_{2\text{H}_2\text{SO}_4} \longrightarrow 2\text{PbSO}_4(s) + 2\text{H}_2\text{O} & E_{\text{cell}}^0 = +2.041 \text{ V}
\end{array}
$$

The decrease in the concentration of sulfuric acid provides an easy method for measuring the degree of discharge, because the density of the solution decreases accordingly. We simply measure the density of the solution with a hydrometer.

One cell creates a potential of about 2 volts. Automobile 12-volt batteries have six cells connected in series. The potential declines only slightly during use, because solid reagents are being consumed. As the cell is used, some H_2SO_4 is consumed, lowering its concentration.

When a potential slightly greater than the potential the battery can generate is imposed across the electrodes, the current flow can be reversed. The battery can then be recharged by reversal of all reactions. The alternator or generator applies this potential when the engine is in operation. The reactions that occur in a lead storage battery are summarized as follows.

A *generator* supplies direct current (dc). An *alternator* supplies alternating current (ac), so a rectifier (an electronic device) is used to convert this to direct current for the battery.

$$\text{Pb} + \text{PbO}_2 + 2[2\text{H}^+ + \text{SO}_4^{2-}] \underset{\text{charge}}{\overset{\text{discharge}}{\rightleftharpoons}} 2\text{PbSO}_4(s) + 2\text{H}_2\text{O}$$

During many repeated charge–discharge cycles, some of the $PbSO_4$ falls to the bottom of the container and the H_2SO_4 concentration remains correspondingly low. Eventually the battery cannot be recharged fully. It can be traded in for a new one, and the lead can be recovered and reused to make new batteries.

This is one of the oldest and most successful examples of recycling.

21-24 THE NICKEL–CADMIUM (NICAD) CELL

To see why a nicad battery produces a constant voltage, write the Nernst equation for its reaction. Look at Q.

The nickel–cadmium (nicad) cell has gained widespread popularity because it can be recharged. It thus has a much longer useful life than ordinary (Leclanché) dry cells. Nicad batteries are used in electronic wristwatches, calculators, and photographic equipment.

The anode is cadmium, and the cathode is nickel(IV) oxide. The electrolytic solution is basic. The "discharge" reactions that occur in a nicad battery are

$$
\begin{array}{lr}
\text{Cd}(s) + 2\text{OH}^-(aq) \longrightarrow \text{Cd(OH)}_2(s) + 2e^- & \text{(anode)} \\
\text{NiO}_2(s) + 2\text{H}_2\text{O}(\ell) + 2e^- \longrightarrow \text{Ni(OH)}_2(s) + 2\text{OH}^-(aq) & \text{(cathode)} \\
\hline
\text{Cd}(s) + \text{NiO}_2(s) + 2\text{H}_2\text{O}(\ell) \longrightarrow \text{Cd(OH)}_2(s) + \text{Ni(OH)}_2(s) & \text{(overall)}
\end{array}
$$

The solid reaction product at each electrode adheres to the electrode surface. Hence, a nicad battery can be recharged by an external source of electricity; that is, the electrode reactions can be reversed. Because no gases are produced by the reactions in a nicad battery, the unit can be sealed. The voltage of a nicad cell is about 1.4 volts, slightly less than that of a Leclanché cell.

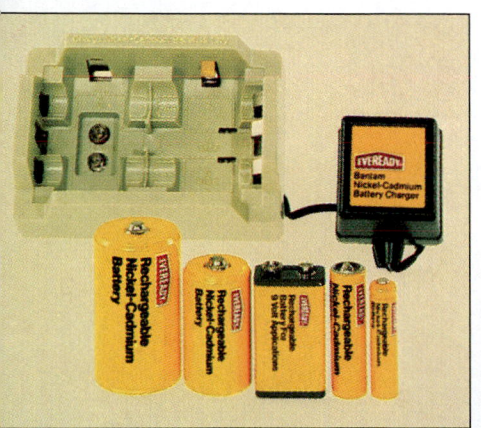

Rechargeable nicad batteries are used to operate many electrical devices.

21-25 THE HYDROGEN–OXYGEN FUEL CELL

Fuel cells are voltaic cells in which the reactants are continuously supplied to the cell and the products are continuously removed. The hydrogen–oxygen fuel cell (Figure 21-18)

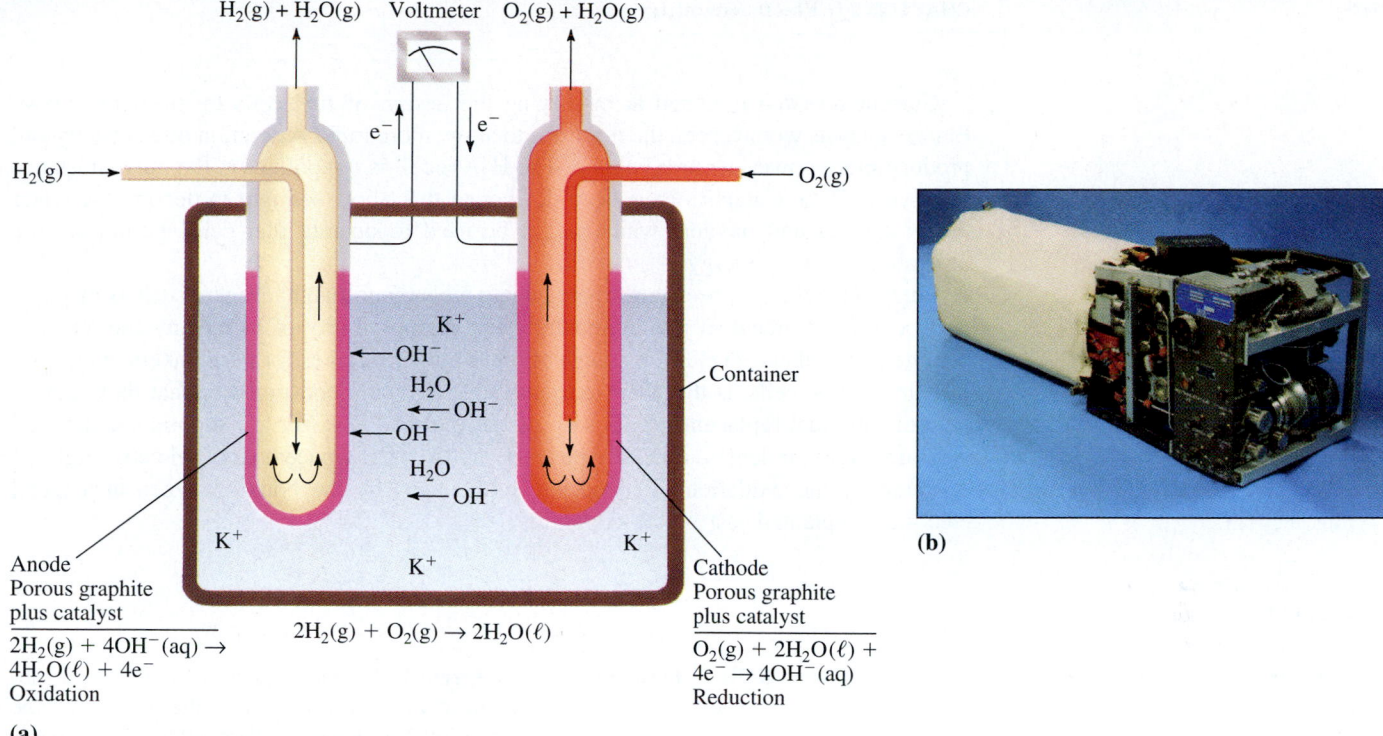

(b)

Figure 21-18 (a) Schematic drawing of a hydrogen–oxygen fuel cell. (b) A hydrogen–oxygen fuel cell that is used in spacecraft.

already has many applications. It is used in spacecraft to supplement the energy obtained from solar cells. Liquid H_2 is carried on board as a propellant. The boiled-off H_2 vapor that ordinarily would be lost is used in a fuel cell to generate electrical power.

Hydrogen (the fuel) is supplied to the anode compartment. Oxygen is fed into the cathode compartment. The diffusion rates of the gases into the cell are carefully regulated for maximum efficiency. Oxygen is reduced at the cathode, which consists of porous carbon impregnated with finely divided Pt or Pd catalyst.

$$O_2 + 2H_2O + 4e^- \xrightarrow{\text{catalyst}} 4OH^- \qquad \text{(cathode)}$$

The OH$^-$ ions migrate to the anode through the electrolyte, an aqueous solution of a base. The anode is also porous carbon containing a small amount of catalyst (Pt, Ag, or CoO). Here H_2 is oxidized to H_2O.

$$H_2 + 2OH^- \longrightarrow 2H_2O + 2e^- \qquad \text{(anode)}$$

The net reaction is obtained from the two half-reactions.

$$
\begin{array}{ll}
O_2 + 2H_2O + 4e^- \longrightarrow 4OH^- & \text{(cathode)} \\
\underline{2(H_2 + 2OH^- \longrightarrow 2H_2O + 2e^-)} & \text{(anode)} \\
2H_2 + O_2 \longrightarrow 2H_2O & \text{(net cell reaction)}
\end{array}
$$

The net reaction is the same as the burning of H_2 in O_2 to form H_2O, but combustion does not actually occur. Most of the chemical energy from the formation of H—O bonds is converted directly into electrical energy, rather than into heat energy as in combustion.

When the H_2/O_2 fuel cell is used aboard spacecraft, it is operated at a high enough temperature that the water evaporates at the same rate as it is produced. The vapor is then condensed to pure water.

The efficiency of energy conversion of the fuel cell operation is 60 to 70% of the theoretical maximum (based on ΔG). This represents about twice the efficiency that can be realized from burning hydrogen in a heat engine coupled to a generator.

Current research is aimed at modifying the design of fuel cells to lower their cost. Better catalysts would speed the reactions to allow more rapid generation of electricity and produce more power per unit volume. The H_2/O_2 cell is nonpolluting; the only substance released is H_2O. Catalysts have been developed that allow sunlight to decompose water into hydrogen and oxygen, which might be used to operate fuel cells, permitting the utilization of solar energy.

Fuel cells have also been constructed using fuels other than hydrogen, such as methane or methanol. Biomedical researchers envision the possibility of using tiny fuel cells to operate pacemakers. The disadvantage of other power supplies for pacemakers, which are primary voltaic cells, is that their reactants are eventually consumed so that they require periodic surgical replacement. As long as the fuel and oxidizer are supplied, a fuel cell can—in theory, at least—operate forever. Eventually, tiny pacemaker fuel cells might be operated by the oxidation of blood sugar (the fuel) by the body's oxygen at a metal electrode implanted just below the skin.

Key Terms

Alkaline cell   A dry cell in which the electrolyte contains KOH.

Ampere   Unit of electrical current; 1 ampere equals 1 coulomb per second.

Anode   The electrode at which oxidation occurs.

Cathode   The electrode at which reduction occurs.

Cathode protection   Protection of a metal (making it a cathode) against corrosion by attaching it to a sacrificial anode of a more easily oxidized metal.

Cell potential   Potential difference, E_{cell}, between reduction and oxidation half-cells; may be at *nonstandard* conditions.

Concentration cell   A voltaic cell in which the two half-cells are composed of the same species but contain different ion concentrations.

Corrosion   Oxidation of metals in the presence of air and moisture.

Coulomb   Unit of electrical charge.

Downs cell   An electrolytic cell for the commercial electrolysis of molten sodium chloride.

Dry cells   Ordinary batteries (voltaic cells) for flashlights, radios, and so on; many are Leclanché cells.

Electrochemistry   The study of the chemical changes produced by electrical current and the production of electricity by chemical reactions.

Electrode potentials   Potentials, E, of half-reactions as reductions versus the standard hydrogen electrode.

Electrodes   Surfaces upon which oxidation and reduction half-reactions occur in electrochemical cells.

Electrolysis   The process that occurs in electrolytic cells.

Electrolytic cell   An electrochemical cell in which electrical energy causes nonspontaneous redox reactions to occur.

Electrolytic conduction   See *Ionic conduction.*

Electroplating   Plating a metal onto a (cathodic) surface by electrolysis.

Faraday   An amount of charge equal to 96,485 coulombs; corresponds to the charge on one mole of electrons, 6.022×10^{23} electrons.

Faraday's Law of Electrolysis   One equivalent weight of a substance is produced at each electrode during the passage of one mole of electrons (96,485 coulombs of charge) through an electrolytic cell.

Fuel cell   A voltaic cell in which the reactants (usually gases) are supplied continuously and products are removed continuously.

Galvanic cell   See *Voltaic cell.*

Half-cell   The compartment in a voltaic cell in which the oxidation or reduction half-reaction occurs.

Hydrogen–oxygen fuel cell   A fuel cell in which hydrogen is the fuel (reducing agent) and oxygen is the oxidizing agent.

Ionic conduction   Conduction of electric current by ions through a solution of pure liquid.

Lead storage battery   A secondary voltaic cell that is used in most automobiles.

Leclanché cell   A common type of dry cell.

Metallic conduction   Conduction of electric current through a metal or along a metallic surface.

Nernst equation   An equation that corrects standard electrode potentials for nonstandard conditions.

Nickel–cadmium cell (nicad battery)   A dry cell in which the anode is Cd, the cathode is NiO_2, and the electrolyte is basic.

Polarization of an electrode   Buildup of a product of oxidation or reduction at an electrode, preventing further reaction.

Primary voltaic cell   A voltaic cell that cannot be recharged; no further chemical reaction is possible once the reactants are consumed.

Sacrificial anode   A more active metal that is attached to a less active metal to protect the less active metal cathode against corrosion.

Salt bridge   A U-shaped tube containing an electrolyte, which connects two half-cells of a voltaic cell.

Secondary voltaic cell   A voltaic cell that can be recharged; the original reactants can be regenerated by reversing the direction of current flow.

Standard cell A cell in which all reactants and products are in their thermodynamic standard states (1 M for dissolved species and 1 atm partial pressure for gases).

Standard cell potential The potential difference, E^0_{cell}, between standard reduction and oxidation half-cells.

Standard electrochemical conditions 1 M concentration for dissolved ions, 1 atm partial pressure for gases, and pure solids and liquids.

Standard electrode A half-cell in which the oxidized and reduced forms of a species are present at unit activity: 1 M solutions of dissolved ions, 1 atm partial pressure of gases, and pure solids and liquids.

Standard electrode potential By convention, the potential (E^0) of a half-reaction as a reduction relative to the standard hydrogen electrode, when all species are present at unit activity.

Standard hydrogen electrode (SHE) An electrode consisting of a platinum electrode that is immersed in a 1 M H$^+$ solution and that has H_2 gas bubbled over it at 1 atmosphere pressure; defined as the reference electrode, with a potential of *exactly* 0.0000 . . . volt.

Voltage Potential difference between two electrodes; a measure of the chemical potential for a redox reaction to occur.

Voltaic cell An electrochemical cell in which spontaneous chemical reactions produce electricity; also called a galvanic cell.

Exercises

Redox Review and General Concepts

1. (a) Define oxidation and reduction in terms of electron gain or loss. (b) What is the relationship between the number of electrons gained and lost in a redox reaction? (c) Why do all electrochemical cells involve redox reactions?

2. Define and illustrate (a) oxidizing agent and (b) reducing agent.

3. For each of the following unbalanced equations, (1) write the half-reactions for oxidation and for reduction, and (2) balance the overall equation in acidic solution using the half-reaction method.
   (a) $Hg^{2+} + Cu \longrightarrow Hg + Cu^{2+}$
   (b) $MnO_2 + Cl^- \longrightarrow Mn^{2+} + Cl_2$
   (c) $Sn^{2+} + O_2 \longrightarrow Sn^{4+} + H_2O$

4. For each of the following unbalanced equations, (1) write the half-reactions for oxidation and reduction, and (2) balance the overall equation using the half-reaction method.
   (a) $FeS + NO_3^- \longrightarrow NO + SO_4^{2-} + Fe^{2+}$ (acidic solution)
   (b) $Cr_2O_7^{2-} + Sn^{2+} \longrightarrow Cr^{3+} + Sn^{4+}$ (acidic solution)
   (c) $S^{2-} + Cl_2 + OH^- \longrightarrow SO_4^{2-} + Cl^- + H_2O$ (basic solution)

5. (a) Compare and contrast ionic conduction and metallic conduction. (b) What is an electrode? (c) What is an inert electrode?

6. Support or refute each of the following statements: (a) In any electrochemical cell the positive electrode is the one toward which the electrons flow through the wire. (b) The cathode is the negative electrode in any electrochemical cell.

Electrolytic Cells—General Concepts

7. (a) Solids such as potassium bromide, KBr, and sodium nitrate, $NaNO_3$, do not conduct electrical current even though they are ionic. Why? Can these substances be electrolyzed as solids? (b) Support or refute the statement that the Gibbs free energy change, ΔG, is negative for any electrolysis reaction.

8. (a) Metallic magnesium cannot be obtained by electrolysis of aqueous magnesium chloride, $MgCl_2$. Why? (b) There are no sodium ions in the overall cell reaction for the electrolysis of aqueous sodium chloride. Why?

9. Consider the electrolysis of molten aluminum oxide, Al_2O_3, dissolved in cryolite, Na_3AlF_6, with inert electrodes. This is the Hall process for commercial production of aluminum (Section 22-6). The following experimental observations can be made when current is supplied:
   (i) Silvery metallic aluminum is produced at one electrode.
   (ii) Oxygen, O_2, bubbles off at the other electrode.
   Diagram the cell, indicating the anode, the cathode, the positive and negative electrodes, the half-reaction occurring at each electrode, the overall cell reaction, and the direction of electron flow through the wire.

10. Do the same as in Exercise 9 for the electrolysis of molten calcium chloride with inert electrodes. The observations are
    (i) Bubbles of pale green chlorine gas, Cl_2, are produced at one electrode.
    (ii) Silvery-white molten metallic calcium is produced at the other electrode.

11. Do the same as in Exercise 9 for the electrolysis of aqueous potassium sulfate, K_2SO_4. The observations are
    (i) Bubbles of gaseous hydrogen are produced at one electrode, and the solution becomes more basic around that electrode.
    (ii) Bubbles of gaseous oxygen are produced at the other electrode, and the solution becomes more acidic around that electrode.

12. Do the same as in Exercise 9 for the electrolysis of an aqueous solution of copper(II) bromide, $CuBr_2$. The observations are
    (i) One electrode becomes coated with copper metal, and the color of the solution around that electrode fades.
    (ii) Around the other electrode, the solution turns brown, as bromine is formed and dissolves in water.

13. (a) Write the equation for the half-reaction when H_2O is reduced in an electrochemical cell. (b) Write the equation for the half-reaction when H_2O is oxidized in an electrochemical cell.

Faraday's Law

14. What are (a) a coulomb, (b) electrical current, (c) an ampere, and (d) a faraday?

15. Calculate the number of electrons that have a total charge of 1 coulomb.

16. Calculate the charge, in coulombs, on a single electron.

17. For each of the following cations, calculate (i) the number of faradays required to produce 1.00 mol of free metal and (ii) the number of coulombs required to produce 1.00 g of free metal. (a) Cd^{2+}, (b) Al^{3+}, (c) K^+.

18. For each of the following cations, calculate (i) the number of faradays required to produce 1.00 mol of free metal and (ii) the number of coulombs required to produce 1.00 g of free metal. (a) Co^{3+}, (b) Hg^{2+}, (c) Hg_2^{2+}.

19. The cells in an automobile battery were charged at a steady current of 5.0 A for exactly 5 hr. What masses of Pb and PbO_2 were formed in each cell? The overall reaction is

$$2PbSO_4(s) + 2H_2O(\ell) \longrightarrow Pb(s) + PbO_2(s) + 2H_2SO_4(aq)$$

*20. The chemical equation for the electrolysis of a fairly concentrated brine solution is

$$2NaCl(aq) + 2H_2O(\ell) \longrightarrow Cl_2(g) + H_2(g) + 2NaOH(aq)$$

What volume of gaseous chlorine would be generated at 735 torr and 5°C if the process were 88% efficient and if a current of 1.5 A flowed for 5.0 hr?

21. A $Cd^{2+}(aq)$ solution is electrolyzed with a current of 0.104 A for a period of 335 s. What mass of Cd(s) is plated out on the cathode?

22. The mass of silver deposited on a spoon during electroplating was 0.775 mg. How much electrical charge passed through the cell?

23. What mass of platinum could be plated onto a ring from the electrolysis of a platinum(II) salt with a 0.345-A current for 125.0 s?

24. What mass of silver could be plated onto a spoon from electrolysis of silver nitrate with a 2.50-A current for 20.0 min?

*25. We pass enough current through a solution to plate out *one* mole of nickel metal from a solution of $NiSO_4$. In other electrolysis cells, this same current plates out *two* moles of silver from $AgNO_3$ solution but liberates only *one-half* mole of O_2 gas. Explain these observations.

*26. A current is passed through 500 mL of a solution of CaI_2. The following electrode reactions occur:

anode: $2I^- \longrightarrow I_2 + 2e^-$
cathode: $2H_2O + 2e^- \longrightarrow H_2 + 2OH^-$

After some time, analysis of the solution shows that 47.7 mmol of I_2 has been formed. (a) How many faradays of charge have passed through the solution? (b) How many coulombs? (c) What volume of dry H_2 at STP has been formed? (d) What is the pH of the solution?

27. An electrolytic cell contains 50.0 mL of a 0.152 *M* solution of $FeCl_3$. A current of 0.775 A is passed through the cell, causing deposition of Fe(s) at the cathode. What is the concentration of $Fe^{3+}(aq)$ in the cell after this current has run for 20.0 min?

28. Suppose 250 mL of a 0.333 *M* solution of $CuCl_2$ is electrolyzed.

How long will a current of 0.75 A have to run in order to reduce the concentration of Cu^{2+} to 0.167 *M*? What mass of Cu(s) will be deposited on the cathode during this time?

29. In a copper-refining plant, a current of 7.50×10^3 A passes through a solution containing Cu^{2+}. What mass, in kilograms, of copper is deposited in 24 hr?

30. The total charge of electricity required to plate out 15.54 g of a metal from a solution of its dipositive ions is 14,475 C. What is the metal?

31. What is the charge on an ion of tin if 7.42 g of metallic tin is plated out by the passage of 24,125 C through a solution containing the ion?

*32. Three electrolytic cells are connected in series; that is, the same current passes through all three, one after another. In the first cell, 1.00 g of Cd is oxidized to Cd^{2+}; in the second, Ag^+ is reduced to Ag; in the third, Fe^{2+} is oxidized to Fe^{3+}. (a) Find the number of faradays passed through the circuit. (b) What mass of Ag is deposited at the cathode in the second cell? (c) What mass of $Fe(NO_3)_3$ could be recovered from the solution in the third cell?

Voltaic Cells—General Concepts

33. (a) What kind of energy is converted into electrical energy in voltaic cells? (b) What kind of process takes place at the cathode of a voltaic cell? At the anode? How does this compare with electrolytic cells?

34. (a) Why must the solutions in a voltaic cell be kept separate and not allowed to mix? (b) What are the functions of a salt bridge?

35. A voltaic cell containing a standard Fe^{3+}/Fe^{2+} electrode and a standard Ga^{3+}/Ga electrode is constructed, and the circuit is closed. Without consulting the table of standard reduction potentials, diagram and completely describe the cell from the following experimental observations. (i) The mass of the gallium electrode decreases, and the gallium ion concentration increases around that electrode. (ii) The ferrous ion, Fe^{2+}, concentration increases in the other electrode solution.

36. Repeat Exercise 35 for a voltaic cell that contains standard Co^{2+}/Co and Au^{3+}/Au electrodes. The observations are: (i) Metallic gold plates out on one electrode, and the gold ion concentration decreases around that electrode. (ii) The mass of the cobalt electrode decreases, and the cobalt(II) ion concentration increases around that electrode.

37. What does voltage measure? How does it vary with time in a primary voltaic cell? Why?

*38. In Section 4-6 we learned how to predict from the activity series (Table 4-13) which metals replace which others from aqueous solutions. From that table, we predict that aluminum will displace silver. The equation for this process is

$$Al(s) + 3Ag^+(aq) \longrightarrow Al^{3+}(aq) + 3Ag(s)$$

Suppose we set up a voltaic cell based on this reaction. (a) What half-reaction would represent the reduction in this cell? (b) What half-reaction would represent the oxidation? (c) Which metal would be the anode? (d) Which metal would be the cathode? (e) Diagram this cell.

39. In a voltaic cell made with metal electrodes, is the more active metal more likely to be the anode or the cathode? Explain.

40. When metallic copper is placed into aqueous silver nitrate, a spontaneous redox reaction occurs. No electricity is produced. Why?

Standard Cell Potentials

41. (a) What are standard electrochemical conditions? (b) Why are we permitted to assign arbitrarily an electrode potential of exactly 0 V to the standard hydrogen electrode?

42. What does the sign of the standard reduction potential of a half-reaction indicate? What does the magnitude indicate?

43. (a) What is the electromotive series? What information does it contain? (b) How was the electromotive series constructed? In general, how do ionization energies and electron affinities of elements vary with their positions in the electromotive series?

44. Standard reduction potentials are 2.9 V for $F_2(g)/F^-$, 0.8 V for $Ag^+/Ag(s)$, 0.5 V for $Cu^+/Cu(s)$, 0.3 V for $Cu^{2+}/Cu(s)$, -0.4 V for $Fe^{2+}/Fe(s)$, -2.7 V for $Na^+/Na(s)$, and -2.9 V for $K^+/K(s)$. (a) Arrange the oxidizing agents in order of increasing strength. (b) Which of these oxidizing agents will oxidize Cu under standard state conditions?

45. Standard reduction potentials are 1.455 V for the $PbO_2(s)/Pb(s)$ couple, 1.360 V for $Cl_2(g)/Cl^-$, 3.06 V for $F_2(g)/HF(aq)$, and 1.77 V for $H_2O_2(aq)/H_2O(\ell)$. Under standard state conditions, (a) which is the strongest oxidizing agent, (b) which oxidizing agent(s) could oxidize lead to lead(IV) oxide, and (c) which oxidizing agent(s) could oxidize fluoride ion in an acidic solution?

46. Arrange the following less commonly encountered metals in an activity series from the most active to the least active: radium [$Ra^{2+}/Ra(s)$, $E^0 = -2.9$ V], rhodium [$Rh^{3+}/Rh(s)$, $E^0 = 0.80$ V], europium [$Eu^{2+}/Eu(s)$, $E^0 = -3.4$ V]. How do these metals compare in reducing ability with the active metal lithium [$Li^+/Li(s)$, $E^0 = -3.0$ V], with hydrogen, and with gold [$Au^{3+}/Au(s)$, $E^0 = 1.5$ V], which is a noble metal and one of the least active of the metals?

47. Arrange the following metals in an activity series from the most active to the least active: nobelium [$No^{3+}/No(s)$, $E^0 = -2.5$ V], cobalt [$Co^{2+}/Co(s)$, $E^0 = -0.28$ V], gallium [$Ga^{3+}/Ga(s)$, $E^0 = -0.53$ V], thallium [$Tl^+/Tl(s)$, $E^0 = -0.34$ V], polonium [$Po^{2+}/Po(s)$, $E^0 = 0.65$ V].

48. Diagram the following cells. For each cell, write the balanced equation for the reaction that occurs spontaneously, and calculate the cell potential. Indicate the direction of electron flow, the anode, the cathode, and the polarity ($+$ or $-$) of each electrode. In each case, assume that the circuit is completed by a wire and a salt bridge.
    (a) A strip of magnesium is immersed in a solution that is 1.0 M in Mg^{2+}, and a strip of silver is immersed in a solution that is 1.0 M in Ag^+.

(b) A strip of chromium is immersed in a solution that is 1.0 M in Cr^{3+}, and a strip of silver is immersed in a solution that is 1.0 M in Ag^+.

49. Repeat Exercise 48 for the following cells.
    (a) A strip of copper is immersed in a solution that is 1.0 M in Cu^{2+}, and a strip of gold is immersed in a solution that is 1.0 M in Au^{3+}.
    (b) A strip of aluminum is immersed in a solution that is 1.0 M in Al^{3+}, and a strip of copper is immersed in a solution that is 1.0 M in Cu^{2+}.

In answering Exercises 50–64, justify your answer by appropriate calculations. Assume that each reaction occurs at standard electrochemical conditions.

50. (a) Will Fe^{3+} oxidize Sn^{2+} to Sn^{4+} in acidic solution? (b) Will dichromate ions oxidize fluoride ions to free fluorine in acidic solution?

51. (a) Will permanganate ions oxidize arsenous acid, H_3AsO_3, to arsenic acid, H_3AsO_4, in acid solution? (b) Will permanganate ions oxidize hydrogen peroxide, H_2O_2, to free oxygen, O_2, in acidic solution?

52. (a) Will dichromate ions oxidize Mn^{2+} to MnO_4^- in acidic solution? (b) Will sulfate ions oxidize arsenous acid, H_3AsO_3, to arsenic acid, H_3AsO_4, in acid solution?

53. Calculate the standard cell potential, E^0_{cell} for the cell described in Exercise 35.

54. Calculate the standard cell potential, E^0_{cell}, for the cell described in Exercise 36.

55. (a) Write the equation for the oxidation of $Zn(s)$ by $Cl_2(g)$. (b) Calculate the potential of this reaction under standard state conditions. (c) Is this a spontaneous reaction?

56. For each of the following cells, (i) write the net reaction in the direction consistent with the way the cell is written; (ii) write the half-reactions for the anode and cathode processes; (iii) find the standard cell potential, E^0_{cell}, at 25°C; and (iv) tell whether the standard cell reaction actually occurs as given or in the reverse direction.
    (a) $Zn|Zn^{2+}||Sn^{2+}|Sn$
    (b) $Ag|Ag^+||Cd^{2+}|Cd$

57. Repeat Exercise 56 for the following cells:
    (a) $Al|Al^{3+}||Ce^{4+}, Ce^{3+}|Pt$
    (b) $Zn|Zn^{2+}||Co^{2+}|Co$

58. Which of the following reactions are spontaneous in voltaic cells under standard conditions?
    (a) $Si(s) + 6OH^-(aq) + 2Cu^{2+}(aq) \longrightarrow$
    $$SiO_3^{2-}(aq) + 3H_2O(\ell) + 2Cu(s)$$
    (b) $Zn(s) + 4CN^-(aq) + Ag_2CrO_4(s) \longrightarrow$
    $$Zn(CN)_4^{2-}(aq) + 2Ag(s) + CrO_4^{2-}(aq)$$
    (c) $MnO_2(s) + 4H^+(aq) + Sr(s) \longrightarrow$
    $$Mn^{2+}(aq) + 2H_2O(\ell) + Sr^{2+}(aq)$$
    (d) $Cl_2(g) + 2H_2O(\ell) + ZnS(s) \longrightarrow$
    $$2HClO(aq) + H_2S(aq) + Zn(s)$$

59. Consult a table of standard reduction potentials and determine which of the following reactions are spontaneous under standard electrochemical conditions.

(a) $Mn(s) + 2H^+(aq) \longrightarrow H_2(s) + Mn^{2+}(aq)$
(b) $2Al^{3+}(aq) + 3H_2(g) \longrightarrow 2Al(s) + 6H^+(aq)$
(c) $H_2(g) \longrightarrow H^+(aq) + H^-(aq)$
(d) $Cl_2(g) + 2Br^-(aq) \longrightarrow Br_2(\ell) + 2Cl^-(aq)$

60. Which of each pair is the stronger reducing agent? (a) Ag or H_2, (b) Sn or Pb, (c) Hg or Au, (d) Cl^- in acidic solution or Cl^- in basic solution, (e) HCl or H_2S, (f) Br^- or Fe^{2+}.

61. Which of each pair is the stronger oxidizing agent? (a) Cu^+ or Ag^+, (b) Sn^{2+} or Sn^{4+}, (c) Fe^{2+} or Fe^{3+}, (d) I_2 or Br_2, (e) MnO_4^- in acidic solution or MnO_4^- in basic solution, (f) Cl_2 or Pb^{2+}

*62. The element ytterbium forms both 2+ and 3+ cations in aqueous solution. $E^0 = -2.797$ V for $Yb^{2+}/Yb(s)$, and -2.267 V for $Yb^{3+}/Yb(s)$. What is the standard state reduction potential for the Yb^{3+}/Yb^{2+} couple?

*63. The standard reduction potential for Cu^+ to $Cu(s)$ is 0.521 V, and for Cu^{2+} to $Cu(s)$ it is 0.337 V. Calculate the E^0 value for the Cu^{2+}/Cu^+ couple.

64. Tarnished silver is coated with a patina of $Ag_2S(s)$. The coating can be removed by boiling the silverware in an aluminum pan, with some baking soda or salt added to make the solution conductive. Explain this from the point of view of electrochemistry.

*65. Describe the process of corrosion. How can corrosion of an easily oxidizable metal be prevented if the metal must be exposed to the weather?

Concentration Effects; Nernst Equation

For cell voltage calculations, assume that the temperature is 25°C unless stated otherwise.

*66. How is the Nernst equation of value in electrochemistry? How would the Nernst equation be modified if we wished to use natural logarithms, ln? What is the value of the constant in the following equation at 25°C?

$$E = E^0 - \frac{\text{constant}}{n} \ln Q$$

67. Identify all of the terms in the Nernst equation. What part of the Nernst equation represents the correction factor for nonstandard electrochemical conditions?

68. By putting the appropriate values into the Nernst equation, show that it predicts that the voltage of a standard half-cell is equal to E^0. Use the Zn^{2+}/Zn reduction half-cell as an illustration.

69. Calculate the potential associated with the following half-reaction when the concentration of the cobalt(II) ion is 1.0×10^{-3} M.

$$Co(s) \longrightarrow Co^{2+} + 2e^-$$

70. Calculate the reduction potential for hydrogen ion in a system having a perchloric acid concentration of 2.00×10^{-4} M and a hydrogen pressure of 3.00 atm. (Recall that $HClO_4$ is a strong acid in aqueous solution.)

71. The standard reduction potentials for the $H^+/H_2(g)$ and $O_2(g)$, $H^+/H_2O(\ell)$ couples are 0.0000 V and 1.229 V, respectively.

(a) Write the half-reactions and the overall reaction, and calculate E^0 for the reaction

$$2H_2(g) + O_2(g) \longrightarrow 2H_2O(\ell)$$

(b) Calculate E for the cell when the pressure of H_2 is 3.00 atm and that of O_2 is 1.25 atm.

72. Consider the cell represented by the notation

$$Zn(s)|ZnCl_2(aq); Cl_2(g, 1 \text{ atm}); Cl^-(aq)|C$$

Calculate (a) E^0 and (b) E for the cell when the concentration of the $ZnCl_2$ is 0.15 mol/L.

73. What is the concentration of Ag^+ in a half-cell if the reduction potential of the Ag^+/Ag couple is observed to be 0.54 V?

74. What must be the pressure of fluorine gas to produce a reduction potential of 2.75 V in a solution that contains 0.40 M F^-?

75. Calculate the cell potential of each of the following electrochemical cells at 25°C.
(a) $Sn(s)|Sn^{2+}(4.5 \times 10^{-3}$ M$)\|Ag^+(0.100$ M$)|Ag(s)$
(b) $Zn(s)|Zn^{2+}(0.500$ M$)\|$
$\qquad Fe^{3+}(7.2 \times 10^{-6}$ M$)$, $Fe^{2+}(0.15$ M$)|Pt$
(c) $Pt|H_2(1 \text{ atm})|HCl(0.00623$ M$)|Cl_2(1 \text{ atm})|Pt$

76. Calculate the cell potential of each of the following electrochemical cells at 25°C.
(a) $Pt|H_2(10.0 \text{ atm})$, $H^+(1.00 \times 10^{-3}$ M$)\|$
$\qquad Ag^+(0.00496$ M$)|Ag(s)$
(b) $Pt|H_2(1.00 \text{ atm})$, $H^+(\text{pH} = 5.97)\|$
$\qquad H^+(\text{pH} = 3.47)$, $H_2(1.00 \text{ atm})|Pt$
(c) $Pt|H_2(0.0361 \text{ atm})$, $H^+(0.0100$ M$)\|$
$\qquad H^+(0.0100$ M$)$, $H_2(5.98 \times 10^{-4} \text{ atm})|Pt$

*77. Find the potential of the cell in which identical iron electrodes are placed into solutions of $FeSO_4$ of concentration 1.5 mol/L and 0.15 mol/L.

*78. We construct a cell in which identical copper electrodes are placed in two solutions. Solution A contains 0.75 M Cu^{2+}. Solution B contains Cu^{2+} at some concentration known to be lower than that in solution A. The potential of the cell is observed to be 0.045 V. What is $[Cu^{2+}]$ in solution B?

*79. We construct a standard copper–cadmium cell, close the circuit, and allow the cell to operate. At some later time, the cell voltage reaches zero, and the cell is "run down." (a) What will be the ratio of $[Cd^{2+}]$ to $[Cu^{2+}]$ at that time? (b) What will be the concentrations?

*80. Repeat Exercise 79 for a standard zinc–nickel cell.

*81. The cell potential for the cell

$$Zn(s) + 2H^+(\underline{?} \text{ M}) \longrightarrow Zn^{2+}(3.0 \text{ M}) + H_2(g) (5.0 \text{ atm})$$

is observed to be 0.450 V. What is the pH in the H^+/H_2 half-cell?

Relationships Among ΔG^0, E^0_{cell}, and K

82. How are the signs and magnitudes of E^0_{cell}, ΔG^0, and K related for a particular reaction? Why is the equilibrium constant K related only to E^0_{cell} and not to E_{cell}?

83. In light of your answer to Exercise 82, how do you explain the fact that ΔG^0 for a redox reaction *does* depend on the number of electrons transferred, according to $\Delta G^0 = -nFE^0_{cell}$?

84. Calculate E^0_{cell} from the tabulated standard reduction potentials for each of the following reactions in aqueous solution. Then calculate ΔG^0 and K at 25°C from E^0_{cell}. Which reactions are spontaneous as written?
    (a) $Sn^{4+} + 2Fe^{2+} \longrightarrow Sn^{2+} + 2Fe^{3+}$
    (b) $2Cu^+ \longrightarrow Cu^{2+} + Cu(s)$
    (c) $3Zn(s) + 2MnO_4^- + 4H_2O \longrightarrow$
        $\qquad\qquad 2MnO_2(s) + 3Zn(OH)_2(s) + 2OH^-$

85. Calculate ΔG^0 (overall) and ΔG^0 per mole of metal for each of the following reactions from E^0 values.
    (a) Zinc dissolves in dilute hydrochloric acid to produce a solution that contains Zn^{2+}, and hydrogen gas is evolved.
    (b) Chromium dissolves in dilute hydrochloric acid to produce a solution that contains Cr^{3+}, and hydrogen gas is evolved.
    (c) Silver dissolves in dilute nitric acid to form a solution that contains Ag^+, and NO is liberated as a gas.
    (d) Lead dissolves in dilute nitric acid to form a solution that contains Pb^{2+}, and NO is liberated as a gas.

86. Use tabulated reduction potentials to calculate the value of the equilibrium constant for the reaction

$$2K(s) + 2H_2O(\ell) \rightleftharpoons 2K^+ + 2OH^- + H_2(g)$$

87. Use tabulated reduction potentials to calculate the equilibrium constant for the reaction

$$2Br^- + Cl_2(g) \rightleftharpoons Br_2(aq) + 2Cl^-$$

88. Using the following half-reactions and E^0 data at 25°C:

$$PbSO_4(s) + 2e^- \longrightarrow Pb(s) + SO_4^{2-} \qquad E^0 = -0.356 \text{ V}$$

$$PbI_2(s) + 2e^- \longrightarrow Pb(s) + 2I^- \qquad E^0 = -0.365 \text{ V}$$

    calculate the equilibrium constant for the reaction

$$PbSO_4(s) + 2I^- \rightleftharpoons PbI_2(s) + SO_4^{2-}$$

89. Calculate the free energy change, ΔG, for each reaction of Exercise 75 under the stated conditions, at 25°C.

90. Calculate ΔG^0 for the half-reaction

$$\tfrac{1}{2}H_2O_2(aq) + H^+ + e^- \longrightarrow H_2O(\ell)$$

    given that $E^0 = 1.77$ V for the $H_2O_2(aq)/H_2O(\ell)$ couple.

Practical Aspects of Electrochemistry

91. Distinguish among (a) primary voltaic cells, (b) secondary voltaic cells, and (c) fuel cells.

92. Sketch and describe the operation of (a) the Leclanché dry cell, (b) the lead storage battery, and (c) the hydrogen–oxygen fuel cell.

93. Why is the dry cell designed so that Zn and MnO_2 do not come into contact? What reaction might occur if they were in contact? How would this reaction affect the usefulness of the cell?

*94. People sometimes try to recharge dry cells, with limited success. (a) What reaction would you expect at the zinc electrode of a

Leclanché cell in an attempt to recharge it? (b) What difficulties would arise from the attempt?

95. Briefly describe how a storage cell operates.

96. How does a fuel cell differ from a dry cell or a storage cell?

97. Does the physical size of a commercial cell govern the potential that it will deliver? What does the size affect?

Mixed Exercises

98. Consider the electrochemical cell represented by $Zn(s)|Zn^{2+}||Fe^{3+}|Fe(s)$. (a) Write the ion–electron equations for the half-reactions and the overall cell equation. (b) The standard reduction potentials for $Zn^{2+}/Zn(s)$ and $Fe^{3+}/Fe(s)$ are -0.763 V and -0.036 V, respectively, at 25°C. Determine the standard potential for the reaction. (c) Determine E for the cell when the concentration of Fe^{3+} is 10.0 mol/L and that of Zn^{2+} is 1.00×10^{-3} mol/L. (d) If 150 mA is to be drawn from this cell for a period of 15.0 min, what is the minimum mass for the zinc electrode?

99. A sample of Al_2O_3 dissolved in a molten fluoride bath is electrolyzed using a current of 1.00 A. (a) What is the rate of production of Al in grams per hour? (b) The oxygen liberated at the positive carbon electrode reacts with the carbon to form CO_2. What mass of CO_2 is produced per hour?

*100. The "life" of a certain voltaic cell is limited by the amount of Cu^{2+} in solution available to be reduced. If the cell contains 25 mL of 0.175 M $CuSO_4$, what is the maximum amount of electrical charge this cell could generate?

101. A magnesium bar weighing 4.0 kg is attached to a buried iron pipe to protect the pipe from corrosion. An average current of 0.025 A flows between the bar and the pipe. (a) What reaction occurs at the surface of the bar? of the pipe? In which direction do electrons flow? (b) How many years will it take for the Mg bar to be entirely consumed (1 year = 3.16×10^7 s)? (c) What reaction(s) will occur if the bar is not replaced after the time calculated in part (b)?

102. The production of uranium metal from purified uranium dioxide ore consists of the following steps:

$$UO_2(s) + 4HF(g) \longrightarrow UF_4(s) + 2H_2O(\ell)$$

$$UF_4(s) + 2Mg(s) \xrightarrow{\Delta} U(s) + 2MgF_2(s)$$

    What is the oxidation number of U in (a) UO_2, (b) UF_4, and (c) U? Identify (d) the reducing agent and (e) the substance reduced. (f) What current could the second reaction produce if 0.500 g of UF_4 reacted each minute? (g) What volume of HF(g) at 25°C and 10.0 atm would be required to produce 0.500 g of U? (h) Would 0.500 g of Mg be enough to produce 0.50 g of U?

103. Which of each pair is the stronger oxidizing agent? (a) H^+ or Cl_2, (b) Ni^{2+} or Se in contact with acidic solution, (c) $Cr_2O_7^{2-}$ or Br_2 (acidic solution).

104. Describe the process of electroplating. (b) Sketch and label an apparatus that a jeweler might use for electroplating silver onto jewelry. (c) A jeweler purchases highly purified silver to use as the anode in an electroplating operation. Is this a wise purchase? Why?

105. The same quantity of electrical charge that deposited 0.583 g of silver was passed through a solution of a gold salt, and 0.355 g of gold was deposited. What is the oxidation state of gold in this salt?

BUILDING YOUR KNOWLEDGE

106. An electrochemical cell was needed in which hydrogen and oxygen would react to form water. (a) Using the following standard reduction potentials for the couples given, determine which combination of half-reactions gives the maximum output potential:

$E^0 = -0.828$ V for $H_2O(\ell)/H_2(g)$, OH^-

$E^0 = 0.0000$ V for $H^+/H_2(g)$

$E^0 = 1.229$ V for $O_2(g)$, $H^+/H_2O(\ell)$

$E^0 = 0.401$ V for $O_2(g)$, $H_2O(\ell)/OH^-$

(b) Write the balanced equation for the overall reaction in (a).

107. (a) Given the following E^0 values at 25°C, calculate K_{sp} for cadmium sulfide, CdS.

$$Cd^{2+}(aq) + 2e^- \longrightarrow Cd(s) \qquad E^0 = -0.403 \text{ V}$$

$$CdS(s) + 2e^- \longrightarrow Cd(s) + S^{2-}(aq) \qquad E^0 = -1.21 \text{ V}$$

(b) Evaluate ΔG^0 at 25°C for the process

$$CdS(s) \rightleftharpoons Cd^{2+}(aq) + S^{2-}(aq)$$

108. Refer to tabulated reduction potentials. (a) Calculate K_{sp} for AgBr(s). (b) Calculate ΔG^0 for the reaction

$$AgBr(s) \rightleftharpoons Ag^+(aq) + Br^-(aq)$$

*109. Under standard state conditions, the following reaction is not spontaneous:

$$Br^- + 2MnO_4^- + H_2O(\ell) \longrightarrow$$
$$BrO_3^- + 2MnO_2(s) + 2OH^- \qquad E^0 = -0.022 \text{ V}$$

The reaction conditions are adjusted so that $E = 0.110$ V by making $[Br^-] = [MnO_4^-] = 1.5$ mol/L and $[BrO_3^-] = 0.5$ mol/L. (a) What is the concentration of hydroxide ions in this cell? (b) What is the pH of the solution in the cell?

110. Show by calculation that $E^0 = -1.662$ V for the reduction of Al^{3+} to Al(s), regardless of whether the equation for the reaction is written

(i) $\frac{1}{3}Al^{3+} + e^- \longrightarrow \frac{1}{3}Al(s) \qquad \Delta G^0 = 160.4$ kJ/mol

or

(ii) $Al^{3+} + 3e^- \longrightarrow Al(s) \qquad \Delta G^0 = 481.2$ kJ/mol

111. We wish to fill a balloon with H_2 at a pressure of 1.05 atm and a temperature of 25°C. The volume of the balloon, when filled, is 750 mL. How long must a current of 2.00 A be passed through the cell in order to produce this amount of H_2 by electrolysis of water?

The mineral rhodochrosite is manganese carbonate, $MnCO_3$. Manganese is an important element in steels that must withstand great shock, such as those used to construct rock crushers.

OBJECTIVES

As you study this chapter, you should learn

• *About major sources of metals*

• *About some pretreatment techniques for ores*

• *About some reduction processes that produce free metals*

• *About some techniques for refining (purifying) metals*

• *The specific metallurgies of five metals: magnesium, aluminum, iron, copper, and gold*

• *The basics of conservation of metals*

METALS

Metals are widely used for structural purposes in buildings, cars, railroads, ships, and aircraft. They also serve as conductors of heat and electricity. Medical and nutritional research during recent decades has provided much insight into important biological functions of metals. The metals Na, K, Ca, and Mg, as well as some nonmetals (C, H, O, N, P, and S), are present in the human body in substantial quantities. Many other metals are present in lesser quantities in the human body, but they are essential to our well-being (see essay "Trace Elements and Life" in Chapter 23). In this chapter, we shall study the occurrence of metals and examine processes for obtaining metals from their ores.

22-1   OCCURRENCE OF THE METALS

In our study of periodicity we learned that metallic character increases toward the left and toward the bottom of the periodic table (Section 4-1) and that oxides of most metals are basic (Section 6-8, part 2). The oxides of some metals (and metalloids) are amphoteric (Section 10-8). In Section 13-17 we described metallic bonding and related the effectiveness of metallic bonding to the characteristic properties of metals.

A cluster of 21 minerals that are present in ores, plus two samples of native ores (free metals, Ag and Bi). These are identified below.

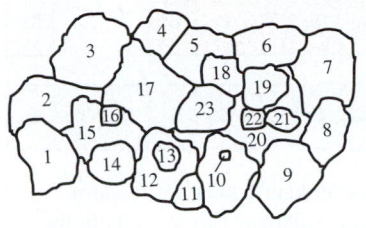

1. Bornite (iridescent)–COPPER
2. Dolomite (pink)–MAGNESIUM
3. Molybdenite (gray)–MOLYBDENUM
4. Skutterudite (gray)–COBALT, NICKEL
5. Zincite (mottled red)–ZINC
6. Chromite (gray)–CHROMIUM
7. Stibnite (*top right*, gray)–ANTIMONY
8. Gummite (yellow)–URANIUM
9. Cassiterite (rust, *bottom right*)–TIN
10. Vanadinite crystal on Goethite (red crystal)–VANADIUM
11. Cinnabar (red)–MERCURY
12. Galena (gray)–LEAD

13. Monazite (white)–RARE EARTHS: cerium, lanthanum, neodymium, thorium
14. Bauxite (gold)–ALUMINUM
15. Strontianite (white, spiny)–STRONTIUM
16. Cobaltite (gray cube)–COBALT
17. Pyrite (gold)–IRON
18. Columbinite (tan, gray stripe)–NIOBIUM, TANTALUM
19. Native BISMUTH (shiny)
20. Rhodochrosite (pink)–MANGANESE
21. Rutile (shiny twin crystal)–TITANIUM
22. Native SILVER (filigree on quartz)
23. Pyrolusite (black, powdery)–MANGANESE

The properties of metals influence the kinds of ores in which they are found, and the metallurgical processes used to extract them from their ores. Metals with negative standard reduction potentials (active metals) are found in nature in the combined state. Those with positive reduction potentials, the less active metals, may occur in the uncombined free state as **native ores.** Examples of native ores are Cu, Ag, Au, and the less abundant Pt, Os, Ir, Ru, Rh, and Pd. Cu, Ag, and Au are also found in the combined state.

Many "insoluble" compounds of the metals are found in the earth's crust. Solids that contain these compounds are the **ores** from which metals are extracted. Ores contain **minerals,** comparatively pure compounds of the metals of interest, mixed with relatively large amounts of **gangue**—sand, soil, clay, rock, and other material. Soluble compounds are found dissolved in the sea or in salt beds in areas where large bodies of water have evaporated. Metal ores can be classified by the anions with which the metal ions are combined (Table 22-1 and Figure 22-1).

Pronounce "gangue" as one syllable with a soft final g.

Table 22-1 *Common Classes of Ores*

Anion	Examples and Names of Minerals
none (native ores)	Au, Ag, Pt, Os, Ir, Ru, Rh, Pd, As, Sb, Bi, Cu
oxide	hematite, Fe_2O_3; magnetite, Fe_3O_4; bauxite, Al_2O_3; cassiterite, SnO_2; periclase, MgO; silica, SiO_2
sulfide	chalcopyrite, $CuFeS_2$; chalcocite, Cu_2S; sphalerite, ZnS; galena, PbS; iron pyrites, FeS_2; cinnabar, HgS
chloride	rock salt, NaCl; sylvite, KCl; carnallite, $KCl \cdot MgCl_2$
carbonate	limestone, $CaCO_3$; magnesite, $MgCO_3$; dolomite, $MgCO_3 \cdot CaCO_3$
sulfate	gypsum, $CaSO_4 \cdot 2H_2O$; epsom salts, $MgSO_4 \cdot 7H_2O$; barite, $BaSO_4$
silicate	beryl, $Be_3Al_2Si_6O_{18}$; kaolinite, $Al_2(Si_2O_8)(OH)_4$; spodumene, $LiAl(SiO_3)_2$

The most widespread minerals are silicates. But extraction of metals from silicates is very difficult. Metals are obtained from silicate minerals only when there is no other more economical alternative.

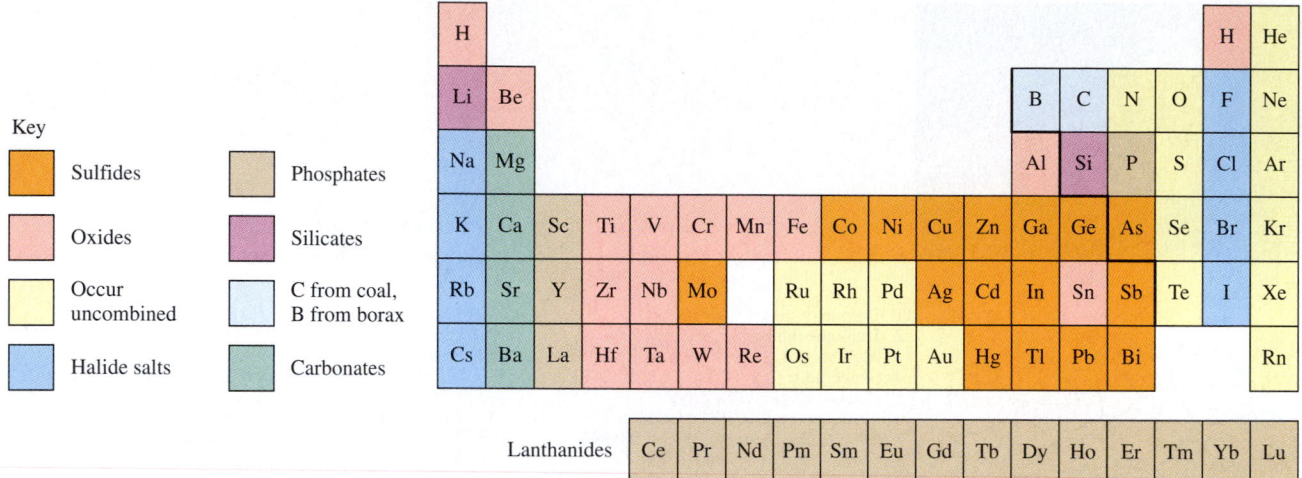

Figure 22-1 Major natural sources of the elements. The soluble halide salts are found in oceans, salt lakes, brine wells, and solid deposits. Most helium is obtained from wells in the United States and Russia. Most of the other noble gases are obtained from air.

METALLURGY

Metallurgy is the commercial extraction of metals from their ores and the preparation of metals for use. It usually includes several steps: (1) mining the ore, (2) pretreatment of the ore, (3) reduction of the ore to the free metal, (4) refining or purifying the metal, and (5) alloying, if necessary.

22-2 PRETREATMENT OF ORES

After being mined, many ores must be concentrated by removal of most of the gangue. Most sulfides have relatively high densities and are more dense than gangue. After pulverization, the lighter gangue particles are removed by a variety of methods. One involves blowing the lighter particles away using a cyclone separator (Figure 22-2a). The lighter particles can be sifted out through layers of vibrating wire mesh or on inclined vibration tables.

The **flotation** method is particularly applicable to sulfides, carbonates, and silicates, which are not "wet" by water or else can be made water-repellent by treatment. Their surfaces are easily covered by layers of oil or other flotation agents. A stream of air is blown through a swirled suspension of such an ore in water and oil (or other agent). Bubbles form on the oil on the mineral particles and cause them to rise to the surface. The bubbles are prevented from breaking and escaping by a layer of oil and emulsifying agent. A frothy ore concentrate forms at the surface. By varying the relative amounts of oil and water, the types of oil additive, the air pressure, and so on, it is even possible to separate one metal sulfide, carbonate, or silicate from another (Figure 22-2b).

Another pretreatment process involves chemical modification. This converts metal compounds to more easily reduced forms. Carbonates and hydroxides may be heated to drive off CO_2 and H_2O, respectively.

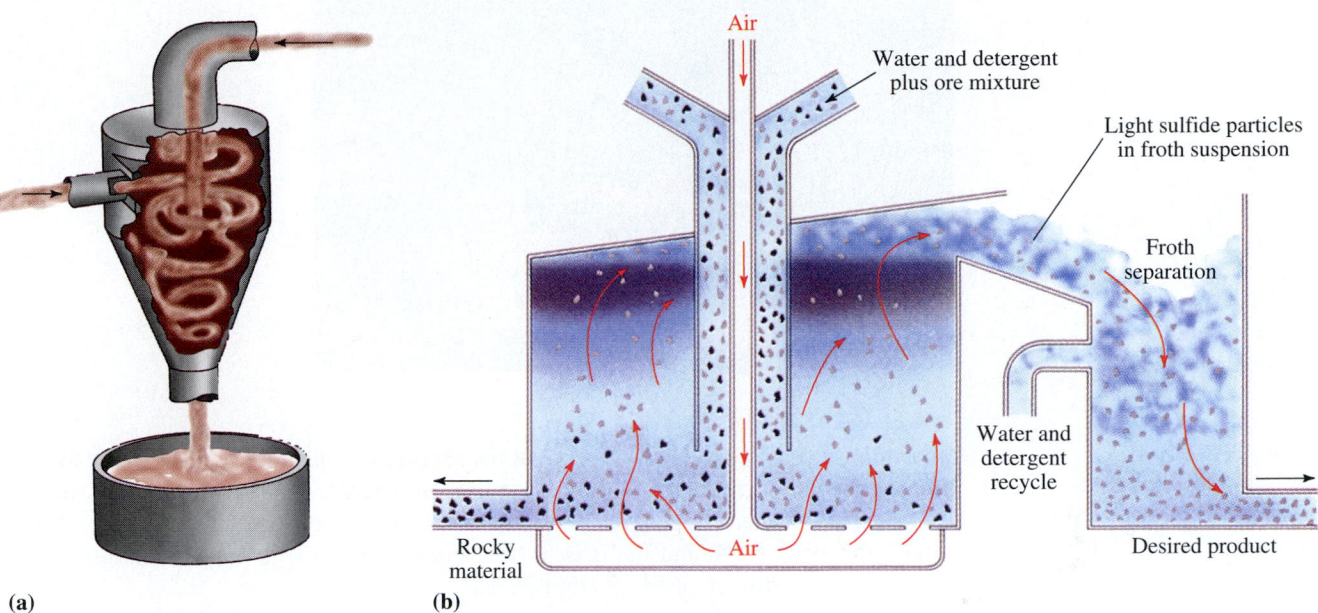

(a) **(b)**

Figure 22-2 (a) The cyclone separator enriches metal ores. Crushed ore is blown in at high velocity. Centrifugal force takes the heavier particles, with higher percentages of metal, to the wall of the separator. These particles spiral down to the collection bin at the bottom. Lighter particles, not as rich in the metal, move into the center. They are carried out the top in the air stream. (b) A representation of the flotation process for enrichment of copper sulfide ore. The relatively light sulfide particles are suspended in the water–oil–detergent mixture and collected as a froth. The denser material sinks to the bottom of the container.

$$CaCO_3(s) \xrightarrow{\Delta} CaO(s) + CO_2(g)$$

$$Mg(OH)_2(s) \xrightarrow{\Delta} MgO(s) + H_2O(g)$$

Some sulfides are converted to oxides by **roasting,** i.e., heating below their melting points in the presence of oxygen from air. For example,

$$2ZnS(s) + 3O_2(g) \xrightarrow{\Delta} 2ZnO(s) + 2SO_2(g)$$

Iron pyrite, FeS_2 (also known as "fool's gold"), at left, and galena, PbS.

Figure 22-3 Trees damaged by acid rain and air pollution in the southwestern United States.

Some of these environmental problems are discussed in the Chemistry in Use essay "Acid Rain" in Chapter 6.

Roasting sulfide ores causes air pollution. Enormous quantities of SO_2 escape into the atmosphere (Section 6-8, part 4), where it causes great environmental damage (Figure 22-3). Federal regulations now require limitation of the amount of SO_2 that escapes with stack gases and fuel gases. Now most of the SO_2 is trapped and used in the manufacture of sulfuric acid (Section 24-12).

22-3 REDUCTION TO THE FREE METALS

The method used for reduction, or smelting, of metal ores to the free metals depends on how strongly the metal ions are bonded to anions. When bonding is stronger, more energy is required to reduce the metals. This makes reduction more expensive. The most active metals usually have the strongest bonding.

The least reactive metals occur in the free state and thus require no reduction. Examples include Au, Ag, and Pt. This is why gold and silver have been used as free metals since prehistoric times. Some less active metals, such as Hg, can be obtained directly from their

Rhodochrosite, $MnCO_3$, is pink.

Rutile contains TiO_2.

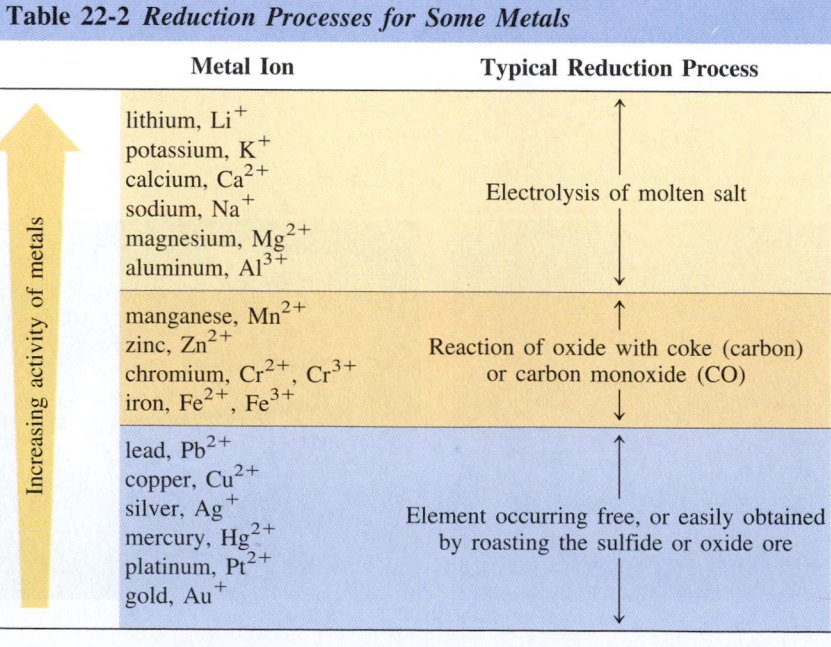

Table 22-2 *Reduction Processes for Some Metals*	
Metal Ion	**Typical Reduction Process**
lithium, Li^+ potassium, K^+ calcium, Ca^{2+} sodium, Na^+ magnesium, Mg^{2+} aluminum, Al^{3+}	Electrolysis of molten salt
manganese, Mn^{2+} zinc, Zn^{2+} chromium, Cr^{2+}, Cr^{3+} iron, Fe^{2+}, Fe^{3+}	Reaction of oxide with coke (carbon) or carbon monoxide (CO)
lead, Pb^{2+} copper, Cu^{2+} silver, Ag^+ mercury, Hg^{2+} platinum, Pt^{2+} gold, Au^+	Element occurring free, or easily obtained by roasting the sulfide or oxide ore

Increasing activity of metals

Table 22-3 *Some Specific Reduction Processes*

Metal	Compound (ore)	Reduction Process	Comments
mercury	HgS (cinnabar)	Roast reduction; heating of ore in air $$HgS + O_2 \xrightarrow{\Delta} Hg + SO_2$$	
copper	sulfides such as Cu_2S (chalcocite)	Blowing of oxygen through purified molten Cu_2S $$Cu_2S + O_2 \xrightarrow{\Delta} 2Cu + SO_2$$	Preliminary ore concentration and purification steps required to remove FeS impurities
zinc	ZnS (sphalerite)	Conversion to oxide and reduction with carbon $$2ZnS + 3O_2 \xrightarrow{\Delta} 2ZnO + 2SO_2$$ $$ZnO + C \xrightarrow{\Delta} Zn + CO$$	Process also used for the production of lead from galena, PbS
iron	Fe_2O_3 (hematite)	Reduction with carbon monoxide $$2C \text{ (coke)} + O_2 \xrightarrow{\Delta} 2CO$$ $$Fe_2O_3 + 3CO \xrightarrow{\Delta} 2Fe + 3CO_2$$	
titanium	TiO_2 (rutile)	Conversion of oxide to halide salt and reduction with an active metal $$TiO_2 + 2Cl_2 + 2C \xrightarrow{\Delta} TiCl_4 + 2CO$$ $$TiCl_4 + 2Mg \xrightarrow{\Delta} Ti + 2MgCl_2$$	Also used for the reduction of UF_4 obtained from UO_2, pitchblende
tungsten	$FeWO_4$ (wolframite)	Reduction with hydrogen $$WO_3 + 3H_2 \xrightarrow{\Delta} W + 3H_2O$$	Used also for molybdenum
aluminum	$Al_2O_3 \cdot nH_2O$ (bauxite)	Electrolytic reduction (electrolysis) in molten cryolite, $Na_3[AlF_6]$, at 1000°C $$2Al_2O_3 \xrightarrow{\Delta} 4Al + 3O_2$$	
sodium	NaCl (sea water)	Electrolysis of molten chlorides $$2NaCl \xrightarrow{\Delta} 2Na + Cl_2$$	Also for calcium, magnesium, and other active metals in Groups IA and IIA

sulfide ores by roasting. This reduces metal ions to the free metals by oxidation of the sulfide ions.

$$HgS(s) + O_2(g) \xrightarrow{\Delta} SO_2(g) + Hg(g)$$
$$\underset{\text{cinnabar}}{} \quad \underset{\text{from air}}{} \qquad \qquad \underset{\substack{\text{obtained as vapor;}\\ \text{later condensed}}}{}$$

Roasting more active metal sulfides produces metal oxides, but no free metals.

$$2NiS(s) + 3O_2(g) \xrightarrow{\Delta} 2NiO(s) + 2SO_2(g)$$

The resulting metal oxides are then reduced to free metals with coke or CO. If C must be avoided, another reducing agent, such as H_2, Fe, or Al, is used.

Coke is impure carbon.

$$SnO_2(s) + 2C(s) \xrightarrow{\Delta} Sn(\ell) + 2CO(g)$$

$$WO_3(s) + 3H_2(g) \xrightarrow{\Delta} W(s) + 3H_2O(g)$$

The very active metals, such as Al and Na, are reduced electrochemically, usually from their anhydrous molten salts. If H_2O is present, it is reduced preferentially. Tables 22-2 and 22-3 summarize reduction processes for some metal ions.

CHEMISTRY IN USE

The Development of Science

Human History Reflects the Activity of Metals

Gold and silver were probably the first two metals discovered by humans. Thousands of years after the discovery of gold and silver, our ancestors probably discovered copper, tin, and iron. Was this order in which we discovered metals one of chance, or does a more fundamental reason exist to explain why metals were found in this particular order? Before we begin to answer such questions, let's review some properties of metals.

One chemical property of metals is their tendency to lose electrons; this tendency is greatest for the metals on the lower left side of the periodic table. The properties of nonmetals are different, i.e., they often gain electrons in chemical reactions. As a result, many metals react with many nonmetals to form compounds.

Very active metals, such as sodium, lose their electrons easily; they are unlikely to be found free in nature. In contrast, noble metals, such as gold, are often found free in nature. Because gold is one of the least active metals, it is usually found free and may have been the first metal discovered by early humans. The larger pieces of free gold near the surface of the earth may have been found during "gold rushes." Interestingly, a prospector found a gold nugget while searching in a Georgia mountain stream in 1991. The nugget weighed a few pounds and was worth tens of thousands of dollars.

There's an interesting flip side to an inactive metal's low reactivity. When heated, compounds of inactive metals yield the free metal. For example, heating mercury(II) oxide produces free mercury and oxygen.

$$2HgO(s) \xrightarrow{\Delta} 2Hg(\ell) + O_2(g)$$

Imagine prehistoric people building a fire on an area rich in mercury(II) oxide and then discovering the liquid mercury, freed by the heat of the fire.

It is easier to extract less active elements, such as mercury, from ores such as mercury(II) oxide than it is to extract more active elements, such as sodium, from ores such as sodium chloride. So we can see why the less active metals were discovered first. A table of standard electrode potentials for common metals can be thought of as a record of metal discovery in reverse, a ladder up which humanity has climbed to discover the metals.

Historians describe periods of time by the primary resources used for making tools, cooking utensils, and weapons. Consequently, the first age was the Stone Age. The Stone Age lasted until about 4000 BC with the discovery of copper, and the Copper Age lasted until about 3000 BC. As humanity climbed higher up the table of electrode potentials, more active metals such as lead, approximately 3500 BC, and tin, approximately 3000 BC, were discovered. When tin is melted with copper, bronze is produced; the Bronze Age began about 3000 BC and lasted until 800 BC.

Iron was discovered around 800 BC, and the Iron Age began. Iron is more active than copper, lead, and tin. We have been able to make stainless steel (12% chromium, 8% nickel, and 80% iron) since 1790, when the even more active metal chromium was discovered. In the early 1800s we discovered very active metals such as potassium and lithium.

Some fear that, with increasing demand, we may run short of less abundant metals such as copper, zinc, and chromium. If we exhaust our supplies, we may have to learn to substitute more abundant iron or aluminum for the depleted metals.

Ronald DeLorenzo
Middle Georgia College

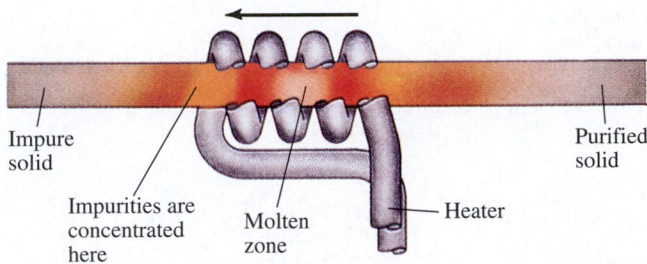

Figure 22-4 A representation of a zone refining apparatus.

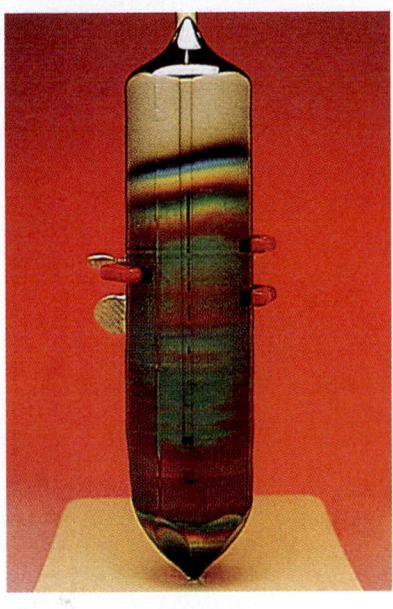

Ultrapure silicon is produced by zone refining.

22-4 REFINING OF METALS

Metals obtained from reduction processes are almost always impure. Further refining (purification) is usually required. This can be accomplished by distillation if the metal is more volatile than its impurities, as in the case of mercury. Among the metals purified electrolytically are Cu, Ag, Au, and Al (Section 21-8). The impure metal is the anode, and a small sample of the pure metal is the cathode. Both are immersed in a solution of the desired metal ion.

Zone refining is often used when extremely pure metals are desired for such applications as solar cells and semiconductors (Section 13-17). An induction heater surrounds a bar of the impure solid and passes slowly from one end to the other (Figure 22-4). As it passes, it melts portions of the bar, which slowly recrystallize as the heating element moves away. The impurity does not fit into the crystal as easily as the element of interest, so most of it is carried along in the molten portion until it reaches the end. Repeated passes of the heating element produce a bar of high purity.

The end containing the impurities can be sliced off and recycled.

After refining, many metals are alloyed, or mixed with other elements, to change their physical and chemical characteristics. In some cases, certain impurities are allowed to remain during refining because their presence improves the properties of the metal. For example, a small amount of carbon in iron greatly enhances its hardness. Examples of alloys include brass, bronze, duralumin, and stainless steel.

Mixtures of metals frequently have properties that are more desirable for a particular purpose than are those of a free metal. In such cases metals are alloyed.

METALLURGIES OF SPECIFIC METALS

The metallurgies of Mg, Al, Fe, Cu, and Au will be discussed as specific examples. The order of increasing standard aqueous electrode potentials of these metals indicates the order of increasing ease of reduction to the free metals.

Mg and Al are active metals, Fe and Cu are moderately active, and Au is relatively inactive.

Reduction Half-Reaction	Standard Reduction Potential E^0, Volts
$Mg^{2+} + 2e^- \longrightarrow Mg$	-2.37
$Al^{3+} + 3e^- \longrightarrow Al$	-1.66
$Fe^{2+} + 2e^- \longrightarrow Fe$	-0.44
$Cu^{2+} + 2e^- \longrightarrow Cu$	$+0.337$
$Au^{3+} + 3e^- \longrightarrow Au$	$+1.50$

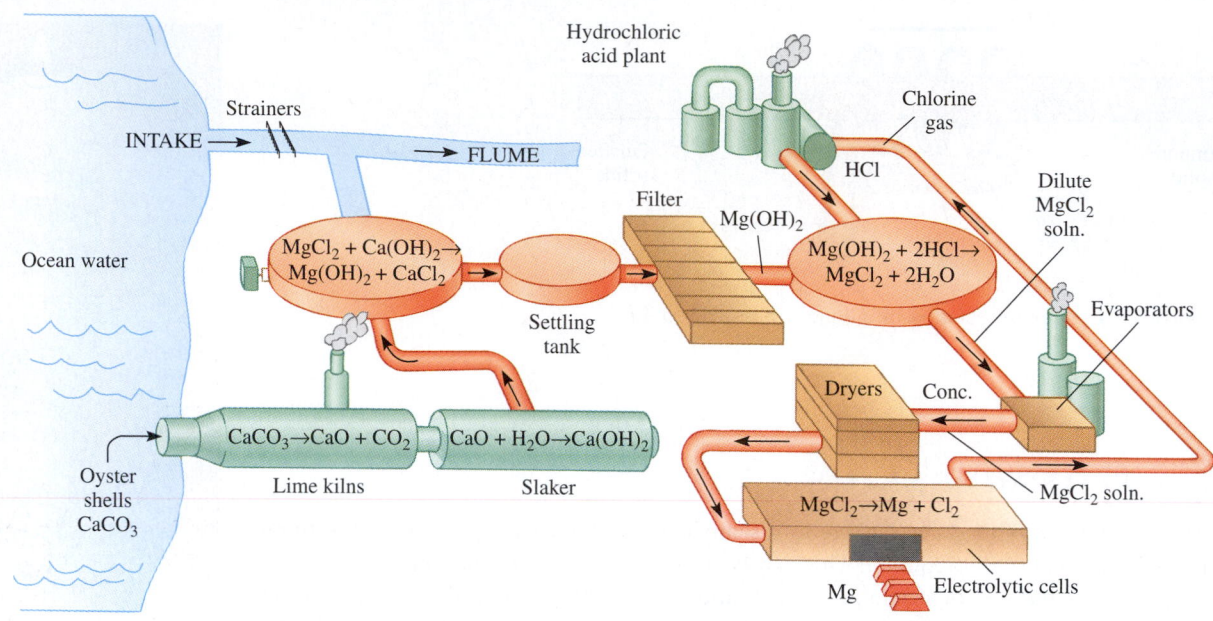

Figure 22-5 Schematic diagram of an industrial plant for the production of magnesium from the Mg^{2+} ions in sea water.

22-5 MAGNESIUM

Magnesium occurs widely in carbonate ores, but most Mg comes from salt brines and from the sea (Figure 22-5). Sea water is 0.13% Mg by mass. Because of its low density (1.74 g/cm^3), Mg is used in lightweight structural alloys for such items as automobile and aircraft parts.

Magnesium ions are precipitated as $Mg(OH)_2$ by addition of $Ca(OH)_2$ (slaked lime) to sea water. The slaked lime is obtained by crushing oyster shells ($CaCO_3$), heating them to produce lime (CaO), and then adding a limited amount of water (slaking).

$$CaCO_3(s) \xrightarrow{\Delta} CaO(s) + CO_2(g) \qquad \text{lime production}$$

$$CaO(s) + H_2O(\ell) \longrightarrow Ca(OH)_2(s) \qquad \text{slaking lime}$$

$$Ca(OH)_2(s) + Mg^{2+}(aq) \longrightarrow Ca^{2+}(aq) + Mg(OH)_2(s) \qquad \text{precipitation}$$

The last reaction occurs because K_{sp} for $Mg(OH)_2$, 1.5×10^{-11}, is much smaller than that for $Ca(OH)_2$, 7.9×10^{-6}. The milky-white suspension of $Mg(OH)_2$ is filtered, and the solid $Mg(OH)_2$ is then neutralized with HCl to produce $MgCl_2$ solution. Evaporation of the H_2O leaves solid $MgCl_2$, which is then melted and electrolyzed (Figure 22-6) under an inert atmosphere to produce molten Mg and gaseous Cl_2. The products are separated as they are formed, to prevent recombination.

$$Mg(OH)_2(s) + 2[H^+(aq) + Cl^-(aq)] \longrightarrow [Mg^{2+}(aq) + 2Cl^-(aq)] + 2H_2O$$

$$\xrightarrow[\text{then melt solid}]{\text{evaporate solution,}} MgCl_2(\ell) \xrightarrow{\text{electrolysis}} Mg(\ell) + Cl_2(g)$$

A bed of limestone ($CaCO_3$), Verde River, Arizona.

Magnesium is cast into ingots or alloyed with other light metals. If no HCl is available, the by-product Cl_2 is used to produce more HCl for neutralization of $Mg(OH)_2$.

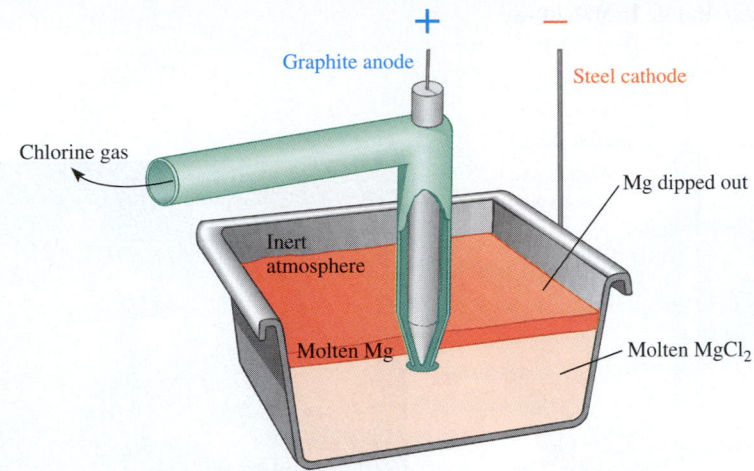

Graphite anode

Steel cathode

Chlorine gas

Mg dipped out

Inert atmosphere

Molten Mg

Molten $MgCl_2$

Figure 22-6 A cell for electrolyzing molten $MgCl_2$. The magnesium metal is formed on the steel cathode and rises to the top, where it is dipped off periodically. Chlorine gas is formed around the graphite anode and is piped off.

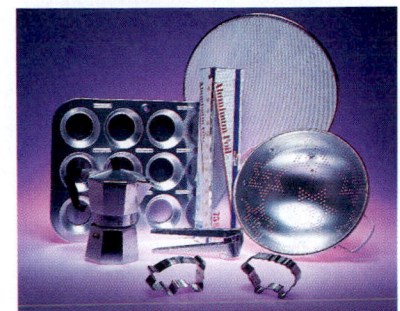

Many consumer items are made of aluminum.

22-6 ALUMINUM

Aluminum is the most commercially important nonferrous metal. Its chemistry and uses will be discussed in Section 23-7. Aluminum is obtained from bauxite, or hydrated aluminum oxide, $Al_2O_3 \cdot xH_2O$. Aluminum ions can be reduced to Al by electrolysis only in the absence of H_2O. First the crushed bauxite is purified by dissolving it in a concentrated solution of NaOH to form soluble $Na[Al(OH)_4]$. Then $Al(OH)_3 \cdot xH_2O$ is precipitated from the filtered solution by blowing in carbon dioxide to neutralize the unreacted NaOH and one OH^- ion per formula unit of $Na[Al(OH)_4]$. Heating the hydrated product dehydrates it to Al_2O_3.

Recall that Al_2O_3 is amphoteric. Impurities such as oxides of iron, which are not amphoteric, are left behind in the crude ore.

$$Al_2O_3(s) + 2NaOH(aq) + 3H_2O(\ell) \longrightarrow 2Na[Al(OH)_4](aq)$$

$$2Na[Al(OH)_4](aq) + CO_2(aq) \longrightarrow 2Al(OH)_3(s) + Na_2CO_3(aq) + H_2O(\ell)$$

$$2Al(OH)_3(s) \xrightarrow{\Delta} Al_2O_3(s) + 3H_2O(g)$$

For clarity, we have not shown waters of hydration.

The melting point of Al_2O_3 is 2045°C; electrolysis of pure molten Al_2O_3 would have to be carried out at or above this temperature, with great expense. However, it can be done at a much lower temperature when Al_2O_3 is mixed with much lower-melting cryolite, a mixture of NaF and AlF_3 often represented as $Na_3[AlF_6]$. The molten mixture can be electrolyzed at 1000°C with carbon electrodes. The cell used industrially for this process, called the **Hall–Heroult process,** is shown in Figure 22-7.

A mixture of compounds typically has a lower melting point than any of the pure compounds (Chapter 14).

The inner surface of the cell is coated with carbon or carbonized iron, which functions as the cathode at which aluminum ions are reduced to the free metal. The graphite anode is oxidized to CO_2 gas and must be replaced frequently. This is one of the chief costs of aluminum production.

(cathode)	$4(Al^{3+} + 3e^- \longrightarrow Al(\ell))$
(anode)	$3(C(s) + 2O^{2-} \longrightarrow CO_2(g) + 4e^-)$
(net reaction)	$4Al^{3+} + 3C(s) + 6O^{2-} \longrightarrow 4Al(\ell) + 3CO_2(g)$

Molten aluminum is more dense than molten cryolite, so it collects in the bottom of the cell until it is drawn off and cooled to a solid.

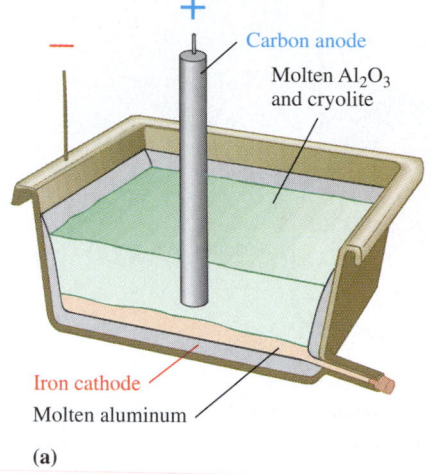

Figure 22-7 (a) Schematic drawing of a cell for producing aluminum by electrolysis of a melt of Al_2O_3 in $Na_3[AlF_6]$. The molten aluminum collects in the container, which acts as the cathode. (b) Casting molten aluminum. Electrolytic cells used in the Hall–Heroult process appear in the background.

(a)

(b)

Recently, a more economical approach, the Alcoa chlorine process, has been developed on a modest commercial scale. The anhydrous bauxite is first converted to $AlCl_3$ by reaction with Cl_2 in the presence of carbon. The $AlCl_3$ is then melted and electrolyzed to give aluminum, and the recovered chlorine is reused in the first step.

$$2Al_2O_3(s) + 3C(coke) + 6Cl_2(g) \longrightarrow 4AlCl_3(s) + 3CO_2(g)$$

$$2AlCl_3(\ell) \longrightarrow 2Al(\ell) + 3Cl_2(g)$$

This process uses only about 30% as much electrical energy as the Hall–Heroult process.

The use of large amounts of electrical energy in electrolysis makes production of aluminum from ores an expensive metallurgy. Methods for recycling used Al use less than 10% of the energy required to make new metal from bauxite by the Hall–Heroult process. Processing of recycled Al now accounts for more than 25% of the production of this metal. This helps to keep down the cost of aluminum.

22-7 IRON

The most desirable iron ores contain hematite, Fe_2O_3, or magnetite, Fe_3O_4. As the available supplies of these high-grade ores have dwindled, taconite, which is magnetite in very hard silica rock, has become an important source of iron. The oxide is reduced in blast furnaces (Figure 22-8) by carbon monoxide. Coke mixed with limestone ($CaCO_3$) and crushed ore is admitted at the top of the furnace as the "charge." A blast of hot air from the bottom burns the coke to carbon monoxide with the evolution of more heat.

$$2C(s) + O_2(g) \xrightarrow{\Delta} 2CO(g) + heat$$

Most of the oxide is reduced to molten iron by carbon monoxide, although some is reduced directly by coke. Several stepwise reductions occur (Figure 22-8), but the overall reactions for Fe_2O_3 can be summarized as follows:

$$Fe_2O_3(s) + 3CO(g) \xrightarrow{\Delta} 2Fe(\ell) + 3CO_2(g) + heat$$

$$Fe_2O_3(s) + 3C(s) \xrightarrow{\Delta} 2Fe(\ell) + 3CO(g) + heat$$

Iron ore is scooped in an open-pit mine.

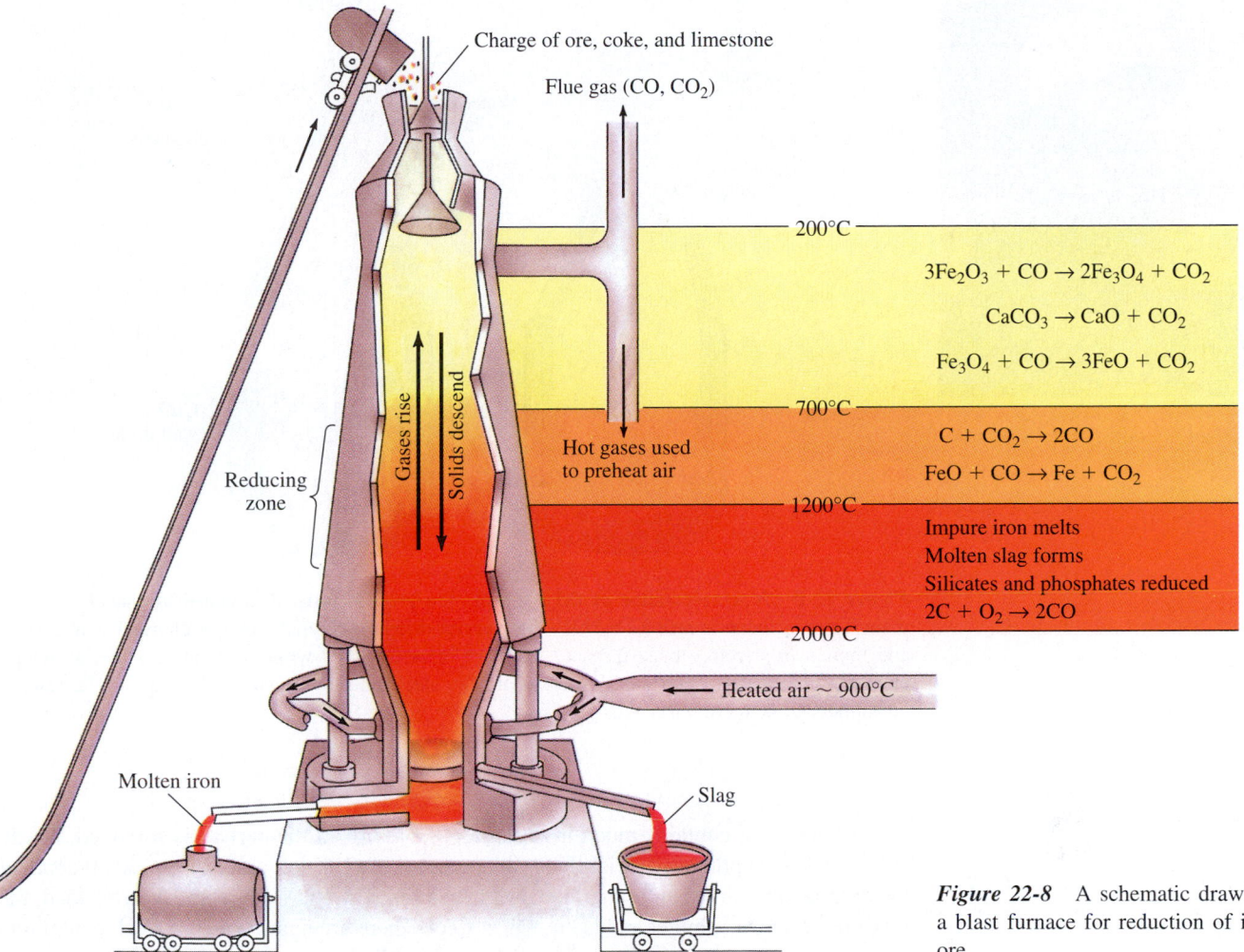

Charge of ore, coke, and limestone

Flue gas (CO, CO$_2$)

Gases rise

Solids descend

Reducing zone

200°C

$3Fe_2O_3 + CO \rightarrow 2Fe_3O_4 + CO_2$

$CaCO_3 \rightarrow CaO + CO_2$

$Fe_3O_4 + CO \rightarrow 3FeO + CO_2$

700°C

$C + CO_2 \rightarrow 2CO$

$FeO + CO \rightarrow Fe + CO_2$

1200°C

Impure iron melts
Molten slag forms
Silicates and phosphates reduced
$2C + O_2 \rightarrow 2CO$

2000°C

Hot gases used to preheat air

Heated air ~ 900°C

Molten iron

Slag

Figure 22-8 A schematic drawing of a blast furnace for reduction of iron ore.

Much of the CO$_2$ reacts with excess coke to produce more CO to reduce the next incoming charge.

$$CO_2(g) + C(s) \xrightarrow{\Delta} 2CO(g)$$

The limestone, called a **flux,** reacts with the silica gangue in the ore to form a molten **slag** of calcium silicate.

$$CaCO_3(s) \xrightarrow{\Delta} CaO(s) + CO_2(g)$$

limestone

$$CaO(s) + SiO_2(s) \xrightarrow{\Delta} CaSiO_3(\ell)$$

gangue   slag

The slag is less dense than molten iron; it floats on the surface of the iron and protects it from atmospheric oxidation. Both are drawn off periodically. Some of the slag is subsequently used in the manufacture of cement.

The iron obtained from the blast furnace contains carbon, among other things. It is called **pig iron.** If it is remelted, run into molds, and cooled, it becomes **cast iron.** This is

Reaction of a metal oxide (basic) with a nonmetal oxide (acidic) forms a salt.

Considerable amounts of slag are also used to neutralize acidic soil. If there were no use for the slag, its disposal would be a serious economic and environmental problem.

Molten steel is poured from a basic oxygen furnace.

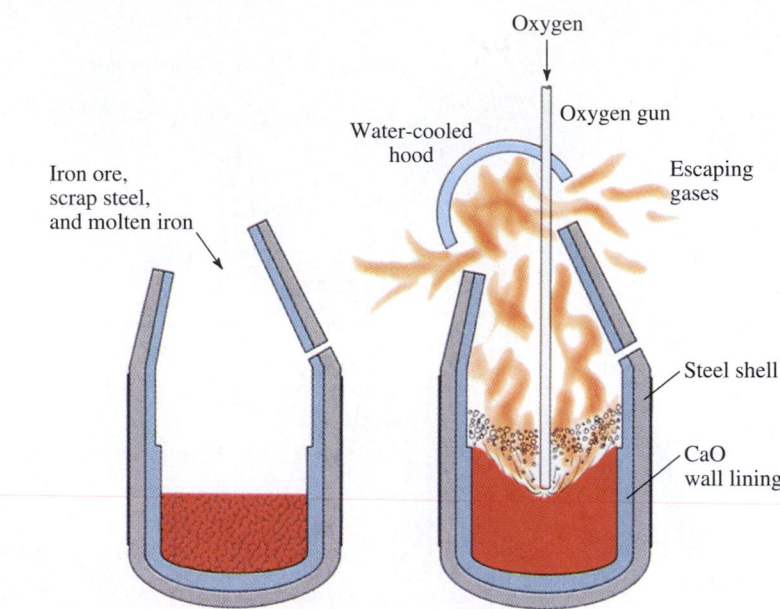

Figure 22-9   A representation of the basic oxygen process furnace. Much of the steel manufactured today is refined by blowing oxygen through a furnace that is charged with scrap and molten iron from a blast furnace. After the refined iron is withdrawn into a ladle, alloying elements are added to give the desired steel. The steel industry is one of the nation's largest consumers of oxygen.

brittle because it contains much iron carbide, Fe_3C. If all the carbon is removed, nearly pure iron can be produced. It is silvery in appearance, quite soft and of little use. If *some* of the carbon is removed and other metals such as Mn, Cr, Ni, W, Mo, and V are added, the mixture becomes stronger and is known as **steel.** There are many types of steel, containing alloyed metals and other elements in various controlled proportions. Stainless steels show high tensile strength and excellent resistance to corrosion. The most common kind contains 14 to 18% chromium and 7 to 9% nickel.

Pig iron can also be converted to steel by burning out most of the carbon with O_2 in a basic oxygen furnace (Figure 22-9). Oxygen is blown through a heat-resistant tube inserted below the surface of the *molten* iron. Carbon burns to CO, which subsequently escapes and burns to CO_2.

22-8 COPPER

Copper is so widely used, especially in its alloys such as bronze (Cu and Sn) and brass (Cu and Zn), that it is becoming very scarce. The U.S. Bureau of Mines estimates that the known worldwide reserves of copper ore will be exhausted during the first half of the next century (Figure 22-10). The two main classes of copper ores are the mixed sulfides of copper and iron—such as chalcopyrite, $CuFeS_2$—and the basic carbonates, such as azurite, $Cu_3(CO_3)_2(OH)_2$, and malachite, $Cu_2CO_3(OH)_2$.

Let's consider $CuFeS_2$ (or $CuS \cdot FeS$). The copper compound is separated from gangue by flotation (Figures 22-2b and 22-11) and then roasted to remove volatile impurities. Enough air is used to convert iron(II) sulfide, but not copper(II) sulfide, to the oxide.

It is now profitable to mine ores containing as little as 0.25% copper. The increased use of fiber optics in place of copper in communications cables may help to lessen the demand for this metal. The use of superconducting materials in electricity transmission lines could eventually provide enormous savings.

Figure 22-10 An open-pit copper mine near Bagdad, Arizona.

Figure 22-11 A copper ore being enriched by flotation.

$$2CuFeS_2(s) + 3O_2(g) \xrightarrow{\Delta} 2FeO(s) + 2CuS(s) + 2SO_2(g)$$

The roasted ore is then mixed with sand (SiO_2), crushed limestone ($CaCO_3$), and some unroasted ore that contains copper(II) sulfide in a reverberatory furnace at 1100°C. CuS is reduced to Cu_2S, which melts. The limestone and silica form a molten calcium silicate glass. This dissolves iron(II) oxide to form a slag less dense than the molten copper(I) sulfide, on which it floats.

$$CaCO_3(s) + SiO_2(s) \xrightarrow{\Delta} CaSiO_3(\ell) + CO_2(g)$$

$$CaSiO_3(\ell) + FeO(s) + SiO_2(s) \xrightarrow{\Delta} CaSiO_3 \cdot FeSiO_3(\ell)$$

The slag is periodically drained off. The molten copper(I) sulfide is drawn off into a Bessemer converter. There it is again heated and treated with air. This oxidizes sulfide ions to SO_2 and reduces copper(I) ions to metallic copper. The overall process is

$$Cu_2S(\ell) + O_2(g) \xrightarrow{\Delta} 2Cu(\ell) + SO_2(g)$$

The impure copper is refined electrolytically (Section 21-8).

A sample that contains two copper-bearing materials. The blue mineral is azurite, $Cu_3(CO_3)_2(OH)_2$ or $2CuCO_3 \cdot Cu(OH)_2$. The green mineral is malachite, $Cu_2CO_3(OH)_2$ or $CuCO_3 \cdot Cu(OH)_2$.

22-9 GOLD

Gold is an inactive metal, so it occurs mostly in the native state. It is sometimes found as gold telluride. Because of its high density, metallic gold can be concentrated by panning. In this operation, gold-bearing sand and gravel are gently swirled with water in a pan. The lighter particles spill over the edge, and the denser nuggets of gold remain. Gold is concentrated by sifting crushed gravel in a stream of water on a slightly inclined shaking table that contains several low barriers. These impede the descent of the heavier gold particles but allow the lighter particles to pass over. The gold is then alloyed with mercury and removed. The mercury is distilled away, leaving behind the pure gold.

Mining low-gold ore in an open-pit mine.

CHEMISTRY IN USE

The Environment

Utilization and Conservation of Metal Resources

Fuels and structural materials are the basis for technological societies such as ours. We are exhausting our supplies of some metal ores, from which we obtain the metals that serve so many structural purposes. The United States is probably the world's richest country in total resources, but our extravagant use of metal ores has forced us to an ever-increasing dependence on foreign resources. The major known world reserves of several important metals are listed in the table.

There are three obvious methods of conserving our metal resources: (1) substitution, (2) recycling, and (3) utilizing lower-grade ores.

Substitution involves replacing some of the metals currently used widely for structural purposes with other more abundant resources, such as glass, granite, wood, and perhaps plastics.

Recycling is possible now for some metal products such as aluminum and steel cans and lead storage batteries (Section 21-23). Unfortunately, it is not presently economical to recycle the steel in automobiles because it is present in so many different alloys, all of which have different properties. Melting down mixtures of the alloys does not produce useful uniform materials. The presence of so many elements in the mixtures results in steel with very weak structural characteristics. It is not possible

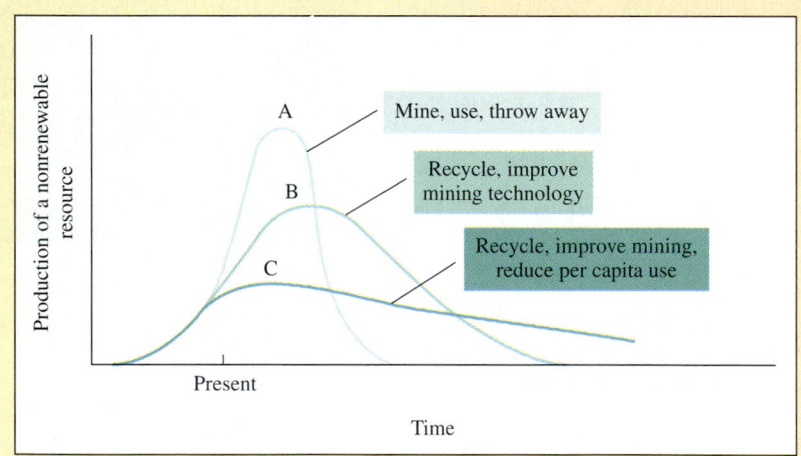

Several possible depletion patterns for a nonrenewable resource.

Because of environmental concerns about mercury toxicity, the cyanide process is increasingly preferred. This is not to suggest that mercury is more toxic than cyanide. The problems due to mercury are greater in that it persists in the environment for a long time, and mercury poisoning is cumulative.

Gold is also recovered from the anode sludge from electrolytic purification of copper (Section 21-8). Gold is so rare that it is also obtained from very low-grade ores by the cyanide process. Air is bubbled through an agitated slurry of the ore mixed with a solution of NaCN. This causes slow oxidation of the metal and the formation of a soluble complex compound.

$$4Au(s) + 8CN^-(aq) + O_2(g) + 2H_2O(\ell) \longrightarrow 4[Au(CN)_2]^-(aq) + 4OH^-(aq)$$

After filtration, free gold can then be regenerated by reduction of $[Au(CN)_2]^-$ with zinc or by electrolytic reduction.

to separate the components of the alloys with reasonable expenditure of energy. This is why there are so many junkyards in which old automobiles are allowed to rust away.

Copper is one of our most important metals whose supply is running low. Copper from the wiring in alternators, starting motors, and elsewhere in automobiles could be easily recycled if most of the wiring were not wound into the frame during assembly. If the construction process were modified, the recycling of copper wire could become economically feasible.

As the sources of ores of a particular metal dwindle, the cost of the metal rises. Consequently, it becomes profitable to extract the metal from *lower-grade ores*. Magnesium is obtained from sea water, and so sea water can be considered an ore. Practically all the known elements can be found in the sea, although most are present in very low concentrations. As the technology of extraction is improved and the prices of the elements rise, we shall probably see more elements being obtained from the sea. There are about 10 million metric tons of gold in the oceans. But it is dissolved in more than a billion cubic kilometers of sea water at a concentration of 7×10^{-9} g/L. It is unlikely that much of it will ever be reclaimed. The seas have such vast volume that they can be considered as a virtually inexhaustible supply of many elements. If we can develop a new, cheap source of energy, it could become economically feasible to do such things as mine the sea for many elements, separate elements from alloys, and obtain aluminum from the very abundant but intractable aluminosilicates that cover the surface of the earth.

In view of the technological advances that must occur before we approach the extremes discussed above, we should concentrate on conserving resources. An estimate of the effects of conservation measures on the depletion pattern of a nonrenewable resource is depicted in the figure. It shows that if we recycle, improve mining techniques, *and* reduce per capita use of a resource, it will be available much longer than if we just mine, use, and throw away.

Reserves of Metals

Metal	Location of Reserves (% of total)
Au	Rep. So. Africa (40)
Hg	Spain (30); Italy (21)
Ag	Russia and former eastern bloc countries (36); U.S. (24)
Sn	Thailand (33); Malaysia (14)
Zn	U.S. (27); Canada (20)
Pb	U.S. (39)
Cu	U.S. (28); Chile (19)
W	China (73)
Mo	U.S. (58); former Soviet Union (20)
Mn	Rep. So. Africa (38); former Soviet Union (25)
Al	Australia (33); Guinea (20)
Co	Rep. Congo (31); Zambia (20)
Pt metals	Rep. So. Africa (47); former Soviet Union (47)
Ni	Cuba (25); New Caledonia (22); Canada (14)
Fe	Former Soviet Union (33); Canada (14)
Cr	Rep. So. Africa (75)

Key Terms

Alloying   Mixing of a metal with other substances (usually other metals) to modify its properties.

Charge   A sample of crushed ore as it is admitted to a furnace for smelting.

Flotation   A method by which hydrophobic (water-repelling) particles of an ore are separated from hydrophilic (water-attracting) particles in a metallurgical pretreatment process.

Flux   A substance added to react with the charge, or a product of its reduction, in metallurgy; usually added to lower a melting point.

Gangue   Sand, rock, and other impurities surrounding the mineral of interest in an ore.

Metallurgy   The overall processes by which metals are extracted from ores.

Native ore   A metal that occurs in an uncombined or free state in nature.

Ore   A natural deposit containing a mineral of an element to be extracted.

Refining   Purifying of a substance.

Roasting   Heating a compound below its melting point in the presence of air.

Slag   Unwanted material produced during smelting.

Smelting   Chemical reduction of a substance at high temperature in metallurgy.

Zone refining   A method of purifying a bar of metal by passing it through an induction heater; this causes impurities to move along in the melted portion.

Exercises

General Concepts

1. List the chemical and physical properties that we usually associate with metals.
2. Define the term "metallurgy." What does the study of metallurgy include?
3. What kinds of metals are most apt to occur in the uncombined (native) state in nature?
4. List the six anions (and their formulas) that are most often combined with metals in ores. Give at least one example of an ore of each kind. What anion is the most commonly encountered?
5. How does an ore differ from a mineral? Name the three general categories of procedures needed to produce pure metals from ores. Describe the purpose of each.
6. Briefly describe one method by which gangue can be separated from the desired mineral during the concentration of an ore.
7. Give the five general steps involved in extracting a metal from its ore and converting the metal to a useful form. Briefly describe the importance of each.
8. Describe the flotation method of ore pretreatment. Are any chemical changes involved?
9. What kinds of ores are roasted? What kinds of compounds are converted to oxides by roasting? What kinds are converted directly to the free metals?
10. Of the following compounds, which would you expect to require electrolysis to obtain the free metals: KCl; $Cr_2(SO_4)_3$; Fe_2O_3; Al_2O_3; Ag_2S; $MgSO_4$? Why?
11. At which electrode is the free metal produced in the electrolysis of a metal compound? Why?
*12. Write the equation that describes the electrolysis of a brine solution to form $NaOH$, Cl_2, and H_2. What mass of each substance will be produced in an electrolysis cell for each mole of electrons passed through the cell? Assume 100% efficiency.
13. The following equations represent reactions used in some important metallurgical processes.

   (a) $Fe_3O_4(s) + CO(g) \longrightarrow Fe(\ell) + CO_2(g)$
   (b) $MgCO_3(s) + SiO_2(s) \longrightarrow MgSiO_3(\ell) + CO_2(g)$
   (c) $Au(s) + CN^- + H_2O(\ell) + O_2(g) \longrightarrow$
   $$[Au(CN)_2]^- + OH^-$$
   Balance the equations. Which one(s) represent reduction to a free metal?

14. Repeat Exercise 13 for

   (a) Al_2O_3 (cryolite solution) $\xrightarrow{\text{electrolysis}}$ $Al(\ell) + O_2(g)$
   (b) $PbSO_4(s) + PbS(s) \longrightarrow Pb(\ell) + SO_2(g)$
   (c) $TaCl_5(g) + Mg(\ell) \longrightarrow Ta(s) + MgCl_2(\ell)$

15. Suggest a method of obtaining manganese from an ore containing manganese(III) oxide, Mn_2O_3. On what basis do you make the suggestion?
16. What is the purpose of utilizing the basic oxygen furnace after the blast furnace in the production of iron?
17. Describe the metallurgy of (a) copper and (b) magnesium.
18. Describe the metallurgy of (a) iron and (b) gold.

Native gold.

19. Briefly describe the Hall–Heroult process for the commercial preparation of aluminum.
20. Name some common minerals that contain iron. Write the chemical formula for the iron compound in each. What is the oxidation number of iron in each substance?
21. What is steel? How does the hardness of iron compare with that of steel?
22. Describe and illustrate the electrolytic refining of Cu.
23. Name the undesirable gaseous product formed during the roasting of copper and other sulfide ores. Why is it undesirable?

BUILDING YOUR KNOWLEDGE

24. The reaction

$$FeO(s) + CO(g) \longrightarrow Fe(s) + CO_2(g)$$

takes place in the blast furnace at a temperature of 800 K. (a) Calculate ΔH^0_{800} for this reaction, using $\Delta H^0_{f, 800} = -268$ kJ/mol for FeO, -111 kJ/mol for CO, and -394 kJ/mol for CO_2. Is this a favorable enthalpy change? (b) Calculate ΔG^0_{800} for this reaction, using $\Delta G^0_{f, 800} = -219$ kJ/mol for FeO, -182 kJ/mol for CO, and -396 kJ/mol for CO_2. Is this a favorable free energy change? (c) Using your values of ΔH^0_{800} and ΔG^0_{800}, calculate ΔS^0_{800}.

25. During the operation of a blast furnace, coke reacts with the oxygen in air to produce carbon monoxide, which, in turn, serves as the reducing agent for the iron ore. Assuming the formula of the iron ore to be Fe_2O_3, calculate the mass of air needed for each ton of iron produced. Assume air to be 21% O_2 by mass and assume that the process is 93% efficient.

26. The following reactions take place during the extraction of copper from copper ore.
(a) $2Cu_2S(\ell) + 3O_2(g) \longrightarrow 2Cu_2O(\ell) + 2SO_2(g)$
(b) $2Cu_2O(\ell) + Cu_2S(\ell) \longrightarrow 6Cu(\ell) + SO_2(g)$
Identify the oxidizing and reducing agents. Show that each equation is correctly balanced by demonstrating that the increase and decrease in oxidation numbers are equal.

*27. Assuming complete recovery of metal, which of the following ores would yield the greater quantity of copper on a mass basis? (a) an ore containing 3.30 mass % azurite, $Cu(OH)_2 \cdot 2CuCO_3$, or (b) an ore containing 4.95 mass % chalcopyrite, $CuFeS_2$.

*28. What mass of copper could be electroplated from a solution of $CuSO_4$, using an electrical current of 3.00 A flowing for 5.00 hr? (Assume 100% efficiency.)

*29. (a) Calculate the weight, in pounds, of sulfur dioxide produced in the roasting of 1 ton of chalcocite ore containing 10.3% Cu_2S, 0.94% Ag_2S, and no other source of sulfur. (b) What weight of sulfuric acid can be prepared from the SO_2 generated, assuming 91% of it can be recovered from stack gases, and 88% of that recovered can be converted to sulfuric acid? (c) How many pounds of pure copper can be obtained, assuming 78% efficient extraction and purification? (d) How many pounds of silver can be produced, assuming 81% of it can be extracted and purified?

*30. Forty-five pounds of Al_2O_3, obtained from bauxite, is mixed with cryolite and electrolyzed. How long would a 0.900-A current have to be passed to convert all the Al^{3+} (from Al_2O_3) to aluminum metal? What volume of oxygen, collected at 785 torr and 125°C, would be produced in the same period of time?

*31. Calculate the percentage of iron in hematite ore containing 62.4% Fe_2O_3 by mass. How many pounds of iron would be contained in 1 ton of the ore?

*32. Find the standard molar enthalpies of formation of Al_2O_3, Fe_2O_3, and HgS in Appendix K. Are the values in line with what might be predicted, in view of the methods by which the metal ions are reduced in extractive metallurgy?

*33. Using data from Appendix K, calculate ΔG^0_{298} for the following reactions.
(a) $Al_2O_3(s) \longrightarrow 2Al(s) + \frac{3}{2}O_2(g)$
(b) $Fe_2O_3(s) \longrightarrow 2Fe(s) + \frac{3}{2}O_2(g)$
(c) $HgS(s) \longrightarrow Hg(\ell) + S(s)$
Are any of the reactions spontaneous? Are the ΔG^0_{298} values in line with what would be predicted based on the relative activities of the metal ions involved? Do increases in temperature favor these reactions?

*34. Sea water contains 0.13 mass % Mg^{2+}. What mass of sea water would have to be processed to yield 1.0 ton of the metal if the recovery process were 82% efficient?

23 Metals II: Properties and Reactions

The tires on the Hummer vehicles used in the Gulf War were made blow-out proof by strong, lightweight magnesium inserts. This is a photo of a civilian model.

OBJECTIVES

As you study this chapter, you should learn

- *About the properties and occurrence of the Group IA metals*

- *Some important reactions of the Group IA metals*

- *Some important uses of the Group IA metals and their compounds*

- *About the properties and occurrence of the Group IIA metals*

- *Some important reactions of the Group IIA metals*

- *Some important uses of the Group IIA metals and their compounds*

- *About the post-transition metals*

- *Some important periodic trends in the properties of Group IIIA metals and some of their compounds*

- *About aluminum and a few of its important compounds*

- *About the d-transition metals and some of their typical compounds*

- *How d-transition metals are classified into subgroups*

- *About the oxides and hydroxides of chromium*

I n this chapter we shall discuss the **representative metals** and some ***d*-transition metals.** The representative elements are those in the A groups of the periodic table. They have valence electrons in their outermost *s* and *p* atomic orbitals. Metallic character increases from top to bottom within groups and from right to left within periods. All the elements in Groups IA (except H) and IIA are metals. The heavier members of Groups IIIA, IVA, and VA are called **post-transition metals.**

Hydrogen is included in IA in the periodic table, but it is *not* a metal.

841

THE ALKALI METALS (GROUP IA)

23-1 GROUP IA METALS: PROPERTIES AND OCCURRENCE

See the discussion of electrolysis of sodium chloride in Section 21-3.

The alkali metals are not found free in nature, because they are so easily oxidized. They are most economically produced by electrolysis of their molten salts. Sodium (2.6% abundance by mass) and potassium (2.4% abundance) are very common in the earth's crust. The other IA metals are quite rare. Francium consists only of short-lived radioactive isotopes formed by alpha-particle emission from actinium (Section 26-4). Both potassium and cesium also have natural radioisotopes. Potassium-40 is important in the potassium–argon radioactive decay method of dating ancient objects (Section 26-11). The properties of the alkali metals vary regularly as the group is descended (Table 23-1).

The free metals, except lithium, are soft, silvery corrosive metals that can be cut with a knife; lithium is harder. Cesium is slightly golden and melts in the hand (wrapped in plastic because it is so corrosive). The relatively low melting and boiling points of the alkali metals result from their fairly weak bonding forces. Each atom can furnish only one

Table 23-1 *Properties of the Group IA Metals*

Property	Li	Na	K	Rb	Cs	Fr
Outer electrons	$2s^1$	$3s^1$	$4s^1$	$5s^1$	$6s^1$	$7s^1$
Melting point (°C)	186	97.8	63.6	38.9	28.5	27
Boiling point (°C)	1347	904	774	688	678	677
Density (g/cm^3)	0.534	0.971	0.862	1.53	1.87	—
Atomic radius (Å)	1.52	1.86	2.31	2.44	2.62	—
Ionic radius, M^+ (Å)	0.60	0.95	1.33	1.48	1.69	—
Electronegativity	1.0	1.0	0.9	0.9	0.8	0.8
E^0 (volts): $M^+(aq) + e^- \longrightarrow M(s)$	−3.05	−2.71	−2.93	−2.93	−2.92	—
Ionization energies (kJ/mol)						
$\quad M(g) \longrightarrow M^+(g) + e^-$	520	496	419	403	376	—
$\quad M^+(g) \longrightarrow M^{2+}(g) + e^-$	7298	4562	3051	2632	2420	—
$\Delta H^0_{hydration}$ (kJ/mol): $M^+(g) + xH_2O \longrightarrow M^+(aq)$	−544	−435	−351	−293	−264	—

electron for metallic bonding (Section 13-17). Because their outer electrons are so loosely held, the metals are excellent electrical and thermal conductors. They ionize when irradiated with low-energy light (the photoelectric effect). These effects become more pronounced with increasing atomic size. Cesium is used in photoelectric cells.

The low ionization energies of the IA metals show that the single electron in the outer shell is very easily removed. In all alkali metal compounds the metals exhibit the +1 oxidation state. Virtually all are ionic. The extremely high second ionization energies show that removal of an electron from a filled shell is impossible by chemical means.

We might expect the standard reduction potentials of the metal ions to reflect the same trends as the first ionization energies. However, the magnitude of the standard reduction potential of Li, -3.05 volts, is unexpectedly large. The first ionization energy is the amount of energy absorbed when a mole of *gaseous* atoms ionize. The standard reduction potential, E^0, indicates the ease with which *aqueous* ions are reduced to the metal (Section 21-15). Thus, hydration energies must also be considered (Section 14-2). Because the Li^+ ion is so small, its charge density (ratio of charge to size) is very high. Therefore, it exerts a stronger attraction for polar H_2O molecules than do the other IA ions. These H_2O molecules must be stripped off during the reduction process. Thus, $\Delta H^0_{hydration}$ for the Li^+ ion and E^0 for the Li^+/Li couple are very negative (Table 23-1).

The high charge density of Li^+ ion accounts for its ability to polarize large anions. This gives a higher degree of covalent character in Li compounds than in other corresponding alkali metal compounds. For example, LiCl is soluble in ethyl alcohol, a less polar solvent than water; NaCl is not. Salts of the alkali metals with small anions are very soluble in water, but salts with large and complex anions, such as silicates and aluminosilicates, are not very soluble.

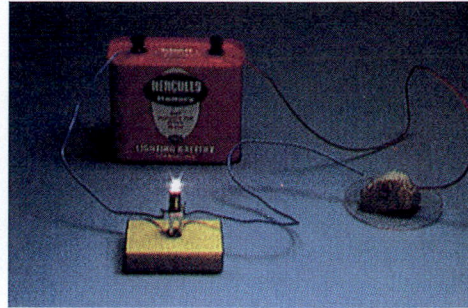
Alkali metals are excellent conductors of electricity.

Polarization of an anion refers to distortion of its electron cloud. The ability of a cation to polarize an anion increases with increasing charge density (ratio of charge/size) of the cation.

23-2 REACTIONS OF THE GROUP IA METALS

Many of the reactions of the alkali metals are summarized in Table 23-2. All are characterized by the loss of one electron per metal atom. These metals are very strong reducing agents. Reactions of the alkali metals with H_2 and O_2 were discussed in Sections 6-7 and 6-8, parts 2; reactions with the halogens in Section 7-2; and reactions with water in Section 4-8, part 2.

Table 23-2 *Some Reactions of the Group IA Metals*

Reaction	Remarks
$4M + O_2 \longrightarrow 2M_2O$	Limited O_2
$4Li + O_2 \longrightarrow 2Li_2O$	Excess O_2 (lithium oxide)
$2Na + O_2 \longrightarrow Na_2O_2$	(sodium peroxide)
$M + O_2 \longrightarrow MO_2$	M = K, Rb, Cs; excess O_2 (superoxides)
$2M + H_2 \longrightarrow 2MH$	Molten metals
$6Li + N_2 \longrightarrow 2Li_3N$	At high temperature
$2M + X_2 \longrightarrow 2MX$	X = halogen (Group VIIA)
$2M + S \longrightarrow M_2S$	Also with Se, Te of Group VIA
$12M + P_4 \longrightarrow 4M_3P$	Also with As, Sb of Group VA
$2M + 2H_2O \longrightarrow 2MOH + H_2$	K, Rb, and Cs react explosively
$2M + 2NH_3 \longrightarrow 2MNH_2 + H_2$	With $NH_3(\ell)$ in presence of catalyst; with $NH_3(g)$ at high temperature (solutions also contain M^+ + solvated e^-)

Sodium reacts vigorously with water.

$$2Na(s) + 2H_2O \longrightarrow$$
$$2[Na^+(aq) + OH(aq)] + H_2(g)$$

The indicator, phenolphthalein, was added to the water. As NaOH forms, the solution turns pink.

The high reactivities of the alkali metals are illustrated by their vigorous reactions with water. Lithium reacts readily; sodium reacts so vigorously that it may ignite; and potassium, rubidium, and cesium burst into flames when dropped into water. The large amounts of heat evolved provide the activation energy to ignite the evolved hydrogen. The elements also react with water vapor in the air or with moisture from the skin.

$$2K + 2H_2O \longrightarrow 2[K^+ + OH^-] + H_2 \qquad \Delta H^0 = -390.8 \text{ kJ/mol rxn}$$

Alkali metals are stored under anhydrous nonpolar liquids such as mineral oil.

As is often true for elements of the second period, Li differs in many ways from the other members of its family. Its ionic charge density and electronegativity are close to those of Mg, so Li compounds resemble those of Mg in some ways. This illustrates the **diagonal similarities** that exist between elements in successive groups near the top of the periodic table.

IA	IIA	IIIA	IVA
Li	Be	B	C
Na	Mg	Al	Si

Lithium is the only IA metal that combines with N_2 to form a nitride, Li_3N. Magnesium readily forms magnesium nitride, Mg_3N_2. Both metals readily combine with carbon to form carbides, whereas the other alkali metals do not react readily with carbon. The solubilities of Li compounds are closer to those of Mg compounds than to those of other IA compounds. The fluorides, phosphates, and carbonates of both Li and Mg are only slightly soluble, but their chlorides, bromides, and iodides are very soluble. Both Li and Mg form normal oxides, Li_2O and MgO, when burned in air at 1 atmosphere pressure. The other alkali metals form peroxides or superoxides.

The IA metal oxides are basic. They react with water to form strong soluble bases.

$$Na_2O(s) + H_2O \longrightarrow 2[Na^+ + OH^-]$$

$$K_2O(s) + H_2O \longrightarrow 2[K^+ + OH^-]$$

The IA cations are derived from strong soluble bases, so they do not hydrolyze.

23-3 USES OF GROUP IA METALS AND THEIR COMPOUNDS

Sodium, Na

Sodium is by far the most widely used alkali metal because it is so abundant. Its salts are essential for life. The metal itself is used as a reducing agent in the manufacture of drugs and dyes and in the metallurgy of metals such as titanium and zirconium.

$$TiCl_4(g) + 4Na(\ell) \xrightarrow{\Delta} 4NaCl(s) + Ti(s)$$

Highway lamps often incorporate Na arcs, which produce a bright yellow glow. A few examples of the uses of sodium compounds are: NaOH, called caustic soda, lye, or soda lye (used for production of rayon, cleansers, textiles, soap, paper, many polymers); Na_2CO_3, called soda or soda ash, and $Na_2CO_3 \cdot 10H_2O$, called washing soda (also used as a substitute for NaOH when a weaker base is acceptable); $NaHCO_3$, called baking soda or bicarbonate of soda (used for baking and other household uses); NaCl (used as table salt

The yellowish glow of some highway lamps is due to a sodium arc. Mercury lamps give a bluish glow.

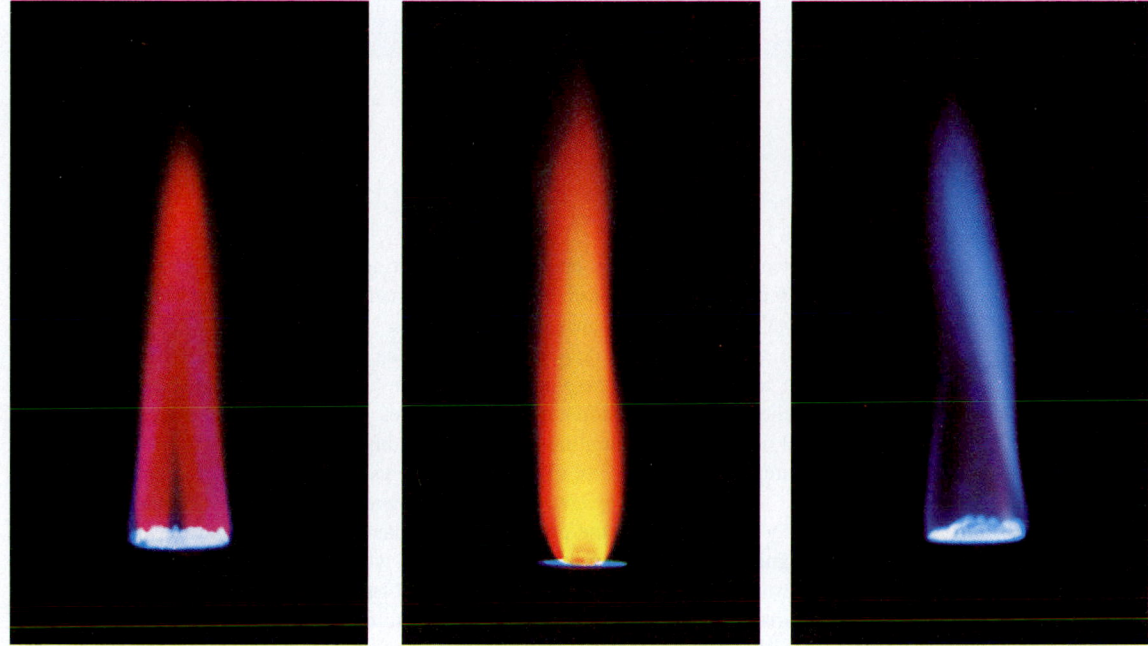

Spacings of energy levels are different for different alkali metals. The salts of the alkali metals impart characteristic colors to flames: lithium (red), sodium (yellow), and potassium (violet).

and as the source of all other compounds of Na and Cl); $NaNO_3$, called Chile saltpeter (a nitrogen fertilizer); Na_2SO_4, called salt cake, a by-product of HCl manufacture (used for production of brown wrapping paper and cardboard boxes); and NaH (used for synthesis of $NaBH_4$, which is used to recover silver and mercury from waste water).

Lithium, Li

Metallic lithium has the highest heat capacity of any element. It is used as a heat transfer medium in experimental nuclear reactors. Extremely lightweight lithium–aluminum alloys are used in aircraft construction. Lithium compounds are used in some lightweight dry cells and storage batteries because they have very long lives, even in extreme temperatures. LiCl and LiBr are very hygroscopic and are used in industrial drying processes and air-conditioning. Lithium compounds are used for the treatment of some types of mental disorders (mainly manic depression).

The highly corrosive nature of both lithium and sodium is a major drawback to applications of the pure metals.

Potassium, K

Like salts of Na (and probably Li), those of potassium are essential for life. KNO_3, commonly known as niter or saltpeter, is used as a potassium and nitrogen fertilizer. Most other major industrial uses for K can be satisfied with the more abundant and cheaper Na.

There are very few practical uses for the rare metals rubidium, cesium, and francium. Cesium is used in some photoelectric cells (Section 5-11).

CHEMISTRY IN USE

The Development of Science

Trace Elements and Life

More than 300 years ago iron was the first trace element shown to be essential in the human diet. An English physician, Thomas Sedenham, soaked "iron and steel filings" in cold Rhenish wine. He used the resulting solution to treat patients suffering from chlorosis, now known to be an iron-deficiency anemia. Nearly 20 trace elements are now believed to be required by humans. The discovery of the biological functions of trace elements is an exciting and controversial area of human nutrition research.

The trace elements can be classified into several categories (see table). In 1989, the National Research Council recognized that iron, iodine, zinc, selenium, copper, chromium, manganese, and molybdenum were dietary essentials for humans. Fluorine is also considered to be valuable for human health, because of its benefits to the teeth and skeleton. These nine trace elements are required by humans and other animals because they are essential components in metalloenzymes and hormones, or because they promote health in a specific tissue (such as fluorine in the teeth and skeleton). The trace elements required by the human body in milligram quantities include iron, zinc, copper, manganese, and fluorine. Trace elements required in microgram (μg) quantities include iodine, selenium, chromium, and molybdenum. Although probably required in μg quantities, no dietary recommendations have been made for arsenic, nickel, silicon, and boron even though there is evidence, primarily from animals,

that they are essential. There is only weak evidence that cadmium, lead, lithium, tin, vanadium, and bromine are essential for humans.

Iron deficiency is one of the most common nutrient deficiencies in the world, occurring in up to 60% of the women, infants, and children of some countries. Anemia, characterized by a low concentration of hemoglobin in the blood or by a low volume of packed red blood cells, is the usual symptom of iron deficiency. Other symptoms include fatigue and cognitive disorders. Up to 1% of the population may have the genetic disease known as hereditary hemochromatosis, which results in excess absorption of dietary iron, and liver and heart damage. Concern about this disease may eventually lead to routine screening so that persons with this disorder can be treated before severe symptoms develop and can avoid foods and supplements with large amounts of iron. The recommended dietary allowance for iron for women ages 23 to 50 years is 50% higher than that for men in the same age group because of the iron lost in menstrual bleeding.

Iodine deficiency remains a major cause of mental retardation and infant mortality and morbidity throughout the world—even though iodine was shown to be essential for human health nearly 100 years ago. More than 1 billion people are believed to be at risk for iodine deficiency. In 1986 the International Council for the Control of Iodine Deficiency Disorders was established in an effort to improve iodine nutrition and alleviate

Many dietary supplements include essential trace elements.

Dietary Trace Elements

Known to Be Essential	Known or Suspected Functions
iron	Hemoglobin, energy metabolism
iodine	Thyroid hormones
zinc	Enzymes, protein synthesis, cell division
copper	Hemoglobin, bone, nerves, vascular system
selenium	Enzymes, protect against oxidant stress
chromium	Insulin action
manganese	Enzymes, bone
molybdenum	Enzymes, sulfur metabolism
fluorine	Bones, teeth

Substantial Evidence for Essentiality	Known or Suspected Functions
arsenic	Amino acid metabolism
boron	Metabolism of calcium, magnesium, hormones
nickel	Not known, suspected in some enzymes
silicon	Bone and connective tissue

Weak Evidence for Essentiality	Known or Suspected Functions
bromine	Not yet known
cadmium	Not yet known
lead	Not yet known
lithium	Not yet known
tin	Not yet known
vanadium	Not yet known

human suffering. This council works closely with the World Health Organization (WHO), the United Nations International Children's Fund (UNICEF), and the United Nations (UN) to alleviate iodine deficiency. Iodine is required for the thyroid hormones, thyroxine and triiodothyronine, that regulate the metabolic rate and O_2 consumption of cells. Iodine is also intimately involved in the control of growth and development, particularly during fetal and infant life.

In the 1930s zinc was discovered to be a dietary essential in animals. Zinc deficiency was recognized as a potential public health problem in the 1960s in Iran where endemic hypogonadism (delayed sexual development) and dwarfism were discovered in adolescents consuming insufficient dietary zinc. More than 200 zinc enzymes have been discovered. Zinc is also important for the structure and function of biomembranes. Loss of zinc from these membranes results in increased susceptibility to oxidative damage, structural strains, and alterations in specific receptor sites and transport systems. Zinc also helps to stabilize

the structures of RNA, DNA, and ribosomes. Several transcription factors contain "Zn fingers" which are needed for the binding of these transcription factors to the DNA. Thus, zinc is absolutely necessary for adequate growth, protein synthesis, and cell division. The best sources of zinc in the human diet are animal foods such as meat, fish, poultry, and dairy products.

Copper was also shown to be essential in the early 1900s. Copper is needed for the absorption and mobilization of iron, so a deficiency of copper causes a type of anemia that is difficult to distinguish from iron deficiency anemia. Copper is also needed for the cardiovascular system, bone, brain, and nervous system. Premature and malnourished infants are particularly susceptible to developing copper deficiency, in part because milk is a poor source of copper. Whole grains, legumes, and nuts are the major dietary sources of copper.

Selenium was first suspected of being a dietary essential in the 1950s. Selenium is considered to be an antioxidant nutrient because it is present in enzymes that help protect against toxic

(Box continues next page.)

CHEMISTRY IN USE (continued)

species of oxygen and free radicals. Selenium deficiency is a major public health problem in certain parts of China where it increases the risk of heart disease, bone and joint disorders, and liver cancer. Selenium is currently under intensive investigation as a possible protector against cancer. The content of selenium in foods is highly variable and dependent on the selenium content of the soil. Generally the best sources of selenium are muscle meats, cereals, and grains.

The 1950s also saw the first evidence that chromium might be a dietary essential. Chromium is believed to promote the action of insulin and thus influences the metabolism of carbohydrates, fats, and proteins. Reports of severe human deficiency of chromium are rare and have been found primarily in people receiving only intravenous feedings for several months or years. Only a few laboratories in the world can accurately measure the amount of chromium in foods and body tissues, because chromium is present in stainless steel which is ubiquitous in analytical laboratories and easily contaminates biological samples.

Manganese and molybdenum are essential for enzymes in humans and other animals, but a dietary deficiency of these minerals is exceedingly rare in humans. Cobalt is essential for vitamin B-12, but the human body cannot make vitamin B-12 from cobalt and thus requires the preformed vitamin from dietary sources. (It is possible to derive some vitamin B-12 from bacterial synthesis in the digestive tract.)

Efforts to discover whether other elements might be essential intensified during the 1970s. Although it is believed that arsenic, nickel, silicon, and boron are probably essential to humans, it has been difficult to determine whether other minerals have specific biological functions in humans or other animals.

There are several reasons why it is difficult to establish the essentiality of trace elements. Some elements such as arsenic and selenium were first recognized for their extreme toxicity, so it has been difficult to convince many health specialists that a toxic element might also be a dietary essential at low levels. Also, most trace elements are present in extremely small amounts in diets and in tissues, and few laboratories are equipped to prevent contamination of samples and to measure these elements with the necessary precision.

Two factors have aided in the discovery of the roles of many trace elements. One is the availability of two highly sensitive analytical techniques, activation analysis and electrothermal atomic absorption spectroscopy, that allow detection of these elements in concentrations of only a few parts per billion. The other is the use of special isolation chambers that allow study of animals under carefully controlled conditions, free of unwanted contaminants. The diets fed to animals and their air supply must be carefully purified to keep out even traces of unwanted elements, and their cages must be made of plastics that contain no metals.

Our understanding of the biological functions of trace elements is changing the way scientists think about diet and health. For example, supplements of manganese, copper, and zinc in combination with calcium have recently been shown to improve human bone health to a greater extent than just calcium alone. Silicon and boron are also believed to be important for bone health. Deficiencies of selenium or copper are suspected by some scientists of increasing the risk of cancer or heart disease. Because chromium, copper, and zinc influence glucose metabolism, future prevention and treatment strategies for diabetes may involve these nutrients.

Mary Ann Johnson
College of Family and Consumer Sciences
University of Georgia

THE ALKALINE EARTH METALS (GROUP IIA)

23-4 GROUP IIA METALS: PROPERTIES AND OCCURRENCE

In Section 8-5 we found that gaseous $BeCl_2$ is linear. However, the Be atoms in $BeCl_2$ molecules act as Lewis acids. In the solid state, they accept shares in electron pairs from Cl atoms in other molecules, to form polymers.

The alkaline earth metals are all silvery white, malleable, ductile, and somewhat harder than their neighbors in Group IA. Activity increases from top to bottom within the group, with Ca, Sr, and Ba being considered quite active. Each has two electrons in its highest occupied energy level. Both electrons are lost in ionic compound formation, though not as easily as the outer electron of an alkali metal. Compare the ionization energies in Tables 23-1 and 23-3. Most IIA compounds are ionic, but those of Be exhibit a great deal of covalent character. This is due to the extremely high charge density of Be^{2+}. Compounds

Table 23-3 *Properties of the Group IIA Metals*

Property	Be	Mg	Ca	Sr	Ba	Ra
Outer electrons	$2s^2$	$3s^2$	$4s^2$	$5s^2$	$6s^2$	$7s^2$
Melting point (°C)	1283	649	839	770	725	700
Boiling point (°C)	2484	1105	1484	1384	1640	1140
Density (g/cm^3)	1.85	1.74	1.55	2.60	3.51	5
Atomic radius (Å)	1.11	1.60	1.97	2.15	2.17	2.20
Ionic radius, M^{2+} (Å)	0.31	0.65	0.99	1.13	1.35	—
Electronegativity	1.5	1.2	1.0	1.0	1.0	1.0
E^0 (volts): $M^{2+}(aq) + 2e^- \longrightarrow 2M(s)$	-1.85	-2.37	-2.87	-2.89	-2.90	-2.92
Ionization energies (kJ/mol)						
$\quad M(g) \longrightarrow M^+(g) + e^-$	899	738	590	549	503	509
$\quad M^+(g) \longrightarrow M^{2+}(g) + e^-$	1757	1451	1145	1064	965	(979)
$\Delta H^0_{hydration}$ (kJ/mol): $M^{2+}(g) \longrightarrow M^{2+}(aq)$	—	-1925	-1650	-1485	-1276	—

of beryllium therefore resemble those of aluminum in Group IIIA (diagonal similarities). The IIA elements exhibit the +2 oxidation state in all their compounds. The tendency to form 2+ ions increases from Be to Ra.

The alkaline earth metals show a wider range of chemical properties than the alkali metals. The IIA metals are not as reactive as the IA metals, but they are much too reactive to occur free in nature. They are obtained by electrolysis of their molten chlorides. Calcium and magnesium are abundant in the earth's crust, especially as carbonates and sulfates. Beryllium, strontium, and barium are less abundant. All known radium isotopes are radioactive and are extremely rare.

23-5 REACTIONS OF THE GROUP IIA METALS

Table 23-4 summarizes some reactions of the alkaline earth metals, which, except for stoichiometry, are similar to the corresponding reactions of the alkali metals. Reactions with hydrogen and oxygen were discussed in Sections 6-7 and 6-8, parts 2.

Table 23-4 *Some Reactions of the Group IIA Metals*

Reaction	Remarks
$2M + O_2 \longrightarrow 2MO$	Very exothermic (except Be)
$Ba + O_2 \longrightarrow BaO_2$	Almost exclusively
$M + H_2 \longrightarrow MH_2$	M = Ca, Sr, Ba at high temperatures
$3M + N_2 \longrightarrow M_3N_2$	At high temperatures
$6M + P_4 \longrightarrow 2M_3P_2$	At high temperatures
$M + X_2 \longrightarrow MX_2$	X = halogen (Group VIIA)
$M + S \longrightarrow MS$	Also with Se, Te of Group VIA
$M + 2H_2O \longrightarrow M(OH)_2 + H_2$	M = Ca, Sr, Ba at 25°C; Mg gives MgO at high temperatures
$M + 2NH_3 \longrightarrow M(NH_2)_2 + H_2$	M = Ca, Sr, Ba in NH$_3$(ℓ) in presence of catalyst; NH$_3$(g) with heat
$3M + 2NH_3(g) \longrightarrow M_3N_2 + 3H_2$	At high temperatures
$Be + 2OH^- + 2H_2O \longrightarrow Be(OH)_4^{2-} + H_2$	Only with Be

Amphoterism is the ability of a substance to react with both acids and bases.

Except for Be, all the alkaline earth metals are oxidized to oxides in air. The IIA oxides (except BeO) are basic and react with water to give hydroxides. Beryllium hydroxide, $Be(OH)_2$, is quite insoluble in water and is amphoteric. Magnesium hydroxide, $Mg(OH)_2$, is only slightly soluble in water. The hydroxides of Ca, Sr, and Ba are strong soluble bases.

Beryllium is at the top of Group IIA. Its oxide is amphoteric, whereas oxides of the heavier members are basic. Metallic character increases from top to bottom within a group and from right to left across a period. This results in increasing basicity and decreasing acidity of the oxides in the same directions, as shown in the following table.

Limestone is mainly calcium carbonate.

Group IA	Group IIA	Group IIIA
Li_2O (basic)	BeO (amphoteric)	B_2O_3 (amphoteric)
Na_2O (basic)	MgO (basic)	Al_2O_3 (amphoteric)
K_2O (basic)	CaO (basic)	Ga_2O_3 (amphoteric)
		In_2O_3 (basic)

(Vertical arrows: Increasing metallic character of elements; Decreasing acidity of oxides; Increasing basicity of oxides)

Increasing metallic character of elements →

Decreasing acidity of oxides →

Increasing basicity of oxides →

Ca, Sr, and Ba react with water at 25°C to form hydroxides and H_2 (see Table 23-4). Magnesium reacts with steam to produce MgO and H_2. Beryllium does not react with pure water even at red heat. It reacts with solutions of strong bases to form the complex ion, $[Be(OH)_4]^{2-}$, and H_2. Group IIA compounds are generally less soluble in water than corresponding IA compounds, but many are quite soluble.

23-6 USES OF GROUP IIA METALS AND THEIR COMPOUNDS

Calcium, Ca

Calcium and its compounds are widely used commercially. The element is used as a reducing agent in the metallurgy of uranium, thorium, and other metals. It is also used as a scavenger to remove dissolved impurities such as oxygen, sulfur, and carbon in molten metals, and to remove residual gases in vacuum tubes. It is a component of many alloys.

Heating limestone produces *quicklime,* CaO, which can then be treated with water to form *slaked lime,* $Ca(OH)_2$, a cheap base for which industry finds many uses. When slaked lime is mixed with sand and exposed to the CO_2 of the air, it hardens to form mortar. Heating gypsum, $CaSO_4 \cdot 2H_2O$, produces plaster of Paris, $2CaSO_4 \cdot H_2O$.

Calcium carbonate and calcium phosphate occur in seashells and animal bones.

Magnesium, Mg

Metallic magnesium burns in air with such a brilliant white light that it is used in photographic flash accessories and fireworks. It is very lightweight and is currently used in many alloys for building materials. Like aluminum, it forms an impervious coating of oxide that protects it from further oxidation. Given its inexhaustible supply in the oceans, it is likely that many more structural uses will be found for it as the reserves of iron ores dwindle.

Beryllium, Be

Because of its rarity, beryllium has only a few practical uses. It occurs mainly as beryl, $Be_3Al_2Si_6O_{18}$, a gemstone which, with appropriate impurities, may be aquamarine (blue) or emerald (green). Because it is transparent to X-rays, "windows" for X-ray tubes are constructed of beryllium. Beryllium compounds are quite toxic.

Strontium, Sr

Strontium salts are used in fireworks and flares, which show the characteristic red glow of strontium in a flame. Strontium chloride is used in some toothpastes for persons with sensitive teeth. The metal itself has no practical uses.

Barium, Ba

Barium is a constituent of alloys that are used for spark plugs because of the ease with which it emits electrons when heated. It is used as a degassing agent for vacuum tubes. A slurry of finely divided barium sulfate from barite, $BaSO_4$, is used to coat the gastrointestinal tract in preparation for X-ray photographs because it absorbs X-rays so well. It is so insoluble that it is not poisonous; all soluble barium salts are very toxic.

A laboratory X-ray tube (left) and a close-up view of one of its windows (right). The windows are made of beryllium metal.

Table 23-5 *Properties of the Group IIIA Elements*

Property	B	Al	Ga	In	Tl
Outer electrons	$2s^2 2p^1$	$3s^2 3p^1$	$4s^2 4p^1$	$5s^2 5p^1$	$6s^2 6p^1$
Physical state (25°C, 1 atm)	solid	solid	solid	solid	solid
Melting point (°C)	2300	660	29.8	156.6	303.5
Boiling point (°C)	2550	2367	2403	2080	1457
Density (g/cm³)	2.34	2.70	5.91	7.31	11.85
Atomic radius (Å)	0.88	1.43	1.22	1.62	1.71
Ionic radius, M^{3+} (Å)	(0.20)*	0.50	0.62	0.81	0.95
Electronegativity	2.0	1.5	1.7	1.6	1.6
E^0 (volts): $M^{3+}(aq) + 3e^- \longrightarrow M(s)$	(−0.90)*	−1.66	−0.53	−0.34	0.916
Oxidation states	−3 to +3	+3	+1, +3	+1, +3	+1, +3
Ionization energies (kJ/mol)					
$\quad M(g) \longrightarrow M^+(g) + e^-$	801	578	576	556	586
$\quad M^+(g) \longrightarrow M^{2+}(g) + e^-$	2427	1817	1971	1813	1961
$\quad M^{2+}(g) \longrightarrow M^{3+}(g) + e^-$	3660	2745	2952	2692	2867
$\Delta H^0_{hydration}$ (kJ/mol): $M^{3+}(g) + xH_2O \longrightarrow M^{3+}(aq)$	—	−4750	−4703	−4159	−4117

For the covalent +3 oxidation state.

THE POST-TRANSITION METALS

There are no true metals in Groups VIA, VIIA, and 0.

The metals below the stepwise division of the periodic table in Groups IIIA through VA are the **post-transition metals.** These include aluminum, gallium, indium, and thallium from Group IIIA; tin and lead from Group IVA; and bismuth from Group VA. Aluminum is the only post-transition metal that is considered very reactive.

23-7 PERIODIC TRENDS—GROUP IIIA

The properties of the elements in Group IIIA (Table 23-5) vary less regularly down the groups than those of the IA and IIA metals. The Group IIIA elements are all solids. Boron, at the top of the group, is a nonmetal. Its melting point, 2300°C, is very high because it crystallizes as a covalent solid. The other elements, aluminum through thallium, form metallic crystals and have considerably lower melting points.

Aluminum, Al

This aluminum honeycomb material is made by bonding aluminum foil sheets to form hexagonal cells. It is used to make sandwich construction panels that have a very high strength-to-weight ratio.

Aluminum is the most reactive of the post-transition metals. It is the most abundant metal in the earth's crust (7.5%) and the third most abundant element. Aluminum is inexpensive compared to most other metals. It is soft and can be readily extruded into wires or rolled, pressed, or cast into shapes.

Because of its relatively low density, aluminum is often used as a lightweight structural metal. It is often alloyed with Mg and some Cu and Si to increase its strength. Many buildings are sheathed in aluminum, which resists corrosion by forming an oxide coating.

Pure aluminum conducts about two thirds as much electrical current per unit volume as copper, but it is only one third as dense (Al, 2.70 g/cm³; Cu, 8.92 g/cm³). As a result, a mass of aluminum can conduct twice as much current as the same mass of copper. Aluminum is now used in electrical transmission lines and has been used in wiring in homes.

CHEMISTRY IN USE

The Development of Science

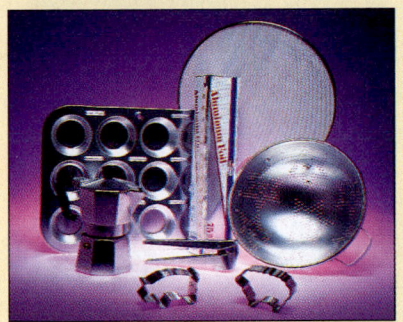

The Most Valuable Metal in the World

Imagine paying eight million dollars for one pound of aluminum! Although aluminum currently costs less than $1.00 a pound, it was considered the most valuable metal in 1827. Aluminum was so cherished by royalty in the early-to-mid 1800s that they alone ate with aluminum spoons and forks while their lower-class guests dined with cheaper gold and silver service. Aluminum is the most abundant metal in the earth's crust (7.5%); why was it originally so expensive?

Aluminum was first prepared by the following displacement reaction.

$$AlCl_3 + 3K \longrightarrow Al + 3KCl$$

Potassium was also expensive because it was made by passing an electric current (from a voltaic cell) through molten KCl. In addition to the great cost of energy required to melt large quantities of KCl, copper and zinc (used in voltaic cells) were also expensive metals in the early 1800s. Thus, the very small amount of aluminum produced by this displacement reaction was extremely expensive.

It was not practical to produce aluminum by passing an electric current through molten Al_2O_3 because it has a high melting point, 2000°C. This high temperature is difficult to achieve and maintain; the components of most voltaic cells melt below this temperature. Zinc melts at 420°C and copper at 1083°C.

The cost of aluminum began to drop as the result of two major advances in the late 1800s. The first came with the invention of the electric generator, which could produce electricity using steam or water. Electricity generated by steam or water was quite inexpensive compared to electricity generated by voltaic cells. Despite this cost reduction, aluminum still cost more than $100,000 a pound. The second advance took place in 1886 when chemists discovered that they could lower the melting point of aluminum oxide by mixing it with complex salts such as $Na_3[AlF_6]$. Since 1886, the price of aluminum has decreased markedly because of lower electrical costs, improved production techniques, and recycling of discarded aluminum products.

Although aluminum is no longer used in table services by royalty, it is of inestimable value in energy conservation. Around our homes we find energy-saving items such as aluminum storm doors and windows, insulation backed with aluminum foil, and aluminum siding. Because vehicle weight significantly affects gas mileage, substituting aluminum for heavier metals in cars, trucks, trains, and aircraft helps preserve our petroleum supplies. Although the cost of aluminum has decreased drastically, it is still a valuable metal because of its ability to help us conserve energy and to improve our standard of living at the same time.

Ronald DeLorenzo
Middle Georgia College

However, the latter use has been implicated as a fire hazard due to the heat that can be generated during high current flow at the junction of the aluminum wire and fixtures of other metals.

Aluminum is a strong reducing agent.

$$Al^{3+}(aq) + 3e^- \longrightarrow Al(s) \qquad E^0 = -1.66 \text{ V}$$

Aluminum is quite reactive, but a thin, transparent film of Al_2O_3 forms when Al comes into contact with air. This protects it from further oxidation. For this reason it is even passive toward nitric acid, HNO_3, a strong oxidizing agent. When the oxide coating is sanded off, Al reacts vigorously with HNO_3.

"Passive toward" means does not react with.

$$Al(s) + 4HNO_3(aq) \longrightarrow Al(NO_3)_3(aq) + NO(g) + 2H_2O(\ell)$$

Figure 23-1   The thermite reaction. A mixture of Fe_2O_3 and aluminum powder was placed in a clay pot with a piece of magnesium ribbon as a fuse. (a) The reaction was initiated by lighting the magnesium fuse. (b) So much heat was produced by the reaction that the iron melted as it was produced. (c) The molten iron dropped out of the clay pot and burned through a sheet of iron that was placed under the pot.

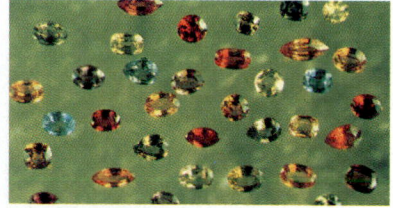

Small amounts of different transition metal ions give different colors to sapphire, which is mostly aluminum oxide, Al_2O_3.

The very negative enthalpy of formation of aluminum oxide makes Al a very strong reducing agent for other metal oxides. The **thermite reaction** is a spectacular example (Figure 23-1). It generates enough heat to produce molten iron for welding steel.

$$2Al(s) + Fe_2O_3(s) \longrightarrow 2Fe(s) + Al_2O_3(s) \qquad \Delta H^0 = -852 \text{ kJ/mol}$$

Anhydrous Al_2O_3 occurs naturally as the extremely hard, high-melting mineral *corundum,* which has a network structure. It is colorless when pure, but becomes colored when transition metal ions replace a few Al^{3+} ions in the crystal. *Sapphire* is usually blue and contains some iron and titanium. *Ruby* is red due to the presence of small amounts of chromium.

Gallium, Ga

Gallium is unusual in that it melts when held in the hand. It has the largest liquid state temperature range of any element (29.8°C to 2403°C). It is used in transistors and high temperature thermometers. Gallium-67 was one of the first artificially produced isotopes to be used in medicine. It concentrates in inflamed areas and in certain melanomas.

Indium, In

Indium is a soft, bluish metal that is used in some alloys with silver and lead to make good heat conductors. Most indium is used in electronics.

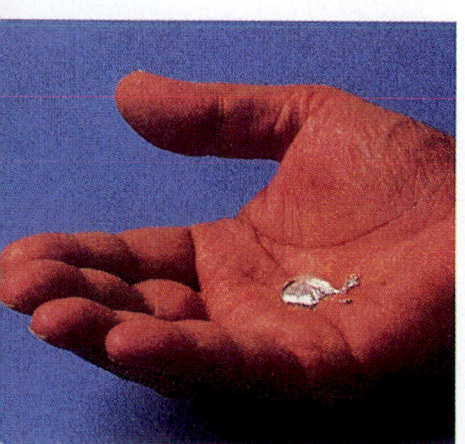

Gallium metal melts below body temperature.

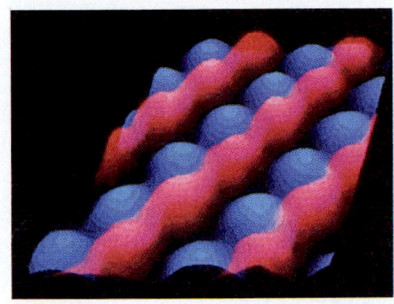

An image of the surface of gallium arsenide, GaAs, produced by a scanning tunneling microscope. This substance is used in microwave generators, lasers, light-emitting diodes, and many other electronic devices.

Thallium, Tl

Thallium is a soft, heavy metal that resembles lead. It is quite toxic and has no important practical uses as a free metal.

The elements at the top of Group IVA—C, Si, and Ge—have relatively high melting points (especially C) because they form covalent crystals. The metals, tin and lead, have lower melting points.

The atomic radii do not increase regularly as Groups IIIA and IVA are descended (Table 6-1). The atomic radius of Ga, 1.22 Å, is *less* than that of Al, 1.43 Å, which is directly above Ga. The transition elements are located between calcium (IIA) and gallium (IIIA), strontium (IIA) and indium (IIIA), and barium (IIA) and thallium (IIIA). The increase in nuclear charge that accompanies filling of the $(n-1)d$ subshell results in the contraction of the size of the atoms. This contraction is caused by the stronger attraction of the more highly charged nuclei for the outer electrons. This causes the radii of Ga, In, and Tl to be smaller than would be predicted from the radii of B and Al. In Group IVA, the atomic radii of Ge, Sn, and Pb are contracted for the same reasons. Atomic radii strongly influence other properties. For example, the densities of Ga, In, and Tl from Group IIIA and Ge, Sn, and Pb from IVA are much higher than those of the elements above them, due to their smaller atomic radii.

> Compare the radii and densities of these elements with those of the IA and IIA metals in the same rows.

The Group IIIA elements have the ns^2np^1 outer electron configuration. Aluminum shows only the +3 oxidation state in its compounds. The heavier metals (Ga, In, Tl) can lose or share either the single p valence electron or the p and both s electrons to exhibit the +1 or +3 oxidation state, respectively. In general the post-transition metals can exhibit oxidation states of $(g-2)+$ and $g+$ where g = periodic group number. As examples, TlCl and TlCl$_3$ both exist, as do SnCl$_2$ and SnCl$_4$. The stability of the lower state increases as the groups are descended. This is called the **inert s-pair effect** because the two s electrons remain nonionized, or unshared, for the $(g-2)+$ oxidation state. To illustrate, AlCl$_3$ exists but not AlCl; TlCl$_3$ is less stable than TlCl. Likewise, in Group IVA GeCl$_4$ is more stable than GeCl$_2$, but PbCl$_4$ is less stable than PbCl$_2$.

> As is generally true, for each pair of compounds, covalent character is greater for the higher (more polarizing) oxidation state of the metal.

> Bi, from Group VA, exhibits the +3 and +5 oxidation states, but the +5 state is quite rare.

THE d-TRANSITION METALS

The term "transition elements" denotes elements in the middle of the periodic table. They provide a transition between the "base formers" on the left and the "acid formers" on the right. The term applies to both the d- and f-transition elements (d and f atomic orbitals are being filled across this part of the periodic table). All are metals. We commonly use the term "transition metals" to refer to the d-transition metals.

> Oxides of most nonmetals are acidic, and oxides of most metals are basic (except those having high oxidation states).

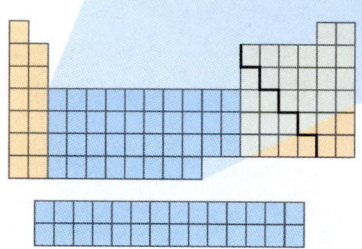

IIIB	IVB	VB	VIB	VIIB	┌──── VIIIB ────┐			IB	IIB
Sc	Ti	V	Cr	Mn	Fe	Co	Ni	Cu	Zn
Y	Zr	Nb	Mo	Tc	Ru	Rh	Pd	Ag	Cd
La	Hf	Ta	W	Re	Os	Ir	Pt	Au	Hg
Ac	Unq	Unp	Unh	Uns					

23-8 GENERAL PROPERTIES

The *d*-transition metals are located between Groups IIA and IIIA in the periodic table. Strictly speaking, a *d*-transition metal must have a partially filled set of *d* orbitals. Zinc, cadmium, and mercury (Group IIB) and their cations have completely filled sets of *d* orbitals. They are *not* *d*-transition metals, but they are often discussed with *d*-transition metals because their properties are similar. All of the other elements in this region have partially filled sets of *d* orbitals, except the IB elements and palladium, which have completely filled sets. Some of the cations of these latter elements have partially filled sets of *d* orbitals.

Some properties of 3*d*-transition metals are listed in Table 23-6. The following are properties of transition elements.

1. All are metals.

2. Most are harder and more brittle and have higher melting points, boiling points, and heats of vaporization than nontransition metals.

Table 23-6 *Properties of Metals in the First Transition Series*

Properties	Sc	Ti	V	Cr	Mn	Fe	Co	Ni	Cu	Zn
Melting point (°C)	1541	1660	1890	1850	1244	1535	1495	1453	1083	420
Boiling point (°C)	2831	3287	3380	2672	1962	2750	2870	2732	2567	907
Density (g/cm^3)	2.99	4.54	6.11	7.18	7.21	7.87	8.9	8.91	8.96	7.13
Atomic radius (Å)	1.62	1.47	1.34	1.25	1.29	1.26	1.25	1.24	1.28	1.34
Ionic radius, M^{2+} (Å)	—	0.94	0.88	0.89	0.80	0.74	0.72	0.69	0.70	0.74
Electronegativity	1.3	1.4	1.5	1.6	1.6	1.7	1.8	1.8	1.8	1.6
E^0 (V) for M^{2+}(aq) + 2e^- ⟶ M(s)	−2.08*	−1.63	−1.2	−0.91	−1.18	−0.44	−0.28	−0.25	+0.34	−0.76
IE (kJ/mol) first	631	658	650	653	717	759	758	737	745	906
second	1235	1310	1414	1592	1509	1561	1646	1753	1958	1733

*For Sc^{3+}(aq) + 3$e^- \to$ Sc(s).

3. Their ions and their compounds are usually colored.

4. They form many complex ions (Chapter 25).

5. With few exceptions, they exhibit multiple oxidation states.

6. Many of them are paramagnetic, as are many of their compounds.

7. Many of the metals and their compounds are effective catalysts.

23-9 ELECTRON CONFIGURATIONS AND OXIDATION STATES

Some of the valence electrons of the d-transition metals are in d orbitals one energy level below the highest occupied level. The properties of these metals vary less dramatically for consecutive elements than do those of representative elements, whose valence electrons are in s and p orbitals in the highest occupied energy level. Electron configurations of the three d-transition series are given in Table 23-7.

Most transition metals exhibit more than one nonzero oxidation state. The *maximum* oxidation state is given by a metal's group number, but this is often not its most stable oxidation state (Table 23-8).

Rhodonite is maganese(II) silicate ($MnSiO_3$), which is pink, mixed with calcium silicate ($CaSiO_3$). The black veins are oxides of manganese. Rhodonite is often cut into cabochons, goblets, vases, and other decorative objects.

Table 23-7 *Ground State Electron Configurations of d-Transition Metals*

Period 4		Period 5		Period 6	
$_{21}$Sc	$[Ar]3d^14s^2$	$_{39}$Y	$[Kr]4d^15s^2$	$_{57}$La	$[Xe]5d^16s^2$
$_{22}$Ti	$[Ar]3d^24s^2$	$_{40}$Zr	$[Kr]4d^25s^2$	$_{72}$Hf	$[Xe]4f^{14}5d^26s^2$
$_{23}$V	$[Ar]3d^34s^2$	$_{41}$Nb	$[Kr]4d^45s^1$	$_{73}$Ta	$[Xe]4f^{14}5d^36s^2$
$_{24}$Cr	$[Ar]3d^54s^1$	$_{42}$Mo	$[Kr]4d^55s^1$	$_{74}$W	$[Xe]4f^{14}5d^46s^2$
$_{25}$Mn	$[Ar]3d^54s^2$	$_{43}$Tc	$[Kr]4d^55s^2$	$_{75}$Re	$[Xe]4f^{14}5d^56s^2$
$_{26}$Fe	$[Ar]3d^64s^2$	$_{44}$Ru	$[Kr]4d^75s^1$	$_{76}$Os	$[Xe]4f^{14}5d^66s^2$
$_{27}$Co	$[Ar]3d^74s^2$	$_{45}$Rh	$[Kr]4d^85s^1$	$_{77}$Ir	$[Xe]4f^{14}5d^76s^2$
$_{28}$Ni	$[Ar]3d^84s^2$	$_{46}$Pd	$[Kr]4d^{10}$	$_{78}$Pt	$[Xe]4f^{14}5d^96s^1$
$_{29}$Cu	$[Ar]3d^{10}4s^1$	$_{47}$Ag	$[Kr]4d^{10}5s^1$	$_{79}$Au	$[Xe]4f^{14}5d^{10}6s^1$
$_{30}$Zn	$[Ar]3d^{10}4s^2$	$_{48}$Cd	$[Kr]4d^{10}5s^2$	$_{80}$Hg	$[Xe]4f^{14}5d^{10}6s^2$

Several of the apparent irregularities in these electron configurations can be explained by the special stability of half-filled and filled sets of d orbitals (Section 5-17).

Table 23-8 *Nonzero Oxidation States of the 3d-Transition Metals**

IIIB	IVB	VB	VIB	VIIB	VIIIB			IB	IIB
Sc	Ti	V	Cr	Mn	Fe	Co	Ni	Cu	Zn
								+1 r	
				+2	+2 r	+2	+2	+2	+2
+3		+3 r	+3		+3	+3 o			
	+4	+4		+4 o					
		+5 o							
			+6 o						
				+7 o					

*o = oxidizing agent; r = reducing agent.

In the "building" of electron configurations by the Aufbau Principle, the outer s orbitals are occupied before the inner d orbitals (Section 5-17).

The outer s electrons lie outside the d electrons and are *always* the first ones lost in ionization. In the first transition series, only scandium and zinc exhibit just one nonzero oxidation state. Scandium loses its two $4s$ electrons and its only $3d$ electron to form Sc^{3+}. Zinc loses its two $4s$ electrons to form Zn^{2+}.

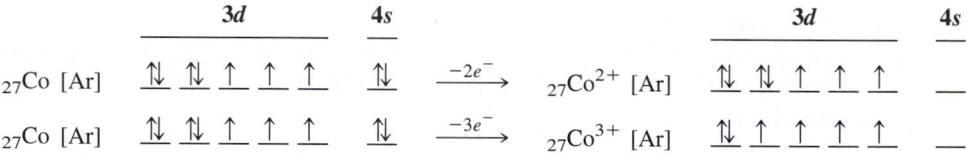

All of the other $3d$-transition metals exhibit at least two oxidation states in their compounds. For example, cobalt can form Co^{2+} and Co^{3+} ions.

The most common oxidation states of the $3d$-transition elements are $+2$ and $+3$. The elements in the middle of each series exhibit more oxidation states than those to the left or right. As one moves down a group, higher oxidation states become more stable and more common (opposite to the trend for representative elements). This is because the d electrons are more effectively shielded from the nucleus as the group is descended and are therefore more easily ionized or more readily available for sharing. For example, cobalt commonly exhibits the $+2$ and $+3$ oxidation states. Rh and Ir are just below Co. Their common oxidation states are $+3$ and $+4$. The $+4$ state is slightly more stable for Ir than for the lighter Rh.

Pentaamminechlorocobalt(III) chloride, $[Co(NH_3)_5(Cl)]Cl_2$, is a compound that contains cobalt in the $+3$ oxidation state (left). Hexaaquacobalt(II) chloride, $[Co(OH_2)_6]Cl_2$, contains cobalt in the $+2$ oxidation state (right).

23-10 CLASSIFICATION INTO SUBGROUPS

The transition metals and Zn, Cd, and Hg are subdivided into eight groups designated by roman numerals (I to VIII) followed by "B." The roman numeral *usually* designates the maximum oxidation number exhibited by members of the group. This does not mean that simple ions with these charges exist; no simple ions of these elements possess a charge greater than $3+$. The elements in corresponding A and B groups form many compounds of similar stoichiometry (Table 23-9). However, their chemical properties are usually dissimilar.

Group VIIIB consists of three columns of three metals each, which have no counterparts among the representative elements. Each horizontal row in VIIIB is called a **triad** and is named after the best-known metal of the row. The three rows are the *iron, palladium,* and *platinum* triads.

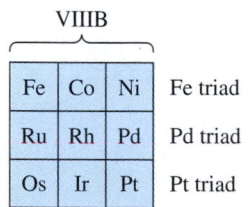

There are differences between the A and B groups, but the transition metals show some of the same trends as the representative elements. For corresponding compounds of metals from a B group in the same oxidation state, covalent character usually decreases and ionic character increases as the group is descended. We observe increasing electrical conductivities of aqueous solutions as well as higher melting and boiling points for the heavier compounds. Consider the metal(V) oxides of Group VB (margin).

Table 23-9 *Typical Compounds and Ions of the A and B Groups*

IA	IIA	IIIA	IVA	VA	VIA	VIIA
NaCl	$MgBr_2$	$Al(NO_3)_3$	CCl_4	$POCl_3$	SO_4^{2-}	Cl_2O_7
KNO_3	$CaCl_2$	$Ga(OH)_3$	PbO_2	PO_4^{3-}	$H_2S_2O_7$	$HClO_4$

IB	IIB	IIIB	IVB	VB	VIB	VIIB
CuCl	$ZnBr_2$	$Sc(NO_3)_3$	$TiCl_4$	$VOCl_3$	CrO_4^{2-}	Mn_2O_7
$AgNO_3$	$CdCl_2$	$Y(OH)_3$	Zro_2	VO_4^{3-}	$H_2Cr_2O_7$	$HMnO_4$

The oxides and hydroxides of lower oxidation states of a given transition metal are basic. Those containing intermediate oxidation states tend to be amphoteric, and those containing high oxidation states tend to be acidic. This is illustrated for the oxides and hydroxides of Cr in the next section.

Oxide	Melting Point	
V_2O_5	690°C	Increasing
Nb_2O_5	1460°C	ionic
Ta_2O_5	1800°C	↓ character

23-11 CHROMIUM OXIDES AND HYDROXIDES

Typical of the metals near the middle of a transition series, chromium shows several oxidation states. The most common are +2, +3, and +6 (Table 23-10).

Oxidation–Reduction

The most stable oxidation state of Cr is +3. Solutions of blue chromium(II) salts are easily air-oxidized to chromium(III).

$$Cr^{3+} + e^- \longrightarrow Cr^{2+} \qquad E^0 = -0.41 \text{ V}$$

Table 23-10 *Some Compounds of Chromium*

Ox. State	Oxide	Hydroxide	Name	Acidic/Basic	Related Salt	Name
+2	CrO black	$Cr(OH)_2$	chromium(II) hydroxide	basic	$CrCl_2$ anhydr. colorless aq. lt. blue	chromium(II) chloride
+3	Cr_2O_3 green	$Cr(OH)_3$	chromium(III) hydroxide	amphoteric	$CrCl_3$ anhydr. violet aq. green $KCrO_2$ green	chromium(III) chloride potassium chromite
+6	CrO_3 dk. red	H_2CrO_4 or $[CrO_2(OH)_2]$	chromic acid	weakly acidic	K_2CrO_4 yellow	potassium chromate
		$H_2Cr_2O_7$ or $[Cr_2O_5(OH)_2]$	dichromic acid	acidic	$K_2Cr_2O_7$ orange	potassium dichromate

Aqueous solutions of some compounds that contain chromium. Left to right: chromium(II) chloride ($CrCl_2$) is blue; chromium(III) chloride ($CrCl_3$) is green; potassium chromate (K_2CrO_4) is yellow; potassium dichromate ($K_2Cr_2O_7$) is orange.

chromate ions, CrO_4^{2-}

dichromate ions, $Cr_2O_7^{2-}$

Chromium(VI) species are oxidizing agents. Basic solutions containing chromate ions, CrO_4^{2-}, are weakly oxidizing. Acidification produces the dichromate ion, $Cr_2O_7^{2-}$, and chromium(VI) oxide, both powerful oxidizing agents.

$$Cr_2O_7^{2-} + 14H^+ + 6e^- \longrightarrow 2Cr^{3+} + 7H_2O \qquad E^0 = +1.33 \text{ V}$$

Chromate–Dichromate Equilibrium

Red chromium(VI) oxide, CrO_3, is the acid anhydride of two acids: chromic acid, H_2CrO_4, and dichromic acid, $H_2Cr_2O_7$. Neither acid has been isolated in pure form, although chromate and dichromate salts are common. CrO_3 reacts with H_2O to produce strongly acidic solutions containing hydrogen ions and (predominantly) dichromate ions.

$$2CrO_3 + H_2O \longrightarrow [2H^+ + Cr_2O_7^{2-}] \qquad \text{dichromic acid (red-orange)}$$

From such solutions orange dichromate salts can be crystallized after adding a stoichiometric amount of base. Addition of excess base produces yellow solutions from which only yellow chromate salts can be obtained. The two anions exist in solution in a pH-dependent equilibrium.

$$\underset{\text{yellow}}{2\overset{+6}{Cr}O_4^{2-} + 2H^+} \rightleftharpoons \underset{\text{orange}}{\overset{+6}{Cr_2}O_7^{2-} + H_2O} \qquad K_c = \frac{[Cr_2O_7^{2-}]}{[CrO_4^{2-}]^2[H^+]^2} = 4.2 \times 10^{14}$$

Adding a strong acid to a solution that contains $CrO_4^{2-}/Cr_2O_7^{2-}$ ions favors the reaction to the right and increases $[Cr_2O_7^{2-}]$. Adding a base favors the reaction to the left and increases $[CrO_4^{2-}]$.

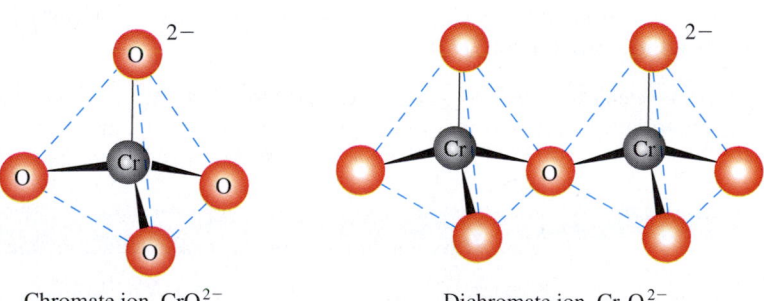

Chromate ion, CrO_4^{2-} Dichromate ion, $Cr_2O_7^{2-}$

This cleaning solution is very dangerous because it is a strong oxidizing agent and is carcinogenic.

Dehydration of chromate or dichromate salts with concentrated H_2SO_4 produces CrO_3. Chromium(VI) oxide is a strong oxidizing agent. A powerful "cleaning solution" used for removing greasy stains and coatings from laboratory glassware is made by adding concentrated H_2SO_4 to a concentrated solution of $K_2Cr_2O_7$. The active ingredients are CrO_3, an oxidizing agent, and H_2SO_4, an excellent solvent.

Chromium(III) hydroxide is amphoteric (Section 10-8).

$$Cr(OH)_3(s) + 3H^+ \longrightarrow Cr^{3+} + 3H_2O \qquad\qquad \text{(rxn. with acids)}$$

$$Cr(OH)_3(s) + OH^- \longrightarrow Cr(OH)_4^- \quad \text{or} \quad CrO_2^- \cdot 2H_2O \qquad \text{(rxn. with bases)}$$

Molybdenum and tungsten are just below chromium in Group VIB. Both MoO_3 and WO_3 are more thermally stable, less acidic, and weaker oxidizing agents than CrO_3.

Key Terms

Alkali metals Group IA metals.

Alkaline earth metals Group IIA metals.

_d_-Transition metals Metals that have partially filled sets of _d_ orbitals; the B Groups except IIB.

Diagonal similarities Chemical similarities of elements of Period 2 to elements of Period 3 one group to the right; especially evident toward the left of the periodic table.

Inert _s_-pair effect The tendency of the two outermost _s_ electrons to remain nonionized or unshared in compounds; characteristic of the post-transition metals.

Post-transition metals Representative metals in the "_p_ block."

Representative metals Metals in the A groups in the periodic table; their outermost electrons are in _s_ and _p_ orbitals.

Triad A horizontal row of three elements in Group VIIIB.

Exercises

1. How do the acidities or basicities of metal oxides vary with oxidation numbers of the same metal?

2. Discuss the general differences in electron configurations of representative elements and _d_-transition metals.

3. Compare the extents to which the properties of successive elements across the periodic table differ for representative elements and _d_-transition metals. Explain.

4. Compare the metals and nonmetals with respect to (a) number of outer shell electrons, (b) electronegativities, (c) standard reduction potentials, and (d) ionization energies.

5. How do the physical properties of metals differ from those of nonmetals?

6. Compare the alkali metals with the alkaline earth metals with respect to (a) atomic radii, (b) densities, (c) first ionization energies, and (d) second ionization energies. Explain the comparisons.

7. Summarize the chemical and physical properties of the alkali metals.

8. Summarize the chemical and physical properties of the alkaline earth metals.

9. Write the general outer electron configurations for atoms of the IA and IIA metals. What oxidation state(s) would you predict for these elements? What types of bonding would you expect in most of the compounds of these elements? Why?

10. Write electron configurations ($\uparrow\downarrow$ notation) for (a) Mg, (b) Mg^{2+}, (c) Na, (d) Na^+, (e) Sn, (f) Sn^{2+}, and (g) Sn^{4+}.

11. Write electron configurations ($\uparrow\downarrow$ notation) for (a) K, (b) K^+, (c) Sr, (d) Sr^{2+}, (e) Al, (f) Al^{3+}, and (g) Ga^{3+}.

12. Are the elements in Groups IA and IIA found in the free state in nature? What are the primary sources for these elements?

13. Describe some uses for (a) lithium and its compounds and (b) sodium and its compounds.

14. Where do the metals of Groups IA and IIA fall with respect to H_2 in the activity series? What does this tell us about their reactivities with water and acids?

15. Write chemical equations describing the reactions of O_2 with each of the alkali and alkaline earth metals. Account for differences within each family.

16. Describe some uses for (a) calcium and its compounds and (b) magnesium and its compounds.

17. Write general equations for reactions of alkali metals with (a) hydrogen, (b) sulfur, and (c) ammonia. Represent the metal as M.

18. Write general equations for reactions of alkali metals with

(a) water, (b) phosphorus, and (c) halogens. Represent the metal as M and the halogen as X.

19. Write general equations for reactions of alkaline earth metals with (a) hydrogen, (b) sulfur, and (c) ammonia. Represent the metal as M.

20. Write general equations for reactions of alkaline earth metals with (a) water, (b) phosphorus, and (c) chlorine. Represent the metal as M.

21. What is meant by the term "diagonal relationships"?

22. Give some illustrations of diagonal relationships in the periodic table, and explain each.

23. What is hydration energy? How does it vary for cations of the alkali metals?

24. How do hydration energies vary for cations of the alkaline earth metals?

25. How do the standard reduction potentials of the alkali metal cations vary? Why?

26. How do the standard reduction potentials of the alkaline earth metal cations vary? Why?

27. Why are the standard reduction potentials of lithium and beryllium out of line with respect to group trends?

*28. Calculate ΔH^0 values at 25°C for the reactions of 1 mol of each of the following metals with stoichiometric quantities of water to form metal hydroxides and hydrogen. (a) Li, (b) K, and (c) Ca. Rationalize the differences in these values.

29. How are the _d_-transition metals distinguished from other elements?

30. What are the general properties of the _d_-transition metals?

31. Why are trends in variations of properties of successive _d_-transition metals less regular than trends among successive representative elements?

32. Write out the electron configurations for the following species: (a) V; (b) Fe; (c) Cu; (d) Zn; (e) Fe^{3+}; (f) Ni^{2+}; (g) Ag; (h) Ag^+.

33. Why do copper and chromium atoms have "unexpected" electron configurations?

34. What are the three triads of the _d_-transition metals?

35. Copper exists in the +1, +2, and +3 oxidation states. Which is the most stable? Which would be expected to be a strong oxidizing agent and which would be expected to be a strong reducing agent?

36. For a given transition metal in different oxidation states, how does the acidic character of its oxides increase? How do ionic and

covalent character vary? Characterize a series of metal oxides as examples.

37. For different transition metals in the same oxidation state in the same group (vertical column) of the periodic table, how do covalent character and acidic character of their oxides vary? Why? Cite evidence for the trends.

38. Chromium(VI) oxide is the acid anhydride of which two acids? Write their formulas. What is the oxidation state of the chromium in these acids?

BUILDING YOUR KNOWLEDGE

*39. Calculate ΔH^0, ΔS^0, and ΔG^0 for the reaction of 1 mol of Na with water to form aqueous NaOH and hydrogen.

*40. Calculate ΔH^0, ΔS^0, and ΔG^0 for the reaction of 1 mol of Rb with water to form aqueous RbOH and hydrogen. Compare the spontaneity of this reaction with that in Exercise 39.

*41. What is the ratio of $[Cr_2O_7^{2-}]$ to $[CrO_4^{2-}]$ at 25°C in a solution prepared by dissolving 1.0×10^{-3} mol of sodium chromate, Na_2CrO_4, in enough of an aqueous solution buffered at pH = 2.00 to produce 200 mL of solution?

*42. How many grams of Co_3O_4 (a mixed oxide, $CoO \cdot Co_2O_3$) must react with excess aluminum to produce 175 g of metallic cobalt, assuming 69.3% yield?

$$3Co_3O_4 + 8Al \xrightarrow{\Delta} 9Co + 4Al_2O_3$$

The tips of "strike anywhere" matches contain tetraphosphorus trisulfide and red phosphorus. Friction converts kinetic energy into heat, which initiates a spontaneous reaction.

$$P_4S_3(s) + 8O_2(g) \longrightarrow P_4O_{10}(s) + 3SO_2(g)$$

OBJECTIVES

As you study this chapter, you should learn about

- *The occurrence and use of the noble gases*

- *Compounds of the noble gases*

- *The occurrence and production of the halogens*

- *Some important reactions and compounds of the halogens*

- *The occurrence and production of sulfur, selenium, and tellurium*

- *Some important reactions and compounds of the heavier Group VIA nonmetals*

- *The occurrence and production of nitrogen and phosphorus*

- *Some important reactions of nitrogen and phosphorus*

- *The occurrence and importance of silicon*

- *A few important compounds of silicon*

Only about 20% of the elements are classified as nonmetals. With the exception of H, they are in the upper right-hand corner of the periodic table. In this chapter we shall consider the chemistry and properties of some nonmetals. These elements best illustrate group trends and individuality of elements within groups of nonmetals.

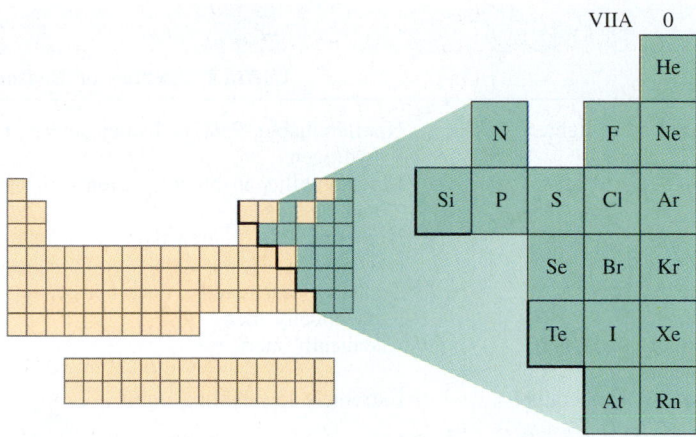

THE NOBLE GASES (GROUP 0)

24-1 OCCURRENCE, USES, AND PROPERTIES

The noble gases are very low-boiling gases. Except for radon, they can be isolated by fractional distillation of liquefied air. Radon is collected from the radioactive disintegration of radium salts. Table 24-1 gives the percentage of each noble gas in the atmosphere.

Helium is produced in the United States from some natural gas fields. This source was discovered in 1905 by H. P. Cady and D. F. McFarland at the University of Kansas, when they were asked to analyze a nonflammable component of natural gas from a Kansas gas well. Uses of the noble gases are summarized in Table 24-2.

The noble gases are colorless, tasteless, and odorless. In the liquid and solid states the only forces of attraction among the atoms are very weak London or van der Waals forces. Polarizability and interatomic interactions increase with increasing atomic size, and so melting and boiling points increase with increasing atomic number. The attractive forces among He atoms are so small that He remains liquid at 1 atmosphere pressure even at a temperature of 0.001 K.

Until the early 1960s, chemists believed that the Group 0 elements would not combine chemically with any elements. In 1962 Neil Bartlett and his research group at the University of British Columbia were studying the powerful oxidizing agent PtF_6. They accidentally prepared and identified $O_2^+PtF_6^-$ by reaction of oxygen with PtF_6. Bartlett reasoned that xenon also should be oxidized by PtF_6 because the first ionization energy of O_2 $(1.31 \times 10^3 \text{ kJ/mol})$ is slightly larger than that of xenon $(1.17 \times 10^3 \text{ kJ/mol})$. He obtained a red crystalline solid initially believed to be $Xe^+PtF_6^-$ but now known to be a more complex compound.

Radon is continually produced in small amounts in the uranium radioactive decay sequence (Section 26-10). Radon gas is so unreactive that it eventually escapes from the soil. Measurable concentrations of radon, a radioactive gas, have been observed in basements of many dwellings.

A pressure of about 26 atmospheres is required to solidify He at 0.001 K.

The noble gases are often called the rare gases. They were formerly called the "inert gases" because it was incorrectly thought that they could not enter into chemical combination.

Table 24-1 *Percentages (by volume) of Noble Gases in the Atmosphere*					
He	**Ne**	**Ar**	**Kr**	**Xe**	**Rn**
0.0005%	0.015%	0.94%	0.00011%	0.000009%	~0%

Table 24-2 *Uses of the Noble Gases*

Noble Gas	Use	Useful Properties or Reasons
helium	1. Filling of observation balloons and other lighter-than-air craft	Nonflammable; 93% of lifting power of flammable hydrogen
	2. He/O_2 mixtures, rather than N_2/O_2, for deep-sea breathing	Low solubility in blood; prevents nitrogen narcosis and "bends"
	3. Diluent for gaseous anesthetics	Nonflammable, nonreactive
	4. He/O_2 mixtures for respiratory patients	Low density, flows easily through restricted passages
	5. Heat transfer medium for nuclear reactors	Transfers heat readily; does not become radioactive; chemically inert
	6. Industrial applications, such as inert atmosphere for welding easily oxidized metals	Chemically inert
	7. Liquid He used to maintain very low temperatures in research (cryogenics)	Extremely low boiling point
neon	Neon signs	Even at low Ne pressure, moderate electric current causes bright orange-red glow; can be modified by colored glass or mixing with Ar or Hg vapor
argon	1. Inert atmosphere for welding	Chemically inert
	2. Filling incandescent light bulbs	Inert; inhibits vaporization of W and blackening of bulbs
krypton	Airport runway and approach lights	Gives longer life to incandescent lights than Ar, but more expensive
xenon	Xe and Kr mixture in high-intensity, short-exposure photographic flash tubes	Both have fast response to electric current
radon	Radiotherapy of cancerous tissues	Radioactive

24-2 XENON COMPOUNDS

Oxygen is second only to fluorine in electronegativity.

Since Bartlett's discovery, many other noble gas compounds have been made. Most are compounds of Xe, and the best characterized compounds are xenon fluorides. Oxygen compounds are also well known. Reaction of Xe with F_2, an extremely strong oxidizing agent, in different stoichiometric ratios produces xenon difluoride, XeF_2; xenon tetrafluoride, XeF_4; and xenon hexafluoride, XeF_6, all colorless crystals (Table 24-3). All involve very electronegative elements.

All the xenon fluorides are formed in exothermic reactions. They are reasonably stable, with Xe—F bond energies of about 125 kJ/mol of bonds. For comparison, strong bond energies range from about 170 to 500 kJ/mol, whereas bond energies of hydrogen bonds are typically less than 40 kJ/mol.

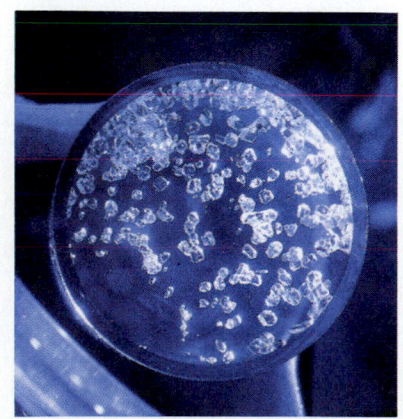

Crystals of the noble gas compound xenon tetrafluoride, XeF_4.

THE HALOGENS (GROUP VIIA)

The elements of Group VIIA are known as **halogens** (Greek, "salt formers"). The term **"halides"** is used to describe their binary compounds. The heaviest halogen, astatine, is an artificially produced element of which only short-lived radioactive isotopes are known.

Table 24-3 *Xenon Fluorides*

Compound	Preparation (Molar ratio Xe:F$_2$)	Reaction Conditions	e^- Pairs Around Xe	Hybridization at Xe	Geometry
XeF$_2$	1:1–3	400°C or irradiation or elec. discharge	5	sp^3d	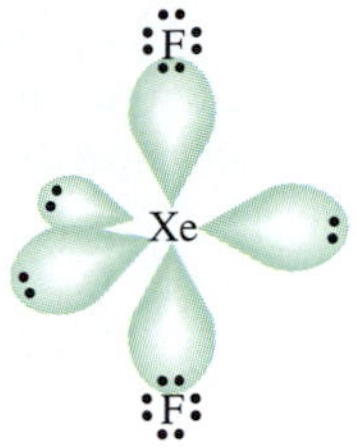
XeF$_4$	1:5	Same as for XeF$_2$	6	sp^3d^2	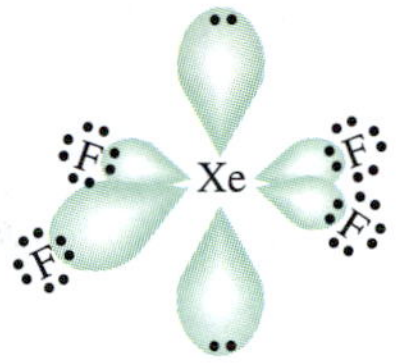
XeF$_6$	1:20	300°C and 60 atm or elec. discharge	7	sp^3d^3(?)	Exact geometry undetermined

24-3 PROPERTIES

The elemental halogens exist as diatomic molecules containing single covalent bonds. Properties of the halogens show obvious trends (Table 24-4). Their high electronegativities indicate that they attract electrons strongly. Many binary compounds that contain a metal and a halogen are ionic.

Table 24-4 *Properties of the Halogens*

Property	F	Cl	Br	I	At
Physical state (25°C, 1 atm)	gas	gas	liquid	solid	solid
Color	pale yellow	yellow-green	red-brown	violet (g); black (s)	—
Atomic radius (Å)	0.64	0.99	1.14	1.33	1.40
Ionic radius (X$^-$) (Å)	1.36	1.81	1.95	2.16	—
Outer shell e^-	$2s^22p^5$	$3s^23p^5$	$4s^24p^5$	$5s^25p^5$	$6s^26p^5$
First ionization energy (kJ/mol)	1681	1251	1140	1008	920
Electronegativity	4.0	3.0	2.8	2.5	2.1
Melting point (°C, 1 atm)	−220	−101	−7.1	114	—
Boiling point (°C, 1 atm)	−188	−35	59	184	—
X—X bond energy (kJ/mol)	158	243	192	151	—

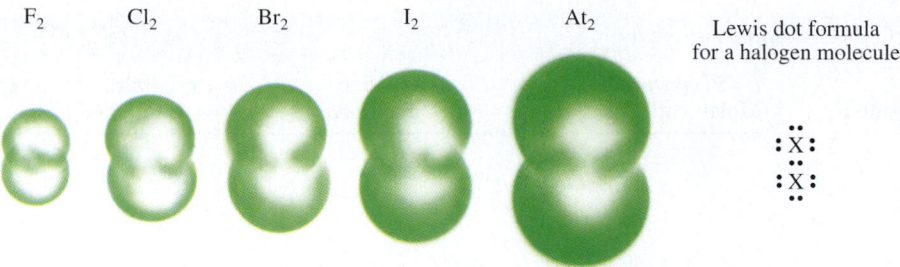

F$_2$ Cl$_2$ Br$_2$ I$_2$ At$_2$

Lewis dot formula
for a halogen molecule

$$:\overset{..}{\underset{..}{X}}:\overset{..}{\underset{..}{X}}:$$

We often represent a halogen atom as X, without specifying a particular halogen.

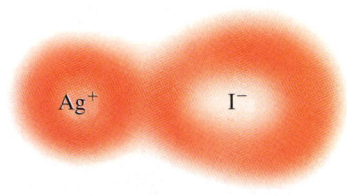

Ag$^+$ I$^-$

The diffuse cloud of the I$^-$ ion is easily polarized by the small Ag$^+$ ion.

The small fluoride ion (radius = 1.36 Å) is not easily polarized (distorted) by cations, whereas the large iodide ion (radius = 2.16 Å) is. As a result, compounds containing I$^-$ ions show greater covalent character than those containing F$^-$ ions. The properties of Cl$^-$ and Br$^-$ ions are intermediate between those of F$^-$ and I$^-$.

The chemical properties of the halogens resemble each other more closely than do those of elements in any other periodic group, with the exception of the noble gases and possibly the Group IA metals. But their physical properties differ significantly. Melting and boiling points of the halogens increase from F$_2$ to I$_2$. This follows their increase in size and increase in ease of polarization of outer shell electrons by adjacent nuclei, resulting in greater intermolecular attractive forces. All halogens except astatine are decidedly nonmetallic. They show the -1 oxidation number in most of their compounds. Except for fluorine, they also exhibit oxidation numbers of $+1$, $+3$, $+5$, and $+7$.

24-4 OCCURRENCE, PRODUCTION, AND USES

The halogens are so reactive that they do not occur free in nature. The most abundant sources of halogens are halide salts. A primary source of iodine is NaIO$_3$. The halogens are obtained by oxidation of the halide ions.

$$2X^- \rightleftharpoons X_2 + 2e^-$$

The order of increasing ease of oxidation is F$^-$ < Cl$^-$ < Br$^-$ < I$^-$ < At$^-$.

Fluorine

Fluorine occurs in large quantities in the minerals *fluorspar* or *fluorite*, CaF$_2$; *cryolite*, Na$_3$AlF$_6$; and *fluoroapatite*, Ca$_5$(PO$_4$)$_3$F. It also occurs in small amounts in sea water, teeth, bones, and blood. F$_2$ is such a strong oxidizing agent that it has not been produced by *direct* chemical oxidation of F$^-$ ions (after 181 years of trying). The pale yellow gas is prepared by electrolysis of a molten mixture of KF + HF, or KHF$_2$, in a Monel metal cell. This must be done under anhydrous conditions because H$_2$O is more readily oxidized than F$^-$.

Monel metal is an alloy of Ni, Cu, Al, and Fe. It is resistant to attack by hydrogen fluoride.

$$2KHF_2 \xrightarrow[\text{melt}]{\text{electrolysis}} F_2(g) + H_2(g) + 2KF(s)$$

In 1986 Carl O. Christe discovered that the strong Lewis acid SbF$_5$ displaces the weaker Lewis acid MnF$_4$ from the hexafluoromanganate(IV) ion, [MnF$_6$]$^{2-}$. MnF$_4$ is thermodynamically unstable, and it decomposes into MnF$_3$ and F$_2$. Christe heated a mixture of potassium hexafluoromanganate(IV) and antimony(V) fluoride in a passivated

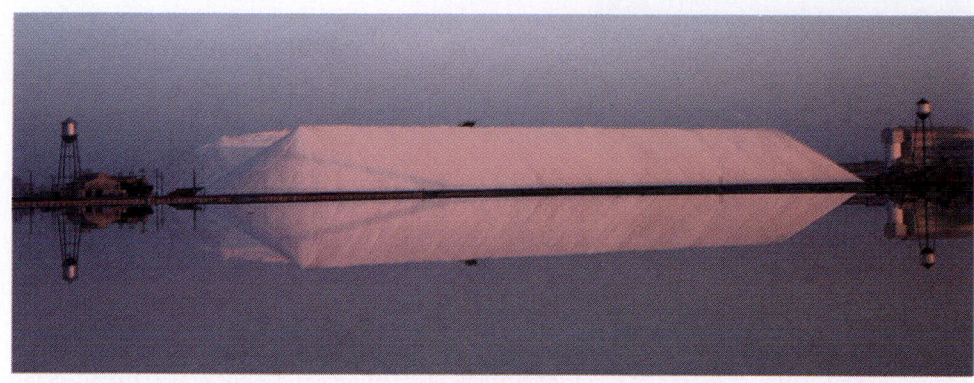

Chlorides occur in salt beds. Twenty-four billion pounds of chlorine were produced in the United States in 1993.

Teflon–stainless steel container at 150°C for 1 hour. He obtained elemental F_2 in better than 40% yield. The overall reaction may be represented as

$$2K_2MnF_6(s) + 4SbF_5(\ell) \longrightarrow 4KSbF_6(s) + 2MnF_3(s) + F_2(g)$$

Fluorine is used as a fluorinating agent. Many fluorinated organic compounds, called fluorocarbons, are stable and nonflammable. They are used as refrigerants, lubricants, plastics (such as Teflon), insecticides, and, until recently, aerosol propellants. Stannous fluoride, SnF_2, is used in toothpaste.

The reactions of F_2 with other elements are dangerous because of the vigor with which F_2 oxidizes other substances. They must be carried out with *extreme* caution!

Chlorine

Chlorine (Greek *chloros*, "green") occurs in abundance in NaCl, KCl, $MgCl_2$, and $CaCl_2$ in salt water and in salt beds. It is also present as HCl in gastric juices. The toxic, yellowish-green gas is prepared commercially by electrolysis of concentrated aqueous NaCl, in which industrially important H_2 and caustic soda (NaOH) are also produced (Section 21-4).

Chlorine is used to produce many commercially important products. Tremendous amounts of it are used in extractive metallurgy and in chlorinating hydrocarbons to produce a variety of compounds (such as polyvinyl chloride, a plastic). Chlorine is present as Cl_2, NaClO, $Ca(ClO)_2$, or Ca(ClO)Cl in household bleaches as well as in bleaches for wood pulp and textiles. Under carefully controlled conditions, Cl_2 is used to kill bacteria in public water supplies.

Bromine

Bromine (Greek *bromos*, "stench") is less abundant than fluorine and chlorine. In the elemental form it is a dense, freely flowing, corrosive, dark-red liquid with a brownish-red vapor at 25°C. It occurs mainly in NaBr, KBr, $MgBr_2$, and $CaBr_2$ in salt water, underground salt brines, and salt beds. The major commercial source for bromine is deep brine wells in Arkansas that contain up to 5000 parts per million (0.5%) of bromide.

Bromine is used in the production of silver bromide for light-sensitive eyeglasses and photographic film; in the production of sodium bromide, a mild sedative; and in methyl bromide, CH_3Br, a soil fumigant that contributes to the destruction of the ozone layer.

Bromine is a dark red liquid.

Iodine reacts with starch (as in this potato) to form a deep-blue complex substance.

Iodine

Iodine (Greek *iodos*, "purple") is a violet-black crystalline solid with a metallic luster. It exists in equilibrium with a violet vapor at 25°C. The element can be obtained from dried seaweed or shellfish or from $NaIO_3$ impurities in Chilean nitrate ($NaNO_3$) deposits. It is contained in the growth-regulating hormone thyroxine, produced by the thyroid gland. "Iodized" table salt is about 0.02% KI, which helps prevent goiter, a condition in which the thyroid enlarges. Iodine is also used as an antiseptic and germicide in the form of tincture of iodine, a solution in alcohol.

More recently available is an aqueous solution of an iodine complex of polyvinylpyrrolidone, or "povidone." It does not sting when applied to open wounds.

The preparation of iodine involves reduction of iodate ion from $NaIO_3$ with sodium hydrogen sulfite, $NaHSO_3$.

$$2IO_3^- + 5HSO_3^- \longrightarrow 3HSO_4^- + 2SO_4^{2-} + H_2O + I_2(s)$$

Iodine is then purified by sublimation (Figure 13-16).

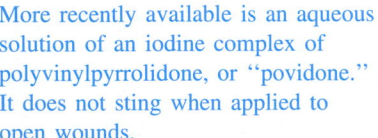

Iron and chlorine react to form iron(III) chloride, $FeCl_3$.

24-5 REACTIONS OF THE FREE HALOGENS

The free halogens react with most other elements and many compounds. For example, all the Group IA metals react with all the halogens to form simple binary ionic compounds (Section 7-2).

General Reaction	Remarks
$nX_2 + 2M \longrightarrow 2MX_n$	All X_2 with most metals (most vigorous reaction with F_2 and Group IA metals)
$X_2 + nX'_2 \longrightarrow 2XX'_n$	Formation of interhalogens (n = 1, 3, 5, or 7); X is larger than X'
$X_2 + H_2 \longrightarrow 2HX$	
$3X_2 + 2P \longrightarrow 2PX_3$	With all X_2, and with As, Sb, Bi replacing P
$5X_2 + 2P \longrightarrow 2PX_5$	Not with I_2; also Sb $\longrightarrow$ SbF_5, $SbCl_5$; As $\longrightarrow$ AsF_5; Bi $\longrightarrow$ BiF_5
$X_2 + H_2S \longrightarrow S + 2HX$	With all X_2
$X'_2 + 2X^- \longrightarrow 2X'^- + X_2$	$F_2 \longrightarrow Cl_2, Br_2, I_2$ $Cl_2 \longrightarrow Br_2, I_2$ $Br_2 \longrightarrow I_2$

Bromine reacts with powdered antimony so vigorously that the flask vibrates.

The most vigorous reactions are those of F_2, which usually oxidizes other species to their highest possible oxidation states. Iodine is only a mild oxidizing agent (I^- is a mild reducing agent) and usually does not oxidize substances to high oxidation states. Consider the following reactions of halogens with two metals that exhibit variable oxidation numbers.

With Fe	With Cu

$$2Fe + 3F_2 \xrightarrow{\overset{(+3)}{}} 2FeF_3 \quad \text{(only)}$$

$$2Fe + 3Cl_2 \text{ (excess)} \xrightarrow{\overset{(+3)}{}} 2FeCl_3 \qquad Cu + X_2 \xrightarrow{\overset{(+2)}{}} CuX_2 \quad (X = F, Cl, Br)$$

$$Fe + Cl_2 \text{ (lim. amt.)} \xrightarrow{\overset{(+2)}{}} FeCl_2$$

$$Fe + I_2 \xrightarrow{\overset{(+2)}{}} FeI_2 \quad \text{(only)} \qquad 2Cu + I_2 \xrightarrow{\overset{(+1)}{}} 2CuI \quad \text{(only)}$$

$$Fe^{3+} + I^- \longrightarrow Fe^{2+} + \tfrac{1}{2}I_2 \qquad Cu^{2+} + 2I^- \xrightarrow{\overset{(+1)}{}} CuI + \tfrac{1}{2}I_2$$

24-6 THE HYDROGEN HALIDES (HYDROHALIC ACIDS)

The hydrogen halides (Figure 24-1) are colorless gases that dissolve in water to give acidic solutions called hydrohalic acids. The gases have piercing, irritating odors. The abnormally high melting and boiling points of HF are due to its very strong hydrogen bonding (Figure 13-5).

For instance, aqueous solutions of hydrogen fluoride are called hydrofluoric acid.

Hydrogen halides can be prepared by combination of the elements.

$$H_2 + X_2 \longrightarrow 2HX(g) \qquad X = F, Cl, Br, I$$

The reaction with F_2 to produce HF is explosive and very dangerous. The reaction producing HCl does not occur significantly in the dark but occurs rapidly by a photochemical **chain reaction** when the mixture is exposed to light. Light energy is absorbed by Cl_2 molecules, which break apart into very reactive chlorine **radicals** (atoms with unpaired electrons). These subsequently attack H_2 molecules and produce HCl molecules, leaving hydrogen atoms (also radicals). The hydrogen radicals, in turn, attack Cl_2 molecules to form HCl molecules and chlorine radicals.

A photochemical reaction is one in which a species (usually a molecule) interacts with radiant energy to produce very reactive species. These then undergo further reaction.

$$Cl_2 \xrightarrow{h\nu} 2 \, \overset{..}{\underset{..}{:}}\overset{..}{Cl} \cdot \qquad \text{initiation step}$$

$$\left. \begin{array}{l} \overset{..}{\underset{..}{:}}\overset{}{Cl} \cdot + H_2 \longrightarrow HCl + H \cdot \\ H \cdot + Cl_2 \longrightarrow HCl + \, \overset{..}{\underset{..}{:}}\overset{}{Cl} \cdot \end{array} \right\} \text{chain propagation steps}$$

This chain reaction continues as long as there is a significant concentration of radicals. **Termination steps** eliminate two radicals and eventually terminate the reaction.

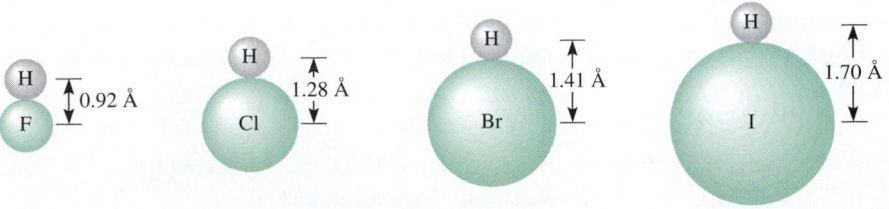

Figure 24-1 Relative sizes (approximate) of the hydrogen halides. The distance between the center of the hydrogen atom and the center of the halogen atom (internuclear distance) is indicated.

$$H \cdot + H \cdot \longrightarrow H_2$$
$$:\ddot{C}l \cdot + :\ddot{C}l \cdot \longrightarrow Cl_2 \quad \Big\}\ \text{termination steps}$$
$$H \cdot + :\ddot{C}l \cdot \longrightarrow HCl$$

The reaction of H_2 with Br_2 is also a photochemical reaction. That of H_2 with I_2 is very slow, even at high temperatures and with illumination.

All hydrogen halides react with H_2O to produce *hydrohalic acids* that ionize.

$$H-\ddot{O}: + H-\ddot{X}: \rightleftharpoons H-\overset{+}{\underset{|}{O}}-H + :\ddot{X}:^-$$
$$\underset{H}{|} \qquad\qquad \underset{H}{|}$$

The reaction is essentially complete for dilute $HCl(aq)$, $HBr(aq)$, and $HI(aq)$. Dilute $HF(aq)$ is a weak acid ($K_a = 7.2 \times 10^{-4}$). In concentrated solutions more acidic dimeric $(HF)_2$ units are present (Figure 24-2). They ionize as follows.

$$(HF)_2(aq) + H_2O(\ell) \rightleftharpoons H_3O^+(aq) + HF_2^-(aq) \qquad K \approx 5$$

The order of increasing acid strengths of the hydrohalic acids is the same as that of the anhydrous hydrogen halides: $HF(aq) \ll HCl(aq) < HBr(aq) < HI(aq)$ (Section 10-5).

The only acid used in industry to a greater extent than HCl is H_2SO_4. Hydrochloric acid is used in the production of metal chlorides, dyes, and many other commercially important products. It is also used on a large scale to dissolve metal oxide coatings from iron and steel prior to galvanizing or enameling.

Hydrofluoric acid is used in the production of fluorine-containing compounds and for etching glass. The acid reacts with silicates, such as calcium silicate, $CaSiO_3$, in the glass to produce a very volatile and thermodynamically stable compound, silicon tetrafluoride, SiF_4.

$$CaSiO_3(s) + 6HF(aq) \longrightarrow CaF_2(s) + SiF_4(g) + 3H_2O(\ell)$$

24-7 THE OXOACIDS (TERNARY ACIDS) OF THE HALOGENS

Table 24-5 lists the known oxoacids of the halogens, their sodium salts, and some trends in properties. Only three oxoacids, $HClO_4$, HIO_3, and H_5IO_6, have been isolated in anhydrous form. The others are known only in aqueous solution. In all these acids the H is bonded through an O.

The Lewis formulas and structures of the chlorine oxoanions are shown in Figure 24-3. The corresponding oxoanions of bromine and iodine have similar structures. Ionization becomes easier as the number of O atoms in the oxoacids increases (Section 10-5). Such behavior is observed for the series of oxoacids of any element, when *all* the H atoms are bonded to O atoms. Concentrated solutions of all of these acids are strong oxidizing agents. Oxidizing power decreases with increasing number of oxygen atoms.

The only oxoacid of fluorine that has been prepared is unstable hypofluorous acid, HOF.

Aqueous *hypohalous acids* (except HOF) can be prepared by reaction of free halogens (Cl_2, Br_2, I_2) with cold water. The smaller the halogen, the farther to the right the equilibrium lies.

$$\overset{0}{X_2} + H_2O \rightleftharpoons \underset{\text{hydrohalic acid}}{\overset{-1}{HX}} + \underset{\text{hypohalous acid}}{\overset{+1}{HOX}} \qquad (X = Cl, Br, I)$$

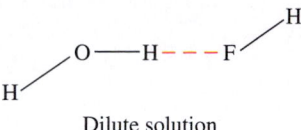

Dilute solution

(HF)$_2$

Concentrated solution

Figure 24-2 Hydrogen bonding (dashed lines) in dilute and concentrated aqueous solutions of hydrofluoric acid.

In HOF the oxidation states are $F = -1$, $H = +1$, $O = 0$.

These reactions all involve *disproportionation* of the halogen.

Table 24-5 *Oxoacids of the Halogens and Their Salts*

Oxidation State	Acid	Name of Acid	Thermal Stability and Acid Strength	Oxidizing Power of Acid	Sodium Salt	Name of Salt	Thermal Stability	Oxidizing Power and Hydrolysis of Anion	Nature of Halogen
+1	HXO (HOX)	hypo*hal*ous acid			NaXO (NaOX)	sodium hypo*hal*ite			X = F, Cl, Br, I
+3	HXO_2	*hal*ous acid			$NaXO_2$	sodium *hal*ite			X = Cl, Br (?)
+5	HXO_3	*hal*ic acid	Increase	Increases	$NaXO_3$	sodium *hal*ate	Increases	Increase	X = Cl, Br, I
+7	HXO_4	per*hal*ic acid			$NaXO_4$	sodium per*hal*ate			X = Cl, Br, I
+7	H_5XO_6	para*per*halic acid			several types	sodium para*per*halates			X = I only

Hypohalite salts can be prepared by reactions of the halogens with *cold* dilute bases.

$$\underset{(0)}{X_2} + 2NaOH \longrightarrow \underset{\substack{(-1)\\ \text{sodium halide}}}{NaX} + \underset{\substack{(+1)\\ \text{sodium hypohalite}}}{NaOX} + H_2O \qquad (X = Cl, Br, I)$$

The hypohalites are used as bleaching agents. Sometimes Cl_2 is used as a bleach or as a disinfectant, as in public water supplies. It reacts slowly with H_2O to form HCl and HOCl. The hypochlorous acid then decomposes into HCl and O radicals, which kill bacteria.

$$Cl_2 + H_2O \rightleftharpoons HCl + HOCl$$

$$HOCl \longrightarrow HCl + :\overset{..}{\underset{..}{O}}\cdot$$

Solid household bleaches are usually Ca(ClO)Cl. This is prepared by reaction of Cl_2 with $Ca(OH)_2$.

$$Ca(OH)_2 + Cl_2 \longrightarrow$$
$$Ca(ClO)Cl + H_2O$$

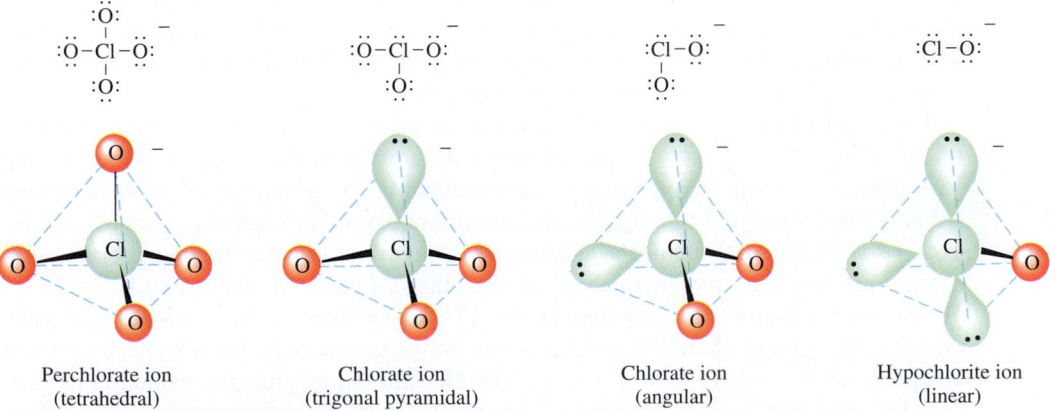

| Perchlorate ion (tetrahedral) | Chlorate ion (trigonal pyramidal) | Chlorate ion (angular) | Hypochlorite ion (linear) |

Figure 24-3 Lewis formulas and structures of the oxoanions of chlorine. All have tetrahedral *electronic geometry,* but only the perchlorate ion, ClO_4^-, has tetrahedral *ionic geometry.* The ionic geometries of the other three are predicted by the (imaginary) successive removal of O atoms.

These oxygen radicals are very strong oxidizing agents. They are the effective bleaching and disinfecting agent in aqueous solutions of Cl_2 or hypochlorite salts.

Anhydrous *halous acids* have not been isolated. Fluorous and iodous acids apparently do not exist, even in aqueous solution. A convenient preparation of aqueous chlorous acid involves reaction of chlorine dioxide and barium hydroxide to produce barium chlorite, a suspension of which is treated with sulfuric acid. The precipitated barium sulfate is removed by filtration, leaving a solution of $HClO_2$.

$$Ba(ClO_2)_2 + H_2SO_4 \longrightarrow BaSO_4(s) + 2HClO_2$$
barium chlorite   chlorous acid

When stoichiometric amounts of reactants are mixed, Ba^{2+} and SO_4^{2-} ions combine to form insoluble $BaSO_4$, which can be separated by filtration. Fairly pure $HClO_3$ and $HBrO_3$ solutions remain.

Aqueous solutions of chloric acid and bromic acids are prepared by reactions of barium chlorate and bromate with sulfuric acid.

$$Ba(XO_3)_2 + H_2SO_4 \longrightarrow BaSO_4(s) + 2HXO_3 \qquad (X = Cl, Br)$$
(precipitates)   halic acid

Treatment of $HClO_4$ with a strong dehydrating agent such as P_4O_{10} produces the explosive anhydride, dichlorine heptoxide, Cl_2O_7. The oxidation state of Cl is +7 in both compounds.

Neither perfluoric acid nor perfluorate ion is known. The other *perhalic acids* and *perhalates* are known. Anhydrous perchloric acid, $HClO_4$, a colorless, oily liquid, distills under reduced pressure (10 to 20 torr) after reaction of a perchlorate salt with nonvolatile, concentrated sulfuric acid.

$$NaClO_4 + H_2SO_4 \longrightarrow NaHSO_4 + HClO_4 \qquad \text{(perchloric acid distills)}$$

Perchloric acid is the strongest of all common acids with respect to ionization, exceeding even HNO_3 and HCl. Hot, concentrated perchloric acid is a very strong oxidizing agent that can explode in the presence of reducing agents. Cold, dilute perchloric acid is only a weak oxidizing agent.

SULFUR, SELENIUM, AND TELLURIUM

24-8 OCCURRENCE, PROPERTIES, AND USES

The Group VIA elements are less electronegative than the halogens. Oxygen and sulfur are clearly nonmetallic, but selenium is less so. Tellurium is usually classified as a metalloid and forms metal-like crystals. Its chemistry is mostly that of a nonmetal. Polonium is a metal. All 29 isotopes of polonium are radioactive.

Irregularities in the properties of elements within a given family increase toward the middle of the periodic table. There are larger differences in the properties of the Group VIA elements than in the properties of the halogens. The properties of elements in the *second period* usually differ significantly from those of other elements in their families, because second-period elements have no low-energy *d* orbitals. So, the properties of oxygen are not very similar to those of the other Group VIA elements (Table 24-6).

The *d* orbitals do not occur until the third energy level.

The outer electron configuration of the VIA elements is ns^2np^4. An atom of each element may gain or share two electrons as it forms compounds. Each forms a covalent compound of the type H_2E in which the VIA element (E) exhibits the oxidation number -2. The maximum number of atoms with which O can bond (coordination number) is four, but S, Se, Te, and probably Po can bond covalently to as many as six other atoms. This is due to the availability of vacant *d* orbitals in the outer shell of each of the VIA elements except O. One or more of the *d* orbitals can accommodate additional electrons to form up to six bonds.

Table 24-6 *Some Properties of Group VIA Elements*

Property	O	S	Se	Te	Po
Physical state (1 atm, 25°C)	gas	solid	solid	solid	solid
Color	colorless (very pale blue)	yellow	red-gray to black	brass-colored, metallic luster	—
Outermost electrons	$2s^2 2p^4$	$3s^2 3p^4$	$4s^2 4p^4$	$5s^2 5p^4$	$6s^2 6p^4$
Melting point (1 atm, °C)	−218	112	217	450	254
Boiling point (1 atm, °C)	−183	444	685	990	962
Electronegativity	3.5	2.5	2.4	2.1	1.9
First ionization energy (kJ/mol)	1314	1000	941	869	812
Atomic radius (Å)	0.66	1.04	1.17	1.37	1.4
Ionic (2−) radius (Å)	1.40	1.84	1.98	2.21	—
Common oxidation states	usually −2	−2, +2, +4, +6	−2, +2, +4, +6	−2, +2, +4, +6	−2, +6

Sulfur

Sulfur makes up about 0.05% of the earth's crust. It was one of the elements known to the ancients. It was used by the Egyptians as a yellow coloring, and it was burned in some religious ceremonies because of the unusual odor it produced; it is the "brimstone" of the Bible. Alchemists tried to incorporate its "yellowness" into other substances to produce gold.

Sulfur occurs as the free element—predominantly S_8 molecules—and as metal sulfides such as galena, PbS; iron pyrite, FeS_2; and cinnabar, HgS. To a lesser extent, it occurs as metal sulfates such as barite, $BaSO_4$, and gypsum, $CaSO_4 \cdot 2H_2O$, and in volcanic gases as H_2S and SO_2.

Sulfur is found in much naturally occurring organic matter, e.g., petroleum and coal. Its presence in fossil fuels causes environmental and health problems because many sulfur-containing compounds burn to produce sulfur dioxide, an air pollutant.

Nearly half of the sulfur used in the United States is recovered from natural gas and oil. Hydrogen sulfide is oxidized to sulfur in the Claus furnace.

$$8H_2S(g) + 4O_2(g) \longrightarrow S_8(\ell) + 8H_2O(g)$$

Elemental sulfur is mined along the U.S. Gulf Coast by the **Frasch process**, or "hot water" process (Figure 24-4). Most of it is used in the production of sulfuric acid, H_2SO_4, the most important of all industrial chemicals. Sulfur is used in the vulcanization of rubber and in the synthesis of many important sulfur-containing organic compounds.

In each of the three physical states, elemental sulfur exists in many forms. The two most stable forms of sulfur, the rhombic (mp 112°C) and monoclinic (mp 119°C) crystalline modifications, consist of different arrangements of S_8 molecules. These are puckered rings containing eight sulfur atoms (Figure 2-3) and all S—S single bonds. Above 150°C, sulfur becomes increasingly viscous and darkens as the S_8 rings break apart into chains that interlock with each other through S—S bonds. The viscosity reaches a maximum at 180°C, at which point sulfur is dark brown. Above 180°C, the liquid thins as the chains are broken down into smaller chains. At 444°C, sulfur boils to give a vapor containing S_8, S_6, S_4, and S_2 molecules.

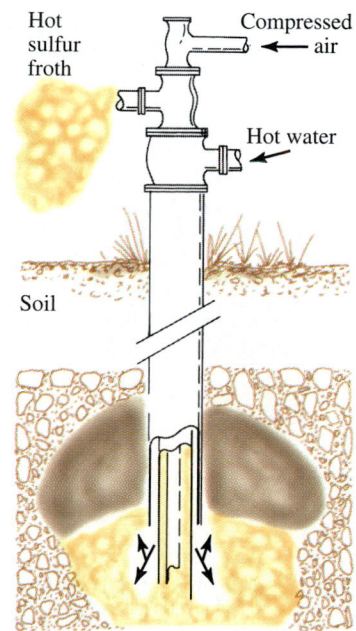

Figure 24-4 The Frasch process for mining sulfur. Three concentric pipes are used. Water at about 170°C and a pressure of 100 lb/in² (7 kg/cm²) is forced down the outermost pipe to melt the sulfur. Hot compressed air is pumped down the innermost pipe. It mixes with the molten sulfur to form a froth, which rises through the third pipe.

Native sulfur. Elemental sulfur is deposited at the edges of some hot springs and geysers. This formation surrounds Emerald Lake in Yosemite National Park.

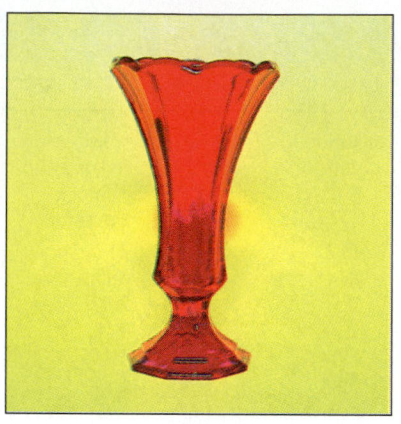

Selenium is often incorporated into glass in the form of Na_2Se. Subsequent heating causes particles of red colloidal selenium to form.

Selenium

Selenium is quite rare ($9 \times 10^{-6}\%$ of the earth's crust). It occurs mainly as an impurity in sulfur, sulfide, and sulfate deposits. It is obtained from the flue dusts that result from roasting sulfide ores and from the "anode mud" formed in the electrolytic refining of copper. It is used as a red coloring in glass. An allotropic form of selenium has an electrical conductivity that is very light-sensitive, so it is used in photocopy machines and in solar cells.

Tellurium

Tellurium is even less abundant ($2 \times 10^{-7}\%$ of the earth's crust) than selenium. It occurs mainly in sulfide ores, especially with copper sulfide, and as the tellurides of gold and silver. It, too, is obtained from the "anode mud" from refining of copper. The element forms brass-colored, shiny, hexagonal crystals having low electrical conductivity. It is added to some metals, particularly lead, to increase electrical resistance and improve resistance to heat, corrosion, mechanical shock, and wear.

24-9 REACTIONS OF GROUP VIA ELEMENTS

Some reactions of the Group VIA elements are summarized below.

General Equation	Remarks
$x\mathrm{E} + y\mathrm{M} \longrightarrow \mathrm{M}_y\mathrm{E}_x$	With many metals
$z\mathrm{E} + \mathrm{M}_x\mathrm{E}_y \longrightarrow \mathrm{M}_x\mathrm{E}_{y+z}$	Especially with S, Se
$\mathrm{E} + \mathrm{H}_2 \longrightarrow \mathrm{H}_2\mathrm{E}$	Decreasingly in the series O_2, S, Se, Te
$\mathrm{E} + 3\mathrm{F}_2 \longrightarrow \mathrm{EF}_6$	With S, Se, Te, and excess F_2
$2\mathrm{E} + \mathrm{Cl}_2 \longrightarrow \mathrm{E}_2\mathrm{Cl}_2$	With S, Se (Te gives $TeCl_2$); also with Br_2
$\mathrm{E}_2\mathrm{Cl}_2 + \mathrm{Cl}_2 \longrightarrow 2\mathrm{ECl}_2$	With S, Se; also with Br_2
$\mathrm{E} + 2\mathrm{Cl}_2 \longrightarrow \mathrm{ECl}_4$	With S, Se, Te, and excess Cl_2; also with Br_2
$\mathrm{E} + \mathrm{O}_2 \longrightarrow \mathrm{EO}_2$	With S (with Se, use $O_2 + NO_2$)

24-10 HYDRIDES OF GROUP VIA ELEMENTS

All the Group VIA elements form covalent compounds of the type H_2E (E = O, S, Se, Te, Po) in which the Group VIA element is in the -2 oxidation state. H_2O is a liquid that is essential for animal and plant life. H_2S, H_2Se, and H_2Te are colorless, noxious, poisonous gases. They are even more toxic than HCN. Egg protein contains sulfur, and its decomposition forms H_2S, giving off the odor of rotten eggs. H_2Se and H_2Te smell even worse. Their odors are usually ample warning of the presence of these poisonous gases.

E represents a Group VIA element.

   Both the melting point and boiling point of water are very much higher than expected by comparison with those of the heavier hydrides (Figure 13-5). This is a consequence of hydrogen bonding in ice and liquid water (Section 13-2) caused by the strongly dipolar nature of water molecules. The electronegativity differences between H and the other VIA elements are much smaller than those between H and O, so no H-bonding occurs in H_2S, H_2Se, or H_2Te.

   Aqueous solutions of hydrogen sulfide, selenide, and telluride are acidic; acid strength increases as the group is descended: $H_2S < H_2Se < H_2Te$. The same trend was observed for increasing acidity of the hydrogen halides. The acid ionization constants are

H_2S is a stronger acid than H_2O. The solubility of H_2S in water is approximately 0.10 mol/L at 25°C.

		H_2S	**H_2Se**	**H_2Te**
$H_2E \rightleftharpoons H^+ + HE^-$	K_1:	1.0×10^{-7}	1.9×10^{-4}	2.3×10^{-3}
$HE^- \rightleftharpoons H^+ + E^{2-}$	K_2:	1.3×10^{-13}	$\sim 10^{-11}$	$\sim 1.6 \times 10^{-11}$

E represents S, Se, or Te.

24-11 GROUP VIA OXIDES

Although others exist, the most important VIA oxides are the dioxides, which are acid anhydrides of H_2SO_3, H_2SeO_3, and H_2TeO_3; and the trioxides, which are anhydrides of H_2SO_4, H_2SeO_4, and H_6TeO_6.

SO_2

Sulfur dioxide is a colorless, poisonous, corrosive gas with a very irritating odor. Even in small quantities, it causes coughing and nose, throat, and lung irritation. It is an angular molecule with trigonal planar electronic geometry, sp^2 hybridization at the S atom, and resonance stabilization.

   Sulfur dioxide is produced in reactions such as the combustion of sulfur-containing fossil fuels and the roasting of sulfide ores.

$$2ZnS + 3O_2 \longrightarrow 2ZnO + 2SO_2$$

SO_2 is a waste product of these operations. In the past, it was released into the atmosphere along with some SO_3 produced by its reaction with O_2. However, efforts are now under way to trap SO_2 and SO_3 and use them to make H_2SO_4. Some coal contains up to 5% sulfur, so both SO_2 and SO_3 are present in the flue gases when coal is burned. No way has been found to remove all the SO_2 from flue gases of power plants. One way of removing most of the SO_2 involves the injection of limestone, $CaCO_3$, into the combustion zone of the furnace. Here $CaCO_3$ decomposes to lime, CaO. This then combines with SO_2 to form calcium sulfite ($CaSO_3$), an ionic solid, which is collected.

If the SO_2 and SO_3 are allowed to escape into the atmosphere, they cause highly acidic rain.

$$CaCO_3 \xrightarrow{\Delta} CaO + CO_2 \qquad \text{followed by} \qquad CaO + SO_2 \longrightarrow CaSO_3$$

+4 Oxidation State

Formula	Name
H_2SO_3	sulfurous acid
H_2SeO_3	selenous acid
H_2TeO_3	tellurous acid

+6 Oxidation State

Formula	Name
H_2SO_4	sulfuric acid
H_2SeO_4	selenic acid
H_6TeO_6	telluric acid

Large amounts of SO_2 and H_2S are released during volcanic eruptions.

This process is called scrubbing. A disadvantage of it is the formation of huge quantities of solid waste ($CaSO_3$, unreacted CaO, and by-products).

Catalytic oxidation is now used by the smelting industry to convert SO_2 into SO_3. This is then dissolved in water to make solutions of H_2SO_4 (up to 80% by mass). The gases containing SO_2 are passed through a series of condensers containing catalysts to speed up the reaction. In some cases the impure H_2SO_4 can be used in other operations in the same plant.

SeO_2 and TeO_2

Selenium and tellurium dioxides can be formed by burning the elements.

$$Se + O_2 \xrightarrow{\Delta} SeO_2 \qquad Te + O_2 \xrightarrow{\Delta} TeO_2$$

Concentrated HNO_3 dissolves Se and Te to give solutions that contain selenous and tellurous acids, respectively.

$$4HNO_3 + Se \longrightarrow H_2SeO_3 + 4NO_2 + H_2O$$

$$4HNO_3 + Te \longrightarrow H_2TeO_3 + 4NO_2 + H_2O$$

When these solutions are evaporated to dryness by heating, the acids decompose to form dioxides.

$$H_2SeO_3 \xrightarrow{\Delta} SeO_2 + H_2O \qquad H_2TeO_3 \xrightarrow{\Delta} TeO_2 + H_2O$$

SO_3

Sulfur trioxide is a liquid that boils at 44.8°C. It is the anhydride of H_2SO_4. It is formed by the reaction of SO_2 with O_2. The reaction is very exothermic, but ordinarily very slow. It is catalyzed commercially in the **contact process** by spongy Pt, SiO_2, or vanadium(V) oxide, V_2O_5, at high temperatures (400 to 700°C).

$$2SO_2(g) + O_2(g) \underset{}{\overset{\text{catalyst}}{\rightleftharpoons}} 2SO_3(g) \qquad \Delta H^0 = -197.6 \text{ kJ/mol} \qquad \Delta S^0 = -188 \text{ J/K} \cdot \text{mol}$$

The high temperature favors SO_2 and O_2 but allows the reaction to proceed much more rapidly, so it is economically advantageous. The SO_3 is then removed from the gaseous reaction mixture by dissolving it in concentrated H_2SO_4 (95% H_2SO_4 by mass) to produce polysulfuric acids—mainly pyrosulfuric acid, $H_2S_2O_7$. This is called oleum, or fuming sulfuric acid. Addition of fuming sulfuric acid to water produces commercial H_2SO_4.

$$SO_3 + H_2SO_4 \longrightarrow H_2S_2O_7 \qquad \text{then} \qquad H_2S_2O_7 + H_2O \longrightarrow 2H_2SO_4$$

In the presence of certain catalysts, sulfur dioxide in polluted air reacts rapidly with O_2 to form SO_3. Particulate matter, or suspended microparticles, such as NH_4NO_3 and elemental S, act as efficient catalysts.

The prefix *pyro* means "heat" or "fire." Pyrosulfuric acid may also be obtained by heating concentrated sulfuric acid, which results in the elimination of one molecule of water from two molecules of sulfuric acid.

$$2H_2SO_4 \longrightarrow H_2S_2O_7 + H_2O$$

SeO_3 and TeO_3

Both SeO_3 and TeO_3 are stronger oxidizing agents than SO_3. They are the anhydrides of selenic and telluric acids, respectively. SeO_3 is prepared by the reaction of potassium selenate, K_2SeO_4, and SO_3.

$$K_2SeO_4 + SO_3 \longrightarrow K_2SO_4 + SeO_3$$

Dehydration of telluric acid, H_6TeO_6, at 300 to 600°C produces water-insoluble TeO_3, which decomposes to Te_2O_5 and O_2 above 400°C.

$$H_6TeO_6 \xrightarrow{\Delta} TeO_3 + 3H_2O$$

24-12 OXOACIDS OF SULFUR

Sulfurous Acid, H_2SO_3

Sulfur dioxide readily dissolves in water to produce solutions of sulfurous acid, H_2SO_3. The acid has not been isolated in anhydrous form.

$$H_2O + SO_2 \rightleftharpoons H_2SO_3$$

The acid ionizes in two steps in water

$$H_2SO_3 \rightleftharpoons H^+ + HSO_3^- \qquad K_1 = 1.2 \times 10^{-2}$$

$$HSO_3^- \rightleftharpoons H^+ + SO_3^{2-} \qquad K_2 = 6.2 \times 10^{-8}$$

When excess SO_2 is bubbled into aqueous NaOH, sodium hydrogen sulfite, $NaHSO_3$, is produced. This acid salt can be neutralized with additional NaOH or Na_2CO_3 to produce sodium sulfite.

$$NaOH + H_2SO_3 \longrightarrow NaHSO_3 + H_2O$$

$$NaOH + NaHSO_3 \longrightarrow Na_2SO_3 + H_2O$$

The sulfite ion is pyramidal and has tetrahedral electronic geometry as predicted by VSEPR theory.

Hydrogen sulfite ion

Sulfite ion

Sulfuric Acid, H_2SO_4

More than 40 million tons of sulfuric acid are produced annually worldwide. The contact process is used for the commercial production of most sulfuric acid. The solution sold commercially as "concentrated sulfuric acid" is 96 to 98% H_2SO_4 by mass, and is about 18 molar H_2SO_4.

Pure H_2SO_4 is a colorless, oily liquid that freezes at 10.4°C and boils at 290 to 317°C while partially decomposing to SO_3 and water. There is some hydrogen bonding in solid and liquid H_2SO_4.

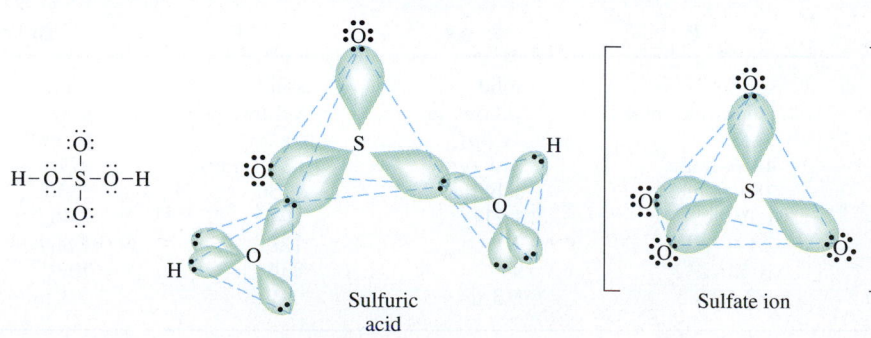

Sulfuric acid

Sulfate ion

Pouring concentrated H_2SO_4 into an equal volume of H_2O liberates a lot of heat, enough to raise the temperature of the resulting solution from room temperature to 121°C.

Tremendous amounts of heat are evolved when concentrated sulfuric acid is diluted. This illustrates the strong affinity of H_2SO_4 for water. H_2SO_4 is often used as a dehydrating agent. Dilutions should always be performed by adding the acid to water to avoid spattering the acid.

Sulfuric acid is a strong acid with respect to the first step of its ionization in water. The second ionization occurs to a lesser extent (Example 18-22).

$$H_2SO_4 \rightleftharpoons H^+ + HSO_4^- \qquad K_1 = \text{very large}$$
$$HSO_4^- \rightleftharpoons H^+ + SO_4^{2-} \qquad K_2 = 1.2 \times 10^{-2}$$

NITROGEN AND PHOSPHORUS

In the nitrogen family, nitrogen and phosphorus are nonmetals, arsenic is predominantly nonmetallic, antimony is more metallic, and bismuth is definitely metallic. Properties of the Group VA elements are listed in Table 24-7.

Oxidation states of the VA elements range from -3 to $+5$. Odd-numbered oxidation states are favored. The VA elements form very few monatomic ions. Ions with a charge of $3-$ occur for N and P, as in Mg_3N_2 and Ca_3P_2.

All of the Group VA elements show the -3 oxidation state in covalent compounds such as NH_3 and PH_3. The $+5$ oxidation state is found only in covalent compounds such as phosphorus pentafluoride, PF_5; nitric acid, HNO_3, and phosphoric acid, H_3PO_4; and in polyatomic ions such as NO_3^- and PO_4^{3-}. Each Group VA element exhibits the $+3$ oxidation state in one of its oxides, e.g., N_2O_3 and P_4O_6. These are acid anhydrides of nitrous acid, HNO_2, and phosphorous acid, H_3PO_3; both are weak acids. No other element exhibits more oxidation states than nitrogen does (Table 24-8).

24-13 OCCURRENCE OF NITROGEN

Every protein contains nitrogen in each of its fundamental amino acid units.

Nitrogen, N_2, is a colorless, odorless, tasteless gas that makes up about 75% by mass and 78% by volume of the atmosphere. Nitrogen compounds form only a minor portion of the earth's crust, but all living matter contains nitrogen. The primary natural inorganic depos-

Table 24-7 *Properties of the Group VA Elements*

Property	N	P	As	Sb	Bi
Physical state (1 atm, 25°C)	gas	solid	solid	solid	solid
Color	colorless	red, white, black	yellow, gray	yellow, gray	gray
Outermost electrons	$2s^2 2p^3$	$3s^2 3p^3$	$4s^2 4p^3$	$5s^2 5p^3$	$6s^2 6p^3$
Melting point (°C)	-210	44 (white)	814 (gray)	631 (gray)	271
Boiling point (°C)	-196	280 (white)	sublimes 613	1750	1560
Atomic radius (Å)	0.70	1.10	1.21	1.41	1.46
Electronegativity	3.0	2.1	2.1	1.9	1.8
First ionization energy (kJ/mol)	1402	1012	947	834	703
Oxidation states	-3 to $+5$	-3 to $+5$	-3 to $+5$	-3 to $+5$	-3 to $+5$

Table 24-8 *Oxidation States of Nitrogen and Examples*

−3	−2	−1	0	+1	+2	+3	+4	+5
NH_3 ammonia	N_2H_4 hydrazine	NH_2OH hydroxylamine	N_2 nitrogen	N_2O dinitrogen oxide	NO nitrogen oxide	N_2O_3 dinitrogen trioxide	NO_2 nitrogen dioxide	N_2O_5 dinitrogen pentoxide
NH_4^+ ammonium ion		NH_2Cl chloramine		$H_2N_2O_2$ hyponitrous acid		HNO_2 nitrous acid	N_2O_4 dinitrogen tetroxide	HNO_3 nitric acid
NH_2^- amide ion						NO_2^- nitrite ion		NO_3^- nitrate ion

its of nitrogen are very localized. They consist mostly of KNO_3 and $NaNO_3$. Most sodium nitrate is mined in Chile.

$:N\equiv N:$

The extreme abundance of N_2 in the atmosphere and the low relative abundance of nitrogen compounds elsewhere are due to the chemical inertness of N_2 molecules. This results from the very high bond energy of the $N\equiv N$ bond (946 kJ/mol).

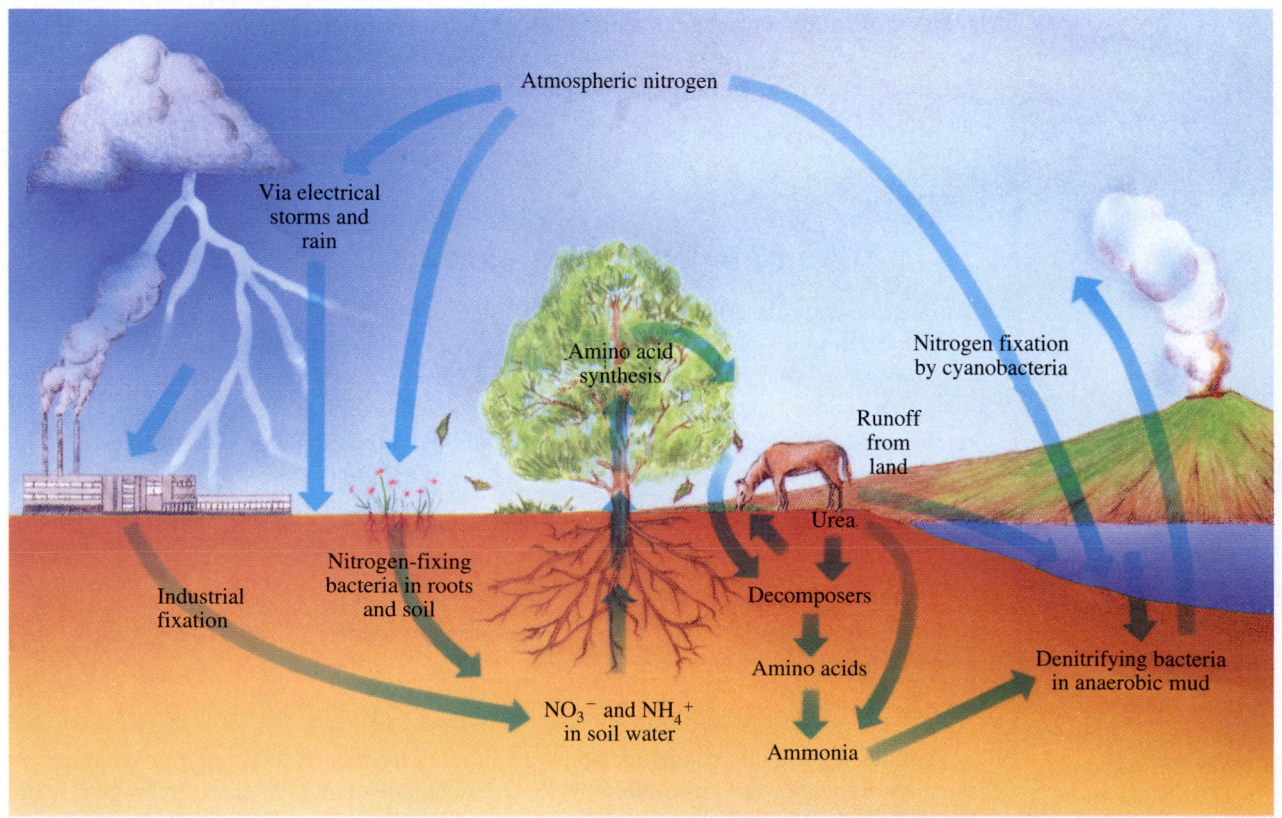

A schematic representation of the nitrogen cycle.

Nitrogen is sold as compressed gas in cylinders. The boiling point of N_2 is $-195.8°C$ ($-320°F$). N_2 is obtained by fractional distillation of liquid air.

Root nodules on soybeans.

Great advances have been made in cattle breeding in recent decades. Semen from superior bulls can be collected and stored in liquid nitrogen for 30 years or more.

Although N_2 molecules are unreactive, nature provides mechanisms by which N atoms are incorporated into proteins, nucleic acids, and other nitrogenous compounds. The **nitrogen cycle** is the complex series of reactions by which nitrogen is slowly but continually recycled in the atmosphere (our nitrogen reservoir), lithosphere (earth), and hydrosphere (water).

When N_2 and O_2 molecules collide near a bolt of lightning, they can absorb enough electrical energy to produce molecules of NO. An NO molecule is quite reactive because it contains one unpaired electron. NO reacts readily with O_2 to form nitrogen dioxide, NO_2. Most NO_2 dissolves in rainwater and falls to the earth's surface (page 238). Bacterial enzymes reduce the nitrogen in a series of reactions in which amino acids and proteins are produced. These are then used by plants, eaten by animals, and metabolized. The metabolic products are excreted as nitrogenous compounds such as urea, $(NH_2)_2CO$, and ammonium salts such as $NaNH_4HPO_4$. These can also be enzymatically converted to ammonia, NH_3, and amino acids.

Nitrogen is converted directly into NH_3 in another way. Members of the class of plants called legumes (including soybeans, alfalfa, and clover) have nodules on their roots. Within the nodules live bacteria that produce an enzyme called nitrogenase. These bacteria extract N_2 directly from air trapped in the soil and convert it into NH_3. The ability of nitrogenase to catalyze this conversion, called **nitrogen fixation**, at usual temperatures and pressures with very high efficiency is a marvel to scientists. They must resort to very extreme and costly conditions to produce NH_3 from nitrogen and hydrogen (the Haber process, Section 17-6).

Ammonia is the source of nitrogen in many fertilizers. Unfortunately, nature does not produce NH_3 and related plant nutrient compounds rapidly enough to provide an adequate food supply for the world's growing population. Commercial synthetic fertilizers have helped to lessen this problem, but at great cost for the energy that is required to produce them.

24-14 HYDROGEN COMPOUNDS OF NITROGEN

We have already described ammonia and some of its reactions. Please review Sections 17-6 and 18-4.

Liquid ammonia (bp $-33.4°C$) is used as a solvent for some chemical reactions. It is hydrogen bonded, just as H_2O is, but NH_3 is a much more basic solvent. Its weak *autoionization* produces the ammonium ion, NH_4^+, and the amide ion, NH_2^-. This is similar to H_2O, which ionizes to produce some H_3O^+ and OH^- ions.

$$NH_3(\ell) + NH_3(\ell) \rightleftharpoons NH_4^+ + NH_2^- \qquad K = 10^{-30}$$

base$_2$ acid$_1$ acid$_2$ base$_1$

Many ammonium salts are known. Most are very soluble in water. They can be prepared by reactions of ammonia with acids.

$$NH_3 + [H^+ + NO_3^-] \longrightarrow [NH_4^+ + NO_3^-] \qquad \text{ammonium nitrate}$$

Amines are organic compounds that are structurally related to ammonia. We think of them as being derived from NH_3 by the replacement of one or more hydrogens with organic groups (Section 27-11). All involve sp^3-hybridized N. All are weak bases because of the unshared pair of electrons on N.

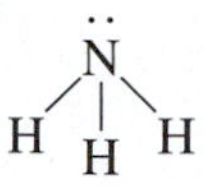

The structures of amines:

- ammonia
- methylamine 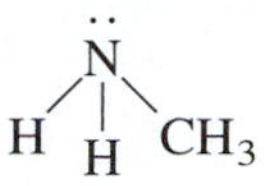
- dimethylamine 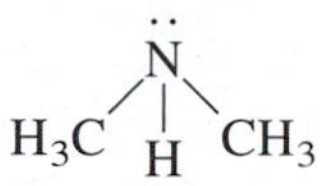
- alanine, a simple amino acid 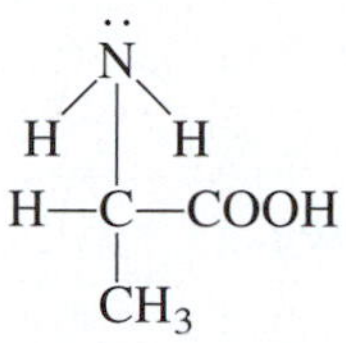

24-15 NITROGEN OXIDES

Nitrogen forms several oxides, in which it exhibits positive oxidation states of 1 to 5 (Table 24-8). All have positive free energies of formation, owing to the high dissociation energy of N_2 and O_2 molecules. All are gases except N_2O_5, a solid that melts at 30.0°C.

Dinitrogen Oxide (+1 Oxidation State)

Molten ammonium nitrate undergoes auto-oxidation–reduction (decomposition) at 170 to 260°C to produce dinitrogen oxide, also called nitrous oxide. At higher temperatures, explosions occur, producing N_2, O_2, and H_2O.

$$\overset{-3}{N}H_4\overset{+5}{N}O_3(s) \xrightarrow{\Delta} \overset{+1}{N_2}O(g) + 2H_2O(g) + \text{heat}$$

Dinitrogen oxide supports combustion because it produces O_2 when heated.

$$2N_2O(g) \xrightarrow{\Delta} 2N_2(g) + O_2(g) + \text{heat}$$

The molecule is linear but unsymmetrical, with a dipole moment of 0.17 D.

Some dentists use N_2O for its mild anesthetic properties. It is also known as laughing gas because of its side effects.

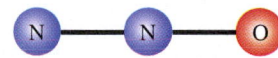

Dinitrogen oxide, or nitrous oxide, N_2O, mp −90.8°C, bp −88.8°C.

Nitrogen Oxide (+2 Oxidation State)

The first step of the Ostwald process (Section 24-16) for producing HNO_3 from NH_3 is used for the commercial preparation of nitrogen oxide, NO.

$$4NH_3 + 5O_2 \xrightarrow[\Delta]{\text{catalyst}} 4NO + 6H_2O$$

NO is not produced in nature under usual conditions. It is formed by direct reaction of N_2 and O_2 in electrical storms.

NO is a colorless gas that condenses at −152°C to a blue liquid. Gaseous NO is paramagnetic and contains one unpaired electron per molecule.

$$:N\!=\!\ddot{O}: \longleftrightarrow :N\!=\!\ddot{O}:$$

Its unpaired electron makes nitric oxide very reactive. Molecules that contain unpaired electrons are called *radicals*. NO reacts with O_2 to form NO_2, a brown, corrosive gas.

$$2NO(g) + O_2(g) \longrightarrow 2NO_2(g)$$

In recent years we have learned that very low concentrations of NO have important roles in our bodies.

Nitrogen oxide, or nitric oxide, NO, mp −163.6°C, bp −151.8°C, bond distance (1.15 Å) intermediate between N≡O (1.06 Å) and N=O (1.20 Å).

Nitrogen Oxides and Photochemical Smog

Nitrogen oxides are produced in the atmosphere by natural processes. Human activities contribute only about 10% of all the oxides of nitrogen (collectively referred to as NO_x) in the atmosphere, but the human contribution occurs mostly in urban areas, where the oxides may be present in concentrations a hundred times greater than in rural areas.

Just as NO is produced naturally by the reaction of N_2 and O_2 in electrical storms, it is also produced by the same reaction at the high temperatures of internal combustion engines and furnaces.

$$N_2(g) + O_2(g) \rightleftharpoons 2NO(g) \qquad \Delta H^0 = 180 \text{ kJ/mol rxn}$$

At ordinary temperatures the reaction does not occur to a significant extent. Because it is endothermic, it is favored by high temperatures. Even in internal combustion engines and furnaces, the equilibrium still lies far to the left, so only small amounts of NO are produced and released into the atmosphere. However, even very small concentrations of nitrogen oxides cause serious problems.

The NO radicals react with O_2 to produce NO_2 radicals. Both NO and NO_2 are quite reactive, and they do considerable damage to plants and animals. NO_2 reacts with H_2O in the air to produce corrosive droplets of HNO_3 and more NO.

$$\overset{+4}{3NO_2} + H_2O \longrightarrow \overset{+2}{NO} + \overset{+5}{2HNO_3} \qquad \text{(nitric acid)}$$

The HNO_3 may be washed out of the air by rainwater (acid rain), or it may react with traces of NH_3 in the air to form solid NH_4NO_3, a *particulate* pollutant.

$$HNO_3 + NH_3 \longrightarrow NH_4NO_3$$

This situation occurs in all urban areas, but the problem is worse in warm, dry climates, which are conducive to light-induced (photochemical) reactions. Here ultraviolet (uv) radiation from the sun produces damaging oxidants. The brownish hazes that often hang over such cities as Los Angeles, Denver, and Mexico City are due to the presence of brown NO_2. Problems begin in the morning rush hour as NO is exhausted into the air. The NO combines with O_2 to form NO_2. Then, as the sun rises higher in the sky, NO_2 absorbs uv radiation and breaks down into NO and oxygen radicals.

$$NO_2 \overset{uv}{\longrightarrow} NO + O$$

The extremely reactive O radicals combine with O_2 to produce O_3 (ozone).

$$O + O_2 \longrightarrow O_3$$

Ozone is a powerful oxidizing agent that damages rubber, plastic materials, and all plant and animal life. It also reacts with

Nitrogen Dioxide and Dinitrogen Tetroxide (+4 Oxidation State)

Nitrogen dioxide is formed by reaction of NO with O_2. It is prepared in the laboratory by heating heavy metal nitrates.

$$2Pb(NO_3)_2(s) \overset{\Delta}{\longrightarrow} 2PbO(s) + 4NO_2(g) + O_2(g)$$

Each NO_2 molecule contains one unpaired electron. NO_2 readily dimerizes to form colorless, diamagnetic dinitrogen tetroxide, N_2O_4, at low temperatures.

$$2NO_2(g) \rightleftharpoons N_2O_4(g) \qquad \Delta H^0 = -57.2 \text{ kJ/mol rxn}$$
$$\text{brown} \qquad\quad \text{colorless}$$

The NO_2 molecule is angular. It is represented by resonance structures.

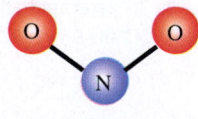

Nitrogen dioxide, NO_2, mp $-11.20°C$, bp $21.2°C$, one unpaired electron, bond length 1.197 Å; brown gas.

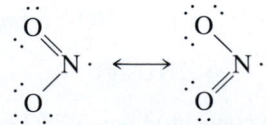

nitrogen dioxide

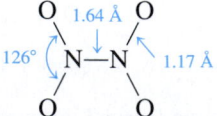

dinitrogen tetroxide

Photochemical smog casts a haze over urban or industrial areas; its severity depends on the weather.

hydrocarbons from automobile exhaust and evaporated gasoline to form secondary organic pollutants such as aldehydes and ketones (Section 27-14). (Obviously, O_3 in the upper atmosphere is not a problem. It is very beneficial because it screens out dangerous uv radiation.) The **peroxyacyl nitrates** (PANs), perhaps the worst of the secondary pollutants, are especially damaging photochemical oxidants that are very irritating to the eyes and throat.

R = hydrocarbon chain or ring

Catalytic converters in automobile exhaust systems reduce emissions of oxides of nitrogen.

24-16 SOME OXOACIDS OF NITROGEN AND THEIR SALTS

The main oxoacids of nitrogen are nitrous acid, HNO_2, and nitric acid, HNO_3.

Nitrogen also forms hyponitrous acid, $H_2N_2O_2$, in which N is in the $+1$ oxidation state, as well as hyponitrite salts such as $Na_2N_2O_2$.

Nitrous Acid (+3 Oxidation State)

Nitrous acid, HNO_2, is unstable and cannot be isolated in pure form. It is prepared as a pale blue solution when H_2SO_4 reacts with cold aqueous sodium nitrite. Nitrous acid is a weak acid ($K_a = 4.5 \times 10^{-4}$). It acts as an oxidizing agent toward strong reducing agents and as a reducing agent toward very strong oxidizing agents.

Lewis formulas for nitrous acid and the nitrite ion follow.

nitrous acid   nitrite ion

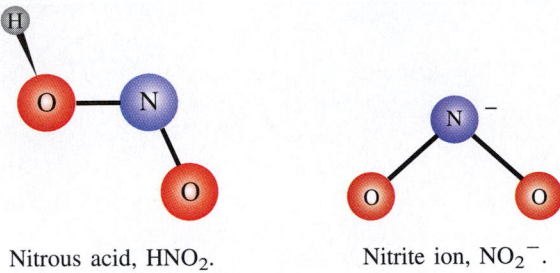

Nitrous acid, HNO_2.   Nitrite ion, NO_2^-.

Nitric Acid (+5 Oxidation State)

Pure nitric acid, HNO_3, is a colorless liquid that boils at 83°C. Light or heat causes it to decompose into NO_2, O_2, and H_2O. The presence of the NO_2 in partially decomposed aqueous HNO_3 causes its yellow or brown tinge. Studies on the vapor phase indicate that the structure of nitric acid is

$$H-\ddot{O}-N\underset{\ddot{\ddot{O}}}{\overset{\ddot{O}}{}} \longleftrightarrow H-\ddot{O}-N\underset{\ddot{O}}{\overset{\ddot{\ddot{O}}}{}}$$

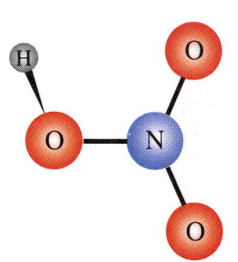

Nitric acid, HNO_3, mp −42°C; bp 83°C; bond lengths N—O (terminal) 1.22 Å, N—O (central) 1.41 Å.

HNO_3 is commercially prepared by the **Ostwald process**. At high temperatures, NH_3 is catalytically converted to NO, which is cooled and then air-oxidized to NO_2. Nitrogen dioxide reacts with H_2O to produce HNO_3 and some NO. The NO produced in the third step is then recycled into the second step. More than 17 billion pounds of HNO_3 were produced in the United States in 1993.

$$4NH_3(g) + 5O_2(g) \xrightarrow[1000°C]{Pt} 4NO(g) + 6H_2O(g)$$

$$2NO(g) + O_2(g) \xrightarrow{cool} 2NO_2(g)$$
$$3NO_2(g) + H_2O(\ell) \longrightarrow 2[H^+ + NO_3^-] + \boxed{NO(g)}$$
$$\text{recycle}$$

Nitric acid is very soluble in water (~16 mol/L). It is a strong acid and a strong oxidizing agent.

Copper (left beaker) and zinc (right beaker) react with concentrated nitric acid.

$NaNO_2$ and $NaNO_3$ as Food Additives

The brown color of "old" meat is the result of oxidation of blood, and is objectionable to many consumers. Nitrites and nitrates are added to food to retard this oxidation and also to prevent growth of botulism bacteria. Nitrate ions, NO_3^-, are reduced to NO_2^- ions, which are then converted to NO. This in turn reacts with the brown oxidized form of the heme in blood. This reaction keeps meat red longer. However, controversy has arisen concerning the possibility that nitrites combine with amines under the acidic conditions in the stomach to produce carcinogenic *nitrosoamines*.

$$\underbrace{\begin{matrix} R' \\ \diagdown \\ N \\ \diagup \\ R \end{matrix}}_{\substack{\text{amine} \\ \text{group}}} - \underbrace{N=O}_{\substack{\text{nitroso} \\ \text{group}}} \qquad \text{(R and R' = organic groups)}$$

24-17 PHOSPHORUS

Phosphorus is always combined in nature. Phosphorus is present in all living organisms—as organophosphates and in calcium phosphates such as hydroxyapatite, $Ca_5(PO_4)_3(OH)$, and fluoroapatite, $Ca_5(PO_4)_3F$, in bones and teeth. It also occurs in these and related compounds in phosphate minerals, which are mined mostly in Florida and North Africa.

Industrially, the element is obtained from phosphate minerals by heating them to 1200 to 1500°C in an electric arc furnace with sand (SiO_2) and coke.

$$\underset{\substack{\text{calcium phosphate} \\ \text{(phosphate rock)}}}{2Ca_3(PO_4)_2} + 6SiO_2 + 10C \overset{\Delta}{\longrightarrow} \underset{\substack{\text{calcium silicate} \\ \text{(slag)}}}{6CaSiO_3} + 10CO + P_4$$

Vaporized phosphorus is condensed to a white solid (mp = 44.2°C, bp = 280.3°C) under H_2O to prevent oxidation. Even when kept under H_2O, white phosphorus slowly converts to the more stable red phosphorus allotrope (mp = 597°C; sublimes at 431°C). Red phosphorus and tetraphosphorus trisulfide, P_4S_3, are used in matches. They do not burn spontaneously, yet they ignite easily when heated by friction. Both white and red phosphorus are insoluble in water.

The largest use of phosphorus is in fertilizers. Phosphorus is an essential nutrient, and nature's phosphorus cycle is very slow owing to the low solubility of most natural phosphates. Therefore, phosphate fertilizers are essential. To increase the solubility of the natural phosphates, they are treated with H_2SO_4 to produce "superphosphate of lime," a mixture of two salts. This solid is pulverized and applied as a powder.

$$\underset{\text{phosphate rock}}{Ca_3(PO_4)_2} + 2H_2SO_4 + 4H_2O \overset{\text{evaporate}}{\longrightarrow} [\underset{\substack{\text{calcium} \\ \text{dihydrogen} \\ \text{phosphate}}}{Ca(H_2PO_4)_2} + \underset{\substack{\text{calcium} \\ \text{sulfate} \\ \text{dihydrate}}}{2(CaSO_4 \cdot 2H_2O)}]$$

superphosphate of lime

This reaction represents the biggest single use of sulfuric acid, the industrial chemical produced in largest quantity.

SILICON

Silicon is a shiny, blue-gray, high-melting, brittle metalloid. It looks like a metal, but it is chemically more like a nonmetal. It is second only to oxygen in abundance in the earth's crust, about 87% of which is composed of silica (SiO_2) and its derivatives, the silicate minerals. The crust is 26% Si, compared to 49.5% O. Silicon does not occur free in nature. Pure silicon crystallizes with a diamond-type structure, but the Si atoms are less closely packed than C atoms. Its density is 2.4 g/cm^3 compared to 3.51 g/cm^3 for diamond.

24-18 SILICON AND THE SILICATES

Elemental silicon is usually prepared by the high-temperature reduction of silica (sand) with coke. Excess SiO_2 prevents the formation of silicon carbide.

$$SiO_2(s, \text{ excess}) + 2C(s) \overset{\Delta}{\longrightarrow} Si(s) + 2CO(g)$$

Reduction of a mixture of silicon and iron oxides with coke produces an alloy of iron and silicon known as *ferrosilicon*. It is used in the production of acid-resistant steel alloys, such as "duriron," and in the "deoxidation" of steel. Aluminum alloys for aircraft are strengthened with silicon.

Pure silicon is used in solar cells to collect energy from the sun.

Elemental silicon is used to make silicone polymers. Its semiconducting properties (Section 13-17) are used in transistors and solar cells.

The biggest chemical differences between silicon and carbon are that (1) silicon does not form stable double bonds, (2) it does not form very stable Si—Si bonds unless the silicon atoms are bonded to very electronegative elements, and (3) it has vacant $3d$ orbitals in its valence shell into which it can accept electrons from donor atoms. The Si—O single bond is the strongest of all silicon bonds and accounts for the stability and prominence of silica and the silicates.

Silicon dioxide (silica) exists in two familiar forms in nature: quartz, small chips of which occur in sand; and flint (Latin *silex*), an uncrystallized amorphous type of silica. Silica is properly represented as $(SiO_2)_n$ because it is a polymeric solid of SiO_4 tetrahedra sharing all oxygens among surrounding tetrahedra (Figure 13-31c). For comparison, solid carbon dioxide (Dry Ice) consists of discrete $O=C=O$ molecules, as does gaseous CO_2.

Some gems and semiprecious stones such as amethyst, opal, agate, and jasper are crystals of quartz with colored impurities.

Most of the crust of the earth is made up of silica and silicates. The natural silicates comprise a large variety of compounds. The structures of all these are based on SiO_4 tetrahedra, with metal ions occupying spaces between the tetrahedra. The extreme stability of the silicates is due to the donation of extra electrons from O into vacant $3d$ orbitals of Si. In many common minerals, called aluminosilicates, Al atoms replace some Si atoms with very little structural change. Because an Al atom has one less positive charge in its nucleus than Si does, it is also necessary to introduce a univalent ion, such as K^+ or Na^+.

The physical characteristics of the silicates are often suggested by the arrangement of the SiO_4 tetrahedra. A single-chain silicate, diopside $[CaMg(SiO_3)_2]_n$, and a double-chain silicate, asbestos $[Ca_2Mg_5(Si_4O_{11})_2(OH_2)]_n$, occur as fibrous or needle-like crystals. Talc, $[Mg_3Si_4O_{10}(OH)_2]_n$, a silicate with a sheet-like structure, is flaky. Micas are sheet-like aluminosilicates with about one of every four Si atoms replaced by Al. Muscovite mica is $[KAl_2(AlSi_3O_{10})(OH)_2]_n$. Micas occur in thin sheets that are easily peeled away from each other.

The clay minerals are silicates and aluminosilicates with sheet-like structures. They result from the weathering of granite and other rocks. The layers have enormous "inner surfaces" that can absorb large amounts of H_2O. Clay mixtures often occur as minute platelets with very large total surface area. When wet, the clays are easily shaped. When heated to high temperatures, they lose H_2O; when fired in a furnace, they become very rigid.

Fused sodium silicate, Na_2SiO_3, and calcium silicate, $CaSiO_3$, are the major components of the glass used in such things as drinking glasses, bottles, and window panes. **Glass** is a hard, brittle material that has no fixed composition or regular structure. Because it has no regular structure, it does not break evenly along crystal planes, but breaks to form rounded surfaces and jagged edges. The basic ingredients are produced by heating a mixture of Na_2CO_3 and $CaCO_3$ with sand until it melts, at about 700°C.

$$[CaCO_3 + SiO_2](\ell) \xrightarrow{\Delta} CaSiO_3(\ell) + CO_2(g)$$

$$[Na_2CO_3 + SiO_2](\ell) \xrightarrow{\Delta} Na_2SiO_3(\ell) + CO_2(g)$$

The resulting "soda–lime" glass is clear and colorless (if all CO_2 bubbles escape and if the amounts of reactants are carefully controlled).

Si is much larger than C. As a result, Si—Si bonds are too long to permit the effective *pi* bonding that is necessary for multiple bonds.

Natural quartz crystals, SiO_2.

"Asbestos" refers to a group of impure magnesium silicate minerals. As you can see, asbestos is a fibrous material. When inhaled these fibers are highly toxic and carcinogenic.

Key Terms

Chain initiation step The first step in a chain reaction; produces reactive species (such as radicals) that then propagate the reaction.

Chain propagation step An intermediate step in a chain reaction; in such a step one or more reactive species is consumed and another reactive species is produced.

Chain reaction A reaction in which reactive species, such as radicals, are produced in more than one step. Consists of an initiation step, one or more propagation steps, and one or more termination steps.

Chain termination step The combination of reactive species (such as radicals) which terminates the chain reaction.

Contact process An industrial process by which sulfur trioxide and sulfuric acid are produced from sulfur dioxide.

Frasch process A method by which elemental sulfur is mined or extracted. Sulfur is melted with superheated water (at 170°C under high pressure) and forced to the surface of the earth as a slurry.

Haber process An industrial process for the catalyzed production of ammonia from N_2 and H_2 at high temperature and pressure.

Halogens Group VIIA elements; F, Cl, Br, I, and At.

Nitrogen cycle The complex series of reactions by which nitrogen is slowly but continually recycled in the atmosphere, lithosphere, and hydrosphere.

Noble gases Group 0 elements; He, Ne, Ar, Kr, Xe, and Rn.

Ostwald process An industrial process for the production of nitrogen oxide and nitric acid from ammonia and oxygen.

PANs Abbreviation for peroxyacyl nitrates, photochemical oxidants in smog.

Particulate matter Finely divided solid particles suspended in polluted air.

Photochemical oxidants Photochemically produced oxidizing agents capable of causing damage to plants and animals.

Photochemical smog A brownish smog occurring in urban areas that receive large amounts of sunlight; caused by photochemical (light-induced) reactions among nitrogen oxides, hydrocarbons, and other components of polluted air that produce photochemical oxidants.

Polymer A large molecule consisting of chains or rings of linked monomer units, usually characterized by high melting and boiling points.

Radical An atom or group of atoms that contains one or more unpaired electrons (usually very reactive species).

Exercises

The Noble Gases

1. Why are the noble gases so unreactive?
2. Why were the noble gases among the last elements to be discovered?
3. List some of the uses of the noble gases and reasons for the uses.
4. Arrange the noble gases in order of increasing (a) atomic radii, (b) melting points, (c) boiling points, (d) densities, and (e) first ionization energies.
5. Explain the order of increasing melting and boiling points of the noble gases in terms of polarizabilities of the atoms and forces of attraction between them.
6. What gave Neil Bartlett the idea that compounds of xenon could be synthesized? Which noble gases are known to form compounds? With which elements are the noble gas atoms bonded?
7. Describe the bonding and geometry in XeF_2, XeF_4, and XeF_6.
*8. Xenon(VI) fluoride can be produced by the combination of xenon(IV) fluoride with fluorine. Write a chemical equation for this reaction. What mass of XeF_6 could be produced from 3.62 g of XeF_4 and excess fluorine?

The Halogens

9. Write the electron configuration for each halogen atom. Write the Lewis symbol for a halogen atom, X. What is the usual oxidation state of the halogens in binary compounds with metals, semiconducting elements, and most nonmetals?
10. Write the Lewis structure of a halogen molecule, X_2. Describe the bonding in the molecule. What is the trend of bond length and strength going down the family from F_2 to I_2?
11. What types of intermolecular forces are found in molecular halogens? What is the trend in these forces going down the group from F_2 to I_2? Describe the physical state of each molecular halogen at room temperature and pressure.
12. List the halogens in order of increasing (a) atomic radii, (b) ionic radii, (c) electronegativities, (d) melting points, (e) boiling points, and (f) standard reduction potentials.
13. Write the equations describing the half-reactions and net reaction for the electrolysis of molten KF/HF mixtures. At which electrodes are the products formed? What is the purpose of the HF?
14. Carl O. Christe's preparation of F_2 did not involve direct chemical oxidation. Explain this statement.
15. Write equations describing general reactions of the free halogens, X_2, with (a) Group IA (alkali) metals, (b) Group IIA (alkaline earth) metals, and (c) Group IIIA metals. Represent the metals as M.
16. Write balanced equations for any reactions that occur in aqueous mixtures of (a) NaI and Cl_2, (b) NaCl and Br_2, (c) NaI and Br_2, (d) NaBr and Cl_2, and (e) NaF and I_2.
*17. An aqueous solution contains either NaBr or a mixture of NaBr and NaI. Using only aqueous solutions of I_2, Br_2, and Cl_2 and a small amount of CH_2Cl_2, describe how you might determine what is in the unknown solution.
*18. Write equations illustrating the tendency of F^- to stabilize high

oxidation states of cations and the tendency of I^- to stabilize low oxidation states. Why is this the case?

19. Why are the free halogens more soluble in water than most non-polar molecules?

20. Distinguish between hydrogen bromide and hydrobromic acid.

21. Refer to Figure 13-5. What is the order of decreasing melting and boiling points of the hydrogen halides? Why is the HF "out of line?"

22. Describe the effect of hydrofluoric acid on glass.

23. What is the acid anhydride of perchloric acid?

24. Name the following compounds: (a) $KBrO_3$; (b) $KOBr$; (c) $NaClO_4$; (d) $NaClO_2$; (e) $HOBr$; (f) $HBrO_3$; (g) HIO_3; (h) $HClO_4$.

25. Write the Lewis formulas and structures of the four ternary acids of chlorine.

26. Write equations describing reactions by which the following compounds can be prepared: (a) hypohalous acids of Cl, Br, and I (in solution with hydrohalic acids); (b) hypohalite salts; (c) chlorous acid; (d) perchloric acid.

27. What is the order of increasing acid strength of the ternary chlorine acids? Explain the order.

28. Choose the strongest acid from each group: (a) HOCl, HOBr, HOI; (b) HOCl, $HClO_2$, $HClO_3$, $HClO_4$; (c) HOI, $HBrO_3$, $HClO_4$. Explain your choices.

Group IIIA Elements

29. Write abbreviated electron configurations for atomic oxygen, selenium, and polonium.

30. Write out the electron configurations of oxide, sulfide, and selenide ions.

31. Characterize the Group VIA elements with respect to color and physical state under normal conditions.

32. The Group VIA elements, except oxygen, can exhibit oxidation states ranging from -2 to $+6$, but not -3 or $+7$. Why?

33. Sulfur, selenium, and tellurium are all capable of forming six-coordinate compounds such as SF_6. Give two reasons why oxygen cannot be the central atom in such six-coordinate molecules.

34. For the following species, draw (1) diagrams that show the hybridization of atomic orbitals and (2) three-dimensional structures that show all hybridized orbitals and outermost electrons. (3) Determine the oxidation state of the Group VIA element (other than oxygen) in each species. (a) H_2S; (b) SF_6; (c) SF_4; (d) SO_2; (e) SO_3.

35. Repeat Exercise 34 for (a) SeF_6, (b) SO_3^{2-}, (c) SO_4^{2-}, (d) HSO_4^-, and (e) thiosulfate ion, $S_2O_3^{2-}$ (one S is central atom).

36. Write equations for the reactions of (a) S, Se, and Te with excess F_2; (b) O_2, S, Se, and Te with H_2; (c) S, Se, and Te with O_2.

37. Write equations for the reactions of (a) S and Te with HNO_3; (b) S and Se with excess Cl_2; (c) S and Se with Na, Ca, and Al.

38. Discuss the acidity of the aqueous Group VIA hydrides, including the relative values of acid ionization constants. What is primarily responsible for the order of increasing acidities in this series?

39. Compare the structures of the dioxides of sulfur, selenium, tellu-

rium, and polonium. How do they relate to the metallic or nonmetallic character of these elements?

40. Write equations for reactions of
(a) NaOH with sulfuric acid (1:1 mole ratio)
(b) NaOH with sulfuric acid (2:1 mole ratio)
(c) NaOH with sulfurous acid (1:1 mole ratio)
(d) NaOH with sulfurous acid (2:1 mole ratio)
(e) NaOH with selenic acid, H_2SeO_4 (1:1 mole ratio)
(f) NaOH with selenic acid, (2:1 mole ratio)
(g) NaOH with tellurium dioxide (1:1 mole ratio)
(h) NaOH with tellurium dioxide (2:1 mole ratio)

*41. How much sulfur dioxide could be produced from complete combustion of 1 ton of coal containing 6.23% sulfur?

42. What mass of H_2SO_4 could be produced in the process given below if 1.00 kg of FeS_2 is used? The *unbalanced* equations for the process are

$$FeS_2(s) + O_2(g) \longrightarrow Fe_2O_3(s) + SO_2(g)$$

$$SO_2(g) + O_2(g) \longrightarrow SO_3(g)$$

$$SO_3(g) + H_2SO_4(\ell) \longrightarrow H_2S_2O_7(\ell)$$

$$H_2S_2O_7(\ell) + H_2O(\ell) \longrightarrow H_2SO_4(aq)$$

43. Common copper ores in western United States contain the mineral chalcopyrite, $CuFeS_2$. Assuming that an average commercially useful ore contains 0.263 mass % Cu and that all the sulfur ultimately appears in the smelter stack gases as SO_2, calculate the mass of sulfur dioxide generated by the conversion of 1.00 kg of the ore.

*44. A gaseous mixture at some temperature in a 1.00 L vessel originally contained 1.00 mol SO_2 and 5.00 mol O_2. Once equilibrium conditions were attained, 78.3% of the SO_2 had been converted to SO_3. What is the value of the equilibrium constant (K_c) for this reaction at this temperature?

Nitrogen and Phosphorus

45. Characterize each of the Group VA elements with respect to normal physical state and color.

46. Write out complete electron configurations for the atoms of the Group VA elements; nitride ion, N^{3-}; and phosphide ion, P^{3-}.

47. Compare and contrast the properties of (a) N_2 and P_4; (b) HNO_3 and H_3PO_4; (c) N_2O_3 and P_4O_6.

48. Describe the natural nitrogen cycle.

49. List natural sources of nitrogen and phosphorus and at least two uses for each.

50. Discuss the effects of temperature, pressure, and catalysts on the Haber process for the production of ammonia. (You may wish to consult Section 17-6.)

51. Determine the oxidation states of nitrogen in the following: (a) NO_3^-; (b) NO_2^-; (c) N_2H_4; (d) NH_3; (e) NH_2^-.

52. Determine the oxidation states of nitrogen in the following: (a) N_2; (b) NO; (c) N_2O_4; (d) HNO_3; (e) HNO_2.

53. Draw three-dimensional structures showing all outer shell elec-

trons, describe molecular and ionic geometries, and indicate hybridization (except for N^{3-}) at the central element, for the following species: (a) N_2; (b) N^{3-}; (c) NH_3; (d) NH_4^+; (e) NH_2^-, amide ion.

54. Draw three-dimensional structures showing all outer shell electrons, describe molecular and ionic geometries, and indicate hybridization at the central element for the following species: (a) NH_2Br, bromamine; (b) HN_3, hydrazoic acid; (c) N_2O_2; (d) $NO_2^+NO_3^-$, solid nitronium nitrate; (e) HNO_3; (f) NO_2^-.

55. Draw three-dimensional structures showing all outer shell electrons for the following species: (a) P_4; (b) P_4O_{10}; (c) As_4O_6; (d) H_3PO_4; (e) AsO_4^{3-}.

56. Write formula unit equations for the following: (a) thermal decomposition of potassium azide, KN_3; (b) reaction of gaseous ammonia with gaseous HCl; (c) reaction of aqueous ammonia with aqueous HCl; (d) thermal decomposition of ammonium nitrate at temperature above 260°C; (e) reaction of ammonia with oxygen in the presence of red hot platinum catalyst; (f) thermal decomposition of nitrous oxide (dinitrogen oxide), N_2O; (g) reaction of NO_2 with water.

57. Write the formula unit equation for the preparation of "superphosphate of lime."

58. Write two equations illustrating the ability of ammonia to function as a Lewis base.

59. In liquid ammonia would sodium amide, $NaNH_2$, be acidic, basic, or neutral? Would ammonium chloride, NH_4Cl, be acidic, basic, or neutral? Why?

60. Which of the following molecules have a nonzero dipole moment—i.e., are polar molecules? (a) NH_3; (b) NH_2Cl; (c) NO; (d) NH_2OH; (e) HNO_3.

61. Describe with equations the Ostwald process for the production of nitrogen oxide, NO, and nitric acid.

62. Why is NO so reactive?

63. Write a Lewis formula for NO_2. Would you predict that it is very reactive? How about N_2O_4 (dimerized NO_2)?

64. At room temperature, a sample of NO_2 gas is brown. Explain why this sample loses its color as it is cooled.

65. Discuss the problem of NO_x emissions with respect to air pollution. Use equations to illustrate the important reactions.

66. What are the acid anhydrides of (a) nitric acid, HNO_3; (b) nitrous acid, HNO_2; (c) phosphoric acid, H_3PO_4; and (d) phosphorous acid, H_3PO_3?

67. Calcium phosphate (phosphate rock) is not applied directly as a phosphorus fertilizer. Why?

68. Discuss the use of sodium nitrite as a meat preservative.

BUILDING YOUR KNOWLEDGE

*69. How many grams of xenon oxide tetrafluoride, $XeOF_4$, and how many liters of HF at STP could be prepared, assuming complete reaction of 6.50 g of xenon tetrafluoride, XeF_4, with a stoichiometric quantity of water according to the equation below?

$$6XeF_4(s) + 8H_2O(\ell) \longrightarrow$$
$$2XeOF_4(\ell) + 4Xe(g) + 16HF(g) + 3O_2(g)$$

*70. Argon crystallizes at −235°C in a face-centered cubic unit cell with $a = 5.43$ Å. Determine the apparent radius of an argon atom in the solid.

*71. A reaction mixture contained 100 g of K_2MnF_6 and 174 g of SbF_5. Fluorine was produced in 38.3% yield. How many grams of F_2 were produced? What volume is this at STP?

*72. Standard enthalpies of formation are −402 kJ/mol for $XeF_6(s)$ and −261.5 kJ/mol for $XeF_4(s)$. Calculate ΔH^0_{rxn} at 25°C for the preparation of XeF_6 from XeF_4 and $F_2(g)$.

73. The average atomic mass of N is 14.0067 amu. There are two isotopes which contribute to this average: $^{14}_7N$ (14.00307 amu) and $^{15}_7N$ (15.00011 amu). Calculate the percentage of $^{15}_7N$ atoms in a sample of naturally occurring nitrogen.

74. The N≡N bond energy is 946 kJ/mol and the N—N bond energy is 159 kJ/mol. Predict whether four gaseous nitrogen atoms would form two gaseous nitrogen molecules or a gaseous tetrahedral molecule similar to P_4, basing your prediction on the amount of energy released as the molecules are formed. Repeat the calculations for phosphorus using 485 kJ/mol for P≡P and 213 kJ/mol for P—P.

*75. Commercial concentrated HNO_3 contains 69.5 mass % HNO_3 and has a density of 1.42 g/mL. What is the molarity of this solution? What volume of the concentrated acid should you use to prepare 10.0 L of dilute HNO_3 solution with a concentration of 6.00 M?

*76. What is the total mass of silicon in the crust of the Earth? Assume that the radius of the Earth is 6400 km, the crust is 50 km thick, the density of the crust is 3.5 g/cm³, and 25.7 mass % of the crust is silicon.

25 Coordination Compounds

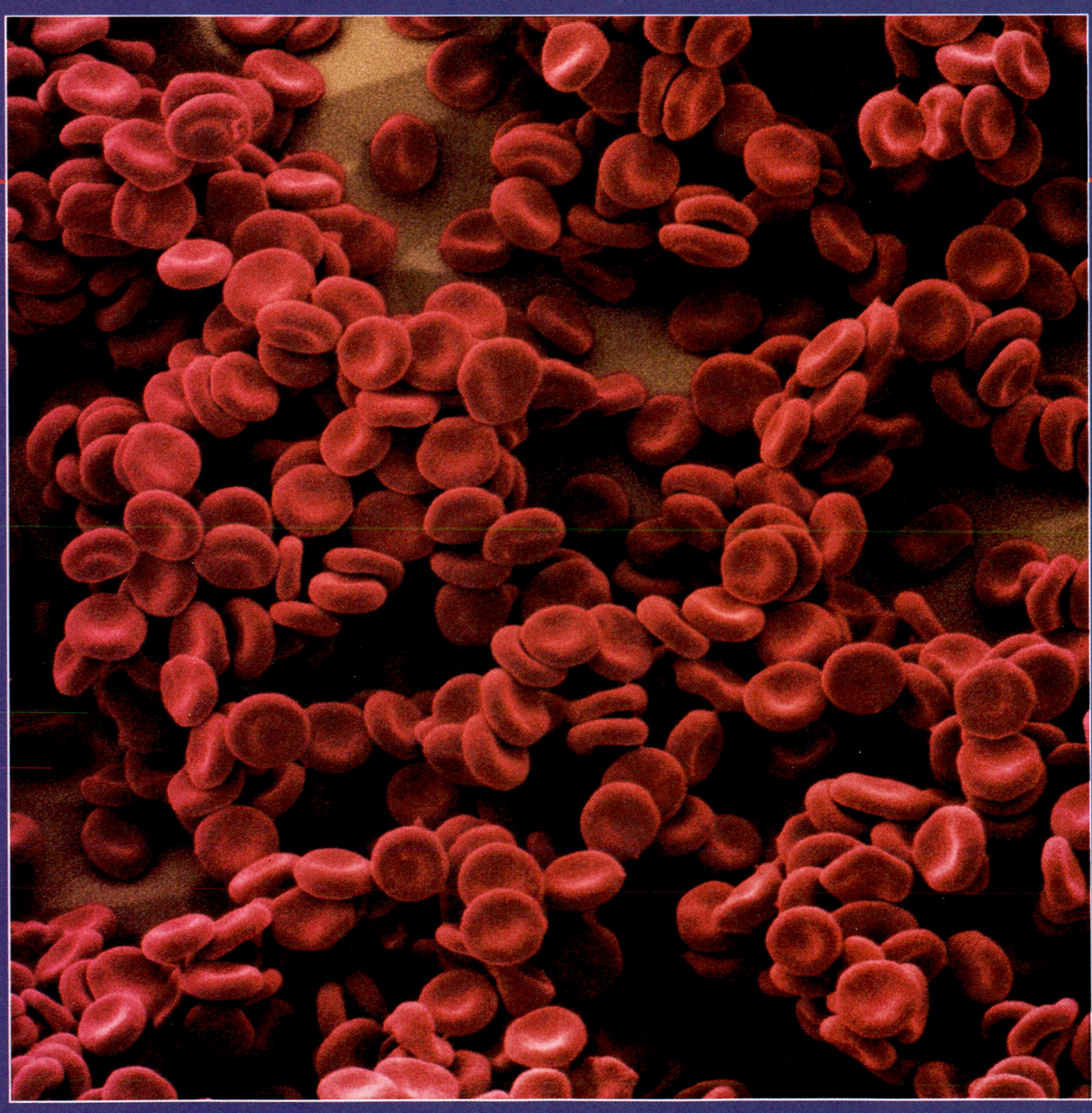

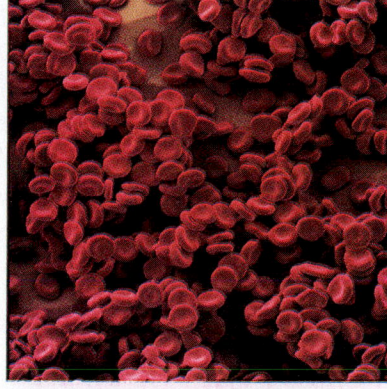

Red blood cells (1200×). The red blood cells that transport O_2 throughout our bodies contain hemoglobin, a coordination compound.

OBJECTIVES

As you study this chapter, you should learn

- *To recognize coordination compounds*

- *The metals that form soluble ammine complexes in aqueous solutions and the formulas for common ammine complexes*

- *About the terminology that describes coordination compounds*

- *The rules for naming coordination compounds*

- *To recognize common structures of coordination compounds*

- *About structural (constitutional) isomers*

- *About stereoisomers*

- *To apply the concepts of valence bond theory to coordination compounds*

- *About the crystal field theory and its advantages*

- *About the origin of color in complex species*

- *About the spectrochemical series*

- *About crystal field stabilization energy (CFSE)*

C oordination compounds are found in many places on the earth's surface. Every living system includes many coordination compounds. They are also important components of everyday products as varied as cleaning materials, medicines, inks, and paints. A list of important coordination compounds appears to be endless because new ones are discovered every year.

25-1 COORDINATION COMPOUNDS

In Section 10-10 we discussed Lewis acid–base reactions. A *base* makes available a share in an electron pair, and an *acid* accepts a share in an electron pair, to form a **coordinate covalent bond.** Such bonds are often represented by arrows.

Covalent bonds in which the shared electron pair is provided by one atom are called *coordinate covalent bonds.*

H—N—H + Cl—B—Cl $\longrightarrow$ H—N$\rightarrow$B—Cl

ammonia,   boron trichloride,
a Lewis base   a Lewis acid

The red arrows represent coordinate covalent bonds. These arrows do not imply that two Sn—Cl bonds are different from the others. Once formed, all the Sn—Cl bonds in the $[SnCl_6]^{2-}$ ion are alike.

2 $:Cl:^-$ + Cl—Sn—Cl $\longrightarrow$ $[SnCl_6]^{2-}$

chloride ion,   tin(IV) chloride,   hexachlorostannate(IV) ion
a Lewis base   a Lewis acid

Most d-transition metal ions have vacant d orbitals that can accept shares in electron pairs. Many act as Lewis acids by forming coordinate covalent bonds in **coordination compounds (coordination complexes, or complex ions).** Complexes of transition metal ions or molecules include $[Cr(OH_2)_6]^{3+}$, $[Co(NH_3)_6]^{3+}$, $[Ni(CN)_4]^{2-}$, $[Fe(CO)_5]$, and $[Ag(NH_3)_2]^+$. Many complexes are very stable, as indicated by their low dissociation constants, K_d (Section 20-6 and Appendix I).

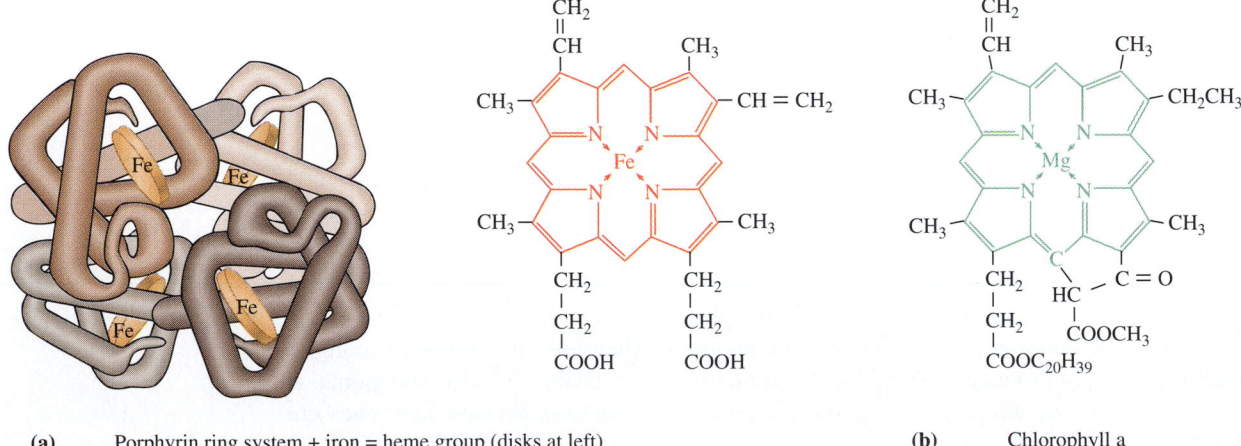

(a)   Porphyrin ring system + iron = heme group (disks at left)

(b)   Chlorophyll a

Figure 25-1 (a) A model of a hemoglobin molecule (MW = 64,500 amu). Individual atoms are not shown. The four heme groups in a hemoglobin molecule are represented by disks. Each heme group contains one Fe^{2+} ion and porphyrin rings. A single red blood cell contains more than 265 million hemoglobin molecules and more than 1 billion Fe^{2+} ions. (b) The structure of chlorophyll a, which also contains a porphyrin ring with a Mg^{2+} ion at its center. Chlorophyll is necessary for photosynthesis. The porphyrin ring is the part of the molecule that absorbs light. The structure of chlorophyll b is slightly different.

We now understand from molecular orbital theory that all substances have some vacant orbitals—they are *potential* Lewis acids. Most substances have unshared pairs of electrons—they are *potential* Lewis bases.

Many important biological substances are coordination compounds. Hemoglobin and chlorophyll are two examples (Figure 25-1). Hemoglobin is a protein that carries O_2 in blood. It contains iron(II) ions bound to large porphyrin rings. The transport of oxygen by hemoglobin involves the coordination and subsequent release of O_2 molecules by the Fe(II) ions. Chlorophyll is necessary for photosynthesis in plants. It contains magnesium ions bound to porphyrin rings. Vitamin B-12 is a large complex of cobalt. Coordination compounds have many practical applications in such areas as water treatment, soil and plant treatment, protection of metal surfaces, analysis of trace amounts of metals, electroplating, and textile dyeing.

Bonding in transition metal complexes was not understood until the pioneering research of Alfred Werner, a Swiss chemist of the 1890s and early 1900s. He received the Nobel Prize in chemistry in 1913. Great advances have been made since in the field of coordination chemistry, but Werner's work remains the most important contribution by a single researcher.

Prior to Werner's work, the formulas of transition metal complexes were written with dots, $CrCl_3 \cdot 6H_2O$, $AgCl \cdot 2NH_3$, just like double salts such as iron(II) ammonium sulfate hexahydrate, $FeSO_4 \cdot (NH_4)_2SO_4 \cdot 6H_2O$. The properties of solutions of double salts are the properties expected for solutions made by mixing the individual salts. However, a solution of $AgCl \cdot 2NH_3$, or more properly $[Ag(NH_3)_2]Cl$, behaves differently from either a solution of (very insoluble) silver chloride or a solution of ammonia. The dots have been called "dots of ignorance," because they signified that the mode of bonding was unknown. Table 25-1 summarizes the types of experiments Werner performed and interpreted to lay the foundations for modern coordination theory.

Werner isolated platinum(IV) compounds with the formulas that appear in the first column of Table 25-1. He added excess $AgNO_3$ to solutions of carefully weighed amounts of the five salts. The precipitated AgCl was collected by filtration, dried, and weighed. He determined the number of moles of AgCl produced. This told him the number of Cl^- ions precipitated per formula unit. The results are in the second column. Werner reasoned that the precipitated Cl^- ions must be free (uncoordinated), whereas the unprecipitated Cl^- ions must be directly bound to Pt so they could not be precipitated by Ag^+ ions. He also measured the conductances of solutions of these compounds of known concentrations. By comparing these with data on solutions of simple electrolytes, he found the number of ions per formula unit. The results are shown in the third column. Piecing the evidence together,

Compounds of the transition metals are often colored, whereas those of A group metals are usually colorless. Aqueous solutions of some nitrate salts (left to right): $Fe(NO_3)_3$, $Co(NO_3)_2$, $Ni(NO_3)_2$, $Cu(NO_3)_2$, and $Zn(NO_3)_2$.

Double salts are ionic solids resulting from the cocrystallization of two salts from the same solution into a single structure. In the example given, the solid is produced from an aqueous solution of iron(II) sulfate, $FeSO_4$, and ammonium sulfate, $(NH_4)_2SO_4$.

The conductance of a solution of an electrolyte is a measure of its ability to conduct electricity. It is related to the number of and the charges on ions in solution.

Table 25-1 *Interpretation of Experimental Data by Werner*

Formula	Moles AgCl Precipitated per Formula Unit	Number of Ions per Formula Unit (based on conductance)	True Formula	Ions/Formula Unit	
$PtCl_4 \cdot 6NH_3$	4	5	$[Pt(NH_3)_6]Cl_4$	$[Pt(NH_3)_6]^{4+}$	$4\ Cl^-$
$PtCl_4 \cdot 5NH_3$	3	4	$[Pt(NH_3)_5Cl]Cl_3$	$[Pt(NH_3)_5Cl]^{3+}$	$3\ Cl^-$
$PtCl_4 \cdot 4NH_3$	2	3	$[Pt(NH_3)_4Cl_2]Cl_2$	$[Pt(NH_3)_4Cl_2]^{2+}$	$2\ Cl^-$
$PtCl_4 \cdot 3NH_3$	1	2	$[Pt(NH_3)_3Cl_3]Cl$	$[Pt(NH_3)_3Cl_3]^+$	Cl^-
$PtCl_4 \cdot 2NH_3$	0	0	$[Pt(NH_3)_2Cl_4]$	no ions	

Colors of coordination compounds depend on which ligands are present. The $[Cu(OH_2)_4]^{2+}$ ion is light blue, whereas the $[Cu(NH_3)_4]^{2+}$ ion is an intense blue.

he concluded that the correct formulas are the ones listed in the last two columns. The NH_3 and Cl^- within the brackets are bonded by coordinate covalent bonds to the Lewis acid, Pt(IV) ion. The charge on a complex is the sum of its constituent charges.

25-2 THE AMMINE COMPLEXES

The **ammine complexes** contain NH_3 molecules bonded to metal ions. Because the ammine complexes are important compounds, we will describe them briefly.

Most metal hydroxides are insoluble in water, and so aqueous NH_3 reacts with nearly all metal ions to form insoluble metal hydroxides, or hydrated oxides. The exceptions are the cations of the strong soluble bases (Group IA cations and the heavier members of Group IIA—Ca^{2+}, Sr^{2+}, and Ba^{2+}).

$$Cu^{2+} + 2NH_3 + 2H_2O \longrightarrow Cu(OH)_2(s) + 2NH_4^+$$

$$Cr^{3+} + 3NH_3 + 3H_2O \longrightarrow Cr(OH)_3(s) + 3NH_4^+$$

In general terms we can represent this reaction as

$$M^{n+} + nNH_3 + nH_2O \longrightarrow M(OH)_n(s) + nNH_4^+$$

where M^{n+} represents all of the common metal ions *except* those of the IA metals and the heavier IIA metals.

The hydroxides of some metals and some metalloids are amphoteric (Section 10-8). Aqueous NH_3 is a weak base ($K_b = 1.8 \times 10^{-5}$), so the $[OH^-]$ is too low to dissolve amphoteric hydroxides to form hydroxo complexes.

However, several metal hydroxides do dissolve in an excess of aqueous NH_3 to form ammine complexes. For example, the hydroxides of copper and cobalt are readily soluble in an excess of *aqueous* ammonia solution.

$$Cu(OH)_2(s) + 4NH_3 \rightleftharpoons [Cu(NH_3)_4]^{2+} + 2OH^-$$

$$Co(OH)_2(s) + 6NH_3 \rightleftharpoons [Co(NH_3)_6]^{2+} + 2OH^-$$

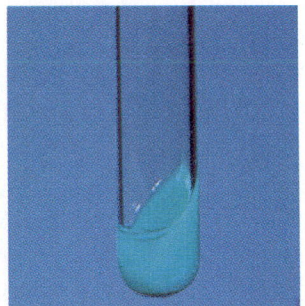

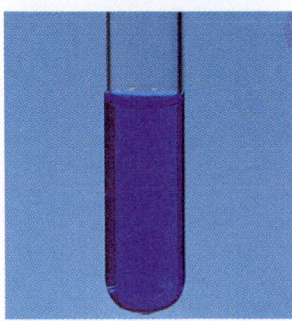

$Cu(OH)_2$ (light blue) dissolves in excess aqueous NH_3 to form $[Cu(NH_3)_4]^{2+}$ ions (deep blue).

Table 25-2 *Common Metal Ions That Form Soluble Complexes with an Excess of Aqueous Ammonia*[a]

Metal Ion	Insoluble Hydroxide Formed by Limited Aq. NH_3	Complex Ion Formed by Excess Aq. NH_3
Co^{2+}	$Co(OH)_2$	$[Co(NH_3)_6]^{2+}$
Co^{3+}	$Co(OH)_3$	$[Co(NH_3)_6]^{3+}$
Ni^{2+}	$Ni(OH)_2$	$[Ni(NH_3)_6]^{2+}$
Cu^+	$CuOH \longrightarrow \frac{1}{2}Cu_2O$[b]	$[Cu(NH_3)_2]^+$
Cu^{2+}	$Cu(OH)_2$	$[Cu(NH_3)_4]^{2+}$
Ag^+	$AgOH \longrightarrow \frac{1}{2}Ag_2O$[b]	$[Ag(NH_3)_2]^+$
Zn^{2+}	$Zn(OH)_2$	$[Zn(NH_3)_4]^{2+}$
Cd^{2+}	$Cd(OH)_2$	$[Cd(NH_3)_4]^{2+}$
Hg^{2+}	$Hg(OH)_2$	$[Hg(NH_3)_4]^{2+}$

[a] *The ions of Rh, Ir, Pd, Pt, and Au show similar behavior.*
[b] *CuOH and AgOH are unstable and decompose to the corresponding oxides.*

Interestingly, the metal hydroxides that exhibit this behavior are derived from the 12 metals of the cobalt, nickel, copper, and zinc families. All the common cations of these metals except Hg_2^{2+} (which disproportionates) form soluble complexes in the presence of excess aqueous ammonia (Table 25-2).

25-3 IMPORTANT TERMS

The Lewis bases in coordination compounds may be molecules, anions, or (rarely) cations. They are called **ligands** (Latin *ligare,* "to bind"). The **donor atoms** of the ligands are the atoms that donate shares in electron pairs to metals. In some cases it is not possible to identify donor atoms, because the bonding electrons are not localized on specific atoms. Some small organic molecules such as ethylene, $H_2C{=}CH_2$, bond to a transition metal through the electrons in their double bonds. Examples of typical simple ligands are listed in Table 25-3.

Ligands that can bond to a metal through only one donor atom at a time are **unidentate** (Latin *dent,* "tooth"). Ligands that can bond simultaneously through more than one donor atom are **polydentate.** Polydentate ligands that bond through two, three, four, five, or six donor atoms are called *bidentate, tridentate, quadridentate, quinquedentate,* and *sexidentate,* respectively. Complexes that consist of a metal atom or ion and polydentate ligands are called **chelate complexes** (Greek *chele,* "claw").

The **coordination number** of a metal atom or ion in a complex is the number of donor atoms to which it is coordinated, not necessarily the number of ligands. The **coordination sphere** includes the metal or metal ion (called the **central atom**) and its ligands, but no uncoordinated counter-ions. For example, the coordination sphere of hexaamminecobalt(III) chloride, $[Co(NH_3)_6]Cl_3$, is the hexaamminecobalt(III) ion, $[Co(NH_3)_6]^{3+}$. These terms are illustrated in Table 25-4.

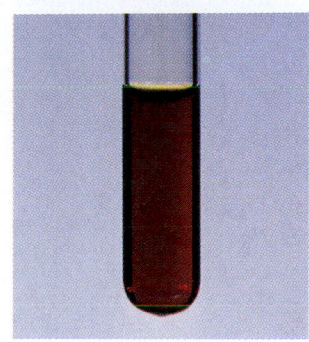

$Co(OH)_2$ (a blue compound that turns gray quickly) dissolves in excess aqueous NH_3 to form $[Co(NH_3)_6]^{2+}$ ions (yellow-orange).

Table 25-3 *Typical Simple Ligands with Their Donor Atoms Shaded*

Molecule	Name	Name as Ligand	Ion	Name	Name as Ligand
:NH_3	ammonia	ammine	:$\ddot{C}l$:⁻	chloride	chloro
:$\ddot{O}H_2$	water	aqua	:$\ddot{F}$:⁻	fluoride	fluoro
:$C{\equiv}O$:	carbon monoxide	carbonyl	:$C{\equiv}N$:⁻	cyanide	cyano[a]
:PH_3	phosphine	phosphine	:$\ddot{O}H^-$	hydroxide	hydroxo
:$\dot{N}{=}\ddot{O}$	nitrogen oxide	nitrosyl	:N (nitrite structure)	nitrite	nitro[b]

[a] Nitrogen atoms can also function as donor atoms, in which case the ligand name is "isocyano."

[b] Oxygen atoms can also function as donor atoms, in which case the ligand name is "nitrito."

Table 25-4 *Some Ligands and Coordination Spheres (complexes)*

Ligand(s)	Classification	Coordination Sphere	Oxidation Number of M	Coordination Number of M
NH_3 ammine	unidentate	$[Co(NH_3)_6]^{3+}$ hexaamminecobalt(III)	+3	6
$H_2N-CH_2-CH_2-NH_2$ (or N⌒N) ethylenediamine (en)	bidentate	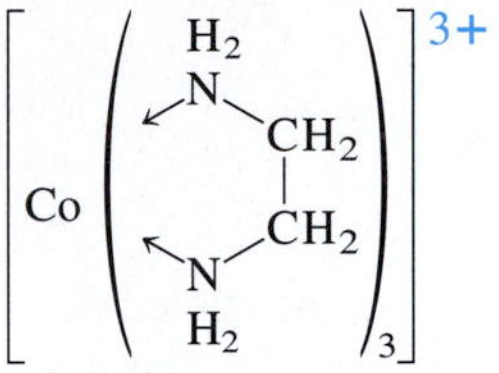 $[Co(en)_3]^{3+}$ tris(ethylenediamine)- cobalt(III) ion	+3	6
Br^- bromo $H_2N-CH_2-CH_2-NH_2$ ethylenediamine (en)	unidentate bidentate	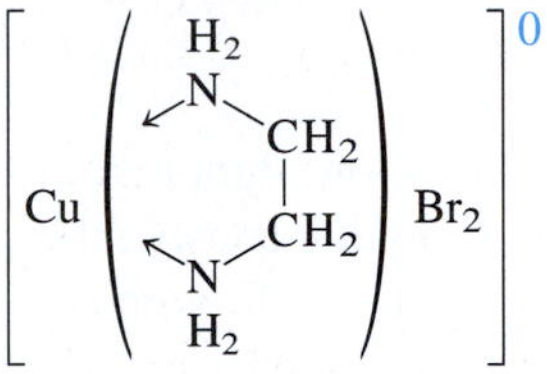 $[Cu(en)Br_2]$ dibromoethylenediamine- copper(II)	+2	4
$H_2N-(CH_2)_2-\overset{H}{N}-(CH_2)_2-NH_2$ (or N⌒N⌒N) diethylenetriamine (dien)	tridentate	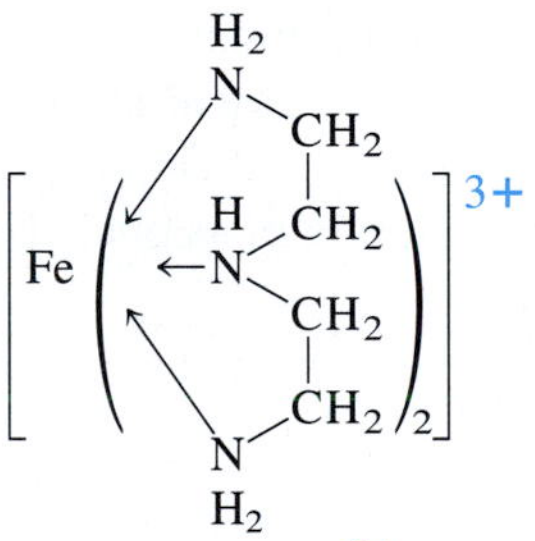 $[Fe(dien)_2]^{3+}$ bis(diethylenetriamine)- iron(III) ion	+3	6
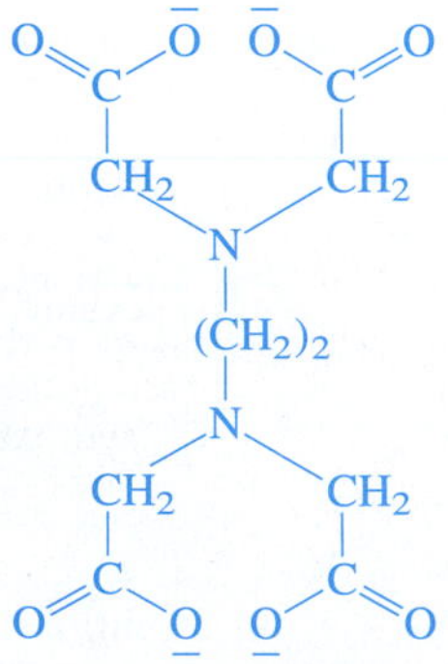 ethylenediaminetetraacetato (edta)	sexidentate	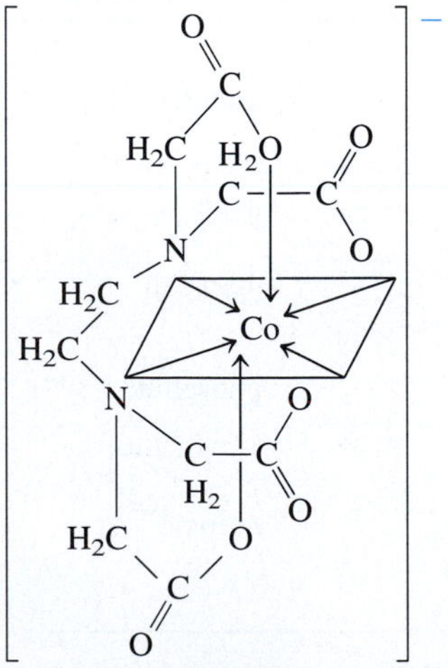 $[Co(edta)]^-$ (ethylenediaminetetraacetato)- cobaltate(III) ion	+3	6

25-4 NOMENCLATURE

The International Union of Pure and Applied Chemistry (IUPAC) has adopted a set of rules for naming coordination compounds. The rules are based on those originally devised by Werner.

1. Cations are always named before anions, with a space between their names.

2. In naming the coordination sphere, ligands are named in alphabetical order. The prefixes di = 2, tri = 3, tetra = 4, penta = 5, hexa = 6, and so on, specify the number of each kind of *simple* (unidentate) ligand. For example, in dichloro, the "di" indicates that two Cl^- act as ligands. For complicated ligands (polydentate chelating agents), other prefixes are used: bis = 2, tris = 3, tetrakis = 4, pentakis = 5, and hexakis = 6. The names of complicated ligands are enclosed in parentheses. The numeric prefixes are not used in alphabetizing. When a prefix denotes the number of substituents on a single ligand, as in dimethylamine, $NH(CH_3)_2$, it *is* used to alphabetize ligands.

3. The names of anionic ligands end in the suffix -o. Examples are F^-, fluoro; OH^-, hydroxo; O^{2-}, oxo; S^{2-}, sulfido; CO_3^{2-}, carbonato; CN^-, cyano; SO_4^{2-}, sulfato; NO_3^-, nitrato; $S_2O_3^{2-}$, thiosulfato.

4. The names of neutral ligands are usually unchanged. Four important exceptions are NH_3, ammine; H_2O, aqua; CO, carbonyl; and NO, nitrosyl.

5. Some metals exhibit variable oxidation states. The oxidation number of such a metal is designated by a Roman numeral in parentheses following the name of the complex ion or molecule.

6. The suffix "-ate" at the end of the name of the complex signifies that it is an anion. If the complex is neutral or cationic, no suffix is used. The English stem is usually used for the metal, but where the naming of an anion is awkward, the Latin stem is substituted. For example, "ferrate" is used rather than "ironate," and "plumbate" rather than "leadate" (Table 25-5).

The following examples illustrate these rules.

$K_2[Cu(CN)_4]$	potassium tetracyanocuprate(II)
$[Ag(NH_3)_2]Cl$	diamminesilver(I) chloride
$[Cr(OH_2)_6](NO_3)_3$	hexaaquachromium(III) nitrate
$[Co(en)_2Br_2]Cl$	dibromobis(ethylenediamine)cobalt(III) chloride
$[Ni(CO)_4]$	tetracarbonylnickel(0)
$[Pt(NH_3)_4][PtCl_6]$	tetraammineplatinum(II) hexachloroplatinate(IV)
$[Cu(NH_3)_2(en)]Br_2$	diammine(ethylenediamine)copper(II) bromide
$Na[Al(OH)_4]$	sodium tetrahydroxoaluminate
$Na_2[Sn(OH)_6]$	sodium hexahydroxostannate(IV)
$[Co(en)_3](NO_3)_3$	tris(ethylenediamine)cobalt(III) nitrate
$K_4[Ni(CN)_2(ox)_2]$	potassium dicyanobis(oxalato)nickelate(II)
$[Co(NH_3)_4(OH_2)Cl]Cl_2$	tetraammineaquachlorocobalt(III) chloride

Table 25-5 *Names for Some Metals in Complex Anions*

Metal	Name* of Metal in Complex Anions
aluminum	aluminate
antimony	antimonate
chromium	chromate
cobalt	cobaltate
copper	*cupr*ate
gold	*aur*ate
iron	*ferr*ate
lead	*plumb*ate
manganese	manganate
nickel	nickelate
platinum	platinate
silver	*argent*ate
tin	*stann*ate
zinc	zincate

Stems derived from Latin names for metals are shown in italics.

The term "ammine" (two m's) signifies the presence of ammonia as a ligand. It is different from the term "amine" (one m), which describes some organic compounds (Section 27-11) that are derived from ammonia.

Water is written as OH_2 rather than H_2O to emphasize that oxygen is the donor atom.

The oxidation state of aluminum is not given because it is always +3.

The abbreviation "ox" represents the oxalate ion $(COO)_2^{2-}$ or $C_2O_4^{2-}$.

25-5 STRUCTURES

The structures of coordination compounds are governed largely by the coordination number of the metal. Most can be predicted by VSEPR theory and have structures similar to

Table 25-6 *Idealized Geometries and Hybridizations for Various Coordination Numbers*

Coordination Number	Geometry	Hybridization	Examples
2	 linear	sp	$[Ag(NH_3)_2]^+$ $[Cu(CN)_2]^-$
4	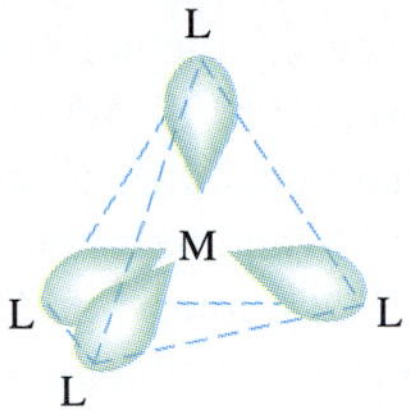 tetrahedral	sp^3	$[Zn(CN)_4]^{2-}$ $[Cd(NH_3)_4]^{2+}$
4	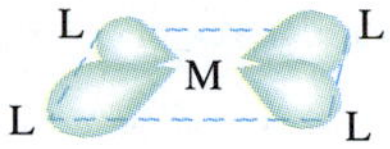 square planar	dsp^2 or sp^2d	$[Ni(CN)_4]^{2-}$ $[Cu(OH_2)_4]^{2+}$ $[Pt(NH_3)_2Cl_2]$
5	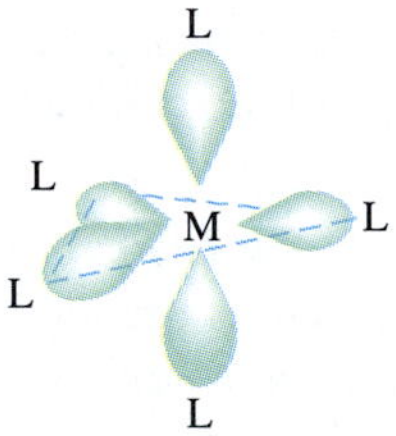 trigonal bipyramidal	dsp^3	$[Fe(CO)_5]$ $[CuCl_5]^{3-}$
5	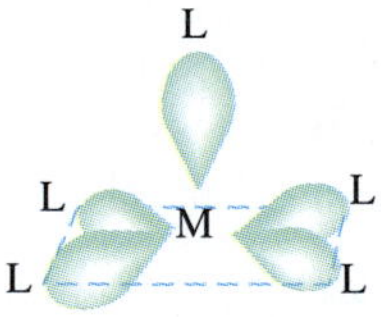 square pyramidal	d^2sp^2	$[Ni(CN)_5]^{3-}$ $[MnCl_5]^{3-}$
6	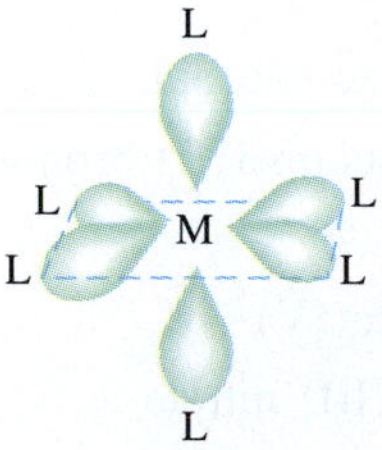 octahedral	d^2sp^3 or sp^3d^2	$[Fe(CN)_6]^{4-}$ $[Fe(OH_2)_6]^{2+}$

the simple molecules we studied in Chapter 8. Unshared pairs of electrons in d orbitals usually have only small influences on geometry because they are not in the outer shell. Table 25-6 summarizes the geometries and hybridizations for common coordination numbers.

Transition metal complexes with coordination numbers as high as 7, 8, and 9 are known. For coordination number 5, the trigonal bipyramidal structure and the square pyramidal structure are both common. The energies associated with these structures are very close. Both tetrahedral and square planar geometries are common for complexes with coordination number 4. The geometries tabulated are ideal geometries. Actual structures are sometimes distorted, especially if the ligands are not all the same. The distortions are due to compensations for the unequal electric fields generated by the different ligands.

ISOMERISM IN COORDINATION COMPOUNDS

Isomers are compounds that have the same number and kinds of atoms arranged differently. The term isomers comes from the Greek word meaning "equal weights." *Because their structures are different, isomers have different properties.* Here we shall restrict our discussion of isomerism to that caused by different arrangements of ligands about central metal ions.

There are two major classes of isomers: structural (constitutional) isomers and stereoisomers. Each can be further subdivided as follows.

Structural Isomers	**Stereoisomers**
1. ionization isomers	1. geometric (positional) isomers
2. hydrate isomers	2. optical isomers
3. coordination isomers	
4. linkage isomers	

Distinctions between simple stereoisomers of coordination compounds involve only one coordination sphere, and the same ligands and donor atoms. Differences between **structural isomers** involve either more than one coordination sphere or different donor atoms on the same ligand. They contain *different atom-to-atom bonding sequences.* Before considering stereoisomers, we shall describe the four types of structural isomers.

25-6 STRUCTURAL (CONSTITUTIONAL) ISOMERS

Ionization (Ion-Ion Exchange) Isomers

These isomers result from the interchange of ions inside and outside the coordination sphere. For example, red-violet $[Co(NH_3)_5Br]SO_4$ and red $[Co(NH_3)_5SO_4]Br$ are ionization isomers.

Some coordination compounds. Starting at the top left and moving clockwise:

$[Cr(CO)_6]$ (white), CO is the ligand

$K_3\{Fe[(COO)_2]_3\}$ (green), $(COO)_2^{2-}$ (oxalate ion) is the ligand

$[Co(H_2N—CH_2—CH_2—NH_2)_3]I_3$ (yellow-orange), ethylenediamine is the ligand

$[Co(NH_3)_5(OH_2)]Cl_3$ (red), NH_3 and H_2O are ligands

$K_3[Fe(CN)_6]$ (red-orange), CN^- is the ligand

A drop of water fell on this sample.

Isomers such as those shown here may or *may not* exist in the same solution in equilibrium. Such isomers are formed by *different* reactions.

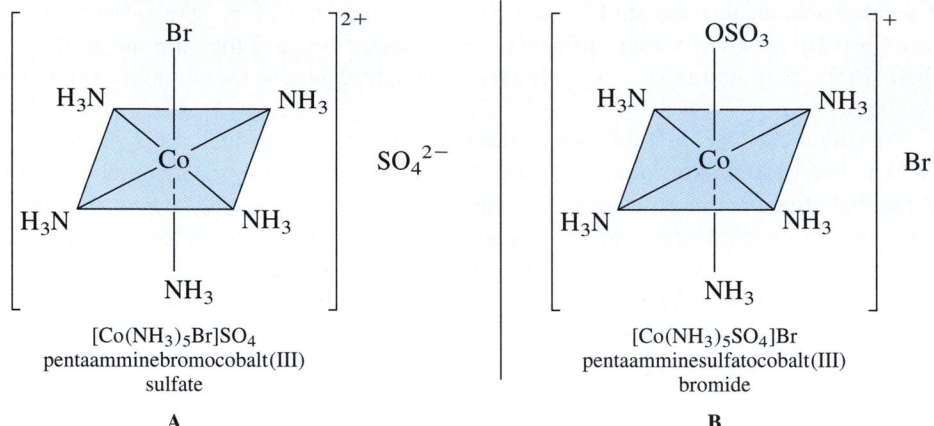

[Co(NH₃)₅Br]SO₄
pentaamminebromocobalt(III)
sulfate

A

[Co(NH₃)₅SO₄]Br
pentaamminesulfatocobalt(III)
bromide

B

Recall that $BaSO_4$ and AgBr are insoluble in H_2O, whereas $BaBr_2$ is soluble and Ag_2SO_4 is moderately soluble.

In A:

$$Ba^{2+} + SO_4^{2-} \longrightarrow BaSO_4(s)$$

$$Ag^+ + SO_4^{2-} \rightarrow \text{no rxn}$$

In B:

$$Ag^+ + Br^- \longrightarrow AgBr(s)$$

$$Ba^{2+} + Br^- \longrightarrow \text{no rxn}$$

In structure A the SO_4^{2-} ion is free and is not bound to the cobalt(III) ion. A solution of A reacts with a solution of barium, $BaCl_2$, to precipitate $BaSO_4$, but does not react with $AgNO_3$. In structure B the SO_4^{2-} ion is bound to the cobalt(III) ion and so it does not react with $BaCl_2$ in aqueous solution. However, the Br^- ion is free, and a solution of B reacts with $AgNO_3$ to precipitate AgBr. *Equimolar* solutions of A and B also have different electrical conductivities. The sulfate solution, A, conducts electric current better because its ions have 2+ and 2− charges rather than 1+ and 1−. Other examples of this type of isomerism include

$[Pt(NH_3)_4Cl_2]Br_2$	and	$[Pt(NH_3)_4Br_2]Cl_2$
$[Pt(NH_3)_4SO_4](OH)_2$	and	$[Pt(NH_3)_4(OH)_2]SO_4$
$[Co(NH_3)_5NO_2]SO_4$	and	$[Co(NH_3)_5SO_4]NO_2$
$[Cr(NH_3)_5SO_4]Br$	and	$[Cr(NH_3)_5Br]SO_4$

Hydrate Isomers

In some crystalline complexes, water can be *inside* and *outside* the coordination sphere. For example, when treated with excess $AgNO_3(aq)$, solutions of the following three hydrate isomers yield three, two, and one mole of AgCl precipitate, respectively, per mole of complex.

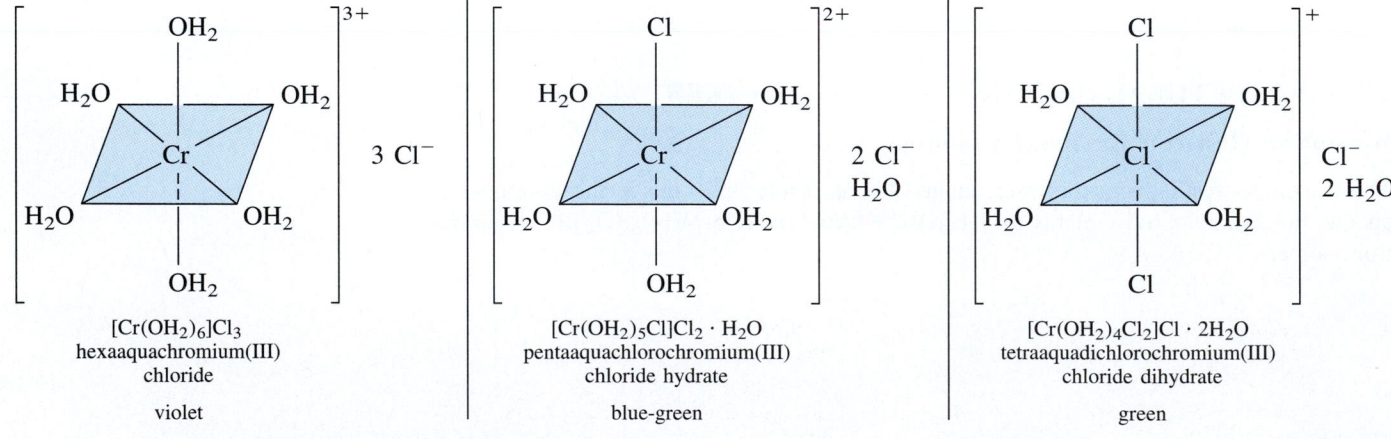

[Cr(OH₂)₆]Cl₃
hexaaquachromium(III)
chloride

violet

[Cr(OH₂)₅Cl]Cl₂ · H₂O
pentaaquachlorochromium(III)
chloride hydrate

blue-green

[Cr(OH₂)₄Cl₂]Cl · 2H₂O
tetraaquadichlorochromium(III)
chloride dihydrate

green

Coordination Isomers

Coordination isomerism can occur in compounds containing both complex cations and complex anions. Such isomers involve exchange of ligands between cation and anion, i.e., between coordination spheres.

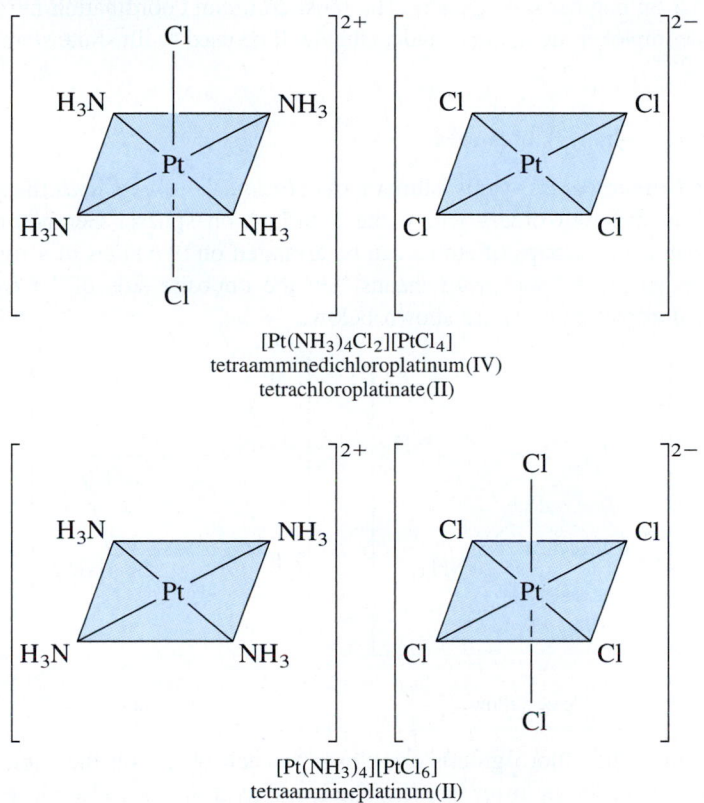

[Pt(NH$_3$)$_4$Cl$_2$][PtCl$_4$]
tetraamminedichloroplatinum(IV)
tetrachloroplatinate(II)

[Pt(NH$_3$)$_4$][PtCl$_6$]
tetraammineplatinum(II)
hexachloroplatinate(IV)

Linkage Isomers

Certain ligands can bind to metal ions in more than one way. Examples of such ligands are cyano, —CN$^-$, and isocyano, —NC$^-$; nitro, —NO$_2$$^-$, and nitrito, —ONO$^-$. The donor atoms are on the left in these representations.

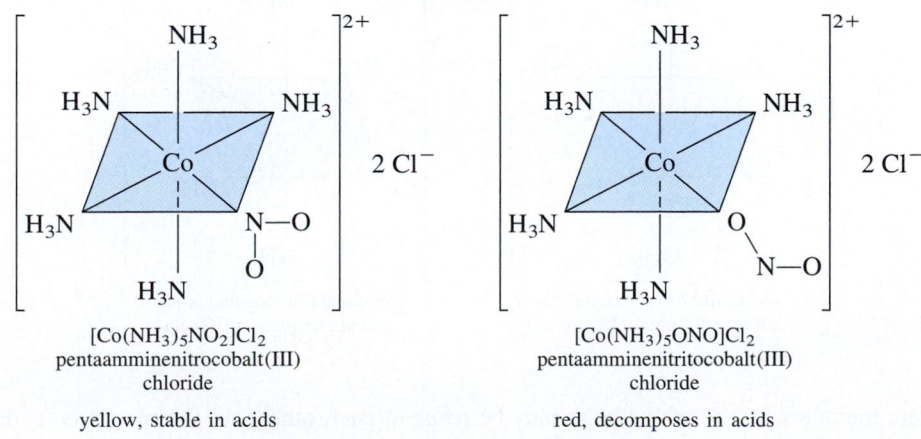

[Co(NH$_3$)$_5$NO$_2$]Cl$_2$
pentaamminenitrocobalt(III)
chloride

yellow, stable in acids

[Co(NH$_3$)$_5$ONO]Cl$_2$
pentaamminenitritocobalt(III)
chloride

red, decomposes in acids

25-7 STEREOISOMERS

A complex with coordination number 2 or 3 that contains only simple *ligands can have only one spatial arrangement. Try building models to see this.*

Compounds that contain the same atoms and the same atom-to-atom bonding sequences, but that differ only in the spatial arrangements of the atoms relative to the central atom, are **stereoisomers.** Complexes with only *simple* ligands can exist as stereoisomers *only if* they have coordination number 4 or greater. The most common coordination numbers among coordination complexes are 4 and 6, and so they will be used to illustrate stereoisomerism.

Geometric (*cis–trans*) Isomers

In **geometric isomers,** or *cis–trans* **isomers,** of *coordination compounds* the same ligands are arranged in different orders within the coordination sphere. Geometric isomerism occurs when atoms or groups of atoms can be arranged on two sides of a rigid structure. *Cis* means "adjacent to" and *trans* means "on the opposite side of." *Cis-* and *trans*-diamminedichloroplatinum(II) are shown below.

The cis isomer has been used in chemotherapy. The trans isomer has no such activity.

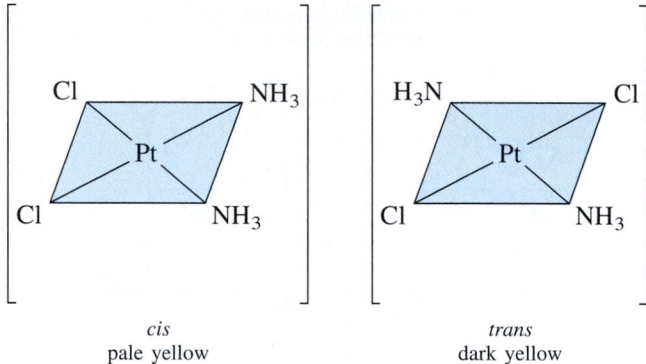

cis
pale yellow

trans
dark yellow

In the *cis* isomer, the chloro groups are closer to each other (on the same side of the square) than they are in the *trans* isomer. The ammine groups are also closer together in the *cis* complex.

Several types of isomerism are possible for octahedral complexes. For example, complexes of the type [MA$_2$B$_2$C$_2$] can exist in several isomeric forms. Consider as an example [Cr(OH$_2$)$_2$(NH$_3$)$_2$Br$_2$]$^+$. First, the members of all three pairs of like ligands may be either *trans* to each other or *cis* to each other.

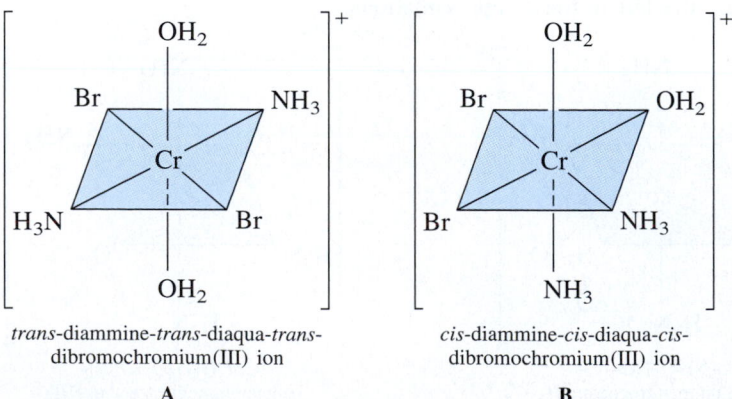

trans-diammine-*trans*-diaqua-*trans*-dibromochromium(III) ion

A

cis-diammine-*cis*-diaqua-*cis*-dibromochromium(III) ion

B

Then, members of one of the pairs may be *trans* to each other, but the members of the other two pairs are *cis*.

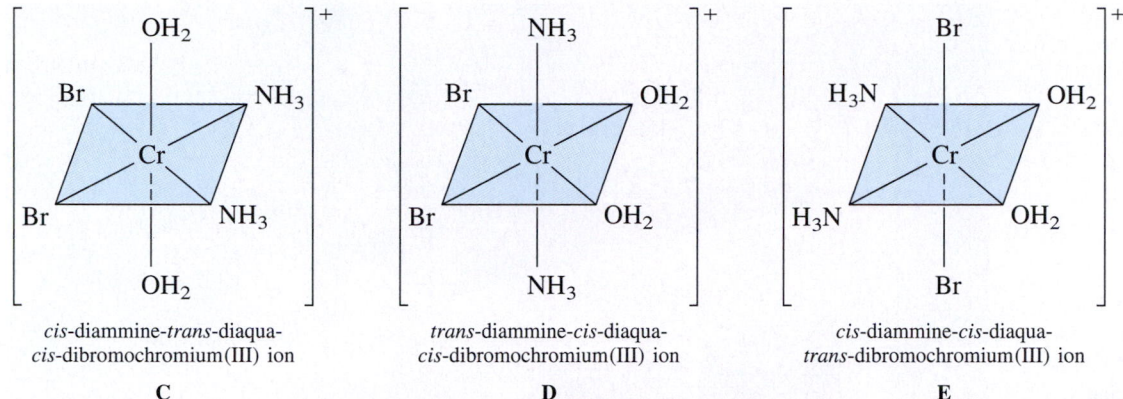

cis-diammine-*trans*-diaqua-   *trans*-diammine-*cis*-diaqua-   *cis*-diammine-*cis*-diaqua-
cis-dibromochromium(III) ion   *cis*-dibromochromium(III) ion   *trans*-dibromochromium(III) ion

     C   **D**   **E**

Further interchange of the positions of the ligands produces no new geometric isomers. However, one of the five geometric isomers (B) can exist in two distinct forms called *optical isomers.*

See whether you can discover why there is no trans-trans-cis isomer.

Optical Isomers

The *cis*-diammine-*cis*-diaqua-*cis*-dibromochromium(III) geometrical isomer (B) exists in two forms that bear the same relationship to each other as left and right hands. They are *nonsuperimposable* mirror images of each other and are called **optical isomers** or **enantiomers.**

An object that is not superimposable with its mirror image is said to be chiral.

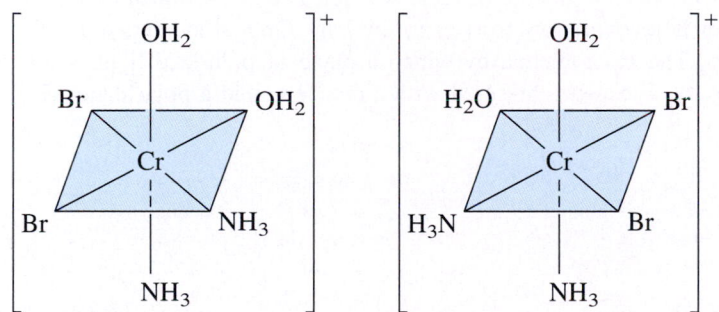

Optical isomers of *cis*-diammine-*cis*-diaqua-*cis*-dibromochromium(III) ion

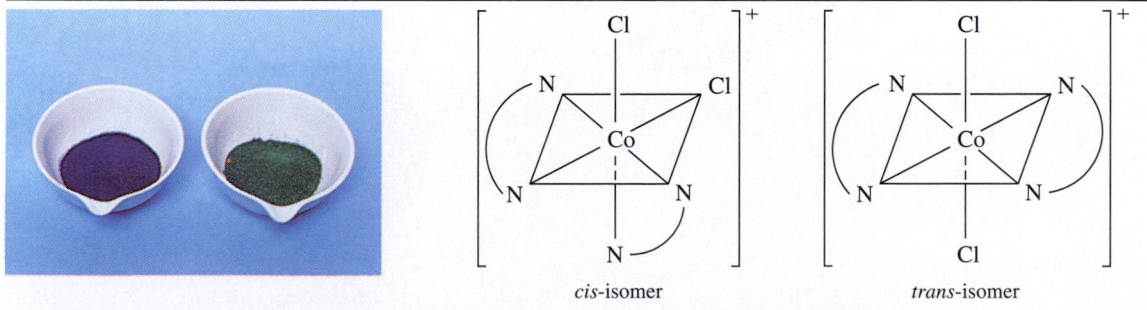

cis-isomer   *trans*-isomer

The dichlorobis(ethylenediamine)cobalt(III) ion, $[Co(en)_2Cl_2]^+$, exists as a pair of *cis–trans* isomers. Ethylenediamine is represented as N⌒N.

Cis-dichlorobis(ethylenediamine)cobalt(III) perchlorate, $[Co(en)_2Cl_2]ClO_4$, is purple.

Trans-dichlorobis(ethylenediamine)cobalt(III) chloride, $[Co(en)_2Cl_2]Cl$, is green.

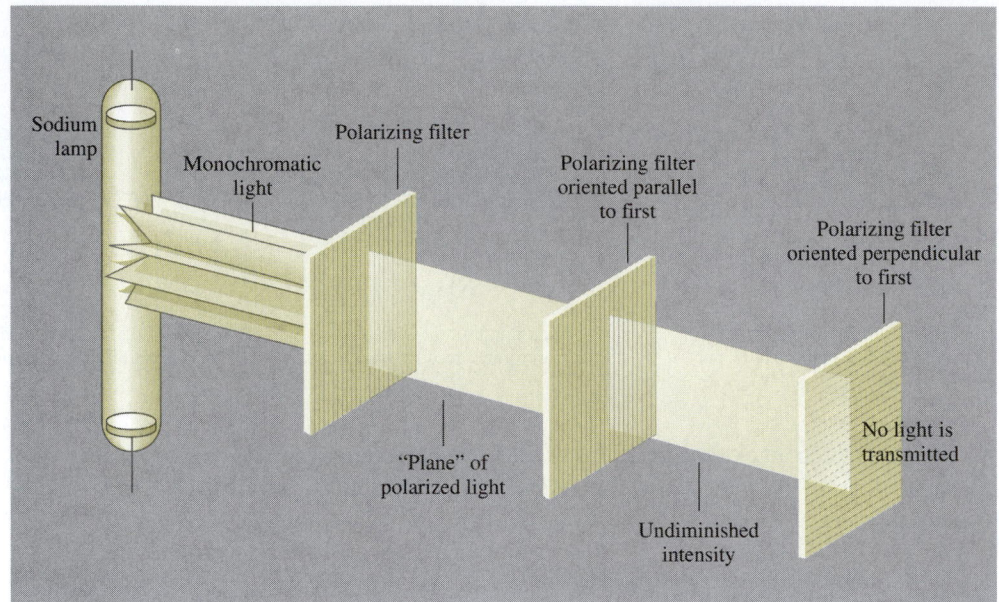

Figure 25-2 Light from a lamp or from the sun consists of electromagnetic waves that vibrate in all directions perpendicular to the direction of travel. Polarizing filters absorb all waves except those that vibrate in a single plane. The third polarizing filter, with a plane of polarization at right angles to the first, absorbs the polarized light completely.

Optical isomers interact with polarized light in different ways. Separate equimolar solutions of each rotate a plane of polarized light (see Figures 25-2 and 25-3) by equal amounts but in opposite directions. One solution is **dextrorotatory** (rotates to the *right*) and the other is **levorotatory** (rotates to the *left*). Optical isomers are called *dextro* and *levo* isomers. The phenomenon by which a plane of polarized light is rotated is called **optical activity.** It can be measured with a device called a polarimeter (Figure 25-3) or

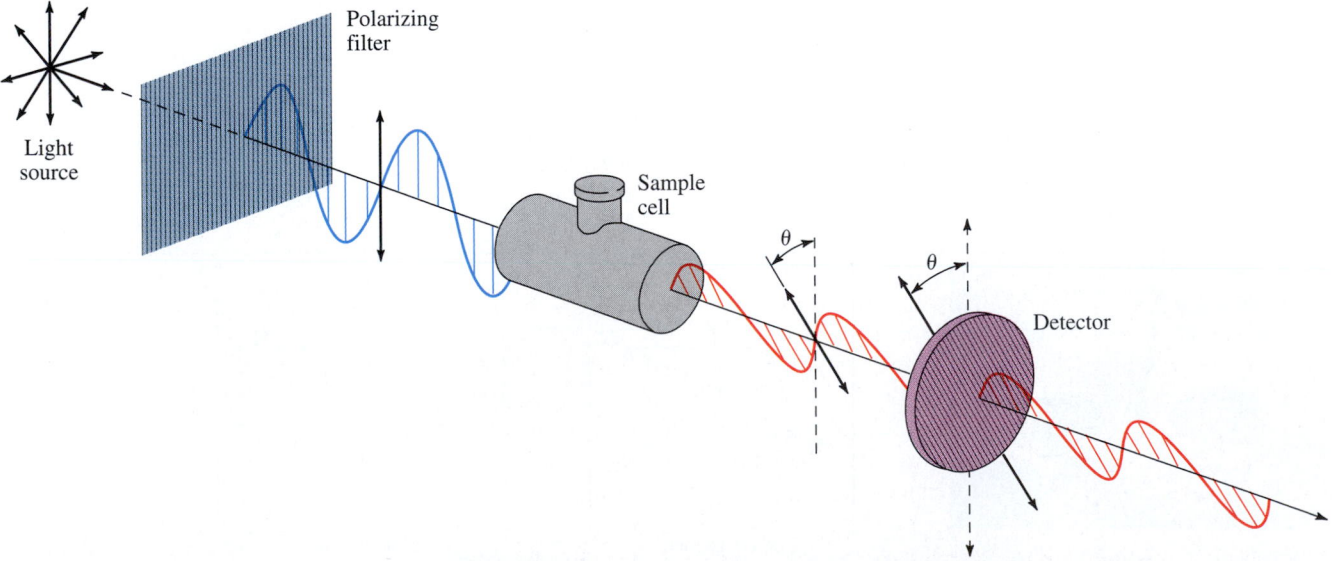

Figure 25-3 The plane of polarization of plane polarized light is rotated through an angle (θ) as it passes through an optically active medium. Species that rotate the plane to the right (clockwise) are dextrorotatory, and those that rotate it to the left are levorotatory.

with more sophisticated instruments. A single solution containing equal amounts of the two isomers is a **racemic mixture.** This solution does not rotate a plane of polarized light. The equal and opposite effects of the two isomers exactly cancel. To exhibit optical activity, the *dextro* and *levo* isomers (sometimes designated as delta, Δ, and lambda, Λ, isomers) must be separated from each other. This is done by one of a number of chemical or physical processes broadly called **optical resolution.** Another pair of optical isomers is shown below. Each contains three molecules of ethylenediamine, a bidentate ligand.

Alfred Werner was also the first person to demonstrate optical activity in an inorganic compound (not of biological origin). This demonstration silenced critics of his theory of coordination compounds; and in his opinion, it was his greatest achievement. Louis Pasteur had demonstrated the phenomenon of optical activity many years earlier in organic compounds of biological origin.

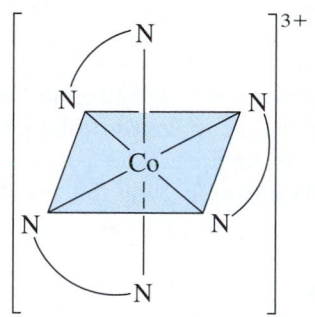

Λ-*tris*(ethylenediamine)cobalt(III) ion

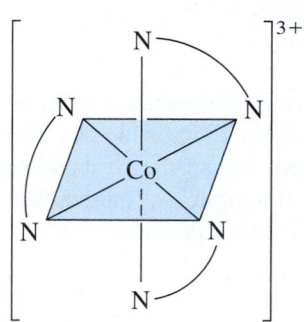

Δ-*tris*(ethylenediamine)cobalt(III) ion

BONDING IN COORDINATION COMPOUNDS

Bonding theories for coordination compounds should be able to account for structural features, colors, and magnetic properties. The earliest accepted theory was the **valence bond theory** (Chapter 8). It can account for structural and magnetic properties. It offers no explanation for the wide range of colors of coordination compounds. However, it has the advantage of being a simple description of bonding that uses the classical picture of the chemical bond. The **crystal field theory** gives quite satisfactory explanations of color as well as of structure and magnetic properties for many coordination compounds.

25-8 VALENCE BOND THEORY

An assumption of the valence bond theory applied to coordination compounds is that the bonds between ligands and metal are entirely covalent. Actually, all ligand-to-metal bonds have some degree of both covalent and ionic character. Because there are so many complex species in which the coordination number of the metal ion is 6, we shall focus our attention on some of them.

Review carefully the shapes of the d orbitals in Figure 5-24 on page 188. The lobes of the $d_{x^2-y^2}$ and d_{z^2} orbitals are directed along the x, y, and z axes. The lobes of the d_{xy}, d_{yz}, and d_{xz} orbitals lie between the axes. The six ligand donor atoms in an octahedral complex are at the corners of an octahedron, two along each of the three axes. Valence bond theory postulates that the donor atoms must donate electrons into a set of octahedrally hybridized d^2sp^3 or sp^3d^2 metal orbitals. Thus, the two metal d orbitals used in the hybridization must be the $d_{x^2-y^2}$ and d_{z^2} orbitals, because they are the only ones directed along the x, y, and z axes (Figure 25-4).

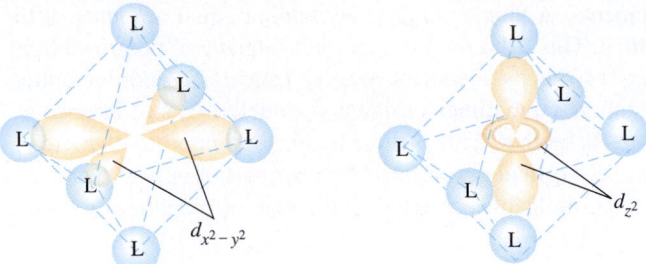

Figure 25-4   Orientation of $d_{x^2-y^2}$ and d_{z^2} orbitals relative to the ligands in an octahedral complex.

A magnetic moment indicates the extent of paramagnetism of a molecule or polyatomic ion.

Let us study some complexes to see how the valence bond theory explains their properties. Hexaaquairon(III) perchlorate, $[Fe(OH_2)_6](ClO_4)_3$, is known to be paramagnetic, with a magnetic moment corresponding to five unpaired electrons per iron atom. The valence bond description of the bonding in the $[Fe(OH_2)_6]^{3+}$ ion follows. Atomic iron has the electron configuration

$$_{26}Fe\ [Ar]\qquad \underset{3d}{\uparrow\downarrow\ \uparrow\ \uparrow\ \uparrow\ \uparrow}\quad \underset{4s}{\uparrow\downarrow}$$

The Fe^{3+} ion is formed by loss of the $4s$ electrons and one of the $3d$ electrons.

$$Fe^{3+}\ [Ar]\qquad \underset{3d}{\uparrow\ \uparrow\ \uparrow\ \uparrow\ \uparrow}\quad \underset{4s}{\underline{}}$$

This is in agreement with Hund's Rule.

To account for the experimentally observed fact that $[Fe(OH_2)_6]^{3+}$ has five unpaired electrons, each $3d$ orbital is assumed to have one unpaired electron in the complex. The vacant $4d_{x^2-y^2}$ and $4d_{z^2}$ orbitals are hybridized with the vacant $4s$ and $4p$ orbitals.

$$Fe^{3+}\ [Ar]\qquad \underset{3d}{\uparrow\ \uparrow\ \uparrow\ \uparrow\ \uparrow}\quad \underset{4s}{\underline{}}\ \underset{4p}{\underline{}\ \underline{}\ \underline{}}\quad \underset{4d}{\underline{}\ \underline{}}$$

$\downarrow$ hybridize

$$Fe^{3+}\ [Ar]\qquad \underset{3d}{\uparrow\ \uparrow\ \uparrow\ \uparrow\ \uparrow}\quad \underset{\text{six } sp^3d^2 \text{ hybrids}}{\underline{}\ \underline{}\ \underline{}\ \underline{}\ \underline{}\ \underline{}}\quad \underset{4d}{\underline{}\ \underline{}\ \underline{}}$$

Each of the six H_2O ligands donates one electron pair into one of the six sp^3d^2 orbitals, forming six coordinate covalent bonds. The electrons originally on the oxygen atoms are represented here as "×" rather than "↑," even though electrons are indistinguishable. These are the only *bonding* electrons; none of the $3d$ electrons of iron is involved in bonding.

All electrons are identical. We use different notations to simplify counting electrons.

$$[Fe(OH_2)_6]^{3+}\ [Ar]\qquad \underset{3d}{\uparrow\ \uparrow\ \uparrow\ \uparrow\ \uparrow}\quad \underset{sp^3d^2}{\times\times\ \times\times\ \times\times\ \times\times\ \times\times\ \times\times}\quad \underset{4d}{\underline{}\ \underline{}\ \underline{}}$$

A set of *outer d* orbitals is used in hybridization, and so $[Fe(OH_2)_6]^{3+}$ is called an **outer orbital complex.**

In deciding whether the hybridization at the central metal ion of octahedral complexes is sp^3d^2 (outer orbital) or d^2sp^3 (inner orbital), *we must know the results of magnetic measurements.* These indicate the number of unpaired electrons.

The hexacyanoferrate(III) ion (ferricyanide ion), $[Fe(CN)_6]^{3-}$, also involves Fe^{3+} (a d^5 ion). But magnetic measurements indicate only one unpaired electron per iron atom. The $3d_{x^2-y^2}$ and $3d_{z^2}$ orbitals (rather than those in the $4d$ shell) are involved in d^2sp^3 hybridization. All but one of the nonbonding $3d$ electrons of Fe^{3+} are paired in this complex.

A convenient way to describe *d*-transition metal ions is to indicate the number of nonbonding electrons in *d* orbitals.

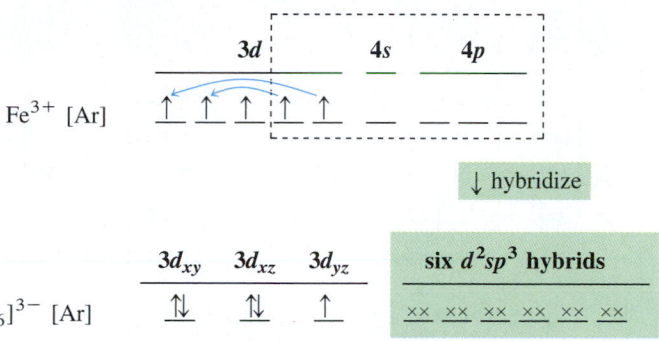

Only inner *d* orbitals are used in hybridization, and so $[Fe(CN)_6]^{3-}$ is called an **inner orbital complex.** As in the previous case, none of the original $3d$ electrons of the Fe^{3+} ion is involved in bonding.

To account for the single (observed) unpaired electron in $[Co(NH_3)_6]^{2+}$ ions (Co^{2+} is a d^7 ion) by valence bond theory, we postulate the *promotion* of a $3d$ electron to a $5s$ orbital as d^2sp^3 hybridization occurs.

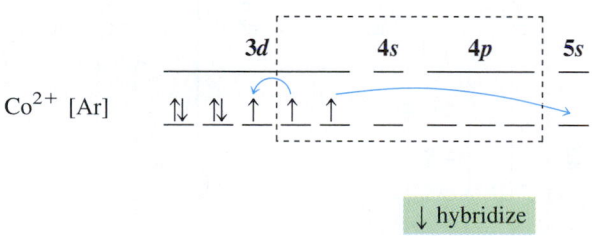

Adding a solution containing Fe^{3+} ions to a solution of potassium hexacyanoferrate(II), $K_4[Fe(CN)_6]$, forms a blue precipitate, $KFe[Fe(CN)_6]$. This compound is called **Turnbull's blue.** Hexacyanoferrate(III) ions react with Fe^{2+} ions to form a blue precipitate known as **Prussian blue.** Prussian blue and Turnbull's blue are thought to be the same compound, $KFe[Fe(CN)_6]$. They were used in making blueprints.

Table 25-7 Bonding and Hybridization in Some Octahedral Complexes

The table presents orbital-box diagrams (with arrows indicating electron spins) for the metal ions and their complex ions; the hybridization and magnetic character are noted.

Metal Ion	Outer Electron Configuration ($3d$, $4s$, $4p$, $4d$)	Complex Ion	Type	Outer Electron Configuration ($3d$, hybrid, $4d$)
Cr^{3+} (d^3)	$3d$: ↑ ↑ ↑	$[Cr(NH_3)_6]^{3+}$	inner orbital (paramagnetic)	$3d$: ↑ ↑ ↑ ; d^2sp^3
Mn^{2+} (d^5)	$3d$: ↑ ↑ ↑ ↑ ↑	$[Mn(CN)_6]^{4-}$	inner orbital (paramagnetic)	$3d$: ⇅ ⇅ ↑ ; d^2sp^3
Fe^{2+} (d^6)	$3d$: ⇅ ↑ ↑ ↑ ↑	$[Fe(CN)_6]^{4-}$	inner orbital (diamagnetic)	$3d$: ⇅ ⇅ ⇅ ; d^2sp^3
Co^{3+} (d^6)	$3d$: ⇅ ↑ ↑ ↑ ↑	$[Co(NH_3)_6]^{3+}$	inner orbital (diamagnetic)	$3d$: ⇅ ⇅ ⇅ ; d^2sp_3
		$[CoF_6]^{3-}$	outer orbital (paramagnetic)	$3d$: ⇅ ↑ ↑ ↑ ↑ ; sp^3d^2
Co^{2+} (d^7)	$3d$: ⇅ ⇅ ↑ ↑ ↑	$[Co(OH_2)_6]^{2+}$	outer orbital (paramagnetic)	$3d$: ⇅ ⇅ ⇅ ↑ ↑ ; sp^3d^2
Ni^{2+} (d^8)	$3d$: ⇅ ⇅ ⇅ ↑ ↑	$[Ni(OH_2)_6]^{2+}$	outer orbital (paramagnetic)	$3d$: ⇅ ⇅ ⇅ ↑ ↑ ; sp^3d^2

The suggested bonding for $[Co(NH_3)_6]^{2+}$ is consistent with the fact that $[Co(NH_3)_6]^{2+}$ is easily oxidized to $[Co(NH_3)_6]^{3+}$, an extremely stable complex ion.

The electron configurations and hybridizations of some octahedral complexes are shown in Table 25-7.

25-9 CRYSTAL FIELD THEORY

Hans Bethe and J. H. van Vleck developed the crystal field theory between 1919 and the early 1930s. It was not widely used until the 1950s. In its pure form it assumes that the bonds between ligand and metal ion are completely ionic. Both ligands and metal ions are treated as infinitesimally small, nonpolarizable point charges. Valence bond theory assumes complete covalence. Modern ligand field theory is an outgrowth of crystal field theory. It attributes partial covalent character and partial ionic character to bonds.

In a metal ion surrounded by other atoms, the d orbitals are at higher energy than they are in an isolated metal ion. If the surrounding electrons were uniformly distributed about the metal ion, the energies of *all* five d orbitals would increase by the same amount (a *spherical crystal field*). Because the ligands approach the metal ion from different directions, they affect different d orbitals in different ways. Here we illustrate the application of these ideas to complexes with coordination number 6 (*octahedral crystal field*).

The $d_{x^2-y^2}$ and d_{z^2} orbitals are directed along a set of mutually perpendicular x, y, and z axes (p. 188). As a group, these orbitals are called e_g **orbitals.** The d_{xy}, d_{yz}, and d_{xz} orbitals, collectively called t_{2g} **orbitals,** lie between the axes. The ligand donor atoms approach the metal ion along the axes to form octahedral complexes. Crystal field theory proposes that the approach of the six donor atoms (point charges) along the axes sets up an electric field (the crystal field). Electrons on the ligands repel electrons in e_g orbitals on the metal ion more strongly than they repel those in t_{2g} orbitals (Figure 25-5). This removes the degeneracy of the set of d orbitals and splits them into two sets, the e_g set at higher energy and the t_{2g} set at lower energy.

> Recall that degenerate orbitals are orbitals of equal energy.

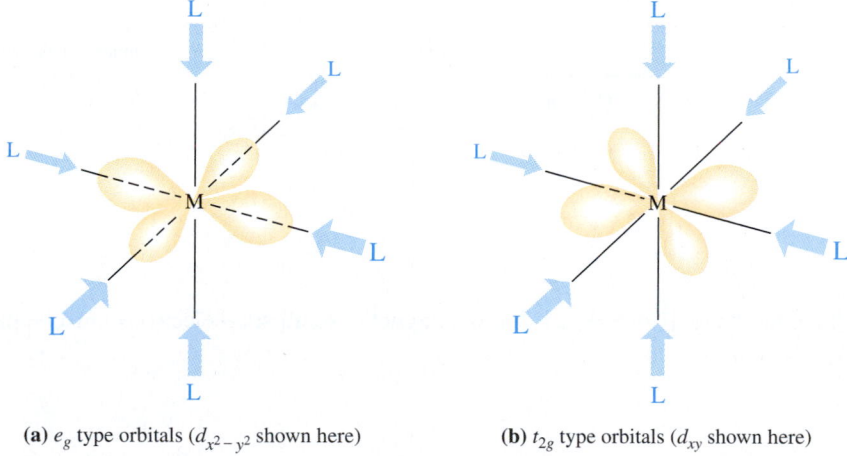

(a) e_g type orbitals ($d_{x^2-y^2}$ shown here)   **(b)** t_{2g} type orbitals (d_{xy} shown here)

Figure 25-5 Effects of the approach of ligands on the energies of d orbitals on the metal ion. In an octahedral complex, the ligands (L) approach the metal ion (M) along the x, y, and z axes, as indicated by the blue arrows. (a) The orbitals of the e_g type—$d_{x^2-y^2}$ (shown here) and d_{z^2}—point directly toward the incoming ligands, so electrons in these orbitals are strongly repelled. (b) The orbitals of the t_{2g} type—d_{xy} (shown here), d_{xz} and d_{yz}—do not point toward the incoming ligands, so electrons in these orbitals are less strongly repelled.

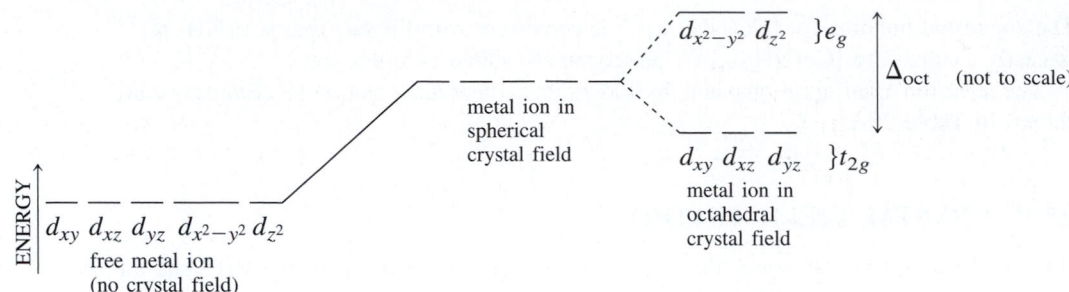

Typical values of Δ_{oct} are between 100 and 400 kJ/mol.

The energy separation between the two sets is called $\Delta_{octahedral}$, or Δ_{oct}. It is proportional to the *crystal field strength* of the ligands—that is, how strongly the ligand electrons repel the electrons on the metal ion.

The d electrons on a metal ion occupy the t_{2g} set in preference to the higher energy e_g set. Electrons that occupy the e_g orbitals are strongly repelled by the relatively close approach of ligands. The occupancy of these orbitals tends to destabilize octahedral complexes.

Let us now describe the hexafluorocobaltate(III) ion, $[CoF_6]^{3-}$, and the hexa-amminecobalt(III) ion, $[Co(NH_3)_6]^{3+}$. Both contain the d^6 Co^{3+} ion; they have already been treated in terms of valence bond theory (Table 25-7). $[CoF_6]^{3-}$ is a paramagnetic outer orbital complex; $[Co(NH_3)_6]^{3+}$ is a diamagnetic inner orbital complex. We will focus our attention on the d electrons.

The Co^{3+} ion has six electrons (four unpaired) in its $3d$ orbitals.

$$\begin{array}{c} \mathbf{3d} \\ \hline \end{array}$$

$$Co^{3+} \ [Ar] \quad \uparrow\downarrow \ \uparrow \ \uparrow \ \uparrow \ \uparrow$$

This is a *high spin* complex.

Magnetic measurements indicate that $[CoF_6]^{3-}$ also has four unpaired electrons per ion. So there must be four electrons in t_{2g} orbitals and two in e_g orbitals.

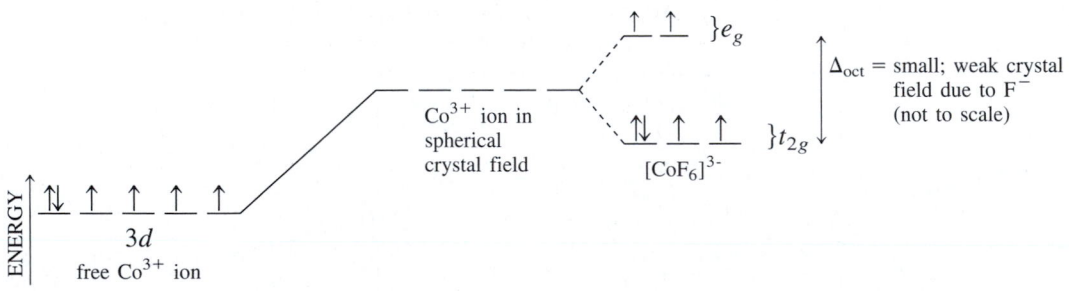

This is a *low spin* complex.

On the other hand, $[Co(NH_3)_6]^{3+}$ is diamagnetic, so all six d electrons must be paired in the t_{2g} orbitals.

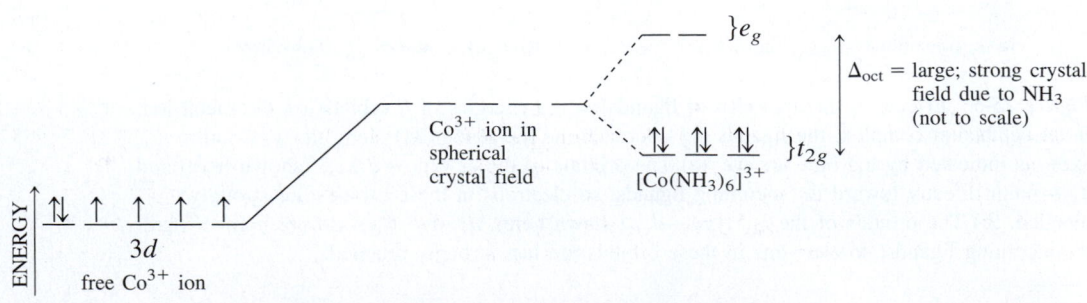

The difference in configurations between $[CoF_6]^{3-}$ and $[Co(NH_3)_6]^{3+}$ is due to the relative magnitudes of the crystal field splitting, Δ_{oct}, caused by the different crystal field strengths of F^- and NH_3. The NH_3 molecule interacts with vacant metal orbitals more strongly than the F^- ion does. As a result, the crystal field splitting generated by the close approach of six NH_3 molecules (strong field ligands) to the metal ion is greater than that generated by the approach of six F^- ions (weak field ligands).

$$\Delta_{oct} \text{ for } [Co(NH_3)_6]^{3+} > \Delta_{oct} \text{ for } [CoF_6]^{3-}$$

The crystal field splitting for $[CoF_6]^{3-}$ is very small. This means that an energetically more favorable situation results if two electrons remain unpaired in the antibonding e_g orbitals. (Hund's Rule requires that electrons singly occupy a set of degenerate orbitals before pairing.) This avoids the expenditure of energy that is necessary to pair electrons by bringing two negatively charged particles into the same region of space. If the approach of a set of ligands removes the d orbital degeneracy but causes a Δ_{oct} less than the electron pairing energy, P, then the electrons will singly occupy the resulting nondegenerate orbitals. After all d orbitals are half-filled, additional electrons will pair with electrons in the t_{2g} set. This is the case for $[CoF_6]^{3-}$, which is called a **high spin complex.**

For $[CoF_6]^{3-}$: F^- is weak field ligand so $\Delta_{oct} < P$; thus, high spin complex

electron pairing energy

A high spin complex in crystal field terminology corresponds to an outer orbital complex in valence bond theory.

In contrast, the $[Co(NH_3)_6]^{3+}$ ion is a **low spin complex.** Δ_{oct} generated by the strong field ligand, NH_3, is greater than the electron pairing energy. So electrons pair in t_{2g} orbitals before any occupy the higher energy e_g orbitals.

$[Co(NH_3)_6]^{3+}$: NH_3 is strong field ligand so $\Delta_{oct} > P$; thus, low spin complex

A low spin complex corresponds to an inner orbital complex.

Low spin configurations exist only for octahedral complexes having metal ions with d^4, d^5, d^6, and d^7 configurations. For d^1–d^3 and d^8–d^{10} ions, only one possibility exists. In these cases, the configuration is designated as high spin. All d^n possibilities are shown in Table 25-8.

$[Co(OH_2)_6]^{2+}$ ions are pink (bottom). A limited amount of aqueous ammonia produces $Co(OH)_2$, a blue compound that quickly turns gray (middle). $Co(OH)_2$ dissolves in excess aqueous ammonia to form $[Co(NH_3)_6]^{2+}$ ions, which are orange-yellow (top).

The electron pairing energy is larger than the crystal field splitting energy.

Table 25-8 High and Low Spin Octahedral Configurations

d^n	Examples	High Spin	Low Spin	d^n	Examples	High Spin	Low Spin
d^1	Ti^{3+}	— — e_g ↑ — — t_{2g}	same as high spin	d^6	Fe^{2+}, Ru^{2+}, Pd^{4+}, Rh^{3+}, Co^{3+}	↑ ↑ e_g ↑↓ ↑ ↑ t_{2g}	— — e_g ↑↓ ↑↓ ↑↓ t_{2g}
d^2	Ti^{2+}, V^{3+}	— — e_g ↑ ↑ — t_{2g}	same as high spin	d^7	Co^{2+}, Rh^{2+}	↑ ↑ e_g ↑↓ ↑↓ ↑ t_{2g}	↑ — e_g ↑↓ ↑↓ ↑↓ t_{2g}
d^3	V^{2+}, Cr^{3+}	— — e_g ↑ ↑ ↑ t_{2g}	same as high spin	d^8	Ni^{2+}, Pt^{2+}	↑ ↑ e_g ↑↓ ↑↓ ↑↓ t_{2g}	same as high spin
d^4	Mn^{3+}, Re^{3+}	↑ — e_g ↑ ↑ ↑ t_{2g}	— — e_g ↑↓ ↑ ↑ t_{2g}	d^9	Cu^{2+}	↑↓ ↑ e_g ↑↓ ↑↓ ↑↓ t_{2g}	same as high spin
d^5	Mn^{2+}, Fe^{3+}, Rn^{3+}	↑ ↑ e_g ↑ ↑ ↑ t_{2g}	— — e_g ↑↓ ↑↓ ↑ t_{2g}	d^{10}	Zn^{2+}, Ag^+, Hg^{2+}	↑↓ ↑↓ e_g ↑↓ ↑↓ ↑↓ t_{2g}	same as high spin

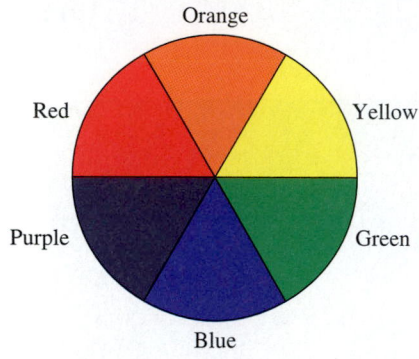

A color wheel shows colors and their complementary colors. For example, green is the complementary color of red. The data in Table 25-9 are given for specific wavelengths. Broad bands of wavelengths are shown in this color wheel.

Table 25-9 *Complementary Colors*

Wavelength Absorbed (Å)	Spectral Color (color absorbed)	Complementary Color (color observed)
4100	violet	lemon yellow
4300	indigo	yellow
4800	blue	orange
5000	blue-green	red
5300	green	purple
5600	lemon yellow	violet
5800	yellow	indigo
6100	orange	blue
6800	red	blue-green

25-10 COLOR AND THE SPECTROCHEMICAL SERIES

A substance appears colored because it absorbs light that corresponds to one or more of the wavelengths in the visible region of the electromagnetic spectrum (4000 to 7000 Å) and transmits or reflects the other wavelengths. Our eyes are detectors for light in the visible region, and so each wavelength in this region appears as a different color. A combination of all wavelengths in the visible region is called "white light"; sunlight is an example. The absence of all wavelengths in the visible region is blackness.

In Table 25-9 we show the relationships among colors absorbed and colors transmitted or reflected in the visible region. The first column displays the wavelengths absorbed. The **spectral color** is the color associated with the wavelengths of light absorbed by the sample. When certain visible wavelengths are absorbed from incoming "white" light, the light *not absorbed* remains visible to us as transmitted or reflected light. For instance, a sample that absorbs orange light appears blue. The **complementary color** is the color associated with the wavelengths that are not absorbed by the sample. The complementary color is seen when the spectral color is removed from white light.

Most transition metal compounds are colored, a characteristic that distinguishes them from most compounds of the representative elements. In transition metal *compounds,* the *d* orbitals in any one energy level of the metals are not degenerate. No longer do all have the same energy, as they do in isolated atoms. They are often split into two sets of orbitals separated by energies, Δ_{oct}, that correspond to wavelengths of light in the visible region. The absorption of visible light causes electronic transitions between orbitals in these sets. Table 25-10 gives the colors of some transition metal nitrates in aqueous solution. Solutions of representative metal nitrates are colorless.

Table 25-10 *Colors of Aqueous Solutions of Some Transition Metal Nitrates*

Transition Metal Ion	Color of Aq. Solution
Cr^{3+}	Deep blue
Mn^{2+}	Pale pink
Fe^{2+}	Pale green
Fe^{3+}	Orchid
Co^{2+}	Pink
Ni^{2+}	Green
Cu^{2+}	Blue

Adding a Co(II) salt to molten glass gives the glass a deep blue color.

914

The colors of complex compounds that contain a given metal depend on the ligands. The yellow compound at the left is a salt that contains $[Co(NH_3)_6]^{3+}$ ions. In the next three compounds, left to right, one NH_3 ligand in $[Co(NH_3)_6]^{3+}$ has been replaced by NCS^- (orange), H_2O (red), and Cl^- (purple). The green compound at the right is a salt that contains $[Co(NH_3)_4Cl_2]^+$ ions.

One transition of a high spin octahedral Co(III) complex is depicted as follows.

Ground State   **Excited State**   **Energy of Light Absorbed**

e_g ↑ ↑ $\xrightarrow[\text{of light}]{\text{Absorption}}$ ↑↓ ↑ ......

t_{2g} ↑↓ ↑ ↑   ↑ ↑ ↑ ......

$\Delta E = h\nu$ depends on Δ_{oct}

Planck's constant is
$h = 6.63 \times 10^{-34}\,\text{J}\cdot\text{s}$.

The frequency (ν), and therefore the wavelength and color, of the light absorbed are related to Δ_{oct}.* This, in turn, depends upon the crystal field strength of the ligands. So the colors and visible absorption spectra of transition metal complexes, as well as their magnetic properties, provide information about the strengths of the ligand–metal interactions.

By interpreting the visible spectra of many complexes, it is possible to arrange common ligands in order of increasing crystal field strengths.

$$I^- < Br^- < Cl^- < F^- < OH^- < H_2O < (COO)_2^{2-} < NH_3 < en < NO_2^- < CN^-$$

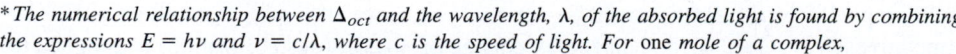

Increasing crystal field strength

This arrangement is called the **spectrochemical series.** Strong field ligands, such as CN^-, usually produce low spin complexes, where possible, and large crystal field splittings. Weak field ligands, such as Cl^-, usually produce high spin complexes and small crystal field splittings. Low spin complexes usually absorb higher-energy (shorter-wavelength) light than do high spin complexes. The colors of several six-coordinate Cr(III) complexes are listed in Table 25-11.

In $[Cr(NH_3)_6]Cl_3$, the Cr(III) is bonded to six ammonia ligands, which produce a relatively high value of Δ_{oct}. This causes the $[Cr(NH_3)_6]^{3+}$ ion to absorb relatively high energy visible light in the blue and violet regions. Thus, we see yellow-orange, the complementary color.

Colors of some copper(II) compounds. Left to right: $CuBr_2$, $CuSO_4 \cdot 5H_2O$, $CuCl_2 \cdot 2H_2O$, $(CuOH)_2CO_3$.

*The numerical relationship between Δ_{oct} and the wavelength, λ, of the absorbed light is found by combining the expressions $E = h\nu$ and $\nu = c/\lambda$, where c is the speed of light. For one mole of a complex,

$$\Delta_{oct} = EN_A = \frac{hcN_A}{\lambda} \qquad \text{where } N_A \text{ is Avogadro's number}$$

Different anions often cause
compounds containing the same
complex cation to have different
colors.

Table 25-11 *Colors of Some Chromium(III) Complexes*

$[Cr(OH_2)_4Br_2]Br$	green		$[Cr(CON_2H_4)_6][SiF_6]_3$	green
$[Cr(OH_2)_6]Br_3$	bluish gray		$[Cr(NH_3)_5Cl]Cl_2$	purple
$[Cr(OH_2)_4Cl_2]Cl$	green		$[Cr(NH_3)_4Cl_2]Cl$	violet
$[Cr(OH_2)_6]Cl_3$	violet		$[Cr(NH_3)_6]Cl_3$	yellow

We see the light that is transmitted
(passes through the sample) or that is
reflected by the sample.

Water is a weaker field ligand than ammonia, and therefore Δ_{oct} is less for $[Cr(OH_2)_6]^{3+}$ than for $[Cr(NH_3)_6]^{3+}$. As a result, $[Cr(OH_2)_6]Br_3$ absorbs lower energy (longer wavelength) light. This causes the reflected and transmitted light to be higher energy bluish gray, the color that describes $[Cr(OH_2)_6]Br_3$.

25-11 CRYSTAL FIELD STABILIZATION ENERGY

Electrons in the t_{2g} orbitals of an octahedral complex are lower in energy, while electrons in the e_g orbitals are higher in energy, than they would be in d orbitals of a metal ion in a spherical field. The lower the total energy of a system, the more stable it is. The **crystal field stabilization energy** (CFSE) of a complex is a measure of the net energy of stabilization (compared to the ion in a spherical field) of a metal ion's electrons. Crystal field stabilization energy is a major factor that contributes to the stability of complex ions with certain electron configurations ($d^4 - d^7$ ions). Here we consider the CFSEs of some octahedral complexes. For a given set of six ligands, the t_{2g} and e_g orbitals are split by a certain amount of energy, Δ_{oct}. Each t_{2g} orbital is $\frac{2}{5}\Delta_{oct}$ *below* the energy of the set of degenerate d orbitals in a spherical (homogeneous) field. Each e_g orbital is $\frac{3}{5}\Delta_{oct}$ *above* the energy of the unsplit d orbitals.

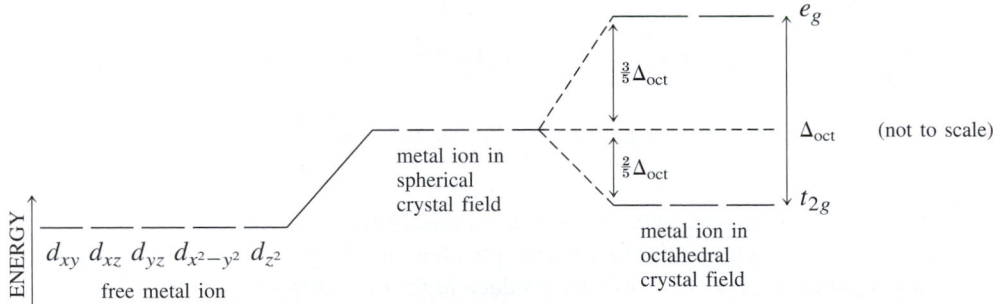

There are three t_{2g} orbitals and two e_g orbitals. The total energy of the sets of t_{2g} and e_g orbitals is the same as the energy of the set of degenerate d orbitals in a spherical field. That is, the change in total energy of a set of d orbitals is zero as a result of the approach of the six ligands along the x, y, and z axes.

$$\Delta E = CFSE = 3(-\tfrac{2}{5}\Delta_{oct}) + 2(\tfrac{3}{5}\Delta_{oct}) = 0$$
$$(t_{2g}) \qquad\quad (e_g)$$

Table 25-8 shows the electron configurations of some high spin and low spin octahedral complexes of d-transition metals. An energy change of $-\tfrac{2}{5}\Delta_{oct}$ (stabilization) is assigned to each t_{2g} electron, and an energy change of $+\tfrac{3}{5}\Delta_{oct}$ (destabilization) is assigned to each e_g electron. The CFSE is the sum of the energies of all the electrons in the metal atom or ion. Table 25-12 shows crystal field stabilization energies for the kinds of octahedral

Table 25-12 *Crystal Field Stabilization Energies for Octahedral d^n Complexes*

d^n	High Spin	Low Spin	d^n	High Spin	Low Spin
d^0	0	same as high spin			
d^1	$-\frac{2}{5}\Delta_{oct}$	same as high spin	d^6	$-\frac{2}{5}\Delta_{oct}$	$-\frac{12}{5}\Delta_{oct}$
d^2	$-\frac{4}{5}\Delta_{oct}$	same as high spin	d^7	$-\frac{4}{5}\Delta_{oct}$	$-\frac{9}{5}\Delta_{oct}$
d^3	$-\frac{6}{5}\Delta_{oct}$	same as high spin	d^8	$-\frac{6}{5}\Delta_{oct}$	same as high spin
d^4	$-\frac{3}{5}\Delta_{oct}$	$-\frac{8}{5}\Delta_{oct}$	d^9	$-\frac{3}{5}\Delta_{oct}$	same as high spin
d^5	0	$-\frac{10}{5}\Delta_{oct}$	d^{10}	0	same as high spin

complexes described in Table 25-8. You should study these two tables together.

For example, for a high spin d^7 complex such as $[Co(OH_2)_6]^{2+}$, the CFSE is $5(-\frac{2}{5}\Delta_{oct}) + 2(+\frac{3}{5}\Delta_{oct}) = -\frac{4}{5}\Delta_{oct}$. So the amount of *energy released* by the octahedral splitting of the occupied orbitals in forming the complex ion is $\frac{4}{5}\Delta_{oct}$. Configurations with the most negative CFSEs in Table 25-12 are the ones for which many stable octahedral complexes are known. No configuration can produce a CFSE greater than zero. That is, no d-transition metal ions should be less stable in an octahedral ligand environment than in a spherical crystal field.

Many properties of complex species of the d-transition elements can be accounted for by the crystal field theory. This is strong evidence for the general validity of the theory. Consider the heats (enthalpies) of hydration for the series of 2+ ions shown in Figure 25-6. All these ions form octahedral, weak field complexes with H_2O, $[M(OH_2)_6]^{2+}$. $\Delta H_{hydration}$ is the amount of energy absorbed when one mole of gaseous ions forms hydrated ions.

$$M^{2+}(g) + 6H_2O \longrightarrow [M(OH_2)_6]^{2+}(aq)$$

The CFSE is zero for weak field (high spin) octahedral d^0, d^5, and d^{10} complexes (Table 25-12). Experimental values for Ca^{2+} (d^0), Mn^{2+} (d^5), and Zn^{2+} (d^{10}) show no CFSE.

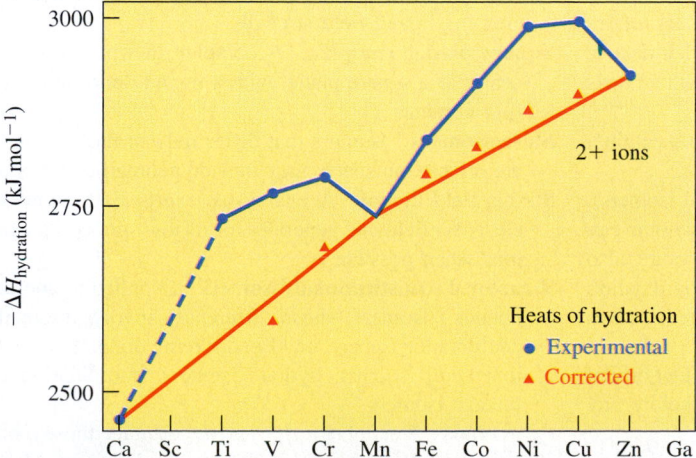

Figure 25-6 Heats of hydration for 2+ transition metal ions. When values of CFSE from spectroscopic Δ_{oct}'s are subtracted from experimental values of $\Delta H_{hydration}$ (●), a plot of the "corrected" values of $\Delta H_{hydrated}$ (▲) versus atomic number is very nearly a straight line.

Key Terms

Ammine complexes   Complex species that contain ammonia molecules bonded to metal ions.

Central atom   The atom or ion to which the ligands are bonded in a complex species.

Chelate   A ligand that utilizes two or more donor atoms in bonding to metals.

cis–trans isomerism   See *Geometric isomerism.*

Complementary color   The color associated with the wavelengths of light that are not absorbed—i.e., the color transmitted or reflected.

Coordinate covalent bond   A covalent bond in which both shared electrons are donated by the same atom; a bond between a Lewis base and a Lewis acid.

Coordination compound or complex   A compound containing coordinate covalent bonds.

Coordination isomers   Isomers involving exchange of ligands between a complex cation and a complex anion of the same compound.

Coordination number   The number of donor atoms coordinated to a metal.

Coordination sphere   The metal ion and its coordinated ligands, but not any uncoordinated counter-ions.

Crystal field stabilization energy   The net energy of stabilization gained by a metal ion's nonbonding d electrons as a result of complex formation.

Crystal field theory   A theory of bonding in transition metal complexes in which ligands and metal ions are treated as point charges; a purely ionic model. Ligand point charges represent the crystal (electrical) field perturbing the metal's d orbitals that contain nonbonding electrons.

Δ_{oct}   The energy separation between e_g and t_{2g} sets of metal d orbitals caused by octahedral complexation of ligands; sometimes called $10Dq$.

Dextrorotatory   Describes an optically active substance that rotates the plane of plane polarized light to the right; also called dextro.

Donor atom   A ligand atom whose electrons are shared with a Lewis acid.

e_g orbitals   A set of $d_{x^2-y^2}$ and d_{z^2} orbitals; those d orbitals within a set with lobes directed along the x, y, and z axes.

Enantiomers   Nonsuperimposable mirror images (optical isomers).

Geometric isomerism   Occurs when atoms or groups of atoms can be arranged in different ways on two sides of a rigid structure; also called *cis–trans* isomerism. In geometric isomers of coordination compounds, the same ligands are arranged in different orders within the coordination sphere.

High spin complex   The crystal field designation for an outer orbital complex; all t_{2g} and e_g orbitals are singly occupied before any pairing occurs.

Hydrate isomers   Isomers of crystalline complexes that differ in terms of the presence of water inside or outside the coordination sphere.

Inner orbital complex   The valence bond designation for a complex in which the metal ion utilizes d orbitals one shell inside the outermost occupied shell in its hybridization.

Ionization isomers   Isomers that result from interchange of ions inside and outside the coordination sphere.

Isomers   Different compounds that have the same formula.

Levorotatory   Refers to an optically active substance that rotates the plane of plane polarized light to the left; also called levo.

Ligand   A Lewis base in a coordination compound.

Linkage isomers   Isomers in which a particular ligand bonds to a metal ion through different donor atoms.

Low spin complex   The crystal field designation for an inner orbital complex; contains electrons paired in t_{2g} orbitals before e_g orbitals are occupied in octahedral complexes.

Optical activity   The rotation of plane polarized light by one of a pair of optical isomers.

Optical isomers   Stereoisomers that differ only by being nonsuperimposable mirror images of each other, like left and right hands; also called enantiomers.

Outer orbital complex   The valence bond designation for a complex in which the metal ion utilizes d orbitals in the outermost (occupied) shell in hybridization.

Pairing energy   The energy required to pair two electrons in the same orbital.

Plane polarized light   Light waves in which all the electric vectors are oscillating in one plane.

Polarimeter   A device used to measure optical activity.

Polydentate   Describes ligands with more than one donor atom.

Racemic mixture   An equimolar mixture of dextro and levo optical isomers that is, therefore, optically inactive.

Spectral color   The color associated with the wavelengths of light that are absorbed.

Spectrochemical series   An arrangement of ligands in order of increasing ligand field strength.

Square planar complex   A complex in which the metal is in the center of a square plane, with a ligand donor atom at each of the four corners.

Stereoisomers   Isomers that differ only in the way in which atoms are oriented in space; they include geometric and optical isomers.

Strong field ligand   A ligand that exerts a strong crystal or ligand electrical field and generally forms low spin complexes with metal ions when possible.

Structural (constitutional) isomers   (Applied to coordination compounds.) Isomers whose differences involve more than a single coordination sphere or else different donor atoms; they include ionization isomers, hydrate isomers, coordination isomers, and linkage isomers.

t_{2g} orbitals   A set of d_{xy}, d_{yz}, and d_{xz} orbitals; those d orbitals within a set with lobes bisecting the x, y, and z axes.

Weak field ligand   A ligand that exerts a weak crystal or ligand field and generally forms high spin complexes with metals.

Exercises

Basic Concepts

1. What property of transition metals allows them to form coordination compounds easily?

2. Suggest more appropriate formulas for $NiSO_4 \cdot 6H_2O$, $Cu(NO_3)_2 \cdot 4NH_3$, and $Ni(NO_3)_2 \cdot 6NH_3$.

3. What are the two constituents of a complex? What type of chemical bonding occurs between these constituents?

4. Define the term "coordination number" for the central atom or ion in a complex. What values of the coordination numbers for metal ions are most common?

5. Describe the experiments of Alfred Werner on the compounds of the general formula $PtCl_4 \cdot nNH_3$ where $n = 2, 3, 4, 5, 6$. What was his interpretation of these experiments?

6. For each of the compounds of Exercise 5, write formulas indicating the species within the coordination sphere. Also indicate the charges on the complex ions.

7. Distinguish among the terms ligands, donor atoms, and chelates.

8. Identify the ligands and give the coordination number and the oxidation number for the central atom or ion in each of the following: (a) $[Co(NH_3)_2(NO_2)_4]^-$; (b) $[Cr(NH_3)_5Cl]Cl_2$; (c) $K_4[Fe(CN)_6]$; (d) $[Pd(NH_3)_4]^{2+}$.

9. Repeat Exercise 8 for (a) $Na[Au(CN)_2]$; (b) $[Ag(NH_3)_2]^+$; (c) $[Pt(NH_3)_2Cl_4]$; (d) $[Co(en)_3]^{3+}$

10. What is the term given to the phenomenon of ring formation by a ligand in a complex? Describe a specific example.

11. Write a structural formula showing the ring(s) formed by a bidentate ligand such as ethylenediamine with a metal ion such as Fe^{3+}. How many atoms are in each ring? The formula for this complex ion is $[Fe(en)_3]^{3+}$.

Ammine Complexes

12. Which of the following insoluble metal hydroxides will dissolve in an excess of aqueous ammonia? (a) $Zn(OH)_2$; (b) $Cr(OH)_3$; (c) $Fe(OH)_2$; (d) $Ni(OH)_2$; (e) $Cd(OH)_2$.

13. Write net ionic equations for the reactions in Exercise 12.

14. Write net ionic equations for reactions of solutions of the following transition metal salts in water with a *limited amount* of aqueous ammonia: (It is not necessary to show the ions as hydrated.) (a) $CuCl_2$; (b) $Zn(NO_3)_2$; (c) $Fe(NO_3)_3$.

15. Write *net ionic* equations for the reactions of the insoluble products of Exercise 14 with an *excess* of aqueous ammonia, if a reaction occurs.

Naming Coordination Compounds

16. Give systematic names for the following compounds.
    (a) $[Ni(CO)_4]$
    (b) $Na_2[Co(OH_2)_2(OH)_4]$
    (c) $[Ag(NH_3)_2]Br$
    (d) $[Cr(en)_3](NO_3)_3$
    (e) $[Pt(NH_3)_4(NO_2)_2]F_2$
    (f) $K_2[Cu(CN)_4]$

17. Name the following substances.
    (a) $Na[Au(CN)_2]$
    (b) $[Pt(NH_3)_4]Cl_2$
    (c) $[CoCl_6]^{3-}$
    (d) $[Co(H_2O)_6]^{3+}$
    (e) $Na_2[Pt(CN)_4]$
    (f) $K[Cr(NH_3)_2(OH)_2Cl_2]$
    (g) $[Ni(NH_3)_4(H_2O)_2](NO_3)_2$
    (h) $Na[Al(H_2O)_2(OH)_4]$
    (i) $[Co(NH_3)_4Cl_2][Cr(C_2O_4)_2]$

18. Write formulas for the following.
    (a) diamminedichlorozinc
    (b) tin(IV) hexacyanoferrate(II)
    (c) tetracyanoplatinate(II) ion
    (d) potassium hexacyanochromate(III)
    (e) tetraammineplatinum(II) ion
    (f) hexaamminenickel(II) bromide
    (g) tetraamminecopper(II) pentacyanohydroxoferrate(III)

19. Write formulas for the following compounds.
    (a) *trans*-diamminedinitroplatinum(II)
    (b) rubidium tetracyanozincate
    (c) triaqua-*cis*-dibromochlorochromium(III)
    (d) pentacarbonyliron(0)
    (e) sodium pentacyanocobaltate(II)
    (f) hexammineruthenium(III) tetrachloronickelate(II)

Structures of Coordination Compounds

20. Write formulas and provide names for three complex cations in each of the following categories.
    (a) cations coordinated to only unidentate ligands
    (b) cations coordinated to only bidentate ligands
    (c) cations coordinated to two bidentate and two unidentate ligands
    (d) cations coordinated to one tridentate ligand, one bidentate ligand, and one unidentate ligand
    (e) cations coordinated to one tridentate ligand and three unidentate ligands

21. Provide formulas and names for three complex anions that fit each description in Exercise 20.

22. How many geometric isomers can be formed by complexes that are (a) octahedral MA_2B_4 and (b) octahedral MA_3B_3? Name any geometric isomers that can exist. Is it possible for any of these isomers to show optical activity (exist as enantiomers)? Explain.

23. Write the structural formulas for (a) two isomers of $[Pt(NH_3)_2Cl_2]$, (b) four isomers (including linkage isomers) of $[Co(NH_3)_3(NO_2)_3]$, and (c) two isomers (including ionization isomers) of $[Pt(NH_3)_3Br]Cl$.

24. Determine the number and types of isomers that would be possible for each of the following complexes.
    (a) tetraamminediaquachromium(III) ion
    (b) triamminetriaquachromium(III) ion
    (c) tris(ethylenediamine) chromium(III) ion
    (d) dichlorobis(ethylenediamine)platinum(IV) chloride
    (e) diamminedibromodichlorochromate(III) ion

*25. Indicate whether the complexes in each pair are identical or isomers.

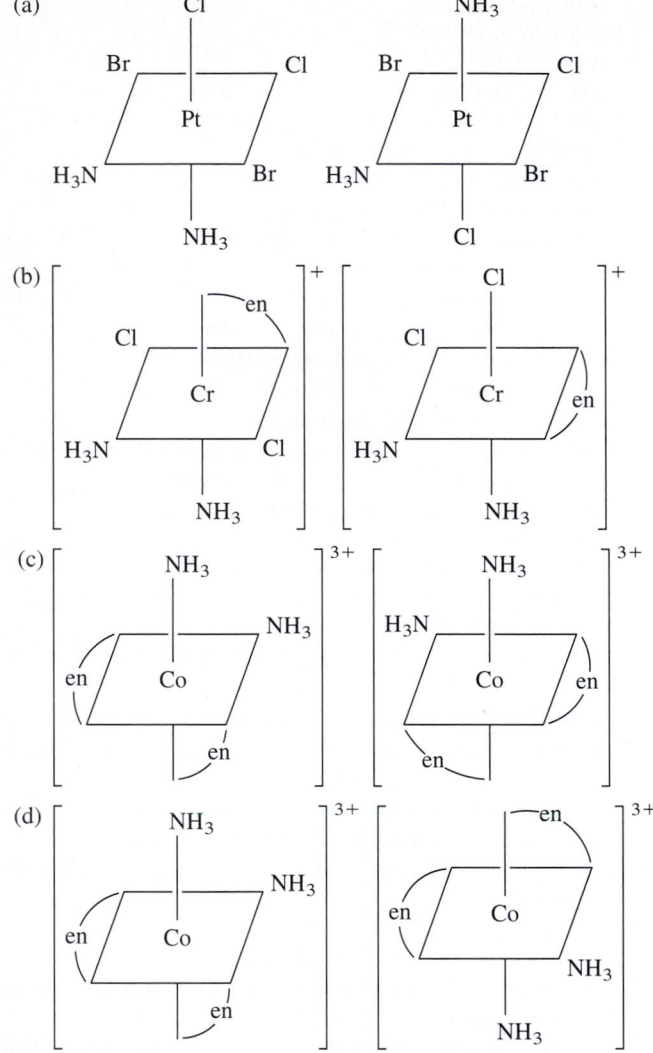

(a)

(b)

(c)

(d)

26. Distinguish between structural isomers and stereoisomers.
27. Distinguish between an optically active complex and a racemic mixture.
28. Write the formula for a potential ionization isomer of each of the following compounds. Name each one.
   (a) $[Cr(NH_3)_4I_2]Br$
   (b) $[Ni(en)_2(NO_2)_2]Cl_2$
   (c) $[Fe(NH_3)_5CN]SO_4$
29. Write the formula for a potential hydrate isomer of each of the following compounds. Name each one.
   (a) $[Cu(OH_2)_4]Cl_2$ (b) $[Ni(OH_2)_5Br]Br \cdot H_2O$
30. Write the formula for a potential coordination isomer of each of the following compounds. Name each one.
   (a) $[Co(NH_3)_6][Cr(CN)_6]$ (b) $[Ni(en)_3][Cu(CN)_4]$
31. Write the formula for a potential linkage isomer of each of the

following compounds. Name each one.
   (a) $[Co(en)_2(NO_2)_2]Cl$ (b) $[Cr(NH_3)_5(CN)](CN)_2$

Valence Bond Theory
32. Describe the hybridization at the metal ion and sketch the structure of the complex part of each of the following. It is not necessary to distinguish between d^2sp^3 and sp^3d^2 hybridization.
   (a) $[Ag(NH_3)_2]Cl$
   (b) $[Fe(en)_3]PO_4$
   (c) $[Co(NH_3)_6]SO_4$
   (d) $[Co(NH_3)_6]_2(SO_4)_3$
   (e) $[Pt(NH_3)_4]Cl_2$
   (f) $(NH_4)_2[PtCl_4]$
33. Repeat Exercise 32 for the following.
   (a) $K_2[PdCl_6]$
   (b) $(NH_4)_2[PtCl_6]$
   (c) $[Co(en)_3]Cl_2$
   (d) $[Co(en)_3]Cl_3$
   (e) $[Cr(en)_2(NH_3)_2](NO_3)_2$
   (f) $[Co(NH_3)_4Cl_2]Cl$
34. On the basis of the spectrochemical series, determine whether each of the following complexes is inner orbital or outer orbital, and diamagnetic or paramagnetic.
   (a) $[Cu(OH_2)_6]^{2+}$
   (b) $[MnF_6]^{3-}$
   (c) $[Co(CN)_6]^{3-}$
   (d) $[Cr(NH_3)_6]^{3+}$
35. On the basis of the spectrochemical series, determine whether each of the following complexes is inner orbital or outer orbital, and diamagnetic or paramagnetic.
   (a) $[CrCl_4Br_2]^{3-}$
   (b) $[Co(en)_3]^{3+}$
   (c) $[Fe(OH_2)_6]^{3+}$
   (d) $[Fe(NO_2)_6]^{3-}$
36. Draw diagrams showing outer electron configurations and hybridizations at the metal ions for each of the complexes in Exercise 35.
37. Using the valence bond theory, describe the bonding in a Co^{3+} octahedral complex with ligands that can (a) occupy only the outer orbitals on the Co^{3+} and (b) occupy inner orbitals on the Co^{3+}. What types of hybridization of the atomic orbitals on Co^{3+} are proposed for each?
38. Predict the number of unpaired electrons in each of the following.
   (a) $[Fe(CN)_6]^{3-}$; (b) $[Fe(OH_2)_6]^{3+}$; (c) $[Mn(OH_2)_6]^{2+}$; (d) $[Co(NH_3)_6]^{3+}$
39. Consider the compound having the formula $[Co(NH_3)_5(H_2O)]^{3+}[Co(NO_2)_6]^{3-}$. In terms of valence bond theory, describe the bonding in each ion. Would you expect this substance to be paramagnetic or diamagnetic?

Crystal Field Theory
40. Describe clearly what Δ_{oct} is. How is Δ_{oct} actually measured experimentally? How is it related to the spectrochemical series?
41. On the basis of the spectrochemical series, determine whether the complexes of Exercise 35 are low spin or high spin.

42. Repeat Exercise 41 for the complexes of Exercise 34.

43. Write out the electron distribution in t_{2g} and e_g orbitals for the following in an octahedral field.

Metal Ions	Ligand Field Strength
V^{2+}	weak
Mn^{2+}	strong
Mn^{2+}	weak
Ni^{2+}	weak
Cu^{2+}	weak
Fe^{3+}	strong
Cu^+	weak
Ru^{3+}	strong

44. Write formulas for two complex ions that would fit into each of the categories of Exercise 43. Name the complex ions you list.

***45.** Describe the relationship among Δ_{oct}, the electron pairing energy, and whether or not a complex is high spin or low spin. Illustrate the relationship with Fe^{2+} in strong and weak octahedral fields.

***46.** What is crystal field stabilization energy? Given the following spectrophotometrically measured values of Δ_{oct}, calculate the CFSEs for the ions. Remember to determine first whether the complex is high spin or low spin.

Complex Ion	Δ_{oct}*
(a) $[Co(NH_3)_6]^{3+}$	22,900 cm^{-1}
(b) $[Ti(OH_2)_6]^{2+}$	20,300 cm^{-1}
(c) $[Cr(OH_2)_6]^{3+}$	17,600 cm^{-1}
(d) $[Co(CN)_6]^{3-}$	33,500 cm^{-1}
(e) $[Co(OH_2)_6]^{2+}$	10,000 cm^{-1}
(f) $[Cr(en)_3]^{3+}$	21,900 cm^{-1}
(g) $[Cu(OH_2)_6]^{2+}$	13,000 cm^{-1}
(h) $[V(OH_2)_6]^{3+}$	18,000 cm^{-1}

*The cm^{-1} is an energy unit; 1 cm^{-1} = 11.96 J/mol.

47. Determine the electron distribution in (a) $[Co(CN)_6]^{3-}$, a low spin complex ion, and (b) $[CoF_6]^{3-}$, a high spin complex ion. Express the crystal field stabilization energy for each complex.

48. Determine the crystal field stabilization energy for (a) $[Mn(H_2O)_6]^{3+}$, a high spin complex; (b) $[Mn(CN)_6]^{3-}$, a low spin complex.

BUILDING YOUR KNOWLEDGE

K_d values are listed in Appendix I.

49. The yellow complex oxalatorhodium compound $K_3[Rh(C_2O_4)_3]$ can be prepared from the wine-red complex compound $K_3[RhCl_6]$ by boiling a concentrated aqueous solution of $K_3[RhCl_6]$ and $K_2C_2O_4$ for two hours and then evaporating the solution until the product crystallizes.

$$K_3[RhCl_6](aq) + 3K_2C_2O_4(aq) \xrightarrow{\Delta}$$
$$K_3[Rh(C_2O_4)_3](s) + 6KCl(aq)$$

What is the theoretical yield of the oxalato complex if 1.00 g of the chloro complex is heated with 4.95 g of $K_2C_2O_4$? In an experiment, the actual yield was 0.88 g. What is the percent yield?

50. Consider the formation of the triiodoargentate(I) ion.

$$Ag^+ + 3I^- \longrightarrow [AgI_3]^{2-}$$

Would you expect an increase or decrease in the entropy of the system as the complex is formed? The standard state absolute entropy at 25°C is 72.68 J/K mol for Ag^+, 111.3 J/K mol for I^-, and 253.1 J/K mol for $[AgI_3]^{2-}$. Calculate ΔS^0 for the reaction and confirm your prediction.

51. Molecular iodine reacts with I^- to form a complex ion.

$$I_2(aq) + I^- \rightleftharpoons [I_3]^-$$

Calculate the equilibrium constant for this reaction given the following data at 25°C.

$$I_2(aq) + 2e^- \longrightarrow 2I^- \qquad E^0 = 0.535 \text{ V}$$
$$[I_3]^- + 2e^- \longrightarrow 3I^- \qquad E^0 = 0.5338 \text{ V}$$

52. Calculate the pH of a solution prepared by dissolving 0.25 mol of tetraamminecopper(II) chloride, $[Cu(NH_3)_4]Cl_2$, in water to give 1.0 L of solution. Ignore hydrolysis of Cu^{2+}.

53. Use the following standard reduction potential data to answer the questions.

$$Co^{3+} + e^- \rightleftharpoons Co^{2+} \qquad E^0 = 1.808 \text{ V}$$
$$Co(OH)_3(s) + e^- \rightleftharpoons Co(OH)_2(s) + OH^-(aq) \qquad E^0 = 0.17 \text{ V}$$
$$[Co(NH_3)_6]^{3+} + e^- \rightleftharpoons [Co(NH_3)_6]^{2+} \qquad E^0 = 0.108 \text{ V}$$
$$[Co(CN)_6]^{3-} + e^- \rightleftharpoons [Co(CN)_5]^{3-} + CN^- \qquad E^0 = -0.83 \text{ V}$$
$$O_2(g) + 4H^+(10^{-7} M) + 4e^- \rightleftharpoons 2H_2O(\ell) \qquad E^0 = 0.815 \text{ V}$$
$$2H_2O(\ell) + 2e^- \rightleftharpoons H_2(g) + 2OH^- \qquad E^0 = -0.828 \text{ V}$$

Which cobalt(III) species among those listed would oxidize water? Which cobalt(II) species among those listed would be oxidized by water? Explain your answers.

54. Calculate (a) the molar solubility of $Zn(OH)_2$ in pure water, (b) the molar solubility of $Zn(OH)_2$ in 0.30 M NaOH solution, and (c) the concentration of $[Zn(OH)_4]^{2-}$ ions in the solution of (b).

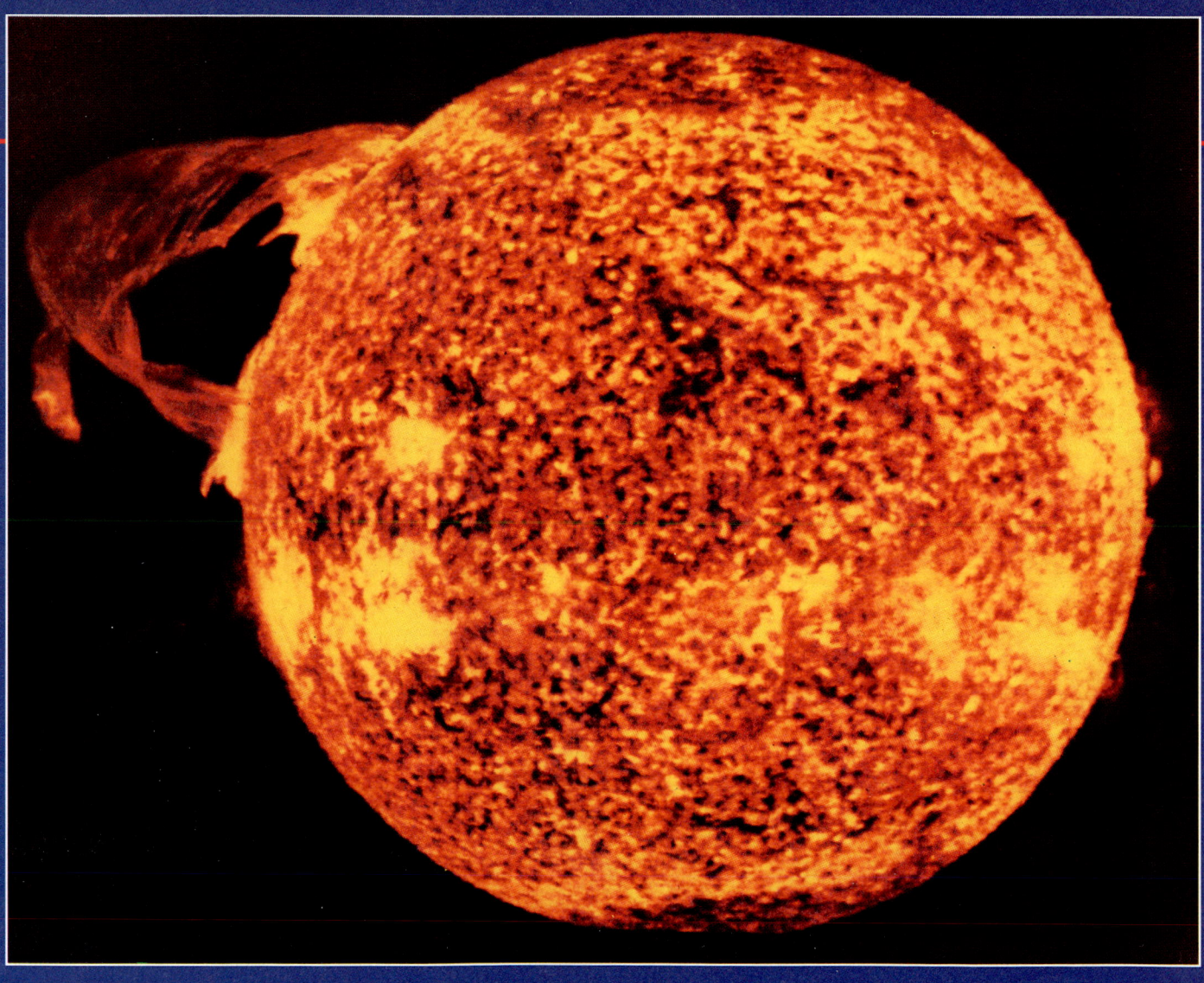

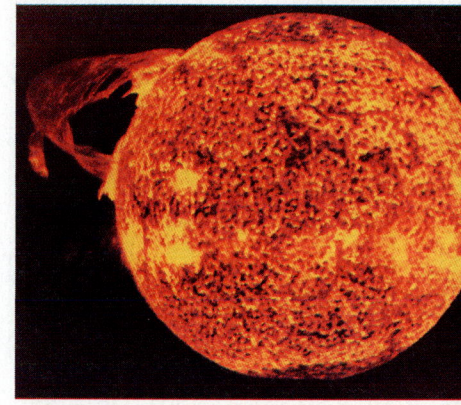

Our sun, like other stars, is a giant nuclear fusion reactor. It supplies energy to the earth from a distance of 93,000,000 miles.

OBJECTIVES

As you study this chapter, you should learn

• *About the makeup of the nucleus*

• *About relationships between neutron–proton ratio and nuclear stability*

• *About the band of stability*

• *To calculate mass deficiency and nuclear binding energy*

• *About common types of radiation emitted when nuclei undergo radioactive decay*

• *To write and balance equations that describe nuclear reactions*

• *About different kinds of nuclear reactions undergone by nuclei, depending on their positions relative to the band of stability*

• *About methods for detecting radiation*

• *To understand half-lives of radioactive elements*

• *To carry out the calculations associated with radioactive decay*

• *About disintegration series*

• *About some uses of radionuclides, including the use of radioactive elements for dating objects*

• *About nuclear reactions that are induced by bombardment of nuclei with particles*

• *About nuclear fission and some of its applications, including the atomic bomb and nuclear reactors*

• *About nuclear fusion and some prospects for and barriers to its use for production of energy*

Marie Sklodowska Curie (1867–1934) is the only person to have been honored with Nobel Prizes in both physics and chemistry. In 1903 Pierre and Marie Curie and Henri Becquerel shared the prize in physics for the discovery of natural radioactivity. Marie Curie also received the 1911 Nobel Prize in chemistry for her discovery of radium and polonium and the compounds of radium. She named polonium for her native Poland. Marie's daughter, Irene Joliot-Curie, and *her* husband, Frederic Joliot, received the 1935 Nobel Prize in chemistry for the first synthesis of a new radioactive element.

In Chapter 5, we represented an atom of a particular isotope by its *nuclide symbol.* Radioisotopes are often called **radionuclides.**

C hemical properties are determined by electron distributions and are only indirectly influenced by atomic nuclei. Up to now, we have discussed ordinary chemical reactions, so we have focused attention on electron configurations. Nuclear reactions involve changes in the composition of nuclei. These extraordinary processes are often accompanied by the release of tremendous amounts of energy and by transmutations of elements. Some differences between nuclear reactions and ordinary chemical reactions follow.

Nuclear Reaction	Ordinary Chemical Reaction
1. Elements may be converted from one to another.	1. No new elements can be produced.
2. Particles within the nucleus are involved.	2. Usually only outermost electrons participate.
3. Tremendous amounts of energy are released or absorbed.	3. Relatively small amounts of energy are released or absorbed.
4. Rate of reaction is not influenced by external factors.	4. Rate of reaction depends on factors such as concentration, temperature, catalyst, and pressure.

Medieval alchemists spent years trying to convert other metals into gold without success. Years of failure and the acceptance of Dalton's atomic theory early in the 19th century convinced scientists that elements are not interconvertible. Then, in 1896 Henri Becquerel discovered "radioactive rays" (**natural radioactivity**) coming from a uranium compound. Ernest Rutherford's study of these rays showed that atoms of one element may indeed be converted into atoms of other elements by spontaneous nuclear disintegrations. Many years later it was shown that nuclear reactions initiated by bombardment of nuclei with accelerated subatomic particles or other nuclei can also transform one element into another—accompanied by the release of radiation (**induced radioactivity**).

Becquerel's discovery led other researchers, including Marie and Pierre Curie, to discover and study new radioactive elements. Many radioactive isotopes, or **radioisotopes,** now have important medical, agricultural, and industrial uses.

Nuclear fission is the splitting of a heavy nucleus into lighter nuclei. **Nuclear fusion** is the combination of light nuclei to produce a heavier nucleus. Huge amounts of energy are released when these processes occur. These processes could satisfy a large portion of our future energy demands. Current research is aimed at surmounting the technological problems associated with safe and efficient use of nuclear fission reactors and with the development of controlled fusion reactors.

26-1 THE NUCLEUS

In Chapter 5 we described the principal subatomic particles (Table 26-1). Recall that the neutrons and protons together constitute the nucleus, with the electrons occupying essentially empty space around the nucleus. The nucleus is only a minute fraction of the total volume of an atom, yet nearly all the mass of an atom resides in the nucleus. Thus, nuclei are extremely dense. It has been shown experimentally that nuclei of all elements have approximately the same density, 2.4×10^{14} g/cm^3.

From an electrostatic point of view, it is amazing that positively charged protons (and uncharged neutrons) can be packed so closely together. Yet nonradioactive nuclei do not spontaneously decompose, so they must be stable. In the early 20th century when Rutherford postulated the nuclear model of the atom, scientists were puzzled by such a situation.

If enough nuclei could be gathered together to occupy one cubic centimeter, the total weight would be about 250 million tons!

Table 26-1 *Fundamental Particles of Matter*

Particle	Mass	Charge
Electron (e^-)	0.00054858 amu	1−
Proton (p or p^+)	1.0073 amu	1+
Neutron (n or n^0)	1.0087 amu	none

Physicists have since detected many very short-lived subatomic particles (in addition to protons, neutrons, and electrons) as products of nuclear reactions. Well over a hundred have been identified. Their functions are not entirely understood, but it is now thought that they help to overcome the proton–proton repulsions and to bind nuclear particles (**nucleons**) together. The attractive forces among nucleons appear to be important over only extremely small distances, about 10^{-13} cm.

26-2 NEUTRON–PROTON RATIO AND NUCLEAR STABILITY

The term **"nuclide"** is used to refer to different atomic forms of all elements. The term "isotope" applies only to different forms of the same element. Most naturally occurring nuclides have even numbers of protons and even numbers of neutrons; 157 nuclides fall into this category. Nuclides with odd numbers of both are least common (there are only four), and those with odd–even combinations are intermediate in abundance (Table 26-2). Furthermore, nuclides with certain "magic numbers" of protons and neutrons are especially stable. Nuclides with a number of protons *or* a number of neutrons *or* a sum of the two equal to 2, 8, 20, 28, 50, 82, or 126 have unusual stability. Examples are $^4_2$He, $^{16}_8$O, $^{42}_{20}$Ca, $^{88}_{38}$Sr, and $^{208}_{82}$Pb. This suggests an energy level (shell) model for the nucleus similar to the shell model of electron configurations.

Figure 26-1 is a plot of the number of neutrons (N) versus number of protons (Z) for the stable nuclides (the **band of stability**). For low atomic numbers, the most stable nuclides have equal numbers of protons and neutrons ($N = Z$). Above atomic number 20, the most stable nuclides have more neutrons than protons. Careful examination reveals an approximately stepwise shape to the plot, due to the stability of nuclides with even numbers of nucleons.

The nuclide symbol for an element (Section 5-7) is

$$^A_Z E$$

where E is the chemical symbol for the element, Z is its atomic number, and A is its mass number.

26-3 NUCLEAR STABILITY AND BINDING ENERGY

Experimentally, we observe that the mass of an atom is always *less* than the sum of the masses of its constituent particles. We now know why this *mass deficiency* occurs. We also know that the mass deficiency is in the nucleus of the atom and has nothing to do with

Table 26-2 *Abundance of Naturally Occurring Nuclides*

Number of protons	even	even	odd	odd
Number of neutrons	even	odd	even	odd
Number of such nuclides	157	52	50	4

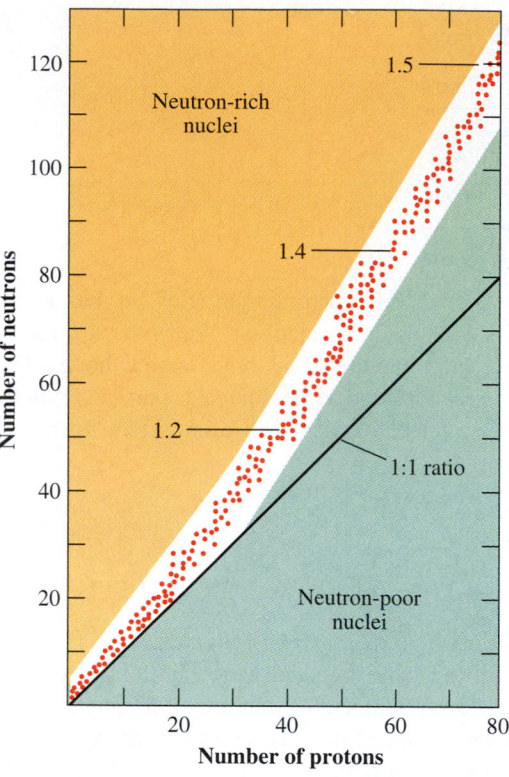

Figure 26-1 A plot of the number of neutrons versus the number of protons in stable nuclei. As atomic number increases, the *N/Z* ratio (the decimal fractions) of the stable nuclei increases. The stable nuclei are located in an area known as the band of stability. Most radioactive nuclei occur outside this band.

the electrons. However, *because tables of masses of isotopes include the electrons, we shall also include them.*

The **mass deficiency, Δm,** for a nucleus is the difference between the sum of the masses of electrons, protons, and neutrons in the atom (calculated mass) and the actual measured mass of the atom.

Do you remember how to find the numbers of protons, neutrons, and electrons in a specified atom? Review Section 5-7.

$$\Delta m = (\text{sum of masses of all } e^-, p^+, \text{ and } n^0) - (\text{actual mass of atom})$$

For most naturally occurring isotopes, the mass deficiency is only about 0.15% or less of the calculated mass of an atom.

EXAMPLE 26-1 *Mass Deficiency*

Calculate the mass deficiency for chlorine-35 atoms in amu/atom and in g/mol atoms. The actual mass of a chlorine-35 atom is 34.9689 amu.

Plan

We first find the numbers of protons, electrons, and neutrons in one atom. Then we determine the "calculated" mass as the sum of the masses of these particles. The mass deficiency is the actual mass subtracted from the calculated mass. This deficiency is commonly expressed either as mass per atom or as mass per mole of atoms.

Solution

Each atom of $^{35}_{17}\text{Cl}$ contains 17 protons, 17 electrons, and $(35 - 17) = 18$ neutrons. First we sum the masses of these particles.

protons:	17×1.0073 amu	$= 17.124$ amu	(masses from Table 26-1)
electrons:	17×0.00054858 amu $=$	0.0093 amu	
neutrons:	18×1.0087 amu	$= 18.157$ amu	

$$\text{sum} = 35.290 \text{ amu} \longleftarrow \text{calculated mass}$$

Then we subtract the actual mass from the "calculated" mass to obtain Δm.

$$\Delta m = 35.290 \text{ amu} - 34.9689 \text{ amu} = \boxed{0.321 \text{ amu}} \qquad \text{mass deficiency (in one atom)}$$

We have calculated the mass deficiency in amu/atom. Recall (Section 5-9) that 1 gram is 6.022×10^{23} amu. We can show that a number expressed in amu/atom is equal to the same number in g/mol of atoms.

$$\frac{?\ \text{g}}{\text{mol}} = \frac{0.321 \text{ amu}}{\text{atom}} \times \frac{1 \text{ g}}{6.022 \times 10^{23} \text{ amu}} \times \frac{6.022 \times 10^{23} \text{ atoms}}{1 \text{ mol } ^{35}\text{Cl atoms}}$$

$$= \boxed{0.321 \text{ g/mol of } ^{35}\text{Cl atoms}} \longleftarrow \text{(mass } \textit{deficiency} \text{ in a mole of Cl atoms)}$$

You should now work Exercises 8a and 10a–b.

What has happened to the mass represented by the mass deficiency? In 1905 Einstein set forth the Theory of Relativity. He stated that matter and energy are equivalent. An obvious corollary is that matter can be transformed into energy and energy into matter. The transformation of matter into energy occurs in the sun and other stars. It happened on Earth when controlled nuclear fission was achieved in 1939 (Section 26-13). The reverse transformation, energy into matter, has not yet been accomplished on a large scale. Einstein's equation, which we encountered in Chapter 1, is $E = mc^2$. E represents the amount of energy released, m the mass of matter transformed into energy, and c the speed of light in a vacuum, 2.997925×10^8 m/s (usually rounded off to 3.00×10^8 m/s).

A mass deficiency represents the amount of matter that would be converted into energy and released if the nucleus were formed from initially separate protons and neutrons. This energy is the **nuclear binding energy, BE.** Thus, we could rewrite the Einstein relationship as

$$BE = (\Delta m)c^2$$

Specifically, if 1 mole of ^{35}Cl nuclei were to be formed from 17 moles of protons and 18 moles of neutrons, the resulting mole of nuclei would weigh 0.321 gram less than the original collection of protons and neutrons (Example 26-1).

Stated differently, the nuclear binding energy for ^{35}Cl is the amount of energy that would be required to separate 1 mole of ^{35}Cl nuclei into 17 moles of protons and 18 moles of neutrons. This has never been done.

Nuclear binding energies may be expressed in many different units, including kilojoules/mole of atoms, kilojoules/gram of atoms, and megaelectron volts/nucleon. Some useful equivalences are

$$1 \text{ megaelectron volt (MeV)} = 1.60 \times 10^{-13} \text{ J} \quad \text{and} \quad 1 \text{ joule (J)} = 1 \text{ kg} \cdot \text{m}^2/\text{s}^2$$

Let's use the value of Δm for ^{35}Cl atoms to calculate their nuclear binding energy.

EXAMPLE 26-2   *Nuclear Binding Energy*

Calculate the nuclear binding energy of ^{35}Cl in (a) kilojoules per mole of Cl atoms, (b) kilojoules per gram of Cl atoms, and (c) megaelectron volts per nucleon.

Plan

The mass deficiency that we calculated in Example 26-1 is related to the binding energy by the Einstein equation.

Solution

The mass deficiency is 0.321 g/mol = 3.21×10^{-4} kg/mol.

(a) $BE = (\Delta m)c^2 = \dfrac{3.21 \times 10^{-4}\ \text{kg}}{\text{mol }^{35}\text{Cl atoms}} \times (3.00 \times 10^8\ \text{m/s})^2 = 2.89 \times 10^{13}\ \dfrac{\text{kg} \cdot \text{m}^2/\text{s}^2}{\text{mol }^{35}\text{Cl atoms}}$

$= 2.89 \times 10^{13}$ J/mol ^{35}Cl atoms = $\boxed{2.89 \times 10^{10} \text{ kJ/mol of }^{35}\text{Cl atoms}}$

(b) From Example 26-1, the actual mass of ^{35}Cl is

$$\dfrac{34.9689\ \text{amu}}{^{35}\text{Cl atom}} \qquad \text{or} \qquad \dfrac{34.9689\ \text{g}}{\text{mol }^{35}\text{Cl atoms}}$$

We use this mass to set up the needed conversion factor.

$$BE = \dfrac{2.89 \times 10^{10}\ \text{kJ}}{\text{mol of }^{35}\text{Cl atoms}} \times \dfrac{1\ \text{mol }^{35}\text{Cl atoms}}{34.9689\ \text{g }^{35}\text{Cl atoms}} = \boxed{8.26 \times 10^8 \text{ kJ/g }^{35}\text{Cl atoms}}$$

The mass number, Z, is equal to the number of nucleons in one atom.

(c) The number of nucleons in *one* atom of ^{35}Cl is 17 protons + 18 neutrons = 35 nucleons.

$$BE = \dfrac{2.89 \times 10^{10}\ \text{kJ}}{\text{mol of }^{35}\text{Cl atoms}} \times \dfrac{1000\ \text{J}}{\text{kJ}} \times \dfrac{1\ \text{MeV}}{1.60 \times 10^{-13}\ \text{J}} \times \dfrac{1\ \text{mol }^{35}\text{Cl atoms}}{6.022 \times 10^{23}\ {^{35}\text{Cl atoms}}}$$

$$\times \dfrac{1\ {^{35}\text{Cl atom}}}{35\ \text{nucleons}} = \boxed{8.57 \text{ MeV/nucleon}}$$

You should now work Exercises 8b and 10c–e.

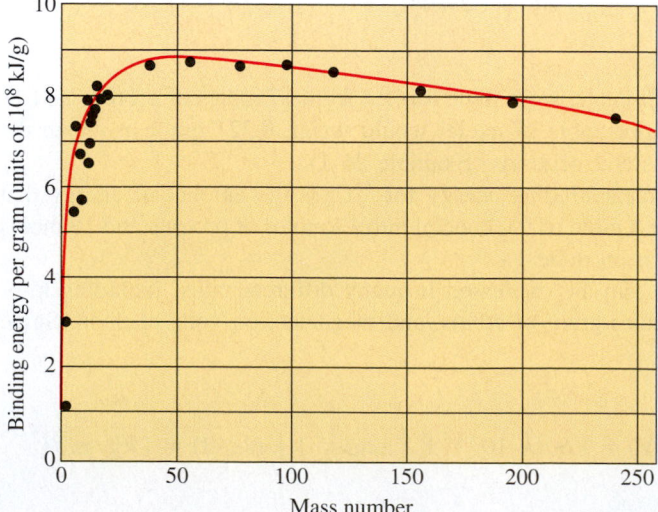

Figure 26-2   Plot of binding energy per gram versus mass number. Very light and very heavy nuclei are relatively unstable.

The nuclear binding energy of a mole of ^{35}Cl nuclei, 2.89×10^{13} J/mol, is an enormous amount of energy—enough to heat 6.9×10^7 kg (~76,000 tons) of water from 0°C to 100°C!

Figure 26-2 is a plot of average binding energy per gram of nuclei versus mass number. It shows that nuclear binding energies (per gram) increase rapidly with increasing mass number, reach a maximum around mass number 50, and then decrease slowly. The nuclei with the highest binding energies (mass numbers 40 to 150) are the most stable. Large amounts of energy would be required to separate these nuclei into their component neutrons and protons. Even though these nuclei are the most stable ones, *all* nuclei are stable with respect to complete decomposition into protons and neutrons because all (except ^1H) nuclei have mass deficiencies. In other words, the energy equivalent of the loss of mass represents an associative force that is present in all nuclei except ^1H. It must be overcome to separate the nuclei completely into their subatomic particles.

Some unstable radioactive nuclei do emit a single proton, a single neutron, or other subatomic particles as they decay in the direction of greater stability. None decomposes entirely into elementary particles.

26-4 RADIOACTIVE DECAY

Nuclei whose neutron-to-proton ratios lie outside the stable region undergo spontaneous radioactive decay by emitting one or more particles and/or electromagnetic rays. The type of decay that occurs usually depends on whether the nucleus is above, below, or to the right of the band of stability (Figure 26-1). Common types of radiation emitted in decay processes are summarized in Table 26-3.

The particles can be emitted at different kinetic energies. These are equal to the energy equivalent of the mass loss of the products relative to reactants (Section 26-3) minus the energy associated with subsequently emitted gamma rays (electromagnetic radiation). Radioactive decay often leaves a nucleus in an excited (high energy) state. Then the decay is followed by gamma ray emission. The energy of the gamma ray ($h\nu$) is equal to the energy difference between the ground and excited nuclear states. This is like the emission of lower energy electromagnetic radiation that occurs as an atom in its excited electronic

Recall that the energy of electromagnetic radiation is $E = h\nu$, where h is Planck's constant and ν is the frequency.

Table 26-3 *Common Types of Radioactive Emissions*

Type and Symbol[a]	Identity	Mass (amu)	Charge	Velocity	Penetration
beta (β^-, $_{-1}^{0}\beta$, $_{-1}^{0}e$)	electron	0.00055	1$-$	≤90% speed of light	low to moderate, depending on energy
positron[b] ($_{+1}^{0}\beta$, $_{+1}^{0}e$)	positively charged electron	0.00055	1+	≤90% speed of light	low to moderate, depending on energy
alpha (α, $_{2}^{4}\alpha$, $_{2}^{4}He$)	helium nucleus	4.0026	2+	≤10% speed of light	low
proton ($_{1}^{1}p$, $_{1}^{1}H$)	proton, hydrogen nucleus	1.0073	1+	≤10% speed of light	low to moderate, depending on energy
neutron ($_{0}^{1}n$)	neutron	1.0087	0	≤10% speed of light	very high
gamma ($_{0}^{0}\gamma$) ray	high energy electromagnetic radiation such as X-rays	0	0	speed of light	high

[a] *The number at the upper left of the symbol is the number of nucleons, and the number at the lower left is the number of positive charges.*

[b] *On the average, a positron exists for only about a nanosecond (1×10^{-9} second) before colliding with an electron and being converted into the corresponding amount of energy.*

A technician cleans lead glass blocks that form part of the giant OPAL particle detector at CERN, the European center for particle physics near Geneva, Switzerland.

state returns to its ground state (Section 5-12). Studies of gamma ray energies strongly suggest that nuclear energy levels are quantized just as are electronic energy levels. This adds further support for a shell model for the nucleus.

$$(\text{excited nucleus}) \longrightarrow {}^{M}_{Z}E^* \longrightarrow {}^{M}_{Z}E + {}^{0}_{0}\gamma$$

The penetrating abilities of the particles and rays are proportional to their energies. Beta particles and positrons are about 100 times more penetrating than the heavier and slower-moving alpha particles. They can be stopped by a $\frac{1}{8}$-inch-thick (0.3 cm) aluminum plate. They can burn skin severely but cannot reach internal organs. Alpha particles have low penetrating ability and cannot damage or penetrate skin. However, they can damage sensitive internal tissue if inhaled. The high energy gamma rays have great penetrating power and severely damage both skin and internal organs. They travel at the speed of light and can be stopped by thick layers of concrete or lead.

Robotics technology is used to manipulate highly radioactive samples safely.

26-5 NEUTRON-RICH NUCLEI (ABOVE THE BAND OF STABILITY)

Nuclei in this region have too high a ratio of neutrons to protons. They undergo decays that *decrease* the ratio. The most common such decay is **beta emission.** A beta particle is an electron ejected *from the nucleus* when a neutron is converted into a proton.

$$\ _{0}^{1}n \longrightarrow \ _{1}^{1}p + \ _{-1}^{0}\beta$$

Beta emission results in a simultaneous increase by one in the number of protons and decrease by one in the number of neutrons. Examples of beta particle emission are

$$\ _{88}^{228}\text{Ra} \longrightarrow \ _{89}^{228}\text{Ac} + \ _{-1}^{0}\beta \quad \text{and} \quad \ _{6}^{14}\text{C} \longrightarrow \ _{7}^{14}\text{N} + \ _{-1}^{0}\beta$$

In the balanced equation for a nuclear reaction,

1. The sums of left-hand superscripts (mass numbers) must be the same on both sides, *and*

2. The sums of left-hand subscripts (atomic numbers) must be the same on both sides.

In all equations for nuclear reactions,

sum of mass numbers of reactants = sum of mass numbers of products
sum of atomic numbers of reactants = sum of atomic numbers of products

26-6 NEUTRON-POOR NUCLEI (BELOW THE BAND OF STABILITY)

Most of these nuclei, especially heavier ones, *increase* their neutron-to-proton ratios by undergoing **alpha emission.** Alpha particles are helium nuclei, $_{2}^{4}\text{He}$—two protons and two neutrons. Alpha emission also results in an increase of the neutron-to-proton ratio. An example is the alpha emission of lead-204.

$$\ _{82}^{204}\text{Pb} \longrightarrow \ _{80}^{200}\text{Hg} + \ _{2}^{4}\alpha$$

α-particles carry a double positive charge, but charge is usually not shown in nuclear reactions.

Two other, less common, types of decay for nuclei below the band of stability are **positron emission** or **electron capture** (*K* capture). Positron emission is most commonly encountered with artificially radioactive nuclei of the lighter elements. Electron capture occurs most often with heavier elements.

A positron has the mass of an electron but a positive charge. Positrons are emitted when protons are converted to neutrons.

$$\ _{1}^{1}p \longrightarrow \ _{0}^{1}n + \ _{+1}^{0}\beta$$

Thus, positron emission results in a *decrease* by one in atomic number and an *increase* by one in the number of neutrons, with *no change* in mass number.

$$\ _{19}^{38}\text{K} \longrightarrow \ _{18}^{38}\text{Ar} + \ _{+1}^{0}\beta \quad \text{and} \quad \ _{8}^{15}\text{O} \longrightarrow \ _{7}^{15}\text{N} + \ _{+1}^{0}\beta$$

The same effect can be accomplished by electron capture (*K* capture), in which an electron from the *K* shell ($n = 1$) is captured by the nucleus.

$$\ _{47}^{106}\text{Ag} + \ _{-1}^{0}e \longrightarrow \ _{46}^{106}\text{Pd} \quad \text{and} \quad \ _{18}^{37}\text{Ar} + \ _{-1}^{0}e \longrightarrow \ _{17}^{37}\text{Cl}$$

Electron capture by the nucleus differs from an atom gaining an electron to form an ion.

Some nuclides, e.g., $_{11}^{22}\text{Na}$, undergo both electron capture and positron emission.

$$\ _{11}^{22}\text{Na} + \ _{-1}^{0}e \longrightarrow \ _{10}^{22}\text{Ne} \ (3\%) \quad \text{and} \quad \ _{11}^{22}\text{Na} \longrightarrow \ _{10}^{22}\text{Ne} + \ _{+1}^{0}\beta \ (97\%)$$

26-7 NUCLEI WITH ATOMIC NUMBER GREATER THAN 83

The only stable nuclide with atomic number 83 is $^{209}_{83}Bi$.

All nuclides with atomic number greater than 83 are beyond the band of stability and are radioactive. Many of these decay by alpha emission.

$$^{226}_{88}Ra \longrightarrow {}^{222}_{86}Rn + {}^{4}_{2}\alpha \qquad \text{and} \qquad {}^{210}_{84}Po \longrightarrow {}^{206}_{82}Pb + {}^{4}_{2}\alpha$$

The decay of radium-226 was originally reported in 1902 by Rutherford and Soddy. It was the first transmutation of an element ever observed. A few heavy nuclides also decay by beta emission, positron emission, and electron capture.

Some isotopes of uranium ($Z = 92$) and elements of higher atomic number, the **transuranium elements,** also decay by nuclear fission. In this process a heavy nuclide splits into nuclides of intermediate mass and neutrons.

$$^{252}_{98}Cf \longrightarrow {}^{142}_{56}Ba + {}^{106}_{42}Mo + 4\,{}^{1}_{0}n$$

26-8 DETECTION OF RADIATIONS

Photographic Detection

Emanations from radioactive substances affect photographic plates just as ordinary visible light does. Becquerel's discovery of radioactivity resulted from the unexpected exposure of such a plate, wrapped in black paper, by a nearby enclosed sample of a uranium-containing compound, potassium uranyl sulfate. After a photographic plate has been developed and fixed, the intensity of the exposed spot is related to the amount of radiation that struck the plate. Quantitative detection of radiation by this method is difficult and tedious.

Detection by Fluorescence

Fluorescent substances can absorb high energy radiation such as gamma rays and subsequently emit visible light. As the radiation is absorbed, the absorbing atoms jump to excited electronic states. The excited electrons return to their ground states through a series of transitions, some of which emit visible light. This method may be used for the quantitative detection of radiation, using an instrument called a **scintillation counter.**

Dry Ice

Figure 26-3 A cloud chamber. The emitter is glued onto a pin stuck into a stopper that is mounted on the chamber wall. The chamber has some volatile liquid in the bottom and rests on dry ice. The cool air near the bottom becomes supersaturated with vapor. When an emission speeds through this vapor, ions are produced. These ions serve as "seeds" about which the vapor condenses, forming tiny droplets, or fog.

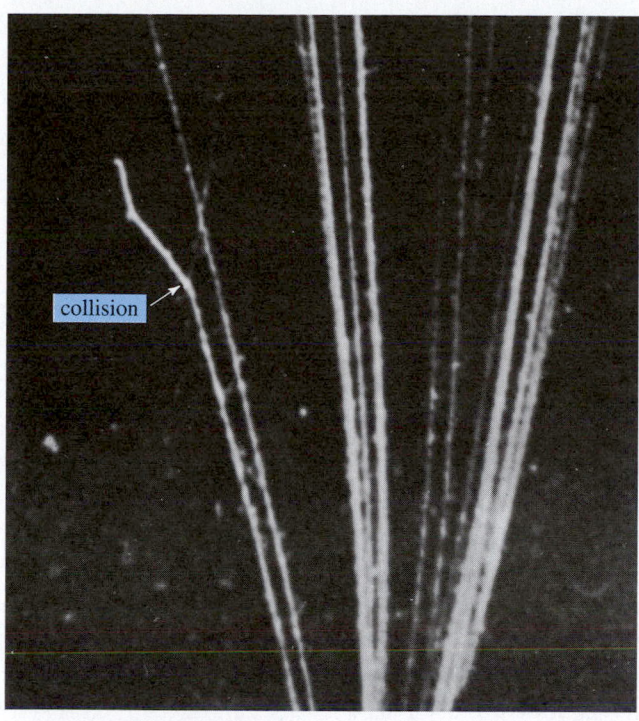

Figure 26-4 A historic cloud chamber photograph of alpha tracks in nitrogen gas. The forked track was shown to be due to a speeding proton (going off to the left) and an isotope of oxygen (going off to the right). It is assumed that the alpha particle struck the nucleus of a nitrogen atom at the point where the track forks.

Cloud Chambers

The original cloud chamber was devised by C. T. R. Wilson in 1911. A chamber contains air saturated with vapor. Particles emitted from a radioactive substance ionize air molecules in the chamber. Cooling the chamber causes droplets of liquid to condense on these ions. The paths of the particles can be followed by observing the fog-like tracks produced. The tracks may be photographed and studied in detail. Figures 26-3 and 26-4 show a cloud chamber and a cloud chamber photograph, respectively.

Gas Ionization Counters

A common gas ionization counter is the **Geiger–Müller counter** (Figure 26-5). Radiation enters the tube through a thin window. Windows of different stopping powers can be used to admit only radiation of certain penetrating powers.

The Geiger counter can detect only β and γ radiation. The α-particles cannot penetrate the walls of the tube.

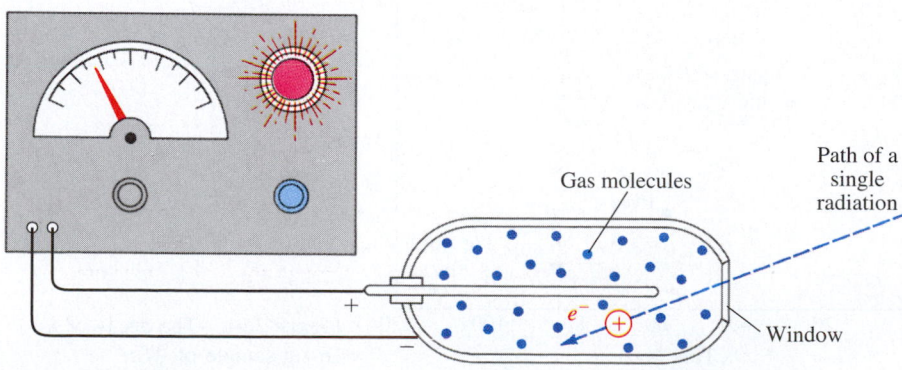

Gas molecules

Path of a single radiation

e^-

$+$

Window

Figure 26-5 The principle of operation of a gas ionization counter. The center wire is positively charged, and the shell of the tube is negatively charged. When radiation enters through the window, it ionizes one or more gas atoms. The electrons are attracted to the central wire, and the positive ions are drawn to the shell. This constitutes a pulse of electric current, which is amplified and displayed on the meter or other readout.

A sample of carnotite, a uranium ore, shown with a Geiger counter.

26-9 RATES OF DECAY AND HALF-LIFE

Radionuclides have different stabilities and decay at different rates. Some decay nearly completely in a fraction of a second and others only after millions of years. The rates of all radioactive decays are independent of temperature and obey *first-order kinetics*. In Section 16-3 we saw that the rate of a first-order process is proportional only to the concentration of one substance.

$$\text{rate of decay} = k[A]$$

The integrated rate equation for a first-order process (Section 16-4) is

$$\ln\left(\frac{A_0}{A}\right) = akt \qquad \text{or} \qquad \log\left(\frac{A_0}{A}\right) = \frac{akt}{2.303}$$

Here A represents the amount of decaying radionuclide of interest remaining at some time t, and A_0 is the amount present at the beginning of the observation. The k is the rate constant, which is different for each radionuclide. Each atom decays independently of the others, so the stoichiometric coefficient a is *always* 1 for radioactive decay. Therefore, we can drop it from the calculations in this chapter. If N represents the number of disintegrations per unit time, a similar relationship holds.

$$\ln\left(\frac{N_0}{N}\right) = kt \qquad \text{or} \qquad \log\left(\frac{N_0}{N}\right) = \frac{kt}{2.303}$$

The half-life, $t_{1/2}$, of a reaction is the amount of time required for half of the original sample to react. For a first-order process, $t_{1/2}$ is given by the equation

$$t_{1/2} = \frac{\ln 2}{k} = \frac{0.693}{k}$$

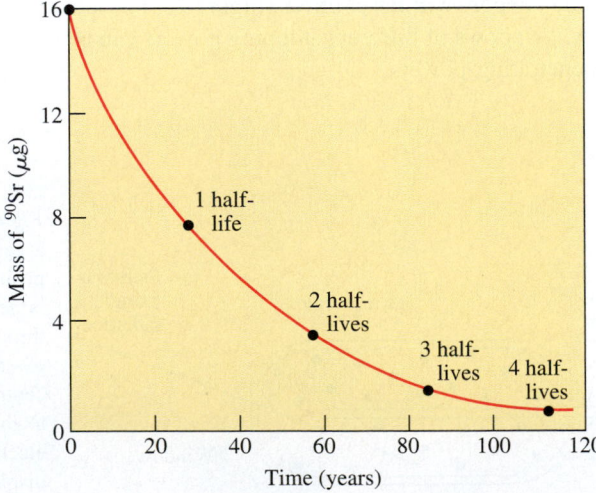

Figure 26-6 The decay of a 16-μg sample of $^{90}_{38}$Sr.

The isotope strontium-90 was introduced into the atmosphere by the atmospheric testing of nuclear weapons. Because of the chemical similarity of strontium to calcium, it now occurs with Ca in measurable quantities in milk, bones, and teeth as a result of its presence in food and water supplies. It is a radionuclide that undergoes beta emission with a half-life of 28 years. It may cause leukemia, bone cancer, and other, related disorders. If we begin with a 16-μg sample of $^{90}_{38}$Sr, 8 μg will remain after one half-life of 28 years. After 56 years, 4 μg will remain; after 84 years, 2 μg; and so on (Figure 26-6).

Similar plots for other radionuclides all show the same shape of **exponential decay curve.** About ten half-lives (280 years for $^{90}_{38}$Sr) must pass for radionuclides to lose 99.9% of their radioactivity.

In 1963 a treaty was signed by the United States, the Soviet Union, and the United Kingdom prohibiting the further testing of nuclear weapons in the atmosphere. Since then, strontium-90 has been disappearing from the air, water, and soil according to the curve in Figure 26-6. So the treaty has largely accomplished its aim up to the present (1995).

EXAMPLE 26-3 *Rate of Radioactive Decay*

The "cobalt treatments" used in medicine to arrest certain types of cancer rely on the ability of gamma rays to destroy cancerous tissues. Cobalt-60 decays with the emission of beta particles and gamma rays, with a half-life of 5.27 years.

$$^{60}_{27}\text{Co} \longrightarrow {}^{60}_{28}\text{Ni} + {}_{-1}^{0}\beta + {}_{0}^{0}\gamma$$

How much of a 3.42-μg sample of cobalt-60 remains after 30.0 years?

Gamma rays destroy both cancerous and normal cells, so the beams of gamma rays must be directed as nearly as possible toward only cancerous tissue.

Plan

We determine the value of the specific rate constant, k, from the given half-life. This value is then used in the first-order integrated rate equation to calculate the amount of cobalt-60 remaining after the specified time.

Solution

We first determine the value of the specific rate constant.

$$t_{1/2} = \frac{0.693}{k} \quad \text{so} \quad k = \frac{0.693}{t_{1/2}} = \frac{0.693}{5.27 \text{ y}} = 0.131 \text{ y}^{-1}$$

This value can now be used to determine the ratio of A_0 to A after 30.0 years.

$$\ln\left(\frac{A_0}{A}\right) = kt = 0.131 \text{ y}^{-1}\,(30.0 \text{ y}) = 3.93$$

Taking the inverse ln of both sides, $\dfrac{A_0}{A} = 51$.

Recall that the stoichiometric coefficient a is always 1 for a nuclear decay process.

$$A_0 = 3.42 \ \mu\text{g, so}$$

$$A = \frac{A_0}{51} = \frac{3.42 \ \mu\text{g}}{51} = \boxed{0.067 \ \mu\text{g} \ {}^{60}_{27}\text{Co}} \text{ remain after 30.0 years.}$$

You should now work Exercise 46.

26-10 DISINTEGRATION SERIES

Many radionuclides cannot attain nuclear stability by only one nuclear reaction. Instead, they decay in a series of disintegrations. A few such series are known to occur in nature. Two begin with isotopes of uranium, ^{238}U and ^{235}U, and one begins with ^{232}Th. All end with a stable isotope of lead ($Z = 82$). Table 26-4 outlines in detail the ^{238}U, ^{235}U, and ^{232}Th disintegration series, showing half-lives. For any particular decay step, the decaying nuclide is called the **parent** nuclide, and the product nuclide is the **daughter.**

Table 26-4 *Emissions and Half-Lives of Members of Natural Radioactive Series**

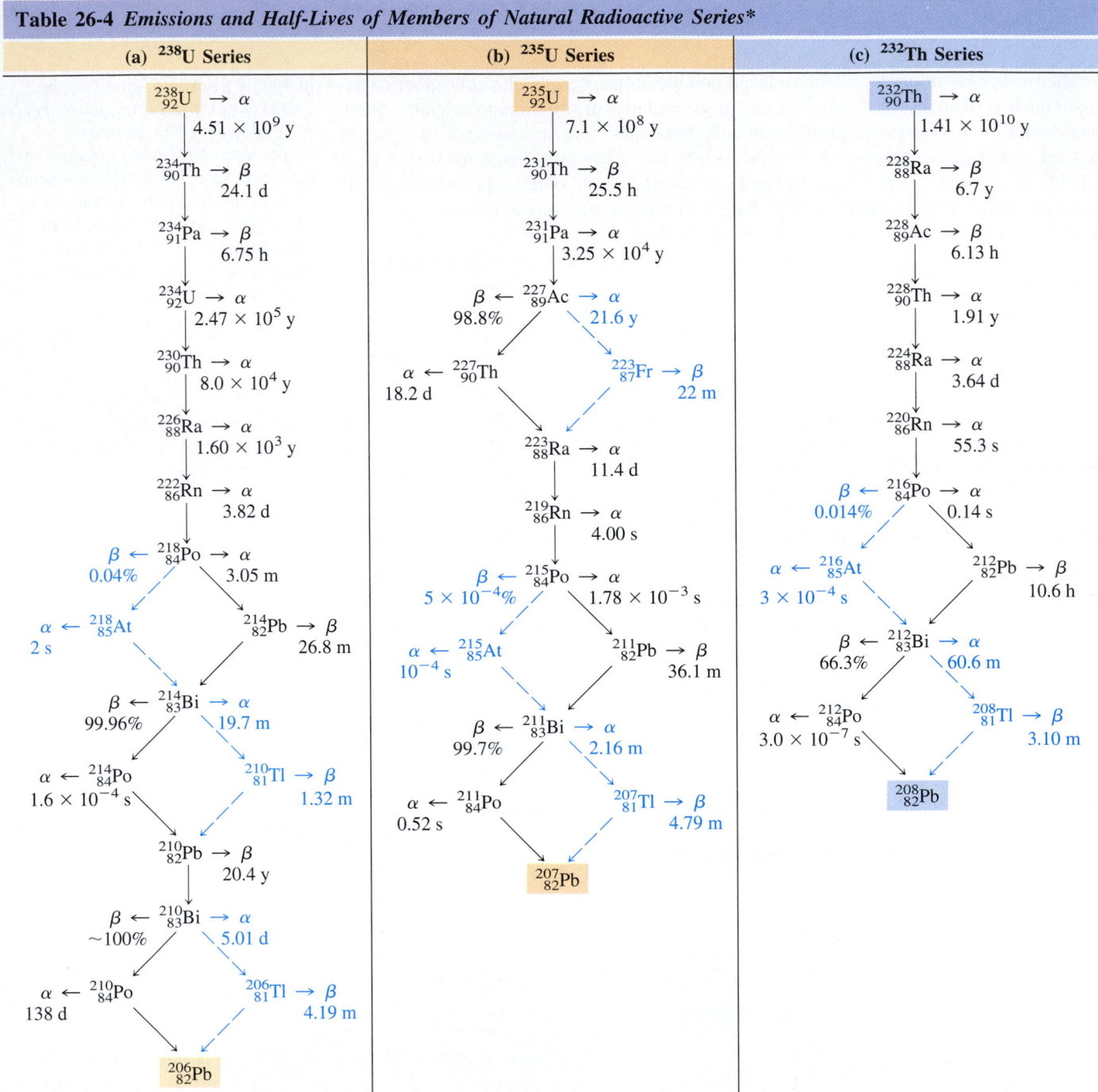

(a) ^{238}U Series	(b) ^{235}U Series	(c) ^{232}Th Series

(a) ^{238}U Series

$^{238}_{92}$U → α
4.51 × 10^9 y

$^{234}_{90}$Th → β
24.1 d

$^{234}_{91}$Pa → β
6.75 h

$^{234}_{92}$U → α
2.47 × 10^5 y

$^{230}_{90}$Th → α
8.0 × 10^4 y

$^{226}_{88}$Ra → α
1.60 × 10^3 y

$^{222}_{86}$Rn → α
3.82 d

β ← $^{218}_{84}$Po → α
0.04%   3.05 m

α ← $^{218}_{85}$At $^{214}_{82}$Pb → β
2 s   26.8 m

β ← $^{214}_{83}$Bi → α
99.96%   19.7 m

α ← $^{214}_{84}$Po $^{210}_{81}$Tl → β
1.6 × 10^{-4} s   1.32 m

$^{210}_{82}$Pb → β
20.4 y

β ← $^{210}_{83}$Bi → α
~100%   5.01 d

α ← $^{210}_{84}$Po $^{206}_{81}$Tl → β
138 d   4.19 m

$^{206}_{82}$Pb

(b) ^{235}U Series

$^{235}_{92}$U → α
7.1 × 10^8 y

$^{231}_{90}$Th → β
25.5 h

$^{231}_{91}$Pa → α
3.25 × 10^4 y

β ← $^{227}_{89}$Ac → α
98.8%   21.6 y

α ← $^{227}_{90}$Th $^{223}_{87}$Fr → β
18.2 d   22 m

$^{223}_{88}$Ra → α
11.4 d

$^{219}_{86}$Rn → α
4.00 s

β ← $^{215}_{84}$Po → α
5 × 10^{-4}%   1.78 × 10^{-3} s

α ← $^{215}_{85}$At $^{211}_{82}$Pb → β
10^{-4} s   36.1 m

β ← $^{211}_{83}$Bi → α
99.7%   2.16 m

α ← $^{211}_{84}$Po $^{207}_{81}$Tl → β
0.52 s   4.79 m

$^{207}_{82}$Pb

(c) ^{232}Th Series

$^{232}_{90}$Th → α
1.41 × 10^{10} y

$^{228}_{88}$Ra → β
6.7 y

$^{228}_{89}$Ac → β
6.13 h

$^{228}_{90}$Th → α
1.91 y

$^{224}_{88}$Ra → α
3.64 d

$^{220}_{86}$Rn → α
55.3 s

β ← $^{216}_{84}$Po → α
0.014%   0.14 s

α ← $^{216}_{85}$At $^{212}_{82}$Pb → β
3 × 10^{-4} s   10.6 h

β ← $^{212}_{83}$Bi → α
66.3%   60.6 m

α ← $^{212}_{84}$Po $^{208}_{81}$Tl → β
3.0 × 10^{-7} s   3.10 m

$^{208}_{82}$Pb

Abbreviations are y, year; d, day; m, minute; and s, second. Less prevalent decay branches are shown in blue.

Uranium-238 decays by alpha emission to thorium-234 in the first step of one series. Thorium-234 subsequently emits a beta particle to produce protactinium-234 in the second step. The series can be summarized as shown in Table 26-4a. The *net* reaction for the ^{238}U series is

$$^{238}_{92}\text{U} \longrightarrow {}^{206}_{82}\text{Pb} + 8\,{}^4_2\text{He} + 6\,{}^0_{-1}\beta$$

"Branchings" are possible at various points in the chain. That is, two successive decays may be replaced by alternate decays, but they always result in the same final product. There are also decay series of varying lengths starting with some of the artificially produced radionuclides (Section 26-12).

26-11 USES OF RADIONUCLIDES

Radionuclides have practical uses because they decay at known rates or, in some cases, because they emit radiation continuously.

Radioactive Dating

The ages of articles of organic origin can be estimated by **radiocarbon dating.** The radioisotope carbon-14 is produced continuously in the upper atmosphere as nitrogen atoms capture cosmic-ray neutrons.

$$^{14}_{7}\text{N} + ^{1}_{0}n \longrightarrow ^{14}_{6}\text{C} + ^{1}_{1}\text{H}$$

The carbon-14 atoms react with oxygen molecules to form $^{14}\text{CO}_2$. This process continually supplies the atmosphere with radioactive $^{14}\text{CO}_2$, which is removed from the atmosphere by photosynthesis. The intensity of cosmic rays is related to the sun's activity. As long as this remains constant, the amount of $^{14}\text{CO}_2$ in the atmosphere remains constant. $^{14}\text{CO}_2$ is incorporated into living organisms just as ordinary $^{12}\text{CO}_2$ is, so a certain fraction of all carbon atoms in living substances is carbon-14. This decays with a half-life of 5730 years.

$$^{14}_{6}\text{C} \longrightarrow ^{14}_{7}\text{N} + ^{0}_{-1}\beta$$

After death, the plant no longer carries out photosynthesis, so it no longer takes up $^{14}\text{CO}_2$. Other organisms that consume plants for food stop doing so at death. The emissions from the ^{14}C in dead tissue then decrease with the passage of time. The activity per gram of carbon is a measure of the length of time elapsed since death. Comparison of ages of ancient trees calculated from ^{14}C activity with those determined by counting rings indicates that cosmic ray intensity has varied somewhat throughout history. The calculated ages can be corrected for these variations. The carbon-14 technique is useful only for dating objects less than 50,000 years old. Older objects have too little activity to be dated accurately.

The **potassium–argon** and **uranium–lead methods** are used for dating older objects. Potassium-40 decays to argon-40 with a half-life of 1.3 billion years.

$$^{40}_{19}\text{K} + ^{0}_{-1}e \longrightarrow ^{40}_{18}\text{Ar}$$

Because of its long half-life, potassium-40 can be used to date objects up to 1 million years old by determination of the ratio of $^{40}_{19}\text{K}$ to $^{40}_{18}\text{Ar}$ in the sample. The uranium–lead method is based on the natural uranium-238 decay series, which ends with the production of stable lead-206. This method is used for dating uranium-containing minerals several billion years old. All the ^{206}Pb in such minerals is assumed to have come from ^{238}U. Because of the very long half-life of $^{238}_{92}\text{U}$, 4.5 billion years, the amounts of intermediate nuclei can be neglected. A meteorite that was 4.6 billion years old fell in Mexico in 1969. Results of $^{238}\text{U}/^{206}\text{Pb}$ studies on such materials of extraterrestrial origin suggest that our solar system was formed several billion years ago.

In 1992, hikers in the Italian Alps found the remains of a man who had been frozen in a glacier for about 4000 years. This discovery is especially important because of the unusual preservation of tissues, garments, and personal belongings. Radiocarbon dating is used to estimate the ages of archaeological finds such as these.

In recent decades atmospheric testing of nuclear warheads has also caused fluctuations in the natural abundance of ^{14}C.

Gaseous argon is easily lost from minerals. Therefore, measurements based on the $^{40}\text{K}/^{40}\text{Ar}$ method may not be as reliable as desired.

In Table 26-4(a) we see that the first step, the decay of ^{238}U, is the slowest step (longest half-life). We learned in Section 16-7 that the slowest step is the rate determining step.

EXAMPLE 26-4 *Radiocarbon Dating*

A piece of wood taken from a cave dwelling in New Mexico is found to have a carbon-14 activity (per gram of carbon) only 0.636 times that of wood cut today. Estimate the age of the wood. The half-life of carbon-14 is 5730 years.

Plan

As we did in Example 26-3, we determine the specific rate constant k from the known half-life. The time required to reach the present fraction of the original activity is then calculated from the first-order decay equation.

Solution

First we find the first-order specific rate constant for ^{14}C.

$$t_{1/2} = \frac{0.693}{k}$$

$$k = \frac{0.693}{t_{1/2}} = \frac{0.693}{5730 \text{ y}} = 1.21 \times 10^{-4} \text{ y}^{-1}$$

The present ^{14}C activity, N (disintegrations per unit time), is 0.636 times the original activity, N_0.

$$N = 0.636 \, N_0$$

We substitute into the first-order decay equation

$$\ln\left(\frac{N_0}{N}\right) = kt$$

$$\ln\left(\frac{N_0}{0.636 \, N_0}\right) = (1.21 \times 10^{-4} \text{ y}^{-1})t$$

We cancel N_0 and solve for t.

$$\ln\left(\frac{1}{0.636}\right) = (1.21 \times 10^{-4} \text{ y}^{-1})t$$

$$0.452 = (1.21 \times 10^{-4} \text{ y}^{-1})t \qquad \text{or} \qquad t = \boxed{3.74 \times 10^3 \text{ y (or 3740 y)}}$$

You should now work Exercises 50 and 52.

A weak radioactive source such as americium is used in some smoke detectors. Radiation from the source ionizes the air to produce a weak current. Smoke particles interrupt the current flow by attracting the ions. This decrease in current triggers the alarm.

EXAMPLE 26-5 *Uranium–Lead Dating*

A sample of uranium ore is found to contain 4.64 mg of ^{238}U and 1.22 mg of ^{206}Pb. Estimate the age of the ore. The half-life of ^{238}U is 4.51×10^9 years.

Plan

The original mass of ^{238}U is equal to the mass of ^{238}U remaining plus the mass of ^{238}U that decayed to produce the present mass of ^{206}Pb. We obtain the specific rate constant, k, from the known half-life. Then we use the ratio of original ^{238}U to remaining ^{238}U to calculate the time elapsed, with the aid of the first-order integrated rate equation.

Solution

First we calculate the amount of ^{238}U that must have decayed to produce 1.22 mg of ^{206}Pb, using the isotopic masses.

$$\underline{?} \text{ mg } ^{238}U = 1.22 \text{ mg } ^{206}Pb \times \frac{238 \text{ mg } ^{238}U}{206 \text{ mg } ^{206}Pb} = 1.41 \text{ mg } ^{238}U$$

Thus, the sample originally contained 4.64 mg + 1.41 mg = 6.05 mg of ^{238}U.

We next evaluate the specific rate (disintegration) constant, k.

$$t_{1/2} = \frac{0.693}{k} \quad \text{so} \quad k = \frac{0.693}{t_{1/2}} = \frac{0.693}{4.51 \times 10^9 \text{ y}} = 1.54 \times 10^{-10} \text{ y}^{-1}$$

Now we calculate the age of the sample, t.

$$\ln\left(\frac{A_0}{A}\right) = kt$$

$$\ln\left(\frac{6.05 \text{ mg}}{4.64 \text{ mg}}\right) = (1.54 \times 10^{-10} \text{ y}^{-1})t$$

$$\ln 1.30 = (1.54 \times 10^{-10} \text{ y}^{-1})t$$

$$\frac{0.262}{1.54 \times 10^{-10} \text{ y}^{-1}} = t \quad \text{or} \quad t = 1.70 \times 10^9 \text{ years}$$

The ore is approximately 1.7 billion years old.

You should now work Exercise 68.

Medical Uses of Radionuclides

The use of cobalt radiation treatments for cancerous tumors was described in Example 26-3. Several other nuclides are used as **radioactive tracers** in medicine. Radioisotopes of an element have the same chemical properties as stable isotopes of the same element, so they can be used to "label" an element in compounds. A radiation detector can be used to follow the path of the element throughout the body. Salt solutions containing ^{24}Na can be injected into the bloodstream to follow the flow of blood and locate obstructions in the circulatory system. Thallium-201 tends to concentrate in healthy heart tissue, whereas technetium-99 concentrates in abnormal heart tissue. The two can be used together to survey damage from heart disease. Iodine-131 concentrates in the thyroid gland, liver, and certain parts of the brain. It is used to monitor goiter and other thyroid problems, as well as liver and brain tumors. The energy produced by the decay of plutonium-238 is converted into electrical energy in heart pacemakers. The relatively long half-life of the isotope allows the device to be used for ten years before replacement.

Research Applications for Radionuclides

The pathways of chemical reactions can be investigated using radioactive tracers. When radioactive ^{35}S^{2-} ions are added to a saturated solution of cobalt sulfide in equilibrium with solid cobalt sulfide, the solid becomes radioactive. This shows that sulfide ion exchange occurs between solid and solution in the solubility equilibrium.

$$\text{CoS(s)} \rightleftharpoons \text{Co}^{2+}\text{(aq)} + \text{S}^{2-}\text{(aq)} \quad K_{sp} = 8.7 \times 10^{-23}$$

Photosynthesis is the process by which the carbon atoms in CO_2 are incorporated into glucose, $C_6H_{12}O_6$, in green plants.

$$6CO_2 + 6H_2O \xrightarrow[\text{chlorophyll}]{\text{sunlight}} C_6H_{12}O_6 + 6O_2$$

The process is more complex than the net equation implies; it actually occurs in many steps and produces a number of intermediate products. By using labeled $^{14}CO_2$, we can identify the intermediate molecules. They contain the radioactive ^{14}C atoms.

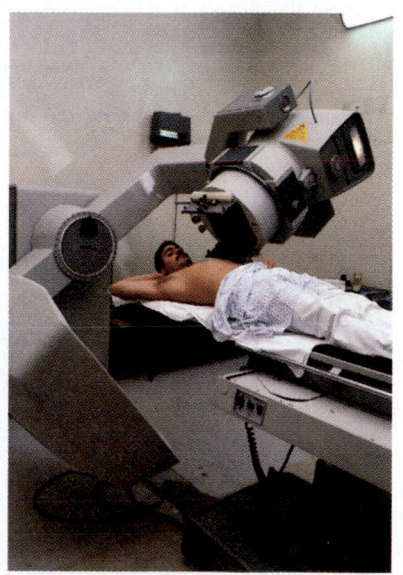

The γ-radiation from ^{60}Co is used to treat cancers near the surface of the body.

Irradiation with gamma rays from radioactive isotopes has kept the strawberries at the right fresh for 15 days, while those at the left are moldy. Such irradiation kills mold spores, but does no damage to the food.

Agricultural Uses of Radionuclides

The pesticide DDT is toxic to humans and animals repeatedly exposed to it. DDT persists in the environment a long time. It concentrates in fatty tissues. The DDT once used to control the screw-worm fly was replaced by a radiologic technique. The irradiation of the male flies with gamma rays alters their reproductive cells, sterilizing them. When great numbers of sterilized males are released in an infested area, they mate with females, who, of course, produce no offspring. This results in the reduction and eventual disappearance of the population.

The procedure works because the female flies mate only once. In an area highly populated with sterile males, the probability of a "productive" mating is very small.

Labeled fertilizers can also be used to study nutrient uptake by plants and to study the growth of crops. Gamma irradiation of some foods allows them to be stored for longer periods. For example, it retards the sprouting of potatoes and onions.

Industrial Uses of Radionuclides

There are many applications of radiochemistry in industry and engineering. Two will be mentioned here. When great precision is required in the manufacture of strips or sheets of metal of definite thicknesses, the penetrating powers of various kinds of radioactive emissions are utilized. The thickness of the metal is correlated with the intensity of radiation passing through it. The flow of a liquid or gas through a pipeline can be monitored by injecting a sample containing a radioactive substance. Leaks in pipelines can also be detected in this way.

26-12 ARTIFICIAL TRANSMUTATIONS OF ELEMENTS

The first artificially induced nuclear reaction was carried out by Rutherford in 1915. He bombarded nitrogen-14 with alpha particles to produce an isotope of oxygen and a proton.

$$^{14}_{7}\text{N} + ^{4}_{2}\alpha \longrightarrow ^{1}_{1}\text{H} + ^{17}_{8}\text{O}$$

Such reactions are often indicated in abbreviated form, with the bombarding particle and emitted subsidiary particles shown parenthetically between the parent and daughter nuclei.

$$^{14}_{7}\text{N} \, (^{4}_{2}\alpha, \, ^{1}_{1}p) \, ^{17}_{8}\text{O}$$

Several thousand artificially induced reactions have been carried out with bombarding particles such as neutrons, protons, deuterons ($^{2}_{1}\text{H}$), alpha particles, and other small nuclei.

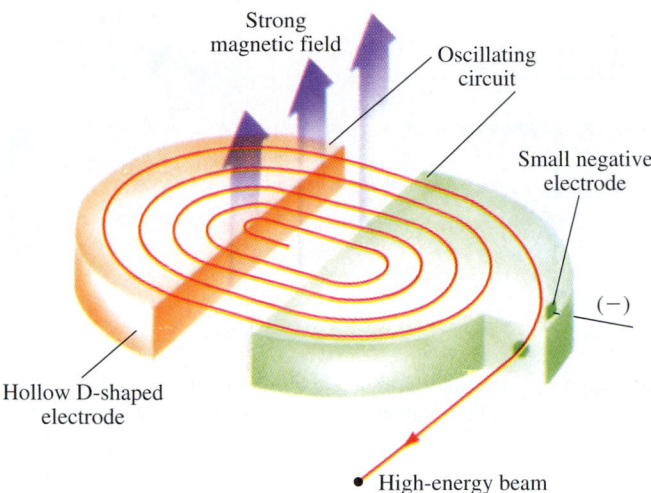

Strong
magnetic field

Oscillating
circuit

Small negative
electrode

(−)

Hollow D-shaped
electrode

High-energy beam

Figure 26-7   Schematic representation of a cyclotron.

Figure 26-8   A beam of protons (bright area) from a cyclotron at the Argonne National Laboratory. Nuclear reactions take place when protons and other atomic particles strike the nuclei of atoms.

Bombardment with Positive Ions

A problem arises with the use of positively charged nuclei as projectiles. For a nuclear reaction to occur, the bombarding nuclei must actually collide with the target nuclei, which are also positively charged. Collisions cannot occur unless the projectiles have sufficient kinetic energy to overcome coulombic repulsion. The required kinetic energies increase with increasing atomic numbers of the target and of the bombarding particle.

Particle accelerators called **cyclotrons** (atom smashers) and **linear accelerators** have overcome the problem of repulsion. A cyclotron (Figure 26-7) consists of two hollow, D-shaped electrodes called "dees." Both dees are in an evacuated enclosure between the poles of an electromagnet. The particles to be accelerated are introduced at the center in the gap between the dees. The dees are connected to a source of high frequency alternating current that keeps them oppositely charged. The positively charged particles are attracted toward the negative dee. The magnetic field causes the path of the charged particles to curve 180 degrees to return to the space between the dees. Then the charges are reversed on the dees, so the particles are repelled by the first dee (now positive) and attracted to the second. This repeated process is synchronized with the motion of the particles. They accelerate along a spiral path and eventually emerge through an exit hole oriented so that the beam hits the target atoms (Figure 26-8).

In a linear accelerator the particles are accelerated through a series of tubes within an evacuated chamber (Figure 26-9). The odd-numbered tubes are at first negatively charged and the even ones positively charged. A positively charged particle is attracted toward the first tube. As it passes through that tube, the charges on the tubes are reversed so that the particle is repelled out of the first tube (now positive) and toward the second (negative) tube. As the particle nears the end of the second tube, the charges are again reversed. As this process is repeated, the particle is accelerated to very high velocities. The polarity is changed at constant frequency, so subsequent tubes are longer to accommodate the in-

The first cyclotron was constructed by E. O. Lawrence and M. S. Livingston at the University of California in 1930.

The path of the particle is initially circular because of the interaction of the particle's charge with the electromagnet's field. As the particle gains energy, the radius of the path increases, and the particle spirals outward.

The first linear accelerator was built in 1928 by a German physicist, Rolf Wideroe.

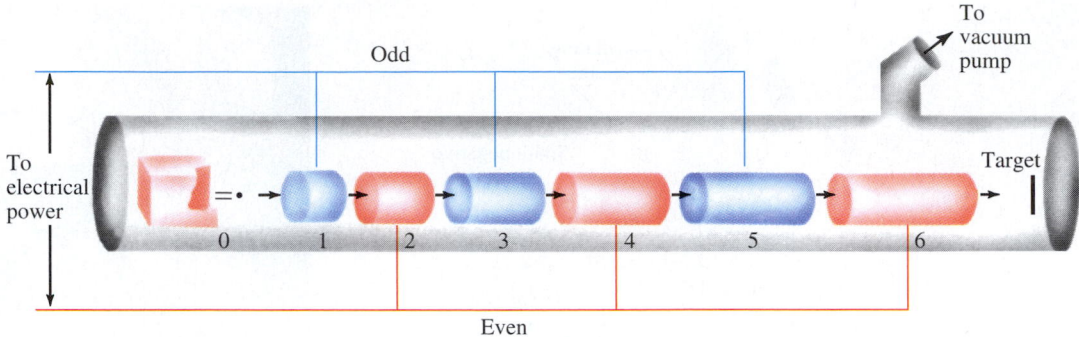

Figure 26-9 Diagram of an early type of linear accelerator. An alpha emitter is placed in the container at the left. Only those alpha particles that happen to be emitted in line with the series of accelerating tubes can escape.

One gigaelectron volt (GeV) = 1 × 10^9 eV = 1.60×10^{-10} J. This is sometimes called 1 billion electron volts (BeV) in the United States.

creased distance traveled by the accelerating particle per unit time. The bombardment target is located outside the last tube. If the initial polarities are reversed, negatively charged particles can also be accelerated. The longest linear accelerator, completed in 1966 at Stanford University, is about 2 miles long. It is capable of accelerating electrons to energies of nearly 20 GeV.

Many nuclear reactions have been induced by such bombardment techniques. At the time of development of particle accelerators, there were a few gaps among the first 92 elements in the periodic table. Particle accelerators were used between 1937 and 1941 to synthesize three of the four "missing" elements: numbers 43 (technetium), 85 (astatine), and 87 (francium).

$$\ce{^{96}_{42}Mo} + \ce{^{2}_{1}H} \longrightarrow \ce{^{97}_{43}Tc} + \ce{^{1}_{0}n}$$

$$\ce{^{209}_{83}Bi} + \ce{^{4}_{2}\alpha} \longrightarrow \ce{^{210}_{85}At} + 3\,\ce{^{1}_{0}n}$$

$$\ce{^{230}_{90}Th} + \ce{^{1}_{1}p} \longrightarrow \ce{^{223}_{87}Fr} + 2\,\ce{^{4}_{2}\alpha}$$

Many hitherto unknown, unstable, artificial isotopes of known elements have also been synthesized so that their nuclear structures and behavior could be studied.

An aerial view of the particle accelerator dedicated in 1978 at the Fermi National Accelerator Laboratory (Fermilab), near Batavia, Illinois. This proton accelerator, 4 miles in circumference, accelerates protons to energies of 1 trillion electron volts. Construction of a vastly larger accelerator known as the superconducting supercollider, or SSC, was begun near Waxahachie, Texas. However, this project was abandoned in 1994 when the U.S. Congress voted to discontinue funding for it.

CHEMISTRY IN USE

The Development of Science

A Dinosaur Extinction Theory

Scientists can use neutron bombardment to determine which elements are present, and how much of each element is present, in a sample. When bombarded with neutrons, atoms of some elements absorb neutrons and are transformed into radioactive atoms. Each radioactive atom emits its own characteristic radiation; by making careful measurements to determine the types of radiation being emitted, we can determine which isotopes are present (*qualitative analysis*). Because the intensity of radiation due to each radioactive element depends only on the amount of that element present, it is also possible to determine how much of each radioactive element is present in a sample (*quantitative analysis*). Neutron bombardment has been used to explain the demise of the dinosaurs.

You may have read the theory that dinosaurs became extinct after a large meteorite struck the earth during the Cretaceous Period, approximately 65 million years ago. The meteorite broke into two pieces that fell in what are now Iowa and the Yucatan Peninsula. Dust from these impacts circled the globe and blocked out the sun for several years. During that time photosynthesis stopped, land and marine plants died, and most animal life vanished. Eventually the dust, very rich in the element iridium, settled on the earth's surface. How did scientists develop this interesting explanation?

Scientists used neutrons to bombard samples from the Cretaceous layer, a sedimentary rock layer that deposited about the time dinosaurs became extinct. These neutrons reacted with iridium in the 65-million-year-old rock layer.

$$^{193}\text{Ir} + {}^{1}_{0}n \longrightarrow {}^{194}\text{Ir}$$

Iridium-193 is not radioactive, but iridium-194 is so radioactive that only billionths of a g/cm^3 in a sample can be detected easily. After bombarding a sample from the Cretaceous layer, the radioactivity from decaying ^{194}Ir atoms was found to be 1.89×10^9 disintegrations per second per cm^3 of soil. When samples from other sedimentary layers were bombarded with neutrons, their rate of radioactive decay was found to be 7.87×10^7 disintegrations per second per cm^3 of soil.

Using the equation,

$$\text{Rate of decay} = kN$$

and knowing that the half-life of iridium-194 is 19 hours, we can calculate the concentration of iridium atoms in g/cm^3 of soil. Performing these calculations on samples from other than the Cretaceous layer, we find

$$\text{Rate} = kN = (0.693/t_{1/2})N$$

$$\frac{7.87 \times 10^7 \; ^{194}\text{Ir atoms}}{\text{s} \cdot \text{cm}^3} = \frac{0.693 \, N}{(19 \text{ h})(3600 \text{ s/h})}$$

Solving for N gives

$$N = 7.77 \times 10^{12} \; ^{194}\text{Ir atoms/cm}^3$$

Next, the mass of iridium-194 present in one cm^3 of soil is calculated.

$$\frac{7.77 \times 10^{12} \; ^{194}\text{Ir atoms}}{\text{cm}^3} \times \frac{194 \text{ g } ^{194}\text{Ir}}{6.02 \times 10^{23} \; ^{194}\text{Ir atoms}}$$

$$= 2.5 \times 10^{-9} \text{ g } ^{194}\text{Ir/cm}^3$$

We can calculate the concentration of iridium-194 atoms in the Cretaceous layer by the same procedure.

$$\text{Rate} = kN = (0.693/t_{1/2})N$$

$$\frac{1.89 \times 10^9 \; ^{194}\text{Ir atoms}}{\text{s/cm}^3} = \frac{0.693 \, N}{(19 \text{ h})(3600 \text{ s/h})}$$

Solving for N gives

$$N = 1.86 \times 10^{14} \; ^{194}\text{Ir atoms/cm}^3$$

Now, the mass of iridium-194 present in one cm^3 of soil can be calculated.

$$\frac{1.86 \times 10^{14} \; ^{194}\text{Ir atom}}{\text{cm}^3} \times \frac{194 \text{ g } ^{194}\text{Ir}}{6.02 \times 10^{23} \; ^{194}\text{Ir atom}}$$

$$= 6.0 \times 10^{-8} \text{ g } ^{194}\text{Ir/cm}^3$$

The percent increase of iridium in the Cretaceous layer is

$$\frac{(6.0 \times 10^{-8} - 2.5 \times 10^{-9}) \text{ g/cm}^3}{2.5 \times 10^{-9} \text{ g/cm}^3} \times 100\%$$

$$= 2300\% \text{ increase in } ^{194}\text{Ir concentration}$$

These calculations show that approximately 65 million years ago, there was a 2300% increase in the concentration of iridium in the earth's crust! Because most of the terrestrial iridium we have found has extraterrestrial origins, the giant meteorite theory appears to offer a satisfactory explanation for the disappearance of dinosaurs.

Ronald DeLorenzo
Middle Georgia College

Neutron Bombardment

Neutrons bear no charge, so they are not repelled by nuclei as positively charged projectiles are. They do not need to be accelerated to produce bombardment reactions. Neutron beams can be generated in several ways. A frequently used method involves bombardment of beryllium-9 with alpha particles.

$$\ce{^{9}_{4}Be} + \ce{^{4}_{2}\alpha} \longrightarrow \ce{^{12}_{6}C} + \ce{^{1}_{0}n}$$

Nuclear reactors (Section 26-15) are also used as neutron sources. Neutrons ejected in nuclear reactions usually possess high kinetic energies and are called **fast neutrons.** When they are used as projectiles they cause reactions, such as (n, p) or (n, α) reactions, in which subsidiary particles are ejected. The fourth "missing" element, number 61 (promethium), was synthesized by fast neutron bombardment of neodymium-142.

$$\ce{^{142}_{60}Nd} + \ce{^{1}_{0}n} \longrightarrow \ce{^{143}_{61}Pm} + \ce{^{0}_{-1}\beta}$$

Fast neutrons move so rapidly that they are likely to pass right through a target nucleus without reacting. Hence, the probability of a reaction is low, even though the neutrons may be very energetic.

Slow neutrons ("thermal" neutrons) are produced when fast neutrons collide with **moderators** such as hydrogen, deuterium, oxygen, or the carbon atoms in paraffin. These neutrons are more likely to be captured by target nuclei. Bombardments with slow neutrons can cause neutron-capture (n, γ) reactions.

$$\ce{^{200}_{80}Hg} + \ce{^{1}_{0}n} \longrightarrow \ce{^{201}_{80}Hg} + \ce{^{0}_{0}\gamma}$$

Slow neutron bombardment also produces the ^{3}H isotope (tritium).

$$\ce{^{6}_{3}Li} + \ce{^{1}_{0}n} \longrightarrow \ce{^{3}_{1}H} + \ce{^{4}_{2}\alpha} \qquad (n, \alpha) \text{ reaction}$$

E. M. McMillan discovered the first transuranium element, neptunium, in 1940 by bombarding uranium-238 with slow neutrons.

$$\ce{^{238}_{92}U} + \ce{^{1}_{0}n} \longrightarrow \ce{^{239}_{92}U} + \ce{^{0}_{0}\gamma}$$

$$\ce{^{239}_{92}U} \longrightarrow \ce{^{239}_{93}Np} + \ce{^{0}_{-1}\beta}$$

Several additional elements have been prepared by neutron bombardment or by bombardment of the nuclei so produced with positively charged particles. Some examples are

$$
\left.
\begin{aligned}
\ce{^{238}_{92}U} + \ce{^{1}_{0}n} &\longrightarrow \ce{^{239}_{92}U} + \ce{^{0}_{0}\gamma} \\
\ce{^{239}_{92}U} &\longrightarrow \ce{^{239}_{93}Np} + \ce{^{0}_{-1}\beta} \\
\ce{^{239}_{93}Np} &\longrightarrow \ce{^{239}_{94}Pu} + \ce{^{0}_{-1}\beta}
\end{aligned}
\right\} \quad \text{plutonium}
$$

$$\ce{^{239}_{94}Pu} + \ce{^{4}_{2}\alpha} \longrightarrow \ce{^{242}_{96}Cm} + \ce{^{1}_{0}n} \qquad \text{curium}$$

$$\ce{^{246}_{96}Cm} + \ce{^{12}_{6}C} \longrightarrow \ce{^{254}_{102}No} + 4\,\ce{^{1}_{0}n} \qquad \text{nobelium}$$

$$\ce{^{252}_{98}Cf} + \ce{^{10}_{5}B} \longrightarrow \ce{^{257}_{103}Lr} + 5\,\ce{^{1}_{0}n} \qquad \text{lawrencium}$$

26-13 NUCLEAR FISSION AND FUSION

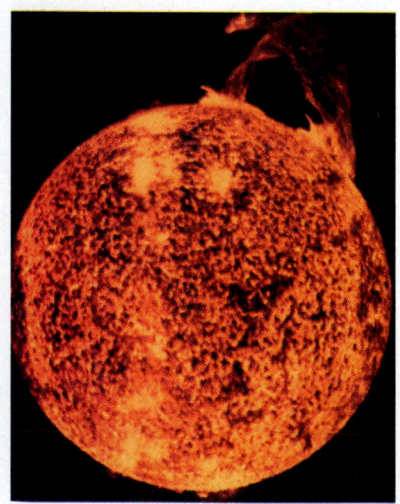

Our sun supplies energy to the earth from a distance of 93,000,000 miles. Like other stars, it is a giant nuclear fusion reactor. Much of its energy comes from the fusion of deuterium, $^{2}_{1}\text{H}$, producing helium, $^{4}_{2}\text{He}$.

Isotopes of some elements with atomic numbers above 80 are capable of undergoing fission in which they split into nuclei of intermediate masses and emit one or more neutrons. Some fissions are spontaneous; others require that the activation energy be supplied by bombardment. A given nucleus can split in many different ways, liberating enormous amounts of energy. Some of the possible fissions that can result from bombard-

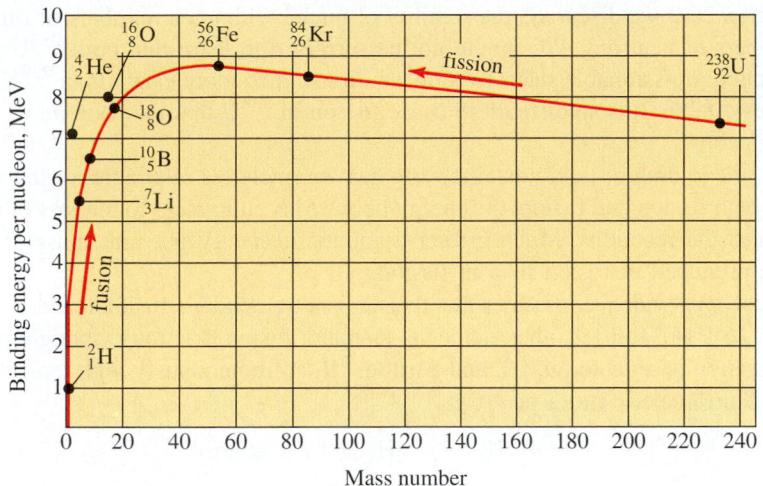

Figure 26-10 Variation in nuclear binding energy with atomic mass. The most stable nucleus is $^{56}_{26}$Fe, with a binding energy of 8.80 MeV per nucleon.

$$1 \text{ MeV} = 1.60 \times 10^{-13} \text{ J}$$

ment of fissionable uranium-235 with fast neutrons follow. The uranium-236 is a short-lived intermediate.

$$^{235}_{92}\text{U} + ^{1}_{0}n \longrightarrow [^{236}_{92}\text{U}] \begin{cases} \nearrow ^{160}_{62}\text{Sm} + ^{72}_{30}\text{Zn} + 4\,^{1}_{0}n + \text{energy} \\ \nearrow ^{146}_{57}\text{La} + ^{87}_{35}\text{Br} + 3\,^{1}_{0}n + \text{energy} \\ \longrightarrow ^{140}_{56}\text{Ba} + ^{93}_{36}\text{Kr} + 3\,^{1}_{0}n + \text{energy} \\ \searrow ^{144}_{55}\text{Cs} + ^{90}_{37}\text{Rb} + 2\,^{1}_{0}n + \text{energy} \\ \searrow ^{144}_{54}\text{Xe} + ^{90}_{38}\text{Sr} + 2\,^{1}_{0}n + \text{energy} \end{cases}$$

Recall that the binding energy is the amount of energy that must be supplied to the nucleus to break it apart into subatomic particles. Figure 26-10 is a plot of binding energy per nucleon versus mass number. It shows that atoms of intermediate mass number have the highest binding energies per nucleon; therefore, they are the most stable. Thus, fission is an energetically favorable process for heavy atoms, because atoms with intermediate masses and greater binding energies per nucleon are formed.

The term "nucleon" refers to a nuclear particle, either a neutron or a proton.

Which isotopes of which elements undergo fission? Experiments with particle accelerators have shown that every element with an atomic number of 80 or more has one or more isotopes capable of undergoing fission, provided they are bombarded at the right energy. Nuclei with atomic numbers between 89 and 98 fission spontaneously with long half-lives of 10^4 to 10^{17} years. Nuclei with atomic numbers of 98 or more fission spontaneously with shorter half-lives of a few milliseconds to 60.5 days. One of the *natural* decay modes of the transuranium elements is via spontaneous fission. In fact, all known nuclides with *mass numbers* greater than 250 do this because they are too big to be stable. Most nuclides with mass numbers between 225 and 250 do not undergo fission spontaneously (except for a few with extremely long half-lives). They can be induced to undergo fission when bombarded with particles of relatively low kinetic energies. Particles that can supply the required activation energy include neutrons, protons, alpha particles, and fast electrons. For nuclei lighter than mass 225, the activation energy required to induce fission rises very rapidly.

In Section 26-2 we discussed the stability of nuclei with even numbers of protons and even numbers of neutrons. We should not be surprised to learn that both ^{233}U and ^{235}U can be excited to fissionable states by slow neutrons much more easily than ^{238}U, because they are less stable. It is so difficult to cause fission in ^{238}U that this isotope is said to be "nonfissionable."

Fusion, the joining of light nuclei to form heavier nuclei, is favorable for the very light atoms. In both fission and fusion, the energy liberated is equivalent to the loss of mass that accompanies the reactions. Much greater amounts of energy per unit mass of *reacting atoms* are produced in fusion than in fission.

Spectroscopic evidence indicates that the sun is a tremendous fusion reactor consisting of 73% H, 26% He, and 1% other elements. Its major fusion reaction is thought to involve the combination of a deuteron, ^2_1H, and a triton, ^3_1H, at tremendously high temperatures to form a helium nucleus and a neutron.

> The deuteron and triton are the nuclei of two isotopes of hydrogen, called deuterium and tritium. Deuterium occurs naturally in water. When the D_2O is purified as "heavy water," it can be used for several types of chemical analysis.

$$^2_1H + ^3_1H \longrightarrow ^4_2He + ^1_0n + \text{energy}$$

Thus, solar energy is actually a form of fusion energy. This is the only unlimited kind of energy available to us. Our fossil fuels are just "leftover" fusion energy.

26-14 THE ATOMIC BOMB (FISSION)

Typically, two or three neutrons are produced per fission reaction. These neutrons can collide with other fissionable atoms to sustain and expand the process. If sufficient fissionable material, the **critical mass,** is contained in a small enough volume, an explosive chain reaction can result. If too few fissionable atoms are present, most of the neutrons escape and no chain reaction occurs. Figure 26-11 depicts a fission chain reaction.

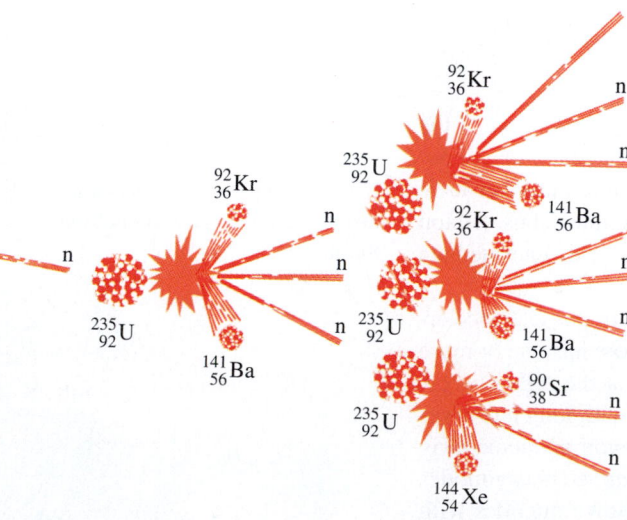

Figure 26-11 A self-propagating nuclear chain reaction. A stray neutron induces a single fission, liberating more neutrons. Each of them induces another fission, each of which is accompanied by release of two or three neutrons. The chain continues to branch in this way, very quickly resulting in an explosive rate of fission.

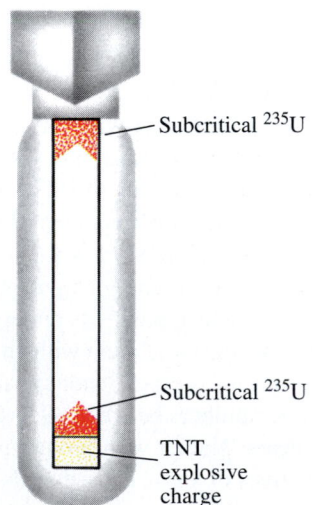

Figure 26-12 One design used in atomic bombs. A conventional explosive is used to bring two subcritical masses together to form a supercritical mass.

CHEMISTRY IN USE

The Development of Science

Chemistry versus Physics

For many years a friendly rivalry has existed between chemists and physicists, with each group claiming to have made the larger contributions to scientific thought. On each opening day of his chemistry class, one chemist brings this rivalry to light when he drops a box of matches. Referring to the falling matches, he says, "That's physics. Physics tells us that the distance the box falls from rest is $\frac{1}{2}gt^2$." He then picks up the box, removes a match, and rubs it against a rough surface. The match bursts into flame. "That's chemistry," he grins. "Chemistry is more striking than physics. Physics doesn't hold a match to chemistry."

Many discoveries are interdisciplinary, i.e., they result from the combined efforts of specialists from many fields. Scientific progress rarely depends exclusively upon one scientific discipline. Historians who write about the evolution of nuclear chemistry and physics appear to have cast physicists into the limelight and neglected some of the important contributions made by chemists. Physicists such as Enrico Fermi played key roles in the development of the first self-sustaining nuclear reaction. This reaction later made possible the development of atomic weapons and nuclear power plants. Fermi's work on radioactivity won him the Nobel Prize for physics in 1938.

However, long before Enrico Fermi's very important discovery, the chemists–physicists Marie and Pierre Curie had done very fundamental work with radioactivity. In 1898, they discovered the elements radium and polonium, and in 1902, they developed the chemical separation techniques used to isolate polonium and radium from the ore pitchblende. For this work, they shared the 1902 Nobel Prize in physics. Nine years later, Marie Curie won the Nobel Prize in chemistry for her work in isolating pure radium.

The distinguished English chemist Frederick Soddy significantly advanced the theory of radioactivity in 1902 and presented the notion of isotopes in 1911. Soddy and others developed techniques for handling minute quantities of radioactive chemicals, techniques that were later indispensable to research on nuclear energy.

The noted German chemist Otto Hahn received the 1944 Nobel Prize in chemistry for discovering the fission of heavy nuclei. He is also credited with the discoveries of thorium and actinium and for distinguishing between regular chemistry and radiochemistry.

Glenn Seaborg, the 1951 Nobel Laureate chemist, is credited with discovering about ten percent of the then-known elements, particularly a very important fissionable element that was later named plutonium. Subsequently, chemical separation techniques were developed and used to isolate fissionable plutonium-239 from uranium-238.

Someday historians may discuss more fully the important roles chemists have played in the development of nuclear energy. However, by that time chemists will probably be occupied with the problems of cleaning up the radioactive wastes left behind by physicists.

Ronald DeLorenzo
Middle Georgia College

One type of atomic bomb (Figure 26-12) contains two subcritical portions of fissionable material. One portion is driven into the other to form a supercritical mass by an ordinary chemical explosive such as trinitrotoluene (TNT). A nuclear fission explosion results. Tremendous amounts of heat energy are released, as well as many radionuclides whose effects are devastating to life and the environment. The radioactive dust and debris are called **fallout.**

The launching of the nuclear submarine *Hyman G. Rickover* into the Thames River in Connecticut (August 27, 1983).

The water that comes near the nuclear core flows in a closed system and is not released to the environment.

26-15 NUCLEAR REACTORS (FISSION)

Controlled fission reactions in nuclear reactors are of great use and even greater potential. The fuel elements of a nuclear reactor have neither the composition nor the extremely compact arrangement of the critical mass of a bomb. Thus, no possibility of nuclear explosion exists. However, various dangers are associated with nuclear energy generation. The possibility of "meltdown" will be discussed with respect to cooling systems in light water reactors. Proper shielding precautions must be taken to ensure that the radionuclides produced are always contained within vessels from which neither they nor their radiations can escape. Long-lived radionuclides from spent fuel must be stored underground in heavy, shock-resistant containers until they have decayed to the point that they are no longer biologically harmful. As examples, strontium-90 ($t_{1/2}$ = 28 years) and plutonium-239 ($t_{1/2}$ = 24,000 years) must be stored for 280 years and 240,000 years, respectively, before they lose 99.9% of their activities. Critics of nuclear energy contend that the containers could corrode over such long periods, or burst as a result of earth tremors, and that transportation and reprocessing accidents could cause environmental contamination with radionuclides. They claim that river water used for cooling is returned to the rivers with too much heat (thermal pollution), thus disrupting marine life. (It should be noted, though, that fossil fuel electric power plants cause the same thermal pollution, for the same amount of electricity generated.) The potential for theft also exists. Plutonium-239, a fissionable material, could be stolen from reprocessing plants and used to construct atomic weapons.

Proponents of the development of nuclear energy argue that the advantages far outweigh the risks. Nuclear energy plants do not pollute the air with oxides of sulfur, nitrogen, carbon, and particulate matter, as fossil fuel electric power plants do. The big advantage of nuclear fuels is the enormous amount of energy liberated per unit mass of fuel. At present, nuclear reactors provide about 17% of the electrical energy consumed in the United States. In some parts of Europe, where natural resources of fossil fuels are scarcer, the utilization of nuclear energy is higher. For instance, in France and Belgium, more than 65% of electrical energy is produced from nuclear reactors. With rapidly declining fossil fuel reserves, it appears likely that nuclear energy and solar energy will become increas-

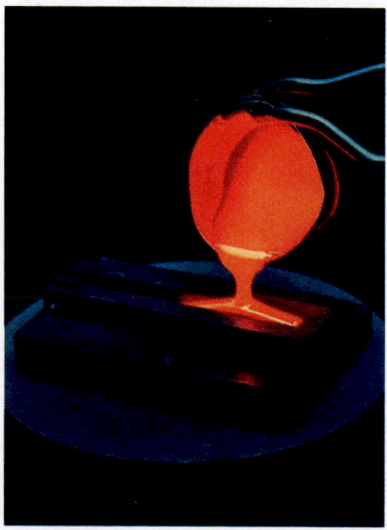

Nuclear waste may take centuries to decompose, so we cannot afford to take risks in its disposal. Suggested approaches include casting it into ceramics, as shown here, to eliminate the possibility of the waste dissolving in ground water. The encapsulated waste could then be deposited in underground salt domes. Located in geologically stable areas, such salt domes have held petroleum and compressed natural gas trapped for millions of years. The political problems of nuclear waste disposal are at least as challenging as the technological ones.

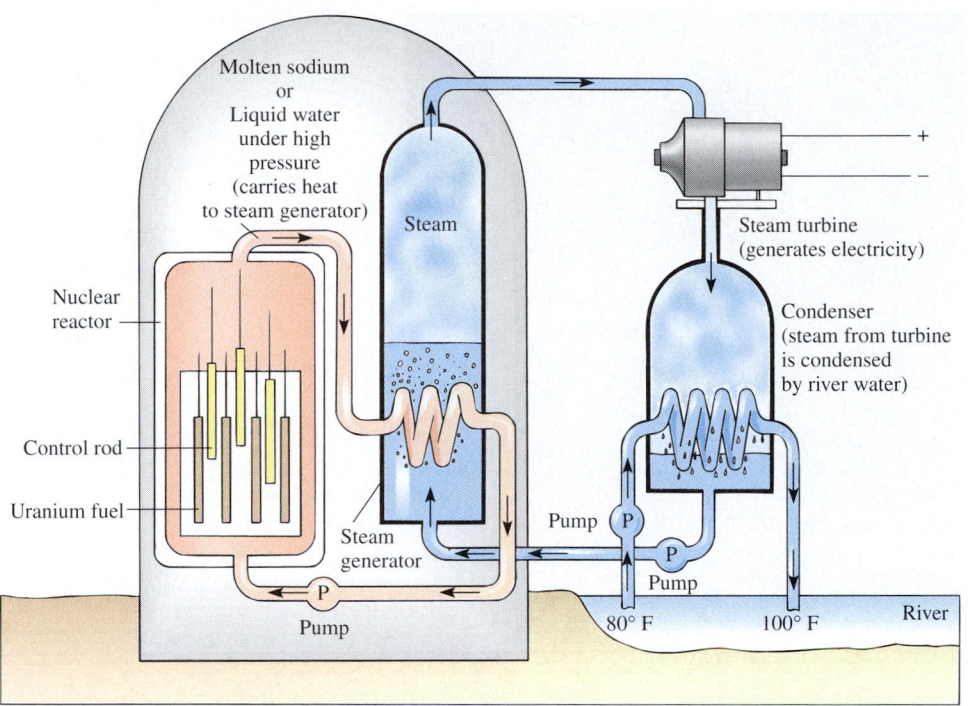

Figure 26-13 A schematic diagram of a light water reactor plant.

ingly important. However, intensifying opposition to nuclear power may mean that further growth in energy production using nuclear power in the United States must await technological developments to overcome the remaining hazards.

Light Water Reactors

Most commercial nuclear power plants in the United States are "light water" reactors, moderated and cooled by ordinary water. Figure 26-13 is a schematic diagram of a light water reactor plant. The reactor core at the left replaces the furnace in which coal, oil, or natural gas is burned in a fossil fuel plant. Such a fission reactor consists of five main components: (1) fuel, (2) moderator, (3) control rods, (4) cooling system, and (5) shielding.

Fuel

Rods of U_3O_8 enriched in uranium-235 serve as the fuel. Unfortunately, uranium ores contain only about 0.7% $^{235}_{92}U$. Most of the rest is nonfissionable $^{238}_{92}U$. The enrichment is done in processing and reprocessing plants by separating gaseous $^{235}UF_6$ from $^{238}UF_6$, prepared from the ore. Separation by diffusion is based on the slower rates of diffusion of heavier gas molecules (Section 12-13). Another separation procedure uses the ultracentrifuge.

A potentially more efficient method of enrichment would involve the use of sophisticated tunable lasers to ionize $^{235}_{92}U$ selectively and not $^{238}_{92}U$. The ionized $^{235}_{92}U$ could then be made to react with negative ions to form another compound, easily separated from the mixture. For this method to work, we must construct lasers capable of producing radiation monochromatic enough to excite one isotope and not the other—a difficult challenge.

Uranium is deposited on the negative electrode in the electrorefining phase of fuel reprocessing. The crystalline mass is about 97% LiCl and KCl. The remaining 3% uranium chloride is responsible for the amethyst color.

Moderator

The most efficient fission reactions occur with slow neutrons. Thus, the fast neutrons ejected during fission must be slowed by collisions with atoms of comparable mass that do not absorb them, called **moderators.** The most commonly used moderator is ordinary water, although graphite is sometimes used. The most efficient moderator is helium, which slows neutrons but does not absorb them all. The next most efficient is "heavy water" (deuterium oxide, $^2_1H_2O$ or $^2_1D_2O$). This is so expensive that it has been used chiefly in research reactors. A Canadian-designed power reactor that uses heavy water is more neutron-efficient than light water reactors.

Control Rods

Cadmium and boron are good neutron absorbers.

$$^{10}_5B + ^1_0n \longrightarrow ^7_3Li + ^4_2\alpha$$

The rate of a fission reaction is controlled by the use of movable control rods, usually made of cadmium or boron steel. They are automatically inserted in or removed from spaces between the fuel rods. The more neutrons absorbed by the control rods, the fewer fissions occur and the less heat is produced. Hence, the heat output is governed by the control system that operates the rods.

Cooling System

Two cooling systems are needed. First, the moderator itself serves as a coolant for the reactor. It transfers fission-generated heat to a steam generator. This converts water to steam. The steam then goes to turbines that drive generators to produce electricity. Another coolant (river water, sea water, or recirculated water) condenses the steam from the turbine, and the condensate is then recycled into the steam generator.

The danger of meltdown arises if a reactor is shut down quickly. The disintegration of radioactive fission products still goes on at a furious rate, fast enough to overheat the fuel elements and to melt them. So it is not enough to shut down the fission reaction. Efficient cooling must be continued until the short-lived isotopes are gone and the heat from their disintegration is dissipated. Only then can the circulation of cooling water be stopped.

The 1979 accident at Three Mile Island, near Harrisburg, Pennsylvania, was due to stopping the water pumps too soon *and* the inoperability of the emergency pumps. A combination of mechanical malfunctions, errors, and carelessness produced the overheating that damaged the fuel assembly. It did not and *could not explode,* although melting of the core material did occur. The 1986 accident at Chernobyl, in the USSR, was far more serious. The effects of that disaster will continue for decades.

Shielding

It is essential that people and the surrounding countryside be adequately shielded from possible exposure to radioactive nuclides. The entire reactor is enclosed in a steel containment vessel. This is housed in a thick-walled concrete building. The operating personnel are further protected by a so-called biological shield, a thick layer of organic material made of compressed wood fibers. This absorbs the neutrons and beta and gamma rays that would otherwise be absorbed in the human body.

The neutrons are the worst problem of radiation. The human body contains a high percentage of H_2O, which absorbs neutrons very efficiently. A new weapon, the neutron bomb, produces massive amounts of neutrons and so is effective against people, but it does not produce the long-lasting radiation of the fission atomic bomb.

26-16 NUCLEAR FUSION (THERMONUCLEAR) ENERGY

Fusion reactions are accompanied by even greater energy production per unit mass of reacting atoms than are fission reactions. However, they can be initiated only by extremely high temperatures. The fusion of ^2_1H and ^3_1H occurs at the lowest temperature of any fusion reaction known, but even this is 40,000,000 K! Such temperatures exist in the sun and other stars, but they are nearly impossible to achieve and contain on earth. **Thermonuclear** bombs (called fusion bombs or hydrogen bombs) of incredible energy have been detonated in tests but, thankfully, never in war. In them the necessary activation energy is supplied by the explosion of a fission bomb.

The explosion of a thermonuclear (hydrogen) bomb releases tremendous amounts of energy. If we could learn how to control this process, we would have nearly limitless amounts of energy.

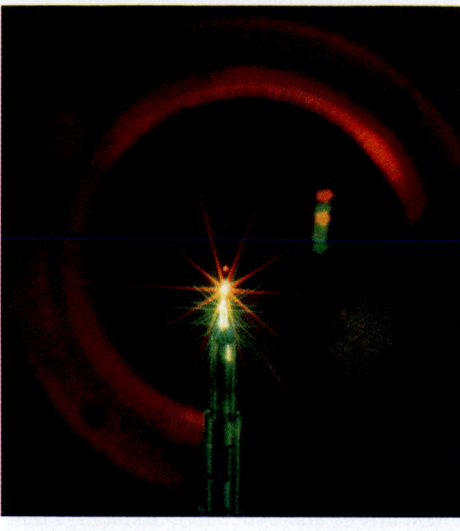

Nuclear fusion provides the energy of our sun and other stars. Development of controlled fusion as a practical source of energy requires methods to initiate and contain the fusion process. Here a very powerful laser beam has initiated a fusion reaction in a 1-mm target capsule that contained deuterium and tritium. In a 0.5-picosecond burst, 10^{13} neutrons were produced by the reaction $_1^2H + _1^3H \rightarrow _2^4He + _0^1n$.

It is hoped that fusion reactions can be harnessed to generate energy for domestic power. Because of the tremendously high temperatures required, no currently known structural material can confine these reactions. At such high temperatures all molecules dissociate and most atoms ionize, resulting in the formation of a new state of matter called a **plasma.** A very high temperature plasma is so hot that it melts and decomposes anything it touches, including structural components of a reactor. The technological innovation required to build a workable fusion reactor probably represents the greatest challenge ever faced by the scientific and engineering community.

Plasmas have been called the fourth state of matter.

Recent attempts at the containment of lower temperature plasmas by external magnetic fields have been successful, and they encourage our hopes. However, fusion as a practical energy source lies far in the future at best. The biggest advantages of its use would be that (1) the deuterium fuel can be found in virtually inexhaustible supply in the oceans; and (2) fusion reactions would produce only radionuclides of very short half-life, primarily tritium ($t_{1/2} = 12.3$ years), so there would be no long-term waste disposal problem. If controlled fusion could be brought about, it could liberate us from dependence on uranium and fossil fuels.

The plasma in a fusion reactor must not touch the walls of its vacuum vessel, which would be vaporized. In the Tokamak fusion test reactor, the plasma is contained within a magnetic field shaped like a doughnut. The magnetic field is generated by D-shaped coils around the vacuum vessel.

Key Terms

Alpha particle (α)   A particle that consists of two protons and two neutrons; identical to a helium nucleus.

Artificial transmutation   An artificially induced nuclear reaction caused by bombardment of a nucleus with subatomic particles or small nuclei.

Band of stability   A band containing stable (nonradioactive) nuclides in a plot of number of neutrons versus number of protons (atomic number).

Beta particle (β)   An electron emitted from the nucleus when a neutron decays to a proton and an electron.

Binding energy (nuclear binding energy)   The energy equivalent ($E = mc^2$) of the mass deficiency of an atom.

Chain reaction   A reaction that, once initiated, sustains itself and expands.

Cloud chamber   A device for observing the paths of speeding particles as vapor molecules condense on them to form fog-like tracks.

Control rods   Rods of materials such as cadmium or boron steel that act as neutron absorbers (not merely moderators), used in nuclear reactors to control neutron fluxes and therefore rates of fission.

Critical mass   The minimum mass of a particular fissionable nuclide, in a given volume, that is required to sustain a nuclear chain reaction.

Cyclotron   A device for accelerating charged particles along a spiral path.

Daughter nuclide   A nuclide that is produced in a nuclear decay.

Electron capture   Absorption of an electron from the first energy level (K shell) by a proton as it is converted to a neutron; also K capture.

Fast neutron   A neutron ejected at high kinetic energy in a nuclear reaction.

Fluorescence   Absorption of high energy radiation by a substance and the subsequent emission of visible light.

Gamma ray (γ)   High energy electromagnetic radiation.

Half-life of a radionuclide   The time required for half of a given sample to undergo radioactive decay.

Heavy water   Water containing deuterium, a heavy isotope of hydrogen, $_1^2$H.

Linear accelerator   A device used for accelerating charged particles along a straight line path.

Mass deficiency   The amount of matter that would be converted into energy if an atom were formed from constituent particles.

Moderator   A substance such as hydrogen, deuterium, oxygen, or paraffin capable of slowing fast neutrons upon collision.

Nuclear binding energy   The energy equivalent of the mass defi-

ciency; energy released in the formation of an atom from subatomic particles.

Nuclear fission   The process in which a heavy nucleus splits into nuclei of intermediate masses and one or more neutrons are emitted.

Nuclear fusion   The combination of light nuclei to produce a heavier nucleus.

Nuclear reaction   A reaction involving a change in the composition of a nucleus; it can evolve or absorb an extraordinarily large amount of energy.

Nuclear reactor   A system in which controlled nuclear fission reactions generate heat energy on a large scale. The heat energy is subsequently converted into electrical energy.

Nucleons   Particles comprising the nucleus; protons and neutrons.

Nuclides   Different atomic forms of all elements (in contrast to isotopes, which are different atomic forms of a single element).

Parent nuclide   A nuclide that undergoes nuclear decay.

Plasma   A physical state of matter that exists at extremely high temperatures, in which all molecules are dissociated and most atoms are ionized.

Positron   A nuclear particle with the mass of an electron but opposite charge.

Radiation   High energy particles or rays emitted in nuclear decay processes.

Radioactive dating   A method of dating ancient objects by determining the ratio of amounts of a parent nuclide and one of its decay products present in an object and relating the ratio to the object's age via half-life calculations.

Radioactive tracer   A small amount of radioisotope that replaces a nonradioactive isotope of the element in a compound whose path (for example, in the body) or whose decomposition products are to be monitored by detection of radioactivity; also called a radioactive label.

Radioactivity   The spontaneous disintegration of atomic nuclei.

Radioisotope   A radioactive isotope of an element.

Radionuclide   A radioactive nuclide.

Scintillation counter   A device used for the quantitative detection of radiation.

Slow neutron   A fast neutron slowed by collision with a moderator.

Thermonuclear energy   Energy from nuclear fusion reactions.

Transuranium elements   The elements with atomic numbers greater than 92 (uranium); none occurs naturally and all must be prepared by nuclear bombardment of other elements.

Exercises

Nuclear Stability and Radioactivity

1. How do nuclear reactions differ from ordinary chemical reactions?

2. What is the equation that relates the equivalence of matter and

energy? What does each term in this equation represent?

3. What is mass deficiency? What is binding energy? How are the two related?

4. What are nucleons? What is the relationship between the number

of protons and the atomic number? What is the relationship among the number of protons, the number of neutrons, and the mass number?

5. Define the term "binding energy per nucleon." How can this quantity be used to compare the stabilities of nuclei?

6. Describe the general shape of the plot of binding energy per nucleon against mass number.

7. (a) Briefly describe a plot of the number of neutrons against the atomic number (for the stable nuclides). Interpret the observation that the plot shows a band with a somewhat step-like shape. (b) Describe what is meant by "magic numbers" of nucleons.

8. The actual mass of a ^{79}Br atom is 78.91834 amu. (a) Calculate the mass deficiency in amu/atom and in g/mol for this isotope. (b) What is the nuclear binding energy in kJ/mol for this isotope?

9. The actual mass of a ^{71}Ga atom is 70.924700 amu. (a) Calculate the mass deficiency in amu/atom and in g/mol for this isotope. (b) What is the nuclear binding energy in kJ/mol for this isotope?

10. Calculate the following for ^{52}Cr (actual mass = 51.94059 amu). (a) mass deficiency in amu/atom; (b) mass deficiency in g/mol; (c) binding energy in J/atom; (d) binding energy in kJ/mol; (e) binding energy in MeV/nucleon.

11. Calculate the following for ^{24}Mg (actual mass = 23.985042 amu). (a) mass deficiency in amu/atom; (b) mass deficiency in g/mol; (c) binding energy in J/atom; (d) binding energy in kJ/mol; (e) binding energy in MeV/nucleon.

12. Calculate the nuclear binding energy in kJ/mol for each of the following. (a) $^{20}_{10}$Ne; (b) $^{87}_{37}$Rb; (c) $^{106}_{46}$Pd. The atomic masses are 19.99244 amu, 86.909187 amu, and 105.9032 amu, respectively.

13. Repeat Exercise 12 for (a) $^{14}_{7}$N, (b) $^{14}_{8}$O, and (c) $^{107}_{47}$Ag. Their respective atomic masses are 14.00307 amu, 14.00860 amu, and 106.905092 amu. Which of these nuclides has the greatest binding energy per nucleon?

14. Compare the behaviors of α, β, and γ radiation (a) in an electrical field, (b) in a magnetic field, and (c) with respect to ability to penetrate various shielding materials, such as a piece of paper and concrete. What is the composition of each type of radiation?

15. Why are α-particles that are absorbed internally by the body particularly dangerous?

16. Name some radionuclides that have medical uses, and give the uses.

17. Describe how radionuclides can be used in (a) research, (b) agriculture, and (c) industry.

18. Name and describe four methods for detection of radiation.

19. Describe how (a) nuclear fission and (b) nuclear fusion generate more stable nuclei.

20. What evidence exists to support the theory that nucleons are arranged in energy levels, or "shells," within the nucleus?

Nuclear Reactions

21. Consider a radioactive nuclide with a neutron/proton ratio that is larger than those for the stable isotopes of that element. What mode(s) of decay might be expected for this nuclide, and why?

22. Repeat Exercise 21 for a nuclide with a neutron/proton ratio that is smaller than those for the stable isotopes.

23. Calculate the neutron/proton ratio for each of the following radioactive nuclides and predict how each of the nuclides might decay: (a) $^{13}_{5}$B (stable mass numbers for B are 10 and 11); (b) $^{92}_{38}$Sr (stable mass numbers for Sr are between 84 and 88); (c) $^{192}_{82}$Pb (stable mass numbers for Pb are between 204 and 208).

24. Repeat Exercise 23 for (a) $^{193}_{79}$Au (stable mass number for Au is 197), (b) $^{189}_{75}$Re (stable mass numbers for Re are 185 and 187), and (c) $^{137}_{59}$Pr (stable mass number for Pr is 141).

25. Write the symbols for the daughter nuclei in the following radioactive decays (β refers to an e^-).

(a) $^{237}_{92}$U $\xrightarrow{-\beta}$ (d) ^{224}Ra $\xrightarrow{-\alpha}$

(b) ^{13}C $\xrightarrow{-n}$ (e) ^{18}F $\xrightarrow{-p}$

(c) ^{11}B $\xrightarrow{-\gamma}$ (f) $^{40}_{19}$K $\xrightarrow{+\beta}$

26. Predict the kind of decays you would expect for the following radionuclides. (a) $^{60}_{27}$Co (n/p ratio too high); (b) $^{20}_{11}$Na (n/p ratio too low); (c) $^{218}_{88}$Rn; (d) $^{67}_{29}$Cu; (e) $^{238}_{92}$U; (f) $^{11}_{6}$C.

27. What are nuclear bombardment reactions? Explain the shorthand notation used to describe bombardment reactions.

28. Fill in the missing symbols in the following nuclear bombardment reactions.

(a) $^{23}_{11}$Na + ? $\longrightarrow$ $^{23}_{12}$Mg + $^{1}_{0}$n

(b) $^{59}_{27}$Co + $^{1}_{0}$n $\longrightarrow$ $^{56}_{25}$Mn + ?

(c) $^{232}_{90}$Th + ? $\longrightarrow$ $^{240}_{96}$Cm + 4 $^{1}_{0}$n

(d) ? + $^{1}_{1}$H $\longrightarrow$ $^{29}_{14}$Si + $^{0}_{0}\gamma$

(e) $^{26}_{?}$Mg + ? $\longrightarrow$ $^{26}_{?}$Al + $^{1}_{0}$n

(f) $^{40}_{18}$Ar + ? $\longrightarrow$ $^{43}_{?}$K + $^{1}_{1}$H

29. Write the symbols for the daughter nuclei in the following nuclear bombardment reactions. (a) $^{60}_{28}$Ni (n, p); (b) $^{98}_{42}$Mo ($^{1}_{0}$n, β); (c) $^{35}_{17}$Cl (p, α); (d) $^{20}_{10}$Ne (α, γ); (e) $^{15}_{7}$N (p, α); (f) $^{10}_{5}$B (n, α).

30. Write the nuclear equation for each of the following bombardment processes. (a) $^{14}_{7}$N (α, p) $^{17}_{8}$O; (b) $^{106}_{46}$Pd (n, p) $^{106}_{45}$Rh; (c) $^{23}_{11}$Na (n, β^-)X. Identify X.

31. Repeat Exercise 30 for the following. (a) $^{113}_{48}$Cd (n, γ) $^{114}_{48}$Cd; (b) $^{6}_{3}$Li (n, α) $^{3}_{1}$H; (c) $^{2}_{1}$H (γ, p)X. Identify X.

32. Write the shorthand notation for each of the following nuclear reactions.

(a) $^{6}_{3}$Li + $^{1}_{0}$n $\longrightarrow$ $^{4}_{2}$He + $^{3}_{1}$H

(b) $^{31}_{15}$P + $^{2}_{1}$H $\longrightarrow$ $^{32}_{15}$P + $^{1}_{1}$H

(c) $^{238}_{92}$U + $^{1}_{0}$n $\longrightarrow$ $^{239}_{93}$Np + $^{0}_{-1}$e

33. Repeat Exercise 32 for the following.

(a) $^{253}_{99}$Es + $^{4}_{2}$He $\longrightarrow$ $^{256}_{101}$Md + $^{1}_{0}$n

(b) $^{27}_{13}$Al + $^{1}_{0}$n $\longrightarrow$ $^{26}_{13}$Al + 2$^{1}_{0}$n

(c) $^{37}_{17}$Cl + $^{1}_{1}$H $\longrightarrow$ $^{1}_{0}$n + $^{37}_{18}$Ar

34. Write the nuclear equations for the following processes. (a) $^{63}_{28}$Ni undergoing β^- emission; (b) two deuterium ions undergoing fusion to give $^{3}_{2}$He and a neutron; (c) a nuclide being bombarded by a neutron to form $^{7}_{3}$Li and an α-particle (identify the unknown nuclide); (d) $^{14}_{7}$N being bombarded by a neutron to form three α-particles and an atom of tritium.

35. Write the nuclear equations for the following processes. (a) $^{220}_{86}$Rn undergoing α decay; (b) $^{110}_{49}$In undergoing positron emission; (c) $^{127}_{53}$I being bombarded by a proton to form $^{121}_{54}$Xe and seven neutrons; (d) tritium and deuterium undergoing fusion to form an

α-particle and a neutron; (e) $^{95}_{42}Mo$ being bombarded by a proton to form $^{95}_{43}Tc$ and radiation (identify this radiation).

36. "Radioactinium" is produced in the actinium series from $^{235}_{92}U$ by the successive emission of an α-particle, a β^--particle, an α-particle, and a β^--particle. What are the symbol, atomic number, and mass number for "radioactinium?"

37. An alkaline earth element (Group IIA) is radioactive. It undergoes decay by emitting three α-particles in succession. In what periodic table group is the resulting element found?

38. A nuclide of element unnilquadium, $^{257}_{104}Unq$, is formed by the nuclear reaction of californium-249 and carbon-12, with the emission of four neutrons. This new nuclide rapidly decays by emitting an α-particle. Write the equation for each of these nuclear reactions.

39. Supply the missing information to each of the following equations.
(a) $^{53}_{24}Cr + ^{4}_{2}He \longrightarrow ^{1}_{0}n + \underline{?}$
(b) $^{187}_{75}Re + \beta \longrightarrow \underline{?}$
(c) $^{243}_{95}Am + ^{1}_{0}n \longrightarrow ^{244}_{?}Cm + \underline{?} + \gamma$
(d) $^{35}_{17}Cl + \underline{?} \longrightarrow ^{32}_{16}S + ^{4}_{2}He$

40. Supply the missing information to each of the following reactions.
(a) $^{14}_{7}N + \underline{?} \longrightarrow ^{17}_{?}O + p$
(b) $^{235}_{92}U + ^{1}_{0}n \longrightarrow ^{137}_{54}? + 2\,^{1}_{0}n + \underline{?}$
(c) $^{241}_{95}Am + ^{12}_{?}C \longrightarrow 4\,^{1}_{0}n + \underline{?}$

41. Describe how (a) cyclotrons and (b) linear accelerators work.

Rates of Decay

42. What does the half-life of a radionuclide represent? How do we compare the relative stabilities of radionuclides in terms of half-lives?

43. Why must all radioactive decays be first order?

44. Describe the process by which steady state (constant) ratios of carbon-14 to (nonradioactive) carbon-12 are attained in living plants and organisms. Describe the method of radiocarbon dating. What factors limit the use of this method?

45. The half-life of $^{19}_{8}O$ is 29 s. What fraction of the isotope originally present would be left after 10.0 s?

46. The half-life of $^{11}_{6}C$ is 20.3 min. How long will it take for 95.0% of a sample to decay? How long will it take for 99.5% of the sample to decay?

47. The activity of a sample of tritium decreased by 5.5% over the period of a year. What is the half-life of $^{3}_{1}H$?

48. A very unstable isotope of beryllium, ^{8}Be, undergoes α emission with a half-life of 0.07 fs. How long does it take for 99.99% of a 1.0-μg sample of ^{8}Be to undergo decay?

49. The $^{14}_{6}C$ activity of an artifact from a tomb (2930 $\pm$ 200 BC) was 8.1/min · g C. The half-life of $^{14}_{6}C$ is 5730 years and the current $^{14}_{6}C$ activity is 15.3/min · g C (that is, 15.3 disintegrations per minute per gram of carbon). How old is the artifact?

50. A piece of wood from a burial site was analyzed using $^{14}_{6}C$ dating and was found to have an activity of 12.7/min · g C. Using the data given in Exercise 49 for $^{14}_{6}C$ dating, determine the age of this piece of wood.

51. Strontium-90 is one of the harmful radionuclides that results from nuclear fission explosions. It decays by beta emission with a half-life of 28 years. How long would it take for 99.99% of a given sample released in an atmospheric test of an atomic bomb to disintegrate?

52. Carbon-14 decays by beta emission with a half-life of 5730 years. Assuming a particular object originally contained 7.50 μg of carbon-14 and now contains 0.72 μg of carbon-14, how old is the object?

Fission and Fusion

53. Briefly describe a nuclear fission process. What are the two most important fissionable materials?

54. What is a chain reaction? Why are nuclear fission processes considered chain reactions? What is the critical mass of a fissionable material?

55. Where have continuous nuclear fusion processes been observed? What is the main reaction that occurs in such sources?

56. The reaction that occurred in the first fusion bomb was $^{7}_{3}Li\,(p, \alpha)$ X. (a) Write the complete equation for the process and identify the product, X. (b) The atomic masses are 1.007825 amu for $^{1}_{1}H$, 4.00260 amu for α, and 7.01600 amu for $^{7}_{3}Li$. Find the energy for the reaction, in kJ/mol.

57. Summarize how an atomic bomb works, including how the nuclear explosion is initiated.

58. Discuss the pros and cons of the use of nuclear energy instead of other, more conventional types of energy based on fossil fuels.

59. Describe and illustrate the essential features of a light water fission reactor.

60. How is fissionable uranium-235 separated from nonfissionable uranium-238?

61. Distinguish between moderators and control rods of nuclear reactors.

62. What are the major advantages and disadvantages of fusion as a potential energy source, compared with fission? What is the major technological problem that must be solved to permit development of a fusion reactor?

Mixed Exercises

*63. Calculate the binding energy, in kJ/mol of nucleons, for the following isotopes. (a) $^{15}_{8}O$ with a mass of 15.00300 amu; (b) $^{16}_{8}O$ with a mass of 15.99491 amu; (c) $^{17}_{8}O$ with a mass of 16.99913 amu; (d) $^{18}_{8}O$ with a mass of 17.99915 amu; (e) $^{19}_{8}O$ with a mass of 19.0035 amu. Which of these would you expect to be most stable?

*64. The first nuclear transformation (discovered by Rutherford) can be represented by the shorthand notation $^{14}_{7}N\,(\alpha, p)\,^{17}_{8}O$. (a) Write the corresponding nuclear equation for this process. The respective atomic masses are 14.00307 amu for $^{14}_{7}N$, 4.00260 amu for $^{4}_{2}He$, 1.007825 amu for $^{1}_{1}H$, and 16.99913 amu for $^{17}_{8}O$. (b) Calculate the energy change of this reaction in kJ/mol.

65. A proposed series of reactions (known as the carbon–nitrogen cycle) that could be important in the very hottest region of the interior of the sun is

$$^{12}C + {}^1H \longrightarrow A + \gamma$$
$$A \longrightarrow B + {}^0_{+1}e$$
$$B + {}^1H \longrightarrow C + \gamma$$
$$C + {}^1H \longrightarrow D + \gamma$$
$$D \longrightarrow E + {}^0_{+1}e$$
$$E + {}^1H \longrightarrow {}^{12}C + F$$

Identify the species labeled A–F.

*66. The ultimate stable product of ^{238}U decay is ^{206}Pb. A certain sample of pitchblende ore was found to contain ^{238}U and ^{206}Pb in the ratio 68.3 atoms ^{238}U:31.7 atoms ^{206}Pb. Assuming that all the ^{206}Pb arose from uranium decay, and that no U or Pb has been lost by weathering, how old is the rock? The half-life of ^{238}U is 4.51×10^9 years.

*67. Potassium-40 decays to ^{40}Ar with a half-life of 1.3×10^9 years.

What is the age of a lunar rock sample that contains these isotopes in the ratio 34.1 atoms ^{40}K:65.9 atoms ^{40}Ar? Assume that no argon was originally in the sample and that none has been lost by weathering.

BUILDING YOUR KNOWLEDGE

68. Show by calculation which reaction produces the larger amount of energy per atomic mass unit of material reacting.

fission: $^{235}_{92}U + {}^1_0n \longrightarrow {}^{94}_{40}Zr + {}^{140}_{58}Ce + 6\,{}^0_{-1}e + 2\,{}^1_0n$

fusion: $2\,{}^2_1H \longrightarrow {}^3_1H + {}^1_1H$

The atomic masses are 235.0439 amu for $^{235}_{92}U$; 93.9061 amu for $^{94}_{40}Zr$; 139.9053 amu for $^{140}_{58}Ce$; 3.01605 amu for ^3_1H; 1.007825 amu for ^1_1H; 2.0140 amu for ^2_1H.

69. What would be the volume of helium, measured at STP, generated from the decay of the ^8Be sample in Exercise 48?

The characteristic flavors of almonds and cinnamon are due to aldehydes.

OBJECTIVES

As you study this chapter, you should learn

- *About saturated hydrocarbons (alkanes and cycloalkanes)—their structures and their nomenclature*

- *About unsaturated hydrocarbons (alkenes and alkynes)—their structures and their nomenclature*

- *About the occurrence of hydrocarbons in petroleum and natural gas*

- *To visualize structural and geometrical isomerism for hydrocarbons*

- *About some aromatic hydrocarbons—benzene, condensed aromatics, and substituted aromatic compounds*

- *About some common functional groups, the resulting classes of organic compounds, and how to name the compounds*
    halides
    alcohols and phenols
    ethers
    amines
    carboxylic acids and some of their derivatives
    aldehydes and ketones

Of the Group IVA elements, carbon is a nonmetal; silicon and germanium are metalloids; and tin and lead are metallic elements.

Organic chemistry is the chemistry of compounds that contain C—C or C—H bonds. Why is one entire branch of chemistry devoted to the behavior of the compounds of just one element? The answer is twofold: (1) There are many more compounds that contain carbon than there are compounds that do not (more than 11 million organic compounds have been identified), and (2) the molecules containing carbon can be so much larger and more complex (a methane molecule contains 5 atoms per molecule and DNA contains tens of billions of atoms per molecule).

Originally the term "organic" was used to describe compounds of plant or animal origin. "Inorganic" was used to describe compounds from minerals. In 1828 Friedrich Wöhler synthesized urea by boiling ammonium cyanate with water.

$$\text{NH}_4\text{OCN} \xrightarrow[\text{boil}]{\text{H}_2\text{O}} \text{H}_2\text{N}-\overset{\displaystyle O}{\overset{\|}{\text{C}}}-\text{NH}_2$$

ammonium cyanate (an inorganic compound)   urea (an organic compound)

Urea, $H_2N-CO-NH_2$, is the principal end product of metabolism of nitrogen-containing compounds in mammals. It is eliminated in the urine. An adult man excretes about 30 grams of urea in 24 hours.

This disproved the theory that held that organic compounds could be made only by living things. Today many organic compounds are manufactured from inorganic materials.

We encounter organic chemistry in every aspect of our lives. All life is based on a complex interrelationship of thousands of organic substances—from simple compounds such as sugars, amino acids, and fats to vastly more complex ones such as the enzymes that catalyze life's chemical reactions and the huge DNA molecules that carry genetic information from one generation to the next. The food we eat (including many additives); the clothes we wear; the plastics and polymers that are everywhere; our life-saving medicines; the paper on which we write; our fuels; many of our poisons, pesticides, carcinogens, dyes, soaps and detergents—all involve organic chemistry.

We normally think of petroleum and natural gas as fuel sources, but most synthetic organic materials are also derived from these two sources. More than half of the top 50 commercial chemicals are organic compounds derived in this way. Indeed, petroleum and natural gas may one day be more valuable as raw materials for organic synthesis than as fuel sources. If so, we may greatly regret our delay in developing alternative energy sources while burning up vast amounts of our petroleum and natural gas deposits as fuels.

A carbon atom has four electrons in its outermost shell with ground state configuration $1s^2 2s^2 2p^2$. The C atom can attain a stable configuration by forming four covalent bonds. As we saw in Chapter 8, each C atom can form single, double, or triple bonds by utilizing various hybridizations. The bonding of carbon is summarized in Table 27-1 using examples that we saw in Chapters 7 and 8. Carbon is unique among the elements in the extent to which it bonds to itself and in the diversity of compounds that are formed. The ability of an element to bond to itself is known as **catenation** ("chain making"). Carbon atoms catenate to form long chains, branched chains, and rings that may also have chains attached to them. A tremendous variety of carbon-containing compounds is known.

Table 27-1 *Hybridization of Carbon in Covalent Bond Formation*

Hybridization and Resulting Geometry	Orbitals Used by Each C Atom	Bonds Formed by Each C Atom	Example	
sp^3, tetrahedral	four sp^3 hybrids	four σ bonds	ethane	$\begin{array}{c} \text{H} \ \ \text{H} \\ \mid \ \ \mid \\ \text{H}-\text{C}-\text{C}-\text{H} \\ \mid \ \ \mid \\ \text{H} \ \ \text{H} \end{array}$
sp^2, trigonal planar	three sp^2 hybrids, one p orbital	three σ bonds, one π bond	ethylene	$\begin{array}{c} \text{H} \quad\quad \text{H} \\ \diagdown \quad\quad \diagup \\ \text{C}=\text{C} \\ \diagup \quad\quad \diagdown \\ \text{H} \quad\quad \text{H} \end{array}$
sp, linear	two sp hybrids, two p orbitals	two σ bonds, two π bonds	acetylene	$\text{H}-\text{C}\equiv\text{C}-\text{H}$

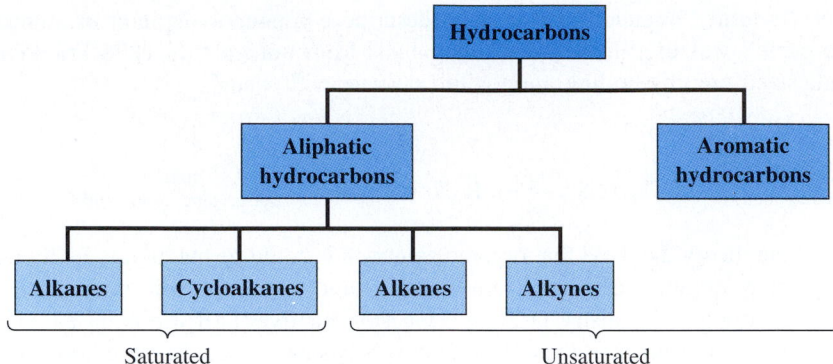

Figure 27-1 Classification of hydrocarbons.

Although millions of organic compounds are known, the elements they contain are very few—C and H; often N, O, S, or a halogen; and sometimes another element. The great number and variety of organic compounds are a result of the many different arrangements of atoms, or *structures,* that are possible. The chemical and physical properties of organic compounds are related to the structures of their molecules. Thus, the basis for organizing and understanding organic chemistry is structural theory.

In this text we shall give only a brief introduction to organic chemistry. In this chapter we organize organic compounds into the most common classes or "families" according to their structural features and learn how to name various types of compounds. Chapter 28 will present a few typical reactions undergone by organic substances.

Organic molecules are based on a framework of carbon–carbon and carbon–hydrogen bonds. Many compounds contain *only* the two elements C and H; they are called **hydrocarbons.** Hydrocarbons that contain delocalized rings such as the benzene ring (Section 9-6) are called **aromatic hydrocarbons.** Those that do not contain such delocalized systems are called **aliphatic hydrocarbons.** Aliphatic hydrocarbons that contain only sigma (σ) bonds (i.e., only single bonds) are called **saturated hydrocarbons.** Those that contain both sigma and pi (π) bonds (i.e., double, triple, or delocalized bonds) are called **unsaturated hydrocarbons.** These classifications are diagrammed in Figure 27-1. The first seven sections of this chapter are devoted to the study of hydrocarbons.

A **functional group** is a special arrangement of atoms within an organic molecule that is responsible for some characteristic chemical behavior of the compound. Different molecules that contain the same functional groups have similar chemical behavior. We shall follow the study of hydrocarbons with a presentation of some important characteristic functional groups.

SATURATED HYDROCARBONS

27-1 ALKANES AND CYCLOALKANES

The term "saturated" comes from early studies in which chemists tried to add hydrogen to various organic substances. Those to which no more hydrogen could be added were called saturated, by analogy with saturated solutions.

The *saturated hydrocarbons,* or **alkanes,** are compounds in which each carbon atom is bonded to four other atoms. Each H atom is bonded to only one C atom. Saturated hydrocarbons contain only single bonds. Petroleum and natural gas are composed mostly of saturated hydrocarbons.

In Section 8-7 we examined the structure of the simplest alkane, *methane,* CH_4. We saw that methane molecules are tetrahedral with sp^3 hybridization at carbon (Figure 27-2).

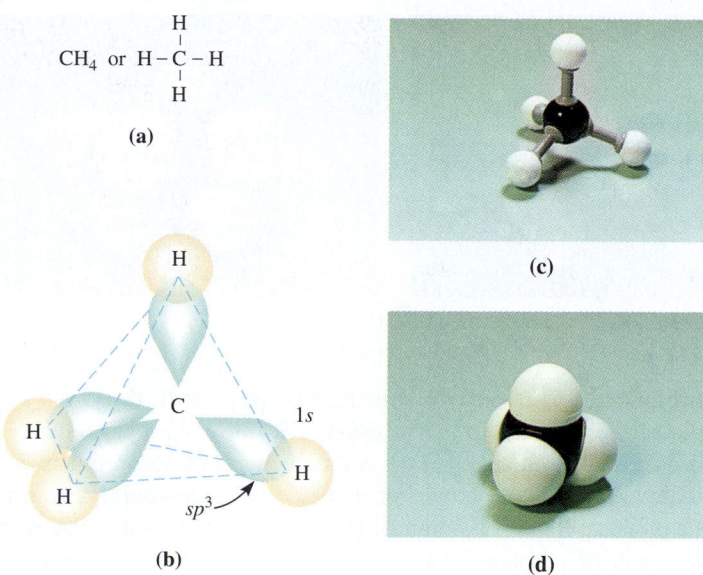

(a)

(b)

(c)

(d)

Figure 27-2 Representations of a molecule of methane, CH_4. (a) The condensed and line formulas for methane. (b) The overlap of the four sp^3 carbon orbitals with the s orbitals of four hydrogen atoms forms a tetrahedral molecule. (c) A ball-and-stick model and (d) a space-filling model of methane.

Ethane, C_2H_6, is the next simplest saturated hydrocarbon. Its structure is quite similar to that of methane. Two carbon atoms share a pair of electrons. Each carbon atom also shares an electron pair with each of three hydrogen atoms. Both carbon atoms are sp^3 hybridized (Figure 27-3). One can visualize the formation of a C_2H_6 molecule from two CH_4 molecules by mentally removing one H atom (and its electron) from each CH_4 molecule and then joining the fragments. *Propane*, C_3H_8, is the next member of the family (Figure 27-4). It can be considered the result of removing one H atom from a methane molecule and one from an ethane molecule, and joining the fragments.

Two different compounds have the formula C_4H_{10} but different structures and, hence, different properties. Such *isomers* result when two molecules contain the same atoms bonded together in different orders. The structures of these two isomeric C_4H_{10} alkanes

Iso = "same"; *mer* = "part." As we saw in Sections 25-5 through 25-7, isomers are substances that have the same numbers and kinds of atoms arranged differently. Isomerism in organic compounds is discussed more systematically in Chapter 28.

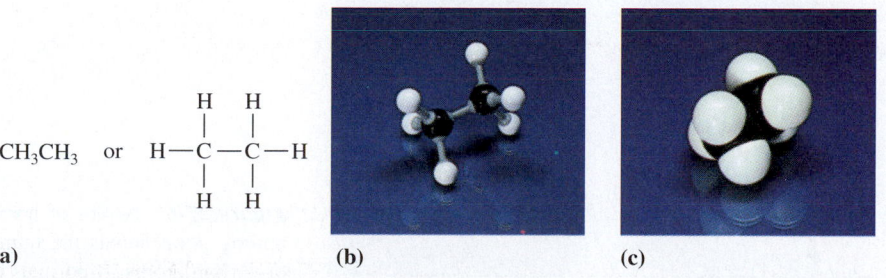

(a)

(b)

(c)

Figure 27-3 Models of ethane, C_2H_6. (a) The condensed and line formulas for ethane. (b) A ball-and-stick model and (c) a space-filling model of ethane.

Figure 27-4 Ball-and-stick and space-filling models of propane, C_3H_8.

are shown in Figure 27-5. These two structures correspond to the two ways in which a hydrogen atom can be removed from a propane molecule and replaced by a —CH_3 group. If a —CH_3 replaces an H on either of the end carbon atoms, the result is *normal butane,* commonly abbreviated as either butane or *n*-butane. Straight-chain, or normal, hydrocarbons are those in which there is no branching. If the —CH_3 group replaces an H from the central carbon atom of propane, however, the result is 2-methylpropane, or *isobutane.* This is the simplest *branched-chain hydrocarbon.*

The formulas of the alkanes can be written in general terms as C_nH_{2n+2}, where *n* is the number of carbon atoms per molecule. The first five members of the series are

Though somewhat misleading, the term "straight-chain" is widely used. The carbon chains are linear only in the structural formulas that we write. They are actually zigzag due to the tetrahedral bond angles at each carbon, and sometimes further kinked or twisted. Think of such a chain of carbon atoms as *continuous.* We can trace a single path from one terminal carbon to the other and pass through every other C atom *without backtracking.*

		CH_4	C_2H_6	C_3H_8	C_4H_{10}	C_5H_{12}
Number of C atoms $= n$ $=$		1	2	3	4	5
Number of H atoms $= 2n + 2 =$		4	6	8	10	12

The formula of each alkane differs from the next by CH_2, a *methylene group.*

A series of compounds in which each member differs from the next by a specific number and kind of atoms is called a **homologous series.** The properties of members of such a series are closely related. The boiling points of the lighter members of the saturated hydrocarbon series are shown in Figure 27-6. As the molecular weights of the normal hydrocarbons increase, their boiling points also increase regularly. Properties such as boiling point depend on the forces between molecules (Chapter 13). Carbon–carbon and

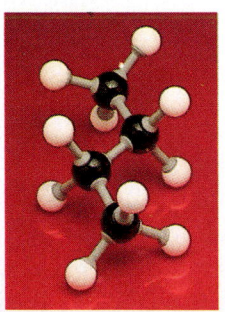

Figure 27-5 Ball-and-stick models of the two isomeric butanes, butane, $CH_3CH_2CH_2CH_3$, and isobutane, $CH_3CH(CH_3)CH_3$.

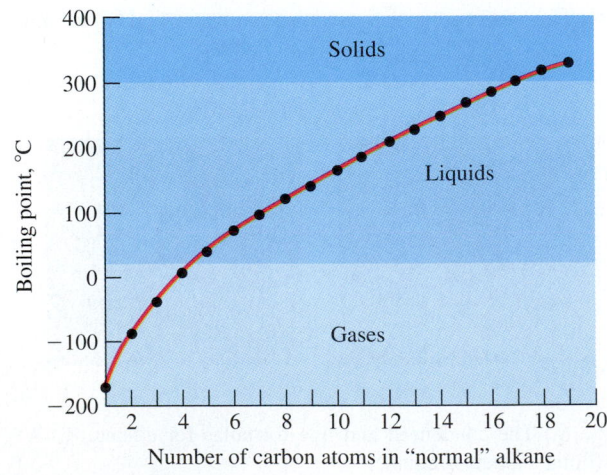

Figure 27-6 A plot of normal boiling point versus the number of carbon atoms in normal, i.e., straight-chain, saturated hydrocarbons.

Molecular Formula	IUPAC Name	Normal bp (°C)	Normal mp (°C)	State at Room Temperature

Table 27-2 *Some Straight-Chain Hydrocarbons (Alkanes)*

Molecular Formula	IUPAC Name	Normal bp (°C)	Normal mp (°C)	State at Room Temperature
CH_4	methane	−161	−184	
C_2H_6	ethane	−88	−183	
C_3H_8	propane	−42	−188	gas
C_4H_{10}	butane	+0.6	−138	
C_5H_{12}	pentane	36	−130	
C_6H_{14}	hexane	69	−94	
C_7H_{16}	heptane	98	−91	
C_8H_{18}	octane	126	−57	
C_9H_{20}	nonane	150	−54	
$C_{10}H_{22}$	decane	174	−30	liquid
$C_{11}H_{24}$	undecane	194.5	−25.6	
$C_{12}H_{26}$	dodecane	214.5	−9.6	
$C_{13}H_{28}$	tridecane	234	−6.2	
$C_{14}H_{30}$	tetradecane	252.5	+5.5	
$C_{15}H_{32}$	pentadecane	270.5	10	
$C_{16}H_{34}$	hexadecane	287.5	18	
$C_{17}H_{36}$	heptadecane	303	22.5	
$C_{18}H_{38}$	octadecane	317	28	solid
$C_{19}H_{40}$	nonadecane	330	32	
$C_{20}H_{42}$	eicosane	205 (at 15 torr)	36.7	

carbon–hydrogen bonds are essentially nonpolar and are arranged tetrahedrally around each C atom. As a result, saturated hydrocarbons are nonpolar molecules, and the only significant intermolecular forces are London forces (Section 13-2). These forces, which are due to induced dipoles, become stronger as the sizes of the molecules and the number of electrons in each molecule increase. Thus, trends such as those depicted in Figure 27-6 are due to the increase in effectiveness of London forces.

Some systematic method for naming compounds is necessary. The system in use today is prescribed by the International Union of Pure and Applied Chemistry (IUPAC). The names of the first 20 straight-chain alkanes are listed in Table 27-2. You should become familiar with at least the first ten. The names of the alkanes starting with pentane have prefixes (from Greek) that give the number of carbon atoms in the molecules. All alkane names have the -*ane* ending.

We have seen that there are two saturated C_4H_{10} hydrocarbons. For the C_5 hydrocarbons, there are three possible arrangements of the atoms. Thus, three different C_5H_{12} alkanes are known.

The number of structural isomers increases rapidly as the number of carbon atoms in saturated hydrocarbons increases. There are five isomeric C_6H_{14} alkanes (Table 27-3). Table 27-4 displays the number of isomers of some saturated hydrocarbons (alkanes). Most of the isomers have not been prepared or isolated. They probably never will be.

As the degree of branching increases for a series of molecules of the same molecular weight, the molecules become more compact (Figure 27-7). A compact molecule can have fewer points of contact with its neighbors than more extended molecules do. As a result, the total induced dipole forces (London forces) are weaker for branched molecules, and the boiling points of such compounds are lower.

Table 27-3 *Isomeric C_6H_{14} Alkanes*

IUPAC Name	Formula	Normal bp (°C)	Normal mp (°C)
hexane	$CH_3CH_2CH_2CH_2CH_2CH_3$	68.7	−94
2-methylpentane	$CH_3CH_2CH_2\underset{\underset{CH_3}{\vert}}{C}HCH_3$	60.3	−153.7
3-methylpentane	$CH_3CH_2\underset{\underset{CH_3}{\vert}}{C}HCH_2CH_3$	63.3	−118
2,2-dimethylbutane	$CH_3CH_2\overset{\overset{CH_3}{\vert}}{\underset{\underset{CH_3}{\vert}}{C}}CH_3$	49.7	−99.7
2,3-dimethylbutane	$CH_3\underset{\underset{CH_3}{\vert}}{C}H-\underset{\underset{CH_3}{\vert}}{C}HCH_3$	58.0	−128.4

Table 27-4 *Numbers of Possible Isomers of Alkanes*

Formula	Isomers
C_7H_{16}	9
C_8H_{18}	18
C_9H_{20}	35
$C_{10}H_{22}$	75
$C_{11}H_{24}$	159
$C_{12}H_{26}$	355
$C_{13}H_{28}$	802
$C_{14}H_{30}$	1,858
$C_{15}H_{32}$	4,347
$C_{20}H_{42}$	366,319
$C_{25}H_{52}$	36,797,588
$C_{30}H_{62}$	4,111,846,763

Cycloalkanes

The cyclic saturated hydrocarbons, or **cycloalkanes,** have the general formula C_nH_{2n}. The cycloalkanes (and other ring compounds that we shall encounter later) are often shown in simplified skeletal form in which each intersection of two lines represents a C atom; we mentally add enough H atoms to give each carbon atom four bonds. The first four unsub-

Figure 27-7 Ball-and-stick models of the three isomeric pentanes. Each contains 5 C atoms and 12 H atoms. The atoms are bonded in a different order in each of the three structural isomers.

stituted cycloalkanes and their simplified representations are

$$
\begin{array}{c}
\text{CH}_2 \\
/ \ \backslash \\
\text{H}_2\text{C}-\text{CH}_2 \\
\text{cyclopropane}
\end{array}
\qquad
\begin{array}{c}
\text{H}_2\text{C}-\text{CH}_2 \\
| \quad | \\
\text{H}_2\text{C}-\text{CH}_2 \\
\text{cyclobutane}
\end{array}
\qquad
\begin{array}{c}
\text{CH}_2 \\
\text{H}_2\text{C} \quad \text{CH}_2 \\
\text{H}_2\text{C}-\text{CH}_2 \\
\text{cyclopentane}
\end{array}
\qquad
\begin{array}{c}
\text{CH}_2 \\
\text{H}_2\text{C} \quad \text{CH}_2 \\
\text{H}_2\text{C} \quad \text{CH}_2 \\
\text{CH}_2 \\
\text{cyclohexane}
\end{array}
$$

In some of these structures, the bond angles are somewhat distorted from the ideal tetrahedral angle of 109.5°, the most severe distortions being 60° in cyclopropane and 90° in cyclobutane. As a result, these rings are said to be "strained," and these two compounds are unusually reactive for saturated hydrocarbons. Cyclopentane is quite stable with a nearly flat ring, because the bond angles in a regular pentagon (108°) are near the tetrahedral angle (109.5°).

Cyclohexane can assume two different geometries that are essentially strain-free, but to do this, the ring "puckers," or becomes nonplanar. The two stable arrangements of cyclohexane are called the *chair* and the *boat* forms (Figure 27-8).

27-2 NAMING SATURATED HYDROCARBONS

It is important to realize that many compounds (and their names) were so familiar to chemists before the development of the IUPAC system (beginning about 1890) that they continued to be called by their common, or "trivial," names. In this and the next chapter, IUPAC names of compounds appear in blue type, and their common alternative names are shown in black type.

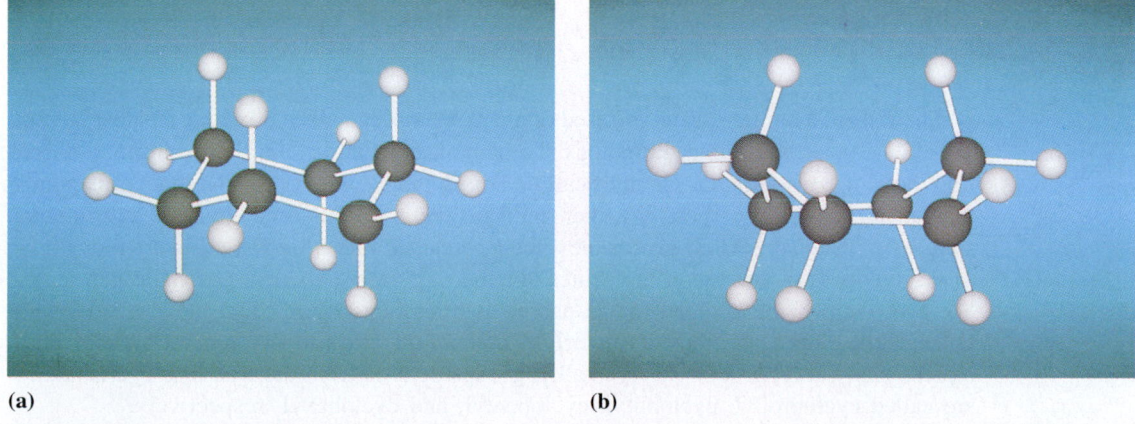

(a)   (b)

Figure 27-8   The two stable forms of the cyclohexane ring: (a) chair and (b) boat. The chair form is more stable because the hydrogens (or other substituents) are, on the average, farther from one another than in the boat form.

Table 27-5 *Some Alkanes and the Related Alkyl Groups*

Parent Hydrocarbon	Alkyl Group, —R	
CH_4, methane	—CH_3, methyl	

$$
\begin{array}{ccc}
H & & \\
| & & \\
H-C-H & or & CH_4 \\
| & & \\
H & &
\end{array}
\qquad
\begin{array}{ccc}
H & & \\
| & & \\
-C-H & or & -CH_3 \\
| & & \\
H & &
\end{array}
$$

| C_2H_6, ethane | —C_2H_5, ethyl | |

$$
\begin{array}{ccc}
H \;\; H & & \\
| \;\; | & & \\
H-C-C-H & or & CH_3CH_3 \\
| \;\; | & & \\
H \;\; H & &
\end{array}
\qquad
\begin{array}{ccc}
H \;\; H & & \\
| \;\; | & & \\
-C-C-H & or & -CH_2CH_3 \\
| \;\; | & & \\
H \;\; H & &
\end{array}
$$

| C_3H_8, propane | —C_3H_7, propyl | —C_3H_7, isopropyl |

$$
\begin{array}{ccc}
H \;\; H \;\; H & & \\
| \;\; | \;\; | & & \\
H-C-C-C-H & or & CH_3CH_2CH_3 \\
| \;\; | \;\; | & & \\
H \;\; H \;\; H & &
\end{array}
$$

$$
\begin{array}{c}
H \;\; H \;\; H \\
| \;\; | \;\; | \\
-C-C-C-H \\
| \;\; | \;\; | \\
H \;\; H \;\; H
\end{array}
\qquad
\begin{array}{c}
H \;\; H \;\; H \\
| \;\; | \;\; | \\
H-C-C-C-H \\
| \;\; \;\; | \\
H \;\;\;\;\;\; H
\end{array}
$$

| C_4H_{10}, butane | —C_4H_9, butyl | —C_4H_9, *sec*-butyl (read as "secondary butyl") |

$$
\begin{array}{ccc}
H \;\; H \;\; H \;\; H & & \\
| \;\; | \;\; | \;\; | & & \\
H-C-C-C-C-H & or & CH_3(CH_2)_2CH_3 \\
| \;\; | \;\; | \;\; | & & \\
H \;\; H \;\; H \;\; H & &
\end{array}
$$

$$
\begin{array}{c}
H \;\; H \;\; H \;\; H \\
| \;\; | \;\; | \;\; | \\
-C-C-C-C-H \\
| \;\; | \;\; | \;\; | \\
H \;\; H \;\; H \;\; H
\end{array}
\qquad
\begin{array}{c}
H \;\; H \;\; H \;\; H \\
| \;\; | \;\; | \;\; | \\
H-C-C-C-C-H \\
| \;\; \;\; | \;\; | \\
H \;\;\;\;\;\; H \;\; H
\end{array}
$$

| C_4H_{10}, methylpropane (common name isobutane) | —C_4H_9, *t*-butyl (read as "tertiary butyl") | —C_4H_9, isobutyl |

$$
\begin{array}{ccc}
CH_3 & & \\
| & & \\
H-C-CH_3 & or & CH(CH_3)_3 \\
| & & \\
CH_3 & &
\end{array}
$$

$$
\begin{array}{c}
CH_3 \\
| \\
-C-CH_3 \\
| \\
CH_3
\end{array}
\qquad
\begin{array}{c}
-CH_2 \\
| \\
H-C-CH_3 \\
| \\
CH_3
\end{array}
$$

The IUPAC naming system is based on the names of the normal hydrocarbons given in Table 27-2 and their higher homologs. To name a branched-chain hydrocarbon, we first find the longest continuous chain of carbon atoms and use the root name that corresponds to that normal hydrocarbon. We then indicate the positions and kinds of *substituents* attached to the chain. **Alkyl group** substituents attached to the longest chain are thought of as fragments of hydrocarbon molecules obtained by the removal of one hydrogen atom. We give them names related to the parent hydrocarbons from which they are derived. Other alkyl groups are named similarly (Table 27-5). We use the general symbol R to represent any alkyl group. The cycloalkyl groups derived from the first four cycloalkanes are called cyclopropyl, cyclobutyl, cyclopentyl, and cyclohexyl, respectively.

Summary of IUPAC Rules for Naming Alkanes

1. Find the longest continuous chain of C atoms. Choose the base name that describes the number of C atoms in this chain, with the ending -*ane* (Table 27-2). The longest chain may not be obvious at first; you may have to "turn corners."

2. Number the C atoms in this longest chain beginning at the end nearest the first branching. If there is branching at equal distances from both ends of the longest chain, begin numbering at the end nearest the second branch. If necessary, go to the third branch closest to an end, and so on, until a difference is located.

3. Assign the name and position number to each substituent. Arrange the substituents in alphabetical order. Hyphenated prefixes, such as *tert* and *sec,* are not used in alphabetization of the substituents.

4. Use the appropriate prefix to group like substituents: *di* = 2, *tri* = 3, *tetra* = 4, *penta* = 5, and so on. Don't consider these prefixes when alphabetizing attached groups.

5. Write the name as a single word. Use hyphens to separate numbers and letters (plus some hyphenated prefixes) and commas to separate numbers. Don't leave any spaces.

Formerly the rule was to name the substituents in order of increasing complexity. This was sometimes difficult to determine. With the use of computers in literature searches, it became necessary to adopt the more definitive alphabetization of the names of substituents.

Let us name the following compound.

$$
\begin{array}{c}
\text{H} \\
| \\
\text{H} \quad \text{H}-\text{C}-\text{H} \quad \text{H} \quad \text{H} \quad \text{H} \quad \text{H} \\
| \qquad | \qquad | \quad | \quad | \quad | \\
\text{H}-\text{C}--\text{C}----\text{C}-\text{C}-\text{C}-\text{C}-\text{H} \qquad or \\
| \qquad | \qquad | \quad | \quad | \quad | \\
\text{H} \qquad \text{H} \qquad \text{H} \quad \text{H} \quad \text{H} \quad \text{H}
\end{array}
$$

$$CH_3CH(CH_3)CH_2CH_2CH_2CH_3$$

We follow Rules 1 and 2 to number the carbon atoms in the longest chain.

$$
\begin{array}{c}
\qquad \text{CH}_3 \\
\qquad | \\
\underset{1}{\text{CH}_3}-\underset{2}{\text{CH}}-\underset{3}{\text{CH}_2}-\underset{4}{\text{CH}_2}-\underset{5}{\text{CH}_2}-\underset{6}{\text{CH}_3}
\end{array}
$$

The methyl group is attached to the *second* carbon atom in a *six-carbon* chain, so the compound is named 2-methylhexane.

The following examples further illustrate the rules of nomenclature.

Parentheses are used to conserve space. Formulas written with parentheses must indicate unambiguously the structure of the compound. The parentheses here indicate that the CH_3 group is attached to the C that precedes it.

It is incorrect to name the compound 5-methylhexane because that violates Rule 2.

EXAMPLE 27-1 *Naming Alkanes*

Name the compound represented by the structural formula

$$
\begin{array}{c}
\qquad \text{CH}_3 \\
\qquad | \\
\text{CH}_3-\text{C}-\text{CH}_2-\text{CH}_2-\text{CH}_3 \qquad or \qquad (CH_3)_3C(CH_2)_2CH_3 \\
\qquad | \\
\qquad \text{CH}_3
\end{array}
$$

Remember that the (dash) formulas indicate which atoms are bonded to each other. They do *not* show molecular geometry. In this molecule, each C atom is tetrahedrally bonded to four other atoms.

Plan

We first find the longest carbon chain and number it to give substituents the smallest possible numbers. Then we name the substituents as in Table 27-5 and specify the number of each as indicated in Rule 4 above.

Solution

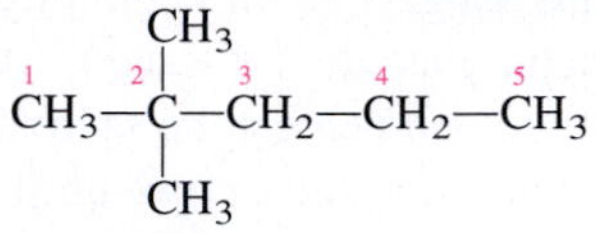

$$\underset{\overset{|}{CH_3}}{\overset{\overset{CH_3}{|}}{\underset{1}{CH_3}-\underset{2}{C}-\underset{3}{CH_2}-\underset{4}{CH_2}-\underset{5}{CH_3}}}$$

The longest chain contains five carbons, so this compound is named as a derivative of pentane. There are two methyl groups, both at carbon number 2. The IUPAC name of this compound is 2,2-dimethylpentane.

EXAMPLE 27-2   *Naming Alkanes*

Name the compound represented by the structural formula

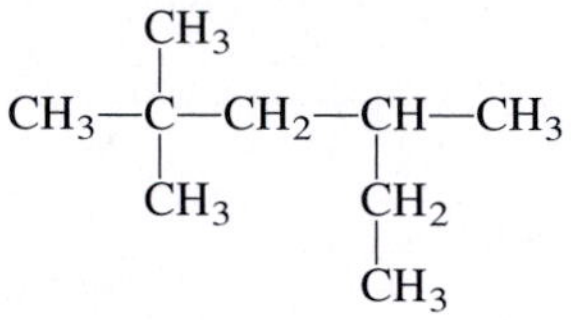

$$\underset{\overset{|}{CH_3}}{\overset{\overset{CH_3}{|}}{CH_3-C-CH_2-CH-CH_3}}\!\!\underset{\overset{|}{CH_3}}{\overset{|}{CH_2}}$$

Plan

The approach is the same as in Example 27-1. We should be aware that the longest continuous carbon chain might not be *written* in a straight line.

Solution

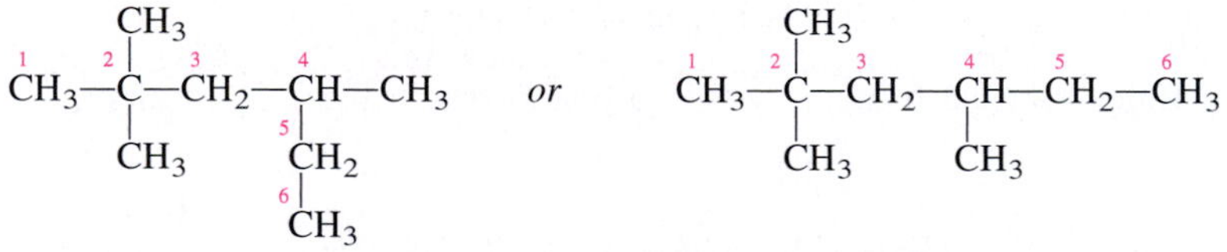

or (written to emphasize better the six-C chain)

The longest chain contains six carbons, so this compound is named as a derivative of hexane. There are three methyl substituents—two at carbon number 2 and one at carbon number 4. The IUPAC name of this compound is 2,2,4-trimethylhexane.

You should now work Exercises 10 and 12.

EXAMPLE 27-3   *From Names to Formulas*

Write the structure for 4-*tert*-butyl-2,5-dimethylheptane.

Plan

The root name "heptane" indicates that there is a seven-carbon chain.

$$C-C-C-C-C-C-C$$

The names and numbers of the substituents tell us where to attach the alkyl groups.

Solution

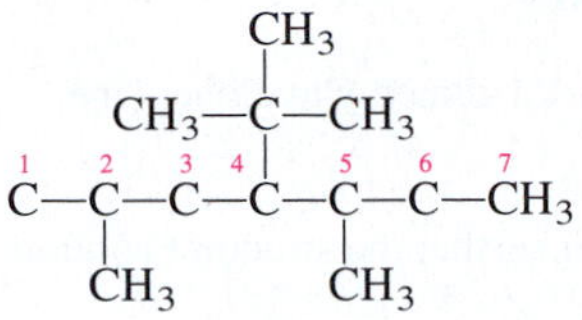

$$CH_3$$
$$CH_3-\overset{|}{C}-CH_3$$
$$\overset{1}{C}-\overset{2}{C}-\overset{3}{C}-\overset{4}{C}-\overset{5}{C}-\overset{6}{C}-\overset{7}{CH_3}$$
$$\overset{|}{CH_3}\quad\quad\overset{|}{CH_3}$$

Then we fill in enough hydrogens to saturate each C atom and arrive at the structure

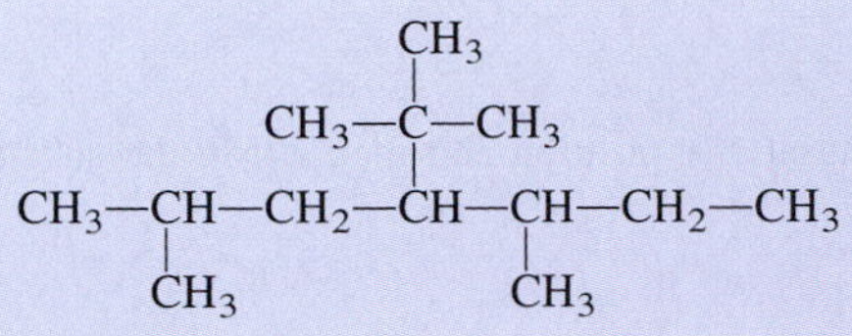

$$CH_3$$
$$CH_3-\overset{|}{C}-CH_3$$
$$CH_3-\overset{|}{CH}-CH_2-\overset{|}{CH}-\overset{|}{CH}-CH_2-CH_3$$
$$\overset{|}{CH_3}\quad\quad\quad\overset{|}{CH_3}$$

You should now work Exercise 13.

EXAMPLE 27-4 *Alkanes with Cycloalkyl Substituents*

Write the structure for the compound 2-cyclopropyl-3-ethylpentane.

Plan
The root name "pentane" tells us that the structure is based on a five-carbon chain. We place the substituents at the positions indicated by the numbers in the name.

Solution

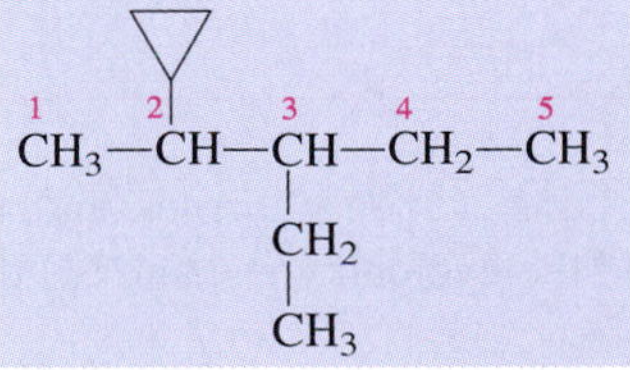

$$\overset{1}{CH_3}-\overset{2}{CH}-\overset{3}{CH}-\overset{4}{CH_2}-\overset{5}{CH_3}$$
$$\overset{|}{CH_2}$$
$$\overset{|}{CH_3}$$

where the symbol  ⋎ represents the cyclopropyl group, $\begin{array}{c} CH_2-CH_2 \\ \diagdown \diagup \\ CH \\ | \end{array}$

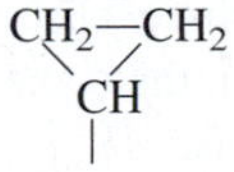

You should now work Exercise 14.

The names of substituted cycloalkanes are derived analogously to those of alkanes. (1) The base name is determined by the number of carbon atoms in the ring using the same base name as the alkane with the addition of cyclo in front. (2) If only one substituent is attached to the ring, no "location number" is required because all positions in a cycloalkane are equivalent. (3) Two or more functional groups on the ring are identified by location numbers, which should be assigned sequentially to the ring carbons in the order that gives the *smallest sum* of location numbers.

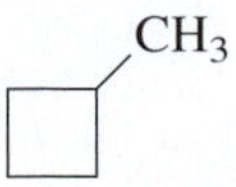

methylcyclobutane

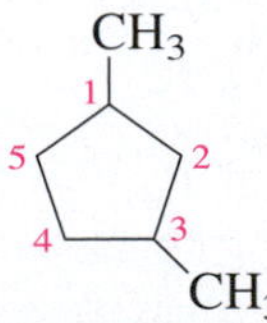

1,3-dimethylcyclopentane
(not 1,4-dimethylcyclopentane)

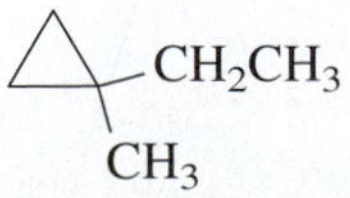

1-ethyl-1-methylcyclopropane

EXAMPLE 27-5   *Cycloalkanes*

Draw the structure for 3-*sec*-butyl-1,1-dimethylcycloheptane.

Plan

The "cycloheptane" base name tells us that the structure contains substituents on a saturated seven-membered ring.

We number this ring starting at any position and then attach the substituents as indicated by the numbers in the name.

Solution

In the simplified notation there is a carbon atom at each intersection and sufficient H atoms to give each C atom four bonds. Thus, in addition to those shown, there are 2 H atoms on each of the C atoms 2, 4, 5, 6, and 7 and 1 H atom on carbon 3.

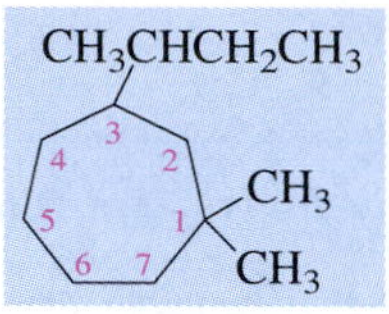

$$CH_3CHCH_2CH_3$$

(ring numbered 1–7 with CH_3 and CH_3 substituents on carbon 1)

You should now work Exercise 16.

UNSATURATED HYDROCARBONS

There are three classes of unsaturated hydrocarbons: (1) the alkenes and their cyclic counterparts, the cycloalkenes; (2) the alkynes and the cycloalkynes; and (3) the aromatic hydrocarbons.

27-3 ALKENES

Recall that the cycloalkanes may also be represented by the general formula C_nH_{2n}.

The simplest **alkenes** contain one carbon–carbon double bond, C=C, per molecule. The general formula for noncyclic alkenes is C_nH_{2n}. The simplest alkene is C_2H_4, which is usually called by its common name, ethylene.

The bonding in ethylene was described in Section 8-13 (see margin). The hybridization (sp^2) and bonding at other double-bonded carbon atoms are similar. Both carbon atoms in C_2H_4 are located at the centers of trigonal planes. Rotation about C=C double bonds does

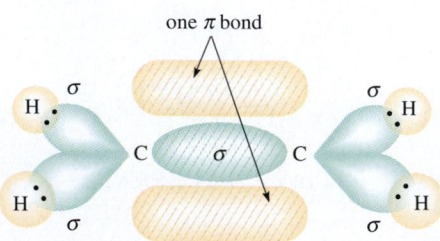

Four C—H σ bonds, one C—C σ bond, and one C—C π bond in the planar C_2H_4 molecule.

CHEMISTRY IN USE

Our Daily Lives

Petroleum

Petroleum refinery towers.

Petroleum, or crude oil, was discovered in the United States (Pennsylvania) in 1859 and in the Middle East (Iran) in 1908. It has been found in many other locations since these initial discoveries, and is now pumped from the ground in many parts of the world. Petroleum consists mainly of hydrocarbons. Small amounts of organic compounds containing nitrogen, sulfur, and oxygen are also present. Each oil field produces petroleum with a particular set of characteristics. Distillation of petroleum produces several fractions, as shown in the table.

Because gasoline is so much in demand, higher hydrocarbons (C_{12} and higher) are "cracked" to increase the amount of gasoline that can be made from a barrel of petroleum. The hydrocarbons are heated, in the absence of air and in the presence of a catalyst, to produce a mixture of smaller alkanes that can be used in gasoline. This process is called *catalytic cracking*.

The **octane number** (rating) of a gasoline indicates how smoothly it burns and how much engine "knock" it produces. (Engine knock is caused by premature detonation of fuel in the combustion chamber.) 2,2,4-trimethylpentane, isooctane, has excellent combustion properties and was arbitrarily assigned an octane number of 100. Normal heptane, $CH_3(CH_2)_5CH_3$, has very poor combustion properties and was assigned an octane number of zero.

$$CH_3-\overset{\overset{\displaystyle CH_3}{|}}{\underset{\underset{\displaystyle CH_3}{|}}{C}}-CH_2-\overset{\overset{\displaystyle CH_3}{|}}{CH}-CH_3$$

isooctane (octane number = 100)

$$CH_3CH_2CH_2CH_2CH_2CH_2CH_3$$

n-heptane (octane number = 0)

Mixtures of these two were prepared and burned in test engines to establish the octane scale. The octane number of such a mixture is the percentage of isooctane in it. Gasolines burned in standard test engines are assigned octane numbers based on the compression ratio at which they begin to knock. A 90-octane fuel produces the same amount of knock as the 90% isooctane–10% *n*-heptane mixture. Branched-chain compounds produce less knock than straight-chain compounds. The octane numbers of two isomeric hexanes are

$$CH_3(CH_2)_4CH_3$$

n-hexane
octane number = 25

$$(CH_3)_3CCH_2CH_3$$

2,2-dimethylbutane
octane number = 92

Petroleum Fractions

Fraction*	Principal Composition	Distillation Range
natural gas	C_1-C_4	below 20°C
bottled gas	C_5-C_6	20–60°
gasoline	C_4-C_{12}	40–200°
kerosene	$C_{10}-C_{16}$	175–275°
fuel oil, diesel oil	$C_{15}-C_{20}$	250–400°
lubricating oils	$C_{18}-C_{22}$	above 300°
paraffin	$C_{23}-C_{29}$	mp 50–60°
asphalt		viscous liquid ("bottoms fraction")
coke		solid

Other descriptions and distillation ranges have been used, but all are similar.

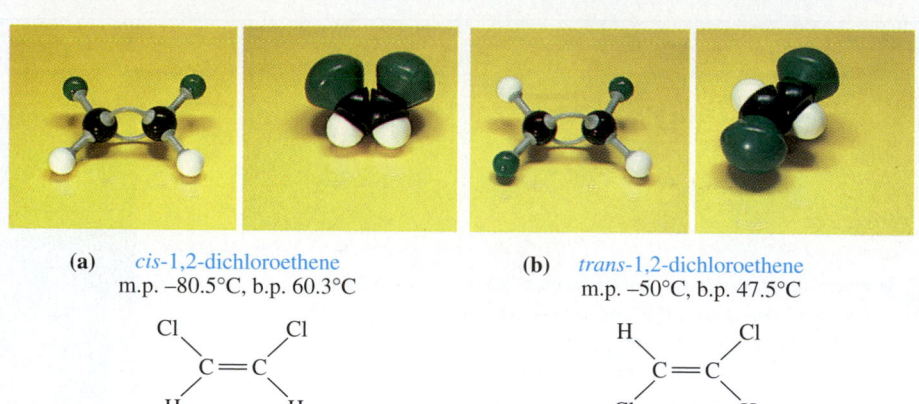

Figure 27-9 Two isomers of 1,2-dichloroethene are possible because rotation about the double bond is restricted. This is an example of *geometric* isomerism. A ball-and-stick model and a space-filling model are shown for each isomer. (a) The *cis* isomer. (b) The *trans* isomer.

(a) *cis*-1,2-dichloroethene
m.p. −80.5°C, b.p. 60.3°C

(b) *trans*-1,2-dichloroethene
m.p. −50°C, b.p. 47.5°C

not occur significantly at room temperature. Therefore, compounds that have the general formula (XY)C=C(XY) exist as a pair of *cis–trans* isomers. Figure 27-9 shows the *cis–trans* isomers of 1,2-dichloroethene. The existence of compounds with different arrangements of groups on the opposite sides of a bond with restricted rotation is called **geometrical isomerism.** This *cis–trans* isomerism can occur across double bonds in alkenes and across single bonds in rings.

Two shared electron pairs draw the atoms closer together than a single electron pair does. Thus, carbon–carbon double bonds are shorter than C—C single bonds, 1.34 Å versus 1.54 Å. The physical properties of the alkenes are similar to those of the alkanes, but their chemical properties are quite different.

The root for the name of each alkene is derived from the alkane having the same number of C atoms as the longest chain containing the double bond. In the trivial (common) system of nomenclature, the suffix *-ylene* is added to the characteristic root. In systematic (IUPAC) nomenclature, the suffix *-ene* is added to the characteristic root.

Summary of IUPAC Rules for Naming Alkenes and Cycloalkenes

1. Locate the C atoms in the *longest* continuous C chain *that contains the double bond*. Use the base name prefix with the ending *-ene*.

2. Number the C atoms of this basic chain sequentially *beginning at the end nearer the double bond*. Insert the number describing the position of the double bond (indicated by its *first* carbon location) before the base name. (This is necessary only for chains of four or more C atoms, because only one position is possible for a double bond in a chain of two or three carbon atoms.)

3. In naming alkenes, the double bond takes positional precedence over substituents on the carbon chain. The double bond is assigned the lowest possible number.

4. To name compounds with possible geometrical isomers, consider the two largest groups within the carbon chain that contains the double bond—these are indicated as part of the base name. The isomer in which the largest groups at each end of the C=C are located on opposite sides is called *trans*. If the largest groups are on the same side, the molecule is referred to as *cis*. Insert the prefix *cis*- or *trans*- just before the number of the double bond to indicate whether the largest groups are on the same or opposite sides, respectively, of the double bond.

5. For cycloalkenes, the double bond is assumed to be between C atoms 1 and 2, so no position number is needed to describe it.

The prefix *trans*- means "across" or "on the other side of." As a reminder of this terminology, think of words such as "transatlantic."

Some illustrations of this naming system follow.

$$\overset{4}{C}H_3 \overset{3}{C}H_2 - \overset{2}{C}H = \overset{1}{C}H_2$$

	CH$_2$=CH$_2$	CH$_3$—CH=CH$_2$	$\overset{4}{C}H_3\overset{3}{C}H_2—\overset{2}{C}H=\overset{1}{C}H_2$
systematic:	ethene	propene	1-butene
trivial:	(ethylene)	(propylene)	(butylene)

$$\overset{1}{C}H_3\overset{2}{C}H=\overset{3}{C}H\overset{4}{C}H_3$$

2-butene

$$CH_3 - C = CH_2$$
$$\quad\quad\ \ |$$
$$\quad\quad CH_3$$

methylpropene
(isobutylene)

There are two isomers of 2-butene, *cis*-2-butene and *trans*-2-butene.

The following two names illustrate the application of Rule 3.

$$\overset{4}{C}H_3 - \overset{3}{C}H_2 - \overset{2}{C} = \overset{1}{C}H_2 \quad\quad \overset{1}{C}H_3 - \overset{2}{C}H = \overset{3}{C}H - \overset{4}{C}H - \overset{5}{C}H_3$$
$$\quad\quad\quad\quad\ |\quad\quad\quad\quad\quad\quad\quad\quad\quad\quad\quad |$$
$$\quad\quad\quad CH_3\quad\quad\quad\quad\quad\quad\quad\quad\quad\quad CH_3$$

2-methyl-1-butene   4-methyl-2-pentene

Some alkenes, called **polyenes,** have two or more carbon–carbon double bonds per molecule. The suffixes *-adiene, -atriene,* and so on are used to indicate the number of (C=C) double bonds in a molecule.

$$\overset{1}{C}H_2=\overset{2}{C}H-\overset{3}{C}H=\overset{4}{C}H_2 \quad\quad \overset{4}{C}H_3-\overset{3}{C}H=\overset{2}{C}=\overset{1}{C}H_2$$

1,3-butadiene   1,2-butadiene

1,3-Butadiene and similar molecules that contain *alternating* single and double bonds are described as having **conjugated double bonds.** Such compounds are of special interest because of their polymerization reactions (Section 28-11).

▼ **PROBLEM-SOLVING TIP**   *When Is the Location of a Substituent or a Multiple Bond Not Included in an IUPAC Name?*

If only one location is possible for a substituent or double bond, the number that indicates the location can be left out of the compound's name. If there is doubt, put the number in the name. For example, "2-methyl-1-propene" is named simply methylpropene. There are no other locations possible for the substituent or the double bond. The name without the numbers is the correct name. However, both numbers are necessary to name 3-methyl-1-butene.

$$CH_3 - C = CH_2 \quad\quad\quad CH_3 - CH - CH = CH_2$$
$$\quad\quad\ \ |\quad\quad\quad\quad\quad\quad\quad\quad\ \ |$$
$$\quad\quad CH_3\quad\quad\quad\quad\quad\quad\quad\ CH_3$$

methylpropene   3-methyl-1-butene

EXAMPLE 27-6   *Naming Alkenes*

Name the following two alkenes.

(a)
$$CH_3 \quad\quad CH_2CH_3$$
$$\quad\ \diagdown\quad\diagup$$
$$\quad\quad C=C$$
$$\quad\diagup\quad\diagdown$$
$$H\quad\quad CH_3$$

(b)
$$CH_3CH_2 \quad\quad CH_2CH_3$$
$$\quad\quad\diagdown\quad\diagup$$
$$\quad\quad\ C=C$$
$$\quad\quad\diagup\quad\diagdown$$
$$H\quad\quad CH_2CH_3$$

Plan

For each compound, we first find the longest chain that includes the double bond, and then number it beginning at the end nearer the double bond (Rules 1 and 2). Then we specify the identities and positions of substituents in the same way we did for alkanes. In (a) we specify the geometrical isomer by locating the two largest groups attached to the double-bonded carbons and then describe their relationship using the *cis–trans* terminology.

Solution

(a)

$$^1CH_3 \qquad ^4CH_2CH_3^5$$
$$\underset{2}{C}=\underset{3}{C}$$
$$H \qquad CH_3$$

The longest such chain contains five atoms and has a double bond beginning at atom 2; thus, the compound is named as a derivative of 2-pentene. Now we must apply Rule 4. The two largest groups in the chain are the terminal —CH_3 (carbon 1) and the —CH_2CH_3 (carbons 4 and 5); these are on the *same side* of the double bond, so we name the compound as a derivative of *cis*-2-pentene. The only substituent is the methyl group at carbon 3. The full name of the compound is

3-methyl-*cis*-2-pentene.

(b) There are two choices for the longest chain, and either one would have an ethyl substituent.

$$^6CH_3CH_2^5 \qquad CH_2CH_3 \qquad\qquad ^6CH_3CH_2^5 \qquad ^2CH_2CH_3^1$$
$$\underset{4}{C}=\underset{3}{C} \qquad\qquad or \qquad\qquad \underset{4}{C}=\underset{3}{C}$$
$$H \qquad ^2CH_2CH_3^1 \qquad\qquad H \qquad CH_2CH_3$$

We could number either chain from the other end and still have the double bond starting at carbon 3; we number from the end that gives the carbon bearing the ethyl group the lowest possible position number, 3. Carbon 3 has two equivalent substituents, so geometrical isomerism is not possible, and we do not use the *cis–trans* terminology. The name is   3-ethyl-3-hexene.

You should now work Exercise 18.

The cycloalkenes are represented by the general formula C_nH_{2n-2}. Two cycloalkenes and their skeletal representations are

$$H_2C\text{——}CH_2$$
$$HC \diagup \quad \diagdown CH_2 \qquad or$$
$$\qquad CH$$

cyclopentene

$$\qquad\qquad CH_2$$
$$H_2C \diagup \quad \diagdown CH_2$$
$$HC \diagdown \quad \diagup CH_2 \qquad or$$
$$\qquad CH$$

cyclohexene

▼ **PROBLEM-SOLVING TIP**   *How to Draw Skeletal Representations*

The rules for drawing skeletal representations follow.

1. A carbon atom is assumed to be at each intersection of two lines and at the end of each line. All carbon–carbon bonds are shown as lines. A carbon atom might be shown for clarity.

2. Because carbon has four bonds, we mentally supply hydrogens to give each carbon its four bonds.

3. Atoms other than hydrogen and carbon are shown. The skeletal representation of 3-chloro-4-ethyl-*cis*-3-heptene is:

and its formula is $C_9H_{17}Cl$.

EXAMPLE 27-7 *Cycloalkenes*

Draw the structure of 3-methylcyclohexene.

Plan

We draw the ring of the specified size with one double bond in the ring. We number the ring so the double bond is between atoms 1 and 2. Then we add the designated substituents at the indicated positions.

Solution

We number the six-membered ring so that the double bond is between atoms 1 and 2.

A methyl group is attached at carbon 3; the correct structure is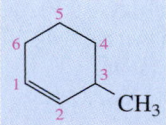

Remember that each intersection of two lines represents a carbon atom; there are enough H atoms at each C atom to make a total of four bonds to carbon.

You should now work Exercise 19.

27-4 ALKYNES

The **alkynes,** or acetylenic hydrocarbons, contain carbon–carbon triple bonds, —C≡C—. The noncyclic alkynes with one triple bond per molecule have the general formula C_nH_{2n-2}. The bonding in all alkynes is similar to that in acetylene (Section 8-14). Triply bonded carbon atoms are *sp* hybridized. The triply bonded atoms and their adjacent atoms lie on a straight line (Figure 27-10).

Alkynes are named like the alkenes except that the suffix *-yne* is added to the characteristic root. The first member of the series is commonly called acetylene. Its molecular formula is C_2H_2. It is thermodynamically unstable, decomposing explosively to C(s) and

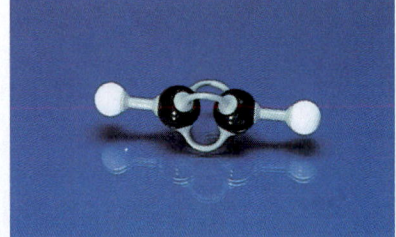
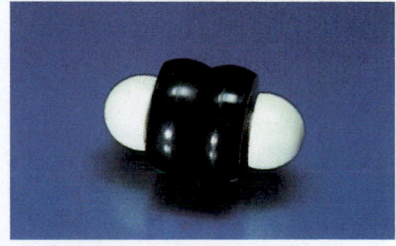

Figure 27-10 Models of acetylene, H—C≡C—H.

$H_2(g)$ at high pressures. It may be converted into ethene and then to ethane by the addition of hydrogen. The formula for acetylene is $H—C\equiv C—H$.

$CH\equiv CH$ $CH_3—C\equiv CH$ $CH_3—CH_2—C\equiv CH$ $CH_3—C\equiv C—CH_3$ $CH_3—CH—C\equiv CH$

ethyne   propyne   1-butyne   2-butyne |
(acetylene) CH_3

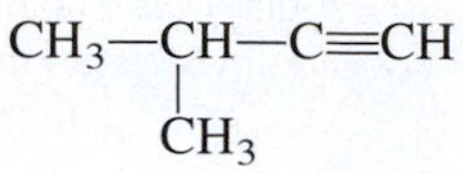

                                                                                                                     3-methyl-1-butyne

The triple bond takes positional precedence over substituents on the carbon chain. It is assigned the lowest possible number in naming.

> **Summary of IUPAC Rules for Naming Alkynes**
>
> Alkynes are named like the alkenes except for the following two points.
>
> 1. The suffix -*yne* is added to the characteristic root.
> 2. Because the linear arrangement about the triple bond does not lead to geometrical isomerism, the prefixes *cis-* and *trans-* are not used.

EXAMPLE 27-8   *Alkynes*

Draw the structure of 5,5-dimethyl-2-heptyne.

Plan

The structure is based on a seven-carbon chain with a triple bond beginning at carbon 2. We add methyl groups at the positions indicated.

Solution

$$\overset{1}{C}—\overset{2}{C}\equiv\overset{3}{C}—\overset{4}{C}—\overset{5}{C}—\overset{6}{C}—\overset{7}{C}$$

There are two methyl groups attached to carbon 5 and sufficient hydrogens to complete the bonding at each C atom.

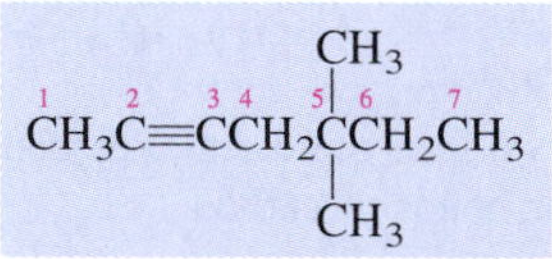

$$\begin{array}{c} CH_3 \\ | \\ \overset{1}{CH_3}\overset{2}{C}\equiv\overset{3}{C}\overset{4}{CH_2}\overset{5}{C}\overset{6}{CH_2}\overset{7}{CH_3} \\ | \\ CH_3 \end{array}$$

The carbon chain is numbered to give the multiple bond the lowest possible numbers even though that results in higher position numbers for the methyl substituents.

You should now work Exercise 26.

EXAMPLE 27-9   *Naming Alkynes*

Name the compound

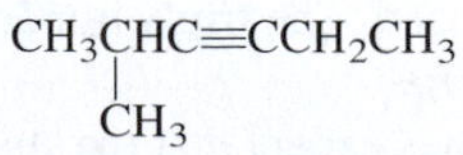

$$\begin{array}{c} CH_3CHC\equiv CCH_2CH_3 \\ | \\ CH_3 \end{array}$$

Plan

We find the longest chain and number it to give the triple bond the lowest possible number. Then we specify the substituent(s) by name and position number.

Solution

The longest continuous carbon chain that includes the triple bond has six C atoms. There are four ways in which we could choose and number such a chain, and in all four the triple bond would be between C atoms 3 and 4.

$$\overset{1}{C}H_3\overset{2}{C}H\overset{3}{C}\equiv\overset{4}{C}\overset{5}{C}H_2\overset{6}{C}H_3 \qquad CH_3\overset{2}{C}H\overset{3}{C}\equiv\overset{4}{C}\overset{5}{C}H_2\overset{6}{C}H_3 \qquad \overset{6}{C}H_3\overset{5}{C}H\overset{4}{C}\equiv\overset{3}{C}\overset{2}{C}H_2\overset{1}{C}H_3 \qquad \overset{5}{C}H_3\overset{4}{C}H\overset{3}{C}\equiv\overset{2}{C}\overset{1}{C}H_2\overset{1}{C}H_3$$
$$\underset{}{\overset{}{C}H_3} \qquad\qquad \underset{1}{CH_3} \qquad\qquad\qquad CH_3 \qquad\qquad\qquad \underset{6}{CH_3}$$

If it does not change the number of the first carbon in the triple bond, we then also want the methyl substituent to have the lowest possible number, so we choose either of the first two possibilities. The name of the compound is 2-methyl-3-hexyne.

You should now work Exercise 28.

AROMATIC HYDROCARBONS

Originally the word "**aromatic**" was applied to pleasant-smelling substances. The word now describes benzene, its derivatives, and certain other compounds that exhibit similar chemical properties. Some have very foul odors because of substituents on the benzene ring. On the other hand, many fragrant compounds do not contain benzene rings.

Steel production requires large amounts of coke. This is prepared by heating bituminous coal to high temperatures in the absence of air. This process also favors production of *coal gas* and *coal tar*. Because of the enormous amount of coal converted to coke, coal tar is produced in large quantities. It serves as a source of aromatic compounds. For each ton of coal converted to coke, about 5 kg of aromatic compounds are obtained. The 19th-century development of the German coal tar industry greatly stimulated the systematic study of organic chemistry.

Early research on the reactions of the aromatic hydrocarbons led to methods for preparing a great variety of dyes, drugs, flavors, perfumes, and explosives. More recently, large numbers of polymeric materials, such as plastics and fabrics, have been prepared from these compounds.

The main components of coal gas are hydrogen (~50%) and methane (~30%).

Distillation of coal tar produces a variety of aromatic compounds.

27-5 BENZENE

Benzene is the simplest aromatic hydrocarbon. By studying its reactions, we can learn a great deal about aromatic hydrocarbons. Benzene was discovered in 1825 by Michael Faraday when he fractionally distilled a by-product oil obtained in the manufacture of illuminating gas from whale oil.

Elemental analysis and determination of its molecular weight showed that the molecular formula for benzene is C_6H_6. The formula suggests that it is highly unsaturated. But its properties are quite different from those of alkenes and alkynes.

The facts that only one monosubstitution product is obtained in many reactions and that no addition products can be prepared show conclusively that benzene has a *symmetrical ring structure*. Stated differently, every H atom is equivalent to every other H atom, and this is possible only in a symmetrical ring structure (a).

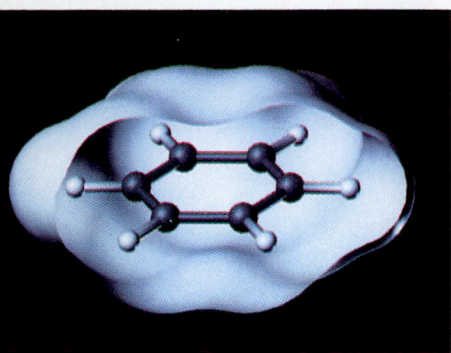

A computer-generated model of a molecule of benzene, C_6H_6. A ball-and-stick model is shown inside a representation of the molecular surface.

(a)

(skeleton only)

(b)

The debate over the structure and bonding in benzene raged for at least 30 years. In 1865 Friedrich Kekulé suggested that the structure of benzene was intermediate between two structures [part (b) above] that we now call resonance structures. We often represent benzene as

or, more simply

The structure of benzene is described in detail in Section 9-6 in terms of MO theory.

All 12 atoms in a benzene molecule lie in a plane. This suggests sp^2 hybridization of each carbon. The six sp^2 hybridized C atoms lie in a plane, and the unhybridized p orbitals extend above and below the plane. Side-by-side overlap of the p orbitals forms pi orbitals (Figure 9-10). The electrons associated with the pi bonds are *delocalized* over the entire benzene ring (Figure 27-11a,b).

27-6 OTHER AROMATIC HYDROCARBONS

Benzene molecules bearing alkyl substituents are called **alkylbenzenes.** The simplest of these is methylbenzene (common name, toluene), shown in Figure 27-11c. The dimethylbenzenes are called xylenes. Three different compounds (Table 27-6) have the formula

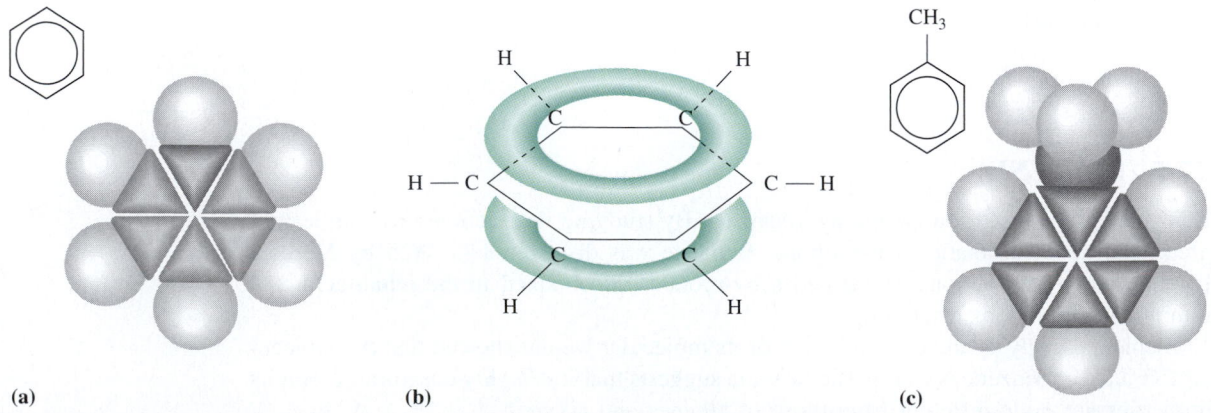

(a)   **(b)**   **(c)**

Figure 27-11 (a) A model of the benzene, C_6H_6, molecule and (b) its electron distribution. (c) A model of toluene, $C_6H_5CH_3$. This is an alkylbenzene, a derivative of benzene in which one H atom has been replaced by a —CH_3 group.

Table 27-6 *Aromatic Hydrocarbons from Coal Tar*

Name	Formula	Normal bp (°C)	Normal mp (°C)	Solubility
benzene	C_6H_6	80	+6	
toluene	$C_6H_5CH_3$	111	−95	
o-xylene	$C_6H_4(CH_3)_2$	144	−27	
m-xylene	$C_6H_4(CH_3)_2$	139	−54	All insoluble
p-xylene	$C_6H_4(CH_3)_2$	138	+13	in water
naphthalene	$C_{10}H_8$	218	+80	
anthracene	$C_{14}H_{10}$	342	+218	
phenanthrene	$C_{14}H_{10}$	340	+101	

$C_6H_4(CH_3)_2$ (see margin). These three xylenes are *structural isomers*. In naming these (as well as other disubstituted benzenes), we use prefixes *ortho-* (abbreviated *o-*), *meta-* (*m-*), or *para-* (*p-*) to refer to relative positions of substituents on the benzene ring. The *ortho-* prefix refers to two substituents located on *adjacent* carbon atoms; e.g., 1,2-dimethylbenzene is *o-*xylene. The *meta-* prefix identifies substituents on C atoms 1 and 3, so 1,3-dimethylbenzene is *m-*xylene. The *para-* prefix refers to substituents on C atoms 1 and 4, so 1,4-dimethylbenzene is *p-*xylene.

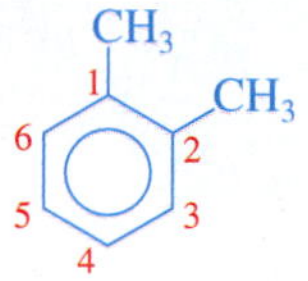

ortho-xylene
bp = 144°C
mp = −27°C

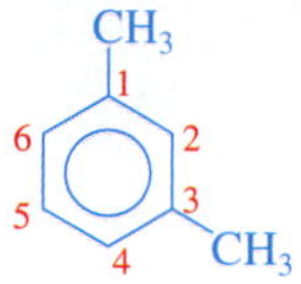

meta-xylene
bp = 139°C
mp = −54°C

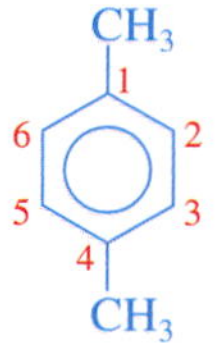

para-xylene
bp = 138°C
mp = 13°C

Summary of Rules for Naming Derivatives of Benzene

1. If there is only one group on the ring, no number is needed to designate its position.

2. If there are two groups on the ring, we use the traditional designations.

> *ortho-* or *o-* for 1,2-disubstitution
> *meta-* or *m-* for 1,3-disubstitution
> *para-* or *p-* for 1,4-disubstitution

3. If there are three or more groups on the ring, location numbers are assigned to give the *minimum sum* of numbers.

Examples are

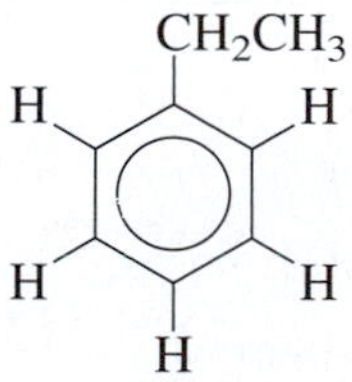

ethylbenzene

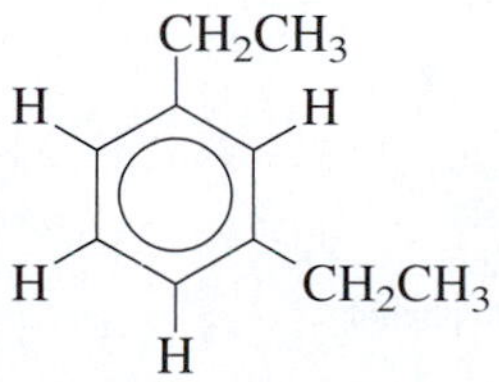

m-diethylbenzene, or
1,3-diethylbenzene

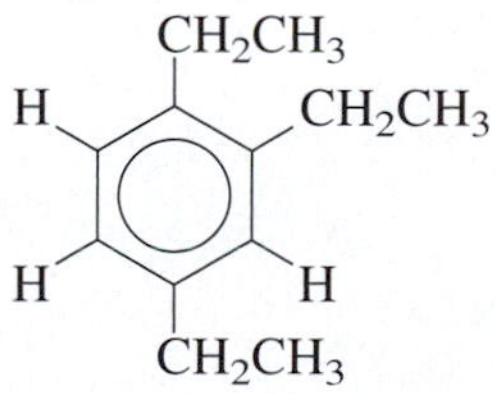

1,2,4-triethylbenzene

When an H atom is removed from a benzene, C_6H_6, molecule, the resulting group, C_6H_5— or 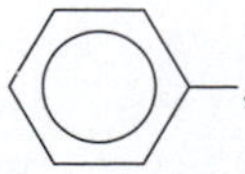 —, is called "phenyl." Sometimes we name mixed alkyl-aromatic hydrocarbons on that basis.

phenylcyclohexane   2-phenyl-*cis*-2-butene

Another class of aromatic hydrocarbons consists of "condensed" or "fused-ring" aromatic systems. The simplest of these are naphthalene, anthracene, and phenanthrene.

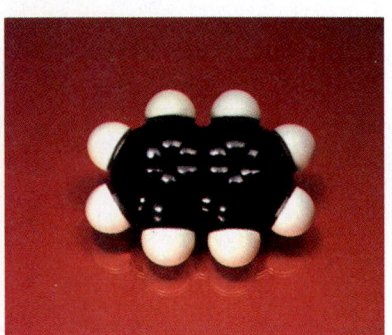

Naphthalene.

naphthalene, $C_{10}H_8$ anthracene, $C_{14}H_{10}$ phenanthrene, $C_{14}H_{10}$

No hydrogen atoms are attached to the carbon atoms that are involved in fusion of aromatic rings, i.e., carbon atoms that are members of two or more aromatic rings.

The traditional name is often used as part of the base name in naming an aromatic hydrocarbon and its derivatives. You should know the names and structures of the fundamental aromatic hydrocarbons discussed thus far: benzene, toluene, the three xylenes, naphthalene, anthracene, and phenanthrene.

Distillation of coal tar provides four volatile fractions as well as the pitch that is used for surfacing roads and in the manufacture of "asphalt" roofing (Figure 27-12). Eight aromatic hydrocarbons are obtained in significant amounts by efficient fractional distillation of the "light oil" fraction (Table 27-6).

Figure 27-12   Fractions obtained from coal tar.

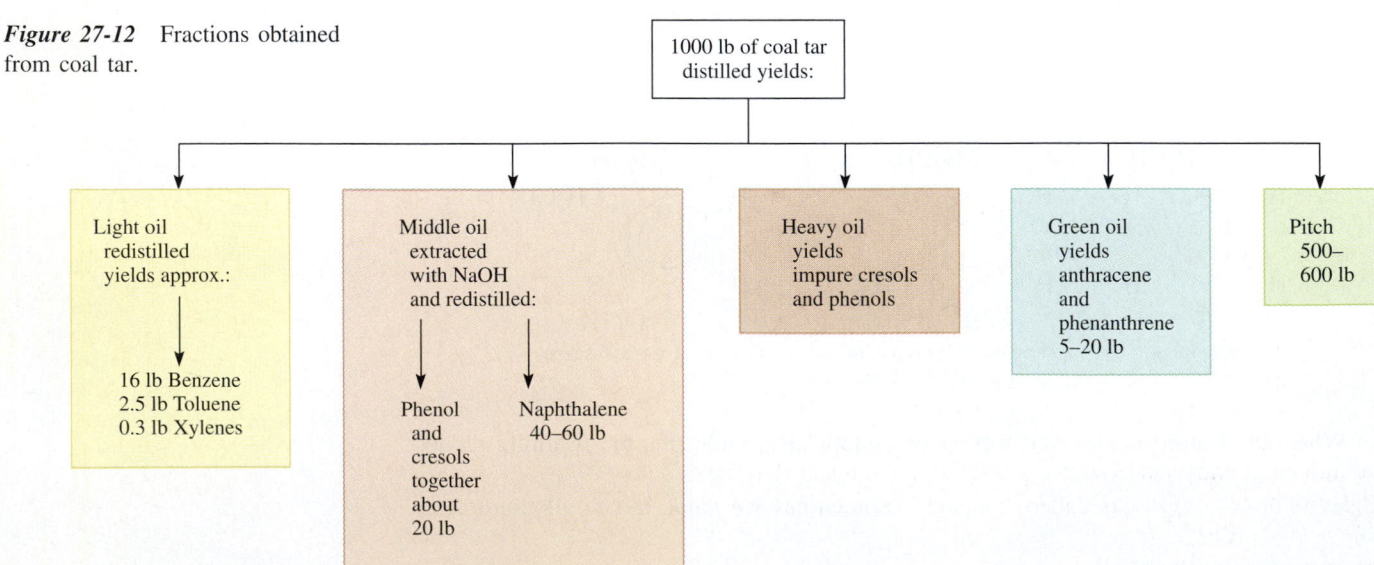

1000 lb of coal tar distilled yields:

Light oil
redistilled
yields approx.:

16 lb Benzene
2.5 lb Toluene
0.3 lb Xylenes

Middle oil
extracted
with NaOH
and redistilled:

Phenol   Naphthalene
and   40–60 lb
cresols
together
about
20 lb

Heavy oil
yields
impure cresols
and phenols

Green oil
yields
anthracene
and
phenanthrene
5–20 lb

Pitch
500–
600 lb

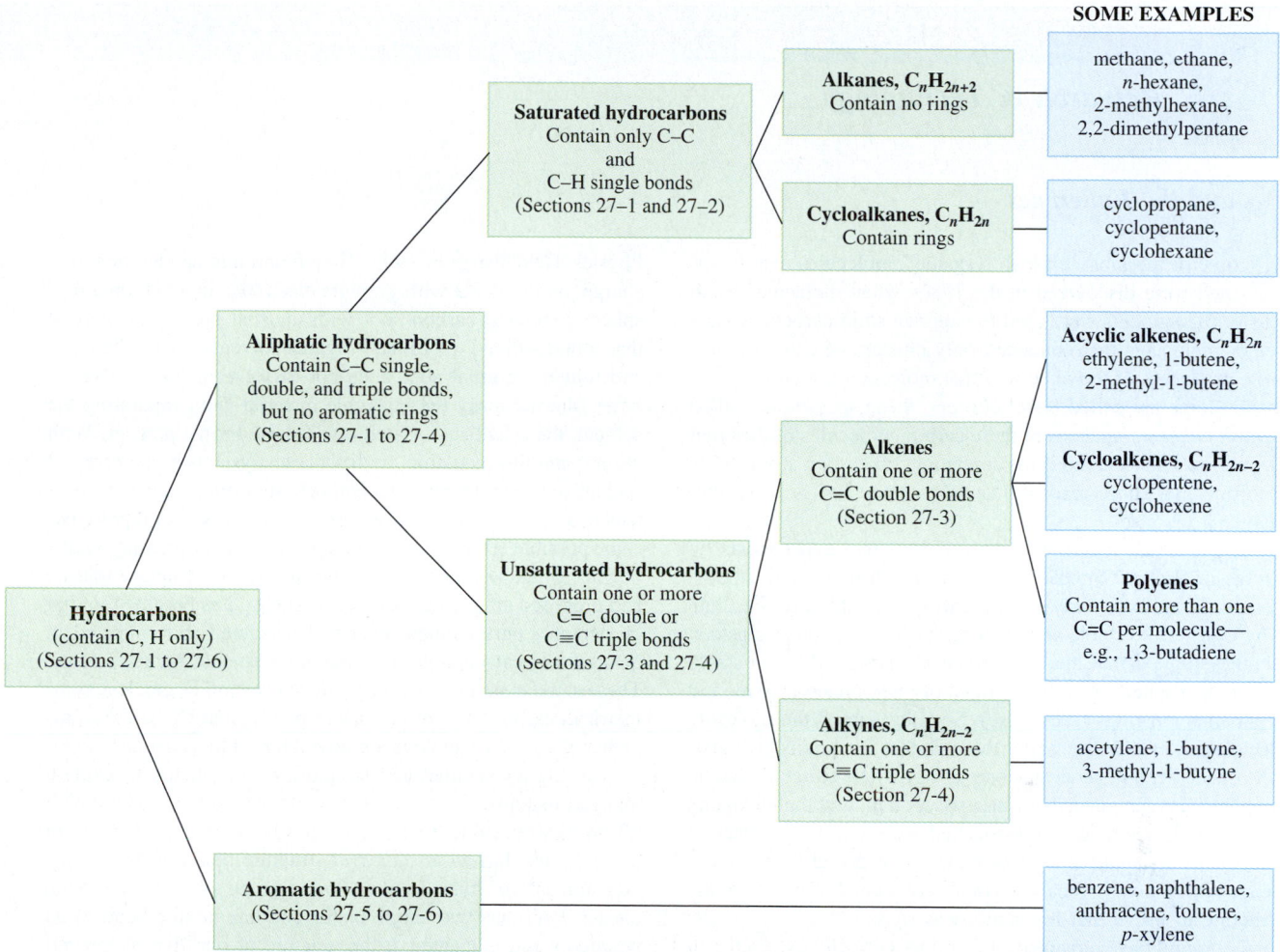

Figure 27-13 A classification of hydrocarbons.

27-7 HYDROCARBONS—A SUMMARY

Hydrocarbons contain only carbon and hydrogen. They can be subdivided into various groups. To assist you in organizing what you have studied up to this point in this chapter, take a few minutes and study Figure 27-13.

FUNCTIONAL GROUPS

The study of organic chemistry is greatly simplified by considering hydrocarbons as parent compounds and describing other compounds as derived from them. In general, an organic molecule consists of a skeleton of carbon atoms with special groups of atoms within or attached to that skeleton. These special groups of atoms are often called **functional groups** because they represent the most common sites of chemical reactivity (func-

CHEMISTRY IN USE

Research & Technology

C_{60} and the Fullerenes

Some of the most unusual "organic" molecules ever imagined were discovered in the 1980s when molecular beam cluster experiments were used to vaporize solid carbon. A variety of molecules that contained only clusters of carbon atoms were produced. In the same way that molecules containing only metal atoms are called *metal clusters,* these species are called *carbon clusters.* Analysis of these carbon molecules of different sizes shows that the molecule containing 60 carbon atoms, C_{60}, is formed far more readily than other sizes and is incredibly stable.

C_{60} was first identified in molecular beam experiments by Professor Richard Smalley and his research team at Rice University. From these early experiments they could only speculate why C_{60} was so stable and so different from the other clusters. Smalley argued that the only structure that could be so stable would be a sphere of carbon formed of interconnected five- and six-membered rings (see figure). Smalley named the molecule "Buckminsterfullerene" after the architect Buckminster Fuller, who specialized in geodesic dome designs. The structure that he proposed has the same shape as a soccer ball, and the nickname "Buckyball" soon became associated with C_{60}. Unfortunately, because of the small amount of material produced in molecular beam experiments, Smalley could not actually measure the structure to prove that his model was right.

C_{60} might have remained a laboratory curiosity except for an exciting discovery by Donald Huffman (University of Arizona) and Wolfgang Krätschmer (Max Planck Institute for Nuclear Physics, Heidelberg) in 1990. They found that an electrical discharge, or arc, made with graphite electrodes in a helium atmosphere generated carbon soot with unusual properties. Part of that soot dissolved in ordinary organic solvents such as benzene and toluene; normal soot does not dissolve in these solvents. After filtering away the insoluble material and evaporating the solvent, the scientists isolated a yellowish brown powder. With gram quantities available, traditional analysis using an array of specialized instrumental techniques (nuclear magnetic resonance and infrared spectroscopy, X-ray crystallography) became possible. To everyone's delight (especially Smalley's), the powder contained mostly C_{60}, and it had the structure that he had proposed almost ten years earlier! C_{60} thus became the first cluster of a pure element ever to be isolated and collected in quantities great enough for traditional chemical experiments. These same experiments also produced several larger, less symmetrical, carbon cage molecules (e.g., C_{70} and C_{84}). Taken together, C_{60} and its analogs are now referred to as the *fullerenes*.

The highly symmetrical structure of C_{60} helps to explain many of its unusual properties. It is almost a perfect sphere with 60 atoms arranged in 20 hexagons and 12 pentagons (32 faces in all). C_{60} also has an *icosahedral* structure. Each of the 60 carbon atoms is sp^2 hybridized and occupies an identical site in the cluster. Each has two single bonds and one double bond to its neighbors and is located at the juncture of one five-membered ring and two six-membered rings. This structure is very different from those of other forms of carbon. For example, diamond

---

tion). The only functional groups that are possible in hydrocarbons are double and triple (i.e., pi) bonds. Atoms other than C and H are called **heteroatoms,** the most common being O, N, S, P, and the halogens. Most functional groups contain one or more heteroatoms.

As you study the following sections, you may wish to refer to the summary in Section 27-15.

In the next several sections we shall introduce some common functional groups that contain heteroatoms and learn a little about the resulting classes of compounds. We shall continue to represent hydrocarbon groups with the symbol R—. We commonly use that symbol to represent either an aliphatic (e.g., alkyl) or an aromatic (e.g., an aryl such as phenyl) group. When we specifically mean an aryl group, we shall use the symbol Ar—.

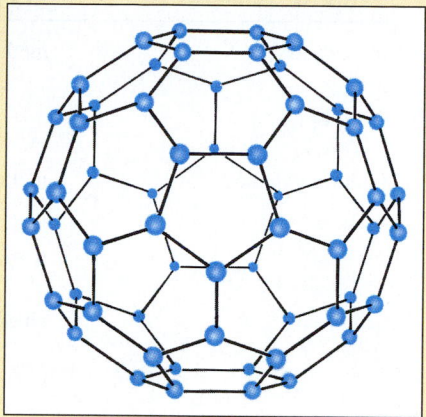

The isosahedral structure of C_{60}.

has tetrahedral bonding to four nearest neighbors. Graphite is more similar to C_{60}, with an infinite array of planar six-membered rings. However, an array of six-membered rings cannot be bent into a closed cage because of the strain on the bonds. Substitution of precisely the correct number of five-membered rings, as occurs in C_{60}, relieves this strain.

There are numerous equivalent ways to draw the chemical bonds in C_{60}. Equivalent bonding arrangements, known as *resonance structures,* are also found for molecules such as benzene and give them their "aromatic" character. C_{60} is therefore also an aromatic molecule. Empirical rules have been developed from chemical bonding theory to predict which other molecules similar to C_{60} might be found. In addition to C_{60}, carbon

molecules with a multiple of sixty atoms, such as the clusters C_{120}, C_{240}, C_{540}, and C_{960}, are also predicted to be stable aromatic molecules with highly symmetrical open-cage structures. Attempts are underway to isolate and characterize these larger fullerenes.

The spherical cage structure makes C_{60} and the other fullerenes fascinating candidates for all kinds of applications in chemistry and for the preparation of new solid materials. For example, crystals of C_{60} have been prepared in which the molecules arrange themselves in a hexagonal close-packed structure. When alkali metal atoms such as cesium or rubidium are added into the gaps between the balls, the resulting compound is a *superconductor.* The open interior of C_{60} is a cavity about 5 Å wide, large enough to contain other atoms—in particular, metals. Numerous research groups are making complexes with encapsulated metal atoms inside C_{60} or in other fullerenes that have larger cavities. The unusual shape of the C_{60} molecule is of special interest in situations in which molecular shapes determine chemical activity, as in biological molecules, pharmaceutical drugs, and polymers. To investigate these kinds of applications, other functional groups or reactive organic systems have already been attached to C_{60}. Long chains of the form (C_{60})—R—(C_{60})—R . . . have also been constructed. The peculiar new molecule, C_{60}, and other members of its family are rapidly emerging from the realm of molecular beams into the mainstream of practical chemistry.

Professor Michael A. Duncan
The University of Georgia

27-8 ORGANIC HALIDES

Almost any hydrogen atom in a hydrocarbon can be replaced by a halogen atom to give a stable compound. Table 27-7 shows some organic halides and their names.

In the IUPAC naming system, the organic halides are named as *halo-* derivatives of the parent hydrocarbons. The prefix *halo-* can be *fluoro-, chloro-, bromo-,* or *iodo-.* Simple alkyl chlorides are sometimes given common names as alkyl derivatives of the hydrogen halides. For instance, the IUPAC name for CH_3CH_2—Cl is chloroethane; it is commonly called ethyl chloride by analogy to H—Cl, hydrogen chloride.

Table 27-7 *Some Organic Halides*

Formula	Structural Formula	Normal bp (°C)	IUPAC Name	Common Name
CH_3Cl	H | H—C—Cl | H	23.8	chloromethane	methyl chloride
CH_2Cl_2	Cl | H—C—Cl | H	40.2	dichloromethane	methylene chloride
$CHCl_3$	Cl | H—C—Cl | Cl	61	trichloromethane	chloroform
CCl_4	Cl | Cl—C—Cl | Cl	76.8	tetrachloromethane	carbon tetrachloride
$CHCl_2Br$	Cl | H—C—Cl | Br	90	bromodichloromethane	—

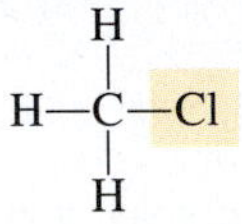

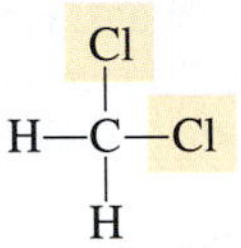

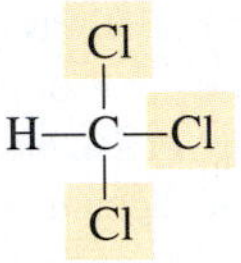

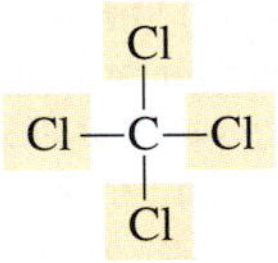

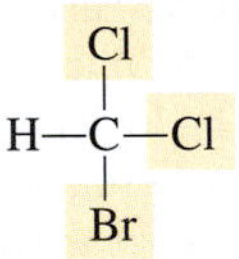

A carbon atom can be bonded to as many as four halogen atoms, so an enormous number of organic halides can exist. Completely fluorinated compounds are known as **fluorocarbons** or sometimes *perfluorocarbons*. The fluorocarbons are even less reactive than hydrocarbons. Saturated compounds in which all H atoms have been replaced by some combination of Cl and F atoms are called *chlorofluorocarbons* or sometimes **freons.** These compounds have been widely used as refrigerants and as propellants in aerosol cans. However, the release of chlorofluorocarbons into the atmosphere has been shown to be quite damaging to the earth's ozone layer. Since January 1978 the use of chlorofluorocarbons in aerosol cans in the United States has been banned, and efforts to develop both controls for existing chlorofluorocarbons and suitable replacements continue. At this time (1995) the production of freons has been banned in several countries. A significant tax has been placed on the sale of freons in the United States.

Freon is a Du Pont trademark for certain chlorofluorocarbons; other companies' related products are known by other names. Typical freons are trichlorofluoromethane, $CFCl_3$ (called Freon-11), and dichlorodifluoromethane, CF_2Cl_2 (called Freon-12).

27-9 ALCOHOLS AND PHENOLS

Alcohols and phenols contain the hydroxyl group (— O—H) as their functional group. **Alcohols** may be considered to be derived from saturated or unsaturated hydrocarbons by the replacement of at least one H atom by a hydroxyl group. The properties of alcohols

Table 27-7 (continued)

Formula	Structural Formula	Normal bp (°C)	IUPAC Name	Common Name
$(CH_3)_2CHI$		89.5	2-iodopropane	isopropyl iodide
$CH_3ClC=CHCH_2CH_2Cl$		40	2,5-dichloro-*cis*-2-pentene; *trans* isomer is also possible	—
C_5H_7Cl		25	3-chlorocyclopentene is shown; other isomers are also possible	—
C_6H_5I		118	iodobenzene	phenyl iodide
C_6H_4ClBr		204	1-bromo-2-chlorobenzene is shown; other isomers are also possible	*o*-bromochlorobenzene

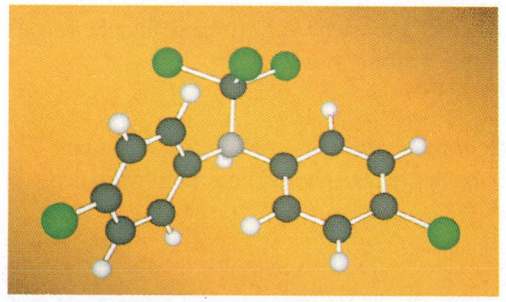

DDT, an organic halide whose full name is 1,1,1-trichloro-2,2-bis(*p*-chlorophenyl)ethane. It was introduced as an insecticide in the 1940s and widely used until the early 1970s. This compound was almost entirely responsible for the virtual eradication of malaria, once the world's most widespread disease, by killing the *Anopheles* mosquito that carried the disease. Unfortunately, this toxic substance stays in the environment for a long time. As a result, it is carried up the food chain and concentrates in the bodies of higher animals. It is especially detrimental in the life cycles of birds. Only about 2% of the DDT in the environment is degraded each year. Even though its use has long been banned, it will be many years before the last traces disappear from the soil.

Figure 27-14 Models of ethanol (also called ethyl alcohol or grain alcohol), CH_3CH_2OH.

result from a hydroxyl group attached to an *aliphatic* carbon atom, —C—O—H. Ethanol (ethyl alcohol) is the most common example (Figure 27-14).

When a hydrogen atom on an aromatic ring is replaced by a hydroxyl group (Figure 27-15), the resulting compound is known as a **phenol.** Such compounds behave more like acids than alcohols. Alternatively, we may view alcohols and phenols as derivatives of water in which one H atom has been replaced by an organic group.

$$H—O—H \qquad \underset{\substack{}}{H—\overset{\displaystyle H}{\underset{\displaystyle H}{C}}—\overset{\displaystyle H}{\underset{\displaystyle H}{C}}—O—H} \qquad \text{⬡—O—H}$$

water   ethanol   phenol

Indeed, this is a better view. The structure of water was discussed in Section 8-9. The hydroxyl group in an alcohol or a phenol is covalently bonded to a carbon atom, but the O—H bond is quite polar. The oxygen atom has two unshared electron pairs, and the C—O—H bond angle is nearly 104.5°.

The presence of a bonded alkyl or aryl group changes the properties of the —OH group. *Alcohols* are so very weakly acidic that they are thought of as neutral compounds. *Phenols* are weakly acidic.

Many properties of alcohols depend on whether the hydroxyl group is attached to a carbon that is bonded to *one, two,* or *three* other carbon atoms.

> **Primary alcohols** contain one R group; **secondary alcohols** contain two R groups; and **tertiary alcohols** contain three R groups bonded to the carbon atom to which the —OH group is attached.

The simplest phenol is called phenol. The most common member of a class of compounds is often called by the class name. Salt, sugar, alcohol, and phenol are examples.

Figure 27-15 Models of phenol, C_6H_5OH.

Representing alkyl groups as R, we can illustrate the three classes of alcohols. The R groups may be the same or different.

$$\begin{array}{ccc} H & R' & R' \\ | & | & | \\ R-C-OH & R-C-OH & R-C-OH \\ | & | & | \\ H & H & R'' \end{array}$$

a primary (1°) alcohol   a secondary (2°) alcohol   a tertiary (3°) alcohol

In writing organic structures, we often use primes when we wish to specify that the alkyl groups might be different, e.g., R, R′, R″.

Naming Alcohols and Phenols

The systematic name of an alcohol consists of the characteristic stem plus an *-ol* ending. A numeric prefix indicates the position of the —OH group on a chain of three or more carbon atoms.

$$CH_3-OH \qquad CH_3-CH_2-OH \qquad CH_3-CH_2-CH_2-OH \qquad \begin{array}{c} OH \\ | \\ CH_3-CH-CH_3 \end{array}$$

methanol   ethanol   1-propanol   2-propanol
methyl alcohol   ethyl alcohol   propyl alcohol   isopropyl alcohol
(wood alcohol)   (grain alcohol)   (a primary alcohol)   (a secondary alcohol)

There are four structural isomers of the saturated acyclic four-carbon alcohols with one —OH per molecule.

Acyclic compounds contain no rings.

1°: $CH_3CH_2CH_2CH_2OH$ 1°: $\begin{array}{c} CH_3-CH-CH_2OH \\ | \\ CH_3 \end{array}$

1-butanol   2-methyl-1-propanol
normal butyl alcohol   isobutyl alcohol

2°: $\begin{array}{c} CH_3CH_2CHCH_3 \\ | \\ OH \end{array}$ 3°: $\begin{array}{c} CH_3 \\ | \\ CH_3-C-OH \\ | \\ CH_3 \end{array}$

2-butanol   2-methyl-2-propanol
secondary butyl alcohol   tertiary butyl alcohol

There are eight structural isomers of the analogous five-carbon alcohols. They are often called "amyl" or "pentyl" alcohols. Two examples are

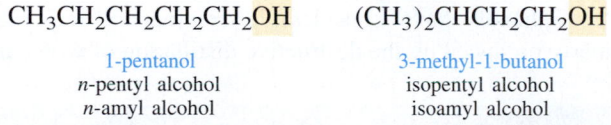

$$CH_3CH_2CH_2CH_2CH_2OH \qquad (CH_3)_2CHCH_2CH_2OH$$

1-pentanol   3-methyl-1-butanol
n-pentyl alcohol   isopentyl alcohol
n-amyl alcohol   isoamyl alcohol

The **polyhydric alcohols** contain more than one —OH group per molecule. Those containing two OH groups per molecule are called **glycols.** Important examples of polyhydric alcohols include

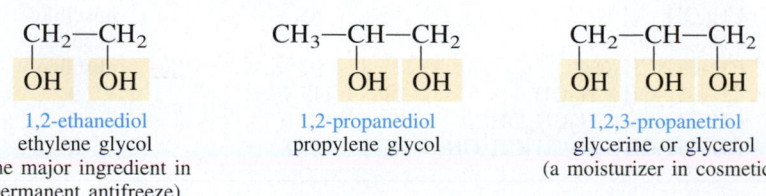

$$\begin{array}{ccc} CH_2-CH_2 & CH_3-CH-CH_2 & CH_2-CH-CH_2 \\ | \quad | & | \quad | & | \quad | \quad | \\ OH \quad OH & OH \quad OH & OH \quad OH \quad OH \end{array}$$

1,2-ethanediol   1,2-propanediol   1,2,3-propanetriol
ethylene glycol   propylene glycol   glycerine or glycerol
(the major ingredient in   (a moisturizer in cosmetics)
permanent antifreeze)

Polyhydric alcohols are used in permanent antifreeze and in cosmetics.

Phenols are usually referred to by their common names. Examples are

The *o*-, *m*-, and *p*- notation was introduced in Section 27-6.

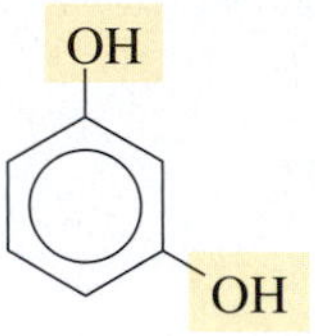

| resorcinol | hydroquinone | *o*-cresol | *m*-cresol | *p*-cresol |

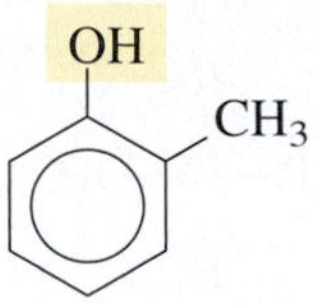

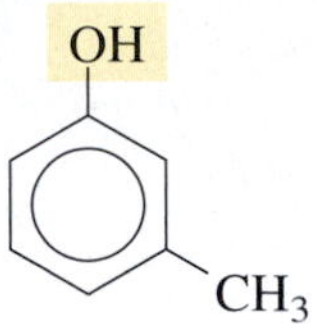

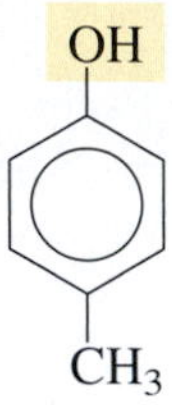

As you might guess, cresols occur in "creosote," a wood preservative.

Physical Properties of Alcohols and Phenols

The hydroxyl group, —OH, is quite polar, whereas alkyl groups, R, are nonpolar. The properties of alcohols depend on two factors: (1) the number of hydroxyl groups per molecule and (2) the size of the nonpolar portion of the molecule.

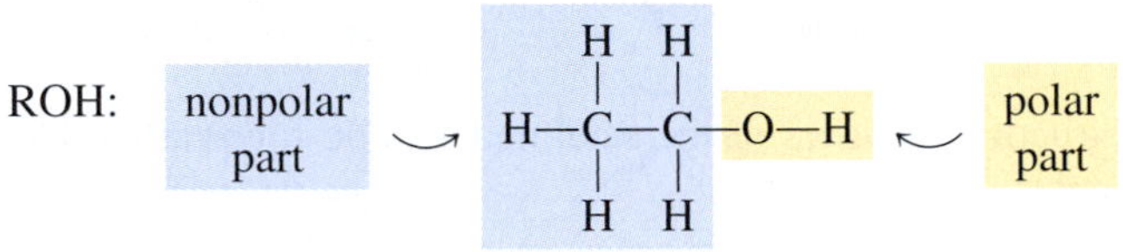

The low-molecular-weight monohydric alcohols are soluble in water in all proportions (miscible). Beginning with the four butyl alcohols, solubility in water decreases rapidly with increasing molecular weight. This is because the nonpolar parts of such molecules are much larger than the polar parts. Many polyhydric alcohols are very soluble in water because they contain two or more polar —OH groups per molecule.

Table 27-8 shows that the boiling points of normal primary alcohols increase, and their solubilities in water decrease, with increasing molecular weight. The boiling points of the alcohols are much higher than those of the corresponding alkanes (Table 27-2) because of the hydrogen bonding of the hydroxyl groups.

Most phenols are solids at 25°C. Phenols are only slightly soluble in water unless they contain other functional groups that interact with water.

Some Uses of Alcohols and Phenols

Many alcohols and phenols have considerable commercial importance. Methanol, CH_3OH, was formerly produced by the destructive distillation of wood and is sometimes

Table 27-8 *Physical Properties of Normal Primary Alcohols*

Name	Formula	Normal bp (°C)	Solubility in H_2O (g/100 g at 20°C)
methanol	CH_3OH	65	miscible
ethanol	CH_3CH_2OH	78.5	miscible
1-propanol	$CH_3CH_2CH_2OH$	97	miscible
1-butanol	$CH_3CH_2CH_2CH_2OH$	117.7	7.9
1-pentanol	$CH_3CH_2CH_2CH_2CH_2OH$	137.9	2.7
1-hexanol	$CH_3CH_2CH_2CH_2CH_2CH_2OH$	155.8	0.59

called wood alcohol. It is now produced in large quantities from carbon monoxide and hydrogen. It is extensively used as a solvent for varnishes and shellacs, as the starting material in the manufacture of formaldehyde (Section 27-14), as a temporary antifreeze (bp = 65°C), and as a fuel additive. It is very toxic and causes permanent blindness when taken internally.

Ethanol, CH_3CH_2OH, also known as ethyl alcohol or grain alcohol, was first prepared by fermentation. The most ancient written literature refers to beverages that were obviously alcoholic! The fermentation of cane sugar is one important source of ethanol. The starches in grains, potatoes, and similar foodstuffs can be converted into sugar by malt; this is followed by fermentation to produce ethanol, which is the most important industrial alcohol. Like the other short-chain alcohols, ethanol participates in hydrogen bonding and is miscible with water.

Fermentation is an enzymatic process carried out by certain kinds of bacteria.

Many simple alcohols are important raw materials in the industrial synthesis of polymers, fibers, explosives, plastics, and pharmaceutical products. Phenols are found in plant products such as flower pigments, tanning agents, and wood. They are widely used in the preparation of plastics and dyes. Dilute aqueous solutions of phenols are used as antiseptics and disinfectants. Some uses of polyhydric alcohols depend on their relatively high boiling points. For instance, glycerine is used as a wetting agent in cosmetic preparations. Ethylene glycol (bp = 197°C), which is miscible with water, is used in commercial permanent antifreeze.

$$H_2C-OH$$
$$|$$
$$H_2C-OH$$
ethylene glycol

27-10 ETHERS

When the word "ether" is mentioned, most people think of the well-known anesthetic, diethyl ether. There are many ethers. Their uses range from artificial flavorings to refrigerants and solvents. An **ether** is a compound in which an O atom is bonded to two organic groups.

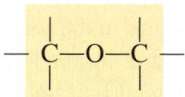

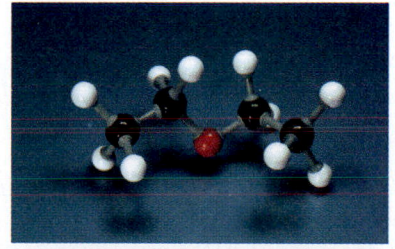

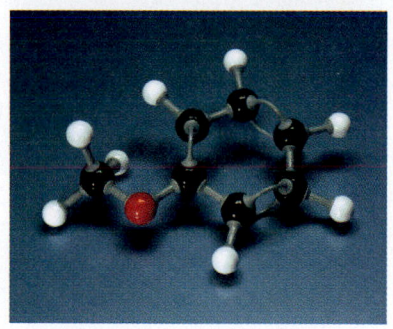

Alcohols are considered derivatives of water in which one H atom has been replaced by an organic group. Ethers may be considered derivatives of water in which both H atoms have been replaced by organic groups.

$$H-O-H \qquad R-O-H \qquad R-O-R'$$
water   alcohol   ether

However, the similarity is only structural because ethers are not very polar and are chemically rather unreactive. (We shall not discuss their reactions here.) In fact, their physical properties are similar to those of the corresponding alkanes; e.g., CH_3OCH_3 is like $CH_3CH_2CH_3$.

Three kinds of ethers are known: (1) aliphatic, (2) aromatic, and (3) mixed. Common names are used for ethers in most cases.

Models of diethyl ether (top) and methyl phenyl ether (bottom).

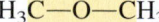

$$H_3C-O-CH_3 \qquad H_3C-O-CH_2CH_3$$

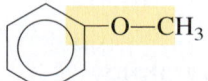 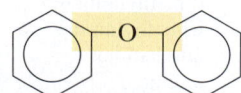

methoxymethane
dimethyl ether
(an aliphatic ether)

methoxyethane
methyl ethyl ether
(an aliphatic ether)

methoxybenzene
methyl phenyl ether
anisole
(a mixed ether)

phenoxybenzene
diphenyl ether
(an aromatic ether)

Diethyl ether is a very-low-boiling liquid (bp = 35°C). Dimethyl ether is a gas that is used as a refrigerant. The aliphatic ethers of higher molecular weights are liquids, and the aromatic ethers are liquids and solids.

Even ethers of low molecular weight are only slightly soluble in water. Diethyl ether is an excellent solvent for organic compounds. It is widely used to extract organic compounds from plants and other natural sources.

Ethers burn readily, and care must be exercised to avoid fires when ethers are used. At room temperature diethyl ether is oxidized by oxygen in the air to a nonvolatile, explosive peroxide. Because of the danger of peroxide explosions, ethereal solutions should never be evaporated to dryness unless proper precautionary steps have been taken to destroy all peroxides in advance.

27-11 AMINES

The **amines** are derivatives of ammonia in which one or more hydrogen atoms have been replaced by alkyl or aryl groups. Many low-molecular-weight amines are gases or low-boiling liquids (Table 27-9). Amines are basic compounds (Table 18-6; Section 28-7). Their basicity differs, depending on the nature of the organic substituents. The aliphatic amines of low molecular weight are soluble in water. Aliphatic diamines of fairly high molecular weight are soluble in water because each molecule contains two highly polar —NH_2 groups that form hydrogen bonds with water.

The odors of amines are quite unpleasant; many of the malodorous compounds that are released as fish decay are simple amines. Amines of high molecular weight are nonvolatile, so they have little odor. One of the materials used to manufacture nylon, hexamethylenediamine, is an aliphatic amine. Many aromatic amines are used to prepare organic dyes that are widely used in industrial societies. Amines are also used to produce many medicinal products, including local anesthetics and sulfa drugs.

Amines are widely distributed in nature in the form of amino acids and proteins, which are found in all higher animal forms, and in alkaloids, which are found in most plants. Some of these substances are fundamental building blocks of animal tissue; and minute amounts of others have dramatic physiological effects, both harmful and beneficial. Countless other biologically important substances, including many vitamins, antibiotics, and drugs, contain amino groups, —NR_2 (where R can represent an H, alkyl, or aryl group).

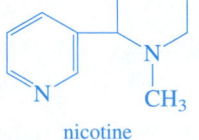

nicotine

strychnine

Ammonia acts as a Lewis base because there is one unshared pair of electrons on the N atom (Section 10-10).

Table 27-9 Boiling Points of Ammonia and Some Amines

Name	Formula	Boiling Point (°C)
ammonia	NH_3	−33.4
methylamine	CH_3NH_2	−6.5
dimethylamine	$(CH_3)_2NH$	7.4
trimethylamine	$(CH_3)_3N$	3.5
ethylamine	$CH_3CH_2NH_2$	16.6
aniline	$C_6H_5NH_2$	184
ethylenediamine	$H_2NCH_2CH_2NH_2$	116.5
pyridine	C_5H_5N	115.3

Structure and Naming of Amines

There are three classes of amines, depending on whether one, two, or three hydrogen atoms have been replaced by organic groups. They are called primary, secondary, and tertiary amines, respectively.

NH_3 RNH_2 R_2NH R_3N

ammonia   methylamine   dimethylamine   trimethylamine
(a primary amine)   (a secondary amine)   (a tertiary amine)

Models of these four molecules are shown in Figure 27-16.

The systematic names of amines are based on consideration of the compounds as derivatives of ammonia. Amines of more complex structure are sometimes named as derivatives of the parent hydrocarbon with the term *amino-* used as a prefix to describe —NH_2.

2-aminobutane or *sec*-butylamine   1-amino-3-ethylcyclohexane

Aniline is the simplest aromatic amine. Many aromatic amines are named as derivatives of aniline.

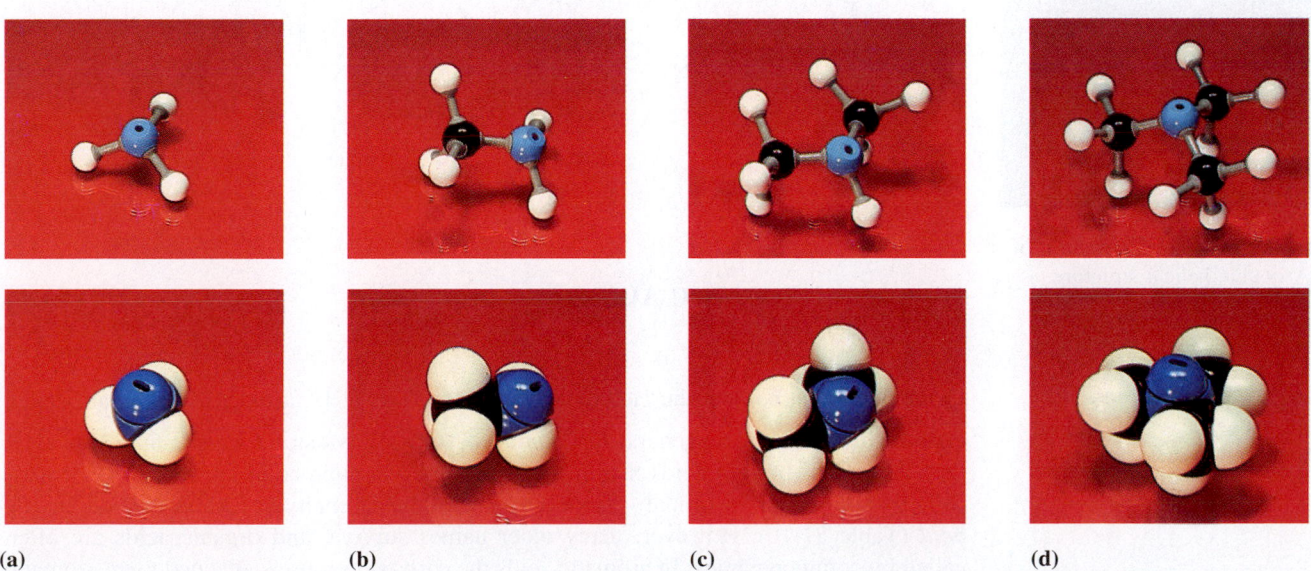

Figure 27-16 Models of (a) ammonia, (b) methylamine, (c) dimethylamine, and (d) trimethylamine.

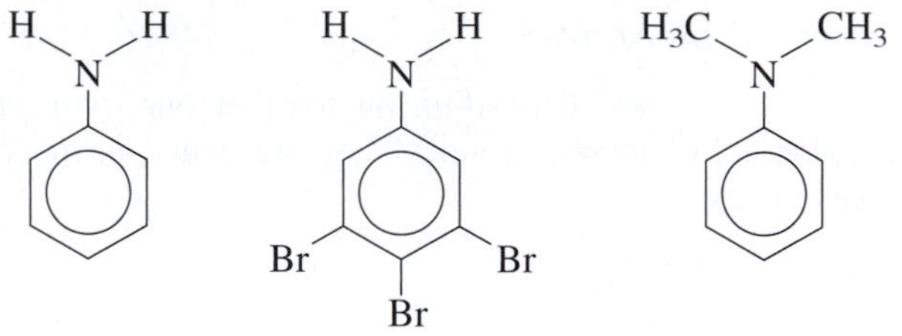

aniline
(primary)

3,4,5-tribromoaniline

N,N-dimethylaniline

Heterocyclic amines contain nitrogen as a part of the ring, bound to two carbon atoms. Many of these amines are found in coal tar and a variety of natural products. Some aromatic and heterocyclic amines are called by their common names.

pyridine
(tertiary)

pyrrole
(secondary)

quinoline
(tertiary)

purine

pyrimidine

Genes, the units of chromosomes that carry hereditary characteristics, are essentially long stretches of double helical deoxyribonucleic acid, or DNA. DNA is composed of four fundamental *nucleotide bases:* adenine, guanine, cytosine, and thymine. The first two are modified purines, and the latter two are modified pyrimidines. The sequence of these building blocks in DNA acts as a code for the order of amino acids in the proteins of an organism. The DNA in each cell of an organism contains the instructions for making the complete organism.

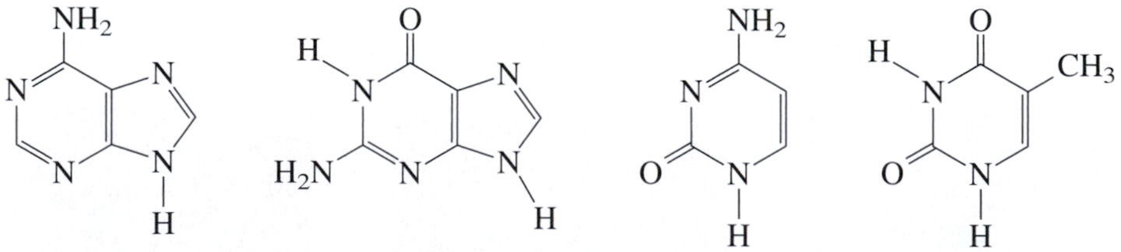

adenine

guanine

cytosine

thymine

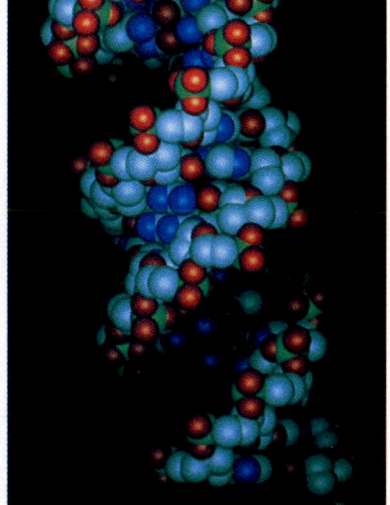

A space-filling model of a portion of the DNA double helical structure.

27-12 CARBOXYLIC ACIDS

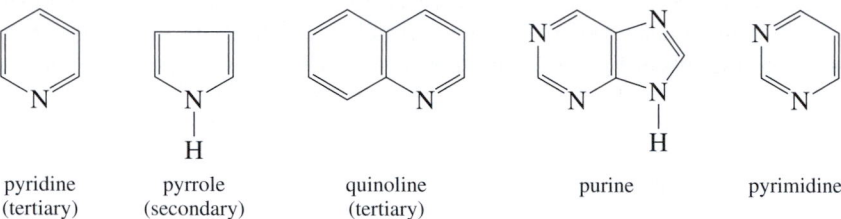

Compounds that contain the **carboxyl group,** $-\overset{\overset{\text{O}}{\|}}{\text{C}}-\text{O}-\text{H}$, are acidic. They are called **carboxylic acids.** Their general formula is R—COOH. Most are *weak acids.* However, they are much stronger acids than most phenols. Carboxylic acids are named systematically by dropping the terminal *-e* from the name of the parent hydrocarbon and adding *-oic acid* (Table 27-10). However, many older names survive, and organic acids are often called by common names. In aromatic acids the carboxyl group is attached to an aromatic ring (Figure 27-17).

Table 27-10 *Some Aliphatic Carboxylic Acids*

Formula	Common Name	IUPAC Name
HCOOH	formic acid	methanoic acid
CH_3COOH	acetic acid	ethanoic acid
CH_3CH_2COOH	propionic acid	propanoic acid
$CH_3CH_2CH_2COOH$	butyric acid	butanoic acid
$CH_3CH_2CH_2CH_2CH_2COOH$	caproic acid	hexanoic acid
$CH_3(CH_2)_{10}COOH$	lauric acid	dodecanoic acid
$CH_3(CH_2)_{14}COOH$	palmitic acid	hexadecanoic acid
$CH_3(CH_2)_{16}COOH$	stearic acid	octadecanoic acid

Formic acid was obtained by distillation of ants (L. *formica,* ant); acetic acid occurs in vinegar (L. *acetum,* vinegar); butyric acid in rancid butter (L. *butyrum,* butter); stearic acid in animal fats (Gr. *stear,* beef suet). *n*-Caproic acid is one of the so-called "goat acids." Its odor is responsible for the name.

Organic acids occur widely in natural products, and many have been known since ancient times. Their common (trivial) names are often derived from a Greek or Latin word that indicates the original source (Table 27-10).

The names of derivatives of carboxylic acids are often derived from the trivial names of the acids. Positions of substituents are sometimes indicated by lowercase Greek letters, beginning with the carbon *adjacent* to the carboxyl carbon, rather than by numbering the carbon chain.

Aliphatic carboxylic acids are sometimes referred to as *fatty acids* because many have been obtained from animal fats.

2-bromopropanoic acid
(α-bromopropionic acid)

3-methylbutanoic acid
(β-methylbutyric acid)

Derivatives of α-hydroxy carboxylic acids are currently used in some cosmetic preparations.

Some carboxylic acid molecules contain more than one —COOH group (Table 27-11). These acids are nearly always called by their common names. Oxalic acid is an aliphatic **dicarboxylic acid,** and phthalic acid is a typical aromatic dicarboxylic acid.

(a) acetic acid
(an aliphatic acid)

(b) benzoic acid
(an aromatic acid)

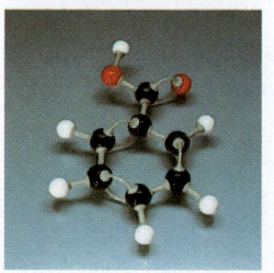

Figure 27-17 (a) Models of acetic acid. (b) Models of benzoic acid.

Table 27-11 *Some Aliphatic Dicarboxylic Acids*

Formula	Name
HOOC—COOH	oxalic acid
HOOC—CH$_2$—COOH	malonic acid
HOOC—CH$_2$CH$_2$—COOH	succinic acid
HOOC—CH$_2$CH$_2$CH$_2$—COOH	glutaric acid
HOOC—CH$_2$CH$_2$CH$_2$CH$_2$—COOH	adipic acid

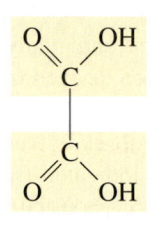

oxalic acid
(an aliphatic acid)

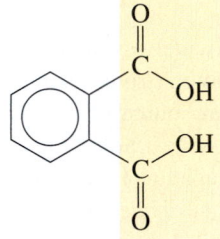

phthalic acid
(an aromatic acid)

Aromatic acids are called by their common names or named as derivatives of benzoic acid, which is considered the "parent" aromatic acid.

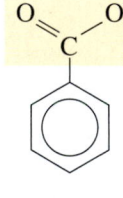

benzoic acid

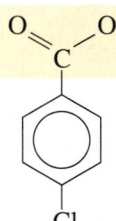

p-chlorobenzoic acid

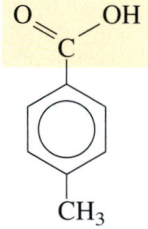

p-toluic acid

Many reactions of carboxylic acids involve displacement of the —OH group by another atom or group of atoms. We find it useful to name the non-OH portions of acid molecules because they occur in many compounds. Such compounds are thought of as derivatives of carboxylic acids.

$$\underset{\substack{\text{an aliphatic}\\\text{carboxylic acid}}}{R-\overset{\overset{\displaystyle O}{\|}}{C}-O-H} \qquad \underset{\substack{\text{an aliphatic}\\\text{acyl group}}}{R-\overset{\overset{\displaystyle O}{\|}}{C}-} \qquad \Big| \qquad \underset{\substack{\text{an aromatic}\\\text{carboxylic acid}}}{Ar-\overset{\overset{\displaystyle O}{\|}}{C}-O-H} \qquad \underset{\substack{\text{an aromatic}\\\text{acyl group}}}{Ar-\overset{\overset{\displaystyle O}{\|}}{C}-}$$

Acyl groups are named as derivatives of the parent acid by dropping *-ic acid* and adding *-yl* to the characteristic stem. Some examples are

$$\underset{\text{acetyl group}}{CH_3-\overset{\overset{\displaystyle O}{\|}}{C}\diagdown} \qquad \underset{\text{propionyl group}}{CH_3CH_2-\overset{\overset{\displaystyle O}{\|}}{C}\diagdown} \qquad \underset{\text{benzoyl group}}{\text{benzoyl group}}$$

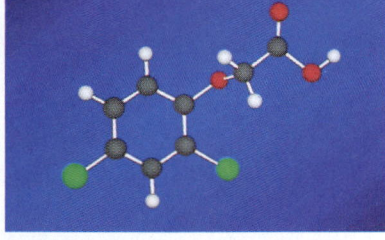

(a) Cl—⬡—O—CH$_2$—$\overset{\overset{\displaystyle O}{\|}}{C}$—OH
     Cl

(b) Cl—⬡—O—CH$_2$—$\overset{\overset{\displaystyle O}{\|}}{C}$—OH
     CH$_3$

These two derivatives of phenoxyacetic acids act as herbicides (weed killers) by overstimulating the plant's growth system.

Although many carboxylic acids occur in the free state in nature, many occur as amides or esters (Section 27-13). Amino acids are substituted carboxylic acids with the general structure

$$R-\underset{\underset{NH_2}{\overset{|}{|}}}{\overset{\overset{H}{\overset{|}{|}}}{C}}-\overset{\overset{O}{\overset{||}{}}}{C}-OH$$

where R can be either an alkyl or an aryl group. Amino acids are the components of proteins, which make up the muscle and tissue of animals. Many other acids are important in the metabolism and synthesis of fats by enzyme systems. Acetic acid (the acid in vinegar) is the end product in the fermentation of most agricultural products. It is the fundamental unit used by living organisms in the biosynthesis of such widely diverse classes of natural products as long-chain fatty acids, natural rubber, and steroid hormones. It is also a powerful solvent and an important reagent in the preparation of pharmaceuticals, plastics, artificial fibers, and coatings. Phthalic acid and adipic acid are used in the production of synthetic polymers that are used as fibers (e.g., Dacron and nylon, Section 28-11).

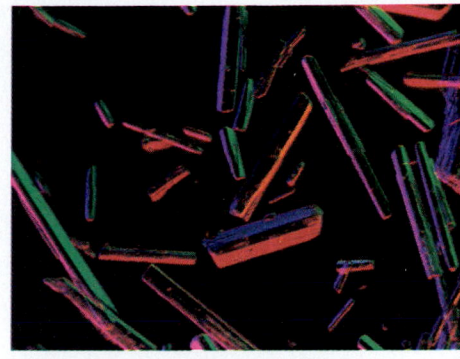

Crystals of glycine viewed under polarized light. Glycine, the simplest amino acid, has the structure shown in the text, with R = H.

27-13 SOME DERIVATIVES OF CARBOXYLIC ACIDS

Four important classes of acid derivatives are formed by the replacement of the hydroxyl group by another atom or group of atoms. Each of these derivatives contains an acyl group.

$$R-\overset{\overset{O}{\overset{||}{}}}{C}-O-\overset{\overset{O}{\overset{||}{}}}{C}-R \qquad R-\overset{\overset{O}{\overset{||}{}}}{C}-Cl \qquad R-\overset{\overset{O}{\overset{||}{}}}{C}-O-R \qquad R-\overset{\overset{O}{\overset{||}{}}}{C}-NH_2$$

| an acid anhydride | an acyl chloride (an acid chloride) | an ester | an amide |

Aromatic compounds of these types (with R = aryl groups) are encountered frequently.

Acid Anhydrides

An acid anhydride can be thought of as the result of removing one molecule of water from two carboxylic acid groups. The structural relationship between monocarboxylic acids and their anhydrides is

$$R-\overset{\overset{O}{\overset{||}{}}}{C}{\overset{}{\diagdown}}{O-H} \atop R-\overset{}{C}{\overset{}{\diagup}}{O-H} \longrightarrow R-\overset{}{C}{\overset{\diagup O}{\diagdown}}{O} \atop R-\overset{}{C}{\diagdown O} \quad + H_2O$$

two molecules of a carboxylic acid   one molecule of acid anhydride

Acid anhydrides are usually prepared by indirect methods. This illustration shows only the structural relationship.

The anhydrides are named by replacing the word "acid" in the name of the parent acid with the word "anhydride." Examples of acids and their anhydrides are

$$CH_3\overset{\overset{\displaystyle O}{\|}}{C}—OH$$

acetic acid

$$CH_3\overset{\overset{\displaystyle O}{\|}}{C}—O—\overset{\overset{\displaystyle O}{\|}}{C}CH_3$$

acetic anhydride

benzoic acid

benzoic anhydride

Acyl Halides (Acid Halides)

The **acyl halides,** sometimes called **acid halides,** are structurally related to carboxylic acids by the replacement of the OH group by a halogen, most often Cl. They are usually named by combining the stems of the common names of the carboxylic acids with the suffix -*yl* and then adding the name of the halide ion. Examples are

$$CH_3—\overset{\overset{\displaystyle O}{\|}}{C}—Cl$$

acetyl chloride

$$CH_3CH_2CH_2—\overset{\overset{\displaystyle O}{\|}}{C}—F$$

butyryl fluoride

benzoyl chloride

Acid halides are very reactive and have not been observed in nature.

Esters

As we shall see in Section 28-9, one method of forming esters involves acid-catalyzed reaction of an alcohol with a carboxylic acid.

Esters can be thought of as the result of removing one molecule of water from a carboxylic acid and an alcohol. Removing a molecule of water from

$$CH_3\overset{\overset{\displaystyle O}{\|}}{C}—OH$$

acetic acid

and $HO—CH_2CH_3$

ethyl alcohol

gives $CH_3\overset{\overset{\displaystyle O}{\|}}{C}—OCH_2CH_3$

ethyl acetate

Models of ethyl acetate, a simple ester, are shown in Figure 27-18.

Figure 27-18 Models of ethyl acetate, $CH_3—\overset{\overset{\displaystyle O}{\|}}{C}—O—CH_2CH_3$, an ester. The $CH_3—\overset{\overset{\displaystyle O}{\|}}{C}—$ fragment is derived from acetic acid, the parent acid; the $O—CH_2CH_3$ fragment is derived from ethanol, the parent alcohol.

Table 27-12 *Some Common Esters*		
Ester	Formula	Odor of
isoamyl acetate	$CH_3COOC_5H_{11}$	bananas
ethyl butyrate	$C_3H_7COOC_2H_5$	pineapples
amyl butyrate	$C_3H_7COOC_5H_{11}$	apricots
octyl acetate	$CH_3COOC_8H_{17}$	oranges
isoamyl isovalerate	$C_4H_9COOC_5H_{11}$	apples
methyl salicylate	$C_6H_4(OH)(COOCH_3)$	oil of wintergreen
methyl anthranilate	$C_6H_4(NH_2)(COOCH_3)$	grapes

Apple blossoms.

Esters are nearly always called by their common names. These consist of, first, the name of the alkyl group in the alcohol, and then the name of the anion derived from the acid.

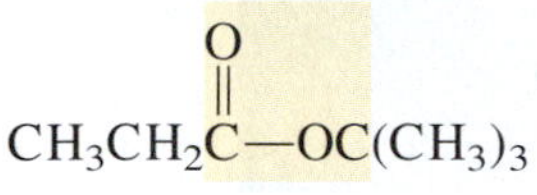

$$CH_3CH_2\overset{\overset{\textstyle O}{\|}}{C}-OC(CH_3)_3 \qquad CH_3\overset{\overset{\textstyle O}{\|}}{C}-O-\bigcirc \qquad \bigcirc-\overset{\overset{\textstyle O}{\|}}{C}-OCH_3$$

t-butyl propionate 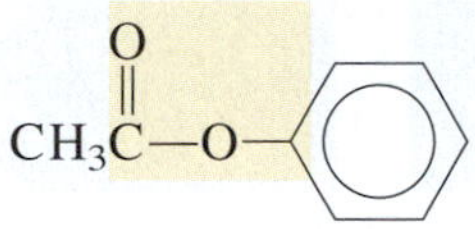   phenyl acetate 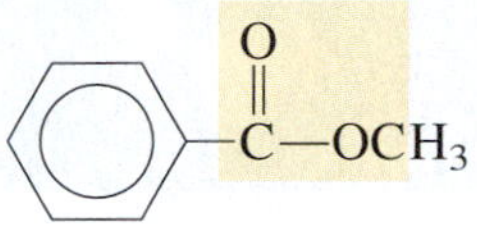   methyl benzoate

Because of their inability to form hydrogen bonds, esters tend to be liquids with boiling points much lower than those of carboxylic acids of similar molecular weights.

Most simple esters are pleasant-smelling substances. They are responsible for the flavors and fragrances of most fruits and flowers and many of the artificial fruit flavorings that are used in cakes, candies, and ice cream (Table 27-12). Esters of low molecular weight are excellent solvents for nonpolar compounds. Ethyl acetate is an excellent solvent that gives many nail polish removers their characteristic odor.

Fats (solids) and **oils** (liquids) are esters of glycerol and aliphatic acids of high molecular weight. "Fatty acids" are all organic acids whose esters occur in fats and oils. Fats and oils have the general formula

Most natural fatty acids contain even numbers of carbon atoms because they are synthesized in the body from two-carbon acetyl groups.

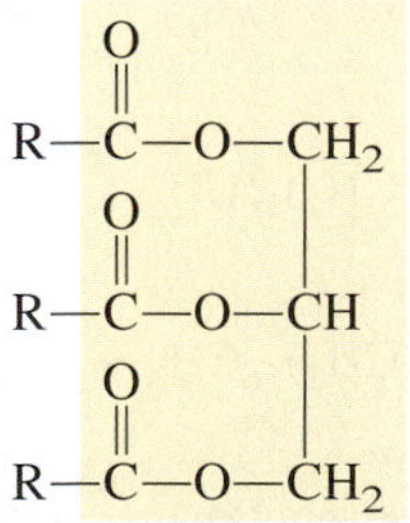

$$R-\overset{\overset{\textstyle O}{\|}}{C}-O-CH_2$$
$$R-\overset{\overset{\textstyle O}{\|}}{C}-O-CH$$
$$R-\overset{\overset{\textstyle O}{\|}}{C}-O-CH_2$$

The fatty acid portions, $R-\overset{\overset{\textstyle O}{\|}}{C}-$, may be saturated or unsaturated. The R's may be the same or different groups.

Fats are solid esters of glycerol and (mostly) saturated acids. Oils are liquid esters that are derived primarily from unsaturated acids and glycerol. The acid portion of a fat usually contains an even number of carbon atoms, often 16 or 18. The acids that occur most frequently in fats and oils are

butyric	$CH_3CH_2CH_2COOH$
lauric	$CH_3(CH_2)_{10}COOH$
myristic	$CH_3(CH_2)_{12}COOH$
palmitic	$CH_3(CH_2)_{14}COOH$
stearic	$CH_3(CH_2)_{16}COOH$

Glycerol is

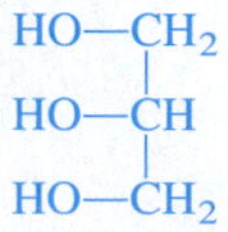

$$HO-CH_2$$
$$HO-CH$$
$$HO-CH_2$$

Figure 27-19 Models of long-chain fatty acids. The saturated fatty acids (a) are linear and tend to pack, like sticks of wood, to form solid masses in blood vessels, thereby constricting them. The *trans* unsaturated fatty acids have a slight Z-shaped kink in the chain, but are also essentially linear molecules. By contrast, *cis* unsaturated fatty acids (b) are bent and so do not pack as well as linear structures and do not collect in blood vessels as readily. Many natural vegetable fats and oils contain esters of *cis* unsaturated fatty acids or polyunsaturated fatty acids. Health problems associated with saturated fatty acids can be decreased by eating less animal fat, butter, and lard. Problems due to *trans* fatty acids are reduced by avoiding processed vegetable fats.

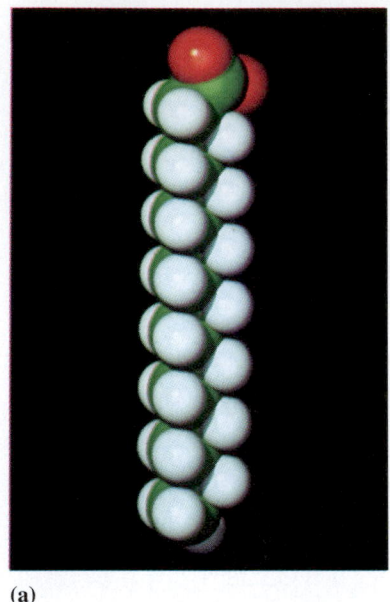

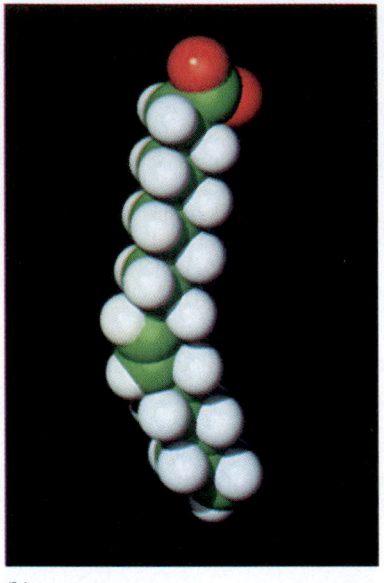

(a)   (b)

oleic	$CH_3(CH_2)_7CH\!=\!CH(CH_2)_7COOH$
linolenic	$CH_3CH_2CH\!=\!CHCH_2CH\!=\!CHCH_2CH\!=\!CH(CH_2)_7COOH$
ricinoleic	$CH_3(CH_2)_5CHOHCH_2CH\!=\!CH(CH_2)_7COOH$

Figure 27-19a shows a model of stearic acid, a long-chain saturated fatty acid.

Naturally occurring fats and oils are mixtures of many different esters. Milk fat, lard, and tallow are familiar, important fats. Soybean oil, cottonseed oil, linseed oil, palm oil, and coconut oil are examples of important oils.

The triesters of glycerol are called glycerides. *Simple glycerides* are esters in which all three R groups are identical. Two examples are

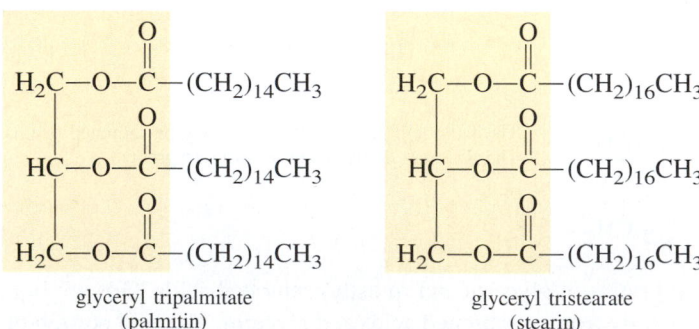

glyceryl tripalmitate   glyceryl tristearate
(palmitin)   (stearin)

Glycerides are frequently called by their common names, indicated in parentheses in the examples above. The common name is the characteristic stem for the parent acid plus an *-in* ending.

Waxes are esters of fatty acids and alcohols other than glycerol. Most are derived from long-chain fatty acids and long-chain monohydric alcohols. Both usually contain even numbers of carbon atoms. Beeswax is largely $C_{15}H_{31}COOC_{30}H_{61}$; carnauba wax contains $C_{25}H_{51}COOC_{30}H_{61}$. Both are esters of myricyl alcohol, $C_{30}H_{61}OH$.

Honeybees produce the wax to build their honeycombs.

Amides

Amides are thought of as derivatives of organic acids and ammonia, primary amines, or secondary amines. Amides contain the $-\overset{\overset{\displaystyle O}{\|}}{C}-N\big<$ grouping of atoms. They are named as derivatives of the corresponding carboxylic acids, the suffix *-amide* being substituted for *-ic acid* or *-oic acid* in the name of the parent acid.

$$CH_3\overset{\overset{\displaystyle O}{\|}}{C}-NH_2 \qquad \langle \bigcirc \rangle -\overset{\overset{\displaystyle O}{\|}}{C}-NH_2$$

acetamide benzamide

The presence of alkyl or aryl substituents attached to nitrogen is designated by prefixing the letter N and the name of the substituent to the name of the unsubstituted amide.

$$CH_3\overset{\overset{\displaystyle O}{\|}}{C}-\overset{\overset{\displaystyle CH_3}{|}}{N}-CH_3 \qquad CH_3\overset{\overset{\displaystyle O}{\|}}{C}-\overset{\overset{\displaystyle H}{|}}{N}-\langle\bigcirc\rangle \qquad CH_3\text{-}\langle\bigcirc\rangle-\overset{\overset{\displaystyle O}{\|}}{C}-N(CH_2CH_3)_2$$

N,N-dimethylacetamide N-phenylacetamide (acetanilide) N,N-diethyl-*m*-toluamide

At room temperature, unsubstituted amides (with the exception of formamide, $HCONH_2$) are crystalline solids with melting and boiling points even higher than those of the carboxylic acids of comparable molecular weight. N,N-dimethylformamide, $HCON(CH_3)_2$, is a good solvent for both polar and nonpolar compounds; it is useful as a reaction medium when such different compounds need to be brought into contact with one another. Acetanilide (sometimes called antifebrin) is the amide of acetic acid and aniline. It is used to treat headaches, neuralgia, and mild fevers. N,N-diethyl-*m*-toluamide, the amide of metatoluic acid and N,N-diethylamine, is the active ingredient in some insect repellents. Proteins are complex amides of high molecular weight. Some synthetic fibers are also polyamides (Section 28-11).

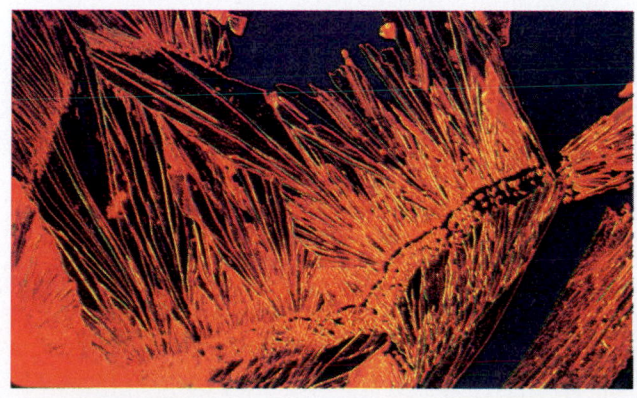

Crystals of acetaminophen (Tylenol) viewed under polarized light. The structure of acetaminophen is

$$CH_3-\overset{\overset{\displaystyle O}{\|}}{C}-\underset{\underset{\displaystyle H}{|}}{N}-\langle\bigcirc\rangle-OH$$

CHEMISTRY IN USE

Our Daily Lives

Butter, Margarine, and trans Fats

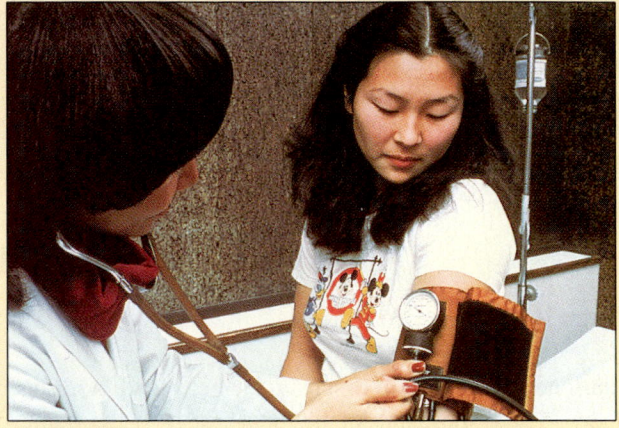
Blood pressure check.

Humans had consumed butter for thousands of years before France introduced the world to margarine in the late 1860s; by 1874 margarine reached the United States. Margarine consumption has increased rapidly, due in part to an increased risk of heart disease that has been associated with consumption of butter.

Both butter and margarine are primarily fats, but they contain different kinds of fats. The main ingredient in butter is cream, the concentrated fat from milk, whereas the main ingredient in margarine is vegetable oil, the concentrated fat from plants. Although it is widely accepted that animal fats pose a greater risk for heart disease than vegetable oils, most vegetable oils used in margarine and processed foods are modified by hydrogenation.

The hydrogenation process adds hydrogen atoms to unsaturated vegetable oils; this causes vegetable oils to solidify, which makes them more creamy, and prolongs their shelf life. Unfortunately, the hydrogenation process changes about 40% of the naturally occurring *cis* molecules of vegetable oil into *trans* isomers; the resulting fat molecules are referred to as *trans* fats. Oils produced in nature contain only *cis* isomers and are considered healthful, but the *trans* fats have been linked to many thousands of deaths due to heart disease. Diets high in hydrogenated vegetable oil may be as unhealthy as those high in saturated animal fats.

To understand the different effects produced by animal fats, vegetable oils, and *trans* fats, we need to know a little about blood cholesterol. Medical specialists recognize several types of cholesterol, two of which are HDL (high density lipoproteins) and LDL (low density lipoproteins). Because high levels of HDL reduce the risk of heart disease, HDL has become known as good cholesterol; because high levels of LDL increase the risk of heart disease, LDL has become known as bad cholesterol. In the average American adult, the total amount of cholesterol (including HDL and LDL) circulating in the blood is about 200 mg per 100 mL of serum.

A diet high in animal fats usually increases blood cholesterol levels beyond 200 mg per 100 mL of serum, thereby increasing the risk of heart disease. However, when animal fats increase the total blood cholesterol level, they increase the levels of both good HDL and bad LDL. In contrast, vegetable oils decrease the total blood cholesterol level, which make them healthier foods than other fats. *Trans* fats (from hydrogenated vegetable oils) also decrease the total blood cholesterol level, giving the appearance of being healthy, but unfortunately they accomplish this reduction by simultaneously reducing the desirable HDL levels and increasing the undesirable LDL levels.

The damage done by hydrogenated vegetable oils is particularly disturbing because margarine is one of the largest sources of calories in the American diet; approximately 40% of margarine fat is the *trans* isomer. Hydrogenated vegetable oils are widely used in shortening, cookies, crackers, chips, and other processed foods. Ironically, most of these foods are touted as being healthy because they contain no cholesterol or saturated fats.

What can we do to protect our health? For many years, nutritionists have been telling us what appears to be the "best solution": reduce your total fat intake to 20% of your daily calories, and reduce your consumption of animal fats by substituting vegetable oils. Whenever possible use vegetable oils that have not been hydrogenated.

Ronald DeLorenzo
Middle Georgia College

27-14 ALDEHYDES AND KETONES

Aldehydes and ketones contain the carbonyl group, $\diagup$C$=$O. In **aldehydes,** at least one H atom is bonded to the carbonyl group. **Ketones** have two alkyl or aryl groups bonded to

(a) **(b)**

Figure 27-20 (a) Models of formaldehyde, HCHO, the simplest aldehyde. (b) Models of acetone, CH_3—CO—CH_3, the simplest ketone.

a carbonyl group. Models of formaldehyde (the simplest aldehyde) and acetone (the simplest ketone) are shown in Figure 27-20.

$$\text{R—}\overset{\overset{\textstyle O}{\|}}{\text{C}}\text{—H} \qquad \text{Ar—}\overset{\overset{\textstyle O}{\|}}{\text{C}}\text{—H} \qquad \text{R—}\overset{\overset{\textstyle O}{\|}}{\text{C}}\text{—R} \qquad \text{Ar—}\overset{\overset{\textstyle O}{\|}}{\text{C}}\text{—Ar} \qquad \text{Ar—}\overset{\overset{\textstyle O}{\|}}{\text{C}}\text{—R}$$

| aliphatic aldehyde | aromatic aldehyde | aliphatic ketone | aromatic ketone | mixed ketone |

The simplest aldehyde, formaldehyde, $H-\overset{\overset{\textstyle O}{\|}}{C}-H$, has two H atoms and no alkyl or aryl groups. Other aldehydes have one alkyl or aryl group and one H atom bonded to the carbonyl group.

Aldehydes are usually called by their common names. These are derived from the name of the acid with the same number of C atoms (Table 27-13). The systematic (IUPAC) name is derived from the name of the parent hydrocarbon. The suffix *-al* is added to the characteristic stem. The carbonyl group takes positional precedence over other substituents.

Table 27-13 *Properties of Some Simple Aldehydes*

Common Name	Formula	Normal bp (°C)
formaldehyde (methanal)	$H-\overset{\overset{\textstyle O}{\|}}{C}-H$	−21
acetaldehyde (ethanal)	$CH_3-\overset{\overset{\textstyle O}{\|}}{C}-H$	20.2
propionaldehyde (propanal)	$CH_3CH_2\overset{\overset{\textstyle O}{\|}}{C}-H$	48.8
benzaldehyde	$\bigcirc\!\!-\overset{\overset{\textstyle O}{\|}}{C}-H$	179.5

Formaldehyde has long been used as a disinfectant and as a preservative for biological specimens (including embalming fluid). Its main use is in the production of certain plastics and in binders for plywood. Many important natural substances are aldehydes and ketones. Examples include sex hormones, some vitamins, camphor, and the flavorings extracted from almonds and cinnamon. Aldehydes contain a carbon–oxygen double bond, so they are very reactive compounds. As a result, they are valuable in organic synthesis, particularly in the construction of carbon chains.

The simplest ketone is called acetone. Other simple, commonly encountered ketones are usually called by their common names. These are derived by naming the alkyl or aryl groups attached to the carbonyl group.

When the benzene ring is a substituent, it is called a phenyl group ($-C_6H_5$).

$$CH_3-\overset{\overset{\displaystyle O}{\|}}{C}-CH_3$$
acetone

$$CH_3-\overset{\overset{\displaystyle O}{\|}}{C}-CH_2CH_3$$
methyl ethyl ketone

$$CH_3CH_2-\overset{\overset{\displaystyle O}{\|}}{C}-CH_2CH_3$$
diethyl ketone

cyclohexanone

acetophenone
(methyl phenyl ketone)

benzophenone
(diphenyl ketone)

The systematic names for ketones are derived from their parent hydrocarbons. The suffix *-one* is added to the characteristic stem.

$$\overset{1}{C}H_3-\overset{\overset{\displaystyle O}{\|}}{\underset{2}{C}}-\overset{3}{C}H_2-\overset{4}{C}H_3$$
2-butanone

$$\overset{1}{C}H_3-\overset{2}{C}H_2-\overset{\overset{\displaystyle O}{\|}}{\underset{3}{C}}-\overset{\overset{\displaystyle CH_3}{|}}{\underset{4}{C}H}-\overset{5}{C}H_2-\overset{6}{C}H_3$$
4-methyl-3-hexanone

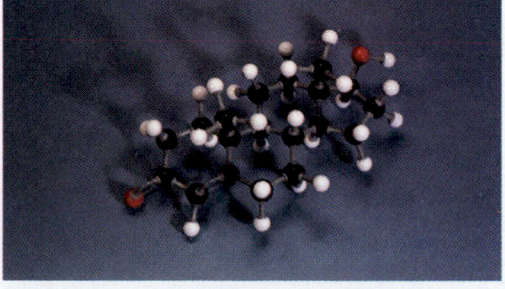

Steroid molecules have similar molecular shapes but different biochemical functions. Progesterone (top), a female sex hormone, and testosterone (bottom), a male sex hormone. Both are ketones.

The ketones are excellent solvents. Acetone is very useful because it dissolves most organic compounds yet is miscible with water. Acetone is widely used as a solvent in the manufacture of lacquers, paint removers, explosives, plastics, drugs, and disinfectants. Some ketones of high molecular weight are used extensively in blending perfumes. Structures of some naturally occurring aldehydes and ketones are

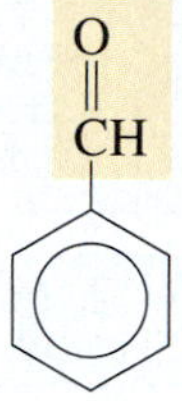

benzaldehyde
(almonds)

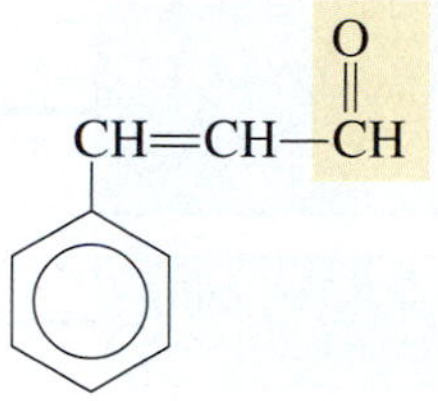

cinnamaldehyde
(cinnamon)

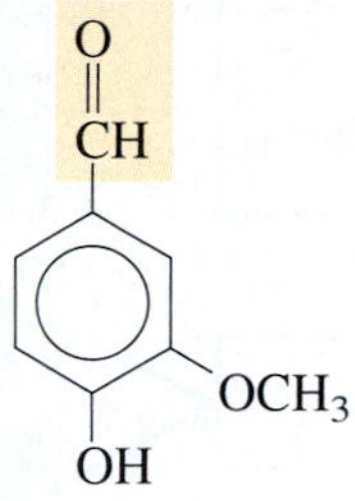

vanillin (vanilla)

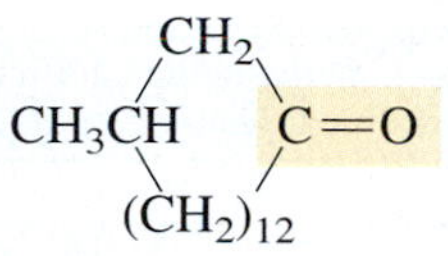

muscone
(musk deer, used
in perfumes)

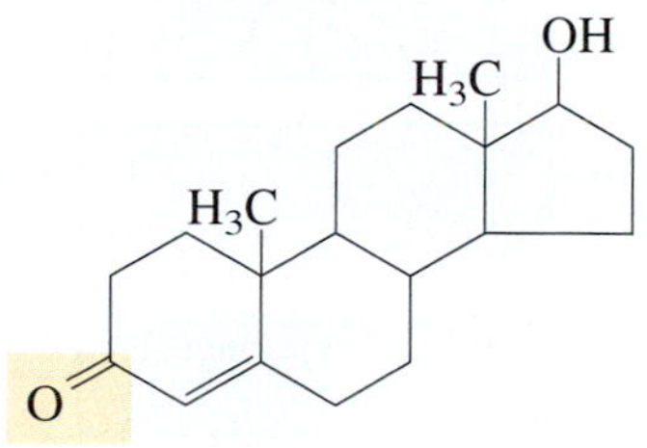

testosterone
(male sex hormone)

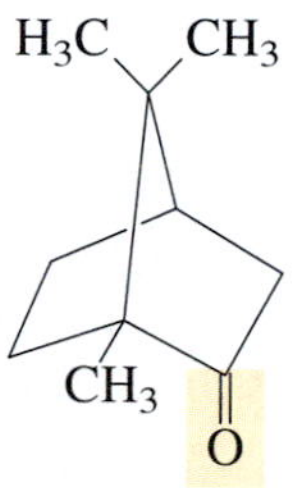

camphor

27-15 SUMMARY OF FUNCTIONAL GROUPS

Some important functional groups and the corresponding classes of related compounds are summarized in Figure 27-21 (on the next page).

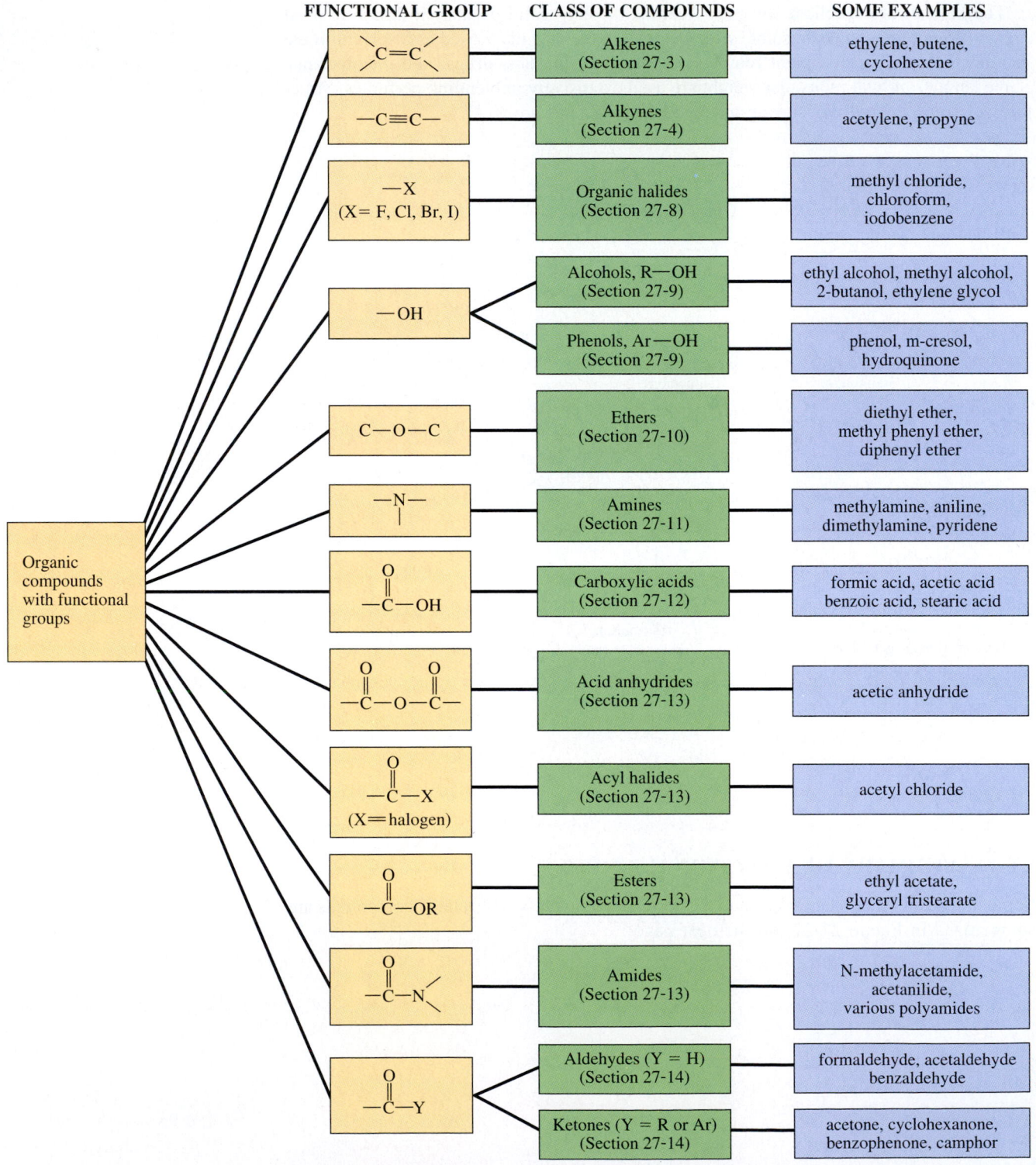

Figure 27-21   Summary of some functional groups and classes of organic compounds.

Key Terms

Acid anhydride A compound that can be thought of as produced by dehydration of a carboxylic acid; general formula is

$$R-\overset{\overset{\displaystyle O}{\|}}{C}-O-\overset{\overset{\displaystyle O}{\|}}{C}-R.$$

Acid halide See *Acyl halide.*

Acyl group The group of atoms remaining after removal of an —OH group of a carboxylic acid.

Acyl halide A compound derived from a carboxylic acid by replacing the —OH group with a halogen (X), usually —Cl; general

formula is $R-\overset{\overset{\displaystyle O}{\|}}{C}-X$; also called acid halide.

Alcohol A hydrocarbon derivative in which an H attached to a carbon atom not in an aromatic ring has been replaced by an —OH group.

Aldehyde A compound in which an alkyl or aryl group and a hydrogen atom are attached to a carbonyl group; general formula is

$R-\overset{\overset{\displaystyle O}{\|}}{C}-H$; R may be H.

Aliphatic hydrocarbons Hydrocarbons that do not contain aromatic rings.

Alkanes See *Saturated hydrocarbons.*

Alkenes Unsaturated hydrocarbons that contain one or more carbon–carbon double bonds.

Alkyl group A group of atoms derived from an alkane by the removal of one hydrogen atom.

Alkylbenzene A compound containing an alkyl group bonded to a benzene ring.

Alkynes Unsaturated hydrocarbons that contain one or more carbon–carbon triple bonds.

Amide A compound containing the $-\overset{\overset{\displaystyle O}{\|}}{C}-N\!\!<$ group.

Amine A compound that can be considered a derivative of ammonia, in which one or more hydrogens are replaced by alkyl or aryl groups.

Amino acid A compound containing both an amino group and a carboxylic acid group.

Amino group The —NH₂ group.

Aromatic hydrocarbons Benzene and similar condensed ring compounds; contain delocalized rings.

Aryl group The group of atoms remaining after a hydrogen atom is removed from an aromatic system.

Carbonyl group The $-\overset{\overset{\displaystyle O}{\|}}{C}-$ group.

Carboxylic acid A compound containing a $-\overset{\overset{\displaystyle O}{\|}}{C}-O-H$ group.

Catenation The ability of an element to bond to itself.

Conjugated double bonds Double bonds that are separated from each other by one single bond, as in C=C—C=C.

Constitutional isomers Compounds that contain the same numbers of the same kinds of atoms but that differ in the order in which their atoms are bonded together. Also known as *Structural isomers.*

Cycloalkanes Cyclic saturated hydrocarbons.

Ester A compound of the general formula $R-\overset{\overset{\displaystyle O}{\|}}{C}-O-R'$ where R and R′ may be the same or different, and may be either aliphatic or aromatic.

Ether A compound in which an oxygen atom is bonded to two alkyl or two aryl groups, or one alkyl and one aryl group.

Fat A solid triester of glycerol and (mostly) saturated fatty acids.

Fatty acid An aliphatic acid; many can be obtained from animal fats.

Functional group A group of atoms that represents a potential reaction site in an organic compound.

Geometrical isomers Compounds with different arrangements of groups on the opposite sides of a bond with restricted rotation, such as a double bond or a single bond in a ring; for example, *cis–trans* isomers of certain alkenes.

Glyceride A triester of glycerol.

Heterocyclic amine An amine in which the nitrogen is part of a ring.

Homologous series A series of compounds in which each member differs from the next by a specific number and kind of atoms.

Hydrocarbon A compound that contains only carbon and hydrogen.

Isomers Different substances that have the same molecular formula.

Ketone A compound in which a carbonyl group is bound to two alkyl or two aryl groups, or to one alkyl and one aryl group.

Oil A liquid triester of glycerol and unsaturated fatty acids.

Organic chemistry The chemistry of substances that contain carbon–carbon or carbon–hydrogen bonds.

Phenol A hydrocarbon derivative that contains an —OH group bound to an aromatic ring.

Pi bonds A chemical bond formed by the side-to-side overlap of atomic orbitals.

Polyene A compound that contains more than one double bond per molecule.

Polyhydric alcohol An alcohol that contains more than one —OH group.

Primary alcohol An alcohol with no or one R group bonded to the carbon bearing the —OH group.

Primary amine An amine in which one H atom of ammonia has been replaced by an organic group.

Saturated hydrocarbons Hydrocarbons that contain only single bonds. They are also called *alkanes* or *paraffin hydrocarbons.*

Secondary alcohol An alcohol with two R groups bonded to the carbon bearing the —OH group.

Secondary amine An amine in which two H atoms of ammonia have been replaced by organic groups.

Sigma bond   A chemical bond formed by the end-to-end overlap of atomic orbitals.

Structural isomers   See *Constitutional isomers*.

Tertiary alcohol   An alcohol with three R groups bonded to the carbon bearing the —OH group.

Tertiary amine   An amine in which three H atoms of ammonia have been replaced by organic groups.

Unsaturated hydrocarbons   Hydrocarbons that contain double or triple carbon–carbon bonds.

Exercises

Basic Ideas

1. (a) What is organic chemistry? (b) What happened to the "vital force" theory?
2. (a) How is carbon's tendency to bond to other carbon atoms unique among the elements? (b) What is catenation?
3. How many "everyday" uses of organic compounds can you think of? List them.
4. (a) What are the principal sources of organic compounds? (b) Some chemists argue that the ultimate source of all naturally occurring organic compounds is carbon dioxide. Could this be possible? *Hint:* Think about the origins of coal, natural gas, and petroleum.

Aliphatic Hydrocarbons

5. (a) What are hydrocarbons? (b) What are saturated hydrocarbons? (c) What are the cycloalkanes? (d) Write the general formula for noncyclic alkanes.
6. Describe the bonding in and the geometry of molecules of the following alkanes: (a) methane, (b) ethane, (c) propane, (d) *n*-butane. How are the formulas for these compounds similar? different?
7. (a) What are "normal" hydrocarbons? (b) What are branched-chain hydrocarbons? (c) Cite three examples of each.
8. (a) What is a homologous series? (b) Provide specific examples of compounds that are members of a homologous series. (c) What is a methylene group? (d) How does each member of a homologous series differ from compounds that come before and after it in the series? (e) Name three homologous series that are also aliphatic hydrocarbons.
9. (a) What are alkyl groups? (b) Draw structures for and write the names of the first five normal alkyl groups. (c) What is the origin of the names for alkyl groups?
10. Could a substance with the molecular formula C_3H_8 be a cycloalkane? Could C_3H_8 be a branched alkane having a methyl group as a substituent attached to the longest chain?
11. Name the following compound by the IUPAC system: $CH_3CH(CH_3)CH_2CH_3$. Draw a constitutional isomer of this compound and give its correct IUPAC name.
12. Give the IUPAC name of each of the following compounds.
    (a) $CH_3CH_2CH_2CH(CH_3)CH_2CH_3$
    (b) $CH_3CH_2CH(CH_2CH_3)CH_2CH_3$
    (c) $CH_3CH(CH_2CH_2CH_3)CH_2CH_3$

(d) $CH_3CH_2CH_2CH(CH_3)CH_3$
(e) $CH_3CH(CH_3)CH_2CH_2CH_3$

13. Draw the structures for the three isomeric saturated hydrocarbons having the molecular formula C_5H_{12}. Name each by the IUPAC system.
14. Draw the structures for 4 constitutional isomers of C_7H_{14} that contain a cyclopropyl substituent. Name each by the IUPAC system.
15. Write the structure for 2,2-dimethylpropane.
16. Write the IUPAC name for each of the following.

17. Draw the structures for 1,1,2-trimethylcyclohexane, isopropylcyclobutane, and *sec*-butylcyclohexane.
18. (a) How does the general formula for the alkenes differ from the general formula for the alkanes? (b) Why are the general formulas identical for alkenes and cycloalkanes that contain the same number of carbon atoms?
19. (a) What are cycloalkenes? (b) What is their general formula? (c) Provide the formulas and names of three examples.
20. Describe the bonding at each carbon atom in (a) ethene, (b) propene, (c) 1-butene, and (d) 2-methyl-2-butene.
21. (a) What are geometric (structural) isomers? (b) Why is rotation around a double bond not possible at room temperature? (c) What do *cis* and *trans* mean? (d) Draw structures for *cis*- and *trans*-3-methyl-3-hexene. How do their melting and boiling points compare?
22. How do carbon–carbon single bond lengths and carbon–carbon double bond lengths compare? Why?
23. (a) What are alkynes? (b) What other name is used to describe them? (c) What is the general formula for alkynes? (d) How does the general formula for alkynes compare with the general formula for cycloalkenes? Why?
24. Describe the bonding and geometry associated with the triple bond of alkynes.
25. Draw the structural formulas of the following compounds: (a) 1-butyne, (b) 2-methylpropene, (c) 2-ethyl-3-methyl-1-butene, (d) 3-methyl-1-butyne.
26. Draw the structural formulas of the following compounds: (a) 3-hexyne, (b) 1,3-pentadiene, (c) 3,3-dimethylcyclobutene, (d) 3,4-diethyl-1-hexyne.

27. Write the IUPAC names for the following compounds.

(a) 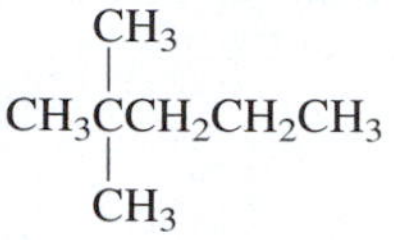

(c) 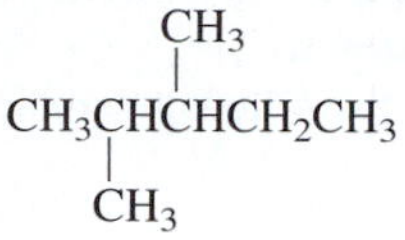

(b) $CH_2{=}C(CH_3)_2$

(d) 

28. Repeat Exercise 27 for

(a) 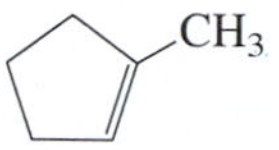

(d) 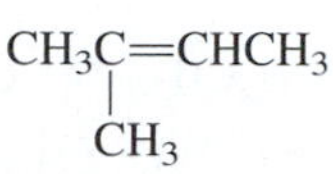

(b) 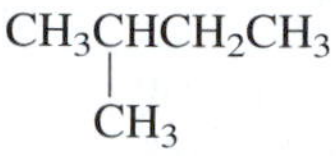

(e) 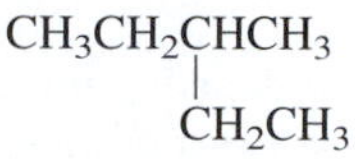

(c) $CH_3C{\equiv}CCH_3$

(f) 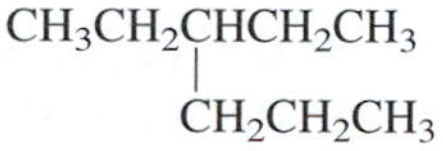

Aromatic Hydrocarbons

29. (a) What are aromatic hydrocarbons? (b) What is the principal source of aromatic hydrocarbons? (c) What is the most common aromatic hydrocarbon?

30. What is a phenyl group? How may isomeric monophenylnaphthalenes are possible?

31. There are only three isomeric trimethylbenzenes. Write their structural formulas and name them.

32. How many isomeric dibromobenzenes are possible? What names are used to designate these isomers?

33. Write the structural formulas for the following compounds: (a) *p*-difluorobenzene, (b) propylbenzene, (c) 1,3,5-tribromobenzene, (d) 1,3-diphenylbutane, (e) *p*-chlorotoluene.

34. Write the IUPAC names for the following compounds.

(a) 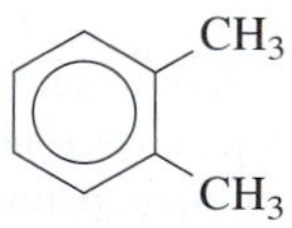

(c) 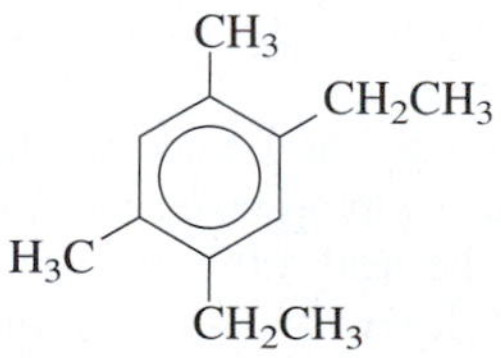

(b) 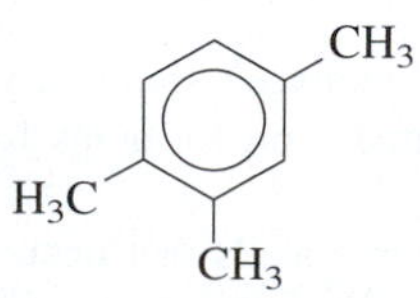

(d) 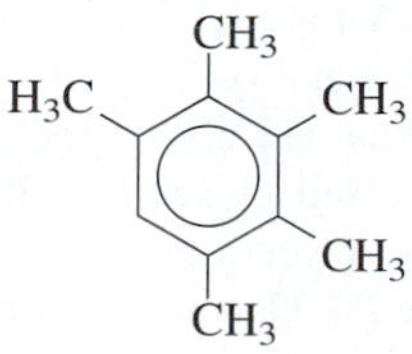

Alkyl and Aryl Halides

35. Write the general representation for the formula of an alkyl halide. How does this differ from the representation for the formula of an aryl halide?

36. Name the following halides.

(a) 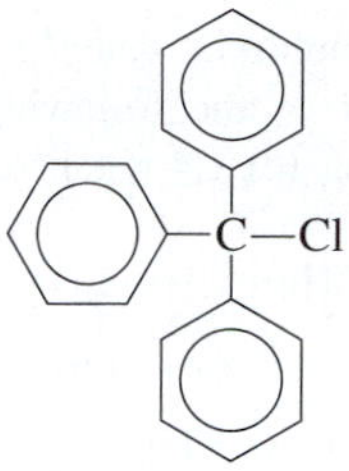

(c) CH_2ClCH_2Cl

(b) 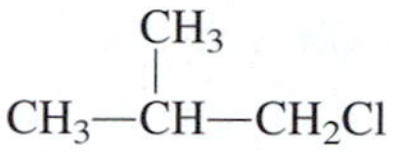

(d) 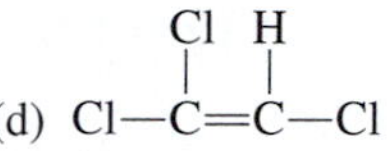

37. Write the structural formulas for the following: (a) 2,2-dichloropentane, (b) 4-bromo-1-butene, (c) 1,2-dichloro-2-fluoropropane, (d) 1,4-dichlorobenzene.

38. Name the following.

(a) 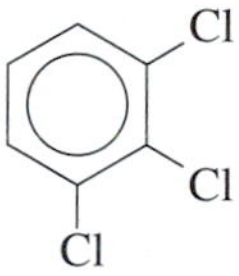

(c) 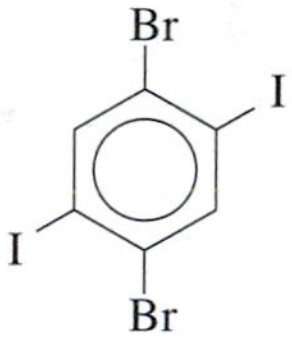

(b) 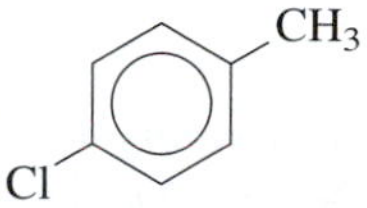

(d) 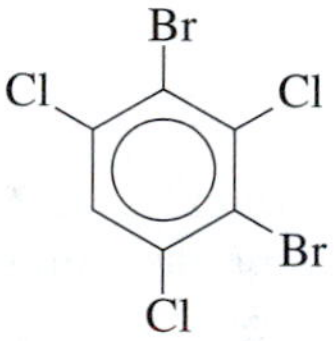

39. The compound 1,2-dibromo-3-chloropropane (DBCP) was used as a pesticide in the 1970s. Recently, agricultural workers have claimed that exposure to DBCP made them sterile. Write the formula for this compound.

Alcohols and Phenols

40. (a) What are alcohols and phenols? (b) How do they differ? (c) Why can alcohols and phenols be viewed as derivatives of hydrocarbons? as derivatives of water?

41. (a) Distinguish among primary, secondary, and tertiary alcohols. (b) Write names and formulas for three alcohols of each type.

42. (a) Draw structural formulas for and write the names of the four (saturated) alcohols that contain four carbon atoms and one —OH group per molecule. (b) Draw structural formulas for and write the names of the eight (saturated) alcohols that contain five carbon atoms and one —OH group per molecule. Which ones may be classified as primary alcohols? secondary alcohols? tertiary alcohols?

43. (a) What are glycols? (b) Draw structures of three examples. (c) Why are glycols more soluble in water than monohydric alcohols that contain the same number of carbon atoms?

44. Refer to Table 27-8 and explain the trends in boiling points and solubilities of alcohols in water.

45. Why are methyl alcohol and ethyl alcohol called wood alcohol and grain alcohol, respectively?

46. Why are most phenols only slightly soluble in water?

47. Write the structural formula for each of the following compounds: (a) 1-butanol, (b) cyclohexanol, (c) 1,4-butanediol.

48. Name the following compounds.

(a) $CH_3CH_2CHCH_2OH$
   |
   CH_3

(c) $CH_3—CH—CH_2$
        | |
        OH   OH

(b) $CH_3—C—CH_2CH_2OH$ (with CH_3 above and CH_3 below the central C)

(d) $CH_3—C—OH$ (with CH_3 above and CH_3 below the central C)

49. Which of the following compounds are phenols? Name each compound.

(a) ⬡—CH_2CH_2OH

(b) cyclohexane with CH_3 and —OH

(c) ⬡ with two —OH groups

(d) bicyclohexyl-substituted ring with —OH

(e) ⬡—OCH_3

50. Write the structural formulas for the following: (a) p-iodophenol, (b) 4-nitrophenol (the nitro group is —NO_2), (c) m-nitrophenol.

Ethers

51. Distinguish among aliphatic ethers, aromatic ethers, and mixed ethers.

52. Briefly describe the bonding around the oxygen atom in dimethyl ether. What intermolecular forces are found in this ether?

53. What determines whether an ether is "symmetrical" or "unsymmetrical"?

*54. Write the structural formulas for the following: (a) methoxymethane, (b) 2-ethoxypropane, (c) 1,3-dimethoxybutane, (d) ethoxybenzene, (e) methoxycyclobutane.

55. Name the following ethers.

(a) $CH_3—O—CH_2CH_2CH_3$

(c) ⬡—$O—CH_2CH_3$

(b) $CH_3—O—CH—CH_3$
                 |
                 CH_3

(d) ⬡—$O—CH_3$

Amines

56. (a) What are amines? (b) Why are amines described as derivatives of ammonia?

57. Write the general representation for the formula for a compound that is (a) a primary amine, (b) a secondary amine, (c) a tertiary amine. Is $(CH_3)_2CHNH_2$ a secondary amine? Give a reason for your answer.

58. Name the following amines.

(a) $CH_3—CH_2$
         |
         NH
         |
    $CH_3—CH_2$

(b) O_2N—⬡—NH_2

(The —NO_2 substituent is called "nitro-.")

(c) ⬠—$NHCH_3$

(d) $CH_3CH_2CH_2CH_2—N—CH_2CH_2CH_2CH_3$
                         |
                     $CH_2CH_2CH_2CH_3$

59. The stench of decaying proteins is due in part to the two compounds whose structures and common names are

$H_2NCH_2CH_2CH_2CH_2NH_2$ putrescine
$H_2NCH_2CH_2CH_2CH_2CH_2NH_2$ cadaverine

Name these compounds as amino-substituted alkanes.

Carboxylic Acids and Their Derivatives

60. (a) What are carboxylic acids? (b) Draw structural formulas for and write the names of five carboxylic acids.

61. (a) Why are aliphatic carboxylic acids sometimes called fatty acids? Cite two examples.

62. (a) What are acyl chlorides, or acid chlorides? (b) Draw structures for four acid chlorides and name them.

63. (a) What are esters? (b) Draw structures for four esters and write their names.

64. (a) What is the general representation for the formula of an acid anhydride? (b) Draw structures for four acid anhydrides and write their names.

65. Write the structural formulas for the following: (a) 2-methylpropanoic acid, (b) 3-bromobutanoic acid, (c) p-nitrobenzoic acid, (d) potassium benzoate, (e) 2-aminopropanoic acid.

66. List six naturally occurring esters and their sources.

67. (a) What are fats? What are oils? (b) Write the general formulas for fats and oils.

68. (a) What are glycerides? Distinguish between simple glycerides and mixed glycerides. (b) Write names and formulas for three simple glycerides.

69. Write the names and formulas for some acids that occur in fats and oils (as esters).

70. What are waxes?

71. Name the following esters.

(a) $CH_3\overset{O}{\overset{\|}{C}}—OCH_2CH_2CH_3$ (b) $CH_3\overset{O}{\overset{\|}{C}}OCH_3$

72. Name the following esters.

(a) [structure: benzene ring—C(=O)—O—benzene ring]

(b) $CH_3CH_2CH_2\overset{\overset{\displaystyle O}{\|}}{C}OCH_2CH_2CH_3$

(b) [structure: benzene ring—CH_2O—$\overset{\overset{\displaystyle O}{\|}}{C}H$]

(e) [structure: benzene ring—$CH_2CH_2\overset{\overset{\displaystyle O}{\|}}{C}Cl$]

(c) [structure: cyclohexane ring—$\overset{\overset{\displaystyle O}{\|}}{C}NH_2$]

Aldehydes and Ketones

73. (a) Distinguish between aldehydes and ketones. (b) Cite three examples (each) of aliphatic and aromatic aldehydes and ketones by drawing structural formulas and naming the compounds.

74. (a) List several naturally occurring aldehydes and ketones. (b) What are their sources? (c) What are some uses of these compounds?

*75. Name the following compounds.

(a) $CH_3CH_2\overset{\overset{\displaystyle }{\underset{\underset{\displaystyle CH_3}{|}}{C}}}HCH_2\overset{\overset{\displaystyle O}{\|}}{C}H$

(c) $H-\overset{\overset{\displaystyle }{\underset{\underset{\displaystyle Br}{|}}{\overset{\overset{\displaystyle Br}{|}}{C}}}}-CH_2-\overset{\overset{\displaystyle O}{\|}}{C}H$

(b) [structure: cyclohexane ring=O]

(d) [structure: benzene ring—$\overset{\overset{\displaystyle O}{\|}}{C}-CH_2-CH_3$]

*76. Write the chemical formulas for the following: (a) 2-methylbutanal, (b) propynal, (c) o-methoxybenzaldehyde, (d) 2-butanone, (e) 1-bromo-2-propanone, (f) 3-heptanone.

Mixed Exercises

*77. Identify the class of organic compounds (ester, ether, ketone, and so on) to which each of the following belongs.

(a) [structure: benzene ring—CH_2CH_2OH]

(d) [structure: benzene ring—O—benzene ring]

(b) [structure: $H_2C-\overset{\overset{\displaystyle O}{\|}}{C}$ and $H_2C-\overset{\underset{\displaystyle O}{\|}}{C}$ bridged by CH_2]

(e) [structure: benzene ring—benzene ring—OH]

(c) $CH_3\overset{\overset{\displaystyle O}{\|}}{C}O\overset{\overset{\displaystyle O}{\|}}{C}C(CH_3)_3$

(f) $CH_3\overset{\overset{\displaystyle O}{\|}}{C}CH_2$—[benzene ring]

*78. Identify the class of organic compounds (ester, ether, ketone, and so on) to which each of the following belongs.

(a) [structure: cyclohexane ring—$\overset{\overset{\displaystyle O}{\|}}{C}OH$]

(d) [structure: benzene ring—$CH\overset{\underset{\displaystyle O}{\diagdown}}{—}CH_2$]

*79. Identify and name the functional groups in each of the following.

(a) $HO-\overset{\overset{\displaystyle O}{\|}}{C}$—[benzene ring]—$\overset{\overset{\displaystyle O}{\|}}{C}-OH$

(b) $CH_2{=}CH-\overset{\overset{\displaystyle O}{\|}}{C}-O-CH_2CH_2CH_2CH_3$

(c) [structure: indane ring system with OH, NO_2, and $O{=}\overset{}{C}-CH_2CH_2OH$]

(d) [structure: benzene ring—$\overset{\overset{\displaystyle }{\underset{\underset{\displaystyle CH_3}{|}}{N}}}-CH_2CH_3$]

80. Identify and name each functional group in the following.

(a) $CH_3CH_2-\overset{\overset{\displaystyle O}{\|}}{C}-CH_2-\overset{\overset{\displaystyle O}{\|}}{C}-CH_2CH_3$

(b) HO—[benzene ring]—$\overset{\overset{\displaystyle OH}{|}}{C}H\underset{\underset{\displaystyle OCH_3}{|}}{CH_2}-\overset{\overset{\displaystyle H}{|}}{N}-CH_3$

(c) dioxane (also known as 1,4-dioxin)

[structure: dioxane ring with two O]

(d) morphine

[structure: morphine skeletal formula with CH_3, N, HO, O, OH]

(e) epinephrine (adrenaline)

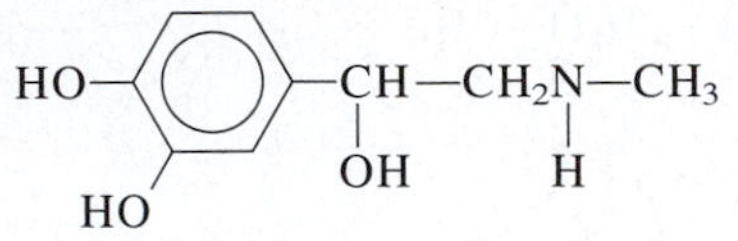

81. Identify and name the functional groups in each of the following.

(a) morpholine

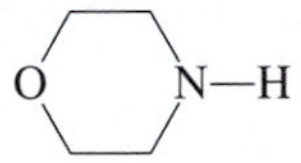

(b) citric acid

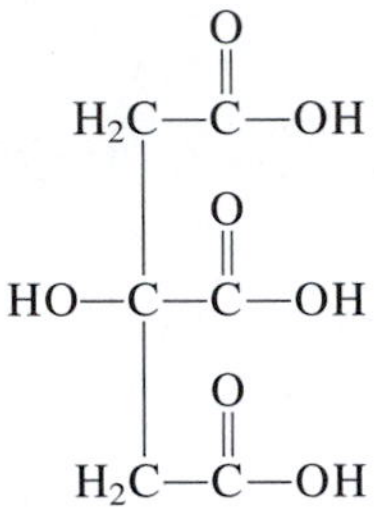

(c) coniine (from the hemlock plant; the poison that Socrates drank)

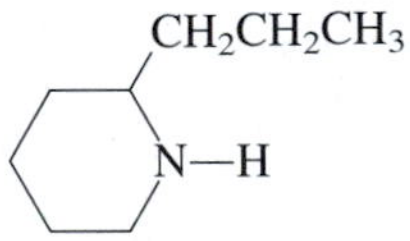

(d) glucose (a simple sugar, also known as dextrose)

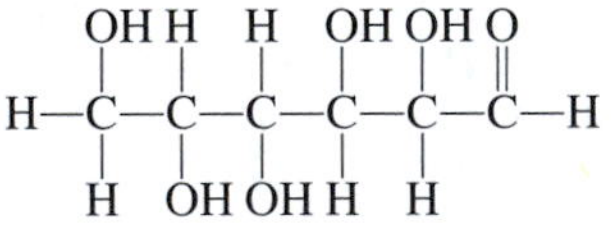

(e) vitamin C (also called ascorbic acid)

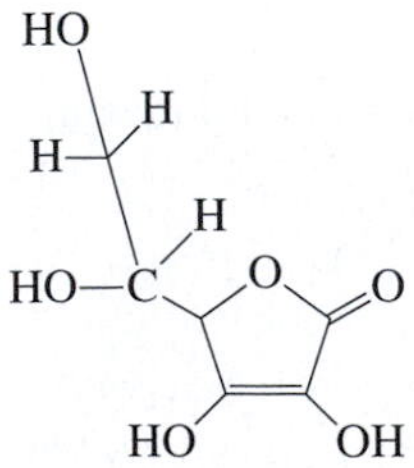

82. Name the following compounds.

(a) $CH_3CH(CH_3)CH_2CH_2OH$

(b)

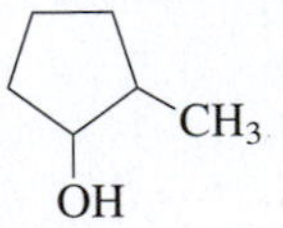

(c) 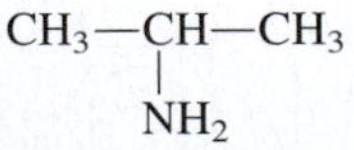 CH_3—CH—CH_3
         |
        NH_2

(d) 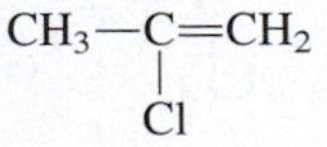 CH_3—C=CH_2
           |
           Cl

(e) 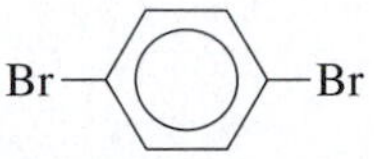 Br—⟨ ⟩—Br

(f) $(CH_3CH_2)_3N$

(g) 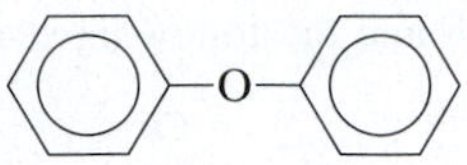 ⟨ ⟩—O—⟨ ⟩

(h)

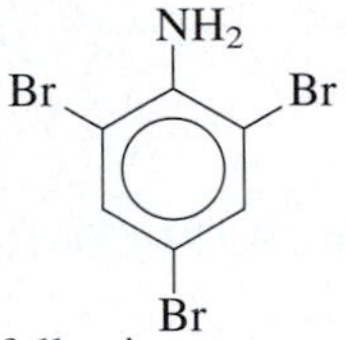

83. Draw the structural formulas for the following compounds: (a) *p*-bromotoluene, (b) cyclohexanol, (c) 2-methoxy-3-methylbutane, (d) diethylamine, (e) *o*-chlorophenol, (f) 1,4-butanediol.

84. Name the following compounds.

(a) 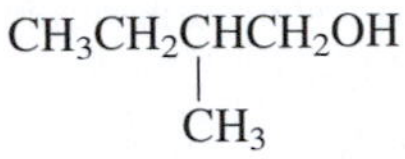 $CH_3CH_2CHCH_2OH$
              |
             CH_3

(b) $CH_3CH_2CH_2CH_2NH_2$

(c) 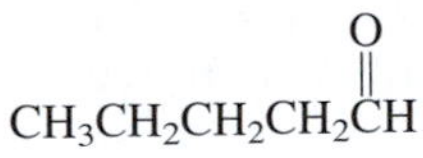 $CH_3CH_2CH_2CH_2\overset{O}{\overset{\|}{C}}H$

(d)

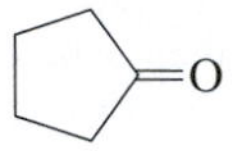

(e) 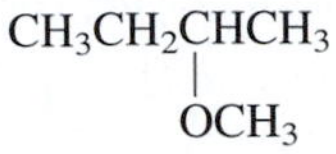 $CH_3CH_2CHCH_3$
              |
             OCH_3

(f) 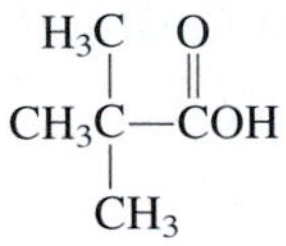 $CH_3\overset{H_3C}{\underset{CH_3}{\overset{\|}{C}}}-\overset{O}{\overset{\|}{C}}OH$

BUILDING YOUR KNOWLEDGE

85. (a) How do the melting points and boiling points of the normal alkanes vary with molecular weight? (b) Do you expect them to vary in this order? (c) Why or why not? Use intermolecular forces to explain your answer.

86. Write a Lewis formula for 2-butanone.

87. (a) What are resonance structures? (b) Draw resonance structures for benzene. (c) What do we mean when we say that the electrons associated with the π bonds in benzene are delocalized over the entire ring?

88. Lidocaine has replaced Novocain as the favored anesthetic in dentistry. What functional group do the two compounds have in common?

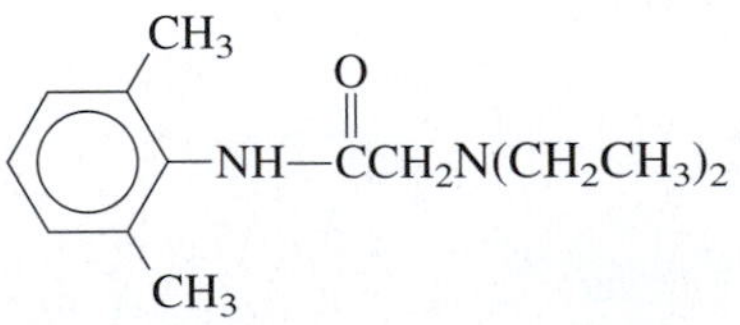

Lidocaine

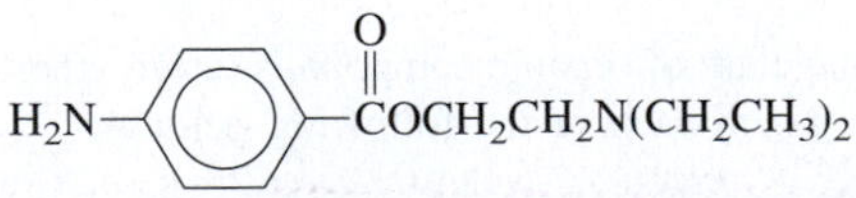

Novocain

89. The K_b for Lidocaine is 7×10^{-6}. What is the pH of a 1.5 percent solution of Lidocaine? The density of the solution is 1.00 g/mL.

Organic Chemistry II: Molecular Geometry and Reactions

28

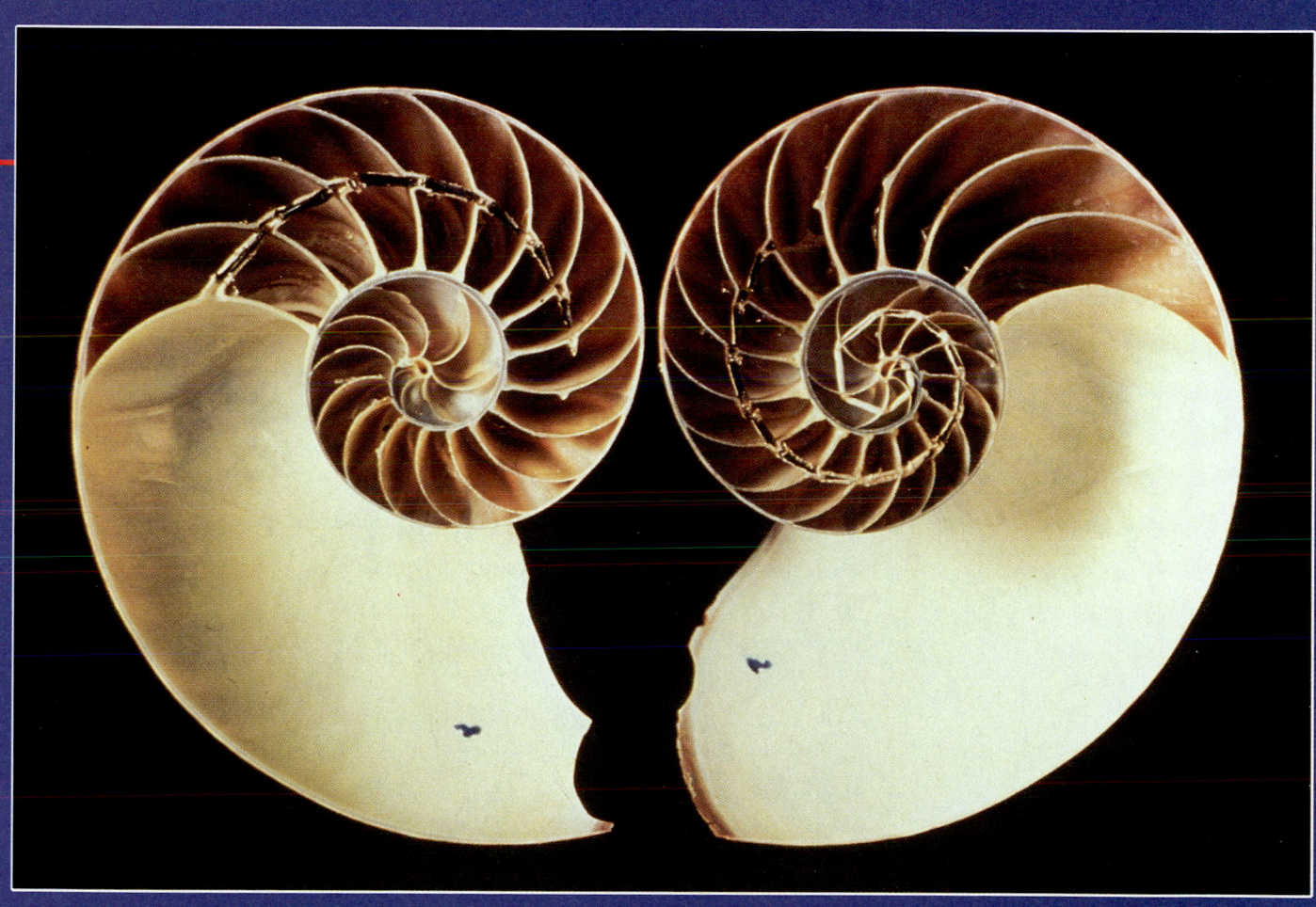

One-half of this bisected nautilus shell has a left-handed spiral. The other half has a right-handed spiral. The two halves of the nautilus shell are nonsuperimposable mirror images.

OBJECTIVES

As you study this chapter, you should learn

- *About the kinds of isomerism that organic compounds can exhibit—constitutional and stereoisomers*

- *About the conformations that molecules can adopt by rotation about single bonds*

- *To recognize and give examples of the three fundamental classes of organic reactions—substitution, addition, and elimination*

- *About some uses of the fundamental classes of organic reactions and some compounds that are prepared by each class of reaction*

- *About some common organic acids and bases and their relative strengths*

- *About oxidations and reductions of organic molecules, and how to recognize each type of transformation*

- *About the relative ease of oxidation of primary, secondary, and tertiary alcohols and the products that are formed*

- *About oxidation and reduction of alkenes and alkynes, the products that are formed, and the uses for such reactions*

- *About reactions in which compounds that contain the carbonyl group are reduced, and the products that are formed*

- *About combustion reactions of organic compounds*

- *About reactions by which esters and amides are formed*

- *About reactions in which esters are hydrolyzed (saponification reactions)*

- *About common polymers and the reactions by which they are formed*

GEOMETRIES OF ORGANIC MOLECULES

As we learned in Chapter 25, the chemical and physical properties of a substance depend strongly on the arrangements, as well as the identities, of its atoms.

> **Isomers** are substances that have the same number and kind of atoms—that is, the same *molecular formula*—but with the atoms arranged differently. *Because their structures are different, isomers have different properties.*

We can also describe isomers as two molecules that cannot be superimposed on one another by any rotation about single bonds.

Isomers can be broadly divided into two major classes: constitutional isomers and stereoisomers. In Chapter 25 we discussed isomerism in coordination compounds, and in Chapter 27 we learned about some isomeric organic compounds. In this chapter we will take a more systematic look at some three-dimensional aspects of organic structures—a subject known as **stereochemistry** ("spatial chemistry").

28-1 CONSTITUTIONAL ISOMERS

> **Constitutional** (or **structural**) **isomers** differ in the *order* in which their atoms are bonded together.

In our studies of hydrocarbons in Sections 27-1 through 27-7, we saw some examples of constitutional isomerism. Recall that there are three isomers of C_5H_{12}.

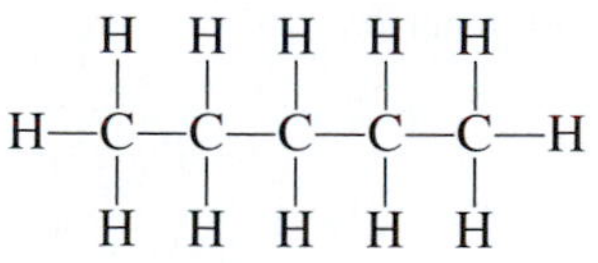

pentane, C_5H_{12} 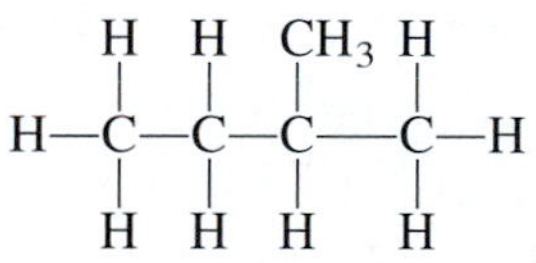   isopentane, C_5H_{12} neopentane, C_5H_{12}
                        methylbutane   dimethylpropane

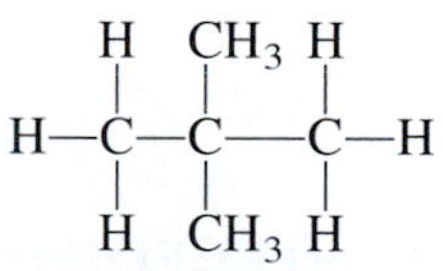

These three isomers differ in the lengths of their base chains but not in the functional groups present (i.e., only alkyl groups are present in this case). As a result, they differ somewhat in their melting and boiling points but differ only very slightly in the reactions they undergo.

In one kind of constitutional isomerism, the compounds have the same *number* and *kind* of functional groups on the *same base chain* or the *same ring* but in different positions. We call such isomers **positional isomers.** These isomers usually have very similar chemical properties differing mainly in physical properties such as melting and boiling points. The following groups of compounds are examples of positional isomers.

$CH_2{=}CHCH{=}CHCH_2CH_3$ $CH_2{=}CHCH_2CH{=}CHCH_3$ $CH_2{=}CHCH_2CH_2CH{=}CH_2$

1,3-hexadiene, C_6H_{10} 1,4-hexadiene, C_6H_{10} 1,5-hexadiene, C_6H_{10}

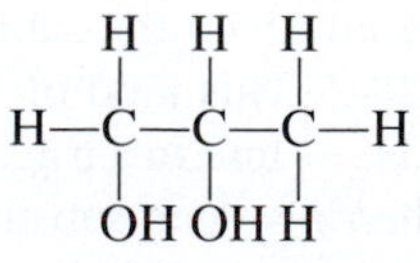

1,2-propanediol, $C_3H_8O_2$ 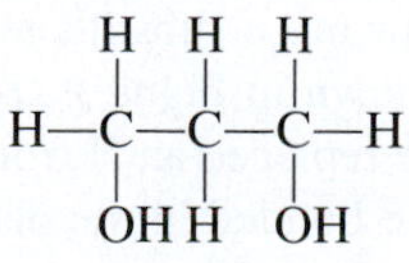   1,3-propanediol, $C_3H_8O_2$

o-dichlorobenzene, $C_6H_4Cl_2$ m-dichlorobenzene, $C_6H_4Cl_2$ p-dichlorobenzene, $C_6H_4Cl_2$

Sometimes the different order of arrangements of atoms results in different functional groups; such isomers are called **functional group isomers.** Some examples of this type of isomerism are

an alcohol and an ether:

ethanol, C_2H_6O dimethyl ether, C_2H_6O

an aldehyde and a ketone:

butanal, C_4H_8O butan-2-one, C_4H_8O
(butyraldehyde)   (methyl ethyl ketone)

28-2 STEREOISOMERS

In **stereoisomers** the atoms are linked together in the same atom-to-atom order, but their arrangements in space are different.

There are two types of stereoisomers.

Geometrical Isomers. Geometrical isomers differ only in the spatial orientation of groups about a plane or direction. Two geometrical isomers have the same molecular formula, the same functional groups, the same base chain or ring, and the same functional groups at the same carbons of the base chain; they differ in orientation either (1) around a double bond or (2) across the ring in a cyclic compound. If comparable groups are on opposite sides of the ring or the double bond, the designation *trans* appears in the name; if they are on the same side, the designation is *cis*. We learned in Section 27-3 about the geometrical isomerism associated with the double bond in alkenes such as the 1,2-dichloroethenes (Figure 27-9). Similarly, two or more substituents can be either on the same side or on opposite sides of the ring, as shown in Figures 28-1 and 28-2. This kind of isomerism is possible when substituents have replaced an H from a —CH_2— unit in a ring. Because substituents on an aromatic ring are bonded in the plane of the ring, such substitutions do not lead to geometrical isomerism.

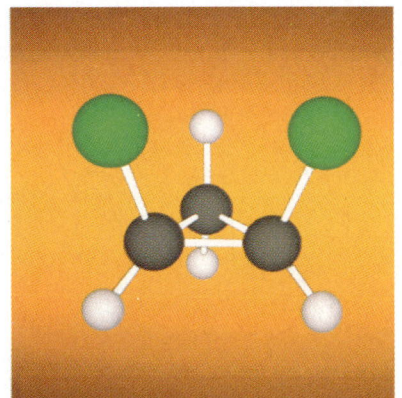

(a)

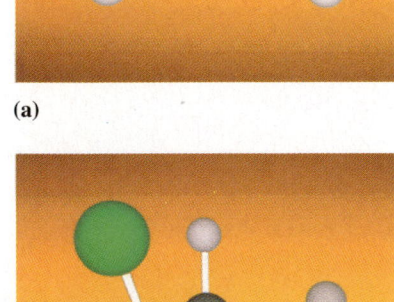

(b)

Figure 28-1 Models of (a) *cis*-dichlorocyclopropane and (b) *trans*-dichlorocyclopropane.

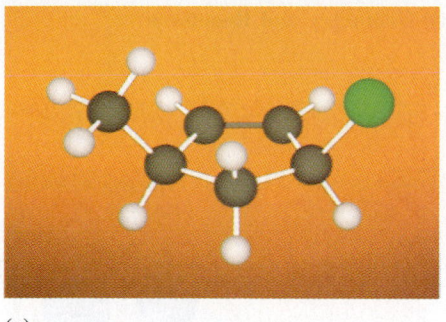
(a)

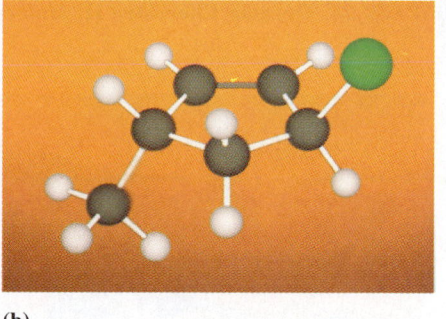
(b)

Figure 28-2 Models of (a) *cis*-3-chloro-5-methylcyclopentene and (b) *trans*-3-chloro-5-methylcyclopentene. In each model, the shaded bond at the back is the double bond.

Optical Isomers. Some macroscopic objects are mirror images of one another but cannot be superimposed. Your two hands are a familiar example of this; each hand is a nonsuperimposable mirror image of the other (Figure 28-3). "Superimposable" means that if one object is placed over the other, the positions of all parts will match.

An object that is *not* superimposable with its mirror image is said to be **chiral** (from the Greek word *cheir,* meaning "hand"); an object that *is* superimposable with its mirror image is said to be **achiral.** Examples of familiar objects that are chiral are a screw, a propeller, a foot, an ear, and a spiral staircase; examples of common objects that are achiral are a plain cup with no decoration, a pair of eyeglasses, and a sock.

We speak of a screw or a propeller as being "right-handed" or "left-handed."

Some molecules can exist in two forms that bear the same relationship to each other as do left and right hands. Such chiral molecules are called **optical isomers** or **enantiomers** of one another.

As an example of this, consider first the two models of bromochloromethane, CH_2BrCl, shown in Figure 28-4. They are mirror images of one another, and they can be superimposed. Thus, this molecule is *achiral* and is not capable of optical isomerism. Now consider bromochloroiodomethane, $CHBrClI$ (Figure 28-5). This molecule is not superimposable with its mirror image, so it is *chiral,* and the two forms are said to be optical isomers of one another. Any compound that contains four different groups bonded to the same carbon atom is chiral; i.e., it exhibits optical isomerism. Such a carbon is said to be

Figure 28-3 Mirror images. Place your left hand in front of a mirror; you will observe that it looks like your right hand. We say that the two hands are mirror images of one another; each hand is in every way the "reverse" of the other. Now try placing one hand directly over the other; they are not identical. Hence, they are nonsuperimposable mirror images. Each hand is a *chiral* object.

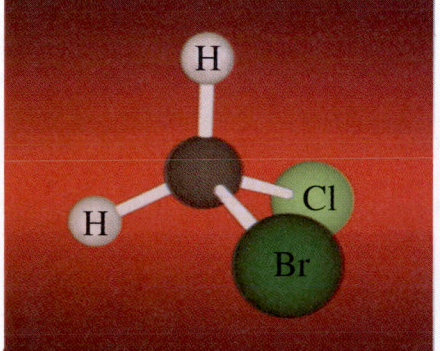

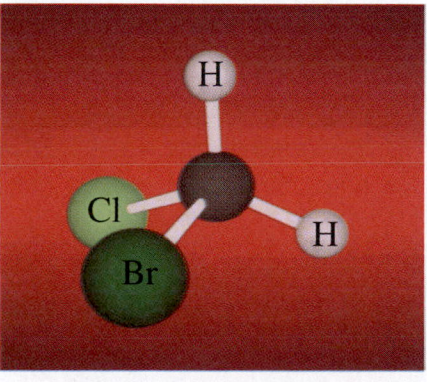

Figure 28-4 Models of two mirror-image forms of bromochloromethane, CH_2BrCl. The two models are the same (superimposable), so they are achiral. CH_2BrCl does *not* exhibit optical isomerism.

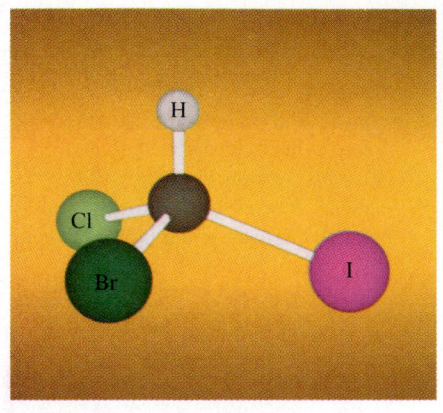

(a)

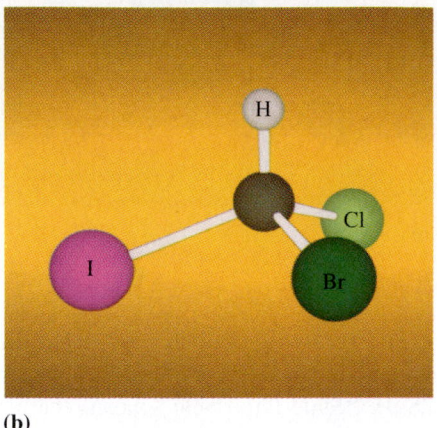

(b)

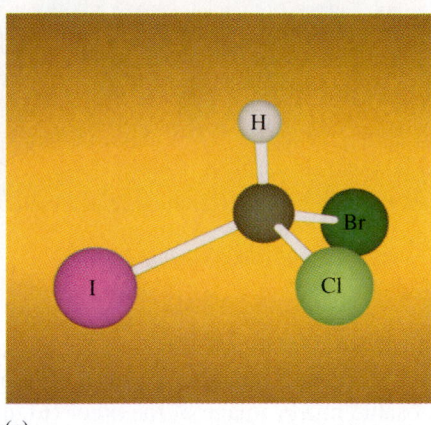

(c)

Figure 28-5   (a, b) Models of the two mirror-image forms of bromochloroiodomethane, CHBrClI. (c) The same model as in (a), turned so that H and I point the same as in (b); however, the Br and Cl atoms are not in the same positions in (b) and (c). The two models in (a) and (b) cannot be superimposed on one another no matter how we rotate them, so they are chiral. These two forms of CHBrClI represent *different compounds* that are optical isomers of one another.

asymmetric (meaning "without symmetry"). Most simple chiral molecules contain at least one asymmetric carbon atom although there are other ways in which molecular chirality can occur.

Optical isomers of a compound have the same type and number of atoms, connected in the same order, but arranged differently in space. In this way they are like geometrical isomers. However, the two types of isomerism differ in that geometrical isomers have *different* physical and chemical properties, whereas optical isomers have *identical* physical properties (such as melting point, boiling point, and density). Optical isomers also undergo the same chemical reactions, except when they interact with other chiral compounds. They often exhibit different solubilities in solvents that are composed of chiral molecules.

Optical isomers do differ from each other in one important physical property; they interact with polarized light in different ways. The main features of this subject were presented in Chapter 25. Separate equimolar solutions of two optical isomers rotate a plane of polarized light (Figures 25-2 and 25-3) by equal amounts but in opposite directions. The solution that rotates the polarized light to the right is **dextrorotatory;** the other, which rotates it to the left, is **levorotatory.** The phenomenon in which a plane of polarized light is rotated is called **optical activity.** It can be measured with a polarimeter (Figure 25-3) or with more sophisticated instruments. A **racemic mixture** is a single sample containing equal amounts of the two optical isomers of a compound. Such a solution does not rotate a plane of polarized light because the equal and opposite effects of the two isomers exactly cancel. The isomers must be separated from each other to exhibit optical activity. The chemical or physical processes that accomplish this are broadly called *optical resolution.*

One very important way in which optical isomers differ chemically from one another is in their biological activities. α-Amino acids have the general structure

$$\text{H}_2\text{N}-\overset{\overset{\displaystyle \text{H}}{|}}{\underset{\underset{\displaystyle \text{R}}{|}}{\text{C}}}-\text{COOH}$$

where R represents any of a number of common substituents. The central carbon atom has four different groups bonded to it. All α-amino acids (except glycine, in which R = H)

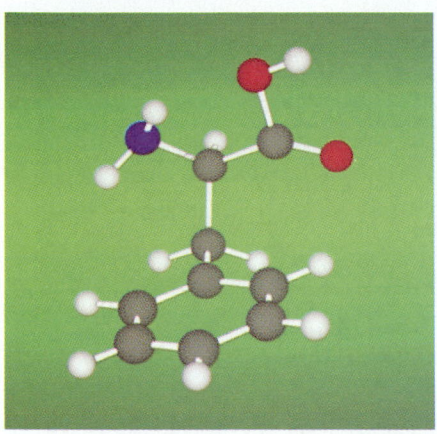
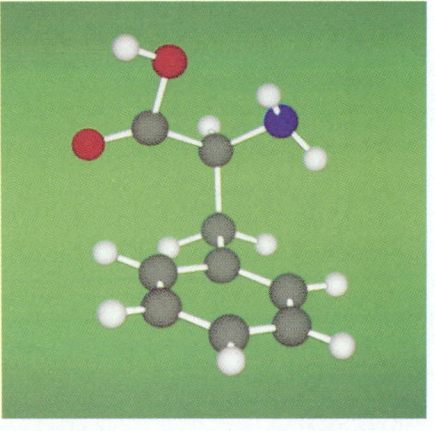

Figure 28-6 Models of the two optical isomers of phenylalanine. The naturally occurring phenylalanine in all living systems is the form shown on the left.

can exist as two optical isomers. Figure 28-6 shows this mirror-image relationship for optical isomers of phenylalanine in which R = —CH₂C₆H₅. All naturally occurring phenylalanine in living systems is of one form. In fact, only one isomer of each of the various optically active amino acids is found in proteins.

28-3 CONFORMATIONS

A *conformation* is one specific geometry of a molecule. The **conformations** of a compound differ from one another in the *extent of rotation about one or more single bonds.* The C—C single bond length, 1.54 Å, is relatively independent of the structure of the rest of the molecule.

Rotation about single C—C bonds is possible; in fact, at room temperature it occurs rapidly. There might appear to be an infinite number of kinds of ethane molecules depending on the rotation of one carbon atom with respect to the other. However, at room temperature ethane molecules possess sufficient thermal energy to cause rapid rotation about the single carbon–carbon bond from one conformation to another. The staggered conformation is slightly more stable than the eclipsed conformation (see Figure 28-7). In the staggered conformation there is less repulsive interaction between H atoms on adjacent C atoms.

As we saw in Section 27-3, rotation does not occur around carbon–carbon double bonds at room temperature.

Staggered
conformation

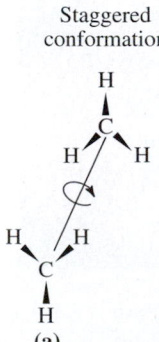

Eclipsed
conformation

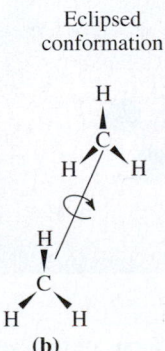
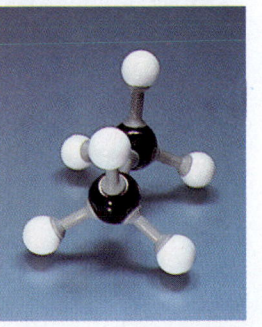

(a)

(b)

Figure 28-7 Two possible conformations of ethane. (a) Staggered. (b) Eclipsed. Rotation of one CH₃ group about the C—C single bond, as shown by the curved arrows, converts one conformation to the other.

Figure 28-8 Two staggered conformations of butane, C_4H_{10}.

Consider two conformations of butane (Figure 28-8). Again, staggered conformations are *slightly* more stable than eclipsed ones. At room temperature many conformations are present in a sample of pure butane.

Take care to distinguish between conformational differences and isomerism. The two forms of butane shown in Figure 28-8 are *not* isomers of one another. Either form can be converted to the other by rotation about a single bond, which is a very easy process that does not involve breaking any bonds. By contrast, at least one chemical bond would have to be broken and then re-formed to convert one isomer to another. This is most obvious with isomerism in which a conversion would change the order of attachment of the atoms. It is also true for geometrical isomers that differ in orientation about a double bond. To convert such a *cis* isomer to a *trans* isomer, it would be necessary to rotate part of the molecule about the double bond. Such a rotation would move the *p* orbitals out of the parallel alignment that is necessary to form the pi component of the double bond (Section 8-13). The breaking of this pi bond is quite costly in terms of energy; it occurs only with the input of energy in the form of heat or light.

We saw in Section 27-1 that cyclohexane exists in two forms called the *chair* and *boat* forms (Figure 28-9). The chair form is the more stable of the two because on the average the hydrogens (or other substituents) are farther from one another than in the boat form. However, chair and boat cyclohexane *are not* different compounds. Either form is easily converted into the other by rotation around single bonds without breaking any bonds, and the two forms cannot be separated. Thus, they are different *conformations* of cyclohexane.

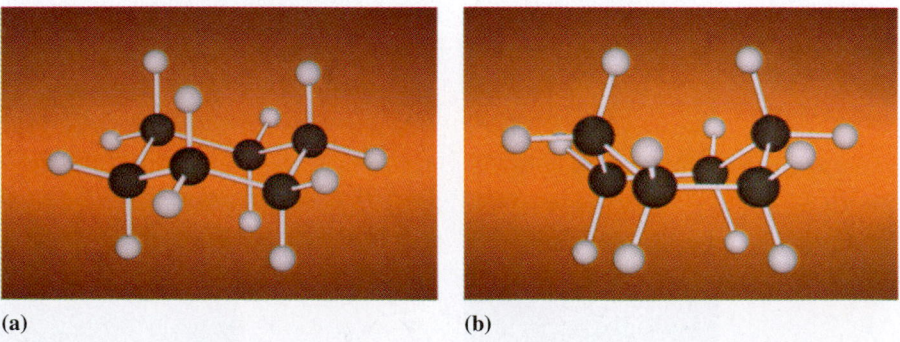

(a)   (b)

Figure 28-9 The two stable forms of the cyclohexane ring: (a) chair and (b) boat.

FUNDAMENTAL CLASSES OF ORGANIC REACTIONS

Organic compounds display very different abilities to react, ranging from the limited reactivity of hydrocarbons and fluorocarbons to the great variety of reactions undergone by the thousands of organic molecules that contain several functional groups. Reactivity depends on structure. We can usually predict the kinds of reactions a compound can undergo by identifying the functional groups it contains. But the electronic and structural features that are *near* a functional group can also affect its reactivity. One of the fascinations of organic chemistry is our ability to "fine-tune" both physical and chemical properties by making small changes in structure. The successes of this approach are innumerable, including the development of fuels and their additives or alternatives, the improvement of pharmaceuticals to enhance their effectiveness and minimize their ill effects, and the development of polymers and plastics with an incredible variety of properties and uses.

In the remainder of this chapter we shall present a few of the kinds of reactions that organic compounds undergo. A topic of such vast scope as reactivity of organic compounds can be made manageable only if we organize it. Nearly all organic transformations involve at least one of three fundamental classes of reactions. The following three sections will address these classes. We shall also look at some reaction sequences that combine several reaction steps from more than one of the fundamental classes.

28-4 SUBSTITUTION REACTIONS

In a **substitution reaction** an atom or a group of atoms attached to a carbon atom is removed and another atom or group of atoms takes its place. No change occurs in the degree of saturation at the reactive carbon atom.

The saturated hydrocarbons (alkanes and cycloalkanes) are chemically rather inert materials. For many years they were known as *paraffin* hydrocarbons because they undergo few reactions. They do not react with such powerful oxidizing agents as potassium permanganate and potassium dichromate. However, they do react with the halogens, with oxygen when ignited, and with concentrated nitric acid. As expected, members of a homologous series (Section 27-1) have similar chemical properties. If we study the chemistry of one of these compounds, we can make predictions about the others with a fair degree of certainty.

The saturated hydrocarbons can react without a big disruption of the molecular structure only by *displacement, or substitution of one atom for another*. At room temperature, chlorine and bromine react very slowly with saturated straight-chain hydrocarbons. At higher temperatures, or in the presence of sunlight or other sources of ultraviolet light, H atoms in the hydrocarbon can be replaced easily by halogen atoms. These substitution reactions are called **halogenation** reactions. The mechanism of reaction of Cl_2 with CH_4 may be represented as

$$: \overset{\cdot\cdot}{\underset{\cdot\cdot}{Cl}} : \overset{\cdot\cdot}{\underset{\cdot\cdot}{Cl}} : \quad \xrightarrow[\text{sunlight}]{\text{heat or}} \quad 2 : \overset{\cdot\cdot}{\underset{\cdot\cdot}{Cl}} \cdot$$

chlorine molecule   chlorine atoms
(radicals)

A radical is an atom or group of atoms that contains at least one unpaired electron. Some carry an electrical charge, others do not.

$$
\begin{array}{cccccc}
& H & & & & H \\
& \ddot{} & & & & \ddot{} \\
H:&C:&H & + & :Cl\cdot & \longrightarrow & H:C\cdot & + & H:Cl: \\
& \ddot{} & & & \ddot{} & & & \ddot{} & \ddot{} \\
& H & & & & H
\end{array}
$$

methane   chlorine   methyl   hydrogen
atom   radical   chloride

$$
\begin{array}{cccc}
H & & & H \\
\ddot{} & & & \ddot{} \\
H:C\cdot + :\ddot{C}l:\ddot{C}l: \longrightarrow H:C:\ddot{C}l: + :\ddot{C}l\cdot \text{ and so on} \\
\ddot{} & & & \ddot{} \\
H & & & H
\end{array}
$$

methyl   chlorine   methyl   chlorine
radical   molecule   chloride   atom

The reaction of Cl_2 with CH_4 is called a **free-radical chain reaction.** Cl_2 molecules absorb light or heat energy and split into very reactive Cl atoms (an *initiation* step). Some Cl atoms attack a methane molecule removing one of the H atoms to form HCl; the methyl radical, in turn, reacts with Cl_2 molecules to form methyl chloride (*propagation* steps). Eventually, Cl atoms may recombine to form Cl_2 molecules. The overall reaction is usually represented as

Note that only one half of the chlorine atoms occur in the organic product. The other half form hydrogen chloride, a commercially valuable compound.

$$
\underset{\text{methane}}{H-\overset{\overset{\textstyle H}{|}}{\underset{\underset{\textstyle H}{|}}{C}}-H} + \underset{\text{chlorine}}{Cl-Cl} \xrightarrow[\text{uv}]{\text{heat or}} \underset{\substack{\text{chloromethane} \\ \text{(methyl chloride)} \\ \text{bp} = 23.8°C}}{H-\overset{\overset{\textstyle H}{|}}{\underset{\underset{\textstyle H}{|}}{C}}-Cl} + HCl
$$

Many organic reactions produce more than a single product. For example, the chlorination of CH_4 may produce several other products in addition to CH_3Cl as the following equations show.

$$
H-\overset{\overset{\textstyle H}{|}}{\underset{\underset{\textstyle H}{|}}{C}}-Cl + Cl-Cl \longrightarrow \underset{\substack{\text{dichloromethane} \\ \text{(methylene chloride)} \\ \text{bp} = 40.2°C}}{Cl-\overset{\overset{\textstyle H}{|}}{\underset{\underset{\textstyle H}{|}}{C}}-Cl} + HCl
$$

$$
Cl-\overset{\overset{\textstyle H}{|}}{\underset{\underset{\textstyle H}{|}}{C}}-Cl + Cl-Cl \longrightarrow \underset{\substack{\text{trichloromethane} \\ \text{(chloroform)} \\ \text{bp} = 61°C}}{Cl-\overset{\overset{\textstyle H}{|}}{\underset{\underset{\textstyle Cl}{|}}{C}}-Cl} + HCl
$$

$$
Cl-\overset{\overset{\textstyle H}{|}}{\underset{\underset{\textstyle Cl}{|}}{C}}-Cl + Cl-Cl \longrightarrow \underset{\substack{\text{tetrachloromethane} \\ \text{(carbon tetrachloride)} \\ \text{bp} = 76.8°C}}{Cl-\overset{\overset{\textstyle Cl}{|}}{\underset{\underset{\textstyle Cl}{|}}{C}}-Cl} + HCl
$$

The mixture of products formed in an organic reaction can be separated by physical methods such as fractional distillation, which relies on differences in the boiling points of different compounds.

When a hydrocarbon has more than one C atom, its reaction with Cl_2 is more complex. The first step in the chlorination of ethane gives the product that contains one Cl atom per molecule.

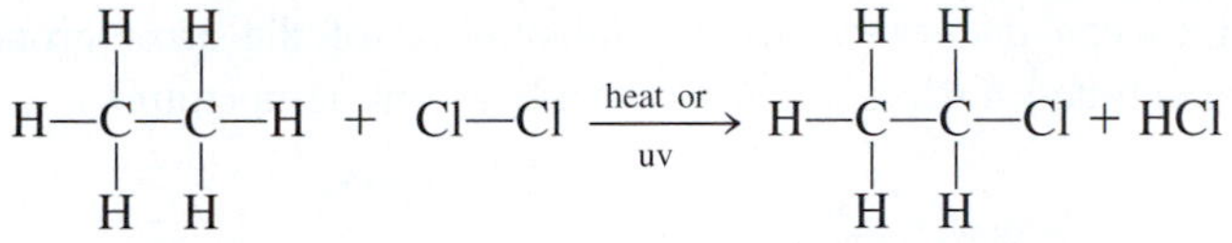

ethane

chloroethane
(ethyl chloride)
bp = 13.1°C

Ethyl chloride is used by athletic trainers as a spray-on pain killer.

When a second hydrogen atom is replaced, a mixture of the two possible products is obtained.

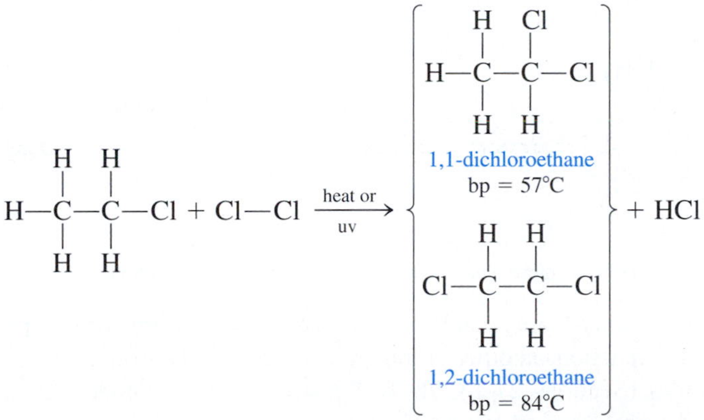

1,1-dichloroethane
bp = 57°C

1,2-dichloroethane
bp = 84°C

The product mixture does not contain equal numbers of moles of the dichloroethanes so we do not show a stoichiometrically balanced equation. Because reactions of saturated hydrocarbons with chlorine can produce many products, the reactions are not always as useful as might be desired.

The reaction of cyclopentane with Cl_2 produces only one monosubstituted product, chlorocyclopentane (cyclopentyl chloride). Multiple substitution also occurs.

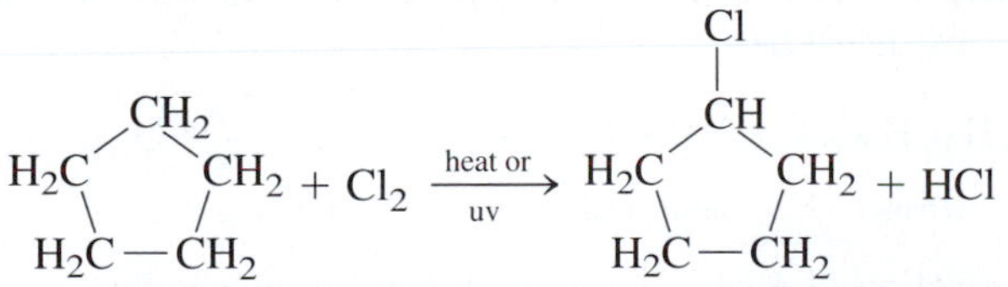

cyclopentane

chlorocyclopentane

Substitution is the most common kind of reaction of the aromatic ring. Halogenation, with chlorine or bromine, occurs readily in the presence of iron or anhydrous iron(III) chloride (a Lewis acid) catalyst.

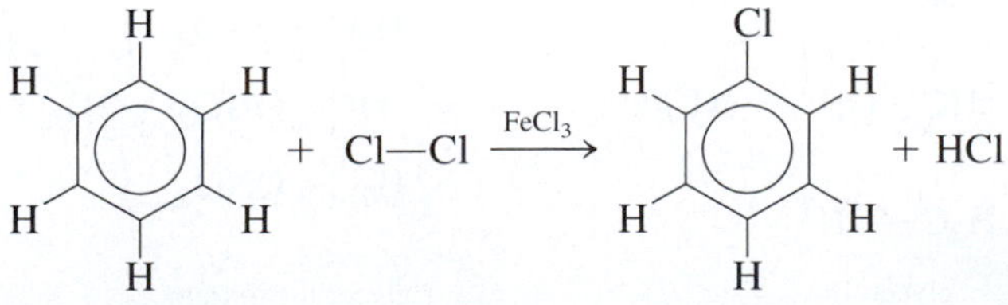

benzene

chlorobenzene

When iron is added as a catalyst, it reacts with chlorine to form iron(III) chloride, which is the true catalyst.

The equation is usually written in condensed form as

Aromatic rings can undergo *nitration,* substitution of the *nitro* group —NO_2, in a mixture of concentrated nitric and sulfuric acids at low temperatures.

nitric acid   nitrobenzene

The H_2SO_4 is both a catalyst and a dehydrating agent. The H in the product H_2O comes from the hydrocarbon; the OH comes from HNO_3.

The explosive TNT (2,4,6-trinitrotoluene) is manufactured by the nitration of toluene in several steps.

toluene   nitric acid   2,4,6-trinitrotoluene (TNT)

Groups other than hydrogen can be substituted by other atoms or groups of atoms. For example, in many reactions another group replaces the —OH group. These reactions form esters and amides (Sections 27-13, 28-9). Substitution of a halogen for the —OH of a carboxylic acid forms an acyl halide (Section 28-9), a type of compound that is widely used in subsequent synthesis.

Alcohols react with common inorganic oxyacids to produce **inorganic esters.** For instance, nitric acid reacts with alcohols to produce nitrates by substitution of nitrate, —ONO_2, for hydroxyl, —OH.

Simple inorganic esters may be thought of as compounds that contain one or more alkyl groups covalently bonded to the anion of a *ternary* inorganic acid. Unless indicated, the term "ester" refers to organic esters.

$$CH_3CH_2OH + HONO_2 \longrightarrow CH_3CH_2—ONO_2 + H_2O$$

ethanol   nitric acid   ethyl nitrate

The substitution reaction of nitric acid with glycerol produces the explosive nitroglycerine. Alfred Nobel's discovery in 1866 that this very sensitive material could be made into a "safe" explosive by absorbing it into diatomaceous earth or wood meal led to his development of dynamite.

Nobel's brother had been killed and his father permanently crippled in a nitroglycerine explosion in 1864. Nobel willed $9,200,000 to establish a fund for annual prizes in physics, chemistry, medicine, literature, and peace. The prizes were first awarded in 1901.

Interestingly, nitroglycerine ("nitro") is taken by some who have heart disease. It acts as a vasodilator (dilates the blood vessels) to decrease arterial tension.

glycerol

glyceryl trinitrate (nitroglycerine)

Cold, concentrated H_2SO_4 reacts with alcohols to form **alkyl hydrogen sulfates.** The reaction with lauryl alcohol is an important industrial reaction.

$$CH_3(CH_2)_{10}CH_2-OH + HOSO_3H \longrightarrow CH_3(CH_2)_{10}CH_2-OSO_3H + H_2O$$

1-dodecanol (lauryl alcohol)   sulfuric   lauryl hydrogen sulfate
                                  acid

The neutralization reaction of an alkyl hydrogen sulfate with NaOH then produces the sodium salt of the alkyl hydrogen sulfate.

$$CH_3(CH_2)_{10}CH_2-OSO_3H + Na^+OH^- \longrightarrow CH_3(CH_2)_{10}CH_2-OSO_3{}^-Na^+ + H_2O$$

sodium lauryl sulfate (a detergent)

Sodium salts of the alkyl hydrogen sulfates that contain about 12 carbon atoms are excellent detergents. They are also biodegradable. (Soaps and detergents were discussed in Section 14-18.)

28-5 ADDITION REACTIONS

An **addition reaction** involves an *increase* in the number of groups attached to carbon. The molecule becomes more nearly saturated.

The principal reactions of alkenes and alkynes (and their cyclic analogs) are addition reactions rather than substitution reactions. For example, contrast the reactions of ethane and ethylene with Cl_2.

ethane: $CH_3-CH_3 + Cl_2 \longrightarrow CH_3-CH_2Cl + HCl$ (substitution, slow)

ethylene: $H_2C{=}CH_2 + Cl_2 \longrightarrow \underset{\underset{Cl}{|}}{CH_2}-\underset{\underset{Cl}{|}}{CH_2}$ (addition, rapid)

Carbon–carbon double bonds are *reaction sites* and so represent *functional groups*. Most addition reactions involving alkenes and alkynes proceed rapidly at room temperature. By contrast, substitution reactions of the alkanes require catalysts and high temperatures.

Bromine adds readily to the alkenes to give dibromides. The reaction with ethylene is

$$H_2C{=}CH_2 + Br_2 \longrightarrow \underset{\underset{Br}{|}}{CH_2}-\underset{\underset{Br}{|}}{CH_2}$$

1,2-dibromoethane
(ethylene dibromide)

The addition of Br_2 to alkenes is used as a simple qualitative test for unsaturation. Bromine, a dark red liquid, is dissolved in a nonpolar solvent. When an alkene is added, the solution becomes colorless as the Br_2 reacts with the alkene to form a colorless compound. This reaction is used to distinguish between alkanes and alkenes.

Hydrogenation is an extremely important addition reaction of the alkenes. Hydrogen adds across double bonds at elevated temperatures, under high pressures, and in the presence of an appropriate catalyst (finely divided Pt, Pd, or Ni).

$$CH_2{=}CH_2 + H_2 \xrightarrow[\text{heat}]{\text{catalyst}} CH_3-CH_3$$

Unsaturated hydrocarbons are converted to saturated hydrocarbons in the manufacture of high-octane gasoline and aviation fuels. Unsaturated vegetable oils can also be converted to solid cooking fats (shortening) by hydrogenation (Figure 28-10). This process produces some *trans* fats.

$$\underset{\text{olein (an oil, liquid)}}{\begin{array}{c} O \\ \| \\ H_2COC(CH_2)_7CH{=}CH(CH_2)_7CH_3 \\ | \\ O \\ \| \\ HCOC(CH_2)_7CH{=}CH(CH_2)_7CH_3 \\ | \\ O \\ \| \\ H_2COC(CH_2)_7CH{=}CH(CH_2)_7CH_3 \end{array}} \xrightarrow[\substack{\text{Ni catalyst} \\ \text{heat}}]{3H_2} \underset{\text{stearin (a fat, solid)}}{\begin{array}{c} O \\ \| \\ H_2COC(CH_2)_{16}CH_3 \\ | \\ O \\ \| \\ HCOC(CH_2)_{16}CH_3 \\ | \\ O \\ \| \\ H_2COC(CH_2)_{16}CH_3 \end{array}}$$

Figure 28-10 Hydrogenation of the olefinic double bonds in a vegetable oil converts it to a solid fat.

The *hydration reaction* (addition of water) is another very important addition reaction of alkenes. It is used commercially for the preparation of a wide variety of alcohols from petroleum by-products. Ethanol, the most important industrial alcohol, is produced industrially by the hydration of ethylene from petroleum, using H_2SO_4 as a catalyst.

Think of H_2O as HOH.

$$H_2C{=}CH_2 + H_2O \xrightarrow{H_2SO_4} H_3C{-}CH_2OH$$

This reaction takes place in two steps.

H_2SO_4, the catalyst, is regenerated in the second reaction.

$$\underset{\substack{\text{ethene} \qquad \text{sulfuric acid}}}{H{-}\overset{\overset{\textstyle H}{|}}{C}{=}\overset{\overset{\textstyle H}{|}}{C}{-}H + HOSO_3H} \xrightarrow{\text{cold}} \underset{\text{ethyl hydrogen sulfate}}{H{-}\overset{\overset{\textstyle H}{|}}{\underset{\underset{\textstyle H}{|}}{C}}{-}\overset{\overset{\textstyle H}{|}}{\underset{\underset{\textstyle H}{|}}{C}}{-}OSO_3H}$$

$$\underset{\text{steam}}{H{-}\overset{\overset{\textstyle H}{|}}{\underset{\underset{\textstyle H}{|}}{C}}{-}\overset{\overset{\textstyle H}{|}}{\underset{\underset{\textstyle H}{|}}{C}}{-}OSO_3H + H{-}OH} \longrightarrow \underset{\text{ethanol}}{H{-}\overset{\overset{\textstyle H}{|}}{\underset{\underset{\textstyle H}{|}}{C}}{-}\overset{\overset{\textstyle H}{|}}{\underset{\underset{\textstyle H}{|}}{C}}{-}OH} + HOSO_3H$$

We see that the first step is an addition reaction in which $H{-}OSO_3H$ adds across the double bond. This is followed by a substitution reaction in which $-OH$ from water (steam) replaces $-OSO_3H$ from ethyl hydrogen sulfate. These reactions amount to the net addition of water to ethene.

Ethylene glycol (Section 27-9) is an important solvent, industrial chemical, and permanent antifreeze. An important commercial synthesis of ethylene glycol from ethylene involves addition followed by substitution. Ethylene reacts with the hypochlorous acid present in chlorine water, and the product is then hydrolyzed in an aqueous solution of sodium carbonate.

Ethylene glycol is miscible with water. It is widely used in permanent antifreeze (bp = 197°C).

$$CH_2{=}CH_2 + \underset{\substack{\text{hypochlorous acid} \\ \text{(chlorine + water)}}}{HOCl} \longrightarrow \underset{\substack{\text{ethylene} \\ \text{chlorohydrin}}}{\underset{\underset{\textstyle OH \quad Cl}{|\qquad|}}{CH_2{-}CH_2}} \xrightarrow[2Na^+CO_3{}^{2-}]{H_2O} \underset{\substack{\text{1,2-ethanediol} \\ \text{(ethylene glycol)}}}{\underset{\underset{\textstyle OH \quad OH}{|\qquad|}}{CH_2{-}CH_2}}$$

One of the commercially most important addition reactions of the alkenes forms *polymers*. This reaction will be discussed in Section 28-11.

The alkynes contain two pi bonds both of which are sources of electrons, and they are more reactive than the alkenes. The most common reaction of the alkynes is addition across the triple bond. The reactions with hydrogen and with bromine are typical.

$$H-C\equiv C-H \xrightarrow{H_2} \begin{array}{c} H \\ \diagdown \\ C=C \\ \diagup \\ H \end{array} \begin{array}{c} H \\ \diagup \\ \diagdown \\ H \end{array} \xrightarrow{H_2} H_3C-CH_3$$

Each sequence may be thought of as a series of stepwise reactions. Stopping either sequence after the first step is difficult because the product still contains a reactive double bond.

$$H-C\equiv C-H \xrightarrow{Br_2} \begin{array}{c} H \\ \diagdown \\ C=C \\ \diagup \\ Br \end{array} \begin{array}{c} Br \\ \diagup \\ \diagdown \\ H \end{array} \xrightarrow{Br_2} H-\overset{\overset{\displaystyle Br}{|}}{\underset{\underset{\displaystyle Br}{|}}{C}}-\overset{\overset{\displaystyle Br}{|}}{\underset{\underset{\displaystyle Br}{|}}{C}}-H$$

1,2-dibromoethene   1,1,2,2-tetrabromoethane

Other unsaturated bonds can also undergo addition reactions. Probably the most impor-

$$\overset{O}{\overset{||}{-C-}}$$

tant example is the carbonyl group, —C—. Because of the availability of unshared pairs of electrons on the oxygen atom, the products can undergo a wide variety of subsequent reactions. For example, HCN adds to the C=O bond of acetone.

$$CH_3-\overset{\overset{\displaystyle O}{||}}{C}-CH_3 + HCN \xrightarrow{NaOH(aq)} CH_3-\overset{\overset{\displaystyle OH}{|}}{\underset{\underset{\displaystyle CN}{|}}{C}}-CH_3$$

This is a key early step in the production of the transparent plastic known as Plexiglas or Lucite.

28-6 ELIMINATION REACTIONS

An **elimination reaction** involves a *decrease* in the number of groups attached to carbon. The degree of unsaturation increases.

One type of elimination reaction takes place when *vicinal* dihalides are treated with a mixture of zinc dust in acetic acid (or ethanol). The two halogen atoms are eliminated from the adjacent carbons, and the single bond between the two carbons is transformed into a double bond. Such a reaction is called a **dehalogenation** reaction. In this reaction the molecule becomes more *unsaturated*.

Vicinal (or *vic*) dihalides are dihalo compounds in which the halogen atoms are situated on adjacent carbon atoms.

$$CH_3-\overset{\overset{}{|}}{\underset{\underset{\displaystyle Cl}{|}}{CH}}-\overset{\overset{}{|}}{\underset{\underset{\displaystyle Cl}{|}}{CH}}CH_3$$

2,3-dichlorobutane (a *vic* dihalide)

The name *geminal* (or *gem*) dihalide is used for dihalides in which both halogen atoms are attached to the same carbon atom.

$$CH_3-\overset{\overset{\displaystyle Cl}{|}}{\underset{\underset{\displaystyle Cl}{|}}{C}}-CH_2CH_3$$

2,2-dichlorobutane (a *gem* dihalide)

$$CH_3-\overset{\overset{}{|}}{\underset{\underset{\displaystyle Br}{|}}{CH}}-\overset{\overset{}{|}}{\underset{\underset{\displaystyle Br}{|}}{CH}}-CH_3 \xrightarrow[\text{in acetic acid or ethanol}]{Zn} \left\{ \begin{array}{c} \begin{array}{cc} CH_3 & H \\ \diagdown & \diagup \\ C=C \\ \diagup & \diagdown \\ H & CH_3 \end{array} \\ \textit{trans-2-butene} \\ \\ + \\ \\ \begin{array}{cc} CH_3 & CH_3 \\ \diagdown & \diagup \\ C=C \\ \diagup & \diagdown \\ H & H \end{array} \\ \textit{cis-2-butene} \end{array} \right\} + ZnBr_2$$

2,3-dibromobutane

Another type of elimination reaction can occur for chloro-, bromo- and iodoalkanes and is sometimes called **dehydrohalogenation.** In such a reaction, the halogen, X, from one C atom and a hydrogen from an adjacent C atom are eliminated. The single bond between two carbon atoms is changed to a double bond; again, the molecule becomes *more unsaturated.* The net reaction is the transformation of an alkyl halide (or haloalkane) into an alkene. Dehydrohalogenation reactions are usually catalyzed by strong bases such as sodium hydroxide, NaOH, and sodium ethoxide, CH_3CH_2ONa.

$$H-\underset{\underset{\displaystyle H}{|}}{\overset{\overset{\displaystyle H}{|}}{C}}-\underset{\underset{\displaystyle Br}{|}}{\overset{\overset{\displaystyle H}{|}}{C}}-H + Na^+OH^- \longrightarrow \underset{\underset{\displaystyle H}{}}{\overset{\overset{\displaystyle H}{}}{C}}{=}\underset{\underset{\displaystyle H}{}}{\overset{\overset{\displaystyle H}{}}{C}} + H_2O + NaBr$$

bromoethane   ethene
(ethylene)

Ethanol, C_2H_5OH, is an even weaker acid than water, HOH. So we know that the ethoxide ion, $C_2H_5O^-$ (from C_2H_5ONa), is an even stronger base than OH^- (from NaOH). The strengths of organic acids and bases are discussed in Section 28-7.

$$CH_3\underset{\underset{\displaystyle Cl}{\underset{\displaystyle |}{\overset{\displaystyle |}{CH_2}}}}{CH}CH_3 + C_2H_5O^-Na^+ \longrightarrow CH_3-\underset{\underset{\displaystyle CH_2}{\displaystyle ||}}{C}-CH_3 + C_2H_5OH + NaCl$$

1-chloro-2-   sodium   2-methylpropene   ethanol
methylpropane   ethoxide

A related reaction is **dehydration** in which an alcohol is converted into an alkene and water by the elimination of —OH and —H from adjacent carbon atoms. The dehydration of an alcohol to form an alkene can be considered the reverse of the hydration of an alkene to form an alcohol (Section 28-5). Dehydration reactions are catalyzed by acids. Dehydration of tertiary alcohols (Section 27-9) is easier than dehydration of secondary alcohols, which in turn is easier than dehydration of primary alcohols. Drastic conditions are required for dehydration of primary alcohols, whereas much gentler conditions can accomplish the change for tertiary alcohols.

3° alcohol:
$$CH_3-\underset{\underset{\displaystyle CH_3}{|}}{\overset{\overset{\displaystyle OH}{|}}{C}}-\underset{\underset{\displaystyle H}{|}}{\overset{\overset{\displaystyle H}{|}}{C}}-H \xrightarrow[85°C]{20\% H_2SO_4} CH_3-\underset{\underset{\displaystyle CH_3}{|}}{C}{=}CH_2 + H_2O$$

2-methyl-2-propanol   2-methylpropene
(*tert*-butyl alcohol)

Recall that H_3PO_4 is a weak acid; here it functions as a dehydrating agent.

2° alcohol:
cyclohexanol $\xrightarrow[165-170°C]{85\% H_3PO_4}$ cyclohexene $+ H_2O$

Concentrated H_2SO_4 is a *very* powerful dehydrating agent.

1° alcohol:
$$H-\underset{\underset{\displaystyle H}{|}}{\overset{\overset{\displaystyle H}{|}}{C}}-\underset{\underset{\displaystyle OH}{|}}{\overset{\overset{\displaystyle H}{|}}{C}}-H \xrightarrow[180°C]{conc. H_2SO_4} \underset{\underset{\displaystyle H}{}}{\overset{\overset{\displaystyle H}{}}{C}}{=}\underset{\underset{\displaystyle H}{}}{\overset{\overset{\displaystyle H}{}}{C}} + H_2O$$

ethanol   ethene
(ethylene)

Concentrated sulfuric acid is an excellent dehydrating agent. Here it removes water from sucrose, a sugar with the formula $C_{12}H_{22}O_{11}$. Dehydration of sucrose produces (mostly) carbon.

Such simple elimination reactions are relatively rare. However, elimination reactions frequently occur as individual steps in more complex reaction sequences. Many more elimination reactions will be encountered in a course in organic chemistry.

SOME OTHER ORGANIC REACTIONS

28-7 REACTIONS OF BRØNSTED–LOWRY ACIDS AND BASES

Many organic compounds can act as weak Brønsted–Lowry acids or bases. Their reactions involve the transfer of H^+ ions, or *protons* (Section 10-4). Like similar reactions of inorganic compounds, these acid–base reactions of organic acids and bases are usually fast and reversible. Consequently, we can discuss the acidic or basic properties of organic compounds in terms of equilibrium constants (Section 18-4).

In the Brønsted–Lowry description, an *acid* is a *proton donor* and a *base* is a *proton acceptor*. Review the terminology of conjugate acid–base pairs in Section 10-4.

Some Organic Acids

The most important organic acids contain carboxyl groups, $-\overset{\overset{\displaystyle O}{\|}}{C}-O-H$. They are called carboxylic acids (Section 27-12). They ionize slightly when dissolved in water, as illustrated with acetic acid.

$$CH_3COOH + H_2O \rightleftharpoons CH_3COO^- + H_3O^+ \qquad K_a = \frac{[CH_3COO^-][H_3O^+]}{[CH_3COOH]} = 1.8 \times 10^{-5}$$

acid$_1$ base$_2$ base$_1$ acid$_2$

Table 28-1 K_a and pK_a Values of Some Carboxylic Acids

Name	Formula	K_a	pK_a
formic acid	HCOOH	1.8×10^{-4}	3.74
acetic acid	CH_3COOH	1.8×10^{-5}	4.74
propionic acid	CH_3CH_2COOH	1.4×10^{-5}	4.85
monochloroacetic acid	$ClCH_2COOH$	1.5×10^{-3}	2.82
dichloroacetic acid	$Cl_2CHCOOH$	5.0×10^{-2}	1.30
trichloroacetic acid	Cl_3CCOOH	2.0×10^{-1}	0.70
benzoic acid	C_6H_5COOH	6.3×10^{-5}	4.20
phenol*	C_6H_5OH	1.3×10^{-10}	9.89
ethanol*	CH_3CH_2OH	$\sim 10^{-18}$	$\sim$18

Phenol and ethanol are not carboxylic acids. Phenol is weakly acidic compared to carboxylic acids, whereas ethanol is even weaker than water.

In Section 18-3 we defined

$$pK_a = -\log K_a$$

When K_a goes down by a factor of 10, pK_a goes up by one unit. We see that the weaker an acid, the higher its pK_a value.

Sources of some naturally occurring carboxylic acids.

This is similar to the reaction of water with active metals.

$$2H—OH + 2Na \longrightarrow$$
$$H_2 + 2[Na^+ + OH^-]$$

Acetic acid is 1.3% ionized in 0.10 M solution. Regardless of the lengths of the chains, the acid strengths of the monocarboxylic acids are approximately the same with K_a values (in water) in the range 10^{-5} to 10^{-4}. Their acid strengths increase dramatically when electronegative substituents are present on the α-carbon atom (K_a values in water range from 10^{-3} to 10^{-1}). Compare acetic acid and the three substituted acetic acids in Table 28-1. There are two main reasons for this increase: (1) The electronegative substituents pull electron density from the carboxylic acid group, and (2) the more electronegative substituents help to stabilize the resulting carboxylate anion by spreading the negative charge over more atoms.

The alcohols are so *very weakly acidic* that they do not react with strong soluble bases. They are much weaker acids than water (see Table 28-1), but some of their reactions are analogous to those of water.

The very reactive metals react with alcohols to form **alkoxides** with the liberation of hydrogen.

$$2CH_3—OH + 2Na \longrightarrow H_2 + 2[Na^+ + CH_3O^-]$$
sodium methoxide
(an alkoxide)

$$2CH_3CH_2—OH + 2Na \longrightarrow H_2 + 2[Na^+ + CH_3CH_2O^-]$$
sodium ethoxide
(an alkoxide)

The low-molecular-weight alkoxides are strong bases that react with water (hydrolyze) to form the parent alcohol and a strong soluble base.

$$[Na^+ + CH_3CH_2O^-] + H—OH \longrightarrow CH_3CH_2OH + [Na^+ + OH^-]$$
sodium ethoxide   ethanol

Phenols react with metallic sodium to produce **phenoxides;** the reactions are analogous to those of alcohols. Because phenols are more acidic than alcohols, their reactions are more vigorous.

$$2 \;\text{(phenol)} + 2Na \longrightarrow H_2 + 2\left[\text{(sodium phenoxide)} + Na^+ \right]$$

phenol   sodium phenoxide

For phenol, $K_a = 10^{-10}$. The phenoxide ion, $C_6H_5O^-$, has considerable resonance stabilization.

Some substituted phenols are much more acidic—for example, nitrophenols and halophenols.

We saw in Section 19-3 that salts derived from strong acids and weak bases react with water (hydrolyze) to give acidic solutions. The example given there involved the acidic character of ammonium ion.

$$\underset{\text{acid}_1}{NH_4^+} + \underset{\text{base}_2}{H_2O} \rightleftharpoons \underset{\text{base}_1}{NH_3} + \underset{\text{acid}_2}{H_3O^+} \qquad K_a = \frac{[NH_3][H_3O^+]}{[NH_4^+]} = 5.6 \times 10^{-10}$$

The weaker a base, the stronger its conjugate acid (Section 10-4).

Similar solvolysis reactions occur with organic ammonium salts. These may be considered analogs of the ammonium ion with one or more of the H atoms replaced by an alkyl or aryl group.

$$\underset{\text{acid}_1}{RNH_3^+} + \underset{\text{base}_2}{H_2O} \rightleftharpoons \underset{\text{base}_1}{RNH_2} + \underset{\text{acid}_2}{H_3O^+} \qquad K_a = \frac{[RNH_2][H_3O^+]}{[RNH_3^+]}$$

We recall from Chapter 19 that the relationship $K_w = K_a K_b$ describes the strengths of any conjugate acid–base pair in aqueous solution. For instance, we can use this relationship for the $CH_3NH_3^+/CH_3NH_2$ pair (we obtain K_b for methylamine, CH_3NH_2, from Appendix G).

$$K_{a\,(CH_3NH_3^+)} = \frac{K_w}{K_{b\,(CH_3NH_2)}} = \frac{1.0 \times 10^{-14}}{5.0 \times 10^{-4}} = 2.0 \times 10^{-11}$$

Because more highly substituted amines are usually weaker bases, their cations are usually stronger acids.

In summary, we can rank the acid strengths of these classes of organic species.

carboxylic acids > phenols > substituted ammonium ions > alcohols

Our discussion has emphasized water solutions of these acids. Many organic compounds are soluble in numerous other solvents. The properties of acids and bases in other solvents depend on solvent properties such as polarity, acidity or basicity, and polarizability.

Some Organic Bases

The most important organic bases are the amines. When dissolved in water, they are partially converted to substituted ammonium ions. This equilibrium is defined as shown in the following equations for the ionization of primary, secondary, and tertiary amines.

Reaction of sodium metal with ethanol gives sodium ethoxide and hydrogen. Ethanol is much less acidic than water, so its reaction with sodium is much less vigorous than the reaction of water with sodium.

$pK_b = -\log K_b$

The weaker a base, the higher its pK_b value.

Table 28-2 *Basicities of Ammonia and Some Amines in Water*

Name	Formula	K_b Values	pK_b
ammonia	NH_3	1.8×10^{-5}	4.74
methylamine	CH_3NH_2	5.0×10^{-4}	3.30
dimethylamine	$(CH_3)_2NH$	7.4×10^{-4}	3.13
trimethylamine	$(CH_3)_3N$	7.4×10^{-5}	4.13
ethylamine	$CH_3CH_2NH_2$	4.7×10^{-4}	3.33
aniline	$C_6H_5NH_2$	4.2×10^{-10}	9.38
ethylenediamine	$H_2NCH_2CH_2NH_2$	8.5×10^{-5}	4.07
pyridine	C_5H_5N	1.5×10^{-9}	8.82

Several mouthwashes contain a pyridinium chloride salt as an antibacterial component.

	base$_1$	acid$_2$	acid$_1$	base$_2$	

primary, 1° $\quad RNH_2 + H_2O \rightleftharpoons RNH_3^+ + OH^- \qquad K_b = \dfrac{[RNH_3^+][OH^-]}{[RNH_2]}$

secondary, 2° $\quad R_2NH + H_2O \rightleftharpoons R_2NH_2^+ + OH^- \qquad K_b = \dfrac{[R_2NH_2^+][OH^-]}{[R_2NH]}$

tertiary, 3° $\quad R_3N + H_2O \rightleftharpoons R_3NH^+ + OH^- \qquad K_b = \dfrac{[R_3NH^+][OH^-]}{[R_3N]}$

Most low-molecular-weight aliphatic amines are somewhat stronger bases than ammonia. Table 28-2 shows that aliphatic amines are much stronger bases than aromatic and heterocyclic amines. The basicities of amines often decrease roughly in the order tertiary > secondary > primary. However, other structural factors and solvation effects may outweigh this tendency, especially with tertiary amines.

The weaker an acid, the stronger its conjugate base (Section 10-4).

Salts derived from weak acids and strong bases react with water to give basic solutions (Section 19-2). The equation $K_w = K_a K_b$ describes this relationship for any conjugate acid–base pair. An important organic example of this is the family of salts that contain the alkoxide ions, RO^-, or phenoxide ions, ArO^-. Alkoxide and phenoxide ions react as bases toward protonic acids (including water).

	base$_1$	acid$_2$	acid$_1$	base$_2$

$$CH_3CH_2O^- + H_2O \rightleftharpoons CH_3CH_2OH + OH^-$$

Phenol and its substituted derivatives have K_a values on the order of 10^{-10}, so phenoxide salts are rather strongly basic; $K_b \approx 10^{-4}$.

Carboxylate salts behave in a similar fashion, but to a lesser extent.

$$RCOO^- + HX \rightleftharpoons RCOOH + X^-$$

$$\underset{\text{base}_1}{RCOO^-} + \underset{\text{acid}_2}{H_2O} \rightleftharpoons \underset{\text{acid}_1}{RCOOH} + \underset{\text{base}_2}{OH^-} \qquad K_b = \frac{[RCOOH][OH^-]}{[RCOO^-]}$$

Applying the relationship $K_w = K_a K_b$ to the K_a values for carboxylic acids in Table 28-1, we see that typical K_b values for unsubstituted carboxylate salts are on the order of 10^{-9}; stronger acids such as the haloacetic acids give salts that are weaker bases.

> In summary, we can rank the base strengths of these common organic bases.
>
> alkoxides > aliphatic amines > phenoxides > carboxylates $\approx$ aromatic amines $\approx$ heterocyclic amines

28-8 OXIDATION–REDUCTION REACTIONS

> **Oxidation** of an organic molecule usually corresponds to *increasing* its *oxygen* content or *decreasing* its *hydrogen* content. **Reduction** of an organic molecule usually corresponds to *decreasing* its *oxygen* content or *increasing* its *hydrogen* content.

We could describe the oxidation and reduction of organic compounds in terms of changes in oxidation numbers, just as we did for inorganic compounds in Sections 4-6 through 4-7. Formal application of oxidation number rules to organic compounds often leads to fractional oxidation numbers for carbon. For organic species, the descriptions in terms of increase or decrease of oxygen or hydrogen are usually easier to apply.

For example, the oxygen content increases when an alkane is converted to an alcohol, so this process is an oxidation.

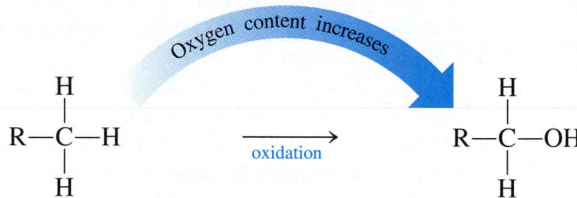

This reaction is not easy to carry out, and alcohols are prepared by many simpler methods. We show it here to illustrate the ideas of organic oxidations and reductions.

Converting a primary alcohol to an aldehyde or a secondary alcohol to a ketone is also an oxidation; the hydrogen content decreases.

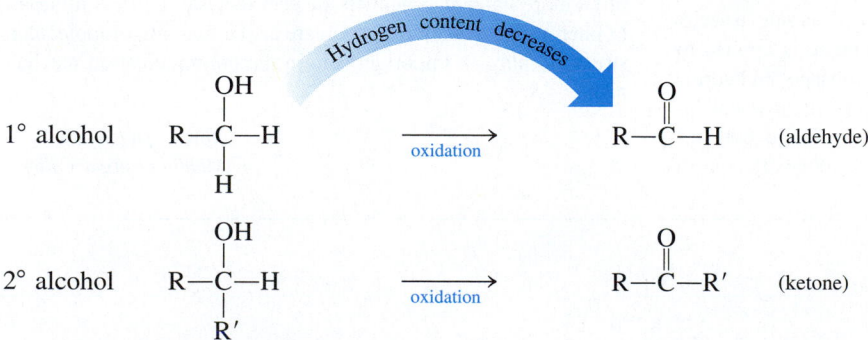

The notation R′ emphasizes that the two R groups may be the same (R = R′, e.g., in the formation of

$$\overset{O}{\underset{\|}{CH_3-C-CH_3}}$$

acetone,) or different (R ≠ R′, e.g., in the formation of methyl ethyl ketone,

$$\overset{O}{\underset{\|}{CH_3-C-CH_2CH_3}}).$$

CHEMISTRY IN USE

Research & Technology

Chemical Communication

The geometries of molecules play important roles in chemical reactivity. Molecular geometry is particularly important in a group of substances known as pheromones. *Pheromones* are chemicals used for communication between members of the same species. Pheromone activity has been observed in many forms of life, from insects to humans, and pheromone research is being done at many scholarly institutions.

If you've ever observed lines of ants moving in opposite directions, you have observed the influence of pheromones on insect behavior. When an ant finds food, it immediately heads toward its nest while secreting 9-oxy-2-decenoic acid from an abdominal gland. When other ants cross this acid trail, they compulsively follow it to the food source and carry the nourishment back to their nest. Soon, many ants will be following the acid trail and reinforcing it with their own 9-oxy-2-decenoic acid secretions. Eventually, the food source becomes exhausted, trail reinforcement stops, and the acid trail evaporates. Ants are so dependent upon the acid trail that if a part of it were wiped away the ants following the trail in both directions would come to a complete stop. They wouldn't know where to go.

Perhaps an even more impressive example than the total dependence upon chemical communication by ants is demonstrated with the so-called death pheromone. Immediately upon the death of an ant, fellow ants continue to groom the dead ant and treat it as if it were still living. This attention continues until the dead ant's body produces the death pheromone, 10-octadecenoic acid. Upon sensing this pheromone, colleagues carry the dead ant to the nearest garbage site. Interestingly, if 10-octadecenoic acid is applied to a living ant, the living ant is similarly dumped into the garbage. The discarded ant will quickly return only to be carried off again, and this process continues until the death pheromone evaporates.

Because pheromones are used by female insects to indicate their state of fertile readiness, pheromones have proven to be an effective weapon in controlling some crop damaging insects. For example, when a specific mating pheromone is applied to crops, male cotton bollworms and female tobacco budworms compulsively mate with one another. Because of physical incompatibilities, their bodies become interlocked and both insects eventually die. Less drastic uses of pheromones to control crop damage involve baiting traps with sex pheromones to lure and trap male insects. Trapping males eventually slows reproduction, and the insect population may decrease to controllable levels. Some of these sex pheromones are so powerful that a single drop has the potential of attracting millions of males. In fact, some male insects can detect a single molecule of female pheromone from a great distance and then successfully seek out and find the female.

Chemical communication is not confined to the insect world. Female dogs secrete the chemical *p*-hydroxybenzoate to attract males. Just like the ants and cotton bollworms who are dependent upon detecting chemicals for their actions, male dogs will attempt to mate with various objects to which *p*-hydroxybenzoate has been applied.

When we examine the molecular structures and functional groups of known pheromones, we find that they have little in common. Some pheromones contain stereoisomers, and some insects can distinguish between the stereoisomers. The structures of pheromones play vital roles in their activity. Part of the structure is an upper limit of about 20 carbon atoms, a limit probably imposed by Graham's Law. Most pheromones must travel through the air; those with low molecular weights are often more volatile. Scientists suspect that the physical motions of pheromone molecules, which are also a function of molecular structure, play an important role in the communication mechanism.

Ronald DeLorenzo
Middle Georgia College

An aldehyde can be oxidized to a carboxylic acid.

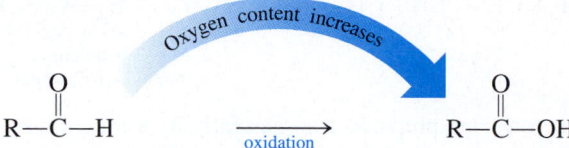

In each of these "oxidation" reactions, something else must act as the oxidizing agent (which is reduced). These oxidizing agents are often inorganic species such as dichromate ions, $Cr_2O_7^{2-}$, or permanganate ions, MnO_4^-. The reverse of each of the above reactions is a reduction of the organic molecule. In this reverse reaction the reducing agent (the substance that is oxidized) is often an inorganic compound.

Let us look at a few important types of organic oxidations and reductions.

Oxidation of Alcohols

Aldehydes can be prepared by the oxidation of *primary* alcohols. The reaction mixture is heated to a temperature slightly above the boiling point of the aldehyde so that the aldehyde distills out as soon as it is formed. Potassium dichromate in the presence of dilute sulfuric acid is the common oxidizing agent.

Aldehydes are easily oxidized to carboxylic acids. Therefore, they must be removed from the reaction mixture as soon as they are formed. Aldehydes have lower boiling points than the alcohols from which they are formed, so the removal of aldehydes is easily accomplished.

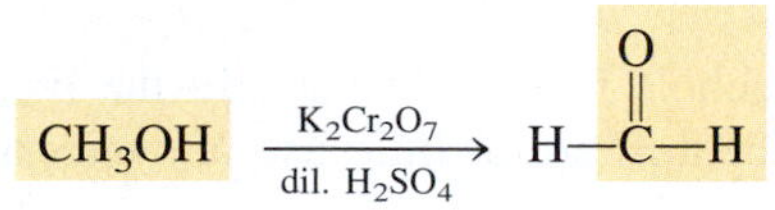

$$CH_3OH \xrightarrow[\text{dil. H}_2SO_4]{K_2Cr_2O_7} H\!-\!\overset{\displaystyle O}{\overset{\|}{C}}\!-\!H$$

methanol
bp = 65°C

methanal
(formaldehyde)
bp = −21°C

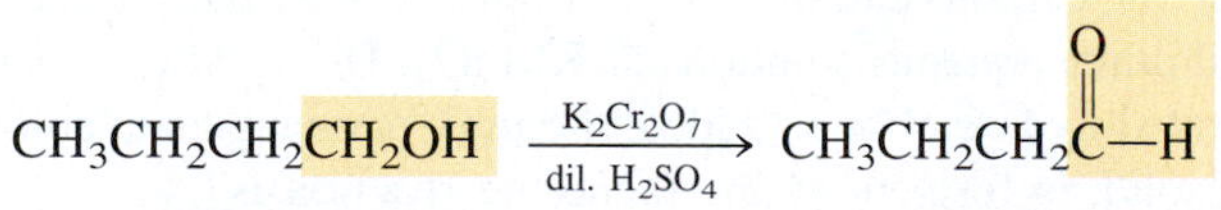

$$CH_3CH_2CH_2CH_2OH \xrightarrow[\text{dil. H}_2SO_4]{K_2Cr_2O_7} CH_3CH_2CH_2\overset{\displaystyle O}{\overset{\|}{C}}\!-\!H$$

1-butanol
bp = 117.5°C

butanal
(butyraldehyde)
bp = 75.7°C

Ketones can be prepared by the oxidation of *secondary* alcohols. Ketones are not as susceptible to oxidation as are aldehydes. Potassium permanganate in alkaline solution may be used as the oxidizing agent.

Ketones are not as easily oxidized as aldehydes, because oxidation of a ketone requires the breaking of a carbon–carbon bond. Thus, it is not as important that they be quickly removed from the reaction mixture.

$$CH_3\!-\!\overset{\displaystyle OH}{\overset{|}{CH}}\!-\!CH_3 \xrightarrow[\text{OH}^-]{KMnO_4} CH_3\!-\!\overset{\displaystyle O}{\overset{\|}{C}}\!-\!CH_3$$

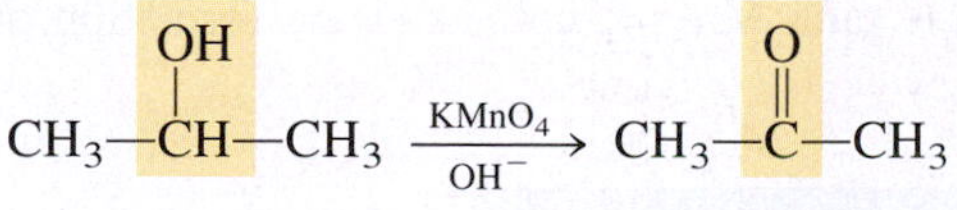

2-propanol
isopropyl alcohol

acetone

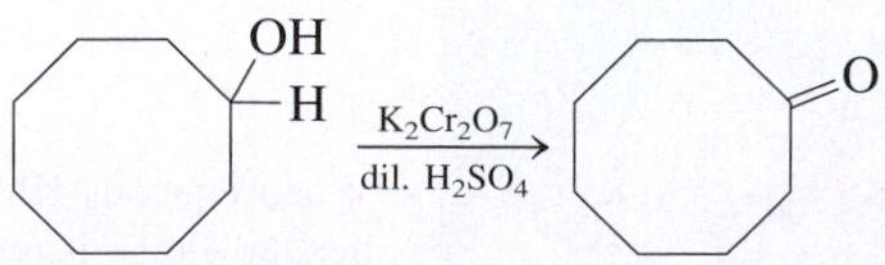

These two reactions can also be described as a type of *elimination* reaction (Section 28-6) called *dehydrogenation*. A molecule of hydrogen, H_2, is eliminated to form a C=O double bond.

cyclooctanol   cyclooctanone

When the carbon bearing the alcohol —OH group also has a hydrogen attached (a primary or secondary alcohol), the oxidation is easy; when it has no hydrogen attached (a tertiary alcohol), the oxidation is difficult.

$$CH_3-CH_2-\underset{\underset{OH}{|}}{CH}-CH_3 \xrightarrow[OH^-]{KMnO_4} CH_3-CH_2-\underset{\overset{\parallel}{O}}{C}-CH_3$$

2-butanol

2-butanone
methyl ethyl ketone

Aldehydes and ketones are prepared commercially by a catalytic process that involves passing alcohol vapors and air over a copper gauze or powder catalyst at approximately 300°C. Here the oxidizing agent is O_2.

$$2CH_3OH + O_2 \xrightarrow[300°C]{Cu} 2H-\underset{\overset{\parallel}{O}}{C}-H + 2H_2O$$

methanol

formaldehyde

Formaldehyde is quite soluble in water; the gaseous compound can be dissolved in water to give a 40% solution.

Acetaldehyde can be prepared by the similar oxidation of ethanol.

$$2CH_3CH_2OH + O_2 \xrightarrow[300°C]{Cu} 2CH_3-\underset{\overset{\parallel}{O}}{C}-H + 2H_2O$$

Oxidation of tertiary alcohols is difficult because the breaking of a carbon–carbon bond is required. Such oxidations are of little use in synthesis.

Oxidation or Reduction of Alkenes and Alkynes

A qualitative test for carbon–carbon unsaturation involves the oxidation of alkenes or alkynes with cold dilute aqueous solutions of $KMnO_4$. Dilute MnO_4^- solutions are pink. When an unsaturated hydrocarbon is added, the pink permanganate color disappears and MnO_2, a brown solid, is formed. With ethene the reaction is

Frequently, only the reactants and products of interest are shown in organic reactions. The reduction product in this reaction is MnO_2.

$$CH_2{=}CH_2 + MnO_4^- \xrightarrow{H_2O} \underset{\underset{OH}{|}}{CH_2}-\underset{\underset{OH}{|}}{CH_2} + MnO_2 + \text{other oxidation products}$$

pink   ethylene glycol   (brown)

Other alkenes undergo similar oxidation reactions. Hot or concentrated $KMnO_4$ solutions oxidize ethene to carbon dioxide and water. Similarly, the unsaturation in alkynes causes the disappearance of the pink MnO_4^- color and the formation of brown MnO_2.

When aqueous $KMnO_4$ is shaken with hexane (left), it retains its pink color. The pink color is removed by reaction with the double bond of 1-hexene (right), and brown solid MnO_2 is formed.

The hydrogenation reactions by which alkenes are converted to alkanes (Section 28-5) increase the hydrogen content so those addition reactions can be classed as *reductions*. Similarly, alkynes can be reduced (hydrogenated) to alkenes.

Hydrogen content increases

$$CH_3CH_2C{\equiv}CCH_2CH_3 + H_2 \xrightarrow[\text{reduction}]{\text{catalyst}}$$

$$\begin{array}{c} CH_3CH_2 \quad\quad CH_2CH_3 \\ \diagdown C{=}C \diagup \\ \diagup \quad\quad \diagdown \\ H \quad\quad\quad H \end{array}$$

3-hexyne   *cis*-3-hexene

Reduction of Carbonyl Compounds

Reduction of a variety of compounds that contain the carbonyl group provides synthetic methods to produce primary and secondary alcohols. A common, very powerful reducing agent is lithium aluminum hydride, $LiAlH_4$; other reducing agents include sodium in alcohol and sodium borohydride, $NaBH_4$.

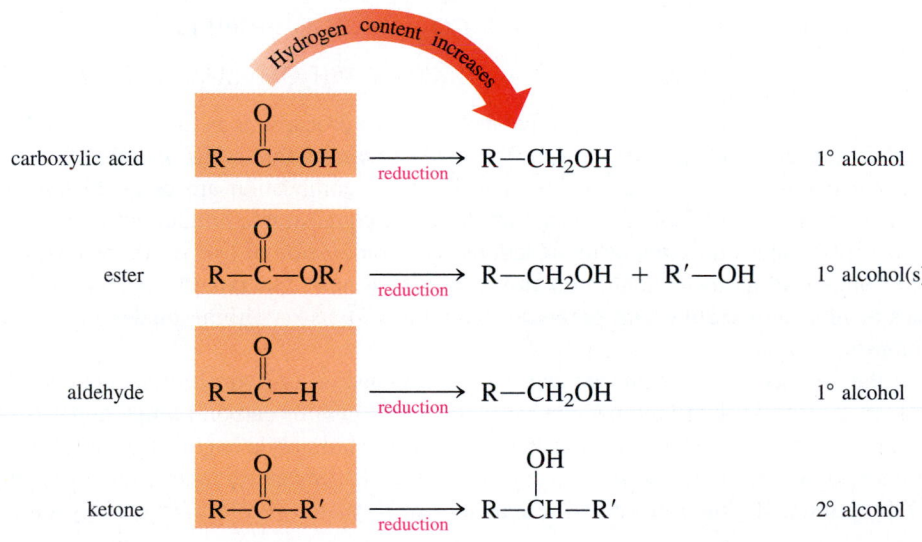

Hydrogen content increases

carboxylic acid	$R{-}\overset{O}{\overset{\|}{C}}{-}OH$ $\xrightarrow{\text{reduction}}$ $R{-}CH_2OH$	1° alcohol
ester	$R{-}\overset{O}{\overset{\|}{C}}{-}OR'$ $\xrightarrow{\text{reduction}}$ $R{-}CH_2OH + R'{-}OH$	1° alcohol(s)
aldehyde	$R{-}\overset{O}{\overset{\|}{C}}{-}H$ $\xrightarrow{\text{reduction}}$ $R{-}CH_2OH$	1° alcohol
ketone	$R{-}\overset{O}{\overset{\|}{C}}{-}R'$ $\xrightarrow{\text{reduction}}$ $R{-}\overset{OH}{\overset{\|}{C}H}{-}R'$	2° alcohol

Organic reactions are sometimes written in extremely abbreviated form. This is often the case when a variety of common oxidizing or reducing agents will accomplish the desired conversion.

Oxidation of Alkylbenzenes

Unsubstituted aromatic hydrocarbons (Sections 27-5, 27-6) are quite resistant to oxidation by chemical oxidizing agents. The reactions of strong oxidizing agents with alkylbenzenes illustrate the stability of the benzene ring system. Heating toluene with a basic solution of $KMnO_4$ results in a nearly 100% yield of benzoic acid. The ring itself remains intact; only the nonaromatic portion of the molecule is oxidized.

$$\text{C}_6\text{H}_5{-}CH_3 \xrightarrow[\text{(2) HCl}]{\text{(1) } \Delta,\ OH^-,\ KMnO_4} \text{C}_6\text{H}_5{-}\overset{O}{\overset{\|}{C}}{-}OH$$

toluene   benzoic acid

The MnO_2 is removed by filtration of the basic solution. Benzoic acid, an insoluble weak acid, is then precipitated by acidification with a strong inorganic acid.

Two such alkyl groups on an aromatic ring are oxidized to give a diprotic acid as the following example illustrates.

Acetylene is produced by the slow addition of water to calcium carbide.

$$CaC_2(s) + 2H_2O(\ell) \longrightarrow$$
$$HC\equiv CH(g) + Ca(OH)_2(s)$$

The light of one kind of headlamp used by miners and cave explorers is given off by the combustion of acetylene.

Terephthalic acid is used to make "polyesters," an important class of polymers (Section 28-11).

Combustion of Organic Compounds

The most extreme oxidation reactions of organic compounds occur when they burn in O_2. Such *combustion reactions* (Section 6-8, part 3) are highly exothermic. When the combustion takes place in excess O_2, the products are CO_2 and H_2O. Examples of alkane combustions are

methane: $CH_4 \;\; + 2O_2 \;\; \longrightarrow \;\; CO_2 \;\; + 2H_2O \;\; + 891 \text{ kJ}$

octane: $2C_8H_{18} + 25O_2 \longrightarrow 16CO_2 + 18H_2O + 1.090 \times 10^4 \text{ kJ}$

Such reactions are the bases for important uses of the hydrocarbons as fuels, e.g., gasoline, diesel fuel, heating oil, and natural gas. The **heat of combustion** is the amount of energy *liberated* per mole of hydrocarbon burned. Heats of combustion are assigned positive values *by convention* (Table 28-3) and are therefore equal in magnitude, but opposite in sign, to ΔH^0 values for combustion reactions. The combustion of hydrocarbons produces large volumes of gases in addition to large amounts of heat. The rapid formation of these gases at high temperature and pressure drives the pistons or turbine blades in internal combustion engines.

Recall that ΔH^0 is negative for an exothermic process.

In the absence of sufficient oxygen, partial combustion of hydrocarbons occurs. The products may be carbon monoxide (a very poisonous gas) or carbon (which deposits on spark plugs, in the cylinder head, and on the pistons of automobile engines). Many modern automobile engines now use microcomputer chips and sensors to control the air supply and to optimize the fuel/O_2 ratio. The reactions of methane with insufficient oxygen are

Table 28-3 *Heats of Combustion of Some Alkanes*

| | | Heat of Combustion | |
Hydrocarbon		*kJ/mol*	*J/g*
methane	CH_4	891	55.7
propane	C_3H_8	2220	50.5
pentane	C_5H_{12}	3507	48.7
octane	C_8H_{18}	5450	47.8
decane	$C_{10}H_{22}$	6737	47.4
ethanol*	C_2H_5OH	1372	29.8

*Not an alkane; included for comparison only.

$$2CH_4 + 3O_2 \longrightarrow 2CO + 4H_2O \quad \text{and} \quad CH_4 + O_2 \longrightarrow C + 2H_2O$$

All hydrocarbons undergo similar reactions.

The alkenes, like the alkanes, burn in *excess* oxygen to form carbon dioxide and water in exothermic reactions.

$$CH_2{=}CH_2 + 3O_2 \text{ (excess)} \longrightarrow 2CO_2 + 2H_2O + 1387 \text{ kJ}$$

When an alkene (or any other unsaturated organic compound) is burned in air, a yellow, luminous flame is observed and considerable soot (unburned carbon) is formed. This reaction provides a qualitative test for unsaturation. Saturated hydrocarbons burn in air without forming significant amounts of soot.

Acetylene lamps are charged with calcium carbide. Very slow addition of water produces acetylene, which is burned as it is produced. Acetylene is also used in the oxyacetylene torch for welding and cutting metals. When acetylene is burned with oxygen, the flame reaches temperatures of about 3000°C.

Like other hydrocarbons, the *complete combustion* of aromatic hydrocarbons, such as benzene, releases large amounts of energy.

$$2C_6H_6 + 15O_2 \longrightarrow 12CO_2 + 6H_2O + 6548 \text{ kJ}$$

Because they are so unsaturated, aromatic hydrocarbons burn *in air* with a yellow, sooty flame.

28-9 FORMATION OF CARBOXYLIC ACID DERIVATIVES

The carboxylic acid derivatives introduced in Section 27-13 can be formed by *substitution* of another group in place of —OH in the carboxyl group. The acyl halides (acid halides) are usually prepared by treating acids with PCl_3, PCl_5, or $SOCl_2$ (thionyl chloride). In general terms, the reaction of acids with PCl_5 may be represented as

$$
\underset{\text{acid}}{R\overset{\overset{\displaystyle O}{\|}}{-}C-OH} + \underset{\substack{\text{phosphorus} \\ \text{pentachloride}}}{PCl_5} \longrightarrow \underset{\substack{\text{an acyl chloride} \\ \text{(an acid chloride)}}}{R\overset{\overset{\displaystyle O}{\|}}{-}C-Cl} + HCl(g) + \underset{\substack{\text{phosphorus} \\ \text{oxychloride}}}{POCl_3}
$$

$$
\underset{\text{acetic acid}}{CH_3\overset{\overset{\displaystyle O}{\|}}{-}C-OH} + PCl_5 \longrightarrow \underset{\text{acetyl chloride}}{CH_3\overset{\overset{\displaystyle O}{\|}}{-}C-Cl} + HCl(g) + POCl_3
$$

The acyl halides are much more reactive than their parent acids. Consequently, they are often used in reactions to introduce an acyl group into another molecule.

When an organic acid is heated with an alcohol, an equilibrium is established with the resulting *ester* and water. The reaction is catalyzed by traces of strong inorganic acids, such as a few drops of concentrated H_2SO_4.

$$
\underset{\text{acetic acid}}{CH_3\overset{\overset{\displaystyle O}{\|}}{-}C-OH} + \underset{\text{ethyl alcohol}}{CH_3CH_2-OH} \underset{}{\overset{H^+, \Delta}{\rightleftharpoons}} \underset{\text{ethyl acetate, an ester}}{CH_3\overset{\overset{\displaystyle O}{\|}}{-}C-O-CH_2CH_3} + H_2O
$$

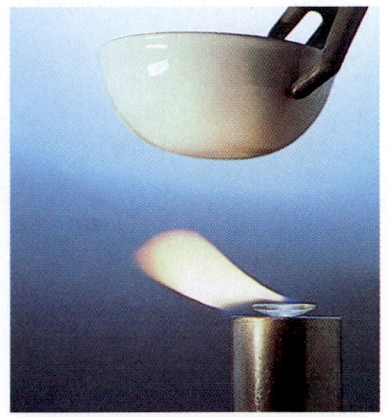

Hexane, C_6H_{14}, an alkane, burns cleanly in air to give CO_2 and H_2O (top). 1-Hexene, C_6H_{12}, an alkene, burns with a flame that contains soot (middle). Burning o-xylene, an aromatic hydrocarbon, produces large amounts of soot (bottom).

In general terms, the reaction of an organic acid and an alcohol may be represented as

(R and R′ may be the same or different groups.)

Reactions between acids and alcohols are usually quite slow and require prolonged boiling (refluxing). However, the reactions between most acyl halides and most alcohols occur very rapidly without requiring the presence of an acid catalyst.

acetyl chloride   ethyl alcohol   ethyl acetate

Amides are usually *not* prepared by the reaction of an amine with an organic acid. Acyl halides and acid anhydrides react readily with primary and secondary amines to produce amides. The reaction of an acyl halide with a primary or secondary amine produces an amide and a salt of the amine.

methylamine   acetyl chloride   N-methylacetamide   methylammonium
(a primary amine)   (an acyl halide)   (an amide)   chloride (a salt)

The reaction of an acid anhydride with a primary or a secondary amine produces an amide and a salt of the amine.

dimethylamine   acetic anhydride   N,N-dimethylacetamide   dimethylammonium
(a secondary amine)   (an acid anhydride)   acetate (a salt)

28-10 HYDROLYSIS OF ESTERS

Because most esters are not very reactive, strong reagents are required for their reactions. Esters can be hydrolyzed by refluxing with solutions of strong bases.

ethyl acetate   sodium acetate   ethanol

The hydrolysis of esters in the presence of strong soluble bases is called **saponification** (soap-making).

American pioneers prepared their soaps by boiling animal fat with an alkaline solution obtained from the ashes of hardwood. The resulting soap could be "salted out" by adding sodium chloride, making use of the fact that soap is less soluble in a salt solution than in water.

In general terms, the hydrolysis of esters may be represented as

$$\underset{\text{ester}}{R-\overset{\overset{\displaystyle O}{\|}}{C}-O-R'} + Na^{+}OH^{-} \xrightarrow{\Delta} \underset{\text{salt of an acid}}{R-\overset{\overset{\displaystyle O}{\|}}{C}-O^{-}Na^{+}} + \underset{\text{alcohol}}{R'OH}$$

Like other esters, fats and oils (Section 27-13) can be hydrolyzed in strongly basic solution to produce salts of the acids and the alcohol glycerol. The resulting sodium salts of long-chain fatty acids are soaps. In Section 14-18 we described the cleansing action of soaps and detergents.

$$
\begin{array}{l}
H_2C-O-\overset{\overset{\displaystyle O}{\|}}{C}-(CH_2)_{16}CH_3 \\[2mm]
\;\;\;| \\[1mm]
HC-O-\overset{\overset{\displaystyle O}{\|}}{C}-(CH_2)_{16}CH_3 \quad + 3Na^{+}OH^{-} \longrightarrow 3CH_3(CH_2)_{16}\overset{\overset{\displaystyle O}{\|}}{C}-O^{-}Na^{+} \;+\;
\begin{array}{l}H_2C-OH\\ | \\ HC-OH\\ | \\ H_2C-OH\end{array} \\[2mm]
\;\;\;| \\[1mm]
H_2C-O-\overset{\overset{\displaystyle O}{\|}}{C}-(CH_2)_{16}CH_3
\end{array}
$$

glyceryl tristearate   sodium stearate   glycerol
(a fat)   (a soap)   (glycerine)

28-11 POLYMERIZATION REACTIONS

A **polymer** is a large molecule that is a high-molecular-weight chain of small molecules. The small molecules that are linked to form polymers are called **monomers**. Typical polymers consist of hundreds or thousands of monomers and have molecular weights up to thousands or millions of grams per mole.

The word fragment *-mer* means "part." Recall that *isomers* are compounds that are composed of the same (*iso*) parts (*mers*). A *monomer* is a "single part"; a large number of monomers combine to form a *polymer,* "many parts."

Polymerization is the combination of many small molecules (monomers) to form large molecules (polymers).

Polymers are divided into two classes—natural and synthetic. Important biological molecules such as proteins, nucleic acids, and polysaccharides (starches and the cellulose in wood and cotton) are natural polymers. Natural rubber and natural fibers such as silk and wool are also natural polymers. Familiar examples of synthetic polymers include plastics such as polyethylene, Teflon, and Lucite (Plexiglas) and synthetic fibers such as Nylon, Orlon, and Dacron. In this section we shall describe some processes by which polymers are formed from organic compounds.

Addition Polymerization

Polymerization is an important addition reaction (Section 28-5) of the alkenes. Polymers formed by this kind of reaction are called **addition polymers.** The formation of polyethylene is an important example. In the presence of appropriate catalysts (a mixture of aluminum trialkyls, R_3Al, and titanium tetrachloride, $TiCl_4$), ethylene polymerizes into chains containing 800 or more carbon atoms.

$$n\text{CH}_2{=}\text{CH}_2 \xrightarrow{\text{catalyst}} {-}(\text{CH}_2{-}\text{CH}_2)_n$$
$$\text{ethylene} \qquad\qquad \text{polyethylene}$$

The polymer may be represented as $CH_3(CH_2{-}CH_2)_nCH_3$, where n is approximately 400. Polyethylene is a tough, flexible plastic. It is widely used as an electrical insulator and for the fabrication of such items as unbreakable refrigerator dishes, plastic cups, and squeeze bottles. Polypropylene is made by polymerizing propylene, $CH_3{-}CH{=}CH_2$, in much the same way. Teflon is made by polymerizing tetrafluoroethylene in a similar reaction.

Teflon is a trade name owned by Du Pont, a company that has developed and manufactured many fluorinated polymers.

$$n\text{CF}_2{=}\text{CF}_2 \xrightarrow[\text{heat}]{\text{catalyst}} {-}(\text{CF}_2{-}\text{CF}_2)_n$$
$$\text{tetrafluoroethylene} \qquad\qquad \text{``Teflon''}$$

The molecular weight of Teflon is about 2×10^6. Approximately 20,000 $CF_2{=}CF_2$ molecules polymerize to form a single giant molecule. Teflon is a very useful polymer. It does *not* react with concentrated acids and bases or with most oxidizing agents. It does not dissolve in most organic solvents.

Natural rubber is obtained from the sap of the rubber tree, a sticky liquid called latex. Rubber is a polymeric hydrocarbon formed in the sap by the combination of about 2000 molecules of 2-methyl-1,3-butadiene, commonly called isoprene. The molecular weight of rubber is about 136,000.

$$2n\text{CH}_2{=}\overset{\overset{\text{CH}_3}{|}}{\text{C}}{-}\text{CH}{=}\text{CH}_2 \longrightarrow {-}(\text{CH}_2{-}\overset{\overset{\text{CH}_3}{|}}{\text{C}}{=}\text{CH}{-}\text{CH}_2{-}\text{CH}_2{-}\overset{\overset{\text{CH}_3}{|}}{\text{C}}{=}\text{CH}{-}\text{CH}_2)_n$$
$$\text{isoprene} \qquad\qquad\qquad \text{natural rubber}$$

When natural rubber is warmed, it flows and becomes sticky. To eliminate this problem, **vulcanization** is used. This is a process in which sulfur is added to rubber and the mixture is heated to approximately 140°C. Sulfur atoms combine with some of the double bonds in the linear polymer molecules to form bridges that bond one rubber molecule to

Many cooking utensils with "nonstick" surfaces are coated with a polymer such as Teflon.

another. This cross-linking by sulfur atoms converts the linear polymer into a three-dimensional polymer. Fillers and reinforcing agents are added during the mixing process to increase the durability of rubber and to form colored rubber. Carbon black is the most common reinforcing agent. Zinc oxide, barium sulfate, titanium dioxide, and antimony(V) sulfide are common fillers.

Some synthetic rubbers are superior to natural rubber in some ways. Neoprene is a synthetic elastomer (an elastic polymer) with properties quite similar to those of natural rubber. The basic structural unit is chloroprene, which differs from isoprene in having a chlorine atom rather than a methyl group as a substituent on the 1,3-butadiene chain.

Numerous other polymers are elastic enough to be called by the generic name "rubber."

$$nCH_2=CH-\underset{\underset{Cl}{|}}{C}=CH_2 \xrightarrow{\text{polymerization}} \left(CH_2-CH=\underset{\underset{Cl}{|}}{C}-CH_2\right)_n$$

chloroprene → neoprene (a synthetic rubber)

Neoprene is less affected by gasoline and oil, and is more elastic than natural rubber. It resists abrasion well and is not swollen or dissolved by hydrocarbons. It is widely used to make hoses for oil and gasoline, electrical insulation, and automobile and refrigerator parts.

When two different monomers are mixed and then polymerized, **copolymers** are formed. Depending on the ratio of the two monomers and the reaction conditions, the order of the units can range from quite regular (e.g., alternating) to completely random. In this way, polymers with a wide variety of properties can be produced. The most important rubber produced in the largest amount in the United States is SBR, a polymer of styrene with butadiene in a 1:3 ratio.

The double bonds in SBR can be cross-linked by vulcanization as described for natural rubber. SBR is used primarily for making tires. Other copolymers are used to make car bumpers, body and chassis parts, wire insulation, sporting goods, sealants, and caulking compounds.

Some addition polymers and their uses are listed in Table 28-4.

Condensation Polymerization

Some polymerization reactions are based on *condensation reactions,* in which two molecules combine by splitting out or eliminating a small molecule. For such a polymer to be formed, each monomer must have two functional groups, one on each end. A polymer formed in this way is called a **condensation polymer.** There are many useful condensation polymers, based on a wide variety of bifunctional molecules.

Table 28-4 *Some Important Addition Polymers*

Polymer Name (some trade names)	Some Uses	Polymer Production, Tons/Year in United States	Monomer Formula	Monomer Name
polyethylene (Polythene)	electrical insulation; toys and molded objects; bags; squeeze bottles	8 million	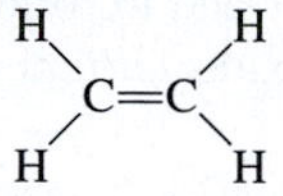 $H_2C=CH_2$	ethylene
polypropylene (Herculon, Vectra)	bottles; films; lab equipment; toys; packaging film; filament for rope, webbing, carpeting; molded auto and appliance parts	2.7 million	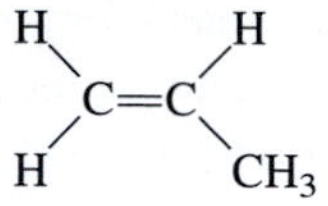 $H_2C=CH(CH_3)$	propylene
polyvinyl chloride (PVC)	pipe, siding, gutters; floor tiles; phonograph records	3.5 million	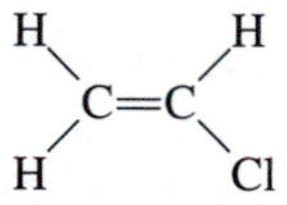 $H_2C=CHCl$	vinyl chloride
polyacrylonitrile (Orlon, Acrilan)	acrylic fibers for carpets, clothing, knitwear	920,000	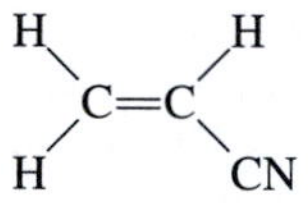 $H_2C=CH(CN)$	acrylonitrile
polystyrene (Styrene, Styrofoam, Styron)	molded toys, dishes, kitchen equipment; insulating foam, e.g., ice chests; rigid foam packaging	2 million	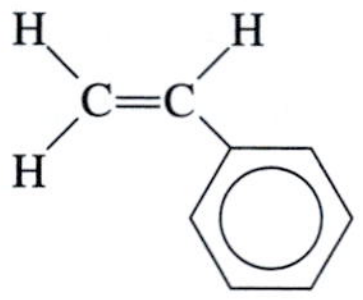 $H_2C=CH(C_6H_5)$	styrene
polyvinylacetate (PVA)	water-based "latex" paint; adhesives; paper and textile coatings	500,000	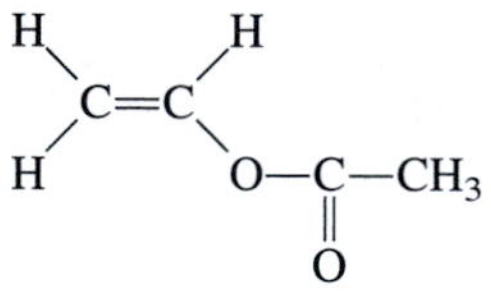 $H_2C=CH-O-C(=O)CH_3$	vinyl acetate
poly(methylmeth-acrylate) (Plexiglas, Lucite)	high-quality transparent objects; water-based paints; contact lenses	450,000	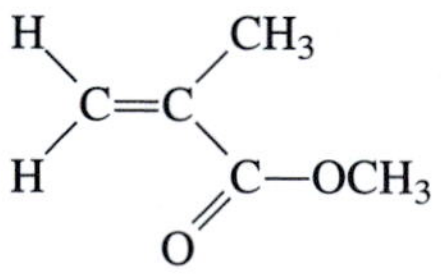 $H_2C=C(CH_3)-C(=O)-OCH_3$	methyl methacrylate
polytetrafluoro-ethylene (Teflon)	gaskets; pan coatings; electrical insulation; bearings	7,000	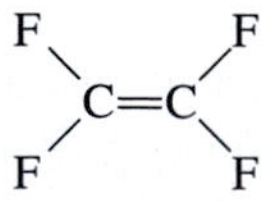 $F_2C=CF_2$	tetrafluoroethylene
polybutadiene	auotmotive tire tread, hoses, belts; metal can coatings	400,000	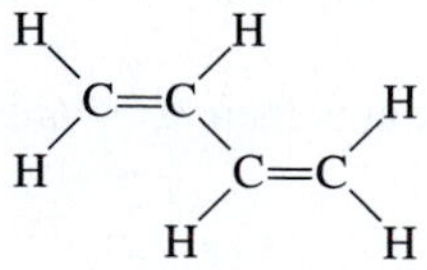 $H_2C=CH-CH=CH_2$	butadiene
ethylene-propylene copolymer	appliance parts; auto hoses, bumpers, body and chassis parts; coated fabrics	150,000	see above	ethylene, propylene
SBR copolymer	tires	1.4 million	see above	styrene, butadiene

Polyesters (short for "*poly*meric *esters*") are condensation polymers that are formed when *dihydric alcohols* react with *dicarboxylic acids*. An ester linkage is formed at each end of each monomer molecule to build up large molecules. A useful polyester is prepared from ethylene glycol and terephthalic acid.

Dihydric alcohols contain two —OH groups per molecule.

$$\text{HO-C-}\bigcirc\text{-C-OH} \quad \text{H-OCH}_2\text{CH}_2\text{O-H} \quad \text{HO-C-}\bigcirc\text{-C-OH} \quad \text{H-OCH}_2\text{CH}_2\text{O-H}$$

terephthalic acid   ethylene glycol   terephthalic acid   ethylene glycol

↓ remove water

$$\text{HOCH}_2\text{CH}_2\text{O}\left[\text{C-}\bigcirc\text{-C-OCH}_2\text{CH}_2\text{O}\right]_n\text{H}$$

polyethylene terephthalate (Dacron, Mylar)

More than 2 million tons of this polymer are produced annually in the United States. Dacron, the fiber produced from this polyester, accounts for approximately 50% of all synthetic fibers. It absorbs very little moisture, and its properties are nearly the same whether it is wet or dry. Additionally, it possesses exceptional elastic recovery properties so it is used to make "permanent-press" fabrics. This polyester can also be made into films of great strength (e.g., Mylar), which can be rolled into sheets 1/30 the thickness of a human hair. Such films can be magnetically coated to make audio and video tapes.

The polymeric amides, **polyamides,** are an especially important class of condensation polymers. **Nylon** is the best known polyamide. It is prepared by heating anhydrous hexa-methylenediamine with anhydrous adipic acid, a dicarboxylic acid. This substance is called Nylon 66 because the parent diamine and dicarboxylic acid each contain six carbon atoms.

The molecular weight of the polymer varies from about 10,000 to about 25,000. It melts at about 260 to 270°C.

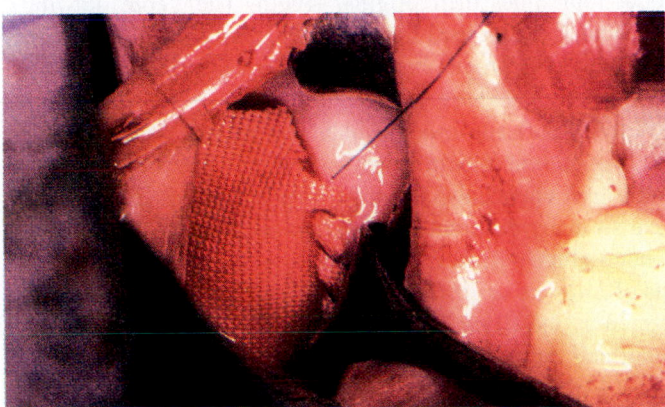

A patch made of Dacron polymer is used to close a defect in a human heart.

Polymers have a wide range of properties and uses. Mylar sheet polymer is used to protect documents, such as this photocopy of a later version of the Declaration of Independence (top). Another kind of polymer coating is used to make shatterproof shields of glass items such as fluorescent light bulbs (bottom).

Nylon is formed at the interface where hexamethylenediamine (in the lower water layer) and adipyl chloride (a derivative of adipic acid, in the upper hexane layer) react. The Nylon can be drawn out and wound on a stirring rod.

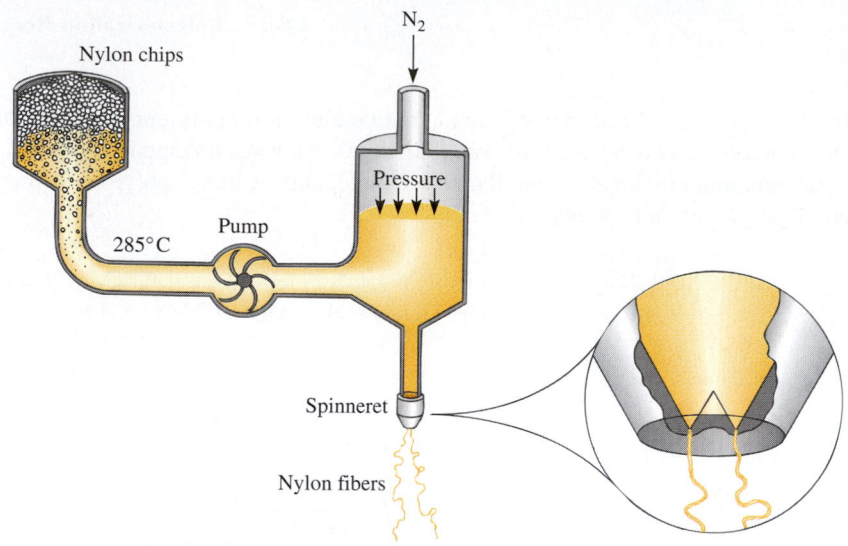

Figure 28-11 Fibers of synthetic polymers are made by extrusion of the molten material through tiny holes, called *spinnerets.* After cooling, Nylon fibers are stretched to about four times their original length to orient the polymer molecules.

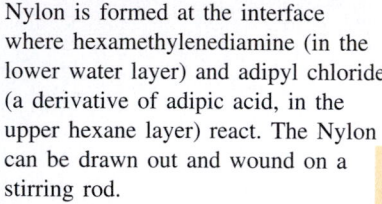

$$HO-\overset{\overset{\displaystyle O}{\|}}{C}-(CH_2)_4-\overset{\overset{\displaystyle O}{\|}}{C}-OH$$

adipic acid

+

$$H_2N-(CH_2)_6-NH_2$$

hexamethylenediamine

$$\xrightarrow[-H_2O]{heat}$$

$$-NH-\left(\overset{\overset{\displaystyle O}{\|}}{C}-(CH_2)_4-\overset{\overset{\displaystyle O}{\|}}{C}-NH-(CH_2)_6-NH\right)_n\overset{\overset{\displaystyle O}{\|}}{C}-$$

Nylon 66
(a polyamide)

Molten Nylon is drawn into threads (Figure 28-11). After cooling to room temperature, these can be stretched to about four times their original length. The "cold drawing" process orients the polymer molecules so that their long axes are parallel to the fiber axis. At regular intervals there are N—H---O hydrogen bonds that *cross-link* adjacent chains to give strength to the fiber.

Petroleum is the ultimate source of both adipic acid and hexamethylenediamine. We do not mean that these compounds are present in petroleum, only that they are made from it. The same is true for many industrial chemicals. The cost of petroleum is an important factor in our economy because so many different products are derived from petroleum.

Some types of natural condensation polymers play crucial roles in living systems. **Proteins** are polymeric chains of *L-amino acids* (Sections 27-12, 28-1) linked by peptide bonds. A **peptide bond** is formed by the elimination of a molecule of water between the amino group of one amino acid and the carboxylic acid group of another.

peptide bond

$$H-\overset{\overset{\displaystyle H}{|}}{\underset{\underset{\displaystyle R}{|}}{N}}-\overset{\overset{\displaystyle O}{\|}}{\underset{}{C}}-OH \;+\; H-\overset{\overset{\displaystyle H}{|}}{\underset{\underset{\displaystyle R'}{|}}{N}}-\overset{\overset{\displaystyle O}{\|}}{\underset{}{C}}-OH \;\longrightarrow\; H-\overset{\overset{\displaystyle H}{|}}{\underset{\underset{\displaystyle R}{|}}{N}}-\overset{\overset{\displaystyle O}{\|}}{\underset{}{C}}-N-\overset{\overset{\displaystyle H}{|}}{\underset{\underset{\displaystyle R'}{|}}{C}}-\overset{\overset{\displaystyle O}{\|}}{\underset{}{C}}-OH \;+\; H_2O$$

Plastics Harder Than Steel

We have learned that vulcanization is used to produce cross-linking between molecules in natural rubber to improve the physical properties of rubber. When cross-linking agents such as sulfur join one linear polymer chain to another, a three-dimensional network is created; this reduces the mobility of the polymer chains. In addition to sulfur, chemists have used oxygen, heat, gamma rays, beta rays, ultraviolet light, and energetic electrons to induce cross-linking.

Chemists have also discovered that high-energy beams of oxygen, carbon, nitrogen, boron, or argon ions have remarkable effects on the surfaces of some polymers. Their surfaces become harder than steel and more resistant to wear and abrasion than steel. Normally, the hardness of untreated polymers is in the range 0.1–0.5 gigapascal, whereas the hardness of stainless steels is in the range 2–3 gigapascals, and the hardness of hardened steels is in the range 8–12 gigapascals. In contrast, some ion-irradiated polymers have a hardness of about 22 gigapascals.

Such polymers could lend themselves to applications such as wear-resistant gears and bearings, artificial hip and knee joints, scratch-resistant plastic sunglasses and airplane windows, and lighter weight materials for automobiles and spacecraft.

In addition to increased hardness and wear resistance, ion-irradiated polymers are also resistant to (do not dissolve in) many solvents and less permeable to gases. The latter property could make them ideal for storing carbonated beverages. Ion-treated polymers also resist oxidation, a serious degrading problem for many polymers; ion-bombardment increases electrical conductivity of polymers.

Chemists have gained enough experience with ion-irradiated polymers to produce several specific kinds of polymer surfaces

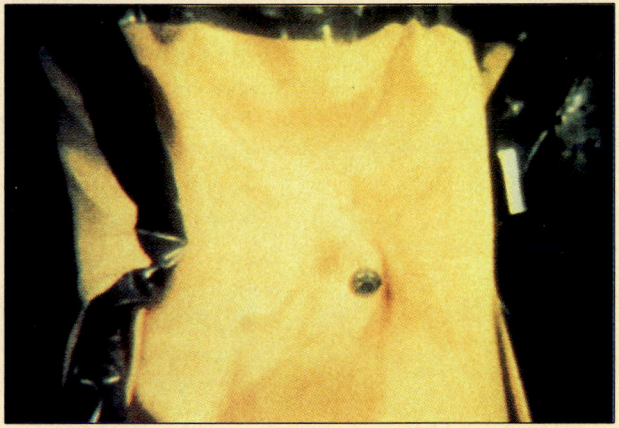

Vest made of Kevlar, a polyamide, has deflected a bullet.

by using different kinds of ions and different ion-bombardment speeds. For example, the polymer's surface hardness is directly proportional to the energy of the ion beam striking the polymer. The degree of ion-penetration and resulting extent of cross-linking vary inversely with the sizes of the ions.

Researchers don't understand fully how ion beams modify polymers to improve their hardness. Cross-linking plays an important role, but many other changes accompany cross-linking. Chemists hope that a better understanding of what happens at the molecular level will lead to even more interesting and exciting research as well as to new polymers.

Even more amazing polymers and potential applications may lie ahead. For example, one laboratory is currently working on a polymer that may be capable of withstanding a simulated nuclear explosion with little damage!

Ronald DeLorenzo
Middle Georgia College

When this process is carried out repeatedly, a large molecule called a **polypeptide** is formed.

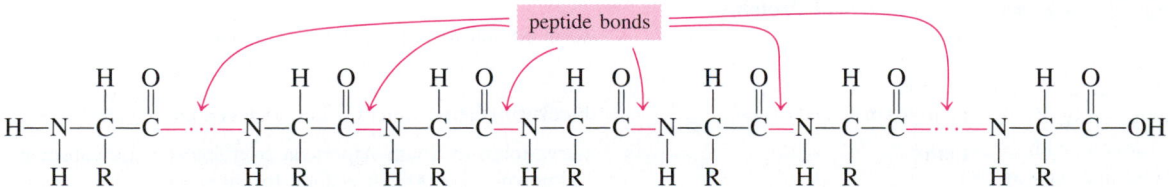

We see that the monomer units of proteins are the amino acids. Most proteins contain 100 to 300 amino acid units. Twenty different amino acids are usually found in proteins. They have in common the general structure indicated in the structure drawn above; but different side chains, R, distinguish the amino acids from one another.

Proteins make up more than 50% of the dry weight of animals and bacteria. They perform many important functions in living organisms, a few of which are indicated in Table 28-5. Each protein carries out a specific biochemical function. Each is a polypeptide with its own unique *sequence* of amino acids. The amino acid sequence of a protein determines exactly how it folds up in a three-dimensional conformation and how it performs its precise biochemical task.

Table 28-5 *Some Functions of Proteins*

Example	Function
Enzymes	
amylase	converts starch to glucose
DNA polymerase I	repairs DNA molecule
trans aminase	transfers amino group from one amino acid to another
Structural Proteins	
viral coat proteins	outer covering of virus
keratin	hair, nails, horns, hoofs
collagen	tendons, cartilage
Hormones	
insulin, glucagon	regulate glucose metabolism
oxytocin	regulates milk production in female mammals
vasopressin	increases retention of water by kidney
Contractile Proteins	
actin	thin contractile filaments in muscle
myosin	thick filaments in muscle
Storage Proteins	
casein	a nutrient protein in milk
ferritin	stores iron in spleen and egg yolk
Transport Proteins	
hemoglobin	carries O_2 in blood
myoglobin	carries O_2 in muscle
serum albumin	carries fatty acids in blood
cytochrome *c*	transfers electrons
Immunological Proteins	
γ-globulins	form complexes with foreign proteins
Toxins	
neurotoxin	blocker of nerve function in cobra venom
ricin	nerve toxin in South American frog (most toxic substance known—0.000005 g is fatal to humans)

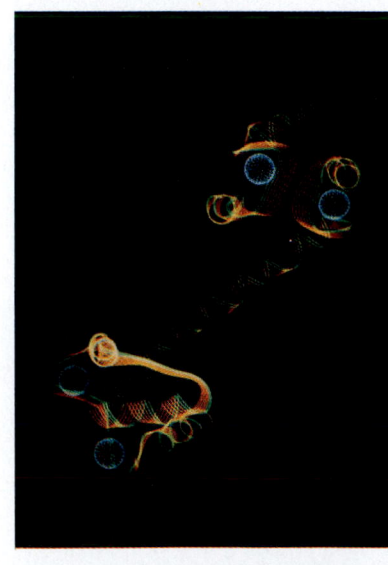

A "ribbon" model of the protein calmodulin. In this type of model, the ribbon represents the polypeptide chain. This protein coordinates with Ca^{2+} ions (white spheres) and aids in transporting them in living systems.

Key Terms

Achiral Describes an object that *can* be superimposed with its mirror image.

Addition reaction A reaction in which two atoms or groups of atoms are added to a molecule, one on each side of a double or triple bond. The number of groups attached to carbon *increases,* and the molecule becomes more nearly saturated.

Brønsted–Lowry acid A species that acts as a proton donor.

Brønsted–Lowry base A species that acts as a proton acceptor.

Chiral Describes an object that *cannot* be superimposed with its mirror image.

Condensation polymer A polymer that is formed by a condensation reaction.

Condensation reaction A reaction in which a small molecule, such as water or hydrogen chloride, is eliminated and two molecules are joined.

Conformation One specific geometry of a molecule. The conformations of a compound differ from one another only by rotation about single bonds.

Constitutional isomers Compounds that contain the same number of the same kinds of atoms but that differ in the order in which their atoms are bonded together. Also known as *structural isomers.*

Copolymer A polymer formed from two different compounds (monomers).

Dehydration The reaction in which H and OH are eliminated from adjacent carbon atoms, to form water and a more unsaturated bond.

Dehalogenation A reaction in which two halogen atoms are eliminated from a *vicinal* dihalide to form a double bond.

Dehydrogenation Elimination of two hydrogen atoms from a compound to form a double bond.

Dehydrohalogenation An elimination reaction in which a hydrogen halide, HX (X = Cl, Br, I), is eliminated from a haloalkane. A C=C double bond is formed.

Elimination reaction A reaction in which the number of groups attached to carbon *decreases.* The degree of unsaturation in the molecule increases.

Enantiomers See *Optical isomers.*

Esterification The reaction in which an alcohol and a carboxylic acid eliminate water and link to form an ester.

Functional group isomers Compounds (isomers) that have different orders of arrangement of atoms resulting in different functional groups being present. A subset of constitutional isomers.

Geometrical isomers Compounds with different arrangements of groups on the opposite sides of a bond with restricted rotation, such as a double bond or a single bond in a ring; for example, *cis–trans* isomers of certain alkenes.

Hydration reaction A reaction in which the elements of water, H and OH, add across a double or triple bond.

Hydrogenation The reaction in which hydrogen adds across a double or triple bond.

Monomers The small molecules from which polymers are formed.

Optical isomers Molecules that are nonsuperimposable mirror images of one another, i.e., that bear the same relationship to each other as do left and right hands; also called *enantiomers.*

Oxidation (as applied to organic compounds) The increase of oxygen content or the decrease of hydrogen content of an organic molecule.

Peptide bond A bond formed by elimination of a molecule of water between the amino group of one amino acid and the carboxylic acid group of another.

Polyamide A polymeric amide.

Polyester A polymeric ester.

Polymerization The combination of many small molecules (monomers) to form large molecules (polymers).

Polymers Large molecules formed by the combination of many small molecules (monomers).

Polypeptide A polymer composed of amino acids linked by peptide bonds.

Positional isomers Constitutional isomers that have the same number and kind of functional groups but in different locations.

Protein A naturally occurring polymeric chain of *L*-amino acids linked together by peptide bonds.

Racemic mixture A single sample containing equal amounts of the two enantiomers (optical isomers) of a compound; does not rotate the plane of polarized light.

Reduction (as applied to organic compounds) The decrease of oxygen content or the increase of hydrogen content of an organic molecule.

Saponification The hydrolysis of esters in the presence of strong soluble bases.

Soap The sodium salt of a long-chain fatty acid.

Stereoisomers Isomers in which the atoms are linked together in the same atom-to-atom order, but with different arrangements in space. See *Geometrical isomers, Optical isomers.*

Structural isomers See *Constitutional isomers.*

Substitution reaction A reaction in which an atom or a group of atoms attached to a carbon atom is replaced by another atom or group of atoms. No change occurs in the degree of saturation at the reactive carbon.

Vulcanization The process in which sulfur is added to rubber and heated to 140°C, to cross-link the linear rubber polymer into a three-dimensional polymer.

Exercises

Geometries of Organic Molecules

1. Name two types of constitutional isomerism. Give two isomers of each type.

2. Name two types of stereoisomerism. Give an example of each type.

3. Which of the following compounds can exist as *cis* and *trans* isomers? Draw them. (a) 2,3-dimethyl-2-butene, (b) 2-chloro-2-butene, (c) dichlorobenzene, (d) 1,1-dichlorocyclobutane.

4. Which of the following compounds can exist as *cis* and *trans* isomers? Draw them. (a) 1-butene, (b) 2-bromo-1-butene, (c) 2-bromo-2-butene, (d) 1,2-dichlorocyclopentane.

5. Distinguish between conformations and isomers.

6. Define optical isomerism. To what do the terms "levorotatory" and "dextrorotatory" refer?

7. Which of the following compounds would exhibit optical isomerism?

(a) $CH_3CHCH_2CH_3$
   |
   Br

(b) CH_3CHCH_3
   |
   OH

(c) HO—⟨◯⟩—C—⟨◯⟩ with H above and Cl below the central C

(d) $CH_3CH{=}CHCH_3$

8. Draw three-dimensional representations of the enantiomeric pairs in Exercise 7.

9. Write formulas and names for the isomers of (a) bromochlorobenzene, (b) trifluorobenzene, and (c) chlorotoluene. What kind of isomerism is illustrated by each of these sets of compounds?

Reactions of Organic Molecules

10. (a) What is a substitution reaction? (b) What is a halogenation reaction? (c) What is a free-radical chain reaction? (d) What is an addition reaction?

11. (a) Describe the reaction of methane with chlorine in ultraviolet light. (b) Write equations that show formulas for all compounds that can be formed by reaction (a). (c) Write names for all compounds in these equations. (d) Why are the halogenation reactions of the larger alkanes of limited value?

12. (a) Describe the reaction of ethane with chlorine in ultraviolet light. (b) Write equations that show formulas for all compounds that can be formed by reaction (a). (c) Write names for all compounds in these equations.

13. Which of the following compounds could undergo addition reactions? (a) propane, (b) 1,3-butadiene, (c) cyclopentene, (d) acetone.

14. Most reactions of the alkanes that do not disrupt the carbon skeleton are substitution reactions, whereas the alkenes are characterized by addition to the double bond. What does this statement mean?

15. How can bromination be used to distinguish between alkenes and alkanes?

16. Write equations for two reactions in which alkenes undergo addition reactions with halogens. Name all compounds.

17. What is the difference between a vegetable oil and a shortening? Illustrate.

18. (a) What is hydrogenation? (b) Why is it important? (c) Write equations for two reactions that involve hydrogenation of alkenes. (d) Name all compounds in part (c).

19. Describe two qualitative tests that can be used to distinguish between alkenes and alkanes. (b) Cite some specific examples. (c) What difference in reactivity is the basis for the qualitative distinction between alkanes and alkenes?

20. (a) Why are alkynes more reactive than alkenes? (b) What is the most common kind of reaction that alkynes undergo? (c) Write equations for three such reactions. (d) Name all compounds in part (c).

21. (a) What is the most common kind of reaction that the benzene ring undergoes? (b) Write equations for the reaction of benzene with chlorine in the presence of an iron catalyst and for the analogous reaction with bromine.

22. Write equations to illustrate both aromatic and aliphatic substitution reactions of toluene using (a) chlorine, (b) bromine, and (c) nitric acid.

23. Write the structural formula of the organic compound formed in each of the following reactions.

(a) CH_3—⟨◯⟩—CH_3 $\xrightarrow[\text{H}_2\text{O, heat}]{\text{excess KMnO}_4}$

(b) $CH_3CH{=}CHCH_3$ $\xrightarrow{\text{HBr}}$

(c) $C_6H_5CHBrCH_2Br$ $\xrightarrow{\text{Zn}}$

24. Write the structural formula for the organic compound or the carbon-containing compound formed in each of the following reactions.

(a) $CH_4 + O_2(\text{excess})$ $\xrightarrow{\Delta}$

(b) $CH_4 + Cl_2$ $\xrightarrow{\text{light}}$

(c) $CH_2{=}CH_2$ $\xrightarrow{\text{H}_2\text{SO}_4}$

25. Repeat Exercise 24 for

(a) $HC{\equiv}CH + O_2(\text{excess})$ $\xrightarrow{\Delta}$

(b) $CH_3CH_2OH + Na$ $\longrightarrow$

(c) $CH_3CH{=}CH_2$ $\xrightarrow{\text{HCl}}$

(d) ⟨◯⟩ $\xrightarrow[\text{Fe}]{\text{Br}_2}$

26. Suppose you have three test tubes. The first contains either hexane or 2-hexene, the second contains either benzene or styrene ($C_6H_5CH{=}CH_2$), and the third contains either cyclohexene or 2-bromopropane. Describe a simple chemical test that would enable you to determine visually which compound is present in each test tube.

27. (a) What are alkyl hydrogen sulfates? (b) How are they prepared? (c) Can alkyl hydrogen sulfates be classified as acids? Why? (d) What are detergents? (e) What is sodium lauryl sulfate? (f) What is the common use of sodium lauryl sulfate?

28. (a) What are inorganic esters? (b) Write equations for the formation of three inorganic esters. (c) Name the inorganic ester formed in each case.

29. (a) What is nitroglycerine? (b) Write the chemical equation that shows the preparation of nitroglycerine. (c) List two important uses for nitroglycerine. Are they similar?

30. Classify each reaction as substitution, addition, or elimination.

(a) $CH_3CH_3 + Br_2 \xrightarrow{\text{light}} CH_3CH_2Br + HBr$

(b) $CH_3CH{=}O + HCN \longrightarrow CH_3\overset{\displaystyle H}{\underset{\displaystyle CN}{C}}{-}OH$

(c) $CH_3CH{=}CH_2 + HOCl \longrightarrow CH_3\underset{\displaystyle OH}{CH}CH_2Cl$

31. Classify each reaction as substitution, addition, or elimination.

(a) $CH_3CH_2Br + CN^- \longrightarrow CH_3CH_2CN + Br^-$

(b) $CH_3\underset{\displaystyle Br}{CH}CH_2Br + Zn \longrightarrow CH_3CH{=}CH_2 + ZnBr_2$

(c) $C_6H_6 + HNO_3 \xrightarrow{H_2SO_4} C_6H_5NO_2 + H_2O$

32. Why are aqueous solutions of amines basic? Show, with equations, how the dissolution of an amine in water is similar to the dissolution of ammonia in water.

33. Show that the reaction of amines with inorganic acids such as HCl are similar to the reactions of ammonia with inorganic acids.

*34. What are the equilibrium concentrations of the species present in a 0.10 M solution of aniline? $K_b = 4.2 \times 10^{-10}$

$$C_6H_5NH_2(aq) + H_2O(\ell) \rightleftharpoons C_6H_5NH_3^+ + OH^-$$

*35. Which solution would be the more acidic: a 0.10 M solution of aniline hydrochloride, $C_6H_5NH_3Cl$ ($K_b = 4.2 \times 10^{-10}$ for aniline, $C_6H_5NH_2$), or a 0.10 M solution of methylamine hydrochloride, CH_3NH_3Cl ($K_b = 5.0 \times 10^{-4}$ for methylamine, CH_3NH_2)? Justify your choice.

36. Choose the compound that is the stronger acid in each set.

(a) $CH_3CH_2CH_2OH$ or   (cyclohexyl)$\overset{\displaystyle O}{-}C{-}OH$

(b) CH_3CH_2OH or   (phenyl)$-OH$

(c) (phenyl)$-OH$ or   (phenyl)$\overset{\displaystyle O}{-}C{-}OH$

(d) (cyclohexyl)$-OH$ or   (phenyl)$-OH$

37. (a) What are alkoxides? (b) What do we mean when we say that the low-molecular-weight alkoxides are strong bases?

38. (a) Write equations for the reactions of three alcohols with metallic sodium. (b) Name all compounds in these equations. (c) Are these reactions similar to the reaction of metallic sodium with water? How?

39. Which physical property of aldehydes is used to advantage in their production from alcohols?

40. Describe a simple test to distinguish between the two isomers 2-pentene and cyclopentane.

41. How are the terms "oxidation" and "reduction" often used in organic chemistry? Classify the following changes as either oxidation or reduction: (a) CH_4 to CH_3OH, (b) $CH_2{=}CH_2$ to $CH_3{-}CH_3$, (c) CH_3CH_2CHO to $CH_3CH_2CH_3$.

(d) (phenyl)$-CH_3$ to   (phenyl)$\overset{\displaystyle O}{-}C{-}OH$

42. Classify the following changes as either oxidation or reduction: (a) CH_3OH to CO_2 and H_2O, (b) CH_2CH_2 to CH_3CHO, (c) CH_3COOH to CH_3CHO, (d) $CH_3CH{=}CH_2$ to $CH_3CH_2CH_3$.

43. Write equations to illustrate the oxidation of the following aromatic hydrocarbons by potassium permanganate in basic solution: (a) toluene, (b) ethylbenzene, (c) 1,2-dimethylbenzene.

44. (a) Do you expect aromatic hydrocarbons to produce soot as they burn? Why? (b) Would you expect the flames to be blue or yellow?

45. Describe the preparation of three aldehydes from alcohols and write appropriate equations. Name all reactants and products.

46. Describe the preparation of three ketones from alcohols and write appropriate equations. Name all reactants and products.

47. Write equations for the formation of three different esters, starting with a different acid chloride and a different alcohol in each case. Name all compounds.

48. Write equations for the formation of three different esters, starting with an acid and an alcohol in each case. Name all compounds.

49. (a) What is saponification? (b) Why is this kind of reaction called saponification?

50. Write equations for the hydrolysis of (a) methyl acetate, (b) ethyl formate, (c) butyl acetate, and (d) octyl acetate. Name all products.

Polymers

51. What is a polymer? What is the term for the smaller molecule that serves as the repeating unit making up a polymer? What are typical molecular weights of polymers?

52. (a) What is polymerization? (b) Write equations for three polymerization reactions.

53. Give an example of a condensation reaction. What is the essential feature of monomers used in condensation polymerizations?

*54. The examples of condensation polymers given in the text are all copolymers, that is, they contain two different monomers. Is it possible for a single monomer to polymerize so as to form a condensation *homo*polymer? If so, suggest an example. If not, explain why not.

55. Poly(vinyl alcohol) has a relatively high melting point, 258°C. How would you explain this behavior? A segment of the polymer is

$$\cdots CH_2CHCH_2CHCH_2CHCH_2CHCH_2CH \cdots$$
$$\qquad | \quad\;\; | \quad\;\; | \quad\;\; | \quad\;\; |$$
$$\qquad OH \;\; OH \;\; OH \;\; OH \;\; OH$$

56. What changes could be made in the structures of polymer molecules that would increase the rigidity of a polymer and raise its melting point?

57. Methyl vinyl ketone, $CH_3\overset{\overset{\displaystyle O}{\|}}{C}CH{=}CH_2$, can be polymerized by addition polymerization. The addition reaction involves only the C=C bond. Write the molecular structure of a four-unit segment of this polymer.

58. (a) What is rubber? (b) What is vulcanization? (c) What is the purpose of vulcanizing rubber? (d) What are fillers and reinforcing agents? (e) What is their purpose?

59. (a) What is an elastomer? (b) Cite a specific example. (c) What are some of the advantages of neoprene compared with natural rubber?

60. What are polyamides? What kind of reaction forms polyamides?

61. (a) What are polyesters? (b) What is Dacron? (c) How is Dacron prepared? (d) What is Mylar? (e) Is it reasonable to assume that a polyester can be made from propylene glycol and terephthalic acid? If so, sketch its structure.

62. Suppose the following diol is used with terephthalic acid to form a polyester. Sketch the structure of the polymer, showing two repeating units.

$$HOCH_2{-}\!\!\bigcirc\!\!{-}CH_2OH$$

63. Give the structural formula of the monomer used in preparation of each of the following polymers.

(a) $\cdots{-}\overset{|}{\underset{|}{C}}{-}CH_2{-}\overset{|}{\underset{|}{C}}{-}CH_2{-}\overset{|}{\underset{|}{C}}{-}CH_2{-}\cdots$
$\qquad\quad CH_3 \qquad CH_3 \qquad CH_3$

(b) $\cdots{-}\overset{CH_3}{\underset{CH_3}{C}}{-}CH_2{-}\overset{CH_3}{\underset{CH_3}{C}}{-}CH_2{-}\overset{CH_3}{\underset{CH_3}{C}}{-}CH_2{-}\cdots$

64. Is it possible to produce a copolymer by addition polymerization? If so, give an example. If not, explain why not.

65. (a) What is Nylon? (b) How is it prepared?

66. Common Nylon is called Nylon 66. (a) What does this mean? (b) Write formulas for two other possible Nylons.

*67. A cellulose polymer has a molecular weight of 750,000 g/mol. Estimate the number of units of the monomer, β-glucose

($C_6H_{12}O_6$) in this polymer. This polymerization reaction can be represented as

$$xC_6H_{12}O_6 \longrightarrow cellulose + (x-1)H_2O$$

68. What is necessary if a molecule is to be capable of polymerization? Name three types of molecules that can polymerize.

69. Describe the structure of a natural amino acid molecule. What kind of isomerism do most amino acids exhibit? Why?

70. How are the amino acid units in a polypeptide joined together? What are the links called?

71. Consider only two amino acids:

$$NH_2{-}\overset{H}{\underset{R'}{C}}{-}COOH \quad and \quad NH_2{-}\overset{H}{\underset{R}{C}}{-}COOH$$

Write the structural formulas for the dipeptides that could be formed containing one molecule of each amino acid.

*72. How many different dipeptides can be formed from the three amino acids A, B, and C? Write the sequence of amino acids in each. Assume that an amino acid could occur more than once in each dipeptide.

*73. How many different tripeptides can be formed from the three amino acids A, B, and C? Write the sequence of amino acids in each. Assume that an amino acid could occur more than once in each tripeptide.

74. Aspartame (trade name NutraSweet) is a methyl ester of a dipeptide:

$$\bigcirc\!\!{-}CH_2{-}\overset{COOCH_3}{\underset{\qquad}{CH}}{-}NH{-}\overset{\overset{\displaystyle O}{\|}}{C}{-}\overset{CH_2COOH}{\underset{\qquad}{CH}}{-}NH_2$$

Write the structural formulas of the two amino acids that are combined to make aspartame (neglecting optical isomerism).

Mixed Exercises

75. Identify the major products of each reaction.

(a) $+\ 2NaOH \longrightarrow$

(b) $\bigcirc\!\!{-}\overset{\overset{\displaystyle O}{\|}}{C}{-}OH + CH_3OH \;\overset{H_2SO_4}{\underset{\Delta}{\longrightarrow}}$

(c) $CH_3\overset{\overset{\displaystyle O}{\|}}{C}H \;\overset{MnO_4{}^-}{\underset{H^+}{\longrightarrow}}$

76. Identify the major products of each reaction.

(a) —CH$_2$OH + Na $\longrightarrow$

(b) CH$_3$CH$_2$$\overset{\overset{\textstyle O}{\|}}{C}OCH_3$ $\xrightarrow[\Delta]{\text{KOH(aq)}}$

(c) $\xrightarrow[\Delta]{\text{NaOH(aq)}}$

77. Why is it hazardous to burn a hydrocarbon fuel in an insufficient supply of oxygen?

78. How does the heat of combustion of ethyl alcohol compare with the heats of combustion of low-molecular-weight saturated hydrocarbons on a per-mole basis and on a per-gram basis?

BUILDING YOUR KNOWLEDGE

79. A laboratory procedure calls for oxidizing 2-propanol to acetone using an *acidic* solution of K$_2$Cr$_2$O$_7$. However, an insufficient amount of K$_2$Cr$_2$O$_7$ is on hand, so the laboratory instructor decides to use an acidic solution of KMnO$_4$ instead. What mass of KMnO$_4$ is required to carry out the same amount of oxidation as 1.00 g of K$_2$Cr$_2$O$_7$?

80. The chemical equation for the water gas reaction is

$$C(s) + H_2O(g) \rightleftharpoons CO(g) + H_2(g)$$

At 1000 K, the value of K_p for this reaction is 3.2. When we treat carbon with steam and allow the reaction to reach equilibrium, the partial pressure of water vapor is observed to be 15.6 atm. What are the partial pressures of CO and H$_2$ under these conditions?

81. (a) In aqueous solution, acetic acid exists mainly in the molecular form ($K_a = 1.8 \times 10^{-5}$). (a) Calculate the freezing point depression for a 0.10 molal aqueous solution of acetic acid, neglecting any ionization of the acid. $K_f = 1.86°C$/molal for water. (b) In nonpolar solvents such as benzene, acetic acid exists mainly as dimers

as a result of hydrogen bonding. Calculate the freezing point depression for a 0.10 molal solution of acetic acid in benzene. $K_f = 5.12°C$/molal for benzene. Assume complete dimer formation.

82. What is the pH of a 0.10 M solution of sodium benzoate? $K_a = 6.3 \times 10^{-5}$ for benzoic acid, C$_6$H$_5$COOH. Would this solution be more or less acidic than a 0.10 M solution of sodium acetate? $K_a = 1.8 \times 10^{-5}$ for acetic acid, CH$_3$COOH.

83. Nylon is decomposed by acids, while polyethylene is not. Suggest an explanation for this difference in behavior.

84. A 2.30-g sample of poly(vinyl alcohol) was dissolved in water to give 101 mL of solution. The osmotic pressure of the solution was 55 torr at 25°C. Estimate the molecular weight of the poly(vinyl alcohol).

85. (a) What is the osmotic pressure of a 1.00% aqueous solution of sucrose, C$_{12}$H$_{22}$O$_{11}$, at 25°C? (b) To what height will the column rise under this pressure? Assume that the density of the solution is 1.00 g/mL.

Qualitative Analysis

Some precipitates observed in qualitative analysis. *Clockwise* from the top: $MgNH_4AsO_4$ (white), CuS (black), Ni(HDMG)$_2$ (bright red), Sb_2S_3 (orange-red), CdS (yellow), and MnS (salmon). Some oxidation of MnS occurs under the photographer's hot lights.

29 Metals in Qualitative Analysis

A photograph of a basic oxygen furnace in operation. Oxygen is blown through molten pig iron (Section 22-7) to oxidize phosphorus, sulfur, and most of the excess carbon. (This is the largest industrial use of oxygen.) The product is called carbon steel; it has several industrial uses. The addition of transition metals to molten carbon steel converts it to stainless steels, which have many more uses.

Many of the metals whose ions are included in the qualitative analysis scheme for cations have already been discussed. We described some important metals and their metallurgy in Chapters 22 and 23. Coordination compounds were covered in Chapter 25. The discussions that follow will refer to the appropriate sections of those chapters.

For each group of metals, we shall (1) tabulate some important properties, (2) discuss some of these properties, (3) indicate naturally occurring sources, (4) briefly describe their metallurgies, and (5) mention a few uses.

29-1 THE METALS OF ANALYTICAL GROUP I

Some properties of the metals of Analytical Group I are tabulated in Table 29-1.

Copper

Copper is a relatively soft, reddish-yellow metal that is an excellent conductor of heat and electricity. Copper was known to many ancient civilizations, and it was probably the first metal used to fabricate tools and utensils.

Table 29-1 *Some Properties of the Metals of Analytical Group I*

Metal	Atomic Number	Periodic Group Number	Atomic Weight	Atomic Radius (Å)	Important Oxidation States	Ionic Radius (Å)	Density at 20°C (g/cm³)	Compounds in Ores
copper	29	IB	63.546	1.28	+1, +2	Cu^{2+}, 0.70	8.9	Cu, Cu_2S, CuS, Cu_2O, CuO, $CuFeS_2$, $Cu(OH)_2 \cdot CuCO_3$
cadmium	48	IIB	112.411	1.54	+2	Cd^{2+}, 0.97	8.7	CdS
bismuth	83	VA	208.9804	1.46	+3, +5	Bi^{3+}, 0.96	9.8	Bi, Bi_2O_3, Bi_2S_3

Four pieces of jewelry made from copper minerals. Clockwise from top: malachite with azurite, malachite, azurite with chalcopyrite, and turquoise, $CuAl_6(PO_4)_4(OH)_8 \cdot 5H_2O$.

This bronze finial comes from Luristan (Persia) and dates from about 1200 B.C.

The fact that copper utensils were used by Native Americans in the southern part of the country is taken as evidence for trade among widely separated tribes.

Copper sometimes occurs as the free metal. The largest known deposits were near Houghton, Michigan. The largest piece found to date weighed more than 400 tons.

The commercially important ores of copper are copper(I) sulfide (Cu_2S, *chalcocite*) and a mixed sulfide ($CuFeS_2$, *chalcopyrite*), which contain less than 10% copper. Most known sources of the high-grade oxide ores—*cuprite*, Cu_2O, and *melaconite*, CuO—have been exhausted. The metallurgy of copper was discussed in Section 22-8. It is refined electrolytically (Section 21-8).

Copper is second in importance to iron, the most widely used metal. Copper has excellent thermal and electrical conductivity, chemical inertness, and usefulness in alloying elements. It is used extensively in alloys such as brass (Cu-Zn), bronze (Cu-Zn-Sn), sterling silver (Cu-Ag), aluminum bronze (Cu-Al), and German silver (Cu-Zn-Ni).

Copper is used in the manufacture of U.S. coins, both pennies and cladded-copper dimes and quarters.

Copper pipes are used in plumbing because copper is easy to "work" and because it does not react with hot or cold water at an appreciable rate. Copper statues and roofs turn brown as a thin adherent film of copper oxide or copper sulfide forms. After long exposure to the atmosphere, they turn green due to the formation of basic copper(II) carbonate, $Cu(OH)_2 \cdot CuCO_3$.

Human need for trace amounts of copper has been demonstrated (Chapter 23). Trace amounts of copper are found in plants that grow near deposits of copper ores. A copper compound (hemocyanin) serves the same oxygen-carrying function in the "blood" of lobsters and oysters that the iron compound hemoglobin serves in the blood of higher animals.

Cadmium sulfide, CdS, is used as a paint pigment.

Cadmium

Cadmium is quite similar to zinc, but it is softer and more malleable and ductile. It is chemically less reactive than zinc.

$$Cd^{2+}(aq) + 2e^- \longrightarrow Cd(s) \qquad E^0 = -0.403 \text{ V}$$

It is used to plate iron when the need for a high-quality coating justifies the cost. Cadmium is used in nicad batteries and as a stabilizer in plastics. Cadmium sulfide is used as a paint pigment.

Cadmium is obtained primarily from zinc smelters and from the sludge produced by the electrolytic refining of zinc. Many zinc ores contain small amounts (<1%) of cadmium sulfide. A rare mineral, *greenockite*, contains a high percentage of CdS.

Cadmium is used in some low-melting alloys such as *Wood's metal* (mp = 65°C), which is 50% Bi, 25% Pb, 12.5% Sn, and 12.5% Cd. It is also used in the manufacture of some antifriction bearings (Cd-Ni) that have higher melting points than Babbit metal (Sb-Sn-Cu) bearings.

Cadmium-plated iron is more resistant to attack by salt water and alkaline solutions than is zinc-plated (galvanized) iron.

Bismuth

Bismuth is the heaviest of the known Group VA elements. Its physical and chemical properties are distinctly metallic. It is a hard, brittle metal with a reddish tint, and it exhibits the unusual property of expanding as it solidifies (as does antimony). It occurs as the free element as well as the oxide, Bi_2O_3, called *bismuth ocher*, and the sulfide, Bi_2S_3, called *bismuth glance*. Because bismuth occurs in many lead ores, its principal commercial source is as a by-product of the refining of lead.

The uses of bismuth are based on the facts that it expands when it solidifies and that some bismuth alloys have very low melting points. Where sharp, well-defined edges of castings are important (in type metals, for example), bismuth and antimony are used as alloying agents. Bismuth alloys are used in automatic sprinkler systems, electrical fuses, and safety plugs for boilers in which low-melting alloys are essential.

Electrical fuses are safety devices based on the low-melting points of bismuth alloys.

29-2 THE METALS OF ANALYTICAL GROUP II

Some properties of the metals of Analytical Group II are tabulated in Table 29-2.

Arsenic

Arsenic and antimony are Group VA metalloids. Arsenic is located just above, and antimony just below, the arbitrary line that separates metals and nonmetals. Arsenic exhibits many properties that are characteristic of nonmetals, but it also exhibits some properties of metals. Arsenic(III) hydroxide (arsenous acid) is amphoteric (Table 10-2). Antimony is decidedly more metallic than arsenic, as is expected from its lower position in the periodic table.

(Text continues on page 1060.)

The mineral orpiment, As_2S_3.

Table 29-2 *Some Properties of the Metals of Analytical Group II*

Metal	Atomic Number	Periodic Group Number	Atomic Weight	Atomic Radius (Å)	Important Oxidation States	Ionic Radius (Å)	Density at 20°C (g/cm³)	Compounds in Ores
arsenic	33	VA	74.9216	1.21	−3, +3, +5	As^{3+}, 0.58	5.7	As_2S_3, As_2S_2, FeAsS
tin	50	IVA	118.710	1.40	+2, +4	Sn^{2+}, 0.93 Sn^{4+}, 0.71	7.3	SnO_2
antimony	51	VA	121.75	1.41	+3, +5	Sb^{3+}, 0.76	6.7	Sb_2S_3

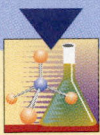

Metal Clusters

One of the frontiers of research in chemistry and physics today is the study of microscopic particles of metals. Particles so small that they contain only a few atoms are referred to as metal atom clusters or simply *metal clusters*. Metal clusters can be indicated with chemical formulas just like conventional molecules. For example, the clusters Ni_3, Fe_5, and Ag_{17} contain 3, 5, and 17 atoms of nickel, iron, and silver, respectively, but no other elements. It is not yet clear whether clusters of metal atoms represent small pieces of solid metal, or whether they are so different that they have a whole family of unusual properties all of their own, like a new class of molecules. Until recently it was pointless to ask such questions because no one could make metal particles this small. Now, however, laboratories around the world can make clusters out of any metal (or other element) in the periodic table. Systematic studies are under way to measure the physical and chemical properties of these unusual species and to determine what their applications might be.

The studies that eventually made cluster research possible were motivated by the need for new solid materials for the chemical and microelectronics industries. In the chemical industry, small metal particles are used as *catalysts* (Section 16-9). When a catalytic metal powder is added to a mixture of chemicals that is about to react, the reactant molecules stick to the metal surface, and the reaction processes can be modified or controlled. Reactions can be made to occur more rapidly, to produce different products, or to minimize unwanted by-products. The activity of a catalyst increases as its particle size decreases. Because they are so small, metal clusters may represent the ultimate metal catalysts. Catalysts are especially important in the petroleum industry; they are used to convert the various components of crude oil into useful products. If catalysis were better understood or could be improved by the use of metal clusters, synthetic chemicals and/or energy would be far less expensive.

In the microelectronics industry, a major goal is to make integrated circuits as small as physically possible. New production techniques are needed to produce "nanoscale" transistors, wires, resistors, and so on, which means that these devices are as small as molecules. They can be made in the right sizes and shapes by "plating," or "depositing," metal clusters onto a supporting surface. However, many questions remain unanswered about such small samples of materials. For example, do microscopically thin wires conduct electricity as easily as larger ones? How small can a piece of semiconductor become without ceasing to function? When electrical current flows, does the heating cause small particles to melt or to move on their support? Investigation of the fundamental properties of metal clusters will provide answers for these and many other technical questions.

Clusters are most often produced by a new technique known as *laser vaporization* (Figure a). In this method, a solid piece of the metal to be studied is mounted in a special holder inside a vacuum chamber. A high-powered pulse of laser light, usually from a Nd:YAG laser, is focused with a lens onto the surface of the metal in much the same way that concentrated sunlight can be focused with a magnifying glass to ignite a piece of paper. However, the laser light is so powerful that it generates a temperature of about 10,000 K, which is enough to vaporize a small amount of the metal. The result is an extremely hot vapor composed of metal atoms. A burst of helium gas at room temperature is squirted through the sample holder to cool the metal vapor. Helium does not react, even with hot metal atoms, but through thousands of collisions it cools the metal vapor to near room temperature. As they cool, metal atoms recombine and condense to form solid material again. It is this recombination of atoms that produces the clusters. If allowed to, the recombining metal vapor would plate out again on the surface from which it came. However, the flowing helium gas pushes the small metal particles out into the vacuum chamber in the form of an expanding gaseous spray. The spray, known as a *molecular beam*, is where cluster experiments are conducted. For example, if the cluster molecular beam is sprayed into a mass spectrometer, the masses, and therefore the sizes, of the clusters formed can be measured. Figure (b) shows a mass spectrum of silver clusters produced by this kind of experiment. The various molecules formed with two, three, four, . . . atoms are called *dimers, trimers, tetramers*, and so on. If the cluster spray is intersected by another laser beam, the absorption spectrum of the clusters can be measured to determine their structures (overall shapes, bond lengths, bond angles). Other experiments with lasers and mass spectrometers measure the energy required to ionize clusters of different sizes or to break the chemical bonds that hold cluster atoms together.

Although clusters of virtually any metal can be formed and studied, these molecules are not "stable" in the usual sense of the word. Aggregated atoms in a cluster, not as stable as more conventional molecules, are still far more stable than the same atoms separated, and so clusters do not spontaneously fly apart. But it is usually not possible to draw simple bonding schemes such as Lewis dot formulas for these kinds of molecules. One exception occurs for dimers of alkali metals, in which each of the atoms has a single *s* electron in its valence shell, just like a hydrogen atom. Like diatomic hydrogen, H_2, the molecules Li_2, Na_2, K_2, and so on are held together by a two-electron covalent single bond. The same is true for the coinage metal dimers, Ag_2, Cu_2, and Au_2, which have a filled *d* electron shell not involved in the bonding and a single *s* electron in the valence shell that forms the covalent bond.

Large particles of sodium, silver, and copper are well-known conductors of electricity, but interestingly the diatomic particles cannot conduct because they contain no free electrons. Transition metals have more complex bonding schemes. For example, a chromium atom has five *d* electrons and one *s* electron in its

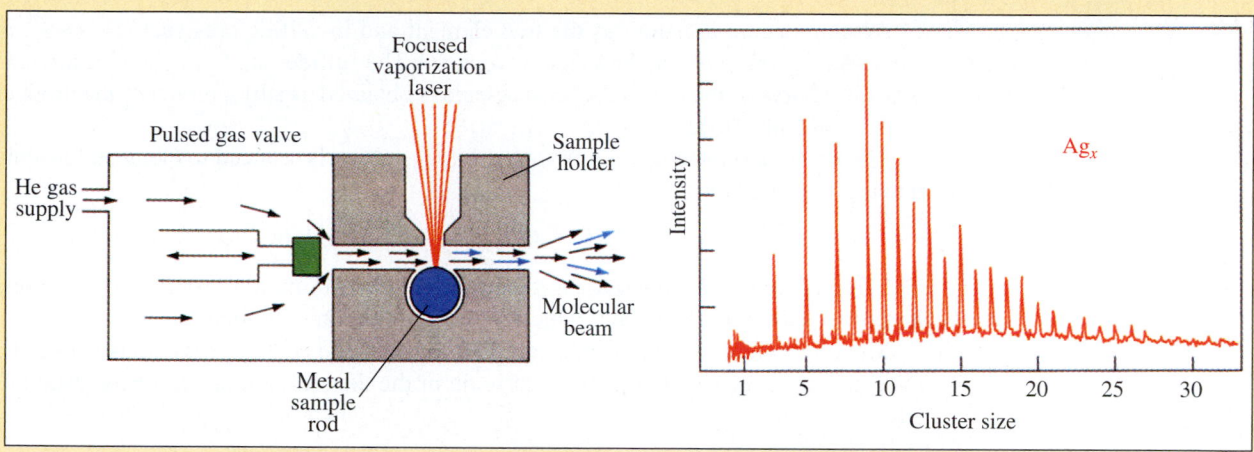

(a) Schematic diagram of the laser vaporization cluster source.

(b) Mass spectrum of silver clusters produced by laser ionization. As shown, not all clusters are produced in equal abundance.

valence shell, and all of these are used in Cr_2, which has a *sextuple* bond! Iron has six *d* electrons and two *s* electrons, but Fe_2 has only a single bond. The "unused" electrons in Fe_2 give it "unsatisfied" bonding capacity, and it is therefore an extremely reactive molecule. When clusters larger than dimers are considered, the chemical bonding becomes too complex for simple description. Some of the electrons are used to hold the atoms together in chemical bonds, and some are delocalized over the surface and volume of the cluster as it begins to "look like" a piece of solid metal. Even with the best computer models available today, it is impossible to predict the chemical bonding scheme in a cluster of about 10 atoms.

Closely related to the chemical bonding in any cluster is its geometric structure. The atoms of solid metals usually are arranged in orderly crystal structures—often close-packed hexagonal or cubic networks in which each atom has 12 nearest neighbors. If these structural patterns were drawn for clusters, they would have sharp edges and flat faces. In reality, effects such as surface tension combine with chemical bonding forces to give clusters a smoother exterior. One especially stable arrangement of cluster atoms that occurs frequently is the 13-atom *icosahedron* shown in Figure (c). This structure resembles close-packed structures in that it has one atom surrounded by 12 nearest neighbors, but its exterior is essentially spherical. Spherical and elliptical structures are commonplace in cluster structures. It is interesting to note that 12 of the 13 atoms in this icosahedral structure are on the surface of the "particle." Even in a cluster with 100 atoms, about 80 are on the surface. This great surface area, where other molecules can stick and react, makes clusters all the more interesting as potential catalysts. One twist is that chemical bonds in clusters are often weak, making the bonding network easy to disrupt. When this begins to occur, cluster bonds may break and re-form rapidly as though the particle were a liquid droplet instead of a solid! This behavior, which is like

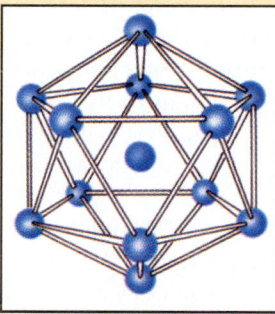

(c) The 13-atom icosahedron structure of many metal clusters.

melting in larger particles, requires a characteristic temperature that depends on the size of the cluster and the stability of the bonding arrangement. If they are heated enough, as occurs with laser excitation, clusters may "evaporate" and lose atoms.

As you can see, many of the simple concepts used in chemistry can be applied to clusters, but because of their small size, the concepts must be modified. Full understanding of cluster properties is only now beginning to emerge. Research is still complicated by the conditions under which clusters are studied. Molecular beam laser vaporization can synthesize virtually any kind of cluster, making it possible to study molecules never before known, but it does not make them in large quantities. In the region of experimentation, there are only about 10^8 clusters per cm^3 (about 10^{-16} moles!). At these concentrations, even the most sensitive measurement techniques available are sometimes not good enough. When clusters can be produced in greater quantities or new techniques to measure them can be developed, more of their unusual properties will be revealed. The strange world of cluster research may then become even more fascinating than it already is.

Professor Michael A. Duncan
The University of Georgia

Arsenic occurs in nature as the free element and in sulfide ores such as As_2S_3, *orpiment*; As_2S_2, *realgar*; and FeAsS, *arsenopyrite*. The sulfides are found in trace amounts in the sulfide ores of many metals. Most arsenic is obtained as a by-product of the production of other metals, notably copper.

The roasting of arsenic ores produces As_4O_6, which is reduced to elemental arsenic by heating with carbon.

$$As_4O_6(s) + 6C(s) \longrightarrow As_4(g) + 6CO(g)$$

The free arsenic is condensed and then purified by sublimation. Native arsenic ores are heated to sublime the arsenic, which is then purified by sublimation.

Most important uses of arsenic compounds are based on their poisonous nature. However, the need for small amounts of arsenic in the diets of animals has been established.

Antimony

Antimony is a brittle, lustrous metalloid with considerably more metallic character than arsenic. It is sometimes found as the free element. Most antimony is obtained from the sulfide ore, *stibnite*, in which Sb_2S_3 occurs as a black solid. The orange-red modification of this compound was used for cosmetic purposes at least 5000 years ago. Antimony exhibits amphoterism in both the +3 and +5 oxidation states. The metallurgy of antimony is similar to that of arsenic.

The principal uses of antimony depend on the facts that it expands when it solidifies (as does bismuth, with which it is used in type metal) and that it is relatively unreactive chemically. The "lead" plates used in lead storage batteries are 94% lead and 6% antimony. The small amount of antimony makes the plates much more resistant to attack by H_2SO_4.

Antimony alloys such as Babbit metal (Sb-Sn-Cu) are used extensively as antifriction machine bearings. Shrapnel is lead that has been hardened by the addition of 10% to 20% antimony.

Tin

Animal need for small amounts of tin has been established (Chapter 23).

Tin is a Group IVA element located near the arbitrary dividing line between metals and nonmetals. It is decidedly metallic in character, but its hydroxides are amphoteric (Table 10-2), which indicates some nonmetallic character.

Tin is relatively soft and ductile. Its excellent resistance to corrosion (reaction with H_2O and O_2) is the basis for its primary uses. Tin vessels have been found in ancient Egyptian tombs. The Romans and Phoenicians used tin that they obtained from the cassiterite deposits in England.

Tin is obtained commercially from *tinstone*, or *cassiterite*, SnO_2. The crushed ore is separated from the lighter rocky material and then roasted to remove arsenic and sulfur as volatile oxides. Oxides of other metals are extracted with hydrochloric acid, in which SnO_2 is insoluble. The remaining ore is then reduced with carbon in a furnace.

$$SnO_2(s) + 2C(s) \longrightarrow Sn(\ell) + 2CO(g)$$

Molten tin is drained off at the bottom of the furnace and cast into blocks. The blocks are then remelted so that the tin flows away from impurities that have higher melting points. Final purification is by electrolysis in a bath of sulfuric acid and hexafluorosilicic acid, $H_2[SiF_6]$. Pure tin cathodes and impure tin anodes are used. The process is similar to the electrolytic refining of copper (Section 21-8).

Tin is used primarily in corrosion protection, i.e., as tin plate for sheet iron. "Tin cans" are protected by a very thin layer of tin. Tin is also used in certain alloys such as bronze (Cu-Zn-Sn) and solder (Sn-Pb).

29-3 THE METALS OF ANALYTICAL GROUP III

Table 29-3 lists some properties of the metals of Analytical Group III.

Aluminum is the only A group metal whose cations occur in Analytical Group III. Therefore, the chemistry of the Analytical Group III cations is largely the chemistry of d-transition metal cations. Most d-transition metal ions form coordination compounds readily.

The metals of Analytical Group III are all classified as "active metals." The production of the free metals requires large amounts of energy. Refer to Appendix J for E^0 values.

Cobalt

Cobalt, like nickel and iron, is magnetic. It looks very much like iron except that it appears slightly pink. It is rendered passive by contact with concentrated nitric acid, as are iron and nickel.

Human need for trace amounts of cobalt has been established (Chapter 23).

There are deposits of cobalt ores containing CoAsS, *cobaltite*, and $CoAs_2$, *smaltite*. Most cobalt is obtained as a by-product of the metallurgy of copper, iron, nickel, silver, and other metals that occur as sulfide ores. After cobalt compounds have been converted to Co_3O_4 and isolated (quite a complex process), the oxide is reduced by metallic aluminum in a *highly* exothermic process.

$$3Co_3O_4(s) + 8Al(s) \xrightarrow{\Delta} 9CO(\ell) + 4Al_2O_3(s) + 4029 \text{ kJ/mol rxn}$$

The metal is purified electrolytically.

Cobalt is alloyed with iron and other metals to produce very hard materials that are used for high-speed cutting tools, surgical instruments, and other tools and instruments that must resist corrosion. Permanent magnets are made of alloys such as Alnico V (8% Al, 14% Ni, 24% Co, 3% Cu, and 51% Fe). Many cobalt compounds are used as contact catalysts.

Table 29-3 *Some Properties of the Metals of Analytical Group III*

Metal	Atomic Number	Periodic Group Number	Atomic Weight	Atomic Radius (Å)	Important Oxidation States	Ionic Radius (Å)	Density at 20°C (g/cm^3)	Compounds in Ores
aluminum	13	IIIA	26.98154	1.43	+3	Al^{3+}, 0.50	2.7	Al_2O_3
chromium	24	VIB	51.9961	1.27	+3, +6	Cr^{3+}, 0.61	7.1	$FeCr_2O_4$
manganese	25	VIIB	54.9380	1.26	+2, +3, +4, +7	Mn^{2+}, 0.80	7.2	MnO_2, Mn_2O_3, Mn_3O_4, mixed oxides
iron	26	VIIIB	55.847	1.26	+2, +3	Fe^{2+}, 0.75 Fe^{3+}, 0.64	7.9	Fe_2O_3, Fe_3O_4, $FeCO_3$, FeS_2
cobalt	27	VIIIB	58.9332	1.25	+2, +3	Co^{2+}, 0.72 Co^{3+}, 0.63	8.7	CoAsS, $CoAs_2$, CoS
nickel	28	VIIIB	58.69	1.24	+2, +4	Ni^{2+}, 0.70	8.9	NiS in mixed sulfides
zinc	30	IIB	65.39	1.38	+2	Zn^{2+}, 0.74	7.1	ZnS, ZnO, $ZnCO_3$

A cluster of 21 minerals that are present in ores, plus 2 samples of native ores (free metals, Ag and Bi). Refer to page 823 for the key.

Nickel

Nickel is a hard, malleable, ductile, silvery-white, highly reflective metal that is resistant to corrosion. Like cobalt, it dissolves very slowly in dilute solutions of nonoxidizing acids such as HCl. It is readily soluble in 6 M HNO$_3$, but is rendered passive by concentrated HNO$_3$.

Metallic nickel and iron are believed to make up most of the core of the earth. Many meteorites are alloys of iron and nickel, which indicates that both metals are present in large quantities in celestial bodies.

Nickel is usually obtained from the ore *pentlandite*, a relatively abundant ore that contains a mixture of Ni, Cu, and Fe sulfides. The (Ni, Cu, Fe)S ore is roasted to convert the mixture of sulfides to oxides.

$$(Ni, Cu, Fe)S + O_2 \longrightarrow (Ni, Cu, Fe)O + SO_2$$

The mixture of oxides is then reduced with carbon to produce a crude alloy known as **Monel metal**.

$$(Ni, Cu, Fe)O + C \longrightarrow (Ni, Cu, Fe) + CO$$

The composition of Monel metal is 60% Ni, 36% Cu, 3.5% Fe and 0.5% Al. It is highly resistant to corrosion and is therefore widely used in industry.

Another preparation of nickel involves separation of the pulverized mixed sulfide ore by selective flotation. The NiS is then roasted and reduced with carbon as indicated above. This process yields metallic nickel that is about 95% pure. It can be purified electrolytically to give a product of better than 99.9% purity.

Alternatively, carbon monoxide can be passed over finely divided impure nickel at about 75°C to produce nickel carbonyl, Ni(CO)$_4$, a volatile (and very toxic) compound that boils at 43°C.

$$Ni(s) + 4CO(g) \xrightarrow{75°C} Ni(CO)_4(g)$$

The nickel carbonyl is condensed to a liquid and then decomposed by heat, in the reverse of the above reaction, to produce very pure nickel.

Because of its resistance to corrosion, nickel is used to plate iron, steel, and copper. Nickel plate offers the additional advantage of a highly reflective surface. *Nichrome* is an alloy (60% Ni, 25% Fe, and 15% Cr) that is used in the heating wires of electric heaters and toasters. There are many other examples of important alloys that contain nickel, such as the stainless steels used in cables, gears, and drive shafts.

Since 1974 we have known that trace amounts of nickel are essential for animal nutrition.

Iron

The two most abundant elements are nonmetals: oxygen (49.5%) and silicon (25.7%). See Table 1-3.

Iron is the second most abundant *metal* (4.7%) in the earth's crust after aluminum (7.5%). It is the most widely used of all metals. Iron was known as early as 4000 BC. It was used by primitive people because of its wide distribution and ease of reduction to the free metal. The red color of many rocks, clays, and soils is due to the presence of Fe$_2$O$_3$, the compound formed when iron rusts.

Iron is a silvery-white metal of high tensile strength that takes a high polish well. It is ductile and soft compared to many other metals. The occurrence and metallurgy of iron were discussed in Section 22-7.

Iron is widely used because its properties can be modified by alloying with a variety of elements. Additionally, it can be protected from corrosion by many different kinds of coatings. For example, iron can be coated with (1) other metals such as Zn, Sn, Cu, Ni, Cr,

Cd, and Pb; (2) ceramic materials such as those used in bathtubs and refrigerators; and (3) organic materials such as paint, lacquer, or asphalt.

Manganese

Manganese is a softer metal than iron. It is gray with a reddish tinge. Common ores of manganese are oxides, with *pyrolusite*, MnO_2, being the most important. Because manganese is used primarily to alloy with iron, the ore is usually reduced with coke in a blast furnace, which also reduces iron oxides. The impure mixture of metals is used to make steel.

$$MnO_2(s) + 2C(s) \longrightarrow Mn(impure) + 2CO(g)$$

Steel that contains 10 to 18% manganese is hard, tough, and resistant to wear. It is used to make rock-crushing machinery, armor plate, railroad rails, and similar items.

Since 1931 we have known that manganese is an essential trace element in the human diet.

Chromium

Chromium is a lustrous, hard metal that is very resistant to corrosion. Like aluminum, it is a strong reducing agent,

$$Cr^{2+}(aq) + 2e^- \longrightarrow Cr(s) \qquad E^0 = -0.91 \text{ V}$$

that forms a thin, tough, protective film of oxide.

The most important ore of chromium is *chromite* (also called chrome iron ore), $FeCr_2O_4$. The ore is reduced by carbon in an electric furnace to form an alloy called *ferrochrome*.

$$FeCr_2O_4(s) + 4C(s) \longrightarrow \underbrace{Fe + 2Cr}_{\text{ferrochrome}} + 4CO(g)$$

Ferrochrome is used in making chromium steels. Stainless steel contains 14 to 18% Cr and is very resistant to corrosion. Chrome-vanadium steel contains 1 to 10% Cr together with about 0.15% vanadium. These very strong steels are used to make axles that must withstand constant strain as well as frequent shock and vibration.

Some chemistry of chromium was described in Section 23-11. The need for trace amounts of chromium in the human diet was established in 1959.

Chromium is plated onto other metals to provide a shiny, protective coating.

Aluminum

Aluminum is the third most abundant element in the earth's crust (Table 1-3), and the most abundant metal. It is a Group IIIA element located just below the dividing line between metals and nonmetals. Not surprisingly, its hydroxide is amphoteric (Table 10-2).

The metallurgy of aluminum was discussed in Section 22-6. Although it is a very reactive metal,

$$Al^{3+}(aq) + 3e^- \longrightarrow Al(s) \qquad E^0 = -1.66 \text{ V}$$

aluminum quickly becomes coated with a thin, tough layer of oxide, Al_2O_3, that protects it from further atmospheric oxidation.

Aluminum is a relatively soft metal that can be extruded, i.e., forced through a die by pressure and thereby formed into the desired shape. Frames for window screens and similar items are made from aluminum that has been shaped by extrusion. Alloys of aluminum that contain *d*-transition metals are much stronger and harder than aluminum itself.

Zinc

Zinc is a Group IIB element, and strictly speaking, it is not a *d*-transition metal. It is a reactive metal, but considerably less so than aluminum.

$$Zn^{2+}(aq) + 2e^- \longrightarrow Zn(s) \qquad E^0 = -0.76 \text{ V}$$

Like aluminum, zinc becomes coated with a protective oxide layer. This is then converted to blue-gray basic carbonate, $Zn(OH)_2 \cdot ZnCO_3$, on exposure to moist air.

Zinc hydroxide is amphoteric (Table 10-2). The metal dissolves in both nonoxidizing acids and strong soluble bases with the evolution of hydrogen.

The main ore of zinc contains zinc sulfide, ZnS, and is called *zinc blende* or *sphalerite*. Cadmium, iron, lead, and arsenic sulfides usually occur in zinc sulfide ores.

Zinc sulfide ores are concentrated by flotation and then roasted.

$$2ZnS(s) + 3O_2(g) \longrightarrow 2ZnO(s) + 2SO_2(g)$$

Then the oxide is reduced by heating it with coal in a clay retort.

$$ZnO(s) + C(s) \longrightarrow Zn(\ell) + CO(g)$$

Zinc is sufficiently volatile (bp = 907°C) that it distills out of the retort as rapidly as it is produced. Impurities, primarily Cd, Fe, Pb, and As, are then separated by careful distillation of the crude zinc.

29-4 THE METALS OF ANALYTICAL GROUP IV

Some properties of the metals of Analytical Group IV are tabulated in Table 29-4.

The cations that occur in Analytical Group IV—Ca^{2+}, Sr^{2+}, and Ba^{2+}—are derived from three metals from Group IIA in the periodic table. These elements exhibit only the +2 oxidation state in their compounds. Sections 23-4, 23-5, and 23-6 contain information on the properties and occurrence, the reactions, and the uses of these metals and their compounds as well as those of magnesium, an Analytical Group V cation.

Table 29-4 *Some Properties of the Metals of Analytical Group IV*

Metal	Atomic Number	Periodic Group Number	Atomic Weight	Atomic Radius (Å)	Important Oxidation States	Ionic Radius (Å)	Density at 20°C (g/cm³)	Compounds in Ores
calcium	20	IIA	40.078	1.97	+2	Ca^{2+}, 0.99	1.6	$CaCO_3$, $CaCO_3 \cdot MgCO_3$, $CaSO_4$, CaF_2, mixed compounds
strontium	38	IIA	87.62	2.15	+2	Sr^{2+}, 1.13	2.6	$SrSO_4$, $SrCO_3$
barium	56	IIA	137.327	2.17	+2	Ba^{2+}, 1.35	3.5	$BaSO_4$, $BaCO_3$

(a)   **(b)**

Very pure calcium carbonate, $CaCO_3$, occurs in two crystalline forms, (a) calcite and (b) aragonite.

Calcium

Calcium makes up about 3.4% of the earth's crust (Table 1-3); it is the fifth most abundant element and the third most abundant metal. Large amounts of calcium occur in *limestone*, a rock that is primarily $CaCO_3$. *Dolomitic limestone* contains large amounts of $MgCO_3$ as well. Calcium carbonate occurs in other natural forms such as marine animal shells and pearls, as well as in deposits of *marl, marble, chalk, calcite* (often as very large crystals that are displayed in museums), and *aragonite*. Calcium sulfate occurs in deposits that contain *gypsum*, $CaSO_4 \cdot 2H_2O$, a hydrated compound, and also as the anhydrous compound *anhydrite*, $CaSO_4$. Calcium fluoride, CaF_2, occurs in a mineral called *fluorite* or *fluorspar*. Calcium occurs in many complex compounds such as *apatite*, which is usually formulated as $Ca_{10}(PO_4)_6(OH,Cl,F)_2$.

Calcium, strontium, and barium are very strong reducing agents.

$$Ca^{2+}(aq) + 2e^- \longrightarrow Ca(s) \qquad E^0 = -2.87 \text{ V}$$
$$Sr^{2+}(aq) + 2e^- \longrightarrow Sr(s) \qquad E^0 = -2.89 \text{ V}$$
$$Ba^{2+}(aq) + 2e^- \longrightarrow Ba(s) \qquad E^0 = -2.90 \text{ V}$$

Standard reduction potentials were determined for reactions that occur in aqueous solutions. Therefore, E^0 values are not strictly applicable to reactions in the absence of water.

They are usually prepared by electrolysis of their molten chlorides.

$$MCl_2(\text{molten}) \xrightarrow{\text{electrolysis}} M(\ell) + Cl_2(g) \qquad (M = Ca, Sr, Ba)$$

Because Ca, Sr, and Ba are such powerful reducing agents, large amounts of energy are required to liberate the metals from their compounds.

Interestingly, calcium does not react with oxygen in room-temperature air at an appreciable rate. Strontium reacts with oxygen quite rapidly, and barium bursts into flame in moist air. Such kinetic differences could *not* be predicted from their E^0 values.

Strontium

Strontium is a rare element, and there are no commercial uses for the free element. Like calcium, it occurs in nature as the carbonate *strontianite*, $SrCO_3$, and as the sulfate *celestite*, $SrSO_4$. Both are quite rare and somewhat less soluble than the corresponding calcium compounds.

Barium

Barium also occurs in nature as the carbonate *witherite*, $BaCO_3$, and as the sulfate *barite*, or *heavy spar*, $BaSO_4$. The solubility of $CaCO_3$ is intermediate between the solubilities of $BaCO_3$ and $SrCO_3$, whereas $BaSO_4$ is much less soluble than $SrSO_4$, which in turn is less soluble than $CaSO_4$.

Barium is produced both by electrolysis of its molten chloride and by reduction of a mixture of barium oxide and barium peroxide with aluminum in a vacuum furnace at 1000 to 1100°C.

$$\left.\begin{array}{c} BaO \\ BaO_2 \end{array}\right\} \xrightarrow{\;Al\;} Ba(g) + Al_2O_3(s)$$

Barium is a stronger reducing agent than aluminum, and we might expect that this reaction would not be possible. However, LeChatelier's Principle comes to the rescue again. The furnace is operated at very low pressures (10^{-3} to 10^{-4} torr) so that barium vaporizes and distills out of the furnace as rapidly as it is formed. Aluminum oxide is thermodynamically a very stable compound ($\Delta H_f^0 = -1676$ kJ/mol, $\Delta G_f^0 = -1562$ kJ/mol). The high thermodynamic stability of Al_2O_3 and the volatility of Ba at 1000 to 1100°C make the process viable.

29-5 THE METALS OF ANALYTICAL GROUP V

Some properties of the metals of Analytical Group V are tabulated in Table 29-5. Magnesium, Mg^{2+}; sodium, Na^+; potassium, K^+; and ammonium, NH_4^+, ions, the so-called soluble group, constitute Analytical Group V. Ammonium ion is a common cation that is not derived from a metal; the group separations place it in Analytical Group V because the solubilities of its compounds are similar to those of the potassium ion. Magnesium is a Group IIA metal; sodium and potassium are found in Group IA.

In addition to Sections 23-4, 23-5, and 23-6, which contain information on the alkaline earth metals, you should read Section 22-5, which describes the commercial recovery of magnesium from seawater and some of its important uses.

Magnesium

Magnesium is the eighth most abundant element in the earth's crust (1.9%) and is less abundant than sodium (2.6%) and potassium (2.4%). Because both $MgCl_2$ and $MgSO_4$ are soluble in water, magnesium ions frequently occur in ground waters as well as in oceans and lakes.

Table 29-5 *Some Properties of the Metals of Analytical Group V*

Metal	Atomic Number	Periodic Group Number	Atomic Weight	Atomic Radius (Å)	Important Oxidation States	Ionic Radius (Å)	Density at 20°C (g/cm³)	Compounds in Ores
magnesium	12	IIA	24.3050	1.60	+2	Mg^{2+}, 0.65	1.7	$MgCl_2$ (brines), $MgCl_2$, $MgSO_4$ (seawater)
sodium	11	IA	22.98977	1.86	+1	Na^+, 0.95	0.97	NaCl, complex compounds
potassium	19	IA	39.0983	2.31	+1	K^+, 1.33	0.86	KCl, complex compounds

In recent years, the problem of magnesium deficiency in the diets of lactating cows has come to light, and some progress has been made in dealing with it. The effects are apparently quite complex and so far are only partially understood. What is known follows. Cows that have young suckling calves and that graze on lush new pasture in the early spring sometimes develop a disorder called *grass tetany*. It is usually fatal unless detected and treated within a few hours. Research has shown that the disease usually develops in apparently healthy animals that have insufficient magnesium in their diets. Two methods have been partially successful in combating this phantom killer of lactating cows. The application of dolomitic limestone ($CaCO_3 \cdot MgCO_3$) to pasturelands significantly reduces the incidence of grass tetany, as does the inclusion of magnesium compounds in the minerals and dry feeds that are fed to cattle.

The costs of beef and dairy products depend on the effectiveness with which agriculture scientists combat problems such as grass tetany.

Sodium and Potassium

Sodium and potassium are widely distributed in nature, being the sixth and seventh most abundant elements (2.6% and 2.4%), respectively, in the earth's crust. The properties, occurrence, and reactions of the alkali metals were discussed in Sections 23-1, 23-2, and 23-3.

Both sodium and potassium can be prepared by electrolysis of anhydrous molten compounds such as NaCl, NaOH, KCl, and KOH. In the production of sodium by the electrolysis of molten NaCl, some $CaCl_2$ is added to the electrolysis mixture to lower its melting point from about 800°C to about 600°C. This significantly reduces the amount of energy required to produce sodium.

Potassium is more expensive than sodium. Because sodium can be used very effectively in most industrial processes, very little potassium is produced today. As an alternative to electrolysis of molten KCl or KOH, potassium is produced by the reaction between molten sodium and molten KCl.

$$Na(\ell) + KCl(\ell) \xrightarrow{\Delta} K(g) + NaCl(\ell)$$

This is another example of a reaction that we might not expect to occur, but that is used in an industrial process. Recall that E^0 values are strictly applicable only to reactions in aqueous solutions. Apparently, the greater volatility of potassium compared with sodium is an important factor (another application of LeChatelier's Principle) in this reaction.

Exercises

Analytical Group I Metals

1. (a) Write the electron configuration (⇅ notation and shorthand $ns^x\ np^y\ nd^z$ notation) for each metal in Analytical Group I. (b) Write the electron configuration for each Group I cation.
2. List the names of the important ores of the Analytical Group I metals and the formulas for the commercially important compound(s) in each ore.
3. Many important ores of the Analytical Group I metals are sulfide ores. What does this suggest about the stability of these compounds?
4. Briefly describe the metallurgy of each metal in Analytical Group I. Write equations for the important reaction(s).
5. List some important uses for each metal (or its compounds) in Analytical Group I.

6. The largest mass of native copper discovered to date is 420 short tons. (It was found in northern Michigan.) The density of copper is 8.65 g/cm^3 at 20°C. (a) If this mass were a perfect cube, how long would each edge be? (b) If it were a perfect sphere, what would its diameter be?

Analytical Group II Metals

7. (a) Write the electron configuration (⇅ notation and shorthand $ns^x\ np^y\ nd^z$ notation) for each metal in Analytical Group II. (b) Write the electron configuration for each Group II cation.
8. List the names of the important ores of the Analytical Group II metals and the formulas for the commercially important compound(s) in each ore.

9. The important ores of the Analytical Group II metals are sulfide or oxide ores. What does this suggest about the stability of these compounds?

10. Briefly describe the metallurgy of each metal in Analytical Group II. Write equations for the important reaction(s).

11. List some important uses for each metal (or its compounds) in Analytical Group II.

Analytical Group III Metals

12. (a) Write the electron configuration ($\uparrow\downarrow$ notation and shorthand $ns^x\ np^y\ nd^z$ notation) for each metal in Analytical Group III. (b) Write the electron configuration for each Group III cation.

13. List the names of the important ores of the Analytical Group III metals and the formulas for the commercially important compound(s) in each ore.

14. The important ores of the Analytical Group III metals are sulfide or oxide ores. What does this suggest about the stability of these compounds?

15. Briefly describe the metallurgy of each metal in Analytical Group III. Write equations for the important reaction(s).

16. List some important uses for each metal (or its compounds) in Analytical Group III.

Analytical Group IV Metals

17. (a) Write the electron configuration ($\uparrow\downarrow$ notation and shorthand $ns^x\ np^y\ nd^z$ notation) for each metal in Analytical Group IV. (b) Write the electron configuration for each Group IV cation.

18. List the names of the important ores of the Analytical Group IV metals and the formulas for the commercially important compound(s) in each ore.

19. The important ores of the Analytical Group IV metals are carbonate and sulfate ores. What does this suggest about the stability of these compounds?

20. (a) List the carbonates of the Analytical Group IV cations in order of increasing solubility in water. (b) List the sulfates of the Analytical Group IV cations in order of increasing solubility in water.

21. Briefly describe the metallurgy of each metal in Analytical Group IV. Write equations for the important reaction(s).

22. List some important uses for each metal (or its compounds) in Analytical Group IV.

23. (a) How is it possible that aluminum, a less active metal, can displace barium, a more active metal, from its molten compounds? (b) If metallic aluminum were placed in an aqueous $BaCl_2$ solution, would you expect metallic barium to be formed? Why?

Analytical Group V Metals

24. (a) Write the electron configuration ($\uparrow\downarrow$ notation and shorthand $ns^x\ np^y\ nd^z$ notation) for each metal in Analytical Group V. (b) Write the electron configuration for each Group V metal cation.

25. List the names of the important ores of the Analytical Group V metals and the formulas for the commercially important compound(s) in each ore.

26. How can large concentrations of NaCl and KCl exist in seawater and brines? (b) Would you expect to find high concentrations of AgCl in seawater or brines? Why?

Introduction to Laboratory Work

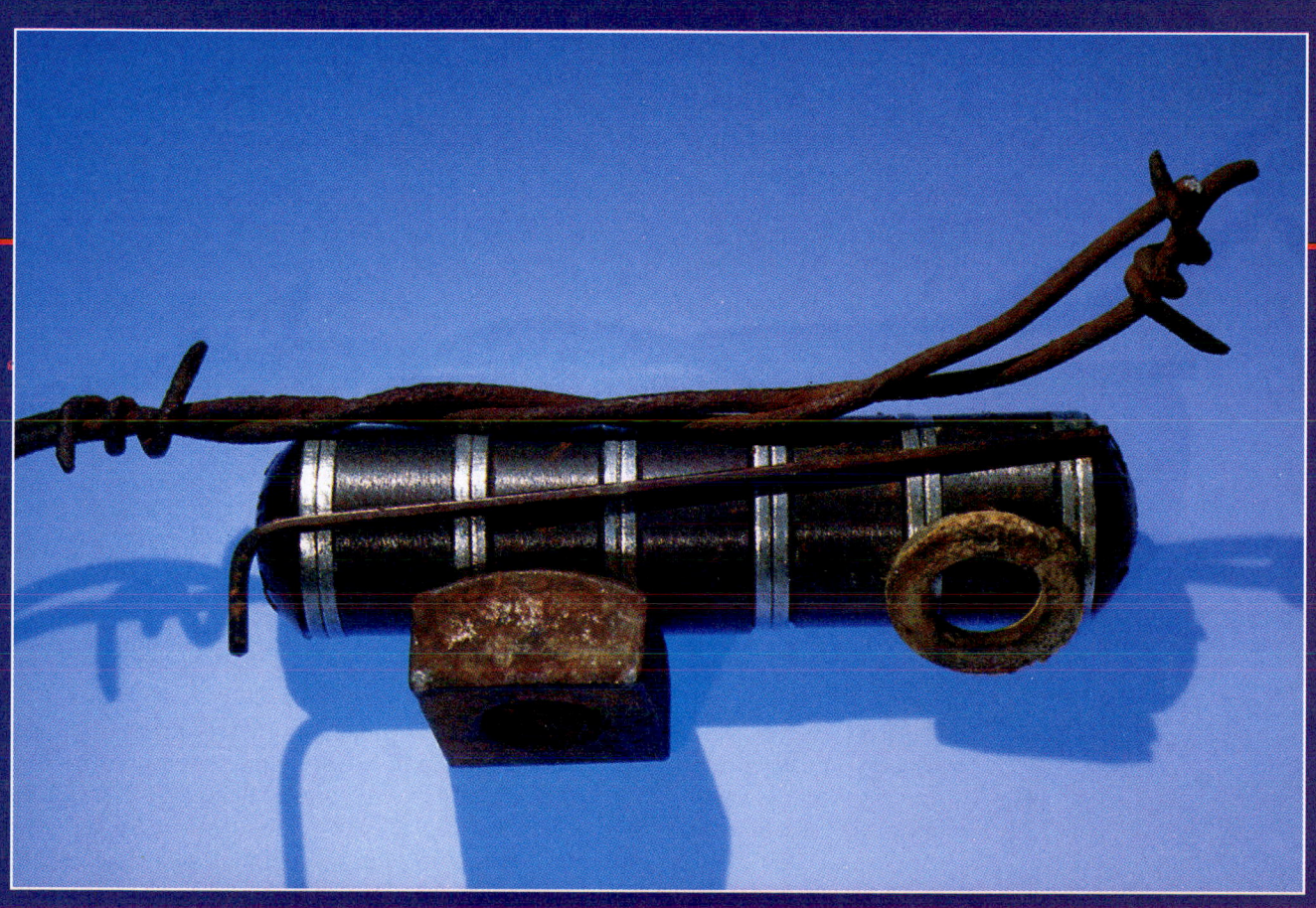

The metals we study in qualitative analysis have a wide variety of practical uses. The finest surgical instruments are made from high quality steels. Artificial hip joints are made of special metal alloys. The "cow magnet" shown in this photograph is made of Alnico, an alloy of aluminum, nickel, and cobalt. The magnet is placed in a cow's first stomach. As she grazes she picks up "scrap iron," which is attracted to the magnet. This prevents "scrap iron" from being carried into the cow's digestive tract where it could cause serious damage.

I n qualitative analysis we learn to make observations accurately and to interpret them logically. These may be the most important skills we develop in the educational process. We shall see the effects of various kinds of equilibria as we observe chemical reactions. We shall observe (1) the precipitation of insoluble compounds and (2) the dissolution of insoluble compounds by complex ion formation, by formation of weak electrolytes, and by oxidation–reduction reactions. We shall also learn careful laboratory manipulations.

30-1 BASIC IDEAS OF QUALITATIVE ANALYSIS

Qualitative analysis is concerned with identification of the ions that are present in a substance. Initially we work with solutions that contain only limited numbers of ions in specific groups. Then we work with more complex mixtures. We separate ions with similar properties into small groups.

We separate the members of the small groups so that we can isolate and identify each one. *Group separations* are based on similar chemical properties, whereas most *identifications* of individual ions are based on properties that are different. The common cations are listed in Figure 30-1, which shows how they can be separated into groups. The order of these operations has been chosen so that the substances added in each step will not interfere in the later steps.

In analyzing *general unknowns* we precipitate Analytical Groups I and II together. Then we dissolve the Group II sulfides in a strongly basic solution (4 *M* NaOH). The Group I sulfides are insoluble in this solution. A *known* solution contains all the cations in a group. An *unknown* solution may contain any or all of the cations in the group.

It is important to recognize that an analytical group number does *not* refer to a group of elements in the periodic table.

30-2 CATIONS OF ANALYTICAL GROUP I

To analyze a solution that contains Group I cations (Bi^{3+}, Cu^{2+}, and Cd^{2+}), we start with a solution that contains either chloride or nitrate salts of these cations. It is made 0.3 molar in HCl and saturated with hydrogen sulfide. Under these conditions the Group I cations

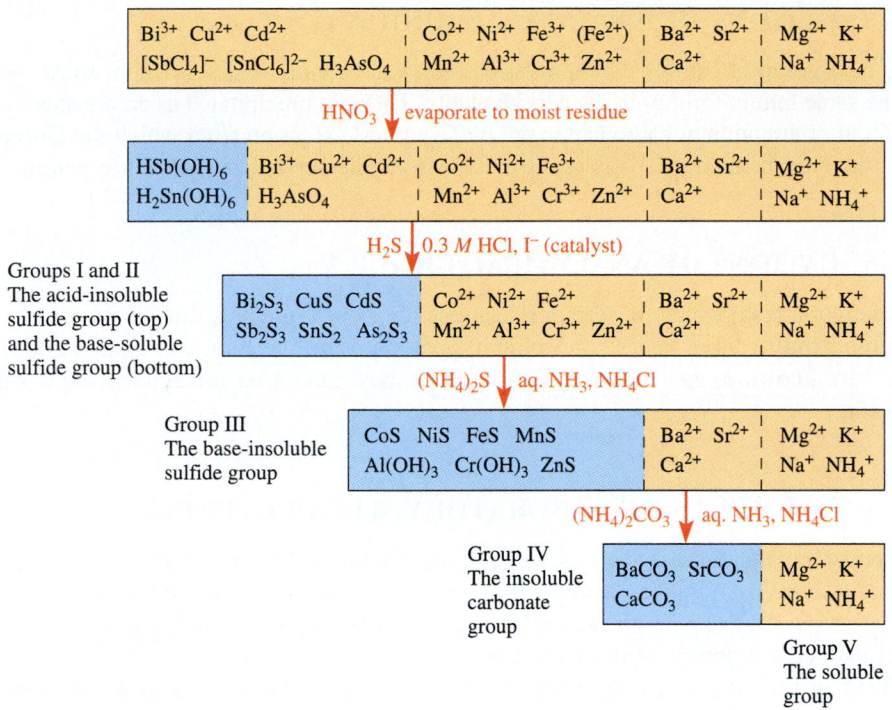

Groups I and II
The acid-insoluble
sulfide group (top)
and the base-soluble
sulfide group (bottom)

Group III
The base-insoluble
sulfide group

Group IV
The insoluble
carbonate
group

Group V
The soluble
group

Figure 30-1 Flow chart for the systematic separation of common cations into groups. Blue areas represent precipitated species.

precipitate as insoluble sulfides. They are bismuth(III), copper(II), and cadmium sulfides. These ions are called the **acid-insoluble sulfide group** to distinguish them from the Group II cations.

30-3 CATIONS OF ANALYTICAL GROUP II

The conditions under which the Group II sulfides are precipitated are similar to those for Group I. (*In general unknowns, Group I and Group II cations will be precipitated together.*) However, antimony and arsenic are both in the +5 oxidation state after the solution has been evaporated with HNO_3. The sulfides As_2S_3 and Sb_2S_3 are easier to precipitate and work with than are As_2S_5 and Sb_2S_5. We add a few drops of NH_4I to reduce both arsenic and antimony to the +3 oxidation state.

An important difference between Group I sulfides and Group II sulfides is the solubility of the Group II sulfides in excess 4 M NaOH solution. The Group I sulfides are insoluble in this solution.

30-4 CATIONS OF ANALYTICAL GROUP III

After the Groups I and II sulfides have been removed, the solution that contains the remaining cations is again saturated with hydrogen sulfide. Then an excess of aqueous ammonia is added. The sulfides of cobalt(II), nickel(II), manganese(II), iron(II), and zinc, together with the hydroxides of aluminum and chromium(III), precipitate from the buffered basic solution that contains ammonium sulfide. These ions are called the **basic-insoluble sulfide group**.

The term "basic-insoluble sulfide group" is not quite accurate because five sulfides and two hydroxides are precipitated in Group III. However, the term is traditional.

30-5 CATIONS OF ANALYTICAL GROUP IV

The next group of cations—barium, strontium, and calcium ions—includes three metals in the same family (group) in the periodic table. They are precipitated as carbonates by the addition of ammonium carbonate to the buffered basic solution from which the Group III ions have been removed. This group is known as the **insoluble carbonate group**.

30-6 CATIONS OF ANALYTICAL GROUP V

The solution from which the Group IV cations have been removed now contains only the **soluble group** cations. These are sodium, potassium, magnesium, and ammonium ions. They are known as the soluble group because they cannot be precipitated by a single reagent.

30-7 COMMENTS ON LABORATORY MANIPULATIONS

In **semimicro** qualitative analysis, we use small volumes in small test tubes so we can separate solids from liquids rapidly with a centrifuge. Small volumes of liquids are conveniently removed from centrifuged solids with capillary pipets. Liquid reagents are added with medicine droppers or dropping bottles.

"Reagent" refers to any substance that is used to analyze or identify another substance. Many are kept in bottles on the laboratory shelf; others must be prepared just before use or even within the solution to be analyzed (*in situ*).

Your instructor will supply a list of laboratory apparatus. When a desk has been assigned, check the apparatus in the desk according to the instructions given by your instructor. Obtain any necessary replacements from the stockroom. Wash all of the equipment thoroughly with detergent solution, rinse it thoroughly with tap water, and finally rinse it with distilled water. All equipment must be *clean* because traces of impurities interfere with many qualitative tests.

30-8 CAPILLARY PIPETS

Liquid reagents may be added with standard medicine droppers or directly from the dropping bottles on the shelves in the laboratory. Ordinary medicine droppers deliver about 1 mL per 20 drops. Capillary pipets, with smaller tips, deliver approximately 1 mL per 40 drops.

Prepare several capillary pipets that will be used to transfer liquids from one test tube to another. (Your instructor may demonstrate such glass-working operations as cutting and fire-polishing.) Choose a 15 to 18-cm length of 8-mm-diameter soft glass tubing, and heat it over the Bunsen flame with rotation until the middle section of the glass softens. Remove the tube from the flame, hold it vertically, and slowly draw it out until the bore (the center opening) is approximately 1 mm across (Figure 30-2).

After the tube has cooled, cut the capillary at the midpoint and fire-polish the capillary ends. Flare the wide ends of the tubes by heating them until they are soft and then quickly pressing them down against a flat metal surface. After the pipets are cool, attach medicine-dropper bulbs to the flared ends.

Figure 30-2   A capillary pipet.

C A U T I O N

Hot glass looks exactly like cold glass! Even experienced lab technicians sometimes burn themselves by picking up hot glassware too soon. If in doubt, use a towel or tongs to pick up glass.

30-9 STIRRING RODS

Use 3-mm soft glass rod (not tubing) to make at least six stirring rods, approximately 12 to 15 cm in length. Fire-polish both ends of each rod.

Only distilled water should be used in the analytical procedures because tap water contains ions such as Ca^{2+}, Mg^{2+}, Fe^{3+}, Fe^{2+}, HCO_3^-, SO_4^{2-}, and Cl^-. These ions are among those tested for in the unknown solutions. Several of them interfere with tests for other ions. Keep a wash bottle (Figure 30-3) filled with distilled water.

30-10 REAGENTS

The small dropping bottles in your desk are used to store commonly used reagents. Remove the droppers, remove the bulbs from the droppers, and wash the bottles and droppers thoroughly in hot detergent solution. Rinse them well with tap water and finally with distilled water. Reassemble the droppers and place them in the bottles. Label the bottles as follows:

1. 1.0 M NH_3 4. 0.1 M HCl   7. 6.0 M H_2SO_4

2. 3.0 M NH_3 5. 6.0 M HCl   8. 4.0 M NaOH

3. 6.0 M NH_3 6. 6.0 M HNO_3 9. 5% thioacetamide

Your instructor will show you where bottles filled with these reagents are located. Use them to fill your dropping bottles. The tenth dropping bottle is filled with distilled water. Label it. You will also be shown other reagents that you will use less frequently. There is another group of bottles filled with *known solutions* on the reagent shelf. Note that each ion is available in an individual bottle, and there are *group knowns* as well.

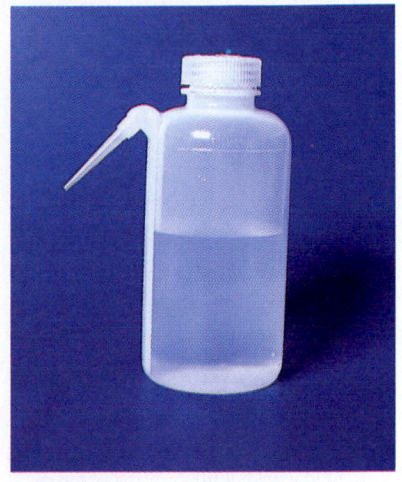

Figure 30-3 A wash bottle should be filled with distilled water. Use it to rinse your glassware.

30-11 PRECIPITATION

We use precipitate formation as the usual method for group separations. Precipitations are carried out in small test tubes. It is imperative that all separations be *as complete as possible.* When a precipitation reaction is carried out, one must always check for *completeness of precipitation.* This is done by centrifuging the reaction mixture at full speed for approximately 1 minute (Section 30-12), adding one drop of the precipitating reagent to the clear solution and observing carefully to see whether additional precipitate forms (Figure 30-4). If the addition of one more drop of the precipitating reagent shows that precipitation is incomplete, the mixture should be stirred thoroughly, centrifuged for approximately 1 minute, and tested for complete precipitation a second time.

Ions from an earlier group that remain in solution may interfere with the analysis of a later group.

> Collision of ions, atoms, or molecules is a necessary condition for reaction.

Because precipitation reactions are carried out in small test tubes, considerable effort is necessary to assure that reactants are mixed intimately. One effective way is to place a stirring rod inside the test tube, hold the top of the test tube between the thumb and index finger, and shake it vigorously sideways so that it strikes the little finger.

Many precipitation reactions result in the formation of **colloidal particles**, particles that are too small to be separated from the liquid phase by filtration or by centrifuging the mixture. Colloidal suspensions can usually be coagulated by heating the mixture in a

This motion imparts rotary motion to the liquid inside the test tube, and the presence of the stirring rod assists in intimate mixing of the contents.

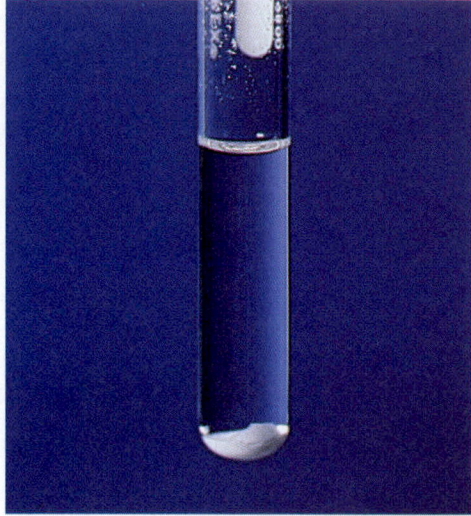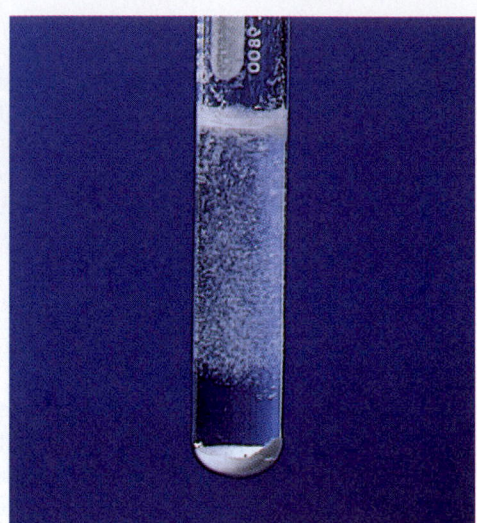

Figure 30-4 Testing for complete precipitation. The solution in the centrifuged mixture should be clear (left). Add one drop of the precipitating reagent. The formation of more precipitate shows that precipitation was not complete (right). Repeat the procedure until no more precipitate forms.

water bath for a few minutes. This process is called **digestion**. As the mixture is warmed, the very small particles dissolve and recrystallize on the surfaces of larger particles. The presence of a colloidal suspension is detected from the opaqueness of the solution; i.e., the solution is not transparent. When a colloidal suspension has coagulated, the solution (the **supernatant liquid**) becomes transparent, or clear (Figure 30-5). "Clear" and "color-less" do not have the same meaning. A *clear* solution is transparent; a *colorless* solution has no color. All true solutions are clear; many are colored.

After a colloidal suspension has coagulated, the mixture is centrifuged and the liquid is withdrawn with a capillary pipet. A slight excess of precipitating reagent is always added to reduce the solubility of the precipitate by the common ion effect. However, a very large excess of precipitating reagent should be avoided because it may form soluble compounds containing complex ions.

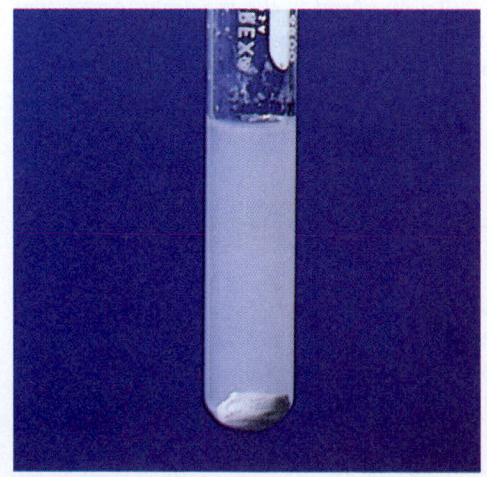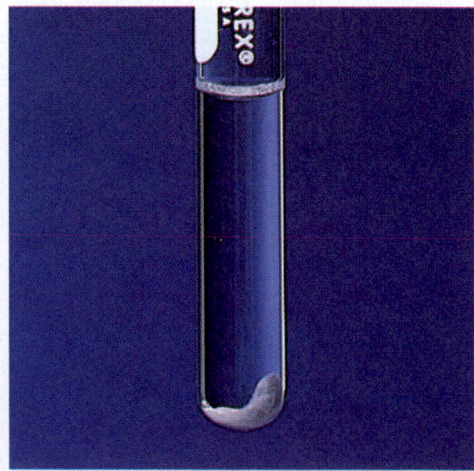

Figure 30-5 A colloidal suspension appears cloudy or opaque (left). A true solution with a precipitate at the bottom of the test tube (right).

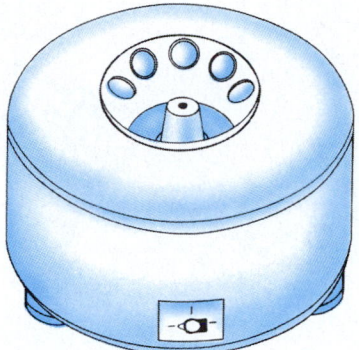

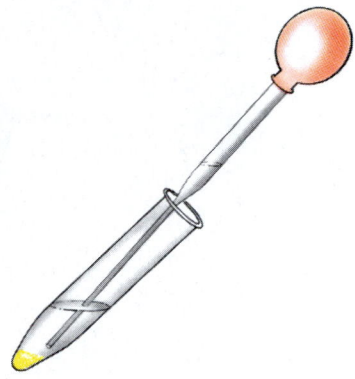

Figure 30-6 A centrifuge.

Figure 30-7 A capillary pipet is used to separate the liquid (centrifugate) from the precipitate.

30-12 CENTRIFUGATION OF PRECIPITATES AND TRANSFER OF THE LIQUID (CENTRIFUGATE)

Precipitates are separated from solutions by centrifugation (Figure 30-6). The mixture is rotated at a very high speed so that the denser precipitate is forced to the bottom of the test tube. A balance tube containing a volume of water equal to the volume of solution to be centrifuged is always placed in the centrifuge exactly opposite the solution of interest. A gummed label should be placed at the top of the balance tube so that it can always be identified easily.

> An unbalanced load in the centrifuge head will cause severe vibration and damage the centrifuge's bearings.

After a mixture has been centrifuged, the tube should be held at an angle so that the liquid (the **centrifugate**) can be easily drawn into a capillary pipet (Figure 30-7). Avoid disturbing the precipitate as the liquid is withdrawn. On occasion, bits of precipitate may be drawn into the capillary pipet with the liquid. To prevent this, wind a very small piece of cotton around the end of the capillary pipet to serve as a filter.

> *Supernate* is often used to describe the clear solution above a precipitate. However, the term *centrifugate* is more descriptive in this case.

30-13 WASHING PRECIPITATES

A precipitate is always wet with the liquid from which it was separated. Precipitates are washed with small amounts of liquid to remove as much of the adhering liquid as possible. Use a stirring rod to break up the precipitate after the wash liquid is added, stir thoroughly, and centrifuge the mixture. When the solution that was initially separated from the precipitate is to be used in subsequent steps, the first wash liquid should be added to that solution. Usually a precipitate is washed twice, and the second wash liquid is discarded.

30-14 DISSOLUTION AND EXTRACTION OF PRECIPITATES

On occasion, it is desirable to dissolve only a part of a precipitate; i.e., certain components of a solid mixture are dissolved while others are not. This process is called **extraction**. It is necessary to break up the precipitate with a stirring rod so that the dissolving liquid comes into intimate contact with the precipitate. Often the mixture is heated, because most substances are more soluble at higher temperatures and most reactions occur more rapidly at higher temperatures. The mixture must always be stirred thoroughly and frequently during extraction. It is then centrifuged and separated.

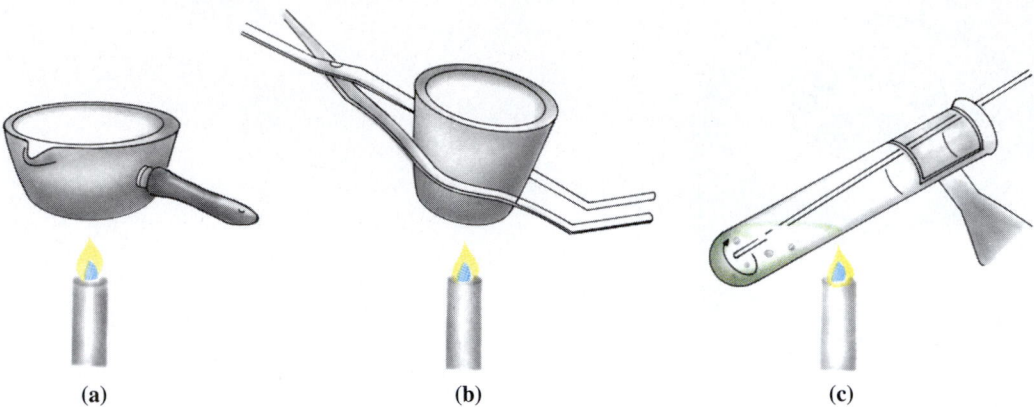

(a)   (b)   (c)

Figure 30-8 Evaporation of small amounts of liquid from (a) a casserole, (b) a crucible, and (c) a test tube.

30-15 HEATING MIXTURES AND SOLUTIONS

Often it is necessary to heat a mixture to cause precipitation or to dissolve or extract precipitates. Keep a water bath gently boiling throughout the laboratory period. When it is necessary to heat a solution, the test tube can be placed in the hot water bath immediately.

30-16 EVAPORATION

Often we wish to reduce the volume of a solution by evaporating some of the liquid, or to evaporate a solution to a moist residue or even to dryness. This may be accomplished in two ways. In the first method the solution is placed in a small evaporating dish or casserole and heated gently until the liquid has evaporated (Figure 30-8a, b). Extreme caution must be exercised in evaporating small amounts of liquid because very little heat is required. Once an evaporating dish or casserole becomes hot, it may remain above the boiling point of water for some time after the flame has been removed. In most cases it is undesirable to bake residues.

   In the second method, evaporations can be done from test tubes (Figure 30-8c), but care is necessary. The tube is held at an angle with the tip of a *very* small flame directed at the upper surface of the liquid. A stirring rod is inserted into the test tube and rotated constantly to break up bubbles of steam. Unless extreme care is used, steam bubbles will cause the solution to "bump" out of the test tube. NEVER POINT A TEST TUBE TOWARD YOURSELF OR ANOTHER PERSON WHILE AN EVAPORATION IS BEING DONE. If the solution does bump, very hot (and possibly caustic or toxic) liquid can spurt over a surprisingly large area.

30-17 KNOWN SOLUTIONS

Remember that known solutions contain all the cations in the group.

Known solutions should be analyzed before unknown solutions are analyzed. This enables you to learn the details of the analytical procedure and to recognize the expected results of tests for ions. The textures of precipitates vary widely. Crystalline precipitates are usually much denser than amorphous precipitates, so there appears to be relatively little of a crystalline precipitate and much more of an amorphous solid when equimolar quantities are present. Such observations are important. Make careful notes as known solutions are analyzed. A "known report" or an "unknown report," should be filled out as each solution is analyzed.

Colors are distinctive, and one must not rely on verbal descriptions of colors. Different people see colors somewhat differently, and therefore it is important to *know* how each colored substance appears *to you*.

The concentrations of cations in both known and unknown solutions are approximately 5 mg cation/mL of solution.

30-18 UNKNOWN SOLUTIONS

After the analysis of a known solution has been completed satisfactorily and the results have been approved by your instructor, ask the instructor to give you an unknown solution. The instructor may wish to ask several questions at this point to make sure you understand the analytical procedures.

The unknown may contain any or all of the ions in the group or groups. There are no unknowns that contain no ions, and no dyes have been added. The color of an unknown may give valuable clues about the presence or absence of certain ions. This is an important, legitimate observation. The results obtained for an unknown must always be consistent with the color of the unknown. Keep in mind that a mixture of colored ions may have a color that is not characteristic of any of the individual ions.

30-19 UNKNOWN REPORTS

Your instructor will indicate the kind of notebook you should use to record the results of the analyses. Observations should be recorded in your notebook *in ink* as soon as they are made. Equations should be written to describe the behavior of each ion as the various tests are run. Your instructor will specify the kind of report form to be used in reporting the analysis of unknowns.

30-20 LABORATORY ASSIGNMENTS

A list of suggested laboratory assignments is given below. Your instructor will indicate the number of unknowns required and the dates they are due.

1. Construct capillary pipets, stirring rods, and a wash bottle if necessary. Fill individual reagent bottles.

Analyze solutions containing the following ions:

Known	**Unknown**
2. The cations of Group I	3. Some Group I cations
4. The cations of Group II	5. Some Group II cations
6. The cations of Groups I and II	7. Some Groups I and II cations
8. The cations of Group III	9. Some Group III cations
10. The cations of Group IV	11. Some Group IV cations
12. The cations of Group V	13. Some Group V cations
	14. Some Cations of Groups I to V

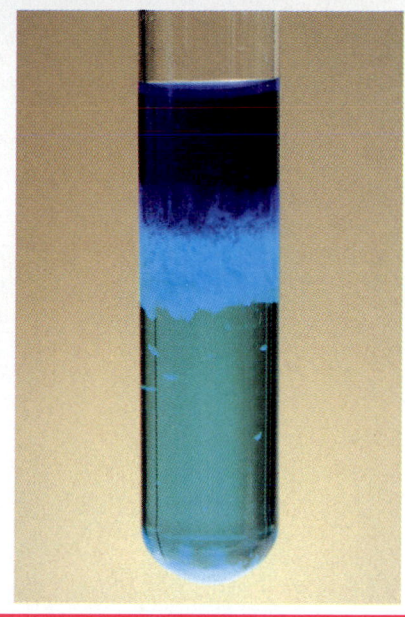

One test for copper(II) ions involves the reaction with an excess of aqueous ammonia. To minimize mixing, concentrated aqueous NH₃ was added slowly to a solution of copper(II) sulfate, CuSO₄. Unreacted blue copper(II) sulfate solution remains in the bottom part of the test tube. The light blue precipitate in the middle is copper(II) hydroxide, Cu(OH)₂. The top layer contains deep blue [Cu(NH₃)₄](OH)₂ which was formed as some Cu(OH)₂ dissolved in excess aqueous NH₃.

31-1 COMMON OXIDATION STATES OF METALS IN CATION GROUP I

Nearly all compounds of bismuth that are stable in aqueous solution contain bismuth in the $+3$ oxidation state. A few compounds of bismuth(V) are known, but Bi(V) is a very powerful oxidizing agent in acidic solutions.

Copper exists in two oxidation states in aqueous solutions, Cu(I) and Cu(II). The following standard reduction potentials show the relative stabilities of simple copper(I) and copper(II) compounds in aqueous solutions.

$$Cu^{+}(aq) + e^{-} \longrightarrow Cu(s) \qquad E^{0} = +0.521 \text{ V}$$

$$Cu^{2+}(aq) + 2e^{-} \longrightarrow Cu(s) \qquad E^{0} = +0.337 \text{ V}$$

$$Cu^{2+}(aq) + e^{-} \longrightarrow Cu^{+}(aq) \qquad E^{0} = +0.153 \text{ V}$$

Combination of the appropriate half-reactions gives

$$2Cu^{+}(aq) \longrightarrow Cu^{2+}(aq) + Cu(s) \qquad E^{0}_{cell} = +0.368 \text{ V}$$

This tells us that simple compounds containing copper(I) are unstable with respect to disproportionation into Cu(II) and metallic copper.

Cadmium exhibits only the $+2$ oxidation state in aqueous solutions.

However, as we shall see, Cu(I) can be stabilized in anionic complexes such as $[Cu(CN)_2]^{-}$.

31-2 INTRODUCTION TO THE ANALYTICAL PROCEDURES

The ions of analytical Group I are referred to as the **acid-insoluble sulfide group** because they are precipitated as sulfides from a saturated solution of H₂S that is also 0.30 M in HCl. These sulfides are insoluble in sodium sulfide solution. Bismuth(III), copper(II), and cadmium ions form sulfides that are insoluble in such solutions.

Thioacetamide serves as a source of hydrogen sulfide. It hydrolyzes in hot acidic solutions to produce hydrogen sulfide and ammonium acetate.

$$\underset{\text{thioacetamide}}{CH_3-\overset{\overset{\textstyle S}{\|}}{C}-NH_2} + 2H_2O \xrightarrow{\Delta} \underset{\substack{\text{usually written as} \\ NH_4CH_3COO}}{\underbrace{CH_3\overset{\overset{\textstyle O}{\|}}{C}-O^{-} + NH_4^{+}}} + H_2S$$

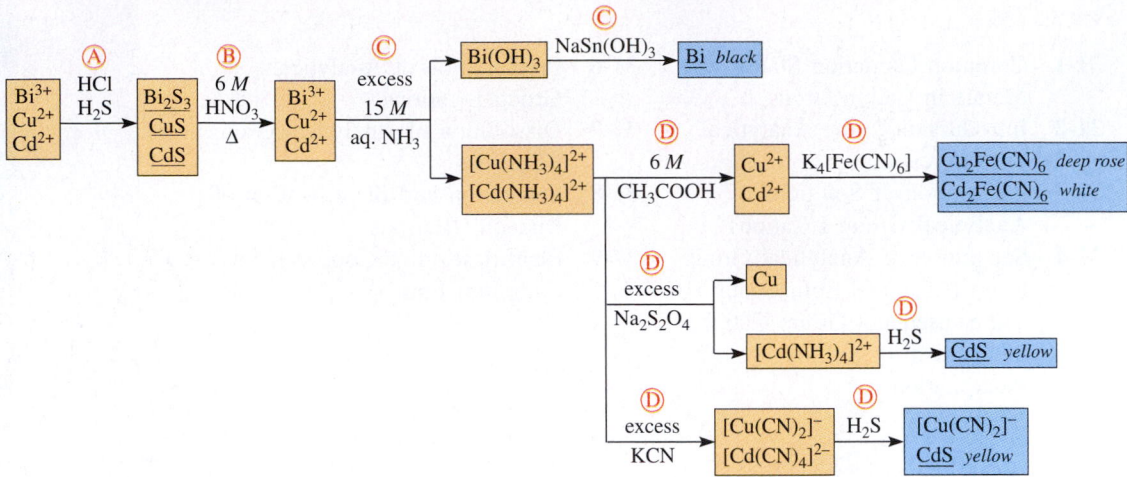

Figure 31-1 Analytical Group I flow chart. Circled letters refer to analytical steps. Species formed in confirmatory tests are shown with blue backgrounds.

Table 31-1 *Solubility Products for Group I Sulfides*	
Sulfide	K_{sp}
Bi_2S_3	1.6×10^{-72}
CuS	8.7×10^{-36}
CdS	3.6×10^{-29}

When directions call for the addition of thioacetamide followed by heating, this amounts to the addition of hydrogen sulfide to the solution. The sulfides of Group I are quite insoluble, as Table 31-1 indicates.

The flow chart in Figure 31-1 outlines schematically the precipitation of the Group I sulfides and the analysis of this group of cations.

31-3 PREPARATION OF SOLUTIONS OF ANALYTICAL GROUP I CATIONS

Solutions of the Group I ions may be prepared by dissolving nitrates in dilute nitric acid to prevent hydrolysis of the metal ions (Section 19-5). Solutions of copper(II) nitrate are blue. Solutions of cadmium and bismuth(III) nitrate are colorless.

Solutions of Cu^{2+}, Cd^{2+}, and Bi^{3+} ions may also be prepared by dissolving their chlorides in dilute hydrochloric acid. Solutions of copper(II) chloride are green; the others are colorless.

31-4 SEPARATION OF ANALYTICAL GROUPS I AND II CATIONS FROM GROUP III AND SUBSEQUENT GROUP CATIONS

Cadmium sulfide is the most soluble of the Group I sulfides, $K_{sp} = 3.6 \times 10^{-29}$ (Table 31-1). Zinc sulfide is the least soluble of the freshly precipitated Group III sulfides, $K_{sp} = 1.1 \times 10^{-21}$ (Table 33-1). Therefore, to effect a clean separation of the Groups I and II from Group III cations, the acidity must be adjusted so that cadmium sulfide is precipitated as completely as possible in Group I and the solubility product for zinc sulfide is not exceeded. The following calculations illustrate this point.

EXAMPLE 31-1   *Equilibria in Sulfide Precipitations*

What is the minimum $[H^+]$ that will prevent the precipitation of zinc sulfide in a solution that is 0.020 M in $Zn(NO_3)_2$ (approximately 1.3 mg Zn^{2+}/mL) and saturated with H_2S?

Plan

In Appendix H we find K_{sp} for ZnS.

$$K_{sp} = [Zn^{2+}][S^{2-}] = 1.1 \times 10^{-21}$$

We are given $[Zn^{2+}]$ and so we solve the K_{sp} expression for the maximum $[S^{2-}]$ that can exist in the solution. Finally, we use the relationship for saturated H_2S solutions that contain a strong acid (Section 36-2) to calculate $[H^+]$.

$$[H^+]^2[S^{2-}] = 1.3 \times 10^{-21}$$

Solution

$$[Zn^{2+}][S^{2-}] = 1.1 \times 10^{-21}$$

$$[S^{2-}] = \frac{1.1 \times 10^{-21}}{[Zn^{2+}]} = \frac{1.1 \times 10^{-21}}{0.020} = 5.5 \times 10^{-20} \, M$$

Therefore, $[S^{2-}]$ cannot exceed $5.5 \times 10^{-20} \, M$ without exceeding K_{sp} for ZnS. In saturated H_2S solutions that contain a strong acid, the following relationship is valid. We can solve for $[H^+]$ because we know $[S^{2-}]$.

$$[H^+]^2[S^{2-}] = 1.3 \times 10^{-21}$$

$$[H^+]^2 = \frac{1.3 \times 10^{-21}}{[S^{2-}]} = \frac{1.3 \times 10^{-21}}{5.5 \times 10^{-20}} = 2.4 \times 10^{-2}$$

$$[H^+] = 0.15 \, M$$

Thus, the concentration of H^+ must be at least 0.15 M to prevent the precipitation of ZnS with the Groups I and II sulfides. The concentration of H^+ is adjusted to 0.3 M to provide some margin of safety.

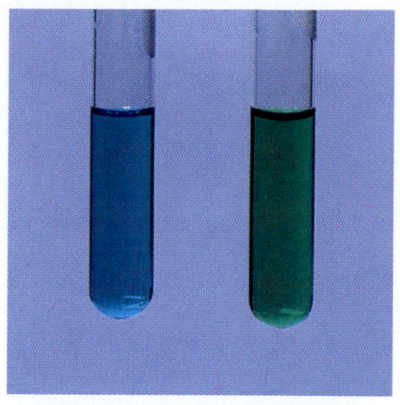

Copper(II) nitrate solutions are blue in excess HNO_3. Copper(II) chloride solutions are green in excess HCl.

Let us now demonstrate that cadmium ions are precipitated essentially completely from a solution under Group I precipitation conditions.

EXAMPLE 31-2   *Equilibria in Sulfide Precipitations*

What is the maximum concentration of Cd^{2+} ions that can exist in a solution that is 0.30 M in H^+ and saturated with H_2S?

Plan

We know the relationship for saturated H_2S solutions that contain a strong acid.

$$[H^+]^2[S^{2-}] = 1.3 \times 10^{-21}$$

We solve it for the maximum $[S^{2-}]$ that can exist in 0.30 M HCl solution. We know K_{sp} for CdS.

$$[Cd^{2+}][S^{2-}] = 3.6 \times 10^{-29}$$

Then we substitute the $[S^{2-}]$ into the K_{sp} expression for CdS to find the maximum $[Cd^{2+}]$ that can exist in this solution.

Solution

$$[H^+]^2[S^{2-}] = 1.3 \times 10^{-21}$$

$$[S^{2-}] = \frac{1.3 \times 10^{-21}}{[H^+]^2} = \frac{1.3 \times 10^{-21}}{(0.30)^2} = 1.4 \times 10^{-20} \, M$$

The maximum concentration of sulfide ions is $1.4 \times 10^{-20} \, M$. Substitution of this value into the solubility product for cadmium sulfide gives

$$[Cd^{2+}][S^{2-}] = 3.4 \times 10^{-29}$$

$$[Cd^{2+}] = \frac{3.4 \times 10^{-29}}{[S^{2-}]} = \frac{3.4 \times 10^{-29}}{1.4 \times 10^{-20}} = \boxed{2.4 \times 10^{-9}}$$

Thus, $2.4 \times 10^{-9} \, M$ is the maximum concentration of cadmium ions that can exist in a solution that is $0.30 \, M$ in H^+ and saturated with H_2S. These calculations show that Cd^{2+} ions are essentially completely precipitated under the Group I precipitation conditions.

31-5 MECHANISM OF SULFIDE PRECIPITATION

The concentration of sulfide ions in a solution saturated with H_2S and $0.3 \, M$ in H^+ ions is $1.4 \times 10^{-20} \, M$, or only about eight S^{2-} ions per milliliter of solution (Example 31-2). Yet the precipitation of many metal sulfides is essentially instantaneous in such solutions. This rapid precipitation of sulfides indicates that the reaction is *not* the combination of a simple metal ion and a simple sulfide ion. The concentration of hydrosulfide ions, HS^-, in saturated ($0.10 \, M$) H_2S solution that is also $0.3 \, M$ in HCl is much greater than the concentration of sulfide ions, as the following calculation shows.

> Strictly speaking, HS^- is the hydrogen sulfide ion. However, it is commonly called hydrosulfide ion. Aqueous solutions of H_2S are usually called hydrogen sulfide, although they are properly named hydrosulfuric acid.

$$H_2S \rightleftharpoons H^+ + HS^- \qquad \frac{[H^+][HS^-]}{[H_2S]} = 1.0 \times 10^{-7}$$

$$[HS^-] = \frac{1.0 \times 10^{-7}[H_2S]}{[H^+]} = \frac{(1.0 \times 10^{-7})(0.10)}{0.30} = 3.3 \times 10^{-8} \, M$$

The concentration of the HS^- ions is about 10^{12} times the concentration of sulfide ions. We assume that a metal ion forms an unstable intermediate hydrosulfide salt that breaks down immediately to form the insoluble metal sulfide and H_2S. This is shown for the precipitation of CuS.

$$Cu^{2+}(aq) + 2HS^- \rightleftharpoons Cu(SH)_2(s) \rightleftharpoons CuS(s) + H_2S(aq)$$

The analogous formation of oxides, through precipitation of hydroxides followed by decomposition into oxides, is well known. For example, when hydroxide ions are added to solutions containing silver ions, an unstable intermediate, silver hydroxide, precipitates. Almost instantly the unstable silver hydroxide decomposes into silver oxide and water.

$$2Ag^+(aq) + 2OH^-(aq) \rightleftharpoons 2AgOH(s) \rightleftharpoons Ag_2O(s) + H_2O$$

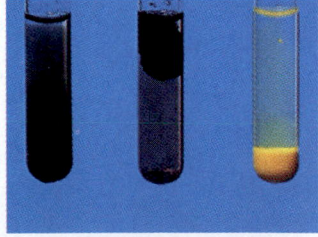

The sulfides of Group I. Left to right: Bi_2S_3 (dark brown), CuS (black), CdS (yellow).

Because equilibrium is concerned only with the final results of chemical reactions, and not with the mechanisms by which they occur, our calculations on the precipitation of sulfides are valid.

The equations for the precipitation of the Group I sulfides in a solution that is 0.30 M in HCl and saturated with H_2S may be written as

$$2Bi^{3+} + 3H_2S \longrightarrow Bi_2S_3(s) + 6H^+$$
dark brown

$$Cu^{2+} + H_2S \longrightarrow CuS(s) + 2H^+$$
black

$$Cd^{2+} + H_2S \longrightarrow CdS(s) + 2H^+$$
yellow

The sulfides of the Group I cations are all colored.

31-6 PRECIPITATION OF ANALYTICAL GROUP I CATIONS

Step A

Precipitation of the Group I sulfides. Use the known or unknown solution that contains some or all of Bi^{3+}, Cu^{2+}, and Cd^{2+}. To 10 drops of the known or unknown solution add 3 M aqueous ammonia until the solution is just basic to litmus. Now add 1 M HCl until the solution is just acidic to litmus. Add 2 drops of 6 M HCl and then dilute the solution to 1.5 mL. Mix well.

Add 12 drops of 5% thioacetamide solution. If directions have been followed carefully, the solution should now be approximately 0.3 M with respect to H^+. Heat the mixture in a hot water bath for 10 minutes to ensure hydrolysis of thioacetamide. Add 1 mL of water and continue to heat the mixture for 5 minutes.

The mixture should be stirred often during the heating process. Centrifuge, and separate the precipitate from the solution. (For general unknowns *only,* save the centrifugate for the analysis of later groups. Add 2 drops of 12 M HCl to the centrifugate and place it in the hot water bath until all H_2S has been expelled. Wash the precipitate twice with 10 drops of 0.1 M HCl. The first 10 drops of wash solution should be added to the original centrifugate and the rest should be discarded.) The precipitate is treated according to Step B.

31-7 DISSOLUTION OF ANALYTICAL GROUP I SULFIDES

Many metal sulfides become colloidal when they are washed with water. A solution of an electrolyte such as ammonium nitrate is used as the wash liquid to help prevent this.

The Group I sulfides are dissolved in hot dilute nitric acid, which oxidizes the sulfide ions to elemental sulfur. This decreases the concentration of sulfide ions so that the solubility products for the metal sulfides are no longer exceeded, $Q_{sp} < K_{sp}$, and so they dissolve. The equations for the dissolution of copper(II) sulfide are

$$CuS(s) \rightleftharpoons Cu^{2+} + S^{2-}$$

$$3S^{2-} + 8H^+ + 2NO_3^- \longrightarrow 3S(s) + 2NO(g) + 4H_2O$$

The equation for the net reaction is the sum of these two equations.

$$3CuS(s) + 8H^+ + 2NO_3^- \longrightarrow 3Cu^{2+} + 3S(s) + 2NO(g) + 4H_2O$$

The equations for the dissolution of the other Group I sulfides are similar. The reduction product from dilute nitric acid is nitric oxide, NO. Sulfur shows up in a variety of forms: sometimes as white flakes, sometimes in "powdery" forms, and sometimes as a "glob." In whatever form it appears, sulfur should be separated from the solution before beginning the next step.

> **Step B**
> Use the residue from Step A, containing CuS, CdS, and Bi_2S_3. Wash the solid sulfides with 16 drops of water to which 4 drops of 1 M NH_4NO_3 have been added, and discard the wash solution. Add 15 drops of 6 M HNO_3 to the residue and heat the mixture in a water bath until the sulfides have dissolved. Stir while the mixture is being heated.

31-8 SEPARATION AND IDENTIFICATION OF BISMUTH(III) IONS

The hydroxides of bismuth(III), copper(II), and cadmium precipitate when NH_3 is added to the solution.

$$Bi^{3+} + 3NH_3 + 2H_2O \longrightarrow Bi(OH)_3(s) + 3NH_4^+$$
$$\text{white}$$

$$Cu^{2+} + 2NH_3 + 2H_2O \longrightarrow Cu(OH)_2(s) + 2NH_4^+$$
$$\text{light blue}$$

$$Cd^{2+} + 2NH_3 + 2H_2O \longrightarrow Cd(OH)_2(s) + 2NH_4^+$$
$$\text{white}$$

Bismuth(III) hydroxide is insoluble in (does not react with) excess aqueous NH_3. As excess aqueous ammonia is added, the hydroxides of copper(II) and cadmium dissolve by forming tetraammine complexes.

$$Cu(OH)_2(s) + 4NH_3 \rightleftharpoons [Cu(NH_3)_4]^{2+} + 2OH^-$$
$$\text{deep blue}$$

$$Cd(OH)_2(s) + 4NH_3 \rightleftharpoons [Cd(NH_3)_4]^{2+} + 2OH^-$$
$$\text{colorless}$$

Bismuth(III) hydroxide is separated from the soluble complexes of copper and cadmium. It is then treated with a powerful reducing agent, sodium stannite, $Na[Sn(OH)_3]$, in the presence of excess NaOH. Bismuth(III) ions are reduced to metallic bismuth, which is black, like most finely divided metals.

The systematic name for $[Sn(OH)_3]^-$ is trihydroxostannate(II) ion.

$$2Bi(OH)_3(s) + 3[Sn(OH)_3]^- + 3OH^- \longrightarrow 2Bi(s) + 3[Sn(OH)_6]^{2-}$$

Because sodium stannite is such a powerful reducing agent, it is not stable in the air. It is prepared just prior to use by treating a solution of tin(II) chloride dissolved in HCl with excess NaOH. The first few drops of NaOH produce a white precipitate, $Sn(OH)_2$, which dissolves in excess NaOH to produce the colorless stannite ion, $[Sn(OH)_3]^-$.

$$[SnCl_3]^- + 2OH^- \longrightarrow Sn(OH)_2(s) + 3Cl^-$$

$$Sn(OH)_2(s) + OH^- \longrightarrow [Sn(OH)_3]^-$$

The stannite ion disproportionates to give black or gray metallic tin on long exposure to air.

$$2[Sn(OH)_3]^- \longrightarrow Sn(s) + [Sn(OH)_6]^{2-}$$

Examine the solution of tin(II) chloride carefully before you prepare sodium stannite. The solution should be clear and there should be pieces of metallic tin in the bottom of the container to keep the tin reduced to the $+2$ oxidation state.

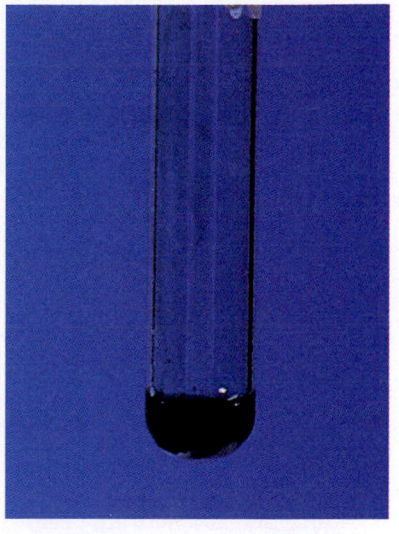

The confirmatory test for bismuth before the mixture is stirred. Some white $Bi(OH)_3$ can be seen in the bottom of the test tube.

Step C

Use the solution from Step B, containing Bi^{3+}, Cu^{2+}, and Cd^{2+}. Add 15 M aqueous NH_3 dropwise until the solution is basic to litmus, and then add 5 more drops. Stir thoroughly. The appearance of a deep blue color shows the presence of copper as tetraamminecopper(II) ions.* The formation of a white precipitate indicates the presence of Bi^{3+} ions. Observe carefully because $Bi(OH)_3$ is a gelatinous precipitate that is difficult to see in the deep-blue-colored solution. Separate the mixture and save the solution for Step D. [Prepare sodium stannite by the dropwise addition of 4 M NaOH to 2 drops of a 1 M solution of tin(II) chloride until a white precipitate forms and then dissolves. Approximately 4 to 6 drops of NaOH will be required.] Add the freshly prepared sodium stannite solution to the white precipitate of $Bi(OH)_3$. Stir thoroughly. The *immediate* formation of a black solid, finely divided metallic bismuth, confirms the presence of bismuth(III) ions.

If the acidity is too low in unknowns that also contain Group III cations, the sulfides of zinc, nickel(II) and cobalt(II) will also precipitate. Nickel(II) ions follow copper(II) ions and give an ammine complex, $[Ni(NH_3)_6]^{2+}$, at this point. Its color is violet, quite different from that of $[Cu(NH_3)_4]^{2+}$.

31-9 IDENTIFICATION OF COPPER(II) AND CADMIUM IONS

The presence of copper(II) ions in the solution is confirmed by the deep blue color of the complex ion, $[Cu(NH_3)_4]^{2+}$. However, if only traces of copper are present, the blue color may be too faint to be seen. A portion of the solution is acidified with acetic acid, and some potassium hexacyanoferrate(II), also called potassium ferrocyanide, $K_4[Fe(CN)_6]$, is added. The formation of a deep-rose-colored precipitate, $Cu_2[Fe(CN)_6]$, is a very sensitive test for copper(II) ions. If copper is absent and cadmium is present, a white precipitate, $Cd_2[Fe(CN)_6]$, should be observed.

$$[Cu(NH_3)_4]^{2+} + 2OH^- + 6CH_3COOH \longrightarrow Cu^{2+} + 4NH_4^+ + 6CH_3COO^- + 2H_2O$$

$$[Cd(NH_3)_4]^{2+} + 2OH^- + 6CH_3COOH \longrightarrow Cd^{2+} + 4NH_4^+ + 6CH_3COO^- + 2H_2O$$

$$2Cu^{2+} + [Fe(CN)_6]^{4-} \longrightarrow Cu_2[Fe(CN)_6](s) \quad \text{(deep rose)}$$

$$2Cd^{2+} + Fe(CN)_6]^{4-} \longrightarrow Cd_2[Fe(CN)_6](s) \quad \text{(white)}$$

Nearly everything is contaminated with traces of iron. Iron(III) ions react with $[Fe(CN)_6]^{4-}$ ions to give a deep-blue compound, $Fe_4[Fe(CN)_6]_3$. The "white" precipitate, $Cd_2[Fe(CN)_6]$, is usually very pale blue owing to traces of iron(III) ions. When copper is absent, this white precipitate serves as a confirmatory test for cadmium ions in Group I unknowns. However, when copper is present, a white precipitate cannot be observed in the presence of a deep-rose-colored precipitate. So an additional test for the presence of cadmium ions is necessary. (In general unknowns, students sometimes precipitate nickel sulfide in Group I. Nickel(II) ions follow copper(II) ions and give a pale blue precipitate, $Ni_2[Fe(CN)_6]$.)

When copper(II) ions are present, the presence of cadmium ions is confirmed by forming yellow CdS. However, before yellow CdS can be observed, copper (II) ions must be removed from the solution because they form a black sulfide, CuS. Two methods are used to detect cadmium ions in the presence of copper(II) ions. Each has some advantages and some disadvantages. Your instructor will indicate the preferred method.

In the first method, a portion of the basic solution that contains $[Cu(NH_3)_4]^{2+}$ and $[Cd(NH_3)_4]^{2+}$ ions is treated with $Na_2S_2O_4$, sodium dithionite, which reduces copper(II) to metallic copper.

$$[Cu(NH_3)_4]^{2+} + S_2O_4^{2-} + 2H_2O \longrightarrow Cu(s) + 2SO_3^{2-} + 4NH_4^{+}$$

However, it does not react with $[Cd(NH_3)_4]^{2+}$ ions. The addition of thioacetamide to the basic solution then results in the precipitation of CdS, a yellow compound, which confirms the presence of cadmium ions.

$$[Cd(NH_3)_4]^{2+} + S^{2-} \longrightarrow CdS(s) + 4NH_3$$

The advantage of the dithionite procedure is that traces of Bi(III) ions, which form a dark sulfide, are reduced to the free metal so that they do not interfere with the test for cadmium ions. The disadvantage of this procedure is that samples of sodium dithionite vary in composition. Some samples contain enough sodium sulfide, as an impurity, to form a (sulfide) precipitate as soon as the $Na_2S_2O_4$ is added. You should be alert to this possibility.

In the second method, also in ammoniacal solution, cyanide ions reduce copper(II) ions to copper(I) ions, which form very stable complex ions, $[Cu(CN)_2]^{-}$, that do not react with sulfide ions. Thus, the addition of excess potassium cyanide to the solution containing the tetraammine complexes of copper(II) and cadmium effectively removes the Cu^{2+} ions so that cadmium sulfide can be precipitated *and* observed.

$$2[Cu(NH_3)_4]^{2+} + 6CN^{-} \longrightarrow 2[Cu(CN)_2]^{-} + (CN)_2 + 8NH_3$$
$$\phantom{2[Cu(NH_3)_4]^{2+} + 6CN^{-} \longrightarrow 2[Cu(CN)_2]^{-}}_{\text{colorless}} _{\text{cyanogen}}$$

$$[Cu(CN)_2]^{-} + S^{2-} \longrightarrow \text{no reaction}$$

Cadmium ions also form complex ions in the presence of excess cyanide ions, but tetracyanocadmate ions are much less stable than dicyanocuprate(I) ions. They react with sulfide ions to form yellow cadmium sulfide.

$$[Cd(NH_3)_4]^{2+} + 4CN^{-} \longrightarrow [Cd(CN)_4]^{2-} + 4NH_3$$
$$_{\text{colorless}} _{\text{colorless}}$$

$$[Cd(CN)_4]^{2-} + S^{2-} \longrightarrow CdS(s) + 4CN^{-}$$
$$\phantom{[Cd(CN)_4]^{2-} + S^{2-} \longrightarrow } _{\text{yellow}}$$

Cyanides are very toxic, and all waste that contains CN^{-} ions should be poured into the cyanide waste container that your instructor designates so that it can be disposed of safely.

The advantage of the cyanide procedure is that it fairly effectively removes Ni^{2+} ions (a problem that occurs only in general unknowns when NiS has been precipitated in Group I rather than in Group III). The disadvantages are the toxicity of cyanides and the fact that any (stray) Bi^{3+} ions interfere with the Cd^{2+} test by forming a dark sulfide.

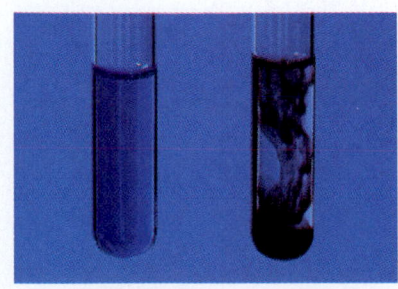

Confirmatory tests for copper. The blue solution contains $[Cu(NH_3)_4]^{2+}$ ions, and the deep-rose precipitate is $Cu_2[Fe(CN)_6]$.

Step D

Use the solution from Step C, containing $[Cu(NH_3)_4]^{2+}$ and $[Cd(NH_3)_4]^{2+}$ ions. A trace of copper can be present even if the solution appears colorless. Place 10 drops of the solution that contains the tetraammine complexes of copper(II) and cadmium in a test tube, add 6 M CH_3COOH until the solution is acidic to litmus, and then add 3 drops of 0.2 M $K_4Fe(CN)_6$. A deep-rose-colored precipitate, $Cu_2[Fe(CN)_6]$, demonstrates the presence of copper. (This was already obvious from the fact that the ammoniacal solution was deep blue.) A pink precipitate, $Cu_2[Fe(CN)_6]$ possibly mixed with $Cd_2[Fe(CN)_6]$, indicates the presence of a very low concentration of Cu^{2+} ions. A white precipitate, $Cd_2[Fe(CN)_6]$, shows the presence of cadmium and the absence of copper. (If NiS has been precipitated in Group I, a pale blue-green precipitate, $Ni_2[Fe(CN)_6]$, may be observed here when copper(II) ions are absent.)

Use either of the following methods, (1) or (2), as directed by your instructor, to detect cadmium ions in the presence of copper(II) ions.

(1) To the remainder of the solution that contains $[Cu(NH_3)_4]^{2+}$ and $[Cd(NH_3)_4]^{2+}$, add an amount of $Na_2S_2O_4$ about the size of four grains of rice. Stir well and heat the mixture for 3 minutes. Centrifuge, and separate the dark precipitate of metallic copper. To the clear solution, add 3 drops of thioacetamide, stir well, and heat the mixture for 3 minutes. A yellow precipitate, CdS, confirms the presence of cadmium ions.

(2) Use the remainder of the solution containing the tetraammine complexes for the following test. In a hood, add 4 drops of 4 M KCN, mix thoroughly, and heat the solution for 1 minute. Then add 2 drops of 5% thioacetamide solution. Heat the mixture in a hot water bath for 3 minutes. The formation of a yellow precipitate,* CdS, confirms the presence of cadmium. When Cu^{2+} ions are absent, the test for Cd^{2+} may be performed as described above by omitting the KCN treatment.

If your precipitate is so dark that you can't tell whether it contains CdS, the interfering dark sulfide Bi_2S_3 must be eliminated. Separate the mixture, discard the liquid into the cyanide waste container, and wash the precipitate twice with 10 drops of water to remove any adhering cyanide ions. Discard the wash liquid into the cyanide waste container. Add 6 drops of water *followed by* 3 drops of 6 M H_2SO_4 to the precipitate, and heat the mixture in the water bath for 2 minutes with constant stirring. Separate and discard any solid residue. Add 2 drops of 6 M aqueous ammonia to the clear, colorless liquid; mix well (the solution should be basic); and then add 2 drops of thioacetamide. Stir and heat for 3 minutes. The formation of a yellow precipitate confirms the presence of cadmium ions.

Traces of bismuth(III) ions interfere with the cadmium test by forming a dark sulfide. If traces of Bi^{3+} ions are present, the "yellow" sulfide precipitate may be olive to black depending on the amount of Bi^{3+} ions present. Because CdS is soluble in 2 M H_2SO_4 (Bi_2S_3 is not), it can easily be separated from the dark-colored sulfide.

Some alloys contain small percentages of copper.

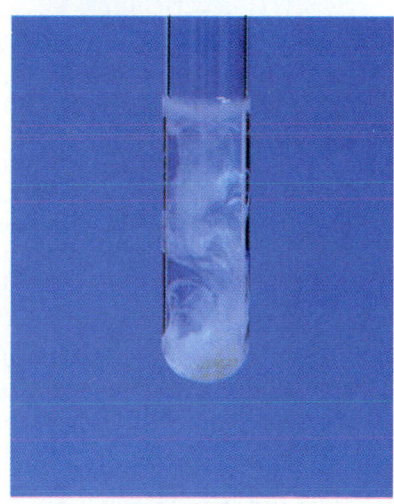

The confirmatory tests for cadmium.

Exercises

General Questions on Group I

1. List the common oxidation states exhibited by the metals whose cations occur in Group I. Indicate those that are usually reducing oxidation states by (R), those that are considered "stable" oxidation states by (S), and those that are oxidizing by (O).

2. The known and unknown solutions contain 5.0 mg of *cation* per milliliter of solution. Calculate the molarity of each *in terms of the Group I metal*. For example, the $Cu(NO_3)_2$ solution contains

5.0 mg Cu/mL, and the molarity should be calculated in terms of mol Cu/L or mmol Cu/mL.

3. (a) What is thioacetamide? (b) For what is it used? Why? (c) Write the equation for the hydrolysis of thioacetamide in hot acidic solution.

4. What is the basis for the separation of the Group I cations from the later groups of cations?

5. Calculate the concentrations of HS^- and S^{2-} ions in saturated H_2S solution that is also 0.30 M in HCl (Section 36-2).

6. Calculate the concentration of S^{2-} in saturated H_2S solutions of (a) pH = 1.00, (b) pH = 2.00, and (c) pH = 3.00.

7. Copper(II) sulfide, $K_{sp} = 8.7 \times 10^{-36}$, is the least soluble of the Group I sulfides; cadmium sulfide, $K_{sp} = 3.4 \times 10^{-29}$, is much more soluble. (a) Calculate the concentrations of Cu^{2+} and Cd^{2+} ions that remain in solution, i.e., unprecipitated, in a solution that is 0.30 M in HCl and saturated with H_2S. (b) What can you conclude about completeness of precipitation of these sulfides in this solution?

8. If the solution described in Exercise 7 were 0.15 M in HCl, what concentrations of Cu^{2+} and Cd^{2+} would remain unprecipitated?

9. (a) Based on your answers to Exercises 7 and 8, is there an obvious relationship between $[H^+]$ and $[Cu^{2+}]$ remaining in solution? What is it? (b) Is there a relationship between $[H^+]$ and $[Cd^{2+}]$ remaining in solution? What is it? (c) Is the relationship between $[H^+]$ and $[Cu^{2+}]$ similar to the relationship between $[H^+]$ and $[Cd^{2+}]$? Why? (d) Would the relationship between $[H^+]$ and $[Bi^{3+}]$ be of the same form? Why?

10. What color is each of the following? Aqueous solutions are indicated by (aq). $Cu(NO_3)_2$(aq); $Bi(NO_3)_3$(aq); $Cd(NO_3)_2$(aq); CuS; Bi_2S_3; CdS.

11. What color is each of the following? $Bi(OH)_3$; Bi; $Cu(OH)_2$; $Cu(NH_3)_4(OH)_2$(aq); $Cu_2[Fe(CN)_6]$; $Cd(OH)_2$; $Cd(NH_3)_4(OH)_2$(aq); $Cd_2[Fe(CN)_6]$.

Group I Reactions

Write balanced net ionic equations for the reactions that occur when the following substances are mixed in aqueous solution. Indicate the colors of all precipitates and complex ions.

Step A

12. Bismuth(III) chloride and hydrogen sulfide.
13. Copper(II) chloride and hydrogen sulfide.
14. Cadmium chloride and hydrogen sulfide.

Step B

15. Solid bismuth(III) sulfide and hot 6 M nitric acid.
16. Solid copper(II) sulfide and hot 6 M nitric acid.
17. Solid cadmium sulfide and hot 6 M nitric acid.

Step C

18. Bismuth(III) sulfate and conc. aq. ammonia.

19. Copper(II) sulfate and a lim. amt. of conc. aq. ammonia.
20. Copper(II) hydroxide and excess conc. aq. ammonia.
21. Cadmium sulfate and a lim. amt. of conc. aq. ammonia.
22. Cadmium hydroxide and excess conc. aq. ammonia.
23. Tin(II) chloride in HCl and a lim. amt. of NaOH.
24. Solid tin(II) hydroxide and excess NaOH.
25. Solid bismuth(III) hydroxide and sodium stannite (in excess NaOH).

Step D

26. Tetraamminecopper(II) hydroxide and acetic acid.
27. Tetraamminecadmium hydroxide and acetic acid.
28. Copper(II) acetate and potassium hexacyanoferrate(II).
29. Cadmium acetate and potassium hexacyanoferrate(II).
30. Tetraamminecopper(II) hydroxide and excess KCN.
31. Tetraamminecadmium hydroxide and excess KCN.
32. Potassium tetracyanocadmate and $(NH_4)_2S$.
33. Tetraamminecopper(II) hydroxide and sodium dithionite.
34. Tetraamminecadmium hydroxide and $(NH_4)_2S$.

Other Questions and Problems

35. Explain the equilibria involved in the dissolution of the Group I sulfides in hot 6 M HNO_3.

36. (a) Which of the Group I cations form hydroxides that are soluble in excess aqueous ammonia? (b) What are the formulas for their ammine complexes?

37. Which of the Group I cations form hydroxides that are soluble in excess NaOH solution?

38. Why is sodium stannite prepared just before it is used in the confirmatory test for bismuth(III) ions?

39. In Step D the ammine complexes of Cu^{2+} and Cd^{2+} are destroyed by adding CH_3COOH before $[Fe(CN)_6]^{4-}$ ions are added. Why?

40. A solution is 0.095 M each in $Cu(NO_3)_2$, $Bi(NO_3)_3$, $Cd(NO_3)_2$, and HCl. If sufficient H_2S is added so that the solution is saturated when precipitation is as complete as possible, what concentrations of Cu^{2+}, Bi^{3+}, and Cd^{2+} remain in solution?

41. Is it possible to increase the acidity of a solution to the point that CdS will not precipitate if the solution is 0.050 M in $Cd(NO_3)_2$ and saturated with H_2S? Justify your answer by appropriate calculations.

42. Aqueous ammonia was added to a Group I unknown. A precipitate formed initially. All of it dissolved and the solution turned deep blue as more aqueous ammonia was added. Which cation(s) could be present?

43. No precipitate was observed in a Group I unknown that was also 6 M in aqueous ammonia. Which cations could be present?

44. A Group I unknown gave a yellow sulfide precipitate in the initial precipitation of the group. Which cations could be present?

45. A Group I unknown gave a black sulfide precipitate in the initial precipitation of the group. Which cations could be present?

Analysis of Cation Group II

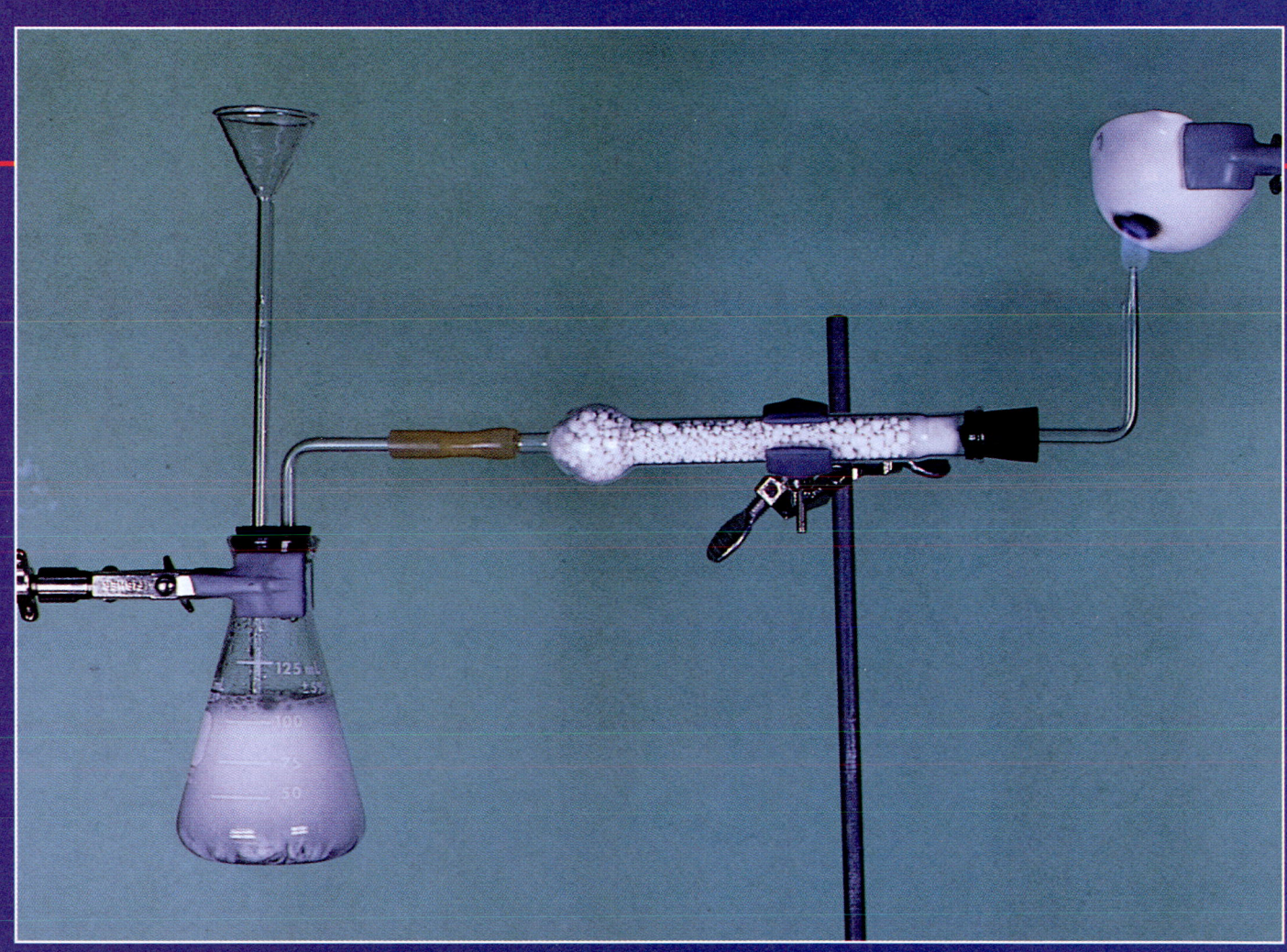

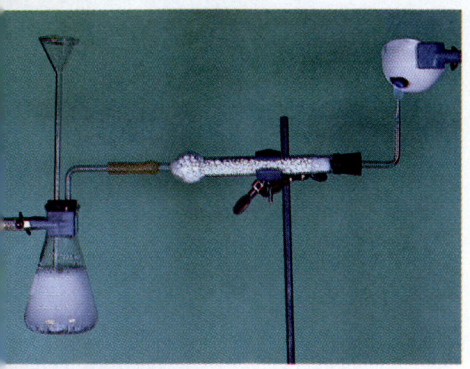

This photo shows a modification of the Marsh test for arsenic, a classic procedure in criminology labs. A solution containing arsenic was added to the flask, which contains Zn and dilute H_2SO_4. As gaseous arsine, AsH_3, is formed, it is carried in the stream of H_2 produced by the reaction of Zn and H_2SO_4. Anhydrous $CaCl_2$ absorbs H_2O vapor in the H_2/AsH_3 gas stream. Hydrogen is ignited as it escapes from the glass tube at the right. Arsine imparts a yellow-green color to the nearly colorless H_2/O_2 flame. The dark spot on the bottom of the evaporating dish is elemental arsenic.

32-1 COMMON OXIDATION STATES OF METALS IN CATION GROUP II

Arsenic is classified as a metalloid. As expected, it exhibits some properties of both metals and nonmetals. In aqueous solutions its important oxidation states are +3 and +5. It can be reduced to the −3 oxidation state by strong reducing agents. As the following half-reaction indicates, arsenic(V) is a mild oxidizing agent.

$$H_3AsO_4(aq) + 2H^+(aq) + 2e^- \longrightarrow H_3AsO_3(aq) + H_2O \qquad E^0 = +0.58 \text{ V}$$

Antimony is decidedly more metallic than arsenic. Compounds containing antimony in the +5 oxidation state are stronger oxidizing agents than are similar compounds of arsenic.

$$[SbCl_6]^-(aq) + 2e^- \longrightarrow [SbCl_4]^-(aq) + 2Cl^-(aq) \qquad E^0 = +0.75 \text{ V}$$

In aqueous solutions the most important oxidation state of Sb is +3.

Tin exhibits two well-defined oxidation states, Sn(II) and Sn(IV). In the +2 oxidation state, tin is a mild reducing agent in acidic solution.

$$Sn^{4+}(aq) + 2e^- \longrightarrow Sn^{2+}(aq) \qquad E^0 = +0.15 \text{ V}$$

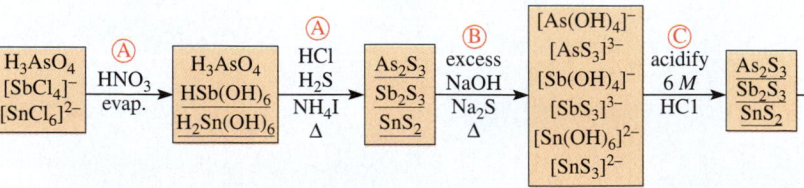

Figure 32-1 Analytical Group II flow chart. Circled letters refer to analytical steps. Species formed in confirmatory tests are shown with blue backgrounds.

Antimony and arsenic, two metalloids, are both semiconducting elements. (Left) A bar of antimony with a sample of one of its ores, stibnite; (right) arsenic.

Metallic tin is added to aqueous solutions of Sn(II) salts to prevent atmospheric oxidation. In basic solutions Sn(II) is such a strong reducing agent that it is unstable.

32-2 INTRODUCTION TO THE ANALYTICAL PROCEDURES

Like the cations of Analytical Group I, the Group II cations form sulfides that are insoluble in dilute acidic solutions. However, sulfides of the Group II cations are soluble in solutions of strong soluble bases, and so the cations of analytical Group II are referred to as the **acid-insoluble, base-soluble sulfide group.** Group II sulfides are quite insoluble in water (Table 32-1).

Exact values of the solubility products for arsenic(III) sulfide, As_2S_3; antimony(III) sulfide, Sb_2S_3; and tin(IV) sulfide, SnS_2, are difficult to determine experimentally. This is because both the cations and the anions of these sulfides hydrolyze extensively in complex reactions. However, many experiments have demonstrated conclusively that these sulfides are quite insoluble.

The flow chart in Figure 32-1 outlines schematically the precipitation of the Group II sulfides and the analysis of this group of cations.

Table 32-1 *Solubility Products for Some Group II Sulfides*

Sulfide	K_{sp}
Sb_2S_3	1.6×10^{-93}
SnS_2	1×10^{-70}

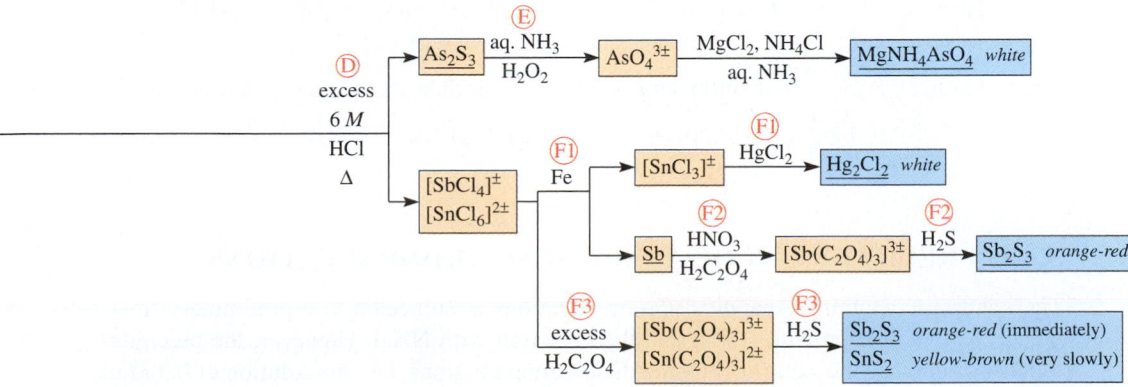

32-3   PREPARATION OF SOLUTIONS OF ANALYTICAL GROUP II CATIONS

Solutions of antimony and tin are usually prepared by dissolving their chlorides in dilute hydrochloric acid to prevent hydrolysis of the metal ions.

$$SbCl_3(s) + [H^+ + Cl^-] \longrightarrow [H^+ + SbCl_4^-] \qquad \text{tetrachloroantimonic(III) acid}$$

$$SnCl_4(\ell) + 2[H^+ + Cl^-] \longrightarrow [2H^+ + SnCl_6^{2-}] \qquad \text{hexachlorostannic(IV) acid}$$

covalent compounds   chloro acids (strong acids)

Solutions containing arsenic are conveniently prepared by dissolving diarsenic pentoxide in hot water.

$$As_2O_5(s) + 3H_2O \xrightarrow{\Delta} 2H_3AsO_4(aq) \qquad \text{arsenic acid (a weak acid)}$$

32-4   EVAPORATION WITH NITRIC ACID

Before the Group II precipitation takes place, the solution is "cleaned up" by evaporation with nitric acid. This converts antimony and tin to insoluble acids in their highest oxidation states, Sb(V) and Sn(IV). Arsenic is already in its highest oxidation state, As(V), in H_3AsO_4.

The formula for $HSb(OH)_6$ has been established. The formula for $H_2Sn(OH)_6$ has not. It is a hydrated oxide, $SnO_2 \cdot xH_2O$, which is written as either H_2SnO_3 or $H_2Sn(OH)_6$. Salts derived from this compound are of the type $Na_2Sn(OH)_6$, so we will represent the acid, or hydrated oxide, as $H_2Sn(OH)_6$.

$$3[SbCl_4]^- + 5H^+ + 2NO_3^- + 14H_2O \xrightarrow[\text{HNO}_3]{\text{evap.}} 3HSb(OH)_6(s) + 2NO + 12HCl(g)$$

antimonic acid
(a white solid)

$$[SnCl_6]^{2-} + 2H^+ + 6H_2O \xrightarrow[\text{HNO}_3]{\text{evap.}} H_2Sn(OH)_6(s) + 6HCl(g)$$

stannic acid
(a white solid)

The precipitation of arsenic(V) sulfide is very slow. Ammonium iodide, NH_4I, is added to reduce arsenic from the +5 to the +3 oxidation state so that it can be precipitated more rapidly as arsenic(III) sulfide.

$$H_3AsO_4(aq) + 2H^+(aq) + 2I^-(aq) \longrightarrow H_3AsO_3(s) + I_2(s) + H_2O$$

arsenic acid   arsenous acid

Antimony is also reduced to the +3 oxidation state by iodide ions.

$$HSb(OH)_6(s) + 2H^+(aq) + 2I^-(aq) \longrightarrow H_3SbO_3(s) + I_2(s) + 3H_2O$$

antimonic acid   antimonous acid

Iodine oxidizes H_2S to free sulfur and reappears as iodide ions in the solution.

$$I_2(s) + H_2S(aq) \longrightarrow 2H^+(aq) + 2I^-(aq) + S(s)$$

32-5   PRECIPITATION OF ANALYTICAL GROUP II CATIONS

The solution containing Analytical Group II cations is subjected to a preliminary treatment, i.e., evaporation with HNO_3 and then reaction with NH_4I. However, the precipitation conditions are the same as for Analytical Group I cations, i.e., the solution of 0.3 M in H^+ and saturated with H_2S. The equations for the precipitation of the Group II sulfides in this solution may be written as

$$2H_3AsO_3(s) + 3H_2S \longrightarrow As_2S_3(s) + 6H_2O$$
$$\text{yellow}$$

$$2H_3SbO_3(s) + 3H_2S \longrightarrow Sb_2S_3(s) + 6H_2O$$
$$\text{orange-red}$$

$$H_2Sn(OH)_6(s) + 2H_2S \longrightarrow SnS_2(s) + 6H_2O$$
$$\text{light tan}$$

The sulfides of the Group II cations are colored.

H_3AsO_3 and H_3SbO_3 may also be written as $As(OH)_3$ and $Sb(OH)_3$ because both are amphoteric. They are also written as $HAsO_2$ and $HSbO_2$, which contain one less H_2O per formula unit.

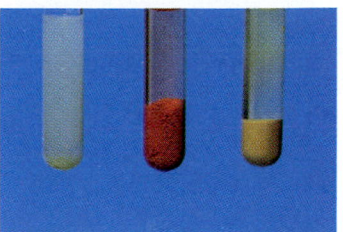

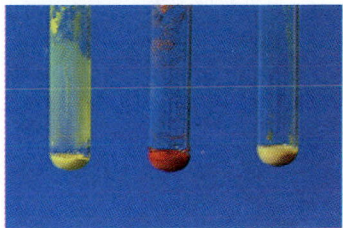

The sulfides of Group II. Left to right: As_2S_3 (pale yellow), Sb_2S_3 (orange-red), SnS_2 (light tan).

Step A

Precipitation of the Analytical Group II sulfides. Use the known or unknown solution that contains some or all of H_3AsO_4, $SbCl_4^-$, and $SnCl_6^{2-}$. Add 4 drops of 6 M HNO_3 to 10 drops of the known or unknown solution and evaporate the solution to a moist residue (Section 30-16). Care should be taken to avoid evaporating the precipitate to dryness. Cool the moist residue and add 10 drops of water, mix well, and add 1 M aqueous ammonia until the solution is just basic to litmus. Now add 1 M HCl until the solution is just acidic to litmus. Add 2 drops of 6 M HCl and 1 drop of 1 M NH_4I, and then dilute the solution to 1.5 mL. Mix well.

Heat the solution in the water bath for 3 minutes. Add 12 drops of 5% thioacetamide solution. If directions have been followed carefully, the solution should now be approximately 0.3 M with respect to H^+. Heat the mixture in a hot water bath for 10 minutes to ensure hydrolysis of thioacetamide. Add 1 mL of water and continue to heat the mixture for 5 minutes.

The mixture should be stirred often during the heating process. Centrifuge, and separate the precipitate from the solution. (For general unknowns *only*, save the centrifugate for the analysis of Groups III to V. Add 2 drops of 12 M HCl to the centrifugate and place it in the hot water bath until all H_2S has been expelled. Wash the precipitate twice with 10 drops of 0.1 M HCl. The first 10 drops of wash solution should be added to the original centrifugate and the rest should be discarded.) The precipitate is treated according to Step B.

32-6 DISSOLUTION OF ANALYTICAL GROUP II SULFIDES IN SODIUM HYDROXIDE SOLUTION

The oxides and hydroxides of arsenic(III), antimony(III), and tin(IV) are amphoteric (Section 10-8). Their sulfides exhibit **thioamphoterism;** that is, the sulfides of these three elements form soluble complex compounds in strongly basic solutions. Sodium hydroxide (NaOH) is the reagent used to dissolve these sulfides. A little thioacetamide is also added so there will be plenty of sulfide ions when we are ready to reprecipitate these sulfides.

$$As_2S_3(s) + 4OH^- \longrightarrow [As(OH)_4]^- + [AsS_3]^{3-} \qquad \text{trithioarsenate(III) ion*}$$

You may wonder if species such as $[As(OH)_2S]^-$ and $[As(OH)S_2]^{2-}$ are formed in this reaction. Yes, they are. Similar species are also formed in the dissolution reactions of Sb_2S_3 and SnS_2. We have represented the reactions in the simplest way that is consistent with accuracy.

$$Sb_2S_3(s) + 4OH^- \longrightarrow [Sb(OH)_4]^- + [SbS_3]^{3-} \qquad \text{trithioantimonate(III) ion}$$

$$3SnS_2(s) + 6OH^- \longrightarrow [Sn(OH)_6]^{2-} + 2[SnS_3]^{2-} \qquad \text{trithiostannate(IV) ion}$$

The prefix thio- *always refers to sulfur.*

The sulfur-containing products of these reactions are called *thio* salts. (The Na^+ ion is the only cation present in significant concentrations in this strongly basic solution.)

C A U T I O N

Failure to dissolve the Group II sulfides in the NaOH solution results in arsenic, antimony, and tin being missed.

As_2S_3, Sb_2S_3, *and* SnS_2 *are soluble in 4* M *NaOH.*

Step B
Use the precipitate from Step A, containing As_2S_3, Sb_2S_3, and SnS_2. Add 12 drops of 4 *M* NaOH and 5 drops of thioacetamide solution to the Group II sulfide precipitate. Heat the mixture in a water bath for 5 minutes, stirring constantly. The precipitate will dissolve completely for unknowns that contain only Group II cations. A precipitate will remain in unknowns that contain *both* Groups I and II. Centrifuge the mixture and treat the clear solution according to Step C. Use the precipitate (Group I sulfides) to begin the analysis for Group I cations with Step B in Group I.

After we have completed the analysis of Group II cations, we shall analyze an unknown solution that contains ions from both Group I and Group II. We use this procedure (Step B), dissolution of Group II sulfides in sodium hydroxide solution, to separate Group I and Group II cations. The sulfides of Group I cations are insoluble in sodium hydroxide solution.

32-7 REPRECIPITATION OF ANALYTICAL GROUP II SULFIDES

The Group II sulfides can be reprecipitated from the strongly basic solution of their hydroxocomplexes and thiosalts by removing the excess S^{2-} and OH^- ions. The addition of an acid decreases the basicity of the solution and destroys the hydroxocomplexes and thiosalts. Sulfide ions then combine with Group II cations to reprecipitate their insoluble sulfides. The equations for the equilibria for the reprecipitation of tin(IV) sulfide from its thiosalt are

$$[SnS_3]^{2-} \rightleftharpoons SnS_2(s) + S^{2-}$$
$$\underline{S^{2-} + 2H^+ \longrightarrow H_2S(g)}$$
$$\text{net reaction} \quad [SnS_3]^{2-} + 2H^+ \longrightarrow SnS_2(s) + H_2S(g)$$

The equilibria for the precipitation of SnS_2 from its hydroxocomplex are

$$[Sn(OH)_6]^{2-} + 6H^+ \longrightarrow Sn^{4+} + 6H_2O$$
$$\underline{Sn^{4+} + 2H_2S \longrightarrow SnS_2(s) + 4H^+}$$
$$\text{net reaction} \quad [Sn(OH)_6]^{2-} + 2H^+ + 2H_2S \longrightarrow SnS_2(s) + 6H_2O$$

These equations show that the addition of a strong acid displaces the equilibrium in the direction of insoluble tin(IV) sulfide, $[Sn^{4+}][S^{2-}]^2 > K_{sp}$, because of the formation of an insoluble compound, SnS_2, and a weak electrolyte, H_2O. Multiplying the first equation by two and combining with the second equation gives

$$2[SnS_3]^{2-} + [Sn(OH)_6]^{2-} + 6H^+ \longrightarrow 3SnS_2(s) + 6H_2O$$

The net ionic equations for the reprecipitation of the Group II sulfides are

Recall from Step B that $[SnS_3]^{2-}$ and $[Sn(OH)_6]^{2-}$ are formed in a $2:1$ mole ratio.

$$[AsS_3]^{3-} + [As(OH)_4]^- + 4H^+ \longrightarrow As_2S_3(s) + 4H_2O$$

$$[SbS_3]^{3-} + [Sb(OH)_4]^- + 4H^+ \longrightarrow Sb_2S_3(s) + 4H_2O$$

$$2[SnS_3]^{2-} + [Sn(OH)_6]^{2-} + 6H^+ \longrightarrow 3SnS_2(s) + 6H_2O$$

Failure to acidify the solution that contains the hydroxocomplexes and the thiosalts of the Group II cations is a common mistake. If the solution is not acidified properly, the II sulfides fail to reprecipitate and are discarded. Ions poured into the waste container will be missed!

Step C
Use the solution from Step B, containing hydroxocomplexes and thiosalts of the Group II ions: $[AsS_3]^{3-}$, $[As(OH)_4]^-$, $[SbS_3]^{3-}$, $[Sb(OH)_4]^-$, $[SnS_3]^{2-}$, and $[Sn(OH)_6]^{2-}$ ions. Add 6 M HCl, with vigorous stirring, until the solution is just acidic to litmus. Heat the mixture in a water bath for 5 minutes. (Because large amounts of H_2S are evolved as the solid sulfides form in the solution, caution must be exercised to determine that the solution is acidic throughout. When you are convinced that the solution is acidic, stir it thoroughly. Then test for acidity again.) Separate the mixture, discard the solution, and save the precipitate of As_2S_3, Sb_2S_3, and SnS_2 for Step D.

32-8 SEPARATION OF ARSENIC FROM ANTIMONY AND TIN

The K_{sp}'s for the Group II sulfides are quite small. To dissolve Sb_2S_3 and SnS_2, while leaving solid As_2S_3 behind, the concentrations of H^+ and S^{2-} ions must be adjusted so that $[Sb^{3+}]^2[S^{2-}]^3 < K_{sp}$ and $[Sn^{4+}][S^{2-}]^2 < K_{sp}$, while $[As^{3+}]^2[S^{2-}]^3 > K_{sp}$. Hot 6 M HCl reduces the concentrations of the ions of antimony(III) and tin(IV) sulfides to values such that these sulfides dissolve.

$$SnS_2(s) + 4H^+ + 6Cl^- \rightleftharpoons [SnCl_6]^{2-} + 2H_2S(g)$$

$$Sb_2S_3(s) + 6H^+ + 8Cl^- \rightleftharpoons 2[SbCl_4]^- + 3H_2S(g)$$

The sulfide of arsenic is not dissolved by 6 M HCl because its solubility product constant is so small. The reactions in which Sb_2S_3 and SnS_2 dissolve in hot 6 M HCl are *reversible*. Reprecipitation of Sb_2S_3 and SnS_2 may occur when the solution cools. If the solution turns orange or yellow *after* it has been drawn into the capillary pipet, the separation has been accomplished. If it turns orange or yellow *before* it is separated from As_2S_3, the mixture must be heated again before separation is possible.

Step D

Use the precipitate from Step C, containing As_2S_3, Sb_2S_3, and SnS_2. Add 1 mL of 6 M HCl to the precipitate, stir the mixture thoroughly, and heat it in a hot water bath for 3 minutes with regular stirring. Separate the residue *as quickly as possible,* and save the centrifugate. Add 1 mL of 6 M HCl to the residue and heat the mixture for 3 minutes, stirring regularly. Separate any residue, and save the combined centrifugates for Step F. (If the HCl solution becomes cool, Sb_2S_3 and SnS_2 may reprecipitate. If the solution turns orange or yellow *inside* the capillary pipet, the separation has been accomplished and there is no need to worry.) The residue is analyzed by Step E. Place the solution containing $[SbCl_4]^-$ and $[SnCl_6]^{2-}$ in the hot water bath so that H_2S will be expelled from it while you are identifying arsenic.

32-9 IDENTIFICATION OF ARSENIC

A saturated solution of arsenic(III) sulfide contains such a low concentration of S^{2-} ions that this sulfide cannot be dissolved completely by the addition of H^+ (from strong acids). The product of $[H^+]^2$ and $[S^{2-}]$ does not exceed the ion product constant for H_2S, even in solutions containing the maximum $[H^+]$ attainable, because $[S^{2-}]$ is so small. Therefore, equilibrium is established before arsenic(III) sulfide dissolves completely.

However, the concentration of sulfide ions can be reduced by *oxidation* so that As_2S_3 dissolves. A mixture of aqueous ammonia and hydrogen peroxide dissolves As_2S_3.

$$As_2S_3(s) + 14H_2O_2 + 12OH^- \longrightarrow 2AsO_4^{3-} + 3SO_4^{2-} + 20H_2O$$

Arsenic(III) ions are oxidized to arsenate ions, AsO_4^{3-}, while sulfide ions are oxidized to sulfate ions. The concentrations of both As^{3+} and S^{2-} are lowered to the point that $[As^{3+}]^2[S^{2-}]^3 < K_{sp}$, so As_2S_3 dissolves.

When magnesia mixture is added to a solution containing arsenate ions, a white crystalline compound, magnesium ammonium arsenate, $MgNH_4AsO_4$, precipitates. Magnesia mixture is a buffered aqueous ammonia–ammonium chloride solution that contains magnesium chloride.

$$Mg^{2+} + NH_4^+ + AsO_4^{3-} \longrightarrow MgNH_4AsO_4(s) \qquad \text{white}$$

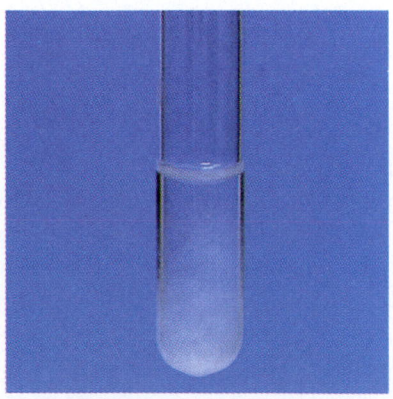

The confirmatory test for arsenic, $MgNH_4AsO_4$.

Step E

Use the residue from Step D, As_2S_3. Add 8 drops of 6 M aqueous ammonia, 4 drops of water, and 6 drops of 6% hydrogen peroxide to the As_2S_3. Stir thoroughly and heat the mixture in a hot water bath with constant stirring for 5 minutes. Separate the mixture. If the solution containing AsO_4^{3-} ions also contains colloidal sulfur, wrap a small piece of cotton around the tip of a capillary pipet and then draw the liquid into the pipet. Most of the colloidal sulfur should be trapped by the cotton. Transfer the liquid to a clean test tube and add 2 drops of 15 M aqueous ammonia and 5 drops of magnesia mixture to the solution. The formation of a white precipitate, $MgNH_4AsO_4$, confirms the presence of arsenate ions. The precipitate may form slowly. Scratching the test tube walls with a stirring rod usually hastens crystallization.

32-10 IDENTIFICATION OF ANTIMONY AND TIN

Recall that antimony(III) and tin(IV) sulfides were dissolved in hot $6\,M$ HCl and are in the form of complex ions, $[SbCl_4]^-$ and $[SnCl_6]^{2-}$. Active metals reduce antimony to the metallic state and tin to the $+2$ oxidation state in hydrochloric acid solution. This reaction provides a convenient method for separating antimony and tin. We use a small iron nail as a source of an active metal.

$$2[SbCl_4]^- + 3Fe(s) \longrightarrow 2Sb(s) + 3Fe^{2+} + 8Cl^-$$

$$[SnCl_6]^{2-} + Fe(s) \longrightarrow [SnCl_3]^- + Fe^{2+} + 3Cl^-$$

Tin is a reducing agent in the $+2$ oxidation state. The reaction of tin(II) chloride with mercury(II) chloride is used as the confirmatory test for tin. The reactions occur in a solution that contains an excess of hydrochloric acid.

$$[SnCl_3]^- + 2[HgCl_4]^{2-} \longrightarrow Hg_2Cl_2(s) + [SnCl_6]^{2-} + 3Cl^-$$

$$Hg_2Cl_2(s) + [SnCl_3]^- + Cl^- \longrightarrow 2Hg(\ell) + [SnCl_6]^{2-}$$

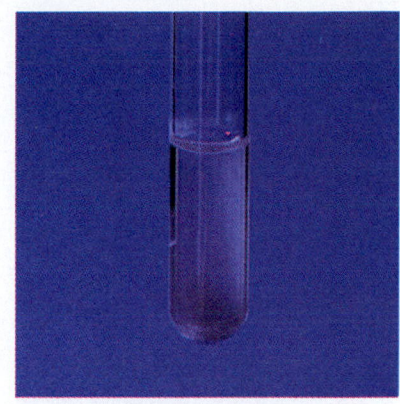

The confirmatory test for tin.

Hg_2Cl_2 is a white, highly crystalline substance as it is formed in this reaction. A little of it is reduced to liquid mercury by $[SnCl_3]^-$ ions.

If black flakes of metallic antimony are formed when the nail is placed in the solution, this provides convincing evidence that antimony was present in the solution. However, we shall perform a confirmatory test. Antimony reacts with dilute nitric acid to form the insoluble oxide, Sb_4O_6. This oxide dissolves in oxalic acid, $H_2C_2O_4$, to form a soluble complex compound, $H_3[Sb(C_2O_4)_3]$.

$$4Sb(s) + 4H^+ + 4NO_3^- \longrightarrow Sb_4O_6(s) + 4NO(g) + 2H_2O$$

$$Sb_4O_6(s) + 12H_2C_2O_4 \longrightarrow 12H^+ + 4[Sb(C_2O_4)_3]^{3-} + 6H_2O$$

These two equations are frequently combined into a single equation

$$Sb(s) + NO_3^- + 3H_2C_2O_4 \longrightarrow [Sb(C_2O_4)_3]^{3-} + 2H^+ + NO(g) + 2H_2O$$

The precipitation of the orange-red sulfide, Sb_2S_3, is the confirmatory test.

$$2[Sb(C_2O_4)_3]^{3-} + 3H_2S + 6H^+ \longrightarrow Sb_2S_3(s) + 6H_2C_2O_4$$

Excess nitric acid remaining in the solution oxidizes hydrogen sulfide to free sulfur and interferes with the precipitation of Sb_2S_3. Therefore, it is important (a) to add no more nitric acid than necessary to dissolve antimony, and (b) to add an excess of thioacetamide.

An alternative test for antimony(III) is available. Both Sb(III) and Sn(IV) ions react with oxalic acid to form soluble complex compounds.

$$[SbCl_4]^- + 3H_2C_2O_4 \longrightarrow [Sb(C_2O_4)_3]^{3-} + 6H^+ + 4Cl^-$$

$$[SnCl_6]^{2-} + 3H_2C_2O_4 \longrightarrow [Sn(C_2O_4)_3]^{2-} + 6H^+ + 6Cl^-$$

$[Sn(C_2O_4)_3]^{2-}$ is a very stable complex ion.

However, the trisoxalatostannate(IV) ion is much more stable than the trisoxalatoantimonate(III) ion. Its stability allows us to detect antimony(III) ions in the presence of Sn(IV) ions. The less stable $[Sb(C_2O_4)_3]^{3-}$ ion reacts with H_2S immediately to give an orange-red precipitate, Sb_2S_3. The $[Sn(C_2O_4)_3]^{2-}$ ion is so stable that it reacts with H_2S only *very slowly* to produce SnS_2, which is light tan.

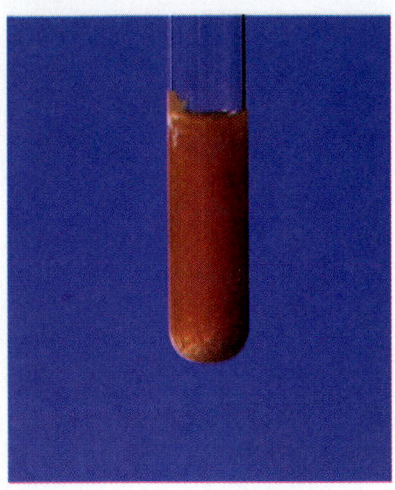

The confirmatory test for antimony, Sb_2S_3.

Step F

Use the solution from Step D, containing $[SbCl_4]^-$ and $[SnCl_6]^{2-}$ ions. All the H_2S should have been expelled by boiling this solution while the test for arsenic was performed.

Perform either (1) and (2) only or (1), (2), and (3) as directed by your instructor. If you perform only (1) and (2), use all of the solution that contains $[SbCl_4]^-$ and $[SnCl_6]^{2-}$ ions for (1). If you perform (1), (2) and (3), use one half of the solution for (1) and one half for (3).

(1) Add 2 drops of 6 *M* HCl to *(one half of)* the solution. Place a clean, small iron nail in the test tube and heat the mixture for 5 minutes. Black flakes of metallic antimony should appear. Withdraw the liquid and place it in a clean test tube. Add 2 drops of 6 *M* HCl to the solution and heat it in a water bath for 3 minutes. Add 5 drops of 0.2 *M* $HgCl_2$ to the warm solution. The formation of a white or gray precipitate confirms the presence of tin.

(2) Remove the iron nail, leaving the black flakes of metallic antimony in the test tube. (Small black pieces of carbon may remain if a significant amount of the nail dissolved. These are usually very small pieces that don't resemble flakes of antimony.) Place the nail in a solid disposal receptacle. Add 1 drop of 6 *M* HNO_3 and an amount of solid $H_2C_2O_4 \cdot 2H_2O$ about the size of a pea to the black flakes. Warm the mixture in a water bath until the black flakes dissolve. (If the black flakes fail to dissolve in 3 minutes, add 1 more drop of 6 *M* HNO_3 and warm the solution.) Add 10 drops of H_2O and 5 drops of thioacetamide to the solution. Stir thoroughly and then place the test tube in a hot water bath for 3 minutes. The formation of an orange-red precipitate, Sb_2S_3, confirms the presence of antimony. (Excess nitric acid oxidizes hydrogen sulfide to free sulfur. Do not confuse free sulfur with an orange-red precipitate of Sb_2S_3.)

(3) *Alternative test for antimony:* Add an amount of solid oxalic acid dihydrate, $H_2C_2O_4 \cdot 2H_2O$, about the size of a pea to *one half* of the solution that contains $[SbCl_4]^-$ and $[SnCl_6]^{2-}$ ions. Place the test tube in a hot water bath and stir it for 3 minutes. Some solid $H_2C_2O_4 \cdot 2H_2O$ should remain—if not, add more and heat for an additional 3 minutes. Separate the clear, colorless liquid and place it in a clean test tube. Add 3 drops of thioacetamide, mix well, and place the test tube in the hot water bath for 3 minutes. The formation of an orange-red precipitate, Sb_2S_3, confirms the presence of antimony. A light tan precipitate that forms very slowly (after a few minutes) is probably tin(IV) sulfide.

Exercises

General Questions on Group II

1. List the common oxidation states exhibited by the metals whose cations occur in Group II. Indicate those that are usually reducing oxidation states by (R), those that are considered "stable" oxidation states by (S), and those that are oxidizing by (O).

2. The known and unknown solutions contain 5.0 mg of *cation* per milliliter of solution. Calculate the molarity of each *in terms of the Group II metal*. For example, the $HSbCl_4$ solution contains 5.0 mg Sb/mL, and the molarity should be calculated in terms of mol Sb/L or mmol Sb/mL.

3. What color is each of the following? $H_3AsO_4(aq)$; $H[SbCl_4](aq)$; $H_2[SnCl_6](aq)$; As_2S_3; Sb_2S_3; $MgNH_4AsO_4$.

Group II Reactions

Write balanced net ionic equations for the reactions that occur when the following substances are mixed in aqueous solution. Indicate the colors of all precipitates and complex ions.

Solution Preparation

4. Antimony(III) chloride dissolves in dil. HCl.

5. Tin(IV) chloride dissolves in dil. HCl.
6. Arsenic(V) oxide dissolves in hot water.

Step A

7. Antimony(III) chloride in dil. HCl and evaporation with HNO_3.
8. Tin(IV) chloride in dil. HCl and evaporation with HNO_3.
9. Arsenic acid and ammonium iodide in excess HCl.
10. Antimonic acid and ammonium iodide in excess HCl.
11. Arsenous acid and hydrogen sulfide.
12. Antimonous acid and hydrogen sulfide.
13. Stannic acid and hydrogen sulfide.

Step B

14. Arsenic(III) sulfide in excess NaOH.
15. Antimony(III) sulfide in excess NaOH.
16. Tin(IV) sulfide in excess NaOH.

Step C

17. Sodium trithioarsenate(III) and sodium tetrahydroxoarsenate(II) plus a lim. amt. of dil. HCl.
18. Sodium trithioantimonate(III) and sodium tetrahydroxoantimonate(III) plus a lim. amt. of dil. HCl.
19. Sodium trithiostannate(IV) and sodium hexahydroxostannate(IV) plus a lim. amt. of dil. HCl.

Step D

20. Tin(IV) sulfide and excess hot 6 *M* HCl.
21. Antimony(III) sulfide and excess hot 6 *M* HCl.

Step E

22. Arsenic(III) sulfide and 6 *M* aq. NH_3 plus 6% H_2O_2.
23. Ammonium arsenate and magnesia mixture.

Step F

24. Tin(IV) chloride in dil. HCl reacts with iron.
25. Tin(II) chloride in dil. HCl and excess mercury(II) chloride in dil. HCl.
26. Antimony(III) chloride in dil. HCl reacts with iron.
27. Antimony and nitric acid plus oxalic acid.
28. A solution of trisoxalatoantimonic(III) acid and hydrogen sulfide.
29. Tin(IV) chloride in dil. HCl and excess oxalic acid.
30. Antimony(III) chloride in dil. HCl and excess oxalic acid.

Other Questions and Problems

31. A Group II unknown gave a yellow sulfide precipitate. Which cation(s) could be present?
32. A Group II unknown gave an orange sulfide precipitate. Which cation(s) could be present?

Aqueous solutions of some compounds that contain chromium. Left to right: chromium(II) chloride ($CrCl_2$) is blue; chromium(III) chloride ($CrCl_3$) is green; potassium chromate (K_2CrO_4) is yellow; potassium dichromate ($K_2Cr_2O_7$) is orange. The last three are observed in the Group III analysis.

33-1 COMMON OXIDATION STATES OF THE METALS IN ANALYTICAL GROUP III

Cobalt exists in aqueous solutions primarily as cobalt(II) compounds. Cobalt(III) is such a strong oxidizing agent that it oxidizes water.

$$Co^{3+}(aq) + e^- \longrightarrow Co^{2+}(aq) \qquad E^0 = +1.82 \text{ V}$$

Only complex compounds containing Co(III) can exist in aqueous solution. In an alternative test for Co, the stable complex ion $[Co(NO_2)_6]^{3-}$ is formed. The hexaamminecobalt(III) ion, $[Co(NH_3)_6]^{3+}$, is stable in aqueous solutions.

Nickel compounds exist in aqueous solution primarily in the +2 oxidation state, although nickel(IV) oxide is precipitated from strongly basic solutions of strong oxidizing agents.

$$NiO_2(s) + 2H_2O + 2e^- \longrightarrow Ni(OH)_2(s) + 2OH^-(aq) \qquad E^0 = +0.49 \text{ V}$$

Nickel(IV) oxide is a powerful oxidizing agent in acidic solutions.

$$NiO_2(s) + 4H^+(aq) + 2e^- \longrightarrow Ni^{2+}(aq) + 2H_2O \qquad E^0 = +1.7 \text{ V}$$

The common oxidation states of iron are iron(II) and iron(III). In the +2 oxidation state, iron is a weak reducing agent. It is a fairly strong oxidizing agent in the +3 oxidation state.

$$Fe^{3+}(aq) + e^- \longrightarrow Fe^{2+}(aq) \qquad E^0 = +0.771 \text{ V}$$

Manganese exhibits a variety of oxidation states in aqueous solutions: +2, +3, +4, +6, and +7. The +2 oxidation state is the most stable. In its higher oxidation states, manganese is a strong oxidizing agent in acidic solutions.

$$MnO_2(s) + 4H^+(aq) + 2e^- \longrightarrow Mn^{2+}(aq) + 2H_2O \qquad E^0 = +1.23 \text{ V}$$

$$MnO_4^-(aq) + 8H^+(aq) + 5e^- \longrightarrow Mn^{2+}(aq) + 4H_2O \qquad E^0 = +1.51 \text{ V}$$

Higher-oxidation-state compounds of manganese are weaker oxidizing agents in basic solutions.

$$MnO_2(s) + 2H_2O + 2e^- \longrightarrow Mn(OH)_2(s) + 2OH^-(aq) \qquad E^0 = -0.05 \text{ V}$$

$$MnO_4^-(aq) + 2H_2O + 3e^- \longrightarrow MnO_2(s) + 4OH^-(aq) \qquad E^0 = +0.588 \text{ V}$$

Solid $MnCl_2 \cdot 4H_2O$ and a concentrated solution of $MnCl_2$ (left). Solid $KMnO_4$ and a dilute solution of $KMnO_4$ (right). The cellulose in a piece of paper towel has reduced $KMnO_4$ to potassium manganate, K_2MnO_4, which is green (center). Under the photographer's hot lights, some K_2MnO_4 has been reduced to MnO_2, which is brown.

Aqueous solutions of some compounds that contain chromium. Left to right: chromium(II) chloride ($CrCl_2$) is blue; chromium(III) chloride ($CrCl_3$) is green; potassium chromate (K_2CrO_4) is yellow; potassium dichromate ($K_2Cr_2O_7$) is orange.

Chromium also exhibits several oxidation states in aqueous solution (+2, +3, and +6). Chromium(II) is such a strong reducing agent that most Cr(II) compounds cannot exist in contact with atmospheric oxygen and water.

$$Cr^{3+}(aq) + e^- \longrightarrow Cr^{2+}(aq) \qquad E^0 = -0.41 \text{ V}$$

By contrast, chromium(VI) is an oxidizing agent in acidic solutions.

$$Cr_2O_7{}^{2-}(aq) + 14H^+(aq) + 6e^- \longrightarrow 2Cr^{3+}(aq) + 7H_2O \qquad E^0 = 1.33 \text{ V}$$

The +3 oxidation state is the most stable oxidation state of chromium in aqueous solutions.

Aluminum and zinc each exhibit only one oxidation state in their compounds in aqueous solutions: +3 for aluminum and +2 for zinc.

The procedures outlined in this chapter may be applied to a Group III known or unknown or to the centrifugate from the Groups I and II sulfide precipitation. The Group III flow chart is given in Figure 33-1.

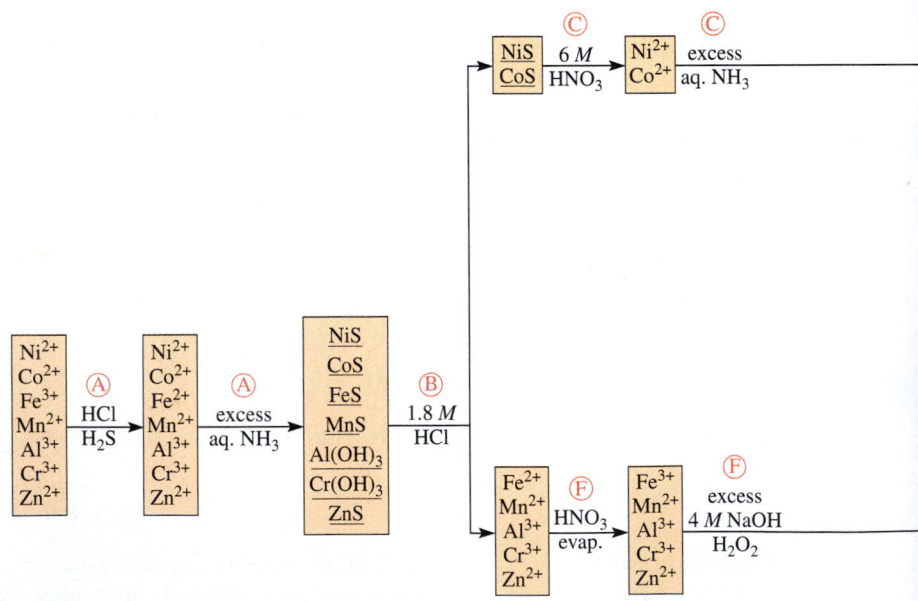

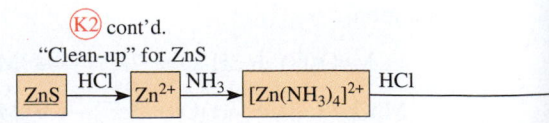

Figure 33-1 Analytical Group III flow chart. Circled letters refer to analytical steps. Species formed in confirmatory tests are shown with blue backgrounds.

33-2 PRECIPITATION OF THE ANALYTICAL GROUP III CATIONS

The Group III cations are nickel, Ni^{2+}; cobalt, Co^{2+}; manganese, Mn^{2+}; iron, Fe^{2+} or Fe^{3+}; aluminum, Al^{3+}; chromium, Cr^{3+}; and zinc, Zn^{2+}. The concentration of sulfide ions in a solution of hydrogen sulfide that is also 0.3 M in hydrogen ions (Groups I and II separations) is too low to precipitate the Group III metal sulfides. However, in a buffered solution of ammonium sulfide, the sulfides of Ni^{2+}, Co^{2+}, Mn^{2+}, Fe^{2+}, and Zn^{2+} do precipitate, whereas Al^{3+} and Cr^{3+} precipitate as hydroxides. The Group III cations are known as the **insoluble basic sulfide group.** Table 33-1 shows solubility products for the freshly precipitated sulfides and hydroxides of Group III.

Zinc sulfide is the least soluble of the freshly precipitated Group III sulfides. We demonstrated that zinc sulfide does not precipitate from solutions that are saturated with H_2S and are 0.30 M in H^+ (Example 31-1). However, when hydrogen sulfide is added to a solution that contains an excess of aqueous ammonia buffered with ammonium nitrate or ammonium chloride, a very high concentration of sulfide ions is produced.

$$2NH_3(aq) + H_2S(aq) \longrightarrow 2NH_4^+(aq) + S^{2-}(aq)$$

Table 33-1 *Solubility Products for Group III Sulfides and Hydroxides*

Group III Compound	K_{sp}
CoS (α)	5.9×10^{-21}
NiS (α)	3.0×10^{-21}
FeS	4.9×10^{-18}
MnS	5.1×10^{-15}
Al(OH)$_3$	1.9×10^{-33}
Cr(OH)$_3$	6.7×10^{-31}
ZnS	1.1×10^{-21}

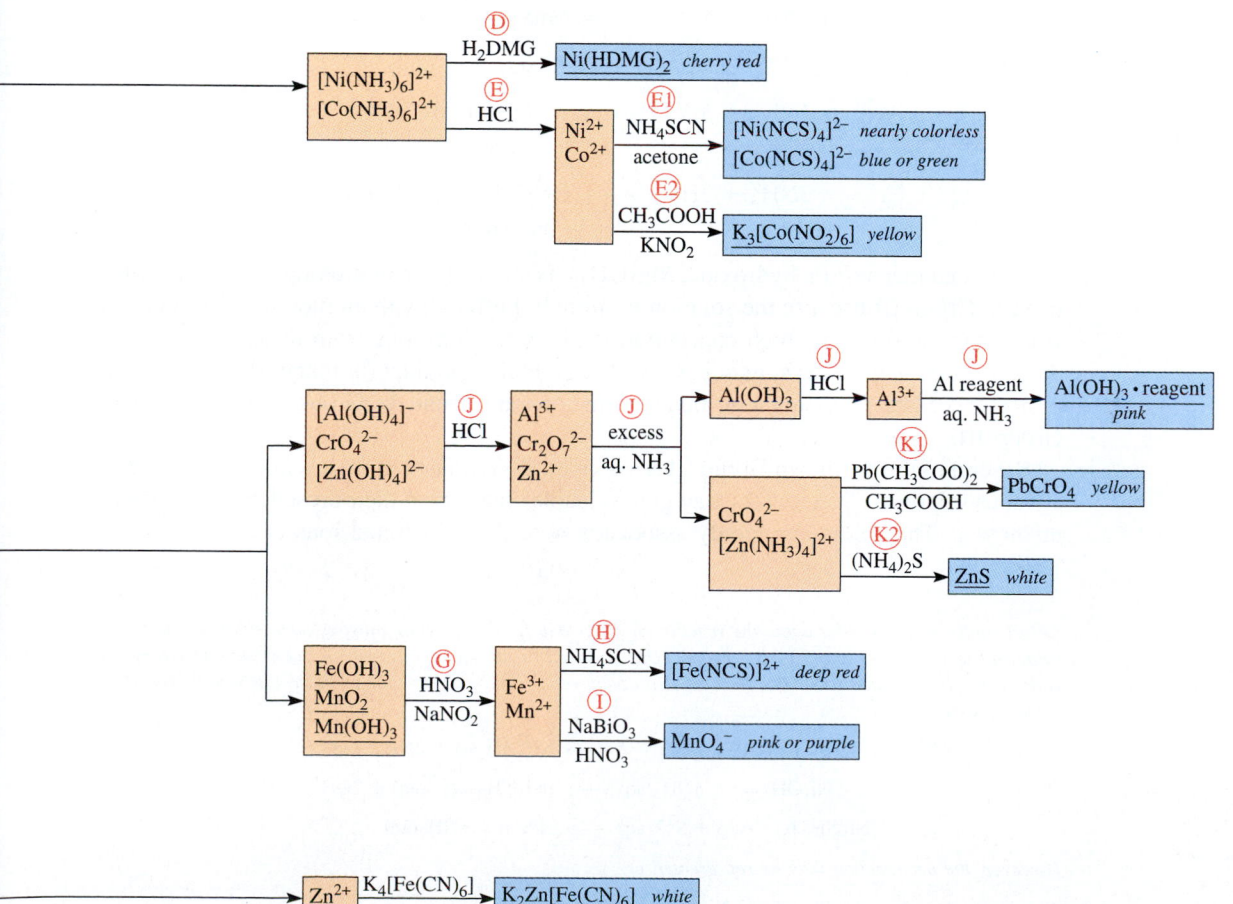

The sulfide ion concentration is sufficiently large that the solubility products for the sulfides of cobalt(II), nickel(II), manganese(II), iron(II), and zinc(II) are exceeded. So those sulfides precipitate from the solution. Similarly, the hydroxide ion concentration of the buffered solution of aqueous ammonia is sufficiently large that the hydroxides of aluminum and chromium(III) precipitate.

The sulfides of aluminum and chromium(III) cannot exist in contact with water because they hydrolyze completely and are converted to the corresponding hydroxides. Aluminum sulfide, Al_2S_3, and chromium(III) sulfide, Cr_2S_3, contain small, highly charged cations that hydrolyze extensively (Section 19-5) as well as an anion that hydrolyzes extensively.

Heating thioacetamide in the acidic solution forms H_2S. In acidic solution the H_2S reduces iron(III) ions to iron(II) ions.

$$2Fe^{3+}(aq) + H_2S(aq) \longrightarrow 2Fe^{2+}(aq) + S(s) + 2H^+(aq)$$

The addition of aqueous ammonia to the acidic solution of hydrogen sulfide produces a high concentration of sulfide ions, and the Group III precipitation reactions occur.*

$$Ni^{2+} + S^{2-} \longrightarrow NiS(s) \qquad \text{black}$$

$$Co^{2+} + S^{2-} \longrightarrow CoS(s) \qquad \text{black}$$

$$Fe^{2+} + S^{2-} \longrightarrow FeS(s) \qquad \text{black}$$

$$Mn^{2+} + S^{2-} \longrightarrow MnS(s) \qquad \text{salmon}$$

$$Zn^{2+} + S^{2-} \longrightarrow ZnS(s) \qquad \text{white}$$

$$Al^{3+} + 3NH_3 + 3H_2O \longrightarrow Al(OH)_3(s) + 3NH_4^+$$
$$\text{white}$$

$$Cr^{3+} + 3NH_3 + 3H_2O \longrightarrow Cr(OH)_3(s) + 3NH_4^+$$
$$\text{gray-green}$$

Although magnesium hydroxide, $Mg(OH)_2$, is an insoluble hydroxide, it is not precipitated in Group III because the solution is strongly buffered with ammonium chloride and ammonium sulfide. The high concentration of ammonium ions from these salts inhibits the ionization of aqueous ammonia so that the solubility product for magnesium hydroxide is not exceeded. Consequently, magnesium ions are found in Group V rather than in Group III.

Several of the cations in Group III are derived from transition metals and are colored in aqueous solutions. Their colors may give valuable clues about their presence or absence in unknowns. The colors commonly associated with these hydrated ions are

---

*Alert students may wonder about the reaction of the Group III cations with aqueous ammonia in the buffered solution (Sections 18-7 and 25-2). Aqueous NH_3 is added to an acidic solution that contains the Group III cations and is saturated with H_2S, an acid. No doubt, other reactions (similar to those illustrated for Ni^{2+}) occur to some extent.

$$Ni^{2+}(aq) + 2NH_3(aq) + 2H_2O \longrightarrow Ni(OH)_2(s) + 2NH_4^+(aq)$$

$$Ni(OH)_2(s) + 6NH_3(aq) \longrightarrow [Ni(NH_3)_6]^{2+}(aq) + 2OH^-(aq)$$

$$[Ni(NH_3)_6]^{2+}(aq) + S^{2-}(aq) \longrightarrow NiS(s) + 6NH_3(aq)$$

However, the net reaction may be represented as

$$Ni^{2+}(aq) + S^{2-}(aq) \longrightarrow NiS(s)$$

We have chosen to represent the Group III precipitation reactions in the simplest way possible.

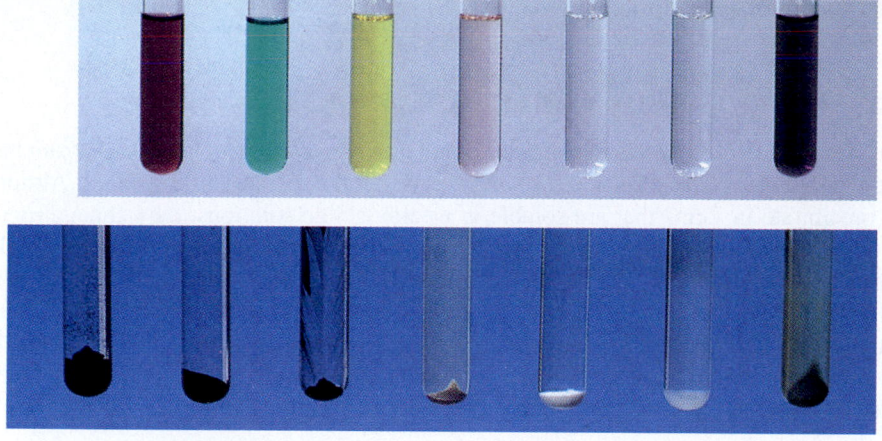

Solutions of Group III nitrates (top). Left to right: Co^{2+}, Ni^{2+}, Fe^{3+}, Mn^{2+}, Al^{3+}, Zn^{2+}, Cr^{3+}. The Group III precipitates (bottom). Left to right: CoS (black), NiS (black), FeS (black), MnS (salmon), ZnS (white), $Al(OH)_3$ (white), $Cr(OH)_3$ (gray-green).

Co^{2+}	red-pink	Mn^{2+}	pale pink
Ni^{2+}	green	Cr^{3+}	deep blue (NO_3^- solutions)
Fe^{2+}	pale green		deep green (Cl^- solutions)
Fe^{3+}	orchid (NO_3^- solutions)		
	yellow (Cl^- solutions)		

The Group III known and unknown solutions are usually prepared by dissolving nitrates of Group III metals in dilute nitric acid solution to suppress hydrolysis of the cations. General unknowns may contain either chlorides or nitrates.

Solutions containing Al^{3+} and Zn^{2+} ions are colorless unless the anions that occur with them impart color. Chromium(III) ion is so intensely colored that its presence usually masks the presence of other colored ions in Group III.

Step A

The solution may contain one or more of Co^{2+}, Ni^{2+}, Fe^{3+}, Mn^{2+}, Al^{3+}, Cr^{3+}, and Zn^{2+}.

(1) If the analysis is to be performed on the centrifugate from the Groups I and II sulfide precipitation, add 8 drops of 5% thioacetamide solution and heat the mixture in a water bath for 3 minutes.

(2) If the analysis is to be performed on a known or unknown solution containing only Group III cations, add two drops of 6 M HCl to 10 drops of the solution, dilute to 1 mL, and add 8 drops of 5% thioacetamide solution. Heat the solution in a hot water bath for 3 minutes.

Add 5 drops of 15 M aqueous ammonia to the solution produced by either treatment (1) or (2), and stir vigorously. Test the solution to determine that it is basic to litmus. If not, continue the dropwise addition of 15 M aqueous ammonia until it is basic. Heat the mixture in a water bath for 5 minutes, stirring frequently. Separate[a] the precipitate, and wash it with a mixture of 10 drops of water and 1 drop of 3 M aqueous ammonia. Save the solution in which the precipitation occurred for the analysis of Groups IV and V if the unknown is a general unknown.[b] Otherwise it may be discarded. The precipitate is treated according to Step B.

[a] *Both cobalt(II) sulfide and nickel(II) sulfide tend to form colloidal suspensions. A colloidal suspension of either or both of these sulfides is usually intensely blue. If a colloidal suspension develops, the mixture should be heated for several minutes until it becomes clear (transparent) before it is centrifuged and separated.*

[b] *Sulfide ions can be oxidized to sulfate ions by atmospheric O_2 in basic solution ($S^{2-} + 2O_2 \rightarrow SO_4^{2-}$). Sulfate ions react with barium ions to form insoluble barium sulfate and cause Ba^{2+} ions to be missed in Group IV. Acidify the solution that contains Groups IV–V with 6 M HCl, and boil it until all the H_2S has been expelled.*

33-3 SEPARATION OF COBALT AND NICKEL

Hydrogen sulfide precipitates the sulfides of cobalt and nickel completely *only* from basic solutions. However, these sulfides are only very slightly soluble in dilute HCl. Although they precipitate in forms that are soluble in weakly acidic solutions, they change rapidly into other crystalline forms that are much less soluble. For example, the solubility product for NiS (α) is 3.0×10^{-21}, that for NiS (β) is 1.0×10^{-26}, and that for NiS (γ) is 2.0×10^{-28}. The fact that CoS and NiS are rapidly converted to much less soluble crystalline forms is the basis for separating them from the other Group III cations. The sulfides of Fe^{2+}, Mn^{2+}, and Zn^{2+} and the hydroxides of Al^{3+} and Cr^{3+} are soluble in 1.8 M HCl; the sulfides of Co^{2+} and Ni^{2+} are not.

$$MnS(s) + 2H^+ \longrightarrow Mn^{2+} + H_2S(g)$$

$$FeS(s) + 2H^+ \longrightarrow Fe^{2+} + H_2S(g)$$

$$Al(OH)_3(s) + 3H^+ \longrightarrow Al^{3+} + 3H_2O$$

$$Cr(OH)_3(s) + 3H^+ \longrightarrow Cr^{3+} + 3H_2O$$

$$ZnS(s) + 2H^+ \longrightarrow Zn^{2+} + H_2S(g)$$

These dissolution reactions involve the formation of covalent compounds, H_2S and H_2O. H_2S is volatile and escapes from the solution as it is formed. This drives the reactions for the dissolution of these sulfides far to the right.

> *Step B*
> Use the precipitate from Step A, containing CoS, NiS, FeS, MnS, $Al(OH)_3$, $Cr(OH)_3$, and ZnS. Add 6 drops of water *followed* by 3 drops of 6 M HCl to the precipitate, and stir vigorously for 2 minutes. Separate the mixture immediately and save the centrifugate for Step F. Add 6 drops of water *followed by* 3 drops of 6 M HCl to the residue, and stir it vigorously for 1 minute. Separate immediately, and add the centrifugate to that obtained in the first extraction. The precipitate is treated according to Step C, and the combined extracts are used in Step F.

About one drop of water remains in the precipitate from Step A, so the HCl concentration is $(3/10) \times 6\ M = 1.8\ M.$

33-4 IDENTIFICATION OF COBALT AND NICKEL

Cobalt(II) and nickel(II) sulfides dissolve in hot nitric acid, which oxidizes the sulfide ions to free sulfur.

$$3CoS(s) + 8H^+ + 2NO_3^- \longrightarrow 3Co^{2+} + 3S(s) + 2NO + 4H_2O$$

$$3NiS(s) + 8H^+ + 2NO_3^- \longrightarrow 3Ni^{2+} + 3S(s) + 2NO + 4H_2O$$

These reactions are similar to the reactions by which the Group I sulfides were dissolved. After the sulfides of cobalt and nickel are dissolved, the solution is boiled for several minutes to remove the oxides of nitrogen. An excess of aqueous ammonia is then added to convert cobalt(II) and nickel(II) ions to hexaammine complexes (Section 25-2).

The reactions of cobalt(II) ions are identical to these.

$$Ni^{2+} + 2NH_3 + 2H_2O \longrightarrow Ni(OH)_2(s) + 2NH_4^+$$
$$\text{lim. amt.}$$

$$Ni(OH)_2(s) + 6NH_3 \longrightarrow [Ni(NH_3)_6]^{2+} + 2OH^-$$
$$\text{excess}$$

Dimethylglyoxime is an organic compound that reacts with nickel(II) ions in aqueous ammonia to form an insoluble, bright cherry-red complex compound. The hydrogen atoms in oximes

$$\text{C}=\text{N}-\text{O}-\text{H}$$

are very weakly acidic. Dimethylglyoxime may be abbreviated H_2DMG.

The arrows indicate coordinate covalent bonds, and the dashed lines indicate hydrogen bonds.

$$2 \quad \begin{array}{l} CH_3-C=N-O-H \\ | \\ CH_3-C=N-O-H \end{array} + [Ni(NH_3)_6]^{2+} \longrightarrow \quad \begin{array}{c} CH_3-C=N \diagdown \diagup N=C-CH_3 \\ | \quad Ni \quad | \\ CH_3-C=N \diagup \diagdown N=C-CH_3 \end{array} + 2NH_4^+ \\ + 4NH_3$$

H_2DMG

$Ni(HDMG)_2(s)$, bright pink-red

Cobalt (II) ions form a brown complex with dimethylglyoxime, but the complex does not interfere with the test for nickel(II) ions.

The simplest confirmatory test for cobalt(II) ions involves their reaction with thiocyanate ions to form tetrathiocyanatocobaltate(II) ions, $[Co(NCS)_4]^{2-}$. The latter are blue or blue-green in 50% H_2O/50% acetone solution.

$$Co^{2+} + 4SCN^- \longrightarrow [Co(NCS)_4]^{2-} \quad \text{blue or blue-green}$$

We have written the formula for the thiocyanate ion as SCN^-, which is the accepted way to write it. However, complex species containing SCN^- ions are usually written like $[Co(NCS)_4]^{2-}$ to emphasize that the nitrogen atom is the donor atom in most complexes of this kind.

If the separation of cobalt(II) and nickel(II) ions from the other Group III cations is less than complete, iron(III) ions interfere with the test for cobalt by forming bright-red complex ions, $[Fe(NCS)]^{2+}$.

$$Fe^{3+} + SCN^- \rightleftharpoons [Fe(NCS)]^{2+} \quad \text{red}$$

This interference may be eliminated by the addition of fluoride ions, which convert the red complex ions into very stable, colorless hexafluoroferrate(III) ions, $[FeF_6]^{3-}$.

$$\underset{\text{red}}{[Fe(NCS)]^{2+}} + 6F^- \rightleftharpoons \underset{\text{colorless}}{[FeF_6]^{3-}} + SCN^-$$

Once the interference by the red $[Fe(NCS)]^{2+}$ ions has been eliminated, the characteristic blue color of the $[Co(NCS)_4]^{2-}$ ions is observed easily.

An alternative test for cobalt(II) ions involves the formation of a golden-yellow precipitate, $K_3[Co(NO_2)_6]$, in mildly acidic solution. Cobalt(II) ions form hexanitrocobaltate(II) ions, $[Co(NO_2)_6]^{4-}$, with excess nitrite ions.

$$Co^{2+} + 6NO_2^- \longrightarrow [Co(NO_2)_6]^{4-}$$

These complex ions are "nitro" complexes, which tells us that the nitrogen atoms are coordinated to cobalt. If coordination were through oxygen atoms, the complex ions would be called "nitrito" complexes.

These are oxidized to hexanitrocobaltate(III) ions, $[Co(NO_2)_6]^{3-}$, by excess NO_2^-.

$$[Co(NO_2)_6]^{4-} + NO_2^- + 2H^+ \longrightarrow [Co(NO_2)_6]^{3-} + NO + H_2O$$

These in turn react with potassium ions in the buffered acetic acid solution to form one of a very few insoluble potassium compounds, potassium hexanitrocobaltate(III).

$$3K^+ + [Co(NO_2)_6]^{3-} \longrightarrow K_3[Co(NO_2)_6](s) \quad \text{golden yellow}$$

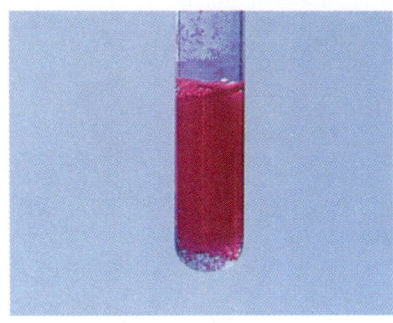

The confirmatory test for nickel, $Ni(HDMG)_2$.

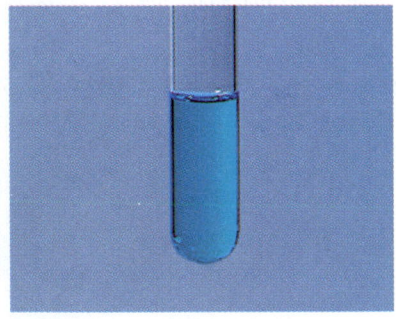

The confirmatory test for cobalt, $[Co(NCS)_4]^{2-}$.

Step C
Use the residue from Step B, containing CoS and NiS. Add 8 drops of 6 *M* HNO_3 to the residue, stir thoroughly, and heat the mixture in a water bath until the black sulfide precipitate has dissolved completely. Stir the solution regularly while it is heated for an additional 3 minutes. Add 10 drops of water and just enough 3 *M* aqueous ammonia to make the solution basic to litmus. Avoid a large excess of aqueous ammonia. Centrifuge the solution and discard any sulfur that remains.

Step D
Test for nickel. Place 6 drops of the solution from Step C in a test tube and add 3 drops of dimethylglyoxime. The formation of a cherry-red precipitate confirms the presence of nickel(II) ions. If cobalt is present and nickel is absent, the solution will turn brown when dimethylglyoxime is added. Don't confuse a brown solution with the bright cherry-red precipitate obtained when nickel is present.

Step E
Test for cobalt. Perform *either* test (1) or (2) as directed by your instructor.

(1) Carefully acidify the remainder of the solution from Step C, which may contain cobalt(II) and nickel(II) ions, with 6 *M* HCl (avoid excess HCl). Then add 5 drops of 4 *M* NH_4SCN solution. Double the volume of solution by adding acetone. Place a clean finger over the top of the test tube and invert it. If cobalt(II) ions are present, the solution will turn blue or blue-green, depending on its acidity. (If the solution is red owing to the incomplete removal of iron, add 4 *M* KF solution dropwise [mix thoroughly after each drop is added] until the red color disappears.)

(2) Transfer the remainder of the solution to a small beaker or evaporating dish, and carefully evaporate it to approximately 1 drop, *but not to dryness.* Be prepared to add a few drops of water to prevent the complete evaporation of liquid. Allow the mixture to cool. Add 3 drops of water, 4 drops of 2 *M* KCH_3COO, and enough 6 *M* CH_3COOH to make the solution acidic to litmus. Mix well, transfer the liquid to a clean test tube, and centrifuge if necessary to remove any solid. Double the volume by adding 6 *M* KNO_2, mix well, and place the solution in the water bath for 3 minutes. Remove the test tube from the water bath, and allow it to stand for 15 minutes. A golden-yellow precipitate of $K_3[Co(NO_2)_6]$, often slow in forming, confirms the presence of cobalt.

33-5 SEPARATION OF IRON AND MANGANESE FROM ALUMINUM, CHROMIUM, AND ZINC

The solution from Step B that contains Fe^{2+}, Mn^{2+}, Al^{3+}, Cr^{3+}, and Zn^{2+} ions also contains H_2S. H_2S must be removed before iron and manganese are separated from aluminum, chromium, and zinc. This is accomplished by evaporating the solution with nitric acid, which oxidizes H_2S to free sulfur and Fe^{2+} ions to Fe^{3+} ions.

$$3H_2S + 2H^+ + 2NO_3^- \longrightarrow 3S(s) + 2NO + 4H_2O$$

$$3Fe^{2+} + 4H^+ + NO_3^- \longrightarrow 3Fe^{3+} + NO + 2H_2O$$

The residue is dissolved in water and treated with 4 *M* NaOH, which results initially in the precipitation of hydroxides of the five metal ions.

$$Fe^{3+} + 3OH^- \longrightarrow Fe(OH)_3(s) \quad \text{red-brown}$$

$$Mn^{2+} + 2OH^- \longrightarrow Mn(OH)_2(s) \quad \text{white}$$

$$Al^{3+} + 3OH^- \longrightarrow Al(OH)_3(s) \quad \text{white}$$

$$Cr^{3+} + 3OH^- \longrightarrow Cr(OH)_3(s) \quad \text{gray-green}$$

$$Zn^{2+} + 2OH^- \longrightarrow Zn(OH)_2(s) \quad \text{white}$$

The continued addition of 4 M NaOH results in the dissolution of the three *amphoteric* hydroxides (Section 10-8).

$$Al(OH)_3(s) + OH^- \longrightarrow [Al(OH)_4]^-$$

$$Cr(OH)_3(s) + OH^- \longrightarrow [Cr(OH)_4]^-$$

$$Zn(OH)_2(s) + 2OH^- \longrightarrow [Zn(OH)_4]^{2-}$$

The strongly basic mixture is then treated with hydrogen peroxide, a strong oxidizing agent in basic solution. The tetrahydroxochromate(III) ions are oxidized to chromate ions (yellow).

$$2[Cr(OH)_4]^- + 3H_2O_2 + 2OH^- \longrightarrow 2CrO_4^{2-} + 8H_2O$$

Manganese(II) hydroxide is oxidized to a mixture of manganese(III) hydroxide (black) and manganese(IV) oxide (dark brown), commonly called manganese dioxide. Both of these compounds are quite insoluble.

$$2Mn(OH)_2(s) + H_2O_2 \longrightarrow 2Mn(OH)_3(s)$$

$$Mn(OH)_2(s) + H_2O_2 \longrightarrow MnO_2(s) + 2H_2O$$

Step F

Use the solution from Step B, containing Mn^{2+}, Fe^{2+}, Al^{3+}, Cr^{3+}, and Zn^{2+}. Add 5 drops of 16 M HNO$_3$ to the solution and evaporate to a moist residue. Cool, add 10 drops of water, mix well, and then add 4 M NaOH dropwise, with constant stirring, until a precipitate forms. Then add another 10 drops of 4 M NaOH. Stir thoroughly, and add 10 drops of 6% hydrogen peroxide to the mixture. Heat it in a water bath for 5 minutes, stirring frequently. Separate the mixture and save the centrifugate for Step J. Wash the residue with 1 drop of 4 M NaOH in 10 drops of water. Treat the precipitate according to Step G.

33-6 IDENTIFICATION OF MANGANESE AND IRON

The residue from Step F contains a mixture of MnO_2, $Mn(OH)_3$, and $Fe(OH)_3$. It is treated with a mixture of HNO_3 and $NaNO_2$. The iron(III) hydroxide dissolves in nitric acid in a typical acid–base reaction.

$$Fe(OH)_3(s) + 3H^+ \longrightarrow Fe^{3+} + 3H_2O$$

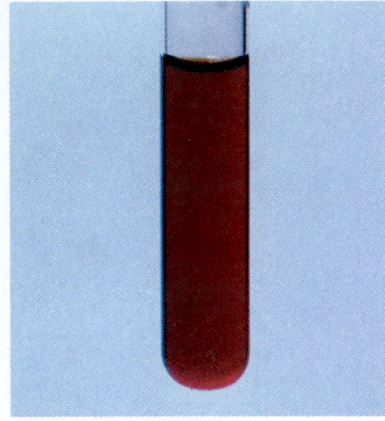

The confirmatory test for iron, $[Fe(NCS)]^{2+}$.

The bismuthate ion, BiO_3^-, is a very powerful oxidizing agent, one of the few that will oxidize Mn^{2+} ions to MnO_4^- ions in acidic solution; the permanganate ion itself is a very strong oxidizing agent.

In acidic solution nitrite ions reduce both manganese(III) and manganese (IV) to manganese(II). Nitrite ions are oxidized to nitrate ions.

$$2Mn(OH)_3(s) + 4H^+ + NO_2^- \longrightarrow 2Mn^{2+} + NO_3^- + 5H_2O$$

$$MnO_2(s) + 2H^+ + NO_2^- \longrightarrow Mn^{2+} + NO_3^- + H_2O$$

The confirmatory test for Fe^{3+} is its reaction with thiocyanate ions, SCN^-, to produce deep-red complex ions, $[Fe(NCS)]^{2+}$. (This was discussed as an interference in the test for cobalt in Section 33-4.) The presence of Mn^{2+} ions does not interfere with the test for Fe^{3+} ions, and Fe^{3+} ions do not interfere with the test for Mn^{2+} ions. Traces of iron contaminate almost everything, and a light pink color may occur even when iron is not present in the unknown. If a light pink solution is obtained in the test for iron(III) ions, a blank test should be run on the reagents, using everything except the solution that contains Fe^{3+} ions, to determine if the pink color is due to impurities in the reagents.

Manganese is detected by converting it to permanganate ions, MnO_4^-, which impart a purple color to the solution. In very dilute solutions permanganate ions appear pink. A powerful oxidizing agent is required to oxidize Mn^{2+} to MnO_4^- ions. Sodium bismuthate, $NaBiO_3$, is commonly used.

$$NaBiO_3(s) + 6H^+(aq) + 2e^- \longrightarrow Bi^{3+}(aq) + Na^+(aq) + 3H_2O \qquad E^0 \approx +1.61 \text{ V}$$

$$2Mn^{2+} + 5BiO_3^- + 14H^+ \longrightarrow 2MnO_4^- + 5Bi^{3+} + 7H_2O$$

Step G

Use the residue from Step F, containing $Fe(OH)_3$, $Mn(OH)_3$, and MnO_2. Add 8 drops of 6 M HNO_3 and two drops of 1 M $NaNO_2$ to the residue. Heat the mixture in a water bath for 6 minutes, stirring frequently. Cool the test tube under running water. Add 10 drops of H_2O. Stir.

Step H

Test for iron. Transfer 10 drops of the solution from Step G to a clean test tube and add 3 drops of 4 M NH_4SCN. A dark-red color develops in the solution if iron is present in the unknown. A faint pink indicates a trace of iron, probably due to impurities, and should be ignored. If a deep red color develops and then fades rapidly, this indicates the presence of unreacted nitrite ions; but the appearance of the deep-red color (even briefly) is taken as confirmation of the presence of iron.

Step I

Test for manganese. To the remainder of the solution from Step G add 2 drops of 6 M HNO_3 and some solid $NaBiO_3$. Sufficient $NaBiO_3$ should be added that the solution is muddy in appearance after it has been stirred thoroughly. Centrifuge the solution and observe the color above the dark brown $NaBiO_3$. The solution should be pink to purple, depending on the amount of manganese present in the unknown.

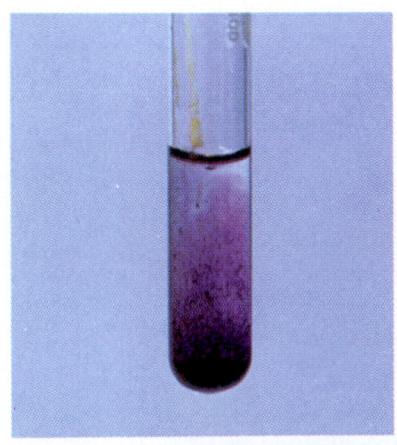

The confirmatory test for manganese, MnO_4^-.

33-7 SEPARATION AND IDENTIFICATION OF ALUMINUM

Hydrochloric acid is added to the basic solution from the H_2O_2 oxidation (Step F) to neutralize the excess sodium hydroxide. It also destroys the hydroxocomplexes of Al and

Zn, converting them first to the insoluble hydroxides

$$[Al(OH)_4]^- + H^+ \longrightarrow Al(OH)_3(s) + H_2O$$

$$[Zn(OH)_4]^{2-} + 2H^+ \longrightarrow Zn(OH)_2(s) + 2H_2O$$

These insoluble hydroxides then dissolve as an excess of HCl is added.

$$Al(OH)_3(s) + 3H^+ \longrightarrow Al^{3+} + 3H_2O$$

$$Zn(OH)_2(s) + 2H^+ \longrightarrow Zn^{2+} + 2H_2O$$

Chromate ions, CrO_4^{2-}, are converted to dichromate ions, $Cr_2O_7^{2-}$, by acids.

$$\underset{\text{yellow}}{2CrO_4^{2-}} + 2H^+ \rightleftharpoons \underset{\text{orange}}{Cr_2O_7^{2-}} + H_2O$$

The addition of aqueous ammonia reprecipitates aluminum hydroxide. It converts $Cr_2O_7^{2-}$ ions back to CrO_4^{2-} (reverses the previous reaction) and forms an ammine complex with zinc ions (Section 25-2).

$$Al^{3+} + 3NH_3 + 3H_2O \longrightarrow Al(OH)_3(s) + 3NH_4^+$$

$$Zn^{2+} + 2NH_3 + 2H_2O \longrightarrow Zn(OH)_2(s) + 2NH_4^+$$

$$Zn(OH)_2(s) + 4NH_3 \longrightarrow [Zn(NH_3)_4]^{2+} + 2OH^-$$

$$Cr_2O_7^{2-} + 2NH_3 + H_2O \rightleftharpoons 2CrO_4^{2-} + 2NH_4^+$$

It is especially important that an excess of aqueous ammonia be added so that all of the zinc hydroxide dissolves and none is left behind with aluminum hydroxide.

In the hydrogen peroxide oxidation (Step F), a strongly basic solution was boiled for several minutes. Strong bases dissolve glass to a slight extent to form soluble silicates. Acidification of solutions that contain silicates produces hydrated silicon dioxide, $SiO_2 \cdot xH_2O$, a nearly colorless gelatinous substance that looks very much like aluminum hydroxide. After the solution has been acidified with HCl, it is mixed thoroughly and centrifuged for a few minutes. Any precipitate of silicon dioxide should be discarded.

To test for aluminum, we first dissolve its hydroxide in hydrochloric acid.

$$Al(OH)_3(s) + 3H^+ \longrightarrow Al^{3+} + 3H_2O$$

Then aluminum hydroxide is precipitated by the addition of aqueous ammonia in the presence of **aluminum reagent,** a red dye that is physically trapped by the gelatinous aluminum hydroxide.

$$Al^{3+} + 3NH_3 + 3H_2O \longrightarrow Al(OH)_3(s) + 3NH_4^+$$

> *Step J*
> Use the solution from Step F, containing $[Al(OH)_4]^-$, CrO_4^{2-}, and $[Zn(OH)_4]^{2-}$. Add enough 6 M HCl to the solution to make it acidic to litmus, and then add an extra 2 drops. Stir vigorously for 1 minute. Centrifuge the solution and observe carefully to see if enough silicon dioxide, SiO_2, has precipitated that it can be separated from the solution. If so, withdraw the solution and discard the silicon dioxide. Add 6 M aqueous ammonia to the solution until it is basic to litmus, then add 8 drops in excess. Stir vigorously for 1 minute. The formation of a white (more

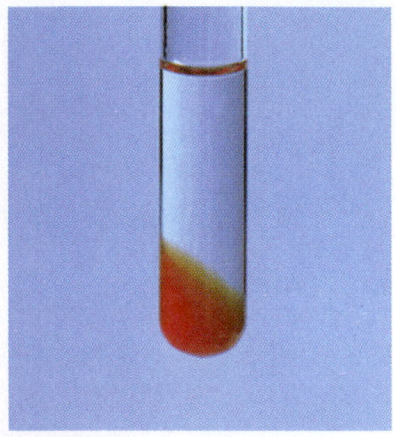

The confirmatory test for aluminum.

accurately, nearly colorless) gelatinous precipitate is likely due to aluminum hydroxide. Separate the mixture and keep the centrifugate for Step K.

To test for aluminum, dissolve the precipitate in 2 drops of 6 M HCl. Add 5 drops of water and 2 drops of aluminum reagent, followed by enough 6 M aqueous ammonia to make the solution basic. The formation of an orange to red precipitate confirms the presence of aluminum.

33-8 IDENTIFICATION OF CHROMIUM AND ZINC

The solution that contains zinc and chromium also contains an excess of aqueous ammonia. This must be neutralized before the solution can be tested for chromate ions. (Zinc ions do not interfere with the test.) We test for chromate ions by adding lead acetate, a covalent compound.

$$Pb(CH_3COO)_2 + CrO_4^{2-} \longrightarrow PbCrO_4(s) + 2CH_3COO^-$$

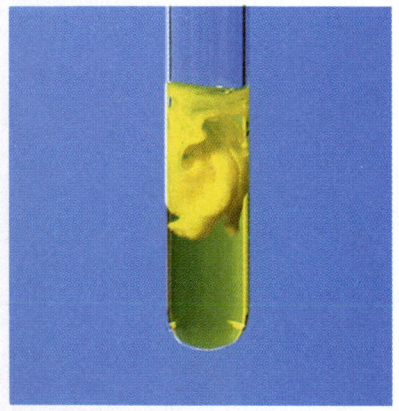

The confirmatory test for chromium, PbCrO$_4$.

Zinc ions are precipitated from the strongly ammoniacal solution by ammonium sulfide, $(NH_4)_2S$. Ammonium sulfide is susceptible to air oxidation. The reagent should be yellow, and there should be no flakes of sulfur floating in it. If there is any question about the reagent, test it by adding 2 drops of the known zinc solution to 10 drops of water and then adding 2 drops of $(NH_4)_2S$. The formation of a large amount of white precipitate, ZnS, indicates that the reagent is all right.

$$[Zn(NH_3)_4]^{2+} + S^{2-} \longrightarrow ZnS(s) + 4NH_3$$

Traces of the Group III ions that form black sulfides—Fe^{3+}, Ni^{2+}, and Co^{2+}—interfere with the test for zinc. They make it impossible to see the white precipitate of zinc sulfide. If the precipitate that should be zinc sulfide is dark, it is then treated with 1.8 M HCl. The clear, colorless solution can be separated from any dark precipitate. The solution is then boiled to expel hydrogen sulfide as completely as possible, the excess HCl is neutralized with aqueous ammonia, and finally the solution is again made slightly acidic with HCl. The addition of potassium hexacyanoferrate(II), $K_4[Fe(CN)_6]$, results in the formation of a white precipitate, $K_2Zn[Fe(CN)_6]$, which proves the presence of zinc.

$$Zn^{2+} + 2K^+ + [Fe(CN)_6]^{4-} \longrightarrow K_2Zn[Fe(CN)_6](s) \qquad \text{white}$$

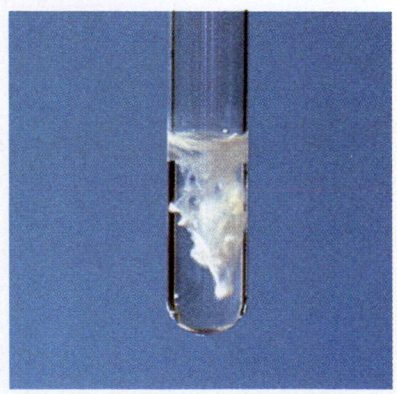

The confirmatory test for zinc, ZnS.

Step K

Use the solution from Step J, containing CrO_4^{2-} and $[Zn(NH_3)_4]^{2+}$.

(1) *Test for chromium.* To half of the solution add 6 M CH$_3$COOH until the solution is acidic to litmus. Add 2 drops of 0.1 M Pb(CH$_3$COO)$_2$. The presence of chromium is verified by the appearance of a bright yellow precipitate of PbCrO$_4$.

(2) *Test for zinc.* To the other half of the solution add 5 drops of 5% $(NH_4)_2S$. The formation of a white precipitate, ZnS, that is soluble in 1.8 M HCl proves that zinc is present. If the precipitate is not white, separate it from the liquid and discard the liquid. Add 7 drops of water followed by 3 drops of 6 M HCl, stir the mixture thoroughly for 1 minute, centrifuge, and withdraw the clear, colorless liquid. Discard any precipitate. Boil the liquid until the H$_2$S has been expelled and then add

sufficient 3 M aqueous ammonia to make it basic to litmus. Now add 10 drops of 1 M HCl, check to determine that the solution is acidic to litmus, and add 5 drops of 0.2 M $K_4[Fe(CN)_6]$. If zinc is present, it precipitates as a white compound, $K_2Zn[Fe(CN)_6]$.

Exercises

General Questions on Analytical Group III

1. List the common oxidation states exhibited by the metals whose cations occur in Group III. Indicate those that are usually reducing oxidation states by (R), those that are considered "stable" oxidation states by (S), and those that are oxidizing by (O).
2. Known and unknown solutions contain 5 milligrams of each cation per milliliter of solution, except that of Al^{3+}, which contains 10 mg/mL. Calculate the molarity of each Group III cation in such solutions.
3. Which of the freshly precipitated Group III sulfides is (a) most soluble? (b) least soluble?
4. Why are Al^{3+} and Cr^{3+} ions precipitated in Group III as hydroxides rather than as sulfides?
5. List all Group III cations that (as NO_3^- or Cl^- salts) fit the following descriptions.
   (a) Colorless solutions.
   (b) Green solutions.
   (c) Soluble in excess aqueous NH_3.
   (d) Insoluble in excess aqueous NH_3.
   (e) Soluble in excess NaOH solution.
   (f) Insoluble in excess NaOH solution.
   (g) Soluble in both excess aqueous NH_3 and excess NaOH solution.
   (h) Insoluble in both excess aqueous NH_3 and excess NaOH solution.
   (i) Soluble in excess aqueous NH_3, but not in excess NaOH solution.
   (j) Soluble in excess NaOH solution, but not in excess aqueous NH_3.
   (k) Pink solution gives a black sulfide.
   (l) Green solution gives a black sulfide.
   (m) Oxidized by H_2O_2.
   (n) Reduced by H_2S.
   (o) Neither oxidized nor reduced in Group III procedures.

Analytical Group III Reactions

Some reactions occur at several points in the analysis scheme for Group III cations. Therefore, these reactions are arranged somewhat differently than earlier group reactions. Write balanced net ionic equations for the reactions that occur when the following substances are mixed in aqueous solutions. Indicate the colors of all precipitates and complex ions.

Reactions with a Limited Amount of NH₃(aq)
6. Cobalt(II) nitrate + lim. amt. of aq. NH_3.
7. Nickel(II) nitrate + lim. amt. of aq. NH_3.
8. Manganese(II) nitrate + lim. amt. of aq. NH_3.
9. Iron(II) nitrate + lim. amt. of aq. NH_3.
10. Aluminum nitrate + lim. amt. of aq. NH_3.
11. Chromium(III) nitrate + lim. amt. of aq. NH_3.
12. Zinc nitrate + lim. amt. of aq. NH_3.

Hydroxides with Excess NH₃(aq)
13. Cobalt(II) hydroxide + excess aq. NH_3.
14. Nickel(II) hydroxide + excess aq. NH_3.
15. Zinc hydroxide + excess aq. NH_3.

Hydroxides with Strong Acids
16. Cobalt(II) hydroxide + nitric acid.
17. Nickel(II) hydroxide + nitric acid.
18. Manganese(II) hydroxide + nitric acid.
19. Iron(III) hydroxide + nitric acid.
20. Aluminum hydroxide + nitric acid.
21. Chromium(III) hydroxide + nitric acid.
22. Zinc hydroxide + nitric acid.

Ammine Complexes with Strong Acids
23. Hexaamminecobalt(II) hydroxide + HCl.
24. Hexaaminenickel(II) hydroxide + HCl.
25. Tetraamminezinc hydroxide + HCl.

Reactions with a Limited Amount of NaOH
26. Cobalt(II) nitrate + lim. amt. of NaOH.
27. Nickel(II) nitrate + lim. amt. of NaOH.
28. Manganese(II) nitrate + lim. amt. of NaOH.
29. Iron(III) nitrate + lim. amt. of NaOH.
30. Aluminum nitrate + lim. amt. of NaOH.
31. Chromium(III) nitrate + lim. amt. of NaOH.
32. Zinc nitrate + lim. amt. of NaOH.

Hydroxides with Excess NaOH (Table 10-2)
33. Cobalt(II) hydroxide + excess NaOH.
34. Aluminum hydroxide + excess NaOH.
35. Chromium(III) hydroxide + excess NaOH.
36. Zinc hydroxide + excess NaOH.

Hydroxocomplexes with Strong Acids

37. Sodium tetrahydroxoaluminate + HCl.
38. Sodium tetrahydroxochromate(III) + HCl.
39. Sodium tetrahydroxozincate + HCl.

Other Reactions

40. Iron(III) nitrate + hydrogen sulfide in acidic solution.
41. Cobalt(II) nitrate + ammonium sulfide in buffered aq. NH_3.
42. Nickel(II) nitrate + ammonium sulfide in buffered aq. NH_3.
43. Iron(II) nitrate + ammonium sulfide in buffered aq. NH_3.
44. Manganese(II) nitrate + ammonium sulfide in buffered aq. NH_3.
45. Aluminum nitrate + ammonium sulfide in buffered aq. NH_3.
46. Chromium(III) nitrate + ammonium sulfide in buffered aq. NH_3.
47. Zinc nitrate + ammonium sulfide in buffered aq. NH_3.
48. Iron(II) sulfide + 1.8 M HCl.
49. Manganese(II) sulfide + 1.8 M HCl.
50. Zinc sulfide + 1.8 M HCl.
51. Cobalt(II) sulfide + hot 6 M HNO_3.
52. Nickel(II) sulfide + hot 6 M HNO_3.
53. Cobalt(II) chloride + ammonium thiocyanate.
54. Cobalt(II) acetate + potassium nitrite (three reactions).
55. Hexaamminenickel(II) hydroxide + dimethylglyoxime (H_2DMG).
56. Iron(II) hydroxide + hydrogen peroxide in excess NaOH.
57. Manganese(II) hydroxide + hydrogen peroxide in excess NaOH (two reactions).
58. Sodium tetrahydroxochromate(III) + hydrogen peroxide in excess NaOH.
59. Manganese(III) hydroxide + sodium nitrite + HNO_3.
60. Manganese(IV) oxide + sodium nitrite + HNO_3.
61. Manganese(II) nitrate + sodium bismuthate + HNO_3.
62. Iron(III) nitrate + ammonium thiocyanate.
63. Sodium chromate + hydrochloric acid.
64. Sodium dichromate + aq. NH_3.
65. Sodium chromate + lead(II) acetate.
66. Tetraamminezinc hydroxide + ammonium sulfide.
67. Zinc chloride + potassium hexacyanoferrate(II).

Other Questions and Problems

68. What is the color of each of the following? Aqueous solutions are indicated by (aq). $Ni(NO_3)_2$(aq); NiS; $Ni(HDMG)_2$; $Co(NO_3)_2$(aq); CoS; $[Co(NCS)_4]^{2-}$(aq); $K_3[Co(NO_2)_6]$; $Fe(NO_3)_3$(aq); $FeCl_3$(aq); FeS; $Fe(OH)_3$; $[Fe(NCS)]^{2+}$(aq); $Mn(NO_3)_2$(aq); MnO_2; MnO_4^-(aq); $Al(NO_3)_3$(aq); $Al(OH)_3$; $Cr(NO_3)_3$(aq); $CrCl_3$(aq); CrO_4^{2-}(aq); $PbCrO_4$; $Zn(NO_3)_2$(aq); ZnS; $K_2Zn[Fe(CN)_6]$.

69. Select reagents that will separate the members of the following pairs in one step, and write a skeletal equation for each separation. Fe^{3+} and Co^{2+}; Ni^{2+} and Al^{3+}; FeS and $Al(OH)_3$; FeS and CoS; MnS and NiS; Fe^{3+} and Cr^{3+}; $Cr(OH)_3$ and $Zn(OH)_2$; $Al(OH)_3$ and $Ni(OH)_2$; $Al(OH)_3$ and $Zn(OH)_2$; $Fe(OH)_3$ and MnO_2; MnO_2 and $Cr(OH)_3$.

70. (a) If a significant amount of iron is found in the test for cobalt, what does this suggest about your technique in Step B? (b) Refer to Table 33-1 and compare solubility products for FeS and ZnS. What is the likely fate of zinc ions?

71. Explain why the addition of an excess of potassium fluoride allows you to detect cobalt even if a large amount of iron is present when you test for cobalt.

*72. Oximes are very weakly acidic. We wrote the formula for dimethylglyoxime in simplified form as H_2DMG to emphasize this fact. The test for nickel is always done in a solution that contains a slight excess of aqueous NH_3. Suggest a reason why the test for nickel fails in solutions that contain a large excess of aqueous NH_3. [*Hint:* The formula for the red precipitate is $Ni(HDMG)_2$.]

73. What is the fate of zinc ions if insufficient aqueous NH_3 is added in Step J?

74. What problem is likely to result from prolonged boiling of the $NaOH/H_2O_2$ solution (Step F) when Al^{3+} ions are absent?

75. What concentration of sulfide ions is necessary to initiate the precipitation of (a) ZnS in a solution that is 0.050 M in $Zn(NO_3)_2$? (b) MnS in a solution that is 0.050 M in $Mn(NO_3)_2$?

76. What is the minimum pH necessary to initiate the precipitation of (a) $Mn(OH)_2$ in a solution that is 0.050 M in $Mn(NO_3)_2$? (b) $Al(OH)_3$ in a solution that is 0.050 M in $Al(NO_3)_3$?

*77. In Section 20-5, Example 20-10, we demonstrated that $Mg(OH)_2$ precipitates from a solution that is 0.10 M in $Mg(NO_3)_2$ and 0.10 M in aqueous ammonia. In Example 20-12, we demonstrated that if the solution is also made 0.15 M in NH_4Cl, the buffering action of NH_4^+ prevents the precipitation of $Mg(OH)_2$. What ratio of $[NH_4^+]/[NH_3]$ is necessary to prevent the precipitation of $Mg(OH)_2$ in Group III? (Magnesium ions occur in Group V.) Recall that the general unknown solution contains 5 mg Mg^{2+}/mL, and 10 drops of unknown are diluted to 25 drops in the Group III precipitation. (*Hint:* Calculate $[Mg^{2+}]$ first.)

Analysis of Cation Group IV

Sea shells are mostly calcium carbonate, CaCO$_3$. Traces of transition metal ions are responsible for their color. Sea shells have a handedness, that is, they are chiral, and virtually all of them are right-handed (Section 25-7). If you cup a shell in your right hand (with your thumb extended), your fingers will follow the curl of the shell as it curls from the outside toward the center. Your thumb will point along the axis of the shell from the narrow end to the wider end.

34-1 INTRODUCTION

The ions of Analytical Group IV are referred to as the **insoluble carbonate group.** They are precipitated as carbonates from a buffered aqueous ammonia solution by the addition of ammonium carbonate.

Group IV contains the ions of calcium, strontium, and barium—three metals in Group IIA in the periodic table. They exhibit the +2 oxidation state in all their common compounds. Because they are derived from metals in the same family of the periodic table, these cations have similar properties. Therefore, separations of the individual ions are more difficult than separations in the earlier groups of cations. Magnesium ions are not precipitated in Group IV because MgCO$_3$ is fairly soluble, and the aqueous ammonia solution is buffered to prevent precipitation of Mg(OH)$_2$, $K_{sp} = 1.5 \times 10^{-11}$ (Section 20-5).

34-2 SOLUBILITIES OF COMPOUNDS OF CALCIUM, STRONTIUM, AND BARIUM

The sulfides and hydroxides of Ca^{2+}, Sr^{2+}, and Ba^{2+} ions are all fairly soluble compounds, so these cations are not precipitated in Group I, II, or III. Because Ca^{2+}, Sr^{2+}, and Ba^{2+} ions have noble gas electron configurations and no partially filled d sublevels, most of their compounds are white.

Most aqueous solutions that contain no cations other than Ca^{2+}, Sr^{2+}, and Ba^{2+} ions are colorless.

Table 34-1 shows that the Group IV carbonates are not nearly as insoluble as most of the compounds formed in earlier group precipitations and that the solubilities of the Group IV carbonates are quite similar. As is often the case with solubilities, the solubilities of these carbonates do not correlate with the positions of the metals in the periodic table.

As the Group IV flow chart in Figure 34-1 shows, Ba^{2+} ions are isolated by precipitating barium chromate, BaCrO$_4$, followed by reprecipitation as BaSO$_4$; Sr^{2+} ions are iso-

Table 34-1 *Solubility Products and Molar Solubilities of Group IV Carbonates*		
Compound	K_{sp}	**Molar Solubility**
CaCO$_3$	4.8×10^{-9}	6.9×10^{-5} M
SrCO$_3$	9.4×10^{-10}	3.1×10^{-5} M
BaCO$_3$	8.1×10^{-9}	9.0×10^{-5} M

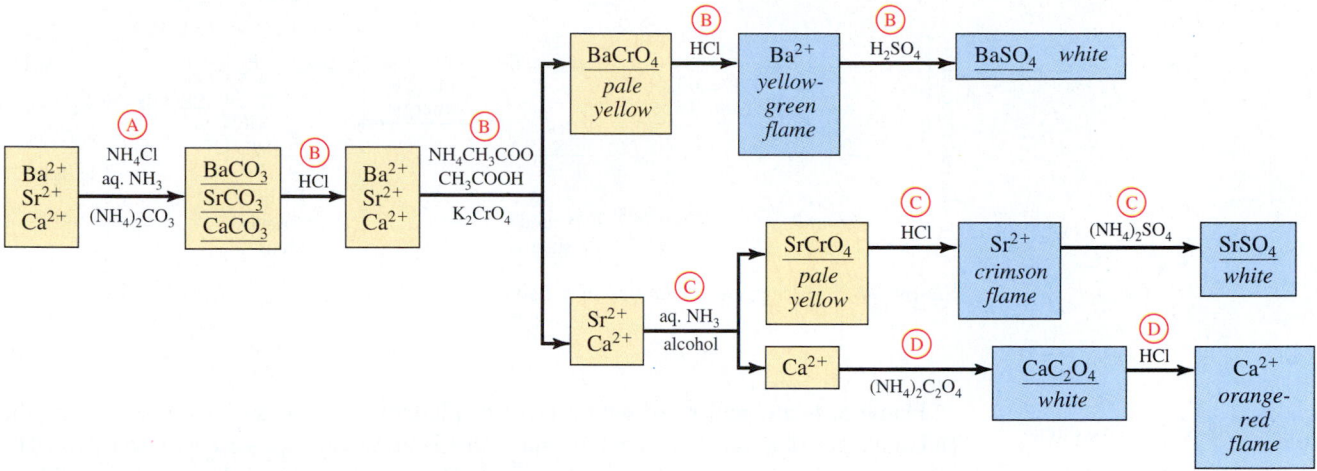

Figure 34-1 Analytical Group IV flow chart. Circled letters refer to analytical steps.

M^{2+}	MCrO$_4$	MSO$_4$	MC$_2$O$_4$
Table 34-2 *Solubility Products for Group IV Chromates, Sulfates, and Oxalates*			
Ca^{2+}	7.1×10^{-4}	2.4×10^{-5}	2.3×10^{-9}
Sr^{2+}	3.6×10^{-5}	2.8×10^{-7}	5.6×10^{-8}
Ba^{2+}	2.0×10^{-10}	1.08×10^{-10}	1.1×10^{-7}

lated as strontium chromate, SrCrO$_4$, followed by reprecipitation as strontium sulfate, SrSO$_4$; Ca^{2+} ions are isolated as calcium oxalate, CaC$_2$O$_4$.

Table 34-2 shows that the solubilities of the chromates, sulfates, and oxalates of the Group IV cations vary little. Therefore, it is relatively easy to precipitate compounds that *should* contain only one kind of cation but that are, in fact, contaminated with another kind of cation. Because these compounds contain one cation and one anion per formula unit, their solubility products can be compared. In each case the molar solubility is $\sqrt{K_{sp}}$.

Many oxalates precipitate as hydrated compounds, CaC$_2$O$_4 \cdot$ H$_2$O, SrC$_2$O$_4 \cdot$ H$_2$O, and BaC$_2$O$_4 \cdot$ 2H$_2$O.

34-3 FLAME TESTS

Laboratory burner flames are hot enough to promote electrons in atoms and ions to higher energy levels. As these excited atoms and ions move out of the hot region of the flame, electrons drop back to their ground state energy levels (Figure 34-2). In some atoms and ions, energy is emitted as visible light (Section 5-12). Such atoms and ions impart characteristic colors to flames. The differences between energy levels are *very specific* and correspond to light of *specific wavelengths in the visible region* (and therefore to specific, characteristic colors). Calcium, strontium, and barium ions impart colors that are used as confirmatory tests for the Group IV cations, as do sodium and potassium ions in Group V. Metal chlorides are more volatile than other common salts, so flame tests are usually done on chloride solutions.

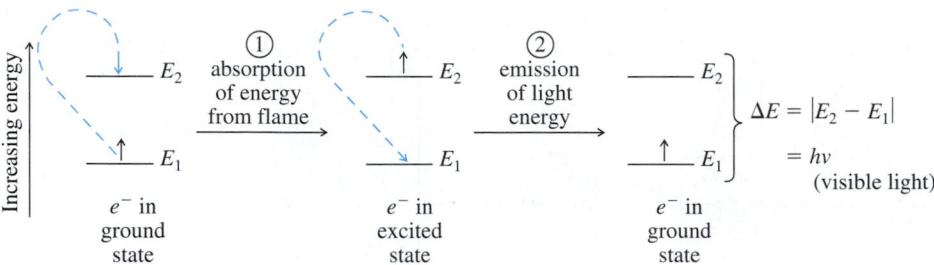

Figure 34-2   Electronic transitions in a flame test.

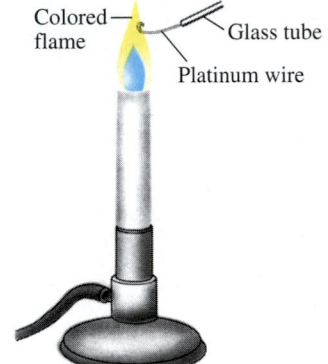

Figure 34-3   An illustration of how a flame test is performed.

Flame tests are performed with a piece of platinum or nichrome wire sealed into the end of a piece of glass tubing or stuck into a small cork. Flame tests are very sensitive. The wire must be clean before a flame test is done. The end of the wire is bent into a small loop so that, when it is dipped into a solution, a film of liquid covers the loop. The burner flame is adjusted so that it is as hot as possible and there is a well-defined blue cone with very little color in the outer part of the flame. The wire loop is dipped in 6 M HCl (in a small test tube), brought slowly up to the outer edge of the blue cone, and held there until the loop is "red hot" (Figure 34-3). The hot loop is dipped in 6 M HCl, and the operation is repeated until the wire imparts no color to the flame.

The clean wire is dipped into the solution to be flame-tested and then brought up to the outer edge of the blue cone. The cations of Groups IV and V color flames as shown in Figure 34-4.

The color imparted to the flame must be observed carefully. Different people see colors somewhat differently. Therefore, it is extremely important that *you* know how each color appears to *you*. Practice with known solutions until you know how each color appears. Never rely on your memory—always flame-test a known solution just before you flame-

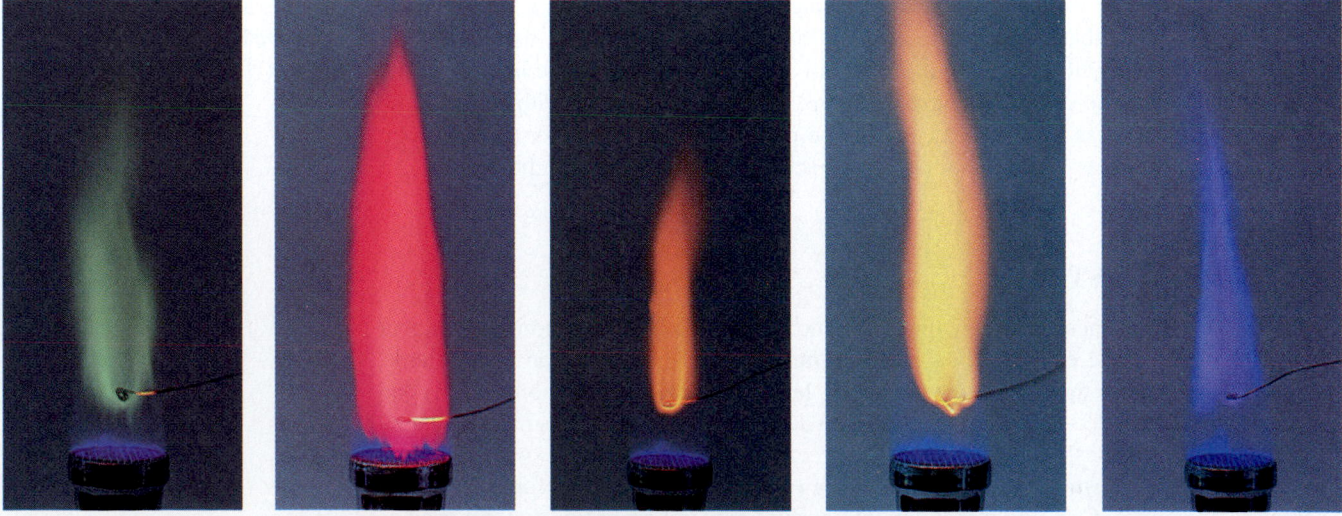

Figure 34-4   Flame colors for Groups IV and V cations. Left to right: barium (apple-green), strontium (crimson), calcium (brick red), sodium (yellow), and potassium (violet).

test an unknown. (Always clean the wire first by dipping it into 6 *M* HCl and then heating the wire.) Precipitates should always be dissolved in a small amount of HCl to provide solutions for flame tests.

One drop of 6 *M* HCl should be added to a few drops of a known or unknown solution before a flame test is performed.

34-4 PRECIPITATION OF ANALYTICAL GROUP IV CATIONS

Known and unknown solutions of the Group IV cations are either nitrates that contain a small amount of HNO_3 or chlorides that contain a little HCl.

The Group IV precipitating reagent, ammonium carbonate, contains the cation of a weak base and the anion of a weak polyprotic acid. Both hydrolyze.

$$NH_4^+ + H_2O \rightleftharpoons NH_3 + H_3O^+ \qquad K_a = 5.6 \times 10^{-10}$$

$$CO_3^{2-} + H_2O \rightleftharpoons HCO_3^- + OH^- \qquad K_b = 2.1 \times 10^{-4}$$

The carbonate ion hydrolyzes to a much greater extent than the ammonium ion, and therefore solutions of $(NH_4)_2CO_3$ do not contain sufficiently high concentrations of CO_3^{2-} ions to precipitate the Group IV carbonates completely. The Group IV precipitation is carried out in a buffered aqueous ammonia solution to suppress hydrolysis of the carbonate ion. The buffering action of NH_4Cl also suppresses the ionization of aqueous ammonia and prevents the precipitation of $Mg(OH)_2$ in Group IV.

Solutions from which Groups I, II, and III have been removed contain high concentrations of ammonium ions because aqueous ammonia was added in earlier procedures. Such solutions must be evaporated to dryness and heated quite strongly to decompose ammonium salts. Ammonium salts undergo thermal decomposition to produce the acid from which they were formed plus ammonia. Oxidizing acids, particularly HNO_3, may oxidize ammonia at elevated temperatures.

Flame tests for strontium (red) and barium (green).

$$NH_4Cl(s) \xrightarrow{\Delta} NH_3(g) + HCl(g)$$

$$NH_4NO_3(s) \xrightarrow{\Delta} NH_3(g) + HNO_3(g)$$

$$NH_3(g) + HNO_3(g) \xrightarrow{\Delta} N_2O(g) + 2H_2O(g)$$

After all ammonium salts have been decomposed, the NH_4^+ concentration can be adjusted to the desired value. One drop of 12 *M* HCl is added and then neutralized by 6 *M* aqueous ammonia,

$$H^+ + Cl^- + NH_3 \longrightarrow NH_4^+ + Cl^-$$

after which the desired excess of aqueous ammonia is added. The precipitating reagent, 1 *M* $(NH_4)_2CO_3$, is then added to the carefully buffered solution.

The reactions by which the Group IV cations are precipitated can be represented as

$$Ca^{2+} + CO_3^{2-} \longrightarrow CaCO_3(s) \qquad \text{white}$$

$$Sr^{2+} + CO_3^{2-} \longrightarrow SrCO_3(s) \qquad \text{white}$$

$$Ba^{2+} + CO_3^{2-} \longrightarrow BaCO_3(s) \qquad \text{white}$$

These carbonates precipitate as dense, fairly crystalline compounds, so there appears to be relatively little of the carbonate precipitate.

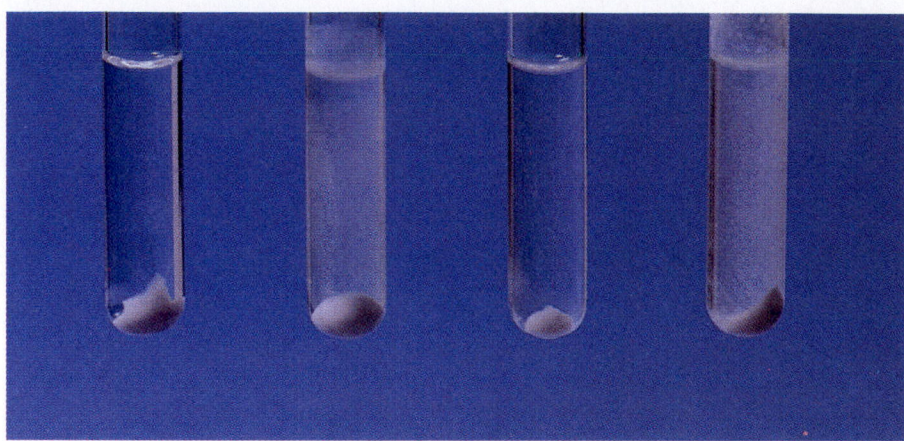

The Group IV precipitates. Left to right: $BaCO_3$, $SrCO_3$, $CaCO_3$, and a mixture of all three.

At temperatures above 80°C, $(NH_4)_2CO_3$ decomposes into ammonia, carbon dioxide, and water.

Step A

Precipitation of the Group IV carbonates: Ba^{2+}, Sr^{2+}, and Ca^{2+}.

(1) If the solution to be analyzed was obtained from the Group III separation, add 3 drops of 16 M HNO_3 and evaporate to dryness in a casserole or crucible. Heat the casserole or crucible red hot for 5 minutes. Allow it to cool, and then dissolve the residue in 12 drops of H_2O and 1 drop of 12 M HCl. Centrifuge, and transfer the solution to a test tube.

(2) If the solution to be analyzed contains only Group IV cations, place 12 drops of the known or unknown in a test tube and add 1 drop of 12 M HCl.

To the solution resulting from either (1) or (2), add 6 M aqueous NH_3 dropwise until the solution is just basic. *Test after each drop of aqueous NH_3 is added.* Now add 1 drop of 6 M aqueous NH_3 in excess and 3 drops of 1 M $(NH_4)_2CO_3$. Mix well and place in the hot water bath at 70 to 80°C for 3 minutes. Cool the mixture and add 1 drop of 1 M $(NH_4)_2CO_3$ to test for complete precipitation. Separate the mixture when precipitation is complete. Treat the residue according to Step B. Save the liquid for the Group V analysis if you are analyzing a general unknown. Otherwise, discard it.

34-5 SEPARATION AND IDENTIFICATION OF BARIUM IONS

Metal carbonates are dissolved by acids stronger than carbonic acid, H_2CO_3 ($K_1 = 4.2 \times 10^{-7}$, $K_2 = 4.8 \times 10^{-11}$). Because the separation of Ba^{2+} ions from Sr^{2+} and Ca^{2+} ions is accomplished best in weakly acidic solutions, the carbonates are dissolved in HCl, the excess HCl is neutralized with aqueous NH_3, and the solution is then made slightly acidic with CH_3COOH. The four equilibria involved in the dissolution may be summarized as follows for calcium carbonate. The equilibria for $SrCO_3$ and $BaCO_3$ are similar.

$$CaCO_3(s) \rightleftharpoons Ca^{2+} + CO_3^{2-}$$

$$H^+ + CO_3^{2-} \rightleftharpoons HCO_3^-$$

$$HCO_3^- + H^+ \rightleftharpoons H_2CO_3$$

$$H_2CO_3 \rightleftharpoons CO_2(g) + H_2O$$

Heating the reaction mixture decreases the solubility of CO_2 in water and shifts these equilibria to the right so that the carbonates dissolve completely. The overall reaction may be represented as

$$2H^+ + CaCO_3(s) \longrightarrow Ca^{2+} + CO_2(g) + H_2O$$

Table 34-2 shows that barium chromate is less soluble than strontium chromate and that strontium chromate is less soluble than calcium chromate. These differences in solubilities provide the basis for separating barium ions from strontium and calcium ions.

$$Ba^{2+} + CrO_4{}^{2-} \longrightarrow BaCrO_4(s) \qquad \text{pale yellow}$$

However, strontium chromate tends to coprecipitate with barium chromate, so the chromate ions are added slowly to a hot solution to minimize this probability. Strontium chromate forms supersaturated solutions readily, and keeping the solution hot increases the probability that $SrCrO_4$ will remain in solution.

Coprecipitation is the phenomenon in which a small amount of a compound precipitates *with* a similar compound whose K_{sp} is exceeded.

The concentration of Sr^{2+} ions is approximately 0.040 M in the solution from which $BaCrO_4$ is precipitated. The maximum concentration of $CrO_4{}^{2-}$ ions that can be present without exceeding K_{sp} for $SrCrO_4$ can be calculated.

$$[Sr^{2+}][CrO_4{}^{2-}] = 3.6 \times 10^{-5}$$

$$[CrO_4{}^{2-}] = \frac{3.6 \times 10^{-5}}{[Sr^{2+}]} = \frac{3.6 \times 10^{-5}}{4.0 \times 10^{-2}} = 9.0 \times 10^{-4} \, M$$

Therefore, it is necessary to keep the concentration of $CrO_4{}^{2-}$ ions below $9.0 \times 10^{-4} \, M$. The concentration of Ba^{2+} ions that remains in such a solution, i.e., unprecipitated, can also be calculated.

$$[Ba^{2+}][CrO_4{}^{2-}] = 2.0 \times 10^{-10}$$

$$[Ba^{2+}] = \frac{2.0 \times 10^{-10}}{[CrO_4{}^{2-}]} = \frac{2.0 \times 10^{-10}}{9.0 \times 10^{-4}} = 2.2 \times 10^{-7} \, M$$

This calculation tells us that Ba^{2+} ions are almost completely precipitated from solutions that are $9.0 \times 10^{-4} \, M$ in $CrO_4{}^{2-}$ ions.

Chromate ions are converted to dichromate ions in acidic solutions.

$$2CrO_4{}^{2-} + 2H^+ \rightleftharpoons Cr_2O_7{}^{2-} + H_2O$$

Therefore, we can control the concentrations of $CrO_4{}^{2-}$ and $Cr_2O_7{}^{2-}$ in solutions by varying the concentration of H^+. The equilibrium constant for the above reaction is

$$\frac{[Cr_2O_7{}^{2-}]}{[CrO_4{}^{2-}]^2[H^+]^2} = 4.2 \times 10^{14}$$

The concentration of $Cr_2O_7{}^{2-}$ is about 0.18 M in the solution from which $BaCrO_4$ is precipitated. We can solve the equilibrium constant expression for the $[H^+]$ that will maintain $[CrO_4{}^{2-}]$ at $9.0 \times 10^{-4} \, M$.

$$[H^+]^2 = \frac{[Cr_2O_7{}^{2-}]}{[CrO_4{}^{2-}]^2(4.2 \times 10^{14})} = \frac{0.18}{(9.0 \times 10^{-4})^2(4.2 \times 10^{14})} = 5.3 \times 10^{-10}$$

$$[H^+] = 2.3 \times 10^{-5} \, M \longleftarrow \text{minimum value for } [H^+]$$

Therefore, to precipitate $BaCrO_4$ almost completely while leaving Sr^{2+} ions in solution, the concentration of H^+ ions should be $2.3 \times 10^{-5} \, M$ or slightly higher (about pH 4.5) to provide some margin of safety.

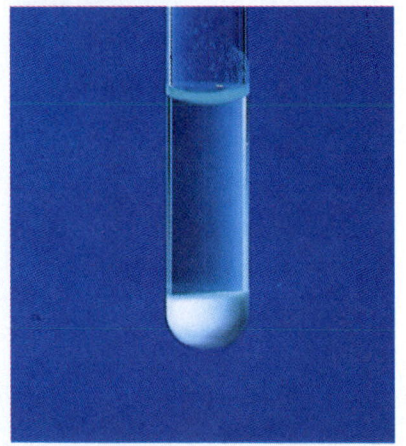

The confirmatory tests for barium: the flame test and precipitated $BaSO_4$.

The ionization constant for CH_3COOH is 1.8×10^{-5}. Buffer solutions (Section 18-7) containing CH_3COOH and NH_4CH_3COO are used to control the pH of the solution in which $BaCrO_4$ is precipitated.

$$\frac{[H^+][CH_3COO^-]}{[CH_3COOH]} = 1.8 \times 10^{-5}$$

$$\frac{[CH_3COO^-]}{[CH_3COOH]} = \frac{1.8 \times 10^{-5}}{[H^+]} = \frac{1.8 \times 10^{-5}}{2.3 \times 10^{-5}} = 0.78$$

The ratio of [salt] to [acid] should be 0.78 or slightly less to provide some margin of safety.

After $BaCrO_4$ has been precipitated and isolated, it must be washed to remove traces of Sr^{2+}, Ca^{2+}, and K^+ ions (from K_2CrO_4) that interfere with the flame test. It is then dissolved in dilute HCl, and a flame test is performed on the resulting solution.

$$2BaCrO_4(s) + 2H^+ \rightleftharpoons 2Ba^{2+} + Cr_2O_7^{2-} + H_2O$$

The high concentration of H^+ reduces the concentration of CrO_4^{2-} to the point that $[Ba^{2+}][CrO_4^{2-}] < K_{sp}$ and $BaCrO_4$ dissolves.

After the flame test has been performed, the solution is treated with H_2SO_4. The formation of a highly crystalline, nearly colorless precipitate, $BaSO_4$, is taken as the final confirmatory test for barium ions.

$$Ba^{2+} + SO_4^{2-} \longrightarrow BaSO_4(s) \qquad \text{white}$$

Step B

Use the precipitate from Step A, containing $BaCO_3$, $CaCO_3$, and $SrCO_3$. Add 10 drops of water and 2 drops of 6 *M* HCl to the precipitate, stir well, and place the test tube in the hot water bath for 3 minutes. All of the white precipitate should dissolve. Stir several times to ensure the removal of CO_2. Add 2 drops of 6 *M* aqueous ammonia, mix well, and then add 2 drops of 6 *M* CH_3COOH. Mix well and place the test tube in the hot water bath for 2 minutes. Use a clean *capillary pipet that delivers small drops* to add 1 *tiny* drop of 0.1 *M* K_2CrO_4 to the hot solution. Mix well and place in the hot water bath for 1 minute. Repeat this step until 10 *tiny* drops of 0.1 *M* K_2CrO_4 have been added, and then heat the mixture for 3 minutes with frequent stirring. A finely divided pale yellow precipitate indicates the presence of barium. Separate the mixture and save the solution for Step C.

Wash the precipitate twice with 3 drops of 3 *M* aqueous ammonia and discard the wash liquid. Dissolve the precipitate in 1 drop of 6 *M* HCl and 2 drops of water, and then perform a flame test on the resulting solution. Barium ions give a short-lived yellow-green color to the flame. Add 1 drop of 6 *M* H_2SO_4 to the solution used for the flame test. A finely divided, nearly colorless precipitate, $BaSO_4$, confirms the presence of barium ions.

34-6 SEPARATION AND IDENTIFICATION OF STRONTIUM IONS

The solubility product for $SrCrO_4$, 3.6×10^{-5}, is fairly large, and so a high concentration of CrO_4^{2-} is required to exceed K_{sp} and precipitate $SrCrO_4$. The addition of aqueous ammonia to the CH_3COOH/NH_4CH_3COO buffer solution neutralizes CH_3COOH and shifts the following equilibrium to the left.

$$2CrO_4^{2-} + 2H^+ \rightleftharpoons Cr_2O_7^{2-} + H_2O$$

When the concentration of CrO_4^{2-} is sufficiently high that K_{sp} for $SrCrO_4$ is exceeded, precipitation should occur. However, $SrCrO_4$ forms supersaturated solutions readily. Precipitation can usually be induced by the addition of ethyl alcohol, which decreases the polarity of the solvent and the solubility of $SrCrO_4$.

$$Sr^{2+} + CrO_4^{2-} \longrightarrow SrCrO_4(s) \qquad \text{pale yellow}$$

After $SrCrO_4$ has been isolated, it must be washed to remove traces of Ca^{2+} and K^+ ions that would interfere with the flame test.

Recall that CrO_4^{2-} ions are yellow and $Cr_2O_7^{2-}$ ions are orange.

Step C

Use the solution from Step B, containing Sr^{2+} and Ca^{2+}. Add 6 M aqueous ammonia dropwise, stirring after each drop, until the orange solution becomes a pretty pastel yellow. No more than 4 drops should be required. Heat the solution in the water bath for 3 minutes, cool it to room temperature, and add enough alcohol to double the volume of the solution. Mix well, and allow the mixture to stand for 3 minutes with frequent stirring. A finely divided yellow precipitate indicates the presence of strontium. Separate the mixture and save the solution for Step D.

Wash the precipitate twice, using a mixture of 3 drops of alcohol and 2 drops of 3 M aqueous ammonia each time. *Add the alcohol first.* Discard the wash solutions. Add 1 drop of 6 M HCl and 2 drops of water to dissolve the precipitate. Flame-test for Sr^{2+}. Strontium ions impart a crimson (pink-red) color to the flame. The flame test is the confirmatory test for Sr^{2+} ions.

If you are unable to decide whether Sr^{2+} ions are present, evaporate the solution that was flame-tested until all the liquid has disappeared. Do not bake. Cool, add 10 drops of water, mix well, and centrifuge. Transfer the liquid to a clean test tube and add 5 drops of 2.0 M $(NH_4)_2SO_4$. Mix well and allow the solution to stand for 5 minutes. A nearly colorless crystalline precipitate, $SrSO_4$, demonstrates the presence of Sr^{2+} *unless* Ba^{2+} ions were incompletely removed earlier.

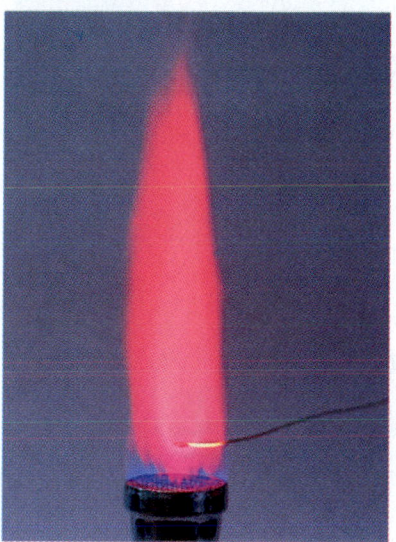

The confirmatory test for strontium.

34-7 SEPARATION AND IDENTIFICATION OF CALCIUM IONS

Calcium oxalate, CaC_2O_4, $K_{sp} = 2.3 \times 10^{-9}$, is precipitated from solutions that contain calcium ions by the addition of ammonium oxalate, $(NH_4)_2C_2O_4$.

$$Ca^{2+} + C_2O_4^{2-} \longrightarrow CaC_2O_4(s) \qquad \text{white}$$

The solution in which CaC_2O_4 is precipitated contains potassium ions (from the addition of K_2CrO_4). The precipitate must be washed to remove potassium ions before a flame test can be performed. The precipitate is dissolved in hydrochloric acid, and the flame test is performed on the resulting solution.

$$CaC_2O_4(s) + H^+ \rightleftharpoons Ca^{2+} + HC_2O_4^-$$

Calcium ions impart an orange-red (brick-red) color to the flame.

The confirmatory test for calcium.

Step D

Use the solution from Step C, containing Ca^{2+}. Place the solution in the hot water bath for 3 minutes, add 3 drops of 0.50 M $(NH_4)_2C_2O_4$, and mix well. If calcium ions are present, a white crystalline precipitate, CaC_2O_4, should form. Because CaC_2O_4 forms supersaturated solutions, it may be necessary to scratch the walls of the test tube with a stirring rod to initiate precipitation. Cool the mixture, separate it, and discard the liquid.

Wash the precipitate twice, using 6 drops of water and 2 drops of 6 M aqueous ammonia each time. Discard the wash liquid. Add 1 drop of 6 M HCl and 2 drops of water to the precipitate and perform a flame test. An orange-red flame confirms the presence of calcium ions.

Now that you have completed the Group IV unknown, flame-test the original solution and compare the results with those obtained using known solutions. Unknowns that contain only a single ion give its characteristic color to the flame. The presence of Ba^{2+} and Sr^{2+}, or Ba^{2+} and Ca^{2+}, in an unknown can be detected easily. Detecting the presence of Sr^{2+} and Ca^{2+} in the same unknown is a little more difficult. The unambiguous detection of all three Group IV cations in a single unknown is even more difficult. However, the results of the tests you have performed for the individual ions must be compatible with a flame test on the original Group IV unknown.

Exercises

General Questions on Group IV

1. In each of the previous groups, we have studied metals that exhibit variable oxidation states in their compounds. The metals in Group IV do not. Why?

2. Why are most compounds of the Group IV cations white in the solid state and colorless in aqueous solutions?

3. What is the solubility rule for metal carbonates? What problems would be caused by the incomplete precipitation of Group III cations in a general unknown?

4. (a) In general unknowns, the centrifugate from the Group III precipitation is always evaporated to dryness with nitric acid. Why? (b) Why is this step unnecessary for unknowns that contain only Group IV cations?

5. (a) For which Group IV carbonate is the molar solubility highest? lowest? (b) What is the ratio (highest molar solubility)/(lowest molar solubility) for the carbonates in (a)? What does this tell you?

6. (a) What is the basis for flame tests? (b) Why are flame tests very specific? (c) Why should one always observe the color of a "known" flame test just before the unknown is flame-tested? (d) List the colors imparted to laboratory burner flames by Group IV cations.

Group IV Reactions

Write balanced net ionic equations for reactions that occur when the following are mixed. Indicate colors of all precipitates.

7. Barium nitrate and ammonium carbonate in buffered aqueous ammonia.

8. Strontium nitrate and ammonium carbonate in buffered aqueous ammonia.

9. Calcium nitrate and ammonium carbonate in buffered aqueous ammonia.

10. Barium carbonate dissolves in 1 M HCl.

11. Strontium carbonate dissolves in 1 M HCl.

12. Calcium carbonate dissolves in 1 M HCl.

13. Barium acetate and potassium chromate (slightly acidic solution).

14. Barium chromate dissolves in 2 M HCl.

15. Dichromic acid reacts with *hot* 2 M HCl.

16. Barium chloride and dilute sulfuric acid.

17. Strontium acetate and potassium chromate.

18. Strontium chromate dissolves in 2 M HCl.

19. Strontium chloride and ammonium sulfate.

20. Calcium acetate and ammonium oxalate.

21. Calcium oxalate dissolves in 2 M HCl.

Other Questions and Problems

22. Why should the solution in which the Group IV precipitation is done be kept below 100°C?

23. Describe and write equations for the important equilibria in buffered aqueous ammonia solutions that contain ammonium carbonate.

24. (a) Describe and write equations for the important equilibria in the dissolution of Group IV carbonates in 1 M HCl. Use $SrCO_3$ as your example. (b) Why do you wait until the carbonates have dissolved before adding aqueous NH_3 in Step B? (c) What is the purpose of adding aqueous NH_3? (d) What are the likely consequences of adding too much aqueous NH_3 in Step B? insufficient aqueous NH_3?

25. (a) Calculate the molar solubilities of the Group IV chromates in pure water. (b) Let M_{El} refer to molar solubilities of these chromates; calculate the following ratios: $(M_{Ba})/(M_{Sr})$, $(M_{Ba})/(M_{Ca})$, $(M_{Sr})/(M_{Ca})$. What do these ratios tell us?

26. The equilibrium constant for the reaction

$$2CrO_4^{2-} + 2H^+ \rightleftharpoons Cr_2O_7^{2-} + H_2O$$

is 4.2×10^{14}.

(a) Write the equilibrium constant expression. (b) What is the value of the equilibrium constant for the reaction

$$Cr_2O_7^{2-} + H_2O \rightleftharpoons 2CrO_4^{2-} + 2H^+?$$

(c) Which ion, CrO_4^{2-} or $Cr_2O_7^{2-}$, is present in greater concentration in a solution (pH = 7.00) in which 0.10 mol/L of $K_2Cr_2O_7$ was dissolved? (d) Repeat (c) for pH = 2.00 and pH = 12.00.

27. Wires made of nichrome, or similar alloys, are used for flame tests in many laboratories. Why do the HCl solutions used to clean nichrome wires turn green? Which metals are suggested by the name "nichrome"?

28. Explain the equilibria involved in the dissolution of $BaCrO_4$ in HCl as you prepare for the flame test.

29. Dichromate ions are strong oxidizing agents in acidic solutions, and they oxidize chloride ions to elemental chlorine in hot solutions. Even if you use a platinum wire for flame tests, the solution that is flame-tested for barium ions turns green rapidly after the hot wire is plunged into it a few times. Why?

30. Suppose you forget to wash the $BaCrO_4$ before you perform a flame test. What unfortunate consequence would you expect? (*Hint:* What reagent was added to precipitate $BaCrO_4$?)

31. Why is aqueous ammonia added to the solution from which $BaCrO_4$ has been removed before $SrCrO_4$ is precipitated?

32. (a) Why is alcohol added to the solution from which $SrCrO_4$ is to be precipitated? (b) What would be the result of adding too little alcohol? Too much alcohol?

33. Refer to the last sentence in Section 34-6. Why do you suppose *unless* is italicized? (Refer to Table 34-2.)

35

Analysis of Cation Group V

Fireworks display colors emitted by excited metal ions.

35-1 INTRODUCTION

Cation Group V is known as the *soluble group* because its cations are not precipitated by the reagents used to precipitate and separate the first four groups or by any other single reagent. These cations are magnesium, Mg^{2+}, sodium, Na^+, potassium, K^+, and ammonium, NH_4^+. None of the metals in this group exhibits variable oxidation states.

The solubility rules (Section 4-2, part 5) tell us that all common *simple* compounds of Na^+, K^+, and NH_4^+ are soluble in water. (In Analytical Group II we precipitated $MgNH_4AsO_4$, an insoluble double salt, in the confirmatory test for arsenic.) Some other complex compounds of these three cations are insoluble. Several compounds of magnesium are insoluble, so magnesium ions are easily separated from the other members of this group.

An Important Note

The test for the ammonium ion, Step F, must always be done on the *original* unknown because aqueous ammonia is added at several points in the analytical procedures. Save some of the original unknown for Step F.

Most of the unknown solutions are acidic, and gaseous ammonia may be absorbed from the air in the laboratory. Therefore, Group V and general unknowns should be kept tightly stoppered.

The solution to be analyzed may contain only Analytical Group V ions, or it may be the centrifugate from the Group IV separation. In general unknowns, care must be exercised before the analysis of Group V is begun to remove any traces of Group IV cations that may have been carried into Group V. Such procedures are not necessary for solutions that contain *only* Group V cations.

Removal of Traces of Barium, Strontium, and Calcium Ions

Even small amounts of the Group IV cations give misleading results in the initial precipitation of magnesium ions. (Traces of Group IV cations fail to precipitate at appropriate points in the Group IV procedures when the acidity is not adjusted properly. This occurs when the concentration of ammonium ions is too high.) As a precautionary measure, a few drops each of $(NH_4)_2C_2O_4$ and $(NH_4)_2SO_4$ solutions are added to the solution to be analyzed. Any precipitate is separated and discarded before the analysis of Group V is begun. The $C_2O_4^{2-}$ ions precipitate traces of Ca^{2+} and Sr^{2+} ions, and the SO_4^{2-} ions precipitate traces of Ba^{2+} ions.

Halite crystals (naturally occurring NaCl).

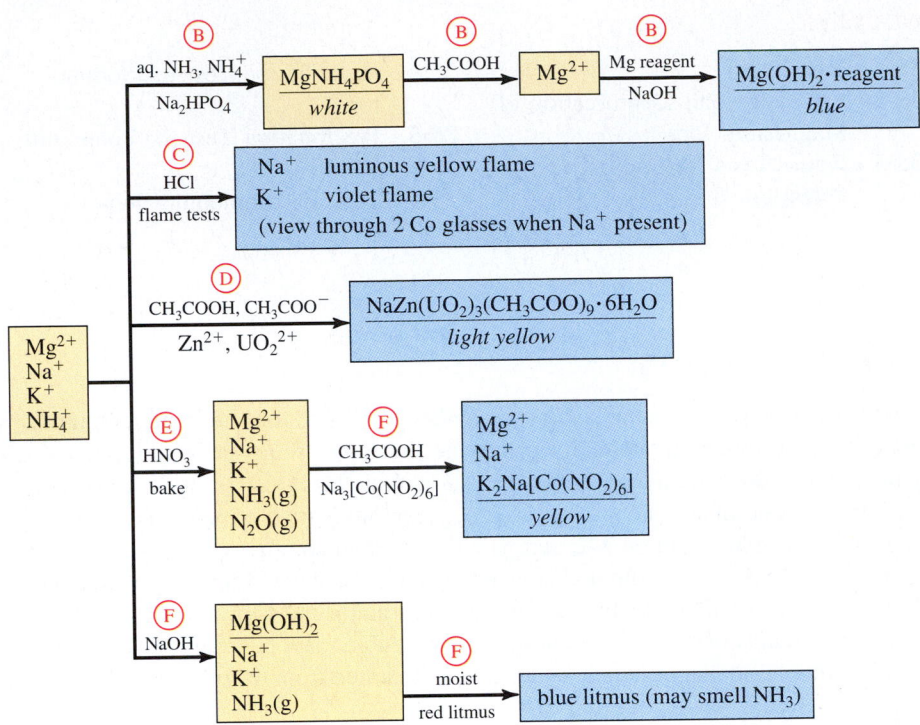

Figure 35-1 Group V flow chart. Circled letters refer to analytical steps.

Step A

Use this procedure for general unknowns only; begin with Step B for knowns and unknowns that contain only Group V cations. Use the solution from the Group IV separation, containing Mg^{2+}, Na^+, K^+, NH_4^+, and possibly traces of Ca^{2+}, Sr^{2+}, and Ba^{2+}. Add 1 drop of 0.5 M $(NH_4)_2C_2O_4$ and 1 drop of 2 M $(NH_4)_2SO_4$ to the solution from which Group IV cations were removed, mix well, and centrifuge the mixture. Wrap a *tiny* piece of cotton around the end of a capillary pipet, and draw the liquid into the pipet. Discard the piece of cotton and any precipitate. Save the solution for Step B.

Although Group V cations cannot be separated individually in a schematic way as was done in earlier groups, each exhibits properties that enable us to test for that ion individually, as Figure 35-1 shows. The reactions indicated in the flow chart are specific for each ion *under carefully controlled conditions*. In the hands of the inexperienced, the precipitation reactions for sodium and potassium ions are less reliable than the flame tests for these ions.

35-2 SEPARATION AND IDENTIFICATION OF MAGNESIUM IONS

A portion of the solution is tested for magnesium ions. The test for arsenate ions, AsO_4^{3-}, in Analytical Group II involved precipitating magnesium ammonium arsenate, $MgNH_4AsO_4$. The formation of a similar compound, magnesium ammonium phosphate,

$MgNH_4PO_4$, is used to separate magnesium ions from the other Group V cations. The reaction of disodium hydrogen phosphate with magnesium ions in buffered aqueous ammonia solutions is

$$Mg^{2+} + NH_3 + HPO_4^{2-} \longrightarrow MgNH_4PO_4(s) \qquad \text{white}$$

The solid $MgNH_4PO_4$ is separated and dissolved in acetic acid, which reverses the above reaction by converting PO_4^{3-} ions to HPO_4^{2-} ions. This shifts the equilibrium to the left. Recall that K_3 for H_3PO_4 is only 3.6×10^{-13} (Section 18-9). Therefore, even the weak acid CH_3COOH ($K_a = 1.8 \times 10^{-5}$) is strong enough to dissolve most phosphates. The concentrations of the ions are reduced so that $[Mg^{2+}][NH_4^+][PO_4^{3-}] < K_{sp}$, and the compound dissolves.

$$MgNH_4PO_4(s) + 2CH_3COOH \rightleftharpoons Mg^{2+} + NH_4^+ + H_2PO_4^- + 2CH_3COO^-$$

The addition of sodium hydroxide solution precipitates magnesium hydroxide, a white compound.

$$Mg^{2+} + 2OH^- \longrightarrow Mg(OH)_2(s)$$

When $Mg(OH)_2$ is precipitated in the presence of an organic dye known as magnesium reagent (*p*-nitrobenzeneazo)resorcinol, the purple dye is adsorbed by the white $Mg(OH)_2$ and the precipitate appears blue. It is sometimes referred to as a "blue lake."

"Lake" is a general term for a coprecipitate of an organic dye with a metallic hydroxide or salt. This kind of reaction can be used to dye cloth.

Step B
Use 6 drops of the solution from Step A or 12 drops of the Group V known or unknown solution, containing Mg^{2+}, Na^+, K^+, and NH_4^+. Test the solution with litmus. If it is not basic, add 6 *M* aqueous ammonia dropwise, with stirring, until it becomes basic and then add 1 drop in excess. Add 2 drops of 1.5 *M* Na_2HPO_4 and mix well, scratching the walls of the test tube with a stirring rod to initiate precipitation if necessary. The appearance of a white crystalline precipitate, $MgNH_4PO_4$, proves the presence of magnesium ions. Centrifuge and separate the mixture. Add 1 drop of 6 *M* CH_3COOH and 4 drops of water to the precipitate. Mix well; after all the precipitate has dissolved, add 1 drop of magnesium reagent (not to be confused with magnesia mixture). Mix well, add 5 drops of 4 *M* NaOH, stir, and centrifuge the mixture. The presence of a *sky-blue precipitate* proves that magnesium ions are present. Observe carefully because the magnesium reagent makes the solution purple.

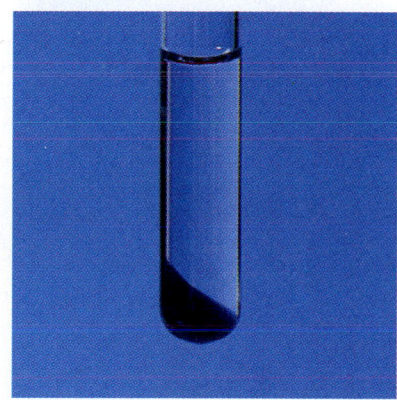

The confirmatory test for magnesium, $MgNH_4AsO_4$.

35-3 FLAME TESTS FOR SODIUM AND POTASSIUM IONS

The most reliable tests for sodium and potassium ions are the flame tests. Both impart characteristic colors to the burner flame. Traces of sodium contaminate almost everything. Use caution in reporting sodium ions.

Flame tests may be performed on the solution obtained in Step A (removal of traces of Group IV cations) or on the original Group V known and unknown solutions. Sodium ions impart an intense yellow color to the burner flame. Flame tests on a known solution should always be compared with those done on unknowns.

Potassium ions impart a characteristic violet color to the flame. However, when sodium ions are also present, their intense yellow color completely masks the violet color pro-

A confirmatory test for sodium.

duced by potassium ions. Two thicknesses of cobalt glass filter out the yellow color produced by sodium ions so that the potassium flame test can be observed even when sodium ions are present. You should practice observing the potassium flame test through two pieces of cobalt glass until you are comfortable with it. First practice on a solution that contains potassium ions, and then use a solution that contains both sodium and potassium ions. Neither magnesium ions nor ammonium ions impart color to laboratory burner flames.

C A U T I O N

Detergents contain large amounts of sodium compounds. Therefore, all test tubes and pipets that are used in the tests for sodium ions must be rinsed thoroughly to remove traces of sodium compounds.

Step C
Use 6 drops of the solution from Step A or 6 drops of the Group V known or unknown, containing Mg^{2+}, Na^+, K^+, and NH_4^+. Add 1 drop of 6 M HCl and flame-test the solution. Observe the flame to determine whether sodium ions are present. If sodium ions are present (even in trace amounts, as they are in almost everything), observe the flame through *two* thicknesses of cobalt glass to determine whether potassium ions are present. Don't confuse the red-hot wire with a flame test for potassium ions. An intense yellow flame and a characteristic violet flame prove the presence of sodium and potassium ions, respectively.

35-4 PRECIPITATION TEST FOR SODIUM IONS

Sodium ions can also be detected by precipitating a crystalline triple salt, sodium zinc uranyl acetate hexahydrate, $NaZn(UO_2)_3(CH_3COO)_9 \cdot 6H_2O$. A saturated solution of zinc acetate, $Zn(CH_3COO)_2$, and uranyl acetate, $UO_2(CH_3COO)_2$, in acetic acid is known as sodium reagent. Because Mg^{2+}, K^+, and NH_4^+ ions do not form precipitates with sodium reagent, the precipitation test may be performed in the presence of these cations.

$$Na^+ + Zn^{2+} + 3UO_2^{2+} + 9CH_3COO^- + 6H_2O \longrightarrow NaZn(UO_2)_3(CH_3COO)_9 \cdot 6H_2O(s)$$
light yellow

The formation of this precipitate is often very slow. This procedure should be done as early in the laboratory period as possible.

Step D
If you use the solution from Step A, perform (a) below and then (c). If you use the original Group V known or unknown, perform (b) below and then (c). The solution contains Mg^{2+}, Na^+, K^+, and NH_4^+.
(a) To 3 drops of the solution from Step A, add 3 M CH$_3$COOH until it is acidic to litmus.
(b) To 3 drops of the Group V known or unknown solution, add 1 M aqueous

ammonia until it is basic to litmus (test after each drop). Then add 1 drop of 3 M CH_3COOH.

(c) Place 10 drops of sodium reagent in a test tube and add 2 drops of the solution prepared in (a) or (b) above. Mix well and allow to stand for an hour if necessary. The appearance of a highly crystalline yellow precipitate indicates that sodium ions are present.

35-5 PRECIPITATION TEST FOR POTASSIUM IONS

After ammonium ions have been removed, potassium ions can be detected in the solution by formation of an insoluble yellow compound, dipotassium sodium hexanitrocobaltate(III), $K_2Na[Co(NO_2)_6]$. A saturated solution of sodium hexanitrocobaltate(III), $Na_3[Co(NO_2)_6]$, is added to a weakly acidic solution that contains potassium ions.

$$2K^+ + Na^+ + [Co(NO_2)_6]^{3-} \longrightarrow K_2Na[Co(NO_2)_6](s) \qquad \text{yellow}$$

However, ammonium ions form a similar precipitate, so they must be removed before the test for potassium ions can be done. Destruction of ammonium ions is accomplished by evaporation with nitric acid followed by ignition to dull red heat, which volatilizes any remaining ammonium salts. These reactions may be summarized in simplified form as

$$NH_4Cl(s) \xrightarrow{\Delta} NH_3(g) + HCl(g)$$

$$NH_4NO_3(s) \xrightarrow{\Delta} NH_3(g) + HNO_3(g)$$

$$NH_3(g) + HNO_3(g) \xrightarrow{\Delta} N_2O(g) + 2H_2O(g)$$

Heat all parts of the dish in which the evaporation and ignition are done so that the expulsion of ammonium compounds is complete.

If NH_4^+ ions are incompletely removed, all is not lost. The test tube that contains $K_2Na[Co(NO_2)_6]$ and/or $(NH_4)_2Na[Co(NO_2)_6]$ can be heated until the solution turns pink. The pink color shows that $[Co(NO_2)_6]^{3-}$ ions have decomposed and that Co(III) has been reduced to Co(II) by nitrite ions in the hot solution. Nitrite ions react with ammonium ions at 100°C and oxidize them to nitrogen and water.

$$NO_2^- + NH_4^+ \longrightarrow N_2 + 2H_2O$$

After any remaining NH_4^+ ions have been destroyed, more $Na_3[Co(NO_2)_6]$ can be added and potassium ions can be detected.

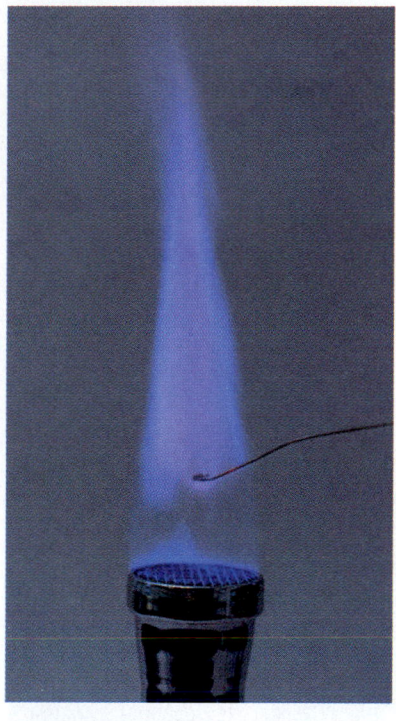

A confirmatory test for potassium.

Step E

Use the remainder of the solution from Step A, containing Mg^{2+}, Na^+, K^+, and NH_4^+, or 10 drops of the original Group V known or unknown. Transfer the solution to a casserole (or a crucible or a small evaporating dish), add 5 drops of 16 M HNO_3, and then carefully evaporate the solution to dryness *under the hood*. Heat the casserole to a dull red heat for a few minutes. Direct the hot part of the flame at all parts of the *outer* surface of the casserole to ensure complete decomposition of ammonium compounds. Allow the casserole to cool to near room temperature, and add 2 drops of 1 M HCl and 6 drops of 6 M CH_3COOH. Swirl the liquid so that it touches all the inner surface of the casserole, and then *carefully* heat the solution to boiling. Add 4 drops of this solution to 10 drops of the saturated $Na_3[Co(NO_2)_6]$ solution and warm gently (50°C) for 5 minutes. A yellow precipitate indicates the presence of potassium ions if ammonium ions were removed completely.

If NH_4^+ ions were incompletely removed, $(NH_4)_2Na[Co(NO_2)_6]$ also precipitates at this point. Heat the solution and yellow precipitate in the hot water bath (100°C) until the solution turns pink. Cool the pink solution and add 3 more drops of $Na_3[Co(NO_2)_6]$. A yellow precipitate indicates conclusively that potassium ions are present.

35-6 TEST FOR AMMONIUM IONS

The test for ammonium ions must always be done on the original known or unknown solution, because aqueous ammonia is added in several procedures. Aqueous ammonia is a weak base, and it ionizes only slightly.

$$NH_3 + H_2O \rightleftharpoons NH_4^+ + OH^-$$

The addition of ammonium ions to a solution of a strong base such as NaOH results in the formation and liberation of gaseous ammonia, i.e., the reverse of the above reaction.

$$NH_4^+ + OH^- \text{ (excess)} \longrightarrow NH_3(g) + H_2O$$

The liberated ammonia can be detected by its characteristic odor or by contact with an acid–base indicator such as litmus.

The confirmatory test for ammonium ions.

Step F
Detection of the ammonium ion. Place a piece of red litmus paper on the convex side of a clean watch glass, and wet the paper with distilled water so that it adheres to the watch glass. Place 20 drops of 4 M NaOH in a small beaker (or a small evaporating dish) supported on a ring stand. Heat the beaker *gently* until it is warm to the touch, but do *not* boil the NaOH solution. Add 5 drops of the *original* known or unknown solution to the warm NaOH. Place the watch glass over the beaker or evaporating dish immediately, and watch for a color change (red litmus turns blue), which proves the presence of ammonium ions. You may wish to remove the watch glass and smell the ammonia.

A word of caution. The NaOH solution must not be hot enough to boil—litmus always turns blue when NaOH spatters onto it!

Step G
A final check. A flame test should be performed on a few drops of the original solution to which 1 drop of 6 M HCl has been added. The results should be consistent with those obtained for the individual ions.

Keep in mind the fact that, in general, the flame tests for unknowns can be observed for any Group IV ions that are present *only if* sodium ions are absent, because sodium ions obscure the flame tests for Group IV cations. Copper(II) ions impart an intense green color to the flame, so the results of a flame test may not be very meaningful in unknowns that contain copper.

Exercises

General Question on Group V

1. (a) List the cations in Group V. (b) Why is Group V called the soluble group?

2. Why are ammonium sulfate and ammonium oxalate added to the solution from which the Group IV cations were removed (Step A) before testing for Group V cations?

3. Why is the test for the ammonium ion always performed on a sample of the original unknown solution?

4. (a) Why does litmus paper often change colors when exposed to the air in the laboratory for long periods of time? (b) What compounds are likely to be responsible for the color changes?

Group V Reactions

Write balanced net ionic equations for reactions that occur when the following are mixed. Indicate colors of all precipitates.

5. Ammonium nitrate + 4 M NaOH.

6. Magnesium nitrate + disodium hydrogen phosphate in buffered aq. NH_3.

7. Magnesium ammonium phosphate + dil. CH_3COOH.

8. Magnesium acetate + 4 M NaOH.

9. The thermal decomposition of ammonium chloride.

10. The thermal decomposition of ammonium nitrate (two reactions).

11. Potassium acetate + sodium hexanitrocobaltate(III) in excess CH_3COOH.

12. Potassium nitrite + ammonium acetate at 100°C in excess CH_3COOH.

13. Sodium nitrate + zinc acetate + uranyl acetate in excess CH_3COOH.

Other Questions and Problems

14. In the precipitation test for potassium (Step E) you are told to heat the solution that contains the yellow precipitate until the solution turns pink. Why?

15. In the alternative confirmatory test for cobalt (Step E(2) in Chapter 33), we represented the insoluble complex compound as $K_3[Co(NO_2)_6]$. In the precipitation test for potassium, we represented the insoluble compound as $K_2Na[Co(NO_2)_6]$. Can you suggest a reason? (*Hint:* What reagent did we add to test for cobalt? In what concentration?)

16. (a) Why are flame tests so distinctive? (b) What is the function of cobalt glass in the flame test for potassium?

17. If you were given a sample of a white solid known to be one of the following, how would you determine which one it is? Answer the question so that you could identify each absolutely. $MgSO_4$, Na_2SO_4, K_2SO_4, $(NH_4)_2SO_4$.

18. Assume that you are given a solution that contains only one Group V cation. How could you determine which cation is present, using *only* two tests? Explain.

36

Ionic Equilibria in Qualitative Analysis

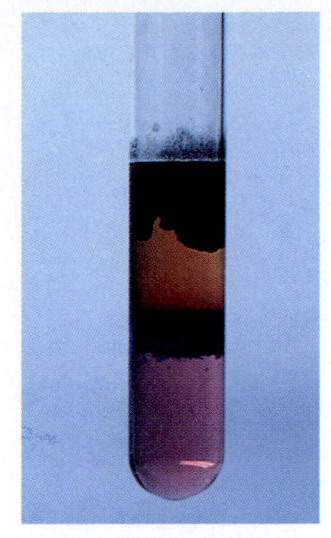

In Chapters 18, 19, and 20 we presented the fundamental ideas about equilibria in aqueous solutions. We now build upon those concepts to explain some of the more sophisticated equilibria encountered in qualitative analysis.

36-1 DISSOLVING PRECIPITATES

In our laboratory work we often find it necessary to dissolve precipitates. A precipitate dissolves when the concentrations of its ions are reduced so that K_{sp} is no longer exceeded, i.e., when $Q_{sp} < K_{sp}$. Precipitates can be dissolved by the three types of reactions discussed in Section 20-6. All involve removing ions from solution. Please review Section 20-6 carefully.

The cations in many slightly soluble compounds can form complex ions. This often results in dissolution of the slightly soluble compound. From the discussion of complex ion formation in Section 20-6, we see (pages 763–765) that the more effectively a ligand competes for a coordination site on the metal ions, the smaller K_d is. This tells us that, in a comparison of complexes with the same number of ligands, the smaller the K_d value, the more stable the complex ion. Some complex ions and their dissociation constants, K_d, are listed in Appendix I.

Some equilibria involving Co^{2+} ions.

Solubility products, like other equilibrium constants, are thermodynamic quantities. They tell us nothing about how fast a given reaction occurs, only that it can, or cannot, occur under specified conditions.

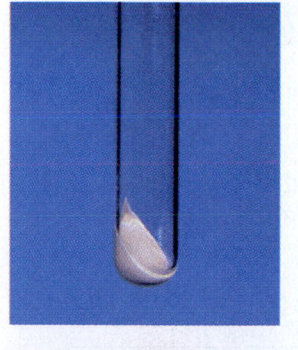
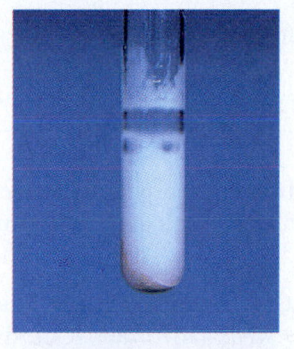
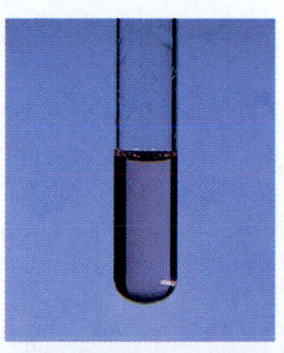

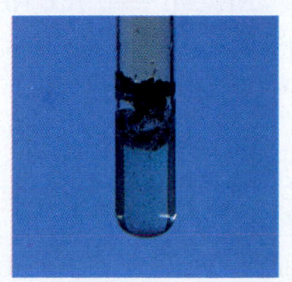

Manganese(II) sulfide, MnS, is salmon-colored. MnS dissolves in 6 M HCl. The resulting solution of $MnCl_2$ is pale pink.

Copper(II) sulfide, CuS, is black. As CuS dissolves in 6 M HNO_3, some NO is oxidized to brown NO_2 by O_2 in the air. The resulting solution of $Cu(NO_3)_2$ is blue.

EXAMPLE 36-1   *Complex Ion Formation*

What are the concentrations of hydrated Cu^{2+}, NH_3, and $[Cu(NH_3)_4]^{2+}$ in a 0.20 M solution of $[Cu(NH_3)_4]SO_4$?

Plan

We write the equation for the *complete dissociation* of the soluble compound that contains the complex ion. This gives us the concentration of the complex ion $[Cu(NH_3)_4]^{2+}$. Then we write the equation for the dissociation of the complex ion, represent the equilibrium concentrations algebraically, and substitute them into the equilibrium constant expression.

Solution

The soluble deep-blue complex salt $[Cu(NH_3)_4]SO_4$ dissociates completely to produce tetraamminecopper(II) ions and sulfate ions.

$$[Cu(NH_3)_4]SO_4(aq) \xrightarrow{100\%} [Cu(NH_3)_4]^{2+}(aq) + SO_4^{2-}(aq)$$
$$0.20\ M \Longrightarrow \quad\quad 0.20\ M \quad\quad\quad 0.20\ M$$

Some of the $[Cu(NH_3)_4]^{2+}$ ions then dissociate. Let x be the concentration of $[Cu(NH_3)_4]^{2+}$ that dissociates.

$$[Cu(NH_3)_4]^{2+} \rightleftharpoons Cu^{2+}(aq) + 4NH_3$$
$$(0.20 - x)\ M \quad\quad x\ M \quad\quad 4x\ M$$

$$K_d = \frac{[Cu^{2+}][NH_3]^4}{[[Cu(NH_3)_4]^{2+}]} = 8.5 \times 10^{-13} = \frac{x(4x)^4}{0.20 - x} = \frac{256\ x^5}{0.20}$$

(assume that $(0.20 - x) \approx 0.20$)

$$x^5 = 6.6 \times 10^{-16}$$

Taking the fifth root of both sides of this equation gives $x = 9.2 \times 10^{-4}$.

$$[Cu^{2+}] = x\ M = 9.2 \times 10^{-4}\ M$$

$$[NH_3] = 4x\ M = 3.7 \times 10^{-3}\ M$$

$$[[Cu(NH_3)_4]^{2+}] = (0.20 - x)\ M \approx 0.20\ M$$

$(0.20 - x) \approx 0.20$ is a valid assumption; 9.2×10^{-4} is much less than 0.20.

Copper(II) hydroxide dissolves in an excess of aqueous NH_3 to form the deep-blue complex ion $[Cu(NH_3)_4]^{2+}$. This decreases the $[Cu^{2+}]$ so that $[Cu^{2+}][OH^-]^2 < K_{sp}$, and so the $Cu(OH)_2$ dissolves.

$$Cu(OH)_2(s) \rightleftharpoons Cu^{2+}(aq) \quad\quad\quad + 2OH^-(aq)$$
$$Cu^{2+}(aq) + 4NH_3(aq) \rightleftharpoons [Cu(NH_3)_4]^{2+}(aq)$$
overall rxn: $\quad Cu(OH)_2(s) + 4NH_3(aq) \rightleftharpoons [Cu(NH_3)_4]^{2+}(aq) + 2OH^-(aq)$

On the other hand, zinc hydroxide is amphoteric (Section 10-8). This tells us that solid $Zn(OH)_2$ dissolves in excess NaOH solution to form the complex ion $[Zn(OH)_4]^{2-}$.

$$Zn(OH)_2(s) + 2OH^-(aq) \rightleftharpoons [Zn(OH)_4]^{2-}(aq)$$

EXAMPLE 36-2   *Complex Ion Equilibria*

Some solid $Zn(OH)_2$ is suspended in a saturated solution of $Zn(OH)_2$. A solution of sodium hydroxide is added until all the $Zn(OH)_2$ just dissolves. The pH of the solution is 11.80. What are the

---

As before, the outer brackets mean molar concentrations. The inner brackets are part of the formula of the complex ion.

Concentrated aqueous NH_3 was added *slowly* to a solution of copper(II) sulfate, $CuSO_4$. Unreacted blue copper(II) sulfate solution remains in the bottom part of the test tube. The light-blue precipitate in the middle is copper(II) hydroxide, $Cu(OH)_2$. The top layer contains deep-blue $[Cu(NH_3)_4]^{2+}$ ions that were formed as some $Cu(OH)_2$ dissolved in excess aqueous NH_3.

concentrations of Zn^{2+} and $[Zn(OH)_4]^{2-}$ ions in the solution? K_{sp} for $Zn(OH)_2 = 4.5 \times 10^{-17}$, and K_d for $[Zn(OH)_4]^{2-} = 3.5 \times 10^{-16}$.

Plan

We write the appropriate chemical equations and equilibrium constants for the two reversible reactions. We see that $[OH^-]$ appears in both of these equilibrium constant expressions. We are given pH, from which we can calculate $[OH^-]$. Once we know $[OH^-]$, we can use K_{sp} for $Zn(OH)_2$ to calculate $[Zn^{2+}]$. Then, knowing both $[OH^-]$ and $[Zn^{2+}]$, we can solve for $[[Zn(OH)_4]^{2-}]$.

Solution

The important equilibria and their equilibrium constant expressions are

$$Zn(OH)_2(s) \rightleftharpoons Zn^{2+}(aq) + 2OH^-(aq) \qquad K_{sp} = [Zn^{2+}][OH^-]^2 = 4.5 \times 10^{-17}$$

$$[Zn(OH)_4]^{2-} \rightleftharpoons Zn^{2+}(aq) + 4OH^-(aq) \qquad K_d = \frac{[Zn^{2+}][OH^-]^4}{[[Zn(OH)_4]^{2-}]} = 3.5 \times 10^{-16}$$

We know the pH, and so we can calculate $[OH^-]$ from pH + pOH = 14.00.

$$pOH = 14.00 - pH = 14.00 - 11.80 = 2.20$$

$$[OH^-] = 10^{-pOH} = 10^{-2.20} = 6.3 \times 10^{-3} \ M = [OH^-]$$

We can use the solubility product expression for $Zn(OH)_2$ to calculate $[Zn^{2+}]$.

$$K_{sp} = [Zn^{2+}][OH^-]^2 = 4.5 \times 10^{-17}$$

$$[Zn^{2+}] = \frac{K_{sp}}{[OH^-]^2} = \frac{4.5 \times 10^{-17}}{(6.3 \times 10^{-3})^2} = \boxed{1.1 \times 10^{-12} \ M \ Zn^{2+}}$$

Both equilibria are established in the same solution, and so the same $[Zn^{2+}]$ and $[OH^-]$ also satisfy the complex ion dissociation equilibrium.

$$K_d = \frac{[Zn^{2+}][OH^-]^4}{[[Zn(OH)_4]^{2-}]} = 3.5 \times 10^{-16}$$

$$[[Zn(OH)_4]^{2-}] = \frac{[Zn^{2+}][OH^-]^4}{3.5 \times 10^{-16}} = \frac{(1.1 \times 10^{-12})(6.3 \times 10^{-3})^4}{3.5 \times 10^{-16}}$$

$$\boxed{[[Zn(OH)_4]^{2-}] = 5.0 \times 10^{-6} \ M}$$

Thus, we see that we are able to shift equilibria (in this case, dissolve $Zn(OH)_2$) by taking advantage of complex ion formation.

36-2 EQUILIBRIA INVOLVING COMPLEX IONS

Dissolving precipitates by complex ion formation is used at several points in qualitative analysis. *By convention, equilibrium constants for complex ions are usually written as dissociation constants.* We have demonstrated that dissociation constants for complex ions can be treated like other equilibrium constants. The dissociation equation and equilibrium constant expression for the diamminesilver ion are

$$[Ag(NH_3)_2]^+ \rightleftharpoons Ag^+ + 2NH_3 \qquad K_d = \frac{[Ag^+][NH_3]^2}{[[Ag(NH_3)_2]^+]} = 6.3 \times 10^{-8}$$

These equilibrium constants *may* also be represented as formation constants.

$$Ag^+ + 2NH_3 \rightleftharpoons [Ag(NH_3)_2]^+$$

$$K_f = \frac{[[Ag(NH_3)_2]^+]}{[Ag^+][NH_3]^2} = \frac{1}{K_d} = \frac{1}{6.3 \times 10^{-8}} = 1.6 \times 10^7$$

The separation of AgCl and Hg_2Cl_2 by dissolving AgCl in an excess of aqueous ammonia is a typical example of the dissolution of a precipitate by forming a soluble complex compound.

$$\left. \begin{array}{l} \underline{AgCl} \\ \underline{Hg_2Cl_2} \end{array} \right\} \xrightarrow[\text{aq. } NH_3]{\text{excess}} \begin{array}{l} [Ag(NH_3)_2]^+ + Cl^- \\ HgNH_2Cl + Hg(\ell) \end{array}$$

Example 36-3 illustrates the quantitative aspects of such equilibria in a somewhat simplified format.

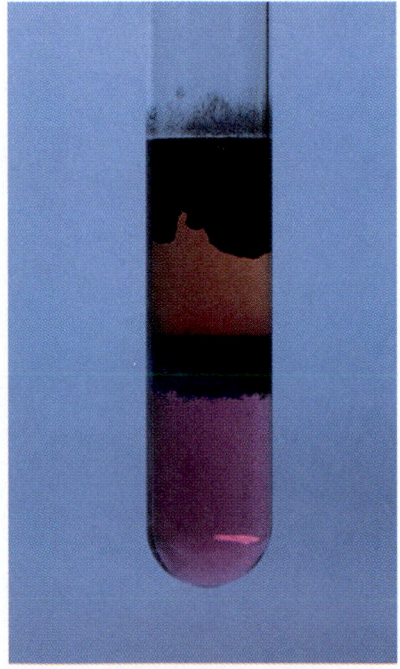

Some equilibria involving Co^{2+} ions.

1. Co^{2+} ions, in the pink solution (bottom), react with aq. NH_3 to form $Co(OH)_2$.
   $K_{sp} = 2.5 \times 10^{-16}$.
2. $Co(OH)_2$, the blue-gray solid (next to bottom layer), reacts with excess aq. NH_3 to form $[Co(NH_3)_6]^{2+}$ ions (tan solution). $K_d = 1.3 \times 10^{-5}$.
3. $[Co(NH_3)_6]^{2+}$ ions, in the tan solution, react with S^{2-} ions to form CoS, the black solid (top), $K_{sp} = 5.9 \times 10^{-21}$.

EXAMPLE 36-3 *Dissolution of a Precipitate*

Calculate the minimum number of moles of gaseous NH_3 required to dissolve 0.020 mole of AgCl in enough water to give 1 liter of solution.

Solution

The dissolution of AgCl in aqueous NH_3 may be represented as

$$\begin{array}{ccccc} AgCl(s) & + & 2NH_3 & \longrightarrow & [Ag(NH_3)_2]^+ + & Cl^- \\ 0.020 \text{ mol} & & 2(0.020 \text{ mol}) \Longrightarrow & & 0.020 \text{ mol} & 0.020 \text{ mol} \end{array}$$

This equation tells us that the dissolution of 0.020 mole of AgCl *requires* 2(0.020) mole of NH_3 and *produces* 0.020 mole of $[Ag(NH_3)_2]^+$ and 0.020 mole of Cl^-. Thus, the *stoichiometry of the reaction requires 0.040 mole of NH_3.* In addition, we must calculate the number of moles of NH_3 in excess of this amount that is *required to satisfy the dissociation constant for the complex ion,* $[Ag(NH_3)_2]^+$.

Two equilibria are involved in the dissolution of AgCl in aqueous NH_3. Therefore, both equilibrium constants must be satisfied. They are

$$AgCl(s) \rightleftharpoons Ag^+ + Cl^- \qquad K_{sp} = [Ag^+][Cl^-] = 1.8 \times 10^{-10}$$

$$[Ag(NH_3)_2]^+ \rightleftharpoons Ag^+ + 2NH_3 \qquad K_d = \frac{[Ag^+][NH_3]^2}{[[Ag(NH_3)_2]^+]} = 6.3 \times 10^{-8}$$

Because Ag^+ ion is the species common to both chemical equations and both equilibrium constants, it provides a "connection" between the two equilibria. The equation for the dissolution of AgCl, a reaction that goes to completion, tells us that the solution contains 0.020 M of Cl^-. Therefore, we can solve the solubility product expression of AgCl for $[Ag^+]$, which gives us a handle on the problem.

$$[Ag^+] = \frac{1.8 \times 10^{-10}}{[Cl^-]} = \frac{1.8 \times 10^{-10}}{0.020} = 9.0 \times 10^{-9} \, M$$

This $[Ag^+]$ is substituted into the dissociation constant expression for $[Ag(NH_3)_2]^+$, which allows us to calculate the *equilibrium concentration of NH_3.* Recall that the stoichiometry of the dissolution reaction tells us that $[[Ag(NH_3)_2]^+] = 0.020 \, M$, so $[NH_3]$ is the only unknown in this expression.

$$[NH_3]^2 = \frac{6.3 \times 10^{-8}[[Ag(NH_3)_2]^+]}{[Ag^+]} = \frac{(6.3 \times 10^{-8})(0.020)}{9.0 \times 10^{-9}}$$

$$[NH_3]^2 = 0.14 \qquad [NH_3] = 0.37 \, M$$

The volume of the solution is one liter. So the total number of moles of NH_3 is the number of moles required by the stoichiometry of the dissolution reaction (0.040 mole) *plus* the number of moles required to satisfy the dissociation constant for $[Ag(NH_3)_2]^+$. That is, in terms of concentration

$$[NH_3]_{total} = 0.040\ M + 0.37\ M = 0.41\ M \qquad \therefore \quad \boxed{0.41 \text{ mole of } NH_3 \text{ is required}}$$

The dissolution of 0.020 mole of AgCl requires one liter of 0.41 M aqueous NH_3. The equilibrium concentration of NH_3 is $0.37/0.04 = 9.3$ times greater than the stoichiometric amount of NH_3.

K_d is fairly large. Therefore, a large excess of NH_3 is required.

Silver cyanide, AgCN, $K_{sp} = 1.2 \times 10^{-16}$, is much less soluble than silver chloride, $K_{sp} = 1.8 \times 10^{-10}$. However, silver cyanide is readily soluble in solutions of soluble cyanides such as potassium cyanide, KCN, because the $[Ag(CN)_2]^-$ ion is so stable ($K_d = 1.8 \times 10^{-19}$).

EXAMPLE 36-4   *Dissolution of a Precipitate*

Calculate the molarity of a KCN solution, one liter of which will just dissolve 0.020 mole of AgCN.

Solution

Again, we have two equilibria and two equilibrium constants to consider.

$$AgCN(s) \rightleftharpoons Ag^+ + CN^- \qquad K_{sp} = [Ag^+][CN^-] = 1.2 \times 10^{-16}$$

$$[Ag(CN)_2]^- \rightleftharpoons Ag^+ + 2CN^- \qquad K_d = \frac{[Ag^+][CN^-]^2}{[[Ag(CN)_2]^-]} = 1.8 \times 10^{-19}$$

The equation for the dissolution of 0.020 mole of AgCN in KCN solution is

$$AgCN(s) + \underset{0.020\ mol}{CN^-} \longrightarrow \underset{0.020\ mol}{[Ag(CN)_2]^-}$$

The stoichiometry of this reaction tells us that 0.020 mole of CN^- is *required* to dissolve 0.020 mole of AgCN, and this *produces* 0.020 mole of $[Ag(CN)_2^-]$.

Examination of the two equilibrium constant expressions reveals that both $[Ag^+]$ and $[CN^-]$ occur in both. Therefore, we solve for one concentration in terms of the other. We solve the solubility product expression for $[Ag^+]$.

$$[Ag^+][CN^-] = 1.2 \times 10^{-16} \qquad \text{or} \qquad [Ag^+] = \frac{1.2 \times 10^{-16}}{[CN^-]}$$

Now we substitute this value into the dissociation constant expression. Recall that $[[Ag(CN)_2]^-] = 0.020\ M$ (from the stoichiometry of the reaction).

$$\frac{[Ag^+][CN^-]^2}{[[Ag(CN)_2]^-]} = 1.8 \times 10^{-19} \qquad \frac{\left(\dfrac{1.2 \times 10^{-16}}{[CN^-]}\right)[CN^-]^2}{(0.020)} = 1.8 \times 10^{-19}$$

This expression contains only one unknown, $[CN^-]$. We solve for $[CN^-]$.

$$[CN^-] = 3.0 \times 10^{-5}\ M$$

The total number of moles of CN^- is the number *required* by the stoichiometry of the dissolution reaction (0.020 mole) *plus* the number required to satisfy the dissociation constant for $[Ag(CN)_2]^-$ (in terms of concentration):

$$\boxed{[CN^-]_{total} = 0.020\ M + 3.0 \times 10^{-5}\ M}$$

K_d is very small. Therefore, only a slight excess of CN^- is required.

From a practical point of view, only slightly more than 0.020 mole per liter of KCN, the stoichiometric amount, would be required to dissolve 0.020 mole of AgCN.

You might like to verify the answer obtained in Example 36-4 by using the equilibrium concentration of CN^- ions just as we used the equilibrium concentration of Cl^- ions in Example 36-3.

Calculations on other complex species are similar to these examples. However, many complex ions contain three, four, or six ligands (Appendix I), and therefore calculations on these involve higher order equations (Appendix A).

36-3 HYDROGEN SULFIDE EQUILIBRIA

Many equilibria in qualitative analysis involve hydrogen sulfide. In Section 18-9 we described the stepwise ionization of polyprotic acids. Let us consider in detail the equilibria in saturated aqueous solutions of H_2S. Hydrosulfuric acid, H_2S, is a *very weak* diprotic acid. Saturated H_2S solutions are 0.10 M.

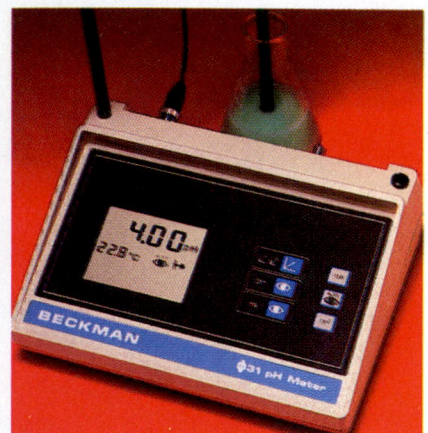

EXAMPLE 36-5 *Solutions of Weak Polyprotic Acids*

Calculate the concentrations of the species in 0.10 M H_2S solution. $K_1 = 1.0 \times 10^{-7}$ and $K_2 = 1.3 \times 10^{-13}$.

Plan

We use the same kind of logic as in Example 18-21.

Solution

Let x = mol/L of H_2S that ionize in the first step.

$$H_2S \quad + H_2O \rightleftharpoons H_3O^+ + HS^-$$
$$(0.10 - x)\ M \qquad\qquad\quad x\ M \quad\ x\ M$$

$$K_1 = \frac{[H_3O^+][HS^-]}{[H_2S]} = \frac{(x)(x)}{(0.10 - x)} = 1.0 \times 10^{-7}$$

Solving with the usual simplifying assumption gives $x = 1.0 \times 10^{-4}$.

$$[H_2S] = (0.10 - x)\ M = \boxed{0.10\ M}$$

$$[HS^-] = [H_3O^+] = x\ M = \boxed{1.0 \times 10^{-4}\ M} \qquad \text{first step}$$

The second step involves ionization of the anion produced in the first step. We represent equilibrium concentrations (where y = mol/L of HS^- that ionizes) as follows.

$$HS^- \quad + H_2O \rightleftharpoons \qquad H_3O^+ \qquad + S^{2-}$$
$$(1.0 \times 10^{-4} - y)\ M \qquad\quad (1.0 \times 10^{-4} + y)\ M \quad y\ M$$

from 1st step   from 2nd step

Substitution into K_2 gives

$$K_2 = \frac{[H_3O^+][S^{2-}]}{[HS^-]} = \frac{(1.0 \times 10^{-4} + y)(y)}{(1.0 \times 10^{-4} - y)} = 1.3 \times 10^{-13}$$

Assume that $(1.0 \times 10^{-4} + y) \approx 1.0 \times 10^{-4}$ and $(1.0 \times 10^{-4} - y) \approx 1.0 \times 10^{-4}$.

$$\frac{(1.0 \times 10^{-4})y}{1.0 \times 10^{-4}} = 1.3 \times 10^{-13} \qquad y = \boxed{1.3 \times 10^{-13}\ M = [HS^-]} = [H_3O^+]_{2nd}$$

Our assumption is valid. Very little HS^- ionizes ($1.3 \times 10^{-13}\ M$).

$$[S^{2-}] = y = \boxed{1.3 \times 10^{-13}\ M}$$

$$[HS^-] = (1.0 \times 10^{-4} - y) = (1.0 \times 10^{-4}) - (1.3 \times 10^{-13}) \approx \boxed{1.0 \times 10^{-4}\ M}$$

$$[H_3O^+] = (1.0 \times 10^{-4} + y) = (1.0 \times 10^{-4}) + (1.3 \times 10^{-13}) \approx \boxed{1.0 \times 10^{-4}\ M}$$

We have already calculated the concentration of nonionized H_2S ($\sim 0.10\ M$).

These concentrations (Example 36-5) tell us that saturated solutions of H_2S (0.10 M) are only slightly acidic. They contain very low concentrations of S^{2-} ions. Recall that for solutions containing only weak polyprotic acids in reasonable concentrations, $[anion^{2-}] = K_2$.

H_2S is a very weak acid. Many experiments have shown that in solutions containing H_2S *and* a strong acid, such as HCl, the acidity is determined by the strong acid. We can derive a simple relationship between $[H^+]$ and $[S^{2-}]$ in such solutions. We multiply K_1 and K_2, the ionization constants for H_2S,

$$K_1K_2 = \frac{[H^+][HS^-]}{[H_2S]} \times \frac{[H^+][S^{2-}]}{[HS^-]} = (1.0 \times 10^{-7})(1.3 \times 10^{-13})$$

to obtain

$$\frac{[H^+]^2[S^{2-}]}{[H_2S]} = 1.3 \times 10^{-20}$$

If we restrict our discussions to *saturated solutions* of H_2S (0.10 M), we can simplify the relationship further.

$$[H^+]^2[S^{2-}] = (1.3 \times 10^{-20})[H_2S] = (1.3 \times 10^{-20})(0.10)$$

$$\boxed{[H^+]^2[S^{2-}] = 1.3 \times 10^{-21}} \qquad \text{valid for saturated } H_2S \text{ solutions that contain a strong acid}$$

This is a very useful relationship for *saturated* H_2S solutions. It tells us that the concentration of sulfide ions *varies inversely with the square of the concentration of hydrogen (hydronium) ions.*

EXAMPLE 36-6 *The Common Ion Effect in H_2S Solutions*

Calculate the concentrations of H^+ and S^{2-} ions in saturated (0.10 M) H_2S solution that is also 0.30 M in HCl, the concentrations used to precipitate Group I sulfides.

Solution

Example 36-5 showed that the maximum $[H^+]$ in saturated H_2S is $1.0 \times 10^{-4}\ M$. Because $[H^+]$ from 0.30 M HCl, a strong acid, is 0.30 M.

$$[H^+]_{HCl} \gg [H^+]_{H_2S}$$

$$0.30\ M \gg 0.00010\ M \qquad \therefore\ \boxed{[H^+] = 0.30\ M}$$

Substitution of $[H^+] = 0.30\ M$ into the relationship we just derived for saturated H_2S solutions gives the value for $[S^{2-}]$.

$$[S^{2-}] = \frac{1.3 \times 10^{-21}}{[H^+]^2} = \frac{1.3 \times 10^{-21}}{(0.30)^2} = \boxed{1.4 \times 10^{-20}\ M}$$

Note that $[S^{2-}] = 1.4 \times 10^{-20}\ M$ in saturated H_2S that is also $0.30\ M$ in HCl. Example 36-5 showed that $[S^{2-}] = 1.3 \times 10^{-13}\ M$ in saturated H_2S. Therefore, the presence of the strong acid has decreased $[S^{2-}]$ by the factor $1.3 \times 10^{-13}/1.4 \times 10^{-20}$, or 9.3×10^6. Stated differently, the concentration of S^{2-} is more than 9 million times greater in saturated H_2S than in saturated H_2S that is *also* $0.30\ M$ in HCl (or any other strong monoprotic acid).

EXAMPLE 36-7   *The Common Ion Effect in H_2S Solutions*

Calculate $[S^{2-}]$ in saturated H_2S that is also $0.60\ M$ in HCl.

Solution
As in Example 36-6, we substitute $[H^+]$, which is $0.60\ M$ in this example, into our useful relationship for saturated H_2S solutions and calculate $[S^{2-}]$.

$$[H^+]^2[S^{2-}] = 1.3 \times 10^{-21}$$

$$[S^{2-}] = \frac{1.3 \times 10^{-21}}{[H^+]^2} = \frac{1.3 \times 10^{-21}}{(0.60)^2} = \boxed{3.6 \times 10^{-21}\ M}$$

We see that this $[S^{2-}]$ is smaller than that in Example 36-6 because the concentration of HCl is twice as great. The ratio of S^{2-} ion concentrations in the two solutions is $1.4 \times 10^{-20}/3.6 \times 10^{-21} = 4$ (within roundoff-error range). Doubling $[H^+]$ decreases $[S^{2-}]$ by a factor of four. Halving $[H^+]$ would increase $[S^{2-}]$ by a factor of four, a typical inverse square relationship.

36-4 PRECIPITATION OF METAL SULFIDES

You may find it helpful to review Section 31-5 carefully.

In discussing the precipitation of the Group I sulfides, we indicated that unstable intermediate compounds containing HS^- ions are probably formed. However, because equilibrium depends on only the reactants and products of chemical reactions, we can describe reactions in which insoluble sulfides precipitate.

EXAMPLE 36-8   *Simultaneous Equilibria in Precipitation Reactions*

Will copper(II) sulfide precipitate in a solution that is $0.10\ M$ in copper(II) nitrate, $0.30\ M$ in HCl, and saturated with H_2S?

Solution
In Example 36-6 we demonstrated that $[S^{2-}] = 1.4 \times 10^{-20}\ M$ in such a solution. Because $Cu(NO_3)_2$ is a soluble ionic compound, we know that $[Cu^{2+}] = 0.10\ M$ in $0.10\ M\ Cu(NO_3)_2$. We calculate Q_{sp} and compare it with K_{sp} for CuS.

$$K_{sp} = [Cu^{2+}][S^{2-}] = 8.7 \times 10^{-36}$$

$$Q_{sp} = (0.10)(1.4 \times 10^{-20}) = 1.4 \times 10^{-21}$$

Because $Q_{sp} \gg K_{sp}$ for CuS, precipitation occurs until $Q_{sp} = K_{sp}$.

We can also calculate the maximum concentration of an ion that can exist in a particular solution (see also Section 31-5).

EXAMPLE 36-9 *Simultaneous Equilibria in Precipitation Reactions*

What is the maximum concentration of Cu^{2+} ions that can exist (without forming a precipitate) in a saturated H_2S solution that is also 0.30 M in HCl?

Solution

Because we know that $[S^{2-}] = 1.4 \times 10^{-20}\,M$ in this solution (from Example 36-6), we can use K_{sp} for CuS to determine the $[Cu^{2+}]$.

$$[Cu^{2+}][S^{2-}] = 8.7 \times 10^{-36}$$

$$[Cu^{2+}] = \frac{8.7 \times 10^{-36}}{[S^{2-}]} = \frac{8.7 \times 10^{-36}}{1.4 \times 10^{-20}} = 6.2 \times 10^{-16}\,M$$

This is only 3.9×10^{-14} g Cu^{2+}/L, and so we conclude that only a tiny trace of Cu^{2+} ions can exist in such solutions.

Note that the statement of Example 36-9 asked for the maximum $[Cu^{2+}]$ that could exist in a particular solution, *not* the $[Cu^{2+}]$ remaining after precipitation. We shall deal with that subject presently.

The equations for the reactions by which the Groups I and II sulfides are precipitated show that many of these reactions produce H^+ ions. As these reactions occur, the acidity of the solution increases (Example 36-10).

EXAMPLE 36-10 *Simultaneous Equilibria in Precipitation Reactions*

Calculate the concentration of Cu^{2+} ions that remains in a solution that was initially 0.10 M in $Cu(NO_3)_2$ and 0.30 M in HCl if sufficient H_2S is added so that the solution is saturated with H_2S when precipitation is as complete as possible.

Solution

The precipitation reaction produces hydrogen ions.

$$Cu^{2+} + H_2S \longrightarrow CuS(s) + 2H^+$$
$$0.10\,M \quad \text{excess} \Longrightarrow \qquad\qquad 0.20\,M$$

The very small K_{sp} for CuS, 8.7×10^{-36}, tells us that CuS is quite insoluble. Therefore, we *assume* that precipitation is essentially complete. We started with 0.10 M $Cu(NO_3)_2$, a soluble ionic compound, so the reaction produces 0.20 M H^+. The total $[H^+]$ is the concentration produced by the reaction *plus* the 0.30 M furnished by 0.30 M HCl.

$$[H^+]_{total} = [H^+]_{reaction} + [H^+]_{HCl} = 0.20\,M + 0.30\,M = 0.50\,M$$

We use the familiar relationship to calculate $[S^{2-}]$ now that $[H^+]_{total}$ is known.

$$[H^+]^2[S^{2-}] = 1.3 \times 10^{-21}$$

$$[S^{2-}] = \frac{1.3 \times 10^{-21}}{[H^+]^2} = \frac{1.3 \times 10^{-21}}{(0.50)^2} = 5.2 \times 10^{-21}\,M$$

We know $[S^{2-}]$ after precipitation has occurred, so we use K_{sp} to calculate $[Cu^{2+}]$.

$$[Cu^{2+}][S^{2-}] = 8.7 \times 10^{-36}$$

$$[Cu^{2+}] = \frac{8.7 \times 10^{-36}}{[S^{2-}]} = \frac{8.7 \times 10^{-36}}{5.2 \times 10^{-21}} = 1.7 \times 10^{-15}\,M$$

The concentration of copper(II) ions remaining in solution is only $1.7 \times 10^{-15}\,M$. Our assumption that precipitation would be essentially complete is valid.

The acidity or basicity of the reaction medium is an extremely important factor in many chemical reactions. By carefully controlling the pH of a solution, it is possible to precipitate some ions nearly completely while others are left in solution. Consider a solution that contains copper(II) nitrate, $Cu(NO_3)_2$; lead(II) nitrate, $Pb(NO_3)_2$; and zinc nitrate, $Zn(NO_3)_2$, all soluble ionic compounds. The three metal ions, Cu^{2+}, Pb^{2+}, and Zn^{2+}, all form insoluble sulfides. Their solubility products indicate significant differences in the solubilities of these "insoluble" compounds. Clearly, CuS is the least soluble and ZnS is the most soluble.

Compound	K_{sp}	Difference (as a ratio)
ZnS	$[Zn^{2+}][S^{2-}] = 1.1 \times 10^{-21}$	1.3×10^6
PbS	$[Pb^{2+}][S^{2-}] = 8.4 \times 10^{-28}$	9.7×10^7
CuS	$[Cu^{2+}][S^{2-}] = 8.7 \times 10^{-36}$	

If a solution is made 0.30 M in HCl and 0.010 M each in $Cu(NO_3)_2$, $Pb(NO_3)_2$, and $Zn(NO_3)_2$, and then saturated with H_2S (0.10 M), will ZnS precipitate? To answer this question, we must determine whether the solubility product for ZnS is exceeded, and so we must calculate both $[Zn^{2+}]$ and $[S^{2-}]$. From Example 36-6, we know that $[S^{2-}] = 1.4 \times 10^{-20} M$ in 0.30 M HCl saturated with H_2S.

The K_{sp} expressions for CuS, PbS, and ZnS are all of the same form, $[M^{2+}][S^{2-}] = K_{sp}$. Because $[S^{2-}]$ is known, we can determine whether any or all of the insoluble sulfides will precipitate from the solution under discussion.

$$K_{sp} = [Zn^{2+}][S^{2-}] = 1.1 \times 10^{-21}$$

In 0.010 M $Zn(NO_3)_2$, we have $[Zn^{2+}] = 0.010 M$, so

$$Q_{sp} = [Zn^{2+}][S^{2-}] = (0.010)(1.4 \times 10^{-20}) = 1.4 \times 10^{-22}$$

$$Q_{sp} = 1.4 \times 10^{-22} < 1.1 \times 10^{-21} = K_{sp}$$

This calculation tells us that ZnS does *not* precipitate from the solution.

Now let's see if PbS and CuS do precipitate from this solution. For lead sulfide we know that

$$K_{sp} = [Pb^{2+}][S^{2-}] = 8.4 \times 10^{-28}$$

In this solution,

$$[Pb^{2+}] = 0.010 M \qquad [S^{2-}] = 1.4 \times 10^{-20} M$$

$$Q_{sp} = [Pb^{2+}][S^{2-}] = (0.010)(1.4 \times 10^{-20}) = 1.4 \times 10^{-22}$$

$$Q_{sp} = 1.4 \times 10^{-22} > 8.4 \times 10^{-28} = K_{sp}$$

This tells us that PbS *does* precipitate from the solution. Because CuS is less soluble than PbS, clearly it will precipitate also.

We have described the reactions that occur in a solution that is

0.30 M in HCl,

0.010 M each in $Cu(NO_3)_2$, $Pb(NO_3)_2$ and $Zn(NO_3)_2$, and

saturated with H_2S

Copper(II) sulfide and lead sulfide precipitate, but zinc sulfide does not.

We might also ask whether the acidity of the solution can be increased enough to prevent the precipitation of PbS by removing S^{2-} from solution to form nonionized H_2S.

To answer the question, we need to know the maximum concentration of H_3O^+ that is possible in HCl solutions.

Concentrated HCl solution is approximately 12 M. However, HCl is not completely ionized in solutions of this concentration. Also, activity coefficients decrease significantly with increasing concentration. Let us *assume* that the concentration of H_3O^+ is about 10 M in concentrated HCl solution.

If $[H_3O^+] = 10\ M$, then we can use the relationship $[H_3O^+]^2[S^{2-}] = 1.3 \times 10^{-21}$ to solve for $[S^{2-}]$ in saturated H_2S solution.

$$[H_3O^+]^2[S^{2-}] = 1.3 \times 10^{-21}$$

$$[S^{2-}] = \frac{1.3 \times 10^{-21}}{[H_3O^+]^2} = \frac{1.3 \times 10^{-21}}{(10)^2}$$

$$[S^{2-}] = 1.3 \times 10^{-23}\ M$$

For PbS we know that

$$K_{sp} = [Pb^{2+}][S^{2-}] = 8.4 \times 10^{-28}$$

Recall that the solution is 0.010 M in Pb^{2+}, so

$$Q_{sp} = [Pb^{2+}][S^{2-}] = (0.010)(1.3 \times 10^{-23}) = 1.3 \times 10^{-25}$$

$$Q_{sp} = 1.3 \times 10^{-25} > 8.4 \times 10^{-28} = K_{sp}$$

Because Q_{sp} is greater than K_{sp}, PbS *does* precipitate from the solution. Therefore, it is not possible to prevent the precipitation of PbS from a solution that is 0.010 M in Pb^{2+} and saturated with H_2S by increasing the acidity of the solution with HCl.

36-5 HYDROLYSIS OF ANIONS OF WEAK POLYPROTIC ACIDS

In Sections 19-2 and 19-4 we described the hydrolysis of anions of weak monoprotic acids. Hydrolysis reactions of anions of weak polyprotic acids are quite similar, *but* they occur in two or more steps.

Consider the reactions that occur when sodium sulfide, Na_2S, the salt of NaOH and H_2S, is placed in water. Sulfide ions hydrolyze to produce hydroxide ions and hydrosulfide ions.

First Step

$$S^{2-} + H_2O \rightleftharpoons HS^- + OH^-$$

$$K_{b(1)} = \frac{[HS^-][OH^-]}{[S^{2-}]} = \frac{K_w}{K_{2(H_2S)}} = \frac{1.0 \times 10^{-14}}{1.3 \times 10^{-13}} = 7.7 \times 10^{-2}$$

The large hydrolysis constant (7.7×10^{-2}) for the first step tells us that sulfide ions hydrolyze extensively. One of the products of the first-step hydrolysis reaction, the HS^- ion, can also hydrolyze to produce more OH^- ions.

Second Step

$$HS^- + H_2O \rightleftharpoons H_2S \times OH^-$$

$$K_{b(2)} = \frac{[H_2S][OH^-]}{[HS^-]} = \frac{K_w}{K_{1(H_2S)}} = \frac{1.0 \times 10^{-14}}{1.0 \times 10^{-7}} = 1.0 \times 10^{-7}$$

When we open a bottle of concentrated HCl, we are greeted by a cloud of gaseous HCl. Clearly, this gaseous HCl is *not* ionized.

$1.3 \times 10^{-23}\ M\ S^{2-}$ corresponds to seven or eight S^{2-} ions per liter of solution.

You may wish to refer to Section 19-2 to refresh your memory on writing expressions for hydrolysis constants.

The much smaller value for the second-step hydrolysis constant tells us that this reaction occurs to a much smaller extent than the first step. In fact, it occurs to an even smaller extent than $K_{b(2)}$ suggests, because the high concentration of OH^- ions from the first step suppresses the second-step reaction.* Let us examine a relatively dilute solution of sodium sulfide, a typical soluble salt of a strong base and a weak polyprotic acid, in detail.

EXAMPLE 36-11   *Hydrolysis Equilibria*

Calculate the pH and percentage of hydrolysis in a 0.00100 M Na_2S solution.

Solution

The equation for the first step of the hydrolysis reaction and the usual algebraic representation of equilibrium concentrations of ions are

$$S^{2-} \qquad + H_2O \rightleftharpoons HS^- + OH^-$$
$$(0.00100 - x)\,M \qquad\qquad x\,M \quad x\,M$$

$$K_{b(1)} = \frac{[HS^-][OH^-]}{[S^{2-}]} = 7.7 \times 10^{-2}$$

Substitution gives a quadratic equation that must be solved by the quadratic formula.

$$\frac{(x)(x)}{0.00100 - x} = 7.7 \times 10^{-2}$$

We clear the fraction and write the equation in standard form.

$$x^2 + (7.7 \times 10^{-2})x - 7.7 \times 10^{-5} = 0$$

The quadratic formula gives two roots, $x = -0.078$ and $x = 0.001$. The negative root is extraneous, and $x = 0.001$ *should be* the root with physical significance.

$$x = [HS^-] = [OH^-] = 0.001\,M$$

For x to be 0.001, $K_{b(1)}$ would have to be infinite. The reaction is nearly complete, but not *that* complete.

However, we cannot accept this value for x because it indicates that the reaction goes to completion, an idea that is incompatible with the fact that $K_{b(1)} = 7.7 \times 10^{-2}$. We conclude that our algebraic representation of the equilibrium is unacceptable, and so we try again.

The large value of $K_{b(1)}$ indicates that the first hydrolysis reaction is nearly complete. Therefore, let us use an algebraic representation in which we let $x = [S^{2-}]$ and $(0.00100 - x) = [HS^-] = [OH^-]$. (To a first approximation we *imagine* that the forward reaction goes to completion. Then we define x as the very small concentrations of HS^- and OH^- that react to form S^{2-} ions to establish equilibrium in the reverse reaction.)

$$S^{2-} + H_2O \rightleftharpoons \qquad HS^- \quad + \quad OH^-$$
$$x\,M \qquad\qquad (0.00100 - x)\,M \quad (0.00100 - x)\,M$$

Substitution into the first hydrolysis constant expression gives

$$\frac{[HS^-][OH^-]}{[S^{2-}]} = \frac{(0.00100 - x)(0.00100 - x)}{x} = 7.7 \times 10^{-2}$$

Because hydrolysis is nearly complete, we make the usual simplifying assumption $[(0.00100 - x) \approx 0.00100]$, which gives

$$\frac{1.0 \times 10^{-6}}{x} = 7.7 \times 10^{-2} \qquad x = 1.3 \times 10^{-5}\,M = [S^{2-}]$$

---

*We encountered a similar situation in Sections 18-9 and 36-3 when we studied the ionization of weak polyprotic acids. The second-step ionization of weak polyprotic acids occurs to a smaller extent than $K_{a(2)}$ would indicate because of the much higher $[H^+]$ produced in the first-step ionization.

Because $[OH^-] = (0.00100 - x) = (0.00100 - 0.000013)M \approx 0.00099\ M$, then $pOH = 3.00$ and $pH = 11.00.$ Considering its concentration, the solution is very basic.

As a reference point, the pH of $0.0010\ M$ NaOH is also 11.00.

We can calculate the percentage of the sulfide ions that are *not hydrolyzed*, because we know the equilibrium $[S^{2-}]$, i.e., the $[S^{2-}]$ not hydrolyzed.

$$\%\ S^{2-}\ (\text{not hydrolyzed}) = \frac{[S^{2-}]_{equil}}{[S^{2-}]_{orig}} \times 100\% = \frac{1.3 \times 10^{-5}\ M}{0.00100\ M} \times 100\%$$

$$= 1.3\%\ S^{2-} \qquad \text{not hydrolyzed}$$

Therefore, 98.7% of the S^{2-} ions hydrolyze. Our assumption that the first-step hydrolysis is nearly complete is valid.

We have subtracted 1.3% from *exactly* 100%. We have not violated the rules for significant figures.

Let us now examine the second-step hydrolysis of sulfide ions. Because the OH^- ions produced in the first step are in the solution, their effect on the second step must be taken into account. The concentrations of HS^- and OH^- are equal. They are $(0.00100 - 1.3 \times 10^{-5})\ M = 0.000987\ M$. We represent the concentration of HS^- that hydrolyzes as $y\ M$.

$$\begin{array}{ccccc} HS^- & + H_2O \rightleftharpoons & H_2S + & OH^- \\ (0.000987 - y)\ M & & y\ M & (0.000987 + y)\ M \end{array}$$

Substitution into the $K_{b(2)}$ expression gives

$$\frac{[H_2S][OH^-]}{[HS^-]} = \frac{(y)(0.000987 + y)}{(0.000987 - y)} = 1.0 \times 10^{-7}$$

We make the usual simplifying assumption, which gives

$$y = 1.0 \times 10^{-7}\ M = [H_2S] = [OH^-] \qquad \text{from 2nd step}$$

$9.9 \times 10^{-4}\ M \gg 1.0 \times 10^{-7}\ M$

The $[OH^-]$ from the second step is so small (equal to that produced by the ionization of pure water) that it does not contribute significantly to the basicity of the solution. The pH of the solution is determined by the first-step hydrolysis of sulfide ions because the percentage hydrolysis in the second step is insignificant compared to that in the first step.

$$\%\ \text{hydrolysis (2nd)} = \frac{[HS^-]_{hydrolyzed}}{[HS^-]_{orig}} \times 100\% = \frac{1.0 \times 10^{-7}\ M}{9.87 \times 10^{-4}\ M} \times 100\% = 0.010\%$$

Let us summarize the results of the calculations we have just done for $0.00100\ M$ Na$_2$S solution.

	$[OH^-]$	pH	$[HS^-]$	$[H_2S]$
first step	$\sim 1.0 \times 10^{-3}\ M$	11.00	$\sim 1.0 \times 10^{-3}\ M$	—
second step	$1.0 \times 10^{-7}\ M$	no change	no change	$1.0 \times 10^{-7}\ M$

The ideas we evolved and used in Example 36-11 are applicable to other soluble salts derived from strong soluble bases and weak polyprotic acids. For example, both sodium carbonate, Na$_2$CO$_3$, and sodium phosphate, Na$_3$PO$_4$, give very basic solutions because both CO_3^{2-} and PO_4^{3-} ions hydrolyze extensively. Appendix F shows that the last-step ionization constants for both carbonic and phosphoric acids are small. For H_2CO_3, $K_2 = 4.8 \times 10^{-11}$; for H_3PO_4, $K_3 = 3.6 \times 10^{-13}$.

36-6 EFFECT OF HYDROLYSIS ON SOLUBILITIES OF SLIGHTLY SOLUBLE COMPOUNDS

In Section 36-5 we showed that anions of weak polyprotic acids hydrolyze extensively. In 0.0010 M Na$_2$S solution, for instance, the percentage hydrolysis is 98.7% (first step).

In Section 33-7 we pointed out the fact that aluminum sulfide, Al$_2$S$_3$, and chromium(III) sulfide, Cr$_2$S$_3$, do not precipitate in Group III. Their hydroxides, Al(OH)$_3$ and Cr(OH)$_3$, precipitate from the buffered aqueous ammonia/ammonium sulfide solution. Both Al^{3+} and Cr^{3+} ions are small, highly charged cations, and both hydrolyze (Section 19-5). Because S^{2-} ions also hydrolyze, Al$_2$S$_3$ and Cr$_2$S$_3$ are unstable in contact with water and decompose into their parent acids and bases.

$$Al_2S_3(s) + 6H_2O \longrightarrow 2Al(OH)_3(s) + 3H_2S(aq)$$

$$Cr_2S_3(s) + 6H_2O \longrightarrow 2Cr(OH)_3(s) + 3H_2S(aq)$$

In fairness, we must emphasize that these "hydroxides" are in fact hydrated oxides. The oxides of both aluminum and chromium(III) are thermodynamically very stable compounds. This stability is the principal "driving force" for these reactions.

Although the insoluble metal sulfides hydrolyze to much smaller extents than Al$_2$S$_3$ and Cr$_2$S$_3$, hydrolysis does affect their solubilities in water.

EXAMPLE 36-12   *Simultaneous Equilibria*

Calculate the molar solubility of copper(II) sulfide in water at 25°C, (a) ignoring hydrolysis and (b) taking hydrolysis into consideration.

Solution

The solubility product for CuS, 8.7×10^{-36}, indicates that it is quite insoluble.

(a) Ignoring hydrolysis, we calculate the molar solubility as we did in Section 20-3.

$$CuS(s) \rightleftharpoons Cu^{2+} + S^{2-}$$
$$x\,mol/L \rightleftharpoons x\,mol/L \quad x\,mol/L$$

$$[Cu^{2+}][S^{2-}] = 8.7 \times 10^{-36} \qquad (x)(x) = 8.7 \times 10^{-36}$$

$$\boxed{x = 2.9 \times 10^{-18}\;mol/L} \quad \longleftarrow \text{ molar solubility ignoring hydrolysis}$$

(b) Let us now calculate the molar solubility of CuS, taking into account the effect of hydrolysis. The hydrolysis constant for Cu^{2+} is quite small (Table 19-2), which indicates that Cu^{2+} hydrolyzes only very slightly.

$$[Cu(OH_2)_4]^{2+} + H_2O \rightleftharpoons [Cu(OH)(OH_2)_3]^{+} + H_3O^{+} \qquad K_a = 1.0 \times 10^{-8}$$

The hydrolysis constant for S^{2-} is quite large (Section 36-5).

$$S^{2-} + H_2O \rightleftharpoons HS^{-} + OH^{-} \qquad K_{b(1)} = 7.7 \times 10^{-2}$$

This tells us that S^{2-} ions hydrolyze to a much greater extent than Cu^{2+} ions. Therefore, we can ignore the hydrolysis of Cu^{2+} ions without introducing serious errors into our calculations. (This also greatly simplifies the arithmetic.) We may represent the dissolution and hydrolysis of CuS in water in three steps.

(1) $CuS(s) \rightleftharpoons Cu^{2+} + S^{2-}$ $\qquad K_{sp} = [Cu^{2+}][S^{2-}] = 8.7 \times 10^{-36}$

(2) $S^{2-} + H_2O \rightleftharpoons HS^{-} + OH^{-}$ $\qquad K_{b(1)} = \dfrac{[HS^{-}][OH^{-}]}{[S^{2-}]} = 7.7 \times 10^{-2}$

(3) $HS^{-} + H_2O \rightleftharpoons H_2S + OH^{-}$ $\qquad K_{b(2)} = \dfrac{[H_2S][OH^{-}]}{[HS^{-}]} = 1.0 \times 10^{-7}$

Because Cu^{2+} ions do not hydrolyze very much, we shall calculate the [Cu^{2+}]. This is equal to the molar solubility of CuS in water at 25°C.

Most of the S^{2-} ions produced in reaction (1) are converted into HS^- ions in reaction (2), and some of the HS^- ions from reaction (2) are converted into H_2S molecules in reaction (3). Our problem is to find a way to calculate the equilibrium concentration of S^{2-} ions that actually satisfies K_{sp} for CuS.

Our first inclination is to ignore reaction (3), because we demonstrated (Example 36-11) that this may be done in some cases. However, in Example 36-11 we worked with a 0.00100 M Na_2S solution. In this example, we have an *extremely* dilute solution of CuS (because K_{sp} is so very small), and there is no basis for ignoring reaction (3). Our calculation in part (a) indicated that the molar solubility of CuS is approximately 2.9×10^{-18} mol/L. The hydrolysis of these S^{2-} ions should produce approximately 2.9×10^{-18} mol/L of OH^- ions, a concentration that is insignificantly small compared to the 1.0×10^{-7} mol/L of OH^- produced by the ionization of water. This gives us a handle on the problem. We *assume* that $[OH^-] = 1.0 \times 10^{-7}$ M and substitute this value into the $K_{b(2)}$ expression for reaction (3).

$$K_{b(2)} = \frac{[HS^-][OH^-]}{[H_2S]} = 1.0 \times 10^{-7}$$

$$\frac{[HS^-]}{[H_2S]} = \frac{1.0 \times 10^{-7}}{[OH^-]} = \frac{1.0 \times 10^{-7}}{1.0 \times 10^{-7}} = 1$$

Thus, for this solution (which is neutral), the ratio tells us that $[HS^-] = [H_2S]$, which gives us two additional bits of information. First, the second-step hydrolysis is important; second, because $[HS^-] = [H_2S]$, exactly half of the HS^- ions produced in reaction (2) are converted into H_2S molecules in reaction (3).

Let us now return to equation (1). It shows that each formula unit of CuS that *dissolves* produces one Cu^{2+} ion and one S^{2-} ion. Stated differently, Cu^{2+} and S^{2-} ions are present in equal concentrations *before* reaction (2) occurs. However, reaction (2) goes essentially to completion in extremely dilute solutions; i.e., each S^{2-} ion produces one HS^- ion. Therefore, we may write

$$[HS^-] = [Cu^{2+}] \qquad \text{before reaction (3) occurs}$$

We have already established that reaction (3) converts one half of the HS^- ions into H_2S molecules ($[HS^-]/[H_2S] = 1$). Therefore, at the final equilibrium,

$$[HS^-] = \tfrac{1}{2}[Cu^{2+}] \qquad \text{final equilibrium}$$

We know that $[OH^-] = 1.0 \times 10^{-7}$ M, so we use $K_{b(1)}$ to obtain the ratio $[HS^-]/[S^{2-}]$.

$$K_{b(1)} = \frac{[HS^-][OH^-]}{[S^{2-}]} = 7.7 \times 10^{-2}$$

$$\frac{[HS^-]}{[S^{2-}]} = \frac{7.7 \times 10^{-2}}{[OH^-]} = \frac{7.7 \times 10^{-2}}{1.0 \times 10^{-7}} = 7.7 \times 10^5$$

This tells us that $[HS^-] = (7.7 \times 10^5)[S^{2-}]$ or $[S^{2-}] = [HS^-]/(7.7 \times 10^5)$. We substitute this value for $[S^{2-}]$ into K_{sp} for CuS.

$$[Cu^{2+}][S^{2-}] = 8.7 \times 10^{-36}$$

$$[Cu^{2+}]\left(\frac{[HS^-]}{7.7 \times 10^5}\right) = 8.7 \times 10^{-36}$$

We have already established a relationship between $[Cu^{2+}]$ and $[HS^-]$, namely $[HS^-] = 0.50[Cu^{2+}]$. Substitution for $[HS^-]$ gives

$$[Cu^{2+}]\left(\frac{0.50[Cu^{2+}]}{7.7 \times 10^5}\right) = 8.7 \times 10^{-36}$$

$$\frac{0.50[Cu^{2+}]^2}{7.7 \times 10^5} = 8.7 \times 10^{-36}$$

$$[Cu^{2+}]^2 = 13 \times 10^{-30}$$

$$[Cu^{2+}] = 3.6 \times 10^{-15} \text{ mol/L} \quad \longleftarrow \text{ molar solubility of CuS}$$

We have shown that the molar solubility of CuS in water, 3.6×10^{-15} mol/L, is more than 1200 times greater than the value calculated in part (a), 2.9×10^{-18} mol/L, when we ignored hydrolysis of the sulfide ions. To simplify the calculation, we ignored the fact that Cu^{2+} ions also hydrolyze to a slight extent, $K_a = 1.0 \times 10^{-8}$. If we had taken this fact into account, we would have obtained a slightly higher molar solubility for CuS.

$$\frac{3.6 \times 10^{-15}}{2.9 \times 10^{-18}} = 1241$$

We should keep in mind that CuS is an *extremely* insoluble compound. Even taking into account the effect of hydrolysis of the S^{2-} ions, the calculated molar solubility, 3.6×10^{-15} mol/L, is still quite small—only 3.4×10^{-13} g CuS/L at 25°C.

The solubility behavior of insoluble carbonates, phosphates, and arsenates is quite similar to the behavior of copper(II) sulfide. The carbonates are derived from carbonic acid (H_2CO_3, $K_2 = 4.8 \times 10^{-11}$), and the phosphates are derived from phosphoric acid (H_3PO_4, $K_3 = 3.6 \times 10^{-13}$). Carbonates and phosphates hydrolyze extensively.

Consider, for a moment, insoluble compounds in which both the cation and anion hydrolyze. The solubility of bismuth(III) sulfide, Bi_2S_3, in water is affected by hydrolysis to a very large extent because both Bi^{3+} ions (Table 19-2) and S^{2-} ions hydrolyze extensively.

$$[Bi(OH_2)_6]^{3+} + H_2O \rightleftharpoons [Bi(OH)(OH_2)_5]^{2+} + H_3O^+ \qquad K_a \quad = 1.0 \times 10^{-2}$$

$$S^{2-} + H_2O \rightleftharpoons HS^- + OH^- \qquad K_{b(1)} = 7.7 \times 10^{-2}$$

Both reactions increase the solubility of Bi_2S_3. Calculations on the effect of hydrolysis on the solubilities of compounds such as Bi_2S_3 are beyond the scope of this text.

Exercises

Dissolution of Precipitates and Complex Ion Formation

1. Explain, by writing appropriate equations, how the following insoluble compounds can be dissolved by the addition of a solution of nitric acid. (Carbonates dissolve in strong acids to form carbon dioxide, which is evolved as a gas, and water.) What is the "driving force" for each reaction? (a) $Cu(OH)_2$; (b) $Al(OH)_3$; (c) $MnCO_3$.

2. Explain, by writing equations, how the following insoluble compounds can be dissolved by the addition of a solution of ammonium nitrate or ammonium chloride. (a) $Mg(OH)_2$; (b) $Mn(OH)_2$; (c) $Ni(OH)_2$.

3. The following insoluble sulfides can be dissolved in 3 M hydrochloric acid. Explain how this is possible and write the appropriate equations. (a) MnS; (b) FeS.

4. The following sulfides are less soluble than those listed in Exercise 3 and can be dissolved in hot 6 M nitric acid, an oxidizing acid. Explain how, and write the appropriate balanced equations. (a) PbS; (b) CuS; (c) Bi_2S_3.

5. Why would MnS be expected to be more soluble in 0.10 M HCl solution than in water? Would the same be true for $Mn(NO_3)_2$?

6. How can most water-insoluble metal hydroxides be dissolved?

Write a chemical equation for the dissolution of $Fe(OH)_3$ in this way.

7. How does the presence of excess H_3O^+ aid in the dissolution of slightly soluble metal carbonates? Write the chemical equation for the dissolution of $MnCO_3$.

8. Find the concentration of Au^{3+} in a solution in which the other equilibrium concentrations are $[Cl^-] = 0.10\ M$ and $[AuCl_4^-] = 0.20\ M$. K_d for $AuCl_4^- = 7.0 \times 10^{-26}$.

Complex Ion Equilibria

9. (a) What is the relationship between the dissociation constant and the formation constant for a complex ion? (b) Illustrate the relationship for $[AgCl_2]^-$, $[Cd(NH_3)_4]^{2+}$, and $[AlF_6]^{3-}$. Appendix I will be helpful.

10. Write equations and equilibrium constant expressions for the equilibria involved in, as well as the overall equation for, the dissolution of (a) AgCl in excess aq. NH_3, (b) $Co(OH)_2$ in excess aq. NH_3, (c) $PbCl_2$ in excess HCl, and (d) AgCl in excess $Na_2S_2O_3$.

11. Calculate the concentrations of complex ion, metal ion, and ammonia in the following solutions. The complex ions are enclosed

in brackets. All these compounds are soluble and ionic.
(a) 0.100 M [Ag(NH$_3$)$_2$]Cl
(b) 0.089 M [Co(NH$_3$)$_6$]SO$_4$
(c) 0.114 M [Co(NH$_3$)$_6$]$_2$(SO$_4$)$_3$

*12. What is the concentration of Ag$^+$ in a solution that is 0.10 M in KSCN and 0.10 M in [Ag(SCN)$_4$]$^{3-}$? Will Ag$_2$SO$_4$ precipitate if [SO$_4{}^{2-}$] = 0.10 M? K_d for [Ag(SCN)$_4$]$^{3-}$ = 2.1 × 10^{-10}.

13. Calculate the minimum number of moles of gaseous NH$_3$ required to dissolve the indicated number of moles of AgCl in enough water to give 1.0 L of solution: (a) 0.15 mol; (b) 0.015 mol; (c) 1.5 × 10^{-3} mol.

14. Calculate the minimum number of moles of the indicated substance required to dissolve 0.025 mol of the indicated silver salt in enough water to give 1.0 L of solution. (a) AgBr in excess HBr; (b) AgCN in excess NaCN; (c) AgBr in excess Na$_2$S$_2$O$_3$.

Hydrogen Sulfide Equilibria

15. Calculate the concentration of each species in saturated H$_2$S solution (0.10 M H$_2$S).

16. Derive the relationship between [H$^+$] and [S^{2-}] in saturated H$_2$S solutions that also contain a strong acid such as HCl.

17. What are the concentrations of H$^+$ and S^{2-} ions in saturated H$_2$S solutions (0.10 M H$_2$S) that are also (a) 0.28 M in HCl and (b) 0.14 M in HCl? What is an inverse square relationship?

18. If a saturated H$_2$S solution is also 0.15 M in HCl and 0.10 M in each of the following metal nitrates, which metal ions would precipitate as insoluble sulfides? Justify your answers by appropriate calculations. (a) Hg(NO$_3$)$_2$; (b) Cd(NO$_3$)$_2$; (c) Zn(NO$_3$)$_2$; (d) Mn(NO$_3$)$_2$.

*19. (a) What is the maximum concentration of each metal ion listed in Exercise 18 that can exist in a solution that is saturated with H$_2$S and also 0.40 M in HCl? What does the answer you obtained for Mn^{2+} indicate? (b) How many grams of each metal per liter of solution is this? Does the answer you obtain for Mn^{2+} "make sense"? Why?

*20. Each of the following solutions is 0.25 M in HCl, and sufficient H$_2$S is added so that the solutions are saturated with H$_2$S when precipitation is as complete as possible. What concentration(s) of metal ion(s) remains(s) in each solution? (a) 0.10 M Hg(NO$_3$)$_2$, (b) 0.10 M Pb(NO$_3$)$_2$, (c) 0.050 M Bi$_2$(SO$_4$)$_3$, (d) 0.10 M Hg(NO$_3$)$_2$ and 0.10 M Pb(NO$_3$)$_2$.

21. Calculate the percentage of each metal ion that is precipitated in (a) and (b) in Exercise 20.

Hydrolysis of Salts Containing Anions of Weak Polyprotic Acids

22. Calculate hydrolysis constants for (a) the first- and second-step hydrolysis reactions for sulfide ions; (b) the first- and second-step hydrolysis reactions for carbonate ions; and (c) the first-, second-, and third-step hydrolysis reactions for arsenate ions. Based on your answers, what observation can you make about the relative values of $K_{b(1)}$ and $K_{b(2)}$ for a given anion? Why?

*23. (a) Calculate the pH of and percentage hydrolysis in 0.0100 M sodium sulfide, Na$_2$S, solution. (b) How do these values compare with those obtained in Example 36-11? (c) The 0.0100 M Na$_2$S solution is decidedly more basic than 0.0010 M Na$_2$S solution

(Example 36-11), but the percentage hydrolysis is less. Why?

*24. (a) Calculate the pH and percentage hydrolysis of 0.0100 M sodium carbonate, Na$_2$CO$_3$, solution. (b) How do these values compare with those obtained in Exercise 23(a)? Why?

*25. Calculate the pH and percentage hydrolysis of 0.0100 M sodium arsenate, Na$_3$AsO$_4$, solution. Compare with Exercises 23 and 24.

Effect of Hydrolysis on Solubility

*26. Calculate the molar solubilities of the following sulfides, (a) ignoring hydrolysis and (b) taking hydrolysis of the sulfide ion into consideration. The hydrolysis constants for Mn^{2+} and Ni^{2+} are sufficiently small that hydrolysis of these cations may be ignored without introducing serious errors.

$$MnS, \quad K_{sp} = 5.1 × 10^{-15}$$
$$NiS, \quad K_{sp} = 3.0 × 10^{-21}$$

(c) What observation can you make about the effect of hydrolysis on the solubilities of MnS and NiS? That is, how do the sizes of the increases in solubilities compare? (d) Compare the values obtained in (c) with the value obtained in Example 36-12.

*27. (a) How does the hydrolysis of carbonate ions, CO$_3{}^{2-}$, affect the solubilities of the insoluble carbonates? (b) Would the effect of hydrolysis be greater for insoluble carbonates or insoluble sulfides? Why?

28. A solution is 0.010 M with respect to Cd^{2+} ions and also 0.010 M with respect to Pd^{2+} ions. Solid KBr is added to the solution (assume negligible change in volume) until [Br$^-$] = 1.0 M. Use the overall dissociation constants of the following complexes to calculate the concentrations of Cd^{2+} ions and Pd^{2+} ions once equilibrium is reached. K_d for [CdBr$_4$]$^{2-}$ = 2.0 × 10^{-4} and for [PdBr$_4$]$^{2-}$ = 7.7 × 10^{-14}.

29. What concentration of Ag$^+$ ions remains in a solution that originally contained 0.10 M Ag$^+$ ions and 1.3 M NH$_3$?

*30. 0.010 mol of solid Zn(OH)$_2$ is suspended in a saturated Zn(OH)$_2$ solution. Some 6.0 M NaOH solution is added and the mixture is stirred vigorously. The volume of the solution is now 400 mL, and its pH is 13.15. (a) Does all the Zn(OH)$_2$ dissolve? (b) What is/was the minimum [OH$^-$] necessary to dissolve Zn(OH)$_2$ completely?

*31. A concentrated, strong acid is added to a solid mixture of 0.010-mol samples of Fe(OH)$_2$ and Cu(OH)$_2$ placed in 1.0 L of water. At what values of pH will the dissolution of each hydroxide be complete? (Assume negligible volume change.)

32. Zinc ion forms a complex ion with EDTA^{4-}, where EDTA^{4-} represents the ethylenediaminetetraacetate ion.

$$[Zn(EDTA)]^{2-} \rightleftharpoons Zn^{2+} + EDTA^{4-} \quad K_d = 2.6 × 10^{-17}$$

Will ZnS form in a solution that was originally 0.10 M in EDTA^{4-}, 0.010 M in S^{2-}, and 0.010 M in Zn^{2+}?

33. A solution is 0.010 M in I$^-$ ions and 0.010 M in Br$^-$ ions. Ag$^+$ ions are introduced to the solution by the addition of solid AgNO$_3$. Determine (a) which compound will precipitate first, AgI or AgBr, and (b) the percentage of the halide ion in the first precipitate that is removed from solution before the precipitation of the second compound begins.

SOME MATHEMATICAL OPERATIONS

In chemistry we frequently use very large or very small numbers. Such numbers are conveniently expressed in *scientific* or *exponential notation.*

A-1 SCIENTIFIC NOTATION

In scientific notation, a number is expressed as the *product of two numbers.* By convention, the first number, called the digit term, is between 1 and 10. The second number, called the *exponential term,* is an integer power of 10. Some examples follow.

$$10000 = 1 \times 10^4 \qquad 24327 = 2.4327 \times 10^4$$
$$1000 = 1 \times 10^3 \qquad 7958 = 7.958 \ \times 10^3$$
$$100 = 1 \times 10^2 \qquad 594 = 5.94 \ \times 10^2$$
$$10 = 1 \times 10^1 \qquad 98 = 9.8 \ \times 10^1$$
$$1 = 1 \times 10^0$$
$$1/10 = 0.1 = 1 \times 10^{-1} \qquad 0.32 = 3.2 \ \times 10^{-1}$$
$$1/100 = 0.01 = 1 \times 10^{-2} \qquad 0.067 = 6.7 \ \times 10^{-2}$$
$$1/1000 = 0.001 = 1 \times 10^{-3} \qquad 0.0049 = 4.9 \ \times 10^{-3}$$
$$1/10000 = 0.0001 = 1 \times 10^{-4} \qquad 0.00017 = 1.7 \ \times 10^{-4}$$

Recall that, by definition, (any base)^0 = 1.

The exponent of 10 is the number of places the decimal point must be shifted to give the number in long form. A *positive exponent* indicates that the decimal point is *shifted right* that number of places. A *negative exponent* indicates that the decimal point is *shifted left.* When numbers are written in *standard scientific notation,* there is one nonzero digit to the left of the decimal point.

$$7.3 \times 10^3 = 73 \times 10^2 \qquad = 730 \times 10^1 \qquad = 7300$$
$$4.36 \times 10^{-2} = 0.436 \times 10^{-1} \quad = 0.0436$$
$$0.00862 = 0.0862 \times 10^{-1} = 0.862 \times 10^{-2} = 8.62 \times 10^{-3}$$

In scientific notation the digit term indicates the number of significant figures in the number. The exponential term merely locates the decimal point and does not represent significant figures.

Addition and Subtraction

In addition and subtraction all numbers are converted to the same power of 10, and the digit terms are added or subtracted.

$$(4.21 \times 10^{-3}) + (1.4 \times 10^{-4}) = (4.21 \times 10^{-3}) + (0.14 \times 10^{-3}) = \underline{4.35 \times 10^{-3}}$$

$$(8.97 \times 10^4) - (2.31 \times 10^3) = (8.97 \times 10^4) - (0.231 \times 10^4) = \underline{8.74 \times 10^4}$$

Multiplication

The digit terms are multiplied in the usual way, the exponents are added algebraically, and the product is written with one nonzero digit to the left of the decimal.

Two significant figures in answer.

$$(4.7 \times 10^7)(1.6 \times 10^2) = (4.7)(1.6) \times 10^{7+2} = 7.52 \times 10^9 = \underline{7.5 \times 10^9}$$

Two significant figures in answer.

$$(8.3 \times 10^4)(9.3 \times 10^{-9}) = (8.3)(9.3) \times 10^{4-9} = 77.19 \times 10^{-5} = \underline{7.7 \times 10^{-4}}$$

Division

The digit term of the numerator is divided by the digit term of the denominator, the exponents are subtracted algebraically, and the quotient is written with one nonzero digit to the left of the decimal.

$$\frac{8.4 \times 10^7}{2.0 \times 10^3} = \frac{8.4}{2.0} \times 10^{7-3} = \underline{4.2 \times 10^4}$$

Three significant figures in answer.

$$\frac{3.81 \times 10^9}{8.412 \times 10^{-3}} = \frac{3.81}{8.412} \times 10^{[9-(-3)]} = 0.45292 \times 10^{12} = \underline{4.53 \times 10^{11}}$$

Powers of Exponentials

The digit term is raised to the indicated power, and the exponent is multiplied by the number that indicates the power.

$$(1.2 \times 10^3)^2 = (1.2)^2 \times 10^{3 \times 2} = 1.44 \times 10^6 = \underline{1.4 \times 10^6}$$

$$(3.0 \times 10^{-3})^4 = (3.0)^4 \times 10^{-3 \times 4} = 81 \times 10^{-12} = \underline{8.1 \times 10^{-11}}$$

These instructions are applicable to most calculators. If your calculator has other notation, consult your calculator instruction booklet.

Electronic Calculators   *To square a number:* (1) enter the number and (2) touch the (x^2) button.

$$(7.3)^2 = 53.29 = \underline{53} \qquad \text{(two sig. figs.)}$$

To raise a number y to power x: (1) enter the number; (2) touch the (y^x) button; (3) enter the power; and (4) touch the (=) button.

$$(7.3)^4 = 2839.8241 = \underline{2.8 \times 10^3} \qquad \text{(two sig. figs.)}$$

$$(7.30 \times 10^2)^5 = 2.0730716 \times 10^{14} = \underline{2.07 \times 10^{14}} \qquad \text{(three sig. figs.)}$$

Roots of Exponentials

The exponent must be divisible by the desired root if a calculator is not used. The root of the digit term is extracted in the usual way, and the exponent is divided by the desired root.

$$\sqrt{2.5 \times 10^5} = \sqrt{25 \times 10^4} = \sqrt{25} \times \sqrt{10^4} = \underline{5.0 \times 10^2}$$

$$\sqrt[3]{2.7 \times 10^{-8}} = \sqrt[3]{27 \times 10^{-9}} = \sqrt[3]{27} \times \sqrt[3]{10^{-9}} = \underline{3.0 \times 10^{-3}}$$

Electronic Calculators   *To extract the square root of a number:* (1) enter the number and (2) touch the ($\sqrt{x}$) button.

$$\sqrt{23} = 4.7958315 = \underline{4.8} \qquad \text{(two sig. figs.)}$$

    To extract some other root: (1) enter the number y; (2) touch the (INV) and then the (y^x) button; (3) enter the root to be extracted, x; and (4) touch the (=) button.

On some models, this function is performed by the $\sqrt[x]{y}$ button.

A-2 LOGARITHMS

The logarithm of a number is the power to which a base must be raised to obtain the number. Two types of logarithms are frequently used in chemistry: (1) common logarithms (abbreviated log), whose base is 10, and (2) natural logarithms (abbreviated ln), whose base is $e = 2.71828$. The general properties of logarithms are the same no matter what base is used. Many equations in science were derived by the use of calculus, and these often involve natural (base e) logarithms. The relationship between $\log x$ and $\ln x$ is as follows.

$$\ln x = 2.303 \log x$$

$\ln 10 = 2.303$

In this text we often present the equation in both terms. To convert from the log form to the ln form, simply drop the factor of 2.303 where it appears.

Finding Logarithms   The common logarithm of a number is the power to which 10 must be raised to obtain the number. The number 10 must be raised to the third power to equal 1000. Therefore, the logarithm of 1000 is 3, written $\log 1000 = 3$. Some examples follow.

Number	Exponential Expression	Logarithm
1000	10^3	3
100	10^2	2
10	10^1	1
1	10^0	0
1/10 = 0.1	10^{-1}	-1
1/100 = 0.01	10^{-2}	-2
1/1000 = 0.001	10^{-3}	-3

    To obtain the logarithm of a number other than an integral power of 10, you must use either a logarithm table or an electronic calculator. On most calculators, you do this by (1) entering the number and (2) pressing the (log) button.

$$\log 7.39 = 0.8686444 = \underline{0.869}$$

$$\log 7.39 \times 10^3 = 3.8686 \qquad = \underline{3.869}$$

$$\log 7.39 \times 10^{-3} = -2.1314 \quad = \underline{-2.131}$$

The number to the left of the decimal point in a logarithm is called the *characteristic,* and the number to the right of the decimal point is called the *mantissa.* The characteristic only locates the decimal point of the number, so it is usually not included when counting significant figures. The mantissa has as many significant figures as the number whose log was found.

To obtain the natural logarithm of a number on an electronic calculator, (1) enter the number and (2) press the (ln) (or ln x) button.

$$\ln 4.45 = 1.4929041 = \underline{1.493}$$

$$\ln 1.27 \times 10^3 = 7.1468 \quad = \underline{7.147}$$

Finding Antilogarithms Sometimes we know the logarithm of a number and must find the number. This is called finding the *antilogarithm* (or *inverse logarithm*). To do this on a calculator, we (1) enter the value of the log; (2) press the (INV) button; and (3) press the (log) button.

On some calculators, the inverse log is found as follows:
1. enter the value of the log
2. press the 2ndF (second function) button
3. press 10^x

$$\log x = 6.131; \quad \text{so } x = \text{inverse log of } 6.131 = \underline{1.352 \times 10^6}$$

$$\log x = -1.562; \quad \text{so } x = \text{inverse log of } -1.562 = \underline{2.74 \times 10^{-2}}$$

To find the inverse natural logarithm, we (1) enter the value of the ln; (2) press the (INV) button; and (3) press the (ln) or (ln x) button.

On some calculators, the inverse natural logarithm is found as follows:
1. enter the value of the ln
2. press the 2ndF (second function) button
3. press e^x

$$\ln x = 3.552; \quad \text{so } x = \text{inverse ln of } 3.552 = \underline{3.49 \times 10^1}$$

$$\ln x = -1.248; \quad \text{so } x = \text{inverse ln of } -1.248 = \underline{2.87 \times 10^{-1}}$$

Calculations Involving Logarithms

Because logarithms are exponents, operations involving them follow the same rules as the use of exponents. The following relationships are useful.

$$\log xy = \log x + \log y \qquad \text{or} \qquad \ln xy = \ln x + \ln y$$

$$\log \frac{x}{y} = \log x - \log y \qquad \text{or} \qquad \ln \frac{x}{y} = \ln x - \ln y$$

$$\log x^y = y \log x \qquad \text{or} \qquad \ln x^y = y \ln x$$

$$\log \sqrt[y]{x} = \log x^{1/y} = \frac{1}{y} \log x \qquad \text{or} \qquad \ln \sqrt[y]{x} = \ln x^{1/y} = \frac{1}{y} \ln x$$

A-3 QUADRATIC EQUATIONS

Algebraic expressions of the form

$$ax^2 + bx + c = 0$$

are called **quadratic equations.** Each of the constant terms (a, b, and c) may be either positive or negative. All quadratic equations may be solved by the **quadratic formula.**

$$x = \frac{-b \pm \sqrt{b^2 - 4ac}}{2a}$$

If we wish to solve the quadratic equation $3x^2 - 4x - 8 = 0$, we use $a = 3$, $b = -4$, and $c = -8$. Substitution of these values into the quadratic formula gives

$$x = \frac{-(-4) \pm \sqrt{(-4)^2 - 4(3)(-8)}}{2(3)} = \frac{4 \pm \sqrt{16 + 96}}{6}$$

$$= \frac{4 \pm \sqrt{112}}{6} = \frac{4 \pm 10.6}{6}$$

The two roots of this quadratic equation are

$$x = 2.4 \qquad \text{and} \qquad x = -1.1$$

As you construct and solve quadratic equations based on the observed behavior of matter, you must decide which root has physical significance. Examination of the *equation that defines* x always gives clues about possible values for x. In this way you can tell which is extraneous (has no physical significance). Negative roots are often extraneous.

When you have solved a quadratic equation, you should always check the values you obtained by substitution into the original equation. In the above example we obtained $x = 2.4$ and $x = -1.1$. Substitution of these values into the original quadratic equation, $3x^2 - 4x - 8 = 0$, shows that both roots are correct. Such substitutions often do not give a perfect check because some round-off error has been introduced.

ELECTRON CONFIGURATIONS OF THE ATOMS OF THE ELEMENTS

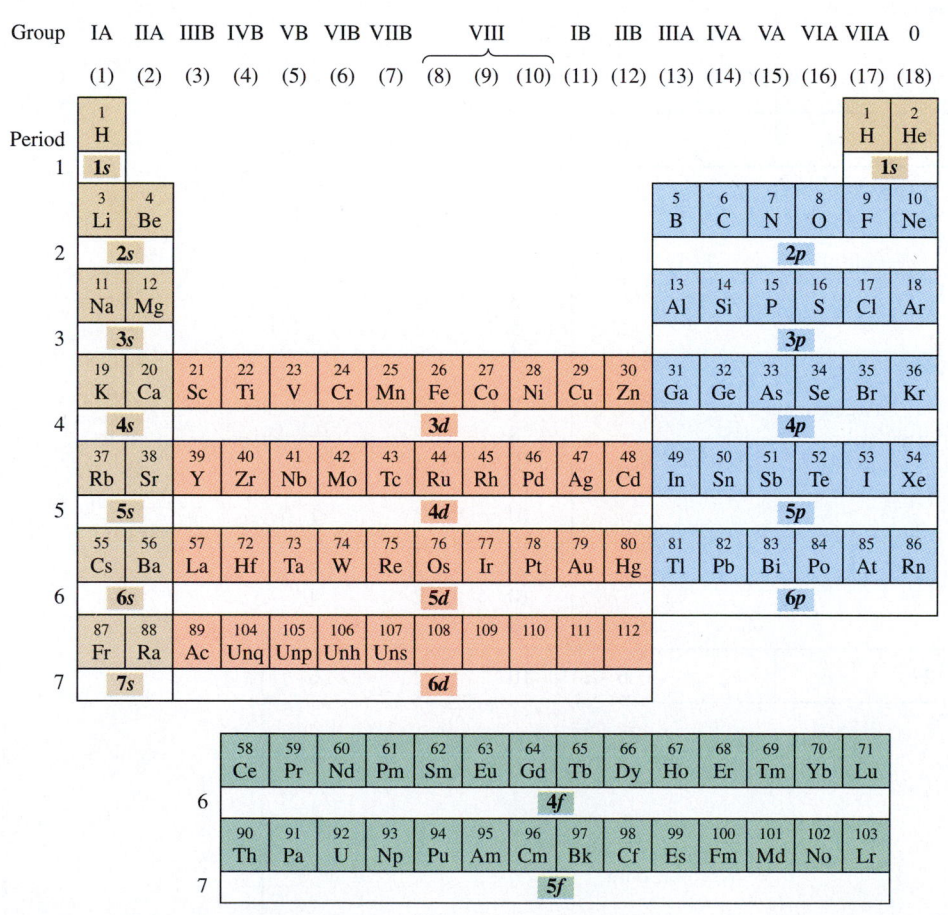

A periodic table colored to show the kinds of atomic orbitals (sublevels) being filled in different parts of the periodic table. The atomic orbitals are given below the symbols of blocks of elements. The electronic structures of the A group and 0 group elements are perfectly regular and can be predicted from their positions in the periodic table, but there are many exceptions in the d and f blocks. The populations of subshells are given in the table on pages A-8 and A-9.

Electron Configurations of the Atoms of the Elements

Element	Atomic Number	Populations of Subshells										
		1s	2s	2p	3s	3p	3d	4s	4p	4d	4f	5s
H	1	1										
He	2	2										
Li	3	2	1									
Be	4	2	2									
B	5	2	2	1								
C	6	2	2	2								
N	7	2	2	3								
O	8	2	2	4								
F	9	2	2	5								
Ne	10	2	2	6								
Na	11	Neon core			1							
Mg	12				2							
Al	13				2	1						
Si	14				2	2						
P	15				2	3						
S	16				2	4						
Cl	17				2	5						
Ar	18	2	2	6	2	6						
K	19	Argon core						1				
Ca	20							2				
Sc	21						1	2				
Ti	22						2	2				
V	23						3	2				
Cr	24						5	1				
Mn	25						5	2				
Fe	26						6	2				
Co	27						7	2				
Ni	28						8	2				
Cu	29						10	1				
Zn	30						10	2				
Ga	31						10	2	1			
Ge	32						10	2	2			
As	33						10	2	3			
Se	34						10	2	4			
Br	35						10	2	5			
Kr	36	2	2	6	2	6	10	2	6			
Rb	37	Krypton core										1
Sr	38											2
Y	39									1		2
Zr	40									2		2
Nb	41									3		2
Mo	42									5		1
Tc	43									5		2
Ru	44									7		1
Rh	45									8		1
Pd	46									10		
Ag	47									10		1
Cd	48									10		2

Element	Atomic Number		Populations of Subshells									
			4d	4f	5s	5p	5d	5f	6s	6p	6d	7s
In	49		10		2	1						
Sn	50		10		2	2						
Sb	51		10		2	3						
Te	52		10		2	4						
I	53		10		2	5						
Xe	54		10		2	6						
Cs	55		10		2	6			1			
Ba	56		10		2	6			2			
La	57		10		2	6	1		2			
Ce	58		10	1	2	6	1		2			
Pr	59		10	3	2	6			2			
Nd	60		10	4	2	6			2			
Pm	61		10	5	2	6			2			
Sm	62		10	6	2	6			2			
Eu	63		10	7	2	6			2			
Gd	64		10	7	2	6	1		2			
Tb	65		10	9	2	6			2			
Dy	66		10	10	2	6			2			
Ho	67		10	11	2	6			2			
Er	68		10	12	2	6			2			
Tm	69	Krypton core	10	13	2	6			2			
Yb	70		10	14	2	6			2			
Lu	71		10	14	2	6	1		2			
Hf	72		10	14	2	6	2		2			
Ta	73		10	14	2	6	3		2			
W	74		10	14	2	6	4		2			
Re	75		10	14	2	6	5		2			
Os	76		10	14	2	6	6		2			
Ir	77		10	14	2	6	7		2			
Pt	78		10	14	2	6	9		1			
Au	79		10	14	2	6	10		1			
Hg	80		10	14	2	6	10		2			
Tl	81		10	14	2	6	10		2	1		
Pb	82		10	14	2	6	10		2	2		
Bi	83		10	14	2	6	10		2	3		
Po	84		10	14	2	6	10		2	4		
At	85		10	14	2	6	10		2	5		
Rn	86		10	14	2	6	10		2	6		
Fr	87		10	14	2	6	10		2	6		1
Ra	88		10	14	2	6	10		2	6		2
Ac	89		10	14	2	6	10		2	6	1	2
Th	90		10	14	2	6	10		2	6	2	2
Pa	91		10	14	2	6	10	2	2	6	1	2
U	92		10	14	2	6	10	3	2	6	1	2
Np	93		10	14	2	6	10	4	2	6	1	2
Pu	94		10	14	2	6	10	6	2	6		2
Am	95		10	14	2	6	10	7	2	6		2
Cm	96		10	14	2	6	10	7	2	6	1	2
Bk	97		10	14	2	6	10	9	2	6		2
Cf	98		10	14	2	6	10	10	2	6		2
Es	99		10	14	2	6	10	11	2	6		2
Fm	100		10	14	2	6	10	12	2	6		2
Md	101		10	14	2	6	10	13	2	6		2
No	102		10	14	2	6	10	14	2	6		2
Lr	103		10	14	2	6	10	14	2	6	1	2
Unq	104		10	14	2	6	10	14	2	6	2	2
Unp	105		10	14	2	6	10	14	2	6	3	2
Unh	106		10	14	2	6	10	14	2	6	4	2

COMMON UNITS, EQUIVALENCES, AND CONVERSION FACTORS

FUNDAMENTAL UNITS OF THE SI SYSTEM

The metric system was implemented by the French National Assembly in 1790 and has been modified many times. The International System of Units, or *le Système International* (SI), represents an extension of the metric system. It was adopted by the 11th General Conference of Weights and Measures in 1960 and has also been modified since. It is constructed from seven base units, each of which represents a particular physical quantity (Table I).

The first five units listed in Table I are particularly useful in general chemistry. They are defined as follows.

1. The *meter* is defined as the distance light travels in a vacuum in 1/299,792,468 second.

2. The *kilogram* represents the mass of a platinum-iridium block kept at the International Bureau of Weights and Measures at Sèvres, France.

3. The *second* was redefined in 1967 as the duration of 9,192,631,770 periods of a certain line in the microwave spectrum of cesium-133.

4. The *kelvin* is 1/273.16 of the temperature interval between absolute zero and the triple point of water.

5. The *mole* is the amount of substance that contains as many entities as there are atoms in exactly 0.012 kg of carbon-12 (12 g of ^{12}C atoms).

Table I *SI Fundamental Units*		
Physical Quantity	**Name of Unit**	**Symbol**
Length	meter	m
Mass	kilogram	kg
Time	second	s
Temperature	kelvin	K
Amount of substance	mole	mol
Electric current	ampere	A
Luminous intensity	candela	cd

Prefixes Used with Metric Units and SI Units

Decimal fractions and multiples of metric and SI units are designated by the prefixes listed in Table II. Those most commonly used in general chemistry are underlined.

Table II *Traditional Metric and SI Prefixes*

Factor	Prefix	Symbol	Factor	Prefix	Symbol
10^{12}	tera	T	10^{-1}	deci	d
10^{9}	giga	G	10^{-2}	centi	c
10^{6}	mega	M	10^{-3}	milli	m
10^{3}	kilo	k	10^{-6}	micro	μ
10^{2}	hecto	h	10^{-9}	nano	n
10^{1}	deka	da	10^{-12}	pico	p
			10^{-15}	femto	f
			10^{-18}	atto	a

DERIVED SI UNITS

In the International System of Units all physical quantities are represented by appropriate combinations of the base units listed in Table I. A list of the derived units frequently used in general chemistry is given in Table III.

Table III *Derived SI Units*

Physical Quantity	Name of Unit	Symbol	Definition
Area	square meter	m^2	
Volume	cubic meter	m^3	
Density	kilogram per cubic meter	kg/m^3	
Force	newton	N	$kg \cdot m/s^2$
Pressure	pascal	Pa	N/m^2
Energy	joule	J	$kg \cdot m^2/s^2$
Electric charge	coulomb	C	$A \cdot s$
Electric potential difference	volt	V	$J/(A \cdot s)$

Common Units of Mass and Weight

1 pound = 453.59 grams

1 pound = 453.59 grams = 0.45359 kilogram
1 kilogram = 1000 grams = 2.205 pounds
1 gram = 10 decigrams = 100 centigrams
        = 1000 milligrams
1 gram = 6.022×10^{23} atomic mass units
1 atomic mass unit = 1.6606×10^{-24} gram
1 short ton = 2000 pounds = 907.2 kilograms
1 long ton = 2240 pounds
1 metric tonne = 1000 kilograms = 2205 pounds

Common Units of Length

1 inch = 2.54 centimeters (exactly)

1 mile = 5280 feet = 1.609 kilometers
1 yard = 36 inches = 0.9144 meter
1 meter = 100 centimeters = 39.37 inches = 3.281 feet
= 1.094 yards
1 kilometer = 1000 meters = 1094 yards = 0.6215 mile
1 Ångstrom = 1.0×10^{-8} centimeter = 0.10 nanometer
= 1.0×10^{-10} meter = 3.937×10^{-9} inch

Common Units of Volume

1 quart = 0.9463 liter
1 liter = 1.056 quarts

1 liter = 1 cubic decimeter = 1000 cubic centimeters
= 0.001 cubic meter
1 milliliter = 1 cubic centimeter = 0.001 liter
= 1.056×10^{-3} quart
1 cubic foot = 28.316 liters = 29.902 quarts
= 7.475 gallons

Common Units of Force* and Pressure

1 atmosphere = 760 millimeters of mercury
= 1.01325×10^{5} pascals
= 14.70 pounds per square inch
1 bar = 10^{5} pascals
1 torr = 1 millimeter of mercury
1 pascal = $1 \text{ kg/m} \cdot \text{s}^2 = 1 \text{ N/m}^2$

Force: 1 newton (N) = 1 kg · m/s², i.e., the force that, when applied for 1 second, gives a 1-kilogram mass a velocity of 1 meter per second.

Common Units of Energy

1 joule = 1×10^{7} ergs

1 thermochemical calorie* = 4.184 joules = 4.184×10^{7} ergs
= 4.129×10^{-2} liter-atmospheres
= 2.612×10^{19} electron volts
1 erg = 1×10^{-7} joule = 2.3901×10^{-8} calorie
1 electron volt = 1.6022×10^{-19} joule = 1.6022×10^{-12} erg = 96.487 kJ/mol[†]
1 liter-atmosphere = 24.217 calories = 101.325 joules = 1.01325×10^{9} ergs
1 British thermal unit = 1055.06 joules = 1.05506×10^{10} ergs = 252.2 calories

The amount of heat required to raise the temperature of one gram of water from 14.5°C to 15.5°C.

[†]*Note that the other units are per particle and must be multiplied by 6.022×10^{23} to be strictly comparable.*

PHYSICAL CONSTANTS

Quantity	Symbol	Traditional Units	SI Units
Acceleration of gravity	g	980.6 cm/s	9.806 m/s
Atomic mass unit (1/12 the mass of ^{12}C atom)	amu or u	1.6606×10^{-24} g	1.6606×10^{-27} kg
Avogadro's number	N	6.0221367×10^{23} particles/mol	6.0221367×10^{23} particles/mol
Bohr radius	a_0	0.52918 Å 5.2918×10^{-9} cm	5.2918×10^{-11} m
Boltzmann constant	k	1.3807×10^{-16} erg/K	1.3807×10^{-23} J/K
Charge-to-mass ratio of electron	e/m	1.75882×10^{8} coulomb/g	1.75882×10^{11} C/kg
Electronic charge	e	1.60218×10^{-19} coulomb 4.8033×10^{-10} esu	1.60218×10^{-19} C
Electron rest mass	m_e	9.10940×10^{-28} g 0.00054858 amu	9.10940×10^{-31} kg
Faraday constant	F	96,485 coulombs/eq 23.06 kcal/volt · eq	96,485 C/mol e$^-$ 96,485 J/V · mol e$^-$
Gas constant	R	$0.08206 \dfrac{\text{L} \cdot \text{atm}}{\text{mol} \cdot \text{K}}$ $1.987 \dfrac{\text{cal}}{\text{mol} \cdot \text{K}}$	$8.3145 \dfrac{\text{kPa} \cdot \text{dm}^3}{\text{mol} \cdot \text{K}}$ 8.3145 J/mol · K
Molar volume (STP)	V_m	22.414 L/mol	22.414×10^{-3} m^3/mol 22.414 dm^3/mol
Neutron rest mass	m_n	1.67495×10^{-24} g 1.008665 amu	1.67495×10^{-27} kg
Planck constant	h	6.6262×10^{-27} erg · s	6.6262×10^{-34} J · s
Proton rest mass	m_p	1.6726×10^{-24} g 1.007277 amu	1.6726×10^{-27} kg
Rydberg constant	R_∞	3.289×10^{15} cycles/s 2.1799×10^{-11} erg	1.0974×10^{7} m^{-1} 2.1799×10^{-18} J
Speed of light (in a vacuum)	c	2.9979×10^{10} cm/s (186,281 miles/second)	2.9979×10^{8} m/s

$\pi = 3.1416$

$e = 2.71828$

$\ln X = 2.303 \log X$

$2.303\,R = 4.576$ cal/mol · K $= 19.15$ J/mol · K

$2.303\,RT$ (at 25°C) $= 1364$ cal/mol $= 5709$ J/mol

SOME PHYSICAL CONSTANTS FOR A FEW COMMON SUBSTANCES

Specific Heats and Heat Capacities for Some Common Substances		
Substance	Specific Heat (J/g · °C)	Molar Heat Capacity (J/mol · °C)
Al(s)	0.900	24.3
Ca(s)	0.653	26.2
Cu(s)	0.385	24.5
Fe(s)	0.444	24.8
Hg(ℓ)	0.138	27.7
H_2O(s), ice	2.09	37.7
H_2O(ℓ), water	4.18	75.3
H_2O(g), steam	2.03	36.4
C_6H_6(ℓ), benzene	1.74	136
C_6H_6(g), benzene	1.04	81.6
C_2H_5OH(ℓ), ethanol	2.46	113
C_2H_5OH(g), ethanol	0.954	420
$(C_2H_5)_2O$(ℓ), diethyl ether	3.74	172
$(C_2H_5)_2O$(g), diethyl ether	2.35	108

Heats of Transformation and Transformation Temperatures of Several Substances						
Substance	mp (°C)	Heat of Fusion (J/g)	ΔH_{fus} (kJ/mol)	bp (°C)	Heat of Vaporization (J/g)	ΔH_{vap} (kJ/mol)
Al	658	395	10.6	2467	10520	284
Ca	851	233	9.33	1487	4030	162
Cu	1083	205	13.0	2595	4790	305
H_2O	0.0	334	6.02	100	2260	40.7
Fe	1530	267	14.9	2735	6340	354
Hg	−39	11	23.3	357	292	58.6
CH_4	−182	58.6	0.92	−164	—	—
C_2H_5OH	−117	109	5.02	78.0	855	39.3
C_6H_6	5.48	127	9.92	80.1	395	30.8
$(C_2H_5)_2O$	−116	97.9	7.66	35	351	26.0

Vapor Pressure of Water at Various Temperatures

Temperature (°C)	Vapor Pressure (torr)	Temperature (°C)	Vapor Pressure (torr)	Temperature (°C)	Vapor Pressure (torr)	Temperature (°C)	Vapor Pressure (torr)
−10	2.1	21	18.7	51	97.2	81	369.7
−9	2.3	22	19.8	52	102.1	82	384.9
−8	2.5	23	21.1	53	107.2	83	400.6
−7	2.7	24	22.4	54	112.5	84	416.8
−6	2.9	25	23.8	55	118.0	85	433.6
−5	3.2	26	25.2	56	123.8	86	450.9
−4	3.4	27	26.7	57	129.8	87	468.7
−3	3.7	28	28.3	58	136.1	88	487.1
−2	4.0	29	30.0	59	142.6	89	506.1
−1	4.3	30	31.8	60	149.4	90	525.8
0	4.6	31	33.7	61	156.4	91	546.1
1	4.9	32	35.7	62	163.8	92	567.0
2	5.3	33	37.7	63	171.4	93	588.6
3	5.7	34	39.9	64	179.3	94	610.9
4	6.1	35	42.2	65	187.5	95	633.9
5	6.5	36	44.6	66	196.1	96	657.6
6	7.0	37	47.1	67	205.0	97	682.1
7	7.5	38	49.7	68	214.2	98	707.3
8	8.0	39	52.4	69	223.7	99	733.2
9	8.6	40	55.3	70	233.7	100	760.0
10	9.2	41	58.3	71	243.9	101	787.6
11	9.8	42	61.5	72	254.6	102	815.9
12	10.5	43	64.8	73	265.7	103	845.1
13	11.2	44	68.3	74	277.2	104	875.1
14	12.0	45	71.9	75	289.1	105	906.1
15	12.8	46	75.7	76	301.4	106	937.9
16	13.6	47	79.6	77	314.1	107	970.6
17	14.5	48	83.7	78	327.3	108	1004.4
18	15.5	49	88.0	79	341.0	109	1038.9
19	16.5	50	92.5	80	355.1	110	1074.6
20	17.5						

IONIZATION CONSTANTS FOR WEAK ACIDS AT 25°C

Acid	Formula and Ionization Equation	K_a
Acetic	$CH_3COOH \rightleftharpoons H^+ + CH_3COO^-$	1.8×10^{-5}
Arsenic	$H_3AsO_4 \rightleftharpoons H^+ + H_2AsO_4^-$	$2.5 \times 10^{-4} = K_1$
	$H_2AsO_4^- \rightleftharpoons H^+ + HAsO_4^{2-}$	$5.6 \times 10^{-8} = K_2$
	$HAsO_4^{2-} \rightleftharpoons H^+ + AsO_4^{3-}$	$3.0 \times 10^{-13} = K_3$
Arsenous	$H_3AsO_3 \rightleftharpoons H^+ + H_2AsO_3^-$	$6.0 \times 10^{-10} = K_1$
	$H_2AsO_3^- \rightleftharpoons H^+ + HAsO_3^{2-}$	$3.0 \times 10^{-14} = K_2$
Benzoic	$C_6H_5COOH \rightleftharpoons H^+ + C_6H_5COO^-$	6.3×10^{-5}
Boric*	$B(OH)_3 \rightleftharpoons H^+ + BO(OH)_2^-$	$7.3 \times 10^{-10} = K_1$
	$BO(OH)_2^- \rightleftharpoons H^+ + BO_2(OH)^{2-}$	$1.8 \times 10^{-13} = K_2$
	$BO_2(OH)^{2-} \rightleftharpoons H^+ + BO_3^{3-}$	$1.6 \times 10^{-14} = K_3$
Carbonic	$H_2CO_3 \rightleftharpoons H^+ + HCO_3^-$	$4.2 \times 10^{-7} = K_1$
	$HCO_3^- \rightleftharpoons H^+ + CO_3^{2-}$	$4.8 \times 10^{-11} = K_2$
Citric	$C_3H_5O(COOH)_3 \rightleftharpoons H^+ + C_4H_5O_3(COOH)_2^-$	$7.4 \times 10^{-3} = K_1$
	$C_4H_5O_3(COOH)_2^- \rightleftharpoons H^+ + C_5H_5O_5COOH^{2-}$	$1.7 \times 10^{-5} = K_2$
	$C_5H_5O_5COOH^{2-} \rightleftharpoons H^+ + C_6H_5O_7^{3-}$	$7.4 \times 10^{-7} = K_3$
Cyanic	$HOCN \rightleftharpoons H^+ + OCN^-$	3.5×10^{-4}
Formic	$HCOOH \rightleftharpoons H^+ + HCOO^-$	1.8×10^{-4}
Hydrazoic	$HN_3 \rightleftharpoons H^+ + N_3^-$	1.9×10^{-5}
Hydrocyanic	$HCN \rightleftharpoons H^+ + CN^-$	4.0×10^{-10}
Hydrofluoric	$HF \rightleftharpoons H^+ + F^-$	7.2×10^{-4}
Hydrogen peroxide	$H_2O_2 \rightleftharpoons H^+ + HO_2^-$	2.4×10^{-12}
Hydrosulfuric	$H_2S \rightleftharpoons H^+ + HS^-$	$1.0 \times 10^{-7} = K_1$
	$HS^- \rightleftharpoons H^+ + S^{2-}$	$1.3 \times 10^{-13} = K_2$
Hypobromous	$HOBr \rightleftharpoons H^+ + OBr^-$	2.5×10^{-9}
Hypochlorous	$HOCl \rightleftharpoons H^+ + OCl^-$	3.5×10^{-8}
Nitrous	$HNO_2 \rightleftharpoons H^+ + NO_2^-$	4.5×10^{-4}
Oxalic	$(COOH)_2 \rightleftharpoons H^+ + COOCOOH^-$	$5.9 \times 10^{-2} = K_1$
	$COOCOOH^- \rightleftharpoons H^+ + (COO)_2^{2-}$	$6.4 \times 10^{-5} = K_2$
Phenol	$HC_6H_5O \rightleftharpoons H^+ + C_6H_5O^-$	1.3×10^{-10}
Phosphoric	$H_3PO_4 \rightleftharpoons H^+ + H_2PO_4^-$	$7.5 \times 10^{-3} = K_1$
	$H_2PO_4^- \rightleftharpoons H^+ + HPO_4^{2-}$	$6.2 \times 10^{-8} = K_2$
	$HPO_4^{2-} \rightleftharpoons H^+ + PO_4^{3-}$	$3.6 \times 10^{-13} = K_3$
Phosphorus	$H_3PO_3 \rightleftharpoons H^+ + H_2PO_3^-$	$1.6 \times 10^{-2} = K_1$
	$H_2PO_3^- \rightleftharpoons H^+ + HPO_3^{2-}$	$7.0 \times 10^{-7} = K_2$
Selenic	$H_2SeO_4 \rightleftharpoons H^+ + HSeO_4^-$	Very large $= K_1$
	$HSeO_4^- \rightleftharpoons H^+ + SeO_4^{2-}$	$1.2 \times 10^{-2} = K_2$
Selenous	$H_2SeO_3 \rightleftharpoons H^+ + HSeO_3^-$	$2.7 \times 10^{-3} = K_1$
	$HSeO_3^- \rightleftharpoons H^+ + SeO_3^{2-}$	$2.5 \times 10^{-7} = K_2$
Sulfuric	$H_2SO_4 \rightleftharpoons H^+ + HSO_4^-$	Very large $= K_1$
	$HSO_4^- \rightleftharpoons H^+ + SO_4^{2-}$	$1.2 \times 10^{-2} = K_2$
Sulfurous	$H_2SO_3 \rightleftharpoons H^+ + HSO_3^-$	$1.2 \times 10^{-2} = K_1$
	$HSO_3^- \rightleftharpoons H^+ + SO_3^{2-}$	$6.2 \times 10^{-8} = K_2$
Tellurous	$H_2TeO_3 \rightleftharpoons H^+ + HTeO_3^-$	$2 \times 10^{-3} = K_1$
	$HTeO_3^- \rightleftharpoons H^+ + TeO_3^{2-}$	$1 \times 10^{-8} = K_2$

*Boric acid acts as a Lewis acid in aqueous solution.

IONIZATION CONSTANTS FOR WEAK BASES AT 25°C

Base	Formula and Ionization Equation			K_b
Ammonia	NH_3	$+ H_2O \rightleftharpoons NH_4^+$	$+ OH^-$	1.8×10^{-5}
Aniline	$C_6H_5NH_2$	$+ H_2O \rightleftharpoons C_6H_5NH_3^+$	$+ OH^-$	4.2×10^{-10}
Dimethylamine	$(CH_3)_2NH$	$+ H_2O \rightleftharpoons (CH_3)_2NH_2^+$	$+ OH^-$	7.4×10^{-4}
Ethylenediamine	$(CH_2)_2(NH_2)_2$	$+ H_2O \rightleftharpoons (CH_2)_2(NH_2)_2H^+$	$+ OH^-$	$8.5 \times 10^{-5} = K_1$
	$(CH_2)_2(NH_2)_2H^+$	$+ H_2O \rightleftharpoons (CH_2)_2(NH_2)_2H_2^{2+}$	$+ OH^-$	$2.7 \times 10^{-8} = K_2$
Hydrazine	N_2H_4	$+ H_2O \rightleftharpoons N_2H_5^+$	$+ OH^-$	$8.5 \times 10^{-7} = K_1$
	$N_2H_5^+$	$+ H_2O \rightleftharpoons N_2H_6^{2+}$	$+ OH^-$	$8.9 \times 10^{-16} = K_2$
Hydroxylamine	NH_2OH	$+ H_2O \rightleftharpoons NH_3OH^+$	$+ OH^-$	6.6×10^{-9}
Methylamine	CH_3NH_2	$+ H_2O \rightleftharpoons CH_3NH_3^+$	$+ OH^-$	5.0×10^{-4}
Pyridine	C_5H_5N	$+ H_2O \rightleftharpoons C_5H_5NH^+$	$+ OH^-$	1.5×10^{-9}
Trimethylamine	$(CH_3)_3N$	$+ H_2O \rightleftharpoons (CH_3)_3NH^+$	$+ OH^-$	7.4×10^{-5}

Appendix H

SOLUBILITY PRODUCT CONSTANTS FOR SOME INORGANIC COMPOUNDS AT 25°C

Substance	K_{sp}	Substance	K_{sp}
Aluminum compounds		**Chromium compounds**	
$AlAsO_4$	1.6×10^{-16}	$CrAsO_4$	7.8×10^{-21}
$Al(OH)_3$	1.9×10^{-33}	$Cr(OH)_3$	6.7×10^{-31}
$AlPO_4$	1.3×10^{-20}	$CrPO_4$	2.4×10^{-23}
Antimony compounds		**Cobalt compounds**	
Sb_2S_3	1.6×10^{-93}	$Co_3(AsO_4)_2$	7.6×10^{-29}
Barium compounds		$CoCO_3$	8.0×10^{-13}
$Ba_3(AsO_4)_2$	1.1×10^{-13}	$Co(OH)_2$	2.5×10^{-16}
$BaCO_3$	8.1×10^{-9}	$CoS\ (\alpha)$	5.9×10^{-21}
$BaC_2O_4 \cdot 2H_2O^*$	1.1×10^{-7}	$CoS\ (\beta)$	8.7×10^{-23}
$BaCrO_4$	2.0×10^{-10}	$Co(OH)_3$	4.0×10^{-45}
BaF_2	1.7×10^{-6}	Co_2S_3	2.6×10^{-124}
$Ba(OH)_2 \cdot 8H_2O^*$	5.0×10^{-3}	**Copper compounds**	
$Ba_3(PO_4)_2$	1.3×10^{-29}	$CuBr$	5.3×10^{-9}
$BaSeO_4$	2.8×10^{-11}	$CuCl$	1.9×10^{-7}
$BaSO_3$	8.0×10^{-7}	$CuCN$	3.2×10^{-20}
$BaSO_4$	1.1×10^{-10}	$Cu_2O\ (Cu^+ + OH^-)^\dagger$	1.0×10^{-14}
Bismuth compounds		CuI	5.1×10^{-12}
$BiOCl$	7.0×10^{-9}	Cu_2S	1.6×10^{-48}
$BiO(OH)$	1.0×10^{-12}	$CuSCN$	1.6×10^{-11}
$Bi(OH)_3$	3.2×10^{-40}	$Cu_3(AsO_4)_2$	7.6×10^{-36}
BiI_3	8.1×10^{-19}	$CuCO_3$	2.5×10^{-10}
$BiPO_4$	1.3×10^{-23}	$Cu_2[Fe(CN)_6]$	1.3×10^{-16}
Bi_2S_3	1.6×10^{-72}	$Cu(OH)_2$	1.6×10^{-19}
Cadmium compounds		CuS	8.7×10^{-36}
$Cd_3(AsO_4)_2$	2.2×10^{-32}	**Gold compounds**	
$CdCO_3$	2.5×10^{-14}	$AuBr$	5.0×10^{-17}
$Cd(CN)_2$	1.0×10^{-8}	$AuCl$	2.0×10^{-13}
$Cd_2[Fe(CN)_6]$	3.2×10^{-17}	AuI	1.6×10^{-23}
$Cd(OH)_2$	1.2×10^{-14}	$AuBr_3$	4.0×10^{-36}
CdS	3.6×10^{-29}	$AuCl_3$	3.2×10^{-25}
Calcium compounds		$Au(OH)_3$	1.0×10^{-53}
$Ca_3(AsO_4)_2$	6.8×10^{-19}	AuI_3	1.0×10^{-46}
$CaCO_3$	4.8×10^{-9}	**Iron compounds**	
$CaCrO_4$	7.1×10^{-4}	$FeCO_3$	3.5×10^{-11}
$CaC_2O_4 \cdot H_2O^*$	2.3×10^{-9}	$Fe(OH)_2$	7.9×10^{-15}
CaF_2	3.9×10^{-11}	FeS	4.9×10^{-18}
$Ca(OH)_2$	7.9×10^{-6}	$Fe_4[Fe(CN)_6]_3$	3.0×10^{-41}
$CaHPO_4$	2.7×10^{-7}	$Fe(OH)_3$	6.3×10^{-38}
$Ca(H_2PO_4)_2$	1.0×10^{-3}	Fe_2S_3	1.4×10^{-88}
$Ca_3(PO_4)_2$	1.0×10^{-25}	**Lead compounds**	
$CaSO_3 \cdot 2H_2O^*$	1.3×10^{-8}	$Pb_3(AsO_4)_2$	4.1×10^{-36}
$CaSO_4 \cdot 2H_2O^*$	2.4×10^{-5}	$PbBr_2$	6.3×10^{-6}

SOLUBILITY PRODUCT CONSTANTS FOR SOME INORGANIC COMPOUNDS AT 25°C (continued)

Substance	K_{sp}	Substance	K_{sp}
Lead compounds (cont.)		**Nickel compounds** (cont.)	
$PbCO_3$	1.5×10^{-13}	NiS (α)	3.0×10^{-21}
$PbCl_2$	1.7×10^{-5}	NiS (β)	1.0×10^{-26}
$PbCrO_4$	1.8×10^{-14}	NiS (γ)	2.0×10^{-28}
PbF_2	3.7×10^{-8}	**Silver compounds**	
$Pb(OH)_2$	2.8×10^{-16}	Ag_3AsO_4	1.1×10^{-20}
PbI_2	8.7×10^{-9}	AgBr	3.3×10^{-13}
$Pb_3(PO_4)_2$	3.0×10^{-44}	Ag_2CO_3	8.1×10^{-12}
$PbSeO_4$	1.5×10^{-7}	AgCl	1.8×10^{-10}
$PbSO_4$	1.8×10^{-8}	Ag_2CrO_4	9.0×10^{-12}
PbS	8.4×10^{-28}	AgCN	1.2×10^{-16}
Magnesium compounds		$Ag_4[Fe(CN)_6]$	1.6×10^{-41}
$Mg_3(AsO_4)_2$	2.1×10^{-20}	Ag_2O ($Ag^+ + OH^-$)†	2.0×10^{-8}
$MgCO_3 \cdot 3H_2O$*	4.0×10^{-5}	AgI	1.5×10^{-16}
MgC_2O_4	8.6×10^{-5}	Ag_3PO_4	1.3×10^{-20}
MgF_2	6.4×10^{-9}	Ag_2SO_3	1.5×10^{-14}
$Mg(OH)_2$	1.5×10^{-11}	Ag_2SO_4	1.7×10^{-5}
$MgNH_4PO_4$	2.5×10^{-12}	Ag_2S	1.0×10^{-49}
Manganese compounds		AgSCN	1.0×10^{-12}
$Mn_3(AsO_4)_2$	1.9×10^{-11}	**Strontium compounds**	
$MnCO_3$	1.8×10^{-11}	$Sr_3(AsO_4)_2$	1.3×10^{-18}
$Mn(OH)_2$	4.6×10^{-14}	$SrCO_3$	9.4×10^{-10}
MnS	5.1×10^{-15}	$SrC_2O_4 \cdot 2H_2O$*	5.6×10^{-8}
$Mn(OH)_3$	$\sim 1.0 \times 10^{-36}$	$SrCrO_4$	3.6×10^{-5}
Mercury compounds		$Sr(OH)_2 \cdot 8H_2O$*	3.2×10^{-4}
Hg_2Br_2	1.3×10^{-22}	$Sr_3(PO_4)_2$	1.0×10^{-31}
Hg_2CO_3	8.9×10^{-17}	$SrSO_3$	4.0×10^{-8}
Hg_2Cl_2	1.1×10^{-18}	$SrSO_4$	2.8×10^{-7}
Hg_2CrO_4	5.0×10^{-9}	**Tin compounds**	
Hg_2I_2	4.5×10^{-29}	$Sn(OH)_2$	2.0×10^{-26}
$Hg_2O \cdot H_2O$* ($Hg_2^{2+} + 2OH^-$)†	1.6×10^{-23}	SnI_2	1.0×10^{-4}
Hg_2SO_4	6.8×10^{-7}	SnS	1.0×10^{-28}
Hg_2S	5.8×10^{-44}	$Sn(OH)_4$	1.0×10^{-57}
$Hg(CN)_2$	3.0×10^{-23}	SnS_2	1.0×10^{-70}
$Hg(OH)_2$	2.5×10^{-26}	**Zinc compounds**	
HgI_2	4.0×10^{-29}	$Zn_3(AsO_4)_2$	1.1×10^{-27}
HgS	3.0×10^{-53}	$ZnCO_3$	1.5×10^{-11}
Nickel compounds		$Zn(CN)_2$	8.0×10^{-12}
$Ni_3(AsO_4)_2$	1.9×10^{-26}	$Zn_2[Fe(CN)_6]$	4.1×10^{-16}
$NiCO_3$	6.6×10^{-9}	$Zn(OH)_2$	4.5×10^{-17}
$Ni(CN)_2$	3.0×10^{-23}	$Zn_3(PO_4)_2$	9.1×10^{-33}
$Ni(OH)_2$	2.8×10^{-16}	ZnS	1.1×10^{-21}

*[H₂O] does not appear in equilibrium constants for equilibria in aqueous solution in general, so it does not appear in the K_{sp} expressions for hydrated solids.

† Very small amounts of oxides dissolve in water to give the ions indicated in parentheses. Solid hydroxides are unstable and decompose to oxides as rapidly as they are formed.

DISSOCIATION CONSTANTS FOR SOME COMPLEX IONS

Dissociation Equilibrium			K_d
$[AgBr_2]^-$	$\rightleftharpoons$	$Ag^+ + 2Br^-$	7.8×10^{-8}
$[AgCl_2]^-$	$\rightleftharpoons$	$Ag^+ + 2Cl^-$	4.0×10^{-6}
$[Ag(CN)_2]^-$	$\rightleftharpoons$	$Ag^+ + 2CN^-$	1.8×10^{-19}
$[Ag(S_2O_3)_2]^{3-}$	$\rightleftharpoons$	$Ag^+ + 2S_2O_3^{2-}$	5.0×10^{-14}
$[Ag(NH_3)_2]^+$	$\rightleftharpoons$	$Ag^+ + 2NH_3$	6.3×10^{-8}
$[Ag(en)]^+$	$\rightleftharpoons$	$Ag^+ + en^*$	1.0×10^{-5}
$[AlF_6]^{3-}$	$\rightleftharpoons$	$Al^{3+} + 6F^-$	2.0×10^{-24}
$[Al(OH)_4]^-$	$\rightleftharpoons$	$Al^{3+} + 4OH^-$	1.3×10^{-34}
$[Au(CN)_2]^-$	$\rightleftharpoons$	$Au^+ + 2CN^-$	5.0×10^{-39}
$[Cd(CN)_4]^{2-}$	$\rightleftharpoons$	$Cd^{2+} + 4CN^-$	7.8×10^{-18}
$[CdCl_4]^{2-}$	$\rightleftharpoons$	$Cd^{2+} + 4Cl^-$	1.0×10^{-4}
$[Cd(NH_3)_4]^{2+}$	$\rightleftharpoons$	$Cd^{2+} + 4NH_3$	1.0×10^{-7}
$[Co(NH_3)_6]^{2+}$	$\rightleftharpoons$	$Co^{2+} + 6NH_3$	1.3×10^{-5}
$[Co(NH_3)_6]^{3+}$	$\rightleftharpoons$	$Co^{3+} + 6NH_3$	2.2×10^{-34}
$[Co(en)_3]^{2+}$	$\rightleftharpoons$	$Co^{2+} + 3en^*$	1.5×10^{-14}
$[Co(en)_3]^{3+}$	$\rightleftharpoons$	$Co^{3+} + 3en^*$	2.0×10^{-49}
$[Cu(CN)_2]^-$	$\rightleftharpoons$	$Cu^+ + 2CN^-$	1.0×10^{-16}
$[CuCl_2]^-$	$\rightleftharpoons$	$Cu^+ + 2Cl^-$	1.0×10^{-5}
$[Cu(NH_3)_2]^+$	$\rightleftharpoons$	$Cu^+ + 2NH_3$	1.4×10^{-11}
$[Cu(NH_3)_4]^{2+}$	$\rightleftharpoons$	$Cu^{2+} + 4NH_3$	8.5×10^{-13}
$[Fe(CN)_6]^{4-}$	$\rightleftharpoons$	$Fe^{2+} + 6CN^-$	1.3×10^{-37}
$[Fe(CN)_6]^{3-}$	$\rightleftharpoons$	$Fe^{3+} + 6CN^-$	1.3×10^{-44}
$[HgCl_4]^{2-}$	$\rightleftharpoons$	$Hg^{2+} + 4Cl^-$	8.3×10^{-16}
$[Ni(CN)_4]^{2-}$	$\rightleftharpoons$	$Ni^{2+} + 4CN^-$	1.0×10^{-31}
$[Ni(NH_3)_6]^{2+}$	$\rightleftharpoons$	$Ni^{2+} + 6NH_3$	1.8×10^{-9}
$[Zn(OH)_4]^{2-}$	$\rightleftharpoons$	$Zn^{2+} + 4OH^-$	3.5×10^{-16}
$[Zn(NH_3)_4]^{2+}$	$\rightleftharpoons$	$Zn^{2+} + 4NH_3$	3.4×10^{-10}

*The abbreviation "en" represents ethylenediamine, $H_2NCH_2CH_2NH_2$.

Appendix J

STANDARD REDUCTION POTENTIALS IN AQUEOUS SOLUTION AT 25°C

Acidic Solution	Standard Reduction Potential, E^0 (volts)
$Li^+(aq) + e^- \longrightarrow Li(s)$	-3.045
$K^+(aq) + e^- \longrightarrow K(s)$	-2.925
$Rb^+(aq) + e^- \longrightarrow Rb(s)$	-2.925
$Ba^{2+}(aq) + 2e^- \longrightarrow Ba(s)$	-2.90
$Sr^{2+}(aq) + 2e^- \longrightarrow Sr(s)$	-2.89
$Ca^{2+}(aq) + 2e^- \longrightarrow Ca(s)$	-2.87
$Na^+(aq) + e^- \longrightarrow Na(s)$	-2.714
$Mg^{2+}(aq) + 2e^- \longrightarrow Mg(s)$	-2.37
$H_2(g) + 2e^- \longrightarrow 2H^-(aq)$	-2.25
$Al^{3+}(aq) + 3e^- \longrightarrow Al(s)$	-1.66
$Zr^{4+}(aq) + 4e^- \longrightarrow Zr(s)$	-1.53
$ZnS(s) + 2e^- \longrightarrow Zn(s) + S^{2-}(aq)$	-1.44
$CdS(s) + 2e^- \longrightarrow Cd(s) + S^{2-}(aq)$	-1.21
$V^{2+}(aq) + 2e^- \longrightarrow V(s)$	-1.18
$Mn^{2+}(aq) + 2e^- \longrightarrow Mn(s)$	-1.18
$FeS(s) + 2e^- \longrightarrow Fe(s) + S^{2-}(aq)$	-1.01
$Cr^{2+}(aq) + 2e^- \longrightarrow Cr(s)$	-0.91
$Zn^{2+}(aq) + 2e^- \longrightarrow Zn(s)$	-0.763
$Cr^{3+}(aq) + 3e^- \longrightarrow Cr(s)$	-0.74
$HgS(s) + 2H^+(aq) + 2e^- \longrightarrow Hg(\ell) + H_2S(g)$	-0.72
$Ga^{3+}(aq) + 3e^- \longrightarrow Ga(s)$	-0.53
$2CO_2(g) + 2H^+(aq) + 2e^- \longrightarrow (COOH)_2(aq)$	-0.49
$Fe^{2+}(aq) + 2e^- \longrightarrow Fe(s)$	-0.44
$Cr^{3+}(aq) + e^- \longrightarrow Cr^{2+}(aq)$	-0.41
$Cd^{2+}(aq) + 2e^- \longrightarrow Cd(s)$	-0.403
$Se(s) + 2H^+(aq) + 2e^- \longrightarrow H_2Se(aq)$	-0.40
$PbSO_4(s) + 2e^- \longrightarrow Pb(s) + SO_4^{2-}(aq)$	-0.356
$Tl^+(aq) + e^- \longrightarrow Tl(s)$	-0.34
$Co^{2+}(aq) + 2e^- \longrightarrow Co(s)$	-0.28
$Ni^{2+}(aq) + 2e^- \longrightarrow Ni(s)$	-0.25
$[SnF_6]^{2-}(aq) + 4e^- \longrightarrow Sn(s) + 6F^-(aq)$	-0.25
$AgI(s) + e^- \longrightarrow Ag(s) + I^-(aq)$	-0.15
$Sn^{2+}(aq) + 2e^- \longrightarrow Sn(s)$	-0.14
$Pb^{2+}(aq) + 2e^- \longrightarrow Pb(s)$	-0.126
$N_2O(g) + 6H^+(aq) + H_2O + 4e^- \longrightarrow 2NH_3OH^+(aq)$	-0.05
$2H^+(aq) + 2e^- \longrightarrow H_2(g)$ (reference electrode)	0.000
$AgBr(s) + e^- \longrightarrow Ag(s) + Br^-(aq)$	0.10
$S(s) + 2H^+(aq) + 2e^- \longrightarrow H_2S(aq)$	0.14
$Sn^{4+}(aq) + 2e^- \longrightarrow Sn^{2+}(aq)$	0.15
$Cu^{2+}(aq) + e^- \longrightarrow Cu^+(aq)$	0.153
$SO_4^{2-}(aq) + 4H^+(aq) + 2e^- \longrightarrow H_2SO_3(aq) + H_2O$	0.17

STANDARD REDUCTION POTENTIALS IN AQUEOUS SOLUTION AT 25°C (continued)

Acidic Solution	Standard Reduction Potential, E^0 (volts)
$SO_4^{2-}(aq) + 4H^+(aq) + 2e^- \longrightarrow SO_2(g) + 2H_2O$	0.20
$AgCl(s) + e^- \longrightarrow Ag(s) + Cl^-(aq)$	0.222
$Hg_2Cl_2(s) + 2e^- \longrightarrow 2Hg(\ell) + 2Cl^-(aq)$	0.27
$Cu^{2+}(aq) + 2e^- \longrightarrow Cu(s)$	0.337
$[RhCl_6]^{3-}(aq) + 3e^- \longrightarrow Rh(s) + 6Cl^-(aq)$	0.44
$Cu^+(aq) + e^- \longrightarrow Cu(s)$	0.521
$TeO_2(s) + 4H^+(aq) + 4e^- \longrightarrow Te(s) + 2H_2O$	0.529
$I_2(s) + 2e^- \longrightarrow 2I^-(aq)$	0.535
$H_3AsO_4(aq) + 2H^+(aq) + 2e^- \longrightarrow H_3AsO_3(aq) + H_2O$	0.58
$[PtCl_6]^{2-}(aq) + 2e^- \longrightarrow [PtCl_4]^{2-}(aq) + 2Cl^-(aq)$	0.68
$O_2(g) + 2H^+(aq) + 2e^- \longrightarrow H_2O_2(aq)$	0.682
$[PtCl_4]^{2-}(aq) + 2e^- \longrightarrow Pt(s) + 4Cl^-(aq)$	0.73
$SbCl_6^-(aq) + 2e^- \longrightarrow SbCl_4^-(aq) + 2Cl^-(aq)$	0.75
$Fe^{3+}(aq) + e^- \longrightarrow Fe^{2+}(aq)$	0.771
$Hg_2^{2+}(aq) + 2e^- \longrightarrow 2Hg(\ell)$	0.789
$Ag^+(aq) + e^- \longrightarrow Ag(s)$	0.7994
$Hg^{2+}(aq) + 2e^- \longrightarrow Hg(\ell)$	0.855
$2Hg^{2+}(aq) + 2e^- \longrightarrow Hg_2^{2+}(aq)$	0.920
$NO_3^-(aq) + 3H^+(aq) + 2e^- \longrightarrow HNO_2(aq) + H_2O$	0.94
$NO_3^-(aq) + 4H^+(aq) + 3e^- \longrightarrow NO(g) + 2H_2O$	0.96
$Pd^{2+}(aq) + 2e^- \longrightarrow Pd(s)$	0.987
$AuCl_4^-(aq) + 3e^- \longrightarrow Au(s) + 4Cl^-(aq)$	1.00
$Br_2(\ell) + 2e^- \longrightarrow 2Br^-(aq)$	1.08
$ClO_4^-(aq) + 2H^+(aq) + 2e^- \longrightarrow ClO_3^-(aq) + H_2O$	1.19
$IO_3^-(aq) + 6H^+(aq) + 5e^- \longrightarrow \frac{1}{2}I_2(aq) + 3H_2O$	1.195
$Pt^{2+}(aq) + 2e^- \longrightarrow Pt(s)$	1.2
$O_2(g) + 4H^+(aq) + 4e^- \longrightarrow 2H_2O$	1.229
$MnO_2(s) + 4H^+(aq) + 2e^- \longrightarrow Mn^{2+}(aq) + 2H_2O$	1.23
$N_2H_5^+(aq) + 3H^+(aq) + 2e^- \longrightarrow 2NH_4^+(aq)$	1.24
$Cr_2O_7^{2-}(aq) + 14H^+(aq) + 6e^- \longrightarrow 2Cr^{3+}(aq) + 7H_2O$	1.33
$Cl_2(g) + 2e^- \longrightarrow 2Cl^-(aq)$	1.360
$BrO_3^-(aq) + 6H^+(aq) + 6e^- \longrightarrow Br^-(aq) + 3H_2O$	1.44
$ClO_3^-(aq) + 6H^+(aq) + 5e^- \longrightarrow \frac{1}{2}Cl_2(g) + 3H_2O$	1.47
$Au^{3+}(aq) + 3e^- \longrightarrow Au(s)$	1.50
$MnO_4^-(aq) + 8H^+(aq) + 5e^- \longrightarrow Mn^{2+}(aq) + 4H_2O$	1.51
$NaBiO_3(s) + 6H^+(aq) + 2e^- \longrightarrow Bi^{3+}(aq) + Na^+(aq) + 3H_2O$	~1.6
$Ce^{4+}(aq) + e^- \longrightarrow Ce^{3+}(aq)$	1.61
$2HClO(aq) + 2H^+(aq) + 2e^- \longrightarrow Cl_2(g) + 2H_2O$	1.63
$Au^+(aq) + e^- \longrightarrow Au(s)$	1.68
$PbO_2(s) + SO_4^{2-}(aq) + 4H^+(aq) + 2e^- \longrightarrow PbSO_4(s) + 2H_2O$	1.685
$NiO_2(s) + 4H^+(aq) + 2e^- \longrightarrow Ni^{2+}(aq) + 2H_2O$	1.7
$H_2O_2(aq) + 2H^+(aq) + 2e^- \longrightarrow 2H_2O$	1.77
$Pb^{4+}(aq) + 2e^- \longrightarrow Pb^{2+}(aq)$	1.8
$Co^{3+}(aq) + e^- \longrightarrow Co^{2+}(aq)$	1.82
$F_2(g) + 2e^- \longrightarrow 2F^-(aq)$	2.87

STANDARD REDUCTION POTENTIALS IN AQUEOUS SOLUTION AT 25°C (continued)

Basic Solution	Standard Reduction Potential, E^0 (volts)
$SiO_3^{2-}(aq) + 3H_2O + 4e^- \longrightarrow Si(s) + 6OH^-(aq)$	−1.70
$Cr(OH)_3(s) + 3e^- \longrightarrow Cr(s) + 3OH^-(aq)$	−1.30
$[Zn(CN)_4]^{2-}(aq) + 2e^- \longrightarrow Zn(s) + 4CN^-(aq)$	−1.26
$Zn(OH)_2(s) + 2e^- \longrightarrow Zn(s) + 2OH^-(aq)$	−1.245
$[Zn(OH)_4]^{2-}(aq) + 2e^- \longrightarrow Zn(s) + 4OH^-(aq)$	−1.22
$N_2(g) + 4H_2O + 4e^- \longrightarrow N_2H_4(aq) + 4OH^-(aq)$	−1.15
$SO_4^{2-}(aq) + H_2O + 2e^- \longrightarrow SO_3^{2-}(aq) + 2OH^-(aq)$	−0.93
$Fe(OH)_2(s) + 2e^- \longrightarrow Fe(s) + 2OH^-(aq)$	−0.877
$2NO_3^-(aq) + 2H_2O + 2e^- \longrightarrow N_2O_4(g) + 4OH^-(aq)$	−0.85
$2H_2O + 2e^- \longrightarrow H_2(g) + 2OH^-(aq)$	−0.8277
$Fe(OH)_3(s) + e^- \longrightarrow Fe(OH)_2(s) + OH^-(aq)$	−0.56
$S(s) + 2e^- \longrightarrow S^{2-}(aq)$	−0.48
$Cu(OH)_2(s) + 2e^- \longrightarrow Cu(s) + 2OH^-(aq)$	−0.36
$CrO_4^{2-}(aq) + 4H_2O + 3e^- \longrightarrow Cr(OH)_3(s) + 5OH^-(aq)$	−0.12
$MnO_2(s) + 2H_2O + 2e^- \longrightarrow Mn(OH)_2(s) + 2OH^-(aq)$	−0.05
$NO_3^-(aq) + H_2O + 2e^- \longrightarrow NO_2^-(aq) + 2OH^-(aq)$	0.01
$O_2(g) + H_2O + 2e^- \longrightarrow OOH^-(aq) + OH^-(aq)$	0.076
$HgO(s) + H_2O + 2e^- \longrightarrow Hg(\ell) + 2OH^-(aq)$	0.0984
$[Co(NH_3)_6]^{3+}(aq) + e^- \longrightarrow [Co(NH_3)_6]^{2+}(aq)$	0.10
$N_2H_4(aq) + 2H_2O + 2e^- \longrightarrow 2NH_3(aq) + 2OH^-(aq)$	0.10
$2NO_2^-(aq) + 3H_2O + 4e^- \longrightarrow N_2O(g) + 6OH^-(aq)$	0.15
$Ag_2O(s) + H_2O + 2e^- \longrightarrow 2Ag(s) + 2OH^-(aq)$	0.34
$ClO_4^-(aq) + H_2O + 2e^- \longrightarrow ClO_3^-(aq) + 2OH^-(aq)$	0.36
$O_2(g) + 2H_2O + 4e^- \longrightarrow 4OH^-(aq)$	0.40
$Ag_2CrO_4(s) + 2e^- \longrightarrow 2Ag(s) + CrO_4^{2-}(aq)$	0.446
$NiO_2(s) + 2H_2O + 2e^- \longrightarrow Ni(OH)_2(s) + 2OH^-(aq)$	0.49
$MnO_4^-(aq) + e^- \longrightarrow MnO_4^{2-}(aq)$	0.564
$MnO_4^-(aq) + 2H_2O + 3e^- \longrightarrow MnO_2(s) + 4OH^-(aq)$	0.588
$ClO_3^-(aq) + 3H_2O + 6e^- \longrightarrow Cl^-(aq) + 6OH^-(aq)$	0.62
$2NH_2OH(aq) + 2e^- \longrightarrow N_2H_4(aq) + 2OH^-(aq)$	0.74
$OOH^-(aq) + H_2O + 2e^- \longrightarrow 3OH^-(aq)$	0.88
$ClO^-(aq) + H_2O + 2e^- \longrightarrow Cl^-(aq) + 2OH^-(aq)$	0.89

SELECTED THERMODYNAMIC VALUES

Species	$\Delta H^0_{f298.15}$ (kJ/mol)	$S^0_{298.15}$ (J/mol·K)	$\Delta G^0_{f298.15}$ (kJ/mol)	Species	$\Delta H^0_{f298.15}$ (kJ/mol)	$S^0_{298.15}$ (J/mol·K)	$\Delta G^0_{f298.15}$ (kJ/mol)
Aluminum				**Cesium**			
Al(s)	0	28.3	0	Cs$^+$(aq)	−248	133	−282.0
AlCl$_3$(s)	−704.2	110.7	−628.9	CsF(aq)	−568.6	123	−558.5
Al$_2$O$_3$(s)	−1676	50.92	−1582	**Chlorine**			
Barium				Cl(g)	121.7	165.1	105.7
BaCl$_2$(s)	−860.1	126	−810.9	Cl$^-$(g)	−226	—	—
BaSO$_4$(s)	−1465	132	−1353	Cl$_2$(g)	0	223.0	0
Beryllium				HCl(g)	−92.31	186.8	−95.30
Be(s)	0	9.54	0	HCl(aq)	−167.4	55.10	−131.2
Be(OH)$_2$(s)	−907.1	—	—	**Chromium**			
Bromine				Cr(s)	0	23.8	0
Br(g)	111.8	174.9	82.4	(NH$_4$)$_2$Cr$_2$O$_7$(s)	−1807	—	—
Br$_2$(ℓ)	0	152.23	0	**Copper**			
Br$_2$(g)	30.91	245.4	3.14	Cu(s)	0	33.15	0
BrF$_3$(g)	−255.6	292.4	−229.5	CuO(s)	−157	42.63	−130
HBr(g)	−36.4	198.59	−53.43	**Fluorine**			
Calcium				F$^-$(g)	−322	—	—
Ca(s)	0	41.6	0	F$^-$(aq)	−332.6	—	−278.8
Ca(g)	192.6	154.8	158.9	F(g)	78.99	158.6	61.92
Ca^{2+}(g)	1920	—	—	F$_2$(g)	0	202.7	0
CaC$_2$(s)	−62.8	70.3	−67.8	HF(g)	−271	173.7	−273
CaCO$_3$(s)	−1207	92.9	−1129	HF(aq)	−320.8	—	−296.8
CaCl$_2$(s)	−795.0	114	−750.2	**Hydrogen**			
CaF$_2$(s)	−1215	68.87	−1162	H(g)	218.0	114.6	203.3
CaH$_2$(s)	−189	42	−150	H$_2$(g)	0	130.6	0
CaO(s)	−635.5	40	−604.2	H$_2$O(ℓ)	−285.8	69.91	−237.2
CaS(s)	−482.4	56.5	−477.4	H$_2$O(g)	−241.8	188.7	−228.6
Ca(OH)$_2$(s)	−986.6	76.1	−896.8	H$_2$O$_2$(ℓ)	−187.8	109.6	−120.4
Ca(OH)$_2$(aq)	−1002.8	76.15	−867.6	**Iodine**			
CaSO$_4$(s)	−1433	107	−1320	I(g)	106.6	180.66	70.16
Carbon				I$_2$(s)	0	116.1	0
C(s, graphite)	0	5.740	0	I$_2$(g)	62.44	260.6	19.36
C(s, diamond)	1.897	2.38	2.900	ICl(g)	17.78	247.4	−5.52
C(g)	716.7	158.0	671.3	HI(g)	26.5	206.5	1.72
CCl$_4$(ℓ)	−135.4	216.4	−65.27	**Iron**			
CCl$_4$(g)	−103	309.7	−60.63	Fe(s)	0	27.3	0
CHCl$_3$(ℓ)	−134.5	202	−73.72	FeO(s)	−272	—	—
CHCl$_3$(g)	−103.1	295.6	−70.37	Fe$_2$O$_3$(s, hematite)	−824.2	87.40	−742.2
CH$_4$(g)	−74.81	186.2	−50.75	Fe$_3$O$_4$(s, magnetite)	−1118	146	−1015
C$_2$H$_2$(g)	226.7	200.8	209.2	FeS$_2$(s)	−177.5	122.2	−166.7
C$_2$H$_4$(g)	52.26	219.5	68.12	Fe(CO)$_5$(ℓ)	−774.0	338	−705.4
C$_2$H$_6$(g)	−84.86	229.5	−32.9	Fe(CO)$_5$(g)	−733.8	445.2	−697.3
C$_3$H$_8$(g)	−103.8	269.9	−23.49	**Lead**			
C$_6$H$_6$(ℓ)	49.03	172.8	124.5	Pb(s)	0	64.81	0
C$_8$H$_{18}$(ℓ)	−268.8	—	—	PbCl$_2$(s)	−359.4	136	−314.1
C$_2$H$_5$OH(ℓ)	−277.7	161	−174.9	PbO(s, yellow)	−217.3	68.70	−187.9
C$_2$H$_5$OH(g)	−235.1	282.6	−168.6	Pb(OH)$_2$(s)	−515.9	88	−420.9
CO(g)	−110.5	197.6	−137.2	PbS(s)	−100.4	91.2	−98.7
CO$_2$(g)	−393.5	213.6	−394.4				
CS$_2$(g)	117.4	237.7	67.15				
COCl$_2$(g)	−223.0	289.2	−210.5				

SELECTED THERMODYNAMIC VALUES (continued)

Species	$\Delta H^0_{f298.15}$ (kJ/mol)	$S^0_{298.15}$ (J/mol·K)	$\Delta G^0_{f298.15}$ (kJ/mol)	Species	$\Delta H^0_{f298.15}$ (kJ/mol)	$S^0_{298.15}$ (J/mol·K)	$\Delta G^0_{f298.15}$ (kJ/mol)
Lithium				**Rubidium**			
Li(s)	0	28.0	0	Rb(s)	0	76.78	0
LiOH(s)	−487.23	50	−443.9	RbOH(aq)	−481.16	110.75	−441.24
LiOH(aq)	−508.4	4	−451.1	**Silicon**			
Magnesium				Si(s)	0	18.8	0
Mg(s)	0	32.5	0	$SiBr_4(\ell)$	−457.3	277.8	−443.9
$MgCl_2(s)$	−641.8	89.5	−592.3	SiC(s)	−65.3	16.6	−62.8
MgO(s)	−601.8	27	−569.6	$SiCl_4(g)$	−657.0	330.6	−617.0
$Mg(OH)_2(s)$	−924.7	63.14	−833.7	$SiH_4(g)$	34	204.5	56.9
MgS(s)	−347	—	—	$SiF_4(g)$	−1615	282.4	−1573
Mercury				$SiI_4(g)$	−132	—	—
$Hg(\ell)$	0	76.02	0	$SiO_2(s)$	−910.9	41.84	−856.7
$HgCl_2(s)$	−224	146	−179	$H_2SiO_3(s)$	−1189	134	−1092
HgO(s, red)	−90.83	70.29	−58.56	$Na_2SiO_3(s)$	−1079	—	—
HgS(s, red)	−58.2	82.4	−50.6	$H_2SiF_6(aq)$	−2331	—	—
Nickel				**Silver**			
Ni(s)	0	30.1	0	Ag(s)	0	42.55	0
$Ni(CO)_4(g)$	−602.9	410.4	−587.3	**Sodium**			
NiO(s)	−244	38.6	−216	Na(s)	0	51.0	0
Nitrogen				Na(g)	108.7	153.6	78.11
$N_2(g)$	0	191.5	0	$Na^+(g)$	601	—	—
N(g)	472.704	153.19	455.579	NaBr(s)	−359.9	—	—
$NH_3(g)$	−46.11	192.3	−16.5	NaCl(s)	−411.0	72.38	−384
$N_2H_4(\ell)$	50.63	121.2	149.2	NaCl(aq)	−407.1	115.5	−393.0
$(NH_4)_3AsO_4(aq)$	−1268	—	—	$Na_2CO_3(s)$	−1131	136	−1048
$NH_4Cl(s)$	−314.4	94.6	−201.5	NaOH(s)	−426.7	—	—
$NH_4Cl(aq)$	−300.2	—	—	NaOH(aq)	−469.6	49.8	−419.2
$NH_4I(s)$	−201.4	117	−113	**Sulfur**			
$NH_4NO_3(s)$	−365.6	151.1	−184.0	S(s, rhombic)	0	31.8	0
NO(g)	90.25	210.7	86.57	S(g)	278.8	167.8	238.3
$NO_2(g)$	33.2	240.0	51.30	$S_2Cl_2(g)$	−18	331	−31.8
$N_2O(g)$	82.05	219.7	104.2	$SF_6(g)$	−1209	291.7	−1105
$N_2O_4(g)$	9.16	304.2	97.82	$H_2S(g)$	−20.6	205.7	−33.6
$N_2O_5(g)$	11	356	115	$SO_2(g)$	−296.8	248.1	−300.2
$N_2O_5(s)$	−43.1	178	114	$SO_3(g)$	−395.6	256.6	−371.1
NOCl(g)	52.59	264	66.36	$SOCl_2(\ell)$	−206	—	—
$HNO_3(\ell)$	−174.1	155.6	−80.79	$SO_2Cl_2(\ell)$	−389	—	—
$HNO_3(g)$	−135.1	266.2	−74.77	$H_2SO_4(\ell)$	−814.0	156.9	−690.1
$HNO_3(aq)$	−206.6	146	−110.5	$H_2SO_4(aq)$	−907.5	17	−742.0
Oxygen				**Tin**			
O(g)	249.2	161.0	231.8	Sn(s, white)	0	51.55	0
$O_2(g)$	0	205.0	0	Sn(s, grey)	−2.09	44.1	0.13
$O_3(g)$	143	238.8	163	$SnCl_2(s)$	−350	—	—
$OF_2(g)$	23	246.6	41	$SnCl_4(\ell)$	−511.3	258.6	−440.2
Phosphorus				$SnCl_4(g)$	−471.5	366	−432.2
P(g)	314.6	163.1	278.3	$SnO_2(g)$	−580.7	52.3	−519.7
$P_4(s, white)$	0	177	0	**Titanium**			
$P_4(s, red)$	−73.6	91.2	−48.5	$TiCl_4(\ell)$	−804.2	252.3	−737.2
$PCl_3(g)$	−306.4	311.7	−286.3	$TiCl_4(g)$	−763.2	354.8	−726.8
$PCl_5(g)$	−398.9	353	−324.6	**Tungsten**			
$PH_3(g)$	5.4	210.1	13	W(s)	0	32.6	0
$P_4O_{10}(s)$	−2984	228.9	−2698	$WO_3(s)$	−842.9	75.90	−764.1
$H_3PO_4(s)$	−1281	110.5	−1119	**Zinc**			
Potassium				ZnO(s)	−348.3	43.64	−318.3
K(s)	0	63.6	0	ZnS(s)	−205.6	57.7	−201.3
KCl(s)	−436.5	82.6	−408.8				
$KClO_3(s)$	−391.2	143.1	−289.9				
KI(s)	−327.9	106.4	−323.0				
KOH(s)	−424.7	78.91	−378.9				
KOH(aq)	−481.2	92.0	−439.6				

ANSWERS TO EVEN-NUMBERED NUMERICAL EXERCISES

When calculations are done by different methods, roundoff errors may give slightly different answers. These differences are larger in calculations with several steps. Usually there is no cause for concern when your answers differ *slightly* from the answers given here.

CHAPTER 1

18. (a) 52600 (b) 0.00000410 (c) 1600. (d) 0.08206
    (e) 9346 (f) 0.009346
22. 1.8390×10^4 pounds
24. (a) 307 in (b) 152 in^3 (c) 6048 m (d) 3.9×10^4 mi/day
26. (a) 1.63×10^{-2} km (b) 1.63×10^4 m (c) 2.47×10^5 g
    (d) 4.32×10^3 mL (e) 8.59 L (f) 7.654×10^6 cm^3
28. (a) 1.5×10^2 km/hr (b) 1.4×10^2 ft/s
30. (a) 24.0 cg (b) 250 cm (c) 0.8 nm is equivalent to 8 Å
    (d) 6.4 m^3
32. (a) 90.9 tons ore (b) 90.9 kg ore
34. 6.3×10^{-3} kg salt
36. 30.63 cents/L
38. (a) 0.045 cm^3/drop; 45 μL/drop (b) 4.4 mm
40. (a) 719 lb (b) 148 kg (c) 9.24×10^5 cg
42. 4.85×10^7 atoms
44. 2.33 g/mL
46. 1.65 g/cm^3
48. 1.2×10^{14} g/cm^3
50. (a) 21.982 cm^3 (b) 7.644 g (c) 21.134 g (d) 21.134 cm^3
    (e) 0.848 cm^3 (f) 9.014 g/cm^3
52. (a) 72.9 cm^3 (b) 4.18 cm (c) 1.65 in
56. (a) -18°C (b) 310.2 K (c) 77°F (d) 74.1°F
58. (c) 285.3°R
60. For Al: 660.4°C; 1221°F   For Ag: 961.9°C; 1763°F
62. 342 J
64. (a) 1.16×10^6 J (b) 12.5°C
68. (a) 7.00×10^3 g (b) 66.8 cm^3
70. 120 mg KCN

72. 9.4542×10^{12} km/yr; 5.8746×10^{12} miles/yr
74. 8.3×10^2 g

CHAPTER 2

16. 3.043
18. 49.3 g; 592 g; 2.48×10^{24} g
20. (a) 40.078 amu; 40.078 g (b) Br; 79.904 g
    (c) 30.9738 amu; 30.9738 g (d) Cr; 51.9961 amu
22. 4.178 mol Ni
24. (a) 253.809 amu (b) 18.0152 amu (c) 183.187 amu
    (d) 261.9676 amu
26. (a) 0.988 mol NH$_3$ (b) 33.2 mol NH$_4$Br
    (c) 0.027 mol PCl$_5$ (d) 1.066 mol Sn
28. 2.516×10^{24} Ni atoms
30. 9.746×10^{-23} g Ni
32. 1.0×10^{16} Ni atoms
34. 2.66×10^{-16} g CH$_4$
36. (a) 3.01×10^{23} molecules CO (b) 3.01×10^{23} molecules
    N$_2$ (c) 6.80×10^{22} molecules P$_4$ (d) 1.36×10^{23}
    molecules P$_2$
38. 8.19×10^{24} H atoms
40. 23.9 mmol H$_2$SO$_4$
44. (a) 74.04% C; 8.699% H; 17.27% N
    (b) 72.21% C; 7.071% H; 4.679% N; 16.03% O
    (c) 63.15% C; 5.300% H; 31.55% O
52. 94.35% C; 5.61% H
56. (b) 131.19 g

60. (a) 3.43 g O (b) 6.86 g O
62. (a) 1.50 g O (b) 2.25 g O
64. 194.0 g Hg
66. 30.2 g $KMnO_4$
68. 75.9 tons Cu_2S
70. 225 g NaCl
72. (a) 395 g $CuSO_4 \cdot H_2O$ (b) 355 g $CuSO_4$
74. (a) 71.40% (b) 51.7%
76. (a) 186 lb $MgCO_3$ (b) 485 lb impurity (c) 53.6 lb Mg
78. (a) 6.00 mol CO_2 (b) 2.00 mol CO_2 (c) 8.00 mol CO_2
80. (a) 1.96 mol O_3 (b) 5.88 mol O (c) 94.0 g O_2
    (d) 62.7 g O_2
82. (a) 1.11 mol Ag (b) 1.11 mol Ag (c) 4.48×10^{-3} mol Ag
    (d) 1.84×10^{-3} mol Ag
88. 15.6 g $CuSO_4 \cdot 5H_2O$
90. 0.300 g Cr_2O_3
96. 406 mL ethanol

CHAPTER 3

8. (b) 600 molecules H_2 (c) 400 molecules N_2
10. (b) 18 mol HCl   (c) 8.8 mol H_2O
12. (a) 15.0 mol O_2 (b) 5.00 mol O_2 (c) 5.00 mol O_2
    (d) 5.00 mol O_2 (e) 20.0 mol O_2
14. (a) 12.5 mol O_2 (b) 10.0 mol NO (c) 15.0 mol H_2O
16. 167 g Cl_2
18. 36.14 g Fe_3O_4
20. 128 g O_2
22. 0.814 g Br_2
24. 40.3 g H_2O
26. 2.68×10^{22} molecules C_3H_8
28. 97.0 g NH_3
30. 735 g superphosphate
32. 78.67 g K
34. 71.6 g PCl_5
36. 96.5%
38. 1.23×10^3 g C_2H_5OBr
40. (a) 4.0×10^5 g or 400 kg CaO (b) 3.8×10^5 g or 380 kg
42. 80.2 g H_2TeO_3
44. 339 g $KClO_3$
46. 7.94 g CH_3CH_3
48. 134 kg Zn
50. (a) 3.21×10^{-3} mol Na_2S (b) 0.250 g Na_2S
    (c) 99.750 g H_2O
52. 84.2 g $(NH_4)_2SO_4$
54. 455 mL soln
56. 2.21 M H_3PO_4
58. 1.34 M H_3PO_4
60. (a) 0.250 M $(CH_3)_2CHOH$ (b) 5.00×10^{-4} mol
    $(CH_3)_2CHOH$
62. (a) 16 g Na_3PO_4
64. (a) 20.0% $CaCl_2$ (b) 2.13 M $CaCl_2$
66. 28.7 M HF

68. 0.0376 M $BaCl_2$
70. 0.675 L conc. HCl soln
72. 270 mL of 0.0600 M $Ba(OH)_2$ soln
74. 4.12 M H_2SO_4
76. 0.0115 L KOH soln
78. 0.0134 L HNO_3 soln
80. 0.00756 M $AlCl_3$ soln
82. 2.74 g AgCl
86. 0.255 L $HClO_4$ soln
88. 0.0354 M NaOH soln
90. 0.118 g $CaCO_3$
92. 76.4 g products; 24.0 g CS_2
94. 5.78% Fe_3O_4 in ore
96. 81.6%
98. Zn: \$65.4/mol H_2; Al: \$36.0/mol H_2
100. 753 g H_3PO_4

CHAPTER 4

96. (a) 0.122 mol O_2 (b) 0.147 mol O_2 (c) 0.0231 mol O_2
98. 14.9 g Zn

CHAPTER 5

6. 1.63×10^{-19} coulombs
8. 2.3×10^{-14}
12. (a) 0.021816% (b) 40.059% (c) 59.918%
24. 7.5%
28. 73.3 amu
30. 52.0 amu
32. 72.64 amu
34. (a) 3.07×10^{14} s^{-1} (b) 6.10×10^{14} s^{-1}
    (c) 6.10×10^9 s^{-1} (d) 6.10×10^{18} s^{-1}
36. (a) 4.47×10^{14} s^{-1} (b) 2.96×10^{-19} J/photon
38. 5.85×10^{-19} J/photon; 3.52×10^5 J/mol or 352 kJ/mol
40. 4.02×10^{-19} J/photon; 25 photons
42. 7.9×10^{26} photons
46. 320 nm (violet/near ultraviolet)
50. (d) 3.20×10^{15} s^{-1}
54. 180 kJ/mol
56. 2.53×10^{18} photons
58. (a) 1.59×10^{-14} m (b) 3.97×10^{-32} m
60. 1.88×10^3 m/s
98. 3.35 m
100. 216 kJ/mol

CHAPTER 6

46. Li^+, 1.8×10^{-19}; Mg^{2+}, 2.7×10^{-19}; Be^{2+}, 2.7×10^{-18};
    Al^{3+}, 9.2×10^{-19}

48. F_2, 0.71 Å; Cl, 0.99 Å; Cl—F, 1.70 Å
84. $1.05 \times 10^{15}\ s^{-1}$
86. 90.06 kJ/g

CHAPTERS 7, 8, 9 AND 10

None

CHAPTER 11

4. 68.2 g $(NH_4)_2SO_4$
6. 5.292 M
8. 1.44 M NaCl
10. 0.828 M Na_3PO_4; 0.37 M NaOH
12. 35.3 mL CH_3COOH soln
14. 0.857 M
16. 2.00 L NaOH soln; 0.333 L H_3PO_4 soln
22. 0.192 M HNO_3
24. 0.1068 M NaOH
28. 23.9% $(COOH)_2 \cdot 2H_2O$
30. 0.114 g $CaCO_3$
32. 1.17 N H_3PO_4
34. 0.276 M H_3AsO_4; 0.828 N H_3AsO_4
36. 0.0732 M $Ba(OH)_2$
38. 0.139 N HCl; 0.139 M HCl
52. 2.67 mL $KMnO_4$ soln
54. 5.00 mL $KMnO_4$ soln
56. (a) 0.1104 M I_2 (b) 0.2765 g As_2O_3
58. (a) 0.0100 L HI soln   (b) 0.118 L HI soln   (c) 0.0311 L HI soln
60. 0.159 M $KMnO_4$
62. 0.1160 M HCl
64. 2.80 mmol HCl; 13.7 mL HCl
66. 379 mL HCl soln
68. 29.3 mL H_2SO_4 soln
70. 19.20 mL NaOH soln
72. 59.36% Fe
74. 80 mmol HCl in stomach; 9.28 mmol HCl
80. 1.2 g $NaHCO_3$

CHAPTER 12

10. (a) 14.6 psi   (b) 75.5 cm Hg   (c) 29.7 in Hg   (d) 101 kPa
    (e) 0.993 atm   (f) 33.7 ft H_2O
12. (a) 854 torr; 1.12 atm   (b) 804 torr; 1.06 atm
14. 2.20×10^3 psi
18. 0.50 atm
20. (a) 8.0 atm   (b) 0.12 L
22. 2200 balloons
28. −173.8°C

30. (a) 1.441 L   (b) 41 cm
32. 3.26 L; 1.30 L; 0.0715 L
36. 3.83 atm
38. 626 K or 353°C
42. 3×10^{14} molecules CO
44. 5.07 g/L
50. 18.5 atm
52. (a) 19.1 K or −254.0°C   (b) 32.1 g Ne/L
54. 1.80×10^8 L Cl_2; 6.34×10^6 ft^3 Cl_2; 40 ft
56. 29.6 g/mol; 2% error
58. 44.0 g/mol
60. 46.7 g/mol
64. 149 atm; 44.5 atm
66. He: 0.310; Ar: 0.368; Xe: 0.322
68. (a) 11.25 atm   (b) 3.75 atm   (c) 3.75 atm
70. 513 mL
72. (a) He: 4.00 atm; N_2: 1.00 atm   (b) 5.00 atm total
    (c) 0.800
74. (a) 3.4×10^{-23} mL   (b) 20. mL   (c) ≈0.089%
78. 1.17
84. 0.695; 43.2 atm
86. (a) 0.821 atm   (b) 0.805 atm
88. 74.8 g NaN_3
90. 0.897 g SO_2
92. 4.33 g $KClO_3$
94. 2.24 g NH_3
96. 190. g KNO_3
98. 25.1% S by mass
100. (a) 2.83 L C_8H_{18} (b) 44.8 min
102. 4000 L air
104. 121 g/mol
106. (a) 6130 g H_2O (b) 6150 mL H_2O
108. 5.10 atm; 2%
116. $P_{CO_2} = 0.294$ atm; $P_{H_2O} = 0.353$ atm
118. 206 g $MgSiO_3$

CHAPTER 13

30. 344.4 torr
32. (a) 43.4 kJ/mol
36. H_2O: 361.8 torr; D_2O: 338.1 torr
38. (c) 3.71×10^4 J/mol   (d) 3.48×10^2 K or 75°C
42. 6.38×10^3 J/mol
44. 2.18×10^5 J
46. 2.33×10^5 J
48. 55.7°C
50. 58.8°C
52. (a) 6.58×10^3 J   (b) 39.9 g
80. 1.00 g/cm^3
82. 206 g/mol; Pb (207.2 g/mol)
84. (d) 1.545 Å   (e) 3.517 g/cm^3
88. 1.542 Å
96. 0.988 atm; 0.982 atm; 0.6%

100. 6.11×10^{-22} g
102. 54.7 min
104. **(a)** Pb(s) = 18.36 cm^3; Pb(ℓ) = 19.87 cm^3; Pb(g) = 1.867 $\times$ 10^5 cm^3 **(b)** 13.5 cm^3 **(c)** Pb(s) = 0.735; Pb(ℓ) = 0.679; Pb(g) = 7.23 $\times$ 10^{-5}

CHAPTER 14

20. at 25°C: 2.4×10^{-4}; at 50°C: 1.7×10^{-4}; decreases
30. 31.5 g NaCl; 179 g H$_2$O
32. **(a)** 20% **(b)** 0.017 **(c)** 0.97 m K$_2$ZrF$_6$
34. 2.67 m C$_6$H$_5$COOH in C$_2$H$_5$OH
36. **(a)** 1.75 m C$_6$H$_{12}$O$_6$
38. C$_2$H$_5$OH: 0.457; H$_2$O: 0.543
40. 0.7763 M K$_2$SO$_4$; 0.8197 m K$_2$SO$_4$; 12.50% K$_2$SO$_4$; 0.9854
44. **(a)** 9.40 torr **(b)** 65.2 torr
46. acetone: 157 torr; chloroform: 161 torr
48. 318 torr; acetone: 0.494; chloroform: 0.506
50. **(a)** chloroform: 55 torr **(b)** acetone: 210 torr **(c)** total: 280 torr
52. 101.10°C
54. −4.00°C
58. 1044°C
60. 32.5 g C$_{10}$H$_8$
62. 120.10 g/mol
64. **(a)** C$_{10}$H$_8$: 70%; C$_{14}$H$_{10}$: 30% **(b)** 80.4°C
70. 0.010 m CaCl$_2$
72. 751 torr
74. 1% ionized
76. i = 1.79; 79% ionized
80. 0.317 atm
82. ΔT_f: 0.0971°C; ΔT_b: 0.0267°C
84. 7.1 atm
86. **(a)** -1.02×10^{-4}°C **(b)** 1.02 torr **(c)** 1000% **(d)** 10% error
94. −843.6 kJ/mol
96. **(a)** 60.5 g/mol **(b)** 117 g/mol
100. 56% lactose by mass
104. 193 g/mol
106. 1.4 atm; 0.062 M

CHAPTER 15

10. **(a)** 2140 kJ evolved **(b)** 52.0 g O$_2$
12. +107 kJ/mol rxn
22. −583 kJ/mol
24. −988.4 kJ/mol rxn
26. +197.8 kJ/mol rxn
28. −293 kJ/mol rxn
30. **(a)** −36.0 kJ/mol rxn **(b)** −1656 kJ/mol rxn **(c)** +624.6 kJ/mol rxn

32. 231 kJ
34. −65.3 kJ/mol SiC(s)
38. **(a)** −96 kJ/mol rxn **(b)** −111 kJ/mol rxn
40. −302 kJ/mol rxn
42. 192 kJ
44. 327.0 kJ
46. 294 kJ/mol
50. 381 J/°C
52. **(a)** 2.1×10^3 J **(b)** -1.0×10^5 J/mol Pb(NO$_3$)$_2$
54. **(b)** −41.8 kJ/g C$_6$H$_6$(ℓ); −3260 kJ/mol C$_6$H$_6$(ℓ)
60. −366 J
64. **(a)** +983 J **(b)** 0 J
76. **(a)** −128.8 J/(mol rxn)·K **(b)** −182 J/(mol rxn)·K **(c)** −24.9 J/(mol rxn)·K
80. −4.78 kJ
82. −51.89 kJ/mol rxn
84. **(a)** ΔH^0: −71.75 kJ/mol rxn; ΔS^0: −268.0 J/(mol rxn)·K; ΔG^0: 8.15 kJ/mol rxn
(b) ΔH^0: 14.7 kJ/mol rxn; ΔS^0: 31.2 J/(mol rxn)·K; ΔG^0: 5.4 kJ/mol rxn
(c) ΔH^0: −367.6 kJ/mol rxn; ΔS^0: −11.62 J/(mol rxn)·K; ΔG^0: −364.1 kJ/mol rxn
88. −106.66 kJ
90. **(a)** ΔH^0: −196.0 kJ/mol rxn; ΔG^0: −233.6 kJ/mol rxn; ΔS^0: +125.6 J/(mol rxn)·K
92. 97°C
98. **(a)** −156 kJ/mol C$_6$H$_{12}$(ℓ) **(b)** −165 kJ/mol C$_6$H$_5$OH(s)
102. q = +8540 J; w = −631 J; ΔE = 7910 J
104. **(a)** 2.88 kJ/°C **(b)** 23.82°C
108. 2.4 hr spent walking
110. 0.128 J/g·°C; 26.5 J/mol·°C
112. **(a)** 30.0 kJ/g **(b)** 7.17 kilocalorie/g **(c)** 35.9 kilocalorie

CHAPTER 16

10. O$_2$: 1.50 M/min; NO: 1.20 M/min; H$_2$O: 1.80 M/min
14. Rate = $(2.5 \times 10^{-2}\ M^{-1} \cdot \text{min}^{-1})$[B][C]
16. **(a)** rate of reaction = $(19\ M^{-2} \cdot \text{s}^{-1})$[NO]^2[O$_2$]
(b) 2.0×10^{-3} M/s
(c) NO: 4.0×10^{-3} M/s; O$_2$: 2.0×10^{-3} M/s; NO$_2$: 4.0×10^{-3} M/s
18. 8
20. **(a)** Rate = $(1.2 \times 10^2\ M^{-2} \cdot \text{s}^{-1})$[ClO$_2$]^2[OH$^-$]
22. Rate = $(2.5 \times 10^{-3}\ M^{-2} \cdot \text{s}^{-1})$[A][B]^2
24. Expt 2: 0.300 M/s; Expt 3: 1.20 M/s
28. 1.76×10^{-1} min or 10.6 s
30. 4.48×10^{-3} mol CH$_3$N=NCH$_3$; 0.0299 mol N$_2$ **(b)** 5.7×10^{-4} g CH$_3$N=NCH$_3$
32. **(a)** 76 yrs **(b)** 54 g NO$_2$ **(c)** 0.91 M
34. 28.0 min
36. 0.023 min^{-1}
38. **(b)** 2.18 mmoles/L
40. **(b)** 8.4 s

46. 83.8 kJ/mol rxn
48. 103 kJ/mol rxn
50. (a) 270. kJ/mol   (b) $7.7 \times 10^{-14}\,s^{-1}$ (c) 730 K
52. 160 kg CO_2/L·hr
54. (a) $0.23\,s^{-1}$ (b) 9°C
64. 64 kJ/mol
66. Rate = $(6.20 \times 10^{-4}\,s^{-1})[N_2O_5]$
68. (a) 3.7×10^{-6} mol N_2O_5 remaining after 1.00 min   (b) 20 s
70. (b) 0.280 L/μmol·s   (c) 0.252 μmol/L·s
76. (a) $-200.$ kJ/mol rxn

CHAPTER 17

12. 1.1×10^{-5}
14. 0.13
16. (a) 8.9×10^5 (b) 1.3×10^{-12} (c) 1.6×10^{-24}
20. (a) 7.4×10^{-5} mol H_2; 1.8×10^{-3} mol HI   (b) 93
22. 0.12
30. 0.024 M HCN
32. 2.50×10^{-2}
34. 6.8 M Cl_2
36. $[SbCl_5] = 6 \times 10^{-4}\,M$; $[SbCl_3] = 5.6 \times 10^{-3}\,M$; $[Cl_2] = 2.8 \times 10^{-3}\,M$
40. 68.7%
52. (a) $K_c = 16$ (b) 0.35 M A
54. (a) $K_c = 0.13$ (b) 0.13 M A and 0.12 M B and C
    (c) 0.70 M A and 0.30 M B and C
56. (a) $[N_2O_4] = 0.0144\,M$; $[NO_2] = 9.18 \times 10^{-3}\,M$
    (b) $[N_2O_4] = 6.4 \times 10^{-3}\,M$; $[NO_2] = 6.12 \times 10^{-3}\,M$
    (c) $[N_2O_4] = 0.031\,M$; $[NO_2] = 0.0135\,M$
60. 1.6×10^{-9}
62. 0.771
64. 7.76
66. 0.108
68. (a) 33.2%
72. $P_{CO} = 0.685$ atm; $P_{CO_2} = 0.315$ atm
76. 6.9×10^{24}
78. 0.18
80. (a) 1.1×10^5 (b) 2.2×10^2 (c) -28.6 kJ/mol rxn; $K_p = 1.0 \times 10^5$
82. (a) -37.9 kJ/mol rxn at 25°C   (b) 4.9×10^{-4} (c) $+68.0$ kJ/mol rxn at 800°C
84. 2.72×10^{22}
86. (a) 7.9×10^{11} (b) 8.9×10^5 (c) 1.3×10^{-12}
    (d) 1.6×10^{-24}
88. 4.32×10^{-3}
90. $\Delta G^0 = -142$ kJ/mol rxn $\Delta G^0 = -371$ kJ/mol
92. 0.13 atm

CHAPTER 18

2. (a) 2.40 M NaCl   (b) 0.975 M H_2SO_4 (c) $9.6 \times 10^{-4}\,M$ C_6H_5OH

4. (a) 0.25 M   (b) 0.055 M   (c) $[Ca^{2+}] = 0.0020\,M$; $[Cl^-] = 0.0040\,M$
6. (a) $[K^+] = 0.030\,M$ and $[OH^-] = 0.030\,M$ (b) $[Ba^{2+}] = 0.017\,M$; $[OH^-] = 0.034\,M$ (c) $[Ca^{2+}] = 0.161\,M$; $[NO_3^-] = 0.322\,M$
14. $1.58 \times 10^{-9}\,M$
16. (a) -4.28 (b) 0.76   (c) -11.24 (d) -6.31
20. (a) 0.301   (b) 1.52   (c) 2.13
22. 10.40
24. $3.0 \times 10^{-4}\,M$
30. (a) 6.812
34. pH = 3.05; $K_a = 9.6 \times 10^{-6}$
36. $K_a = 1.6 \times 10^{-5}$
38. $[C_6H_5COOH] = 0.35\,M$; $[H_3O^+] = [C_6H_5COO^-] = 4.7 \times 10^{-3}\,M$; $[OH^-] = 2.1 \times 10^{-12}\,M$
40. 5.8%
42. benzoic acid: 4.20; hydrocyanic acid: 9.40
46. 8.82
48. 5.0×10^{-4}
50. (a) $[OH^-] = 1.3 \times 10^{-3}\,M$; % ionization = 1.3%; pH = 11.11
    (b) $[OH^-] = 6.8 \times 10^{-3}\,M$; % ionization = 6.8%; pH = 11.83
52. (a) $2.5 \times 10^{-3}\,M$ (b) $1.0 \times 10^{-3}\,M$
56. 8.2
58. $4.5 \times 10^{-4}\,M$ $HCOO^-$
60. (a) 3.36   (b) 4.74
62. (a) pH = 9.43   (b) pH = 8.83
68. pH = 1.00
70. The pH decreases by 0.43 units.
72. (a) 9.34   (b) 9.26   (c) 1.00
74. $[CH_2BrCOOH] = 5.7 \times 10^{-2}\,M$; $[NaCH_2BrCOO] = 0.14\,M$
76. 0.0020 M
78. 0.19 M
80. $V_B = 0.34$ L; $V_A = 0.66$ L
82. (a) 0.625 M   (b) 0.250 M   (c) 0.500 M   (d) $2.3 \times 10^{-5}\,M$
    (e) 4.65
84. $1.1 \times 10^{-2}\,M$
86.

0.200 M H_3AsO_4 Solution		0.100 M H_3PO_4 Solution	
Species	Concentration (M)	Species	Concentration (M)
H_3AsO_4	0.193	H_3PO_4	0.076
H_3O^+	0.0071	H_3O^+	0.024
$H_2AsO_4^-$	0.0071	$H_2PO_4^-$	0.024
$HAsO_4^{2-}$	5.6×10^{-8}	HPO_4^{2-}	6.2×10^{-8}
OH^-	1.4×10^{-12}	OH^-	4.2×10^{-13}
AsO_4^{3-}	2.4×10^{-18}	PO_4^{3-}	9.3×10^{-19}

88. $[H_3O^+] = 0.11\,M$; $[HSeO_4^-] = 0.09\,M$; $[OH^-] = 9.11 \times 10^{-14}\,M$; $[SeO_4^{2-}] = 0.01\,M$
90. (a) 1.07   (b) $6.4 \times 10^{-5}\,M$

92. 4.19
94. 2.07
96. $[H_3O^+] = 2.8 \times 10^{-3}$ M; pH = 2.55

CHAPTER 19

12. 5.3×10^{-10}
14. (a) 2.2×10^{-11} (b) 2.9×10^{-7} (c) 5.6×10^{-11}
16. (a) $1.2 \times 10^{-3}\%$ (b) 0.14% (c) $1.9 \times 10^{-3}\%$
18. pH = 8.16
22. (a) 5.04 (b) 5.77 (c) 2.72
30. (a) pH = 2.89; 0.87% (b) pH = 5.21; $8.2 \times 10^{-3}\%$
(c) pH = 6.17; $4.5 \times 10^{-4}\%$
34. (a) 0.903 (b) 1.06 (c) 1.300 (d) 1.900 (e) pH = 7
(f) 12.00
38. (a) pH = 8.93 (b) pH = 8.76 (c) pH = 8.36
50. pH = 5.96

CHAPTER 20

8. (a) 3.5×10^{-5} (b) 7.7×10^{-19} (c) 6.9×10^{-15}
(d) 1.00×10^{-4}
12. (a) 2.3×10^{-6} mol CuI/L; 4.4×10^{-4} g/L
(b) 6.5×10^{-7} mol $Ba_3(PO_4)_2$/L; 3.9×10^{-4} g/L
(c) 2.1×10^{-3} mol PbF_2/L; 0.51 g/L
(d) 7.7×10^{-10} mol $Pb_3(PO_4)_2$/L; 6.2×10^{-7} g/L
16. 5.3×10^{-3} mol Ag_2SO_4/L 0.15 M K_2SO_4 soln
18. (a) 6.7×10^{-8}mol $Mg(OH)_2$/L 0.015 M NaOH soln
(b) 1.6×10^{-5} mol $Mg(OH)_2$/L 0.015 M $MgCl_2$ soln
26. (a)(i) 2.0×10^{-11} M (ii) 1.0×10^{-8} M (iii) 7.3×10^{-19} M
(b)(i) 1.7×10^{-7} M (ii) 1.0×10^{-8} M (iii) 7.3×10^{-19} M
28. 0.013%
32. (b) 1.1×10^{-4} M; 99.89% (c) $[Au^+] = 1.1 \times 10^{-7}$ M;
$[Ag^+] = 9.5 \times 10^{-5}$ M
34. (b) 3.6×10^{-13} M (c) 3.6×10^{-7} M (d) 0.050 M;
5.0×10^{-8} M
36. 7.61
38. 2.5×10^{-7} mol CaF_2/L
42. 4.30
44. (a) 9.65 (b) 2.0×10^{-4} g/100 mL
54. (b) 99.955%
56. 0.6 g/L
58. 29%

CHAPTER 21

16. 1.602×10^{-19} C/e^-
18. (ii)(a) 4.91×10^3 C (b) 962 C (c) 481 C
20. 2.9 L Cl_2

22. 0.693 C
24. 3.35 g Ag
26. (a) 0.0954 faradays (b) 9.21×10^3 C (c) 1.07 L_{STP} H_2
(d) 13.281
28. 1.07×10^4 s; 2.97 h; 2.64 g Cu
32. (a) 0.0178 faradays (b) 1.92 g Ag (c) 4.31 g $Fe(NO_3)_3$
48. (a) +3.17 V (b) +1.54 V
54. +1.78 V
56. (a) +0.62 V (b) −1.202 V
62. −1.207 V
70. −0.233 V
72. (a) +2.123 V (b) +2.178 V
74. 1.2×10^{-5} atm
76. (a) 0.870 V (b) 0.148 V (c) 0.0527 V
78. 0.023 M
80. (a) 2×10^{17} (b) $[Ni^{2+}] = 1 \times 10^{-17}$ M; $[Zn^{2+}] = 2.00$ M
84. (a) −0.62 V; 120 kJ/mol rxn; 10^{-21}
(b) +0.368 V; −35.5 kJ/mol rxn; 1.6×10^6
(c) +1.833 V; −1061 kJ/mol rxn; 10^{186}
86. 7.7×10^{70}
88. 2
90. −171 kJ/mol rxn
98. (b) +0.727 V (c) +0.836 V (d) 0.0457 g Zn
100. 840 C
102. (f) 10.2 C/s (g) 0.0206 L HF(g)
108. (a) 10^{-12} (b) +68 kJ/mol rxn
110. (i) −1.662 V (ii) −1.662 V

CHAPTER 22

12. 39.997 g NaOH; 1.008 g H_2; 35.45 g Cl_2
24. (a) −15 kJ/mol rxn (b) +5 kJ/mol rxn
(c) −25 J/(mol rxn)·K
28. 17.8 g Cu
30. 4.09 years; 9.49×10^3 L O_2
34. 9.4×10^2 tons seawater

CHAPTER 23

28. (a) −201.4 kJ/mol rxn (b) −138.9 kJ/mol rxn
(c) −415.0 kJ/mol rxn
40. $\Delta H^0 = -195.4$ kJ/mol Rb(s); $\Delta S^0 = 29.4$ J/K per mol Rb(s);
$\Delta G^0 = -204.0$ kJ/mol Rb(s)
42. 345 g Co_3O_4

CHAPTER 24

8. 4.28 g XeF_6
42. 1.64 kg H_2SO_4
44. 2.8

70. 1.92 Å
72. -140 kJ/mol rxn
76. 2.2×10^{25} g Si

CHAPTER 25

46. (a) -657 kJ/mol (b) -194 kJ/mol (c) -253 kJ/mol
  (d) -962 kJ/mol (e) -95.7 kJ/mol (f) -314 kJ/mol
  (g) -93.3 kJ/mol (h) -172 kJ/mol
48. (a) $-3/5\ \Delta_{oct}$ (b) $-8/5\ \Delta_{oct}$
50. -153.5 J/(mol rxn)$\cdot$K
52. 10.40
54. (a) 2.2×10^{-6} mol $Zn(OH)_2$/L (b) ≈ 0.010 mol $Zn(OH)_2$/L
  (c) $0.010\ M$

CHAPTER 26

8. (a) 0.740 amu/atom or 0.740 g/mol (b) 6.66×10^{10} kJ/mol of ^{79}Br atoms
10. (a) 0.491 amu/atom (b) 0.491 g/mol (c) 7.34×10^{-11} J/atom (d) 4.42×10^{10} kJ/mol (e) 8.82 MeV/nucleon
12. (a) 1.56×10^{10} kJ/mol of ^{20}Ne atoms (b) 7.34×10^{10} kJ/mol of ^{87}Rb atoms (c) 8.82×10^{10} kJ/mol of ^{106}Pd atoms
46. 87.9 min; 155 min
48. 0.93 fs
50. 1540 yr
52. 1.94×10^4 yr
56. -1.68×10^9 kJ/mol rxn
64. (b) 1.15×10^8 kJ/mol rxn
66. 2.48×10^9 yr

CHAPTER 27

None

CHAPTER 28

34. $[C_6H_5NH_2] = 0.100\ M$: $[C_6H_5NH_3^+] = 6.5 \times 10^{-6}\ M$;
  $[OH^-] = 6.5 \times 10^{-6}\ M$; $[H_3O^+] = 1.5 \times 10^{-9}\ M$
80. 7.1 atm
82. 8.60
84. 7.7×10^3 g/mol

QUALITATIVE ANALYSIS

CHAPTER 29

12. (a) 353 cm (b) 438 cm

CHAPTER 30

None

CHAPTER 31

2. 0.024 mmol Bi/mL, 0.079 mmol Cu/mL, 0.045 mmol Cd/mL
6. (a) $1.3 \times 10^{-19}\ M$ (b) $1.3 \times 10^{-17}\ M$ (c) $1.3 \times 10^{-15}\ M$
8. $[Cu^{2+}] = 1.5 \times 10^{-16}\ M$; $[Pb^{2+}] = 1.4 \times 10^{-8}\ M$
40. $[Cu^{2+}] = 4.0 \times 10^{-15}\ M$; $[Bi^{3+}] = 1.2 \times 10^{-5}\ M$; $[Cd^{2+}] = 1.6 \times 10^{-8}\ M$

CHAPTER 32

2. 0.067 mmol As/mL, 0.041 mmol Sb/mL, 0.042 mmol Sn/mL

CHAPTER 33

2. $[Cr^{3+}] = 0.096\ M$, $[Mn^{2+}] = 0.091\ M$, $[Fe^{3+}] = 0.089\ M$,
  $[Co^{2+}] = 0.085\ M$, $[Ni^{2+}] = 0.085\ M$, $[Zn^{2+}] = 0.077\ M$,
  $[Al^{3+}] = 0.37\ M$
76. (a) pH = 7.98 (b) pH = 3.53

CHAPTER 34

26. (b) $K' = 2.4 \times 10^{-15}$ (c) $Cr_2O_7^{2-}$ (d) $Cr_2O_7^{2-}$ at pH = 2.00, $Cr_2O_4^{2-}$ at pH = 12.00

CHAPTER 35

None

CHAPTER 36

8. $1.4 \times 10^{-22}\ M$
12. $2.1 \times 10^{-7}\ M$; $Q_{sp} < K_{sp}$, so Ag_2SO_4 will not ppt.
14. (a) $5.9 \times 10^6\ M$ (impossible) (b) $(0.025 + 3.5 \times 10^{-5})\ M$ NaCN (c) $0.060\ M\ Na_2S_2O_3$
18. (a) HgS: $Q_{sp} = 5.8 \times 10^{-21} > 3.1 \times 10^{-53} = K_{sp}$, HgS ppts.
  (b) CdS: $Q_{sp} = 5.8 \times 10^{-21} > 3.6 \times 10^{-29} = K_{sp}$, CdS ppts.
  (c) ZnS: $Q_{sp} = 5.8 \times 10^{-21} > 1.1 \times 10^{-21} = K_{sp}$, ZnS ppts., but probably not enough to be seen.
  (d) MnS: $Q_{sp} = 5.8 \times 10^{-21} > 5.2 \times 10^{-15} = K_{sp}$, MnS does not ppt.
20. (a) $4.7 \times 10^{-33}\ M$ (b) $1.3 \times 10^{-7}\ M$ (c) $4.5 \times 10^{-6}\ M$
  (d) $[Hg^{2+}] = 9.7 \times 10^{-33}\ M$, $[Pb^{2+}] = 2.7 \times 10^{-7}\ M$

22. (a) $K_{b(1)} = 7.7 \times 10^{-2}$, $K_{b(2)} = 1.0 \times 10^{-7}$ (b) $K_{b(1)} = 2.1 \times 10^{-4}$, $K_{b(2)} = 2.4 \times 10^{-8}$ (c) $K_{b(1)} = 4.0 \times 10^{-11}$, $K_{b(2)} = 1.8 \times 10^{-7}$; $K_{b(3)} = 3.3 \times 10^{-2}$

24. (a) pH = 11.11, 0.13% hydrolysis (b) This solution is decidedly less basic, and the percentage hydrolysis is less because CO_3^{2-} ions hydrolyze to a lesser extent than S^{2-} ions.

26. (a) MnS, 7.1×10^{-8} M; NiS, 5.5×10^{-11} M (b) MnS, 4.4×10^{-5} M; NiS, 6.8×10^{-8} M (c) Each is about 10^3 times more soluble than the (incorrect) calculation in part (a) would indicate. (d) Within roundoff error range, the results obtained here agree with those in Example 36-12 (10^3 times greater).

28. $[Cd^{2+}] = 2.0 \times 10^{-6}$ M, $[Pd^{2+}] = 7.7 \times 10^{-10}$ M

30. (a) no (b) 0.44 M

32. $Q_{sp} > K_{sp}$, so ZnS would ppt.

Illustration and Table Credits

Chapter 1

Chapter opener: Milton Heiberg/Photo Researchers, Inc.; **1-2(a–c), 1-4(a–d), 1-6(a, b), 1-10, 1-14(a), unnum. figures pp. 15, 28:** Charles Steele; **unnum. figure p. 3:** from Petit Format/Nestle/Photo Researchers, Inc.; **1-1(a), 1-13, unnum. figure p. 11:** Charles D. Winters; **1-1(b, c), 1-12:** James Morgenthaler; **1-11:** Ohaus Corporation; **1-14(b):** from BioPhoto Associates, N.H.P.A.; **unnum. figure p. 30:** Lawrence Livermore National Laboratory.

Chapter 2

Chapter opener: © 1990 Peter Menzel; **2-4(parts 1 and 2), 2-6, 2-8, 2-9, 2-10, unnum. figures pp. 46, 64, 67, 68:** Charles Steele; **2-4(parts 3 and 4), unnum. figures pp. 49, 58:** Charles D. Winters; **2-5:** John Ozcomert, Michael Trenary; **unnum. figure p. 73:** Dennis Drenner.

Chapter 3

Chapter opener, unnum. figures pp. 90, 94(top), 95, 96, 101(2 parts), 104(3 parts), 108: Charles Steele; **3-2(a–c), 3-3(a–c), 3-4(a–d):** Charles D. Winters; **unnum. figure p. 81:** from Atlanta Gas Light Co.; **unnum. figure p. 94(bottom):** Tom McHugh/Photo Researchers, Inc.; **unnum. figure p. 98:** Dennis Drenner.

Chapter 4

Chapter opener: Yoav Levy/Phototake NYC; **4-2(a–c), unnum. figures pp. 123, 134, 142:** Charles D. Winters; **4-3(a, b), 4-4, 4-5, unnum. figures pp. 117(top left and right), 122, 136, 138, 141 (bottom), 150:** Charles Steele; **unnum. figures pp. 139(top and bottom), 141(top):** J. Morgenthaler; **unnum. figure p. 117(bottom):** Jeff Smith/The Image Bank; **unnum. figure p. 118:** AT&T Bell Laboratories; **unnum. figure p. 119:** from J. W. van Spronsen, *The Periodic System of Chemical Elements: A History of the First Hundred Years,* Elsevier, 1969; **unnum. figure p. 120:** Andrea Pistolesi/The Image Bank; **unnum. figure p. 129:** Kevin Schafer/Tom Stack & Associates; **Table 4-10:** from *Chemical & Engineering News,* 4 July 1994.

Chapter 5

Chapter opener: Steve Elmore/Tom Stack & Associates; **5-10(a):** photo by Dave Pierce, Finnigan MAT; **5-10(b):** Charles Steele; **unnum. figures pp. 158, 176:** AIP Emilio Segrè Visual Archives; **unnum. figure p. 161:** Gawthron Inst., Nelson, New Zealand/courtesy AIP E. Segrè Visual Archives; **unnum. figure p. 162:** University of Oxford, Museum of History of Science, courtesy of AIP Niels Bohr Library; **unnum. figure p. 169:** Alfred Pasieda/Peter Arnold, Inc.; **unnum. figure p. 170(bottom):** National Portrait Gallery, London; **unnum. figure p. 170(top):** Charles D. Winters; **unnum. figure p. 173:** OM-4T Black, product of Olympus America, Inc.; **unnum. figure p. 175:** Comstock; **unnum. figure p. 180:** R. Creighton, Sandia Laboratories; **unnum. figure p. 182:** Anglo-Australian Observatory, photography by David Malin; **unnum. figure p. 181(top):** Philips Electronic Instruments, Inc.; **unnum. figure p. 181(bottom):** David Scharf/Peter Arnold, Inc.; **unnum. figure p. 193:** Runk/Schoenberger from Grant Heilman; **unnum. figure p. 194:** Paul Silverman/Fundamental Photographs, New York; **unnum. figure p. 204(left):** courtesy of James Mauseth; **unnum. figure p. 204(right):** from W. L. Masterton, E. J. Slowinski, C. L. Stanitski: *Chemical Principles,* 6th ed., Saunders College Publishing, 1985; **unnum. figure p. 205:** Peter Arnold, Inc., by Dagmar Hailer-Hamann.

Chapter 6

Chapter opener: Richard Megna, Fundamental Photographs, NYC; **6-5, 6-7, 6-9, unnum. figure p. 209:** Charles D. Winters; **6-6(a, b):** Grant Heilman for Grant Heilman; **6-10:** Georgia Power Company by Max Fundom; **6-11:** k10234(2) courtesy Department of Library Services, American Museum of Natural History; **6-12:** National Center for Atmospheric Research/National Science Foundation; **unnum. figure p. 208:** AT&T Bell Laboratories; **unnum. figures pp. 210, 230:** Charles Steele; **unnum. figures p. 236:** EPA; **unnum. figures p. 237(top left and right):** Dean and Chapter of Lincoln; **unnum. figure p. 237(bottom):** Ohio Edison.

Chapter 7

Chapter opener, unnum. figures pp. 246(top), 260, 264: Charles D. Winters; **unnum. figure p. 246(bottom):** Hal Levin; **unnum. figures pp. 248, 273:** Charles Steele; **unnum. figure p. 258:** Ralph Earlandson/ Imagery; **unnum. figure p. 272:** National Center for Atmospheric Research/National Science Foundation.

Chapter 8

Chapter opener, unnum. figure p. 306: Charles D. Winters; **unnum. figure p. 279:** Alfred Pasieka/Peter Arnold, Inc.; **unnum. figure p. 283:** Bob Evans/Peter Arnold, Inc.; **unnum. figures pp. 284, 287, 289, 294, 296, 298, 302:** Charles Steele; **unnum. figure p. 308:** courtesy of Bethlehem Steel.

Chapter 9
Chapter opener: © Daniel Quat/Phototake NYC; **unnum. figure p. 318:** Leon Lewandowski; **unnum. figure p. 325:** © 1991 Peter Menzel.

Chapter 10
Chapter opener, unnum. figures pp. 348, 351(top), 358: Charles D. Winters; **unnum. figure p. 337(top left and right):** Marna G. Clarke; **unnum. figures pp. 337(bottom), 351(bottom), 352:** James W. Morgenthaler; **unnum. figure p. 338:** Larry Lefever/Grant Heilman, Inc.; **unnum. figure p. 347:** Martin Dohrn/Science Photo Library/Photo Researchers, Inc.; **unnum. figure p. 349:** Charles Steele.

Chapter 11
Chapter opener: Leon Lewandowski; **11-1, unnum. figures p. 387(left, bottom right):** Charles D. Winters; **unnum. figure p. 361:** Brinkman Instruments; **unnum. figures pp. 365, 366, 378(top and bottom):** James W. Morganthaler; **unnum. figure p. 368:** Richard Megna/Fundamental Photographs, New York; **unnum. figures pp. 372, 373, 375(top), 382:** Charles Steele; **unnum. figure p. 375(bottom):** Robert E. Daemmrich/Tony Stone Images; **unnum. figure p. 380:** NASA; **unnum. figures pp. 386, 387(top right):** Dennis Drenner.

Chapter 12
Chapter opener, 12-2(c), 12-15(b), unnum. figure p. 430: Charles Steele; **12-2(a), unnum. figures pp. 391, 400, 426(bottom):** Charles D. Winters; **12-2(b):** Taylor Scientific Instruments; **12-8:** Marna G. Clarke; **unnum. figure p. 393:** Jacques Jangoux/Peter Arnold, Inc.; **unnum. figure p. 399:** The Granger Collection, New York; **unnum. figure p. 407:** National Center for Atmospheric Research/National Science Foundation; **unnum. figure p. 418:** Paul Cherfils/Tony Stone Images; **unnum. figure p. 429:** courtesy of Saab Automobiles; **unnum. figure p. 434:** NASA/JPL; **unnum. figure p. 436:** David M. Dennis/Tom Stack & Associates; **unnum. figure p. 437:** courtesy of Fisher Scientific.

Chapter 13
Chapter opener: Michael Lustbader/Photo Researchers, Inc.; **13-9, 13-14, 13-31(b), unnum. figures pp. 448, 463(bottom right), 464:** Charles D. Winters; **13-11(c), 13-31(c), unnum. figures pp. 450(bottom left), 457, 463(top), 483, 485:** Charles Steele; **13-16:** Marna G. Clarke; **13-18:** from Marvin L. Hackert; **13-31(a):** Gemological Institute of America; **unnum. figure p. 449:** David Parker/Science Photo Library/Photo Researchers, Inc.; **unnum. figure p. 450(middle left):** Manfred Daneger/Peter Arnold, Inc.; **unnum. figure p. 450(bottom middle and right):** Charles Steele/from Rain X; **unnum. figure p. 463(bottom left):** Telegraph Colour Library/FPG International; **unnum. figure p. 465:** from Joesten: *The World of Chemistry* Program 8, ''Chemical Bonds''; **unnum. figure p. 481(left):** Chem. Design/Science Photo Library/Photo Researchers, Inc.; **unnum. figure p. 481(right):** Masato Murakimi/Istec/© 1991 Discover Magazine; **unnum. figure p. 487:** R. E. Davis.

Chapter 14
Chapter opener: courtesy of Ashland Oil, Inc.; **unnum. figure p. 501:** Richard Megna/Fundamental Photographs; **14-4, 14-5, 14-6, unnum.**

figures pp. 502, 507(left), 509, 510, 535, 540: Charles Steele; **unnum. figure p. 504:** Kip Peticolas, Fundamental Photographs; **unnum. figure p. 505:** Petit Format/Institut Pasteur/Charles Dauget/Photo Researchers, Inc.; **unnum. figure p. 506:** Bill Wood/Bruce Coleman Ltd.; **14-20, unnum. figures pp. 507(right), 511:** © Robert W. Metz; **unnum. figure p. 508:** Leon Lewandowski; **14-14(a), unnum. figure p. 533:** James W. Morgenthaler; **14-14(b):** Union Carbide Industrial Gases, Linde Division; **unnum. figure p. 520:** Bethlehem Steel; **unnum. figure p. 521:** Union Carbide Corporation; **unnum. figures p. 527, 539(bottom left):** Charles D. Winters; **unnum. figure p. 529:** courtesy of R. F. Baker, University of Southern California Medical School; **unnum. figure p. 530:** courtesy of E. I. DuPont de Nemours and Company; **unnum. figure p. 531:** SIU School of Medicine/Bruce Coleman, Inc., New York; **unnum. figure p. 532:** from NALCO Chemical Company; **unnum. figure p. 536:** Paolo Koch, Photo Researchers, Inc.; **unnum. figures p. 539(top left), 541;** Dennis Drenner.

Chapter 15
Chapter opener: Comstock; **unnum. figure p. 545:** Courtesy of Niagara Mohawk Power Corp./Frank Warner; **15-2, unnum. figures pp. 551, 581, 585, 586, 589:** Charles D. Winters; **unnum. figure p. 570:** Charles Steele; **unnum. figure p. 574:** NASA; **unnum. figures p. 576:** Dennis Drenner; **unnum. figure p. 583:** Courtesy of Department of Energy; **unnum. figure p. 584:** NASA; **unnum. figure p. 591:** Atlanta Gas Light Co.

Chapter 16
Chapter opener, unnum. figure p. 625: NASA; **unnum. figure p. 598:** Robert Herko/The Image Bank; **unnum. figures pp. 599(top right), 634:** Charles D. Winters; **unnum. figures pp. 599(bottom right), 617, 626, 635(top right):** Charles Steele; **unnum. figure p. 602:** Leon Lewandowski; **unnum. figures pp. 627, 630, 635(bottom right):** J. Morgenthaler; **unnum. figure p. 632:** from Harshaw/Fitrol Partnership; **16-17(b):** General Motors.

Chapter 17
Chapter opener: Dresser Industries/Kellogg Company; **unnum. figures pp. 658, 660, 670:** Charles Steele; **unnum. figure p. 659:** Charles D. Winters; **unnum. figure p. 663:** courtesy of M. W. Kellogg Company.

Chapter 18
Chapter opener: Leon Lewandowski; **18-1, unnum. figures pp. 699, 702(top and bottom), 719:** Beckman Instruments; **18-2, unnum. figures pp. 713, 723:** Charles D. Winters; **unnum. figure p. 692:** Lester Lefkowitz/Telegraph Colour Library/FPG International; **18-3, unnum. figure p. 710:** Marna G. Clarke; **unnum. figures pp. 703, 705, 709:** J. Morgenthaler; **18-4, unnum. figures pp. 688, 694, 716:** Charles Steele.

Chapter 19
Chapter opener, unnum. figures pp. 738, 744: Charles D. Winters; **unnum. figures pp. 730, 731:** Beckman Instruments; **unnum. figure p. 735:** Science Museum Library, London; **unnum. figure p. 736:** Charles Steele; **unnum. figure p. 745:** Jan Cobb/The Image Bank.

Chapter 20

Chapter opener, unnum. figures pp. 763(top and bottom): James Morgenthaler; **unnum. figures pp. 747(top), 764:** Charles D. Winters; **unnum. figure p. 747(bottom):** M. Timothy O'Keefe/Tom Stack & Associates; **unnum. figure p. 748(top left):** CNRI/Science Photo Library/Photo Researchers, Inc.; **unnum. figure p. 748(bottom left and right):** Brian Parker/Tom Stack & Associates; **unnum. figures pp. 752, 755, 757:** Charles Steele.

Chapter 21

Chapter opener: Nathan S. Lewis/California Institute of Technology; **unnum. figure p. 771:** Yoav Levy/Phototake NYC; **unnum. figure p. 775:** Oesper Collection in the History of Chemistry, University of Cincinnati; **unnum. figure p. 778:** Gordon Garradd/Science Photo Library/Photo Researchers, Inc.; **21-5(c):** ASARCO, Inc.; **21-6(b), 21-12(b), unnum. figures pp. 774, 795(bottom left):** Charles D. Winters; **unnum. figures pp. 782, 783:** J. Morgenthaler; **unnum. figure p. 785:** courtesy of Donald M. West; **unnum. figure p. 795(top):** M. D. Ippolito; **unnum. figure p. 795(bottom right):** Tom Stack & Associates; **unnum. figure p. 796:** Patricia Caufield/Photo Researchers, Inc.; **unnum. figure p. 797:** Culver Pictures, Inc.; **21-14(b):** credit unknown; **21-16(b):** Duracell, Inc.; **unnum. figures pp. 810(right), 812:** courtesy of Eveready Battery Company; **21-18(b):** United Technologies.

Chapter 22

Chapter opener, unnum. figures p. 826(top), 835(top): Brian Parker/Tom Stack & Associates; **unnum. figure p. 823:** Paul Silverman, Fundamental Photographs New York; **unnum. figure p. 825(left):** Wards Natural Science Establishment; **unnum. figures pp. 825(right), 829, 831:** Charles D. Winters; **22-3:** © Ray Pfortner/Peter Arnold Inc.; **unnum. figure p. 826(bottom):** Allen B. Smith/Tom Stack & Associates; **unnum. figure p. 828:** North Wind Picture Archives; **22-7(b):** Aluminum Association of America; **unnum. figures pp. 832, 834:** courtesy of Bethlehem Steel; **22-10, unnum. figure p. 830:** James Cowlin; **22-11:** from H. C. Metcalfe, J. E. Williams, J. F. Costka, *Modern Chemistry,* Holt, Rinehart, and Winston, 1986; **unnum. figure p. 835(bottom):** courtesy of E. I. DuPont de Nemours & Company; **unnum. figure p. 838:** © Gemological Institute of America.

Chapter 23

Chapter opener: Courtesy of AM General Corporation; **unnum. figures pp. 843, 844(top), 846, 850, 851, 860:** Charles Steele; **unnum. figure p. 844(bottom):** Steve Elmore/Tom Stack & Associates; **unnum. figure p. 845:** from Masterton, Slowinski, and Stanitski: *Chemical Principles,* 6th edition, Saunders College Publishing; **unnum. figure p. 850(top):** Dick George/Tom Stack & Associates; **unnum. figure p. 850(bottom):** Kevin Schafer/Tom Stack & Associates; **unnum. figure p. 852:** courtesy of American Cyanamid Company; **23-1(a–c), unnum. figure p. 853:** Charles D. Winters; **unnum. figure p. 854(top):** © Gemological Institute of America; **unnum. figure p. 854(bottom):** Leon Lewandowski; **unnum. figure p. 855:** courtesy of IBM; **unnum. figures pp. 857, 858:** J. Morgenthaler.

Chapter 24

Chapter opener, unnum. figures pp. 869(bottom), 870(top, middle, and bottom), 880, 887(top and middle), 888(top): Charles Steele; **unnum.** **figure p. 866:** from Argonne National Laboratory; **unnum. figure p. 869(top):** Kevin Schafer/Tom Stack & Associates; **unnum. figure p. 876(left):** Don and Pat Valenti/Tom Stack & Associates; **unnum. figure p. 876(right):** J. Morgenthaler; **unnum. figure p. 878:** William Felger/Grant Heilman; **unnum. figures pp. 882(top), 886:** Metcalfe, Williams, Castka, and Scott; **unnum. figure p. 882(bottom):** American Breeders Service; **unnum. figure p. 885:** NASA; **unnum. figure p. 886:** Brian Parker/Tom Stack & Associates, **unnum. figure p. 887:** Standard Oil Company, Ohio; **unnum. figure p. 888(bottom):** Particulate Mineralogy Unit/Avondale Research Center/U.S. Bureau of Mines.

Chapter 25

Chapter opener: David Scharf/Peter Arnold, Inc.; **unnum. figures pp. 895, 896(bottom), 897, 905, 913, 915(bottom):** J. Morgenthaler; **unnum. figures pp. 896(top), 901:** Charles D. Winters; **unnum. figure p. 909:** Charles Steele; **unnum. figure p. 914(bottom):** Runk/Schoenberger/Grant Heilman, Inc.; **unnum. figure p. 915(top):** from Masterton, Slowinski, Stanitski, *Chemical Principles,* 6th ed., Saunders College Publishing.

Chapter 26

Chapter opener, unnum. figures pp. 933, 944: NASA; **unnum. figure p. 924:** AIP Niels Bohr Library; **unnum. figure p. 930(top):** Philippe Plailly/Science Photo Library/Photo Researchers, Inc.; **unnum. figure p. 930(bottom):** Oak Ridge National Laboratory; **26-4:** from Brescia, Mehlman, Pellegrini, and Stambler: *Chemistry: A Modern Introduction 2/e,* Saunders College Publishing, 1978; **unnum. figure p. 934:** Dave Davidson/Tom Stack and Associates; **unnum. figure p. 937:** Paul Hanny/Gamma Liaison International; **unnum. figure p. 938:** Marna G. Clarke; **unnum. figure p. 939:** Stand Levy/Photo Researchers, Inc.; **unnum. figure p. 940:** from the International Atomic Energy Agency; **26-8:** Argonne National Laboratory; **unnum. figure p. 942:** courtesy of Fermilab; **unnum. figure p. 943:** © 1992 The Field Museum/Photo by John Weinstein/Neg.#GEO-CK-13.Tc; **unnum. figure p. 947:** Yoav Levy/Phototake NYC; **unnum. figure p. 948(top):** courtesy of E. I. DuPont deNemours and Company; **unnum. figure p. 948(bottom):** Science Photo Library/Photo Researchers, Inc.; **unnum. figure p. 950:** Argonne National Laboratory; **unnum. figure p. 951:** Courtesy of the National Atomic Museum; **unnum. figure p. 952(top):** Lawrence Livermore Laboratory; **unnum. figure p. 952(bottom):** courtesy of Princeton Plasma Physics Laboratory.

Chapter 27

Chapter opener, 27-5, 27-20, unnum. figure p. 998: Charles D. Winters; **27-2(c, d), 27-3, 27-4, 27-7, 27-9, 27-10, 27-14, 27-15, 27-16, 27-17, 27-18, unnum. figures pp. 980, 987, 989, 1002:** Charles Steele; **27-8, unnum. figures pp. 985, 994:** R. E. Davis; **unnum. figure p. 971:** courtesy of The American Petroleum Institute; **unnum. figure p. 978:** J. Weber, University of Geneva, Switzerland; **unnum. figure p. 992:** illustration copyright by Irving Geis; **unnum. figure p. 995:** © Herb Charles Ohlmeyer/Fran Heyl Associates; **unnum. figure p. 997:** Biophoto Associates; **27-19(a, b):** Leonard Lessin; **unnum. figure p. 999:** Phillip A. Harrington/Fran Heyl Associates, **unnum. figure p. 1000:** Science VU-WHC/Visuals Unlimited.

Chapter 28

Chapter opener: J. Kirk Cochran; **28-1, 28-2, 28-4, 28-5, 28-6, 28-9:** R. E. Davis; **28-3:** Ray Ellis/Photo Researchers, Inc.; **28-7, 28-8, unnum.**

figures pp. 1034, 1037: Charles Steele; 28-10: James Morgenthaler; unnum. figures pp. 1027, 1028, 1029, 1030, 1044: Charles D. Winters; unnum. figure p. 1032: © 1990 Trygve Steen; unnum. figure p. 1036: photo by Charles D. Winters, spelunker—John Reynolds; unnum. figure p. 1039: North Wind Picture Archives; unnum. figure p. 1040: credit unknown; unnum. figure p. 1043(left): courtesy of Drs. James L. Monro and Gerald Shore and the Wolfe Medical Publications, London, England; unnum. figures pp. 1043(right top, right bottom), 1045: courtesy of E. I. duPont de Nemours and Company; unnum. figure p. 1046: Evans and Sutherland.

Qualitative Analysis
Unit opener: James Morgenthaler.

Chapter 29
Chapter opener: Bethlehem Steel; unnum. figures pp. 1056(top), 1065: Leon Lewandowski; unnum. figures pp. 1056(margin), 1057(top): J. Morgenthaler; unnum. figure p. 1057(bottom): Charles D. Winters; unnum. figure p. 1062: Paul Silverman, Fundamental Photographs, New York; unnum. figure p. 1063: Tom Stack & Associates.

Chapter 30
Chapter opener: Charles Steele; 30-3, 30-4, 30-5: J. Morgenthaler.

Chapter 31
Chapter opener: Charles D. Winters; unnum. figures pp. 1081, 1082, 1085, 1086, 1087: J. Morgenthaler.

Chapter 32
Chapter opener: James Morgenthaler; unnum. figure p. 1091(left): Paul Silverman/Fundamental Photographs, New York; unnum. figure p. 1091(right): Runk/Schoenberger/Grant Heilman; unnum. figures pp. 1093, 1096, 1097, 1098: J. Morgenthaler.

Chapter 33
Chapter opener, unnum. figure p. 1102: Charles Steele; unnum. figures pp. 1101, 1105, 1108, 1110, 1112: J. Morgenthaler.

Chapter 34
Chapter opener: Charles D. Winters; 34-4, unnum. figures pp. 1119, 1120, 1122, 1123, 1124: J. Morgenthaler.

Chapter 35
Chapter opener: Charles Steele; unnum. figure p. 1127: from Levin; unnum. figures pp. 1129, 1130, 1131, 1132: J. Morgenthaler.

Chapter 36
Chapter opener, unnum. figures pp. 1135, 1138: J. Morgenthaler; unnum. figure p. 1136: Charles D. Winters; unnum. figure p. 1140: from Beckman Instruments.

Entries in *italics* indicate illustrations; page numbers followed by *t* indicate tables. Glossary terms, printed in **boldface**, are defined here as well as in the text and in Key Terms at the ends of appropriate chapters. Some terms used generally in the text are defined here without giving specific page references.

LOCATION OF COMMONLY USED INFORMATION